B/2A

Supplement C2
The chemistry of
triple-bonded functional groups

THE CHEMISTRY OF FUNCTIONAL GROUPS

A series of advanced treatises under the general editorship of
Professors Saul Patai and Zvi Rappoport

$$-C{\equiv}C-, \quad -C{\equiv}N, \quad -\overset{+}{N}{\equiv}N$$

Supplement C2
The chemistry of
triple-bonded functional groups

Edited by

SAUL PATAI

The Hebrew University, Jerusalem

1994

JOHN WILEY & SONS

CHICHESTER–NEW YORK–BRISBANE–TORONTO–SINGAPORE

An Interscience® Publication

Copyright © 1994 by John Wiley & Sons Ltd,
Baffins Lane, Chichester,
West Sussex, PO19 1UD, England
Telephone: National   Chichester (0243) 779 777
International   + 44 243 779 777

Other Wiley Editorial Offices

John Wiley & Sons, Inc., 605 Third Avenue,
New York, NY 10158-0012, USA

Jacaranda Wiley Ltd, 33 Park Road, Milton,
Queensland 4064, Australia

John Wiley & Sons (Canada) Ltd, 22 Worcester Road,
Rexdale, Ontario M9W 1L1, Canada

John Wiley & Sons (SEA) Pte Ltd, 37 Jalan Pemimpin #05-04,
Block B, Union Industrial Building, Singapore 2057

Library of Congress Cataloging-in-Publication Data

The Chemistry of triple-bonded functional groups   /   edited by Saul
  Patai.
        p.   cm.—(The Chemistry of functional groups.   Supplement;
  C2)
    An Interscience publication.
    Includes bibliographical references and index.
    ISBN 0-471-93559-X (v. 2)
    1. Chemistry, Physical organic.   2. Functional groups.   I. Patai,
Saul.   II. Series.
QD476.C535   1994   93-21238
547.13—dc20   CIP

British Library Cataloguing in Publication Data

A Catalogue record for this book is available from the British Library

ISBN 0 471 93559 X

Typeset in Times 9/10 pt by Thomson Press (I) Ltd, New Delhi
Printed and bound in Great Britain by Biddles, Guilford, Surrey

Contributing authors

Frank H. Allen — Cambridge Crystallographic Data Centre, 12 Union Road, Cambridge, CB2 IEZ, UK

Derek V. Banthorpe — Department of Chemistry, University College London, 20 Gordon Street, London, WC1H OAH, UK

Harold Basch — Department of Chemistry, Bar Ilan University, Ramat Gan 52900, Israel

Zvi Ben-Zvi — Department of Clinical Pharmacology, Ben-Gurion University of the Negev and Soroka Medical Center, Beer Sheva, Israel

Sárka Beranová — Department of Chemistry, The University of Akron, Akron, OH 44325–3601, USA

G. V. Boyd — Department of Organic Chemistry, The Hebrew University of Jerusalem, Jerusalem 91904, Israel

C. Chatgilialoglu — I. Co. C.E.A., Consiglio Nazionale delle Richerche, Via Gobetti 101, 40129 Bologna, Italy

Abraham Danon — Department of Clinical Pharmacology, Ben-Gurion University of the Negev and Soroka Medical Center, Beer Sheva, Israel

Wiendelt Drenth — Ovidiuslaan 4, 3584 AW Utrecht, The Netherlands

Renate Dworczak — Karl-Franzens University of Graz, Heinrichstrasse 28, A-8010 Graz, Austria

C. Ferreri — Dipartimento di Chimica Organica e Biologica, Universitá di Napoli, Via Mezzocannone 16, 80134 Napoli, Italy

Stephanie E. Garner — Cambridge Crystallographic Data Centre, 12 Union Road, Cambridge, CB2 IEZ, UK

Rolf Gleiter — Organisch-Chemisches Institut der Universität Heidelberg, Im Neuenheimer Feld 270, D-69120 Heidelberg, Germany

Harold Hart — Department of Chemistry, Michigan State University, 320 Chemistry Building, East Lansing, MI 48824–1322, USA

William M. Horspool — Department of Chemistry, The University of Dundee, DD1 4HN, Scotland, UK

Tova Hoz — Department of Chemistry, Bar Ilan University, Ramat Gan 52900, Israel

Hans Junek — Karl-Franzens University of Graz, Heinrichstrasse 28, A-8010 Graz, Austria

T. Kasprzycka-Guttman — Department of Chemistry, University of Warsaw, 02–093 Warsaw, Pasteura 1, Poland

Toshio Masuda	Department of Polymer Chemistry, Kyoto University, Kyoto 606–01, Japan
Vera L. Pardini	Instituto de Química, Universidade de São Paulo, 01498 São Paulo, Brazil
Michael J. Polce	Department of Chemistry, The University of Akron, Akron, OH 44325–3601, USA
Wolfgang Schäfer	Boehringer Mannheim GmbH, Sandhofer Str. 116, D-68305 Mannheim, Germany
Hideki Shirakawa	Institute of Materials Science, University of Tsukuba, Tsukuba, Ibaraki 305, Japan
John Shorter	School of Chemistry, University of Hull, Hull, HU6 7RX, UK
Peter J. Smith	Department of Chemistry and Chemical Engineering, University of Saskatchewan, Saskatoon, Saskatchewan, Canada
Peter J. Stang	Department of Chemistry, The University of Utah, Salt Lake City, UT 84112, USA
Kenji Takeda	Research Center, Japan Synthetic Rubber Co., Ltd, 25 Miyukigaoka, Tsukuba, Ibaraki 305, Japan
Reinaldo R. Vargas	Instituto de Química, Universidade de São Paulo, 01498 São Paulo, Brazil
Hans Viertler	Instituto de Química, Universidade de São Paulo, 01498 São Paulo, Brazil
Chrys Wesdemiotis	Department of Chemistry, The University of Akron, Akron, OH 44325–3601, USA
Kenneth C. Westaway	Department of Chemistry, Laurentian University, Ramsey Lake Road, Sudbury, Ontario P3E 2C6, Canada
D. Whittaker	Department of Chemistry, University of Liverpool, Liverpool, L69 3BX, UK
Viktor V. Zhdankin	Department of Chemistry, University of Minnesota-Duluth, MN 55812, USA
Jacob Zabicky	The Institutes for Applied Research, Ben-Gurion University of the Negev, Beer-Sheva 84110, Israel

Foreword

The first supplementary volume dealing with the chemistry of triple-bonded functional groups ($-C\equiv C-$, $-C\equiv N$, $-\overset{+}{N}\equiv N$ and related groups) was published in 1983. Since that time, the subject has advanced considerably, and the chapters contained in the present Supplement C2 volume treat almost exclusively material which was reported in the last fifteen years. In most cases the literature was covered up to the middle or end of 1992 and less often up to the beginning of 1993.

Some chapters intended for the present volume did not materialize. These were on 'Chiroptical Properties', 'Acidity, Basicity, H-bonding and Complex Formation' and on 'Free Radical Reactions Involving the Diazonium Group'.

I will be most obliged to readers who will call my attention to omissions or mistakes relating to this and other volumes of the series'.

Jerusalem
June, 1994

SAUL PATAI

The Chemistry of Functional Groups
Preface to the series

The series 'The Chemistry of Functional Groups' was originally planned to cover in each volume all aspects of the chemistry of one of the important functional groups in organic chemistry. The emphasis is laid on the preparation, properties and reactions of the functional group treated and on the effects which it exerts both in the immediate vicinity of the group in question and in the whole molecule.

A voluntary restriction on the treatment of the various functional groups in these volumes is that material included is easily and generally available in secondary or tertiary sources, such as Chemical Reviews, Quarterly Reviews, Organic Reactions, various 'Advances' and 'Progress' series and in textbooks (i.e. in books which are usually found in the chemical libraries of most universities and research institutes), should not, as a rule, be repeated in detail, unless it is necessary for the balanced treatment of the topic. Therefore each of the authors is asked not to give an encyclopaedic coverage of his subject, but to concentrate on the most important recent developments and mainly on material that has not been adequately covered by reviews or other secondary sources by the time of writing of the chapter, and to address himself to a reader who is assumed to be at a fairly advanced postgraduate level.

It is realized that no plan can be devised for a volume that would give a complete coverage of the field with no overlap between chapters, while at the same time preserving the readability of the text. The Editors set themselves the goal of attaining reasonable coverage with moderate overlap, with a minimum of cross-references between the chapters. In this manner, sufficient freedom is given to the authors to produce readable quasi-monographic chapters.

The general plan of each volume includes the following main sections:

(a) An introductory chapter deals with the general and theoretical aspects of the group.

(b) Chapters discuss the characterization and characteristics of the functional groups. i.e. qualitative and quantitative methods of determination including chemical and physical methods, MS, UV, IR, NMR, ESR and PES—as well as activating and directive effects exerted by the group, and its basicity, acidity and complex-forming ability.

(c) One or more chapters deal with the formation of the functional group in question, either from other groups already present in the molecule or by introducing the new group directly or indirectly. This is usually followed by a description of the synthetic uses of the group, including its reactions, transformations and rearrangements.

(d) Additional chapters deal with special topics such as electrochemistry, photo-chemistry, radiation chemistry, thermochemistry, syntheses and uses of isotopically labelled compounds, as well as with biochemistry, pharmacology and toxicology. Whenever applicable, unique chapters relevant only to single functional groups are also included (e.g. 'Polyethers', 'Tetraaminoethylenes' or 'Siloxanes').

This plan entails that the breadth, depth and thought-provoking nature of each chapter will differ with the views and inclinations of the authors and the presentation will necessarily be somewhat uneven. Moreover, a serious problem is caused by authors who deliver their manuscript late or not at all. In order to overcome this problem at least to some extent, some volumes may be published without giving consideration to the originally planned logical order of the chapters.

Since the beginning of the Series in 1964, two main developments have occurred. The first of these is the publication of supplementary volumes which contain material relating to several kindred functional groups (Supplements A, B, C, D, E and F). The second ramification is the publication of a series of 'Updates', which contain in each volume selected and related chapters, reprinted in the original form in which they were published, together with an extensive updating of the subjects, if possible, by the authors of the original chapters. A complete list of all above mentioned volumes published to date will be found on the page opposite the inner title page of this book.

Advice or criticism regarding the plan and execution of this series will be welcomed by the Editors.

The publication of this series would never have been started, let alone continued, without the support of many persons in Israel and overseas, including colleagues, friends and family. The efficient and patient co-operation of staff-members of the publisher also rendered us invaluable aid. Our sincere thanks are due to all of them.

The Hebrew University   SAUL PATAI
Jerusalem, Israel   ZVI RAPPOPORT

Contents

xi

Contents

List of abbreviations used

Ac	acetyl (MeCO)
acac	acetylacetone
Ad	adamantyl
AIBN	azoisobutyronitrile
Alk	alkyl
All	allyl
An	anisyl
Ar	aryl
Bz	benzoyl (C_6H_5CO)
Bu	butyl (also t-Bu or Bu^t)
CD	circular dichroism
CI	chemical ionization
CIDNP	chemically induced dynamic nuclear polarization
CNDO	complete neglect of differential overlap
Cp	η^5-cyclopentadienyl
Cp*	η^5-pentamethylcyclopentadienyl
DABCO	1,4-diazabicyclo[2.2.2]octane
DBN	1,5-diazabicyclo[4.3.0]non-5-ene
DBU	1,8-diazabicyclo[5.4.0]undec-7-ene
DIBAH	diisobutylaluminium hydride
DME	1,2-dimethoxyethane
DMF	N,N-dimethylformamide
DMSO	dimethyl sulphoxide
ee	enantiomeric excess
EI	electron impact
ESCA	electron spectroscopy for chemical analysis
ESR	electron spin resonance
Et	ethyl
eV	electron volt
FC	ferrocenyl
FD	field desorption
FI	field ionization

FT	Fourier transform
Fu	furyl (OC_4H_3)
GLC	gas–liquid chromatography
Hex	hexyl (C_6H_{13})
c-Hex	cyclohexyl (C_6H_{11})
HMPA	hexamethylphosphortriamide
HOMO	highest occupied molecular orbital
HPLC	high performance liquid chromatography
i-	iso
Ip	ionization potential
IR	infrared
ICR	ion cyclotron resonance
LAH	lithium aluminium hydride
LCAO	linear combination of atomic orbitals
LDA	lithium diisopropylamide
LUMO	lowest unoccupied molecular orbital
M	metal
M	parent molecule
MCPBA	*m*-choloroperbenzoic acid
Me	methyl
MNDO	modified neglect of diatomic overlap
MS	mass spectrum
n	normal
Naph	naphthyl
NBS	*N*-bromosuccinimide
NCS	*N*-chlorosuccinimide
NMR	nuclear magnetic resonance
Pc	phthalocyanine
Pen	pentyl (C_5H_{11})
Pip	piperidyl ($C_5H_{10}N$)
Ph	phenyl
ppm	parts per million
Pr	propyl (also *i*-Pr or Pr^i)
PTC	phase transfer catalysis or phase transfer conditions
Pyr	pyridyl (C_5H_4N)
R	any radical
RT	room temperature
s-	secondary
SET	single electron transfer
SOMO	singly occupied molecular orbital
t-	tertiary
TCNE	tetracyanoethylene

TFA	trifluoroacetic acid
THF	tetrahydrofuran
Thi	thienyl (SC_4H_3)
TLC	thin layer chromatography
TMEDA	tetramethylethylene diamine
TMS	trimethylsilyl or tetramethylsilane
Tol	tolyl (MeC_6H_4)
Tos or Ts	tosyl (*p*-toluenesulphonyl)
Trityl	triphenylmethyl (Ph_3C)
Xyl	xylyl ($Me_2C_6H_3$)

In addition, entries in the 'List of Radical Names' in *IUPAC Nomenclature of Organic Chemistry*, 1979 Edition. Pergamon Press, Oxford, 1979, p. 305–322, will also be used in their unabbreviated forms, both in the text and in formulae, instead of explicitly drawn structures.

CHAPTER **1**

The nature of the triple bond

HAROLD BASCH and TOVA HOZ

Department of Chemistry, Bar Ilan University, Ramat Gan 52900, Israel

LIST OF ABBREVIATIONS

AE	all electron
BE	binding energy
BMA	bond midpoint axis
BSSE	basis set superposition error
CEP	compact effective potentials
CI	configuration interaction
CIS	configuration interaction singles method
CT	charge transfer
DF	Dirac–Fock
EA	electron affinity
IP	ionization potential

Supplement C2: The chemistry of triple-bonded functional groups
Edited by S. Patai © 1994 John Wiley & Sons Ltd

L	ligand
M^+	metal ion
MP2	Moeller–Plesset to the second order
PA	proton affinity
PV	combined P(main) + V(metal) basis set
Q	quadratic
QCISD	a CI method including all single and double excitation configurations, in which all additional terms are quadratic in the configuration coefficients
QCISD(T)	QCISD with the added contribution of triple (T) substitutions
RCEP	relativistic compact effective potentials
RHF	restricted Hartree–Fock
ZPE	zero-point energy

I. INTRODUCTION

The nature of the triple bond, whether $-C\equiv C-$, $-C\equiv N$ or $=N\equiv N$ and their higher row analogues, has been reviewed several times in this series[1–4], in theoretical chapters. The nature of a chemical entity is defined by its geometric and electronic structural properties. The triple bond is characterized by short bond distances, local cylindrical symmetry, relatively high electron density prependicular to the bond axis and relatively large polarizability parallel to the bond axis. Its description in terms of σ and π bonding is classic[5], and qualitatively explains the characteristic properties of the triple bond.

A more quantitive measure of the properties of a system is obtained by observing the results of perturbing effects. The true nature of a bond reveals itself in the way that it reacts or adapts to interactions with other species (intermolecular), or through substituent atoms or groups attached to one or both of the triply bonded atoms (intramolecular). Both geometric and electronic structural properties are affected by such perturbation, and their changes, as a function of the strength of the intermolecular interaction, allow an analysis of the bonding situation in the unperturbed substrate. Of course, the interactions can be sufficiently strong that they disrupt the bonding situation in the precursor and cause existing bonds to be broken and new bonds to be formed. Such processes usually give little information about the original substrate. The focus here is on relatively mild perturbations where the existing bonding patterns in the precursor substrate are essentially maintained.

The significance of measuring the properties of a chemical system from the effect of substituents or perturbers is the possibility that new insight and information can be obtained even for well known substrates. An outstanding example of such a discovery can be found in the role that the chemically inert nitrogen molecule was found to play as a ligand in transition metal complexes only some 30 years ago[6]. On the other hand, certain triply bonded species, such as fulminic acid ($H-C\equiv N=O$), have not been satisfactorily characterized from a theoretical point of view with regard to their geometric and electronic structural properties[7]. A description of these properties for the isolated molecule is, apparently, difficult to achieve. A reasonable expectation is that, by examining the results of interactions with perturbers and the effect of substituents, its intrinsic properties can be revealed.

In the realm of intermolecular interactions, the collision of metal ions, M^+, with single or multiple ligands has developed into a very active research area, both experimentally[8–17] and theoretically[18–21]. Certainly, a very energetic ion, both kinetically and/or electronically, can break any chemical bond. Although the resultant fragmentation pattern and consequent metal ion fragment complexes that are formed may give information on bond strengths, very little is learned about the parent substrate and about the direct metal ion–ligand interactions. On the other hand, the association of metal ions having relatively low electron affinities (for an ion) with organic and inorganic substrates leads to long-lived coordination complexes that essentially maintain the integrity of the ligand. These type metal ions are then excellent probes of the geometric and electronic structural properties of the triple bond and from them much can also be learned about the nature of the metal ion–substrate bond.

From a procedural point of view, there have been many recent advances in theory, methodology and computer technology that have allowed ever-widening use of computational chemistry as a useful and practical tool for problem solving in the field of molecular structure. The study of gas-phase metal ion–ligand complexes by the various spectrometric and spectroscopic techniques referenced above does not give direct or detailed information on the geometric structure of the complexes, or indicate preferred site attachment and orientation where the substrate offers the possibility of several coordination sites and conformational possibilities, as is often the case. Computational methods, on the other hand, automatically give a superabundance of numbers whose significance and accuracy need to be understood. Experiments then give few numbers that have to be cleverly expanded into useful results, while theory gives many numbers that have to be cleverly condensed into meaningful information.

We have here chosen to review the properties of the triple bonds by surveying their 1:1 coordination complexes with the Cu^+, Ag^+ and Au^+ metal ions. These metal ions are relatively weak perturbers because of their low electron affinities (for ions) (7.724, 7.574 and 9.22 eV, respectively[22, 23]) and spherical atomic symmetry. A comparison of bonding capabilities and their effect on a specific ligand among different rows of transition metals in the Periodic Table is also very enlightening[20, 24–26]. The triply bonded systems studied here include $N\equiv N$, $HC\equiv N$, $HC\equiv CH$, $N\equiv N=O$, $FC\equiv N$, $HC\equiv N=O$, $HN=N\equiv N$, $HC\equiv CF$, $CH_2=N\equiv N$, $CH_3C\equiv N$, $FC\equiv CF$, $FC\equiv N=O$ and $CH_3C\equiv N=O$. This substrate set includes different types of triple bonds and substituent effects, and examines both classical ($N\equiv N$, $-C\equiv N$, $-C\equiv N-$) and hypervalent ($N\equiv N=O$, $-N=N\equiv N$, $-C\equiv N=O$) structures. These latter are defined by their higher formal bond coordination number for a nitrogen atom. For comparison purposes, parallel calculations were carried out on the corresponding ethylene systems. Besides metal ion coordination, the protonated substrates are also treated. These allow a comparison with the complexed ligands for the properties studied here and the progressive change in these properties from bare substrate to metal ion complex to protonated species. Properties of interest include equilibrium geometries, competing site attachments, binding energies, atomic charges, atomic orbital populations, harmonic vibrational frequencies and a description of excited state energies and their natures, all as a function of cation, ligand and binding site location. These properties combine to give a picture of the chemical bonding situation in these triply-bonded systems.

Besides reviewing the literature, a calculational survey was also carried out to obtain the properties of these systems within a uniform and consistent theoretical framework for comparison purposes. The computational and theoretical methodology used to generate these results are similar to those used previously[20,21] with several refinements. The geometries of the bare substrates and cation–ligand complexes (including protonated species) were gradient optimized at the MP2 (Moeller-Plesset to second order) level[27, 28] using compact effective potentials (CEP) for the main group elements (C, N, O and F)[29] and their relativistic analogues (RCEP) for the transition metal ions (Cu^+, Ag^+ and Au^+)[30]. The single electronic configuration restricted Hartree–Fock[27–31] (RHF) method was also used to generate optimized geometries, but these served for comparison purposes only, and all subsequent calculations, beyond the MP2 level, used only the MP2 optimized geometries. All calculations used the GAUSSIAN90[32] and GAUSSIAN92[33] systems of computer programs.

The basis sets were taken from the respective CEP[29] and RCEP[30] tabulations. For the main group elements a CEP-211G split of the published CEP-31G valence sp atomic orbital basis set[29] was augmented by a double set of polarization d-type basis functions (5 components). The single gaussian exponent values for the set of d functions were taken from the internally stored values of the GAUSSIAN programs[32-34]. For the hydrogen atom, the 311G triple-zeta set with a single gaussian p-type polarization function was taken, also from GAUSSIAN[32-34]. All told, the basis functions for the main group elements (including hydrogen) are designated CEP-N11G(2d,p), with $N = 2$ for the carbon, nitrogen, oxygen and fluorine atoms, and $N = 3$ for hydrogen. These are polarized valence triple-zeta basis sets which should be capable of describing the contraction of the valence atomic orbital shells due to the overall positive charge on each substrate species, for all degrees of ligand to metal (L-M) charge transfer dative bonding. The double set of d-type polarization functions for the main group elements is designed to satisfy the need for both angular correlation of the inner parts of the valence atomic orbitals and polarization in the more diffuse bonding region, as described by Magnusson[35]. These different requirements in terms of spatial extent leads to the double-zeta set of d-type functions for the main group elements.

The transition metal ions, Cu^+, Ag^+ and Au^+, all have a $nd^{10}(^1S)$ electronic state-configuration, with $n = 3$, 4 and 5, respectively. The RCEP used here were generated from Dirac–Fock (DF) all electron (AE) relativistic atomic orbitals[36], and therefore implicitly include the indirect relativistic effects of the core electron on the valence electrons[37], which in these metal ion systems are the major radial scaling effect. In these RCEP the ns^2np^6 subshells are included in the valence orbital space together with the nd, $(n + 1)s$ and $(n + 1)p$ atomic orbitals and all must be adequately represented by basis functions. The need for such 'semi-core' or 'semi-valence' electrons to be treated explicitly together with the traditional valence orbitals for the heavier elements has been adequately documented[38, 39]. The gaussian function basis set on each metal atom consists of the published[30] $4sp3d$ distribution which is double-zeta each in the nsp and $(n + 1)sp$ orbital space, and triple-zeta for the nd electrons.

In standard notation[27] this basis set might be labelled RCEP-5121/411G for Cu, RCEP-4121/311G for Ag and RCEP-4111/311G for Au to show explicitly the gaussian primitive distribution in each basis function. For brevity, this basis set is designated $RCEP[4^{sp}3^d]$ which shows the number of contracted basis functions, and is called the valence (V) metal basis. A larger basis set for the metal ions was formed by augmenting the $RCEP[4^{sp}3^d]$ functions by a double set of single primitive f-type functions (7 components). The coarse grid optimized exponents for the f-type functions were obtained by minimizing the MP2 energy of the metal ions in their ^1S ground electronic states. Initial calculations on Ag^+ showed an optimum exponent ratio of about 3 and this ratio was adopted for all three metal ions. The resultant exponents are shown in Table 1. The need for f-type functions to correlate the ns, np and nd electrons in the transition metal atoms is well documented[40, 41]. The angular correlation f-orbital needs to be more contracted while the polarization f-

TABLE 1. Double set of f-type functions for metal (M^+) ions[a]

M^+	f_1	f_2
Cu^+	5.10	1.70
Ag^+	2.70	0.90
Au^+	1.80	0.60

[a]Double set of single gaussian f-type functions (seven components) with exponents f_1 and f_2.

gaussian needs to be more concentrated in the valence region. Radial correlation of the important valence d orbitals is accomplished by using a triple-zeta d set. Radial and angular correlation of the valence $n$spd shell of electrons increases both the $(n+1)$s and nd ionization potentials in the neutral atoms and affects their spatial distribution. This, in turn, enhances the electron attraction properties of the monocations, and affects the metal ion–ligand interaction energy and optimal bond distance. This basis set for the metals, labelled RCEP[$4^s$$3^d$$2^f$], is called the polarization (P) metal basis. The only basis set used for the main group elements, the polarization CEP-N11G(2d,p) or CEP[$3^s$$2^d$/$3^s$$1^p$], is designated P.

All RHF and MP2 geometry optimizations were carried out in the combined P(main) + V(metal) basis set, called PV for short. The MP2 derived structures were then used to calculate the QCISD(T)[42] energies as single point calculations in both the PV and P(main) + P(metal) = PP combined basis sets. QCISD is a configuration interaction (CI) method that includes all single and double excitation configurations from the occupied to the virtual (unoccupied) space, and all additional terms that are quadratic (Q) in the configuration coefficients. To this is added the contribution of triple (T) substitutions as a perturbation. QCISD and QCISD(T) are more extensive methods than MP2 for calculating correlation energy within the single reference electronic configuration framework. MP2, QCISD and QCISD(T) each share the feature of being size-consistent methods[43], meaning that the energy of non-interacting units calculated together is equal to the energy sum of the individual units calculated separately, all within that method. This property is important for the proper calculation of dissociation energies, where the theoretical method is appropriately balanced between the complex and the component fragments. Thus, a comparison of RHF/PV optimized, MP2/PV optimized, QCISD//MP2/PV, QCISD(T)//MP2/PV, MP2/PP//MP2/PV, QCISD/PP//MP2/PV and QCISD(T)/PP//MP2/PV energies and properties will reveal the sensitivity of the results to basis set and level of theory.

The MP2 optimized geometries were also used to calculate stationary state geometry force constants and harmonic vibrational frequencies. These results serve two purposes. Firstly, they test the stability of the calculated equilibrium geometry[44]. As is well known, a true energy minimum has all real frequencies which are the eigenvalues of the force constant matrix in a normal coordinate analysis. A saddle point or transition state is characterized by having a single imaginary frequency. It is therefore possible to distinguish between the two types of stationary points, the minimum and the saddle point, by noting the number of imaginary calculated frequencies. Secondly, comparing the calculated vibrational frequencies between the free ligand and its metal ion complexes (and as a function of coordination site) offers another mode of studying the effect of complexation on the substrate. The vibrational frequencies of the additional modes due to the metal ion in the complex also give a measure of the strength of the metal–substrate interaction. These comparisons also include the protonated species.

Atomic charges and orbital populations have always been part of the language used by chemists in analysing experimental results and understanding chemical structure and reactivity. Quantitatively, in theoretical methods that make use of basis functions to represent a wave function as in the methods used here, a partitioning of the basis function space in the wave function charge distribution is used to define atomic charges and populations. The oldest and most widely used of these methods is the Mulliken population analysis[45]. Since these quantities are not observables, any definition is arbitrary. The Mulliken definition has been shown to suffer from being basis set dependent and to sometimes give unreasonable results with diffuse basis functions[46]. This latter defect can manifest itself as a negative electron population. However, for a series of compounds in a common basis set with no especially diffuse basis functions, trends in atomic charges and orbital populations are expected to be reliable and useful. We have therefore here calculated both the RHF and MP2 densities in the PV basis set, and the MP2 densities also in the PP basis set. It should be noted that the MP2 densities used here are the relaxed densities calculated as energy derivatives[47]. They therefore contain both the first-order and orbital relaxation contributions to the MP2 density.

One of the advantages of using the CEP and RCEP for the core electrons is to reduce the basis set superposition error (BSSE), which is due to the basis functions on one fragment in a complex correcting for basis set deficiencies on the other fragment. Such contributions exaggerate the complex binding energy and, if large enough, can also affect the equilibrium geometry. Core electron energy has been shown to be a large part of BSSE[48] at the RHF level. These are not absent from the calculated results presented here because of the explicit inclusion of the $n$sp subshell of electrons on the metal ions. The mutual energy improvement in the valence region between the metal ions and the ligands also has to be considered. To evaluate the possible BSSE contribution to the calculated binding energies the function counterpoise correction method[49] has been applied selectively to a number of metal ion–substrate complexes at the MP2 level. In this procedure, each single fragment energy is calculated both alone and in the combined fragments basis set and the energy differences for the two fragments are used to correct the calculated binding energy of the complex relative to asymptotes. This method has been shown[50] to give good estimates of the BSSE effect.

Another property monitored here is the hierarchy of excited electronic state energies and their electric dipole intensities in optical absorption for the metal ion complexes (M^+L). In particular, the focus is on the ligand-to-metal charge transfer (CT) excited state to formally form the ML^+ charge distribution, which is expected to be a low-lying energy excited state[51]. The method used here is the CI singles (CIS) method most recently described by Foresman and coworkers[52]. Although not very accurate or reliable in an absolute sense, CIS is simple to implement and can be useful for following trends in the CT transition as a function both of metal ion and nature of the substrate. In this type application, the CIS method can be helpful. For practical reasons only the first ten or twenty excited states each for spin triplet and singlet excited states (from the closed-shell spin singlet ground state) are usually listed in the Tables. In comparing different metal ions (M^+) and spin states for a given ligand (L) this can lead to different energy ranges at the upper end of the calculated spectrum. Intensities are measured by their calculated oscillator strength (f). The excited states of gas-phase transition metal ion complexes are currently being studied experimentally[13, 53, 54] and discussed theoretically[55].

Before entering discussions on specific ligands and complexes the difficulty of describing triple bonded systems with the MP2 level of theory should be addressed. It is well known[34, 56] that the MPn series oscillates for triply-bonded N_2, for example, in calculating property values such as bond dissociation energy, equilibrium bond distance and harmonic vibrational frequency, in going RHF $\rightarrow$ MP2 $\rightarrow$ MP3 $\rightarrow$ MP4. Accurate values of these properties require f-type functions and perhaps higher in the basis set[34, 57] and a multiconfiguration reference description[56–58]. However, the metal ion-substrate interactions surveyed here lead to what can be described as weakly bound complexes[59]. The usual molecular orbital bonding theory language[5] to describe this interaction competes with the perturbation theory description of electrostatic, exchange-overlap repulsion, polarization, charge transfer and dispersion (also called correlation)[60]. Since the ligand geometries are only mildly perturbed by coordination with the metal ion, changes in properties should be adequately described. These aspects will be discussed on a case by case basis. The relevant energies of the ligand substrate, protonated forms and bare metal ions are tabulated in Tables 2–4, respectively.

The ethylene system is discussed first. Its metal ion complexes are better known than those of the triple bond systems. Much of the detailed background material for the discussion of properties in contained in this section.

TABLE 2. Energies (in a.u.) of ligand substrates[a]

Ligand	BF[b]	RHF	MP2	QCISD	QCISD(T)
C_2H_4	68	−13.288018			
		−13.287532	−13.585348	−13.614005	−13.626186
N_2	44	−19.421992			
		−19.414904	−19.757232	−19.759985	−19.776216
HCN	50	−15.739907			
		−15.735506	−16.051236	−16.059845	−16.075415
HCCH	56	−12.073892			
		−12.072016	−12.353469	−12.369675	−12.383729
NNO	66	−34.994844			
		−34.979906	−35.568367	−35.554979	−35.586530
FCN	66	−38.990824			
		−38.984375	−39.505112	−39.510524	−39.532868
HCNO[c, e]	72	−31.331874			
		−31.324686	−31.871259	−31.870430	−31.898405
HCNO[d, f]		−31.319113	−31.872218	−31.869424	−31.898762
HNNN	72	−29.555277			
		−29.545665	−30.103650	−30.100941	−30.131228
HCCF	72	−35.326784			
		−35.323860	−35.808241	−35.822114	−35.842279
CH_2NN[c, e]	78	−25.944896			
		−25.940698	−26.451191	−26.463249	−26.489397
CH_2NN[d, f]		−25.939142	−26.451326	−26.462647	−26.489047
CH_3CN	84	−22.416665			
		−22.412224	−22.879021	−22.899387	−22.920838
FCCF	88	−58.570865			
		−58.566766	−58.252671	−59.265275	−59.291394
FCNO[c, e]	88	−54.561897			
		−54.553172	−55.297719	−55.297592	−55.331463
FCNO[d, f]		−54.544486	−55.308207	−55.304184	−55.342893
CH_3CNO[c, e]	106	−38.005370			
		−37.998072	−38.693958	−38.706298	−38.739735
CH_3CNO[d, f]		−37.994582	−38.694124	−38.705349	−38.739470

[a]The first line for each molecule is for the RHF/CEP [3^{s}P2^d/3^{s}1^P] optimized geometry and subsequent lines are for MP2/CEP [3^{s}P2^d/3^{s}1^P] optimized geometries.
[b]Number of basis functions (five d-type components).
[c]Linear or planar.
[d]Bent or non-planar.
[e]MP2 transition state.
[f]Equilibrium geometry.

TABLE 3. Energies (in a.u.) of protonated ligand substrates[a]

Ligand	BF[b]	RHF	MP2	QCISD	QCISD(T)
$C_2H_5^{+\ e,f}$	74	−13.568434			
		(176.0)			
		−13.568096	−13.852927	−13.884806	−13.895189
		(176.1)	(167.9)	(169.9)	(168.8)
$HN_2^{+\ d,e}$	50	−19.539092			
		−19.533231	−19.870650	−19.879308	−19.895928
$HNN^{+\ c,f}$		−19.612531			
		(119.6)			
		−19.604822	−19.949763	−19.955716	−19.972102
		(119.2)	(120.8)	(122.8)	(122.9)
$H_2CN^{+\ d,g}$	56	−15.920441			
		−15.919756	−16.192052	−16.226621	−16.239456
$HCNH^{+\ f,h}$		−16.024770			
		(178.8)			
		−16.021572	−16.327217	−16.337906	−16.354363
		(179.5)	(173.2)	(174.5)	(175.0)
$C_2H_3^{+\ c,d}$	50	−12.340318			
		−12.340022	−12.589303	−12.620136	−12.631150
$C_2H_3^{+\ e,f}$		−12.333584			
		(162.9)			
		−12.332291	−12.600921	−12.622530	−12.635114
		(163.3)	(155.3)	(158.7)	(157.7)
$HNNO^{+\ f,h}$	72	−35.201891			
		(129.9)			
		−35.190332	−35.788842	−35.774469	−35.808389
		(132.0)	(138.3)	(137.7)	(139.2)
$NNOH^{+\ f,i}$		−35.238733			
		(153.0)			
		−35.226865	−35.786968	−35.790847	−35.817507
		(155.0)	(137.2)	(148.0)	(144.9)
$FCNH^{+\ f,h}$	72	−39.267909			
		(173.9)			
		−39.263781	−39.773031	−39.782371	−39.802878
		(175.3)	(168.1)	(170.6)	(169.4)
$H_2CNO^{+\ f,g}$	78	−31.612384			
		(176.0)			
		−31.608774	−32.133823	−32.150478	−32.176287
		(178.3)	(164.2)	(176.4)	(174.1)
$HCNOH^{+\ f,i}$		−31.643592			
		(200.1)			
		−31.638729	−32.153704	−32.167799	−32.190569
		(200.6)	(176.6)	(187.2)	(183.1)
$HNNNH^{+\ f,j}$	78	−29.780967			
		−29.769454	−30.353694	−30.340516	−30.375937
$HNNNH^{+\ f,p}$		−29.809879			
		(159.8)			
		−29.804029	−30.354253	−30.362213	−30.392871
		(162.1)	(157.3)	(163.9)	(164.2)
$H_2NNN^{+\ d,k}$		−29.869458			
		−29.860824	−30.394704	−30.404020	−30.429446
$H_2NNN^{+\ f,l}$		−29.872482			
		(199.0)			
		−29.862465	−30.397112	−30.406388	−30.431967
		(198.8)	(184.1)	(191.7)	(188.7)
$H_2CCFH^{+\ f,m}$	78	−35.602109			
		(172.8)			

(*continued*)

TABLE 3. (*continued*)

Ligand	BF[b]	RHF	MP2	QCISD	QCISD(T)
		−35.601302 (174.1)	−36.064274 (160.7)	−36.089152 (167.6)	−36.107406 (166.4)
CH_2NNH^+ [f,h]	84	−26.245794 (188.8)			
		−26.243343 (189.9)	−26.734749 (177.9)	−26.760776 (186.7)	−26.785067 (185.5)
$CH_3N_2^+$ [f,g]		−26.320814 (235.9)			
		−26.311418 (232.6)	−26.805676 (222.4)	−26.823342 (226.0)	−26.845742 (223.6)
CH_3CNH^+ [f,h]	90	−22.727912 (195.3)			
		−22.724773 (196.1)	−23.180337 (189.1)	−23.205350 (192.0)	−23.225220 (191.0)
$HC_2F_2^+$ [d,e]	94	−58.800353			
		−58.797301	−59.475396	−59.492175	−59.517837
$HC_2F_2^+$ [c,f]		−58.833024 (164.5)			
		−58.830878 (165.7)	−59.493259 (151.0)	−59.518160 (158.7)	−59.542432 (157.5)
$FCNOH^+$ [f,n]	94	−54.745529 (115.2)			
		−54.736540 (120.5)	−55.475903 (105.2)	−55.498032 (121.6)	−55.536178 (121.3)
$FCNOH^+$ [f,g]		−54.854663 (183.7)			
		−54.848831 (191.0)	−55.573851 (166.7)	−55.589936 (179.3)	−55.622021 (175.2)
$FCNOH^+$ [f,i]		−54.879363 (199.2)			
		−54.873351 (206.4)	−55.590701 (177.3)	−55.602079 (186.9)	−55.631507 (181.1)
CH_3CNOH^+ [f,o]	112	−38.295720 (182.2)			
		−38.292316 (186.8)	−38.968785 (172.3)	−38.997455 (183.3)	−38.092527 (182.0)
CH_3CNOH^+ [f,i]		−38.342124 (211.3)			
		−38.337334 (215.1)	−39.002214 (193.3)	−39.028490 (202.8)	−39.057183 (199.4)

[a] The first line for each molecule is for the RHF/CEP $[3^s2^p2^d/3^s1^p]$ optimized geometry. The calculated proton affinity (in kcal mol^{-1}) is in parentheses. The second line is for the MP2/CEP $[3^s2^p2^d/3^s1^p]$ optimized geometry.
[b] Number of basis functions (five d-type components).
[c] Asymmetric structure.
[d] Transition state.
[e] Bridged symmetric structure.
[f] Equilibrium state.
[g] Protonated at carbon.
[h] Protonated at nitrogen.
[i] Protonated at oxygen.
[j] HNNNH in a *W* or *syn* conformation.
[k] Planar geometry.
[l] Pyramidal NH_2 group.
[m] Protonated at the C(H).
[n] Protonated at fluorine.
[o] Protonated at the middle C.
[p] *Gauche* conformation.

TABLE 4. Energies (in a.u.) of metal (M) ions (^1S state) and atoms (^2S state)

M	BF[a]	HF[b]	$<S^2>$	MP2	QCISD	QCISD(T)
Cu+ (^1S)	31	−195.048650	0.	−195.488305	−195.420387	−195.423138
	45	−195.048650	0.	−195.738093	−195.659848	−195.668303
Cu (^2S)	31	−195.290944	0.752	−195.752421	−195.682332	−195.686813
		(6.59)[c]		(7.19)	(7.13)	(7.17)
	45	−195.290944	0.752	−196.010257	−195.927784	−195.939207
		(6.59)		(7.41)	(7.29)	(7.37)
Ag+ (^1S)	31	−145.176067	0.	−145.356341	−145.352778	−145.354710
	45	−145.176067	0.	−145.710204	−145.682748	−145.692491
Ag (^2S)	31	−145.406045	0.752	−145.601939	−145.600003	−145.603139
		(6.26)		(6.68)	(6.73)	(6.76)
	45	−145.406045	0.752	−145.974337	−145.942367	−145.954576
		(6.26)		(7.19)	(7.06)	(7.13)
Au+ (^1S)	31	−134.877405	0.	−134.999461	−135.001615	−135.003899
	45	−134.877405	0.	−135.317117	−135.292833	−135.301902
Au (^2S)	31	−135.155426	0.753	−135.298239	−135.302297	−135.305830
		(7.56)		(8.13)	(8.18)	(8.22)
	45	−135.155426	0.753	−135.643535	−135.610989	−135.622559
		(7.56)		(8.88)	(8.66)	(8.73)

[a]Number of basis functions. The smaller number is for the RCEP[$4^{sp}3^d$] basis with five d-type components, and the larger is for the RCEP[$4^{sp}3^d2^f$] basis with five d-type and seven f-type components.
[b]RHF etc. for closed shell, and UHF etc. for open shell.
[c]Values in parentheses are ionization potentials in eV.

II. C_2H_4

The calculated properties of C_2H_4, protonated ethylene (ethyl cation) and its metal ion complexes are shown in Tables 2, 3 and 5–8. The MP2 calculated geometry of ethylene is shown in Table 6. This structure is well known experimentally[61]. The calculated (experimental) structure parameters are $R(C—C) = 1.337$ Å (1.339 Å), $R(C—H) = 1.082$ Å (1.086 Å) and $\angle HCC = 121.4°$ (121.2°). The agreement is very good and is better than will be found for the triply bonded substrates, as anticipated. The calculated vibrational frequencies for C_2H_4 are also tabulated in Table 7 for comparison purposes with the metal ion complexes. The protonated species, $C_2H_5^+$ (Tables 3, 6 and 7), has both classical ($CH_3CH_2^+$) and non-classical bridged stationary state structures[62]. The bridged form calculated here is a real equilibrium structure, as shown by its having all real MP2 computed vibrational frequencies. Based both on theory and experiments[63, 64] it has been concluded that the ethyl cation has a bridged structure. The asymmetric classical geometry is calculated to have a single imaginary frequency (401 cm^{-1}). This is a transition state structure on the path to interchanging hydrogen atoms in the bridged cation. The C—C bond length increases (Table 6) and harmonic stretch frequency decreases (Table 7) upon protonation, reflecting the decrease in degree of multiple bonding.

The calculated proton affinity (PA) of ethylene to form the $C_2H_5^+$ bridged structure, as a function of theoretical level, is shown in Table 3. Characteristically, the RHF value is too large because it is more accurate for the more saturated bonding system. The correlation methods show a damped mild oscillation with improved level of theory [MP2 → QCISD → QCISD(T)]. The best value obtained here of 168.8 kcal mol^{-1} is only the electronic energy contribution and to this must be added vibrational, rotational, translational (temperature) and work ($RT\Delta n$) energy differences. The largest correction term is usually for the vibrational zero-point energy (ZPE) differences and the MP2 calculated ZPE values are

32.2 and 38.8 kcalmol^{-1}, respectively, for C_2H_4 and $C_2H_5^+$ (Table 7). Substracting the MP2 ZPE difference from the QCISD(T) calculated PA in Table 3 gives 162.2 kcalmol^{-1}. The best experimental value is 162.6 kcalmol^{-1} [65,66]. The exact agreement is almost certainly serendipitous, although PA is usually a relatively easy property to calculate close to experiment using a large basis set and a reasonably high correlation level treatment. More importantly, we see here the expected converging oscillatory pattern of calculated values where the QCISD(T) electronic PA energy is somewhat larger than the experimental value. These patterns can serve as a model for ensuing discussions, where their applicability to the triply bonded systems will be examined.

The results of the $MC_2H_4^+$ metal ion complex calculations are shown in Tables 5–8. The geometry studied here was the non-classical π bridging structure with C_{2v} symmetry, as previously [18]. Side-on coordination of a metal ion opposite two hydrogen atoms, either along the $C=C$ bond axis or in an in-plane bridging position, may also give stable structures due to the higher polarizability [67] of the ligand in those directions relative to the π bridging structure. A larger polarizability gives an enhanced charge-induced dipole interaction [68,69] which, in the absence of a permanent dipole moment, is expected to be the leading electrostatic interaction term between the metal ion and ethylene. Such complexes have recently been considered for alkanes [70,71]. However, ethylene has a positive quadrupole moment [67] and the charge–quadrupole interaction favours the π bridging geometry [68].

The trends in metal ion–ethylene binding energies in Table 5 can be analysed as functions of basis set, theoretical level and metal ion. In contrast to the proton affinity calculation described above where zero-point energy difference contributed more than 6 kcalmol^{-1} to the final binding energy (BE) value, the vibrational frequencies in Table 7 show that the electronic energy part of the complex coordination energy will be reduced only by about

TABLE 5. Energies (in a.u.) of $MC_2H_4^+$ complexes[a]

M^+	BF[b]	RHF	MP2	QCISD	QCISD(T)
Cu^+	99[c, e]	– 208.383344 (29.3)			
	99[d,e]	– 208.371975 (22.5)	– 209.149404 (47.5)	– 209.105190 (44.4)	– 209.107851 (36.7)
	113[f]	– 208.372270 (22.6)	– 209.409060 (53.7)	– 209.349699 (47.6)	– 209.363311 (43.2)
Ag^+	99[c, e]	– 158.497995 (21.3)			
	99[d, e]	– 158.494092 (18.8)	– 158.988609 (29.4)	– 159.010877 (27.7)	– 159.026099 (28.4)
	113[f]	– 158.494776 (19.6)	– 158.355608 (37.7)	– 159.347709 (32.0)	– 159.372587 (33.8)
Au^+	99[c, e]	– 148.221513 (35.2)			
	99[d, e]	– 148.212952 (29.8)	– 148.673240 (55.5)	– 148.699569 (52.7)	– 148.715968 (53.8)
	113[f]	– 148.214478 (31.1)	– 149.015930 (71.2)	– 149.001392 (59.3)	– 149.027690 (62.5)

[a] Numbers in parentheses are metal ion binding energies in kcal mol^{-1}.
[b] Number of basis functions; CEP[3^{s}2^d/3^{s}1^p] basis set with five d-type components for ethylene.
[c] RHF optimized.
[d] MP2 optimized.
[e] RCEP[4^{s}3^d] basis with five d-type components for M^+.
[f] RCEP[4^{s}3^{d}2^f] basis with five d-type and seven f-type components for M^+. Complex geometry taken from 99 basis function MP2 optimization.

1.3 (Ag^+) to 2.0 (Au^+) kcal mol^{-1} for ZPE differences, which is to be applied to all the binding energy values in Table 7. The other thermodynamic correction terms mentioned above should largely cancel this small ZPE term.

A significant change in BE results in going from the single configuration RHF level to the correlated level methods of MP2 and QCISD (Table 5). The calculated metal ion dissociation energies for all three metal complexes increases substantially in going RHF $\rightarrow$ MP2, from where they show a mildly damped oscillatory behaviour in going MP2$\rightarrow$ QCISD$\rightarrow$QCISD(T) for Ag^+ and Au^+, but decreasing values for Cu^+. The f-type functions on the metal ions, which aren't involved in the metal ligand bonding and don't mix with the valence nd and (n + 1)sp atomic orbitals at the RHF level, affect the properties of the complexes only on the correlation levels. Some of the reasons for the jump in binding energy values in going RHF $\rightarrow$ MP2 can be understood on the basis of improvement of the individual metal ion and substrate properties. For example, the electron affinity of the metal ions into the (n + 1)s atomic orbital (M^+, $^1S \rightarrow M^0, ^2S$) improves from 6.59, 6.26 and 7.56 eV for Cu^+, Ag^+ and Au^+, respectively, at the SCF level to 7.19, 6.68 and 8.13 eV at the MP2 level (Table 4) using the valence metal ion RCEP[$4^{sp}3^d$] basis set[72]. The corresponding experimental numbers are 7.72, 7.57 and 9.22 eV[22, 23], respectively. Adding f-type functions to form the polarization metal ion RCEP[$4^{sp}3^d2^f$] basis set improves the MP2 calculated electron affinities even more, to 7.41, 7.19 and 8.88 eV, respectively. Analogously, in the CEP[$3^{sp}2^d/3^s1^p$] basis set the ionization potential of ethylene improves from a Hartree–Fock calculated value of 8.73 eV to a MP2 value of 10.31 eV, where the experimental number is 10.51 eV[73]. Other properties may not be as well converged. Reasonably accurate values of these quantities, which determine the energetic and spatial wave function properties of the HOMOs and LUMOs of the interacting fragment units, are expected to be important for a proper description of the bonding situation in the complexes.

The increased binding energy with improved level of basis set and theoretical treatment brings the calculated values into good agreement with the recently measured binding energy for the $AgC_2H_4^+$ complex of 33.7 kcal mol^{-1}[74]. The best calculated value here is the QCISD(T)/PP//MP2/PV 33.8 kcalmol^{-1}, which becomes 32.5 kcalmol^{-1} after correction for ZPE difference (Table 7). This very good agreement must have a measure of error cancellation since the calculations can still improve at both the basis set and configuration interaction levels. As noted above, the calculated properties of the metal ions and ethylene still haven't completely converged to their experimental values. On the other hand, BSSE is probably present here, to a certain extent, as will be shown in subsequent systems. Apparently, these two deficiencies give rise to errors of opposite sign which cancel to a large degree in these calculations for the $AgC_2H_4^+$ complex, and give a near-experimental binding energy.

Recent experiments[75,76] have also set a lower limit to the $CuC_2H_4^+$ complex metal ion binding energy and *ab initio* electronic structure calculations[77] have also been carried out to determine its value more precisely. The latter calculate a binding energy (uncorrected for ZPE differences) of 36.0 kcalmol^{-1} and estimate that correcting for deficiencies in their theoretical treatment should add 3–5 kcalmol^{-1} to that number. The calculated 43.2 kcal mol^{-1} binding energy shown in Table 5 at the QCISD(T)/PP//MP2/PV level is about 3 kcal mol^{-1} larger than that estimate[77]. Both calculated values are significantly larger than the experimental 26–28 kcalmol^{-1} lower bound to the true finding energy[75, 76]. If the calculated binding energy in Table 5 is reduced by an estimated 3 kcalmol^{-1} for BSSE, then it comes into close agreement with the adjusted value of Sodupe and collaborators[77] at a dissociation energy of about 40 kcalmol^{-1} for $CuC_2H_4^+$. This size BSSE adjustment is consistent with what will be shown is appropriate for strong Cu^+ coordination to the triply bonded substrates. Ag^+ complexes show a smaller calculated BSSE than that for Cu^+, which is consistent with the above comparison of calculated and experimental $AgC_2H_4^+$ binding energies.

TABLE 6. Mulliken populations and geometric parameters for $MC_2H_4^{+a}$

| M⁺ | Mulliken | | | | | | | Bond lengths (Å) | | | Angles (deg) | |
| | M⁺ populations | | | | charges | | | | | | | |
	s	p	d	f	M⁺	C	H	M–C	C–C	H–C	HCC	HCCM
—	—	—	—	—	—	−0.24	0.12	—	1.320	1.077	121.6	—
—	—	—	—	—	—	−0.25	0.12	—	1.337	1.082	121.4	—
H⁺ᵇ	—	—	—	—	0.25	−0.14	0.25	1.320	1.377	1.077	120.6	90.4
	—	—	—	—	0.29	−0.17	0.26	1.313	1.386	1.085	120.4	90.0
Cu⁺ᵇ	2.22	6.12	10.00	—	0.67	−0.24	0.21	2.336	1.344	1.077	121.2	93.6
	2.41	6.17	9.86	—	0.56	−0.22	0.22	2.054	1.379	1.085	120.6	95.6
	2.44	6.17	9.83	0.03	0.53	−0.21	0.22					
Ag⁺ᵇ	2.16	6.08	9.98	—	0.77	−0.29	0.20	2.596	1.341	1.077	121.3	93.6
	2.24	6.11	9.93	—	0.73	−0.30	0.22	2.396	1.365	1.085	120.9	94.5
	2.27	6.10	9.84	0.09	0.69	−0.28	0.22					
Au⁺ᵇ	2.38	6.10	9.88	—	0.64	−0.26	0.22	2.376	1.363	1.077	120.9	95.4
	2.58	6.10	9.67	—	0.65	−0.29	0.23	2.165	1.404	1.084	120.1	98.4
	2.65	6.10	9.58	0.11	0.56	−0.25	0.24					

ᵃBasis sets, energies and levels of theory for the neutral and protonated C_2H_4, and complexes are described in Tables 2, 3 and 5.
ᵇBridged structure.

The trend in going down the column, $Cu^+ \rightarrow Ag^+ \rightarrow Au^+$, at all levels of basis set and theory is for first decreased and then increased binding energy with ethylene in Table 5, with Au^+ having the largest dissociation energy. This general trend has also been found for other ligands[24,70,78] in going from the first- to the second-row transition metal ions and has been explained by the larger radius of the Ag^+ compared to Cu^+, for example[78]. For electrostatic binding, where the charge-induced dipole and charge–quadrupole interactions are dominant, the distance dependence of the interaction energy is strong[68,69]. However, Au^+ does not fit this trend, since although its radius is larger than that of Ag^+[79] and has larger Au^+–ethylene equilibrium bond distances, as is seen in Table 6, the Au^+–C_2H_4 binding energy is calculated to be largest of the three metal ion complexes.

A possible explanation for the calculated trend in the metal ion binding energies to ethylene, which will be discussed for the triply bonded substrates, could lie in a consideration of the Dewar–Chatt–Duncanson[80,81] donor–acceptor model for bridging-type metal–olefin complexes. Their proposed two-way interaction involves mixing of the olefin π electrons with a metal $(n + 1)sp$ σ hybrid atomic orbital (L $\rightarrow$ M, for short) and simultaneous 'back donation' (M $\rightarrow$ L) of metal nd electrons of appropriate symmetry into the olefin π^* molecular orbital MO. For the monocation metal ions the latter-type interaction should be less favourable due to stabilizaion of the nd electrons by the charge on the metal. L $\rightarrow$ M should be favoured for the same reason; stabilization of the $(n + 1)s$ and $(n + 1)p$ orbitals by the $+1$ charge.

These two oppositely directed but reinforcing trends can be quantified in the asymptotic (individual) fragment units by combining the values of the relevant ionization potentials and electron affinities. As has been noted above, the electron affinities (EA) of Cu^+, Ag^+ and Au^+, respectively, into the $(n + 1)s$ orbital are 7.72, 7.57 and 9.22 eV[22, 23]. Thus, the L $\rightarrow$ M electron transfer energies, noted in Table 8, as the differences between the ethylene

TABLE 7. MP2 calculated harmonic vibrational frequencies (in cm^{-1}) for $MC_2H_4^{+a}$

Exp.[b] C_2H_4	Calculated				
	C_2H_4	$CuC_2H_4^{+d}$	$AgC_2H_4^{+d}$	$AuC_2H_4^{+d}$	$C_2H_5^{+d}$
		253	224	371	748
		341	234	350	844
		514	423	667	1103
843	832	819	835	832	1144
959	924	1011	1011	1073	1156
969	974	1057	1067	1099	1264
1044	1066	1057	1081	1053	1307
1245	1239	1225	1240	1232	1358
1370	1382	1306	1342	1260	1499
1473	1491	1481	1491	1488	1585
1655	1668	1591	1616	1575	2204
3147	3177	3156	3163	3167	3171
3153	3195	3165	3173	3174	3175
3232	3267	3259	3263	3272	3284
3234	3293	3278	3285	3289	3301
31.9[c]	32.2[c]	33.6[c]	33.5[c]	34.2[c]	38.8[c]

[a] MP2 optimized geometry; CEP[$3^sP2^d/3^s1^P$] basis set for C_2H_4 and RCEP[4^sP3^d] for M^+, with five d-type components.
[b] Harmonic frequencies from Reference 27.
[c] Zero-point energy; in $kcal\,mol^{-1}$.
[d] Bridged structure.

ionization potential of 10.51 eV[73] and the respective metal ion EA, are 2.79, 2.94 and 1.29 eV. Here, the lower energy for the L → M dative bond interaction for Au^+ relative to the other metal ions correlates with its larger complex binding energy.

Ag^+ with the largest charge transfer (CT) energy also has the lowest binding energy. In addition, M → L back donation should vary as the nd ionization potential (IP) for the three metal atoms. Looking at the $nd^{10}(n+1)s^1$ (^2S) → $nd^9(n+1)s^1$ (^3D) ionization process (averaged over J multiplets[22, 23]), for example, gives 10.53, 12.61 and 11.51 eV, respectively, for Cu, Ag and Au. Here, again, Ag has the highest value, although Cu then will have the lowest M → L CT energy for a given ligand species. The unfavourable position of the $AgC_2H_4^+$ complex relative to the other two cation complexes correlates with these asymptotic unit or fragment properties.

The higher binding energy of the Au^+ complex relative to Cu^+ could then be due to a combination of L → M σ and M → L π effects in the following manner. The σ dative bond is stronger in the Au^+ complex than in the Cu^+ complex. This dative CT reduces the effective charge on Au^+ more than for Cu^+, which in turn reduces the nd atomic orbital ionization potential on Au^+ relative to Cu^+ in their respective complexes. As a result, the M → L π interaction is strengthened preferentially for Au^+ relative to Cu^+. The language used here in terms of the pure asymptotes is, of course, not exact but the same effects come out of considering MO mixing between the appropriate fragment orbitals, when the influence of partial charge on the position of the asymptotic unit orbitals in the complexes is taken into account.

The geometric structures of the $MC_2H_4^+$ complexes and Mulliken populations are shown in Table 6. Generally, the trends in these properties follow the order of the binding energies; the $AuC_2H_4^+$ complex shows the largest change from $Au^+ + C_2H_4$ and $AgC_2H_4^+$ has the smallest changes relative to its own asymptotes. In terms of bonding interactions the $(n+1)s$ metal ion atomic orbital is most occupied in Au^+ and least occupied in Ag^+ in their respective complexes. This, presumably, represents the strength of the L → M dative σ bond, but perhaps also some BSSE effects. The question of the M → L π dative bonding interaction can perhaps be addressed by noting the symmetry of the metal ion nd atomic orbitals that are most depleted in the complexes. A detailed examination of the metal orbital populations (not shown in Table 6) reveals a differential depletion (δ) from those orbitals that have the appropriate symmetry to interact with the ethylene π^* MO. This effect is largest for Au^+ ($\delta = 0.12e$), smallest for Ag^+ ($\delta = 0.03e$) and is in between for Cu^+ ($\delta = 0.06e$) at the optimized MP2/PV levels. Thus, the reduction in the nd orbital populaion divides between isotropic correlation excitation into the $(n+1)s$, $(n+1)p$ and f-type (at the MP2/PP level) orbitals and an anisotropic contribution to the metal ion–ethylene bonding. As anticipated from a consideration of the metal atom properties, the M → L π bond strength is weakest in the Ag^+ complex.

The calculated Cu—C bond distance of 2.336 Å in Table 6 at the optimized RHF/PV level translates to a metal–BMP (C—C bond midpoint) distance of 2.24 Å. This can be compared directly with the 2.25 Å distance calculated by Sodupe and coworkers[77] also at the RHF level. Correlation effects, as shown by the MP2/PV optimized structures, have a strong influence on this distance in all three metal ion complexes. The reduction in the Cu—C distance in going RHF → MP2 is relatively constant at 0.18–0.21 Å with only small increases in going down the Periodic Table. As Table 7 shows, the vibrational frequencies associated with the metal ion–C_2H_4 motions are at low energy. This gives a flat energy surface for these motions and wide amplitude displacements from equilibrium. A change of about 0.20 Å in the M^+–C distance will contribute only from 2–4 kcal mol^{-1} to the binding energy, depending on the metal ion.

The calculated equilibrium C—C distance shows the same trends with metal ion (Cu^+ → Ag^+ → Au^+) as the binding energies and M—C bond lengths, except here the C—C distance increases with correlation, as expected. As noted above, the experimental C—C bond length for ethylene is 1.338 Å[82], which compares well with the MP2/CEP-211G(2d,p) optimized value in Table 6. The displacement of the olefin hydrogen atoms from planarity (HCCM dihedral angle) also is largest for the Au[+] complex and smallest for Ag[+], reflecting the degree of perturbation of the C=C double bond. In this context, the planarity of the C_2H_4 group in the bridged ethyl cation shows that protonation is not just a perturbation of the double bond but a breaking of the olefin π bond and the creation of a new 3-center bond involving the carbon and bridging hydrogen atoms[83].

The calculated harmonic vibrational frequencies for ethylene and the metal ion complexes are shown in Table 7, together with the estimated experimental harmonic frequencies for ethylene[84]. The calculated MP2 frequencies agree well with the experimental values. Of particular interest is the harmonic C—C stretch frequency which agrees with experiment to within 13 cm[-1]. Upon complexation with the metal ions the C—C stretch mode frequency decreases, as is generally observed experimentally[85, 86]. MP2 level calculated harmonic stretch frequencies in an extended basis set have been found to be reliable[87]. The largest calculated decrease is for Au[+] (–93 cm[-1]), next is Cu[+] (– 77 cm[-1]) and the smallest shift is for Ag[+] (– 52 cm[-1]). The four C—H stretch frequencies in the 3200 cm[-1] region are also affected.

The three metal ion–ethylene vibrational frequencies are listed at the top of each column in Table 7 and form corresponding modes from the normal coordinate analyses, even though they are not in strictly increasing energy order. The first frequency represents a rocking motion of the ethylene unit relative to the metal ion–bond midpoint axis, where the metal ion–methylene group distances alternately shorten and lengthen. This motion takes the complex from a bridging to a classical structure. The second frequency is the metal ion–ethylene unit pseudo-diatomic stretch mode. These first two modes involve essentially the same force constants but different reduced masses. The third mode is approximately a rotation of the ethylene unit in place about the C—C bond axis. This would convert the observed π bridging complex to the side-on bridging complex discussed for alkanes[70,71], and is the highest calculated metal ion–ethylene vibrational frequency. The height of the rotation barrier was not investigated.

The calculated vertical transition energies to the excited electronic states of the $MC_2H_4^+$ complexes are presented in Table 8. Only the lowest ten excitations for each spin state have been considered. The notation used is to relate each excited state to a one-electron excitation. Thus, for example, a promotion of an electron from orbital a to orbital b (a → b) triplet coupled becomes the ^{3}(a, b) excited state. The excitation energy spectrum is expected to be composed of a combination of complex perturbed intra-unit (M^+ and C_2H_4) transitions, plus L → M and M → L charge transfer transitions. The excited states of ethylene have been discussed extensively and recently reviewed, both experimentally and theoretically[88, 89]. Below 8 eV are found two Rydberg states and the famous $\pi \to \pi^*$ transitions (triplet and singlet). The Rydberg excited states are ignored here, since they are expected to be shifted out of range to higher energies due to the positive charge on the complex. It should also be noted that no Rydberg-type gaussian functions were used here in the ethylene basis set. This leaves the $^3(\pi, \pi^*)$ (T) and strongly electric dipole allowed $^1(\pi, \pi^*)$ (V) states at experimental vertical excitation energies of 4.36 eV and 7.60 eV, respectively[88, 89]. The electron affinities of the metal ion (n + 1)s orbital are sufficiently large that the energy of the highest σ MO in ethylene could also be relevant to L → M CT transitions. From the experimental photoelectron spectrum it is known that the σ(C—H) MO is 2.34 eV higher in energy than the π MO[73].

The lowest transition energies of the bare (^1S) metal ions correspond to the $nd \to (n + 1)$s, excitation to the ^3D (2.81 eV, 5.03 eV and 2.29 eV) and ^1D (3.26 eV, 5.71 eV and 3.67 eV) excited states, respectively, for Cu[+], Ag[+] and Au[+] [22,23], averaged over J multiplets. The (n + 1)p orbital of the metal ions are also low lying; their calculated electron affinities ($^1S \to {}^2P$) are 3.91 eV, 3.83 eV and 4.27 eV for Cu[+], Ag[+] and Au[+], respectively, averaged

TABLE 8. CIS calculated vertical excited state energies (in eV) for $MC_2H_4^{+a}$

Assignment		$CuC_2H_4^+$			$AgC_2H_4^+$			$AuC_2H_4^+$		
		Asymptote		Complex	Asymptote		Complex	Asymptote		Complex
		Exp.	Calc.	Calc.	Exp.	Calc.	Calc.	Exp.	Calc.	Calc.
$^3(\pi,\pi^*)$	T	4.36^b	3.61	3.95	4.36^b	3.95	4.01	4.36^b	3.95	4.46
$^3[\pi,(n+1)s]$	$^3CT_1$	2.79^c	3.67	4.82	2.94^c	4.01	5.10	1.29^c	2.83	3.73
$^1[\pi,(n+1)s]$	$^1CT_1$	2.79^c	3.67	5.63	2.94^c	4.01	6.06	1.29^c	2.83	4.74
				(0.018)			(0.070)			(0.035)
$^3[nd,(n+1)s]$	^3D	2.81^d	4.12^h	5.57–7.48^h	5.03^d	5.72^h	6.56–7.58^h	2.29^d	2.83^h	4.87–7.04^h
$^1[nd,(n+1)s]$	^1D	3.26^d	4.81^h	6.14^i	5.71^d	6.52^h	7.27^i	3.67^d	3.93^h	5.51^i
				(0.003)			(0.000)			(0.000)
$^3[\pi,(n+1)p]$	$^3CT_2$	6.60^e	6.88^g	—	6.68^e	7.13^g	8.25^g	6.24^e	6.92^g	—
$^1[\pi,(n+1)p]$	$^1CT_2$	6.60^e	6.88^g	—	6.68^e	7.13^g	8.84^g	6.24^e	6.92^g	—
							(0.118)			
$[\sigma,(n+1)s]$	CT_3	5.13^f	7.13	—	5.28^f	7.67	—	5.44^f	6.48	—
$^1(\pi,\pi^*)$	V	7.60^b	8.06	7.26	7.60^b	8.06	7.58	7.60^b	8.06	6.86
		(0.300)	(0.582)	(0.288)	(0.300)	(0.582)	(0.421)	(0.300)	(0.582)	(0.236)
$^3(nd,\pi^*)$	$^3CT_4$	—	—	7.49^i	—	—	—	—	—	7.07^i
$^1(nd,\pi^*)$	$^1CT_4$	—	—	8.08^i	—	—	—	—	—	7.78^i
				(0.005)						(0.006)

^{a}MP2 optimized ground state geometries. Values in parentheses are oscillator strengths. For M = Cu, $n = 3$; M = Ag, $n = 5$; M = Au, $n = 5$.
^{b}See Reference 88.
^{c}Based on the first IP (10.51 eV) of ethylene (Reference 73) and electron affinity of M^+ into the $(n+1)s$ atomic orbital (References 22 and 23).
^{d}From References 22 and 23.
^{e}Based on the first IP (10.51 eV) of ethylene (Reference 73) and electron affinity of M^+ into the $(n+1)p$ atomic orbital (References 22 and 23).
^{f}See Reference 73 for the energy difference between the highest π and σ molecular orbitals in ethylene.
^{g}Three components.
^{h}Five components.
^{i}First of multiple components.

over J multiplets. Thus, the relevant orbitals for L → M CT are the $\sigma(C—H)$ and π MO on ethylene, and the $(n+1)s$ and $(n+1)p$ atomic orbitals on the metal ions. For possible M → L $(nd \to \pi^*)$ CT in the separated units, an ionization potential $(M^{1+} \to M^{2+})$ from the nd atomic orbital $(^1S \to {}^2D)$ is just too high in energy to register on this energy scale[22, 23]. However, in the complex, charge flow compensation can shift the M → L CT transition to lower energy.

The results, tabulated in Table 8, compare calculated with experimental in the individual M^+ and C_2H_4 units, for the asymptotic charge transfer energies, and with the calculated vertical excitation energies for the equilibrium geometry complexes. (For each complex, in Table 8 and in all other Tables showing CIS energies, the values under the heading 'Asymptote' show the experimental and calculated values for the dissociated complex with no bond between the metal ion and the ligand.) The comparison for the asymptotic units gives a measure of the accuracy of the calculated values. The M → L CT states are degenerate for the spin singlet and triplet states of a given one-electron transfer and like the L → M CT states have no intensity at the zero overlap asymptotic limits. The bare metal ion $^1S \to {}^1D$ transitions in the spherical metal ion symmetry are electric dipole forbidden. In the C_{2v} symmetry of the complexes only excitations to A_2 excited states are electric dipole forbidden, and ^1D

and CT transitions will have allowed components. All spin triplet states carry no intensity in the absence of spin–orbit coupling, which is not considered here.

Of particular interest is the location and nature of the CT transitions in the complexes. As anticipated, the first such inter-unit transition, $^1CT_1$, is $\pi \to (n+1)$s. Focusing on the singlet states, the calculated vertical excitation energies are at 5.63 eV, 6.06 eV and 4.74 eV, respectively, in the Cu^+, Ag^+ and Au^+ complexes. These are the lowest energy transitions in the electronic spectrum but are calculated to be of low intensity in the complexes. If these energy values are corrected by the difference between calculated and experimental in the asymptotic region, then the vertical $^1CT_1$ energies in the complexes are estimated to be 4.75 eV, 4.99 eV and 3.20 eV, respectively, for Cu^+, Ag^+ and Au^+.

This latter value for Au^+ is the only one for the three metal ion complexes that is smaller than the combined binding energy calculated for the complex (Table 5) plus experimental asymptotic $^1CT_1$ energy, which together give the energy of the (CT) excited state M + $C_2H_4^+$ asymptote relative to the ground state equilibrium complex. This relationship has recently been discussed for the metal ion–benzene complexes[54, 55], where the CT state was found to be dissociative for Cu^+ and Ag^+. The $^1CT_1$ energy curve for the $MC_2H_4^+$ complexes was not probed here but, from the estimated vertical excitation energies relative to the asymptotic M + $C_2H_4^+$ energies, the calculation above shows that even if the $^1CT_1$ state is dissociative for Cu and Ag it may be found for Au. The neutral MC_2H_4 (M = Cu, Ag and Au) complexes, which correspond to the CT_1 state without the ion-induced dipole stabilization of the positive charge on C_2H_4, have been observed in low-temperature matrix isolation[90].

Other states of interest in the complexes are the ^1D and V states. The former are predicted to split into five components, shift to higher energy and acquire a small amount of intenstiy from configuration mixing with other (allowed) transitions in the lower symmetry of the complexes. The strongly allowed ethylene V state shifts to lower energy upon complexation and is somewhat reduced in intensity by MO and configuration mixing. Completing the picture for excited state energies below 8 eV are the $nd \to \pi^*$, M $\to$ L CT transitions in Cu^+ and Au^+, shifted to lower energy in the complexes relative to asymptotic $M^+ + C_2H_4$ by the reduced charge on the metals and partial positive charge on ethylene. These excitations appear in the Cu^+ and Au^+ complexes below 8 eV but not in the Ag^+ complex, which is further evidence of the greater importance of $nd \to \pi^*$ dative bonding in the Cu^+ and Au^+ complexes relative to the Ag^+ complex. Table 8 lists only those calculated excitations below ca 8 eV.

III. N₂

The calculated results for N_2, protonated N_2 and the metal ion complexes of N_2 are presented in Tables 2, 3 and 9–12. The experimental N–N equilibrium distance of 1.098 Å[91] is between the RHF and MP2 optimized values for bare N_2. Very accurate determination of the ground state properties of triply bonded N_2 has been found to require larger basis sets, including f-type functions[34] and extensive configuration interaction[27, 92]. The error in the bond length at the MP2 level (Table 10) is 0.02 Å. The calculated harmonic stretch frequency is 2,153 cm^{-1} (Table 11), compared to the derived experimental value of 2,360 cm^{-1} [27,91], a difference of 207 cm^{-1}. These are the reference quantities for subsequent discussions of the N≡N triple bond.

One test of the ability of these methods to accurately follow changes in the properties is to look at the proton affinity. The calculated value (Table 3) for the linear HNN^+ structure at the QCISD(T)/PP//MP2/PV level is 122.9 kcalmol^{-1}. The ZPE difference between N_2 (3.1 kcalmol^{-1}) and linear HN_2^+ (9.8 kcalmol^{-1}) from the MP2 frequencies is 6.7 kcal mol^{-1} (Table 11). This gives an adjusted QCISD(T) result of 116.2 kcalmol^{-1}, to be compared with the experimental value of 118.2 kcalmol^{-1} [65, 66, 93]. The direct MP2 calculated PA, corrected for ZPE differences, is only 114.1 kcalmol^{-1}. The best calculated value here is still smaller than experiment. This could be due to the calculated N≡N bond distance in the linear HNN$^+$ geometry actually being somewhat smaller than in free N$_2$, indicating a stronger triple bond.

The geometric structure of protonated N$_2$ has recently been reviewed[94, 95]. The calculated energy results in Tables 3 show that the linear HNN$^+$ form is the equilibrium structure. The symmetric bridging structure is a transition state, characterized by one MP2 calculated imaginary vibrational frequency (704 cm^{-1}). The electronic barrier to rearrangement is QCISD(T)//MP2 calculated to be 47.8 kcalmol^{-1}. The experimentally determined equilibrium structure of HNN^{+} [96] shows an N—N distance of 1.093 Å and a H—N bond length of 1.034 Å. The corresponding MP2 calculated quantities are 1.112 Å and 1.039 Å (Table 10), respectively. The 0.02 Å difference between calculation and experiment for the N—N distance is maintained in HNN$^+$ relative to N$_2$. The comparison of harmonic vibrational frequencies in Table 11 for HNN$^+$ shows that the calculated N—N stretch frequency is higher than the experimental value, while the calculated H—N frequency is lower than experiment. Both differences are about 150 cm^{-1} and are difficult to interpret, since the two motions are somewhat coupled in the normal coordinate analysis. The exact coincidence of the 687 cm^{-1} doubly-degenerate bending mode in Table 11 between calculated and experiment is not a typographical error.

The metal ion–N$_2$ complexes show the same preferred linear structures as the protonated species. The linear structure of these weakly bound complexes is favoured both by the larger polarizability of N$_2$ along the bond axis and by its negative quadrupole moment[97]. This has recently been discussed for the CoN$_2^+$ complex[98]. For CuN$_2^+$ the enthalpic binding energy has been measured at about 6 kcal mol^{-1} in the gas phase[99]. This low value seems to be inconsistent with the experimental dissociation energy of CrN$_2^+$ at 14.1 kcal mol^{-1} [100], since CuL$^+$ binding energies are usually larger than those of CrL$^+$ for ligand (L) types that bind to metal ions similarly to N$_2$. Combining experimental[100,101] and theoretical[98,102,103] binding energies for ML$^+$ with M = Na, Cr and Cu and L = N$_2$, CO and Ar gives an internally consistent dissociation energy estimate for linear CuN$_2^+$ of close to 22 kcal mol^{-1}.

The calculated binding energies are shown in Table 9. For linear CuN$_2^+$ the QCISD(T)/PP//MP2/PV level result is 22.9 kcalmol^{-1}. The BSSE correction[49] for the MN$_2^+$ systems has been computed at the MP2/PV level and amounts to 7.0 kcalmol^{-1} in the linear geometry. Reducing the QCISD(T)/PP binding energy for this effect leaves a net dissociation energy of 15.9 kcalmol^{-1}. For linear AgN$_2^+$, adjusting the 14.7 kcalmol^{-1} QCISD(T)/PP BE by the 2.0 kcalmol^{-1} MP2/PV BSSE effect (Table 9) gives 12.7 kcalmol^{-1}. For linear AuN$_2^+$ the same exercise (2.3 kcalmol^{-1} BSSE) results in a net calculated 21.6 kcalmol^{-1} dissociation energy.

There are several points to be discussed here with regard to these numerical exercises. Firstly, the steady decrease of the QCISD(T) energies in Table 9 for linear CuN$_2^+$ relative to QCISD and MP2 is similar to that found for CuC$_2$H$_4^+$ (Table 5). This result must be due to a slow convergence of the perturbation series for Cu$^+$ with increased level of excitation to the extent carried here. Interestingly enough, the MP2 → QCISD difference isn't large. The bridging Cu$^+$ complex shows the same decrease in binding energy with correlation level. The geometric and electronic structures of the linear MN$_2^+$ series don't seem to show any obvious break in properties between M = Cu$^+$ and M = Ag$^+$ or Au$^+$, where the energy convergencies [MP2 → QCISD → QCISD(T)] are reasonable for both the linear and bridging structures. In any event, the alternating nature expected for such convergencies means that the next energy contribution term for linear or bridged CuN$_2^+$ will probably increase their calculated binding energies.

All the metal ion complex binding energies calculated here (after BSSE correction) will probably increase with a larger basis set and higher level of correlation treatment. In addition, core-valence correlation will also increase the well depth of the equilibrium

TABLE 9. Energies (in a.u.) of MN_2^+ complexes[a]

M^+	BF^b	RHF	MP2	QCISD	QCISD(T)
Cu^{+g}	$75^{c,e}$	−214.488375			
		(11.1)			
	$75^{d,e}$	−214.472264	−215.288891	−215.222102	−215.228923
		(5.5)	(27.2,20.2)	(26.2)	(18.6)
	89^f	−214.472500	−215.545195	−215.465751	−215.481062
		(5.6)	(31.3)	(28.8)	(22.9)
Cu^{+h}	$75^{c,e}$	−214.474983			
		(2.7)			
	$75^{d,e}$	−214.461555	−215.258741	−215.193095	−215.207975
		(−1.3)	(8.3,5.2)	(8.0)	(5.4)
	89^f	−214.461647	−215.512203	−215.434814	−215.456539
		(−1.2)	(10.6)	(9.4)	(7.5)
Ag^{+g}	$75^{c,e}$	−164.608319			
		(6.4)			
	$75^{d,e}$	−164.598668	−165.132462	−165.131159	−165.150161
		(4.8)	(11.9,9.9)	(11.5)	(12.1)
	89^f	−164.599033	−165.491883	−165.464503	−165.492064
		(5.1)	(15.3)	(13.7)	(14.7)
Ag^{+h}	$75^{c,e}$	−164.599868			
		(1.1)			
	$75^{d,e}$	−164.591504	−165.119186	−165.118195	−165.136992
		(0.3)	(3.5,2.0)	(3.4)	(3.8)
	89^f	−164.591595	−165.476226	−165.449879	−165.476923
		(0.4)	(5.5)	(4.5)	(5.2)
Au^{+g}	$75^{c,e}$	−154.311426			
		(7.5)			
	$75^{d,e}$	−154.297775	−154.787198	−154.791018	−154.811248
		(3.4)	(19.1,16.8)	(18.5)	(19.5)
	89^f	−154.299000	−155.115451	−155.087965	−155.116148
		(4.2)	(25.8)	(22.1)	(23.9)
Au^{+h}	$75^{c,e}$	−154.301358			
		(5.7)			
	$75^{d,e}$	−154.290304	−154.766224	−154.770473	−154.790281
		(−1.3)	(6.0,3.7)	(5.6)	(6.4)
	89^f	−154.290857	−155.090998	−155.065493	−154.092915
		(−0.9)	(10.4)	(8.0)	(9.3)

[a]Numbers in parentheses are metal ion binding energies in kcalmol^{-1}; a second number in the parentheses is after BSSE correction.
[b]Number of basis functions; CEP [3^{sp}2^d/3^{s}1^p] basis set with five d-type components for N_2.
[c]RHF optimized.
[d]MP2 optimized.
[e]RCEP [4^{sp}3^d] basis with five d-type components for M^+.
[f]RCEP [4^{sp}3^{d}2^f] basis with five d-type and seven f-type components for M^+. Complex geometry taken from 75 basis function MP2 optimization.
[g]MNN$^+$ linear geometry.
[h]Perpendicular (bridged) geometry.

complex. This has been shown[104] using a core polarization potential[105]. Therefore, the BSSE correction will be offset by these missing terms and the QCISD(T)/PP binding energies presented here are probably close to the experimental values, as was found for $AgC_2H_4^+$ above and estimated for linear CuN_2^+. Dependence on cancellation of errors is necessary for those systems for which the more exacting calculation is difficult and to test levels of approximation for larger systems.

The bridged MN_2^+ complexes are also transition states, like the protonated species, showing one imaginary vibrational frequency (Table 11). They are only very weakly bound and the electronic barrier to metal ion transfer between nitrogen atoms is calculated to be between 8–15 kcalmol^{-1}, depending on the metal ion. The bridged complexes are probably not bound at the RHF level, except perhaps for Au$^+$, and almost all the binding comes at the correlation level description. This result is in contrast to the bridging $MC_2H_4^+$ complexes where there is substantial binding even at the RHF level (Table 5). The difference can probably be attributed mainly to the much higher ionization potential of N_2 (15.58 eV from the $3\sigma_g$ MO[91]) which results in much less L $\rightarrow$ M dative bonding, and which,

TABLE 10. Mulliken populations and geometric parameters for MN_2^{+a}

M+	Mulliken									Angle (deg) MNN
	M$^+$populations				charges			Bond lengths (Å)		
	s	p	d	f	M$^+$	N	N	M—N	N—N	
—	—	—	—	—	—	0	0	—	1.071	—
—	—	—	—	—	—	0	0	—	1.118	—
H^{+b}	—	—	—	—	0.47	0.28	0.25	1.025	1.064	180
	—	—	—	—	0.47	0.28	0.25	1.039	1.112	180
H^{+c}	—	—	—	—	0.52	0.24	0.24	1.277	1.095	64.6
	—	—	—	—	0.52	0.24	0.24	1.288	1.142	63.7
Cu^{+b}	2.05	6.05	10.00	—	0.91	0.01	0.08	2.182	1.069	180
	2.20	6.12	9.89	—	0.78	0.07	0.15	1.867	1.116	180
	2.21	6.13	9.87	0.03	0.77	0.08	0.15			
Cu^{+c}	2.03	6.04	10.01	—	0.92	0.04	0.04	2.642	1.074	78.3
	2.11	6.10	9.97	—	0.82	0.09	0.09	2.181	1.128	75.4
	2.12	6.10	9.95	0.03	0.81	0.09	0.09			
Ag^{+b}	2.02	6.02	10.00	—	0.96	0	0.04	2.577	1.069	180
	2.07	6.05	9.97	—	0.91	−0.01	0.09	2.316	1.117	180
	2.08	6.05	9.89	0.09	0.90	0	0.10			
Ag^{+c}	2.02	6.02	10.00	—	0.96	0.02	0.02	2.863	1.072	79.9
	2.04	6.05	9.99	—	0.92	0.04	0.04	2.685	1.122	78.2
	2.04	6.04	9.91	0.09	0.92	0.04	0.04			
Au^{+b}	2.06	6.03	9.97	—	0.95	0	0.05	2.434	1.069	180
	2.20	6.06	9.84	—	0.90	−0.02	0.12	2.099	1.117	180
	2.23	6.07	9.74	0.10	0.85	0.03	0.12			
Au^{+c}	2.02	6.02	10.00	—	0.95	0.02	0.02	2.790	1.073	79.8
	2.09	6.07	9.96	—	0.89	0.06	0.06	2.526	1.126	77.3
	2.10	6.05	9.87	0.10	0.87	0.07	0.07			

[a]Basis sets, energies and levels of theory for the neutral and protonated N_2, and complexes are described in Tables 2, 3 and 9.
[b]Linear geometry.
[c]Perpendicular (bridged) geometry.

consequently, also discourages M → L back delocalization. The result is a system that depends crucially on an accurate description of the constituent units which can be obtained only at the correlation level due to the inherent defects of the RHF method, for both the metal ions and N_2.

The calculated equilibrium geometries and Mulliken populations are shown in Table 10. The M—N distance is substantially reduced at the MP2 level relative to RHF. Going RHF→ MP2 enhances the L → M dative bond interaction, as indicated by the reduced charge on the metals. BSSE may contribute to give a too short M—N distance, but the other contributions mentioned above will tend to decrease the M—L bond distance which here is mimicked by the BSSE effect. The optimized N—N distances in the linear complexes are very similar to those in the bare N_2 unit and in the protonated species. For the bridging metal ion complexes the N—N bond lengths increase by at most 0.01 Å relative to free N_2. Thus, the triple bond character of the N_2 bond is essentially maintained in the metal ion complexes. This is consistent with the linear coordination of the complexes, where the major dative bonding MO on N_2 is a σ non-bonding lone pair of electrons. The cylindrical π electrons can also donate to the metal $(n + 1)p$ π system and the metal nd can delocalize into the N_2 π^* system, but a close examination of the Mulliken populations shows that the σ interaction is dominant, as expected. For the weakly bound bridging π complexes the triple bond is barely perturbed. The calculated harmonic vibrational frequencies for the metal ion complexes shown in Table 11 are consistent with the above discussion.

The CIS calculated vertical excited state energies for the linear metal ion–N_2 complexes are shown in Table 12, and compared with experiment in the asymptote, where possible. Only the first ten excited states of each spin state are listed. The spectrum is composed of complex perturbed N_2 and metal ion intra-unit excitations, and L → M CT transitions. None of the intra-unit (asymptote) transitions listed in Table 12 are electric dipole allowed. The first vertical ionization potential of N_2 is at 15.58 eV[91] from the $3\sigma_g$ MO (called σ in Table 12) and the second vertical ionization energy is at 16.95 eV from the doubly degenerate $1\pi_u$ MO[106, 107] (labelled π in Table 12). The reverse ordering of MO ionization potentials (π below σ) is obtained at the RHF level, both within the Koopmans' theorem[108] (frozen orbital) approximation and with direct variational calculation of each ion state separately[109]. The single configuration electronic structure description of N_2 incorrectly predicts that the first ionization energy is from the $1\pi_u$ MO. This also affects the CIS calculated L → M charge transfer energies. The LUMO in N_2 is the doubly degenerate $1\pi_g$ (called π^* in Table 12), and this is the terminating MO for the internal N_2 and L → M CT transitions.

TABLE 11. MP2 calculated harmonic vibrational frequencies (in cm^{-1}) for MN_2^{+a}

N_2	CuN_2^+		AgN_2^+		AuN_2^+		Linear HNN^+	
	Linear	Bridged	Linear	Bridged	Linear	Bridged	Calc.	Exp.[e]
	303^b	191^c	184^b	146^c	223^b	163^c	687^b	687^b
	369	180	182	107	252	138	2104	2258
2153	2176	2075	2166	2124	2162	2091	3381	3234
3.1^d	4.5^d	—	3.9^d	—	4.1^d	—	9.8^d	9.8^d

[a]MP2 optimized geometry; CEP[$3^{sp}2^d/3^s1^p$] basis set for N_2 and RCEP[$4^{sp}3^d$] for M^+, with five d-type components.
[b]Doubly degenerate.
[c]Imaginary frequency.
[d]Zero-point energy, in kcal mol^{-1}.
[e]Fundamental vibrational frequencies from ref. 96.

TABLE 12. CIS calculated vertical excited state energies (in eV) for linear MNN^{+a}

Assignment		CuNN⁺			AgNN⁺			AuNN⁺		
		Asymptote		Complex	Asymptote		Complex	Asymptote		Complex
		Exp.	Calc.	Calc.	Exp.	Calc.	Calc.	Exp.	Calc.	Calc.
$^3[nd,(n+1)s]$	^3D	2.81^d	4.12^h	$4.55-5.16^h$	5.03	5.72^h	$5.83-6.23^h$	2.29	2.83^h	$3.11-3.40^h$
$^1[nd,(n+1)s]$	^1D	3.26^d	4.81^h	$5.46-5.86^{h,k}$	5.71	6.52^h	$6.68-6.97^{h,k}$	3.67	3.93^h	$4.46-5.01^{h,k}$
$^3(\pi,\pi^*)$	$^3\Sigma_u^+$	6.22^b	5.83	5.67	6.22^b	5.83	5.70	6.22^b	5.83	5.67
$^3(\sigma,\pi^*)$	$^3\Pi_g$	7.39^b	7.78^i	7.64^i	7.39^b	7.78^i	7.77^i	7.39^b	7.78^i	7.57^i
$^3(\pi,\pi^*)$	$^3\Delta_u$	7.41^b	6.93^i	6.75^i	7.41^b	6.93^i	6.84^i	7.41^b	6.93^i	6.77^i
$^3[\sigma,(n+1)s]$	$^3CT_1$	7.86^c	—	—	8.01^c	11.00	—	6.36^c	—	—
$^1[\sigma,(n+1)s]$	$^1CT_1$	7.86^c	—	—	8.01^c	11.00	—	6.36^c	9.81	—
$^3(\pi,\pi^*)$	$^3\Sigma_u^-$	8.22^b	8.11	7.48	8.22^b	8.11	8.01	8.22b	8.11	7.90
$^3[nd,(n+1)p]$	^3P	8.33^d	9.67^f	—	10.16^d	11.01^f	10.82^j	—	9.12^f	8.43^j
$^1(\pi,\pi^*)$	$^1\Sigma_u^-$	8.45^b	8.17	7.87	8.45^b	8.11	8.01	8.45^b	8.11	7.90
$^1(\sigma,\pi^*)$	$^1\Pi_g$	8.59^b	9.81^i	—	8.59^b	9.81^i	—	8.59^b	9.81^i	—
$^1[nd,(n+1)p]$	^1F	8.92^d	10.30^g	9.73^j (0.000)	11.05^d	11.32^g	9.76^j (0.000)	—	9.94^g	8.78^j (0.041)
$^1(\pi,\pi^*)$	$^1\Delta_u$	8.94^b	8.74^i	8.38^i	8.94^b	8.69^i	8.58^i	8.94^b	8.74^i	8.39^i
$^3[\pi,(n+1)s]$	$^3CT_2$	9.23^e	9.94^i	—	8.38^e	10.29^i	—	7.72^e	9.10^i	—
$^1[\pi,(n+1)s]$	$^1CT_2$	9.23^e	9.94^i	—	8.38^e	10.29^i	—	7.72^e	9.10^i	—

^{a}MP2 optimized ground state geometries. For M = Cu, n = 3; M = Ag, n = 4; M = Au, n = 5.
^{b}See Reference 91.
^{c}Based on the first IP (15.58 eV) of N_2 (Reference 91) and EA of M^+ into the $(n + 1)$s atomic orbital (References 22 and 23).
^{d}From References 22 and 23.
^{e}See Reference 106 for the energy difference between the highest σ and π molecular orbitals in N_2.
^{f}Three components.
^{g}Seven components.
^{h}Five components.
^{i}Doubly degenerate.
^{j}First of multiple components.
^{k}f is no larger than 0.003 for any component.

Because of the high energy first IP of N_2, the lowest energy electronic transitions in its metal ion complexes are to the metal ion ^3D and ^1D states. These are predicted to be appropriately split in the linear MNN^+ symmetry and mildly shifted to higher energy in the complexes relative to the free ions. Several triplet $\pi \to \pi^*$ and $\sigma \to \pi^*$ excited states of N_2 are predicted to come next. The triplet $\sigma,\pi \to \pi^*$ and lower energy $^1(\pi,\pi^*)$ N_2 transitions are all calculated to within 0.5 eV of experiment in the bare molecule and are predicted to shift by less than 0.5 eV in the complexes. However, the $\sigma \to (n + 1)$s CT transitions are calculated at more than 3 eV too high for the asymptotic units and are correspondingly shifted in the complexes. If we borrow from the experience with $MC_2H_4^+$, where the L $\to$ M charge transfer states were calculated about 1 eV too high in the asymptotic regime and predicted to be blue shifted in the complexes (Table 8), the $\sigma \to (n + 1)$s CT states in Table 12 for MNN^+ are not expected below 9 eV. This would still place them energetically above the $nd \to (n + 1)$s, $\pi \to \pi^*$ and $\sigma \to \pi^*$ states discussed above. In that same region above 9 eV several more excited state transitions are predicted, including $\pi \to \pi^*(^1\Sigma_u^+)$, $\sigma \to \pi^*$ ($^1\Pi_g$), $nd \to (n + 1)$p and $\pi \to (n + 1)$s. Although excitations arising from the N_2 σ MO are calculated too high in energy relative to experiment, the attachment of the metal ions in a linear fashion through the σ framework of N_2 by itself does not seem to shift these type transitions to higher energy, as evidenced by the $^3\Pi_g$ state. Finally, the sum of the asymptotic CT energy (7.9 eV) plus the largest metal ion binding energy (ca 1 eV) will be smaller than the estimated CT energy in the complexes (>9 eV) so that the latter are expected to be dissociative.

IV. HCN

The data on the HCN, protonated HCN and metal ion–HCN complexes are contained in Tables 2, 3 and 13–16. A recent analysis[110] of the experimental results on the geometric structure of gaseous HCN gives an equilibrium bond distance of 1.153 Å for C—N and 1.065 Å for C—H. Again, as in the case of N_2, the calculated lengths at the RHF and MP2 optimized levels bracket these values, with the MP2 error being 0.02 Å for the triple bond. The measured dipole moment of HCN is $2.940D$[111], compared to the MP2/CEP[$3^{sp}2^d/3^s1^p$] calculated value of $2.72D$. The calculated harmonic vibrational frequencies in Table 15 for HCN are within 28 cm^{-1} of the derived experimental harmonic modes[27] for all three motions; C—H and C—N stretches, and the bending mode.

The structure of protonated hydrocyanic acid (HCN) has recently been reviewed[94, 112]. As shown in Table 3, two protonated structures are obtained: one from attachment at the

TABLE 13. Energies (in a.u.) of HCNM$^+$ complexes[a]

M$^+$	BF[b]	RHF	MP2	QCISD	QCISD(T)
Cu$^+$	81[c, e]	−210.847834			
		(37.2)			
	81[d, e]	−210.837815	−211.624940	−211.565127	−211.560165
		(33.7)	(53.6,48.4)	(43.3)	(38.7)
	95[f]	−210.838285	−211.882507	−211.809401	−211.815649
		(34.0)	(58.5)	(56.3)	(45.1)
Ag$^+$	81[c, e]	−160.959833			
		(27.5)			
	81[d, e]	−160.954459	−161.460840	−161.465456	−161.483216
		(26.9)	(33.4,32.2)	(33.2)	(33.3)
	95[f]	−160.955362	−161.822832	−161.800387	−161.827008
		(27.5)	(38.5)	(36.3)	(37.1)
Au$^+$	81[c, e]	−150.669691			
		(32.9)			
	81[d, e]	−150.662227	−150.124905	−151.134776	−151.153455
		(30.9)	(46.6,45.1)	(46.0)	(46.5)
	95[f]	−150.664308	−151.456360	−151.433401	−151.460230
		(32.3)	(55.2)	(50.7)	(52.0)

[a]Numbers in parentheses are metal ion binding energies in kcal mol^{-1}; a second number in the parentheses is after BSSE correction.
[b]Number of basis functions; CEP[$3^{sp}2^d/3^s1^p$] basis set with five d-type components for HCN.
[c]RHF optimized.
[d]MP2 optimized.
[e]RCEP[$4^{sp}3^d$] with five d-type components for M$^+$.
[f]RCEP[$4^{sp}3^d2^f$] with five d-type and seven f-type components for M$^+$. Complex geometry taken from 81 basis function MP2 optimization.

nitrogen atom ($HCNH^+$) and one at the carbon atom (H_2CN^+). A frequency calculation of the vibrational energy modes at the MP2 level shows one imaginary frequency (604 cm^{-1}) for the H_2CN^+ isomer with a normal coordinate mode motion that will swing one hydrogen atom towards the nitrogen and the other hydrogen atom towards alignment with the C—N bond. The H_2CN^+ form is then a transition state for the interchange of hydrogen atoms in the $HCNH^+$ structure. From the QCISD(T) calculated proton affinities in Table 3, the electronic energy difference between the two isomers is calculated to be approximately 72 kcal mol^{-1}. The equilibrium C—N bond length in H_2CN^+ (Table 14) is somewhat lengthened relative to HCN, as expected from the large increase in charge on the nitrogen atom.

$HCNH^+$ is predicted to be linear, as is also found experimentally[113]. Its geometric structure and vibrational frequencies were essentially predicted theoretically and helped experimentalists identify the species and derive its structural and spectroscopic properties[94,112]. The measured C—N bond distance is 1.1345 Å[113] compared to the MP2 calculated 1.146 Å (Table 14). Analogously, the experimental (theoretical) values for C—H and N—H bond lengths are 1.078 Å (1.080 Å) and 1.010 Å (1.018 Å), respectively. Agreement is respectably good. The vibrational frequencies for $HCNH^+$ are also shown in Table 15, comparing theory with experiment. The (degenerate) bending modes and C—N stretch (at 2100–2200 cm^{-1}) are close but both the calculated N—H and C—H stretch motions are somewhat far from experiment. The comparison here is with the real (anharmonic) experimental values[113] and similar differences have been found by others[114, 115]

TABLE 14. Mulliken populations and geometric parameters for HCNM^{+a}

M$^+$		Mulliken												
	M$^+$ populations				charges				Bond lengths (Å)			Angles (deg)	
	s	p	d	f	M$^+$	H	C	N	M—N	C—H	C—N	MNC	HCN
—	—	—	—	—	—	0.23	0.02	−0.24	—	1.058	1.128	—	180
—	—	—	—	—	—	0.22	−0.07	−0.14	—	1.067	1.170	—	180
H^{+b}	—	—	—	—	0.43	0.36	0.41	−0.21	1.003	1.073	1.113	180	180
	—	—	—	—	0.36	0.44	0.30	−0.10	1.018	1.080	1.146	180	180
H^{+c}	—	—	—	—	0.34	0.34	0.11	0.21	1.098	1.098	1.205	117.7	117.7
	—	—	—	—	0.37	0.37	0.03	0.22	1.119	1.119	1.198	118.8	118.8
Cu$^+$	2.11	6.08	9.98	—	0.82	0.31	0.20	−0.33	2.028	1.065	1.124	180	180
	2.27	6.15	9.88	—	0.70	0.31	0.13	−0.14	1.827	1.073	1.159	180	180
	2.29	6.15	9.85	0.03	0.68	0.31	0.13	−0.12					
Ag$^+$	2.07	6.05	9.98	—	0.90	0.30	0.17	−0.36	2.334	1.064	1.125	180	180
	2.12	6.08	9.95	—	0.85	0.30	0.08	−0.23	2.194	1.073	1.162	180	180
	2.14	6.07	9.87	0.09	0.83	0.30	0.09	−0.22					
Au$^+$	2.17	6.06	9.91	—	0.86	0.31	0.20	−0.37	2.191	1.065	1.123	180	180
	2.32	6.08	9.79	—	0.81	0.31	0.14	−0.26	2.024	1.074	1.158	180	180
	2.36	6.09	9.69	0.11	0.75	0.32	0.14	−0.21					

^{a}Basis sets, energies and levels of theory for the neutral and protonated HCN, and complexes are described in Tables 2, 3 and 13.
^{b}HNCH$^+$.
^{c}H$_2$CN$^+$.

for $HCNH^+$. Harmonic stretch frequencies are usually somewhat larger than anharmonic and this adjustment should cover part of the gap between calculated and experimental for the hydrogen stretch modes in Table 15. In HCN, the difference between calculated harmonic and measured fundamental frequencies is 131 cm^{-1} for C—H but only 32 cm^{-1} for C—N[27].

The calculated PA for $HCNH^+$ is [QCISD(T)//MP2 in Table 3] 175.0 kcal mol^{-1}. Subtracting the ZPE difference (Table 15) of 7.1 kcal mol^{-1} leaves 167.9 kcal mol^{-1}, relative to the experimental value of 168–171 kcal mol^{-1} [65, 66]. This is good agreement, and, again, the calculated value is smaller than experiment. In geometry, the C—H bond length increases and the C—N distance decreases (in both calculated and experiment) in going from HCN to $HCNH^+$. A similar decrease is observed for the N—N bond length in protonated N_2. Part of the decrease is possibly due to the delocalization of the positive charge which has a general contracting effect on the atomic orbitals, although the C—H bond length doesn't decrease. However, protonation doesn't seem to adversely affect the C—N triple bond properties, as also evidenced by the preferred linear geometric structure of $HCNH^+$.

The metal ion–HCN coordinated complexes, attached at the nitrogen atom ($HCNM^+$), are also calculated to be linear (Table 14). Attempts to form the carbon coordinated M^+ complexes led to spontaneous rearrangement to the linear nitrogen bound complex. The MP2 calculated equilibrium C—N bond lengths in the complexes are also found to be close to the bare HCN value and slightly smaller (Table 14). This is similar to the result found for the linear MNN^+ complexes (Table 10). Apparently, coordination through the σ framework to ostensibly lone pair electrons hardly affects the bond, as expected, even though (M $\rightarrow$ L) backbonding into the ligand π^* orbitals is still possible in the linear geometry. This, of course, is true for the metal ions but not for the proton, yet the geometry changes show the same trends. Also, as in protonation, the C—H bond is lengthened upon complex formation.

The calculated coordination binding energies for the metal ions to HCN, shown in Table 13, are larger than for N_2, as expected, due to the permanent dipole moment of HCN and its larger polarizability[97]. BSSE calculated at the MP2/PV level is 5.2, 1.2 and 1.5 kcal

TABLE 15. MP2 calculated harmonic vibrational frequencies (in cm^{-1}) for $HCNM^{+a}$

Exp[b] HCN	Calculated					Exp[e] HCNH$^+$
	HCN	HCNCu$^+$	HCNAg$^+$	HCNAu$^+$	HCNH$^+$	
		256[d]	178[d]	227[d]	631[d]	646[d]
		431	263	341	818[d]	802[d]
727[d]	752[d]	813[d]	797[d]	791[d]	2119	2156
2129	2003	2114	2085	2120	3340	3188
3442	3470	3416	3418	3409	3626	3483
10.0[c]	10.0[c]	11.6[c]	11.0[c]	11.3[c]	17.1[c]	16.8[c]

[a] MP2 optimized geometry; CEP[3^{s}P2^d/3^{s}1^P] basis set for HCN and RCEP[4^{s}P3^d] for M$^+$, with five d-type components.
[b] Harmonic frequencies from Reference 27.
[c] Zero-point energy, in kcal mol.
[d] Doubly degenerate.
[e] Experimental fundamental frequencies from Reference 113.

mol^{-1}, respectively, for Cu^+, Ag^+ and Au^+. The ZPE difference between the complexes and free HCN (Table 13) is 1.6, 1.0 and 1.3 kcal mol^{-1} for the three metal ions in that order. Subtracting these two adjustments from the QCISD(T)/PP//MP2/PV calculated binding energies results in net metal ion–HCN dissociation energies of 38.3, 34.9 and 49.2

TABLE 16. CIS calculated vertical excited state energies (in eV) for linear HCNM^{+a}

Assignment		HCNCu$^+$			HCNAg$^+$			HCNAu$^+$		
		Asymptote		Complex	Asymptote		Complex	Asymptote		Complex
		Exp.	Calc.	Calc.	Exp.	Calc.	Calc.	Exp.	Calc.	Calc.
$^3[nd,(n+1)s]$	^3D	2.81^d	4.12^h	4.13–4.99^h	5.03^d	5.72^h	5.48–6.23^h	2.29^d	2.83^h	2.93–4.03^h
$^1[nd,(n+1)s]$	^1D	3.26^d	4.81^h	5.08^k 5.50^i (0.006) 6.65^{k,i}	5.71^d	6.52^h	6.38^k 6.83^i (0.008) 6.91^{k,i}	3.67^d	3.93^h	4.27^k 4.71^{k,i} 4.93^i (0.009)
$^3(\pi,\pi^*)$	$^3\Sigma^+$	—	5.92	5.29	—	5.92	5.23	—	5.92	5.31
$^3(\pi,\pi^*)$	$^3\Delta$	—	6.95^i	6.30^i	—	6.95^i	6.29^i	—	6.95^i	6.34^i
$^3(\pi,\pi^*)$	$^3\Sigma^-$	—	7.96	7.26	—	7.96^i	7.28	—	7.96	7.29
$^3[\pi,(n+1)s]$	$^3CT_1$	6.09^c	7.26^i	?^{e}	6.24^c	7.61^t	9.69^i	4.59^c	6.42^i	9.55^i
$^1[\pi,(n+1)s]$	$^1CT_1$	6.09^c	7.26^i	10.6^i (0.021)	6.24^c	7.61^i	9.98 (0.060)	4.59^c	6.42^i	10.0^i (0.001)
$^3(\sigma,\pi^*)$	$^3\Pi$	—	8.77^i	—	—	8.77^i	—	—	8.77^i	—
$^1(\pi,\pi^*)$	$^1\Sigma^-$	6.48^b	7.96	7.26	6.48^b	7.96	7.28	6.48^b	7.96	7.29
$^1(\pi,\pi^*)$	$^1\Delta$	7.48^b	8.39	7.66	7.48^b	8.39	7.72	7.48^b	8.39	7.68
$^1[nd,(n+1)p]$	^1F	8.92^d	10.30^g	8.96^i (0.062)^j	11.05^d	11.32^g	10.13^i (0.015)	—	9.94^g	8.60^i (0.095)^j
$^3[nd,(n+1)p]$	^3P	8.33^d	9.67^f	8.25^i 9.10	10.16^d	10.61^f	9.23^i 10.08	—	9.12^f	7.67^i 8.76
$^1(\sigma,\pi^*)$	$^1\Pi$	ca9^b	10.47 (0.024)	—	ca9^b	10.47 (0.024)	—	ca9^b	10.47 (0.024)	—
$^3[\sigma,(n+1)s]$	$^3CT_2$	6.28^c	9.20	—	6.43^c	9.54	—	4.78^c	8.36	—
$^1[\sigma,(n+1)s]$	$^1CT_2$	6.28^c	9.20	—	6.43^c	9.54	—	4.78^c	8.36	—

^{a}MP2 optimized ground state geometries. Metal ion attachment is at the nitrogen atom. For M = Cu, $n = 3$; M = Ag, $n = 4$; M = Au, $n = 5$.
^{b}See Reference 118.
^{c}Based on the IP (13.81 and 14.00 eV) of HCN (Reference 116) and EA of M$^+$ into the $(n + 1)$s atomic orbital (References 22 and 23).
^{d}From References 22 and 23.
^{e}Strong mixing with other transitions.
^{f}Three components.
^{g}Seven components.
^{h}Five components.
^{i}Doubly degenerate.
^{j}First of multiple components.
$^k f = 0.000$.

kcalmol^{-1}, respectively, for Cu^+, Ag^+ and Au^+. As in the N_2 case, the ordering of the magnitudes of the BSSE error within the HCN complex is according to the metal–N distance; the shorter the coordinating bond length the larger the BSSE. This comparison does not extend to different ligands of the same metal ion. The metal ion–N equilibrium bond distance are shorter and the degree of L → M CT (as measured by the Mulliken populations) is larger in HCN relative to N_2 as ligands, as expected, but the BSSE is smaller. The calculated property trends in Table 14 as a function of metal ion generally follow the $Au^+>Cu^+>Ag^+$ QCISD(T) binding energy order. However, the RHF, MP2 and QCISD binding energies in Table 13 for both the V and P metal ion basis sets show a $Cu^+>Au^+>Ag^+$ trend. The problem seems to be with the Cu^+ complex energies.

The calculated harmonic vibrational frequencies in Table 15 for the metal ion complexes at the MP2 level are consistent with the geometry changes. The C—H stretch (3400–3500 cm^{-1}) is hardly affected, the C—N stretch (ca 2100 cm^{-1}) increases in parallel with the decreased C—N bond length and the HCN bending mode 750–800 cm^{-1}) increases in energy. The HCN—M stretch frequency is calculated to be lowest for M = Ag (263 cm^{-1}), as expected, but HCN—Cu is larger than HCN—Au, in accord with the MP2 binding energy order and due, at least in part, to the difference in reduced mass.

The CIS calculated vertical excited state energies for MHCN$^+$ are shown in Table 16. The photoelectron spectrum of HCN shows two ionization processes very close to each other at the lower end of the energy spectrum[116]. These have been interpreted in a one-electron model as corresponding to the removal of a σ MO electron (13.61 eV) and a doubly degenerate π MO electron (14.00 eV). Thus, as in the case of N_2 the first ionization potential is out of a σ orbital, but in HCN the σ–π energy gap is much smaller. Since the σ MO is found to be preferentially stabilized by the metal ion in a linear coordination geometry, the valence π MO should be the one involved in the lowest energy L → M charge transfer transitions in the complexes. The calculated $\pi \rightarrow (n + 1)$s electron transfer between the isolated units (HCN → M$^+$, CT$_1$) is calculated too high by 1.17 eV, whereas the $\sigma \rightarrow (n + 1)$s electron transfer is too high by almost 3 eV relative to experiment. The $\pi \rightarrow (n + 1)$s transitions (both spin triplet and singlet) in the complexes are difficult to identify as such because of strong mixing with the corresponding spin $nd \rightarrow (n + 1)$s transitions, both on the orbital and configurational levels. Those that are identifiable are predicted to shift to higher energy by over 3 eV relative to the isolated unit electron transfer values. They are still predicted to fall lower in energy than the $\sigma \rightarrow (n + 1)$s transitions.

The intra-unit HCN electronic transitions parallel those of isoelectronic N_2 with the lowest energy excited states arising from excitations to the $\pi \rightarrow \pi^*$ and $\sigma \rightarrow \pi^*$ multiplet excited states. However, in HCN the $^1\Pi$ state is dipole allowed. The assignment of the measured optical absorption spectrum of HCN is complicated by bending in the excited state[61] and vibronic coupling[117]. Two spin singlet $\pi \rightarrow \pi^*$ states are assigned experimentally as the lowest energy transitions observed[61, 118] and are predicted to shift to lower energy upon metal ion complexation. The third HCN excited state in the singlet manifold is $^1(\sigma, \pi^*)$, with a vertical excitation energy of about 9 eV[118]. The intra-atomic electronic transitions for the spherical metal ions split into the lower order linear symmetry degeneracies. For Cu^+ and Au^+ the $^{3,1}D$ excited states are calculated to shift to higher energy upon complexation, but for Ag^+ the atomic ion value brackets the split components in the linear complex with HCN. An intense transition is the $^1S \rightarrow ^1P$ to the $^1[nd, (n + 1)p]$ state, but this is just beyond the energy range in Table 16. Intensity borrowing by the ^1D transitions is not calculated to be substantial. However, the lowest energy component of the linear symmetry split ^1F excited state gains considerable intensity and competes with the $^1CT_1$ state as to which is lower. The symmetries of the internal HCN excited states do not change upon linear complexation of the metal ions, so the selection rules don't change and the dipole forbiddeness of $\pi \rightarrow \pi^*$ transitions in Table 16 remains.

V. C₂H₂

The calculated data on the C_2H_2, $C_2H_3^+$ and metal ion–C_2H_2 complexes are contained in Tables 2, 3 and 17–20. The calculated equilibrium structure of acetylene[61] has C—C and C—H bond distances of 1.208 Å and 1.058 Å, respectively. The corresponding MP2/CEP[$3^{sp}2^d/3^s1^p$] optimized geometry (Table 18) has a C—C distance of 1.213 Å and a C—H length of 1.063 Å. The corresponding RHF values are 1.184 Å and 1.055 Å, respectively. Again, as with HCN, the experimental values are bracketed by the RHF and MP2 numbers, with the C—C distance closer to the MP2 determined bond length. With the acetylene system we now have reference property values for all three types of triple bonds; N≡N, C≡N and C≡C.

Protonated acetylene ($C_2H_3^+$) has been studied extensively, both experimentally[119, 120] and theoretically[121-123]. Both theory and experiment predict the non-classical bridged structure to be more stable than the classical asymmetrically protonated geometry. As seen in Table 3, the classical structure is more stable at the RHF level but at the correlation method level the bridged structure is more stable[123]. The classical geometry is a transition state with one MP2 calculated imaginary frequency. The energy difference at the QCISD(T)//MP2 level between the symmetric and asymmetric structures is only 2.5 kcalmol⁻¹. The ZPE of the classical structure is calculated to be 21.5 kcalmol⁻¹ at the MP2 level not counting the imaginary frequency, and the ZPE for the bridging structure is 21.8 kcalmol⁻¹ (Table 19). Subtracting the ZPE difference of 0.3 kcalmol⁻¹ from the 2.5 kcalmol⁻¹ gives the

TABLE 17. Energies (in a.u.) of $MC_2H_2^+$ complexes[a]

M⁺	BF[b]	RHF	MP2	QCISD	QCISD(T)
Cu⁺	87[c, e]	−207.163998 (26.0)			
	87[d, e]	−207.149576 (18.1)	−207.909488 (42.5,35.6)	−207.852622 (39.3)	−207.857847 (32.0)
	101[f]	−207.149875 (18.3)	−208.168567 (48.3)	−208.096808 (42.2)	−208.112604 (38.0)
Ag⁺	87[c, e]	−157.278770 (18.1)			
	87[d, e]	−157.274069 (16.3)	−157.749131 (24.7,22.1)	−157.759821 (23.4)	−157.776995 (24.2)
	101[f]	−157.274718 (16.7)	−158.113963 (31.6)	−158.095558 (27.1)	−158.122139 (28.8)
Au⁺	87[c, e]	−146.995794 (27.9)			
	87[d, e]	−146.985917 (22.9)	−147.424943 (45.2,39.2)	−147.439741 (43.0)	−147.458395 (44.4)
	101[f]	−146.987716 (24.0)	−147.764417 (58.9)	−147.740329 (48.8)	−147.768637 (52.1)

[a] Numbers in Parentheses are metal ion binding energies in kcalmol⁻¹; a second number in the parentheses is after BSSE correction.
[b] Number of basis functions; CEP[$3^{sp}2^d/3^s1^p$] basis set with five d-type components for acetylene.
[c] RHF optimized.
[d] MP2 optimized.
[e] RCEP[$4^{sp}3^d$] with five d-type components for M⁺.
[f] RCEP [$4^{sp}3^d2^f$] with five d-type and seven f-type components for M⁺. Complex geometry taken from 87 basis function MP2 optimization.

TABLE 18. MP2 Mulliken populations and optimized geometric parameters for $MC_2H_2^+$[a]

| M^+ | Mulliken | | | | | | | | | | | |
| | M^+ populations | | | | Charges | | | Bond lengths (Å) | | | Angles (deg) | |
	s	p	d	f	M^+	H	C	M—C	H—C	C—C	MCC	HCC
	—	—	—	—	—	0.20	−0.20	—	1.055	1.184	—	180.0
	—	—	—	—	—	0.19	−0.19	—	1.063	1.213	—	180.0
H^+[b]	—	—	—	—	0.29	0.29	−0.11	1.086	1.086	1.260	119.7[d]	179.8
						0.33	0.20		1.072			
	—	—	—	—	0.31	0.31	−0.15	1.099	1.099	1.260	120.2[d]	179.8
						0.33	0.20		1.082			
H^+[c]	—	—	—	—	0.32	0.31	0.03	1.278	1.070	1.207	61.8	179.1
	—	—	—	—	0.35	0.32	0.01	1.279	1.078	1.231	61.2	179.4
Cu^+	2.16	6.11	10.00	—	0.74	0.26	−0.13	2.315	1.062	1.195	75.0	174.1
	2.35	6.18	9.85	—	0.63	0.27	−0.09	2.020	1.074	1.238	72.2	169.0
	2.37	6.18	9.82	0.03	0.60	0.27	−0.07					
Ag^+	2.11	6.07	9.99	—	0.84	0.26	−0.18	2.605	1.061	1.193	76.8	174.4
	2.17	6.10	9.94	—	0.79	0.27	−0.16	2.417	1.071	1.226	75.3	173.0
	2.19	6.09	9.86	0.09	0.76	0.27	−0.15					
Au^+	2.28	6.08	9.88	—	0.75	0.26	−0.14	2.568	1.063	1.203	75.4	171.6
	2.46	6.12	8.69	—	0.73	0.28	−0.14	2.169	1.074	1.247	73.3	166.1
	2.52	6.11	9.60	0.11	0.65	0.29	−0.11					

[a]Basis sets, energies and levels of theory for the neutral and protonated C_2H_2, and complexes are described in Tables 2, 3 and 17, respectively.
[b]Classical (asymmetric) structure.
[c]Non-classical (bridging) structure.
[d]Averaged.

bridged geometry more stable by 2.2 kcalmol^{-1}. Other theoretical estimates are in the 1–4 kcalmol^{-1} range[121–123]. The equilibrium geometry of the non-classical $C_2H_3^+$ structure shows slightly longer C—C (0.02 Å) and C—H (0.015 Å) bond lengths relative to C_2H_2. The HCC angles are very close to 180° (Table 18).

The proton affinity of C_2H_2 is estimated experimentally at 151.7[124]–154.3[65] kcalmol^{-1}. The QCISD(T)//MP2 calculated value (Table 3) is 157.7 kcalmol^{-1}. The ZPE energy difference between $C_2H_3^+$ and C_2H_2 is 5.4 kcalmol^{-1}. When subtracted from the total electronic energy difference this gives a PA of 152.3 kcalmol^{-1}, which is close to the experimental numbers.

The MP2 calculated harmonic vibrational frequencies for $C_2H_3^+$ are shown in Table 19 and compared with experimentally derived harmonic frequencies[33]. The theoretical values for the C—H stretch are only 36–40 cm^{-1} larger than experiment. For the non-classical bridged structure the C—H stretch frequencies are predicted to decrease by about 120 cm^{-1} relative to C_2H_2. This is also the expected direction[120].

The metal ion–acetylene complexes are calculated in the bridging structure geometry. No asymmetric structures were examined. Theoretical studies on symmetric Cu^+– and Ag^+–acetylene complexes have recently been reported[125, 126]. The only experimental value of the dissociation energy for the three systems studied here is for $CuC_2H_2^+$[76], where an upper bound of 30.5 kcalmol^{-1} has been estimated. The QCISD(T)/PP//MP2/PV calculated binding energy in Table 17 is 38.0 kcalmol^{-1}. The ZPE difference of 0.6 kcalmol^{-1} (Table 19) reduces this to 37.4 kcalmol^{-1}. Previously calculated Cu$^+$–ethylene dissociation energies gave values of 26.7[125] and 32.8[126] kcalmol^{-1}, with a projected 3.5 kcalmol^{-1} error (underestimation) in the latter case. The adjusted Table 17 value probably includes some BSSE. However, the above results on N$_2$ complexes shows that π complexes have less BSSE than the σ complexes. It should be noted that increasing the basis set size and level of correlation will also increase the calculated binding energies and tend to somewhat cancel the BSSE. Thus, the theoretical dissociation energies obtained here and previously[126] for CuC$_2$H$_2^+$ converge on similar values which are somewhat above the experimentally estimated upper limit[76]. Theoretical estimates of 14.8[125], 23.6[126] and 28.8 (Table 17) kcal mol^{-1} are also given for the AgC$_2$H$_2^+$ dissociation energies. Here again, after consideration of the effects discussed above, the latter two values converge to being very close. This result is similar to that obtained for the AgC$_2$H$_4^+$ complex discussed above.

The AuC$_2$H$_2^+$ complex is calculated to have the largest binding energy independent of basis set size or a theoretical level. The binding energy trends for the acetylene complexes as a function of metal ion closely parallel those of the ethylene complexes (Table 5), except that they are uniformly smaller. This result is obtained even though the metal ion–carbon distances in the two cases are very similar, as is seen from a comparison of Tables 5 and 18, with the Cu—C bond length being larger for the ethylene than the acetylene complex. A purely electrostatic interaction would predict a direct dependence of the binding energy on metal–atom distance. However, as noted by Sodupe and coworkers[77], the polarizability of these two ligands in the perpendicular (π) direction is larger for ethylene than for acetylene. Thus, the charge-induced dipole interaction will be larger for the metal ion–ethylene complexes. This comparison can be roughly extended to the values of the metal ion–ligand stretch frequencies in Tables 7 and 19, ethylene and acetylene, respectively. In the former case, the calculated harmonic frequencies are 341, 234 and 350 cm^{-1}, respectively, for Cu$^+$, Ag$^+$ and Au$^+$. The corresponding values for the acetylene substrate are 345, 227 and 333 cm^{-1}.

The MP2 calculated C—C bond distances and HCC angles in the complexes (Table 18) both follow the trend showed by the binding energy; the perturbation relative to free C$_2$H$_2$ is largest for Au$^+$ and smallest for Ag$^+$ coordination. The Mulliken populations and charges at the MP2 level also follow this order and are uniformly smaller than for the corresponding ethylene complexes. Acetylene has the same type of σ L $\rightarrow$ M dative bond between the in-plane ligand π electrons and the metal ion ($n + 1$)sp atomic orbitals and π M $\rightarrow$ L dative bond between the in-plane $nd\pi$ metal ion electrons and the empty π^* MO as in ethylene. However, for acetylene there is the added possibility of a L $\rightarrow$ M interaction between the C$_2$H$_2$ π electrons perpendicular to the MCC plane and corresponding metal ion ($n + 1$)p orbital, and a M $\rightarrow$ L dative interaction between the perpendicular $nd\pi$ MO that lies parallel to the C—C axis and the acetylene π^* MO. These latter two interactions are of a δ type, where each of the participating orbitals is perpendicular to the metal–bond centre axis and is, consequently, expected to be very weak and contribute little to the bonding.

The impression obtained from the Mulliken populations and atomic charges in Tables 6 and 18 is that L $\rightarrow$ M charge transfer is larger for the ethylene ligand than for acetylene, while the M $\rightarrow$ L back bonding is not very different for the two substrates. The fact that the ionization potential of acetylene at 11.41 eV[61] is 0.9 eV larger than for ethylene explains the larger L $\rightarrow$ M dative bonding of ethylene and, consequently, the resultant larger metal ion binding energy for the ethylene complexes relative to acetylene.

The excited electronic states of acetylene have been discussed extensively, both experimentally[61, 127, 128] and theoretically[129–132]; the latter also review the measured data. All the low-lying excited states of acetylene, both triplet and singlet, are either known or suspected of being either *cis* or *trans* bent to some degree. In the discussion here of the acetylene and

TABLE 19. MP2 calculated harmonic vibrational frequencies (in cm^{-1}) for $MC_2H_2^{+a}$

Exp[b] C_2H_2	Calculated				
	C_2H_2	$CuC_2H_2^+$	$AgC_2H_2^+$	$AuC_2H_2^+$	$C_2H_3^{+e}$
		280	217	330	581
		345	227	333	661
624[d]	515[d]	541	580	608	790
		589	623	736	936
747[d]	777[d]	780	784	771	1317
		790	841	831	1911
2011	1962	1830	1897	1796	2377
3415	3451	3338	3371	3337	3300
3497	3537	3418	3457	3431	3397
16.8[c]	16.4[c]	17.0[c]	17.1[c]	17.4[c]	21.8[c]

[a]MP2 optimized geometry; CEP[3s2d/3s1p] basis set for C_2H_2 and RCEP[4s p/3d] for M$^+$, with five d-type components.
[b]Experimental harmonic frequencies from Reference 27.
[c]Zero-point energy, in kcal mol^{-1}.
[d]Doubly degenerate.
[e]Non-classical bridged structure.

complexed acetylene transitions (Table 20) only vertical excitation energies are calculated, so that the linear acetylene symmetry designations are used. The pure acetylene low-lying excited states arise from a one-electron $\pi_u \rightarrow \pi_g^*$ excitation. As with N_2 and HCN (without the g,u inversion symmetry) this gives rise to three distinct excited states of Σ_u^+, Δ_u and Σ_u^- symmetries, both triplets and singlets. In the lower symmetry of the metal ion complexes the Δ_u states split into non-degenerate components, besides the energy shift they experience from coordination of the metal ion. The first excited state of substantial intensity in C_2H_2 is the $^1\Pi_u$[61, 130, 131] which apparently is calculated beyond the range of energies treated here. However, lower-lying triplet and singlet $\pi_u \rightarrow \sigma_u^*$ (Π_g) transitions were found where the σ_u^* MO may not be completely valence-like.

The purely metal ion electronic transitions [$nd \rightarrow (n + 1)$s, to give the ^3D and ^1D states] have been mentioned above. An atomic state of D symmetry is five-fold degenerate and splits into five non-degenerate components in the lower symmetry of the bridging metal ion complexes. The degree of energy splitting depends on the different interaction strengths of each of the five angular momentum components with the ligand. Not all the five components of the D states could be identified in the energy range covered in Table 5. The lowest energy $nd \rightarrow (n + 1)$p is ^3P with three components, but the corresponding lowest-energy singlet state is ^1F[22, 23].

The charge transfer transitions are expected to be mainly L $\rightarrow$ M: $\pi_u \rightarrow (n + 1)$s and $\pi_u \rightarrow (n + 1)$p. The experimental ionization potential of acetylene is 11.41 eV[61], both vertical and adiabatic since the first vibronic band in the photoelectron spectrum is also the most intense[133]. The asymptotic charge transfer energies are then the difference between this IP and appropriate EA of each metal ion, and are the same for both triplet and singlet electronic states. These are tabulated in Table 20.

The effect of complexation on the different asymptotic excitation energies is measured by the shift in the calculated values. The location of the triplet $\pi \rightarrow \pi^*$ excited states in Table 20 for C_2H_2 is from electron impact studies[127] and is approximate. These states have been much discussed lately[128, 132]. The location of the singlet states are better characterized but also not free from uncertainty[129, 130]. The purpose of the calculations here is not to

TABLE 20. CIS calculated vertical excited state energies (in eV) for $MC_2H_2^{+a}$

Assignment		$CuC_2H_2^+$			$AgC_2H_2^+$			$AuC_2H_2^+$		
		Asymptote		Complex	Asymptote		Complex	Asymptote		Complex
		Exp.	Calc.	Calc.	Exp.	Calc.	Calc.	Exp.	Calc.	Calc.
$^3[\pi,(n+1)s]$	$^3CT_1$	3.69^b	4.54^c	4.87	3.84^b	4.88^c	5.35	2.19^b	3.70^c	3.64
				5.17			5.93			3.91
$^1[\pi,(n+1)s]$	$^1CT_1$	3.69^b	4.54^c	5.76	3.84^b	4.88^c	6.27	2.19^b	3.70^c	4.72
				(0.020)			(0.028)			(0.016)
				8.25			6.45			4.77
				(0.026)			(0.029)			(0.020)
$^3[nd,(n+1)s]$	^3D	2.81^d	4.12^e	5.56–	5.03^d	5.72^e	6.50–	2.29^d	2.83^e	4.82–
				8.41^e			7.96^e			7.95^e
$^1[nd,(n+1)s]$	^1D	3.26^d	4.81^e	5.69^f	5.71^d	6.52^e	7.16^f	3.67^d	3.93^e	5.42^f
				(0.007)			(0.000)			(0.000)
$^3(\pi,\pi^*)$	$^3\Sigma_u^-$	5.2^g	4.46	4.47	5.2^g	4.46	4.47	5.2^g	4.46	4.44
$^3(\pi,\pi^*)$	$^3\Delta_u$	5.9–	5.43^c	4.65	5.9–	5.43^c	5.18	5.9–	5.43^c	4.73
		6.1^g		5.30	6.1^g		5.49	6.1^g		5.41
$^3(\pi,\pi^*)$	$^3\Sigma_u^+$	—	6.30	6.98	—	6.30	6.76	—	6.30	7.12
$^1(\pi,\pi^*)$	$^1\Sigma_u^-$	5.23^g	6.30	5.28	5.23^g	6.30	5.88	5.23^g	6.30	5.31
$^1(\pi,\pi^*)$	$^1\Delta_u$	6.71^g	6.66^c	6.46	6.71^g	6.65^c	6.68	6.71^g	6.66^c	6.38
				(0.009)			(0.005)			(0.016)
				7.57			7.28			7.66
$^3[\pi,(n+1)p]$	$^3CT_2$	7.50^h	7.75^i	9.08^f	7.58^h	8.00^i	9.05^f	7.14^h	7.79^i	8.25^f
$^1[\pi,(n+1)p]$	$^1CT_2$	7.50^h	7.75^i	9.83^f	7.58^h	8.00^i	9.49^f	7.14^h	7.79^i	9.34^f
				(0.120)			(0.238)			(0.255)
$^3[nd,(n+1)p]$	^3P	8.33^d	9.67^j	7.57^f	10.61^d	—	9.60^f	—	—	7.54^f
$^1[nd,(n+1)p]$	^1F	8.92^d	—	8.28^f	11.05^d	—	10.33^f	—	—	8.07^f
				(0.066)			(0.026)			(0.027)
$^3(\pi,\sigma^*)$	$^3\Pi_g$	g	8.97^c	—	g	8.97^c	—	g	8.97^c	—
$^1(\pi,\sigma^*)$	$^1\Pi_g$	g	9.61^c	—	g	9.61^c	—	g	9.61^c	—

^{a}MP2 optimized ground state geometries. Values in parentheses are oscillator strengths. For M = Cu, $n = 3$; M = Ag, $n = 4$; M = Au, $n = 5$.
^{b}Based on the IP (11.41 eV) of acetylene (Reference 61) and EA of M^+ into the $(n + 1)$s atomic orbital (References 22 and 23).
^{c}Doubly degenerate.
^{d}From References 22 and 23.
^{e}Five components.
^{f}First of multiple components.
^{g}Adiabatic; from References 127 and 129; see text.
^{h}Based on the IP (11.41 eV) of acetylene (Reference 61) and EA of M^+ into the $(n + 1)$p atomic orbital (References 22 and 23).
^{i}Six components.
^{j}Three components.

reproduce experiment for the free acetylene molecule, but to trace the excited state energies from free molecule to complex. All the free acetylene $\pi \to \pi^*$ transitions in Table 20 are either spin or dipole forbidden. Only the lower-energy component of the $^1\Delta_u$ state acquires some intensity in the complex. For a given state, the shifts are consistent in their direction for all three metal ions. The Σ states display all three possible shift directions upon complexation; higher energy, lower energy or unchanged. This, of course, depends on the orbital natures of the different states, in their interaction with the metal ion in the complex. The Δ states split off one component to lower energy and one remains essentially unchanged from free acetylene.

Both the pure metal ion D states and the $\pi \to (n + 1)$sp charge transfer states are predicted to shift to higher energy upon complexation. This is the usual result. If the calculated CT energies in the complex are adjusted by the error in the asymptote CT energies, then the first low-intensity singlet CT transitions in the complexes ($^1CT_1$) are predicted to fall at 4.91 eV, 5.23 eV and 3.21 eV, respectively, for Cu^+, Ag^+ and Au^+. The lowest-energy ^1D state component in each complex is not far behind. As noted above, the pure acetylene $^1\Pi_u$ state, which has substantial Rydberg character, is not shown in Table 20. Experimentally[134] this is a vertical transition at about 8.2 eV with a high absorption intensity. The first strong ($f > 0.03$) absorption in Table 20 is for the $^1CT_2$, $\pi \to (n + 1)$p CT state, which is predicted to shift by over 2 eV to higher energy upon complexation. This is likely to be actually the lowest energy intense transition, since the acetylene $^1\Pi_u$ state's Rydberg character should also induce a considerable blue shift upon complexation.

VI. N₂O

The $N\equiv N\equiv O$ **(I)** molecule has the first example in this series of a central nitrogen atom which, in order to satisfy all the usual valencies of the atoms bonded to it, would have to be described as a five-coordinate nitrogen atom. The proper description of such molecular systems, which are commonly called 1,3 dipoles because of their chemistry of 1,3 addition to double bonds, has been discussed extensively[135, 136]. Consensus opinion is that NNO, for example, has both a fully formed double bond and a fully formed triple bond. Specifically, the central nitrogen atom participates in five fairly normal two-electron covalent bonds. The central nitrogen atom in N_2O can therefore legitimately be called hypervalent.

Actually, a number of resonance hybrid structures of NNO can also be written. In addition to the hypervalent configuration **I** there are three relevant zwitterionic resonance structures:

$$\bar{N}\!=\!\overset{+}{N}\!=\!O \longleftrightarrow N\!\equiv\!\overset{+}{N}\!-\!\bar{O} \longleftrightarrow \overset{+}{N}\!=\!N\!-\!\bar{O}$$

$$\text{(II)} \qquad\qquad \text{(III)} \qquad\qquad \text{(IV)}$$

Structure **IV** is responsible for the 1,3 dipole behaviour of N_2O and does not have a hypervalent nitrogen atom. Almost each resonance structure above reduces the order of an atom–atom bond relative to **I**. Thus, the $N\equiv N$ bond distance is expected to be somewhat longer than the ordinary triple N—N bond length if structures **II** and **IV** are important and the $N\!=\!O$ distance should be somewhat elongated relative to a normal N—O double bond length if structures **III** and **IV** contribute strongly. As will be shown subsequently, these considerations are useful in correlating the geometric and electronic structural properties of these systems.

The data on N_2O, HN_2O^+ and the metal ion-N_2O complexes are contained in Tables 2, 3 and 21–24. The calculated equilibrium geometry of N_2O[82] shows a linear structure with measured N—N and N—O bond distances of 1.128 Å and 1.184 Å, respectively. The experimental values are again bracketed by the RHF and MP2 optimized bond lengths (Table 22) with the latter being about 0.03 Å longer than measured. Here, with a hypervalent nitrogen atom, the difference between the RHF and MP2 results for N—N is larger than for N_2, although the measured increase of 0.04 Å in the N≡N bond length

TABLE 21. Energies (in a. u.) of $MNNO^+$ complexes[a]

M^+	BF^b	X^g	RHF	MP2	QCISD	QCISD(T)
Cu^+	$97^{c,e}$	O	−230.073134			
			(18.6)			
	$97^{d,e}$	O	−230.056900	−231.090540	−231.016996	−231.029118
			(17.8)	(21.3)	(26.1)	(12.2)
	111^f	O	−230.057185	−231.344863	−231.259774	−231.280328
			(18.0)	(24.1)	(28.2)	(16.0)
	$97^{c,e}$	N	−230.065776			
			(14.0)			
	$97^{d,e}$	N	−230.045149	−231.115503	−231.028409	−231.050481
			(10.4)	(36.9)	(33.3)	(25.6)
	111^f	N	−230.045427	−231.372348	−231.272466	−231.303348
			(10.6)	(41.3)	(36.2)	(30.4)
Ag^+	$97^{c,e}$	O	−180.191276			
			(12.8)			
	$97^{d,e}$	O	−180.176504	−180.943253	−180.930588	−180.096281
			(12.9)	(11.6)	(14.3)	(13.5)
	111^f	O	−180.176847	−181.301576	−181.263295	−181.303837
			(13.1)	(14.4)	(16.0)	(15.6)
	$97^{c,e}$	N	−180.183851			
			(8.1)			
	$97^{d,e}$	N	−180.167094	−180.954808	−180.933918	−180.969416
			(7.0)	(18.9)	(16.4)	(17.7)
	111^f	N	−180.167618	−181.315720	−181.268094	−181.312339
			(7.3)	(23.3)	(19.1)	(20.9)
Au^+	$97^{c,e}$	O	−169.877707			
			(13.3)			
	$97^{d,e}$	O	−169.877073	−170.591989	−170.585648	−170.618502
			(12.8)	(15.2)	(18.2)	(17.6)
	111^f	O	−169.878716	−170.917696	−170.836623	−170.922236
			(13.4)	(20.2)	(21.4)	(21.2)
	$97^{c,e}$	N	−169.887375			
			(9.5)			
	$97^{d,e}$	N	−169.868179	−170.609865	−170.593829	−170.630618
			(6.8)	(26.4)	(23.4)	(25.2)
	111^f	N	−169.869541	−170.938732	−170.890886	−170.935697
			(7.7)	(33.4)	(27.0)	(29.7)

[a] Numbers in parentheses are metal ion binding energies in $kcal\,mol^{-1}$.
[b] Number of basis functions; $CEP[3^sP2^d/3^s1^P]$ basis set with five d-type components for NNO.
[c] RHF optimized.
[d] MP2 optimized.
[e] $RCEP[4^sP3^d]$ with five d-type components for M^+.
[f] $RCEP[4^sP3^d2^f]$ with five d-type and seven f-type components for M^+. Complex geometry taken from 97 basis function MP2 optimization.
[g] Atom to which M^+ is attached. For X = N attachments is to the terminal atom.

$N_2 \rightarrow N_2O$) is reproduced in the MP2 calculated geometries. This increase does not bring the N—N distance in N_2O down to double-bond range. H_2N_2, for example, has an equilibrium N=N distance of 1.252 Å[82]. The experimental N—N bond length of 1.128 Å for NNO is certainly closer to the unperturbed triple bond N_2 distance of 1.098 Å[91] than to double-bonded H_2N_2. Thus the N≡N bond distance in NNO is slightly elongated relative to a 'normal' N—N triple-bond length, in accord with structure **II** and **IV** above. The experimental N—O bond length of NNO at 1.184 Å is difficult to compare with a standard N=O bond length because N=O containing compounds usually have a number of possible resonance hybrid structures just like NNO. In addition, the hypervalency of the central nitrogen atom should contract its valence shell somewhat and naturally lead to shorter bond lengths, which will tend to offset the effect of the single-bonded structures. Perhaps the best comparison is with *trans*-HONO, where the N—O bond length is measured at 1.170 Å[82], 0.014 Å shorter than in NNO, as expected on the basis of the greater importance of structures **III** and **IV** above. The calculated N—O bond length of 1.190 Å at the MP2 level is in good agreement with experiment.

The measured dipole moment of N_2O is a small $0.161D$[111], with polarity +NNO–[137]. The RHF calculated value is $0.675D$ and the RHF wave function at the MP2 geometry gives $0.457D$, both with the observed polarity. However, the MP2 density gives a dipole moment of $0.182D$, but with a reversed –NNO+ polarity. This difference between the RHF and MP2 results is reflected in the atomic charges in Table 22, which shows a relative reduction of negative charge on the oxygen atom in going RHF → MP2. In general, the multi-configuration representation tends to reduce atomic charges compared to RHF because it usually reduces the ionicity of the molecular wave function[138]. In any event, the calculated atomic charges show a –+– (NNO) distribution of charges obtained by combining structures **II** and **III**. The small absolute value for the dipole moment of N_2O and the question of its sign recalls an analogous situation with the CO molecule. Here, the experimental polarity is –CO+ and the correct sign can be calculated only by taking into account subtle configuration interaction effects[139, 140] which are not included in the MP2 method. In spite of the small value of its dipole moment, CO coordinates to metal ions almost exclusively at the carbon atom.

The calculated harmonic vibrational frequencies for N_2O are shown in Table 23. The derived experimental values are also quoted[141]. As expected, the MP2 N—N (2159 cm^{-1}) and N—O (1251 cm^{-1}) stretches are underestimated, with the triple-bond harmonic mode having the larger calculated–experiment gap. The observed fundamental vibrational frequencies[142] (2224 and 1285 cm^{-1}, respectively) for these two modes are naturally smaller than the derived harmonic values, and are therefore closer to the calculated stretch frequencies. The bending mode is very well calculated, despite the absence of f-type functions from the basis set[143], as has been shown necessary for acetylene.

Protonated N_2O has been extensively discussed recently[112, 142, 144, 145]. The experimental proton affinity has been reported at 138.8 kcalmol^{-1}[65] and 137.3 kcalmol^{-1}[146], where the exact attachment site of the proton is not known directly but is inferred from a match between a calculated and observed rotational constant[147]. Protonation at the oxygen atom is predicted to be the more stable isomer. The calculated QCISD(T)//MP2 values in Table 3 favour the oxygen-bound structure with a purely electronic energy PA of 144.9 kcalmol^{-1}. The ZPE difference is 6.2 kcalmol^{-1}, which gives an adjusted PA of 138.7 kcalmol^{-1}. Other thermodynamic corrections to bring the conditions of the calculated value to those of the measured quantity have been estimated[144] to add about 0.4 kcalmol^{-1} to give 139.1 kcalmol^{-1} for the PA. There is thus good agreement of the QCISD(T)//MP2 calculated value with experiment. The MP2 calculated PA (Table 3) is only 0.9 kcalmol^{-1} smaller.

The energy difference between the NNOH$^+$ and HNNO$^+$ structures is QCISD(T)//MP2 calculated to be 5.7 kcalmol^{-1} (Table 3), compared to an estimated experimental value of

TABLE 22. Mulliken populations and geometric parameters for MNNO^{+a}

| M$^+$ | Mulliken | | | | | | | | Bond lengths (Å) | | | Angles (deg) | |
| | M$^+$ populations | | | | Charges | | | | | | | | |
	s	p	d	f	M$^+$	N	N	O	M–X^d	N–N	N–O	MXN^d	NNO
H^{+b}	—	—	—	—	—	−0.21	0.38	−0.17	—	1.085	1.179	—	180
	—	—	—	—	—	−0.19	0.21	−0.02	—	1.161	1.190	—	180
H^{+c}	—	—	—	—	0.46	−0.14	0.47	0.21	1.021	1.117	1.103	131.2	173.2
	—	—	—	—	0.45	−0.07	0.32	0.30	1.032	1.163	1.152	135.9	169.2
	—	—	—	—	0.48	0.06	0.53	−0.06	0.972	1.067	1.277	108.1	174.0
	—	—	—	—	0.48	0.08	0.45	−0.01	0.996	1.130	1.275	108.0	172.5
Cu^{+b}	2.05	6.06	9.99	—	0.89	−0.35	0.43	0.03	2.120	1.093	1.142	179.6	179.9
	2.20	6.14	9.90	—	0.76	−0.30	0.35	0.20	1.851	1.157	1.169	180.	180.
	2.22	6.14	9.87	0.03	0.74	−0.29	0.35	0.20					
Cu^{+c}	2.04	6.04	10.00	—	0.92	−0.13	0.42	−0.20	2.106	1.075	1.209	140.4	178.6
	2.11	6.08	9.95	—	0.86	−0.09	0.30	−0.07	1.950	1.146	1.213	137.1	177.5
	2.12	6.08	9.93	0.03	0.85	−0.09	0.31	−0.07					
Ag^{+b}	2.03	6.03	9.99	—	0.95	−0.27	0.33	−0.01	2.499	1.092	1.149	179.3	179.9
	2.10	6.06	9.96	—	0.88	−0.29	0.25	0.17	2.233	1.160	1.171	180.	180.
	2.10	6.06	9.88	0.09	0.87	−0.29	0.25	0.17					
Ag^{+c}	2.02	6.02	10.00	—	0.96	−0.15	0.39	−0.20	2.444	1.077	1.202	149.9	179.2
	2.04	6.04	9.99	—	0.93	−0.12	0.25	−0.07	2.343	1.149	1.207	145.6	178.5
	2.04	6.03	9.90	0.09	0.93	−0.12	0.25	−0.07					
Au^{+b}	2.06	6.03	9.96	—	0.95	−0.30	0.34	0.01	2.370	1.092	1.145	179.4	180
	2.20	6.07	9.84	—	0.89	−0.33	0.25	0.20	2.075	1.157	1.170	179.9	179.6
	2.23	6.08	9.75	0.10	0.84	−0.27	0.23	0.20					
Au^{+c}	2.05	6.03	9.99	—	0.94	−0.14	0.41	−0.20	2.410	1.077	1.208	136.9	178.7
	2.12	6.05	9.94	—	0.89	−0.09	0.29	−0.09	2.257	1.148	1.215	129.5	177.4
	2.14	6.05	9.84	0.10	0.86	−0.09	0.30	−0.07					

^{a}Basis sets, energies and levels of theory for the neutral and protonated NNO, and complexes are described in Tables 2, 3 and 13, respectively.
^{b}M$^+$ attached to terminal N.
^{c}M$^+$ attached to O.
^{d}The atom X is defined in Table 21.

6.2 ± 0.5 kcalmol^{-1} [148]. The theoretically preferred isomer in Table 3 oscillates with the level of theory and for MP2, the level at which the geometries were optimized, the HNNO$^+$ structure is marginally more stable. This alternation has been noted previously[140, 144, 145]. Since the PA to form the NNOH$^+$ structure is apparently calculated correctly, the problem has been attributed to the difficulty in calculating the HNNO$^+$ structure at the MP2 level[145]. Both isomers are equilibrium structures as shown by their all-real harmonic vibrational frequencies in Table 23.

The geometries shown in Table 22 for the protonated structures are very similar to previous corresponding level calculational results[140, 144, 145]. For the HNNO$^+$ structure, the MP2 bond length for N—N changes very little from NNO while the N—O distance shortens by 0.04 Å. The latter is probably due to the increased importance of structure **II** in the N-protonated species. The expected lengthening of the N—N bond may be offset by the general cation contraction. For NNOH$^+$ the N—N distance decreases by 0.03 Å (Table 22), possibly due to enhancement of the N≡N triple-bond form according to structure **III**, which is emphasized in the O-protonated isomer. Consistent with this interpretation is the large (0.085 Å) increase in the N—O bond length upon O-protonation. The N—O harmonic stretch frequency (Table 23) decreases to 1049 cm^{-1} from 1251 cm^{-1} in NNOH$^+$.

An outstanding feature of the protonated structures is their non-linearity, although NNO is only mildy bent. Both isomer species have the *trans* geometry. The H—O—N angle in NNOH$^+$ (*ca* 108°) is normal for the angle between two single bonds to oxygen, as would be predicted by structures **III** and **IV**. The MP2 optimized HNN angle in HNNO$^+$ at 135.9° is larger than normal, which could be taken as the measured HN═N angles in HN$_3$ (114.1°) and H$_2$N$_2$ (116.9°)[82]. This large angle is somewhat reduced by geometry optimization at higher levels of theory[144] but remains well above 109°. The larger angle could be taken as reflecting the persistence of the degree of N≡N triple-bond character upon N-protonation, which would prefer a linear H—N—N group. This explanation is consistent with the small change calculated for the N—N distance upon N-protonation.

The calculated properties of the metal ion coordinated N$_2$O are shown in Tables 21–23. As with protonation, metal ion coordination can take place at either the terminal nitrogen or oxygen ends. All the resonating valence bond structures for N$_2$O (I–IV) predict the middle nitrogen atom to be neutral or positively charged and, therefore not conducive to binding a cation. Pre-positioning a proton near the middle nitrogen atom with RHF or MP2 optimization gave the HNNO$^+$ structure with no stationary state for middle nitrogen atom attachment. Such a structure has actually been identified as a transition state connecting HNNO$^+$ and NNOH$^+$ [144].

However, contrary to the result for protonation, metal ion coordination favours terminal N-atom attachment over O-atom attachment (Table 21). This preferred stabilization is found at every level of correlation (MP2 and both QCISD). Only at the RHF level is oxygen coordination of Cu$^+$, Ag$^+$ or Au$^+$ energetically more stable than the corresponding nitrogen-bound complex. This must be the simple electrostatic result as has been shown for Li$^+$ interaction with N$_2$O, even at correlated levels[149]. The total energy differences between MNNO$^+$ and NNOM$^+$ for a given M$^+$ show a damped, converging oscillation with level of theory. If the alternating stabilities are averaged, the results give MNNO$^+$ more stable than NNOM$^+$ by about 12, 5 and 8 kcalmol^{-1} for M = Cu, Ag and Au, respectively. The actual bond dissociation energy for NNOCu$^+$ is difficult to arrive at because of the apparent lack of convergence of the correlation level treatment used here. For Au$^+$ and Ag$^+$ the calculated QCISD(T)//MP2 energies in Table 21 seem to be converging to metal ion binding energies of 29 and 20 kcalmol^{-1}, respectively, for MNNO$^+$.

The preferred nitrogen atom coordination site for these metal ions is reflected in the relative M$^+$—N and M$^+$—O calculated equilibrium distances at the MP2 level. Here, contrary to expectation, the M$^+$—N bond length is from 0.10 Å (Cu$^+$) to 0.18 Å (Au$^+$) shorter than M$^+$—O. The opposite (and expected) result is obtained at the RHF level for

TABLE 23. MP2 calculated harmonic vibrational frequencies (in cm^{-1}) for MN_2O^{+a}

N_2O Exp.[b]		Calculated							
	N_2O	CuN_2O^+		AgN_2O^+		AuN_2O^+		HN_2O^+	
		O^e	N^f	O^e	N^f	O^e	N^f	O^e	N^f
		101	173[d]	59	129[d]	87	104[d]	449	587
		291	332	178	187	226	228	479	636
596[d]	594[d]	548	629	560	624	550	591	1049	751
		570	645	569	633	574	595	1384	1343
1298	1251	1236	1347	1244	1315	1217	1336	2143	2402
2282	2159	2104	2368	2110	2317	2080	2363	3477	3427
6.8[c]	6.6[c]	6.9[c]	8.1[c]	6.7[c]	7.6[c]	6.8[c]	7.7[c]	12.8[c]	13.1[c]

[a]MP2 optimized geometry; CEP[$3^{sp}2^d/3^s$] basis set for NNO and RCEP[$4^{sp}3^d$] for M^+, with five d-type components.
[b]Experimental harmonic frequencies from Reference 141.
[c]Zero-point energy, in kcalmol^{-1}.
[d]Doubly degenerate.
[e]M^+ attachment at O.
[f]M^+ attachment at N.

Cu^+ and Ag^+ which have shorter calculated equilibrium M^+—O lengths relative to M^+—N, at similar geometries for the given coordination site. For a given basis set type and theoretical level, the calculated $MNNO^+$ binding energies are generally smaller than those for $HCNM^+$ (Table 13) and larger than for MNN^+ (Table 9), and the M^+—N bond lengths vary accordingly (Tables 22, 14 and 10).

The variation in the MP2 optimized N—N and N—O bond lengths follows the discussion on the protonated structures above in terms of ionic resonance structures **II** to **IV**. These equilibrium bond lengths for the metal ion complexes are bounded by their values in free N_2O and in the corresponding attachment site protonated species. Thus, for N-coordination the N≡N distance is virtually unchaged, as appropriate to the linear geometry which preserves the triple bond. The N—O bond length decreases by ca 0.02 Å as structure **II** is more important relative to **III** and **IV** in $MNNO^+$ compared to free NNO. For $NNOM^+$, structures **III** and **IV** seem more important as N—N decreases in length by a little over 0.01 Å, and the N—O distance gets larger by ca 0.02 Å.

The $MNNO^+$ structure is calculated to be linear, at both the RHF and the MP2 levels. The linear geometry would be expected on the basis of the alignment of the metal ion with the direction having the larger NNO polarizability[67] for a ligand with an almost zero dipole moment. In a model using localized MO distributed polarizabilities[150] the larger interaction energy of a metal ion with the nitrogen end of NNO is easy to envision. Alternately, the highest occupied σ MO on NNO has a large component on the terminal nitrogen atom (σ_N), as do both the HOMO and LUMO π molecular orbitals[133]. These distributions are convenient for both $L \rightarrow M [\sigma \rightarrow (n+1)s, \sigma \rightarrow (n+1)p]$ and $M \rightarrow L (nd \rightarrow \pi^*)$ dative bonding. The M^+ orbital populations in Table 22 show moderate amounts of each type of bonding for the linear $MNNO^+$ complexes.

These σ_N HOMO and π LUMO orbital characters are less convenient for a linear $NNOM^+$ geometry and the metal ion apparently prefers a bent structure in order to induce HOMO $\pi \rightarrow (n+1)s$ $L \rightarrow M$ dative bonding to supplement the weaker $\sigma \rightarrow (n+1)s$ interaction. The M^+ orbital populations in Table 22 show less dative bonding for $NNOM^+$ than for $MNNO^+$ for all types of orbital interactions. In accord with the bent $NNOM^+$ structure and increased N—O distance, Table 23 shows a decrease in the calculated N—O stretch frequency in the 1250 cm^{-1} region upon metal ion coordination. The N—N

TABLE 24. CIS calculated excited state energies (in eV) for MNNO^{+a}

Assignment		CuNNO+ Asymptote Exp.	CuNNO+ Asymptote Calc.	CuNNO+ Complex Calc.	AgNNO+ Asymptote Exp.	AgNNO+ Asymptote Calc.	AgNNO+ Complex Calc.	AuNNO+ Asymptote Exp.	AuNNO+ Asymptote Calc.	AuNNO+ Complex Calc.
$^3[\pi,(n+1)s]$	$^3CT_1$	5.17^b	6.71^c	8.95^c	5.32^b	7.06	8.57^c	3.67^b	5.87^c	8.21^c
$^1[\pi,(n+1)s]$	$^1CT_1$	5.17^b	6.71^c	9.52^c (0.000)	5.32^b	7.06^c	9.02^c (0.004)	3.67^b	5.87^c	8.66^c (0.006)
$^3[nd,(n+1)s]$	^3D	2.81^d	4.12^e	4.44– 5.03^e	5.03^d	5.72^e	5.76– 6.29^e	2.29^d	2.83^e	3.05– 3.87^e
$^1[nd,(n+1)s]$	^1D	3.26^d	4.81^e	5.33 5.57^c (0.005) 5.73^c	5.71^d	6.52^e	6.63 6.88^c (0.006) 6.97^c	3.67^d	3.93^e	4.40 4.65^c 4.81^c (0.006)
$^3(\pi,\pi^*)$	$^3\Sigma_u^+$	4.28^g	4.39	4.25	4.28^g	4.39	4.27	4.28	4.39	4.29
$^3(\pi,\pi^*)$	$^3\Delta_u$	—	5.26^c	5.06^c	—	5.26^c	5.09^c	—	5.26^c	5.09^c
$^3(\pi,\pi^*)$	$^3\Sigma_u^-$	—	6.13	5.84	—	6.13	5.89	—	6.13	5.88
$^1(\pi,\pi^*)$	$^1\Sigma_u^-$	4.54^g	6.13	5.84	4.54^g	6.13	5.89	4.54^g	6.13	5.88
$^1(\pi,\pi^*)$	$^1\Delta_u$	6.81^g	6.53^c	6.19^c	6.81^g	6.53^c	6.27^c	6.81^g	6.53^c	6.23^c
$^3[nd,(n+1)p]$	^3P	8.33^d	9.67^i	8.68^f	10.16^d	—	9.93^f	—	9.12^j	8.12^f
$^1[nd,(n+1)p]$	^1F	8.92^d	10.30^i	9.30^f (0.051)	11.05^d	11.32^j	—	—	9.94^j	9.01^f (0.061)
$^3[\pi,(n+1)p]$	$^3CT_2$	8.98^h	9.93^k	—	9.06^h	10.18^k	—	8.62^h	—	—
$^1[\pi,(n+1)p]$	$^1CT_2$	8.98^h	9.93^k	—	9.06^h	10.18^k	10.58^f (0.046)	8.62^h	—	—

^{a}MP2 optimized ground state geometries. Values in parentheses are oscillator strengths. For M = Cu, $n = 3$; M = Ag, $n = 4$; M = Au, $n = 5$. Metal ion attachment is at the terminal N atom.

^{b}Based on the IP (12.89 eV) of NNO (Reference 133) and EA of M$^+$ into the $(n + 1)$s atomic orbital (References 22 and 23).

^{c}Doubly degenerate.

^{d}From References 22 and 23.

^{e}Five components.

^{f}First of multiple components.

^{g}From Reference 150.

^{h}Based on the IP (12.89 eV) of NNO (Reference 133) and EA of M$^+$ into the $(n + 1)$p atomic orbital (References 22 and 23).

frequency (ca 2160 cm^{-1}) also decreases. For MNNO$^+$, on the other hand, both the N—O and N—N stretch modes increase in energy upon metal ion complexation. The significant difference in N—N stretch frequency between N coordination and O coordination could be used as a means of chemical identification of these different species, for both metal ions and protonation.

The electronic absorption spectrum of bare N$_2$O has been summarized and interpreted in recent tabulations[61, 151]. The ground state electronic structure of NNO relevant to this discussion can be written as $\pi_1^4\,\sigma_N^2\,\pi_2^4\,(\pi_3^*)^0$. The lower energy part of the excitation spectrum is dominated by spin triplet and spin singlet $\pi_2 \to \pi_3^*$ ($\pi \to \pi^*$) transitions. The only dipole allowed excitation of this group is to the $^1\Sigma_u^+$ state with a vertical transition energy of 9.66 eV and an oscillator strength of 0.36[151]. This assigned state was not found among the lowest twenty vertical excited states generated for Table 24. A lower energy transition observed at 8.52 eV and assigned as $\pi_2 \to \sigma^*$ is calculated here at 10.44 eV ($f = 0.001$). The corresponding triplet state is predicted at 9.91 eV and a $^3(\sigma_N, \pi^*)$ excitation is calculated at 9.91 eV. None of these transitions is found in the complexes in the energy range covered here and they have not been listed in Table 24. Where calculated can be compared to observed, the differences are small, except for the $^1\Sigma_u^-$ state which is ca 1.6 eV too high in energy.

The electronic excitation spectrum of the bare metal ion has been discussed above. The asymptotic energy values of L $\to$ M electron transfer are based on the vertical ionization potential of N$_2$O from the π_2 MO which is reported at 12.89 eV[133]. The next ionization process is from the σ_N MO at 16.38 eV. Because of its high energy, the L $\to$ M CT spectrum is not expected to involve this σ MO even though the metal ion bonding is in a σ configuration. Thus, the electronic excitations in the asymptotic region (metal ions + N$_2$O) have the $nd \to (n + 1)$s triplet state at lowest energy, followed by either the CT or $\pi_2 \to \pi_3^*$ transitions, depending on the metal ion. The ordering of the singlet states among these three types of excitations is metal ion dependent. None of the asymptote excited states in Table 24 carry electric dipole intensity.

The effect of complexation between the metal ions and N$_2$O at the terminal nitrogen atom is also shown in Table 24. The N$_2$O $\pi_2 \to \pi_3^*$ states seem to be only mildly affected and generally move to slightly lower excitation energies. The metal ion based transitions, ^3D, ^3P, ^1D and ^1F, are all spin and/or dipole forbidden from the totally symmetric (^1S) electronic ground states. Upon attachment to the ligand they are found to split into their multiplet components and generally shift to higher energies relative to asymptote. The singlet excited states also pick up some intensity through mixing with allowed monocentre and CT transitions in the lower symmetry of the complex relative to the metal ion. The $\pi_2 \to (n + 1)$s L $\to$ M CT transitions are consistently shifted to higher energy upon complexation, by 1.5–2.8 eV relative to asymptote, independent of metal ion. The singlet states are calculated to be particularly shifted because they interact strongly with the $nd \to (n + 1)$s excitation of the same symmetry. These latter are shown in Table 24 as the second (Cu$^+$ and Ag$^+$) or third (Au$^+$) line for the ^1D states in the complexes with a small oscillator strength. The assignments here would benefit greatly from a natural orbital analysis of the CIS wave function[152]. In any event, since the dipole forbidden NNO $^1(\pi, \pi^*)$ states cannot acquire intensity in the same-symmetry linear MNNO$^+$ complex, the lowest-energy dipole allowed transition will be a component of the complex perturbed ^1D state mixed with the ^{1}CT$_1$ state.

VII. FCN

FCN is the lightest member of the cyanogen halides. Although a gas at room temperature, it polymerizes rapidly and is difficult to maintain as an isolated monomer unit at room temperature[153]. The presently calculated data for FCN, FCNH$^+$ and FCNM$^+$ (M = Cu, Ag and Au) are summarized in Tables 2, 3 and 25–28. FCN is linear and the experimental geometry[82] shows an equilibrium C—F distance of 1.262 Å and a C≡N distance of 1.159

TABLE 25. Energies (in a.u.) of MFCN$^+$ complexes[a]

M$^+$	BF[b]	RHF	MP2	QCISD	QCISD(T)
Cu$^+$	97[c,e]	− 234.094934 (34.8)			
	97[d,e]	− 234.083926 (31.9)	− 235.073851 (50.5)	− 235.010242 (49.8)	− 235.013619 (36.2)
	111[f]	− 234.084370 (32.2)	− 235.331280 (55.3)	− 235.254556 (52.8)	− 235.268596 (42.3)
Ag$^+$	97[c,e]	− 184.207378 (25.4)			
	97[d,e]	− 184.200673 (25.2)	− 184.910660 (30.9)	− 184.911760 (30.4)	− 184.936248 (30.5)
	111[f]	− 184.201523 (25.8)	− 184.272470 (35.9)	− 185.246542 (33.4)	− 185.279868 (34.2)
Au$^+$	97[c,e]	− 173.916411 (30.2)			
	97[d,e]	− 173.907922 (29.0)	− 174.573087 (43.0)	− 174.579493 (42.3)	− 174.604868 (42.7)
	111[f]	− 173.909846 (30.2)	− 174.903884 (51.2)	− 174.877748 (46.7)	− 174.911205 (48.0)

[a]Numbers in parentheses are binding energies in kcalmol^{-1}.
[b]Number of basis functions; CEP[3^{s}P2^d/3^{s}1^P] basis set with five d-type components for FCN.
[c]RHF optimized.
[d]MP2 optimized.
[e]RCEP[4^{s}P3^d] with five d-type components for M$^+$.
[f] RCEP[4^{s}P3^{d}2^f] with five d-type and seven f-type components for M$^+$. Complex geometry taken from 97 basis function MP2 optimization.

Å. Again, the observed bond lengths are bracketed by the RHF and MP2 optimized values, but closer to the latter. The MP2 errors in the C—F and C≡N distances are 0.017 Å and 0.015 Å, respectively. The measured dipole moment of FCN is 2.12D[154] compared to the MP2 calculated value of 2.18D (+FCN–). The MP2 calculated harmonic vibrational frequencies are shown in Table 27. The observed C≡N fundamental stretch frequency is 2319 cm^{-1} [155] compared to the calculated harmonic vibrational energy of 2227 cm^{-1}. The calculated harmonic frequency is expected to be larger than the observed fundamental value, so that the error here is probably larger than the 92 cm^{-1} difference between the above two numbers. The measured F—C fundamental frequency is 1076 cm^{-1} [155] compared to the MP2 calculated harmonic value of 1037 cm^{-1}. Again, the calculated harmonic value should be larger than the observed fundamental frequency because of anharmonicity effects. The bending mode is observed at 451 cm^{-1} [155] and is calculated at 445 cm^{-1}. This is the unusually good agreement for this mode found here for simple linear ligand systems.

FCN protonates preferentially at the nitrogen atom to form a linear geometry. The structural parameters are shown in Table 26. Attempts to locate a prependicular protonated structure by initially positioning the proton off the carbon atom leads spontaneously to the linear FCNH$^+$ structure in the geometry optimization. In FCNH$^+$ both the F—C and C≡N bond distances are seen to contract upon protonation of FCN. Accordingly, both the calculated C—F and C≡N harmonic vibrational frequencies shift to higher energy upon N-protonation. The largest change in the atomic charges is on the carbon atom, which becomes substantially more positive with the proton attached to the nitrogen atom.

TABLE 26. Mulliken populations and geometric parameters for FCNM⁺ᵃ

M⁺	Mulliken								Bond Lengths (Å)			Angles (deg)	
	M⁺ populations				Charges								
	s	p	d	f	M⁺	F	C	N	M–N	C≡N	C–F	MNC	FCN
	—	—	—	—	—	-0.12	0.45	-0.33	—	1.127	1.249	—	180
	—	—	—	—	—	-0.07	0.32	-0.25	—	1.174	1.279	—	180
H⁺	—	—	—	—	0.44	0.03	0.76	-0.24	1.004	1.113	1.206	180	180
	—	—	—	—	0.44	0.10	0.63	-0.17	1.017	1.146	1.231	180	180
Cu⁺	2.10	6.08	9.98	—	0.83	-0.04	0.58	-0.38	2.031	1.124	1.225	180	180
	2.26	6.14	9.89	—	0.71	0.03	0.47	-0.22	1.830	1.162	1.248	180	180
	2.27	6.15	9.86	0.03	0.69	0.03	0.47	-0.20					
Ag⁺	2.06	6.05	9.98	—	0.91	-0.05	0.56	-0.42	2.340	1.125	1.228	180	180
	2.12	6.07	9.95	—	0.86	0.01	0.44	-0.31	2.197	1.165	1.252	180	180
	2.13	6.07	9.87	0.09	0.84	0.01	0.44	-0.30					
Au⁺	2.15	6.05	9.92	—	0.87	-0.04	0.59	-0.42	2.204	1.123	1.225	180	180
	2.30	6.08	9.80	—	0.82	0.03	0.48	-0.33	2.035	1.161	1.248	180	180
	2.33	6.09	9.71	0.11	0.74	0.03	0.49	-0.29					

ᵃBasis sets, energies and levels of theory for the neutral and protonated FCN, and complexes are described in Tables 2, 3 and 25, respectively.

TABLE 27. MP2 calculated harmonic vibrational frequencies (in cm^{-1}) for MFCN^{+} [a]

Exp[d] FCN	Calculated				
	FCN	FCNCu^{+}	FCNAg^{+}	FCNAu^{+}	FCNH^{+}
		158[b]	107[b]	135[b]	425[b]
		351	210	260	550[b]
451[b]	445[b]	496[b]	486[b]	484[b]	1156
1076	1037	1171	1133	1161	2425
2319	2227	2374	2334	2383	3618
6.1[c]	5.9[c]	7.4[c]	7.0[c]	7.2[c]	13.1[c]

[a] MP2 optimized geometry; CEP[3^{sp}2^{d}/3^{s}1^{p}] basis set for FCN and RCEP[4^{sp}3^{d}] for M^{+}, with five d-type components.
[b] Doubly degenerate.
[c] Zero-point energy, in kcal mol^{-1}.
[d] Experimental fundamental vibrational frequencies from Reference 155.

Valence shell contraction on carbon could account for the shortened F—C and C≡N bond lengths. For HCNH^{+} (Table 14) the C≡N bond also shortens relative to HCN while the atomic charge on carbon increases, but the H—C bond length increases. A general explanation of bond strengthening upon protonation of the less electronegative base in bidentate bases has recently been given[156].

The proton affinity of FCN has been estimated at 151 kcal mol^{-1} [65, 157]. The QCISD (T)/ /MP2 calculated value in Table 3 is 169.4 kcal mol^{-1} which, after subtracting the 7.2   kcal mol^{-1} ZPE difference between FCNH^{+} and FCN from Table 27, gives an adjusted PA of 162.2 kcal mol^{-1}. Other theoretical estimates have also placed the proton affinity of   FCN at about 5 kcal mol^{-1} less than for HCN[158]. As described above, the difference in the (ZPE adjusted) QCISD(T) calculated (167.9 kcal mol^{-1}) and measured[65, 66] PA of HCN is very small. The 162.2 kcal mol^{-1} estimate of the PA of FCN is also expected to be reason-ably close to experiment, as were all the other PA calculated here. The reported 151 kcal mol^{-1} value should therefore be re-examined.

Only the linear geometry for the metal ion complexes coordinated to the nitrogen atom was examined. Presumably it should be possible to obtain fluorine complexed (or proto-nated) structures but these weren't probed here. For Al^{+} coordination to FCN, only the N coordinated linear structure was found[159]. The QCISD(T)/PP//MP2/PV calculated metal ion binding energies for the metal ion complexes, presented in Table 25, show a Au^{+}>Cu^{+}>Ag^{+} trend in going down the Periodic Table, although all the other theoretical levels predict Cu^{+}>Au^{+}. Similar to all protonation energy difference, the metal ion dissociation energy is 3–4 kcal mol^{-1} smaller than for the corresponding HCNM^{+} complex at the QCISD(T) level. ZPE corrections are in the 1.1–1.5 kcal mol^{-1} range (Table 27).

The metal ion complexed geometries at the MP2 level in Table 26 show the same bond shortening trend for the F—C and C≡N bonds relative to FCN found for the protonated species, except to a lesser extent, as expected. The metal ion atomic orbital populations in the FCN complexes are very similar to those for the HCN complexes except that the $(n+1)$s orbital occupation numbers are smaller in the FCN complexes. This correlates with their smaller metal ion binding energies, which can be attributed to the higher ionization potential of FCN[160] relative to HCN. Substituent (X) effects on the charge distribution of the C≡N bond in XCN and the effect of cations have recently been examined[161]. The calculated harmonic frequencies in Table 27 have the metal–ligand stretch vibration as the second entry in each column for the metal ion complex, and the double degenerate bending mode as the top entry. These normal mode frequencies are difficult to compare between

TABLE 28. CIS calculated excited state energies (in eV) for FCNM^{+a}

Assignment		FCNCu$^+$			FCNAg$^+$			FCNAu$^+$	
	Asymptote		Complex	Asymptote		Complex	Asymptote		Complex
	Exp.	Calc.	Calc.	Exp.	Calc.	Calc.	Exp.	Calc.	Calc.
$^3[nd, (n+1)s]$ ^3D	2.81^d	4.12^e	4.15–4.96^e	5.03^d	5.72^e	5.49–6.22^e	2.29^d	2.83^e	2.93–3.98^e
$^1[nd, (n+1)s]$ ^1D	3.26^d	4.81^e	5.09–5.63^e	5.71^d	6.52^e	6.40^f 6.82^c (0.009)	3.67^d	3.93^e	4.27–4.88^e
$^3[\pi, (n+1)s]$ $^3CT_1$	5.93^b	7.02^c	8.77^c	6.08^b	7.36^c	10.21^c	4.43^b	6.18^c	8.54^c
$^1[\pi, (n+1)s]$ ^1CT	5.93^b	7.02^c	9.81^c (0.047)	6.08^b	7.36^c	10.58^c (0.020)	4.43^b	6.18^c	9.53^c (0.020)
$^3(\pi, \pi^*)$ $^3\Sigma^+$	g	5.63	6.04	g	5.63	6.00	g	5.63	6.06
$^3(\pi, \pi^*)$ $^3\Delta$	g	6.75^c	7.09^c	g	6.75^c	7.08^c	g	6.75^c	7.13^c
$^3(\pi, \pi^*)$ $^3\Sigma^-$	g	7.78	8.01	g	7.78	8.05	g	7.78	8.05
$^3(\pi, \sigma^*)$ $^3\Pi$	g	7.61^c	—	g	7.61^c	—	g	7.61^c	—
$^3(\sigma, \pi^*)$ $^3\Pi$	g	9.40^c	—	g	9.40^c	—	g	9.40^c	—
$^1(\pi, \pi^*)$ $^1\Sigma^-$	g	7.78	8.01	g	7.78	8.05	g	7.78	8.05
$^1(\pi, \pi^*)$ $^1\Delta$	g	8.24^c	8.35^c	g	8.24^c	8.46^c	g	8.24^c	—
$^1(\pi, \sigma^*)$ $^1\Pi$	g	8.89^c (0.071)	—	g	8.89^c (0.071)	—	g	8.89^c (0.071)	—
$^3[\sigma, (n+1)s]$ $^3CT_2$	6.84^k	9.82	—	6.99^k	10.17	—	5.34^k	8.98	—
$^1[\sigma, (n+1)s]$ $^1CT_2$	6.84^k	9.82	—	6.99^k	10.17	—	5.34^k	8.98	—
$^3[nd, (n+1)p]$ ^3P	8.33^d	9.67^i	8.43^f	10.61^d	—	9.62^f	—	9.12^i	7.84–8.80^i

(*continued*)

TABLE 28. (continued)

Assignment		FCNCu$^+$			FCNAg$^+$			FCNAu$^+$		
		Asymptote		Complex	Asymptote		Complex	Asymptote		Complex
		Exp.	Calc.	Calc.	Exp.	Calc.	Calc.	Exp.	Calc.	Calc.
^{1}[nd, (n + 1)p]	^1F	8.92^d	10.30^j	9.13^{c,f} (0.062)	11.05^d	11.32^j	10.25^{c,f} (0.054)	—	9.94^j	8.37^{c,f} 8.78^c (0.097)
^{3}[π, (n + 1)p]	$^3CT_3$	9.74^h	—	9.17^f	9.82^h	10.481	8.50^f	9.38^h	—	—
^{1}[π, (n + 1)p]	$^1CT_3$	9.74^h	10.23^l	9.88^f (0.283)	9.82^h	10.481	9.10^c (0.048)	9.38^h	—	9.63^{c,f}

^{a} MP2 optimized ground state geometries. Values in parentheses are oscillator strengths. For M = Cu, n = 3; M = Ag, n = 4; M = Au, n = 5. Metal ion attachment is at the N atom.
^{b} Based on the IP (13.65) of FCN (Reference 160) and EA of M$^+$ into the (n + 1)s atomic orbital (References 22 and 23).
^{c} Doubly degenerate.
^{d} From References 22 and 23.
^{e} Five components.
^{f} First of multiple components.
^{g} See Text.
^{h} Based on the IP (13.65) of FCN (Reference 160) and EA of M$^+$ into the (n + 1)p atomic orbital (References 22 and 23).
^{i} Three components.
^{j} Seven components.
^{k} Based on the IP (14.56) of FCN (Reference 160) and EA of M$^+$ into the (n + 1)s atomic orbital (References 22 and 23).
^{l} Six components.

different substituents and different cations because of the different degrees of mode mixing in each case. Comparison with Table 15, however, shows that the $C\equiv N$ vibrational energy increases more for FCN than for HCN complexation, for a given metal ion.

The CIS calculated vertical excitation energies for the FCN metal ion complexes is given in Table 28 using the MP2 optimized geometries. The isolated FCN unit electronic absorption spectrum is not well known but has been extensively studied for the other cyanogen halide (XCN, X = Cl, Br and I)[151, 162]. Like the isoelectronic NNO system, the valence excited states are expected to be dominated by the $\pi_2 \to \pi_3^*$ transitions. However, whereas in NNO the energy gap between the HOMO π and σ is 3.49 eV[133], in FCN this gap is only 0.91 eV[160] so that the σ HOMO, which is mainly localized on the nitrogen atom, can be expected to be involved in the same energy region, in generating both intramolecular and CT excited states. In addition, the inductive effect of the very electronegative fluorine atom is expected to stabilize the σ^* LUMO, so that it may also contribute excitations (from the HOMO π and σ) in the valence electron excited state spectrum of both FCN and the metal ion complexes.

This is shown in Table 28 where, in addition to the usual $\pi \to \pi^*$ multiplet distribution of spin triplet and singlet transitions, the σ HOMO and σ^* LUMO give rise to $\sigma \leftrightarrow \pi$ and $\sigma \to (n + 1)$s excitations. However, as described above also for the HCN complexes, the σ HOMO in FCN is strongly stabilized by its interaction with the metal ion centre in FCNM$^+$. The σ^* LUMO is apparently strongly mixed with the metal ion $(n + 1)$sp hybrid MO and repelled to higher energies. The net result is that all the electronic excitations in the complexes involving the σ HOMO and σ^* LUMO are shifted to higher energies, out of the (lowest twenty excitations of each spin type) range of Table 28 relative to the asymptotic units. The $\pi \to \pi^*$ transitions are found to be mildly shifted to higher energy upon metal ion coordination, as was found also in N_2O. They cannot acquire electric dipole intensity in the linear complex geometry.

The metal ion based excitations, $nd \to (n + 1)$s and $nd \to (n + 1)$p, are also affected by complexation. The former are mildly shifted to higher energy and the singlet states can acquire some intensity through mixing with charge transfer transitions and allowed $nd \to (n + 1)$p state. The latter are stabilized in the complex (ca1 eV) and, although the lowest energy excited state of this type in the free metal ion is the dipole forbidden ^1F state[22, 23], some intensity is acquired through mixing with allowed states in the lower point group symmetry of the complex.

Two types of CT transitions are listed in the energy range of Table 28. The $\pi \to (n + 1)$s excitation is calculated to be 1.1–1.7 eV too high in the asymptote and to be shifted to higher energy in the metal ion complexes by 1.7–3.4 eV, depending on the metal ion and spin state. Adjusting the singlet state excitation energies for the asymptotic error gives transition energies of 7.8, 8.7 and 9.6 eV for Au$^+$, Cu$^+$ and Ag$^+$, respectively, with $f\sim$0.02–0.05 intensity. The $\pi \to (n + 1)$p excitations, which are higher lying in the asymptotes, are stabilized by complexation and could be located only about 1 eV above the lower energy CT states.

VIII. HCNO

Fulminic acid (HCNO) is one of six possible non-ring species with the empirical formula CHNO. These and other type structures have been enumerated and studied theoretically[163]. Like NNO, H—$C\equiv N$=O has a hypervalent central nitrogen atom[135, 136] and bond structure types like I–IV for NNO are appropriate also to HCNO, where H—C replaces the terminal nitrogen atom. The experimental aspects of HCNO, also called formonitrile oxide to emphasize the HCN^+O^- valence bond structure[136], has recently been reviewed[7]. Despite extensive experimental structure and spectroscopic studies[164] and application of advanced *ab initio* computational methods to this problem[165, 166], the

TABLE 29. Energies (in a.u.) of the MHCNO$^+$ complexes[a]

M$^+$	BF[b]	X[g]	RHF	MP2	QCISD	QCISD(T)
Cu$^+$	103[c,e]	O	− 226.441945 (38.5)			
	103[d,e]	O	− 226.434090 (41.6)	− 227.426589 (41.5)	− 227.367657 (48.8)	− 227.352198 (19.0)
	117[f]	O	− 226.434599 (41.9)	− 227.682122 (45.1)	− 227.610950 (51.3)	− 227.610406 (27.2)
	103[c,e]	C	− 226.402498 (13.8)			
	103[d,e]	C	− 226.392320 (15.4)	− 226.412540 (32.6)	− 227.344295 (34.2)	− 227.368336 (29.1)
	117[f]	C	− 226.792320 (15.6)	− 227.412540 (38.0)	− 227.344295 (37.4)	− 227.621402 (34.1)
Ag$^+$	103[c,e]	O	− 176.554472 (29.2)			
	103[d,e]	O	− 176.548008 (33.1)	− 177.271777 (27.1)	− 177.272031 (31.3)	− 177.300224 (29.3)
	117[f]	O	− 176.548808 (33.6)	− 177.631973 (31.1)	− 177.606108 (33.8)	− 177.642789 (32.4)
	103[c,e]	C	− 176.5163511 (5.3)			
	103[d,e]	C	− 176.509423 (8.9)	− 177.254737 (16.4)	− 177.250082 (17.5)	− 177.283432 (18.8)
	117[f]	C	− 176.510036 (9.3)	− 177.618843 (22.9)	− 177.585654 (21.0)	− 177.628183 (23.2)
Au$^+$	103[c,e]	O	− 166.260473 (32.1)			
	103[d,e]	O	− 166.253372 (35.7)	− 166.926756 (34.6)	− 166.933785 (39.4)	− 166.962523 (37.6)
	117[f]	O	− 166.255238 (36.8)	− 167.255715 (41.7)	− 167.232540 (44.1)	− 167.268733 (42.7)
	103[c,e]	C	− 166.237566 (17.8)			
	103[d,e]	C	− 166.230150 (21.0)	− 166.928123 (34.4)	− 166.933480 (39.2)	− 166.967411 (40.6)
	117[f]	C	− 166.231781 (22.1)	− 167.265694 (47.9)	− 167.234278 (45.2)	− 167.276995 (47.9)

[a]Numbers in parentheses are binding energies in kcalmol^{-1}; a second number in parentheses is after BSSE correction.
[b]Number of basis functions; CEP[3^{s}2^d/3^{s}1^p] basis set with five d-type components for HCNO.
[c]RHF optimized.
[d]MP2 optimized.
[e]RCEP[4^{s}3^d] with five d-type components for M$^+$.
[f]RCEP[4^{s}3^{d}2^f] with five d-type and seven f-type components for M$^+$. Complex geometry taken from 103 basis function MP2 optimization.
[g]Atom to which M$^+$ is attached.

question is still debated as to whether HCNO is statically linear or bent. Experimentally, fulminic acid is considered to be a classic example of a quasi-linear molecule where the large amplitude and anharmonicity of the HCN bending mode must be taken into account for an accurate fit of the rovibrational energies[167, 168]. The best description proposes that the HCN chain is moderately bent (relative to linear CNO) with a barrier height to inversion at linearity that is below the zero-point energy level of the bending mode. The result is that *ab initio* methods alternate between linear and bent equilibrium structures, depending on the theoretical model and level. All methods agree that the bending potentials is very flat[164–166].

The results obtained here for the neutral, protonated and metal ion complexed HCNO are shown in Tables 2, 3 and 29–32, respectively. The lowest-energy MP2 optimized geometry is bent (by 0.60 kcal mol^{-1} relative to optimized linear) but the RHF method predicts a linear structure. At the optimized bent and linear MP2 geometries the QCISD energy is lower for the linear conformation (by 0.63 kcalmol^{-1}) while QCISD(T) prefers the bent structure (by 0.22 kcalmol^{-1}). This alternation with method and level emphasizes the difficulties encountered in predicting the equilibrium geometry of fulminic acid. In these comparisons both the linear and the bent geometries are individually MP2 optimized. The MP2 optimized linear geometry shows one calculated imaginary harmonic frequency of 326 cm^{-1} (doubly degenerate) in the HCN bending mode. Similar results have been obtained previously[165, 166]. The MP2 calculated dipole moment is 2.56D at the bent geometry and 2.97D at the (also MP2 optimized) linear geometry, compared to the experimental value of 3.10D[169]. This property is better described by the static linear structure.

The bent geometry also has a slightly non-linear CNO angle, where the HCNO conformation is *trans* with respect to the C—N bond. The C—N bond distance lengthens by *ca* 0.01 Å in the bending process. This shows the complicated coupling between all the geometrical aspects of this molecule in its ground state potential energy surface. The MP2 calculated (bent geometry) N=O bond length (Table 30) is 0.01 Å larger than in NNO (Table 22) and the equilibrium C—N distance is 0.02 Å larger than in HCN (Table 14) and FCN (Table 26). These increases reflect a small loss of multiple bonding character in HCNO for these bonds in the bent geometry. However, since the bending motion is apparently more dynamic than static, the fixed bent geometry may exaggerate these effects. The MP2 calculated harmonic frequencies for the bent geometry are shown in Table 31 and compared with experiment[170]. The latter are interpreted in terms of a linear geometry, so that the HCN bending mode cannot be compared between methods. The CNO bending mode (*ca* 570 cm^{-1}) is split in the fixed bent structure and calculated too high compared to a linear bending mode. The two CNO stretch modes with measured fundamental frequencies of 1253 cm^{-1} and 2196 cm^{-1} are close to the calculated harmonic values of 1281 cm^{-1} and 2185 cm^{-1}, respectively. Finally, the C—H harmonic stretch is calculated above experiment, as expected. Clearly, more work is needed to elucidate the dynamic aspects of the ground state potential properties of HCNO.

The protonation of HCNO has recently been treated both experimentally and theoretically[171]. Three possible isomers were probed here (Tables 3, 30 and 31), each having the proton attached to one of the heavy atoms. The initial geometry placing the proton near the nitrogen atom rearranged spontaneously in the geometry optimization process to the H_2CNO^+ form. In previous work[171] the nitrogen attached protonated species was obtained only in a triplet electronic state, while all the structures reported here have a closed-shell singlet electronic ground state. Of the other two isomers obtained, Table 3 shows that the O-attached $HCNOH^+$ structure is more stable than the C-attached H_2CNO^+. The measured proton affinity of HCNO has been reported to be 181.2 kcalmol^{-1} [171]. The QCISD (T) //MP2 value for proton attachment to the oxygen atom is 183.1 kcalmol^{-1}. Subtracting the ZPE difference with the neutral parent HCNO (Table 31) of 7.6 kcalmol^{-1} gives an adjusted theoretical value of 175.5 kcalmol^{-1}, about 6 kcal mol^{-1} below the reported experimental value. This is a larger difference than has been found for most of the already discussed triple-bonded systems above, even for the hypervalent NNO where agreement with experiment was good. It may be that unusual dynamical effects even in protonated HCNO will contribute more than usual to converting the (ZPE adjusted) electronic energy

TABLE 30. Mulliken populations and geometric parameters for MHCNO[+] [a]

| M+ | Mulliken | | | | | | | | | Bond lengths (Å) | | | | Angles (deg) | | |
| | M+ populations | | | | Charges | | | | | | | | | | | |
	s	p	d	f	M+	H	C	N	O	M–X[b]	H–C	C–N	N–O	MXN[b]	HCN	CNO
c	—	—	—	—	—	0.27	0.10	−0.06	−0.31	—	1.054	1.127	1.202	—	180	180
d	—	—	—	—	—	0.26	−0.04	−0.06	−0.16	—	1.061	1.181	1.206	—	180	180
d	—	—	—	—	—	0.23	−0.08	−0.03	−0.13	—	1.064	1.193	1.202	—	149.6	172.1
H+	—	—	—	—	0.45	0.36	0.42	−0.09	−0.14	0.962	1.070	1.111	1.290	107.9	179.4	173.3
	—	—	—	—	0.45	0.36	0.29	0.00	−0.10	0.986	1.077	1.151	1.297	107.0	178.1	171.0
H+	—	—	—	—	0.29	0.29	0.06	0.17	0.20	1.080	1.080	1.233	1.111	117.1	117.1	180.
	—	—	—	—	0.30	0.30	−0.05	0.19	0.27	1.091	1.091	1.257	1.146	116.5	116.5	180.
Cu+	2.07	6.06	9.99	—	0.88	0.33	0.28	−0.15	−0.34	2.007	1.062	1.117	1.239	135.5	180	177.4
	2.16	6.10	9.93	—	0.80	0.33	0.16	−0.08	−0.21	1.891	1.070	1.162	1.241	132.8	179.4	176.0
	2.17	6.10	9.91	0.03	0.79	0.33	0.16	−0.08	−0.20							
Cu+	2.17	6.10	9.99	—	0.74	0.27	−0.10	0.07	0.02	2.125	1.081	1.198	1.133	120.0	117.4	175.5
	2.34	6.15	9.89	—	0.62	0.29	−0.07	0.03	0.12	1.905	1.090	1.220	1.167	119.3	120.9	174.2
	2.36	6.16	9.86	0.03	0.59	0.29	−0.05	0.04	0.13							

Ag+	2.04	6.04	10.00	—	0.92	0.32	0.27	−0.17	−0.34	2.307	1.061	1.118	1.233	142.1	179.6	178.3
	2.07	6.06	9.98	—	0.89	0.32	0.12	−0.11	−0.22	2.244	1.068	1.166	1.234	136.3	179.5	177.5
	2.08	6.05	9.90	0.09	0.88	0.32	0.13	−0.10	−0.21							
Ag+	2.11	6.06	9.98	—	0.86	0.27	−0.18	0.07	−0.03	2.418	1.077	1.187	1.140	117.3	121.7	174.6
	2.15	6.08	9.94	—	0.82	0.29	−0.18	0.01	0.06	2.279	1.080	1.206	1.177	112.9	132.3	172.9
	2.18	6.07	9.86	0.09	0.79	0.29	−0.17	0.01	0.07							
Au+	2.12	6.04	9.95	—	0.89	0.32	0.27	−0.13	−0.36	2.248	1.062	1.117	1.241	129.3	179.9	177.2
	2.22	6.07	9.89	—	0.82	0.33	0.14	−0.05	−0.24	2.151	1.070	1.163	1.246	123.9	179.7	175.6
	2.25	6.07	9.79	0.10	0.78	0.33	0.15	−0.05	−0.20							
Au+	2.31	6.07	9.87	—	0.75	0.27	−0.15	0.08	0.05	2.176	1.080	1.205	1.130	120.9	116.5	176.5
	2.43	6.08	9.76	—	0.73	0.29	−0.20	0.05	0.14	2.042	1.088	1.223	1.167	117.9	122.3	175.7
	2.50	6.08	9.66	0.11	0.65	0.29	−0.15	0.06	0.15							

[a] Basis sets, energies and levels of theory for the neutral and protonated HCNO, and complexes are described in Tables 2, 3 and 29, respectively.
[b] X is defined in Table 29.
[c] RHF.
[d] MP2.

difference to a thermodynamic quantity. The value of the PA of HCNO needs further investigation.

The two protonated structures show some interesting geometric properties (Table 30). Protonation at the carbon atom gives a linear structure with two equivalent C—H bonds. The C—N bond lengthens considerably (ca 0.06 Å at the MP2 level) relative to the (bent) parent HCNO and the N—O distance shortens by about the same amount. Both changes correlate with the C—N bond going down from triple towards double bond character and the N—O bond character increasing from a double towards a triple bond type. In the latter case, lone pair electron density is pulled from oxygen into the N=O region to stabilize the $H_2C=\overset{+}{N}=O$ valence bond structure having the positive charge on nitrogen. The result is a bond structure that looks more like $H_2C=N\equiv\overset{+}{O}$, with the positive charge on oxygen. In fact, in Table 30, the oxygen atom undergoes the largest depletion of atomic charge (from – to +) upon protonation at the carbon atom. The carbon protonated isomer is the less stable form at all levels of theory (Table 3).

Protonation at the oxygen atom produces a non-linear, zig-zag planar structure with the opposite geometric changes. Here, the C—N bond decreases in length by ca 0.04 Å and the N—O bond length increases by 0.095 Å. These type changes have already been explained in the NNOH$^+$ case as being due to the enhanced importance of the $H—C\equiv N^+—O^-$ dipolar structure which increases the triple bond character of the C—N bond (over the $H—C^-=N^+=O$ form, with which it is mixed in the parent HCNO), and decreases the degree of double bond character in the N—O bond.

The calculated harmonic vibrational frequencies for both protonated fulminic acids are shown in Table 31. The C—N and N—O stretch modes are mixed in HCNO (2185 cm^{-1} and 1281 cm^{-1}) and in the potonated forms. The higher energy mode increases for O-protonation (2252 cm^{-1}) and decreases for C-protonation (2057 cm^{-1}) as expected on the basis of the C—N bond length changes. However, the lower energy mode decreases in both cases (1040 cm^{-1} and 1180 cm^{-1}, respectively, not in accord with the N—O bond length changes.

TABLE 31. MP2 calculated harmonic vibrational frequencies (in cm^{-1}) for HCNOM^{+a}

HCNO		HCNOCu$^+$		HCNOAg$^+$		HCNOAu$^+$		HCNO$^+$	
Experiment^e	Calculated	O^c	C^d	O^c	C^d	O^c	C^d	O^c	C^d
		113	119	83	96	113	121	371	433
		344	395	240	249	288	377	404	491
		409	595	374	422	403	517	637	740
224^f	486	468	608	406	605	474	569	720	1050
537^f	567	529	621	521	610	519	645	1040	1180
	573	556	821	537	781	566	887	1430	1367
1253	1281	1213	1319	1221	1337	1175	1300	2252	2057
2196	2185	2202	2130	2189	2226	2178	2092	3374	3088
3336	3465	3434	3176	3450	3281	3432	3194	3604	3232
11.9^b	12.2^b	13.2^b	14.0^b	12.9^b	13.7^b	13.1^b	13.9^b	19.8^b	19.5^b

^{a}MP2 optimized geometry; CEP[3^{s}p2^d/3^{s}1^P] basis set for NNO and RCEP[4^{s}p3^d] for M$^+$, with five d-type components.
^{b}Zero-point energy, in kcalmol^{-1}.
^{c}M$^+$ attachment at O.
^{d}M$^+$ attachment at C.
^{e}Experimental fundamental frequencies from Reference 170.
^{f}Doubly degenerate.

The calculated dissociation energies of metal ion–HCNO, complexed at either the carbon or oxygen atoms, are shown in Table 29. The preferred coordination site is seen to be a function of both the theoretical level and the metal ion. For Cu^+ attachment RHF, MP2 and QCISD//MP2 favour the oxygen atom independent of the basis set, while QCISD(T)//MP2 predicts the carbon atom as more stable. The latter result is due to a large negative contribution of the triples to the QCISD, whose size is basis set dependent. This seems to indicate a severe non-convergence of the perturbation correction for $HCNOCu^+$ and this result may be unreliable. The metal ion binding energy seems to behave reasonably for Cu^+ coordination to the carbon atom. For complexation of Ag^+, all levels of theory predict the oxygen atom site to be preferred [by ca 10 kcalmol^{-1} at the QCISD(T)//MP2 level for both the PV and PP basis sets] and the triples perturbation corrections to the QCISD energies is moderate for attachment to either atom site. For protonated HCNO the oxygen site is more stable by 8.7 kcalmol^{-1} at the same level of theory. Au^+ coordination, on the other hand, is uniformly predicted by all the correlation methods to be more stable at the carbon atom site. At the ZPE difference-adjusted QCISD(T)//MP2 level the energy difference with the oxygen complexed site is 4.6 kcalmol^{-1}. The behaviour of the triples correction here seems reasonable for both site attachments. Only at the RHF level is the oxygen site preferred for Au^+ coordination. Au^+ is predicted to have the largest binding energy to HCNO at the QCISD(T)//PP and MP2//PP levels but the QCISD energies predict Cu^+ attached to oxygen to be more stable.

The MP2 optimized geometric structures of both the oxygen and carbon atom bound complexes are shown in Table 30. For a given atom coordination of the metal ions the conformations are very similar. For metal ion attachment to the oxygen atom the HCNO skelton is essentially linear, more so than for protonation. The MON angle is larger for metal ion coordination compared to protonation. This result correlates with the gradual change of bonding from purely electrostatic, which would prefer a 180° angle for maximum alignment with the dipole moment direction, to purely covalent in the protonated species, with a HON angle characteristic of covalent XOY binding. Ag^+ with the smallest binding energy has the largest MON angle while Cu^+ and Au^+, with the larger binding energies of the metal ions and the larger covalent components, have the smaller angles. For carbon atom attachment the metal ion complexes have a skeletal geometric structure very similar to of the protonated species.

The bond lengths in the metal ion complexes change relative to isolated fulminic acid in the same way as for the corresponding protonated species, but to an intermediate degree. Thus, for coordination at the oxygen atom the C—N bond length decreases and the N—O bond distance increases. This has been interpreted as a preferential stabilization of the zwitterion structure $H—C≡N^+—O^-$ by cation attachment to oxygen. On the other hand, for the carbon site attachment complex the C—N bond length increases and the N—O bond distance gets smaller. Here, the preferentially stabilized dipolar structure is $H—C^-=N^+=O$, where the N=O bond may even take on partial triple bond character. The C—H bond length, of course, increases due to repulsion with the cation. The intermediate nature of these changes result from the larger M—O and M—C equilibrium bond lengths and smaller binding energies for the metal ions relative to protonation.

The calculated harmonic vibrational frequencies for the metal ion complexes are listed in Table 31. The normal coordinate analysis does not show a single low energy mode that can be unambiguously correlated with the pure metal–ligand stretching motion, either for the oxygen or carbon attachment complexes. The skeletal modes are all mixed, as was noted above for the protonated species. However, the C—H stretch fundamental is easily identified as the highest energy mode and from the normal coordinate analysis. The C—H frequency is here seen to be considerably shifted to lower energy by coordination to the carbon atom, while in the oxygen attached complex it stays almost unchanged. This large

TABLE 32. CIS calculated vertical excited state energies (in eV) for HCNOM^{+a}

Assignment		HCNOCu$^+$			HCNOAg$^+$			HCNOAu$^+$		
		Asymptote		Complex	Asymptote		Complex	Asymptote		Complex
		Exp.	Calc.	Calc.	Exp.	Calc.	Calc.	Exp.	Calc.	Calc.
^{3}[nd, (n + 1)s]	^3D	2.81^d	4.12^e	3.77– 4.49^e	5.03^d	5.72^e	5.20– 5.97^e	2.29^d	2.83^e	2.53– 3.50^e
^{1}[nd, (n + 1)s]	^1D	3.26^d	4.82^e	4.65^f 4.91^e (0.006) 5.06 (0.004)	5.71^d	6.52^e	6.24^f (0.006) 6.60 (0.006)	3.67^d	3.93^e	4.19^f (0.006)
^{3}[π, (n + 1)s]	^{3}CT$_1$	3.11^b	4.48^c	7.35 7.71	3.16^b	4.82^c	6.95 7.11	1.61^b	3.64^c	6.42 6.55
^{1}[π, (n + 1)s]	^{1}CT$_1$	3.11^b	4.48^c	7.85 (0.001) 8.13 (0.045)	3.16^b	4.82^c	5.98 (0.008) 7.72 (0.064)	1.61^b	3.64^c	3.83 (0.004) 7.00 (0.000)
^{3}(π, π*)	$^3\Sigma^+$	g	4.09	4.82	g	4.09	4.68	g	4.09	4.79
^{3}(π, π*)	$^3\Delta$	g	4.91^c	5.74 5.75	g	4.91^c	5.57 5.60	g	4.91^c	5.69 5.71
^{3}(π, π*)	$^3\Sigma^-$	g	5.65	6.56	g	5.65	6.39	g	5.65	6.53
^{1}(π, π*)	$^1\Sigma^-$	4.70^g	5.65	6.53	g	5.65	6.41	g	5.65	6.44
^{1}(π, π*)	$^1\Delta$	g	5.97^c	6.90 6.93 (0.000)	g	5.97^c	6.73 6.74 (0.000)	g	5.97^c	6.84 6.85 (0.004)

		$M = Cu$			$M = Ag$			$M = Au$		
$^3[nd, (n+1)p]$	^3P	8.33[d]	—	9.03[f]	10.16[d]	—	10.28[f]	—	—	8.63[f]
$^1[nd, (n+1)p]$	^1F	8.92[d]	—	9.74[c,f] (0.021)	11.05[d]	—	10.84[c,f] (0.009)	—	—	9.33[c,f] (0.006)
$^3[\pi, (n+1)p]$	$^3CT_3$	6.92[h]	7.69[i]	8.46[f]	7.00[h]	7.94[i]	9.07[f]	6.56[h]	7.73[i]	7.89[f]
$^1[\pi, (n+1)p]$	$^1CT_2$	6.92[h]	7.69[i]	8.98[f] (0.023)	7.00[h]	7.94[i]	9.48[f] (0.030)	6.56[h]	7.73[i]	8.62[f] (0.047)
$^3(\pi, \sigma^*)$	$^3\Pi$	g	8.21[c]	—	g	8.21[c] 10.12	9.98	g	8.21[c]	—
$^1(\pi, \sigma^*)$	$^1\Pi$	g	9.10[c] (0.001)	—	g	9.10[c] (0.001) 10.77 (0.073)	10.76 (0.009)	g	9.10[c] (0.001)	—

[a] MP2 optimized ground state geometries. Values in parentheses are oscillator strengths. For M = Cu, n = 3; M = Ag, n = 4; M = Au, n = 5. Metal ion attachment is at the N atom.
[b] Based on the IP (10.83 eV) of HCNO (Reference 172) and EA of M^+ into the $(n + 1)s$ atomic orbital (References 22 and 23).
[c] Doubly degenerate.
[d] From References 22 and 23.
[e] Five components.
[f] First of multiple components.
[g] From Reference 173; see text.
[h] Based on the IP (10.83 eV) of HCNO (Reference 172) and EA of M^+ into the $(n + 1)p$ atomic orbital (References 22 and 23).
[i] Six components.

difference in behaviour can be the basis of spectroscopic identification of the type of metal ion complex formed with HCNO, with regard to attachment site.

The valence electron photoelectron spectrum of HCNO has been measured and discussed[172], based on a linear or quasi-linear structure. The electronic structure description of linear HCNO is the same as that for NNO discussed above, except for the ordering of the occupied π and σ MOs. The lowest energy vertical ionization processes are both out of π molecular orbitals with energies of 10.83 (π_2) and 15.92 (π_1) eV. The highest occupied σ MO is next at 17.79 eV. Therefore, in the low energy electronic excitation spectrum only transitions from the HOMO π MO (π_2) are expected to be important. The gas-phase absorption spectrum of fulminic acid has been observed[173, 174] above 2400 Å and only a single band system is observed with a vibrational progression in the HCN bending mode. This excitation has been assigned as a $\pi \rightarrow \pi^*$ transition to a bent HCNO with an average transition energy of ca 4.7 eV. At higher energies a rising absorption continuum is found which probably is related to the photoisomerization of HCNO to a more stable isomer structure[162, 166].

As has been detailed previously[151], the lowest energy $\pi \rightarrow \pi^*$ transitions in these linear 16 valence electron systems give rise to three possible singlet excited states: $^1\Sigma^-$, $^1\Delta$ and $^1\Sigma^+$, and corresponding triplets. Comparison with NNO and later with HN$_3$ favours the $^1\Sigma^-$ assignment for the observed ca 4.7 eV band. In any event, the calculated vertical electronic excitation spectrum in Table 32 shows that only transitions from the π HOMO fall in the lower energy region. The $^1\Sigma^+$ state, as usual, is not calculated in this energy range. The effect of metal ion complexation at the oxygen atom (Table 32) is seen to generally shift the $\pi \rightarrow \pi^*$ type transitions to higher energy by 0.7–1.0 eV. Triplet and singlet $\pi \rightarrow \sigma^*$ excitations are also predicted to appear in HCNO above the 8-eV region, although the accuracy of these numbers is not known. In any event, they are expected (and for Ag$^+$ found) to be shifted to considerably higher energies upon complexation. All degeneracies are removed in the non-linear HCNOM$^+$ complex structures.

The two types of L $\rightarrow$ M charge transfer processes, $\pi \rightarrow (n + 1)$s and $\pi \rightarrow (n + 1)$p, are also predicted to shift to higher energies upon metal ion coordination. If the energy of the ^{1}CT$_1$ transition is adjusted (downward) by the error between calculated and experiment in the asymptotic region, then the lowest energy L $\rightarrow$ M CT bands should fall at 6.5 eV, 4.4 eV and 1.6 eV, respectively, for the Cu$^+$, Ag$^+$ and Au$^+$ complexes. The high value for Cu$^+$ results from strong mixing at the configuration level between the 3d $\rightarrow$ 4s and $\pi \rightarrow$ 4s excitations. A more accurate treatment may be needed to resolve this interaction. The $\pi \rightarrow (n + 1)$p excitations are also calculated to blue shift in the higher energy region of the spectrum. As usual, the nd $\rightarrow (n + 1)$s excitations split without much shifting. The nd $\rightarrow (n + 1)$p transitions in the pure metal ions are calculated too high and don't appear in Table 28 in the 'Asymptote' category. However, they are calculated to red shift upon metal ion complexation and therefore do appear in the 'Complex' columns.

IX. HN$_3$

Hydrazoic acid (HN$_3$) is another member of the set of 16 valence electron 1,3 dipoles. As with both NNO and HCNO, HNNN has a hypervalent central nitrogen atom that is both doubly and triply bound to its neighbouring atoms. The calculated properties of the neutral, protonated and metal ion complexed hydrazoic acid are tabulated in Tables 2, 3 and 33–36. The photochemistry and detonation properties of HN$_3$ have been of particular interest.

The RHF and MP2 optimized geometries of HN$_3$ are shown in Table 34. Unlike fulminic acid, hydrazoic acid is statically bent with a *trans* conformation. The measured HNN angle is 108.8° and the NNN angle is 171.3°[175]. The corresponding MP2 calculated quantities are 108.6° and 171.1°. The experimental (MP2 calculated) bond lengths are H—N = 1.015 Å

(1.022 Å), N=N = 1.243 Å (1.255 Å) and N≡N = 1.134 Å (1.151 Å). Again, the MP2 error in thr triple bond length is *ca* 0.02 Å. The N=N calculated bond length of 1.255 Å corresponds about exactly to the measured N=N double bond length in diimide[82] of 1.252 Å. The calculated N≡N triple bond length of 1.151 Å in hydrazoic acid is even slightly

TABLE 33. Energies (in a. u.) of the MHN_3^+ complexes[a]

M^+	BF^b	X^g	RHF	MP2	QCISD	QCISD(T)
Cu^+	103[c,e]	N^h	−224.636503			
			(20.4)			
	103[d,e]	N^h	−224.621823	−225.663871	−225.588279	−225.605024
			(17.3)	(45.1)	(42.0)	(31.8)
	117[f]	N^h	−224.622167	−225.921197	−225.832561	−225.859257
			(41.9)	(49.9)	(51.3)	(27.2)
	103[c,e]	N^i	−224.660414			
			(35.4)			
	103[d,e]	N^i	−224.645958	−225.665457	−225.598761	−225.602619
			(32.4)	(46.1)	(48.6)	(30.3)
	117[f]	N^i	−224.646475	−225.922845	−225.843134	−225.858919
			(32.7)	(50.9)	(51.7)	(37.3)
Ag^+	103[c, e]	N^h	−174.752345			
			(13.2)			
	103[d,e]	N^h	−174.740919	−175.499808	−175.490137	−175.524582
			(12.0)	(25.0)	(22.9)	(24.2)
	117[f]	N^h	−174.741620	−175.861738	−175.824989	−175.868324
			(12.5)	(30.0)	(25.9)	(28.0)
	103[c,e]	N^l	−174.7715711			
			(25.2)			
	103[d,e]	N^i	−174.760526	−175.504287	−175.500261	−175.531445
			(24.3)	(27.8)	(29.2)	(28.6)
	117[f]	N^i	−174.761490	−175.866821	−175.835613	−175.875650
			(24.9)	(33.2)	(32.6)	(32.6)
Au^+	103[c,e]	N^h	−164.457233			
			(15.4)			
	103[d,e]	N^h	−164.443249	−165.157351	−165.152271	−165.188136
			(12.7)	(34.0)	(31.2)	(33.3)
	117[f]	N^h	−164.444923	−164.487400	−164.449970	−164.493977
			(13.7)	(41.8)	(35.3)	(38.2)
	103[c,e]	N^i	−164.483908			
			(32.1)			
	103[d,e]	N^i	−164.471514	−165.169059	−165.171196	−165.203016
			(30.4)	(41.4)	(43.1)	(42.6)
	117[f]	N^i	−164.473755	−165.502243	−165.471414	−165.511304
			(31.8)	(51.1)	(48.7)	(49.1)

[a]Numbers in parentheses are metal ion binding energies in kcalmol^{-1}.
[b]Number of basis functions; CEP[$3^{sp}2^d/3^s1^p$] basis set with five d-type components for HN_3.
[c]RHF optimized.
[d]MP2 optimized.
[e]RCEP[$4^{sp}3^d$] with five d-type components for M^+.
[f]RCEP[$4^{sp}3^d2^f$] with five d-type and seven f-type components for M^+. Complex geometry taken from 103 basis function MP2 optimization.
[g]Atom to which M^+ is attached.
[h]Terminal N atom.
[i]N(H).

TABLE 34. Mulliken populations and geometric parameters for MHN_3^{+a}.

M⁺	Mulliken									Bond lengths (Å)				Angles (deg)		
	M⁺ populations				Charges											
	s	p	d	f	M⁺	H	Nᵇ	Nᶜ	Nᵈ	M–Nᵉ	H–N	N–Nᶜ	N–Nᵈ	MNN	HNN	NNNᵈ
H⁺ᶠ	—	—	—	—	—	0.38	−0.01	0.48	0.15	—	1.006	1.245	1.091	—	107.4	173.9
	—	—	—	—	—	0.29	−0.37	0.22	−0.14	—	1.022	1.255	1.151	—	108.6	171.1
H⁺ᵍ	—	—	—	—	0.40	0.40	−0.11	0.43	−0.11	1.022	1.022	1.153	1.153	115.4	115.4	171.0
	—	—	—	—	0.41	0.41	−0.09	0.37	−0.09	1.034	1.034	1.178	1.180	121.3	121.2	166.4
	—	—	—	—	0.37	0.37	−0.33	0.45	0.14	1.002	1.002	1.285	1.073	116.7	116.7	180.
	—	—	—	—	0.38	0.38	−0.28	0.38	0.14	1.018	1.018	1.289	1.127	116.9	116.9	180.
Cu⁺	2.07	6.08	9.98	—	0.87	0.33	−0.37	0.44	−0.27	2.073	1.010	1.196	1.104	176.9	112.0	173.1
	2.23	6.16	9.89	—	0.72	0.36	−0.24	0.41	−0.26	1.840	1.024	1.212	1.153	177.2	117.4	170.5
	2.24	6.16	9.87	0.03	0.70	0.36	−0.24	0.41	−0.24							
Cu⁺	2.12	6.09	9.98	—	0.81	0.32	−0.58	0.47	−0.01	2.040	1.005	1.278	1.079	127.2	107.4	176.6
	2.26	6.14	9.90	—	0.70	0.34	−0.41	0.36	0.02	1.872	1.023	1.279	1.138	127.9	107.1	175.6
	2.28	6.14	9.88	0.03	0.68	0.34	−0.40	0.36	0.02							

Ag$^+$	2.05	6.03	9.99	—	0.93	0.32	−0.41	0.32	−0.17	2.421	1.009	1.204	1.102	175.4	111.2	173.2
	2.12	6.08	9.95	—	0.84	0.35	−0.27	0.32	−0.24	2.191	1.023	1.219	1.154	176.8	115.7	170.4
	2.13	6.08	9.87	0.09	0.83	0.35	−0.27	0.33	−0.24							
Ag$^+$	2.08	6.05	9.98	—	0.88	0.31	−0.62	0.45	−0.03	2.341	1.006	1.273	1.080	128.1	107.2	175.9
	2.13	6.08	9.96	—	0.84	0.33	−0.50	0.33	−0.01	2.233	1.022	1.273	1.140	126.4	108.0	174.6
	2.14	6.07	9.87	0.09	0.82	0.33	−0.48	0.33	0.00							
Au$^+$	2.08	6.04	9.95	—	0.93	0.33	−0.40	0.32	−0.18	2.293	1.010	1.199	1.102	175.7	111.7	173.1
	2.23	6.08	9.82	—	0.86	0.35	−0.25	0.30	−0.27	2.043	1.024	1.215	1.152	176.0	116.8	169.9
	2.27	6.09	9.73	0.11	0.81	0.36	−0.25	0.27	−0.19							
Au$^+$	2.19	6.06	9.91	—	0.84	0.33	−0.63	0.47	−0.01	2.206	1.004	1.274	1.079	126.1	108.6	176.5
	2.32	6.08	9.81	—	0.78	0.34	−0.51	0.36	0.02	2.075	1.021	1.275	1.138	124.8	109.8	176.2
	2.37	6.09	9.72	0.11	0.71	0.35	−0.44	0.35	0.03							

[a] Basis sets, energies and levels of theory for the neutral and protonated HNNN, and metal ion complexes are described in Tables 2, 3 and 33, respectively.
[b] N(H).
[c] Middle N atom.
[d] Terminal N atom.
[e] The first set of entries for each M$^+$ is for attachment to the terminal N atom and the second set is for attachment to N(H).
[f] HNNNH$^+$ gauche conformation.
[g] Pyramidal H$_2$NN conformation.

smaller than the N≡N triple bond length in N_2O (Table 22). The total dipole moment of HN_3 is apparently not known, but a recent extensive basis set and multi-configuration study gave a best theoretical result of $1.839D$[176] at the experimental geometry. The MP2 density value for the optimized geometry in Table 34 is $1.856D$. The measured fundamental vibrational frequencies[177] for HN_3 are compared with the MP2 calculated harmonic frequencies in Table 35. The N—H stretching mode (3519 cm^{-1}) is calculated too high, as expected, as is the NNN antisymmetric stretch (2257 cm^{-1}). The corresponding symmetric NNN stretch (1131 cm^{-1}) and the bending modes are all closer to experiment. A general study of HN_3 and all the halogen azides has recently been reported[178].

The properties of protonated hydrazoic acid have recently been reviewed[179, 180]. As in the case of fulminic acid there are basically two types of protonated structures, resulting from protonation at one or the other of the terminal nitrogen atoms. Protonation at the bare terminal nitrogen atoms gives a HNNNH$^+$ arrangement that can adopt one of several conformations, including all-*anti* (W form), *gauche* and *syn–anti*. The energies of the all-*anti* and *gauche* forms are reported in Table 3 and the geometry of the *gauche* conformer (dihedral HN...NH angle of 90°) only is listed in Table 34. Protonation at the N(H) end gives the H_2NNN^+ structure, which can have the H_2NN fragment either planar or non-planar (pyramidal). All theoretical results, including those presented here, predict that the H_2NNN^+ (aminodiazonium ion) isomer is more stable than the HNNNH$^+$ (iminodiazenium ion) form.

The proton affinity of HN_3 has been reported at 179 kcalmol^{-1} [157] and 176.6 kcalmol^{-1} [179]. The QCISD(T)//MP2 level PA value in Table 3 to form H_2NNN^+ is 188.7 kcalmol^{-1}, from which is subtracted the ZPE difference with the neutral parent (Table 35) of 7.9 kcalmol^{-1}. This gives an adjusted PA of 180.8 kcalmol^{-1}, which is in good agreement with both reported measured values. It has recently been claimed[180b] that the true experimental PA of HN_3 is somewhat (<9 kcalmol^{-1}) higher than reported experimentally[179] from proton transfer equilibria. The diazenium cations is QCISD(T) calculated to be 24.5 kcal mol^{-1} less stable than the diazonium ion (Table 3).

TABLE 35. MP2 calculated harmonic vibrational frequencies (in cm^{-1}) for MHN$_3^{+a}$

HN$_3$		CuHN$_3^+$		AgHN$_3^+$		AuHN$_3^+$		H$_2$NNN^{+f}	HNNNH^{+g}
Exp.^{e}	Calc.	N^c	N^d	N^c	N^d	N^c	N^d		
		176	149	143	115	96	138	411	558
		181	385	156	270	114	192	471	583
		351	395	212	277	256	344	655	743
527	546	631	532	618	540	575	527	1077	844
588	576	692	566	669	560	603	593	1330	852
1168	1131	941	1163	974	1161	938	1168	1630	1336
1273	1237	1342	1371	1306	1347	1327	1359	2206	2173
2150	2257	2376	2220	2346	2225	2384	2235	3428	3376
3324	3519	3499	3499	3508	3509	3495	3516	3528	3405
12.9^b	13.2^b	14.6^b	14.7^b	14.2^b	14.3^b	14.0^b	14.4^b	21.1^b	19.8^b

^{a} MP2 optimized geometry; CEP[3^{s}P2^d/3^{s}1^p] basis set for HNNN and RCEP[4^{s}P3^d] for M$^+$, with five d-type components.
^{b} Zero-point energy, in kcalmol^{-1}.
^{c} M$^+$ attachment at terminal N.
^{d} M$^+$ attachment at N(H).
^{e} Experimental fundamental vibrational frequencies; from Reference 177.
^{f} Pyramidal conformation.
^{g} Gauche conformation.

TABLE 36. CIS calculated vertical excited state energies (in eV) for HNNNM[+a]

Assignment[c]		HNNNCu+			HNNNAg+			HNNNAu+		
		Asymptote		Complex	Asymptote		Complex	Asymptote		Complex
		Exp.	Calc.	Calc.	Exp.	Calc.	Calc.	Exp.	Calc.	Calc.
$^3[nd, (n+1)s]$	^3D	2.81[d]	4.12[e]	4.30–4.97[e]	5.03[d]	5.72[e]	5.65–6.25[e]	2.29[d]	2.83[e]	3.00–3.91[e]
$^1[nd, (n+1)s]$	^1D	3.26[d]	4.82[e]	5.21–5.65[e]	5.71[d]	6.52[e]	6.54–7.59[e,j]	3.67[d]	3.93[e]	4.34–4.86[e,j]
$^3[\pi, (n+1)s]$	$^3CT_1$	3.02[b]	4.40	7.45	3.17[b]	4.75	7.12	1.52[b]	3.56	6.68
		4.53	6.28	8.99	4.53	6.63	8.65	3.03	5.44	8.22
$^1[\pi, (n+1)s]$	$^1CT_1$	3.02[b]	4.40	7.81	3.17[b]	4.75	6.41	1.52[b]	3.56	7.12
				(0.004)			(0.010)			(0.000)
		4.53	6.28	9.29	4.53	6.63	8.90	3.03	5.44	8.48
				(0.028)			(0.009)			(0.042)
$^3(\pi, \pi^*)$	$^3\Sigma^+$	g	3.95	3.54	g	3.95	3.58	g	3.95	3.62
$^3(\pi, \pi^*)$	$^3\Delta$	g	4.05	3.79	g	4.05	3.82	g	4.05	3.78
$^3(\pi, \pi^*)$	$^3\Sigma^-$	4.47	5.07	5.07	4.56	5.07	5.07	4.51	5.07	—
$^3(\pi, \sigma^*)$	$^3\Pi$	g	6.68	5.98	g	6.68	6.11	g	6.68	6.03
		g	7.44	—	g	7.44	8.62	g	7.44	—
$^3(\sigma, \pi^*)$	$^3\Pi$	—	8.54	9.11	9.90	8.54	9.11	—	8.54	—
				8.19						
$^1(\pi, \pi^*)$	$^1\Sigma$	4.70[g]	4.60	4.14	4.70[g]	4.60	4.20	4.70[g]	4.60	4.12
		(0.0006)	(0.0002)	(0.0002)	k	k	(0.0002)	k	k	(0.)
$^1(\pi, \pi^*)$	$^1\Delta$	6.20[g]	6.34	5.84	6.20[g]	6.34	5.93	6.20[g]	6.34	5.85
		(0.003)	(0.016)	(0.003)	k	k	(0.005)	k	k	(0.003)
		6.56	7.39	6.72	6.56	7.39	6.85	6.56	7.39	6.76
		(0.015)	(0.003)	(0.003)	k	k	(0.004)	k	k	(0.003)

(continued)

TABLE 36. (continued)

Assignment		HNNNCu⁺ Asymptote Exp.	HNNNCu⁺ Asymptote Calc.	HNNNCu⁺ Complex Calc.	HNNNAg⁺ Asymptote Exp.	HNNNAg⁺ Asymptote Calc.	HNNNAg⁺ Complex Calc.	HNNNAu⁺ Asymptote Exp.	HNNNAu⁺ Asymptote Calc.	HNNNAu⁺ Complex Calc.
$1(\pi, \sigma^*)$	$^1\Pi$	7.29[g] (0.01) 7.94 (0.3)	8.07 (0.010) 9.87 (0.170)	—	7.29[g] [k] 7.94 [k]	8.07 [k] 9.87 [k]	9.15 (0.003) —	7.29[g] [k] 7.94 [k]	8.07 [k] 9.87 [k]	—
[nd, (n + 1)p]	^3P	8.33[d]	—	8.86[f]	10.16[d]	9.11[f]	10.17[f] (0.063)	—	9.12[i]	7.72[f]
[nd, (n + 1)p]	^1F	8.92[d]	—	8.85[f] (0.055)	11.05[d]	—	—	—	9.94[f]	8.61[f] (0.047)
[π, (n + 1)p]	$^3CT_2$	6.83[h] 8.34	7.62[i] 9.49[i]	8.79[f]	6.91[h] 8.42	7.87[i] 9.75[i]	9.37[f]	6.47[h] 7.98	7.66[i] 9.53[i]	8.67[f]
[π, (n + 1)p]	$^1CT_2$	6.83[h] 8.34	7.62[i] 9.49[i]	9.20[f] (0.958)	6.47[h] 8.42	7.87[i] 9.75[i]	9.39[f] (0.761)	6.47[h] 7.98	7.66[i] 9.53[i]	8.97[f] (0.856)

[a] MP2 optimized ground state geometries. Values in parentheses are oscillator strengths. For M = Cu, $n = 3$; M = Ag, $n = 4$; M = Au, $n = 5$. Metal ion attachment is at the N atom.

[b] Based on the first two IP (10.74 eV and 12.25 eV) of HNNN (Reference 172) and EA of M⁺ into the $(n + 1)$s atomic orbital (References 22 and 23).

[c] π molecular orbitals are symmetry split.

[d] From References 22 and 23.

[e] Five components.

[f] First of multiple components.

[g] From Reference 151; see text.

[h] Based on the first two IP (10.74 eV and 12.25 eV) of HNNN (Reference 172) and EA of M⁺ into the $(n + 1)$p atomic orbital (References 22 and 23).

[i] Three components.

[j] Some components have low ($f \sim$ 0.005–0.009) calculated intensity.

[k] Same as for Cu⁺.

Two conformations of the aminodiazonium ion were studied here. The energies of both are given in Table 3 but only the geometry of the lower energy, H_2NN pyramidal form, is given in Table 34. The vibrational frequencies of the pyramidal form of this isomer are tabulated in Table 35. The higher energy conformer is planar and a frequency calculation at its optimized geometry gives one imaginary harmonic vibrational energy (508 cm^{-1}) with the normal mode coordinates showing the two hydrogen atoms bending out of plane synchronously. The electronic energy barrier to inversion at the QCISD(T)//MP2 level is 1.6 kcalmol^{-1}. The barrier at the RHF level is 1.9 kcalmol^{-1}.

The bond distances in pyramidal H_2NNN^+ show the middle $N{=}N$ bond distance increasing while the end $N{\equiv}N$ bond length decreases relative to HN_3. These trends indicate an emphasis on the $H{-}N^-{-}N^+{\equiv}N$ structure upon protonation. In accordance with this interpretation, Table 34 shows that the atomic charge on the central nitrogen atom becomes more positive while N(H) remains substantially negative upon protonation to form H_2NNN^+. On the other hand, in forming $HNNNH^+$ the precursor $(H)N{=}N$ bond decreases in length and the end bond $N{\equiv}N$ distance lengthens, in accord with the emphasized $HN{=}N^+{=}NH$ valence bonding structure, which gives the symmetric geometry and charge distribution shown in Table 34.

The calculated harmonic vibrational frequencies in Table 35 show that the NNN antisymmetric stretch at 2257 cm^{-1} decreases to 2206 cm^{-1} in H_2NNN^+, while the symmetric stretch shifts from 1131 cm^{-1} to 1077 cm^{-1}. It is difficult to interpret these moderate changes upon protonation in terms of bonding and bond length changes, because the normal modes are not pure linear motions of only the heavy atoms. In $HNNNH^+$ the symmetric stretch increases to 1336 cm^{-1} relative to HN_3 and the antisymmetric stretch decreases to 2173 cm^{-1}. The intensity of the symmetric stretch mode should decrease substantially for this isomer because of the quasi-inversion symmetry of the cation which is shown, for example, by a zero coefficient for the middle nitrogen atom in the normal mode analysis vector.

The optimized geometries of the metal ion complexes resulting from coordination at the N(H) to give $MN(H)NN^+$ or at the opposite end terminal nitrogen atom ($HNNNM^+$) are shown in Table 34, and their energies in Table 33. No attempt was made to probe the large number of possible conformations for either structure type and it is possible that one or more lower energy conformations exist. Only the nature of each calculated stationary state was tested with a frequency calculation. These were all found to be real and the calculated harmonic vibrational fundamentals for the MP2 geometries are shown in Table 35. Based on the results for the protonated hydrazoic acid[179, 180] the energy differences between these metal ion complex geometries and any energetically lower-lying conformations are expected to be small.

Cu^+, Ag^+ or Au^+ attached to the terminal nitrogen atom of HNNN give complexes that have almost linear MNN bond angles. The NNN angle is also almost linear and the MNNN group forms a *syn* or shallow arc conformation. The N—H bond points out of the arc. Focusing on the QCISD(T)/PP//MP2/PV results in Table 33, the $HNNNM^+$ structure is generally the less stable of the two site-attachment isomers considered here for the metal ion complexes, for all the metal ions. However, the MP2 optimized metal ion–N bond distances are consistently shorter for the higher energy imine complex ($HNNNM^+$) compared to the amine analogue [$MN(H)NN^+$] complexes, although the RHF shows the expected ordering of amine<imine. In $HNNNM^+$ the middle $N{=}N$ bond length decreases and the terminal $N{\equiv}N$ bond distance is essentially unchanged relative to free hydrazoic acid at the MP2 level. The almost linear MNN bond angle also shows that the terminal $N{\equiv}N$ bond in HN_3 is virtually undisturbed in the complexes, unlike the symmetric geometry of the $HNNNH^+$ protonated species where an allene-type resonance has been proposed as partly stabilizing the di-imine structure[180b]. In the calculated harmonic vibrational frequencies for the imine-type complexes the metal–ligand stretch mode can be

identified as the third largest energy in Table 35 (351, 212 and 256 cm^{-1} for Cu$^+$, Ag$^+$ and Au$^+$, respectively). For the other normal modes, the N—H (*ca* 3500 cm^{-1}) stretch seems to be virtually unaffected by metal ion complexation at the other end of the molecule, while the (calculated) 1131 cm^{-1} NNN skeletal motion in HN$_3$ is strongly reduced in the imine complexes.

As shown in Table 33, the amine-analogue complexes are more stable at almost every level of theory for each metal ion. The presence of the bonded hydrogen atom makes this nitrogen atom more negative and more attractive to cations than the terminal N atom. The N=N bond increases in length and the terminal N≡N bond decreases in length upon complexation in accord with what is found for the corresponding protonated species, and for the same reasons. The amine-analogue complexes found here are planar even at the MP2 level and the NNN bond angle is nearly 180° (Table 34). Comparing the metal ion atomic orbital populations between the amine- and imine-types of metal ion complexes shows that the L → M charge transfer, mainly into the (n + 1)s orbital, seems to be slightly larger for the amine-type complex. This result is consistent with expectations in terms of the greater relative availability of an amine lone pair of electrons for such an interaction compared to a relatively more stable σ MO on a terminal nitrogen atom, like in N$_2$.

The electronic absorption spectrum of HN$_3$ has been widely investigated[151, 181], mainly to study its photodissociation[182-184]. Interestingly enough, the thermodynamic bond dissociation energy of HN—N$_2$ to form ground state products is only 0.48 eV[182]. Since the photon absorption that leads to this dissociation is to a singlet electronic state and the dissociated products are NH in its ground triplet state, an intersystem crossing must take place. Interest has focussed on the nature(s) of the absorbing state at various excitation wavelengths and the behaviour of the different excited electronic states in the dissociation process.

In the bent ground state equilibrium geometry of HN$_3$ all the degeneracies of the quasi-linear or actually linear 16 valence electron systems like HCNO and NNO are broken. The almost linear NNN frame of the hydrazoic acid means that the π MOs retain their identity and character, more or less[184]. The degenerate π MO pairs are split energetically since the in-plane component interacts with the angled H—N bond while the out-of-plane component retains its pristine π character. This is true of both the occupied and virtual MOs.

The CIS calculated vertical excited states of the HN$_3$ metal ion complexes are shown in Table 36. The assignment designations use linear symmetry term symbols for the HN$_3$ intravalent excited states for convenience and show the symmetry split components in the metal ion complexes. The imine-analogue (HNNNM$^+$) geometry has been adopted also for convenience and by analogy to the HCNO and NNO complexes. The asymptotic energies are independent of the metal ion complex isomer form, in any case. The photoelectron spectrum to give the ionization potentials has been studied and assigned[172]. The first two ionization processes correspond to the vertical ejection of an electron from each of the symmetry split components of the doubly degenerate π HOMO for a linear system at 10.74 eV and 12.25 eV. The next ionization process (>3 eV higher) is, presumably, too high to be relevant to the lower-energy electronic absorption spectrum.

The assignment of the pure HN$_3$ spectrum in terms of symmetry split π → π* and π → σ* singlet states is from McGlynn and coworkers[151, 181], although questions have been raised by the photodissociation experiments as to the number and location of states in the long-wavelength region[183]. The calculated results seem to fit this assignment. A triplet σ → π* excitation is calculated to fall in the higher energy region of the Table for uncomplexed HN$_3$. As expected, however, this transition shifts to higher energies beyond the range of the Table upon complexation. The π → π* excitations all shift to lower energy while π → σ* is also calculated to shift to higher energy upon complexation. The symmetry splittings of some of the 'degenerate' Π and Δ states is sometimes considerable. The lower C_s symmetry of HN$_3$ induces some intensity into the (linear symmetry) forbidden $^1\Sigma^-$ and $^1\Delta$ states. The strongly dipole allowed $^1\Sigma^+$ ($\pi \to \pi^*$) state is not calculated in this energy range but mixes strongly with the $\pi \to (n + 1)$p $^1CT_2$ state and enhances its intensity considerably.

The pure metal ion transitions originate from the nd atomic orbital and their direction of shift upon ligand attachment depends on the terminating orbital. Excitations to the $(n + 1)$s atomic orbital are calculated to shift to higher energies and transitions to the higher-lying $(n + 1)$p orbital are red shifted. Some of the intermediate energy components of the symmetry split ^1D state in the complexes acquire moderate intensity ($f\sim0.008$). The L $\to$ M charge transfer transitions are calculated to be blue shifted, irrespective of the metal ion terminating orbital. These results are similar to those found in the previous systems discussed here, as is natural from their valence electronic structure similarity. It should be noted that both the orbital and configurational mixing in the complexes is considerable and, as has been noted above, the assignments would benefit greatly from a natural orbital analysis[152]. The asymptote error adjusted $^1CT_1$ states in the metal ion complexes (HNNNM$^+$) are predicted to fall at 6.4 eV, 4.8 eV and 5.1 eV, respectively, for Cu$^+$, Ag$^+$ and Au$^+$.

X. HCCF

Monofluoroacetylene (HCCF) or fluoroethyne is a member of the 16 valence electron set of molecules that is not a 1,3 dipole and does not have a hypervalent central nitrogen atom. The properties of HCCF, protonated HCCF and the metal ion complexes are shown in Tables 2, 3 and 37–40. The effect of the fluorine atom substitution on the C—C triple bond can be studied in several ways. The equilibrium goemetry of HCCF is linear and,

TABLE 37. Energies (in a.u.) of MC$_2$HF$^+$ complexes[a]

M$^+$	BF[b]	RHF	MP2	QCISD	QCISD(T)
Cu$^+$	103[c,e]	−230.408170 (20.5)			
	103[d,e]	−230.388426 (10.0)	−231.356298 (37.6)	−231.295998 (33.6)	−231.308373 (27.0)
	117[f]	−230.388713 (10.2)	−231.615647 (43.5)	−231.540355 (36.6)	−231.563332 (33.1)
Ag$^+$	103[c,e]	−180.523873 (25.4)			
	103[d,e]	−180.518291 (11.5)	−181.196852 (20.2)	−181.205007 (18.9)	−181.228766 (19.9)
	117[f]	−180.518881 (11.9)	−181.560836 (26.6)	−181.540483 (22.4)	−181.573496 (24.3)
Au$^+$	103[c, e]	−170.239067 (21.9)			
	103[d, e]	−170.222387 (13.3)	−170.872578 (40.7)	−170.883873 (37.7)	−170.909798 (39.9)
	117[f]	−170.224374 (14.5)	−171.212736 (54.8)	−171.184470 (43.6)	−171.220249 (47.7)

[a]Numbers in parentheses are metal ion binding energies in kcalmol^{-1}.
[b]Number of basis functions; CEP[3^{sp}2^d/3^{s}1^p] basis set with five d-type components for HCCF.
[c]RHF optimized.
[d]MP2 optimized.
[e]RCEP[4^{sp}3^d] with five d-type components for M$^+$.
[f]RCEP[4^{sp}3^{d}2^f] with five d-type and seven f-type components for M$^+$. Complex geometry taken from 103 basis function MP2 optimization.

TABLE 38. Mulliken populations and geometric parameters for MC_2HF^+[a]

M^+	Mulliken M^+ populations				Mulliken Charges					Bond lengths (Å)				Angle (deg)		
	s	p	d	f	M^+	H	C	C	F	M–C[b]	H–C	C–C	C–F	MCC	HCC	CCF
	—	—	—	—	—	0.21	-0.14	0.10	-0.17	—	1.054	1.175	1.264	—	179.9	180
	—	—	—	—	—	0.22	-0.16	0.11	-0.17	—	1.061	1.206	1.291	—	179.9	180
H^+	—	—	—	—	0.30	0.30	-0.14	0.49	0.05	1.083	1.083	1.260	1.197	118.9	118.9	180
	—	—	—	—	0.31	0.31	-0.19	0.43	0.14	1.093	1.093	1.270	1.216	119.0	119.0	180
Cu^+	2.14	6.10	10.00	—	0.75	0.28	-0.19	0.25	-0.09	2.250	1.062	1.190	1.244	88.0	165.0	179.7
	2.37	6.19	9.83	—	0.61	0.29	-0.10	0.22	-0.02	1.951[c]	1.074	1.239	1.271	78.5	161.3	171.1
	2.39	6.19	9.81	0.03	0.58	0.29	-0.08	0.23	-0.02							
Ag^+	2.09	6.06	9.99	—	0.86	0.28	-0.22	0.19	-0.10	2.555	1.060	1.187	1.248	88.9	167.0	179.6
	2.15	6.09	9.95	—	0.81	0.29	-0.22	0.14	-0.02	2.331[d]	1.071	1.224	1.268	92.8	158.7	178.0
	2.17	6.08	9.87	0.09	0.78	0.29	-0.20	0.15	-0.02							
Au^+	2.25	6.08	9.91	—	0.76	0.29	-0.22	0.25	-0.07	2.309	1.063	1.200	1.239	90.4	158.8	179.7
	2.52	6.12	9.63	—	0.73	0.30	-0.15	0.15	-0.02	2.106[e]	1.074	1.253	1.272	75.8	161.7	164.0
	2.58	6.12	9.54	0.11	0.64	0.30	-0.10	0.16	-0.01							

[a]Basis sets, energies and levels of theory for the neutral and protonated HCCF, and complexes are described in Tables 2, 3 and 37, respectively.
[b]Attached to C(H).
[c]Cu—C(F) = 2.092 Å.
[d]Ag—C(F) = 2.685 Å.
[e]Au—C(F) = 2.151 Å.

experimentally, the C—C and C—H bond lengths seem to change by only a few thousandths of an Å compared to the parent acetylene[82, 185]. The MP2 optimized structure (Table 38) shows a slight increase in the C—C bond length compared to C_2H_2 (Table 18). The MP2 calculated C—C bond length is, at most, 0.008 Å larger than experiment. The C—F bond length is reported to be in the 1.273–1.281 Å range[82, 185], which is bracketed by the RHF and MP2 optimized values. The measured dipole moment is reported in a compilation[186] at $0.721 D$. The MP2 calculated value is $0.897 D$.

The vibrational frequencies of HCCF have been calculated with a quartic force field[187] and the measured spectrum analysed using a vibration–rotation Hamiltonian[188]. The fitted harmonic frequencies[188] are shown in Table 39, together with the MP2 calculated values. The C—H (ca 3500 cm^{-1}), C—F (ca 1060 cm^{-1}) and C≡C (ca 2250 cm^{-1}) modes show good agreement between experiment and theory, of the same degree as was found also for C—H and C≡C in acetylene (Table 19). Even the (doubly degenerate) bending modes are well described, although this is more than is expected within the theory and basis set level used here[187]. A high-level theoretical study of HCCF has recently appeared[110b].

Protonated HCCF has been examined by several groups[157, 189, 190] who all agree that the most stable equilibrium geometry is the classical structure resulting from protonation at the carbon(H) atom (H_2CCF^+). The MP2 optimized geometry for this structure is given in Table 38. An attempt to generate the MP2 optimized geometry with the proton at C(F) by prepositioning the proton there resulted in spontaneous migration of the proton to the C(H) position. The C(F) protonated structure has been reported at the RHF level[189]. Compared to the parent H—C≡C—F the C—C bond length increases by 0.064 Å and the C—F bond distance decreases by 0.073 Å in going to H_2CCF^+, or what looks almost like $H_2C=C^+=F$. Comparison with the $C_2H_2 \rightarrow C_2H_3^+$ protonation is difficult since the latter has a non-classical (bridging) equilibrium structure. The proton affinity of HCCF has been reported[65] at 165 kcalmol^{-1} and estimated[157] to be 161 kcalmol^{-1}. The QCISD(T)//MP2 value in Table 3 is 166.4 kcalmol^{-1}, from which the ZPE difference between HCCF and H_2CCF^+ of 5.9 kcalmol^{-1} (Table 39) is subtracted to give the adjusted calculated PA of 160.5 kcalmol^{-1}. The HCC(F)H$^+$ structure has been calculated to be ca 18 kcalmol^{-1} less stable[189].

TABLE 39. MP2 calculated harmonic vibrational frequencies (in cm^{-1}) for MC$_2$HF^{+a}.

HCCF ^{b}		CuC$_2$HF$^+$	AgC$_2$HF$^+$	AuC$_2$HF$^+$	H$_2$CCF$^+$
Exp^d	Calc.				
	145	81	174	300	
	326	249	358	398	
	392	385	378	741	
375 ^{b}	375 ^{b}	438	416	462	856
597 ^{b}	616 ^{b}	627	595	601	1077
	658	680	742	1277	
1077	1044	1041	1073	1040	2073
2277	2231	2088	2168	2028	3081
3501	3511	3361	3402	3362	
12.6 ^{c}	12.5 ^{c}	13.0 ^{c}	12.9 ^{c}	13.1 ^{c}	18.6^c

^{a} MP2 optimized geometry; CEP[3^{sp}2^d/3^{s}1^p] basis set for HCCF and RCEP[4^{sp}3^d] for M$^+$, with five d-type components.
^{b} Double degenerate.
^{c} Zero-point energy, in kcal as mol^{-1}.
^{d} Experimental harmonic frequencies from References 188.

TABLE 40. CIS calculated vertical excited state energies (in eV) for MC$_2$HF^{+a}

Assignment		CuC$_2$HF$^+$			AgC$_2$HF$^+$			AuC$_2$HF$^+$		
		Asymptote		Complex	Asymptote		Complex	Asymptote		Complex
		Exp.	Calc.	Calc.	Exp.	Calc.	Calc.	Exp.	Calc.	Calc.
^{3}[nd,(n + 1)s]	^3D	2.81^d	4.12^e	5.62–8.07^e	5.03^d	5.72^e	6.05–7.49^e	2.29^d	2.83^e	5.02–7.78^e
^{1}[nd,(n + 1)s]	^1D	3.26^d	4.82^e	6.20– (0.003) 8.00^e (0.026)	5.71^d	6.52^e	6.74– (0.017) 7.64^e (0.011)	3.67^d	3.93^e	5.60– (0.000) 8.48^e (0.080)
^{3}[π,(n + 1)s]	^{3}CT$_1$	3.78^b	4.78^c	4.84 5.18	3.93^b	5.13^c	4.76 5.14	2.28^b	3.94^c	3.76 3.93
^{1}[π,(n + 1)s]	^{1}CT$_1$	3.78^b	4.78^c	5.69 (0.009) 5.75 (0.022)	3.93^b	5.13^c	5.73 (0.013) 6.08 (0.048)	2.28^b	3.94^c	4.75 (0.022) 4.81 (0.020)
	$^3\Sigma^*$	g	5.14	4.45	g	5.14	5.35	g	5.14	4.23
	$^3\Delta$	g	6.20^c	4.55 5.56	g	6.20^c	6.00 7.29	g	6.20^c	4.53 5.24
	$^3\Sigma^-$ $^3\Pi$	g g	7.09 7.30^c	7.57 9.36 10.08	g g	7.09 7.30^c	— 9.36 —	g g	7.09 7.30^c	7.61 9.28 —
	$^1\Sigma^-$	g	7.09	5.20 (0.000)	g	7.09	7.19 (0.003)	g	7.09	4.93 (0.000)

$^1(\pi,\pi^*)$	7.44^c	g	6.50 (0.027) 8.11 (0.002)	g	7.44^c	7.88 (0.007) —	g	7.44^c	6.17 (0.041) 8.20 (0.002) (0.417)
$^1(\pi,\sigma^*)$	8.54^c (0.066)	g	—	g	8.54^c (0.066)	10.03 (0.008)	g	8.54^c (0.066)	9.91 (0.071) 9.97
$^3[nd,(n+1)p]$	9.67^i	8.33^d	7.45^f	10.16^d	—	10.07^f	—	9.12^i	7.33^f
$^1[nd,(n+1)p]$	—	8.92^d	8.27^f (0.013)	11.05^d	—	—	—	9.94^j	7.94^f (0.007)
$^3[\pi,(n+1)p]$	8.00^k	7.59^h	9.01^f	7.67^h	8.25^k	8.65^f	7.23^h	8.04^k	8.20^f
$^1[\pi,(n+1)p]$	8.00^k	7.59^h	9.13^f (0.086)	7.67^h	8.25^k	8.48^f (0.158)	7.23^h	8.04^k	9.91^f (0.071)

[a] MP2 optimized ground state geometries. Values in parentheses are oscillator strengths. For M = Cu, n = 3; M = Ag, n = 4; M = Au, n = 5. Metal ion attachment is at the N atom.
[b] Based on the IP (11.50 eV) of C_2HF (Reference 193) and EA of M^+ into the $(n+1)$s atomic orbital (References 22 and 23).
[c] Doubly degenerate.
[d] From References 22 and 23.
[e] Five components.
[f] First of multiple components.
[g] See text.
[h] Based on the IP (11.50 eV) of C_2HF (Reference 193) and EA of M^+ into the $(n+1)$p atomic orbital (References 22 and 23).
[i] Three components.
[j] Seven components.
[k] Six components.

The MP2 calculated harmonic vibrational frequencies for H_2CCF^+ are also tabulated in Table 39. As expected, the C—F stretch increases (1044 $cm^{-1} \to 1277$ cm^{-1}), C—H decreases (3511$cm^{-1} \to 3081, 3180$ cm^{-1}) and C≡C decreases (2231 $cm^{-1} \to 2073$ cm^{-1}) upon protonation of HCCF to give the classical H_2CCF^+ structure with equivalent C—H bonds.

The calculated energy and structural results for the metal ion complexes are shown in Tables 37 and 38. The calculated equilibrium geometries at the MP2 level for the MC_2HF^+ complexes show essentially a bridged (non-classical) structure where the metal ion is somewhat closer to C(H) than to C(F). These geometries are similar to those found for the corresponding $MC_2H_2^+$ complexes. Table 38 shows that the difference in the two bond lengths in MC_2HF^+ in the MP2 optimized geometries ranges from only 0.065 Å (Au^+) to 0.354 Å (Ag^+). The HCCF group is almost linear with the CCF angle most linear for Ag^+ and most bent for Au^+, while HCC is most bent for the Ag^+ complex. The bending directions for both H(C) and F(C) are always away from the metal ion, except for F(C) with Ag^+. These small geometry differences correlate with the calculated binding strengths of the metal ion complexes. The weaker the binding energy (Ag^+, Table 37) the more asymmetric the complex. The calculated C—C and C—F bond distances also change in accord with the binding strengths of the complex and these changes relative to free HCCF are much milder than for the protonated species.

The calculated binding energies are shown in Table 37. As expected, these values are somewhat smaller than the corresponding metal ion binding energies to acetylene (Table 17) as the fluorine atom reduces the electron density on its carbon atom. The MC_2HF^+ binding energies follow the usual trend with Au^+ largest and Ag^+ smallest, independent of basis set and theoretical level, and seem to be relatively well converged as a function of theoretical level [MP2 $\to$ QCISD $\to$ QCISD(T)]. The RHF values are more than 50% of the QCISD(T) binding energies and, according to the M^+ atomic orbital populations in Table 38, the correlation levels enhance both the L $\to$ M and M $\to$ L dative bond interactions. The weakest of the metal–ligand bonds, with Ag^+, seems to have a larger electrostatic component. This would be a reasonable interpretation for its having the largest geometry asymmetry which puts it closer to the negatively charged C(H).

The MP2 calculated harmonic vibrational frequencies in Table 39 show all the trends discussed above for protonated fluoroethyne in the C—H, C—C and C—F stretch frequencies. The lowest energy mode (81 cm^{-1} for Ag^+, for example) connects between a more bridging to a more asymmetric structure. The low value for this swing motion in the Ag^+ complex shows how relatively easy such a conversion is for this complex. The next line of vibrational frequencies lists essentially the metal–ligand stretch mode. Here, the Au^+ complex has the largest energy.

The calculated vertical excited state energies for HCCF are shown in Table 40. Both the intra-ligand and L $\to$ M charge transfer transitions depend on the ligand ionization potential. The effect of fluorine substitution on molecular electronic energy levels has been widely noted[191, 192]. Generally, the π ionization energy stays almost unchanged when fluorine atoms substitute hydrogen in the σ framework of a planar molecule[191]. On the other hand, the σ electrons are strongly stabilized, depending on the degree of fluorine substitution. The π^* MO also seems to be unshifted upon fluorine replacing hydrogen in π electron systems, at least for monofluorination, since the position of the V state in partially substituted ethylenes remains essentially unchanged[192]. The σ^* MOs seem to be stabilized like their bonding counterparts[192], but to a greater extent.

These effects have actually also been observed here in the photoionization and electronic excitation spectrum of HCN and FCN (Tables 16 and 28, respectively). Thus the first IP in both molecules is out of a π MO and hardly changes from 13.61 eV to 13.65 eV when F replaces H. On the other hand, the first σ MO (non-bonding on nitrogen) ionization energy goes from 14.00 eV in HCN to 14.56 in FCN[116, 160]. The calculated spectra at the asymptotes (Tables 16 and 28) also show the $\pi \to \sigma^*$ excitation energies decreasing from HCN to FCN by 1–2 eV. The expectation then for HCCF relative to HCCH is for the $\pi \to \pi^*$ excitations to stay at about the same energy, $\pi \to \sigma^*$ to shift to lower energies, $\sigma \to \pi^*$ higher and L $\to$ M CT to stay about the same. The measured IP of C_2H_2 and C_2HF are 11.41 eV[61] and 11.50 eV[193], so previous experience with fluorine substitution[191, 192] appears to be applicable here also.

The CIS calculated vertical excited state energies are shown in Table 40. The bare monofluoroacetylene lower energy excitations include $\pi \to \pi^*$ and $\pi \to \sigma^*$ transitions. The $\sigma \to \pi^*$ excited states don't fall in the energy range of the Table. The double degenerate Δ ($\pi \to \pi^*$) and Π ($\pi \to \sigma^*$) states are split in the lower symmetry of the metal ion complexes. The $^{3,1}(\pi, \pi^*)$ excited states in bare HCCF are calculated to shift uniformly to higher energies by 0.7–0.8 eV relative to acetylene (Table 20). The $\pi \to \pi^*$ states are calculated to mainly (but not always) shift to lower energy upon metal ion complexation of fluoroacetylene and the dipole forbidden spin singlet states acquire some intensity through mixing with the strongly allowed transitions . The $\pi \to \sigma^*$ states are substantially (ca 1.5–2.0 eV) shifted to higher energy upon metal ion coordination, as has been noted in previously discussed systems.

The metal ion $nd \to (n + 1)$s and $nd \to (n + 1)$p transitions behave as usual. The former are calculated to increase in energy and the latter shift to lower energies upon complexation. Both type excited states split into their degeneracy number components and , in the former case, the first four members cluster together while the highest energy state is strongly blue shifted by configuration mixing. It should be noted that some of the assignments in this Table are arbitrary due to strong mixing both on the orbital and configurational levels. Thus, the second $^1\Pi$ state in AuC_2HF^+ at 9.97 eV has a calculated oscillator strength of 0.417. This degree of intensity bears no resemblance to the calculated intensity in the bare ligand ($f = 0.066$) and most certainly is mixed with the more strongly allowed $nd \to (n + 1)$p ^1P state (not shown in Table 40).

The lowest-energy electric dipole allowed charge transfer state $^1CT_1$, arising from the $\pi \to (n + 1)$s transition, is calculated to shift to somewhat higher energies in the metal ion complexes relative to asymptotes. The calculated minus experiment asymptote adjusted $^1CT_1$ energies are 4.7 eV, 4.5 eV and 3.1 eV, respectively, in the Cu^+, Ag^+ and Au^+ complexes. It remains to be investigated whether these states are bound or dissociative[53–55]. If the sum of the calculated binding energies [QCISD(T)/PP/MP2/PV in Table 37] and experimental $^1CT_1$ asymptote energies are compared to the adjusted $^1CT_1$ excitation energies in the complexes, then the $^1CT_1$ states would seem to be dissociative for all three metal ions.

XI. CH$_2$N$_2$

Diazomethane (H_2CNN) is a 16 valence electron 1,3 dipole with a hypervalent central nitrogen atom[135, 136, 194]. The relevant calculated results for the parent, protonated and metal ion complexed species are shown in Tables 2, 3 and 41–44. The precise geometric structure of diazomethane is apparently difficult to determine[195]. The RHF optimized geometry in Tables 2 and 42 shows a planar structure with C=N and N≡N equilibrium bond distances of 1.291 Å and 1.106 Å, respectively. MP2 optimization of an initially planar structure maintains the planarity of the geometry and the linearity of the CNN group. However, a MP2 calculation of the harmonic vibrational frequencies gives one imaginary mode of 243 cm^{-1} corresponding to an out-of-plane motion of the hydrogen atoms. Subsequent MP2 optimization of the non-planar structure results in a geometry that is mildly pyramidal about the carbon atom with a slightly bent CNN group in a trans conformation to the planar CH$_2$ group relative to C=N. The HCH angle bisector makes a 154.1° angle with the C=N bond and C_s symmetry is maintained with equivalent hydrogen atoms. This geometry shows all real harmonic frequencies (Table 43). However, the QCISD energies at the MP2 optimized non-planar geometry are both above the corresponding planar geometry energies (Table 2). These results recall a similar situation with the isoelectronic fulminic acid (Tables 2 and 30) where the static MP2 structure is bent but the molecule has a dynamic quasi-linear geometry (Section VIII). Diazomethane, apparently, has a similar situation with respect to planarity. Higher-level theoretical methods will be necessary to treat adequately these effects. Previous *ab initio* studies of diazomethane have not considered a non-planar structure[135, 136, 194, 196–199], although the H_2C^-— $N^+\equiv N$ valence bond structure[136, 194] suggests such a geometry. The MP2 calculated dipole moments for diazomethane are $1.33D$ (planar) and $1.30D$ (pyramidal), compared to a quoted experimental value of $1.50D$[186]. The direction of the dipole places the negative end on the terminal nitrogen atom, as in the $H_2C= N^+=N^-$ valence bonding structure.

The calculated harmonic vibrational spectrum of (non-planar) diazomethane is compared with the experimental fundamental frequencies in Table 43[200]. The lowest energy mode, as expected, is the CH_2 out-of-plane deformation motion (measured $= 406\,cm^{-1}$) and is calculated too low, appropriate to the above discussion. The next two higher-energy modes, observed at $421\,cm^{-1}$ and $564\,cm^{-1}$, correspond to CNN bending in-plane and out-of-plane, respectively. Here, the calculated values are close to experiment. At the upper end of the energy spectrum the calculated harmonic C—H stretches are larger than experiment, as is the N—N stretch measured at $2102\,cm^{-1}$. This is similar to that found for HN_3 (Table 35). Anharmonicity effects will reduce the harmonic stretch frequencies towards the experimental values[197, 200]. If the tendency towards a pyramidal H_2CN group is real, its effects should be discernable in high-resolution vibration–rotation spectra of diazomethane.

Diazomethane can be protonated either at the carbon atom or at the terminal nitrogen atom. In the former case the methyl diazonium cation, CH_3NN^+, is formed which is assumed to be the lower energy form[157, 180c, 201]. Table 3 shows preferred protonation at the carbon atom to be the result obtained here, also, by a substantial $38.1\,kcalmol^{-1}$ at the QCISD(T)//MP2 level compared to CH_2NNH^+. This is similar to protonation at N(H) being the preferred site for HN_3, but by a smaller energy difference relative to the competing site at the bare nitrogen atom end. For NNO, protonation at the oxygen atom is only slightly more stable (Table 3) than the N-attachment site. The calculated proton affinity $(CH_2NN \rightarrow CH_3NN^+)$ at the QCISD(T)//MP2 level is $223.6\,kcalmol^{-1}$. Subtracting the ZPE difference (Table 43) of $8.5\,kcalmol^{-1}$ gives an adjusted PA of $215.1\,kcalmol^{-1}$. This value is higher than the reported $212\,kcalmol^{-1}$[202], the $201–207\,kcalmol^{-1}$ estimated from core binding energies[157] and the $205\,kcalmol^{-1}$ calculated from enthalpies of formation[65]. Both the calculated and experimental PA values for carbon atom attachment are large compared, for example, to formaldehyde $(H_2C=O)$ which has a proton affinity of 171.7 $kcalmol^{-1}$[65]. A precise experimental determination of the proton affinity of diazomethane is needed.

The geometry of CH_3NN^+ is linear with equivalent C—H bonds[201]. The $N\equiv N$ bond distance in CH_3NN^+ (Table 42) is calculated to increase by at least $0.155\,\text{Å}$ at the MP2 level relative to CH_2NN, which is much larger than the increase in $H_2N=N$ bond length for H_2NNN^+ relative to HN_3 (Table 34). The C—N bond in CH_3NN^+ is single bond length. The $N\equiv N$ bond distance, however, shortens, as expected, in forming the diazonium cation. CH_2NNH^+, the alternate protonated diazomethane, is calculated to be (ZPE corrected) $32.0\,kcalmol^{-1}$ less stable than CH_3NN^+ (Table 3) at the QCISD(T)//MP2 level. In the N-protonated isomer the CNN backbone is slightly bent (Table 42) and both the H—N bond and the CH_2 bond midpoint axis (BMA) are bent *anti* to the CNN bend direction. The N—C—BMA angle is only $178.1°$. The cation maintains over-all C_s symmetry. The equilibrium C=N and $N\equiv N$ bond lengths are calculated to decrease and increase, respectively, in going from CH_2NN to CH_2NNH^+, in accord with the enhanced $H_2C=N^+=N^-$ structure type in the cation at the expense of the CH_2^-—$N^+\equiv N$ form. The

TABLE 41. Energies (in a.u.) of $MCH_2N_2^+$ complexes[a]

M^+	BF^b	X^g	RHF	MP2	QCISD	QCISD(T)
Cu^+	$109^{c,e}$	N	−221.038079 (27.9)			
	$109^{d,e}$	N	−221.028665 (25.6)	−222.015774 (47.8)	−221.960896 (48.9)	−221.967638 (34.8)
	123^f	N	−221.029086 (25.9)	−222.273401 (52.7)	−222.205112 (51.8)	−222.223216 (41.3)
	$109^{c,e}$	C	−221.065744 (45.3)			
	$109^{d,e}$	C	−221.050543 (39.4)	−222.038463 (62.0)	−221.978805 (60.1)	−221.984368 (45.3)
	123^f	C	−221.051108 (39.7)	−222.298269 (68.3)	−222.224323 (63.9)	−222.242538 (53.5)
Ag^+	$109^{c,e}$	N	−171.151269 (19.2)			
	$109^{d,e}$	N	−171.146077 (19.4)	−171.850313 (26.8)	−171.859834 (27.9)	−171.888901 (28.3)
	123^f	N	−171.146891 (19.9)	−172.221237 (37.5)	−172.194765 (31.0)	−172.232737 (32.1)
	$109^{c,e}$	C	−171.174052 (33.3)			
	$109^{d,e}$	C	−171.162177 (29.5)	−171.871828 (40.3)	−171.876868 (38.6)	−171.905238 (38.6)
	123^f	C	−171.163353 (30.2)	−172.239663 (49.0)	−172.215268 (43.8)	−172.253181 (45.0)
Au^+	$109^{c,e}$	N	−160.864846 (26.7)			
	$109^{d,e}$	N	−160.858361 (26.2)	−161.515602 (40.7)	−161.532164 (42.6)	−161.562845 (43.9)
	123^f	N	−160.860334 (27.5)	−161.849533 (50.9)	−161.832665 (48.4)	−161.871876 (50.8)
	$109^{c,e}$	C	−160.909216 (54.5)			
	$109^{d,e}$	C	−160.896041 (49.9)	−160.563619 (70.8)	−161.575188 (69.6)	−161.604178 (69.8)
	123^f	C	−160.897952 (51.1)	−161.906254 (86.5)	−161.879500 (77.8)	−161.917412 (79.4)

[a] Numbers in parentheses are metal ion binding energies in kcalmol⁻¹.
[b] Number of basis functions; CEP[$3^sP2^d/3^s1^P$] basis set with five d-type components for CH_2N_2.
[c] RHF optimized.
[d] MP2 optimized.
[e] RCEP [4^sP3^d] with five d-type components for M^+.
[f] RCEP[$4^sP3^d2^f$] with five d-type and seven f-type components for M^+. Complex geometry taken from 109 basis function MP2 optimization.
[g] Atom to which M^+ is attached.

CH_2NNH^+ geometry in Table 42 shows all real calculated harmonic vibrational frequencies, as appropriate to a genuine equilibrium state. The calculated harmonic vibrational frequencies for both isomeric protonated forms are shown in Table 43. An outstanding feature of these energies is the large decrease in the C—N stretch mode from 1119 cm⁻¹ in CH_2N_2 to 720 cm⁻¹ in $CH_3N_2^+$ in accord with the change in C—N bond length and character discussed above.

TABLE 42. Mulliken populations and geometric parameters for $MCH_2N_2^{+}$[a]

M⁺	Mulliken M⁺ populations				Charges					Bond Lengths (Å)				Angles (deg)		
	s	p	d	f	M⁺	H	C	N[b]	N[c]	M–X[d]	H–C	C–N	N–N	MXY[e]	HCN	CNN
f	—	—	—	—	—	0.17	−0.37	0.08	−0.04	—	1.067	1.291	1.106	—	117.2	180
g	—	—	—	—	—	0.17	−0.37	0.08	−0.04	—	1.074	1.319	1.143	—	116.6	180
g	—	—	—	—	—	0.16	−0.38	0.08	−0.04	—	1.076	1.324	1.143	—	114.9	176.6
H⁺	—	—	—	—	0.36	0.26	0.04	0.06	0.01	1.023	1.078	1.237	1.169	112.4	118.2	172.2
	—	—	—	—	0.37	0.27	−0.05	0.13	0.01	1.041	1.087	1.260	1.196	111.9	117.4	171.2
H⁺	—	—	—	—	0.25	0.25	−0.17	0.20	0.22	1.079	1.079	1.535	1.066	104.1	104.1	180
	—	—	—	—	0.26	0.26	−0.26	0.24	0.23	1.089	1.089	1.479	1.118	105.4	105.4	179.9
Cu⁺	2.08	6.09	9.99	—	0.85	0.21	−0.16	−0.01	−0.10	2.035	1.071	1.259	1.124	179.8	117.8	180
	2.24	6.17	9.88	—	0.71	0.23	−0.22	0.11	−0.06	1.825	1.079	1.290	1.148	179.9	116.7	179.8
	2.25	6.24	9.86	0.03	0.69	0.23	−0.22	0.11	−0.04							
Cu⁺	2.26	6.12	9.97	—	0.65	0.21	−0.39	0.16	0.15	2.077	1.082	1.482	1.069	113.8	102.7	177.1
	2.47	6.15	9.87	—	0.51	0.23	−0.29	0.16	0.16	1.906	1.091	1.424	1.125	114.3	104.8	178.3
	2.50	6.15	9.84	0.03	0.47	0.23	−0.26	0.16	0.16							
Ag⁺	2.05	6.04	9.99	—	0.92	0.20	−0.20	−0.10	−0.03	2.363	1.070	1.263	1.123	180	117.7	180
	2.12	6.08	9.95	—	0.85	0.22	−0.26	0.01	−0.04	2.187	1.078	1.296	1.148	180	116.6	180
	2.13	6.08	9.87	0.09	0.83	0.22	−0.25	0.02	−0.03							

M[+]															
Ag[+]	2.19	6.07	9.96	—	0.77	0.22	−0.09	0.18	0.14	2.330	1.081	1.462	1.071	114.6	103.5	176.5
	2.28	6.09	9.92	—	0.71	0.24	−0.50	0.16	0.14	2.217	1.089	1.402	1.129	114.1	105.9	176.9
	2.31	6.09	9.83	0.09	0.67	0.24	−0.47	0.16	0.15							
Au[+]	2.21	6.05	9.91	—	0.83	0.22	−0.13	0.07	−0.20	2.220	1.072	1.256	1.135	129.6	117.9	177.1
	2.37	6.08	9.80	—	0.76	0.23	−0.18	0.11	−0.15	2.058	1.081	1.287	1.163	130.4	116.9	175.4
	2.43	6.09	9.70	0.11	0.68	0.24	−0.17	0.10	−0.08							
Au[+]	2.47	6.07	9.83	—	0.64	0.23	−0.44	0.17	0.17	2.137	1.080	1.503	1.068	111.8	102.5	178.2
	2.63	6.07	9.70	—	0.59	0.25	−0.46	0.18	0.18	2.030	1.090	1.440	1.124	111.1	104.7	179.4
	2.72	6.07	9.60	0.11	0.50	0.26	−0.40	0.20	0.19							

[a]Basis sets, energies and levels of theory for the neutral and protonated CH_2N_2, and complexes are described in Tables 2, 3 and 41, respectively.
[b]Middle N atom.
[c]Terminal N atom.
[d]The first set of entries for each M[+] is for attachment to the terminal N atom and the second set is for attachment to $C(H_2)$.
[e]For attachment to terminal N the MXY angle is MNN; for attachment to $C(H_2)$ the MXY angle is MCN.
[f]RHF.
[g]MP2.

Metal ion coordination to diazomethane can also take place at either the methylene carbon atom or at the terminal nitrogen atom. Table 41 shows that the carbon atom is the overwhelmingly preferred site, energetically, for all the metal ions and at all levels of theory, even though it is not at the negative end of the CH_2N_2 dipole moment. The geometric structures of the MCH_2NN^+ complex are described in Table 42. CNN is almost linear and the metal ion approximately takes the place of the proton in the $CH_3N_2^+$ structure, giving close to C_s symmetry. The $C=N$ bonds lengthen and $N\equiv N$ bonds decrease in distance upon metal ion complexation at $C(H_2)$, as expected. The metal ion binding energies (Table 41) follow the usual trend of $Au^+ > Cu^+ > Ag^+$ independent of basis set and theoretical level. The Au^+ dissociation energy is a substantial 79.4 kcalmol^{-1} at the calculated QCISD(T)/ PP//MP2/PV level and appears to be converging with correlation level [MP2 $\rightarrow$ QCISD $\rightarrow$ QCISD(T)]. Such size binding energies are indicative of almost regular covalent bonding, similar to the protonated species. As noted above, both the H^+ and M^+ binding energies to $C(H_2)$ in diazomethane are unusually large.

The other site coordination for the metal ions, at the terminal nitrogen atom to give CH_2NNM^+, is 13–29 kcalmol^{-1} less stable than MCH_2NN^+ at the highest correlation level, depending on the metal ion (Table 41). Here, however, an unusual conformational situation is encountered. CH_2NNCu^+ and CH_2NNAg^+ both have linear CNNM groups while CH_2NNAu^+ has a bent (ca 130° at the MP2 level) AuNN angle (Table 42). The MP2 geometry optimizations were all initiated from bent structures (117° for CuNN, for example) so that the final linearity of the CuNN and AgNN groups is reliable. The MP2 calculated harmonic vibrational frequencies (Table 43) are all real for each of the three metal ion complexes in their respective MP2 optimized geometries. Thus both the linear conformer for Cu^+ and Ag^+, and the bent conformer for Au^+ are real equilibrium geometric

TABLE 43. MP2 calculated harmonic vibrational frequencies (in cm^{-1}) for $MCH_2N_2^{+a}$

CH₂N₂		CuCH₂N₂⁺		AgCH₂N₂⁺		AuCH₂N₂⁺		HCH₂N₂⁺	
Exp.[e]	Calc.	N[c]	C[d]	N[c]	C[d]	N[c]	C[d]	N[c]	C[d]
		114	149	88	123	124	143	405	341[f]
		181	357	141	317	309	362	439	
		361	400	214	382	330	384	905	720
406	321	483	454	440	506	473	606	1046	1157[f]
421	439	487	587	478	542	544	675	1109	
564	560	592	818	592	860	662	768	1177	1439
1109	1119	1109	1109	1116	1047	1113	1162	1222	1479[f]
1170	1145	1215	1275	1187	1252	1187	1265	1451	
1413	1445	1426	1448	1428	1460	1437	1450	2008	2174
2102	2302	2270	2121	2265	2114	2110	2109	3134	3099
3077	3229	3196	3109	3205	3125	3184	3117	3276	3229
3185	3366	3345	3192	3354	3210	3332	3201	3292	3231
19.2[b]	19.9[b]	21.1[b]	21.5[b]	20.7[b]	21.4[b]	21.2[b]	21.8[b]	27.8[b]	28.4[b]

[a]MP2 optimized geometry; CEP[3ˢ2ᵈ/3ˢ1ᵖ] basis set for CH_2N_2 and RCEP[4ˢᵖ3ᵈ] for M^+, with five d-type components.
[b]Zero-point energy, in kcalmol^{-1}.
[c]M^+ attachment at terminal N.
[d]M^+ attachment at $C(H_2)$.
[e]Experimental fundamental vibrational frequencies; from Reference 200.
[f]Doubly degenerate.

structures. In this regard, the Au^+ complex behaves like the same site protonated species, while the Cu^+ and Ag^+ ions align with the diazomethane dipole moment direction. As noted for previous systems, this seems to indicate a transition to a more covalent type bonding situation for Au^+ relative to Cu^+ and, of course, Ag^+. The change in $C=N$ and $N\equiv N$ bond lengths upon metal ion complexation at the terminal nitrogen atom follows the same trends as in the same-site protonated species (Table 42).

The calculated harmonic vibrational frequencies in Table 43 can, in principle, be assigned to specific diatomic or pseudo-diatomic stretch and bending modes from their normal coordinates. The calculated values of the force constants also help the indentification. In practice, these motions are sometimes too mixed to make such an assignment, unambiguously. Thus, the metal ion–substrate pseudo-diatomic stretch mode is found at 361 cm^{-1} (CH_2NNCu^+), 400 cm^{-1} ($CuCH_2NN^+$), 214 cm^{-1} (CH_2NNAg^+) and 384 cm^{-1} ($AuCH_2NN^+$) but cannot be clearly identified in $AgCH_2NN^+$ and CH_2NNAu^+. The C—N stretch, on the other hand, calculated in CH_2NN at 1119 cm^{-1}, is found to shift to slightly higher energies in CH_2NNCu^+ (1215 cm^{-1}), CH_2NNAg^+ (1187 cm^{-1}) and CH_2NNAu^+ (1187 cm^{-1}). In the MCH_2NN^+ complexes, however, this mode moves to much lower energy (818 cm^{-1} for Cu^+, 860 cm^{-1} for Ag^+ and 768 cm^{-1} for Au^+), appropriate to the strong reduction in C—N bond order and increased bond length. As noted above, in CH_3NN^+ the C—N stretch is calculated even lower at 720 cm^{-1}. The $N\equiv N$ stretch mode, calculated at 2302 cm^{-1} in CH_2NN, changes much less upon metal ion coordination or protonation (2174 cm^{-1} in CH_3NN^+) but always shifts to lower energy, irrespective of attachment site. For the MCH_2NH^+ complexes, where the equilibrium $N\equiv N$ bond lengths shorten, an opposite direction shift might have been expected.

The low-lying excited electronic states of diazomethane have been surveyed[61] and assigned theoretically[151, 203, 204]. The photoelectron spectrum[172] shows the two lowest-lying vertical ionizations at 9.00 eV and 14.00 eV, corresponding to the (planar geometry) C_{2v} symmetry splitting of the highest occupied π MO which is doubly degenerate in the linear symmetry of NNO and HCNO. This π MO is also split in HN_3, but to a much smaller extent (10.74 eV and 12.25 eV), as noted in Section IX. The coordinate axis convention used here identifies linear CNN as the z axis, and CH_2 as lying in the yz plane, with the x axis perpendicular to the molecular plane. In C_{2v} symmetry the irreducible representations transform as $a_1(z)$, $b_1(x)$, $b_2(y)$ and $a_2(xy)$. Only electronic transitions to overall $^1A_2$ state symmetry are formally electric dipole forbidden from the $^1A_1$ electronic ground state. Within this framework the purely linear geometry π systems (as in NNO) break into two sets in H_2CNN; three b_1 MOs, two of which are doubly occupied ($1b_1$ and $2b_1$) and one empty (b) and two b_2 MOs (b_2, doubly occupied and b_2^* unoccupied). There is another b_2 set of MOs comprised of the antisymmetric combination of C—H bonds. These play no role here. The two HOMO assigned in the photoelectron spectrum are $2b_1$ (9.00 eV) and b_2 (14.13 eV).

The CIS calculated vertical excited state energies are shown in Table 44. The bare diazomethane electronic spectrum at lower energies can be assigned to one electron transition between the $1b_1$, b_2 and $2b_1$ doubly occupied MOs and the virtual b_2^* and b_1^* π based MO, but also to $a_1\sigma \rightarrow b_2^*$ and $2b_1 \rightarrow a_1\sigma^*$ excitations involving σ and σ^* MOs. Electronic states of the same spin and spatial symmetry mix and tend to confute the simple one-electron transition assignments. All previous work agrees[151, 203, 204] in assigning the first two observed transitions in bare diazomethane to the $^1A_2$ ($2b_1 \rightarrow b_2^*$) and $^1A_1$ ($2b_1 \rightarrow b_1^*$) excited states at 3.14 eV and 5.70 eV, respectively. The calculated intensity for the latter excitation is too high compared to experiment, apparently due to exaggerated intensity borrowing from the $^1A_1$ ($b_2 \rightarrow b_2^*$) transition. The latter has been assigned[151] to an observed 7.06 eV absorption band but is calculated here to be above 10 eV. The observed 6.53 eV transition assigned as $^1B_1$ ($2b_1 \rightarrow a_1\sigma^*$) is calculated at 7.17 eV, in good agreement with experiment, as is the observed (calculated) 3.14 eV (3.18 eV) $2b_1 \rightarrow b_2^*$ excitation.

TABLE 44. CIS calculated vertical excited state energies (in eV) for MCH₂N₂⁺[a]

Assignment		CuCH₂N₂⁺ Asymptote		CuCH₂N₂⁺ Complex	AgCH₂N₂⁺ Asymptote		AgCH₂N₂⁺ Complex	AuCH₂N₂⁺ Asymptote		AuCH₂N₂⁺ Complex
		Exp.	Calc.	Calc.	Exp.	Calc.	Calc.	Exp.	Calc.	Calc.
$^3[2b_1,(n+1)s]$	$^3CT_1$	1.28^b	2.37	3.77	1.43^b	2.71	4.12	-0.22^b	1.53	3.00
$^1[2b_1,(n+1)s]$	$^1CT_1$	1.28^b	2.37	4.69 (0.032)	1.43^b	2.71	5.32 (0.120)	-0.22^b	1.53	4.07 (0.058)
$^3[nd,(n+1)s]$	^3D	2.81^d	4.12^e	5.05–7.35^e	5.03^d	5.72^e	5.68^f	2.29^d	2.83^e	4.46–7.62^e
$^1[nd,(n+1)s]$	^1D	3.26^d	4.82^e	5.66–(0.005) 8.36^e (0.081)	5.71^d	6.52^e	6.98–(0.006) 9.28^e (0.019)	3.67^d	3.93^e	5.12–(0.000) 8.68^e (0.033)
$^3(2b_2,b_2^*)$	$^3A_2$	g	2.52	5.41	g	2.52	4.94	g	2.52	5.62
$^3(2b_1,b_1^*)$	$^3A_1$	g	3.22	4.94	g	3.22	4.87	g	3.22	5.10
$^1(2b_1,b_2^*)$	$^1A_2$	3.14^g	3.18	5.99 (0.001)	3.14^g	3.18	5.48 (0.001)	3.14^g	3.18	6.17 (0.001)
$^3(b_2,b_2^*)$	$^3A_1$	g	5.88	6.12	g	5.88	6.00	g	5.88	6.10
$^3[2b_1,(n+1)p]$	$^3CT_2$	5.09^h	5.58^i	6.88^f	5.17^h	5.83^i	7.11^f	4.73^h	5.62^i	6.45^f
$^1(2b_1,b_1^*)$	$^1A_1$	5.70^g	6.70 (0.345)	7.20 (0.193)	5.70^g	6.70 (0.345)	7.02 (0.111)	5.70^g	6.70 (0.345)	7.20 (0.226)
$^3(b_2,b_2^*)$	$^3A_2$	g	8.85	7.54	g	8.85	—	g	8.85	7.42
$^1[2b_1,(n+1)p]^l$	CT_2	5.09^h	5.58^i	7.83^f (0.101)	5.17^h	5.83^i	7.90^f (0.127)	4.73^h	5.62^i	7.67^f (0.101)

$^3(2b_1,a_1\sigma^*)$	$^3B_1$	g	6.63	g	g	6.63	—	g	6.63	—
$^1(2b_1,a1s^*)$	$^1B_1$	6.53[g]	7.17 (0.022)	6.53[g]	6.53[g]	7.17 (0.022)	—	6.53[g]	7.17 (0.022)	—
$^3[b_2,(n+1)s]$	$^3CT_3$	6.41[b]	8.26	6.56[b]	g	8.61	—	4.91[b]	7.42	—
$^1[b_2,(n+1)s]$	$^1CT_3$	6.41[b]	8.26	6.56[b]	g	8.61	—	4.91[b]	7.42	—
$^3(2b_1,b_2\sigma^*)$	$^3A_2$	g	8.42	g	g	8.42	—	g	8.42	—
$^3(2b_1,a_1\sigma^*)$	$^3B_1$	g	8.56	g	g	8.56	—	g	8.56	—
$^1(2b_1,b_2\sigma^*)$	$^1A_2$	g	8.83	g	g	8.83	—	g	8.83	—
$^1(2b_1,a_1\sigma^*)$	$^1B_1$	g	9.25 (0.019)	g	g	9.25 (0.019)	—	g	9.25 (0.019)	—
$^3(1b_1,b_2^*)$	$^3A_2$	g	9.69	g	g	9.69	8.07	g	9.69	8.13
$^1(b_2,b_1^*)$	$^1A_2$	g	9.45	g	g	9.45	8.28 (0.037)	g	9.45	8.18 (0.009)
$^3(1b_1,b_1^*)$	$^3A_1$	g	9.97	g	g	9.97	8.65	g	9.97	8.51
$^3(a_1\sigma,b_1^*)$	$^3B_1$	g	—	g	g	—	9.26	g	—	—
$^3[nd,(n+1)p]$	^3P	8.33[d]	9.67[i]	8.78[f]	10.16[d]	—	9.81[f]	—	9.12[i]	8.30[f]
$^1[nd,(n+1)p]$	^1F	8.92[d]	10.30[c]	9.23[f] (0.000)	11.05[d]	—	10.33[f] (0.001)	—	9.94[c] (0.016)	8.76[f] (0.016)

[a] MP2 optimized ground state geometries. Values in parentheses are oscillator strengths. For M = Cu, $n = 3$; M = Ag, $n = 4$; M = Au, $n = 5$. Metal ion attachment is at the C atom.
[b] Based on the IP (9.00 eV and 14.13 eV) of CH_2N_2 (Reference 172) and EA of M^+ into the $(n + 1)s$ atomic orbital (References 22 and 23).
[c] Seven components.
[d] From References 22 and 23.
[e] Five components.
[f] First of multiple components.
[g] From Reference 151; see text.
[h] Based on the IP (9.00 eV) of CH_2N_2 (Reference 172) and EA of M^+ into the $(n + 1)p$ atomic orbital (References 22 and 23).
[i] Three components.

Upon complexation at the carbon atom the lower-energy $\pi \to \pi^*$ type transitions are calculated to shift to somewhat higher energies. The $^1A_2$ excited states are dipole allowed in the lower symmetry of the complex and gain some intensity. As has been found previously, electronic transitions that involve σ either as originating or terminating MOs are shifted out of the range of Table 44 to much higher energies upon metal ion coordination. The occupied σ MO involved in the asymptotic excitations consists essentially of the non-bonding pair of electrons on the terminal nitrogen atom and is assigned to the 15.13 eV vertical ionization in the photoelectron spectrum[172].

The intra-metal ion transitions to the $^{3,1}D$, ^3P and ^1F excited states show the usual behaviour. Upon metal ion complexation the $nd \to (n+1)s$ excitations shift to higher energy and split, with the highest energy member appearing far from the others due to configuration mixing with other low-lying transitions. Only the lowest and highest energy components are shown in Table 44 together with their calculated intensities for the singlet states. The $nd \to (n+1)p$ excitations are stabilized by metal ion coordination. Although the lowest energy such transition from the bare metal ions is to the dipole forbidden ^1F state[22, 23], the strongly absorbing ^1P state is not far above and can influence the calculated spectrum both in energy and intensity through configuration mixing in the lower symmetry of the complexes.

The relevant charge transfer states are $2b_1 \to (n+1)s$, $b_2 \to (n+1)s$ and $2b_1 \to (n+1)p$. In fact, for the Au^+ complex, the difference between the ionization potential of diazomethane (9.00 eV)[172] and the electron affinity of Au^+ (9.22 eV)[22, 23] is negative, so that asymptotically the electron transfer state $(Au^\circ + CH_2N_2^+)$ is more stable than $Au^+ + CH_2N_2$. It is only in the complex that the metal cation state is stabilized, by about 2.5 eV for $^1CT_1$ according to the calculation. In general, the $2b_1 \to (n+1)s$ excitations shift to higher energy upon metal ion complexation. The $2b_1 \to (n+1)p$ transitions are also calculated to shift to higher (ca 2 eV) energy upon metal ion attachment to the carbon atom. These transitions have substantial electric dipole intensity in the xz plane and magnetic dipole intensity in the y direction. The CT_3 excitations, $b_2 \to (n+1)s$, show up in Table 44 only in the asymptote. They are apparently shifted out of energy range upon metal ion complexation.

XII. CH₃CN

Acetonitrile is a 16 valence electron molecule with a C—N triple bond and no hypervalent central nitrogen atom. It is to be compared with HCN (Section IV) and FCN (Section VII). The properties of CH_3CN, CH_3CNH^+ and CH_3CHM^+ are detailed in Tables 2, 3 and 45–48. Substituent effects on the charge distribution and bonding of the C—N group have recently been discussed[161]. A review of the known structural data on acetonitrile from microwave spectroscopy[205] gives a C—C bond distance of 1.462 Å, C—N = 1.157 Å and C—H = 1.095 Å. The corresponding calculated MP2 values are 1.466 Å, 1.172 and 1.089 Å, respectively. The measured C—N bond distance is bracketed by the RHF and MP2 values. The observed dipole moment is 3.913 D[205] and the MP2 calculated value in the CEP[3ˢp2ᵈ/3ˢ1ᵖ] basis set is 4.272D. As can be seen from a comparison of Tables 14, 26 and 46, the C—N bond length is hardly affected by the nature of the substituent X in X—C≡N for these three cases and is calculated at a constant value of 1.17 Å.

The harmonic vibrational frequencies for acetonitrile, both experimental[206] and MP2 calculated, are shown in Table 47. Both the C—H stretch frequencies calculated at 3107 cm⁻¹ and 3197 cm⁻¹ and the C—C mode (1437 cm⁻¹) are close to the experimentally derived harmonic frequencies[206], while the C—N stretch differs by over 100 cm⁻¹, as appropriate to the more difficult bond type. The measured fundamental C—H frequencies are about 123 cm⁻¹ below the derived harmonic values, as expected[206]. The CCN bend mode at 360 cm⁻¹ in Table 47 is particularly well reproduced. This may also represent a cancellation of errors at this particular level of basis set (tendency to overestimate) and theory (tendency to underestimate) for the lowest-energy bending mode of a pseudo-linear molecule.

The first spectroscopic identification of protonated acetonitrile, CH_3CNH^+, has only very recently been reported[207], by observation of the N—H fundamental stretch band at 3527 cm^{-1}. The MP2 calculated harmonic value in Table 47 is 3664 cm^{-1}. If the difference between the observed fundamental and calculated harmonic C—H stretch frequency of ca 123 cm^{-1} for acetonitrile is arbitrarily subtracted from the calculated 3664 cm^{-1} N—H frequency of CH_3CNH^+, the resulting extrapolated 'fundamental' frequency is 3541 cm^{-1}, which is close to the measured value.

The MP2 calculated geometry of CH_3CNH^+ is described in Table 46. All previous theoretical work[207-209] has predicted the N-protonated form to be the most stable and that the CNH group is linear. These are the results also obtained here for CH_3CN, as well as for HCN (Table 14) and FCN (Table 26). Efforts to generate a C(N) carbon protonated form by initially positioning a proton near an appropriately bent CC^*N group produced only the dissociated CH_3^+ + HCN. As was found also for $HCNH^+$ and $FCNH^+$, formation of CH_3CNH^+ reduces both the X—C (X = F, C) and C—N bond distances relative to the parent neutral.

The proton affinity of CH_3CN has been reported at 188.4[65], 186.4[66], 183[157] and 188.6[210] kcal mol^{-1}. The QCISD(T)//MP2 value in Table 3 is 191.0 kcalmol^{-1}, from which the ZPE difference of 6.7 kcalmol^{-1} (Table 47) is subtracted to give an adjusted PA of 184.2 kcalmol^{-1}. Additional correction terms to match standard thermodynamic conditions are positive and may amount to ca 2 kcalmol^{-1} [27]. This brings the value calculated here to match that

TABLE 45. Energies (in a.u.) of CH_3CNM^+ complexes[a]

M^+	BF[b]	RHF	MP2	QCISD	QCISD(T)
Cu^+	115[c, e]	− 217.537643 (45.4)			
	115[d, e]	− 217.528067 (42.2)	− 218.467960 (63.1)	− 218.420064 (62.9)	− 218.415554 (44.9)
	129[f]	− 217.528626 (42.5)	− 218.726031 (68.3)	− 218.664586 (66.1)	− 218.673098 (52.7)
Ag^+	115[c, e]	− 167.647555 (34.0)			
	115[d, e]	− 167.642215 (33.8)	− 168.300949 (41.2)	− 168.317196 (40.8)	− 168.340997 (41.4)
	129[f]	− 167.643340 (34.5)	− 168.663893 (46.9)	− 168.652744 (44.3)	− 168.685522 (45.3)
Au^+	115[c, e]	− 157.359622 (41.1)			
	115[d, e]	− 157.352450 (39.4)	− 157.967988 (56.2)	− 157.989477 (55.5)	− 158.014255 (56.2)
	129[f]	− 157.354758 (40.9)	− 158.300372 (65.4)	− 158.288591 (60.5)	− 158.321570 (62.0)

[a] Numbers in parentheses are metal ion binding energies in kcalmol^{-1}.
[b] Number of basis functions; CEP[$3^sp2^d/3^s1^p$] basis set with five d-type components for CH_3CN.
[c] RHF optimized.
[d] MP2 optimized.
[e] RCEP[4^sp3^d] with five d-type components for M^+.
[f] RCEP[$4^sp3^d2^f$] with five d-type and seven f-type components for M^+. Complex geometry taken from 115 basis function MP2 optimization.

TABLE 46. Mulliken populations and geometric parameters for CH_3CNM^{+a}

M^+	Mulliken									Bond lengths (Å)				Angles (deg)		
	M^+ populations				Charges					$M-N$	$H-C^b$	$C-C$	$C-N$	MNC	HCC	CCN
	s	p	d	f	M^+	H	C^b	C	N							
	—	—	—	—	—	0.16	-0.30	0.11	-0.30	—	1.083	1.472	1.130	—	109.4	180
	—	—	—	—	—	0.17	-0.36	0.01	-0.18	—	1.089	1.466	1.172	—	109.7	180
H^+	—	—	—	—	0.42	0.23	-0.23	0.47	-0.34	0.998	1.084	1.467	1.120	180	108.0	180
	—	—	—	—	0.42	0.24	-0.30	0.37	-0.22	1.013	1.091	1.454	1.152	180	108.4	180
Cu^+	2.12	6.10	9.98	—	0.80	0.20	-0.29	0.34	-0.45	1.999	1.082	1.471	1.128	180	108.6	180
	2.30	6.15	9.88	—	0.66	0.22	-0.36	0.28	-0.25	1.815	1.089	1.457	1.164	180	109.1	180
	2.32	6.16	9.85	0.03	0.64	0.22	-0.36	0.28	-0.22							
Ag^+	2.08	6.06	9.98	—	0.88	0.20	-0.29	0.30	-0.48	2.291	1.082	1.472	1.129	180	108.8	180
	2.15	6.09	9.94	—	0.82	0.21	-0.35	0.22	-0.33	2.161	1.089	1.461	1.166	180	109.2	180
	2.16	6.09	9.86	0.09	0.80	0.21	-0.35	0.22	-0.31							
Au^+	2.20	6.07	9.90	—	0.84	0.20	-0.29	0.36	-0.52	2.156	1.083	1.471	1.127	180	108.7	179.9
	2.36	6.08	9.78	—	0.78	0.22	-0.35	0.30	-0.38	2.006	1.090	1.459	1.163	179.7	109.1	179.8
	2.40	6.10	9.68	0.08	0.72	0.22	-0.35	0.29	-0.33							

[a]Basis sets, energies and level of theory for the neutral and protonated CH_3CN, and complexes are described in Tables 2, 3 and 45, respectively.
[b]Methyl carbon atom.

obtained by the theoretical G2 method[66], but still *ca* 2 kcalmol^{-1} below the best purely experimental PA[62, 210]. The proton affinity of CH_3CN is calculated to be substantially larger (by *ca* 16 kcalmol^{-1}) than that of HCN. This has been attributed to larger electrostatic, polarization and charge transfer contributions[209]. The PA of HCN is predicted to be only 5–6 kcalmol^{-1} larger than for FCN.

The metal ion binding energies to CH_3CN are shown in Table 45. These follow the familiar trend of $Au^+>Cu^+>Ag^+$ at the QCISD(T)/PP//MP2/PV level but have the order of Au^+ and Cu^+ reversed at all other levels. Again, the large negative (T) correction for Cu^+ is the primary cause of this change. The Mulliken metal ion populations and atomic charges seem to support the $Au^+>Cu^+$ binding strength ordering, but the CH_3CN bond length changes upon metal ion complexation show the opposite trend. Both sets of properties are calculated at the MP2 level of theory. At a given level of theory and basis set the metal ion binding energies also follow the expected order of $CH_3CN>HCN>FCN$, although the differences are not as large as for the corresponding proton affinities. The acetonitrile–metal ion complexes are predicted to be N atom coordinated with a linear MNCC backbone (Table 46). Attempts to generate C(N) attached complexes led to geometry rearrangement to N-site coordination with no barrier. The calculated property values in Tables 14 (HCN), 26 (FCN) and 46 (CH_3CN) are generally consistent with the trends discussed above for protonation and metal ion complexation.

The calculated harmonic vibrational frequencies for the metal ion complexes are shown in Table 47. The degenerate modes involve bending motions, except for the C—H stretch at 3100–3200 cm^{-1}. The metal–ligand stretch mode frequencies are in the second line of the metal ion complex columns: 375 cm^{-1} (Cu^+), 238 cm^{-1} (Ag^+) and 292 cm^{-1} (Au^+). The low-energy CCN bend mode (360 cm^{-1}) is seen to be strengthened by metal ion complexation (Cu^+, 415 cm^{-1}; Ag^+, 394 cm^{-1}; Au^+, 405 cm^{-1}) but less so by protonation (378 cm^{-1}). The HNC bending mode in the latter is at 563 cm^{-1}. The C—N stretch (2183 cm^{-1}) in acetonitrile is seen to increase in energy upon attachment of a metal ion, or protonation. The C—C stretch (1437 cm^{-1}) stays relatively unchanged, as do the C—H modes. The changes, or lack thereof, are all essentially consistent with the general bonding picture and properties discussed above.

TABLE 47. MP2 calculated harmonic vibrational frequencies (in cm^{-1}) for CH_3CNM^{+a}

Exp^d CH_3CN	Calculated				
	CH_3CN	CH_3CNCu^+	CH_3CNAg^+	CH_3CNAu^+	CH_3CNH^+
		152^b	108^b	132^b	378^b
		375	238	292	563^b
365^b	360^b	415^b	394^b	405^b	888
929	911	960	929	947	1060^b
1062^b	1076^b	1066^b	1068^b	1066^b	1415^b
1418	1437	1427	1429	1426	1474^b
1478^b	1516^b	1491^b	1496^b	1492^b	2275
2294	2183	2271	2246	2279	3088
3044	3107	3101	3106	3100	3190^b
3135^b	3197^b	3198^b	3203^b	3199^b	3664
28.3^c	28.5^c	29.7^c	29.3^c	29.5^c	35.3^c

^{a}MP2 optimized geometry; CEP[3^{s}P2^d/3^{s}1^P] basis set for CH_3CN and RCEP[4^{s}P3^d] for M^+, with five d-type components.
^{b}Doubly degenerate.
^{c}Zero-point energy, in kcalmol^{-1}.
^{d}Experimental frequencies from Reference 206.

TABLE 48. CIS calculated vertical excited state energies (in eV) for CH_3CNM^{+a}

Assignment		CH_3CNCu^+			CH_3CNAg^+			CH_3CNAu^+		
		Asymptote		Complex	Asymptote		Complex	Asymptote		Complex
		Exp.	Calc.	Calc.	Exp.	Calc.	Calc.	Exp.	Calc.	Calc.
$^3[nd, (n+1)s]$	^3D	2.81^d	4.12^e	3.99–4.91^e	5.03^d	5.72^e	5.31^f	2.29^d	2.83^e	2.86–5.83^e
$^1[nd, (n+1)s]$	^1D	3.26^d	4.82^e	4.94 (0.000) 5.38^c (0.006) 5.56^c (0.000)	5.71^d	6.52^e	6.25 (0.000) 6.74^c (0.010) 6.87^c (0.000)	3.67^d	3.93^e	4.19 (0.001) 4.68^c (0.000) 9.33^c (0.001)
$^3[\pi, (n+1)s]$	$^3CT_1$	4.49^b	5.90^c	9.52^c	4.64^b	6.25^c	9.03^c	2.99^b	5.06^c	3.79^c
$^1[\pi, (n+1)s]$	$^1CT_1$	4.49^b	5.90^c	9.84^c (0.009)	4.64^b	6.25^c	9.33^c (0.013)	2.99^b	5.06^c	4.85^c (0.010)
$^3(\pi, \pi^*)$	$^3\Sigma^+$	5.5^g	5.21	5.60	5.5^g	5.21	5.58	5.5^g	5.21	5.62
$^3(\pi, \pi^*)$	$^3\Delta$	6.7^g	6.23^c	6.57^c	6.7^g	6.23^c	6.57^c	6.7^g	6.23^c	6.61^c
$^3(\pi, \pi^*)$	$^3\Sigma^-$	7.7^g	7.22	7.47	7.7^g	7.22	7.51	7.7^g	7.22	7.51
$^1(\pi, \pi^*)$	$^1\Sigma^-$	g	7.22	7.47	g	7.22	7.51	g	7.22	7.51
$^1(\pi, \pi^*)$	$^1\Delta$	8.45^g	7.65^c (0.001)	7.81^c (0.001)	8.45^g	7.65^c (0.001)	7.90^c (0.001)	8.45^g	7.65^c (0.001)	7.84^c (0.001)
$^3[\sigma, (n+1)s]$	$^3CT_2$	5.42^h	8.57	—	5.54^h	8.91	—	3.92^h	7.73	—
$^1[\sigma, (n+1)s]$	$^1CT_2$	5.42^h	8.57	—	5.54^h	8.91	—	3.92^h	7.73	—
$^3(\sigma, \pi^*)$	$^3\Pi$	g	8.60^c	—	g	8.60^c	—	g	8.60^c	—

State	Term	Cu			Ag			Au		
³[π, (n + 1)p]	³CT₃	8.30[k]	9.11[l]	—	8.38[k]	9.37[l]	—	7.94[k]	9.16[l]	8.67[f]
¹[π, (n + 1)p]	¹CT₃	8.30[k]	9.11[l]	9.66[f] (0.462)	8.38[k]	9.37[l]	10.53[f] (0.675)	7.94[k]	9.16[l]	9.46[f] (0.625)
¹(σ, π*)	¹Π	9.0[g]	10.49[c] (0.021)	—	9.0[g]	10.49[c] (0.021)	—	9.0[g]	10.49[c] (0.021)	—
¹(π, σ*)	¹Π	g	10.18[c] (0.009)	—	g	10.18[c] (0.009)	—	g	10.18[c] (0.009)	—
³[nd, (n + 1)p]	³P	8.33[d]	—	8.22[f]	10.16[d]	—	9.27[f]	—	9.12[i]	7.60[f]
¹[nd, (n + 1)p]	¹F	8.92[d]	10.30[j]	8.92[f] (0.066)	11.05[d]	—	9.97[f] (0.063)	—	9.94[f]	8.56[f] (0.104)

[a] MP2 optimized ground state geometries. Values in parentheses are oscillator strengths. For M = Cu, n = 3; M = Ag, n = 4; M = Au, n = 5. Metal ion attachment is at the N atom.
[b] Based on the first IP (12.21 eV) of CH_3CN (Reference 133) and EA of M^+ into the (n + 1)s atomic orbital (References 22 and 23).
[c] Doubly degenerate.
[d] From References 22 and 23.
[e] Five components.
[f] First of multiple components.
[g] From Reference 213; see text.
[h] Based on the second IP (13.14 eV) of CH_3CN (Reference 133) and EA of M^+ into the (n + 1)s atomic orbital (References 22 and 23).
[i] Three components.
[j] Seven components.
[k] Based on the first IP (12.21 eV) of CH_3CN (Reference 133) and EA of M^+ into the (n + 1)p atomic orbital (References 22 and 23).
[l] Six components.

The valence electronic structure of acetonitrile is similar to that of HCN and FCN. The HOMO π ionization energy is 12.21 eV and the σ HOMO is not much higher at 13.14 eV[133, 211]. Unlike fluorine atom substitution, the methyl group affects both π and σ electron ionization energies substantially, reducing both IP from their 13.81 eV and 14.00 eV for HCN. Although CH_3CN has C_{3v} symmetry we will continue to use the homologous linear notation for the equivalent MOs and electronic states, for consistency with HCN and FCN. Lower symmetry equilibrium geometries in the excited electronic states are not considered here. The relevant electronic structure description of acetonitrile for a discussion of valence excited states is, $\sigma^2\pi^4\pi^{*0}$. The σ MO consists mainly of the non-bonding lone pair of electrons on the nitrogen atom.

The electronic absorption spectrum of CH_3CN has been recorded and discussed extensively[61, 88, 192, 212, 213]. Many transitions to Rydberg excited states have been observed and classified, and many of the valence $\pi \rightarrow \pi^*$ and $\sigma \rightarrow \pi^*$ electronic transitions have been assigned[212, 213]. The assignments are not definitive and rely on the expected ordering of excited states and the angular dependence of electron impact spectra. The CIS calculated vertical excited state energies for acetonitrile are shown in Table 48. In C_{3v} symmetry the spin singlet $A_1(\Sigma^+)$ and $E(\Pi, \Delta)$ are dipole allowed excited states from the $^1A_1$ ground state in optical absorption. $^1\Sigma^-$ is still dipole forbidden. In electron impact spectra, triplet states are also observed since the electron scattering process involves a temporary negative ion with spin $(S) = 1/2$, which can then lose an electron to give either $S = 0$ (singlet) or $S = 1$ (triplet) excited states. The assigned excited states for bare CH_3CN in Table 48 show good agreement between the calculated $\pi \rightarrow \pi^*$ energies and the observed spectral features. These excitation energies are predicted not to shift by any significant amount upon metal ion complexation. The $\sigma \rightarrow \pi^*$ and $\pi \rightarrow \sigma^*$ excitations are calculated to shift out of energy range in the complex, as usual. The latter type excitation is probably intimately mixed with the Rydberg-type transitions, which aren't treated here.

The L $\rightarrow$ M charge transfer transitions, $\pi \rightarrow (n+1)$s,p and $\sigma \rightarrow (n+1)$s, are low-lying. The calculations do poorly in reproducing the asymptotic energy value for the latter process. Metal ion complexation is predicted to shift those CT excitations terminating in the primarily $(n+1)$s MO to much higher energy. These transitions mix with the symmetry split $nd \rightarrow (n+1)$s excitations with which, asymptotically, they are close in energy. The latter gain in intensity from the mixing and the singlet states could appear as weak bands in the lower energy part of the optical absorption spectrum. The very intense valence transitions are expected just below 9 eV from a combination of the $\pi \rightarrow (n+1)$p and $nd \rightarrow (n+1)$p (^1F and ^1P) transitions.

XIII. C_2F_2

The calculated properties of C_2F_2, $HC_2F_2^+$ and $MC_2F_2^+$ are listed in Tables 2, 3 and 49–52. Difluoroethyne has a MP2 calculated $C\equiv C$ triple bond length of 1.195 Å (Table 50), which decreases progressively from C_2H_2 (1.213 Å, Table 18) and C_2HF (1.206 Å, Table 37). At the same time, the calculated harmonic C—C stretch frequency increases from 1962 cm^{-1} for C_2H_2 (Table 19), to 2277 cm^{-1} in C_2HF (Table 39) to 2464 cm^{-1} (Table 51). The best explanation for these trends is simply that the increased positive charge on the carbon atom valence electron radial distribution, shortens the C—C bond length and tightens the vibrational potential.

Difluoroethyne is a short-lived molecule at room temperature and decomposes to yield a polymer and other reactive species. Its ground state geometric structure and vibrational spectrum have recently been investigated[214, 215]. The best combined experimental and theoretical estimate of the equilibrium bond lengths gives 1.19 Å for C—C and 1.28–1.29 Å for C—F in a linear conformation, to be compared with the calculated numbers in Table 50. The measured fundamental vibrational frequencies[215] are listed in Table 51 and are very close to the MP2 calculated harmonic frequencies. In general, a comparison of calculated harmonic with experiment, whether derived harmonic or measured fundamental frequencies, shows good agreement also for C_2H_2 (Table 19) and C_2HF (Table 39). The anharmonicity corrections for this series are apparently not very large[124].

Two structures for protonated difluoroethyne ($HC_2F_2^+$) were generated, by analogy with protonated acetylene: the classical asymmetric form and the non-classical (bridged) geometry. As can be seen from Table 3, in contrast to the result for $C_2H_3^+$, the asymmetric classical structure is the more stable of the two isomeric forms, both at the RHF and MP2 optimization levels. $HC_2F_2^+$ is thus like $C_2H_2F^+$. The symmetric bridged geometry shows one imaginary harmonic frequency (450 cm^{-1}) calculated at the MP2 level, with a normal mode motion that shows it to be a transition state connecting the two possible asymmetric structures. The electronic energy difference between the classical equilibrium structure and bridged stationary state at the QCISD(T)//MP2 level is 15.4 kcalmol^{-1}. The proton affinity of C_2F_2 to give the asymmetric $HC_2F_2^+$ at this theory level is 157.5 kcalmol^{-1}. Subtracting the ZPE difference of 6.8 kcalmol^{-1} (Table 51) gives an adjusted PA of 150.7 kcalmol^{-1}. The PA to C_2H_2, C_2HF and C_2F_2 are predicted to decrease by a regular interval (Table 3) of ca 9 kcalmol^{-1} for each substitution of fluorine for a hydrogen atom.

The MP2 calculated C—C bond length in the classical $HC_2F_2^+$ structure of 1.305 Å is almost double-bond length. In general, the C—C bond distance increases for the asymmetric protonated acetylenic structure with increased fluorine substitution (Tables 18, 39 and

TABLE 49. Energies (in a.u.) of $MC_2F_2^+$ complexes[a]

M^+	BF^b	RHF	MP2	QCISD	QCISD(T)
Cu^{+g}	$119^{c,e}$	-253.638964 (12.2)			
	$119^{d,e}$	-253.606139 (-5.8)	-254.791501 (31.7)	-254.725605 (25.1)	-254.746684 (20.2)
	134^f	-253.606457 (-6.8)	-255.052154 (38.5)	-254.970231 (28.3)	-255.002476 (26.8)
Ag^{+g}	$119^{c,e}$	-203.757313 (6.5)			
	$119^{d,e}$	-203.749442 (4.1)	-204.629550 (12.9)	-204.636631 (11.7)	-204.666497 (12.8)
	134^f	-203.749890 (4.4)	-204.992498 (18.6)	-204.971665 (14.8)	-205.010660 (16.8)
Au^{+g}	$119^{c,e}$	-193.468382 (12.6)			
	$119^{d,e}$	-193.445561 (0.9)	-194.311500 (37.3)	-194.319503 (33.0)	-194.353118 (36.3)
	134^f	-193.448041 (2.4)	-194.652271 (51.8)	-194.619423 (38.5)	-194.663279 (43.9)

[a] Numbers in parentheses are metal ion binding energies in kcalmol^{-1}; a second number in parentheses is after BSSE correction.
[b] Number of basis functions; CEP[$3^{sp}2^d/3^s1^p$] basis set with five d-type components for C_2F_2.
[c] RHF optimized.
[d] MP2 optimized.
[e] RCEP[$4^{sp}3^d$] with five d-type components for M^+.
[f] RCEP[$4^{sp}3^d2^f$] with five d-type and seven f-type components for M^+. Complex geometry taken from 119 basis function MP2 optimization.
[g] Symmetric bridged structure.

50). The two fluorine atoms in protonated difluoroethyne are *trans* to each other across the C—C bond and the F*CC group [in F*CC(F)H$^+$] is only mildly bent. The geometric structure is perfectly planar. The terminal F*—C bond shortens considerably (*ca* 0.08 Å) upon asymmetric protonation at the other carbon site, while the C—F bond length at the protonated carbon atom hardly changes. The composite picture from these bond distance changes is a bonding structure that looks almost like F$^+$=C=C(H)F, which in the perpendicular direction has the allene π configuration. This valence bonding configuration may contribute to the preferred stability of the asymmetric protonated structure. The same decrease in calculated C—F bond length upon C(H) protonation is found in CHF$_2$ (Table 38) and the same F$^+$=C=CH$_2$ π structure is possible. The MP2 calculated harmonic vibrational frequencies for HC$_2$F$_2^+$ are also shown in Table 51. The primarily symmetric C—F stretch at 765 cm^{-1} (with zero IR intensity in C$_2$F$_2$) is seen to shift down to 659 cm^{-1} in the protonated species. The C—C stretch also decreases substantially from 2464 cm^{-1} (IR forbidden) to 2039 cm^{-1}. Both vibrational modes should be observed in HC$_2$F$_2^+$.

The calculated properties of the complexes resulting from metal ions coordinating to C$_2$F$_2$ are shown only for the symmetric bridged structure in Tables 49–52. This geometry is expected to be the most stable by analogy to the MC$_2$HF$^+$ structures, which were approximately bridging despite the intrinsic asymmetry of the ligand. An attempt to

TABLE 50. Mulliken populations and geometric parameters for MC$_2$F$_2^{+a}$

M$^+$			Mulliken									
	M$^+$ populations				Charges			Bond lengths (Å)			Angles (deg)	
	s	p	d	f	M$^+$	F	C	M—C	F—C	C—C	MCC	FCC
—	—	—	—	—	—	−0.17	0.17	—	1.272	1.163	—	180
—	—	—	—	—	—	−0.12	0.12	—	1.299	1.195	—	180
H^{+b}	—	—	—	—	0.28	−0.03	0.39	1.299	1.234	1.198	62.6	176.9
	—	—	—	—	0.31	0.04	0.30	1.307	1.258	1.225	62.0	177.6
H^{+c}	—	—	—	—	0.25	−0.12	0.35	1.082	1.281	1.291	116.7	122.8
						0.06	0.47		1.197			175.9
	—	—	—	—	0.26	−0.07	0.25	1.097	1.307	1.305	113.7	125.2
						0.15	0.40		1.216			173.3
Cu^{+b}	2.12	6.09	10.01	—	0.78	−0.11	0.22	2.426	1.261	1.177	76.0	175.5
	2.43	6.22	9.74	—	0.61	−0.04	0.24	1.957	1.288	1.243	71.5	163.2
	2.46	6.22	9.72	0.03	0.57	−0.04	0.26					
Ag^{+b}	2.06	6.05	10.00	—	0.89	−0.13	0.18	2.782	1.264	1.173	77.8	176.9
	2.14	6.09	9.96	—	0.81	−0.05	0.14	2.506	1.288	1.212	76.0	173.4
	2.15	6.08	9.88	0.09	0.79	−0.05	0.15					
Au^{+b}	2.23	6.08	9.93	—	0.76	−0.10	0.22	2.474	1.258	1.186	76.1	172.1
	2.60	6.13	9.51	—	0.76	−0.04	0.17	2.080	1.284	1.260	72.4	159.0
	2.67	6.13	9.43	0.11	0.66	−0.03	0.20					

^{a}Basis sets, energies and levels of theory for the neutral and protonated C$_2$F$_2$, and complexes are described in Tables 2, 3 and 49.
^{b}Symmetric bridged structure.
^{c}Classical structure. First line refers to protonated carbon atom. The two fluorine atoms are *trans*.

generate the asymmetric $AuC_2F_2^+$ complex geometry by initializing a MP2 geometry optimization in an asymmetric bent geometry led to the bridging structure without a barrier. The metal ions are calculated to be relatively weakly bound, continuing the trend of decreasing metal ion dissociation energies from C_2H_2 (Table 17) and C_2HF (Table 37). Generally, the metal ion binding energy is due mainly to correlation effects, which is characteristic of the π bridged ligands. The metal–carbon equilibrium bond distance decreases substantially from RHF to MP2. This shows the bonding to be sensitive to the details of the metal ion and ligand properties, which is characteristic of a ligand with no permanent dipole moment and low polarizability in the bonding direction. The metal ion binding energy order of $Au^+ > Cu^+ > Ag^+$ is obtained independent of basis set or theoretical level of calculation.

The $AgC_2F_2^+$ complex is particularly weak and shows a very low harmonic vibrational frequency (54 cm^{-1}) in Table 51. This normal mode motion converts the symmetric geometry to an asymmetric structure and in $AgC_2F_2^+$ this is a very soft mode. The corresponding frequencies in the Cu^+ and Au^+ complexes are 204 cm^{-1} and 206 cm^{-1}, respectively. This tendency towards an $AgC_2F_2^+$ asymmetric structure parallels what was found also for MC_2HF^+ where the Ag^+ complex was calculated to be the most asymmetric of the three metal ions. The metal ion–carbon equilibrium bond distances (Table 50) both in RHF and MP2 optimized geometries are particularly long compared to the $MC_2H_2^+$ and MC_2HF^+ complexes. The changes in C—C bond distances, harmonic stretch frequencies, and the deviation of the FCC bond angles from linearity are proportional to the metal ion binding energies. The C—C bonds, however, don't reach the near-double bond length of the asymmetric $HC_2F_2^+$ structure. The C—F bond length in the complexes stays within 0.015 Å of the bare C_2F_2 value. The Au^+ atomic orbital populations show particularly lower 5d occupancy for the C_2F_2 ligand complex than for C_2HF or C_2H_2. This could be interpreted as enhanced $M \rightarrow L$ ($nd \rightarrow \pi^*$) dative bonding promoted by the increased positive charge on the carbon atoms with increased fluorine substitution.

TABLE 51. MP2 calculated harmonic vibrational frequencies (in cm^{-1}) for $MC_2F_2^{+a}$

Experimental[d]	Calculated			
C_2F_2	C_2F_2	$CuC_2F_2^{+e}$	$AgC_2F_2^{+e}$	$AuC_2F_2^{+e}$	$HC_2F_2^{+f}$
		204	54	206	177
		217	133	214	180
258[b]	260[b]	222	191	217	659
		308	201	307	718
282[b]	281[b]	373	284	402	956
		404	293	514	1198
795	765	776	779	803	1393
1349	1334	1313	1340	1318	2039
2450	2464	2215	2382	2145	3097
8.1[c]	8.1[c]	8.6[c]	8.1[c]	8.8[c]	14.9[c]

[a]MP2 optimized geometry; CEP[3s2p2d/3s1p] basis set for C_2F_2 and RCEP[4s4p3d] for M$^+$, with five d-type components.
[b]Doubly degenerate.
[c]Zero-point energy, in kcalmol^{-1}.
[d]Experimental fundamental vibrational frequencies from Reference 215.
[e]Symmetric bridged structure.
[f]Asymmetric geometry.

TABLE 52. CIS calculated vertical excited state energies (in eV) for $MC_2F_2^{+}$[a]

Assignment	$CuC_2F_2^+$			$AgC_2F_2^+$			$AuC_2F_2^+$		
	Asymptote		Complex	Asymptote		Complex	Asymptote		Complex
	Exp.	Calc.	Calc.	Exp.	Calc.	Calc.	Exp.	Calc.	Calc.
$^3[nd, (n+1)s]$ ^3D	2.81[d]	4.12[e]	5.20– 7.45[e]	5.03[d]	5.72[e]	6.39– 6.25[e]	2.29[d]	2.83[e]	5.45– 7.67[e]
$^1[nd, (n+1)s]$ ^1D	3.26[d]	4.82[e]	6.62– (0.002) 8.03[e] (0.019)	5.71[d]	6.52[e]	7.08– (0.031) 7.43[e] (0.003)	3.67[d]	3.93[e]	6.00– (0.000) 7.86[e] (0.007)
$^3[\pi, (n+1)s]$ $^3CT_1$	3.88[b]	5.06[c]	5.35 7.92	4.03[b]	5.41[c]	5.12 5.73	2.08[b]	4.22[c]	3.93 4.00
$^1[\pi, (n+1)s]$ $^1CT_1$	3.88[b]	5.06[c]	5.87 (0.028) 6.06 (0.019)	4.03[b]	5.41[c]	6.08 (0.048) 6.16 (0.031)	2.08[b]	4.22[c]	4.69 (0.022) 5.00 (0.028)
$^3(\pi, \pi^*)$ $^3\Sigma_u^+$	[g]	5.89	3.50	[g]	5.89	5.02	[g]	5.89	3.78
$^3(\pi, \sigma^*)$ $^3\Pi_g$	[g]	6.49[c]	—	[g]	6.49[c]	8.73 8.78	[g]	6.49[c]	—
$^3(\pi, \pi^*)$ $^3\Delta_u$	[g]	7.12[c]	3.16 5.89	[g]	7.12[c]	5.15 6.37	[g]	2.12	3.08 5.43
$^3(\pi, \pi^*)$ $^3\Sigma_u^-$	[g]	8.04	8.05	[g]	8.04	7.88	[g]	8.04	8.27
$^3(\pi, \sigma^*)$ $^3\Pi_u$	[g]	7.95[c]	—	[g]	7.95[c]	—	[g]	7.95[c]	—
$^3(nd, \pi^*)$ $^3CT_2$	[g]	—	6.05[f]	[g]	—	—	[g]	—	6.69[f]
$^1(nd, \pi^*)$ $^1CT_2$	[g]	—	7.49[f] (0.010)	[g]	—	—	[g]	—	7.27[f] (0.017)

		Cu			Ag			Au		
$^1(\pi, \pi^*)$	$^1\Sigma_u^-$	g	8.04	4.16	g	8.04	6.30	g	8.04	3.99
	$^1\Pi_g$	g	8.20^c	—	g	8.20^c	9.67	g	8.20^c	—
							(0.000)			
$^1(\pi, \sigma^*)$	$^1\Pi_u$	g	8.84^c	—	g	8.84^c	—	g	8.84^c	—
			(0.127)			(0.127)			(0.127)	
$^1(\pi, \pi^*)$	$^1\Delta_u$	g	8.39^c	5.62	g	8.39^c	7.08	g	8.39^c	5.41
				(0.051)			(0.031)			(0.058)
				8.66			8.63			8.80
				(0.000)			(0.000)			(0.000)
$^1(\pi, \pi^*)$	$^3\Sigma_u^+$	g	—	10.15	g	—	9.48	g	—	10.09
				(0.267)			(0.080)			(0.281)
$^3[\pi, (n+1)p]$	$^3CT_3$	7.69^h	8.28^i	9.01^f	7.77^h	8.53^i	9.08^f	7.33^h	8.32^i	8.20^f
$^1[\pi, (n+1)p]$	$^1CT_3$	7.69^h	8.28^i	9.01^f	7.77^h	8.53^i	8.16^f	7.33^h	8.32^i	8.66^f
				(0.102)			(0.166)			(0.000)

^{a} MP2 optimized symmetric ground state geometries. Values in parentheses are oscillator strengths. For M = Cu, $n = 3$; M = Ag, $n = 4$; M = Au, $n = 5$.

^{b} Based on the IP (11.60 eV) of C_2F_2 (Reference 193) and EA of M^+ into the $(n + 1)s$ atomic orbital (References 22 and 23).

^{c} Doubly degenerate.

^{d} From References 22 and 23.

^{e} Five components.

^{f} First of multiple components.

^{g} See text.

^{h} Based on the IP (11.60 eV) of C_2F_2 (Reference 193) and EA of M^+ into the $(n + 1)p$ atomic orbital (References 22 and 23).

^{i} Six components.

H. Basch and T. Hoz

TABLE 53. Energies (in a.u.) of MFCNO$^+$ complexes[a]

M$^+$	BF[b]	X[g]	RHF	MP2	QCISD	QCISD(T)
Cu$^+$	119[c,e]	O	−249.674155 (39.9)			
	119[d,e]	O	−249.664973 (45.1)	−250.857610 (38.3)	−250.798164 (46.1)	−250.786007 (12.5)
	133[f]	O	−249.665507 (45.4)	−251.111330 (40.8)	−251.041556 (48.6)	−251.045155 (21.3)
	119[c,e]	C	−249.640058 (18.5)			
	119[d,e]	C	−249.626635 (21.0)	−250.845048 (30.5)	−250.783622 (37.1)	−250.810130 (27.7)
	133[f]	C	−249.627019 (21.3)	−251.104072 (36.3)	−251.028194 (40.3)	−251.064688 (33.6)
Ag$^+$	119[c,e]	O	−199.786226 (30.3)			
	119[d,e]	O	−199.769789 (30.9)	−200.702000 (23.5)	−200.699568 (26.7)	−200.735674 (23.9)
	133[d,e]	C	−199.770583 (31.4)	−201.062236 (27.5)	−201.033710 (29.4)	−201.078300 (26.9)
	119[c,e]	C	−199.7521821 (8.9)			
	119[d,e]	C	−199.739264 (11.7)	−200.683345 (11.8)	−200.683947 (16.9)	−200.724553 (16.9)
	133[d,e]	C	−199.739918 (12.2)	−201.047641 (18.3)	−201.019422 (20.4)	−201.069300 (21.3)
Au$^+$	119[c,e]	O	−189.492776 (33.6)			
	119[d,e]	O	−189.484024 (39.0)	−189.357988 (31.6)	−190.364238 (36.7)	−190.399363 (33.0)
	133[f]	O	−189.485996 (40.2)	−190.687407 (39.0)	−190.663460 (41.7)	−190.706048 (38.4)
	119[c,e]	C	−189.476279 (23.2)			
	119[d,e]	C	−189.465682 (27.5)	−190.363351 (34.9)	−190.373452 (42.5)	−190.413608 (41.9)
	133[f]	C	−189.467384 (28.5)	−190.701372 (47.7)	−190.673856 (48.2)	−190.722972 (49.1)

[a] Numbers in parentheses are binding energies in kcalmol^{-1}.
[b] Number of basis functions; CEP[3^{sp}2^d/3^{s}1^p] basis set with five d-type components for FCNO.
[c] RHF optimized.
[d] MP2 optimized.
[e] RCEP[4^{sp}3^d] with five d-type components for M$^+$.
[f] RCEP[4^{sp}3^{d}2^f] with five d-type and seven f-type components for M$^+$. Complex geometry taken from 119 basis function MP2 optimization.
[g] Atom to which M$^+$ is attached.

The CIS calculated vertical electronic excitation energies are shown in Table 52. The excited states of C_2F_2 are classified exactly like those of C_2H_2 with full inversion symmetry and the doubly degenerate π HOMO (π_u) and π^* LUMO (π_g). In addition, fluorine substitution stabilizes the valence σ^* MOs[192], so that in the series, $C_2H_2 \rightarrow C_2HF \rightarrow C_2F_2$, $\pi \rightarrow \sigma^*$ transitions are progressively lowered in energy. However, the resultant Π_g and Π_u

states hardly appear in the metal ion complexes in the energy range of Table 52. Perfluorination in ethylene has the effect of shifting the $\pi \rightarrow \pi^*$ transition to higher energy[192] and the same effect is observed here in comparing C_2H_2 (Table 20) and C_2HF (Table 40) to C_2F_2. The first vertical IP of C_2F_2 is 11.60 eV[193] so that, compared with C_2HF (11.50 eV[193]) and C_2H_2 (11.40 eV[133]), the L $\rightarrow$ M CT energies in the asymptote are found to hardly change.

Upon complexation with a metal ion in the non-classical bridged geometry the various asymptote excitation energies are affected differently. Besides splitting into their non-degenerate components, the $\pi \rightarrow \pi^*$ transitions are generally shifted to lower energy, to a much greater extent than is found in the C_2H_2 and C_2HF systems. It is tempting to attribute this excited state stabilization to a differential stabilization of the π^* MO in the metal ion complexes. Of course, a simple, uniform one-electron model cannot explain multi-electron effects which are undoubtedly involved here. However, Table 52 does show the $nd \rightarrow \pi^*$ CT transitions being in the energy range discussed here in the complexes but not in the asymptote. The $nd \rightarrow \pi^*$ transitions appear both in the calculated C_2H_2 and C_2HF spectra very heavily mixed with the $nd \rightarrow (n+1)\pi$ transitions in their respective complexes. In Tables 20 and 40 these excitations are classified as $nd \rightarrow (n+1)p$, because that seems to be their major component and because these metal ion transitions appear in the asymptote list among the first 20 excitations. In C_2F_2 the major component is $nd \rightarrow \pi^*$ and the primarily $nd \rightarrow (n+1)p$ excitations don't fall among the lowest twenty asymptote excited states. As has been expressed several times above, these assignments would benefit greatly from a natural orbital analysis[152] and a higher-level theoretical treatment. In any event, tracing the combined $nd \rightarrow (n+1)p$, π^* transitions from C_2H_2 to C_2F_2 shows their steady shift to lower energies with increased fluorination. Another manifestation of the stabilization of $\pi \rightarrow \pi^*$ in the complexes is the appearance of the only $\pi \rightarrow \pi^*$ electric dipole allowed state in bare C_2F_2: $^1\Sigma_u^+$, in Table 52. On the other hand, $\pi \rightarrow \sigma^*$, as usual, seems to be shifted out of energy range in the complexes.

The $\pi \rightarrow (n+1)s$ CT transitions are calculated to generally (but not always) shift to somewhat higher energies in the complexes. The changes are about the same for C_2F_2 as for C_2HF and C_2H_2. The same trends are found for the $\pi \rightarrow (n+1)p$ CT and the energies slowly decreases with increased fluorination. The respective asymptote error adjusted ^1CT and ^1D electronic state energies in the metal ion complexes predict the CT state to be above the lowest-energy ^1D derived multiplet for Cu^+, but to be the lowest-energy state in the Ag^+ and Au^+ complexes.

XIV. FCNO

Fluorformonitrile oxide (FCNO) is predicted to be one of the less stable forms of the isomeric set of CFNO compounds and to have a triplet electronic ground state[216]. These results have to be tested at higher levels of basis set and theory. Only the F—$C\equiv N$=O isomer with the hypervalent N atom has been treated here, in a closed-shell ground state electronic configuration. The results for FCNO, protonated FCNO and metal ion coordinated FCNO are shown in Tables 2, 3 and 53–55.

The ground state geometric structure of bare FCNO was optimized in both linear and non-linear geometries at the MP2 level, by analogy to HCNO. RHF gave only the linear symmetry. As shown in Tables 2 and 54, however, the bent geometry is more stable at all levels of correlation. MP2 harmonic vibrational analysis of the linear geometry shows one imaginary frequency of 411 cm^{-1} corresponding to bending or kinking about the carbon atom. The non-linear structure is planar, having a *trans* configuration with respect to the C—N bond. The MP2 calculated harmonic vibrational frequencies for the bent structure are all real, as shown in Table 55. The most easily identified mode is the C—N stretch motion at 2321 cm^{-1}. This is larger than the 2185 cm^{-1} energy calculated for the same mode in HCNO (Table 31).

TABLE 54. Mulliken populations and geometric parameters for MFCNO⁺ᵃ.

M⁺	Mulliken									Bond lengths (Å)				Angles (deg)		
	M⁺ populations					Charges										
	s	p	d	f	M⁺	F	C	N	O	M–Xᵇ	F–C	C–N	N–O	MXNᵇ	FCN	CNO
h	—	—	—	—	—	-0.11	0.58	-0.12	-0.35	—	1.260	1.115	1.229	—	180	180
i	—	—	—	—	—	-0.07	0.41	-0.13	-0.20	—	1.296	1.168	1.222	—	180	180
i	—	—	—	—	—	-0.11	0.32	-0.07	-0.14	—	1.312	1.205	1.211	—	140.2	160.8
H⁺	—	—	—	—	0.44	0.02	0.78	-0.10	-0.14	0.962	1.216	1.108	1.306	107.2	178.8	171.6
	—	—	—	—	0.45	0.02	0.76	-0.09	-0.14	0.984	1.242	1.148	1.315	105.9	175.8	165.4
H⁺	—	—	—	—	0.27	-0.09	0.44	0.15	0.22	1.079	1.267	1.250	1.120	118.7	119.1	178.2
	—	—	—	—	0.28	-0.04	0.31	0.18	0.27	1.088	1.298	1.282	1.158	117.8	118.6	178.1
Cu⁺	2.08	6.06	9.99	—	0.87	-0.04	0.70	-0.17	-0.36	2.006	1.236	1.110	1.261	133.8	179.5	176.7
	2.17	6.11	9.93	—	0.79	0.02	0.55	-0.11	-0.24	1.887	1.264	1.154	1.261	129.9	175.3	172.3
	2.18	6.11	9.90	0.03	0.78	0.02	0.55	-0.11	-0.23							
Cu⁺	2.20	6.13	9.98	—	0.69	-0.17	0.31	0.09	0.08	2.104	1.302	1.244	1.133	124.1	115.0	172.2ᶜ
	2.40	6.17	9.86	—	0.56	-0.11	0.32	0.08	0.15	1.883	1.334	1.255	1.174	120.6	118.7	168.4ᶜ
	2.43	6.18	9.83	0.03	0.53	-0.11	0.34	0.09	0.16							

Ag$^+$	2.04	6.04	9.99	—	0.92	−0.05	0.69	−0.19	−0.37	2.299	1.240	1.110	1.256	140.5	179.6	177.5
	2.08	6.06	9.88	—	0.88	0.02	0.50	−0.12	−0.25	2.244	1.274	1.174	1.243	135.2	156.4	165.7
	2.08	6.06	9.90	0.09	0.87	0.02	0.50	−0.12	−0.24							
Ag$^+$	2.13	6.07	9.98	—	0.82	−0.17	0.20	0.10	0.06	2.389	1.303	1.238	1.136	124.5	116.0	170.8^d
	2.17	6.09	9.94	—	0.81	−0.09	0.20	0.02	0.06	2.277	1.320	1.219	1.189	114.5	130.5	165.5^e
	2.19	6.08	9.86	0.09	0.78	−0.09	0.22	0.03	0.06							
Au$^+$	2.13	6.05	9.95	—	0.88	−0.04	0.69	−0.15	−0.38	2.240	1.236	1.110	1.265	128.1	179.7	176.3
	2.24	6.07	9.88	—	0.81	0.01	0.54	−0.09	−0.27	2.139	1.263	1.156	1.266	122.3	173.5	170.6
	2.28	6.08	9.79	0.10	0.76	0.02	0.54	−0.09	−0.23							
Au$^+$	2.36	6.09	9.85	—	0.71	−0.17	0.27	0.09	0.10	2.148	1.295	1.245	1.132	122.5	115.3	173.3^f
	2.53	6.09	9.69	—	0.69	−0.11	0.17	0.08	0.17	1.997	1.325	1.258	1.174	118.4	118.9	170.7^g
	2.60	6.10	9.59	0.11	0.61	−0.09	0.21	0.09	0.18							

^{a}Basis sets, energies and levels of theory for the neutral and protonated FCNO, and complexes are described in Tables 2, 3 and 53, respectively.
^{b}X is defined in Table 53.
^{c}Angle MCF is 120.8°.
^{d}Angle MCF is 119.4°.
^{e}Angle MCF is 115.0°.
^{f}Angle MCF is 122.2°.
^{g}Angle MCF is 122.7°.
^{h}RHF.
^{i}MP2.

The MP2 calculated F—C bond length in FCNO (1.312 Å, Table 54) is longer than in FCN (1.279 Å, Table 26), FCCH (1.291 Å, Table 38) and FCCF (1.299 Å, Table 50). This trend is consistent with the argument that fluorine destabilizes the CNO group[216]. The C—N bond length increases from 1.067 Å in HCN (Table 14) to 1.174 Å in FCN (Table 26), possibly because of the $F^+=C=N^-$ bonding structure component in the over-all valence bond electronic structure description. This $F^+=C=N—O^-$ type bonding arrangement will be less favourable in the nitrile oxide, and is inconsistent with the large calculated F—C bond distance. From HCNO to FCNO the C—N equilibrium distance increases by ca 0.01 Å, again showing the absence of stabilization by the fluorine substituent. The N—O bond increases ca 0.01 Å for each transition from NNO (Table 22) to HCNO to FCNO. Thus, all the bond distances in FCNO are larger than any combination of structural parameters from appropriate reference molecules.

FCNO can be protonated at any of its four atoms. Protonation at the nitrogen leads to

$$HNC^+\big\backslash^{\nearrow O}_{F}$$

a structure which shows one imaginary harmonic vibrational frequency at the MP2 level, and will not be discussed further. The fluorine atom can also be protonated and the energies of the resultant HFCNO$^+$ cation are shown in Table 3. The C—F bond is elongated to 1.581 Å and the MP2 harmonic vibrational frequencies (not shown in Table 55) are all real. HFCNO$^+$ is therefore the equilibrium structure of a weakly bound molecular complex. This cationic species will not be described here in any more detail. Similarly to HCNO, protonation at the carbon or at oxygen atoms leads to stable, planar structures. Their geometries are described in Table 54, energies in Table 3 and harmonic vibrational frequencies in Table 55.

Carbon protonated FC(H)NO$^+$ has a QCISD(T)//MP2 calculated proton affinity of 175.2 kcalmol^{-1} which is larger than the 174.1 kcalmol^{-1} value for C-protonation of HCNO. After adjustment for the ZPE difference these binding energies become 168.5 kcalmol^{-1} and 166.8 kcalmol^{-1}, respectively. The F—C bond length in FC(H)NO$^+$ decreases by ca 0.01 Å relative to bent FCNO. These last two trends, which are counter-intuitive, are perhaps due, at least in part, to a F...H stabilizing interaction. The MP2 calculated HCF angle in FC(H)NO$^+$ is 123.6° while the HCH angle in H$_2$CNO$^+$ (Table 30) is larger at 127.0°. The $X—C^-=N^+=O$ precursor bonding structure is also important to the C-protonated cation. The C—N bond length increases (by ca 0.08 Å) and N—O decreases (by ca 0.05 Å) upon protonation at the carbon atom, as expected, due to the

$$^{F\diagdown}_{H\diagup}C\overset{+}{=}N=O$$

structure. The calculated C—N stretch frequency (1972 cm^{-1}) in Table 55 is a substantial decrease from bent FCNO (2321 cm^{-1}). These trends were noted also for H$_2$CNO$^+$. The slight in-plane CNO bend in FC(H)NO$^+$ is toward the hydrogen atom.

The most stable isomer, FCNOH$^+$, has a QCISD(T)//MP2 PA of 181.1 kcalmol^{-1} which becomes 174.0 kcalmol^{-1} after deducting the ZPE difference. The corresponding HCNOH$^+$ proton binding energy is 175.5 kcalmol^{-1}. Fluorine substitution for hydrogen decreases the PA at the remote oxygen site. The N—O bond length increases by ca 0.10 Å and C—N decreases by ca 0.06 Å in going to FCNOH$^+$ from bent FCNO. In accord with the strengthened C≡N bond in FCNOH$^+$, the FCNO skeleton becomes more linear. These changes are best explained by the precursor $F—C≡N^+—O$ structure favoured by the O-protonated cation. The atom charges show fluorine and especially carbon becoming more positive, and the F—C bond length decreasing by ca 0.07 Å in forming FCNOH$^+$. These latter changes indicate a contraction of valence shell electron density about the carbon atom that tends to shorten all the carbon–X (X = F,N) bond distances. Perhaps the

TABLE 55. MP2 calculated harmonic vibrational frequencies (in cm^{-1}) for MFCNO^{+a}

FCNO	CuFCNO$^+$		AgFCNO$^+$		AuFCNO$^+$		HFCNO$^+$	
	O^c	C^d	O^c	C^d	O^c	C^d	O^c	C^d
	89	123	61	95	80	127	241	240
	126	242	98	164	104	246	280	264
	190	329	169	249	196	321	323	672
335	283	389	213	274	244	364	409	717
446	316	422	346	458	314	374	816	1088
558	438	671	389	593	444	671	1298	1213
957	881	1044	894	970	862	1070	1489	1436
1460	1376	1295	1409	1399	1347	1311	2545	1972
2321	2497	1896	2336	2044	2475	1896	3626	3203
8.7^b	8.9^b	9.2^b	8.5^b	8.9^b	8.5^b	9.1^b	15.8^b	15.4^b

^{a}MP2 optimized geometry; CEP[3^{s}P2^d/3^{s}1^p] basis set for FCNO and RCEP[4^{s}P/3^d] for M$^+$, with five d-type components.
^{b}Zero-point energy. in kcalmol^{-1}.
^{c}M$^+$ attachment at O.
^{d}M$^+$ attachment at C.

F$^+$=C=N—O—H bonding structure also contributes. The differences in vibrational frequencies calculated for FCNOH$^+$ relative to FC(H)NO$^+$ and their shifts relative to bent FCNO shown in Table 55 are sufficiently large to serve as a basis of separate identification of each of the two cation isomers.

The energies of the metal ion bound FCNO complexes show a similar pattern of relationships as the HCNO metal ion complexes (Table 29). For both substrates the oxygen-bound Ag$^+$ complex is consistently calculated to be more stable than the carbon-attached compound, irrespective of basis set and correlation level. For Cu$^+$ coordination, the MP2 level calculations show the oxygen site attachment to be more stable. This result persists at the QCISD level but the triples correction, which is negative and much larger for FCNOCu$^+$ than for FC(Cu$^+$)NO$^+$, reverses the order of Cu$^+$ binding energies. For Au$^+$ coordination, all levels of basis set and correlation show the carbon-bound complex to be more stable than the oxygen-bound complex and, like for Ag$^+$, the triples correction to the QCISD energy is reasonable. Discounting the QCISD(T) results for Cu$^+$, the general result is that the preferred binding site of the metal ions for FCNO changes from oxygen (Cu$^+$, Ag$^+$) to carbon (Au$^+$). Further, only for the Au$^+$ complex is the binding energy to FC*NO larger than to HC*NO.

The MP2 optimized geometric structures for the FCNOM$^+$ and FC(M)NO$^+$ complexes are described in Table 54. All three metal ions were also found to attach to the nitrogen atom

$$MNC\begin{smallmatrix}\nearrow O \\ \searrow F\end{smallmatrix}^+$$

to form the MNC$^+$ rearranged cation, like the protonated species described above.

Metal ion coordination to the backside of the fluorine atom was also found to give stable (MP2) complexes. However, like the protonated species, the metal ion binding energies to fluorine were calculated to be much less than for FCNOM$^+$ and FC(M)NO$^+$ and will not be discussed further. FCNOM$^+$ has a planar structure for M$^+$ = Cu$^+$ and Au$^+$, and is slightly kinked out of plane for M = Ag⁺. All these complexes have a zig-zag pattern for the FCNO chain. FC(M)NO⁺ is planar for all the metal ions and N—O is *syn* to C—M. The trends in bond atom distances from bare FCNO to each of the complex ion isomer forms are usually similar and of intermediate degree to those for the corresponding site protonated species. Deviations from the trends found for the protonated FCNO, such as the larger C—F bond lengths in the FC(M)NO⁺ complexes, can probably be attributed to the size of the metal ions. Analogously, the calculated harmonic stretch frequencies for the different isomer forms show that vibrational spectra can be used to distinguish between the two possible coordination sites.

XV. CH₃CNO

Acetonitrile N-oxide (CH₃CNO) is unstable at room temperature and undergoes dimerization. All experimental results are obtained either at lower temperatures for the pure substance, or else in solution or in matrix isolation. CH₃CNO is a 1,3 dipole with a hypervalent central nitrogen atom. Of the possible methyl–CNO chain (non-branched) isomers, acetonitrile N-oxide is predicted to be the least stable at the RHF level in a small basis set[216]. However, CH₃—C≡N=O is a model compound for higher alkylated RCNO compounds which are known experimentally[7]. The calculated properties for bare, protonated and metal ion coordinated CH₃CNO are listed in Tables 2, 3, and 56–58. An RHF study of the space partitioned charge density of CH₃CNO has recently been published[217].

The microwave spectrum of CH₃CNO[218] gives measured bond distances of 1.441 Å for C—C, 1.167 Å for C—N and 1.219 Å for N=O, and a dipole moment of 4.49D. The MP2 level calculation on CH₃CNO repeats the experience with HCNO and FCNO. The linear CCNO optimized geometry (Table 57) has one imaginary harmonic vibrational frequency with normal coordinate eigenvectors that indicate a bending about the central carbon atom. Following this mode in geometry optimization gives an equilibrium structure also described in Table 57. In this non-linear geometry all the calculated harmonic vibrational frequencies are real (Table 58). The symmetry is C_s with the planar HCCNO skeleton in a zig-zag configuration. The MP2 calculated bond distances are larger than experiment by 0.03 Å (C—C), 0.02 Å (C≡N) and 0.01 Å (N=O). The bent CH₃CNO structure has a MP2 dipole moment of 3.85D and the linear geometry has a MP2 calculated dipole moment of 4.09D. Again, the linear structure has the dipole moment closer to experiment. RHF optimization gives only the linear structure with an exaggerated dipole moment of 5.59D. QCISD energies at the MP2 optimized linear and bent conformations favour the linear structure. The very low harmonic vibrational mode (34 cm⁻¹) for the bent conformation (Table 58), and a CNN angle that is larger than HCN in HCNO (Table 30) and FCN in FCNO (Table 54) indicate the great likelihood that the CCNO group is linear. However, the tendency for the nitrile oxides to bend in a *trans* conformation, as has been found here for HCNO, FCNO and CH₃CNO, has been used to interpret the mechanism of nucleophilic attack at the 1,3 (C,O) positions and the conformation of the adduct[7, 219–221]. Several vibrational bands of acetonitrile N-oxide have been identified in matrix isolation[219] and can be compared with the MP2 calculated harmonic frequencies in Table 58. The fundamental C—N stretch is observed at 2309 cm⁻¹ and calculated at 2256 cm⁻¹. The other measured fundamental frequencies are more mixed-mode types. The mainly N—O stretch is observed at 1332 cm⁻¹ and calculated to be 1423 cm⁻¹, while the primarily C—C stretch motion is reported at 780 cm⁻¹ and calculated at 790 cm⁻¹.

CH₃CNO protonated at the nitrogen atom rearranged spontaneously to give CH₃C(H)NO⁺. Protonation at the middle carbon atom gives a MP2 optimized C_s structure with the in-plane (terminal) C—H bond in a *syn* conformation with C—N and *anti* to the (middle) C—H bond across the C—C bond. The CNO group is essentially linear (Table 57). All the harmonic vibrational frequencies are real for this conformation. The eclipsed C—H

TABLE 56. Energies (in a.u.) of MCH$_3$CNO$^+$ complexes[a]

M$^+$	BF[b]	X[g]	RHF	MP2	QCISD	QCISD(T)
Cu$^+$	137[c,e]	O	−233.127977			
			(46.4)			
	137[d,e]	O	−233.120254	−234.263624	−234.218866	−234.193501
			(46.1)	(51.1)	(57.8)	(19.2)
	151[f]	O	−233.120861	−234.519660		
			(46.5)	(55.0)		
	137[c,e]	C	−233.073507			
			(4.8)			
	137[d,e]	C	−233.062613	−234.241271		
			(9.3)	(37.0)		
	151[f]	C	−223.062993	−234.500467		
			(10.2)	(42.9)		
Ag$^+$	137[c,e]	O	−183.238519			
			(35.8)			
	137[d,e]	O	−183.232257	−184.105826	−184.119167	−184.152955
			(36.5)	(34.8)	(37.7)	(36.7)
	151[f]	O	−183.233282	−184.466834		
			(37.1)	(39.3)		
	137[c,e]	C	h			
	137[d,e]	C	−183.180339	−184.081005		
			(5.6)	(19.3)		
	151[f]	C	−183.180983	−184.445910		
			(4.3)	(26.2)		
Au$^+$	137[c,e]	O	−172.946490			
			(40.0)			
	137[d,e]	O	−172.939615	−173.763954	−173.784207	−173.818644
			(40.2)	(44.2)	(47.9)	(47.1)
	151[f]	O	−172.941830	−174.094217		
			(41.6)	(52.2)		
	137[c,e]	C	−172.909420			
			(21.3)			
	137[d,e]	C	−172.901392	−173.756470		
			(16.3)	(39.6)		
	151[f]	C	−172.903174	−174.096225		
			(17.4)	(53.4)		

[a]Numbers in parentheses are metal ion binding energies in kcalmol^{-1}.
[b]Number of basis functions; CEP[3^{s}p2^d/3^{s}1^p] basis set with five d-type components for CH$_3$CNO.
[c]RHF optimized.
[d]MP2 optimized.
[e]RCEP[4^{s}p3^d] with five d-type components for M$^+$.
[f]RCEP[4^{s}p3^{d}2^f] with five d-type and seven f-type components for M$^+$. Complex geometry taken from 137 basis function MP2 optimization.
[g]Atom to which M$^+$ is attached.
[h]Rearranged spontaneously to the oxygen attached metal ion complex.

bond conformer was not investigated. The C—C bond length increased by 0.06 Å, C—N increases by 0.07 Å and N—O decreases by 0.06 Å in going from CH$_3$CNO to CH$_3$C(H)NO$^+$. These are the same trends as for the corresponding C-protonation of FCNO, except for the magnitude of the changes which are smaller for FCNO. The changes in C—N and N—O

TABLE 57. Mulliken populations and geometric parameters for MCH$_3$CNO$^+$[a].

M$^+$	Mulliken										Bond lengths (Å)					Angles (deg)		
	M$^+$ populations				Charges													
	s	p	d	f	M$^+$	H[c]	C[k]	C	N	O	M–X[b]	H–C[k]	C–C	C–N	N–O	MXN[b]	CCN	CNO
H$^+$[l]	—	—	—	—	—	0.16	−0.24	0.26	−0.12	−0.38	—	1.083	1.475	1.126	1.218	—	180	180
	—	—	—	—	—	0.17	−0.29	0.12	−0.12	−0.21	—	1.090	1.468	1.180	1.215	—	180	180
H$^+$[d,m]	—	—	—	—	—	0.16	−0.29	0.12	−0.12	−0.20	—	1.091	1.472	1.186	1.212	—	161.2	174.6
[e]	—	—	—	—	0.44	0.22	−0.23	0.49	−0.19	−0.17	0.960	1.083[c]	1.476	1.116	1.300	107.5	179.4	172.5
	—	—	—	—	0.43	0.23	−0.28	0.36	−0.10	−0.13	0.983	1.090[c]	1.461	1.155	1.310	106.3	177.8	169.2
	—	—	—	—	0.27	0.20	−0.32	0.21	0.08	0.15	1.082	1.082[c]	1.520	1.234	1.120	124.4	121.9	179.2
	—	—	—	—	0.28	0.22	−0.37	0.12	0.10	0.23	1.092	1.089[c]	1.525	1.253	1.156	113.4	120.8	178.5
Cu$^+$	2.09	6.07	9.99	—	0.86	0.20	−0.23	0.41	−0.24	−0.38	1.986	1.082[c]	1.477	1.119	1.255	133.3	179.6	176.9
	2.19	6.11	9.93	—	0.77	0.21	−0.28	0.28	−0.16	−0.25	1.871	1.089[c]	1.467	1.163	1.255	131.3	179.9	175.6
Cu$^+$[f]	2.10	6.11	9.90	—	0.76	0.18	−0.28	0.28	−0.16	−0.24	2.142	1.081[c]	1.534	1.190	1.146	109.7	126.2	176.1
[g]	2.19	6.11	9.99	0.03	0.71	0.19	−0.29	0.10	−0.01	−0.05	1.904	1.087[c]	1.535	1.218	1.177	109.3	126.7	176.0
	2.38	6.16	9.89	—	0.57	0.19	−0.34	0.15	−0.04	0.07								
	2.41	6.16	9.86	0.03	0.54	0.20	−0.34	0.17	−0.04	0.08								
Ag$^+$	2.05	6.04	9.99	—	0.91	0.19	−0.23	0.39	−0.25	−0.39	2.274	1.082[c]	1.478	1.120	1.250	138.5	179.5	177.7
	2.09	6.08	9.97	—	0.86	0.20	−0.27	0.25	−0.18	−0.27	2.211	1.089[c]	1.468	1.166	1.249	133.7	179.6	176.7
Ag$^+$[h]	2.10	6.06	9.89	0.09	0.85	0.20	−0.28	0.26	−0.18	−0.26	2.291	1.088[c]	1.513	1.207	1.188	103.2	138.6	174.5
	2.17	6.08	9.95	—	0.80	0.19	−0.32	0.03	−0.08	0.00								
	2.20	6.08	9.86	0.09	0.77	0.20	−0.32	0.04	−0.08	0.01								
Au$^+$	2.15	6.05	9.94	—	0.86	0.20	−0.23	0.39	−0.21	−0.41	2.212	1.082[c]	1.477	1.119	1.259	127.0	179.6	176.5
	2.27	6.07	9.87	—	0.79	0.21	−0.28	0.27	−0.12	−0.29	2.121	1.089[c]	1.467	1.164	1.263	121.8	179.8	174.5
	2.31	6.08	9.77	0.10	0.73	0.21	−0.28	0.27	−0.13	−0.24								
Au$^+$[j,i]	2.35	6.07	9.87	—	0.71	0.18	−0.30	0.05	0.00	0.00	2.177	1.081[c]	1.536	1.203	1.139	113.6	121.9	177.9
	2.50	6.09	9.75	—	0.67	0.20	−0.36	0.02	−0.02	0.10	2.042	1.088[c]	1.530	1.222	1.176	109.7	127.1	178.0
	2.56	6.09	9.65	0.11	0.59	0.20	−0.36	0.06	−0.02	0.11								

[a] Basis sets, energies and levels of theory for the neutral and protonated CH$_3$CNO, and complexes are described in Tables 2, 3 and 56, respectively.
[b] X is defined in Table 56.
[c] Averaged.
[d] Angle MCC = 124.4 °.
[e] Angle MCC = 125.7 °.
[f] Angle MCC = 124.1 °.
[g] Angle MCC = 124.0 °.
[h] No RHF structure; MP2 angle MCC = 118.2 °.
[i] Angle MCC = 124.5 °.
[j] Angle MCC = 123.2 °.
[k] Methyl carbon.
[l] CH$_3$C(H)NO$^+$.
[m] CH$_3$CNOH$^+$.

TABLE 58. MP2 calculated harmonic vibrational frequencies (in cm^{-1}) for CH$_3$CNOM^{+a}

CH$_3$CNO	CuCH$_3$CNO$^+$		AgCH$_3$CNO$^+$		AuCH$_3$CNO$^+$		H$_3$CNOH$^+$	
	O^c	C^d	O^c	C^d	O^c	C^d	O^c	C^d
	16	79	−26	76	20	110	10	142
	85	114	63	94	82	119	223	203
	208	179	184	153	202	186	237	457
34	213	298	190	209	204	263	376	621
116	351	377	267	267	319	348	380	625
472	403	592	418	549	400	532	728	872
478	476	639	450	591	488	625	1055	1042
790	779	891	770	836	761	897	1060	1092
1066	1062	1055	1063	1051	1061	1056	1168	1149
1066	1064	1057	1065	1069	1064	1061	1419	1320
1423	1309	1366	1318	1411	1269	1350	1450	1426
1449	1433	1422	1435	1429	1431	1422	1480	1494
1517	1500	1505	1504	1507	1498	1503	1484	1508
1526	1501	1518	1505	1523	1500	1517	2387	2018
2256	2336	2165	2317	2299	2315	2104	3092	3099
3088	3100	3113	3102	3105	3099	3109	3189	3144
3168	3195	3214	3197	3202	3195	3212	3199	3200
3176	3197	3221	3198	3205	3198	3218	3641	3221
30.9^b	31.8^b	32.6^b	31.5^b	32.3^b	31.6^b	32.4^b	38.0^b	38.1^b

^{a}MP2 optimized geometry; CEP[3^{s}p2^d/3^{s}1^p] basis set for CH$_3$CNO and RCEP[4^{s}p3^d] for M$^+$, with five d-type components.
^{b}Zero-point energy, in kcal mol^{-1}.
^{c}M$^+$ attachment at O.
^{d}M$^+$ attachment at C(N).

bond distances, found also in H$_2$CNO$^+$ (Table 30), are explained by emphasis of the CH$_3$—C(H)=N$^+$=O bonding structure, although it is the oxygen atom which shows the largest increase in charge (Table 57). As noted for HCNO, these changes imply a role of a >C=N≡O$^+$ bonding structure. Consistent with the bond length change, the C—N stretch frequency decreases from 2256 cm^{-1} to 2018 cm^{-1}. The (H$_3$)CCH angle is 125.7° and the increase in C—C length might be partially due to steric repulsion between CH$_3$ and H. In CH$_3$CN (Table 46) the C—C bond distance is the same as in CH$_3$CNO.

The more stable protonated structure is CH$_3$CNOH$^+$ with a QCISD(T)//MP2 calculated proton affinity of 199.4 kcalmol^{-1} at the oxygen atom. This is larger than the PA found for both HCNO and FCNO for the same site attachment, and larger than C-attachment to acetonitrile N-oxide by an electronic energy difference of 17.4 kcalmol^{-1}. The ZPE difference adjusted PA for O-attachment is 192.3 kcalmol^{-1}. CH$_3$CNOH$^+$ has approximate C_s symmetry with an almost linear CCNO chain in a flat zig-zag arrangement. The C—H and O—H bonds are *trans* to each other. The C—C bond length is essentially unaffected, C—N decreases and N—O increases upon O-protonation in accord with the stabilized CH$_3$—C≡N$^+$—OH structure. The calculated harmonic frequencies (Table 58) show a very low energy coupled H$_3$C...OH rotation mode, indicating pseudo-cylindrical symmetry for CH$_3$CNOH$^+$. The C—N stretch can be identified at 2387 cm^{-1} and N—O at 1168 cm^{-1}. These vibrational frequencies are shifted higher and lower, respectively, relative to CH$_3$CNO, in accord with the specific changes in bond length.

Metal ion–acetonitrile N-oxide complexes attached at the CNO carbon atom look very much like their protonated counterparts. The CNO group is slightly bent towards the metal ion and the complexes have essentially C_s symmetry. The metal ion binding energies at the MP2 level are larger for CH_3CNO than for FCNO or HCNO, with the ordering, $CH_3 > H \geqslant F$. At the higher correlation levels the trend is less clear. The bond length changes upon metal ion coordination are intermediate between the bare CH_3CNO and the carbon atom protonated species, as are the corresponding harmonic stretch frequencies. For the Ag^+ complex, which is calculated to be the weakest bound of the three, as usual, the RHF optimized geometry could not be obtained. Initial positioning of Ag^+ near the carbon atom gave the CH_3CNOAg^+ complex in the geometry optimization.

Metal ion complexation at the oxygen is mainly calculated to be the energetically preferred site. Here, the metal ion binding energies are larger for CH_3CNO than for FCNO or HCNO for all the metal ions, basis set and level of theory. The symmetry of the three metal ion complexes is almost C_s with the CH_3 group rotated out of planar alignment. This rotation is reflected in the low-energy smallest harmonic vibrational frequencies for the CH_3CNOM^+ series which involve mainly cylindrical CH_3 motion about the C—C bond. For CH_3CNOAg^+ this frequency is even negative at the MP2 optimized geometry which is taken as indicative of a transition state. However, the methyl rotation is almost completely unhindered with a calculated force constant of 0.0005 mdyne/Å. This is shown also by the very low positive frequencies for the Cu^+ and Au^+ oxygen site complexes. The energy gradient is calculated to be very small for any CH_3 rotation angle and the exact equilibrium geometry of CH_3CNOAg^+ was not pursued further. The CCNO group is bent in a shallow arc and the metal ion sits outside the sense of the arc. As usual, the CH_3CNO bond lengths and corresponding harmonic stretch frequencies are intermediate between the free ligand and the O-protonated form.

The behaviour of the perturbative triples correction (T) to the QCISD energy for the Cu^+ complexes of CH_3CNO is anomalously large and negative (Table 57), as has also been found for many of the other copper ion complexes in this study. The magnitude of the (T) energy term is variable, depending on the ligand and on the basis set. The PP basis set consistently shows the smaller effect. These results apparently confirm the conclusions of a recent analysis of excitation level contributions[222] that the QCISD(T) method exaggerates the triples contribution in those cases where triples level excitation effects are important. It must therefore be that those Cu^+ complexes showing unusually large (T) energy correction terms need to be treated at an appropriately higher level of correlation than the corresponding Ag^+ and Au^+ complexes for a comparable level of accuracy.

XVI. REFERENCES

1. E. Clementi, in *The Chemistry of the Cyano Group* (Ed. Z. Rappoport), Wiley, London, 1970, p. 1.
2. J. B. Moffat, in *The Chemistry of Diazonium and Diazo Groups* (Ed. S. Patai), Wiley, Chichester, 1978, p. 1.
3. M. Simonetta and A. Gavezzotti, in *The Chemistry of the Triple Bond* (Ed. S. Patai), Wiley, Chichester, 1978, p. 1.
4. J. B. Moffat, in *The Chemistry of Triple-Bonded Functional Groups, Supplement C* (Ed. S. Patai), Wiley, Chichester, 1983, pp. 1015, 1305.
5. C. A. Coulson, *Valence,* Oxford University Press, London, 1961.
6. F. A. Cotton and G. Wilkinson, *Advanced Inorganic Chemistry, fourth edition,* Wiley, New York, 1980, p. 86; see also, M. R. A. Blomberg and P. E. M. Seigbahn, *J. Am. Chem. Soc.,* **115**, 6908 (1993).
7. G. Bianchi, R. Gandolfi and P. Grunanger, Reference 4, p. 737.
8. J. A. M. Simoes and J. L. Beauchamp, *Chem. Rev.,* **90**, 629 (1990).
9. P. B. Armentrout, *Ann. Rev. Phys. Chem.,* **41**, 313 (1990).

10. L. M. Lech and S. Freiser, *Organometallics*, **7**, 1948 (1988).
11. K. Eller and H. Schwartz, *Chem. Rev.*, **91**, 1121 (1991).
12. A. W. Castleman, J. K. Weil, S. W. Sigworth, R. E. Leuchtner and R. G. Keesee, *J. Chem. Phys.*, **86**, 3829 (1987).
13. K. F. Willey, C. S. Yeh, D. L. Robbins and M. A. Duncan, *Chem. Phys. Lett.*, **192**, 179 (1992).
14. F. Bouchard, J. W. Hepburn and T. B. McMahon, *J. Am. Chem. Soc.*, **111**, 8934 (1989).
15. H. Higashide, T. Kaya, M. Kobayashi, H. Shinchara and H. Sato, *Chem. Phys. Lett.*, **171**, 297 (1990).
16. M. S. El-Shall, K. E. Schriver, R. L. Whetten and M. Meot-Ner, *J. Phys. Chem.*, **93**, 7969 (1989).
17. K. X. He, T. D. Hammond, C. B. Winstead, J. L. Gole and D. A. Dixon, *J. Chem. Phys.*, **95**, 7183 (1991).
18. H. Basch, *J. Chem. Phys.*, **56**, 441 (1972).
19. See, for example, C. W. Bauschlicher, Jr., H. Partridge and S. R. Langhoff, *J. Phys. Chem.*, **96**, 3273 (1992); M. Sodupe, C. W. Baushlicher Jr. and H. Partridge, *Chem. Phys. Lett.*, **192**, 185 (1992).
20. H. Basch and T. Hoz, in *Supplement B: The Chemistry of Acid Derivatives, Vol. 2* (Ed. S. Patai), Wiley, Chichester, 1992, p. 1.
21. T. Hoz and H. Basch, *Supplement S: The Chemistry of Sulphur Containing Functional Groups* (Eds. S. Patai and Z. Rappoport), Wiley, Chichester, 1993.
22. C. E. Moore, Atomic Energy Levels, NSRDS-NBS 35/V.II, U.S. Government Printing Office, Washington, D. C. 20402, 1971.
23. C. E. Moore, Atomic Energy Levels, NSRDS-NBS 35/V.III, U.S. Government Printing Office, Washington, D. C. 20402, 1971.
24. L. A. Barnes, M. Rosi and C. W. Bauschlicher, Jr., *J. Chem. Phys.*, **93**, 609 (1990).
25. C. W. Bauschlicher, Jr., H. Partridge, J. A. Sheehy, S. R. Langhoff and M. Rosi, *J. Phys. Chem.*, **96**, 6969 (1992).
26. R. Akesson, L. G. M. Pettersson, M. Sandstrom, P. E. M. Siegbahn and U. Wahlgren, *J. Phys. Chem.*, **96**, 10773 (1992).
27. W. J. Hehre, L. Radom, P. v. R. Schleyer and J. A. Pople, *Ab Initio Molecular Orbital Theory*, Wiley-Interscience, New York, 1986.
28. C. Moller and M. S. Plesset, *Phys. Rev.*, **46**, 618 (1934).
29. W. J. Stevens, H. Basch and M. Krauss, *J. Chem. Phys.*, **81**, 6026 (1984).
30. W. J. Stevens, M. Krauss, H. Basch and P. G. Jasien, *Can. J. Chem.*, **70**, 612 (1992).
31. C. C. J. Roothaan, *Rev. Mod. Phys.*, **23**, 69 (1951).
32. Gaussian 90. M. J. Frisch, M. Head-Gordon, J. B. Foresman, Trucks, K. Raghavachari, H. B. Schlegel, M. Robb, J. S. Binkley, C. Gonzales, D. J. DeFrees, D. J. Fox, R. A. Whiteside, R. Seeger, C. F. Melius, J. Baker, L. R. Kahn, J. J. P. Stewart, E. M. fluder, S. Topiol and J. A. Pople, Gaussian, Inc., Pittsburgh, PA 15213, 1990.
33. Gaussian 92, Revision A. M. J. Frisch, G. W. Trucks, M. Head-Gordon, P. M. W. Gill, W. Wong, J. B. Foresman, H. B. Schlegel, M. A. Robb, E. S. Replogle, R. Gomperts, J. L. Andres, K. Raghavachari, J. S. Binkley, C. Gonzales, R. L. Martin, D. J. Fox, D. J. DeFrees, J. Baker, J. J. P. Stewart and J. A. Pople, Gaussian, Inc., Pittsburgh PA 15213, 1992.
34. M. J. Frisch, J. A. Pople and J. S. Binkley, *J. Chem. Phys.*, **80**, 3265 (1984).
35. E. Magnusson, *J. Am. Chem. Soc.*, **115**, 1051 (1993).
36. J. P. Desclaux, *Comput. Phys. Commun.* **9**, 31 (1975).
37. See, for example, K. G. Dyall, *J. Chem. Phys.*, **98**, 2191 (1993).
38. L. G. M. Petterson, U. Wahlgren and O. Gropen, *Chem. Phys.*, **80**, 7 (1983).
39. M. M. Hurley, L. Fernandez, P. A. Pacios, R. B. Christiansen, R. B. Ross and W. C. Ermler, *J. Chem. Phys.*, **84**, 6840 (1986).
40. C. W. Bauschlicher, Jr., S. P. Walch and H. Partridge, *J. Chem. Phys.*, **76**, 1033 (1982); see also, A. W. Ehlers, M. Bohme, S. Dapprich, A. Gobbi, A. Hollwarth, V. Jonas, K. F. Kohler, R. Stegmann, A. Veldkamp and G. Frenking, *Chem. Phys. Lett.*, **208**, 111 (1993).
41. K. Raghavachari and G. W. Trucks, *J. Chem. Phys.*, **91**, 1062, 2467 (1989).
42. J. A. Pople, M. Head-Gordon and K. Raghavachari, *J. Chem. Phys.*, **87**, 5968 (1987).
43. J. A. Pople, R. Seeger and R. Krishnan, *Int. J. Quantum Chem., Symp.*, **11**, 149 (1977).
44. H. B. Schlegel, *Adv. Chem. Phys.*, Vol. LXVII (Ed. K. P. Lawley), Wiley, Chichester, 1987, p. 249.
45. R. S. Mulliken, *J. Chem. Phys.*, **55**, 3428 (1962).

46. R. S. Mulliken and P. Politzer, *J. Chem. Phys.*, **55**, 5135 (1971).
47. K. B. Wiberg, C. M. Hadad, T. J. LePage, C. M. Breneman and M. J. Frisch, *J. Phys. Chem.*, **96**, 671 (1992).
48. P. G. Jasien and W. J. Stevens, *J. Chem. Phys.*, **84**, 3271 (1986).
49. S. F. Boys and F. Bernardi, *Mol. Phys.*, **19**, 553 (1970).
50. T. F. M. Tao and Y. K. Pan, *J. Phys. Chem.*, **95**, 3582 (1991).
51. R. S. Mulliken and W. P. Person, *Molecular Complexes*, Wiley-Interscience, New York, 1969.
52. J. B. Foresman, M. Head-Gordon, J. A. Pople and M. J. Frisch, *J. Phys. Chem.*, **96**, 135 (1992).
53. D. E. Lessen, R. L. Asher and P. J. Brucat, *J. Chem. Phys.*, **95**, 1414 (1991).
54. K. F. Willey, C. S. Yeh, D. L. Robbins and M. A. Duncan, *J. Phys. Chem.*, **96**, 9106 (1990).
55. P. B. Armentrout and J. Simons, *J. Am. Chem. Soc.*, **114**, 8627 (1992).
56. W. D. Laidig, P. Saxe and R. J. Bartlett, *J. Chem. Phys.*, **86**, 887 (1986).
57. S. R. Langhoff, C. W. Bauschlicher, Jr., and P. R. Taylor, *Chem. Phys. Lett.*, **180**, 88 (1991).
58. H. Basch, S. Hoz, M. Goldberg and L. Gamess, *Israel J. Chem.*, **31**, 335 (1991).
59. J. H. van Lenthe, J. G. C. M. van Duijneveldt and F. B. van Duijneveldt, *Adv. Chem. Phys.*, Vol. LXIX (Ed. K. P. Lawley), Wiley, Chichester, 1987, p. 521.
60. W. J. Stevens and W. H. Fink, *Chem. Phys. Lett.*, **139**, 15 (1987).
61. G. Herzberg, *Electronic Spectra and Electronic Structure of Polyatomic Molecules*, D. Van Nostrand, Inc., New York, 1966.
62. See, for example, G. Trinquier, *J. Am. Chem. Soc.*, **114**, 6807 (1992).
63. B. Ruscic, J. Berkowitz, L. A. Curtiss and J. A. Pople, *J. Chem. Phys.*, **91**, 114 (1989).
64. W. Klopper and W. Kutzelnigg, *J. Phys. Chem.*, **94**, 5625 (1990).
65. S. G. Lias, J. E. Bartmess, J. F. Liebman, J. L. Holmes, R. D. Levin and W. G. Mallard, *J. Phys. Chem. Ref. Data*, **17**, Supplement No. 1 (1988).
66. B. J. Smith and L. Radom, *J. Am. Chem. Soc.*, **115**, 4885 (1993).
67. M. A. Spackman, *J. Phys. Chem.*, **93**, 7594 (1989).
68. A. D. Buckingham, *Adv. Chem. Phys.*, **12**, 107 (1967).
69. P. D. Dacre, *J. Chem. Phys.*, **80**, 5677 (1984).
70. M. Rosi, C. W. Bauschlicher, Jr., S. R. Langhoff and H. Partridge, *J. Phys. Chem.*, **94**, 8656 (1990).
71. J. K. Perry, G. Ohanessian and W. A. Goddard III, *J. Phys. Chem.*, **97**, 5238 (1993).
72. M. Guse, N. S. Ostland and G. D. Blyholder, *Chem. Phys. Lett.*, **61**, 526 (1979).
73. C. R. Brundle, M. B. Robin, N. A. Kuebler and H. Basch, *J. Am. Chem. Soc.*, **94**, 1451 (1972).
74. B. C. Guo and A. W. Castleman, Jr., *Chem. Phys. Lett.*, **181**, 16 (1991).
75. R. C. Burnier, G. D. Byard and B. S. Freiser, *Anal. Chem.*, **52**, 1641 (1980).
76. E. R. Fisher and P. B. Armentrout, *J. Phys. Chem.*, **94**, 1674 (1990).
77. M. Sodupe, C. W. Bauschlicher, Jr., S. R. Langhoff and H. Partridge, *J. Phys. Chem.*, **96**, 2118 (1992); Erratum: *J. Phys. Chem.*, **96**, 5670 (1992).
78. C. W. Bauschlicher, Jr., H. Partridge and S. R. Langhoff, *J. Phys. Chem.*, **96**, 3273 (1992).
79. J. P. Desclaux, *At. Data Nucl. Data Tables*, **12**, 311 (1973).
80. M. J. S. Dewar, *Bull. Soc. Chim. Fr.*, **18**, C79 (1951).
81. J. Chatt and L. A. Duncanson, *J. Chem. Soc.*, 2939 (1953).
82. Landolt-Börnstein, *Numerical Data and Functional Relationships in Science and Technology*, New Series (Ed. K.-H. Hellwege), Group II, Vol. 7, *Structure Data of Free Polyatomic Molecules*, J. H. Callomon, E. Hirota, K. Kuchitsu, W. F. Lafferty, A. G. Maki and C. S. Pote (Eds. K.-H. Hellwege and A. M. Hellwege); Springer-Verlag, Berlin, 1976.
83. B. Zurawaki, R. Ahlrichs and W. Kutzelnigg, *Chem. Phys. Lett.*, **14**, 385 (1973).
84. B. G. Johnson, P. M. W. Gill and J. A. Pople, *J. Chem. Phys.*, **98**, 5612 (1993).
85. H. W. Quinn, J. S. McIntyre and D. J. Peterson, *Can. J. Chem.*, **43**, 2896 (1965).
86. Ch. Elschenbroich and A Salzer, *Organometallics*, VCH Publishers, Weinheim, Germany, 1989.
87. See, for example, J. F. Stanton, J. Gauss and R. J. Bartlett, *Chem. Phys. Lett.*, **195**, 194 (1992).
88. M. B. Robin, *Higher Excited States of Polyatomic Molecules*, Vol. III, Academic Press, New York, 1985.
89. K. B. Wiberg, C. M. Hadad, J. B. Foresman and W. A. Chupka, *J. Phys. Chem.*, **96**, 10756 (1992); L. Serrano-Andres, M. Merchan, I. Nebot-Gil, R. Lindh and B. O. Roos, *J. Chem. Phys.*, **98**, 3151 (1993).
90. P. H. Kasai, D. McLeod Jr. and W. Watanabe, *J. Am. Chem. Soc.*, **102**, 179 (1980); P. H. Kasai,

J. Am. Chem. Soc., **105**, 6704 (1983); J. A. Howard, H. A. Joly and B. Mile, *J. Phys. Chem.*, **94**, 1275, 6627 (1990).

91. K. Huber and G. Herzberg, *Constants of Diatomic Molecules*, Van Nostrand Reinhold, New York, 1979.
92. T. J. Lee, R. B. Remington, Y. Yamaguchi and H. F. Schaefer III, *J. Chem. Phys.*, **89**, 408 (1988).
93. B. Ruscic and J. Berkowitz, *J. Chem. Phys.*, **95**, 4378 (1991).
94. P. Botschwina, *Ion and Cluster Ion Spectroscopy and Structure*, Elsevier Science Publishers B. V., Amsterdam, 1989.
95. R. Glaser, C. J. Horan and P. E. Haney, *J. Phys. Chem.*, **97**, 1835 (1993).
96. J. C. Owrutsky, C. S. Gudeman, C. C. Martner, L. M. Tack, N. H. Rosenbaum and R. J. Saykally, *J. Chem. Phys.*, **84**, 605 (1986).
97. P. W. Fowler and G. H. F. Dierksen, *Chem. Phys. Lett.*, **167**, 105 (1990).
98. C. W. Bauschlicher, H. Partridge and S. R. Langhoff, *J. Phys. Chem.*, **96**, 2475 (1992).
99. M. S. El-Shall, K. E. Schriver, R. L. Whetten and M. Meot-Ner, *J. Phys. Chem.*, **93**, 7969 (1989).
100. D. E. Lessen, R. L. Asher and P. J. Brucat, *Chem. Phys. Lett.*, **177**, 380 (1991).
101. R. G. Keesee and A. W. Castleman Jr., *J. Phys. Chem. Ref. Data*, **15**, 1011 (1986).
102. L. A. Barnes, M. Rosi and C. W. Bauschlicher Jr., *J. Chem. Phys.*, **93**, 609 (1990).
103. H. Partridge, C. W. Bauschlicher Jr. and S. R. Langhoff, *J. Phys. Chem.*, **96**, 5350 (1992).
104. L. G. M. Petterson and H. Akeby, *J. Chem. Phys.*, **94**, 2968 (1991).
105. W. Muller, J. Flesch and W. Meyer, *J. Chem. Phys.*, **80**, 3297 (1984).
106. J. H. D. Eland, *Photoelectron Spectroscopy*, Wiley, New York, 1974.
107. A. Lofthus and P. H. Krupenie, *J. Phys. Chem. Ref. Data*, **6**, 113, (1977).
108. T. Koopmans, *Physica*, **1**, 104 (1933).
109. G. Verhegen, W. G. Richards and C. M. Moser, *J. Chem. Phys.*, **47**, 2595 (1967).
110. (a) S. Carter, I. M. Mills and N. C. Handy, *J. Chem. Phys.*, **97**, 1606 (1992).
(b) See also recent work in J. R. Thomas, B. J. De Leeuw, G. Vacek, T. D. Crawford, Y. Yamaguchi and H. F. Schaefer III, *J. Chem. Phys.*, **99**, 403 (1993).
111. Landolt-Börnstein, *Numerical Data and Functional Relationships in Science and Technology*, New Series (Ed. K.-H. Hellwege), Group II, Vol. 6, *Molecular Constants*, J. Demaison, W. Huttner, B. Stark, I. Buck, R. Tischer and M. Winnewisser (Eds. K.-H. Hellwege and A. M. Hellwege), Springer-Verlag, Berlin, (1974).
112. H. F. Schaefer III, in *Ion and Cluster Ion Spectroscopy and Structure*, Elsevier Science Publishers B. V., Amsterdam, Netherlands, 1989.
113. (a) D.-J. Liu, S.-T. Lee and T. Oka, *J. Mol. Spectrpsc.*, **127**, 275 (1988).
(b). M. W. Crofton, M.-F. Jagod, B. D. Rehfuss and T. Oka, *J. Chem. Phys.*, **91**, 5139 (1989).
114. T. J. Lee and H. F. Schaefer III, *J. Chem. Phys.*, **82**, 2977 (1984).
115. P. Botschwina, *Chem. Phys. Lett.*, **124**, 382 (1986).
116. D. C. Frost, S. T. Lee and C. A. McDowell, *Chem. Phys. Lett.*, **23**, 472 (1973).
117. H. Koppel, L. S. Cederbaum, W. Domcke and W. von Niessen, *Chem. Phys.*, **37**, 303 (1979).
118. S. Peric, S. D. Peyerimhoff and R. J. Buenker, *Can. J. Chem.*, **55**, 3664 (1977); *Mol. Phys.*, **62**, 1339 (1987); M. Peric, H. Dohmann, S. D. Peyerimhoff and R. J. Buenker, *Z. Phys. D*, **5**, 65 (1987).
119. Z. Vager, R. Naaman and E. P. Kanter, in *Ion and Cluster Ion Spectroscopy and Structure*, Elsevier Science Publishers B. V., Amsterdam, 1989.
120. T. Oka, *Phil. Trans. Roy. Soc.* London Ser. *A*, **324**, 81 (1988); M. W. Crofton, M. F. Jagod, B. D. Rehfuss and T. Oka, *J. Chem. Phys.*, **91**, 5139 (1989).
121. R. Lindh, B. O. Roos and W. P. Kraemer, *Chem. Phys. Lett.*, **139**, 407 (1987).
122. T. J. Lee and H. F. Schaeffer III, *J. Chem. Phys.*, **85**, 3437 (1986).
123. L. A. Curtiss and J. A. Pople, *J. Chem. Phys.*, **88**, 7405 (1988).
124. J. Berkowitz, C. A. Mayhew and B. Ruscic, *J. Chem. Phys.*, **88**, 7396 (1988).
125. J. Miralles-Sabater, M. Merchan, I. Nebot-Gil and P. Virula-Martin, *J. Phys. Chem.*, **92**, 4853 (1988).
126. M. Sodupe and C. W. Bauschlicher, *J. Phys. Chem.*, **95**, 8640 (1991).
127. D. F. Dance and I. C. Walker, *Chem. Phys. Lett.*, **18**, 601 (1973).
128. P. Dupre, R. Jost, M. Lombardi, P. G. Green, E. Abramson and R. W. Field, *Chem. Phys.*, **152**, 293 (1991).
129. H. Lischka and A Karpfen, *Chem. Phys.*, **102**, 77 (1986).
130. (a) M. Peric, S. D. Peyerimhoff and R. J. Buenker, *Mol. Phys.*, **62**, 1339 (1987).

    (b) M. Peric, H. Dohmann, S. D. Peyerimhoff and R. J. Buenker, *Z. Phys. D*, **5**, 65 (1987).
131. J. O. Jensen, G. F. Adams and C. F. Chabalowski, *Chem. Phys. Lett.*, **172**, 379 (1990).
132. G. Vacek, J. R. Thomas, B. J. DeLeeuw, Y. Yamaguchi and H. F. Schaefer III,   *J. Chem. Phys.*, **98**, 4766 (1993); J. K. Lundberg, R. W. Field, C. D. Sherrill, E. T. Seidl, Y. Xie and H. F. Schaefer III, *J. Chem. Phys.*, **98**, 8384 (1993); Y. Yamaguchi, G. Vacek and H. F. Schaefer III, *Theor. Chim. Acta*, **86**, 97 (1993).
133. K. Kimura, S. Katsumata, Y.Achiba, T. Yamazaki and S. Iwata, *Handbook of HeI Photoelectron Spectra of Fundamental Organic Molecules*, Halstead Press, New York, 1981.
134. A. Gedanken and O. Schnepp, *Chem. Phys. Lett.*, **37**, 373 (1973).
135. S. D. Kahn, W. J. Hehre and J. A. Pople, *J. Am. Chem. Soc.*, **109**, 1871 (1987).
136. D. L. Cooper, J. Gerratt and M. Rainmondi, *J. Chem. Soc., Perkin Trans. 2*, 1187 (1987).
137. H. Jalink, D. H. Parker and S. Stolte, *J. Mol. Spectrosc.*, **121**, 236 (1987).
138. J. C. Slater, *Quantum Theory of Molecules and Solids*, Volume 1, McGraw-Hill, New York, 1963.
139. first noted by F. Grimaldi, A. Lecourt and C. Moser, *Int. J. Quantum Chem., Symp.*, **1**, 153 (1967).
140. M. J. Frisch and J. E. Del Bene, *Int. J. Quantum Chem., Symp.*, **23**, 363 (1989).
141. J.-L. Teffo and A. Chedin, *J. Mol. Spectrosc.*, **135**, 389 (1989).
142. F. Melen and M. Herman, *J. Phys. Chem. Ref. Data*, **21**, 831 (1992).
143. E. D. Simandiras, J. E. Rice, T. J. Lee, R. D. Amos and N. C. Handy, *J. Chem. Phys.*, **88**, 3187 (1988).
144. J. M. L. Martin and T. J. Lee, *J. Chem. Phys.*, **98**, 7951 (1993).
145. S. P. Ekern, A. Illies and M. L. McKee, *Mol. Phys.*, **78**, 263 (1993).
146. N. G. Adams, D. Smith, M. Tichy, G. Javahery, N. D. Twiddy and E. E. Ferguson, *J. Chem. Phys.*, **91**, 4037 (1989).
147. J. E. Rice, T. J. Lee and H. F. Schaefer III, *Chem. Phys. Lett.*, **130**, 333 (1986).
148. G. Javahery, J. Glosik, N. D. Twiddy and E. E. Ferguson, *Int. J. Mass. Spectrom. Ion Proc.*, **98**, 225 (1990).
149. J. E. Del Bene and M. J. Frisch, *Int. J. Quantum Chem., Symp.*, **23**, 371 (1989).
150. D. R. Garmer and W. J. Stevens, *J. Phys. Chem.*, **93**, 8263 (1989).
151. J. W. Rabalais, J. M. McDonald, V. Scherr and S. P. McGlynn, *Chem. Rev.*, **71**, 73 (1971).
152. E. R. Davidson, *Rev. Mod. Phys.*, **44**, 451 (1972).
153. B. Corain, *Coord. Chem. Rev.*, **47**, 165 (1982).
154. A. G. Maki and S. M. Freund, *J. Mol. Spectrosc.*, **66**, 493 (1977).
155. V. K. Wang and J. Overend, *Spectrochim. Acta*, **A29**, 1623 (1973).
156. M. Alcami, O. Mo, M. Yanez, J. L. M. Abbound and J. Elguero, *Chem. Phys. Lett.*, **172**, 471 (1990).
157. D. B. Beach, C. J. Eyerman, S. P. Smit, S. F. Xiang and W. L. Jolly, *J. Am. Chem. Soc.*, **106**, 536 (1984).
158. S. Marriott, R. D. Topsom, C. B. Lebrilla, I. Koppel, M. Mishima and R. W. Taft, *Theochem*, **137**, 133 (1986).
159. M. Alcami, O. Mo and M. Yanez, *Theochem*, **234**, 357 (1991).
160. L. Asbrink, A. Svensson, W. Von Neissen and G. Bieri, *J. Electron Spectrosc. Relat. Phenom.*, **24**, 293 (1981).
161. Y. Aray, J. Murgich and M. A. Luna, *J. Am. Chem. Soc.*, **113**, 7135 (1991).
162. G. W. King and A. W. Richardson, *J. Mol. Spectrosc.*, **21**, 339 (1966).
163. D. Poppinger, L. Radom and J. A. Pople, *J. Am. Chem. Soc.*, **99**, 7806 (1977).
164. J. Teles, G. Maier, B. A. Hess, L. J. Schaad, M. Winnewisser and B. P. Winnewisser, *Chem. Ber.*, **122**, 732 (1989).
165. M. T. Nguyen, K. Pierloot and L. G. Vanquickenborne, *Chem. Phys. Lett.*, **181**, 83 (1991); A. P. Rendell, T. J. Lee and R. Lindh, *Chem. Phys. Lett.*, **194**, 84 (1992).
166. N. Pinnavaia, M. J. Bramley, Ming-Der Su, W. H. Green and N. C. Handy, *Mol. Phys.*, **78**, 319 (1993); K. Yokoyama, S.-Y. Takane and T. Fueno, *Bull. Chem. Soc. Jpn.*, **64**, 2230 (1991).
167. P. R. Buenker, B. M. Landsberg and B. P. Winnewisser, *J. Mol. Spectrosc.*, **74**, 9 (1979).
168. F. Iachello, N. Manini and S. Oss, *J. Mol. Spectrosc.*, **156**, 190 (1992).
169. R. Takasi, K. Tanaka and T. Tanaka, *J. Mol. Spectrosc.*, **138**, 450 (1989).
170. Reported in, M. E. Jacox, *J. Phys. Chem. Ref. Data*, **19**, 1387 (1990).
171. C. E. C. A. Hop, J. L. Holmes, P. J. A. Ruttink, G. Schaftenaar and J. K. Terlouw, *Chem. Phys. Lett.*, **156**, 251 (1989).

172. J. Bastide and J. P. Maier, *Chem. Phys.*, **12**, 177 (1976).
173. W. D. Sheasley and C. W. Mathews, *J. Mol. Spectrosc.*, **43**, 467 (1972).
174. V. E. Bondybey, J. H. English, C. W. Mathews and R. J. Contolini, *J. Mol. Spectrosc.*, **92**, 431 (1982).
175. B. P. Winnewisser, *J. Mol. Spectrosc.*, **82**, 220 (1980).
176. D. P. Chong. *Chem. Phys. Lett.*, **175**, 525 (1990).
177. C. B. Moore and K. J. Rosengreen, *J. Chem. Phys.*, **40**, 2461 (1964).
178. M. Otto, S. D. Lotz and G. Frenking, *Inorg. Chem.*, **31**, 3647 (1992).
179. F. Cacace, M. Attina, G. De Petris, F. Grandinetti and M. Speranza, *Gazz. Chim. Ital.*, **120**, 691 (1990).
180. (a) R. Glaser and G. S.-C. Choy, *J. Phys. Chem.*, **95**, 7682 (1991).
     (b) R. Glaser and G. S.-C. Choy, *J. Org. Chem.*, **57**, 4976 (1992).
     (c) R. Glaser and G. S.-C. Choy, *J. Am. Chem. Soc.*, **115**, 2340 (1993).
181. J. R. McDonald, J. W. Rabalais and S. P. McGlynn, *J. Chem. Phys.*, **52**, 1332 (1970).
182. H. Okabe, *J. Chem. Phys.*, **49**, 2726 (1968).
183. A. P. Baronavski, R. G. Miller and J. R. McDonald, *Chem. Phys.*, **30**, 119 (1978).
184. U. Meier and V. Staemmler, *J. Phys. Chem.*, **95**, 6111 (1991).
185. J. K. Holland, D. A. Newmham and I. M. Mills, *Mol. Phys.*, **70**, 319 (1990).
186. D. R. Lide (Ed.), *CRC Handbook of Chemistry and Physics*, 72nd edition (1991–1992), The Chemical Rubber Co., Cleveland, Ohio.
187. W. H. Green, D. Jayatilaka, A. Willetts, R. D. Amos and N. C. Handy, *J. Chem. Phys.*, **93**, 4965 (1990). See also P. Botschwina, M. Oswald, J. Flugge, A. Heyl and R. Oswald, *Chem. Phys. Lett.*, **209**, 117 (1993).
188. J.K. Holland, D. A. Newnham, I. M. Mills and M. Herman, *J. Mol. Spectrosc.*, **151**, 346 (1992).
189. P. S. Martin, K. Yates and I. G. Csizmadia, *Can. J. Chem.*, **67**, 2178 (1989).
190. D. I. Stams, T. D. Thomas, D. C. MacLaren, D. Ji and T. H. Morton, *J. Am. Chem. Soc.*, **112**, 1427 (1990).
191. C. R. Brundle, M. B. Robin, N. A. Kuebler and H. Basch, *J. Am. Chem. Soc.*, **94**, 1451 (1972).
192. M. B. Robin, *Higher Excited States of Polyatomic Molecules*, Vol. II, Academic Press, New York, 1975.
193. G. Bieri, L. Asbrink and W. von Niessen, *J. Electron Spectrosc. Relat. Phenom.*, **23**, 281 (1981).
194. D. L. Cooper, J. Gerratt, M. Raimondi and S. C. Wright, *Chem. Phys. Lett.*, **138**, 296 (1987).
195. C. B. Moore and G. C. Pimentel, *J. Chem. Phys.*, **40**, 329 (1964); A. P. Cox, L. F. Thomas and J. Sheridan, *Nature (London)*, **181**, 1000 (1958).
196. C. Guimon, S. Khayar, F. Gracian, M. Begtrup and G. Pfister-Guillouzo, *Chem. Phys.*, **138**, 157 (1989).
197. L. Nemes, J. Vogt and M. Winnewisser, *J. Mol. Struct.*, **218**, 219 (1990).
198. F. Wang, M. A. Winnik, M. R. Peterson and I. G. Csizmadia, *J. Mol. Struct.*, **232**, 203 (1991).
199. M. A. McAllister and T. T. Tidwell, *J. Am. Chem. Soc.*, **114**, 6553 (1992).
200. J. Vogt, M. Winnewisser, K. Yamada and G. Winnewisser, *Chem. Phys.*, **83**, 309 (1984).
201. R. Glaser, G. S.-C. Choy and M. K. Hall, *J. Am. Chem. Soc.*, **113**, 1109 (1991); see also, T. Neuheuser, M. von Armim and S. D. Peyerimhoff. *Theor. Chim. Acta*, **83**, 123 (1992).
202. M. S. Foster and J. L. Beauchamp, *J. Am. Chem. Soc.*, **94**, 2425 (1972).
203. P. Ertl and J. Leska, *J. Mol. Struct.*, **165**, 1 (1988).
204. M. Rittby, S. Pal and R. J. Bartlett, *J. Chem. Phys.*, **90**, 3214 (1989).
205. D. Boucher, J. Burie, A. Bauer, A. Dubrulle and J. Demaison, *J. Phys. Chem. Ref. Data*, **9**, 659 (1980).
206. J. L. Duncan, D. C. McKean, F. Tullini, G. D. Nivellini and J. P. Pena, *J. Mol. Spectrosc.*, **69**, 123 (1978); J. L. Duncan, *Spectrochim. Acta*, **20**, 1197 (1964).
207. T. Amano, *J. Mol. Spectrosc.*, **153**, 654 (1992).
208. C. A. Deakyne, M. Meot-Ner, C. L. Campbell, M. G. Hughes and S. P. Murphy, *J. Chem. Phys.*, **84**, 4958 (1986).
209. C. A. Deekyne, M. Meot-Ner, T. J. Buckley and R. Metz, *J. Chem. Phys.*, **86**, 2334 (1987).
210. M. Meot-Ner and L. W. Sieck, *J. Am. Chem. Soc.*, **113**, 4448 (1991).
211. D. C. Frost, F. G. Herring, C. A. McDowell and L. A. Stenhouse, *Chem. Phys. Lett.*, **4**, 533 (1970).
212. C. Fridh, *J. Chem. Soc., Faraday Trans. 2*, **74**, 2193 (1978).

213. R. Rianda, R. P. Frueholz and A. Kuppermann, *J. Chem. Phys.*, **80**, 4035 (1984).
214. H. Burger, W. Schneider, S. Sommer, W. Thiel and H. Willner, *J. Chem. Phys.*, **95**, 5660 (1991).
215. J. Breidung, W. Schneider, W. Thiel and T. J. Lee, *J. Chem. Phys.*, **97**, 3498 (1992).
216. D. Poppinger and L. Radom, *J. Am. Chem. Soc.*, **100**, 3674 (1978).
217. K. B. Wiberg and C. M. Breneman, *J. Am. Chem. Soc.*, **112**, 8765 (1990).
218. P. B. Blackburn, R. D. Brown, F. R. Burden, J. G. Crofts and I. R. Gillard, *Chem. Phys. Lett.*, **7**, 102 (1970).
219. Z. Mielke, M. Hawkins and L. Andrews, *J. Phys. Chem.*, **93**, 558 (1989).
220. M. T. Nguyen, M. Sana, G. Leroy, K. J. Dignam and A. F. Hegarty, *J. Am. Chem. Soc.*, **102**, 573 (1980).
221. K. K. Sharma and A. K. Aggarwal, *Indian J. Chem.*, *Sect. A*, **28A**, 63 (1989) and references cited therein.
222. Z. He and D. Cremer, *Theor. Chim. Acta*, **85**, 305 (1993) and references cited therein.

CHAPTER **2**

Structural chemistry of triple-bonded groups

FRANK H. ALLEN and STEPHANIE E. GARNER

Cambridge Crystallographic Data Centre, 12 Union Road, Cambridge CB2 1EZ, England

Supplement C2: The chemistry of triple-bonded functional groups
Edited by S. Patai © 1994 John Wiley & Sons Ltd

I. INTRODUCTION

Earlier volumes in this series have contained surveys[1–5] of the structural chemistry of triple-bonded groups from both a theoretical and experimental standpoint, with coverage of the literature through *ca* 1977[1–3] and 1982[4,5]. In the experimental area, these articles provide excellent coverage of results obtained by electron diffraction (ED), microwave (MW) and other spectroscopic (S) methods. By contrast, available data from X-ray (X) and neutron (N) diffraction studies were rather limited at that time (see Figure 1), thus restricting discussion to some of the individual structures then available.

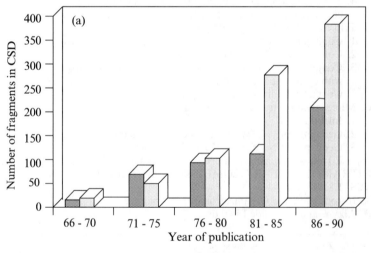

FIGURE 1.

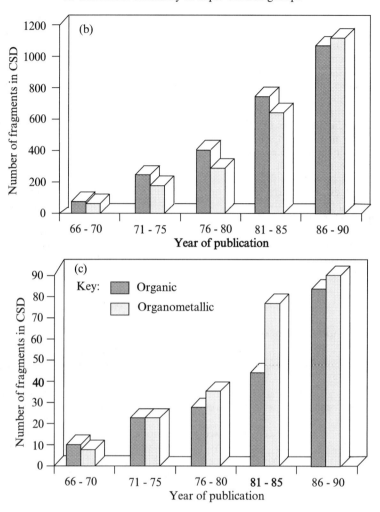

FIGURE 1. Number of of fragments in CSD over the years (a) number of C≡C containing entries; (b) number of C≡N containing entries; (c) number of N≡N, C=N=N and N=N=N containing entries

Over the past fifteen years or so, crystal structure analysis has become a method of choice rather than a method of last resort. As a result, the number of crystal structures containing the acetylene (yne), cyano, isocyano, diazo and azido groups has risen considerably, particularly through the involvement of these groups as ligands in organometallics and metal complexes. Further, the past fifteen years have seen the crystallographic structural databases[6] arrive at some degree of maturity. This development not only permits rapid access to the relevant literature references, but also to the three-dimensional (3D) atomic coordinates and molecular geometry that are the fundamental numerical results of a crystal structure analysis.

In this article, we use the Cambridge Structural Database (CSD)[7] to generate a systematic analysis of the geometrical characteristics of the acetyleno, cyano, isocyano, diazo and azido functional groups and of their immediate chemical environments. The CSD is comprehensive for all organic, organometallic and metal complex small molecular structures determined by X or N methods. Because of this broad chemical coverage, we are able to present typical geometries of groups and group environments drawn from both organic and metallo-organic species. Further, because of the crystallographic origin of the data, we also present a brief survey of the molecular recognition properties of triple-bonded functional groups derived from their non-bonded interactions in the solid state.

The vast majority of results used in this survey arise from X-ray diffraction studies: neutron studies comprise less than 1% of CSD content. However, wherever possible we cite relevant neutron and spectroscopic results and discuss (and attempt to quantify) the numerical differences that exist between N-, S- and X-geometries.

Given the plethora of X-ray studies that are relevant to this area of chemistry, it is obviously impossible to reference each study individually, as might be expected in a classical review format. Indeed, this referencing is a basic function of the database itself. Instead we allow each individual study to make its own contribution to the structural scene through careful selection and averaging as described in Section II and highlight, through direct citation, those structures which differ from the norm either numerically or chemically. In this sense, the words 'review' and 'survey' are synonymous in this chapter and we thank those many crystallographers whose careful work has contributed to results presented below.

II. METHODOLOGY OF DATABASE USE

A. Information Content of the CSD

Each crystal structure of an organic or metallo-organic compound published in the open literature generates an *entry* in the CSD. Each entry is identified by a reference code (*refcode*) and contains three categories of information: (i) bibliographic information: authors names, literature citations, (ii) chemical information: compound name(s), molecular formula, chemical class(es) and the 2D chemical structural diagram as a connection table complete with 2D display coordinates, (iii) structural information: precision indicators, unit-cell and space-group symmetry, 3D atomic coordinates determined by X or N analysis. The CSD is fully retrospective (earliest reference 1935) and is maintained on a current basis. The Version 5 CSD of April 1993, used to generate results presented here, contained 109,992 entries abstracted from 639 literature sources. The CSD also acts as a depository for unpublished atomic coordinate data for more than 20 major journals.

B. The CSD Software System

A fully interactive menu-driven software system has been written[7, 8] to search, retrieve, display and analyse the information held in the CSD. The program QUEST3D permits searches based on all text and numeric fields, substructure searches based on the 2D chemical connectivity information, and 3D (geometric) searches and tabulations of user-defined geometrical parameters derived from the 3D coordinate data. The program will display 2D (chemical) and 3D (structural) diagrams of entries that satisfy search criteria. It will also generate histograms and scattergrams of selected geometrical parameters and provide summary statistics (mean values, estimated standard deviations, maximum and minimum values etc.) for individual parameter distributions. QUEST3D will tabulate, search and analyse both intramolecular and intermolecular geometry. The program GSTAT[7,9,10] is also available to perform more complex analyses of molecular geometry.

C. Generation of Results

Detailed surveys of substructure search and data analysis methodologies applied to the CSD have been published elsewhere[7,11–13]. Here we present only a brief summary of those methods that are relevant to the generation of the results presented in this review. In general, these methods are closely similar to those employed earlier[14,15] in generating standard bond lengths in a broad range of organic and metallo-organic species.

A typical substructure, e.g. the α-ketoacetylene illustrated in Figure 2, is sketched via the CSD menu system. The substituent carbon C_1 is constrained to be an sp^3 hybrid, to avoid further conjugative interactions with the triple bond, and any substitution is permitted at C_4. Further secondary search constraints are also applied to restrict the substructure search to entries which are: (i) error-free, i.e. have passed the CSD check and evaluation procedures, (ii) do not exhibit disorder in their crystal structures, (iii) have a crystallographic R-factor (*RFAC in Figure 2b) ≤ 0.05, or (iv) have an R-factor in the range 0.051–0.070 and a mean estimated standard deviation for CC bond lengths ≤ 0.010 Å; this corresponds to an average sigma (or AS) flag (*SIGF, Figure 2b) of 1 or 2 in the CSD.

For each instance of every fragment that passes tests (i)–(iv) we calculate an appropriate set of geometrical parameters: the set selected for the α-ketoacetylenes is illustrated in Figure 2a. After all information has been accumulated, we obtain a data matrix containing n lines where n is the number of fragments counted, each line containing the ordered list of (chosen) geometrical parameters. This matrix may now be analysed using a variety of graphical, statistical or numerical techniques[13]. For this review, simple histograms and descriptive statistics for distributions of individual variables (e.g. D1, D2 etc. in Figure 2b) have been used extensively to assemble the numerical results. The Tables presented below include some or all of the following quantities for each cited parameter.

(a) Typical substructure search fragment and definition of geometrical parameters (see text).

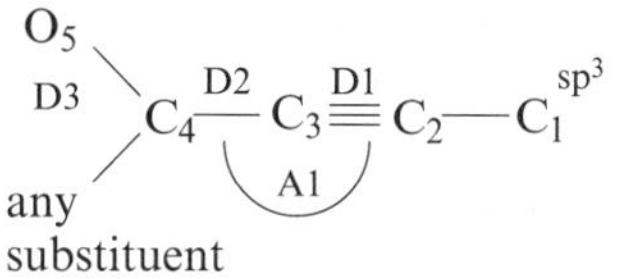

(b) Fragment-by-fragment tabulation of molecular geometry and simple descriptive statistics for each parameter distribution. The structural precision indicators Rfact and AS are discussed in the text.

Nfrag	Refcode	*RFAC	*SIGF	D1	D2	D3	A1
1	AMTETA	0.056	1	1.186	1.487	1.237	176.318
2	AMTETA	0.056	1	1.186	1.487	1.237	176.318
3	FIVGOG	0.043	1	1.200	1.460	1.216	177.843
4	MTROLA10	0.050	1	1.187	1.454	1.222	178.332
5	SAZKIN	0.053	1	1.178	1.450	1.180	177.610
6	SAZKIN	0.053	1	1.178	1.450	1.180	177.610
7	TETROL01	0.040	0	1.178	1.441	1.204	176.440
	Nobs	7	7	7	7	7	7
	Mean	0.050	0.857	1.185	1.461	1.211	177.210
	SDSample	0.006	0.378	0.008	0.018	0.024	0.831
	SDMean	0.002	0.143	0.003	0.007	0.009	0.314
	Minimum	0.040	0.000	1.178	1.441	1.180	176.318
	Maximum	0.056	1.000	1.200	1.487	1.237	178.332

FIGURE 2. Typical substructure search fragment for an α-ketoacetylene (a) and relevant data (b)

(a) The unweighted sample mean, d for a bond length, θ for a valence angle, τ for a torsion angle, where, for example,

$$d = \sum_{i=1}^{n} d_i / n$$

Here, d_i is the ith value in a sample of n observations. It has been shown[16-18] that the unweighted mean is an acceptable, even preferable, alternative to a weighted mean for structural data. This is especially true when the structures used have been pre-screened so that they are all of comparable precision.

(b) The sample standard deviation, denoted as σ_s throughout this work, where, for example,

$$\sigma_s = \sum_{i=1}^{n} [(d_i - d)^2 / (n-1)]^{1/2}$$

(c) The standard deviation of the mean, denoted as σ, is calculated as

$$\sigma = \sigma_s / \sqrt{n}$$

(d) The minimum and maximum values of the parameter in that distribution, denoted as e.g. d_{min}, d_{max}, etc.

(e) The total number, n, of observations in the sample.

Outlying observations will occur in any study of this type. These are observations which differ very significantly from the mean value for the distribution. Outliers can arise for a number of reasons, but principally (i) due to some (unknown) systematic error(s) in a particular crystal structure analysis, or (ii) due to some very specific chemical effect(s) within a particular fragment. In this work, we have defined outliers as being more than $4\sigma_s$ from the initial mean value. In these (very few) cases, the mean is recomputed for inclusion in the final tables with any outliers excluded.

D. Summary of Abbreviations and Table Layout

The parameters described in the previous sections are collected for convenience in Table 1, which also summarizes the abbreviations used throughout this review for the various methods of structure determination. In most of the tabulations in this review, mean values are presented in the form, e.g., $d(\sigma, n)$. Thus the value 1.474 (1, 88) indicates a mean bond length of 1.474 Å and a standard deviation of 0.001 Å obtained from a distribution of 88 observations. In some cases, the value of n is constant for all entries in a given line of a table and here, to conserve space, mean values are given only in the form, e.g., $d(\sigma)$, with the value of n in a separate column.

E. Organization of the Review

The numerical data are organized in three main sections. Section III is concerned with the intramolecular geometries of triple-bonded groups as observed in molecules that are generally regarded as being organic: chemical classes 1–65 and 70 of the CSD System. Section IV is entirely concerned with the structural chemistry of triple-bonded groups in metallo-organic species: chemical classes 71–86 of the CSD. Here, we are particularly concerned with the varied modes of bonding of triple-bonded groups at metal centres. Finally, in Section V we present a brief overview of available studies of the types and patterns of intermolecular non-bonded interactions involving triple-bonded functional groups in extended crystal structures.

TABLE 1. Glossary of abbreviations used in text and tables

Geometrical parameters

d_i	individual bond lengths or non-bonded distance in Å
θ_i	individual valence angle in degrees
τ_i	individual torsion angle in degrees
d	mean bond length or non-bonded distance in Å
θ	mean valence angle in degrees
τ	mean torsion angle in degrees
σ_s	standard deviation of sample of parameters
σ	standard deviation of mean of a sample
n	number of contributors to an averaged sample
min	minimum value of cited geometrical parameters
max	maximum value of cited geometrical parameters
e	error estimates from non-crystallographic studies

Experimental methods

X	X-ray diffraction
N	neutron diffraction (Np = neutron powder diffraction)
ED	electron diffraction (gas)
MW	microwave spectroscopy
S	spectroscopic
T	temperature in K

Other

CSD	Cambridge Structural Database
2D	two-dimensional
3D	three-dimensional
m	methyl
p	primary
s	secondary
t	tertiary
Cp	cyclopropyl
E_s	strain energy

III. INTRAMOLECULAR GEOMETRY IN ORGANIC SYSTEMS

A. Parent Triple-bonded Systems

1. Acetylenic compounds

Bond lengths obtained from individual structure determinations of parent acetylenes R^1—C≡C—R^2 (R=H, D, CH$_3$) are collected in Table 2. Gas-phase results for C≡C, for which an early mean value of 1.204 Å has been cited[19], are spread over a very narrow range from 1.202 to 1.214 Å as a measure of the internuclear C≡C separation. By contrast the solid state (primarily neutron diffraction) results show a much wider spread, from 1.136(5) to 1.193(6) Å. Crystalline acetylene and deutero-acetylene undergo a phase transition at *ca* 130 K and 140 K, respectively (see Reference 25 for a discussion) from a cubic phase of space group Pa3 to an orthorhombic phase of space group Acam. It is reported[25] that the high-temperature cubic modification exhibits large rigid-body librational motion about the molecular centre of mass, thus leading to an apparent foreshortening of the observed bond lengths. This effect is minimal in the low-temperature orthorhombic modification and bond-length values recorded here[25, 26] are consistent with each other and with the gas-phase data within 3σ (neutron). The C≡C bond length of 1.193(6) Å is probably the most accurate available from neutron crystal structure analysis.

TABLE 2.   Bond lengths, given as $d_i(e_i$ or $\sigma_i)$, in parent acetylenes R^1—C≡C—R^2 (R^1, R^2 = H,D,Csp^3) obtained from individual structure determinations

R^1	R^2	C≡C	C—R^1		C—R^2	Method	Reference
H	H	1.2120(4)	—		—	ED(r$_g$)	20^a
H	H	1.20257(9)	1.06215(17)			RAMAN	21^a
H	H	1.2033(2)	1.0605(3)			IR	22^a
H	H	1.2088(—)	—			ED(rs)	23^a
H	H	1.178(2)	1.043(—)			X(T = 141)	24
H	H	1.177(6)	1.024(10)			N(T = 141)	25
H	H	1.186(4)	1.016(8)			N(T = 131)	25
D	D	1.193(6)	1.070(5)			N(T = 15)	25^b
D	D	1.138(5)	1.036(3)			N(T = 143)	25
D	D	1.180(6)	1.061(5)			N$_p$(T = 4.2)	26^b
D	D	1.136(5)	—			N$_p$(T = 77)	26
D	D	1.162(5)	—			N$_p$(T = 109)	26
CH$_3$	H	1.2062(—)	1.4589(—)		1.0562(—)	D(r$_s$)	27^a
CH$_3$	H	1.2073(—)	1.4577(—)		1.0602(—)	ED(r$_o$)	27^a
CH$_3$	CH3	1.2135(13)		1.4675(16)		ED(r$_g$)	20^a

^{a} = Mean values for these data are in Table 3.
^{b} = Low-temperature orthorthombic phase (see text).

   The X-ray C≡C distance in crystalline acetylene of 1.178(2) Å^{24} is consistent with values obtained from the CSD (Table 3) of 1.172(1, 94), 1.184(1, 44) and 1.186(1, 399) for Csp^3—C≡C—H, Csp^3—C≡C—Csp^3 and A—C≡C—A substructures, respectively. However, the X-ray values are considerably foreshortened by comparison with both individual (Table 2) and averaged (Table 3) gas-phase and neutron results. This is, of course, an artefact of the X-ray method which provides separations between electron density maxima, rather than true internuclear separations. Normally, X—N and X—G differences are not large, but they are always appreciable for bonds to hydrogen atoms and for bonds in electron-rich systems. For hydrogen, the spherically symmetric charge density distribution is subject to a relatively large distortion on formation of an A—H bond^{28}. The H position obtained from X-ray data, and using the normal spherical scattering factor, is displaced towards A along the A—H vector yielding bond lengths that are considerably shorter than those obtained by N or G experiments or by theoretical calculations. For example, X—N shifts for C—H and O—H bonds in sucrose were found to be – 0.13(1) and – 0.18(1) Å, respectively^{29}.
   Detectable asphericity shifts are not, however, restricted to H and can be significant^{30} for elements of the first row of the periodic table. Recently^{31} a systematic pairwise analysis of X and N bond lengths showed X—N differences across a wide range of bonding situations: – 0.0096(9) for C=C (ethylenic), – 0.0052(9) for C=C (aromatic), – 0.0078(16) for C≡N, etc. The data of Table 3a show clearly that a similar effect operates for C≡C as well. For Csp^3—C≡C—Csp^3 it is reasonable to assign a Csp^3—Csp^3 separation of $2 \times 1.461 + 1.209 = 4.131$ Å in the gas phase. This separation is $2 \times 1.474 + 1.184 = 4.132$ Å across 44 X-ray crystal structures. The principal geometrical difference between these two cases lies in an inward movement of the two Csp^1 atoms by – 0.0125 Å each. This X—G value is entirely consistent with the X—N differences described earlier^{31}.
   Table 3a also reports Csp^1—Csp^3 distances for Csp^3 = methyl and for primary, secondary and tertiary Csp^3 centres. There is a small but typical^{14} increase of ca 0.01 Å in this bond length as Csp^3 substitution increases. Further comparative data are presented in Table 4a and are fully discussed in Section III.A.3.

TABLE 3. Mean geometry for parent acetylenes and cyanides R^1—C≡C—R^2, R^1—C≡N (R^1=H or Csp^3)^a

(a) Acetylenes

R^1	R^2	C≡C	C—R^1	C—R^2	R—C≡C	Method	n
H	H	1.208(2)	1.061(—)		—	G	5
CH_3	H	1.209(2)	1.461(3)	1.074(15)	—	G	3
overall (G)		1.208(1)	—	—	—	G	8
Csp^3	H	1.172(1)	1.476(1)	—	175.4(6)	X	94
CH_3(m)	Csp^3,H	1.174(2)	1.470(2)	—	179.0(2)	X	6
CH_2—C(p)	Csp^3,H	1.180(2)	1.467(7)	—	174.6(5)	X	53
CH—C_2(s)	Csp^3,H	1.177(4)	1.473(3)	—	176.6(7)	C	19
C—C_3(t)	Csp^3,H	1.177(1)	1.479(1)	—	176.9(2)	X	99
Csp^3	Csp^3	1.184(1,44)	1.474(1,88)		177.4(2,88)	X	
A^b	A^b	1.186(1)	—		—	X	399

(b) Cyanides

R^2	C≡N	C—R^1	R^1—C≡N	Method	n
H, halogen	1.158(1)	1.063(—) [H]	—	G	5
CH_3	1.156(1)	1.461(1)	—	G	5
overall (G)	1.157(1)	—	—	G	10
Csp^3	1.152(2)	—	—	N	5
CH_3(m)	1.128(3)	1.437(4)	177.7(3)	X	77
CH_2—C(p)	1.131(1)	1.462(2)	178.4(1)	X	58
CH—C_2(s)	1.137(1)	1.468(1)	177.8(1)	X	122
C—C_3(t)	1.136(1)	1.470(1)	177.4(1)	X	329
Csp^3	1.134(1)	1.464(1)	177.6(2)	X	586
Csp^3—m^c	1.136(1)	1.469(1)	177.6(2)	X	509

^{a} Mean gas-phase (G) values (see text) are derived from data in Table 2 for acetylenes, and from data given by Moffatt (Reference 4, Tables 1 and 5) for cyanides. Mean X-ray results are derived from the CSD as described in Section II.
^{b} From organic molecules only, X is any substituent including H or D.
^{c} CH_3—C≡N excluded, see text.

One individual structure, that of 3,3-bistrimethylsilylethyny1-5-trimethylsilylpenta-1,4-diyne[32], appeared just as this review was being completed and is worthy of mention. The structure is the first to contain a tetraethynylmethyl group in which the mean C≡C distance is foreshortened to 1.152 Å (range 1.142 to 1.161 Å) and the mean Csp^3—Csp^1 distance is elongated to 1.492 Å (range 1.479 to 1.502 Å).

2. Cyano compounds

By comparison with parent acetylenes, the Csp^3-substituted cyano group is ubiquitous in the X-ray literature and is also represented by 5 neutron examples of acceptable structural precision. Examination of the data presented by Moffatt[4] shows that gas-phase observations of the C≡N bond length are tightly clustered and the mean value of 1.157(1) of Table 3b is representative. As expected, the neutron mean is close to this value at 1.152(2) with d_{min}, d_{max} = 1.146, 1.157. By comparison, the overall X-ray mean value of 1.136(1) shows X—G, X—N shortenings of − 0.021, − 0.016 Å, respectively. These values are directly comparable to the acetylenic case and, again, they have their origins in X-ray asphericity effects[31]. For C≡N, the Csp^3—N linear separation is 2.617 in the gas phase and 2.605 for X-ray studies (acetonitriles excluded as discussed below). However, here the terminal N must be assumed to remain static since there is an almost exact balance between the triple-bond density on the one hand and the N lone-pair density on the other[33]. Thus most of the X—G C≡N discrepancy must be attributed to a movement of the Csp^1 atom towards nitrogen. We will return to this point in discussing the cyanocyclopropyl fragment in Section III.B.2.

Table 3b also lists mean Csp^3—Csp^1 bond lengths in cyano compounds subdivided according to the degree of substitution of the Csp^3 atom. As with the acetyleno equivalents, there is a very small increase in this bond length in changing from primary through secondary to tertiary Csp^3 centres. However, by contrast with the acetylenes, the value for Csp^3 = methyl is no longer approximately equal to the $Csp^3(p)$—Csp^1 bond length and an apparent shortening of 0.025 Å is observed. The reason for this shortening is directly related to the reason why so many (77) observations of the parent acetonitrile structure are available in the X-ray literature. This molecule is a commonly occurring molecule of crystallization and such isolated species are stabilized in the extended crystal structure only by hydrogen bonding or by even weaker intermolecular forces. Their atoms almost always exhibit higher thermal motion than their counterparts in larger and more rigid molecular frameworks, leading to librational shortening of the uncorrected bond lengths that have been used to assemble the data of Table 3. Hence, we observe the highly significant shortening of CH_3—Csp^1 in X-ray observations of acetonitrile and a smaller shortening (of *ca* 0.01 Å) in the C≡N distance itself. It is for this reason that we present mean Csp^3—C≡N geometry in Table 3b, with $CH_3C≡N$ values omitted and use this C≡N distance in the discussion of the previous paragraph. It is worth noting that, if stabilizing electrostatic interactions do occur in acetonitrile solvates, then these interactions most likely involve the C≡N group rather than the C—CH_3 group.

3. Diazo and azido compounds

The vast majority of diazo (generically C—N≡N) compounds studied by X-ray diffraction involve extensive delocalization through unsaturated groups involving the C atom. Very few diazoalkanes or azidoalkanes have been reported and even here the canonical forms contributing to the effective resonance hybrids involve considerable electron delocalization. For these reasons, the geometries of diazo and azido compounds for all types of substitution are covered in Section III.B.4.

4. The Csp^1—Csp^3 bond length in acetylenes and cyanides: comparison with related systems

Table 4 quantifies the well-known shortening of Csp^1—Csp^3 single bonds [1.474(1) for C≡C, 1.469(1) for C≡N] by comparison with their Csp^2—Csp^3 [1.507(1) for ethylenic C, 1.513(1) for aromatic C^{14}] and Csp^3—Csp^3 [1.530(1)[14]] counterparts. It is also well known[34] that the length of a C—C bond is dependent on its degree of substitution and Table 3 documents this effect for Csp^1—Csp^3 bonds. This effect has also been carefully documented in Reference 14 for the corresponding Csp^2 and Csp^3 systems and all available results are collected in Table 4.

For Csp^3—Csp^3 bonds, the overall arithmetic means, D, represent a 'best attempt' to quantify the length of typical Csp^3—Csp^3(m,p,s,t) bonds in organic systems. If we accept these estimates [which may be low for Csp^3(t) due to the paucity of Csp^3(t)—Csp^3(t) single bonds in unstrained systems], then Table 4 gives rise to the following observations:

(a) Bond lengths of the type Csp^K—CH_3 differ little from their Csp^K—$Csp^3(p)$ counterparts for $K = 3$, 2 and for the acetylenes. This would confirm that the CH_3—Csp^1 bond length in acetonitriles is indeed atypically short for reasons noted earlier.

TABLE 4. Comparative lengths of Csp^3—Csp^K (K = 1,2,3) single bonds, taken from this work and from Reference 14, showing the effect of degree of substitution (S) at Csp^3 [a]

k	S	m	p	s	t	D
3	m	—	1.513(1,192)	1.524(1,226)	1.534(1,825)	1.524
3	p	1.513(1,192)	1.524(1,2459)	1.531(1,217)	1.538(1,330)	1.527
3	s	1.524(1,226)	1.531(1,1217)	1.542(1,321)	1.556(1,215)	1.538
3	t	1.534(1,825)	1.538(1,330)	1.556(1,215)	1.588(4,21)	1.554
3	overall	1.528(1,1243)	1.527(1,4198)	1.535(1,1979)	1.540(1,1391)	
2(ethylenic)		1.503(1,215)	1.502(1,483)	1.510(1,564)	1.522(1,193)	1.509
Δ(23)		0.025	0.025	0.025	0.018	
2(aromatic)		1.506(1,454)	1.510(1,674)	1.515(1,363)	1.527(1,308)	1.515
Δ(23)		0.022	0.017	0.020	0.013	
1(acetylenic)		1.470(2,6)	1.467(1,53)	1.473(3,19)	1.479(1,99)	1.473
Δ(13)		0.058	0.060	0.062	0.061	
1(cyano)		[1.437(4,77)]	1.462(2,58)	1.468(1,122)	1.470(1,329)	1.467
Δ(13)		[0.091]	0.055	0.067	0.070	

[a]In this table, S defines Csp^3 as a m (methyl), or as a p (primary), s (secondary) or t (tertiary) centre. Bond lengths (Å) are given as $d(\sigma,n)$ where n is the number of observations contributing to each mean value. The overall mean value for each row (D) is the mean of means cited in that row. Δ(k3) is the difference between the Csp^3—Csp^3 bond length and the Csp^3—Csp^k bond length for k = 1, 2.

(b) In moving from Csp^3(p) to Csp^3(t) the C—C single-bond lengthening varies from 0.008 to 0.020 Å (mean = 0.014 Å). Values of 0.012 Å (acetylenes) and 0.008 Å (cyano) are at the lower end of this range.

(c) In moving from Csp^3 to Csp^2 the Δ values for Csp^3—Csp^K bond lengths are very consistent for the m, p, s, t subgroups. The mean shortening of a Csp^3—Csp^2 single bond by comparison with its Csp^3—Csp^3 counterpart is 0.021(2) Å over the 8 Csp^2 contributors. The corresponding mean shortening of a Csp^3—Csp^1 single bond relative to Csp^3—Csp^3 values is 0.062(2) Å (acetonitriles omitted). Hence the overall shortening, Δ(12), in moving from Csp^2 to Csp^1 participation in the Csp^3—Csp^K single bond is 0.041 Å, almost double the Δ(23) values deduced from Table 4.

(d) The D-values of Table 4 can be used in a simple derivation of covalent radii (r) for the various C hybridization states[35]. If we take the mean D-value for Csp^3—Csp^3 bonds (1.536 Å) as being typical of organic systems, then $r(sp^3)$ = 0.768 Å. Subtracting this value from $D(Csp^2, Csp^1)$ values then yields $r(Csp^2$ aromatic) = 0.747, $r(Csp^2$ ethylenic) = 0.741, $r(Csp^1$ C≡C) = 0.705 and $r(Csp^1$ C≡N) = 0.699 Å.

B. Triple Bonds in Conjugated Systems

1. Conjugated acetylenes

The geometrical characteristics, determined by X-ray diffraction, for systems R^1—C≡C—R^2 in which acetylene is directly bonded to R = C≡C (ethylenic), C=C (aromatic), C=O (ketones and carboxylic acids), C≡N and to other C≡C bonds are collected in Table 5. In Table 6, we summarize the results of accurate ED studies of dimethylacetylene[20], vinylacetylene[36], propynal[37] and diacetylene[38] which were designed to study the effects of conjugation in these systems.

The gas-phase results of Table 6 show that the C≡C bonds in vinylacetylene and in propynal are very slightly elongated by comparison with C≡C in parent acetylene (Tables 2 and 3). However, the C≡C bond length in diacetylene [1.2176(6)[38]] is significantly

TABLE 5. Acetylenes: Mean X-ray bond lengths in conjugated systems $R^3 \sim R^1-C\equiv C-R^2 \sim R^4$ where $\sim$ indicates a multiple bond[a].

$R^3 \sim R^1$	$R^2 \sim R^4$	$C\equiv C$	$C-R^1$	$R^1 \sim R^3$	$R^1-C\equiv C$	n
$C=C$	H	1.179(6)	1.432(8)	1.333(7)	—	7
C(ar)	H	1.175(4)	1.438(4)	—	177.5(6)	4
Csp^2	H	1.178(5)	1.434(5)	—	—	11
$C=C$	Csp^3	1.188(3)	1.427(3)	1.336(2)	—	30
C(ar)	Csp^3	1.190(3)	1.438(3)	—	176.0(5)	14
$C=O$	Csp^3	1.186(3)	1.459(6)	1.211(7)	—	6
Csp^2	Csp^3	1.190(1)	1.433(2)	—	—	49
$C=C$	Csp^2	1.195(2)	1.423(3)	1.359(8)	—	16
C(ar)	C(ar)	1.192(2)	1.434(1)[b]	—	—	26
$C=O$	Csp^2	1.193(3)	1.459(4)	1.227(6)	—	19
Csp^2	Csp^2	1.193(1)	1.438(1)[b]	—	176.3(6)[b]	56
$C\equiv C$	Csp^3	1.191(1)	1.379(1)	—	—	81
$N\equiv C$	Csp^3	1.199(19)	1.384(6)	1.145(19)	—	4
$C\equiv C$	$C\equiv C$	1.198(5)	1.359(9)[b]	—	—	4

[a]In some cases R^2 represents a single atom, e.g. H, Csp^3, Csp^2 etc., while in most cases of multiple conjugation $R^3 \sim R^1 = R^2 \sim R^4$ and only a single mean is given for these symmetric systems.
[b]Symmetrical fragment $R^1 \sim R^3 \equiv R^2 \sim R^4$: mean values derived from distribution of $2n$ observations.

TABLE 6. Bond lengths (ED, r_g values) determined by Kuchitsu and co-workers[20,36–38] for dimethylacetylene, vinylacetylene, propynal and diacetylene using ED methods

	$C\equiv C$	$C-C$	Reference
$CH_3-C\equiv C-CH_3$	1.213(1)	1.467(1)	20
$CH_2=CH-C\equiv CH$	1.215(3)	1.434(3)	36
$O=CH-C\equiv CH$	1.211(2)	1.453(1)	37
$HC\equiv C-C\equiv CH$	1.2176(6)	1.3837(8)	38

longer than the parent value. The X-ray results for the few $Csp^2-C\equiv CH$ systems yield a mean $C\equiv C$ bond length [1.178(5, 11), Table 5] that differs very little from individual (Table 2) or mean (Table 3) $C\equiv C$ distances in parent acetylenes. However, the mean $C\equiv C$ bond lengths in $Csp^2-C\equiv C-Csp^3$, $Csp^2-C\equiv C-Csp^2$, $Csp^1-C\equiv C-Csp^3$ and $C\equiv C-C\equiv C$ systems [1.190(1, 49), 1.193(1, 56), 1.191(1, 81), 1.198(5, 4), respectively: Table 5] do appear to mirror the small elongations apparent in the gas-phase data, albeit at the limits of statistical significance.

The joint effects of conjugation and of hybridization changes in $C(R^1)$ are clearly revealed in the $\equiv C-R^1$ bond lengths of Tables 5 and 6. Indeed, the trends in $C-R^1$ in Tables 6 [1.467(CH_3), 1.453($C=O$), 1.434($C=C$), 1.384($C\equiv C$)] are closely mirrored by the comparable X-ray results of Tables 4 and 5 [1.474(CH_3), 1.459($C=O$), 1.427($C=C$) and 1.379($C\equiv C$)]. Table 7 represents a simple attempt to separate out the hybridization and conjugative effects on $C-C$ bonds in $C\sim C-C\sim C$ systems ($C\sim C$ represents a multiple bond) in a manner analogous to that used by Allen[35]. Here we use the covalent radii deduced in Section III.A.4 for Csp^3, Csp^2 (aromatic), Csp^2 (ethylenic) and Csp^1 (acetylenic) atoms to predict (D_{calc}) the lengths of Csp^K-Csp^K ($K = 1, 2$) single bonds taking account only of hybridization effects. For Csp^2-Csp^2 single bonds, Reference 14 provides data for both conjugated and unconjugated systems and, for the latter, Table 7

TABLE 7. The effects of conjugation and changes in hybridization on C—C bond lengths in C~C—C~C systems where C~C represents a single or multiple bond[a]

Substructure	u or c	D_{obs}	D_{calc}	$D_{obs} - D_{calc}$
Csp^3—Csp^3	u	1.536	1.536	0.0
Csp^3—Csp^2 (ar)	u	1.515	1.515	0.0
Csp^3—Csp^2(e)	u	1.509	1.509	0.0
Csp^3—Csp^1	u	1.473	1.473	0.0
Csp^2(ar)—Csp^2(ar)	u	1.488	1.494	−0.006
Csp^2(ar)—Csp^2(ar)	c	1.470	1.494	−0.024
Csp^2(e)—Csp^2(e)	u	1.478	1.482	−0.004
Csp^2(e)—Csp^2(e)	c	1.455	1.482	−0.027
Csp^2(ar)—Csp^1	c	1.438	1.452	−0.014
Csp^2(e)—Csp^1	c	1.427	1.446	−0.019
Csp^1—Csp^1	c	1.379	1.410	−0.031

[a]The D_{obs} values are taken from the X-ray data of Tables 4 and 5 and from Reference 14 . The D_{calc} values are obtained using the r (Csp^n) values obtained in Section III.A.4. Csp^2(e) indicates ethylenic carbon, Csp^2(ar) represents carbon in aromatic systems. The designators u and c indicate unconjugated and conjugated systems, where c for Csp^2=C—C≡Csp^2 systems denotes a C—C—C—C torsion angle within ± 20° of 0 or 180°.

shows very small differences between the observed and predicted C—C single-bond lengths. However, in substructures where conjugative interactions are in effect, the difference D_{obs}—D_{calc} now indicates an extra conjugative foreshortening of the C—C bond that varies from – 0.014 Å to – 0.031 Å. Despite the simplicity of these calculations, it is interesting to note that the conjugative foreshortening of a Csp^2—Csp^1 single bond is quite small (*ca* – 0.0165 Å over both Csp^2 atom types) by comparison to the –0.031 Å shortening of the Csp^1—Csp^1 single bond. We note that this result is inversely correlated to the elongations of the C≡C bond noted in the previous paragraph.

2. Conjugated cyano groups

The geometrical characteristics of R—C≡N systems in which R is a conjugative grouping are collected in Table 8 (X-ray results) and in Table 9 (relevant gas-phase data). The mean C≡N bond lengths obtained from available X-ray data for conjugated systems do not differ significantly (within 3σ) from the X-ray mean of 1.136(1) Å obtained for 'parent' cyanides (Table 3), with the exception of C≡N distances in tetracyano-quinodimethane (TCNQ). Here two cyano groups are attached to a 'C=C' bond in a highly delocalized system: one in which the length of the C=C bond itself is considerably

TABLE 8. Cyano compounds: Mean X-ray bond lengths in conjugated systems R^2 ~R^1—C≡N where ~ indicates a multiple bond[a]

R^2 ~R^1	C≡N	C—R^1	(C—R^1)$_{calc}$	R^2 ~ R^1	R^1 —C≡N	n
C=C	1.139(2)	1.426(3)	1.440	1.326(6)	177.4(8)	11
C(ar)	1.138(1)	1.442(1)	1.446	1.390(1)	178.3(2)	172
C(cp)	1.138(1)	1.446(1)	1.449[b]	—	177.3(3)	70
C=C (TCNQ)	1.144(1)	1.425(1)	—	1.395(1)	178.3(1)	570
C≡C	1.145(19)	1.384(6)	1.404	1.199(19)	—	4

[a]In some cases R^1 represents a single atom e.g. Csp^2 , or a ring atom, e.g. C(ar) phenyl, C(cp)—cyclopropyl. (C—R^1)$_{calc}$ is obtained as described in Table 7.
[b]Calculated using the covalent radius of 0.750 for C(cp)[35].

TABLE 9. Cyano compounds: Relevant gas-phase data for conjugated systems $R^2 \sim R^1 —C≡N$ (~ indicates a multiple bond, Cp indicates cyclopropyl)

Compound	C≡N	C—C	Reference
$CH_2=CH—C≡N$	1.163	1.425	39
$CH_2=CH—C≡N$	1.167	1.438	40
$Cp—C≡N$	1.161(4)	1.420(6)	41
$HC≡C—C≡N$	1.159	1.378	42
$N≡C—C≡N$	1.1634(19)	1.3925(22)	43

elongated by comparison with its formal ethylenic counterpart (see Table 8). We note that, although the mean C≡N bond in —C≡C—C≡N systems is apparently elongated at 1.145(19), there are only 4 contributors to this mean and individual distances span a very broad range from 1.115 to 1.200 Å. This fact is reflected in the very high σ-value for the cited mean. Similarly, none of the gas-phase data of Table 9 can be interpreted in terms of any systematic elongation of the C≡N bond over the mean value of 1.157 cited in Table 3. Certainly, the obvious elongation of the C≡C triple bond seen in both X-ray and gas-phase studies of diacetylene is not apparent in cyanoacetylene or in cyanogen.

The C—C single bonds in conjugated R—C≡N systems are comparable to those in R—C≡C systems and are in the order expected from the hybridization of the C(R) atom. However, if we accept the C covalent radius in C≡N of 0.699 Å (Section III.A.4), which is 0.006 Å smaller than for C≡C, then we obtain the $(C—R)_{calc}$ values of Table 8 predicted on the basis of hybridization effects alone. By comparison with the C≡C case (Table 7) the differences (Δ) between $(C—R)_{obs}$ and $(C—R)_{calc}$ are systematically much smaller, although ordered in the same way. Thus $\Delta(C≡N)$, $\Delta(C≡C)$ are (— 0.004, — 0.014), (— 0.014, — 0.019) and (— 0.020, — 0.031) for R = C (aromatic), C (ethylenic) and Csp^1, respectively.

Although the bond-length differences cited here and in the previous section are small, they are derived by careful averaging over many examples of substructures in the X-ray literature. Further, the results appear to follow expected trends as far as the changes in C hybridization are concerned. Hence, the analysis of C≡N and C—C bonds in conjugated systems would confirm that electron delocalization from the C≡N bond into the adjacent C—C bond is inhibited in comparison to conjugated C≡C systems, due to the increased electronegativity of the N atom in the C≡N bond[5].

3. Conjugative interactions of acetylene and cyano groups with cyclopropane rings

It has long been recognized that the strained hydrocarbon cyclopropane has many properties that are analogous to those of ethylene[44]. In particular, the ability of the cyclopropyl group (Cp) to conjugate with adjacent π-acceptor groups, e.g. —C≡C, —C≡O, —C≡N etc., has long been a subject of both theoretical[45–47] and experimental[35,41,48] interest, since these interactions lead to observable bond-length asymmetry in

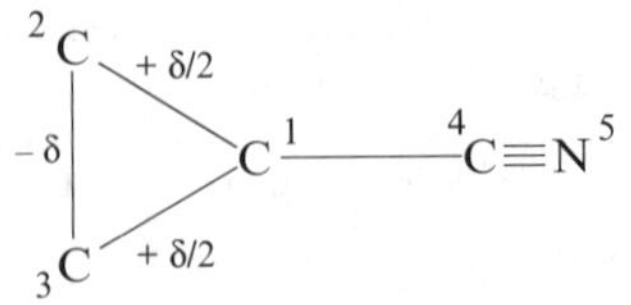

FIGURE 3. Designation of the atoms in cyanocyclopropane

the Cp ring. The highest occupied Cp orbital with the correct symmetry to interact with π-acceptors[45,46] is the 3e' orbital[49]. Maximum overlap of Cp-3e' with low-lying unoccupied acceptor orbitals occurs when the two orbital systems are parallel. The interaction is conformation-dependent for double-bond acceptors but is always operative for, e.g., $C\equiv N$, $C\equiv C$. Since Cp-3e' has bonding character for the Cp bonds vicinal to the π-acceptor (1–2, 1–3, Figure 3) but has antibonding character for the distal (2–3) bond, then transfer of electron density from 3e' to the π-system will strengthen (shorten) the 2–3 bond (say by $-\delta$ Å) and weaken (lengthen) the 1–2, 1–3 bonds by $+\delta/2$ Å in each case, as illustrated in Figure 3. In cases of multiple substitution of a Cp ring by the π-acceptor groups, it is suggested[47] that net ring asymmetry should be a summation of individual effects.

Early gas-phase studies (reviewed in Reference 50) and X-ray analyses of cyclopropanecarbohydrazide[51], 2,5-dimethyl-7,7-dicyanonorcaradiene[52] and cyclopropane-1,1-dicarboxylic acid[53] provided experimental evidence for the predicted bond-length asymmetry in Cp and for the approximate validity of the additivity principle. The magnitude of δ was indicated to be *ca* 0.015-0.025 Å for $-C\equiv O$, $-C\equiv N$. A more rigorous attempt[48] to quantify δ for various substituents made use of 299 relevant substructures available in the X-ray literature in 1980. This work yielded δ-values of 0.026(5), 0.022(4) and 0.017(2) for $-C\equiv O$, $-C\equiv C$, and $-C\equiv N$ substituents, respectively, and clearly implicated a number of other groups in effective conjugative interactions.

More recently, accurate microwave studies of Cp—$C\equiv N$ and Cp—$C\equiv C$ have been reported[41] (see Table 10). The mean Cp ring bond length is 1.519 Å in each case, from which δ-values of 0.019 and 0.016 are obtained for $-C\equiv N$ and $-C\equiv C$ substituents, respectively. X-ray structural activity in this area has remained high and there are now 57 precise (R $\leq$ 0.070) entries in the CSD containing the Cp—$C\equiv N$ fragment; to date Cp—$C\equiv C$ remains uncharacterized by the X-ray method. In particular, the structure of *trans*-1,2,3-tricyanocyclopropane has been determined[54] and compared to results for the *cis* isomer[55] (Table 10). The expected equality of the Cp C—C bonds (δ cancel in all bonds) is observed in the *cis* isomer. However, the slight asymmetry ($\delta = 0.008$ Å) in the *trans* isomer is ascribed to differences in non-bonded interactions between the *cis*-2,3-cyano groups and the *trans*-1,2- and *trans*-1,3-cyano groups. This observation indicates some limitations in the additivity principle and suggests that further detailed work in this area may soon be possible.

Bearing in mind the possible limitations on δ-additivities, we have selected 22 of the available X-ray observations of the Cp—$C\equiv N$ fragment, excluding substructures that are part of a more complex strained ring system and also those with more than four substituents of any type on the Cp ring. The —$C\equiv N$ δ-values deduced from these selected substructures range from 0.004 to 0.033 Å with a mean δ of 0.020(2). If we exclude only five observations that differ from this mean by more than 4 σ(0.008 Å), then we obtain a mean $\delta(C\equiv N)$ of 0.019(1) Å. Both of these means agree with the gas-phase result[41].

TABLE 10. Acetylene and cyano compounds: Geometrical effects of conjugative interactions with cyclopropane rings (atom numbering from Figure 3)

Compound	1–2	1–3	2–3	1–4	4–5	Reference
Cp—$C\equiv N$	1.529	1.529	1.500	1.420	1.161	41
Cp—$C\equiv C$	1.527	1.527	1.503	1.422	1.211	41
cis-Cp—$(C\equiv N)_3$	1.518	1.518	1.518	1.449	1.144	55
trans-Cp—$(C\equiv N)_3$	1.508	1.508	1.521	1.446	1.137	54

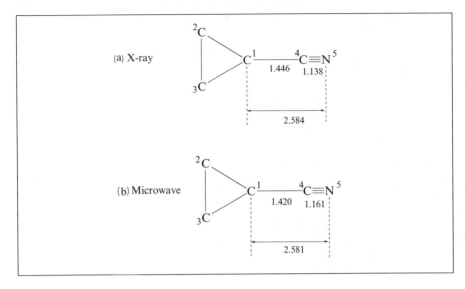

FIGURE 4. Bond lengths (Å) of cyanocyclopane obtained by (a) X-ray and (b) microwave

One parameter in Cp-C≡N for which the X-ray (Table 8) and gas-phase results do disagree very significantly[41] is the C_1—C_4 single bond length. The numerical situation is illustrated in Figure 4 and is a clear example of the effect of atomic asphericity[31] on X-ray bond lengths: there is a clear and substantial 'movement' of C_4 towards the triple-bond density, while N_5 remains static due to the 'balancing' effect[33] of the N lone pair.

4. Geometry of diazo and azido groups

This section is primarily concerned with the geometrical characteristics of the diazo group obtained by X-ray diffraction studies of benzenediazonium compounds (1), α-diazoketones (resonance hybrids 2a,b,c) and diazoalkanes (3a,b). The section concludes with some geometrical details for the analogous azido group (4a,b). There are 50 crystal-

(1)

(2a)   (2b)   (2c)

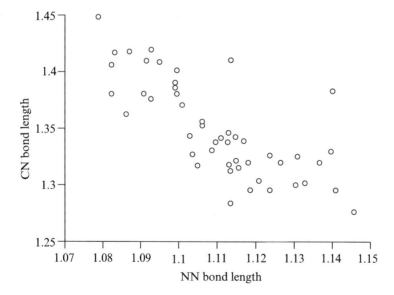

TABLE 11. Diazo and azido compounds: Mean geometrical parameters for **1–4**[a]

	d_{12}	d_{23}	d_{34}	d_{45}	θ_{123}	θ_3	n
1	1.093(2)	1.401(5)	1.394(6)[b]	—	176.1(9)	125.7(6)	17
2	1.115(2)	1.329(5)	1.437(3)	1.219(3)	177.3(3)	124.2(5)[c]	29[d]
3	1.135(5)	1.306(11)	—	—	178.9(8)		4
4	1.127(2)	1.224(3)	1.470(3)	—	172.3(2)	115.5(5)	95
DZME	1.130	1.280	—	—	—	—	—
DIN2	1.097	—	—	—	—	—	—

[a]The angle θ_3 is 5–3–4 in **1,3**, 3–4–5 in **2** and 2–3–4 in **4**. DZME refers to gas-phase results for parent diazomethane[56], DIN2 refers to the N≡N bond in dinitrogen[57].
[b]Mean taken over d_{34} and d_{35} in **1** (34 values).
[c]The mean N2—C3—C4 angle is 116.7(7,29)°.
[d]The N2—C3—C4=O5 torsion angle is within −10 to +10° for 25 of these examples and between 143 and −165° for the other 4.

FIGURE 5. Plot of the N≡N bond length versus the C—N bond length for diazo compounds

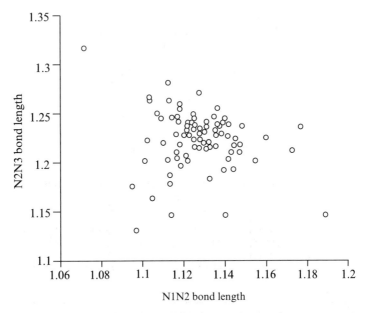

FIGURE 6. Plot of the N1N2 bond length versus the N2N3 bond length for azido compounds

lographic observations of the diazo group in structures having $R \leq 0.10$: 17 observations of **1**, 29 observations of **2** and 4 observations of **3**. There are 95 observations of **4** with R = C. Mean geometries for **1–4** are collected in Table 11. A scattergram of the $N_1 \sim N_2$ and $N_2 \sim C_3$ bond lengths for all diazo compounds is in Figure 5, while a comparable plot of $N_1 \sim N_2$ versus $N_2 \sim N_3$ for azido compounds is in Figure 6.

The mean geometry of the benzenediazonium subset (Table 11) is entirely consistent with resonance structure **1** as the predominant form. The $N{\equiv}N$ distances are the shortest observed and span a very narrow range from 1.078 to 1.114 Å. The mean value of 1.093(2) Å is directly comparable to the $N{\equiv}N$ distance in dinitrogen (1.097[57]). The mean C—N distance of 1.401(5) Å (range 1.375 to 1.450 Å) can be assigned as an almost pure Csp^1—Nsp^1 single bond through its comparison to Csp^2 (aromatic)—Csp^1 values of 1.438(3,14) and 1.442(1,172) in acetylenes and cyanides, respectively. These values are *ca* 0.04 Å longer than C—N in **1**, a difference which is similar to that between Csp^3—Csp^3 and Csp^3—Nsp^3 bond lengths[14]. The phenyl ring bond lengths d_{34}, d_{35} in **1** are unremarkable, but we note that the intra-annular valence angle at the point of substitution is consistently enlarged (122.8–131.3°) to yield a mean value of 125.7(6)°. This is probably a reflection of the large -*I* inductive effect of the diazonium group, and is in accord with a systematic study of substituent induced ring deformations in benzene[58].

In moving to the α-diazoketones (**2**), we see a significant increase in the $N{\equiv}N$ distance and a concomitant shortening of C—N. The mean C—N bond length of 1.329(5) Å is longer than that in a C=N double bond [1.279(1) Å[14]], but is comparable to C ~ N in delocalized systems[14] such a pyrazole [1.329(4) Å], imidazole [1.313(2) Å] and the C ~ N^+—H [1.335(2) Å] or C ~ N^+—Csp^3 [1.346(1) Å] units of pyridinium cations. By contrast, the C3—C4 distance at 1.437(3) is very much shorter than the 1.464(1) Å normally expected in conjugated O=C—C=C systems, while the mean angle at C3 is 116.5(7)°. The C=O distance, 1.219(3) Å, is typical of O=C—C=C systems [1.222(1)

$(i\text{-}Pr)_2N$ $N(i\text{-}Pr)_2$

$(i\text{-}Pr)_2N$ P $\overset{4}{}$ C$_3$ P $\overset{4'}{}$ $N(i\text{-}Pr)_2$

N$_2$

N$_1$

(5)

(6)

Si(Ph)$_3$

(7)

Å^{14}]. The MO calculations of Csizmadia and coworkers[59] agree well with these experimental observations. We may conclude that resonance hybrid **2a** plays a major role in α-diazoketone structures with contributions from **2b** and, to a much smaller extent, **2c**.

There are very few structures in the X-ray literature that can be classified as diazoalkanes (**3**). Table 11 lists mean geometry for four examples only: structure **5**[60] and three examples of the diazafluorene unit (**6**). In **5** the N≡N distance is the longest observed at 1.146 Å, while the C—N distance is the shortest at 1.278 Å. These data are reinforced by results for phenyl(triphenylsilyl) diazomethane[61] (**7**, crystallographic R=0.117, cf 0.052 for **5**) where N≡N=1.130 Å and C—N=1.280 Å. However, C$_3$ in **7** is bonded to a single delocalized system which can interact with C=N=N; this is not the case in **5**[60]. In structures of type **6**, C$_3$ is connected to two delocalized systems and N≡N, C—N in these three examples are (1.141, 1.297 Å)[62], (1.124, 1.326 Å)[63], (1.127, 1.371 Å)[63]. It is clear that **5** and possibly **7** are best represented by a very large contribution from resonance hybrid **3a** and their geometry agrees well with data for the parent diazomethane. However, in structures of type **6** we can observe an increased contribution from **3b**, although not to the extent that **2b** contributes to the α-diazoketone structure.

The results of Table 11, then, demonstrate the reduction in N≡N triple-bond character and the increase in C—N double-bond character in moving from **1** to **3**. The linearity of this effect over all 50 diazo-substructures is clearly visualized in the C—N vs N≡N plot of Figure 5. Even within the two major subgroups, **1** (short N≡N, long C—N) and **2** (longer N≡N, shorter C—N), this approximate linearity is preserved.

Data for the azido group are much less susceptible to chemical subdivision. The N≡N vs N—N scatterplot shows a large coherent cluster of observations together with a few outliers. The mean N≡N bond lies between the values observed for α-diazoketones (**2**) and the diazoalkanes (**3**), while the N—N bond is somewhat shorter [1.224(3) Å] than that in Csp^3—N=N—Csp^3 substructures [1.240(2) Å^{14}]. We conclude that the contribution of resonance hybrid **4b** is significant, However, it seems that further detailed work, outside the scope of this review, is required to analyse fully the available data for both diazo and azido compounds.

C. Miscellaneous Structural Topics in Organic Systems

1. Introduction

Sections III.A and III.B are primarily concerned with the structural chemistry of triple bonds that are bonded to the carbon atoms of various organic functional groups. There exist, however, in the crystallographic literature a growing number of structures involving —C≡C—X and N≡C—X systems in which the symbol X excludes metals (see Section IV), carbon, hydrogen or deuterium. The geometrical characteristics of these systems are briefly summarized in Sections III.C.2 and III.C.3 below. Finally, in this survey of intramolecular geometry in organic systems, we summarize geometrical data for simple cyclic systems containing the C≡C triple bond. The observed ring conformations and the geometrical deformations of the —C≡C— system are discussed in Section III.C.4.

2. Geometry of —C≡C—X (X ≠ C, H, D, metal)

Table 12 lists available mean geometry in —C≡C—X systems. The only major grouping is for X = Si(sp^3) where the Csp^1—Si distance of 1.840(2) Å may be compared with Csp^3—Si = 1.888(2) Å[14] and C$_{ar}$—Si = 1.868 Å[14]. The decrease of 0.020 Å in moving from Csp^3 to Csp^2 is typical of data in Table 4 for C≡C—C systems. However, the further decrease of 0.028 Å from Csp^2—Si to Csp^1—Si is smaller than for the carbon case (ca 0.040 Å, see Table 4). The mean C≡C distance in the Si dataset [1.201(3) Å] is somewhat longer than the typical X-ray value of 1.186(1) of Table 3. This elongation of C≡C is also apparent in other entries in Table 12. However, given the very small number of contributors to many of the cited means within an increased R-factor limit of 0.10, more detailed data analysis, and probably more crystal structures, will be needed to confirm (or not) these observations.

3. Geometry of N≡C—X (X ≠ C, H, D, metal)

The only major groupings under this heading (Table 13) involve X=S and X=N. The Csp^1—S distance for divalent sulphur [1.681(5) Å] is very considerably shorter than for

TABLE 12. Geometrical characteristics of C≡C—X (X ≠ C,H,D, metal) systems from X-ray data[a]

System	C—X	C≡C	C≡C—X	n
C≡C—Si(4)	1.840(2)	1.201(3)	173.7(8)	53
C≡C—P(3)	1.770(3)	1.203(3)	171.7(10)	9
C≡C—P(4)=	1.780(7)	1.181(5)	178.0(4)	6
C≡C—P$^+$(4)	1.708(10)	1.201(9)	171.7(27)	3
C≡C—P(5)	1.807(—)	1.202(—)	178.5(—)	1
C≡C—B(3)	1.547(6)	1.213(3)	176.8(6)	9
C≡C—N(3)	1.322(9)	1.205(9)	177.0(5)	6
C≡C—O(2)	1.355(9)	1.156(4)	176.9(12)	3
C≡C—Cl(1)	1.635(4)	1.176(2)	178.0(7)	4
C≡C—Br(1)	1.759(—)	1.184(—)	176.1(—)	1
C≡C—I(1)	2.029(15)	1.174(4)	177.4(5)	8
C≡C—Se(2)	1.836(—)	1.175(—)	179.2(—)	1
C≡C—Te(2)	2.037(—)	1.210(—)	178.2(—)	1

[a] X(i) indicates the number (i) of chemical bonds formed by atom X; X(i)= indicates that one of these bonds is a double bond. The crystallographic R-factor limit is R ≤ 0.10.

TABLE 13. Geometrical characteristics of N≡C—X (X ≠ C,H,D, metal) systems from X-ray data[a]

System	C—X	C≡N	N≡C—X	n
N≡C—S—C	1.681(5)	1.147(4)	176.1(6)	26
N≡C—S⁻	1.632(8)	1.148(6)	177.6(6)	14
N≡C—N(2)=	1.334(3)	1.146(2)	172.9(3)	37
N≡C—N(3)	1.323(6)	1.147(8)	177.5(3)	16
N≡C—N(all)	1.330(4)	1.146(5)	174.3(3)	53
N≡C—O(2)	1.280(7)	1.129(5)	176.5(4)	4
N≡C—O⁻	1.210(7)	1.08(6)	177(3)	2
N≡C—P(3)	1.767(9)	1.134(4)	173.8(11)	4
N≡C—P(4)	1.813(7)	1.133(7)	175.5(6)	6
N≡C—P(5)	1.792(8)	1.138(12)	179.4(7)	2
N≡C—B(3)	1.577(4)	1.142(5)	176.7(9)	7
N≡C—Se(2)	1.818(7)	1.142(4)	177.6(6)	12
N≡C—Te(any)	2.097(17)	1.115(11)	176.6(9)	6

[a] X(i) indicates the number of chemical bonds (i) formed by atom X; X(i)= indicates that one of these bonds is a double bond. The crystallographic R-factor limit is $R \leq 0.10$.

TABLE 14. Effect of cyclization on X—C≡C—X geometry[a]

r	X	C≡C	X—C≡C	C—X	E_s
6	C	1.341(9, 8)	122.4(8, 16)	1.432(10, 16)	
	Si	1.199(—, 1)	148.7(—, 2)	1.843(—, 2)	
7	C	1.259(16, 8)	140.8(19, 16)	1.498(8, 16)	31.1
	Si	1.213(—, 1)	160.9(—, 2)	1.823(—, 2)	
8	C	1.194(3, 7)	156.0(8, 14)	1.437(6, 14)	20.8
	Si	1.192(14, 3)	165.8(8, 6)	1.840(6, 6)	
9	C	1.184(—, 1)	166.1(—, 2)	1.455(—, 2)	16.4
10	C	1.191(3, 32)	168.2(5, 32)	1.458(6, 32)	9.9

[a] X-ray results for ring size (r) of 6–10 atoms (see text for chemical selection criteria) from structures with an R-factor ≤ 0.12. E_s is the strain energy of the hydrocarbon ring.

Csp^3—S [1.819(1) Å[14]] or Csp^2—S [1.751(2) Å[14]]. It is very similar to values for a variety of C=S double bonds [1.660–1.681[14]]. The Csp^1—S⁻ distance in thiocyanides [1.632(8) Å] is, as expected, considerably shorter than for divalent S. Within the statistical limits of the averaging process, it is not possible to distinguish Csp^1—N= and Csp^1—Nsp^3 bond lengths. The data given in Table 13, together with the short Csp^1—Csp^3 distance in Table 12 imply considerable delocalization into C—N bonds in these systems.

As with the C≡C bond, there is an apparent elongation of C≡N in some of the substructures of Table 13 over the values cited in Tables 3b and 8. However, none of these elongations is statistically significant.

4. Effects of cyclization on —C≡C—geometry

The valence-angle constraints imposed by cyclization lead to very high strain energies (E_s) in the small- and medium-ring cycloalkynes. Members of the series below ring size 6 are too highly strained to be isolable, while cyclohexyne and cycloheptyne ($E_s = 31.1$ kcal mol^{-1}) are only stabilized through π-complexation to transition metals. Even at ring size 10, a strain energy of 9.9 kcal mole^{-1} still exists. Major decreases in the two C—C≡C valence angles from 180° also lead to a decreased p-orbital participation in the C≡C π-system and this should lead to a gradual increase in the C≡C bond length with decreasing ring size.

These effects are explored as far as possible in Table 14. All 6,7-membered rings except two in these small datasets are π-complexed to various metal atoms. For 6-rings, there is only one genuine cyclohexyne, four benzynes and two cyclohexenynes. The 7-ring dataset is dominated by 1-thiacyclohept-4-yne rings. The two silicon-containing 6- and 7-rings are a tetrasilacyclohexyne derivative[64] and decamethyl-pentasilacycloheptyne[65]. The 8,9,10-rings are all uncomplexed carbocycles containing up to two —C≡C—units and, in some cases, other unsaturated bonds. A small subset of 8-rings containing Si—C≡C—Si units is also reported. Thus, the C—X distances for X=C in Table 14 cover a variety of C hybridization states.

Despite the small size of the available datasets, Table 14 quantifies the relationship between C—C≡C angles and the strain energy E_s. It also shows that the presence of ring heteroatoms, that effectively enlarge the overall size of the ring, permits a significant increase in X—C≡C angles, thus explaining the isolation of the Si-containing rings without metal π-complexation of the triple bond[64,65]. Table 14 would also indicate an increase in the C≡C bond length for the smaller rings of size ≤ 9 atoms. However, the π-complexation of all 6,7-rings must be taken into account (see Section IV.B) before this correlation can be quantified.

IV. INTRAMOLECULAR GEOMETRY IN METALLO-ORGANIC SYSTEMS

A. Introduction and Organization of Material

This section discusses the structural chemistry of triple-bonded ligands bonded to metal centres in the following subsections: (B) alkynes and acetylides, (C) dinitrogen and azides, (D) cyanides and isocyanides. Mean geometrical descriptors are generated from the CSD as described in Section II. Searches were restricted to d- and f-block metals, and metal–metal bonds in substructure definitions were constrained to be single bonds in all cases. Ligand geometries in metallo-organic species are less precise than for the parent organic systems considered in Section III. This is exemplified in a recent systematic compilation of bond lengths in these systems[15] where much higher σ for mean values, drawn from much wider numerical ranges, are the norm. There are two basic reasons for reduced precision in mean geometry. Firstly, lighter atoms such as C,N,O cannot be located as precisely in crystal structures that contain heavy elements as they can in wholly organic structures. Secondly, and most importantly, bonding modes in metallo-organic systems are less well understood, making it more difficult to categorize datasets in chemical terms. This problem is compounded by the very wide range of metals that must be treated in any systematic study.

However, if we are to understand and classify the varied bonding models in metallo-organics, then the systematic analysis of crystallographic results for carefully chosen classes and subclasses of these compounds represents an essential starting point. Indeed, crystal structure analysis has underpinned the rapid developments in metallo-organic chemistry, particularly over the past 10–15 years, to the extent that some 48% of CSD entries now fall into this category. For these reasons, no modern review of triple-bonded functional groups would be complete without some consideration of their interaction with metal centres.

Given the crystallographic and chemical complexities stated above, the organization and layout of key material differs slightly from that adopted in Section III in two major respects:

(a) *Chemical diagrams* are shown for each substructure for which geometrical data are tabulated. These formulae are clearly identified (in bold type) by codes of the form **MAi**, **MNi** and **MCi** for alkynes/acetylides, dinitrogen/azides and cyanides/isocyanides, respectively; i is a sequence number. Other chemical diagrams, used for explanatory purposes, continue in the bold numbered sequence already established in Section III.

(b) *Numerical tabulations* use the chemical fragment identifier described above and now contain one line per geometrical parameter. The line contains the parameter identifier such as C≡C, M—C≡C, etc., the mean values such as $d(\sigma)$, $\theta(\sigma)$, etc., the minimum and maximum values of that parameter and the number of observations n. We feel that the inclusion of d_{min}, d_{max}, θ_{min}, θ_{max} in this section provides valuable additional information.

B. Alkyne and Acetylide Ligands

1. Overview

There are strong similarities between alkyne (R—C≡C—R) and acetylide (C≡C—R) bonding modes in organometallic complexes, hence these two ligands are considered together. A recent extensive review[66] describes the great structural diversity in their interactions with transition metals, and there have been a number of other reviews on alkyne and acetylide ligands[67–70]. Here, we summarize structural data for the main bonding modes, subdivided according to the number (one to four) of metal atoms involved in the interaction(s). We note that this structural summary is not comprehensive, especially for higher-order metal complexes.

The large number of bonding modes is due to the fact that the alkyne and acetylide ligands[71] have two mutually perpendicular π-orbitals, which are available for bonding to metal centres, and also two mutually perpendicular empty π^*-orbitals available for π-back-bonding from the metal atoms. Bonding in monometal alkynes is normally interpreted in terms of the Dewar–Chatt–Duncanson model[72,73], and the orbital interactions are depicted in Reference 66. Alkyne ligands may donate between 2 and 4 electrons to a cluster, depending upon the mode of coordination[66]. With this large variation in bonding modes, the C≡C bond lengths extend over a wide range (1.2 to 1.45 Å) and, with coordination, the CCR bond angles deviate strongly from 180°.

2. One-metal fragments

The mean C≡C bond of the end-on acetylide ligand, **MA1**, is comparable to that in parent acetylene and in other organic systems (Section III) and the CCR angle displays only a small deviation from 180°. Of the 50 **MA1** fragments, 17 have R=Csp^3, 27 have R=Csp^2, and there is one example[74] of R=Csp^1 (see Table 15). The four remaining fragments have R=H. The Csp^1—Csp^R bond-length shortening here is directly comparable to that in organic systems (Section III.A.4).

The side-on bonding of **MA2** is more complicated as it is possible for the alkyne ligand to donate 2,3 or 4 electrons, depending on whether one or two π-bonds are used for donation to the metal atom. This is reflected in the wide range of C≡C bond lengths and CCR angles. It has been shown[15] that as the number of electrons donated by the alkyne ligand to the metal atom increases, the C≡C bond increases, as one would expect. A graph of C≡C distance versus CCR bend-back angle for the 173 fragments shows a negative correlation of -0.65. This illustrates that as the π-back-donation from the metal to the alkyne ligand increases, there is a reduction of the triple-bond character of the CC

TABLE 15. Mean geometric parameters for the alkyne and acetylide substructures **MA1** and **MA2**

Ligand	Parameter	Mean(σ)	Minimum	Maximum	n
MA1	C≡C	1.193(3)	1.150	1.226	50
	C≡C—R	175.9(3)	170.3	180.0	50
	M—C≡C	175.9(4)	168.3	180.0	50
MA1	C≡C	1.192(6)	1.150	1.226	17
(R = Csp^3)	C—R	1.477(6)	1.418	1.524	17
MA1	C≡C	1.195(3)	1.156	1.215	27
(R = Csp^2)	C—R	1.445(3)	1.421	1.508	27
MA1	C≡C	1.198, 1.207	—	—	2
(R = Csp^1)	C—R	1.351, 1.362	—	—	2
MA2	C≡C	1.280(3)	1.188	1.379	173
	Av. C≡C—R	144.1(5)	120.2	178.2	346

$$\begin{array}{cc} & \underset{\displaystyle\nwarrow}{R}\qquad\underset{\displaystyle\nearrow}{R} \\ M-C\equiv C-R & C\equiv C \\ & \downarrow \\ & M \end{array}$$

(MA1)	(MA2)
R = C or H	R = C or H

bond (see also Section IV.B.6). Although there is a large variation in the average CCR angle for **MA2**, 152 of the fragments lie in the range of 131 to 158°. Four fragments have a small average CCR angle (121 to 127°), and these are benzyne and cyclohexyne fragments (see also Table 14), and thus the constraints of the ring cause the very small CCR angles. 17 fragments have an average CCR angle greater than 160°, and of these 9 are tribenzocyclododeca-1,5,9-triene-3,7,11-triyne fragments, where again the constraints of the ring cause the CCR angles to remain close to linear (see Table 14).

As noted by Raithby and Rosales[66] **MA1**- and **MA2**-type bonding to one metal is not observed in cluster complexes. In the database study, most of the complexes contain only one metal, a few are dimetallic complexes, and for **MA2** there is one trimetallic structure[75] and one tetrametallic complex[76] which also has an **MA8**-type alkyne ligand. The lack of cluster complexes containing terminal acetylides can be explained by the fact that the alkyne and acetylide ligands are capable of acting as strong π-acceptors, and can maximise back-bonding by coordinating to two, three or even more metals.

3. Two-metal fragments

The μ-η^2 bridging acetylide ligand of fragment **MA3** is formally a 3-electron donor, whilst for **MA4** the acetylide is a 1-electron donor. Table 16 shows that the mean C≡C bond is longer for the π-bridging acetylide of **MA3** than for **MA1**and **MA2** (Table 15). As with many of these organometallic fragments there is large variation in the bond angles of the ligand, where for some fragments the M—C≡C—R unit only shows small deviations from linearity, but for others the MCC and CCR angles lie between 150° and 160°. In all cases the M—C sigma bond is shorter than the two M—C π-bonds, and also the M—C_α π-bond is shorter than the M—C_β π-bond. In some fragments the M—C_β bond may be quite long (> 2.6 Å). The geometry of the bound acetylide ligand, **MA3,** for some

TABLE 16. Mean geometric parameters for the alkyne and acetylide substructures **MA3**, **MA4**, **MA5** and **MA6**

Ligand	Parameter	Mean(σ)	Minimum	Maximum	n
MA3	C≡C	1.221(6)	1.174	1.262	16
	CCR	164.0(20)	151.9	179.3	16
	M1C$_\alpha$C$_\beta$	166.9(4)	156.5	175.4	16
	VANG	43.2(23)	32.7	62.9	16
MA4	C≡C	1.193(10)	1.169	1.199	5
	CCR	174.2(27)	163.3	177.9	5
	Av.MCC	137.3(66)	91.1	173.5	10
	MCM	83.9(41)	74.6	96.5	5
MA5	C≡C	1.343(3)	1.285	1.409	65
	Av. CCR	139.7(9)	125.6	155.8	130
	VANG	90.5(2)	85.7	96.2	64[a]
MA6	C≡C	1.317(4)	1.278	1.368	22
	Av. CCR	124.7(8)	117.3	133.3	44
	VANG	4.3(9)	0.8	14.7	21[b]

[a]All **MA5** fragments, except for Ref. 82.
[b]All **MA6** fragments, except for the *trans* structures[80].

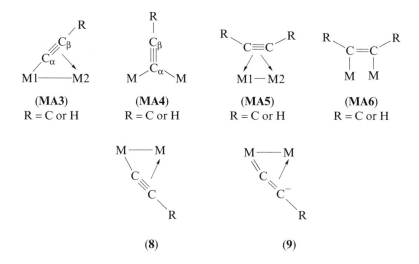

(**MA3**)   (**MA4**)   (**MA5**)   (**MA6**)
R = C or H   R = C or H   R = C or H   R = C or H

(8)   (9)

structures may suggest a contribution from canonical form **9**, as well as **8**[77]. Under the conditions of the database search, only 5 **MA4** fragments were obtained, but their bond lengths are comparable to **MA3** and **MA1** bonding modes.

For the alkyne ligand two bonding modes are possible for dimetallic complexes, **MA5** and **MA6**, which are known as perpendicular and parallel acetylide complexes, respectively. The bonding mode of **MA5** is the most common (Table 16). These two modes have been studied in detail by Hoffmann and coworkers[78,79]. The perpendicular **MA5** alkyne is seen as a 4-electron donor of 2 π-electrons to each metal, whilst the parallel alkyne ligand donates 2 electrons via σ bonds. At the time of publication of References 78 and 79 (1982) only a few dimetallic alkyne crystal structures had been determined. However, Hoffmann predicted from MO considerations that the parallel arrangement (**MA6**) should have smaller CCR angles, and shorter CC bond lengths, than the perpendicular arrangement (**MA5**). Table 16 confirms this prediction. The M_2C_2 unit of **MA5** is tetrahedral, whilst **MA6** may be viewed as an alkene with two substituents replaced by metal atoms where the mean angles at the alkyne carbon atoms and the CC bond length suggest that the carbon atoms are of sp^2 character. For **MA6**, no bond between the two metals was coded and, in all cases except for one[80], the two metals are mutually *cis*, with the CC bond lying approximately parallel to the MM vector [τ(M—C—C—M) = 0 to 22°]. For the *trans* structure[80], τ(M—C—C—M) is 171°.

The angle between the CC and MM vectors, VANG, is shown in Table 16 for **MA3**, **MA5** and **MA6**. For the acetylide fragment **MA3** the acetylide is not perfectly coplanar with or parallel to the M—M vector[81], whilst for **MA5** all the fragments show only small deviations from the perpendicular geometry, except for one example[82] (see Table 16), in which the alkyne ligand has been twisted from the perpendicular by 35°. The twisting of alkyne ligands has been explained by Hoffmann and Calhorda[83] on the basis of extended Hückel calculations. The result of the twisting is a large difference between the M—C bond lengths. For the twisted ditungsten complex[82,84], the M1—C3 and M2—C4 bond lengths are on average 0.49 Å shorter than the M2—C3 and M1—C4 bond lengths. (**10–12**), whilst for the other fragments the difference in bond lengths is between 0.0 and 0.09 Å. Therefore, it is suggested that the μ-alkyne ligand can be better described by the two valence bond descriptions of **11** and **12** shown below, rather than that of **10**[84]. Molecular orbital studies have also been carried out on μ-alkyne metal alkoxide complexes[85].

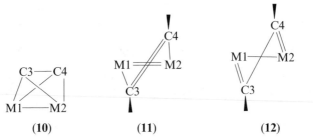

(10)   (11)   (12)

For all four of these bridging ligands, the majority of structures are dimetallic, with only a few structures containing more than two metals and, as for the mononuclear complexes, the bond lengths and angles of the alkyne ligands show a wide variation.

4. Three-metal fragments

For trimetallic complexes the bonding mode of acetylide ligands is **MA7**, whilst for alkyne ligands two modes, **MA8** and **MA9**, are possible. As with dinuclear complexes, the

(MA7)   (MA8)   (MA9)
R = C or H   R = C or H   R = C or H

TABLE 17. Mean geometric parameters for the alkyne and acetylide substructures **MA7**, **MA8** and **MA9**

Ligand	Parameter	Mean(σ)	Minimum	Maximum	n
MA7	C≡C	1.303(6)	1.241	1.364	21
	CCR	142.1(8)	135.6	154.3	21
	MCC	154.6(6)	150.2	162.5	21
MA8	C≡C	1.391(4)	1.311	1.444	54
	Av. CCR	123.6(5)	104.6	137.6	108
MA9	C≡C	1.391			1
	MCC	126.3			1
	Av.CCR	132.0			2

acetylide ligand has a shorter CC multiple bond length than the alkyne ligands, although the acetylide of **MA7** has a high formal electron donation, being a 5-electron donor (1 σ- and 2 π-bonds)[66]. The alkyne ligands of **MA8** and **MA9** are 4- and 5-electron donors, respectively.

In all 21 acetylide fragments of **MA7** the metal–carbon σ-bond is significantly shorter than the metal–carbon π-distances by 0.15 to 0.31 Å. Also, the mean M—C_α π-bond distances are shorter than the mean M—C_β values, except for two triosmium fragments[86,87]. As in the **MA5** bonding mode, the CC multiple bond of **MA7** lies approximately perpendicular to a M—M bond (M2—M3), with the angle between the CC and MM vectors being in the range 83–96°. Therefore, this ligand is often described as μ_3–($\eta_2 - \perp$).

Similar to the dimetallic fragments, there are two possible bonding modes for the R—C≡C—R ligand of **MA8** and **MA9**. These are distinguished by whether the CC bond lies parallel or perpendicular to the MM bond, respectively, i.e., μ_3–($\eta_2 - \parallel$) (**MA8**) or μ_3–($\eta_2 - \perp$) (**MA9**)[88]. The angle between the M1—M2 and CC vectors for the **MA8** fragments lies between 0 and 10°. These alkyne fragments may also be described in terms of cluster polyhedra, where the alkyne carbons are taken to be an integral part of the M_3C_2 5-vertex cluster. Based on this description, **MA8** is a nido cluster, whilst **MA9** is a closo structure[89,90]. Table 17 shows that bonding mode **MA8** is almost exclusive: **MA9** has only one structure under the constraints of the CSD search. If the search is carried out with no constraints on R-factor or AS flag (Section II), then 6 **MA9** fragments are obtained.

The alkyne ligand or **MA8** is a 4-electron donor, coordinated to the 3 metal atoms via 2 σ-bonds to 2 metal atoms and 1 π-bond to the third metal. The mean CC bond length of 1.391 Å is long, showing the ligand is highly reduced, and the CCR angles are correspondingly small. In most cases, as one would expect, the mean M—C σ-bonds are shorter than the mean M—C π-bond by 0.1 to 0.2 Å. A number of the fragments are heteronuclear clusters, and the fluxionality in such systems has been observed in solution[89,91]. Only one isomer is usually observed in the solid state, but there have been X-ray studies of 2 isomers of a ditungsten osmium complex[91,92]. The M—M edge which is parallel to the alkyne ligand has been studied by MO theory[89], with the conclusion that the orientation may be correlated with electron-accepting properties of the trimetallic fragment. The preferred isomer tends to have the more electron-attracting metal groups at positions 1 or 2, rather than 3, in **MA8**[89,90].

5. Four-metal fragments

As one might expect, the number of fragments available in the CSD decreases as the number of coordinated metals increases. For the acetylide bonding mode **MA10** only 2

$$C_\alpha \equiv C_\beta$$

(MA10)
R = C or H

(MA11)
R = C or H

(MA12)
R = C or H

fragments were obtained under the strict search constraints, however the CC multiple bond length is longer than for the trimetallic acetylide ligand, and the CCR angle is smaller. The acetylide of **MA10** is a 5-electron donor with 2 metal–carbon σ-bonds and two metal–carbon π-bonds.

For the alkyne ligand, two bonding modes are again possible, **MA11** and **MA12**, where the alkyne is a 4-electron donor in both cases. The alkyne ligand for the **MA11** mode bonds to the four-metal cluster by two σ-bonds to opposite metals, and two π-bonds to the other two metals. For bonding mode **MA12** the alkyne ligand binds to each metal via a σ-bond, similar to the **MA6** mode, the carbons of the multiple bond may be viewed as being of sp^3 character with two metal atoms acting as two of the substituents. **MA11** is the most common mode (28 fragments), whilst there is only one fragment of bonding mode **MA12**, and only two more **MA12** fragments were found in the database if the R-factor and AS restrictions (Section II) are removed.

For bonding mode **MA12** the 4 metal atoms need to be essentially planar, whilst for **MA11** the 4 metal atoms may be either planar, or have a butterfly geometry where there is a M1—M3 bond, in which the CC multiple bond lies parallel to this hinge metal–metal bond[93]. The planarity of the M$_4$ unit can be measured by the torsion angle, τ(M2—M1—M3—M4). For the 28 **MA11** fragments, 18 have a τ value between 106 and 122° with a M1—M3 bond distance between 2.4 and 3.1 Å, and thus have a butterfly configuration, 8 fragments have τ in the range of 153 to 164° with a M1M3 distance between 3.9 and 4.1 Å and represent an essentially planar M$_4$ cluster, and two structures have an intermediate τ value of 141° with M1M3 distances of 3.8 and 3.5 Å[94,95]. For all structures the CC multiple bond lies approximately parallel to the M1—M3 vector (the angle between C—C and M1—M3 vectors lies in the range 0 to 6°). Also, the M1M3RCCR unit is essentially planar [τ(R—C—C—R) = 0 to 19°, τ(M1—C—C—M3) = 0 to 9°].

The **MA11** and **MA12** bonding modes show the longest CC multiple bonds and greatest bend back of the CCR angles, where the carbon atoms are of sp^2 character (Table 18). The CC bond lengths display a wide range from 1.395 to 1.489 Å, the longest bond length being from an iridium complex[96]. The CCR angle also shows wide variation from 108 to 129°, however there is no strong correlation between CC bond length and CCR angle. However, a graph of the mean CCR angle versus τ(M2—M1—M3—M4), the degree of pucker of the M$_4$ cluster, shows a strong correlation of -0.834. This means that as the M$_4$ unit becomes more puckered, and thus adopts a butterfly geometry, the CCR angles of the alkyne ligand become smaller, hence there is greater bend back of the alkyne substituents. There is not such a strong correlation between the CC bond length and τ (correlation coefficient $= -0.51$), although it appears that only the butterfly-type structures have the long CC bonds, whereas the CC bond range for planar structures is 1.39 to 1.44 Å. The smaller CC bond range for planar structures, however, may simply be a reflection of the small number of structures. The σ metal–carbon distances are generally shorter than the π metal–carbon distances as expected. A number of the **MA11** complexes are heterometallic, and for the butterfly structures it is possible to have metal hinge–wing isomerism[93].

TABLE 18. Mean geometric parameters for the alkyne and acetylide substructures **MA10**, **MA11** and **MA12**

Ligand	Parameter	Mean(σ)	Minimum	Maximum	n
MA10	C≡C	1.341, 1.348			2
	CCR	141.4, 143.4			2
	Av. MCC	140.0, 139.0			2
MA11	C≡C	1.432(5)	1.395	1.489	28
	Av. CCR	122.3(6)	108.6	128.7	56
MA12	C≡C	1.446			1
	Av.CCR	119.4			2

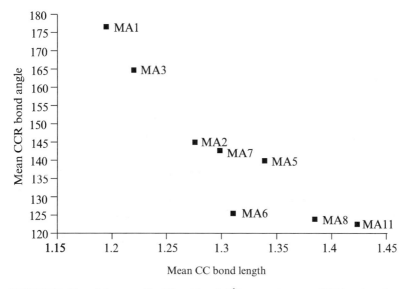

FIGURE 7. Plot of the mean C≡C bond length (Å) versus the mean CCR bond angle (°) for the acetylide and alkyne fragments

6. Summary

Tables 15–18 show that, for all the bonding modes, the C≡C bond lengths and CCR bond angles cover wide ranges. However, there are some general trends: (a) the alkyne or acetylide ligands generally coordinate to as many metal atoms as possible; (b) for a particular metal cluster the mean C≡C bonds of acetylide ligands are always smaller than those for alkyne ligands; (c) as the number of coordinated metals increases, i.e., as the π-donation to the metal atom(s) and the π-back-donation to the alkyne or acetylide ligand increases, the CC multiple bond becomes longer and the CCR angle becomes smaller[97]. Point (c) is illustrated in Figure 7, where the mean CC bond length is plotted against mean CCR angle for all the main bonding modes that have a population greater than 5. Stanghellini and coworkers[98] have also illustrated the correlation between C≡C distance and CCR bond angle by a plot of C≡C versus CCR for 70 π-bonded alkyne metal complexes. Their plot divided into reasonably well-separated sectors, dependent on the number of metal atoms to which the alkyne is coordinated, and showed that neither the type of metal nor the geometry of the complex have a significant effect on the $C\equiv C$ distance and CCR angle.

C. Dinitrogen and Azide Ligands

1. The dinitrogen ligand

The first dinitrogen complex, $[Ru(NH_3)_5(N_2)]^{2+}$, was discovered only in 1965 by Allen and Senoff[99]. The most common bonding mode of N_2 is 'end-on' coordination, although 'side-on' coordination is possible. For 'end-on' coordination the dinitrogen ligand can bond to one metal atom (**MN1**) or to two metal atoms (**MN2**). The end-on bonding mode of N_2 is similar to carbonyl coordination, where there is σ-donation from the filled lone pair of the N_2 ligand to a vacant d–p hybrid metal orbital, and π-back-donation from filled metal d–π orbitals to empty π^*g orbitals of N_2. However, the N_2 ligand has lower stability compared to the carbonyl ligand, being a poorer π-acceptor, and can readily undergo replacement by other 2-electron donor ligands. Generally, molecular dinitrogen complexes contain 'electron-rich' metals in low oxidation states. The bonding and structures of dinitrogen complexes are adequately reviewed[100–104].

$$M-N\equiv N \qquad\qquad M-N\equiv N-M$$
$$\textbf{(MN1)} \qquad\qquad\qquad \textbf{(MN2)}$$

For the end-on bonding mode of **MN1**, the mean NN bond length of 1.110 Å (Table 19) is only slightly longer than 'free' N_2 (1.0976 Å[57]), although the longest bond length is observed at 1.135 Å. This geometry exemplifies the poorer coordination of N_2 compared to CO, and shows that a bond order of 3 is retained for the NN bond. **MN1** coordination is nearly linear, with a mean MNN angle of 177.4°.

Table 19 shows that the NN bond lengths for the 7 fragments of mode **MN2** in the CSD are significantly longer than those found for **MN1**, and that the **MN2** bond lengths cover a wide range from 1.182 to 1.298 Å. The 7 fragments can be split into two groups. Three of the fragments are ditungsten or ditantalum complexes, in which the NN bond is 1.298, 1.292 and 1.272 Å (longer than a NN double bond) and the M—N bond lengths are relatively short. In these cases bonding of the **MN2** fragment can be seen as 2 d-electrons from each metal being donated to N_2 to produce a 'hydrazido (4-)' ligand, i.e., N_2^{4-} oxidizing the two metal centres[102,105–107]. In terms of valence bond theory, the coordinated NN ligand may be represented by the contributors **13** and **14**.

For the other 4 fragments the NN bond length is between 1.182 and 1.236 Å, where the coordinated metal atoms are Ti, Mo, Zr and W. In general it is thought that W and Ta are

TABLE 19. Mean geometric parameters for the dinitrogen substructures **MN1** and **MN2**

Ligand	Parameter	Mean(σ)	Minimum	Maximum	n
MN1	$N\equiv N$	1.110(6)	1.018	1.135	19
	MNN	177.4(4)	173.9	180.0	19
MN2	$N\equiv N$	1.239(19)	1.182	1.298	7
	Av. MNN	173.8(8)	168.2	178.1	14

$$M-N\equiv N-M \qquad\qquad M=N-N=M$$

(13) (14)

$$\overset{-}{\underset{M}{\overset{..}{N}}}-\overset{+}{N}\equiv N \qquad\qquad \overset{..}{\underset{M}{N}}=\overset{+}{N}=\overset{..}{N}{}^{-}$$

(15) (16)

TABLE 20. Mean geometric parameters for the azide substructures **MN3**, **MN4** and **MN5**

Ligand	Parameter	Mean(σ)	Minimum	Maximum	n	n for M = Cu
MN3	$N_\beta N_\gamma$	1.149(2)	1.076	1.187	76	30
	$N_\alpha N_\beta$	1.186(2)	1.131	1.226	76	
	NNN	176.4(2)	167.2	179.9	76	
	MNN	126.2(7)	115.9	164.0	76	
MN4	$N_\beta N_\gamma$	1.141(4)	1.119	1.161	10[a]	12[b]
	$N_\alpha N_\beta$	1.207(4)	1.177	1.222	10	
	NNN	177.2(8)	171.4	180.0	10	
	Av. MNN	124.3(13)	109.6	134.9	20	
	MNM	103.5(24)	96.4	123.9	10	
MN5	Av NN	1.169(3)	1.147	1.197	12	4
	NNN	177.3(4)	175.9	178.6	6	
	Av. MNN	128.4(18)	120.4	139.5	12	

[a]4 fragments removed as long N_β—N_γ bonds: 1.241, 1.23, 1.329, 1.27 Å.
[b]This is including the four fragments mentioned in note a.

better at reducing dinitrogen, but for one tungsten structure[108] the NN bond has remained relatively short (1.189 Å), and correspondingly the W—N bonds are longer than for the hydrazido-type complexes[107,109]. For all 7 fragments the M—NN—M unit is relatively linear.

The 'side-on' bonding mode of dinitrogen remains very rare and there appear to be no structures recorded in the April 1993 version of the CSD. The side-on bonding mode of N_2 is similar to the bonding mode of ethene, where the π-bond of N_2 donates electrons to the metal atom, and back-donation is from metal d-type orbitals to empty N_2 π^*-orbitals[100].

2. The azide ligand

The bonding of the azide ligand is described in an early review[110]. There is a π-interaction between the p-orbital on N_α and the d-orbitals on the metal atom, and it is assumed that the larger the π-interaction, the greater the difference between the 2 N—N bond lengths, although there is also the effect of the N—N σ-bonds to be taken into consideration. In terms of valence bond theory the coordinated azide ligand, **MN3**, may be seen to have two contributors of **15** and **16**. These are similar to the valence bond structures for organic azides (see Section III.B.4).

The bond lengths for the azide **MN3** fragment cover a wide range, as shown in Table 20, and are of a similar range to the organic azide fragments. The mean bond lengths are not equal, N_α—N_β = 1.186 Å, N_β—N_γ = 1.149 Å, and in general the NN bond coordinated to the metal is the longest. For the organic azides there is a much larger difference between the two NN bond lengths. For the free azide ion the NN bond lengths are equal at 1.154(15) Å[110], and thus the NN mean values for the **MN3** fragment do not differ significantly from this value. Of the 76 **MN3** azide fragments, only 4 have the N_α—N_β bond significantly shorter than the N_β—N_γ bond. The largest difference between the two NN bond lengths is found for a copper complex, where (N_α—N_β) – (N_β—N_γ) = 0.12 Å. The NNN and MNN angles of the azide ligand also show a degree of flexibility. The MNN bond angle lies between 116 and 135° in all the fragments, except for two outliers[111,112] having MNN = 141°, 164°. The MNN angle may be influenced[112] by hydrogen bonding from N_γ of the azide ligand to a coordinated water molecule. The MNN values between 116 and 135° are in the expected range for a trigonally hybridized nitrogen atom[110].

For dimetallic azide complexes there are two modes of bonding: (a) end-on bonding where one nitrogen atom is coordinated to both metal atoms (**MN4**), and (b) end-to-end bonding, denoted as bonding mode **MN5**. Both of these binuclear azide bonding modes are not as common as the mononuclear coordinated azides of **MN3**. The mean NN bond lengths for **MN4** and **MN5** are not significantly different from those of **MN3**, although **MN4** shows the largest variation in the two NN bond lengths. As shown in Table 20, 4 fragments have not been included in the calculation of the mean values for **MN4** as the NN bond lengths for these fragments are very long compared to the other fragments, and

(MN3)   (MN4)   (MN5)

it appears that the atomic displacement (thermal) parameters are quite large, which may affect the accuracy of the bond lengths. For a clearer description of the bond lengths more structures are needed. Of the 10 **MN4** fragments the NNN bond angle shows only small deviation from linearity. For all 14 fragments the MNN bond angles lie in the range 109 to 135°, which is similar to MNN angles of **MN3**, and the MNM bond angle is between 91 and 106°, except for one dirhodium complex in which MNM is 123.9°[113]. Summation of the bond angles at N_α (SUM) show that of the 14 fragments, 10 have a planar N_α atom (SUM = 355 to 360°), whilst for the other 4 structures SUM = 321 to 345° and the geometry of N_α is pyramidal. Additionally the two MN distances are quite similar, except for 3 fragments, where one MN distance is relatively long (2.54, 2.42 and 2.33 Å). Therefore the **MN4** bridging azide shows a large degree of overall flexibility in its structural geometry.

From a CSD search with strict crystallographic constraints (Section II), only 6 azide fragments of bonding mode **MN5** were obtained. The NNN bond angle is linear (Table 20), and the MNN bond angles (120 to 140°) are similar to those for **MN3** and **MN4**. The two MN bond lengths of **MN5** are similar, except for one fragment[114] (1.98 Å, 2.46 Å). As discussed[115] for **MN3** azide complexes there may be weak interactions between the N_γ nitrogen atom and another metal, which again shows the large structural diversity of the azide ligand.

Azide coordinated complexes, particularly copper complexes, have been extensively studied in recent years to try and understand the enormous variation of magnetic spin exchange interactions. This is of importance in the study of active sites in copper-containing proteins. For studies of particular magnetic exchange interactions it is important to determine X-ray structures of small binuclear and mononuclear azide-containing copper complexes. In Table 20 we indicate the number of fragments in which M = Cu for the 3 azide-type bonding modes, since this illustrates the interest in copper azide complexes, particularly those which are dinuclear.

It has been found that for the two types of bridging azides, the bonding mode **MN4** favours a triplet ground state (a ferromagnetic interaction), whilst bonding mode **MN5** favours a singlet ground state (an antiferromagnetic interaction)[116]. The magnetic properties of azide copper(II) complexes will not be discussed here, but are treated in papers[115,117–120] which describe crystal structure determinations with particular emphasis on the magnetic properties of these complexes.

D. Cyanide and Isocyanide Ligands

1. The cyanide ligand

Table 21 lists the geometry of cyanide ligands for the terminal and bridging bonding modes, **MC1** and **MC2**, respectively. The bonding of terminal cyanide, which is the most

$$M—C \equiv N \qquad M—C \equiv N—M$$
$$\text{(MC1)} \qquad\qquad \text{(MC2)}$$

common, is similar to that of carbonyl ligands, i.e. σ-donation to the metal atom and π-acceptance from the metal to empty π^*-orbitals, although it is thought that the negative charge of the CN^- ligand decreases its π-acceptor ability compared to CO. The cyanide ligand, however, can form complexes in low and high oxidation states and, due to its charge, is a strong nucleophile.

For bonding mode **MC1** the mean CN bond length of 1.140 Å is similar to that of organic cyanides (Section III.A.2), thus illustrating the weak π-back-donation. The cyanide ligand shows some deviation from linearity, but again is similar to the organic cyanides.

In the bridging cyanide of **MC2** the lone pair of the nitrogen atom is used to coordinate to the second metal. All of the fragments found in the C SD have a polymeric structure, except for one[121]. It is known that cyanides, such as AuCN and $Zn(CN)_2$, are polymeric with bridging cyanide ligands, and that hexacyanides are able to enclathrate small molecules like water and ammonia because of their polymeric structure which has extensive cyanide bridging[122]. Table 21 shows that the CN bond is similar to the terminal cyanides, and that the MCN bond angle only deviates slightly from linearity, whilst the

TABLE 21. Mean geometric parameters for the cyanide substructures **MC1** and **MC2**

Ligand	Parameter	Mean(σ)	Minimum	Maximum	n
MC1	C≡N	1.140(1)	1.083	1.196	156
	MCN	176.9(2)	170.0	180.0	156
MC2	C≡N	1.144(3)	1.126	1.165	18
	MCN	177.2(4)	172.8	179.4	18
	CNM	170.3(19)	155.3	179.7	18

MNC angle covers a much wider range from 155 to 180°. For all fragments the M—C bond length is shorter than the M—N bond. In some cases this difference is small, but for 7 fragments the difference is greater than 0.35 Å, thus reflecting the polymeric nature of these structures.

2. The isocyanide ligand

Isocyanide appears to be a stronger σ-donor than CO, and also a stronger π-acceptor than CN⁻ and is capable of forming complexes in high and low oxidation states. From Tables 21 and 22 one can see that the number of terminal isocyanides is nearly double that of terminal cyanides. As for cyanides, the terminal bonding mode **MC3** of isocyanides is far more common than the two bridging modes of **MC4** and **MC5**.

$$M—C\equiv N—C \qquad \qquad \qquad$$

(MC3)   (MC4)   (MC5)

Table 22 shows that for the terminal isocyanides, **MC3**, the CN bond length is longer than that found in the cyanide ligands, and also in the organic cyanides, although not significantly so. This, however, could suggest that there is a greater degree of π-back-

TABLE 22. Mean geometric parameters for the isocyanide substructures **MC3**, **MC4** and **MC5**

Ligand	Parameter	Mean(σ)	Minimum	Maximum	n
MC3	C≡N	1.152(1)	1.094	1.203	300
	N—C	1.434(2)	1.325	1.549	300
	CNC	173.3(3)	139.5	180.0	300
	MCN	175.3(2)	160.6	180.0	300
MC4	C≡N	1.234(11)	1.174	1.273	8
	CNC	134.4(11)	130.2	137.4	8
	M1CN	167.9(17)	158.6	173.8	8
MC5	C≡N	1.228(7)	1.161	1.297	29
	CNC	133.9(28)	120.2	175.2	29
	Av. MCN	138.6(9)	120.3	155.0	58

TABLE 23. Mean geometric parameters for the parent isocyanide substructures, M—C≡N—R [R = t-Bu, C(C₂H), C(CH₂), CH₃, Ph]

MCNR	C≡N	N—C	MCN	CNC	n
MCNBu-t	1.149(2)	1.458(2)	175.4(3)	173.7(5)	94
MCNCC₂H	1.145(2)	1.464(5)	174.7(10)	173.8(10)	14
MCNCCH₂	1.153(3)	1.439(6)	176.5(4)	170.8(11)	16
MCNCH₃	1.145(2)	1.440(3)	175.6(4)	175.6(5)	43
MCNPh	1.155(2)	1.402(2)	175.1(3)	172.7(5)	110

bonding in these complexes. The MCN angle shows some deviation from 180°, and the CNC angle extends over a very wide range from 140 to 180°. Due to the large variation in values the fragments were split into subsets depending on the substitution at C (see Table 23), in a similar way to the organic cyanides. For the 5 sub-groups the mean C≡N bond lengths do not differ significantly. However, there are differences in the mean N—C bond lengths due to the increasing substitution at C, as observed in the organic cyanides. For

$$\overset{-}{Mo}-C\equiv\overset{+}{N}-R \qquad\qquad Mo=C=\overset{..}{N}\diagdown R$$

(17)   (18)

MCNC(C_2H) the NC bond is longer than in butyl isocyanide, but the number of fragments is small. The phenyl isocyanide has a much shorter NC bond length of 1.402 Å, which is due to greater delocalization through the system. Under the crystallographic constraints of the search one structure with 6 trifluorocarbon isocyanide ligands, $Cr(CNCF_3)_6$[123], was found, which has long C≡N bond lengths between 1.17 and 1.2 Å. One would expect the strongly electronegative CF_3 group to cause stronger π-back-donation, and the N—C bonds are relatively short between 1.36 and 1.41 Å. The majority of MCN and CNC bond angles are between 170 and 180°. Some of the bond length and bond angle outliers could be due to large thermal parameters for the isocyanide ligand. However, for one tertiary butyl isocyanide[124], which has the smallest CNC angle of 140° of the dataset, and a relatively long C≡N bond of 1.193 Å, and a short Mo—C bond of 1.996 Å, the structure is said to be due to a strong contribution from canonical form **18** rather than that of **17**[124]. The electron-rich molybdenum of this structure means that there is extensive π-back-bonding to the isocyanide ligand. Although, a molecular orbital study of $Cp_2MoCNMe$, which looked at the bending of the isocyanide ligand, shows that electronic effects do not clearly favour the bending of the isocyanide ligand, but that bending may originate from other effects like crystal packing forces.

The isocyanide ligand can also bridge two metal atoms, either by a σ + π coordination (**MC4**), or by 2 σ-bonds via the carbon atom (**MC5**). Neither of these bridging modes are as common as the terminal isocyanide, **MC3** (Table 22). The mean C≡N bond for **MC4** is longer than for the terminal isocynide, showing a reduction in the multiplicity of the bond, although it is not as long as a C=N double bond[14]. For all 8 **MC4** fragments the M—N π-bond is shorter than the M—C π-bond, and, as one would expect, the M—C σ-bond is shorter than the M—C/N π-bond distances. The MCN bond angle remains relatively linear, however the CNC bond angle lies between 130 and 137°, thus indicating a change in hybridization at the nitrogen atom, away from the linear sp^1 which is found for the majority of terminal isocyanides (**MC3**). The bonding mode of **MC5** shows a much wider variation in geometry than **MC4**. The C≡N bond length is generally long (1.22 to 1.30 Å), except for four fragments (C≡N = 1.16 to 1.19 Å)[125,126]. A scatterplot of C≡N bond length versus CNC angle shows a correlation of -0.77, where the four fragments with short C≡N bonds also have a nearly linear CNC angle (158 to 175°), whilst the remainder of the fragments have CNC angles between 120 and 140°. Therefore 25 of the **MC5** fragments have a geometry similar to **MC4**, where there is a change in hybridization of the nitrogen atom from sp^1, and four fragments have a geometry similar to **MC3**. There appears to be no obvious steric reason for the linearity of the **MC5** bonding mode, and an electronic explanation is being investigated[125].

3. Other cyanide derivatives

There are a large number of cyanide ligands in which coordination to the metal is via the nitrogen atom (**MC6** bonding mode), instead of via the carbon atom (**MC1**). In fact from this search (see Section II), there are nearly twice the number of **MC6** fragments (292) compared to **MC1** (156 fragments). The mean C≡N bond lengths for **MC1** and **MC6** are very similar (Tables 22 and 24), but there is much wider variation of the MNC bond angle for MC6, compared to the MCN angle for **MC1**, which remains close to linear. For **MC6** the MNC bond angle varies from 140 to 180°, however only 13 fragments have a bond angle less than 160°. The smallest bond angle of 140° is for a copper TCNQ complex[127].

For the cyanate ligand, **MC7**, the NC bond lengths and MNC bond angles show a large variation. It has been suggested that there is a correlation between these two parameters, where for small MNC angles (120°) the CN distance is long, and tends towards a C=N bond, whilst for MNC of 180° the CN distance is short and indicative of a C≡N bond[128]. A plot is given in Reference 128 of MNC versus CN, showing the correlation. However, for the 31 fragments given here a strong correlation is not present, although the structures with the smallest MNC angles do have the longer CN distances. The mean CN distance is quite short, and is comparable to **MC1** and **MC6**. In comparison to the MNC angle, the

$$M-N{\equiv}C-C \qquad M-N{=}C{=}O$$

$$\begin{array}{c} O \\ \| \\ C \\ \| \\ N \\ \diagup \quad \diagdown \\ M \qquad M \end{array}$$

(MC6)   (MC7)   (MC8)

TABLE 24. Mean geometric parameters for the cyanide derivative substructures **MC6, MC7, MC8, MC9** and **MC10**

Ligand	Parameter	Mean (σ)	Minimum	Maximum	n
MC6	N≡C	1.130(1)	1.058	1.204	292
	NCC	177.8(1)	170.8	180.0	292
	MNC	173.0(3)	140.5	180.0	292
MC7	N≡C	1.131(7)	1.029	1.214	31
	CO	1.195(5)	1.113	1.245	31
	NCO	177.6(3)	169.5	179.8	31
	MNC	161.9(22)	127.8	180.0	31
MC8	NC	1.157(6)	1.133	1.189	7
	CO	1.185(3)	1.173	1.199	7
	NCO	178.3(4)	176.5	179.8	7
	Av. MNC	131.9(11)	125.8	145.4	14
MC9	N≡C	1.152(1)	1.107	1.189	272
	C–S	1.622(1)	1.562	1.663	272
	NCS	178.5(1)	174.0	180.0	272
	MNC	165.4(5)	138.2	180.0	272
MC10	N≡C	1.149(4)	1.113	1.182	18
	CS	1.641(4)	1.610	1.669	18
	NCS	178.0(3)	175.4	179.7	18
	MNC	162.1(23)	142.8	177.1	18
	CSM	98.2(13)	83.7	108.1	18

NCO angle remains relatively linear. The mean $C\equiv N$ bond length for the bridging cyanate ligand **MC8** is slightly longer than for the terminally bonded cyanate ligand **MC7**.

$$M-N\equiv C-S \qquad\qquad M-N\equiv C-S$$

(MC9)   (MC10)   M

The NCO angle is close to 180°, and the MNC angle is in the expected range of 125 to 145°.

The geometry of the thiocyanate ligand for **MC9** is similar to that of the NCO ligand of **MC7**, where the MNC shows a wide range of angles (although the minimum value for **MC7** is 10° less than that for **MC9**), and the NCS and NCO bond angles remain nearly linear. The mean CN bond is longer for **MC9**, although not significantly so. For **MC9** there is no correlation between CN distances and MNC angles.

The thiocyanate ligand can also bridge two metals via the nitrogen and sulphur atoms, although from the CSD search (Section II), only 18 fragments were obtained. The mean $C\equiv N$ bond length is the same as for the terminally bound ligand **MC9**, and the mean C—S bond length is slightly longer for **MC10**. For **MC9** and **MC10** the mean C—S bond length is similar to that found in $N\equiv C-S^-$ and is significantly shorter than the mean S—C bond length in $N\equiv C-S-C^{14}$ (see Table 13).

V. INTERMOLECULAR INTERACTIONS OF TRIPLE-BONDED GROUPS

A. Introduction

The crystallographic method is unique in providing precise geometrical descriptions of non-bonded interactions as they occur in extended crystal structures. The shorter, and usually more directional, of these non-bonded contacts govern molecular aggregation in the solid state and it is reasonable to infer that these interactions are heavily implicated in molecular recognition phenomena. An understanding of molecular recognition is increasingly important in modern chemistry, in studies of protein folding, protein ligand interactions, supramolecular chemistry and in crystal engineering. Crystal structure data have always been a primary source of information on hydrogen bonding[129–131] and, in recent years, detailed analyses using the CSD[132] have become more common. A particular success of database analysis has been the clear characterization[133] of the C—H---O hydrogen bond as an important interaction in organic and biological systems. More recent analyses[134,135] have concentrated on the ability of more general non-bonded interactions, not mediated by hydrogen, to stabilize extended crystal structures through the formation of recognizable molecular motifs. In all of those studies, the directionality of short non-bonded contacts, i.e. their geometrical relationship to putative lone-pair positions or to areas of increased electron density (as in the π-systems), has been a key feature of the analysis. In this short section, we indicate a number of recent analyses in which the non-bonded contacts formed by acetylene and cyano groups have been the focus of interest.

B. Non-bonded Interactions of Acetylene

The acetylenic proton is well known to be more acidic than its alkyl or alkenyl analogues. It should, therefore[133], be more susceptible to H-bond formation with suitable acceptor atoms, e.g. C—H---O. Indeed, the result of theoretical studies on the acetylene–water system[136,137] indicated that acetylene does form H-bonds to water oxygens. This was confirmed by microwave and radio-frequency spectroscopy[138] which yielded an H---O distance of 2.229 Å and a stretching force constant of 0.065 mdyn Å^{-1}.

A systematic study of crystallographic C≡C—H---O contacts was carried out by Desiraju and Murty[139]. These workers selected nine crystal structures (out of 51 containing C≡C—H and oxygen) for which IR acetylenic C—H stretches were also available from measurements in the solid state. The normal v_{C-H} in solution is 3300 cm^{-1}, but for 8 of the selected compounds v_{C-H} in the solid state ranged from 3190 to 3280 cm^{-1}, indicative of a weakening of the C---H bond and probable H-bond formation to ether or carbonyl oxygens. The C---O distances ranged from 3.20 to 3.38 Å, entirely consistent with the H---O distance of 2.229 cited above, and were strongly correlated with the v_{C-H} values. In the ninth compound, C---O > 4.0 Å and v_{C-H} = 3300 cm^{-1} is identical to the normal solution value. The authors remark that 'the acetylenic hydrogen closely approaches an oxygen atom along a well-defined and anticipated geometrical direction', i.e. H approaches along an O–lone-pair vector.

Molecular packing in the low-temperature orthorhombic phase of perdeutero-acetylene[25,26] shows a layer structure with a T-shaped interaction between pairs of molecules within a layer. The D---C distances form an isosceles triangle with edges of 2.738, 2.737 Å[25] and a D to C≡C midpoint distance of 2.672 Å. The D---C contacts are some 0.2 Å shorter than those expected from van der Waals radii considerations. The geometrical arrangement clearly implies a donor–acceptor interaction between D and the triple-bond π-electrons, a situation described as long ago as 1972[140] to account for close C≡C---I interactions in crystals of diiodo-acetylene. More recently[141] five O—H---C≡C and five NH---C≡C π-interactions with mean H---C distances of 2.69 (for O—H) and 2.61 (for N—H) have been located in the CSD.

This T-shaped geometry is also identified in a microwave study of the C_2H_2.HF complex[142] where the H---C≡C midpoint distance is short at 2.19 Å consistent with the higher acidity of HF by comparison with acetylene[25]. Microwave and infrared studies of acetylene dimer also reveal a T-shape deviating by 27° from the fully perpendicular arrangement[143]. The distance between the centres of mass of the two monomer units is given as 4.38 Å.

C. Non-bonded Interactions Involving Cyano Groups

Much early work on the intermolecular interactions of cyano groups was reviewed by Britton[144] with particular reference to the probable importance of C≡N---halogen interactions. These interactions were first postulated by Hassell[145] and, although weak, appear to be the crucial stabilizing influence in structures of halogenophenyl cyanides[144]. A systematic database study of C≡N---Cl and C≡N---Br contacts was carried out by Desiraju and Harlow[146] in 1989. In both cases the histogram of N---halogen distances showed a bimodal distribution. The longer-distance peak (ca 3.40 Å for Cl, 3.70 Å for Br) corresponds to normal van der Waals separations, but a shorter-distance peak (ca 3.00 to 3.30 Å for Cl, ca 3.20 Å for Br) is also observed. These shorter distances appear to correspond to the largest (>130°) C≡N---halogen angles and the approximate equality of the N---Cl, N---Br distances is ascribed to polarization effects. These authors conclude that these weak but directional interactions play a decisive role in crystal packing in the absence of any stronger interactions. More recently[147], Desiraju and coworkers have used the weak C≡N---Cl interaction to engineer linear molecular arrays in crystals of 1,4-dimethoxy-2,3-dichloro-5,6-dicyanobenzene and in 1-hydroxy-4-octyloxy-2,3-dichloro-5,6-dicyanobenzene. Pairs of molecules are stabilized by two C≡N---Cl interactions to form infinite motifs described by these authors as 'linear tapes'.

The most obvious example of the importance of non-bonded interactions in organic cyano compounds lies in the role played by tetracyanoquinodimethane in complexes that exhibit metallic conductivity. There is now a vast literature in this area and many excellent reviews are already available[148–150].

The —C≡N group is, of course, a highly suitable H-bond acceptor due to the (linear) lone pair on nitrogen. As long ago as 1951, the crystal structure of HCN[151] was shown to consist of linear ---H—C≡N---H—C≡N--- chains with N---C = 3.2 Å and N---H = 2.2 Å in agreement with v_{C-H} (vapour) = 3312 cm^{-1} and v_{C-H}(crystal) = 3132 cm^{-1}. Although —C≡N---H—X hydrogen bonds have been observed and commented on in many individual crystal structures, we can find no systematic review of their geometric properties. A very rapid survey of the CSD isolated 47 —C≡N---H—O bonds in organic structures having R-factor ≤ 0.10, and N---O ≤ 3.2 Å. The mean d(N---O) and d(N---H) are 2.93(2) and 2.26(7), respectively. Interestingly, the C≡N---H angle appears to adopt two discrete values: 24 instances have θ(C≡N---H) in the range 145–180° as expected, 21 have θ(C≡N---H) in the range 105–135°; there are three outlying observations. These results are directly comparable to a similar 'overview' survey of the CSD reported by Sarma and collaborators[152] carried out as part of a developing strategy for the design of non-centrosymmetric structures. The effectiveness of even weaker C—H---N≡C— interactions is graphically illustrated in the 1:1 molecular complex of 1,3,5-tricyanobenzene and hexamethylbenzene[153]. There is no doubt, however, that much more detailed analysis is required to provide a clear picture of —C≡N---H—X bonding and, indeed, of other intermolecular interactions involving triple-bonded groups.

VI. SUMMARY AND CONCLUSIONS

This review has presented a survey of the structural chemistry of triple-bonded groups as observed (primarily) in several hundreds of crystal structure analyses. There will be areas that we have overemphasized and *vice versa*, as in all reviews. However, we have attempted to provide characteristic geometries for a large number of substructures involving triple bonds and, where relevant, to compare and contrast these geometries. For organic substructures in particular, it has been possible to make inferences based upon very small geometrical differences (of *ca* 0.02 Å in bond lengths) by comparing *mean* values drawn from *large distributions*. In many cases, it would not be possible to make these comparisons with any statistical certainty on the basis of bond lengths etc. drawn from individual crystal structure determinations. It is only when many geometrical observations of the same substructure all say the same thing that we can perform comparisons at or beyond the precision levels of normal experimental observations.

Despite the large body of information used to assemble this review, it has been frequently necessary to comment that 'further work is necessary to clarify' the structural chemistry in a number of areas. This 'further work' is of two forms: (a) more detailed and interpretative studies of the crystal structure data that are already available, and (b) the need for more individual crystal structures to expand the available information pool to a point where it is suitable for systematic analysis. It is hoped that, if nothing else, this review will provoke activity in both of these important areas.

VII. REFERENCES

1. M. Simonetta and A. Gavezzotti in *The Chemistry of the C≡C Triple Bond* (Ed. S. Patai), Wiley, Chichester; 1979, pp. 1–55.
2. J. L. Hencher, in *The Chemistry of the C≡C Triple Bond* (Ed. S. Patai), Wiley, Chichester; 1978, pp. 57–67.
3. S. Sorriso, in *The Chemistry of Diazonium and Diazo Groups* (Ed. S. Patai), Wiley, Chichester; 1978, pp. 96–135.
4. J. B. Moffatt, in *The Chemistry of Functional Groups, Supplement C* (Eds. S. Patai and Z. Rappoport), Wiley, Chichester; 1983, pp. 1016–1034.

148   F. H. Allen and S. E. Garner

5. J. B. Moffatt, in *The Chemistry of Functional Groups, Supplement C* (Eds. S. Patai and Z. Rappoport), Wiley, Chichester; 1983, pp. 1305–1343.
6. F. H. Allen, G. Bergerhoff and R. Sievers (Eds.), *Crystallographic Databases*, International Union of Crystallography, Chester, UK, 1987.
7. F. H. Allen, J. E. Davies, J. J. Galloy, O. Johnson, O. Kennard, C. F. Macrae, E. M. Mitchell, G. F. Mitchell, J. M. Smith and D. G. Watson, *J. Chem. Inf. Comp. Sci.*, **31**, 187 (1991).
8. F. H. Allen and O. Kennard, *Chemical Design Automation News*, **8**, 1 (1993).
9. P. Murray-Rust and J. Raftery, *J. Mol. Graphics*, **3**, 50 (1985).
10. P. Murray-Rust and J. Raftery, *J. Mol. Graphics*, **3**, 60 (1985).
11. F. H. Allen, O. Kennard, J. J. Galloy, O. Johnson and D. G. Watson, in *Chemical Structures 2* (Ed. W. A. Warr), Springer-Verlag, Berlin, 1993.
12. F. H. Allen, O. Kennard and D. G. Watson, in *Structure Correlation* (Eds. H. B. Bürgi and J. D. Dunitz), VCH Publishers, Weinheim, 1993.
13. R. Taylor and F. H. Allen, in *Structure Correlation* (Eds. H. B. Bürgi and J. D. Dunitz), VCH Publishers, Weinheim, 1993.
14. F. H. Allen, O. Kennard, D. G. Watson, L. Brammer, A. G. Orpen and R. Taylor, *J. Chem. Soc., Perkin Trans. 2*, S1–S19 (1987).
15. A. G. Orpen, L. Brammer, F. H. Allen, O. Kennard, D. G. Watson and R. Taylor, *J. Chem. Soc., Dalton Trans.*, S1–S83 (1989).
16. R. Taylor and O. Kennard, *Acta Crystallogr.*, **B39**, 517 (1983).
17. R. Taylor and O. Kennard, *Acta Crystallogr.*, **A41**, 85 (1985).
18. R. Taylor and O. Kennard, *J. Chem. Inf. Comp. Sci.*, **26**, 28 (1986).
19. J. Dale, in *The Chemistry of Acetylenes* (Ed. H. G. Viehe), Marcel Dekker, New York, 1969.
20. M. Tanimoto, K. Kuchitsu and Y. Morino, *Bull. Chem. Soc. Jpn.*, **42**, 2519 (1969).
21. A. Baldacci, S. Ghersetti, S. C. Hurlock and K. N. Rao, *J. Mol. Spectrosc.*, **59**, 116 (1976).
22. H. Fast and H. L. Welsh, *J. Mol. Spectrosc.*, **41**, 203 (1972).
23. W. J. Lafferty, E. K. Plyler and E. D. Tidwell, *J. Chem. Phys.*, **37**, 1981 (1962).
24. G. J. H. van Nes and F. van Bolhuis, *Acta Crystallogr.*, **B35**, 2580 (1979).
25. R. K. McMullan, A. Kvick and P. Popelier, *Acta Crystallogr.*, **B48**, 726 (1992); **B49**, 145 (1993).
26. H. K. Koski, *Acta Crystallogr.*, **B31**, 933 (1975).
27. C. C. Costain, *J. Chem. Phys.*, **29**, 824 (1958).
28. R. F. Stewart, E. R. Davidson and W. T. Simpson, *J. Chem. Phys.*, **42**, 3175 (1965).
29. J. C. Hanson, L. C. Sieker and L. H. Jensen, *Acta Crystallogr.*, **B29**, 797 (1973).
30. B. Dawson, *Acta Crystallogr.*, **17**, 990 (1964).
31. F. H. Allen, *Acta Crystallogr.*, **B42**, 515 (1986).
32. K. S. Feldman, C. M. Kraebel and M. Parvez, *J. Am. Chem. Soc.*, **115**, 3846 (1993).
33. P. Coppens, *Acta Crystallogr.*, **B30**, 255 (1974).
34. B. P. Stoicheff, *Tetrahedron*, **17**, 135 (1962).
35. F. H. Allen, *Acta Crystallogr.*, **B37**, 890 (1981).
36. T. Fukuyama, K. Kuchitsu and Y. Morino, *Bull. Chem. Soc. Jpn.*, **42**, 379 (1969).
37. M. Sugie, T. Fukuyama and K. Kuchitsu, *J. Mol. Struct.*, **14**, 333 (1972).
38. M. Tanimoto, K. Kuchitsu and Y. Morino, *Bull. Chem. Soc. Jpn.*, **44**, 386 (1971).
39. C. C. Costain and B. P. Stoicheff, *J. Chem. Phys.*, **30**, 777 (1958).
40. T. Fukuyama and K. Kuchitsu, *J. Mol. Struct.*, **51**, 131 (1970).
41. M. D. Harmony, R. N. Nandi, J. V. Tietz, J. I. Choe, S. J. Getty and S. W. Staley, *J. Am. Chem. Soc.*, **105**, 3947 (1983).
42. J. K. Tyler and J. Sheridan, *Trans Faraday Soc.*, **59**, 2661 (1963).
43. Y. Morino, K. Kuchitsu, Y. Hori and M. Tanimoto, *Bull. Chem. Soc. Jpn.*, **41**, 2349 (1968).
44. M. Charton, in *The Chemistry of Alkenes Vol. 2* (Ed. J. Zabicky), Wiley, London, 1970, pp. 511–610.
45. R. Hoffmann, *Tetrahedron Lett.*, 2907 (1970).
46. R. Hoffmann and W. D. Stohrer, *J. Am. Chem. Soc.*, **93**, 6941 (1971).
47. R. Hoffmann and R. B. Davidson, *J. Am. Chem. Soc.*, **93**, 5699 (1971).
48. F. H. Allen, *Acta Crystallogr.*, **B36**, 81 (1980).
49. W. F. Jorgensen and L. Salem, *The Organic Chemists' Book of Orbitals*, Academic Press, New York, 1973.
50. R. E. Penn and J. E. Boggs, *J. Chem. Soc., Chem. Commun.*, 666 (1972).
51. D. B. Chesnut and R. E. Marsh, *Acta Crystallogr.*, **11**, 413 (1958).

52. C. J. Fritchie, *Acta Crystallogr.*, **20**, 27 (1966).
53. M. A. M. Meester, H. Schenk and C. H. MacGillavry, *Acta Crystallogr.*, **B27**, 630 (1971).
54. P. G. Jones and G. Schrumpf, *Acta Crystallogr.*, **C43**, 1179 (1987).
55. A. Hartman and F. L. Hirshfeld, *Acta Crystallogr.*, **20**, 80 (1966).
56. J. Sheridan, *Advances in Molecular Spectroscopy*, Pergamon, Oxford, 1962, p. 139.
57. B. P. Stoicheff, *Can. J. Phys.*, **82**, 630 (1954).
58. A. Domenicano and P. Murray-Rust, *Tetrahedron Lett.*, 2283 (1979).
59. I. G. Csizmadia, S. A. Houlden, O. Meresz and P. Yates, *Tetrahedron*, **25**, 2121 (1969).
60. M. J. Menu, M. Dartiguenave, Y. Dartiguenave, J. J. Bonnet, G. Bertrand and A. Baceiredo, *J. Organomet. Chem.*, **372**, 201 (1989).
61. C. Glidewell and G. M. Sheldrick, *J. Chem. Soc., Dalton Trans.*, 2409 (1972).
62. A. Miyazaki, A. Izuoka, T. Sugawara, D. Bethell and P. Gallagher, *Acta Crystallogr.*, **C47**, 1054 (1991).
63. T. H. Tulip, P. W. R. Corfield and J. A. Ibers, *Acta Crystallogr.*, **B34**, 1549 (1978).
64. Y. Pang, A. Schneider, T. J. Barton, M. S. Gordon and M. R. Carroll, *J. Am. Chem. Soc.*, **114**, 4920 (1992).
65. W. Ando, N. Nakayama, Y. Kabe and T. Shimizu, *Tetrahedron Lett.*, **31**, 3597 (1990).
66. P. R. Raithby and M. J. Rosales, *Adv. Inorg. Chem. Radiochem.*, **29**, 169 (1985).
67. E. Sappa, A. Tiripicchio and P. Braunstein, *Chem. Rev.*, **83**, 203 (1983).
68. A. J. Carty, *Pure Apple. Chem.*, **54**, 113 (1982).
69. R. S. Dickson and P. J. Fraser, *Adv. Organometal. Chem.*, **12**, 323 (1974).
70. E. Sappa, A. Tiripicchio and P. Braunstein, *Coord. Chem. Rev.*, **65**, 219 (1985).
71. A. A. Cherkas, A. J. Carty, E. Sappa, M. A. Pellinghelli and A. Tripicchio, *Inorg. Chem.*, **26**, 3201 (1987).
72. M. J. S. Dewar, *Bull. Soc. Chim. Fr.*, **18**, C71 (1951).
73. J. Chatt and L. A. Duncanson, *J. Chem. Soc.*, 2939 (1953).
74. R. Kergoat, L. C. Gomes de Lima, C. Segat, N. Le Berre, M. M. Kubicki, J. E. Guerchais and P. L'Haridon, *J. Organomet. Chem.*, **389**, 71 (1990).
75. J. C. Jeffery and M. J. Went, *J. Chem. Soc., Chem. Commun.*, 1766 (1987).
76. Hong Chen, B. F. G. Johnson, J. Lewis, D. Braga, F. Grepioni and P. Sabatino, *J. Organomet. Chem.*, **405**, C22 (1991).
77. S. F. T. Froom, M. Green, R. J. Mercer, K. E. Nagle, A. G. Orpen and R. A. Rodrigues, *J. Chem. Soc., Dalton Trans.*, 3171 (1991).
78. D. M. Hoffman, R. Hoffmann and C. R. Fisel, *J. Am. Chem. Soc.*, **104**, 3858 (1982).
79. D. M. Hoffman and R. Hoffmann, *J. Chem. Soc., Dalton Trans.*, 1471 (1982).
80. J. Breimair, M. Steinmann, B. Wagner and W. Beck, *Chem. Ber.*, **123**, 7, 1990.
81. A. A. Cherkas, N. Hadj-Bagheri, A. J. Carty, E. Sappa, M. A. Pellinghelli and A. Tiripicchio, *Organometallics*, **9**, 1887, (1990).
82. M. H. Chisholm, B. W. Eichhorn and J. C. Huffman, *Organometallics*, **8**, 80 (1989).
83. M. J. Calhorda and R. Hoffmann, *Organometallics*, **5**, 2181 (1986).
84. K. J. Ahmed, M. H. Chisholm, K. Folting and J. C. Huffman, *Organometallics*, **5**, 2171 (1986).
85. M. H. Chisholm, B. K. Conroy, D. l. Clark and J. C. Huffman, *Polyhedron*, **7**, 903 (1988).
86. K. I. Hardcastle, T. McPhillips, A. J. Acre, Y. de Sanctis, A. J. Deeming and N. I. Powell, *J. Organomet. Chem.*, **389**, 361 (1990).
87. R. D. Adams and Gong Chen, *Organometallics*, **10**, 3020 (1991).
88. M. G. Thomas, E. L. Muetterties, R. O. Day and V. W. Day, *J. Am. Chem. Soc.*, **98**, 4645 (1976).
89. J. -F. Halet, J. -Y. Saillard, R. Lissillour, M. J. McGlinchey and G. Jaouen, *Inorg. Chem.*, **24**, 218 (1985).
90. F. W. B. Einstein, K. G. Tyers, A. S. Tracey and D. Sutton, *Inorg. Chem.*, **25**, 1631 (1986).
91. L. Busetto, M. Green, B. Hessner, J. A. K. Howard, J. C. Jeffery and F. G. A. Stone, *J. Chem. Soc., Dalton Trans.*, 519 (1983).
92. M. R. Churchill, C. Bueno and H. J. Wasserman, *Inorg. Chem.*, **21**, 640 (1982).
93. E. Sappa, A. Tiripicchio, A. J. Carty and G. E. Toogood, *Prog. Inorg. Chem.*, **35**, 437 (1987).
94. S. Aime, G. Nicola, D. Osella, A. M. M. Lanfredi and A. Tiripicchio, *Inorg. Chim. Acta*, **85**, 161 (1984).
95. E. Sappa, A. Tiripicchio and M. Tiripicchio Camellini, *J. Organomet. Chem.*, **199**, 243 (1980).
96. G. F. Stuntz, J. R. Shapley and C. G. Pierpont, *Inorg. Chem.*, **17**, 2596 (1978).

150					F. H. Allen and S. E. Garner

97. S. A. MacLaughlin, N. J. Taylor and A. J. Carty, *Organometallics*, **2**, 1194 (1983).
98. G. Gervasio, R. Rossetti and P. L. Stanghellini, *Organometallics*, **4**, 1612 (1985).
99. J. Chatt and G. J. Leigh, *Chem. Soc. Rev.*, **1**, 121 (1972). (The first synthesis of these complexes is ascribed here (p. 128), and in other secondary publications, to Allen and Senoff, but no original literature citation to these workers is provided in any of these sources.)
100. H. M. Colquhoun, *Acc. Chem. Res.*, **17**, 23 (1984).
101. J. Chatt, J. R. Dilworth and R. L. Richards, *Chem. Rev.*, **78**, 589 (1978).
102. G. J. Leigh, *Trans. Met. Chem.*, **11**, 118 (1986).
103. R. A. Henderson, G. J. Leigh and C. J. Pickett, *Adv. Inorg. Chem. Radiochem.*, **27**, 197 (1983).
104. P. Pelikán and R. Boca, *Coord. Chem. Rev.*, **55**, 55 (1984).
105. R. R. Schrock, R. M. Kolodziej, A. H. Liu, W. M. Davis and M. G. Vale, *J. Am. Chem. Soc.*, **112**, 4338 (1990).
106. R. D. Sanner, J. M. Manriquez, R. E. Marsh and J. E. Bercaw, *J. Am. Chem. Soc.*, **98**, 8351 (1976).
107. M. B. O'Regan, A. H. Liu, W. C. Finch, R. R. Schrock and W. M. Davis, *J. Am. Chem. Soc.*, **112**, 4331 (1990).
108. C. J. Harlan, R. A. Jones, S. U. Koschmieder and C. M. Nunn, *Polyhedron*, **9**, 669 (1990).
109. M. R. Churchill, Yong-Ji Li, K. H. Theopold and R. R. Schrock, *Inorg. Chem.*, **23**, 4472 (1984).
110. Z. Dori and R. F. Ziolo, *Chem. Rev.*, **73**, 247 (1973).
111. P. Leoni, M. Pasquali, D. Braga and P. Sabatino, *J. Chem. Soc., Dalton Trans.*, 959 (1989).
112. F. A. Mautner and M. A. S. Goher, *Cryst. Res. Technol.*, **25**, 1271 (1990).
113. W. Rigby, P. M. Bailey, J. A. McCleverty and P. M. Maitlis, *J. Chem. Soc., Dalton Trans.*, 371 (1979).
114. I. Bkouche-Waksman, S. Sikorav and O. Kahn, *J. Cryst. Spectrosc.*, **13**, 303 (1983).
115. P. Chaudhuri, K. Oder, K. Weighardt, B. Nuber and J. Weiss, *Inorg. Chem.*, **25**, 2818 (1986).
116. M.-L. Boillot, O. Kahn, C. J. O'Connor, J. Gouteron, S. Jeannin and Y. Jeannin, *J. Chem. Soc., Chem. Commun.*, 178 (1985).
117. K. Matsumoto, S. Ooi, K. Nakatsuka, W. Mori, S. Suzuki, A. Nakahara and Y. Nakao, *J. Chem. Soc., Dalton Trans.*, 2095 (1985).
118. J. Comarmond, P. Plumeré, J.-M. Lehn, Y. Agnus, R. Louis, R. Weiss, O. Kahn and I. Morgenstern-Badarau, *J. Am. Chem. Soc.*, **104**, 6330 (1982).
119. Y. Agnus, R. Louis, J.-P. Grisselbrecht and R. Weiss, *J. Am. Chem. Soc.*, **106**, 93 (1984).
120. O. Kahn, T. Mallah, J. Gouteron, S. Jeannin and Y. Jaeannin, *J. Chem. Soc., Dalton Trans.*, 1117 (1989).
121. A. Christofides, N. G. Connelly, H. J. Lawson, A. C. Loyns, A. G. Orpen, M. O. Simmonds and G. H. Worth, *J. Chem. Soc., Dalton Trans.*, 1595 (1991).
122. D. F. Mullica, D. B. Tippin and E. L. Sappenfield, *Inorg. Chim. Acta*, **174**, 129 (1990).
123. D. Lentz, *J. Organomet. Chem.*, **381**, 205 (1990).
124. A. M. Martins, M. J. Calhorda, C. C. Romão, C. Völkl, P. Kiprof and A. C. Filippou, *J. Organomet. Chem.*, **423**, 367 (1992).
125. A. D. Burrows, H. Fleischer and D. M. P. Mingos, *J. Organomet. Chem.*, **433**, 311 (1992).
126. Y. Yamamoto and H. Yamazaki, *Inorg. Chem.*, **25**, 3327 (1986).
127. J. P. Cornelissen, J. H. van Diemen, L. R. Groeneveld, J. G. Haasnoot, A. L. Spek and J. Reedijk, *Inorg. Chem.*, **31**, 198 (1992).
128. R. Cortés, M. I. Arriortua, T. Rojo, J. L. Mesa, X. Solans and D. Beltran, *Acta Crystallogr.*, **C44**, 986 (1988).
129. G. C. Pimental and A. L. McClellan, *The Hydrogen Bond*, Freeman, San Francisco, 1960.
130. W. C. Hamilton and J. A. Ibers, *Hydrogen Bonding in Solids*, Benjamin, New York, 1968.
131. G. A. Jeffrey and W. Saenger, *Hydrogen Bonding in Biological Structures*, Springer-Verlag, Berlin, 1991.
132. R. Taylor and O. Kennard, *Acc. Chem. Res.*, **17**, 320 (1984).
133. R. Taylor and O. Kennard, *J. Am. Chem. Soc.*, **104**, 5063 (1982).
134. G. R. Desiraju, *Crystal Engineering: The Design of Organic Solids*, Elsevier, Amsterdam, 1989.
135. R. Taylor, A. Mullaley and G. W. Mullier, *Pestic. Sci.*, **29**, 197 (1990).
136. A. C. Hopkinson, M. H. Lein, K. Yates, P. G. Mezey and I. Csizmadia, *J. Chem. Phys.*, **67**, 517 (1977).
137. M. J. Frisch, J. A. Pople and J. E. Del Bene, *J. Chem. Phys.*, **78**, 4063 (1983).
138. K. I. Peterson and W. Klemperer, *J. Chem. Phys.*, **81**, 3842 (1984).

139. G. R. Desiraju and B. N. Murty, *Chem. Phys. Lett.*, **139**, 360 (1987).
140. J. D. Dunitz, A. Gehrer and D. Britton, *Acta Crystallogr.*, **A28**, 1899 (1972).
141. M. A. Viswamitra, R. Radhakrishnan, J. Bandekar and G. R. Desiraju, *J. Am. Chem. Soc.*, **115**, 4868 (1993).
142. W. G. Read and W. H. Flygare, *J. Chem. Phys.*, **76**, 2238 (1982).
143. D. G. Pritchard, R. N. Nandi and J. S. Muenter, *J. Chem. Phys.*, **89**, 115 (1988).
144. D. Britton, in *Perspectives in Structural Chemistry Vol. 1* (Eds. J. D. Dunitz and J. A. Ibers), Wiley, New York, 1967, pp. 190–171.
145. O. Hassell, *Mol. Phys.*, **1**, 241 (1958).
146. G. R. Desiraju and R. L. Harlow, *J. Am. Chem. Soc.*, **111**, 6757 (1989).
147. D. S. Reddy, K. Panneerselvam, T. Pilati and G. R. Desiraju, *J. Chem. Soc., Chem. Commun.*, 661 (1993).
148. J. B. Torrance, *Acc. Chem. Res.*, **12**, 79 (1979).
149. J. M. Williams, M. A. Beno, H. H. Wang, P. C. W. Leung, T. J. Emge, U. Geiser and K. D. Carlson, *Acc. Chem. Res.*, **18**, 261 (1985).
150. D. J. Sandman and G. P. Ceasar, *Isr. J. Chem.*, **27**, 293 (1986).
151. W. J. Dulmage and W. N. Lipscomb, *Acta Crystallogr.*, **4**, 330 (1951).
152. J. A. R. P. Sarma, M. S. K. Dhurjati, K. Bhanuprakash and K. Ravikumar, *J. Chem. Soc., Chem. Commun.*, 440 (1993).
153. D. S. Reddy, B. S. Goud, K. Panneerselvam and G. R. Desiraju, *J. Chem. Soc., Chem. Commun.*, 661 (1993).

CHAPTER **3**

Photoelectron spectra of systems with triple bonds between main group atoms—recent studies

ROLF GLEITER

Organisch-Chemisches Institut der Universität Heidelberg, Im Neuenheimer Feld 270, D-69120 Heidelberg, Germany

and

WOLFGANG SCHÄFER

Boehringer Mannheim GmbH, Sandhofer Str. 116, D-68305 Mannheim, Germany

Supplement C2: The chemistry of triple-bonded functional groups
Edited by S. Patai   © 1994 John Wiley & Sons Ltd

I. INTRODUCTION AND SCOPE

Photoelectron (PE) spectroscopy using a He(I) or He(II) light source has been used quite frequently to help to elucidate the electronic structure of molecules. The principles of this type of spectroscopy have been discussed extensively in the literature[1-5], so we will provide only a very brief introduction.

In a conventional ultraviolet PE-spectrometer a neutral molecule M is irradiated in the gas phase with monochromatic light, leading to a radical cation M^+ and a free electron with kinetic energy E_{kin}:

$$M + h\nu \rightarrow M^{+\cdot} + e \qquad (1)$$

Energy conservation requires that

$$E(M) + h\nu = E(M^{+\cdot}) + E_{kin}(e) \qquad (2)$$

or

$$E(M^{+\cdot}) - E(M) = IE = h\nu - E_{kin}(e) \qquad (3)$$

The left side of equation 3 is, by definition, identical to the ionization energy (IE) of the molecule M. Since the radiation energy $h\nu$ is known, equation 3 allows determination of the ionization energies by measuring the kinetic energies, $E_{kin}(e)$, of the emitted electrons.

Experimentally, a sample of the compound is irradiated in the gas phase by an ultraviolet light source. The emitted electrons are sorted by means of electric fields according to their kinetic energy and counted. The photoelectron spectrum is then obtained by plotting this count rate (intensity of the photoelectric current) against the energy difference $h\nu - E_{kin}(e)$ of equation 3. This is depicted schematically in Figure 1.

From the definition of the aforementioned energy difference it is clear that only ionization events up to the energy of the radiation source ($h\nu$) can be obtained. In the case of organic compounds, usually He(Iα) radiation of 21.21 eV [sometimes He(IIα) radiation of 40.80 eV] is used since the energy range covered by these light sources allows investigation of energetically low-lying electronic states that are involved in chemical bonding. Especially, the π-systems including the molecules reported below containing triple bonds can easily be investigated by this method, since ionization events associated with these systems usually occur in the range of 7 to 12 eV.

Koopmans' Theorem. In contrast to many other spectroscopic methods, interpretation of photoelectron spectra requires the support of calculations. According the equation 3 it is in principle necessary to calculate the total energy of both the ground state and the various cationic states of a molecule. In most cases—for exceptions, see Section III—it is sufficient to calculate the electronic structure of the ground state of the molecule within the self-consistent-field (SCF) method.

The most important approximation of this procedure is to replace the exact interaction terms between the individual electrons by an average potential. The Schrödinger equation can then be replaced by a matrix equation

$$FC = C\varepsilon \qquad (4)$$

(in an orthogonalized basis; F is the Fock operator, C the set of canonical molecular orbitals and ε the orbital energies).

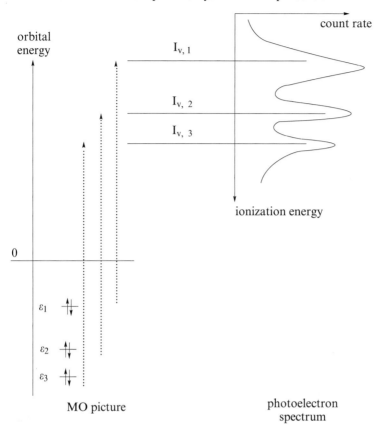

FIGURE 1. Relation between energy of light source (length of arrow), orbital energies ε_j and ionization energies, $I_{v,j}$, assuming Koopmans' theorem

Koopmans[6] has shown that the canonical orbital energy ε_j is equal to the negative value of the vertical ionization energy, $I_{v,j}$, resulting from the ejection of an electron from orbital j.

$$I_{v,j} = -\varepsilon_j \qquad (5)$$

Therefore, a single SCF calculation is sufficient. Many investigations have shown that, for the large majority of organic compounds, this approximation can be applied. The success of Koopmans' theorem is due to the fact that electron relaxation effects as well as correlation effects nearly cancel each other. Nevertheless, it must be emphasized that MOs cannot be observed; they are one-electron wave functions which merely serve as convenient model building blocks in the theoretical construction of molecular many-electron states[7]. It should also be kept in mind that this simple picture may occasionally break down, particularly for compounds with small electronic excitation energies[8,9].

Vibrational Structure. The time scale of the electron emission is extremely short compared to the motions of the nuclei. Therefore, the geometry of the molecule does not change during ionization (vertical ionization). In Figure 2 we distinguish two cases:

R. Gleiter and W. Schäfer

(a) If the electron is ejected from a nonbonding or a weakly bonding orbital, the potential surfaces of M and M^+ should be similar. According to the Franck–Condon principle, the transition to the vibrational ground state of the radical cation should have the highest intensity (case 1 in Figure 2). The PE spectrum shows peaks of the form (a).

(b) If an electron is ejected from a strongly binding orbital, transitions to higher vibrational levels become more probable (case 2) which leads to peaks of type (b).

The onset of the peaks in the PE spectrum is usually referred to as adiabatic ionization energy, I_a.

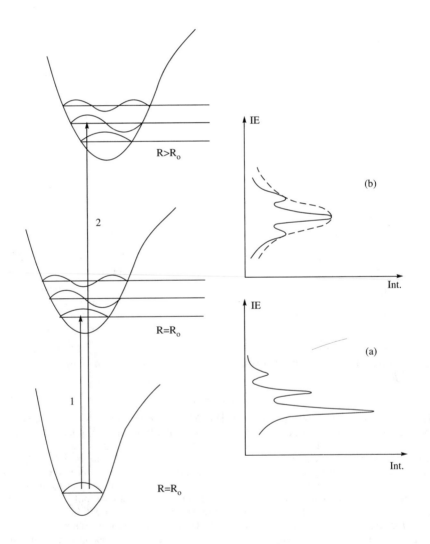

FIGURE 2. Qualitative explanation of band shapes on the basis of the Franck–Condon principle

In many PE spectra, especially in compounds with triple bonds, such a vibrational fine structure can be observed. The observed energy differences between the vibrational states of the radical cation can be compared with IR frequencies of the ground state and may be helpful in the assignment of the PE spectrum. In many cases vibrational fine structures of 0.25 eV can be observed. This value corresponds to the characteristic triple-bond IR frequencies around 2000 cm^{-1}. As an example we show the first PE bands of tetraacetylene (octatetrayne)[10] (figure 3). All four bands of tetraacetylene show vibrational fine structure with the following spacing: v (cm^{-1}): (1) 2100, (2) 1950, (3) 1960, (4) $\approx$ 400.

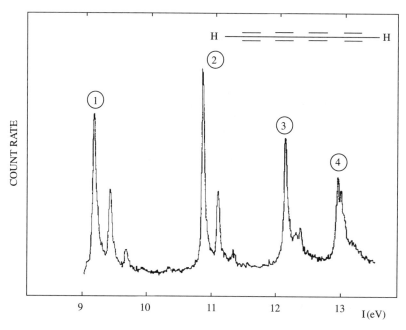

FIGURE 3. PE spectrum of octatetrayne[10]

Scope. The scope of this review covers recent results of the PE spectra of molecules containing triple bonds between main group elements. We will not consider work on the emission spectra and refer the reader to reviews by Maier [11a]. Our citation of the work on electron transmission spectroscopy is incomplete and will be mentioned only in passing [11b]. We will concentrate on work published after 1977 for CC triple bonds and 1983 for CN triple bonds and refer to the reviews by Heilbronner and coworkers[12] and by Stafast and Bock[13], respectively.

II. PHOTOELECTRON SPECTRA OF ACETYLENES

A. Monoacetylenes

The photoelectron spectra of acetylene (1) and its dideutero derivative were first reported by Baker and Turner[14]. A vibrational fine structure of 1830 cm^{-1} is found for 1$^+$. The first ionization energy at 11.4 eV is assigned to the ejection of electrons from the $e_{2u}(\pi)$ levels. Later on many derivatives, such as 2 and 3, were investigated[15-19]. As anticipated, the perturbation by the alkyl group(s) shifts the first ionization energy towards lower energy. A PE study on 11 monosubstituted and 25 disubstituted aliphatic alkynes revealed a correlation between the first ionization energy with Taft's polar σ^* constant[15]. As a consequence of the symmetry of these molecules, the π-orbitals associated with these ionizations remain degenerate.

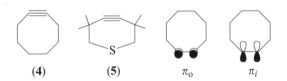

$$H\!\!=\!\!\!=\!\!H \qquad R\!\!=\!\!\!=\!\!H \qquad R\!\!=\!\!\!=\!\!R'$$

$$\textbf{(1)} \qquad\qquad \textbf{(2)} \qquad\qquad \textbf{(3)}$$

$$R, R' = \text{alkyl groups, SiMe}_3, \text{GeMe}_3, \quad \text{CH}_3$$

In strongly bent alkynes, such as cyclooctyne (**4**) and 3,3,6,6-tetramethyl-1-thiacycloheptyne (**5**), one expects a removal of the degeneracy of the π-MOs, since the interactions of the 'in plane' MO π_i and the 'out of plane' MO π_o with the σ-framework become different. This expectation is observed in the PE spectrum of **5**[20] where π_i is lowered by 0.31 eV as compared to π_o. In **4** this energy splitting is much less pronounced (0.1 eV)[21]. However, these effects show up much stronger in electron transmission (ET) spectroscopy[22].

$$\textbf{(4)} \qquad\qquad \textbf{(5)} \qquad\qquad \pi_o \qquad\qquad \pi_i$$

The PE spectra of mono- and dihaloacetylenes **6–9** have been investigated in great detail by Heilbronner's group[12, 23]. These molecules are ideal model compounds for PE studies

$$H\!\!=\!\!\!=\!\!X \qquad X\!\!=\!\!\!=\!\!Y \qquad X, Y = F, Cl, Br, I$$

$$\textbf{(6)} \qquad\qquad \textbf{(7)} \qquad\qquad \textbf{(a) (b) (c) (d)}$$

$$R\!\!=\!\!\!=\!\!X \qquad X = F, Cl, Br, I$$

$$R = Me \quad R = SiMe_3 \qquad \textbf{(a) (b) (c) (d)}$$

$$\textbf{(8)} \qquad \textbf{(9)}$$

for several reasons: They have a rigid molecular skeleton possessing high-frequency normal modes, mainly of the $C\equiv C$ stretching type. The $^2\pi_\Omega$ states of the radical cations have the same configuration but belong to different total angular momentum numbers $\Omega = 3/2$ and 1/2. Thus the two $^2\pi_\Omega$ states are split by spin–orbit coupling(SOC), Δ_{SOC} of the $^2\pi_{3/2}$ state of a given configuration being more stable than that of the $^2\pi_{1/2}$ state. The energy split due to SOC is given by

$$\Delta_{SOC,j} = E_j(^2\pi_{3/2}) - E_j(^2\pi_{1/2}) = I_{v,j,3/2} - I_{v,j,1/2} \qquad (6)$$

This split obeys simple sum rules and therefore provides additional information concerning the electronic structure of a particular state, especially for heavy atoms such as Br and I[24, 30, 32]. We will return to these features in the next section on polyacetylenes.

For the haloacetylenes as well as for the dihaloacetylenes, a correlation between the ionization energies and the valence state ionization energies of the free halogen atoms has been found[12, 23]. This is demonstrated for the symmetric dihaloacetylenes in Figure 4. In the case of fluoro compounds such as **6a–9a**, one finds that the ionization energies of the π-bands are very similar to the parent compounds (X = H)[23]. However, if one or two H

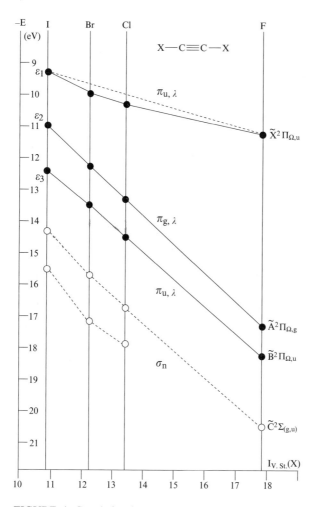

FIGURE 4. Correlation between the valence state ionization energies of the free halogen atoms $[I_{V.\ St.}(X)]$ and the ionization energies (ε) of symmetric dihalocetylenes X—C≡C—X with X = F, Cl, Br and I[12, 24]

atoms are replaced by CF_3, a substantial shift towards higher energy is observed. Typical examples are **10** and **11**.

$$F_3C \!\!=\!\!=\!\! H \qquad F_3C \!\!=\!\!=\!\! X \qquad X = Cl, Br, I$$
$$\qquad\quad \textbf{(10)} \qquad\qquad\quad \textbf{(11)}$$

B. Polyacetylenes

The PE spectra of di-, tri-, tetra- and pentaalkynes have been studied extensively[1,10,12,14,23–29]. Since the linearity of these compounds leads to a separation of the σ-

$$R \!\!=\!\!=\!\!=\!\! R' \qquad\qquad R \!\!=\!\!=\!\!=\!\!=\!\! R$$
$$R, R' = H,\ Me,\ Et,\ n\text{-Pr} \qquad R = H,\ t\text{-Bu}$$
$$\textbf{(a) (b) (c) (d)} \qquad\qquad \textbf{(a) (d)}$$
$$\textbf{(12)} \qquad\qquad\qquad \textbf{(14)}$$

$$R \!\!=\!\!=\!\!=\!\! R \qquad t\text{-Bu} \!\!=\!\!=\!\!=\!\!=\!\!=\!\! t\text{-Bu}$$
$$R = H,\ t\text{-Bu}$$
$$\textbf{(13)} \qquad\qquad\qquad \textbf{(15)}$$

and π-orbitals, they are ideal cases to study the interaction between conjugated π-systems. The PE spectra of **12a–14a** consist of one, two, three and four well-separated π-bands, respectively, where the vibrational fine structure is well resolved. As an example the PE spectrum of **14a** is shown in Figure 3.

For the interpretation of the PE spectra MO theoretical calculations have been carried out[1,14,30,31] using different degrees of sophistication. However, as Heilbronner has pointed out[10,12,25], sophisticated models are not needed for the interpretation of the π-bands due to their simplicity. In Figure 5 a correlation diagram is shown for the π-band positions of acetylene, di-, tri- and tetraacetylene. The orbital energies in this diagram are derived by a simple model using linear combinations of semilocalized bond orbitals (LCBO)[10,12,25]. For the $^2\Sigma^+$ states (ejection of electrons from σ-orbitals) the situation is more complex.

$$H \!\!=\!\!=\!\! X \qquad\qquad X = F,\ Cl,\ Br,\ I$$
$$\textbf{(16)} \qquad\qquad\qquad \textbf{(a) (b) (c) (d)}$$

$$X \!\!=\!\!=\!\! X \qquad\qquad X = F,\ Cl,\ Br,\ I$$
$$\textbf{(17)} \qquad\qquad\qquad \textbf{(a) (b) (c) (d)}$$

$$I \!\!=\!\!=\!\!=\!\! I \qquad\qquad F_3C \!\!=\!\!=\!\! F$$
$$\textbf{(18d)} \qquad\qquad\qquad \textbf{(19)}$$

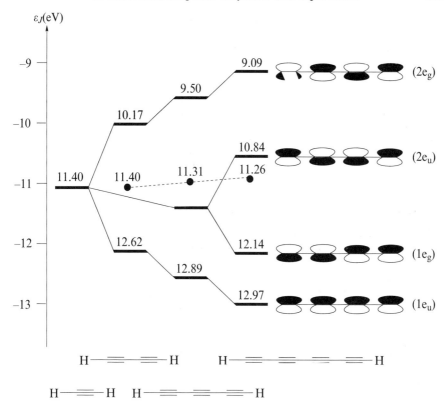

FIGURE 5. Left side: Correlation between the π-orbital energies of mono-, di-, tri- and tetraacetylene. Right side: Schematic drawing of the highest occupied π-MOs for octatetrayne[12, 25]

Of the various halogen derivatives of the polyacetylenes structures **16–19** have been investigated by PE spectroscopy. As already stated in the previous section, there is a correlation found between the 'observed' orbital energies of the symmetric dihalodiacetylenes **17** and the valence state ionization energies of the free halogen atoms[12].

Compounds **16–18** have been shown to be excellent models to aid in the study of spin–orbit coupling[12, 23, 29, 30, 32]. According to the simple first-order perturbation treatment, the size of the spin–orbit coupling split $\Delta_{SOC, j}$ is determined by the square of the AO coefficient $C^2_{X, j}$ in the π-orbital Ψ_j from which the electron is ejected, and by the size of the spin–orbit coupling coefficient $\xi(X)$ of atom X. Thus,

$$\Delta_{SOC, j} = C^2_{X, j} \, \xi(X) \tag{7}$$

for acetylenes of type $H(C{\equiv}C)_n X$ or $X(C{\equiv}C)_n X$, and

$$\Delta_{SOC, j} = C^2_{X, j} \, \xi(X) + C^2_{Y, j} \, \xi(Y) \tag{8}$$

for acetylenes of type $X(C{\equiv}C)_n Y$.

From $\Sigma_C{}^2 = 1$ in the simple LCBO approach, it follows that

$$\Sigma \, \Delta_{SOC,j} = \xi(X) \tag{9}$$

$$\Sigma \, \Delta_{SOC,j} = \xi(X) + \xi(Y) \tag{10}$$

That this is indeed the case has been demonstrated for several haloacetylenes[12, 23, 24]. A typical example is provided by diiodoacetylene (**7d**), diiododiacetylene (**17d**) and diiodotriacetylene (**18d**). The sum of the observed splits $\Sigma_j \Delta_{SOC,j}$ due to spin–orbit coupling is equal to 1.26 eV, which follows also from equation 10 for $\xi(I) = 0.63$ eV.

The PE spectra of difluorodiacetylene (**17a**) and perfluoro-1,3-pentadiyne (**19**) have also been recorded[23]. In the case of **17a** as well as **7a** the assignment is supported by a calculation using the many-body Green's function approach[23].

C. Diynes Separated by One or More Carbon Atoms

In this section we will treat the PE spectra of diynes in which the triple bonds are separated by one, two or more CR_2 groups. Examples are shown in **20–37**[33–37]. In the case

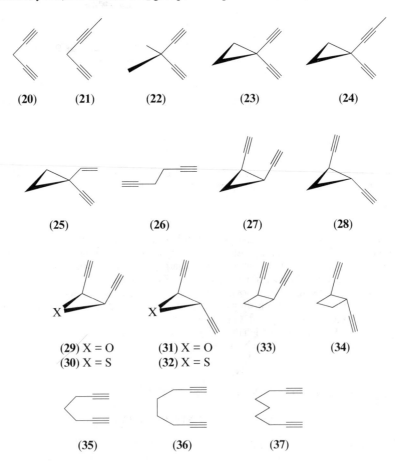

| (20) | (21) | (22) | (23) | (24) |

| (25) | (26) | (27) | (28) |

| (29) X = O | (31) X = O | (33) | (34) |
| (30) X = S | (32) X = S |

| (35) | (36) | (37) |

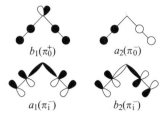

FIGURE 6. Schematic drawing of the highest occupied MOs of diethynylmethane (**20**)

where the triple bonds are separated by only one C-atom, such as **20–24**, we expect some homoconjugative interaction. A split of 0.6 eV is encountered for the first PE bands of **20–22**. In Figure 6 we show a schematic drawing of the four resulting π-MOs of **20**. It is seen that π_o^+ as well as π_i^- and π_i^+ interact considerably with the σ-system. In addition, a through-space interaction, especially for the π_i set, must be taken into account. In the case of **23** the antisymmetric Walsh e-orbital[38] will destabilize b_1 (π_o^+) considerably and thus the split of the π-bands is enlarged (1.2 eV) (see Figure 7)[39].

The analysis of the PE spectrum of 1,5-hexadiyne **26**[33, 35] is difficult because the first three bands (10.5 eV) overlap strongly. MO calculations on **26** suggest[33,35] that the antiplanar conformation is dominant in the gas phase. In figure 8 we have shown schematically the highest occupied MOs of **26** in the antiplanar conformation. One encounters a strong interaction either with the central C—C σ-bond [$9a_g(\pi_i^+)$] (through-bond interaction) or with the adjacent C—H σ-bond [$2a_u(\pi_o^+)$ and $2b_g(\pi_o^-)$] (hyperconjugation).

The PE spectra of **27** and **28** show[33] three peaks in the low-energy region (9–11 eV). This splitting is due to a strong interaction between Walsh-type MOs of the three-membered ring[40, 41] and the π-MOs of the ethynyl fragments. In the case of **29–32** a peak from the 2p(3p) lone pair of the heteroatom is added.

Similar to the PE spectra of **27** and **28** are those of **33** and **34** (see Figure 9). The splitting of the π-MOs is due to a σ/π interaction between the Walsh-type σ-MOs of the four-membered ring[40] and the triple bonds. For the PE spectra of **35** to **37** no splitting of the first peak around 10.1 eV is observed. The half-height width of the π-ionization decreases with the length of the alkyl chain, indicating that some interaction in the early members of the series does occur.

Special cases are **39** and **41** in which two diyne moieties are separated by a bicyclic framework. In both compounds the sharp and easily recognizable PE bands of the triple bonds can be used to investigate long-range interactions. In **39** the interaction between both acetylenic moieties was found to be small[41]. In **41**, however, a splitting of 0.5 eV has been detected[42] (see Figure 10). It has been traced back to a strong interaction between the valence orbitals of the bicyclo[1.1.1]pentane moiety and the π-MOs of the triple bonds[42,43].

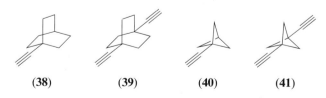

    (38)   **(39)**   **(40)**   **(41)**

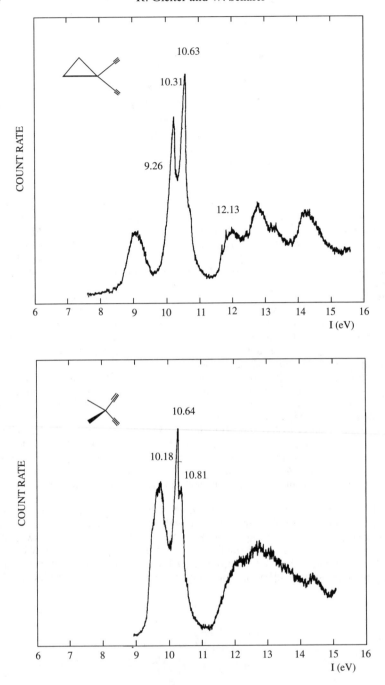

FIGURE 7. PE spectra of 3,3-dimethyl-1,4-pentadiyne and 1,1-diethynylcyclopropane[36b]

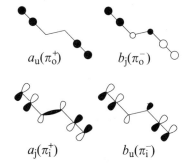

$$a_u(\pi_o^+) \qquad b_j(\pi_o^-)$$

$$a_j(\pi_i^+) \qquad b_u(\pi_i^-)$$

FIGURE 8. Schematic drawing of the highest occupied MOs of 1,5-hexadiyne

D. Cyclic Nonconjugated Diynes and Polyynes

1. Simple diynes

Recently, quite a number of nonconjugated diynes of the type **42** have been synthesized and investigated by PE spectroscopy[21, 44, 45]. The most interesting cases are those with n, $m = 2$ and n, $m = 3$, i.e. 1,5-cyclooctadiyne (**42a**)[21] and 1,6-cyclodecadiyne (**42b**)[44, 45], respectively. The first PE bands of these compounds have been interpreted in terms of the ionization from the symmetric and antisymmetric linear combinations of the 'in plane' (π_i) and 'out of plane' (π_o) orbitals. All four are shown in Figure 11.

$$(CH_2)_n \qquad (CH_2)_m$$

(42)

	n, m		n, m
(a)	2, 2	(f)	3, 4
(b)	3, 3	(g)	3, 5
(c)	2, 3	(h)	4, 4
(d)	2, 4	(i)	5, 5
(e)	2, 5		

For reasons of symmetry, the interaction of the two π_i combinations with the σ-framework depends on the number of CH_2 groups that connect the triple bonds. The π_o linear combinations are prone to interact with the adjacent C—H σ-orbitals (see also figure 8). In Figure 12 we have compared the first PE bands of **42a** with those of **42b**. Although in **42b** both triple bonds are more separated (3.0 Å) than in **42a** (2.65 Å), the bands are more spread apart in **42b**. This is due to the fact that π_i^- in **42a** does not interact with the σ-frame while in **42b** it shows a strong interaction. As a consequence of this different interaction the HOMOs of both species are separated considerably. As a corollary, a different reactivity of 1,6-cyclodecadiyne as compared to 1,5-cyclooctadiyne was predicted[46]. This led to the investigation of cyclic diynes towards electrophiles and transition metal fragments. As a result, new syntheses of superphanes resulted[47]. In Figure 13 we have compared the first PE

$$\begin{array}{c} \diagdown \qquad \diagup \\ {-}Si \equiv\!\!\equiv\!\!\equiv Si_{\diagdown} \\ | \qquad\qquad | \\ {}_{\diagdown}Si \!\!\equiv\!\!\equiv\!\!\equiv Si^{\diagup} \\ \diagup \qquad \diagdown \end{array}$$

(43)

R. Gleiter and W. Schäfer

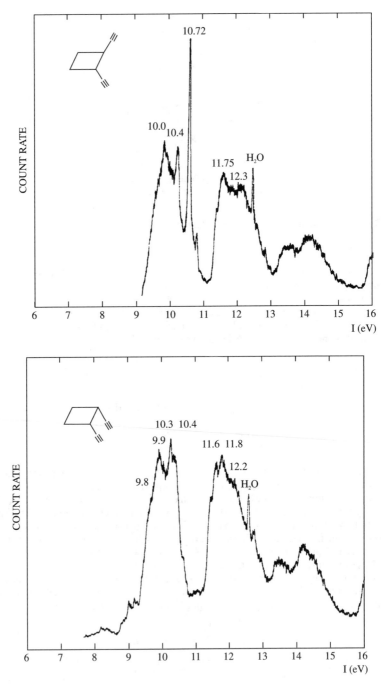

FIGURE 9. PE spectra of *cis*- and *trans*-1,2-diethynylcyclobutane[39]

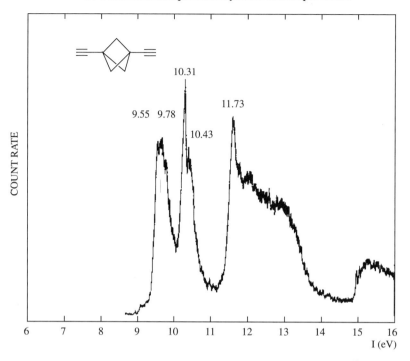

FIGURE 10. PE spectrum of 1,3-diethynylbicyclo[1.1.1]pentane[42]

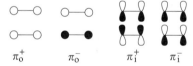

$\pi_o^+ \qquad \pi_o^- \qquad \pi_i^+ \qquad \pi_i^-$

FIGURE 11. Schematic drawing of the 'in-plane' (π_i) and 'out-of-plane' (π_o) π-MOs of two triple bonds

bands of **42b, 42c, 42f, 42g** and **42i**. A maximum splitting of the π-bands is encountered in **42b** as pointed out above (see also Figure 12).

A further example that the σ-bonds can influence the π-splitting drastically can be seen by comparing the first four bands of the PE spectra of 1,5-cyclooctadiyne (**42a**) with those of 3,3,4,4,7,7,8,8-octamethyl-3,4-7,8-tetrasilacycloocta-1,5-diyne (**43**)[48]. The PE spectrum of **43** is shown in Figure 14a.

A comparison with the first bands of the PE spectrum of **42a** (Figures 12 and 14b) reveals a strong destabilization of the band which has been assigned to $a_g(\pi_i^+)$.

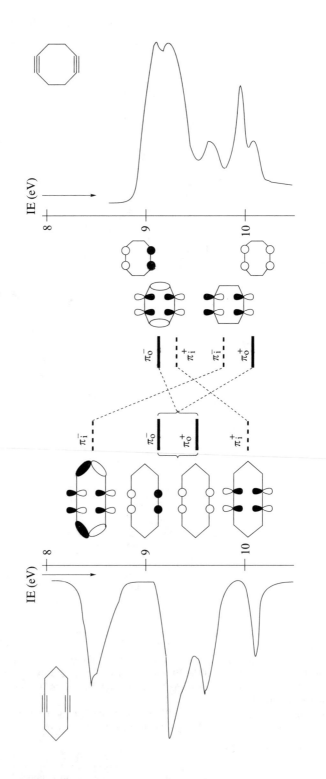

FIGURE 12. Comparison between the first PE bands of 1,6-cyclodecadiyne and 1,5-cyclooctadiyne[44]

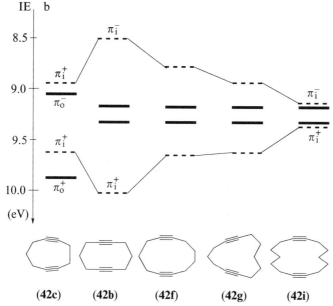

(42c) **(42b)** **(42f)** **(42g)** **(42i)**

FIGURE 13. Comparison of the first PE bands of **42b**, **42c**, **42f**, **42g** and **42i**[45]

2. Skipped enediynes and dienediynes

In skipped enediynes and dienediynes of type **44–54** a homoconjugative interaction between the double bond(s) and the triple bonds is possible. A PE study of **44–53** provides evidence for such an interaction[49]. This can best be seen by comparing the first PE bands

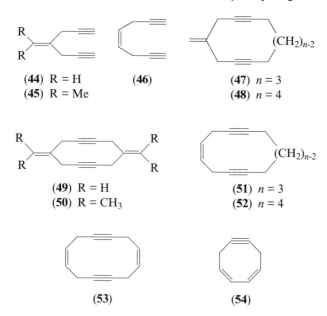

(44) R = H   **(46)**   **(47)** $n = 3$
(45) R = Me   **(48)** $n = 4$

(49) R = H   **(51)** $n = 3$
(50) R = CH$_3$ **(52)** $n = 4$

(53)   **(54)**

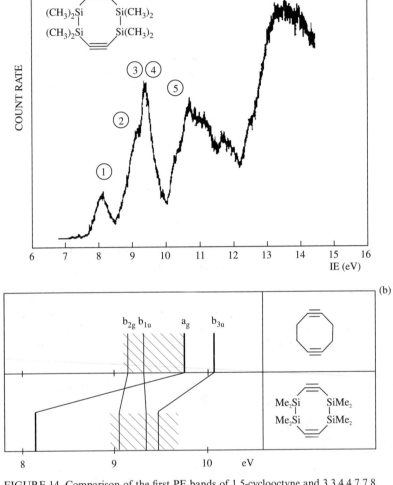

FIGURE 14. Comparison of the first PE bands of 1,5-cyclooctyne and 3,3,4,4,7,7,8, 8-octamethyl-3,4,7,8-tetrasilacycloocta-1,5-diyne[48]

of **42b** with those of **49** (Figure 15). The π_i^+ linear combination is clearly influenced by the exomethylene double bond.

An explanation for the homoconjugation between double and triple bonds in **47–49** as well as in **51–53** was given by assuming a very similar basis orbital energy of the π-MOs of the double and triple bonds. Confirmation of this explanation is found when looking at the PE spectrum of **50** (*cf* Figure 15). In this case the basis orbital energy of the isopropylidene double bonds are shifted by about 1 eV, due to the CH_3 groups, towards lower energy as compared to the exomethylene groups in **49**.

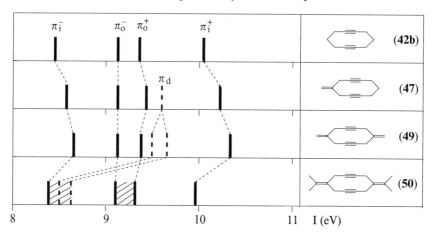

FIGURE 15. Correlation between the first PE bands of 1,6-cyclodecadiyne (**42b**) with **47**, **49** and **50** to demonstrate the homoconjugation between double and triple bonds[49]

Another example where no homoconjugation has been observed due to different basis orbital energies is **54**, whose PE spectrum has been reported[50].

3. Heterocyclic diynes

Recently, the PE spectra of a number of 1,6-dihetero-cyclodeca-3,8-diynes have been investigated. The splitting of the four π-bands π_i^+, π_i^-, π_o^+ and π_o^- has been found to occur in the range of 1.2–1.6 eV. In figure 16 we show a correlation of the first PE bands of **55**–**60**[51]. As anticipated, in the PE spectrum of dipropynyl sulfide (**61**) the signals for the triple bonds are hardly split[52].

X, Y	
(**55**)	O, O
(**56**)	O, S
(**57**)	S, S
(**58**)	CH$_2$, S
(**59**)	S, Se
(**60**)	Se, Se

(**61**) X = S
(**62**) X = O

In addition to these spectra with O, S and Se as heteroatoms, also some aza-compounds like **63**–**65** have been studied[53]. In the case of the cyclic species the splitting of the π-bands is similar as for **55**–**60**; for **65**, a similar broad peak for the triple bonds as for **61** has been reported[54].

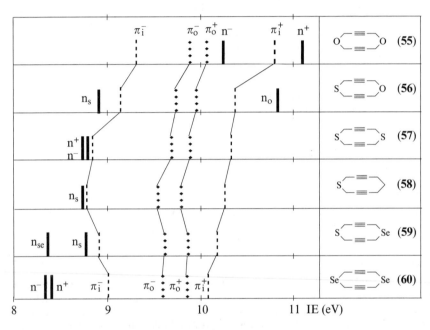

(63)   (64)   (65)

R = H   Me   Et   *i*-Pr   *t*-Bu

(a)   (b)   (c)   (d)   (e)

FIGURE 16. Correlation between the first PE bands of the heterosubstituted 1,6-cyclodecadiynes **55–60**[51]

4. Pericyclenes and related compounds

The PE spectra of **66–68** have been reported[48, 55]. A weak homoconjugative interaction in the cationic state of these polyynes has been detected. The electronic structure of 1,5,9-cyclododecatriyne (**69**) and the related silicon compound **70** have been discussed, based on PE data[48, 56]. In the PE spectrum of **69** there is one broad band between 9.2 and 9.9 eV, which encompasses all of the acetylenic π-ionization. Due to a strong interaction of in-plane σ-MOs with the energetically high-lying Si—Si σ-bond, the bands of **70** are spread out much more[48]. The comparison of **69** and **70** reminds one very much of that between **42a** and **43**. So far the PE spectra of only two cyclic tetraynes (**71, 72**)[57, 58] have been investigated. Both show two sharp bands between 8 and 9 eV, each of which has been assigned to two ionic states.

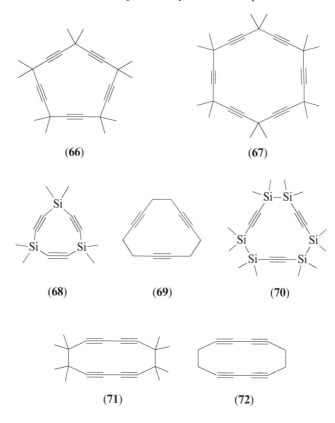

(66) (67)

(68) (69) (70)

(71) (72)

E. Conjugated Systems

In this section we will summarize the PE spectra of compounds in which one or several acetylenic unit(s) are directly connected to a π-system[33, 50, 54, 58–62] or to a heteroatom which provides an occupied p-type lone-pair orbital. Prominent examples are vinyl- (73)[58] and

(73) (74) (75) (76)

(77) (78) (79) (80)

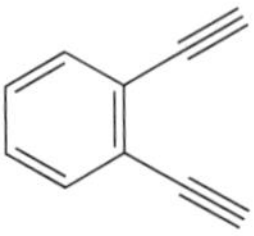

(81) 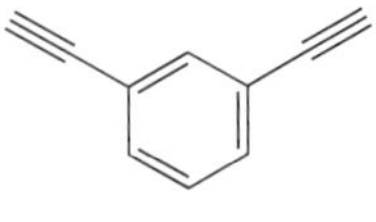   (82)    (83) 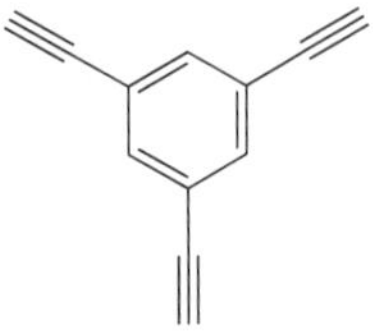   (84)

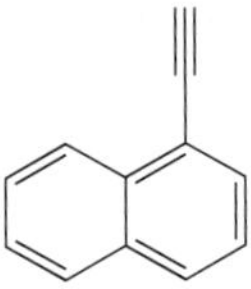

CH₃ CH₃

(85) 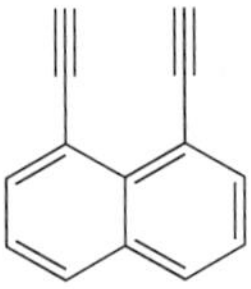   (86) 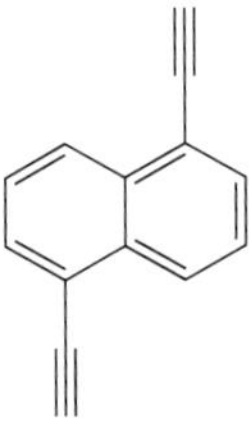   (87) 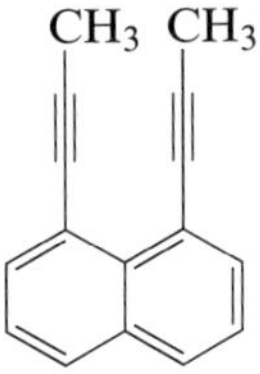   (88)

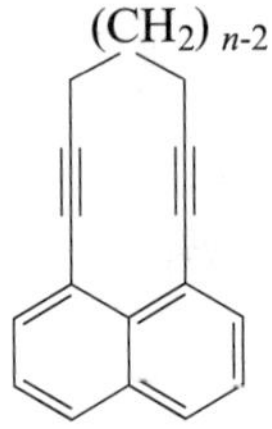

(CH₂)ₙ₋₂

(89)   (90)   (91)

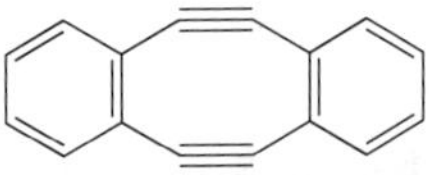

n = 3, 4, 5

(a) (b) (c)

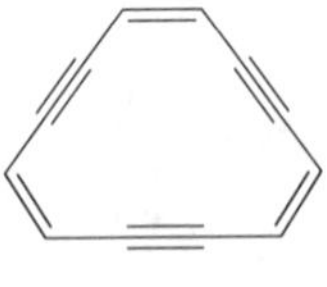

(92) 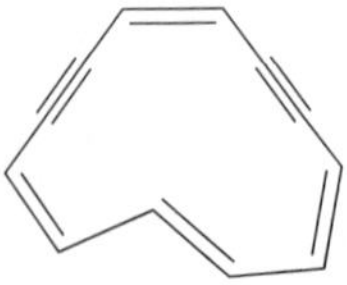   (93)    (94)    (95)

divinylacetylene (74)[33], cis-75 and trans-76 diethynylethylene[33], ethynylallene (77)[54], ethynylbenzene (78)[33] and some related structures such as 79–94. From our point of view the lesson learned from these investigations is twofold: (a) the π-system of the triple bond

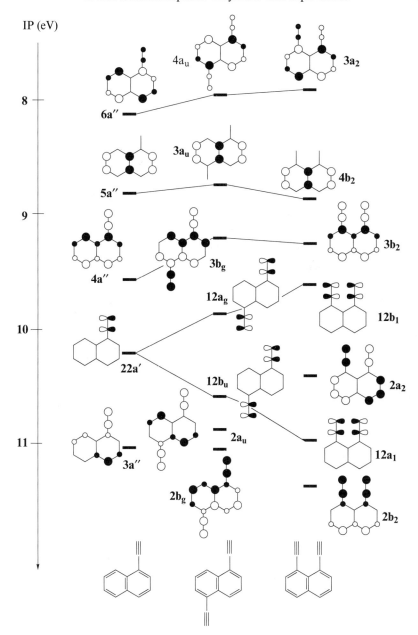

FIGURE 17. Comparison of the first PE bands of 1-ethynylnaphthalene, 1,5- and 1,8-diethynylnaphthalene[61]

interacts strongly with the adjacent π-system, and (b) in **86–89** there is considerable through-space interaction between the 'in-plane' π-MOs[61, 62].

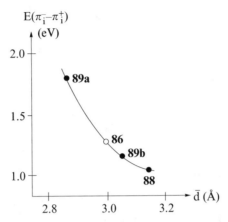

$E(\pi_i^- - \pi_i^+)$
(eV)

2.0

● **89a**

1.5

○ **86**
● **89b**

1.0

88

$\bar{d}$ (Å)

2.8   3.0   3.2

FIGURE 18. Plot of the energy difference E between π_i^- and π_i^+ against the distance in **86, 88, 89a** and **89b**. Full circles: X-ray structures; open circle: structure calculated by MINDO/3[62]

Using **73–76, 78, 81–84, 91** and **94** as models it has been shown[33, 60] that the first PE bands can be reproduced very well by a LCBO model. In the case of **73–76** and **91–94** the diagonal terms are: $A_d^o = \langle \pi_d/H/\pi_d \rangle = -10.1$ eV and $A_t^o = \langle \pi_t/H/\pi_t \rangle = -11.4$ eV, where π_d stands for a double bond and π_t for a triple bond. The higher value for A_t^o as compared to A_d^o indicates that the π-electrons of the triple bond are more localized than the double bond.

R—X≡

X = O, S, Se
(a) (b) (c)
(96)

R—X≡≡—X—R

R = CH$_3$
(97)

$\begin{array}{c} R \\ \diagdown \\ R \end{array} N ≡≡ N \begin{array}{c} R \\ \diagup \\ R \end{array}$

(98)

≡≡≡—N O

(99)

$≡≡≡ \overset{O}{\underset{Y}{\diagup}}$

Y = H, F, Cl, OH
(100)

$≡≡ \overset{S}{\underset{H}{\diagup}}$

(101)

To demonstrate the strong through-space interaction of the 'in-plane' π-MOs (π_i^+, π_i^-) of two triple bonds in peri-positions, we have compared the first PE bands of **85–87**[61] in Figure 17. In Figure 18 we have plotted the energy difference ΔE between π_i^- and π_i^+ of the PE spectra of **86**, **88**, **89a** and **89b** as a function of the average distance between the triple bonds. From this figure a strong dependence of the interaction between the triple bonds upon their distance can be derived[62].

Finally, we should mention that very recently the PE spectrum of *o*-benzyne (**95**) was recorded[63]. A simple Koopmans' theorem interpretation of the time-of-flight PE spectrum (10.49 eV) of **95** yields 9.03 ± 0.05 eV for the highest π-MO (out-of-plane) and 9.77 ± 0.03 eV for the in-plane MO.

In this section also substituted alkynes like **96–100** should be mentioned[64–68]. In the donor-substituted mono- and diynes the n-type lone pair(s) at the O, S, Se and N, respectively, interacts with the π-system of the triple bond[64–67]. The interaction of the triple bonds with the acceptor groups in **100** and **101** leads to the anticipated rise in the ionization energies of the π-bands[52, 68]. The compounds **1–101** whose PE spectra have been discussed so far are listed in Table 1 together with the corresponding references of the reported PE data.

TABLE 1. References to the He(I) PE spectra of molecules with carbon–carbon triple bonds (the term 'hetero systems' refers to molecules where the hetero atom is not involved in the triple bond; for these compounds see Table 2)

Structure	Name	References
	1. Open-chain monoacetylenes	
	1.1. Hydrocarbons	
1	acetylene	14
2, 3	mono- and dialkylacetylenes	15, 16, 17, 18
	1.2. Systems containing hetero atoms	
2, 3	mono- and disilyl-germyl acetylenes	17,28
6	monohaloacetylenes	32
7	dihaloacetylenes	23, 24
8	1-halo-propynes	12
9	trimethylsilylhaloacetylenes	18
10	trifluromethylacetylene	12
11	trifluromethyl-haloacetylenes	12
96, 97	methoxy-, methylthio-, methylselenoacetylenes	64, 67
98	bis(dialkylamino)acetylene	65
	2. Acyclic conjugated polyynes	
	2.1. Hydrocarbons	
12	butadiynes	16, 25, 26, 27, 28, 33
13	mono- and dialkyl-hexatriynes	16, 25, 27, 28, 33
14	mono- and dialkyl-octatetraynes	10, 27, 28
15	di-*t*-butyl-decapentayne	27, 28
99	*N*-butadiynyl-morpholine	67
	2.2. Systems containing hetero atoms	
16	monohalo-butadiynes	29
17	dihalo-butadiynes	23, 29
18d	diiodo-hexatriyne	29
19	perfluoro-1,3-pentadiyne	23

(*continued*)

TABLE 1. (*continued*)

Structure	Name	References

3. Open-chain diynes

3.1. Hydrocarbons

20	1,4-pentadiyne	16, 33
21	1,4-hexadiyne	16, 33
22	3,3-dimethyl-1,4-pentadiyne	36b
26	1,5-hexadiyne	16, 33, 35
35	1,6-heptadiyne	37
36	1,7-octadiyne	37
37	1,8-nonadiyne	37
44	4-methylene-1,6-heptadiyne	49
45	4-isopropylidene-1,6-heptadiyne	49
46	*cis*-oct-4-ene-1,7-diyne	33
75	*cis*-1,2-diethynylethylene	33
76	*trans*-1,2-diethynylethylene	33

3.2. Systems containing hetero atoms

61	dipropargylsulfide	52
62	dipropargylether	52
65	dipropargylamine	54

4. Cycloalkylacetylenes

4.1. Triple bond(s) attached to a monocyclic ring

23	1,1-diethynyl-cyclopropane	34
24	1-ethynyl-1-propyne-1-yl-cyclopropane	34
25	1-ethynyl-1-vinyl-cyclopropane	34
27	*cis*-1,2-diethynyl-cyclopropane	33
28	*trans*-1,2-diethynyl-cyclopropane	33
29	*cis*-2,3-diethynyl-oxirane	33
30	*cis*-2,3-diethynyl-thiirane	33
31	*trans*-2,3-diethynyl-oxirane	33
32	*trans*-2,3-diethynyl-thiirane	33
33	*cis*-1,2-diethynyl-cyclobutane	39
34	*trans*-1,2-diethnyl-cyclobutane	39

4.2 Triple bond(s) attached to a bicyclic system

38	1-ethynylbicyclo[2.2.2]octane	41
39	1,4-diethynylbicyclo[2.2.2]octane	41
40	1-ethynylbicyclo[1.1.1]pentane	42
41	1,3-diethynylbicyclo[1.1.1]pentane	42

5. Cyclic acetylenes

5.1. Monoacetylenes

4	cyclooctyne	21
5	3,3,6,6-tetramethyl-1-thia-cycloheptyne	20
54	1,3-cyclooctadiene-6-yne	50
90	1,5-cyclooctadiene-3-yne	50

5.2. Cyclic diynes

5. 2. 1. Hydrocarbons

42a	1,5-cyclooctadiyne	21
42	cycloalkadiynes	44, 45

(continued)

TABLE 1. (*continued*)

Structure	Name	References
47	4-methylene-1,6-cyclodecadiyne	49
48	4-methylene-1,6-cycloundecadiyne	49
49	4,9-dimethylene-1,6-cyclodecadiyne	49
50	4,9-diisopropylidene-1,6-cyclodecadiyne	49
51	4-cycloundecene-1,7-diyne	49
52	4-cyclododecene-1,7-diyne	49
53	1,7-cyclododecadiene-4,10-diyne	49
91	dibenzo-cycloocta-1,5-diene-diene-3,7-diyne	60
93	1,5,7,9-cyclododecatetraene-3,11-diyne (1,2,5,6-tetradehydro[12]annulene)	60

5.2.2. Systems containing hetero atoms

43	1,2,5,6-tetrasila-cycloocta-3,7-diyne	48
55	1,6-dioxa-cyclodeca-3,8-diyne	51
56	6-oxa-1-thia-cyclodeca-3,8-diyne	51
57	1,6-dithia-cyclodeca-3,8-diyne	21
58	1-thia-cyclodeca-3,8-diyne	51
59	1-selena-6-thia-cyclodeca-3,8-diyne	51
60	1,6-diselena-cyclodeca-3,8-diyne	51
63	6-aza-1-thia-cyclodeca-3,8-diynes	60
64	1,6-diazacyclodeca-3,8-diynes	60

5.3. Cyclic acetylenes with three or more triple bonds

66	permethyl-1,4,7,10,13-pentadecapentayne ([5]pericyclyne)	55
67	permethyl-1,4,7,10,13,16-cyclooctadecahexayne ([6]pericyclyne)	55
68	permethyl-1,4,7-trisila-2,5,8-cyclononatriyne	48
69	1,5,9-cyclododecatriyne	56
70	permethyl-1,2,6,9,l0-octasila- 3,7,10-cyclododecatriyne	48
71	permethyl-1,3,7, 9-cyclododecatetrayne	57
72	1,3,7,9-cyclododecatetrayne	58
92	1,5,9-cyclododecatriene-3,7,11-triyne	60
94	tribenzocyclododeca-1,5,9-triene-3,7,11-triyne	60

6. Acetylenes, conjugated to double bonds

73	vinylacetylene	16, 33
74	divinylacetylene	33
75	*cis*-1,2-diethynylethylene	33
76	*trans*-1,2-diethynylethylene	33
77	ethynylallene	54
90	1,5-cyclooctadiene-3-yne	50
92	1,5,9-cyclododecatriene-3,7,11-triyne	60
93	cyclododeca-1,3,5,9-tetraene-7,11-diyne	60
95	benzyne	63
100	2-butynal and related oxyalkynes	68
101	propynethial	52

7. Acetylenes, conjugated to aromatic systems

78	ethynylbenzene	33
79	phenyl-cyclopropyl-acetylene	59
80	1-methyl-1-(phenylethyny)cyclopropane	59
81	1,2-diethynylbenzene	33
82	1,3-diethynylbenzene	33

(*continued*)

TABLE 1. (*continued*)

Structure	Name	References
83	1,4-diethynylbenzene	33
84	1,3,5-triethynylbenzene	33
85	1-ethynylnaphthalene	61
86	1,8-diethynylnaphthalene	61
87	1,5-diethynylnaphthalene	61
88	1,8-dipropyn-1-ylnaphthalene	62
89	1,8-naphtho-cycloalkadiynes	62
91	dibenzo-cycloocta-1,5-diene-3,7-diyne	60
94	tribenzo-cyclododeca-1,5,9-triene-9,7,11- triyne	60

III. HETERO TRIPLE BONDS

A. PE Spectra of N_2, CO, CS, PN and P_2

The PE spectra of two-atomic molecules are of interest to spectroscopists and theoreticians. The PE spectra of N_2 (**102**) and CO (**103**) have been known since the early days of

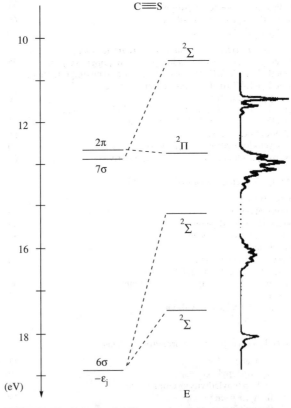

FIGURE 19. Calculated MO energies (left) and ionic states (center) for CS. The shape of the PE bands is indicated at the right

PE spectroscopy[1-5]. For the interpretation of the bands one has to go beyond the independent particle picture. Corrections are necessary because electron correlations and relaxation effects must be considered (see Section I). Calculations applied to interpret the PE spectra of N_2[9, 70] and CO[9, 69, 70] are based on the Green's function method[9, 70] or the configuration interaction technique[9, 69].

The He(I) PE spectrum of the short-lived species CS (**104**)[71], PN (**105**)[72] and P_2 (**106**)[73] have been reported by several groups[71-73]. Calculations on these species using the Green's function approach show that the inclusion of many-body corrections to Koopmans' theorem is necessary to obtain the correct ordering of states[74, 75]. To demonstrate this we have compared in Figure 19 the calculated energies of the highest occupied MOs of CS with the results of a treatment going beyond Koopmans' theorem[74] and the experiment. It is seen that in this molecule Koopmans' theorem fails to predict the sequence of the two lowest states. An even more serious breakdown of Koopmans' theorem is found for the third MO (6σ). The analysis[74] shows that the corresponding Koopmans' configuration $(6\sigma)^{-1}$ contributes to two cationic states with about equal weights and thus the orbital picture is no longer valid. There is no unique correspondence between PE peaks and MOs (Figure 19).

B. PE Spectra of NSF, NSCl, NSF₃, RCP, RNSi, XBS, RBO and RBNR′

The PE spectrum of NSF (**107**) shows a pronounced fine structure which has been analyzed[76]. The first bands of this spectrum have been assigned by semiempirical calculations under the assumption of the validity of Koopmans' theorem and by correlation with the PE bands of NSCl (**108**) and NSF₃ (**109**)[77].

More recently, the PE spectra of methylidenephosphanes (**110**) have been investigated, mainly by Kroto and coworkers[78-84]. Most of these species have been generated in the gas phase by pyrolysis. The PE studies show that the first band (9.5–11 eV) is due to the ionization from the C(π) MO located at the CP triple bond. Well separated from this band one finds the band which stems from the ionization of the lone pair on P [a_1(n)]. In Figure 20 we have compared the first PE bands of some RCP species with each other.

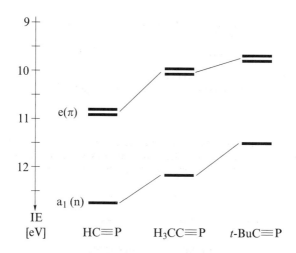

FIGURE 20. Correlation of the first PE bands of HCP, CH₃CP and t-C₄H₇CP

$$R \longrightarrow \equiv P \qquad R = H, Me, \textit{t}\text{-Bu}$$

$$(110)$$

$$Cl_2P \longrightarrow \equiv PCl_2 \longrightarrow P \equiv \equiv P$$

$$(111) \qquad\qquad (112)$$

Recently, evidence has accumulated that the reaction of **111** with a $MgCl_2/MgO/SiO_2$ catalyst produces minor amounts of **112**[85]. The phenylsilaisocyanide **114** has been generated by heating the azide **113**[86]. Its PE spectrum has been investigated[86].

$$(113) \qquad\qquad\qquad (114)$$

The PE spectra of XBS (X = H, F, Cl, Br) (**115**) have been studied mainly by Kroto and coworkers[78, 87]. The first ionization energy (10.4–11.1 eV) is assigned to the $e(\pi)$ orbitals of the triple bond between boron and sulfur. More recently, methylboroxide CH_3BO (**116**) has been generated in the gas phase. The PE data obtained[88] fit very well those of **115**.

$$X - B \equiv S \qquad X = H, F, Cl, Br$$

$$(115)$$

The iminoborane system is also isoelectronic to a CC triple bond.

$$R - N \equiv B - R \qquad R = Me, \textit{t}\text{-Bu}$$

$$(119) \quad (117)$$

$$N_3BMe_2 \xrightarrow{600\text{–}700\,°C} MeN \equiv BMe$$

$$(118) \qquad\qquad (119)$$

The PE spectrum of **117** has been recorded at room temperature[89a], that of **119** at 700 °C. It was generated in the gas phase from azidodimethylborane **118**[89b].

C. Nitriles and Isonitriles

The PE spectra of nitriles have been summarized in a very comprehensive review by Stafast and Bock[13]. We only mention here those papers which are not referred to in this reference. First we should mention papers which were concerned with the calculation of

$$CH_3CN \qquad CH_2(CN)_2 \qquad CH(CN)_3 \qquad C(CN)_4$$

$$(120) \qquad\quad (121) \qquad\qquad (122) \qquad\quad (123)$$

$$BrC(CN)_3 \qquad PhC(CN)_3$$

$$(124) \qquad\quad (125)$$

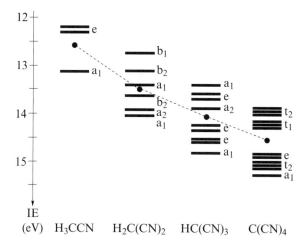

IE
(eV)   H₃CCN   H₂C(CN)₂   HC(CN)₃   C(CN)₄

FIGURE 21. Correlation between the π- and n-bands of CH₃CN, CH₂(CN)₂, CH(CN)₃ and C(CN)₄; the center of gravity is marked by a dot

ionization energies of nitriles[90–93], and the investigation of the high-resolution PE spectra of BrCN[94] and ICN[94]. Bock and Dammel have compared the PE spectra of the cyanomethanes **120–125**[95, 96]. In Figure 21 we have summarized the PE bands of **120–123**. It is seen that the energy difference between the bands originating from the lone pairs on nitrogen and those arising from the ionization from the π-MOs of the triple bond(s) remains essentially the same. The center of gravity of these bands, however, is shifted towards higher energy in going from **120** to **123**. This result is also confirmed by MO calculations[96].

Further small molecules with CN groups whose PE spectra were published lately are azocarbodinitrile (**126**)[96], N-cyanomethaneimine (**127**)[96, 97] and iminoacetonitrile (**128**)[98]. The latter two compounds are of interest as potential interstellar and prebiotic molecules.

NC–N=N–CN (126)   H₂C=N–CN (127)   H–N=C(CN)H (128)

(126)   (127)   (128)

The PE spectra of the dicyanobicycloalkadienes **129** and **130** as well as the dicyanobarrelenes **131** and **132** have been studied[99] to understand the ground- and excited-state properties of these species.

The PE spectra of alkyl- and arylisonitriles (**133**) were reported some time ago[100–103]. The first ionization energies are assigned to the ejection of electrons from the lone pair at carbon (ca 11 eV) and the π-electrons from the triple bond (ca 12 eV). This assignment is in line with the band shape criteria discussed in Section I. The second band in the PE spectrum of methylisocyanide [12.46 eV. e(π)][100] correlates well with the first band in the PE spectrum of acetonitrile [12.46 eV. e(π)][104]. As anticipated, the two lone-pair bands are about 2 eV apart (see figure 22).

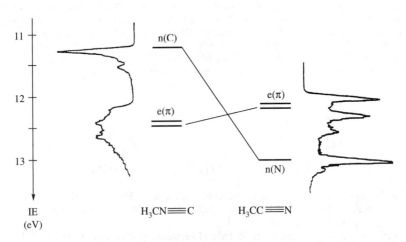

n = 1, 2, 3   n = 1, 2, 3
(129)   **(130)**

(131)   **(132)**

R—NC   R = Me, Et, *i*-Pr, *n*-Bu, *t*-Bu
(133) C_6H_{11}, Ph, CH_2Ph

F_3C—NC $\longrightarrow$ F_3C-CN
(134)   **(135)**

Figure 22. Comparison between the PE spectrum of acetonitrile and methylisocyanide

The thermal rearrangement of trifluoromethylisocyanide (**134**) to trifluoromethylcyanide (**135**) has been monitored by PE spectroscopy[105]. Another interesting isonitrile is acetyleneisocyanide (**136**)[106] which, like other simple linear molecules, is a candidate to be detected in outer space. Related to **136** is diisocyanogen (**137**) or its isomer isocyanogen

TABLE 2. References to the He(I) PE spectra of molecules where heteroatoms are involved in the triple bond

Structure	Name	References
	1. Heteroanalogs of acetylene	
102	N_2	1–5
103	CO	1–5
104	CS	71
105	PN	72
106	P_2	73
	2. Substituted systems	
107	NSF	76
108	NSCl	77
109	NSF_3	77
110	RCP	78–84
112	PCCP	85
114	RNSi	86
115	RBS	78, 87
116	RBO	88
117,119	RBNR	89
	3. Cyano compounds	
120–125	cyanomethanes	95, 96, 104
126	azocarbodinitrile	96
127	N-cyanomethaneimine	96, 97
128	iminoacetonitrile	98
129–132	dicyanobarrelenes and -bicycloalkadienes	99
135	trifluoroacetonitrile	105
	4. Isocyano compounds	
133	RNC (isonitriles)	100, 101, 102
134	CF_3 NC (trifluoromethylisonitrile)	105
136	HCCNC (ethinyl isocyanide)	106
137	CN—NC (diisocyanogen)	107
138	CNCN (isocyanogen)	107
	5. Others	
139	nitrile imines	109
140	fulminic acid	110

(138). A recent report on the PE spectrum of **137**[107] has been questioned[108] on the basis of calculations using Green's function theory. A comparison of the experimental spectrum with the results of a calculation for **137** and **138** clearly favors **138**.

$$HC \equiv C-N \equiv C \qquad C \equiv N-N \equiv C \qquad N \equiv C-N \equiv C$$
$$\textbf{(136)} \qquad\qquad \textbf{(137)} \qquad\qquad \textbf{(138)}$$

Finally, to conclude this chapter two investigations should be mentioned: nitrile imine **(139)**[109] and fulminic acid **(140)**[110]. Both species are of interest as being 1,3-dipoles. The compounds **102–140** whose PE spectra have been discussed in the preceding chapter are listed in Table 2 together with the corresponding references of the reported PE data.

$$R-C\equiv\overset{+}{N}-\overset{=}{N}-SiMe_3 \qquad H-C\equiv\overset{+}{N}-\overset{=}{\underline{O}}\,|$$

$$R = Me, Ph$$

$$(139) \qquad\qquad (140)$$

IV. CONCLUDING REMARKS

This review shows that the electronic structure of molecules with triple bonds between main group atoms can be elucidated by means of PE spectroscopy. The triple bond(s) in a molecule can be used as a probe to study conjugation, homoconjugation and the effect of π/σ interaction. PE spectroscopy not only helps to unravel the interactions in molecules, but can also contribute towards understanding their similarity and differences as well as their reactivities towards electrophiles.

V. ACKNOWLEDGMENTS

We are grateful to all coworkers and colleagues whose names appear in the references. We thank Prof. E. Heilbronner for permission to reproduce figures 3–5. We acknowledge continuous financial support from the Deutsche Forschungsgemeinschaft, the Fonds der Chemischen Industrie and BASF Aktiengesellschaft.

VI. REFERENCES

1. D. W. Turner, C. Baker, A. D. Baker and C. R. Brundle, *Molecular Photoelectron Spectroscopy*, Wiley, London, 1970.
2. C. R. Brundle and M. B. Robbin, in *Determination of Organic Structures by Physical Methods*, Vol.3 (Eds. F. C. Nachod and J. J. Zuckerman) Academic Press, New York, 1971.
3. J. H. D. Eland, *Photoelectron Spectroscopy*, Butterworth, London, 1974.
4. J. W. Rabalais, *Principles of Ultraviolet Spectroscopy*, Wiley, New York, 1977.
5. C. R. Brundle and A. D. Baker (Eds.), *Electron Spectroscopy, Theory, Techniques and Applications*, Academic Press, London, 1977.
6. T. Koopmans, *Physica* **1**, 104 (1934).
7. See, for example, K. Wittel and S. P. McGlynn, *Chem. Rev.*, **77**, 745 (1977).
8. J. Spanget-Larsen, *Croat. Chim. Acta*, **57**, 991(1984).
9. L. S. Cederbaum, W. Domcke, J. Schirmer and W. v. Niessen, *Adv. Chem. Phys.*, **65**, 115 (1986).
10. M. Allan, E. Heilbronner, E. Kloster-Jensen and J. P. Maier, *Chem. Phys. Lett.*, **41**, 228 (1976).
11. (a) J. P. Maier, *Acc. Chem. Res.*, **15**, 18 (1982); *Angew. Chem.*, **93**, 649 (1981) and references cited therein.
    (b) V. Balaji, L. Ng, K. D. Jordan, M. N. Paddon-Row, H. K. Patney, *J. Am. Chem. Soc.*, **109**, 6957 (1987), and references therein.
12. G. Bieri, E. Heilbronner, T. B. Jones, E. Kloster-Jensen and J. P. Maier, *Phys. Scr.*, **16**, 202 (1977) and references cited therein.
13. H. Stafast and H. Bock, in *The Chemistry of Functional Groups, Supplement C* (Eds. S. Patai and Z. Rappoport), Wiley, *J. Chem. Soc.*, Chichester, 1983, p. 137 and references cited therein.
14. C. Baker and D. W. Turner, *Chem. Commun.*, 797 (1967); *Proc. R. Soc. London Ser, A.* **308**, 19 (1968).
15. P. Carlier, J. E. Dubois, P. Masclet and G. Mouvier, *J. Electron Spectrosc. Relat. Phenom.*, **7**, 55 (1975).
16. G. Bieri, F. Burger, E. Heilbronner and J. P. Maier, *Helv. Chim. Acta*, **60**, 2213 (1977).
17. W. Ensslin, H. Bock and G. Becker, *J. Am. Chem. Soc.*, **96**, 2757 (1974); G. Bieri, F. Brogli, E. Heilbronner and E. Kloster-Jensen, *J. Electron Spectrosc. Relat. Phenom.*, **1**, 67 (1972); N.

Villem, *Tartu Riikliku,* **558**, 133 (1981).
18. I. A. Boyarskaya, V. N. Baidin and I. N. Domnin, *Zh. Org. Khim.,* **26**, 2303 (1990).
19. M. V. Andreocci, P. Bitchev, P. Carusi and A. Furlani, *J. Electron Spectrosc. Relat. Phenom.,* **16**, 25 (1979).
20. H. Schmidt, A. Schweig and A. Krebs, *Tetrahedron Lett.,* 1471 (1974).
21. G. Bieri, E. Heilbronner, E. Kloster-Jensen, A. Schmelzer and J. Wirz, *Helv. Chim. Acta,* **57**, 1265 (1974).
22. L. Ng, K. D. Jordan, A. Krebs and W. Rüger, *J. Am. Chem. Soc.,* **104**, 7414 (1982).
23. G. Bieri, E. Heilbronner, J.-P. Stadelmann, J. Vogt and W. v. Niessen, *J. Am. Chem. Soc.,* **99**, 6832 (1977).
24. E. Heilbronner, V. Hornung and E. Kloster-Jensen, *Helv. Chim. Acta,* **53**, 331 (1970).
25. F. Brogli, E. Heilbronner, V. Hornung and E. Kloster-Jensen, *Helv. Chim. Acta,* **56**, 2171 (1973).
26. E. Heilbronner, T. B. Jones and J. P. Maier, *Helv. Chim. Acta,* **60**, 1697 (1977).
27. E. Heilbronner, T. B. Jones, E. Kloster-Jensen and J. P. Maier, *Helv. Chim. Acta,* **61**, 2040 (1978).
28. E. Haselbach, U. Klemm, U. Buser, R. Gschwind, M. Jungen, E. Kloster-Jensen, J. P. Maier, O. Marthaler, H. Christen and P. Baertschi, *Helv. Chim. Acta,* **64**, 823 (1981).
29. E. Heilbronner, V. Hornung, J. P. Maier and E. Kloster-Jensen, *J. Am. Chem. Soc.,* **96**, 4252 (1974).
30. E. Heilbronner, K. A. Muszkat and J. Schäublin, *Helv. Chim. Acta,* **54**, 58 (1971).
31. M. J. S. Dewar, G. P. Ford and H. S. Rzepa, *Chem. Phys. Lett.,* **50**, 262 (1977).
32. H. J. Haink, E. Heilbronner, V. Hornung and E. Kloster-Jensen, *Helv. Chim. Acta,* **53**, 1073 (1970).
33. F. Brogli, E. Heilbronner, J. Wirz, E. Kloster-Jensen, R. G. Bergman, K. P. C. Vollhardt and A. J. Ashe III, *Helv. Chim. Acta,* **58**, 2620 (1975).
34. M. Eckert-Maksic, R. Gleiter, N. S. Zefirov, S. I. Kozhushkov and T. S. Kuznetsova, *Chem. Ber.,* **124**, 371 (1991).
35. P. Bischof, R. Gleiter, H. Hopf and F. T. Lenich, *J. Am. Chem. Soc.,* **97**, 5467 (1975).
36. (a) R. Gleiter, *Top. Curr. Chem.,* **86**, 197 (1979).
    (b) R. Gleiter and J. Spanget-Larsen, in *Advances in Strain in Organic Chemistry* (Ed. B. Halton), Vol. 2, JAI Press, 1992, p. 143 and references cited therein.
37. E. Nagy-Felsobuki, J. B. Peel and G. D. Willett, *Chem. Phys. Lett.,* **68**, 523 (1979).
38. A. D. Walsh, *Nature (London),* **159**, 167 (1947); *Trans. Faraday Soc.,* 179 (1949); E. Honegger, E. Heilbronner and A. Schmelzer, *Nouv. J. Chim.,* **6**, 519 (1982) and references cited therein.
39. R. Gleiter and H. Musso, unpublished results.
40. L. Salem and J. S. Wright, *J. Am. Chem. Soc.,* **91**, 5947 (1969); R. Hoffmann and R. B. Davidson, *J. Am. Chem. Soc.,* **93**, 5699 (1971).
41. E. Honegger, E. Heilbronner, N. Hess and H.-D. Martin, *Chem. Ber.,* **120**, 187 (1987).
42. R. Gleiter, K. –H. Pfeifer, G. Szeimies and U. Bunz, *Angew. Chem., Int. Ed. Engl.,* **29**, 413 (1990).
43. M. N. Paddon-Row and K. D. Jordan, *J. Am. Chem. Soc.,* **115**, 2952 (1993).
44. R. Gleiter, M. Karcher, R. Jahn and H. Irngartinger, *Chem. Ber.,* **121**, 735 (1988).
45. R. Gleiter, D. Kratz, W. Schäfer and V. Schehlmann, *J. Am. Chem. Soc.,* **113**, 9258 (1991).
46. R. Gleiter, M. Karcher, D. Kratz, S. Rittinger and V. Schehlmann, in *Organometallics in Organic Synthesis 2* (Eds. H. Werner and G. Erker), Springer-Verlag, Berlin, Heidelberg, 1989, p. 109.
47. R. Gleiter, *Angew. Chem.,* **104**, 29 (1992); *Angew. Chem., Int. Ed. Engl.,* **31**, 27 (1992); R. Gleiter and D. Kratz, *Acc. Chem. Res.,* **26**, 311 (1993) and references cited therein.
48. R. Gleiter, W. Schäfer and H. Sakurai, *J. Am. Chem. Soc.,* **107**, 3046 (1985).
49. R. Gleiter, R. Merger and H. Irngartinger, *J. Am. Chem. Soc.,* **114**, 8927 (1992).
50. H. Meier, P. König, T. Molz, R. Gleiter and W. Schäfer, *Chem. Ber.,* **118**, 210 (1985).
51. R. Gleiter, S. Rittinger and H. Irngartinger, *Chem. Ber.,* **124**, 365 (1991).
52. O. M. Nefedov, V. A. Korolev, L. Zanathy, B. Solouki and H. Bock, *Mendeleev Commun.,* 67 (1992).
53. R. Gleiter and J. Ritter, unpublished results.
54. O. I. Osman, D. McNaughton, R. J. Suffolk, J. D. Watts and H. W. Kroto, *J. Chem. Soc., Perkin Trans. 2,* 683 (1987).
55. L. T. Scott, *Pure Appl. Chem.,* **58**, 105 (1986).
56. K. N. Houk, R. W. Strozier, C. Santiago, R. W. Gandour and K. P. C. Vollhardt, *J. Am. Chem. Soc.,* **101**, 5183 (1979).

57. C. Santiago, K. N. Houk, G. J. DeCicco and L. T. Scott, *J. Am. Chem. Soc.*, **100**, 692 (1978).
58. R. Gleiter and J. Chavez, unpublished results
59. V. V. Plemenkov, O. Y. Butenko, M. V. Kleimenov, V. I. Kleimenov, A. V. Ignatchenko and I. G. Boselov, *Zh. Obshch. Khim.*, **61**, 1896 (1991).
60. J. Wirz, *Helv. Chim. Acta*, **59**, 1647 (1976).
61. R. Gleiter, W. Schäfer and M. Eckert-Maksic, *Chem. Ber.*, **114**, 2309 (1981).
62. R. Gleiter, W. Schäfer and A. flatow, *J. Org. Chem.*, **49**, 372 (1984).
63. X. Zhang and P. Chen, *J. Am. Chem. Soc.*, **114**, 3147 (1992).
64. H. Bock, W. Ried and U. Stein, *Chem. Ber.*, **114**, 673 (1981).
65. H. Bock, W. Kaim, M. Kira, L. Réné and H. –G. Viehe, *Z. Naturforsch.*, **39b**, 763 (1984).
66. A. A. Bredikhin, *Izv. Akad. Nauk SSSR, Ser. Khim.*, 1789 (1991).
67. P. Rademacher, M. Woydt, M. Barz and G. Himbert, *Chem. Ber.*, **120**, 1441 (1987).
68. D. Klapstein and R. T. O'Brien, *Can. J. Chem.*, **66**, 143 (1988).
69. P. S. Bagus and W.-K. Viinikka, *Phys. Rev. A*, **15**, 1486 (1977).
70. J. Schirmer, L. S. Cederbaum, W. Domcke and W. v. Niessen, *Chem. Phys.*, **26**, 149 (1977).
71. G. H. King, H. W. Kroto and R. J. Suffolk, *Chem. Phys.Lett.*, **13**, 457 (1972); N. Jonathan, A. Morris, M. Okuda, D. J. Smith and K. J. Ross, *Chem. Phys. Lett.*, **13**, 334 (1972); N. Jonathan, A. Morris, M. Okuda, K. J. Ross and D. J. Smith, *Discuss. Faraday Soc.*, **54**, 48 (1972); D. C. Frost, S. T. Lee and C. A. McDowell, *Chem. Phys. Lett.*, **17**, 153 (1972).
72. D. K. Bulgin, J. M. Dyke and A. Morris, *J. Chem. Soc., Faraday Trans. 2*, **73**, 983 (1977).
73. D. K. Bulgin, J. M. Dyke and A. Morris, *J. Chem. Soc., Faraday Trans. 2*, **72**, 2225 (1976); A. W. Potts, K. G. Glenn, and W. C. Price, *J. Chem. Soc., Faraday Discuss.* 65 (1972); H. Bock and H. Müller, *Inorg. Chem.*, **23**, 4365 (1984).
74. W. Domcke, L. S. Cederbaum, W. v. Niessen and W. P. Kraemer, *Chem. Phys. Lett.*, **43**, 258 (1976).
75. J. Schirmer, W. Domcke, L.S. Cederbaum and W. v. Niessen, *J. Phys. B: Atom. Molec. Phys.*, **11**, 1901 (1978).
76. (a) D. O. Cowan, R. Gleiter, O. Glemser, E. Heilbronner and J. Schäublin, *Helv. Chim. Acta*, **54**, 1559 (1971).
    (b) R. N. Dixon, G. Duxbury, G. R. Fleming and J. M. V. Hugo, cited in Reference 77.
77. D. O. Cowan, R. Gleiter, O. Glemser and E. Heilbronner, *Helv. Chim. Acta*, **55**, 2418 (1972).
78. H. W. Kroto, *Chem. Soc. Rev.*, **11**, 435 (1982) and references cited therein.
79. J. K. Tyler, *J. Chem. Phys.*, **40**, 1170 (1964); D. C. Frost, S. T. Lee and C. A. McDowell, *Chem. Phys. Lett.*, **23**, 472 (1973).
80. H. W. Kroto, J. F. Nixon and N. P. C. Simmons, *J. Mol. Spectrosc.*, **77**, 270 (1979); N. P. C. Westwood, H. W. Kroto, J. F. Nixon and N. P. C. Simmons, *J. Chem. Soc., Dalton Trans.*, 1405 (1979).
81. J.C.T.R. Burckett-St.Laurent, M.A. King, H.W. Kroto, J.F. Nixon and R.J. Suffolk, *J. Chem. Soc., Dalton Trans.*, 755 (1983).
82. B. Solouki, H. Bock, R. Appel, A. Westerhaus, G. Becker and G. Uhl, *Chem. Ber.*, **115**, 3747 (1982).
83. H. W. Kroto, J. F. Nixon and N. P. C. Simmons, *J. Mol. Spectrosc.*, **82**, 185 (1980); H. W. Kroto, J. F. Nixon, N. P. C. Simmons and N. P. C. Westwood, *J. Am. Chem. Soc.*, **100**, 446 (1978).
84. T. J. Dennis, S. Firth, H. W. Kroto, G. Y. Matti, C.-Y. Mok, R. J. Suffolk and D. R. M. Walton, *J. Chem. Soc., Faraday Trans.*, **87**, 917 (1991); S. Lacombe, G. Pfister-Guillouzo, J. C. Guillemin and J. M. Denis, *J. Chem. Soc., Chem. Commun.*, 403 (1991).
85. H. Bock and M. Bankmann, *Phosporus, Sulfur and Silicon*, **53**, 167 (1990).
86. H. Bock and R. Dammel, *Angew. Chem. Int. Ed. Engl.*, **24**, 111 (1985).
87. H. W. Kroto, R. J. Suffolk and N. P. C. Westwood, *Chem. Phys. Lett.*, **22**, 495 (1973); T. P. Fehlner and D. W. Turner, *J. Am. Chem. Soc.*, **95**, 7175 (1973).
88. H. Bock, L. S. Cederbaum, W. v. Niessen, P. Paetzold, P. Rosmus and B. Solouki, *Angew. Chem.*, **101**, 77 (1989).
89. (a) P. Paetzold, C. v. Plotho, G. Schmid, R. Boese, B. Schrader, D. Bougeard, U. Pfeiffer, R. Gleiter and W. Schäfer, *Chem. Ber.*, **117**, 1089 (1984).
    (b) P. Paetzold, E. Eleftheriadis, R. Minkwitz, V. Wölfel, R. Gleiter, P. Bischof and G. Friedrich, *Chem. Ber.*, **121**, 61 (1988).
90. W. v. Niessen and R. Cambi, *Chem. Phys.*, **103**, 11 (1986).

91. R. Cambi and W. v. Niessen, *J. Electron Spectrosc. Relat. Phenom.* **42**, 245 (1987).
92. J. Mirek and A. Buda, *Z. Naturforsch.*, **39a**, 386 (1984).
93. O. Walter, L. S. Cederbaum, B. Solouki, P. Rosmus and H. Bock, *J. Phys. Chem.*, **89**, 1384 (1985).
94. B. Kovač, *J. Phys. Chem.*, **91**, 4231 (1987).
95. R. Dammel and H. Bock, *J. Electron Spectrosc. Relat. Phenom.*, **41**, 311 (1986); H. Bock and R. Dammel, *Z. Naturforsch.*, **42b**, 315 (1987).
96. H. Bock, R. Dammel, P. Lorenčak and C. Wentrup, *Z. Naturforsch.*, **45b**, 59 (1990).
97. I. B'Shary, C. Guimon, M. Grimaud, G. Pfister-Guillouzo and D. Liotard, *Can. J. Chem.*, **66**, 2123 (1988).
98. R. A. Evans, S. M. Lacombe, M. J. Simon, G. Pfister-Guillouzo and C. Wentrup, *J. Phys. Chem.*, **96**, 4801 (1992).
99. T. Kobayashi, Z. Yoshida, Y. Asako, S. Miki and S. Kato, *J. Am. Chem. Soc.*, **109**, 5103 (1987); T. Kobayashi, S. Miki, Z. Yoshida, Y. Asako and C. Kajimoto, *J. Am. Chem. Soc.*, **110**, 5622 (1988); T. Kobayashi, H. Yamochi, K. Nakasuji and I. Murata, *J. Am. Chem. Soc.*, **110**, 6696 (1988).
100. R. F. Lake and H. W. Thompson, *Spectrochim. Acta*, **27A**, 783 (1971).
101. V. Y. Young and K. L Cheng, *J. Electron Spectrosc. Relat. Phenom.*, **9**, 317 (1976).
102. H. U. van Piggelen and C. Worrell, *Spectrosc. Lett.*, **10**, 79 (1977).
103. F. P. Colonna, G. Distefano, D. Jones, A. Modelli and V. Galasso, *Z. Naturforsch.*, **36a**, 385 (1981).
104. M. Gochel-Dupuis, J. Delwiche, M.-J. Hubin-Franskin and J. E. Collin, *Chem. Phys. Lett.*, **193**, 41 (1992).
105. H. Bock, R. Dammel and D. Lentz, *Inorg. Chem.*, **23**, 1535 (1984).
106. L. Zanathy, H. Bock, D. Lentz, D. Preugschat and P. Botschwina, *J. Chem. Soc., Chem. Commun.*, 403 (1992).
107. O. Grabandt, C. A. de Lange, R. Mooyman, T. van der Does and F. Bickelhaupt, *Chem. Phys. Lett.*, **155**, 221 (1989).
108. L. S. Cederbaum, F. Tarantelli, H.-G. Weikert, M. Scheller and H. Köppel, *Angew. Chem.*, **101**, 770 (1989); *Angew. Chem., Int. Ed. Engl.*, **28**, 761 (1989).
109. H. Bock, R. Dammel, S. Fischer and C. Wentrup, *Tetrahedron Lett.*, **28**, 617 (1987).
110. J. Bastide and J. P. Maier, *Chem. Phys.*, **12**, 177 (1976).

CHAPTER **4**

Analytical aspects of alkynes, nitriles, cyanates, thiocyanates, diazonium compounds and their isomeric functional groups

JACOB ZABICKY

The Institutes for Applied Research, Ben-Gurion University of the Negev, Beer-Sheva, Israel

Supplement C2: The chemistry of triple-bonded functional groups
Edited by S. Patai © 1994 John Wiley & Sons Ltd

I. INTRODUCTION

A. General Remarks

The triple bonded organic functional groups considered in the present chapter are $-CC-$, $-CN$, $-CNO$ (nitrile oxide), $-OCN$, $-SCN$ and $-N_2^+$. Some isomeric functions such as $-NC$, $-NCO$ and $-NCS$, will also be included, as they sometimes occur together with their isomers. The general aspects and specific methods for the analysis of alkynes[1], nitriles[2], diazonium compounds[3], cyanates and thiocyanates[4], and other related functional groups were adequately described in previous books of the series *The Chemistry of Functional Groups*. Consideration was made there of detection and determination by means of chemical reactions and application of characteristic spectroscopic properties of the groups.

The present chapter is supplementary to the reviews mentioned above, and will therefore avoid fundamental aspects of the analytical methods. It will deal briefly with some advancements of analytical methods that appeared in the last few years, it will consider new possible schemes for derivatising the functional groups both as organic compounds and as complexes with transition metals and will list compounds that have found application in industry, as many pertinent analytical methods have been developed for quality control, occupational safety and monitoring of environmental problems.

B. Anal1ytical Aspects

1. Synthesis of derivatives

It is worthwhile to recall briefly the requirements for a good derivatising scheme that were discussed elsewhere[5], because such schemes will be an important part of the present chapter.

    i. Functional selectivity. Only the functional group of interest reacts in a predetermined way, while the rest of the molecule remains untouched.

    ii. Analytical compatibility. The properties to be measured should be enhanced in the derivative.

    iii. Stability. Derivatives should be stable under ordinary laboratory conditions and those involved in the analytical process.

    iv. Ease of handling. The derivatizing process should be easy to perform within a reasonable time.

2. Complexes with transition metals

The electronic structure of the functional groups considered in this chapter enables the formation of complexes with transition metals. Several advantages of analytical interest could be achieved through complex formation:

    1. Selectivity for a specific functional group may be high, or it may be enhanced by applying the appropriate conditions of reaction.

    2. Stable crystalline complexes may be isolated for further characterization of the organic compound. This allows structural analysis, including determination of absolute configurations.

3. Stable dissolved complexes can be separated more easily from other components of mixtures by chromatographic methods.

4. Large gravimetric amplification factors can be attained on complex formation due to the high molecular weight of the metal surrounded by other ligands.

5. Yields of stable dissolved complexes may be quantitative, even if yields reported for the isolated crystals are usually less than 100%.

6. Complexes frequently are reactive intermediates from which pure organic molecules of analytical interest may be derived.

7. Very effective universally available instrumental methods are useful for the analysis of complexes. The methods reported for the examples of the present chapter include IR and UV-visible spectrophotometry, ^{1}H-NMR, ^{13}C-NMR and metal-NMR for dissolved complexes and X-ray crystallography of single crystals. Electrochemical methods, EPR and thermal analysis were also used occasionally.

Thousands of complexes of transition metals with organic compounds bearing the functions relevant to this chapter have been reported in the literature, however, only part of them have potential analytical interest. The complexes listed in this chapter appeared in works published during the last few years. These sources are intended to serve as leading references for various types of complexes. The selection was made applying the following criteria:

1. Complexes are stable at room temperature.

2. Crystals are obtained in reasonable yields—say over 30%, but usually much higher.

3. Complex formation takes place in a reasonable time—usually a few minutes to overnight at most.

4. Workup of the crystals is not very laborious and does not involve unusual operations.

Other important criteria from the analytical point of view, such ease of development of the most appropriate measuring methods, availability or ease of synthesis of the complexing agent, toxicity and costs, are left to the reader for evaluation, as the main objective of this chapter is to awaken interest in the analytical potential of complexes.

3. Elemental analysis of heteroatoms

All functional groups of the present chapter except for the alkynes, involve nitrogen and some of them also oxygen or sulphur. Determination of these heteroatoms may serve as means for detecting or determining the presence of certain compounds in samples. Elemental analysis of nitrogen[3,6], automatic elemental analysers[5] and element-specific detectors for gas chromatography[5] were discussed in other volumes of the *Functional Groups* series.

A brief recapitulation of elemental analysers is in place because of their importance in modern analytical laboratories and their relevance to our subject. The chemical processes on which these instruments are based are similar to the Pregl determination of C and H, by conversion to CO_2 and H_2O, Dumas' conversion of organic N to N_2, the bomb method for S, converting to SO_2 and SO_3, and Schultze's method for O, by converting into CO. The main difference between the various models on the market resides on the methods for separation and determination of these products. For example, in one simultaneous CHN analyser[7] the combustion gases are cleaned from other elements on a silver gauze, nitrogen oxides are converted to N_2, H_2O and CO_2 are adsorbed on silica gel columns, N_2 is determined with a thermal conductivity detector, followed by desorption and determination of H_2O and CO_2 with the same detector. In a simultaneous CHNS analyser[8], after catalytic conversion of nitrogen oxides to N_2 and SO_3 to SO_2, the combustion products are separated by chromatography and the elements determined with a thermal conductivity detector. Oxygen analysers are usually available as adaptation kits for CHN analysers. For example, in one model[7], after burning the sample in hydrogen, passing the gases over a hot carbon bed and removing impurities over silver wool and a soda lime bed, CO is measured with a non-dispersive IR detector.

The elemental analysis of the transition metal and other heteroatoms present in complex derivatives is beyond the scope of this book. A modern method of choice could be ICP spectrometry[9].

C. Occupational and Environmental Aspects of Chemicals

One may consider several stages during the lifetime of chemicals:

1. At the *industrial stage*, chemicals are being produced or used within systems where professional supervision and official regulation *can* be enforced. Such systems include manufacturing industries of all types, modern agricultural operations and medical services in clinics and hospitals. It is fortunate that most chemicals remain in this stage, and especially within the realm of the manufacturing industries; nevertheless, enough slips out of control, either by design or for lack of proper care, to make official agencies and the public anxious.

2. At the *transit stage*, chemicals are transported from one station to another for whatever purpose. This is usually carried out by trained personnel with dedicated vessels. Nevertheless, a small but significant volume of dangerous chemicals moves as general cargo. Accidents happen occasionally at this stage with deleterious effects.

3. At the *consumer stage*, chemicals are in the hands of the public at large, as a result of acquisition in the free market. This takes place with virtually no regulation. Frequently, the consumer has very scant knowledge about the nature of such materials and the effects of improper handling.

4. At the *expiration stage*, chemicals cease to fulfill their designed functions. In such cases they remain on site in the original form or a degraded one, until disposal. Chemicals have to be disposed of by methods adequate to each specific case, lest the products become harmful on site or cause environmental pollution.

Handling of chemicals involves occupational and environmental dangers at all stages of their lifetime. In industrially developed countries, various official agencies deal with regulation for the use of chemicals, to minimize such dangers. Proper development of a safety or environmental protection standard for each chemical must include specifications for its monitoring, including instructions for sampling, detection and determination. In the USA, a very large effort has been made in this field by the Environmental Protection Agency (EPA), National Institute for Occupational Safety and Health (NIOSH), Occupational Safety and Health Administration (OSHA), Food and Drug Administration (FDA) and others. Of special interest is the Registry of Toxic Effects of Chemical Substances (RTECS), which includes many industrial chemicals bearing the functional groups treated here. The specifications developed by these agencies for certain chemicals may be used as a basis for development of analytical methods for analogous substances. Since access to official documents of the USA may be rather involved in academic establishments outside that country, the widely available publications cited in References 10 and 11 may serve as a source for leading references. Compilations of occupational and environmental documents on chemicals began to appear in compact-disk format for personal computers, and it is expected that this trend will grow in the next few years.

II. ALKYNES

A. General

Mixtures of alkenes and alkynes can be analysed with detection limits in the nanogram range, by means of Fe(I) chemical ionization using tandem sector MS or GC-FTMS. The

TABLE 1. Acetylenic compounds of industrial application[a]

Compound and CAS registry number	NIOSH/OSHA RTECS[b]
Acetylene [74-86-2]	AO 9600000
Acetylenedicarboxamide [543-21-5]	AO 9900000
Acetylenedicarboxylic acid monopotassium salt [928-04-1][c]	AP 0700000
Adipic acid di-2-propynyl ester [6900-06-7]	AV 1750000
1,4-Bis(diethylamino)-2-butyne [105-18-0]	ES 1270000
Bis(dimethylthallium) acetylide	
1,1-Bis(4-fluorophenyl)-2-propynyl N-cycloheptylcarbamate [20929-99-1]	GU 3150000
1,1-Bis(4-fluorophenyl)-2-propynyl N-cyclooctylcarbamate [20930-00-1]	GX 96250000
Bis(triethyltin)acetylene [994-99-0]	
(E) (1RS,4RS)-Bornan-2-one O-2-propynyloxime [73886-28-2][c]	
Bromoacetylene [593-61-3]	
1-Bromocyclotrideca-2,12-dien-4,8,10-triyne	
1-Bromo-2-propyne [106-96-7]	UK 4375000
1-Buten-3-yne [689-97-4]	
1-Butyne [107-00-6]	ER 9553000
Butoxyacetylene [3329-56-4]	
t-Butylnitroacetylene	
2-(4-t-Butylphenoxy)cyclohexyl 2-propyn-1-yl sulphite [2312-35-8][c]	
Chloroactylene [593-63-5]	
3-Chlorocarbanilic acid 1-methyl-2-propynyl ester [1967-16-4][c]	FD 8575000
3-Chloropropynenitrile [2003-31-8]	
3-Chloro-1-propyne [624-65-7]	
Dibromoacetylene [624-61-3]	
Dichloroacetylene [7572-29-4]	AP 1080000
1,4-Dichloro-2-butyne [821-10-3]	ER 9600000
1,6-Dichloro-2,4-hexadiyne [16260-59-6]	
1-(Dichloromethyldimethylsilyl)-1-hexyn-3-ol [25898-71-9]	MR 0182000
Dicyanoacetylene [1071-98-3]	
1-Diethylamino-1-buten-3-yne [1809-53-6]	
1,4-Dipyrrolidinyl-2-butyne [51-73-0]	UY 0175000
Empentrin [54406-48-3][c]	
Ethoxyacetylene [927-80-0]	
1-Ethoxy-2-propyne [628-33-1]	
4-Ethynyl-1,2,2,6,6-pentamethyltetrahydropyridine hydrochloride [63905-91-9]	UT 8350000
Ethynyl vinyl selenide	
Fluoroacetylene [2713-09-9]	
2-Ethynylfuran [18649-64-4]	
6-Fluoro-1-hexyne [373-16-0]	MR 0180000
9-Fluoro-1-nonyne [463-19-4]	RB 3400000
8-Fluoro-1-octyne [408-15-1]	RI 2700000
Flupropadine [81613-59-4][c]	
1,6-Heptadiyne [2396-63-6]	MI 5600000
1,5-Hexadien-3-yne [821-08-9]	
4,5-Hexadien-2-yn-1-ol [2749-79-3]	
3-Hexyne-2,5-diol [3031-66-1]	MR 0175000
1-Hexyn-3-ol [105-31-7]	MR 0181000
1-Iodo-3-penten-1-yne	
1-Iodo-3-phenyl-2-propyne [73513-15-2]	
Lithium chloroacetylide	
Methoxyacetylene	
3-Methoxypropyne [627-41-8]	

(continued)

TABLE 1. (*continued*)

Compound and CAS registry number	NIOSH/OSHA RTECS[b]
2-Methyl-1-buten-3-yne [78-80-8]	EN 0960000
3-Methyl-3-buten-1-ynyltriethyllead	
2-Methyl-3-butyn-2-ol [115-19-5]	ES 0810000
2-Methyl-1,5-hexadiene-3-yne [820-54-2]	MM 3470000
3-Methyl-3-pentene-1-yn-3-ol [323-06-1]	SB 4950000
2-Nonene-4,6,8-triyn-1-al	
1,3-Pentadiyne [4911-55-1]	
1,3-Pentadiyn-1-ylcopper	
1,3-Pentadiyn-1-ylsilver	
1-Pentyne [627-19-0]	
1-Pentyn-3-one [7299-55-0]	SC 5250000
Phenylacetylene [536-74-3]	DA 0780000
Prallethrin [23031-36-9][c]	
Propargyl bromide [106-96-7]	UK 4375000
α-Propargylmercaptoisobutyramide [64037-68-9]	ES 4460000
Propyne [74-99-1]	UK 4250000
2-Propyn-1-ol [107-19-7]	UK 5075000
2 Propyn-1-thiol [27846-30-6]	
2-(Propynoylamino)fluorene [60550-90-5]	UD 9276000
2-Propynylamine [2450-71-7]	UK 5250000
1-(2-Propynyl)cyclohexyl carbamate [358-52-1]	FD 1150000
2-Propynyl ether [6921-27-3]	UK 5950000
2-(Propynylmercapto)propane [67465-99-0]	UE 4528000
2-Propynyl 3,7,11-trimethyl-2(*E*),4(*E*)-dodecadienoate [37882-31-8]	
2-Propynyl vinyl sulphide	
Silver acetylide [7659-31-6]	
Sodium ethoxyacetylide	
Sodium phenylacetylide [1004-22-4]	
Tetraethynylgermanium [4531-35-5]	
Tetraethynyltin [16413-88-0]	
Trichlorosilylacetylene [75-94-5]	VV 6125000
Triphenylstannyl propiolate [57410-20-2]	WH 5702000

[a] See also Reference 11.
[b] *Registry of Toxic Effects of Chemical Substances* of National Institute for Occupational Safety and Health/ Occupational Safety and Helth Administration.
[c] See also Reference 21.

source for Fe(I) is $Fe(CO)_5$. Alkynes with more than 12 carbon atoms are problematic because Fe(I) is too reactive, leading to consecutive reactions with big molecules, particularly in the low-pressure regime of the MS instruments[12].

Terminal alkynes can be determined by potentiometric titration with silver nitrate solution, in aqueous or non-aqueous media, using an ion-selective silver sulphide electrode[13].

The possible effects on the atmospheric ozone layer of volatile organic pollutants, inclusive of acetylene, were assessed by measuring the reaction rate with OH radical at room temperature, in the presence of oxygen; pollutants were classified as reactive or non-reactive according to US-EPA[14].

Acetylene and methane can be detected in remote locations using diode lasers coupled with optical fibres[15]. Environmental measurements of acetylene are possible based on the overtone transition at $\lambda \sim 789$ nm, using a stabilized AlGaAs diode laser, operating in an external optical cavity configuration. The detection limit is 0.2 ppm/km for acetylene[16].

Column densities of ethane and acetylene in the atmosphere were determined by ground measurements of the solar spectrum in the λ 8–10 μm range[17].

Adsorption of ethylene and acetylene on platinum in the presence of hydrogen was investigated using elastic recoil detection analysis (ERDA). Below 200 K no change was observed for acetylene, while at room temperature C_2H_3 radicals were formed[18]. The geometry of acetylene adsorbed on the Ni (100) face was determined by LEED (low energy electron diffraction) intensity analysis[19].

Very complex alkynes (1, 2) were isolated from the roots of *Ferula communis* and their structure was deduced from ^{1}H-NMR, ^{13}C-NMR 1D and 2D spectra (one bond and long range interactions)[20].

Table 1 lists acetylenic compounds of industrial interests.

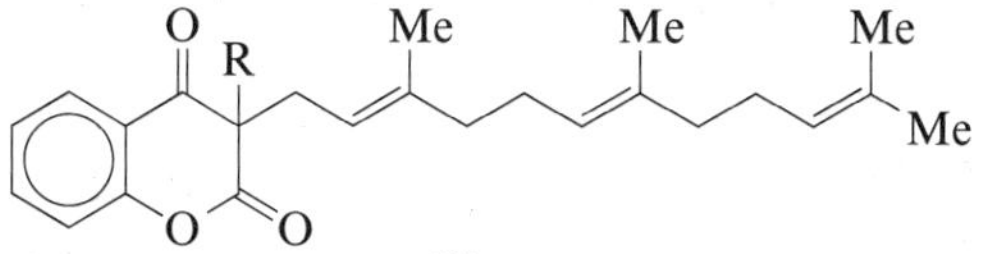

(1)

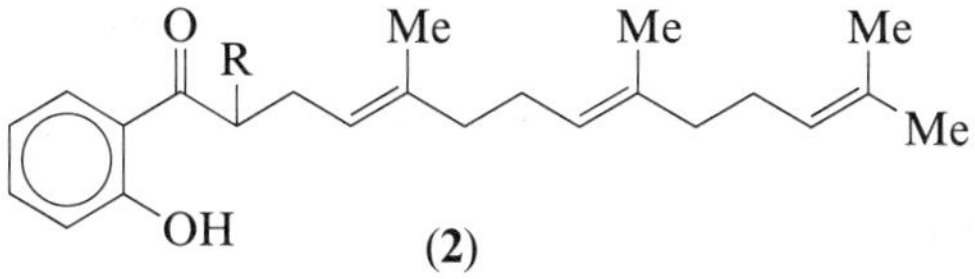

(2)

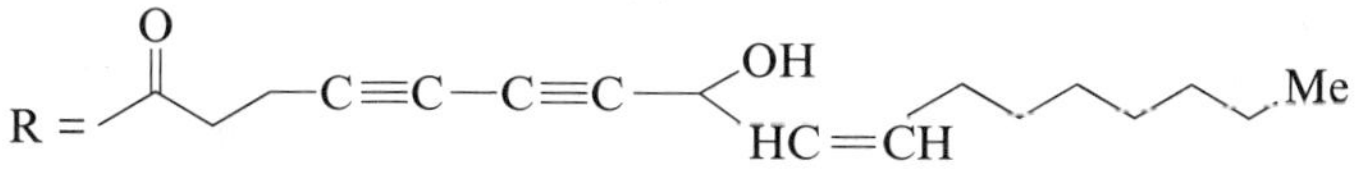

$$R =$$

B. Functional Derivatives

The position of the triple bond in alkynes and alkynecarboxylic esters can be established by ozonolysis in methanol, as in reaction 1. Ozonolysis yields are in the 90–99% range and no degradation of the corresponding carboxylic acid is observed. The resulting carboxylic

$$RC\equiv CR' \xrightarrow{\text{O}_3/\text{MeOH}} RCO_2H \xrightarrow{\text{N}_2CH_2} RCO_2Me \qquad (1)$$

R = Alkyl; R' = Alkyl, CO_2Alkyl

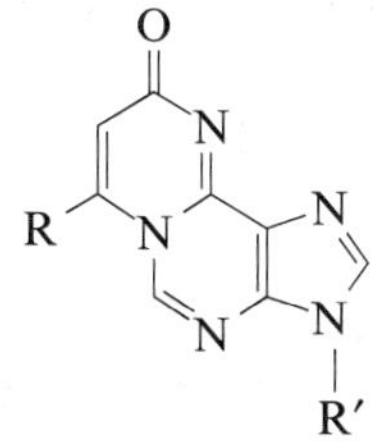

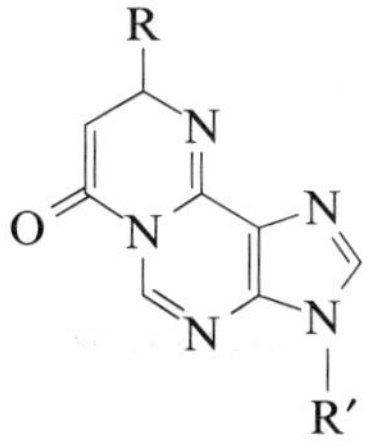

(R' = CH_2Cl,CH_2I,H,CO_2Me)

(3)   (4)

acid is methylated with diazomethane and determined by GLC[22,23]. An alternative to ozonolysis is treatment with a perruthenate salt; this, however, is a longer process. Treatment with ruthenium tetroxide leads to degraded carboxylic acids[22].

Reactions of alkynes and other unsaturated hydrocarbons, catalysed by early transition metal complexes, were reviewed[24].

Activated acetylenic compounds, for example esters $RC{\equiv}CCO_2R''$ undergo cycloadditions to adenosine derivatives to yield two possible products (**3**, **4**), depending on the conditions or reaction[25]. The derivatives of 2-alkynoic acids undergo a regiospecific and stereospecific hydroacetoxylation reaction, catalysed by palladium acetate, as shown in reaction 2[26].

A rule was proposed for assignment of the absolute configuration of chiral terminal alkynes with a hydroxyl group in the propargyl position. Configuration assignment depends on the selectivity of acetal formation with a lactol of type **5**. Diasteroisomer **6** is

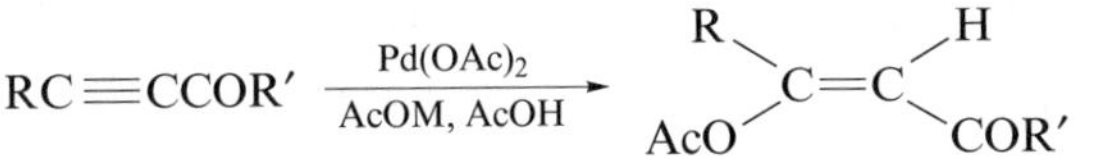

$$RC{\equiv}CCOR' \xrightarrow[\text{AcOM, AcOH}]{\text{Pd(OAc)}_2} \qquad R=Me, \textit{n}\text{-Bu, Ph} \qquad R'=OMe, OPh, NH_2 \qquad (2)$$

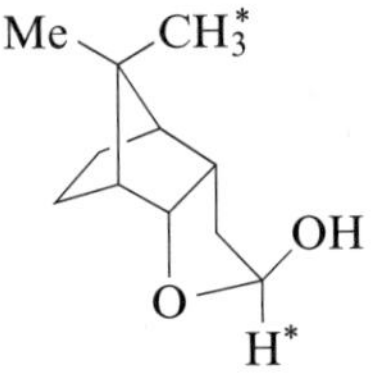

(**5**)

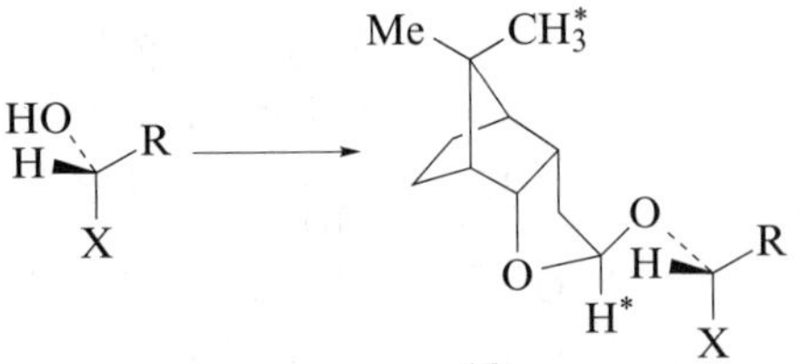

(**6**)

R = Me, Et, PhCH$_2$OCH$_2$, Ph

X = C$\equiv$CH, C$\equiv$N, CH=O, Ph

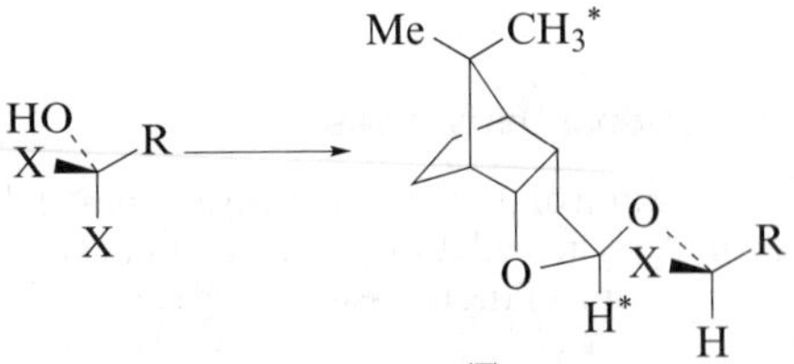

(**7**)

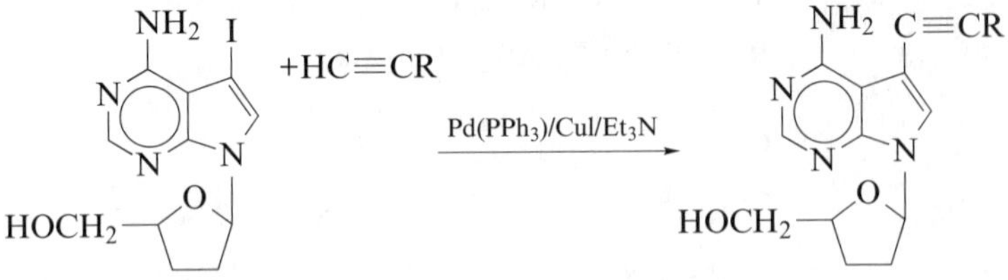

$$\text{(3)}$$

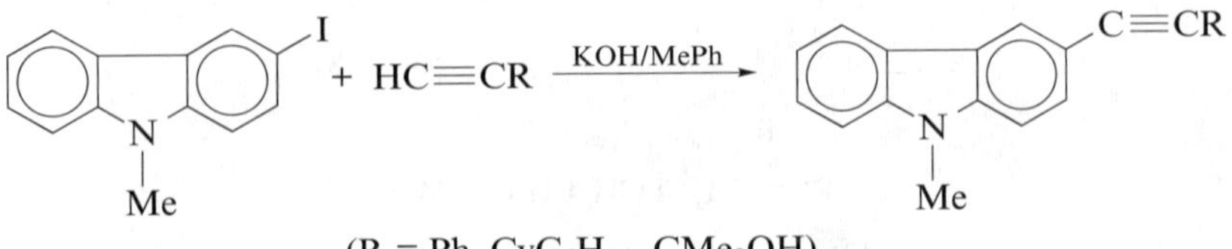

$$\xrightarrow{\text{Pd(PPh}_3)/\text{Cul/Et}_3\text{N}}$$

$$\xrightarrow{\text{KOH/MePh}} \qquad (4)$$

(R = Ph, CyC$_6$H$_{11}$, CMe$_2$OH)

obtained in higher yield than **7**. Long range interactions between the X group and the protons denoted by * affect characteristic chemical shifts of the diasteroisomers **6** and **7**; thus, they are also helpful in the assignment of absolute configuration[27].

Terminal alkynes undergo insertion into iodonucleosides[28] and iodo-9-methylcarbazole[29], as shown in reactions 3 and 4, respectively.

Terminal alkynes undergo Markonikov hydroiodination catalysed by boron triiodide, as shown in reaction 5[30].

$$RC\equiv CH \xrightarrow{\text{BI}_3\text{(PhNEt}_2\text{)/AcOH}} \underset{\underset{I}{|}}{\overset{\overset{R}{\diagdown}}{C}}=CH_2 \qquad (5)$$

Terminal alkynes can be conveniently silanized as shown in reaction 6[31]. They undergo platinum-catalysed hyrosilylation and hydrogermylation. The germylation reaction 7 is almost selective for the terminal carbon [32].

$$RC\equiv CH + ClSi\underset{R'}{\overset{R'}{|}}-\bigcirc-\bigcirc \longrightarrow RC\equiv C-Si\underset{R'}{\overset{R'}{|}}-\bigcirc-\bigcirc \qquad (6)$$

$$RC\equiv CH + \underset{S}{\boxed{}}-MHMe_2 \xrightarrow{H_2PtCl_6}$$

$$\underset{S}{\boxed{}}-MMe_2 \underset{H}{\overset{}{\diagdown}}C=C\underset{R}{\overset{H}{\diagup}} + \underset{S}{\boxed{}}-MMe_2 \underset{R}{\overset{}{\diagdown}}C=C\underset{H}{\overset{H}{\diagup}} \qquad (7)$$

R = 1-Ad, *t*-Bu, Ph, CH_2OMe, CO_2Me
M = Si, Ge

Terminal alkyne complexes with ruthenium react with carbon dioxide and amines to yield urethanes, as shown in reaction 8[33].

$$RC\equiv CH + HNR'_2 + CO_2 \xrightarrow{RuCl_3.3H_2O} RCH=CHOCONR'_2 \qquad (8)$$

Acetylenes yield acrylanilides by insertion of aryl isocyanate into an intermediate tantalum complex, as shown in reaction 9[34].

$$RC\equiv CR' \xrightarrow{\text{TaCl}_5,\text{Zn}} \underset{TaL_n}{\overset{R\diagdown\quad\diagup R'}{\diagup\diagdown}} \xrightarrow{PhNCO} \underset{CONHPh}{\overset{R\diagdown\quad\diagup R}{C=C}} \qquad (9)$$

Cycloaddition reactions of alkynes aided by transition metals were reviewed[35]. Various trimerization processes of acetylenic compounds have been reported. Titanium chloride catalyses the trimerization of acetylenic compounds, by way of intermediate complexes that can be isolated and characterized. This is shown in Table 2 for $TiCl_4$ and 2-butyne[36]. Acetylenes activated by ether groups in the propargyl position undergo trimerization catalysed by $NiBr_2/Mg$. Acetylenes without activation also undergo the same reaction, but with lower yields[37]. Iron π-complexes can catalyse stepwise polymerization of alkynes[38].

Acetylenic compounds form adducts with acrylic acid derivatives, catalysed by a ruthenium complex with 1,5-cyclooctadiene (cod) and 1,3,5-cyclooctatriene (cot), as shown in reaction 10[3].

$$RC\equiv CR' + CH_2{=}CHX \xrightarrow{\text{Ru(cod)(cot)}} \underset{\underset{H}{|}}{R}\diagdown\underset{}{C}{\diagup}\!\!\!\diagdown\underset{}{C}\diagdown\cdots\quad (10)$$

$$X = CO_2Me, CONMe_2$$

α, ω-Alkyne nitriles undergo the cobalt-catalysed cycloaddition reaction 11[40].

Acetylenes undergo the platinum-catalysed cycloaddition to compound **8** appearing in reaction 12, yielding solid derivatives[41].

$$2HC\equiv C(CH_2)_nCN \xrightarrow{\text{Co}^{2+}\text{--AlR}_3} (CH_2)_n\!\!\overbrace{\qquad}^{}\!\!\!\diagup\!\!\!{-}(CH_2)_nCN \quad (11)$$

(8)

In Table 2 appear complexes of transition metals with acetylenic compounds. A study was made of the structure and interconversion of acetylene complexes with binuclear transition metals, where acetylene lies parallel or perpendicular to the metal–metal bond[42,43]. Carbonyl alkyne complexes with binuclear iron give good separations in reverse-phase HPLC[44].

TABLE 2. Complexes of acetylenic compounds with transition metals

Complexing reagents	Products	Ref.	Notes
$Co_2(CO)_8$		45	
$CoM(Cp)(CO)_7$; (M = Mo, W)	$CoM(Cp)(CO)_7(R—C\equiv C—R')_2$	46	a
$Co_2Rh_2(CO)_{12}$	$Co_2Rh_2(CO)_{10}(HC\equiv C—R)$	47	b
$Cr(CO)_5(—NPh=N^+=NHPh) + CO$	(9) and (10)	48	
		49	c, d
$[Cu(NCMe)_4][X] + bpy$ (X = PF_6^-, SbF_6^-)	$[Cu(bpy)(R—C\equiv C—R')][X]$	50	e
$[Cu(O_2CF_3)]_4 \cdot 2C_6H_6$		51	
$[Et_3NH]^+$		52	
$FeH(L)(ppp)$ (L = H_2, N_2)	$CH_2=CHR$	53	c, f
$IrCl(CO)(tpp)_2$	$IrCl(CO)(tpp)_2(OTf)(—C\equiv C—R)$	54	g–i

(*continued*)

TABLE 2. (*continued.*)

Complexing reagents	Products	Ref.	Notes
Ir(OTf)(CO)(tpp)$_2$	[Ir(⬡O)(CO)(tpp)$_2$]$^+$[OTf]$^-$ 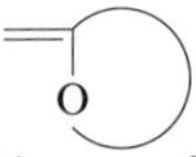	55	*g, h, j*
Mo(II) reagents	a wide range of Mo(II) complexes with acetylenic compounds is reviewed	56	
MoCo(Cp)(CO)$_7$;	CoMo(Cp)(CO)$_7$(R—C≡C—R')$_2$	46	*a*
Mo(Cp)(Cl)$_2$(=NR)	Mo(Cp)(Cl)(=NR)(R'C≡CR'')	57	*a*
(R = H, *t*-Bu)	(R' = H; R'' = *t*-Bu, Ph)		
NbCl$_3$(dme) + thf	NbCl$_3$(thf)$_2$(R—C≡C—R')	58	*k, l*
Ni(L)$_2$(C$_2$H$_4$)	Ni(L)$_2$(R—C≡C—R')	59	*g, m–o*
Ni(cod)$_2$ + 2L	Ni(L)$_2$(R—C≡C—R')		
(L = tpp, tcp, ttop)	(R = Me, Ph; R' = Me, Et, *i*-Pr *t*-Bu, SiMe$_3$, SiPh$_3$)		
Os(CO)$_{10}$(Si(OMe)$_3$)H(NCMe)	Os(CO)$_{10}$(Si(OMe)$_3$)(—CH=CHR) 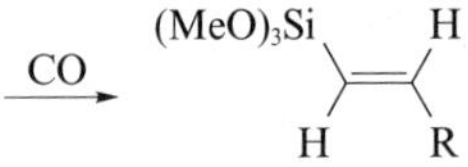	60	*c*
[Pd$_3$(CO)(tfa)(dppm)$_3$]$^+$[tfa]	[Pd$_3$(XC≡CY)(tfa)(dppm)$_3$]$^+$[tfa] (X, Y = CO$_2$Alkyl)	61	*p–r*
Pt(OTf)(Me)(tpp)	[Pt(⬡O)(Me)(tpp)]$^+$[OTf]$^-$ 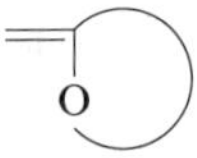	55	*g, h, j*
[Re(CO)$_5$]$^-$ Na$^+$		62	*p*
	(R = Me, CO$_2$Me)		
ReH$_5$(Cyttp)	ReH$_3$(Cyttp)(RO$_2$CCH=CHCO$_2$R)	63	*p, s*
Co$_2$Rh$_2$(CO)$_{12}$	CoRh$_2$(CO)$_{10}$(HC≡C—R)	47	*b*
RhCl(CO)(tpp)$_2$	RhCl(CO)(tpp)$_2$(OTf)(—C≡C—R)	54	*g–i*
RuCl$_2$(L)(hmb) + ROH	[RuCl(L)(hmb)(=C(OH)CH$_2$R')]$^-$	64	*c, t*
(L = PMe$_3$, PMe$_2$Ph, PPh$_3$; R = Me, *i*-Pr)	(R' = *t*-Bu, Me, Ph)		
Ru(CO)Cl(tpp)$_2$(Me$_2$Hpz)(H)	Ru(CO)Cl(tpp)$_2$(Me$_2$Hpz)	65	*c, u*
(R = *t*-Bu, C$_8$H$_{17}$, Ph, SiMe$_3$)	(—CH=CHR)		
Ti(Cp*)$_2$(Cl)$_2$ + Mg	Ti(Cp*)$_2$(RC≡CR')	66	*v*
Ti(Cp)$_2$(tmp)$_2$	Ti(Cp)$_2$(RC≡CR) 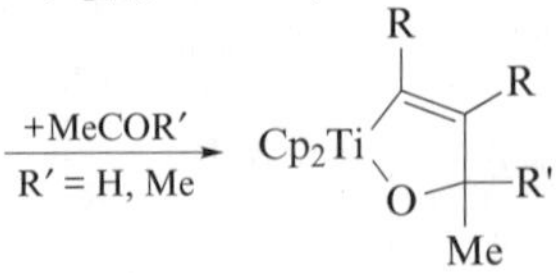	67	*w*
TiCl$_4$ + 3MeC≡CMe	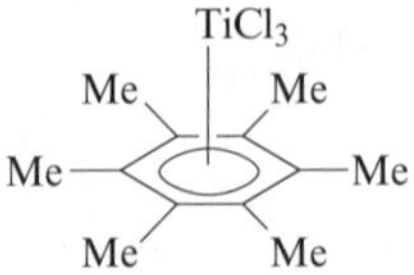 TiCl$_3$	36	

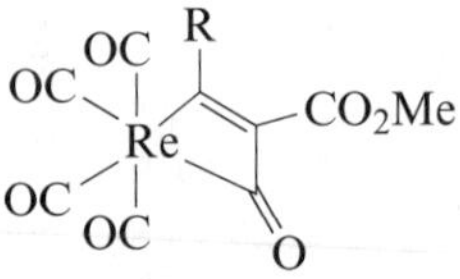

(*continued*)

TABLE 2. (*continued*)

Complexing reagents	Products	Ref.	Notes
W(II) reagents	a wide range of W(II) complexes with acetylenic compounds was reviewed		
WCl₄, Et₂O	WCl₄(R—C≡C—R).Et₂O	68	
CoW(Cp)(CO)₇; (M = Mo, W)	CoW(Cp)(CO)₇(R—C≡C—R')₂	46	a
W(Br)(I)(CO)(NCMe)₂ (R = R' = CH₂Cl, Ph; R = Me, R' = Ph)		69	x
W(Cp*)(X)₄ + Na-Hg (R = PhO⁻, Cl⁻)	W(Cp*)(RC≡CR')₂ (R, R' = H, Me, Ph, SiMe₃)	70	v

^{a}Cp = cyclopentadienyl anion.
^{b}Quantitative yields are claimed; on further treatment of the product in solution with triphenylphosphine (tpp), Co₂Rh₂(CO)₈(tpp)₂(HC≡C—R) is obtained.
^{c}For terminal alkynes.
^{d}Regioselective and diastereoselective reaction.
^{e}bpy – 2,2'-bispyridine.
^{f}ppp = P(CH₂CH₂PPh₂)₃.
^{g}tpp = triphenylphosphine.
^{h}OTf = triflate anion, CF₃SO₃⁻.
^{i}For terminal acetylenes via the iodonium intermediate [RC≡C—I—Ph]⁺[OTf]⁻.
^{j}For alkynols with triple bond at terminal position.
^{k}dme = 1,2-dimethoxyethane.
^{l}thf = tetrahydrofuran
^{m}cod = 1,5-cyclooctadiene.
^{n}tcp = tris(cyclopentadienyl)phophine.
^{o}ttop = tris(o-tolyloxy)phosphine.
^{p}for activated alkynes.
^{q}tfa = trifluoroacetate anion.
^{r}dppm = Ph₂PCH₂PPh₂.
^{s}Cyttp = PhP[CH₂CH₂CH₂P(cyclo-C₆H₁₁)₂]₂.
^{t}hmb = hexamethylbenzene.
^{u}Me₂Hpz = 3,5-dimethylpyrazole.
^{v}Cp* = pentamethylcyclopentadienyl anion.
^{w}tmp = trimethylphosphine.
^{x}Complex is formed in two stages. In the first one a dimeric complex is formed of formula [W(Br)(I)(CO)(NCMe) (RC≡CR')]₂, which reacts further with RC≡CR' to yield the final complex.

III. NITRILES AND ISONITRILES

A. General

Nitriles and isonitriles can interfere with various methods for the determination of cyanide in water and wastewaters, similarly to the interference known for thiocyanates[71].

Non-fluorescent compounds such as nitriles in the gas phase can be detected by selective photolysis and the fragments recognized by dye laser-induced fluorescence[72].

Gas chromatography columns for interplanetary exploration should be very efficient, due to the stringent demands imposed on payload. Porous-layer open tubular (PLOT) columns loaded with a styrene-divinyl benzene copolymer showed better performance than Cromosorb 103 or Porapak Q columns in the analysis of low molecular weight hydrocarbons and nitriles[73]. Potassium chloride-deactivated alumina PLOT columns, on the other hand, were not good for the intended purpose, because low molecular weight nitriles were difficult to elute[74]. Wall coated open tubular (WCOT) columns with a stationary chemically bonded phase of dimethyl siloxane have the mechanical resilience to endure the conditions of extraterrestrial exploration and separated efficiently C_1–C_4 nitriles[75]. Permanent gases were analysed in the presence of hydrocarbons and nitriles with a PLOT capillary column[76]. The equilibrium constant for the interaction of nitrile groups with various solutes was determined by GLC[77].

Aliphatic nitriles in aqueous solution were converted to the corresponding amines by electrolytic reduction with a Pd/Ni cathode. The amines were treated with reagent 11, separated on Lichosorb RP-8 with aqueous methanol, and determined using a fluorescence detector, with detection limits of 20 mg/L of aliphatic nitriles[78]. The following procedure was used for screening rape products in search for nitriles: rape seed meal was incubated with myrosinase in the presence of Fe(II) sulphate, lyophilized, extracted with chloroform, the extract hydrolysed with concentrated alkali and the liberated ammonia was titrated[79].

(11)   (12)

Aldicarb nitrile (12), a residual degradation product of the carbamate pesticide Aldicarb, can be analysed by GC using a short column and MS detection, with a detection limit of 0.15 ng. The short column is required to avoid thermal degradation of Aldicarb to 12 that could distort the analysis[80].

Determination of ppm concentrations of nitriles in light naphtha was carried out by capillary GC, by either of the following procedures. Method 1: Clean-up and preconcentration by medium pressure LC removed the hydrocarbons and increased the nitrile concentration ten fold; the samples were analysed using an FI detector. Method 2: Direct injection into a capillary GC using a chemiluminescence nitrogen detector[81].

Pyrethroid insecticides containing nitrile groups can be detected by TLC on silica gel plates, with a detection limit of ca 1 μg, by successive treatment with alkali and cupric acetate solutions. The liberated cyanide ions give a blue colour with phosphomolybdic acid-o-tolidine reagent. No interference was shown by organochlorine, organophosphorus or carbamate insecticides[82].

Determination of σ constants of anionic substituents of benzonitrile is impossible by the usual spectrophotometric methods. Thus, the constants σ^+, σ° and $\Delta\sigma_R^+$ were determined for such substituents by measuring the frequency (cm^{-1}) and integrated intensity (L mol^{-1} cm^{-2}) of the C≡N stretching frequency, in HMPA and DMSO solvents[83].

Tables 3 and 4 list nitriles and isonitriles of industrial interest.

TABLE 3. Nitriles of industrial application[a]

Compounds and CAS registry number	NIOSH/OSHA RTECS[b]
Acetonitrile [75-05-8]	AL 7700000
2-Acetoxyacrylonitrile [3061-65-2]	AG 4550000
(Acetoxy)methylmalononitrile [7790-71-4]	WW 9100000
Acrylonitrile [107-13-1][c]	AT 5250000
Adiponitrile [111-69-3]	AV 2625000
Alkylnitrile[d]	AZ 9375500
3-Allyloxypropionitrile [3088-44-6]	UG 0175000
Aminoacetonitrile [540-61-4]	MC 1400000
Aminoacetonitrile sulphate [5466-22-8]	MC 2100000
2-Aminopropionitrile [2134-48-7]	
3-Aminopropionitrile [151-18-8]	UG 0350000
3-Aminopropionitrile fumarate [2079-89-2]	UG 0700000
Anthranilonitrile [1885-29-6]	CB 4575000
1-Aziridinepropionitrile [1072-66-8]	CM 8750000
Azobisisobutyronitrile [78-67-1]	UG 0800000
Benz(c)acridine-7-carbonitrile [3123-27-1]	CU 3050000
Benzalmalononitrile [2700-22-3]	
Benz(a)anthracene-7-acetonitrile [63018-69-9]	CV 8975000
1,3-Benzenedicarbonitrile [626-17-5]	CZ 1900000
1,4-Benzenedicarbonitrile [623-26-7]	CZ 1925000
2-Benzimidazoleacetonitrile [4414-88-4]	DD 5650000
2-Benzofuranacetonitrile [6149-69-9]	DF 6550000
Benzonitrile [100-47-0]	DI 2450000
4,4′-Biphenyldicarbonitrile [1591-30-6]	DV 2625000
Bis(acrylonitrile)nickel(0) [12266-58-9]	
β-(Bis(β-chloroethyl)amino) benzylidenemalononitrile [4213-30-3]	OO 3600000
p-(Bis(β-chloroethyl)amino)propionitrile [63815-37-2]	UG 0875000
Bis(β-cyanoethyl)amine [111-94-4]	UG 2975000
N,N-Bis(β-cyanoethyl)benzenesulphonamide	DB 0525000
Bis(β-cyanoethyl)methylamine [1555-58-4]	UG 3675000
1,4-Bis(dicyanomethylene)cyclohexane [1518-15-6]	GU 9700000
2,3-Bis(4-hydroxyphenyl)valeronitrile [65-14-5]	YV 8200000
α-Bromophenylacetonitrile [5798-79-8]	AL 8050000
4-Bromophenylacetonitrile [16532-79-9]	AL 8050000
3-Bromopropionitrile [2417-90-5]	UG 1050000
3-Butenenitrile [109-75-1]	EM 8050000
3-Butoxypropionitrile [6959-71-3]	UG1225000
Butyronitrile [109-74-0]	ET 8750000
Cacium cyanamide [156-62-7]	GS 6000000
N-(1-Carbamoyl-4-(nitrosocyanamido)butyl)benzamide	CV 2443700
Carbonyl cyanide 3-chlorophenylhydrazone [555-60-2]	FG 5600000
Chloroacetonitrile [107-14-2]	AL 8225000
2-Chloroacrylonitrile [920-37-6]	AT 5525000
2-Chlorobenzonitrile [873-32-5]	DI 2625000
4-Chlorobenzonitrile [623-03-0]	DI 2800000
Chlorocyanoacetylene [2003-31-8]	
3-Chloro-6-cyano-2-norbornanone O-methylcarbamoyloxime [15271-41-7]	RB 7700000
3-Chlorolactonitrile [33965-80-9]	OD 8575000
3-Chloropropionitrile [542-76-7]	UG 1400000
Cinnamonitrile [4360-47-8]	UD 1440000
Citronellyl nitrile [51566-62-2]	RH 2250000

(continued)

TABLE 3. (*continued*)

Compounds and CAS registry number	NIOSH/OSHA RTECS[b]
Cyanazine [21725-46-2][e]	
Cyanoacetamide [107-91-5]	AB 5950000
Cyanoacetic acid hydrazide [104-87-4]	AG 4200000
Cyanoacetic acid [372-09-8]	AG 3675000
N-Cyanoacetyl ethyl carbamate [6629-04-5]	EZ 3480000
1-Cyano-3-*t*-amylguanidine [1113-10-6]	MF 0175000
4-Cyanobenzaldehyde [105-07-7]	CU 5250000
10-Cyano-1,2-benzanthracene [7476-08-6]	CW 1050000
N-Cyano-2-bromoethylbutylamine [87049-67-0]	
N-Cyano-2-bromoethylcyclohexylamine	
5-Cyano-10,11-dihydro-5-(3-dimethylaminopropyl)-	HO 8050000
5H-dibenzo(*a,d*)cycloheptene hydrochloride [1769-99-9]	
5-Cyano-10,11-dihydro-5-(3-methylaminopropyl)-5H-	HO 8225000
dibenzo(*a,d*)cycloheptene hydrochloride [60763-63-5]	
Cyanodimethylarsine [683-45-4]	CH 2100000
5-Cyano-9,10-dimethyl-1,2-benzanthracene [63018-68-8]	CW 1225000
4-Cyanoethoxy-2-methyl-2-pentanol [10141-15-8]	SA 5950000
N-(Cyanoethyl)diethylenetriamine [65216-94-6]	IE 1380000
β-Cyanoethylmercaptan [1001-58-7]	KJ 0430000
2-Cyanoethyltrichlorosilane [1071-22-3]	UG 4900000
2-Cyanoethyltriethoxysilane [919-31-3]	VV 2750000
2-Cyano-2′-fluorodiethyl ether [353-18-4]	UG 2330000
Cyanoformic acid methyl ester [17640-15-2]	LQ 7525000
6-Cyanohexanoic acid ethyl ester [10137-65-2]	MO 7200000
Cyanomethyl acetate [1001-55-4]	MC 7700000
o-Cyanonitrobenzene [612-24-8]	DI 4903000
4-Cyanophenyl isothiocyanate [2719-32-6]	NX 8479000
Cyanophos [2636-26-2][e]	
2-Cyano-2-propanol [75-86-5]	OD 9275000
1-Cyanopropene [4786-20-3]	GC 6322000
1-Cyano-2-propen-1-ol [600-17-9]	
Cycloprothin [63935-38-6][e]	
Cyfluthrin [68359-37-5][e]	
Cyhalothrin [68085-85-8][e]	
λ-Cyhalothrin [91465-08-6][e]	
Cymoxanil [57966-95-7][e]	
Cypermethrin [52315-07-8][e]	
Detamethrin [52918-63-5][e]	
Diallylcyanamide [538-08-9]	GS 6125000
Diazidomalononitrile	
Diazidomethylenecyanamide	
Dibromoacetonitrile [3252-43-5]	AL 8450000
Dibromocyanoacetamide [10222-01-2]	AB 5956000
2,6-Dibromo-4-cyanophenyl octanoate [1689-99-2]	DI 3325000
3,5-Dibromo-4-hydroxybenzonitrile [1689-84-5][e]	DI 315000
(3,5-Di-*t*-butyl-4-hydroxybenzylidene)malononitrile [10537-47-0]	OO 3737000
2,6-Dichlorobenzonitrile [1194-65-6][e]	
2,3-Dichloro-5,6-dicyanobenzoquinone [84-58-2]	GU 4825000
4-(2,3-Dichlorophenyl)-1H-pyrrole-3-carbonitrile [74738-17-3][e]	
Dicyanoacetylene [1071-98-3]	
1,4-Dicyano-2-butene [1119-85-3]	MP 6750000

(*continued*)

TABLE 3. (*continued*)

Compounds and CAS registry number	NIOSH/OSHA RTECS[b]
β,β'-Dicyanodiethyl sulfide [111-97-7]	UG 4375000
2,3-Dicyano-1,4-dithiaanthraquinone [3347-22-6][e]	
3,3-Dicyclopropyl-2-(ethoxycarbonyl)acrylonitrile [5232-99-5]	AT 6200000
N,N-Diethylaminoacetonitrile [3010-02-4]	AI 8575000
4-Diethylamino-2-isopropyl-2-phenylvaleronitrile	YV 8520000
2,4-Dihydroxy-3,3-dimethylbutyronitrile [10232-92-5]	ET 9625000
3,5-Diiodo-4-hydroxybenzonitrile [2961-61-7]	DI 4200000
Dimethylaminoacetonitrile [926-64-7]	AL 9450000
3-Dimethylaminopropionitrile [1738-25-6]	UG 1575000
α-[(1,3-Dioxolan-2-yl)methoxyimino]benzeneacetonitrile [74782-23-3]	
Diphenylacetonitrile [86-29-3]	AL 9800000
2,3-Diphenylacrylonitrile [2510-95-4]	AT 6225000
Dodecanenitrile [2437-25-4]	JR 2600000
1,2-Epoxybutyronitrile [6509-08-6]	ET 9900000
3,4-Epoxycyclohexanecarbonitrile [141-40-2]	RN 7350000
Ethenetetracarbonitrile [670-54-2]	KM 7300000
Ethoxymethylenemalononitrile [123-06-8]	OO 3850000
3-Ethoxypropionitrile [14631-45-9]	UG 1800000
3-(2-Ethylbutoxy)propionitrile [10232-91-4]	UG 1925000
Ethyl cyanoacetate [105-56-6]	AG 4110000
Ethylcyanocyclohexyl acetate [1331-45-9]	GU 7700000
N-Ethyl-N-(2-cyanoethyl)aniline [148-87-8]	BY 0100000
Ethylenebis(tris(2-cyanoethyl)phosphonium bromide) [10310-38-0]	TA 2305000
(Ethylenedinitrilo)tetraacetonitrile [5766-67-6]	AM 0100000
3-(2-Ethylhexyloxy)propionitrile [10231-75-9]	UG 2100000
Fluoroacetonitrile [503-20-8]	AM0175000
4-Fluorobutyronitrile [407-83-0]	EU 0175000
12-Fluorododecanonitrile [334-71-4]	JR 4375000
7-Fluoroheptanonitrile [334-44-1]	MJ 7875000
4-Fluorophenylacetonitrile [459-22-3]	AM 0210000
5-Fluorovaleronitrile [353-13-9]	YV 8575000
Fumaronitrile [764-42-1]	LT 2300000
Geranylnitrile [5585-39-7]	RG 5370000
Glutaronitrile [544-13-8]	YI 3500000
Heptanedinitrile [646-20-8]	MI 9020000
Hexanenitrile [628-73-9]	MO 3900000
3-Hexenedinitrile [1119-85-3]	MP 6750000
Hydroxyacetonitrile [107-16-4]	AM 0350000
4-Hydroxybenzonitrile [767-00-0]	DI 4375000
2-Hydroxy-3-butenenitrile [5809-59-6]	EM 8225000
1-Hydroxycycloheptanecarbonitrile [931-97-5]	GU 7710000
4-Hydroxy-3,5-diiodobenzonitrile [1689-83-4][e]	DI 4025000
4-Hydroxy-3-iodo-5-nitrobenzonitrile [1689-89-0]	DI 4600000
3-Indolylacetonitrile [771-51-7]	AM 0700000
Isobutyronitrile [78-82-0]	TZ 4900000
Isovaleronitrile [625-28-5]	NY 1750000
Lactonitrile [78-97-7]	OD 8225000
Malononitrile [109-77-3]	OO 3150000
Mandelonitrile [532-28-5]	OO 840000
4-Methoxyphenylacetonitrile [104-47-2]	AM 0810000
2-(4-methoxyphenyl)-3,3-diphenylacrylonitrile [16143-89-8]	AT 6900000

(*continued*)

TABLE 3. (continued)

Compounds and CAS registry number	NIOSH/OSHA RTECS[b]
3-Methoxypropionitrile [110-67-8]	UG 3150000
Methylacrylonitrile [126-98-7]	UD 1400000
7-Methylbenz(a)anthracene-8-carbonitrile [63018-70-2]	CW 0730000
9-Methylbenz(a)anthracene-10-carbonitrile [63020-25-7]	CW 1400000
Methyl cyanoacetate [105-34-0]	AG 4375000
2-Methyllactonitrile [75-86-5]	OD 9275000
N-Methyl-N-nitrosoaminoacetonitrile [3684-97-7]	AM 0950000
4-Methylpentanenitrile [542-54-1]	YV 8588000
Methylphosphothionic acid O-4-cyanophenyl ester O-phenyl ester [5954-90-5]	TB 1700000
4-Morpholinoacetonitrile [5807-02-3]	QD 6650000
Morpholinocarbonylacetonitrile [15029-32-0]	AM 1230000
1-Naphthylacetonitrile [132-75-2]	AM 1240000
2-Nitrobenzonitrile [612-24-8]	
3-Nitrobenzonitrile [619-24-9]	DI 4900000
3-Nitrobenzylidenemalononitrile [2826-32-6]	OO 4375000
N-Nitrosoiminodiacetonitrile [16339-18-7]	AM 1300000
3,3'-Oxydipropionitrile [1656-48-1]	UG 4200000
α-Phenylacetoacetoacetonitrile [4468-48-8]	AL 7708000
Phenylacetonitrile [140-29-4]	AM 1400000
Phenyl-2-(2-piperidinoethyl)-3-methylvaleronitrile [1228-02-0]	YV 8800000
Phthalonitrile [91-15-6]	TI 8575000
1-Piperidineacetonitrile [3010-03-5]	AM 1925000
1-Piperidinepropionitrile [3088-41-3]	TN 3714000
Pivalonitrile [630-18-2]	
Propionitrile [107-12-0]	UF 9625000
Sebaconitrile [1871-96-1]	VS 1500000
Succinonitrile [110-61-2]	WN 3850000
Succinylnitrile [63979-84-0]	WN 5800000
Tetraacrylonitrilecopper(I) perchlorate	
Tetraacrylonitrilecopper(II) perchlorate	
Tetrachloroisophthalonitrile [1897-45-6][e]	NT 2600000
Tetracyanoethylene [670-54-2]	KM 7300000
Tetramethylsuccinonitrile [3333-52-6]	WN 4025000
m-Tolunitrile [620-22-4]	XV 0525000
o-Tolunitrile [529-19-1]	XV 0600000
p-Tolunitrile [104-84-8]	XV 0700000
Tributylstannanecarbonitrile [681-99-2]	WH 8583500
Trichloroacetonitrile [545-06-2]	AM 2450000
Tridecanenitrile [629-60-7]	YD 3500000
Tridecanenitrile (mixed isomers) [1070-01-5]	YD 3675000
3-Trifluoromethylphenylacetonitrile [2338-76-3]	CY 1820000
Trinitroacetonitrile [630-72-8]	
3,3,3-Triphenylacrylonitrile [6304-33-2]	AT 7100000

[a] See also Reference 11.
[b] Registry of Toxic Effects of Chemical Substances of National Institute for Occupational Safety and Health, Occupational Safety and Health Administration.
[c] For polymers derived from the acrylonitrile monomer, see NIOSH specifications AT 6970000, 6975000, 6976000, 6978000.
[d] A mixture of nitriles, $C_9H_{19}CN$ to $C_{17}H_{35}CN$.
[e] See also Reference 21.

TABLE 4. Industrial organic isocyanides[a]

Compound and CAS registry number	NIOSHOSHA RTECS[b]
t-Butyl isocyanide [7188-38-7]	EQ 7102000
1,4-Di-p-oxyphenyl-2,3-diisonitrilo-1,3-butadiene [580-74-5]	KX 1625000
Ethyl isocyanide [624-79-3]	FH 4025000
Methyl isocyanide [593-75-9]	

[a] See also Reference 11.
[b] *Registry of Toxic Effects of Chemical Substances* of National Institute for Occupational Safety and Health/Occupational Safety and Health Administration.

B. Functional Derivatives

Activated nitriles form adducts with conjugated azoalkenes, as shown for example in reaction 13[84].

$$Me$$
$$MeO_2CCH=CN=NCO_2Me + RCOCH_2CN \xrightarrow{MeONa}$$

$$Me$$
$$MeO_2CCH-C=NNHCO_2Me \quad (13)$$
$$RCO-CH-CN$$

Nitriles with active α-hydrogen atoms can effect nucleophilic displacement of aromatic halogen, in the presence of complex bases, as depicted in reaction 14[85].

$$(Z)_n \quad Br \quad \xrightarrow[\text{2. RCH}_2\text{CN}]{1.\ NaNH_2/EtOCH_2CH_2OCH_2CH_2OH/(MeO)_2CH_2} \quad (Z)_n \quad CHCN$$
$$(Z = H, OMe, Me; R = H, Me, Et, Ph; n = 1,2,3) \quad R \quad (14)$$

A rule was proposed for assignment of the absolute configuration of chiral nitriles with a hydroxyl group in the α-position. Assignment depends on the selectivity of acetal formation with a lactol of type **5**. Diastereoisomer **6** is obtained in higher yield than **7**. The long range interaction between the X group and the protons denoted by * on characteristic chemical shifts of the diastereoisomers **6** and **7** are also helpful in the assignment of absolute configuration[27].

$$\text{Li}^+ \quad Pr\text{-}i \quad \text{Li}^+ \quad Pr\text{-}i \quad NHPr\text{-}i$$
$$+ RCN \longrightarrow \xrightarrow{NH_4Cl} \quad (15)$$

TABLE 5. Complexes of nitriles with transition metals

Complexing reagents	Products	Refs.	Notes
[Co(NH₃)₅(OTf)](OTf)₂	[Co(NH₃)₅(NC—R)](X)₃ (X = OTf, ClO₄⁻; R = NH2, NMe₂, Me, CH₂=CH, Ph, o-O₂N-C₆H₄, p-F-C₆H₄)	88	a
Cr(CO)₅(thf)	Cr(CO)₅(NC-COR) + thf	89	b, c
Fe(CO)₅ (catalyst: dbu; solvent: 1-ClC₁₀H₇)	**13**	90	d, e
[FeH(H₂)(depp)₄][BX₄] (X = F, Ph)	[FeH(NC—R)(depp)₄][BX₄]	91	f
[Mn(CO)₃(C₆H₆)]⁺[PF₆]⁻	(see structure) Mn(CO)₃ CR₂CN H	92	g
Rh(ClO₄)(CO)(tpp)₂	Rh(ClO₄)(CO)(tpp)₂(NC—R) (R = Me, CH₂=CH, CH₂=CHCH₂, CH₂=CCH₃, CH₃CH=CH)	93	h
Ru(—C=C—Ph / (NC)₂C—C(CF₃)₂)(tpp)(Cp) (NCMe)	Ru(—C=C—Ph / (NC)₂C—C(CF₃)₂)(tpp)(Cp) (NCR)	94	h–j
SnCl₄	SnCl₄(NC—R) (R = Me, t-Bu, CH₂=CH, CN—CH₂, Ph, o-NC-C₆H₄)	95, 96	
(dipp)₃Ta(=Ph / —Ph)	(dipp)₃Ta(=Ph / —Ph) RC≡N	97	k
	R = R'CH₂ → (dipp)₃ Ta (see structure with Ph, Ph, N, H, CHR')		
V(mes)₃(thf)	t-Bu—N=mes / mes—V—N—t-Bu / N / t-Bu mes	98	b, l
	O₂, H₂O → 3t-BuNHCO-mes		
ZnCl₂ + HCl	[ZnCl₃]₂⁻ [ArCH₂CNH]⁺ [ArCH₂C(Cl)=NH₂]	99	

ᵃ OTf = triflate anion, CF₃⁺SO₃⁻.
ᵇ thf = tetrahydrofuran.
ᶜ For acyl cyanides.
ᵈ For aromatic ortho-dinitriles.
ᵉ dbu = 1,8-diazabiclyclo[5,4,0]undec-7-ene (**14**).
ᶠ depp = diethoxyphenylphosphine.
ᵍ For nitriles with active α-H.
ʰ tpp = triphenylphosphine.
ⁱ Another nitrile displaces MeCN of the complexing agent.
ʲ Cp = cyclopentadienyl anion.
ᵏ dipp = 2,6-diisopropylphenoxide anion.
ˡ mes = 2,4,6-trimethylphenyl.

(13)

(14)

α,ω-Alkyne nitriles undergo the cobalt-catalysed cyaloaddition reaction 11[40].
Nitriles give good yields of 4-aminopyridine derivatives with γ-dianions, as shown in reaction 15. Very hindered nitriles, such as t-BuCN, do not undergo this reaction[86].
Nitriles afford ylides with diazomethane. These reactive intermediates may be scavenged by labile compounds, as shown in reaction 16[87].

$$CH_2N_2 \xrightarrow{hv} CH_2 \xrightarrow{MeCN} Me-C\equiv N^+CH_2^- \xrightarrow{NCCH=CHCN} \quad (16)$$

In Tables 5 and 6 appear complexes of transition metals with organic compounds containing nitrile and isonitrile functional groups, respectively.

TABLE 6. Complexes of organic isocyanides with transition metals

Complexing reagents	Products	Ref.	Notes
K[Au(CN)$_2$]		100	
Co(Cp)(CH$_2$=CH$_2$)$_2$	Co(Cp)(CN—R)$_2$ + 2CH$_2$=CH$_2$	101	a
Cr(CO)$_6$ (catalyst: CoCl$_2$, PdO)	Cr(CO)$_{6-n}$(CN—R)$_n$ + nCO	102	
[Cr(CO)$_5$I]$^-$[NEt$_4$]$^+$	Cr(CO)$_5$(CN—Ar)	103	
Cr(CO)$_5$	Cr(CO)$_5$	104	
Fe(CO)5 (catalyst: CoCl$_2$)	Fe(CO)$_{5-n}$(CN—R)$_n$ + nCO	105	
FeHCl(dppe)$_2$ + Tl[A] (A = PF$_6$, BF$_4$)	[FeH(CN—R)(dppe)$_2$] [A] + TlCl	106	b
[FeHCl(dppe)$_2$(N$_2$)][A] (A = PF$_6$, BF$_4$)	[FeH(CN—R)(dppe)$_2$][A] + N$_2$	106	b
[Fe(L)][ClO$_4$] (L = tetraphenylporphyrin and its o-, m- and p-Me derivatives) (solvent: CH$_2$Cl$_2$)	[Fe(L)(CN—R)][ClO$_4$]	107	
[FeH(H$_2$)(depp)$_4$][BX$_4$] (X = F, Ph)	[FeH(CN—Ar)(depp)$_4$][BX$_4$]	91	c

(*continued*)

TABLE 6. (*continued*)

Complexing reagents	Products	Ref.	Notes
Hf(Cp*)(dmbd)(Cl)		108	*d, e*
Mo(CO)$_6$ (catalyst: CoCl$_2$, PdO)	Mo(CO)$_{6-n}$(CN—R)$_n$ + nCO	102	
Mo(Cp)$_2$HI	[Mo(Cp)$_2$H(CN—R)]$^+$I$^-$	109	*a*
[Re$_3$H$_4$(CO)$_9$(MeCN)]$^-$[NEt$_4$]$^+$	[Re$_3$H$_4$(CO)$_9$(CN—R)]$^-$[NEt$_4$]$^+$	110	
[Rh(tmpp)$_2$][BF$_4$]$_2$	[Rh(tmpp)$_2$(CN—R)$_2$][BF$_4$]$_2$	111	*f*
Ru(CO)Cl(tpp)$_2$(—CH=CHR) (R = *t*-Bu, C$_8$H$_{17}$, Ph, SiMe$_3$)		112	*g*
Ru(CO)(L)Cl(L)$_2$(—CH=CHR) (L = Py, Me$_2$Hpz) (R = t-Bu, C$_8$H$_{17}$, Ph, SiMe$_3$)		112	*h, i*
Ru(ttp)(CO)	Ru(ttp)(CN—R)$_2$	113	*j*
Ti (Cp*)(dmbd)(Cl)		108	*d, e*
V$_2$(Cp)$_2$(CO)$_4$(SMe)$_2$	V$_2$(Cp)$_2$(CO)$_3$(CN R)(SMe)$_2$ V$_2$(Cp)$_2$(CO)$_2$(CN—R)$_2$(SMe)$_2$	114	*a*
W(CO)6 (catalyst: CoCl$_2$, PdO)	W(CO)$_{6-n}$(CN—R)$_n$ + nCO	102	
[W(CO)$_5$I]$^-$[NEt$_4$]$^+$	W(CO)$_5$(CN—Ar)	103	
		104	

^{a}Cp = cyclopentadienyl anion.
^{b}dppe = Ph$_2$PCH$_2$CH$_2$PPh$_2$.
^{c}depp = diethoxyphenylphosphine.
^{d}dmbd = 2,3-dimethyl-1,3-butadiene.
^{e}Cp* = pentamethylcyclopentadienyl anion.
^{f}tmpp = tris(2,4,6-trimethoxyphenyl)phosphine.
^{g}tpp = triphenylphosphine.
^{h}Py = pyridine.
^{i}Me$_2$Hpz = 3,5-dimethylpyrazole.
^{j}tpp = *meso*-tetraphenylporphyrin.

IV. NITRILE OXIDES

A. General

Nitrile oxides, with mesomeric structure **15**, are very reactive intermediates, as illustrated in Section IV. B. Table 7 summarizes various compounds of industrial interest, including one nitrile oxide.

$$-\ddot{C}-\ddot{N}\longrightarrow O \longleftrightarrow -C\equiv N \rightarrow O$$

(15)

TABLE 7. Organic cyanates and a nitrile oxide that found industrial application[a]

Compound and CAS registry number	NIOSH/OSHA RTECS[b]
Cyanatotributyltin [4027-17-2]	WH 6871000
Cyclohexyl tricyanatoacetate [5349-27-9]	AJ 6125000
Dicyanogen N,N'-dioxide [4331-98-0]	

[a] See also Reference 11.
[b] *Registry of Toxic Effects of Chemical Substances* of National Institute for Occupational Safety and Helth/ Occupational Safety and Health Administration.

B. Functional Derivatives

Nitrile oxides undergo dipolar [3 + 2] cycloadditions to carbon–carbon double bonds and triple bonds, for example, as in reactions 17–19 with an olefin[115], a uracil derivative[116] and a 1,4-dihydropyridine derivative[117].

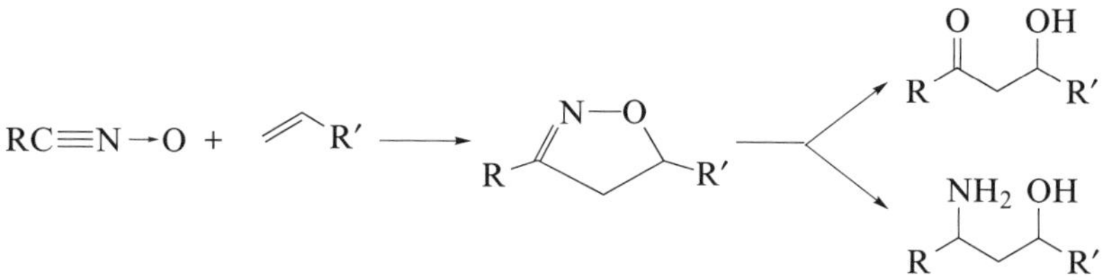

(17)

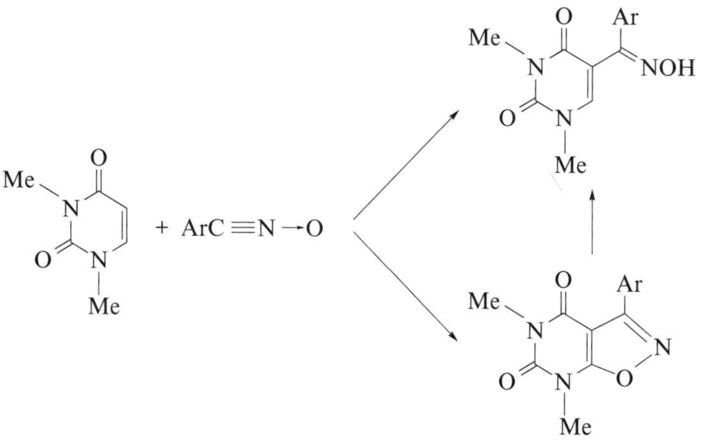

(18)

$$RC\equiv N^+ - O^-$$

R = CO₂Et, R' = Me (labeled on arrows)

(19)

In Table 8 appear complexes of rhodium with nitrile oxides, as well as some other related complexes with thiocyanates, isocyanates and isothiocyanates.

TABLE 8. Complex of nitrile oxides, organic thiocyanates, isocyanates and isothiocyanates with transition metals

Complexing reagents	Products	Ref.	Notes
a. *Nitrile oxides* Rh(Cp)(tip)(CO)		118	a, b
b. *Organic thiocyanates* AgNO₃, NH₃, NaOH HgCl₂, KI, NaOH	RSAg (RS)₂Hg	119 119	c c
c. *Organic isocyanates* Co(tmp)(Cp)(CN—R)		120	b, d
d. *Organic isothiocyanates* Co(tmp)(Cp)(CN—R)		120	b, d

a tip = triisopropylphosphin.
b Cp = cyclopentadienyl anion.
c For unsaturated thiocyanates.
d tmp = trimethylphosphine.

V. DIAZONIUM COMPOUNDS

A. General

The rate of decomposition of arenediazonium compounds catalysed by Fe(II)–EDTA could be measured by EPR under conditions of continuous flow[121]. No scrambling of N atoms was detected, using ^{15}N-labeling and ^{15}N-NMR spectroscopy[122]. Decamethylferrocene and ferrocyanide were used as one-electron reducing agents for substituted benzenediazonium compounds. The reaction is of second order kinetics, first order in each of the reacting species[123].

Triazenes $ArN\!\!=\!\!NNHAr$ are formed as by-products of the diazotization reaction of primary aromatic amines. They can be determined by liberating the amine $ArNH_2$ in strong acid, followed by potentiometric titration with nitrite in hydrobromic acid, using a gold electrode and a standard calomel electrode[124].

The formation of genotoxic substances, such as benzenediazonium ions, was shown to take place in nitrosated smoked fish and meat products[125]. A cytotoxic, antineoplastic, antibiotic agent—agaricin (16)—was isolated from the basidomycete *Agaricus xanthodermus Genevier*, after sodium sulphite extraction and gel chromatography. Its formula is that of the tautomeric 4-hydroxybenzenediazonium sulphite or the corresponding diazosulphonate. Agaricin yields hydroquinone on heating in acid solution[126,127].

$$p\text{-}HOC_6H_4\,-\!N_2^+\,SO_3^- \;\;\rightleftharpoons\;\; p\text{-}HOC_6H_4\,-\!N\!=\!N\!-\!SO_3^-$$

$$(16)$$

The electrochemical reduction mechanism of arenediazonium tetrafluoroborates in aprotic medium, shown in reactions 20 and 21, was studied using many variations of the polarographic technique. A mechanism involving two one-electron steps was proposed for the reaction leading to nitrogen evolution. The basicity of the solvent affects the reduction mechanism. A satisfactory linear correlation was obtained in DMF between the reduction potential of the absorption prewave (equation 20) and the substituent constants σ_p^{+}[128,129].

$$ArN_2^+ + e \;\;\rightleftharpoons\;\; (ArN_2)^{\cdot} \qquad (20)$$

$$(ArN_2)^{\cdot} + 3e + 3H^+ \;\;\rightleftharpoons\;\; ArNHNH_2 \qquad (21)$$

Table 9 summarizes diazonium compounds of industrial interest.

B. Functional Derivatives

Arenediazonium salts undergo coupling with various classes of compounds, for example, reagent 17. There is evidence for nitrogen exchange with N_2 at high pressure[130]. Spectrophotometric determination of the coupling product in the UV-visible range is the measuring method of choice. The coupling reaction can be applied using a diazonium compound as analytical reagent. For example, reagent 18 was used for the determination of non-enzymatic glycosylation products of lysine[131]. Arenediazonium compounds served as photoaffinity labels for proteins[132]. For example, after binding benzenediazonium compounds *p*-substituted with electron-donating groups and light activation, acetylcholinesterase activity was inhibited irreversibly[133]. Care should be taken when using diazonium reagent that were stored for a long time, as it was found that solid arenediazonium

TABLE 9. Industrial diazonium compounds[a]

Compound and CAS registry number
4-Aminobenzenediazonium perchlorate
Benzenediazonium-2-carboxylate [1608-42-0]
Benzenediazonium chloride [100-34-5]
Benzenediazonium nitrate [619-97-6]
Benzenediazonium-4-sulphonate [2154-66-7]
Bis(5-chlorotoluenediazonium) zinc tetrachloride [89453-69-0]
o- and m-Chlorobenzenediazonium salts (e.g. [89-90-7], [100-77-6])
4-Chloro-2-methylbenzenediazonium salts (e.g. [94-95-1])
Di(benzenediazonium) zinc tetrachloride [69092-42-8]
3,6-Dimethylbenzenediazonium-2-carboxylate [68596-88-3]
4,6-Dinitrobenzenediazonium-2-olate
4-Hydroxybenzenediazonium-2-carboxylate
4-Iodobenzenediazonium-2-carboxylate [68596-90-7]
2-Naphthalenediazonium mercury trichloride [68448-47-5]
4-Nitrobenzenediazonium azide
4-Nitrobenzenediazonium nitrate [42238-29-9]
3-Nitrobenzenediazonium perchlorate [22721-24-2]
o-Toluenediazonium perchlorate [68597-04-6]
3,4,5-Triiodobenzediazonium nitrate [68596-99-6]

[a] See also Reference 11.

compounds undergo S_N-type substitutions of *ortho*- and *para*-chloro and nitro substituents in the solid state[134].

(17)   (18)

$4\text{-}HO_3SC_6H_4N_2^+$

Arenediazonium ions were titrated with sodium tetraphenylborate, using ion-selective PVC membrane electrodes[135]. Conversely, aromatic hydroxy compounds, amines and active methylene compounds were titrated with various arenediazonium compounds, using the same indicating electrodes[136]. Compound **19** was found to be a convenient titrating reagent[137]. Ampholytic diazonium compounds, for example, those containing hydroxy or carboxylic acid functions, showed unclear end-points for the tetraphenylborate titration. If a sulphonic acid was present, no direct titration was possible. An alternative determination procedure proposed was to treat these problematic diazonium compounds with excess 3-methyl-1-phenyl-5-pyrazolone (**20**), followed by back-titration with reagent **19**[138].

(19)   (20)

α-Substituted phosphorylacetaldehydes undergo coupling as shown in reaction 22[139], however, the course (equation 23) is different when R' = H. Reactions were followed by UV spectrophotometry[140].

$$4\text{-}YC_6H_4N_2X \ + \ (RO)_2P(O)CHR'CH=O$$

$$\xrightarrow{H_2O} \ 4\text{-}YC_6H_4NHN = CR'P(O)(OR)_2 \quad (22)$$

$$(Y = NO_2, SO_2NH_2; X = Cl, BF_4; R = Et, i\text{-}Pr; R' = Cl, Me, Ph)$$

$$4\text{-}YC_6H_4N_2X \ + \ (MeO)_2P(O)CH_2CH=O \xrightarrow{H_2O}$$

$$4\text{-}YC_6H_4NHN = C[P(O)(OMe)_2]CH=O \quad (23)$$

$$(Y = o\text{-}NO_2, p\text{-}NO_2, p\text{-}F, p\text{-}Cl, p\text{-}Br; X = BF_4$$

Some arenediazonium compounds serve as derivatising reagents, undergoing coupling reactions with active hydrogen compounds. The azo dyes resulting from the reaction can be determined spectrophotometrically. In some cases colour change is observed due to instability of the product, as, for example, phenylbutazone, which is determined after derivatizing according to reaction 24[141].

$$(24)$$

Arenediazonium salts are converted to the corresponding fluoroarenes in the presence of boron trifluoride–diethyl ether complex[142].

Arenediazonium compounds displace trimethylsilyl groups at the allylic position of olefinic substrates, as shown in reactions 25 and 26. If the point of insertion of the diazo group bears a hydrogen atom, a further rearrangement takes place, as in reaction 26[143].

$$(25)$$

$$(26)$$

Arenediazonium salts yield sulphone derivatives 21 when treated with benzenesulphinates, or diazonitriles 22 when treated with cyanides. In the latter case a crown compound is required for transfer into the methylene chloride solution[144].

$$ArN\!=\!NSO_2Ph \qquad ArN\!=\!N\!-\!CN$$

$$(21) \qquad\qquad (22)$$

Thiocarboxylate anions[145] and halogeno or azido groups in Si-substituted trimethyl-silanes[146] displace the diazonium group in aprotic solvents, as shown in reactions 27 and 28 respectively. Treatment with chlorotrimethylsilane and then with water leads to hydrogenation at the substitution site, as shown in reaction 29[146].

$$ArN_2^+ + RCOS^- \xrightarrow{\text{DMSO}} RCOS\!-\!Ar + N_2 \qquad (27)$$

$$ArN_2^+BF_4^- + XSiMe_3 \xrightarrow[X = I, N_3]{\text{DMF, etc.}} Ar\!-\!X + N_2 \qquad (28)$$

$$ArN_2^+BF_4^- \xrightarrow[\text{b. }H_2O]{\substack{\text{a. } Cl-SiMe_3}} ArH \qquad (29)$$

Arenediazonium compounds yield acylated arenes according to reaction 30[147].

$$ArN_2X + CO + SnR_4 \xrightarrow{Pd(OAc)_2} ArCOR \qquad (30)$$

$$(X = BF_4, PF_6; R = Me, Et, Ph)$$

Heterocyclic diazonium compounds undergo cycloaddition with compounds containing reactive CH bonds, as shown, for example, in reaction 31[148,149]. See also reaction 24.

$$(31)$$

Alkyldiazonium salts are very unstable intermediates that undergo nitrogen elimination through carbonium ion intermediates, leading to various products, depending on the nature of the organic radical and the chemical environment[150,151]. Indirect kinetic evidence was adduced for a secondary alkyldiazonium intermediate in the decomposition of (E)-2-butanediazoic acid[152].

Non-aromatic diazonium cations decompose in the presence of nucleophiles with nitrogen elimination, and do not effect coupling reactions. However, some heterocyclic diazonium compounds were found to have reactivity similar to that of arenediazonium compounds, including production of coupling derivatives. Such is the case of 1,3,4-thiadiazolinediazonium cations (reaction 32)[153] and diazonium betaines (reaction 33)[154]

$$(32)$$

$$(33)$$

In Table 10 appear complexes of transition metals with diazonium compounds.

TABLE 10. Complexes of diazonium compounds with various metals

Complexing reagents	Products	Ref.	Notes
[FeH(H$_2$)(depp)$_4$][BX$_4$] (X = F, Ph)	[FeH(N$_2$—Ar)(depp)$_3$][BX$_4$]	91	a
Pt$_2$Cl$_2$(dppm)$_2$	[Pt$_2$Cl$_2$(dppm)$_2$(N$_2$C$_6$H$_4$X-p)]$^+$[BF$_4$]$^-$ (X = NO$_2$, F, OMe)	155	b, c
[SbCl$_6$]$^-$		156	d

^{a} depp = diethoxyphenylphosphine.
^{b} For diazonium fluoroborates.
^{c} dppm = Ph$_2$PCH$_2$PPh$_2$.
^{d} Vinyldiazonium compounds are very unstable and they become stabilized by the chlorostibnate anion.

VI. ORGANIC CYANATES AND THIOCYANATES

A. General

Separations on cyanogen bromide-activated polysaccharide resins (Sepharose, Sephadex) yield imidocarbamates and cyanates. The former are determined after selective liberation of ammonia and a modification of the ninhydrin reaction. Organic cyanates can be detected and determined spectrophotometrically in nanomolar quantities, by treatment with a reagent consisting of pyridine and dimethylbarbituric acid (23) as depicted in reactions 34 and 35[157,158].

$$(34)$$

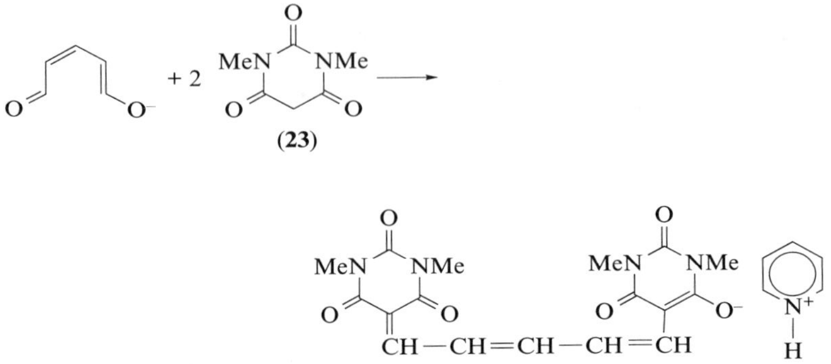

(23)

(35)

ε 157,000; λ 588 nm

Methylene dithiocyanate (see Table 11) can be analysed in water samples and industrial materials using reverse-phase HPLC with UV detector at 254 nm. The detection limit is 2 ppm[159].

In Tables 7 and 11 appear organic cyanates and thiocyanates of industrial interest.

B. Functional Derivatives

Allyl cyanates undergo a [3,3] sigmatropic rearrangement to allyl isocyanates, as depicted in reaction 36[160].

(36)

Aromatic cyanates undergo catalytic trimerisationon heating, leading to triaryl cyanurates, as shown in reaction 37[161].

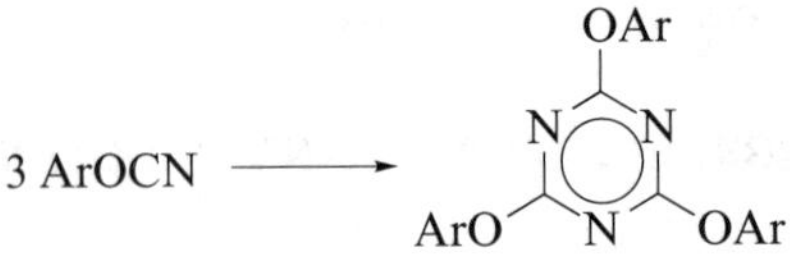

$$3 \text{ ArOCN} \longrightarrow$$

(37)

Thiocyanates undergo condensation with active methylene compounds in the presence of nickel acetylacetonate, as depicted in reaction 38[162].

$$\text{RSCN} + \text{MeCOCH}_2\text{COR}' \longrightarrow \text{MeCO}-\underset{\underset{\text{R'CO}}{|}}{\text{C}}=\underset{\underset{\text{SR}}{|}}{\text{C}}-\text{NH}_2$$

(38)

(R = Me, Et, Pr, Ph, p-Tol; R′ = Me, Ph, OEt)

In Table 8 appear derivatives from unsaturated thiocyanates and silver and mercury salts.

TABLE 11. Industrial organic thiocyanates[a]

Compound and CAS registry number	NIOSH/OSHA RTECS[b]
Allyl thiocyanate [764-49-8]	XL 2000000
Amyl thiocyanate [32446-40-5]	XL 1750000
Benzyl thiocyanate [3012-37-1][c]	XK 8155000
Bicyclo(2.2.1)hept-5-en-2-ylmethyl thiocyanatoacetate [63906-51-4]	AJ 5950000
2-(2-Butoxyethoxy)ethyl thiocyanate [112-56-1]	XK 8400000
n-Butyl thiocyanate [628-83-1]	XK 8500000
2-Chloroethyl thiocyanate [928-57-4]	XK 8750000
2,4-Dinitrophenyl thiocyanate [1594-56-5]	XK 9450000
n-Dodecyl thiocyanate [765-15-1]	XK 9625000
Ethyl thiocyanate [542-90-5]	XK 9900000
5-Fluoroamyl thiocyanate [661-18-7]	XL 0950000
2-Fluoroethyl thiocyanate [505-13-5]	XL 0700000
6-Fluorohexyl thiocyanate [408-28-6]	XL 0875000
3-Fluoropropyl thiocyanate	
Isobornyl thiocyanatoacetate [115-31-1]	AJ 6300000
Methylene dithiocyanate [6317-18-6]	XL 1560000
Methyl thiocyanate [556-64-9]	XL 1575000
4-(2-Nitro-1-propenyl)phenyl thiocyanate [950-00-5]	XL 1642000
Octyl thiocyanate [19942-78-0]	XL 1650000
2-Oxo-2-(phenylamino)ethyl selenocyanate [63981-21-5]	VT 0875000
Phenyl thiocyanate [5285-87-0]	XL 1850000
4-Thiocyanatoaniline [2987-46-4]	XK 7700000
3-Thiocyanatobenzoic acid [16671-87-7]	DH 6865000
4-Thiocyanatobenzoic acid [16671-88-8]	DH 6905000
4-Thiocyanato-N,N-dimethylaniline [7152-80-9]	BX 8050000
Thiocyanic acid ester with 2,2'-(methylamino)diethanol hydrochloride [63915-59-3]	XK 9800000
Trichloromethyl thiocyanate [20233-04-9]	XL 2450000
Triphenylstannyl thiocyanate [7224-23-9]	WH 8900000

[a] See also Reference 11.
[b] *Registry of Toxic Effects of Chemical Substances* of National Institute for Occupational Safety and Health/ Occupational Safety and Health Administration.
[c] See also Reference 21.

VII. ORGANIC ISOCYANATES AND ISOTHIOCYANATES

A. General

A dithiocarbamate antifilarial agent and its active isothiocyanate derivative in biological fluids were analysed by HPLC[163].

Concentrations of the pesticide methyl isothiocyanate (MITC, see Table 13) in air were measured around fumigated fields and compared with a calculated model[164]. Traces of MITC in water could be determined with a detection limit of 1 μg/L, by a large volume injection of water samples in a reverse-phase LC column. The detection limits were lowered to 100 pg/L by a fast extraction[165]. An alternative procedure for MITC in water, with the same detection limits, is HPLC using an octadecyl bonded column, with a mobile phase containing hexadecyltrimethylammonium bromide[166]. Methyl isothiocyanate was determined in wines by static headspace GC[167].

Primary amines in biological samples (plasma, urine, cerebrospinal fluid and tissues) were converted to isothiocyanates by treatment with carbon disulphide, and were determined by GC using a nitrogen detector[168].

Various alkyl isothiocyanates formed on enzymatic hydrolysis of non-indolyl glucosinolates of rapeseed were determined by GC-MS[169,170].

Carbamoyl isothiocyanates (24) in solution exist in equilibrium with thiocarbamoyl cyanates (25). This equilibrium is well established at 80–90 °C. The species have distinct absorption bands in the IR region at 1980 and 2240 cm^{-1} for 24 and 25, respectively[171].

$$
\underset{\underset{O}{\|}}{RR'NC} - N = C = S \rightleftharpoons \underset{\underset{S}{\|}}{RR'NC} - O - C \equiv N
$$

(24)   (25)

In Tables 12 and 13 appear organic isocyanates and isothiocyanates of industrial interest.

TABLE 12. Industrial organic isocyanates^a

Compound and CAS registry number	NIOSH/OSHA RTECS^b
Allyl isocyanate [1476-23-9]	NQ 8175000
1,2-Benzanthryl-10-isocyanate [63018-56-4]	NO 8225000
Benzene-1,3-diisocyanate [123-61-5]	NR 0150000
Butyl isocyanate [111-36-4]	NQ 8250000
6-Chlorohexyl isocyanate [13654-91-6]	NQ 8500000
3-Chlorophenyl isocyanate [2909-38-8]	NQ 8560000
4-Chlorophenyl isocyanate [104-12-1]	NQ 8575000
Bianisidine diisocyanate [91-93-0]	NQ 8800000
1,6-Diisocyanatohexane [822-06-0]	MO 1740000
(Diisocyanatomethyl)benzene [1321-38-6]	CZ 6200000
2,4-Diisocyanatotoluene [584-84-9]	CZ 6300000
2,6-Diisocyanatotoluene [91-08-7]	CZ 6310000
3,3'-Dimethyl-4,4'-biphenylene diisocyanate [91-97-4]	DV 3960000
Diphenylmethane diisocyanate [26447-40-5; 101-68-8]	NQ 9350000
Durene isocyanate [58149-28-3]	DA 3700000
Ethyl isocyanate [109-90-0]	NQ 8825000
Isophorone diisocyanate [4098-71-9]	NQ 5400000
5,5'-Methylenebis(2-isocyanatotoluene) [139-25-3]	NQ 8820000
Methyl isocyanate [624-83-9]	NQ 9450000
1,5-Naphthalene diisocyanate [3173-72-6]	
3-Nitro-3-azapentane 1,5-diisocyanate [7046-71-9]	NR 0100000
Octadecyl isocyanate [112-96-9]	NR 0140000
Phenyl isocyanate [103-71-9]	DA 3675000
Propyl isocyanate [110-78-1]	NR 0190000
Tributylstannyl isocyanate [681-99-2]	WH 8583500
m-Trifluoromethylphenyl isocyanate [329-01-1]	NR 0200000
m-Xylene α,α'-diisocyanate [3634-83-1]	ZE 4375000

^{a} See also Reference 11.
^{b} Registry of Toxic Effects of Chemical Substances of National Institute for Occupational Safety and Health/ Occupational Safety and Health Administration.

TABLE 13. Industrial organic isothiocyanates[a]

Compound and CAS registry number	NIOSH/OSHA RTECS[b]
3-Acetamidophenyl isothiocyanate [3137-83-5]	NX 8220000
Allyl isothiocyanate [57-06-7]	NX 8225000
Benzyl isothiocyanate [622-78-6]	NX 8250000
4-Chlorobenzyl isothiocyanate [3694-45-9]	NX 8470000
4-Chlorophenyl isothiocyanate [2131-55-7]	NX 8473000
4-Cyanophenyl isothiocyanate [2719-32-6]	NX 8479000
Cyclohexyl isothiocyanate [1122-82-3]	NX 8480000
Ethylene diisothiocyanate [3688-08-2]	NX 8580000
3-Fluorophenyl isothiocyanate [404-72-8]	NX 8750000
4-Fluorophenyl isothiocyanate [1544-68-9]	NX 8760000
Methyl isothiocyanate [556-61-1][c]	PA 9625000
1-Naphthyl isothiocyanate [551-06-4]	NX 9100000
3-Nitrophenyl isothiocyanate [3529-82-6]	NX 9110000
2-Phenylethyl isothiocyanate [2257-09-2]	NX 9115000
Phenyl isothiocyanate [103-72-0]	NX 9275000
Trimethylstannyl isothiocyanate [15597-43-0]	WH 8583600
Tripropylstannyl isothiocyanate [31709-32-7]	WH 8583700

[a] See also Reference 11.
[b] *Registry of Toxic Effects of Chemical Substances* of National Institute of Occupational Safety and Health/ Occupational Safety and Health Administration.
[c] See also Reference 21.

B. Functional Derivatives

A general scheme for synthesis of isocyanate and isothiocyanate derivatives is the reaction with amines (equation 39). This yields derivatives of urea or thiourea, however, the methods should be developed in accordance with the intended application as an analytical tool.

$$RN{=}C{=}X + R'R''NH \longrightarrow RNHCXNR'R''$$ (39)

$$(X = O, S; R, R', R'' = alkyl, aryl, H)$$

Isocyanates can be determined by indirect titration of dibutylamine[172] or by GC determination of this reagent[173], after reaction 39 (X = O; R' = R'' = Bu) goes to completion. A variation of this procedure for isothiocyanate insecticides and other formulations consists of treating a solution in acetonitrile with excess primary amine, scavenging the excess amine with carbon disulphide, yielding a dithiocarbamate and determining the concentration by a 3-step titration with Cu(II) in acetonitrile[174].

A colourimetric method was developed for determination of allyl isothiocyanate, with a detection limit of *ca* 1 μg S. It is based on alkaline hydrolysis to yield sulphide ions, which subsequently react (as shown in equation 40) to yield methylene blue, which in turn is measured at 665 nm. The following sulphur-containing compounds were tested as possible interference with the method: Thioacetamide, thiourea, cysteine, methionine, cysteic acid,

(40)

thioacetamide, benzenesulphonamide, glucose-6-sulphate and sinigrin. None showed interference in the determination when their S content was less than ten fold that of allyl isothiocyanate[175].

Organic isothiocyanates have been isolated from marine fauna[176] and plants[177,178]. Alkyl, allyl and benzyl isothiocyanates occur in nature as glucosinolates, in plants such as garden cress, horseradish and mustard. Their principal metabolite is a mercapturic acid (**26**) present in urine. It is analysed by decomposing at pH 5 according to reaction 41; the isothiocyanate is treated with *n*-butylamine (reaction 39, X = S) and the resulting thiourea is extracted and analysed by HPLC[179].

$$RNHC(=S)-SCH_2CH(NHAc)CO_2H \longrightarrow RNCS + HSCH_2CH(NHAc)CO_2H \quad (41)$$

(**26**) R = PhCH$_2$, Allyl, Alkyl

Isocyanates are derivatised with tryptamine (**27**) and determined with a dual detection system based on fluorescence and amperometric oxidation[180]. Tryptamine and *N*-ω-methyltryptamine (**28**) were found to be the most promising scavenging reagents for airborne organic isocyanates, before analysis by HPLC with reverse-phase columns and UV detection[181]. A personal sampler for airborne isocyanates in workplaces was developed based on **27**[182].

(**27**)   (**28**)

Isothiocyanates may be converted into derivatives of thiourea or 4-imidazoline-2-thione as shown in reactions 42 and 43, respectively[183].

$$RNCS + PhCOCH_2NH_2 \longrightarrow RNHCSNHCH_2COPh \quad (42)$$

$$RNCS + MeCOCH_2NH_2 \longrightarrow R-N \underset{\underset{S}{\parallel}}{\overset{Me}{\diagup}} NH \quad (43)$$

Table 8 shows cobalt complexes derived from organic isothiocyanates.

Some isothiocyanates are important analytical tools in protein research. The Edman method for sequence analysis of proteins consists of attaching phenyl isothiocianate to the amino end of the protein chain, and converting the thiourea into a 5-substituted 3-phenyl-2-thiohydantoin, as shown in reactions 44 and 45. After hydrolysis of the peptide bonds the end amino acid is labelled accordingly[184]. The method has been investigated in automatic sequencing at sub-picomole levels, using HPLC[185]. Tyrosine *O*-sulphate was identified in proteins, after hydrolysis with barium hydroxide, chromatography on Dowex AG 50 × 8, precolumn derivatizing with phenyl isothiocyanate and reversed-phase HPLC[186]. The

$$NH_2CHRCONH\text{-protein} + PhN=C=S \xrightarrow{OH^-}$$

$$PhNHCSNHCHRCONH\text{-protein} \quad (44)$$

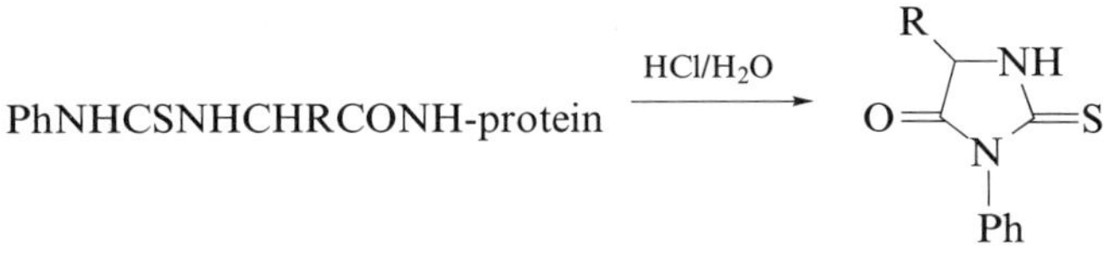

PhNHCSNHCHRCONH-protein $\xrightarrow{\text{HCl/H}_2\text{O}}$

(45)

4-Me$_2$NC$_6$H$_4$—N=N—C$_6$H$_4$—N=C=S-4

(29)

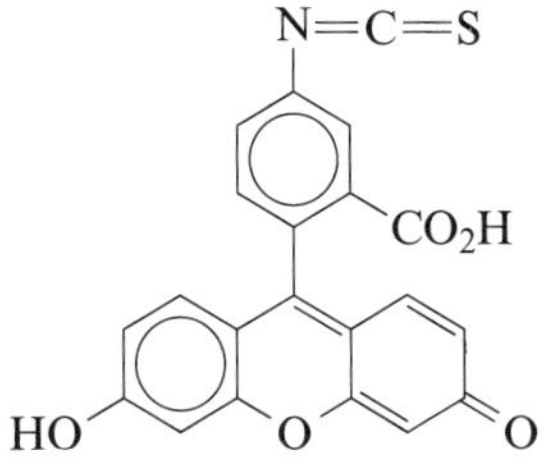

HO

(30)

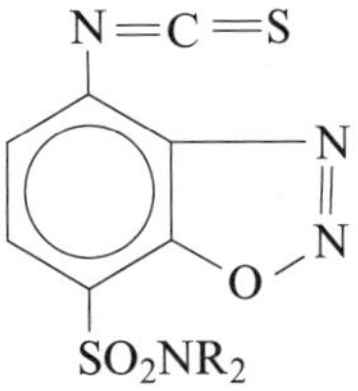

SO$_2$NR$_2$

(31) R$_2$ = H$_2$, Me$_2$

SO$_2$(CH$_2$)$_3$—N=C=S

NMe$_2$

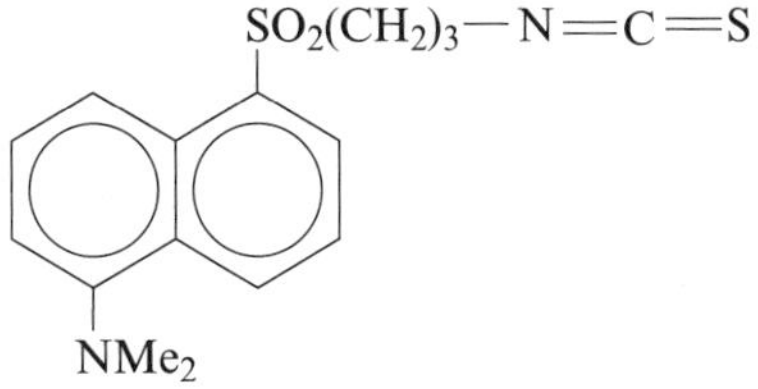

(32)

activity of β-aspartylpeptidase was determined by precolumn derivatizing with phenyl isothiocyanate, and HPLC[187]. Derivatising with phenyl isothiocyanate was proposed for fast HPLC determination of free amino acids in biological samples[188]. Other isothiocyanates, such as dimethylaminoazobenzene isothiocyanate **(29)**[189–197], fluorescein isothiocyanate **(30)**[198–211] and the benzoxadiazolyl derivatives **31**[212], were also proposed as aids in protein sequence analysis and as fluorescent labels in biological and biochemical studies. Reagent **32** was recommended for fluorescent labeling with a spacer between the bound amino group and the fluorophore[213]. Aminofluorescein, a possible impurity of reagent **30**, can be determined spectrofluorometrically[214].

Amino sugars and amino sugar alcohols can be derivatized with phenyl isothiocyanate and determined by HPLC. Some amino sugars rendered also cyclic compounds, after the hemiacetal form opened and reacted further[215].

Ethylenediaminephosphoramidate derivatives of nucleotides have been prepared and further derivatised with **30**[216].

2,3,4,6-Tetra-O-acetyl-β-D-glucopyranosyl isothiocyanate (TAGI, **33**) is an example of a chiral reagent that takes advantage of reaction 39 to attain derivatives of asymmetric amines before chromatographic analysis[217–221]. TAGI is capable of reacting with amino groups even in aqueous media. It was used for HPLC separation of enantiomers of drugs[222], thiols[223], amino acids, β-adrenergic blockers, alkyoxiranes[224], cyclic amino acids

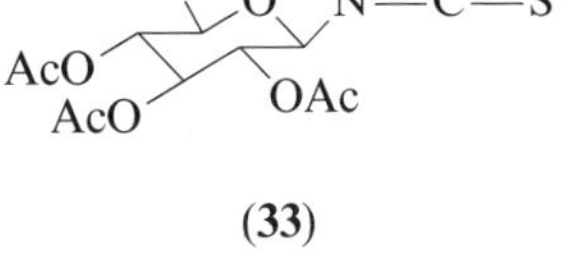

AcOCH$_2$

AcO

AcO   OAc

(33)

Ph—C···Me

H

N=C=S

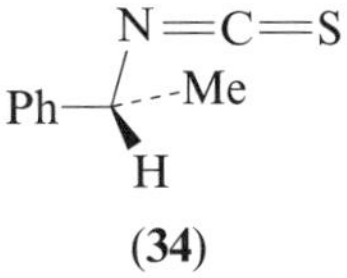

(34)

and β-amino acids[225]. R-α-Methylbenzyl isothiocyanate (**34**) was proposed as an alternative to **33** for chiral amines, and resolution is performed on reverse-phase HPLC[226]. Reagents **33** and **34** are commercially available.

Isothiocyanatobenzyl-EDTA is a chelating agent for the determination of the yield of antibody radiolabeling[227].

VIII. REFERENCES

1. K. A. Connors, in *The Chemistry of Carbon–Carbon Triple Bonds* (Ed. S. Patai), Wiley-Interscience, Chichester, 1978, pp. 137ff; see also *Supplement C, The Chemistry of Triple-Bonded Functional groups* (Eds. S. Patai and Z. Rappoport), Wiley-Interscience, Chichester, 1983.
2. D. J. Curran and S. Siggia, in *The Chemistry of the Cyano Group* (Ed. Z. Rappoport), Wiley-Interscience, London, 1970, pp. 167ff.
3. D. A. Ben-Efraim, in *The Chemistry of Diazonium and Diazo Groups* (Ed. S. Patai), Wiley-Interscience, Chichester, 1978, pp. 149ff.
4. D. A. Ben-Efraim, in *The Chemistry of Cyanates and their Thio Derivatives* (Ed. S. Patai), Wiley-Interscience, Chichester, 1978, pp. 191ff.
5. J. Zabicky, in *The Chemistry of Sulphenic Acids and their Derivatives* (Ed. S. Patai), Wiley-Interscience, Chichester, 1990, pp. 83ff.
6. J. Zabicky, in *The Chemistry of the Amino Group* (Ed. S. Patai), Wiley-Interscience, London, 1968, pp. 79ff.
7. *Elemental Analyzer CHN-O Rapid, Operating Instructions*, W. C. Heraeus GmbH, Hanau, Germany, 1986.
8. B. Colombo, M. Baccant and J. Theobald, *Int. Labmate*, **13**, [7] (1988).
9. P. W. J. M. Boumans, *Inductively Coupled Plasma Emission Spectroscopy*, Parts I and II, Wiley-Interscience, New York, 1987.
10. R. E. Lenga (Ed.), *The Sigma-Aldrich Library of Chemical Safety Data*, 2nd edn., Sigma-Aldrich Corp., Milwaukee, 1988.
11. N. I. Sax, *Dangerous Properties of Industrial Materials*, 6th edn., Van Nostrand Reinhold, New York, 1984.
12. D. A. Peake, M. L. Gross and S. K. Huang, *Anal. Chem.*, **59**, 1557 (1987).
13. M. M. Buzlanova, R. Shahid, I. I. Karandi, S. I. Obtemperanskaya and N. N. Godovikov, *J. Anal. Chem. USSR*, **43**, 1530 (1988).
14. R. R. Arnts, R. L. Seila and J. J. Bufalini, *Proc. APCA 80th Am. Mat.*, Vol. 4, 87/71.4 (1987); *JAPCA—J. Air Waste Manag. Assoc.*, **39**, 453 (1989).
15. H. Tai, K. Yamamoto, M. Uchida, S. Osawa and K. Uehara, *IEEE Photonics Tech. Lett.*, **4**, 804 (1992).
16. F. S. Pavone, F. Marin, M. Inguscio, K. Ernst and G. Dilonardo, *Appl. Optics*, **32**, 259 (1993).
17. R. Zander, J. Brault and G. Stokes, *C. R. Acad. Sci., Ser. 2*, 295 (1982).
18. R. Yu and T. Gustafsson, *J. Vacuum Sci. Technol. A, Vacuum Surf. Films*, **5**, 814 (1987).
19. G. Casalone, M. G. Cattania and M. Simonetta, *Surf. Sci.*, **103**, L121 (1981).
20. J. D. Teresa, M. Grande, J. M. Hernandez, J. R. Moran, J. G. Urones and M. A. Villaseco, *Planta Medica*, 458 (1986).
21. C. R. Worthing (Ed.), *The Pesticide Manual*, British Crop Protection Council, 1991.
22. L. S. Silbert and T. A. Foglia, *J. Am. Oil Chem. Soc.*, **61**, 696 (1984).
23. L. S. Silbert and T. A. Foglia, *Anal. Chem.*, **57**, 1404 (1985).
24. H. Yasuda and A. Nakamura, *Angew. Chem., Int. Ed. Engl.*, **26**, 723 (1987).
25. P. Roques, J. Y. Legall, L. Lacombe and M. Olomucki, *J. Org. Chem.*, **57**, 1579 (1992).
26. X. Y. Lu, G. X. Zhu and S. M. Ma, *Tetrahedron Lett.*, **33**, 7205 (1992).
27. C. R. Noe, M. Knollmuller, B. Oberhauser, G. Steibauer and E. Wagner, *Chem. Ber.*, **119**, 729 (1986).
28. M. J. Robins, R. S. Vinayak and S. G. Wood, *Tetrahedron Lett.*, **31**, 3731 (1990).
29. S. F. Vasilevskii, M. V. Nikitin and M. S. Shvartsberg, *Bull. Acad. Sci. USSR, Div. Chem. Sci.*, **40**, 2306 (1991).
30. C. K. Reddy and M. Periasamy, *Tetrahedron Lett.*, **31**, 1919 (1990).
31. J. Anthony and F. Diederich, *Tetrahedron Lett.*, **32**, 3787 (1991).
32. E. Lukevits, R. Y. Strukovich and O. A. Pudova, *Zh. Obshch. Khim.*, **58**, 815 (1988).

4. Analytical aspects of triple-bonded functional groups227

33. P. H. Dixneuf, *Pure Appl. Chem.*, **61**, 1763 (1989).
34. K. Takai, Y. Kataoka, K. Toshizumi, Y. Oguchi and K. Utimoto, *Chem. Lett.*, 1479 (1991).
35. N. E. Schore, *Chem. Rev.*, **88**, 1081 (1988).
36. E. Solari, C. Floriani, A. Chiesivilla and C. Guastini, *J. Chem. Soc., Chem. Commun.*, 1747 (1989).
37. P. Alphonse, F. Moyen and P. Mazerolles, *J. Organomet. Chem.*, **345**, 209 (1988).
38. A. Funhoff, H. Schaufele and U. Zenneck, *J. Organomet. Chem.*, **345**, 331 (1988).
39. T. Mitsudo, S. W. Zhang, M. Nagao and Y. Watanabe, *J. Chem. Soc., Chem. Commun.*, 598 (1991).
40. U. M. Dzhemilev, V. R. Khafizov and F. A. Selimov, *Bull. Acad. Sci. USSR, Div. Chem. Sci.*, **35**, 1478 (1986).
41. M. Tanaka, Y. Ichimaru and H. J. Lautenschlager, *Organometallics*, **10**, 16 (1991).
42. D. M. Hoffman, C. R. Fisel and R. Hoffmann, *J. Am. Chem. Soc.*, **104**, 3858 (1982).
43. D. M. Hoffman and R. Hoffmann, *J. Chem. Soc., Dalton Trans.*, 1471 (1982).
44. A. Mangia, G. Predieri and E. Sappa, *Anal. Chim. Acta*, **149**, 349 (1983).
45. A. Devasagayaraj and M. Periasamy, *Tetrahedron Lett.*, **30**, 595 (1989).
46. R. Yáñez, N. Lugan and R. Mathieu, *Organometallics*, **9**, 2998 (1990).
47. I. Ojima, N. Clos, R. J. Donovan and P. Ingallina, *Organometallics*, **10**, 3211 (1991).
48. R. Aumann and H. Heinen, *J. Organomet Chem.*, **389**, C1 (1990).
49. K. H. Dotz, H. G. Erben and K. Harms, *J. Chem. Soc., Chem. Commun.*, 692 (1989).
50. D. L. Reger and M. F. Huff, *Organometallics*, **11**, 69 (1992).
51. D. L. Reger, M. F. Huff, T. A. Wolfe and R. D. Adams, *Organometallics*, **8**, 848 (1989).
52. D. Seyferth, J. B. Hoke and G. B. Womack, *Organometallics*, **9**, 2662 (1990).
53. C. Banchini, A. Meli, M. Peruzzini, P. Frediani, C. Bohanna, M. A. Esteruelas and L. A. Oro, *Organometallics*, **11**, 138 (1992).
54. P. J. Stang and C. M. Crittell, *Organometallics*, **9**, 3191 (1990).
55. P. J. Stand and Y. H. Huang, *J. Organomet. Chem.*, **431**, 247 (1992).
56. J. L. Templeton, *Adv. Organomet. Chem.*, **29**, 1 (1989).
57. M. L. H. Green, P. C. Konidaris, P. Mountford and S. J. Simpson, *J. Chem. Soc., Chem. Commun.*, 256 (1992).
58. J. B. Hartung and S. F. Pedersen, *Organometallics*, **9**, 1414 (1990).
59. T. Batik, B. Happ, M. Iglewsky, H. Bandmann, R. Boese, P. Heimbach, T. Hoffmann and E. Wenschuh, *Organometallics*, **11**, 1235 (1992).
60. R. D. Adams, J. E. Cortopassi and M. P. Pompeo, *Organometallics*, **11**, 1 (1992).
61. M. Rashidi, G. Schoettel, J. J. Vittal and R. J. Puddephatt, *Organometallics*, **11**, 2424 (1992).
62. L. L. Padolik, J. Gallucci and A. Wojcicki, *J. Organomet. Chem.*, **383**, 383 (1990).
63. Y. Kim, J. Gallucci and A. Wojcicki, *Organometallics*, **11**, 1963 (1992).
64. H. Lebozec, K. Ouzzine and P. H. Dixneuf, *Organometallics*, **10**, 2768 (1991).
65. A. Romero, A. Santos and A. Vegas, *Organometallics*, **7**, 1988 (1988).
66. U. Rosenthal, H. Gorls, V. V. Burlakov, V. B. Shur and M. E. Volpin, *J. Organomet. Chem.*, **426**, C53 (1992).
67. H. G. Alt and G. S. Herrmann, *J. Organomet. Chem.*, **390**, 159 (1990).
68. K. H. Theopold, S. J. Holmes and R. R. Schrock, *Angew. Chem., Int. Ed. Engl.*, **22**, 1010 (1983).
69. P. K. Baker, M. J. Morris, P. F. Reinisch, M. McPartlin and H. R. Powell, *Polyhedron*, **8**, 2587 (1989).
70. M. B. Oregan, M. G. Vale, J. F. Payack and R. R. Schrock, *Inorg. Chem.*, **31**, 1112 (1992).
71. R. Rubio, M. T. Galceran and G. Rauret, *Analyst*, **115**, 959 (1990).
72. S. C. Lee, B. J. Stanton, B. A. Eldridge and E. L. Wehry, *Anal. Chem.*, **64**, 268 (1992).
73. L. Do and F. Raulin, *J. Chromatogr.*, **481**, 45 (1989).
74. L. Do and F. Raulin, *J. Chromatogr.*, **514**, 65 (1990).
75. L. Do and F. Raulin, *J. Chromatogr.*, **591**, 297 (1992).
76. E. Devanssay, P. Capilla, D. Coscia, L. Do, R. Sternberg and F. Raulin, *J. Chromatogr.*, **639**, 255 (1993).
77. M. Lafosse and M. Dreux, *J. Chromatogr.*, **193**, 9 (1980).
78. Y. Suzuki and T. Inoue, *Bunseki Kagaku*, **35**, 614 (1986); *Chem. Abstr.*, **105**, 183141 (1986).
79. R. V. Kononova, *Ind. Lab. USSR*, **52**, 796 (1986).
80. M. L. Trehy, J. J. McCreary and R. A. Yost, *Anal. Chem.*, **56**, 1281 (1984).
81. U. Ehrmann, L. Carbognani and C. Ceccarelli, *J. High Resolut. Chromatogr.*, **15**, 467 (1992).
82. V. B. Patil, M. T. Sevalkar and S. V. Padalikar, *Analyst*, **117**, 75 (1992).

83. I. G. Binev, I. N. Juchnovski, J. Kanety and R. B. Kuzmanova, *J. Chem. Soc., Perkin Trans. 2*, 1533 (1982).
84. O. A. Attanasi, L. Decrescentini, S. Santeusanio, F. Serrazanetti, A. Mckillop and Z. Y. Liao, *J. Chem. Soc., Perkin Trans. 1*, 1009 (1992).
85. M. C. Carre, A.S. Ezzinadi, M. A. Zouaoui, P. Geoffroy and P. Caubere, *Synth. Commun.*, **19**, 3323 (1989).
86. G. Bartoli, M. Bosco, C. Cimarelli, R. Dalpozzo, G. Demunno, G. Guercio and G. Palmieri, *J. Org. Chem.*, **57**, 6020 (1992).
87. J. A. Lavilla and J. L. Goodman, *Tetrahedron Lett.*, **29**, 2623 (1988).
88. D. P. Fairlie and W. G. Jackson, *Inorg. Chem.*, **29**, 140 (1990).
89. A. A. Ismail, I. S. Butler, G. Jaouen, R. J. Angelici and G. N. Glavee, *Inorg. Synth.*, **26**, 31 (1989).
90. M. Hanack, G. Renz, J. Strahle and S. Schmid, *J. Org. Chem.*, **56**, 3501 (1991).
91. G. Albertin, S. Antoniutti and E. Bordignon, *J. Am. Chem. Soc.*, **111**, 2072 (1989).
92. F. Balssa, K. Aniss and F. Rosemunch, *Tetrahedron Lett.*, **33**, 1901 (1992).
93. C. S. Chin, S. N. Choi and K. D. Kim, *Inorg. Chem.*, **29**, 145 (1990).
94. M. I. Bruce, T. W. Hambley, M. J. Liddell, A. G. Swincer and E. R. T. Tiekink, *Organometallics*, **9**, 2886 (1990).
95. P. G. Huggett, R. J. Lynch, T. C. Waddington and K. Wade, *J. Chem. Soc., Dalton Trans.*, 1164 (1980).
96. M. Mishima and T. Okuda, *Bull. Chem. Soc. Jpn.*, **63**, 1206 (1990).
97. J. R. Strickler and D. E. Wigley, *Organometallics*, **9**, 1665 (1990).
98. M. Vivanco, J. Ruiz, C. Floriani, A. Chiesivilla and C. Guastini, *Organometallics*, **9**, 2185 (1990).
99. A. V. Sereda, I. E. Sukhov, B. M. Zolotarev, I. V. Yartseva and O. N. Tolkarev, *Tetrahedron Lett.*, **33**, 4205 (1992).
100. C. M. Che, W. T. Wong, T. F. Lai and H. L. Kwong, *J. Chem. Soc., Chem. Commun.*, 243 (1989).
101. I. Beaumont and A. H. Wright, *J. Organomet. Chem.*, **425**, C11 (1992).
102. M. O. Albers and N. J. Coville, *Inorg. Synth.*, **28**, 140 (1990).
103. R. A. Michelin, G. Facchin and P. Ugliagliati, *Inorg. Chem.*, **23**, 961 (1984).
104. R. Aumann, *Chem. Ber.*, **125**, 1141 (1992).
105. M. O. Albers, E. Singleton and N. J. Coville, *Inorg. Synth.*, **28**, 179 (1990).
106. M. B. Baptista, M. A. N. D. A. Lemos, J. J. R. F. Dasilva and A. J. L. Pombeiro, *J. Organomet. Chem.*, **424**, 49 (1992).
107. G. Simonneaux, F. Hindre and M. Leplouzennec, *Inorg. Chem.*, **28**, 823 (1989).
108. B. Hessen, J. Blenkers, J. H. Teuben, G. Helgesson and S. Jagner, *Organometallics*, **8**, 830 (1989).
109. A. M. Martins, M. J. Calhorda, C. C. Romao, C. Volkl, P. Kiprof and A. C. Filippou, *J. Organomet. Chem.*, **423**, 367 (1992).
110. T. Beringhelli, G. Dalfonso, M. Freni, G. Ciani, M. Moret and A. Sironi, *J. Organomet. Chem.*, **399**, 291 (1990).
111. K. R. Dunbar and S. C. Haefner, *Organometallics*, **11**, 1431 (1992).
112. J. Montoya, A. Santos, J. Lopez, A. M. Echavarren, J. Ros and A. Romero, *J. Organomet. Chem.*, **426**, 383 (1992).
113. C. Geze, N. Legrand, A. Bondon and G. Simonneaux, *Inorg. Chim. Acta*, **192**, 73 (1992).
114. F. Y. Petillon, P. Schollhammer and J. Talarmin, *J. Organomet. Chem.*, **411**, 159 (1991).
115. T. Legall, J. P. Lellouche, L. Toupet and J. P. Beaucourt, *Tetrahedron Lett.*, **30**, 6517 (1989).
116. J. N. Kim and E. K. Ryu, *J. Org. Chem.*, **57**, 1088 (1992).
117. M. D. Taylor, R. J. Himmelsbach, B. E. Kornberg, J. Quin, E. Lunney and A. Michel, *J. Org. Chem.*, **54**, 5585 (1989).
118. H. Werner, U. Brekau, O. Nurnberg and B. Zeier, *J. Organomet. Chem.*, **440**, 389 (1992).
119. M. Giffard and I. Leaute, *J. Chem. Res. (S)*, 320 (1990).
120. H. Werner and B. Stecker, *J. Organomet. Chem.*, **413**, 379 (1991).
121. B. C. Gilbert, P. Hanson, J. R. Jones, A. C. Whitwood and A. W. Timms, *J. Chem. Soc., Perkin Trans. 2*, 1715 (1992).
122. L. M. Anderson, A. R. Butler and A. S. McIntosh, *J. Chem. Soc., Perkin Trans. 2*, 1239 (1987).
123. J. R. Pladziewicz and M. P. Doyle, *J. Imaging Sci.*, **33**, 57 (1989).
124. J. Kubias and J. Lakomy, *Chem. Listy*, **80**, 400 (1986).
125. H. Oshima, M. Friesen, C. Malaveille, I. Brouet, A. Hautefeuille and H. Bartsch, *Food Chem. Technol.*, **27**, 193 (1989).
126. K. Dornberger, W. Ihn, L. Radics, W. Schade, D. Tresselt and A. Zureck, *Tetrahedron Lett.*, **27**, 559 (1986).

127. K. Dornberger, W.Ihn, L. Radics, W. Schade, D. Tresselt and A. Zureck, *Pharmazie*, **42**, 212 (1987).
128. P. Janderka and I. Cejpek, *Collect. Czech. Chem. Commun.*, **54**, 1496 (1989).
129. P. Janderka and I. Cejpek, *Collect. Czech. Chem. Commun.*, **54**, 3135 (1989).
130. M. D. Ravenscroft, K. Tagaki, B. Weiss and H. Zollinger, *Gazz. Chim. Ital.*, **117**, 353 (1987).
131. G. Candiano, G. Delfino, G. M. Ghiggeri, F. Ginevri and C. Queirolo, *Anal. Chim. Acta*, **184**, 323 (1986).
132. F. Kotzybahibert, J. Langenbuchcachat, M. Goeldner and J. Jaganthen, *FEBS Lett.*, **182**, 297 (1985).
133. B. L. Kieffer, M. P. Goeldner and C. G. Hirth, *J. Chem. Soc., Chem. Commun.*, 398 (1981).
134. R. W. Trimmer, L. R. Stover and A. C. Skjold, *J. Org. Chem.*, **50**, 3612 (1985).
135. K. Vytras, M. Remes and H. Kubesova-Svoboda, *Anal. Chim. Acta*, **124**, 91 (1981).
136. K. Vytras, J. Kalous and T. Capoun, *Anal. Chim. Acta*, **162**, 141 (1984).
137. K. Vytras, J. Latinak, T. Capoun and H. Svoboda, *Chem. Prum.*, **32**, 81 (1982); *Chem. Abstr.*, **97**, 57105 (1982).
138. K. Vytras, J. Kalous and M. Vosmanska, *Anal. Chim. Acta*, **175**, 313 (1985).
139. M. P. Sokolov, G. V. Mavrin, I. G. Gazizov, V. N. Ivanova and T. A. Zyablikova, *Zh. Obshch. Khim.*, **59**, 53 (1989).
140. M. P. Sokolov, I. G. Gazizov, V. N. Ivanova, B. G. Liorber, G. V. Mavrin and V. A. Pavlov, *Zh. Obshch. Khim.*, **57**, 1249 (1987).
141. G. Waldheim, H. Mohrle and S. Rudiger, *Pharmazie*, **42**, 11 (1987).
142. K. Shinhama, S. Aki, T. Furuta and J. I. Minamikawa, *Synth. Commun.*, **23**, 1577 (1993).
143. H. Mayr and K. Grimm, *J. Org. Chem.*, **57**, 1057 (1992).
144. M. F. Ahern, J. R. Beadle, G. W. Gokel and A. Leopold, *J. Am. Chem. Soc.*, **104**, 548 (1982).
145. G. Petrillo, M. Novi, G. Garbarino and M. Filiberti, *Tetrahedron*, **45**, 7411 (1989).
146. T. Keumi, T. Umeda, Y. Inoue and H. Kitajima, *Bull. Chem. Soc. Jpn.*, **62**, 89 (1989).
147. K. Kikukawa, K. Kono, T. Matsuda and F. Wada, *Chem. Lett.*, 1 (1982).
148. H. S. Elkashef, H. H. Alnima, M. H. Elnagdi and K. U. Sadek, *J. Chem. Eng. Data*, 103 (1982).
149. M. K. A. Ibrahim, A. H. Elghandour and K. Abouhadeed, *Phosphorus Sulfur Silicon Relat. Elem.*, **60**, 119 (1991).
150. W. Kirmse, J. Rode and K. Kirmse, *Chem. Ber.*, **119**, 3672 (1986).
151. W. Kirmse and J. Rode, *Chem. Ber.*, **119**, 3694 (1986).
152. J. I. Finneman, J. Ho and J. C. Fishbein, *J. Am. Chem. Soc.*, **115**, 3016 (1993).
153. V. I. Rabinov and M. V. Gorelik, *Zh. Org. Khim.*, **23**, 663 (1987).
154. R. Neidlein and A. A. Johmann, *Monatsh. Chem.*, **122**, 215 (1991).
155. F. Neve, M. Ghedini, G. Demunno and A. Crispini, *Inorg. Chem.*, **31**, 2979 (1992).
156. R. Glaser, G. S. Chen and C. L. Barnes, *Angew. Chem., Int. Ed. Engl.*, **31**, 740 (1992).
157. J. Kohn and M. Wilchek, *Anal. Biochem.*, **115**, 375 (1981).
158. J. Kohn, E. C. Albert, R. Langer and M. Wilchek, *Anal. Chem.*, **58**, 3184 (1986).
159. H. Shustina and J. H. Lesser, *J. Liquid Chromatogr.*, **7**, 2653 (1984).
160. Y. Ichikawa, *Synlett*, 238 (1991).
161. P. Penczek and W. Kaminska, *Adv. Polym. Sci.*, **97**, 41 (1990).
162. V. A. Dorokhov, M. F. Gordeev, E. M. Shashkova, A. V. Komkov and V. S. Bogdanov, *Bull. Acad. Sci. USSR, Div. Chem. Sci.*, **40**, 2274 (1991).
163. S. C. Bhatia, S. N. Revankar, K. J. Doshi, N. D. Desai, E. D. Bharucha and C. G. Sahajwalla, *J. Chromatogr., Biomed. Appl.*, **434**, 296 (1988).
164. F. Vandenberg, *Atmos. Environ. Part A, Gen. Topics*, **27**, 63 (1993).
165. E. A. Hogendoorn, C. Verschraagen, U. A. T. Brinkman and P. Vanzoonen, *Anal. Chim. Acta*, **268**, 205 (1992).
166. F. G. P. Mullins and G. F. Kirkbright, *Analyst*, **112**, 701 (1987).
167. M. B. Mariani, G. Vinci and M. R. Milana, *Ann. Chim.*, **82**, 589 (1992).
168. N. Narasimhachari and R. O. Friedel, *Clin. Chim. Acta*, **110**, 235 (1981).
169. O. Suzuki and H. Hattori, *Biomed. Mass Spectrom.*, **10**, 430 (1983).
170. M. N. Radwan and B. C. Y. Lu, *J. Am. Chem. Oil Soc.*, **63**, 1442 (1986).
171. J. Goerdeler and H. J. Bartsch, *Chem. Ber.*, **118**, 2294 (1985).
172. *ASTM Method* D-2572-87, vol. 6.02, p. 374 (1990).
173. C. B. Fanska, T. J. Byerley and J. D. Eick, *J. Chromatogr.*, **537**, 357 (1991).
174. B. C. C. Verma, S. B. Kalia, V. S. Jamwal, S. Kumar, D. K. Sharma and A. Sud, *Talanta*, **38**, 217 (1991).

175. Y. M. Chae and M. A. Tabatabai, *Anal. Lett., A, Chem. Anal.*, **16**, 1197 (1983).
176. D. L. Burgoyne, E. J. Dumdei and R. J. Andersen, *Tetrahedron*, **49**, 4503 (1993).
177. S. Kosemura, S. Yamamura and K. Hasegawa, *Tetrahedron Lett.*, **34**, 481 (1993).
178. W. Dorsch, O. Adam, T. Ziegeltrum and J. Weber, *Eur. J. Pharmacol.*, **107**, 17 (1984).
179. W. H. Menicke, T. Kral, G. Krumbiegel and N. Rittmann, *J. Chromatogr., Biomed. Appl.*, **414**, 19 (1987).
180. W. S. Wu, R. E. Stoyanoff, R. S. Szklar, V. S. Gaind and M. Rakanovic, *Analyst*, **115**, 801 (1990).
181. W. S. Wu, R. E. Stoyanoff and V. S. Gaind, *Analyst*, **116**, 21 (1991).
182. W. S. Wu and V. S. Gaind, *Analyst*, **117**, 9 (1992).
183. J. Fuentes, W. Moreda, C. Ortiz, I. Robina and C. Welsh, *Tetrahedron*, **48**, 6413 (1992).
184. P. Edman, *Acta Chem. Scand.*, **4**, 277 (1950).
185. P. Tempst and L. Riviere, *Anal. Biochem.*, **183**, 290 (1989).
186. D. L. Christie, P. M. Barling, R. M. Hill and K. Isakow, *Anal. Biochem.*, **154**, 92 (1986).
187. F. R. Vanderleij and G. W. Welling, *J. Chromatogr.*, **383**, 35 (1986).
188. V. Fierabracci, P. Masielo, M. Novelli and E. Bergamini, *J. Chromatogr., Biomed. Appl.*, **570**, 285 (1991).
189. J. Y. Chang, *Anal. Biochem.*, **102**, 384 (1980).
190. J. Y. Chang, *Biochem. J.*, **199**, 537 (1981).
191. H. Vonbahrlindstrom, J. Hempel and F. H. Woeller, *J. Protein Chem.*, **1**, 257 (1982).
192. J. Y. Chang, *Methods Enzymol.*, **91**, 79 (1983).
193. J. Y. Chang, *Methods Enzymol.*, **91**, 455 (1983).
194. G. Winkler, F. X. Heinz and C. Kunz, *J. Chromatogr.*, **297**, 63 (1984).
195. M. Bauer, L. Mailhe and L. Nguyen, *J. Chromatogr.*, **292**, 468 (1984).
196. M. Salva and F. X. Aviles, *Anal. Biochem.*, **180**, 374 (1989).
197. N. Bergenhem and U. Carlsson, *J. Chromatogr.*, **472**, 318 (1989).
198. R. Cotrufo, M. R. Monsurro, G. Delfino and G. Geraci, *Anal. Biochem.*, **134**, 313 (1983).
199. D. L. Reynolds and L. A. Pachla, *J. Pharm. Sci.*, **74**, 1091 (1985).
200. B. Houston and D. Peddie, *Anal. Biochem.*, **177**, 263 (1989).
201. S. L. Wu and N. J. Dovichi, *J. Chromatogr.*, **480**, 141 (1989).
202. S. Hanau, F. Dallocchio and M. Rippa, *Biochem. Mol. Biol. Int.*, **29**, 837 (1993).
203. I. N. Smirnova and L. D. Faller, *Biochemistry*, **32**, 5957 (1993).
204. I. N. Smirnova and L. D. Faller, *FASEB J.*, **7**, A1145 (1993).
205. S. H. Lin and L. D. Faller, *FASEB J.*, **7**, A1145 (1993).
206. K. Nakamura, Y. Mimura and K. Takeo, *Electrophoresis*, **14**, 81 (1993).
207. E. K. Koepf and L. D. Burtnick, *Eur. J. Biochem.*, **212**, 713 (1993).
208. I. Orr, C. Martin, R. Ashley and V. Shoshanbarmatz, *J. Biol. Chem.*, **268**, 1376 (1993).
209. P. Champeil, *Biophys. J.*, **64**, A353 (1993).
210. R. Pei, M. Arjomandshamsai, C. T. Deng, A. Cesbron, J. D. Bignon and J. H. Lee, *Human Immunol.*, **36**, 55 (1993).
211. T. Oshita and N. Katunuma, *J. Chromatogr.*, **633**, 281 (1993).
212. K. Imai, S. Uzu, K. Nakashima and S. Akiyama, *Biomed. Cromatogr.*, **7**, 56 (1993).
213. W. Hallenbach and L. Horner, *Synthesis*, 791 (1985).
214. A. G. Stepanova and N. V. Abrosimova, *J. Anal. Chem. USSR*, **37**, 245 (1982).
215. K. R. Anumula and P. B. Taylor, *Anal. Biochem.*, **197**, 113 (1991).
216. A. N. Al-Deen, D. C. Cecchini, S. Abdel-Baky, N. M. A. Moneam and R. W. Giese, *J. Chromatogr.*, **512**, 409 (1990).
217. N. Nimura, T. Kinoshita and H. Ogura, *J. Chromatogr.*, **202**, 375 (1980).
218. T. Kinoshita, Y. Kasahara and N. Nimura, *J. Chromatogr.*, **210**, 77 (1981).
219. J. Gal and R. C. Murphy, *J. Liquid Chromatogr.*, **7**, 2307 (1984).
220. M. Ahnoff, K. Balmer and Y. Lindman, *J. Chromatogr.*, **592**, 323 (1992).
221. D. M. Desai and J. Gal, *J. Chromatogr., Biomed. Appl.*, **579**, 165 (1992).
222. H. Nishi, N. Fujimura, H. Yamaguchi and T. Fukuyama, *J. Chromatogr.*, **539**, 71 (1991).
223. S. Ito, A. Ota, K. Yamamoto and Y. Kawashima, *J. Chromatogr.*, **626**, 187 (1992).
224. M. Lobell and M. P. Schneider, *J. Chromatogr.*, **633**, 287 (1993).
225. T. Miyazawa, H. Iwanaga, T. Yamada and S. Kuwata, *Anal. Lett.*, **26**, 367 (1993).
226. J. Gal and A. J. Sedman, *J. Chromatogr.*, **314**, 275 (1984).
227. C. F. Meares, C. I. Diamanti, D. A. Goodwin, D. I. Reardan and M. J. McCall, *J. Labeled Compd. Radiopharm.*, **21**, 1031 (1984).

CHAPTER 5

Electronic effects of cyano, isocyano, acetylenic and diazonio groups

JOHN SHORTER

School of Chemistry, University of Hull, Hull HU6 7RX, UK

Supplement C2: The chemistry of triple-bonded functional groups
Edited by S. Patai © 1994 John Wiley & Sons Ltd

I. INTRODUCTION

A. The Scope of this Chapter

Previous articles by the present contributor in *The Chemistry of Functional Groups* series have dealt with the electronic effects of the sulphonio group[1], of the sulphinyl and sulphonyl groups[2], of SOOH and related groups[3], of amidino, guanidino and related groups[4] and of ether and hydroxyl groups[5]. In the first two cases[1,2] there was copious information in the literature from which to draw, but fairly comprehensive surveys were practicable. In the third and fourth contributions[3,4] the amount of information available was very restricted. In the fifth case[5] the amount of available material was enormous and the treatment was necessarily highly selective both in the topics covered and in the illustrative examples provided. This will be the situation for the present chapter. Treatment will be restricted to the triply-bonded groups cyano CN, isocyano NC, acetylenic RC≡C and diazonio N_2^+. Of these there is far more material available for the cyano group than for the others. While cyano tends not to be so commonly used as nitro, many studies of substituent effects include it, so by far the larger part of this chapter is devoted to CN.

It is appropriate that the present Introduction should contain a specifically historical section, particularly in relation to the cyano group, whose electronic effect played a distinctive role in organic chemistry long before it was recognized as such. In this section and in later sections references are often given to classical papers and texts, whose importance has been overlaid by more recent work.

The quantitative study of the electronic effects of these groups is naturally much concerned with the Hammett equation and its extensions. The next main section therefore contains a summary of the salient features of the Hammett equation and cognate linear free-energy relationships, along the general lines of corresponding sections in certain of the contributor's previous articles in the series[1,2,5]. This prepares the ground for a discussion of the electronic effects of the cyano group on the strengths of carboxylic and other acids: alicyclic, aliphatic and aromatic systems are covered. The discussion of aromatic systems leads to a further main section on the *ortho*-effect of CN, in which electronic effects are moderated by steric and other influences. The emphasis of much of the chapter is deliberately 'chemical', but in the next section the application of modern spectroscopic and theoretical techniques to the cyano group is explored. The following section deals with several important areas for the manifestation of the electronic effects of CN: the formation and stabilization of carbanions, nucleophilic substitution (both aliphatic and aromatic), electrophilic aromatic substitution and liquid crystals. Most of the treatment thus far has concerned CN as a substituent directly attached to the molecular skeleton of interest. A further section now deals with various substituents containing one or more cyano groups.

In a short section the electronic effect of the isocyano group is discussed. The possibility of including unipolar substituents in Hammett and similar treatments has long been a matter of controversy and this problem is discussed with special reference to the N_2^+ substituent. The effects of acetylenic groups are examined in some detail and various unsatisfactory aspects are uncovered. Multiparameter treatments such as the Yukawa–Tsuno equation and the dual substituent-parameter equation have long been important and further treatments have been devised in recent years. A final section is devoted to some of these, with an indication of the place of CN, and to a lesser extent of the other groups of interest, in these treatments.

This chapter is dedicated to the memory of Bartholomeus Meindert Wepster (1920–1992). The contributor's personal indebtedness to Bart Wepster over a period of more than thirty years is very great. The present chapter contains much discussion of Wepster's work in linear free-energy relationships: see Sections II.B, III.B, III.C, and IX.C.

B. Historical Introduction

Hydrogen cyanide was discovered by Scheele in 1782 and cyanogen by Gay-Lussac in 1815. The study of these compounds and their derivatives, both inorganic and organic, was pursued energetically throughout the 19th century. Many cyano compounds were made and a great deal of information about their behaviour was accumulated[6], much of which was difficult to understand before the theory of valency and the structural theory of organic chemistry were developed in the third quarter of the century. By 1870 it was known that there were two types of organic cyanides: the *nitriles* in which a radical R was attached to the carbon of CN, and the *isonitriles* (carbamines) in which R was attached to the nitrogen, RCN and RNC, respectively.

Nitriles had been known since the eighteen-thirties. CN played an important part in the development of the concept of the 'compound radical', and benzoyl cyanide was one of the compounds prepared by Liebig and Wöhler[7] in 1832 during their work which led to the recognition of the benzoyl radical. The first simple aliphatic nitrile to be made was propionitrile by Pelouze[8] in 1834 and benzonitrile was first made by Fehling[9] in 1844. Isonitriles were not obtained until somewhat later. Hofmann is usually credited with their proper recognition: in 1867 he made an isonitrile, more or less accidentally, by heating a primary amine with chloroform and alcoholic potash, and he then prepared several others more purposefully[10]. According to H.E. Armstrong[11], Hofmann had been unaware that one of these nauseous compounds had already been obtained by Gautier, who had failed to publish his work properly. Also, "... E. Meyer, many years before, had allowed the ethyl compound to pass through his hands unrecognised, as by heating silver cyanide with ethyl iodide and distilling the product he had obtained a liquid having an overwhelming odour and yielding ethylamine on hydrolysis"[11]. In fact, too, some eight years before Hofmann's serendipitous work, Lieke had made allyl carbamine by the action of silver cyanide on allyl iodide[12].

The existence of both nitriles and isonitriles stimulated interest in the structure of hydrogen cyanide and its possible tautomerism, and this was discussed for many years to come[13]. There was also interest in the way some metallic cyanides reacted with alkyl halides to give nitriles, while others gave isonitriles. The latter certainly excited great interest from the standpoint of valency theory. While nitriles were easily formulated with tetravalent carbon bound to R by a single bond and to N by a triple bond, the formulation of isonitriles posed a problem. The most obvious arrangement involved a single bond from N to R, and a double bond from N to C, i.e. R—N=C. This preserved the trivalency of nitrogen, but carbon was apparently divalent. In those days pentavalent nitrogen was acceptable and a structure R—N≡C would have preserved the tetravalency of carbon, but the concept of a quadruple bond could not easily be fitted into the developing ideas of stereochemistry. Thus the isocyanides, like carbon monoxide, were regarded as exceptional molecules involving divalent carbon and their characteristic reactions were interpreted in terms of that carbon atom possessing two free valencies[14].

In the second half of the 19th century the CN group of nitriles (but not of isonitriles) began to acquire importance in preparative organic chemistry. This was not only because of its useful characteristic reactions, but also because of its power of activating the hydrogen atom(s) on the carbon to which the CN is attached. It was not found to be unique in this respect; other groups such as NO_2 and COOEt were found to show similar behaviour. Indeed this area of organic chemistry really began with the discovery of acetoacetic ester (ethyl acetoacetate) by Geuther in 1863[15], which was shortly followed by Frankland and Duppa's investigations of the acetoacetic ester condensation and of the reactions of acetoacetic ester[16,17]. The details of this chemistry took many years to elucidate, with important contributions being made by Claisen[18]. (This is why his name is often associated with the acetoacetic ester condensation.) The overall stoichiometry (i.e. omitting the part played by the reagent sodium) of the acetoacetic ester condensation may be expressed by the equation:

$$CH_3COOEt + CH_3COOEt \longrightarrow CH_3COCH_2COOEt + EtOH$$

However, Claisen found in 1889 that a second molecule of ethyl acetate (with CH_3 hydrogen atoms activated by COOEt) was not the only possible partner for a condensation reaction[19]. It could be replaced by a variety of other molecules, such as aldehydes and ketones (activation by CO group) and methyl cyanide and its derivatives (activation by CN group). Many condensation reactions of esters with nitriles were discovered, e.g. diethyl oxalate reacts with methyl cyanide in the presence of sodium ethoxide (Fleischhauer, 1893[20]):

$$\begin{array}{c} COOEt \\ | \\ COOEt \end{array} + CH_3CN \longrightarrow \begin{array}{c} COOEt \\ | \\ COCH_2CN \end{array} + EtOH$$

Other condensations depending on the activating effect of CN were also discovered. Thus acetone and other ketones and ketonic esters condense with cyanoacetic ester (Perkin and Haworth, 1908[21]):

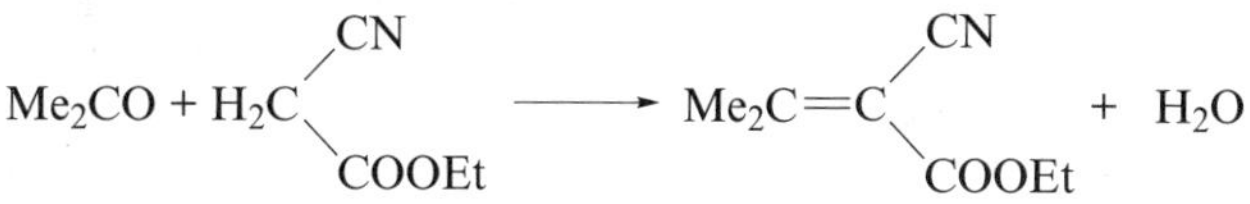

$$Me_2CO + H_2C\underset{COOEt}{\overset{CN}{\diagup}} \longrightarrow Me_2C{=}C\underset{COOEt}{\overset{CN}{\diagup}} + H_2O$$

Aldehydes condense with cyanoacetamide (Gabriel, 1903[22]):

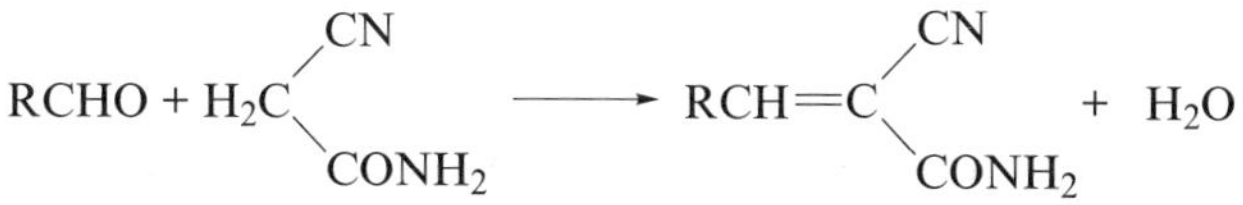

$$RCHO + H_2C\underset{CONH_2}{\overset{CN}{\diagup}} \longrightarrow RCH{=}C\underset{CONH_2}{\overset{CN}{\diagup}} + H_2O$$

Frankland and Duppa[16] suggested that in the acetoacetic ester condensation, hydrogen in the second molecule of ethyl acetate was replaced by sodium, to form an intermediate compound $CH_2NaCOOEt$. Thus almost from the start of this area of organic chemistry the activation of CH was believed to be associated with incipient acidity. Over the course of time this was formulated in various ways and the tautomerism of acetoacetic ester and related compounds was also the subject of much discussion. However, the essential features of the mechanisms of such reactions as those shown above, that are accepted today, were put forward by Lapworth early in the 20th century[23]. That is, that the partner in a condensation reaction is activated through the removal of a hydrogen ion by a molecule of base, thereby producing a highly reactive carbanion. For the reactions set out above, the intermediate species would be: $^-CH_2COOEt$, $^-CH_2CN$, $^-CH(CN)COOEt$ and $^-CH(CN)$ $CONH_2$. The carbanion then attacks the carbon of ester, ketone or aldehyde function.

Although these ideas were first propounded in 1901, it was only much later that they were generally accepted by organic chemists. For a long time authors appear to have been reluctant to accept mechanisms involving carbanions. Thus in the early nineteen-twenties Cohen[13] at no point clearly introduces carbanionic intermediates, even though he makes much reference to Lapworth's work on reaction mechanisms. However, in the nineteen-thirties formulations involving carbanions became more widely used[24, 25], although the actual term carbanion was not rapidly adopted. Hammett's book (1940)[26] appears to have been one of the first texts to use the term freely. By the nineteen-thirties a fair number of 'stable' carbanions, as opposed to highly reactive intermediates, were recognized, and this no doubt helped to make the idea of carbanionic intermediates more acceptable. Thus it was known that malononitrile was sufficiently acidic to give a pK_a value of about 11 in water, while cyanoform $HC(CN)_3$ was quite a strong acid.

Returning to the beginning of the 20th century, it should be mentioned that the effect of CN and other groups on the addition of bromine to a $C{=}C$ double bond was studied by Bauer[27]. It was found that the presence of CN attached to the double bond in stilbene and related compounds considerably *retarded* the addition of bromine. The retarding effect of various groups increased in the same order as they raised the dissociation constant of acetic acid. (Cyanoacetic acid was included in the wide selection of organic acids whose dissociation constants were measured by Ostwald around 1889[28].) By early in the century it was also appreciated that CN, like NO_2, was *meta* directing and retarding in aromatic substitution, i.e. nitration, bromination, etc.[29]. Thus these substituent effects of CN are, so to say, the other side of the coin from the activating effects described earlier.

In the nineteen-twenties the cyano group played a minor role in the development of the electronic theory of organic reactions. In Robinson's version of the theory[30], the CN group was a possible component of a 'katio-enoid' system (see **1**) in which the C=C double bond acquired 'kationoid' reactivity, in contrast to its usual 'anionoid' reactivity, i.e. it would react with 'anionoid' reagents such as CN^- or NH_3, rather than with 'kationoid' reagents such as Br_2. This was a result of the conjugative polarization indicated by the curly arrows in **1**. As Ingold's version of the electronic theory developed[31], CN was classified, like NO_2, as $-I$ and $-T$ in electronic effects, signifying electron attraction by both the *Inductive Effect* and the *Tautomeric Effect*. Later, the latter was subdivided into the *Mesomeric Effect M* (polarization) and *Electromeric Effect E* (polarizability).

$$\underset{}{\diagup}C{=}\overset{|}{C}{-}C{\equiv}N$$

(1)

The electronic theory of valency provided the opportunity to clarify the nature of the bonding in isocyanides. There was, of course, no problem in dealing with nitriles in terms of the theory: the C atom was envisaged as achieving its octet through forming a single bond and a triple bond, while the N atom achieved its octet through forming a triple bond. The traditional formulation of isocyanides in terms of divalent carbon left that atom with only a sextet of electrons. However, it was realized that in such a structure the N atom had a lone pair of electrons which could advantageously be shared with the C atom in the formation of a coordinate bond, with accompanying charge separation, thereby forming a rather polar triple bond: $-\overset{+}{N}{\equiv}\overset{-}{C}$: and bringing both nitrogen and carbon to the octet condition. In agreement with this formulation it was shown that phenyl isocyanide had the substantial dipole moment of about 3.5 D, with the negative end of the dipole on the isocyanide carbon[32]. Much other evidence was adduced for the above formulation, including parachor, heats of formation, vibrational frequencies and bond lengths[33]. With regard to the last mentioned, it was found that the carbon to nitrogen distances in cyanides and isocyanides are all about the same, in accord with the bond being essentially a triple bond in both. The observed dipole moments of isocyanides are not large enough to correspond to a full separation of unit charges to the bond distance. This would give a dipole moment of about 5.5 D. Doubtless the charge separation is reduced by the electronegativity of nitrogen being greater than that of carbon. The difference in electronegativity is also responsible for the dipole moments of nitriles, e.g. the value for benzonitrile is about 3.9 D.

Cyano and isocyano groups also played some part in the elucidation of substituent effects through dipole moment measurements in the nineteen-thirties. There was great interest in comparing the dipole moments of corresponding aliphatic and aromatic compounds, e.g. MeCN and PhCN. It was found that these frequently differed somewhat, in a direction which appeared to confirm the occurrence in aromatic compounds of a movement of electrons corresponding to the mesomeric effect that had been postulated in the electronic theory of organic chemistry. This led to the concept of 'mesomeric moment' and various ways of estimating this from observed dipole moments were devised. The actual values depended on the method of estimation employed, but various methods applied to a series of substituents gave similar trends[34]. Thus comparison of the dipole moments of MeCN and PhCN in benzene gives a mesomeric moment of $- 0.46$, the minus sign indicating mesomeric electron withdrawal by CN in the aromatic system. A comparable value for NO_2 is $- 0.95$ and for COMe is $- 0.17$. These values contrast with positive values for substituents such as halo and methoxy, indicating mesomeric electron release. A similar comparison of the dipole moments of EtNC and PhNC[33b] gives a mesomeric moment of

-0.06 for NC. This may be regarded as effectively zero within the limitations of the treatment. The isocyano group is not usually regarded as capable of exerting a mesomeric effect in either direction. Much of the way in which dipole moment measurements were interpreted in the nineteen-thirties is now regarded as over-simplified[35], but there was great interest at the time in the relationship of the signs of the mesomeric moments to the *ortho/para* or *meta* orienting influence of the groups.

Not much needs to be said about historical aspects of the substituent effects of the other groups with which this chapter is concerned, namely acetylenic groups and the unipolar diazonium group.

It has long been known that acetylene and related compounds RC≡CH easily form metallic derivatives, i.e. hydrogen attached to triply-bonded carbon shows incipient acidity to a much greater extent than the hydrogen in alkenes or alkanes[36]. Thus C≡C has a pronounced electronegative character, which is also displayed by its influence on the strengths of carboxylic acids. Ostwald showed in 1889[28] that phenylpropiolic acid PhC≡CCOOH was a much stronger acid than phenylpropionic acid, the dissociation constants being about 10^{-2} and 2.5×10^{-5} respectively. Acetylenic groups RC≡C played little part in the development of the electronic theory of organic chemistry, but were identified by Ingold[31] as $-I$ in character. In his article in *Chemical Reviews*[31] he does not appear to consider their possible mesomeric effects. Such electron movement could, however, conceivably be in either direction with respect to an attached benzene ring, so acetylenic groups should be classified as $\pm M$.

Since aromatic diazo compounds were discovered by Griess in 1858[37] and then exhaustively investigated by him, they have been of great interest because of their numerous reactions and their application to the preparation of azo-dyes[38]. However, the diazonium group was for a long time of little interest as a substituent for influencing the behaviour of other functional groups. The structure of aromatic diazo compounds was a matter of dispute for many years, until their nature as salts was properly appreciated and they were correctly formulated as $[\text{Ar—N≡N}]^+[\text{X}]^{-}$[39]. The substituent effect of a diazonium group is therefore (usually) essentially that of the unipole N_2^+. In the development of the electronic theory of organic chemistry, much attention was devoted to the substituent effects of unipoles such as NMe_3^+, SMe_2^+, O^-, etc., but little attention was given to N_2^+. Ingold did, however, classify the group as $-I, -M$ in electronic effects[31].

II. THE HAMMETT EQUATION[40]

A. Introduction

The Hammett equation is the best-known example of a linear free-energy relationship (LFER), that is, an equation which implies a linear relationship between free energies (Gibbs energies) of reaction or activation for two related processes[41]. It describes the influence of polar *meta*- or *para*- substituents on reactivity for side-chain reactions of benzene derivatives.

The Hammett equation (1937)[42–47] takes the form of equation 1 or 2:

$$\log k = \log k^0 + \rho\sigma \tag{1}$$

$$\log K = \log K^0 + \rho\sigma \tag{2}$$

The symbol k or K is the rate or equilibrium constant, respectively, for a side-chain reaction of a *meta*- or *para*-substituted benzene derivative, and k^0 or K^0 denotes the statistical quantity (intercept term) approximating to k or K for the 'parent' or 'unsubstituted' compound. The *substituent constant* σ measures the polar (electronic) effect of replacing H by a given substituent (in the *meta-* or *para-*position) and is, in principle, independent of the nature of the reaction. The *reaction constant* ρ depends on the nature of the reaction (including conditions such as solvent and temperature) and measures the susceptibility of the reaction to polar effects. Hammett chose the ionization of benzoic acids in water at 25 °C as a standard process. For this, ρ is defined as 1.000, and the value of σ for a given substituent is then $\log (K_a/K_a^0)$, where K_a is the ionization constant of the substituted benzoic acid and K_a^0 that of benzoic acid itself. Selected values of σ for well-known substituents are given in Table 1. They are readily interpreted qualitatively in simple electronic terms, i.e. through the inductive (I) effect and the resonance or conjugative (R) effect.

TABLE 1. Selected values [a] of σ, σ^+ and σ^- contants

Substituent	σ_m	σ_p	σ_p^+	σ_p^-
Me	-0.07	-0.17	-0.31	—
OMe	0.12	-0.27	-0.78	—
SMe	0.15	0.00	-0.60	0.21
OH	0.12	-0.37	-0.92	—
SH	0.25	0.15	—	—
NMe_2	-0.15	-0.63	-1.7	—
F	0.34	0.06	-0.07	—
Cl	0.37	0.23	0.11	—
CF_3	0.43	0.54	—	0.65
CN	0.61	0.65	—	0.88
NO_2	0.71	0.78	—	1.24
CO_2H	0.37	0.45	—	0.73

[a] These values, drawn from various sources, are presented solely for illustration. The table should not itself be used uncritically as a source of σ values for correlations. See rather References 44 and 48. The values for CN will be discussed later in this chapter.

Jaffé (1953)[49] showed that while many rate or equilibrium data conform well to the Hammett equation (as indicated by correlation coefficient), many such data are outside the scope of the equation in its original form and mode of application. Deviations are commonly shown by *para*-substituents with considerable $+R$ or $-R$ effect[50]. Hammett himself found that p-NO_2 ($+R$) showed deviations in the correlation of reactions of anilines or phenols. The deviations were systematic in that a σ value of *ca* 1.27 seemed to apply, compared with 0.78 based on the ionization of p-nitrobenzoic acid. Other examples were soon discovered and it became conventional to treat them similarly in terms of a 'duality of substituent constants'.

When σ values based on the ionization of benzoic acids are used, deviations may occur with $+R$ *para*-substituents for reactions involving $-R$ electron-rich reaction centres, and with $-R$ *para*-substituents for reactions involving $+R$ electron-poor reaction centres. The explanation of these deviations is in terms of 'cross-conjugation', i.e. conjugation involving substituent and reaction centre.

In the ionization of the p-nitroanilinium ion, the free base is stabilized by delocalization of electrons involving the canonical structure **2**.

An analogous structure is not possible for the p-nitroanilinium ion. In the ionization of p-nitrophenol, analogous delocalization is possible in both phenol and phenate species, but is more marked in the ion. Thus, in both the aniline and the phenol system p-NO_2 is

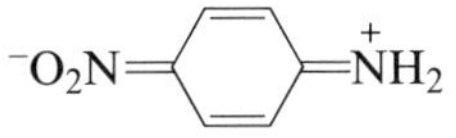

$$^-O_2N =\!\!\left\langle \underline{\quad}\right\rangle\!\!= \overset{+}{N}H_2 \qquad\qquad O_2N\!-\!\!\left\langle \underline{\quad}\right\rangle\!\!= \overset{+}{C}O_2^-H$$

(2) 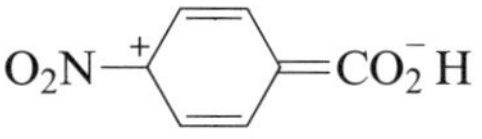   (3)

effectively more electron-attracting than in the ionization of benzoic acid, where the reaction centre is incapable of a $-R$ effect, and indeed shows a small $+R$ effect (3).

An example of a reaction series in which large deviations are shown by $-R$ *para*-substituents is provided by the rate constants for the solvolysis of substituted *t*-cumyl chlorides, $ArCMe_2Cl^{51}$. This reaction follows an S_N1 mechanism, with intermediate formation of the cation $ArCMe_2^+$. A $-R$ *para*-substituent such as OMe may stabilize the activated complex, which resembles the carbocation–chloride ion pair, through delocalization involving structure **4**. Such delocalization will clearly be more pronounced than in the species involved in the ionization of *p*-methoxybenzoic acid, which has a reaction centre of feeble $+R$ type (5). The effective σ value for *p*-OMe in the solvolysis of *t*-cumyl chloride is thus -0.78, compared with the value of -0.27 based on the ionization of benzoic acids.

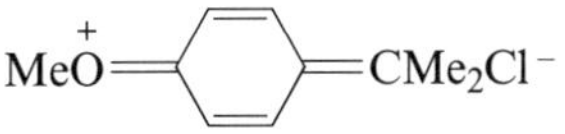

$$MeO\overset{+}{=}\!\!\left\langle \underline{\quad}\right\rangle\!\!=\!\overset{-}{C}Me_2Cl^- \qquad\qquad MeO\overset{+}{-}\!\!\left\langle \underline{\quad}\right\rangle\!\!=\!\overset{-}{C}O_2^-H$$

(4) 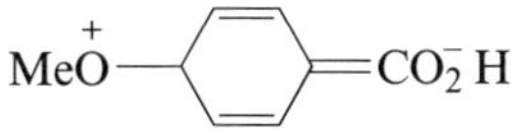   (5)

The special substituent constants for $+R$ *para*-substituents are denoted by σ^-, and those for $-R$ *para*-substituents are denoted by σ^{+51}. They are based respectively on the reaction series discussed above. Selected values are given in Table 1. Characteristic σ^- or σ^+ values are sometimes distinguished for *meta*-substituents also, but only for a minority of substituents which show very marked $+R$ or $-R$ effects do these differ significantly from ordinary σ values. The range of applicability of the Hammett equation is greatly extended by means of σ^- and σ^+, notably to nucleophilic (by σ^-) and to electrophilic (by σ^+) aromatic substitution.

However, the 'duality of substituent constants' and the attempt to deal with cross-conjugation by selecting σ^+, σ or σ^- in any given case is somewhat artificial. The contribution of the resonance effect of a substituent relative to its inductive effect must in principle vary continuously as the electron-demanding quality of the reaction centre is varied, i.e. whether it is electron-rich or electron-poor. A 'sliding scale' of substituent constants would be expected for each substituent having a resonance effect and not just a pair of discrete values: σ^+ and σ for $-R$, or σ^- and σ for $+R$ substituents[52].

B. Multiparameter Extensions[46,53,54]

There are two main types of treatment, both involving multiparameter extensions of the Hammett equation, which essentially express the 'sliding scale' idea.

In the Yukawa–Tsuno equation (1959)[55] (equation 3), the sliding scale is provided by multiple regression on σ and $(\sigma^+-\sigma)$ or $(\sigma^--\sigma)$, depending on whether the reaction is more or is less electron-demanding than the ionization of benzoic acid. (There is a corresponding form of the equation for equilibria.) The quantity $r^\pm$ gives the contribution of the enhanced $\pm R$ effect in a given reaction. (The equation was modified in 1966[56] to use σ^0 instead of σ values, see below, but the essential principles are unaltered.)

$$\log k = \log k^0 + \rho[\sigma + r^{\pm}(\sigma^{\pm} - \sigma)] \tag{3}$$

In the form of treatment developed by Taft and his colleagues since 1956[57-59], the Hammett constants are analysed into inductive and resonance parameters, and the sliding scale is then provided by multiple regression on these. Equations 4 and 5 show the basic relationships, the suffix BA signifying benzoic acid. The σ_I scale is based on alicyclic and

$$\sigma_m = \sigma_I + 0.33\sigma_R(\text{BA}) \tag{4}$$

$$\sigma_p = \sigma_I + \sigma_R(\text{BA}) \tag{5}$$

aliphatic reactivities (see below), and the factor 0.33 in equation 4 is the value of a 'relay coefficient', α, giving the indirect contribution of the resonance effect to σ_m. However, the ionization of benzoic acids is not regarded as an entirely satisfactory standard process, since it is subject to some slight effect of cross-conjugation (see structure 5 above). Consideration of 'insulated series', not subject to this effect, e.g. the ionization of phenylacetic acids, is used as the basis of a σ^0 scale, which can be analysed by equations 6 and 7[60]. (Note the different value of α.) By a different procedure Wepster and colleagues[52] devised an analogous σ^n scale (n = normal, i.e. free from the effects of cross-conjugation). Analysis of σ^+ and σ^- constants correspondingly involves σ_R^+ and σ_R^-.

$$\sigma_m^0 = \sigma_I + 0.5\sigma_R^0 \tag{6}$$

$$\sigma_p^0 = \sigma_I + \sigma_R^0 \tag{7}$$

Multiple regression on σ_I and σ_R-type parameters employs the 'dual substituent-parameter equation', which may be written as in equation 8[61]. (The combining of the k and k^0 terms implies that there is no intercept allowed, and k^0 is now the actual value for the parent system,

$$\log(k/k^0) = \rho_I\sigma_I + \rho_R\sigma_R \tag{8}$$

cf below.) For any given reaction series the equation is applied to meta- and para-substituents separately, and so values of ρ_I and ρ_R characteristic both of reaction and of substituent position are obtained. The various σ_R-type scales are linearly related to each other only approximately. In any given application the scale which gives the best correlations must be found[62].

Values of σ^0, σ_I and σ_R-type parameters for certain substituents are given in Table 2. It should be mentioned that Exner has developed a slightly different procedure for analysing sigma values[63] into inductive and resonance components[46,47,64].

TABLE 2. Selected values[a] of σ^0, σ_I and σ_R-type constants

Substituent	σ_m^0	σ_p^0	σ_I	$\sigma_R(\text{BA})$	σ_R^0	σ_R^+	σ_R^-
Me	−0.07	−0.15	−0.05	−0.12	−0.10	−0.25	—
OMe	0.06	−0.16	0.26	−0.53	−0.41	−1.02	—
NO$_2$	0.70	0.82	0.63	0.15	0.19	—	0.61
F	0.35	0.17	0.52	−0.46	−0.35	−0.57	—
Cl	0.37	0.27	0.47	−0.24	−0.20	−0.36	—

[a] See footnote to Table 1.

A slightly different procedure for carrying out multiple regression on $\sigma_I\delta$ and σ_R-type parameters employs the 'extended Hammett equation' of Charton[65], which may be written

$$Q = \alpha\sigma_{I,X} + \beta\sigma_{R,X} + h \tag{9}$$

as in equation 9. For the substituent X, Q is the absolute value of the property to be correlated (log k or log K in the case of reactivity), i.e. not expressed relative to X = H, h is introduced as the appropriate intercept term, and the regression coefficients are α and β. (Charton has used various symbols at various times.)

The correlation analysis of spectroscopic properties in terms of σ_I and σ_R-type parameters has been very important. Substituent effects on ^{19}F NMR shielding in fluorobenzenes have been studied in great detail by Taft and colleagues[60, 66, 67]. For δ_m^F linear regression on σ_I is on the whole satisfactory, but a term in σ_R^0 with a small coefficient is sometimes introduced. The correlation analysis of δ_p^F, however, requires terms in both σ_I- and σ_R- type parameters, with σ_R^0 being widely applicable. Many new values of these parameters have been assigned from fluorine chemical shifts. In recent years there has also been extensive use of correlation analysis of ^{13}C NMR data[68,69].

The correlation analysis of infrared data has been much examined by Katritzky, Topsom and colleagues[70,71]. Thus the intensities of the ν_{16} ring-stretching bands of mono- and di-substituted benzenes may be correlated with the σ_R^0 values of the substituents and these correlations may be used to find new σ_R^0 values.

Finally, in this account of multiparameter extensions of the Hammett equation, we comment briefly on the origins of the σ_I scale. This had its beginnings around 1956[59] in the σ' scale of Roberts and Moreland[72] for substituents X in the reactions of 4-X-bicyclo[2.2.2]octane-1 derivatives. However, at that time few values of σ' were available. A more practical basis for a scale of inductive substituent constants lay in the σ^* values for XCH_2 groups derived from Taft's analysis of the reactivities of aliphatic esters into polar, steric and resonance effects[59,73–75]. For the few σ' values available it was shown that σ' for X was related to σ^* for XCH_2 by the equation $\sigma' = 0.45\,\sigma^*$. Thereafter the factor 0.45 was used to calculate σ_I values of X from σ^* values of XCH_2[76]. These matters will be referred to again later in this chapter, and other methods of determining σ_I values will also be mentioned. Taft's analysis of ester reactivities was also important because it led to the definition of the E_s scale of substituent steric parameters, thereby permitting the development of multiparameter extensions of the Hammett equation involving steric as well as electronic terms.

III. ELECTRONIC EFFECTS OF CN ON THE STRENGTHS OF CARBOXYLIC AND OTHER ACIDS[77]

A. Alicyclic and Aliphatic Systems

The simplest indicator of the electronic effect of a substituent X is its influence on the ionization constant of an organic acid into which it is substituted. For the least complicated behaviour, the group should not be conjugated with the molecular skeleton and should not be too close to the acidic centre. The change in acid strength produced by X is conveniently expressed as ΔpK_a, defined as $(pK_a)_H - (pK_a)_X$, so that an increase in acid strength is associated with a positive value of ΔpK_a. In Table 3 the ΔpK_a value of 0.539 for 4-cyanobicyclo[2.2.2]octane-1-carboxylic acid (6) and of 3.04 for the 4-cyanoquinuclidinium ion (7) (in water at 25 °C) are clear indications of the electronegative or electron-attracting nature of CN. The influence of this reaches the acidic centre either by induction through the bonds of the molecular skeleton or through the electric field of the substituent as moderated by the dielectric behaviour of the molecular cavity and the solvent. The respective roles of these two modes of transmission have long been a matter of controversy[78]. Both are 'inductive' in the most general meaning of this term in physics and we shall continue the traditional practice of describing them collectively as the 'inductive effect'.

TABLE 3. The influence of the cyano group on the strengths of alicyclic and aliphatic acids[77]

Acid	Solvent	Temp. (°C)	pK_a X = H	pK_a X = CN	$\Delta pK_a{}^a$	$\sigma_I{}^b$ (calc)
1. XCH$_2$COOH	H$_2$O	25	4.756	2.47	2.286	0.57^c
2. XCH$_2$COOH	50% w/w EtOH-H$_2$O	25	5.84	3.39	2.45	0.592
3. 4-X-Bicyclo [2. 2. 2.] octane-1-carboxylic acid	H$_2$O	25	5.084	4.545	0.539	0.559
4. 4-X-Bicyclo [2. 2. 2.] octane-1-carboxylic acid	50% v/v^d EtOH-H$_2$O	25	6.75	5.90	0.85	0.576
5. 4-X-Quinuclidinium ion	H$_2$O	25	11.12	8.08	3.04	0.559
6. trans-4-X-Cyclohexane-1-carboxylic acid	H$_2$O	24.91	4.90	4.48	0.42	0.562
7. trans-4-X-Cyclohexane-1-carboxylic acid	50% v/v^d EtOH-H$_2$O	24.91	6.40	5.74	0.66	0.582

^{a} $\Delta pK_a = (pK_a)_H - (pK_a)_{CN}$.
^{b} From the appropriate regression equations in Charton's review[81].
^{c} By definition, see text.
^{d} i.e. a solvent made up from equal volumes of ethanol and water.
See Sections III.B and III.C and Note 107.

(6)   (7)

The data in Table 3 for the behaviour of CN in the 4-position of cyclohexane-1-carboxylic acid also demonstrate the electron-attracting inductive effect, although there may be some acid-strengthening contribution from converting the parent acid which contains appreciable ax-COOH in $eq \rightleftharpoons ax$ equilibrium into the largely diequatorial trans-4-cyanocyclohexane-1-carboxylic acid[79]. The pK_a values of acetic acid and cyanoacetic acid (Table 3) also show the electron-attracting effect of CN, although with the substituent so close to the acidic centre, some contribution of a steric effect cannot be ruled out. However, the behaviour of substituted acetic acids containing very bulky alkyl groups shows that a steric effect in that situation is acid-weakening and involves the inhibition of solvation of the carboxylate ion[80].

Data for the bicyclooctane system in 50% w/w EtOH–H$_2$O are the basis for primary σ_I values according to Charton[81], calculated through equation 10[82].

$$\sigma_I = \Delta pK_a / 1.56 \qquad (10)$$

However, the pK_a value for the cyano-substituted acid in this solvent was not available, so the σ_I value for CN was calculated by substituting the pK_a value for cyanoacetic acid in water at 25 °C (2.47) in the regression equation 11:

$$pK_a = -4.05 \, \sigma_I + 4.791 \qquad (11)$$

which had been established by using data for substituents whose σ_I values were available from equation 10. This procedure gave the value 0.573 for σ_I of CN, which was rounded to 0.57. The available values of σ_I (including that for CN as 0.57) were used by Charton to establish regression equations of the general form of equation 12:

$$pK_a = L\sigma_I + h \tag{12}$$

for systems 2 to 7 in Table 3[83]. The back-calculated values of σ_I for CN are shown in the last column of the Table. The range of values from about 0.56 to 0.59 indicates that the value of 0.57 is reasonably applicable throughout these systems. It should be noted, however, that the data points for CN exert a strong influence on the regressions, since at $\sigma_I = 0.57$ CN is at one extreme of the scale for the substituents involved. For comparison we mention σ_I values as follows: CF₃, 0.40; Cl, 0.47; NO₂, 0.67. The value for CN of 0.57 is in good agreement with the value of 0.58 for the equivalent σ' constant which had been derived by Roberts and Moreland in 1953[72]. Taft's analysis of substituent effects in aliphatic ester reactions[59] yielded $\sigma^* = 1.30$ for CH₂CN, which gave $\sigma_I = 0.59$ by application of the equation $\sigma_I = 0.450^*$ (see Section II.B). Other estimates of σ_I for CN, determined in various ways, will be mentioned later in this chapter.

Data are available for the effect of CN on acid strength in a variety of alicyclic carboxylic acids. Systems treated by Charton[81] include the following (back-calculated value of σ_I given in parenthesis): 4-substituted bicyclo[2.2.2]oct-2-ene-1-carboxylic acids in 50% w/w EtOH–H₂O, 25 °C (0.613); 4-substituted bicyclo[2.2.1]heptane-1-carboxylic acids in H₂O, 25 °C (0.580); 6-substituted spiro[3.3]heptane-2-carboxylic acids in 50% w/w EtOH–H₂O, 25 °C (0.603); 4-substituted cubane-1-carboxylic acids in 50% w/w EtOH-H₂O, 24.95 °C (0.552). (It must be remembered that the points for CN were included in the regressions.) It appears that overall the value of 0.57 for CN is reasonably applicable, although the back-calculated values are not as concordant as one might have hoped. pK_a data (H₂O, 25 °C) are available for trans-2-substituted cyclopropane-1-carboxylic acids and the corresponding derivatives of the fused ring structures bicyclo[2.1.0]pentane and bicyclo[3.1.0]hexane[84]. It is of interest that the ΔpK_a values for CN substitution are very close together: 1.11, 1.115 and 1.163, respectively. On the other hand, ΔpK_a for the related trans-2-cyanocyclohexane-1-carboxylic acid is slightly lower, 1.035[77a]. This acid probably contains some diaxial conformer in equilibrium with the predominant diequatorial conformer.

Charton[81] also presents regressions on σ_I for the pK_a values of 3-(substituted methyl)benzoic acids and 4-(substituted methyl)benzoic acids in 50% v/v EtOH–H₂O, 25 °C, and of substituted methylammonium ions in H₂O, 25 °C. The back-calculated values of σ_I for CN are 0.576, 0.552 and 0.560, respectively. Thus the value of 0.57 seems again to be fairly satisfactory as σ_I for CN, this time in structures very different from those with which we have been concerned in the earlier parts of this section.

The bimolecular reaction of carboxylic acids with diazodiphenylmethane (DDM) involves a slow proton transfer to form a diphenylmethanediazonium carboxylate ion-pair. The reaction has been greatly used during the past 40 years as a kinetic probe of acidity[85]. The value $\sigma_I = 0.57$ for CN seems also to express fairly well the effect of this substituent on log k (second-order rate coefficient) for the reaction of acetic acid with DDM at 30 °C in MeOH, EtOH, Pr^{i}OH, Bu^{t}OH, MeO(CH₂)₂OH or BuO(CH₂)₂OH as solvents[86]. These data were considered by Charton[81] in his review, and back-calculation from the regressions[87] gives value of σ_I for CN of 0.541, 0.550, 0.555, 0.574, 0.550 and 0.530, respectively. All these values except one are below 0.57 and there is a consistent trend with branching of the alcohol at the α-carbon atom. There is thus some contrast with what was found above for pK_a values in water or aqueous ethanol in the various systems, in so far as back-calculated values of σ_I in those cases are distributed fairly equally above and below 0.57. However, it cannot be claimed that there is any clear indication of a solvent effect on σ_I for

CN, cf the behaviour of OMe[88]. Nevertheless it seems possible that hydrogen-bonding in protic solvents could be a factor enhancing the electron-attracting effect of CN; see structure **8**.

$$X-O-H\cdots:N\equiv C-CH_2-COOH$$

$$(X = H \text{ or alkyl})$$

$$(8)$$

It might be hoped, as in the case of OMe[88], that the reaction of DDM with substituted acetic acids in aprotic solvents would provide some clearer indications. For this reaction in dimethyl sulphoxide and ethyl acetate at 30 °C, Charton's regressions give back-calculated values of σ_I for CN of 0.537 and 0.575, respectively[89]. These values provide no support for the view that hydrogen-bonding to protic solvents is a factor enhancing the electron-attracting effect of CN. It is, however, possible that such an effect might be revealed by carefully designed experimental studies of aliphatic and alicyclic carboxylic acids, coupled with a penetrating analysis of the results.

Results are available for rate coefficients of the reactions of *trans*-4-substituted cyclohexane-1-carboxylic acids with DDM in various alcohols[79,90]. The data were analysed in terms of the σ' values of Roberts and Moreland[72], a value of 0.579 being used for CN. Conformity of $\log k$ to linear regression on σ' was only rather modest, probably because of substituent-dependent variations in the relative amount of *eq-eq* and *ax-ax* conformers in the various solvents. While these *trans* acids should be mainly in the *eq-eq* conformation, the occurrence of a small amount of *ax-ax* is probably encouraged when the 4-substituent is rather small, as is the case with CN.

pK_a data are available for the acids $NC(CH_2)_n COCH$ in water at 25 °C, where $n = 1, 2$ or 3. The values of n, pK_a are: 1, 2.47; 2, 3.99; 3, 4.44. These data show clearly the increasing 'damping' effect of the $(CH_2)_n$ chain as n is increased. The pK_a value of butanoic acid is 4.82, which is not far short of the limiting value of pK_a for a long-chain fatty acid. A decremental factor of about 2.7 per CH_2 has been suggested for the effect on the σ_I values of polar substituents[59]. On the basis of $\sigma_I = 0.57$ for CN and the above factor, σ_I for NCCH$_2$ should be about $0.57/2.7 = 0.21$ and σ_I for $NC(CH_2)_2$ should be about $0.57/(2.7)^2 = 0.08$. The values calculated from the above pK_a values and Charton's regression[81] are 0.20 and 0.09, respectively. ΔpK_a data are also available for the effect of introducing CN into the α-position of several aliphatic acids. The values are as follows (H$_2$O, 25 °C): acetic, 2.286; propanoic, 2.506; 2-methylpropanoic, 2.44; 3-methylbutanoic, 2.38; cyclohexylacetic, 2.434. The introduction of CN into acetic acid does seem to have a slightly smaller effect than that shown by acids which are somewhat substituted on the α-carbon atom.

B. Aromatic Systems

The pK_a values in water at 25 °C of *m*-cyanobenzoic acid and *p*-cyanobenzoic acid were first measured accurately by Briegleb and Bieber in 1951 by an electrometric method (H$_2$/Pt electrode)[91]. The values obtained were 3.598 and 3.551, respectively, compared with 4.213 for benzoic acid itself. (Measurements were also made at several other temperatures.) On the basis of these values the Hammett σ constant[43] (Section II.A) is 0.615 for *m*-CN and 0.662 for *p*-CN. Much more recently Hoefnagel and Wepster[92] have determined these σ values as 0.64 and 0.68, respectively, by measurements with a glass electrode. (The value for *m*-CN was based on an extrapolation from measurements in 10, 4 and 2% v/v EtOH–H$_2$O mixtures, while that for *p*-CN was based directly on measurements in water.) Detailed consideration of the work of these various authors leads to weighted mean values of 0.62 for *m*-CN and 0.67 for *p*-CN[93]. We shall regard these values as the most reliable at present available (1993).

It seems to be the case, however, that the determination of the pK_a values of the cyanobenzoic acids in water is not altogether straightforward. Widequist[94] reported that *o*-cyanobenzoic acid undergoes rapid hydrolysis in water, which is easily understandable as acidic hydrolysis of CN catalysed by neighbouring COO*H* (see Section IV.C). Contrary to what is stated in Reference 77a, he did not report rapid hydrolysis for the *meta* acid, and the pK_a value which he found, 3.602, agrees closely with that determined by Briegleb and Bieber[91]. Very recently Ludwig and coworkers[95] have found a value of 3.59 for this acid, but they found 3.50 for the *para* isomer, somewhat lower than the value determined by Briegleb and Bieber[91]. It is also of interest that in 1973 Hoefnagel and Wepster[96] reported a value of 0.70 for σ_p of CN, based on unpublished measurements by Boers and Hoefnagel on benzoic acids in water at 25 °C, i.e. slightly higher than their more recent value of 0.68 referred to above, or that based on the work of Briegleb and Bieber. Thus carefully determined values of σ_p for CN vary between 0.66 and 0.71.

In the tabulation of substituent constants by Hammett (1940)[97], a value of 0.678 for *m*-CN was based on the application of the $\rho\sigma$ equation to the reactions of phenolate ions with ethylene oxide or propylene oxide, while a value of 1.00 for *p*-CN was based on the ionization of phenols. Hammett recognized that the latter value would probably be valid only for the reactions of phenol and aniline derivatives. He gave no such *caveat* for the value of 0.678 for *m*-CN, but it is evident that this also is a value enhanced by the nature of the reaction on which it is based (Sections II.A and III.D), albeit less dramatically than in the case of *p*-CN. There were, in fact, pK_a values available for *m*- and *p*-cyanobenzoic acids in the late 1930s which Hammett might have used. For the *meta* acid there was a concentration-based value of 3.70, which had been measured by Ostwald in 1889[28] (conductivity method), compared with 4.22 for benzoic acid itself (temperature unspecified). These values would give a σ constant of 0.52 for *m*-CN, very much on the low side. Through taking this value (among others) into consideration, McDaniel and Brown[98] arrived at 0.56 for σ_m, a value which is still sometimes quoted[99]. For the *para* acid there was a pK_a value of 3.51 dating from 1930[100] (H_2/Pt electrometric method), compared with 4.175 for benzoic acid itself. These values give σ_p as 0.665, fairly close to the values derived above from more modern work.

The marked acid-strengthening effect of *p*-CN is usually attributed to the influence of the electron-attracting inductive effect ($+I$), augmented by a small electron-attracting mesomeric or resonance effect ($+R$). The smaller acid-strengthening effect of *m*-CN is explained as the resultant of the inductive effect and a small 'relayed' influence of the resonance effect. If σ_p is regarded simply as a sum of σ_I and σ_R (Section II.B) and σ_I is taken as 0.57 (Section II.A), a value of $0.67 - 0.57 = 0.10$ is indicated for σ_R. The relay factor of 0.33 for the resonance effect accounts reasonably well for the value of σ_m as $\sigma_I + 0.33\sigma_R = 0.57 + 0.03 = 0.60$; *cf* 0.62 above.

The σ values of 0.62 and 0.67 for *m*-CN and *p*-CN, respectively, appear to be reasonably well applicable to reactions of benzoic acid and its derivatives in aqueous organic solvents. However, the application of the simple classical Hammett equation should always be approached with circumspection. The possibility of solvent effects on σ values should be borne in mind and also the possible intervention of hydrophobic effects of substituents, which may mimic electronic effects. The importance of hydrophobic effects in influencing reactivity has only been realized in recent years. Such effects have been discussed in detail by Hoefnagel and Wepster[92] and have been expressed in terms of Hansch's hydrophobic substituent constant π[101]. The π value of CN is modest at -0.57, *cf* Cl, 0.71; CF$_3$, 0.88; Bu^t, 1.98; and under most circumstances substituent hydrophobic effects will be of minor significance in the case of CN. (For a brief introduction to substituent hydrophobic effects on reactivity, see the account of OPh in a previous article in this series[102].)

Fairly recently Pytela and coworkers[103] have obtained indications of a solvent effect on σ_m and/or σ_p for CN. This research group (Ludwig and coworkers[95]) determined the apparent pK_a values for a large number of substituted benzoic acids in one-component organic solvents. Their results for m- and p-cyanobenzoic acids are most easily discussed in terms of the relative order of the pK_a values. In MeOH or EtOH the pK_a values are in the order m-CN > p-CN, as in water, but in DMF or acetone the order is m-CN < p-CN, while in sulpholane m-CN ≈ p-CN. These relationships suggest that there may be a reversal of the order of the σ values in aprotic solvents, which would probably be connected with the absence of hydrogen bonding of the solvent to the N of CN. This could be seem as diminishing the $+R$ effect. Unfortunately, however, the pK_a values determined in DMF or acetone are not very reproducible, and the mean errors quoted are of such a size as to cast doubt on the reality of the reversal of the pK_a values. Further, the reversal does not occur with acetonitrile as solvent, but again the mean errors quoted are rather large. It is interesting, however, that reversals are also found in certain solvents when the substituent is SO_2Me or NO_2.

In certain respects the electronic effects of CN and NO_2 are very similar. The values of σ_I are 0.57 and 0.63, respectively (Tables 2 and 3), while the value of σ_R is 0.10 for CN (above) and 0.15 for NO_2 (Table 2). However, the $+R$ effects of the two groups differ in one important respect: that of NO_2 is subject to steric inhibition of resonance, while that of CN is not. This is shown most simply by the differing effects of inserting flanking 3- and 5-methyl groups into 4-cyano- and 4-nitro-benzoic acids. ΔpK_a values of 4-cyano-, 3-methyl- and 4-cyano-3,5-dimethyl-benzoic acids in 44.1% w/w EtOH–H_2O, 25 °C, are respectively as follows: 1.05, –0.17 and 0.81. If the effect of the groups in the 4-cyano-3,5-dimethyl acid is strictly additive, its ΔpK_a value should be $1.05 - 2 \times 0.17 = 0.71$. The observed effect 0.81 indicates that, if anything, the acid-strengthening effect of 4-CN is enhanced by the flanking methyls. The ΔpK_a values of 4-nitro-, 3-methyl- and 4-nitro-3,5-dimethyl-benzoic acids in 44.1% w/w EtOH–H_2O, 25 °C, are respectively as follows: 1.25, –0.17 and 0.80. If the effect of the groups in the 4-nitro-3,5-dimethyl acid is strictly additive, its ΔpK_a value should be $1.25 - 2 \times 0.17 = 0.91$. The observed effect 0.80 indicates that the acid-strengthening effect of 4-NO_2 is appreciably reduced by the flanking methyl groups. This is usually attributed to an inhibition of the resonance effect of the NO_2 group by the methyl groups twisting it out of the plane of the benzene ring, thus making the $p\pi$ orbital overlap of NO_2 and ring less effective. In contrast, the corresponding orbital overlap in the case of the cylindrically-symmetrical CN group is not subject to interference by twisting as a result of steric interaction with the methyl groups. However, the interpretation of the above results for NO_2 in terms of steric inhibition of resonance is not everywhere accepted[104,105].

Hammett σ values are available for many substituents containing the CN moiety. Some of these will be discussed later in this chapter (Section VII.C). At this stage we will deal only with one of these: the CH_2CN group. From pK_a measurements in 80% w/w 2-MeO$(CH_2)_2$OH–H_2O and in 1:1 v/v EtOH–H_2O, Exner[64] obtained σ values of 0.16 for m-CH_2CN and 0.18 for p-CH_2CN. As mentioned in Section III.A, the value of σ_I for CH_2CN is about 0.20. In terms of the analysis of σ into contribution from σ_I and σ_R, it appears that the $+I$ effect of CH_2CN is opposed by a small $-R$ effect arising from the conjugation of the CH_2 moiety with the benzene ring; cf CH_3 for which σ_R is about -0.13. However, a quantitative analysis cannot be done since the order $\sigma_m < \sigma_p$ for CH_2CN and $\sigma_I = 0.20$ are obviously incompatible with equations 4 and 5. If equations 4 and 5 are regarded as simultaneous equations in σ_I and σ_R as unknowns, the insertion of the above values for σ_m and σ_p leads to $\sigma_I = 0.15$ and $\sigma_R = 0.03$, but such values are not readily interpretable.

C. Acids of the Type Ph–G–COOH

It was mentioned in Section II.B that the ionization of benzoic acids is not regarded as an entirely satisfactory standard process, since in the case of $-R$ substituents, such as OMe, it is subject to some slight effect of cross-conjugation (see structure **5**). Consideration of 'insulated series', not subject to this effect, e.g. the ionization of phenylacetic acids, is used as the basis of the σ^0 scale. For the sake of uniformity σ^0 values for $+R$ substituents have also been based on such systems. Wepster and colleagues[96,106], however, have criticized the use of systems in which the substituent is 'insulated' by methylene groups from the reaction centre for its tendency to lead to slightly exalted values of σ^0 for $+R$ substituents, i.e. the supposed insulation is not 100% effective. It is suggested that CH_2 is capable of a slight $-R$ interaction with the benzene ring through its hyperconjugation and this can lead to cross-conjugation with a $+R$ substituent. They see an analogy to the very pronounced exaltations that occur in the effects of $+R$ substituents on the ionization of Ph—G—COOH with G = $NHCH_2$, OCH_2 or SCH_2. If this view be accepted, for $+R$ substituents it is better simply to assume that σ values based on benzoic acid ionization are effectively values of σ^0, since COOH shows a $+R$ resonance effect and there can be no cross-conjugation. This has been done explicitly in a recent compilation by Exner[47].

The above points will now be illustrated with respect to the cyano group. The most convenient data for this purpose are for ionization in 50% v/v EtOH–H_2O[96,106,107]. (Data for other solvent composition are also in the references[96,106].) ΔpK_a values for the effects of m–CN and p–CN on the various acids Ph—G—COOH, along with the corresponding Hammett ρ values (Section II.A), are shown in Table 4. The values of σ_m (calc) and σ_p (calc) are obtained as σ (calc) = $\Delta pK_a/\rho$.

TABLE 4. The influence of the cyano group on the strengths of acids of the type Ph—G—COOH in 50%v/v EtOH–H_2O, 25 °C[96,106,107]

Acid	$\Delta pK_a{}^a$ m–CN	$\Delta pK_a{}^a$ p–CN	ρ	σ_m (calc)	σ_p (calc)
PhCH$_2$COOH	0.45	0.52	0.71	0.63	0.73
PhCH$_2$CH$_2$COOH	0.23	0.26	0.36	0.64	0.72
PhCMe$_2$CH$_2$COOH	—	0.31	0.38	—	0.82
PhNHCH$_2$COOH	—	0.35	0.38	—	0.92
PhOCH$_2$COOH	—	0.43	0.52	—	0.83

$^a\Delta pK_a = (pK_a)_H - (pK_a)_{CN}$.

The data for the phenylacetic and 3-phenylpropanoic acid systems give values of σ_m (calc) very close to 0.62, considered to be the most reliable value for σ_m of CN (Section III.B). However, the values of σ_p (calc) for the effects of CN in these acids are appreciably greater than 0.67, considered to be the most reliable value for σ_p of CN (Section III.B). This supports the views of Wepster and colleagues[96,106] that cross-conjugation between a $+R$ substituent and methylene, as in structure **9**, may enhance the electron-attracting effect of

$$N\equiv C-\!\!\!\left\langle\;\right\rangle\!\!\!-CH_2COOH \;\longleftrightarrow\; N\equiv C=\!\!\!\left\langle\;\right\rangle\!\!\!=\overset{H^+}{CHCOOH}$$

(9)

the substituent. While it is possible that such enhancement might be due to a specific influence of solvent, no enhancement is apparent in the corresponding data for p-cyanobenzoic acid: $\Delta pK_a = 1.03$, $\rho = 1.50$; therefore σ_p (calc) = 0.69[92]. The behaviour of p-CN in 3-methyl-3-phenylbutanoic acid indicates further enhancement of the electronic effect, presumably through C—C hyperconjugation involving the methyl groups. The enhancement is even more marked in the acids involving NH or O as part of G, conjugation of the lone pair electrons of N or O doubtless being involved, e.g. structure **10**. Data also

$$N\equiv C-\!\!\!\left\langle\!\!\!\bigcirc\!\!\!\right\rangle\!\!\!-NHCH_2COOH \quad\longleftrightarrow\quad N\!\!=\!\!C\!\!=\!\!\left\langle\!\!\!\bigcirc\!\!\!\right\rangle\!\!=\!\!\overset{+}{N}HCH_2COOH$$

(10)

exist for the ionization of phenoxy- and phenylthio-acetic acids in water, which yield a σ_p (calc) value for CN of about 0.8 and 1.0, respectively. The σ_p (calc) values for CN in cross-conjugation with NH, O or S are thus not very different from the σ_p^- value for the group, which lies in the range 0.9 to 1.0 (Section III.D).

D. Phenol and Anilinium Ion

The ionization of phenol and anilinium ion are both processes which are greatly facilitated by $+R$ para-substituents such as NO_2, SO_2Me, CN, etc. (Section II.A). The ρ values are most reliably determined by linear regression of $-pK_a$ on σ for the meta-substituted substrates only, and the following equations 13 and 14 are typical[108]. (pK_a values are for solutions in water at 25 °C.)

Phenols

$$-pK_a = -9.936 + 2.205\sigma$$
$$(\pm 0.078)$$
$$n = 9, r = 0.9965, s = 0.0579, \psi = 0.105$$

(13)

Anilinium ions

$$-pK_a = -4.567 + 2.847\sigma$$
$$(\pm 0.079)$$
$$n = 11, r = 0.9965, s = 0.0603, \psi = 0.092$$

(14)

(In these regression n = number of data points, r = correlation coefficient, s = standard deviation of estimate, ψ = Exner's statistic of goodness of fit[109,110].) The insertion of observed pK_a values for the meta-cyano-substituted compounds (8.61 and 2.75, respectively) into the above expressions gives apparent values of σ_m for CN as 0.601 and 0.638, respectively. These values are close to the most reliable value of 0.62 for σ_m of CN (Section III.B). However, the corresponding insertion of pK_a values (7.97 and 1.74, respectively) for the para-cyano-substituted compounds gives apparent values of σ_p for CN as 0.892 and 0.993, respectively. These values are considerably enhanced from the most reliable value of 0.67. The explanation of this enhancement is in terms of cross-conjugation, as given in Section II.A for the corresponding behaviour of the nitro group.

Such enhanced sigma values for $+R$ groups are commonly designated as σ^- values. In the case of CN, Exner[48] tabulates σ_p^- as 0.88 from phenol ionization and gives values of 0.99 and 1.02 based on two different studies of anilinium ionization. The discrepancy between the phenol- and anilinium-based values indicates that the choice of standard systems for the determination of σ^- values presents some problems. The ionization of substituted phenols and the ionization of substituted anilinium ions are usually regarded as being essentially equivalent as bases for the definition of σ^- values. This view probably arose because the σ_p^- values for the nitro-group (on whose behaviour the 'duality of substituent constants' was largely founded, see Section II.A) on the bases of the two systems agree fairly closely, 1.28 and 1.25 respectively. An inspection of Exner's compilation[48], however, reveals a number of groups for which discrepancies exist as between the phenol and anilinium scales. The most significant are usually in the sense that the enhancement of the $+R$ effect is smaller for the ionization of phenol than for the ionization of the anilinium ion, e.g. CO_2R, 0.64 and

0.74; PO(OEt)$_2$, 0.75 and 0.84; SCF$_3$, 0.57 and 0.64; SO$_2$Me, 0.98 and 1.14; SO$_2$CF$_3$, 1.36 and 1.65, respectively. There are other groups for which there is good agreement or a small difference in the opposite direction. This provides a warning that the values based on the two systems should not be mixed in correlations. Ideally only one of these systems should be chosen as the basis for the σ^- scale. The other should be regarded as a system for treatment by the Yukawa–Tsuno equation[55,56] or other multiparameter extensions[53] of the Hammett equation (Section II.B).

Further light is shed on the behaviour of $+R$ substituents in phenol by studies of gas-phase acidity. Fujio, McIver and Taft[111] measured the gas-phase acidities, relative to phenol, of 38 *meta*- or *para*-substituted phenols by the ion cyclotron resonance (ICR) equilibrium constant method. The results were treated by linear free-energy relationships and comparisons were made with behaviour in aqueous solution. The present author has summarized elsewhere the salient features of this work[112]. We will restrict the present discussion to $+R$ substituents. For CN the apparent value of σ_m in the gas-phase is 0.65, which the authors regard as essentially the same as the value in aqueous solution, which they take as 0.61. However, σ_p^- (g) is 0.83, compared with 0.99 quoted as the value of σ_p^- (aq). For several other $+R$ substituents, e.g. NO$_2$ and SO$_2$Me, σ_p^- (g) is also considerably lower than σ_m^-(aq). The most important inference from this situation is that[111] "the previously generally held view that σ_p^-(aq) values represent the inherent internal π-electron-acceptor ability of $+R$ substituents must be incorrect. Instead, σ_m^-(aq) values are shown to involve a complex composite of field/inductive, internally enhanced π-electron delocalization, and specific substituent HBA solvation assisted resonance effects." Thus, while the enhancement of the electron-attracting effect of CN by hydrogen bonding to water is practically nil in the *meta* position, it is substantial in the *para* position because of the delocalization of charge from O$^-$ into the substituent in the anion. The situation may be represented schematically as in structure **11**.

(11)

The difference between CN and NO$_2$ in respect of the influence of flanking methyl groups which was noted above (Section III.B) for the benzoic acid system is more marked in phenol ionization. The ΔpK_a values of 4-cyano- and 3,5-dimethyl-phenol (compared with phenol itself) are 2.04 and –0.19, respectively (water, 25 °C)[113]. Thus the ΔpK_a (calc) value for 4-cyano-3,5-dimethylphenol, assuming strict additivity, is 1.85. The value of ΔpK_a(obs) is 1.78, in reasonable agreement with ΔpK_a (calc). The corresponding data[113] for 4-NO$_2$ give ΔpK_a (calc) = 2.60, compared with ΔpK_a (obs) = 1.75, indicating marked steric inhibition of resonance through twisting the NO$_2$ out of the plane of the ring by the methyl groups.

The effects of both 4-CN and 4-NO$_2$ on the acidity of phenol are somewhat enhanced by the introduction of methyl groups in the 2,6 positions[113]. Thus ΔpK_a values for 4-cyano- and 2,6-dimethyl-phenol are 2.04 and – 0.61, respectively, giving ΔpK_a (calc) for 4-cyano-2,6-dimethylphenol as 1.43. ΔpK_a (obs) is in fact 1.72. The corresponding calculation for the nitro group gives ΔpK_a (calc) = 2.20 and ΔpK_a (obs) = 2.80. This enhancement is no doubt due to steric inhibition of the solvation of the O$^-$ in the phenate ion, which increases sensitivity to the electronic effects of the substituents, i.e. the ρ value. This effect is even more marked in 2,6-di-*tert*-butylphenol, for which the ΔpK_a value (relative to phenol) is –3.06

(1:1 v/v EtOH–H$_2$O, 25 °C)[114]. In this system ΔpK_a values for 4-CN and 4-NO$_2$ compounds are 4.07 and 6.73, respectively, relative to 2,6-di-*tert*-butylphenol, corresponding to a ρ value of about 4, compared with about 2.9 for the ionization of phenol itself in the same solvent.

IV. THE *ORTHO*-EFFECT OF CN

A. Introduction

The term *ortho*-effect has long been used to cover the peculiar influence of a substituent in the position *ortho* to a reaction centre, which often differs markedly from that of the same substituent in the *meta*- or *para*-position[74,115,116]. Steric phenomena have long been recognized as playing a major part in the *ortho*-effect. Primary steric effects of various kinds, including steric hindrance to the approach of the reagent or to solvation, and secondary steric effects have been invoked. In certain systems hydrogen-bonding and other intramolecular interactions have been postulated.

One of the main difficulties in understanding the *ortho*-effect, however, lies in adequately specifying the electronic effects of *ortho*-substituents. The relative contributions of *I* and *R* effects to the influence of *ortho*-substituents are liable to be very different from those operating at the *meta*- or *para*-position. There have been many attempts to develop scales of 'sigma-*ortho*' constants analogous to σ, σ^0, σ^+, σ^-. etc. (Section II) for the *meta*- and *para*-position, but such scales are never found to be of very general application[74,116]. The composition of the electronic influence of *ortho*-substituents with respect to *I* and *R* effects seems greatly subject to variation with the nature of the reaction, the side-chain, the solvent, etc. The inductive effect of an *ortho*-substituent operates at much shorter range than that of a *meta*- or *para*-substituent, but the orientations of substituent dipoles with respect to the reaction centre are very different from those of *meta*- or *para*-substituents. It is sometimes supposed that the resonance effect of an *ortho*-substituent tends to be inherently weaker than that of the same substituent in the *para*-position, because *ortho*-quinonoid instead of *para*-quinonoid structures may be involved in its operation. However, the resonance effect also is being delivered at rather short-range from the *ortho*-position.

The most fruitful treatment of the electronic effects of *ortho*-substituents involves the use of the same σ_I and σ_R-type constants as may be employed in correlation analysis for *meta*- and *para*-substituents by means of the 'dual substituent-parameter equation'[61] or the 'extended Hammett equation'[65] (Section II.B). Obviously it is a considerable assumption that these are valid for *ortho*-substituents and the implication is that in the correlation analysis any peculiarities in the electronic effects may be adequately expressed through the coefficients of the inductive and resonance terms. Really satisfactory correlation analysis for any given reaction system requires a large amount of data and can only rarely be accomplished.

In Section IV.B we will discuss the *ortho*-effect of CN as manifested in the ionization of carboxylic and other acids and (in Section IV.C) in the reactions of substituted benzoic acids with diazodiphenylmethane (DDM). Only in the case of the latter system can really satisfactory correlation analysis be taken as the basis for discussion. For most of the other systems discussion will have to be qualitative or, at best, semi-quantitative.

B. Ionization of Carboxylic and Other Acids

ortho-substituted benzoic acids involving electron-attracting substituents tend to be considerably stronger than their *para* isomers. The pK$_a$ values (water, 25 °C) for some *p*-X, *o*-X pairs are respectively as follows: NO$_2$, 3.44, 2.17; Cl, 3.98, 2.92: F, 4.14, 3.27

SO_2Me, 3.53, 2.53[117]; $COOMe$, 3.75[96], 3.32 (30 °C). Thus a decrease in pK_a of about one unit is fairly typical for the effect of moving such a substituent from *para*- to *ortho*-position. Doubtless this is considerably due to the increased inductive effect from the *ortho*-position. However, some contribution will also be made by the substituent twisting the carboxyl group out of the plane of the benzene ring, thereby reducing the extent of conjugation of the ring with the side-chain. This has the result of destabilizing the undissociated form of the acid relative to the carboxylate ion, thereby enhancing acid strength. The deconjugation effect is shown very clearly in *ortho-t*-butylbenzoic acid, whose pK_a value is 3.54; *cf* the acid-weakening effect of the group in the *para*-position, $pK_a = 4.40$. The deconjugation effect must play some part in the case of the more bulky polar groups, for example SO_2Me, but not much in the case of F. Another factor in the case of some substituents is hydrogen-bonding. Thus the very large increase in acidity as between *p*-hydroxy- and *o*-hydroxy-benzoic acid, pK_a values 4.58 and 3.03 respectively, is attributed to the stabilization of the anion of salicylic acid by internal hydrogen-bonding[118].

The pK_a value of *o*-cyanobenzoic acid is 3.14. It has already been mentioned (Section III.B) that Widequist[94] found this acid to be rapidly hydrolysed in water (initially to phthalamic acid). He seems to have taken great care to minimize the effects of this on the conductivity measurements which were used to determine the dissociation constant, but it must be supposed that the above value is not of the highest accuracy possible for pK_a determinations. The pK_a value for the *para* acid is about 3.55. Thus the general pattern for the effect of moving an electron-attracting group from *para*- to *ortho*-position is followed, but the increase in acidity is not so marked as in most of the cases mentioned above. To decide whether the increase in acidity is reasonable requires correlation analysis by the extended Hammett equation (Section II.B) in the form of equation 15:

$$- pK_a = \alpha\sigma_I + \beta\sigma_R + \varphi\upsilon + h \qquad (15)$$

where σ_I and σ_R are, respectively, the inductive and resonance constants of Taft's analysis of ordinary Hammett σ constants (see Section II.B), and υ is the steric substituent constant developed by Charton[119-121]. There are plenty of data for the ionization of *ortho*-substituted benzoic acids in water at 25 °C, but unfortunately correlation analysis proves not to be altogether straightforward. The establishment of the regression equation cannot be done unambiguously through the use of 'well-behaved' substituents (*cf* Section IV.C), and in the various equations that may be derived by taking various selections of substituents, the regression coefficients and the intercept term vary quite a lot. It would not be appropriate in this chapter to go into great detail. Suffice it to say that a fairly satisfactory equation may be based on 16 substituents: H, Me, Et, Pr^i, Bu^t, F, Cl, Br, I, OMe, OPh, SMe, NH_2, SH, NO_2, Ph. Charton's values of σ_I and σ_R have been used (these were derived originally in connection with the correlation analyses outlined in Section IV.C) and NO_2 and Ph are assumed to lie orthogonal to the benzene ring. (The conformation affects the appropriate values of σ_R and υ.) Regression equation 16 was obtained:

$$- pK_a = 2.478\,\sigma_I + 1.803\,\sigma_R + 0.763\,\upsilon - 4.093 \qquad (16)$$
$$(\pm 0.132)\quad (\pm 0.130)\quad (\pm 0.109)$$
$$n = 16,\ R = 0.988,\ s = 0.111,\ \psi = 0.178$$

R = multiple correlation coefficient; the other symbols have the same meanings as in connection with equations 13 and 14 in Section III.D.)

A rather striking feature of this equation is that the intercept 4.093 does not correspond closely to the observed pK_a value of 4.205 for benzoic acid itself, i.e. this acid is not acting as the true parent of the series of *ortho*-substituted benzoic acids. Examination of the data and the results of the correlation analysis suggests that all the *ortho*-substituted acids receive an initial increment in acidity caused in some way by the replacement of o-H by any other substituent. The regression was therefore repeated without the benzoic acid point, and equation 17 was the result:

$$-pK_a = 2.360\ \sigma_I + 1.910\ \sigma_R + 0.588\ \upsilon - 3.920 \qquad (17)$$
$$(\pm 0.140)\quad (\pm 0.135)\quad (\pm 0.142)$$
$$n = 15,\ R = 0.990,\ s = 0.103,\ \psi = 0.166$$

From this equation it appears that the effective parent acid is at a pK_a value of 3.92. If the characteristic constants of CN ($\sigma_I = 0.57$, $\sigma_R = 0.08$, $\upsilon = 0.40$) are inserted in equation 17, pK_a(calc) is 2.19, i.e. almost one pK_a unit lower than pK_a(obs) of 3.14. The discrepancy might be due to great experimental error in the observed value, but we have already commented on the relatively small difference in pK_a values for the *para* and *ortho* acids, compared with other pairs. It seems possible that o-cyanobenzoic acid is stabilized in the undissociated form by internal hydrogen-bonding involving the carboxylic proton and either the lone-pair electrons of the nitrogen atom or the π-electrons of the C$\equiv$N triple bond. It must be admitted, however, that there is no sign of the role of such bonding in influencing reactivity towards DDM in solution in alcohols (Section IV.C). This is in stark contrast to the effect of hydrogen-bonding in stabilizing the anion of o-hydroxybenzoic acid; this can be seen not only in the dissociation of the acid in water, but also clearly in the reactivity towards DDM in alcohols[122,123]. At present the explanation of the marked discrepancy between pK_a(calc) and pK_a(obs) for o-cyanobenzoic acid must be regarded as unclear. As already mentioned, various regression equations may be obtained for the effect of *ortho*-substituents on the ionization of benzoic acid, depending on the selection of substituents used, but all the equations tested agree in giving pK_a(calc) for the o-CN acid in the range 2.30 ± 0.15, well below pK_a(obs).

Data for the effect of o-CN in acids of the general type Ph—G—COOH are restricted to phenoxyacetic acid. The pK_a values (water, 25 °C) for some p-X, o-X pairs are as follows: CN, 2.93, 2.98; NO$_2$, 2.89, 2.90; F, 3.13, 3.09; Cl, 3.10, 3.05; Br, 3.13, 3.12; I, 3.16, 3.17; OMe, 3.21, 3.23; cf the pK_a value for phenoxyacetic acid itself, 3.17. Thus the rather small effects of substituents on the acidity of this acid differ little as between the *para*- and the *ortho*-position. No steric inhibition by the *ortho*-substituent of conjugation involving the carboxyl group can occur in this system. Presumably any increase in the transmission of the inductive effect of the substituent in moving from the *para*- to the *ortho*-position is more or less cancelled out by changes in the orientation of substituent dipoles relative to the reaction centre. The behaviour of CN accords with that of the other electron-attracting substituents. It is not worth speculating about the variations in the pattern of pK_a values, $o < p$ or $o > p$, involving a few units in the second place of decimals.

In phenol ionization, pK_a values (water, 25 °C) for some p-X, o-X pairs are as follows: CN, 7.95, 6.90; F, 9.91, 8.71; Cl 9.42, 8.53; Br, 9.36, 8.44; NO$_2$, 7.15, 7.23. (Various different values may be found in the literature[77c] for some of these substituted phenols. As far as possible the individual values quoted for the members of each pair have been determined by the same authors.) Here the pattern for CN is similar to those for F, Cl and Br, but different from that shown by NO$_2$, for which the isomeric phenols differ little in acidity. The increase in acidity when an electron-attracting group is moved from *para*- to *ortho*-position in phenol is doubtless due to an increased transmission of the inductive effect to the reaction centre, which more than outweighs any unfavourable effect of change in orientation. The anomalously low acidity of o-nitrophenol is usually attributed to stabilization of the undissociated form by hydrogen-bonding between OH and ONO.

For the ionization of the anilinium ion, pK_a values (water, 25 °C) for some p-X, o-X pairs are as follows: CN, 1.75, 0.95; NO$_2$, 1.13, −0.05; F, 4.65, 3.20; Cl, 4.05, 2,71; Br, 3.95, 2.55 Thus in every case a marked increase in acidity occurs when the substituent is moved from the *para*- to the *ortho*-position, although the change for CN is the smallest quoted above, with NO_2 the next to smallest. It might have been expected that at least with *o*-NO_2 hydrogen-bonding would play a part in stabilizing the protonated form relative to the neutral form, but there is no clear indication of this. Such an effect is apparent for the dimethylaniline system: NO_2, 0.61, 2.92. Here no internal hydrogen-bonding is possible in the unprotonated form and the stabilization of the protonated form by hydrogen bonding between NH^+ and ONO leads to a very marked decrease in acidity.

C. The Reactions of *ortho*-Substituted Benzoic Acids with Diazodiphenylmethane (DDM)

The point has already been made (Section IV.A) that discussion of the *ortho*-effect is frequently hampered by a lack of data for the wide range of *ortho*-substituted compounds necessary for satisfactory correlation analysis by means of an appropriate form of extended Hammett equation[65]. This situation was remedied some years ago by the author and his colleagues[123,124] for the reactions of *ortho*-substituted benzoic acids with DDM. Rate coefficients (1 mol^{-1} min^{-1}) at 30 °C were measured for the reactions of benzoic acid and 32 *ortho*-substituted benzoic acids in 11 alcohols (including 2-methoxyethanol) as solvents[123]. The reaction involves a rate-determining proton transfer from the carboxylic acid to the DDM to form a diphenylmethanediazonium carboxylate ion-pair. Subsequent fast-product governing stages have been variously formulated[123]. A more restricted study was carried out for reactions at 30 °C of the substituted benzoic acids with DDM in 7 aprotic solvents[124], in which the proton transfer is believed to be rate-limiting rather than rate-determining.

The correlation analysis employed the extended Hammett equation in the form of equation 18:

$$\log k = \alpha\sigma_I + \beta\sigma_R + \varphi\upsilon + h \tag{18}$$

(The symbols were defined in Section IV.B.) A full discussion of the *ortho*-effect as revealed in this work would be inappropriate here. We must restrict ourselves to the more limited task of indicating the role of *o*-CN. We discuss first the work involving alcohols as solvents. To apply the extended Hammett equation, i.e. to determine the regression coefficients α, β and φ and the intercept term h, it is first necessary to select a set of substituents which can be expected to be 'well-behaved'. Particular problems for σ_R and υ may be caused by conformational effects, and internal hydrogen-bonding may occur as a further factor governing reactivity, for which parametrization is not included in equation 18.

Nine substituents (Set A: H, Me, Bu^t, F, Cl, Br, I, CF$_3$ and CN) were selected as a basic set for the following qualities[123]: (i) symmetry and simplicity, (ii) freedom from conformational effects, (iii) lack of a large resonance effect, (iv) lack of any marked tendency to form hydrogen-bonds with the reaction centre. The inclusion of CN is obviously in accord with criteria (i) to (iii) and, *a priori*, it would be expected that CN would also fulfil criterion (iv). In view of what emerged in Section IV.B about the effect of *o*-CN on the ionization of benzoic acid in water, one cannot be so sure about criterion (iv) but, as we shall see, confidence in the suitability of CN as a member of the basic set of well-behaved substituents may be justified *a posteriori*. It proved possible to expand the list from 9 to 18 by making reasonable assumptions about the conformations of certain substituents, thus enabling these to be placed on the σ_R and υ scales [Set B: Set A + Et, Pr^i, OMe, OEt, OPh, SMe, SO$_2$Me, CH$_2$Ph, (CH$_2$)$_2$Ph]. Correlations based on Set B turned out to be superior to those based on Set A.

The regression equations were established for data in 11 alcohols as solvents and were then used to assess the peculiar behaviour of another 15 *ortho*-substituents in respect of conformational effects and intra-molecular hydrogen-bonding[123,125]. Here we are concerned with assessing whether confidence in o-CN as a well-behaved substituent was justified. We first give as an example the regression for 2-methoxyethanol as solvent:

$$\log k = 1.624\,\sigma_I + 0.964\,\sigma_R + 0.346\,\upsilon + -0.305 \qquad (19)$$
$$(\pm 0.074)\quad(\pm 0.082)\quad(\pm 0.060)$$
$$n = 18,\ R = 0.990,\ s = 0.070$$

The regression coefficients are positive for the σ_I and σ_R terms because electron-attracting groups accelerate the reaction and electron-releasing groups retard it. The positive regression coefficient for the υ term corresponds to the reaction being subject to steric acceleration by $ortho$-substituents through deconjugation of COOH with the benzene ring; cf Section IV.B.

If the parameters for CN are inserted into equation 19 (that is $\sigma_I = 0.57$, $\sigma_R = 0.08$ and $\upsilon = 0.40$), we obtain $\log k(\text{calc}) = 0.836$, compared with $\log k(\text{obs}) = 0.834$. Without quoting the regression equations[123,126], we give the results of similar calculations for the reactions in various alcohols [solvent, $\log k(\text{calc})$, $\log k(\text{obs})$]: methanol, 1.265, 1.238; butan-2-ol, 0.877, 0.883; 2-methylbutan-2-ol, 0.399, 0.408; benzyl alcohol, 1.881, 1.811; cyclopentanol, 0.877, 0.857. In every case the data point for o-CN conforms well to the regression equation; cf its apparent behaviour in the ionization of benzoic acid in water (Section IV.B).

We turn now to the studies in aprotic solvents[124]. Here the series of acids used could not be complete for every solvent, for reasons of insolubility and other factors. (There were also other complications in this part of the work[124].) Thus some substituents of Set B could not be studied in every solvent used and some ceased to be well-behaved, because they were found to develop a capacity for internal hydrogen-bonding in the non-hydroxylic media. The o-CN acid could be studied in acetone, ethyl acetate, DMA, DMF and DMSO, but not in nitrobenzene or chlorobenzene. In most of the five solvents o-CN continued to be well-behaved, $\log k(\text{calc})$ deviating from $\log k(\text{obs})$ by about the standard deviation of the estimate or considerably less than this. Only for ethyl acetate was there an indication of o-CN deviating from the regression equation.

Finally, it must be mentioned that solid samples of o-cyanobenzoic acid deteriorated noticeably over the course of 2 to 3 months, i.e. a solution made from such a sample showed a significant decrease in the rate coefficient for reaction with DDM. Solutions of the acid in some of the solvents deteriorated much more rapidly; cf Widequist's finding for aqueous solutions[94].

V. SUBSTITUENT CONSTANTS OF CN FROM THE APPLICATION OF MODERN EXPERIMENTAL AND THEORETICAL TECHNIQUES

A. Experimental Techniques

The emphasis in the foregoing parts of this chapter has been deliberately 'chemical'. We have tried to explore the role of substituent constants in relation to understanding the effect of structure on reaction rates and equilibria, with particular reference to the CN group as a substituent. This chemical emphasis will continue in the later parts of the chapter, for CN and for the other substituents with which we are concerned, but in the present section there will be a change. In Section II.B brief reference was made to the use of substituent constants in the correlation analysis of spectroscopic data, particularly ^{19}F and ^{13}C NMR substituent chemical shifts and infrared frequencies and intensities. These matters must now be explored in greater detail.

Attempts were made to apply Hammett benzoic acid-based σ_m and σ_p constants to the correlation analysis of spectroscopic data. Some significant correlations were obtained, but many of the correlations were rather poor, trends rather than precise relationships. Success in this area was found to involve the separation of inductive and resonance effects and the application of the dual substituent parameter (DSP) equation (Section II.B). Indeed the development of the DSP equation became closely connected with the correlation analysis of ^{19}F NMR shielding of substituted fluorobenzenes at an early stage, around 1957[76]. σ_I and σ_R^0 were applied extensively to ^{19}F NMR data[60], and within a few years the correlations were being used to investigate 'the effect of solvent on the inductive order'[66], and 'the effect of structure and solvent on resonance effects'[67]. New σ_I and σ_R^0 values were based on the correlations. What happened with ^{19}F NMR set a pattern which was followed by later work. Established σ_I and σ_R^0 values for substituents which were expected to be 'well-behaved' were used to set up regression equations. In the very early days the established substituent constants were all based on chemical reactivity (rate constants or equilibrium constants). ^{19}F NMR data for groups for which no appropriate substituent constants were available were then substituted into the regression equations to obtain '^{19}F NMR-based values' of the substituent constants. Further, for the substituents which had been used to establish the regression equations, back-calculation from the NMR data led to '^{19}F NMR-based values' for those substituents as well. Thus for many substituents both 'reactivity-based' and '^{19}F NMR-based' values of σ_I and σ_R^0 became available. For certain substituents there was a proliferation of values based on reactivity under various conditions or on ^{19}F NMR in different solvents. Slightly later, correlation analysis of infrared data led in particular to new σ_R^0 values and, to a lesser extent, new σ_I values, which were described as 'IR-based'[70,71]. Somewhat later the same development occurred in connection with ^{13}C NMR, leading to '^{13}C NMR-based' values[68,69].

The CN group played a part in these developments. Thus CN was characterized by Taft[57,59] with a σ_I value of 0.59, based on the appropriate $\sigma*$ value in his analysis of reactivities in ester reactions. The application of ^{19}F NMR found a σ_I value for CN of 0.48 in 'normal solvents', 0.43 in dioxan, 0.53 in 'weakly protonic solvents' and 0.74 in trifluoroacetic acid[66]. The last-mentioned high value was attributed to the effect of hydrogen-bonding of CN to CF$_3$COOH, but with this exception all the ^{19}F NMR-based values lie considerably below the reactivity value of 0.59, which agrees well with Charton's reactivity value of 0.57 (Section III.A)[81]. 'Normal solvents' included cyclohexane, benzene, carbon tetrachloride and other more polar solvents, while the principal 'weakly protonic solvents' were methanol and formic acid. The above figures appear to give some indication of slight enhancement of the $+I$ effect of CN by hydrogen-bonding to 'weakly protonic solvents'. It will be recalled, however, that reactivity studies (Section III.A) failed to give any clear sign of a solvent-dependent σ_I value for CN. The very low value for σ_I in dioxan solution finds a parallel in the behaviour of σ_I for other electron-attracting groups (e.g. NO$_2$ and COCH$_3$) and was attributed to Lewis acid-bonding of dioxan to the substituent[66]. A recent compilation of substituent constants gives a ^{19}F NMR-based value for CN of 0.53 from a chemical shift determined in dilute solution in hydrocarbon solvent[99]; cf 0.48 above. The increase of 0.05 may be due to using a slightly different regression equation. (The application of this equation to the chemical shift data for 'weakly protonic solvents' would give a σ_I value for CN of 0.58, in good agreement with the reactivity-based value.) However, in the recent compilation[99] the value of 0.53 for CN is put in parentheses to indicate 'magnetic or other complications". (Values for a few other substituents are similarly distinguished, including those for F, I and NMe$_2$.) Thus it seems that exponents of correlation analysis involving ^{19}F NMR data now have doubts about the validity of basing a σ_I value for CN on ^{19}F NMR data. The σ_I value recommended in 1973 for CN by Ehrenson, Brownlee and Taft[62] was 0.56, while a recent tabulation[127] gives the equivalent σ_F value (see below) as 0.60.

A σ_R^0 value of 0.13 was given by Ehrenson, Brownlee and Taft[62]. This seems to have been derived by substracting a σ_I value of 0.56[62] from the σ_p^0 value of 0.69 given by Taft[128], which is apparently a reactivity-based value. Slightly different values of σ_p^0 for CN were obtained by Yukawa, Tsuno and Sawada[129], but by using these or by using slightly different values of σ_I in the range 0.57 to 0.59, very similar values of σ_R^0 would be obtained. The value 0.13 for σ_R^0 of CN may be taken as typical of those based on the reactivity of 'insulated' reaction series (Section II.B). It is slightly higher than any value based on the assumption that for CN there is no valid distinction between σ_R^0 and $\sigma_R(BA)$; see Section III.C, e.g. Charton's value 0.08 for σ_R, which was calculated by subtracting $\sigma_I = 0.57$ from $\sigma_p = 0.65^{81}$. ^{19}F NMR studies shed no direct light on σ_R^0 for CN, because cross-conjugation between F and p-CN in p-cyanoflurobenzene leads to enhancement of the $+R$ effect of CN (Section II.A). It is only for $-R$ substituents that ^{19}F NMR studies yield σ_R^0 values. The application of the ^{19}F NMR method to $+R$ substituents yields 'apparent' values, which may be designated $\bar{\sigma}_R^{58}$. For CN the value of $\bar{\sigma}_R$ is about 0.2 from measurements in a wide range of supposedly indifferent solvents[67]. As in the case of σ_I, $\bar{\sigma}_R$ is increased by certain solvents. Thus in trifluoroacetic acid $\bar{\sigma}_R \approx 0.30$. Presumably hydrogen-bonding enhances both $+I$ and $+R$ effects. Measurements of infrared intensities do, however, give information about σ_R^0 constants for $+R$ substituents. Katritzky and colleagues[130] give a σ_R^0 value of 0.085 for CN, based on measurements with a solution of cyanobenzene in carbon tetrachloride. (Note: the sign of σ_R^0 is not actually given by the correlation of infrared intensities, because the square of σ_R^0 is involved. The sign can usually be inferred from other knowledge about the substituent.) This agrees nicely with the reactivity-based value of 0.08[81].

There is a continual tendency for the values of σ_I and σ_R^0 (and other σ_R–type constants) to be adjusted in the light of new measurements. Thus measurements in 1979[131] of $para$ ^{13}C substituent chemical shifts for a series of mono-substituted benzenes in very dilute solution in cyclohexane, carbon tetrachloride or deuteriochloroform were the basis for a redefinition of the σ_R^0 scale and some amendment of σ_R^0 values. For most substituents the changes compared with the values recommended earlier[62] were very small, but σ_R^0 for CN came out at 0.08 and the previous value of 0.13[62] was by implication abandoned.

Reynolds and coworkers[132] based a similar operation on ^{13}C substituent chemical shifts of $meta$- and $para$-substituted styrenes. Iterative multiple regression was used for the redefinition of the σ_I and σ_R^0 scales. The authors also took the opportunity to replace the symbol σ_I by σ_F, having become convinced that the so-called inductive effect was entirely a field effect (see the present author's discussion of this matter[46]). The authors presented an extensive table in which their values of the substituent parameters are compared with those obtained by other authors. Their σ_R^0 value of 0.071 for CN essentially confirmed the various values of around 0.08 already mentioned, but their σ_I value of 0.599 extended the upper limit of the range of values recorded.

Happer[133] determined ^{13}C substituent chemical shifts for $meta$- and $para$-substituted styrenes in seven different solvents. Data for the side-chain carbons, and in the $meta$ series for the ring carbon $para$ to the substituent, were analysed as a basis for assessing solvent effects on σ_I, σ_R^0, $\sigma_R(BA)$ and $\bar{\sigma}_R$. The σ_I values for CN varied from 0.52 in DMSO to 0.64 in CDCl$_3$; in EtOH the value was 0.60. The σ_R^0 values for CN varied from 0.04 in DMSO to 0.09 in benzene; in EtOH the value was 0.06. Some doubts were, however, expressed regarding the reliability of the σ_R^0 values for CN, NO$_2$ and CF$_3$.

The influence of solvent on the inductive order of substituents was studied by Laurence and collaborators through infrared measurements on 4-substituted camphors[134]. From these Laurence[135] has tabulated new σ_F values applicable to solutions in carbon tetrachloride or other solvents of low dielectric constant. CN comes out at 0.57.

Mention must also be made of the use of studies of chemical reactions in the gas phase as a means of determining substituent constants. The investigation of substituent effect and linear free-energy relationships in the gas phase has become an enormous subject with which we can deal only briefly. Part of this subject was established a long time ago and consists in the study of such reactions as the pyrolysis of esters by the techniques of gas kinetics (see the review by Smith and Kelly[136]). One purpose of such work is to see how far substituent constants based on processes in solution may be applied successfully in the gas phase. This leads to the possibility of determining substituent constants in the complete absence of solvent. Work of this nature continues today; see the recent review by Holbrook in this Series[137], which updates the earlier review by Taylor[138]. CN frequently features as a substituent in gas phase studies and there are many examples of its inclusion in Hammett or Taft correlations[137–139].

However, the major activity in gas phase studies now depends on the use of modern techniques such as ion cyclotron resonance (ICR). Thus, as already mentioned (Section III.D), Fujio, MacIver and Taft[111] measured the gas phase acidities, relative to phenol, of 38 *meta*- or *para*-substituted phenols by the ICR equilibrium constant method, and their results for $+R$ substituents led them to suggest that such substituents in aqueous solution exerted solvation-assisted resonance effects. It was later[140] shown by comparison of gas phase acidities of phenols with acidities of phenols in solution in DMSO that solvation-assisted resonance effects could also occur even when the solvent did not have hydrogen-bond donor properties. Indeed for p-CN and certain other substituents these effects appeared to be larger than in aqueous solution.

Taft and Topsom[127] have recently written an extensive review of the electronic effects of substituents in the gas phase. This article includes a tabulation of substituent inductive and resonance parameters. The inductive parameters (designated σ_F) are based upon measured substituent effects on spectroscopic properties in either the gas phase or in hydrocarbon or similar solvents. The resonance parameters were arrived at through the treatment of 38 gas phase reactivity series by iterative multiple regression, using the σ_R^0 values of Bromilow and coworkers[131] as the starting point. The σ_F value for CN was found to be 0.60 (quoted above), while the resonance parameter is given as 0.10. The column heading in Taft and Topsom's table is simply σ_R, but inspection shows that the values must be regarded as being those of σ_R^0, when the distinction matters, i.e. with $-R$ substituents.

B. Theoretical Techniques

The application of *ab initio* molecular orbital theory to suitable model systems has led to theoretical scales of substituent parameters, which may be compared with the experimental scales. Calculations (3-21G or 4-31G level) of energies or electron populations were made by Marriott and Topsom in 1984[141]. The results are well correlated with σ_F (i.e. σ_I) for a small number of substituents whose σ_F values on the various experimental scales (gas phase, non-polar solvents, polar solvents) are concordant. The regression equations are the basis of theoretical σ_F values for about fifty substituents. Unfortunately, the position of CN in the treatment is anomalous. Its various experimental σ_F values are deemed concordant, but it is not possible to include CN in the selection of substituents used in establishing the regression equations, for the theoretical value of σ_F comes out at 0.47 or 0.45, according to the method of calculation employed. This is well below the experimental values in the range 0.57 to 0.60. An explanation is provided as follows: "It may well be that the higher values in the literature reflect not only the effect of the simple dipole related to the cyano substitution but also the charge transfer (hyperconjugation) that can occur between cyano group and CH or CC bonds if the atom of attachment is sp^3 hybridized".

A theoretical scale of substituent resonance effects was based on calculations of electron populations in substituted ethylenes[142]. A suitable regression equation was again set up by using standard substituents, but in this case the quantum-mechanical quantity was correlated with infrared-based σ_R^0 values. The equation was the basis of theoretical σ_R^0 values for more than forty substituents. In this case it was possible to include the data for CN in the setting up of the regression equation, and the theoretical value for σ_R^0 came out in exact agreement with the experimental value which had been used, namely 0.09. A further redefinition of the theoretical scale was made recently[143] as a result of a change of views as to the most suitable level of MO approximation. In the latest version σ_R^0 comes out at 0.13, the literature value of 0.09 having been used in setting up the regression equation. Thus the new theoretical value for σ_R^0 is not an improvement as far as CN is concerned. Other recent theoretical papers by Topsom in which CN is featured are on a scale of variable π-electron transfer[144], the influence of water on substituent field effects[145], the influence of water on substituent resonance effects[146], acidities of *ortho*-substituted phenols[147] and theoretical studies of the effects of hydration on organic equilibria[148]. There is also an extensive review of theoretical studies of electronic substituent effects[149]. It seems rash, however, to regard theoretical substituent parameters as in any way replacing those founded on experimental results.

There have been various other theoretical treatments of substituent effects, e.g. the correlation analysis of substituent effects on the acidity of benzoic acid by the AM1 method[150] and direct prediction of linear free-energy substituent effects from 3D structures using comparative molecular field analysis, the relevant data set being 49 substituted benzoic acids[151].

VI. ELECTRONIC EFFECTS OF CN ON VARIOUS SYSTEMS AND PROCESSES

A. Formation and Stabilization of Carbanions

1. Introduction

The Historical Introduction (Section I.B) showed how the activation of the hydrogen atom(s) on the carbon to which CN is attached was recognized at an early stage to be associated with incipient acidity. Reaction mechanisms essentially involving carbanions were proposed by Lapworth[23] early in the 20th century and were gradually accepted by organic chemists over the next 30 to 40 years. By the 1930s some 'stable' carbanions, as opposed to highly reactive intermediates, were recognized, e.g. the behaviour of malononitrile and cyanoform as acids was known.

In the last half-century the pK_a values of thousands of carbon acids have been measured in a variety of solvents. In many such acids the ionization is stimulated by the CN group, sometimes in association with other $+I$, $+R$ groups such as NO_2, COR or COOR. There have also been many studies of the kinetic acidities of carbon acids and of the mechanisms of reactions involving carbanion intermediates; again, in many cases CN has played an important role. Clearly a comprehensive survey of CN-promoted C—H acidity is not feasible here, and it will be necessary to concentrate on outlining a few main topics. The role of CN in increasing C—H equilibrium acidity will be illustrated with a few structural series, some recent studies of kinetic acidity will be outlined, and developments in the correlation analysis of C—H acidity and in understanding of the stabilization of carbanions by CN and related groups will be discussed.

2. Equilibrium acidities of cyanocarbon acids

Many pK_a values for cyanocarbon acids are in the compilations of Palm and colleagues[152]. A comprehensive study of the ionization of many carbon acids in DMSO has been made by Bordwell and colleagues[153] and many of them are cyano compounds.

The pK_a value of methane in water has been estimated as about 40, and that of CH_3CN as about 25. Further substitution of H by CN leads to pK_a values of 11.2 for $CH_2(CN)_2$ and − 5.1 for $CH(CN)_3$. Considering the enormous changes involved and the mutual proximity of the cyano groups, the effects of successive substitution seem to show an extraordinary degree of additivity, ΔpK_a for each substitution averaging about 15 ± 2. (Strictly speaking, statistical correction should be made for the different numbers of ionizable hydrogen atoms, but is hardly worthwhile in view of the uncertainties in the data for methane and methyl cyanide.) This is probably connected with the small size of CN and the linearity of C—C≡N; the corresponding series for NO_2 does not show additivity.

There are many other examples of increased acidity produced by successive substitution of H by CN in a given molecular skeleton. Thus for CN substitution in cyclopentadiene, pK_a values (around ambient temperature in water) are as follows[154,155]: cyclopentadiene, 15; 1-cyanocyclopentadiene, 9.78; 2,5-dicyanocyclopentadiene, 2.52: 3,4-dicyanocyclopentadiene, 1.11. To study the effect of further substitution, it is necessary to change to acetonitrile as solvent: 3,4-dicyanocyclopentadiene, 10.17; 1,3,5-tricyanocyclopentadiene, 3.00; 2,3,4,5-tetracyanocyclopentadiene, 0.0; 1,2,3,4,5-pentacyanocyclopentadiene, < -2. (The last-mentioned corresponds to < -11 on the water scale.)

Various series of acids can be devised to demonstrate the relative effects of CN and other substituents in promoting acidity of neighbouring C—H. As might be expected, CN is inherently not quite so effective as NO_2. Thus the pK_a value of CH_3NO_2 is 10.2 in water; cf about 25 for CH_3CN. The series RCH_2NO_2 presents some interesting results for different groups R: H, 10.2; Me, 8.6; Cl, 7.2; COOMe, 5.65; CN, 4.9; NO_2, 3.6; Ph, 6.9. The position of the 'parent' H point in this series is decidedly anomalous. The relatively small difference between the effects of CN and NO_2 in this series, compared with that noted above, is presumably due to mutual interference of the NO_2 groups in $CH_2(NO_2)_2$.

CN may also feature as a substituent influencing C—H acidity from a more remote site. Thus for ring substitution in the series $PhCH(NO_2)_2$, the pK_a values for the m- and p-cyano compounds are 2.86 and 2.73, respectively, compared with 3.89 for the unsubstituted compound (water, 20 °C). The ρ value for this reaction series is about 1.7, and these results for m- and p-CN appear to indicate that ordinary Hammett σ values are applicable, whereas σ^- values might have been expected to be more relevant. Presumably this means that the delocalization of the negative charge in the carbanion is dominated by the NO_2 groups, so that delocalization into the ring plays a minor role.

In the series $RCH(NO_2)_2$, data are available for the effect of substituting CN in an alkyl chain, pK_a values for various groups R being as follows: Et, 5.6; Bu, 5.45; CH_2CN, 2.34; $(CH_2)_2CN$, 3.5; $(CH_2)_3CN$, 4.34; $(CH_2)_4CN$, 5.0 (all values in water at 20 °C), i.e. a butyl chain damps out most of the effect of CN.

3. Kinetic acidities of cyanocarbon acids

Various types of reaction of carbon acids are believed to proceed via dehydronation (commonly referred to as deprotonation)[156], notably halogenation in the presence of bases, and the rates of such processes are a measure of the kinetic acidities of the carbon acids. Extensive data of this nature exist, including some for cyanocarbon acids[152]. In recent years kinetic acidity has also been studied through the detritiation of appropriately tritiated compounds, e.g. benzyl cyanide and substituted benzyl cyanides, tritiated in the side-chain. The existence also of equilibrium acidity data for the substrates under the same conditions permits rate–equilibrium correlations of the Brønsted type[157], as well as Hammett treatment. Studies of the rates of reaction of a given substrate with a series of bases also generate Brønsted correlations, and Hammett correlations in appropriate cases.

Thus Bowden and colleagues[158] have studied the kinetic acidities of $meta$-substituted benzyl cyanides through detritiation by secondary aliphatic amines and guanidines as bases in DMSO. Brønsted correlations for seven benzyl cyanides reacting with the series of bases give almost identical β values of about 0.92. The α values for variation of carbon acid with a given base are not quite so constant, lying mainly between 0.68 and 0.77, with one outlier at 0.46. There is no evidence for a reactivity–selectivity[159] relationship in these studies. The authors comment, "The structure of the transition state appears to be remarkably invariant with simple polar changes in both base and substrate, but steric factors can affect this state markedly". The results of a related study[160] of a series of substituted benzyl cyanides undergoing detritiation by a series of substituted benzylamines as bases were subjected to both Brønsted and Hammett treatments. Again no reactivity–selectivity relationship was apparent.

There have also been studies of kinetic and equilibrium acidities of 4-nitro- and 2,4-dinitro-benzyl cyanide and 9-cyanofluorene in aqueous DMSO[161,162].

4. Correlation analysis of C—H acidity and the origin of the stabilization of carbanions by CN

It is commonly supposed that the stabilization of carbanions by $+R$ groups is due to delocalization of negative charge away from the carbanionic carbon. In the case of CN, resonance structure as in 12 are written[155]:

$$R^1\!-\!\overset{..}{\underset{\underset{R^2}{|}}{C}}\!-\!C\!\equiv\!N\!: \longleftrightarrow R^1\!-\!\underset{\underset{R^2}{|}}{C}\!=\!C\!=\!\overset{..}{N}\!\bar{:}$$

(12)

with the additional possibility that some groups R^1 and/or R^2 may also participate in the delocalization. It might therefore be expected that correlation analysis of the effects of C—H acidity of the substituent X in X—C—H would require electron delocalization parameters for X, perhaps related in some way to σ^- constants. In fact the problem of dealing with proximate substituents is much more difficult than treating the effect of a remote p-X on the acidities of phenols or anilinium ions (Sections II. A and III.D). It is much more analogous to dealing with the *ortho*-effect (Section IV).

Attempts at correlation analysis for X—C—H acidity have been made for a long time. Bowden and colleagues[163] examined the problem in 1970 for the pK_a values of 9-X-fluorenes and related series. For a limited selection of substituents X in 9-X-fluorenes, a correlation with Taft's σ^* values for X was found[59]. A more general correlation, albeit with considerable scatter, was found with ΔM, a parameter based on LCAO–MO calculations. 9-Cyanofluorene fitted the line quite well, but malononitrile deviated strongly. More extensive correlation with σ^* values was found for the effect of X in the series $XCH(NO_2)_2$[155,164], but the substituents X did not include any of strong $+R$ effect such as CN or NO_2.

In recent years Pagani and coworkers have made detailed studies of the problem. In the space available we can only outline their work and interested readers should consult the very detailed papers. These authors have developed special scales of substituent constants for dealing with 'contiguous functionalities'[165]. These new substituent constants are σ_C^- (which seems to be fairly closely related to the ordinary σ^-), σ_{IB} (which bears some relationship to σ_I, but not that close), and σ_{R-}, a special delocalization parameter. It is claimed[166] that "these scales are appropriate for describing interactions between contiguous functionalities, as opposed to literature values which account for remote interactions". Various C—H acidities in gas phase and in solution are successfully correlated by means of multiple regressions on σ_{IB} and σ_{R-}. The same parameters are used in the correlation of ^{13}C chemical shifts for the central carbon in the carbanions.

The most recent paper (1993)[167] is devoted largely to the behaviour of cyano carbanions and there is an estimation of the transfer of π-charge from the carbanionic carbon to the CN. It is concluded that the charge demand of CN is relatively modest. The description of $[RCHCN]^-$ as essentially $RCH\!=\!C\!=\!N^-$ is held to be incorrect. ("Breakdown of a Myth" is the expression used.) It is, however, held that strong π interaction between the carbanionic C and the CN C (shown by structural investigations) gives the CC linkage a high double-bond character, but this apparently does not imply the reduction of the CN bond from triple to double. This is taken as a sign of the 'inadequacy' of the Valence Bond approach to electron delocalization.

It may be recalled that for some time there have been doubts about the role of π(pd) bonding in the stabilization of carbanions by sulphinyl or sulphonyl groups[2].

Whatever the explanation of the stabilization of carbanions by CN may be, new manifestations of this continue to be found, e.g. the behaviour of $C(CN)_3^-$ and $C(CN)_2NO_2^-$ as the leaving groups in the generation of the t-cumyl and t-butyl carbocations, respectively, by C—C bond fission[168–170].

B. Nucleophilic Substitution

CN can exert important effects on reactivity both as a substituent in the substrate and in the nucleophile, and both in aliphatic[171] and in aromatic[172] substitution. We shall illustrate all the relevant features of the effects of CN in this respect.

The normal effect of CN on S_N1 solvolysis of substrates such as benzhydryl chloride is to retard reaction. Thus in the ethanolysis of $XC_6H_4CHPhCl$, log k values (first-order rate coefficients, s^{-1}) are as follows (25 °C): H, -4.27; m-CN, -6.80; p-CN, -7.05 (approximate value by extrapolation from higher temperatures)[173]. The ρ value for this reaction is about -4.0, so the log k values for the m-CN and p-CN derivatives correspond fairly closely to the recommended σ values (Section III.B) of 0.62 and 0.67, respectively. This reaction is, of course, strongly accelerated by $-R$ para-substituents through cross-conjugation in the carbocationic transition state (Section II.A).

Although the effect of CN as a remote substituent on S_N1 reactions is exerted through $+I$ and $+R$ electronic effects, there is evidence that CN and certain other groups which normally π-electron acceptors may function as π-electron donors under extreme electron demand. Molecular orbital calculations support this idea[144,174] and there is now a fair amount of experimental evidence[175–178]. This is mainly of the nature that when CN is actually attached to the reaction centre for S_N1 solvolysis, it is not quite so retarding as might be expected from the operation of its inductive effect. The π-electron donor effect of CN in stabilizing a carbocation may be expressed in terms of resonance involving structure 13.

$$N\equiv C-\overset{+}{\underset{R^2}{\overset{R^1}{C}}} \longleftrightarrow \overset{+}{N}=C=\overset{R^1}{\underset{R^2}{C}}$$

(13)

In contrast to the usual behaviour of the cyano group as a remote substituent in systems undergoing S_N1 reactions. o-CN and p-CN are activating in a substrate undergoing aromatic nucleophilic substitution, although not so strongly as o-NO_2 and p-NO_2. The attack of chlorobenzene by EtO^- in ethanol provides a suitable example. Chlorobenzene itself is not appreciably attacked below 150 °C, but the reactions of o-nitrochlorobenzene and p–nitrochlorobenzene with this reagent can be conveniently studied around 60 °C. Values of log k (second-order rate coefficients, $dm^3\ mol^{-1}s^{-1}$) extrapolated to 25 °C are -6.45 and -6.05, respectively. The reaction of 2,4-dinitrochlorobenzene with EtO^- in EtOH can be studied at room temperature and log k at 25 °C is -1.31. It does not appear to be possible to study the activating effect of CN by itself as a substituent in this system, but for 2-cyano-4-nitrochlorobenzene and 2-nitro-4-cyanochlorobenzene log k at 25 °C has the value -3.15 and -3.11, respectively, i.e. replacing NO_2 by CN decrease log k by 1.84 and 1.80, respectively. Thus the reactivity decreases by a factor of about 70.

The activating effect of NO_2 or CN is due to the electron-attracting effect of these groups, and in particular the $+R$ component, assisting the delocalization of negative charge away from the reaction centre. This may be represented for the intermediate complex in terms of resonance involving structures such as **14**. The relevant transition states are similarly affected[172]. A heterocyclic N atom is also able to assist the delocalization of negative charge

(14)

away from the reaction centre. The reaction of 2-chloro- or 4-chloro-pyridine with EtO$^-$ in EtOH can be studied at around 100 °C. Log k values extrapolated to 25 °C are -8.4 and -6.8, respectively[179], rather lower than those for the corresponding chloronitrobenzenes. However, the presence of two heterocyclic N atoms exerts a powerful influence, e.g. log k for 2-chloropyrimidine at 25 °C is -2.60[179].

The reaction of EtO$^-$ with chloro compounds in ethanol is actually not an entirely satisfactory system for studying the activating effect of CN on aromatic nucleophilic substitution because there is a tendency for CN itself to react, forming the conjugate base of iminoether, $-C(OEt)=NH$[180].

The activating effect of CN has been much studied in association with the activating effect of NO_2 and heterocyclic N in the reactions of 2-chloro-3-cyano-5-nitropyridine (and related compounds) with aniline and substituted anilines and other amine nucleophiles in various solvents[181–183]. 2-Chloro-3-cyanopyridine is too unreactive for study, but the reaction of 2-chloro-5-nitropyridine with aniline in methanol gives a log k value of -5.31 at 40 °C[181]. For the corresponding reaction of 2-chloro-3-cyano-5-nitropyridine log k is -1.68, i.e. the activation by CN is by a factor of over 4000[183]. The kinetics of the reactions of 2-fluoro-3-cyano-5-nitropyridine have also been studied[182].

It might be expected that the activating effect of p-CN would be related to the σ_p^- value of the substituent. From a study of nucleophilic aromatic substitution, Miller and Parker[184] obtained a σ_p^- value for CN of 1.00, very close to the values based on anilinium ionization (Section III.D).

It should also be mentioned that CN is sometimes used as one of the electron-attracting substituents necessary to stabilize Meisenheimer-type complexes. See, for example, a recent study of the reaction of MeO$^-$ with 2,6-dinitro-4-cyanoanisole[185].

The introduction of CN into the aniline molecule would be expected to reduce the nucleophilicity of NH_2 and from the *para* position this should be more or less according to the σ_p^- of CN. There is plenty of information regarding the analogous effect of p-NO_2, but examples involving CN are hard to find. It is easy, however, to find an example for the reduction of amine nucleophilicity via the $+I$ effect of CN. Grob and Schlageter[186] have presented extensive results for the quaternization of 4-X-substituted quinuclidines (*cf* structure **7**) with methyl iodide in methanol at 10.0 °C. According to Charton[81], the ρ_I constant for this reaction is -1.12, and the log k values for X = H and CN are -2.35 and -3.00, respectively[186]. These correspond well to a σ_I value of 0.57 for CN (Section III.A). Grob and Schlageter[186] carried out an analogous correlation in terms of σ_I^q, an inductive parameter based directly on the pK_a values of 4-X-substituted quinuclidinium ions (*cf* structure **7**) as a standard system[187], for which ρ_I is defined as 1.000. The σ_I^q value of CN is 3.04. Grob has made extensive use of this σ_I^q in correlating the effects of substituents on $\log k$ for various reactions, including the solvolyses of 1-substituted 3-bromoadamantanes[188], 4-substituted 2-chloro-2-methylbutanes[189] and 4-substituted bicyclo[2.2.2]octyl *p*-nitrobenzenesulphonates[190]. CN features as a substituent in all of these, which therefore present examples of the retarding effect of remote CN in the S_N1 reactions of saturated aliphatic and carbocyclic systems[191].

Certain nucleophilic substitutions of course depend on the CN group actually to generate the nucleophile, i.e. the nucleophile is a carbanion from a cyanocarbon acid. For example, kinetic and equilibrium measurements have been made for nucleophilic attack at unsubstituted ring position of trinitro-substituted aromatics by various carbanions, including those derived from ethyl cyanoacetate and 4-nitro-, 4-cyano- and 2-cyano-benzyl cyanides, the solvent being methanol[192]. The results were used to determine 'intrinsic reactivities' for the carbanions in these σ-adduct forming processes. The reactions of 2-halobenzonitriles with carbanions derived from phenylacetonitriles (pK_a values between 19 and 23) have also been studied. In these reactions CN both activates the substrate and generates the nucleophile; the halogen is displaced and (2-cyanoaryl)arylacetonitriles are formed[193].

C. Electrophilic Aromatic Substitution

Like the NO_2 group, CN is deactivating in electrophilic aromatic substitution. There is much less literature to deal with than there was for the article on ether and hydroxyl groups[5]. As in the case of the previous article, however, Roger Taylor's excellent monograph[194] is a good guide to what is available and references to particular parts of this book will be appropriate.

TABLE 5. Isomer distributions (%) for nitration or chlorination of PhX, with X = CN or NO_2[195]

X	Reagent	Temp. (°C)	o	m	p	o:p	m:p
Nitration							
CN	HNO₃–oleum	0	13.8	85.0	1.5	4.6	28.3
	HNO₃	0	16.8	80.8	1.95	4.3	20.7
	HNO₃–CF₃COOH	75	25	70	4.4	2.8	8.0
NO₂	HNO₃	0	6.4	93.2	0.3	10.7	155
	HNO₃–H₂SO₄	25	6.1	91.8	2.1	1.5	21.9
	HNO₃–CF₃COOH	75	10.5	84.7	4.8	1.1	8.8
Chlorination							
CN	HOCl–H⁺	?	23.2	73.9	2.9	4.0	12.7
NO₂	HOCl–H⁺	?	17.6	80.9	1.5	5.9	27.0

Table 5 displays data for the isomer proportions formed in the nitration and halogenation of benzonitrile, with data for the corresponding reactions of nitrobenzene for comparison[195]. As with all $+I$, $+R$ substituents, the formation of the *meta* isomer predominates, with CN being slightly less *meta*-directing than NO_2, as might be expected. The reactions are far slower than those of benzene itself. Partial rate factors for the influence of CN are not available, but for the reaction of nitrobenzene with HNO_3–H_2SO_4 they are as follows: f_o, 1.08×10^{-8}; f_m, 16.2×10^{-8}; f_p, 7.26×10^{-9} (25 °C)[195]. In the formation of the minor products, *ortho* substitution tends to be favoured relative to *para*, as shown by the values of the ratio $\frac{1}{2}o:p$. At one time special interactions between the reagent and the substituent were often invoked to account for the ortho preference, which is now known to be widespread for other $+R$ substituents and other electrophilic reagents. It is more likely that a general greater deactivation of the *para* position is responsible. The pattern of $\frac{1}{2}m:p$ ratios (Table 5) ia also in accord with this. It has been suggested that the greater deactivation of the *para* position is due to it carrying a slightly greater share of the positive charge than either *ortho* position in the Wheland intermediate[195], and there is quantum-mechanical support for this idea.

In highly acidic media the actual state of substituents such as NO_2 or CN with regard to protonation or hydrogen-bonding is somewhat problematical. The slightly reduced *meta* preference for reactions in HNO_3–CF_3COOH (Table 5) suggests some difference in the state of the substituents in this medium, but it should be noted that these reactions were at a rather higher temperature.

Partial rate factors have been measured for the bromination of the mesitylenes **15** (*meta* position) and the durenes **16** (*para* position). The values for X = CN are 8.7×10^{-7} and 3.1×10^{-6}, respectively; *cf* for X = Cl, 4.9×10^{-4} and 0.145, respectively; and for X = OMe, 0.19 and 1.64×10^{5}, respectively[196].

(15)   **(16)**

In Taylor's tabulation of sigma values [197], σ_p^+ for CN is given as 0.66, very close to the value of 0.675 given for σ_p. No distinction between these two types of substituent constant is to be expected for CN in connection with electrophilic aromatic substitution. However, a distinctive value for σ_m^+ is given as 0.56; *cf* 0.61 for σ_m of CN. This value of σ_m^+ is attributed to Stock and Brown[198] and may have been influenced by the use of the erroneous σ_m value of 0.56 for CN in the relevant correlation (Section III.B)[98]. The substituent $CNAlCl_3$ is given $\sigma_p^+ = 1.2$, and a considerable increase in electron-attracting power compared with uncoordinated CN is reasonable.

For CH_2CN a value of 0.12 for σ_p^+ is quoted, compared with 0.18 for σ_p[197]. Values are also given of σ_m^+ and σ_p^+ for the unipole $CHCN^-$ as -1.1 and -5.4, respectively, the source being Juchnovski and Binev[199]. These large negative values undoubtedly reflect the powerful $-I$ and $-R$ effects of the negative unipole, but one must have reservations about their quantitative generality. This applies to the sigma values of all unipoles (Section IX).

D. Liquid Crystals

During the last 20 years or so, the cyano group has played a key role in the development of liquid crystal materials for display technology. For this purpose a 'nematic' phase is required, i.e. a state of one-dimensional order. The formation of a nematic phase is a property of certain rod-like molecules and substituents may exert considerable effects on the characteristics of such a phase. The cyano group is particularly important in controlling the temperature range over which the nematic phase exists and also its dielectric anisotropy. The structural features of CN that are significant in this connection are its high bond dipole moment, its ability to conjugate with aromatic systems, its cylindrical shape and small bulk, and the linearity of $C-C\equiv N$. Nitro compounds may also form liquid crystals, but these are often 'smectic' phases of two-dimensional order. Further, the colour and relatively low stability of nitro compounds makes them useless in display technology, for which cyano compounds have the advantages of being colourless and very stable.

The relationship between molecular structure and properties of liquid crystals is a complicated subject which we cannot go into here. The present discussion must be confined to indicating the types of cyano compound that are of interest.

The compounds which first found applications in display devices were the 4-alkyl-4'-cyanobiphenyls **17** and the 4-alkoxy-4'-cyanobiphenyls **18** (Gray and colleagues, 1973)[200]. It has subsequently proved possible to incorporate other structural features and thereby modify the liquid crystal characteristics in interesting and sometimes useful ways. For

(**17**) X = alkyl
(**18**) X = alkoxy

(**19**) X = alkyl

example, heterocycles can be introduced, as in **19**, but it is not necessary for the rings to be aromatic. Thus one or both of the rings may be saturated carbocycles, as in the 4-[*trans*-4-alkylcyclohexyl] benzonitriles **20**[201], or the *trans, trans*-4'-alkylbicyclohexyl-4-carbonitriles **21**[202]. The compounds **20** form nematic phases with higher clearing points and wider

(**20**)   (**21**)

temperature intervals than those of **17**, while compounds **21** show nematic phases whose transition temperatures to the isotropic phase are higher than those of **20**. Some members of **20** and **21** have now achieved considerable importance in display technology. Compounds analogous to **21**, but with *axial* CN groups instead of *equatorial*, also show liquid crystal behaviour[203]. (The CN is forced into the axial disposition by having a large group R' attached instead of H to the same carbon atom.) These compounds also have high transition temperatures. Further information about the liquid crystalline behaviour of cyano compounds is in a review[204]. (The present author thanks members of the Liquid Crystal Research Group of the University of Hull for assistance in connection with Section VI.D, particularly Professor John Goodby and Dr. Ken Toyne.)

VII. SUBSTITUENT CONSTANTS OF CN AND OF GROUPS CONTAINING IT: RECAPITULATION AND SOME EXTENSION

A. Introduction

At the end of this long account of the electronic effects of CN, it seems useful to bring together and summarize some of the material which is scattered throughout it. Also, there are some relevant substituents which have not so far been mentioned, including polycyano groups and various substituents in which CN is attached to some atom or group X, which is in turn bonded to the molecular skeleton.

B. Recapitulation

For CN itself the recommended values of σ_m and σ_p (benzoic acid scale) are 0.62 and 0.67, respectively (Section III.B). When these values are used for correlations of processes taking place in other than highly aqueous media, the possibility of specific solvent effects should be borne in mind. The fact that the σ values of CN are rather at the upper end of the scale for commonly used substituents means that they exert a strong influence in regression analysis and there is the danger of their biasing a correlation unduly.

For correlations in which σ^0 values are considered relevant, the ordinary σ_m and σ_p values of CN should be used (Section III. C). Similarly, for correlations of electron-demanding processes in which the use of σ^+ values is appropriate for $-R$ substituents, the ordinary σ_m and σ_p values of CN should be used (Sections VI. C and VI. D). When the use of σ_p^- values is appropriate for $+R$ substituents, the value 0.88 for CN may be used in association with other values based on phenol ionization, while a value of 1.00 may be used in association with other values based on anilinium ionization (Section III. D).

As to σ_I, the value of 0.57 for CN seems to be of wide application (Sections III. C and V.A), although here also the possibility of solvent effects and the dangers of bias in correlations must be borne in mind. It is probably best to make no distinction between σ_R(BA) and σ_R^0 for CN and to use a value of 0.08 for both (Section V.A.)

For the CH_2CN group σ_m and σ_p may be taken as 0.16 and 0.18, respectively (Section III. B), and σ_I as 0.20 (Section III. A).

C. The Effects of Substituents Containing CN

Various compilations of sigma values provide information for about thirty substituents containing CN[48,81,99,101]. Detailed examination of all the available data would raise problems of discrepancies, reliability of particular items, etc., and would be too lengthy to undertake here, so a more limited objective is necessary. Hansch, Leo and Taft[99] have compiled data for the ^{19}F substituent chemical shifts of *meta-* and *para-* substituted fluorobenzenes relating to 269 substituents, all the experimental work having been done by Taft and coworkers or by their associates in other laboratories. The solvents used were mainly of the type classified earlier as 'normal' (Section V.A), e.g. cyclohexane, benzene, CCl_4, CH_2Cl_2, $CFCl_3$, etc. The experimental results were analysed to give values of σ_I (these authors currently prefer the symbol σ_F[132]) and σ_R. The exact significance of σ_R depends on whether the substituent is of the $-R$ or $+R$ type. For $-R$ substituents the σ_R values tabulated may be regarded as values of σ_R^0, but for $+R$ substituents the σ_R values tabulated must be regarded as slightly enhanced from σ_R^0 by cross-conjugation of F with the $+R$ *para-* substituent. While certain *minutiae* of their Table may perhaps be subjected to criticism, this large body of data is highly self-consistent in the experimental method and procedures and in the treatment of the results. It contains information on some 16 substituents involving CN (including CN itself) and is very suitable for discussing the various structural effects, including those of polysubstitution by CN or of separating CN from the benzene ring by another moiety. The relevant data have been abstracted to form Table 6. These data can be supplemented by some information regarding values of σ_m and σ_p, but for many of the substituents σ_m and σ_p values can only be based on appropriately combining σ_I and σ_R (Section II.B), the dissociation constants of the substituted benzoic acids or properties of other suitable systems not having been measured.

Entry 1 for CN itself is regarded by Hansch, Leo and Taft[99] as of doubtful significance because of magnetic or other complications. Entries 2 to 4 show the effect of successive

TABLE 6. σ_I and σ_R values for substituents containing CN[99]

No.	Substituent	σ_I	$\sigma_R{}^a$	No.	Substituent	σ_I	$\sigma_R{}^a$
1	CN	$(0.54)^b$	$(0.18)^b$	9	$CNBCl_3$	1.17	0.43
2	CH_2CN	0.32	-0.06	10	COCN	0.62	0.31
3	$CH(CN)_2$	0.61	-0.02	11	NHCN	0.44	-0.30
4	$C(CN)_3$	0.95	0.02	12	$N(CN)_2$	0.94	-0.21
5	$CMe(CN)_2$	0.69	-0.04	13	$P(CN)_2$	0.80	0.14
6	$CH{=}CHCN$ (*trans*)	0.33	0.10	14	OCN	0.86	-0.25
7	$CH{=}C(CN)_2$	0.47	0.25	15	SCN	0.50	0.05^c
8	$C(CN){=}C(CN)_2$	0.72	0.26	16	SO_2CN	0.99	0.27

^{a} For $-R$ substituents σ_R is $\sigma_R{}^0$, but for $+R$ substituents σ_R is slightly exalted by cross-conjugation.
^{b} Of doubtful significance because of magnetic or other complications.
^{c} Sign doubtful; see text.

replacement by CN of the H atoms of CH_3, for which $\sigma_I = -0.01$ and $\sigma_R^0 = -0.13$. There is a fairly linear increase in the $+I$ effect (*cf* the influence of these groups on C—H acidity already noted, Section VI.A), while the $-R$ effect is rapidly reduced effectively to zero. The enhancement of the $+I$ effect by the replacement of H in $CH(CN)_2$ by Me to give $CMe(CN)_2$ (entry 5) is not in accord with the traditional veiw of alkyl groups as electron-releasing. Polysubstitution in the olefin side-chain $CH{=}CH_2$ (for which $\sigma_I = 0.07$ and $\sigma_R^0 = -0.01$), entries 6 to 8, likewise increases the $+I$ effect, but not uniformly. The introduction of the second CN at the β position has less effect than the first (the second is in the *cis* disposition with respect to the benzene ring), but it nevertheless augments the $+R$ effect considerably. The final CN, on the α position, naturally has a marked effect on σ_I, but is not able to augment the $+R$ effect directly, since it is not conjugated with the benzene ring.

The coordination of CN to the Lewis acid BCl_3 would be expected to enhance the $+I$ and $+R$ effects (entry 9). These σ_I and σ_R values would combine to give σ_m and σ_p values of about 1.4 and 1.6, respectively. There are experimental values for these of 0.95 and 0.86, respectively[99, 205], so there are discrepancies to be resolved. (Values of σ_m and σ_p also exist for $CNBBr_3$ of 0.61 and 0.48, respectively, and for $CNBF_3$ of 0.72 and 0.66, respectively[99, 205].)

The σ_I and σ_R values for COCN (entry 10) may be compared with those for CHO, 0.35 and 0.23, respectively. Thus the replacement of H by CN increases σ_I by 0.27 and σ_R by 0.08. The latter must correspond to an indirect enhancement of the $+R$ effect of the carbonyl group, since the CN is not itself conjugated with the benzene ring. The entries so far examined in Table 6 appear to agree that the replacement of α-H by CN leads to an increase in σ_I roughly in the range 0.25 to 0.35, and this may usefully be borne in mind when considering the remaining entries. This corresponds to a damping factor of about 0.5 for the effect of the α-carbon atom; *cf* the damping factor of about $1/2.7 \approx 0.4$ per CH_2 group in aliphatic chains (Section III. A)[59]. It is interesting that the replacement of the first β-H in $CH{=}CH_2$ increases σ_I by 0.25, entry 6 discussed above; a rather smaller increase, perhaps 0.15, influenced by further damping, might have been expected. Probably the greater polarizability of C=C compared with C—C facilitates the transmission of the inductive effect. The fact that the Hammett ρ values for the ionization of phenylacetic and *trans*-cinnamic acids are about the same is in accord with this [206, 207].

The stepwise introduction of CN into NH_2 (for which $\sigma_I = 0.09$ and $\sigma_R^0 = -0.48$), entries 11 and 12, and the introduction of two cyano groups into PH_2 (for which $\sigma_I = 0.07$ and $\sigma_R^0 = -0.02$[208]) lead to slightly larger increases in σ_I per CN than those considered above, i.e. they are near or beyond the upper limit of 0.35. This probably indicates strong interaction between CN and the lone pair of electrons on N or P. In agreement with this the $-R$ effect is strongly reduced and conjugative interaction with the benzene ring actually changes to $+R$ in the case of P. By convention this would be regarded as $\pi(\mathrm{pd})$ conjugation.

The introduction of CN into OH (for which $\sigma_I = 0.32$ and $\sigma_R^0 = -0.43$), entry 14, leads to an even more marked increase in the $+I$ effect, accompanied by reduction in the $-R$ effect. Strong interaction between CN and the lone pairs of electrons on O is indicated. The conversion of SH (for which $\sigma_I = 0.27$ and $\sigma_R^0 = -0.13$) into SCN, entry 15, is accompanied by a not so marked increase in σ_I; maybe the conjugative interaction of the lone pairs with CN is not so strong. However, the $-R$ effect is reduced and, according to the σ_R value given, the resonance effect actually changes sign, corresponding to $\pi(\mathrm{pd})$ conjugation. In the treatment adopted in the paper of Hansch, Leo and Taft[99] there is a dubiety in the signs of very small resonance effects, when on structural grounds an effect of either sign is conceivable. A σ_R^0 value of -0.05 for SCN has previously been given [209]. For SO_2CN, entry 16, we must use SO_2Me ($\sigma_I = 0.61$ and $\sigma_R = 0.14$) for comparison. As with SCN there is an increase in σ_I around the lower limit of 0.25 mentioned above; the situation rather resembles that for COCN, entry 10, and there is similarly a marked (indirect) enhancement of the $+R$ effect.

Hansch, Leo and Taft[99] tabulate σ_I and σ_R values based on[19] F NMR for eight substituents HgX involving attachment of Hg to the benzene ring, but these do not include HgCN. However, such values are available from other authors[210], the values being 0.23 and 0.11, respectively. For comparison we take $\sigma_I = 0.12$ and $\sigma_R = 0.01$ for HgMe from Hansch, Leo and Taft[99]. The increase in σ_I on replacing Me by CN is fairly modest compared with the changes produced by some rather less electron-attracting groups, as indicated in the original Table[99], but the increase in σ_R is exceptionally large. The σ_R values for the eight HgX substituents given are all in the range 0.01 to 0.03. No immediate explanation of the peculiar behaviour of HgCN as a substituent comes to mind. Because of the weakness of the Hg—C bond (bond energy ca 15 kcal mol^{-1}) the behaviour of organomercury compounds is highly individualistic, as has long been recognized[211].

VIII. ELECTRONIC EFFECTS OF NC, THE ISOCYANO GROUP

There is only very little information about the electronic behaviour of this group. No doubt the objectionable smell of isonitriles acts as a disincentive to working with them.

The electronic structure of this substituent (see Section I. B.) suggests that the group should exert a strong $+I$ effect, with little or no R effect. Exner (1978)[48] quotes a σ_I value for NC of 0.67, said to be based on the pK_a value of isocyanoacetic acid and communicated to him personally by Charton. However, Charton (1981) states the value as 0.63[81], based on a pK_a value for isocyanoacetic acid of 2.23 (water, 25 °C), determined by Ashworth and Coller[212]; cf Charton's value of 0.57 for the σ_I constant of CN[81] (Section III.A). In contrast, ^{19}F NMR studies of m- and p-fluoroisocyanobenzene find the considerably lower value of 0.47 for σ_I of NC, along with the (expected) very small σ_R^0 value of 0.02[213] (CCl_4 or $CCCl_3F$ as solvent). We recall that the σ_I value for CN found by ^{19}F NMR studies in 'normal' solvents was 0.48 (Section V.A)[66].

No one appears to have measured the pK_a values of isocyano-substituted benzoic acids or acids of the type Ph—G—COOH. Exner[48] suggests values for σ_m^0 and σ_p^0 as 0.48 and 0.49, respectively, on the basis of $\sigma_I = 0.47$ and $\sigma_R^0 = 0.02$, by applying the appropriate relations (equations 6 and 7 in Section II.B).

Ashworth and Coller[212] have provided another item of quantitative information about the relative reactivity of substrates substituted with CN or NC. Second-order rate coefficients for the alkaline hydrolysis of esters $CH_2XCOOEt$ with X = CN or NC in water at 25 °C were found to be 21.1 and 33.7 $\mathrm{dm}^3\ \mathrm{mol}\ \mathrm{s}^{-1}$, respectively; cf about 0.11 for X= H. This is a reaction which is strongly accelerated by electron-attracting groups X, but is also subject to some degree of steric hindrance by CH_2X. Assuming, however, that CH_2CN and CH_2NC are of roughly the same bulk, the above results indicate that CH_2NC is slightly more electron-attracting than CH_2CN. This analysis may be done more quantitatively by correlating the $\log k$ values for a suitable set of esters RCOOEt with the σ_I values and Taft steric parameters E_s[59, 73-75]. The regression equation 20 is typical:

$$\log k = -1.275 + 15.27\ \sigma_I + 0.549\ E_s$$
$$(\pm 0.75) \qquad (\pm 0.165) \qquad\qquad (20)$$
$$n = 15,\ R = 0.987,\ s = 0.329,\ \psi = 0.177$$

If we assume that CH_2NC has the same value of E_s as CH_2CN, the σ_I value of CH_2NC may be calculated from the above regression to be 0.017 units greater than that of CH_2CN. If the decremental factor per methylene group is taken as 2.7 (Section III.B)[59], this difference in σ_I values for CH_2NC and CH_2CN corresponds to a difference of 0.046 in the σ_I values of NC and CN. This fits in fairly well with the pK_a-based values of σ_I for NC and CN as 0.63 and 0.57, respectively.

When the carbon atom of NC is attached to O or S in NCO or NCS, the electronic effects are considerably modified. The σ_I value is reduced to 0.36 and 0.42, respectively, and an appreciable $-R$ substituent effect is developed, the σ_R^0 values being -0.17 and -0.07, respectively (^{19}F NMR studies)[48]. These changes may be ascribed to the character of these groups as resonance hybrids of structures of the types **22**[214]:

$$-\ddot{N}{=}C{=}X \longleftrightarrow -\overset{+}{N}{\equiv}C{-}\overset{-}{X}$$

(22)

The reduced $+I$ effect and the $-R$ effect result from the contribution of the structure involving double bonds, with a lone-pair of electrons on the nitrogen atom. This leads also to the order $\sigma_m > \sigma_p$, the values being 0.30 and 0.24, respectively, for NCO and 0.48 and 0.38, respectively, for NCS. (According to Exner[48], values from different sources show some discrepancies.)

IX. ELECTRONIC EFFECTS OF THE UNIPOLE N_2^+

A. Introduction

Various sigma values for N_2^+ may be found in the literature. Before these are quoted here, it is necessary to say that sigma values for unipolar substituents are to be regarded as no more than a rough indication of their effects on reactivity or other properties in given systems under given conditions. The Hammett equation and its usual extensions (Section II) are essentially based on dipolar substituents and are not really applicable to unipolar substituents. This matter will be elaborated in Section IX.B. In Section IX.C it will be shown how an attempt has been made to modify the Hammett equation to incorporate a valid treatment of unipolar substituents.

Lewis and Johnson[215] measured pK_a values for benzoic acid substituted with m-N_2^+ or p-N_2^+, the gegen ion being BF_4^-. They calculated values of σ_m and σ_p as 1.76 and 1.91, respectively. Apparent σ values for positively charged substituents, such as NR_3^+ are often in the region of 1.0, so the values for N_2^+ are at first sight unexpectedly high. However, N_2^+ may be expected to behave as a strong $+R$ substituent, so considerably larger σ values than those characteristic of NR_3 unipoles are not unreasonable, although a greater difference between σ_m and σ_p might have been expected.

The same authors[215] also measured the pK_a value of phenylacetic acid substituted by p-N_2^+ and derived a value of σ_p^0 as 2.18. The enhanced value compared with σ_p possibly reflects the imperfections of phenylacetic acid for the determination of σ_p^0 values for $+R$ substituents (see discussion for CN in Section III.C). They also measured the pK_a values of phenol and of anilinium ion substituted by $p\text{-}N_2^+$ and calculated values of σ_p^- as 3.04 and 3.43, respectively, indicating considerable enhancement of the $+R$ effect (cf Section III.D regarding the tendency of the anilinium system to give higher values of σ^- than the phenol system). In contrast, a value of σ_p^- of 1.87 (i.e. actually less than the value of σ_p given above) was based on a study of nucleophilic aromatic substitution[184]. This will serve to emphasize the above warning regarding the significance of sigma values for unipolar substituents.

Values purporting to be of σ_p, σ_p^+ and σ_R^0, 1.83, 1.88 and 0.15, respectively, were based on a ^{13}C NMR study[101, 216].

Values of σ_R^0 were determined through infrared studies as ± 0.30[130] (D_2O as solvent) and ± 0.29)[217] (DMSO as solvent). (Signs are not determined by the infrared method, see Section V.A.) These values seem rather small. A ^{19}F NMR study found σ_I to be 1.34 and σ_R^0 to be 0.64[213], values which seem fairly in harmony with the magnitudes of σ_m and σ_p as determined by Lewis and Johnson[215].

B. The Behaviour of Unipolar Substituents with Respect to the Hammett Equation

The point was made in Section IX.A that the simple Hammett equation is not really applicable to unipolar substituents. For the latter it is not possible to define sigma values on the basis of suitable reference systems, and then to apply these widely and successfully to many different processes.

The Hammett treatment is essentially based on dipolar substituents. In the original presentation of the Hammett equation (1937)[42,218], substituent constants were not tabulated for any unipolar substituent. When Jaffé (1953)[49] reviewed the state of the Hammett equation, he tabulated σ values for several 'cationic' or 'anionic' substituents, e.g. m- and $p\text{-}NMe_3^+$, $m\text{-}NH_3^+$, m- and $p\text{-}CO_2^-$, $p\text{-}SO_3^-$, m- and $p\text{-}O^-$, and others. He seemed fairly optimistic that the extension of the Hammett equation to embrace ionic substituents would prove a successful development. He commented[219]: "Since the available data indicate no greater uncertainty for substituent constants of ionic substituents than for those of neutral groups, the Hammett equation also appears to be applicable to substituents which carry an integral charge". However, he also pointed out that ionic substituents will interact strongly with polar solvents, and their substituents constants might be expected to be particularly solvent-dependent.

In fact, problems that may be expected to arise in the attempt to include unipolar substituents within the scope of the Hammett equation may readily be envisaged. A given reaction of a substrate containing a unipolar substituent will be of a charge type different from that of the same reaction of a substrate containing a dipolar substituent. This difference has immediate consequences. The reactions of substrates of different charge type will respond differently to changes in ionic strength and in dielectric constants of the medium, i.e. whether equilibria or rates are involved, reactants of different charge type will show different primary salt effects and different solvent effects.

Suppose we establish a Hammett relation for a given reaction series involving dipolar substituents under given conditions. We may then use appropriate data for unipolar substituents to measure apparent sigma constants for those substituents. If we now consider another set of data for the same reaction but under markedly different conditions of ionic strength and/or dielectric constant of the medium, we may derive a new Hammett relation for the dipolar substituents and also try to include the unipolar substituents on the basis of the sigma values previously derived. The general result will be that the latter will not be found applicable under the new conditions, because the substrates involving unipolar substituents, on the one hand, and those involving dipolar substituents, on the other, have responded differently to the changes in the medium. Thus any simple definition of sigma values for unipolar substituents to be used in a treatment which is dominated by dipolar substituents is in principle impossible. The inclusion of dipolar and unipolar substituents in one treatment can only be done by including specific consideration of the effects of ionic strength and dielectric constant in an elaboration of the Hammett equation. In essence this is what Wepster tried to do[220] (Section IX.C).

It is not surprising that the attempt during the past forty years to include unipolar substituents in simple Hammett treatment has produced what the present author has described elsewhere as a "long history of anomalies, failures, and warnings"[4,46]. Thus authors have persistently found that sigma values of unipolar substituents show marked dependence on ionic strength and dielectric constant of medium and that values based on behaviour in a given reaction under given conditions prove totally inapplicable to another reaction under different conditions, for no immediately obvious reason. The present author has summarized this history in various publications[1,46]. In his chapter in a very recent volume in this series[5] the history was exemplified by showing the sigma values for m- and p-O^- which have been proposed at various times; there is a great deal of such data. For N_2^+ there is much less data and the display of these data and the comments made in Section IX.A must suffice.

C. Wepster's Treatment of Unipolar Substituents

The essence of this treatment lies in considering a 'Bjerrum field effect term' as a separate component of the Hammett equation. Wepster's[220] paper occupies some 25 pages of the *Journal of Organic Chemistry* and contains extensive tabulation of experimental results (partly Wepster's own work, but mainly from the literature, with many references) and of the results of applying the new treatment. The experimental results are for the effects of both unipolar and dipolar substituents in a variety of reactions and they have been analysed in terms of the classical ideas developed by G. N. Lewis and Bjerrum and revived (now some 20 years ago) by Palm and his coworkers[221]. The outcome is that the Hammett equation needs only a simple extension to cover the effects of both dipoles and unipoles. The present author has previously written[46] a fairly lengthy summary of Wepster's paper, while recommending that seriously interested readers should consult the original paper. Here, however, only a brief summary will be attempted, with an indication of application to N_2^+.

The attempt to correct experimental data to zero ionic strength is fundamental to the treatment, even though often this can be done only approximately. The model of non-conjugative substituent effects which is used is a combination of Lewis's model of the inductive effect as a through-bonds displacement of electrons, and the electrostatic model of the field effect as devised by Bjerrum in his treatment of the first and second ionization constants of aliphatic dicarboxylic acids in water. As Wepster acknowledges, these models have their limitations, but he claims that their combination has nevertheless led to a very successful treatment.

The Hammett equation is rewritten as

$$\Delta = \rho\sigma \tag{21}$$

where Δ is the substituent effect (log K – log K_H or log k-log k_H; note the use of the actual reactivity for the parent system), $\rho \equiv \rho_m$ is the reaction constant as obtained by using standard σ_m values[52], and σ is the substituent constant which may either be Wepster's own σ^n in the absence of 'mesomeric *para* interaction' (as he terms cross-conjugation), or an exalted value, perhaps σ^+ or σ^-. From the rewritten Hammett equation and the above models, Wepster writes equations 22 and 23:

$$\Delta = \delta^L + \delta^B \tag{22}$$

$$\Delta = \rho^L\sigma^L + \delta^B \tag{23}$$

δ^L signifies the effect through the molecule (the superscript standing for Lewis) and δ^B is the electrostatic effect (the superscript standing for Bjerrum); σ^L is a sigma value free from the Bjerrum field effect and ρ^L is a reaction constant based on such (*meta*) σ^L values. The Bjerrum term is based on the well-known expressions in electrostatics for the interaction energies of two point charges, in the case of a unipolar substituent, or of point charge and dipole, in the case of a dipolar substituent, the interactions being moderated by the dielectric constant of the solvent.

The extension of the Hammett equation to unipolar substituents formally requires a redefinition of the σ values of dipolar substituents as σ^L values, but the changes involved are very small and in practice $\sigma^L \approx \sigma^n$ for such substituents, and $\rho^L = \rho_m$ as defined above. σ^L values (strictly σ^{Ln}) for many unipolar substituents were derived from experimental results on various acid–base equilibria and applied to other reactions through equation 23. The general success of the treatment was shown graphically and by statistical tabulation. Individual σ^{Ln} values and those for structural series were specially discussed.

The values of σ_m^{Ln} and σ_p^{Ln} for N_2^+ were found to be 1.11 and 1.32, respectively. These values were not so securely established as those of some unipolar substituents, since the data available were limited to those obtained by E. S. Lewis and Johnson[215] for the effects of N_2^+ as a substituent in benzoic acid, as discussed in Section IX.A. The σ^{Ln} values for N_2^+ are easily the largest for any positive pole; compare with 0.59 and 0.53 for m-NMe$_3^+$ and p-NMe$_3^+$, respectively, and 0.76 and 0.77 for m-SMe$_2^+$ and p-SMe$_2^+$, respectively. No doubt this is an indication of the strong $+R$ effect of N_2^+. Wepster also gives a value of 2.76 for σ^L_{-p} of N_2^+, based on Lewis and Johnson's[215] determination of the pK_a values for the substituted phenol, and of 3.00, based on their determination of the pK_a value for the substituted anilinium ion. N_2^+ is almost alone in showing such enhancements, since most unipolar substituents do not exert strong $+R$ effects.

X. THE ELECTRONIC EFFECTS OF ACETYLENIC GROUPS

A. The Characteristics of the Ethynyl Group

In the Historical Introduction (Section I.B) it was mentioned that the pronounced electronegative character of the carbon–carbon triple bond has long been recognized. It is indicated by the acidity of terminal acetylenes, RC≡CH, and by the effect of C≡C on the strengths of carboxylic acids. This electronegative or electron-attracting character is commonly attributed to the sp hybridization of the carbon atoms involved. The electronegativity of a carbon atom increases with the percentage of s character in the hybrid orbitals, thus in the order: sp^3<sp^2 < sp, the percentage of s character being 25%, 33% and

$$R-\overset{..}{\underset{}{C}}=C=\!\!\!\left\langle\!\!\bigcirc\!\!\right\rangle \quad + \qquad R-\overset{+}{C}=C=\!\!\!\left\langle\!\!\bigcirc\!\!\right\rangle\!\!:^-$$

$+R$ effect $-R$ effect

(23)   (24)

50%, respectively. It was also mentioned in the Historical Introduction that Ingold[31] (1934) classified acetylenic groups as –I in character (his sign convention[50]). He appears not then to have considered their possible mesomeric effects, but such movement of electrons could conceivably be in either direction with respect to a benzene ring (see 23 and 24), and so acetylenic groups should be classified as $\pm M$ in Ingold's symbolism. For the present chapter we classify acetylenic groups as $+I, \pm R$. We shall try to assess quantitatively the electronic effects of the ethynyl group HC≡C. Unfortunately this will prove to be not entirely straightforward.

The $+I$ character of the group is most simply shown by its effect on the acidity of acetic acid. The apparent pK_a value of ethynylacetic acid at an ionic strength $I = 0.1$ (NaCl) is 3.32 (water, 25 °C)[222]. This is a so-called 'mixed' constant[77b], involving activity of hydrogen ions, but concentrations of other species. For comparison with the proper thermodynamic pK_a value of acetic acid, a correction is required. This[223] amounts to: $0.5I^{1/2} (1 + I^{1/2})^{-1}$, which for $I = 0.1$ will be 0.12, so the corrected pK_a value of ethynylacetic acid is 3.44, the pK_a value of acetic acid being 4.756, $\Delta pK_a = 1.316$ (see Section III.A for definition). This may be compared with $\Delta pK_a = -0.064$ for butanoic acid and 0.404 for 3-butenoic acid, indicating the considerably greater electron-attracting inductive effect of HC≡C. From Charton's equation[81] relating pK_a values of substituted acetic acids to the σ_I values of the substituents (our equation 11 in Section III.A), σ_I for HC≡C may be calculated as 0.33. Charton[81] actually prefers to base the σ_I value for the ethynyl group on its effect on the ionization of the quinuclidinium ion[187] (Section III.A, Table 3), and obtains the value 0.29. Charton's earlier value (1964)[224] was 0.35, and was based on the pK_a value of ethynylacetic acid as above[222], but without applying the ionic strength correction and with a slightly different regression equation.

Landgrebe and Rynbrandt[225] measured the pK_a values of m- and p-ethynylbenzoic acids in 1:1 EtOH–H_2O at 25 °C as 5.45 and 5.41, respectively, compared with their own value of 5.75 for benzoic acid. From these results and the appropriate ρ value, values of σ_m and σ_p were calculated as 0.21 and 0.23, respectively. In a recent re-appraisal of pK_a data measured in 1:1 EtOH–H_2O and σ values based thereon (in connection with a IUPAC project[93]), the present author has arrived at essentially the same σ values for HC≡C. If we regard the two equations for Taft's analysis of σ values into inductive and resonance components as a pair of simultaneous equations with σ_I and $\sigma_R(BA)$ as the unknowns (equations 4 and 5 in Section II.B), values of σ_I and $\sigma_R(BA)$ may be calculated from the above values of σ_m and σ_p as 0.20 and 0.03, respectively, i.e. σ_I is calculated to be much lower than indicated above at ca 0.3. Charton has implicitly recognized this problem in Table 27 of his article[81], where he obtains a value for σ_R [effectively $\sigma_R(BA)$] of -0.04 by way of applying the extended Hammett equation to the pK_a values of $trans$-3-XCH≡CHCOOH. He then calculates σ_p for HC≡C as $\sigma_I + \sigma_R = 0.29 -0.04 = 0.25$, and σ_m simply as 0.29, apparently not considering the correction for the relayed resonance effect as worth applying, since it amounts to only $0.33\sigma_R = -0.01$ (our equation 4). There is thus considerable uncertainty as to the σ_m and σ_p values for HC≡C, beyond the probability that they are both in the range 0.20 to 0.30. It even seems uncertain whether $\sigma_m < \sigma_p$ or $\sigma_p > \sigma_m$.

At various times many years ago, Charton obtained other values of σ_p for HC≡C by interpolation in sundry correlations; 0.27[226], 0.22[227] and 0.28[228]. Thus we may regard the range 0.22 to 0.28 for σ_p as providing a point of reference for assessing the behaviour of this substituent in processes in which it might be expected to reveal more clearly the dichotomy of its resonance effect, i.e. $\pm R$. Landgrebe and Rynbrandt[225] in fact provide data on the solvolysis of benzyl chlorides in 1:1 EtOH–H_2O, in which enhanced $-R$ effects may be observed, and they give a σ_p^+ value of 0.18 for HC≡C, certainly less positive than σ_p. Unfortunately the rate coefficient which they obtained for m-ethynylbenzyl chloride requires a distinctive value of σ_m^+ as 0.33, a result which is not easy to explain. As regards enhanced $+R$ character, Eaborn and coworkers[229] based a value of σ_p^- for the ethynyl group on the alkaline cleavage of arylstannanes. They found σ_p^- to be 0.52, certainly indicating an increased $+R$ effect.

The evidence from spectroscopic studies is likewise not completely concordant. Values of σ_I and σ_R from ^{19}F NMR substituent chemical shifts are given as 0.15 and 0.08, respectively[99], but these values are considered to be of doubtful significance because of magnetic or other complications. Laurence and colleagues[135] have determined σ_I for the ethynyl group as 0.26 from their infrared camphor model, which must be considered in reasonable agreement with Charton's[81] preferred value of 0.29. However, they quote in parentheses a value of 0.23, which is described as a 'statistical value', and give a reference to Taft and Topsom (1987)[127] and to 'R. W. Taft, personal communication'. Reference 127 certainly tabulates a value of 0.23 for σ_I (they say σ_F), but the basis for this is not explained. The infrared intensity method[130] gives σ_R^0 as ± 0.072 for the ethynyl group, the sign not being determined by the method.

We turn now to the part played by the ethynyl group in Topsom's papers on theoretical scales of substituent field and resonance parameters[141-143]. In the first of these[141] σ_F values for HC≡C are quoted as 0.20 in the gas phase[111] and 0.29[81] in polar solvents. This substituent accordingly does not meet the criteria for inclusion in the regression (see Section V.B). A theoretical σ_F value of 0.17 for the ethynyl group comes out of the treatment and presumably this is regarded as being in reasonable agreement with the gas-phase value of 0.20. At the first stage [142] of the development of the theoretical scale of σ_R^0, HC≡C acquires a theoretical value of $- 0.01$, which is compared with an experimental value of $- 0.19$ for non-polar solvents[230]. At the second stage[143] the theoretical value of σ_R^0 for the ethynyl group comes out at 0.05, the experimental value still being quoted as $- 0.19$. In Reference 127 no value of σ_R^0 is tabulated to go with the value of 0.23 (see above) for σ_F.

The situation regarding substituent parameters for the ethynyl group seems at present to be somewhat confused. The well-known compilations of sigma values[44,81,99,101] give limited data for other groups containing C≡C, but such data should often be regarded with some reservations.

B. Structural Effects in Aliphatic Acids Containing C≡C

In Section X.A the considerable acid-strengthening effect of C≡C was shown for the series: butanoic acid, 3-butenoic acid, 3-butynoic acid. This effect is revealed even more dramatically by the pK_a values (water, 25 °C) in the series: propanoic acid, 4.88; propenoic acid, 4.25; propynoic acid, 1.96. (This is Mansfield and Whiting's value[222], corrected for the effect of ionic strength; see Section X.A.). The $- R$ effect of HC≡C may be a factor tending to weaken the acidity of HC≡C—COOH by stabilizing the undissociated acid relative to the ionized form (cf the $- R$ effect of Ph in benzoic acid), but it is certainly completely swamped by the $+I$ effect.

Structural series for acids containing C≡C are in Table 7. Most of the data come from Mansfield and Whiting and, for the present discussion, we leave their pK_a values determined at $I = 0.1$ uncorrected for ionic strength.

Series 1 shows the damping effect of introducing successive methylene groups on the acid strengthening effect of the ethynyl group. This seems to be rather more severe than would correspond to the decremental factor of about 2.7 which often applies to the inductive effects of polar groups (Section III.A)[59]. The effect of the HC≡C group seems largely to be damped out by three or four methylene groups.

Series 1 shows that extending the C≡C conjugated system increases the acidity, but successive C≡C units are of diminishing effect.

In series 3 we see the effect of replacing the ethynyl H of HC≡CCOOH by various substituents. The most remarkable feature is the fairly strong acid-weakening effect of alkyl substitution, while substitution by halogen has apparently little acid-strengthening effect. Substitution by MeCH=CH appears to be more acid-weakening than alkyl substitution, substitution by Ph rather less so. Even substitution by MeC≡C is acid-weakening; cf the effect of extending the C≡C conjugated system in series 2. It appears that there is some effect, acid-weakening in nature, of replacing H by any other substituent, which reinforces the inherent acid-weakening effect of alkyl groups and greatly reduces the normal acid-strengthening effect of halogen substituents. Indications of such an effect of replacing may be seen in the results of a treatment involving the extended Hammett equation carried out by Charton twenty years ago[232]. In a correlation analysis of the acidities

TABLE 7. The pK_a values of various acids containing $C{\equiv}C^a$

Series	pK_a	Series	pK_a
1. $HC{\equiv}C(CH_2)_nCOOH$		4. $XC{\equiv}CCOOH$	
n = 0	1.84	X = H	1.84
1	3.32	Me	2.59
2	4.205	Et	2.605
3	4.595	Pr	2.60
4	4.575	t-Bu	2.655
2. $Et(C{\equiv}C)_nCOOH$		neo-Pen	2.53
n = 1	2.605	$MeCH{=}CH$	2.665
2	1.90	Ph	2.23
3	1.67	$MeC{\equiv}C$	1.94
3. $XC{\equiv}CCH_2COOH$		Cl	1.795
X = H	3.32	Br	1.805
Me	3.59		
Ph	3.435		
$CH_2{=}CH$	3.39		
$MeC{\equiv}C$	3.23		

^{a} Data mainly from Mansfield and Whiting[222] for I = 0.1, uncorrected for effect of ionic strength. Data for X = Cl and Br in series 4 from Guillème and Wojtkowiak[231], corrected to the same scale as data from Mansfield and Whiting.

$XC{\equiv}CCOOH$, the point for X = H shows a large deviation from the regression equation, i.e. the intercept term does not correspond to the experimental value for X = H.

The pattern of results in series 4 seems to suggest that in $XC{\equiv}CCH_2COOH$ an effect similar to that discussed in the previous paragraph is operating, but may be not so marked.

For further information regarding pK_a values of acids containing $C{\equiv}C$, see the paper by Pethybridge[233].

C. The Transmission of Substituent Effects through $C{\equiv}C$

$C{\equiv}C$ may be used as the connective G in the acids PhGCOOH, and so its transmission of the effects of substituents in Ph may be studied. When G = CH_2 or $(CH_2)_2$ the Hammett ρ values for ionization in water at 25 °C are 0.56 and 0.24, respectively, compared to 1.00 for benzoic acid[206]. Transmission by $CH{=}CH$ is, however, greater than that by $CH_2{-}CH_2$, the ρ value for the ionization of *trans*-cinnamic acids being 0.42. In a simple way it may be imagined that the π-bond component makes the $C{=}C$ double bond more polarizable than the $C{-}C$ single bond and this facilitates the transmission of substituent effects The ρ values for the rate coefficients of reaction of the acids with diazodiphenylmethane in ethanol at 30 °C show much the same pattern: CH_2, 0.40; $(CH_2)_2$, 0.22; $CH{=}CH$, 0.42, compared with 0.95 for benzoic acid. It might be predicted that $C{\equiv}C$ would show better transmission than $CH{=}CH$, but this is not the case. In the DDM reaction the ρ value for $C{\equiv}C$ is 0.33. The ionization of phenylpropynoic acids cannot be studied in water, but in 1:1 EtOH–H_2O at 25 °C, the ρ value is 0.42, compared with 0.70 for *trans*-cinnamic acids. Similar relationship hold for the alkaline hydrolysis of the ethyl esters[234], although earlier data seemed to suggest better transmission through $C{\equiv}C$ than through $CH{=}CH$. Supporting evidence regarding the relative transmitting powers of these moieties has also been obtained by studies of ^{13}C NMR substituent chemical shifts[235].

Bowden[206] and Bowden, Chapman and Shorter[207] tried to develop a simple theory of the relative transmitting powers of various groups G. The theory achieved a modest success, but no real explanation of the inferior transmission by $C{\equiv}C$ was offered. It was simply suggested that C≡C might be less polarizable than C=C, and reference was made to a statement by Eliel[236]. Modern calculations of the polarizabilities of $H_2C=CH$ and $HC≡C$ as substituents appear to give no support to this idea (see Section XI.D). It does seem possible, however, that the enhanced electronegativity of C≡C (arising from the sp hybridization of the carbon atoms, see Section X.A), which is responsible for the electron-attracting behaviour of RC≡C, is also responsible for firmly holding the π-electrons in such a way that transmission of polar substituent effects is reduced compared with CH=CH.

XI. SOME FURTHER MULTIPARAMETER TREATMENTS OF SUBSTITUENT EFFECTS

A. Introduction

Earlier sections of this chapter contain accounts of the Yukawa–Tsuno equation[55,56], the Dual Substituent-Parameter (DSP) equation[61,62] and Extended Hammett (EH) equation[65] (see Section II.B), with the particular intention of showing how these may be applied to data sets involving the substituents of particular interest for this chapter. These equations are not now the only possibilities for multiparameter treatment. In this section we shall give accounts of some of the other approaches. The accounts will necessarily be brief, but key references will be given, with indications as to how the substituents of interest for this chapter fit into the various treatments.

B. Exner's Analysis

This is essentially a method of providing an alternative set of σ_I and σ_R parameters for use in the DSP equation or EH equation. In the mid-1960's Exner[64] found evidence that the inductive effect from the *para* position of benzoic acid was stronger than that from the *meta* position by a factor of 1.14. He also suggested that σ_I values current at that time and based on alicyclic and aliphatic reactivities were out of scale with σ_m and σ_p by a factor of 1.10, and should be multiplied by this to introduce the π-inductive component. This led Exner to a revised analysis of σ_m and σ_p in terms of inductive and resonance components. He calculated revised σ_I values by multiplying the alicyclic/aliphatic values by 1.10, and then multiplying these further by 1.14 before subtraction from σ_p values to obtain revised values of σ_R.

The most dramatic changes were for some $+R$ substituents, such as NO_2 and CN, whose σ_R values dropped to zero. The implication of this is that such substituents are normally not conjugated with the benzene ring and only become so in the presence of a $-R$ *para* substituent with which cross-conjugation is possible (Section II.A). Exner's re-calculation of σ_R values imposes less dramatic changes on $-R$ substituents, although these are still appreciable.

The status of Exner's revised σ_I and σ_R values has been debated for more than a quarter of a century. A number of prominent workers in the field are rather critical of Exner's approach. For a recent appraisal of the situation, see an article by the present author[46]. Exner has continued to propagate his view on this matter in his recently published book[47]. Some of his papers in the past few years indicate that he is developing further criticisms of aspects of the 'traditional' separation of inductive and resonance effects and of the ways in which correlation analysis of substituent effects is generally carried out[237–239].

C. C. G. Swain's Treatments

These began with a paper by Swain and Lupton in 1968[240]. The approach was slightly modified and greatly extended by Hansch's group in 1973[241]. During the first 15 years of so of its life, the Swain–Lupton treatment was applied extensively, but was also severely criticized. A revised version appeared in 1983 in a paper by Swain and coworkers[242]. This revised version was in its turn severally criticized, but also applied. The Swain–Lupton treatment was reviewed by the present author in 1978[53] and again more briefly in 1982[54]. A more recent review[46] covers also the revised version and an account of a mini-symposium in print in which several of Swain's critics set forth their views, and Swain replied[243–246].

The Swain–Lupton treatment[240] was a reaction against the proliferation of scales of polar substituent constants. The authors maintained that the polar effect of any given substituent could be adequately expressed in terms of just two basic characteristics: a field constant $\mathfrak{F}$ and a fixed resonance constant $\mathfrak{R}$. Swain and Lupton maintained that the correlation analysis of chemical reactivity data and spectroscopic data of aromatic systems could be carried out satisfactorily in terms of $\mathfrak{F}$ and $\mathfrak{R}$. (cf the four σ_R-type parameters introduced for the DSP equation), meta and para series being dealt with separately, as in the case of the DSP equation. The assumptions involved in establishing the $\mathfrak{F}$ and $\mathfrak{R}$. scales provoked much criticism. Nevertheless, the treatment achieved fair success when applied to chemical reactivity data and some spectroscopic data, particularly NMR[53,54]. The most notable success, however, was in the correlation analysis of biological activity data[247].

The revised version[242] developed new scales of field and resonance parameters, the awkward symbols $\mathfrak{F}$ and $\mathfrak{R}$ being replaced by the more straightforward $\mathfrak{F}$ and $\mathfrak{R}$. Some of the criticism made of the earlier form of the treatment had been met by the modifications, but the critics were still not satisfied[243–245].

A compilation of F and R constants as revised by Hansch, for numerous substituents, appeared in a book by Hansch and Leo[101]. A more recent compilation of substituent constants includes F and R values, revised again by Hansch[99]. Values are provided for CN, N_2^+ and other substituents of interest in this chapter.

D. The Poly Substituent-Parameter (PSP) Equation

This equation is an elaboration of the dual substituent-parameter (DSP) equation. Its development has been relatively recent, but Taft and Topsom, who have been closely associated with it, have already written a long review article[127] involving the equation, and this article will probably acquire the status in respect of the PSP equation that the article of Ehrenson, Brownlee and Taft[62] has in connection with the DSP equation. The name Poly Substituent-Parameter Equation was devised by the present author in a short account thereof[46]. Hopefully, that account and the present briefer one will encourage study of Taft and Topsom's article[127].

The new treatment had its origins partly in ab initio molecular orbital calculations of substituent effects and partly in extensive studies of gas-phase proton transfer reactions from about 1980 (Section V.A). Various aspects of this work essentially drew attention to the importance of substituent polarizability. In 1986 Taft, Topsom and their colleagues[248] developed a scale of 'directional substituent polarizability parameters', σ_α, by ab initio calculations of directional electrostatic polarization potentials at the 3-21G//3-31G level for a large set of CH_3X molecules. The σ_α values were shown to be useful in the correlation analysis of gas-phase acidities of several series of substrates[248], and such work has subsequently been extended by Taft and Topsom[127].

Values of σ_α are available for over thirty substituents. H is the standard at 0.00 and the values range from $+0.13$ for F to -0.81 for Ph. The values for CN, NC and HC≡C are -0.46, -0.33 and -0.60, respectively. To set these values in context we mention that the σ_α values for NH_2, NO_2, Me, Cl and H_2C=CH are -0.16, -0.26, -0.35, -0.43 and -0.50, respectively. (Note HC≡C is more polarizable than H_2C=CH; cf Section X.C)

The PSP equation is written by Taft and Topsom[127] in various forms. Equation 24 is a convenient form with which to begin this discussion:

$$- \delta \Delta G^\circ = \rho_F \sigma_F + \rho_R \sigma_R + \rho_\alpha \sigma_\alpha + \rho_\chi \sigma_\chi \qquad (24)$$

The equation is written in terms of Gibbs energy changes, rather than $\log K$ or $\log k$, because much of its application initially was to gas-phase reactions for which the use of Gibbs energies is conventional. Corresponding equations in terms of $- \delta \Delta E^\circ$ or $-\delta \Delta H^\circ$ have also been used. The negative sign is introduced to make the signs of ρ values correspond to the conventions of the Hammett equation. σ_F is Taft and Topsom's preferred symbol for the inductive constant σ_I (see Section V.A), σ_R is a resonance constant closely related to σ_R^0, σ_α the substituent polarizability parameter as above and σ_χ is the substituent electronegativity parameter.

The inclusion of σ_χ is to deal with the possibility that consideration of electronegativity may be helpful in understanding substituent effects. Values of σ_χ come from *ab initio* calculations. On this scale H is taken as a standard at $\sigma_\chi = 0.00$ and the values range from $- 0.15$ for SMe to $+ 0.70$ for F. CN and HC≡C are at 0.30 and 0.12, respectively. To set these values in context we mention that the σ_χ values for NH_2, NO_2, CH_3 and Cl are at 0.33, 0.46, 0.00 and 0.16, respectively. However, except at very short range, electronegativity effects of substituents are found not to be important, and the PSP equation may be simplified to equation 25:

$$- \delta \Delta G^\circ = \rho_F \sigma_F + \rho_R \sigma_R + \rho_\alpha \sigma_\alpha \qquad (25)$$

Taft and Topsom's article[127] and also Topsom's[149] should be consulted for details of the setting up of the scales of substituent parameters. The equation has been applied to a wide range of gas-phase reactivities. (In the multiple regressions an intercept term is often permitted, but usually this turns out to be indistinguishable from zero, as it should be if equation 25 is valid.) For aliphatic and alicyclic saturated systems the resonance term is duly negligible. The roles of field, resonance and polarizability effects are discussed and the interpretation of the various ρ values is attempted.

When the equation is applied to reactions in solution, it is found that polarizability effiects tend to be much smaller than in the gas phase, but the PSP equation has to be adapted to include Substituent Solvation Assisted Resonance (SSAR). (See Section III.D.) The PSP equation then assumes the form of equation 26:

$$- \delta \Delta G^\circ \text{(soln.)} = \rho_F \sigma_F + \rho_R \sigma_R + \rho_S \Delta \sigma_R \qquad (26)$$

where $\Delta \sigma_R$ is the SSAR parameter. A scale of $\Delta \sigma_R$ values has been established. It is also necessary to use special σ_F(aq.) values for some hydrogen-bond acceptor substituents in aqueous solution.

The SSAR phenomenon affects only $+R$ substituents. The $\Delta \sigma_R$ value of H is 0.00. Values for several $+R$ substituents are as follows[127]: SO_2Me, 0.02; CN, 0.07; COMe, 0.10; NO_2 0.18. Several of the substituents for which enhanced σ_F(aq.) values are tabulated are $+R$ in nature, but they do not include CN.

A recent study applied the PSP equation to good effect in discussing the gas-phase and aqueous solution basicities of about fifty 2-, 3- or 4-substituted pyridines and some 2,6-disubstituted compounds[249]. The substituents studied included 2-, 3- and 4-CN, and these conformed fairly well to various relations and graphical plots. CN also features extensively among the $+R$ substituents in a recent paper on the inherent dependence of resonance effects of strongly conjugated substituents on electron demand[250].

E. Charton's LDR Equation

This has been developed since 1986. The title letters stand for *L*ocalized *D*elocalized *R*esponse. The Localized effect is Charton's preferred name for the inductive effect and delocalized effect is his preferred name for the resonance effect. Indeed, he would like to change the usual symbols from σ_I to σ_L and σ_R to σ_D for the purposes of the Extended Hammett (EH or LD) equation[81]. The response referred to is that of the substituent to the electronic demand of the site (i.e. reaction site in the correlation analysis of reactivity). Thus this equation, like the PSP equation, is concerned with the parametrization of substituent polarizability.

We shall describe the treatment only rather briefly, because a detailed article[251] and a useful introductory account[252] have already appeared. (The latter includes a table of substituent constants for about thirty common substituents.)

The LDR equation may be written as in equation 27:

$$Q_X = L\sigma_{l,x} + D\sigma_{d,x} + R\sigma_{e,x} + h \tag{27}$$

where Q_X is the property influenced by the substituent X, σ_l is the localized effect parameter, identical to σ_I, σ_d is the intrinsic delocalized effect parameter for minimal electronic demand of the active site and σ_e gives the sensitivity of X to changes in electronic demand of the active site; h is the intercept term. Quantities σ_d and σ_e are defined by equation 28:

$$\sigma_D = \sigma_e \eta + \sigma_d \tag{28}$$

where η expresses the electronic demand of the active site and σ_D (i.e. σ_R) is the relevant delocalized electronic parameter which would be used in the EH treatment of the system, i.e. a σ_R-type quantity. The main article mentioned above[251] should be consulted for the methods whereby the substituent parameter scales were established. Several hundred data sets have now been treated by means of the LDR equation, and the various sigma parameters have been tabulated for more than 120 substituents.

As already mentioned, the σ_l values correspond closely to those of σ_I as derived by Charton[81], while the values of σ_d are broadly similar to Charton's values of σ_R[81]. However, individual values may sometimes differ by a few units in the second place of decimals, consequent upon σ_d being derived from σ_D (i.e. σ_R) in equation 28 by subtracting an electronic response term. Thus for CN and HC≡C, σ_d values are 0.12 and −0.02, respectively; cf 0.08 and −0.04, respectively, for Charton's σ_R values[81]. H is the standard for σ_e at 0.00, and the scale runs from +0.041 for F to −0.29 for PPh$_2$. The values for CN and HC≡C are −0.055 and −0.10, respectively. To set these values in context we mention the values for a few selected groups: NH$_2$, −0.13; NO$_2$, −0.077; OMe, −0.064; Me, −0.030, Cl, −0.011.

The electronic demand parameter η, characteristic of a given process, is equal to the ratio of the coefficients R/D and has been shown to depend on the nature of the active site, skeletal group and medium. Contrary to the general view, electronic demand is roughly the same in magnitude for σ_R (based on benzoic acid ionization) and σ_R^0 scales, but is positive for the former and negative for the latter.

It is claimed that, "The LDR equation is the first sucessful model for electronic effects of substituents bonded to carbon in all substrates"[253].

XII. REFERENCES AND NOTES

1. J. Shorter, in *The Chemistry of the Sulphonium Group* (Eds. C. J. M. Stirling and S. Patai), Chap. 9, Wiley, Chichester, 1981.
2. J. Shorter, in *The Chemistry of Sulphones and Sulphoxides* (Eds. S. Patai, Z. Rappoport and C. J. M. Stirling), Chap. 10, Wiley, Chichester, 1988.
3. J. Shorter, in *The Chemistry of Sulphinic Acids, Esters and their Derivatives* (Ed. S. Patai), Chap. 17, Wiley, Chichester, 1990.

4. J. Shorter, in *The Chemistry of Amidines and Imidates*, Vol. 2 (Eds. S. Patai and Z. Rappoport), Chap. 13, Wiley, Chichester, 1991.
5. J. Shorter, in *The Chemistry of Hydroxyl, Ether and Peroxide Groups*, Supplement E2: (Ed. S. Patai), Chap. 9, Wiley, Chichester, 1993.
6. See, for example, *Watts' Dictionary of Chemistry*, 2nd ed. (Eds. M. M. Pattison Muir and H. Forster Morley), Longmans, Green and Co., London, in four vols., 1890–94; especially the entries under Cyan... in Vol. II, pp. 296–360.
7. J. Liebig and F. Wöhler, *Liebigs Ann.*, **3**, 267 (1832).
8. T. J. Pelouze, *Liebigs Ann.*, **10**, 249 (1834).
9. H. Fehling, *Liebigs Ann.*, **49**, 91 (1844).
10. A. W. Hofmann, *Liebigs Ann.*, **144**, 114 (1867).
11. H. E. Armstrong, in *Memorial Lectures Delivered before The Chemical Society 1893–1900*, Gurney and Jackson, London, 1901, pp. 707–710.
12. W. Lieke, *Liebigs Ann.*, **112**, 316 (1859).
13. J. B. Cohen, *Organic Chemistry for Advanced Students, Part I, Reactions*, 4th ed., Edward Arnold, London, 1923, pp. 70–73. (This book and its companions: *Part II, Structure* and *Part III, Synthesis* reveal the state of the art at the time when electronic theory of organic reactions was just coming in. In Part I, Chap. I, "Historical Introduction" and Chap. II, "The Valency of Carbon" there are extended accounts of the development of theoretical organic chemistry. A very unusual feature for an organic textbook of those times is the long Chap. IV, "Dynamic of Organic Reactions".)
14. Reference 13, pp. 66–70.
15. A. Geuther, *Jahresber.*, **16**, 323 (1863).
16. E. Frankland and B. F. Duppa, *Phil. Trans. Roy. Soc.*, **156**, 37 (1866).
17. An extended account of early work on the acetoacetic ester condensation is given in Reference 13, pp.260–273.
18. L. Claisen, *Chem. Ber.*, **20**, 647 (1887); **36**, 3674 (1903); **38**, 709 (1905); **41**, 1260 (1908).
19. L. Claisen, *Chem. Ber.*, **22**, 1009 (1889).
20. H. Fleischhauer, *J. Prakt. Chem.*, **47**, 375 (1893).
21. W. H. Perkin and W. N. Haworth, *Trans. Chem. Soc.*, **93**, 1944 (1908); **95**, 480 (1909).
22. S. Gabriel, *Chem. Ber.*, **36**, 570 (1903).
23. A. Lapworth, *Trans. Chem. Soc.*, **79**, 1269 (1901).
24. F. Arndt and B. Eistert, *Chem. Ber.*, **69**, 2381 (1936).
25. H. B. Watson, *Modern Theories of Organic Chemistry*, 2nd ed., Chap. 11, Oxford University Press, Oxford, 1941.
26. L. P. Hammett, *Physical Organic Chemistry*, McGraw-Hill, New York, 1940.
27. H. Bauer, *Chem. Ber.*, **37**, 3317 (1904); *J. Prakt. Chem.*, **72**, 201 (1905).
28. W. Ostwald, *Z. Phys. Chem.*, **3**, 170, 241, 369 (1889).
29. The historical introduction in Reference 5 summarizes some early work on directing effects on substituents in aromatic substitution.
30. R. Robinson, *Outline of an Electrochemical (Electronic) Theory of the Course of Organic Reactions*, Institute of Chemistry, London, 1932, 52 pp.
31. C. K. Ingold, *Chem. Rev.*, **15**, 225(1934).
32. D. Ll. Hammick, R. G. A. New, N. V. Sidgwick and L. E. Sutton, *J. Chem. Soc.*, 1876 (1930).
33. Summaries in (a) Reference 25, Chap. 8, and (b) N. V. Sidgwick, *The Chemical Elements and Their Compounds*, Clarendron Press, Oxford, 1950, pp. 546, 672.
34. Reference 25, pp. 59–63.
35. O. Exner, *Dipole Moments in Organic Chemistry*, Chap. 6, Georg Thieme, Stuttgart, 1975.
36. See entry under Acetylene in Reference 6, Vol. I, p. 41.
37. P. Griess, *Liebigs Ann.*, **106**, 123 (1858).
38. See entry under Diazo Compounds in Reference 6, Vol. I, p. 397.
39. This complicated history is summarized by N. V. Sidgwick, in *The Organic Chemistry of Nitrogen* (Eds. T. W. J. Taylor and W. Baker), Clarendon Press, Oxford, 1937, pp. 413–425.
40. This section is largely based on previous accounts in References 1, 2 and 5, in which certain material was adapted (by kind permission of Oxford University Press) from J. Shorter, *Correlation Analysis in Organic Chemistry; An Introduction to Linear Free-Energy Relationships*, Chap. 2, Oxford Chemistry Series, 1973.
41. J. Shorter, *Correlation Analysis of Organic Reactivity*, Chap. 1, Research Studies Press, Wiley, Chichester, 1982.

42. Reference 26, Chap. 7.
43. L. P. Hammett, *Physical Organic Chemistry*, 2nd edn., Chap. 11, McGraw-Hill, New York, 1970.
44. O. Exner, in *Advances in Linear Free Energy Relationships* (Eds. N.B. Chapman and J. Shorter), Chap. 1, Plenum Press, London, 1972.
45. Reference 41, Chap. 3.
46. J. Shorter, in *Similarity Models in Organic Chemistry, Biochemistry, and Related Fields* (Eds. R.I. Zalewski, T.M. Krygowski and J. Shorter), Chap. 2, Elsevier, Amsterdam, 1991.
47. O. Exner, *Correlation Analysis of Chemical Data*, Plenum, New York and SNTL, Prague, 1988.
48. O. Exner, in *Correlation Analysis in Chemistry: Recent Advances* (Eds. N.B. Chapman and J. Shorter), Chap. 10, Plenum Press, New York, 1978.
49. H. H. Jaffé, *Chem. Rev.*, **53**, 191 (1953).
50. The symbol and sign conventions used for substituent effects in this chapter are those most frequently used by writers on correlation analysis in organic chemistry. I or R effects which withdraw electrons from the ring are regarded as positive. The sign convention is the opposite of that used by Ingold[31], and which was observed in Section I.B. See Reference 41, pp. 229–230 for a more detailed consideration of symbol and sign conventions.
51. H. C. Brown and Y. Okamoto, *J. Am. Chem. Soc.*, **80**, 4979 (1958).
52. H. van Bekkum, P.E. Verkade and B.M. Wepster, *Recl. Trav. Chim. Pays-Bas*, **78**, 815 (1959).
53. J. Shorter, in Reference 48, Chap. 4.
54. Reference 41, Chap. 3.
55. Y. Yukawa and Y. Tsuno, *Bull. Chem. Soc. Jpn.*, **32**, 971 (1959).
56. Y. Yukawa, Y. Tsuno and M. Sawada, *Bull. Chem. Soc. Jpn.*, **39**, 2274 (1966).
57. R. W. Taft and I.C. Lewis, *J. Am. Chem. Soc.*, **80**, 2436 (1958).
58. R. W. Taft and I.C. Lewis, *J. Am. Chem. Soc.*, **81**, 5343 (1959).
59. R. W. Taft, in *Steric Effects in Organic Chemistry* (Ed. M.S. Newman), Chap. 13, Wiley, New York, 1956.
60. R. W. Taft, S. Ehrenson, I.C. Lewis and R.E. Glick, *J. Am. Chem. Soc.*, **81**, 5352 (1959).
61. S. Ehrenson, *Prog. Phys. Org. Chem.*, **2**, 195 (1964).
62. S. Ehrenson, R.T.C. Brownlee and R.W. Taft, *Prog. Phys. Org. Chem.*, **10**, 1 (1973).
63. It should be mentioned that the expression 'σ value' or 'σ constant' has acquired both a specialized and a more general meaning. The former denotes substituent constants based on the ionization of benzoic acids; the latter signifies polar (electronic) substituent constants in general. Thus the meaning of σ often has to be understood in context, but expressions such as 'sigma value' or 'σ-type constant' are preferable when the more general meaning is intended.
64. O. Exner, *Collect. Czech. Chem. Commun.*, **31**, 65 (1966).
65. See, for example, M. Charton, *J. Am. Chem. Soc.*, **91**, 6649 (1969).
66. R. W. Taft, E. Price, I.R. Fox, I.C. Lewis, K.K. Andersen and G.T. Davis, *J. Am. Chem. Soc.*, **85**, 709 (1963).
67. R. W. Taft, E. Price, I.R. Fox, I.C. Lewis, K.K. Andersen and G.T. Davis, *J. Am. Chem. Soc.*, **85**, 3146 (1963).
68. D. F. Ewing, in Reference 48, Chap. 8.
69. D. J. Craik and R.T.C. Brownlee, *Prog. Phys. Org. Chem.*, **14**, 1 (1983).
70. A. R. Katritzky and R.D. Topsom, in Reference 44, Chap. 3.
71. G. P. Ford, A.R. Katritzky and R.D. Topsom, in Reference 48, Chap. 3.
72. J. D. Roberts and W.T. Moreland, *J. Am. Chem. Soc.*, **75**, 2167 (1953).
73. Reference 41, Chap. 4.
74. J. Shorter, in Reference 44, Chap. 2.
75. Reference 40, Chap. 3.
76. R. W. Taft, *J. Am. Chem. Soc.*, **79**, 1045 (1957).
77. Except where a more specific reference is given, pK_a values quoted explicitly, or quoted implicitly through correlation equations, in this chapter have been taken from the following compilations:
(a) G. Kortüm, W. Vogel and K. Andrussow, *Dissociation Constants of Organic Acids in Aqueous Solution*, Butterworths, London, 1961 (for I.U.P.A.C.). This is a reprint from *Pure and Applied Chemistry*, Vol. 1, Nos. 2 and 3.
(b) E.P. Serjeant and B. Dempsey, *Ionisation Constants of Organic Acids in Aqueous Solution*, Pergamon Press, Oxford, 1979 (I.U.P.A.C. Chemical Data Series, No. 23).
(c) V.A. Palm (Ed.), and several compilers (From the Laboratory of Chemical Kinetics and

Catalysis, Tartu State University, Estonia), *Tables of Rate and Equilibrium Constants of Heterolytic Organic Reactions*, Moscow, 1975–79, in 5 volumes, each being in two parts (10 books in all). Six supplementary volumes were published in 1984–90. Information about pK_a values may be found in particular in the first volume of each series. These volumes are especially useful for pK_a values determined in aqueous organic or purely organic solvents. Detailed references may be found in the above sources. Some use has also been made of Tables in Charton's review[81], where detailed references are also given. Reference 77c has also been used as a source of rate constants in the present chapter, where no more specific reference is given.

78. For a recent review of the 'through-bonds' versus 'through-space' controversy, see J. Shorter in Reference 46, especially pp. 117–120. See also K. Bowden and E.J. Grubbs, *Prog. Phys. Org. Chem.*, **19**, 183 (1993).
79. N. B. Chapman, J. Shorter and K.J. Toyne, *J. Chem. Soc.*, 1077 (1964).
80. K. Bowden, N.B. Chapman and J. Shorter, *J. Chem. Soc.*, 3370 (1964).
81. M. Charton, *Prog. Phys. Org. Chem.*, **13**, 119 (1981).
82. Equation 58 on p. 137 of Reference 81.
83. Equation 36 on p. 131 of Reference 81, with slightly different symbols. The correlation equations which are relevant to Table 3 are nos. 6, 53, 21, 18, 36, 33 and 35 in Charton's Table 9, for our entries 1 to 7, respectively.
84. Reference 77c, supplementary volume 1.
85. The practice began with J.D. Roberts about 1950. See, for example, J.D. Roberts and W. Watanabe, *J, Am. Chem. Soc.*, **72**, 4869 (1950).
86. K. Bowden, M. Hardy and D.C. Parkin, *Can. J. Chem.*, **46**, 2929 (1968).
87. The relevant correlations in Charton's review[81] are nos. 55–60 of his Table 9, respectively.
88. Reference 5, p. 417.
89. The relevant correlations in Charton's review[81] are nos. 6 and 7, respectively, of his Table 15.
90. N .B. Chapman, A. Ehsan, J. Shorter and K.J. Toyne, *J. Chem. Soc.(B)*, 256 (1967).
91. G. Briegleb and A. Bieber, *Z. Elektrochem.*, **55**, 250 (1951).
92. A.J. Hoefnagel and B.M. Wepster, *J. Chem. Soc., Perkin Trans. 2*, 977 (1989).
93. The author is the chairman of a Working Party on Structure-Reactivity Parameters and Equations under the auspices of the IUPAC Commission for Physical Organic Chemistry. At present (1993) a document on σ values which are based directly on the pK_a values of substituted benzoic acids in water is in course of publication in *Pure and Applied Chemistry*. The σ values for CN are from this document. Other material in this chapter is also from material being considered by the Working Party.
94. S. Widequist, *Arkiv. Kemi Mineral. Geol.*, **2**, Nr. 25, 383 (1950), cited in Reference 77a.
95. M. Ludwig, V. Baron, K. Kalfus, O. Pytela and M. Večeřa, *Collect. Czech. Chem. Commun.*, **51**, 2135 (1986).
96. A.J. Hoefnagel and B.M. Wepster, *J. Am. Chem. Soc.*, **95**, 5357 (1973).
97. Reference 26, Chap. 7.
98. D. H. McDaniel and H.C. Brown, *J. Org. Chem.*, **23**, 420 (1958).
99. C. Hansch, A. Leo and R.W. Taft, *Chem. Rev.*, **91**, 165 (1991).
100. H. P. Kirschman, B. Wingfield and H.J. Lucas, *J. Am. Chem. Soc.*, **52**, 23 (1930).
101. C. Hansch and A.J. Leo, *Substituent Constants for Correlation Analysis in Chemistry and Biology*, Wiley, New York, 1979.
102. Reference 5, p. 423.
103. O. Pytela, M. Ludwig and M. Večeřa, *Collect. Czech. Chem. Commun.*, **51**, 2143 (1986).
104. V. Všetečka and O. Exner, *Collect. Czech. Chem. Commun.*, **39**, 1140 (1974).
105. This matter is very much tied in with Exner's approach to the analysis of σ values into inductive and resonance components. See References 45 and 64, and an appraisal of the situation by the present author in Reference 46. See Section XI.B.
106. A.J. Hoefnagel, J.C. Monshouwer, E.C.G. Snorn and B.M. Wepster, *J. Am. Chem. Soc.*, **95**, 5350 (1973).
107. 50% v/v EtOH–H_2O is different from 1:1 v/v EtOH–H_2O mentioned in Section III.B. The latter is made by mixing equal volumes of ethanol and water, whereas the former is made by taking V cm^3 of ethanol and making up to $2V$ cm^3 with water. 1:1 v/v EtOH–H_2O is sometimes referred to by those who use it as "50% ethanol", so careful reading of a paper may be needed to discover what solvent composition has actually been used.
108. Reference 5, p. 427.

109. O. Exner, *Collect. Czech. Chem. Commun.*, **31**, 3222 (1966).
110. Reference 41, Chapter 7.
111. M. Fujio, R.T. McIver and R.W. Taft, *J. Am. Chem. Soc.*, **103**, 4017 (1981).
112. Reference 46, p. 102–104.
113. G. W. Wheland, R.M. Brownell and E.C. Mayo, *J. Am. Chem. Soc.*, **70**, 2492 (1948).
114. L. A. Cohen and W.M. Jones, *J. Am. Chem. Soc.*, **85**, 3397 (1963).
115. Reference 25, p. 241.
116. Reference 41, Chapter 4.
117. M. Hojo, M. Utaka and Z. Yoshida, *Tetrahedron*, **27**, 4031 (1971).
118. G. E. K. Branch and D.L. Yabroff, *J. Am. Chem. Soc.*, **56**, 2568 (1934).
119. M. Charton and B.I. Charton, *J. Org. Chem.*, **43**, 1161 (1978) and earlier papers referred to therein.
120. M. Charton, in *Topics in Current Chemistry 114: Steric Effects in Drug Design* (Eds. M. Charton and I. Motoc), Springer-Verlag, Berlin, 1983, pp. 57–91.
121. M. Charton, in Reference 46, Chap. 11.
122. Reference 5, p. 435.
123. M. H. Aslam, A.G. Burden, N.B. Chapman, J. Shorter and M. Charton, *J. Chem. Soc., Perkin Trans. 2*, 500 (1981).
124. M. H. Aslam, N.B. Chapman, J. Shorter and M. Charton, *J. Chem. Soc., Perkin Trans. 2*, 720 (1981).
125. Reference 5, p. 433–436.
126. M. H. Aslam, Ph.D. Thesis, University of Hull, 1978. Also deposited material associated with References 123 and 124.
127. R. W. Taft and R.D. Topsom, *Prog. Phys. Org. Chem.*, **16**, 1 (1987).
128. R. W. Taft, *J. Phys. Chem.*, **64**, 1805 (1960).
129. Y. Yukawa, Y. Tsuno and M. Sawada, *Bull. Chem. Soc. Jpn.*, **45**, 1198 (1972).
130. R. T. C. Brownlee, R.E.J. Hutchinson, A.R. Katritzky, T.T. Tidwell and R.D. Topsom, *J. Am. Chem. Soc.*, **90**, 1757 (1968).
131. J. Bromilow, R.T.C. Brownlee, V. O. Lopez and R.W. Taft *J. Org. Chem.*, **44**, 4766 (1979).
132. W. F. Reynolds, A. Gomes, A. Maron, D.W. MacIntyre, A. Tanin, G.K. Hamer and I. R. Peat, *Can. J. Chem.*, **61**, 2376 (1983).
133. D. A. R. Happer, *J. Chem. Soc., Perkin Trans. 2*, 1673 (1984).
134. C. Laurence, M. Berthelot, M. Luçon, M. Helbert, D.G. Morris and J.–F. Gal, *J. Chem. Soc., Perkin Trans. 2*, 705 (1984).
135. C. Laurence in Reference 46, Chap. 5.
136. G. G. Smith and F.W. Kelly, *Prog, Phys. Org. Chem.*, **8**, 75 (1971).
137. K. A. Holbrook, in *Supplement B: The Chemistry of Acid Derivatives* (Ed. S. Patai), Chap. 12, Wiley, Chichester, 1992.
138. R. Taylor, in *Supplement B: The Chemistry of Acid Derivatives* (Ed. S. Patai), Wiley, Chichester, 1979, p. 860.
139. See, for example, I. Martin, G. Chuchani, I. Avila, A. Rotinov and R. Olmos, *J, Phys. Chem.*, **84**, 9 (1980).
140. M. Mashima, R.T.McIver, R.W. Taft, F.G. Bordwell and W.N. Olmstead, *J. Am. Chem. Soc.*, **106**, 2717 (1984).
141. S. Marriott and R. D. Topsom, *J. Am. Chem. Soc.*, **106**, 7 (1984).
142. S. Marriott and R. D. Topsom, *J. Chem. Soc., Perkin Trans. 2*, 1045 (1985).
143. S. Marriott, A. Silvestro and R. D.Topsom, *J. Chem. Soc., Perkin Trans. 2*, 457 (1988).
144. S. Marriott, A. Silvestro and R. D.Topsom, *J. Molec. Struct.* (*Theochem*), **184**, 23 (1989).
145. Cai Jinfeng and R.D. Topsom, *J. Molec. Struct.* (*Theochem*), **204**, 353 (1990).
146. Jinfeng Cai and R.D. Topsom, *J. Molec. Struct.* (*Theochem*), **228**, 181 (1991).
147. Tony Silvestro and R.D. Topsom, *J. Molec. Struct.* (*Theochem*), **206**, 309 (1990).
148. R. D. Topsom, *Prog. Phys. Org. Chem.*, **17**, 107 (1990).
149. R. D. Topsom, *Prog. Phys. Org. Chem.*, **16**, 125 (1987).
150. T. Sotomatsu, Y. Murata and T. Fujita, *J. Comput. Chem.*, **10**, 94 (1989).
151. K. H. Kim and Y. C. Martin, *J. Org. Chem.*, **56**, 2723 (1991).
152. Reference 77c, Volume II, Part 1 and Supplementary Volume I, Issue 5.
153. F. G. Bordwell, *Acc. Chem. Res.*, **21**, 456 (1988) and references cited therein.
154. O. W. Webster, *J. Am. Chem. Soc.*, **88**, 3046 (1966).

284   J. Shorter

155. J. R. Jones. *The Ionisation of Carbon Acids*, Chap.5, Academic Press, London, 1973.
156. D. J. Cram, *Fundamentals of Carbanion Chemistry*, Academic Press, New York, 1965.
157. R. P. Bell, *The Proton in Chemistry*, 2nd edn., Chapman and Hall, London, 1973.
158. K. Bowden, N. S. Nadvi and R.J. Ranson, *J. Chem. Res. (S)*, 299 (1990); *(M)*, 2474–2496 (1990).
159. C. D. Johnson, *Chem. Rev.*, **75**, 755 (1975).
160. K. Bowden and S.I.J. Hirani, *J. Chem. Soc., Perkin Trans. 2*, 1889 (1990)
161. C. F. Bernasconi and S. A. Hibdon, *J. Am. Chem. Soc.*, **105**, 4343 (1983).
162. C. F. Bernasconi and F. Terrier, *J. Am. Chem. Soc.*, **109**, 7115 (1987).
163. K. Bowden, A. F. Cockerill and J. R. Gilbert, *J. Chem. Soc. (B)*, 179 (1970).
164. I. V. Tselinskii, A.S. Kosmynina, V. N. Dronov and I.N. Shokhor, *Reakts. Spos. Org. Soedinenii* **7**, 50 (1970).
165. S. Bradamente and G.A. Pagani, *J. Org. Chem.*, **45**, 105, 114 (1980).
166. S. Bradamente and G.A. Pagani, *J. Chem. Soc., Perkin Trans. 2*, 1047 (1986).
167. A. Abbotto, S. Bradamente and G. A. Pagani, *J. Org. Chem.*, **58**, 449 (1993).
168. T. Mitsuhashi, *J. Am. Chem. Soc.*, **108**, 2394 (1986).
169. H. Hirota and T. Mitsuhashi, *Chem. Lett.*, 803 (1990).
170. T. Mitsuhashi and H. Hirota, *J. Chem. Soc., Chem. Commun.*, 324 (1990)
171. C. A. Bunton, *Nucleophilic Substitution at a Saturated Carbon Atom*, Elsevier, Amsterdam, 1963.
172. J. Miller, *Aromatic Nucleophilic Substitution*, Elsevier, Amsterdam, 1968.
173. Unless a more specific reference is given, rate coefficients in Section VI.B have been taken from the volumes of Reference 77c: Volume 3, Part I (first order nucleophilic substitution); Volume 4, Part II (aromatic nucleophilic substitution).
174. W. F. Reynolds, P. Dais, D. W. MacIntyre, R. D. Topsom, S. Marriott, E. von Nagy-Felsobuki and R.W. Taft, *J. Am. Chem. Soc.*, **105**, 378 (1983).
175. P. G. Gassman and J. J. Talley, *J. Am. Chem. Soc.*, **102**, 1214 (1980).
176. P. G. Gassman and T. T. Tidwell, *Acc. Chem. Res.*, **16**, 279 (1983).
177. T. T. Tidwell, *Angew. Chem., Int. Ed. Engl.*, **23**, 20 (1984).
178. A. D. Allen, V. M. Kanagasabapathy and T. T. Tidwell, *J. Am. Chem. Soc.*, **108**, 3470 (1986).
179. N. B. Chapman and D. A. Russell-Hill, *J. Chem. Soc.*, 1563 (1956).
180. Reference 172, p. 23.
181. N. B. Chapman, D. K. Chaudhury and J. Shorter, *J. Chem. Soc.*, 1975 (1962).
182. D. M. Brewis, N.B. Chapman, J. S. Paine, J. Shorter and D.J. Wright, *J. Chem. Soc., Perkin Trans. 2*, 1787 (1974).
183. D. M. Brewis, N. B. Chapman, J.S. Paine and J. Shorter, *J. Chem. Soc., Perkin Trans. 2*, 1802 (1974).
184. J. Miller and A. J. Parker, *Aust. J. Chem.*, **11**, 302 (1958).
185. P. C. M. F. Castilho, M. R. Crampton and J. Yarwood, *J. Chem. Res. (S)*, 370 (1989).
186. C. A. Grob and M. G. Schlageter, *Helv. Chim. Acta*, **60**, 1884 (1977).
187. C. A. Grob and M. G. Schlageter, *Helv. Chim. Acta*, **59**, 264 (1976).
188. W. Fischer and C. A. Grob, *Helv. Chim. Acta*, **61**, 1588 (1978).
189. C. A. Grob and A. Waldner, *Helv. Chim. Acta*, **62**, 1736 (1979).
190. C. A. Grob and R. Rich, *Helv. Chim. Acta*, **62**, 2793 (1979).
191. C. A. Grob, *Angew. Chem., Int. Ed. Engl.*, **15**, 569 (1976).
192. M. R. Crampton and J. A. Stevens, *J. Chem. Soc., Perkin Trans. 2*, 1715 (1991).
193. M. B. Sommer, M. Begtrup and K. P. Bogeso, *J. Org. Chem.*, **55**, 4817 (1990).
194. R. Taylor, *Electrophilic Aromatic Substitution*, Wiley, Chichester, 1990.
195. Reference 194, pp. 308, 395 and 485–487.
196. Reference 194, pp. 398–399.
197. Reference 194, pp. 458–463.
198. L. M. Stock and H. C. Brown, *Adv. Phys. Org. Chem.*, **1**, 35 (1963).
199. I. N. Juchnovski and I. G. Binev, *Tetrahedron*, **33**, 2993 (1977).
200. G. W. Gray, K. J. Harrison and J. A. Nash, *Electron. Lett.*, **9**, 130 (1973).
201. R. Eidenschink, D. Erdmann, J. Krause and L. Pohl, *Angew. Chem., Int. Ed. Engl.*, **16**, 100 (1977).
202. R. Eidenschink, D. Erdmann, J. Krause and L. Pohl, *Angew. Chem., Int. Ed. Engl.*, **17**, 133 (1978).
203. R. Eidenschink, G. Haas, M. Romer and B. S. Scheuble, *Angew. Chem., Int. Ed. Engl.*, **23**, 147 (1984).
204. R. Eidenschink, *Mol. Cryst. Liq. Cryst.*, **123**, 57 (1985).
205. M. G. Hogben and W. A. G. Graham, *J. Am. Chem. Soc.*, **91**, 283 (1969).

206. K. Bowden, *Can. J. Chem.*, **41**, 2781 (1963).
207. K. Bowden, N. B. Chapman and J. Shorter, *Can. J. Chem.*, **42**, 1979 (1964).
208. W. Prikoszovich and H. Schindlbauer, *Chem. Ber.*, **102**, 2922 (1969).
209. W. A. Sheppard and R. W. Taft, *J. Am. Chem. Soc.*, **94**, 1919 (1972).
210. L. M. Yagupol' skii, A. Ya. Il'chenko and N. B. Kondratenko, *Uspekhi Khim.*, **43**, 64 (1974); *Russ. Chem. Rev.*, **43**, 32 (1974).
211. Reference 33b, pp. 299–316.
212. J. Ashworth and B. A. W. Coller, *Trans. Faraday Soc.*, **67**, 1069 (1971).
213. L. G. Vaughan and W. A. Sheppard, *J. Am. Chem. Soc.*, **91**, 6151 (1969).
214. Reference 33b, pp. 673–676.
215. E. S. Lewis and M. D. Johnson, *J. Am. Chem. Soc.*, **81**, 2070 (1959).
216. Yu. A. Ustynyuk, O. A. Subbotin, L. M. Buchneva and L. A. Kazitsyna, *Dokl. Akad. Nauk SSR*, **227**, 101 (1976); English version, p. 175.
217. N. C. Cutress, T. B. Grindley, A. R. Katritzky, M. V. Sinnott and R. D. Topsom, *J. Chem. Soc., Perkin Trans. 2*, 2255 (1972).
218. L. P. Hammett, *J. Am. Chem. Soc.*, **59**, 96 (1937).
219. Reference 49, p. 239.
220. A. J. Hoefnagel, M. E. Hoefnagel and B. M. Wepster, *J. Org. Chem.*, **43**, 4720 (1978).
221. I. A. Koppel, M. M. Karelson and V. A. Palm, *Org. React. (Tartu)*, **11**, 101 (1974).
222. G. H. Mansfield and M. C. Whiting, *J. Chem. Soc.*, 4761 (1956).
223. R. P. Bell, *Acids and Bases: Their Quantitative Behaviour*, Methuen, London, 1952.
224. M. Charton, *J. Org. Chem.*, **29**, 1222 (1964).
225. J. A. Landgrebe and R. H. Rynbrandt, *J. Org. Chem.*, **31**, 2585 (1966).
226. M. Charton, *J. Chem.*, **30**, 552 (1965).
227. M. Charton, *J. Chem. Soc.*, 1205 (1964).
228. M. Charton, and H. Meislich, *J. Am. Chem. Soc.*, **80**, 5940 (1958).
229. C. Eaborn, A. R. Thompson and D. R. M. Walton, *J. Chem. Soc. (B)*, 859 (1969).
230. A. R. Katritzky and R. D. Topsom, *Chem. Rev.*, **77**, 639 (1977).
231. J. Guillème and B, Wojtkowiak, *Bull. Soc. Chim. France*, 3007 (1969).
232. M. Charton, *Prog. Phys. Org. Chem.*, **10**, 81 (1973). The material of interest is on pp. 152, 153 and 202.
233. A. D. Pethybridge, *J. Chem. Soc., Perkin Trans. 2*, 102 (1973).
234. R. Fuchs, *J. Org. Chem.*, **28**, 3209 (1963).
235. G. Butt and R. D. Topsom, *Spectrochim. Acta*, **38**A, 649 (1982).
236. E. Eliel, *Stereochemistry of Carbon Compounds*, Chap. 6, McGraw-Hill, New York, 1962.
237. O. Exner and M. Budĕšinský, *Magn. Reson. Chem.*, **27**, 27 (1989).
238. M. Budĕšinský and O. Exner, *Magn. Reson. Chem.*, **27**, 585 (1989).
239. M. Ludwig, S. Wold and O. Exner, *Acta Chem. Scand.*, **46**, 549 (1992).
240. C. G. Swain and E. C. Lupton, *J. Am. Chem. Soc.*, **90**, 4328 (1968).
241. C. Hansch, A. Leo, S. H. Unger, K. H. Kim, D. Nakaitani and E. J. Lien, *J. Med. Chem.*, **16**, 1207 (1973).
242. C. G. Swain, S. H. Unger, N. R. Rosenquist and M. S. Swain, *J. Am. Chem. Soc.*, **105**, 492 (1983).
243. W. F. Reynolds and R. D. Topsom, *J. Org. Chem.*, **49**, 1989 (1984).
244. A. J. Hoefnagel, W. Oosterbeek and B. M. Wepster, *J. Org. Chem.*, **49**, 1993 (1984).
245. M. Charton, *J. Org. Chem.*, **49**, 1997 (1984).
246. C. G. Swain, *J. Org. Chem.*, **49**, 2005 (1984).
247. C. Hansch, in Reference 48, Chap. 9.
248. W. J. Hehre, C. -F. Pau, A. D. Headley, R. W. Taft and R. D. Topsom, *J. Am. Chem. Soc.*, **108**, 1711 (1986).
249. J.-L. M. Abboud, J. Catalan, J. Elguero and R. W. Taft, *J. Org. Chem.*, **53**, 1137 (1988).
250. R. W. Taft, J. L. M. Abboud, F. Anvia, M. Berthelot, M. Fujio, J. -F. Gal, A. D. Headley, W. G. Henderson, I. Koppel, J. H. Qian, M. Mishima, M. Taagepera and S. Ueji, *J. Am. Chem. Soc.*, **110**, 1797 (1988).
251. M. Charton, *Prog. Phys. Org. Chem.*, **16**, 287 (1987).
252. M. Charton and B. Charton, *Bull. Soc. Chim. France*, 199 (1988).
253. Reference 251, p.238.

CHAPTER **6**

Advances in acetylene chemistry*

G. V. BOYD

Department of Organic Chemistry, The Hebrew University of Jerusalem, Jerusalem 91904, Israel

---

*It had originally been intended to include in this chapter sections on cyanides, isocyanides and diazonium compounds but the idea was abandoned when it was found that there had not been sufficiently important developments in the chemistry of these classes since the appearance of previous volumes in this series (*The Chemistry of Diazonium and Diazo Groups*, 1978 and *Supplement C: The Chemistry of Triple-bonded Functional Groups*, 1983).

Supplement C2: The chemistry of triple-bonded functional groups
Edited by S. Patai © 1994 John Wiley & Sons Ltd.

I. INTRODUCTION

This chapter presents a selective account of new developments in the chemistry of acetylenes since the appearance of previous reviews in this series[1]. Three outstanding features of recent acetylene chemistry are the use of metal derivatives, the synthesis of large carbon rings containing several triple bonds and studies in the field of enediyne anti-cancer antibiotics, e.g. Dynemicin A(1), which were discovered only six years ago[2a-c].

(1)

II. SYNTHESIS OF ACETYLENES

A. From Alkenes

Perfluoroalkylacetylenes 4 are prepared from olefins 2 (R_F = perfluoroalkyl) by exhaustive chlorination under ultraviolet irradiation, followed by treatment with zinc. The resulting acetylenic zinc compounds 3 are decomposed by dilute hydrochloric acid[3]. 1,1-Dibromoalk-1-enes 5 (R^1 = Me, cyclohexyl etc.; R^2 = H or Ph) are converted into the rearranged acetylenes 7 by the action of samarium(II) iodide in benzene containing 10% HMPA; alkylidenecarbenes 6 are presumed to be intermediates in this process[4].

$$R_FCH{=}CH_2 \longrightarrow R_FCCl{=}CCl_2 \longrightarrow R_FC{\equiv}CZnCl \longrightarrow R_FC{\equiv}CH$$

(2)   (3)   (4)

$$R^1R^2C{=}CBr_2 \longrightarrow R^1R^2C{=}C{:} \longrightarrow R^1C{\equiv}CR^2$$

(5)   (6)   (7)

A number of novel acetylene syntheses from allenes has been described. 1-Bromoallenes 8 (R^1 = alkyl; R^2 = H or alkyl) are transformed into 3-phenyl-1-alkynes 9 on treatment with the complex organometallic reagent [PhCu. MgBr$_2$ LiBr]. Optically active bromoallenes, e.g. 1-bromo-1,2,3-butatriene (**8**; R^1 = Me; R^2 = H), yield chiral products in this reaction[5]. Grignard alkylation of methoxyallene (**10**) gives the acetylenes **11** (R = 3-methylbutyl or 4-methylhexyl)[6]. The reaction of allenylstannanes **12** (R^1 = H or pentyl) with methyllithium results in an equilibrium mixture of metal derivatives, **13** and **14**, of allenes and acetylenes. Addition of an aldehyde or a ketone R^2 COR^3 yields a mixture of allenic and acetylenic alcohols, **15** and **16**, in which the latter predominate[7]. Allenic stannanes **17** (R = Et, *i*-Pr or Bu) react with isoquinoline in the presence of methyl chloroformate to give the rearranged acetylenic dihydroisoquinolines **18**. Quinoline reacts analogously[8]. Conjugate addition of allenic stannanes **19** (R = H or Me) to α,β-enones in the presence of titanium(IV) chloride leads to rearranged β-propargyl ketones; thus cyclohex-2-enone gives **20**. Nitroolefins react similarly, β-nitrostyrene affording the acetylenes **21**[9].

$$R^1 \backslash C=C=C \diagup^H_{Br} \quad \longrightarrow \quad R^1-\overset{Ph}{\underset{R^2}{C}}-C\equiv CH$$

$$\text{(8)} \hspace{6cm} \text{(9)}$$

$$H_2C=CHOMe \quad \xrightarrow{RMgBr} \quad RCH_2C\equiv CH$$

$$\text{(10)} \hspace{5cm} \text{(11)}$$

$$H_2C=C=CR^1SnBu_3 \quad \longrightarrow \quad H_2C=C=CR^1Li \;+\; LiH_2CC\equiv CR^1$$

$$\text{(12)} \hspace{4cm} \text{(13)} \hspace{3cm} \text{(14)}$$

$$H_2C=C=CR^1-\overset{R^2}{\underset{OH}{C}}\!\diagdown^{R^3} \quad + \quad \overset{R^2}{\underset{R^3}{C}}-C\equiv CR^1$$

$$\text{(15)} \hspace{5cm} OH \quad \text{(16)}$$

$$H_2C=C=CRSnBu_3 + \quad \text{(isoquinoline)} \quad \xrightarrow{ClCO_2Me} \quad \text{(dihydroisoquinoline, } N-CO_2Me, CH_2-C\equiv CR)$$

$$\text{(17)} \hspace{7cm} \text{(18)}$$

$$RHC=C=CHSnMe_3 + \quad \text{(cyclohex-2-enone)} \quad \longrightarrow \quad \text{(cyclohexanone, } CHR-C\equiv CH)$$

$$\text{(19)} \hspace{7cm} \text{(20)}$$

$$PhHC=CHNO_2 \quad \xrightarrow{19} \quad PhHCCH_2NO_2$$

$$\hspace{6cm} \underset{CHR-C\equiv CH}{|}$$

$$\text{(21)}$$

Treatment of the trichloroenediyne **22** with butyllithium, followed by morpholine, results in 1,6-dimorpholinohexa-1,3,5-triyne (**23**)[10]. The 'push–pull' triacetylene **24** is similarly produced from compound **22**, butyllithium and methyl chloroformate. The corresponding tetraacetylene **25** has also been reported. These compounds have low thermal stability[11].

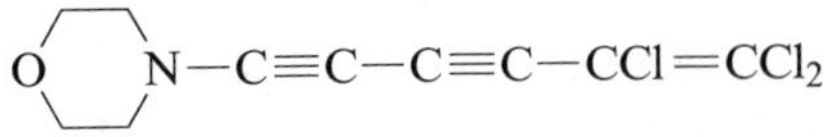

$$O \diagdown N-C\equiv C-C\equiv C-CCl=CCl_2$$

(22)

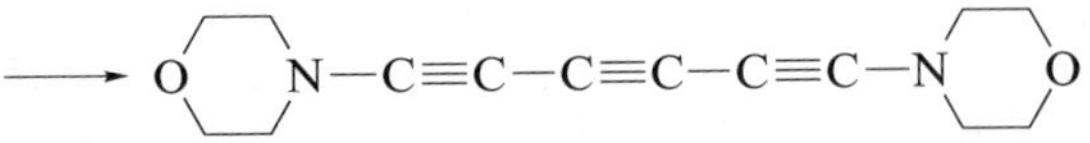

$$\longrightarrow O \diagdown N-C\equiv C-C\equiv C-C\equiv C-N \diagup O$$

(23)

$$O \diagdown N-C\equiv C-C\equiv C-C\equiv C-CO_2Me$$

$$O \diagdown N+C\equiv C\frac{}{4}CO_2Me$$

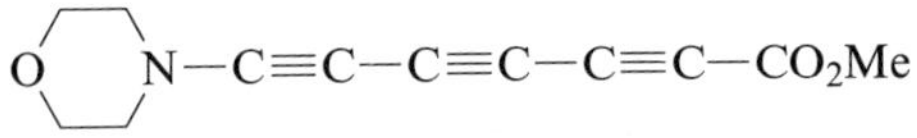

(24) 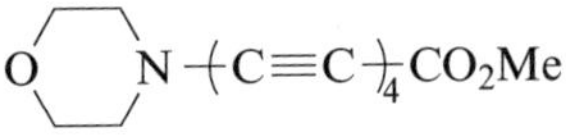   **(25)**

B. From Simpler Acetylenes

An alkynylxenonium salt, the tetrafluoroborate **27**, has been obtained by the low-temperature reaction of the lithium acetylide **26** with xenon trifluoride and boron trifluoride[12].

$$Bu^t-C\equiv CLi \longrightarrow Bu^t-C\equiv C-Xe^+$$
$$BF_4^-$$

(26)   **(27)**

Several new methods for the preparation of 1-haloalkynes have been described. High yields of bromo compounds, e.g. **28**, are obtained by treatment of alkynes with triphenylphosphine/carbon tetrabromide[13], or with a concentrated aqueous solution of potassium hypobromite and potassium hydroxide[14] (equation 1). 1-Iodoalkynes are produced from terminal alkynes and bis(pyridine)iodine(I) tetrafluoroborate in methanol in the presence of sodium methoxide (equation 2)[15] or from alkynes with a mixture of iodine, potassium carbonate, copper(I) iodide and tetrabutylammonium chloride under phase-transfer catalysis[16]. Lithium acetylides **29** (R = Ph, t-Bu, HOCH$_2$ etc.) react with zinc iodide and bis(trimethylsilyl) peroxide to yield 1-iodoalkynes[17]. The method has been

$$PhOH_2C-C\equiv CH \longrightarrow PhOH_2C-C\equiv CBr$$

(28)

$$RC\equiv CH \longrightarrow RC\equiv CBr \qquad (1)$$

R = Me, Et, Pr or Bu

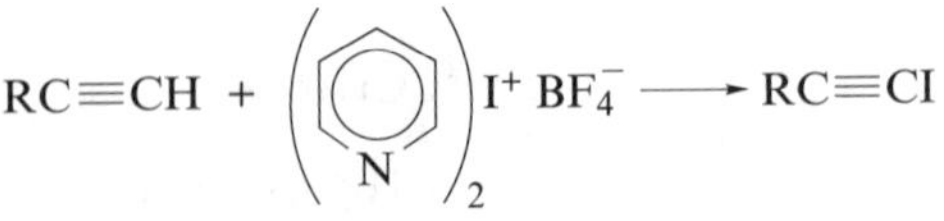

$$RC\equiv CH + \left(\bigcirc_N\right)_2 I^+ BF_4^- \longrightarrow RC\equiv CI \qquad (2)$$

R = alkyl, aryl etc.

extended to the preparation of chloro- or bromoalkynes by using bis(trimethylsilyl) peroxide and copper(I) chloride or bromide, respectively[18].

$$RC{\equiv}CLi + ZnI_2 \xrightarrow{(Me_3Si)_2 O_2} RC{\equiv}CI$$

(29)

A number of syntheses of di- and polyacetylenes has been reported. 1-Iodo-1-alkynes couple with terminal acetylenes under palladium–copper catalysis to give 1,3-diynes; thus γ-iodopropargyl alcohol and phenylacetylene afford compound **30**[19]. Oxidative coupling of 1-alkynes to yield symmetrical 1,3-diynes is brought about by air and copper(I) chloride in the presence of N, N'-tetramethylethylenediamine (equation 3)[20,21]. Trialkylsilyl substituents serve as protecting groups in this reaction; the silyl groups can be removed with aqueous methanolic alkali, see e.g. equation 4[22]. A series of polyacetylenes **31** ($n = 4,5,6,7$, 9,10 or 12), **32** ($n = 2,6$ or 8), **33** ($n = 2,3,4$ or 5), **34** ($n = 3,4$ or 5) and **35** has been prepared by this method[23,24].

$$HOH_2C-C{\equiv}CI + HC{\equiv}CPh \longrightarrow HOH_2C-C{\equiv}C-C{\equiv}CPh$$

(30)

$$2RC{\equiv}CH + [O] \longrightarrow R-C{\equiv}C-C{\equiv}C-R \qquad (3)$$

$$Et_3SiC{\equiv}CH \longrightarrow Et_3Si-C{\equiv}C-C{\equiv}C-SiEt_3 \longrightarrow HC{\equiv}C-C{\equiv}CH \qquad (4)$$

$$H{\left(C{\equiv}C\right)}_n$$

(31)

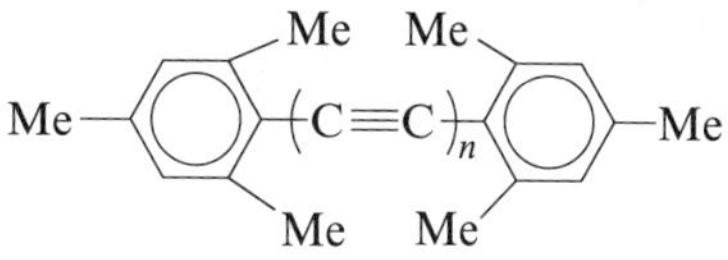

(32)

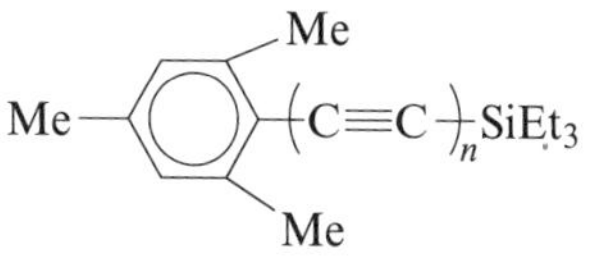

(33)

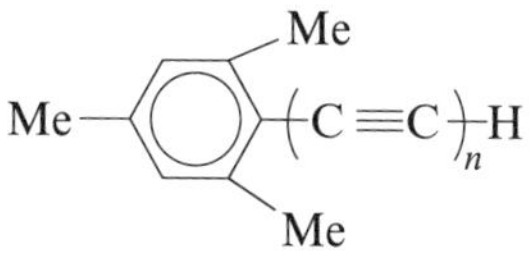

(34)

$$Bu^t{\left(C{\equiv}C\right)}_{12}Bu^t$$

(35)

Terminal acetylenes dimerize to conjugated enynes in the presence of catalytic amounts of palladium(II) acetate and the hindered phosphine tris(2,6-dimethoxyphenyl)phosphine. 1-Octyne, for instance, affords the enyne **36**[25]. Palladium-catalysed condensations of 1-alkynes $RC{\equiv}CH$ ($R = Ph$ or CH_2OH) with vinyl bromide, (E)-2-bromostyrene, iodobenzene and 2-bromopyridine to yield **37**, **38**, **39**, and **40**, respectively (Scheme 1), have been reported[26] (see also References 27 and 28). The palladium[tetrakis(triphenylphosphine)]-catalysed alkynylation of 1,1-dichloroethene with terminal alkynes **41** (R = pentyl, 3-chloropropyl, Ph or SiMe_3) results in the enynes **42**[29].

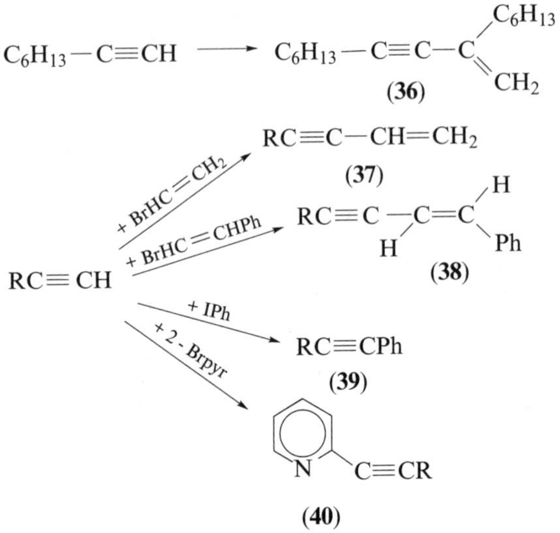

$$C_6H_{13}-C{\equiv}CH \longrightarrow C_6H_{13}-C{\equiv}C-C\underset{CH_2}{\overset{C_6H_{13}}{<}}$$

(36)

$$RC{\equiv}C-CH{=}CH_2$$

(37)

$$RC{\equiv}C-\underset{H}{\overset{}{C}}{=}C\underset{Ph}{\overset{H}{<}}$$

(38)

$$RC{\equiv}CH \quad \xrightarrow{+\ BrHC{\equiv}CH_2} \quad \xrightarrow{+\ BrHC{\equiv}CHPh}$$

$$\xrightarrow{+\ IPh} RC{\equiv}CPh$$

(39)

$$\xrightarrow{+\ 2\text{-}Br_{pyr}}$$

$$\underset{N}{\overset{}{\bigcirc}}\ C{\equiv}CR$$

(40)

SCHEME 1

$$RC{\equiv}CH + Cl_2C{=}CH_2 \longrightarrow RC{\equiv}C-CCl{=}CH_2$$

(41)   (42)

A general method for the preparation of 1,3-diynes is exemplified by the synthesis of compound **45**: condensation of the acetylene **43** with *cis*-1,2-dichloroethene in the presence of Pd(PPh₃)₄.CuI gives **44**, which is dehydrochlorinated by the action of tetra-butylammonium fluoride in THF[30]. Conjugated (*E*)-enynes **48** are obtained from terminal acetylenes **46** (R = Bu, Ph or Me₃Si) and the (*Z*)-bromo ester **47** under catalysis by (Ph₃P)₂PdCl₂. CuI and triethylamine[31]. The same catalyst system promotes the condensa-

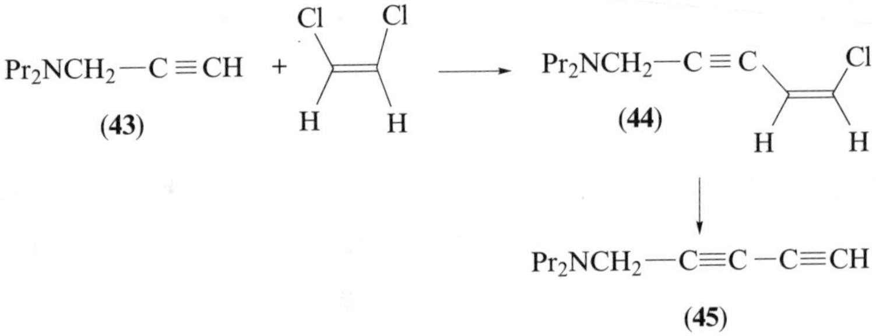

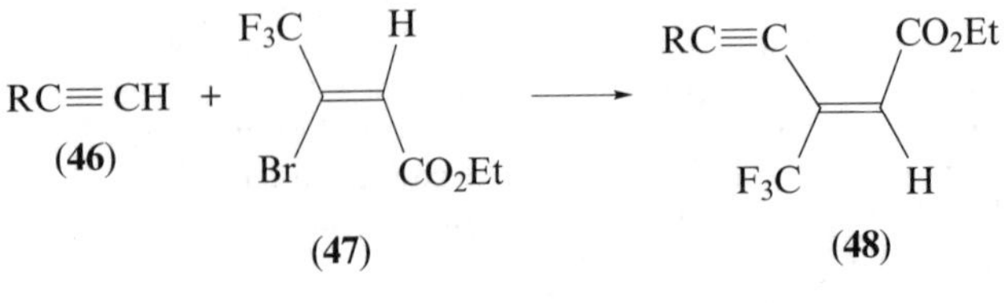

$$Pr_2NCH_2-C{\equiv}CH \ + \ \underset{H\quad H}{\overset{Cl\quad Cl}{>{=}<}} \longrightarrow Pr_2NCH_2-C{\equiv}C\underset{H\quad H}{\overset{Cl}{>{=}<}}$$

(43)   (44)

$$Pr_2NCH_2-C{\equiv}C-C{\equiv}CH$$

(45)

$$RC{\equiv}CH \ + \ \underset{Br\quad CO_2Et}{\overset{F_3C\quad H}{>{=}<}} \longrightarrow \underset{F_3C\quad H}{\overset{RC{\equiv}C\quad CO_2Et}{>{=}<}}$$

(46)   (47)   (48)

tion of terminal alkynes **49** (R = alkyl or Ph) with fluorinated vinyl iodides, e.g. **50**, to yield the (E)-enynes **51**[32–34]. Analogous palladium-catalysed reactions of alkynes with N-arylimidoyl iodides **52** (equation 5) have been reported[35].

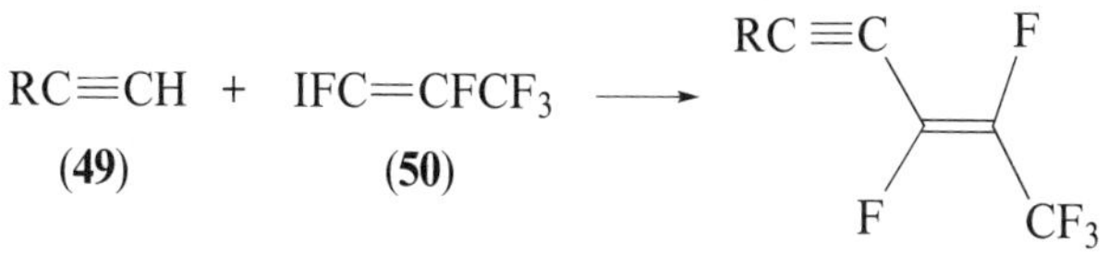

$$RC{\equiv}CH \; + \; IFC{=}CFCF_3 \;\longrightarrow$$

(49)   **(50)**

$$\begin{array}{c} RC{\equiv}C \qquad F \\ \diagdown \qquad \diagup \\ C{=}C \\ \diagup \qquad \diagdown \\ F \qquad \quad CF_3 \end{array}$$

(51)

$$RC{\equiv}CH \; + \; \begin{array}{c} I \\ \diagdown \\ C{=}NAr \\ \diagup \\ F_3C \end{array} \;\longrightarrow\; \begin{array}{c} RC{\equiv}C \\ \diagdown \\ C{=}NAr \\ \diagup \\ F_3C \end{array} \qquad (5)$$

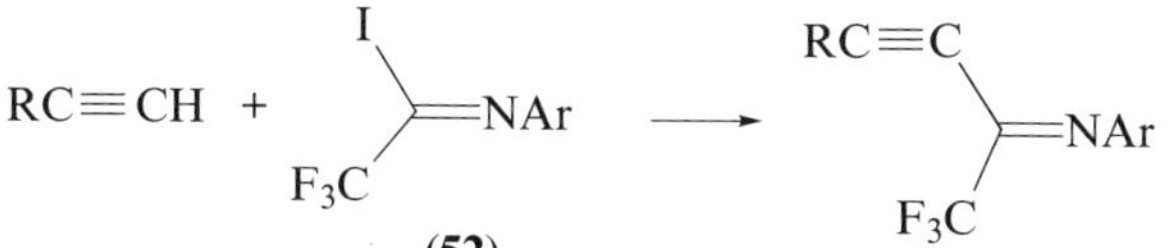

(52)

Aniline, acetanilide, benzyl alcohol or methyl benzoate containing a halogen atom in the ortho-position react with 1-alkynes in the presence of bis(triphenylphosphine)palladium(II) chloride and copper(I) iodide to give the corresponding o-alkynylbenzenes (e.g. equation 6). Some of the products can be cyclized: the aniline derivative **53**, for instance, is converted into 2-phenylindole in 99% yield on treatment with a catalytic amount of copper(I) iodide[36].

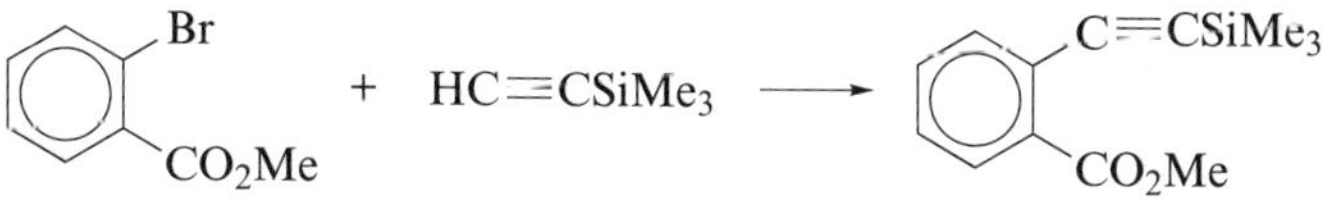

$$\begin{array}{c} Br \\ \diagup \\ \diagdown \\ CO_2Me \end{array} \; + \; HC{\equiv}CSiMe_3 \;\longrightarrow\; \begin{array}{c} C{\equiv}CSiMe_3 \\ \diagup \\ \diagdown \\ CO_2Me \end{array} \qquad (6)$$

$$\begin{array}{c} C{\equiv}CPh \\ \diagup \\ \diagdown \\ NH_2 \end{array} \;\longrightarrow\; \begin{array}{c} \diagup \diagdown \\ \diagdown N \!-\! Ph \\ H \end{array}$$

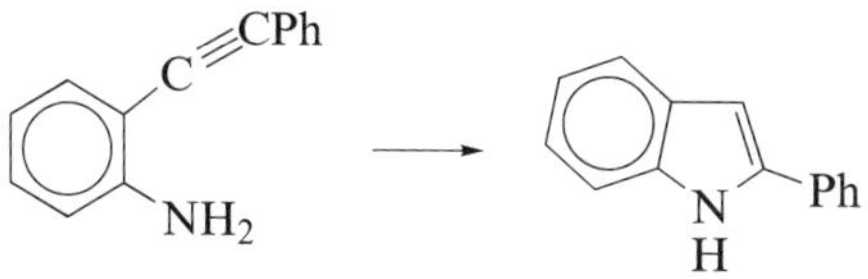

(53)

Derivatives of tetrafluoroiodobenzene react with terminal alkynes in the presence of a palladium–copper(I) iodide catalyst by displacement of the iodine atom (e.g. equation 7)[37]. m-Methoxyphenol condenses with phenylacetylene in the presence of the sulphonyl fluoride **54** and bis(triphenylphosphine)palladium(II) chloride to give m-methoxy-diphenylacetylene (equation 8)[38].

The 'crowded' diarylacetylene **56** has been obtained from the trifluoromethanesulphonate **55** and 2,6-dimethoxyphenylacetylene[39]. The palladium-catalysed reaction of the

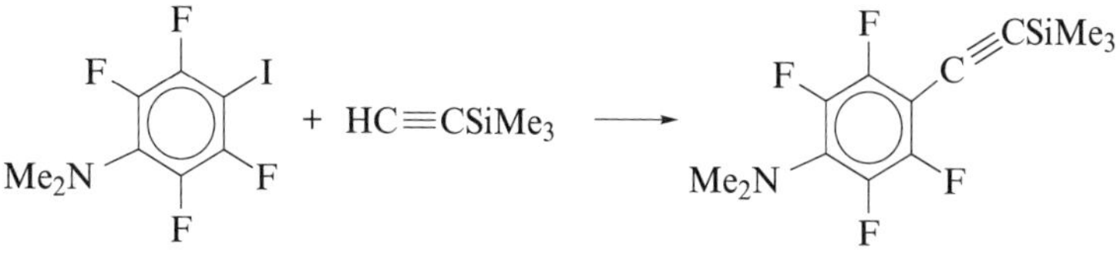

$$\begin{array}{c} F \\ F \diagup \diagdown I \\ \diagdown \diagup \\ Me_2N \diagup \diagdown F \\ F \end{array} \; + \; HC{\equiv}CSiMe_3 \;\longrightarrow\; \begin{array}{c} F \quad C{\equiv}CSiMe_3 \\ F \diagup \diagdown C \\ \diagdown \diagup \\ Me_2N \diagup \diagdown F \\ F \end{array} \qquad (7)$$

$$(8)$$

(55)   (56)

(57)   (58)   (59)

bromomethano[10]annulene **57** with bis(trimethylsilyl)acetylene gave **58**, which was desilylated to the alkyne **59**[40]. 4-Bromo- and 5-bromothiazoles and 4-bromo- and 5-bromooxazoles condense with terminal acetylenes under palladium catalysis to give good yields of ethynyl derivatives[41]. A novel synthesis of symmetrical or unsymmetrical diarylbutadiynes **62** is the reaction of aromatic or heteroaromatic bromo or iodo compounds **60** with the propargyl alcohols **61** in the presence of (PPh$_3$)$_2$PdCl$_2$. CuI and benzyltrimethylammonium chloride in benzene–aqueous sodium hydroxide under phase-transfer conditions[42].

$$R^1X + HOMe_2CC\equiv C-C\equiv CR^2 \longrightarrow R^1C\equiv C-C\equiv CR^2$$

(60)   (61)   (62)

$$R^1, R^2 = \text{aryl, 2-pyridyl, 2-quinolyl or 2-benzothiazolyl}$$

The action of terminal alkynes **63** (R = Bu, CH$_2$OH, Ph or SiMe$_3$) on 2,5-dibromopyridine in triethylamine containing bis(triphenylphosphine)palladium dichloride and copper(I) iodide results in the 2-alkynyl compounds **64** regiospecifically[43].

(64)

Three alkynyl groups are introduced by the palladium-catalysed reaction of 3,4,5-triiodobenzoic acid with phenylacetylene (equation 9); 3,4,5-triiodonitrobenzene reacts analogously[44]. The hexaethynylation of hexabromobenzene with (trimethylsilyl)acetylene, followed by desilylation, to give **65** has been reported[45] and the syntheses of the hexakis (butadiynyl)benzenes) **66** (R = t-Bu or SiPr$_3$) have been achieved[46].

The action of alkynyl Grignard reagents on the pyridinium salt **67** affords solely 1,2-dihydropyridine derivative **68**, whereas alkyl Grignard reagents yield mixtures of 1,2- and 1,4-dihydropyridines (equation 10). The contrasting behaviour of the two nucleophiles was explained by means of the Principle of Hard and Soft Acids and Bases[47]. Application of the reaction to the action of the acetylenic Grignard compound **69** on 1-methoxycarbonyl-2-methylpyridinium chloride gave the dihydropyridine derivative **70**, which was transformed into (±) -monomorine **71**, a train pheromone of the pharao ant[48].

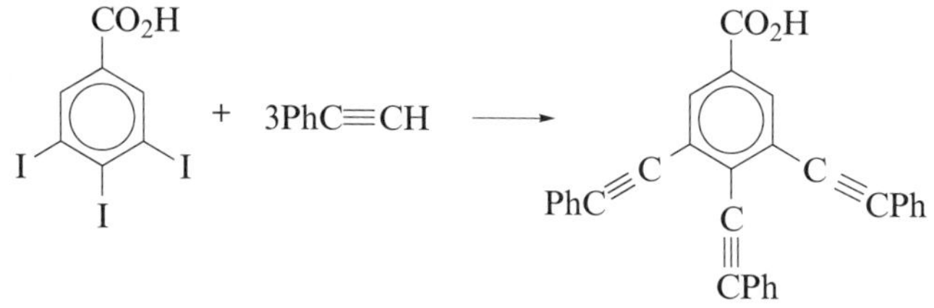

$$CO_2H + 3PhC{\equiv}CH \longrightarrow \qquad\qquad (9)$$

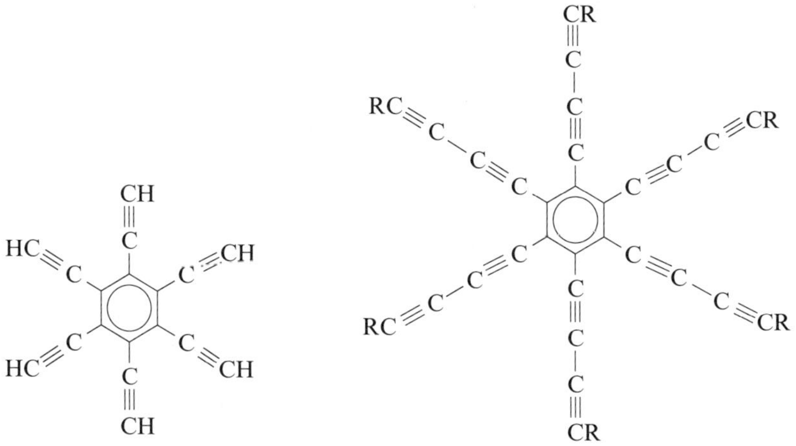

(65)   (66)

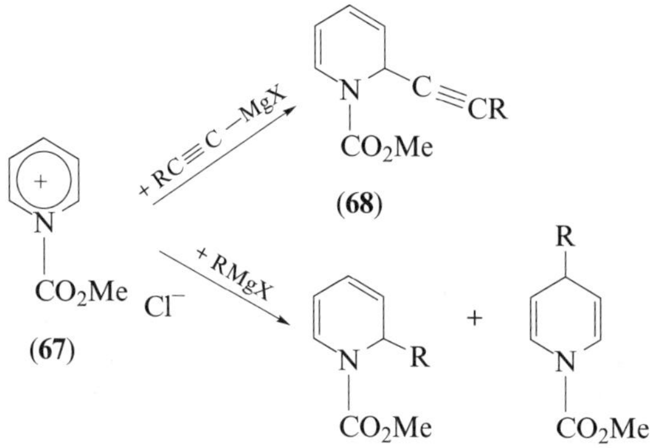

(10)

(67)   (68)

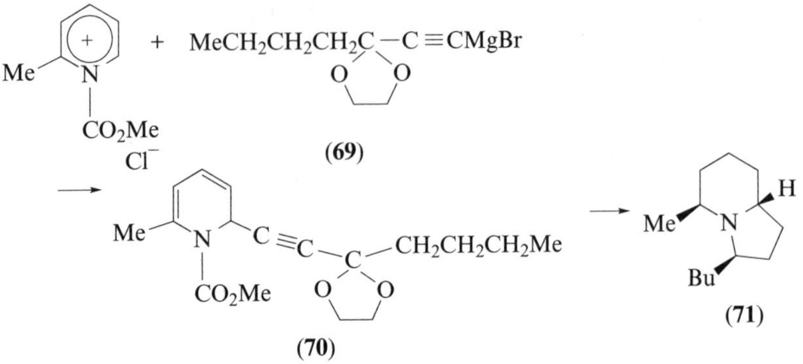

(69)

(70)

(71)

N-Benzoyloxypyridinium chloride (72), prepared from pyridine oxide and benzoyl chloride, reacts with silver phenylacetylide selectively at position 2 to afford 2-(phenylethynyl)pyridine (73)[49]. The bromine atom of the tetrahydro-1,4-oxazin-2-one 74 is replaced by an alkynyl group on treatment with stannanes 75 (R = hexyl or Ph); the products 76 are transformed into (S)-amino acids 77 by catalytic hydrogenation[50].

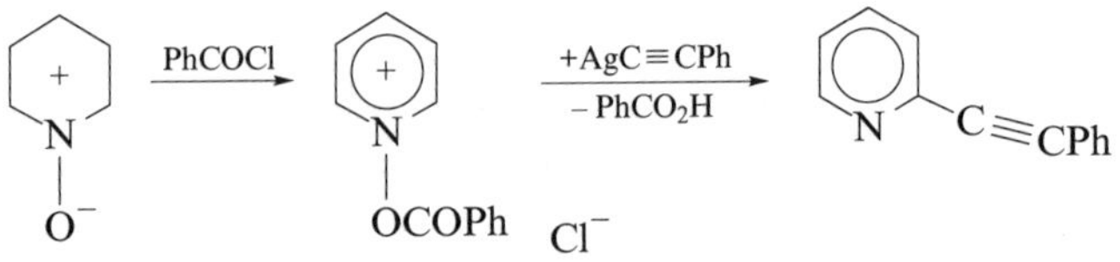

(72)   (73)

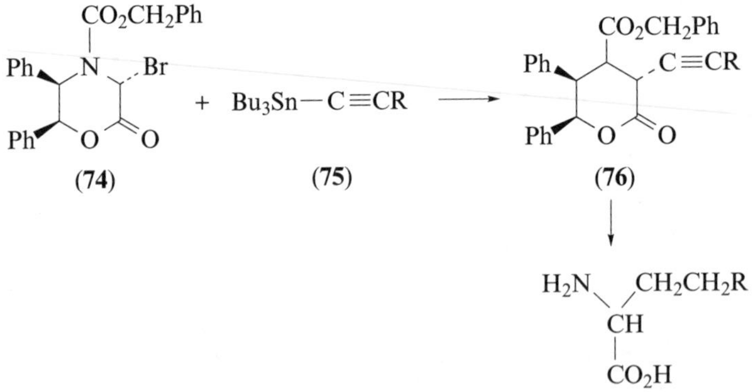

(74)   (75)   (76)

(77)

An *ortho* amide 78 of propiolic acid has been prepared; it readily condenses with acetylacetone (equation 11)[51]. Novel N,N-bis(trimethylsilyl)ynamines 80 are obtained by electrophilic amination of lithium acetylides with the hydroxylamine derivative 79[52a,b].

The action of lead tetraacetate on di(alk-1-ynyl)mercury compounds or alk-1-ynyltrimethylstannanes gives highly unstable alk-1-ynyllead triacetates 81 (R = H hexyl, Me_3Si, Ph or Ar), which can be employed for the C-alkynylation of β-dicarbonyl compounds, e.g. equation 12[53-55].

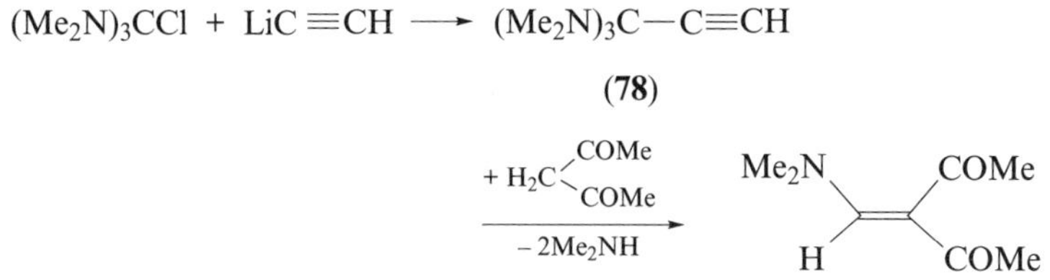

$$(Me_2N)_3CCl + LiC\equiv CH \longrightarrow (Me_2N)_3C-C\equiv CH$$

$$\textbf{(78)}$$

$$(11)$$

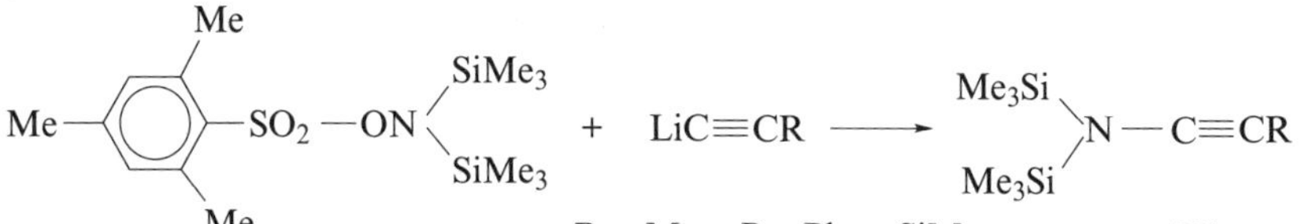

$$R = Me, \textit{t}-Bu, Ph \text{ or } SiMe_3 \qquad \textbf{(80)}$$

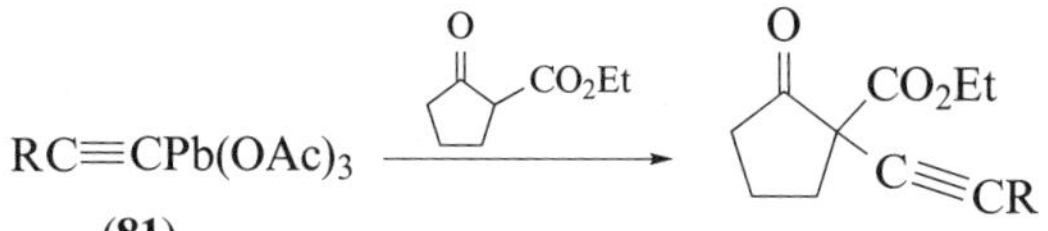

$$\textbf{(79)}$$

$$(12)$$

$$\textbf{(81)}$$

$$Me_3SiC\equiv CMgCl$$

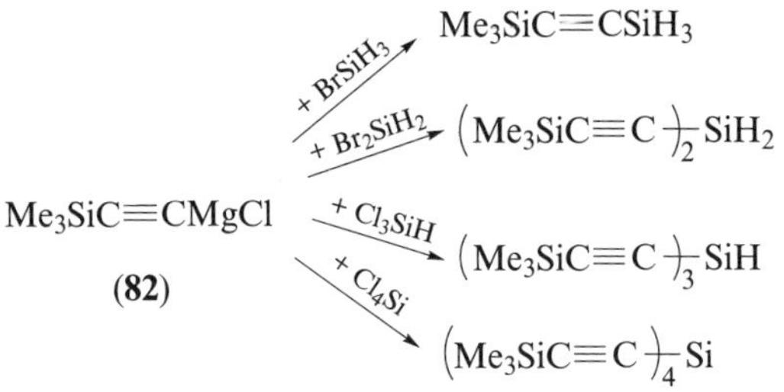

$$\textbf{(82)}$$

SCHEME 2

A series of acetylenic silanes has been prepared from the Grignard reagent **82** and various silicon halides (Scheme 2)[56]. Aldehyde acetals **83** (R^1 = $PhCH_2CH_2$ or octyl, R^2 = Me or $PhCH_2$) react with 1-(trimethylsilyl)acetylenes **84** (R^3 = H, Me or pentyl) in the presence of catalytic amounts of tin(IV) chloride and zinc chloride to give propargyl ethers **85**[57,58]. Methods for preparing selenoalkynes $R^1SeC\equiv CR^2$ have been reviewed[59]. New procedures are treatment of Grignard or lithium compounds with elemental selenium, followed by 1-bromoalkynes (e.g. equation 13) and the action of dialkyl or diaryl diselenides on lithium acetylides (equation 14)[59]. Symmetrical dialk-1-yn-1-yl tellurides **86** (R = Bu, $SiMe_3$, Ph or

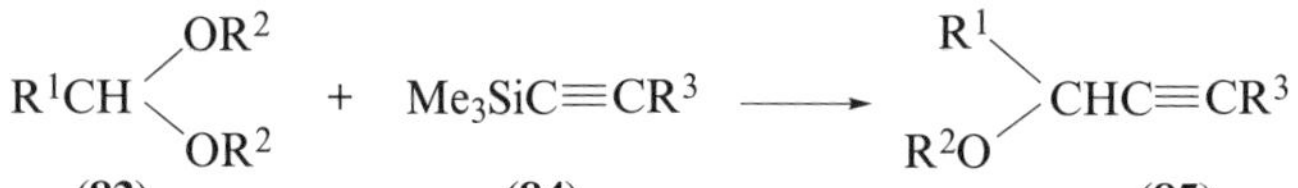

$$\textbf{(83)} \qquad\qquad \textbf{(84)} \qquad\qquad\qquad \textbf{(85)}$$

Ar) are obtained from tellurium tetrachloride by the action of LDA, followed by a terminal alkyne[60].

$$BuLi + Se \longrightarrow \left[BuSeLi\right] \xrightarrow{Br-C\equiv C-Ph} BuSeC\equiv CPh \qquad (13)$$

$$R^1_2Se_2 + LiC\equiv CR^2 \longrightarrow R^1SeC\equiv CR^2 \qquad (14)$$

$$TeCl_4 \xrightarrow[\text{(ii) HC}\equiv\text{CR}]{\text{(i) LDA}} RC\equiv C-Te-C\equiv CR$$

(86)

A variety of nitrogen-containing acetylenes has been described. The aminals **87** (R = Ph or 2-thienyl) react with phenylacetylene in the presence of copper(I) salts to yield 3-piperidinoalkynes **88**[61]. The action of propargyl bromide on Schiff's bases has been investigated. Whereas the reaction with N-(benzylidene)benzylamine in aqueous sodium hydroxide/DMSO/dichloromethane containing the phase-transfer catalyst tetrabutylammonium bromide resulted in a low yield of the primary amine **89**, treatment of the lithium derivative **90** of N-(diphenylmethylene)benzylamine with propargyl bromide gave 70% of the imine **91**[62]. Immonium salts **92** (R = Ph, 2-furyl or 2-thienyl) react with silver or copper(I) phenylacetylide to yield the tertiary amines **93**[63]. The action of terminal acetylenes **94** (R³ = alkyl or aryl) on imidoyl chlorides **95** (R¹ = alkyl or aryl, R² = aryl) in triethylamine in the presence of bis(triphenylphosphine)palladium(II) chloride–copper(I) iodide results in the acetylenic imines **96**[64]. N-Propargylic amides **99**, together with 2-oxazolines **100** and 4H-1,3-oxazines **101**, are produced from the imines **97** (R¹ = Ph, Ar or t-Bu) and alkynes **98** (R² = H, Ph, Me₃Si, COMe or CO₂Et)[65].

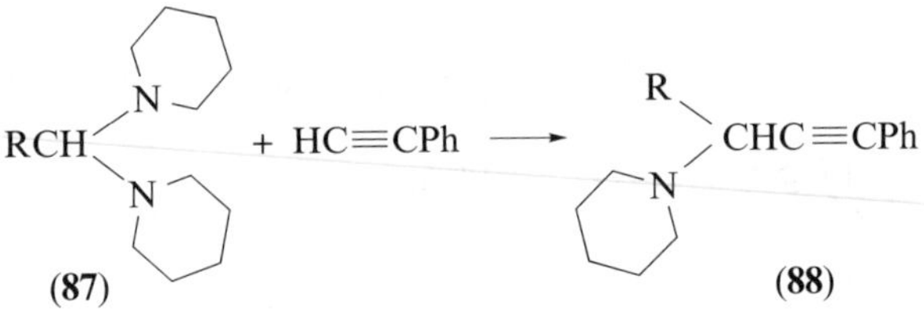

(87)   + HC≡CPh ⟶   CHC≡CPh

(88)

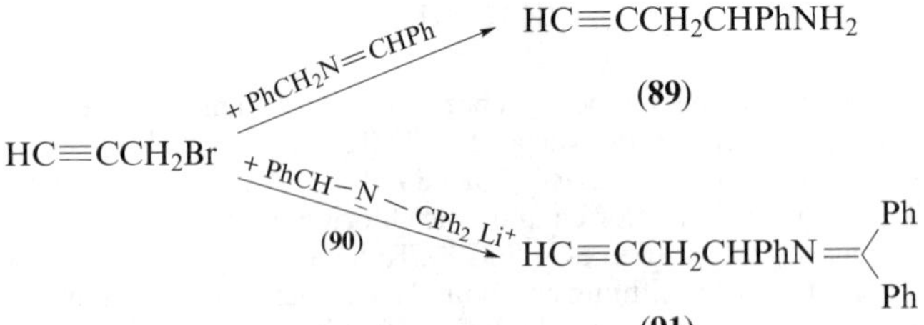

$$HC\equiv CCH_2Br$$

$$\xrightarrow{+ PhCH_2N=CHPh} HC\equiv CCH_2CHPhNH_2$$

(89)

$$\xrightarrow{+ PhCH-N=CPh_2\ Li^+} HC\equiv CCH_2CHPhN=\begin{cases}Ph\\Ph\end{cases}$$

(90)

(91)

$$Me_2\overset{+}{N}=CHR\ \ X^- + AgC\equiv CPh \longrightarrow Me_2NCHRC\equiv CPh$$

(92)   **(93)**

$$R^3C{\equiv}CH + \underset{R^2}{\overset{Cl}{>}}{=}NR^1 \longrightarrow R^3C{\equiv}C-CR^2{=}NR^1$$

$$(94) \qquad (95) \qquad\qquad (96)$$

$$R^1\underset{\underset{O}{\|}}{\overset{}{C}}-N{=}\underset{CF_3}{\overset{CF_3}{<}} + HC{\equiv}CR^2$$

$$(97) \qquad\qquad (98)$$

$$\longrightarrow R^1\underset{\underset{O}{\|}}{\overset{}{C}}-N\overset{H}{\underset{F_3C}{-}}C\underset{CF_3}{-}C{\equiv}CR^2 + \underset{R^1}{\overset{CF_3}{N\underset{O}{\overset{}{<}}}}\underset{CHR^2}{\overset{CF_3}{<}} + \underset{R^1}{\overset{F_3C\ \ CF_3}{N}}\underset{O\qquad R^2}{}$$

$$(99) \qquad\qquad (100) \qquad\qquad (101)$$

The enantioselective addition of dialkylzinc reagents to alkynyl aldehydes catalysed by (S)-(+)-(1-methylpyrrolidin-2-yl)diphenylmethanol has been described. Thus the laevorotatory alcohol **102** was obtained in 78% enantiomeric excess (equation 15)[66]. The action of diethylaluminium chloride on alkynyllithium compounds yields ether-free diethylalkynylalanes **103**, which undergo conjugate addition to nitroalkenes (equation 16)[67]. Conjugate addition also occurs in the reactions of (1-alkynyl)diisopropoxyboranes **104** (R = Bu, t-Bu or Ph) with α,β-unsaturated ketones **105** (R^2 = H, Me or Ph; R^3 = Me, hexyl or Ph) (equation 17)[68].

$$Me_3SiC{\equiv}CCHO + Et_2Zn \xrightarrow{\underset{Me}{\overset{}{N}}\text{-}CPh_2OH} Me_3SiC{\equiv}C-\underset{OH}{\overset{H}{\underset{|}{C}}}\text{-}Et \qquad (15)$$

$$(102)$$

$$RC{\equiv}CLi + Et_2AlCl \longrightarrow RC{\equiv}CAlEt_2$$

$$\xrightarrow{\underset{Me}{\overset{Me}{>}}C{=}C\underset{Me}{\overset{NO_2}{<}}} RC{\equiv}C-\underset{Me}{\overset{Me}{\underset{|}{C}}}\text{-}\underset{Me}{\overset{NO_2}{\underset{|}{C}}}H \qquad (16)$$

$$R = \text{Bu or Ph}$$

$$R^1C{\equiv}C-B\underset{OPr^i}{\overset{OPr^i}{<}} + \underset{R^2}{\overset{H}{>}}C{=}C\underset{H}{\overset{COR^3}{<}} \longrightarrow R^1C{\equiv}CCHR^2CH_2COR^3 \qquad (17)$$

$$(104) \qquad\qquad (105)$$

$$R^1CO_2R^2 + LiC{\equiv}CR^3 \longrightarrow R^1COC{\equiv}CR^3 \qquad (18)$$

$$(106) \qquad (107) \qquad\qquad (108)$$

Esters **106** (R^1 = Me, Et or Pr; R^2 = Et, Pr, t-Bu or PhCH$_2$) of aliphatic carboxylic acids react with lithium acetylides **107** (R^3 = H, C$_5$H$_{11}$ or Ph) in the presence of boron trifluoride etherate in THF to give acetylenic ketones **108** (equation 18)[69]. Palladium-[tetrakis(triphenylphosphine)]–copper(I) iodide catalyses the oxidative addition–decarboxylation of propargyl methyl carbonates, e.g. **109**, with terminal alkynes to yield 1,2-dien-4-ynes (allenylacetylenes) **110**[70]. The regiochemistry of the palladium-catalyzed addition of phenylacetylene to the allenic ester **111** depends on the nature of the catalyst used: palladium(III) acetate–triphenylphosphine yields a 81:19 mixture of adducts **112** and **113**, while in the presence of tetrakis(carbomethoxy)palladacyclopentadiene–tris(2,4,6-trimethoxyphenyl)phosphine the ratio is reversed to 9 : 91[71].

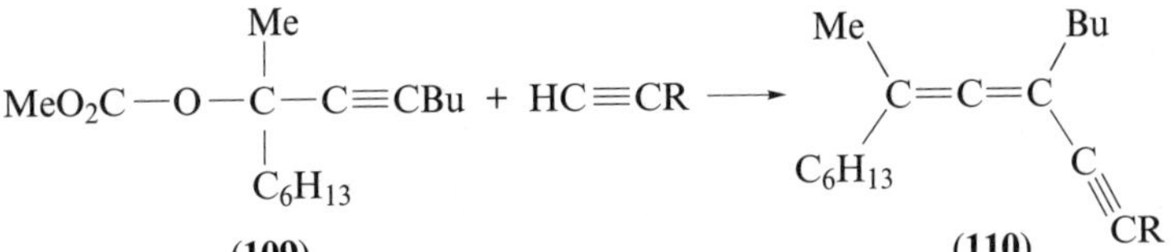

$$\text{MeO}_2\text{C}-\text{O}-\underset{\underset{\text{C}_6\text{H}_{13}}{|}}{\overset{\overset{\text{Me}}{|}}{\text{C}}}-\text{C}\equiv\text{CBu} + \text{HC}\equiv\text{CR} \longrightarrow$$

(109)

(110)

$$\text{PhC}\equiv\text{CH} +$$

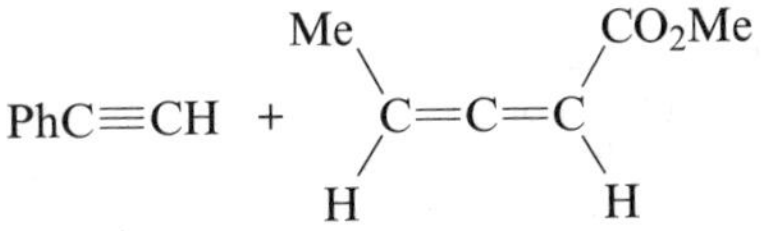

(111)

$$\longrightarrow$$

$$+$$

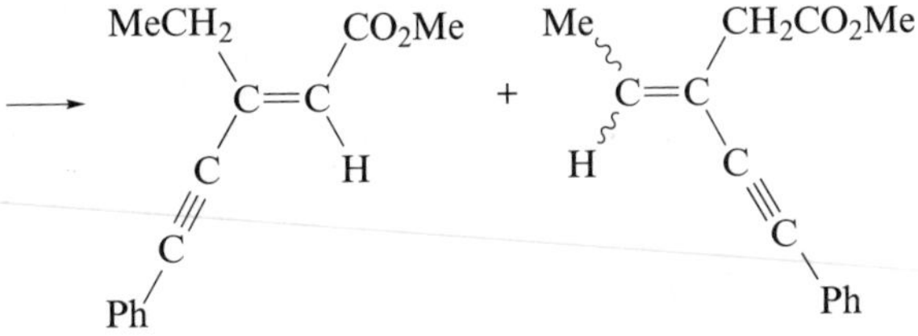

(112)   (113)

Joint catalysis by tin(IV) chloride and zinc chloride induces coupling of the acetal **114** with the alkyne **115** to give the propargyl ether **116**[72]. Allyl halides condense with terminal alkynes in DMF in the presence of copper(I) iodide, potassium carbonate and tetrabutylammonium chloride under phase-transfer conditions to afford eneynes, e.g. equation 19[73]. Palladium-catalysed coupling reactions of 1-bromoallenes with terminal acetylenes have been reported for the first time: thus acetylene and the allenes **117** (R^1, R^2 = Me or Et) in triethylamine in the presence of tetrakis(triphenylphosphine)palladium and

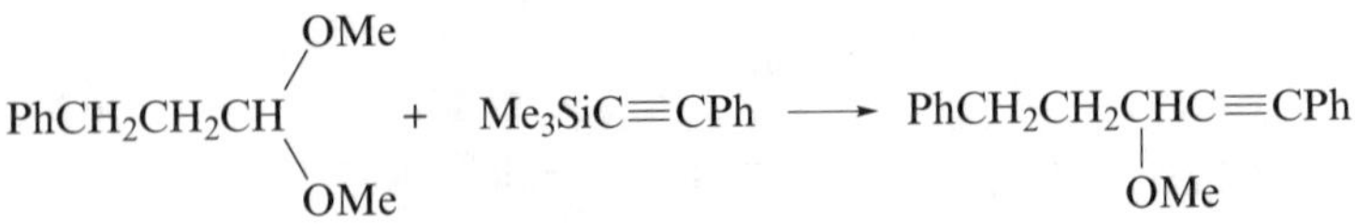

$$\text{PhCH}_2\text{CH}_2\text{CH}\overset{\overset{\text{OMe}}{/}}{\underset{\underset{\text{OMe}}{\backslash}}{}} + \text{Me}_3\text{SiC}\equiv\text{CPh} \longrightarrow \text{PhCH}_2\text{CH}_2\underset{\underset{\text{OMe}}{|}}{\text{CHC}}\equiv\text{CPh}$$

(114)   (115)   (116)

copper(I) iodide afforded 1,2,6,7-octatetraen-4-ynes **118** and a similar reaction of **117** with diacetylene gave 1,2,8,9-decatetraen-4,6-diynes **119**[74].

$$H_2C{=}CMeCH_2Cl + HC{\equiv}CCMe_2OH \longrightarrow H_2C{=}CMeCH_2C{\equiv}CCMe_2OH \quad (19)$$

(117)

(118)   **(119)**

A synthesis of dienediynes from 2-(hydroxymethylene)cyclopentanone involves the dienol ditriflate [bis(trifluoromethanesulphonate)] **120**, which reacts with an alkyne in the presence of a palladium catalyst to give **121**, accompanied by only small amounts of the regioisomer **122**. The major product on treatment with the same or different alkyne yields the dienediyne **123**[175.] 2-(Alkyn-1-yl)-1-methylpyrrolidines **125** (R = Me, pentyl, Me$_3$Si or Ph) have been obtained from the thiolactam **124** by sequential alkylation, treatment with a lithium acetylide and reduction with LAH[76]. The phenylthio group in α-(phenylthio)-lactams is displaced by an alkynyl group by the action of an alkyn-1-ylzinc chloride or of di(alkyn-1-yl)zinc compounds. Thus the β-lactam **126** gave **127** and the azaprostacyclin II **128** gave **129**[77].

(123)   **(121)**   **(122)**

1-Alkynes add to aldehydes in the presence of tin(II)trifluoromethanesulphonate and 1,8-bis(dimethylamino)naphthalene to yield secondary acetylenic alcohols (e.g. equation

(124) → (125)

Structure (124): 1-methyl-2-thioxopyrrolidine (N–Me, C=S)

Structure (125): N-Me pyrrolidine with C≡CR group

(126) → MeC≡CZnCl → (127)

Structure (126): azetidinone with Me–OSi–Bu^t (Me, Me), O=, SPh, NH

Structure (127): azetidinone with Me–OSi–Bu^t (Me, Me), O=, C≡CMe, NH

(128) → Me$_3$SiC≡CZnCl → (129)

Structure (128): bicyclic lactam, Ph–Bu^t–SiO–CH$_2$, Ph–OSi–Bu^t, Ph, SPh, NH, O

Structure (129): bicyclic lactam, Ph–Bu^t–SiOCH$_2$, Ph, Ph–OSi–Bu^t, Ph, C≡CSiMe$_3$, NH, O

$$\text{cyclohexyl-CHO} + \text{PhC}\equiv\text{CH} \longrightarrow \text{cyclohexyl-C(H)(OH)-C}\equiv\text{CPh} \qquad (20)$$

20)[78]. The stereoselectivity of the reaction of 1-alkynylzinc bromides 130 (R^1 = heptyl or Ph) with various chiral α-benzyloxy aldehydes 131 (R^2 = Me, i-Pr or PhCH$_2$OCH$_2$) to yield the alcohols 132 depends on the nature of the substituents R^1 and R^2 [79]. Regiospecific propargylation with acylsilanes is exemplified by the reaction of the organozinc bromide 133 (from 2-octynyl bromide and zinc dust in THF) with benzoyltrimethylsilane, followed by desilylation, to yield only the alcohol 134[80].

$$R^1C\equiv CZnBr + \underset{(131)}{R^2\text{-CH(OCH}_2\text{Ph)-CHO}} \longrightarrow \underset{(132)}{R^2\text{-CH(OCH}_2\text{Ph)-CH(OH)-C}\equiv CR^1}$$

(130)   (131)   (132)

$$\text{Me(CH}_2)_4\text{C}\equiv\text{CCH}_2\text{ZnBr} + \text{PhCOSiMe}_3 \longrightarrow \text{Me(CH}_2)_4\text{C}\equiv\text{CCH}_2\text{CHPhOH}$$

(133)   (134)

1,2-Dien-4-ynes **136** (R^1–R^4 = alkyl) are produced from propargylic carbonates **135** and terminal alkynes in the presence of a palladium–phosphine complex and copper(I) iodide[81]. The linear co-dimerization of terminal acetylenes and 1,3-dienes is catalyzed by ruthenium(cyclooctadiene)(cyclooctatriene)(trialkylphosphine) (alkyl = Et, Bu or octyl); thus 1-hexyne and methyl penta-2,4-dienoate give a mixture of the eneynes **137** and **138**[82]. Coupling of octa-1,7-diyne (**139**) with the acetylenic bromo acid **140** in aqueous THF–methanol containing butylamine, hydroxylamine hydrochloride and copper(I) chloride gave a mixture of the triynyl acids **141** and **142**[83].

$$MeO_2COCR^1R^2C\equiv CR^3 + HC\equiv CR^4 \longrightarrow R^1R^2C=C=CR^3C\equiv CR^4$$

(135)   **(136)**

$$BuC\equiv CH + H_2C=CHCH=CHCO_2Me \longrightarrow$$

(137)   **(138)**

$$HC\equiv C(CH_2)_4C\equiv CH + BrC\equiv C(CH_2)_8CO_2H \longrightarrow$$

(139)   **(140)**

$$HC\equiv C(CH_2)_4C\equiv CC\equiv C(CH_2)_8CO_2H + MeC\equiv C(CH_2)_3C\equiv CC\equiv C(CH_2)_8CO_2H$$

(141)   **(142)**

(21)

Highly unsaturated macrocyclic silahydrocarbons have been produced from double acetylenic Grignard reagents and fluorosilanes as exemplified by equation 21[84].

Alken-1-en-4-yn-3-ones are formed by the palladium-catalyzed reaction of vinyl trifluoromethanesulphonates with 1-alkynes and carbon monoxide, e.g. equation 22[85].

$$(22)$$

Syntheses of alkynyl-substituted *p*- and *o*-benzoquinones have been described. The action of lithium acetylides on 4,5-dimethoxy-*o*-benzoquinone (**143**) gives the adducts **144** (R = H, Bu, Ph or CO$_2$Et), which are converted into the *p*-quinones **145** by treatment with dilute sulphuric acid. An analogous sequence, starting with the *p*-benzoquinone **146**, yields alkynyl-*o*-quinones **147**[86].

Palladium-catalysed coupling of alkenyl iodides with alkynylstannanes is stereospecific to afford high yields of conjugated enynes, as shown by the three examples in Scheme 3[87].

Introduction of the chloroethynyl group is accomplished by the reaction of dichloroacetylene with the lithium enolates of ketones or of esters of α, α-dialkyl carboxylic acids (Scheme 4)[88].

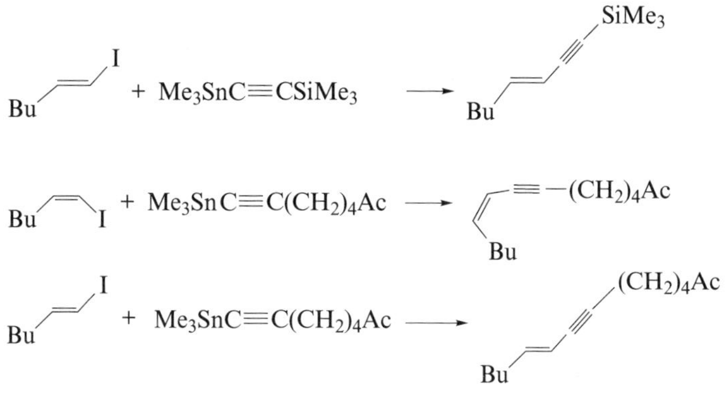

SCHEME 3

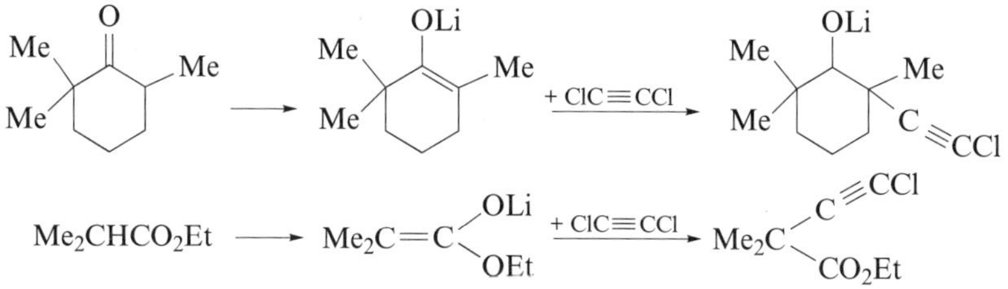

SCHEME 4

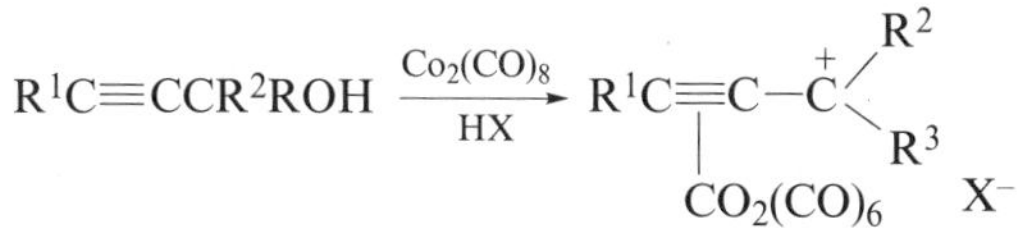

(148)

The reaction of propargyl alcohols with dicobalt octacarbonyl to give the complex salts **148** (X = BF₄ or PF₆) and synthetic uses of the latter have been reviewed[89]. The salts react with electron-rich aromatic compounds ArH, such as anisole, phenol or N,N-dimethylaniline, to yield substitution products **149** after oxidative demetallation with an iron(III) or cerium(IV) salt[90]; with β-diketones or β-keto esters the corresponding propargyl-substituted compounds **150** are obtained[91]. Acetone reacts in an analogous fashion to give **151**[92]. The action of the cobalt complexes **148** on allylsilanes **152** leads to enynes **153**[93]. Indole reacts with the complex **148** (R¹ = H; R² = R³ = Me) in the presence of boron trifluoride etherate to give **154**, which was converted into **155** by the action of iron(III) nitrate[94].

$$148 + ArH \longrightarrow \longrightarrow R^1C\equiv CCR^2R^3Ar$$

$$(149)$$

$$148 + H_2C \underset{COR^5}{\overset{COR^4}{\diagdown}} \longrightarrow \longrightarrow R^1C\equiv CCR^2R^3CH \underset{COR^5}{\overset{COR^4}{\diagdown}}$$

$$(150)$$

$$148 + MeCOMe \longrightarrow \longrightarrow R^1C\equiv CCR^2R^3CH_2COMe$$

$$(151)$$

$$148 + R^4R^5C=CHCHR^6SiMe_3 \longrightarrow \longrightarrow R^1C\equiv CCR^2R^3CR^4R^5CH = C \underset{H}{\overset{R^6}{\diagdown}}$$

$$(152) \qquad\qquad\qquad\qquad\qquad (153)$$

(154)   (155)

C. From Aldehydes and Ketones

Aliphatic and aromatic aldehydes react with carbon tetrabromide and triphenylphosphine to yield 1,1-dibromoalkenes **156**, which are converted into alkynes **157** by the action of butyllithium or lithium amalgam[95]. A convenient modification of the second step is the use of magnesium metal in boiling THF[96]. 1-Chloro-1-alkynes **159** (R = Bu, hexyl, heptyl etc.) are produced from aldehydes and carbon tetrachloride/triphenylphosphine/magnesium, followed by dehydrochlorination of the products **158** with potassium hydroxide in the presence of the phase-transfer agent Aliquat 336[97].

$$RCHO \longrightarrow RCH=CBr_2 \longrightarrow RC\equiv CH$$

$$(156) \qquad\qquad (157)$$

$$RCHO \longrightarrow RCH=CCl_2 \longrightarrow RC\equiv CCl$$

$$(158) \qquad\qquad (159)$$

α-Methylene ketones **160** (R^1 = Me or Ph; R^2 = Ph, COPh or CO_2Et) are readily dehydrated to the acetylenes **161** by a reagent prepared *in situ* from triphenylphosphine oxide and trifluoromethanesulphonic anhydride[98]. The preparation of terminal acetylenes from methyl ketones is accomplished by successive treatment with LDA or lithium 2,2,6,6-tetramethylpiperidide, diethyl chlorophosphate and again the lithium compound (equation 23)[99]. The application of the method to β-ionone (**162**) has been described in detail[100].

Palladium-catalysed reactions of iodocubane with terminal alkynes yielded the corresponding alkynylcyclooctatetraenes rather than the desired alkynylcubanes. An alkynylcubane, compound **164**, was obtained from methyl 4-iodocubyl ketone (**163**) by the procedure just described[101].

$$R^1COCH_2R^2 \xrightarrow{-H_2O} R^1C \equiv CR^2$$
$$\text{(160)} \qquad\qquad \text{(161)}$$

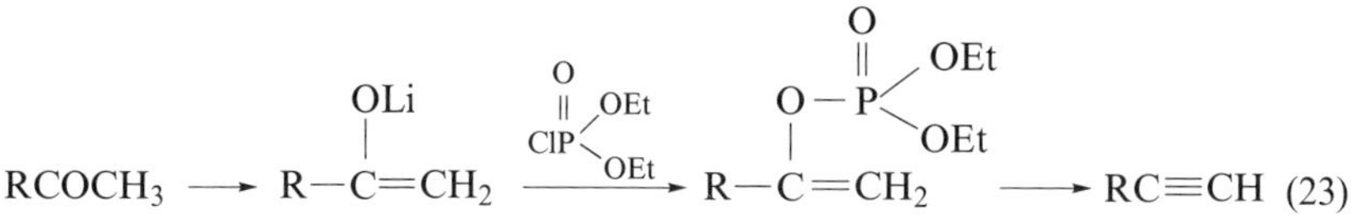

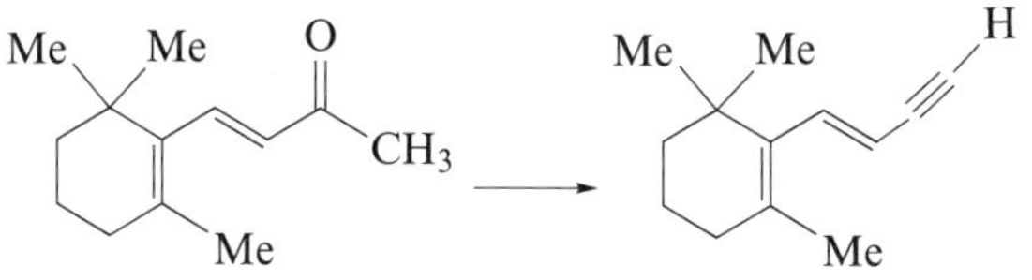

$$RCOCH_3 \longrightarrow R-\underset{\underset{\text{OLi}}{|}}{C}=CH_2 \xrightarrow[\text{OEt}]{\overset{O}{\underset{||}{ClP}}\diagdown\text{OEt}} R-\underset{\underset{O-P\diagdown\text{OEt}}{\overset{O}{|}}}{C}=CH_2 \longrightarrow RC\equiv CH \quad (23)$$

(162)

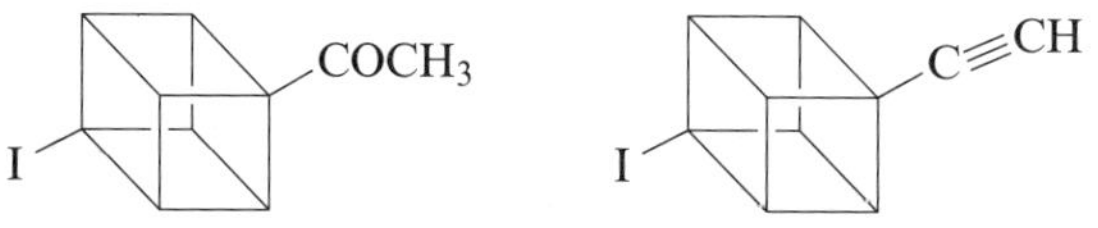

(163) **(164)**

Aromatic ketones **165** (R = Me or *i*-Pr) react with phosphorus pentachloride to yield trichloro compounds **166**, which are converted into mixtures of (*Z*)- and (*E*)-dichloroalkenes **167** by heating or by the action of 1,5- diazabicyclo[4.3.0]non-5-ene. The dichloroalkenes afford acetylenes **168** in 77–96% yields on treatment with magnesium[102].

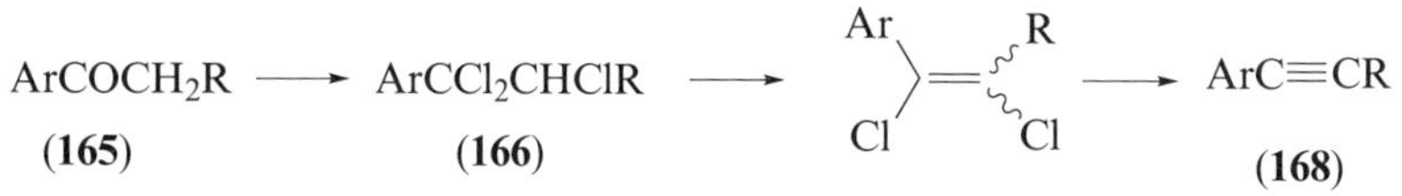

$$ArCOCH_2R \longrightarrow ArCCl_2CHClR \longrightarrow \underset{\text{(167)}}{\underset{Cl}{\overset{Ar}{\diagup}}=\underset{Cl}{\overset{R}{\diagdown}}} \longrightarrow ArC\equiv CR$$
$$\text{(165)} \qquad\qquad \text{(166)} \qquad\qquad\qquad\qquad \text{(168)}$$

D. From Heterocyclic Compounds

3,4-Disubstituted 5-isoxazolones **169** (R^1, R^2 = alkyl or aryl) are converted into acetylenes **170** by the combined action of sodium nitrite and iron(II) sulphate in aqueous scetic acid; a plausible mechanism is shown in equation 24[103]. Aminoacetylene (**172**) has been produced by flash-vacuum pyrolysis of the isoxazolones **171** (R = Me or Ph); it was observed by low-temperature IR spectroscopy and by its collision activation mass spectrum (CAMS)[104]. Aminoacetylene is also formed by pyrolysis of the derivative **173** of Meldrum's acid; the product, an equilibrium mixture of the ketenes **174** and **175**, decomposes above 580°C to aminoacetylene[104].

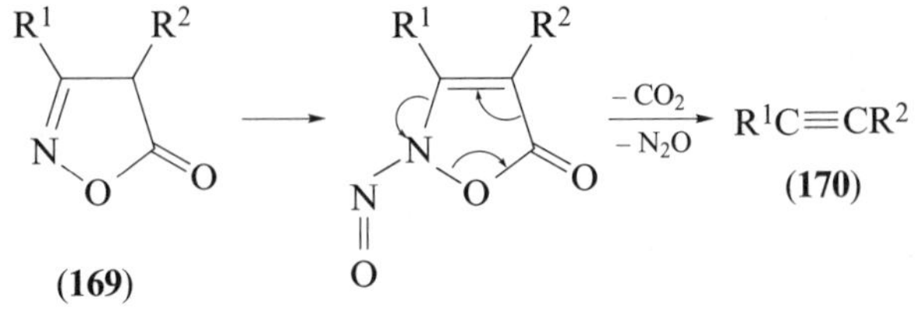

(24)

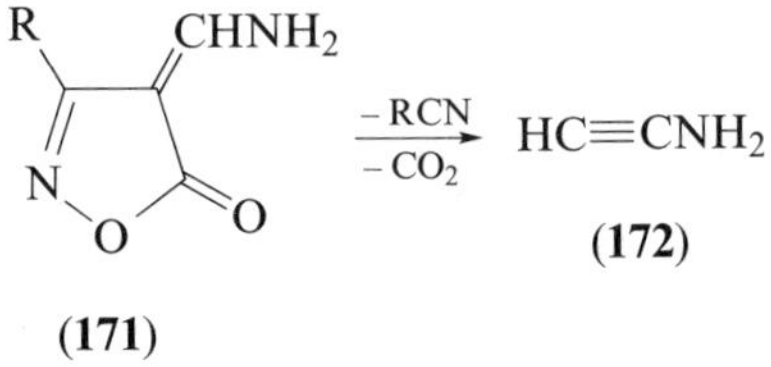

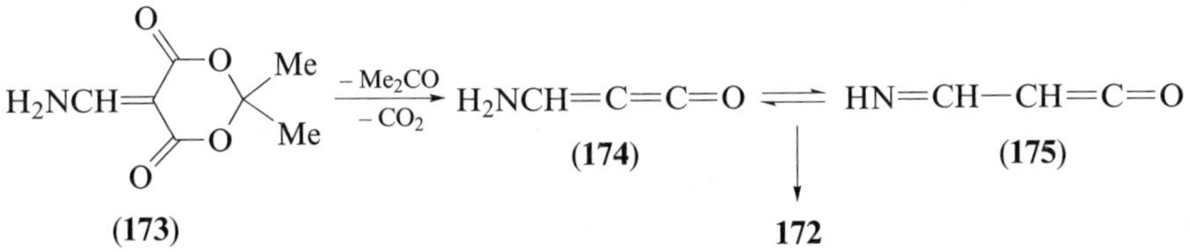

1,2,3-Selenadiazoles **176**, which are readily formed by the action of selenium dioxide on semicarbazones of α-methylene ketones[105], decompose thermally to acetylenes (equation 25)[106]. This, the Lalezari reaction, is widely applicable. Recent uses are illustrated by the synthesis of four medium-sized cycloalkadiynes: cyclonona-1,5-diyne (**177**), cyclodeca-1,5-diyne (**178**), cycloundeca-1,5-diyne (**179**) and cycloundeca-1,6-diyne (**180**) (Scheme 5)[107]. Strained cycloalkynes with additional double bonds, e.g. the cyclododeca-3,7,11-triene-1-yne **181**, have also been obtained by the Lalezari method. Compound **181** isomerizes spontaneously to the bicyclic hydrocarbon **182**[108].

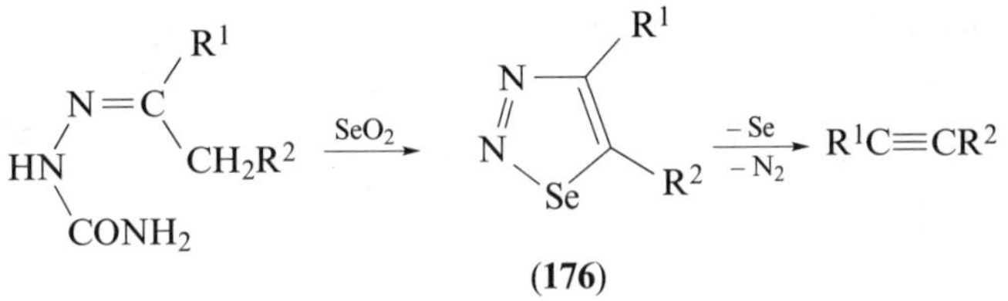

(25)

E. From Phosphoranes

Aliphatic or aromatic aldehydes react with (ethoxycarbonyliodomethyl)triphenyl-phosphorane (**183**) in the presence of potassium carbonate in a two-phase liquid–solid system to give acetylenic esters **184**[109]. Pyrolysis of α-halophosphoranes **185** (X = Cl or Br; R = Ar or *t*-Bu) results in 1-haloalkynes **186**[110]. Vacuum pyrolysis of the betaine **188**, formed from the phenoxymethylenephosphorane **187**, yields the acetylenic ether **189**[111]. Flash-vacuum pyrolysis of the phosphorane **190** at 750°C gives triphenylphosphine oxide and phenylacetylene with elimination of the ethoxycarbonyl group[112].

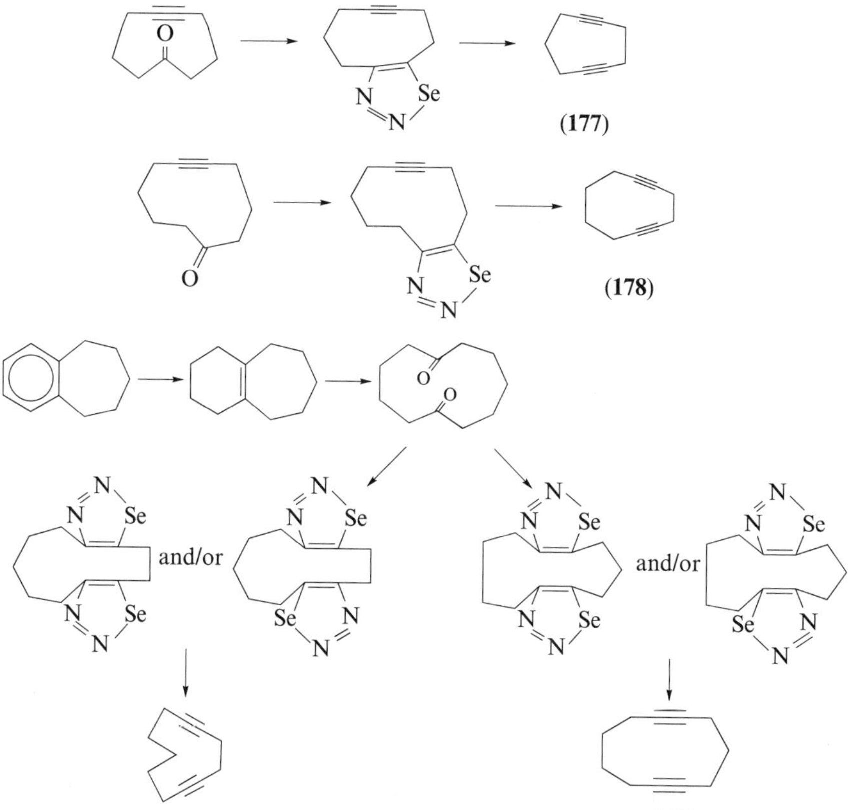

(177)

(178)

and/or

and/or

(180)

(179)

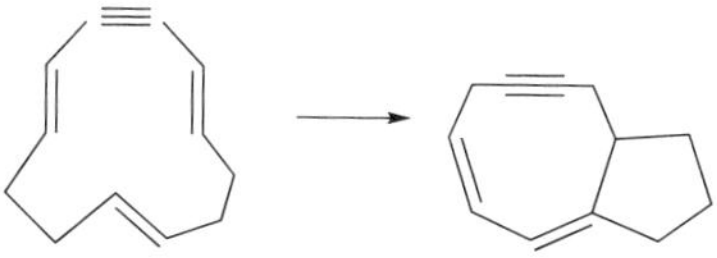

SCHEME 5

(181)   **(182)**

$$RCHO + \underset{I}{\overset{EtO_2C}{\diagdown}}C{=}PPh_3 \longrightarrow RC{\equiv}CCO_2Et$$

(183)   **(184)**

$$Ph_3P{=}CXCOR \longrightarrow XC{\equiv}CR$$

(185)   **(186)**

$$\text{Ph}_3\text{P}{=}\text{CHOPh} + \text{C}_2\text{F}_5\text{COCl} \longrightarrow \underset{\underset{(188)}{\text{OPh}}}{\overset{\text{O}^-}{\text{Ph}_3\overset{+}{\text{P}}{-}\text{C}\overset{\diagup\text{C}\diagdown}{\diagdown}\text{C}_2\text{F}_5}} \longrightarrow \text{PhOC}{\equiv}\text{CC}_2\text{F}_5$$

(187)   (189)

$$\text{Ph}_3\text{P}{=}\text{CHCO}_2\text{Et} + \text{PhCOCl} \longrightarrow \underset{\underset{(190)}{\text{COPh}}}{\text{Ph}_3\text{P}{=}\text{C}\overset{\diagup\text{CO}_2\text{Et}}{\diagdown}} \longrightarrow \text{PhC}{\equiv}\text{CH}$$

F. From Miscellaneous Precursors

Irradiation at 12 K of matrix-isolated ketene under argon with 308-nm pulses from a XeCl excimer laser yields pure hydroxyethyne (191)[113]. Treatment of chloroacetaldehyde diethyl acetal (192) with sodamide in liquid ammonia, followed by an electrophile E$^+$, leads to ethoxyacetylenes (equation 26). Ethyl bromide, for instance, gives 1-ethoxypropyne (193), acetone gives the alcohol 194 and water gives 1-ethoxyethyne[114]. 1-Aryl-3,3,3-trifluoropropynes 196 are produced by the action of potassium hydroxide on the aldehydes 195[115]. N-Nitroso-β-acetoxyamides 198 (R = Ph or hexyl), obtained from the amides 197 and dinitrogen tetroxide, are converted into acetylenes 199 in boiling THF (equation 27)[116a,b].

$$\text{H}_2\text{C}{=}\text{C}{=}\text{O} \longrightarrow \text{HC}{\equiv}\text{COH}$$

(191)

$$\underset{(192)}{\text{ClCH}_2\text{CH}\overset{\diagup\text{OEt}}{\diagdown\text{OEt}}} \longrightarrow \left[\text{Na}\overset{+}{\text{C}}{\equiv}\overset{-}{\text{C}}\text{OEt} \right] \overset{\text{E}^+}{\longrightarrow} \text{EC}{\equiv}\text{COEt} \qquad (26)$$

$$\underset{(193)}{\text{EtC}{\equiv}\text{COEt}} \qquad\qquad \underset{\underset{(194)}{\text{HO}}}{\overset{\text{Me}\diagdown}{\text{Me}{-}\text{C}{-}\text{C}{\equiv}\text{COEt}}}$$

$$\underset{(195)}{\overset{\text{F}_3\text{C}\diagdown}{\underset{\text{F}_3\text{C}\diagup}{\text{C}{=}\text{C}}}\overset{\diagup\text{Ar}}{\diagdown\text{CHO}}} \longrightarrow \underset{(196)}{\text{F}_3\text{CC}{\equiv}\text{CAr}}$$

$$\underset{(197)}{\underset{\underset{\text{Ac}}{\text{NH}}}{\overset{\text{OAc}}{R}}} \longrightarrow \underset{(198)}{\underset{\underset{\text{Ac}\diagdown\text{NO}}{\text{N}}}{\overset{\text{OAc}}{R}}} \longrightarrow \underset{(199)}{\overset{\text{OAc}}{R}\overset{\overset{+}{\text{N}_2}}{\diagup}} \longrightarrow \text{RC}{\equiv}\text{CH} \qquad (27)$$

III. REACTIONS OF ACETYLENES

A. Formation of Open-chain Compounds

1. Formation of alkenes

A review on additions of organometallic reagents to alkynes to form vinyl metal compounds has appeared[117]. The stereospecific *cis*-addition of hydrogen to a wide variety of alkynes is effected in ethanol by zinc dust activated by 1,2-dibromoethane[118]. A combination of the transition metal catalyst $Co(PPh_3)_4Cl_2$ and samarium(II) iodide and a proton donor, such as isopropyl alcohol, reduces acetylenes selectively to (Z)-olefins[119]. (Z)-Olefins are also obtained from alkynes by the combined action of zinc and niobium(V) chloride or tantalum(V) chloride[120]. The copper hydride reagent $[(Ph_3P)CuH]_6$ hydrogenates terminal acetylenes at room temperature; internal acetylenes yield *cis*-alkenes at elevated temperatures[121]. *cis*-Addition to the triple bond of alkynes and enynes occurs to the extent of 85% or more by hydrogenation in the presence of an interlamellar montmorillonite-anchored bis(triphenylphosphine)palladium(II) complex[122].

ω-Hydroxyalkynes (**200**, R = Pr, Bu, pentyl or heptyl; n =1,2,3 or 4) rearrange in the presence of alkyllithium compounds and N,N,N',N'-tetramethylethylenediamine to give the allenes **201**, accompanied by smaller amounts of the isomeric allenes **202**[123]. A procedure for the conversion of 1,3-diynes into (Z)-3-en-1-ynes is outlined in equation 28. The diyne is treated with ethylmagnesium bromide, followed by triethylsilyl chloride, to give the silylated diyne, which is reduced to the silylated (Z)-enyne by catalytic hydrogenation in the presence of palladium or by means of activated zinc. The triethylsilyl group is finally removed with methanolic potassium hydroxide to yield the product.[124-126]

$$HO(CH_2)_nCH_2C{\equiv}CCH_2R \longrightarrow$$

$$(\textbf{200})$$

$$HO(CH_2)_nCH{=}C{=}CHCH_2R \ + \ HO(CH_2)_nCH_2CH{=}C{=}CHR$$

$$(\textbf{201}) \qquad\qquad\qquad (\textbf{202})$$

$$RC{\equiv}C{-}C{\equiv}CH \longrightarrow RC{\equiv}C{-}C{\equiv}CSiMe_3 \longrightarrow \underset{H}{\overset{R}{\diagdown}}C{=}C\underset{H}{\overset{C{\equiv}CSiMe_3}{\diagup}}$$

$$\underset{H}{\overset{R}{\diagdown}}C{=}C\underset{H}{\overset{C{\equiv}CH}{\diagup}} \qquad (28)$$

Terminal acetylenes are selectively hydrogenated to alkenes in the presence of the iron(II) precursors $[P(CH_2CH_2PPh_3)\ FeH(X_2)\ BPh_4]$ (X = N or H) under mild conditions in THF[127]. *Cis*-Addition of hydrogen to 1-iodoalkynes is brought about by reaction with diimide, N_2H_2[128].

Palladium-catalysed hydroarylations and hydrovinylations of olefins and acetylenes have been reviewed[129]. Vinyl iodide, for instance, reacts with internal acetylenes in the presence of $(PPh_3)_2Pd(OAc)_2$, triethylamine and formic acid to yield the alkenes **204** stereospecifically via the intermediates **203**[130]. Iodobenzene similarly adds to 1-(trimethylsilyl)alkynes to give β-(trimethylsilyl)styrenes **205** regioselectively[131]. The reaction of vinyl or aryl iodides R^3I with the acetylenic esters **206** leads to unsaturated γ-lactones **207**[132].

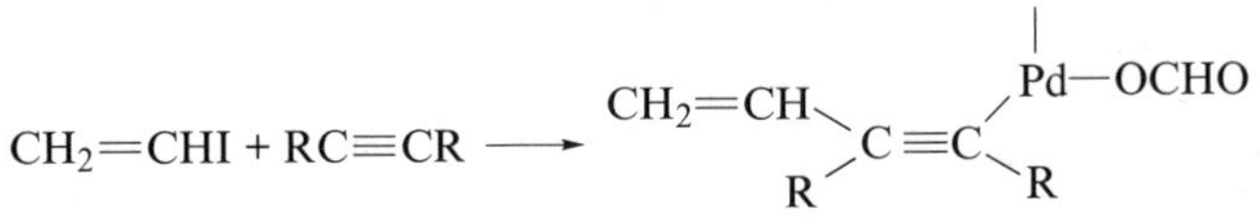

$$CH_2=CHI + RC\equiv CR \longrightarrow \underset{R}{CH_2=CH}\!\!\diagdown\!\!\underset{R}{C\equiv C}\!\!\diagup\!\!Pd-OCHO$$

(203)

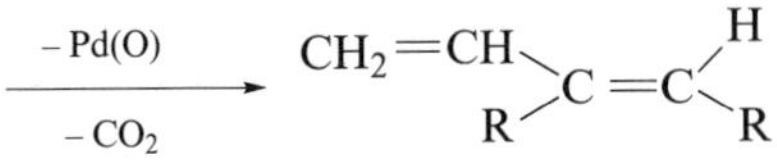

$$\xrightarrow[-CO_2]{-Pd(O)} \quad \underset{R}{CH_2=CH}\!\!\diagdown\!\!C=C\!\!\diagup\!\!\underset{R}{H}$$

(204)

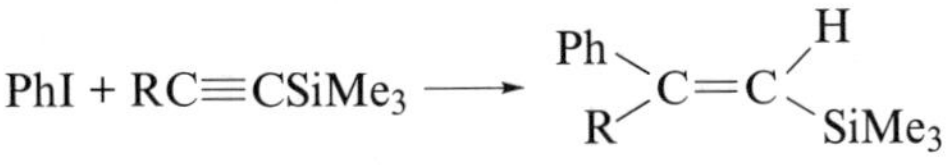

$$PhI + RC\equiv CSiMe_3 \longrightarrow \underset{R}{Ph}\!\!\diagdown\!\!C=C\!\!\diagup\!\!\underset{SiMe_3}{H}$$

(205)

$$R^3I + \underset{R^2}{\overset{R^1}{\diagdown}}\!\!\underset{OH}{\overset{|}{C}}\!\!-C\equiv C-CO_2Et \xrightarrow{Pd(0)} \underset{R^2}{\overset{R^1}{\diagdown}}\!\!\underset{\underset{OH}{\overset{|}{C}}}{\overset{-Pd-}{C=C}}\!\!\diagup\!\!\underset{CO_2Et}{\overset{R^3}{\diagup}} \longrightarrow R^1\!\!\diagdown\!\!\underset{R^2}{\underset{O}{\diagdown}}\!\!\overset{R^3}{\underset{O}{\diagup}}$$

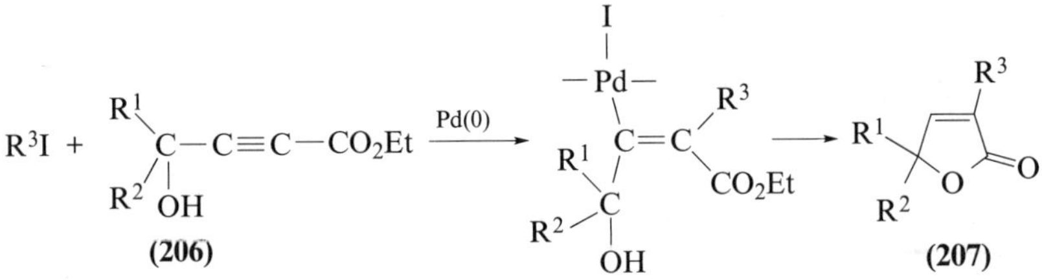

(206)   **(207)**

Hydroboration of terminal alkynes, e.g. 1-hexyne, 1-octyne or cyclohexylacetylene, with a dialkylborane, such as bis(1,2-dimethylpropyl)borane, followed by copper(I)cyanide and copper(II) acetate in HMPA containing a trace of water, gives isomerically pure (E)-1-cyanoalk-1-enes (equation 29)[133]. Successive treatment of 1-bromo-1-alkynes with dialkylboranes and sodium methoxide results in the borinate esters **208**, which are converted into (E)-alkenes of greater than 99% isomeric purity by protonolysis. The action of alkaline hydrogen peroxide on the borinates produces ketones (equation 30)[134].

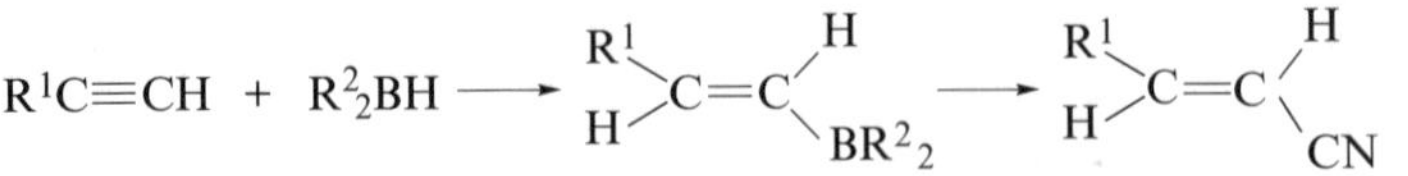

$$R^1C\equiv CH + R^2_2BH \longrightarrow \underset{H}{\overset{R^1}{\diagdown}}\!\!C=C\!\!\diagup\!\!\underset{BR^2_2}{\overset{H}{\diagup}} \longrightarrow \underset{H}{\overset{R^1}{\diagdown}}\!\!C=C\!\!\diagup\!\!\underset{CN}{\overset{H}{\diagup}} \qquad (29)$$

2. Formation of ketones

p-Methoxybenzenetellurinic acid anhydride, $[p\text{-MeOC}_6H_4Te(O)]_2O$, catalyses the hydration of terminal acetylenes to yield methyl ketones (equation 31)[135]. Unactivated alkynes **209** (R^1 = H or alkyl, R^2 = alkyl or Ph) are converted into ketones in refluxing aqueous

$$R^1C\equiv CBr \xrightarrow{R^2_2BH} \underset{H}{\overset{R^1}{>}}C=C\underset{BR^2_2}{\overset{Br}{<}}$$

$$\xrightarrow{NaOMe} \underset{H}{\overset{R^1}{>}}C=C\underset{R^2}{\overset{BR^2OMe}{<}} \xrightarrow{H^+} \underset{H}{\overset{R^1}{>}}C=C\underset{R^2}{\overset{H}{<}}$$

$$\xrightarrow{H_2O_2} R^1CH_2COR^2 \tag{30}$$

$$(208)$$

methanol in the presence of NaAuCl$_4$[136]. Zero-valent nickel, generated from nickel(cyclooctadiene) and tri(n-octyl)phosphine, effects the stereoselective hydroacylation of 4-octyne with aldehydes RCHO (R = Pr, i-Pr or Ph) to the (E)-adducts 210. In the case of unsymmetrical alkynes, the hydrogen atom of the aldehyde tends to add to the carbon atom bearing the bulkier group (e.g. equation 32)[137].

$$RC\equiv CH + H_2O \longrightarrow RCOCH_3 \tag{31}$$

$$R^1C\equiv CR^2 \longrightarrow R^1CH_2COR^2$$
$$(209)$$

$$PrC\equiv CPr + RCHO \longrightarrow \underset{RCO}{\overset{Pr}{>}}C=C\underset{H}{\overset{Pr}{<}}$$
$$(210)$$

$$MeC\equiv CBu^t + RCHO \longrightarrow \underset{RCO}{\overset{Me}{>}}C=C\underset{H}{\overset{Bu^t}{<}} \tag{32}$$

Ruthenium-catalysed additions to alkynes have been reviewed[138]. Terminal acetylenes add allyl alcohol in the presence of a ruthenium catalyst to give enones. Phenylacetylene affords a mixture of conjugated and unconjugated enones, 211 and 212, respectively, in this reaction, but 1-tridecyne gives only the conjugated enone 213[139a,b]. Treatment of internal acetylenes with a mixture of 35% aqueous hydrogen peroxide, 12-tungstophosphoric acid and cetylpyridinium chloride under phase-transfer conditions results in mixtures of α,β-

$$PhC\equiv CH + HOCH_2-CH=CH_2 \longrightarrow$$

$$(211) \qquad (212)$$

$$(213)$$

unsaturated ketones and α,β-epoxy ketones; 4-octyne, for example, is converted into the enone **214** (15%) and the epoxide **215** (62%)[140]. α,α-Dichloroketones are obtained from acetylenes by the action of hydrogen chloride and 'Oxone' (potassium monoperoxysulphate) in DMF (equation 33)[141].

$$PrC\equiv CPr \longrightarrow \underset{(214)}{\text{(enone structure)}} + \underset{(215)}{\text{(epoxide structure)}}$$

$$R^1C\equiv CR^2 \longrightarrow R^1\underset{O}{\underset{\|}{C}}CCl_2R^2 \qquad (33)$$

The phase-transfer-assisted permanganate oxidation of alkynes and alkenes has been reviewed[142]. Terminal and internal alkynes are oxidized to 1,2-dicarbonyl compounds by the combined action of diphenyl disulphide, ammonium peroxidisulphate and water[143] or by sodium periodate in the presence of ruthenium dioxide[144] (equation 34). Other reagents for the conversion of acetylenes into 1,2-dicarbonyl compounds are hydrogen peroxide in the presence of (2,6-dicarboxylatopyridine)iron(II)[145], the complex oxo(N,N'-ethylenebissalicylideneiminato)chromium(V) trifluoromethanesulphonate (**216**)[146] and ruthenium tetroxide as a mediator in electrooxidation[147]. 1-Acetoxyalkan-2-ones **217** are obtained by the oxidation of terminal acetylenes with sodium perborate and mercury(II) acetate in acetic acid[148]. Terminal alkynes give α-ketoaldehydes **218** on treatment with dilute hydrogen peroxide, combined with mercury(II) acetate and sodium molybdate or sodium tungstate under phase-transfer conditions[149].

$$R^1C\equiv CR^2 \longrightarrow R^1COCOR^2 \qquad (34)$$

$$R-\underset{O}{\underset{\|}{C}}-CH_2OAc$$
$$(217)$$

(**216**) $CF_3SO_3^-$

$$RC\equiv CH \longrightarrow R\underset{O}{\underset{\|}{C}}CHO$$
$$(218)$$

α-Keto esters **220** are produced by the oxidation of (trimethylsilyl)acetylenes **219** (R pentyl, hexyl or Ph) with t-butyl hydroperoxide in methanol in the presence of osmiur tetroxide[150]. The action of PdCl$_2$(MeCN)$_2$ on the propargyl acetates **221** results in th rearranged acetoxy ketones **222**, which afford dicarbonyl compounds **223** on alkalin hydrolysis[151]. The cobalt(II)-catalysed oxygenation of p-tolylacetylene in the presence c isopropyl alcohol yields the ketone **224**, accompanied by minor amounts of the esters **22** and **226**[152].

$$RC\equiv CSiMe_3 \longrightarrow RCOCO_2Me$$
$$\mathbf{(219)} \qquad\qquad \mathbf{(220)}$$

$$R^1C\equiv CCHR^2OAc \longrightarrow R^1C-C=CHR^2 \longrightarrow R^1COCOCH_2R^2$$
$$\mathbf{(221)} \qquad\qquad\quad \underset{\substack{\| \quad | \\ O \quad OAc}}{} \qquad\qquad \mathbf{(223)}$$
$$\mathbf{(222)}$$
$$R^1, R^2 = C_6H_{13} \text{ or } Ph$$

$$p\text{-TolC}\equiv CH \longrightarrow p\text{-TolCOCH}_3 + p\text{-TolCH(OH)CO}_2Pr^i + p\text{-TolCOCO}_2Pr^i$$
$$\mathbf{(224)} \qquad\qquad \mathbf{(225)} \qquad\qquad \mathbf{(226)}$$

3. Formation of halogen compounds

Electrophilic acetylenes 227 ($R^1 = EtO_2C$, CHO or CN; $R^2 = EtCO_2$, Ph or $(EtO)_2CH$) add the elements of hydrogen fluoride on treatment with caesium fluoride under phase-transfer catalysis to yield mainly (Z)-fluoroalkenes 228[153]; dimethyl acetylenedicarboxylate affords a 85:15 mixture of (Z)- and (E)-adducts under these conditions. No addition occurs in homogeneous solutions[154].

$$R^1C\equiv CR^2 \longrightarrow \underset{H}{\overset{R^1}{}}C=C\underset{R^2}{\overset{F}{}}$$
$$\mathbf{(227)} \qquad\qquad \mathbf{(228)}$$

$$RC\equiv CM + I_2 \longrightarrow \underset{H}{\overset{R}{}}C=C\underset{I}{\overset{H}{}}$$

$$M = metal$$

Metallated acetylenes, obtained by hydrozirconation[155], hydroalumination[156] or hydroboration[157], react with elemental iodine to give (E)-iodoalkene[158]. Terminal alkynes 229 (R = octyl or decyl) add hydrogen iodide, generated from the boron triiodide/N,N–diethylaniline complex and acetic acid, in a Markovnikov sense to afford the iodoalkenes 230[159]. cis-Addition of hydrogen iodide, produced in situ from trimethylsilyl chloride and aqueous sodium iodide, to a number of internal alkynes has been reported[160].

$$RC\equiv CH + HI \longrightarrow \underset{I}{\overset{R}{}}C=CH_2$$
$$\mathbf{(229)} \qquad\qquad \mathbf{(230)}$$

$$HC\equiv CCO_2Me + Bu^tI \longrightarrow \underset{H}{\overset{Bu^t}{}}C=C\underset{I}{\overset{CO_2Me}{}}$$
$$\mathbf{(231)}$$

Electron-deficient alkynes undergo an 'iodine atom-transfer addition reaction' with secondary or tertiary alkyl iodides in the presence of hexabutyldistannane and under sunlamp irradiation to give iodoalkenes; methyl propiolate and t-butyl iodide, for instance, afford a 4:1 mixture of the (E)- and (Z)-adducts 231[161]. Analogous reactions with simple alkynes have been described for α-iodo ketones, α-iodo carboxylic esters and α-iodonitriles. Thus 1-heptyne and iodoacetonitrile furnished the adducts 232 [(E) : (Z) = 1 : 2.6][162].

$$C_5H_{11}C\equiv CH \ + \ NCCH_2I \ \longrightarrow \ \underset{I}{\overset{C_5H_{11}}{\diagdown}}C=C\underset{CH_2CN}{\overset{H}{\diagup}}$$

(232)

Acetylenes react with perfluoroalkyl iodides in hexane under catalysis by tetrakis(triphenylphosphine)palladium(0) to give good yields of perfluoroalkylated alkenyl iodides[163]. The analogous reaction of 1-hexyne with compound 233 in the presence of magnesium gives a mixture of (Z)- and (E)- adducts 234[164]. Heptafluoroisopropyl iodide and terminal acetylenes 235 (R = H, Ph or CH$_2$OAc) afford only (E)-adducts 236[165].

$$BuC\equiv CH \ + \ I(CF_2)_4Cl \ \longrightarrow \ \underset{I}{\overset{Bu}{\diagdown}}C=C\underset{(CF_2)_4Cl}{\overset{H}{\diagup}}$$

(233)   (234)

$$RC\equiv CH \ + \ \underset{I}{\overset{F}{\diagdown}}C\underset{CF_3}{\overset{CF_3}{\diagup}} \ \longrightarrow \ \underset{I}{\overset{R}{\diagdown}}C=C\underset{H}{\overset{CF\diagup CF_3}{\diagup}}$$

(235)   (236)

Analogous *trans*-additions of perfluoroalkyl iodides to terminal or internal alkynes in the presence of sodium dithionite and mediated by ultrasonic irradiation[166] or catalysed by triethylborane[167] have been reported. 1,2–Dibromotetrafluoroethane reacts with terminal acetylenes in DMF in the presence of the redox system (NH$_4$)$_2$S$_2$O$_8$/HCO$_2$Na.2H$_2$O to yield a mixture of mainly (E)- and some (Z)-olefins 237, the initial adducts having undergone reductive debromination under the reaction conditions[168].

$$RC\equiv CH \ + \ F_2BrCCBrF_2 \ \longrightarrow \ \underset{H}{\overset{R}{\diagdown}}C=C\underset{CF_2CBrCF_2}{\overset{H}{\diagup}}$$

(237)

R = Bu, SiMe$_3$ or CH$_2$OH

β-Iodostyrenes 241 are produced by the action of alkynes 238 (R^1 = Bu or Ph; R^2 = H or Me) on aromatic compounds 239 (benzene, toluene, p-xylene, naphthalene or bromobenzene) in the presence of (dipyridine)iodonium tetrafluoroborate (240)[169].

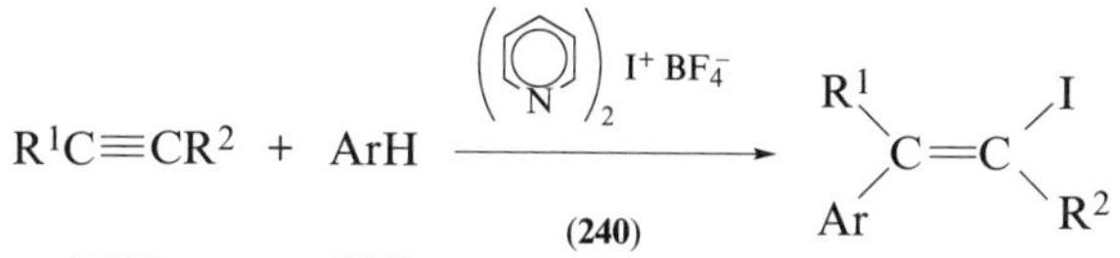

$$R^1C\equiv CR^2 \;+\; ArH \xrightarrow[\quad(240)\quad]{\left(\bigcirc_N\right)_2 I^+ BF_4^-} \begin{array}{c} R^1 \\ \diagdown \\ Ar \end{array} C=C \begin{array}{c} I \\ \diagup \\ R^2 \end{array}$$

(238)   (239)   (241)

The iodonium salt also mediates the reaction of nuleophiles with alkynes to yield iodoalkenes. Thus 3-hexyne and lithium bromide give the bromoiodo compound **242**[170] and, generally, internal or terminal acetylenes add hydrogen chloride, hydrogen bromide, formic acid and acetic acid in the presence of the reagent to give iodoolefins **243** as single regio- and stereoisomers[171]. The reaction has been extended to the preparation of 1-bromo-1-iodoalpenes **244,** which are formed regio- and stereospecifically from 1-bromo-1-alkynes, the iodonium salt and nucleophiles[172].

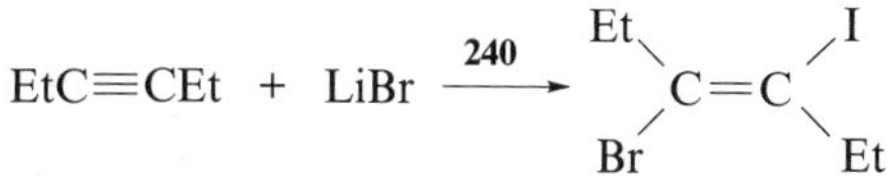

$$EtC\equiv CEt \;+\; LiBr \xrightarrow{240} \begin{array}{c} Et \\ \diagdown \\ Br \end{array} C=C \begin{array}{c} I \\ \diagup \\ Et \end{array}$$

(242)

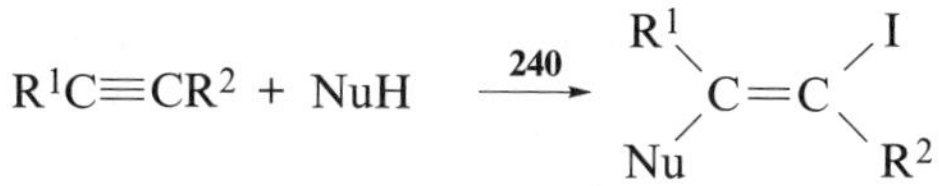

$$R^1C\equiv CR^2 \;+\; NuH \xrightarrow{240} \begin{array}{c} R^1 \\ \diagdown \\ Nu \end{array} C=C \begin{array}{c} I \\ \diagup \\ R^2 \end{array}$$

(243)

Nu = Cl, Br, HCO_2 or AcO

$$RC\equiv CBr \;+\; NuH \xrightarrow{240} \begin{array}{c} R \\ \diagdown \\ Nu \end{array} C=C \begin{array}{c} I \\ \diagup \\ Br \end{array}$$

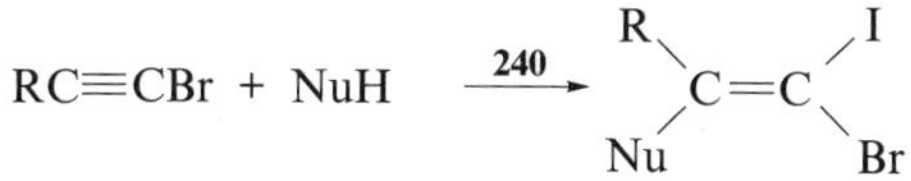

(244)

Nu = Cl, Br, HCO_2 or AcO

Bromine adsorbed on graphite in dichloromethane reacts with diverse acetylenes to give (E)- dibromoalkenes exclusively. No (E)- → (Z)-isomerization, which is usually catalysed by bromine, occurs in the presence of graphite[173]. Alkynes adsorbed on alumina add iodine stereoselectively; e.g. 3-hexyne gives 97% (E)-3,4-diiodo-3-hexene (**245**)[174]; a similar *trans-*addition of iodine occurs in the presence of Florisil[175]. The reaction of alkynes with tetrabutylammonium dichlorobromate or bromine chloride is non-regiospecific but stereospecific. Thus 1-phenylpropyne (**246**) affords a mixture of (E)-bromochloroalkenes **247** and **248**[176]. The joint action of iodine and mercury compounds HgX_2 (X = Cl, OAc, OTs, SCN or SCH_2Ph) on internal or terminal alkynes give stereoselectively (E)- adducts **249**; unsymmetrical alkynes yield Markovnikov products[177].

$$EtC\equiv CEt \;+\; I_2 \longrightarrow \begin{array}{c} Et \\ \diagdown \\ I \end{array} C=C \begin{array}{c} I \\ \diagup \\ Et \end{array}$$

(245)

$$\text{PhC}\equiv\text{CMe} \longrightarrow \underset{\text{Cl}}{\overset{\text{Ph}}{}}\text{C}=\text{C}\underset{\text{Me}}{\overset{\text{Br}}{}} \quad + \quad \underset{\text{Br}}{\overset{\text{Ph}}{}}\text{C}=\text{C}\underset{\text{Me}}{\overset{\text{Cl}}{}}$$

(246)   (247)   (248)

$$\text{R}^1\text{C}\equiv\text{CR}^2 \longrightarrow \underset{\text{X}}{\overset{\text{R}^1}{}}\text{C}=\text{C}\underset{\text{R}^2}{\overset{\text{I}}{}}$$

(249)

4. Formation of enol esters

The [ruthenium(cyclooctadienyl)$_2$ (tributylphosphine)]- catalysed addition of carboxylic acids **250** (R^1 = cyclohexyl, $PhCH_2$ or CH_2 = CMe) to terminal acetylenes **251** (R^2 = Bu or Ph) results in enol esters **252**[178-183]; similarly, benzoic acid adds to propyne in the presence of a rhodium catalyst to yield 2-(benzoyloxy)propene (**253**) regiospecifically[184].

Enol formates, which are formylating agents under neutral conditions, are obtained from 1-alkynes and formic acid by arene-ruthenium(II) catalysis[185]. Heating acetylenes (phenylacetylene, diphenylacetylene, 1-octyne or 4-octyne) with formic acid to 100°C produces ketones and carbon monoxide. It was shown by NMR spectroscopy that the process involves the intermediate formation of enol formates (equation 35)[186]. Treatment of terminal alkynes with carbon dioxide and secondary amines in the presence of a ruthenium catalyst affords vinyl carbamates, e.g. equation 36[187,188]; for reviews, see References 189 and 190.

$$\text{R}^1\text{CO}_2\text{H} + \text{R}^2\text{C}\equiv\text{CH} \longrightarrow \underset{\text{R}^1\text{CO}_2}{\overset{\text{R}^2}{}}\text{C}=\text{CH}_2$$

(250)   (251)   (252)

$$\text{PhCO}_2\text{H} + \text{MeC}\equiv\text{CH} \longrightarrow \underset{\text{PhCO}_2}{\overset{\text{Me}}{}}\text{C}=\text{CH}_2$$

(253)

$$\text{R}^1\text{C}\equiv\text{CR}^2 + \text{HCO}_2\text{H} \longrightarrow \underset{\underset{\text{O}}{\overset{\|}{\text{HC}-\text{O}}}}{\overset{\text{R}^1}{}}\text{C}=\text{C}\underset{\text{R}^2}{\overset{\text{H}}{}} \xrightarrow{-\text{CO}} \text{R}^1\text{COCH}_2\text{R}^2 \quad (35$$

$$\text{BuC}\equiv\text{CH} + \text{HNEt}_2 \xrightarrow{\text{CO}_2} \underset{\text{H}}{\overset{\text{Bu}}{}}\text{C}=\text{C}\underset{\text{O}-\text{CONEt}_2}{\overset{\text{H}}{}} \quad (36$$

5. Formation of carboxylic acids and derivatives

Phase-transfer hydrocarboxylation of terminal acetylenes **254** (R = Bu, pentyl, Ph or PhCH$_2$CH$_2$) with carbon monoxide, catalysed by nickel(II) cyanide, affords the alkenoic acids **255**[191]. Similarly, phenylacetylene reacts with carbon monoxide and alcohols in the presence of palladium(dibenzylideneacetone)$_2$ to give esters of unsaturated acids, e.g. **256**[192].

$$HC \equiv CR \xrightarrow{CO} H_2C = CRCO_2H$$

$$(254) \qquad\qquad (255)$$

$$HC \equiv CPh + CO \xrightarrow{Bu^tOH} H_2C = CPhCO_2Bu^t$$

$$(256)$$

Analogous high-pressure carbonylations of terminal as well as internal alkynes in the presence of PtH(SnCl$_3$)(PPh$_3$)$_2$ have been described[193]. Treatment of phenylacetylene with carbon monoxide in the presence of cobalt(II) chloride, potassium cyanide and nickel(II) cyanide under phase-transfer conditions gave a mixture of the acids **257** and **258**, in which the former predominated[194].

$$HC \equiv CPh \longrightarrow \underset{\underset{CO_2H}{|}}{H_3CCHPh} + \underset{\underset{CO_2H}{|}}{H_2CCH_2Ph}$$

$$(257) \qquad\qquad (258)$$

α-Haloalkynes react with carbon monoxide in the presence of nickel(II) cyanide, aqueous sodium hydroxide, 4-methyl-2-pentanone (methyl isobutyl ketone) and tetrabutylammonium bromide to give allenic monocarboxylic acids, together with unsaturated dicarboxylic acids. 3-Chloro-3-methylpent-1-yne (**259**), for instance, yielded a mixture containing 87% of the allenic acid **260** and 13% of the alkylidenesuccinic acid **261**[195].

$$\underset{\underset{Et}{|}}{\overset{\overset{Me}{|}}{Cl-C-C \equiv CH}} + CO \longrightarrow$$

$$(259)$$

$$\underset{Et}{\overset{Me}{>}}C = C = C \overset{H}{\underset{CO_2H}{<}} \quad + \quad \underset{Et}{\overset{Me}{>}}C = C \overset{CH_2CO_2H}{\underset{CO_2H}{<}}$$

$$(260) \qquad\qquad\qquad (261)$$

Combined catalysis by palladium(II) chloride and copper(II) chloride induces the reaction of 1-alkynes with oxygen and carbon monoxide at room temperature to give a mixture of the anhydrides **262** and the corresponding maleic and fumaric acids (equation 37)[196].

$$RC\equiv CH \longrightarrow \underset{\textbf{(262)}}{\underset{O \quad O \quad O}{\overset{R}{\diagdown}}} + \underset{HO_2C}{\overset{R}{\diagdown}}\underset{CO_2H}{\overset{H}{\diagup}} + \underset{HO_2C}{\overset{R}{\diagdown}}\underset{H}{\overset{CO_2H}{\diagup}} \qquad (37)$$

Oxidative cleavage of acetylenes $RC\equiv CH$ (R = Ph, Bu, hexyl, heptyl, cyclopentyl or $PhCH_2CH_2$) with bis[trifluoroacetoxy)iodo]pentafluorobenzene in wet benzene gives carboxylic acids RCO_2H[197]. Alkynes **263** (R^1, R^2 = H, Me, Et, Pr, Bu or Ph) undergo an oxidative rearrangement by the action of [hydroxy(tosyloxy)iodo] benzene in methanol to yield the esters **264**[198].

$$R^1C\equiv CR^2 \longrightarrow R^1R^2CHCO_2Me$$

$$\textbf{(263)} \qquad\qquad\qquad \textbf{(264)}$$

6. Formation of silanes

The lithium organosilylcuprate **265** adds to hex-3-yne to yield a copper derivative **266**, which can be converted into a variety of alkenylsilanes by treatment with electrophiles (e.g. equation 38). 1-Hexyne forms the analogue **267** regioselectively[199,200].

$$EtC\equiv CEt + (PhMe_2Si)_2CuLi\cdot LiCN \longrightarrow \underset{Et}{\overset{(Cu)}{\diagdown}}\underset{Et}{\overset{SiMe_2Ph}{\diagup}}$$

$$\textbf{(265)} \qquad\qquad\qquad\qquad \textbf{(266)}$$

$$\overset{MeI}{\longrightarrow} \underset{Et}{\overset{Me}{\diagdown}}\underset{Et}{\overset{SiMe_2Ph}{\diagup}} \qquad (38)$$

$$BuC\equiv CH \longrightarrow \underset{Bu}{\overset{(Cu)}{\diagdown}}\underset{H}{\overset{SiMe_2Ph}{\diagup}}$$

$$\textbf{(267)}$$

Terminal acetylenes **268** (R = Bu, t-Bu or Ph) react with the lithium silylcuprate **269** by cis-addition to afford, after work-up with aqueous ammonium chloride, the silylated olefins **270**[201]. A similar reaction of 1-dodecyne with the reagent prepared from methylmagnesium iodide and lithium (dimethyl)phenylsilane and a trace of tris(tributylphosphine)platinum(II) chloride gave solely the (E)-alkene **271**. In contrast, in the presence of bis(tri-o-tolylphosphine)palladium(II) chloride the regioisomer **27** was the main product[202].

The outcome of the reaction of triethylsilane with terminal acetylenes in the presence of rhodium(II) perfluorobutyrate depends on the order in which the reagents are added. Thus adding triethlsilane to the catalyst, followed by 1-octyne, gives the rearranged silane **273** as a mixture of (E)- and (Z)- isomers. In contrast, when triethylsilane is added last, a mixture of **274, 275** and **276** is produced[203].

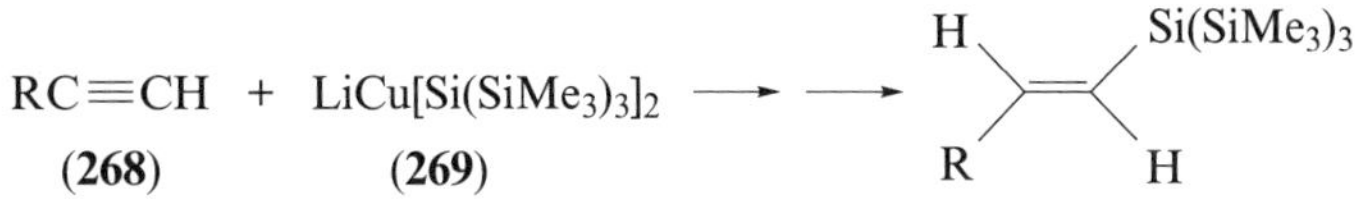

$$RC\equiv CH \ + \ LiCu[Si(SiMe_3)_3]_2 \longrightarrow \longrightarrow$$
(**268**)   (**269**)

(**270**)

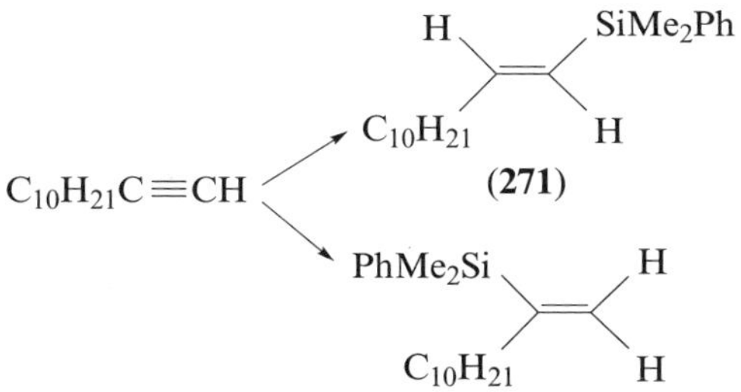

$$C_{10}H_{21}C\equiv CH$$

(**271**)

(**272**)

$$C_6H_{13}-C\equiv CH$$

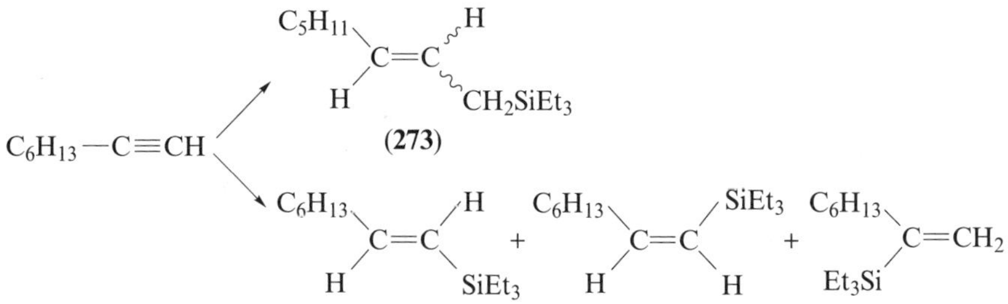

(**273**)

(**274**)   (**275**)   (**276**)

Hydrosilylation of phenylacetylene with triethylsilane catalysed by *trans*-chloro(carbonyl)bis(triphenylphosphine)iridium yields the *trans*-adduct **277**, accompanied by traces of the *cis*-isomer **278** and of the regioisomer **279**[204].

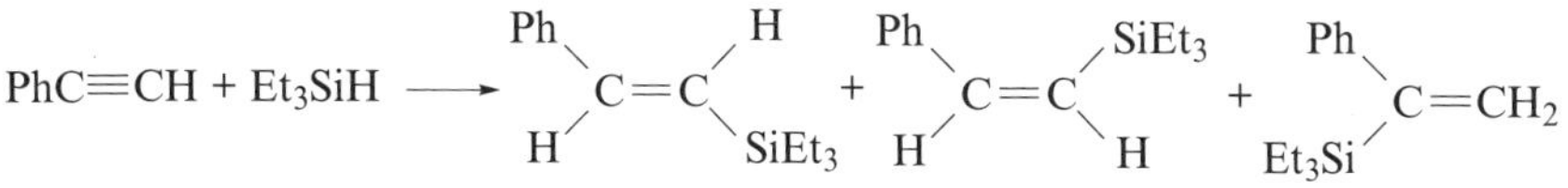

$$PhC\equiv CH + Et_3SiH \longrightarrow$$

(**277**)   (**278**)   (**279**)

Whereas terminal alkynes add metal silanes, e.g. PhMe$_2$SiM (M= metal), to form solely 2-metallo-1-silylalkenes (**267**), the reverse mode of addition has been observed with manganese reagents. Thus, treatment of lithium phenyldimethylsilane with methylmagnesium iodide, followed by manganese (II) chloride and a 1-alkyne, gave a mixture of (E)- and (Z)- metallated olefins **280**[205]. The reagent formed from lithium trimethylsilane, methylmagnesium iodide and manganese(II) chloride adds to bis(trimethylsilyl)acetylene to yield, after aqueous work-up, the highly strained tetrakis(trimethylsilyl)ethene **281**[205].

Products **283** of *cis*-addition are obtained regio-and stereospecifically from terminal acetylenes **282** (R = Ph or substituted Ph) and cyanotrimethylsilane in the presence of a catalytic amount of the palladium(II) chloride–pyridine complex[206a,b]. Alkynes **284** (R^1 = H, Me, Ph, or $HOCH_2$; R^2 = H, Me or CO_2Me) undergo 'silylformylation' with (dimethyl)phenylsilane under carbon monoxide pressure in the presence of tetrarhodium dodecacarbonyl to give α,β-unsaturated aldehydes **285**. In the case of terminal alkynes the silicon atom attaches itself specifically to C(1)[207].

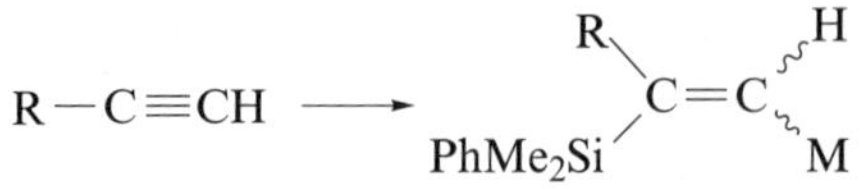

$$R-C\equiv CH \longrightarrow \underset{PhMe_2Si}{\overset{R}{>}}C=C\underset{M}{\overset{H}{<}}$$

(280)

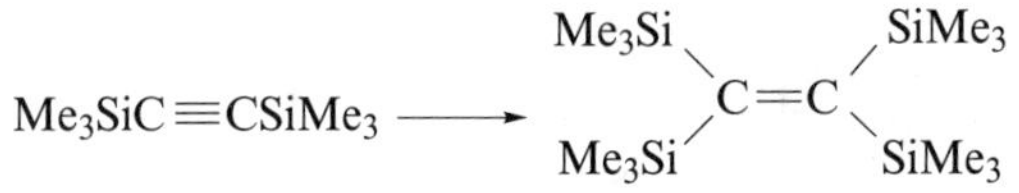

$$Me_3SiC\equiv CSiMe_3 \longrightarrow \underset{Me_3Si}{\overset{Me_3Si}{>}}C=C\underset{SiMe_3}{\overset{SiMe_3}{<}}$$

(281)

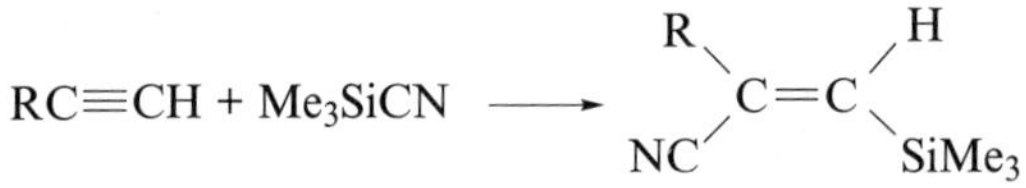

$$RC\equiv CH + Me_3SiCN \longrightarrow \underset{NC}{\overset{R}{>}}C=C\underset{SiMe_3}{\overset{H}{<}}$$

(282)   **(283)**

$$R^1C\equiv CR^2 + HSiMe_2Ph \xrightarrow{CO} \underset{OHC}{\overset{R^1}{>}}C=C\underset{SiMe_2Ph}{\overset{R^2}{<}}$$

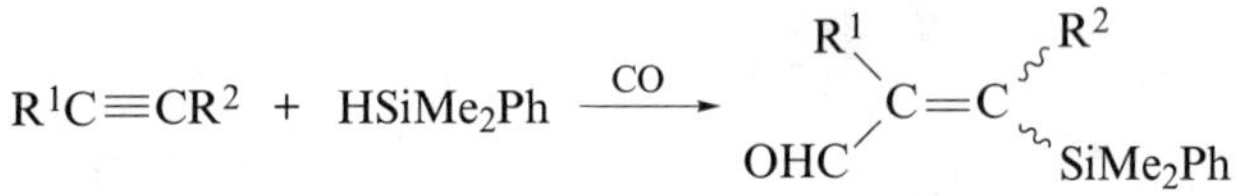

(284)   **(285)**

Symmetrical tetramethylidisilane adds to dimethyl acetylenedicarboxylate in the presence of bis(triethylphosphine)palladium(II) chloride to yield the *cis*-adduct **286**[208]. The palladium-catalysed disilylation of terminal acetylenes proceeds poorly when hexamethyldisilane is used; in contrast, the disilane **287** affords good yields of the *cis*-adducts **288**[209].

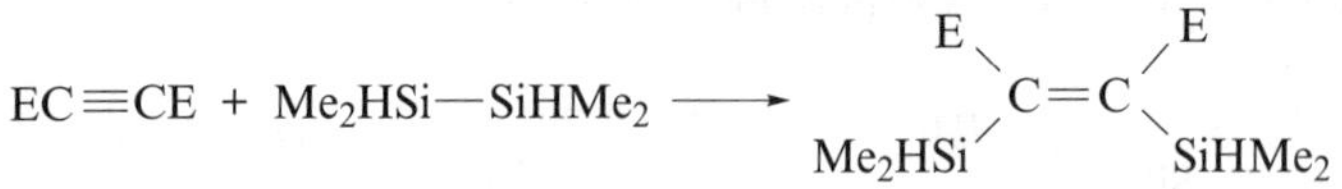

$$EC\equiv CE + Me_2HSi-SiHMe_2 \longrightarrow \underset{Me_2HSi}{\overset{E}{>}}C=C\underset{SiHMe_2}{\overset{E}{<}}$$

(286)

E = CO_2Me

$$RC\equiv CH + MeOMe_2Si-SiMe_2OMe \longrightarrow \underset{MeOMe_2Si}{\overset{R}{>}}C=C\underset{SiMe_2OMe}{\overset{H}{<}}$$

(287)   **(288)**

Good yields are obtained in the disilylation of alkynes with hexamethylsilane and the catalyst system $Pd(OAc)_2$ –$Me_3CCH_2CMe_2NC$ in toluene; 2-nonyne, for example, gave 95% of the (Z)-olefin **289**[210]. The difluorosilane **290** undergoes *cis*-addition to a variety of alkynes in the presence of bis(triphenylphosphine)palladium(II) chloride (equation 39)[211].

A cyclic unsaturated disilane, the disilacycloheptene **292**, was prepared from acetylene and the disilane **291** with the same catalyst[212].

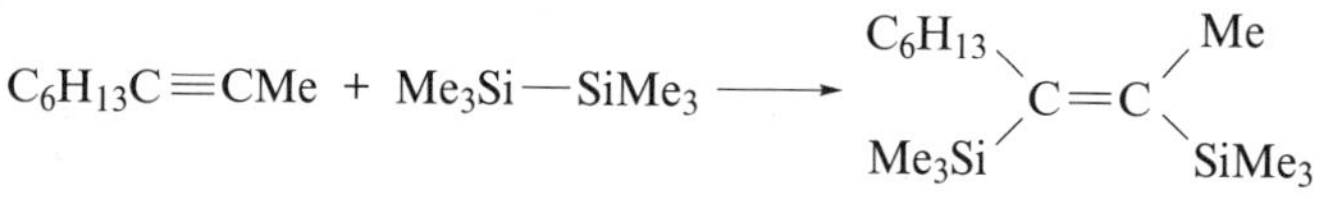

$$C_6H_{13}C \equiv CMe + Me_3Si-SiMe_3 \longrightarrow \begin{matrix} C_6H_{13} \\ Me_3Si \end{matrix} C = C \begin{matrix} Me \\ SiMe_3 \end{matrix}$$

(289)

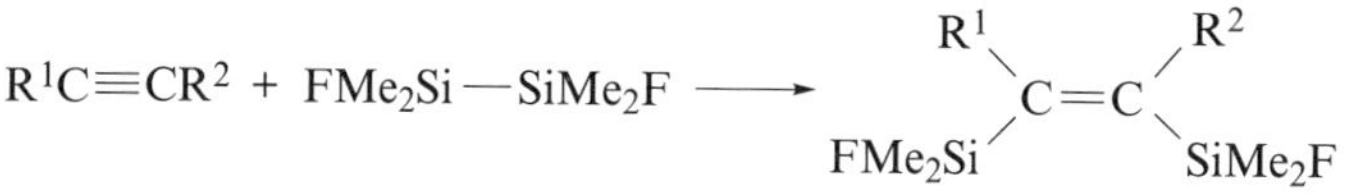

$$R^1C \equiv CR^2 + FMe_2Si-SiMe_2F \longrightarrow \begin{matrix} R^1 \\ FMe_2Si \end{matrix} C = C \begin{matrix} R^2 \\ SiMe_2F \end{matrix} \qquad (39)$$

(290)

$$HC \equiv CH + \begin{matrix} Me_2Si-SiMe_2 \\ \diagdown \diagup \end{matrix} \longrightarrow Me_2Si \begin{matrix} === \\ \end{matrix} SiMe_2$$

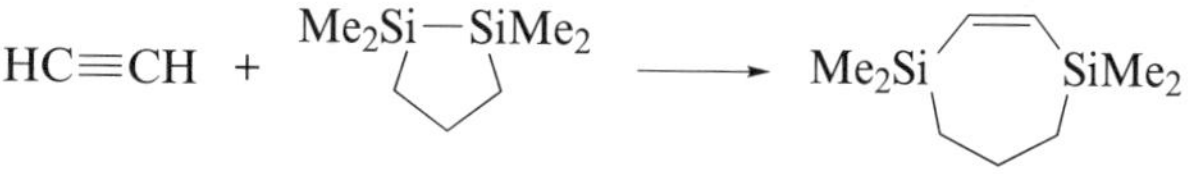

(291)   **(292)**

7. Fomation of organotin and organogermanium compounds

The orientation and steric course of the hydrostannylation of alkynes to yield alkenylstannanes is determined by the nature of the reagent and the catalyst used. The trimethylstannylcopper reagent $Me_3SnCu \cdot Me_2S$ adds regioselectively to terminal acetylenes **293** ($n = 1,2,3,4$ or 6, X = OH, Cl, $OSiMe_2Bu^t$ or 2-tetrahydropyranyl) to give, after treatment with methanol, the stannanes **294**[213].

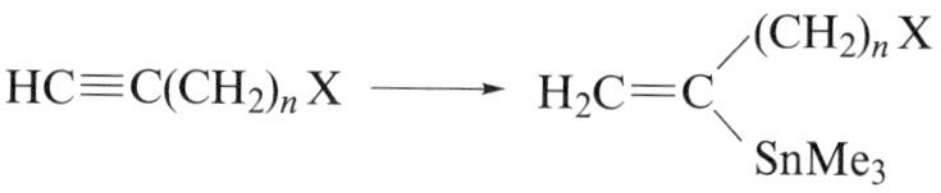

$$HC \equiv C(CH_2)_n X \longrightarrow H_2C = C \begin{matrix} (CH_2)_n X \\ SnMe_3 \end{matrix}$$

(293)   **(294)**

The tetrakis(triphenylphosphine)palladium(0)-catalysed reaction of tributylstannane with phenylacetylene results in a 1:1 mixture of the adducts **295** and **296**[214].

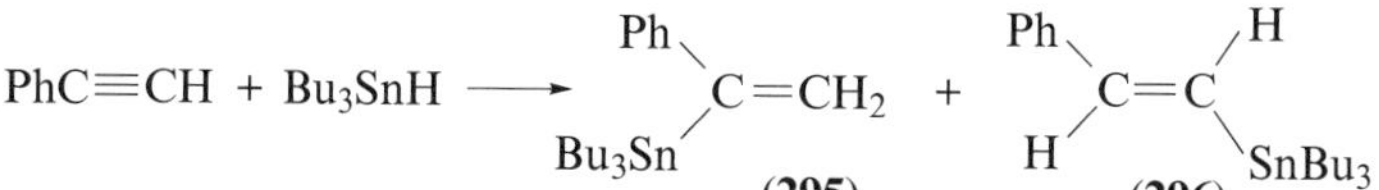

$$PhC \equiv CH + Bu_3SnH \longrightarrow \begin{matrix} Ph \\ Bu_3Sn \end{matrix} C = CH_2 + \begin{matrix} Ph \\ H \end{matrix} C = C \begin{matrix} H \\ SnBu_3 \end{matrix}$$

(295)   **(296)**

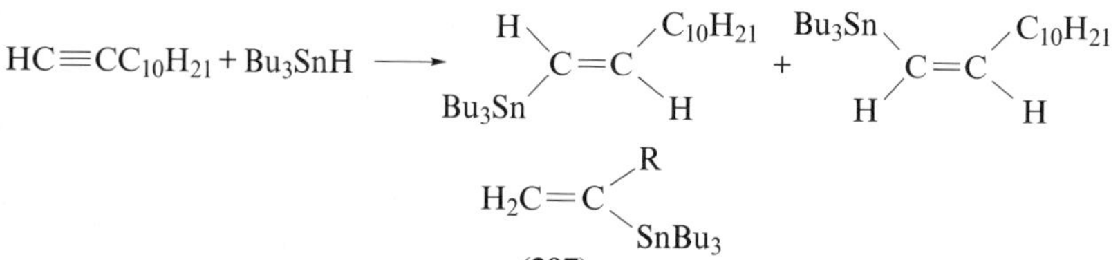

$$HC \equiv CC_{10}H_{21} + Bu_3SnH \longrightarrow \begin{matrix} H \\ Bu_3Sn \end{matrix} C = C \begin{matrix} C_{10}H_{21} \\ H \end{matrix} + \begin{matrix} Bu_3Sn \\ H \end{matrix} C = C \begin{matrix} C_{10}H_{21} \\ H \end{matrix} \qquad (40)$$

$$H_2C = C \begin{matrix} R \\ SnBu_3 \end{matrix}$$

(297)

Tributyltin hydride adds to terminal acetylenes in the presence of triethylborane to give mixtures of (E)- and (Z)- alkenes, in which the tin atom is attached to C(1), e.g. equation 40[215,216]. The reverse regiochemistry is observed in the reaction of tributylstannane with terminal acetylenes in the presence of rhodium complexes, e.g. [RhCl(1,5-cyclooctadiene)]₂, to yield exclusively the adducts **297**[217].

The tetrakis(triphenylphosphine)palladium - catalysed hydrostannation of hex-3-yn-2-one (**298**) with trimethylstannane proceeds with high regio- and stereoselectivity to afford **299**[218]. Stannylalkenes **301** (R = Bu, Ph, HOCH₂CH₂ or Me₂NCH₂) are produced regioselectively from the silicon–tin compound **300** and terminal acetylenes under catalysis by palladiumtetrakis(triphenylphosphine), followed by desilylation with tetrabutylammonium fluoride[219].

$$\text{EtC}\equiv\text{CCOMe} + \text{Me}_3\text{SnH} \longrightarrow \underset{\text{H}}{\overset{\text{Et}}{>}}\text{C}=\text{C}\underset{\text{SnMe}_3}{\overset{\text{COMe}}{<}}$$

(**298**)   (**299**)

$$\text{RC}\equiv\text{CH} + \text{Bu}_3\text{Sn}-\text{SiMe}_2\text{Ph} \longrightarrow \underset{\text{Bu}_3\text{Sn}}{\overset{\text{R}}{>}}\text{C}=\text{CH}_2$$

(**300**)   (**301**)

1-Decyne reacts with tributyltin-diethylaluminium in the presence of copper(I) cyanide in THF to yield a mixture of **302** and **303** in the ratio 91:9; in contrast, when the reaction is conducted in THF–HMPA the ratio is reversed to 6:94[220].

$$\text{HC}\equiv\text{CC}_8\text{H}_{17} + \text{Bu}_3\text{Sn}-\text{AlEt}_2 \longrightarrow \text{H}_2\text{C}=\text{C}\underset{\text{SnBu}_3}{\overset{\text{C}_8\text{H}_{17}}{<}} \quad + \quad \underset{\text{Bu}_3\text{Sn}}{\overset{\text{H}}{>}}\text{C}=\text{C}\underset{\text{H}}{\overset{\text{C}_8\text{H}_{17}}{<}}$$

(**302**)   (**303**)

The steric course of the addition of various (trimethylstannyl)cuprate reagents to ethyl 2-butynoate (**304**) to give (Z)- or (E)- isomers of **305** depends on the constitution of the reagents, as indicated in equation 41, which gives the main products of the two reactions[221].

$$\text{MeC}\equiv\text{CCO}_2\text{Et} \quad \xrightarrow[\text{Me}_3\text{SnCu·LiBr·Me}_2\text{S}]{[\text{Me}_3\text{SnCuSPh}]\text{Li}} \quad \begin{array}{c} \underset{\text{Me}_3\text{Sn}}{\overset{\text{Me}}{>}}\text{C}=\text{C}\underset{\text{CO}_2\text{Et}}{\overset{\text{H}}{<}} \\ (Z)\text{-}(\mathbf{305}) \\ \\ \underset{\text{Me}_3\text{Sn}}{\overset{\text{Me}}{>}}\text{C}=\text{C}\underset{\text{H}}{\overset{\text{CO}_2\text{Et}}{<}} \end{array}$$

(**304**)

(41)

(E)-(**305**)

The action of triethylborane on lithium acetylides **306** (R¹ = hexyl, Ph or Me₃SiCH₂) leads to the salts **307**, which react with trimethyltin chloride to form the stannanes **308**. The latter are useful precursors for alkenylstannanes **309**, which are obtained by the sequential addition of butyllithium, the copper(I) bromide–dimethylsulphide complex and an alkyl halide $R^2X^{222a,b}$.

$$LiC\equiv CR^1 \xrightarrow{BEt_3} Li^+Et_3\bar{B}-C\equiv CR^1$$

(306)   **(307)**

$$\xrightarrow{Me_3SnCl} \underset{Et_2B}{\overset{Et}{}}C=C\underset{SnMe_3}{\overset{R^1}{}} \longrightarrow \underset{R^2}{\overset{Et}{}}C=C\underset{SnMe_3}{\overset{R^1}{}}$$

(308)   **(309)**

$$R^2 = Me \text{ or } CH_2{=}CHCH_2$$

Treatment of acetylene with the salt $Bu_3Sn(Me)CuCNLi_2$, followed by an electrophile, gives (Z)-vinylstannanes. Thus trimethylsilyl chloride affords **310**, bromine the bromoalkenylstannane **311**, and cyclohex-2-enone the conjugate adduct **312**[223].

$$Me_3Si\diagdown_{=}\diagup SnBu_3 \qquad Br\diagdown_{=}\diagup SnBu_3$$

(310)   **(311)**   **(312)**

(Z)-Distannylalkenes **314** are obtained by the tetrakis(triphenylphosphine)palladium–catalysed *cis*-addition of hexamethyldistannane to terminal acetylenes **313** ($R = H$, Bu, Ph, $PhCH_2$, $HOCH_2$ or CO_2Me)[224a,b]. A similar reaction of terminal alkynes with the tin–silicon compound $Me_3Sn–SiMe_2Bu^t$ results in the (Z)-adducts **315** regiospecifically[225]. $Me_3Sn–SiMe_3$ behaves analogously towards terminal acetylenes; with ethyl phenylpropiolate only compound **316** is produced[226].

$$RC\equiv CH + Me_2Sn-SnMe_3 \longrightarrow \underset{Me_3Sn}{\overset{R}{}}C=C\underset{SnMe_3}{\overset{H}{}}$$

(313)   **(314)**

$$\underset{Me_3Sn}{\overset{R}{}}C=C\underset{SiMe_2Bu^t}{\overset{H}{}} \qquad \underset{Me_3Sn}{\overset{Ph}{}}C=C\underset{SiMe_3}{\overset{CO_2Et}{}}$$

(315)   **(316)**

The palladium–catalysed hydrostannylation of the thioacetylenes **317** (R = alkyl or $HOCH_2$) with tributyltin hydride affords selectively the (E)-alkenes **318**[227]. Allenylstannanes **320** ($R^1 = Me$ or Bu; $R^2 = H$, Et, $PhCH_2$ or $PhCH_2CH_2$) are formed in the reaction of lithium tributyltin with the acetylenes **319**[228].

$$\text{PhSC} \equiv \text{CR} \longrightarrow \underset{\text{Bu}_3\text{Sn}}{\overset{\text{PhS}}{\diagdown}} \text{C} = \text{C} \underset{\text{H}}{\overset{\text{R}}{\diagup}}$$

(317)   (318)

$$\text{R}^1\text{—C} \equiv \text{C—CHR}^2\text{OEt} \longrightarrow \underset{\text{Bu}_3\text{Sn}}{\overset{\text{R}^1}{\diagdown}} \text{C} = \text{C} = \text{C} \underset{\text{OEt}}{\overset{\text{R}^2}{\diagup}}$$

(319)   (320)

Additions of trihalogermanes to alkynes, alkenes and carbonyl compounds have been reviewed[229]. Hydrogermylation of phenylacetylene with triphenylgermane in the presence of Pd(PPh$_3$)$_4$ gives a mixture of three vinylgermanes (equation 42)[230].

PhC$\equiv$CH + Ph$_3$GeH

$$\longrightarrow \underset{\text{Ph}_3\text{Ge}}{\overset{\text{Ph}}{\diagdown}} \text{C} = \text{CH}_2 + \underset{\text{H}}{\overset{\text{Ph}}{\diagdown}} \text{C} = \text{C} \underset{\text{H}}{\overset{\text{GePh}_3}{\diagup}} + \underset{\text{H}}{\overset{\text{Ph}}{\diagdown}} \text{C} = \text{C} \underset{\text{GePh}_3}{\overset{\text{H}}{\diagup}} \quad (42)$$

The triethylborane–catalysed addition of triphenylgermane to 1-dodecyne at $-78°$C affords solely the (Z)-product **321**; at 60 °C the (E)-isomer **322** is obtained, which is also produced when the (Z)-isomer is heated in benzene in the presence of traces of triethylborane and triphenylgermane[231].

$$\text{HC} \equiv \text{CC}_{10}\text{H}_{21} + \text{Ph}_3\text{GeH} \longrightarrow \underset{\text{H}}{\overset{\text{Ph}_3\text{Ge}}{\diagdown}} \text{C} = \text{C} \underset{\text{H}}{\overset{\text{C}_{10}\text{H}_{21}}{\diagup}} \longrightarrow \underset{\text{H}}{\overset{\text{Ph}_3\text{Ge}}{\diagdown}} \text{C} = \text{C} \underset{\text{C}_{10}\text{H}_{21}}{\overset{\text{H}}{\diagup}}$$

(321)   (322)

Dimethyl(2-thienyl)germane (**323**) adds to terminal acetylenes **324** (R = t-Bu, Ph, MeOCH$_2$ or CO$_2$Me) in the presence of H$_2$PtCl$_6$ to yield mixtures of regioisomeric adducts (equation 43). Compound **323** is less reactive than its silicon analogue[232]. Palladium(II) chloride catalyses the reaction of (cyano)trimethylgermane with phenylacetylene to give (Z)-3-(trimethylgermyl)-2-phenylprop-2-enonitrile (**325**)[233].

$$\underset{\text{S}}{⟨⟩}\text{—GeMe}_2\text{H} + \text{HC} \equiv \text{CR} \longrightarrow \underset{\text{S}}{⟨⟩}\text{—GeMe}_2 \underset{\text{H}}{\overset{}{}} \underset{\text{R}}{\overset{\text{H}}{}}\text{C}=\text{C} + \underset{\text{S}}{⟨⟩}\text{—GeMe}_2 \underset{\text{R}}{\overset{}{}}\text{C}=\text{CH}_2 \quad (43)$$

(323)   (324)

$$\text{HC} \equiv \text{CPh} + \text{NCGeMe}_3 \longrightarrow \underset{\text{Me}_3\text{Ge}}{\overset{\text{H}}{\diagdown}} \text{C} = \text{C} \underset{\text{CN}}{\overset{\text{Ph}}{\diagup}}$$

(325)

8. Formation of sulphur, selenium and tellurium compounds

The reactions of acetylenes with sulphenyl halides and related compounds have been reviewed[234]. N-(Benzenesulphenyl)-p-nitroaniline (326) reacts with 1-pentyne in acetonitrile in the presence of boron trifluoride etherate to yield solely the 2-azabutadiene derivative 327[235].

$$\text{PhS—NHAr} + \text{HC}\equiv\text{CPr} \xrightarrow{\text{MeCN}} \underset{\substack{\text{H} \\ \quad }}{\overset{\text{PhS}}{\diagup}}\text{C}=\text{C}\underset{\text{N}=\text{C}}{\overset{\text{Ph}}{\diagup}}\text{NHAr}$$

(326)

$\text{Ar} = p\text{-O}_2\text{NC}_6\text{H}_4$ (327)

Treatment of (4-nitrobenzenesulphenyl)aniline (328) with internal acetylenes 329 (R^1, R^2 = alkyl or Ph) in acetic acid under the same conditions leads to mixtures of regioisomeric acetates 331 and 332. It is suggested that the reactions proceed by way of thiirenium ions 330[236]. The action of the sulphenylaniline 326 on alkynes in the presence of tetrabutylammonium tetrafluoroborate results in compounds which represent adducts of phenylsulphenyl fluoride to the triple bond. Thus 1-pentyne affords solely the (E)-product 333[237a,b].

$$\text{PhNH—SAr} \longrightarrow \text{Ph}-\overset{\overset{\displaystyle BF_3^-}{\underset{\displaystyle |}{+}}}{\underset{\displaystyle H}{N}}-\text{SAr} \xrightarrow[(329)]{R^1C\equiv CR^2} \underset{R^2}{\overset{R^1}{\diagup}}\overset{+}{\underset{\quad}{\triangleright}}\overset{+}{S}-\text{Ar}$$

(328)   (330)   PhNHBF$_3^-$

$$\xrightarrow{\text{AcOH}} \underset{\text{ArS}}{\overset{R^1}{\diagup}}\text{C}\underset{R^2}{\overset{\text{OAc}}{}}\text{C} \quad + \quad \underset{R^2}{\overset{\text{ArS}}{\diagup}}\text{C}\underset{\text{OAc}}{\overset{R^1}{}}\text{C}$$

(331)   (332)

$\text{Ar} = p\text{-O}_2\text{NC}_6\text{H}_4$

$$\text{326} + \text{HC}\equiv\text{CPr} \xrightarrow{\text{Bu}_4\text{N}^+\text{BF}_4^-} \underset{\text{H}}{\overset{\text{PhS}}{\diagup}}\text{C}=\text{C}\underset{\text{F}}{\overset{\text{Pr}}{\diagup}}$$

(333)

Phthalimidosulphenyl chloride (334), a stable solid, adds to terminal alkynes 335 (R = Bu or Bu^t) in a $trans$-fashion to give the chloroalkenes 336, accompanied by up to 15% of the other regioisomers. But-2-yne gives the analogue 337, which is transformed into 338 by the action of t-butyllithium. Phthalimidosulphenyl chloride has thus acted as a synthon for the inaccessible t-butylsulphenyl chloride[238].

Propargyl alcohols 339 (R^1 = H, alkyl or Ph; R^2 = H or Me; R^3 = H, D or alkyl) react with 4-morpholinesulphenyl chloride (340) to yield the rearranged allenic sulphinamides 341, which are converted into the acetylenes 342 by acid-catalysed hydrolysis[239].

'Iodosulphonization' of 1-alkynes with sodium tosylate/iodine results in (E)-2-iodo-1-tosyl-1-alkenes **343**, hydrogenation of which in the presence of palladium on charcoal and quinoline yields the pure (Z)-vinyl sulphones **344**[240].

Olefinic betaines, e.g. **345**, are formed from alkynes and the dimethylsulphide–sulphur trioxide complex[241a,b]. The surprising formation of the ketene **347** on heating the perfluorinated acetylene **346** with sulphur trioxide has been reported[242].

The stereoselective free-radical addition of diphenyl diselenide to the alkynes **348** (R = pentyl, HOCH$_2$, SiMe$_3$ or Ph) gives mixtures of (E)- and (Z)-olefins **349**, in which the former predominate. Application of the reaction to allyl propargyl ether resulted in 10% of the adduct **350** and 65% of the tetrahydrofuran derivative **351**[243].

$$RC\equiv CH + Ph_2Se_2 \longrightarrow$$
(348)

(349)

$$H_2C=CHCH_2OCH_2C\equiv CH \longrightarrow$$

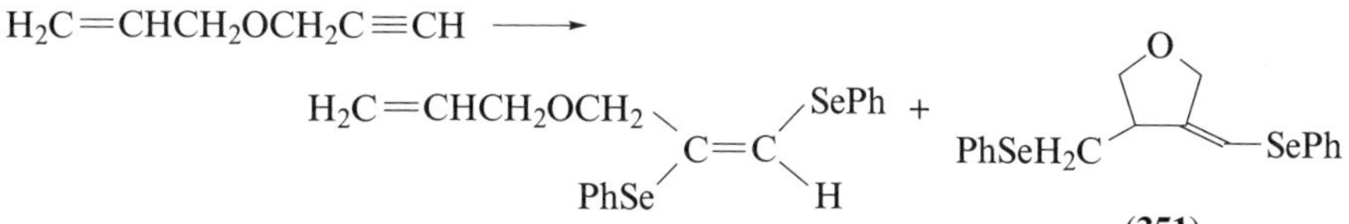

(350)

(351)

Terminal acetylenes **352** (R = i-Bu, Me$_3$Si, Ph, HOCH$_2$ or HOCMe$_2$) react with diphenyl diselenide in the presence of tetrakis(triphenylphosphine)palladium to give exclusively (Z)-adducts **353**; in the presence of carbon monoxide S-phenyl esters **354** of unsaturated carboselenoic acids are formed. Diaryl disulphides undergo analogous reactions[244].

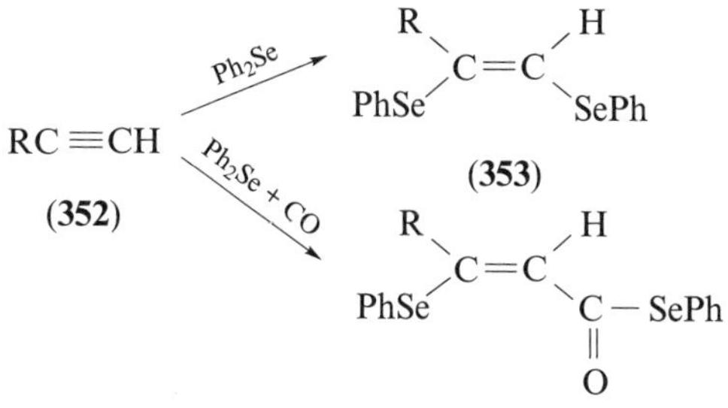

$$RC\equiv CH$$
(352)

(353)

(354)

Benzeneselenyl tosylate (**355**), generated from phenylselenyl chloride and silver tosylate, adds to alkynes **356** (R^1, R^2 = H, alkyl, Ph or CO$_2$Me) in a *trans*-manner to form mixtures of regioisomers **357** and **358**[245a,b].

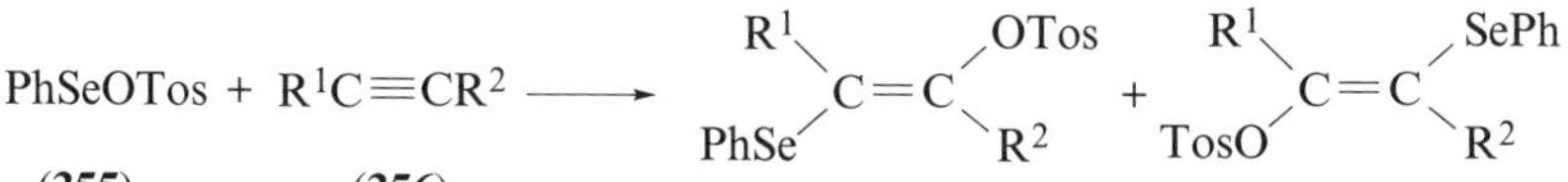

$$PhSeOTos + R^1C\equiv CR^2 \longrightarrow$$
(355) **(356)**

(357) **(358)**

'Phenylselenofluoration' of acetylenes **359** (R = Pr or Bu) to give **360** is accomplished by the joint action of N-(phenylseleno)phthalimide and Et$_3$N·3HF[246]. Free-radical selenosulphonation of conjugated enynes by PhSeTos under photochemical conditions or in the presence of azoisobutyronitrile yields mixtures of stereoisomers, e.g. equation 44[247].

The action of p-methoxyphenyltellurium trichloride on terminal alkynes RC≡CH (R = cyclohexyl, HOCH$_2$ or Ar) results in cis-adducts **361** (Ar = p-MeOC$_6$H$_4$)[248]. Sodium benzenetellurolate, PhTeNa, formed in situ from diphenyl ditelluride and sodium borohydride in ethanol, reacts with phenylacetylene, 1-octyne or propargyl alcohol by *trans*-addition to afford mainly the regioisomers **362**[249]. Bis(phenyltelluro)alkenes **363** (R = hexyl, Ph or HOCH$_2$) are obtained as (*E*)-isomers when mixtures of diphenyltelluride and terminal acetylenes are exposed to visible light[250].

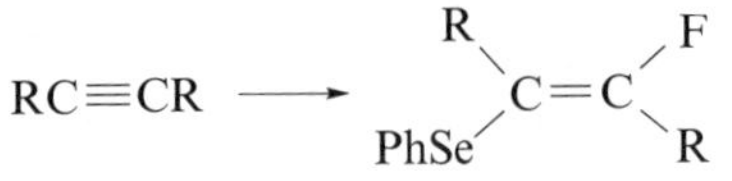

$$RC{\equiv}CR \longrightarrow \underset{PhSe}{\overset{R}{\diagdown}}C=C\underset{R}{\overset{F}{\diagup}}$$

(359)   **(360)**

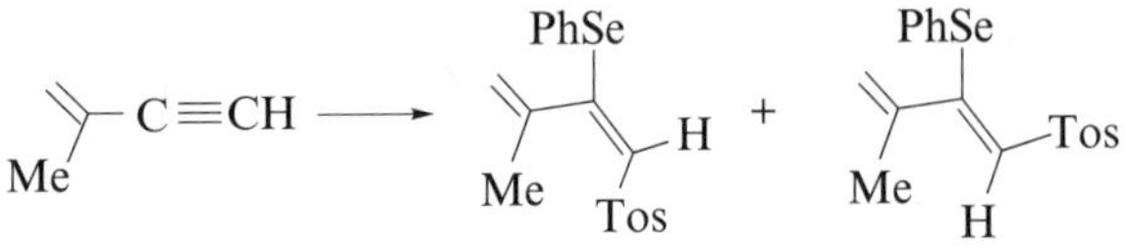

$$\underset{Me}{\diagup}C{\equiv}CH \longrightarrow \underset{Me}{\overset{PhSe}{\diagup}}C=C\underset{Tos}{\overset{H}{}} + \underset{Me}{\overset{PhSe}{\diagup}}C=C\underset{H}{\overset{Tos}{}} \qquad (44)$$

$$\underset{Cl}{\overset{R}{\diagdown}}C=C\underset{TeArCl_2}{\overset{H}{\diagup}}$$

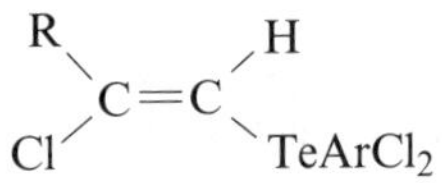

(361)

$$\underset{H}{\overset{R}{\diagdown}}C=C\underset{TePh}{\overset{H}{\diagup}} \qquad \underset{PhTe}{\overset{R}{\diagdown}}C=C\underset{H}{\overset{TePh}{\diagup}}$$

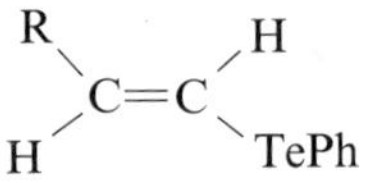

(362) 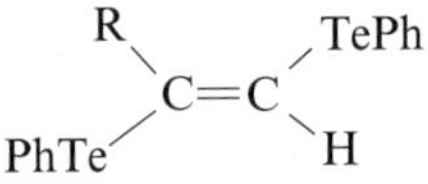   **(363)**

9. Miscellaneous reactions

The base-catalysed reaction of phenol with the acetylene **364** affords solely the (*Z*)-adduct **365**, whereas thiophenol gives the (*E*)-analogue **366**. The contrasting behaviour of the reagents was rationalized in terms of frontier–orbital interactions[251].

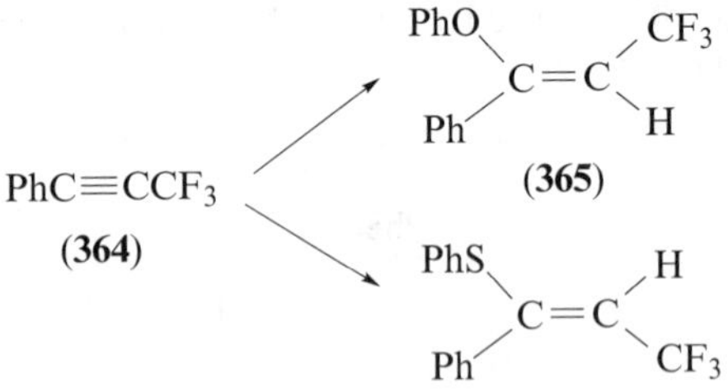

$$PhC{\equiv}CCF_3 \quad \nearrow \quad \underset{Ph}{\overset{PhO}{\diagdown}}C=C\underset{H}{\overset{CF_3}{\diagup}}$$

(364)   **(365)**

$$\searrow \quad \underset{Ph}{\overset{PhS}{\diagdown}}C=C\underset{CF_3}{\overset{H}{\diagup}}$$

(366)

cis-Addition of hydrogen cyanide to a veriety of terminal and internal acetylenes in the Markovnikov sense is accomplished by the combined action of potassium cyanide, K$_2$[Ni(CN)$_4$] and sodium borohydride or zinc dust[252]. A twofold Michael reaction of methyl propiolate with nitroethane in the presence of potassium fluoride and tetrabutylammonium fluoride results in **367** as a mixture of geometrical isomers[253].

Vinylalanes **368**, formed from terminal acetylenes and trimethylaluminium in the presence of zirconocene dichloride, react with conjugated enones in the presence of CuCN·2LiCl to give mixtures of stereoisomeric adducts (equation 45)[254]. A similar reaction of 1-octyne with $Cp_2Zr(H)Cl$, MeLi and $Me_2Cu(CN)Li_2$ affords the vinylic cuprate $(C_6H_{13}CH=CH)MeCu(CN)Li_2$, which forms solely the (E)-adduct **369** with cyclohex-2-enone[255].

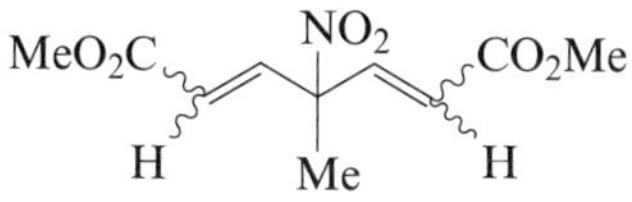

$$MeO_2C \diagdown \diagup \underset{Me}{\overset{NO_2}{|}} \diagup \diagdown CO_2Me$$

(**367**)

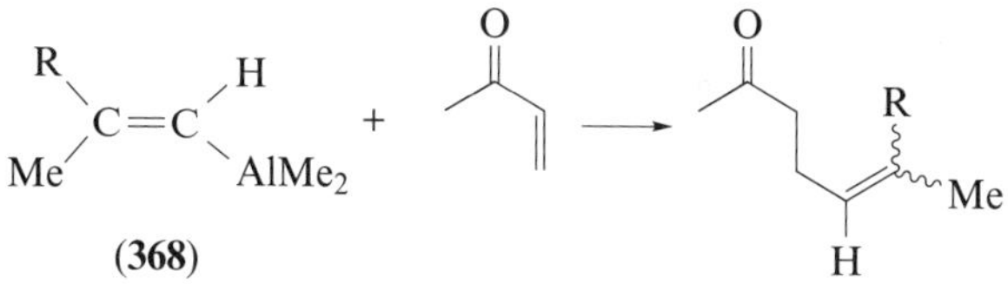

$$\underset{Me}{\overset{R}{>}} C = C \underset{AlMe_2}{\overset{H}{<}} \quad + \quad \text{(enone)} \quad \longrightarrow \quad \text{(adduct)} \qquad (45)$$

(**368**)

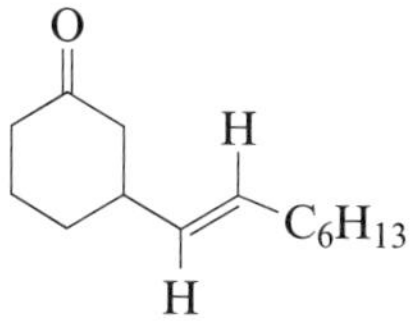

(**369**)

The combined action of zinc and tantalum(V) chloride on alkynes generates low-valent tantalum–alkyne complexes, which react with carbonyl compounds to yield (E)-allylic alcohols stereoselectively. The tantalum–6-dodecyne complex, for instance, and β-phenyl-propionaldehyde afford the alcohol **370** in 96% yield[256]. Mixtures of structurally isomeric α,β-unsaturated (E)-amides **371** and **372** (R^1, R^2 = alkyl or Me_3Si) are obtained from the tantalum complexes of alkynes $R^1C≡CR^2$ and isocyanates R^3NCO (R^3 = Bu or Ph)[257].

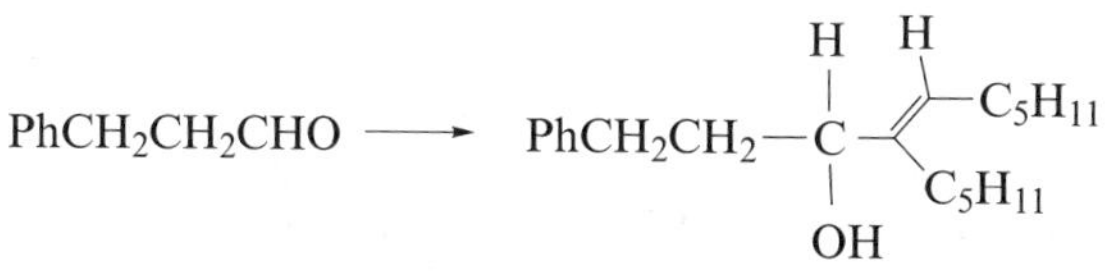

$$PhCH_2CH_2CHO \longrightarrow PhCH_2CH_2 - \underset{\underset{OH}{|}}{\overset{\overset{H}{|}}{C}} - C \underset{C_5H_{11}}{\overset{H}{\diagup}} \diagdown C_5H_{11}$$

(**370**)

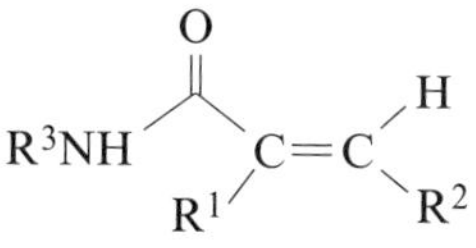

(**371**)

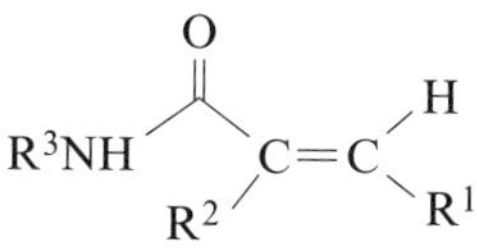

(**372**)

The action of methanol on terminal acetylenes 373 (R = hexyl, Ph, PhOCH$_2$, CH$_2$=CHCH$_2$OCH$_2$ etc.) in the presence of triethylamine and a catalytic amount of mercury(II) chloride furnishes the adducts 374 regiospecifically[258].

$$RC\equiv CH \longrightarrow \underset{MeO}{\overset{R}{\diagdown}}C=CH_2$$

(373)   (374)

'Methoxymercuration' of diphenylacetylene by mercury(II) acetate in methanol results in a mixture of geometrical isomers 375 and 376. Internal aliphatic acetylenes yield mercurated ketones; 2-nonyne, for example, gives a mixture of 377 and 378; 2-heptyne behaves in an analogous fashion[259]. The reaction of mercury(II) thiocyanate with terminal alkynes affords 379 (R = Pr, Bu, t-Bu, hexyl or Ph), which are converted into unsaturated isothiocyanates 380 by acidolysis[260].

$$PhC\equiv CPh \xrightarrow[Hg(OAc)_2]{MeOH} \underset{MeO}{\overset{Ph}{\diagdown}}C=C\overset{Ph}{\underset{HgOAc}{\diagup}} + \underset{MeO}{\overset{Ph}{\diagdown}}C=C\overset{HgOAc}{\underset{Ph}{\diagup}}$$

(375)   (376)

$$MeC\equiv CC_6H_{13} \longrightarrow \underset{AcOHg}{\overset{Me}{\diagdown}}C-\underset{H}{\underset{O}{C}}-C_6H_{13} + Me-\underset{O}{C}-C\overset{C_6H_{13}}{\underset{HgOAc}{\diagup}}$$

(377)   (378)

$$RC\equiv CH \xrightarrow{Hg(SCN)_2} \underset{N}{\overset{R}{\diagdown}}C=C\overset{H}{\underset{HgSCN}{\diagup}} \longrightarrow \underset{N}{\overset{R}{\diagdown}}C=CH_2$$

(379)   (380)

The free-radical reaction of diphenylphosphine with terminal acetylenes 381 (R = Pr, t-Bu or Ph) in the presence of the initiator azoisobutyronitrile or under UV irradiation furnishes the (Z)-alkenylphosphines 382[261]. 'Hydrophosphorylation' of propargyl alcohol 383 (R = HOCH$_2$) and its derivatives 383 (R = BuOCH$_2$, HOCMe$_2$ or PhCO$_2$CH$_2$) with Ph$_2$POH results in the adducts 384, which may add a further molecule of the reagent to yield 385[262].

$$RC\equiv CH \xrightarrow{Ph_2PH} \underset{H}{\overset{R}{\diagdown}}C=C\overset{PPh_2}{\underset{H}{\diagup}}$$

(381)   (382)

$$RC{\equiv}CH \xrightarrow{Ph_2POH} \underset{(384)}{\begin{array}{c} R \\ \diagdown \\ H \end{array} C{=}C \begin{array}{c} \diagup \\ \diagdown \\ H \end{array}} \xrightarrow{Ph_2POH} \underset{(385)}{RCH{-}CH_2{-}P}$$

(383)

B. Formation of Cyclic Compounds

Several reviews on transition-metal-mediated cycloaddition reactions of alkynes leading to carbocycles and heterocycles of various ring sizes have appeared[263–265].

1. Formation of three-membered rings

The Fischer carbene-complex 386 undergoes tricyclization to 387 in boiling benzene[266]. Chlorofluorocarbene adds to alkynes much more readily than dichlorocarbene: thus the sterically shielded acetylene 388, which does not react with dichlorocarbene, forms an adduct with fluorocarbene which was converted into the cyclopropenone 389 by hydrolysis *in situ*[267].

(386)   (387)

(388)   (389)

Cycloalkadiynes 390 (n = 3, 4 or 5) react with ethyl diazoacetate in the presence of rhodium(II) acetate to yield the dicyclopropenes 391 as mixtures of *syn*- and *anti*-isomers, which were transformed into dicyclopropenylium salts 392[268].

(390)   (391)   (392)

The action of diphenylacetylene on the dimethylaminocarbene complex **393** affords **394**, which reacts with benzylidene-*N*-methylamine *in situ* to give the bicyclic lactam **395**[269].

(393)   (394)   (395)

Mixtures of 3-phthalimidoazirines **398** and **399** are produced from *N*-aminophthalimide and acetylenic esters **396** (R^1 = H, R^2 = $(CH_2)_8CO_2Me$; R^1 = Me, R^2 = $(CH_2)_7CO_2Me$; R^1 = octyl, R^2 = $(CH_2)_7CO_2Me$) in the presence of palladium(IV) acetate. It is suggested that the process involves the formation of intermediate *N*-phthalimidoazirines **397**[270].

(396)   (397)

(398)   (399)

The aluminium trichloride–catalysed reaction of dichlorophosphines $RPCl_2$ (R = Me, *t*-Bu or Ph) with 3-hexyne affords the 1-chlorophosphirenium tetrachloroaluminates **400**.

(400)

(401)   (402)

3-Hexyne reacts with phosphorus trichloride to give the phosphirenium salt **402** by insertion of a second molecule of the alkyne into the phosphorus–chlorine bond of the intermeazdiate **401**[271].

2. Formation of four-membered ring

a. Homocyclic compounds. Successive treatment of lithium(dimethyl) phenylsilane with methylmagnesium iodide, copper(I) iodide and 5-tosyloxy-1-pentyne (**403**) results in the (silylmethylene)cyclobutane **404**; the analogous derivative **405** of 1-butyne affords a mixture of the methylenecyclopropane **406** and the silylated cyclobutene **407** in this reaction[272].

Cyclobutenones are obtained by the cycloaddition of dichloroketene (from trichloroacetyl chloride and zinc–copper couple) to alkynes, followed by dechlorination of the resulting dichlorocyclobutenones with zinc. The preparation of 3-butylcyclobutenone from 1-hexyne by this method (equation 46) has been described in detail[273].

The tungsten carbene complex **408** reacts with phenylacetylene to give the phenol **409**, accompanied by the methanol trapping product **410** and the cyclobutenone **411**[274]. Cycloaddition of 1,1-dimethylallene to ethyl propiolate affords the cyclobutene **412**[275].

$$\underset{Me}{\overset{Me}{>}}C{=}C{=}CH_2 \ + \ HC{\equiv}CCO_2Et \ \longrightarrow \ \underset{Me}{\overset{Me}{>}} \hspace{-0.3em}\square\hspace{-0.3em}-CO_2Et$$

(412)

The kinetically stabilized cyclobutadiene **413** adds methyl propiolate in two senses to produce a mixture of the Dewar benzenes **414** and **415**, but addition of a number of terminal alkynes $RC{\equiv}CH$ [R = CO_2Me, Tos or $Ph_2P(O)$] to the *t*-butyl ester **416** gave only the former type of cycloadduct, i.e., **417**[276].

(413)	(414)	(415)

(416)	(417)

Norbornene reacts with dimethyl acetylenedicarboxylate in the presence of ruthenium(cyclooctadiene)(cyclooctatriene) to form the *exo*-cycloadduct **418**. The ability to activate norbornene is unique to low-valent ruthenium[277a,b]. The photochemical addition of symmetrical acetylenes $RC{\equiv}CR$ (R = H, Me, Ph or $SiMe_3$) to 3,6-

E = CO_2Me (418)

(419)	(420)	(421)

dihydrophthalic anhydride (419) results in a mixture of products 420 and 421, the proportion of which depends on the nature of the substituents R[278].

[1.1.1]Propellane (422) reacts with dimethyl acetylenedicarboxylate to form the 1:1 adduct 423, together with the 2:1 adducts 424 and 425[279]. Treatment of the acetylenes 426 ($NR_2=NMe_2$, NEt_2 or piperidino) with iron pentacarbonyl induces head-to-tail dimerization to yield yellow air-stable iron–cyclobutadiene complexes 427[280].

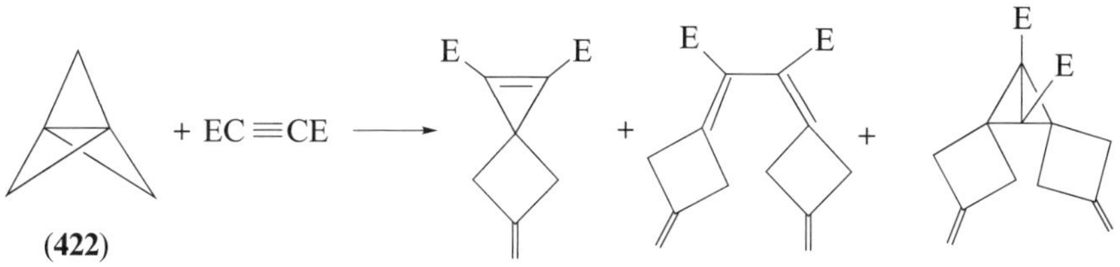

(422)   + EC≡CE ⟶

E = CO_2Me   (423)   (424)   (425)

$Me_3SiC \equiv CNR_2$ ⟶

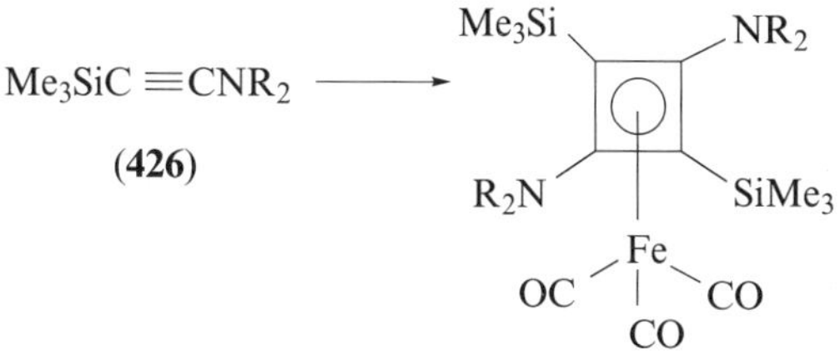

(426)

(427)

b. Heterocyclic compounds. The stable disilene 428 (Ar = $2,4,6\text{-}Me_3C_6H_2$) adds terminal acetylenes 429 (R = Ph, EtO, CO_2Me or $SiMe_3$) to yield approximately equal amounts of the stereoisomers 430 and 431, consistent with a stepwise mechanism of the cycloaddition[281]. The action of sulphur in the presence of antimony(V) chloride, or of disulphur dichloride in the presence of aluminium(III) chloride, on symmetrical acetylenes RC≡CR (R = H, Me, Et, Pr, Bu, t-Bu or Ph) at 250 K gives 1,2-dithiete radical cations 432, which on warming to 300 K are converted into dithiin radical cations 433[282].

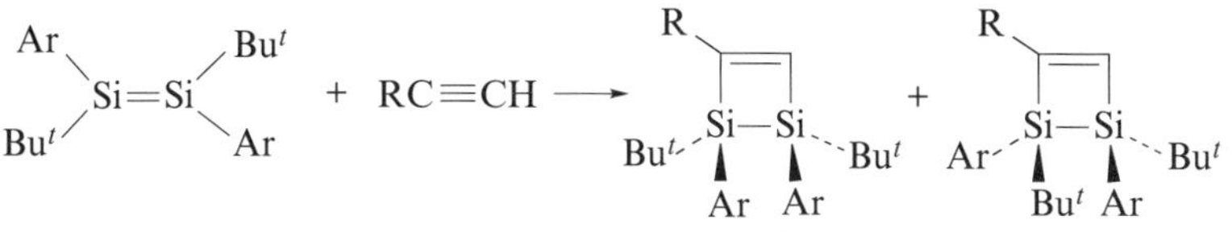

(428)   (429)   + RC≡CH ⟶   (430)   +   (431)

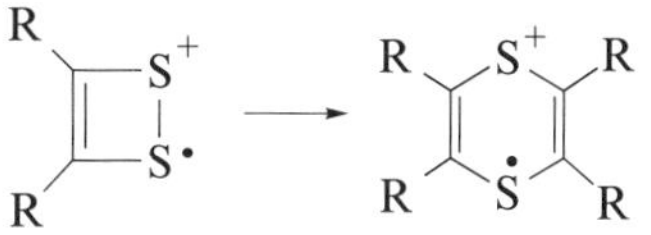

(432)   (433)

The 6-thia-1-aza-5λ^5-phosphabicyclo[3.2.0]hept-3-ene 435 is produced from the 1,3,2-diazaphosphete 434 and dimethyl acetylenedicarboxylate[283]. The reaction of the sulphur diimide **436** with the ynamine **437** at low temperatures leads to a mixture of the thiazetine **438** and its open-chain valence isomer **439**; at higher temperatures the diazabutadiene **440** is formed[284].

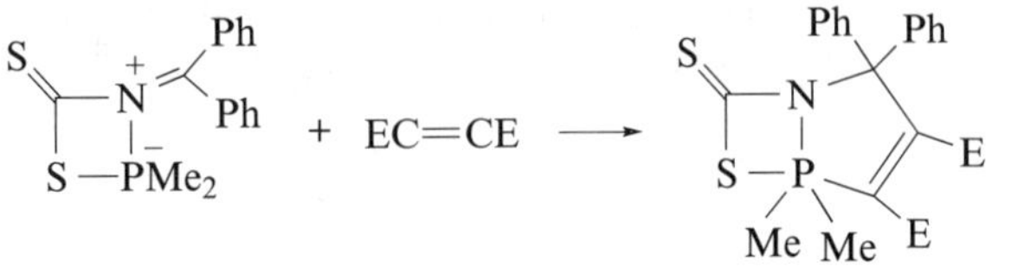

(**434**)   E = CO₂Me   (**435**)

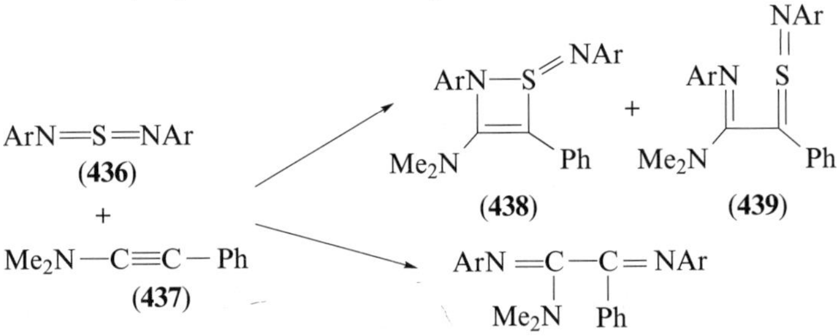

ArN=S=NAr
(**436**)

+

Me₂N—C≡C—Ph
(**437**)

(**438**)   (**439**)

ArN=C—C=NAr
      | |
   Me₂N Ph

Ar = p-O₂NC₆H₄   (**440**)

3. Formation of five-membered rings

a. Homocyclic compounds. Thermal reactions of alkynes by flash-vacuum pyrolysis have been reviewed[285]. *o*-Tolylacetylene rearranges at 740 °C to indene, presumably by way of the (arylmethylene)carbene shown in equation 47[286]. The ethynyl ketones **441** (R = H, D, Me or SiMe₃) undergo a related reaction on thermolysis to give mixtures of bicyclic and spiro-cyclopentenones, **442** and **443**[287]. The action of 1-pentyne on the chelated chromium complex **444** leads to the cyclopentadiene derivative **445**[288].

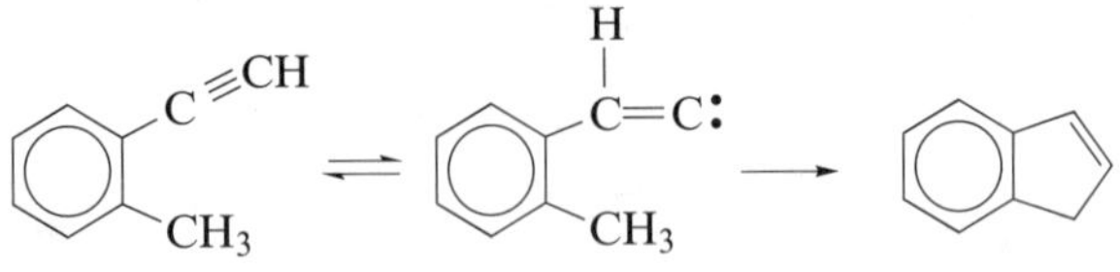

(47)

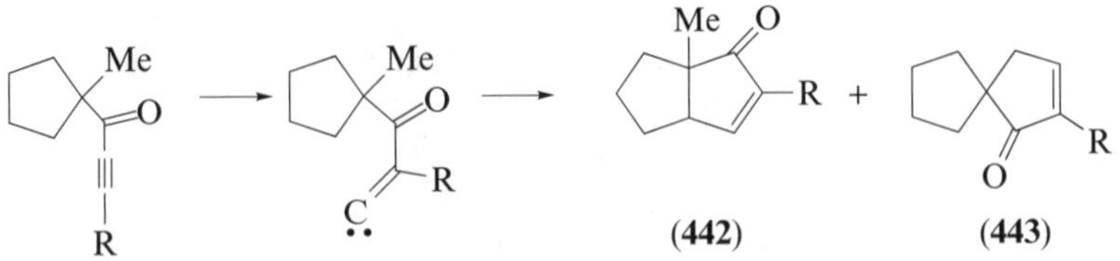

(**441**)   (**442**)   (**443**)

Alkynyl(phenyl)iodonium salts react with certain carbon nucleophiles to yield products which are converted into carbenes by reductive elimination of iodobenzene. The carbene may finally cyclize to derivatives of cyclopentene. Thus dec-1-yn-1-yl(phenyl)iodonium fluoroborate and the anion of methyldimedone afford the cyclopentene shown in equation 48[289].

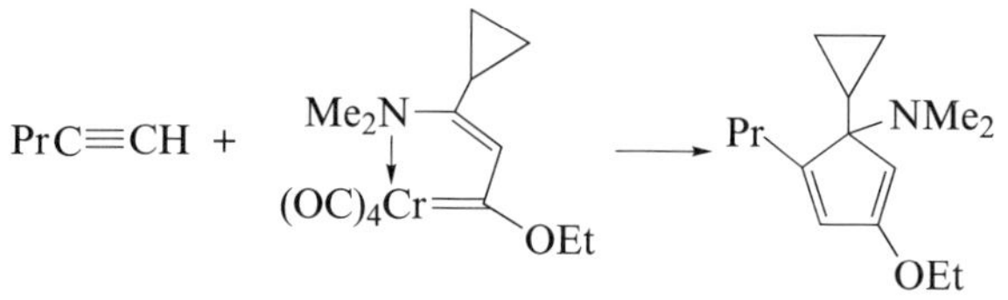

(444)   (445)

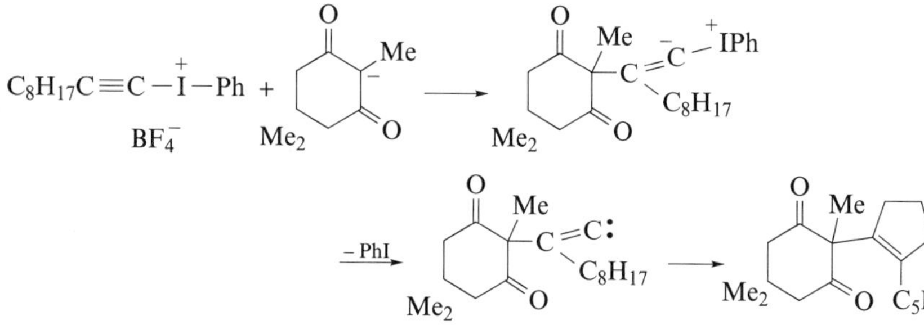

(48)

The silyloxyketene **447**, generated by a retro-Diels–Alder reaction of the bridged dihydroanthracene **446**, reacts *in situ* with diphenylacetylene to yield the methylenecyclopentenedione **451**. The process is thought to involve formation of the cyclobutenone **448**, electrocyclic ring-opening to **449**, recyclization to the diradical **450** and, finally, rearrangement to the product (equation 49). The reaction of **446** with 5-decyne gave, in addition to the analogue **452**, the quinone **454**, which is presumably formed by an

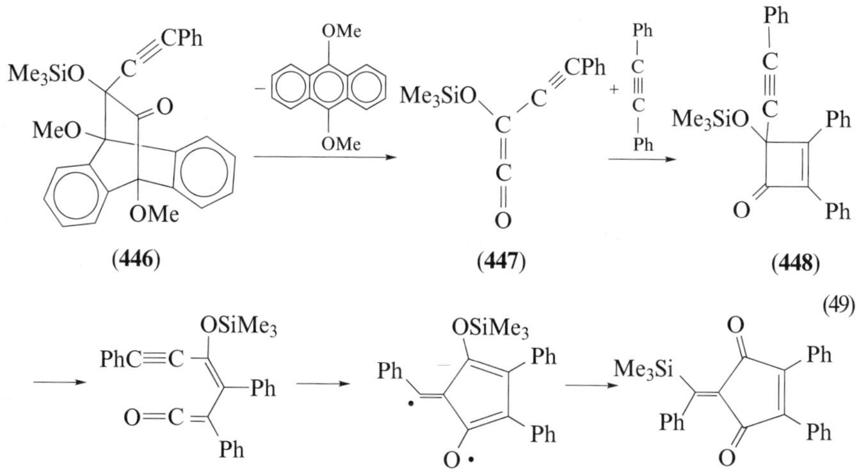

(446)   (447)   (448)

(49)

(449)   (450)   (451)

alternative cyclization of the intermediate **453** (equation 50). Formation of yet another type of product was observed when the cyano(silyloxy)ketene **456** was generated from the dihydroanthracene **455** in the presence of diphenylacetylene: only the furan **457** was isolated. It is believed to arise from an intermediate silylated oxonium betaine (equation 51)[290].

A new fulvene synthesis is the reaction of the iodostyrene **458** with two molecules of ethynyltrimethylsilane in the presence of $Pd(CH_3CN)_2Cl_2$ (equation 52)[291]. The palladium(II) acetate-triphenylphosphine-catalysed reaction of (*E*)- or (*Z*)-vinyl bromide **459** (R^1 = Ph or EtO) with diphenylacetylene or 3-hexyne results in the fulvenes **460** (R^1 = Ph or EtO, R^2 = Ph or Bu)[292].

R¹, R², C=C, H, Br **(459)** + R²–C≡C–R² **(460)** → R¹, R², R², R², R²

The 2-bromododeca-1,11-dien-6-yne derivative **461** undergoes a triple cyclization, followed by rearrangement, under the influence of a palladium catalyst to yield **462**[293]. Cross-coupling of the iodo compound **463** with (*E*)-1-hexen-1-ylzinc chloride in the presence of tetrakis(triphenylphosphine)palladium yields 68% of the indane **464** and 19% of the uncyclized product **465**[294].

(461)   E = CO₂Me   **(462)**

(463) + ClZn, H, Bu, H → **(464)** + **(465)**

The *ortho*-manganated aryl ketone complex **466** forms 3-hydroxyindenes **467** by the action of alkynes; in contrast, the complexed amide **468** reacts with diphenylacetylene to afford the diphenylindenone **469**[295].

(466) + R–C≡C–R → **(467)**

R = H or Ph

(468) + Ph–C≡C–Ph → **(469)**

Oxidation of the complex prepared from acetophenone and benzylmanganese pentacarbonyl with trimethylamine oxide, followed by the addition of an alkyne $R^1C{\equiv}CR^2$ (R^1 = H, alkyl, EtO or Ph; R^2 = H, alkyl, Ph, Me$_3$Si or CO$_2$Et), affords indenols **470** in which the bulkier substituent of the alkyne is attached to C(2). Analogous products **471** are obtained from α-tetralone[296].

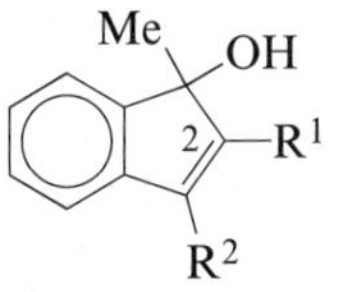

(470) 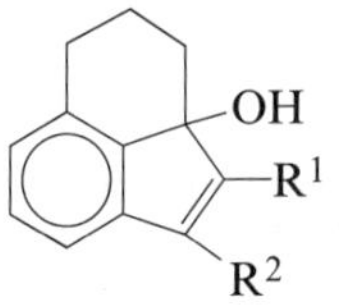   **(471)**

Several reactions of metal–carbene complexes with alkynes leading to five-membered ring compounds have been described. The action of acetylenes on the chromium phenyl(pyrrolidino)carbene complex **472** results in mixtures of indanones **473** and indenes **474**[297]. Terminal alkynes (pent-1-yne or hex-1-yne) react with the molybdenum carbene complex **475** to afford, after oxidative work-up, indanones **476**; in contrast, trimethylsilylacetylene gave only the naphthoquinone **477**[298].

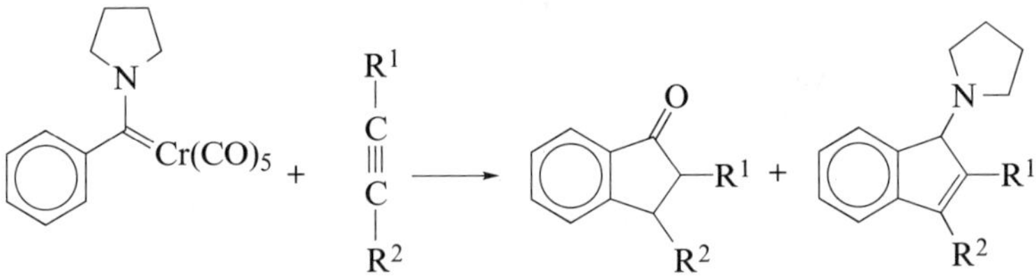

(472)   **(473)**   **(474)**

$$R^1 = H, R^2 = Bu$$
$$R^1 = R^2 = Et, Bu \text{ or } Ph$$
$$R^1 = Me, R^2 = Ph$$

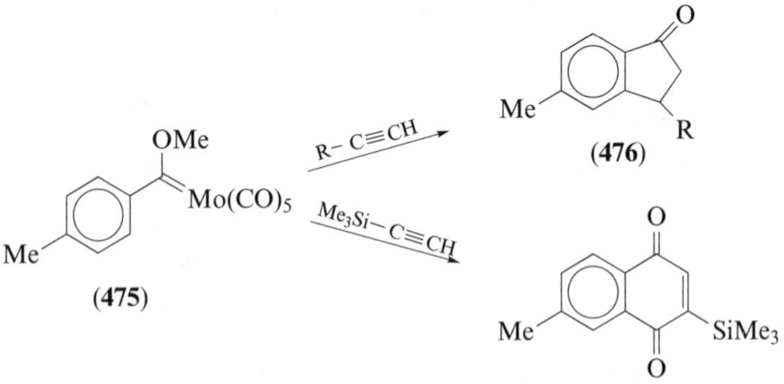

(475)

(476)

(477)

A mixture of products **479–481** is produced from the carbene complex **478** and phenylacetylene[299]. Under conditions of high dilution, terminal acetylenes from mixture of isomeric cyclopentenones, e.g. **483** and **484**, on treatment with the (cyclopropyl) (methoxy)chromium carbene complex **482**[300].

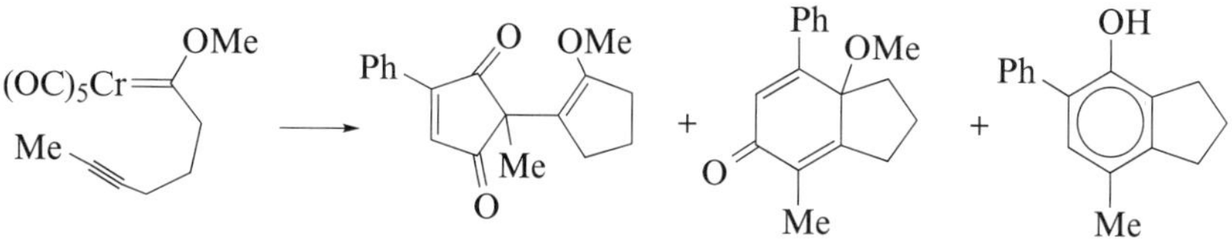

(**478**)   54%   20%   16%

(**479**)   (**480**)   (**481**)

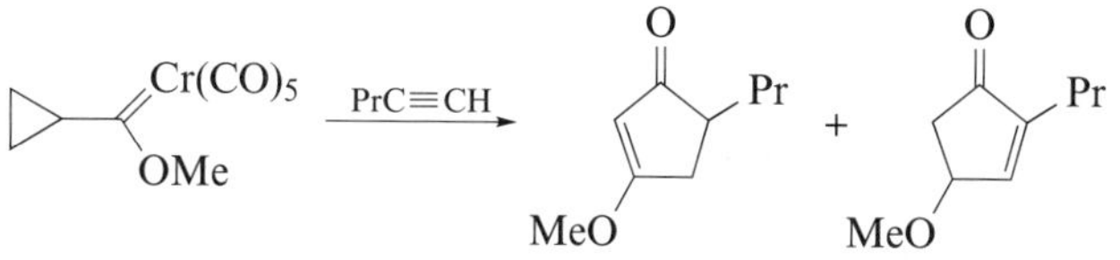

(**482**)   (**483**)   (**484**)

The Pauson–Khand reaction[301] is the formation of cyclopentenones **486** by the action of dicobalt octacarbonyl on an alkyne, followed by an alkene. The process involves the intermediacy of cobalt complexes **485** (equation 53). (For reviews, see References 302 and 303.) Terminal alkynes, including acetylene and arylacetylenes, give better yields than internal acetylenes. An example of an intramolecular Pauson–Khand reaction is the conversion of the enyne **487** into the cyclopentanocyclopentenone **488**[304].

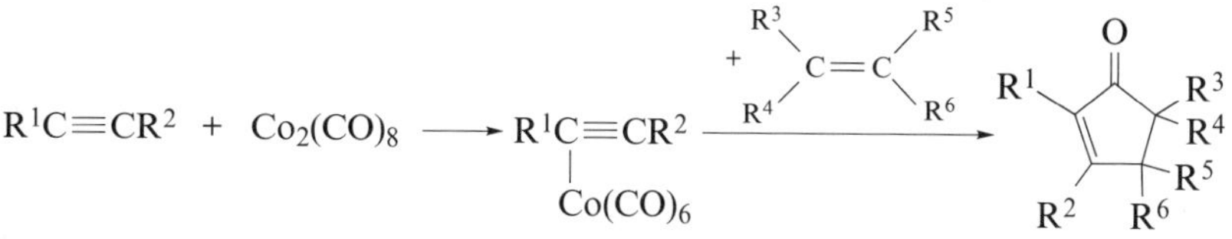

(**485**)   (**486**)

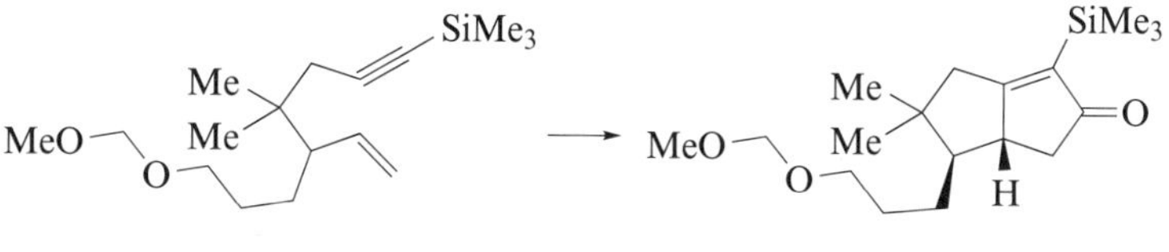

(**487**)   (**488**)

Alkyne–dicobalt hexacarbonyl complexes, formed *in situ* from alkynes RC≡CH (R = octyl or Ph), cobalt(II) bromide, zinc dust and carbon monoxide in THF, give mixtures of **489** and **490** by the action of norbornene[305].

'Directed Pauson–Khand reactions', giving only one of two possible regioisomers, occur when the carbon chain of the alkene is attached to a nitrogen or sulphur atom. Thus treatment of the propyne–dicobalthexaarbonyl complex with the thioether **491** gave solely the cyclopentenone **492**[306].

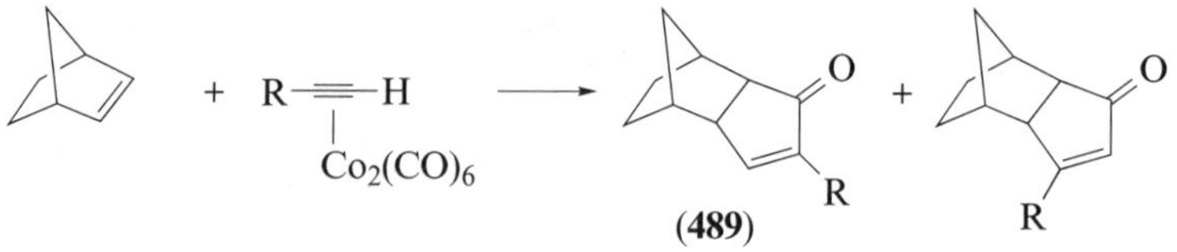

(489)   (490)

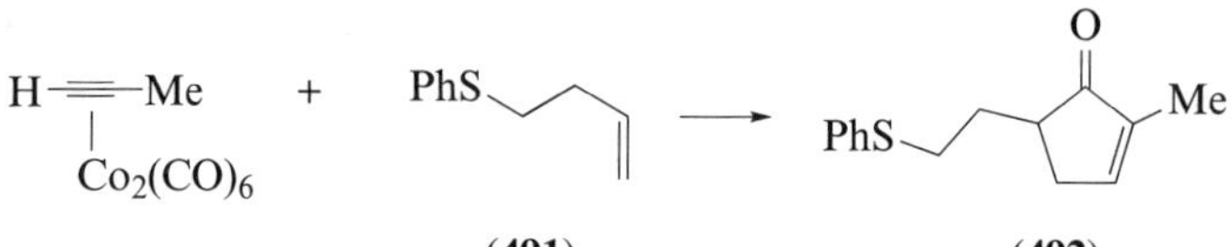

(491)   (492)

'Carbonylative' cycloaddition of allyl bromides and acetylenes in methanol in the presence of nickel tetracarbonyl results in the formation of cyclopentenones. For instance, 3-bromocyclopentene and methyl but-2-ynoate afford the diester **493**, the halogen atom being replaced by the methoxycarbonyl group[307]. Nickel tetracarbonyl also catalyses the reaction of 1-(bromomethyl) cycloalkenes **494** of ring size 5–8 with methyl but-2-ynoate in methanol to yield the spiro-compounds **495** (equation 54)[308].

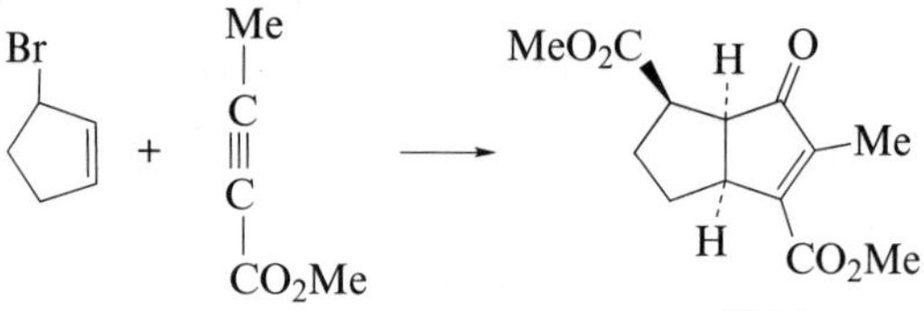

(493)

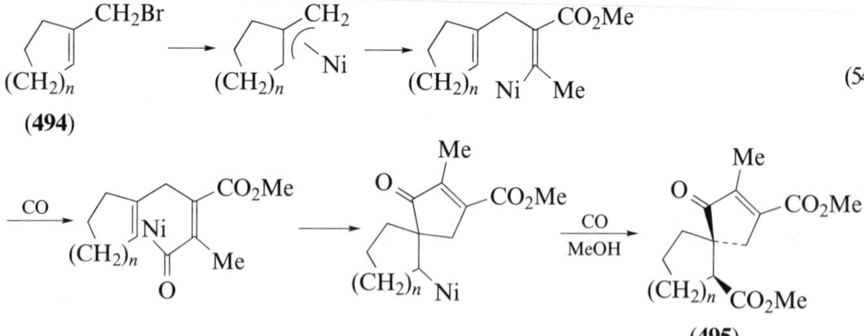

(494)   (54)

(495)

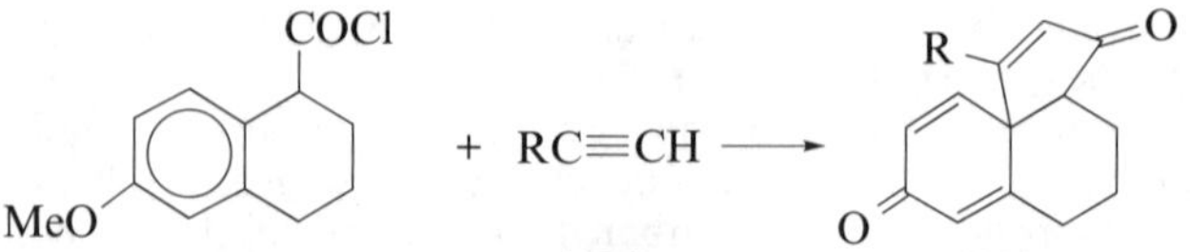

(496)   (497)   (498)

The action of aluminium trichloride on a mixture of 1,2,3,4-tetrahydro-6-methoxynaphthoyl chloride (**496**) and terminal alkynes **497** (R = Pr, pentyl or Ph) produces tricyclospirodienones **498**[309].

A general cyclization reaction of δ-lithiated (trimethylsilyl)alkynes is exemplified by the conversion of the iodo compound **499** into the methylenecyclopentane **500** by treatment with *t*-butyllithium[310]. Heating a mixture of the acetylenic silane **501** and 2-(benzyloxymethyl)allylzinc bromide **502** resulted in the diene **503**, which cyclized to the methylenecyclopentene **504** in the presence of tetrakis(triphenylphosphine)palladium[311].

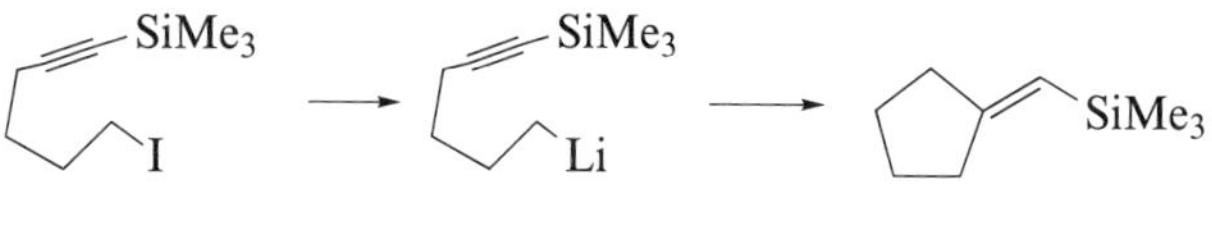

(**499**)   (**500**)

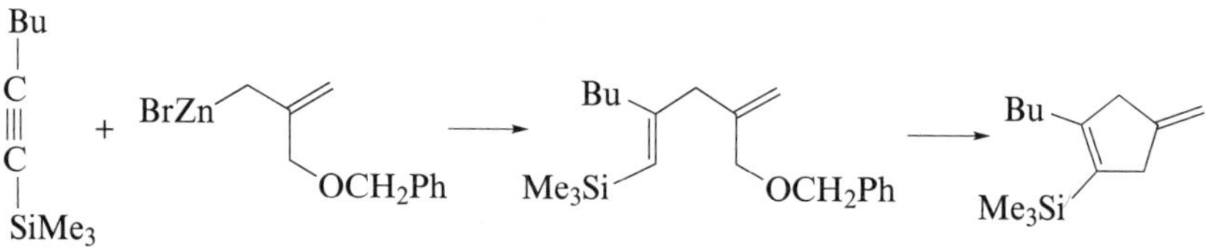

(**501**)   (**502**)   (**503**)   (**504**)

Acetylenes with one bulky substituent are transformed into symmetrically or unsymmetrically substituted cyclopentadienones by the combined action of dicobalt octacarbonyl and carbon monoxide; see, for example, equation 55[312].

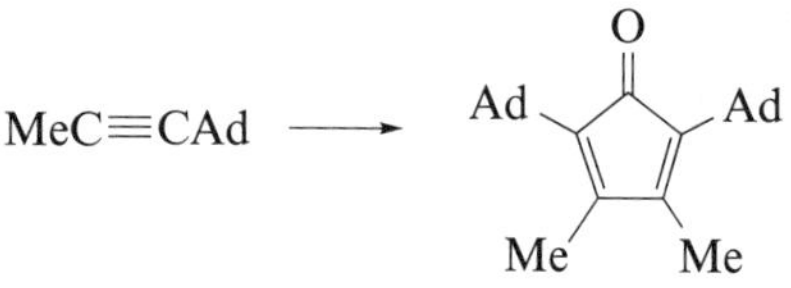

$$MeC{\equiv}CAd \longrightarrow$$

(55)

Ad = 1- adamantyl

Reductive cyclization of 1,6-diynes in the presence of triethylsilane and a catalyst prepared *in situ* from palladium(0) and acetic acid leads to dialkylidenecyclopentanes. Thus the diester **505** gave compound **506**[313].

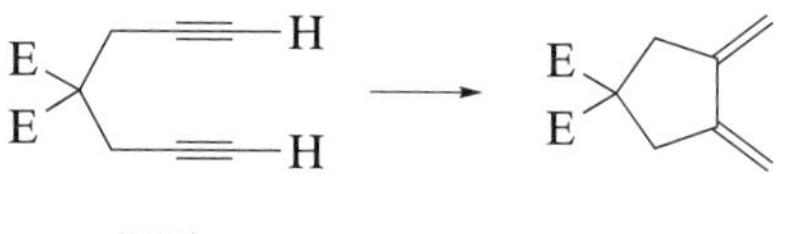

(**505**)   (**506**)

E = CO$_2$Me

Hept-5-yn-1-yl iodide (**507**) cyclizes to the iodoethylidenecyclopentane **508** by a free-radical atom-transfer process initiated by tributyltin radicals[314]. The thermal iodine-transfer-addition reaction of the dicyanopropargyl iodide **509** with (*E*)-oct-4-ene gives **510**, which in the presence of tributyltin hydride undergoes a second such reaction to yield the methylenecyclopentane **511** stereoselectively[315].

(507)   **(508)**

(509)   **(510)**   **(511)**

Cobalt(II) iodide–triphenylphosphine–zinc systems catalyse the homo-Diels-Alder re-action of norbornadiene with alkynes $R^1C\equiv CR^2$ (R^1 = Bu, Ph or SiMe$_3$; R^2 = H, Me or Et) to give the tetracyclic adducts **512**[316]. Asymmetric induction has been reported for the cobalt acetylacetonate-catalysed homo-Diels-Alder addition of 1-hexyne to norbornadiene in the presence of the diphosphine (*R*)-(+)-Ph$_2$PCH$_2$CHMePPh$_2$ to yield the dextrorota-tory product **513** in 78% enantiomeric excess. Six new stereocentres are created in this reaction[317].

R^1 R^2 **(513)**

(512)

b. Heterocyclic compounds.   The dicobalt octacarbonyl-catalysed reaction of the bulky 1-adamantylacetylene with carbon monoxide affords the spiro-compound **514**, as well as the 'normal' products: the benzene derivative **515** and the cyclopentadienone **516**[318]. Furans **518** result from the action of arylacetylenes on the diazo compounds **517** (R^1 = Me or EtO; R^2 = Ac or CO$_2$Et) in the presence of rhodium(II) acetate[319].

(514)   **(515)**   **(516)**

(517)   (518)

The bismuthonium ylide **519** is converted into the annelated furans **522** on treatment with terminal alkynes in the presence of copper(I) chloride. It is suggested that the process involves the carbene **520** and the diradical **521**[320]. Intramolecular [2 + 2 + 2] cycloaddition of the triyne **523** mediated by tris(triphenylphosphine)rhodium(I) chloride gives the tetrahydrofuranobenzofuran **524**[321].

(519)   (520)

(520) +

(521)   (522)

(523)   (524)

5-*t*-Butyl-2-iodophenol furnishes benzofurans **526** by the action of terminal alkynes **525** (R = Bu, HOCH₂ or CO₂Et) in pyridine containing copper(I) oxide[322]. Similarly, treatment of *o*-iodophenol **527** (X = CH) or the corresponding 3-pyridinol **527** (X = N) with a variety of terminal alkynes in the presence of piperidine, copper(I) iodide and bis(triphenylphosphine)palladium diacetate gives benzofurans **528** (X = CH) or furanopyridines **528** (X = N), respectively[323].

(525)   (526)

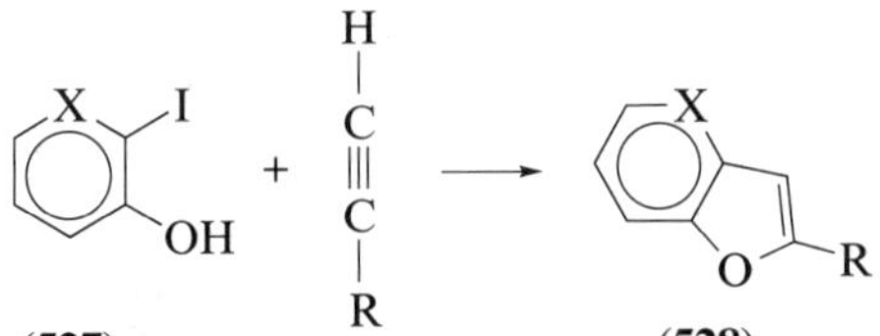

(527)   (528)

Pyrroles **531** are formed from the chromium complex **529** and alkynes **530** (R^1 = H, Me or Ph; R^2 = Me, Ph or NEt_2)[324]. The dicobaltoctacarbonyl-catalysed reaction of cyanotrimethylsilane with a variety of acetylenes $R^1C{\equiv}CR^2$ (R^1, R^2 = alkyl or Ph) furnishes pyrroles **532**, in which the bulkier of the two substituents of unsymmetrical internal acetylenes appears at the position marked with an asterisk[325a,b]. An indole synthesis from o-iodo-aniline and alkynes $R^1C{\equiv}CR^2$ (R^1, R^2 = alkyl or Ph) in the presence of palladium(II) acetate, triphenylphosphine, lithium chloride and potassium carbonate has been described (equation 56). In the case of unsymmetrical alkynes, the bulkier substituent tends to be in position 2 of the indole[326].

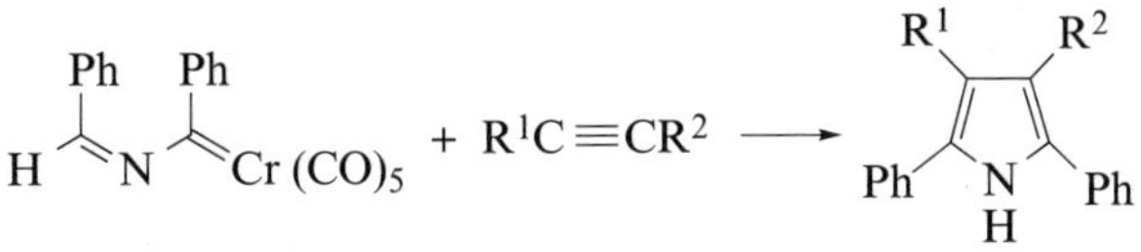

(529)   (530)   (531)

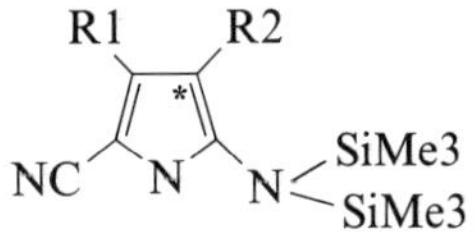

(532)

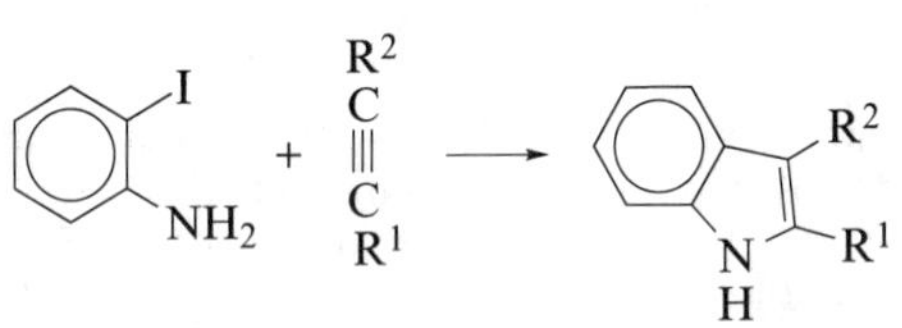

(56)

The action of potassium trithiocarbonate on the acetylenes **533** (R = H, Me or i-Pr X = Cl, Br or TosO) results in the dithiolethiones **534**, which afford butatrienes **535** on desulphurization with nickel[327].

A 'one-pot' procedure for the preparation of 5-substituted 3-chloroisoxazoles **539** has been described: the oxime **(536)** of glyoxylic acid is chlorinated and the product is treated with a mixture of a terminal alkyne **538** (R = Me, pentyl, Ph, $HOCH_2$, HOCHMe or $BrCH_2$) and potassium hydrogen carbonate. The nitrile oxide **537**, which is thus generated, undergoes a 1,3-dipolar cycloaddition to the alkyne to give the product regiospecifically[328]. Similarly, treatment of dibromoformaldoxime **(540)** with diverse terminal alkynes in the presence of a base affords the isoxazoles **541**[329].

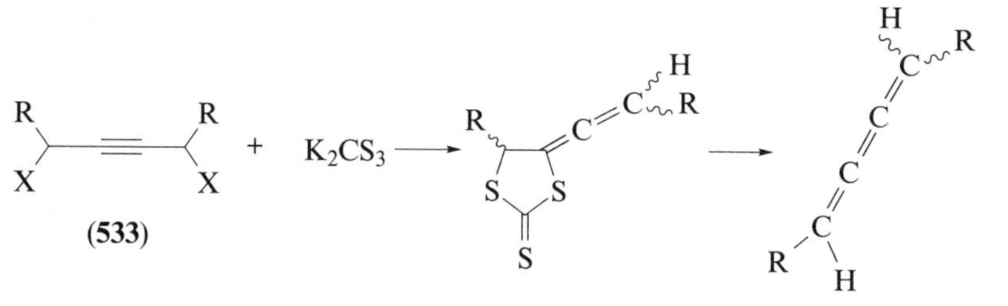

(533)   + K₂CS₃ → (534)   → (535)

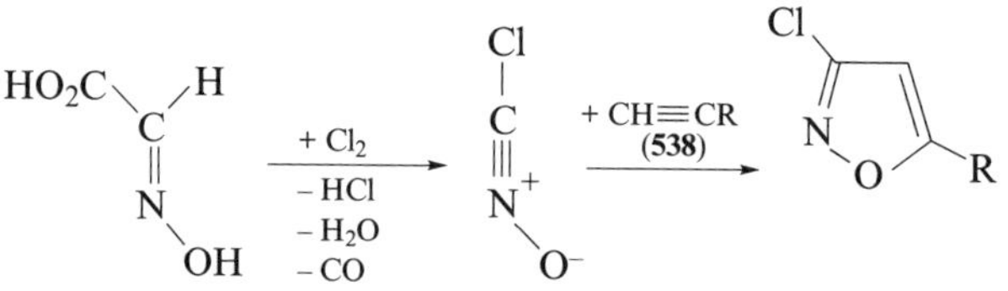

(536)   (537)   (539)

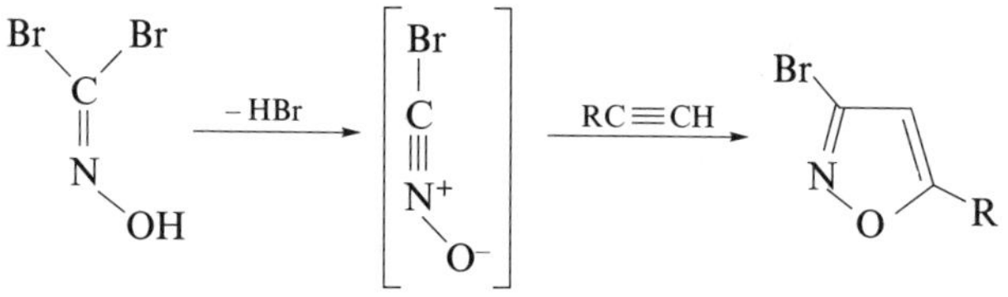

(540)   (541)

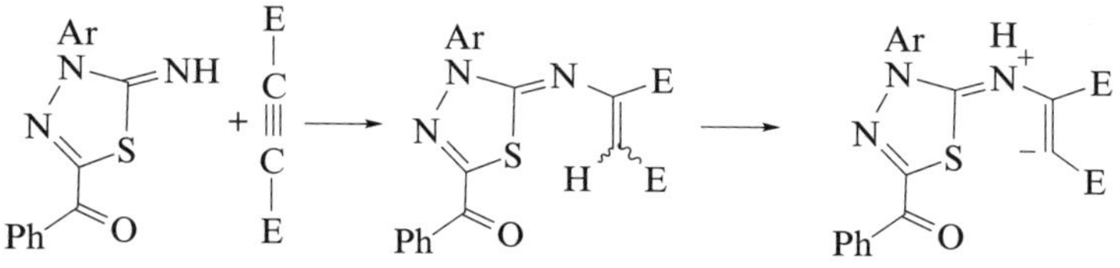

(542)   (543)   (544)

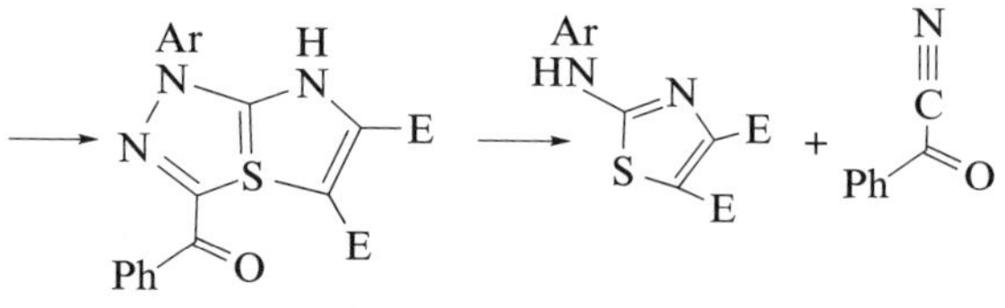

(545)   (546)

E = CO₂Me

Dimethyl 2-(arylamino)thiazole-4,5-dicarboxylates **546** are produced from 3-aryl-5-benzoyl-2-imino-1,3,4-thiadiazolines **542** and dimethyl acetylenedicarboxylate. It is suggested that the initial adducts **543** rearrange to the betaines **544** by a proton shift and that the betaines cyclize to the hypervalent sulphur compounds **545**. Subsequent loss of benzoyl cyanide yields the products[330].

Addition of dimethyl acetylenedicarboxylate to the mesoionic dehydrodithizone (**547**) yields the stable betaine **549** by ring-cleavage of the initial dipolar cycloadduct **548**, which is destabilized by the presence of eight π-electrons in the tetrazoline ring[331]. Dichlorodithionitronium hexafluoroarsenate $CISNSCl^+ AsF_6^-$ reacts with acetylene in liquid sulphur dioxide to give the 1,3,2-dithiazolium salt **550**[332].

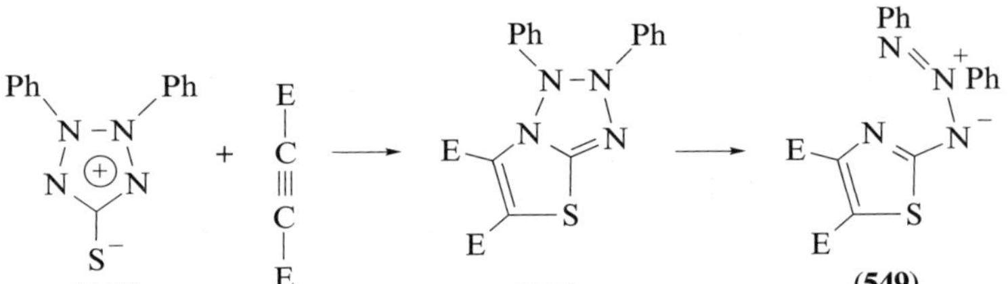

$E = CO_2Me$

(**550**)

4. Formation of six-membered rings

a. Homocyclic compounds. The cyclocodimerization of 1,3-butadienes with non-activated terminal acetylenes in the presence of a rhodium(I) complex yields mixtures of cyclohexadienes; e.g. 2-methylbutadiene and phenylacetylene afford the isomers shown in equation 57[333].

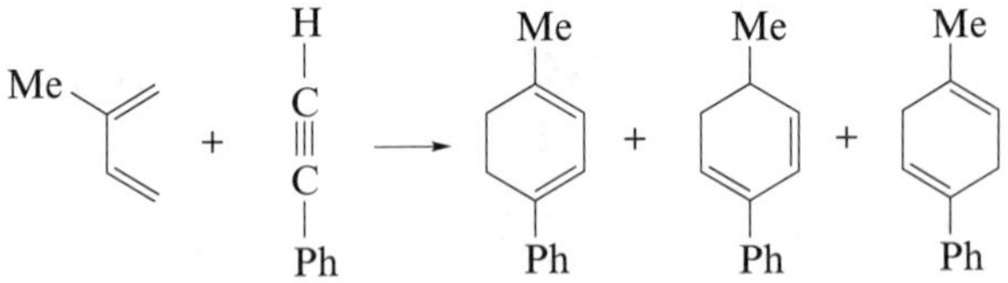

(57)

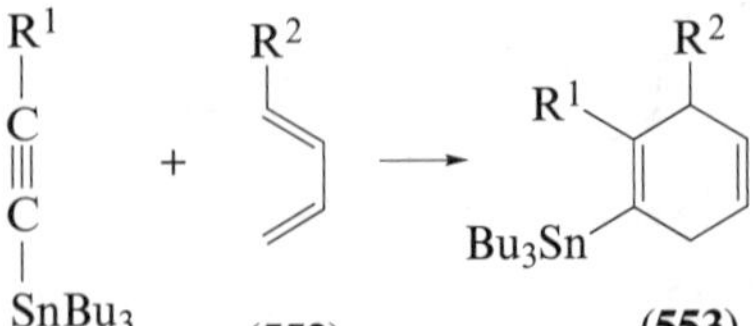

$R^1 = Ac, CO_2Me$ or CN
$R^2 = Me$ or Et

The regiospecific Diels–Alder addition of alkynylstannanes **551** to 1-substituted buta-1,3-dienes **552** gives solely the 1,4-cyclohexadienes **553**[334]. The enediyne **554** does not function as a diene towards maleic anhydride or tetracyanoethylene; with 2,3-dimethylbutadiene it reacts as a dienophile to give the cyclohexadiene **555**[335].

(554)   **(555)**

The action of pent-1-yne on the indolylcarbene complex **556** results in the formation of the tricyclic compound **557**; the isomeric complex **558** yields the analogue **559**[336].

(556)   **(557)**

(558)   **(559)**

Bis(ethene)(toluene)iron catalyses the cyclotrimerization of alkynes to derivatives of benzene; terminal alkynes $RC\equiv CH$ (R = t-Bu, Ph or CO_2Me) yield mainly 1,2,4-trisubstituted benzenes **560**, while dimethyl acetylenedicarboxylate furnishes the iron complex **561**[337].

(560)   **(561)**   E = CO_2Me

The niobium complex **562** is formed by successive treatment of $CpNbCl_4$ with magnesium metal and phenylacetylene. The complex yields a mixture of 1,2,4-and 1,3,5-triphenylbenzene on standing (equation 58). The catalytic system $CpNbCl_4/Mg(0)$ mediates the reaction of the diyne **563** with hex-3-yne to give the benzene derivative **564**[338].

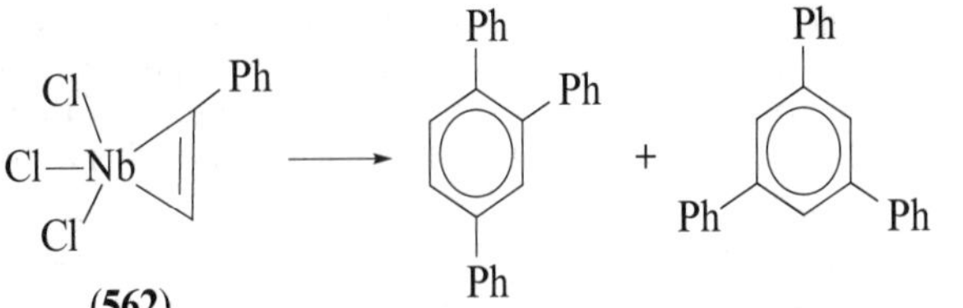

(58)

(562)

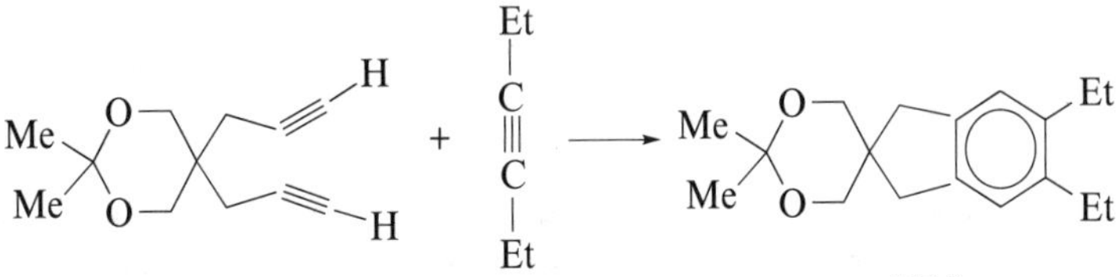

(563)   (564)

Trimerization of alkynes to derivatives of benzene is promoted by $[Co(pyridine)_6]^+$ BPh_4^-. In the presence of hydrogen the catalyst induces the formation of a mixture of the dienes 565–567 and the tributylbenzenes 568 and 569 from 1-hexyne; the composition of the mixture depends on the hydrogen pressure[339].

The action of the complex 570 on various alkynes gives diverse products: ethyl propiolate affords a mixture of all three possible triethyl benzenetricarboxylates, phenylacetylene yields mainly the *trans*-enyne 571; 3-hexyne behaves similarly, but 1-octyne isomerizes to 2-octyne[340]. The palladacyclopentadiene 572 catalyses the co-trimerization of hex-3-yne

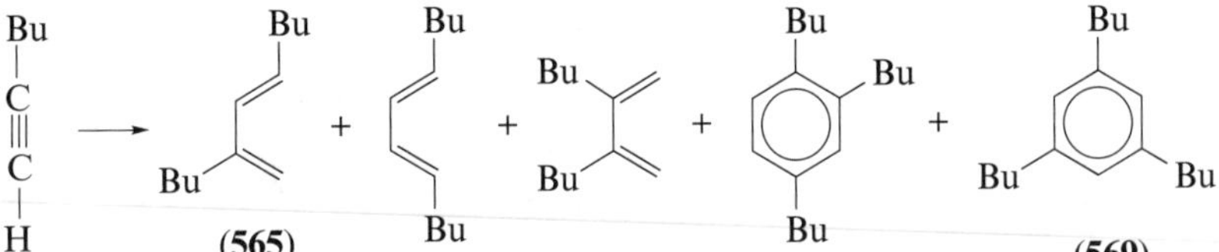

(565)   (566)   (567)   (568)   (569)

$[CoH(N_2)(PPh_3)_3]$

(570)

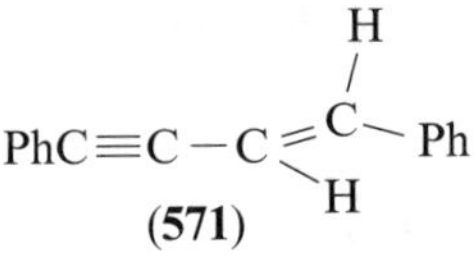

(571)

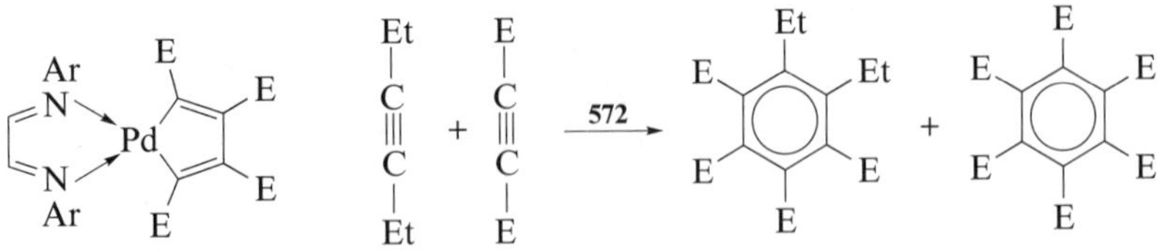

(572)   (573)   (574)

$Ar = $, $E = CO_2Me$

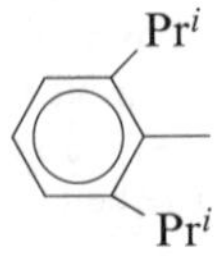

with dimethyl acetylenedicarboxylate to give the tetraester **573**, accompanied by hexamethyl benzenehexacarboxylate (**574**)[341].

The phase-transfer-catalysed trimerization of various acetylenes PhC≡CR (R = Me, Et, Ph or Ac) in the presence of rhodium(III) chloride and Aliquat 336 in 1,1,2,2-tetrachloroethane – water to yield mainly the benzene derivatives **575** has been reported[342]. Biphenyldiazonium salts **576** (X = H, Cl or MeO) are converted into phenanthrenes **578** by the action of terminal acetylenes **577** (R = Ph, SiMe₃ or CO₂Et) in pyridine. 2-Phenyl-1-naphthyldiazonium tetrafluoroborate (**579**) similarly affords the substituted chrysenes **580**[343].

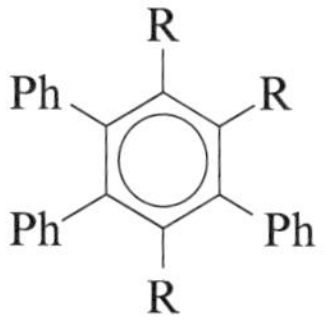

(**575**)

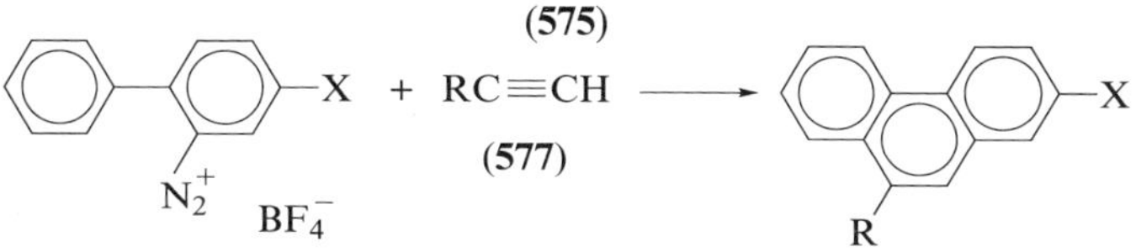

(**576**)   +   RC≡CH   ⟶   (**578**)

(**577**)

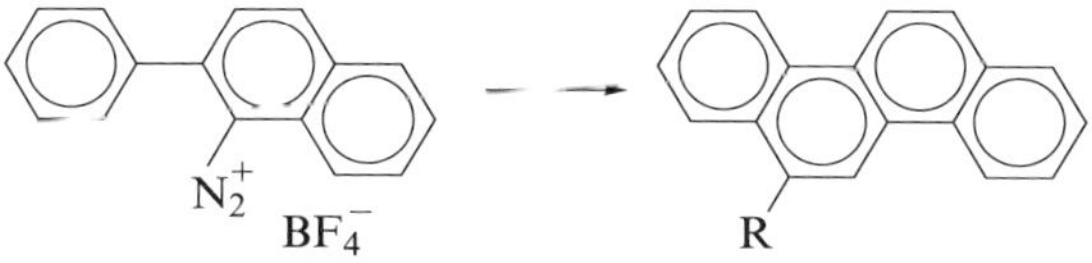

(**579**)   − − →   (**580**)

The Diels–Alder reaction of α-pyrone with bis(trimethylsilyl)acetylene, obtained from dilithium acetylide and trimethylsilyl chloride, gives *o*-bis(trimethylsilyl)benzene by spontaneous decarboxylation of the initial adduct (equation 59)[344].

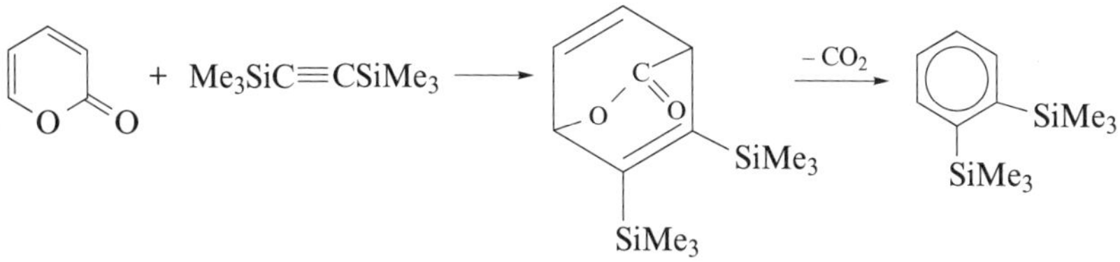

$$+ \ Me_3SiC\equiv CSiMe_3 \ \longrightarrow \ \xrightarrow{-CO_2} \qquad (59)$$

Acid-catalysed cyclization of the alcohol **581** yields the dihydro-γ-pyrone **582**, which is converted into the benzene derivatives **586** by sequential treatment with trimethylsilyl chloride and acetylenes **584** (R¹ = H, Me or CO₂Me; R² = Bz or CO₂Me). The process involves formation of the pyran **583**, Diels–Alder addition to the alkynes and extrusion of acetone from the adducts **585**[345].

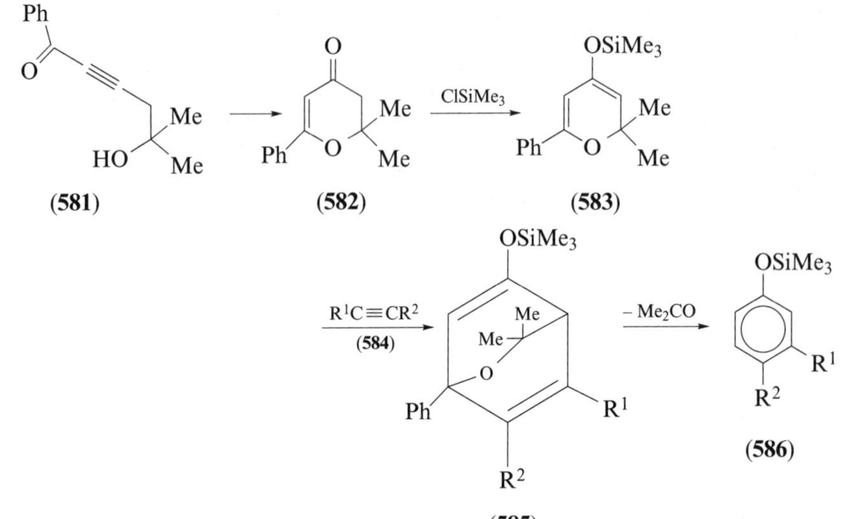

(581)   (582)   (583)

(585)

(586)

A method for the synthesis of highly substituted aromatic and heteroaromatic hydroxy compounds is the photochemical Wolff rearrangement of an unsaturated α-diazo ketone in the presence of an alkyne. The product, an alkenylcyclobutenone, undergoes ring-opening and recyclization to a phenol (equation 60). Three examples of the reaction are the formation of the naphthols **587** and **588** and that of the hydroxybenzofuran **589**[346]. Complexes **590** (R^1 = alkyl or aryl; R^2 *t*-alkyl or Me_3Si), produced from THF, alkynes and niobium trichloride. 1,2-dimethoxyethane, react regiospecifically with phthaladehyde to yield derivatives **591** of 1-naphthol. Similarly, treatment of the 4-octyne complex with *o*-formylacetophenone gives the naphthol **592**[347].

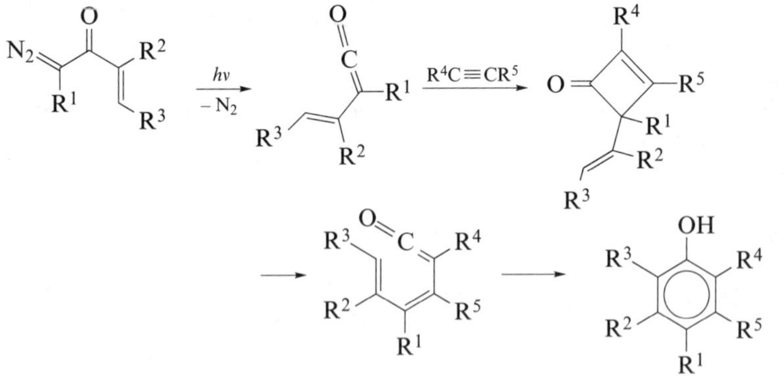

$$\text{(60)}$$

Complexes formed from tantalum(V) chloride or niobium(V) chloride, alkynes and zinc undergo analogous reactions[348]. Mixtures of phenols are obtained from cyclobutenones and alkynes in the presence of nickel(cyclooctadiene)$_2$ at 0 °C. Thus 4-methyl-3-phenylcyclobut-2-enone and 4-methylpent-2-yne yield **593** and **594**[349].

The formation of benzoquinones from alkynes and the cobalt complex **595** proceeds with moderate regioselectivity; e.g. 1-hexyne affords the isomers **596** and **597** in the ratio 4:1[350].

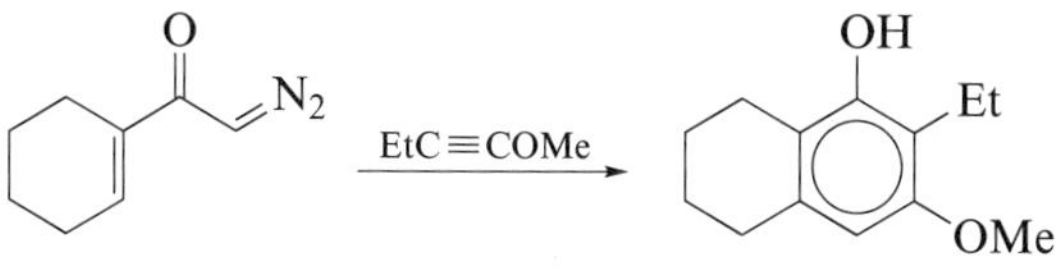

(587)

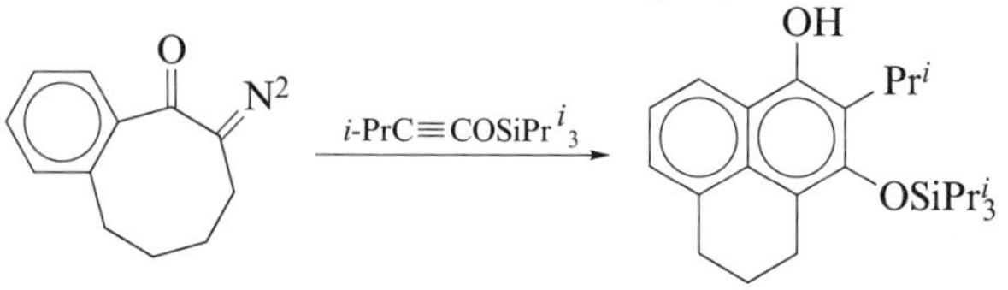

(588)

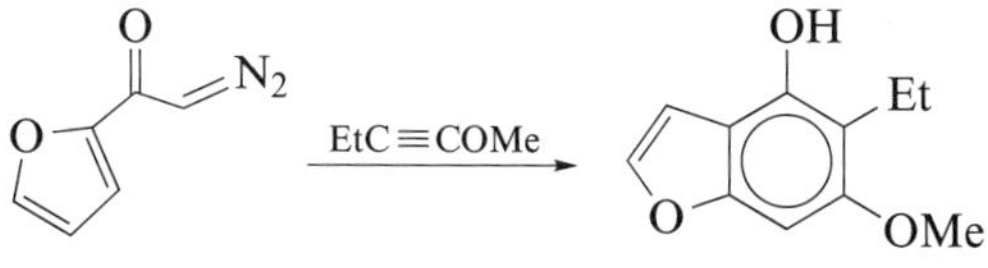

(589)

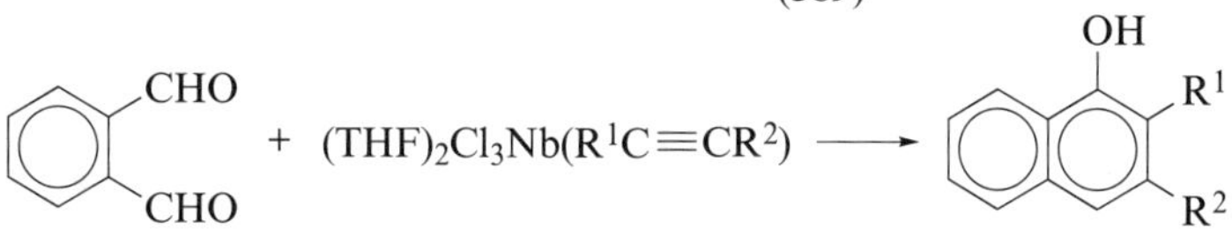

(590)   (591)

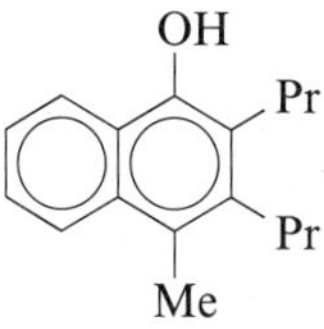

(592)

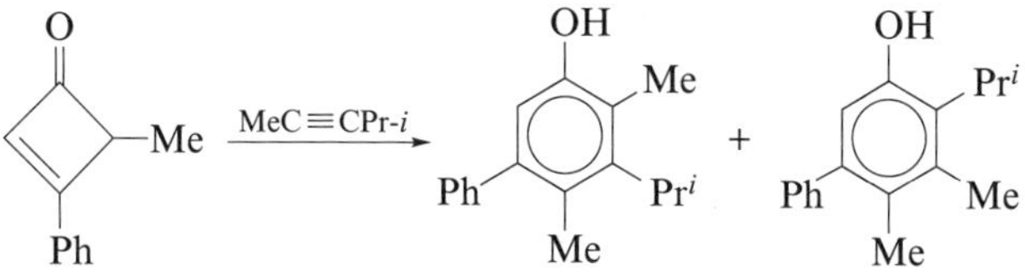

(593)   (594)

In the presence of tin(IV) chloride the selectivity is enhanced to 10:1; and in the case of ethyl but-2-ynoate the proportion of **598** is twenty times that of its regioisomer[351].

b. Heterocyclic compounds. The unstable cyclopropylidenedimedone **599** has been trapped as its Diels-Alder adduct **600** with 1-(diethylamino)propyne[352]. Heating the dialkynylazidobenzoquinone **601** (Tol = *o*-tolyl) generates the cyanoketene **602**, which

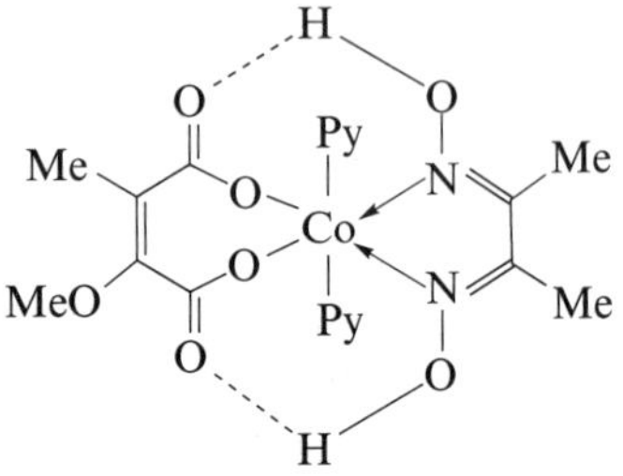

(595)

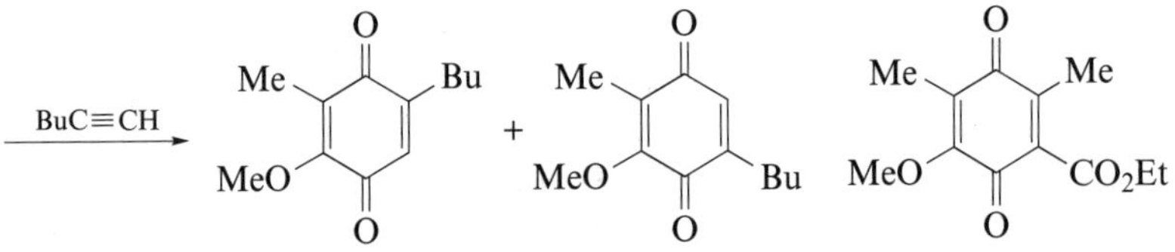

$BuC\equiv CH$ →

(596) (597) (598)

py = pyridine

$MeC\equiv CNEt_2$ →

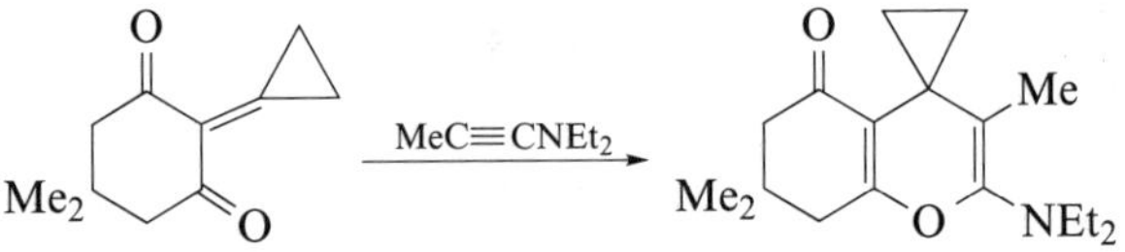

(599) (600)

reacts with diphenylacetylene to yield the dibenzopyran 606. It is suggested that the ketene adds the acetylene to form the cyclobutenone 603, which is converted into the diradical 604 by ring-opening, followed by recyclization in an alternative way. The diradical 604 rearranges to the oxygen–benzyl diradical 605 by hydrogen transfer; cyclization completes the process[353].

2-Substituted chromones are formed by heating terminal acetylenes with o-iodophenols under carbon monoxide pressure in the presence of palladium[bis(diphenylphosphino-ferrocene)] dichloride; e.g. o-iodophenol and phenylacetylene afford 2-phenylchromone (equation 61)[354]. Acetylenic aldiminium salts undergo cyclization to six-membered nitrogen heterocycles in polar aprotic solvents. Thus heating solutions of 5-(butylamino) pent-1-yne (607), an aldehyde RCHO (R = Me, Et, Ph or CO_2Et) and tetrabutylammonium iodide in acetonitrile yields the piperidine derivatives 608 and similar reactions of 4-(benzylamino)but-1-yne lead to tetrahydropyridines 609[355].

Cobalt metal catalyses the co-trimerization of one molecule of acetonitrile with two of 1-pentyne or 1-hexyne to yield mixtures of the pyridines 610 and 611 (R = Pr or Bu)[356]. The regioselctivity of the reaction of nitriles with terminal acetylenes has been investigated. It was found that, in general, mixtures of 2,4,6- and 2,3,6-trisubstituted pyridines are produced. The selectivity depends on the size of the substituents attached to the triple bond and the cyanide group. If either of these is large, only 2,4,6-substituted pyridines are obtained. Ethyl propiolate and aliphatic or aromatic nitriles (RCN) yield, in addition to the usual type of product, i.e. 612 and 613, the 2,3,5-isomer 614[357]. Pyridines 617 result from the action of alkynes $RC\equiv CH$ (R = Ac, PrCO or CO_2Me) on the electron-rich

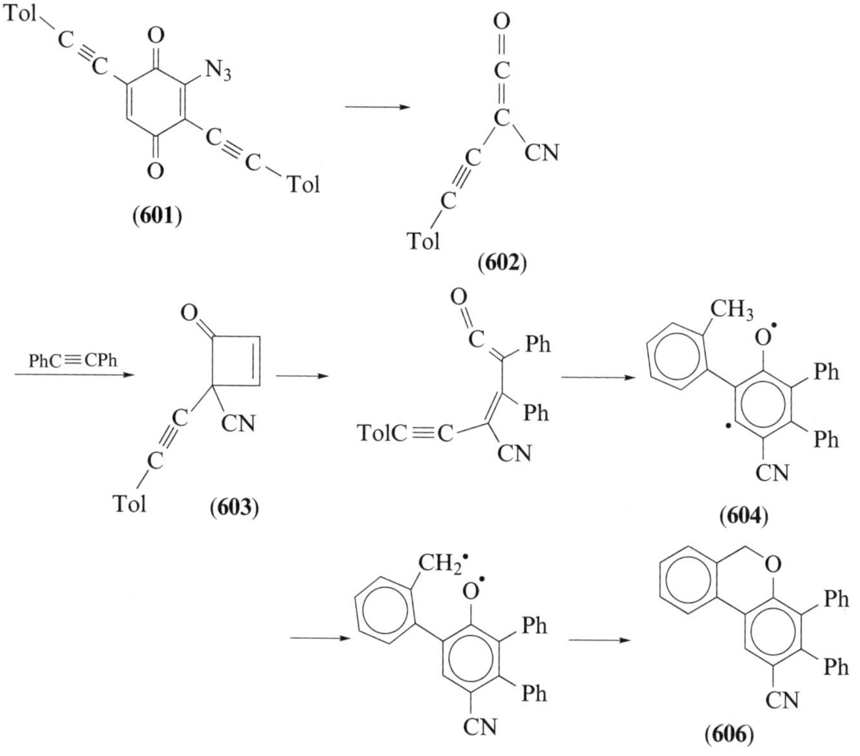

(601) → (602)

PhC≡CPh → (603) → (604) → (605) → (606)

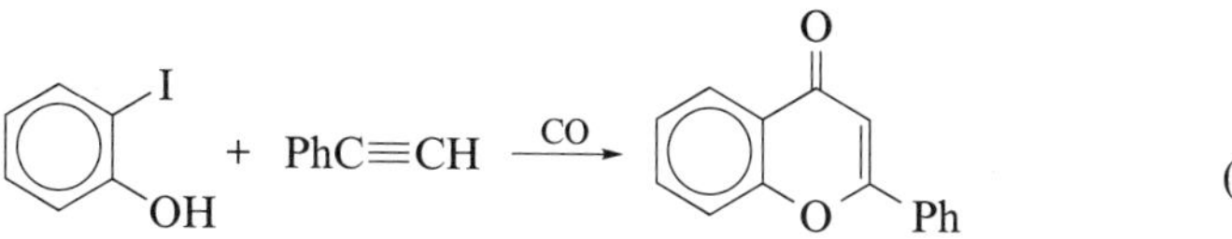

$$\text{(61)}$$

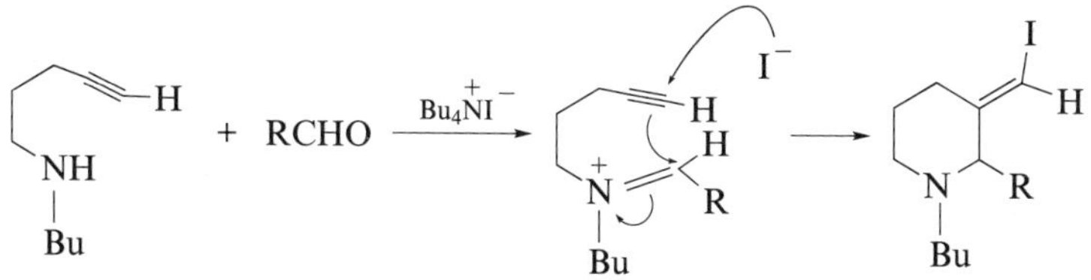

(607) + RCHO $\xrightarrow{\text{Bu}_4\overset{+}{\text{N}}\text{I}^-}$ → (608)

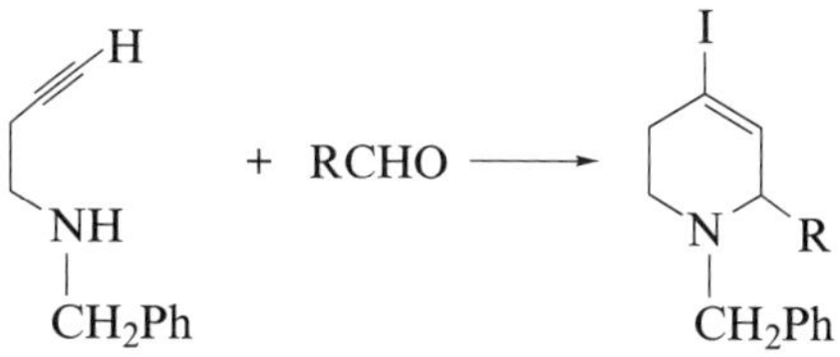

(609)

(610)   (611)

(612)   (613)   (614)   E = CO₂Et

dihydropyrimidine **615**. The initial cycloadducts **616** undergo a retro-Diels-Alder reaction to give acetone imine and the products[358].

(615)   (616)   (617)

   Treatment of *para*-substituted benzylideneanilines **618** (R = Me, Cl or MeO) with phenylacetylene in the presence of diisopropyl peroxydicarbonate results in mixtures of 6- and 7-substituted 2,4-diphenylquinolines, **621** and **623**. It is thought that the process is initiated by the formation of arylimidoyl radicals **619**, which add to phenylacetylene to form the radicals **620**. The latter cyclize to the 6-substituted quinolines by Route A. An alternative mode of cyclization (Route B) proceeds via the spiro-radicals **622**, which eventually yield the 7-isomers[359].

(618)   (619)   (620)

(623)   (622)   (621)

The 'cyclopalladated' Schiff's base **624** reacts with symmetrical alkynes RC≡CR [R = Et, CO₂Me or (EtO)₂CH] to give isoquinolinium salts **625**[360]. 2-Phenylcinnolinium salts **627** are similarly formed from the palladium complex **626** and acetylenes[361]. Palladium-catalysed reactions of *o*-bromo- or *o*-iodoaniline with terminal alkynes under carbon monoxide pressure give good yields of 4-quinolones (equation 62)[362].

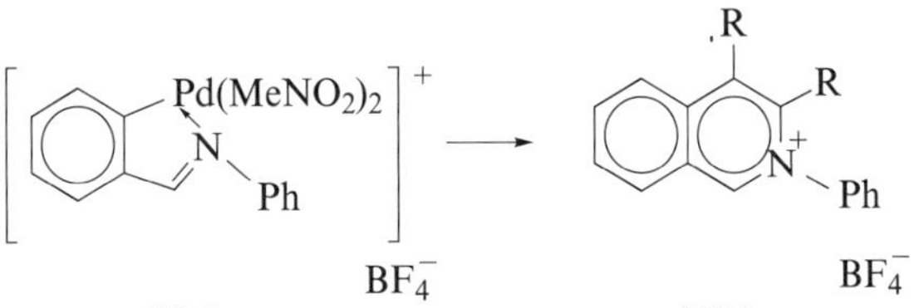

(624) → (625)

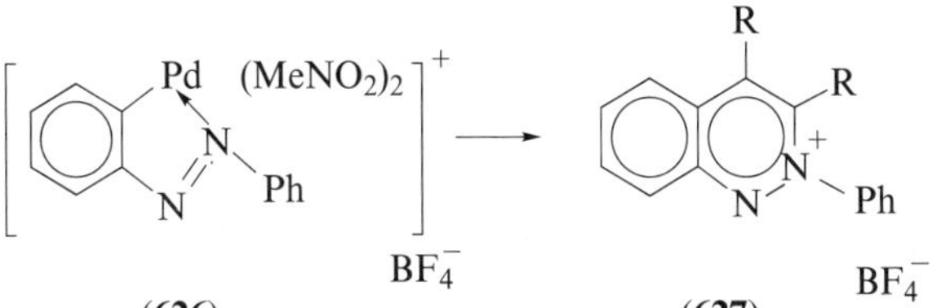

(626) → (627)

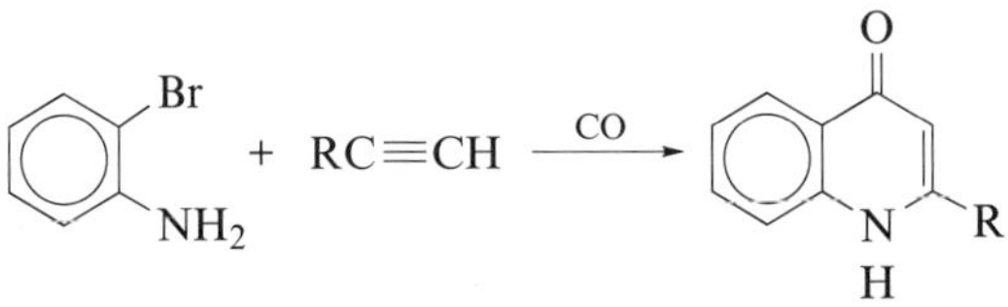

$$\text{(62)}$$

R = alkyl or aryl

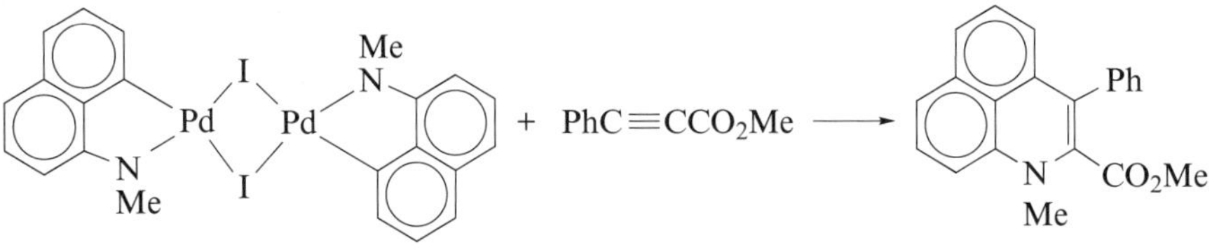

(628) + PhC≡CCO₂Me → (629)

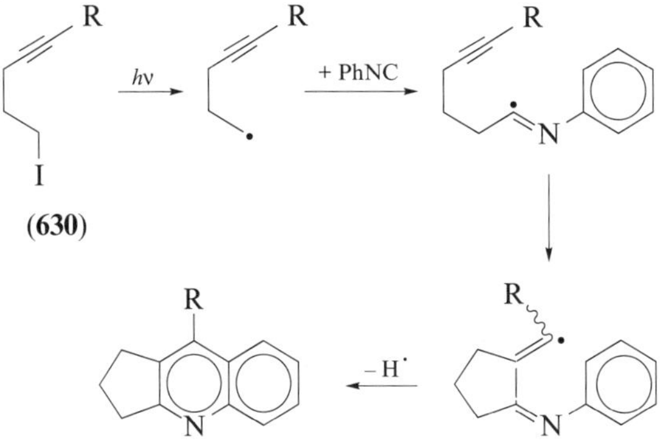

(630)

$$\text{(63)}$$

The complex **628** is converted into the benzo[*de*]quinoline **629** by the action of methyl phenylpropiolate[363]. Irradiation of mixtures of the ω-iodoalkynes **630** (R = H, Me, Ph or PhCH$_2$) and phenyl isocyanide produces cyclopentano[*b*]quinolines by way of a radical addition to the isocyanide, followed by two cyclizations (equation 63)[364].

Intramolecular Diels–Alder reactions of the pyrazine derivatives **631** (X = O, S or NAc; R = SiMe$_3$ or HOCH$_2$) yield mixtures of the tricyclic heterocycles **633** and **634**. Both arise from the intermediate cycloadducts **632**, which can lose the elements of hydrogen cyanide in two ways, as indicated[365].

The disylanylstannane **635** adds terminal alkynes regio- and stereoselectively to form (Z)-alkenes **636**, which undergo a tetrakis(triphenylphosphine)palladium–catalysed reaction with phenylacetylene to give 1-sila-4-stannacyclohexa-2,5-dienes **637**[366]. 1,4-Dithianes **639** (*n* = 1) and 1,4-dithiepanes **639** (*n* = 2) are obtained by the azoisobutyronitrile–induced homolytic cycloaddition of alkynes R^{1}C≡CR^2 (R^1 = Pr, Bu or CMe$_2$OH; R^2 = H or Pr) to the dithols **638**[367a,b].

'Dithioannulation' of various terminal or internal acetylenes by bis(dithiobenzil)nickel in pyridine yields 1,4-dithiins (equation 64)[368].

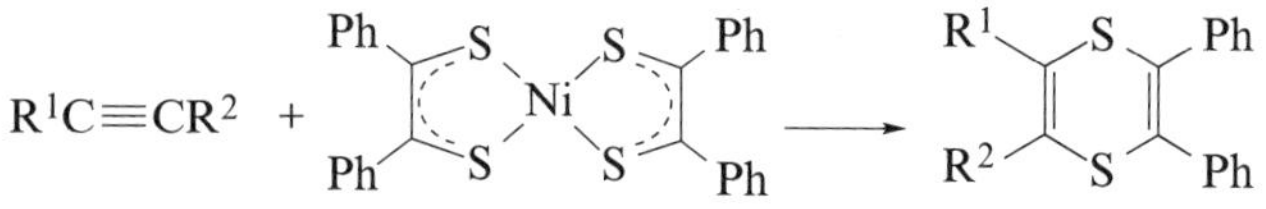

$$R^1C{\equiv}CR^2 \; + \; \text{(structure)} \longrightarrow \text{(structure)} \tag{64}$$

1,1,3,3-tetrakis(dimethylamino)-1λ^5,3λ^5- diphosphete (**640**) adds phenylacetylene to give a mixture of the 1λ^5,3λ^5-diphosphorines **641** and **642** (R = Ph); in contrast, 3,3,3-trifluoropropyne yields only **641** (R = CF$_3$)[369]. 2,4,6-Tripheny-3-aza arsinine **643** reacts with alkynes **644** (R^1 = Ph, R^2 = H or CHO; R^1 = Ac, R^2 = Ph; R^1 = CO$_2$Me, R^2 = H) to give λ^3-arsinines **645** by the cycloaddition–cycloreversion sequence shown[370].

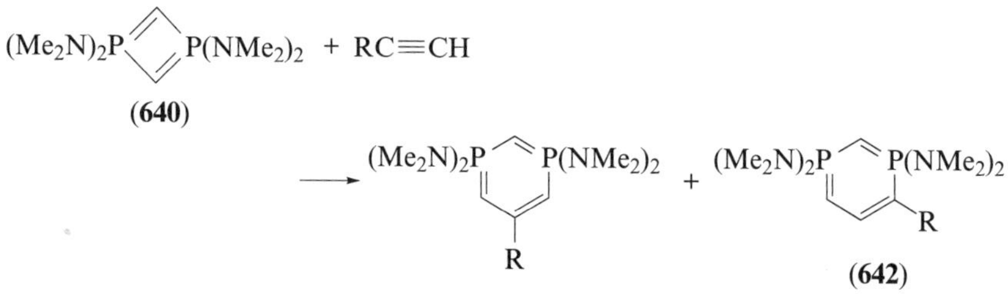

$$(Me_2N)_2P{<}\text{(structure)}P(NMe_2)_2 \; + \; RC{\equiv}CH$$

(**640**)

$$\longrightarrow \; (Me_2N)_2P\text{(structure)}P(NMe_2)_2 \quad + \quad (Me_2N)_2P\text{(structure)}P(NMe_2)_2$$

(**641**)   (**642**)

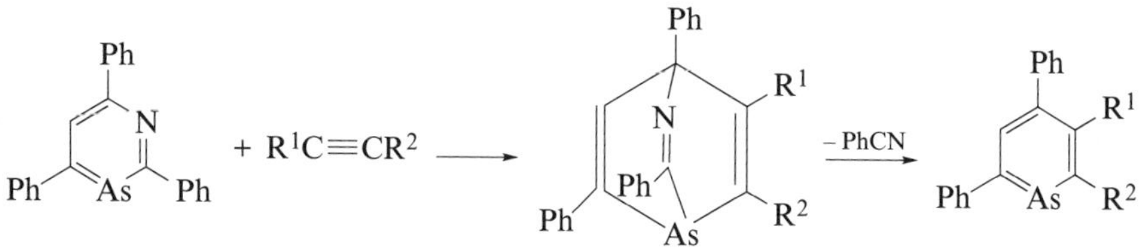

(**643**)   (**644**)   (**645**)

5. Formation of large rings

Reviews on cyclic diacetylenes[371] and on transition-metal-catalysed cyclizations of α,ω-diynes **646** (e.g. equation 65)[372] have appeared.

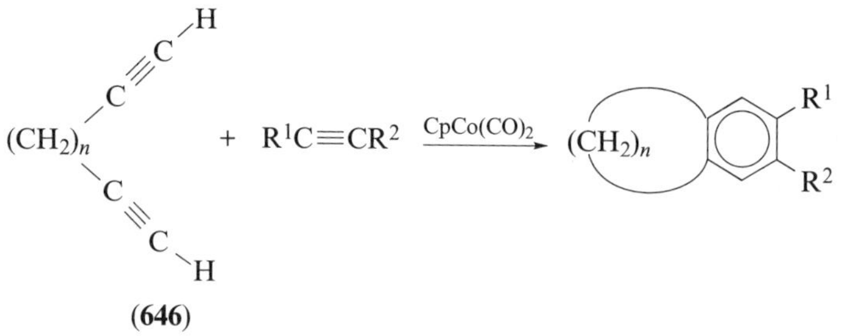

$$\text{(CH}_2)_n\text{(structure)} \; + \; R^1C{\equiv}CR^2 \xrightarrow{\text{CpCo(CO)}_2} \text{(CH}_2)_n\text{(structure)} \tag{65}$$

(**646**)

A mixture of the ten-membered cyclic *cis*-diol **648** and its *trans*-isomer is obtained from the enediyne dialdehyde **647**[373]. A new family of hydrocarbon rings, the three 'skipped' cyclic $C_{12}H_{12}$ dienediynes **649 – 651**, has been described. These compounds were prepared from dilithium derivatives of enediynes and diiodo- or dibromoalkenes as shown in Scheme 6. Treatment of **651** with CpCo(CO)$_2$ gave the cobalt sandwich complex **652**[374].

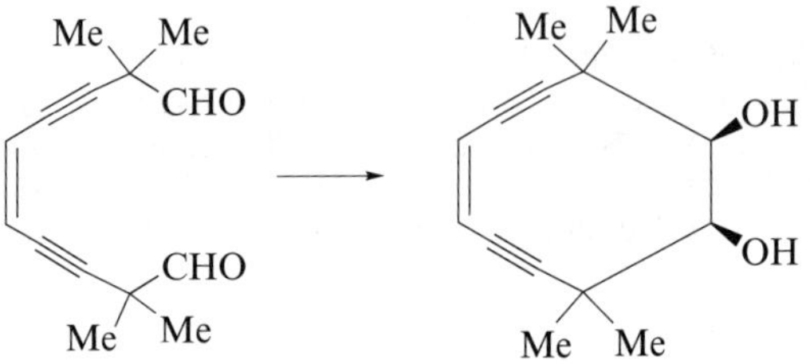

SCHEME 6

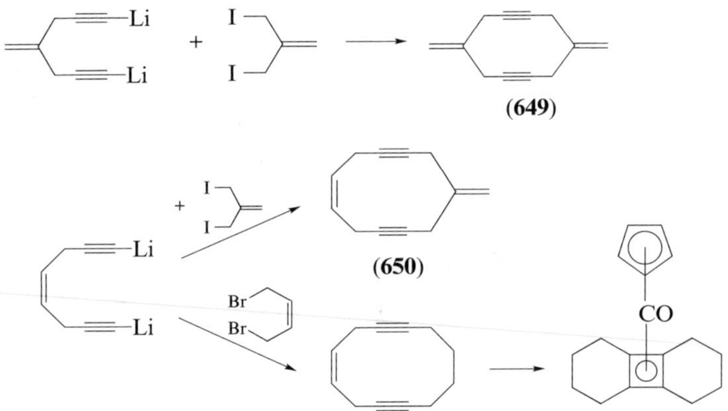

The black–purple tetramethyloctadehydrotridecapentadecafulvalene **654** was obtained in 22% yield by the oxidative ring-closure of **653** mediated by copper(II) acetate in pyridine – ether – methanol[375].

A series of cyclic homoconjugated polyacetylenes of ring-sizes 15, 20, 25 and 30, **655 – 658**, the 'cyclenes', has been reported. The general method of synthesis is exemplified by the preparation of **657**. The penta-1,4-diyne **661** was obtained from the chloro compound **659** and the Grignard reagent **660**, followed by removal of the trimethylsilyl groups. The diyne was converted into its dicopper derivative and the latter was coupled with two molecules of the protected bromodiyne **662** to yield the hexayne **663**. Deprotection, followed by renewed coupling with two molecules of **662**, gave the decayne **664**. The synthesis was completed by removal of the two trimethylsilyl groups and cyclization of the resulting

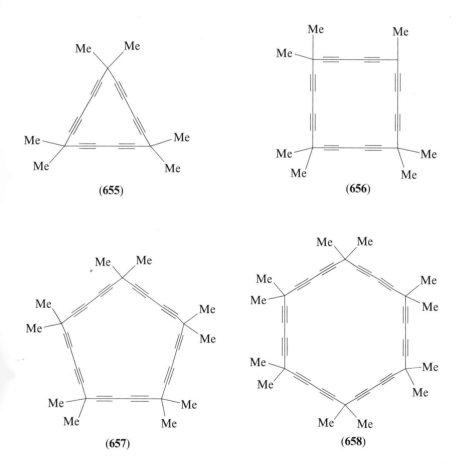

(653)   (654)

double-terminal acetylene (Scheme 7). The UV spectra of the four cyclenes give evidence of homoconjugation; the effect is particularly pronounced in the 15-membered ring compound[376].

(655)   (656)

(657)   (658)

The colourless conjugated macrocycle **666**, which consists of twelve benzene rings and twelve acetylene units, has been obtained by palladium-catalysed intramolecular coupling of the iodo compound **665**[377].

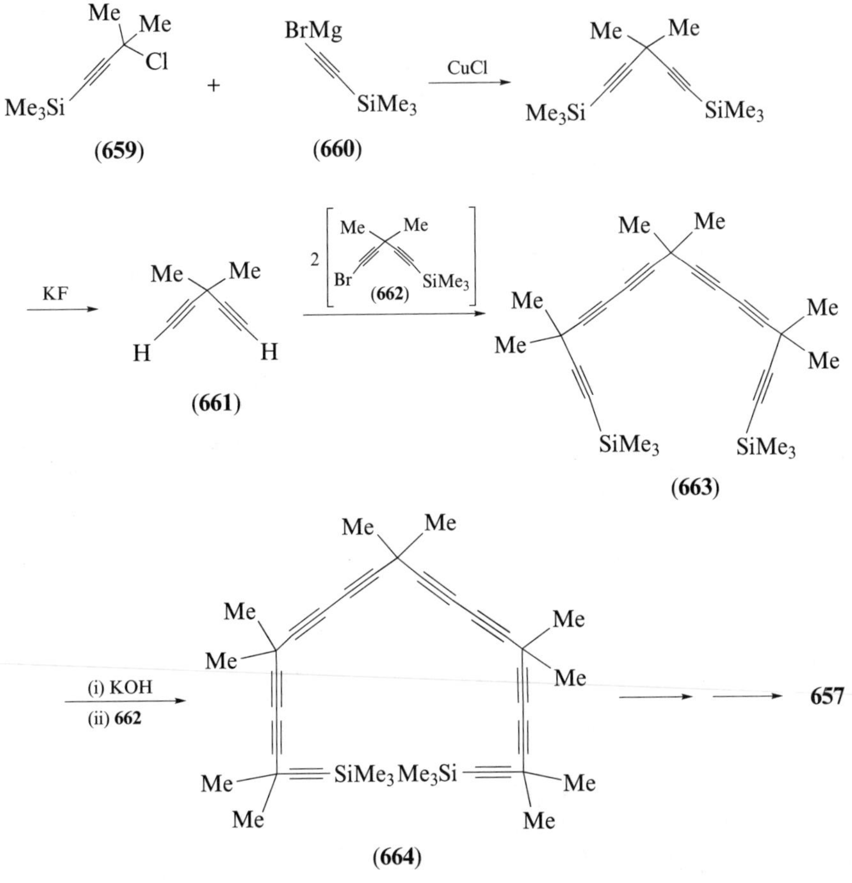

SCHEME 7

Treatment of *o*-dibromobenzene with propargyl alcohol in the presence of tetrakis-(triphenylphosphine)palladium(0) yielded the bromo alcohol **667**, which condensed with a second molecule of propargyl alcohol to give the diyne **668**. The dimethanesulphonate of the latter was added to propylamine over a period of 20–30 hours to afford the 18-membered nitrogen heterocycle **669**[378].

(665)

(666)

Br → C≡CCH₂OH → C≡CCH₂OH

Br Br C≡CCH₂OH

(667) (668)

N Pr ← Prnh₂ C≡CCH₂O₃SMe

Pr N C≡CCH₂O₃SMe

(669)

IV. REFERENCES

1. (a) P. F. Hudrlik and A. M. Hudrlik, in *The Chemistry of the Carbon–Carbon Triple Bond* (Ed. S. Patai), Wiley, Chichester, 1978, p. 199.
   (b) D. A. Ben-Efraim, in *The Chemistry of the Carbon–Carbon Triple Bond* (Ed. S. Patai), Wiley, Chichester, 1978, p. 755.
   (c) K. Friedrich, in *The Chemistry of Triple-bonded Functional Groups* (Eds. S. Patai and Z. Rappoport), Wiley, Chichester, 1983, p. 1345.
2. (a) I. H. Goldberg, *Acc. Chem. Res.*, **24**, 191 (1991).
   (b) M. D. Lee, G. A. Ellestad and D. B. Borders, *Acc. Chem. Res.*, **24**, 235 (1991).
   (c) K. C. Nicolaou and W. M. Dai, *Angew. Chem., Int. Ed. Engl.*, **30**, 1387 (1991).
3. D. J. Burton and T. D. Spawn, *J. Fluorine Chem.*, **38**, 119 (1988).
4. M. Kunishima, K. Hioki, T. Ohara and S. Tani, *J. Chem. Soc., Chem. Commun.*, 219 (1992).
5. A. M. Caporusso, C. Polizzi and L. Lardicci, *J. Org. Chem.*, **52**, 3920 (1987).
6. N. A. Urdaneta, M. V. Mavrov and C. H. Nguyen, *Izv. Akad. Nauk SSSR, Ser. Khim.*, 2384 (1987); *Chem. Abstr.*, **109**, 54256a (1988).
7. M. Suzuki, Y. Morita and R. Noyori, *J. Org. Chem.*, **55**, 441 (1990).
8. R. Yamaguchi, M. Moriyasu, I. Takase, M. Kawanisi and S. Kozima, *Chem. Lett.*, 1519 (1987).
9. J. Haruta, K. Nishi, S. Matsuda, S. Akai, Y. Tamura and Y. Kita, *J. Org. Chem.*, **55**, 4853 (1990).
10. D. Faul, E. Leber and G. Himbert, *Synthesis*, 73 (1987).
11. A. Bartlome, U. Stämpfli and M. Neuenschwander, *Helv. Chim. Acta*, **74**, 1264 (1991).
12. V. V. Zhdankin, P. J. Stang and N. S. Zefirov, *J. Chem. Soc., Chem. Commun.*, 578 (1992).
13. A. Wagner, M. P. Heitz and C. Mioskowski, *Tetrahedron Lett.*, **31**, 3141 (1990).
14. L. Brandsma and H. D. Verkruijsse, *Synthesis*, 984 (1990).
15. J. Barluenga, J. M. Gonzalez, M. A. Rodriguez, P. J. Campos and G. Asensio, *Synthesis*, 661 (1987).
16. T. Jeffery, *J. Chem. Soc., Chem. Commun.*, 909 (1988).
17. A. Ricci, M. Taddei, P. Dembech, A. Guerrini and G. Seconi, *Synthesis*, 461 (1989).
18. A. Casarini, P. Dembech, G. Reginato, A. Ricci and G. Seconi, *Tetrahedron Lett.*, **32**, 2169 (1991).
19. J. Wityak and J. B. Chan, *Synth. Commun.*, **21**, 977 (1991).
20. A. S. Hay, *J. Org. Chem.*, **27**, 3320 (1962).
21. D. O'Krongly, S. R. Denmeade, M. Y. Chiang and R. Breslow, *J. Am. Chem. Soc.*, **107**, 5544 (1985).
22. R. Eastmond and D.R. M. Walton, *J. Chem. Soc., Chem. Commun.*, 204 (1968).
23. R. Eastmond, T. R. Johnson and D. R. M. Walton, *Tetrahedron*, **28**, 4601 (1972).
24. T. R. Johnson and D. R. M. Walton, *Tetrahedron*, **28**, 5221 (1972).
25. B. M. Trost, C. Chan and G. Ruhter, *J. Am. Chem. Soc.*, **109**, 3486 (1987).
26. D. Sonogashira, Y. Tohda and N. Hagihara, *Tetrahedron Lett.*, 4467 (1975).
27. L. Cassar, *J. Organomet. Chem.*, **93**, 253 (1975).

28. H. A. Dieck and F. R. Heck, *J. Organomet. Chem.*, **93**, 259 (1975).
29. V. Ratovelomanana, A. Hammoud and G. Linstrumelle, *Tetrahedron Lett.*, **28**, 1649 (1987).
30. A. S. Kende and C. A. Smith, *J. Org. Chem.*, **53**, 2655 (1988).
31. W. Y. Huang and L. Lu, *Chin. Chem. Lett.*, **2**, 769 (1991); *Chem. Abstr.*, **116**, 173169d (1992).
32. Z. Y. Yang and D. J. Burton, *Tetrahedron Lett.*, **31**, 1369 (1990).
33. Z. Y. Yang and D. J. Burton, *J. Fluorine Chem.*, **53**, 307 (1991).
34. S. Eddarir, C. Francesch, H. Mestdagh and C. Rolando, *Tetrahedron Lett.*, **31**, 4449 (1990).
35. K. Uneyama and H. Watanabe, *Tetrahedron Lett.*, **32**, 1459 (1991).
36. D. Villemin and D. Goussu, *Heterocycles*, **29**, 1255 (1989).
37. B. V. Nguyen, Z. Y. Yang and D. J. Burton, *J. Fluorine Chem.*, **50**, 265 (1990).
38. Q. Chen and Y. He, *Synthesis*, 896 (1988).
39. K. L. Evans, P. Prince, E. T. Huang, K. R. Boss and R. D. Gandour, *Tetrahedron Lett.*, **31**, 6753 (1990).
40. R. Neidlein and R. Winkler, *Monatsh. Chem.*, **122**, 177 (1991).
41. T. Sakamoto, H. Nagata, Y. Kondo, M. Shiraiwa and H. Yamanaka, *Chem. Pharm. Bull.*, **35**, 823 (1987).
42. S. A. Nye and K. T. Potts, *Synthesis*, 375 (1988).
43. J. W. Tilley and S. Zawoiski, *J. Org. Chem.*, **53**, 386 (1988).
44. W. Tao, S. Nesbitt and R. F. Heck, *J. Org. Chem.*, **55**, 63 (1990).
45. R. Diercks, J. S. Armstrong, R. Boese and K. P. C. Vollhardt, *Angew. Chem.*, *Int. Ed. Engl.*, **25**, 268 (1986).
46. R. Boese, J. R. Green, J. Mittendorf, D. L. Mohler and K. P. C. Vollhardt, *Angew. Chem.*, *Int. Ed. Engl.*, **31**, 1643 (1992).
47. R. Yamaguchi, Y. Nakazono, T. Matsuki, E. Hata and M. Kawanisi, *Bull. Chem. Soc. Jpn.*, **60**, 215 (1987).
48. R. Yamaguchi, E. Hata, T. Matsuki and M. Kawanisi, *J. Org. Chem.*, **52**, 2094 (1987).
49. N. Nishiwaki, S. Minakata, M. Komatsu and Y. Ohshiro, *Chem. Lett.*, 773 (1989).
50. D. Zhai, W. Zhai and R. M. Williams, *J. Am. Chem. Soc.*, **110**, 2501 (1988).
51. W. Kantlehner, P. Speh, H. Lehmann, E. Haug and W. W. Mergen, *Chem.-Ztg.*, **114**, 176 (1990).
52. (a) R. H. Weigmann and E. U. Würthwein, *Tetrahedron Lett.*, **30**, 6147 (1989).
   (b) R. H. Weigmann and E. U. Würthwein, *Tetrahedron Lett.*, **31**, 3082 (1990).
53. M. G. Moloney, J. T. Pinhey and E. G. Roche, *Tetrahedron Lett.*, **27**, 5025 (1986).
54. M. G. Moloney, J. T. Pinhey and E. G. Roche, *J. Chem. Soc.*, *Perkin Trans. 1*, 333 (1989).
55. Review: T. Pinhey, *Aust. J. Chem.*, **44**, 1353 (1991).
56. H. Schmidbaur, J. Ebenhöch and G. Müller, *Z. Naturforsch.*, *B: Chem. Sci.*, **43**, 49 (1988).
57. M. Hayashi, A. Inubushi and T. Mukaiyama, *Chem. Lett.*, 1975 (1987).
58. M. Hayashi, A. Inubushi and T. Mukaiyama, *Bull. Chem. Soc. Jpn.*, **61**, 4037 (1988).
59. H. Lang, H. Keller, W. Imhof and S. Martin, *Chem. Ber.*, **123**, 417 (1990).
60. T. Murai, K. Imaeda, S. Kajita, H. Ishihara and S. Kato, *J. Chem. Soc.*, *Chem. Commun.*, 832 (1991).
61. V. N. Komissarov, L. Yu. Ukhin, Zh. I. Orlova and D. A. Tokarskaya, *Zh. Org. Khim.*, **23**, 1325 (1987); *Chem. Abstr.*, **108**, 204183v (1988).
62. L. A. Remizova, T. A. Shustrova and I. A. Favorskaya, *Zh. Org. Khim.*, **22**, 992 (1986); *Chem. Abstr.*, **106**, 213506p (1987).
63. L. Yu. Ukhin, V. N. Komissarov, Zh. I. Orlova, D. A. Tokarskaya, A. I. Yanovskii and Yu. T. Struchkov, *Zh. Org. Khim.*, **23**, 1323 (1987); *Chem. Abstr.*, **108**, 204182u (1988).
64. S. Lin, H. Sheng and Y. Huang, *Synthesis*, 235 (1991).
65. K. Burger, N. Sewald, E. Huber and R. Ottlinger, *Z. Naturforsch.*, *B: Chem. Sci.*, **44**, 1298 (1989).
66. S. Niwa and K. Soai, *J. Chem. Soc.*, *Perkin Trans. 1*, 937 (1990).
67. A. Pecunioso and R. Menicagli, *J. Org. Chem.*, **54**, 2391 (1989).
68. H. Fujishima, E. Takada, S. Hara and A. Suzuki, *Chem. Lett.*, 695 (1992).
69. M. Yamaguchi, K. Shibato, S. Fujiwara and I. Hirao, *Synthesis*, 421 (1986).
70. T. Mandai, H. Murayama, T. Nakata, H. Yamaoki, M. Ogawa, M. Kawada and J. Tsuji, *J. Organomet. Chem.*, **417**, 305 (1991).
71. B. M. Trost and G. Kottirsch, *J. Am. Chem. Soc.*, **112**, 2816 (1990).
72. M. Hayashi, A. Inubushi and T. Mukaiyama, *Bull. Chem. Soc. Jpn.*, **61**, 4037 (1988).
73. T. Jeffery, *Tetrahedron Lett.*, **30**, 2225 (1989).
74. G. Märkl, P. Attenberger and J. Kellner, *Tetrahedron Lett.*, **29**, 3651 (1988).

368   G. V. Boyd

75. S. W. Scheuplein, K. Harms , R. Brückner and J. Suffert, *Chem. Ber.*, **125**, 271 (1992).
76. H. Takahata, K. Takahashi, E. C. Wang and T. Yamazaki, *J. Chem. Soc.*, *Perkin Trans. 1*, 1211 (1989).
77. S. Mori, H. Iwakura and S. Takechi, *Tetrahedron Lett.*, **29**, 5391 (1988).
78. M. Yamaguchi, A. Hayashi and T. Minami, *J. Org. Chem.*, **56**, 4091 (1991).
79. K. T. Mead, *Tetrahedron Lett.*, **28**, 1019 (1987).
80. A. Yanagisawa, S. Habaue and H. Yamamoto, *J. Org. Chem.*, **54**, 5198 (1989).
81. T. Mandai, T. Nakata, H. Murayama, H. Yamaoki, M. Ogawa, M. Kawada and J. Tsuji, *Tetrahedron Lett.*, **31**, 7179 (1990).
82. T. Mitsudo, Y. Hori and Y. Watanabe, *Bull. Chem. Soc. Jpn.*, **59**, 3201 (1986).
83. T. V. Kulik and A. Ya. Ilchenko, *Zh. Org. Khim.*, **25**, 728 (1989); *Chem. Abstr.*, **112**, 178019g (1990).
84. M. G. Voronkov, O. G. Yarosh, G. Yu. Turkina, V. Yu. Vitkovskii and A. I. Albanov, *Metalloorg. Khim.*, **2**, 511 (1989); *Chem. Abstr.*, **112**, 118916n (1990).
85. P. G. Ciattini, E. Morera and G. Ortar, *Tetrahedron Lett.*, **32**, 6449 (1991).
86. H. W. Moore, K. F. West, U. Wriede, K. Chow, M. Fernandez and N. V. Nguyen, *J. Org. Chem.*, **52**, 2537 (1987).
87. J. K. Stille and J. H. Simpson, *J. Am. Chem. Soc.*, **109**, 2138 (1987).
88. A. S. Kende and P. Fludzinski, *Tetrahedron Lett.*, **23**, 2373 (1982).
89. K. M. Nicholas, *Acc. Chem. Res.*, **20**, 207 (1987).
90. R. F. Lockwood and K. M. Nicholas, *Tetrahedron Lett.*, 4163 (1977).
91. H. D. Hodes and K. M. Nicholas, *Tetrahedron Lett.*, 4349 (1978).
92. K. M. Nicholas, M. Mulvaney and M. Baeyer, *J. Am. Chem. Soc.*, **102**, 2508 (1980).
93. J. E. O'Boyle and K. M. Nicholas, *Tetrahedron Lett.*, **21**, 1595 (1980).
94. M. Nakagawa, J. Ma and T. Hino, *Heterocycles*, **30**, 451 (1990).
95. E. J. Corey and P. L. Fuchs, *Tetrahedron Lett.*, 3769 (1972).
96. L. Van Hijfte, M. Kolb and P. Witz, *Tetrahedron Lett.*, **30**, 3655 (1989).
97. P. Vinczer, S. Struhar, L. Novak and C. Szantay, *Tetrahedron Lett.*, **33**, 683 (1992).
98. J. B. Hendrickson and M. S. Hussoin, *Synthesis*, 217 (1989).
99. E. Negishi, A. O. King, W. L. Klima, W. Patterson and A. Silveira, Jr., *J. Org. Chem.*, **45**, 2526 (1980).
100. E. Negishi, A. O. King and J. M. Tour, *Org. Synth.*, Coll. Vol. **7**, 63 (1990).
101. P. E. Eaton and D. Stoessel, *J. Org. Chem.*, **56**, 5138 (1991).
102. T. A. Engler, K. D. Combrink and J. E. Ray, *Synth. Commun.*, **19**, 1735 (1989).
103. J. Boivin, L. Elkaim, P. G. Ferro and S. Z. Zard, *Trtrahedron Lett.*, **32**, 5321 (1991).
104. C. Wentrup, H. Briehl, P. Lorencak, U. J. Vogelbacher, H. W. Winter, A. Maquestiau and R. Flammang, *J. Am. Chem. Soc.*, **110**, 1337 (1988).
105. I. Lalezari, A. Shafiee and M. Yalpani, *Tetrahedron Lett.*, 5105 (1969).
106. I. Lalezari, A. Shafiee and M. Yalpani, *Angew. Chem.*, *Int. Ed. Engl.*, **9**, 464 (1970).
107. R. Gleiter, D. Kratz and V. Schehlmann, *Tetrahedron Lett.*, **29**, 2813 (1988).
108. H. Meier, N. Hanold, T. Molz, H. J. Bissinger, H. Kolshorn and J. Zountsas, *Tetrahedron*, **42**, 1711 (1986).
109. J. Chenault and J. F. E. Dupin, *Synthesis*, 498 (1987).
110. A. L. Braga and J. V. Comasseto, *Synth. Commun.*, **19**, 2877 (1989).
111. Y. Shen, W. Cen and Y. Huang, *Synthesis*, 626 (1987).
112. R. A. Aitken and S. Seth, *Synlett.*, 211 (1990).
113. R. Hochstrasser and J. Wirz, *Angew. Chem.*, *Int. Ed. Engl.*, **29**, 411 (1990).
114. W. M. Stalick, R. N. Hazlett and R. E. Morris, *Synthesis*, 287 (1988).
115. A. Laurent, I. LeDrean and A. Selmi, *Tetrahedron Lett.*, **32**, 3071 (1991).
116. (a) A. G. Godfrey and B. Ganem, *J. Am. Chem. Soc.*, **112**, 3717 (1990).
     (b) M. E. Jung, *Chemtracts: Org. Chem.*, **3**, 392 (1990).
117. E. Negishi, *Acc. Chem. Res.*, **20**, 65 (1987).
118. M. H. P. J. Aerssens, R. van der Heiden, M. Heus and L. Brandsma, *Synth. Commun.*, **20**, 342 (1990).
119. J. Inanaga, Y. Yokoyama, Y. Baba and M. Yamaguchi, *Tetrahedron Lett.*, **32**, 5559 (1991).
120. Y. Kataoka, K. Takai, K. Oshima and K. Otimoto, *Tetrahedron Lett.*, **31**, 365 (1990).
121. J. F. Daeuble, C. McGettigan and J. M. Stryker, *Tetrahedron Lett.*, **31**, 2397 (1990).
122. G. V. M. Sharma, B. M. Choudary, M. R. Sarma and K. K. Rao, *J. Org. Chem.*, **54**, 2991 (1989).

123. M. Enomoto, T. Katsuki and M. Yamaguchi, *Tetrahedron Lett.*, **27**, 4599 (1986).
124. B. G. Shakovskoi, M. D. Stadnichuk and A. A. Petrov, *J. Gen. Chem. USSR*, 2646 (1964).
125. M. H. P. J. Aerssens and L. Brandsma, *J. Chem. Soc., Chem., Commun.*, 735 (1984).
126. M. H. P. J. Aerssens, R. van der Heiden, M. Heus and L. Brandsma, *Synth. Commun.*, **20**, 344 (1990).
127. C. Bianchini, A. Meli, M. Peruzzini, F. Vizza, F. Zanobini and P. Frediani, *Organometallics*, **8**, 2080 (1989).
128. H. A. Dieck and F. R. Heck, *J. Org. Chem.*, **40**, 1083 (1975).
129. S. Cacchi, *Pure Appl. Chem.*, **62**, 713 (1990).
130. S. Cacchi, M. Felici and B. Pietroni, *Tetrahedron Lett.*, **25**, 3137 (1984).
131. A. Arcadi, S. Cacchi and F. Marinelli, *Tetrahedron Lett.*, **27**, 6397 (1986).
132. A. Arcadi, E. Bernocchi, A. Burini, S. Cacchi, F. Marinelli and B. Pietroni, *Tetrahedron Lett.*, **30**, 3465 (1989).
133. Y. Masuda, M. Hoshi and A. Arase, *J. Chem. Soc., Chem. Commun.*, 748 (1991).
134. H. C. Brown, D. Basavaiah, S. U. Kulkarni, H. D. Lee, E. Negishi and J. J. Katz, *J. Org. Chem.*, **51**, 5270 (1986).
135. N. X. Hu, Y. Aso, T. Otsubo and F. Ogura, *Tetrahedron Lett.*, **27**, 6099 (1986).
136. Y. Fukuda and K. Utimoto, *J. Org. Chem.*, **56**, 3729 (1991).
137. T. Tsuda, T. Kiyoi and T. Saegusa, *J. Org. Chem.*, **55**, 2554 (1990).
138. C. Bruneau, M. Neveux, Z. Kabouche, C. Ruppin and P. H. Dixneuf, *Synlett.*, 755 (1991).
139. (a) B. M. Trost, G. Dyker and R. J. Kulawiec, *J. Am. Chem. Soc.*, **112**, 7809 (1990).
    (b) K. Narasaka and N. Iwasawa, *Chemtracts: Org. Chem.*, **4**, 393 (1991).
140. Y. Ishii and Y. Sakata, *J. Org. Chem.*, **55**, 5545 (1990).
141. K. K. Kim, J. N. Kim, K. M. Kim, H. R. Kim. and E. K. Ryu, *Chem. Lett.*, 603 (1992).
142. D. G. Lee, E.J. Lee and K. C. Brown, *ACS Symp. Ser. 326 (Phase Transfer. Catal. : New Chem., Catal. Appl.)*, 82 (1987); *Chem. Abstr.*, **107**, 39111q (1987).
143. M. Tiecco, L. Testaferri, M. Tingoli, D. Chianelli and D. Bartoli, *J. Org. Chem.*, **56**, 4529 (1991).
144. R. Zibuck and D. Seebach, *Helv. Chim. Acta*, **71**, 237 (1988).
145. C. Sheu, S. A. Richert, P. Cofre, B. Ross, Jr., A. Sobkowiak, D. T. Sawyer and J. R. Kanofsky, *J. Am. Chem. Soc.*, **112**, 1936 (1990).
146. B. Rihter and J. Masnovi, *J. Chem. Soc., Chem. Commun.*, 35 (1988).
147. S. Torii, T. Inokuchi and Y. Hirata, *Synthesis*, 377 (1987).
148. K. L. Reed, J. T. Gupton and K. L. McFarlane, *Synth. Commun.*, **19**, 2595 (1989).
149. F. P. Ballistreri, S. Failla and G. A. Tomaselli, *J. Org. Chem.*, **53**, 830 (1988).
150. P. C. Page and S. Rosenthal, *Tetrahedron Lett.*, **27**, 1947 (1986).
151. R. Mahrwald and H. Schick, *Angew. Chem., Int. Ed. Engl.*, **30**, 593 (1991).
152. A. Nishinaga, K. Maruyams, K. Yoda and H. Okamoto, *J. Chem. Soc., Chem. Commun.*, 876 (1990).
153. A. Gorgues, D. Stephan and J. Cousseau, *Janssen Chim. Acta*, **7**, 3 (1989); *Chem. Abstr.*, **114**, 23390p (1991).
154. A. Gorgues, D. Stephan and J. Cousseau, *J. Chem. Soc., Chem. Commun.*, 1493 (1989).
155. J. Schwartz, *J. Organomet. Chem. Lib. 1*, 468 (1977).
156. G. Zweifel and C. C. Whitney, *J. Am. Chem. Soc.*, **89**, 2753 (1967).
157. H. C. Brown, T. Hamaoka and N. Ravindran, *J. Am. Chem. Soc.*, **95**, 5786 (1973).
158. M. E. Jung and L. A. Light, *Tetrahedron Lett.*, **23**, 3851 (1982).
159. C. K. Reddy and M. Periasamy, *Tetrahedron Lett.*, **31**, 1919 (1990).
160. N. Kamiya, Y. Chikami and Y. Ishii, *Synlett.*, 675 (1990).
161. D. P. Curran and D. Kim, *Tetrahedron*, **47**, 6171 (1991).
162. D. P. Curran, D. Kim and C. Ziegler, *Tetrahedron*, **47**, 6189 (1991).
163. T. Ishihara, M. Kuroboshi and Y. Okada, *Chem. Lett.*, 1895 (1986).
164. Q. Chen, Z. Qiu and Z. Yang, *J. Fluorine Chem.*, **36**, 149 (1987).
165. S. I. Pletnev, S. M. Igumnov, I. N. Rozhkov, G. D. Rempel, V. I. Ponomarev, L. E. Deev and V. S. Shaidurov, *Izv. Akad. Nauk SSSR, Ser. Khim.*, 2057 (1989); *Chem. Abstr.*, **112**, 197549w (1990).
166. G. Rong and R. Keese, *Tetrahedron Lett.*, **31**, 5615 (1990).
167. Y. Takeyama, Y. Ichinose, K. Oshima and K. Utimoto, *Tetrahedron Lett.*, **30**, 3159 (1989).
168. C. Hu, Y. Qiu and F. Qing, *J. Fluorine Chem.*, **51**, 295 (1991).
169. J. Barluenga, M. A. Rodriguez, J. M. Gonzalez and P. J. Campos, *Tetrahedron Lett.*, **31**, 4207

370   G. V. Boyd

(1990).
170. J. Barluenga, M. A. Rodriguez, J. M. Gonzalez, P. J. Campos and G. Asensio, *Tetrahedron Lett.*, **27**, 3303 (1986).
171. J. Barluenga, M. A. Rodriguez and P. J. Campos, *J. Org. Chem.*, **55**, 3104 (1990).
172. J. Barluenga, M. A. Rodriguez and P. J. Campos, *Synthesis*, 270 (1992).
173. M. Kodomari, T. Sakamoto and S. Yoshitomi, *Bull. Chem. Soc. Jpn.*, **62**, 4053 (1989).
174. S. Larson, T. Luidhardt, G. W. Kabalka and R. M. Pagni, *Tetrahedron Lett.*, **29**, 35 (1988).
175. G. J. S. Doad, D. J. Austin and T. C. Owen, *J. Chem. Res.*, *Synop.*, 313 (1989).
176. T. Negoro and Y. Ikeda, *Bull. Chem. Soc. Jpn.*, **59**, 3515 (1986).
177. J. Barluenga, J. M. Martinez-Gallo, C. Najera and M. Yus, *J. Chem. Soc.*, *Perkin Trans.*, 1, 1017 (1987).
178. T. Mitsudo, Y. Hori, Y. Yamakawa and Y. Watanabe. *J. Org. Chem.*, **52**, 2230 (1987).
179. T. Mitsudo, Y. Hori and Y. Watanabe. *J. Org. Chem.*, **50**, 1566 (1985).
180. T. Mitsudo, Y. Hori, Y. Yamakawa and Y. Watanabe, *Tetrahedron Lett.*, **27**, 2152 (1986).
181. Y. Hori, T. Mitsudo and Y. Watanabe, *Tetrahedron Lett.*, **27**, 5389 (1986).
182. Y. Hori, T. Mitsudo and Y. Watanabe, *J. Organomet. Chem.*, **321**, 397 (1987).
183. M. Rotem and Y. Shvo, *Organometallics*, **2**, 1689 (1983).
184. C. Bianchini, A. Meli, M. Peruzzini, F. Zanobini, C. Bruneau and P. H. Dixneuf, *Organometallics*, **9**, 1155 (1990).
185. M. Neveux, C. Bruneau and P. H. Dixneuf, *J. Chem. Soc.*, *Perkin Trans.* 1, 1197 (1991).
186. N. Menashe, D. Reshef and Y. Shvo, *J. Org. Chem.*, **56**, 2912 (1991).
187. Y. Sasaki and P. H. Dixneuf, *J. Chem. Soc.*, *Chem. Commun.*, 790 (1986).
188. R. Mahe, Y. Sasaki, C. Bruneau and P. H. Dixneuf, *J. Org. Chem.*, **54**, 1518 (1989).
189. P. H. Dixneuf, *Pure Appl. Chem.*, **61**, 1763 (1989).
190. P. H. Dixneuf, C. Bruneau and J. Fournier, *NATO ASI Ser.*, *Ser. C, 314 (Enzym. Model Carbo-xylation Reduct. React. Carbon Dioxide Util.*), 65 (1990); *Chem. Abstr.*, **114**, 61957m (1991).
191. I. Amer and H. Alper, *J. Organomet. Chem.*, **383**, 573 (1990).
192. B. E. Ali and H. Alper, *J. Mol. Catal.*, **67**, 29 (1991).
193. Y. Tsuji, T. Kondo and Y. Watanabe, *J. Mol. Catal.*, **40**, 295 (1987).
194. J. T. Lee and H. Alper, *Tetrahedron Lett.*, **32**, 1769 (1991).
195. H. Arzoumanian, F. Cochini, D. Nuel, J. F. Petrignani and N. Rosas, *Organometallics*, **11**, 49⁻ (1992).
196. D. Zargarian and H. Alper, *Organometallics*, **10**, 2914 (1991).
197. R. M. Moriarty, R. Penmasta, A. K. Awasthi and I. Prakash, *J. Org. Chem.*, **53**, 6124 (1988).
198. R. M. Moriarty, R. K. Vaid, M. P. Duncan and B. K. Vaid, *Tetrahedron Lett.*, **28**, 2845 (1987).
199. I. Fleming, T. W. Newton and F. Roessler, *J. Chem. Soc.*, *Perkin Trans.* 1, 2527 (1981).
200. I. Fleming and T. W. Newton, *J. Chem. Soc.*, *Perkin Trans.* 1, 1805 (1984).
201. H. M. Chen and J. P. Oliver, *J. Organomet. Chem.*, **316**, 255 (1986).
202. H. Hayami, M. Sato, S. Kanemoto, Y. Morizawa, K. Oshima and H. Nozaki, *J. Am. Chem. Soc*, **105**, 4491 (1983).
203. M. P. Doyle, K. G. High, C. L. Nesloney, T. W. Clayton, Jr. and J. Lin, *Organometallics*, **10**, 122 (1991).
204. L. I. Kopylova, V. B. Pukhnarevich and M. G. Voronkov, *Zh. Obshch. Khim.*, **61**, 2606 (1991 *Chem. Abstr.*, **116**, 194409x (1992).
205. K. Fugami, J. Hibino, S. Nakatsukasa, S. Matsubara, K. Oshima, K. Utimoto and H. Nozak *Tetrahedron*, **44**, 4277 (1988).
206. (a) N. Chatani and T. Hanafusa, *J. Chem. Soc.*, *Chem. Commun.*, 838 (1985).
     (b) N. Chatani, T. Takeyasu, N. Horiuchi and T. Hanafusa, *J. Org. Chem.*, **53**, 3539 (1988).
207. I. Matsuda, A. Ogiso, S. Sato and Y. Izumi, *J. Am. Chem. Soc.*, **111**, 2332 (1989).
208. H. Okinoshima, K. Yamamoto and M. Kumada, *J. Organomet. Chem.*, **86**, C27 (1975).
209. H. Watanabe, M. Kobayashi, M. Saito and Y. Nagai, *J. Organomet. Chem.*, **216**, 149 (1981).
210. Y. Ito, M. Suginome and M. Murakami, *J. Org. Chem.*, **56**, 1948 (1991).
211. K. Tamao, M. Hayashi and M. Kumada, *J. Organomet. Chem.*, **114**, C19 (1976).
212. H. Sakurai, Y. Kamiyama and Y. Nakadaira, *J. Am. Chem. Soc.*, **97**, 931 (1975).
213. E. Piers and J.M. Chong, *J. Chem. Soc.*, *Chem. Commun.*, 934 (1983).
214. H. Miyake and K. Yamamura, *Chem. Lett.*, 981 (1989).
215. K. Nozaki, K. Oshima and K. Utimoto, *J. Am. Chem. Soc.*, **109**, 2547 (1987).
216. K. Nozaki, K. Oshima and K. Utimoto, *Tetrahedron*, **45**, 923 (1989).

217. K. Kikukawa, H. Umekawa, F. Wada and T. Matsuda, *Chem. Lett.*, 881 (1988).
218. J.C. Cochran, B.S. Bronk, K.M. Terrence and H.K. Phillips, *Tetrahedron Lett.*, **31**, 6621 (1990).
219. K. Ritter, *Synthesis*, 218 (1989).
220. S. Sharma and A.C. Oehlschlager, *Tetrahedron Lett.*, **27**, 6161 (1986).
221. E. Piers, J.M. Chong and H.E. Morton, *Tetrahedron Lett.*, **22**, 4905 (1981).
222. (a) K. K. Wang and K. H. Chu, *J. Org. Chem.*, **51**, 767 (1986).
    (b) K. K. Wang, K. H. Chu, Y. Lin and J. H. Chen, *Tetrahedron*, **45**, 1105 (1989).
223. A. Barbero, P. Cuadrado, I. Fleming, A. M. Gonzalez and F. J. Pulido, *J. Chem. Soc., Chem. Commun.*, 351 (1992).
224. (a) T. N. Mitchell, A. Amamria, H. Killing and D. Rutschow, *J. Organomet., Chem.*, **241**, C45 (1983).
    (b) T. N. Mitchell, A. Amamria, H. Killing and D. Rutschow, *J. Organomet. Chem.*, **304**, 257 (1986).
225. B. L. Chenard, E. D. Laganis, F. Davidson and T. V. RajanBabu, *J. Org. Chem.*, **50**, 3666 (1985).
226. T. N. Mitchell, R. Wickenkamp, A. Amamria, R. Dicke and U. Schneider, *J. Org. Chem.*, **52**, 4868 (1987).
227. P. A. Magriotis, J. T. Brown and M. E. Scott, *Tetrahedron Lett.*, **32**, 5047 (1991).
228. T. Takeda, H. Ohshima, M. Inoue, A. Togo and T. Fujiwara, *Chem. Lett.*, 1345 (1987).
229. W. Wolfsberger, *Chem.-Ztg.*, **115**, 161 (1991).
230. Y. Ichinose, H. Oda, K. Oshima and K. Utimoto, *Bull. Chem. Soc. Jpn.*, **60**, 3468 (1987).
231. Y. Ichinose, K. Nozaki, K. Wakamatsu, K. Oshima and K. Utimoto, *Tetrahedron Lett.*, **28**, 3709 (1987).
232. E. Lukevics, R. Ya. Sturkovich and O. A. Pudova, *Zh. Obshch. Khim.*, **58**, 815 (1988); *Chem. Abstr.*, **110**, 75677z (1989).
233. N. Chatani, N. Horiuchi and T. Hanafusa, *J. Org. Chem.*, **55**, 3393 (1990).
234. G. Capozzi, G. Modena and L. Pasquato, in *The Chemistry of Sulphenic Acids and their Derivatives* (Ed. S. Patai). Wiley, Chichester, 1990, p. 403.
235. L. Benati, P. C. Montevecchi and P. Spagnolo, *J. Chem. Soc., Perkin Trans. 1*, 1105 (1989).
236. L. Benati, D. Casarini, P. C. Montevecchi and P. Spagnolo, *J. Chem. Soc., Perkin Trans. 1*, 1113 (1989).
237. (a) L. Benati, P. C. Montevecchi and P. Spagnolo, *Gazz. Chim. Ital.*, **119**, 609 (1989).
    (b) L. Benati, P. C. Montevecchi and P. Spagnolo, *J. Chem. Soc., Perkin Trans. 1*, 1691 (1990).
238. G. Capozzi, L. Gori and S. Menichetti, *Tetrahedron Lett.*, **31**, 6212 (1990).
239. J. B. Baudin, S. A. Julia and Y. Wang, *Tetrahedron Lett.*, **30**, 4965 (1989).
240. T. Kobayashi, Y. Tanaka, T. Ohtani, H. Kinoshita, K. Inomata and H. Kotake, *Chem. Lett.*, 1987, 1209.
241. (a) T. V. Popkova, A. V. Shastin, E. I. Lazhko and E. S. Balenkova, *Zh. Org. Khim.*, **22**, 2460 (1986); *Chem. Abstr.*, **107**, 157007x (1987).
    (b) A. V. Shastin, T. V. Popkova, E. S. Balenkova, V. K. Belskii and E. I. Lazhko, *Zh. Org. Khim.*, **23**, 2313 (1987); *Chem. Abstr.*, **109**, 109500h (1988).
242. M. V. Galakhov, V. F. Cherstkov, S. R. Sterlin and L. S. German, *Izv. Akad. Nauk SSSR, Ser. Khim.*, 958 (1987); *Chem. Abstr.*, **108**, 111779d (1988).
243. A. Ogawa, H. Yokoyama, K. Yokoyama, T. Masawaki, N. Kambe and N. Sonoda, *J. Org. Chem.*, **56**, 5721 (1991).
244. H. Kuniyasu, A. Ogawa, S. Miyazaki, I. Ryu, N. Kambe and N. Sonoda, *J. Am. Chem. Soc.*, **113**, 9796 (1991).
245. (a) T. G. Back and K. R. Muralidharan, *Tetrahedron Lett.*, **31**, 1957 (1990).
    (b) T. G. Back and K. R. Muralidharan, *J. Org. Chem.*, **56**, 2781 (1991).
246. C. Saluzzo, G. Alvernhe, D. Anker and G. Haufe, *Tetrahedron Lett.*, **31**, 2127 (1990).
247. T. G. Back, E. K. Y. Lai and K. R. Muralidharan, *J. Org. Chem.*, **55**, 4595 (1990).
248. J. V. Comasseto, H. A. Stefani, A. Chieffi and J. Zukerman-Schpector, *Organometallics*, **10**, 845 (1991).
249. H. Takahashi, K. Ohe, S. Uemura and N. Sugita, *Nippon Kagaku Kaishi*, 1508 (1987); *Chem. Abstr.*, **108**, 74916y (1988).
250. A. Ogawa, K. Yokoyama, H. Yokoyama, R. Obayashi, N. Kambe and N. Sonoda, *J. Chem. Soc., Chem. Commun.*, 1748 (1991).
251. C. L. Baumgardner, J. E. Bunch and M. H. Whangbo, *Tetrahedron Lett.*, **27**, 1883 (1986).
252. T. Funabiki, H. Sato, N. Tanaka, Y. Yamazaki and S. Yoshida, *J. Mol. Catal.*, **62**, 157 (1990).

372 G. V. Boyd

253. D. A. Anderson and J. R. Hwu, *J. Chem. Soc., Perkin Trans. 1*, 1694 (1989).
254. B. H. Lipshutz and S. H. Dimock, *J. Org. Chem.*, **56**, 5761 (1991).
255. B. H. Lipshutz and E. L. Ellsworth, *J. Am. Chem. Soc.*, **112**, 7440 (1990).
256. K. Takai, Y. Kataoka and K. Utimoto, *J. Org. Chem.*, **55**, 1707 (1990).
257. K. Takai, Y. Kataoka, K. Yoshizumi, Y. Uguchi and K. Utimoto, *Chem. Lett.*, 1479 (1991).
258. J. Barluenga, F. Aznar and M. Bayod, *Synthesis*, 1544 (1988).
259. M. Bassetti, B. Floris and G. Spadafora, *J. Org. Chem.*, **54**, 5934 (1989).
260. M. Giffard, J. Cousseau, L. Gouin and M.R. Crahe, *Tetrahedron*, **42**, 2243 (1986).
261. T. N. Mitchell and K. Heesche, *J. Organomet. Chem.*, **409**, 163 (1991).
262. E. E. Nifantev, L. A. Solovetskaya, V. I. Maslennikova and N. M. Sergeev, *Zh. Obshch. Khim.*, **57**, 514 (1987); *Chem. Abstr.*, **108**, 167564x (1988).
263. N. E. Schore, *Chem. Rev.*, **88**, 1081 (1988).
264. P. M. Maitlis, *J. Organomet. Chem.*, **200**, 161 (1980).
265. M. Pfeffer, *Recl. Trav. Chim. Pays-Bas*, **109**, 567 (1990).
266. P. F. Korkowski, T. R. Hoye and D. R. Rydberg, *J. Am. Chem. Soc.*, **110**, 2676 (1988).
267. E. V. Dehmlow and A. Winterfeldt, *Tedrahedron*, **45**, 2925 (1989).
268. R. Gleiter and M. Merger, *Tetrahedron Lett.*, **33**, 3473 (1992).
269. L. S. Hegedus and D. B. Miller, Jr., *J. Org. Chem.*, **54**, 1241 (1989).
270. M. H. Ansari, F. Ahmad and M. Ahmad, *JAOCS, J. Am. Oil Chem. Soc.*, **63**, 92 (1986).
271. S. Lochschmidt, F. Mathey and A. Schmidpeter, *Tetrahedron Lett.*, **27**, 2635 (1986).
272. Y. Okuda, Y. Morizawa, K. Oshima and H. Nozaki, *Tetrahedron Lett.*, **25**, 2483 (1984).
273. R. L. Danheiser, S. Savariar and D. D. Cha, *Org. Synth.*, **68**, 32 (1990).
274. D. W. Wulff and Y. C. Xu, *Tetrahedron Lett.*, **29**, 415 (1988).
275. D. J. Pasto and W. Kong, *J. Org. Chem.*, **53**, 4807 (1988).
276. H. Wingert, H. Irngartinger, D. Kallfass and M. Regitz, *Chem Ber.*, **120**, 825 (1987).
277. (a) T. Mitsudo, K. Kokuryo, T. Shinsugi, Y. Nakagawa, Y. Watanabe and Y. Takegami, *J. Org. Chem.*, **44**, 4492 (1979).
    (b) T. Mitsudo, Y. Hori and Y. Watanabe, *J. Organomet. Chem.*, **334**, 157 (1987).
278. R. Askani and J. Hoffmann, *Chem. Ber.*, **124**, 2307 (1991).
279. K. B. Wiberg and S. T. Waddell, *J. Am. Chem. Soc.*, **112**, 2194 (1990).
280. R. B. King, R. M. Murray, R. E. Davis and P. K. Ross, *J. Organomet. Chem.*, **330**, 115 (1987).
281. D. J. De Young and R. West, *Chem. Lett.*, 883 (1986).
282. H. Bock, P. Rittmeyer and U. Stein, *Chem. Ber.*, **119**, 3766 (1986).
283. M. Fulde, W. Ried and J.W. Bats, *Helv. Chim. Acta*, **72**, 139 (1989).
284. H. Gotthardt, T. Loehr and D.J. Brauer, *Chem. Ber.*, **120**, 751 (1987).
285. R. F. C. Brown, *Recl. Trav. Chim. Pays-Bas*, **107**, 655 (1988).
286. R. F. C. Brown, F. W. Eastwood, K. J. Harrington and G. L. McMullen, *Aust. J. Chem.*, **30**, 175? (1977).
287. M. Karpf and A. S. Dreiding, *Helv. Chim. Acta*, **62**, 852 (1979).
288. M. Dütsch, R. Lackmann, F. Stein and A. De Meijere, *Synlett.*, 324 (1991).
289. M. Ochiai, M. Kunishima, Y. Nagao, K. Fuji, M. Shiro and E. Fujita, *J. Am. Chem. Soc.*, **108**, 8281 (1986).
290. D. J. Pollart and H. W. Moore, *J. Org. Chem.*, **54**, 5444 (1989).
291. G. C. M. Lee, B. Tobias, J. M. Holmes, D. A. Harcourt and M. E. Garst, *J. Am. Chem. Soc.*, **112**, 9330 (1990).
292. L. J. Silverberg, G. Wu, A. L. Rheingold and R. F. Heck, *J. Organomet. Chem.*, **409**, 411 (1991).
293. F. E. Meyer, J. Brandenburg, P. J. Parsons and A. De Meijere, *J. Chem. Soc., Chem. Commun*, 390 (1992).
294. E. Negishi, Y. Noda, F. Lamaty and E. J. Vawter, *Tetrahedron Lett.*, **31**, 4393 (1990).
295. N. P. Robinson, L. Main and B. K. Nicholson, *J. Organomet Chem.*, **364**, C37 (1989).
296. L. S. Liebeskind, J. R. Gasdaska and J. S. McCallum, *J. Org. Chem.*, **54**, 669 (1989).
297. A. Yamashita, *Tetrahedron Lett.*, **27**, 5915 (1986).
298. K. H. Doetz and H. Larbig, *J. Organomet. Chem.*, **405**, C38 (1991).
299. Y. C. Xu, C. A. Challener, V. Dragisich, T. A. Brandvold, G. A. Peterson, W. D. Wulff and P. C. Williard, *J. Am. Chem. Soc.*, **111**, 7269 (1989).
300. J. W. Herndon, S. U. Tumer and W. F. K. Schnatter, *J. Am. Chem. Soc.*, **110**, 3334 (1988).
301. I. U. Khand, G. R. Knox, P. L. Pauson and W. E. Watts, *J. Chem. Soc., Chem. Commun.*, ? (1971).

302. P. L. Pauson, *Organomet.. Org. Synth. [Proc. Symp. 'Org. Synth. Organomet,'], 1st*, 233 (1987).
303. N. E. Schore, *Org. React.*, **40**, 1 (1991).
304. C. Exon and P. Magnus, *J. Am. Chem. Soc.*, **105**, 2477 (1983).
305. A Davasagayaraij and M. Periasamy, *Tetrahedron Lett.*, **30**, 595 (1989).
306. M. E. Krafft, C. A. Juliano, I. L. Scott, C. Wright and M. D. McEachin, *J. Am. Chem. Soc.*, **113**, 1693 (1991).
207. F. Camps, A. Llebaria, J. M. Moreto and L. Pages, *Tetrahedron Lett.*, **33**, 109 (1992).
308. F. Camps, A. Llebaria, J. M. Moreto and L. Pages, *Tetrahedron Lett.*, **33**, 113 (1992).
309. F. T. Boyle, Z. S. Matusiak, O. Hares and D. A. Whiting, *J. Chem Soc., Chem. Commun.*, 518 (1990).
310. G. Wu, F. E. Cederbaum and E. Negishi, *Tetrahedron Lett.*, **31**, 493 (1990).
311. J. Van der Louw, J. L. Van der Baan, F. Bickelhaupt and G. W. Klumpp, *Tetrahedron Lett.*, **28**, 2889 (1987).
312. E. V. Dehmlow and A. Winterfeldt, *Z. Naturforsch., B: Chem. Sci.*, **44**, 455 (1989).
313. B. M. Trost and D. C. Lee, *J. Am. Chem. Soc.*, **110**, 7255 (1988).
314. D. P. Curran, M. H. Chen and D. Kim, *J. Am. Chem. Soc.*, **111**, 6265 (1989).
315. D. P. Curran and C. M. Seong, *J. Am. Chem. Soc.*, **122**, 9401 (1990).
316. I. F. Duan, C. H. Cheng, J. S. Shaw, S. S. Cheng and K. F. Liou, *J. Chem. Soc., Chem. Commun.*, 1347 (1991).
317. M. Lautens, J. C. Lautens and A. C. Smith, *J. Am. Chem. Soc.*, **112**, 5627 (1990).
318. E. V. Dehmlow, A. Winterfeldt and J. Pickardt, *J. Organomet Chem.*, **363**, 223 (1989).
319. H. M. L. Davies and K. R. Romines, *Tetrahedron*, **44**, 3343 (1988).
320. T. Ogawa, T. Murafuji, K. Iwata and H. Suzuki, *Chem. Lett.*, 325 (1989).
321. R. Grigg, R. Scott and P. Stevenson, *J. Chem. Soc., Perkin Trans 1*, 1357 (1988).
322. J. S. Gurinder, J. A. Barltrop, C. M. Petty and T. C. Owen, *Tetrahedron Lett.*, **30**, 1597 (1989).
323. A. Arcadi, F. Marinelli and S. Cacchi, *Synthesis*, 749 (1986).
324. R. Aumann and H. Heinen, *J. Organomet. Chem.*, **391**, C7 (1990).
325. (a) N. Chatani and T. Hanafusa, *Tetrahedron Lett.*, **27**, 4201 (1986).
     (b) N. Chatani and T. Hanafusa, *J. Org. Chem.*, **56**, 2166 (1991).
326. R. C. Larock and E. K. Yum, *J. Am. Chem. Soc.*, **113**, 6689 (1991).
327. R. Herges and C. Hoock, *Angew. Chem., Int. Ed. Engl.*, **31**, 1611 (1992).
328. D. Chiarino, M. Napoletano and A. Sala, *Synth. Commun.*, **18**, 1171 (1988).
329. D. Chiarino, M. Napoletano and A. Sala, *J. Heterocycl. Chem.*, **24**, 43 (1987).
330. Y. Yamamoto, T. Tsuchiya, M. Ochiumi, S. Arai, N. Inamoto and K. Akiba, *Bull. Chem. Soc. Jpn.*, **62**, 211 (1989).
331. G. V. Boyd, T. Norris, P. F. Lindley and M. M. Mahmoud, *J. Chem. Soc., Perkin Trans. 1*, 1612 (1977).
332. S. Parsons, J. Passmore, M. J. Schriver and P. S. White, *Can. J. Chem.*, **68**, 852 (1990).
333. I. Matsuda, M. Shibata, S. Sato and Y. Izumi, *Tetrahedron Lett.*, **28**, 3361 (1987).
334. B. Jousseaume and P. Villeneuve, *Tetrahedron*, **45**, 1145 (1989).
335. H. Hopf and M. Kreutzer, *Angew. Chem., Int. Ed. Engl.*, **29**, 393 (1990).
336. W. E. Bauta, W. D. Wulff, S.F. Pavkovic and E.J. Zaluec, *J. Org. Chem.*, **54**, 3249 (1989).
337. A. Funhoff, H. Schäufele and U. Zenneck, *J. Organomet. Chem.*, **345**, 331 (1988).
338. A.C. Williams, P. Sheffels, D. Sheehan and T. Livinghouse, *Organometallics*, **8**, 1566 (1989).
339. P. Biagini, A. M. Coporusso, T. Funaioli and G. Fachinetti, *Angew. Chem., Int. Ed. Engl.*, **28**, 1009 (1989).
340. R. Herrmann and A. J. Pombeiro, *Monatsh. Chem.*, **119**, 583 (1988).
341. H. tom Dieck, C. Munz and C. Müller, *J. Organomet. Chem.*, **384**, 243 (1990).
342. I. Amer, J. Blum and K. P. C. Vollhardt, *J. Mol. Catal.*, **60**, 323 (1990).
343. R. Leardini, D. Nanni, A. Tundo and G. Zanardi, *Synthesis*, 333 (1988).
344. P. R. Jones, T. E. Albanesi, R. D. Gillespie, P. C. Jones and S. W. Ng, *Appl. Organomet. Chem.*, **1**, 521 (1987); *Chem. Abstr.*, **110**, 154365z (1989).
345. D. Obrecht, *Helv. Chim. Acta*, **74**, 27 (1991).
346. R. L. Danheiser, R. G. Brisbois, J. J. Kowalczyk and R. F. Miller, *J. Am. Chem. Soc.*, **112**, 3093 (1990).
347. J. B. Hartung, Jr. and S. E. Pedersen, *J. Am. Chem. Soc.*, **111**, 5468 (1989).
348. Y. Kataoka, J. Miyai, M. Tezuka, K. Takai, K. Oshima and K. Utimoto, *Tetrahedron Lett.*, **31**, 369 (1990).

349. M. A. Huffman and L. S. Liebeskind, *J. Am. Chem. Soc.*, **133**, 2771 (1991).
350. L. S. Liebeskind, J. P. Leeds, S. L. Baysdon and S. Iyer, *J. Am. Chem. Soc.*, **106**, 6451 (1984).
351. S. Iyer and L. S. Liebeskind, *J. Am. Chem. Soc.*, **109**, 2759 (1987).
352. E. Vilsmaier and R. Baumheier, *Chem. Ber.*, **122**, 1285 (1989).
353. K. Chow, N. V. Nguyen and H. W. Moore, *J. Org. Chem.*, **55**, 3876 (1990).
354. V. N. Kalinin, M. V. Shostakovskii and A. B. Ponomarev, *Tetrahedron Lett.*, **31**, 4073 (1990).
355. L. E. Overman and A. K. Sarkar, *Tetrahedron Lett.*, **33**, 4103 (1992).
356. G. Vitulli, S. Bertozzi, R. Lazzarone and P. Salvadori, *J. Organomet. Chem.*, **307**, C35 (1986).
357. P. Diversi, G. Ingrosso, A. Lucherini and D. Vanacore, *J. Mol. Catal.*, **41**, 261 (1987).
358. C. Kashima, M. Shimizu and Y. Omote, *Chem. Pharm. Bull.*, **35**, 2694 (1987).
359. R. Leardini, D. Nanni, G. F. Pedulli, A. Tundo and G. Zanardi, *J. Chem. Soc., Chem. Commun.*, 1591 (1986).
360. G. Wu, S. J. Geib, A. L. Rheingold and R. F. Heck, *J. Org. Chem.*, **53**, 3238 (1988).
361. G. Wu, A. L. Rheingold and R. F. Heck, *Organometallics*, **6**, 2386 (1987).
362. S. Torii, H. Okumoto and L. H. Xu, *Tetrahedron Lett.*, **32**, 237 (1991).
363. F. Maassarani, M. Pfeffer and G. LeBorgne, *J. Chem. Soc., Chem. Commun.*, 565 (1987).
364. D. P. Curran and H. Liu, *J. Am. Chem. Soc.*, **113**, 2127 (1991).
365. N. Haider and H. C. Van der Plas, *Tetrahedron*, **46**, 3641 (1990).
366. M. Murakami, Y. Morita and Y. Ito, *J. Chem. Soc., Chem. Commun.*, 428 (1990).
367. (a) D. V. Demchuk, E. I. Troyanskii and G. I. Nikishin, *Izvest. Akad. Nauk SSSR. Ser. Khim.*, 1443 (1989); *Chem. Abstr.*, **112**, 77140j (1990).
    (b) D. V. Demchuk, A. I. Lutsenko, E. I. Troyanskii and G. I. Nikishin, *Izv. Akad. Nauk, SSSR, Ser. Khim.*, 2801 (1990); *Chem. Abstr.*, **114**, 207218v (1991).
368. S. Ghosal, M. Nirmal, J. C. Medina and K. S. Kyler, *Synth. Commun.*, **17**, 1683 (1987).
369. E. Fluck, W. Plass and G. Heckmann, *Z. Anorg, Allg. Chem.*, **588**, 181 (1990).
370. G. Märkl and S. Dietl, *Tetrahedron Lett.*, **29**, 539 (1988).
371. R. Gleiter, *Angew. Chem., Int. Ed. Engl.*, **31**, 27 (1992).
372. K. P. C. Vollhardt, *Angew. Chem., Int. Ed. Engl.*, **23**, 539 (1984).
373. K. C. Nicolaou, E. J. Sorensen, R. Discordia, C.-K. Hwang, R. E. Minto, K. N. Bharucha and R. G. Bergman, *Angew. Chem., Int. Ed. Engl.*, **31**, 1044 (1992).
374. R. Gleiter, R. Merger and B. Huber, *J. Am. Chem. Soc.*, **114**, 8921 (1992).
375. H. Higuchi, K. Kitamura, J. Ojima, K. Yamamoto and G. Yamamoto, *J. Chem. Soc., Perkin Trans. 1*, 1343 (1992).
376. L. T. Scott, M. J. Cooney and D. Johnels, *J. Am. Chem. Soc.*, **112**, 4054 (1990).
377. J. S. Moore and J. Zhang, *Angew. Chem., Int. Ed. Engl.*, **31**, 922 (1992).
378. G. Just and R. Singh, *Tetrahedron Lett.*, **28**, 5981 (1987).

CHAPTER **7**

Photochemistry and radiation chemistry

WILLIAM M. HORSPOOL

Department of Chemistry, The University of Dundee, Dundee, DD1 4HN, Scotland, UK

Supplement C2: The chemistry of triple-bonded functional groups
Edited by S. Patai © 1994 John Wiley & Sons Ltd

I. INTRODUCTION

The areas covered by this chapter deal with the photochemistry and radiation chemistry of triply bonded systems. The definition of a system with a triple bond is fairly loose and the review will deal with compounds which contain the following functionalities: alkynes nitriles, diazo compounds, diazonium salts and azides. The photochemical reactivity of these compounds has been reviewed extensively[1]. This review is not intended to be encyclopaedic but will highlight some of the areas currently of interest.

II. PHOTOCHEMICAL REACTIONS OF ALKYNES

In addition to the reviews mentioned above, a short review, in Japanese, has also been published[2]. Since 1978 many new processes involving compounds within which there is an alkyne functional group have been published.

A. Free Radical Reactions

Some photochemical processes have been described where the primary photochemical event involves the fission of a bond between a substituent and the alkyne terminal carbon atom. Thus Inoue and coworkers[3] report that the irradiation of 1-bromo and 1-iodohex

1-yne leads to fission of the halogen–C bond and the formation of products derived from free radical paths. A free radical path is also involved on irradiation of the alkyne **1**. Here C—S bond fission affords radicals which add to norbornene to yield the *trans*-adduct **2**[4]. Hydrogen abstraction is also a reaction that can be observed on irradiation of alkynes. Typical of this reaction mode is the irradiation of dec-1-yne which undergoes photochemical reduction to dec-1-ene on irradiation at 185 nm in pentane[5]. Dec-5-yne behaves similarly but yields a mixture of *cis*- and *trans*-dec-5-ene as well as the totally reduced species decane[6]. Intermolecular hydrogen abstraction also dominates the photoreaction of the alkyne **3** affording the alkene **4**[7]. In the case of the dienyne **5** intermolecular hydrogen abstraction from the solvent affords the triene **5**, again a conventional reduction process. In contrast, the irradiation of the more highly strained enyne **7** results in bond fission and isomerization to yield the cyclohexene derivative **8**[8]. One example has been described, the irradiation of cyclononyne, where hydrogen abstraction occurs intramolecularly to afford bicyclononene **9** via the biradical **10**, which cyclizes to the final product[6].

$$Ph-C\equiv C-SO_2Ph$$
(1)

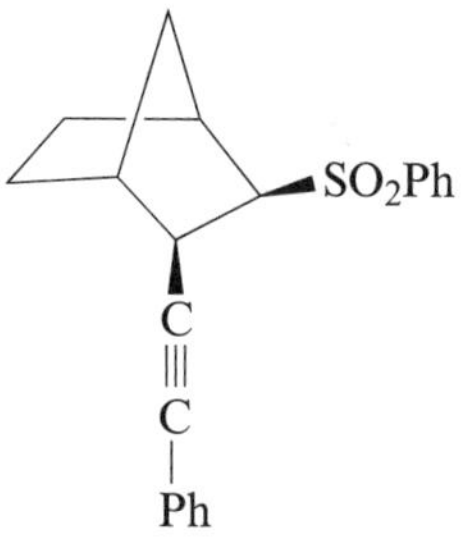

(2)

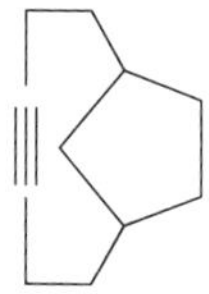

(3)

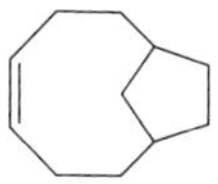

(4)

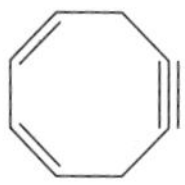

(5)

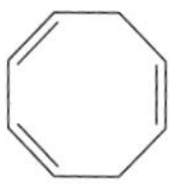

(6)

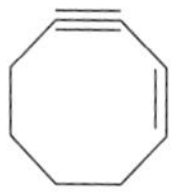

(7)

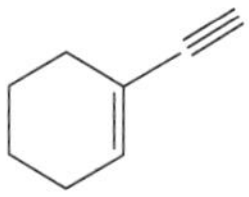

(8)

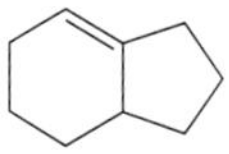

(9)

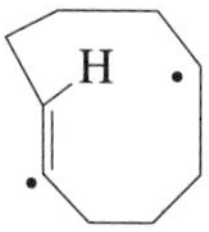

(10)

B. Photo-hydration of Alkynes and Related Processes

Excitation of alkynes in protic media leads to an enhancement of the basicity of the alkyne moiety. This process, which leads to protonation and addition of solvent, is a well known reaction in alkenes[9].

A study of the photohydration of the alkynes **11** and **12** in sulphuric acid has shown that only the normal Markownikov products, in this instance in the keto form **13**, are formed[10]. The mechanism, as proposed by Yates and collaborators[10] involves a rate-limiting protonation of the S_1 state of the alkyne. This leads to the cation **14**, solvation of which affords the enol of the final product. Associated with this, and providing spectroscopic substantiation, is the report of the direct observation of the enol of acetophenone during a study of the photohydration of phenylacetylene[11]. The introduction of nitro substituents as in **15** leads to abnormal Markownikov addition products. Excitation in these systems results in a reversal of polarity in the excited state, consistent with the

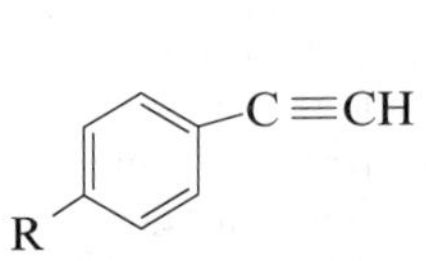

(11) R = H or Me 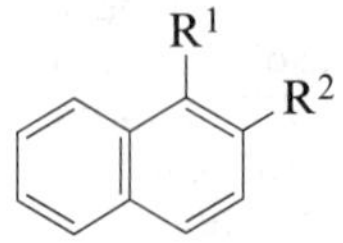   **(12)** R^1 = H, R^2 = C≡CH or CH=CH$_2$ **(13)**
                               R^1 = C≡CH, R^2 = H

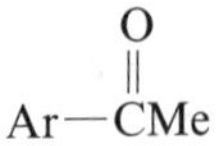

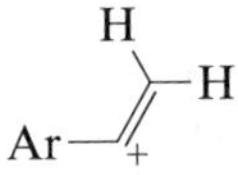

(14) 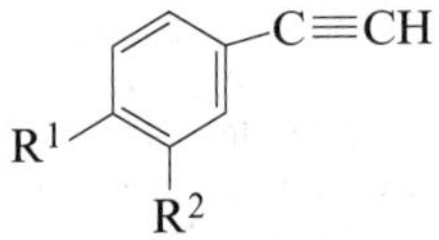   **(15)** R^1 = NO$_2$, R^2 = H   **(16)**   **(17)**
                               R^1 = H, R^2 = NO$_2$

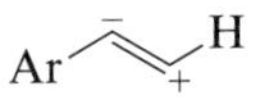

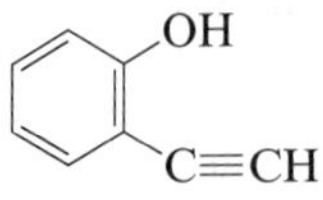

enhanced electron-withdrawing properties of the nitro group in the excited state, leading
to the zwitterionic excited state **(16)** which protonates on the carbon adjacent to the aryl
ring[10]. The influence of *m*- and *p*-substituted groups on the photohydration of
phenylacetylene derivatives has been studied with a view to establishing a linear free-
energy relationship for such photohydrations[12]. With *ortho*-substituents, as in **17**, the
mechanism of photohydration has implicated an intramolecular proton transfer[13].

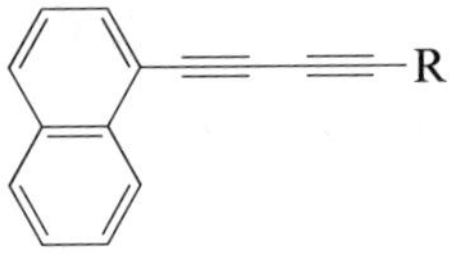

(18) R = Ph, Me or Bu^t 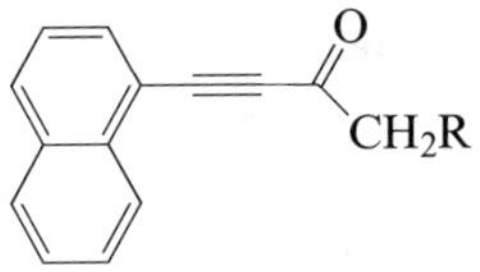   **(19)** 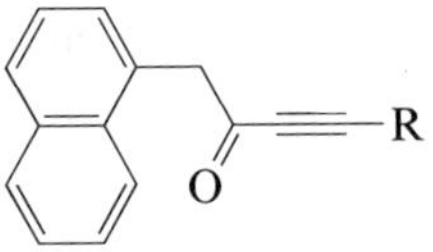   **(20)**

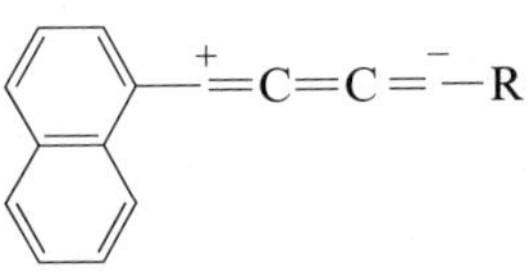

(21) 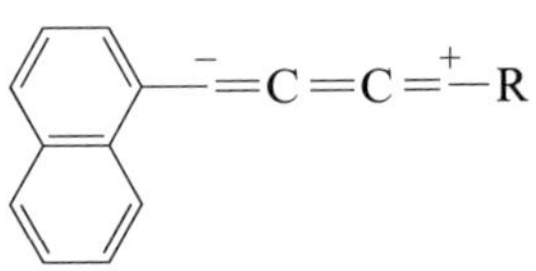   **(22)**

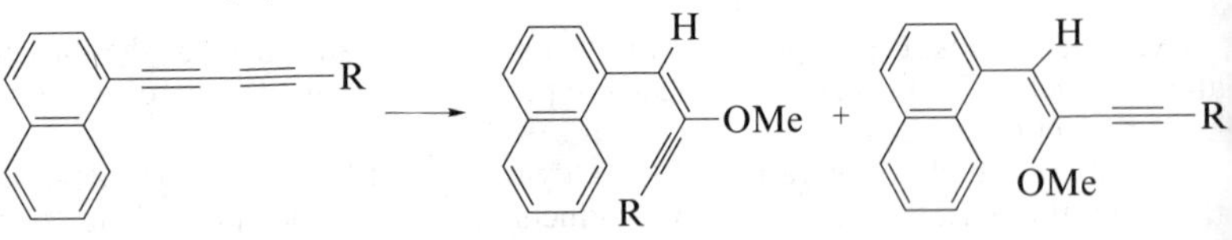

R = Ph	7%	—
R = Bu^t	10%	11%
R = Me	12%	19%
R = TMS	14%	—

SCHEME 1

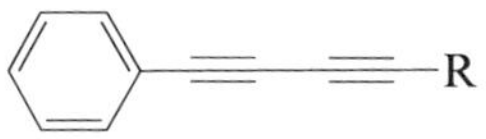

(23) (a) R = Ph
     (b) R = Bu^t

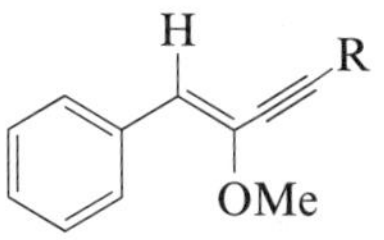

(24) (a) 23%
     (b) 27%

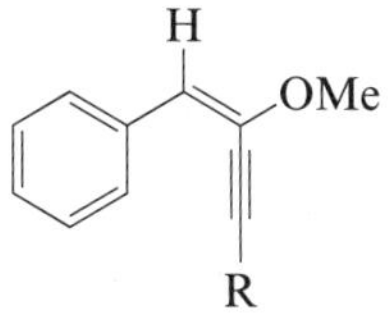

(25) (a) 24%
     (b) 28%

$$HC\equiv CSiR_2R'$$

(26)

R = Me or Et

R'= Me, Et, OEt, OPr, or F

$$(R_2R'SiC=CH)_2S$$

(27)

Hydration of the diynes **18** takes place via the singlet excited state. Irradiation of these compounds in acetonitrile/water results in the formation of the two ketones **19** and **20**. These are formed via hydration of the two charge separated states **21** and **22** with the former **19**, a Markownikov product, predominant. Interestingly, substituents do influence the excited states involved and when the substituent R is alkyl in **18** the ketonic product **19** is formed from the triplet state of the diyne[14]. Other solvents also undergo addition to the same alkynes. Thus in methanol, addition of the solvent takes place affording the products shown with the yields in Scheme 1. It is suggested, as in the previous example, that the addition involves a polarized excited state[15] but both the singlet and triplet excited states are involved. An analogous addition reaction was encountered with the phenyl derivatives **23** affording the adducts **24** and **25**[15,16]. A photostationary mixture of the alkenes is obtained on prolonged irradiation. Photochemical addition of hydrogen sulphide to the alkynes **26** has been reported to be a route to the alkenes **27**[17].

C. Cycloaddition Reactions of Alkynes

Several reviews have been devoted to the subject of, among other cycloaddition reactions, (2 + 2)-cycloaddition reactions to enones. Within these articles the additions of alkynes has been dealt with[18]. Specifically, a short review has discussed the cycloaddition reactions of enones with alkynes[19].

1. Open-chain systems

An interesting cycloaddition reaction of the (2 + 4)-category has been described for the addition of dicyanoacetylene and of the acetylenes **28** to the divinylamine **29**. This affords the adducts **30** and **31**, respectively[20,21].

$$R'O_2CC\equiv CCO_2R'$$

(28)

R'= Me, Et, Pr, Pr^i, Bu, CHMeEt,
    Bu^t, cyclopentyl, or cyclohexyl

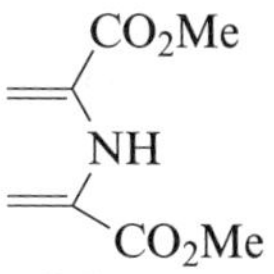

(29)

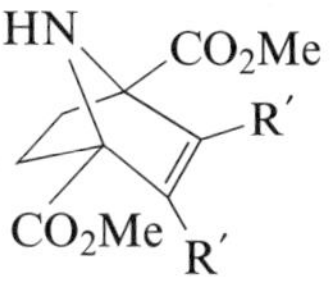

(30) R' = CN

(31) R' = CO$_2$R'

2. Additions to cyclopentenones

More conventional (2 + 2)-cycloaddition reactions are encountered with cyclic enones and related systems. Thus both 3-methylbut-1-yne and 3,3-dimethylbut-1-yne undergo photochemical addition to cyclopentenone to yield the two adducts **32** and **33**. The latter of these is photochemically reactive and irradiation converts it into the 1,3-migration product **34**[22]. Substituted cyclopentenones also undergo addition to alkynes and **35a** adds acetylene to give indan-1-one in 7% yield. This product is formed by a double-addition, (2 + 2)-cycloaddition of acetylene, followed by ring opening of the resultant cyclobutene and elimination of HCN to yield the final product. Addition of acetylene to **35b** also gives the indanone, but in a better yield (23%). Several other products were obtained among which **36** (4.5%), **37** (13%) and the double addition product **38** (3.1%) were identified. Compound **36** is formed by a straightforward (2 + 2)-cycloaddition and **38** by a second (2 + 2)-addition to **36**, while **10** arises by an oxa-di-π-methane rearrangement[23] of **36**[24]. The position of the substituent on the cyclopentenone does not affect the outcome of the cycloaddition and cyclopentenone **39** undergoes photoaddition to hex-1-yne to yield the (2 + 2)-cycloadduct **40** in good yield. The same starting materials, using Pyrex-filtered light, lead to a mixture of the tropovalene **41** (25%) and the bicyclic compound **40** (38%). Independent irradiation of **41** yields the tropone **42** in 83% yield. The same product can be obtained by irradiation of **43**, formed by elimination of acetic acid from **40**[25]. The same cyclopentenone **39** and the 2-methoxy substituted derivative **44** also undergo addition of ethyne and but-1-yne using left-circularly polarized light[26]. The resultant cyclobutene derivatives are readily converted to the optically active enones **45**. Pentyne adds photochemically to the sterloidal enone **46** to give the isomeric products **47** and **48** in a ratio of 10 : 1[27].

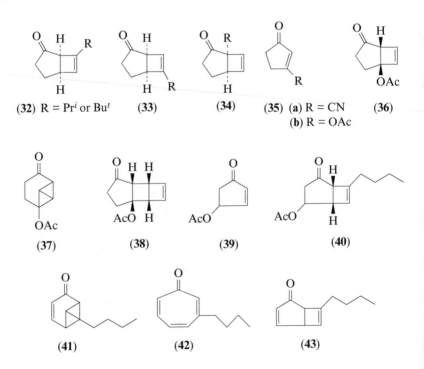

(32) R = Pr^i or Bu^t **(33)** **(34)** **(35) (a)** R = CN **(36)**
(b) R = OAc

(37) **(38)** **(39)** **(40)**

(41) **(42)** **(43)**

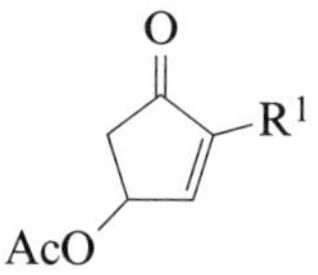

(44) R^1 = MeO 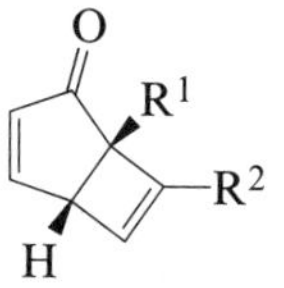   **(45)** R^2 = H or Bu 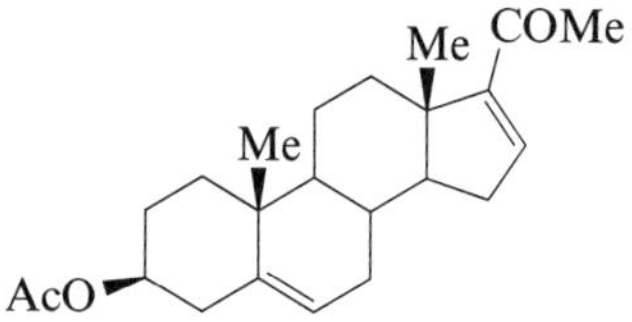   **(46)**

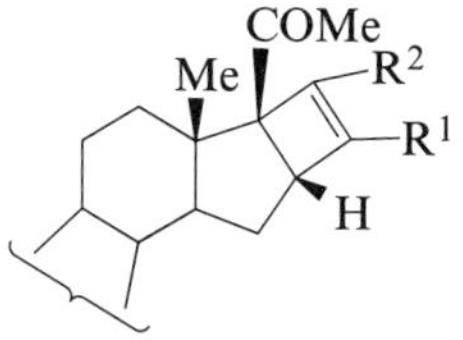

(47) R^1= H, R^2 = Pr
(48) R^1 = Pr, R^2 = H

Lactone **51** is also a good substrate for cycloaddition reactions and affords a mixture of head-to-head and head-to-tail isomers, **49** and **50**, respectively, on irradiation in the presence of the 2-prop-2-yn-1-ol[28] and hex-1-yne to yield **52**[29]. This latter addition reaction also yields the adduct **53** which suggests that the cycloaddition occurs by a two-step mechanism affording biradical **54** from the head-to-tail addition mode which is

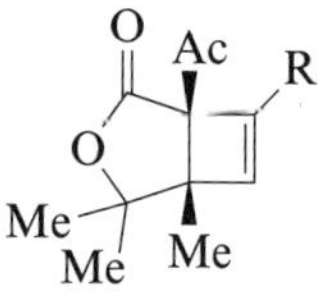

(49) R = CH$_2$OH   **(50)** 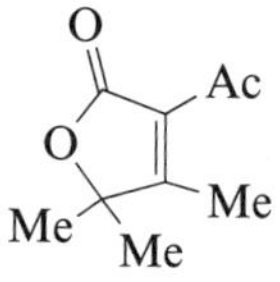   **(51)** 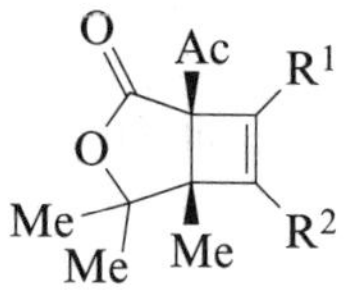   **(52)** R^1 = pentyl, R^2 = H

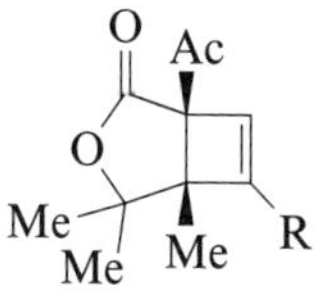

or *vice versa*

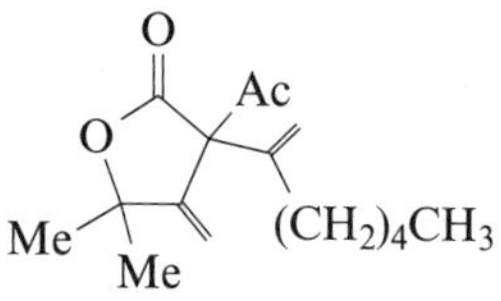

(53) 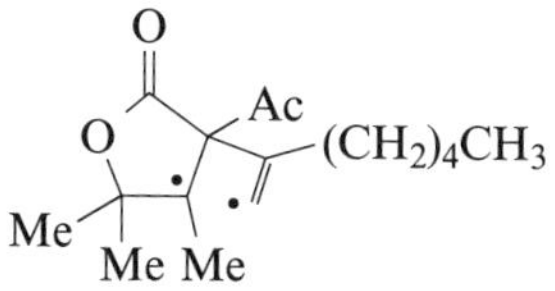   **(54)**    **(55)**

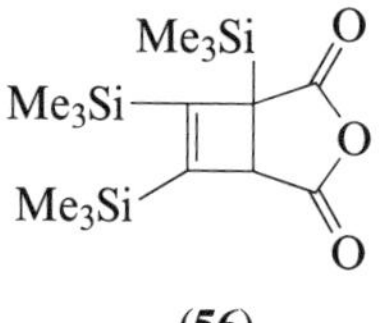

(56)

capable of undergoing a hydrogen abstraction to afford the final product. Trimethylsilyl substituted maleic anhydride photochemically adds the disilylacetylene **55** to afford the photolabile (2 + 2)-cycloadduct **56**[30]. Many years ago Askani[31] described the photochemical addition of alkynes to the anhydride **57** whereby either a cyclobutene **58** or a pentacyclic product **59** were obtained. In a recent re-investigation the scope of the reaction has been studied. Apparently the outcome of the reaction is both substituent and

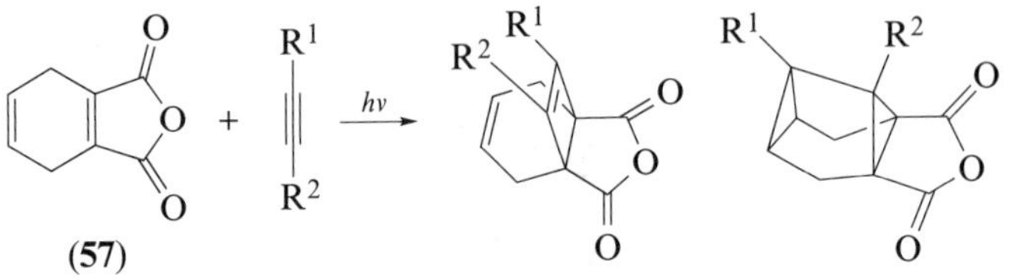

(57)   (58)   (59)

R[1]	R[2]			Solvent
H	H	2%	98%	acetone
Me	Me	8	92	acetone
Me	Me	22	78	diethyl ether
Si(Me)$_3$	Si(Me)$_3$	100	—	diethyl ether
CH$_2$OMe	CH$_2$OMe	—	100	diethyl ether

SCHEME 2

temperature dependent. A selection of the results obtained are shown in Scheme 2[32]. (2 + 2)-Cycloaddition of alkynes such as phenylacetylene to the enediones **60** results in the formation of cyclobutenes. The one shown (**61**) is formed from phenylacetylene and unsubstituted **60** in 40% yield. The adducts obtained can be converted into azatropolones, e.g. **62** by thermolysis or photolysis in yields around 60%[33].

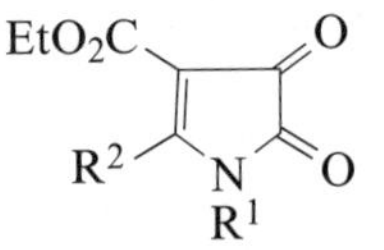

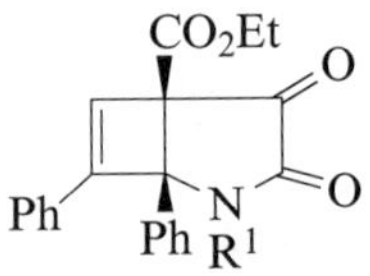

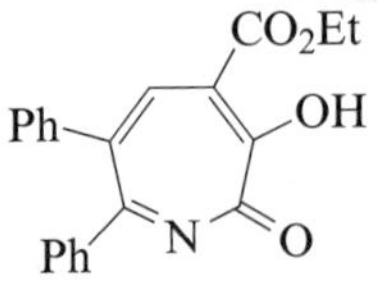

(60) R[1] = H or Me; R$_2$ = Ph or 4-BrC$_6$H$_4$ (61)   (62)

3. Additions to cyclohexenones

a. Intermolecular additions. Acetylene undergoes non-stereoselective photoaddition to the enone **63** to afford the adduct **64**[34]. Diphenylacetylene adds photochemically to the enone **65** to yield the adducts **66**. This can be converted thermally into the enone **67**, which can be transformed subsequently into 1,8-diphenylcyclo-octatetraene[35]. Earlier (2 + 4)-cycloadditions were discussed with open-chain systems. Such a process is also reported for the addition of dimethylacetylene dicarboxylate with the pyrone **68** to yield a thermally unstable adduct **69** that can be transformed into a benzenoid derivative by decarboxylation[36].

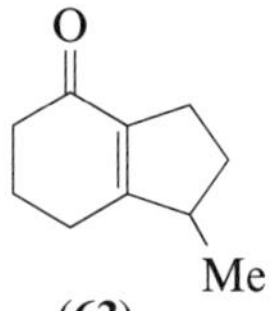

(63) 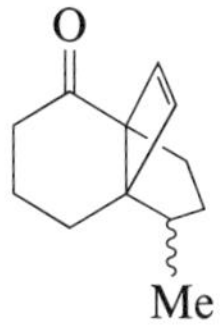 (64) 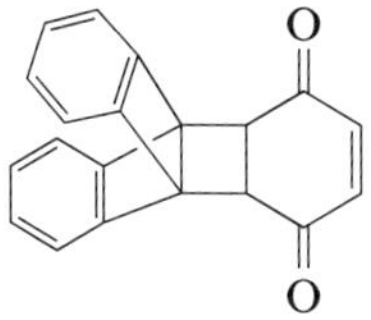 (65) 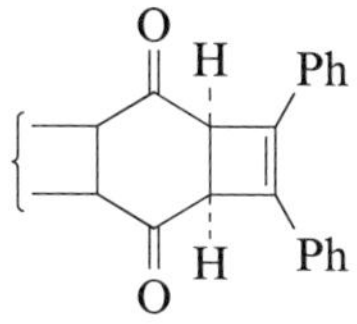 (66)

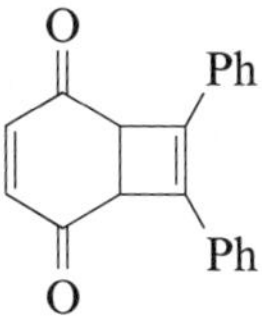

(67) 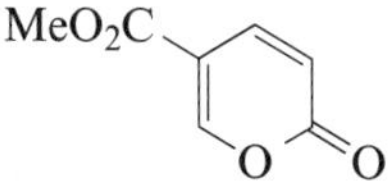 (68) 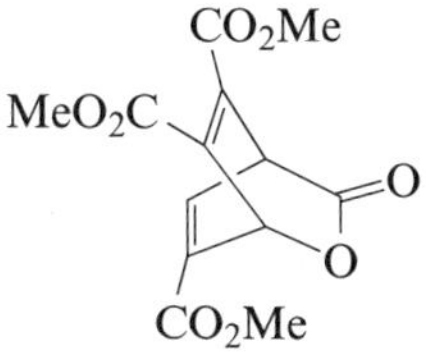 (69)

Two reports have dealt with the result of the photoaddition of ethyne to the hexenulose **70**. This reaction occurs stereo-specifically to afford the adduct **71**. The adduct can be readily reduced to the corresponding cyclobutane derivative[37] or can be elaborated into optically active tricothecene derivatives[38]. Acetone-sensitized irradiation of the pyrone **72** in the presence of ethyne affords the adduct **73** and the rearrangement product **74**[39]. The formation of this latter adduct presumably involves cyclobutene ring-opening followed by crossed addition of the diene. Cycloaddition of but-1-yne to the quinolone **75** affords the head-to-tail (2 + 2)-cycloadduct **76**[40] while addition to hex-3-yne yields **77**[41]. The route using the adducts **76** has been developed as a synthetic path to 3-substituted quinolones.

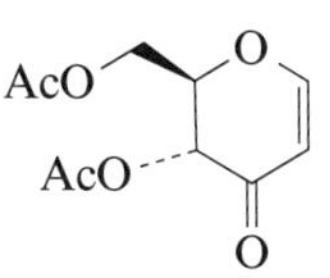

(70) 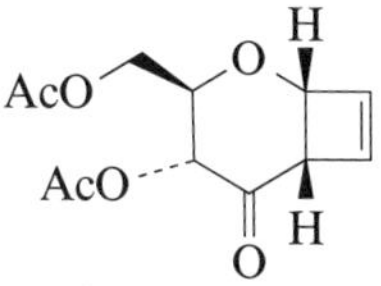 (71) 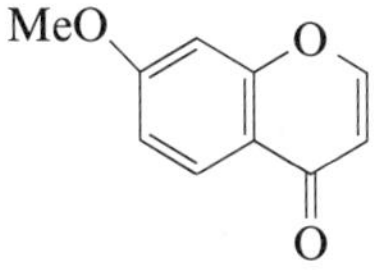 (72)

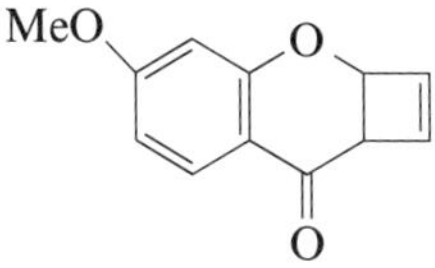

(73) 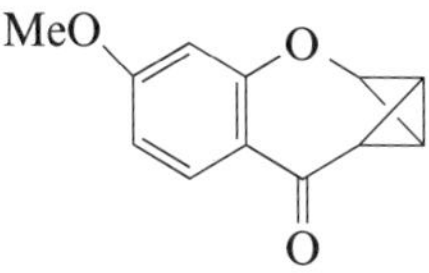 (74) 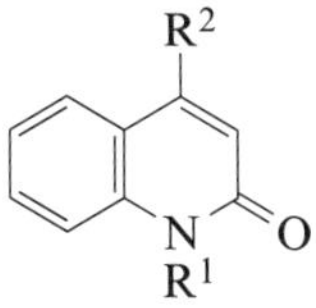 (75) (a) R^1 = H, R^2 = OMe
                        (b) R^1 = Me, R^2 = OMe
                        (c) R^1 = R^2 = Me
                        (d) R^1 = H, R^2 = OAc

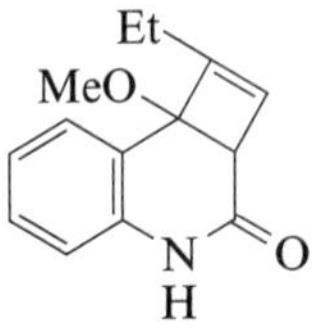

(76) 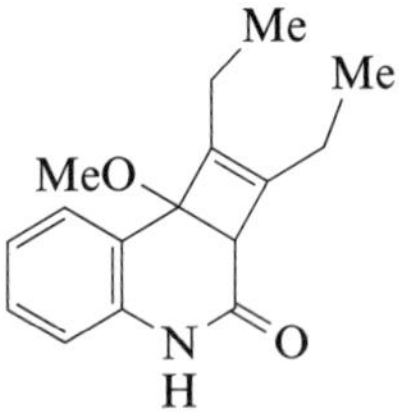   **(77)**

b. Intramolecular additions. A head-to-head adduct is formed in the photochemical (2 + 2)-cycloaddition of the enone **78** yielding **79**. This product was subsequently transformed by a series of thermal reactions into the methylene cyclo-octatriene **80**[42]. Head-to-head addition is also involved in the formation of the unstable adduct **81** obtained on intramolecular cyclization of the enone **82**[43].

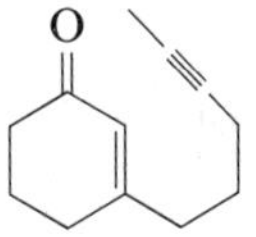

(78) 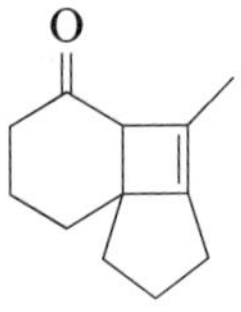   **(79)** 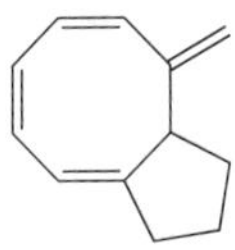   **(80)** 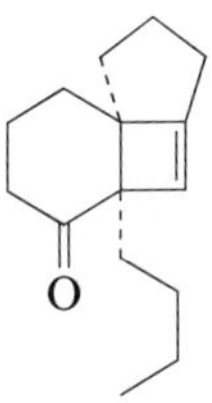   **(81)** 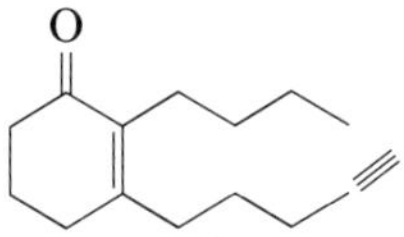   **(82)**

The influence of different attachment points for the alkynyl chain on the enone has also been studied. Thus intramolecular cyclization of the enone **83a** affords a 2 : 1 mixture of the two adducts **84** and **85** in a total yield of 55%. The enone **83b** behaves similarly and

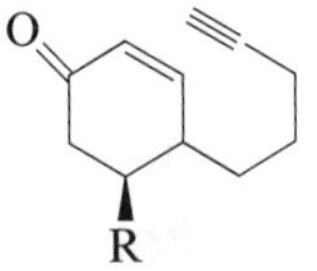

(83) (a) R = H   **(84)**   **(85)**   **(86)**
      (b) R = Pr^i

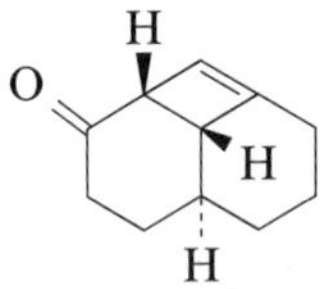

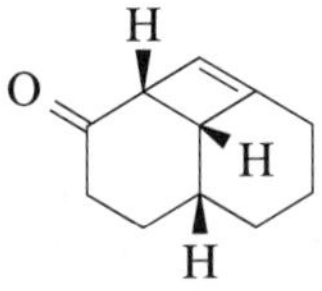

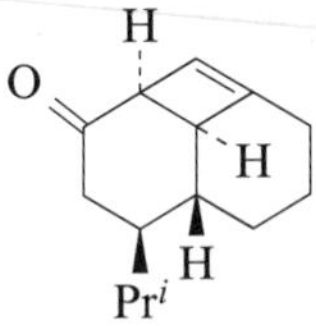

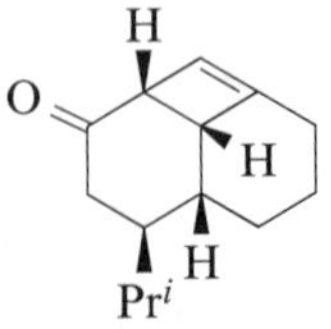

(87) 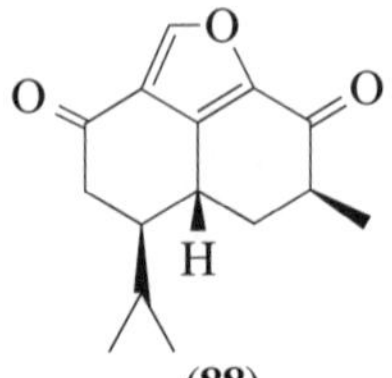   **(88)**

yields **86** and **87** in a ratio of 1.5 : 1 and a total yield of 60%[44,45]. This cycloaddition path has been used as a synthetic route to hibiscone C **88**.

Direct irradiation of the pyridone derivatives **89** brings about conversion to the valence isomers **90** in the yields shown. Benzophenone-sensitized irradiation of **89b** follows a different path and yields the (2 + 2)-cycloadduct **91** in 25% yield as well as the corresponding valence isomer[46].

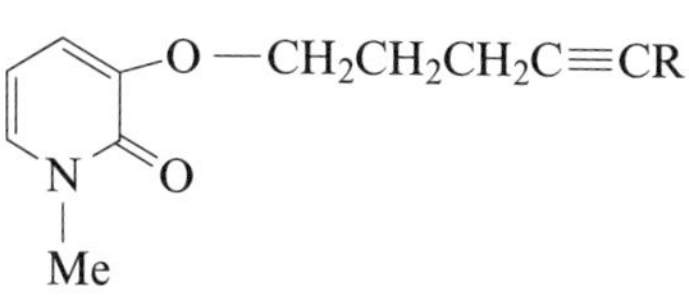

(89) (a) R = H
(b) R = CO$_2$Et

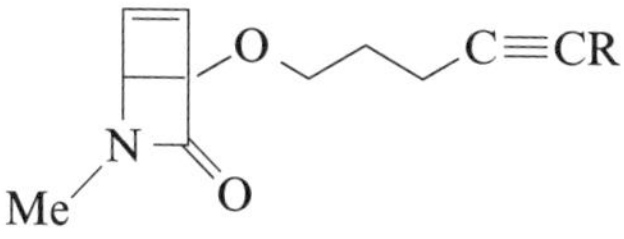

(90) (a) 57%
(b) 50%

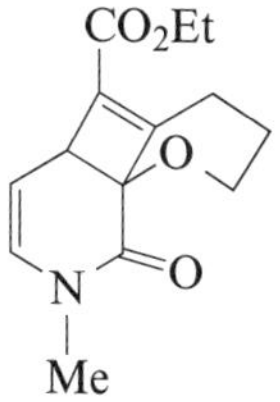

(91)

Irradiation of the alkynyltropones **92** in acidic methanol affords the barbaralones **93** and **94** as an equilibrium mixture[47]. The products are obtained by secondary irradiation, a di-π-methane rearrangement, of the initially formed, but undetected, enones **95**.

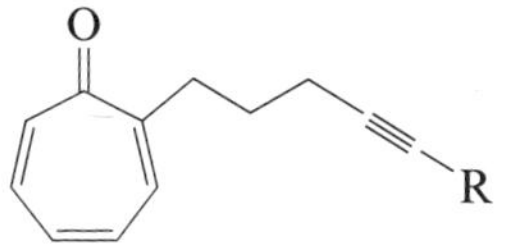

(92) R = Me, SiMe$_3$, H 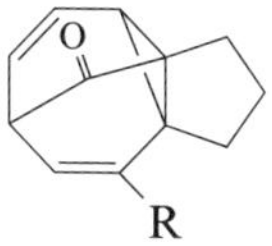   (93) 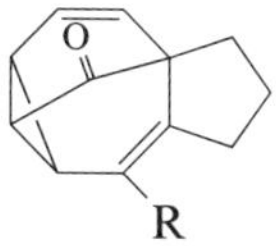   (94) 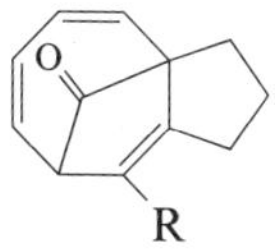   (95)

D. Additions to Aromatic Systems

1. Intermolecular additions

Acetylenes commonly undergo a (2 + 2)-cycloaddition to arenes. The addition of the alkyne **96** to benzene is typical of this reactivity and affords the (2 + 2)-adduct **97**[48]. The original discovery of this type of addition was associated with the reactions of dimethylacetylene dicarboxylate and benzene. Bryce-Smith and his coworkers have re-examined this addition particularly to assess the influence of acid on the outcome of the reaction. As always, the addition affords the cyclobutene **98**. It is this intermediate which usually ring opens to afford the standard product, dimethyl cyclo-octatetraene dicarboxylate. However, the presence of acid interrupts the normal ring opening and the cyclobutene protonates and ring opens to afford an allene, from which the two esters shown in Scheme 3 are produced[49]. (2 + 2)-Cycloaddition also occurs with pentafluoropyridine when the adducts **99** are obtained with the alkynes 1-phenylprop-1-yne and 1-phenyl-3,3-dimethyl but-1-yne[50]. Both 1 : 1 and 2 : 1 photoadducts have been obtained from the photoaddition of pentafluoropyridine and but-2-yne[51]. The cycloaddition of phenylacetylene to hexafluorobenzene has been described[52]. Photocycloaddition of ethoxyethyne to 2-phenylbenzothiazole (**100**) yields the products **101** and **102**. Ethoxypropyne adds to the same substrate in an analogous fashion but yields a more complicated reaction mixture. The products obtained from this are shown in Scheme 4[53].

$Bu^t—C≡C—CO_2Me$

(96)

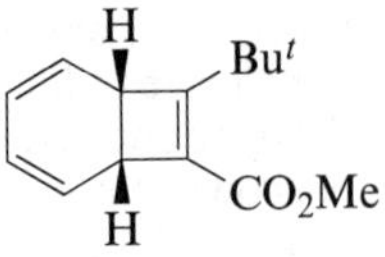

(97)

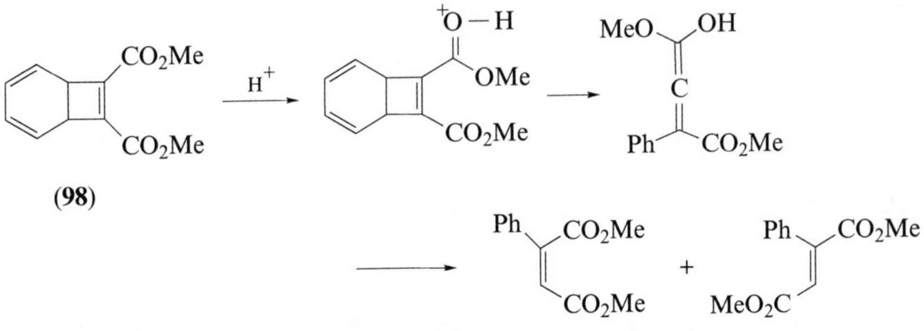

(98)

SCHEME 3

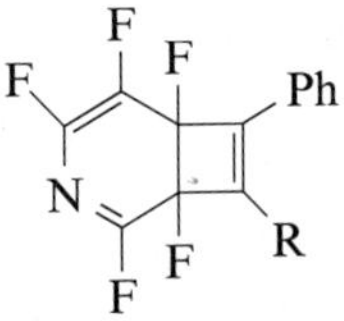

(99) R = Me, or Bu^t

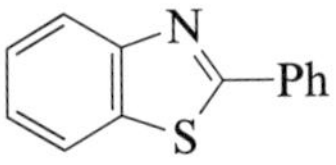

(100)

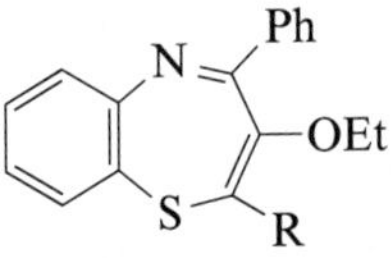

(101) (a) R = H (25%)
      (b) R = Me

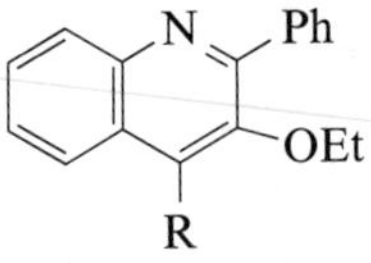

(102) (a) R = H (13%)
      (b) R = Me

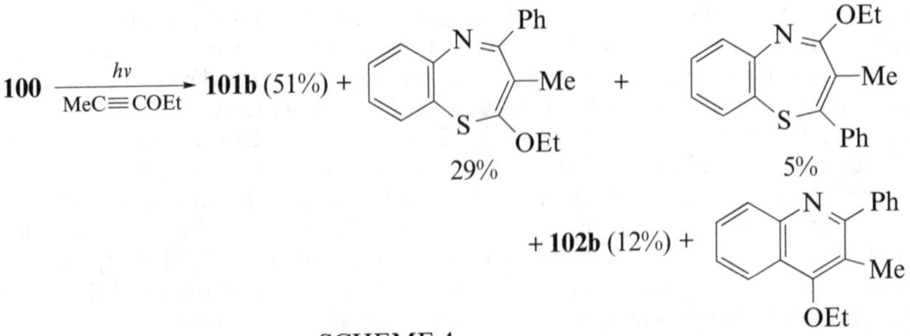

$$100 \xrightarrow[MeC≡COEt]{hv} 101b \ (51\%) +$$

29%

5%

+ **102b** (12%) +

SCHEME 4

Interestingly, ring expansion of the thiazole ring does not occur when the thiazole **103** is irradiated in the presence of dimethyl acetylenedicarboxylate. Instead, the azepine **104** is formed on photoaddition of 2 moles of the alkyne[54].

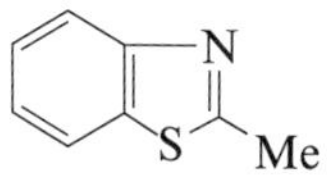

(103) 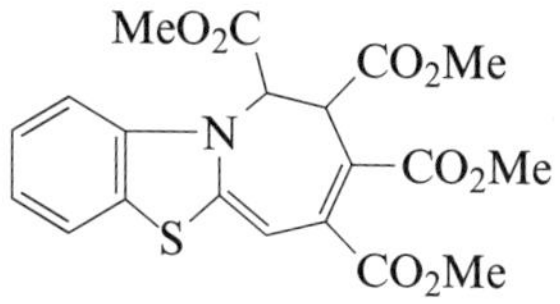　　　　**(104)**

2. Intramolecular additions

Pirrung[55] has reported that the silyl substituted alkyne group can undergo efficient intramolecular cycloaddition in the arene **105**. The cyclobutene adduct that results undergoes facile ring opening to yield the cyclo-octatetraene **106** in 46% yield. Alkynes can also undergo cyclizations of the *cis*-stilbene type six-electron cyclizations. Thus the acetylene derivative **107**, formed by elimination of HCl from the alkene **108**, undergoes

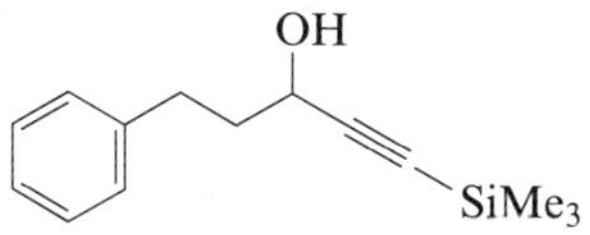

(105) 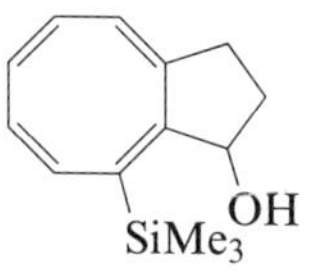　　　　**(106)**

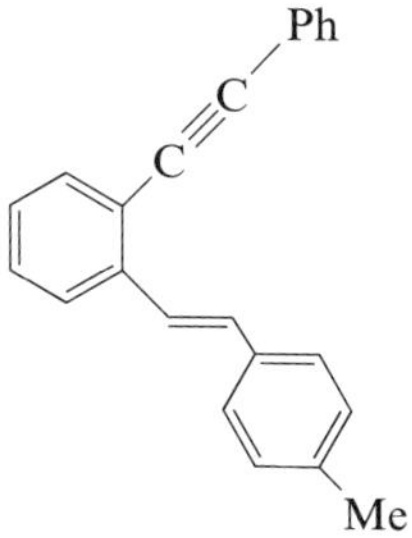

(107) 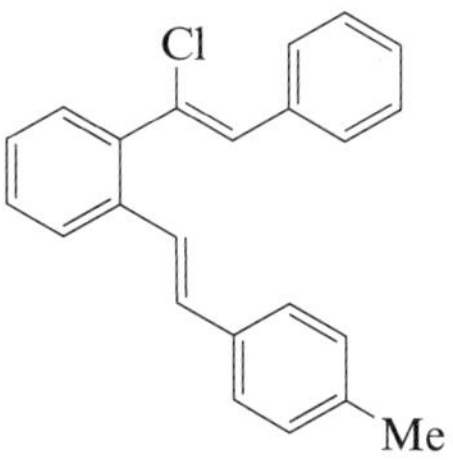　　　　**(108)**

cyclization to yield the naphthalene derivative **109**[56]. *cis*-Stilbene-type cyclization is also seen from the singlet state of the alkyne system **110**. In an aprotic solvent containing oxygen, this treatment affords the two products **111** and the oxygenated product **112**[57]. The formation of the second product is thought to involve the intermediate cycloallene **113**. This, conceivably, could yield **112** via a 1,2-dioxetane.

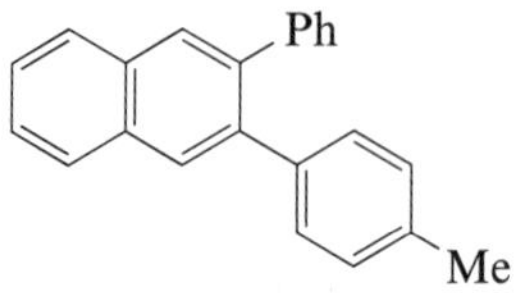

(109)

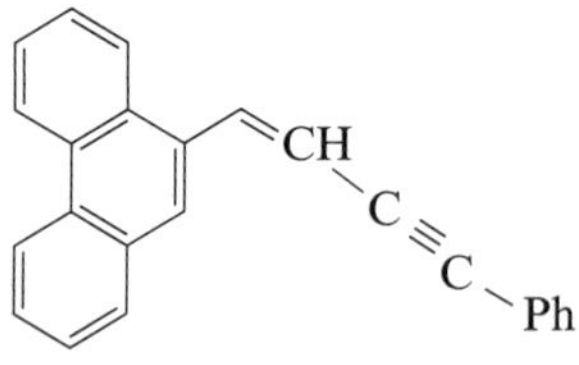

(110)

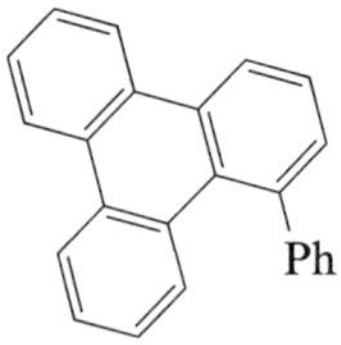

(111)

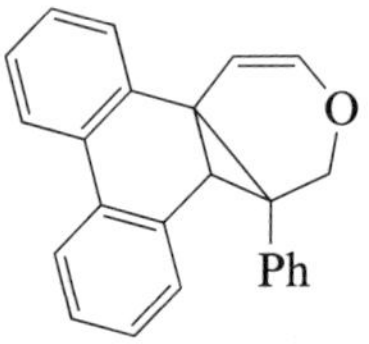

(112)

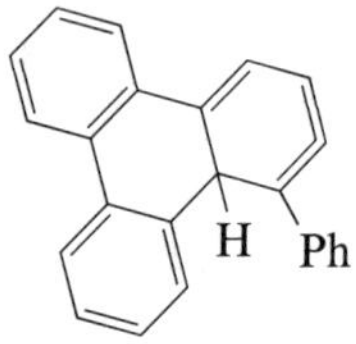

(113)

E. Oxetane Formation

Oxetane formation is reported following the irradiation of the ynone **114a** with 2,3-dimethylbut-2-ene. This affords the products **115** and **116a**. In addition to the foregoing, a product **117** was obtained perhaps by a hydrogen abstraction path. The related ynone **114b** also yields an oxetane **116b**, but the major product from this reaction was the dihydrofuran **118** (68%)[58]. A dihydrofuran derivative **119** is also obtained on irradiation of the decyn-2-one **114c** in the presence of 2,3-dimethylbut-2-ene. An oxetane is also obtained by (2 + 2)-cycloaddition in this reaction system[59]. Oxetes are formed when 1,4-diphenylbuta-1,3-diyne undergoes addition to p-benzoquinone and tetrachloro-p-benzoquinone. However, this product is unstable and the quinomethanes **120** are the isolated products[60].

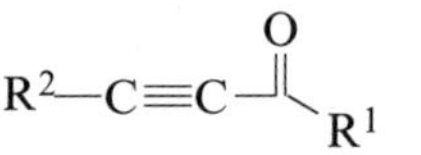

(114) (a) $R^1 = R^2 = Bu^t$
(b) $R^1 = Bu^t$, $R^2 = SiMe_3$
(c) $R^1 = CF_3$, $R^2 = C_6H_{13}$

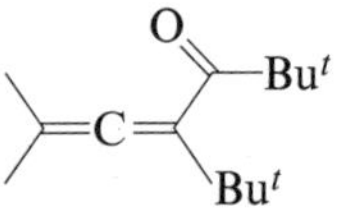

(115)

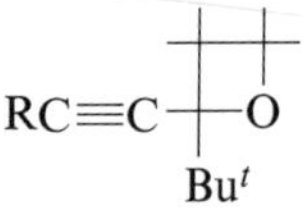

(116) (a) R = Bu^t
(b) R = $SiMe_3$

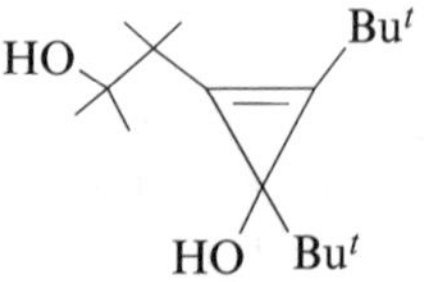

(117)

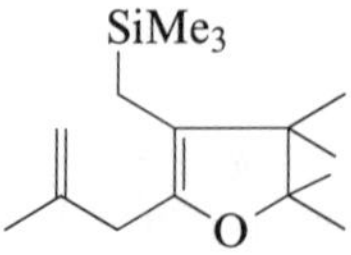

(118)

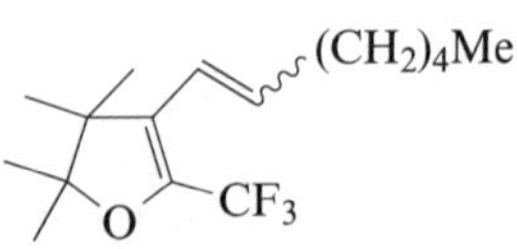

(119)

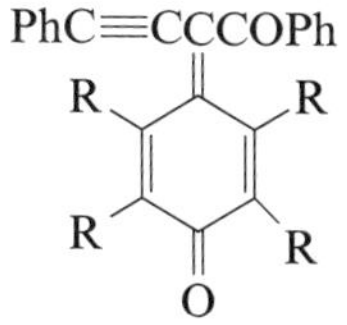

PhC≡CCCOPh

(120)

R = H or Cl

Alkynes can add to thiocarbonyl groups to afford stable thietes. A reaction typical of this behaviour is the addition of hex-3-yne or diphenylacetylene to the monothio phthalimide **121** to yield the thietes **122**[61,62]. In some instances the resultant thietes are unstable and undergo ring opening to a thioester as with the addition of dithiomethoxyacetylene to the same phthalimide **121**. The product from this reaction is the thioester **123**[61,62]. Interestingly the dithiophthalimide is unreactive under these conditions. Additions of this type have also been reported for xanthenethione and thioxanthenethione[63]. The formation of thioesters **124** occurs on irradiation of the thiones **125** with dithio-*t*-butoxyacetylene. The reaction involves the formation of an unstable thiete which ring opens to afford the final products[64].

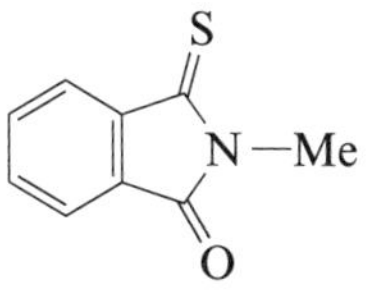

(121) 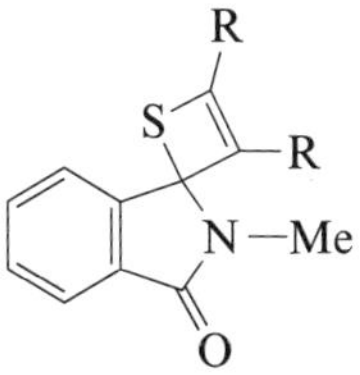   (122) R = Et or Ph 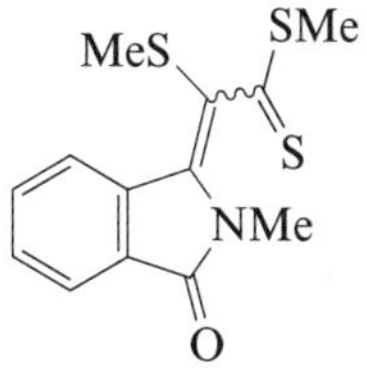   (123)

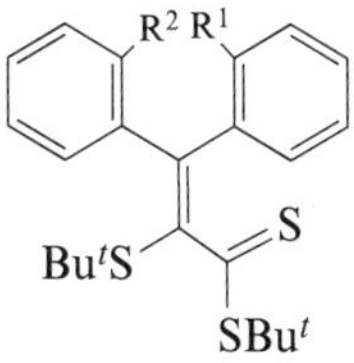

(124) 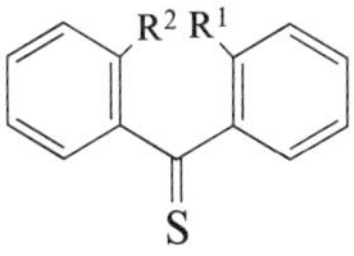   (125)

R¹–R² = O or S

R¹ = R² = H

F. Miscellaneous Alkyne Addition Reactions

As well as undergoing addition to enones, arenes, carbonyl groups and thiocarbonyl groups, alkynes also undergo additions to alkenes. Thus tolan affords cyclobutenes on photoaddition to the tetrahydronaphthalene **126**[65]. The triplet-sensitized irradiation of vinylacetylene in the presence of 1,1-dichloroethene yields the (2 + 2)-cycloadduct **127** as well as the dimers **128** and **129**. Similar behaviour is observed when dienes are the reaction substrate. In both cases the addition occurs exclusively at the vinyl group. In the case of buta-1,3-diyne, however, the cycloaddition with vinylacetylene affords the adduct **130**[66,67]. (2 + 2)-Cycloaddition is also reported to occur on irradiation of 1,4-diphenylbuta-1,3-diyne. This affords the *bis* adduct **131** when irradiated in the presence of 2,3-dimethylbut-2-ene. It is thought that this reaction involves a biradical intermediate. When cyclohexa-1,4-diene is used as the substrate a *bis* adduct **132** is formed again and this is used to support the postulate of a biradical intermediate[68]. Cycloaddition of this type does not occur with electron-withdrawn alkenes. Cycloaddition to 1-phenylpenta-1,3-diyne (in

(126)   **(127)**   **(128)**   **(129)**   **(130)**

(131)   **(132)**   **(133)**

(134) R = OEt (5%)   **(135)** $Ph—C≡C—C≡C—C≡C—Me$
R = CN (14%)   **(136)**

(137)

(a) R^1 = $(CH_2)_8CO_2Me$; R^2 = H; R^3 = PhC≡C—C≡C—; R^4 = Me
(b) R^1 = $(CH_2)_8CO_2Me$; R^2 = H; R^3 = PhC≡C—; R^4 = —C≡C—Me
(c) R^1 = H; R^2 = $(CH_2)_8CO_2Me$; R^3 = PhC≡C—; R^4 = —C≡C—Me
(d) R^1 = H; R^2 = $(CH_2)_8CO_2Me$; R^3 = Ph; R^4 = —C≡C—C≡C—Me

either the singlet or the triplet excited state) affords a 1 : 1 adduct **133** in 10% yield when it is irradiated in the presence of 2,3-dimethylbut-2-ene[69]. Addition to ethyl vinyl ether and acrylonitrile also affords (2 + 2)-cycloadducts **134** in a regio- and site-selective fashion. 1 : 2-Adducts **135** are also obtained[69,70]. The triyne **136** undergoes photochemical (2 + 2)-cycloaddition to methyl undec-10-enoate to afford a mixture of the four cyclobutenes **137**[71].

Other modes of addition are also possible with diynes and triynes. thus irradiation of 1,4-diphenylbuta-1,3-diyne at 350 nm populates the triplet state which undergoes addition to 2,3-dimethylbut-2-ene, affording the 2 : 1 adduct **138**[72]. This same diyne also adds

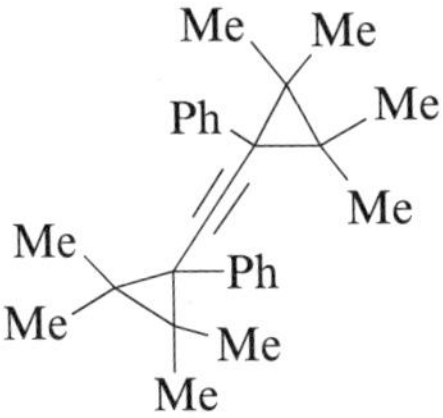

(138)

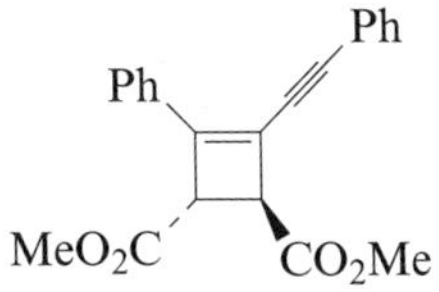

(139)

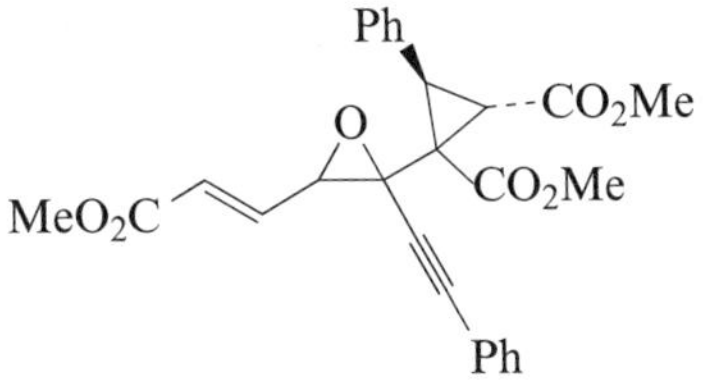

(140)

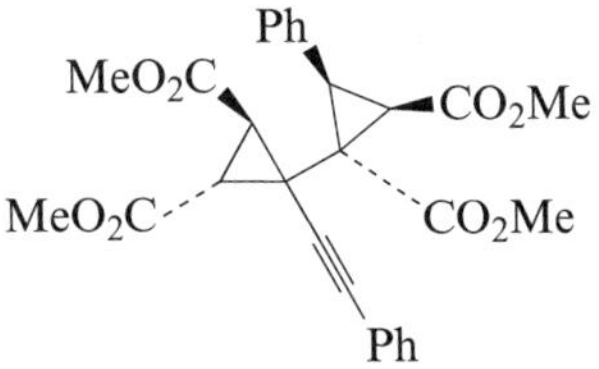

(141)

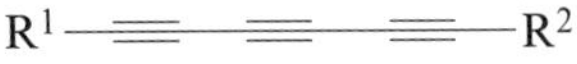

$R^1 \longequal \longequal \longequal R^2$

(142) (a) $R^1 = Ph$, $R^2 = H$
(b) $R^1 = Ph$, $R^2 = Bu^tSiMe_2$
(c) $R^1 = R^2 = Ph$

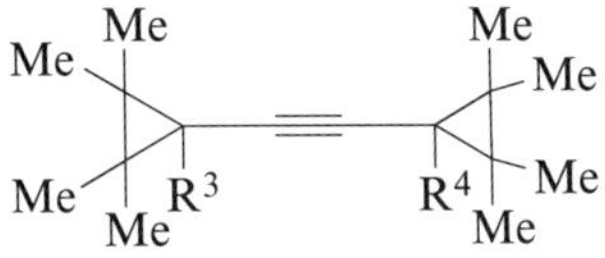

(143) (a) $R^3 = PhC \equiv C$, $R^4 = H$
(b) $R^3 = PhC \equiv C$, $R^4 = Bu^tSiMe_2$

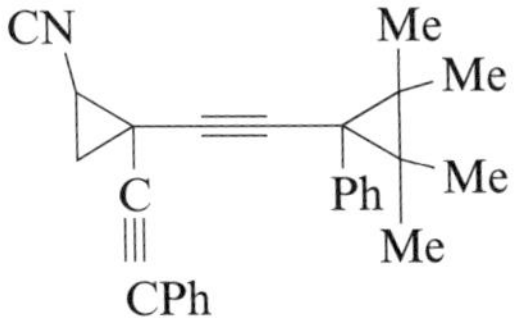

(144)

dimethyl fumarate photochemically to yield the (2 + 2)-adduct **139** via the triplet state of the diyne. Subsequent irradiation of **139** in the presence of excess ester brings about its conversion into two new products, **140** and **141**[73]. The triynes **142a** and **b** add to 2,3-dimethylbut-2-ene to afford the cyclopropyl derivatives **143** on irradiation at 350 nm in neat degassed alkene. The triyne **142c** was also irradiated with the same olefin in the presence of acrylonitrile when a mixture of adducts **143a** and **144** was obtained[74–76].

G. Di-π-methane Reactivity

The photochemical reactivity of triaryl systems such as **145** has been studied[77]. The reactions which compounds of this type undergo are of the di-π-methane type but involve only phenyl–phenyl interactions. The key reaction is the formation of a carbene by the extrusion of biphenyl following the di-π-methane process (Scheme 5)[77–80]. The presence of the alkyne substituent appears not to influence the outcome of the reaction. This is not the case with the boron substituted alkynes, which also undergo the di-π-methane type processes and have been under study for the past ten years. Among the earliest examples of this reaction is the photochemical formation of the borirene **146** from the complex **147** on irradiation in benzene[81]. Eisch and coworkers[82] have also reported the di-π-methane rearrangement of the ethynylborane derivative **148**. The irradiation of this compound in benzene-pyridine using 300-nm light brings about its conversion to the borirene **149** by a 1,2-migration of the mesityl group from boron to carbon, a process typical of the di-π-methane type reaction. Further examples of the di-π-borate process, notably by Schuster and his coworkers[83,84], have been described. Thus, for example, the conversion of **150** affords **151** in reasonable yields[85,86].

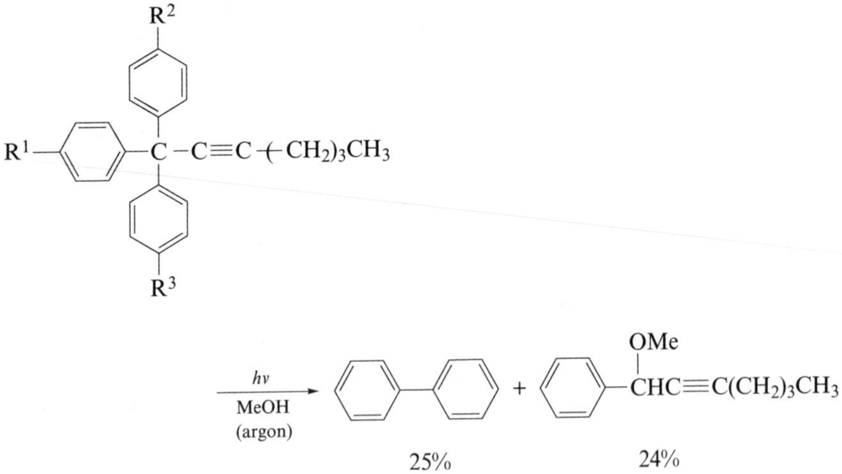

$$R^1-\text{Ar}-C-C\equiv C+(CH_2)_3CH_3$$

(with R² on top aryl and R³ on bottom aryl)

$$\xrightarrow[\substack{MeOH \\ (argon)}]{hv} \quad \text{(biphenyl)} \quad + \quad \text{Ph}-CHC\equiv C(CH_2)_3CH_3$$

25%   24%

(**145**) (a) $R^1 = R^2 = R^3 = H$
        (b) $R^1 = R^2 = H, R^3 = Me$
        (c) $R^1 = R^2 = R^3 = Me$
        (d) $R^1 = R^2 = H, R^3 = OMe$
        (e) $R^1 = R^2 = R^3 = OMe$

SCHEME 5

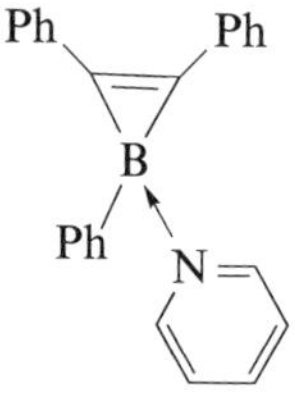

Ph⟍   ⟋Ph
    ⟍ ⟋
     C=C
      |
      B
   ⟋ ↑
 Ph   N=
      ⟨ ⟩

(146)

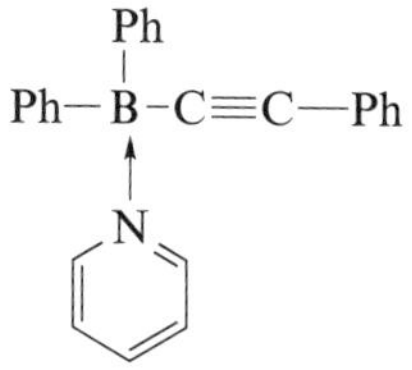

Ph
 |
Ph—B-C≡C—Ph
      ↑
      N
   ⟨ ⟩

(147)

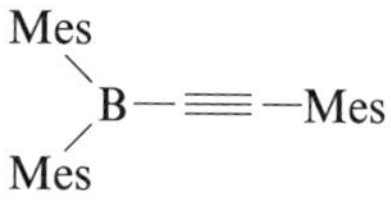

Mes
  ⟍
   B—≡—Mes
  ⟋
Mes

(148)

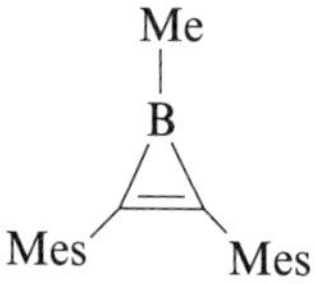

Me
 |
 B
△
Mes⟋ ⟍Mes

(149)

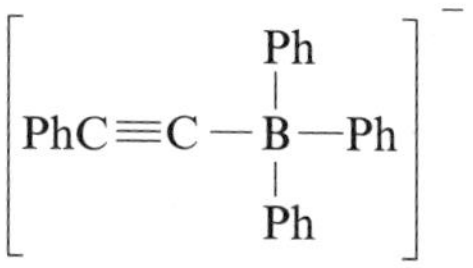

⎡   Ph   ⎤⁻
⎢ | ⎥
⎢ PhC≡C — B —Ph ⎥
⎢ | ⎥
⎣   Ph   ⎦

(150)

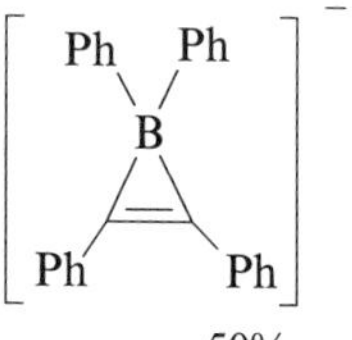

⎡ Ph Ph ⎤⁻
⎢   \ /   ⎥
⎢   B   ⎥
⎢   △   ⎥
⎢ Ph   Ph ⎥
⎣   ⎦
   50%

(151)

Mes = 2,4,6-trimethylphenyl

H. Rearrangements and Cyclizations of Alkynes

Some alkyne systems, with comparatively labile bonds, undergo rearrangement processes. Thus photorearrangement of the silylacetylene **152** 1,3-silyl migration affording the allene **153**[87].

Trimerizations of alkynes are also of interest and Ferris and Guillemin[88] have demonstrated that irradiation of the alkyne **154** at 185 or 206 nm results in the formation of 1,3,5-tricyanobenzene. This product is also formed on irradiation at 254 nm, but then it is accompanied by 1,2,4-tricyanobenzene and tetracyanocyclo-octatetraene.

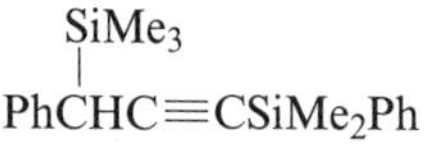

SiMe₃
 |
PhCHC≡CSiMe₂Ph

(152)

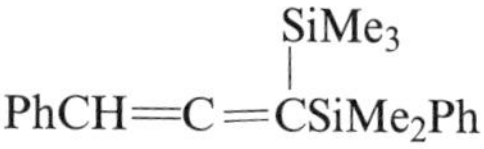

SiMe₃
 |
PhCH=C=CSiMe₂Ph

(153)

H——≡—CN

(154)

Several reports have dealt with the reductive cyclization of enynes where the alkyne component undergoes addition to an adjacent carbonyl group. Typical of this is irradiation of **155**, in the presence of the SET sensitizer triethylamine, which affords the bicyclic product **156**, a piperidone, that was used in a synthesis of iso-oxyskyanthine[89]. Unsaturated aldehydes such as **157** also undergo this photo-electron transfer-induced cyclization to afford a mixture of alcohols **158** (Scheme 6)[90]. Pete and his coworkers have extended their study of this reductive cyclization of unsaturated ketones in a synthesis of hirsutene **159**. The reaction involves the irradiation of the ketone **160** at 254 nm in the presence of triethylamine, whereby the tricyclic compound **161** is formed in 58% yield. This is subsequently transformed into hirsutene[91,92].

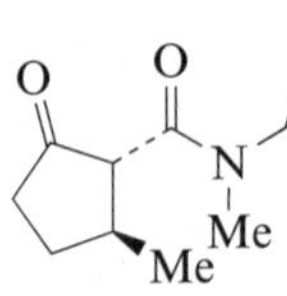

(155) 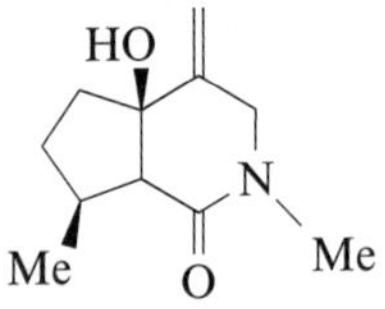   (156)

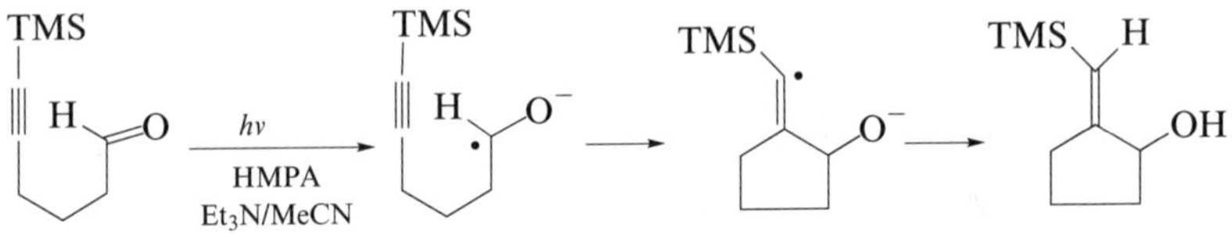

(157)

$$\xrightarrow[\substack{HMPA \\ Et_3N/MeCN}]{hv}$$

(158)

SCHEME 6

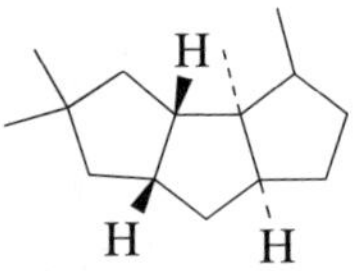

(159) 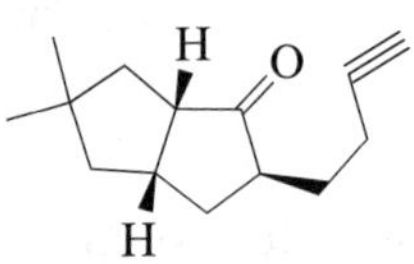   (160) 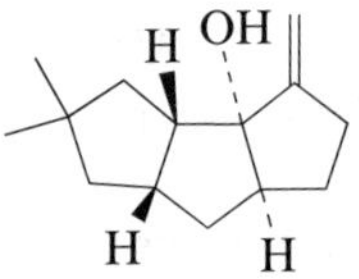   (161)

I. Miscellaneous Reactions of Ynones

A flash photochemical study of the behaviour of 1-phenyl-4,4-dimethylpent-2-yne-1-one has been reported. Apparently the triplet state of this ynone is long-lived. Decomposition of the ynone takes place on irradiation in cyclohexane, but in benzene the compound is unreactive[93]. Reductive ring closure of the ynone 162 affords the indanone 163

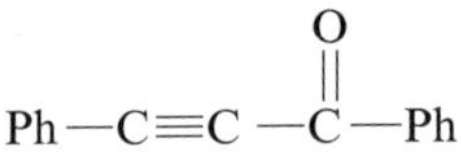

$$Ph-C\equiv C-\overset{O}{\underset{\|}{C}}-Ph$$

(162)

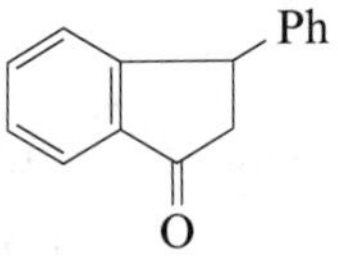

(163)

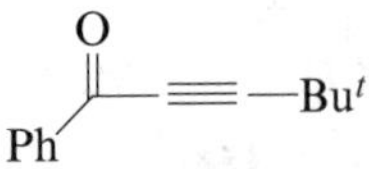

(164)

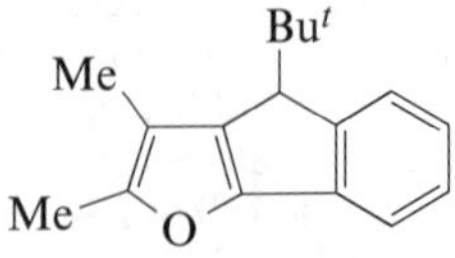

(165) 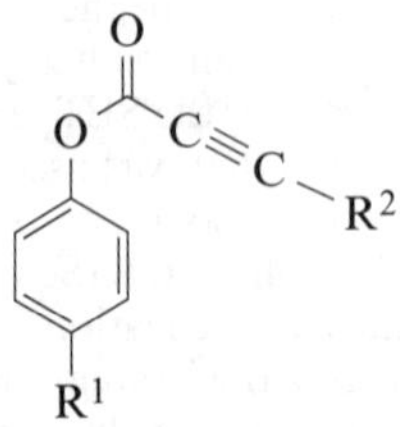   (166)

on irradiation[94]. Biradical species are presumably involved in the conversion of the ynone **164** into the furan **165** (60%) on irradiation in the presence of but-2-yne[95]. Radicals are also involved in the photo-Fries reactivity of the ynone **166**, which has been studied as part of a synthetic strategy for the synthesis of chromones[96].

Agosta and his coworkers have examined the photochemical reactivity of the enynones such as **167**. The principal reaction encountered is Norrish Type II hydrogen abstraction to yield a 1,4-biradical as shown in Scheme 7. Cyclization within this biradical affords a carbene from which the products shown are formed by hydrogen migration reactions[97]. Hydrogen abstraction reactions are also dominant with the isomeric enynones **168** as shown in Scheme 8. Again the biradical transforms to a carbene from which the alkenes are formed by hydrogen migration. Interception by solvent to afford ether products is also observed[98–100].

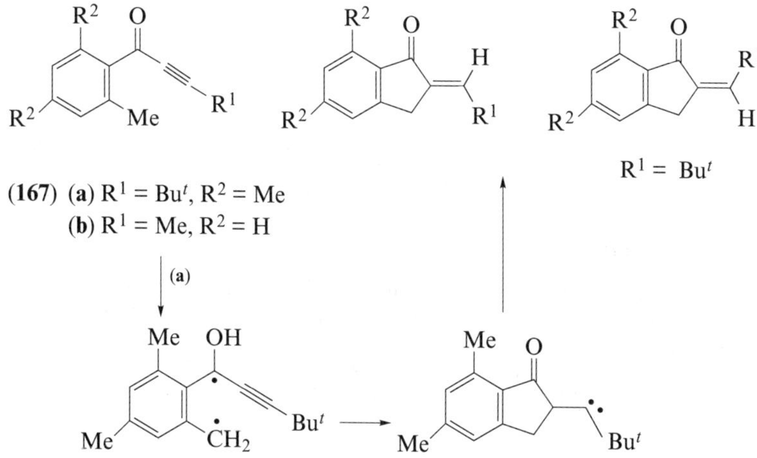

(**167**) (a) $R^1 = Bu^t$, $R^2 = Me$
(b) $R^1 = Me$, $R^2 = H$

$R^1 = Bu^t$

SCHEME 7

(**168**) (a) $R = Me$
(b) $R = Bu^t$

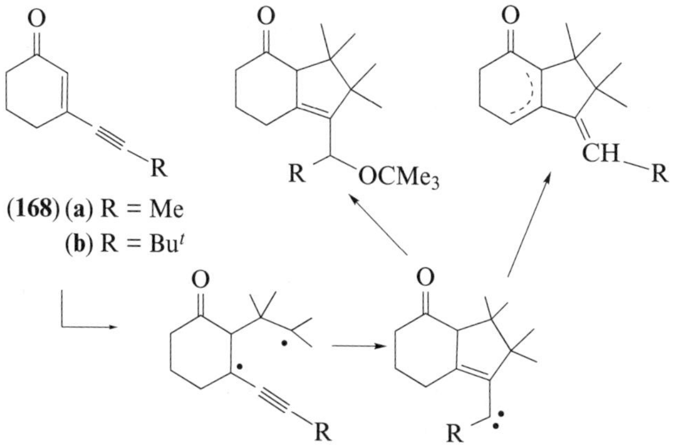

SCHEME 8

III. PHOTOCHEMISTRY OF COMPOUNDS CONTAINING NITRILE GROUPS

As with the photochemistry of alkynes described in the previous section, there is little photochemistry that is related solely to the nitrile group. However, the presence of such a functionality often plays a major role in determining the outcome of the reaction in terms of regiochemistry, stereochemistry and excited state involved.

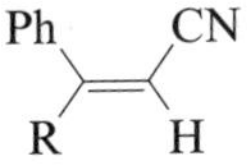

(169)

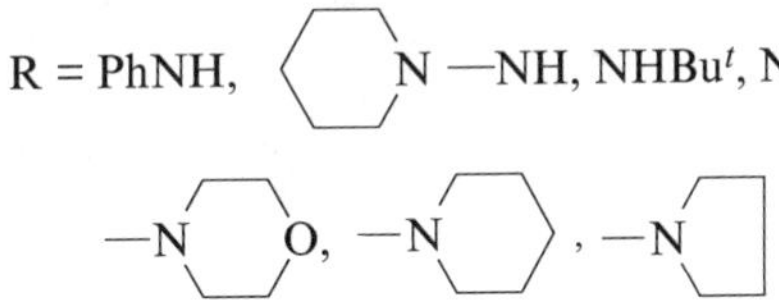

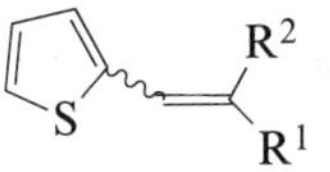

$R = PhNH,$ [structures] N —NH, NHBu^t, NH$_2$

$-N$ O, $-N$, $-N$

(170) (a) $R^1 = CSNH_2, R^2 = H$
       (b) $R^1 = CN, R^2 = Ph$

A. cis,trans-Isomerization

As with a normal alkene, a cyano substituent on the alkene does not alter the ability of the system to undergo E,Z-isomerization. Thus a study of the photochemical isomerization of the cyano alkenes (169) has shown that heavy atom additives and quenchers have no effect on the process and that the isomerization arises from a singlet state[101]. Other systems such as the vinylthiophenes (170) also undergo isomerization[102,103].

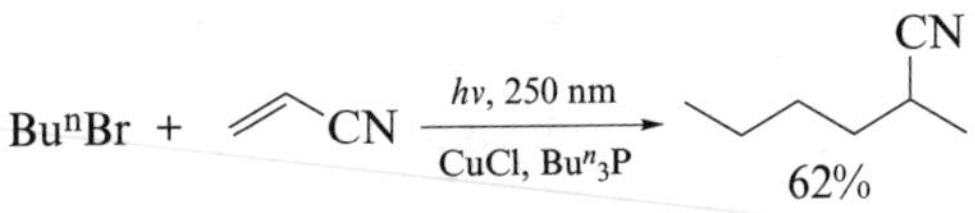

$$Bu^nBr + \ \diagup\!\!\diagdown CN \xrightarrow[\text{CuCl, Bu}^n_3\text{P}]{hv, 250 \text{ nm}} \ \text{[product]} \quad 62\%$$

SCHEME 9

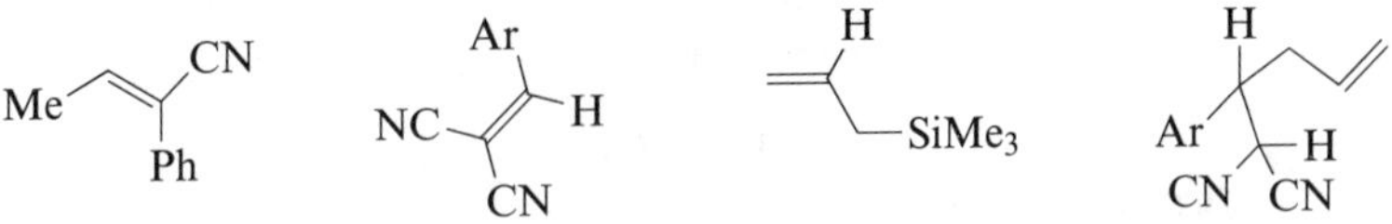

(171)   (172)   (173)   (174)

Ar = Ph, p-CNC$_6$H$_4$, p-ClC$_6$H$_4$, p-MeC$_6$H$_4$ or p-MeOC$_6$H$_4$

B. Free Radical Addition Reactions to Cyanoalkenes

Mitani and coworkers[104] report that alkyl halides can be induced to undergo addition to acrylonitrile. Typical results from this reaction, which is catalysed by Cu(I) complexes, are shown in Scheme 9. The reaction is brought about by irradiation at 251 nm in the

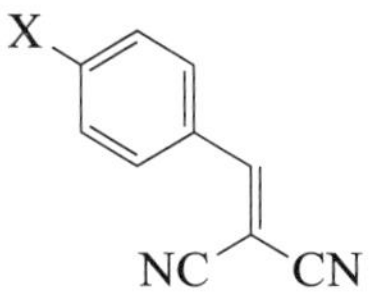

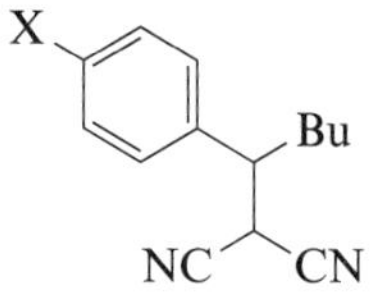

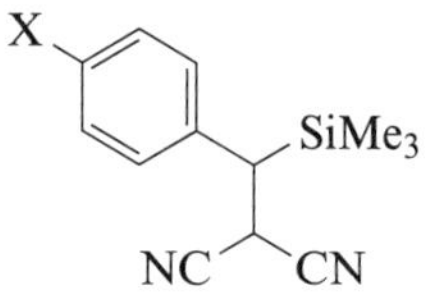

(175) (a) X = H
(b) X = OMe
(c) X = Me
(d) X = Cl
(e) X = CN

(176) X = H   (a) 85%
X = MeO (b) 20%
X = Me   (c) 78%
X = Cl   (d) 66%
X = CN   (e) 61%

(177) 75%
10
80
54
84

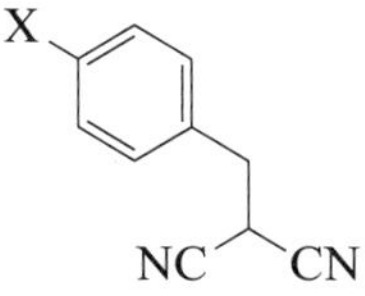

(178)   %
10
trace
8
32
6

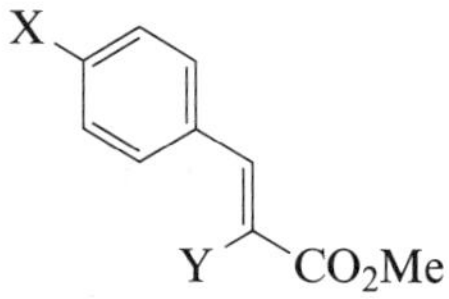

(179)

X	Y
H	CN
X	CO$_2$Me
Cl	CO$_2$Me
CN	CO$_2$Me

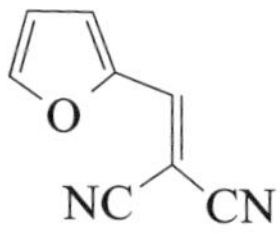

(180) 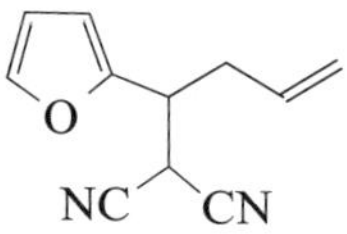   (181)

presence of a phosphine. A short review of the photochemical behaviour of cyanoalkenes such as 171 has been published[105]. The formation of radical cations of such systems is a common occurrence and can be carried out by irradiation in the presence of a suitable electron-donating sensitizer. Thus the generation of the radical anion of the alkene 172 can be brought about on irradiation with phenanthrene as the electron donor. When this is done in the presence of the silanes 173, the final products are the allylated compounds 174[106]. Extension of the scope of the reaction has used the alkenes 175 in propionitrile as solvent with tetrabutyl stannane. SET generates a butyl radical and the radical anion of the alkene. Combination of these affords the observed products 176 in yields ranging from good to excellent. It is clear that substitution of the aryl ring can play a major role, as shown by the methoxy derivative where yields of product are poor[107]. With phenanthrene and hexamethyl disilane the same set of alkenes provides another example of this regioselective addition affording the adducts 177 and the alkanes 178. Again, the yields of products illustrate that the reaction is less efficient when the aryl group is substituted with an electron-donating moiety. The reaction has also been extended to the esters 179[108]. Interestingly, the reaction can be extended to other aryl systems as in the reaction of the furan derivatives 180 to afford 181, again exhibiting the same regiospecificity[109].

Cyano-substituted alkenes can also act as traps for biradical intermediates. This behaviour is exemplified by the formation of the three products **182**–**184** that are obtained on the irradiation of homobenzvalene **185**[110].

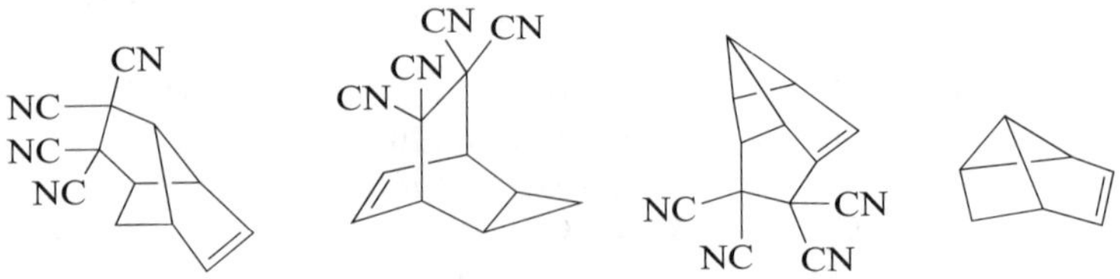

(182)   (183)   (184)   (185)

C. Photoreactions of Cyanocyclopropanes and Synthesis of Cyanocyclopropanes

SET, using triethylamine as the electron donor, can be used to bring about ring opening of cyanocyclopropanes as in the conversion of 1,1-diphenyl-2,2,3,3-tetracyanocyclopropane into 1,1-diphenyl-2,2-dicyanoethene in 57% yield[111]. Ring opening is also involved in the rearrangement of the cyclopropanonaphthalene **186**[112]. The reaction, to afford the products shown in Scheme 10, is solvent dependent and non-polar solvents such as benzene promote a faster reaction rate. The isomeric naphthalene derivative **187**, a product of the original irradiation, yields 1-cyanomethylnaphthalene as the principal product on irradiation. Ring opening is also observed for the cyanocyclopropane **188** to yield the photoreactive cyanovinyl derivative **189**. Subsequent cis-stilbene cyclization and oxidation affords the final product **190**[113].

A patent has been lodged dealing with the photochemical trans–cis isomerization of the cyanocyclopropane **191**. The photostationary state reached in these photolyses consists of approximately 65% cis[114]. Both direct and sensitized irradiation of the cyanocyclopropane **192** leads to the formation of the cis-isomer **193**. Two rearrangement products, **194** and **195**, are also formed in this reaction. The biradical intermediate **196** is proposed as the

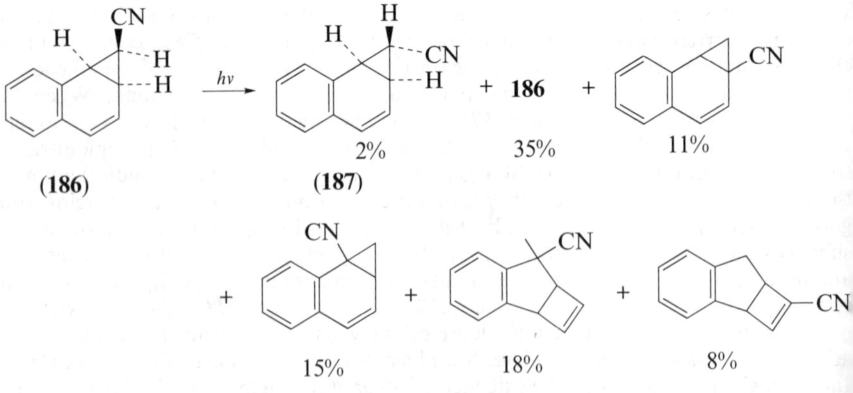

(186)   (187)
                 2%   35%   11%

15%   18%   8%

SCHEME 10

(188) **(189)** **(190)**

precursor to the alkenes[115]. The ring opening of the anion **197** to the allyl anion **198** was reported some time ago. Calculations concerning the substituent effects and the strength of the carbon–metal bond have been carried out[116]. Ring opening[116] is also reported for the cyanobicyclobutane **199** which undergoes C—C bond fission to afford a biradical that adds to the cyclic aza-compounds to yield the adducts **200**[117].

(191) **(192)** **(193)** **(194)**

(195) **(196)** **(197)** **(198)**

(199) **(200)**

Some paths are available for the synthesis of cyanocyclopropyl systems from cyanoalkenes. An interesting example of this is the photoisomerization of 2-aminopropenenitrile to yield the aziridine **201**. Products of fragmentation such as HCN and acetonitrile are also obtained[118]. The N-acylimine **202** is converted on irradiation into the bicyclic products **203** and **204**. The route to products involves bond formation by attack of the carbonyl oxygen atom on the styryl moiety. This affords the biradical intermediate **205**, where radical stabilization by the cyano group is important. Cyclization within this yields the two bicyclic compounds. The same reaction path is followed on irradiation of **202** in the crystalline phase. Here, however, this path is minor (30%) and the major path affords a (2 + 2)-cycloadduct[119].

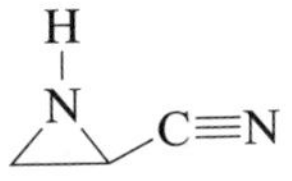

(201) 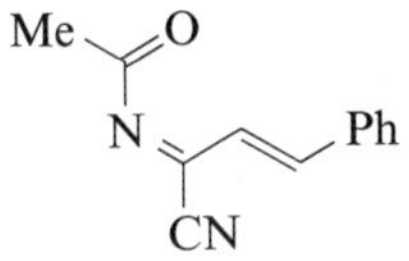 (202) 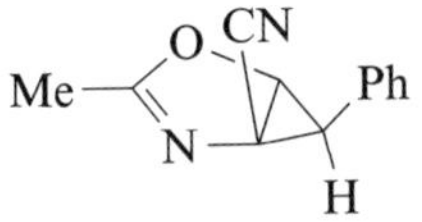 (203)

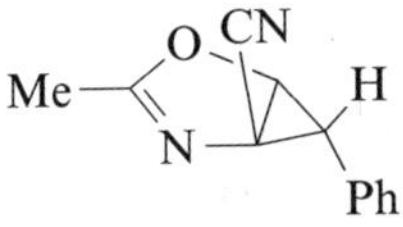

(204) 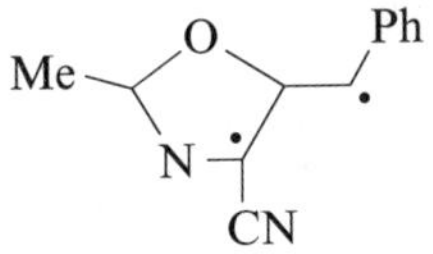 (205)

A $\pi\pi^*$ singlet state is involved in the conversion of the cyanoalkenes **206** into the cyclopropane derivatives **207**. Irradiation produces the biradical (Scheme 11) within which 1,2-hydrogen transfer affords the transient, which is proposed to exhibit ionic behaviour. This subsequently ring closes to the cyclopropane **207**. Quantum yield determinations were carried out and a slight solvent dependency was detected[120,121]. Details of the photochemical rearrangements of the cycloheptatriene **208** have been obtained[122]. This compound ring closes to the cyanocyclopropane **209** (33%) as the principal reaction path.

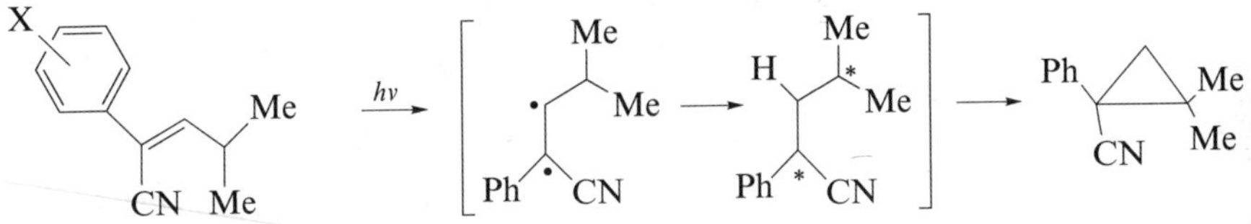

(206)

X = p-MeO
X = m-MeO
X = p-Me
X = p-Cl
X = p-CN

SCHEME 11

(207)

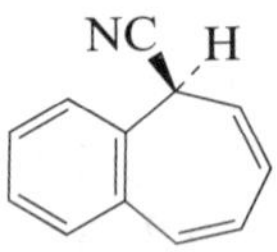

(208) 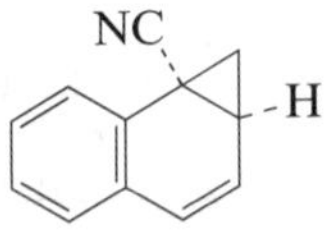 (209)

Di-π-methane processes and related studies. The ubiquitous di-π-methane process does occur even with cyano substituents. Thus Ferreira and Salisbury[123] have examined this with the cyanoalkene **120**. This compound reacts via a singlet state and affords the two isomeric cyclopropanes, **211** and **212**, while irradiation of the alkene under sensitized conditions only brings about *trans,cis*-isomerization. Zimmerman and his coworkers[124–126] have also addressed the di-π-methane reactivity of cyano substituted 1,4-dienes, but in the solid state. The photoreactivity of the diene **213** has been shown to be different from the

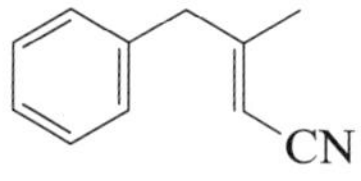

(210)

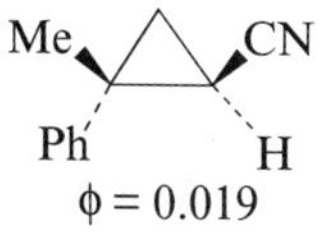

Me, CN
Ph H
φ = 0.019
(211)

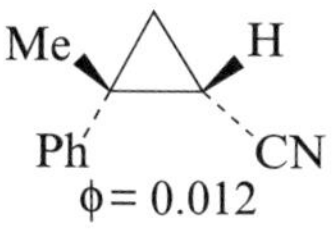

Me, H
Ph CN
φ = 0.012
(212)

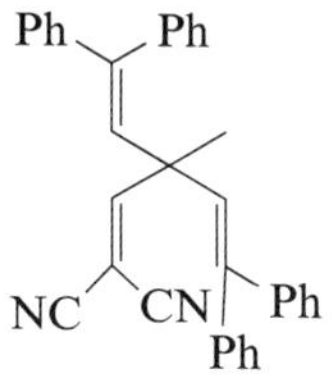

Ph Ph
NC CN Ph
Ph
(213)

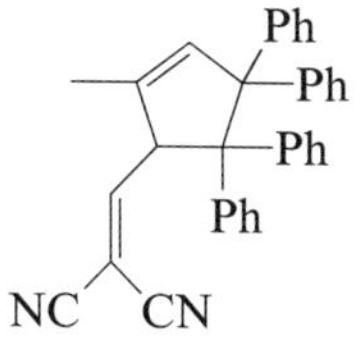

Ph
Ph
Ph
Ph
NC CN
(214)

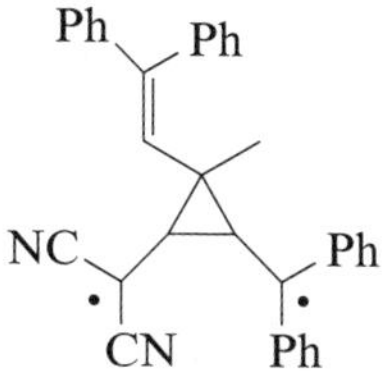

Ph Ph
NC CN Ph
CN Ph
(215)

Pr^i Pr^i
Ph Ph CN
CN
(216)

$\xrightarrow[\text{CH}_3\text{CN}]{h\nu}$

Pr^i Pr^i
Ph -CN
Ph CN
(217)

H
Ph Ph CN
CN

+

Ph H
Ph, CN
CN

+

Ph

+

CN
CN

+

CN
CN
Ph

+

CN
CN
Ph

+

CN
CN
Ph Ph

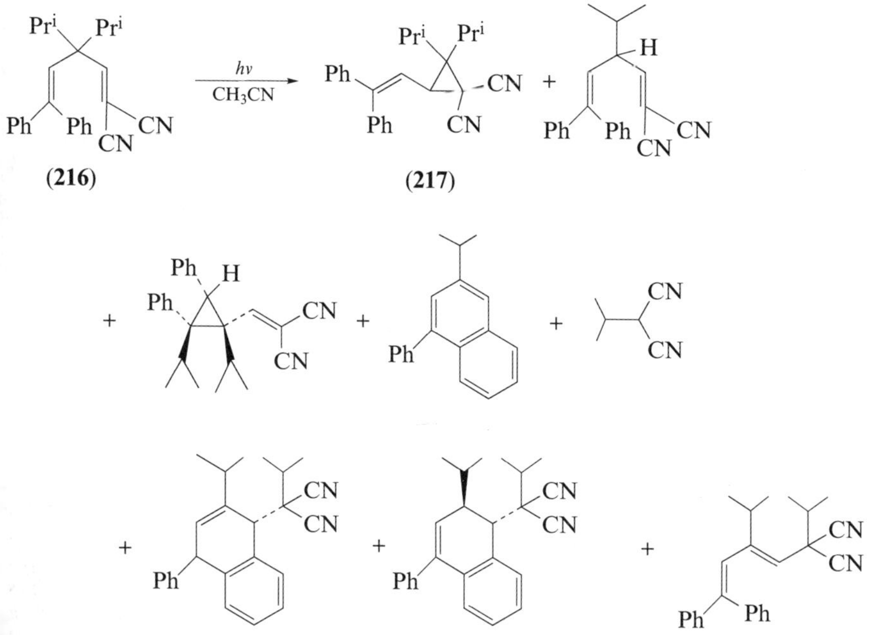

SCHEME 12

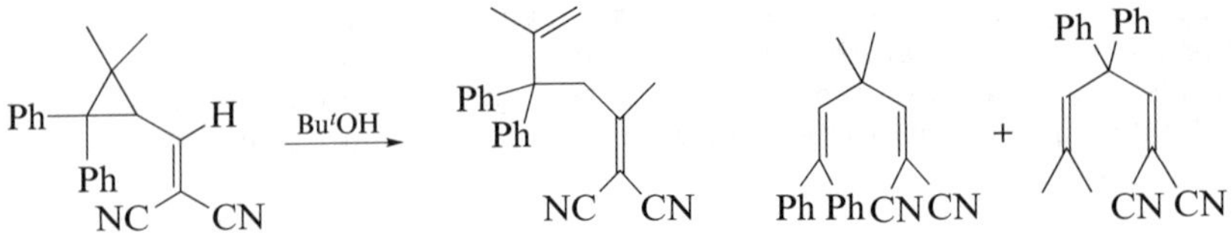

(218)   (219)   (220)

SCHEME 13

photochemical reactivity in solution. The quantum efficiency in the crystalline phase is, however, much lower than the quantum efficiency in solution phase. The dicyanodiene **213** yields only the cyclopentene **214** via the biradical **215**, the result of vinyl–cyanovinyl bridging, in the crystalline phase. Apparently, the path to **214** is preferred to the more conventional path to a cyclopropane. Acetophenone-sensitized irradiation of the diene **216** also follows the same bridging interaction in solution phase and affords the cyclopropane **217** in 35% yield. The reaction is reasonably efficient with a quantum yield of 0.041[127]. Direct irradiation in acetonitrile is more complex and, in addition to the di-π-methane rearrangement product **217**, six other products are formed (Scheme 12). A detailed study of the irradiation of the vinyl cyclopropane **218** has been carried out. A variety of processes occur such as Griffin fragmentation and single-bond fission processes. Sensitized irradiation follows a different path in which phenyl migrations occur. The vinyl cyclopropane **218** is also reactive on direct irradiation, affording the products shown in Scheme 13. Here a reverse di-π-methane reaction affords the dienes **219** and **220**[128].

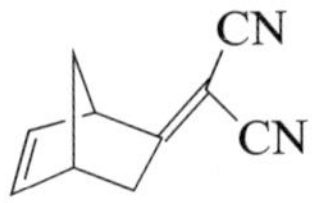

(221) 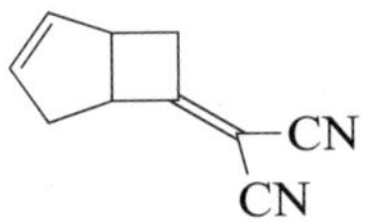   (222) 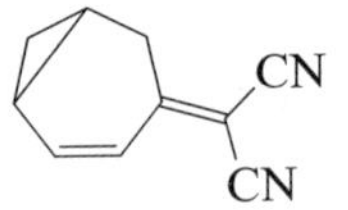   (223)    (224)

(225) 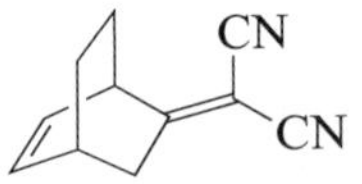   (226) 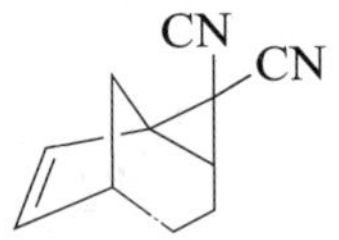   (227) 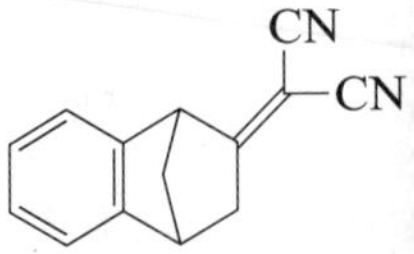   (228)

Di-π-methane reactivity is also seen with cyclic cyano-substituted 1,4-dienes. Thus the dicyanobicycloheptene **221** undergoes photochemical rearrangement by 1,3-carbon migrations to yield products **222** and **223** and a di-π-methane process to the non-isolated intermediate **224**, which transforms into the isolated product **225**[129]. The authors suggest that the cyano substituents play a major part in controlling the reaction, but they suggest that a charge transfer mechanism is not operative. The diene **226** also undergoes a di-π-methane transformation into **227**. The interest in such compounds relates to energy storage, and it is of importance in this respect that **227** can be converted thermally to the starting material[130]. Di-π-methane reactivity is also reported for the conversion of **228** into **229**, by way of the thermally unstable intermediate **230**[131], and for the tetraene **231**[132]. Considerable interest has been shown in the di-π-methane rearrangement of norbornadiene systems. The influence of cyano groups on this rearrangement has been studied notably by Paquette and his coworkers. Thus the introduction of a cyano group at

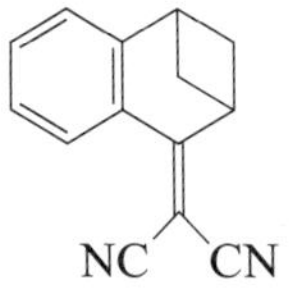

(229) 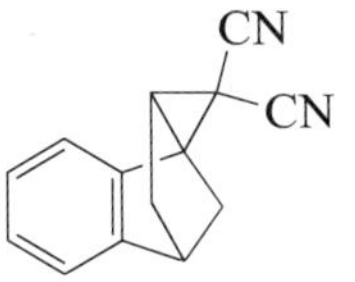   (230) 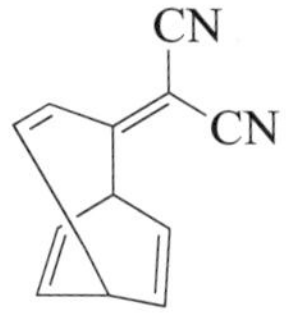   (231)

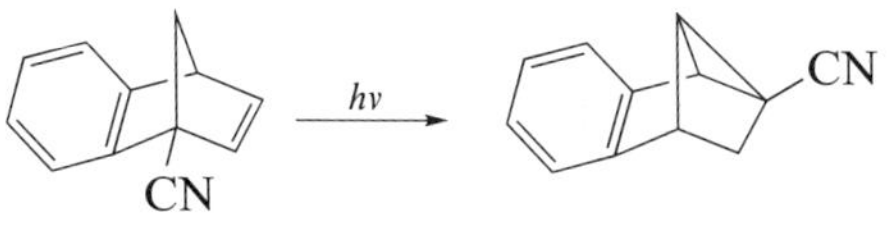

(232)   (233) 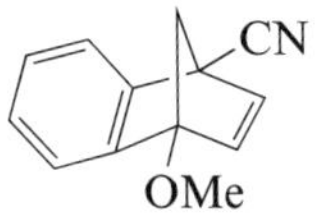   (234)

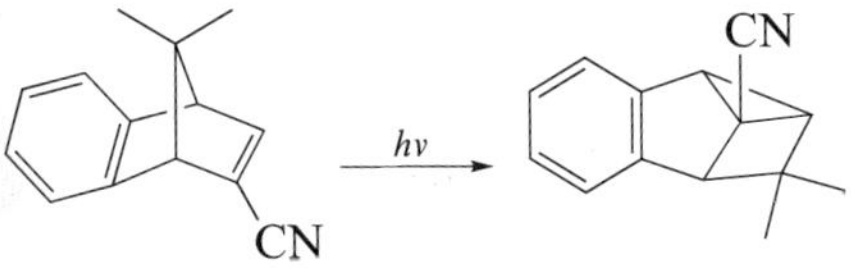

100%   100%

(235)   (236)

SCHEME 14   SCHEME 15

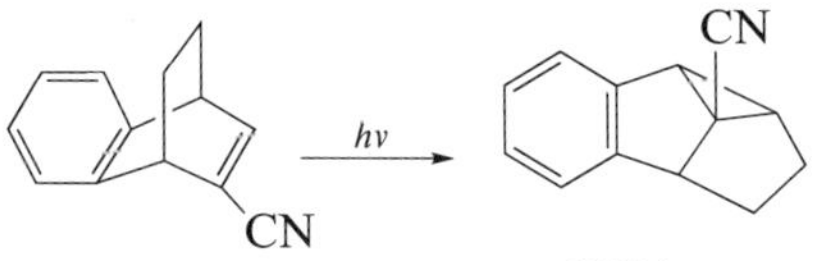

the bridgehead in **232** controls the cyclization, so that **233** is formed exclusively[133-135]. Similar control is observed with the methoxy cyano-substituted derivative **234**[136]. Other workers[137] have also studied the influence of a cyano substituent and also the influence of ring size on the outcome of the di-π-methane process in systems such as **235** and **236**. Again the control of the cyano substituent affords exclusively the products shown in Schemes 14 and 15. Both acetone and acetophenone sensitization are effective in bringing about the transformation, indicating that the triplet state is operative. Interestingly, the inclusion of a second cyano group reduced the reactivity considerably. Shi and his coworkers have reported the di-π-methane reactivity, by phenyl–phenyl interaction of the cyanotriarylmethane derivatives[138].

Bender and his coworkers[139-141] have examined the influence of the position of a cyano substituent on the outcome of the reactions of the naphthalene derivatives **237**, **238** and **239**. Typical of the reactions is the direct irradiation ($\lambda > 290$ nm) of **237**, resulting in its conversion into the three products shown in Scheme 16. The semibullvalene **240** is also photochemically reactive and is converted into the cyclo-octatetraene in 75% yield on further irradiation. This cyclo-octatetraene is the sole product obtained on the sensitized irradiation of the starting material **237**. Labelling studies have shown that the semibullvalene **240** is formed by two paths, the major of which is a di-π-methane process. The sensitized irradiation of **238** does not involve a di-π-methane process and the fission of the internal bond of the cyclobutane affords the product directly[139]. The direct irradiation of **238** also yields a semibullvalene **241** (71%) as the principal product by a di-π-methane path[140] as does the irradiation of **239**, which yields **242** (55%) on direct irradiation and the same compound in 78% yield on sensitization[141]. Interestingly, the naphthalene derivative **243** does not rearrange by a di-π-methane path[141].

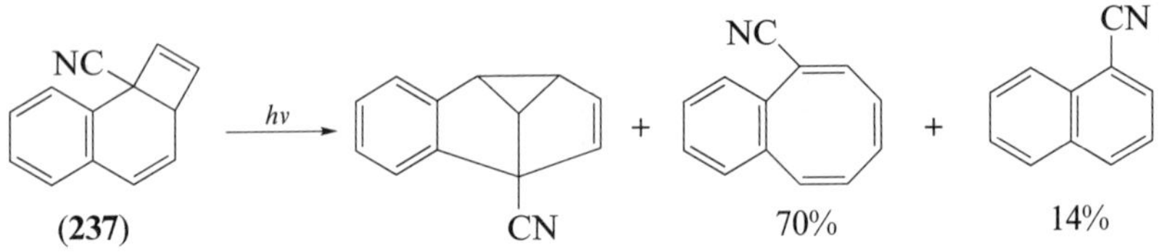

(237)   CN   70%   14%
                               (240) 5%

SCHEME 16

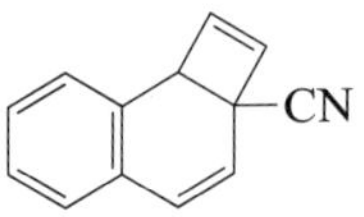

(238) 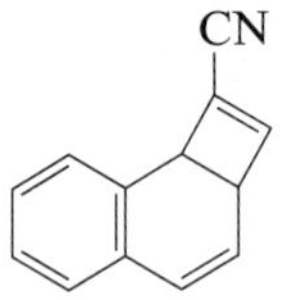   (239) 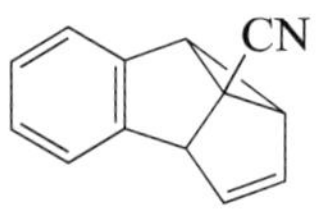   (241)

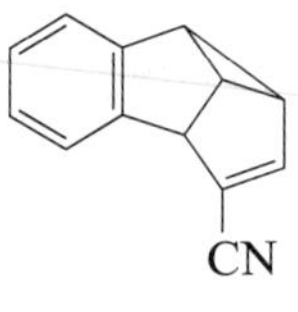

(242) 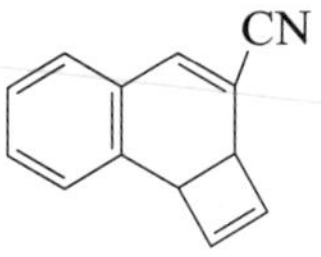   (243)

D. Cyclobutane Formation by Addition to Cyanoalkene Systems

Many examples of cyclobutane formation involving cyano-substituted alkenes or enone components have been reported over the years. Some of these reactions arise as a result of SET to an electron-accepting sensitizer such as 1,4-dicyanonaphthalene. One such reaction among the many reported involves the photoreaction of the diene **244** in benzene solution. Under these conditions the (2 + 2)-cycloadduct **245** is formed in reasonable yields (*ca* 70%). In acetonitrile, however, a different reaction occurs leading to the allylation of the sensitizer affording **246**, which undergoes (2 + 2)-cycloaddition to yield the cyclobutane **247**[143]. Other intramolecular cycloadditions have also been reported such as the formation of the cage compound **248** from irradiation of the pentaene **249**[144]

CN

Me₂Si — (244)    (245) SiMe₂   (246) 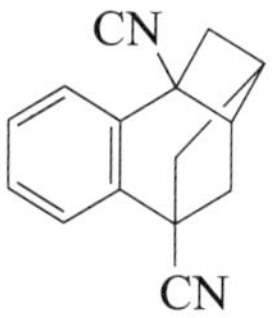   (247)

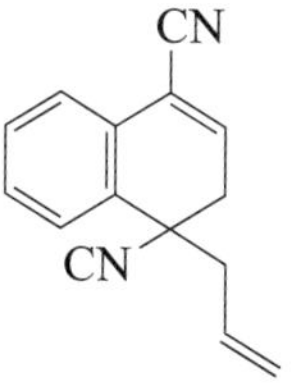

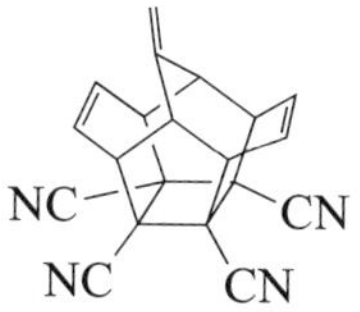

(248) 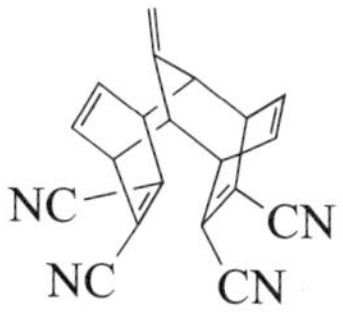   (249)

Cyano-substituted quadricyclanes are also of interest because of the use that such substrates have in solar energy conversion projects. Thus the derivative **250** is formed on direct irradiation (at 366 nm) of the norbornadiene **251**. The quantum yield for the cyclization is high at 0.68. However, triplet sensitization is found to be less efficient ($\phi = 0.1$) and the authors[145] suggest that the lowest singlet state of the norbornadiene is more reactive than the triplet. SET using phenanthrene, anthracene, pyrene and *N*-methylcarbazole as the electron-donating sensitizers can also be used to synthesize quadricyclanes, and the dicyano derivative **252** is readily obtained from **253**[146–148]. Interest has also been shown in the ring-opening reactions of quadricyclane derivatives such as **254** which open to the biradical **255** on irradiation at 300 nm in hexane solution[149,150].

(250) 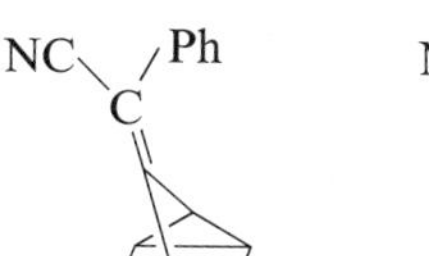   (251) 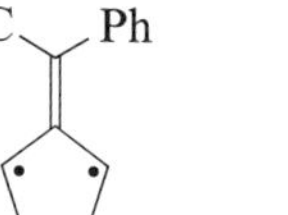   (252)

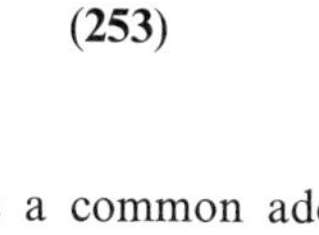

(253) 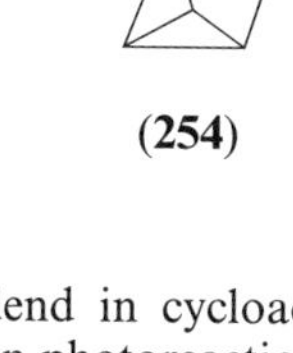   (254) 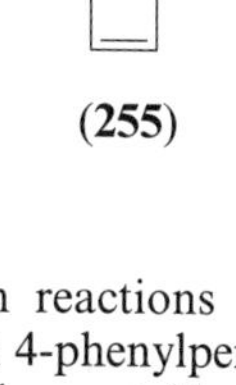   (255)

Acrylonitrile is a common addend in cycloaddition reactions and several (2+2)-cycloadducts have been reported on photoreaction with 4-phenylpenta-2,4-diyne[151,152], naphthalene[153], mesityl oxide[154], enone **256** affording the unstable adduct **257**[155], 4,4-

dimethylcyclohexenone[156] and 1,4-naphthoquinone to yield the cyclobutane **258** in 5% yield[157]. This last process is a good example of the influence that a cyano group can exert on a reaction, since acrylonitrile is an alkene with a high oxidation potential. Under these conditions the addition takes place to form a cyclobutane, while with an electron-rich alkene the addition affords a spiro-oxetane.

(256)   (257)   (258)

Acrylonitrile also undergoes dimerization to afford *trans*-fused cyclobutanes by using ethylene-*bis*-triphenyl phosphine nickel or *bis*-triphenylphosphine nickelacyclopentane as catalysts[158]. Dimers are obtained on irradiation (λ > 400 nm) of the crystalline stilbene derivative **259**, which at room temperature yields the (2 + 2)-cycloadduct **260**. Oligomers are also produced from this reaction. The reactivity is temperature dependent, however, and irradiation at −78°C using λ > 300 nm yields the double (2 + 2)-product **261** in 27% yield[159]. Interestingly, the cycloadducts are formed only by cyanoalkene–alkene interactions. A re-investigation of the solid-phase dimerization of **262** has identified **263** as the

(259)   (260)

(261)   (262)   (263)

(264)   (265)

Ar = *m*-Br, *m*-Cl, *p*-Cl, *m*-OMe,
      p-OMe, *m*-NO$_2$ and *p*-NO$_2$C$_6$H$_4$

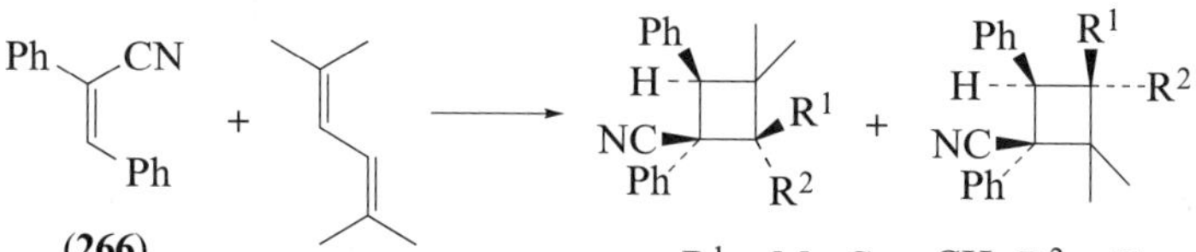

$$R^1 = Me_2C = CH, R^2 = H$$
$$R^1 = H, R^2 = Me_2C = CH$$

SCHEME 17

product of the cycloaddition[160]. The involvement of a singlet exciplex is proposed to account for the selective photocycloaddition of 2,3-dimethylbut-2-ene to the *trans*-nitrile **264** affording the adducts **265**[161]. Lewis and coworkers[162] have also reported the involvement of a singlet exciplex in the reaction of the cinnamonitrile (**266**)–hexa-1,3-diene system. The process affords (2 + 2)-cycloaddition products when carried out in non-polar solvents (Scheme 17). With polar solvents such as acetonitrile, a change in the reaction is observed and products of hydrogen abstraction, such as **267** , accompany the (2 + 2)-cycloadducts. The unusual formation of the 2-amino-1-dicyanopropene from the irradiation of tetracyanoethylene in acetonitrile is thought to involve a SET. This leads to a cycloadduct **268**, which undergoes ring opening and hydrolysis. Apparently, the addition of a drop of water speeds up the reaction remarkably and is probably responsible for the hydrolysis process[163]. 1,5-Bonding has been reported by Oku and collaborators[164] on irradiation of the aryl diene **269**, which yields the *endo–exo*- **270** and the *exo–exo*- **271** intramolecular adducts.

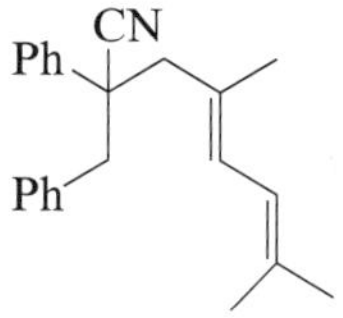

(267) 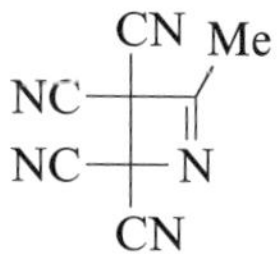   (268)

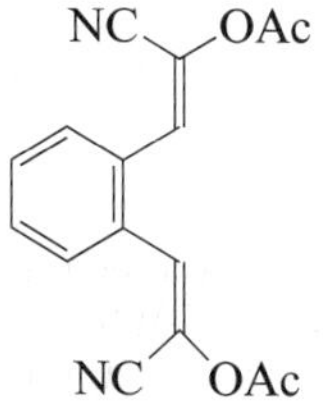

(269) 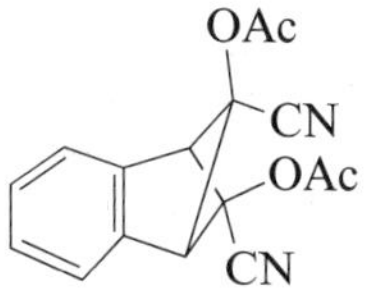   (270) 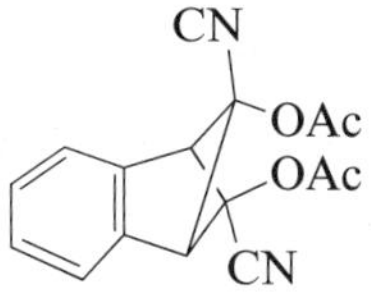   (271)

Rearrangement reactions occasionally have been reported in cyclic cyano-substituted systems. Thus the lactam **272** undergoes *cis,trans*-isomerization to afford **273**[165]. A much more extensive rearrangement occurs on irradiation of the substituted 4*H*-pyrans **274**. The irradiation of these compounds through Pyrex in methylene dichloride solution gives the cyclobutenes **275** in about 30% yield. These products are accompanied by the fragmentation products 1-cyanophenylacetylene and the amide **276**, which are produced by secondary irradiation of the cyclobutene. The reaction affording the cyclobutenes is presumed to involve bridging (possibly following intramolecular SET) of the type illustrated in **277**. Collapse of this intermediate as illustrated yields the observed products. A full report of this work has been published with the results of experiments to substantiate the proposed electron-transfer mechanism[166,167]. The isomeric pyran derivatives **278** also undergo photoreaction involving a reversible C—O bond fission[168].

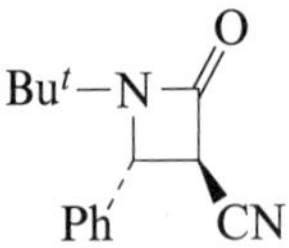

(272) 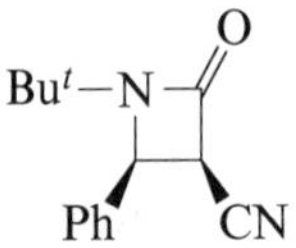   (273) 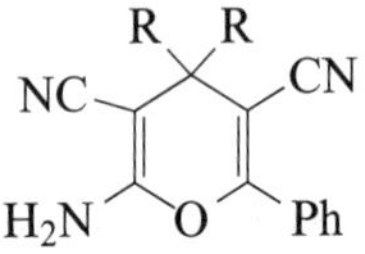   (274)

R–R = (CH₂)₃ or (CH₂)₄
R = Me

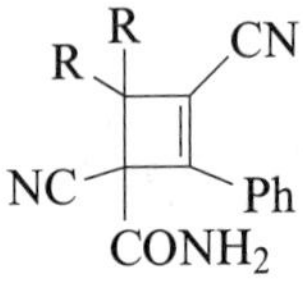

(275) 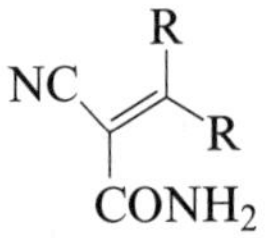   (276) 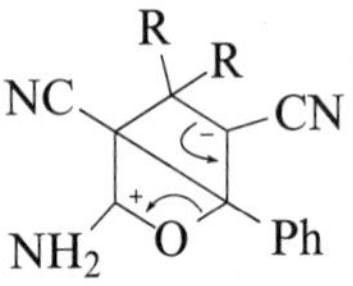   (277)

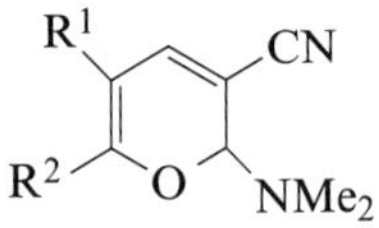

(278)

R¹ = Me, Ph or OMe
R² = CO₂Me, CO₂Et or COMe

There is little doubt that the addition of acrylonitrile and related compounds to a variety of substrates provides a useful synthetic path for the formation of many bicyclic compounds. Typical of this is the photoaddition of malononitrile or fumaronitrile to 3-methylcyclohex-2-enone which provides four adducts in a ratio of 29 : 15 : 21 : 35[169]. The structure of the predominant isomer was determined by X-ray crystallography and was established to be the cyclobutane **279**. The ratio of the products obtained from these experiments was shown to be independent of the dose of irradiation. Quantum yield

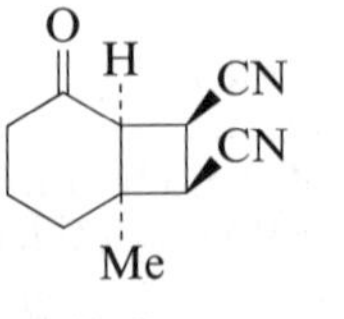

(279) 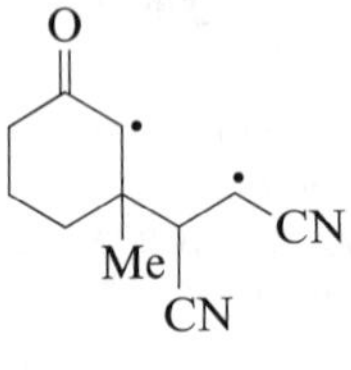   (280) 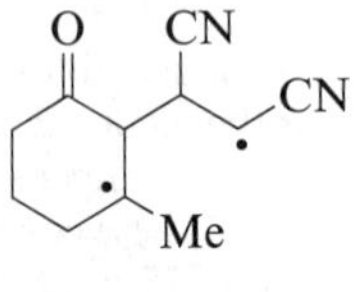   (281)

studies have shown that capture of the triplet enone is almost unity (96%). The evidence presented by the authors[169] suggests, in accord with the work of Hastings and Weedon[170] on the (2 + 2)-cycloadditions to cyclopentenone, that no exciplex is involved in these cycloadditions and that the triplet biradicals **280** and **281** are involved. The authors[169] concede that the exciplex mechanism, while useful for the prediction of the outcome of the cycloaddition, should be treated with scepticism.

Cycloaddition of acrylonitrile also occurs to the nicotinamide **282**. This process takes place stereospecifically and affords the regioisomeric adducts **283** and **284**[171–173]. Addition of tetracyanoethylene to the pyridone **285** does not yield a stable cyclobutane adduct and ring opening affording the dienes **286** is obtained. Several examples of this reaction with nicotinamides having different N-substituents were reported[174]. Other nitrogen

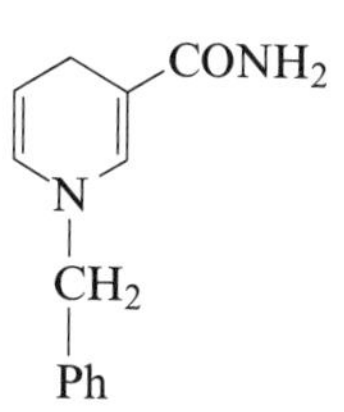

(**282**) 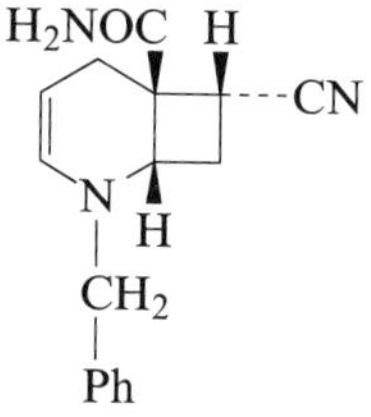 (**283**) 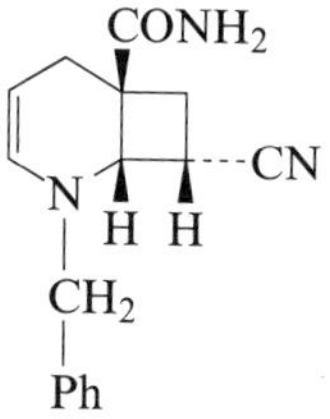 (**284**)

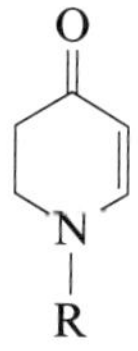

(**285**) 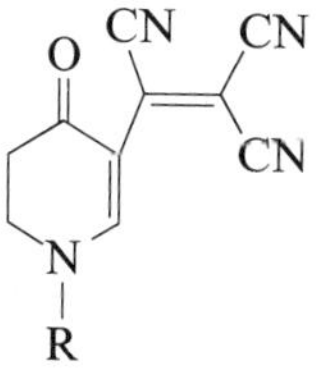 (**286**) (**287**) (**288**)

(a) R = Me
(b) R = PhCH$_2$
(c) R = Ph$_2$CH
(d) R = CH$_2$=CH(CH$_2$)$_2$
(e) R = CH$_2$=CH(CH$_2$)$_3$

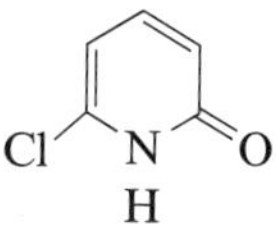

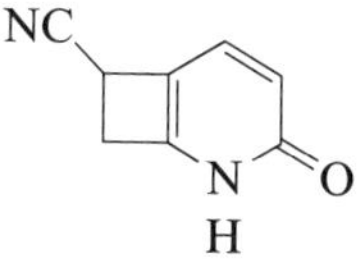

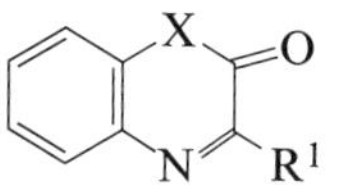

(**289**) (a) X = NH, R^1 = C$_4$H$_9$ (**290**) (**291**) R^2 = Me, R^3 = CN
(b) X = O, R^1 = Me, Bu or CH$_2$CH$_2$Pr^i R^2 = CN, R^3 = Me

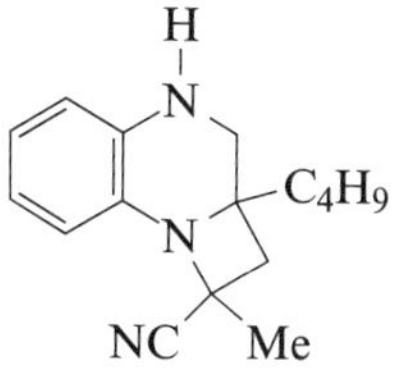

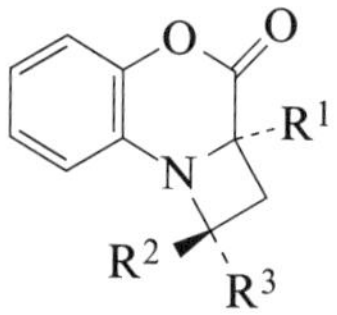

heterocycles also undergo addition of cyanoalkenes on irradiation. Thus a patent has been lodged dealing with, among other addends, the addition of acrylonitrile to the pyridone **287** to afford a cyclobutane derivative, which eliminates HCl to yield the adduct **288**[175]. Quinoxalin-2-ones **289a**[176,177] and 1,4-benzoxazin-2-ones **289b** undergo a non-stereoselective but regioselective cycloaddition reaction on irradiation in the presence of 2-cyanopropene. This affords the adducts **290** and **291**, respectively.

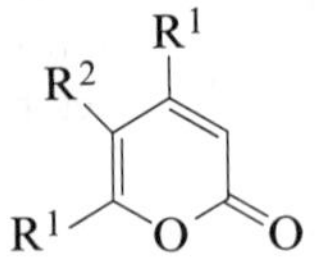

(292) (a) R^1 = H, R^2 = CO$_2$Me
(b) R^1 = Me, R^2 = H

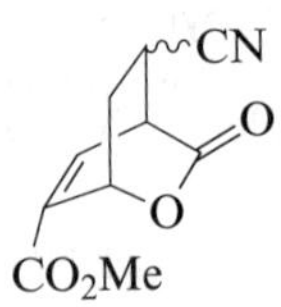

(293)

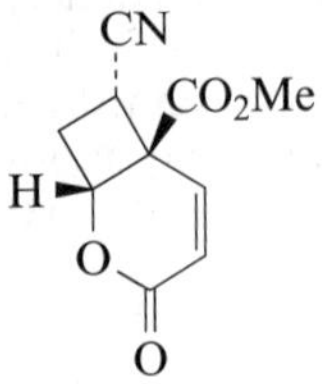

(294)

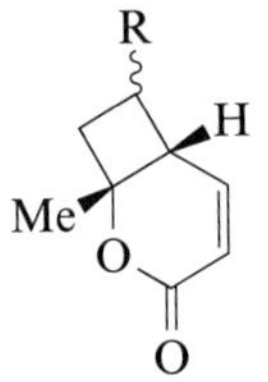

(295)   R = α-CN or β-CN

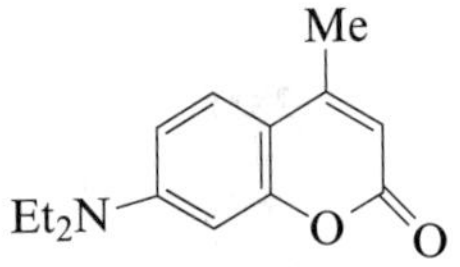

(296)

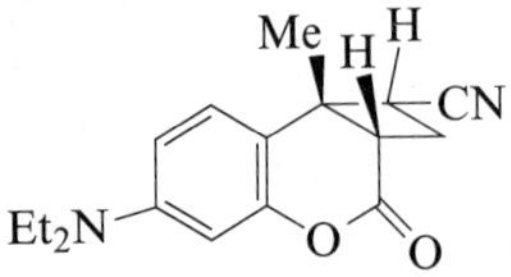

(297)

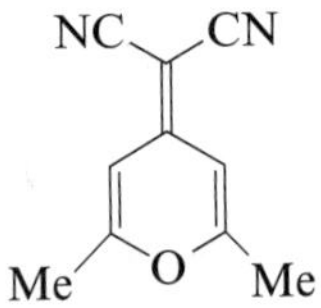

(298)

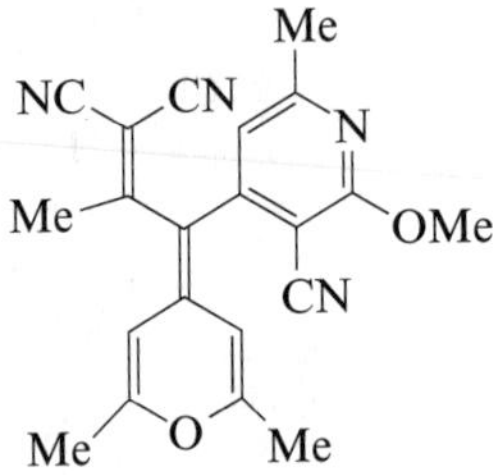

(299)

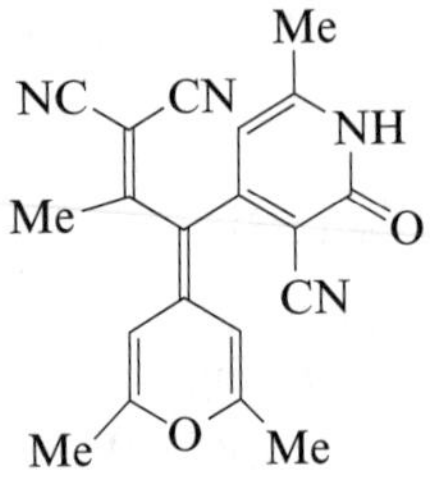

(300)

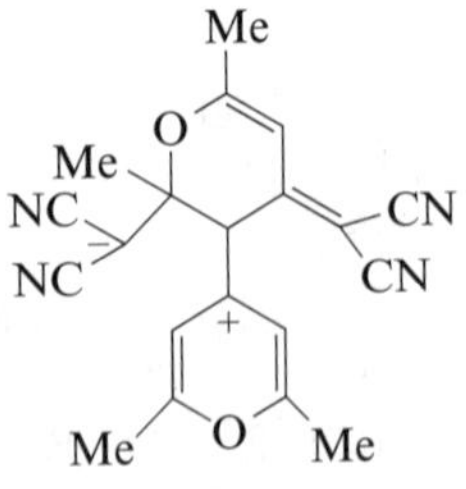

(301)

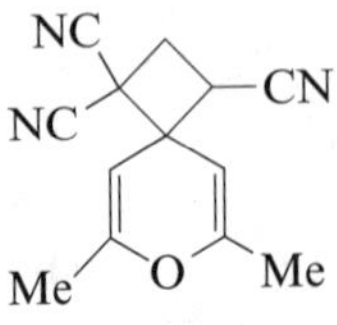

(302)

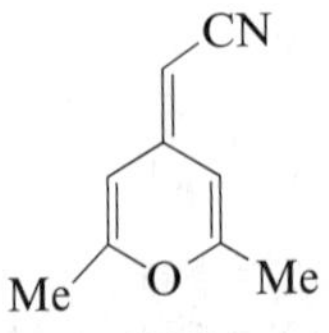

(303)

Acrylonitrile has also been reported to undergo addition reactions with pyrones. Thus, pyrone **292a** affords **293** as the major product accompanied by a low yield of the (2 + 2)-cycloadduct **294**, while addition to **292b** affords the (2 + 2)-cycloadducts **295** in a regiospecific manner[179,180]. (2 + 2)-Cycloaddition is reported for the benzopyrone **296** in its singlet state to afford the adduct **297**[181]. Samuel and his coworkers[182] report that the irradiation of the pyrone analogue **298** in methanol does not bring about rearrangement of the ring system. The products obtained from the reaction are the dimers **299** and **300**. These compounds arise by the photochemical addition of the exocyclic double bond of the pyrone analogue to a ring double-bond affording a cyclobutane. This step is followed by ring opening to the zwitterion **301** from which the final products are obtained. Cycloaddition reactions are also encountered between the same pyrone and acrylonitrile. Again, the addition encountered involves the exocyclic double-bond to yield an unstable cyclobutane **302**. This undergoes ring opening to a zwitterion which, in this case, fragments to yield **303**.

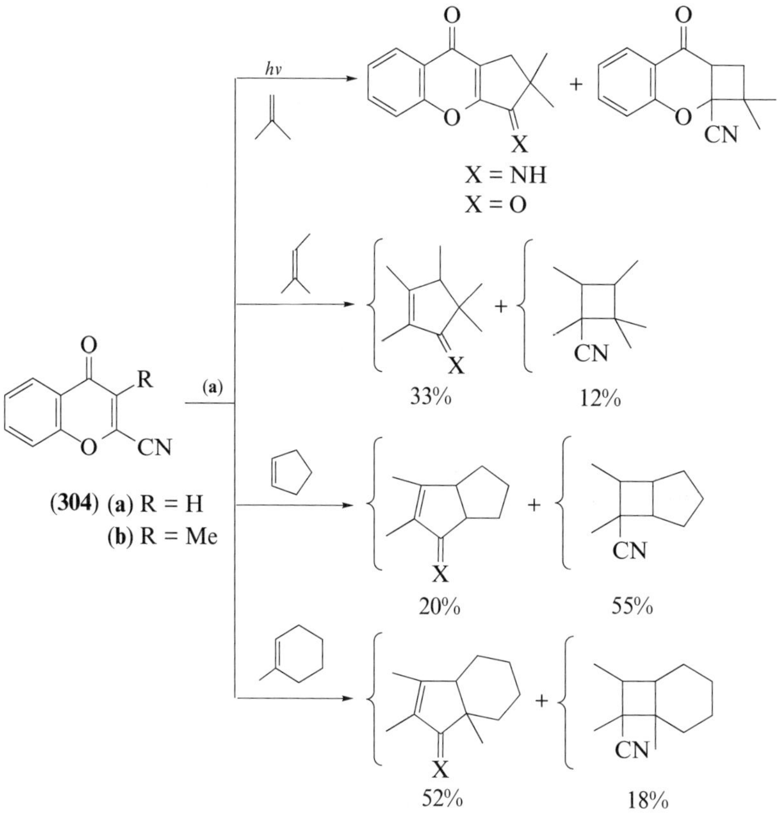

SCHEME 18

The γ-pyrone (**304a**) undergoes photoreaction when irradiated with alkenes. Two types of adducts are obtained, the normal cyclobutane adduct in 12% yield when using

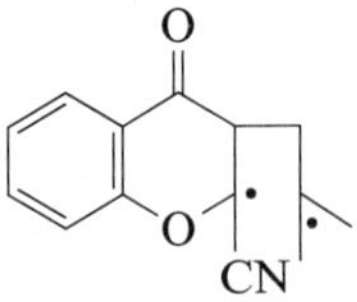

(305)

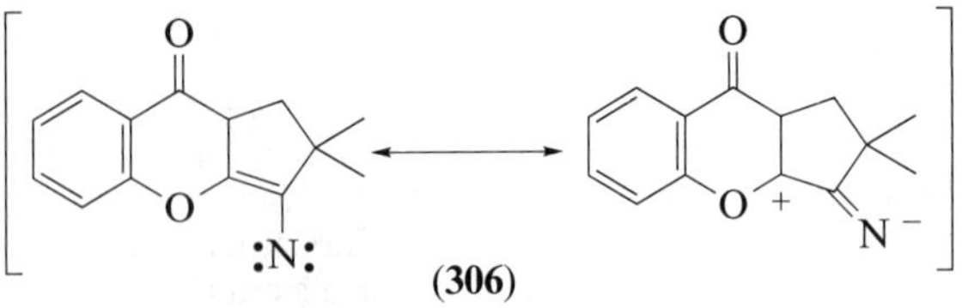

(306)

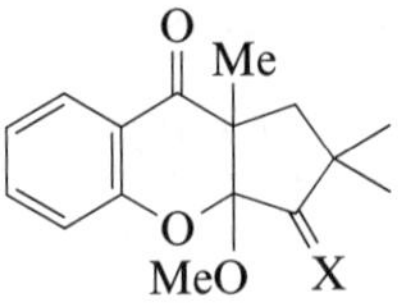

(307) X = NH or O

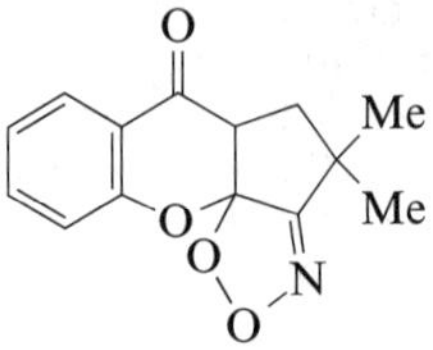

(308)

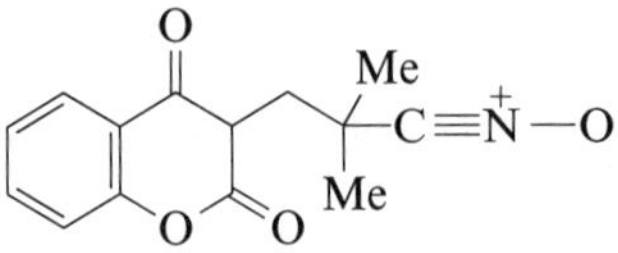

(309)

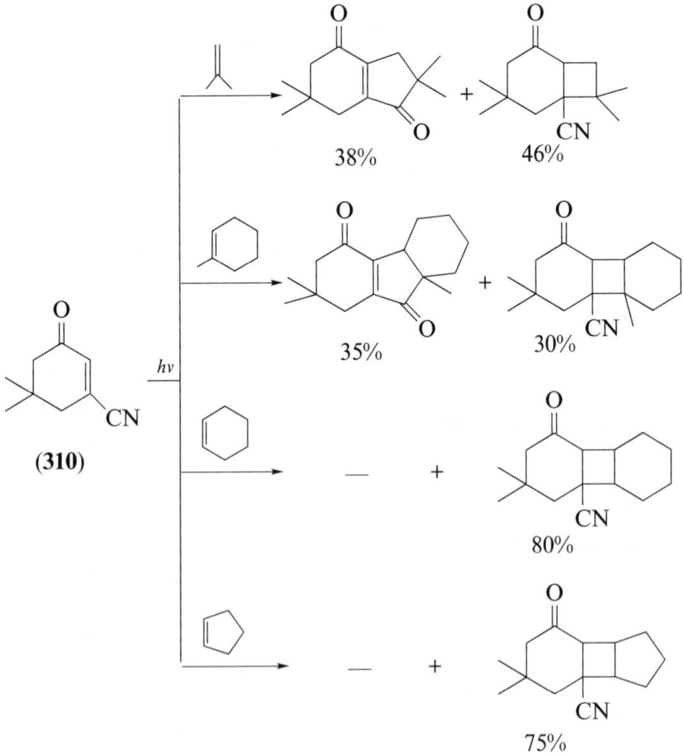

SCHEME 19

2-methylpropene as the addend and the ring-expanded compound in 81% from the same alkene. A sample of the results are shown in Scheme 18[183,184]. The route to product follows the normal addition mode and affords a biradical **305**, which either cyclizes to the cyclobutane or else undergoes addition to the nitrile function affording a nitrene **306** which subsequently forms an imine or, on hydrolysis, a ketone. It is of interest that the nitrene generated from enone **304b** can be trapped by methanol as the adduct **307**. Substantiation for the involvement of a biradical comes from trapping reactions with oxygen which show that a species such as **306** affords an oxygen adduct **308**, which ring opens to yield the nitrile oxide **309**[185]. A similar reaction mode is proposed for the enone **310**, the yields from which and some typical examples are shown in Scheme 19[186]. Reactions with cyano-uracils have also been studied[187,188].

Gelas-Mialhe and coworkers[189] report the efficient cyclization of the cyano substituted enones **311** to afford the cyclized compounds **312** and **313**. Conrotatory cyclization within **311** yields the ylide **314**. This is followed by a 1,4-hydrogen migration.

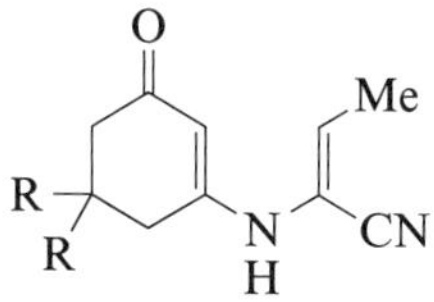

(**311**) R = H or Me

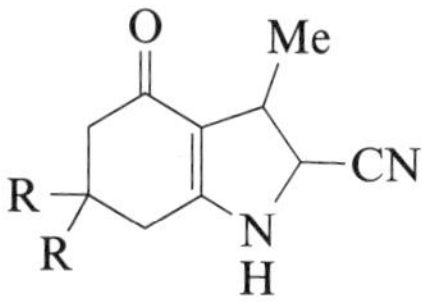

(**312**)

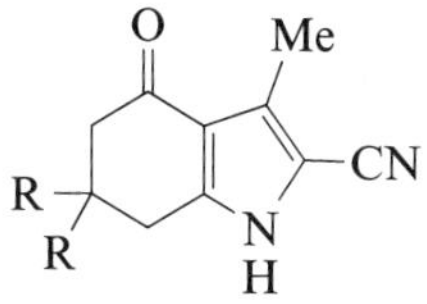

(**313**) 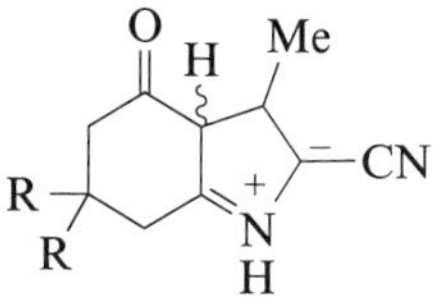 (**314**)

E. Addition to Cyanoarenes

Cyclobutane formation does occur in some cases by addition of alkenes or cyano groups to aromatic systems. An example of this is the addition of 2,3-dimethylbut-2-ene to 9-cyanophenanthrene[190]. An exciplex is thought to be involved in the photoaddition of phenoxyethylene to the cyanophenanthridene derivative **315** to yield the azocine **316** in ethanol solution. Cycloaddition to the same phenanthridene with the alkene 1-(*p*-methoxyphenyl)propene affords the cyclobutane adduct **317** which, interestingly, does not undergo ring opening[191].

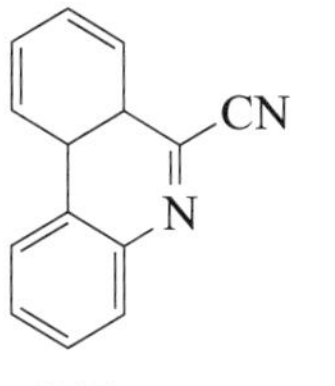

(**315**) 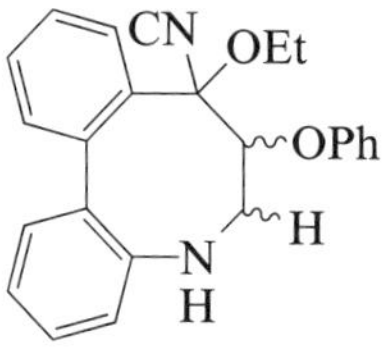 (**316**) 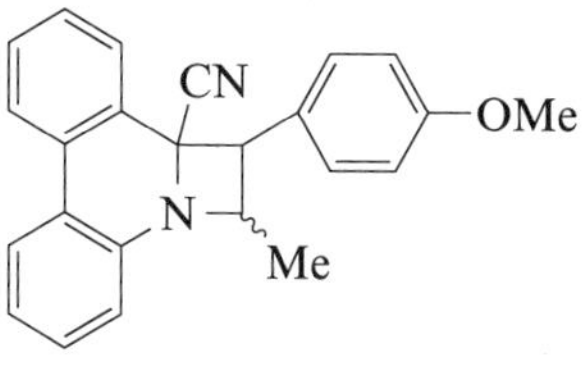 (**317**)

The outcome of cycloaddition reactions to benzene derivatives can often depend on the nature of the substituent in the ring. Apparently, the nature of the substituent can dictate the addition path that is followed. Thus *ortho* addition appears to be favoured over other paths such as *meta* addition if polar groups are present. This reactivity apparently correlates with the magnitude of the charge transfer interactions between the alkene and the arene. Thus polar groups, which would favour charge transfer, drive the reaction towards the *ortho* addition mode. This is seen with cyano substituents in the photoaddition of benzonitrile to 2,3-dihydrofuran which yields the *ortho*-adduct 318[192]. Another 1,2-addition is observed in the formation of the *ortho*-adduct **319a** as the major product following cycloaddition of acrylonitrile to anisole, or adduct **319b** on addition to 1,4-dimethoxy benzene[193]. The photoaddition of acrylonitrile to the *p*-cyanophenol presumably affords a cyclobutane, which ring opens to yield the product **320**[194]. The 1,2-addition mode followed by ring opening of the cyclobutane originally formed has been developed as a new synthetic route to azocine-2-ones **321**. This route involves the photoaddition of the nitrile group of benzonitrile or naphthonitriles to phenols. Initially, addition of the cyano group occurs to the phenol ring to yield the **322**, which ring opens to afford the isolated products[195].

(318)   (319) (a) R = H   (320)
                     (b) R = OMe

(321) Ar = Ph or 1-naphthyl   (322)

F. Photochromism

Electron cyclizations of the *cis*-stilbene type can lead to highly coloured compounds that can be of some industrial value. A few examples of photochromism of this type with

(323)   (324)   (325)

R = 9-anthryl, *p*-NO$_2$C$_6$H$_4$
p-Me$_2$NC$_6$H$_4$, *o*-NO$_2$C$_6$H$_4$

(326)   (327)   (328)

cyano substituents have been reported. Typical example of this is the work of Daub and collaborators[196–198]. They have reported studies of the photochromic properties of **323** and its conversion into **324**, and also the photobehaviour of the closely related tetraenes **325** which ring open to the pentaenes **326**[197,198]. Another example is the more typical cyclization of the thiophene derivative **327**, which affords the coloured product **328**[199].

G. Cyano-oxiranes

The presence of electron-withdrawing cyano groups in a molecule obviously changes the ionization potential. This, in many instances, can make the molecule a useful target for SET processes. This is seen with the cyano-oxiranes (**329**) which undergo SET induced ring-opening, a mode of reaction that was studied by Griffin and his coworkers some time ago[200,201]. A laser-flash study of the epoxide **330** has also been reported and the ylide formed by this process has been studied[202]. Such ylides can be of some synthetic values as has been reported by Brokatzky-Geiger and Eberbach[203]. Thus photochemical ring opening of **331** to the corresponding ylide is followed by intramolecular 1,3-dipolar addition to afford the tetrahydrofuran derivative **332**.

(329)   (330)   (331)   (332)

$R^1 = R^3 = CN$, $R^2 = R^4 = $ *p*-anisyl, *p*-tolyl or Ph

$R^1 = R^4 = CO_2Me$, $R^2 = R^3 = $ *p*-Ar, *p*-tolyl or Ph

$R^1 = R^3 = CO_2Me$, $R^2 = R^4 = $ *p*-Ar or *p*-tolyl

Oxirane systems with cyanovinyl groups have also attracted attention as in the ring opening and rearrangement reactions of **333**[204]. Irradiation brings about conversion into the products **334–339** either by the intermediacy of an ylide (**340**) or a carbene (**341**). The corresponding cyclopropane derivative **342** also undergoes ring opening on direct irradiation in an analogous fashion to afford an ylide that gives **343** as the major product. This is accompanied by the cyclopropenylnitrile **344**, presumably formed via a carbene intermediate. Triplet sensitized irradiation of **343** brings about formation of the Z-isomer[205]. Irradiation of the unsaturated oxiranes **345** in the presence of triethylamine results in the formation of the zwitterion **346**, presumably by a SET process. Intramolecular trapping of this system by the pendant hydroxy group affords the spiroacetals **347** and **348** in the yields shown[206].

(**333**)   (**334**)   (**335**)   (**336**)   (**337**)

(**338**)   (**339**)   (**340**)   (**341**)

(**342**)   (**343**)   (**344**)   (**345**) $n = 1$ or 2

(**346**)

(**347**) $n = 1$ 7%   (**348**) 29%
         $n = 2$ 7%   19%

H. Oxetane Formation

Cyanoalkenes are also substrates used for oxetane formation. Thus benzil (in its lowest triplet state) adds to the alkene **349**, to yield the oxetane **350** (in 54% yield) as the main product[207]. A detailed study of the photochemical formation of oxetanes by the addition of *E*-1,2-dicyanoethene and *Z*-1,2-diethoxyethene to the adamantanones **351** has been reported[208]. In addition Turro, Le Noble and coworkers[209] have studied the photochem cal behaviour of **351** on irradiation in acetonitrile and in aqueous solutions of β-cyclodextrin with fumaronitrile. The influence of environment on the formation of the *syn*- and *anti*-oxetanes **352** and **353**, respectively, was studied.

Carbonyl compounds **354** also photo-add to furans in yields that often can be high

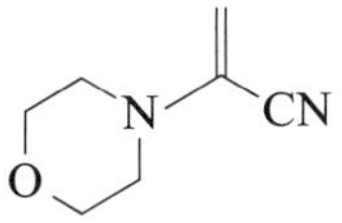

(349) 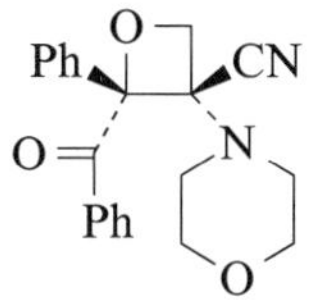   (350) 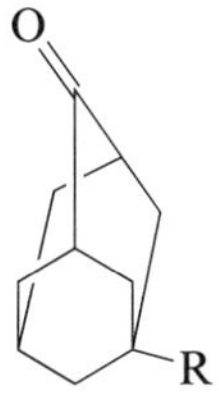   (351)

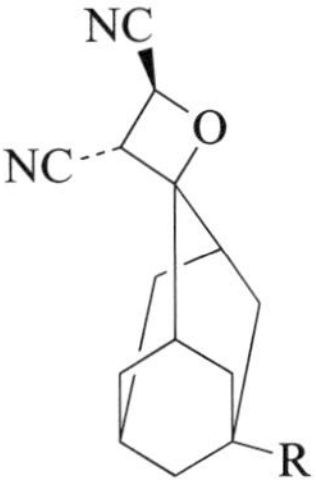

(352) *syn* 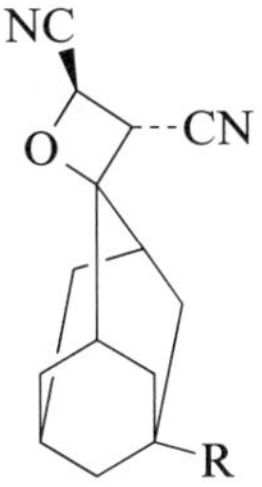   (353) *anti*

R = F, Cl, Br, OH, Ph, Bu^t

(Scheme 20). In this series the *endo–exo* ratio has been shown to be temperature dependent. However, in this system, when optically active addends were employed, only a low diastereoisomeric excess was observed[210]. Other studies have highlighted the control that

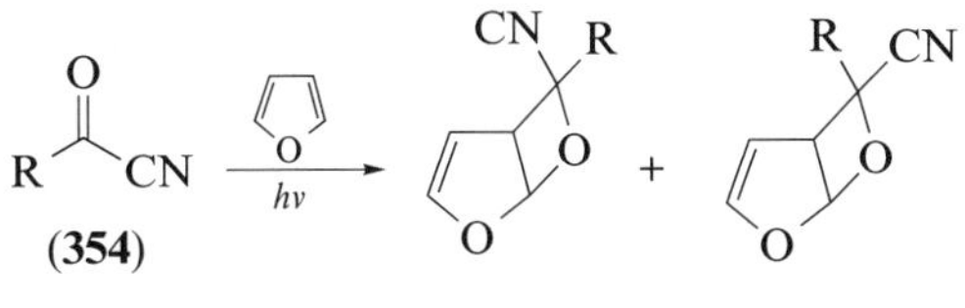

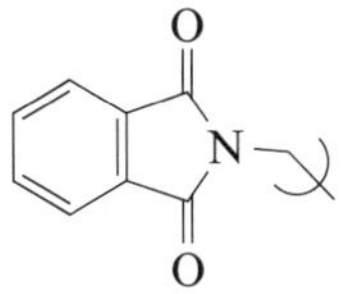

(354)

R =	30%	ratio	3.5:1	at – 55 °C
R = Bu^t	86%	ratio	8.9:1	at ambient
	89%	ratio	9.3:1	at – 55 °C
R =Ph	77%	ratio	3.7:1	at ambient
	95%	ratio	5.3:1	at – 55°C

SCHEME 20

can be exercised by the nature of the substituent on the furan. Thus the addition of 3-fluorobenzaldehyde to 2-cyanofuran affords a mixture of products **355** and **356** in a ratio of > 20 : 1[211]. This is to be contrasted with the results from the addition of benzaldehyde to 2-methylfuran which yields a 1.3 : 1 mixture of the adducts.

(355)   (356)

   Photoaddition of cyanoalkenes (acrylonitrile and 2-methylacrylonitrile) to the indoline thiones **357** affords the isomeric derivatives **358**. These are produced by ring opening of the thio-oxetane **359** formed as the key intermediate[212]. The two thio-oxetanes **360** and **361** are obtained on irradiation of the thione **362** in the presence of acrylonitrile. The cycloaddition involves an upper excited state of the thiocarbonyl group[213,214].

(357) (a) R^1 = Me, Bu, $PhCH_2$ or Ph   (358) R^4 = Me   (359)
      (b) $R^2 - R^3$ = Me or Ph $R^4 = CH_2SH$
      (c) R^1 = Me, $R^2 = C_5H_{11}$
      (d) $R^1 - R^2 = (CH_2)_n$, $(CH_2)_2O(CH_2)_2$
          n = 4 or 5

(360)   (361)   (362)

I. Fragmentations of Cyanoalkenes and Related Compounds

   The alkene **363** is photochemically reactive on irradiation in a matrix at low temperature. An analysis of the IR spectrum of the photolysate showed that the ketene **364** and acetaldehyde are formed[215]. Extended irradiation of **363** in argon led to the formation of a new compound identified as 1,1-dicyano-2-methyloxirane, produced by the addition of dicyanocarbene, formed by secondary irradiation of the ketene, to acetaldehyde.

(363) (364)

The dicyanophenylbutene **365** on irradiation at 254 nm affords two products identified as **366** and **367**. The former of these is produced by a 1,3-benzyl migration. The product ratio is solvent dependent and in cyclohexane, a non-polar solvent, the migration product is formed exclusively. However, in polar solvents such as acetonitrile, the ratio of products is 0.34 : 1. The authors[216] reasoned that the indene **367** must arise from a zwitterionic intermediate. Again the ability of cyano groups to stabilize such species is important. A 1,2-bond migration is involved in the transformation of the cyanonorbornene **368** into the major product **369** (95%). A minor product **370** arises by a 1,3-migration path[217]. Bond fission to afford the biradical intermediate **371** is involved in the transformation of the cyanoalkene **372** into the two products **373** and **374**[218]. Mukai and his coworkers[219] have also reported on the photochemical behaviour of **375**. Again 1,3-migration processes are the route to the products **376–378** on irradiation at 254 nm in cyclohexane. A dependence for product formation on the multiplicity of the reaction has been demonstrated for the conversion of the diene **379**. Irradiation of this diene at 254 nm affords the single product **380**, while acetone sensitization follows a different path and yields **381**. The mechanism for the formation of these products involves fission of the initial adduct into cycloheptatriene and tetracyanoethylene followed by a thermal re-addition of the ethene to afford the two

(365) (a) R = H   (366)   (367)   (368)
(b) R = Me

(369)   (370)   (371)   (372)

(373)   (374)   (375)   (376)

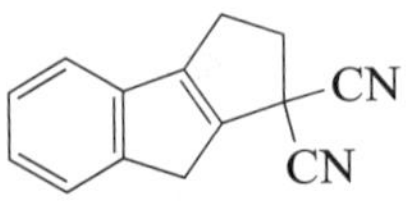

(377)

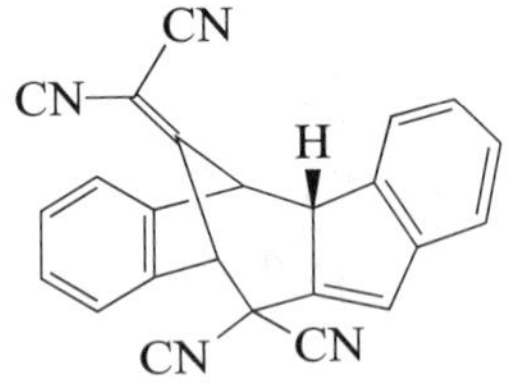

(378)

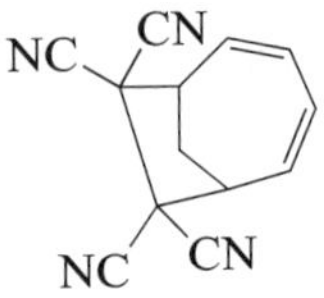

(379)

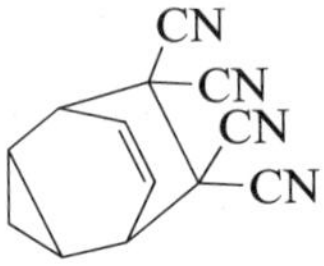

(380)

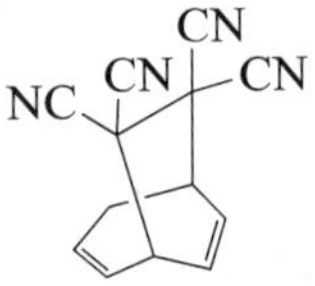

(381)

products. The authors suggest that the singlet and triplet states of **379** do not decay along a common path and that polarization of excited states is important[220].

J. Reduction Processes of Systems Containing Cyano Groups

Two competing processes have been reported for the photoreaction of 2-cyanopyridine in the presence of methanol and ethanol. The products formed are the ethers **382** by attack on the pyridine ring and the hydroxyalkyl derivatives **383** by reaction at the cyano group[221]. Reductive decyanation occurs on irradiation of the pyrazines **384** in the presence of secondary or tertiary amines. It is likely that a SET process is involved[222]. SET processes are also involved in the reactions of 9-cyanophenanthrene in the presence of amines. Using non-polar solvents affords the diethylaminyl radical exclusively and the products **385** (principal product), **386** and **387** are obtained. In acetonitrile, however, the

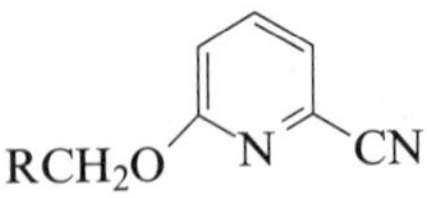

(382)

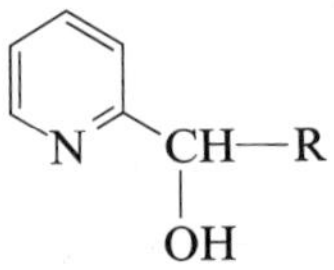

(383) R = H or Me

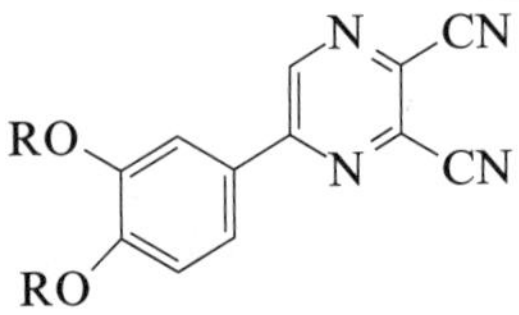

(384) R = Me

R–R = $O(CH_2CH_2OCH_2CH_2)_2$

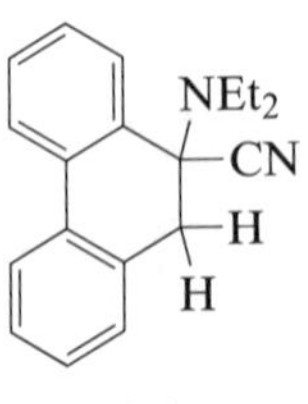

(385)

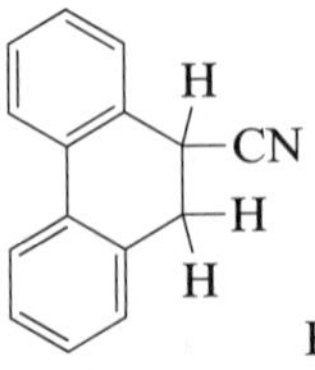

(386)

EtN$=$CHCH$_3$

(387)

diethylaminyl radical is accompanied by the 1-(ethylamino)ethyl radical and the products formed are **386** and **387**[223].

The Norrish Type II activity of ketones does not seem to be adversely affected by the presence of a cyano group as evidenced by the conversion of the cyanoketone **388** in benzene solution and affords the photo-enol **389**, which can be trapped by a variety of dienophiles[224].

(**388**)   (**389**)

IV. PHOTOCHEMISTRY OF COMPOUNDS CONTAINING A DIAZO GROUP

This section deals with the photochemical loss of nitrogen from compounds containing a diazo group. As a result of the scale of the work some material has had to be excluded. Within this section paths to the synthesis of diazo compounds from hydrazones have been omitted. Much of the use of diazo compounds and the photochemical (and thermal) elimination of nitrogen is in the generation of carbenes, from which a variety of reactions of synthetic value can occur. This short section is aimed mainly at that area, although some reference will be made to the elegant low-temperature studies that have been carried out.

A. Synthesis of Cyclopropanes

One of the common reaction paths for carbenes generated by irradiation of the corresponding diazo compounds is trapping with alkenes to provide a convenient route to cyclopropanes. A reaction typical of this is the decomposition of diaryldiazomethanes in the presence of 1,1-diarylalkenes to yield the corresponding 1,1,2,2-tetraarylcyclopropanes[225]. Related to this type of reactivity is the report of a laser-jet study of the reaction of diphenylcarbene with 1,1-di-*p*-methoxyphenylethene where a 1,3-biradical, obtained by a non-concerted addition via the triplet state, is detected presumably on the path to the cyclopropane[226]. The non-concerted addition of diphenylcarbene and of triplet fluorenylidene to fumarate esters has also been demonstrated[227]. In general it has been shown that diphenylcarbene reacts more readily with electron-rich and with electron-poor styrenes than with the parent[228]. A description of the reaction of diphenylcarbene with benzene has been published[229]. Photodecomposition of 4-pyridyldiazomethane in the presence of 1,1-dimethylallene yields the cyclopropane **390**[230]. Two separate studies have examined the carbenes derived from diazasuberones. Thus the carbene from the diazasuberone **391a** adds to styrene to yield the corresponding spirocyclopropanes **392**[231]. The isomeric species from **391b** also affords spirocyclopropanes **393** by a reaction involving the singlet state carbene[232]. The carbene **394** can be formed in the presence of styrene by irradiation of the corresponding diazo compound. The ratio of the mixture of *cis*- and *trans*-cyclopropanes **395** formed was dependent on temperature as illustrated[233]. A cyclopropane **396** is also formed in good yield on addition of the carbene **397** to 1,2-ditrimethylsilylethyne[234]. Interestingly, direct IR observation of the addition of the cyclopentadienylidene from diazocyclopentadiene to ethene in an argon matrix has been reported[235,236]. Cycloaddition to 2-methylacrylonitrile occurs with the carbene **398**. Here, two modes of addition were observed. One yields the cyclopropane **399** while alternatively, via the ylide, **400**, involving trapping of a sort which has been mentioned previously, a further mole of the alkene adds to yield **401**[237].

(**390**)   (**391**) (**a**) Y = H, X = N   (**392**)   (**393**)
                (**b**) Y = N, X = H

(**394**)   (**395**)

Temp.	cis : trans Ratio
RT	10 : 1
−78 °C	2.8 : 1
−196 °C	1.5 : 1

(**396**)   (**397**)   (**398**)   (**399**)   (**400**)

Intramolecular addition of carbenes to pendant alkenes also provides a cyclization path as e.g., the irradiation of the diazoketone **402** in benzene. No photo-Wolff reaction, rearrangements of the type described in the next section, occurs and the two products

(**401**)   (**402**)   (**403**)

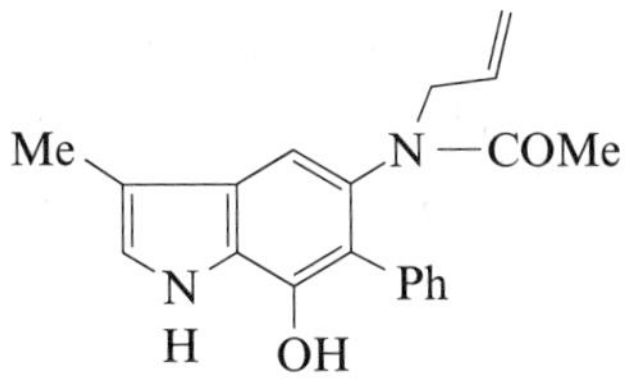

(404)

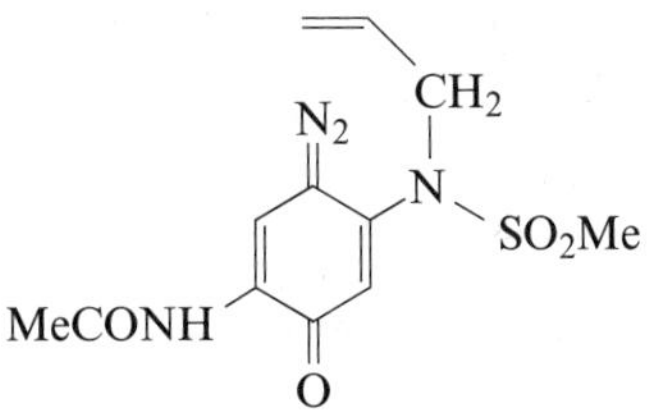

(405)

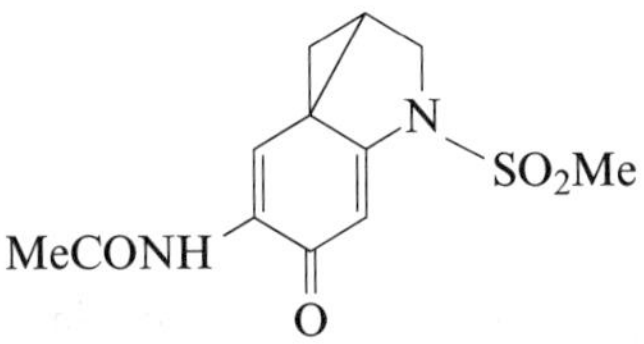

(406)

obtained result from addition of the carbene to the double bond affording **403** or attack on solvent to yield **404**[238]. The quinone diazide **405** also shows this cyclization mode and yields **406**[239].

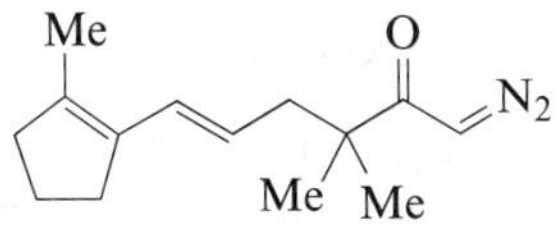

(407)

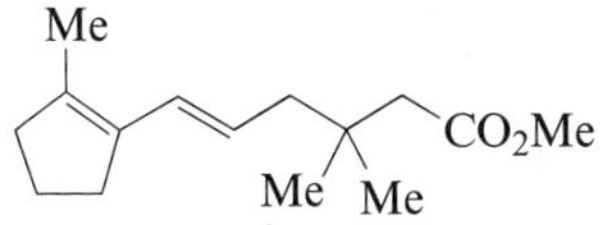

(408)

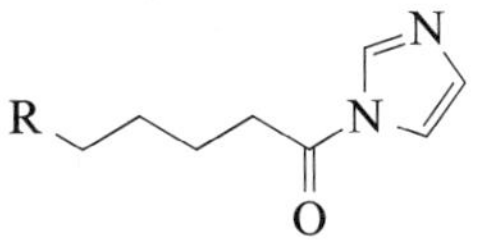

(409)

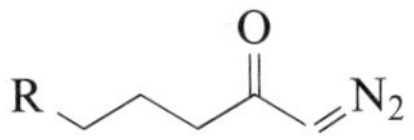

(410)

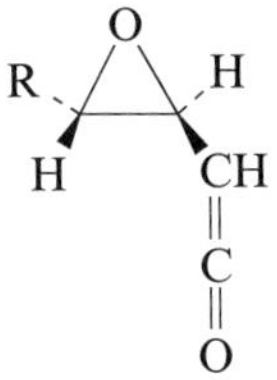

(411)

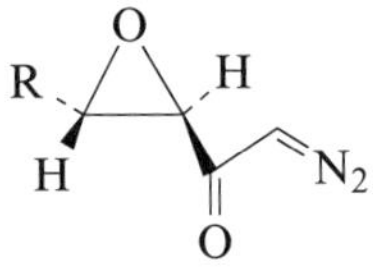

(412)

B. Photo-Wolff reactivity

The conversion of α-diazocarbonyl compounds into the corresponding ketenes is referred to as the photo-Wolff reaction and it has its thermal counterpart in the Arndt–Eistert process of α-diazoketones. The photo-reaction involves the singlet state of the diazo compound. The reaction is useful synthetically and has been studied and used frequently. A new mechanism for the reaction has been proposed, following results from MINDO/3 molecular orbital calculations, which suggest that a one-step process is involved[240]. The reaction can involve the migration of a variety of groups from the carbonyl function to the developing carbene. Alkyl migrations to afford ketenes have been reported for the conversion of **407** into **408**. The ketene in this instance was trapped by methanol[241]. Trapping by an amine is used in the preparation of **409** from the diazoketone **410**, again involving alkyl group migration. The product is prone to further photoisomerization[242]. A ketene **411**, involving the migration of an oxiranyl group, is also obtained on the decomposition of the oxirane derivative **412**. This reaction is followed by trapping with ethanol and this provides a stereospecific path to the ester **413**[243,244]. Simple alkyl migration affords ketene **414** from **415**[245].

(413)   (414)   (415)

Conversion to amides is reported to be an efficient path involving alkyl group migration following irradiation of the diazodiketone **416** and its derivatives in the presence of amines[246]. However, when given a choice both aryl and alkyl groups are reported to undergo migration on irradiation of the diazodiketone **417** in methanol affording the two esters **418** and **419**[247]. Alkyl group migrations are also observed in the conversion of **420** into **421**[248] while the migration of an aryl group accounts for the formation of the ketene responsible for the formation of the ester **422** from **423**[249]. Interestingly, a carbomethoxy group is the migrating species in the formation of the ketene **424** from **425**. Again this ketene is trapped readily by alcohols and can be used as a synthetic path to mixed malonate esters[250]. The production of ketenes in the gas phase has also been examined[251]. α,β-Unsaturated ketenes can be suitable substrates for Diels-Alder additions. This approach has been used in a synthetic strategy aimed at the synthesis of the host defence stimulant *maesanin*. The route involves the irradiation of the diazoketone **426** to afford

(416)   (417)   (418)

(419)   (420)   (421)

(422)   (423)   (424)   (425)

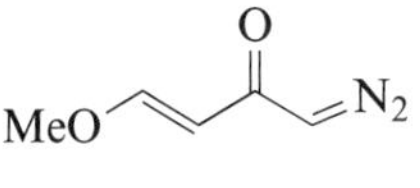

$$\text{MeO}\diagup\diagdown\overset{\overset{\displaystyle O}{\|}}{\diagdown}\diagup\diagdown=\text{N}_2$$

(426)

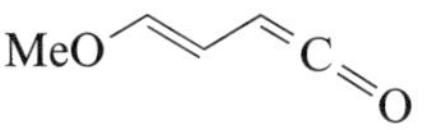

$$\text{MeO}\diagup\diagdown=\diagup\diagdown\text{C}=\text{O}$$

(427)

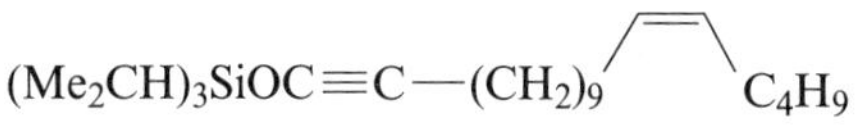

$$(\text{Me}_2\text{CH})_3\text{SiOC}\equiv\text{C}-(\text{CH}_2)_9\diagup\diagdown\text{C}_4\text{H}_9$$

(428)

OMe
OH
(CH$_2$)$_9$ C$_4$H$_9$
OSi(Pr^i)$_3$

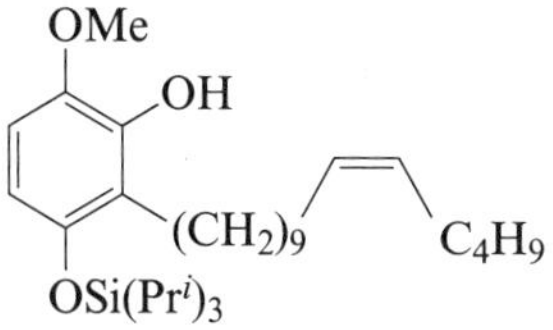

(429)

the ketene **427**, which is trapped by the substituted alkyne **428** ultimately affording the aromatic derivative **429**[252].

Irradiation of diazoketones in cyclic systems has been used frequently as a path to contraction of the ring system. Typically, the cyclobutane system **430** undergoes loss of nitrogen and photo-Wolff rearrangement to yield the ketene **431**[253]. This species can be trapped by (2 + 2)-cycloaddition to 1-dimethylaminopropyne, to yield the spiro compound **432**. Ring contraction from C$_5$ to C$_4$ systems is much more common and many examples abound in the literature. Thus irradiation of **433a** yields the ketene from which the ester **434a** is obtained if the reaction is carried out in methanol[255,256]. Similar ring contractions to cyclobutanes were observed for **433b, c** to **434b, c**[256]. 1,2-Diazetidinones (aza-β-lactams) can also be synthesized in this manner from the pyrazolidine derivatives **435**[256-258]. The ketene produced in this ring contraction can be trapped by alcohols, amines or water. The previous examples all involved the migration of a carbon centre to the carbene. In some cases migration of sulphur takes place and the thietane **436** is formed, while the ketene **437** is obtained by alkyl group migration in the photo-Wolff reaction of **438**. The former product undergoes secondary photolysis, while in methanol the ketene is trapped as the corresponding ester[259]. Ring contraction by migration of a

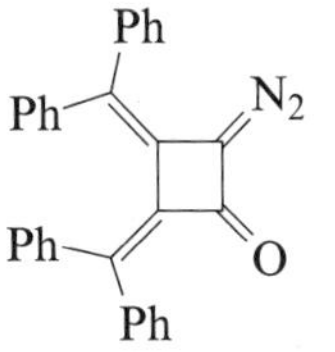

(430)

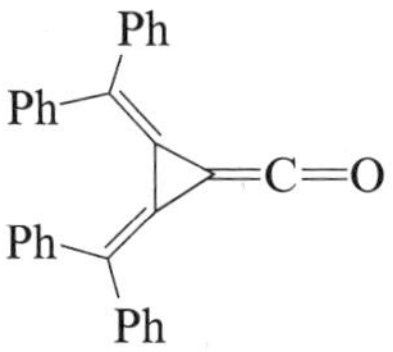

(431)

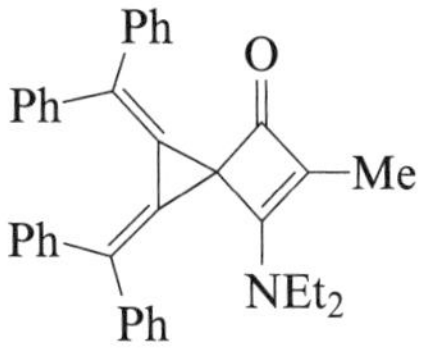

(432)

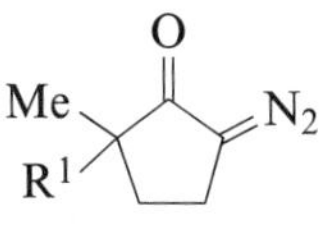

(433) (a) R^1 = Me
(b) R^1 = Ph
(c) R^1 = CH$_2$CH$_2$OMe or CH=CH$_2$

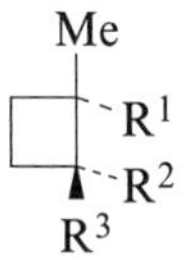

(434) (a) R^2 = CO$_2$Me, R^3 = H
(b) R^2= CO$_2$Me, R^3 = H
(c) R^2 = R^3= H or CO$_2$Me

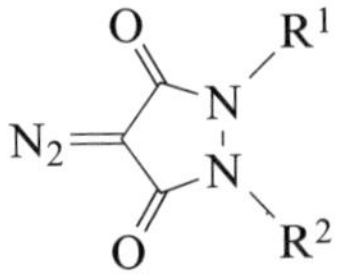

(435) 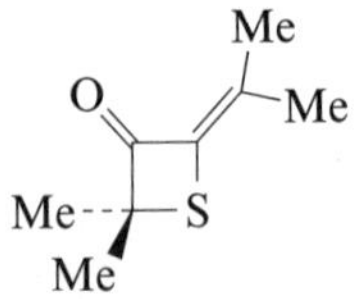 (436) 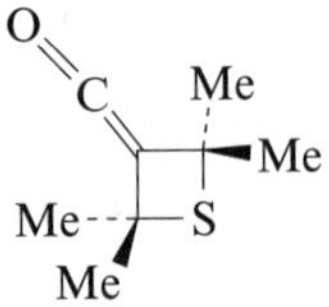 (437) 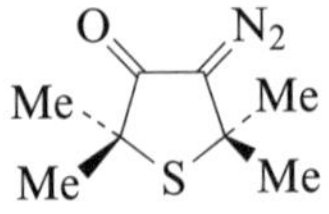 (438)

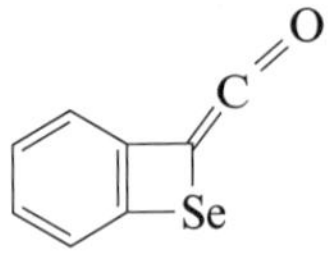

(439) 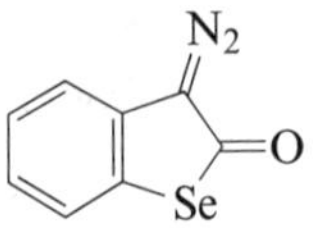 (440) 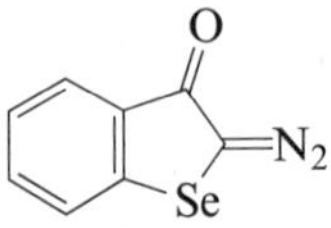 (441) 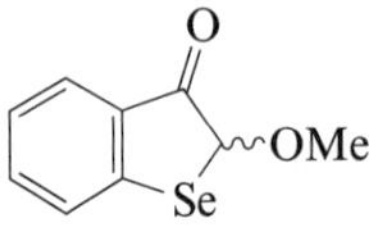 (442)

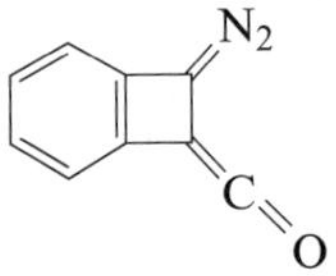

(443) 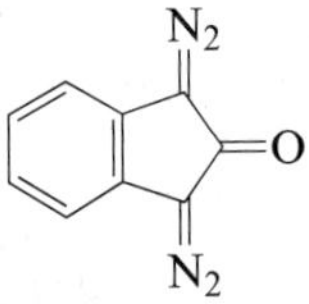 (444) 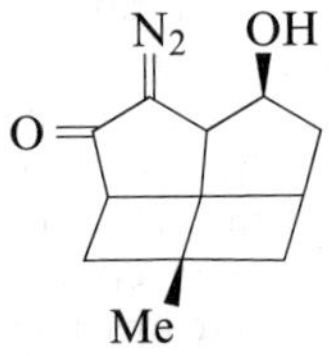 (445) 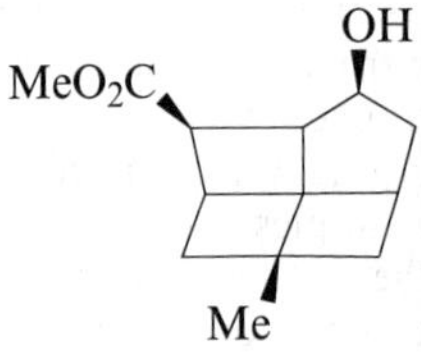 (446)

selenium function is responsible for the formation of the ketene **439** that can be trapped as an ester from the diazoketone **440**. In contrast, the isomeric diazoketone **441** affords a carbene that undergoes insertion into the solvent to yield the ether **442**[260]. Ring contraction to a diazoketene **443** is the first step in the irradiative conversion of the didiazoketone **444** both in solution and in a matrix[261,262]. Ring contraction dominates the reaction even in highly strained environments such as **445**. Irradiation of this in methanol yields the ring-contracted fenestrane ester **446**[263].

Ring contraction from cyclohexane to cyclopentane systems also features in the photochemists armoury of reactions. Thus irradiation of the diazodiketone **447** yields the ketoesters **448** when an optically active ester is used as the ketene trap[263]. The ketene generated by the irradiation of **447** has also been examined in a low-temperature matrix[264]. Several variants on the systems that can undergo ring contraction have been studied. Thus ring contraction occurs with **449** affording **450**[265], while **451** is obtained from **452**[266]. Other pyrimidines such as **453** also undergo ring contraction into a ketene that can be trapped as the ester **454**[267]. A new route to the novel carbapenem ring system

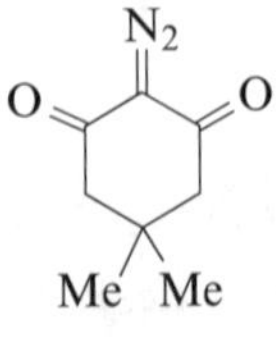

(447) 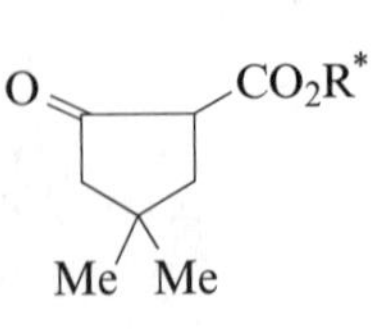 (448) 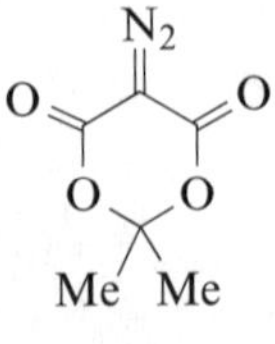 (449) 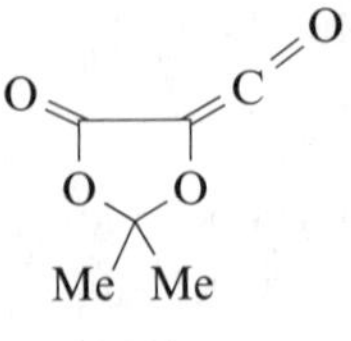 (450)

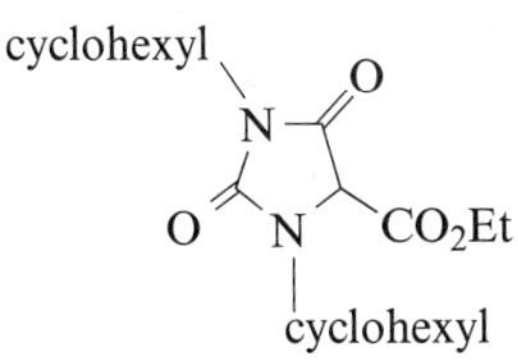

(451) 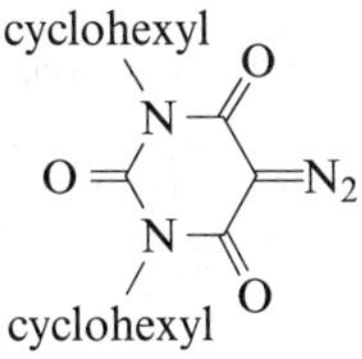 (452) 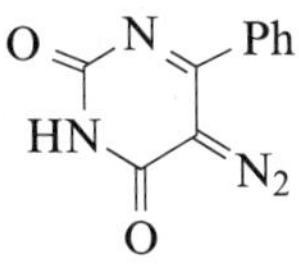 (453) 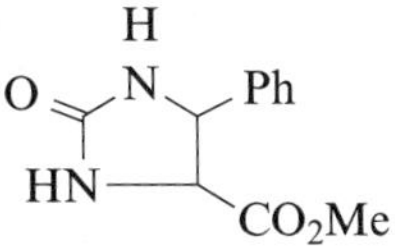 (454)

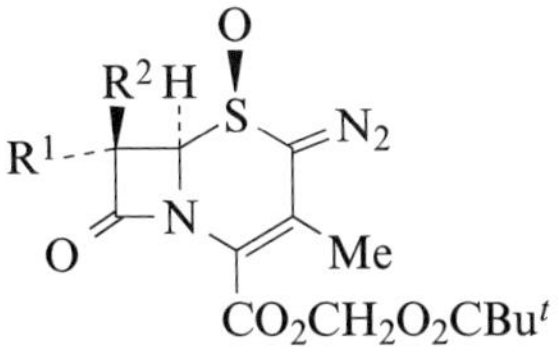

(455) 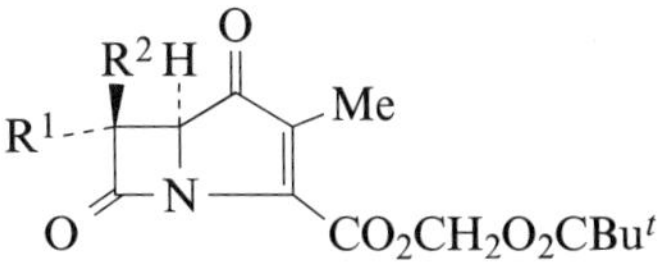 (456)

has been described following the ring contraction of diazosulphoxide derivative **455**. This affords the carbapenem derivative **456**[268].

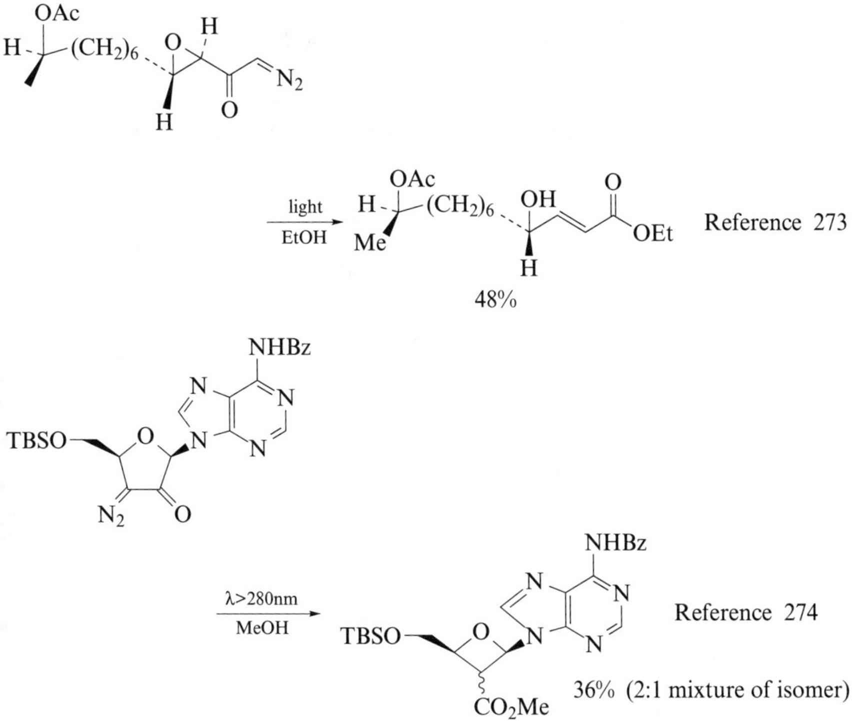

Reference 273

48%

Reference 274

36% (2:1 mixture of isomer)

SCHEME 21

Ring contraction in more complicated ring systems still appears to be a viable reaction path. Thus the irradiation of **457** yields **458**[269] while **459** also undergoes ring contraction[270]. An approach to the paddlane skeleton has also be described using the double contraction to the diketone **460** and brought about by irradiation of **461**[271]. Other examples of ring contraction have been used as key steps in the synthesis of (–)-aspicilin[272], patulolide C[273] and (–)-oxetanocin (Scheme 21)[274].

(457)   (458)   (459)

(460)   (461)

Interest in the photochemical decomposition of diazoquinones has continued. Many of these compounds have become of interest in the expanding market of photo-resist

(462)   (463)   (464)

R = H or p-MeC$_6$H$_4$SO$_3^-$

(465)   (466)   (467)   (468)

production[275,276]. Thus the *o*-quinone derivative **462**[277] and related sulphonate derivatives[278,279] ring-contract to **463** while **464** affords **465** on irradiation[280]. The phenan-
throquinone system **466** is also reactive in this mode[281, 282]. *p*-Quinone systems are also of
interest since they readily undergo decomposition to carbenes. This is typified by the
irradiation of **467** in carbon tetrachloride where the adduct **468** is obtained[283].

C. Hydrogen Abstraction and Insertion Reactions

Hydrogen abstraction reactions in carbene systems can take place intra- or intermolecularly. For example, the irradiation of the diazoester **469** in cyclohexane does not
undergo the usual photo-Wolff reaction discussed above but instead affords the cyclohexane addition product **470**[284]. Both N—H yielding **471** and C—H giving **472** insertion
reactions occur when the aryldiazomethanes are irradiated in diethylamine[285]. Observations of substituent effects have been reported for the carbenes generated by irradiation of
the *para*-substituted and disubstituted diaryl diazomethanes. Here, the reaction with
methanol was studied and ether formation was enhanced by electron-donating groups
and retarded by electron-withdrawing groups[286]. The influence of a 4-nitro group is seen
with the carbene generated from 4-nitrophenyldiazomethane, where only hydrogen ab-

(469)   **(470)** $ArCH_2NEt_2$
                                     (471)

$ArCH_2CH\overset{Me}{\underset{NHEt}{}}$

(472)   **(473)**   **(474)**   **(475)** (a) R = C_6H_{11}
                                                  (b) R = C_6H_5

(476)   **(477)**

straction is observed[285]. Singlet state reactivity has also been described for the carbene from 1-naphthyldiazomethane[287] in its reactions with cyclohexane or toluene and also for fluorenylidene in contrast to the reactions of diphenylcarbene, where triplet state reactivity is encountered[288]. Steric factors do play a part in the selectivity shown by a carbene. Thus fluorenylidene is indiscriminate compared to the highly congested carbene **473**[289]. In some systems the carbenes have been described as electrophilic, such as **474** generated by loss of nitrogen from the corresponding diazo compound. This species reacts by insertion reactions and affords **475a** from cyclohexane and **475b** from reaction with benzene[290]. Insertion is also found with the carbene **476**, which does not undergo ring contraction by the photo-Wolff reaction; instead the product **477** is obtained on reaction with cyclooctane[291].

Intramolecular hydrogen abstraction processes have also been observed, but usually when the carbene is generated in a rigid glass. Thus a 1,4-hydrogen shift affords the singlet *o*-xylylene on irradiation[292] of **478** at 4.6 K which initially affords the triplet carbene. Interestingly, the bis diazo compound **479** affords 1,3-didehydrobenzoquinodimethane on irradiation at 22 K. This transforms to triplet *m*-xylylene on annealing to 77 K in deuteriated ethanol[293]. Intramolecular hydrogen transfer also accounts for the formation of the triplet biradical **480** following generation of the carbene **481** obtained by irradiation of the corresponding diazo compound in a matrix[294]. Intramolecular hydrogen abstraction is also the key to the formation of hydrocarbon products from the carbene **482**. A 1,5-hydrogen transfer yields the *o*-xylylene **483** from which the products are obtained[295].

(478)   **(479)**   **(480)**

                                              R = H, Cl, Br, I, S Ph or PhSO$_2$

(481)   **(482)** R = mesityl   **(483)**

D. Ring Opening, Ring Formation and Fragmentations

In some instances when the carbene centre is adjacent to a ring, opening or fragmentation takes place. This is exemplified in one of the attempts to synthesize tetrahedrane from the carbene produced on irradiation of **484**. Here, cyclization did not occur and instead the formation of two alkynes **485** and **486** results[296,297]. Such ring opening appears to be the result of the generation of a triplet carbene. The singlet path is followed by **487** and the

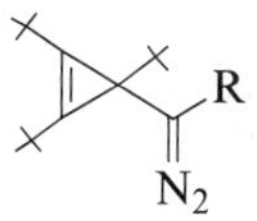

(484) R = Bu^t or SiMe$_3$
(487) R = $+$CO$_2$
(490) R = Bu^t

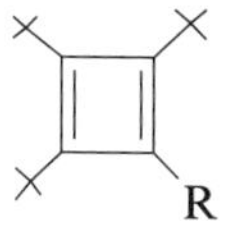

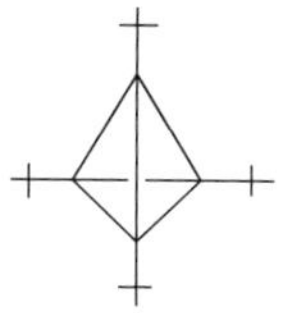

Bu^{t}C≡CBu^t
(485)

Bu^{t}C≡C—R
(486) R = Bu^t or SiMe$_3$

(488) R = CO$_2+$ **(491)**
(489) R = Bu^t

cyclobutadiene **488** is formed[298]. Cyclization to a cyclobutadiene **489** has also been reported for the carbene **490**. Further irradiation of the cyclobutadiene **489** provides a route to the tetrahedrane **491**[299].

Interesting behaviour, dependent on the wavelength used, has been demonstrated for the bis diazo compound **492**. Conventional irradiation at 254 nm leads to consecutive loss of nitrogen and cyclization to yield the mixture of isomers **493**[300,301]. Using pulsed-laser irradiation at 248 nm affords the bis carbene **494**, which dimerizes to yield the cyclo-octatriene **495**.

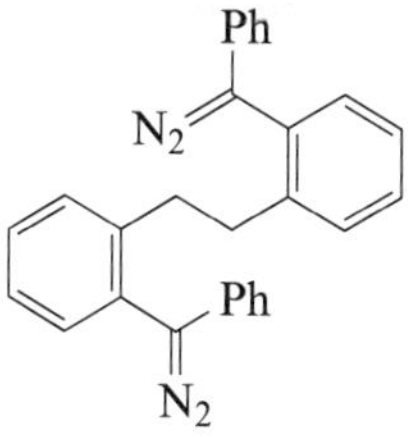

(492) 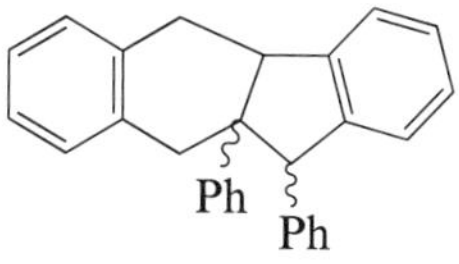   **(493)**

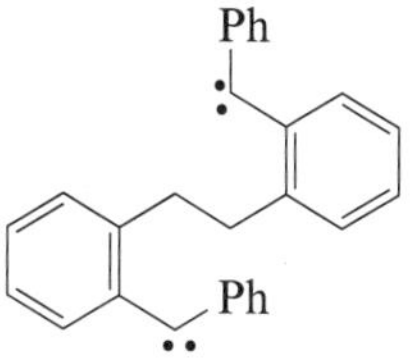

(494) 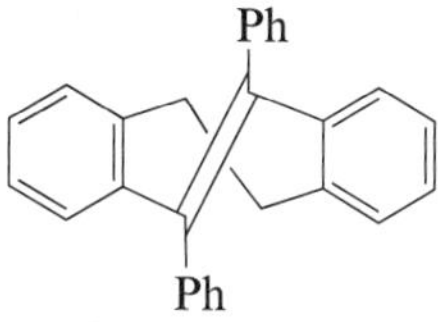   **(495)**

Ring opening is reported to take place on irradiation of **496**, where formation of the carbene is followed by fission of the four-membered ring to yield cycloheptatetraene[302].

Some research has addressed the question of oxacarbene–oxirene equilibria (see later for further examples). One example of this, accompanied by a ring-opening process, has been observed for the ketocarbene developed from **497** in toluene. This reaction affords the ring-opened ynone **498** and the benzofuran **499**[303].

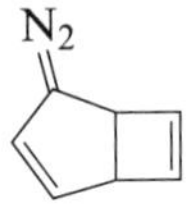

(496)

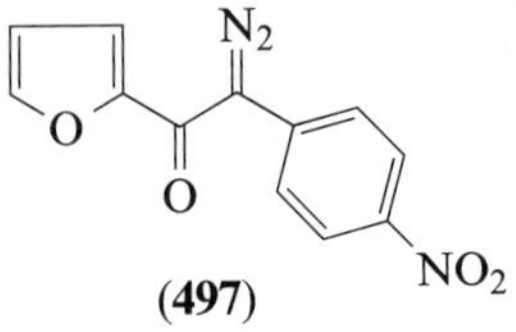

(497)

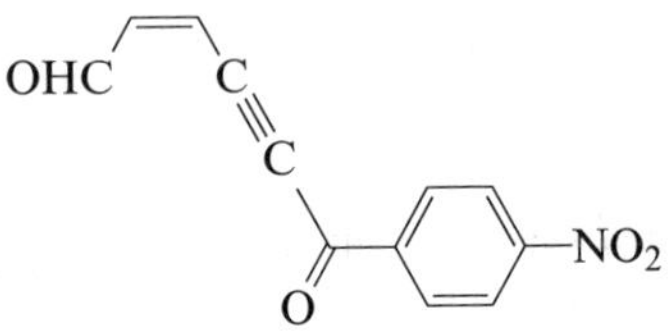

(498)

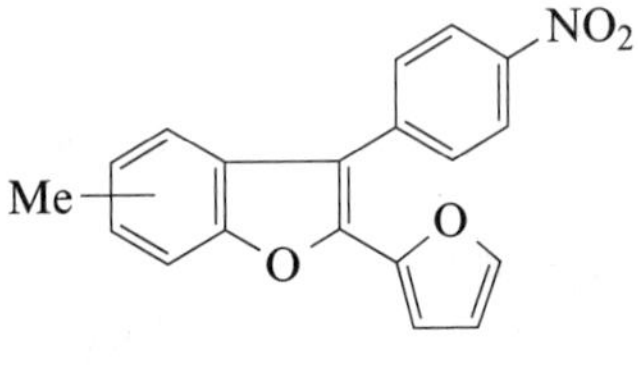

(499)

The incorporation of nitrogen into a ring system containing a carbene can confer instability on the resultant system. This point is exemplified by the formation of the α-diazonitriles **500** from ring opening of the carbene produced by irradiation of **501**[304]. Another example of such behaviour provides a synthetic route to the esters **502** by ring opening of the carbene from **503**[305]. Complete fragmentation to C_4O_2 occurs on irradiation of the bisdiazo compound **504** in an argon matrix[306]. An unusual loss of a carbon atom by an unknown path has been described for the compound **505**, which affords triptycene on irradiation[307].

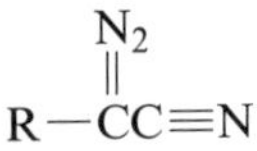

(500) 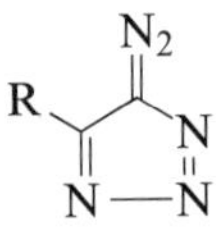   **(501)** 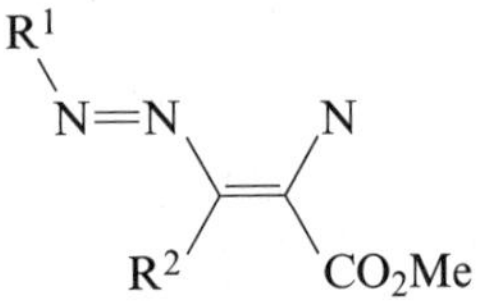   **(502)**

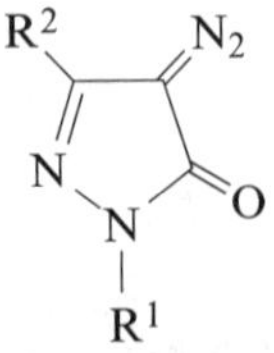

(503) 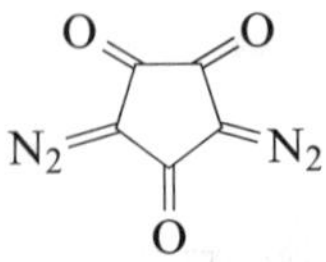   **(504)** 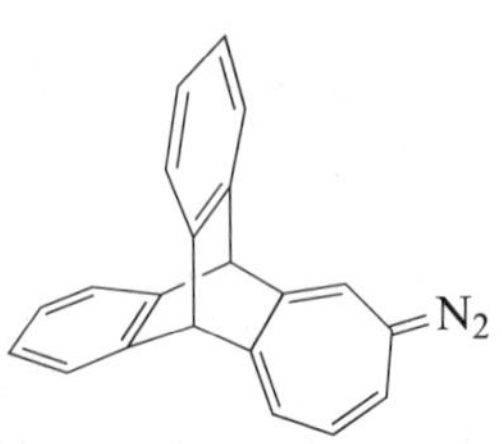   **(505)**

E. Other Carbenes

1. Phosphorus substitution

Most of the reactions encountered with the carbon-substituted carbenes are found with phosphorus substitution. Typically, the irradiation of α-diazobenzylphosphonate in the presence of 2-methylbut-2-ene results in the formation of a cyclopropane[308]. Group migration is also observed as in the irradiation of the diazo compound **506**, which in benzene yields the product **507** formed by the 1,2-migration of an amino group from phosphorus to carbon, while in methanol the adduct **508** is formed[309]. A 1,2-phenyl migration from phosphorus to carbene occurs on irradiation of **509** to yield **510**[310].

The carbenes formed from such systems can also be trapped prior to any rearrangement as in the irradiation of **511** in the presence of *t*-butylisonitrile that affords the adduct **512**[311]. An interesting reaction is described for the carbene **513**, which in the presence of dimethylamine affords **514** with loss of the silyl group[312].

2. Silicon substitution

The photochemical reactions that the silyldiazoalkanes undergo involve silylcarbene chemistry. In line with reactions described for the more conventional carbenes, a variety of processes are common to both types. Thus irradiation brings about 1,2-alkyl shifts, for example, when **515** is photolysed in alcohols to afford a carbene, which transforms into a silene. Four products **516–519** are obtained in good yield[313]. The formation of **519** confirms that the carbene generated from **515** affords a silene via 1,2-methyl migration. An additional migration path, that of 1,2-ethoxy migration, is operative in this system and involves the formation of a trimethylsilyl ethoxy ketene. Other examples of silene formation are seen on irradiation of the polysilylated diazomethanes **520**[314], **521**[315], **522**[316] as well as in the diazo compounds **523**[317] and **524**[318]. In these cases the reactions involve the migration of silyl groups from silicon to carbon. In this last example a 1,2-ethoxy migration occurs in the silene to yield the ketene **525**. Proof of formation of a silene via 1,2-migration has been demonstrated spectroscopically by matrix isolation in 3-

$$\underset{\text{(515)}}{\overset{\overset{\text{N}_2}{\|}}{\text{Me}_3\text{SiCCO}_2\text{Et}}} \qquad \underset{\text{(516)}}{\overset{\text{Me}_3\text{SiCHCO}_2\text{R}}{\underset{\text{OEt}}{|}}} \qquad \underset{\text{(517)}}{\overset{\text{Me}_3\text{SiCHCO}_2\text{R}}{\underset{\text{OR}}{|}}} \qquad \underset{\text{(518)}}{\overset{\text{Me}_2\text{SiCHCO}_2\text{Et}}{\underset{\text{OR}}{|}}}$$

$$\underset{\text{(519)}}{\overset{\text{Me}_2\text{Si}-\text{CHCO}_2\text{Et}}{\underset{\text{RO}\quad\text{Me}}{|\qquad|}}} \qquad \underset{\text{(520)}}{\text{Me}_3\text{Si}-\overset{\overset{\text{Me}}{|}}{\underset{\overset{|}{\text{Me}}}{\text{Si}}}-\overset{}{\underset{\overset{\|}{\text{N}_2}}{\text{C}}}-\overset{}{\underset{\overset{\|}{\text{O}}}{\text{C}}}\text{+}} \qquad \underset{\text{(521)}}{\text{Me}_3\text{Si}-\overset{\overset{\text{Me}}{|}}{\underset{\overset{|}{\text{Me}}}{\text{Si}}}-\overset{\overset{\text{N}_2}{\|}}{\text{C}}-\overset{\overset{\text{O}}{\|}}{\text{C}}\text{R}}$$

R = Me, Pr^i, Ph, p-Tolyl or adamantyl

$$\underset{\text{(522)}}{\text{Me}_3\text{Si}-\overset{\overset{\text{Me}}{|}}{\underset{\overset{|}{\text{Me}}}{\text{Si}}}-\overset{}{\underset{\overset{\|}{\text{N}_2}}{\text{C}}}\diagdown^{\text{CO}_2\text{Et}}} \qquad \underset{\text{(523)}}{\text{Me}_3\text{Si}-\overset{\overset{\text{Me}}{|}}{\underset{\overset{|}{\text{Me}}}{\text{Si}}}\diagdown^{\text{H}}_{\text{N}_2}} \qquad \underset{\text{(524)}}{\text{Me}_3\text{SiSiMe}_2-\overset{}{\underset{\overset{\|}{\text{N}_2}}{\text{C}}}-\text{CO}_2\text{Et}}$$

$$\underset{\text{(525)}}{\overset{\text{Me}_3\text{Si}}{\underset{\underset{\text{OEt}}{\text{Me}_2\text{Si}}}{}}\diagdown\diagup^{\text{C}=\text{C}=\text{O}}}$$

(525)   (526)   (527)   (528)

methylpentane[319]. An interesting example of a silene, again obtained by the migration o a silyl group, is the formation of the silabenzene **526**. This is obtained by the formation o the carbene **527** within which migration occurs. The silabenzene is stable at – 100 °C bu will add methanol, to yield the final product **528**[320, 321]. Other derivatives of silaarene have also been recorded in the ring-expansion reactions of the silacyclopentadiene deriva tives **529**[322,323]. Trapping experiments again confirm the existence of the silenes. Silene are also obtained from the irradiation of the diazo compounds **530**. In these example where there is conjugation with an acyl group, silenes are formed again by 1,2-sily migrations as is the case in the irradiation of diazo derivative **530** to yield **531**[324,325].

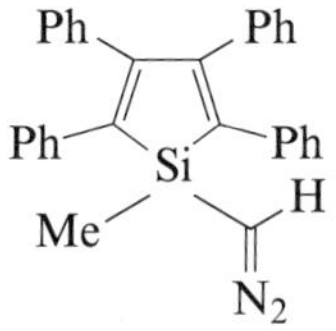

(529)

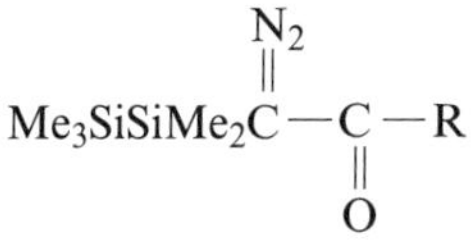

(530) R = *t*-Bu, *i*-Pr, Me
or 1-adamantyl

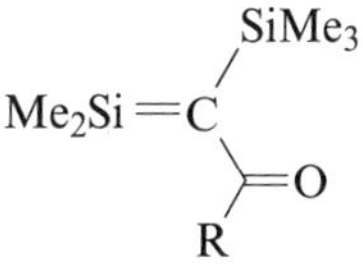

(531)

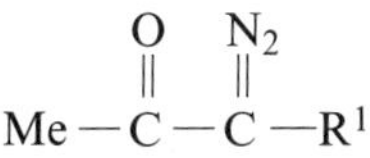

(532) R^1 = SiEt$_3$ or SiPr$^i_3$

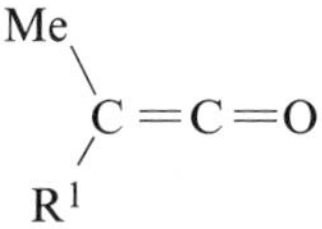

(533)

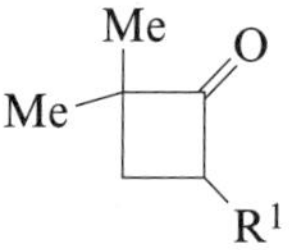

(534)

Occasionally, different reaction paths have been described following the generation of a sila carbene. This is seen in the competing paths in the reaction of the carbene formed from diazo compound **532**. Here, either a ketene **533** is formed or else insertion affords the cyclobutanones **534**[326]. A surprising result is reported in the irradiation of the bis diazo compound **535** when a low yield of the remarkable sila bicyclo compound **536** is obtained[327].

More complex systems such as **537** have been examined recently. For instance, bis silene **538** is obtained which, in the absence of trapping agents, leads to the bicyclic compound **539**[328].

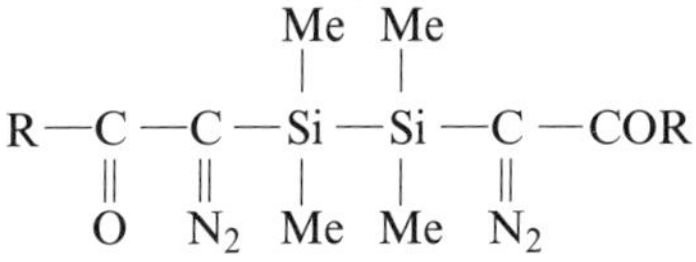

(535)

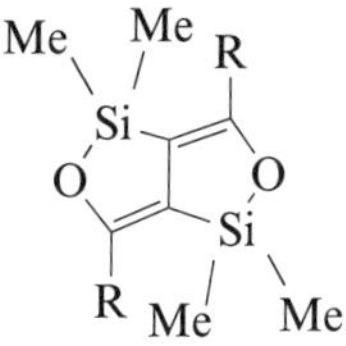

(536)

(537)

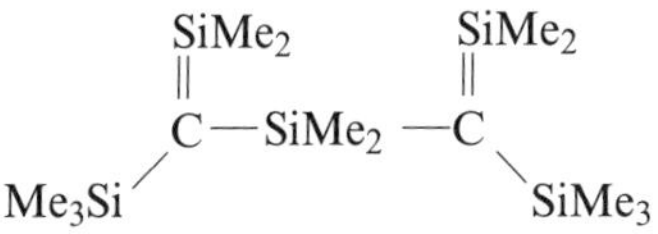

(538)

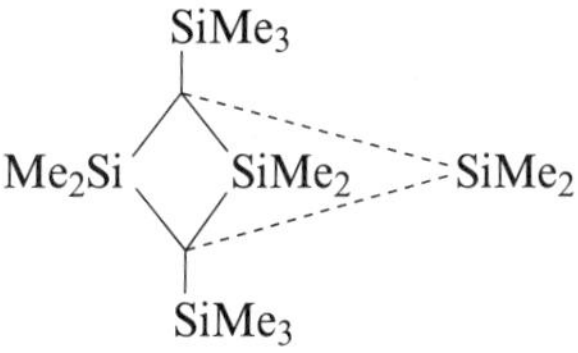

(539)

F. Spectroscopic and Related Studies

A considerable amount of information has been accumulated on the spectroscopic details of carbenes. Often, experiments are carried out in low-temperature matrices, as in the direct observation of di-t-butylcarbene produced from the corresponding diazo compound by irradiation at 254 nm at 14 K[329]. However, solution-phase studies are also important and a study of diadamantyldiazomethane has shown that the derived carbene behaves as a triplet state in solution. This behaviour is reported to be most unusual for a dialkyl carbene[330].

Several reports have discussed the ring-expansion reactions of phenylcarbene when it is produced in a low-temperature matrix. Under these conditions, ring expansion to cyclohepta-1,2,4,6-tetraene results[331–333]. The formation of this species occurs at temperatures as low as 10 K. The tetraene is extremely reactive and undergoes thermal (2 + 2)-cycloaddition followed by ring opening of the cyclobutane to yield the final product **540**. Surprisingly, the naphthyl analogue **541** does not undergo ring expansion to a cycloheptatetraene and instead affords the tricyclic hydrocarbon **542**[334].

(540)   **(541)**   **(542)**

The low-temperature irradiation of the series of quinoline diazomethanes **543** has shown that these are good precursors of triplet carbenes. These carbenes have been shown to be present as two rotamers[335]. The cyclopentadienylcarbene derivative **544**[336] and fluorenylidene[337] have also been shown to be triplets. It is clear that structural factors play an important role in determining the reactivity of carbenes. Typical of this is the behaviour of the dianthrylcarbene **545**, which reacts differently from other diaryl carbenes[338].

R[1]	R[2]	R[3]
CHN$_2$	H	H
H	CHN$_2$	H
H	H	CHN$_2$

(543)   **(544)**   **(545)**

A review has been published in Japanese dealing with low-temperature matrix isolation techniques applied to carbene systems[339]. Low-temperature studies have examined the trapping of carbene with a variety of trapping agents. Thus with oxygen benzophenone C oxide has been observed spectroscopically when diphenyldiazomethane is irradiated at 515 nm[340]. Interestingly, phenyldiazomethane does not react similarly[341]. However, other carbenes do react in a similar fashion to diphenylcarbene and the ylide **546** is observed on irradiation of diazocyclopentadiene[342]. This study has shown that the ylide is accompanied by the isomeric dioxirane **547**. The related ylides **548** and **549** have also been observed by trapping the corresponding carbenes with oxygen[343]. Furthermore, the first report of a carbonyl oxide from a π-electron-donating ring system has been made and deals with the formation of **550** on irradiation of the corresponding diazo compound in an oxygen-doped matrix[344]. The carbene generated from 1,4-benzoquinone diazide can also be trapped by oxygen as the corresponding carbonyl ylide[345]. Experiments have also been described of the irradiation of alkyl diazo compounds in the presence of oxygen[346].

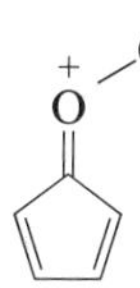

(546)    (547) 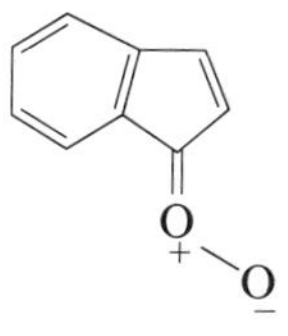   (548) 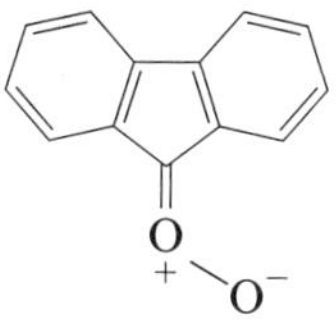   (549) 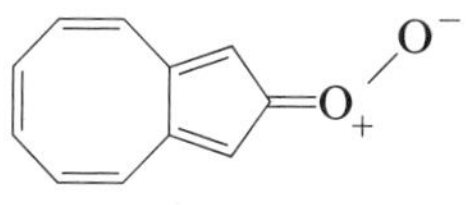   (550)

Other trapping agents have also been added to the matrix in which carbenes have been produced by irradiation. Thus the formation of diphenylcarbene in frozen alcoholic matrices has shown that the triplet carbene produced is trapped as the corresponding ether[347]. A similar observation has been carried out with S-(+)-butan-2-ol where an enantiomerically pure ether is obtained again from the triplet state carbene, but since this is produced in the frozen alcohol no rotational movement is possible[348]. Interestingly, when reactions of diphenyl carbene are carried out in fluid solutions of alcohols, the singlet state of the carbene is proposed as the forerunner to the ethers[349]. Carbenes can also be trapped intramolecularly by the presence of a neighbouring alcohol functional group. This is seen with the irradiation of the diazo compound **551**, which affords the oxirane **552** and the hydroxyketone **553** when irradiation is carried out in cyclohexane[350].

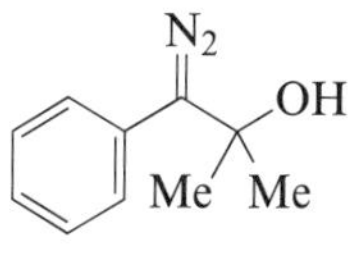

(551) 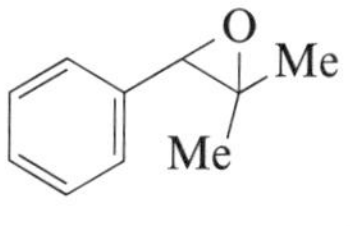   (552) 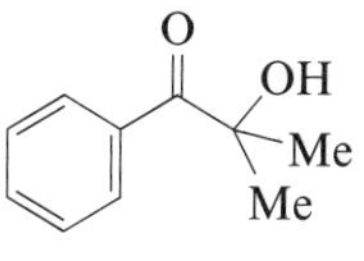   (553)

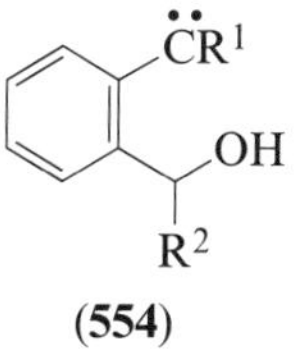

(a) R^1 = H, R^2 = H
(b) R^1 = Ph, R^2 = H
(c) R^1 = H, R^2 = Me

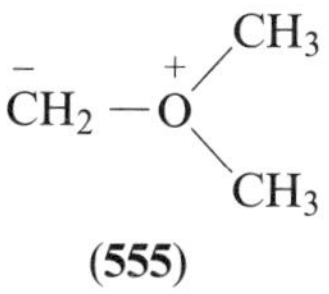

(554)

(555)

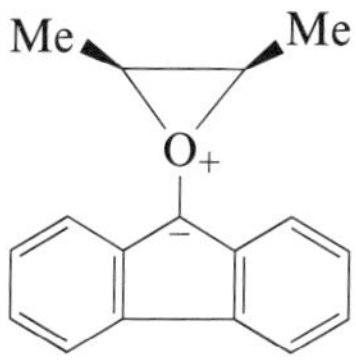

(556)

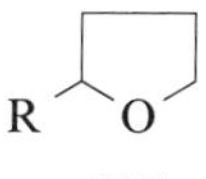

R = H, Ph or CO$_2$Et

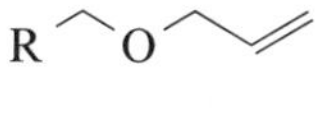

(557)   (558)

Others have also observed both an intra- and inter-molecular reactivity with the carbenes 554[351,352]. Ethers are also reported to act as traps for carbenes and singlet methylene is reported to form the ylide 555 on reaction with dimethyl ether[353]. Indeed, the selectivity of reaction shown by some carbenes can be influenced by the presence of 1,4-dioxane, since this compound is capable of stabilizing singlet carbenes by complexation[354]. Analogous to such reactivity is the report that the reaction of singlet fluorenylidene with oxiranes provides a stereospecific path for the removal of oxygen via the complex 556[355]. Other cyclic ethers also react with carbenes, as can be seen from the Stevens-type reaction of carbenes derived from diazomethane, phenyldiazomethane and ethyl diazoacetate with oxetane in methanol to yield either the ring-expansion product 557 or else the fragmentation compound 558[356]. Stereospecificity of reaction can also be observed in such systems when optically active oxetanes are used as the substrate[357].

Addition of singlet carbenes to other trapping agents, such as aldehydes and ketones, also has some synthetic value. Thus singlet methylene can be trapped as an ylide on its production in the presence of formaldehyde[358]. The singlet carbene, from the diazodiester 559 in the presence of acetaldehyde, also affords an ylide 560 prior to a second addition yielding the final product 561[359]. Room-temperature formation of fluorenylidene in acetone gives rise to ylide 562 on the path which leads to the oxirane 563[360]. Di-t-butylthioketone also affords an ylide under the same conditions. In this case the sulphur ylide was shown to be considerably longer lived than the oxygen counterpart[361].

Carbenes can also be trapped by thioamides as in the reactions of the carbenes obtained by irradiation of 564. The possibility is that a sulphur ylide is formed in this system and the subsequent intramolecular reaction with the amide nitrogen affords the final product 565[363]. Even in simple systems an ylide can be formed, as in the reaction of singlet methylene with acetone[363].

$R^1 = Ph$, p-Anisyl p-ClC$_6$H$_4$ or p-O$_2$NC$_6$H$_4$

Singlet methylene has also been shown to undergo reaction with acetonitrile to yield the nitrile ylide **566**[364,365]. Both singlet and triplet diphenylcarbene can be trapped with diphenyldiazomethane to afford an azine[366]. A similar outcome is observed with the fluoroalkyl diazo compound **567**[367].

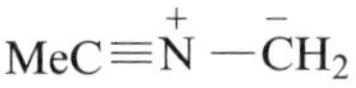

$$Mec \equiv \overset{+}{N} - \overset{-}{CH_2}$$

(566)

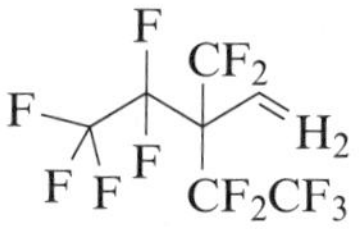

(567)

Intramolecular trapping has been reported for the ester **568**. Here, irradiation in a matrix at 10 K yields the photochemically reactive ylide **569**. This can be transformed to the isomeric oxirane on irradiation[368,369].

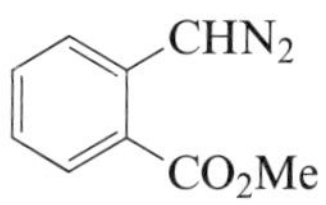

(568)

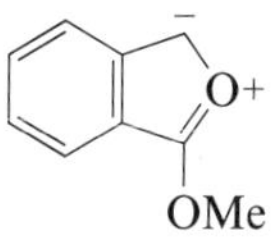

(569)

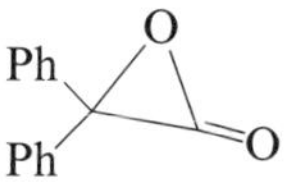

(570)

Various reports have shown that diphenylcarbene and arylcarbenes can be trapped by a variety of materials such as silicon, tin or germanium hydrides[370]. Furthermore, both CO[371] and CO_2[372] can be used to trap diphenylcarbene. The latter example is carried out at 35 K in a carbon dioxide-doped argon or xenon matrix. This treatment permits the formation of matrix-trapped oxiranone **570**.

There is no doubt that matrix isolation experiments have provided many examples of great interest in this and other areas where reactive intermediates are produced. α-Diazoketones were discussed in an earlier section since the singlet state reaction undergoes the well-known photo-Wolff process. Other studies on these compounds in matrices at 77 K have shown that the diazoketones **571** and **572** undergo equilibration via an oxirene intermediate. Again this is an example of intramolecular trapping[373,374]. Others have also observed the formation of oxirenes and the existence of a keto-carbene/oxirene equilibrium[375,376] in a variety of systems, such as the diazoquinone **573**[377].

(571)

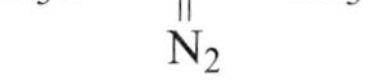

(572)

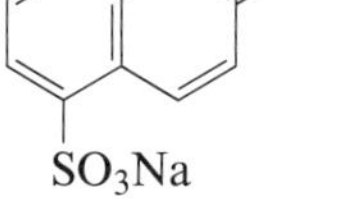

(573)

V. IRRADIATION OF AROMATIC DIAZONIUM SALTS

Irradiation of aromatic diazonium salts results in an efficient elimination of nitrogen and the formation of substitution products. To some extent the mechanism of the reaction is dependent upon the conditions under which the reaction is being carried out and whether sensitization is being employed. Typically, the irradiation of the aryl diazonium salts **574** in trifluoromethyl sulphonic acid affords an efficient path to the aryl triflates **575**. This behaviour is typical of the loss of nitrogen with the formation of an aryl carbocation[378]. However, in cases where sensitization (using benzophenone, xanthene, thioxanthene or benzil) is employed, e.g. with **576**, the loss of nitrogen affords a radical which yields the corresponding arene. CIDNP studies have demonstrated that triplet precursors are involved and that a SET process, rather than energy transfer, is taking place[379]. Further evidence for SET processes in loss of nitrogen from aryl diazonium salts comes from the studies on **576b** using xanthene dyes as the sensitizers[380]. The results from the irradiation of **557a** suggest that in these cases an intramolecular charge or electron transfer may be involved. The quantum yields for the loss of nitrogen are in the range 0.125–0.005 with electron-donating groups enhancing the efficiency[381]. A similar study was carried out with the stilbene derivative **577b**[382].

(574) R = MeO or CF$_3$	**(575)** R = MeO; 82% R = CF$_3$; 81%

(576) (a) R = H, Cl, Me or MeO
     (b) R = OMe or O-dodecyl

(577) (a) Ar = R—⟨ ⟩
     R = Cl, Me, Me$_3$N$^+$, H, MeO,
     Me$_2$N or Et$_2$N

     (b) Ar = MeS⟨S⟩

Heterocyclic diazonium salts have also been studied in some detail. Thus the irradiation of **578a** in aqueous HBF$_4$ provides a path to carbocations and the corresponding hydroxy compound **578b** is obtained. When the nitrogen substituent is phenyl as in **578c**, irradiation in dioxane, toluene or THF affords the radical product **578d**[383]. Irradiation of the heterocyclic compound **579a** provides an efficient path to the corresponding fluoro

(578) (a) R^1 = N$_2^+$, R^2 = H, Me or CH$_2$ CH$_2$OH
     (b) R^1 = HO, R^2 = H, Me or CH$_2$CH$_2$OH
     (c) R^1 = N$_2^+$, R^2 = Ph
     (d) R^1 = H, R^2 = Ph

derivatives **579b**[384]. The conversion of an amine group to a hydroxy function has been demonstrated in the conversion of the purine derivative **580a** into the corresponding hydroxy compound **580b**[383].

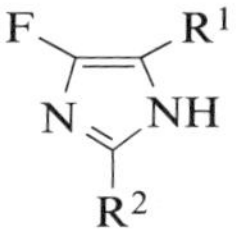

(579) (a) $R^1 = CO_2Et$ or $CONH_2$
$R^2 = N_2^+ BF_4^-$
(b) $R^1 = CO_2Et$ (53%) or $CONH_2$ (28%)
$R^2 = F$

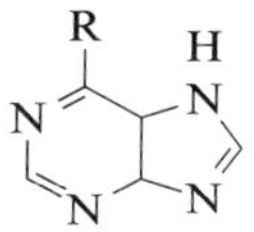

(580) (a) $R = N_2^+$
(b) $R = OH$

The ability of producing a reactive species, either a radical or a cation, from the photochemical denitrogenation of diazonium salts has been used in the photo-Pschorr synthesis as an approach to polycyclic compounds. Thus the irradiation of the diazonium salt **581** affords the azafluorenones **582**[385].

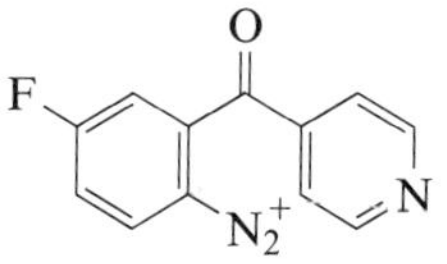

(581) 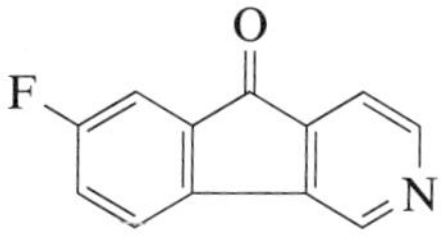   **(582)**

VI. IRRADIATION OF AZIDES

A. Alkyl Azides

The typical reaction of alkyl azides is loss of nitrogen and reaction of the nitrene so generated. Typically, the nitrenes obtained either undergo 1,2-hydrogen or 1,2-alkyl migration with the formation of an imine or undergo C—H insertion or insertion into neighbouring double bonds to yield cyclic compounds.

Several examples of the formation of imines by 1,2-hydrogen migration have been reported. A typical example of this is observed in the irradiation of azide **583a** isolated in an argon matrix at 11 K where a mixture of the corresponding isomeric imines is obtained[386]. The reaction also functions on a preparative scale as observed by the efficient

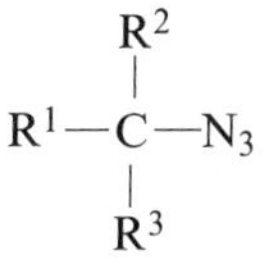

(583) (a) $R^1 = CN$, $R^2 = R^3 = H$
(b) $R^1 = $ alkyl, $R^2 - R^3 = NH$
(c) $R^1 = $ alkyl, $R^2 = H$, $R^3 = OMe$

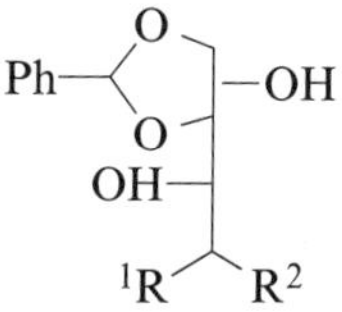

(584) (a) $R^1 = N_3$, $R^2 = H$
(b) $R^1 - R^2 = NH$ quantitative

formation of the imines **583b** from irradiation of the azido alkanes **583c**[387]. Hydrogen atom migration is also reported to take place in the conversion of the carbohydrate derivative **584a** into the corresponding imine **584b**[388]. This reaction path was used in a synthesis of L-lyxose from L-arabinitol.

Alkyl groups also undergo migration as in the conversion of **585** into **586** by methyl migration[389] or conversion of **587** into **588** by the exclusive 1,2-migration of an *iso*-butyl group[390]. In a related example, the transformation of **589** into **590**, a methoxy carbonyl group undergoes the migration[390]. Alkyl or acyl migration can also occur on formation of the nitrene by irradiation of 2-alkyl-2-azido-3-oxobutanoic esters or amides[391]. In some cases aryl groups are the migrating species affording the imines **591** by migration of the group R in **592** or yielding **593** by migration of one of the xanthenyl aryl groups[392].

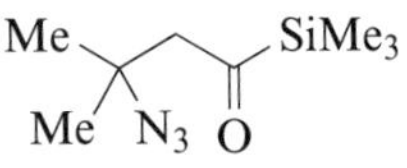

(585) 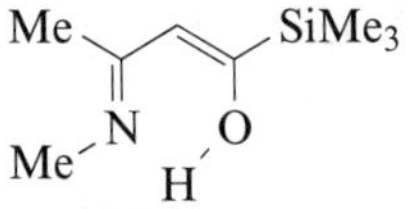   (586) 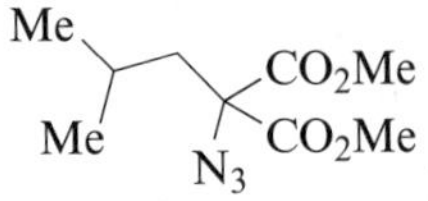   (587) 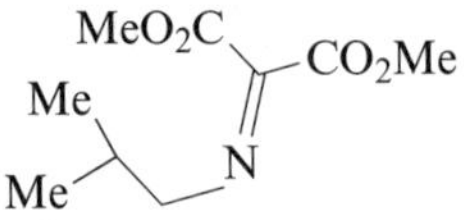   (588)

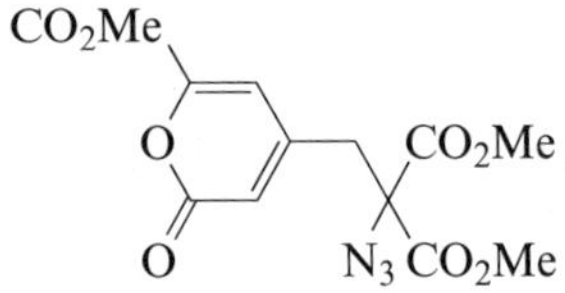

(589)

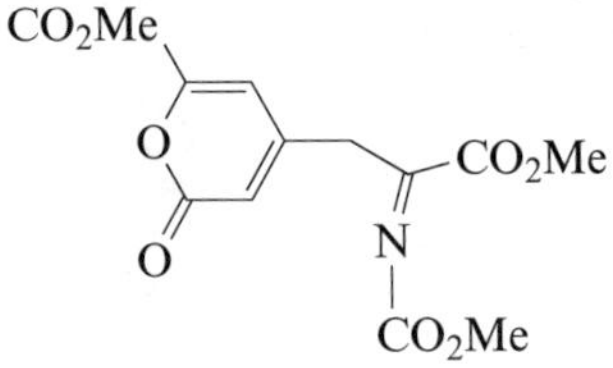

(590) (24%) 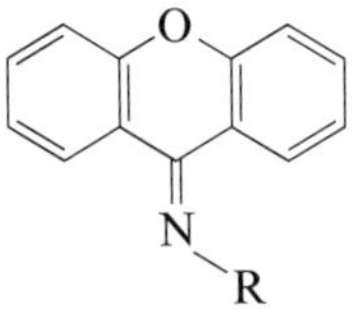   (591)

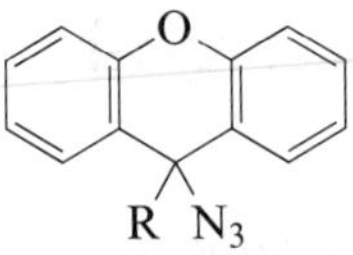

(592)   (593)

R = Ph, *p*-MeOC$_6$H$_4$, *p*-ClC$_6$H$_4$
or *p*-FC$_6$H$_4$

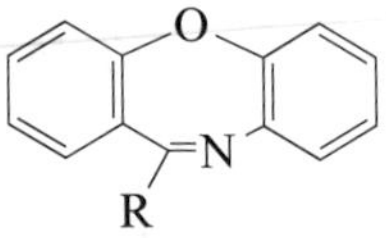

The migration of ring bonds can also be categorized as migration of an alkyl group. This is illustrated in the ring-expansion reactions of the nitrene generated by irradiation of the azide **594**. Again a 1,2-alkyl migration would afford an imine such as **595**. However, in this instance the imine could not be detected even at 77 K in a hydrocarbon medium. When the reaction is run in protic solvents such as methanol, the adduct **596** was obtained as proof of the existence of the imine[393]. Interestingly, it is reported that bridgehead imines have been identified from the irradiation of the azide **597**[394]. Interest in the photochemical decomposition of the 3-azidonoradamantane **598** has shown that 2-azaadamantan-1-ene (**599**) and 4-azaprotoadamantan-3-ene (**600**) can be formed[395]. These arise by the migration of the two bonds adjacent to the nitrene generated by irradiation of **598**. The former imine **599** is reported to be one of the more highly twisted bridgehead imines that has been isolated by irradiation of an azide in a matrix[396]. The imines formed by the ring-expansion processes can be trapped by reaction with methanol[395], ethanol or butyl amine[397]. Bridgehead imines **601** which undergo dimerization can also be generated from the azidotribenzobarralenes **602**[398,399]. Again, these imines can be trapped success-fully by alcohols.

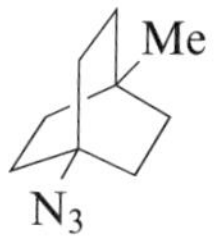

(594)

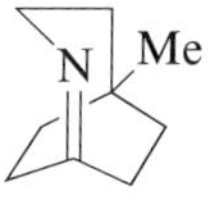

(595)

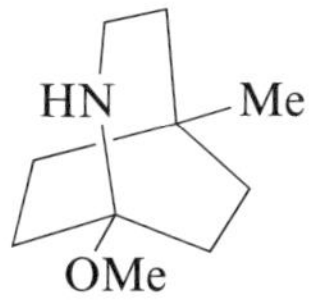

(596)

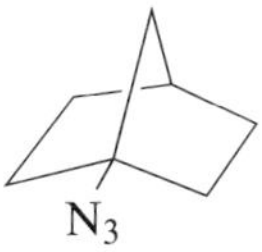

(597)

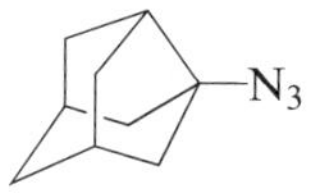

(598)

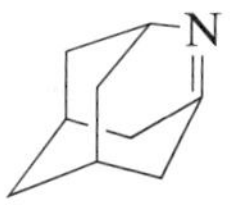

(599)
(38% as MeOH adduct)

(600)
(43% as MeOH adduct)

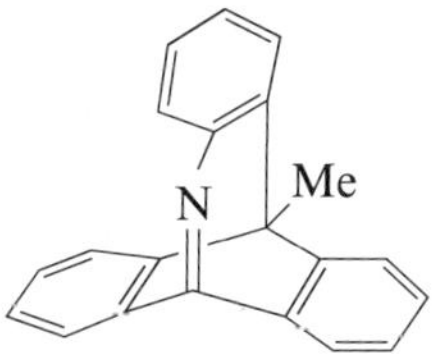

(601)

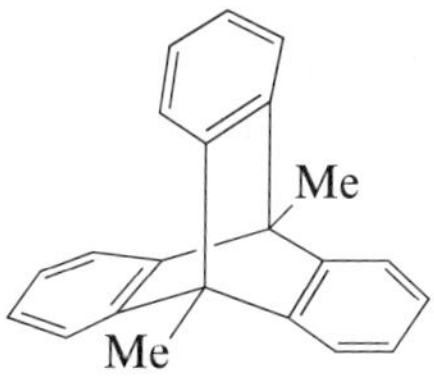

(602)

Some examples of the synthesis of triaziridine systems have been reported following the irradiation of azides with a proximate N—N double bond. Thus the irradiation of **603** affords the triaziridine **604**[400] or the unusually stable derivative **605** from irradiation of **606**[401,402]. Addition of a nitrene to a double or triple bond is not an uncommon reaction path to aziridine derivatives. Such behaviour has been reported even for the gas-phase decomposition of HN₃ in the presence of ethyne, when addition to yield 1*H*-azirine occurs in preference to insertion reactions[403].

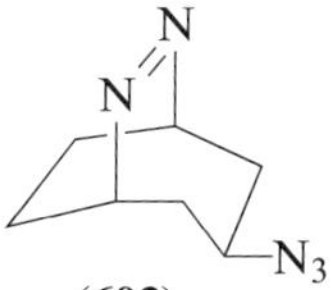

(603)

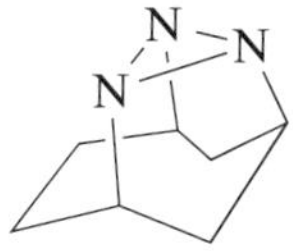

(604)

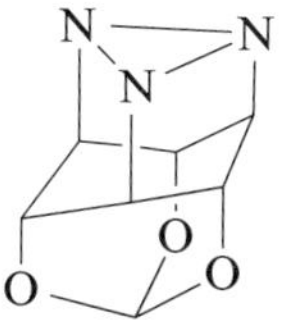

(605) 65%

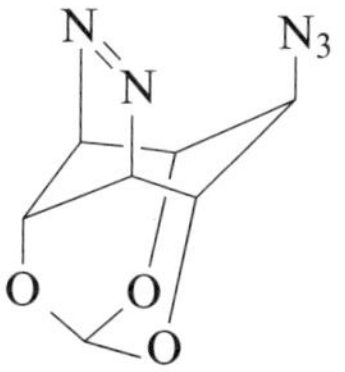

(606)

When the nitrene is generated in strained systems, ring fission can be the result. This reaction mode is seen on irradiation of the azide **607**. The resultant nitrene undergoes fission to yield the tricyclic product **608** even when the reaction is carried out at 12 K[404]. An interesting fragmentation and loss of a side chain is observed on irradiation of the bisazide **609** in the presence of oxygen[405]. Whether carbenes are involved in this process, as they are in the decomposition of 1,1-diazido carbohydrate derivatives[406], remains to be proven.

(607)   (608)   (609)

Fluoro groups also undergo 1,2-migration with the formation of fluoroimines on irradiation of fluoroalkyl azides[407].

B. Acyl Azides

Photolysis of acyl azides also results in the formation of a nitrene, or else rearrangement by a photo-Curtius route yields an isocyanate. The nitrenes produced by the irradiation can be either in the singlet or in the triplet state. In the singlet state the acyl nitrenes undergo a stereoselective addition to double bonds to produce aziridines. Alternatively, the singlet can undergo insertion reactions into C—H bonds. Interestingly, this latter path is the only reaction which the triplet-state nitrene undergoes and it can be used as a synthetic path to amides. In some cases the mode of excitation can also be important, as seen with the behaviour of 4-acetylbenzoyl azide which, on $\pi\pi^*$ excitation, affords both types of reaction, namely the production of an acylnitrene and an isocyanate by the Curtius rearrangement, while $n\pi^*$ excitation yields only the acylnitrene by decomposition of the triplet acyl azide[408]. The type of substituent could be important in determining the outcome of the reaction, since both direct irradiation and sensitized irradiation of β-naphthoyl azide give products of singlet acyl nitrene behaviour[409].

Curtius-type behaviour is followed on irradiation of the azide **610** at 10 K when methoxyisocyanate is obtained by migration of the methoxy group from carbon to nitrogen[410]. Carbamoyl azide is also photolabile under similar conditions and provides a route to aminonitrene[411]. The substituted amide derivative **611** also undergoes conversion to the corresponding isocyanate and, if this is brought about in the presence of carbodiimides, cycloaddition results in the formation of triazoles[412].

(610)   (611)

Considerable synthetic use has been made of ethyl azidoformate. Typical reactions of this are the additions to double bonds as a path to aziridines. Thus addition of ethoxycarbonyl nitrene to 1-ethoxy-2-methyl-1-trimethylsilyloxypropene affords the aziridine **612** that can be ring opened to yield (75%) the diester **613**[413]. Addition of the same nitrene to siloxydienes has also been reported[414]. One of the common reactions of ethoxycarbonyl nitrene is addition to arenes as a route to azepines. An example of this process is seen in the synthesis of the hexafluoroazepine **614** by addition of the nitrene to hexafluorobenzene[415]. In some instances acyl nitrenes do not yield the aziridine but instead undergo 1,3-addition to an alkene to yield oxazoline derivatives. This reaction type is reported for the reaction of arylcarbonyl nitrenes to 1-methoxypropene[416].

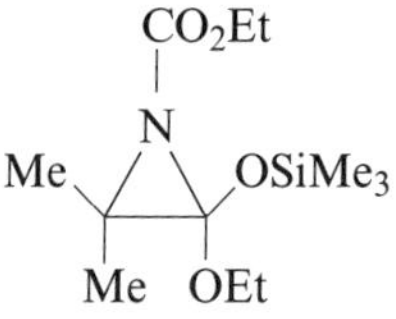

(612) 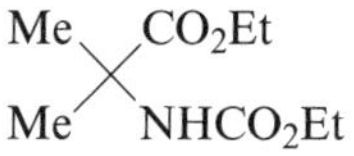   (613) 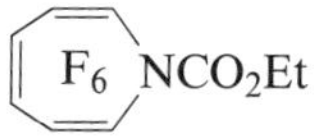   (614)

Insertion into C—H bonds is also a reaction type common to nitrene chemistry. This is observed in the formation of the cyclopropylamine **615** from reaction with 1-ethoxy-1-trimethylsilyloxycyclopropane[417]. Details have been reported of the use that can be made of the addition of the same nitrene to the chiral silyloxyalkene **616**. This leads to an asymmetric synthesis of the amine **617**[418]. Insertion of a nitrene into C—H bonds is also reported for the photodecomposition of p-trimethylsilylbenzoyl azide in hydrocarbons[415]. Reports have been made concerning the non-specificity of this particular nitrene. Apprently, some control can be exercised on its reactivity by using a micellar system. Under conditions of this type, using N, N-dimethyloctylammonium propionate as the micellar material, only the formamidine **618** is obtained[420,421].

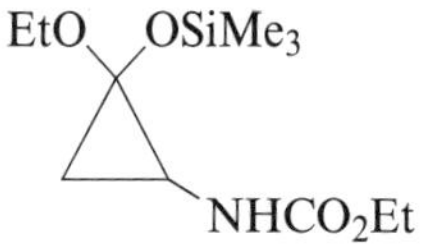

(615) 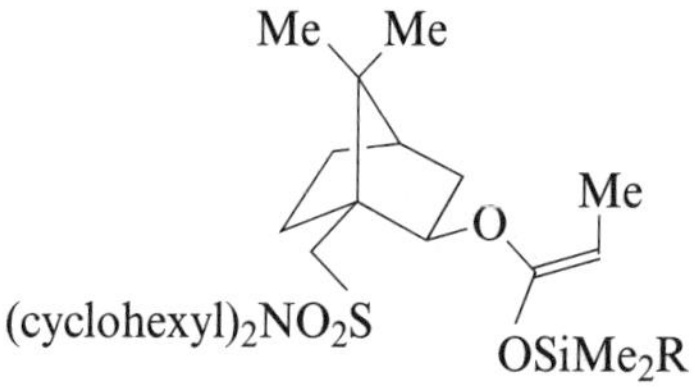   (616)

R	yield (%)	de
t-Bu	77	70
Me	14	50

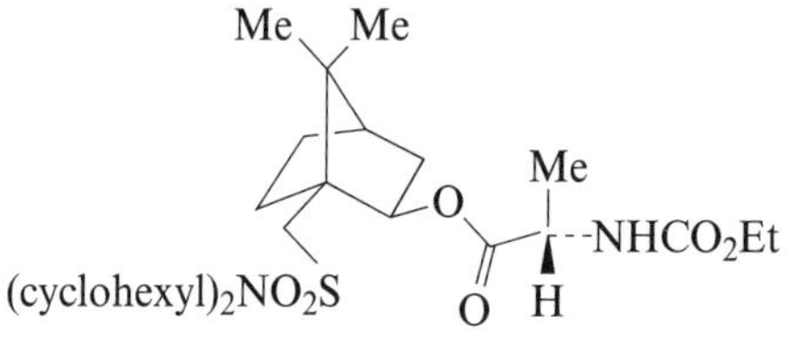

(617)

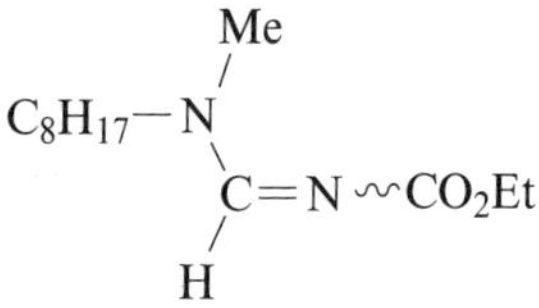

(618)

Intramolecular trapping of the singlet nitrene by C—H insertion reactions also features n synthetic routes to a variety of skeletous. Thus the azide **619** readily loses nitrogen on rradiation and yields the two products **620** and **621** by insertion into a C—H or a C—D bond[422]. An interesting variant of such intramolecular trapping is seen in the reaction of the azide **622**. Reaction of this affords the aldehydes **623** by a path involving insertion, ring opening and hydrolysis (Scheme 22)[423].

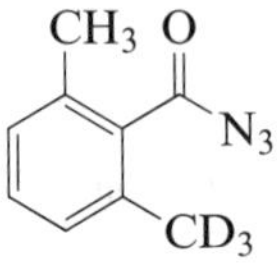

(619)

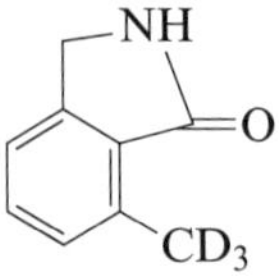

(620)

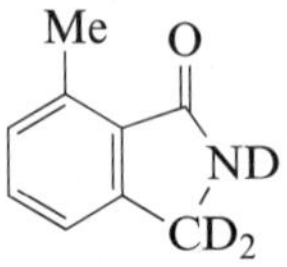

(621)

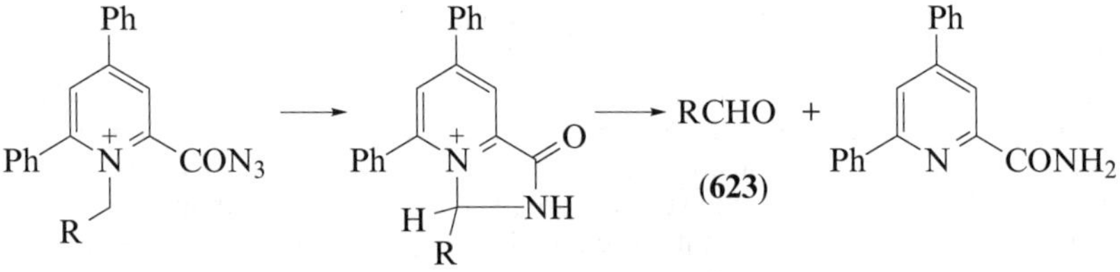

(622) R = Ph, *p*-ClC_6H_4, *p*-Tolyl

SCHEME 22

C. Irradiation of Vinyl Azides

The usual photoreaction of a vinyl azide is intramolecular cyclization to yield a 2*H*-azirine. Although this process is a well-established path to these compounds, the details of the mechanism by which the reaction occurs is unresolved. In some instances the formation of the azirine is thought not to involve the nitrene intermediate. However, other studies have shown that azirines and vinyl nitrenes are in thermal equilibrium and thus it seems likely that a vinylnitrene is the key intermediate in such reactions.

A typical high-yield synthesis of a vinyl substituted 2*H*-azirine is shown by the conversion of the azide **624** into **625**[424]. Similar behaviour is seen for the isomeric azidoeneynes **626** and **627** that are converted into the alkynyl substituted azirine **628**[425] while irradiation of the azidoallylphosphonates **629** yields the corresponding azirines **630**[426] and **631** is transformed into **632** at 0 °C[427]. This route to azirines has been extended to the synthesis of the bis-2*H*-azirines **633** and **634**, which are formed by irradiation of the bis azides **635**[428] and **636**[429], respectively. The formation of these products arises by sequential loss of nitrogen. Thus the bis azide **635** is converted to **637** prior to the ultimate formation of the final product **633**. Methylene azirines such as **638** are formed on photoelimination of nitrogen from the azides **639**. In this instance, however, the presence of the azirine derivative was established by trapping experiments with HCN when the adducts **640** are formed[430].

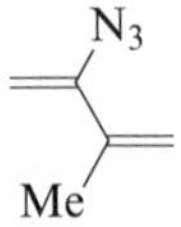

(624)

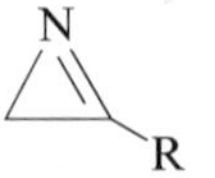

(625) R = C(Me)=CH_2

(628) R = C≡CH

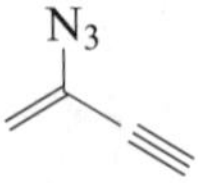

(626)

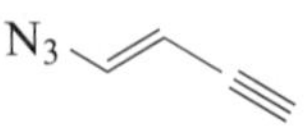

(627)

$$(629) \qquad (630) \qquad (631)$$

$$R^1 = Me, R^2 = Me \text{ or } Ph$$

$$(632)$$

$$(633) \ R = H \qquad (635) \ R = H \qquad (637)$$
$$(634) \ R = Me \qquad (636) \ R = Me$$

$$(638) \qquad (639) \qquad (640)$$
$$R = Me, CH_2OMe, CH_2Cl$$

The azirine formed from the conversion of vinyl azides can often be sufficiently labile, as with **638** above, to undergo further reactions. Thus the azirine **641**, obtained from irradiation of **642**, undergoes trimerization on further irradiation[431]. Ring opening of the azirines **643** to yield nitrile ylides can be used as a synthetic path to oxazoles[432]. Cyclization to yield an azirine is also proposed to account for the formation of **644** from **645** on irradiation in morpholine[433]. An azirine of this type (**646**) has also been proposed as the key intermediate in the formation of **647** from irradiation of the uracil azide **648**. This reaction is carried out in methylamine and ring expansion of the azirine occurs on attack by the amine[434]. When the reaction is carried out in methanol, the azirine does not undergo ring expansion and instead ring opening affords the acetal derivative **649**. Instead of a 1,3-dipole, which is the more usual intermediate obtained on ring opening of an azirine, irradiation of the vinyl azide **650** affords a carbene **651**, which undergoes addition to the adjacent double bond to yield the cyclopropane derivative **652**[435]. Other examples have shown that this type of behaviour can be a useful synthetic strategy[436].

In some instances the vinyl azide does not undergo the conversion to the 2*H*-azirine and instead yields a nitrene, which adds to alkenes and dienes. This is seen with the azide **653** which, on irradiation through Pyrex, affords the corresponding nitrene that adds to alkenes stereospecifically to yield aziridines[437]. Stereospecific addition to dienes is exemplified by the irradiation of the quinone azide **654** in the presence of *trans,trans*-hexa-2,4-diene. Aziridines do not result from this process and instead the stereoisomeric indole derivatives **655** and **656** are obtained[438]. This process has been developed as a route to

(641)   **(642)**   **(643)**

R = Ph, *o*-Tolyl, *o*-PhCH₂C₆H₄
or fluoren-1-yl

(644)   **(645)**   **(646)**

(647)   **(648)**   **(649)**

(650)   **(651)**   **(652)** $n = 1, 95\%$
$n = 1$ or 2 $n = 2, 67\%$

mitomycin precursors[439]. An alternative approach to mitomycin analogues has made use of insertion reactions of nitrenes generated from the quinone azides **657** which, on irradiation in methanol, are converted into the pyrrole derivatives **658**[440]. Insertion reactions can be of considerable synthetic value as in the reactions encountered with the azidocoumarins[441] or with the azidouridines **659** that are converted into **660**[442].

(653)   **(654)**

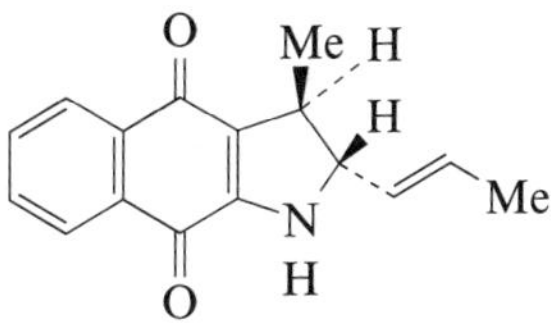

(655) 60% yield 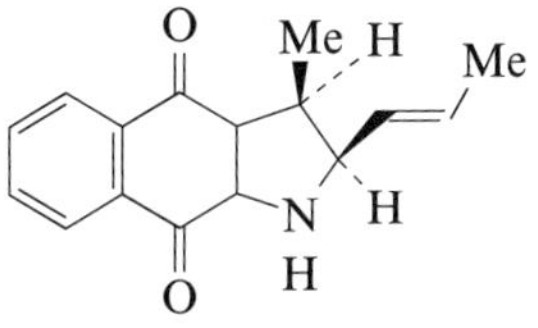 **(656)** 18% yield

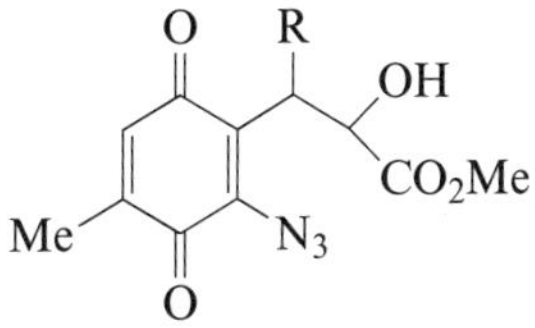

(657) 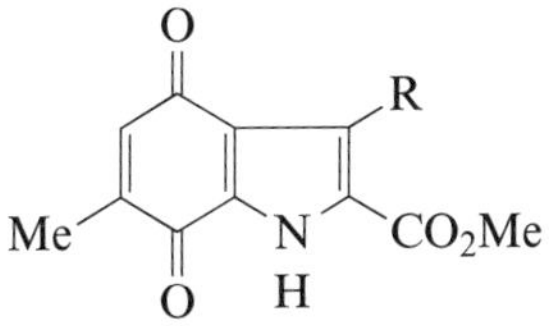 **(658)**

R = H, Me, Et, *n*-Pr, CO$_2$Me, (CH$_2$)$_2$CO$_2$Me
yield (%) 42, 62, 61, 60, 40, 64

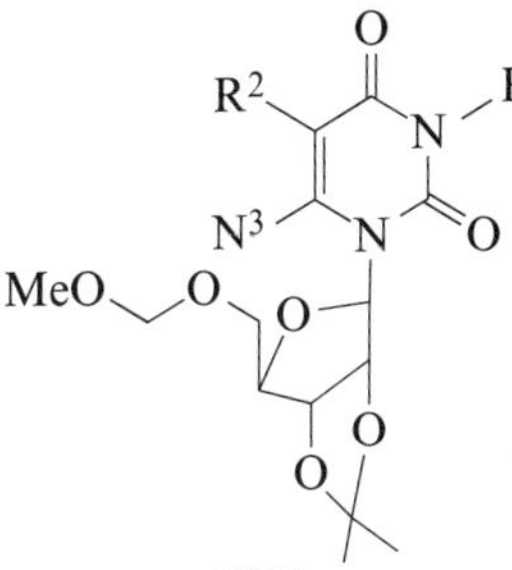

(659) 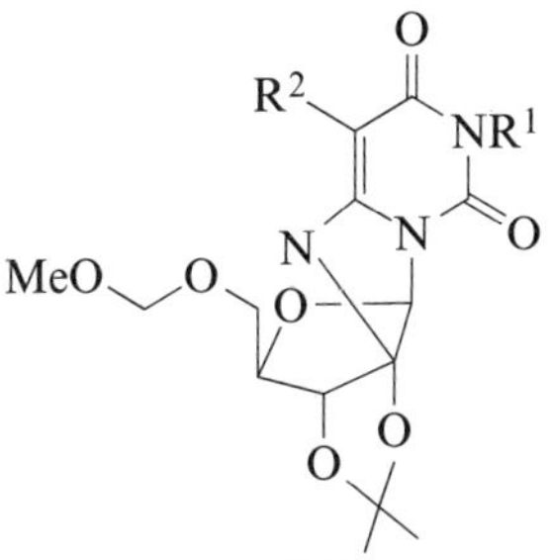 **(660)**

R^1 = R^2 = H or Me
R^1 = H, R^2 = Me
R^1 = Me, R^2 = H

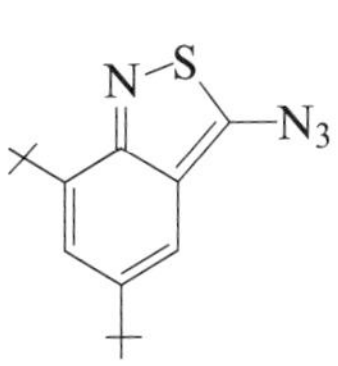

(661) 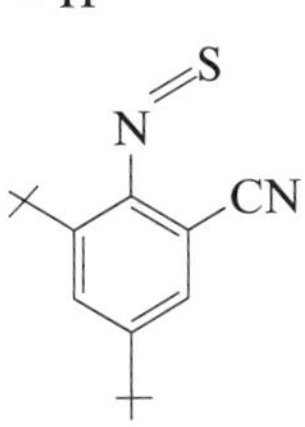 **(662)**

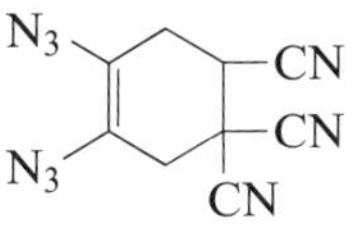

(663) 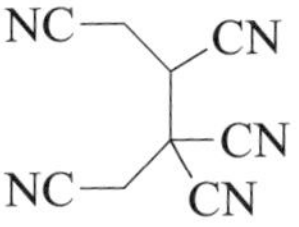 **(664)** 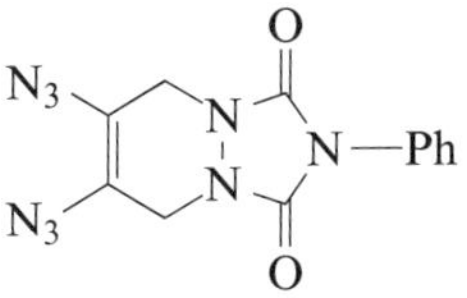 **(665)** 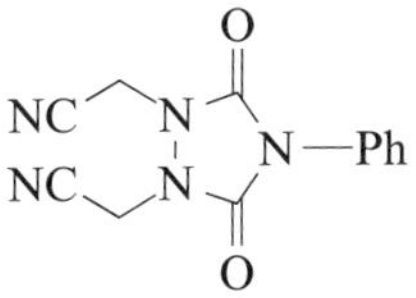 **(666)**

With cyclic azido compounds, ring-cleavage reactions can occur following irradiation and loss of nitrogen as with the generation of the nitrene from the azide **661** at 12 K which ring opens to the highly reactive thionitrosobenzene derivative **662**[443]. Ring opening has also been reported at higher temperatures following irradiation of the vicinal diazide **663**, which yields the pentanitrile **664**[444] or conversion of **665** into **666**[445]. Ring opening has also provided a route to nitriles, as in the fragmentation of the colchicine derivative **667** to yield **668**[446] or of **669** to give **670**[447].

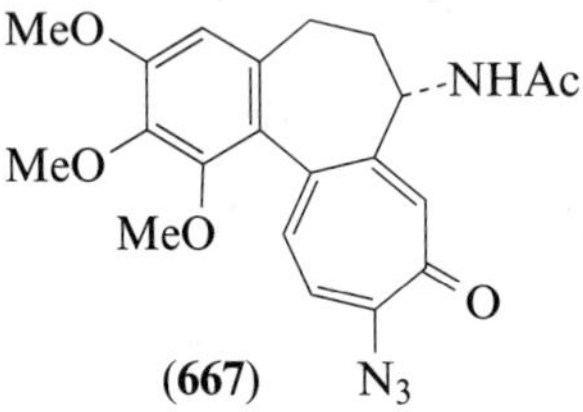

Meo .--NHAc
Meo
Meo
O
(667)   N₃

Meo .---NHAc
Meo
Meo CH=C=O
CN
(668)

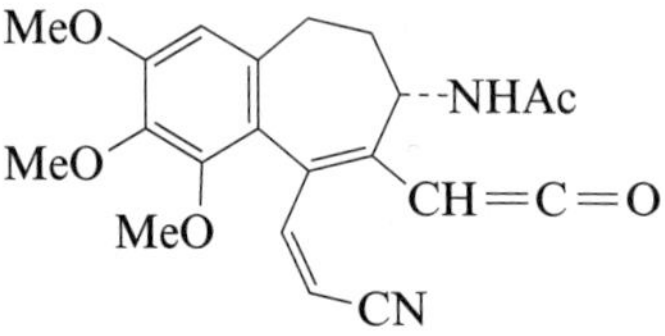

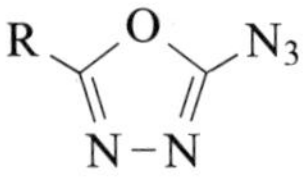

R   O   N₃
N–N
(669)

O
‖
RCC≡N
(670)

D. Miscellaneous Azides

A variety of other azides, where the azido group is bonded to an atom other than carbon, have also been studied. The processes which the nitrenes produced by irradiation of these compounds undergo are fundamentally the same as those for the azides described previously. Thus 1,2-group migration is important. This is seen in the 1,2-hydrogen migration yielding **671** following the irradiation at 254 nm of azidosilane in a low-temperature matrix[448]. Under similar conditions trimesitylsilyl azide also undergoes group migration, a mesityl group in this instance, to yield **673**. This product of group migration is accompanied by the insertion product **674**[449,450]. When the dimesityl(trimethylsilyl)silyl azide **675** is irradiated, a preference for migration of the trimethylsilyl group affords the silanimine **676**[451].

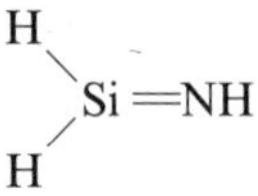

H
\
  Si=NH
/
H
(671)

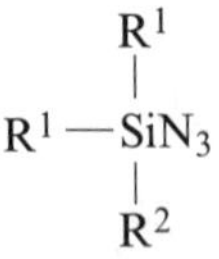

R¹
|
R¹—SiN₃
|
R²
(672) R¹ = R² = mesityl
(675) R¹ = mesityl, R² = SiMe₃

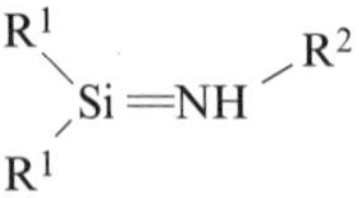

R¹   R²
\   /
  Si=NH
/
R¹
(673) R¹ = R² = mesityl
(676) R¹ = mesityl, R² = SiMe₃

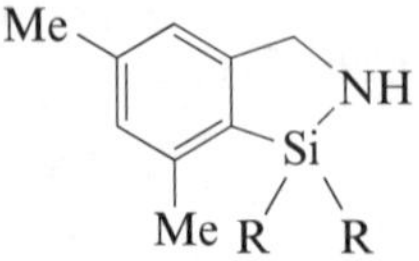

Me
NH
Si
Me R   R
(674)

Silyl bisazides such as **677** has also been studied under low-temperature matrix isolation conditions. The irradiation brings about complete loss of nitrogen to yield di-*t*-butylsilylene as a ground-state singlet[452]. A silylene is also obtained on the irradiation of the bisazide **678**[453]. A slightly different reaction path is observed for the low-temperature (77 K) irradiation of the bisazide **679**. Here, sequential loss of nitrogen followed by migration of a silyl group affords the first reported example of a silanediimine **680**[454].

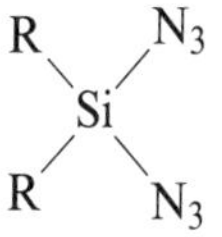

$$Me_3Si-N=Si=N-SiMe_3$$

(680)

R—Si(R)(N$_3$)(N$_3$)

(677) R = Bu^t
(678) R = Me
(679) R = SiMe$_3$

In the phosphorus series, group migrations are also predominant. Thus a 1,2-methyl migration is observed in the irradiative conversion of **681** into the transient phosphine imide **682**[455]. This species is not isolated but undergoes (2 + 2)-dimerization at the P—N double bonds. When given a choice of migrating groups as in **683**, a preference is shown for alkyl group migration rather than aryl[456]. The preference for aryl group migration and the influence of substituents on the aryl group has been studied in some detail as in **684**[457,458]. Migration of amino groups has also been reported, for example, in the conversion of **685** into **686** in methanol. The phosphoimine **687** is formed initially and is trapped by the solvent[459]. A transient phosphonitrile **688** can be obtained by irradiation of the azide **689**. In the absence of trapping agents this species undergoes dimerization, while in methanol the phosphoimine **690** is obtained[460].

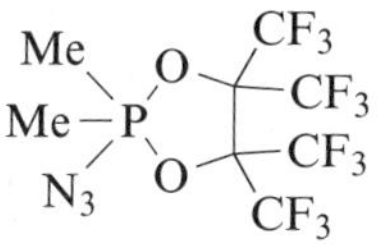

(681) 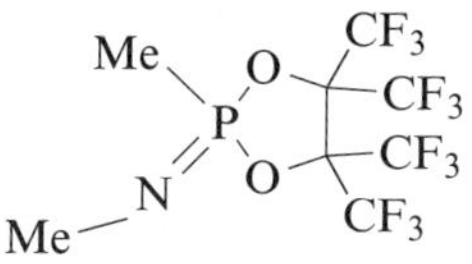   **(682)** 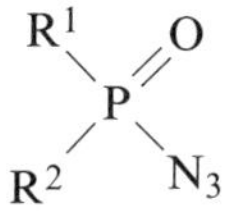   **(683)** R^1 = Ph, R^2 = alkyl
                                      (684) R^1 = R^2 = aryl
                                      (685) R^1 = R^2 = Pr$^i_2$N

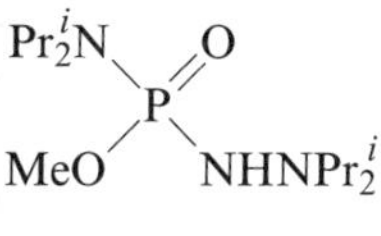

(686) 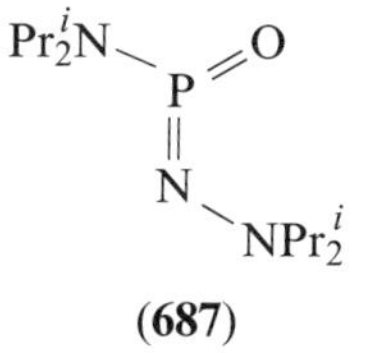   **(687)** 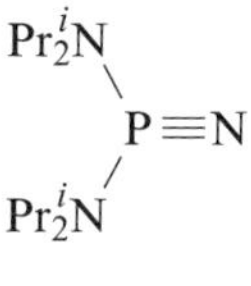   **(688)** 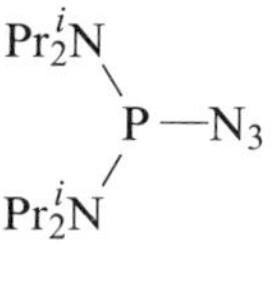   **(689)** 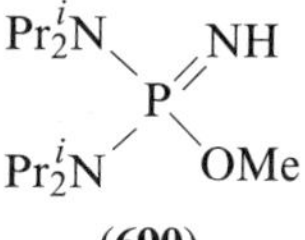   **(690)**

Germanium systems have also been studied. Here again insertion reactions and group migrations are observed. Thus the irradiation of **691** in cyclohexane affords the nitrene **692**. In a matrix, however, irradiation is followed by a 1,2-mesityl migration to yield the germanimine 692[461]. A germanium-centred carbene has also been prepared by the irradiation of the bisazidodimethyl germane in a low-temperature matrix[462].

$$R_3GeN_3 \qquad \overset{R}{\underset{R}{\diagdown}}Ge=N-R$$

(691) R = mesityl   (692)

E. Aryl Azides

A variety of spectroscopic studies on the ring-expansion path of arylnitrenes to didehydroazepines have been reported. Spectroscopic evidence for the formation of 693 by irradiation of phenyl azide in solution at room temperature shows the presence of 3H-azepines prior to reaction with nucleophilic reagents[463]. Additional evidence has been presented that indicates the involvement of the singlet nitrene in the ring-expansion process[464,465] and similar observations have been made for the 1- and 2-naphthyl azide systems[466,367]. Interestingly, other studies have examined the triplet-state reactivity of phenyl nitrene at 12 K in an argon matrix and have suggested that the triplet state also yields the didehydroazepine[468]. A flash photolytic examination of the reaction of 694 has supported the view that dihydroazepine is a key intermediate in the formation of products[469].

(693)   (694)

Some studies have shown that the generation of the nitrene from pentafluorophenyl azide[470] or from 2,6-dimethylphenyl azide[471], from irradiation in a low-temperature matrix, permits its trapping by carbon monoxide and the formation of the corresponding isocyanate. No evidence for the ring-expansion process was seen in these experiments. Other low-temperature studies in the absence of CO trap and using ^{15}N-labelled phenyl azide have shown that even at 12 K there is evidence for the ring expansion to the didehydroazepine[472].

1. Ring-expansion reactions

The ring expansion of phenyl azides on irradiation and the concomitant ring expansion to a didehydroazepine and subsequent trapping provides a useful synthetic route to a variety of azepine systems. Thus irradiation of 695, with a variety of p-substituents, occurs readily in THF/water to yield the azepinones 696[473]. Interest in the influence of other substituents on this process has been recorded in the study of the reactions of 3- and 4-nitrophenyl azides[474]. This work has illustrated the profound effect that substituents can have on the reaction. Ring expansion is also observed for a tyrosine derivative[475]. Irradiation of 697 in excess diethylamine brings above the exclusive conversion into 698[476].

Ring expansion of the quinoline derivatives 699[477] or 700[478] is also facile. Clean and efficient ring expansion of 701 into 702 occurs in basic methanol solution[479]. However, the outcome of the reaction is dependent on the reaction conditions and irradiation in the

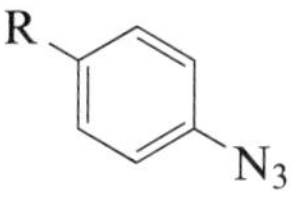

(695) 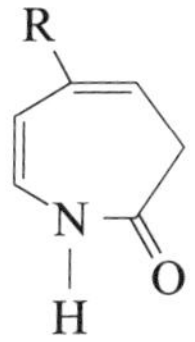　　　(696) 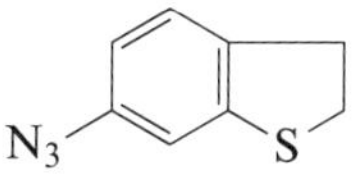　　　　(697) 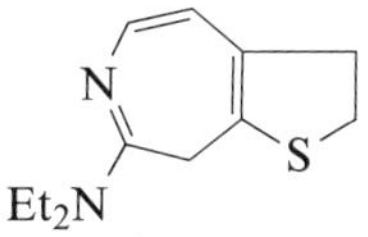　　　　　(698)

R = CO$_2$Me, CN, CF$_3$, SO$_2$NPh, CO$_2$H

yield (%)　　　45, 60, 60,　　32,　　36

presence of an acid, such as HCl, affords a mixture of the two phenazines **703** and **704**. In neutral methanol only **703** is obtained. This influence of solvent change also plays a part in the formation of 3-amino-4-chloropyridine obtained on irradiation of 4-azidopyridine in HCl (6 M)[480]. This phenazine derivative **705** is also reactive and on irradiation in dipropylamine the two ring-expanded products **706** and **707** are produced[481].

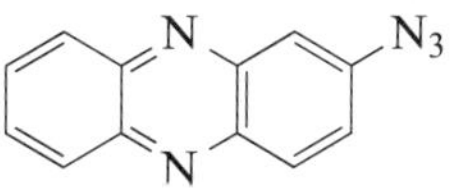

(701)

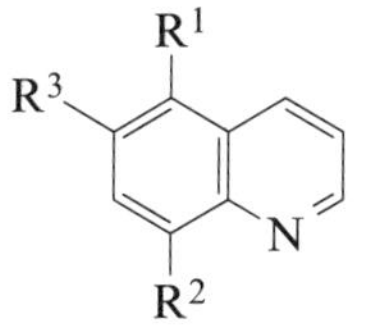

R^1	R^2	R^3
N$_3$	H	H
H	N$_3$	H
H	MeO	N$_3$
H	N$_3$	MeO

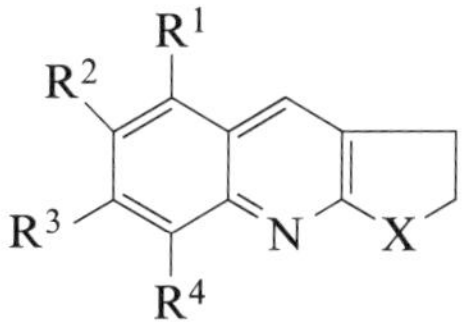

(699)

(700)　　X = S or O

R^1	R^2	R^3	R^4
N$_3$	H	H	H
H	N$_3$	H	H
H	H	N$_3$	H
H	H	H	N$_3$

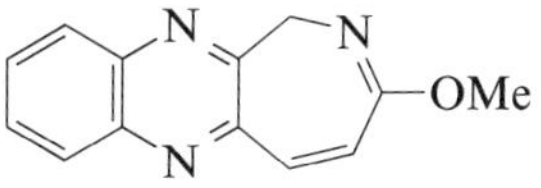

(702) 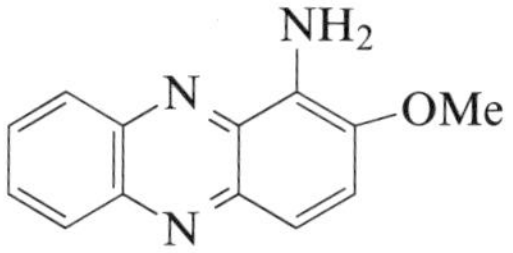　　　　　　(703) 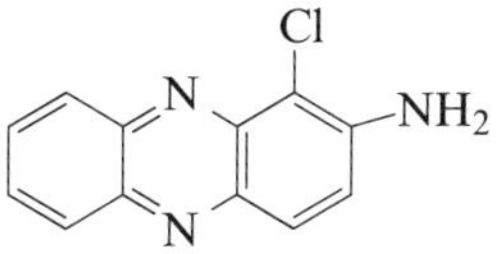　　　　　　(704)

The behaviour of 2-azidobiphenyl also involves ring expansion. However, in the presence of tetracyanoethylene an adduct **708** is obtained which is thought to arise by trapping of a carbene[482]. The reactivity of such azidobiphenyls can be influenced by the use of a hydrophobic cage[483]. A novel diazepine **709** can be obtained by the irradiation of the 1,7-diazidonaphthalene in the presence of sodium methoxide[484].

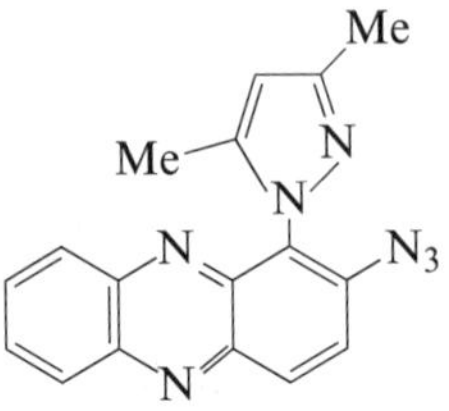

(705) 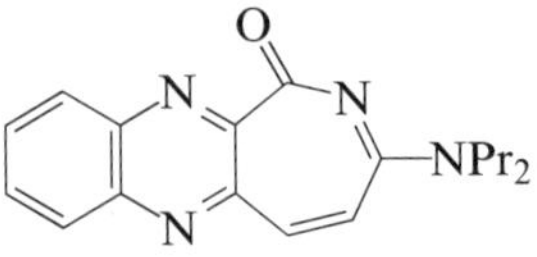　　　　　　(706) 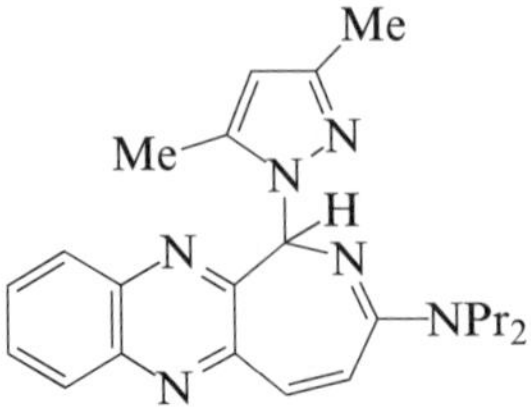　　　　　　(707)

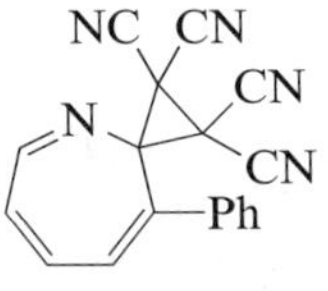

(708) 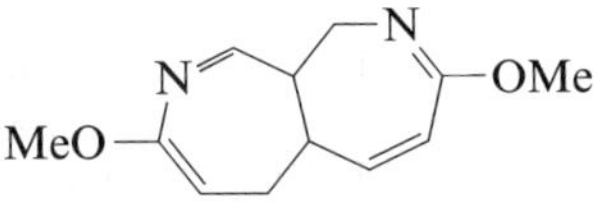　　　　　　(709)

2. Diazepines and triazepines

Ring expansion of azidopyridines and related compounds provides an efficient route to diazepines and triazepines. Typically, the 3-azidopyridines 710 undergo ring expansion, predominantly to the unsubstituted position, in solvents such as methanol with sodium methoxide to yield the diazepines 711[485,486]. 1,4-Diazepines, e.g. 712, can also be prepared starting from 4-azidopyridines[487–490]. 3- and 4-Azidopyridine N-oxides do not undergo ring expansion to the corresponding diazepines on reaction in nucleophilic solvents[491]. Quinolines behave analogously opening the route to benzo-1,4-diazepines regardless of whether the 3-azido- or 4-azido-quinoline starting material is used. Thus 713 affords 714 in methanol solution[492] while irradiation of 715 yields 716[493,494]. Interestingly, change of the nucleophilic species to ethanethiol alters the reactivity of the 3-azidoquinoline and no ring expansion occurs and the product obtained, perhaps by a free radical path, is 3-imino-4-thioethoxyquinoline[495]. With the isoquinolines 717 1,3-diazepines 718 result on ring expansion in methanol[496,497].

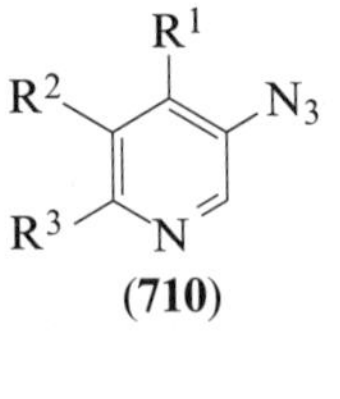

(710) 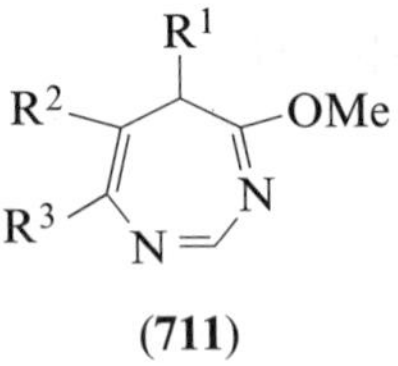　　　　　　(711) 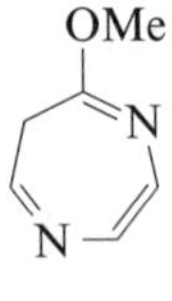　　　　　　(712)

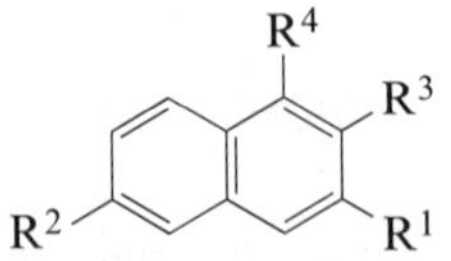

(713) $R^4 = R^2 = H$, $R^3 = N_3$, $R^1 = H$, Me or Ph
(715) $R^3 = H$, $R^4 = N_3$, $R^1 = R^2 = H$; $R^1 = H$, $R^2 = Cl$;
　　　　$R^1 = $ Me or Ph, $R^2 = H$

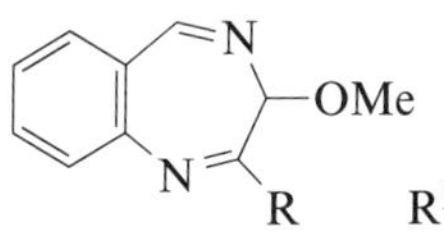

(714) 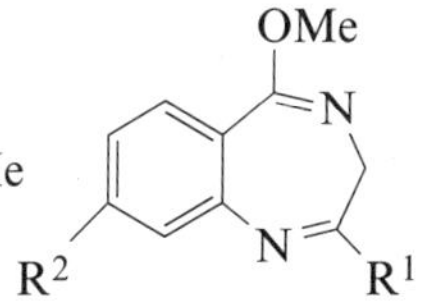   (716) 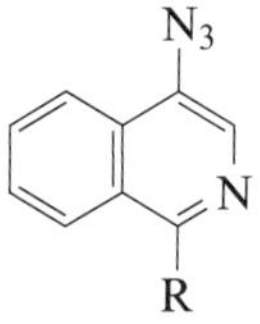   (717)   (718)

R = H, Me, Ph, OMe, Cl

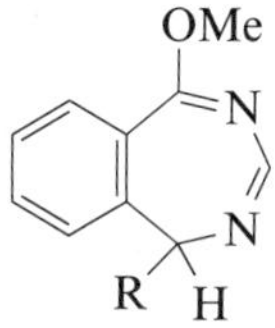

The synthesis of triazepines can be approached by the ring expansion of azidopyridazines **719** in sodium methoxide or diethylamine. This affords the products **720**[498,499]. The first fully unsaturated 1,3,5-triazepine has been reported[500].

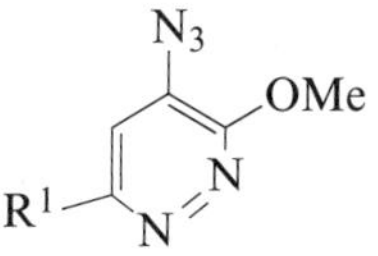

(719)   (720)
R^1 = H, Me or   R^3 = NEt$_2$ or MeO
MeO

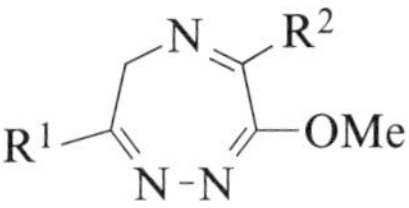

F. Intramolecular Cyclizations

The formation of a nitrene adjacent to a double bond can be used in the synthesis of novel heterocyclic systems. For example, the azide **721** converts smoothly into the tricyclic product **722** on irradiation[501] and the same *modus operandi* is used for the synthesis of pyrrolo[2,3-*d*]pyrimidines from substituted 4-azidopyrimidines[502]. Aziridine formation is also encountered in the cyclization of **723** to the tetrahydroquinoline derivatives **724**[503]. Other cyclizations provide routes to imidazoindoles **725**[504] or thienopyrroles **726**[505,506].

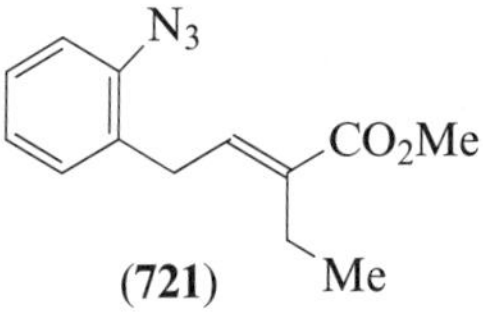

(721) 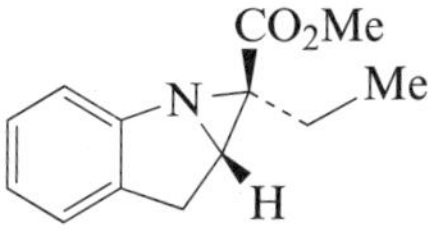   (722) 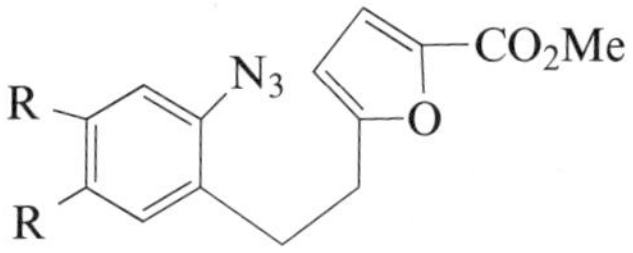   (723)

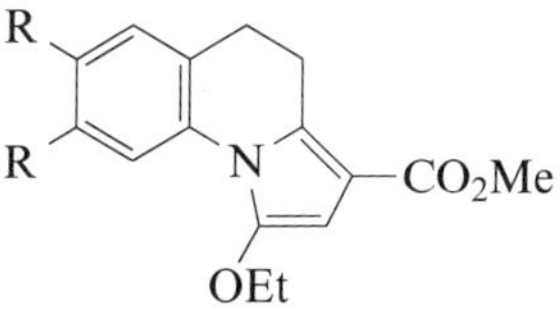

(724) 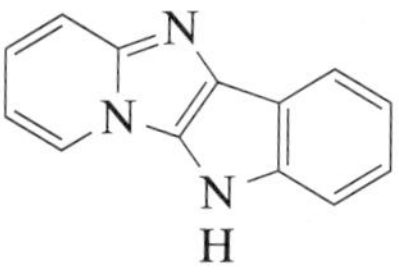   (725) 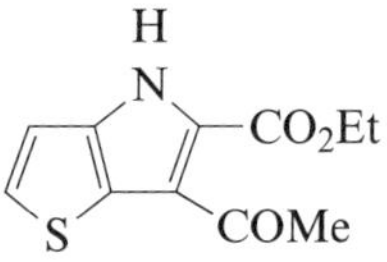   (726)

Other interesting cyclizations also occur involving the *peri* positions of naphthalene derivatives. Thus the irradiation of **727** produces a nitrene that is trapped intramolecularly to yield the product **728**[507]. A few reports have dealt with the evidence for the formation of **729** from the photon pulsed laser irradiation of 1,8-diazidonaphthalene in a low-temperature matrix[508,509]. Similar behaviour is observed with **730** when **731** is formed[510] and for the formation of a low yield of the cinnoline **732** from **733**[511]. A mixture of singlet and triplet reactivity accounts for the formation of products by intramolecular cyclization following the irradiation of the quinone azide **734**[512].

(727)   (728)   (729)   (730)

(731)   (732)   (733)   (734)

G. Photoaffinity Labelling

Interest in the insertion reactions of aryl nitrenes as a means of labelling proteins, etc. has led to considerable interest in such C—H and N—H insertion processes. Pentahalophenyl nitrenes have been studied in relation to this[513]. Pentafluorophenyl nitrene, for example, undergoes insertion reactions into the C—H bonds of cyclopentane[514] and toluene[515]. The nitrene from **735** undergoes both C—H and N—H insertion reactions when it is generated in a cyclohexane solution of diethylamine[516,517]. Iodoaryl azides have also been examined for the purpose of photo-affinity labelling. In this system two reactions are observed providing a triplet nitrene, and therefore aniline as the product or else loss of iodine from the aryl ring[518]. Interest in related areas such as intracellular distribution has attracted attention via the decomposition of polychlorobiphenyl azides[519]. Many examples have been published dealing with the use of aryl azides as a means of labelling proteins, etc. by C—H or N—H insertion reactions[520–524].

(735)

VII. RADIATION CHEMISTRY

In a review of this sort no attempt has been made to be encyclopaedic. However, it is interesting to note that the material available on this subject, since the last review articles in this series[1], where the radiation chemistry of the functional groups are considered specifically, is fairly sparse. As with the previous sections the material that has been selected for inclusion has focused on processes which affect the functional groups (alkyne, nitrile, diazo, azide and diazonium salt) directly. Often, reactions are observed where the functional group plays no part in determining the outcome of the reaction and in these cases no references have been included. The exclusions include radiolysis of the azide ion, photographic processes and polymerization in general.

A. Alkynes

Radiation provides a convenient method for the production of radical cations. Thus the radiolysis of propynoic acid brings about loss of CO_2 and yields the cyclobutadiene radical cation. Protection of the acid function by esterification suppresses the decarboxylation. Thus methyl propynoate undergoes loss of an electron from the ester group on radiolysis to yield the radical cation **736**[525]. At lower temperatures (77 K) radiolysis of but-2-ynoic acid affords the corresponding radical cation without undergoing decarboxylation, and in addition the radical **737** is formed by deprotonation of the radical cation[526]. A pulsed radiolysis study has examined the pairing of radical cations derived from the diyne **738**[527]. The formation of radical cations of alkynes (propyne and but-2-yne) by radiolysis has also been reported[528]. Other radiative sources have also been examined, such as the synchrotron bombardment of acetylene as a method for the determination of C—H bond strength[529] and X-ray irradiation of hex-1-yne and hex-3-yne has provided evidence for the generation of anions[530]. Polymerization by radiolysis is a vast field dealing with primary polymerization and also with grafting processes. Typical of the work encountered is the polymerization of p-nitrophenylacetylene in both the solid and solution phase. The rate of polymerization is proportional to the γ-ray intensity and, in line with the examples above, the radiation produces cationic species which are responsible for the polymerization[531]. Degradation of polyacetylenes such as poly(1-phenylprop-1-yne) can also be induced by γ-irradiation. The results show that radiation in the presence of air induces loss in the molecular weight of the polymer. Radiation under vacuum does not cause chain fission. A study of silicon-containing polyacetylenes has also been reported[532].

$$\left(HC\equiv C\overset{O}{\overset{\|}{C}}-OCH_3 \right)^{\overset{\bullet}{+}}$$

$$\overset{\bullet}{C}H_2C\equiv CCO_2H$$

$$Ph-C\equiv C-C\equiv\!\!=\!\!-Ph$$

(736)   (737)   (738)

B. Nitriles

Radiolysis at 77 K of t-butyl nitrile in trichlorofluoromethane has shown that the radical cation is produced[533]. γ-Radiolysis of 1-cyanoadamantane affords the corresponding 1-cyanoadamantyl radical which isomerizes to the 1-adamantaneiminyl radical[534]. It was also reported that α-radiolysis of nitriles **739** affords radicals[535] and electron-impact induced formation of cyano radicals from acetonitrile, fluoroacetonitrile and chloroacetonitrile has been carried out[536]. Pulsed radiolysis studies have also been reported and interest shown in the study of change in absorption bands brought about by phase effects on the pulse radiolysis of the liquid crystalline material **740**[537]. Several studies of radiation-induced decomposition of nitrile-group-containing compounds have been published. Thus γ-irradiation of bromoxynil **741** has been shown to bring about a first-order decomposition[538]. γ-Irradiation of aqueous solutions of nitriles such as acetonitrile provides many oxidation products such as di- and tri-carboxylic acids[539]. Other studies have observed the oxidation of succinonitrile in aqueous solution[540] and α-radiolysis has been used as a means of eliminating cyanides from the waste water from acrylonitrile manufacture[541].

(739) n = 1 or 2

(740)

(741)

Much of the study on nitrile-substituted species has dealt with the radiation-induced grafting reactions. Thus γ-irradiation of acrylonitrile on filter paper has shown that grafting to the cellulose occurs and that increasing the dose gives enhanced grafting[542]. The grafting of acrylonitrile to polypropylene[543] and to poly(trimethylvinylsilane)[544] can also be achieved with γ-irradiation and reasonable levels of crosslinking attained. The latter example was presented as a means of fabricating membranes for the separation of He and N$_2$. Other studies have examined the effect of irradiation of polyacrylonitrile[545,546], acrylonitrile–styrene copolymers[547–549] and nitrile rubbers[550].

C. Diazonium Salts

Radiolysis of aryldiazonium salt solutions has been studied as a path to heterolysis and the formation of the corresponding cation by loss of nitrogen. Under radiolysis conditions there is little effect due to the presence of aryl substituents[551]. Another study has examined the behaviour of benzenediazonium 2-carboxylate on irradiation in benzene in the presence of ^{18}F hydrofluoric acid. This has shown that reasonable yields of the corresponding labelled 2-fluorobenzoic acid can be obtained[552].

D. Azides

γ-Radiolysis of the azide **742** at 77 K in fluorotrichloromethane yields the radical cation. The electron density within this is unevenly spread over the three nitrogens of the azido group. The spectroscopic study shows that the highest electron density occurs on the nitrogen adjacent to the silicon atom[553]. A study of the radiolytic behaviour of the azide derivatives **743** in the gas phase (in methane at 760 torr and at 310 K) has been carried out. The results show, in the protonated form, that the azide derivatives undergo nitrogen loss by displacement with the formation of diaryl amines[554].

Me$_3$SiN$_3$

(742)   **(743)** R = F or Me

VIII. REFERENCES

1. Several reviews have dealt with the photochemical reactions of the types of compounds included in this chapter. Alkynes, J. D. Coyle in *The Chemistry of the Carbon–Carbon Triple Bond* (Ed. S. Patai), Wiley, Chichester, 1978, p. 523; Nitriles, in *The Chemistry of the Cyano Group*, diazonium salts and diazo compounds by W. Ando, in *The Chemistry of Diazonium and Diazo Groups* (Ed. S. Patai), Wiley, Chichester, 1978, p. 341; and radiation chemistry by Z. B. Alfassi, in *The Chemistry of Functional Groups, Supplement C* (Eds. S. Patai and Z. Rappoport), Wiley, Chichester, 1983, p. 187. Azides have been the subject of another extensive review [E. F. Scriven and K. Turnbull, *Chem. Rev.*, **88**, 298 (1988)]. In addition to these references, extensive information on the photochemical reactivity of such groups is to be found in the appropriate chapters in Volumes 1–24 of *Photochemistry* published by the Royal Society of Chemistry, London, from 1968–1993.

2. Y. Inoue and T. Hakushi, *Kagaku To Kogyo (Osaka)*, **56**, 165 (1982); *Chem. Abstr.*, **97**, 109200 (1982).

3. Y. Inoue, T. Fukunaga and T. Hakushi, *J. Org. Chem.*, **48**, 1732 (1983).

4. O. De Lucchi, G. Licini, L. Pasquato and M. Senta, *J. Chem. Soc., Chem. Commun.*, 1597 (1985).

5. P. S. Mariano, J. L. Stavinoha, G. Pepe and E. F. Meyer, jun., *J. Am. Chem. Soc.*, **100**, 7114 (1978).

6. Y. Inoue, Y. Ueda and T. Hakushi, *J. Am. Chem. Soc.*, **103**, 1806 (1981).

7. C. Chulz-Popitz and H. Meier, *Nouv. J. Chim.*, **7**, 395 (1983); *Chem. Abstr.*, **100**, 5910 (1984).

8. H. Meier and P. Koenig, *Nouv. J. Chim.*, **10**, 437 (1986); *Chem. Abstr.*, **107**, 22991 (1987).

9. P. J. Kropp, *Org. Photochem.*, **4**, 1 (1979).

10. P. Wan, S. Culshaw and K. Yates, *J. Am. Chem. Soc.*, **104**, 2509 (1982); J. McEwen and K. Yates, *J. Phys. Org. Chem.*, **4**, 193 (1991).

11. Y. Chiang, A. J. Kresge, M. Capponi and J. Wirz, *Helv. Chim. Acta*, **69**, 1331 (1986).

12. J. McEwen and K. Yates, *J. Am. Chem. Soc.*, **109**, 5800 (1987).

13. M. Isaks, K. Yates and P. Kalanderopoulos, *J. Am. Chem. Soc.*, **106**, 2728 (1984).

14. S. C. Shim and T. S. Lee, *J. Chem. Soc., Perkin Trans. 2*, 1739 (1990).

15. S. C. Shim and T. S. Lee, *J. Photochem. Photobiol., A*, **53**, 323 (1990).

16. T. S. Lee, S. C. Shim and S. S. Kim, *Bull. Korean Chem. Soc.*, **7**, 116 (1986), (1989); *Chem. Abstr.*, **106**, 6569 (1987).

17. M. G. Voronkov, N. N. Vlasova, G. Yu. Zhila, O. G. Yarosh, E. I. Brodskaya, A. I. Albanov and V. Yu. Vitkovskii, *Zh. Obshch. Khim.*, **54**, 868 (1984); *Chem. Abstr.*, **101**, 91043 (1984).

18. A. C. Weedon, in *Synthetic Organic Photochemistry* (Ed. W. M. Horspool), Plenum Press, New York, 1984, p. 61.

19. E. P. Serebryakov, *Izv. Akad. Nauk SSSR, Ser. Khim.*, 136 (1984); *Chem. Abstr.*, **100**, 155887 (1984).

20. T. Zaima, C. Matsuno, Y. Matsunaga and K. Mitsuhashi, *Nippon Kagaku Kaishi*, 1293 (1984); *Chem. Abstr.*, **101**, 230340 (1984).

21. T. Zaima, Y. Matsunaga and K. Mitsuhashi, *J. Heterocycl. Chem.*, **20**, 1 (1983); *Chem. Abstr.*, **98**, 179144 (1983).

22. S. Hussain, D. I. Schuster and K. El-Bayoumy, *Tetrahedron Lett.*, **23**, 153 (1982).

23. M. Demuth, *Org. Photochem.*, **11**, 37 (1991).

24. M. Cavazza, G. Guella and F. Pietra, *Helv. Chim. Acta*, **71**, 1608 (1988).

25. M. Cavazza and F. Pietra, *J. Chem. Soc., Perkin Trans. 1*, 2283 (1985).

26. M. Cavazza and M. Zandomeneghi, *Gazz. Chim. Ital.*, **117**, 17 (1987); *Chem. Abstr.*, **107**, 175533 (1987).

27. A. V. Kamernitskii, V. N. Ignatov, I. S. Levina, E. P. Serebryakov, V. S. Bogdanov and E. G. Cherepanova, *Izv. Akad. Nauk SSSR, Ser. Khim.*, 1184 (1986); *Chem. Abstr.*, **106**, 196671 (1987).

28. A. A. Avetisyan, A. Kh. Margaryan, G. Nalbandyan and T. V. Avetisyan, *Zh. Org. Kim.*, **25**, 530 (1989); *Chem. Abstr.*, **111**, 214333 (1989).

29. A. A. Avetisyan, A. Kh. Magaryan, G. K. Nalbandyan and T. V. Avetisyan, *Khim. Geterotsikl. Soedin.*, 1315 (1986); *Chem. Abstr.*, **107**, 96524 (1987).

30. G. Maier, H. W. Lage and H. P. Reisenauer, *Angew. Chem., Int. Ed. Engl.*, **20**, 976 (1981).

31. R. Askani, *Chem. Ber.*, **98**, 3618 (1965).

32. R. Askani and J. Hoffmann, *Chem. Ber.*, **124**, 2307 (1991).
33. T. Sano, Y. Horiguchi and T. Tsuda, *Chem. Pharm. Bull.*, **38**, 3283 (1990); *Chem. Abstr.*, **114**, 228702 (1991).
34. A. B. Smith and P. J. Jerris, *tert.*, *J. Org. Chem.*, **47**, 1845 (1982).
35. G. I. Fray, G. R. Green and N. A. Whiteside, *Synthesis*, 956 (1982).
36. T. Shimo, K. Somekawa, M. Sato and S. Kumamoto, *Nippon Kagaku Kaishi*, 1927 (1984); *Chem. Abstr.*, **102**, 149041 (1985).
37. T. Matsui, Y. Kawano and M. Nakayama, *Chem. Express*, **5**, 697 (1990); *Chem. Abstr.*, **114**, 82281 (1991).
38. M. Fetizon, D. D. Khac and N. D. Tho, *Tetrahedron Lett.*, **27**, 1777 (1986).
39. T. Matsui and M. Nakayama, *Bull. Chem. Soc. Jpn.*, **56**, 3531 (1983).
40. T. Naito and C. Kaneko, *Chem. Pharm. Bull.*, **31**, 366 (1983); *Chem. Abstr.*, **99**, 53565 (1983).
41. T. Naito and C. Kaneko, *Chem. Pharm. Bull.*, **31**, 366 (1983).
42. T. Wang and L. A. Paquette, *J. Org. Chem.*, **51**, 5232 (1986).
43. E. R. Koft and A. B. Smith, *tert.*, *J. Org. Chem.*, **49**, 832 (1984).
44. E. R. Koft and A. B. Smith, *tert.*, *J. Am. Chem. Soc.*, **104**, 5568 (1982).
45. E. R. Koft and A. B. Smith, *tert.*, *J. Am. Chem. Soc.*, **106**, 2115 (1984).
46. K. Somekawa, H. Oda and T. Shimo, *Chem. Lett.*, 2077 (1991).
47. K. S. Feldman, J. H. Come, G. J. Fegley, B. D. Smith and M. Parvez, *Tetrahedron Lett.*, **28**, 607 (1987).
48. Y. Hanzawa and L. A. Paquette, *Synthesis*, 661 (1982).
49. D. Bryce-Smith, A. Gilbert, N. Al-Jalal, R. R. Deshpande, J. Grzonka, M. A. Hems and P. Yianni, *Z. Naturforsch., B: Anorg. Chem. Org. Chem.*, **38**, 1101 (1983).
50. B. Sket and M. Zupan, *Tetrahedron*, **45**, 1755 (1989).
51. M. G. Barlow, D. E. Brown and R. N. Haszeldine, *J. Fluorine Chem.*, **20**, 745 (1982).
52. B. Sket, N. Zupancic and M. Zupan, *Tetrahedron*, **40**, 3795 (1984).
53. M. Sindler-Kulyk and D. C. Neckers, *J. Org. Chem.*, **47**, 4914 (1982).
54. T. Itoh, K. Matsuzaki, S. Suzuki, K. Kisjida, H. Ogura, N. Kawahara and T. Nakjima, *Heterocycles*, **20**, 1321 (1983).
55. M. C. Pirrung, *J. Org. Chem.*, **52**, 1635 (1987).
56. P. M. op den Brouw and W. H. Laarhoven, *J. Org. Chem.*, **47**, 1546 (1982).
57. R. J. F. M. Van Arendonk and W. H. Laarhoven, *Recl.: J. R. Neth. Chem. Soc.*, **100**, 263 (1981).
58. S. Wolff and W. C. Agosta, *J. Am. Chem. Soc.*, **106**, 2363 (1984).
59. P. Margaretha, C. Schroeder, S. Wolff and W. C. Agosta, *J. Fluorine Chem.*, **30**, 429 (1986); *Chem. Abstr.*, **106**, 4776 (1987).
60. S. S. Kim, D. Y. Yoo, A. E. Rhan, I. H. Cho and S. C. Shim, *Bull. Korean Chem. Soc.*, **10**, 66 (1989); *Chem. Abstr.*, **111**, 153293 (1989).
61. J. D. Coyle, P. A. Rapley, J. Kamphuis and H. J. T. Bos, *J. Chem. Soc.*, *Perkin Trans. 1*, 2173 (1986).
62. J. D. Coyle, P. A. Rapley, J. Kamphuis and H. J. T. Bos, *Tetrahedron Lett.*, **26**, 2249 (1985).
63. A. C. Brouwer and H. J. T. Bos, *Recl.: J. R. Neth. Chem. Soc.*, **103**, 152 (1984).
64. A. C. Brouwer, A. V. E. George and H. J. T. Bos, *Recl.: J. R. Neth. Chem. Soc.*, **102**, 83 (1983); *Chem. Abstr.*, **99**, 5481 (1983).
65. G. Kaupp and E. Jostkleigrewe, *Angew. Chem. Suppl.*, 1100 (1982).
66. H. Siegel, L. Eisenhuth and H. Hopf, *Chem. Ber.*, **118**, 597 (1985).
67. L. Eisenhuth, H. Siegel and H. Hopf, *Chem. Ber.*, **114**, 3772 (1981).
68. S. C. Shim and S. S. Kim, *Tetrahedron Lett.*, **26**, 765 (1985).
69. J. H. Kwan, S. J. Lee and S. C. Shim, *Tetrahedron Lett.*, **32**, 6719 (1991).
70. S. C. Shim, S. J. Lee and J. H. Kwan, *Chem. Lett.*, 1767 (1991).
71. C. S. Lee and S. C. Shim, *Photochem. Photobiol.*, **55**, 323 (1992); *Chem. Abstr.*, **116**, 173648 (1992).
72. T. S. Lee, S. J. Lee and S. C. Shim, *J. Org. Chem.*, **55**, 4544 (1990).
73. S. J. Lee and S. C. Shim, *Tetrahedron Lett.*, **31**, 6197 (1990).
74. S. C. Shim and T. S. Lee, *J. Org. Chem.*, **53**, 2410 (1988).
75. S. C. Shim and T. S. Lee, *Chem. Lett.*, 1075 (1986).
76. S. C. Shim and T. S. Suk, *Bull. Korean Chem. Soc.*, **7**, 304 (1986); *Chem. Abstr.*, **107**, 11528: (1987).
77. M. Shi, Y. Okamoto and S. Takamuku, *J. Chem. Soc.*, *Perkin Trans. 1*, 2391 (1991).

78. M. Shi, Y. Okamoto and S. Takamuku, *Tetrahedron Lett.*, **30**, 6709 (1989).
79. M. Shi, Y. Okamoto and S. Takamuku, *J. Org. Chem.*, **55**, 3821 (1990).
80. M. Shi, Y. Okamoto and S. Takamuku, *Chem. Lett.*, 1297 (1989).
81. J. J. Eisch, F. Shen and K. Tamao, *Heterocycles*, **18**, 245 (1982).
82. J. L. Eisch, B. Shafii and A. L. Rheingold, *J. Am. Chem. Soc.*, **109**, 2526 (1987).
83. J. D. Wilkey and G. B. Schuster, *J. Am. Chem. Soc.*, **110**, 7569 (1988).
84. M. A. Kropp and G. B. Schuster, *J. Am. Chem. Soc.*, **111**, 2316 (1989).
85. M. A. Kropp, M. Baillargeon, K. M. Park, K. Bhamidapaty and G. B. Schuster, *J. Am. Chem. Soc.*, **113**, 2155 (1991).
86. K. M. Park and G. B. Schuster, *J. Org. Chem.*, **57**, 2502 (1992).
87. L. Fabry and P. Gomory, *Acta Chim. Acad. Sci. Hung.*, **106**, 291 (1981); *Chem. Abstr.*, **95**, 149532 (1981).
88. J. P. Ferris and J. C. Guillemin, *J. Org. Chem.*, **55**, 5601 (1991).
89. J. Cossy and C. Leblanc, *Tetrahedron Lett.*, **32**, 3051 (1991).
90. J. Cossy, J.-P. Pete and C. Portella, *Tetrahedron Lett.*, **30**, 7361 (1989).
91. J. Cossy, D. Belotti and J.-P. Pete, *Tetrahedron*, **46**, 1859 (1990).
92. J. Cossy, D. Belotti and J.-P. Pete, *Tetrahedron Lett.*, **28**, 4547 (1987).
93. B. Guerin and L. J. Johnston, *J. Org. Chem.*, **54**, 3176 (1989).
94. K. Yamamoto, *Kenkyu Kiyo-Konan Joshi Daigaku*, **25**, 89 (1988); *Chem. Abstr.*, **112**, 55187 (1990).
95. V. B. Rao, C. Schroeder, P. Margaretha, S. Wolff and W. C. Agosta, *J. Org. Chem.*, **50**, 3881 (1985).
96. M. Alvaro, H. Garcia, S. Iborra, M. A. Miranda and J. Primo, *Tetrahedron*, **43**, 143 (1987).
97. V. B. Rao, S. Wolff and W. C. Agosta, *J. Am. Chem. Soc.*, **107**, 521 (1985).
98. P. Margaretha, H. J. Rathjen, S. Wolff and W. C. Agosta, *J. Chem. Soc., Chem. Commun.*, 841 (1988).
99. H.-J. Rathjen, P. Margaretha, S. Woll and W. C. Agosta, *J. Am. Chem. Soc.*, **113**, 3904 (1991).
100. S. Hussain and W. C. Agosta, *Tetrahedron*, **38**, 1132 (1982).
101. U. Chiaccio, G. Musumarra and G. Purello, *J. Chem. Soc., Perkin Trans. 2*, 1591 (1988).
102. L. L. Costanzo, S. Giuffrida, S. Pistara and G. Condoelli, *J. Photochem.*, **18**, 307 (1982); *Chem. Abstr.*, **97**, 91413 (1982).
103. L. L. Costanzo, S. Giuffrida, S. Pistara, G. Scarlata and M. Torre, *J. Photochem.*, **18**, 317 (1982); *Chem. Abstr.*, **97**, 91414 (1982).
104. M. Mitani, I. Kato and K. Kogama, *J. Am. Chem. Soc.*, **105**, 6719 (1983).
105. T. Mukai, T. Kumagai, T. Segawa and N. Tanaka, *Kokagaku*, **11**, 75 (1987); *Chem. Abstr.*, **109**, 229740 (1988).
106. K. Mizuno, M. Ikeda and Y. Otsuji, *Chem. Lett.*, 1507 (1988).
107. K. Mizuno, N. Nakanishi, A. Tachibana and Y. Otsuji, *J. Chem. Soc., Chem. Commun.*, 344 (1991).
108. K. Mizuno, K. Nakanishi, J. Chosa, T. Nguyen and Y. Otsuji, *Tetrahedron Lett.*, **30**, 3689 (1989).
109. K. Mizuno, S. Toda and Y. Otsuji, *Chem. Lett.*, 203 (1987).
110. E. Kim, M. Christl and J. K. Kochi, *Chem. Ber.*, **123**, 1209 (1990).
111. H. Tomioka and M. Kanda, *Chem. Lett.*, 2223 (1990).
112. H. Kobayashi, M. Kato and T. Miwa, *Bull. Chem. Soc. Jpn.*, **58**, 490 (1985).
113. G. Mitchell and C. W. Rees, *J. Chem. Soc., Perkin Trans. 1*, 403 (1987).
114. K. Salisbury, *Eur. Pat. Appl. EP 70 608*; *Chem. Abstr.*, **99**, 22242 (1983).
115. H. E. Zimmerman and R. W. Binkley, *Tetrahedron Lett.*, **26**, 5859 (1985).
116. M. A. Fox, *J. Am. Chem. Soc.*, **101**, 4008 (1979); M. A. Fox, C.-C. Chess and K. A. Campbell, *J. Org. Chem.*, **48**, 321 (1983).
117. R. L. Amey and B. E. Smart, *J. Org. Chem.*, **46**, 4090 (1981).
118. S. Drenkard, J. Ferris and A. Eschenmoser, *Helv. Chim. Acta*, **73**, 1373 (1990).
119. M. Teng, J. W. Lauher and F. W. Fowler, *J. Org. Chem.*, **56**, 6840 (1991).
120. T. Kumagai, T. Segawa, Z. Endo and T. Mukai, *Tetrahedron Lett.*, **27**, 6225 (1986).
121. T. Kumagai, T. Segawa, T. Miyashi and T. Mukai, *Chem. Lett.*, 475 (1989).
122. H. Kobayashi, K. Takatoku, M. Kato and T. Miwa, *Bull. Chem. Soc. Jpn.*, **56**, 3449 (1983).
123. A. B. B. Ferreira and K. Salisbury, *J. Chem. Soc., Perkin Trans. 2*, 25 (1982).
124. H. E. Zimmerman and M. J. Zuraw, *J. Am. Chem. Soc.*, **111**, 2358 (1989).

125. H. E. Zimmerman and M. J. Zuraw, *J. Am. Chem. Soc.*, **111**, 7974 (1989).
126. H. E. Zimmerman and M. J. Zuraw, *J. Am. Chem. Soc.*, **111**, 2358 (1989).
127. H. E. Zimmerman and J. M. Cassel, *J. Org. Chem.*, **54**, 3800 (1989).
128. H. E. Zimmerman, F. L. Oakes and P. Campos, *J. Am. Chem. Soc.*, **111**, 1007 (1989).
129. T. Kumagai, H. Honda and T. Mukai, *Tetrahedron Lett.*, **26**, 5771 (1985).
130. T. Kumagai, K. Murukami, H. Hotta and T. Mukai, *Tetrahedron Lett.*, **23**, 4705 (1982).
131. T. Kumagai, T. Nittono, N. Tanaka and T. Mukai, *Tetrahedron Lett.*, **26**, 6093 (1985).
132. G. R. Tian, S. Sugiyama, A. Mori and H. Takeshita, *Bull. Chem. Soc. Jpn.*, **62**, 614 (1989).
133. L. A. Paquette and E. Bay, *J. Org. Chem.*, **47**, 4597 (1982).
134. L. A. Paquette and E. Bay, *J. Am. Chem. Soc.*, **106**, 6693 (1984).
135. L. A. Paquette, A. Varadarajan and L. D. Burke, *J. Am. Chem. Soc.*, **108**, 8032 (1986).
136. L. A. Paquette, A. Varadarajan and E. Bay, *J. Am. Chem. Soc.*, **106**, 6702 (1984).
137. H. Hemetsberger and M. Nobbe, *Tetrahedron*, **44**, 67 (1988).
138. M. Shi, Y. Okamoto and S. Takamuku, *J. Chem. Res. (S)*, 131 (1990); *Tetrahedron Lett.*, **30**, 6709 (1989); *J. Org. Chem.*, **55**, 3821 (1990).
139. C. O. Bender, D. S. Clyne and D. Dolman, *Can. J. Chem.*, **69**, 70 (1991).
140. C. O. Bender, D. L. Bengston, D. Dolman and S. F. O'Shea, *Can. J. Chem.*, **64**, 237 (1986).
141. C. O. Bender, D. Dolman and G. K. Murphy, *Can. J. Chem.*, **66**, 1656 (1988).
142. C. O. Bender, and D. Dolman, *Can. J. Chem.*, **67**, 82 (1989).
143. K. Nakanishi, K. Mizuno and Y. Otsuji, *J. Chem. Soc., Chem. Commun.*, 90 (1991).
144. F.-G. Klarner, B. M. J. Dogan, R. Weider, D. Ginsburg and E. Vogel, *Angew. Chem., Int. Ed. Engl.*, **25**, 346 (1986).
145. H. Izekawa, C. Kutal, K. Yasufuku and H. Yamazaki, *J. Am. Chem. Soc.*, **108**, 1589 (1986).
146. Q. H. Wu, B. W. Zhang, Y. F. Ming and Y. Cao, *J. Photochem. Photobiol. A: Chem.*, **61**, 53 (1991).
147. Q. Wu, B. Zhang and Y. Cao, *Youji Huaxue*, **11**, 590 (1991); *Chem. Abstr.*, **116**, 106021 (1992).
148. Y. Ming, Q. Wu and Y. Cao, *Chin. Chem. Lett.*, **1**, 161 (1990); *Chem. Abstr.*, **115**, 207286 (1991).
149. T. Hirano, T. Kumagai, T. Kumagai, T. Miyashi, K. Akiyama and Y. Ikegama, *J. Org. Chem.*, **56**, 1907 (1991).
150. T. Hirano, T. Kumagai, T. Miyashi, K. Akiyama and Y. Ikegami, *J. Org. Chem.*, **57**, 876 (1992).
151. J. H. Kwan, S. J. Lee and S. C. Shim, *Tetrahedron Lett.*, **32**, 6719 (1991).
152. S. C. Shim, S. J. Lee and J. H. Kwan, *Chem. Lett.*, 1767 (1991).
153. R. M. Bowman, T. R. Chamberlain, C.-W. Huang and J. J. McCullough, *J. Am. Chem. Soc.*, **96**, 692 (1974).
154. F. N. Putilin, V. V. Krotov, S. E. Nolde, V. E. Shevchenko and I. G. Bolesov, *Vestn. Mosk. Univ., Ser. 2: Khim.*, **32**, 583 (1991); *Chem. Abstr.*, **116**, 234844 (1992).
155. L. F. Tietze, A. Bergmann and K. Brueggemann, *Vortragstag. Fachgruppe Photochem.*, 101 (1983); *Chem. Abstr.*, **102**, 5358 (1985).
156. J. M. Rao and D. I. Schuster, Indian J. Chem., Sect. B, **22**, 114 (1983); *Chem. Abstr.*, **99**, 70262 (1983).
157. D. Bryce-Smith, E. H. Evans, A. Gilbert and H. S. McNeill, *J. Chem. Soc., Perkin Trans. 1*, 485 (1992).
158. A. Miyashita, S. Ikezu and H. Nohira, *Chem. Lett.*, 1235 (1985).
159. C. Chung, F. Nakamura, Y. Hashimoto and M. Hasegawa, *Chem. Lett.*, 779 (1991).
160. S. Chimichi, P. Sarti-Fantoni, G. Coppini, F. Perghem and G. Renzi, *J. Org. Chem.*, **52**, 5124 (1987).
161. Y. M. Chae, S. K. Yoon and S. C. Shim, *Bull. Korean Chem. Soc.*, **6**, 95 (1985); *Chem. Abstr.*, **104**, 33667 (1986).
162. F. D. Lewis, R. J. DeVoe and D. B. MacBlane, *J. Org. Chem.*, **47**, 1392 (1982).
163. K. Tsujimoto, T. Fujimori and M. Ohashi, *J. Chem. Soc., Chem. Commun.*, 304 (1986).
164. A. Oku, S. Urano, T. Nakaji, G. Qing and M. Abe, *J. Org. Chem.*, **57**, 2263 (1992).
165. A. Padwa, K. F. Koehler and A. Rodriguez, *J. Am. Chem. Soc.*, **103**, 4947 (1981).
166. D. Armesto, W. M. Horspool, N. Martin, A. Ramos and C. Seoane, *J. Org. Chem.*, **54**, 3069 (1989).
167. D. Armesto, W. M. Horspool, N. Martin, A. Ramos and C. Seoane, *J. Chem. Soc., Chem Commun.*, 1231 (1987).
168. Zh. A. Krasnaya, T. S. Stytsenko, V. S. Bogdanov and A. S. Dvornikov, *Izv. Akad. Nauk SSSR Ser. Khim.*, 1323 (1989); *Chem. Abstr.*, **112**, 76100 (1990).

169. D. I. Schuster, G. E. Heibel and J. Woning, *Angew. Chem., Int. Ed. Engl.*, **30**, 1345 (1991).
170. D. J. Hastings and A. C. Weedon, *J. Am. Chem. Soc.*, **113**, 8525 (1991).
171. G. Adembri, D. Donati, S. Fusi and F. Ponticelli, *Heterocycles*, **23**, 2885 (1985).
172. G. Adembri, D. Donati, S. Fusi, and F. Ponticelli, *Heterocycles*, **26**, 3221 (1987); *Chem. Abstr.*, **109**, 128872 (1988).
173. G. Adembri, D. Donati, S. Fusi and F. Ponticelli, *J. Chem. Soc., Perkin Trans. 1*, 2033 (1992).
174. P. Guerry and R. Neier, *Chimia*, **41**, 341 (1987); *Chem. Abstr.*, **109**, 128783 (1988).
175. Seitetsu Kagaku Co. Ltd., Jpn. Kokai Tokkyo JP 82 192 368; *Chem. Abstr.*, **98**, 198032 (1983).
176. T. Nishio, *J. Chem. Soc., Perkin Trans. 1*, 565 (1990).
177. T. Nishio, *J. Org. Chem.*, **49**, 827 (1984).
178. T. Nishio and Y. Omote, *J. Org. Chem.*, **50**, 1370 (1985).
179. T. Shimo, K. Somekawa, M. Sato and S. Kumamoto, *Nippon Kagaku Kaishi*, 1927 (1984); *Chem. Abstr.*, **102**, 149041 (1985).
180. T. Shimo, K. Somekawa and S. Kumamoto, *Nippon Kagaku Kaishi*, 394 (1983); *Chem. Abstr.*, **99**, 53539 (1983).
181. M. A. Kirpichenok, L. M. Mel'nikova, L. K. Denisov and I. I. Grandberg, *Khim. Geterosikl. Soedin.*, 1169 (1988); *Chem. Abstr.*, **111**, 77774 (1989).
182. C. J. Samuel, H. G. Beaton, K. Davey and C. J. Reader, *J. Photochem. Photobiol. A: Chem.*, **58**, 307 (1991).
183. I. Saito, K. Shimozono and T. Matsuura, *Tetrahedron Lett.*, **23**, 5439 (1982).
184. I. Saito, K. Shimozono and T. Matsuura, *J. Am. Chem. Soc.*, **102**, 3948 (1980); I. Saito, K. Kanehira, K. Shimozono and T. Matsuura, *Tetrahedron Lett.*, **21**, 2939 (1980); I. Saito, K. Shimozono and T. Matsuura, *J. Org. Chem.*, **47**, 4356 (1982).
185. I. Saito, K. Shimozono and T. Matsuura, *Tetrahedron Lett.*, **24**, 2195 (1983).
186. I. Saito, K. Shimozono and T. Matsuura, *J. Org. Chem.*, **47**, 4356 (1982).
187. I. Saito, K. Shimozono and T. Matsuura, *J. Am. Chem. Soc.*, **105**, 963 (1983); I. Saito, K. Shimozono and T. Matsuura, *J. Am. Chem. Soc.*, **102**, 3948 (1980).
188. I. Saito, F. Kubota, K. Shimozono and T. Matsuura, *Angew. Chem., Int. Ed. Engl.*, **22**, 629 (1983).
189. Y. Gelas-Mialhe, G. Mabiala and R. Vessirere, *J. Org. Chem.*, **52**, 5395 (1987).
190. F. D. Lewis and R. J. DeVoe, *Tetrahedron*, **38**, 1069 (1982).
191. S. Futamura, H. Ohta and Y. Kamiya, *Bull. Chem. Soc. Jpn.*, **55**, 2190 (1982).
192. J. Mattay, J. Runsink, R. Heckendorn and T. Winkler, *Tetrahedron*, **43**, 5781 (1987).
193. N. Al-Jalal and A. Gilbert, *J. Chem. Res. (S)*, 266 (1983).
194. N. Al-Jalal, *Gazz. Chim. Ital.*, **119**, 569 (1989).
195. N. Al-Jalal, *J. Photochem. Photobiol., A* **54**, 99 (1990); *J. Heterocycl. Chem.*, **27**, 1323 (1990).
196. J. Daub, J. Salbeck, T. Knochel, C. Fischer, H. Kunkely and K. M. Rapp, *Angew. Chem., Int. Ed. Engl.*, **28**, 1494 (1989).
197. S. Gierisch, W. Bauer, T. Burgemeister and J. Daub, *Chem. Ber.*, **122**, 2341 (1989).
198. S. Gierisch and J. Daub, *Chem. Ber.*, **122**, 69 (1989).
199. M. Irie and M. Mohri, *J. Org. Chem.*, **53**, 803 (1988).
200. P. K. Das, A. J. Muller, G. W. Griffin, I. R. Gould, C. H. Tung and N. J. Turro, *Photochem. Photobiol.*, **39**, 281 (1984); *Chem. Abstr.*, **101**, 71978 (1984).
201. P. K. Das, A. J. Muller and G. W. Griffin, *J. Org. Chem.*, **49**, 1977 (1984).
202. P. K. Das and G. W. Griffin, *J. Photochem.*, **27**, 317 (1984); *Chem. Abstr.*, **102**, 166261 (1985).
203. J. Brokatzky-Geiger and W. Eberbach, *Heterocycles*, **20**, 1519 (1983); *Chem. Abstr.*, **100**, 33896 (1984).
204. K. Ishii and M. Sakamoto, *Chem. Lett.*, 1107 (1985).
205. K. Ishii, M. Abe and M. Sakamoto, *J. Chem. Soc., Perkin Trans. 1*, 1937 (1987).
206. K. Ishii, T. Nakano, T. Zenko, M. Kotera and M. Sakamoto, *J. Chem. Soc., Perkin Trans. 1*, 2057 (1991).
207. D. Döpp, H. R. Memarian, M. A. Fischer, A. M. J. van Eijk and C. A. G. O. Varma, *Chem. Ber.*, **125**, 983 (1992).
208. W.-S. Chung, N. J. Turro, S. Srivastava and W. J. Le Noble, *J. Org. Chem.*, **56**, 5020 (1991).
209. W.-S. Chung, N. J. Turro, J. Silver and W. J. Le Noble, *J. Am. Chem. Soc.*, **112**, 1202 (1990).
210. C. Zagar and H.-D. Scharf, *Chem. Ber.*, **124**, 967 (1991).
211. H. A. J. Carless and A. F. E. Halfhide, *J. Chem. Soc., Perkin Trans. 1*, 1081 (1992).
212. T. Nishio, N. Okuda, and Y. Omote, *J. Chem. Soc., Perkin Trans. 1*, 1663 (1988).

213. V. P. Rao and V. Ramamurthy, *J. Org. Chem.*, **50**, 5009 (1985).
214. V. P. Rao and V. Ramamurthy, *J. Org. Chem.*, **53**, 332 (1988).
215. J. E. Gano and S. E. Rowan, *J. Org. Chem.*, **52**, 4608 (1987).
216. R. C. Cookson, D. E. Sadler and K. Salisbury, *J. Chem. Soc., Perkin Trans. 2*, 774 (1981).
217. I. A. Akhtar, J. J. McCullough, S. Vaitekunas, R. Faggiani and C. J. L. Lock, *Can. J. Chem.*, **60**, 1657 (1982).
218. T. Kumagai, K. Murukami, H. Hotta and T. Mukai, *Chem. Lett.*, 257 (1982); *Chem. Abstr.*, **96**, 180426 (1992).
219. N. Tanaka, T. Kumagai and T. Mukai, *Tetrahedron Lett.*, **27**, 6221 (1986).
220. Z. Goldschmidt and E. Genizi, *Tetrahedron Lett.*, **28**, 4867 (1987).
221. A. Sugimori, T. Furihata, S. Mikayama, M. Yoshida and Y. Nakanishi, *Bull. Chem. Soc. Jpn.*, **55**, 2906 (1982).
222. M. Tada, H. Hamazaki and H. Hirano, *Bull. Chem. Soc. Jpn.*, **55**, 3865 (1982).
223. F. D. Lewis and P. E. Correa, *J. Am. Chem. Soc.*, **103**, 7347 (1981).
224. R. C. Munjal and T. Durst, *Tetrahedron Lett.*, **33**, 7277 (1992).
225. D. R. Arnold, D. D. M. Wayner and M. Yoshida, *Can. J. Chem.*, **60**, 2313 (1982).
226. R. M. Wilson and K. A. Schnapp, *J. Am. Chem. Soc.*, **110**, 982 (1988).
227. L. M. Tolbert and M. B. Ali, *J. Am. Chem. Soc.*, **107**, 4589 (1985).
228. H. Tomioka, K. Ohno, Y. Izawa, R. A. Moss and R. C. Munjal, *Tetrahedron Lett.*, **25**, 5415 (1984).
229. K. Hannemann, *Angew. Chem., Int. Ed. Engl.*, **27**, 284 (1988).
230. X. Creary and M. E. Mehrsheikh-Mohammadi, *Tetrahedron Lett.*, **29**, 749 (1988).
231. S. H. Doss, A. A. Abdel-Wahab, E. M. Fruhof, H. Dürr, I. R. Gould and N. J. Turro, *J. Org. Chem.*, **52**, 434 (1987).
232. O. S. Mohamed, H. Dürr, M. T. Ismail and A. A. Abdel-Wahab, *Tetrahedron Lett.*, **30**, 1935 (1989).
233. H. Tomioka, Y. Ozaki, Y. Koyabu and Y. Izawa, *Tetrahedron Lett.*, **23**, 1917 (1982).
234. G. Maier and B. Wolf, *Synthesis*, 871 (1985).
235. A. K. Mal'tsev, P. S. Zuev, Yu. V. Tomilov and O. M. Nefedov, *Izv. Akad. Nauk SSSR, Ser. Khim.*, 2202 (1987); *Chem. Abstr.*, **108**, 221098 (1988).
236. O. M. Nefedov, P. S. Zuev, A. K. Mal'tsev and Yu. V. Tomilov, *Tetrahedron Lett.*, **30**, 763 (1989).
237. A. S. Kende, P. Hebeisen, P. J. Sanfilippo and B. H. Toder, *J. Am. Chem. Soc.*, **104**, 4244 (1982).
238. R. J. Sundberg, E. W. Baxter, W. J. Pitts, R. Ahmed-Schofield and T. Nishiguchi, *J. Org. Chem.*, **53**, 5097 (1988).
239. R. J. Sundberg and E. W. Baxter, *Tetrahedron Lett.*, **27**, 2687 (1986).
240. M. Tsuda and S. Oikawa, *J. Photopolym. Sci. Technol.*, **2**, 325 (1989).
241. T. Hudlicky, D. C. Banu, S. M. Naqvi and A. Srnak, *J. Org. Chem.*, **50**, 123 (1985).
242. R. Pelliciari, B. Natalini, S. Cecchetti and S. Santucci, *Tetrahedron Lett.*, **25**, 3103 (1984).
243. L. Thijs, F. J. Dommerholt, F. M. C. Leemuis and B. Zwanenburg, *Tetrahedron Lett.*, **31**, 6589 (1990).
244. F. J. Dommerholt, L. Thijs and B. Zwanenburg, *Tetrahedron Lett.*, **32**, 1495 (1991).
245. S. Ghosh, I. Dutta, R. Chakrabaory, T. K. Das, J. Sengupta and D. C. Sarkar, *Tetrahedron*, **45**, 1441 (1989).
246. J. Cossy, D. Belotti, A. Thelland and J.-P. Pete, *Synthesis*, 720 (1988).
247. H. Tomioka, N. Hayashi, T. Asano and Y. Izawa, *Bull. Chem. Soc. Jpn.*, **56**, 758 (1983).
248. U. Burger and D. Zellweger, *Helv. Chim. Acta*, **69**, 676 (1986).
249. A. Padwa, S. P. Carter, H. Nimmesgern and P. D. Stull, *J. Am. Chem. Soc.*, **110**, 2894 (1988).
250. R. R. Gallucci and M. Jones, *J. Org. Chem.*, **50**, 4404 (1985).
251. C. Marfisi, P. Verlaque, G. Davidovics, J. Pourcin, L. Pizzala, J.-P. Aycard and H. Bodot, *J. Org. Chem.*, **48**, 533 (1983).
252. R. L. Danheiser and D. D. Cha, *Tetrahedron Lett.*, **31**, 1527 (1990).
253. H. Nickels, H. Durr and F. Toda, *Chem. Ber.*, **119**, 2249 (1986).
254. A. Ghosh, U. K. Banerjee and R. V. Venkateswaran, *Tetrahedron*, **46**, 3077 (1990).
255. U. K. Banerjee and R. V. Venkateswaran, *Tetrahedron Lett.*, **24**, 423 (1983).
256. G. Lawton, C. J. Moody and C. J. Pearson, *J. Chem. Soc., Perkin Trans. 1*, 877 (1987).
257. G. Lawton, C. J. Moody, C. J. Pearson and D. J. Williams, *J. Chem. Soc., Perkin Trans. 1*, 88 (1987).

258. C. J. Moody, C. J. Pearson and G. Lawton, *Tetrahedron Lett.*, **26**, 3167 (1985).
259. J. M. Bolster and R. M. Kellogg, *J. Org. Chem.*, **47**, 4429 (1982).
260. S. Yamazaki, K. Kohgami, M. Okazaki, S. Yamabe and T. Arai, *J. Org. Chem.*, **54**, 240 (1989).
261. S. Murata, T. Yamamoto, H. Tomioka, H. K. Lee, H. K. Kim and A. Yabe, *J. Chem. Soc.*, *Chem. Commun.*, 1258 (1990).
262. H. K. Lee, H. K. Kim and H. Tomioka, *Bull. Korean Chem. Soc.*, **9**, 399 (1988); *Chem. Abstr.*, **111**, 133742 (1989).
263. (a) S. Wolff, B. R. Venepalli, C. F. George and W. C. Agosta, *J. Am. Chem. Soc.*, **110**, 6785 (1988).
     (b) F. Kunisch, K. Hobert and P. Welzel, *Tetrahedron Lett.*, **26**, 5433 (1985).
264. M. Ulbricht, J. U. Thurner, M. Siegmund and G. Tomaschewski, *Z. Chem.*, **28**, 102 (1988).
265. V. A. Nikolaev, N. N. Khimich and I. K. Korobitsyna, *Khim. Geterosikl. Soedin.*, 321 (1985); *Chem. Abstr.*, **103**, 87818 (1985).
266. M. Ulbricht, G. Tomaschewski and J. U. Thurner, *J. Prakt. Chem.*, **331**, 873 (1989).
267. W. Klotzer, G. Dorler, B. Stanovnik and M. Tisler, *Heterocycles*, **22**, 1763 (1984).
268. R. L. Rosati, L. V. Kapili, P. Morrissey and J. A. Retsema, *J. Med. Chem.*, **33**, 2910 (1990); R. L. Rosati, L. V. Kapili, P. Morrissey, J. Bordner and E. Subramanian, *J. Am. Chem. Soc.*, **104**, 4262 (1982).
269. R. Yamaguchi, K. Honda and M. Kawanisi, *J. Chem. Soc.*, *Chem. Commun.*, 83 (1987).
270. M. Banciu, *Rev. Roum. Cheim.*, **27**, 769 (1982); *Chem. Abstr.*, **97**, 71208 (1982).
271. P. E. Eaton and B. D. Liepzig, *J. Am. Chem. Soc.*, **105**, 1657 (1983).
272. P. P. Waanders, L. Thijs and B. Zwanenburg, *Tetrahedron Lett.*, **28**, 2409 (1987).
273. L. Thijs, D. M. Egenberger and B. Zwanenburg, *Tetrahedron Lett.*, **30**, 2153 (1989).
274. D. W. Norbeck and J. B. Kramer, *J. Am. Chem. Soc.*, **110**, 7217 (1988).
275. B. Baumbach, J. Bendig and T. Rampke, *Z. Chem.*, **30**, 28 (1990).
276. G. G. Lazarev, V. L. Kushov and Ya. S. Lebedev, *Chem. Phys. Lett.*, **170**, 94 (1990).
277. K. Tanigaki, T. Honda and T. W. Ebbesen, *J. Photopolym. Sci. Technol.*, **2**, 341 (1989).
278. H. Pasch, H. Much and T. Buchheim, *J. Prakt. Chem.*, **330**, 634 (1988).
279. Yu. I. Kol'tsov, V. I. Yudina and G. V. Solomonenko, *Zh. Prikl. Khim. (Leningrad)*, **62**, 191 (1989); *Chem. Abstr.*, **111**, 114816 (1989).
280. T. N. Chikirisova, D. B. Askerov, N. A. Sakharova, V. F. Tarasov, T. M. Filippova and R. D. Erlikh, *Zh. Org. Khim.*, **24**, 1714 (1988); *Chem. Abstr*, **110**, 154122 (1989).
281. H. Boettcher, J. Marx, J. Lucas and B. Strehmel, *J. Prakt. Chem.*, **324**, 237 (1982).
282. V. A. Kalibabchuk, G. L. Linnikova and V. D. Romanenko, *Zh. Fiz. Khim.*, **56**, 2271 (1982); *Chem. Abstr.*, **97**, 215313 (1982).
283. A. Z. Yankelevich, S. V. Rykov and G. A. Nikoforov, *Izv. Akad. Nauk SSSR, Ser. Khim.*, 2325 (1987); *Chem. Abstr.*, **109**, 54173 (1988).
284. T. Terasawa, N. Ikekawa and M. Morisaki, *Chem. Pharm. Bull.*, **34**, 935 (1986).
285. H. Tomioka, K. Tabayashi and Y. Izawa, *J. Chem. Soc.*, *Chem. Commun.*, 906 (1985).
286. L. M. Hadel, V. M. Maloney, M. S. Platz, W. G. McGimpsey and J. C. Scaiano, *J. Phys. Chem.*, **90**, 2488 (1986).
287. G. W. Griffin and K. A. Horn, *J. Am. Chem. Soc.*, **109**, 4919 (1987).
288. T. G. Savino, V. P. Sentilnathan and M. S. Platz, *Tetrahedron*, **42**, 2167 (1986).
289. D. Griller, A. S. Nazran and J. C. Scaiano, *Acc. Chem. Res.*, **17**, 283 (1984).
290. T. J. Amick and H. Schechter, *Tetrahedron Lett.*, **27**, 901 (1986).
291. S.-J. Chang, B. K. R. Shankar and H. Schechter, *J. Org. Chem.*, **47**, 4226 (1982).
292. R. J. McMahon and O. L. Chapman, *J. Am. Chem. Soc.*, **109**, 683 (1987).
293. B. B. Wright and M. S. Platz, *J. Am. Chem. Soc.*, **105**, 628 (1983).
294. M. J. Fritz, F. L. Ramos and M. S. Platz, *J. Org. Chem.*, **50**, 3523 (1985).
295. D. W. Jones and A. Pomfret, *J. Chem. Soc.*, *Chem. Commun.*, 991 (1982).
296. G. Maier, K. A. Reuter, L. Franz and H. P. Reisenauer, *Tetrahedron Lett.*, **26**, 1845 (1985).
297. G. Maier and D. Born, *Angew. Chem.*, *Int. Ed. Engl.*, **28**, 1050 (1989).
298. P. Eisenbarth and M. Regitz, *Chem. Ber.*, **115**, 3796 (1982).
299. G. Maier and F. Fieischer, *Tetrahedron Lett.*, **32**, 57 (1991).
300. K. Hannemann, J. Wirz and A. Riesen, *Helv. Chim. Acta*, **71**, 1841 (1988).
301. K. Hannemann and J. Wirz, *Angew. Chem.*, *Int. Ed. Engl.*, **27**, 853 (1988).
302. O. L. Chapman and C. J. Abelt, *J. Org. Chem.*, **52**, 1218 (1987).
303. W. Ruhl, F. Bolsing and E. Hofer, *Z. Naturforsch.*, *B: Chem. Sci.*, **42**, 1487 (1987).

304. H. K.-W. Hui and H. Schechter, *Tetrahedron Lett.*, **23**, 5115 (1982).
305. P. Umrigar, G. W. Griffin, S. N. Ege, A. D. Adams and P. K. Das, *Can. J. Chem.*, **62**, 2456 (1984).
306. G. Maier, H. P. Reisenauer, H. Balli, W. Brandt and R. Janoschek, *Angew. Chem.*, **102**, 920 (1990).
307. S. Tivakornpannarai and E. E. Waali, *J. Am. Chem. Soc.*, **108**, 6058 (1986).
308. H. Tomioka and K. Hirai, *J. Chem. Soc., Chem. Commun.*, 362 (1989).
309. A. Bsceidero, A. Igau, G. Bertrand, M. J. Menu, Y. Dartiguenave and J. J. Bonet, *J. Am. Chem. Soc.*, **108**, 7868 (1986).
310. H. Tomioka and K. Hirai, *J. Chem. Soc., Chem. Commun.*, 1611 (1990).
311. G. R. Gillette, A. Baceiredo and G. Bertrand, *Angew. Chem., Int. Ed. Engl.*, **29**, 1429 (1990).
312. A. Baceiredo, G. Bertrand and G. Sicard, *J. Am. Chem. Soc.*, **107**, 4781 (1985).
313. W. Ando, A. Sekiguchi, T. Hagiwara, T. Migita, V. Chowdhry, F. H. Westheimer, S. L. Kammula, M. Green and M. Jones, Jr., *J. Am. Chem. Soc.*, **101**, 6393 (1979).
314. G. Maas, M. Alt, K. Schneider and A. Fronda, *Chem. Ber.*, **124**, 1295 (1991).
315. K. Schneider, B. Daucher, A. Fronda and G. Maas, *Chem. Ber.*, **123**, 589 (1990).
316. A. Sekiguchi, T. Sato and W. Ando, *Organometallics*, **6**, 2337 (1988).
317. A. Sekiguchi and W. Ando, *Organometallics*, **6**, 1857 (1988).
318. A. Sekiguchi, W. Ando and K. Honda, *Tetrahedron Lett.*, **26**, 2337 (1985).
319. A. Sekiguchi and W. Ando, *Chem. Lett.*, 2025 (1986).
320. G. Markl and W. Schlosser, *Angew. Chem., Int. Ed. Engl.*, **27**, 963 (1988).
321. G. Markl, W. Schlosser and W. S. Sheldrick, *Tetrahedron Lett.*, **29**, 467 (1988).
322. W. Ando, H. Tanikawa and A. Sekiguchi, *Tetrahedron Lett.*, **24**, 4245 (1983).
323. A. Sekiguchi, H. Tanikawa and W. Ando, *Organometallics*, **4**, 584 (1985).
324. A. Sekiguchi and W. Ando, *J. Am. Chem. Soc.*, **106**, 1486 (1984).
325. G. Maas, K. Schneider and W. Ando, *J. Chem. Soc., Chem. Commun.*, 72 (1988).
326. R. Bruckmann, K. Schnieder and G. Maas, *Tetrahedron*, **45**, 5517 (1989).
327. G. Maas and A. Fronda, *J. Organomet. Chem.*, **398**, 229 (1990).
328. W. Ando, H. Yoshida, K. Kurishima and M. Sugiyama, *J. Am. Chem. Soc.*, **113**, 7790 (1991).
329. J. E. Gano, R. H. Wettach, M. S. Platz and V. P. Senthilnathan, *J. Am. Chem. Soc.*, **104**, 2326 (1982).
330. D. R. Meyers, V. P. Senthilnathan, M. S. Platz and M. Jones, *J. Am. Chem. Soc.*, **108**, 4232 (1986).
331. P. R. West, O. L. Chapman and J.-P. LeRoux, *J. Am. Chem. Soc.*, **104**, 1779 (1982).
332. R. J. McMahon, C. J. Abelt, O. L. Chapman, J. W. Johnson, C. L. Kreil, J.-P. LeRoux, A. M. Mooring and P. R. West, *J. Am. Chem. Soc.*, **109**, 2456 (1987).
333. R. J. McMahon and O. L. Chapman, *J. Am. Chem. Soc.*, **108**, 1713 (1986).
334. P. R. West, A. M. Mooring, R. J. McMahon and O. L. Chapman, *J. Org. Chem.*, **51**, 1316 (1986).
335. R. S. Hutton, H. D. Roth, M. L. M. Schilling and J. W. Suggs, *J. Am. Chem. Soc.*, **103**, 5147 (1981).
336. J. U. Thurner, R. Stessre, B. Hinzmann and G. Tomaschewski, *Z. Chem.*, **28**, 62 (1988).
337. P. B. Grasse, J. J. Zupancic, S. C. Lapin, M. P. Hendrich and G. B. Schuster, *J. Org. Chem.*, **50**, 2352 (1985).
338. D. J. Astles, M. Girard, D. Griller, R. J. Kolt and D. D. M. Wayner, *J. Org. Chem.*, **53**, 6053 (1988).
339. H. Tomioka and Y. Izawa, *Yuki Gosei Kagaku Kyokaishi*, **43**, 344 (1985); *Chem. Abstr.*, **102**, 220160 (1985).
340. W. Sander, *Angew. Chem., Int. Ed. Engl.*, **25**, 255 (1986).
341. W. Sander, *Angew. Chem., Int. Ed. Engl.*, **24**, 988 (1985).
342. I. R. Dunkin and C. J. Shields, *J. Chem. Soc., Chem. Commun.*, 154 (1986).
343. I. R. Dunkin and G. A. Bell, *Tetrahedron*, **41**, 339 (1985).
344. S. Murata, H. Tomika, T. Kawase and M. Oda, *J. Org. Chem.*, **55**, 4502 (1990).
345. W. W. Sander, *J. Org. Chem.*, **53**, 2091 (1988).
346. R. Stoesser, J. U. Thurner, W. Warncke, G. Tomaschewski and T. Taplick, *Z. Chem.*, **29**, 67 (1989).
347. E. Leyva, R. L. Barcus and M. S. Platz, *J. Am. Chem. Soc.*, **108**, 7786 (1986).
348. J. Zayasa and M. S. Platz, *J. Am. Chem. Soc.*, **107**, 7065 (1985).

349. N. J. Turro, Y. Cha and I. R. Gould, *Tetrahedron Lett.*, **26**, 5951 (1985).
350. H. Tomioka and Y. Nunome, *J. Chem. Soc., Chem. Commun.*, 1243 (1990).
351. W. Kirmse and K. Kund, *J. Org. Chem.*, **55**, 2325 (1990).
352. W. Kirmse, K. Kund, E. Ritzer, A. E. Dorigo and K. N. Houk, *J. Am. Chem. Soc.*, **108**, 6045 (1986).
353. G. A. Olah, H. Doggweiler and J. D. Felberg, *J. Org. Chem.*, **49**, 2116 (1984).
354. H. Tomioka, Y. Ozaki and Y. Izawa, *Tetrahedron*, **41**, 4987 (1985).
355. C. J. Shields and G. B. Schuster, *Tetrahedron Lett.*, **28**, 853 (1987).
356. K. Friedrich, U. Jansen and W. Kirmse, *Tetrahedron Lett.*, **26**, 193 (1985).
357. W. Kirmse, P. V. Chiem and V. Schurig, *Tetrahedron Lett.*, **26**, 197 (1985).
358. G. K. S. Prakash, R. W. Ellis, J. D. Felberg and G. A. Olah, *J. Am. Chem. Soc.*, **108**, 1341 (1986).
359. R. P. L'Esperance, T. M. Ford and M. Jones, *J. Am. Chem. Soc.*, **110**, 209 (1988).
360. J. C. Scaiano, W. G. McGimpsey and H. L. Casal, *J. Am. Chem. Soc.*, **107**, 7204 (1985).
361. W. G. McGimpsey and J. C. Scaiano, *Tetrahedron Lett.*, **27**, 547 (1986).
362. L. Capuano, G. Bolz, R. Burger, V. Burckhardt and V. Huch, *Justus Liebigs Ann. Chem.*, 239 (1990).
363. N. J. Turro and Y. Cha, *Tetrahedron Lett.*, **28**, 1723 (1987).
364. N. J. Turro, Y. Cha, I. R. Gould, A. Padwa, J. R. Gasdaska and M. Tomas, *J. Org. Chem.*, **50**, 4415 (1985).
365. A. Padwa, J. R. Gasdaska, M. Tomas, N. J. Turro, Y. Cha and I. R. Gould, *J. Am. Chem. Soc.*, **108**, 6739 (1986).
366. D. Griller, M. Majewski, W. G. McGimpsey, A. S. Nazran and J. C. Scaiano, *J. Org. Chem.*, **53**, 1550 (1988).
367. P. L. Coe, M. I. Cook, N. J. Goodchild and P. N. Edwards, *J. Fluorine Chem.*, **34**, 191 (1986).
368. S. Murata, Y. Ohtawa and H. Tomioka, *Chem. Lett.*, 853 (1989).
369. H. Tomioka, Y. Ohtawa and S. Murata, *Nippon Kagaku Kaishi*, 1413 (1989); *Chem. Abstr*, **112**, 118034 (1990).
370. A. Alberti, *Gazz. Chim. Ital.*, **119**, 541 (1989).
371 G. A. Bell and I. R. Dunkin, *J. Chem. Soc., Faraday Trans. 2*, **81**, 725 (1985).
372. W. W. Sander, *J. Org. Chem.*, **54**, 4265 (1989).
373. M. Torres, J. L. Bourdelande, A. Clement and O. P. Strausz, *J. Am. Chem. Soc.*, **105**, 1698 (1983).
374. P. G. Mahaffy, D. Visser, M. Torres, J. L. Bourdelande and O. P. Strausz, *J. Org. Chem.*, **52**, 2680 (1987).
375. C. Bachmann, T. Y. N'Guessan, F. Debu, M. Monnier, J. Pourcin, J.-P. Aycard and H. Bodot, *J. Am. Chem. Soc.*, **112**, 7488 (1990).
376. W. Ruhl, F. Bolsing, E. Hofer, D. Speer and M. Jansen, *Z. Naturforsch., B: Anorg. Chem., Org. Chem.*, **41B**, 772 (1986).
377. K. Tanigaki and T. W. Ebbesen, *J. Am. Chem. Soc.*, **109**, 5883 (1987).
378. N. Yoneda, T. Fukuhara, T. Mizokami and S. Susuki, *Chem. Lett.*, 459 (1991).
379. H. Baumann, H. G. O. Becker, K. P. Kronfeld, D. Pfeifer and H. J. Timpe, *J. Photochem.*, **28**, 393 (1985).
380. W. Ortmann and E. Fanghaenel, *Z. Chem.*, **25**, 109 (1985).
381. F. Walkow and H. Kress, *Z. Chem.*, **30**, 27 (1990).
382. F. Walkow, *Z. Chem.*, **30**, 62 (1990).
383. M. Spasova and R. Zakharieva, *Collect. Czech. Chem. Commun.*, **54**, 196 (1989).
384. T. Takahashi, K. L. Kirk and L. A. Cohen, *J. Org. Chem.*, **49**, 1951 (1984).
385. E. P. Kyba, S.-T. Liu, K. Chocklingam and B. R. Reddy, *J. Org. Chem.*, **53**, 3513 (1988).
386. P. Lorencak, G. Raabe, J. J. Radziszewski and C. Wentrup, *J. Chem. Soc., Chem. Commun.*, 916 (1986).
387. A. Hassner, R. Fibiger and A. S. Amarasekara, *J. Org. Chem.*, **53**, 22 (1988).
388. M. Petrusova, M. Matulova, M. Fedoronko and L. Petrus, *Synthesis*, 209 (1991).
389. D. I. Gasking and G. H. Whitham, *J. Chem. Soc., Perkin Trans. 1*, 409 (1985).
390. J. Frank, G. Stoll and H. Musso, *Justus Liebigs Ann. Chem.*, 1990 (1986).
391. O. E. Edwards and W. Rank, *Can. J. Chem.*, **68**, 1425 (1990).
392. P. Coombes, A. Goosen and B. Taljaard, *Heterocycles*, **28**, 559 (1989).
393. H. Quast and B. Seiferling, *Justus Liebigs Ann. Chem.*, 1553 (1982).

394. J. G. Radziszewski, J. W. Downing, C. Wentrup, P. Kaszynski, M. Jawdosiuk, P. Kovacic and J. Michl, *J. Am. Chem. Soc.*, **107**, 2799 (1985).
395. T. Sasaki, S. Eguchi and T. Okano, *Tetrahedron Lett.*, **23**, 4969 (1982).
396. J. G. Radziszewski, J. W. Downing, C. Wentrup, P. Kaszynski, M. Jawdosiuk, P. Kovacic and J. Michl, *J. Am. Chem. Soc.*, **106**, 7996 (1984).
397. M. Jawdosiuk and P. Kovacic, *J. Chem. Soc., Perkin Trans. 1*, 2583 (1984).
398. H. Quast, P. Eckert and B. Seiferling, *Chem. Ber.*, **118**, 3535 (1985).
399. H. Quast, P. Eckert and B. Seiferling, *Justus Liebigs Ann. Chem.*, 696 (1985).
400. W. Marterer, O. Klinger, R. Thiergardt, E. Beckmann, H. Fritz and H. Prinzbach, *Chem. Ber.*, **124**, 621 (1991).
401. O. Klinger and H. Prinzbach, *Angew. Chem., Int. Ed. Engl.*, **26**, 566 (1987).
402. W. Marterer, H. Fritza and H. Prinzbach, *Tetrahedron Lett.*, **28**, 5497 (1987).
403. S. Kodama, *J. Phys. Chem.*, **92**, 5019 (1988).
404. P. E. Eaton and R. E. Hormann, *J. Am. Chem. Soc.*, **109**, 1268 (1987).
405. C. O. Kappe and G. Faerber, *J. Chem. Soc., Perkin Trans. 1*, 1342 (1991).
406. J.-P. Praly, Z. El Kharraf and G. Descotes, *Tetrahedron Lett.*, **31**, 4441 (1990).
407. C. G. Crespan, *J. Org. Chem.*, **51**, 332 (1986).
408. M. E. Sigman, T. Autrey and G. B. Schuster, *J. Am. Chem.Soc.*, **110**, 4297 (1988).
409. T. Autrey and G. B. Schuster, *J. Am. Chem. Soc.*, **109**, 5814 (1987).
410. J. H. Teles and G. Maier, *Chem. Ber.*, **122**, 745 (1989).
411. J. H. Teles, G. Maier, B. A. Hess and L. J. Schaad, *Chem. Ber.*, **122**,749 (1989).
412. W. Lwowski, S. Kanemasa, R. A. Murray, V. T. Ramakrishnan, T. V. Thiruvengadam, K. Yoshida and A. Subbaraj, *J. Org.Chem.*, **51**, 1719 (1986).
413. M. Mitani, O. Tachizawa, H. Takeuchi and K. Koyama, *Chem. Lett.*, 1029 (1987).
414. M. A. Loreto, L. Pellacani and P. A. Tardella, *Tetrahedron Lett.*, **30**, 5025 (1989).
415. M. G. Barlow, G. M. Harrison, R. N. Haszeldine, W. D. Morton, P. Shaw-Luckman and M. D. Ward, *J. Chem. Soc., Perkin Trans. 1*, 2101 (1982).
416. V. P. Semenov, A. N. Studenikov, B. B. Ivanov and K. A. Ogloblin, *Zh. Org. Khim.*, **26**, 331 (1990); *Chem. Abstr.*, **113**, 59001 (1990).
417. M. Mitani, O. Tachizawa, H. Takenchi and K. Koyama, *J. Org. Chem.*, **54**, 5397 (1989).
418. M. A. Loreto, L. Pellacani and P. A. Tardella, *Tetrahedron Lett.*, **30**, 2975 (1989).
419. W. Abraham, S. Siegert and D. Kreysig, *J. Prakt. Chem.*, **331**, 177 (1989).
420. E. Kozlowska-Gramsz and W. E. Hahn, *Pol. J.Chem.*, **59**, 493 (1985).
421. R. Bertolaccini, M. A. Loreto, L. Pellacani, P. A. Tardella and G. Cerichelli, *J.Org.Chem.*, **52**, 1859 (1987).
422. E. Eibler, J. Kasbauer, H. Pohl and J. Sauer, *Tetrahedron Lett.*, **28**, 1097 (1987).
423. A. R. Katritzky and T. Siddiqui, *J.Chem.Soc., Perkin Trans. 1*, 2953 (1982).
424. K. Banert, *Angew. Chem., Int. Ed. Engl.*, **24**, 216 (1985).
425. K. Banert, *Chem. Ber.*, **122**, 1175 (1989).
426. R. A. Abramovitch, M. Konieczny, W. Pennington, S. Kanamathareddy and M. Vedachalam, *J.Chem.Soc., Chem. Commun.*, 269 (1990).
427. K. Isomura, M. Sakurai, T. Komura, M. Saruwatari and H. Taniguchi, *Chem. Lett.*, 883 (1987).
428. K. Banert, *Tetrahedron Lett.*, **26**, 5261 (1985).
429. K. Banert, *Chem. Ber.*, **120**, 1891 (1987).
430. K. Banert and M. Hagedorn, *Angew, Chem., Int. Ed.Engl.*, **29**, 103 (1990).
431. D. M. B. Hickey, C. J. Moody and C. W. Rees, *J.Chem. Soc., Perkin Trans. 1*, 1119 (1986).
432. R. W. Saalfrank, E. Ackermann, M. Fischer, U. Wirth and H. Zimmermann, *Chem. Ber.*, **123**, 115 (1990).
433. H. Suschitzky, W. Kramer, R. Neidlein and H. Uhl, *J.Chem. Soc., Perkin Trans. 1*, 983 (1988).
434. K. Hirota, K. Maruhashi, N. Kitamura, T.Asao and S. Senda, *J.Chem. Soc., Perkin Trans. 1*, 1719 (1984).
435. C. J. Moody and G. J. Warrellow, *J.Chem. Soc., Perkin Trans. 1*, 1123 (1986).
436. C. J. Moody and G. J. Warrellow, *J.Chem. Soc., Perkin Trans.* 1913 (1987).
437. A. Subbaraj, O. S. Rao and W. Lowowsji, *J.Org.Chem.*, **54**, 3945 (1989).
438. Y. Naruta, T. Yokota, N. Nagai and K. Maruyama, *J.Chem. Soc., Chem. Commun.*, 972 (1986).
439. Y. Naruta, N. Nagai, T. Yokota and K. Maruyama, *Chem. Lett.*, 1185 (1986).
440. T. Kozuka, *Bull. Chem. Soc. Jpn.*, **55**, 2922 (1982).

441. W. Stadlbauer, *Monatsh. Chem.*, **118**, 1297 (1987).
442. T. Miyasaka, H. Tanaka, K. Satoh, M. Imahashi, K. Yamaguchi and Y. Iitaka, *J.Heterocycl. Chem.*, **24**, 873 (1987).
443. R. Okazaki, M. Takahashi, N. Inamot, T. Sugawara and H. Iwamura, *Chem. Lett.*, 2083 (1989).
444. K. Banert, *Chem. Ber.*, **122**, 123 (1989).
445. K. Banert, *Angew. Chem.*, *Int. Ed.Engl.*, **26**, 879 (1987).
446. M. E. Staretz and S. B. Hastie, *J.Org. Chem.*, **56**, 428 (1991).
447. P. N. Congalone and R. B. Woodward, *J.Am. Chem. Soc.*, **105**, 902 (1983).
448. G. Maier, J. Glatthaar and H. P. Reisenauer, *Chem. Ber.*, **122**, 2403 (1989).
449. S. S. Zigler, L. M. Johnson and R. West, *J.Organomet. Chem.*, **341**, 187 (1988).
450. S. S. Zigler, R. West and J. Michl, *Chem.Lett.*, 1025 (1986).
451. A. Sekiguchi, W. Ando and K. Honda, *Chem. Lett.*, 1029 (1986).
452. K. M. Welsh, J. Michl and R. West, *J.Am. Chem. Soc.*, **110**, 6689 (1988).
453. G. Raabe, H. Vancik, R. West and J. Michl, *J.Am.Chem.Soc.***108**, 671 (1986).
454. S. S. Zigler, K. M. Welsh and R. West, *J.Am. Chem. Soc.*, **109**, 4392 (1987).
455. A. Baceiredo, G. Bertrand, J. P. Majoral, U. Wermuth and R. Schmutzler, *J.Am.Chem.Soc.*, **106**, 7065 (1984).
456. M. J. P. Harger and S. Westlake, *Tetrahedron* **38**, 3073 (1982).
457. M. J. P. Harger and S. Westlake, *Tetrahedron*, **38**, 1511 (1982).
458. M. J. P. Harger and S. Westlake, *J.Chem. Soc.*, *Perkin Trans. 1*, 2351 (1984).
459. A. Baceiredo, G. Bertrand, J. P. Majoral, F. E. Anba and G. Manuel, *J. Am. Chem. Soc.*, **107**, 3945 (1985).
460. G. Sicard, A. Baceiredo, G. Bertrand and J. P. Magoral, *Angew. Chem.*, *Int. Ed. Engl.*, **23**, 459 (1984).
461. T. Tsumuaya and W. Ando, *Chem. Lett.*, 1043 (1989).
462. J. Barrau, D. L. Bean, K. M. Welsh, R. West and J. Michl, *Organometallics*, **8**, 2606 (1989).
463. Y.-Z. Li, J. P. Kirby, M. W. George, M. Poliakoff and G. B. Schuster, *J. Am. Chem. Soc.*, **110**, 8092 (1988).
464. E. Leyva, M. S. Platz, G. Persy and J. Wirz, *J. Am. Chem. Soc.*, **108**, 3783 (1986).
465. A. K. Schrock and G. B. Schuster, *J. Am. Chem. Soc.*, **106**, 5228 (1984).
466. E . Leyva and M. S. Platz, *Tetrahedron Lett.*, **28**, 11 (1987).
467. A. K. Schrock and G. B. Schuster, *J. Am. Chem. Soc.*, **106**, 5234 (1984).
468. J. C. Hayes and R. S. Sheridan, *J. Am. Chem. Soc.*, **112**, 5879 (1990).
469. C. J. Shields, D. R. Chrisope, G. B. Schuster, A. J. Dixon, M. Poliakoff and J. J. Turner, *J. Am. Chem. Soc.*, **109**, 4723 (1987).
470. I. R. Dunkin and P. C. P. Thomson, *J. Chem. Soc.*, *Chem. Commun.*, 1192 (1982).
471. I. R. Dunkin, T. Donnelly and T. S. Lockhart, *Tetrahedron Lett.*, **26**, 359 (1985).
472. T. Donnelly, I. R. Dunkin, D. S. D. Norwood, A. Prentice, C. J. Shields and P. C. P. Thomson, *J. Chem. Soc.*, *Perkin Trans. 2*, 307 (1985).
473. K. Lamara and R. K. Smalley, *Tetrahedron*, **47**, 2277 (1991).
474. T. Y. Liang and G. B. Schuster, *J. Am. Chem. Soc.*, **109**, 7803 (1987).
475. U. Henriksen and O. Buchardt, *Tetrahedron Lett.*, **31**, 2443 (1990).
476. C. M. Daly, B. Iddon, H. Suschitzky, U. Jordis and F. Sauter, *J. Chem. Soc.*, *Perkin Trans. 1*, 1933 (1988).
477. D. I. Patel, E. F. V. Scriven, R. K. Smalley, H. Suschitsky and D. I. C. Scopes, *J. Chem. Soc. Perkin Trans. 1*, 1911 (1985).
478. R. Hayes and R. K. Smalley, *J. Chem. Res. (S)*, 14 (1988).
479. A. Albini, G. F. Bettinetti, E. Fasani and S. Pietra, *Gazz. Chim. Ital.*, **112**, 13 (1982).
480. H. Sawanishi, T. Hirai and T. Tsuchiya, *Heterocycles*, **19**, 1043 (1982).
481. A. Albini, G. Bettinetti, G. Minoli and S. Pietra, *Gazz. Chim. Ital.*, **118**, 61 (1988).
482. S. Murata, T. Sugara and H. Iwamura, *J. Chem. Soc.*, *Chem. Commun.*, 1198 (1984).
483. J. Kikuchi, *J. Appl. Chem.*, **60**, 549 (1988).
484. H. Sawanishi, H. Muramatsu and T. Tsuchiya, *J. Chem. Soc.*, *Chem. Commun.*, 628 (1990).
485. H. Sawanishi, K. Tajima and T. Tsuchiya, *Chem. Pharm. Bull.*, **35**, 4101 (1987).
486. H. Sawanishi and T. Tsuchiya, *Chem. Pharm. Bull.*, **33**, 5603 (1985).
487. H. Sawanishi, K. Tajima and T. Tsuchiya, *Chem. Pharm. Bull.*, **35**, 3175 (1987).
488. H. Sawanishi, K. Tajima, M. Osada and T. Tsuchiya, *Chem. Pharm. Bull.*, **32**, 4694 (1984).

489. H. Sashida, A. Fujii and T. Tsuchiya, Chem. Pharm. Bull., 35, 3182 (1987).
490. H. Sashida, M. Kaname and T. Tsuchiya, Chem. Pharm. Bull., 35, 4676 (1987).
491. R. A. Abramovitch, B. Bachowska and P. Tomasik, Pol. J. Chem., 58, 805 (1984).
492. H. Sashida, A. Fujii and T. Tsuchiya, Chem. Pharm. Bull., 35, 4110 (1987).
493. H. Sashida, A. Fujii, H. Sawanishi and T. Tsuchiya, Heterocycles, 24, 2147 (1986).
494. H. Sashida, M. Kaname and T. Tsuchiya, Chem. Pharm. Bull., 38, 2919 (1990).
495. H. Sawanishi, T. Hirai and T. Tsuchiya, Heterocycles, 22, 1501 (1984).
496. H. Sawanishi, H. Sashida and T. Tsuchiya, Chem. Pharm. Bull., 33, 4564 (1985).
497. H. Sawanishi and T. Tsuchiya, Heterocycles, 22, 2725 (1984).
498. H. Sawanishi, S. Saito and T. Tsuchiya, Chem. Pharm. Bull., 38, 2992 (1990).
499. H. Sawanishi, S. Saito and T. Tsuchiya, Chem. Pharm. Bull., 36, 4240 (1988).
500. H. Sawanishi and T. Tsuchiya, J. Chem. Soc., Chem. Commun., 723 (1990).
501. Y. Morimoto, F. Matsuda and H. Shirahama, Tetrahedron Lett., 31, 6031 (1990).
502. Y. Kondo, R. Watanabe, T. Sakamoto and H. Yamanaka, Chem. Pharm. Bull., 37, 2933 (1989).
503. K. Yakushijin, T. Tsuruta and H. Furukawa, Chem. Pharm. Bull., 30, 140 (1982).
504. J. C. Teulade, A. Gueiffier, H. Viols, J. P. Chapat, G. Grassy, B. Perly and G. Dauphin, J. Chem. Soc., Perkin Trans. 1, 1895 (1989).
505. R. S. Gairns, C. J. Moody and C. W. Rees, J. Chem. Soc., Chem. Commun., 1818 (1985).
506. R. S. Gairns, C. J. Moody and C. W. Rees, J. Chem. Soc., Perkin Trans. 1, 501 (1986).
507. P. C. Montevecchi and P. Spagnolo, J. Org. Chem., 47, 1996 (1982).
508. A. Yabe, K. Honda, H. Nakanishi, R. A. Hayes and O. L. Chapman, J. Am. Chem. Soc., 110, 4441 (1988).
509. A. Yabe, A. Ochi and H. Moriyama, J. Chem. Soc., Chem. Commun., 1744 (1987).
510. A. Yabe, K. Honda, H. Nakanishi and K. Someno, Chem. Lett., 1407 (1984).
511. A. Yabe, A. Ochi, H. Moriyama and E. Masuda, Jpn. Kokai Tokkyo Koho JP 63, 225, 361; Chem. Abstr., 110, 212841 (1989).
512. A. V. El'tsov, F. M. Dmitiev, L. M. Gornostaev and N. I. Rtishchev, Zh. Org. Khim., 22, 2361 (1986); Chem. Abstr., 107, 58268 (1987).
513. B. Ganem, Chemtracts: Org. Chem., 3, 367 (1990).
514. M. J. T. Young and M. S. Platz, Tetrahedron Lett., 30, 2199 (1989).
515. R. Poe, J. Grayzar, M. J. T. Young, E. Leyva, K. A. Schnapp and M. S. Platz, J. Am. Chem. Soc., 113, 3209 (1991).
516. S. X. Cai and J. F. W. Keana, Tetrahedron Lett., 30, 5409 (1989).
517. J. F. W. Keana and S. X. Kai, J. Org. Chem., 55, 3640 (1990).
518. D. S. Watt, K. Kawada, E. Leyva and M. S. Platz, Tetrahedron Lett., 30, 899(1989).
519. A. Bruendl, E. Clausen and K. Buff, Z. Naturforsch., B: Chem. Sci., 45, 1072 (1990).
520. T. Melvin and G. B. Schuster, Photochem. Photobiol., 51, 155 (1990).
521. N. Soundararajan and M. S. Platz, J. Org. Chem., 55, 2034 (1990).
522. E. Levya, D. Munoz and M. S. Platz, J. Org. Chem., 54, 5938 (1989).
523. K. W. Haider, E. Migridicyan, M. S. Platz, N. Soundararajan and A. Depres, Mol. Cryst. Liq. Cryst., 176, 85 (1989).
524. M. W. Shaffer and M. S. Platz, Tetrahedron Lett., 30, 6465 (1989).
525. C. J. Rhodes and M. C. R. Symons, J. Chem. Res. (S), 28 (1988).
526. C. J. Rhodes, J. Chem. Res. (S), 170 (1989).
527. S. Yamamoto, Y. Yamamoto and K. Hayashi, Bull. Chem. Soc. Jpn., 64, 346 (1991).
528. H. Tachikawa, M. Shiotani, O. Masaru and K. Ohta, J. Chem. Phys., 96, 164 (1992); Chem Abstr., 116, 40785 (1992).
529. H. Shiromaru, Y. Achiba, K. Kimura and Y. T. Lee, J. Chem. Phys., 91, 17 (1987); Chem Abstr., 106, 32294 (1987).
530. K. Matsuua and H. Muto, J. Chem. Phys., 94, 4078 (1991); Chem. Abstr., 115, 63079 (1991).
531. H. Yamakita, M. Tazawa and K. Hayakawa, Radioisotopes, 34, 374 (1985); Chem. Abstr., 105, 115459 (1986).
532. H. Yamaoka, T. Matsuyama, B. Z. Tang, T. Matsuda and T. Higashimura, Proc. Tihany Symp Radiat. Chem., 511 (1987); Chem. Abstr., 108, 56732 (1988).
533. J. Rideout and M. R. C. Symons, J. Chem. Res. (S), 268 (1984); Chem. Abstr., 102, 4537: (1985).
534. J. P. Michaut, J. J. Pinvidic, J. Roncin and H. Szwarc, Chem. Phys. Lett., 120, 555 (1985); Chem Abstr., 104, 185471 (1986).

535. J. T. Guthrie and L. J. Squires, *Polym., Photochem.*, 135 (1984); *Chem. Abstr.*, **101**, 6429 (1984).
536. J. Hacaloglu, A. S. S. Gokmen and E. Illenberger, *Turk. Kim. Derg.*, **14**, 133 (1990); *Chem. Abstr.*, **114**, 100907 (1991).
537. N. Kato, Y. Kawai, M. Matsushima, T. Miyazaki and K. Fueki, *Radiat. Phys. Chem.*, **27**, 13 (1986).
538. D. Luther, H. C. Abendorth and H. G. Koennecke, *Z. Chem.*, **25**, 221 (1985); *Chem. Abstr.*, **103**, 137074 (1985).
539. A. Negron-Mendoza, Z. D. Draganic, R. Navarro-Gonzalez and I. G. Draganic, *Radiat. Res.*, **95**, 248 (1983); *Chem. Abstr.*, **99**, 118481 (1983).
540. G. Albarran, C. Juarez and A. Negron-Mendoza, *Proc. Tihany Symp. Radiat. Chem.*, **7**, 397 (1991); *Chem. Abstr.*, **115**, 250650 (1991).
541. E. P. Petryaev, L. A. Kiseleva, A. M. Kovalevskaya and V. G. Shlyk, *Khim. Promst. (Moscow)*, 23 (1984); *Chem. Abstr.*, **100**, 144469 (1984).
542. M. L. Sagu and K. K. Bhattacharyya, *J. Macromol. Sci., Chem.*, **A23**, 1099 (1986); *Chem. Abstr.*, **105**, 62493 (1986).
543. M. H. Rao and K. N. Rao, *Radiat. Phys. Chem.*, **26**, 669 (1985); *Chem. Abstr.*, **104**, 51939 (1986).
544. S. G. Durgar'yan, V. V. Teplyakov, I. N. Kozhikhova and L. E. Starannikova, *Vysokomol. Soedin., Ser. A*, **28**, 67 (1986); *Chem. Abstr.*, **104**, 149901 (1986).
545. D. N. Aneli and M. I. Topchiashvili, *Izv. Akad. Nauk Gruz. SSR, Ser. Khim.*, **12**, 56 (1986); *Chem. Abstr.*, **104**, 225722 (1986).
546. L. I. Nantanzon and V. A. Tarasenko, *Vysokomol. Soedin., Ser. B*, **27**, 835 (1985); *Chem. Abstr.*, **104**, 131294 (1986).
547. A. K. Fritzsche, *J. Appl. Polym. Sci.*, **32**, 3541 (1986); *Chem. Abstr.*, **105**, 135036 (1986).
548. M. Vandermarliere, G. Groeninckx, H. Reynaers, C. Riekel and M. H. J. Koch, *Morphol. Polym., Proc., Europhys. Conf. Macromol. Phys., 17th.* (Ed. B. Sedlacek), 421 (1986); *Chem. Abstr.*, **105**, 191911 (1986).
549. V. Ya. Kabanov and R. E. Aliev, *Radiat. Phys. Chem.*, **26**, 697 (1985); *Chem. Abstr.*, **104**, 110310 (1986).
550. A. H. A. Roediger and T. A. Du Plessis, *Radiat. Phys. Chem.*, **27**, 461 (1986); *Chem. Abstr.*, **105**, 98321 (1986).
551. H. B. Ambroz, T. J. Kemp, G. K. Przybytniak and T. Wronka, *Radiat. Phys. Chem.*, **36**, 207 (1990).
552. A. D. Strouphauer, C. L. Liotta and R. W. Fink, *Int. J. Appl. Radiat. Isot.*, **35**, 787 (1984); *Chem. Abstr.*, **102**, 24197 (1985).
553. C. J. Rhodes, *J. Chem. Res. (S)*, 28 (1989); *Chem. Abstr.*, **111**, 97344 (1989).
554. M. Atina and A. Ricci, *Int. J. Mass Spectrom. Ion Processes*, **115**, 89 (1992); *Chem. Abstr.*, **117**, 25684 (1992).

CHAPTER **8**

The thermochemistry of the C≡C bond

T. KASPRZYCKA-GUTTMAN

Department of Chemistry, University of Warsaw, 02-093 Warsaw, Pasteura 1, Poland

I. INTRODUCTION

Measurement of the enthalpies of hydrogenation of alkynes is not an easy thermochemical problem. There are three phases present during the hydrogenation reaction: liquid (solvent + alkyne), gaseous (H_2) and solid (catalysator). Participation of these three phases requires one to include some correction factors in the measured data in order to reach reasonable final results. Values of the enthalpies of hydrogenation have been collected and improved many times, involving a great number of arduous experiments until reliable results were obtained. The calculation methods are here of very great importance, e.g. the additivity method to estimate heats of formation, entropies and heat capacities by Benson and Buss[1–3]. Very helpful are also the methods of molecular mechanics[4,5]. Calculated results obtained by numerical methods are usually compared with those of experiment, and are often found to be very accurate. Differences between the calculated results and the experimental data engender very constructive analysis of both methods, resulting in improvement of the final results. The review is devoted to the experimental studies regarding the heats of hydrogenation of linear alkynes and phenylalkynes[6,7] and to the determination of the enthalpy of mixing of alkyne either with an alkane or an alkene and alkynes as well as that of the mixing of some other solvents[8,9].

Supplement C2: The chemistry of triple-bonded functional groups
Edited by S. Patai © 1994 John Wiley & Sons Ltd

II. HEATS OF FORMATION AND HYDROGENATION OF LINEAR ALKYNES

The heats of formation of alkynes, generally for the ideal gas state at 298.15 K, were investigated mainly by Wyatt and Stafford[11], Williams and Smith[12] and Okabe and Dibeler[13].

The work of Benson and coworkers[1] was fundamental for the development of additivity methods for estimation of the heats of formation, heat capacities and entropies of organic compounds in general, not only alkynes. They proposed the two simplest methods: atom and bond additivity. These methods are often sufficient for estimation of the thermochemical properties of ideal gases, including also acetylenes.

Shaw[2] extended the principles of additivity to the heat capacities of organic liquids including but-2-yne. This was very important, because most alkynes are liquid in the standard state and the higher homologues are not even very volatile.

In 1977 Luria and Benson[3] extended group additivity for liquid alkynes as a function of temperature between 150 K and 280 K for but-1-yne and between 250 K and 290 K for but-2-yne.

Rogers and coworkers[6] studied the hydrogenation of alkynes in hexane solution, with the reaction producing an essentially indefinitely diluted solution of reaction product in hexane. This experimental method may be used for hydrocarbons of more than five carbon atoms. The reaction

$$C_nH_{2n-2}(7\% \text{ in hexane}) + 2H_2(g) = C_nH_{2n+2}(\text{dil})$$

is not thermochemically different from the gas-phase reaction

$$C_nH_{2n-2}(g) + 2H_2(g) = C_nH_{2n+2}(g)$$

The above reactions are correct in the absence of the necessary heats: the heat of solution and the heat of vaporization. This would be also qualitatively acceptable if the separation of alkyne molecules from the pure liquids to the gas is energetically comparable to separation of alkyne molecules by an inert solvent to form the diluted reactant solution, and if vaporization of the alkane is energetically comparable to separation of product alkane molecules to form the very diluted product solution. All this is possible in theory alone. The only substantial energetic effect which might be superimposed on the heat of hydrogenation is the heat of dilution of the 7% alkyne solution in hexane to a very diluted solution of the hydrogenation product, also in hexane. Because of its use as a thermochemical standard we must make the same assumptions with regard to the hydrogenation of hexene as we make regarding the alkyne reaction. The authors[6] suggest that there is no substantial energetic difference between the reactions in diluted hexane solution and the gas-phase reaction. They believe that 'these assumptions regarding the reaction' in hexane solvent v the gas phase are good ones, and that even the little error which does result will be largely cancelled out when the alkyne is compared with the standard 1-hexene. Rogers and coworkers[6] compared the heats of hydrogenation of some terminal alkenes, determined by direct hydrogenation, with the corresponding values determined indirectly from heats of combustion.

The heat of hydrogenation of 1-hexene equals -126.65 kJ mol^{-1} in the gas phase; the heats of hydrogenation of the C_6 to C_{10} terminal alkenes[15] measured directly equal respectively -127.76, -126.77, -125.77, -127.07 and -126.98 kJ mol^{-1}.

The heats of hydrogenation were determined using a calorimeter according to the procedure described previously[6,15]. All alkynes were 99% pure or better, only the 5-decyne was 98% or better.

The heats of hydrogenation of linear alkynes from C_2 to C_{10} are listed in Table 1, which shows that the terminal alkynes are much more energetic than their other isomers. All C to C_{10} terminal alkynes are about equally exothermic on hydrogenation; the enthalpy

TABLE 1. Comparison of experimental heats of hydrogenation of linear alkynes with those calculated by molecular mechanics in kJ mol^{-1} at 298 K[a]

Compound	Location of triple bond in chain				
	1	2	3	4	5
Ethyne	−314.26				
	(0.01)[26]				
	−309.95				
Propyne	−291.82				
	(0.001)[26]				
	−291.53				
Butyne	−292.45	−274.57			
	(0.01)	(0.01)[26]			
	−290.98	−274.40			
Pentyne	−290.98	−275.49			
	(0.02)[27]	(0.03)[27]			
	−289.98	−273.40			
Hexyne	−289.56	−274.86	−272.60		
	(0.005)	(0.02)	(0.01)		
	−289.52	−272.48	−272.60		
Heptyne	−291.61	−272.60	−270.59		
	(0.02)	(0.01)	(0.01)		
	−289.52	−271.93	−271.60		
Octyne	−289.56	−272.56	−271.30	−268.88	
	(0.03)	(0.005)	(0.01)	(0.01)	
	−288.81	−271.39	−271.05	−270.59	
Nonyne	−291.15	−272.48	−270.84	−270.84	
	(0.02)	(0.02)	(0.01)	(0.02)	
	−288.81	−271.18	−270.55	−270.09	
Decyne	−291.61	−273.27	−271.56	−289.59	−268.37
	(0.02)	(0.02)	(0.02)	(0.02)	(0.02)
	−288.76	−271.18	−270.30	−269.59	−269.50

[a]The first value is the experimental value, the number in parentheses indicates the 95% confidence limit on both experimental and literature data, while the last value is the calculated value.

the triple bond is about the same for all of them. There is a sharp drop in the magnitude of the heat of hydrogenation of a triple bond in the 1 position to one in the 2 position, followed by a small consistent decrease as we proceed from the 2 position to the 5 position in decyne or the 4 position in nonyne and octyne.

From the heptyne series through the decyne series there are some clear generalizations that can be drawn from the observed heats of hydrogenation without smaller acetylene molecules. The 1-alkynes (except acetylene) have a heat of hydrogenation which is 290.56 ± 1.25 kJ mol^{-1}; the 2-alkynes for the same group have heats of hydrogenation which are 273.81 ± 1.25 kJ mol^{-1}.

Turner[16] investigated the heat of hydrogenation of 4-octyne in glacial acetic acid. The value found for this alkyne was 262.94 kJ mol^{-1}. This value is less exothermic than those obtained by other investigators. Rogers and coworkers[6] suggest what is very probable, that alkenes or alkynes being Lewis bases react with glacial acetic acid exothermically, but alkanes do not. This acid–base pair is broken up during hydrogenation of an alkene or an alkyne to the corresponding alkane.

Williams[17], Skinner[18] and Rogers[19] determined the heat of hydrogenation with the catalyst present as a slurry in the acid, then corrected the results for the heat of solution of the reaction product determined in a separate experiment. The heat of solution of octane in acetic acid according to Skinner is 6.28 kJ mol^{-1}; Rogers obtained 5.44 kJ mol^{-1}. If these endothermic correction factors are substracted from Turner's value for H_{H2} of 4-octyne, they obtain $- 268.53$ and $- 269.37$ kJ mol^{-1}: these fit the values for hydrogenation in hexane solutions (Table 1). It has been suggested that the heats of solution of alkynes in acetate acid might be near zero, but this seems to be improbable. Rogers obtained 3.97 kJ mol^{-1} for the heat of solution of 4-octyne and 6.99 kJ mol^{-1} for octane and corrected the heat of hydrogenation value of Turner to $- 265.86$ kJ mol^{-1}. This value is lower than the result obtained by Rogers. The difference between the values in Table 1 and Skinner's value for the heat of hydrogenation of 1,7-octadiyne[18] corrected for the heat of solution of octane, which is $- 584.89$ kJ mol^{-1}, is negligible within uncertainties.

Rogers and coworkers[6] did not anticipate conjugative or steric influence on a linear diyne with distant triple bonds and thus one-half of Skinner's value should be equal to the experimental data for 1-octyne. The 2-alkynes for the same groups of compounds have heats of hydrogenation which are 273.81 kJ mol^{-1}. The bulk of the difference, 16.73 kJ mol^{-1}, is clearly due to a substituent effect, replacement of a hydrogen by an alkyl.

The 3-alkynes have heats of hydrogenation in the range 271.30 ± 1.25 kJ mol^{-1}. Rogers and coworkers[6] found that the 4-alkynes have heats of hydrogenation of 269.62 ± 1.25 kJ mol^{-1} and the 5-alkyne has a heat of hydrogenation of 268.37 ± 2.93 kJ mol^{-1}. There is a trend that the heat of hydrogenation of the triple bond is monotonically diminished as the bond moves from the end of the chain toward the middle of the chain.

The same calorimeter was used to determine the enthalpy of hydrogenation of phenylalkynes[7]. In this case the calorimetric standard was allylbenzene, used in order to minimize the difference in reaction medium interactions for the reactant and for the standard. The kinetics of hexene hydrogenation was progressively slowed by the presence of accumulated aromatic product in the calorimeter. When allylbenzene was used as the thermochemical standard this kinetic effect did not occur.

The enthalpies of hydrogenation for phenylacetylene, 1-phenyl-1-propyne, 1-phenyl-1-butyne, 1-phenyl-1-hexyne, diphenylacetylene and diphenylbutadiyne are listed on Table 2. The authors[7] used a 10% (without stating whether this is by volume or by weight) standard solution of allylbenzene in hexane which was injected into a 25 ml calorimeter, containing a slurry of Pd catalyst on carbon support, also in hexane. Following this a 10% solution of alkyne in hexane was injected into the same calorimeter. Taking the enthalpy

TABLE 2. The heat of hydrogenation and formation of phenylalkynes in kJ mol^{-1}

Compound	H_h	H_h (lit)	H_f	H_f (lit)
Phenylacetylene	$- 276.74$ ± 1.67	$- 296.00^{20}$ $- 270.88^{29}$	306.76 ± 1.71	327.72^2
1-Phenyl-1-propyne	$- 260.42$ ± 2.09		268.33 ± 2.18	
1-Phenyl-1-butyne	$- 262.51$ ± 0.83		248.77 ± 1.05	
1-Phenyl-1-hexyne	$- 262.09$ ± 1.25		226.07^a	
Diphenylacetylene	$- 249.53$ ± 2.51	$- 268.37^{30}$	385.18 ± 2.67	
Diphenylbutadiyne	$- 496.13^a$ ± 4.18			

^{a} Approximate value.

TABLE 3. Comparison of experimental heats of formation of linear alkynes with those calculated by molecular mechanics, in kJ mol^{-1}, at 298 K[a]

Compound	Calculated		Experimental	
	$h_f^°$	$h_f^°$(gas)	$h_f^°$	$h_f^°$(gas)
Acetylene	228.81	225.13[10]	227.51(0.79)	227.14 ± 0.79[10]
Propyne	187.82	185.59[10]	185.85(0.88)	185.55 ± 0.88[10]
1-Butyne	166.01	165.82[10]	165.34(0.88)	165.07 ± 0.88[10]
2-Butyne	149.26	146.05[10]	145.32(0.83)	145.09 ± 0.84[10]
1-Pentyne	143.86		144.02(3.56)	
2-Pentyne	126.27		128.53(3.56)	
1-Hexyne	122.17		122.38(1.21)	
2-Hexyne	105.13		107.72(2.43)	
3-Hexyne	105.25		105.46(1.92)	
1-Heptyne	100.48		103.83(2.55)	
2-Heptyne	83.40		84.82(2.22)	
3-Heptyne	83.06		82.81(2.42)	
1-Octyne	79.04		80.76(3.64)	
2-Octyne	61.63		63.81(1.51)	
3-Octyne	61.29		62.55(1.82)	
4-Octyne	60.83		60.12(2.13)	
1-Nonyne	57.86		62.29(2.97)	
2-Nonyne	40.23		43.62(3.14)	
3-Nonyne	39.61		41.99(2.47)	
4-Nonyne	39.14		41.99(2.84)	
1-Decyne	36.59		41.90(3.43)	
2-Decyne	19.01		23.57(3.42)	
3-Decyne	18.13		21.85(3.26)	
4-Decyne	17.42		19.88(3.14)	
5-Decyne	17.33		18.67(3.26)	

[a]Numbers in parentheses indicate 95% confidence limits.

of hydrogenation of a terminal, unconjugated double bond[14] to be − 126.65 kJ mol^{-1}, one can calculate. H_h of the alkyne from the ratios of the heats produced.

The enthalpy of hydrogenation approximates the standard state value. For example, the enthalpy of hydrogenation for phenylacetylene obtained by the author[2] is − 276.83 ± 0.25 kJ mol^{-1}. The earlier value by Rogers and coworkers[29] is − 270.88 kJ mol^{-1}.

The heat of hydrogenation by Flitcroft and Skinner was − 293.51 kJ mol^{-1}. They used ethanol as the calorimeter fluid. After correction for the enthalpy of solution of ethylbenzene in ethanol, they found H_h = − 296.02 ± 4.19 kJ mol^{-1}, for the pure liquid to liquid hydrogenation[20]. The authors suggested that a catalyst interaction and enthalpies of vaporization are included in the earlier values, making them more exothermic than they should be by about 20.93 kJ mol^{-1}, which seems to be very probable.

Rogers calculated the heat of formation for linear alkynes by the method of molecular mechanics, as described previously[2,22]. The values of the heat of formation for linear alkynes are listed in Table 3. These values are very similar to values of Benson and coworkers[1], Cox and Pilcher[10] or Stull and coworkers[23].

The heats of formation of phenylalkynes were obtained from combustion by Rossini's group[24,28]. For ethylbenzene they used H_f = 29.94 ± 0.79 kJ mol^{-1} and obtained the enthalpy of formation for phenylacetylene equal to 306.76 ± 1.71 kJ mol^{-1}. Using the combustion results of Rossini's group[24,28], for n-propylbenzene 7.91 ± 0.79 J mol^{-1} Rogers[7] obtained the enthalpy of formation for 1-phenyl-1-propyne 268.33 ± 2.18 kJ mol^{-1} The same method used for 1-phenyl-1-butyne, assuming that H_f for n-butylbenzene equals

-13.73 ± 0.79 kJ mol^{-1}, gave the result $H_f = 248.77 \pm 1.05$ kJ mol^{-1}. The value of H_f for 1-phenyl-1-hexyne has not been determined precisely because the value for the enthalpy of formation of n-hexylbenzene is not known. Although the authors[7] tried to extrapolate the latter from values for phenylacetylene, phenylpropyne and phenylbutyne, obtaining the value $H_f = -35.59$ kJ mol^{-1}, still the final value H_f of 1-phenyl-1-hexyne is an estimate only.

The results obtained by Coops and coworkers[25] on diphenylbutadiyne, and its hydrogenation product in the crystalline state, are not comparable to the present results, because the appropriate enthalpies of sublimation are not known.

III. HEATS OF MIXING OF SOME ALKYNES

The heats of mixing of alkynes confirm earlier observations, namely that the description of the triple bond as two orthogonal ethylene-like π bonds is inadequate on thermochemical grounds. The sum of the strength of the two carbon–hydrogen bonds in acetylene can be calculated[47] using the value for the heat of formation of C_2 which equals[31] 836.8 ± 3.8 kJ mol^{-1}, as well as that for the hydrogen atom, which is[31] 217.778 ± 0.004 kJ mol^{-1}, and that for acetylene[10] (227.14 ± 0.79 kJ mol^{-1}). The value obtained for the sum of the strengths of the two carbon–hydrogen bonds is 1045.3 ± 3.9 kJ mol^{-1}.

The heat of mixing for some alkynes with n-alkanes or n-alkenes was investigated by Woycicki[8]. Changes in the observed values were caused by alteration of the position of the unsaturated bond in the alkyne or by varying its chain length even when using mixtures with the same saturated hydrocarbon.

Woycicki[8] investigated binary mixtures, i.e. n-hexane with 1-n-hexyne or with 3-n-hexyne as well as n-heptane with n-heptyne and also a second series, namely tetrachloromethane with 1-n-hexyne, 3-n-hexyne or 1-n-heptyne. Molar excess heats of mixing were measured in a continuous titration calorimeter of the Becker–Kiefer type[32, 33] over a wide concentration range at 298.15 K. The method has been described elsewhere[32,33]. The purities of all substances were determined by g.l.c. and were better than 99.31%.

Experimental excess heat of mixing (H^E) values are given in Table 4, where $(1-x)$ means mole fractions of n-hexane or n-heptane or tetrachloromethane. In the mixtures of n-alkanes with n-alkynes the excess heat of mixing values diminish as the chain length is increased. Similar results were found for mixtures of n-alkanes + n-alkanes with n-alkenes[34]. The H^E_{max} values are about 10 times as large as those in the n-alkanes + n-alkene mixtures. We can say that the interactions of the π bond doublets in the triple bonds between the alkyne molecules are disproportionally stronger than those of the single π bonds between the alkene molecules. However, since cis–$trans$ isomerism does not occur in alkynes, the effect of shift of the triple bond towards the centre of the molecule cannot be correctly compared with that observed in alkane + alkene mixtures. This effect manifests itself in alkynes much more strongly—a drop of about 180 J mol^{-1}—than it does in alkane–alkene mixtures. The result of the interaction of the alkyne bond with tetrachloromethane is a very small positive excess heat of mixing in the case of tetrachloromethane with 1-n-hexyne, while there occur both positive and negative excess heats of mixing in the case of tetrachloromethane with 1-n-heptyne, and strongly negative heats of mixing in the case of tetrachloromethane with 3-n-hexyne.

Transition from tetrachloromethane + 1-n-hexyne to tetrachloromethane + 1-n-heptyne is accompanied by a decrease in the excess heat of mixing by about 30 J mol^{-1} in the middle range of concentrations. It is interesting that, in the same direction and by about the same magnitude, the excess heat of mixing varies from n-hexane + 1-n-hexyne to n-heptane + 1 n-heptyne.

The enthalpy of mixing observed is assumed to be a sum of several contributions[35]. The decrease in excess heat of mixing in the tetrachloromethane mixtures appears to be

TABLE 4. Excess heats of mixing: n-alkane + n-alkyne, tetrachloromethane + n-alkyne at 298.15 K

$1-x$	H^E (J mol^{-1})	$1-x$	H^E (J mol^{-1})	$1-x$	H^E (J mol^{-1})
$(1-x)C_6H_{14} + xCH{\equiv}CC_4H_9$		$(1-x)C_6H_{14} + xC_2H_5C{\equiv}CC_2H_5$		$(1-x)C_7H_{16} + xCH{\equiv}CC_5H_{11}$	
0.05	109	0.05	79	0.05	105
0.10	208	0.10	149	0.10	198
0.15	294	0.15	212	0.15	280
0.20	371	0.20	267	0.20	351
0.25	436	0.25	314	0.25	411
0.30	489	0.30	352	0.30	461
0.35	531	0.35	382	0.35	501
0.40	562	0.40	403	0.40	530
0.45	583	0.45	416[a]	0.45	549
0.50	592	0.45	394[b]	0.50	557[a]
0.55	589	0.50	421[a]	0.50	567[b]
0.60	576[a]	0.50	397[b]	0.55	554[a]
0.65	585[b]	0.55	418[a]	0.55	562[b]
0.70	558	0.55	392[b]	0.60	541[a]
0.75	518	0.60	406[a]	0.60	546[b]
0.80	466	0.60	380[b]	0.65	520
0.85	401	0.65	359	0.70	482
0.90	322	0.70	332	0.75	432
0.95	230	0.75	297	0.80	371
	123	0.80	253	0.85	297
		0.85	202	0.90	211
		0.90	143	0.95	112
		0.95	76		
$(1-x)CCl_4 + xCH{\equiv}CC_4H_9$		$(1-x)CCl_4 + xC_2H_5C{\equiv}CC_2H_5$		$(1-x)CCl_4 + xCH{\equiv}CC_5H_{11}$	
0.05	0	0.05	−85	0.05	−5
0.10	0	0.10	−165	0.10	−8
0.15	1	0.15	−237	0.15	−12
0.20	3	0.20	−301	0.20	−14
0.25	4	0.25	−359	0.25	−15
0.30	6	0.30	−408	0.30	−16
0.35	8	0.35	−449	0.35	−16
0.40	11	0.40	−480	0.40	−16
0.45	13	0.45	−501	0.45	−14
0.50	15	0.50	−512[a]	0.50	−13
0.55	17	0.50	−514[b]	0.55	−10
0.60	19[a]	0.55	−514[a]	0.60	−8[a]
0.65	23[b]	0.55	−513[b]	0.60	−6[b]
0.70	23	0.60	−506[a]	0.65	−5[a]
0.75	22	0.60	−501[b]	0.65	−4[b]
0.80	20	0.65	−479	0.70	−2
0.85	17	0.70	−445	0.75	0
0.90	13	0.75	−400	0.80	2
0.95	7	0.80	−343	0.85	3
		0.85	−273	0.90	3
		0.90	−193	0.95	2
		0.95	−96		

First component added to n-alkyne.
n-Alkyne added to the first component.

primarily accounted for by contributions due to differences in π–π interactions in 1-n-hexyne and 1-n-heptyne. For tetrachloromethane with alkenes the excess heat of mixing increases as the chain length is increased, whereas for tetrachloromethane with alkynes it is decreased. The nature of this interaction continues to be undecided: Ghassemi, Grolier and Kiehiaian's suggestion[36] of an n–π type interaction seems to be correct. The large negative excess heat of mixing for tetrachloromethane with 3-n-hexyne surprises, but similar values were found by Wilhelm[37]. The differences between the values of heat of mixing for tetrachloromethane + 1-n-hexyne and 3-n-hexyne, about 530 J mol^{-1}, cannot be explained in terms of the differences of thermal contributions due to π–π interactions, because [{H^E (0.5 CH≡CC$_4$H$_9$ + 0.5 C$_6$H$_{14}$) – H^E (0.5 C$_2$H$_5$C≡CC$_2$H$_5$ + 0.5 C$_6$H$_{14}$)} $\cong$ 80 J mol^{-1}]. The authors[8] suggest that this rise in excess enthalpy must be associated with a considerable increase of the contributions due to π–π interactions or due to interactions of more than one tetrachloromethane molecule with one 3-n-hexyne molecule. The π–π bond, which is very exposed in the structurally simple 3-n-hexyne molecule, could make the last-mentioned mode of interaction possible. The slight shift of H^E_{max} towards higher mole fractions of tetrachloromethane can hardly bear out this suggestion.

Letcher and coworkers[38–45] investigated the excess heat of mixing of alkynes with different solvents; such as cycloalkanes(cyclopentane, cyclohexane, cycloheptane, cyclooctane, cyclodecane) with an n-alkyne (n-hexyne or n-heptyne or n-octyne)[38]; bicyclic compounds[decahydronaphthalene (C$_{10}$H$_{18}$), bicyclohexyl (C$_{12}$H$_{22}$), 1,2,3,4-tetrahydronaphthalene (C$_{10}$H$_{12}$), cyclohexylbenzene (C$_{12}$H$_{16}$)] with n-hexyne or n-heptyne[39]; benzene with n-hexyne[40], n-heptyne or n-octyne[41]; and finally, 1,3,5-trimethylbenzene with n-hexyne n-heptyne or n-octyne[9].

For all these series excess heats of mixing were measured using an LKB 2107 type

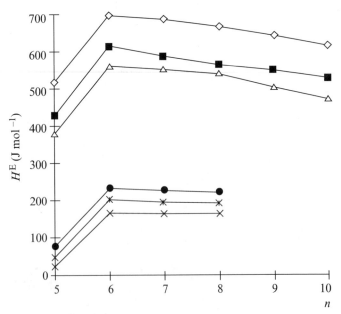

FIGURE 1. Heat of mixing H^E(J mol^{-1}), vs the cycloalkane carbon number n for $x_1 = x_2 = 0.5$ at 298.15 K: c–C$_n$H$_{2n}$ with $\diamond$ n–C$_6$H$_{10}$, $\square$ n–C$_7$H$_{12}$, $\triangle$ n–C$_8$H$_{14}$; c–C$_n$H$_{2n}$ with × n–C$_6$H$_{14}$, * n–C$_7$H$_{16}$, ● n–C$_8$H$_{18}$

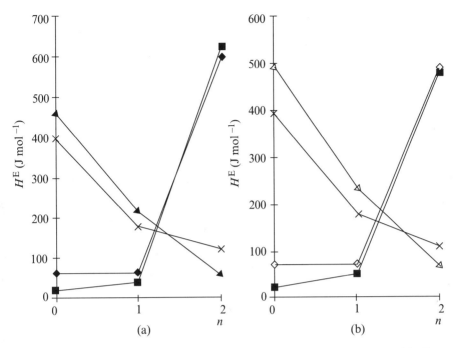

FIGURE 2. Heat of mixing H^E (J mol^{-1}) vs number of bonds n in the functional group of the C_6 or C_7 compounds, for $x_1 = x_2 = 0.5$: (a) ◆ $C_{10}H_{18}$ + n-hexane or n-hexene or n-hexyne, ■ $C_{12}H_{22}$ + n-hexane or n-hexene or n-hexyne, ▲ $C_{10}H_{12}$ + n-hexane or n-hexene or n-hexyne, × $C_{12}H_{16}$ + n-hexane or n-hexene or n-hexyne; (b) ◆ $C_{10}H_{18}$ + n-heptane or n-heptene or n-heptyne, ■ $C_{12}H_{22}$ + n-heptane or n-heptene or n-heptyne, ▲ $C_{10}H_{12}$ + n-heptane or n-heptene or n-heptyne, × $C_{12}H_{16}$ + n-heptane or n-heptene or n-heptyne

calorimeter[42]. The method has been described elsewhere[42]. All substances used had a purity of 98.8 to 99 moles per cent, as shown by g.l.c.

Figure 1 illustrates the changes in the magnitude of the heat of mixing for $x_1 = x_2 = 0.5$ of cycloalkanes c-C_nH_{2n} with both alkynes and alkanes[43–46] where n is the carbon number of the cycloalkane. In the binary mixtures c-C_nH_{2n} + C_nH_{2n-2} when the length of chain of the n-alkynes increases, the values of the excess enthalpy of mixing decrease. The more positive H^E values for cycloalkane + n-alkyne mixtures are probably caused by the dissociation of π–π or H–π bonding between the alkyne molecules on mixing with the cycloalkanes. In the case of bicyclic compounds with n-hexyne or n-heptyne, Figure 2 shows the change in H^E values as a function of the number n of the bonds in the functional group of the n-hexyne, n-hexene, n-hexane or n-heptyne, n-heptene, n-heptane molecules.

Two groups of values of H^E (decalin with n-hexyne or n-heptyne, and bicyclohexyl with n-hexyne or n-heptyne) give similar patterns (Figure 2). In the case of tetralin with n-hexyne or n-heptyne and cyclohexylbenzene with n-hexyne or n-heptyne the values of H^E are smaller, and are even negative for tetraline with n-heptyne. These results show that the π–π interactions between alkynes and an aromatic bicyclic compound are stronger than the same with n-alkenes or n-alkanes.

Figure 3 compares the values of heat of mixing of benzene with an n-alkane (n-hexane, n-heptane, n-octane) and those of benzene with an n-alkyne (n-hexyne, n-heptyne, n-octyne). The values of H^E for $x_1 = x_2 = 0.5$ for benzene with n-alkanes are positive and higher

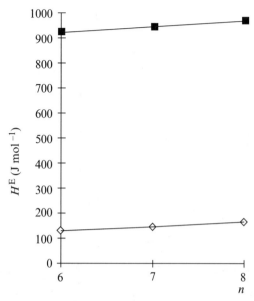

FIGURE 3. Heat of mixing H^E (J mol^{-1}) vs number of carbons in chain n at 298.15 K: ■ benzene with an n-alkane, ◇ benzene with an n-alkyne.

than the values for benzene with n-alkynes. The authors concluded that the very much smaller H^E value for the mixtures of benzene with n-alkynes (C_6 to C_8) is probably due to the stronger π–π interactions between the benzene and the n-alkynes than between benzene and the n-alkanes.

For the mixtures 1,3,5-trimethylbenzene + n-hexyne the values of H^E are positive and small ($H^E_{max} = 32.5$ J mol^{-1}), but for the mixture of 1,3,5-trimethylbenzene with n-heptyne $H^E_{max} = -156.2$ J mol^{-1}; for a mixture of 1,3,5-trimethylbenzene with n-octyne $H^E_{max} = 231$ J mol^{-1}. For the last two mixtures the excess heat of mixing is negative. This effect could well be due to the inductive effect of the substituted methyl groups increasing the electron density in the aromatic ring and resulting in either an enhanced π–π interaction or an enhanced π (aromatic ring)–H interaction.

All authors who investigated alkynes with different series of solvents gave only simple interpretations of the results, because it is not possible to explain fully the data based on one method of measurement. Further investigations involving, e.g., solid–liquid phase equilibria seem to be worthwhile.

IV. REFERENCES

1. S. W. Benson and J. H. Buss., *J. Chem. Phys.*, **29**, 546 (1958).
2. R. Shaw, *J. Chem. Eng. Data*, **14**, 461 (1969).
3. M. Luria and S. W. Benson, *J. Chem. Eng. Data*, **22**, 90 (1977).
4. U. Burkert and L. N. Allinger, *Molecular Mechanics*, ACS Monograph 177, American Chemical Society, Washington, D.C., 1982.
5. U. A. Szibajew and M. I. Ulianowskaja and B. Puzanow, *Zurnal Fizyczeskoj Chimii*, **64**, 52 (1990).
6. D. W. Rogers, O. A. Dagdagan and N. L. Allinger, *J. Am. Chem. Soc.*, **101**, 671 (1979).

7. H. E. Davis, N. L. Allinger and D. W. Rogers, *J. Org. Chem.*, **50**, 3601 (1985).
8. W. Woycicki and P. Rhensius, *J. Chem. Thermodyn.*, **11**, 153 (1979).
9. M. T. Letcher and A. Koteswari Prasad, *J. Chem. Thermodyn.*, **23**, 643 (1991).
10. J. D. Cox and G. Pilcher, *Thermochemistry of Organic and Organometallic Compounds*, Academic Press, New York, 1970.
11. J. R. Wyatt and F. E. Stafford, *J. Phys. Chem.*, **76**, 1913 (1972).
12. A. Williams and D. B. Smith, *Chem. Rev.*, **70**, 267 (1970).
13. H. Okabe and V. H. Dibeler, *J. Chem. Phys.*, **59**, 2430 (1973).
14. H. A. Skinner and A. Snelson, *Trans. Faraday Soc.*, **55**, 404 (1958).
15. D. W. Rogers and S. Skanupong, J. Phys. Chem., **78**, 569 (1974).
16. R. B. Turner, A. D. Jarrett, P. Goebel and B. J. Mallon, *J. Am. Chem. Soc.*, **95**, 790 (1973).
17. R. B. Williams, *J. Am. Chem. Soc.*, **64**, 1395 (1942).
18. T. Flitcroft, H. A. Skinner and M. C. Whiting, *Trans. Faraday Soc.*, **54**, 47 (1958).
19. E. Bretschneider and D. W. Rogers, *Microchim. Acta*, 482 (1970).
20. T. Flitcroft, H. A. Skinner, *Trans. Faraday Soc.*, **54**, 47 (1958).
21. N. L. Allinger and A. Y. Meyer, Tetrahedron, **31**, 1807 (1975).
22. N. L. Allinger, *J. Am. Chem. Soc.*, **99**, 8127 (1977).
23. D. R. Stull, E. F. Westrum and G. C. Sinke, *The Chemical Thermodynamics of Organic Compounds*, Wiley, New York, 1969.
24. J. E. Prossen, R. Gilmont and D. F. Rossini, *J. Res. Natl. Bur. Stand.*, **65**, 34 (1945); **36**, 455 (1946).
25. J. Coops, J. G. Hoijtink and T. Kramer, *Recl. Trav. Chim. Pays-Bas*, **72**, 981 (1963).
26. J. B. Conn, G. B. Kistiakowsky and E. A. Smith, *J. Am. Chem. Soc.*, **61**, 1868 (1939).
27. F. D. Rossini, K. S. Pilzer and D. D. Wagman, *Selected Values of Physical and Thermodynamic Properties of Hydrocarbons and Related Compounds*, Carnege Press, Pittsburgh, Pa., 1953.
28. D. D. Wagman, J. E. Kilpatrick, K. S. Pitzer and F. D. Rossini, *J. Res. Natl. Bur. Stand.*, **35**, 467 (1945).
29. W. D. Rogers and J. F. McLafferty, *Tetrahedron*, **27**, 3765 (1971).
30. T. Flitcroft, A. H. Skinner and C. M. Whiting, *Trans. Faraday Soc.*, **63**, 784 (1957).
31. D. R. Stull (Ed.), *JANAF, Thermochemical Tables*, Dow Chemical Company, Midland, Michigan, 1969.
32. F. Becker and M. Kiefer, *Z. Naturforsch*, **24a**, 7 (1969).
33. P. Rhensius, Diplomarbeit, University of Frankfurt, 1975.
34. W. Woycicki, *J. Chem. Thermodyn.*, **7**, 77 (1975).
35. W. Woycicki, *J. Chem. Thermodyn.*, **7**, 1007 (1975).
36. M. H. Korbalai Ghassemi, J. P. E. Grolier and H. V. Kiehiaian, *J. Chim. Phys.*, **73**, 923 (1976).
37. E. Wilhelm, University of Vienna, personal communication.
38. M.T. Letcher, S. Taylor and R. Baxter, *J. Chem. Thermodyn.*, **20**, 1265 (1988).
39. M. T. Letcher and A. Lucas, *Fluid Phase Equilibr.*, **8**, 301 (1982).
40. M. T. Letcher, W. L. Spiteri and H. W. B. Scoones, *J. Solution Chem.*, **11**, 423 (1982).
41. M. T. Letcher, F. E. Z. Schoonbaert and J. Mercer-Chalmers, *J. Chem. Thermodyn.*, **22**, 815 (1990).
42. M. T. Letcher and H. W. B. Scoones, *J. Chem. Thermodyn.*, **14**, 703 (1982).
43. W. Spateri and M. T. Letcher, *J. Chem. Thermodyn.*, **14**, 1047 (1982).
44. W. Spateri and M. T. Letcher, *Thermochimica Acta*, **57**, 73 (1982).
45. M. T. Letcher, C. Heyward and L. W. Spiteri, *J. Chem. Thermodyn.*, **15**, 395 (1983).
46. C. C. Benson, *Int. Data Ser.*, A. 1., 19 (1971).
47. R. Shaw, in *The Chemistry of the Carbon–Carbon Triple Bond* (Ed. S. Patai), Wiley, Chichester, 1978, p. 70.

CHAPTER 9

Rearrangements involving triple bonded groups

D. WHITTAKER

Department of Chemistry, University of Liverpool, UK

Supplement C2: The chemistry of triple-bonded functional groups
Edited by S. Patai   © 1994 John Wiley & Sons Ltd

I. INTRODUCTION

Chapters on the rearrangements of cyanides, diazo and diazonium compounds, and acetylenes have appeared in books in this series in 1970 and 1978. Chapters on arynes (1983) and isocyanides (1983) have also appeared. This chapter aims to update these earlier works to the end of 1992.

The problem of defining a rearrangement remains. In an earlier chapter, a rearrangement was regarded as a reaction changing the carbon skeleton of a molecule. This definition has had to be widened to include isomerization reactions such as acetylene–allene inteconversions, so the term is now difficult to define and this chapter probably includes some reactions where the rearrangement element is dubious, but the reaction was too interesting to leave out.

This chapter attempts to summarize all the major trends in the study of rearrangements of triple bonded molecules in the last two decades. It makes no attempt to be comprehensive. Examples have been chosen for their illustrative value, and no consideration has been given to priority of publication.

II. REARRANGEMENTS INVOLVING ACETYLENES

A. Alkyne–Allene Interconversions

1. Rearrangements of hydrocarbons

Isomerization of 1-alkynes into 2-alkynes on heating with alcoholic potassium hydroxide solution was first observed by Favorski[1]. Investigation[2] of the isomerization of 1-pentyne (1), 2-pentyne (2) or 1,2-pentadiene (3) with 4M potassium hydroxide solution at 175 °C showed the equilibrium mixture to be as follows:

(1) $HC{\equiv}C{-}CH_2{-}CH_2{-}CH_3$ 1.3%

$\Updownarrow$

(2) $H_3C{-}C{\equiv}C{-}CH_2{-}CH_3$ 3.5%

$\Updownarrow$

(3) $CH_2{=}C{=}CH{-}CH_2{-}CH_3$ 92.5%

When the reaction is carried out with metallic sodium[3] then the 1-alkyne is the major product, since it is withdrawn from the equilibrium mixture by formation of the acetylide. Similar results are obtained using an excess of sodium in liquid ammonia[4].

The mechanism of this rearrangement has been investigated by a number of workers[5], showing it to be a carbanion mechanism similar to that found in other prototropic rearrangements.

$$\bar{B} + -CH_2-C{\equiv}C- \longrightarrow BH + \left[-\bar{C}H-C{\equiv}C- \longleftrightarrow -CH{=}C{=}\bar{C}- \right]$$

$$\longrightarrow -CH{=}C{=}CH- + \bar{B}$$

Investigations by Cram and his coworkers[6] showed the intramolecularity of the reaction to vary from 88% to 18%, depending on substrate and solvent, and labelling experiments[7] demonstrated the incorporation of deuterium. More recently[8] this has been developed into a useful method of deuteration of alkynes.

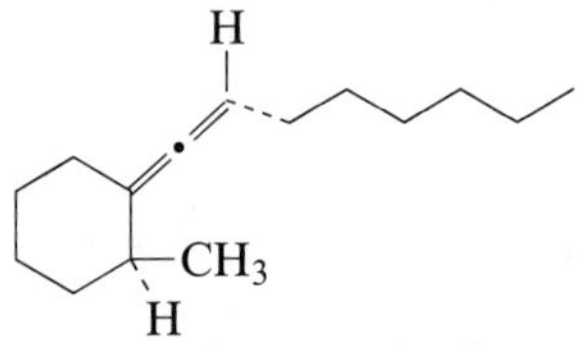

(4)

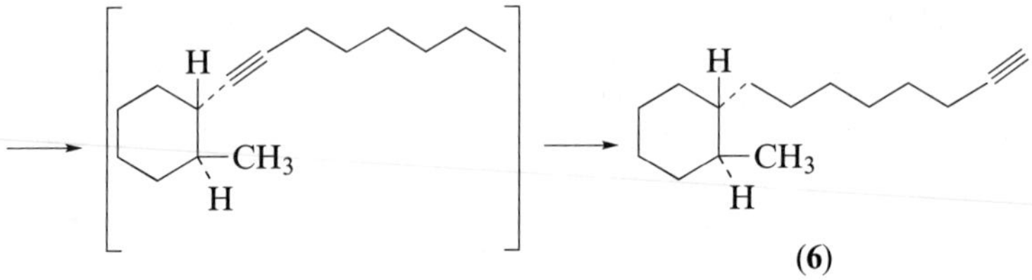

(6)

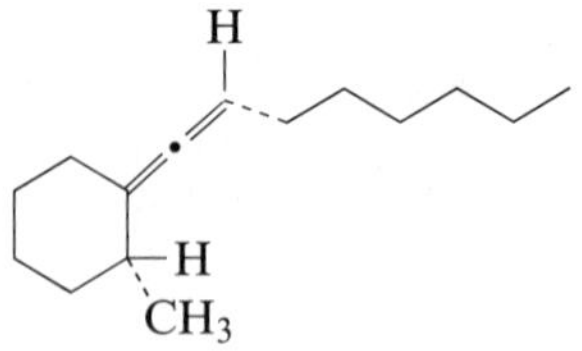

(5)

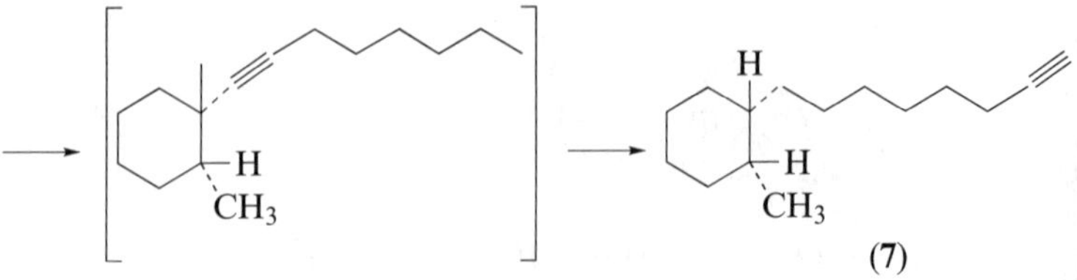

(7)

During the last few years, more efficient reagents for transforming alkynes into allenes have been developed[9,10] and even an enzyme catalysed reaction has been investigated[11]. The main mechanistic development has been the solution to the problem of whether rearrangements of an alkyne into an allene (and *vice versa*) involve an intermediate carbanion, or if the reaction can occur via a concerted mechanism in which a proton is donated from the solvent synchronously with the abstraction of a proton from the substrate by a base.

The approach used[12] was to study the stereochemical integrity of the reaction. Two substrates, **4** and **5**, were used, differing only in orientation of the allenic and ring hydrogens, as shown. On reaction with the mildest of the 1,3-diaminopropane-based isomerization reagents, the lithium amide, a concerted reaction would preserve the sterochemical integrity of the systems, so that **4** would yield only **6**, and **5** would yield only **7**, while reaction via a discrete anionic intermediate would yield a mixture of both **6** and **7** from **4** and from **5**. The experiment gave identical mixtures of **6** and **7** in the ratio 3:1, clearly showing that the main reaction proceeded via a discrete carbanion, through not completely ruling out the existence of a concerted bicyclic mechanism.

Interconversion of alkynes and allenes can be induced electrochemically[13]; it probably involves deprotonation of the parent compound by the electrogenerated radical anion of the parent species.

2. Rearrangements of lithium alkynes

Substituted prop-2-ynyl lithium compounds are being increasingly used in organic synthesis, though they frequently give mixtures of acetylenic and allenic isomers. The reaction[14] of electrophiles such as aldehydes and ketones with the allenyllithium reagent **8** has shown that both allenic (**9**) and acetylenic (**10**) alcohols are formed. Overall, aldehydes favour formation of acetylenic alcohols, while ketones favour formation of allenic alcohols, suggesting that the direction of reaction is under steric control, though it is unlikely that this is the sole directing factor. West[15] has suggested that the principle of hard and soft acids and bases may be relevant, as the propargyl site of the dilithiated allenes is characterized as soft, having high p character, while the sp hybridized acetylide end is classified as hard.

Reactions of substituted prop-2-ynyl lithium compounds can similarly give acetylenic and allenic products. Huynh and Linstrumelle[16] prepared the lithium acetylide **11** and confirmed its structure by NMR spectroscopy. Reaction with propylene oxide then gave a mixture of **12** (45%) and **13** (55%), showing that both organo-lithium species can react to give either allenic or acetylenic products.

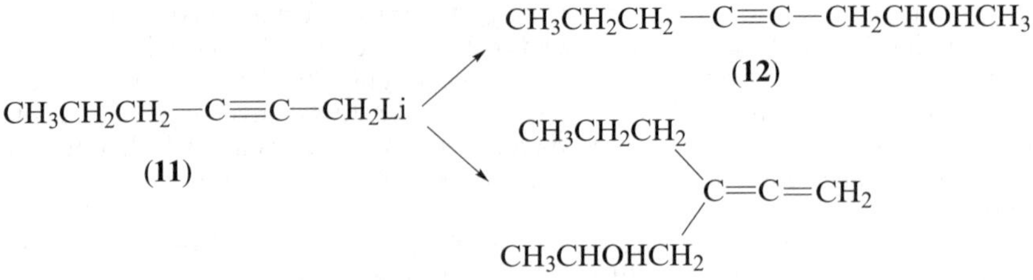

$$CH_3CH_2CH_2-C\equiv C-CH_2CHOHCH_3$$

(12)

$$CH_3CH_2CH_2-C\equiv C-CH_2Li$$

(11)

$$CH_3CH_2CH_2$$
$$\diagdown$$
$$C=C=CH_2$$
$$\diagup$$
$$CH_3CHOHCH_2$$

(13)

Huynh and Linestrumelle then kept **11** in the presence of hexamethylphosphoramide at –75 °C, and showed that it rearranged to the allenic lithium compound **14**. They suggest the reaction takes place through an ion pair. It is not clear whether or not the reaction is reversible, since the lithium allene is thermodynamically stable, though it has been claimed[14] that the rearrangement of a lithium allene to a lithium acetylide is a forbidden concerted process.

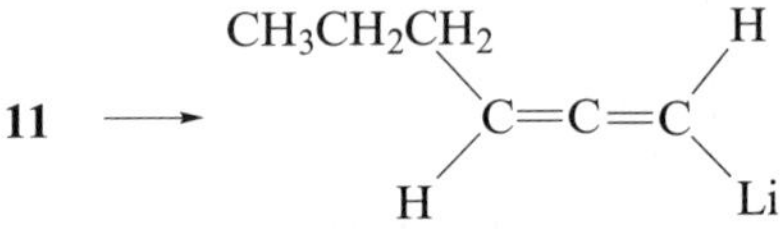

$$CH_3CH_2CH_2 \qquad H$$
$$\diagdown \qquad \diagup$$
11 $\longrightarrow$ $\qquad C=C=C$
$$\diagup \qquad \diagdown$$
$$H \qquad Li$$

(14)

Another interesting rearrangement of an alkynyllithium compound has been reported by Grovenstein[17]. He prepared the alkynyllithium **15**, in tetrahydrofuran, and established its identity by carbonation with CO_2. He then allowed the solution to warm to 0 °C for 4 hours, before showing (again by carbonation) that it had rearranged to **16**. A trapping experiment with lithium *t*-butyl acetylide indicated that the reaction was intramolecular rather than proceeding via an elimination–re-addition mechanism.

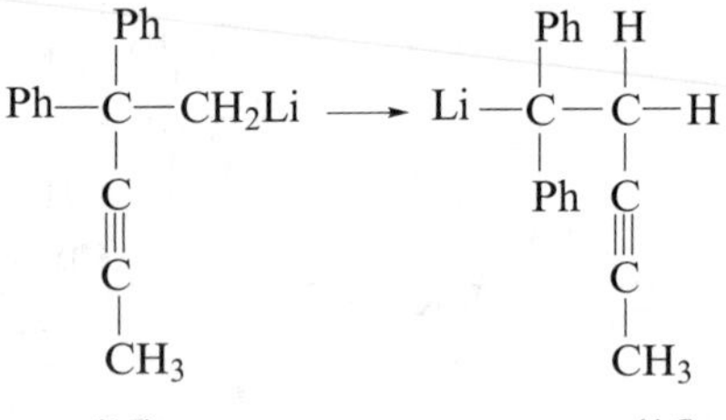

$$
\begin{array}{ccc}
\text{Ph} & & \text{Ph} \quad \text{H} \\
| & & | \quad\; | \\
\text{Ph}-\text{C}-\text{CH}_2\text{Li} \longrightarrow \text{Li}-\text{C}-\text{C}-\text{H} \\
| & & | \quad\; | \\
\text{C} & & \text{Ph} \quad \text{C} \\
\parallel\!\parallel & & \qquad \parallel\!\parallel \\
\text{C} & & \qquad \text{C} \\
| & & \qquad | \\
\text{CH}_3 & & \qquad \text{CH}_3 \\
\end{array}
$$

(15)   (16)

A number of rearrangements of alkyne ethers and sulphides on treatment with alkyllithium reagents are known[18], such as

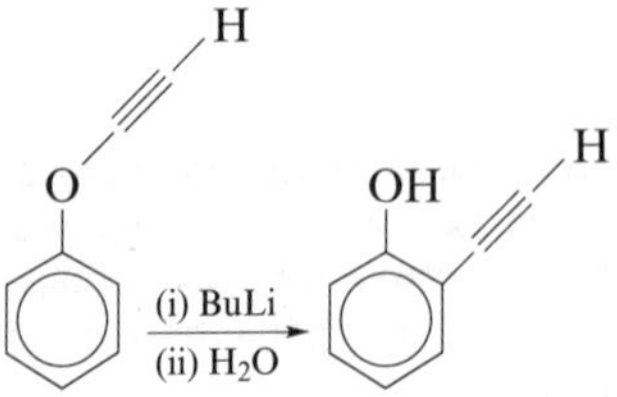

The reaction is believed to involve initial anionic intramolecular attack of an o-lithio group on a lithioethynyl group, the reaction possibly proceeding through a 2,3-dilithiobenzofuran. If the terminal alkyne is replaced by a substituted group, then direct reaction of the lithium species with the alkyne leads to the bicyclic anion, which can be quenched to yield the benzofuran.

3. Rearrangements of tin alkynes

Tin alkynes rearrange to yield allene products in much the same way as do lithium alkynes, except that the reaction involves a radical mechanism. It is very similar to the reaction of allyl stannanes with alkyl halides[19], which substitutes the allyl group. Similar reactions are reported for allyl derivatives of cobalt, rhodium and iridium, but this work has not been extended to alkyne derivatives.

Reaction[20,21] of a radical with propargyltriphenylstannane (17) yields an allene, together with some rearranged allenyltriphenylstannane (18). However, irradiation of 17 at 40 °C, or reaction with azobis(isobutyronitrile) at 70 °C (but not thermally at 70 °C) gave 18 as the sole product. The authors[21] do not consider this reaction to be involved in the rearrangement, since 2-butynylstanane (19) does not undergo rearrangement under similar conditions, but does yield an allenic product on reaction with radicals.

4. Rearrangements of alkynyltrimethylsilanes

Propargyltrimethylsilanes react with electrophiles[22,23] such as Br_2, SO_3, $MeSO_3H$ and H^+ to give allenes of high purity in good yield. The acid cleavage of propargylsilanes cannot be strictly compared to the reactions discussed previously, since in this case we have the extra factor of facilitation of the reaction by β-silicon stabilization of a cationic transition state. On refluxing with base, **20** yields 2-nonyne and 1,2-nonadiene in the ratio of 3:2.

5. Rearrangements of mercury alkynes

The propargylic and allenic mercurials[24] both rearrange readily when treated with halogens or with acylating agents ($AlCl_3$—CH_2Cl_2 plus an acid chloride) giving good yields of rearranged products of high purity, making the rearrangement a valuable synthetic route.

$$CH_3C{\equiv}CCH_2HgI \xrightarrow{CH_3CH_2CH_2COCl} \underset{H_3C}{\overset{CH_3CH_2CH_2\overset{\displaystyle O}{\overset{\|}{C}}}{>}}C{=}C{=}CH_2$$

$$CH_3CH{=}C{=}CHHgCl \xrightarrow{CH_3CH_2CH_2COCl} CH_3CH_2CH_2\overset{\displaystyle O}{\overset{\|}{C}}{-}\underset{CH_3}{\overset{\displaystyle |}{C}H}{-}C{\equiv}CH$$

6. Rearrangements of alkynyl selenoxides

Propargyl selenides[25], such as phenyl propargyl selenide, are readily deprotonated by LDA at $-78\,^\circ$C to give a stable dilithium reagent which is of synthetic value in preparing disubstituted propargyl selenides.

However, on oxidation to the selenoxide[25, 26] rearrangement occurs at $-35\,^\circ$C. The sole produce of rearrangement of **21** is **23**, but the reaction is believed to proceed via **22**.

$$\underset{Ph}{\overset{\displaystyle O}{\overset{\|}{Se}}}{\diagdown}CH_2{-}C{\equiv}CH \longrightarrow \left[\underset{H}{\overset{PhSe}{H_2C{=}C{=}C{\diagup}}}{\diagup}O \right] \longrightarrow \underset{CH_2}{\overset{PhSe{\diagdown}\overset{\displaystyle O}{\overset{\|}{C}}{\diagup}H}{\overset{\displaystyle \|}{C}}}$$

$$\quad\quad (21) \quad\quad\quad\quad\quad\quad\quad (22) \quad\quad\quad\quad\quad (23)$$

A further experiment revealed that the reaction is partially intermolecular. Reaction of a mixture of **21** with **24** yielded 30% of the undeuterated product **25**.

$$m\text{-}CF_3C_6H_4{-}\overset{\displaystyle O}{\overset{\|}{Se}}{-}CH_2{-}C{\equiv}CD \longrightarrow \underset{CH_2}{\overset{m\text{-}CF_3C_6H_4{-}Se{\diagdown}\overset{\displaystyle O}{\overset{\|}{C}}{\diagdown}H(D)}{\overset{\displaystyle \|}{C}}}$$

$$\quad\quad\quad\quad (24) \quad\quad\quad\quad\quad\quad\quad\quad (25)$$

7. Rearrangements of alkynyl phosphines

Primary ethyn-1-yl phosphines (26) rearrange[26] readily under mild conditions in the presence of a Lewis base such as Et_3N to yield phosphaalkynes (27). The reaction probably takes place via the phosphaallene. Other rearrangements of alkynyl phosphines to phosphaallenes have been reported[28,29].

$$H-C\equiv C-P\begin{matrix}H\\\\H\end{matrix} \longrightarrow \left[\begin{matrix}H\\H\end{matrix}C=C=P\begin{matrix}H\\H\end{matrix}\right] \longrightarrow H_3C-C\equiv P$$

(26) (27)

All the organometallic alkynes rearrange along similar lines, both during reaction and, in some cases, spontaneously. The alkyne-to-allene interconversion is a consistent reaction of alkynes over a very wide range of substituents.

B. Alkyne Dimerizations

The dimerization of terminal acetylenes by transition metal catalysts has provided a highly attractive route to unsaturated C-4 units. Recent developments in this field include head-to-tail coupling by Ti or Pd catalysts[30] to give 2,4-disubstituted enynes and head-to-head dimerization of ethynylsilanes by Pd or Rh complexes[31] or by iridium complexes[32] to give 1,4-disubstituted enynes[31].

A study[33] of dimerization of alkynes catalysed by $RhCl(PMe_3)_3$ showed the reaction to yield a mixture of straight-chain and branched products. The authors were able to identify a number of rhodium/alkyne complexes along the reaction route, and conclude that the reaction may be summarized by the scheme presented here. Extrusion of the metal then gave rise to products.

$$R-C\equiv CH + M$$

$$\downarrow$$

$$R-C\equiv C-M-H$$

$$\Big| R'-C\equiv CH$$

$$\begin{matrix}H-C\equiv C-R'\\|\\R-C\equiv C-M-H\end{matrix}$$

$R-C\equiv C-CH=C-MH$ $\quad\quad$ $R-C\equiv C-M-CH=CH-R'$
$\qquad\quad |$
$\qquad\quad R'$

$+$ $\qquad\qquad\qquad\qquad\qquad\qquad +$

$R-C\equiv C-C=CH-MH$ $\quad\quad$ $R-C\equiv C-M-C=CH_2$
$\qquad\quad |$ $\qquad\qquad\qquad\qquad\qquad\qquad\quad |$
$\qquad\quad R'$ $\qquad\qquad\qquad\qquad\qquad\qquad\quad R'$

A more selective reaction than this was observed[34a] for the ruthenium catalysis of dimerization of t-butylacetylene (28) to (Z)-1,4-di-t-butylbutatriene (29). The authors were able to isolate a number of complexes along the reaction pathway which were themselves capable of catalysing the dimerization, and are probably involved in the reaction pathway.

$$
\begin{array}{ccc}
 & \underset{\underset{CH_3}{|}}{\overset{\overset{CH_3}{|}}{CH_3-C-C{\equiv}CH}} & \longrightarrow & \underset{H}{\overset{H_3C}{\underset{H_3C}{>}C{=}}}C{=}C{=}C\underset{}{\overset{CH_3}{<}}{\overset{CH_3}{CH_3}} \\
 & (28) & & (29)
\end{array}
$$

A different addition reaction has been used successfully[34b] to produce pentatetraenes. In this case, reaction of a 2,4-pentadiynyl ester with an alkyl silver compound gave the pentatetraene, 30.

$$
\underset{\underset{O}{\parallel}}{H-C{\equiv}C-C{\equiv}C-\underset{\underset{OSCH_3}{|}}{C}-C_3H_7\text{-}i} \xrightarrow{t\text{-}C_4H_9Ag} t\text{-}C_4H_9CH{=}C{=}C{=}C{=}C\underset{i\text{-}C_3H_7}{\overset{H}{<}}
$$

$$(30)$$

Unlike the trienes, the pentatetraenes are thermally unstable, though they can be characterized by usual spectroscopic techniques.

C. Acetylene–Vinylidene Interconversion

Vinylidene (the $H_2C{=}C$: molecule) is the simplest possible vinyl carbene, and as such has aroused the interest of chemists. It can rearrange rapidly to acetylene, the reaction having a very low energy barrier[35].

$$H_2C{=}C: \longrightarrow HC{\equiv}CH$$

It can be generated by pyrolysis of acetylene[36], and detected by addition to benzene:

$$H_2C{=}C: + C_6H_6 \longrightarrow$$

Catalysis by alkali metal ions has recently been reported as an alternative route[37]. In an argon matrix, acetylene forms a π complex with the metal. On irradiation, it isomerizes to the vinylidene form, $M:C{=}CH_2$. When complexed with metals, vinylidene is much more stable, in the same way that metal carbenoids are generally much more stable than carbenes, and rearrangement of a tungsten alkyne complex to a tungsten vinylidene complex has been reported[38].

Vinylidene complexes can also be decomposed to recover acetylenes, as has been shown by Bullock[39]. When $HC{\equiv}CMe$ is bubbled through a methanol solution of $(C_5H_5)(PMe_3)_2RuCl$, containing NH_4PF_6, then the complex $(C_5H_5)(PMe_3)_2R$ $(HC{\equiv}CMe)^+PF_5^-$ is precipitated as a yellow powder. The complex rearranges at room temperature to $(C_5H_5)(PMe_3)_2Ru\!=\!C\!=\!CHMe^+PF_5^-$, which on heating with acetonitrile at 80–110 °C regenerates $HC\!\equiv\!CMe$ in high yield. Again the double-bonded structure is more stable than the triple-bonded, even in rather unusual circumstances.

D. Rearrangements of Alkyne–Metal Complexes

Most of the metal complexes of alkynes discussed so far are stabilized by proximity of an sp hybrid orbital on the alkyne carbon; the alternative approach to metal complex formation is donation of the triple-bond electrons to the vacant orbitals of a metal. This yields a vast range of complexes, many of which are stable and can be used in synthesis, but is some cases the alkyne can undergo rearrangements which are not observed in the free organic species.

1. Triple-bond cleavage

Although the $C\!\equiv\!C$ bond is known to be strong, examples of its cleavage to yield two carbyne fragments stabilized by bonding to an organometallic triangular framework are known.

The reaction[40] of $CpCo(CO)_2$ with $Me_3Si\!-\!C\!\equiv\!C\!-\!SiMe_3$ at 137 °C provides a simple thermal example of such a cleavage. The reaction also involves dimerization of the substrate, and it seems likely that it may proceed via bis (trimethylsilyl)butadiyne.

Compounds 33 and 34 are readily formed from 31 by direct reaction with $CpCo(CO)_2$. A possible reaction sequence is formation of the triene, 31, from alkyne dimerization followed by reaction with the cobalt species to give the three complexes. Reaction of 34 with alkynes yielded only cyclobutadiene complexes; alkyne metathesis was not observed, probably since the carbon-to-metal bonds are too strong.

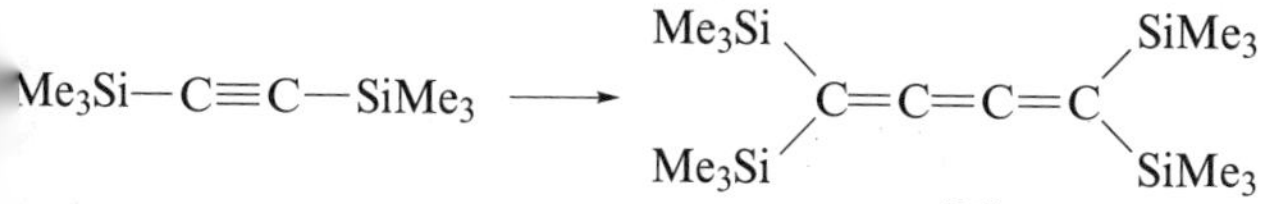

$$Me_3Si\!-\!C\!\equiv\!C\!-\!SiMe_3 \longrightarrow$$

(31)

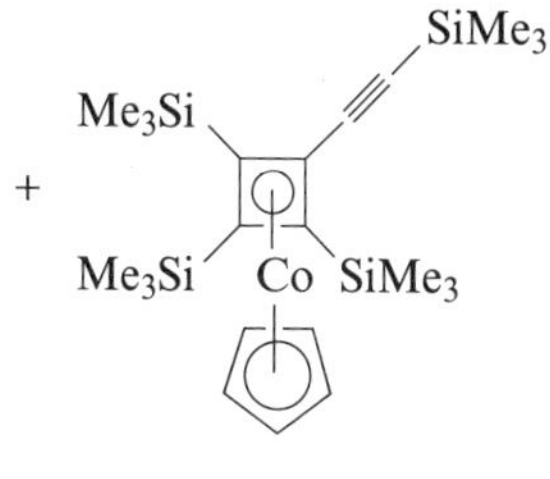

(32)

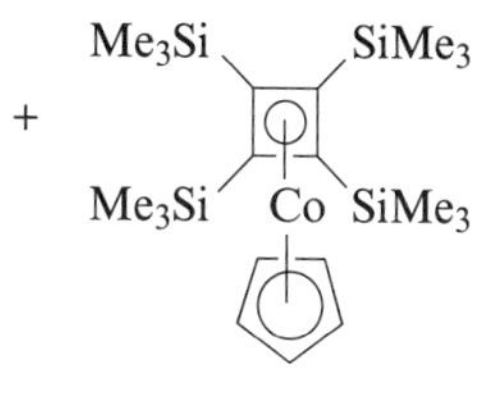

(33)

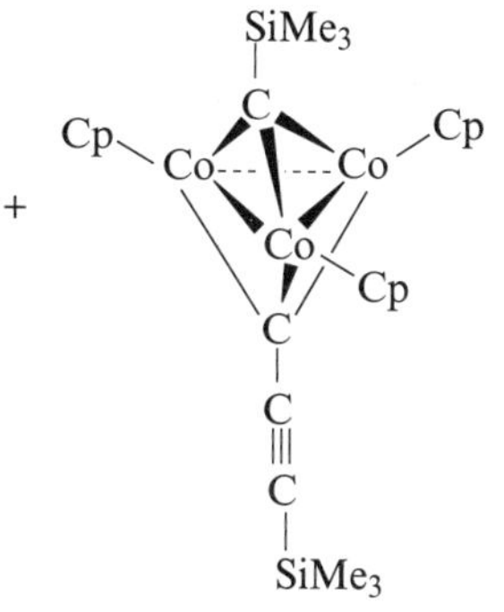

(34)

A similar reaction has been reported[41] to take place at room temperature in the presence of a catalyst, sodium diphenylketyl. The mode of action of this catalyst is not clear.

$$
\text{MeC}\equiv\text{CNEt}_2 \xrightarrow{\text{Fe}_2(\text{CO})_9} (\text{CO})_3\text{Fe}
\begin{array}{c}
\text{Me} \\
| \\
\text{C} \\
\diagdown \diagup \\
\text{Fe(CO)}_3 \, \text{Fe(CO)}_3 \\
\diagup \diagdown \\
\text{C} \\
| \\
\text{NEt}_2
\end{array}
$$

2. Alkyne metathesis

Alkyne metathesis, paralleling olefin metathesis, is the process of conversion of an unsymmetrical alkyne into a mixture of itself and the two possible symmetrical alkynes, e.g

$$\text{PhC}\equiv\text{CMe} \longrightarrow \text{PhC}\equiv\text{CPh} + \text{MeC}\equiv\text{CMe}$$

All three alkynes are present at equilibrium, the ratio depending on thermodynamic stabilities. While the process clearly involves triple-bond fission, it is not related to that described above, as a stable bond to the metal is not formed. The process can be illustrated

$$
\text{M}\equiv\text{CR}^1 + \text{R}^2\text{C}\equiv\text{CR}^3 \longrightarrow
\begin{array}{c}
\text{R}^1\text{C}=\text{M} \\
| \quad | \\
\text{R}^2\text{C}=\text{CR}^3
\end{array}
\longrightarrow
\begin{array}{c}
\text{R}^1\text{C}-\text{M} \\
\| \quad \| \\
\text{R}^2\text{C}-\text{CR}^3
\end{array}
\longrightarrow \text{R}^1\text{C}\equiv\text{CR}^2 + \text{M}\equiv\text{CR}^3
$$

A number of authors have reported successful alkyne metathesis reactions catalysed by either tungsten[42–44] molybdenum[45] complexes.

The catalysts used are usually alkylidyne complexes, such as **35**. This compound can react[46] with diphenylacetylene in stoichiometric quantities to give $\text{PhC}\equiv\text{CCMe}_3$ in 95%

$$
\begin{array}{c}
\text{CH}_3 \quad\quad \text{OPEt}_3 \\
| \quad\quad \diagup \\
\text{H}_3\text{C}-\text{C}-\text{C}\equiv\text{W}-\text{Cl} \\
| \quad\quad | \diagdown \text{Cl} \\
\text{CH}_3 \quad \text{Cl}
\end{array}
\qquad\qquad
\begin{array}{c}
\quad\quad \text{OPEt}_3 \\
\quad\quad \diagup \\
\text{Ph}-\text{C}\equiv\text{W}-\text{Cl} \\
\quad\quad | \diagdown \text{Cl} \\
\quad\quad \text{Cl}
\end{array}
$$

$$\qquad\qquad (\textbf{35}) \qquad\qquad\qquad\qquad\qquad (\textbf{36})$$

$$
\begin{array}{c}
\text{CH}_3 \\
| \\
\text{H}_3\text{C}-\text{C}-\text{C}\equiv\text{W(OCMe}_3)_3 \\
| \\
\text{CH}_3
\end{array}
$$

$$(\textbf{37})$$

yield, together with the complex **36**. The presence of the phosphine oxide ligand slows the reaction sufficiently to demonstrate this reaction; the metathesis[46] reaction uses catalysts such as **37**. Other catalysts, such as [Mo(CO)$_6$]–PhOH, are believed to act by initial formation of an alkylidene complex[47].

The evidence presented so far does not exclude the possibility that the reaction proceeds by exchange of substituents while the triple bond remains intact. This, however, has been excluded by ^{13}C-labelling experiments[47,48].

$$Ph^{13}C\equiv CC_6H_4Me\text{-}p \longrightarrow Ph^{13}C\equiv{}^{13}CPh + p\text{-}MeC_6H_4C\equiv CC_6H_4Me\text{-}p$$

The reaction cannot be used to metathesize terminal alkynes; it is suggested[49] that this is because a proton can readily be lost to the catalyst.

3. Insertion into metal–carbon bonds

This reaction is often assumed to be of importance in alkyne polymerization. The first clues to the mode of action came from the work of Thorn and Hoffmann[50] on the insertion of alkenes into Pt—H bonds, which they suggested involved a four-coordinate intermediate in which the alkane replaces a ligand and achieves a coordination site *cis* to the hydride.

This was supported[51] by a study of the kinetics of cyclization of **38** and analogues with extra methylene groups inserted between the oxygen and alkyne groups to form analogues of **39**. Overall, the reaction may be written as in the accompanying scheme. Insertion of alkynes into aminocarbene complexes of chromium is of potential use in synthesizing heterocycles[52].

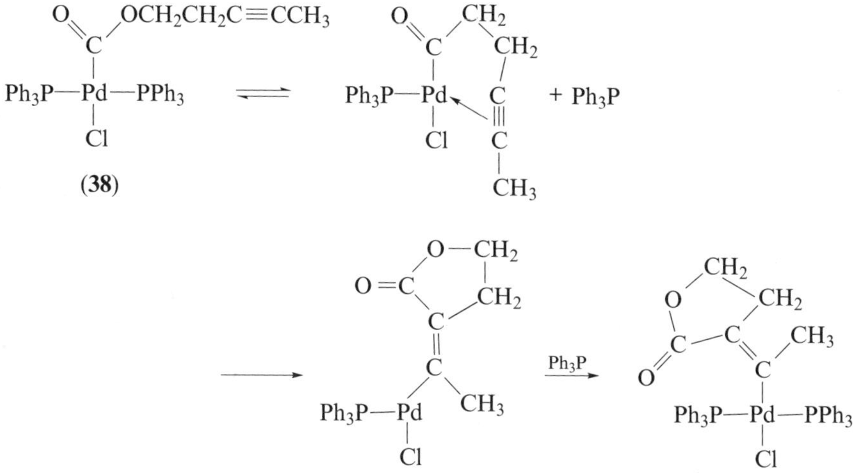

4. Polymerization of alkynes

Alkyne metathesis is catalysed by alkylidene complexes of tungsten and molybdenum, but not by all alkylidene complexes of these metals. Complexes which do not catalyse metathesis catalyse polymerization of acetylenes to give polypropynes.

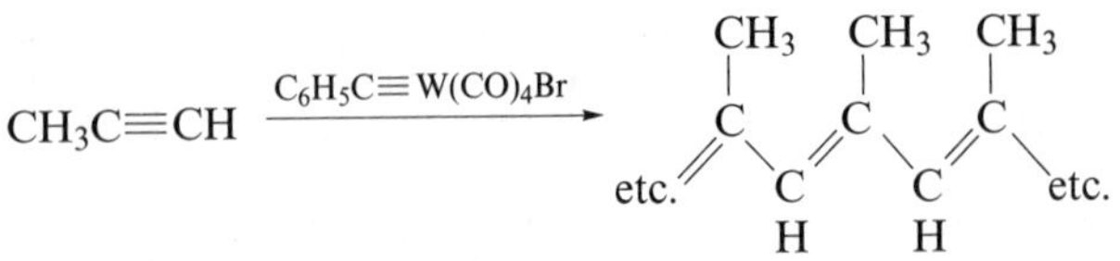

$$CH_3C\equiv CH \xrightarrow{\ C_6H_5C\equiv W(CO)_4Br\ }$$

$$
\begin{array}{ccccc}
 & CH_3 & CH_3 & CH_3 & \\
 & | & | & | & \\
\text{etc.} & C & C & C & \text{etc.} \\
 & \diagup\diagdown & \diagup\diagdown & \diagup\diagdown & \\
 & C & & C & \\
 & H & & H & \\
\end{array}
$$

Katz and coworkers[53] suggested that the polymerization of acetylenes in this way results from catalysis by metal carbenes. The authors speculate that metal carbynes may well be the source of metal carbenes; if the metal carbyne is coordinatively unsaturated, then metathesis can occur, but if not, it must acquire coordinative unsaturation in order to react with the alkyne, which it can do only by transforming into a metal carbene and catalysing polymerization.

A number of metal complexes catalyses specific alkyne polymerizations, giving rise to four-, six- or eight-membered carbocyclic rings. The first work in this area was the nickel-catalysed formation of cyclooctatetraene (40) from acetylene by the group of Reppe[54], but since then formation of cyclic systems from acetylenes has been found to be also catalysed by molybdenum[55], cobalt[56], iridium[57] and tantalum[58].

The mechanism of the Reppe reaction has been studied by Colborn and Vollhardt[59]. They used ^{13}C-labelled acetylene and studied the labelling pattern of the cyclooctatetraene. Their results were not consistent with reaction through cyclobutadiene or by recombination of carbyne units resulting from triple-bond cleavage, but were consistent with either the concerted or stepwise mechanisms outlined below.

Concerted

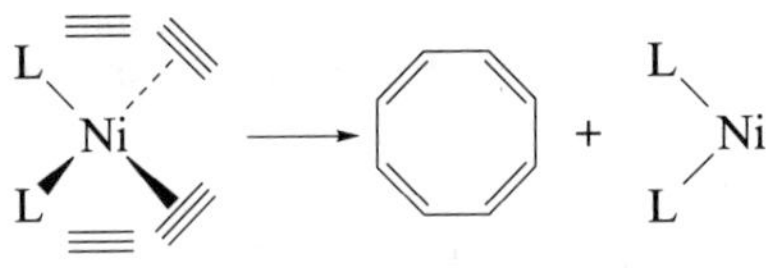

(40)

Stepwise

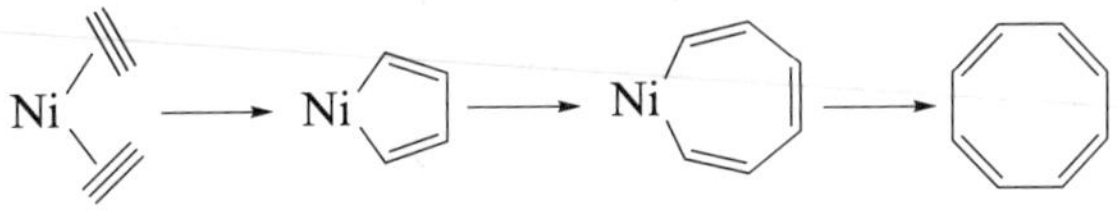

(40)

Each kind of species involved in the stepwise mechanism is well documented[60].

A similar scheme is suggested by Bergman and collaborators[56] for the trimerization of alkynes. The intermediate 41 has been isolated, characterized and shown to display catalytic activity. The authors suggest that the intermediate between 41 and hexamethylbenzene may well be the Diels-Alder adduct, 42.

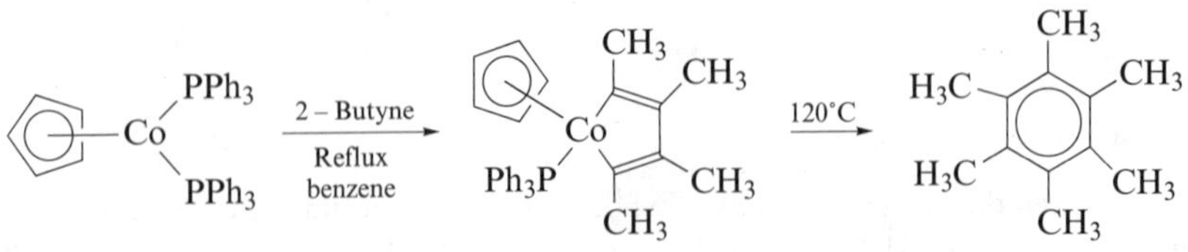

(41)

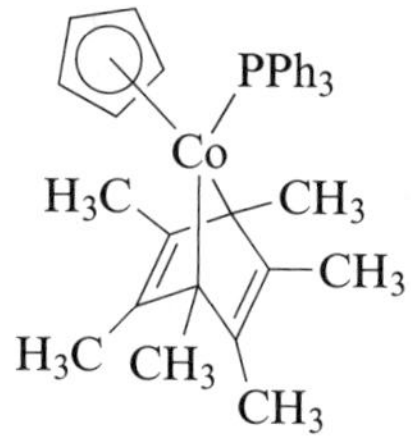

(42)

Addition of an alkyne to an already complexed alkyne to give a cyclobutadiene complex has been observed, though the reaction proceeds only under harsh conditions[61]; intramolecular reaction of the acetylene groups of a cyclic diyne has also been recorded yielding a complex of a cyclobutadiene[62] (43). Using palladium as a catalyst gives a more controlled reaction, in particular by varying the degree of intramolecularity of the reaction[63].

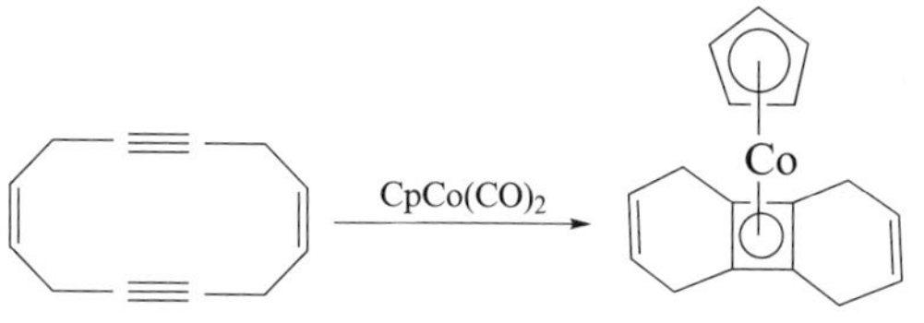

(43)

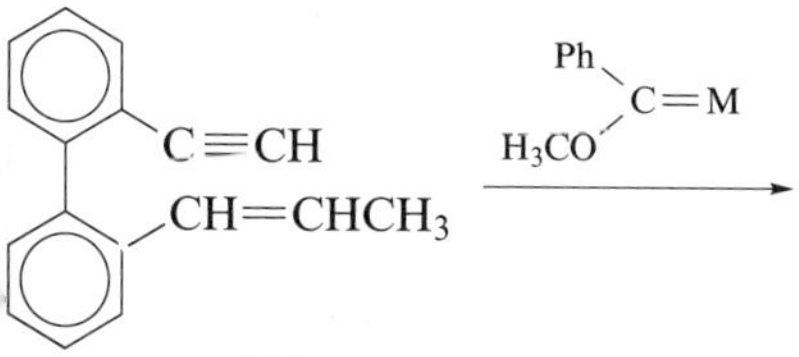

(44)

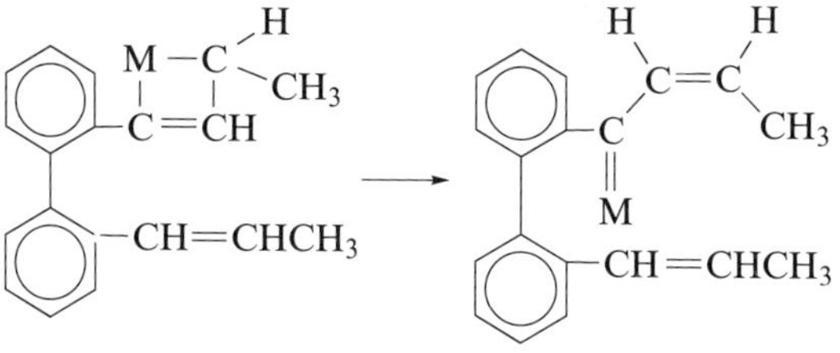

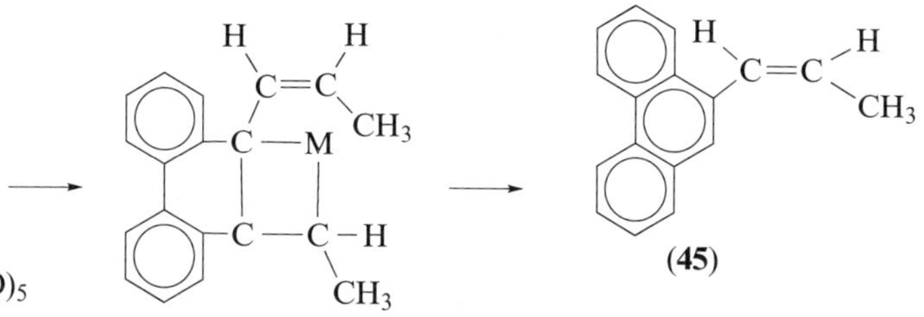

M = W(CO)$_5$

(45)

Alkyne reactions like these are of obvious use in organic synthesis, and a number of examples of their use have appeared. The synthesis of **45** from **44** has been carried out with a catalytic amount of a tungsten complex[64]. The reaction involves the cycle shown nearby. The initial run through the cycle exchange the substituents on the carbon of the C=W bond; then the new catalyst carries out the reaction, as shown.

Reaction of an alkyne with a distant alkene group can be brought about using a palladium(0) catalyst[65].

An ingenious use of reactions of this type is to generate a heterocyclic ring[66]. More commonly, the reaction is used to generate aromatic rings[67,68].

E. Rearrangements Involving Reaction of Alkynes with Electrophiles

The reaction of alkynes with fluorosulphonic acid at −170 °C has been found[69] to be instantaneous and quantitative. With terminal alkynes, the first step is production of a vinyl cation–fluorosulphonate an ion pair (**46**). This ion pair subsequently collapses to give the vinyl fluorosulphonate. Experiments with FSO_3D showed that the addition is dominantly *syn*; it is probable that collapse of the ion pair gives a *syn* product, while separation and recombination yields a *syn–anti* mixture.

(46)

With disubstituted alkynes[70] the reaction appears to involve the free vinyl cations (i.e vinyl cations which do not form part of an ion pair, though they are obviously solvated)

Reaction of the vinyl cation is with another molecule of alkyne, giving the cyclobutenyl cation (47).

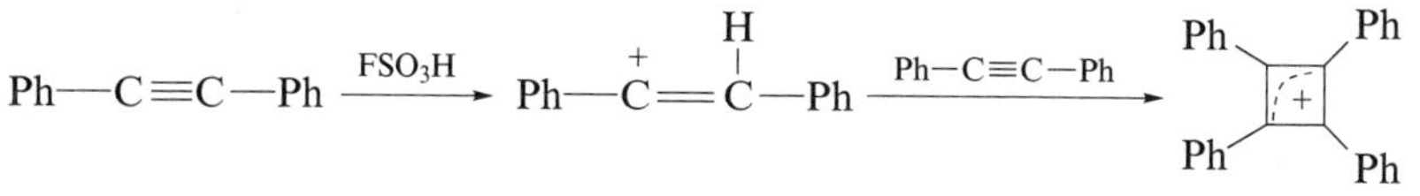

$$Ph-C\equiv C-Ph \xrightarrow{FSO_3H} Ph-\overset{+}{C}=\overset{H}{\underset{}{C}}-Ph \xrightarrow{Ph-C\equiv C-Ph}$$

(47)

The vinyl cation is also proposed as an intermediate in the acid-catalysed hydration of alkynes[71] which proceeds to the corresponding aldehyde.

A similar reaction has been observed for the hydration of α-ethynyl alcohols[72]. Under normal Rupe rearrangement conditions (acidic media) the reaction proceeds as follows:

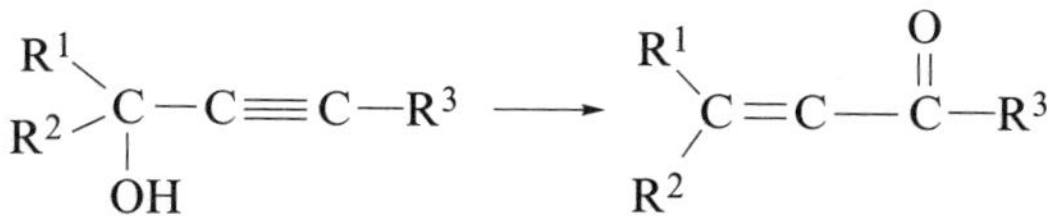

$$\underset{OH}{\overset{R^1}{\underset{R^2}{\diagdown}}C-C\equiv C-R^3} \longrightarrow \overset{R^1}{\underset{R^2}{\diagdown}}C=C-\overset{O}{\overset{\|}{C}}-R^3$$

Using the fluorinated α-ethynyl alcohol (48), the reaction under Rupe conditions was hydration of the triple bond to 49; on prolonged reaction this gave rise to an acyloin condensation to yield 50.

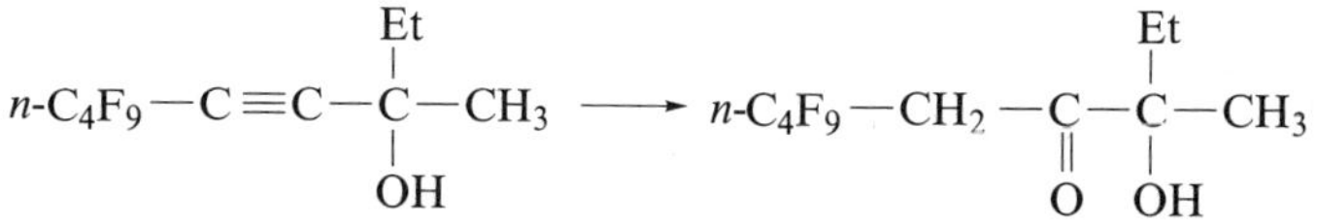

$$n\text{-}C_4F_9-C\equiv\overset{Et}{\underset{OH}{C}}-CH_3 \longrightarrow n\text{-}C_4F_9-CH_2-\overset{Et}{\underset{O\ \ OH}{C-C}}-CH_3$$

(48)   (49)

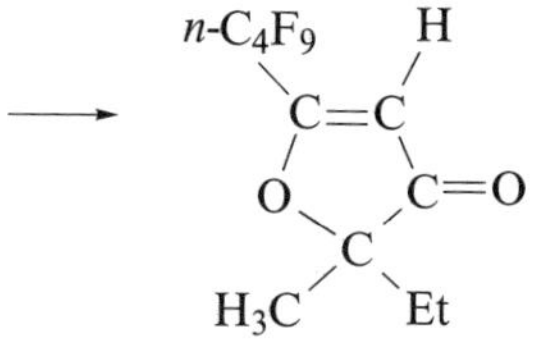

(50)

In practice, most hydration reactions of alkynes are carried out using mercuric acetate as a Lewis acid catalyst because of the milder conditions required. Under favourable circumstances[73] even this reaction is not immune to the tendency of alkynes to rearrange, and provides a useful route to benzofurans (51).

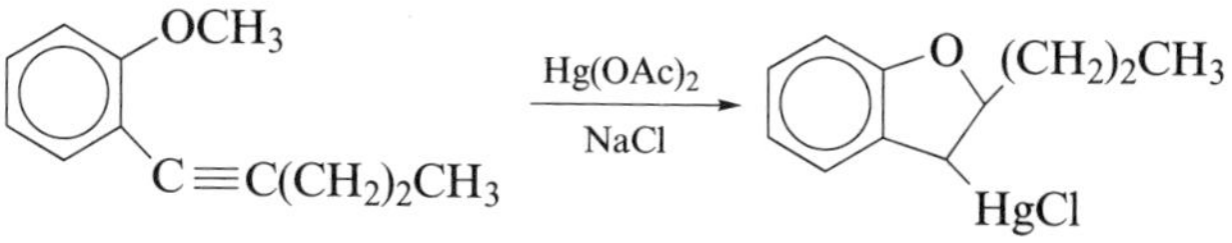

(51)

The reaction works equally well with the oxygen atom of the substrate replaced by nitrogen.

F. Triple-bond Participation

The triple bond, being electron-rich, is readily attacked by electrophiles. In cases where the electrophile and the triple bond are part of the same molecule, this leads to cyclization of the molecule, or, when the electrophilic centre and the triple bond are close together, to direct interaction.

1. Homopropargylic rearrangements

A triple bond can participate in formation of a carbocation when it is separated from the developing carbocation centre by two or more CH_2 groups, so that interaction proceeds through space instead of along the bond.

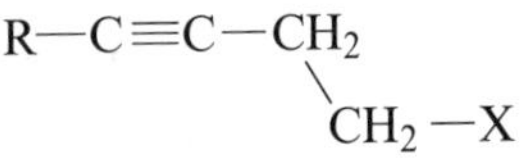

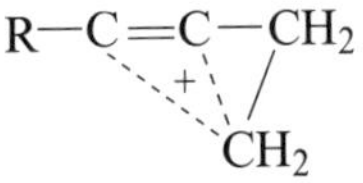

The products of this reaction are usually the cyclopropyl (52) and cyclobutyl (53) derivatives.

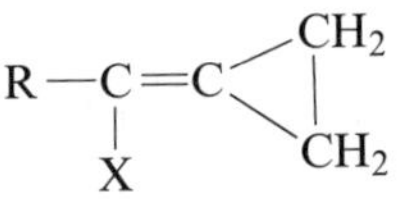

(52) 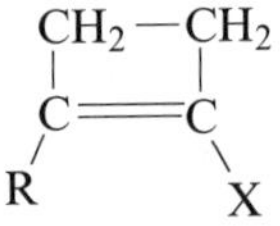   (53)

Participation by the triple bond produces a small rate acceleration of a factor of six relative to the saturated ester[74].

The work of Harding and Hanack[75] on this system has confirmed the view that the reaction is observed only in solvents of low nucleophilicity; in highly nucleophilic solvents, a bimolecular reaction proceeds without rearrangement. A series of isotope effect measurements confirms that in solvents of low nucleophilicity and good ionizing power, the homopropargylic assistance is of importance in the solvolysis reaction.

2. Remote triple-bond participation

a. Interaction with ketones. The reaction[75] of 5-cyclodecynone (54) with HCl in methanol or BF_3 in aprotic solvents yields only bicyclo-[4,4,0]-1(6)-decene-2-one (55).

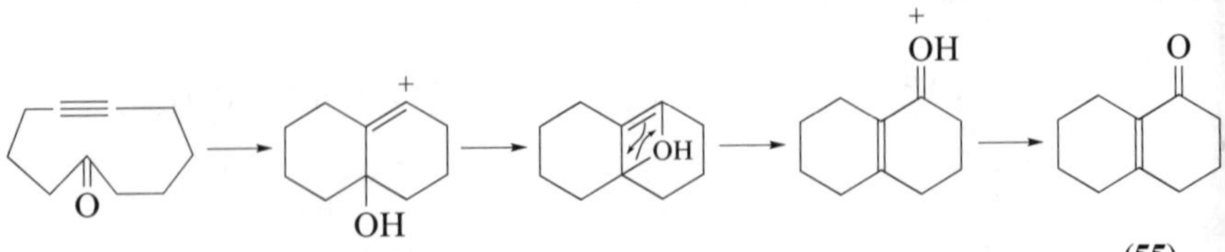

(54)   (55)

A recent report[76] suggests the 'oxete' mechanism outlined above. Alternatively the scheme shown below, involving enol intermediates, could operate.

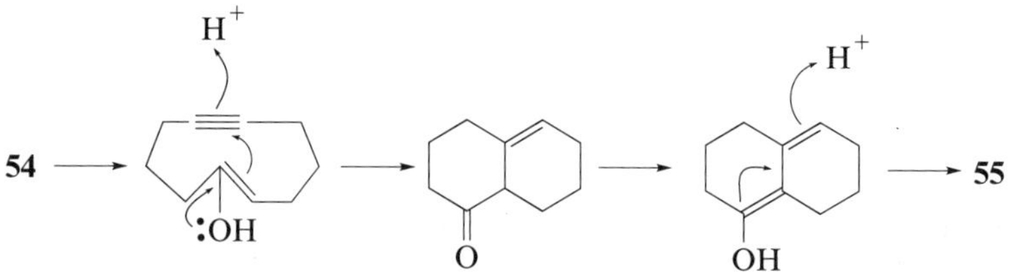

To resolve the situation, Harding and King[77] prepared 6-octyn-2-one (**56**) and isomerized it with ethanolic HCl and (separately) with BF_3 etherate. Both reactions yielded 55% of **57** and 45% of **58**. The authors showed that the reaction did not involve oxygen exchange with O^{18}-enriched solvent.

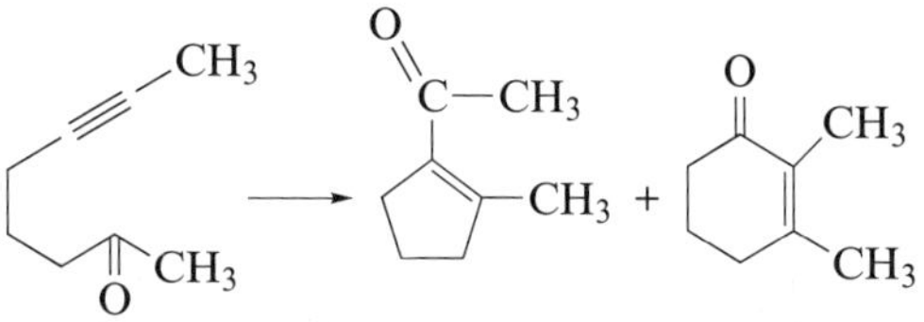

(**56**)   (**57**)   (**58**)

Both mechanisms can give rise to **57**, but the enol intermediate mechanism would yield **59** rather than **58**.

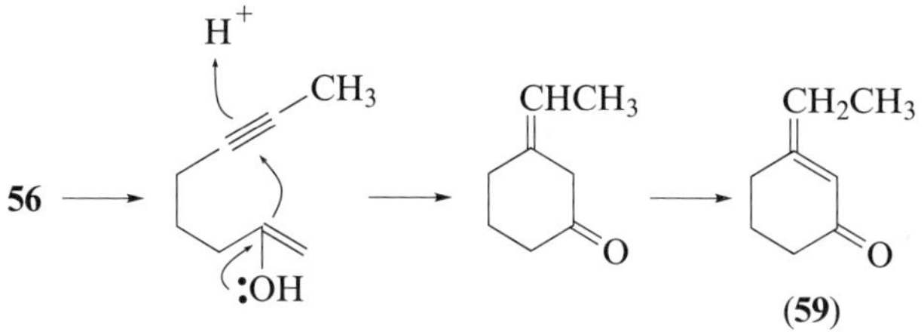

On the other hand, the 'oxete' mechanism provides a route to **58**:

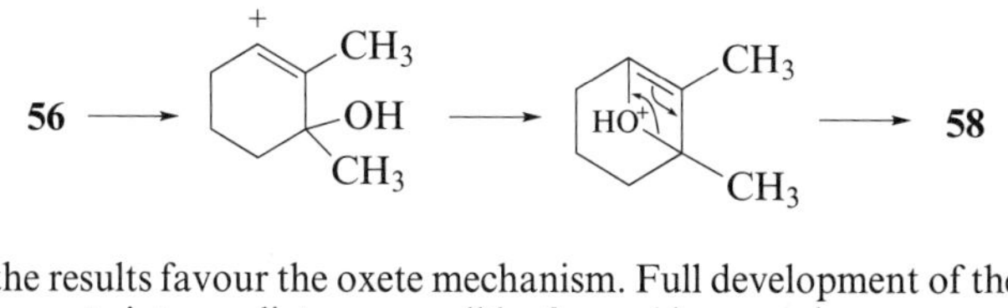

Clearly, the results favour the oxete mechanism. Full development of the vinyl cation is unlikely; the oxete intermediate may well be formed in a synchronous process.

b. Interaction with cations. Through-space interaction of a cation with an unsaturated system is a useful cyclization reaction and works well with alkynes. The simplest possible interaction of this type has been demonstrated by Hanack and collaborators[78]. They solvolysed hept-6-yn-1-yl trifluoroacetate to give the product of internal return of the ion pair in 50% yield even in water as solvent.

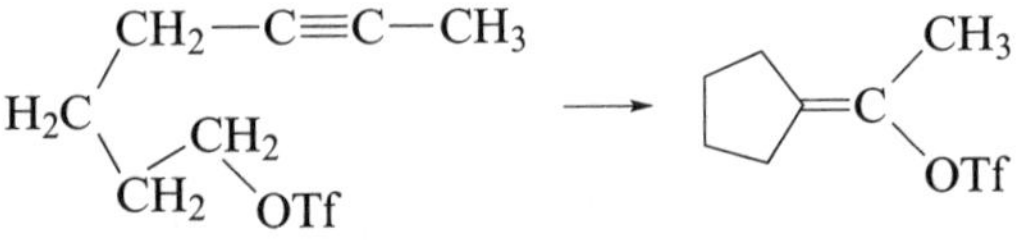

The participation is probably not synchronous with ion formation; indeed, in 80% ethanol–water the main product is the unrearranged ethyl ether.

The method is of general use, and can be used in synthesis[79], such as the synthesis of decalone (**60**) from cyclization of **61**.

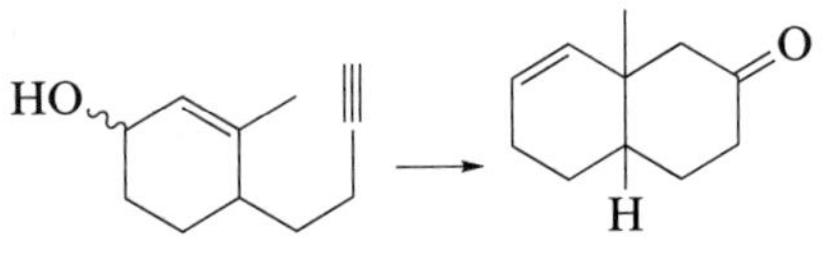

(**61**)   (**60**)

In recent years, the method has been used for the synthesis of a number of heterocyclic systems. Heimstra and Speckamp[80] cyclized a propargysilane (**62**) with trifluoroacetic acid at 0 °C.

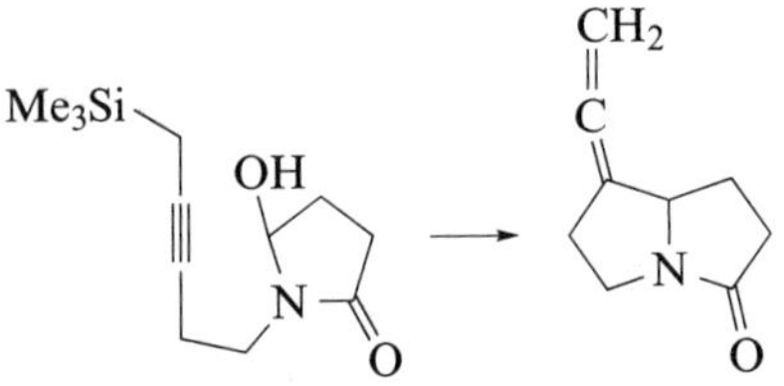

(**62**)

Weinreb and Scola[81] used the method to synthesize 1,3-oxazines (**63**), using BF$_3$ etherate as catalyst.

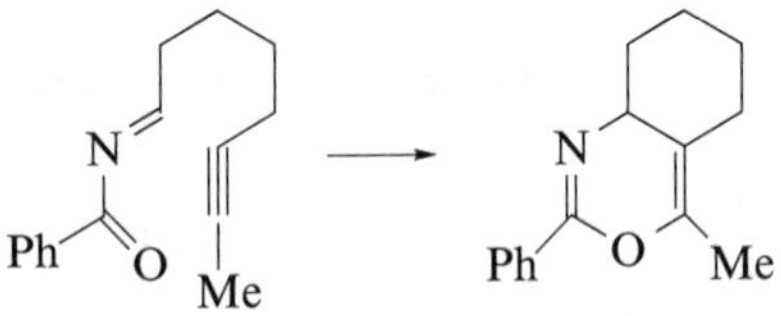

(**63**)

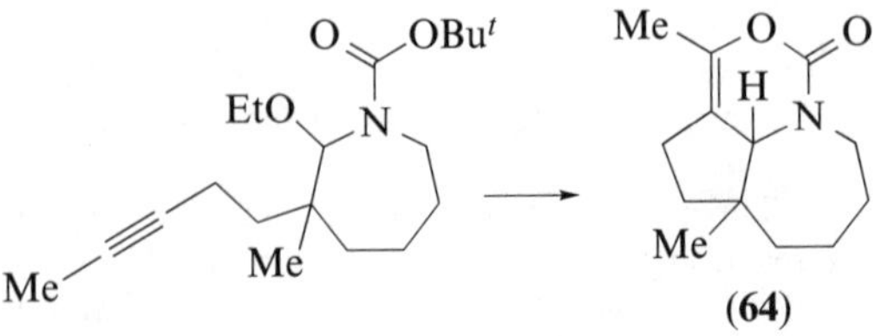

(**64**)

Fisher and Overman[82] used $SnCl_4$ as catalyst when cyclizing the carbonate to the enol carbonate (64), a reaction which proceeded in 69% yield.

c. Interaction with radicals. The understanding of factors controlling relative stereo-chemistry in radical cyclization reactions has progressed rapidly[83,84] and the research frontier is now in the area of controlling acyclic stereochemistry in intermolecular radical reactions[85–87].

Recently, Curran and coworkers[88] had demonstrated that chiral radicals derived from camphor sultam give high levels of asymmetric addition in cyclization reactions with

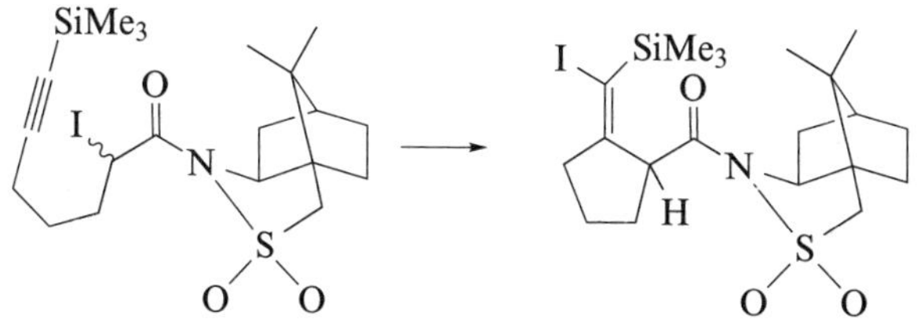

(65)   (66)

achiral alkynes. Thus 65, on treatment with $Bu_3SnSnBu_3$ in benzene at 80 °C under irradiation, gave an E/Z mixture of 66. Removal of the I and SiMe groups showed 66 to be a 90:10 mixture of diastereoisomers, the main component being as shown. The stereoselectivity of the reaction appears to depend on the nature of the substituents involved[89].

The radicals can also be trapped efficiently by oxime ethers; this is a useful approach to intramolecular cyclization[90].

A variation of this type of reaction involves interaction of two alkyne groups via radical formation. The mechanism proposed by Garratt[91] involves initial base-catalysed isomer-ization to the bis allene, which then undergoes an intramolecular allene dimerization to give the bis methylene diradical (67).

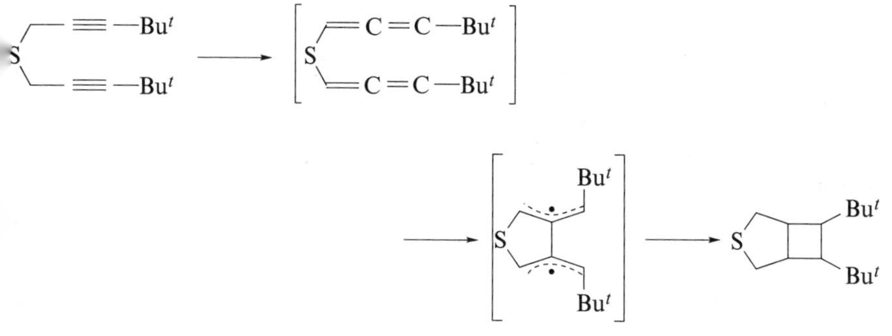

(67)   (68)

The main product is 68, but with some combinations of the heteroatom and substituent groups, the radical 67 can dimerize.

Other examples of this type of reaction are known[92]; in some cases, there is no evidence of radical involvement, and reaction could be a simple valence isomerization.

Thermal reaction of two alkyne groups forms the basis of the Bergmann[93] rearrangement, in which 1,5-diyne-3-enes form arene-1,4-diyls (69). The reaction rate is influenced by the radical trapping agent used[94].

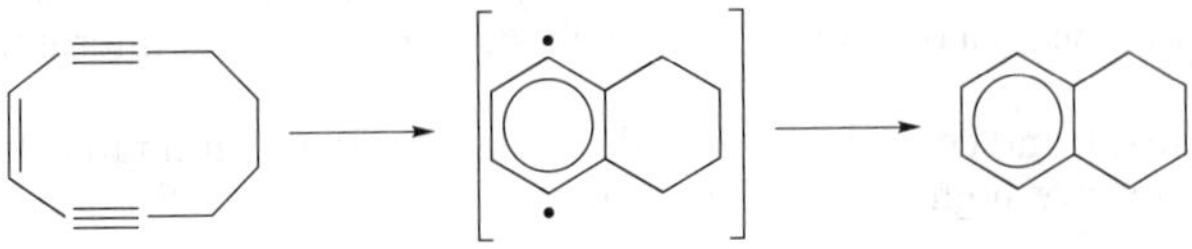

(69)

d. Interaction with organometallic catalysts.   Organocopper reagents can be added to acetylenes in a synthetically useful manner[95]; this reaction has proved valuable as a general cyclization method[96] in which an alicyclic organocuprate is formed from a halide, and then cyclizes.

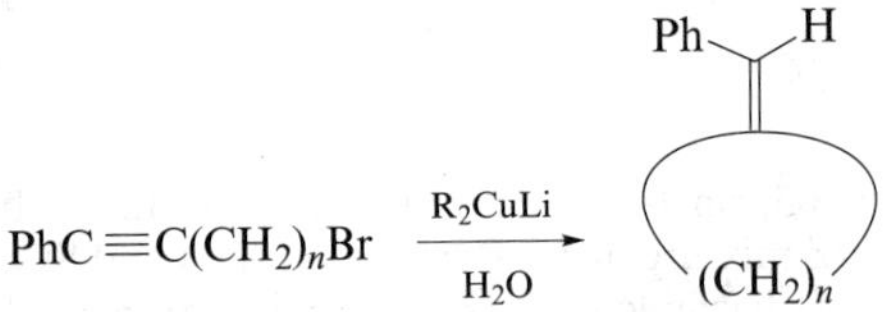

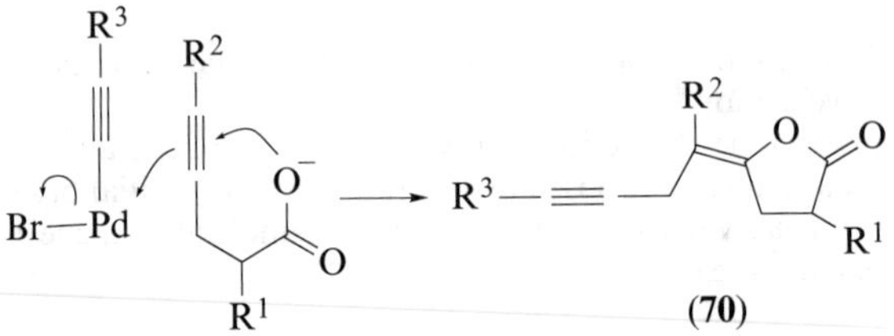

$$PhC \equiv C(CH_2)_nBr \xrightarrow[H_2O]{R_2CuLi}$$

(70)

When n = 4, the yield is close to 80%; when n = 3, this is lowered to 40%. Quenching wit' D_2O instead of H_2O leads to incorporation of D instead of H attached to the double bond The complexation process also activates the acetylenic bond. Formation of a palladium complex[97] has been used to react a γ-acetylenic carboxylate with a 1-bromo-1-alkyne t give an ynenol lactone (70). Palladium has also been used to cyclize a zinc substitute alkyne[98].

Propargyl alcohols add to allylic indium sesquihalides, but the mechanism of the process is not known[99]. Molybdenum complexes have been used to cyclize enynes and dienynes[100,10]

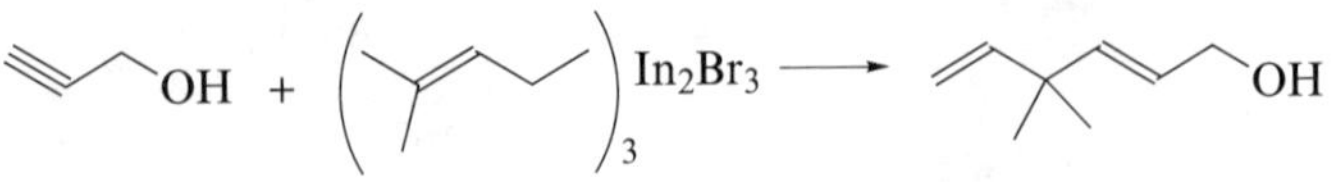

G. Cycloaddition Reactions of Alkynes

Alkynes react as dienophiles in [4+2] cycloadditions, particularly when activated groups such as sulphones[102]. Recently, use of a bis-sulphone has been reported to give : effective dienophile[103].

Alkynyl cyanoketenes readily undergo [2+2] cycloadditions with alkenes, and the reaction has now been extended to alkynes[104]. The initial reaction is conventional, **71** forming **72**. However, **72** can then rearrange to form the dibenzopyran, **74**.

When the starting material **71** was unsubstituted, this reaction route is unavailable. Reaction of **75** proceeds via **76**, a similar intermediate to **73**, but then adds another molecule of PhC≡CCH₃ to give the bicyclo[4.2.0]octatriene **77**.

The reaction has also been used to make cyclopentenes by adding silylated propargyl cations to alkenes[105]; addition to a titanium imide complex has also been reported[106].

The photochemical [2+2] cycloaddition of α,β-unsaturated carbonyl compounds to alkynes has been accomplished, and proved to be of synthetic value[107].

The photochemical addition of α,β-acetylenic ketones with alkenes yields acid-sensitive products[108–110]. In the absence of acids, reaction of 3-pentyn-2-one with tetramethylethylene **78** yields **79**, **80** and **81**. In the presence of acids, **80** and **81** rearrange to give **82**. The reaction probably proceeds through a biradical.

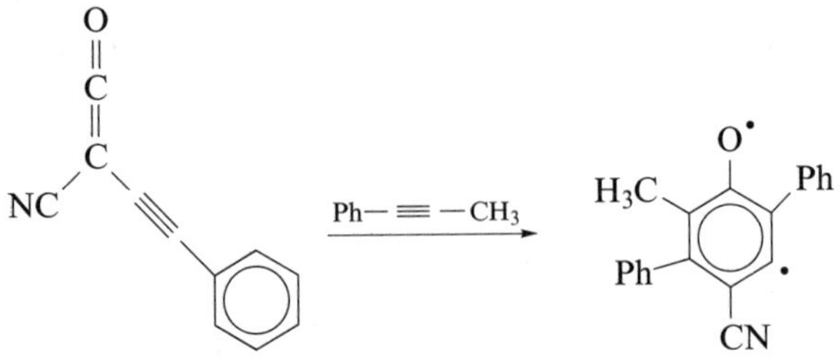

(75)   (76)

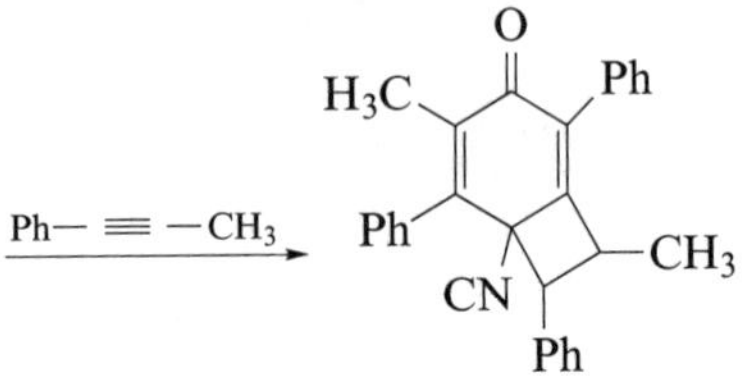

(77)

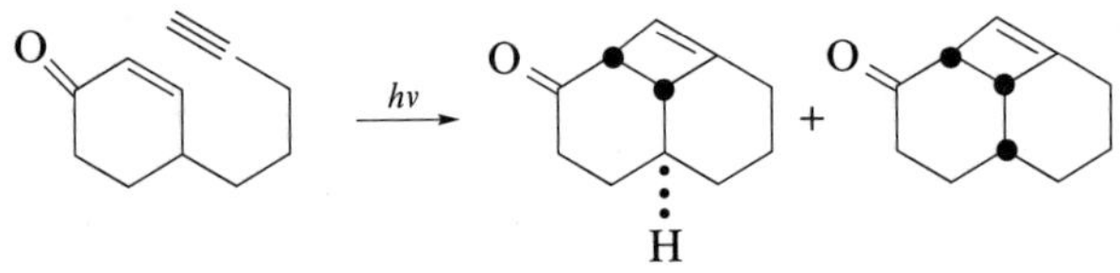

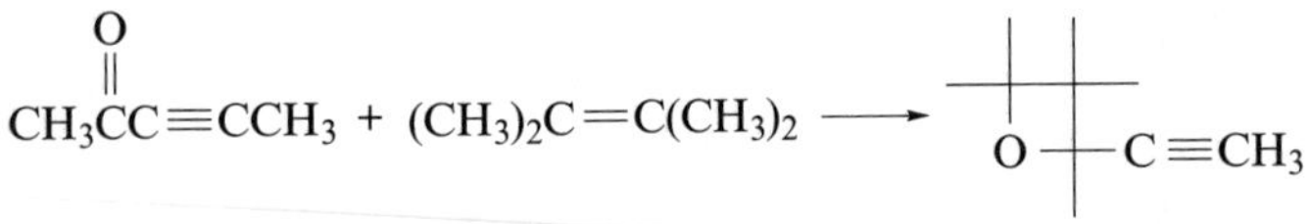

(78)   (79)

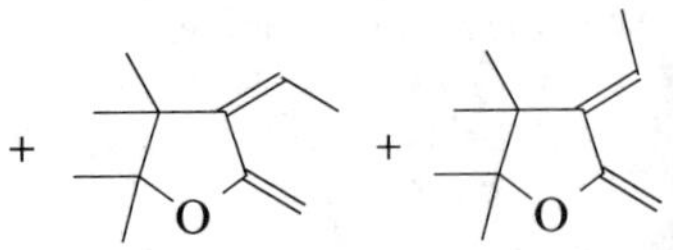

(80)   (81)

The carbene has been trapped with methanol.

Complexation with organometallic molecules has proved a valuable method of activating enynes to undergo cyclization. Complexation[111] with $Co_2(CO)_8$ was used to activate **8** to cyclize to **84** in a Khand reaction. Propargylsilanes have also been used in a [3+2] annulation[112].

Alkynyl complexes have been reported to undergo [2+2] and [2+4] cycloaddition reactions with ketenes[113]. Alkynes have been reacted with cyclopropylcarbene–chromium complexes[114], and with vinylketene–tricarbonyl iron(0) complexes, the latter in a highly regioselective reaction[115].

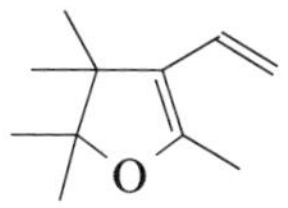

(82) 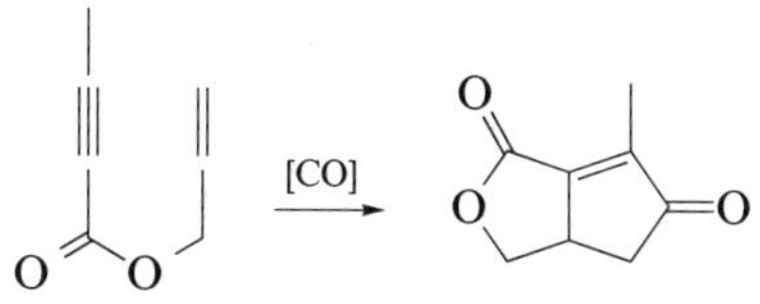 (83) (84)

H. Sigmatropic Rearrangements of Alkynes

The simplest possible sigmatropic reaction is a [1,2] shift within a cation, radical or anion.

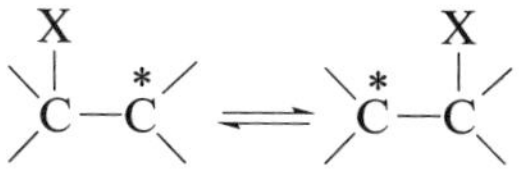

When the reaction is synchronous, i.e. bond formation and bond breaking are concerted, then the orbital symmetry rules apply to the process.

Alkyne rearrangements generally involve [2,3] or [3,3] sigmatropic processes. The latter involve the Claisen and Cope rearrangements. The Cope rearrangement of 1-hexene-5-yne (**85**) proceeds by formation of the alkatriene **86**, which cyclizes.

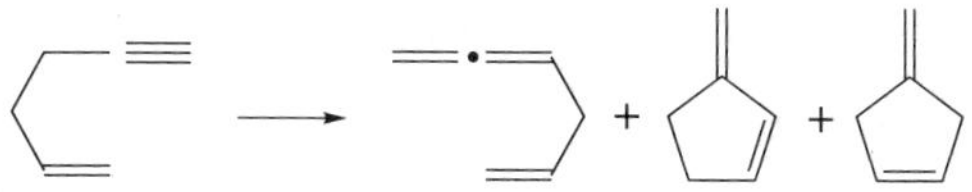

(85) (86)

A biradical intermediate has been suggested for the interconversion of **85** and **86**; however, recent work[116] has suggested that while the biradical **87** may well be a transition state for the conversion of **88** into **89**, it is unlikely to be an intermediate.

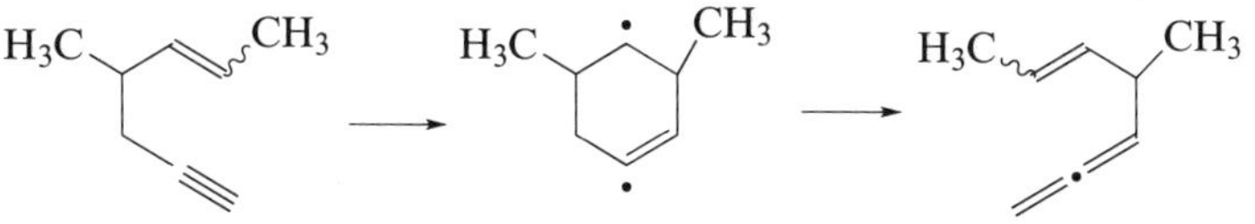

(88) (87) (89)

A Cope-type rearrangement in a radical cation has been reported, when 1,5-hexadiyne **90**) is irradiated in a Freon matrix. The intermediate (**91**) can also be generated by treatment of the diallene under similar conditions.

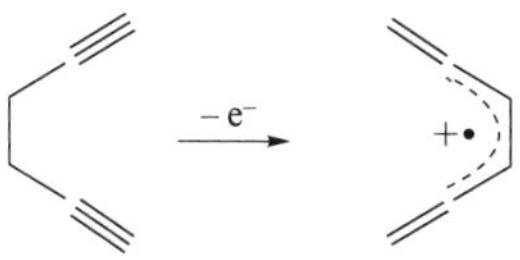

(90) (91)

The [2,3] Wittig rearrangement is a useful reaction involving a [2,3] sigmatropic shift, and has been used to obtain homopropargylic alcohols from allylic propargylic ethers with high stereoselectivity[117]. The reaction has been modified by Marshall[118] to provide a useful route to optically active allenic alcohols.

These reactions have found use in syntheses as different as construction of a 14-membered carbocyclic ring[119] and preparation of optically active 2,5-dihydrofurans[120].

I. Thermal and Photochemical Rearrangements of Alkynes

1. Concerted rearrangements

1-Alkynyl ethers bearing hydrogen atoms β to the oxygen atom fragment readily into an alkene and a ketene at moderate temperatures. Semi-empirical calculations are consistent with the reaction having a highly synchronous character, and fit the observed isotope effects[121].

The reaction has proved to be of synthetic value in the synthesis of silylketenes[122].

Aluminium bromide is known[123] to catalyse formation of cyclobutadiene from alkynes at low temperatures.

2. Formation of radical cations

If a mixture of a dialkylalkyne and aluminium chloride in dichloromethane at low temperature is irradiated with U.V. light, the E.S.R. spectrum of the corresponding tetra alkylcyclobutadiene radical cation can be observed[124].

$$R-C\equiv C-R \xrightarrow[hv]{AlCl_3} R-C\overset{\bullet}{\underset{}{\equiv}}\overset{+}{C}-R \xrightarrow{R-C\equiv C-R}$$

Using arylalkynes[125] the reaction yields the corresponding azulene on a quantitative scale.

The conversion of diphenylacetylene into 1,2,3-triphenylazulene under catalytic conditions[126] is well documented, but it has not previously been observed for the radical cation.

Photogenerated radical cations derived from alkynes by irradiation in methanol with 1, 4-dicyanonaphthalene have been shown to provide another route to alkyne–

(92)   (93)

llene interconversion[127]. The reaction also yields small quantities of the ethers **92** and **3**. Carrying out the reaction in MeOD resulted in 80% labelling of the H substituent, ndicating that the reaction is mainly intermolecular.

III. REARRANGEMENTS INVOLVING CYANIDES

A. Rearrangements to Isocyanides and Ketenimines

The isocyanide–cyanide (isonitrile–nitrile) rearrangement was discovered in 1873 by Weith[128] and is best described as a cationotropic 1,2-shift.

$$CH_3—N\equiv C \rightleftharpoons CH_3—C\equiv N$$

(94)

The equilibrium[129] favours the more stable cyanide **94** by 15 kcal mol^{-1}. The interconversion[130] reaction of the cyanide is first order, and proceeds readily at around 200 °C. It is believed to proceed through a symmetrical intermediate such as **95**. The reaction proceeds with retention of optical activity[131].

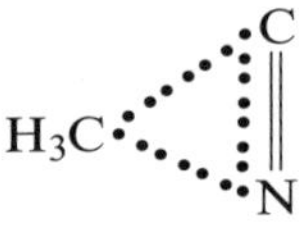

$$H_3C\cdots\overset{\displaystyle C}{\underset{\displaystyle N}{\triangle}}$$

(95)

The reaction is now of principal interest for testing theories of chemical kinetics, and it has been used to test the Rice–Ramsperger–Kassel–Marcus theory[132] which calculates rates on the basis of such parameters as the transition-state vibrational frequencies and moments of inertia. The results suggested that during the reaction, neither the C≡N triple-bond length nor the H—C—H bond angle in the migrating methyl group is seriously changed. This leads to the suggestion that neither steric nor electronic effects on the rate of the reaction are likely to be important, and therefore no distinct structure–reactivity relationship is to be expected[133]. A survey of published data showed that 19 primary, secondary, tertiary, cyclic bicyclic, bridgehead, benzyl, substituted benzyl, α-carbomethoxymethyl and triphenylmethyl isocyanides varied in reaction rate by only a factor of 67, and in ΔG by only ±2 kcal mol^{-2}. A sole exception was 9-triptycyl isocyanide (**96**), which reacts more slowly than a simple alkyl system by a factor of 3000, probably due

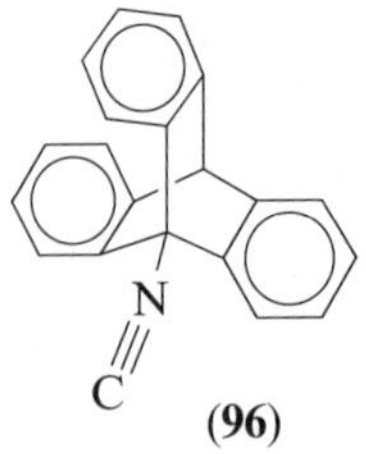

(96)

to interference with the transition state by the three *peri* hydrogen in **96**. Aromat isocyanides isomerize about ten times faster, independent of polar *para* substituents ar bulky *ortho* substituents. These data all favour a tight three-membered cyclic transitic state, such as **95**.

Cyanides which possess one or more hydrogen atoms attached to the α-carbon atom are potentially capable of tautomeric equilibrium[134] between the cyanide (97) and the ketenimine (98), a reaction similar to the alkyne–allene interconversion discussed earlier.

$$\begin{array}{c} R \\ \diagdown \\ \hspace{1em} CH-C\equiv N \\ R^1 \diagup \\ \hspace{1em} (97) \end{array} \rightleftharpoons \begin{array}{c} R \\ \diagdown \\ \hspace{1em} C=C=NH \\ R^1 \diagup \\ \hspace{1em} (98) \end{array}$$

The α-hydrogen of the cyanide 97 is acidic, and the carbanion produced is a resonance hybrid, to which a structure similar to 98 contributes substantially.

$$\begin{array}{c} R \\ \diagdown \\ \hspace{1em} \overset{\cdot\cdot}{C}{}^{-}-C\equiv N\colon \\ R^1 \diagup \end{array} \rightleftharpoons \begin{array}{c} R \\ \diagdown \\ \hspace{1em} C=C=\overset{\cdot\cdot}{N}\colon{}^{-} \\ R^1 \diagup \end{array}$$

The synthesis of N-t-butylketenimines can be accomplished readily[135] at –80 °C. On warming to – 40 °C the ketenimines dimerized to 2-iminoazetidines (99); it is suggested that the reaction is initiated by protonation of the ketenimine.

$$\begin{array}{c}
CH_3CH=C=\overset{+}{N}H-Bu^t \\
\\
CH_3CH=C=\overset{\cdot\cdot}{N}-Bu^t
\end{array}
\quad
\begin{array}{c}
CH_3CH=C-NHBu^t \\
\hspace{1.5em}| \\
\hspace{1.5em}\overset{+}{N}\,Bu^t \\
\hspace{1.5em}\| \\
\hspace{1.5em}C \\
\hspace{1.5em}\| \\
\hspace{1em}H_3CCH
\end{array}
\longrightarrow
\begin{array}{c}
\hspace{3em}H \\
H_3C-\!\!\!-\!\!\!\!=NBu^t \\
HC=\!\!\!-\!\!\!\!NBu^t \\
\hspace{2em}| \\
\hspace{2em}CH_3 \\
\hspace{2em}(99)
\end{array}$$

$$-C\equiv C-N\diagup\diagdown \rightleftharpoons \diagdown\diagup C=C=N \rightleftharpoons -\overset{|}{\underset{|}{C}}-C\equiv N$$

$$(100)$$

A further tautomer[134] is the ynamine structure (100); only tertiary ynamines are stable. The tautomeric equilibrium strongly favours the cyanide form.

Closely related to these tautomers is cyanocarbene (101), which is believed to exist in an equilibrium with the allene (102).

$$-\overset{\cdot\cdot}{C}-C\equiv N\colon \rightleftharpoons -\overset{+}{C}=C=\overset{-}{N}\colon$$

$$\hspace{1em}(101) \hspace{4em} (102)$$

This was originally assumed to be linear, until the work of Wasserman[136] suggested that carbenes had a substantially bent triplet state, even though experimental evidence to extend this view to cyanocarbene could not be obtained at the time. Recent theoretical treatments[137] suggest that HCCN is a quasi-linear molecule, with an HCC angle of 138°C, and a barrier to linearity of only about 2 kcal mol^{-1}.

B. Rearrangements Involving Metal Complexes

Organic cyanides do not form transition metal complexes as readily as do isocyanides[138] or alkynes. A number of metal complexes catalyse reactions of cyanides, and these are considered when discussing the appropriate reactions.

Cyanides do form metal complexes, and are important as linking groups between metals[139]. An important reaction of cyanides is with a molecule already complexed to a metal. This reaction is quite similar to the reaction reported by Hogeveen[140] between aluminium halide σ complexes of cyclobutadiene and cyanides, yielding a mixture of pyridines.

A series of similar reactions, in which a tungsten complex of an alkyne (103) can undergo reaction with a cyanide, has been reported[141]. The reaction yields cyclic products (104) usually involving five- or seven-membered rings.

$$W(OBu-t)_6(\mu\text{-}C_2H_2)(py) + PhC \equiv N \longrightarrow$$

(103)   (py = pyridine)   (104)

The reaction is believed to be a direct addition, and not involve the cleavage of the cyanide bond by tungsten complexes reported earlier[142].

$$(Me_3CO)_3W \equiv W(OCMe_3)_3 + PhCN \longrightarrow (Me_3CO)_3W \equiv CPh + (Me_3CO)_3W \equiv N$$

Unsaturated cyanides can undergo a reaction which is basically alkene metathesis. Long-chain unsaturated cyanides[143] react with the catalyst system WCl_6–$Me_3Al_2Cl_2$, while more general metathesis has been achieved[144] using the catalyst system WCl_6–Me_4Sn. The reaction is inevitably more complex than an alkene metathesis, since an alkene and an unsaturated cyanide must yield a mixture of four products.

$$2CH_2{=}CH[CH_2]_nCN$$

$$+$$

$$2CH_3CH_2CH{=}CH[CH_2]_2CH_3$$

$$\rightleftharpoons$$

$$CH_2{=}CHCH_2CH_3$$
$$+ CH_2{=}CH[CH_2]_2CH_3$$
$$+ CH_3CH_2CH{=}CH[CH_2]_nCN$$
$$+ CH_3[CH_2]_2CH{=}CH[CH_2]_nCN$$

C. Rearrangements Involving Cations

1. Effects of the cyano group on neighbouring carbocations

The cyano group is electron withdrawing, so should destabilize a carbocation; hence, α- and β-substituted carbocations would be expected to form at greatly reduced rates relative to their H analogues. However, a β-cyano group is more rate retarding[145-147], $k_H/k_{\beta\text{-CN}} = 10^5$ to 10^7, than is an α-cyano group, $k_H/k_{\alpha\text{-CN}} = 10^3$ to 10^4. To explain this result, it was suggested that α-cyano groups are less destabilizing than would be expected on the basis of the inductive effect, since the group may also display conjugative stabilization, giving an intermediate with more nitrenium ion character.

Evidence that the α-cyano carbocations did display some nitrenium ion character was obtained by Olah and coworkers[148-150]. A study of the ion (**105**) generated from benzophenone cyanhydrin indicated from ^{15}N chemical shift data that the ion included a substantial contribution from the nitrenium ion form (**106**).

$$\underset{(105)}{\overset{Ph}{\underset{Ph}{\diagdown}}\overset{+}{C}-C\equiv N} \quad \longleftrightarrow \quad \underset{(106)}{\overset{Ph}{\underset{Ph}{\diagdown}}C=C=\overset{+}{N}}$$

A study of the products of decomposition of cyano-substituted carbocations revealed a more confusing picture[151]. Comparison of the rates of solvolysis of the substrates **107** and **108** showed the unsaturated system to react faster by a factor of 10^8, compared to 10^{10} when

(107) (108)

the cyano group is replaced by hydrogen. In both **107** and **108**, the products were formed from a rearranged ion reacting with solvent at the carbon atom β to the cyano group.

A similar observation was made by Farcasiu[152] from a study of the reactions of adamantyl systems; he attributed the rearrangement to steric strain relief on the basis of force-field calculations.

The question has been further investigated[153] in an ingenious experiment in which the equilibration of the carbocations **109** and **110** was studied.

(109) (110)

The reaction was difficult to study, since attempts to solvolyse the ester **111** yielded almost exclusively the ester **112** via tight ion pairs. This problem was overcome by studying the decomposition of the appropriate diazonium ions to give **109** and **110**. Then, both reactions gave a mixture of products derived from **109** and **110** in which products obtained from the β-cyano carbocation predominated.

The authors[153,154] were able to show, however, that the trifluoroacetic acid esters, **111** and **112**, differed in stability by approximately 10 kcal mol^{-1}, with the β-cyano ester **112** being the more stable. Clearly, in this case the rate of formation of a carbocation is not a good guide to the stability of the carbocation, as the energy difference between the esters exceeds that between the carbocations.

(111)   (112)

2. Through-space interactions of carbocations with cyanides

Reaction of a carbocation with a cyanide yields an amide in a reaction known as the Ritter reaction.

$$RC{\equiv}N + R'^{+} \rightleftharpoons \left[RC{\equiv}\overset{+}{N}R' \longleftrightarrow R\overset{+}{C}{\equiv}\overset{..}{N}R' \right] \xrightarrow{H_2O} RCONHR'$$

The reaction is believed to involve the nitrilium ion; evidence for the existence of this ion in fluorosulphonic acid has been obtained by Olah[155].

The reaction is of value in carrying out cyclizations[156] but then an alternative reaction pathway is available[157]. Protonation of the nitrile is followed by transannular double-bond participation to produce the 3-exo-hydroxyadamantanone **113**.

(113)

Examples of nucleophilic attack on the nitrilium ion, $-C{\equiv}\overset{+}{N}-$, are known in other fields, where the ions are more commonly generated by halide solvolysis[158]. The reaction has since been extended into a more general process for the synthesis of heterocyclic molecules by Bishop[159].

Recent research into the Ritter reaction has concentrated on alternative means of generating the carbocation. Radiolytically formed carbocations[160] have been reacted in the gas phase with aromatic and aliphatic cyanides yielding the corresponding nitrilium ions, which undergo condensation with water to yield N-alkylamides.

The development of carbocations of organometallic complexes of transition metals[161] offers an alternative method of carrying out the Ritter reaction with primary and secondary alcohols, which normally give poor yields. Reaction of the complex **114** with concentrated sulphuric acid at −15 °C in the presence of acetonitrile gives the amide complex in good yield[162].

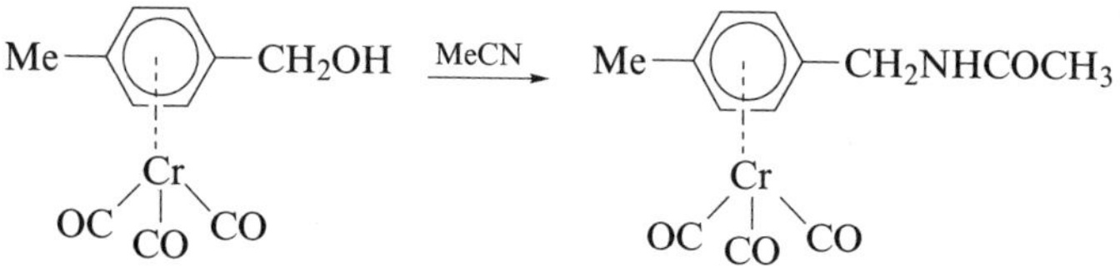
Me
CH2OH
MeCN
Me
CH2NHCOCH3
Cr
OC
CO
CO
Cr
OC
CO
CO

(114)

The reaction has also been used to stabilize the carbocation from propargyl alcohols.

The same principles have been applied to activate a double bond to react with a cyanide[163]. The method involves the use of the highly electrophilic complex $Pd(CH_3CN)_4(BF_4)_2$, which interacts strongly with alkenes to give intermediates showing some of the properties of carbocations, which can add to the cyano group to give a nitrilium ion. In the presence of water, this can complete a conventional Ritter reaction, but alternatively it can react readily with available nucleophiles to produce more complex products.

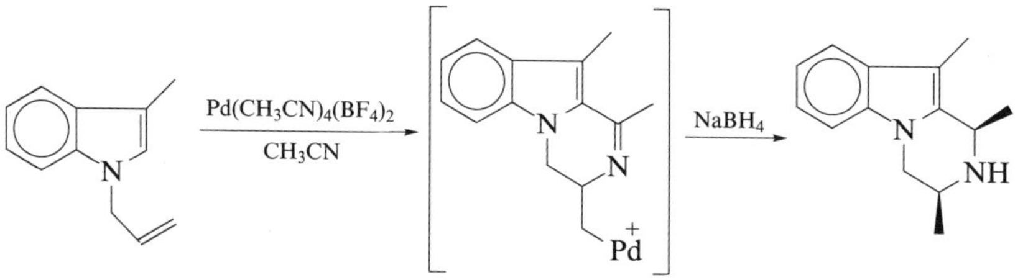
N
Pd(CH3CN)4(BF4)2
CH3CN
N
N
Pd
NaBH4
N
NH

D. Rearrangements Involving Carbenes and Nitrenes

Cyanocarbenes ($\ddot{C}$—C≡N) and cyanonitrene (:N—C≡N) have been known for some time, and display the expected addition and related reactions. New approaches to obtaining both species have appeared during the last decade. Cyanonitrene has been obtained by an α-elimination process[164] as outlined below. Evidence that the cyanonitrene is obtained is

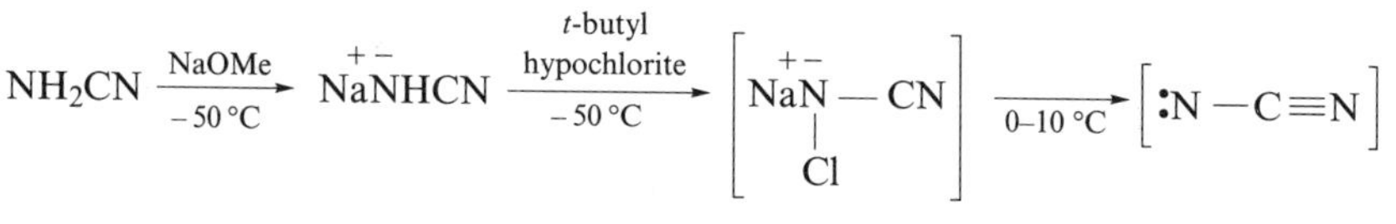
NH2CN
NaOMe
- 50 °C
NaNHCN
t-butyl
hypochlorite
- 50 °C
NaN — CN
Cl
0–10 °C
:N — C≡N

provided by reaction of the nitrene with an amine to yield an aminimide, a reaction which does not take place below the 0–10 °C range, and formation of the explosive dicyanodiazine **115**, a known dimerization product of cyanonitrene[165].

$$[:N{-}C{\equiv}N] + (CH_3)_3N \longrightarrow (CH_3)_3\overset{+}{N}\overset{-}{N}CN$$

$$N{\equiv}C{-}N{=}N{-}C{\equiv}N$$

(115)

A simple route to a cyanocarbene has been found by warming dimethylaminomalonitrile **116** to 60 °C under nitrogen, when it eliminates hydrogen cyanide and radicals are formed, as shown by the intense E.S.R. spectrum of the dimethylglycinonitrile radical, **117**. The rather complex product mixture is consistent with formation of **117**; elimination of HCN from 116 is a symmetry-allowed $[\sigma^{2s} + \sigma^{2a}]$ non-linear cheletropic fragmentation.

(116)   (117)

E. Rearrangements Involving Radicals

Radicals interact with cyano groups in much the same way as they interact with alkynes; when both groups are in the same molecule, this leads to cyclization[166]. The cyclic radical **118** formed is unstable, and undergoes ring opening to give the rearranged radical, **119**. The

(118)   (119)

overall reaction involved in this case is thus a 1,4-shift of the cyano group. The reaction is general to a variety of unsaturated groups[167]. Evidence[168] for the existence of an intermediate such as **118** has been obtained by generating **120** by thermal decomposition of **121**; the intermediate decomposes immediately to yield the radical **122**. A rare example of generation of a stable cyclic compound has been reported[169]; the lactone **123** yields **124** with tributyltin hydride.

(121)   (120)   (122)

(123)   (124)

Coupling of radicals obtained from cyanides has provided a useful route to diamines[170]. The radical is generated by addition of $NbCl_4(THF)_2$ and is probably stabilized by the complex formed.

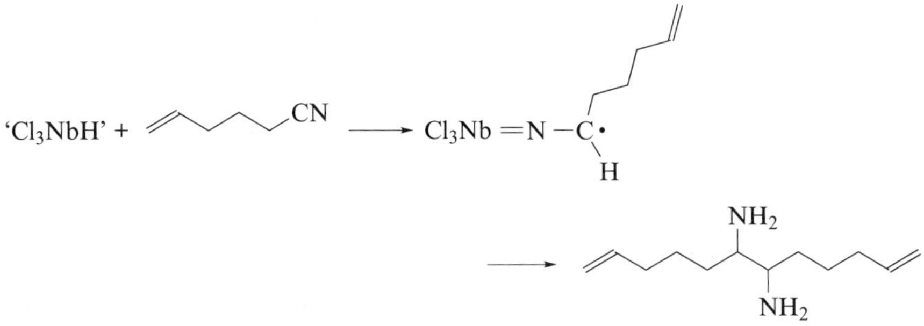

'Cl₃NbH' + (structure with CN) ⟶ Cl₃Nb =N—C•—H (with alkenyl group)

⟶ (structure with two NH₂ groups on chain with two terminal alkenes)

Methanesulphonyl cyanide or *p*-toluenulphonyl cyanide readily release the CN radical, which can be used as a radical trap. The method provides a useful route to the synthesis of cyanides[171].

F. Photochemical Rearrangements

A well-known rearrangement of α,β-unsaturated ketones is the 'Type A' process[172] shown below:

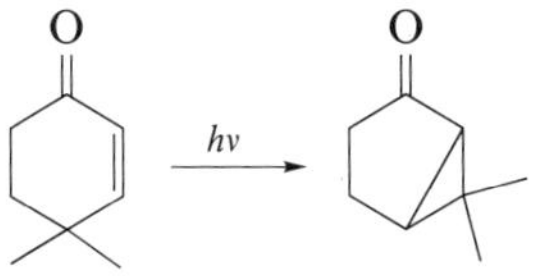

(structure of 4,4-dimethylcyclohex-2-enone) —hv→ (bicyclic ketone product)

A rearrangement of 1-cyanocyclohexene which formally resembles this process has been observed[173].

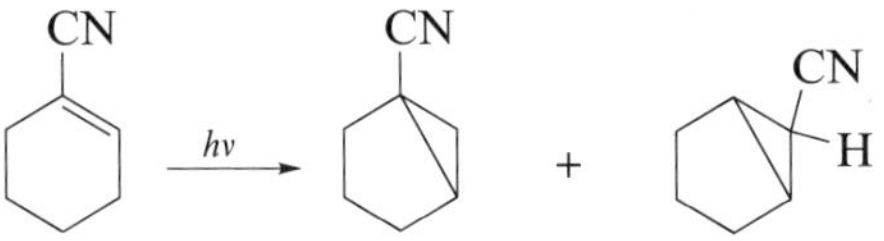

(1-cyanocyclohexene) —hv→ (bicyclic CN product) + (bicyclic CN—H product)

Similarly,

(substituted cyanocyclohexene) —hv→ (bicyclic CN product) + (CN H bicyclic product)

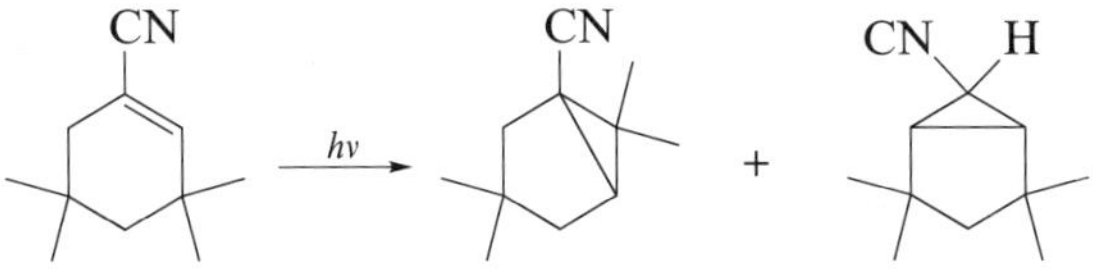

Another Type A rearrangement was studied by Zimmerman and Pasteris[174], who showed that **125** rearranged to **126** on irradiation.

(125)   (126)

G. Cycloaddition Reactions of Cyanides

The addition reactions of cyanoketenes with alkynes have already been discussed. Cyanoketenes, however, also undergo reaction with ketenes, the reaction pathway depending on the structure of the ketene[175].

In contrast to the above[176], the reaction of *t*-butylcyanoketene (127) with isocyanides follows a different pathway, shown nearby.

Acrylonitrile[177] undergoes a [2+2] cycloaddition reaction with ketene alkylsilyl acetals. Like cyanoketenes, cyanoacetylenes readily undergo cycloaddition reactions, and research was given an additional stimulus by their observation in the atmosphere of Titan[178].

The parent cyanoacetylene was reacted with [2.2] paracyclophane[179], giving products which were explained on the basis of a preliminary dimerization of cyanoacetylene to 1,3-cyclobutadiene-1,2-dicarbonitrile, which exists as an equilibrium of valence tautomers, **128**. These then react with [2.2] paracyclophane to give the [2 + 4] adducts. The key adduct

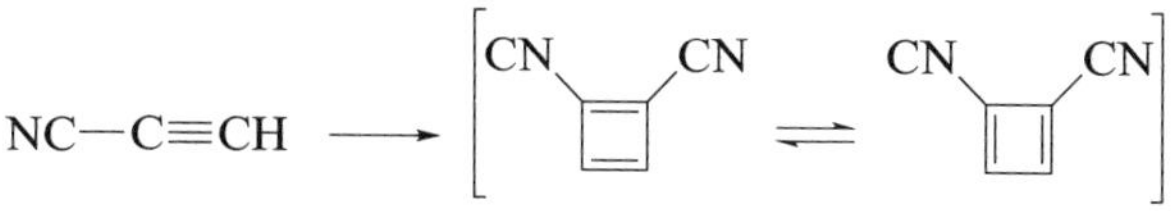

(128)

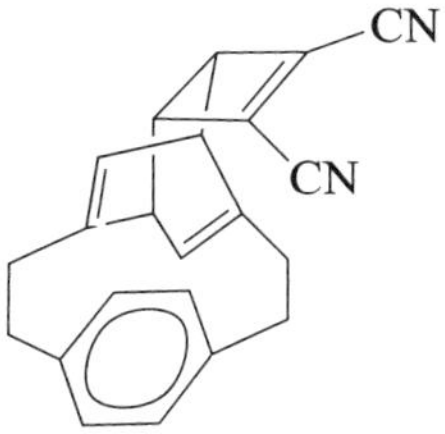

(129)

in the mechanistic scheme is **129**; this product cannot be formed by Diels–Alder addition of cyanoacetylene followed by a [2 + 2] cycloaddition. The photochemical polymerization[180] of cyanoacetylene proceeds via **130** and also via the 1,3-dicyanocyclobutadiene **131**. Both dicyanocyclobutadienes can react by addition of a further molecule of cyanoacetylene to give the aromatic structures shown, or can dimerize to yield tetracyano-cyclooctatetraenes.

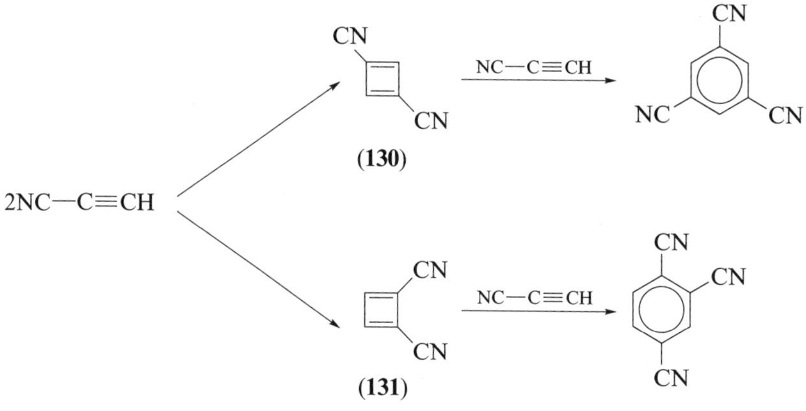

(130)

(131)

With ethylene, irradiation of cyanoacetylene yields 1-cyanocyclobutene[181].

4. Sigmatropic Rearrangements of Cyanides

Cyanides undergo sigmatropic rearrangements in which the cyanide group is directly involved in the reaction, an example being the [2,3] sigmatropic rearrangement of cyanosulphonium ylides to ketenimines[182].

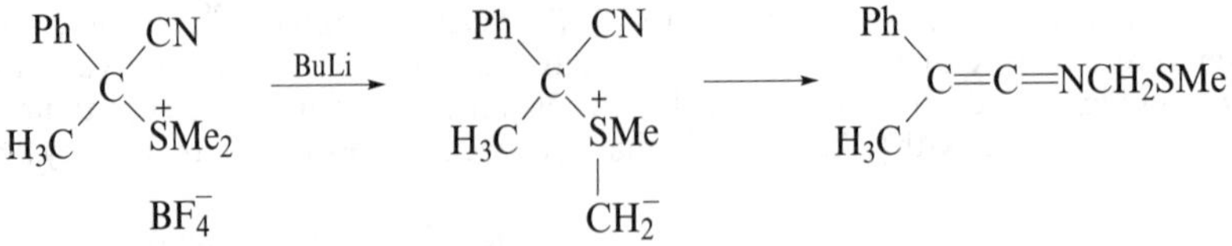

Even when the cyanide group is not directly involved in the reaction, it can have a considerable influence on the reaction mechanism. A study of kinetic deuterium isotope effects on a series of Cope rearrangements[183] showed that rearrangement of the 3,3-dicyanohexa-1,5-diene (**132**) had the k_H/k_D ratio at C(4) (i.e. bond breaking) greater than the k_D/k_H ratio at C(6) bond making) rather than the reverse, which was observed with other substituents. A theoretical study[184] of the reaction indicated that this reaction is

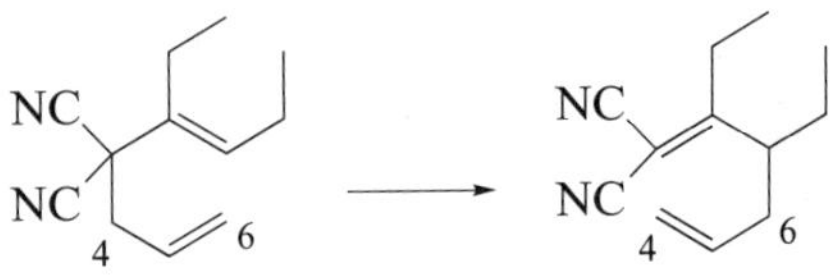

(132)

probably taking place via an aromatic transition state, typical of a pericyclic rearrangement, rather than a biradical-type intermediate, probably as a result of the change in hybridization of the atoms forming the transition state as a result of substitution.

An interesting reaction, which may take place via a simple [1,3] sigmatropic shift, though other processes are possible, is the rearrangement of the cyanothiophenes. Methyl cyanothiophenes rearrange[185] their substituents on heating or irradiation—a typical pattern is shown below.

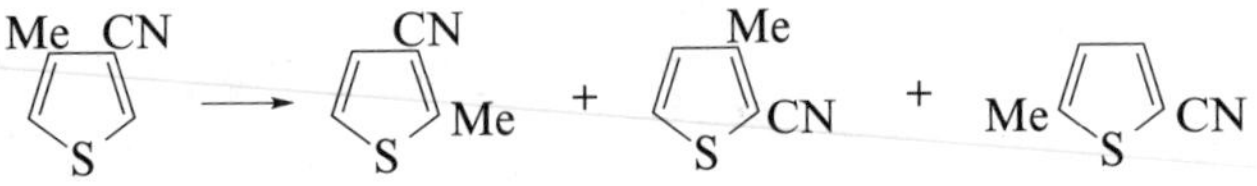

The sequential order of the ring carbons is preserved, leading to the suggestion that the reaction proceeds via a 'sulphur walk'. In support, a small amount of the cyanothiophene valence tautomer 2-cyano-3-methyl-5-thiabicyclo[2.1.0]pent-2-ene (**133**) was isolated from

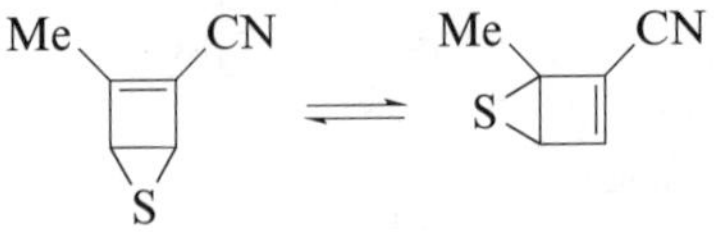

(133)   (134)

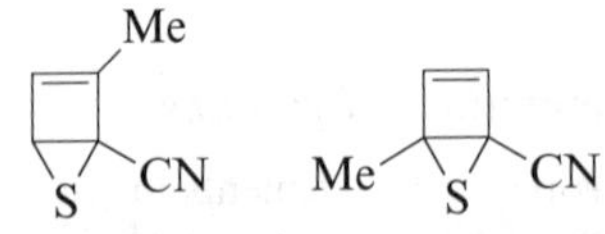

(135)   (136)

the reaction mixture[186]. On forming a furan adduct[187] a mixture was obtained, indicating that **133** existed in dynamic equilibrium with its valence tautomer **134**. The authors found no evidence for the existence of the other possible tautomers, **135** and **136**, but they may well be present only in very small amounts.

IV. REARRANGEMENTS INVOLVING ISOCYANIDES

Isocyanides are linear molecules, with an electron distribution which makes them 'carbenoid' in character[188].

$$\overset{\cdot\cdot}{R\overset{}{N}}{=}C\colon \longleftrightarrow R\overset{+}{N}{\equiv}\overset{-}{C}\colon$$

However, calculation of the gross atomic charges indicates that the nitrogen is electron-rich relative to both the isocyano carbon and the carbon attached to the isocyano group, so that the gross dipole moment of the isocyano group is directed from the terminal carbon towards the nitrogen; this results from the dipole caused by σ charge distribution making a greater overall contribution than does π delocalization. Thus, the isocyano group can stabilize a neighbouring carbocation.

A. Insertion Reactions of Isocyanides

Isocyanides are stable nucleophilic carbenes, well able to insert into carbon–hydrogen, carbon–halogen[189] or heteroatom–hydrogen[190] bonds. The reactions may proceed thermally, or can be catalysed by acids or metal salts. Thus[191] heating the isocyanonaphthalene **137** to 235 °C results in isomerization into the benzocycloheptindole **138** and the benzophenanthridene **139**, in addition to isomerization into the cyanide.

(137)   (138)   (139)

Insertion into carbon-sulphur bonds[192] can take place at room temperature. Thus, *t*-butyl isocyanide inserts into the C—S bond of α-cyano sulphides, such as **140**, giving thioimidates **141** which then rearrange further to a mixture of *E* and *Z* isomers of *N*-vinyl carbamates **142**.

Isocyanides also insert into Zr—H bonds[193] and Pd—C bonds[194] of metal complexes.

(140)   (141)   (142)

Successive insertions into a Li—Si bond are believed to form the basis of a novel dimerization of isocyanides[195] Reaction of a range of isocyanides with phenyldimethylsilyllithium, followed by treatment with trialkylchlorosilane, offered silylated dimers of the isocyanide.

$$2RNC + PhMe_2SiLi \longrightarrow PhMe_2Si-\overset{\displaystyle \underset{RN}{\|}}{C}-\overset{\displaystyle \underset{NR}{\|}}{C}-Li$$

B. Rearrangements Involving Carbocations

Calculations[196] of the stabilities of isocyano-substituted carbocations suggest that the charge is spread over the two carbon atoms.

$$\underset{R}{\overset{R}{>}}\overset{+}{C}-N\equiv C \quad \longleftrightarrow \quad \underset{R}{\overset{R}{>}}C=N=\overset{+}{C}$$

Stabilization over the unsubstituted carbocation varies from 25 kcal mol^{-1} for R = H to 10 kcal mol^{-1} for R = CH$_3$.

Isocyanides, being nucleophilic, and having a greater proton affinity than the corresponding cyanides, add readily to carbocations, though the acid sensitivity of isocyanides limits the reaction to those carbocations[197] which can be prepared under non-acidic conditions. The complex of cyclobutadiene with AlCl$_3$ provides a useful example of this type of reaction[198]. The isocyanide can also insert into the C—Cl bond of **143**.

(143)

Diethylaluminium chloride promotes 1,4-addition of α,β-unsaturated carbonyl compounds with isocyanides, a reaction of value in the synthesis of γ-butyrolactones[199].

With $TiCl_4$, however, the reaction gives only the cyanide.

It is probable that the isocyanate adds to the activated enone, and then undergoes β-elimination to give the β-cyano enolate plus the t-butyl carbocation[200]. Evidence in support of this has been obtained[201] by adding a second molecule of an isocyanide to the complex formed by adding isocyanide to acetals in the presence of $TiCl_4$.

C. Reactions with Radicals

Isocyanides react readily with radicals. In the case of phosphorus or silicon based radicals[202] addition gives a new radical species (**144**) which loses an alkyl radical.

$$(CH_3)_3CN = C + \overset{\bullet}{S}iR_3 \longrightarrow (CH_3)_3CN = \overset{\bullet}{C}SiR_3 \longrightarrow (CH_3)_3\overset{\bullet}{C} + N \equiv CSiR_3$$

$$(144)$$

However, the radical species such as **144** can be trapped in suitable conditions, so the reaction can be used to carry out cyclizations[203].

The addition of perfluoroalkyl iodides to isocyanides to give perfluoroalkylimidol iodides has been shown to be catalysed by copper powder[204].

D. Rearrangements Involving Metal Complexes

Isocyanides react readily with organometallic reagents; if the isocyanide possesses an α-hydrogen atom, then this will be abstracted to give an α-metalated isocyanide. If not, then α-addition to the isocyanide yields a metalloaldimine. Both reactions are valuable in synthetic studies.

Rearrangements are less common, since the complexes are usually stable. However, if a cobalt–acetylene complex is reacted with an isocyanide, a metallacycle (**145**) is formed[205].

Replacement of a phenyl group with an electron-withdrawing group shows that the latter occupies the β-position (i.e. diametrically across from the cobalt atom), suggesting that the regioselectivity observed when unsymmetrical alkynes are used is electronic in nature.

E. Cycloaddition Reactions of Isocyanides

Isocyanides readily undergo cycloaddition reactions, and these are very valuable in the formation of heterocyclic rings. Reaction[206] of β-nitrostyrene with an alkyl isocyanide gives a hydroxy indole (**146**). Reaction proceeds even more readily[207a] between tosylmethyl isocyanide (**147**), in which the methylene is activated, and aryldiazonium compounds. With ketenes, isocyanides give imino lactones[207b]. However, with t-butylcyanoketene, the reaction follows a different pathway[208] involving the carbonyl bond of the ketene, to yield **148**. A [1 + 3] cycloaddition of an isocyanide to a 1,3-dipole has been used to prepare azetidines[209]. The method has been used for synthesis of a number of azetidines[210].

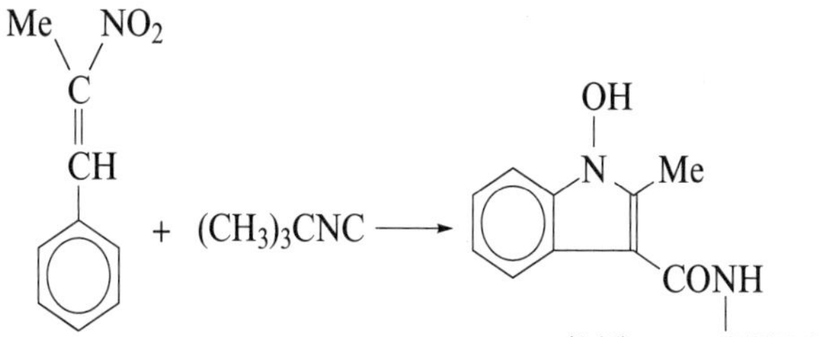

(146)

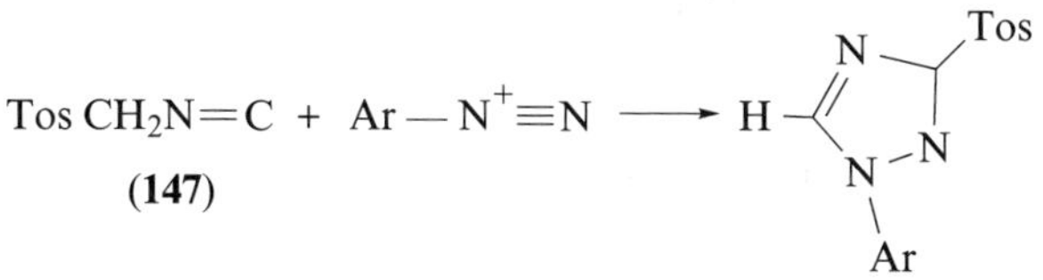

(147)

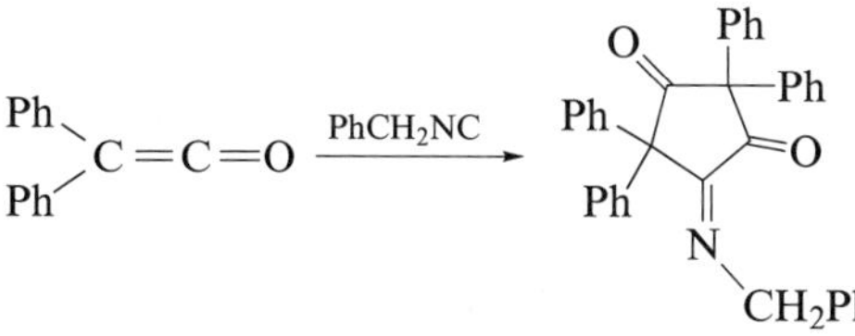

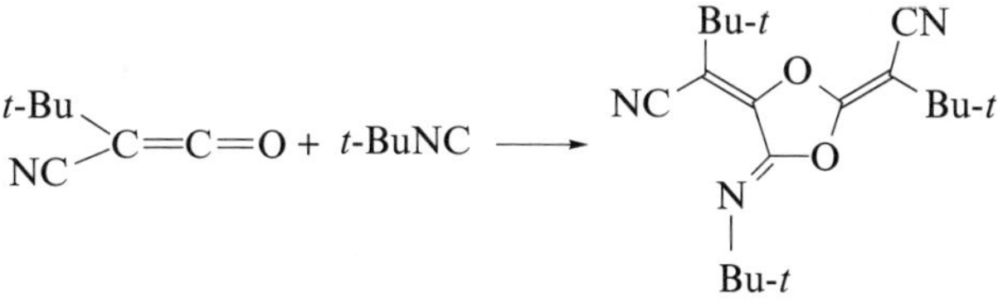

(148)

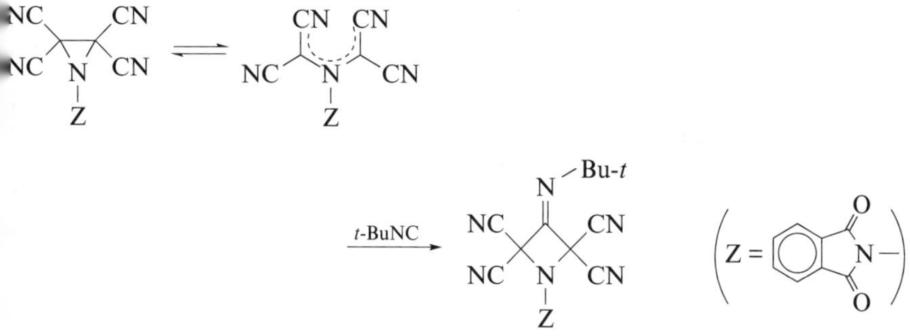

A deceptively simple synthesis of cyclopenta-fused quinolines involves initial radical addition to an isocyanide, followed by a [4 + 1] radical annulation[211].

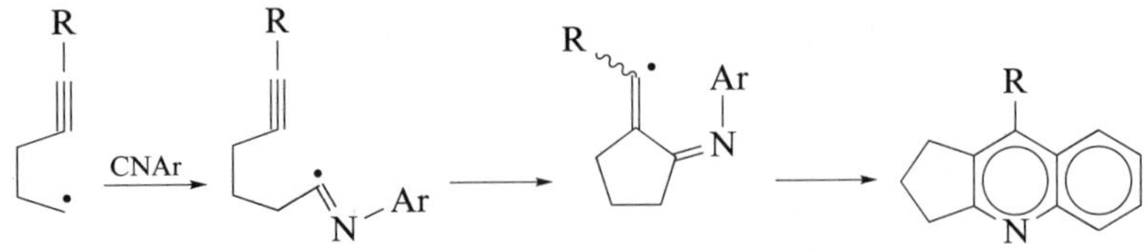

Recently[212] isocyanides have been shown to undergo benzannulation to amines with chromium carbene complexes such as **149**.

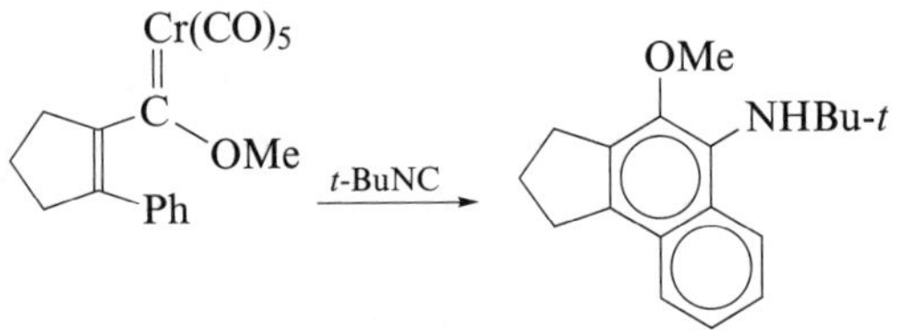

(149)

V. REARRANGEMENTS INVOLVING ARYNES

Many rearrangements occur via reactive intermediates, but arynes must be among those least prone to rearrange. The generation and reaction of an aryne does involve some measure of rearrangement, which provided strong evidence for the existence of arynes.

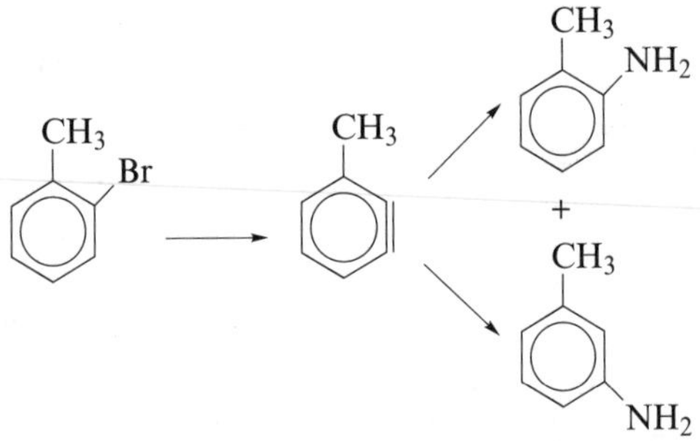

The extra election pair involved in the aryne does not affect the aromaticity of the species. The original sextet still functions as a closed ring, and the two additional electrons are merely located in a π orbital that covers only two carbon atoms.

(150)

The type of reaction shown above can take place intramolecularly with a suitable substrate, and yield cyclic systems **(150)**.

Although arynes are reluctant to rearrange, examples of ring contraction of an aryne have been recorded. An example of an aryne rearrangement to a carbene[213] has been recorded; it is probable that the aryne and carbene are in equilibrium under the vigorous conditions (flash vacuum pyrolysis at 960 °C) of the reaction[214,215]. A similar ring contraction in a heterocyclic system probably proceeds via a similar mechanism[216].

The arynes also undergo cycloaddition reactions very readily; they even add to fullerenes[217]. An excellent review of these reactions has appeared[218].
An unusual reaction of arynes is with thiirane to yield phenyl vinyl sulphides[219].

Benzyne also reacts with oxiranes and aziridines[220].
The arynes discussed above are entirely of the o-aryne type; that is to say, the hydrogen atoms 'removed' from benzene are *ortho* to each other. It is not necessary for them to be

(151)   (152)

ortho; 1,4-dehydrobenzene exists, probably in the form **151**, and the hydrogens can even be removed from different rings (**152**).

The representation of these molecules as diradicals is probably correct, though in **152** concerted addition of an alkene is possible.

The possibility that 1,4-dehydrobenzene, or a molecule of similar symmetry, is an intermediate in a rearrangement has been explored by Bergman[221,222]. He has shown that the dialkynes **153** and **154** interconvert; the probable pathway is through 1,4-dehydrobenzene (**151**).

$$\text{(153)} \quad\rightleftharpoons\quad \text{(151)} \quad\rightleftharpoons\quad \text{(154)}$$

A double cycloaromatization[223] to 2,6-didehydronaphthalene has been reported.

A number of other examples of formation of 1,4-biradicals of this type by electrocyclization from enediynes have been reported[224]. The reactions take place under surprisingly mild conditions.

An alternative representation of the structure of 1,4-dehydrobenzene is the butalene structure **155**. Evidence that this may be a contributor to the stability of 1,4-dehydrobenzene

(**155**)

is again obtained from trapping experiments[225]. On treatment of the chloride **156** with LiNMe$_2$ in HNMe$_2$ – THF at 0 °C, dimethylaniline is formed; the labelling pattern when formed from labelled **156** is as expected. However, if diphenylisobenzofuran is added to the

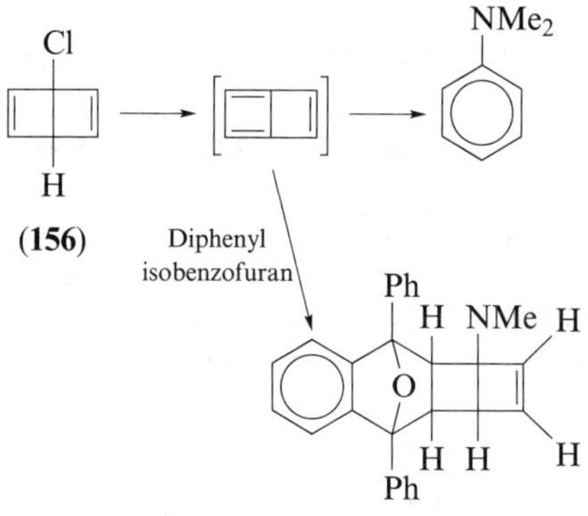

(156)   Diphenyl isobenzofuran

(157)

reaction, then the adduct **157** is obtained, consistent with reactions proceeding via the butalene structure **155**. Direct synthesis of an aromatic species provides further evidence of a biradical structure for 1,4-dehydro benzene[226]. Reaction of **158** in the gas phase at 320 °C yields a mixture of the three products shown.

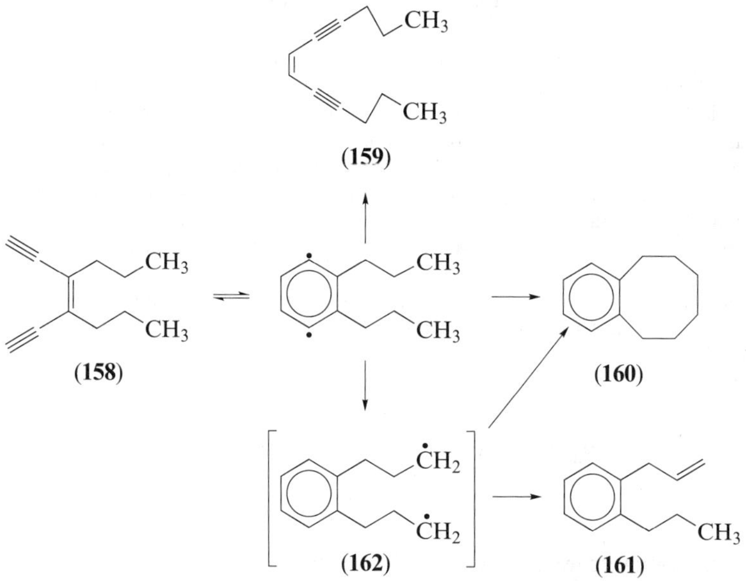

(159)

(158)   **(160)**

(162)   **(161)**

In the presence of a good hydrogen atom donor (1,4-cyclohexadiene) the reduced product **163** is then obtained; using a deuterium donor, it was shown that most (66%) of **163** comes from the biradical.

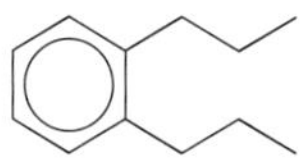

(163)

Other biradical structures have been obtained by rearrangement of non-aromatic precursors, including α-3-dehydrotoluene[227] and 1,3-dehydronaphthalene.

VI. REARRANGEMENTS INVOLVING DIAZO AND DIAZONIUM GROUPS

Diazo compounds and diazonium compounds differ from the triple-bonded molecules discussed so far in that they readily lose nitrogen, forming carbenes and carbocations, respectively. This opens up two large fields of rearrangements, but on the other hand reduces the number of rearrangements involving reaction of the group with a reactive intermediate, and restricts the scope of addition reactions.

A. Isomerization of Diazo Compounds

Diazo compounds do not readily undergo isomerization reactions, although diazomethane itself is unique among small molecules in its number of theoretically possible structural

$$
\begin{array}{ccc}
\mathrm{H} & & \mathrm{H}\quad\mathrm{N} \\
\diagdown\mathrm{C}{=}\overset{+}{\mathrm{N}}{=}\overset{-}{\mathrm{N}} & \rightleftharpoons & \diagdown\mathrm{C}\overset{\|}{\diagup} \\
\diagup & & \diagup\quad\| \\
\mathrm{H} & & \mathrm{H}\quad\mathrm{N}
\end{array}
$$

$$\textbf{(164)} \qquad\qquad \textbf{(165)}$$

isomers[228]. The only isomer with which diazomethane (164) can be interconverted is diazirine (165), a cyclic structure[229]. Involvement of diazomethane in the photolysis of diazirine has been demonstrated for both photochemical[230] and thermal[231] reactions. Loss of diazirine, measured by U.V. spectroscopy, is slower than nitrogen production by a factor of three. More recently, the work of Liu and coworkers[232] has suggested a diazirine decomposition mechanism in which a carbene–nitrogen complex is formed as the first intermediate. This can either collapse to the diazo compound or lose nitrogen to yield the free carbene. Isomerization to the diazo compound is thus not necessarily a general reaction of diazirines.

The isomerization of diazo compounds into diazirines is less common, but the photochemical isomerization of α-diazoamides into the corresponding diazirinylamides has been reported[233].

$$
\mathrm{N_2CHCONC_5H_{10}} \longrightarrow
\begin{array}{cc}
\mathrm{N} & \mathrm{H} \\
\|\diagdown\mathrm{C}\diagup \\
\mathrm{N} & \diagdown\mathrm{CONC_5H_{10}}
\end{array}
$$

B. Rearrangements Involving Metal Complexes

Like cyanides, diazo compounds do not form complexes with metals as readily as do alkynes. Reaction of diphenyldiazomethane with a metal-to-metal triple bond to give

$$
\mathrm{Cp(CO)_2Mo{\equiv}Mo(CO)_2Cp + Ph_2CN_2} \longrightarrow
\begin{array}{cc}
\mathrm{Ph} & \mathrm{N} \\
\diagdown\diagup\| \\
\mathrm{Ph{-}C} & \mathrm{N} \\
| & | \\
\end{array}
\ \mathrm{Cp(CO)_2Mo{=}Mo(CO)_2Cp}
$$

a metal complex has been reported[234]. An unusual electrochemical coupling of diazomethane ligands can occur to give a ligand which bridges two metal complexes[235].

$$2\left[WF(N_2CH_2)(dppe)_2\right]^+ BF_4^- \longrightarrow \left[WF(dppe)_2N_2CH_2CH_2N_2WF(dppe)_2\right]$$

$$dppe = Ph_2PCH_2CH_2PPh_2$$

C. Addition Reactions of Diazo Compounds

1. Nucleophilic addition

Some of the canonical forms contributing to the stability of diazomethane place a negative charge on the carbon atom.

$$CH_2\overset{+}{=}N\overset{-}{=}N \longleftrightarrow \overset{-}{C}H_2-\overset{+}{N}\equiv N \longleftrightarrow \overset{-}{C}H_2-N\overset{+}{=}N$$

Consequently, diazomethane shows some carbanion character, which the basis of several useful reactions. An unusual use of this property is the reaction of germylene, Ph_2Ge, with diazoalkanes[236].

(166)

The double-bonded germanium compound, **166** undergoes rapid oligomerization, and has not been isolated.

2. Cycloaddition

Diazoalkanes are 1,3-dipoles, and undergo the many cycloaddition reactions expected for compounds of this class[237]. However, the nitrogen involved in the rings which are generated is easily lost, and in many cases the addition compound is not observed, so that some slight doubt remains as to the mechanism of some reactions.

An example of the ready reactivity of diazoalkanes is provided by the addition of liquid ketene to liquid diazomethane at −145 °C, which yields cyclopropanone[238].

Even at this temperature, there is no sign of a heterocyclic intermediate, so proof that this is a cycloaddition reaction has not been obtained.

$$Ph_2C = S + CH_2 = \overset{+}{N} = \overset{-}{N} \longrightarrow$$

(167)

By using a more stable system it is possible to isolate the initial reaction product, and study its decomposition[239]. The reaction of diazomethane with thiobenzophenone at −78 °C (the Schönberg reaction) yielded 2,2-diphenyl-1,3,4-thiodiazoline (167). The product was obtained as colourless crystals which, in the words of the authors, went 'pfft' around −20 °C. At −30 °C in tetrahydrofuran solution the reaction proceeds more smoothly to give 2,2,3,3-tetraphenyl-1,4-dithiane (168) in a reaction which is probably dimerization of the ylide 169. Decomposition of the ylide can also proceed via 170 to 171, or it can be trapped by dipolarophiles.

$$167 \longrightarrow (169) \quad \overset{(\times 2)}{\longrightarrow} \quad (168)$$

$$(170) \longrightarrow Ph_2C = CH_2 \quad (171)$$

Generation of ylides by reaction of allylic substrates with diazo compounds, catalysed by Rh(OAc)₄ and Rh₆(CO)₁₆, has been reported[240]. In this case, a heterocyclic intermediate is not observed and the reaction could proceed directly.

The reaction works well between allyl acetals and diazo esters, yielding 2,5-dialkoxy-4-alkenoates[241].

$$\underset{CH(OMe)_2}{\overset{\diagup}{\diagdown}} \quad + \quad N_2CHCOOEt$$

$$\xrightarrow{Rh(OAc)_4} \quad \underset{OMe}{\overset{MeO}{\diagdown}}\!\!\!\diagup\!\!=\!\!\diagdown\!\!\!\diagup^{COOEt} \quad + \quad \underset{EtOOC}{\overset{H}{\diagdown}}\!\!\triangle\!\!\diagdown_{CH(OMe)_2}$$

Cycloaddition can proceed intramolecularly, as in the reaction of allyldiazomethanes[242], which react via a nitrene-type 1,1-cycloaddition to give 1,2-diazabicyclo[3.1.0]hex-2-enes (172). On further heating 172 can rearrange to give the dihydropyridazine 173.

$$\underset{(172)}{\overset{\overset{N_2}{\|}}{PhCCH_2CH}\!=\!CHPh} \quad \rightleftharpoons \quad \text{(172)}$$

$$172 \longrightarrow \text{(173)} \longrightarrow \text{(173)}$$

(173)

The authors suggest that the reason for this pathway being followed rather than a concerted 1,3-dipolar addition taking place lies in the constraints imposed on the system. For concerted 1,3-dipolar cycloaddition to take place, the atoms of the dipolarophile should be arranged so as to have their p-orbitals in a plane parallel to the plane of the 1,3-dipole. This is prevented by the constraints of the system. Instead, attack of the terminal nitrogen atom of the diazo group on the neighbouring double bond occurs to generate the 1,2-diazabicyclohexane ring system. The result is a highly stereoselective reaction which parallels the stereospecific addition of singlet nitrenes, and the intramolecular 1,1-cyclo-addition of nitrogen ylides and nitrile imines. The reaction provides evidence of the electrophilic nature of the terminal nitrogen.

(174)   (175)

A further example of electrophilic behaviour of the terminal nitrogen of the diazo group is provided by the cyclization of the 1-aryl-3-diazoalkene (174) to give the benzo-1,2-diazepine system (175).

A mechanistic investigation[243] has shown that this reaction is also reversible. At high temperatures, nitrogen is lost and an irreversible cyclization proceeds via a carbene.

Cycloaddition reactions of diazo compounds can be also promoted by a suitable rhodium catalyst. Addition of diazo dimedone to furan has been carried out in this way[244].

D. Rearrangements Involving Diazonium Ions

A good review of this area has been published by Kirmse[245].

The simplest possible route to the diazonium ion is by protonation of the diazo compound.

$$Ph_2CN_2 \xrightleftharpoons{H^+} Ph_2CHN_2^+$$

The diazonium ion is much less stable than the diazo compound. This route is rarely used, as the diazonium ion is easily prepared from the more accessible amine.

1. Formation of aliphatic diazonium ions

Several methods are in general use.

a. Nitrosation of amines. A full review of nitrosation has been published by Williams[246]. This is still the most popular method. The amine is treated with nitrous acid, generated *in situ* from sodium nitrite and an acid. The reagent is believed to be NO^+. The

$$RNH_2 + HNO_2 \longrightarrow RN\overset{NO}{\overset{|}{H}} \longrightarrow RNNOH \xrightarrow{Acid} RN_2^+$$

(176)

reaction can be used only with primary amines, as secondary nitrosamines cannot tautomerize to the diazotic acid (176). The decomposition of the diazotic acid is probably catalysed by the acid used to generate the nitrous acid, so the diazonium ion is formed as an ion pair, which also includes a water molecule.

The reaction can be carried out using NOCl as a source of NO^+, when the reaction can be stopped at the nitrosamine stage[247], and by using the powerful nitrosating reagent acetyl nitrite[248].

b. Thermal decomposition of N-alkyl-N-nitrosamides. In this reaction, the amine is converted first into an amide, then into a nitrosamide and finally decomposed in a suitable solvent[249,250]. In this case the ion pair formed does not include a water molecule. Strong

$$RNH_2 \longrightarrow RN\overset{H}{\overset{|}{\underset{\overset{||}{O}}{C}}}R^1 \longrightarrow RN\overset{N=O}{\overset{|}{\underset{\overset{||}{O}}{C}}}R^1 \longrightarrow \left[RN=NOCR^1\overset{O}{\overset{||}{}}\right] \longrightarrow \left[RN_2^+ \bar{O}OCR^1\right]$$

nucleophiles such as tertiary amine oxides change the mechanism of the reaction; rearrangement of the n-alkyl-n-nitrosamide is promoted, and ester formation (from the ion pair) is suppressed[251].

c. The triazine reaction. In this reaction, the aliphatic amine in reacted with a diazonium salt to yield a triazine, which then undergoes acid-catalysed decomposition in a suitable solvent[252]. The mechanism resembles that for decomposition of the N-alkyl-N-nitrosamide.

$$
\overset{\overset{\displaystyle H}{|}}{RN}-N=NAr \xrightarrow{HX} RN=N-\overset{\overset{\displaystyle H}{|}}{N}-NAr \xrightarrow{HX} RN=\overset{+}{N}\;\;\underset{\underset{\displaystyle H}{|}}{\overset{\overset{\displaystyle H}{|}}{N}}-Ar
$$
$$\overset{-}{X}$$

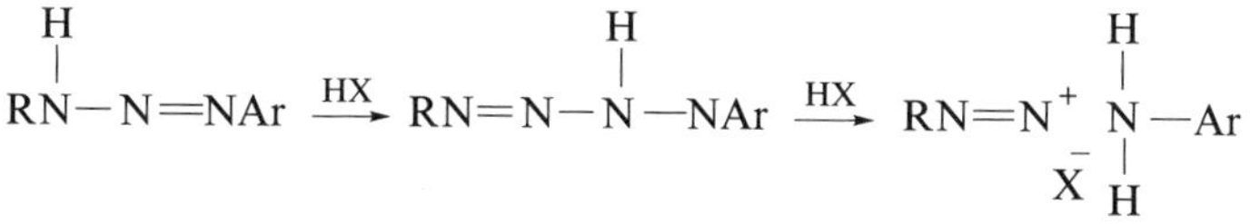

Again, the ion pair formed does not include a water molecule.
An improved synthesis of trialkyltriazines has been reported[253]. A study of the reaction confirmed the mechanism shown above.

d. Nitrosation with alkyl nitrites. Alkyl nitrites are effective nitrosating reagents[254] and can be used for the nitrosation of amines. In water, the reaction is kinetically zero order in the substrate, consistent with rate-limiting NO formation[255].

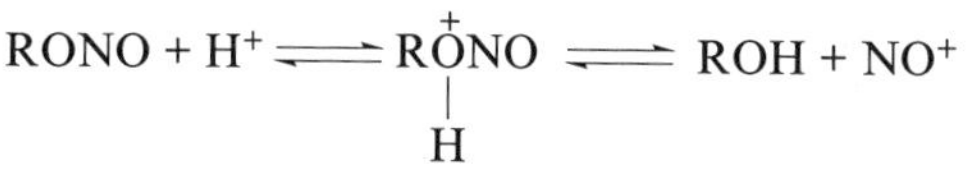

$$
RONO + H^+ \rightleftharpoons R\overset{+}{\underset{\underset{\displaystyle H}{|}}{O}}NO \rightleftharpoons ROH + NO^+
$$

Propyl nitrite has been shown to be a poor reagent for nitrosation of aniline derivatives, but the reaction can be catalysed by halide ions; the probable mechanism involves rapid equilibrium formation of the nitrosyl halide, which then reacts with aniline[256]. A similar mechanism has been proposed when alkyl nitrites and titanium tetrahalides are used as a nitrosating reagent[257].

e. Nitrosation with ruthenium nitrosyl complexes. Aromatic amines have been nitrosated with the ruthenium nitrosyl complex, [Ru(bpy)$_2$(NO)Cl]$^{2+}$ (bpy is 2,2'-pibyridine), to give diazonium complexes[258]. The properties of the complex are consistent with its possessing

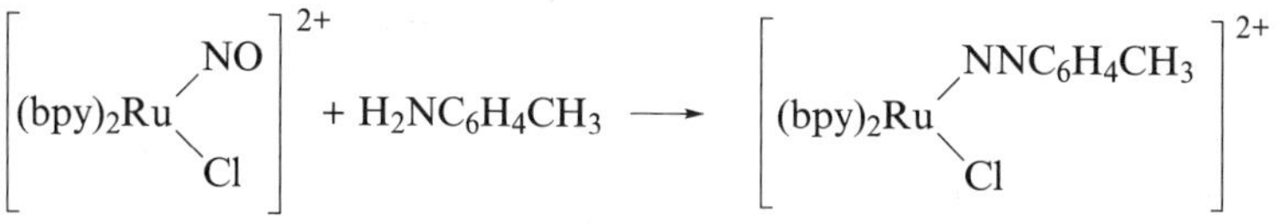

$$
\left[(bpy)_2Ru\overset{\displaystyle NO}{\underset{\displaystyle Cl}{<}}\right]^{2+} + H_2NC_6H_4CH_3 \longrightarrow \left[(bpy)_2Ru\overset{\displaystyle NNC_6H_4CH_3}{\underset{\displaystyle Cl}{<}}\right]^{2+}
$$

a considerable degree of diazonium ion character, and undergoing such reactions as diazo coupling with β-naphthol.
A review of metal-nitrosyl compounds has been published[259].

f. Nitrosation with sodium nitroprusside. Sodium nitroprusside, Na$_2$Fe(CN)$_5$NO, has been found to be a useful reagent for effecting diazotization under basic conditions[260]. Under these conditions, the diazonium ion reacted by nucleophilic attack, rather than by loss of a proton to yield the diazo compound.

2. Reactions of diazonium ions

Most diazonium ions are based on alkyl or aryl groups, but the reactions of aminodiazonium ions[261] and cyanodiazonium ions[262] with aromatic species make them valuable aminating and cyanating reagents.

The fluorine diazonium ion has been prepared, but it has little fluorinating power, probably because it does not yield F^+.

$$NH_2\overset{+}{N_2}\,AlCl_4^- + \bigcirc \longrightarrow \bigcirc\!-\!NH_2$$

$$H_2NC \equiv N + N\overset{+}{O}\,BF_4^- \longrightarrow \left[N \equiv C - \overset{+}{N} \equiv N\,BF_4^-\right] \overset{\bigcirc}{\longrightarrow} \bigcirc\!-\!CN$$

3. Rearrangement of aliphatic diazonium ions

Alkyl diazonium ions are very reactive intermediates, but have none the less been observed in superacid solution[263]. Dissolving diazomethane in fluorosulphonic acid at $-120\,°C$ yielded the methyl diazonium ion (**177**) and the methylene diazenium ion (**178**) in the ratio 4 to 1.

$$CH_2N_2 + FSO_3H \longrightarrow CH_3-\overset{+}{N}\equiv N + CH_2=\overset{+}{N}=N-H$$
$$\qquad\qquad\qquad\qquad\qquad (\mathbf{177}) \qquad\qquad (\mathbf{178})$$

On the basis of theoretical studies, it has been suggested that alkyl diazonium ions should undergo a rearrangement similar to the cyanide–isocyanide interconversion[264]; this reaction, which could only be detected by shift of labelled nitrogen, has not been observed.

Reaction of a diazonium ion proceeds usually by one of four routes.

(i) Deprotonation to a diazo compound. Incorporation of deuterium from the solvent during deamination indicates that diazo–diazonium interconversion occurs[265].

(ii) Reaction with nucleophiles[262]. This reaction takes place with complete inversion of configuration[266].

(iii) Elimination to yield alkenes.

(iv) Carbocation formation.

The serious problem of disentangling diazonium ion reactions from carbocation reactions makes diazonium ion rearrangements difficult to study. It seems likely that products of primary aliphatic amine nitrosation are derived from the diazonium ion rather than a primary aliphatic carbocation formed by spontaneous loss of nitrogen, since primary carbocations are unknown in solution. Products from secondary aliphatic amine nitrosation are formed from either the diazonium ion or the carbocation.

On the basis of this assumption, rearrangements of primary diazonium ions must involve a bond shift from within the aliphatic part of the molecule to expel the leaving nitrogen. Many such examples are known.

(i) Ring expansions[267]

$$\bigcirc\!-\!CH_2NH_2 \longrightarrow \bigcirc\!-\!OH$$

(ii) Ring contractions[268]

It is probable that expulsion of nitrogen involves formation of a secondary carbocation. Secondary aliphatic diazonium ions have these routes available, but can also spontaneously lose nitrogen to give an unrearranged carbocation.

Diazotization of α-aminonitriles has been reported to generate free radicals[269], presumably from the diazonium ion. The cyanide group is clearly important, as this reaction is usually observed only with cyano diazonium ions. Diazotization of 2-amino-2-methylpropanenitrile (**179**) yields 2-nitrosopropanenitrile (**180**) which can be trapped to give **181**, presumably by reaction of the radical **182** with NO. Further evidence of the existance of **182** is provided by observation of its dimerization product.

(179)   (180)   (181)   (182)

4. Formation of carbocations from aliphatic diazonium ions

Deamination reactions frequently involve extensive rearrangement of the substrate, and in an attempt to explain this two theories have been put forward:

(i) The aliphatic diazonium ion has a very short lifetime (much less than the time of rotation of a C—C bond[270]) and loses nitrogen in a low-energy process, which leaves a high-energy carbocation which, in turn, can undergo rearrangements not available to a normal carbocation.

(ii) Rearrangement of the diazonium ion rather than a carbocation is responsible for many of the rearranged products.

It was pointed out[271] that the first explanation raises the problem of how this extra energy is accommodated in the molecule. No satisfactory explanation of this point has been provided. A modification of the high-energy carbocation theory was put forward by Kirmse and Voigt[272] who suggested that, since loss of nitrogen from the diazonium ion is a low-energy reaction, the carbocation was formed early on the reaction profile, and hence differed geometrically from ions formed later on the reaction profile of solvolysis reactions.

The alternative view, that reaction of the diazonium ion was important in deamination, was put forward by Streitwieser[273] who pointed out that on the high-energy carbocation hypothesis, diastereoisomeric cyclic amines should yield similar products on nitrous acid decomposition, whereas different products were obtained. He therefore favoured the diazonium ion as the branching point for competing reactions.

Recent research has shed light on this problem and clearly favours the diazonium ion reaction theory. Observation of the diazonium ion in superacid[263] shows that it can have a long lifetime; while superacid conditions do not compare with normal conditions a close relationship between reaction pathways has been demonstrated[274], and the observation of the diazonium ion in conditions where primary aliphatic carbocations have not been observed does not favour the theory of an extremely rapid reaction.

The observation by Friedman and coworkers[265] of incorporation of deuterium from solvent during the reaction of 1-butylamine to yield 1-butanol provides clear evidence that the diazonium ion is involved in the deamination reaction pathway, and has sufficient lifetime to equilibrate with the diazo compound. A carbocation, if it existed, would not incorporate deuterium without isomerizing to give 2-butyl products.

An important observation has been made recently by Brosch and Kirmse[266]. They showed that the nitrous acid deamination of optically active [1-^{2}H] butylamine and [1-^{2}H] methyl propylamine both proceed with complete ($\pm 2\%$) inversion of configuration. Reaction via a carbocation is unlikely to proceed with complete inversion of configuration; reaction of a diazonium ion with a nucleophile must proceed with complete inversion of configuration. Although ion pairs would be involved in the carbocation reaction[275] it is unlikely that they would give a completely inverted product.

This leaves the question of why diazonium ion reactions give such a wide variety of rearranged products. The answer may well be because the C—N bond of a primary aliphatic diazonium ion does *not* break spontaneously to yield the primary carbocation. It can form a carbocation only by nucleophilic displacement by an intramolecular bond shift displacing the nitrogen molecule and leaving a rearranged secondary or tertiary carbocation. In a non-nucleophilic solvent, this becomes the only available pathway, so extensive rearrangement is observed.

When we turn to a secondary aliphatic diazonium ion, then loss of nitrogen to yield an unrearranged carbocation is an available pathway, so we get products of both diazonium ion decomposition and carbocation decomposition. We should thus expect to get less rearrangement from secondary than from primary aliphatic diazonium ions because this lower-energy pathway is available to the diazonium ion.

5. Rearrangements during amine nitrosation

The study of amine nitrosation has made progress as analytical methods have increased in sophistication, since complex mixtures of products in various stages of racemization are often obtained.

The work of Whiting[275] has shown the importance of ion pairing of the reactive intermediates. The deamination of trans-4-t-butylcyclohexylamine in acetic acid was shown to yield the alcohol by reaction with water produced during the reaction; this water was retained close to the reacting intermediate, and the reaction proceeded mainly with retention of configuration. Whiting was able to show that reaction with a nucleophile generated during reaction gave different amounts of retention and inversion from reaction with an external nucleophile.

In further pursuit of the importance of ion pairing in deamination, the reactions of 1-amino-n-octane and 4-amino-n-octane were compared[276]. The reactions were carried out in a variety of ways and it was shown that the primary alkyl diazonium ion had properties independent of its method of preparation, in that the alkyl cation generally does not capture the leaving group which accompanies its formation, but reacts with solvent to give a constant set of products. The secondary alkylamine behaves differently. In this case, the leaving group which is generated during reaction is captured by the intermediate to varying degrees, after as well as before rearrangement. This points to the primary alkyl diazonium ion having a much longer lifetime than the secondary, and thus allowing the ion pair to diffuse away. A similar experiment[248] using acetyl nitrite as nitrosating reagent in propionic acid gave only traces of the acetate with both primary and secondary amines; however, acetyl nitrite could equilibrate in propionic acid with propionyl nitrite, so the actual reagent involved in uncertain.

Whiting's results differ from those of the groups of White[277,278], who studied the decomposition of butyl N-nitrocarbonates in ethanol, and found that the primary alkyl intermediate was more likely to trap the counterion than the secondary or tertiary. However, Whiting[276] pointed out that the intermediate in this case is the species R—N=N=O. Certainly, a comparison of the products of this reaction with amine nitrosation showed overall similarities, but considerable differences in detail[279].

The importance of ion pairing in amine nitrosation has been further demonstrated by Kirmse and Siegfried[280]. They studied the nitrosation of the optically active 2-norbornamines, and confirmed that the endo isomer **183** yields a mixture of 90% racemic exo alcohol **184** and 10% racemic endo alcohol **185**. Formation of **184** is easily explained by formation of the

| | | |
| (183) | (184) | (185) |

delocalized carbocation, but nucleophilic attack on the diazonium ion derived from **183** would proceed with inversion of configuration to give optically active exo alcohol, which is not observed. However, loss of the diazonium group from the endo isomer cannot be synchronous with delocalization; the active endo alcohol probably arises from a water molecule generated within the diazonium ion pair capturing the carbocation in competition with delocalization.

A similar picture emerged from a study[281] of the diazonium ion **186** obtained from 5-norbornen-2-ylamine. A small amount of product only comes from the diazonium ion, the bulk from a delocalized ion.

| | | |
| NNHTs | (186) | (187) |

This reaction has been used[282] in the synthesis of exo-tetracyclo[4.2.1.0³,⁵0⁴,⁹] non-7-en-2-ol (**188**) by deamination of exo-2-nortriquinacenylamine (**189**). In contrast, the endo isomer gave a mixture of **190** and **191**, with acetates and alcohols in the approximate ratio of 2 to 1.

| | |
| (189) | (188) |

Presumably delocalization of the double bond cannot assist loss of nitrogen from the *endo* ion.

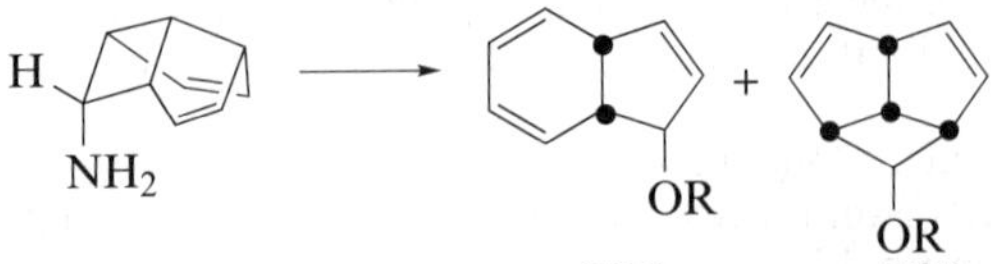

(190)   (191)

6. Aryldiazonium ions

Aryldiazonium ions are much more stable than their aliphatic counterparts, and are therefore much less likely to be involved in rearrangements. They are readily prepared by nitrosation of phenylamines. No evidence[283] of scrambling of the nitrogen atoms in solution has been detected, so that rearrangements of the diazonium ion **192** to a cyclic structure **193** can be ruled out.

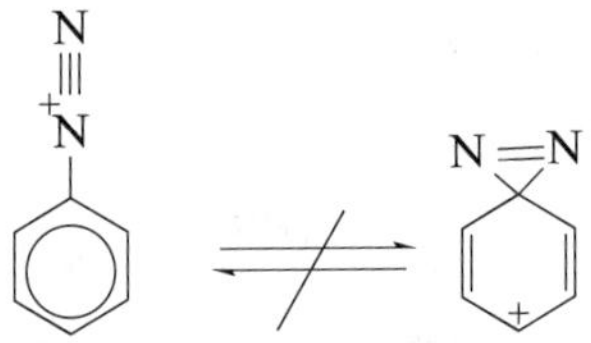

(192)   (193)

Scrambling of the nitrogen atoms during decomposition of aryldiazonium ions has been observed; a study of the reaction[284] in which secondary deuterium isotope measurements were carried out excluded the possibility of a benzene spirodiazirine similar to **193**, and favoured a carbocation with closely associated nitrogen, which can collapse to reform the diazonium ion.

7. Addition reactions of aryldiazonium ions

Being much more stable than their aliphatic counterparts, aryldiazonium ions can undergo addition reactions in which nitrogen is not lost. Reactions of aryldiazonium salts

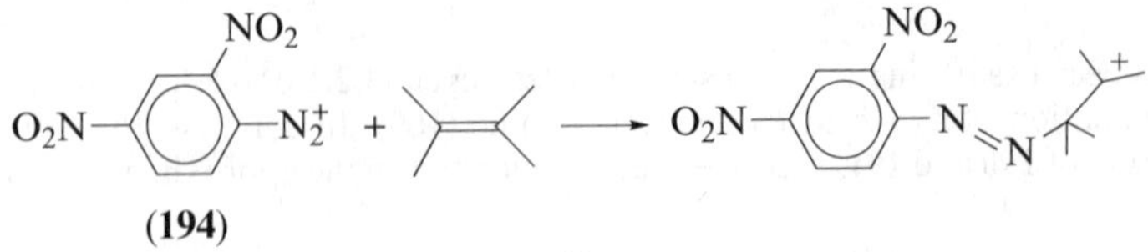

(194)

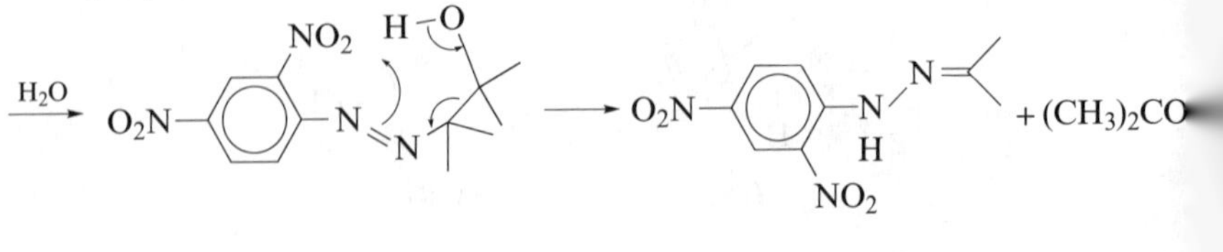

as electrophiles have been used in the dye industry since 1858, and new reactions of this type are still being discovered.

The 2,4-dinitrobenzenediazonium ion (194) is sufficiently reactive to undergo addition to an ordinary double bond[285].

The reaction also takes place with enol ethers[286], enamines[287] and allylsilanes[288].

When carried out with an isonitrile, the initial coupling reaction is followed by cyclization[207].

Free radicals also add to diazonium ions, yielding azo compounds[289].

$$t\text{-Bu}^\bullet + \text{ArN}_2^+ \longrightarrow t\text{-BuN}=\text{N}^\bullet \overset{+}{\text{Ar}} \longrightarrow t\text{-BuN}=\text{NAr}$$

The other important reaction of aryldiazo compounds is cycloaddition. Their dienophilic activity was demonstrated[290] by reaction of 4-nitrobenzenediazonium chloride with 2,3-dimethylbutadiene. The reaction has been shown to take place with a number of dienes[291]. It yields the thermodynamic equilibrium product.

With cyclopentadiene[292] the reaction follows a different course.

The authors suggest that this azo coupling reaction of 1,3-dienes affords a competing pathway only in cases where the Diels-Alder reaction leads to an unstable product.

8. Substitution reactions of aryldiazonium ions

Replacement of the nitrogen of an aryldiazonium ion by a substituent usually proceeds without rearrangement, though there is evidence of small-scale interconversion of *ortho*, *meta* and *para* isomers during replacement of the diazo group with halide ions[293]. They probably result from the spreading of the charge of the diazonium group around the ring, which can be demonstrated by ^{13}C nuclear magnetic resonance spectroscopy[294]. The substitution reaction has been reviewed by Zollinger[295].

Such rearrangements as do take place during reactions of aryldiazonium ions occur when an electrophilic species can attack a neighbouring aromatic ring. This may be a simple electrophilic addition of the diazonium ion, or generation of a reactive intermediate by loss of nitrogen, which then undergoes reaction. If the first process then loses nitrogen, it is difficult to distinguish the pathways.

An example of such a reaction is the Pschorr reaction[296], in which 2-amino-α-arylcinnamic acid (**195**) yields a phenanthrene (**196**). The reaction proceeds in 10–20% yield on irradia-

(195)   (196)

tion. The main product is an acetamide derivative formed from reaction with solvent acetonitrile; the authors suggest that this is formed via a cation, while **196** results from concurrent radical formation.

In the presence of the catalyst tris-(2,2'-bipyridyl)ruthenium(II), the yield of **196** is increased to almost 100%. The authors suggest that a radical mechanism is again involved; irradiation of the ruthenium complex produces an excited state which is quenched by electron transfer from N. Loss of nitrogen leaves a radical, which attacks the neighbouring aryl ring to yield the phenanthrene.

A novel cyclization[297,298] occurs during nitrosation of 4-amino[2.2](1,4)naphthaleno-paracyclophane (**197**). The mechanism of the reaction is uncertain. One possibility is transannular attack by an aryl cation obtained from the diazonium ion from the upper to the lower ring, giving an intermediate which is cleaved in the aqueous acid present to give the product, **199**. Alternatively, a transannular attack by the electrophilic diazonium ion could yield the intermediate **198** having an azo linkage between the rings, and this could lose nitrogen and add water to yield **199**. Deuterium tracer experiments suggest that this is the

(197)   (198)   (199)

correct pathway, and suggest that decomposition of **198** may proceed by generation of an aryne in the naphthalene ring which then adds water.

A similar reaction in which an aryne has been postulated as intermediate[299] is the formation of benzobarrelene (**200**) from 5-amino-[3,3]paracyclophane (**201**).

(**201**)   (**200**)

If the benzobarrelene has been prepared from an aryne generated from 5-amino-[3,3]paracyclophane[300], this may well be a route to arynes.

E. Rearrangements Involving Carbenes and Carbenoids

1. Formation of carbenes

Carbenes can be obtained by pyrolysis or photolysis of diazo compound.

$$PhCN_2 \longrightarrow Ph_2C\colon + N_2$$

In many cases the diazo compounds are sufficiently stable to be isolated and characterized, but less stable diazo compounds can be prepared from toluene-*p*-sulphonyl hydrazones and decomposed *in situ*.

$$\underset{R}{\overset{R^1}{>}}C{=}N{-}NH{-}SO_2Tol\text{-}p \longrightarrow \underset{R}{\overset{R^1}{>}}C{=}N{-}\bar{N}{-}SO_2Tol\text{-}p$$

$$\longrightarrow \underset{R}{\overset{R^1}{>}}C{=}N_2 \longrightarrow \underset{R}{\overset{R^1}{>}}C\colon + N_2$$

Carbenes can also be generated by decomposition of diazirines, though diazo compounds have been implicated in some of the decompositions, both thermal[230] and photolytic[231]. There is, however, at least one case of the reverse of this[301]; photolysis of 9-diazo-1,8-diazafluorene (**202**) is believed to proceed via the diazirine (**203**).

(**202**)   (**203**)

Evidence[302] that carbenes are genuine intermediates, as distinct from energy maxima or points on the energy slope, has been obtained. Dimethyl carbene has been shown to have a lifetime of between 1 and 100 nanoseconds.

However, a study[303] of phenyl carbene generated directly and generated by a carbene-to-carbene rearrangement shows significant differences in product composition, so it is likely that at least some of the products reported as products of carbene reactions are formed directly from the carbene precursor, in a reaction which does not involve a carbene.

The rate of carbene formation by pyrolysis of diazo compounds shows unexpected substituent effects. The unimolecular rate constant for thermolysis of diphenyldiazomethane is increased by any single *para* substituent, be it electron-donating or electron-withdrawing. If both rings are substituted, one by an electron-donating and the other by an electron-withdrawing substituent, then this disubstituted compound decomposes faster than a compound monosubstituted with either substituent. However, symmetrically disubstituted diphenyldiazomethanes decompose more slowly than do the compounds monosubstituted with the same substituents.

These observations have been rationalized[304,305] by the suggestion that there is extensive interaction of each phenyl group with an electron-deficient or an electron-rich orbital developing within a tetrahedral-like transition state (**204**).

(**204**)

Early in the study of carbenes, it was realized that carbenes could exist with the non-bonding electrons paired, in the singlet form, or unpaired, in the triplet or biradical form. Recently, this hypothesis has been supported by spectroscopic data[306].

$$Ph_2C\!\downharpoonright\qquad Ph_2C\!\upharpoonright\upharpoonright$$
$$\text{Singlet}\qquad\text{Triplet}$$

An important characteristic reaction of carbenes is addition to an alkene to yield a cyclopropane[307]. This reaction allows distinction between singlet and triplet forms, since addition of the singlet form is stereospecific while addition of the triplet form is not[308]. However, rates of reaction of carbenes and alkenes varied widely[309] so the reaction was a useful rather than definitive diagnostic test.

It was soon established that singlet and triplet forms of carbenes could interconvert[310]. The singlet form underwent reactions such as reacting with water to yield alcohols and with alcohols to yield ethers[311], while the triplet form lead to dimerization reactions such as azine and dimeric alkene formation.

For arylcarbenes, low-temperature EPR spectroscopy has shown that the ground state is a bent triplet with two orthogonal singly occupied orbitals[312,313]. The first excited state is an electrophilic singlet and, in the case of diphenylcarbene, the free energy difference between these two states has been estimated[314] to be 5.1 ± 1 kcal mol^{-1}.

Direct generation of triplet carbene from the diazo compound is usually by irradiation with a triplet photosensitizer. Although the distinctions between reactions of singlet carbenes and triplet carbenes can be clearly made[315,316], product studies are always subject to the limitation that the final product depends on the relative speeds of singlet–triplet interconversion and the rates of the reactions of these species to yield the observed products.

In 1976, Closs and Rabinow[317] made the first measurement of rate constants for the reaction of a carbene. They used a flash photolysis technique. A brief flash of radiation generated carbene intermediates, and their decay was then monitored spectrophotometrically. In this way the rate constant for reaction of diphenylcarbene with 1,1-diphenylethylene was found to be 4.8×10^5 M^{-1} s^{-1}. Using this technique, measurements were possible on a microsecond time scale. With the advent of laser flash photolysis techniques, the resolution time was reduced to nanoseconds, causing a resurgence of interest in the kinetics of carbene reactions. The early results of use of this technique have been discussed by Griller and coworkers[318].

Using laser techniques, the rate of decomposition of the triplet carbene in the fluorenylidene system has been shown to be 3.6×10^9 s^{-1} at room temperature[319]. This triplet state is very much more reactive than that of diphenylcarbene[320]; it has been suggested that the latter may be stabilized by the ability of the phenyl rings to rotate, so that the unpaired electrons that are in orthogonal orbitals may be more efficiently delocalized.

Along with the development of laser photolysis for the study of carbenes, the development of low-temperature studies of carbenes in polycrystalline solids and glasses has provided a valuable tool for carbene studies. In the cases of carbenes which have a triplet ground state, important differences have been observed between behaviour at room temperature and at the temperature of liquid nitrogen. Partly, the different product distribution results from different activation parameters for different processes, and for different rates of interconversion of singlet and triplet species. In cases where the carbene has a triplet ground state, a large increase in triplet reactions over singlet reactions is observed. Part of the change must, however, be attributed to the control that the matrix exerts over the reactions taking place by restricting movement of the reactants. For example[321], generation of diphenylcarbene at room temperature by a number of routes produced similar products but, at -196 °C, very different chemistry was observed. The authors comment that: 'The fate of carbenes generated in a matrix may be predestined by site preference imposed upon the precursor'. Thus, 1,2-hydrogen shifts of singlet carbenes in a perfluorinated matrix proceed without appreciable isotope effect[322]. The method has found much use for generation of predominantly triplet species.

Recently, carbenes with a boron atom next to the carbene carbon atom have been prepared, and are believed to be linear ground state triplets[323]. Silylcarbenes, in which the carbene carbon atom is also next to a less electronegative atom, also exist as linear triplets[324].

2. Rearrangements of carbenes

Until the last decade, product studies formed the main evidence for carbene formation; singlet carbenes formed cyclopropanes from alkenes stereospecifically, while triplet carbenes formed cyclopropanes non-stereospecifically. Formation of a cyclopropane (though not by addition to an alkene) via a carbocation route was demonstrated[325] and, more recently[326], it has been shown that ρ values for insertion–addition selectivity and for cyclopropanation stereoselectivity vary as to photochemical or thermal generation of the carbene. The authors of this latter study suggest that a ground state diazo compound could be masquerading as a carbene in its thermal reaction with olefins, possibly[327] by electrocyclic addition followed by loss of nitrogen.

Carbenes give rise to products by rearrangement, addition or insertion, and can lead to complex product mixtures[328]. The reactions involved are best considered separately.

a. Insertion into a C̶H bond. Consideration of spin conservation leads to the conclusion that the insertion of a singlet carbene into a C—H bond is a concerted process involving a three-centre transition state (**205**).

(205)

In contrast, the insertion of a triplet carbene is a two-step process involving radical formation and combination.

Evidence in support of these mechanisms has been obtained from CIDNP experiments[329]. Photolysis of diazomethane in toluene does not yield polarized ethylbenzene, so does not involve a radical pair, while benzophenone-sensitized decomposition of diazomethane in toluene does involve a radical pair so presumably reacts as a triplet[330].

The problem of facile singlet–triplet interconversion causes difficulties in interpretation of data. A different approach to the problem has been to observe the ESR spectrum of the

(205)   (206)

biradical which, together with product studies, offers convincing evidence of reaction pathway[331]. Thus photolysis of 1-(diazomethyl)-8-methylnaphthalene (205) proceeds via a biradical to give acenaphthalene (206).

A valuable boost to the study of reactions of triplet species came from the observation by Moss[332] that the chemistry of some carbenes is very sensitive to temperature. Photolysis of 9-diazofluorene (207) in isobutylene at 273 K produced the cyclopropane (208) characteristic of a singlet carbene reaction, but at 133 K addition took place to yield 209. The change in products was shown to occur at the temperature of change in phase of the solution.

Use of isobutylene with the CH_2 carbon labelled with ^{13}C showed the label to be scrambled in the product [209], suggesting the reaction proceeds via a radical pair[333].

The reaction has been shown to proceed in a variety of organic glasses, but at very low temperatures the classical H-atom abstraction mechanism changes to one involving quantum mechanical tunnelling[334].

b. *Insertion into a C—Cl bond.* Insertion into C—Cl bonds gives rise to a number of interesting rearrangments[335], the best known of which is the reaction of methylene with carbon tetrachloride to yield the tetrachloride 210.

By use of CIDNP techniques[329] it has been shown that chlorine abstraction is a major reaction of singlet methylene. Triplet methylene, though occasionally involved in chlorine abstraction[336], shows a strong preference for hydrogen abstraction. Thus the direct photolysis of diazirine (165) in deuteriotrichloromethane gave 211 with a strong CIDNP signal; the photosensitized reaction gave 212 with a strong CIDNP signal.

In the case of singlet phenylchlorocarbene and triplet diphenylcarbene, the abstraction of chlorine has been shown to be accelerated by the presence of electron-donating substituents on the aryl carbene, by increasing ease of reduction of the chlorine donor and by an increase in solvent polarity. This suggests that the transition state probably has considerable carbene carbon to chlorine bond formation and charge development[337].

$$DCH_2 \cdot\cdot CCl_3 \longrightarrow DCH_2CCl_3$$
(212)

$$ClCH_2 \cdot\cdot CDCl_2 \longrightarrow ClCH_2CDCl_2$$
(211)

(165)

c. Alkene-forming insertions. These reactions are of the general type.

$$R^2-\overset{\overset{\displaystyle R^1}{|}}{\underset{\underset{\displaystyle R^3}{|}}{C}}-\overset{\displaystyle \cdot\cdot}{C}\diagdown_{R^4} \longrightarrow \overset{\displaystyle R^2}{\underset{\displaystyle R^3}{\diagup}}C=C\overset{\displaystyle R^1}{\underset{\displaystyle R^4}{\diagdown}}$$

The reaction consists of an intramolecular insertion of the carbene into one of the bonds to the α-carbon atom. These reactions are frequently described as '1,2-shift' reactions by comparison with carbocation rearrangements. The mechanism of the reaction has been studied[338] by observation of isotope effects in the reaction of 1-phenyl-2-diazopropane-1-d (213). The alkenes 214 and 215 are obtained as both the *cis-* and *trans-*isomers. The isotope effects,

$$Ph-\overset{\overset{\displaystyle D}{|}}{\underset{\underset{\displaystyle H}{|}}{C}}-\overset{\overset{\displaystyle N_2}{\parallel}}{C}-CH_3 \overset{\Delta}{\longrightarrow} \overset{\displaystyle D}{\underset{\displaystyle Ph}{\diagup}}C=C\overset{\displaystyle H}{\underset{\displaystyle CH_3}{\diagdown}} + \overset{\displaystyle H}{\underset{\displaystyle Ph}{\diagup}}C=C\overset{\displaystyle D}{\underset{\displaystyle CH_3}{\diagdown}} + Ph-\overset{\overset{\displaystyle D}{|}}{\underset{\underset{\displaystyle H}{|}}{C}}-\overset{\overset{\displaystyle H}{|}}{C}=CH_2$$

(213) (214) (215)

k_H/k_D, vary from 1.2 to 1.5, increasing with electron-withdrawing substituents on the phenyl group for the formation of both *cis-* and *trans-β*-methylstyrenes. The authors suggest the results are consistent with a 'pull–push' mechanism, which can be regarded as electrophilic attack on the C—H bond by the phantom p orbital of the carbene occurring simultaneously with backside nucleophilic attack by the carbene unshared electron pair to push the H away and form the π bond.

Consistent with this, it has been shown[339] that the alignment of the carbene electrons and the bond into which they are inserted is of great importance in predicting the direction of insertion. In a rigid system, such as *endo*-4-phenyl-5-brexanone tosylhydrazone (216),

(216) (217)

formation of 4-phenyl-4-brexene (217) is favoured by a factor of 1000. Laser flash photolysis[340] of diazirines has allowed rates of 1,2-hydrogen shifts to be measured. A Hammett plot gave a ρ value of -1.0, consistent with a hydride-like shift to the carbene centre. However, laser flash photolysis of 3-neopentyl-3-fluorodiazirine[341] yielded the alkene with a kinetic isotope effect of approximately 5. This isotope effect is much larger than any previously reported, but there is at present no explanation of this unexpected result.

The course of a 1,2-hydrogen shift reaction is strongly affected if it is carried out in a matrix[342]. Thus, at room temperature, 1,2-diphenyl-1-diazopropane (218) yields, on irradiation in methyl cyclohexane, a mixture of alkenes 219 and 220, which result from H migration, and 221, which results from phenyl migration, in roughly equal amounts. When the temperature of the reaction is decreased, then the product of phenyl migration increases steadily as long as the solvent remains fluid, but is completely suppressed once the environment becomes solid. This is consistent with insertion of the singlet carbene into the C—H bond, and phenyl migration in the triplet, but then the restricted mobility imposed by the solid matrix places an energy premium on the triplet reaction, favouring the singlet.

(218)   (219)   (220)   (221)

The rate of the insertion reaction has been measured by laser flash photolysis[342]. Photolysis of chlorocyclopropyldiazirine (222) yielded the carbene 223 which, at 20 °C, decayed with $k_1 = 8.5 \times 10^5$ s^{-1} with an Arrhenius energy $E_a = 2.7$ kcal mol^{-1}. The reaction is of a rate comparable to addition to an alkene, but slower than formation of an ylide by reaction with pyridine.

(222)   (223)

d. Ylide formation.   Carbenes form ylides by reaction with electron donors such as pyridine[342], methacrylonitrile[343], aldehydes[344] and acetonitrile[345].
   Laser flash photolysis[322] of chlorocyclopropyldiazirine (222) with acetonitrile in pyridine-isooctane formed the ylide with $k_2 = 3.6 \times 10^8$ M^{-1} s^{-1}, a somewhat faster reaction than addition to an alkene. The ylide itself has various tautomeric forms (224), and is relatively stable, having a lifetime of several hundred microseconds.

$$\underset{\text{(224)}}{\overset{\overset{\displaystyle Cl}{\diagup}}{\triangleleft}\overset{-}{C}-\overset{+}{N}=C-CH_3 \quad \rightleftharpoons \quad \overset{\overset{\displaystyle Cl}{\diagup}}{\triangleleft}\overset{-}{C}=\overset{+}{N}=\overset{-}{C}_{\diagdown CH_3} \quad \rightleftharpoons \quad \overset{\overset{\displaystyle Cl}{\diagup}}{\triangleleft}C=N-\overset{..}{C}-CH_3}$$

Ylides can be quenched[345] with electron-deficient alkenes to yield Δ'-pyrrolines, **225**. They also react, albeit more slowly, with less activated alkenes such as diethyl maleate or 2,3-dimethyl-2-butene. In the absence of reactants, the ylide forms a 2*H* azirine (**226**), but on irradiation[346] this reverts to the ylide, **227**.

$$Ph-\overset{-}{C}H-\overset{+}{N}\equiv C-CH_3 \quad \xrightarrow{\overset{\nearrow CN}{}} \quad \underset{\text{(225)}}{\underset{\overset{\displaystyle H}{|}}{\overset{\displaystyle Ph}{\diagdown}}\overset{\displaystyle N}{\underset{\diagdown CN}{C}}\overset{\diagup\diagdown}{}\overset{\displaystyle \nwarrow}{C}-CH_3}$$

Phenylchlorocarbene, generated by laser flash photolysis of phenylchlorodiazirine, reacts readily with γ,γ-dimethylallyl methyl sulphide to yield the s-ylide **228**. This then undergoes a [2,3] sigmatropic rearrangement to **229**, which hydrolyses to the ketone, **230**[347].

$$\underset{\text{(227)}}{} \qquad \underset{\text{(226)}}{}$$

$$\underset{\overset{\displaystyle Ph}{}}{\overset{\displaystyle Cl}{\diagdown}}C\colon \;+\; CH_3SCH_2CH=C(CH_3)_2 \;\longrightarrow\; CH_3\overset{+}{S}CH_2CH=C(CH_3)_2$$
$$\underset{\text{(228)}}{\qquad\qquad\qquad\qquad\qquad\qquad\qquad\quad \overset{\displaystyle \overset{Ph}{\diagdown}\overset{-}{C}\overset{Cl}{\diagup}}{|}}$$

$$\underset{\text{(229)}}{\overset{\overset{\displaystyle H_3CS \;\; CH_3}{|\;\;\;\;\;\;|}}{Ph-\underset{\underset{\displaystyle Cl\;\;\;CH_3}{|\;\;\;\;\;\;|}}{C}-C-CH=CH_2}} \quad \xrightarrow{H_2O} \quad \underset{\text{(230)}}{\overset{\overset{\displaystyle O\;\;\;CH_3}{\|\;\;\;\;\;|}}{Ph-C-\underset{\underset{\displaystyle CH_3}{|}}{C}-CH=CH_2}}$$

e. Insertion into OH bonds. Ylide formation is one of the possible mechanisms of reaction of a carbene with an OH bond[348]. Others are one-step insertion into the OH bond.

and protonation of the carbene to give a carbocation pair. Kirmse and coworkers[349] were able to show that the electrophilic or nucleophilic character of the carbene played a major role in selecting the mechanism. Thus, cyclopentadienylidene (**231**) follows the ylide path, while cycloheptatrienylidene (**232**) favours the carbocation path. The reaction is also

(**231**)

(**232**)

solvent-dependent; the group of Kirmse[350] studied the reaction of diphenylcarbene with 2,2,2-trifluoroethanol in acetonitrile (the reaction with the more nucleophilic water in acetonitrile was then too fast to follow) using laser flash photolysis techniques and was able to observe the spectrum of the diphenylmethyl carbocation. Using improved laser flash equipment[351], the diphenylmethyl carbocation was observed in aqueous acetonitrile. This contrasts with isotope effect studies on thermally generated carbenes[348]; that work used lower concentrations of water in acetonitrile, so the ylide mechanism may take over when the protonating power of the solvent is reduced. The reaction of carbenes with cyclopropanols involves ring opening of the latter. Reaction via a radical pair of cyclopropyloxy radical and a carbene-derived substituted methyl radical has been suggested[352].

Photolysis of diphenyldiazomethane in solid (S)–(+)-2-butanol at 137 K yields the alcohol **233**, enantiomerically pure, by reaction of the triplet diphenylcarbene with the alcohol to give a radical pair, which collapses. The solid state matrix directs the collapse to proceed with complete retention of configuration[353].

$$\text{Ph}_2\text{C}-\underset{\underset{\text{CH}_2\text{CH}_3}{|}}{\overset{\overset{\text{OH}}{|}}{\text{C}}}-\text{CH}_3$$

(**233**)

f. Insertion into BH bonds. Insertion of methylene, generated by irradiation of diazomethane, into a B—H bond of carborane has been reported by Yuan and Jones[354]. The mechanism of the process is unknown.

g. Cyclopropane formation. Addition of a carbene to an alkene to yield a cyclopropane is the most characteristic reaction of a carbene, and one that is often used as evidence of reaction via a carbene mechanism. It is generally accepted that addition of a singlet carbene to an alkene proceeds stereospecifically, while triplet carbenes give non-stereospecific addition. The variation of selectivity of carbene has been discussed[355] and has been shown to be consistent with the predictions of frontier molecular orbital theory.

In cases where we have addition to an allene, so that the carbene can add to either of the double bonds, then the competition affects the overall course of the reaction[316]. In general, the singlet carbene is believed to add to the more substituted double bond. Diphenylcarbene, generated photochemically, added to the less substituted double bond.

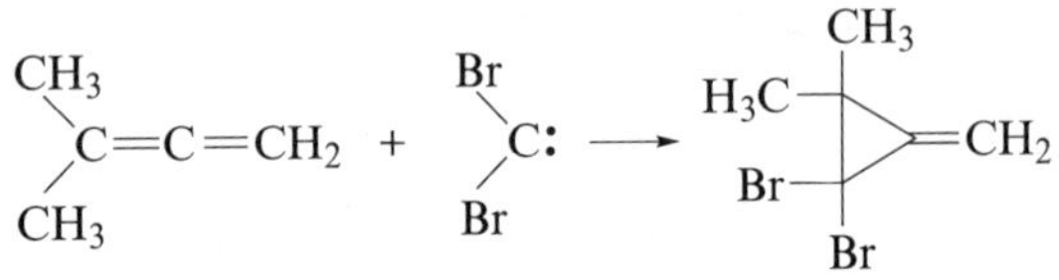

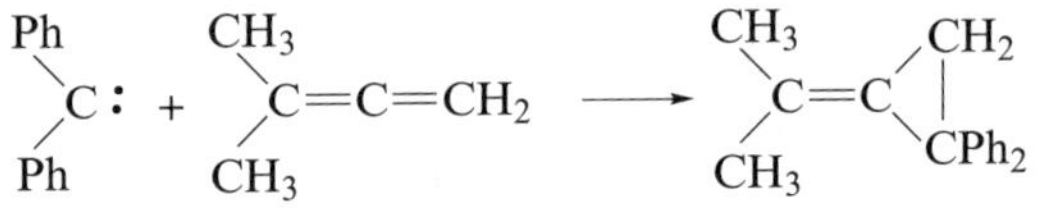

It was suggested that this reaction proceeds through a triplet species. The problem of singlet-triplet interconversion, however, makes clear conclusions doubtful from product studies of this type, though the selectivity is clearly significant. Triplet diphenylcarbene has been shown[320] to be unreactive relative to triplet fluorenylidene by two or three orders of magnitude.

Laser flash photolysis has, as usual, illuminated the problem. Jones and Rettig[356] photodecomposed 9-diazofluorene (188) in hexafluorobenzene and cis-4-methyl-2-pentene mixtures, and showed that the degree of stereoselectivity in the cyclopropane products depended on the concentration of the alkene. Laser flash photolysis[357] showed that the first detectable intermediate in the photolysis reaction is the triplet carbene, and suggests that the product studies are consistent with initial formation of a singlet fluorenylidene which has an extremely short life (less than 5 ns) before forming the triplet. The singlet can be trapped only by high alkene concentrations while the more stable triplet is easily trapped.

h. Fragmentation reactions. Generation of a carbene in a labile system may cause fragmentation of the molecule. A simple example of this is the reaction shown below. The authors[358] were able to prove the presence of both diazo compound and carbene in the reaction sequence by trapping experiments.

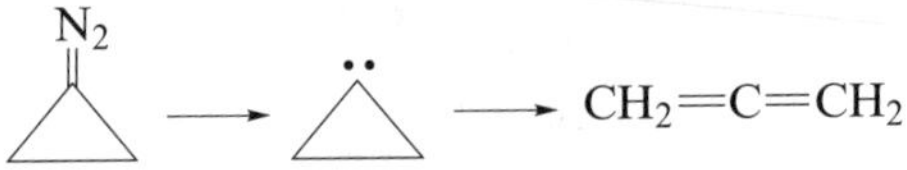

Another example[359,360] of fragmenting carbenes is provided by the 3H-pyrazolylidenes. The 3H-diazo-pyrazole (234), on pyrolysis, yielded the carbene 235 probably stabilized as shown below. Although the carbene can undergo typical carbene reactions, such as insertion into a C—H bond, ultimately yielding 236 with cyclohexane, it can also decompose to the 2H-azirine 237. The methyl cyanocarbene subsequently yields acrylonitrile.

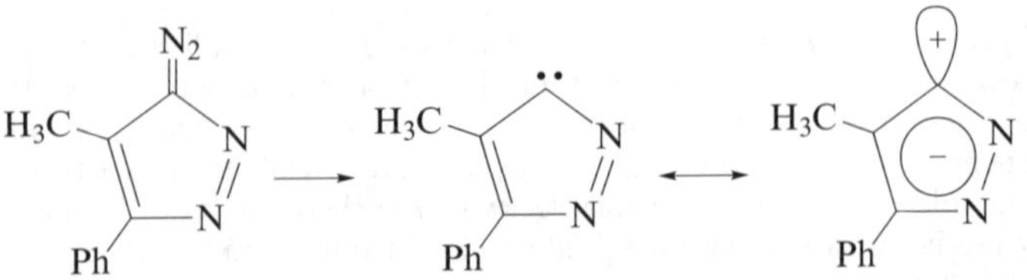

(234)   (235)

i. Carbene-to-carbene rearrangements. Carbene-to-carbene rearrangements are relatively uncommon[361], since carbenes are highly reactive species with a variety of reaction pathways available to them. When they occur, they often result from intramolecular addition to form a cyclopropane, which subsequently decomposes to yield a different carbene[362], such as

Successive ring expansion and contraction can cause a substituent to 'walk' round an aromatic ring[363]. The carbene is eventually trapped intramolecularly to yield an ylide, which gives the observed products.

If the carbene centre is next to a heterocyclic ring carrying a nitrogen atom, nitrene formation takes place. Nitrene formation from carbenes is favoured by the greater stability of the nitrene, even when a competing carbene route is available[364].

When the ring is substituted with oxygen or sulphur, this route is not available[365] and the main observed reaction is ring opening of the carbene. Formation of cyclopentylidenes from cyclopropylidenes is known but, surprisingly[366], 2-vinylcyclobutylidene does not behave in a similar fashion, yielding cyclopropyl, cyclobutyl and cyclopentyl products, but not cyclohexyl. A number of carbene rearrangements of vinyl carbenes and alkynyl carbenes are discussed below.

3. Vinyl carbenes

Vinyl carbenes are readily generated from vinyl diazo compounds as well as by thermal ring-opening of cyclopropenes[367]. Their main reaction pathway is by interaction of the carbene with the vinylic double bond, as observed in many carbene-to-carbene rearrangements. Thus, bicyclo[3.2.1]-octa-2,6-dien-4-ylidene (**238**) rearranges to styrene[368]. The carbene (**238**) can be trapped with cyclohexane to give an addition compound. Putting in an extra double bond gives a more complex system, and hence gives scope to a more complicated series of carbene-to-carbene rearrangements[369]. The related carbene bicyclo[3.2.0]hepta-2,6-dien-4-ylidene (**239**), has been shown to rearrange in a similar way[370] yielding cycloheptatetraene (**240**). The parent carbene, vinylmethylene, does not form cyclopropane[371] but, when generated in carbon tetrachloride, abstracts a chlorine atom to give an ion pair. A strong nuclear spin polarization signal was observed when the reaction was carried out in the N.M.R. spectrometer.

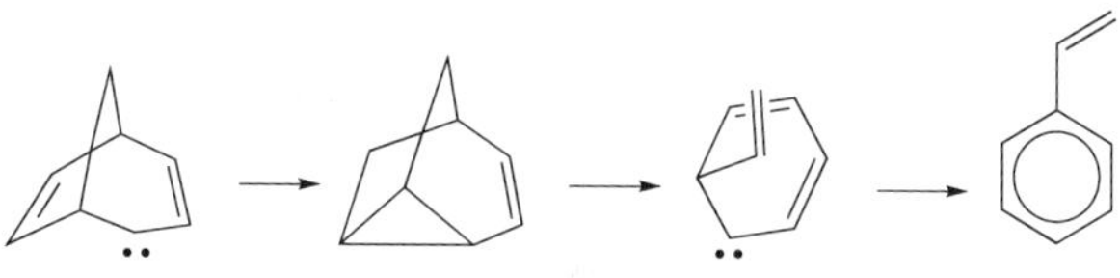

(238)

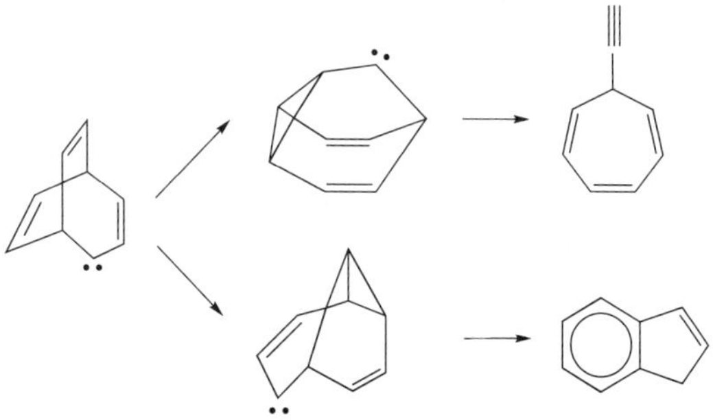

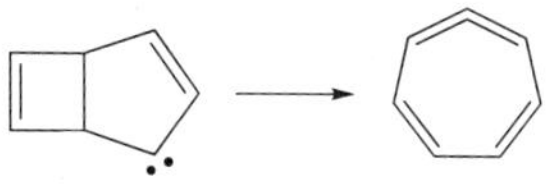

(239)   (240)

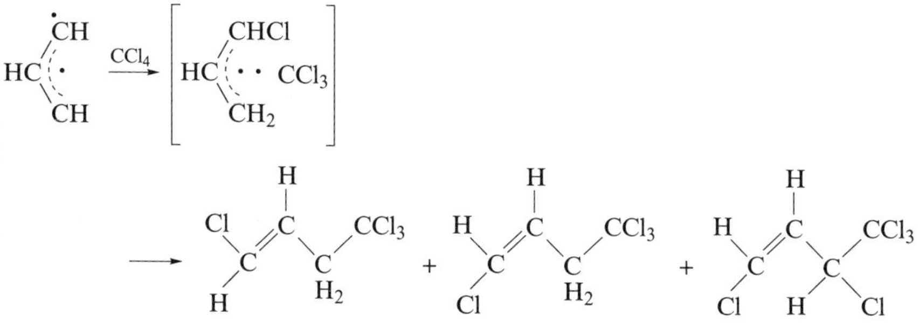

4. Alkynyl carbenes

Thermolysis of diazoalkynes yields alkynyl carbenes; like vinyl carbenes, these undergo carbene-to-carbene rearrangements very readily[372]. Propargylene (241) was prepared in 1971, and it was found that the singlet form reacted with alkenes only at C_1, the position vacated by nitrogen, but the triplet form reacted at either C_1 or C_3. Propargylene and its methyl and phenyl derivatives showed electron paramagnetic resonance[373], and the predicted structure[374] is of C_2 symmetry, best described in terms of a diradical valence structure.

A similar reaction with alcohols has been observed[375]. With methanol, 242 and 243 are formed in the ratio of 2 to 3, but with isopropanol, the ratio is 5 to 1 in favour of 242.

$$RC_{(3)} \equiv C_{(2)} \ddot{C}_{(1)}H \rightleftharpoons R\ddot{C}_{(3)}C_{(2)} \equiv C_{(1)}H$$

(241)

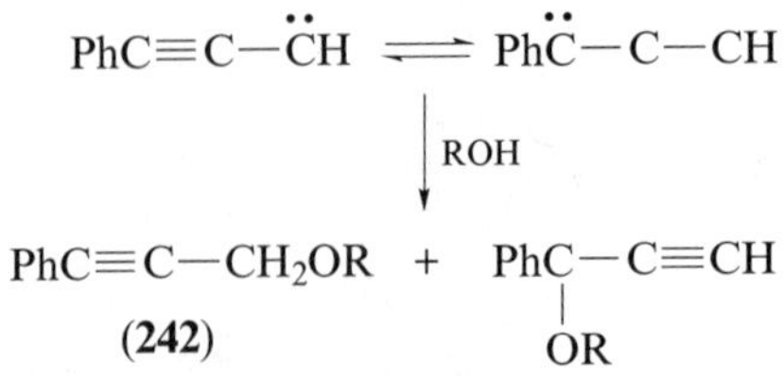

$$PhC\equiv C-\overset{..}{C}H \rightleftharpoons Ph\overset{..}{C}-C-CH$$

$$\downarrow ROH$$

$$PhC\equiv C-CH_2OR \quad + \quad PhC-C\equiv CH$$
$$\underset{(242)}{} \qquad\qquad \underset{OR}{|}$$
$$(243)$$

When the carbene centre and alkyne groups are more remote, the system behaved as a conventional carbene[376] except that cyclization by interaction between the carbene and the alkyne group yielded a rearranged carbene. Attempts to demonstrate that the rearrangement of **244** to **245** was reversible were not successful.

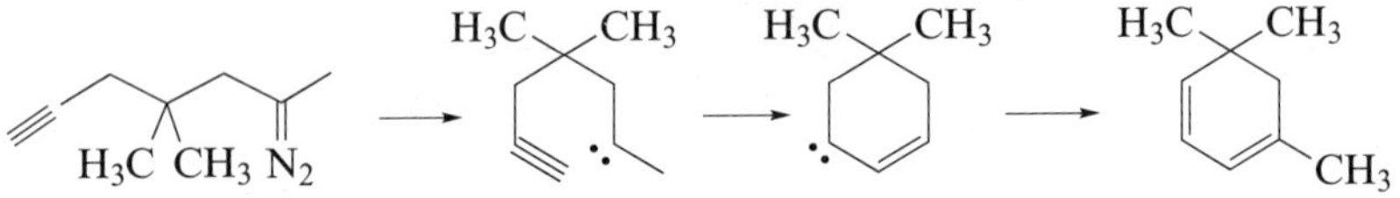

(244)   (245)

5. Phosphinocarbenes

Phosphoryl carbenes can be prepared from the corresponding diazo compounds; they rearrange readily to methylene phosphene oxides[377,378].

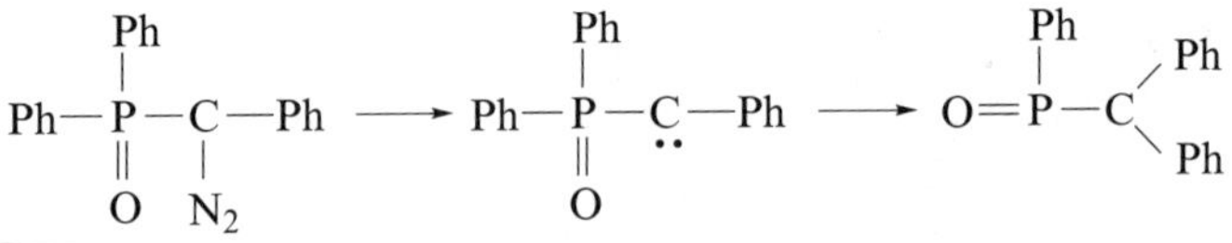

More recently, α-diazophosphines have been prepared[379]; on irradiation α-phosphinocarbenes are formed[380]. Although phosphinocarbenes always possess multiple bond character, they can rearrange either as carbenes or as ylides.

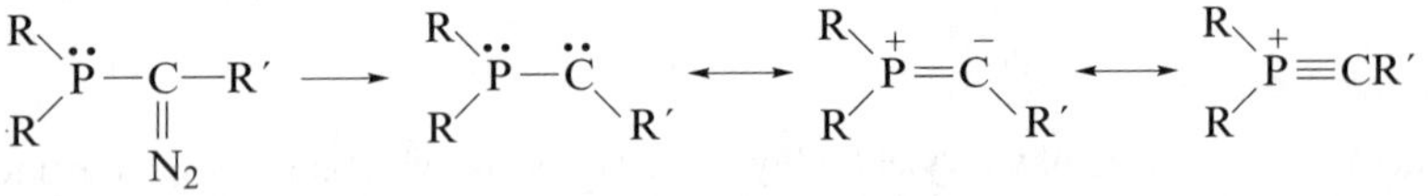

These carbenes have not been trapped by alkenes. Presumably the presence of an electron-rich phosphorus atom next to an electron-deficient carbon atom results in some delocalization of the phosphine non-bonded electron pair, tending to produce a dipolar species as shown above, and reducing the carbene character of the intermediate.

6. Germylenes

Preparation of a germanium-substituted diazo compound was accomplished by the tosylhydrazone route. On flash vacuum pyrolysis of the diazo compound[381] the major product was germaindane (**246**). The probable route to its formation involves initial production of the carbene **247** followed by a carbene-to-carbene rearrangement to **24** which can cyclize. Germylenes have been shown to insert into the Si—H bond[382]. The early development of germylenes has been reviewed[383].

(247)   (248)   (246)

7. Silylenes

Silylenes are very similar to germylenes, and can be prepared in the same way[384]. The reaction involves a carbene-to-carbene rearrangement, as does the similar germylene reaction.

The product of decomposition of an aliphatic silyl diazo compound is a silylene, which then dimerizes[385].

$$(Me_3Si)_2CN_2 \xrightarrow{400\ ^\circ C} Me_2Si{=}C\overset{Me}{\underset{SiMe_3}{\diagup}}$$

$$\underset{Me_3Si\overset{\overset{\displaystyle N_2}{\|}}{C}GeMe_3}{} \longrightarrow Me_2Si{=}C\overset{GeMe_3}{\underset{Me}{\diagup}}$$

Formation of a mixed silicon/germanium diazo compound allows competition between formation of a silylene and a germylene; the silylene is favoured by a factor of four.

$$:SiH_2 + SiH_4 \longrightarrow Si_2H_6$$

Silylene insertion into the S—H bond of silanes has been reported[386].

8. Carbenoid rearrangements

Carbenoids can be defined as: 'Intermediates which exhibit reactions qualitatively similar to those of carbenes without necessarily being free divalent carbon species'. Usually, the carbene is associated with a metal species. It is difficult to distinguish between a carbene and a carbenoid; methods used include reactivity with alkenes and a study of the stereochemistry of reaction of the carbene(oid). A recent study[387] has shown that the ρ value for the reactions of the alkylidene carbene derived from 1-diazo-2-methyl-1-propene is similar to that of the same species from α-elimination of the corresponding vinyl triflate; this may well develop into a valuable technique. Chemists carrying out syntheses have long made use of carbenoids formed when a diazoalkane is decomposed in the presence of a copper salt; the carbenoid is more likely to undergo intermolecular reaction than would a photolytically generated carbene from the same diazoalkane.

The advent of rhodium(II) salts as superior catalysts to generate electrophilic metal carbenes, even from α-diazo ketones, has generated a great deal of interest[388-390].

Unusual uses of the method include the formation of nitrocyclopropanes from nitrodiazocompounds and alkynes[391]:

Catalysis of insertion into a C—H bond has been observed using this reagent[392].

An important factor in the efficiency of this reaction may well be complexation between the rhodium carbene and the bond into which carbene insertion is to take place; certainly rhodium mediated C—H bond activation is known to take place[393]. The addition to alkenes can also be catalysed[394]; rearrangement of the product provides a useful route to cyclopentane rings.

By a process of cyclopropane formation followed by a Cope rearrangement, rhodium(II) catalysed decomposition of vinyldiazomethanes in the presence of dienes has been used to generate 1,4-cycloheptadienes[395].

Reaction with benzene follows a similar pathway, yielding a bicyclo[3.2.2]nonatriene structure[396]. Vinylcarbenoids also react with pyrroles to give tropanes via a cyclopropanation–Cope rearrangement route[397]. The direct addition of carbenes to acetylenes does not give satisfactory yields of cyclopropanes, but the rhodium carboxylate catalysed reaction of diazo compounds with acetylenes is a useful source of cyclopropanes[398]. Carbenoids can also attack a carbonyl oxygen atom, giving rise to a zwitterion (249). An excellent review of intramolecular carbenoid reactions has appeared[399].

(249)

F. Rearrangements of Diazoketones

Diazoketones undergo a number of interesting reactions, such as intramolecular insertions and additions and intramolecular additions, but are best known for the Wolff rearrangement.

1. The Wolff rearrangement

Thermal or photolytic decomposition of a diazo compound which has a carbonyl group α to the diazo group often yields a ketene, which can be trapped by conventional techniques, such as reaction with water to give a carboxylic acid[400].

The Wolff rearrangement is often accompanied by formation of a ketocarbene, and the competition between the two reactions has aroused much interest. The investigations of Tomioka and coworkers[401] have shown that Wolff rearrangement to form the ketene takes place directly from the excited singlet state of the s-Z conformer, whereas the excited state of the s-E conformer loses nitrogen to yield the singlet carbonyl carbene, which can either undergo characteristic carbene reactions or rearrange to a ketene. Thus[402] the di-butyl substituted diazoketone (250) is predominantly in an s-E conformation, so gives mainly carbene products on decomposition, while the diazocyclohexanone (251), which is locked

(250)   (251)

in the *s-Z* conformation, forms exclusively the ring-contracted ketene. When a molecule undergoes the Wolff rearrangement directly, loss of nitrogen is concerted with migration of the carbon–carbon bond[403]. Stereochemical[404] and CIDNP[329,404] evidence favours this mechanism, as do calculations on a retro-Wolff rearrangement[405].

In most other cases, the Wolff rearrangement proceeds through the singlet state. The few α-ketocarbenes which have been studied spectroscopically have triplet ground states[406], but in the main these do not undergo the Wolff rearrangement. An exception[407] to this general rule is the α-diazoketone (**252**) which, on irradiation in an argon matrix at 10–15 K, gave Wolff rearrangement to the ketone **253**, though it does not give this product in solution on either thermal or photochemical decomposition. Further experiments showed that the initial product in matrix conditions was the triplet α-ketocarbene, and this underwent a photochemical rearrangement to the ketene, probably[408] as a result of photolysis of the initially produced triplet α-keto carbene to yield the singlet.

(**252**)   (**253**)

This work clearly establishes that a ketene is formed from an α-ketocarbene, so demonstrating a stepwise Wolff rearrangement. The mechanism of reaction of the ketocarbene has also aroused interest; reaction via an oxirene has been suggested as a possible route. Evidence of formation of oxirenes from α-ketocarbenes has been obtained by labelling experiments[409]. It could be argued, however, that since it has been demonstrated that a ketene can arise from an α-ketocarbene, the oxirene may be involved in a ketocarbene-to-ketocarbene rearrangement rather than in the formation of ketene, and laser flash photolysis suggests that this could the case[410]. Evidence of involvement of oxirenes in the Wolff rearrangement has been obtained spectroscopically. Laser flash photolysis[411] of sodium 1-oxo-2-diazonaphthoquinone-5-sulphonate (**254**) yielded the ketene **255**. The reaction involved two detectable intermediates, the second of which was

(**254**)   (**255**)

the ketene: evidence that the first was the oxirene rather than the ketocarbene was provided by the failure to trap this intermediate with oxygen. Firmer evidence[412] that oxirenes are involved in the Wolff rearrangement has been obtained from irradiation of α-diazoketones in frozen gas matrices; photolysis of $CH_3COCN_2CH_3$ yielded oxirenes as minor products, stable at temperatures below 25 K, but which isomerized to ketenes on irradiation. Kinetic data showed the rates of conversion of diazo compound to ketocarbene, ketocarbene to ketene and ketene to oxirene were of the same order of magnitude. This work does not, however, establish that the oxirene is *necessarily* involved in the conversion of ketocarbene to ketene.

Other attempts to explore the intermediacy of oxirenes in the Wolff rearrangement are inconclusive; evidence in favour[413] and evidence against[414] their involvement has been obtained.

The Wolff rearrangement has also been observed in the gas phase[415] induced by an electron beam[416] and by collision with helium molecules[417]. It can also take place where the carbonyl group forms part of an ester or thioester; a competitive experiment[418] showed that migration of the sulphur entity is favoured over the oxygen.

A selenium substituent next to the diazo group blocked the Wolff rearrangement[419].

If however, the selenium was next to the oxygen, the reaction proceeded, yielding **256** in a sequence which probably involves Wolff rearrangement, ring opening and dimerization.

The Wolff rearrangement forms the basis of two reactions important in synthesis:
(a) It is an essential part of the Arndt–Eistert synthesis, used to lengthen the chain of a carboxylic acid.
(b) It is a valuable method of contracting a ring[420].

Recently[421] a new use of the Wolff rearrangement in synthesis has been demonstrated. Irradiation of a suitable α-diazo ketone induces a Wolff rearrangement, which produces an aryl or vinyl ketene, which then combines with an acetylene in a regiospecific [2+2] cycloaddition. Further irradiation or thermolysis then brings about 4π electrocyclic cleavage of the resulting 4-substituted cyclobutenone, producing a dienylketene which undergoes 6π electrocyclization to a 2,4-cyclohexadiene which can tautomerize to an aromatic system. The method clearly has great scope for development as a useful synthetic route.

2. The vinylogous Wolff rearrangement

In a series of papers[422–426] Smith and his coworkers investigated the mechanism of a rearrangement which can take place when a rearranged acid **258** is formed from an unsaturated diazo ketone **257** during the Arndt–Eistert reaction. The reaction was promoted by $CuSO_4$, $Cu(AcAc)_2$ or $Cu(OTf)_2$ in the presence of various alcohols.

The reaction is believed to proceed by initial formation of a ketocarbene, which can insert into the double bond to give a bicyclic ketone (**259**). This then yields the β,γ-unsaturated ketene **260**, which adds alcohol to give an ester. The reaction appears to be applicable to a wide range of β,γ unsaturated diazo ketones[425]; the stereochemistry of the reaction is consistent with the proposed mechanism.

G. Rearrangements of Ketocarbenes

While ketocarbenes can readily undergo the Wolff rearrangement, this section concentrates on reactions which do not follow this pathway. The best way to block the Wolff rearrangement is to carry out the reaction in a highly strained cyclic system so that the increase in strain resulting from a ring contraction renders the reaction energetically unfavourable. This route was used[427] to prepare some strained ketocarbenes in a matrix, and show that they existed as triplets. Even these strained, triplets on irradiation, underwent a stepwise Wolff rearrangement.

(261)   (262)

In solution[428], **261** does not undergo a Wolff rearrangement, and reactions of the triplet ketocarbene can be studied. With 2-propanol, hydrogen abstraction from the alcohol yields propan-2-one. With cyclooctane, insertion into a C—H bond was observed, yielding 2-cyclooctylacenaphthenone. Decomposition of **261** in benzonitrile yields **263**. Ketocarbene-to-ketocarbene rearrangements can take place, possibly through oxirene intermediates. Evidence of an intermediate of similar symmetry to an oxirene in the Wolff rearrangement has long been available from labelling experiments[429], but evidence of a ketocarbene-to-ketocarbene rearrangement, possibly through an oxirene, has been obtained[430]. Photolysis of **264** gives a mixture of Wolff rearrangement and OH insertion products; under similar conditions, the isomeric diazoketone **265** gives similar products in similar yields.

(263)

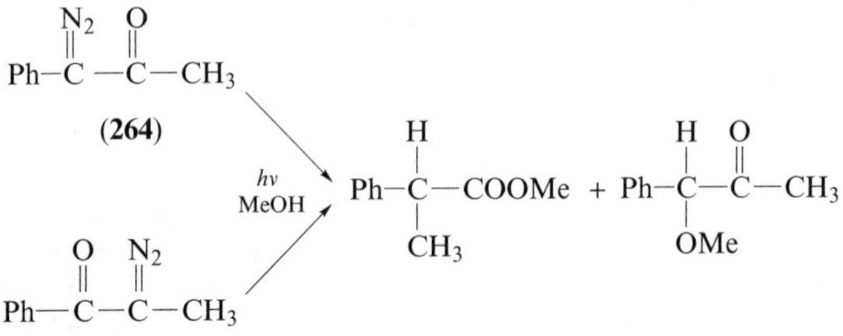

$$\underset{\substack{\| \\ Ph-C}}{\overset{\substack{N_2 \quad O \\ \| \quad \|}}{}} \underset{}{-C-CH_3}$$

(264)

$$\xrightarrow[\text{MeOH}]{hv}$$

$$\underset{\substack{| \\ CH_3}}{\overset{H}{Ph-C}}-COOMe \; + \; \underset{\substack{| \\ OMe}}{\overset{H \quad O}{Ph-C-C}}-CH_3$$

$$\underset{\substack{\| \quad \| \\ Ph-C-C-CH_3}}{\overset{O \quad N_2}{}}$$

(265)

Gas-phase photolysis of hexafluoro-3-diazo-2-butanone (266) in an excess of hexafluoro-2-butyne (267) yields the furan 268.

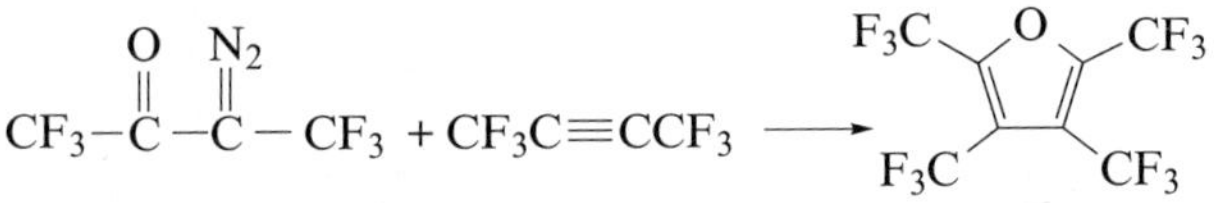

$$\underset{\substack{\| \quad \| \\ CF_3-C-C-CF_3}}{\overset{O \quad N_2}{}} + CF_3C\equiv CCF_3 \longrightarrow \underset{\substack{F_3C \qquad CF_3}}{\overset{F_3C \diagup O \diagdown CF_3}{}}$$

(266)   (267)   (268)

This may result from a 1,3-addition reaction, but formation of a pyrazolene followed by loss of nitrogen cannot be excluded[431].

Reaction of a di(diazo)ketone has provided a route to transient formation of a strained alkyne[432]. Irradiation of 269 proceeds via the diazoketene 270 to the cyclopropenone 271. Continued irradiation gives carbon monoxide and the allene 272, which is probably formed via the cyclopentyne 273.

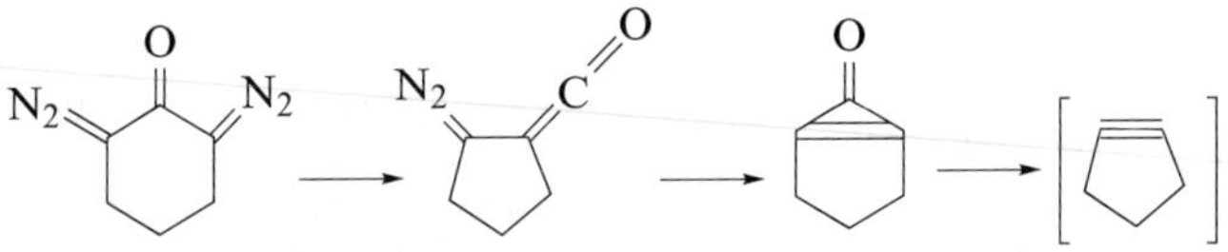

$$N_2 \diagdown \overset{O}{\diagup} \diagdown N_2$$

(269)   (270)   (271)   (272)

$$\longrightarrow H_2C=C=C \overset{CH_2}{\underset{CH_2}{\diagup |}}$$

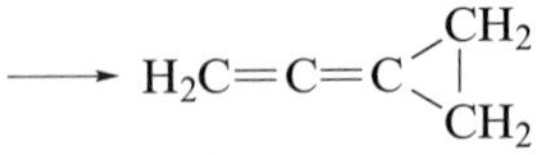

(273)

Photolysis of α-diazo-β-diketones showed initial formation of an α-oxo-ketene, which decarboxylated to an oxocarbene, then underwent a conventional Wolff rearrangement[433]. α-Diazocarbonyl compounds in which the carbonyl group is part of an ester function can undergo the Wolff rearrangement, but conventional carbene reactions can compete more readily.

Thus, dimethyl diazomalonate (274) yields bis(methoxycarbonyl)carbene (275) in a thermal decomposition. While 275 does undergo a Wolff rearrangement, it is also reported[434] to add to naphthalene to yield the cyclopropane 276 in 30–40% yield. When the α-diazocarbonyl compound has a double bond in a suitable position, cyclization completely replaces the Wolff rearrangement[435].

H. Acid-catalysed Rearrangements of α-Diazoketones

The presence of an α-keto group would be expected to increase the stability of a diazo group, and in fact the acid-catalysed decomposition of an α-diazoketone is slower than that of the corresponding diazo compound. Kinetic studies have shown that primary α-diazoketones react in acid via a rapid pre-equilibrium protonation[436], whereas secondary α-diazoketones[437,438] (and a few primary[439]) react via a rate-determining protonation. The position of protonation is uncertain. In superacids it has been shown to be almost entirely on oxygen[440], but reaction would require an S_N2 displacement at an sp^2 centre, so the possibility remains that the oxygen-protonated intermediate does not undergo reaction, and that reaction proceeds via a small amount of material protonated on nitrogen.

Since the carbocation would be destabilized by the presence of an α-keto group, it seems very likely that rearrangement would occur by displacement of nitrogen from the diazonium ion. Evidence of this[441] comes from reaction of the α-diazoketone **277**. Formation of

278 and **279** are consistent with *exo* protonation of **277** to give the diazonium ion **280**, which loses nitrogen to give **281**. A comprehensive review of the area has been published by Smith and Dieter[442].

Reaction of α-diazoketones can also be catalysed by Lewis acids, and cyclization of a β,γ-unsaturated diazoketone has been used extensively in synthesis. The mechanism of the reaction has been investigated[443–446]; it is suggested that the Lewis acid adds to the carbonyl group, to give an intermediate such as **282**, which cyclizes to **283** reaction is also catalysed by copper trifluoroacetate[447]. Lewis acid catalysis[448] has also been used to improve yields in cycloaddition reactions with nitriles, yielding oxazoles.

(282)   (283)

The reaction also takes place with a wide range of catalysts, or thermally[449].

I. Rearrangements of α-Keto Carbenoids

Rhodium catalysts have again gained in importance during the last decade[450]. Reaction of α-diazoketones with salts of copper or rhodium leads to formation of carbenoids[389]. These can undergo the usual carbene reactions plus the Wolff rearrangement (in a few cases, but more generally with silver salts). Chiral Rh(II) carboxylates have been shown to generate high levels of enantioselectivity in asymmetric synthesis[451–454].

1. Insertion into a C—H bond

Rhodium salts of strong organic acids have been used to catalyse the insertion of diazoesters into the C—H bonds of alkanes[455]. The rearrangement is much more selective and hence of greater use, when used to form cyclic systems[355,456,457].

2. Insertion into an O—H bond

Rhodium acetate has proved an effective catalyst for generation of a carbenoid which will insert into an OH bond[458]. It is also effective as a catalyst for insertion into N—H and S—H bonds.

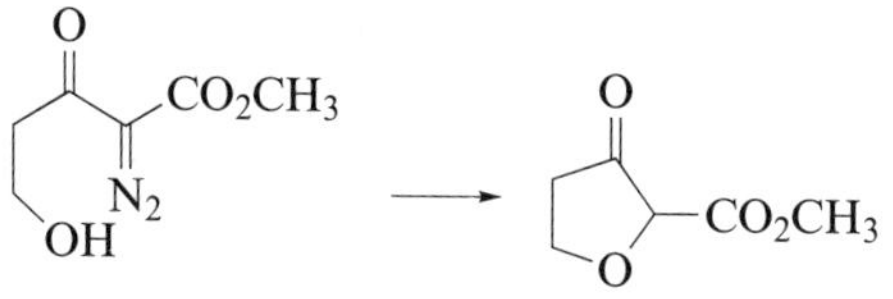

The method has been of considerable value in developing syntheses[459,460].

3. Addition to multiple bonds

Rhodium acetate is the usual catalyst for cyclopropane formation from the reaction of α-diazoketones with alkenes[461], though the Lewis acid $(\eta^5\text{-}C_5H_5)Fe^+(CO)_2(THF)BF_4^-$ has also been used[462].
The reaction is effective with aromatic systems[463], proceeding via a hexatriene.

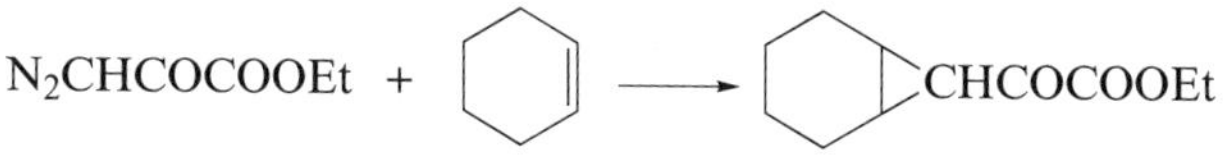

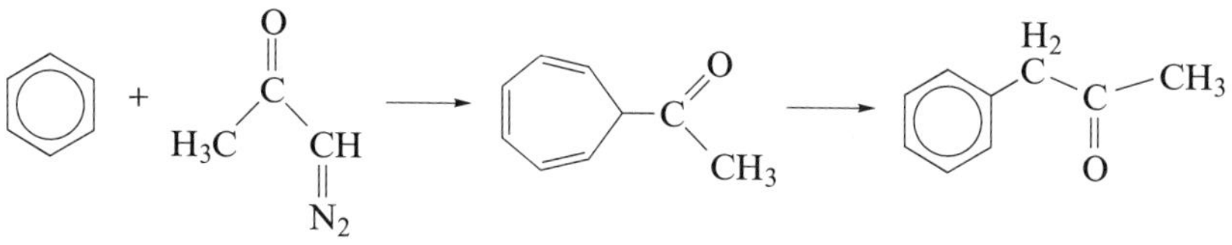

$N_2CHCOCOOEt$ +

Intramolecular insertion has been observed in reactions of homoallylic diazoacetates[464]. With acetylenes, reaction can either consist of an insertion into the C—C bond to the acetylene[465] or an addition reaction forming a cyclopropene, which spontaneously rearranges[466] to a vinyl carbene intermediate (**284**). In a polyunsaturated system, this carbene can undergo further addition[467].

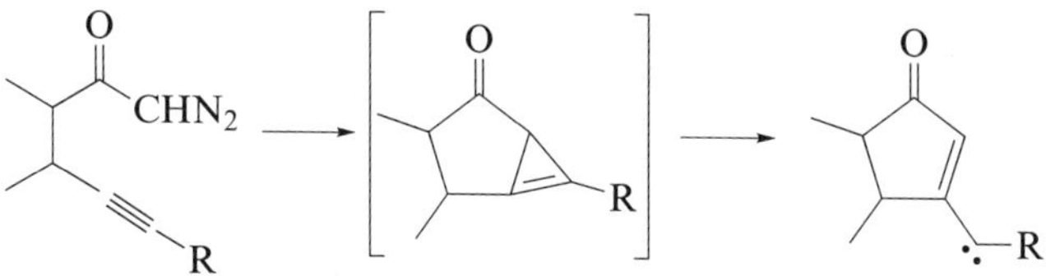

(284)

Insertion reactions probably involve formation of a rearranged carbenoid. This can be trapped with a suitable alkyne[468].

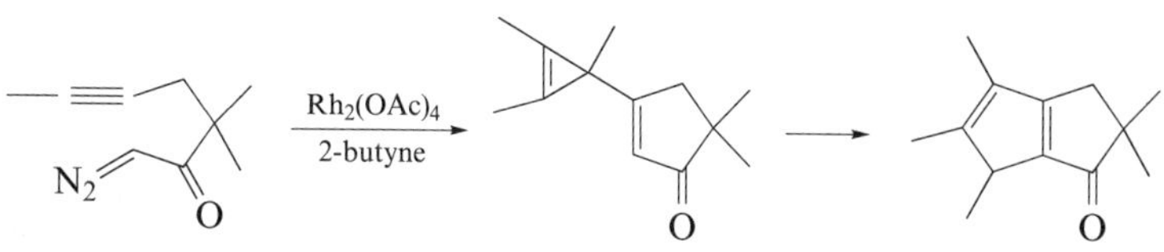

Alternatively, reaction with two alkyne molecules can give rise to a phenol[464].

Carbenoids can insert into the C≡N bond of cyanides, yielding oxazoles[469]. The reaction has been carried out successfully with diazomalonates[470] and with diazosulphones[471] catalysed by rhodium(II) acetate.

4. Formation of ylides

Carbenoids, like carbenes, can react with heteroatoms carrying a lone pair of electrons to give ylides, which can then undergo cycloaddition reactions[472,473].

Other workers[474] have found that ylide formation is favoured by using rhodium(II) perfluorobutyrate as the catalyst.

In cases where the molecule has a suitable double bond, ylide formation can be followed by an intramolecular cycloaddition[475,476] to yield a tetrahydrofuran (286).

$$CH_2=CH(CH_2)_4COCH_2CH_2COCHN_2$$

(285)   (286)

The ylide (285) can be trapped with dimethyl acetylenedicarboxylate.

5. Diazophosphonate reactions

The reactions of α-diazo-β-keto alkylphosphonates with rhodium(II) lead to carbenoid intermediates which can insert into C—H and OH bonds. Insertion into an OH bond yields ethers, leaving the phosphonate group available for further reaction. The method has been used to make cyclic ethers[477,478].

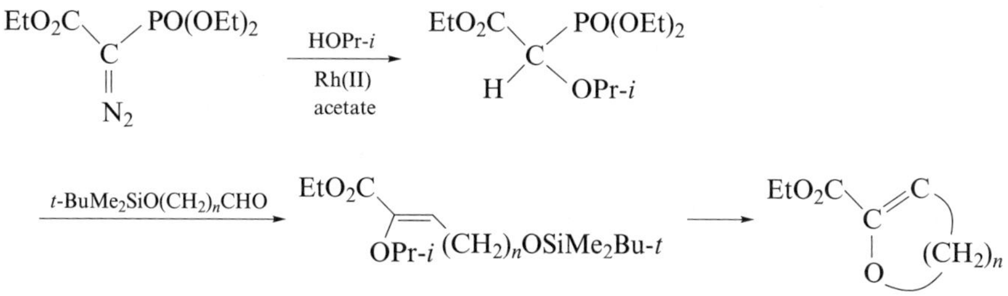

EtO$_2$C   PO(OEt)$_2$
     C
     ‖
     N$_2$
$\xrightarrow[\text{acetate}]{\begin{array}{c}\text{HOPr-}i\\ \text{Rh(II)}\end{array}}$
EtO$_2$C   PO(OEt)$_2$
     C
  H   OPr-i

$\xrightarrow{t\text{-BuMe}_2\text{SiO(CH}_2)_n\text{CHO}}$
EtO$_2$C
       OPr-i (CH$_2$)$_n$OSiMe$_2$Bu-t
$\longrightarrow$
EtO$_2$C   C
    C
    O   (CH$_2$)$_n$

Insertion into C—H bonds has also been reported[479] using rhodium(II) catalysis, but this reaction competes with a Wolff rearrangement.

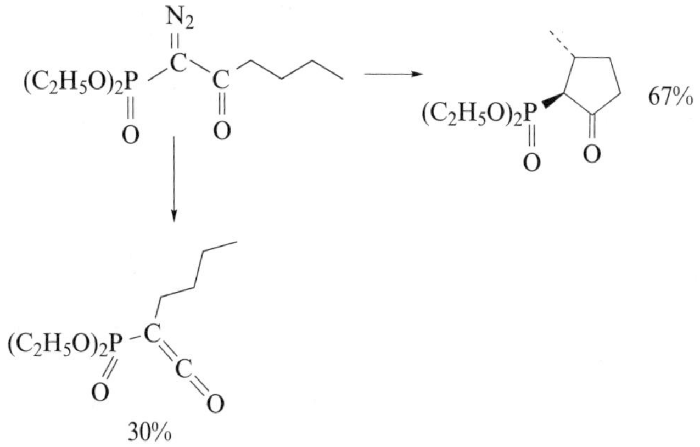

$$(C_2H_5O)_2P\underset{\overset{\|}{O}}{} \overset{\overset{N_2}{\|}{C}}{} C\underset{\overset{\|}{O}}{} \text{(chain)} \longrightarrow (C_2H_5O)_2P\underset{\overset{\|}{O}}{}\text{(cyclopentanone)} \quad 67\%$$

$$\downarrow$$

$$(C_2H_5O)_2P\underset{\overset{\|}{O}}{}\overset{}{C}\underset{\overset{}{C}=O}{} \\ 30\%$$

J. Addition Reactions of α-Diazoketones

α-Diazocarbonyl compounds undergo cycloaddition to a number of electron-poor alkynes[480]. Alkynyl(phenyl)iodinium salts (**287**) represent an extreme example of electron-poor alkynes, and readily undergo 1,3-dipolar cycloaddition[481].

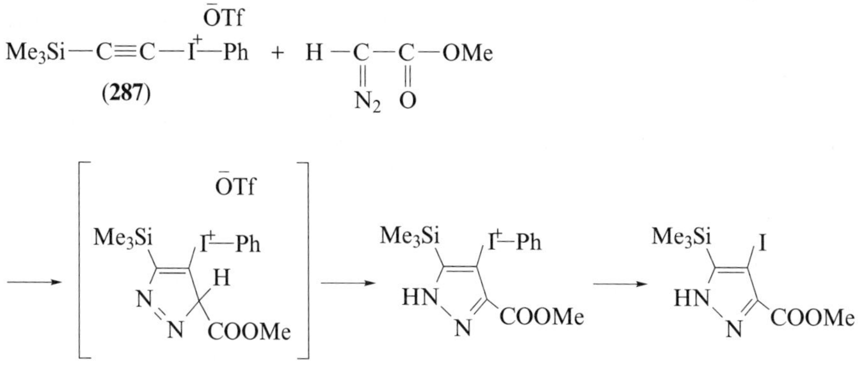

$$\overset{\overline{\text{O}}\text{Tf}}{Me_3Si-C\equiv C-\overset{+}{I}-Ph} \quad + \quad H-\underset{\overset{\|}{N_2}}{C}-\underset{\overset{\|}{O}}{C}-OMe$$
$$(287)$$

$$\longrightarrow \left[\begin{array}{c} \overset{\overline{\text{O}}\text{Tf}}{Me_3Si \quad \overset{+}{I}-Ph} \\ \quad\quad H \\ N \quad\quad\quad N \quad COOMe \end{array} \right] \longrightarrow \begin{array}{c} Me_3Si \quad \overset{+}{I}-Ph \\ HN \\ \quad N \quad COOMe \end{array} \longrightarrow \begin{array}{c} Me_3Si \quad I \\ HN \\ \quad N \quad COOMe \end{array}$$

VII. REFERENCES

1. A. E. Favorski, *J. Prakt. Chim.*, **37**, 417 (1888).
2. T. L. Jacobs, R. Awakie and R. C. Cooper, *J. Am. Chem. Soc.*, **73**, 1273 (1951).
3. A. E. Favorski and Z. I. Iotsich, *J. Prakt. Chim.*, **29**, 30 (1897).
4. M. D. Carr, L. H. Gan and I. Reid, *J. Chem. Soc., Perkin Trans. 2*, 672 (1973).
5. F. Theron, M. Verny and R. Vessiere, in *The Chemistry of the Carbon–Carbon Triple Bond* (Ed. S. Patai), Wiley, Chichester, 1978, p. 389.
6. D. J. Cram, F. Willey, H. P. Fischer, H. M. Relles and D. A. Scott, *J. Am. Chem. Soc.*, **88**, 2759 (1964).
7. R. J. Bushby and G. H. Whitham, *J. Chem. Soc. (B)*, 67 (1969).
8. S. R. Abrams, *J. Org. Chem.*, **49**, 3587 (1984).
9. P. Crabbe, H. Fillion, D. Andre and J-L Luche, *J. Chem. Soc., Chem. Commun.*, 859 (1979).
10. T. Kimmel and D. Becker, *J. Org. Chem.*, **49**, 2494 (1984).
11. J. M. Schwab and D. C. T. Lin, *J. Am. Chem. Soc.*, **107**, 6046 (1985).
12. S. R. Abrams and A. C. Shaw, *J. Org. Chem.*, **52**, 1835 (1987).
13. J. Z. Stemple and D. G. Peters, *J. Org. Chem.*, **54**, 5318 (1989).
14. X. Creary, *J. Am. Chem. Soc.*, **99**, 7632 (1977).
15. W. Priester and R. West, *J. Am. Chem. Soc.*, **98**, 8421, 8426 (1976).
16. C. Huynh and G. Linstrumelle, *J. Chem. Soc., Chem. Commun.*, 1133 (1983).
17. E. Grovenstein Jr., K.-W Chiu and B. B. Patel, *J. Am. Chem. Soc.*, **102**, 5848 (1980).
18. F. Johnson and R. Subramaniam, *J. Org. Chem.*, **50**, 5430 (1985); **51**, 5040 (1986).
19. M. Kosugi, K. Kurino, K. Takamaya and T. Migita, *J. Organomet. Chem.*, **56**, C11 (1973).
20. J. E. Baldwin, R. M. Adlington and R. M. Basak, *J. Chem. Soc., Chem. Commun.*, 1284 (1984).
21. G. A. Russell and L. L. Herold, *J. Org. Chem.*, **50**, 1037 (1985).
22. T. Flood and P. E. Peterson, *J. Org. Chem.*, **45**, 5006 (1980).
23. J. Parnet, *Tetrahedron Lett.*, **21**, 2049 (1980).
24. R. C. Larock, M-S. Chow and S. J. Smith, *J. Org. Chem.*, **51**, 2623 (1986).
25. H. J. Reich and S. K. Shah, *J. Am. Chem. Soc.*, **99**, 263 (1977).
26. H. J. Reich, S. K. Shah, P. M. Gold and R. E. Olson, *J. Am. Chem. Soc.*, **103**, 3112 (1981).
27. J.-C. Guillemin, T. Janati and J.-M. Dennis, *J. Chem. Soc., Chem. Commun.*, 415 (1992).
28. G. Markl and S. Reitinger, *Tetrahedron Lett.*, **29**, 463 (1988).
29. G. Markl and S. Reitinger, *Angew. Chem., Int. Ed. Engl.*, **27**, 1360 (1988).
30. M. Akita, H. Yasuda and A. Nakamura, *Bull. Chem. Jpn.*, **57**, 480 (1984); B. M. Trost, C. Chan and G. Runter, *J. Am. Chem. Soc.*, **109**, 3486 (1987).
31. M. Ishikawa, J. Oshita and A. Minato, *J. Organomet. Chem.*, **346**, C58 (1988); J. Oshita, K. Furumori, A. Matsuguchi and M. J. Ishikawa, *J. Org. Chem.*, **55**, 3277 (1990).
32. C.-H. Jun, Z. Lu and R. H. Crabtree, *Tetrahedron. Lett.*, **33**, 7119 (1992).
33. I. P. Kovalev, K. V. Yevdakov, Yu. A. Strelenko, M. G. Vinogradov and G. I. Nikishim *J. Organomet. Chem.*, **386**, 139 (1990).
34. (a) Y. Wakatsuki, H. Yamazaki, K. Kumegawa, T. Satoh and J. Y. Satoh, *J. Am. Chem. Soc.*, **113**, 9604, (1991).
    (b) E. A. Oostveen, C. J. Elsevier, J. Meijer and P. Vermeer, *J. Org. Chem.*, **47**, 371 (1982).
35. Y. O. Samura, H. F. Schaefer, III, S. K. Gray and W. H. Miller, *J. Am. Chem. Soc.*, **103**, 1904, (1981).
36. R. P. Duran, V. T. Amorebieta and A. J. Colussi, *J. Am. Chem. Soc.*, **109**, 3154 (1987).
37. P. H. Kasai, *J. Am. Chem. Soc.*, **114**, 3299 (1992).
38. K. R. Birdwhistell, T. L. Tonker and J. L. Templeton, *J. Am. Chem. Soc.*, **107**, 4474 (1985).
39. R. M. Bullock, *J. Chem. Soc., Chem. Commun.*, 165 (1987).
40. J. R. Fritch, K. P. C. Vollhardt, M. R. Thompson and V. W. Day, *J. Am. Chem. Soc.*, **101**, 2768 (1979).
41. E. Cabera, J. C. Daran and Y. Jeannin, *J. Chem. Soc., Chem. Commun.*, 607, (1988).
42. L. G. McCullough and R. R. Schrock, *J. Am. Chem. Soc.*, **106**, 4067 (1984).
43. G. A. Koutsantonis and J. P. Selegue, *J. Am. Chem. Soc.*, **113**, 2316 (1991).
44. L. M. Atagi, S. C. Critchlow and J. M. Mayer, *J. Am. Chem. Soc.*, **114**, 9223 (1992).
45. M. Petit, A. Mortreux and F. Petit, *J. Chem. Soc., Chem. Commun.*, 1385 (1982).
46. J. H. Wengrovius, J. Sancho and R. R. Schrock, *J. Am. Chem. Soc.*, **103**, 3932 (1981).
47. G. J. Leigh, M. T. Rahman and D. R. M. Walton, *J. Chem. Soc., Chem. Commun.*, 541 (1982).

48. A. Mortreux, F. Petit and M. Blanchard, *Tetrahedron Lett.*, **19**, 4967 (1978).
49. L. G. McCullough, M. L. Listemann, R. R. Schrock, M. R. Churchill and *J. Am. Chem. Soc.*, **105**, 6729 (1983).
50. D. L. Thorn and R. Hoffmann, *J. Am. Chem. Soc.*, **100**, 2079 (1978).
51. E. G. Samsel and J. R. Norton, *J. Am. Chem. Soc.*, **106**, 5505 (1984).
52. E. Chelain, R. Goumont, L. Hamon, A. Parlier, M. Rudler, H. Rudler, J.-C. Daran and J. Vaissermann, *J. Am. Chem. Soc.*, **114**, 8088 (1992).
53. T. J. Katz, T. H. Ho, N.-Y. Shih, Y.-C. Ying and V. I. W. Stuart, *J. Am. Chem. Soc.*, **106**, 2659 (1984).
54. W. Reppe, O. Schlichting, K. Klager and T. Toepel, *Justus Liebigs Ann. Chem.*, **560**, 1 (1948).
55. M. Green, N. C. Norman and A. G. Orpen, *J. Am. Chem. Soc.*, **103**, 1269 (1981).
56. D. R. McAlister, J. E. Bercaw and R. G. Bergman, *J. Am. Chem. Soc.*, **99**, 1666 (1977).
57. B. J. Rappoli, M. R. Churchill, T. S. Janik, W. M. Rees and J. D. Atwood, *J. Am. Chem. Soc.*, **109**, 5145 (1987).
58. M. D. Curtis and J. Real, *J. Am. Chem. Soc.*, **108**, 4668 (1986).
59. R. E. Colborn and K. P. C. Vollhardt, *J. Am. Chem. Soc.*, **108**, 5470 (1986).
60. K. P. C. Vollhardt, Acc. Chem. Res., **10**, 1 (1977); *Angew. Chem., Int. Ed. Engl.*, **23**, 539 (1984).
61. W.-Y. Yeh and L. K. Liu, *J. Am. Chem. Soc.*, **114**, 2267 (1992).
62. R. Gleiter and R. Merger, *Tetrahedron Lett.*, **31**, 1845 (1990).
63. E. Neglishi, L. S. Harring, Z. Owezarczyk, M. M. Mohamud and M. Ay, *Tetrahedron Lett.*, **33**, 3253 (1992).
64. T. J. Katz and T. M. Sivavec, *J. Am. Chem. Soc.*, **107**, 737 (1985).
65. B. M. Trost, W. P. Frengle, H. Urabe and J. Dumas, *J. Am. Chem. Soc.*, **114**, 1923 (1992).
66. D. J. Brien, A. Naima and K. P. C. Vollhardt, *J. Chem. Soc., Chem. Commun.*, 133 (1982).
67. R. L. Funk and K. P. C. Vollhardt, *J. Am. Chem. Soc.*, **101**, 215 (1979).
68. W. D. Wulff, K.-S. Chan and P.-C. Tang, *J. Org. Chem.*, **49**, 2293 (1984).
69. G. A. Olah and R. J. Spear, *J. Am. Chem. Soc.*, **97**, 1845 (1975).
70. A. E. Lodder, H. M. Buck and L. J. Oosterhoff, *Recl. Trav. Chim. Pays-Bas*, **89**, 1229 (1970).
71. V. Lucchini and G. Modena, *J. Am. Chem. Soc.*, **112**, 6291 (1990).
72. L. Gomez, P. Calas and A. Commeyras, *J. Chem. Soc., Chem. Commun.*, 1493 (1985).
73. R. C. Larcock and L. W. Harrison, *J. Am. Chem. Soc.*, **106**, 4218 (1984).
74. J. W. Wilson, *J. Am. Chem. Soc.*, **91**, 3238 (1969).
75. L. E. Harding and M. Hanack, *Tetrahedron Lett.*, 1253 (1971).
76. C. E. Harding and G. R. Stanford, *J. Org. Chem.*, **54**, 3054 (1989).
77. C. E. Harding and S. L. King, *J. Org. Chem.*, **57**, 883 (1992).
78. M. Hanack, K. A. Fuchs and C. J. Collins, *J. Am. Chem. Soc.*, **105**, 4008 (1983).
79. K. E. Harding, J. L. Cooper and P. M. Puckett, *J. Am. Chem. Soc.*, **100**, 993 (1978).
80. H. Heimstra and W. N. Speckamp, *Tetrahedron Lett.*, **24**, 1407 (1983).
81. S. M. Weinreb and P. M. Scola, *J. Org. Chem.*, **51**, 3248 (1986).
82. M. J. Fisher and L. E. Overman, *J. Org. Chem.*, **55**, 1447 (1990).
83. D. P. Curran, *Synthesis*, 417, 489 (1988).
84. B. Giese, *Radicals, in Organic Synthesis: Formation of Carbon–Carbon Bonds*, Pergamon Press, Oxford, 1986.
85. B. Giese, *Angew, Chem., Int. Ed. Engl.*, **28**, 969 (1989).
86. M. Journet, W. Smadja and M. Malacria, *Syn. Lett.*, 320 (1990).
87. M. Journet and M. Malacria, *J. Org. Chem.*, **57**, 3085 (1992).
88. D. P. Curran, W. Shen, J. Zhang and T. A. Heffner, *J. Am. Chem. Soc.*, **112**, 6738 (1990).
89. M. Journet and M. Malacria, *Tetrahedron Lett.*, **33**, 1893 (1992).
90. E. J. Enholm, J. A. Burroff and L. M. Jaramillo, *Tetrahedron Lett.*, **31**, 3727 (1990).
91. P. J. Garratt and S. B. Neoh, *J. Am. Chem Soc.*, **97**, 3256 (1975).
92. A. Huth, H. Straub and E. Müller, *Tetrahedron Lett.*, 2845 (1973).
93. R. G. Bergman, *Acc. Chem. Res.*, **6**, 25 (1973); K. C. Nicolau, G. Zuccarello, Y. Orgawa, E. J. Schweigo and T. Kumazawa, *J. Am. Chem. Soc.*, **110**, 4866, (1988).
94. M. F. Semmelhack, T. Neu and F. Foubelo, *Tetrahedron Lett.*, **33**, 3277 (1992).
95. J. F. Normant, G. Caihez, C. Chuit and J. Villeras, *J. Organomet. Chem.*, **77**, 281 (1974).
96. J. K. Crandall, P. Battioni, J. T. Wehlacz and R. Bindra, *J. Am. Chem. Soc.*, **97**, 7171 (1975).
97. D. Bouyssi, J. Gore and G. Balme, *Tetrahedron Lett.*, **33**, 2811 (1992).
98. J. v. d. Louw, J. L. v. d. Baan, C. M. D. Komen, A. Knol, F. J. J. de Kanter, F. Bickelhaupt and G. W. Klumpp, *Tetrahedron*, **48**, 6105 (1992).

574   D. Whittaker

99. S. Araki, A. Imai, K. Shimizu and Y. Butsugan, *Tetrahedron Lett.*, **33**, 2581 (1992).
100. D. F. Harvey and M. F. Brown, *J. Org. Chem.*, **57**, 5559, (1992).
101. D. F. Harvey, K. P. Lund and D. A. Neil, *J. Am. Chem. Soc.*, **114**, 8424 (1992).
102. O. De Lucchi and L. Pasquate, *Tetrahedron*, **44**, 6755 (1988).
103. A. Riera, M. Marti, A. Moyano, M. Pericas and J. Santamaria, *Tetrahedron Lett.*, **31**, 2173 (1990).
104. K. Chow, N. V. Nguyen and H. W. Moore, *J. Org. Chem.*, **55**, 3876 (1990).
105. H. Mayr, E. Bauml, G. Cibura and R. Koschinsky, *J. Org. Chem.*, **57**, 768 (1992).
106. P. L. McGrane and T. Livinghouse, *J. Org. Chem.*, **57**, 1323 (1992).
107. E. R. Koft and A. B. Smith III, *J. Am. Chem. Soc.*, **104**, 5568 (1982).
108. S. Hussain and W. C. Agosta, *Tetrahedron*, **37**, 3301 (1981).
109. S. Saba, S. Wolff, C. Schröder, P. Margaretha and W. C. Agosta, *J. Am. Chem. Soc.*, **105**, 6902 (1983).
110. V. B. Rao, S. Wolff and W. C. Agosta, *J. Am. Chem. Soc.*, **107**, 521 (1985).
111. M. E. Krafft, R. H. Romero and I. L. Scott, *J. Org. Chem.*, **57**, 5277 (1992).
112. R. L. Danheiser, B. R. Dixon and R. W. Gleason, *J. Org. Chem.*, **57**, 6094 (1992).
113. A. G. M. Barrett, N. E. Carpenter, J. Mortier and M. Sabat, *Organometallics*, **9**, 151 (1990).
114. S. U. Tumer, J. W. Herndon and L. A. McMullen, *J. Am. Chem. Soc.*, **114**, 8394 (1992).
115. K. G. Morris, S. P. Saberi, A. M. Z. Slawin, S. E. Thomss and D. J. Williams, *J. Chem. Soc., Chem. Commun.*, 1788 (1992).
116. K. A. Owens and J.A. Berson, *J. Am. Chem. Soc.*, **110**, 627 (1988).
117. T. Nakai and K. Mikami, *Chem. Rev.*, **86**, 885 (1986).
118. J. A. Marshall, E. D. Robinson and A. Zapata, *J. Org. Chem.*, **54**, 5854 (1989).
119. J. A. Marshall, T. M. Jenson and B. S. DeHoff, *J. Org. Chem.*, **51**, 4316 (1986).
120. J. A. Marshall and X. Wang, *J. Org. Chem.*, **55**, 2995 (1990).
121. A. Moyano, M. A. Pericàs, F. Serratosa and E. Valentí, *J. Org. Chem.*, **52**, 5532, (1987).
122. E. Valenti, M. A. Pericàs and F. Serratosa, *J. Org. Chem.*, **55**, 395 (1990).
123. H. Hogeween and D. M. Kok, *Tetrahedron Lett.*, **21**, 659 (1980).
124. J. L. Courtneidge, A. G. Davies, S. M. Tollerfield, J. Rideout and M.R. C. Symons, *J. Chem. Soc., Chem. Commun.*, 1092 (1985).
125. C. J. Cooksey, J. L. Courtneidge, A. G. Davies, P. S. Gregory, J. C. Evans and C. C. Rowlands, *J. Chem. Soc., Perkin Trans 2*, 807 (1988).
126. A. Alberola and M. C. Borque, *An. Quim.*, **73**, 872 (1977).
127. M. W. Klett and R. P. Johnson, *J. Am. Chem. Soc.*, **107**, 6615 (1985).
128. W. Weith, *Ber. Dtsch. Chem. Ges.*, 210 (1873).
129. M. Hunt, J. A. Kerr and A. F. Trotman-Dickenson, *J. Chem. Soc.*, 5074 (1965).
130. J. Casanova Jr., N. D. Werner and R. E. Schuster, *J. Org. Chem.*, **31**, 3473 (1966).
131. M. Meier and C. Rüchardt, *Chem. Ber.*, **120**, 1 (1987).
132. P. Saxe, Y. Yamaguchi, P. Pulay and H. F. Schaefer III, *J. Am. Chem. Soc.*, **102**, 3718 (1980).
133. M. Meier, B. Müller and C. Rüchardt, *J. Org. Chem.*, **52**, 648 (1987).
134. J.Casanova, Jr., in *The Chemistry of the Cyano Group* (Ed. Z. Rappoport), Wiley, London, 1970, p. 885.
135. R. Fuks, D. Baudoux, C. Piccinni-Leopardi, J-P. De Clercq and M. Van Meerssche, *J. Org. Chem.*, **53**, 18 (1988).
136. E. Wasserman, W. A. Yager and V. J. Kuck, *Chem. Phys. Lett.*, **7**, 409 (1970).
137. K. S. Kim, H. F. Schaefer III, L. Radom, J. A. Pople and J. S. Binkley, *J. Am. Chem. Soc.*, **105**, 4148 (1983).
138. H. Latka, *Z. Anorg. Allg. Chem.*, **353**, 243 (1967).
139. Y. Lei, T. Buranda and J. F. Endicott, *J. Am. Chem. Soc.*, **112**, 8820 (1990).
140. H. Hogeween, R. F. Kingma and D. M. Kok, *J. Org. Chem.*, **47**, 989 (1982).
141. M. H. Chisholm, D. M. Hoffman and J. C. Huffman, *J. Am. Chem. Soc.*, **106**, 6815 (1984).
142. R. R. Schrock, M. L. Listemann and L. G. Sturgeoff, *J. Am. Chem. Soc.*, **104**, 4291 (1982).
143. R. Nakamura, S. Matsumoto and E. Echigoya, *Chem. Lett.*, 1019 (1976).
144. R. H. A. Bosma, A. P. Kouwenhoven and J. C. Mol, *J. Chem. Soc., Chem. Commun.*, 1081 (1981).
145. P. G. Grassman, K. Saito and J. J. Tally, *J. Am. Chem. Soc.*, **102**, 7613 (1980).
146. P. G. Gassman and K. Saito, *Tetrahedron Lett.*, **22**, 1311 (1981).
147. P. G. Gassman and J. J. Talley, *J. Am. Chem. Soc.*, **102**, 1214 (1980).

148. G. A. Olah, G. K. Surya Prakash and M. Arvanaghi, *J. Am. Chem. Soc.*, **102**, 6641 (1980).
149. G. A. Olah, M. Arvanaghi and G. K. Surya Prakash, *J. Am. Chem. Soc.*, **104**, 1628 (1982).
150. V. V. Krishnamurthy, G. K. Surya Prakash, P. S. Iyer and G. A. Olah, *J. Am. Chem. Soc.*, **108**, 1575 (1986).
151. P. G. Gassman and J. J. Talley, *J. Am. Chem. Soc.*, **102**, 4138 (1980).
152. D. Farcasiu, *J. Org. Chem.*, **43**, 3878 (1978).
153. W. Kirmse and B. Goer, *J. Am. Chem. Soc.*, **112**, 4556 (1990).
154. Y.-D. Wu, W. Kirmse and K. N. Houk, *J. Am. Chem. Soc.*, **112**, 4557 (1990).
155. G. A. Olah and T. E. Kiovsky, *J. Am. Chem. Soc.*, **90**, 4666, (1968).
156. R. Bishop and G. Burgess, *Tetrahedron Lett.*, **28**,1585 (1987).
157. A. Hassner, T. K. Morgan, Jr. and A. R. McLaughlin, *J. Org. Chem.*, **44**, 1999 (1979).
158. M. T. McCormack and A. F. Hegarty, *J. Chem. Soc., Perkin Trans. 1*, 1701 (1976).
159. R. Bishop, S. C. Hawkins and I. C. Ibana, *J. Org. Chem.*, **53**, 427 (1988).
160. F. Cacace, G. Ciranni and P. Giacomello, *J. Am. Chem. Soc.*, **104**, 2258 (1982).
161. D. Seyferth, J. S. Merola and C. E. Eschbach, *J. Am. Chem. Soc.*, **100**, 4124 (1978).
162. S. Top and G. Jaouen, *J. Org. Chem.*, **46**, 78 (1981).
163. L. S. Hegedus, T. A. Mulhernand and H. Asada, *J. Am. Chem. Soc.*, **108**, 6224 (1986).
164. M. G. K. Hutchings and D. Swern, *J. Org. Chem.*, **47**, 4847 (1982).
165. F. E. Marsh and M. E. Hermes, *J. Am. Chem. Soc.*, **87**, 1819 (1965).
166. A. L. J. Beckwith, D. M. O'Shea, S. Gerba and S. W. Westwood, *J. Chem. Soc., Chem. Commun.*, 666 (1987).
167. A. L. J. Beckwith, D. M. O'Shea and S. W. Westwood, *J. Am. Chem. Soc.*, **110**, 2565 (1988).
168. B. P. Roberts and J. N. Winter, *J. Chem. Soc., Perkin Trans. 2*, 1353 (1979).
169. B. Chernera, C.-P. Chuang, D. J. Hart and L.-Y. Hsu, *J. Org. Chem.*, **50**, 5409 (1985).
170. E. J. Roskamp and S. F. Pedersen, *J. Am. Chem. Soc.*, **109**, 3152 (1987).
171. D. H. R. Barton, J. C. Jaszberenyi and E. A. Theodorakis, *Tetrahedron Lett.*, **32**, 3321, (1991).
172. O. L. Chapman, T. A. Rettig, A. I. Dutton and P. Fitton, *Tetrahedron Lett.*, 2049 (1963).
173. C. Manning and J. J. McCullough, *J. Chem. Soc., Chem. Commun.*, 75 (1977).
174. H. Zimmerman and R. J. Pasteris, *J. Org. Chem.*, **45**, 4864 (1980).
175. H. W. Moore and D. S. Wilbur, *J. Org. Chem.*, **45**, 4483 (1980).
176. H. W. Moore and C. C. Yu, *J. Org. Chem.*, **46**, 4935 (1981).
177. G. Rousseau and A. Quendo, *Tetrahedron*, **48**, 6361 (1992).
178. V. G. Kunde, A. C. Aikin, R. A. Hanel, D. E. Jennings, W. C. Maguire and R. E. Samuelson, *Nature*, **292**, 686 (1981).
179. B. Witulski, L. Ernst, P. G. Jones and H. Hopf, *Angew. Chem., Int. Ed. Engl.*, **28**, 1279 (1989).
180. J. P. Ferris and J. C. Guillemin, *J. Org. Chem.*, **55**, 5601 (1990).
181. R. L. Cobb and J. E. Mahon, *J. Org. Chem.*, **42**, 2597 (1977).
182. G. Morel, S. Khamsitthideth and A. Foucaud, *J. Chem. Soc., Chem. Commun.*, 274 (1978).
183. J. J. Grajewski and N. D. Conrad, *J. Am. Chem. Soc.*, **101**, 6693 (1979).
184. M. S. Dewar and C. Jic, *J. Chem. Soc., Chem. Commun.*, 18 (1989).
185. J. A. Barltrop, A. C. Day and E. Irving, *J. Chem. Soc., Chem. Commun.*, 881 (1979).
186. J. A. Barltrop, A. C. Day and E. Irving, *J. Chem. Soc., Chem. Commun.*, 966 (1979).
187. B. Jaques and G. R. Wallace, *Tetrahedron*, **33**, 581 (1977).
188. H. M. Walborsky and M. P. Periasamy, *The Chemistry of the Functional Groups, Supplement C* (Eds. S. Patai and Z. Rappoport), Wiley, Chichester, 1983, p. 835.
189. M. Tordeux and C. Wakselman, *Tetrahedron*, **37**, 315 (1981).
190. A. F. Hegarty and A. Chandler, *Tetrahedron Lett.*, **21**, 885 (1980); D. Marmet, P. Boullanger and G. Descotes, *Tetrahedron Lett.*, **21**, 1459 (1980).
191. J. Boyer and J. R. Patel, *J. Chem. Soc., Chem. Commun.*, 855 (1977).
192. G. Morel, E. Marchand, K. H. Nguyen Thi and A. Foucaud, *Tetrahedron*, **40**, 1075 (1984).
193. J. R. Bocarsly, C. Floriani, A. Chiesi-Villa and C. Guastino, *Organometallics*, **5**, 2380 (1986).
194. J. Dupont and M. Pfeffer, *J. Chem. Soc., Dalton Trans.*, 3193 (1990).
195. Y. Ito, T. Matsuura, S. Nishimura and M. Ishikawa, *Tetrahedron Lett.*, **27**, 3261 (1986).
196. J. B. Moffat, *Tetrahedron Lett.*, **22**, 1001 (1981).
197. M. Meot-Ner, Z. Karpas and C. A. Deakyne, *J. Am. Chem. Soc.*, **108**, 2913 (1986).
198. P. B. J. Driessen and H. Hogeveen, *Tetrahedron Lett.*, 271 (1979).
199. Y. Ito, H. Kato and T. Saegusa, *J. Org. Chem.*, **47**, 741 (1982).
200. Y. Ito, H. Kato, H. Imai and T. Saegusa, *J. Am. Chem. Soc.*, **104**, 6449 (1982).

201. H. Pellissier, A. Meou and G. Gil, *Tetrahedron Lett.*, **27**, 3505 (1986).
202. J. A. Baban and B. P. Roberts, *J. Chem. Soc., Perkin Trans. 2*, 1607 (1986).
203. C. Chatgilialoglu, B. Giese and B. Kopping, *Tetrahedron Lett.*, **31**, 6013 (1990).
204. M. Tordeaux and C. Wakselman, *Tetrahedron*, **37**, 315 (1981).
205. Y. Wakatsuki, Sin-ya Miya, S. Ikuta and H. Yamazaki, *J. Chem. Soc., Chem. Commun.*, 35 (1985).
206. H. Person, M. DelAguila Pardo and A. Foucaud, *Tetrahedron Lett.*, **21**, 281 (1980).
207. (a) A. M. van Leusen, B. E. Hogenboom and H. A. Houwing, *J. Org. Chem.*, **41**, 711 (1976).
    (b) I. Ugi and K. Rosendahl, *Chem. Ber.*, **94**, 2233 (1961).
208. H. W. Moore and C.-C. Yu, *J. Org. Chem.*, **46**, 4935 (1981).
209. J. Charrier, A. Foucaud, H. Person and E. Loukakou, *J. Org. Chem.*, **48**, 481 (1983).
210. K. Burger, G. Marschke and F. Manz, *J. Heterocycl. Chem.*, **19**, 1315 (1982).
211. D. P. Curran and H. Liu., *J. Am. Chem. Soc.*, **113**, 2127 (1991).
212. C. A. Merlic, E. G. Burns, D. Xu and S. Y. Chen, *J. Am. Chem. Soc*, **114**, 8722 (1992).
213. M. R. Anderson, R. F. C. Brown, K. J. Coulston, F. W. Eastwood and A. Ward, *Aust. J. Chem.*, **43**, 1137 (1990).
214. M. Barry, R. F. C. Brown, F. W. Eastwood, D. A. Gunawardana and C. Vogel, *Aust. J. Chem.*, **37**, 1643 (1984).
215. R. F. C. Brown, K. J. Coulston, F. W. Eastwood and C. Vogel, *Aust. J. Chem.*, **41**, 1687 (1988).
216. R. F. C. Brown, K. J. Coulston, F. W. Eastwood and M. R. Moffat, *Tetrahedron Lett.*, **32**, 801, (1991).
217. S. H. Hoke, II, J. Molstad, D. Dilettato, J. Jay, D. Carlson, B. Kahr and R. G. Cooks, *J. Org. Chem.*, **57**, 5069 (1992).
218. T. L. Gilchrist, in *Supplement C: The Chemistry of Triple-bonded Functional Groups* (Eds. S. Patai and Z. Rappoport), Wiley, Chichester, 1983, p. 383.
219. J. Nakayama, S. Takeue and M. Hoshino, *Tetrahedron Lett.*, **25**, 2679 (1984).
220. R. M. Bryce and J. J. Bernon, *Adv. Heterocycl. Chem.*, **28**, 183 (1981).
221. R. R. Jones and R. G. Bergman, *J. Am. Chem. Soc.*, **94**, 660 (1972).
222. R. G. Bergman, *Acc. Chem. Res.*, **6**, 25 (1973).
223. K. N. Barucha, R. M. Marsh, R. E. Minto and R. G. Bergman, *J. Am. Chem. Soc.*, **114**, 3120 (1992).
224. A. G. Myers, P. S. Dragovich and E. Y. Kuo, *J. Am. Chem. Soc.*, **114**, 9369 (1992).
225. R. Breslow, J. Napierski and T. C. Clarke, *J. Am. Chem. Soc.*, **97**, 6275 (1975).
226. T. P. Lockhart, C. B. Mallon and R. G. Bergman, *J. Am. Chem. Soc.*, **102**, 5976 (1980).
227. A. G. Myers, E. Y. Kuo and N. S. Finney, *J. Am. Chem. Soc.*, **111**, 8057 (1989).
228. B. T. Hart, *Aust. J. Chem.*, **26**, 461 (1973).
229. T. S. Cameron, P. K. Bakshi, B. Borecka and M. T. H. Liu, *J. Am. Chem. Soc.*, **114**, 1889 (1992).
230. R. A. Smith and J. R. Knowles, *J. Chem. Soc., Perkin Trans. 2*, 686 (1975).
231. B. M. Jennings and M. T. H. Liu, *J. Am. Chem. Soc.*, **98**, 6416 (1976).
232. M. T. H. Liu, M. Tencer and I. D. R. Stevens, *J. Chem. Soc., Perkin Trans. 2*, 211 (1986).
233. G. Lowe and J. Parker, *J. Chem. Soc., Chem. Commun.*, 1135 (1971).
234. L. Messerle and M. D. Curtis, *J. Am. Chem. Soc.*, **102**, 7789 (1980).
235. C. J. Pickett, J. E. Tolhurst, A. Coppenhaver, T. A. George and R. Lester, *J. Chem. Soc., Chem. Commun.*, 1071 (1982).
236. P. Riviere, A. Castel and J. Satge, *J. Am. Chem. Soc.*, **102**, 5413 (1980).
237. D. Whittaker, in *The Chemistry of Diazonium and Diazo Groups* (Ed. S. Patai), Wiley, Chichester, 1978, p. 596.
238. E. F. Rothgery, R. J. Holt and H. A. McGee, Jr., *J. Am. Chem. Soc.*, **97**, 4971 (1975).
239. I. Kalwinsch, L. Xingya, J. Gottstein and R. Huisgen, *J. Am. Chem. Soc.*, **103**, 7032 (1981).
240. M. P. Doyle, W. H. Tamblyn and V. Bagheri, *J. Org. Chem.*, **46**, 5094 (1981).
241. M. P. Doyle, J. H. Griffin, M. S. Chinn and D. van Leusen, *J. Org. Chem.*, **49**, 1917 (1984).
242. A. Padwa, A. Rodriguez, M. Tohidi and T. Fukunaga, *J. Am. Chem. Soc.*, **105**, 933 (1983).
243. T. K. Miller, J. T. Sharpe, H. R. Sood and E. Stefaniuk, *J. Chem. Soc., Perkin Trans. 2*, 823 (1984).
244. M. C. Pirrung and J. Zhang, *Tetrahedron Lett.*, **33**, 5987 (1992).
245. W. Kirmse, *Angew. Chem., Int. Ed. Engl.*, **15**, 251 (1976).
246. D. L. H. Williams, in *Advances in Physical Organic Chemistry* (Eds. D. Bethell and V. Gold), Vol 19, Academic Press, London, 1983, p. 381.
247. E. Mueller and H. Haiss, *Chem. Ber.*, **96**, 570 (1963).

248. A. B. Kyte, R. Jones-Parry and D. Whittaker, *J. Chem. Soc., Chem. Commun.*, **74**, (1982).
249. R. Huisgen and H. Reimlinger, *Ann. Chem.*, **599**, 161, 183 (1956).
250. E. H. White and C. Aufdermarsh, Jr., *J. Am. Chem. Soc.*, **83**, 1179 (1961).
251. N. Nikolaides, A. G. Godfrey and B. Ganem, *Tetrahedron Lett.*, **31**, 6009, (1990).
252. E. H. White and M. Scherrer, *Tetrahedron Lett.*, 758 (1961).
253. D. H. Sieh, D. J. Wilbur and C. Michejda, *J. Am. Chem. Soc.*, **102**, 3883 (1980).
254. M. P. Doyle, J. W. Terpstra, R. A. Pickering and D. M. Le Poire, *J. Org. Chem.*, **48**, 3379 (1983).
255. M. J. Crooks and D. L. H. Williams, *J. Chem. Soc., Chem. Commun.*, 571, (1988).
256. S. E. Aldred and D. L. H. Williams, *J. Chem. Soc., Perkin Trans, 2*, 1021 (1981).
257. M. P. Doyle, R. J. Bosch and P. G. Seites, *J. Org. Chem.*, **43**, 4120, (1978).
258. W. L. Bowden, W. F. Little and T. J. Meyer, *J. Am. Chem. Soc.*, **99**, 4340 (1977).
259. J. A. McCleverty, *Chem. Rev.*, **79**, 53 (1979).
260. G. J. McGarvy and M. Kimura, *J. Org. Chem.*, **51**, 3915 (1986).
261. A. Mertens, K. Lammerstma, M. Arranaghi and G. A. Olah, *J. Am. Chem. Soc.*, **105**, 5657 (1983).
262. G. A. Olah, K. Laali, M. Farnia, J. Shih, B. P. Singh, C. J. Schack and K. C. Christie, *J. Org. Chem.*, **50**, 1339 (1985).
263. D. Berner and J. F. McGarrity, *J. Am. Chem. Soc.*, **101**, 3135 (1979).
264. G. W. Van Dine and R. Hoffman, *J. Am. Chem. Soc.*, **90**, 3227 (1968).
265. L. Friedman, A. T. Jurewicz and J. H. Bayless, *J. Am. Chem. Soc.*, **91**, 1795 (1969).
266. W. Brosch and W. Kirmse, *J. Org. Chem.*, **56**, 907 (1991).
267. P. A. S. Smith and D. R. Baer, *J. Am. Chem. Soc.*, **74**, 6135 (1952).
268. D. V. Nightingale, J. D. Kerr, J. A. Gallacher and M. Maienthal, *J. Org. Chem.*, **17**, 1017 (1952).
269. M. Bunse, D. Jodicke and W. Kirmse, *J. Chem. Soc., Chem. Commun.*, 47 (1992).
270. D. J. Cram and J. E. McCarty, *J. Am. Chem. Soc.*, **79**, 2866 (1957).
271. E. J. Corey, J. Casanova, Jr., P. A. Vatakencherry and R. Winter, *J. Am. Chem. Soc.*, **85**, 169 (1963).
272. W. Kirmse and G. Voigt, *J. Am. Chem. Soc.*, **96**, 7598 (1974).
273. A. Streitwieser, *J. Org. Chem.*, **22**, 861 (1957).
274. E. M. Arnett and C. Petro, *J. Am. Chem. Soc.*, **100**, 2563, 5402, 5408 (1978).
275. H. Maskill and M. C. Whiting, *J. Chem. Soc., Perkin Trans. 2*, 1462 (1976).
276. R. M. Southam and M. C. Whiting, *J. Chem. Soc., Perkin Trans. 2*, 597 (1982).
277. E. H. White and K. W. Field, *J. Am. Chem. Soc.*, **97**, 2148 (1975).
278. F. H. White, K. W. Field, W. H. Hendrickson, P. Dzadzic, D. F. Roswell, S. Palk and P. W. Mullen, *J. Am. Chem. Soc.*, **114**, 8023 (1992).
279. H. Maskill and A. A. Wilson, *J. Chem. Soc., Perkin Trans. 2*, 1369 (1984).
280. W. Kirmse and R. Siegfried, *J. Am. Chem. Soc.*, **105**, 950 (1983).
281. W. Kirmse and N. Knöpfel, *J. Am. Chem. Soc.*, **98**, 4672 (1976).
282. H. C. Berk, C. R. Degenhardt and L. A. Paquette, *J. Org. Chem.*, **43**, 4516 (1978).
283. L. M. Anderson, A. R. Butler and S. A. McIntosh, *J. Chem. Soc., Perkin Trans. 2*, 1239 (1987).
284. I. Szele. and H. Zollinger, *Helv. Chim. Acta*, **64**, 2728 (1981).
285. H. Marxmeier and E. Pfeil, *Chem. Ber.*, **97**, 53 (1964).
286. A. P. Terent'ev and V. Jagorsvkii, *Zh, Obshch. Khim.*, **26**, 211 (1956).
287. J. W. Crary, O. R. Quayle and C. T. Lester, *J. Am. Chem. Soc.*, **78**, 5584 (1956).
288. H. Mayr and K. Grimm, *J. Org. Chem.*, **57**, 1057 (1992).
289. F. Minisci, F. Coppa, F. Fontana, G. Pianese, and L. Zhao, *J. Org. Chem.*, **57**, 3929, (1992).
290. B. A. Carlson, W. A. Sheppard and O. W. Webster, *J. Am. Chem. Soc.*, **97**, 5291 (1975).
291. F. Bronberger and R. Huisgen, *Tetrahedron Lett.*, **25**, 57 (1984).
292. R. Huisgen and F. Bronberger, *Tetrahedron Lett.*, **25**, 61 (1984).
293. G. A. Olah and J. Welch, *J. Am. Chem. Soc.*, **97**, 208 (1975).
294. G. A. Olah and J. L. Grant, *J. Am. Chem. Soc.*, **97**, 1546 (1975).
295. H. Zollinger, *Angew. Chem., Int. Ed. Engl.*, **17**, 141 (1978).
296. H. Cano-Yelo and A. Deronzier, *J. Chem. Soc., Perkin Trans. 2*, 1093 (1984).
297. N. Mori and T. Tachibana, *J. Am. Chem. Soc.*, **106**, 6115 (1984).
298. Y. Takada, K. Tsuchiya, S. Takahashi and N. Mori, *J. Chem. Soc., Perkin Trans. 2*, 2141 (1990).
299. N. Mori, T. Takemura and K. Tsuchiya, *J. Chem. Soc., Chem. Commun.*, 575 (1988).
300. D. T. Longone and J. A. Gladysz, *Tetrahedron Lett.*, 4559 (1976).
301. Y. Z. Li and G. B. Schuster, *J. Org. Chem.*, **51**, 3804 (1986).
302. D. Modarelli and M. S. Platz, *J. Am. Chem. Soc.*, **113**, 8985, (1991).

578   D. Whittaker

303. J. M. Fox, J. E. G. Scacheri, K. G. L. Jones, M. Jones, Jr., P. B. Shevlin, B. Armstrong and R. Sztyrbicka, *Tetrahedron Lett.*, **33**, 5021 (1992).
304. R. J. Miller, L. S. Yang and H. Shechter, *J. Am. Chem. Soc.*, **99**, 938 (1977).
305. R. S. Miller and H. Shechter, *J. Am. Chem Soc.*, **100**, 7920 (1978).
306. J. R. Ammann, R. Subramanian and R. S. Sheridan, *J. Am. Chem. Soc.*, **114**, 7592 (1992).
307. W. v. E. Doering and A. K. Hoffmann, *J. Am. Chem. Soc.*, **76**, 6162 (1954).
308. P. S. Skell and R. C. Woodworth, *J. Am. Chem. Soc.*, **78**, 4496 (1956).
309. R. A. Moss, *Acc. Chem. Res.*, **13**, 58 (1980).
310. D. Bethell, D. Whittaker and J. D. Callister, *J. Chem. Soc.*, 2466 (1965).
311. W. Kirmse, *Justus Liebigs Ann. Chem.*, **666**, 9 (1963).
312. A. M. Trozzolo, R. W. Murray and E. Wasserman, *J. Am. Chem. Soc.*, **84**, 4990 (1962).
313. I. Moritani, S. I. Murahashi, K. Yoshinaga and H. Ashitaka, *Bull. Chem. Soc. Jpn.*, **40**, 1506 (1967).
314. K. B. Eisenthal, N. Turro, M. A. Aikawa, J. A. Butcher Jr., C. Dupuy, W. Hetherington, G. M. Korenowski and M. J. McAuliffe, *J. Am. Chem. Soc.*, **102**, 6563 (1980).
315. J. J. Havel, *J. Org. Chem.*, **41**, 1464 (1976).
316. X. Creary, *J. Am. Chem. Soc.*, **102**, 1611 (1980).
317. G. L. Closs and B. E. Rabinow, *J. Am. Chem. Soc.*, **98**, 8190 (1976).
318. D. Griller, A. S. Nazran and J. C. Scaiano, *Acc. Chem. Res.*, **17**, 283 (1984).
319. P. B. Grasse, B. E. Brauer, J. J. Zupanic, K. J. Kaufmann, and G. B. Schuster, *J. Am. Chem. Soc.*, **105**, 6833 (1983).
320. L. M. Hadel, M. S. Platz and J. C. Scanio, *J. Am. Chem. Soc.*, **106**, 283 (1984).
321. H. Tomioka, G. W. Griffin and K. Nishiyama, *J. Am. Chem. Soc.*, **101**, 6009 (1979).
322. H. Tomioka, N. Hayashi, Y. Izawa, V. P. Senthilnathan and M. S. Platz, *J. Am. Chem. Soc.*, **105**, 5053 (1983).
323. R. J. Blanche, J. Li, L. C. Bush and M. Jones, Jr., *J. Am. Chem. Soc.*, **114**, 9236 (1992).
324. M. R. Chedekel, M. S. Keglund, R. L. Kreeger and H. Schechter, *J. Am. Chem. Soc.*, **98**, 7846 (1976).
325. A. H. Jurewicz and L. Friedman, *J. Am. Chem. Soc.*, **89**, 149 (1967).
326. H. Tomoika, S. Suzuki and Y. Izawa, *J. Am. Chem. Soc.*, **104**, 1047 (1982).
327. C. Ho, R. T. Conlin and P. P. Gasper, *J. Am. Chem. Soc.*, **96**, 8109 (1974).
328. J. J. Havel, *J. Org. Chem.*, **41**, 1464 (1976).
329. H. D. Roth, *Acc. Chem. Res.*, **10**, 85 (1977).
330. H. D. Roth, *J. Am. Chem. Soc.*, **94**, 1761 (1972).
331. M. C. Biewer, M. S. Platz, M. Roth and J. Wirz, *J. Am. Chem. Soc.*, **113**, 8069 (1991).
332. R. A. Moss and U. H. Dolling, *J. Am. Chem. Soc.*, **93**, 954 (1971).
333. R. A. Moss and M. A. Joyce, *J. Am. Chem. Soc.*, **100**, 4475 (1978).
334. J. Ruzicka, E. Leyva and M. S. Platz, *J. Am. Chem. Soc.*, **114**, 897 (1992).
335. W. H. Urry and J. R. Eizner, *J. Am. Chem. Soc.*, **73**, 2977 (1951).
336. H. D. Roth, *J. Am. Chem. Soc.*, **93**, 4935 (1971).
337. M. B. Jones, V. M. Maloney and M. S. Platz, *J. Am. Chem. Soc.*, **114**, 2163 (1992).
338. D. T. T. Su and E. R. Thornton, *J. Am. Chem. Soc.*, **100**, 1872 (1978).
339. A. Nickon and J. K. Bronfenbrenner, *J. Am. Chem. Soc.*, **104**, 2022 (1982).
340. M. T. H. Liuard and R. Bonneau, *J. Am. Chem. Soc.*, **114**, 3604 (1992).
341. R. A. Moss, G. J. Ho, W. Liu and C. Sierakowski, *Tetrahedron Lett.*, **33**, 4287 (1992).
342. R. A. Moss, G. J. Ho, S. Shen and K. Krogh-Jespersen, *J. Am. Chem. Soc.*, **112**, 1638 (1990).
343. A. S. Kende, P. Hebeisen, P. J. San Filippo and B. H. Toder, *J. Am. Chem. Soc.*, **104**, 4244 (1982).
344. P. de March and R. Huisgen, *J. Am. Chem. Soc.*, **104**, 4952 (1982).
345. R. L. Barcus, L. M. Hadel, L. J. Johnstone, M. S. Platz, T. G. Savino and J. C. Scaiano, *J. Am. Chem. Soc.*, **108**, 3928 (1986).
346. P. B. Grasse, B. E. Brauer, J. J. Zupancic, K. J. Kaufmann and G. B. Schuster, *J. Am. Chem. Soc.*, **105**, 6833 (1983).
347. R. A. Moss, G. J. Ho and C. Sievakowski, *J. Am. Chem. Soc.*, **114**, 3128 (1992).
348. D. Bethell, A. R. Newall, G. Stevens and D. Whittaker, *J. Chem. Soc. (B)*, 749 (1969); D. Bethell, A. R. Newall and D. Whittaker, *J. Chem. Soc. (B)*, 23 (1971).
349. W. Kirmse, K. Loosen and H. D. Sluma, *J. Am. Chem. Soc.*, **103**, 5935 (1981).
350. W. Kirmse, J. Kilian and S. Steenken, *J. Am. Chem. Soc.*, **112**, 6399 (1990).
351. J. E. Chateauneuf, *J. Chem. Soc., Chem. Commun.*, 1437 (1991).

352. A. O. Ku, M. Iwamoto, K. Sanada and M. Abe, *Tetrahedron Lett.*, **33**, 7169 (1992).
353. J. Zayas and M. S. Platz, *J. Am. Chem. Soc.*, **107**, 7065 (1985).
354. K. Yuan and M. Jones Jr., *Tetrahedron Lett.*, **33**, 7481 (1992).
355. R. A. Moss, *Acc. Chem. Res.*, **13**, 58 (1980).
356. M. Jones and K. R. Rettig, *J. Am. Chem. Soc.*, **87**, 4013 (1965).
357. D. Griller, C. R. Montgomery, J. C. Scanio, M. S. Platz and L. Hadel, *J. Am. Chem. Soc.*, **104**, 6813 (1982).
358. W. M. Jones and M. H. Grasley, *Tetrahedron Lett.*, **4**, 927 (1962); W. M. Jones and J. M. Walbrick, *J. Org. Chem.*, **34**, 2217 (1969).
359. W. L. Magee and H. Schechter, *J. Am. Chem. Soc.*, **99**, 633 (1977).
360. W. L. Magee and H. Schechter, *Tetrahedron Lett.*, **20**, 4697 (1979).
361. W. M. Jones, *Acc. Chem. Res*, **10**, 353 (1977).
362. S. Chari, G. K. Agopian and M. Jones Jr., *J. Am. Chem. Soc.*, **101**, 6125 (1979).
363. H. Tomioka, N. Kobayashi, S. Murata and Y. Ohtawa, *J. Am. Chem. Soc.*, **113**, 8771 (1991).
364. C. Mayor and C. Wentrup, *J. Am. Chem. Soc.*, **97**, 7467 (1975).
365. R. V. Hoffman, G. G. Orphanides and H. Schechter, *J. Am. Chem. Soc.*, **100**, 7927 (1978).
366. U. H. Brinker and L. Konig, *J. Am. Chem. Soc.*, **101**, 4738 (1979).
367. B. Halton and M. G. Banwell, 'Cyclopropenes' in *The Chemistry of Functional Groups, Cyclopropanes*, Vol. 2 (Ed. Z. Rappoport), Wiley, Chichester, 1988.
368. S. Murahashi, K. Okumura, T. Naota and S. Nagase, *J. Am. Chem. Soc.*, **104**, 2466 (1982).
369. P. K. Freeman and K. E. Swenson, *J. Org. Chem.*, **47**, 2040 (1982).
370. O. L. Chapman and C. J. Abelt, *J. Org. Chem.*, **52**, 1218 (1987).
371. M. L. Manion and H. D. Roth, *J. Am. Chem. Soc.*, **97**, 6919 (1975).
372. R. Selvarajan and J. H. Boyer, *J. Org. Chem.*, **36**, 1679 (1971).
373. D. J. De Frees and A. D. Mclean, *Astrophys. J.*, **308** (1986).
374. W. J. Hehre, J. A. Pople, W. A. Latham, L. Radom, E. Wasserman and Z. R. Wasserman, *J. Am. Chem. Soc.*, **98**, 4378 (1976).
375. A. Padwa, Y. Gareau and S. L. Xu., *Tetrahedron Lett.*, **32**, 983 (1991).
376. P. K. Freeman, J. C. Danino, B. K. Stevenson and G. E. Clapp, *J. Org. Chem.*, **55**, 3867 (1990).
377. M. Regitz, A. Liedheger, W. Anschutz and H. Eckes, *Chem. Ber.*, **104**, 2177 (1971).
378. M. Regitz, *Diazoalkane*, George Thieme Verlag, Stuttgart, 1977; M. Regitz and G. Maas, *Top. Curr. Chem.*, **97**, 71 (1981).
379. H. Keller, G. Maas and M. Regitz, *Tetrahedron Lett.*, **27**, 1903 (1986).
380. A. Baceiredo, A. Igau, G. Bertrand, M. J. Menu, Y. Dartiguenave and J. J. Bonnet, *J. Am. Chem. Soc.*, **108**, 7868 (1986).
381. E. B. Norsoph, B. Coleman and M. Jones Jr., *J. Am. Chem. Soc.*, **100**, 994 (1978).
382. K. M. Baines, J. A. Cooke and J. J. Vittal, *J. Chem. Soc., Chem. Commun.*, 1484 (1992).
383. J. Satge, M. Massol and P. Riviere, *J. Organomet. Chem.*, **56**, 1 (1973).
384. W. Ando, A. Sekiguchi, A. J. Rothschild, R. R. Galluci, M. Jones Jr., T. J. Barton and J. A. Kilgour, *J. Am. Chem. Soc.*, **99**, 6995 (1977).
385. T. J. Barton and S. K. Hockman, *J. Am. Chem. Soc.*, **102**, 1584 (1980).
386. R. Becerra, H. M. Frey, B. P. Mason, R. Walsh and M. S. Gordon, *J. Am. Chem. Soc.*, **114**, 2751 (1992).
387. J. C. Gilbert and D. H. Giamalva, *J. Org. Chem.*, **57**, 4185 (1992).
388. M. P. Doyle, *Acc. Chem. Res.*, **19**, 348 (1986).
389. M. P. Doyle, *Chem. Rev.*, **86**, 919 (1986).
390. M. Gerhard, *Top. Curr. Chem.*, **137**, 77 (1987).
391. P. E. O'Bannon and W. P. Dailey, *J. Org. Chem.*, **56**, 2258 (1991).
392. D. F. Taber and R. E. Ruckle, Jr., *J. Am. Chem. Soc.*, **108**, 7686 (1986).
393. W. D. Jones and F. Feher, *J. Am. Chem. Soc.*, **107**, 620, (1985).
394. H. M. L. Davies and B. Hu, *Tetrahedron Lett.*, **33**, 453 (1992).
395. H. M. L. Davies, T. J. Clark and H. D. Smith, *J. Org. Chem.*, **56**, 3817 (1991).
396. H. M. L. Davies, H. D. Smith, B. Hu, S. M. Klenzak and F. J. Hegner, *J. Org. Chem.*, **57**, 6900 (1992).
397. H. W. Davies and N. J. S. Huby, *Tetrahedron Lett.*, **33**, 6935 (1992).
398. D. L. Hertzog, D. J. Austin, W. R. Nadler and A. Padwa, *Tetrahedron Lett.*, **33**, 4731, (1992).
399. A. Padwa and K. E. Krumpe, *Tetrahedron*, **48**, 5385 (1992).
400. E. G. Lewars, *Chem. Rev.*, **83**, 519 (1983).

401. H. Tomioka, H. Okuno and Y. Izawa, *J. Org. Chem.*, **45**, 5278 (1980).
402. F. Kaplan and M. L. Mitchell, *Tetrahedron Lett.*, **20**, 759 (1979).
403. H. Meier and K.-P. Zeller, *Angew. Chem., Int. Ed. Engl.*, **14**, 32 (1975).
404. H. D. Roth and N. L. Mannion, *J. Am. Chem. Soc.*, **98**, 3392 (1976).
405. M. T. Nguyen, M. R. Hajnal, T. K. Ha, L. G. Vanquickenbourne and C. Wentrup, *J. Am. Chem. Soc.*, **114**, 4387 (1992).
406. A. M. Trozzolo, *Acc. Chem. Res.*, **1**, 329 (1968).
407. R. A. Hayes, T. C. Hess, R. J. McMahon and O. L. Chapman, *J. Am. Chem. Soc.*, **105**, 7786 (1983).
408. R. J. McMahon, O. L. Chapman, R. A. Hayes, T. C. Hess and H. P. Krimmer, *J. Am. Chem. Soc.*, **107**, 7597 (1985).
409. J. Fenwick, G. Frater, K. Ogi and O. P. Strausz, *J. Am. Chem. Soc.*, **95**, 124, (1973).
410. M. Barra, T. A. Fisher, G. J. Cernigliaro, R. Sinta and J. C. Scanio, *J. Am. Chem. Soc.*, **114**, 2630 (1992).
411. K. Tanigaki and T. W. Ebbesen, *J. Am. Chem. Soc.*, **109**, 5883 (1987).
412. C. Bachmann, T. Y. N'Guessan, F. Debû, M. Monnier, J. Pourein, J. P. Aycard and H. Bodot, *J. Am. Chem. Soc.*, **112**, 7488 (1990).
413. E. Lewars and S. Siddiqi, *J. Org. Chem.*, **50**, 135 (1985).
414. Y. Ogata, Y. Sawaki and T. Ohno, *J. Am. Chem. Soc.*, **104**, 216 (1982).
415. C. Marfisi, P. Verlaque, G. Davidovics, J. Pourcin, L. Pizzala, J. P. Aycard and H. Bodot, *J. Org. Chem.*, **48**, 533 (1983).
416. J. Pacansky and H. Coufal, *J. Am. Chem. Soc.*, **102**, 410 (1980).
417. A. T. Lebedev, R. N. Hayes and J. H. Bowie, *J. Chem. Soc., Perkin Trans. 2*, 1127 (1991).
418. V. Georgian, S. K. Boyer and B. Edwards, *J. Org. Chem.*, **45**, 1686 (1980).
419. S. Yamazaki, K. Kohgami, M. Okazaki, S. Yamabe and T. Arai, *J. Org. Chem.*, **54**, 240, (1989).
420. K. B. Wiberg, B. L. Furteke and L. K. Olli, *J. Am. Chem. Soc.*, **101**, 7675 (1979).
421. R. L Danheiser, R. G. Brisbois, J. J. Kowalczyk and R. F. Miller, *J. Am. Chem. Soc.*, **112**, 3093 (1990).
422. A. B. Smith, III, *J. Chem. Soc., Chem. Commun.*, 695 (1974).
423. A. B. Smith, III, B. H. Toder and S. J. Branca, *J. Am. Chem. Soc.*, **98**, 7456 (1976).
424. S. J. Branca, R. Lock and A. B. Smith, III, *J. Org. Chem.*, **42**, 3165 (1977).
425. A. B. Smith, III, B. H. Toder and S. J. Branca, *J. Am. Chem. Soc.*, **106**, 3995 (1984).
426. A. B. Smith, III, B. H. Toder, R. E. Richmond and S. J. Branca, *J. Am. Chem. Soc.*, **106**, 4001 (1984).
427. R. J. McMahon, O. L. Chapman, R. A. Hayes, T. C. Hess and H. D. Krimmer, *J. Am. Chem. Soc.*, **107**, 7597 (1985).
428. S. J. Chang, B. K. Ravi Shankar and H. Shechter, *J. Org. Chem.*, **47**, 4226 (1982).
429. J. Fenwick, G. Frater, K. Ogi, and O. P. Strausz, *J. Am. Chem. Soc.*, **95**, 124 (1973).
430. H. Tomioka, H. Okuno, S. Kondo and Y. Izawa, *J. Am. Chem. Soc.*, **102**, 7123 (1980).
431. P. G. Mahaffy, D. Visser, M. Torres, J. L. Bourdelande and O. P. Strausz, *J. Org. Chem.*, **52**, 2680 (1987).
432. O. L. Chapman, J. Gano, P. R. West, M. Regitz and G. Maas, *J. Am. Chem. Soc.*, **103**, 7033 (1981).
433. R. Leung-Toung and C. Wentrup, *J. Org. Chem.*, **57**, 4850 (1992).
434. M. Pomerantz and M. Levanon, *Tetrahedron Lett.*, **32**, 995 (1991).
435. S. R. Wilson, A. M. Venkatesan, C. E. Augelli-Szafran and A. Yasmin, *Tetrahedron Lett.*, **32**, 2339 (1991).
436. R. A. More O'Ferral, *Advan. Phys. Org. Chem.*, **5**, 331 (1967).
437. H. Dahn, H. Gold, M. Ballenegger, J. Lenoir, G. Diderich and R. Malherbe, *Helv. Chim. Acta*, **51**, 2065 (1968).
438. H. Dahn and M. Ballenegger, *Helv. Chim. Acta*, **52**, 2417 (1969).
439. W. Jugelt and L. Berseck, *Tetrahedron Lett.*, **10**, 2659 (1968).
440. C. Wentrup and H. Dahn, *Helv. Chim. Acta*, **53**, 1637 (1970).
441. P. Yates and J. D. Fenwick, *J. Am. Chem. Soc.*, **93**, 4618 (1971).
442. A. B. Smith, III and R. K. Dieter, *Tetrahedron*, **37**, 2407 (1981).
443. A. B. Smith, III, *J. Chem. Soc., Chem Common.*, 274 (1975).
444. A. B. Smith, III, B. H. Toder, S. J. Branca and R. K. Dieter, *J. Am. Chem. Soc.*, **103**, 1996 (1981).
445. A. B. Smith, III and R. K. Dieter, *J. Am. Chem. Soc.*, **103**, 2009 (1981).

446. A. B. Smith, III and R. K. Dieter, *J. Am. Chem. Soc.*, **103**, 2017 (1981).
447. M. P. Doyle and M. L. Trudell, *J. Org. Chem.*, **49**, 1196 (1984).
448. M. P. Doyle, W. E. Buhro, J. G. Davidson, R. C. Elliott, J. W. Hoekstra and M. Oppenhuizen, *J. Org. Chem.*, **45**, 3657 (1980).
449. I. J. Turchik and M. J. S. Dewar, *Chem. Rev.*, **75**, 389 (1975).
450. G. Maas, *Top. Curr. Chem.*, **137**, 75 (1987).
451. M. Kennedy, M. A. McKervey, A. R. Maguire and G. H. P. Roos, *J. Chem. Soc., Chem. Comm.*, 361 (1990).
452. M. P. Doyle, A. V. Oeveren, L. J. Westrum, M. N. Protopopova, and T. W. Clayton, Jr., *J. Am. Chem. Soc.*, **113**, 8982 (1991).
453. M. A. McKervey and T. Ye, *J. Chem. Soc., Chem. Commun.*, 823 (1992).
454. N. McCarthy, M. A. McKervey, T. Ye, M. McCann, E. Murphy and M. P. Doyle, *Tetrahedron Lett.*, **33**, 5983 (1992).
455. A. Demonceau, A. F. Noels, A. J. Hubert and P. Teyssie, *J. Chem. Soc., Chem. Commun.*, 688 (1981).
456. H. R. Sonawane, N. S. Bellur, J. R. Ahuja and D. G. Kulkarni, *J. Org. Chem.*, **56**, 1434 (1991).
457. S. Hashimoto, N. Watanabe and S. Ikegami, *Tetrahedron Lett.*, **31**, 5173 (1990).
458. M. P. Moyer, P. L. Feldman and H. Rapoport, *J. Org. Chem.*, **50**, 5223 (1985).
459. M. J. Davies, C. J. Moody and R. J. Taylor, *J. Chem. Soc., Perkin Trans. 1*, 1 (1991).
460. M. J. Davies and C. J. Moody, *J. Chem. Soc., Perkin Trans. 1*, 4 (1991).
461. S. Bien and Y. Segal, *J. Org. Chem.*, **42**, 1685 (1977).
462. W. J. Seitz, A. K. Saha, D. Caspar, and M. Hossain, *Tetrahedron Lett.*, **33**, 7755 (1992).
463. M. A. McKervey, D. N. Russell and M. F. Twohig, *J. Chem. Soc., Chem. Commun.*, 491 (1985).
464. A. Padwa and S. L. Xu, *J. Am. Chem. Soc.*, **114**, 5881 (1992).
465. S. F. Martin, C. J. Oalmann and S. Liras, *Tetrahedron Lett.*, **33**, 6727 (1992).
466. A. Padwa, U. Chiacchio, Y. Garveau, J. M. Kassir, K. E. Krumpe and A. M. Schoffstall, *J. Org. Chem.*, **55**, 414 (1990).
467. A. Padwa, D. J. Austin and S. L. Xu *J. Org. Chem.*, **57**, 1330 (1992).
468. T. R. Hoye and C. J. Dinsmore, *Tetrahedron Lett.*, **32**, 3755 (1991).
469. K. J. Doyle and C. J. Moody, *Tetrahedron Lett.*, **33**, 7769 (1992).
470. A. Padwa, K. E. Krumpe and J. M. Kassir, *J. Org. Chem.*, **57**, 4940 (1992).
471. S. Yoo, *Tetrahedron Lett.*, **33**, 2159 (1992).
472. A. Padwa, S. P. Carter and H. Nimmesgern, *J. Org. Chem.*, **51**, 1157 (1986).
473. A. Padwa, Y. S. Kulkarni and Z. Zhang, *J. Org. Chem.*, **55**, 4144 (1990).
474. M. P. Doyle, V. Bagheri and E. E. Claxton, *J. Chem. Soc., Chem. Commun.*, 46, (1990).
475. A. Padwa, S. F. Hornbuckle, G. E. Fryxell and Z. J. Zhang, *J. Org. Chem.*, **57**, 5747 (1992).
476. J. S. Clark, *Tetrahedron Lett.*, **33**, 6193 (1992).
477. C. J. Moody, E. R. H. B. Sie and J. J. Kulagowski, *Tetrahedron Lett.*, **32**, 6947 (1991).
478. C. J. Moody, E. R. H. B. Sie and J. J. Kulagowski, *Tetrahedron.*, **48**, 3991 (1992).
479. B. Corbel, D. Hernot, J. P. Haelters and G. Sturtz, *Tetrahedron Lett.*, **28**, 6605 (1987).
480. M. Regitz and H. Heydt, in *1,3-Dipolar Cycloaddition Chemistry* (Ed. A. Padwa), Vol. 1, Chap. 4, Wiley, New York, 1984.
481. G. Maas, M. Regitz, U. Moll, R. Rahm, F. Krebs, R. Hector, P. J. Stang, C. M. Crittell and B. L. Williamson, *Tetrahedron*, **48**, 3527 (1992).

CHAPTER **10**

The electrochemistry of the triple bond

HANS VIERTLER, VERA L. PARDINI and REINALDO R. VARGAS

Instituto de Química, Universidade de São Paulo, São Paulo, Brazil

I. CARBON–CARBON TRIPLE BOND

A. Introduction

Cathodic reduction or anodic oxidation of carbon–carbon triple bonds are not accomplished easily even when appropriate substitution lowers the required potentials. Therefore, if electron transfer occurs the products formed are likely to undergo electrochemical reduction or oxidation[1].

Most of the electrochemical processes studied in the last decade with substrates containing carbon–carbon triple bonds, were reductions. In many cases electrolytic generation of intermediates which attack the triple bond is involved rather than direct charge transfer. The latter requires the use of aprotic solvents, but isolated triple bonds generally are reduced/oxidized beyond the accessible potential range of the more common electrolytic solvents and only conveniently substituted alkynes may undergo direct electron transfer involving the triple bond.

Supplement C2: The chemistry of triple-bonded functional groups
Edited by S. Patai © 1994 John Wiley & Sons Ltd

B. Cathodic Reduction

The cathodic hydrodimerization of activated alkynes was reported by Kern and Schäfer[2] at a Hg electrode. The products were formed by competing hydrodimerization, hydrogenation and nucleophilic addition to the triple bond, and their distribution depends strongly on the electrolyte composition. Table 1 shows the products obtained for the substrates investigated. The data make it clear that, besides experimental conditions, minor structural differences in the substrates can also lead to a different product distribution. Thus while ethyl propynate, when electrolysed in aqueous DMF, yielded predominantly two hydrodimers besides vinyl ether, the analogous methyl derivative gave under the same experimental conditions the vinyl ether as the major product (entries 2 and 4). When reduction of ethyl propynate was performed in dry DMF (entry 3) triethyl 1,3,5-benzenetricarboxylate was the only isolated product. Substituting water for tertiary amine salts, the 3-dialkylaminoacrylates were formed in the methyl propynate electrolyses as the sole product (entries 5 and 6). Dimer formation occurred mainly through coupling at the 3-position of the examined substrates except for 4-phenyl-3-butyn-2-one (entry 9), where coupling involved the carbonyl group. No 1,3-dienes, the primary products of the hydrodimerization, were isolated because they are readily reduced to alkenes at the potentials used for the triple bond reduction. If hydrodimerization of alkynes is considered for dimer preparation, it is likely to be necessary to perform several trial runs to establish the best experimental conditions.

Peters and coworkers[3–6] investigated the electrochemical reduction of several haloalkynes on a Hg cathode in DMF solutions. Their results are summarized in Table 2. The interest in the electrochemistry of haloalkynes is due to the possibility that certain acetylenic halides may undergo intramolecular cyclizations, a reaction which has been observed when this type of compounds were treated with organo-metallic reducing agents[3]. The halogens play an important role in these electrochemical reductions since 1-phenyl-1-hexyne (entry 1) under similar conditions afforded only hydrogenation products. For 6-chloro-1-phenyl-hexyne[5] (entry 2) the controlled potential electrolysis in the region of the first polarographic wave, which was attributed to the reduction of the carbon–carbon triple bond, afforded benzylidene-cyclopentane as the major product by an intramolecular cyclization of a radical anion formed through one-electron transfer to the triple bond. When the substrate concentration was increased (entry 3) the acetylene isomerized during electrolysis to an allene, 6-chloro-1-phenyl-1,2-hexadiene, which was reduced at a more positive potential ($E_{1/2} = -2.25$ V). It is worth mentioning that this kind of isomerization was also observed for several alkynes bearing sulphonyl, sulphinyl and sulphonium groups[7], where the corresponding allenes also showed more positive $E_{1/2}$ potentials. The electrochemical behaviour of the bromo- and iodo-derivatives[6] was markedly different, because their first polarographic wave corresponds to the reduction of the carbon–halogen bond. In both cases the product distribution obtained in their electrolyses was potential dependent (entries 4–7). Electrochemical cleavage of the carbon–halogen bonds was the only reaction occurring at less negative potentials, but some direct electroreduction involving the carbon–carbon triple bond cannot be ruled out when conducting the experiments at more negative potentials. Although the major reduction products present essentially the same trend as a function of potential, there are several significant differences regarding their absolute yields. Higher yields of both 1-phenyl-1-hexyne and the N-methylformamide adduct as well as a lower one for benzylidene cyclopentane are found with the bromo compound.

1-Bromo- and 1-iodo-5-decyne[3] were studied because they do not possess electro-active carbon–carbon triple bonds and therefore the electrolytic cleavage of the carbon–halogen bonds can be investigated over a relatively wide range of potentials. The reduction product distribution proved to be dependent upon the potentials employed during the

TABLE 1. Electrochemical reduction of activated alkynes at mercury cathode[2]

Compound	Solvent/Potential (V vs SCE)/F mol^{-1}	Main Products (%)
(1) MeO$_2$CC $\equiv$ CCO$_2$Me	DMSO–H$_2$O(3:1v/v)/–1.05/1.6	MeO$_2$CCH$_2$C(CO$_2$Me)=C(CO$_2$Me)CH$_2$CO$_2$Me, $cis,trans$ 1:1 (46); MeO$_2$CCH=CHCO$_2$Me $cis,trans$ 1:1 + MeO$_2$C(CH$_2$)$_2$CO$_2$Me (13); MeO$_2$CCH$_2$CH(OAc)CO$_2$Me (21); polymers (20)
(2) CH$\equiv$CCO$_2$Et	DMF–H$_2$O(96:4v/v)/–1.80/1.01	EtO$_2$CCH$_2$CH=CHCH$_2$CO$_2$Et-$cis,trans$ + EtO$_2$CC(Me)=CH-CH$_2$CO$_2$Et-$cis,trans$ + EtO$_2$CCH=CH–O–CH=CHCO$_2$Et-$trans,trans$ 1:0.7:0.4 (50)
(3) CH$\equiv$CCO$_2$Et	DMF/–1.80.1.01	1,3,5-C$_6$H$_3$(CO$_2$Et)$_3$ (60)
(4) CH$\equiv$CCO$_2$Me	DMF–H$_2$O(96:4v/v)/–1.80/1.01	$trans,trans$-MeO$_2$CCH=CH–O–CH=CHCO$_2$Me (66)
(5) CH$\equiv$CCO$_2$Me	DMF–Et$_3$NHClO$_4$/–1.80/1.01	Et$_2$NCH=CHCO$_2$Me (80)
(6) CH$\equiv$CCO$_2$Me	DMF–Bu$_3$NHClO$_4$/–1.80/1.01	Bu$_2$NCH=CHCO$_2$Me (83)
(7) MeC$\equiv$CCO$_2$Et	DMF–H$_2$O(95:5v/v)/–2.2/1.6	(E,Z)-EtO$_2$CCH$_2$CH$_2$CH(Me)C(Me)=CHCO$_2$Et + (E,Z)-EtO$_2$CCH$_2$CH$_2$C(Me)=C(Me)CH$_2$CO$_2$Et (46)
(8) PhC$\equiv$CCO$_2$Et	DMF/–1.9/1.1	(E or Z)-EtO$_2$CCH$_2$CCH$_2$CPh=CPhCH$_2$CO$_2$Et (29), (E,Z)-EtO$_2$CCH$_2$CCH$_2$CPh=C(CO$_2$Et)CH$_2$Ph (13)
(9) PhC$\equiv$CCOMe	DMF–H$_2$O(9:1v/v)/–1.3/1.44	PhC$\equiv$CC(OH)(Me)C(OH)(Me)C$\equiv$CPh (38) + other dimers

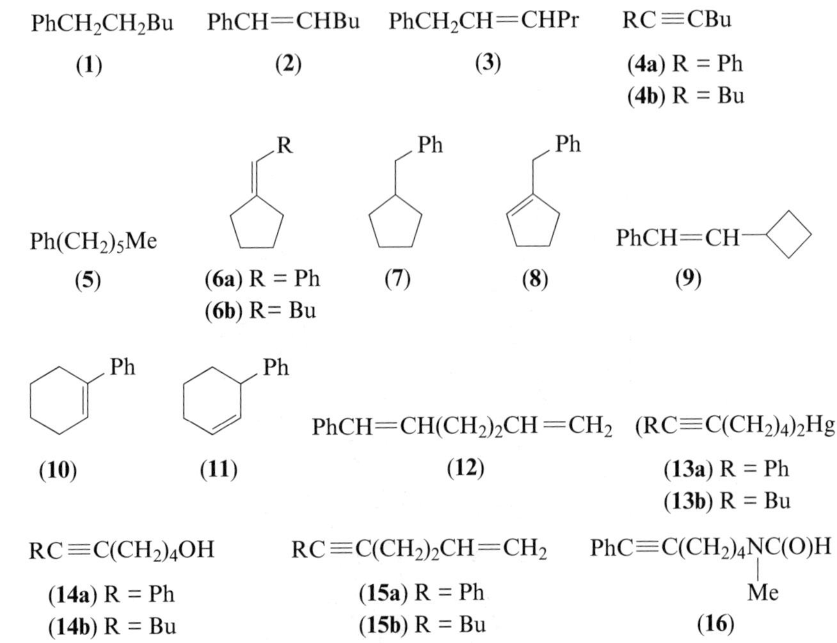

$PhCH_2CH_2Bu$　　　$PhCH=CHBu$　　　$PhCH_2CH=CHPr$　　　$RC\equiv CBu$

(1)　　　　　　　　(2)　　　　　　　　(3)　　　　　　　　(4a) R = Ph

　　　　　　　　　　　　　　　　　　　　　　　　　　　　(4b) R = Bu

$Ph(CH_2)_5Me$

(5)　　　　(6a) R = Ph　　(7)　　　　(8)　　　　(9)

　　　　　　(6b) R = Bu

(10)　　　　　(11)

$PhCH=CH(CH_2)_2CH=CH_2$　　$(RC\equiv C(CH_2)_4)_2Hg$

(12)　　　　　(13a) R = Ph

　　　　　　　(13b) R = Bu

$RC\equiv C(CH_2)_4OH$　　　$RC\equiv C(CH_2)_2CH=CH_2$　　$PhC\equiv C(CH_2)_4NC(O)H$

(14a) R = Ph　　　　(15a) R = Ph　　　　　　　　　　Me

(14b) R = Bu　　　　(15b) R = Bu　　　　　　　　　　(16)

FIGURE 1. Electrolysis products of haloalkynes (see Table 2)

TABLE 2. Electrochemical reduction of haloalkynes in DMF at Hg cathode[a]

Compound		Potential (V vs SCE)	Products (%)[b]	Reference
(1)	$PhC\equiv CBu$	-2.82^c	1(22), cis/trans-2(61), trans-3(9)	4
(2)	$PhC\equiv C(CH_2)_4Cl$	-2.50^d	4a(4), cis/trans-2(5), trans-3(2), 5(1), 6a(81), 7(6)	5
(3)	$PhC\equiv C(CH_2)_4Cl$	-2.50	8(36), 6a(30), 7(13), trans-9(8), 10(4), trans-3(3), trans-12(2), 11(1), trans-2(3)	5
(4)	$PhC\equiv C(CH_2)_4Br$	-2.25	13a(43), 4a(35), 14a(12), 16(9), 15a(4), 6a(1)	6
(5)	$PhC\equiv C(CH_2)_4Br$	-2.55	4a(32), 16(17), 6a(11), 15a(6), 13a(3)	6
(6)	$PhC\equiv C(CH_2)_4I$	-2.05	13a(63), 6a(24), 14a(4), 15a(2)	6
(7)	$PhC\equiv C(CH_2)_4I$	-2.45	4a(18), 6a(11), 16(14), 14a(2), 15a(3), 13a(9)	6
(8)	$BuC\equiv C(CH_2)_4Br$	-2.05	13b(72), 4b(12), 6b(3), 15b(2), 14b(1)	3
(9)	$BuC\equiv C(CH_2)_4I$	-2.45	4b(42), 15b(21), 13b(13), 14b(2), 6b(1)	3
(10)	$BuC\equiv C(CH_2)_4Br$	-2.00	13b(65), 4b(17), 15b(2), 6b(1)	3
(11)	$BuC\equiv C(CH_2)_4Br$	-2.45	4b(50), 15b(24), 13b(2), 14b(1)	3

[a]Hg/DMF containing Bu_4NClO_4 or Me_4NClO_4; substrate concentration 2.5 mM.
[b]See Figure 1 for the structures of the compounds.
[c]Substrate concentration 5 mM.
[d]Substrate concentration 0.25 mM.

electrolyses, but no major differences were observed between 1-bromo- and 1-iodo-5-decynes (entries 8–11).

It was observed[8] that phenylacetylene presented two 2-electron polarographic waves at -1.937 and -2.030 V (vs Hg pool), the latter with polarographic characteristics of styrene.

TABLE 3. Polarographic data for the electrochemical reduction of aryl phenylethynyl chalcogenides[8,9]

4-RC$_6$H$_4$EC≡CC$_6$H$_5$ E	R	$-E_{1/2}$ (V vs Hg pool)	n	$-E_{1/2}$ (V vs Hg pool)	n
S	MeO	1.618	2.0	1.930	1.5
Se	MeO	1.450	2.0	1.948	0.94
S	Me	1.598	1.8	1.930	1.4
Se	Me	1.436	1.92	1.946	0.93
Te	Me	0.837	1.96	1.973	3.95
S	H	1.568	1.9	1.930	1.5
Se	H	1.426	1.91	1.950	0.78
Te	H	0.816	1.97	1.992	3.6
S	F	1.544	2.0	1.928	1.6
Se	F	1.374	2.0	1.943	0.9
Te	F	0.778	2.0	2.006	4.0
S	Cl	1.525	2.1	1.929	1.6
Se	Cl	1.343	2.0	1.957	0.97
Te	Cl	0.771	1.98	1.998	4.7
S	Br	1.566	1.7	1.959	1.5
Te	Br	0.743	1.99	1.998	6.0

The two waves merged into one 4-electron wave in the presence of proton donors. Substituting H by phenylchalcogen groups[8,9] (4-RC$_6$H$_4$E, where E = S, Se, Te) the potential values for the first wave decreased, possessing the order for the anodic shifts Te>Se>S (Table 3) following the order of C$_{sp}$—E bond strengths. This reduction process led to the formation of an arenechalcogenolate anion and an arylacetylene as a result of the cleavage of the C$_{sp}$—E bond by an ECE (Electron transfer–Chemical reaction–Electron transfer) mechanism. The second wave was due to the triple bond of the formed phenylacetylene.

When the polarographic reduction of phenyl(4-methylphenyltelluro)acetylene[10] was examined in MeCN in the presence of benzoic acid, the first reduction wave corresponding to the 2-electron electrochemical cleavage of the C$_{sp}$—Te bond was progressively superseded by the 4-electron process of the hydrogenation of the triple bond. This reduction has been interpreted as involving an electron transfer to the adsorbed substrate yielding an anion radical, which after protonation and further reduction would form the saturated telluride (Scheme 1).

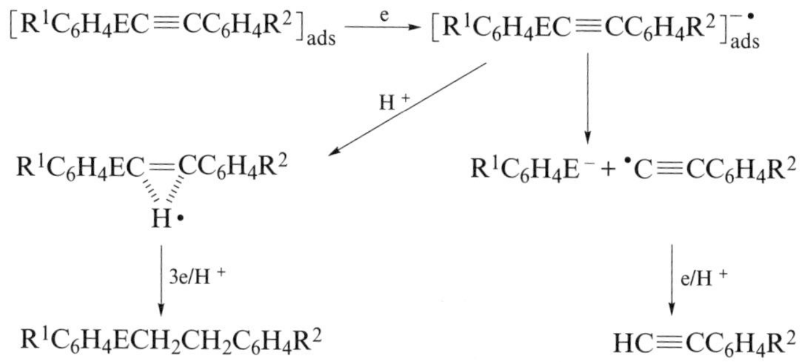

$$[R^1C_6H_4EC≡CC_6H_4R^2]_{ads} \xrightarrow{e} [R^1C_6H_4EC≡CC_6H_4R^2]^{-\bullet}_{ads}$$

$$R^1C_6H_4EC=CC_6H_4R^2 \qquad R^1C_6H_4E^- + {}^\bullet C≡CC_6H_4R^2$$
$$H\bullet$$

$$\Big| 3e/H^+ \qquad\qquad\qquad \Big| e/H^+$$

$$R^1C_6H_4ECH_2CH_2C_6H_4R^2 \qquad\qquad HC≡CC_6H_4R^2$$

SCHEME 1

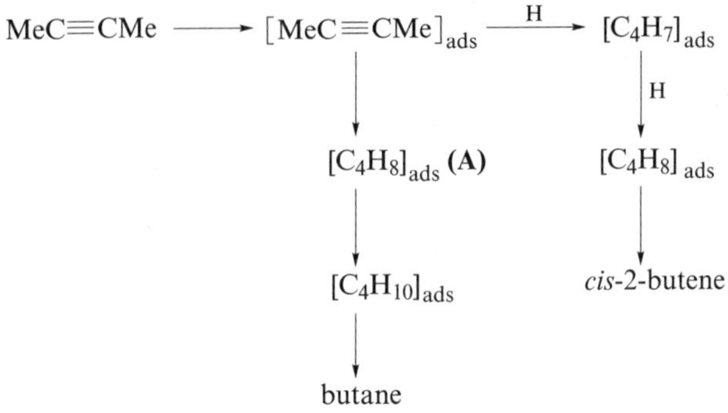

$$MeC\equiv CMe \longrightarrow [MeC\equiv CMe]_{ads} \xrightarrow{\text{H}} [C_4H_7]_{ads}$$

$$\downarrow \qquad\qquad\qquad\qquad \downarrow \text{H}$$

$$[C_4H_8]_{ads} \text{ (A)} \qquad\qquad [C_4H_8]_{ads}$$

$$\downarrow \qquad\qquad\qquad\qquad \downarrow$$

$$[C_4H_{10}]_{ads} \qquad\qquad cis\text{-2-butene}$$

$$\downarrow$$

$$\text{butane}$$

SCHEME 2

The electrochemical reduction of 2-butyne on platinized Pt electrodes in aqueous H_2SO_4 solutions yielded butane and cis/trans-2-butene as major products[11]. The product distribution was dependent on the applied potential and was related to the adsorbed hydrogen on the electrode. When the potential was 0.15 V (vs NHE), cis-2-butene was the major product and its formation was explained as occurring through two sequential hydrogen addition steps to the adsorbed butyne. At potentials smaller than 0.05 V, butane was the main product and the results of the electroreduction replacing H_2SO_4 by D_2SO_4 showed that butane formation did not involve cis-2-butene once formed, but probably further reduction of the 1-butene-like intermediate (A) adsorbed on the electrode surface (Scheme 2).

Hydrogen electrogenerated and adsorbed on a Raney Ni electrode consisting of a Ni plate covered with Ra/Ni catalyst reduced the carbon–carbon triple bond of 1-heptyne and diphenylacetylene mainly to the corresponding alkanes[12] (Table 4). Diphenylacetylene was also hydrogenated on Pd black and Pt black cathodes[13] and the product distributions were compared with the catalytic hydrogenation using the same catalyst. Except for Pt black in basic solution, no significant differences were observed. Under those conditions 82% yield of stilbene was obtained against only 20% for the catalytic reaction. Dimethyl acetylene-dicarboxylate afforded mainly dimethyl succinate when the reduction was conducted using a Pd black cathode in both acidic and alkaline media. Comparing the cis/trans ratios of the unsaturated compounds formed by cathodic reduction and catalytic hydrogenation, one concludes that the former are smaller than the latter. Both substrates were also electrolysed on a Hg pool cathode in both media and only the diester was significantly reduced. It is noteworthy that only trans-isomer was formed under these conditions (Table 4, entries 10–11) suggesting that in this case a direct reduction via radical anion is involved, contrary to the metal black electrodes where adsorbed hydrogen is the reducing species.

An interesting method to synthesize methylene cyclopentanols was developed by Shono and coworkers[14] electrolysing γ-ethynylketones with terminal triple bonds in anhydrous DMF containing tetraethylammonium tosylate at a carbon rod electrode (Table 4). When an alkyl group was introduced at the terminal C atom of the triple bond, a decrease in the yield of cyclic product was observed and acetylenic alcohols [$RC\equiv C(CH_2)_3C(OH)(Me)Et$, R = Me or Et] were formed as by-products. Since both the carbonyl group and the triple bond are only reduced at very negative potentials, it seems reasonable to assume that an electrocatalytically generated 'tetraethylammonium amalgam' is involved in this reaction as was suggested for the electrochemical cyclization of keto-olefins[15].

The use of electrogenerated bases was reported for the deconjugative alkylations of α,β-acetylenic esters by electrolysing mixtures of the esters and alkyl iodides at a Pt cathode in HMPA or DMF solutions containing tetrabutylammonium iodide[16,17]. Some examples

TABLE 4. Electrochemical hydrogenation of alkynes

Compound	Electrode/solvent	Products (%)	Reference
(1) MeC≡CMe	Pt/H_2SO_4–H_2O	cis-MeCH=CHMe;C_4H_{10}	11
(2) HexC≡CH	Ra-Ni/MeOH–MeONa	HexCH=CH_2 (11); n-C_8H_{18} (39)	12
(3) PhC≡CPh	Ra-Ni/MeOH–MeONa	cis-PhCH=CHPh (22); $PhCH_2CH_2Ph$ (60)	12
(4) PhC≡CPh	Pd black/MeOH–1M H_2SO_4	PhCH=CHPh (68) cis/trans: 6.1; $PhCH_2CH_2Ph$ (20)	13
(5) PhC≡CPh	Pd black/MeOH–2M NaOH	PhCH=CHPh (60) cis/trans: 4.8; $PhCH_2CH_2Ph$ (9)	13
(6) PhC≡CPh	Pt black/MeOH–1M H_2SO_4	PhCH=CHPh (27) cis/trans: 9; $PhCH_2CH_2Ph$ (32)	13
(7) PhC≡CPh	Pt black/MeOH–2M NaOH	PhCH=CHPh (82) cis/trans: 16; $PhCH_2CH_2Ph$ (2)	13
(8) MeO_2CC≡CCO_2Me	Pd black/MeOH–1M H_2SO_4	MeO_2CCH=$CHCO_2Me$ (13) cis/trans: 1.4; $MeO_2CCH_2CH_2CO_2Me$ (35)	13
(9) MeO_2CC≡CCO_2Me	Pd black/MeOH–2M NaOH	MeO_2CCH=$CHCO_2Me$ (trace); $MeO_2CCH_2CH_2CO_2Me$ (13)	13
(10) MeO_2CC≡CCO_2Me	Hg pool/MeOH–1M H_2SO_4	trans-MeO_2CCH=$CHCO_2Me$ (32); $MeO_2CCH_2CH_2CO_2Me$ (30)	13
(11) MeO_2CC≡CCO_2Me	Hg pool/MeOH–2M NaOH	trans-MeO_2CCH=$CHCO_2Me$ (27); (85–95)	13
(12) HC≡$CCH_2CR^1CH_2COR^2$	C/DMF–Et_4NTos		14

R^1 = H, Me; R^2 = Me, Et, i-Pr, Bu

TABLE 5. Electrochemical alkylation of α,β-acetylenic esters[16,17]

Compound	Solvent/RI	Products (%)^a
(1) $HC{\equiv}CCO_2Et$	DMF/MeI	$MeC{\equiv}CCO_2Et$ (15); $H_2C{=}C{=}CMeCO_2Et$ (14); $HC{\equiv}CCMe_2CO_2Et$ (15); $MeC{\equiv}CCMe_2CO_2Et$ (37)
(2) $HC{\equiv}CCO_2Et$	HMPA/MeI	$MeC{\equiv}CCO_2Et$ (17); $HC{\equiv}CCMe_2CO_2Et$ (17); $MeC{\equiv}CCMe_2CO_2Et$ (50)
(3) $MeC{\equiv}CCO_2Et$	HMPA/MeI	$HC{\equiv}CCMe_2CO_2Et$ (19); $MeC{\equiv}CCMe_2CO_2Et$ (37);
(4) $EtC{\equiv}CCO_2Et$	DMF/MeI	$MeC{\equiv}CCMe_2CO_2Et$ (71)
(5) $EtC{\equiv}CCO_2Et$	DMF/EtI	$MeC{\equiv}CCEt_2CO_2Et$ (59)
(6) $EtC{\equiv}CCO_2Et$	HMPA/BuI	$MeC{\equiv}CCBu_2CO_2Et$ (89)
(7) $EtC{\equiv}CCO_2Et$	DMF/All I	$MeC{\equiv}CC(All)_2CO_2Et$ (74)
(8) $BuC{\equiv}CCO_2Et$	DMF/MeI	$PrC{\equiv}CCMe_2CO_2Et$ (72)

^{a}Based on consumed acetylenic ester, divided cell, constant current electrolysis, 0.1 A cm^{-2}, Pt electrode, ratio acetylenic ester/RI 1:10.

are presented in Table 5. Voltammetric studies showed that the half-wave reduction potentials of the esters are more negative than that of methyl iodide on both dropping Hg and Pt electrodes. Removal of an acetylenic hydrogen by the electrogenerated base (B) formed in the reduction of the iodide would give the acetylide anion which reacted with alkyl halide affording the products (Scheme 3).

Tokuda and coworkers[18] also reported the electrochemical alkylation of phenylacetylene with alkyl halides (MeI, EtBr, EtI, n-BuI) in HMPA/Bu$_4$NI, in good yields. Secondary halides and benzyl and allyl bromides did not give the expected products. It was suggested

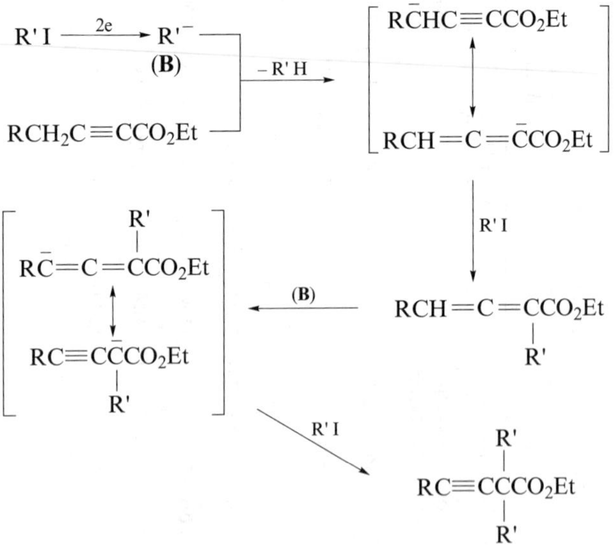

SCHEME 3

Initiation: $\qquad$ $C_4F_9I \xrightarrow{\text{e}} C_4F_9{}^{\bullet} + I^-$

Propagation:

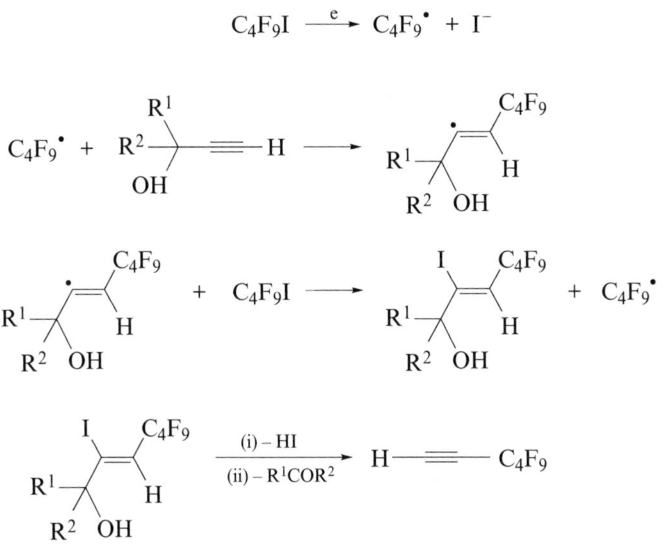

$R^1 = H$ or Me, $R^2 = H$, Me or Et

(i) KOH, MeOH, room temp., (ii) NaOH, heat

SCHEME 4

that this cathodic reaction probably proceeds via the acetylide anion, which may be formed by 1-electron reduction of the acetylene. Although several experimental evidences are given supporting the direct reduction, the quoted reduction potential of –0.95 V vs SCE for phenylacetylene in THF seems unlikely[1,19], meaning that an electrogenerated base could be involved[17].

An efficient electrochemical route to (perfluoroalkyl)alkynes, in which an electrocatalytic addition of perfluoroalkyliodides ($R = C_4F_9$, C_6F_{13}) to 3-hydroxy-3-alkylbut-1-ynes to produce iodoalkenes was the key step, has been described by Calas and coworkers[20]. The reaction was performed on a carbon fiber cathode at constant current in an emulsion of the

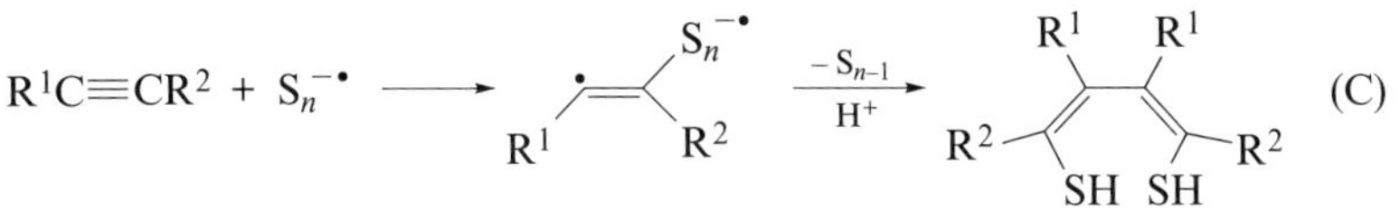

R^1	R^2	$D\,(\%)$
CO_2Et	H	10
CO_2Me	Ph	85
CHO	Ph	82
CN	Ph	80
CO_2Me	CO_2Me	32 (+ 53% of C)

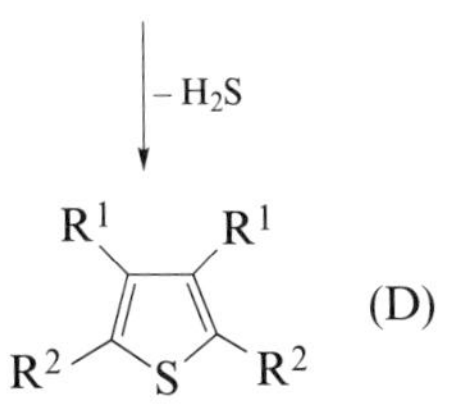

SCHEME 5

iodide, alkynol and water containing KCl. The addition involves a radical chain mechanism[21], which was suggested based on preparative scale electrolyses and cyclic voltammetry experiments (Scheme 4).

Activated acetylenes, when reduced at the cathode consisting of an intimate mixture of sulphur and carbon, afforded substituted thiophenes in good yields[19]. The potential applied (–0.9V vs SCE) was less negative than the reduction potentials of the substrates (–1.3 to –2.0V vs SCE), making a direct electron transfer to the substrates highly unlikely. Therefore, it was suggested that sulphur would be the initial electroactive species and that the reaction was initiated by attack of a sulphur radical anion on the acetylene (Scheme 5).

Bromoacetals containing carbon–carbon triple bonds were cyclized to 5-membered heterocyclic rings (5-*exo*-exclusively) using [chloropyridine-bis(dimethylglioximato)] cobalt(I) (cobaloxime I) generated from cobaloxime III as an electron transfer reagent (mediator) in alkaline methanol solution at constant current[22]. The most efficient molar substrate/mediator ratio was 2:1 with 70–87% yields of isolated products. The proposed

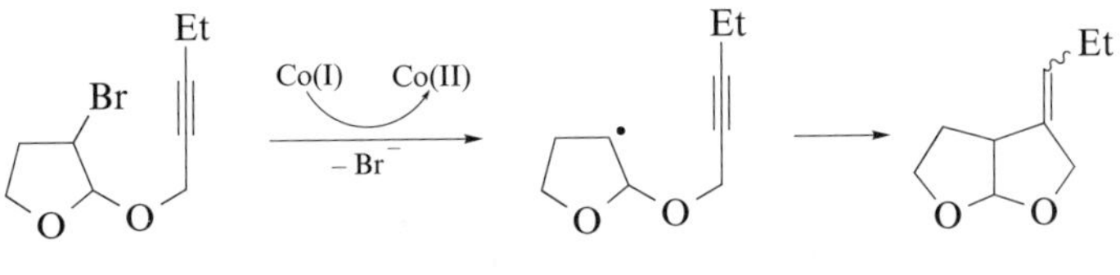

SCHEME 6

mechanism involves an electroreductive cleavage of the carbon–bromine bond by cobaloxime I, generating a radical which cyclizes and a Co(II) species which completes the catalytic cycle by cathodic reduction to cobaloxime (I). An example of this reaction is in Scheme 6.

It is noteworthy that this mediated type of cyclization is very effective and therefore could be a convenient alternative route for direct electroreductive cyclization for haloalkynes studied by Peters and coworkers[3–6].

The direct electrolysis of diphenylacetylene in the presence of CO_2 gave both diphenylmaleic anhydride and diphenylfumaric acid in low yields and, although this incorporation of CO_2 represents an interesting possibility for the formation of C–C bonds, the yields observed were discouraging[1]. The limitation presented by the direct reduction

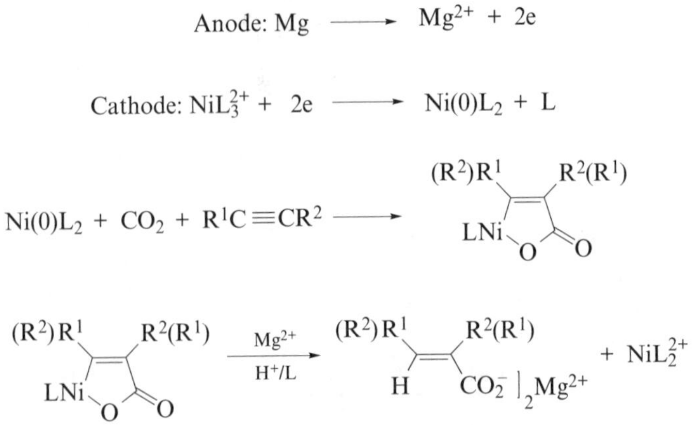

$$\text{Anode: Mg} \longrightarrow \text{Mg}^{2+} + 2e$$

$$\text{Cathode: NiL}_3^{2+} + 2e \longrightarrow \text{Ni(0)L}_2 + L$$

SCHEME 7

TABLE 6. Ni(0) catalysed electrochemical carboxylation of alkynes[a,b]

$R^1C{\equiv}CR^2$		Temp. (°C)	Products (%)[c]		
R^1	R^2		$HO_2CCR^1{=}CHR^2$	$R^1CH{=}CR^2CO_2H$	Others
(1) Pr	H	45	38	7	—
(2) Hex	H	20	54	6	—
(3) c-Pent	H	50	45.5	4.5	—
(4) $NC(CH_2)_4$	H	50	36	4	—
(5) $Cl(CH_2)_4$	H	40	30	16	—
(6) $AcO(CH_2)_3$	H	70	32.5	32.5	—
(7) Ph	H	20	25	10	$Ph(CH_2)_2CO_2H$ (6)
(8) Pr	Pr	20	75(E)/11(Z)		$PrCH(CO_2H)CH(CO_2H)Pr$ (8)
(9) Ph	Ph	80	21(E,Z)		$PhCH_2CH(CO_2H)Ph$ (42) $PhCH(CO_2H)CH(CO_2H)Ph$ (6) (E,Z)$PhCH{=}CHPh$ (25)
(10) Ph	Me	20	25	40	$PhCH(CO_2H)CH(CO_2H)Me$ (7)
(11) Ph	CO_2Et	5	20		$PhCH(CO_2H)CH(CO_2H)CO_2Et$ (55)

[a] Reference 23a–c.
[b] Conditions: $Ni(bpy)_3(BF_4)_2$, CO_2 at atmospheric pressure.
[c] Yields based on converted alkynes.

was circumvented by the use of electrogenerated Ni(0) complexes causing a simultaneous activation of CO_2 and alkynes[23], diynes[24] and 1,3-enynes[25]. The electrolyses were run in a one-compartment cell using a carbon fibre cathode and a sacrificial magnesium anode in DMF solutions containing Bu_4NBF_4 under controlled-current conditions. Carbon dioxide was bubbled into the solution at atmospheric pressure or at 5 atm and the solution temperature varied between 20–65 °C. The substrate/Ni complex ratio was 10:1 and, among several complexes studied, two deserved more attention: $Ni(bpy)_3(BF_4)_2$ (bpy = 2,2'-bipyridine) and $NiBr_2$, DME + 2 PMDTA (PMDTA = N,N,N',N'',N''-pentamethyl-diethylene triamine). Based on cyclic voltammetry and electrolysis studies, a mechanistic cycle was proposed (Scheme 7), where the presence of the magnesium salt was shown to be important for the hydrolysis of 1-oxa-2-nickelocyclopentenone. The difference in stability of the intermediate metallocycles is probably responsible for the observed regioselectivity of the carboxylation. Several examples of the electrochemical carboxylation of alkynes are presented in Table 6. A high regioselectivity was observed for the carboxylation of terminal alkynes when the triple bond was substituted by alkyl groups (entries 1–3). The presence of functional groups in the alkyl chains may cause a decrease in regioselectivity (entries 4–6). The regiocontrol was less effective with phenylacetylene (entry 7), probably due to an electronic effect of the phenyl ring, and some reduction of the carbon–carbon double bond of the carboxylated products was observed (see also entry 9).

A selective cis-carboxylation was obtained with a dialkyl substituted acetylene (entry 8). With a phenyl and an alkyl substituent (entry 10) preferential carboxylation occurred at the alkylated carbon. Electron-deficient alkynes led to the saturated dicarboxylic acids (entry

TABLE 7. Electrocarboxylation of some diynes in the presence of Ni–PMDTA complex[a,b]

$R^1C{\equiv}C(CH_2)_nC{\equiv}CR^2$			Major products (%)[c]
R^1	R^2	n	
(1) Pen	Pen	0	(E)-PenC≡CC(CO₂H)=CHPen (58); (E)-PenC≡CCH=C(CO₂H)Pen (25)
(2) c-Pen	c-Pen	0	(E)-c-PenC≡CC(CO₂H)CHc-Pen (86)
(3) Ph	Ph	0	(E)-PhC≡CC(CO₂H)=CHPh (40)
(4) PhOCH₂	PhOCH₂	0	(E)-PhOCH₂C≡CC(CO₂H)=CHCH₂OPh (88)
(5) Pen	H	1	PenC≡CCH₂C(CO₂H)=CH₂ (63)
(6) Bu	H	2	BuC≡C(CH₂)₂C(CO₂H)=CH₂ (77); (E)-BuC≡C(CH₂)₂CH=CHCO₂H (8)
(7) Bu	H	8	BuC≡C(CH₂)₈C(CO₂H)=CH₂ (39); (E)-BuC≡C(CH₂)₈CH=CHCO₂H (19)
(8) Me	Me	2	(E)-MeC≡C(CH₂)₂C(CO₂H)=CHMe (36); (E)-MeC≡C(CH₂)₂CH=C(CO₂H)Me (54)
(9) H	H	4	HC≡C(CH₂)₄C(CO₂H)=CH₂ (59)

(6)

| (10)[d] H | H | 4 | HC≡C(CH₂)₄C(CO₂H)=CH₂ (24); |

[a]Reference 24a–b.
[b]CO₂–5 atm, 20 °C.
[c]Yields based on converted diynes.
[d]CO₂–1 atm, 65 °C.

11). If the metallocycles are intermediates in the electrocarboxylation, one would expect stereospecific *cis*-addition to the triple bond, but it was shown that a *cis/trans* isomerization of the addition products occurred in the presence of Ni(0) species[23a].

1,3-Diynes were carboxylated with high regioselectivity and complete stereoselectivity (Table 7, entries 2–5). The presence of one terminal triple bond in non-conjugated diynes made the carboxylation a totally chemospecific reaction with decreasing regioselectivity as the number of methylene groups between the triple bonds increases (entries 5–7). Experimental conditions may have a striking influence on the course of the reaction as can be seen for 1,7-octadiyne (entries 9–10). Thus changing the CO_2 pressure and the solution temperature, a decrease of the carboxylic acid was observed and a dimer (see Table 7, entry 10) became an important product. It is worthwhile mentioning that the nature of the ligand

TABLE 8. Electrochemical carboxylation of 1,3-enynes[a,b]

Compound	Complex[c]	Products (%)[d]
(1) RC≡CH	1	RC(CO$_2$H)=CH$_2$ (84) RCH=CHCO$_2$H (9)
(2) RC≡CH	2	RC(CO$_2$H)=CH$_2$ (68) RCH=CHCO$_2$H (7) 1,3,5-R$_3$C$_6$H$_5$ (25)
(3) CH$_2$=CMeC≡CBu	1	(*E*)-CH$_2$=CMeC(CO$_2$H)=CHBu (43); (*E*)-CH$_2$=CMeCH=C(CO$_2$H)Bu (43)
(4) CH$_2$=CMeC≡CBu	2	(*E*)-CH$_2$=CMeC(CO$_2$H)=CHBu (36); (*E*)-CH$_2$=CMeCH=C(CO$_2$H)Bu (36); (*E,Z*)-CH$_2$=CMeC(CO$_2$H)=C(CO$_2$H)Bu (7)
(5) (*Z*)-PenCH=CHC≡CPen	1	PenCH=CHC(CO$_2$H)=CHPen (52); (*Z*) (*E*) PenCH=CHCH=C(CO$_2$H)Pen (22) (*Z*) (*E*)
(6) (*Z*)-PenCH=CHC≡CPen	2	PenCH=CHC(CO$_2$H)=CHPen (38); (*Z*) (*E*) PenCH=CHCH=C(CO$_2$H)Pen (25); (*Z*) (*E*) Pen-CH=CHC(CO$_2$H)=C(CO$_2$H)Pen (27) (*Z*) (*E,Z*)
(7) (*E*)-PenCH=CHC≡CPen	1	PenCH=CHC(CO$_2$H)=CHPen (40); (*E*) (*E*) PenCH=CHCH=C(CO$_2$H)Pen (40); (*E*) (*E*)
(8) (*E*)-PenCH=CHC≡CPen	2	PenCH=CHC(CO$_2$H)=CHPen (26); (*E*) (*E*) PenCH=CHCH=C(CO$_2$H)Pen (47); (*E*) (*E*) PenCH=CHC(CO$_2$H)=C(CO$_2$H)Pen (18) (*E*) (*E,Z*)

[a] Reference 25.
[b] CO_2–5 atm, 20°C.
[c] (1) Ni–PMDTA; (2) Ni(bpy)$_3$.
[d] Yields based on converted enynes.

in the nickel complex has an important influence on the course of this reaction[24a]. When PMDTA was substituted by Ph_3P and the reaction conducted at 1 atm of CO_2 and 65 °C, 1,7-octadiyne afforded the aforementioned dimer in 70% yield.

The nickel catalysed electrochemical carboxylation was also applied to 1,3-enynes[25], being chemoselective in relation to the triple bond and occurring through a stereoselective cis-addition. PMDTA and bpy again behaved as efficient ligands for the incorporation of CO_2, yielding monocarboxylic acids as main products. When bpy was the ligand, dicarboxylation occurred to some extent. Table 8 summarizes the results. As was observed with other alkynes, high regioselectivity was obtained with a terminal triple bond (entries 1–2) whereas it was lost for internal triple-bond carboxylation (entries 3–4). The configuration of the conjugated double bond was maintained during the stereocontrolled addition on the triple bond and slightly affected the regioselectivity, which was shown to be different for the two nickel complexes used (entries 6–8). The regiochemical course was examined at the metallocycle stage and considered to be mainly determined by steric factors.

A promising electrochemical hydrogenation of diphenylacetylene to 1,2-diphenylethane was described by Coche and Moutet[26] using a catalytic cathode consisting of a surface-confined pyrrole–viologen polymer on carbon felt impregnated with Pd. The substrate was hydrogenated with electrolytically generated hydrogen in 98% yield.

C. Anodic Oxidation

Alkynes are more difficult to oxidize electrochemically than alkenes[1]. Most of the reported examples of preparatively significant electrochemical oxidations of acetylenes do not involve initial electron transfer from the triple bond.

The anodic oxidation of acetylene at a gold anode in aqueous solution, although having been studied in great detail[1], was reinvestigated by voltammetry and controlled potential electrolysis[27]. In addition to CO_2, acetic acid, glyoxal, glycolaldehyde, formaldehyde and acetaldehyde were detected and yields determined. It was shown that acetaldehyde was formed mainly by an electrochemical route and not via an acid-catalysed hydration. A solid deposited on the anode which previously had been identified as an organic polymer[28], was actually a mixture containing gold oxides and salts, and it was suggested that formation of these could have occurred through an intermediate complex like gold acetylide. Based on experimental evidences a mechanism involving the oxidation of the metal was proposed (Scheme 8).

$$(C_2H_2Au)_{ads} \xrightarrow[-H^+]{-e} (C_2HAu)_{ads} \longrightarrow ? \longrightarrow Au^0, Au^{3+}, MeCO_2H$$

$$+ \text{ other products}$$

SCHEME 8

The catalytic behaviour of Pt–Ru anodes in aqueous solution was examined with respect to oxidation of acetylene[29]. It was suggested that the electrocatalytic activity observed may be due to oxygen or oxides formed on the electrodes facilitating the oxidation of the intermediate species formed during the initial stage of the acetylene oxidation.

The direct anodic oxidations of phenylacetylene, vinylacetylene and 1,4-diphenylbutadiyne were compared with the oxidation of the respective alkenes in MeCN at a carbon anode[30]. The measured $E_{1/2}$ vs SCE were ca 300–600 mV more positive than those of the corresponding alkenes and the oxidation was completely irreversible for phenyl- and vinylacetylenes but only nearly reversible for 1,4-diphenylbutadiyne. The difference in $E_{1/2}$ for the former two alkynes, determined on several anode materials, was interpreted as involving adsorbed substrates for the first charge transfer of an ECEC mechanism yielding an intermediate

$$PhC\equiv C-C\equiv CPh \xrightarrow[-e]{MeOH} PhC(OMe)_2C\equiv CC(OMe)_2Ph \quad (80\%)$$

$$PhC\equiv CH \xrightarrow[-e]{MeOH} PhC(OMe)_2CH=CHC(OMe)_2Ph$$

$$+ \; PhC(OMe)=CHCH(OMe)C(OMe)_2Ph \; + \; PhC(OMe)_2CH(OMe)_2$$

$$MeOH \Big| -e$$

$$PhC(OMe)_3 \; + \; HC(OMe)_3$$

SCHEME 9

radical cation. It was suggested that the chemical reaction of this radical cation would be responsible for the final composition of the reaction mixture. HMO calculations did not provide a precise prediction of the reactivity of the intermediate radical cation except for the highly conjugated diphenylbutadiyne. When the oxidation was conducted in methanol several products, shown in Scheme 9, were identified.

The oxydation of acetylenic and related steroid hormones was examined in aqueous MeOH solution by several electrochemical techniques at Hg electrodes[31]. The electrochemical behaviour was strongly affected by reactant and product adsorption and a mercury acetylide compound was formed during the controlled potential electrolysis.

A convenient method for chlorinating the triple bonds of alkynes was developed by Simonet and coworkers[32] oxidizing chloride ion in the presence of acetylenic substrates in DMF or MeCN at a polished Pt anode. Through cyclic voltammetry experiments the authors observed that the peak corresponding to the chloride oxidation was not modified in the presence of the acetylenes but the chlorine reduction peak, although still present, became smaller with decreasing sweep rates. At $10 \, mV \, s^{-1}$ a clear splitting of the substrate oxidation peak (phenylacetylene) was observed with the more anodic one identified as corresponding to the oxidation of dichlorostilbene. Table 9 summarizes the results for the anodic chlorination under controlled potential conditions of the studied alkynes. The formate obtained in the electrolyses of phenyl- and diphenylacetylenes (entries 1–2) was explained in terms of the solvent leading to an intermediate iminium salt which hydrolysed during the work-up procedure, whereas for the chlorinated ketones (entries 1 and 2) the intermediate was a diiminium salt as shown in Scheme 10 for phenylacetylene. The product in entry 6 was a result of migration of the phenyl group of an intermediate vinyl carbocation.

This chlorination is an example of an electrochemical reaction involving electrogenerated reagents. As the anodic oxidation of alkynes occurs only at more positive potentials than

TABLE 9. Anodic chlorination of alkynes in DMF^{32}

Compound	Charge (F mol^{-1})	Products (%)
(1) PhC≡CH	4	PhC(OCHO)=CHCl (20); PhCOCHCl$_2$ (54); PhC≡CCl (26)
(2) PhC≡CPh	2	PhC(OCHO)=CClPh (84)
(3) PhC≡CPh	4	PhCCl$_2$COPh (68)
(4) HOCH$_2$C≡CH	2–4	HOCH$_2$CCl=CHCl (50)
(5) PhCH(OH)C≡CH	2–4	(E)-PhCH(OH)CCl=CHCl (70)
(6) Ph$_2$C(OH)C≡CH	4	PhCOCPh=CHCl (50)

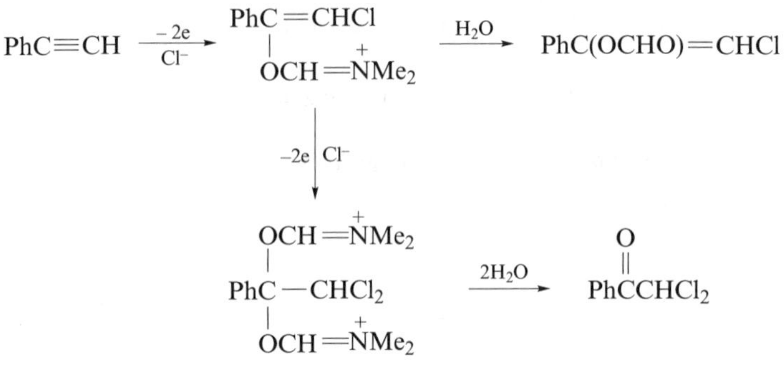

$$PhC\equiv CH \xrightarrow[Cl^-]{-2e} \underset{OCH=\overset{+}{N}Me_2}{\overset{PhC=CHCl}{|}} \xrightarrow{H_2O} PhC(OCHO)=CHCl$$

$$\downarrow -2e \mid Cl^-$$

$$\underset{OCH=\overset{+}{N}Me_2}{\overset{OCH=\overset{+}{N}Me_2}{\underset{|}{\overset{|}{PhC-CHCl_2}}}} \xrightarrow{2H_2O} \overset{O}{\overset{\|}{PhCCHCl_2}}$$

SCHEME 10

the chloride ion oxidation, the latter was oxidized preferentially to Cl_2, the electrophile of the observed addition reaction.

α-Arylseleno-α,β-unsaturated aldehydes and ketones[33] were formed in 62–94% yields by electrochemical oxidation of 4,4'-dichlorodiphenyl diselenide in the presence of α-hydroxy alkynes in MeCN/H_2O solutions at a Pt anode. The hydrolysis of an intermediate selenirenium-type cation formed by the reaction of anodically generated 4-chlorophenyl selenenic acid with the alkyne would afford an enol which, by dehydration, gave the final products (Scheme 11). The regioselective addition was ascribed to a strict control by the alkyne hydroxyl group.

A further example[34] of an indirect electrooxidation of alkynes is their transformation into 1,2-diones when catalytic amounts of RuO_4 are generated anodically from RuO_2 in a two-phase system (CCl_4-saturated aqueous NaCl solution). Overoxidation to carboxylic acids is a minor reaction for disubstituted acetylenes, but terminal triple bonds are cleaved

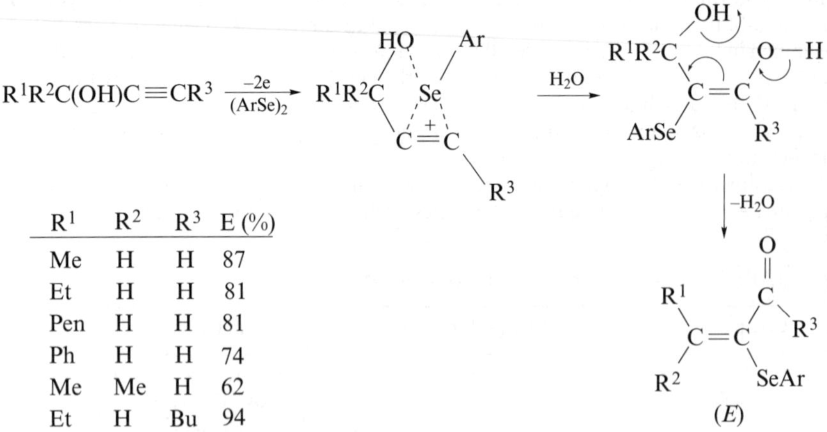

$$R^1R^2C(OH)C\equiv CR^3 \xrightarrow[(ArSe)_2]{-2e} R^1R^2\underset{\overset{+}{C}=C}{\overset{/}{C}}\overset{HO\quad Ar}{\underset{\diagdown R^3}{\overset{/}{Se}}} \xrightarrow{H_2O} \underset{ArSe}{\overset{R^1R^2C}{\overset{/}{C}=\overset{/}{C}}}\overset{OH}{\overset{O-H}{\underset{R^3}{}}}$$

R^1	R^2	R^3	E (%)
Me	H	H	87
Et	H	H	81
Pen	H	H	81
Ph	H	H	74
Me	Me	H	62
Et	H	Bu	94
Et	H	Ph	90
Me	Me	Ph	65

$$\downarrow -H_2O$$

$$\underset{R^2}{\overset{R^1}{\overset{}{C}}}=\underset{SeAr}{\overset{C}{\overset{\|}{\underset{}{C}}}}\overset{O}{\underset{R^3}{}}$$

$$(E)$$

SCHEME 11

$$R^1C{\equiv}CR^2 \xrightarrow[\text{RuO}_2]{-2e} \overset{\overset{\displaystyle O}{\displaystyle \|}}{R^1C} - \overset{\overset{\displaystyle O}{\displaystyle \|}}{CR^2} + R^1CO_2H$$

R^1	R^2	Yield (%)	
Bu	Bu	75	7
Hex	Hex	83	7
Ph	Et	73	10
Pen	CH$_2$OAc	69	29
Hex	CH$_2$OAc	80	11
Ph	H	—	59
Hex	H	—	76

SCHEME 12

and carboxylic acids are formed exclusively. No mechanistic rationalization was given. Some examples are shown in Scheme 12.

II. CARBON–NITROGEN TRIPLE BOND

A. Cyano Group

1. Introduction

Oxidations and reductions involving electron transfer from or to the cyano group in organic molecules are very difficult and are strongly dependent on the experimental conditions and substrate structures. Due to its electron-withdrawing ability the presence of the cyano group facilitates reduction and hinders oxidation, as already pointed out by Yoshida[35].

2. Cathodic reduction

Several possible pathways for the electrochemical reduction of cyano derivatives are depicted in Scheme 13. The mechanism to be followed is greatly influenced by the structure of R and the electrolysis conditions.

Electrochemical methods for the reduction of nitriles to the corresponding amines may be considered as an alternative to catalytic hydrogenation and therefore have received some attention. Cathode material plays an important role and, in many cases, an electrocatalytical hydrogenation involving adsorbed hydrogen seems to be the mechanistic pathway.

The electrochemical transformations of nitriles into primary amines was reviewed by Antonova and coworkers[36], giving special attention to the influence of the cathode material, temperature and current densities on the yield of the product in aqueous solution.

Raney-Ni powder as cathode material was used by Chiba and coworkers[12] to electrohydrogenate pentanenitrile, heptanenitrile, adiponitrile, benzonitrile and phenylacetonitrile in alkaline methanol to the corresponding amines in ca 70% yield. The same electrode was used for the electrochemical synthesis[37] of aminonitriles starting with adiponitrile and azelanitrile. In the case of adiponitrile a through study of experimental conditions revealed that significantly lower current densities than those used previously[12],

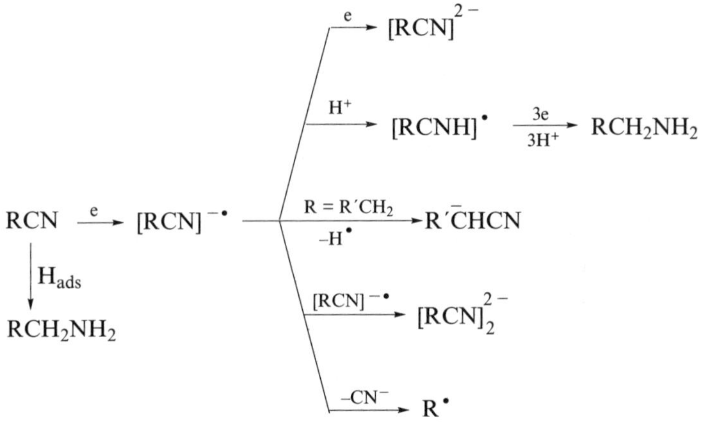

$$RCN \xrightarrow{e} [RCN]^{-\bullet}$$

$$\xrightarrow{e} [RCN]^{2-}$$

$$\xrightarrow{H^+} [RCNH]^{\bullet} \xrightarrow[3H^+]{3e} RCH_2NH_2$$

$$\xrightarrow[-H^{\bullet}]{R = R'CH_2} R'\bar{C}HCN$$

$$\xrightarrow{[RCN]^{-\bullet}} [RCN]_2^{2-}$$

$$\xrightarrow{-CN^-} R^{\bullet}$$

$$RCN \downarrow H_{ads}$$

$$RCH_2NH_2$$

SCHEME 13

in ethanol/water–NH$_4$OAc electrolyte at 35–45 °C, afforded 6-aminohexanenitrile with selectivities in the range of 79–97%. Two by-products, hexamethylenediamine and hexamethyleneimine, were observed. For azelanitrile, the corresponding 9-aminononanenitrile was obtained with similar selectivities. Attempts to reduce selectively adiponitrile to the diamine were made using deposited Ni black on graphite in aqueous acidic solutions[38], but all experimental conditions gave 6-aminohexanenitrile as a by-product. When the same electrode was employed to reduce p-tolunitrile[39] on aqueous ethanolic ammonium sulphate medium, p-methylbenzylamine was obtained in 74% yield. Better yields (84%) were achieved using a Pd black electrode in ethanolic hydrochloric acid.

Electrocatalytic hydrogenation of several nitriles was investigated using Ra-Ni or Pd/C powders placed onto a Ni plate[40] and Table 10 summarizes some of these results. On Ra-Ni, the reduction occurred more efficiently than on Pd/C. With the latter, higher selectivity for nitro group reduction in the presence of cyano group was achieved (entries 5–6).

Antonova and coworkers[41] reported the electroreduction of 4-amino-5-cyano-2-methylpyrimidine to the corresponding amine in acidic aqueous medium. A competitive pathway due to hydrolysis of the intermediate aldimine may occur yielding an aldehyde or an alcohol (Scheme 14). A Ni cathode in HCl solution containing PdCl$_2$ and substrate at 20 °C and current densities of 2A dm^{-2} gave the maximum yield (84%) of the expected amine. It is noteworthy that Pd is electrodeposited on the electrode at the beginning of the

TABLE 10. Electrocatalytic hydrogenation of nitriles[40]

Compound	Conditions	Products (%)
(1) PhCN	1 M NaOMe/MeOH/NH$_3$ /Ra/Ni	PhCH$_2$NH$_2$ (97)
(2) PenCN	0.5 M NaOMe/MeOH/NH$_3$ /Ra-Ni	PenCH$_2$NH$_2$ (91)
(3) PenCN	0.5 M NaOMe/MeOH/NH$_3$ /Pd/C	PenCH$_2$NH$_2$ (39)
(4) t-BuCN	1 M NaOMe/MeOH/NH$_3$ /Ra-Ni	t-BuCH$_2$NH$_2$ (96)
(5) 4-O$_2$NC$_6$H$_4$CN	0.5 M NaClO$_4$/H$_2$O–THF(1:9) t-BuCO$_2$H/Ra-Ni	4-H$_2$NC$_6$H$_4$CN (66)
(6) 4-O$_2$NC$_6$H$_4$CN	0.5 M NaClO$_4$/H$_2$O–THF(1:9) t-BuCO$_2$H/Pd/C	4-H$_2$NC$_6$H$_4$CN (77)

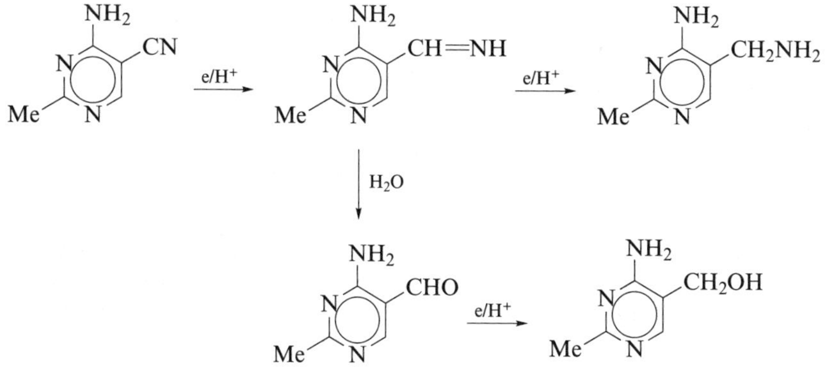

SCHEME 14

electrolysis, thus acting as a surface catalyst which has a longer useful lifetime than the films electrodeposited from $HCl/PdCl_2$ aqueous solution, before adding the substrate.

Glassy carbon felt electrodes modified by electrodeposited poly(pyrrole-viologen) films containing electroprecipitated microparticles of precious metals like Pt, Pd, Rh or Ru have been shown to be suitable for the electrocatalytical hydrogenation of several organic substrates including benzonitrile, in acidic aqueous solution[42]. Pd exhibited the highest current efficiency and yields for benzylamine formation when compared with Pt and Rh.

Phthalonitrile was reduced at controlled potential in alcoholic solution at a C cathode in a two-compartment cell in which cathode and anode were separated by a Nafion cation-exchange membrane[43]. The anolyte contained metallic salts such as $CuSO_4$, $NiSO_4$, etc. The products formed in the cathode compartment were metal phthalocyanine complexes. Low yields were obtained when the metal ions were placed into the cathode compartment. The phthalonitrile radical anion formed by a 1-electron transfer was invoked as being important in the cyclization process leading to the metal complexes. Several intermediates

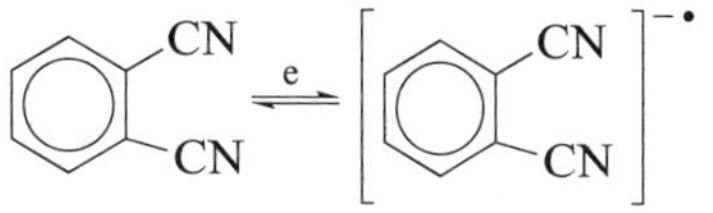

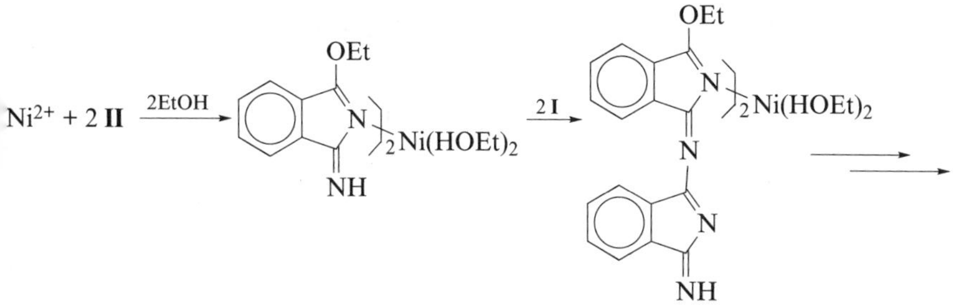

SCHEME 15

towards nickel phthalocyaninate formation were separated and characterized[44] mainly by mass spectrometry. The alcohols used as solvents play an important role allowing higher yields of the Ni complex with fewer by-products (Scheme 15).

The cathodic behaviour of aliphatic nitriles was examined in aprotic media using Fe and Pt electrodes and tetraalkylammonium salts as supporting electrolytes[45]. Cyclic voltammetric experiments made with Bu_4NBF_4/MeCN solutions at Pt showed that MeCN was reduced to the cyanomethyl anion without participation of the supporting electrolyte. The formation of the anion was confirmed by preparative, controlled potential electrolyses using an Fe or Pt cathode and Mg anode in a one-compartment cell. The insoluble magnesium derivative of MeCN formed during the electrolysis when treated with MeI afforded propanenitrile. Similar results were obtained with propanenitrile at an Fe cathode leading to 2-methylbutanenitrile after treatment with EtI. Phenylacetonitrile and malononitrile were also reduced to the respective carbanions, which were then alkylated.

Solvated electrons generated electrochemically in HMPA[46] were used to reduce several nitriles to amines. Under these conditions acetonitrile and acrylonitrile gave ethylamine and propylamine, respectively, but with dinitriles (adiponitrile and malononitrile) mixtures of diamines and cyanoamines were obtained and their ratios depended upon the excess of solvated electrons produced.

The electrode reduction mechanism of benzenodicarbonitrile isomers was examined by polarography, cyclic voltammetry and controlled potential electrolysis in DMF solutions at a Hg cathode[47]. 1,2- and 1,4-dicyanobenzenes were reduced in two successive steps under polarographic conditions, where the first step corresponds to a quasi-reversible one-electron transfer. Cyclic voltammetric experiments provided more information on the electrode reduction mechanism and allowed one to suggest the mechanistic scheme for 1,2- and 1,4-dicyanobenzenes shown in Scheme 16.

$$
\begin{array}{ccc}
 & -E_{1,2}^0 \;(\text{V } vs \text{ Ag/AgCl}) & -E_{1,4}^0 \\
C_6H_4(CN)_2 \xrightleftharpoons{e} \left[C_6H_4(CN)_2\right]^{-\bullet} & 1.32 & 1.25 \\
\left[C_6H_4(CN)_2\right]^{-\bullet} \xrightleftharpoons{e} \left[C_6H_4(CN)_2\right]^{2-} & 2.35 & 2.20
\end{array}
$$

$$\left[C_6H_4(CN)_2\right]^{2-} + HA \longrightarrow C_6H_5CN + CN^- + A^-$$

SCHEME 16

Controlled potential electrolysis at a Hg pool cathode in the absence of proton donors at potentials corresponding to the first electron transfer consumed 1e/molecule, yielding the radical anion with no C—CN bond breaking. When proton donors were added under otherwise similar conditions, 1,2-dicyanobenzene gave benzonitrile and cyanide ion (2e/

$$1,2\text{-}C_6H_4(CN)_2^{-\bullet} + HA \longrightarrow 1,2\text{-}C_6H_4(CN)_2H^{\bullet} + A^-$$

$$\downarrow 1,2\text{-}C_6H_4(CN)_2^{-\bullet}$$

$$1,2\text{-}C_6H_4(CN)_2H^- + 1,2\text{-}C_6H_4(CN)_2$$

$$\downarrow$$

$$C_6H_5CN + CN^-$$

SCHEME 17

molecule), whereas the 1,4-isomer consumed 4e/molecule but C—CN bond cleavage was prevented. An explanation was offered based on the difference of basicities of the radical anions obtained from both nitriles, where the radical anion derived from the 1,2-isomer is more basic. These results can be accounted for by the reactions in Scheme 17.

When the reductions were performed in the presence of H_2O at potentials corresponding to the second electron transfer, both nitriles consumed 4e/molecule giving benzene and cyanide ion, from benzonitrile which was the first reduction product. For 1,3-dicyano-benzene, the existence of a single quasi-reversible one-electron transfer step yielding a radical anion was observed which, through a radical-radical coupling reaction, underwent a fast dimerization[47,48]. Under controlled potential electrolysis conditions in the presence of up to 1 M H_2O, 1e/molecule was required but neither benzonitrile nor cyanide were found among the products. When phenol or acetic acid were used as proton donors, 2e/molecule were consumed and C—CN bond cleavage was observed. In MeCN solutions[49] the electrochemical reduction of the three benzenedicarbonitrile isomers was essentially the one observed for DMF solutions.

Rotating ring-disk electrodes made of amalgamated gold were used to examine the electroreduction of benzonitrile, p-tolunitrile, 1,2-, 1,3- and 1,4-dicyanobenzenes and 1,2,4,5-tetracyanobenzene, in DMSO solution containing Bu_4NBF_4 as supporting electrolyte[50]. The radical anions formed at the first cathodic wave by a one-electron transfer were detected on the ring for all nitriles except for 1,3-dicyanobenzene, in which case cyanide ion or a dianion obtained via fast coupling of a radical anion was suggested. When benzonitrile was used as a ligand in bis(arene)chromium complexes, $(C_6H_5CN)Cr(C_6H_5X)$ where X = H, Me, MeO, CN, F, Cl, CF_3, COMe, its behaviour changed completely during the cathodic reduction. The first one-electron polarographic wave changed into a two-electron wave, which was interpreted as a reversible transfer of the first electron followed by chemical reaction of the radical anion, whose products were reduced at the same potential. CN^- was detected at the ring electrode, unless the substituent of the other ligand could be cleaved preferentially. The observed destabilization of the nitrile radical anion in the

Initiation:

$$PhCN \xrightarrow{e} PhCN^{-\bullet}$$

Propagation:

$$PhCN^{-\bullet} + PhBr \longrightarrow PhCN + PhBr^{-\bullet}$$

$$PhBr^{-\bullet} \longrightarrow Br^- + Ph^\bullet$$

$$Ph^\bullet + PhS^- \longrightarrow Ph_2S^{-\bullet}$$

$$Ph_2S^{-\bullet} + PhBr \longrightarrow Ph_2S + PhBr^{-\bullet}$$

and/or

$$Ph_2S^{-\bullet} + PhCN \longrightarrow Ph_2S + PhCN^{-\bullet}$$

Termination:

$$Ph^\bullet + HA \longrightarrow PhH + A^\bullet$$

(HA: DMSO or BuN_4^+)

SCHEME 18

complex as compared to the free one was explained by different spin density distributions.

The radical anion derived from benzonitrile was used as a redox catalyst for electrochemical initiation of $S_{NR}1$ reactions between bromobenzene and Bu$_4$NSPh in DMSO[51]. Good yields of diphenyl sulphide were obtained and the low consumption of charge/mol of bromobenzene was indicative of a chain reaction. It is noteworthy that benzonitrile is reversibly reduced by one electron at *ca* 500 mV less negative potentials than bromo benzene (Scheme 18).

Polarographic techniques were used to examine the electrochemical reduction of 4-cyano-1-methylpyridinium ion in MeCN[52]. Two reduction waves were observed: the first, with $E_{1/2}$ of +0.066 V (*vs* Ag/Ag$^+$), corresponds to a reversible one-electron process whereas the second one at –1.669 V, equal in height to the first, was in accord with a reversible

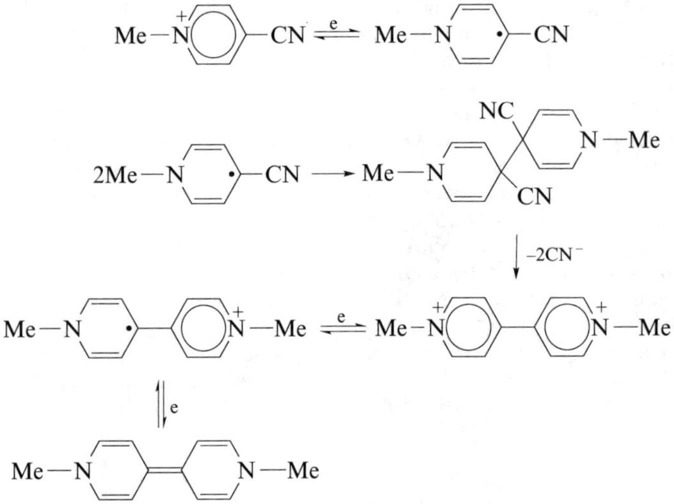

SCHEME 19

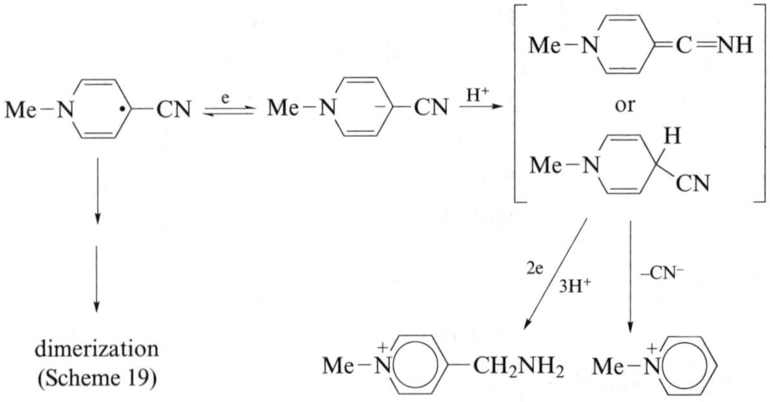

SCHEME 20

electron transfer followed by a fast chemical reaction. Controlled potential coulometry[53] at -1.05 V (1.09 F mol^{-1}) reduced the 1,1'-dimethyl-4,4'-bispyridinium dication to a radical cation in 71% yield as determined by UV spectroscopy, but a sharp decrease in yield was observed at -1.8 V. This was attributed to a further reduction of the radical cation. As a consequence of time scale differences in polarographic measurements and controlled potential electrolysis, the proposed reduction mechanisms for both techniques were diverse. The radical formed in the first electron transfer step, although stable under polarographic conditions, dimerized during the electrolysis and, after loss of cyanide ions, produced the 4,4'-bispyridinium dication which was further reduced under the experimental conditions (Scheme 19).

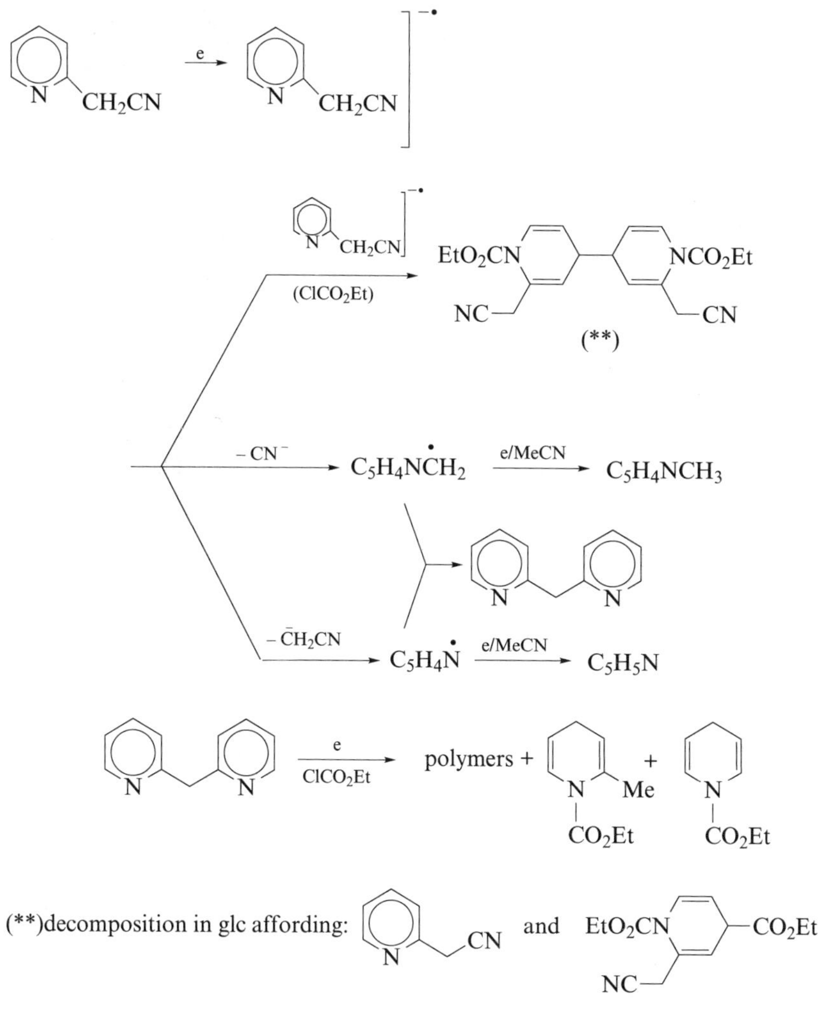

SCHEME 21

In aqueous solution, the electrochemical behaviour of 4-cyano-1-methylpyridinium ion was dependent on pH[54]. Polarographic measurements exhibited one well-defined wave (4e) at pH < 7 which split into two waves at higher pH. Electrolyses of solutions buffered at pH 2–6, involving a four-electron process, gave the 4-amino-1-methylpyridinium ion. When the pH was in the range 8–10 the reduction at the potential of the first wave consumed between 1–1.5 electrons/molecule, and 1,1'-dimethyl-4,4'-bispyridinium radical cation was the product, whereas at the second wave involving 2–3.5 electrons/molecule 4-aminomethyl-1-methylpyridinium and 1-methylpyridinium ions were obtained. The formation of the radical cation was explained as shown in Scheme 19, but for the other products a further reduction of the intermediate radical was proposed (Scheme 20). Similar behaviour was observed for the 2-cyano-1-methylpyridinium ion[55].

2-, 3-, and 4-cyanomethylpyridines were also reduced in MeCN containing Et$_4$NCl at a Hg cathode[56]. Several products were characterized after reaction with ethyl chloroformate, and their formation was rationalized as in Scheme 21 for the 2-cyano derivative.

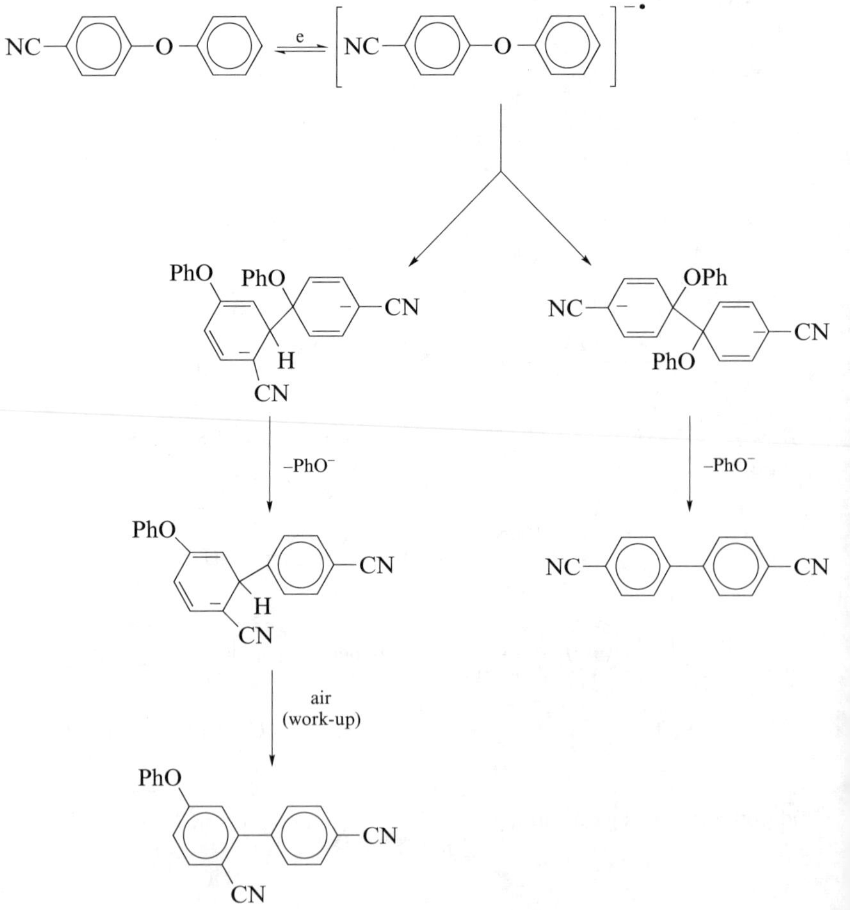

SCHEME 22

Radical anions and dianions were also observed in the electroreduction of 2- and 4-cyanopyridines in liquid ammonia at a Pt cathode[57]. Under similar conditions the 3-cyano derivative afforded a radical anion which underwent a fast dimerization.

Cyano substituents have important influences on the electrochemical reductive cleavage of C—O bonds of 2-, 3- and 4-cyanoanisoles in DMF[58]. Under cyclic voltammetric conditions reversible electron transfer, generating radical anions, was observed for 2- and 3-cyanoanisoles, but for the 4-isomer the wave was chemically irreversible at low sweep rates. C—O bond breaking was observed during controlled potential electrolyses of the 2- and 4-compounds resulting in cyanophenols and methane in good yields in a two-electron process. On the contrary, decyanation and C—O bond cleavage occurred with low efficiency for 3-cyanoanisole. It was shown that the electrochemically generated radical anions had reacted by different reaction pathways.

The investigation on C—O bond cleavage was followed up with the reduction of 2- and 4-cyanodiphenyl ethers in DMF[59]. 4-Cyanodiphenyl ether was reduced to an unstable radical anion and then transformed by a concentration-dependent pathway. Electrolyses gave phenol, 4,4'-dicyanobiphenyl and 2,4'-dicyano-5-phenoxybiphenyl as major products. These results were rationalized by 4-4 and 2-4 couplings of the radical anion generated by electron transfer (Scheme 22).

The radical anion derived from 2-cyanodiphenyl ether underwent reversible dimerization resulting in a dianion relatively resistant towards further chemical reaction. The products observed after electrolysis were mainly phenol, diphenyl ether and 2',4-dicyano-3-phenoxybiphenyl. The structure of the isolated biphenyl suggested that the radical anion dimerization had occurred via 2-4 coupling.

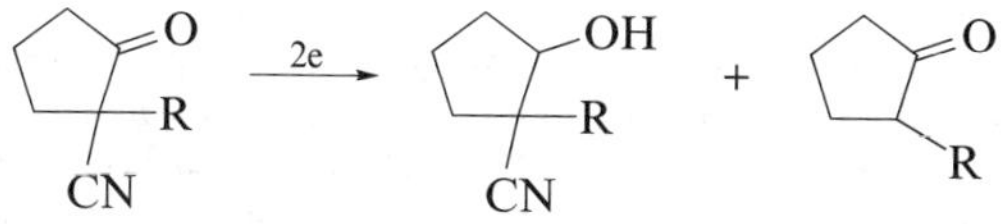

R		Yields (%)	
Me	LiCl	47	45
	Et$_4$NI	10	87
HC≡CCH$_2$	Et$_4$NI	20	65

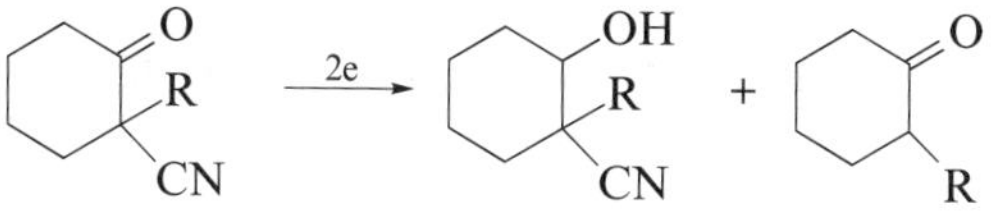

R		Yields (%)	
Me	LiCl	40	53
	Et$_4$NI	15	75
H$_2$C=CHCH$_2$	LiCl	48	30
	Et$_4$NI	10	75

SCHEME 23

The reactivity of cyanide ion with aryl radicals generated electrochemically in liquid ammonia from 2-, 3- and 4-chlorobenzonitriles was reported by Amatore and coworkers[60]. The coupling between the aryl radicals and the nucleophile appears to be the key step of a $S_{RN}1$ reaction. In these reactions the C—Cl bond was cleaved in the radical anion formed by electroreduction of the chlorobenzonitrile, giving a radical that then reacted with cyanide to yield dicyanobenzenes. The steps involved in the reaction sequence are described in Yoshida's review[35].

Cleavage of the CN group was also observed in aliphatic compounds[61]. 2-Alkyl-2-cyanocycloalkanones were reduced at a Hg cathode in aqueous/ethanol solution and, besides the expected 2-hydroxy nitriles, 2-alkylcycloalkanones were obtained. It was observed that the bond cleavage did not depend on the working potential but on the nature of the supporting electrolyte (Scheme 23). It should be emphasized that this cleavage was not observed in the absence of the 2-alkyl substituent. The mechanism suggested is presented in Scheme 24.

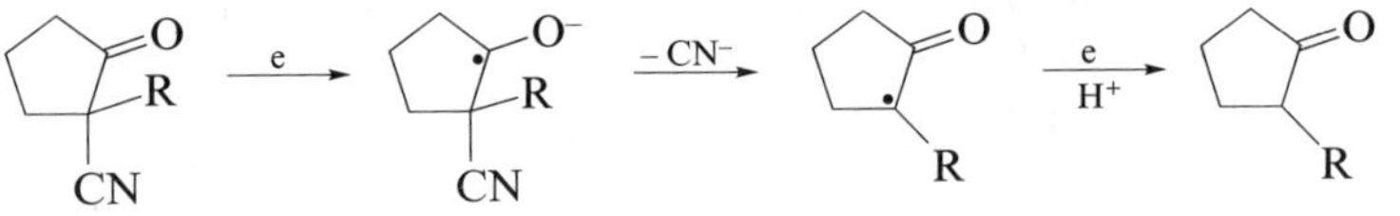

SCHEME 24

A variety of aliphatic nitriles were decyanated by electrochemical reduction[62]. Zn cathodes in DMF containing Et_4NTos as supporting electrolyte were shown to afford the best experimental condition to conduct this reaction. Similar results were obtained with Mg or W electrodes, but with Pt or Pb a sharp decrease in yields was observed. Substituting LiCl for Et_4NTos caused the electrodeposition of Li metal on the cathode and, since the starting material was recovered, this electroreduction is not the same as the Birch-type reduction. α-Amino and α-alkoxynitriles were also decyanated. Their ease of reduction as well as that of benzyl nitriles was explained in terms of stabilization of a radical intermediate by the adjacent alkoxy, amino or phenyl group. The proposed mechanism is illustrated in Scheme 25 and some selected examples of this reduction are given in Table 11. It should be mentioned that due to the very negative reduction potentials of nitriles, direct electron transfer from the electrode to the substrate may not be involved and an alternative route should be considered[15].

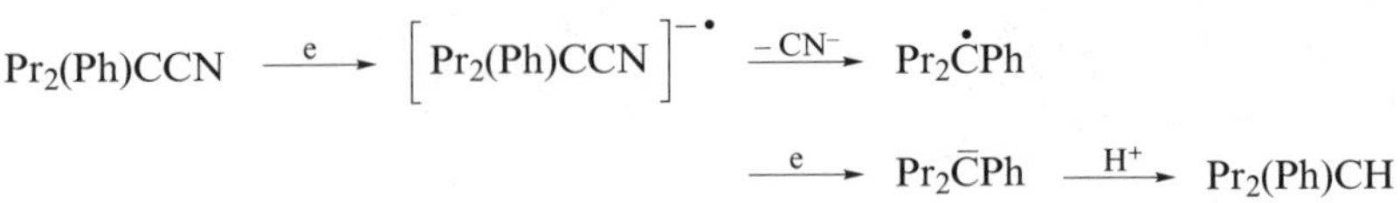

$$Pr_2(Ph)CCN \xrightarrow{e} \left[Pr_2(Ph)CCN \right]^{-\bullet} \xrightarrow{-CN^-} Pr_2\overset{\bullet}{C}Ph$$

$$\xrightarrow{e} Pr_2\overset{-}{C}Ph \xrightarrow{H^+} Pr_2(Ph)CH$$

SCHEME 25

The electroreductive behaviour of nitriles containing functional groups has received some attention[63]. Several bromoalkanenitriles were examined at Hg or vitreous C cathode in DMF solutions containing Et_4NClO_4, by polarography, cyclic voltammetry and controlled potential electrolysis. Depending on the position of Br in relation to the cyano group, cyclic nitriles could result from the electrochemical reduction. Based on polarographic and voltammetric data as well as electrolysis product analysis (Table 12), it was shown that electroreductive cleavage of the C—Br bond occurred yielding a carbanion

TABLE 11. Some electroreductive decyanations of nitriles[62]

Compound	Product (%)[a,b]
(1) $C_{11}H_{23}CN$	$C_{11}H_{24}$ (50[c])
(2) $C_{10}H_{21}CEt_2CN$	$C_{10}H_{21}CHEt_2$ (80)
(3) $Pr_2(Ph)CCN$	$PhCHPr_2$ (85)

(4) → $N-CO_2Me$ (75[c])

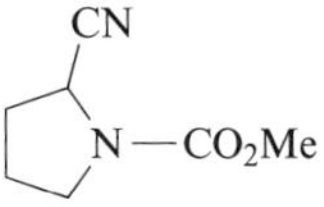

(5) → $N-CO_2Me$ (89[c])

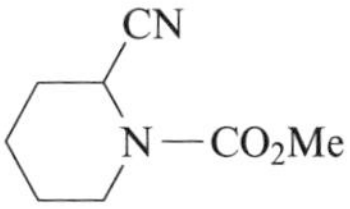

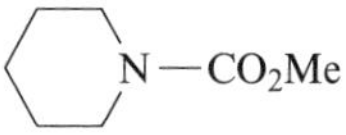

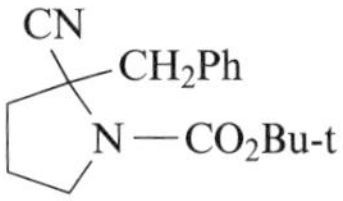

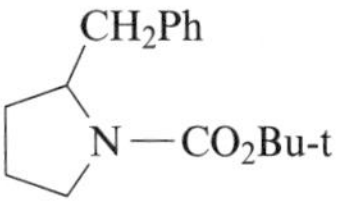

(6[d]) → $N-CO_2Bu$-t (85)

(7) → $N-CO_2Me$ (100[c])

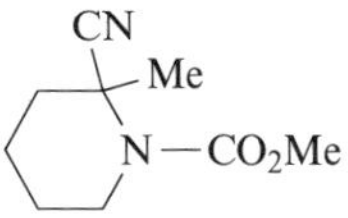

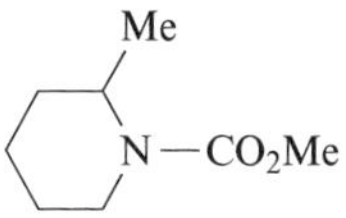

(8[d]) → (57)

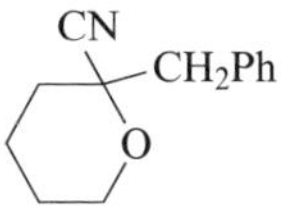

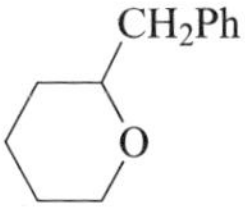

[a]Conditions: Zn cathode/DMF–Et$_4$NTos.
[b]Isolated yield.
[c]Determined by glc.
[d]MeCN.

which was involved in further chemical reactions. The debromination pathway was the major route observed, but three- and four-membered ring products were also formed in low yields.

TABLE 12. Electrochemical reduction of bromonitriles[63]

Compound	Cathode	Products (%)
(1) Me_2CBrCN	Hg	Me_2CHCN (35); $NCCMe_2CMe_2CN$ (16); polymers (30)
(2) Me_2CBrCN	C	Me_2CHCN (45); $CH_2=CMeCN$ (45)
(3) $Br(CH_2)_2CN$	Hg	$MeCH_2CN$ (36); $NC(CH_2)_4CN$ (5); polymers (45)
(4) $Br(CH_2)_2CN$	C	$MeCH_2CN$ (45); $CH_2=CHCN$ (45)
(5) $Br(CH_2)_3CN$	Hg	$Me(CH_2)_2CN$ (50); c-PrCN (35)
(6) $Br(CH_2)_4CN$	Hg	$Me(CH_2)_3CN$ (45); c-BuCN (15)

TABLE 13.   Electroreductive cyclization of 2-[2- and 3-cyano-alkyl] alkanones and cycloalkanones[64,65]

Compounds	Temperature (°C)	Products (%)
(1)	25	(77)
(2)	65	(64)
(3)	25	(76, *cis/trans*: 67/33)
(4)	65	(71)
(5)	25	(60)
(6)	25	(69, *cis/trans*: 67/33)
(7)	65	(60)
(8)	25	(60)
(9)	25	(62, *cis/trans*: 75/25)

Shono and coworkers[64,65] achieved electroreductive intra- and intermolecular couplings of ketones and nitriles in 2-propanol solutions containing Et_4NTos using Sn cathodes at controlled potential (–2.8 V vs SCE). Intramolecular coupling of cyclic γ- and δ-cyano ketones, besides good to excellent yields, proved to be cis stereoselective when α-hydroxy ketones with bicyclo[3.3.0] or [4.3.0] skeletons were formed. When the reactions were carried out at 65 °C instead of 25 °C, dehydration of hydroxy ketones occurred and the corresponding α,β-unsaturated ketones were obtained. The presence of alkyl or 2-ethoxycarbonyl substituents did not hinder the cyclization. In Table 13 some representative examples are shown.

Following the observation that only the carbonyl group was electrochemically reduced under the experimental conditions used, a radical formed via protonation of the ketyl radical anion resulting from a one-electron transfer was suggested as the key intermediate responsible for the coupling (Scheme 26).

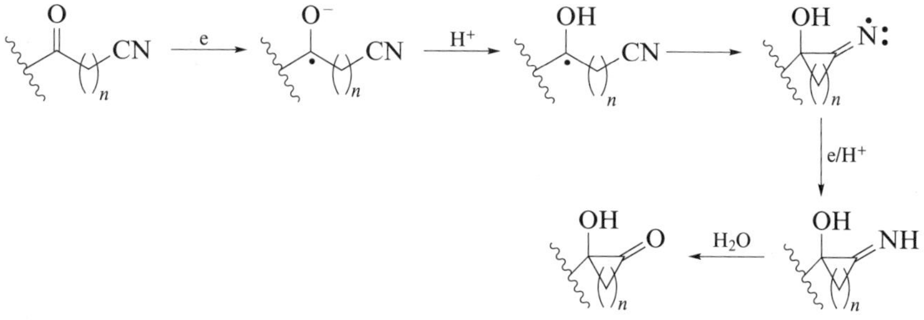

SCHEME 26

Attempted intermolecular[64] coupling of ketones and nitriles under conditions similar to those used for intramolecular coupling led to mixtures of two types of ketone–nitrile coupling products and alcohols resulting from ketone electroreduction. Product selectivity could be changed altering nitrile/solvent (2-propanol or ethanol) composition. Some results for cyclohexanone/acetonitrile reductions are shown in Scheme 27.

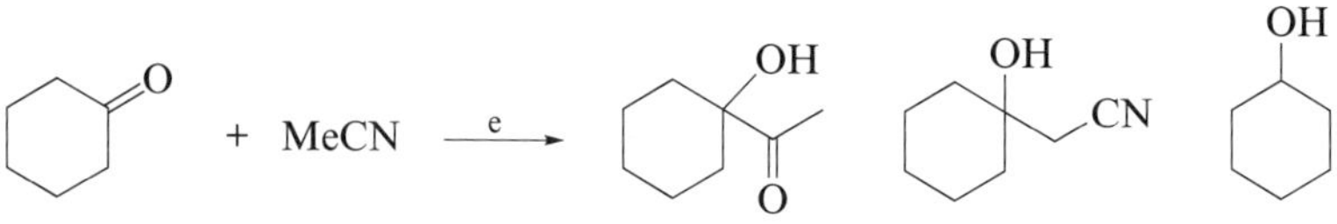

Solvent	Yields (%)		
MeCN	—	90	—
MeCN/i-PrOH 5/1	57	25	11
MeCN/i-PrOH 2/1	67	8	18
MeCN/EtOH 10/1	65	3	19

SCHEME 27

Electroreduction of α,β-unsaturated nitriles[66] $trans$-PhCH=CHCN, PhCMe=CHCN and CH_2=CHCN was investigated in acetonitrile solutions at a Hg cathode, to establish the most favourable experimental conditions for cyanomethylation to occur at the expense of hydrodimerization and reductive hydrogenation. Table 14 shows some of the results obtained. For cinnamonitrile (entry 1), high temperature, low current densities and low concentrations gave best yields of 3-phenylglutaronitrile. In this case cyanomethylation proved to be a catalytic chain process competing efficiently with hydrodimerization (Scheme 28).

TABLE 14. Electroreduction of α,β-unsaturated nitriles[66]

Compound	Conditions[a]	Products (%)
(1) PhCH=CHCN	10 mM, 79–81 °C, 10 mA	PhCH(CH$_2$CN)$_2$ (80)
(2) PhCMe=CHCN	0.1 M, 16°C, –3.0 V vs Ag/Ag$^+$	PhCHMeCH$_2$CN (10)
		PhCMe(CH$_2$CN)$_2$ (17)
		NCCH$_2$CMePhCMePhCH$_2$CN (50; d/l: meso 6.6:1)[b]
(3) CH$_2$=CHCN	12 mM, 21–24 °C, 10 mA	NC(CH$_2$)$_3$CN(45); NC(CH$_2$)$_4$CN (4)
(4) CH$_2$=CHCN	39 mM, azobenzene 75 mM, –2.0 V vs Ag/Ag$^+$	NC(CH$_2$)$_3$CN (26.5) + polymers

[a]Hg cathode, 0.1 M Et$_4$NBF$_4$/MeCN.
[b]Determined as diastereomeric mixture of 1-amino-2-cyano-3,4-dimethyl-3,4-diphenylcyclopent-1-ene.

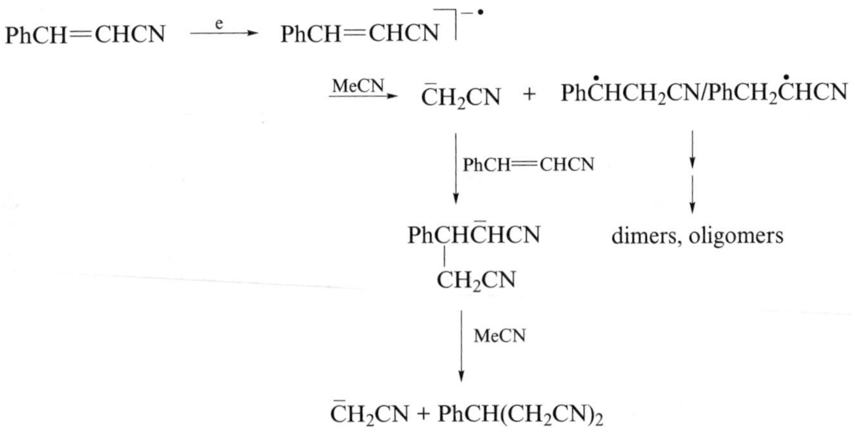

$$PhCH=CHCN \xrightarrow{\ e\ } PhCH=CHCN \rceil^{-\bullet}$$

$$\xrightarrow{MeCN} \bar{C}H_2CN \ + \ Ph\overset{\bullet}{C}HCH_2CN/PhCH_2\overset{\bullet}{C}HCN$$

$$\Big\downarrow PhCH=CHCN \qquad\qquad \Big\downarrow$$

$$Ph\overset{|}{CH}\bar{C}HCN \qquad\qquad dimers, oligomers$$
$$\overset{|}{C}H_2CN$$

$$\Big\downarrow MeCN$$

$$\bar{C}H_2CN + PhCH(CH_2CN)_2$$

SCHEME 28

With 3-methylcinnamonitrile (entry 2), hydrodimerization was the main pathway followed, whereas acrylonitrile (entry 3) afforded the best yields of cyanomethylated product under conditions similar to those observed for cinnamonitrile. Efforts to accomplish the Michael addition of $^-$CH$_2$CN to acrylonitrile, with the anionic reagent obtained via deprotonation of MeCN by electroreduced azobenzene (entry 4), gave glutaronitrile in low yields but no adiponitrile (hydrodimerization product). Anionic polymerization of acrylonitrile was the main reaction pathway.

When a mixture of phenylsulphonylacetonitrile and acrylonitrile in DMF was reduced at a less negative potential than the reduction potential of acrylonitrile, the products obtained were acetonitrile, propanenitrile, glutaronitrile and adiponitrile. It is noteworthy that glutaronitrile formation continued after all acrylonitrile had been consumed. This was accounted for by a series of chemical and electrochemical reactions initiated by the electroreduction of phenylsulphonylacetonitrile.

Simonet and coworkers[67] described the formation of d,l-hydrodimers and cyclic enamines when 3-(2-furyl)propenenitriles were electrochemically reduced at a Hg cathode in aqueous EtOH or DMF, under controlled potential (Table 15). The stereoselective formation of the cis-cyclopentene derivative by exclusive cyclization of the $meso$-hydrodimer and the lack of reactivity toward cyclization of the d,l-hydrodimer were best explained by a preferential adsorption on the cathode of the former. It is worth mentioning that the $trans$-enamine was obtained when the d,l-hydrodimer was treated with sodium t-butoxide in t-butanol.

TABLE 15. Preparative electroreduction[a] of 3-(2-furyl)propenenitriles[67]

Compounds	Products (%)		
R—CH=C(Ph)CN X	(RCHCH(Ph)CN)$_2$ |		
R = X—furyl—O			
(1)[b] H	48		32
(2) Me	60		20
(3) PhCH$_2$	45		30
(4)[c] Ph(CH$_2$)$_2$	42		18

[a]Conditions: Hg cathode, –1.8 V vs SCE, Et$_4$NI/H$_2$O–DMF.
[b]H$_2$O–DMF–EtOH.
[c]Bu$_4$NI.

TABLE 16. Preparative electrolyses of 2-alkoxy-3-phenylpropenenitriles[68]

Compound	Conditions (V vs SCE)	Products (%)
(1) PhCH=C(OMe)CN	MeCN/H$_2$O/LiCl/–1.90	PhCH$_2$CH$_2$CN (30) PhCH$_2$CH(OMe)CN (70)
(2) PhCH=CH(OBu-t)CN	DMF/H$_2$O/LiCl/–1.90	PhCH$_2$CH(OBu-t)CN (100)
(3) PhCH=CH(OCH$_2$Ph)CN	MeCN/H$_2$O/LiCl/–1.82	PhCH$_2$CH$_2$CN (93) PhCH$_2$CH(OCH$_2$Ph)CN (7)

Another example of an electrochemical reduction of α,β-unsaturated nitriles was reported later by Simonet and coworkers[68]. (E)- and (Z)-2-alkoxy-3-phenylpropenenitriles, PhCH=C(OR)CN, where R = Me, Et, t-Bu, Ph or PhCH$_2$, showed two reduction waves in the cyclic voltammetry experiments made in DMF at a Hg cathode. The first wave was reversible for R = Me, Et and Ph whereas the second, of much higher current, was irreversible. Presence of increasing amounts of phenol more than doubled the current of the first wave with a simultaneous decrease of the second. Cathodic behaviour of E and Z stereoisomers was very similar. Preparative electrolyses performed in DMF, DMF/H$_2$O or MeCN/H$_2$O are shown in Table 16. Two pathways may be followed during reduction, both leading to a saturated compound but one involving a C—O bond cleavage (Scheme 29).

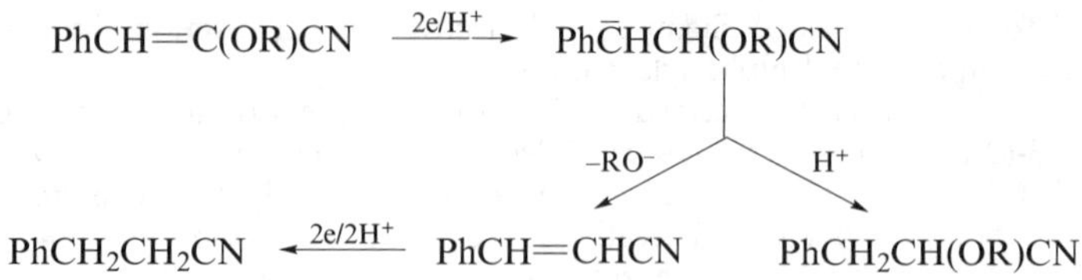

$$PhCH{=}C(OR)CN \xrightarrow{2e/H^+} Ph\bar{C}HCH(OR)CN$$

$$PhCH_2CH_2CN \xleftarrow{2e/2H^+} PhCH{=}CHCN \qquad PhCH_2CH(OR)CN$$

SCHEME 29

TABLE 17. Indirect electroreduction[a] of α,β-epoxynitriles[69]

Compound	Mediator (mol%)	Products (%)
(1)	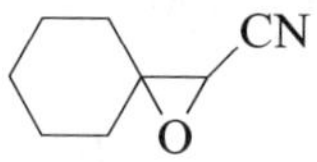 Ph_2Se_2 (60)[b]	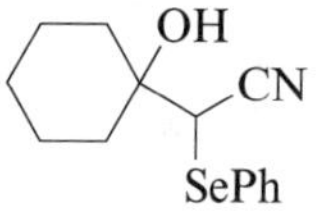 (60)
(2)	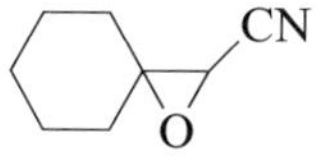 Ph_2Se_2 (2)	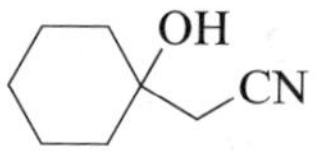 (82)
(3)	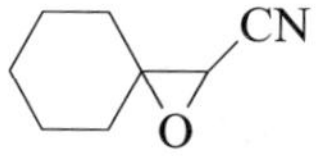 Ph_2Te_2 (60)[b]	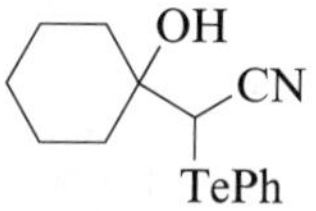 (81)
(4)	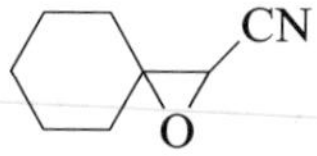 Ph_2Te_2 (2)	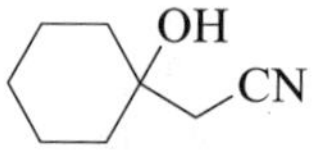 (85)
(5)	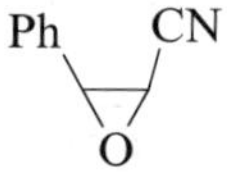 Ph_2Se_2 (2)	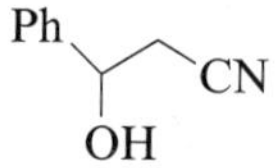 (87)
(6)	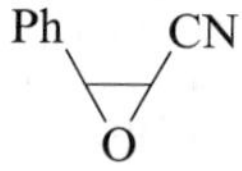 Ph_2Te_2 (2)	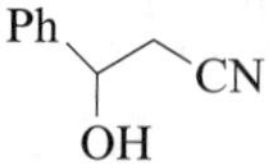 (91)
(7)	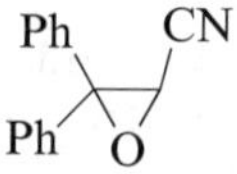 Ph_2Se_2 (2)	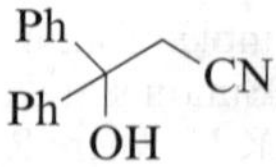 (93)
(8)	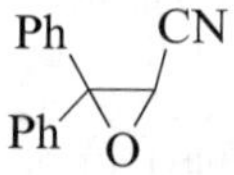 Ph_2Te_2 (2)	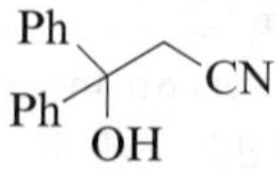 (95)

[a]Conditions: MeOH / 0.2 M $NaClO_4$, Pt cathode, 50 °C,
[b]Room temperature

α,β-Epoxynitriles were electrolysed in the presence of diphenyl diselenide or diphenyl ditelluride in methanolic solutions, at a Pt cathode, affording 3-hydroxynitriles and 2-phenylseleno- or 2-phenyltelluro-3-hydroxynitriles[69] (Table 17). This electroreductive ring opening was suggested to proceed by an initial attack of electrogenerated chalcogenide anions at the α-position of the cyano group. In the presence of catalytic amounts of diselenide/ditelluride the reaction led to 3-hydroxynitriles. However, with larger amounts of dichalcogenides and at room temperature (entries 1 and 3) the products were the corresponding α-substituted hydroxynitriles. These compounds were stable under the electrolysis conditions in the absence of the dichalcogenides but in their presence transformed into the 3-hydroxynitriles. This result indicated that the reduction occurred in two consecutive steps.

Polarographic investigations[70] of a number of substituted nitro- and m-dinitrobenzonitriles in DMF solution revealed stable radical anion formation at the first polarographic wave by a reversible 1-electron transfer. A cyano group at the 5-position of nitrobenzene and 1,3-dinitrobenzene proved to exert an important influence on the half-wave potentials of these benzonitrile derivatives and on the spin densities of the radical anions which were formed. The latter effect was confirmed by ESR measurements.

Molecules containing dicyanomethylene units were examined by cyclic voltammetry[71] in acetonitrile solution and two reduction waves, corresponding to radical anion and dianion formation, were observed. In view of the existing linear relationships between orbital energies and oxidation potentials, attempts to correlate LUMO energies with the reduction potentials were made. Large scatter was observed using first reduction potentials but a smooth curve was obtained with second reduction potentials. No explanation was given for this type of behaviour.

Dialkyl N-cyanoimidodithiocarbonates, $NCN{=}C(SR)SMe$ or $(NCN{=}C(SR)S)_2CH_2$, were electrochemically reduced at a vitreous C cathode in DMF[72a]. An irreversible one-electron process giving a radical anion followed by a fast C—S bond cleavage into a thiolate and alkyl radical was suggested (Scheme 30). Electroreduction of carbon–carbon double bonds becomes easier with increasing number of cyano substituents. Three polarographic waves were observed[72b] for 1,2-dicyano-1,2-dialkylthio-ethylenes in DMF solution, the first at ca –1.0 V (vs SCE). Controlled potential electrolysis in MeOH/HOAc/NaOAc solution at a Hg pool cathode produced the hydrogenated double-bond derivatives $(NC(SR)CHCH(SR)CN)$.

$$NCN{=}C(SMe)SR \xrightarrow{\ e\ } \left[NCN{=}C(SMe)SR \right]^{-\bullet} \longrightarrow NCNC(SMe)S^- + R^\bullet$$

SCHEME 30

3. Anodic oxidation

Since nitriles show great resistance towards electrochemical oxidation, few reports on nitrile oxidation are available in the literature. As a matter of fact the direct oxidation of the cyano group is not observed.

Methyl substituted benzonitriles were examined at a Pt anode in $FSO_3H/HOAc$, at –76 °C, using cyclic voltammetry[73]. At least two oxidation peaks were observed, the first corresponding to a 1-electron transfer leading to radical cations. This step was completely reversible for tetra- and pentamethyl derivatives and deprotonation only occurred at the second oxidation peak. However, with three or less methyl groups deprotonation was observed to occur at the first peak, generating a radical which was further oxidized to the corresponding cation that reacted with the solvent (Scheme 31). Cyclic voltammetry behaviour was also shown to be dependent on the existence of the protonated benzonitrile derivatives.

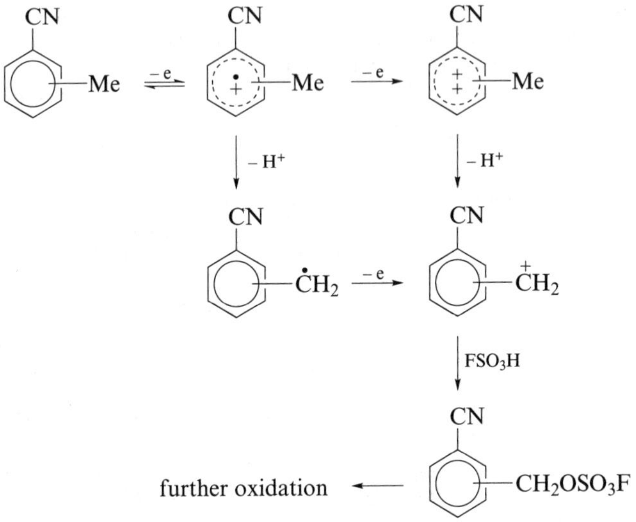

SCHEME 31

1,5-Dimethyl-2-pyrrolecarbonitrile[74] was oxidized in $H_2O/MeOH/NaCN$ solution, at a Pt anode under controlled potential conditions, giving 5-cyanomethyl-1-methyl-2-pyrrolecarbonitrile (61%, based on converted material) as shown in Scheme 32.

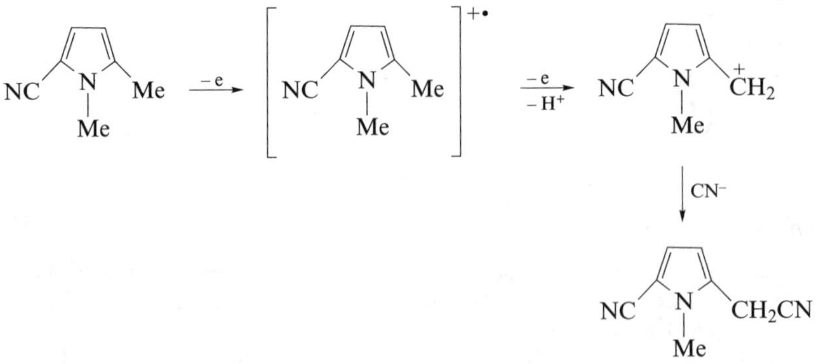

SCHEME 32

Electrogenerated cationic species can be trapped by nitriles yielding amides, through a Ritter-type reaction. This kind of reaction was observed when a variety of *ortho-*

toluenesulphonamides[75] were oxidized at a Pt anode in nitrile solution (MeCN, EtCN, *i*-PrCN, PhCN) (Scheme 33). Some of the results are in Table 18.

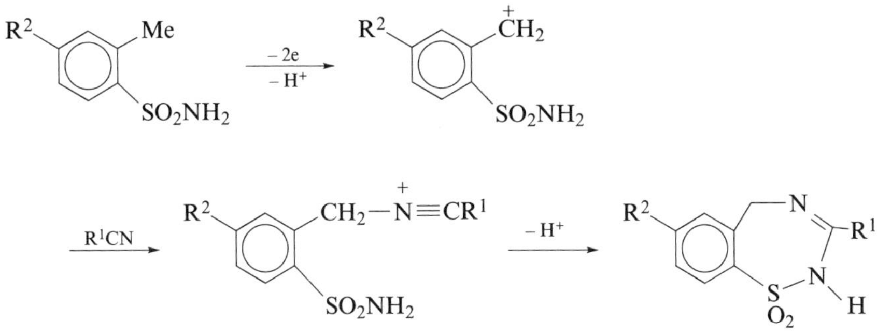

SCHEME 33

TABLE 18. Anodic oxidation of *ortho*-toluenesulphonamides[74]

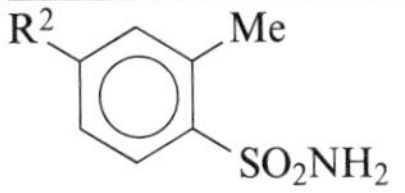

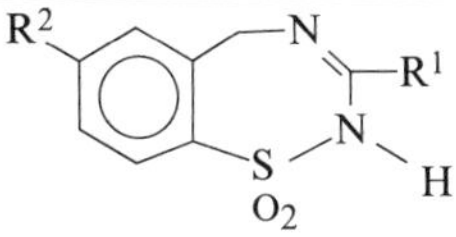

R^2	R^1	(%)
H	Me	87
H	Et	71
H	Pr	59
H	*i*-Pr	63
F	Me	65
Me	Me	61
Cl	Me	79
Br	Me	72
NO$_2$	Me	55

The anodic behaviour of E,Z-3-alkoxy-propenenitriles was investigated by Simonet and coworkers[76] at Pt and glassy C electrodes, in MeCN solution. The presence of both electron-donating and electron-withdrawing groups at the double bond was expected to cause some specific reactivity under electrochemical conditions. The compounds were oxidized in one irreversible step and coulometric measurements yielded n-values of 1.3–1.6 electrons/molecule. Only small differences could be noted when the isolated E- and Z-isomers were oxidized anodically. Preparative electrolysis at a Pt anode in dry MeCN afforded complex mixtures and polymerization occurred extensively. Table 19 summarizes these results. As can be seen, γ-diketones were the major products and their yields increased when high temperatures were used. Considering that starting materials are acid-sensitive and may undergo solvolysis in contact with electrogenerated acids, 2,6-lutidine or solid Na$_2$CO$_3$ were added to the electrolysis solution. In the presence of these bases the electrolysis product composition changed for some of the enol ethers, but still remained complex.

It was shown later[77] that 2-cyanoketones, the hydrolysis products of enol ethers, could be anodically oxidized under similar conditions affording dimeric products. The formation

TABLE 19. Electroreduction of some 3-alkoxy-propenenitriles[76]

Compound		Temp. (°C)	Products (%)		
NC(Ph)C=C(OR^2)R^1			NC(Ph)CHCOR^1	(NC(Ph)CCOR^1)$_2$	Others
R^1	R^2				
(1) H	Et	20	15	15	20^a
(2) Me	Me	20	18	40	8^b
(3) Me	Me	70	10	60	—
(4) Et	Et	20	8	28	7^b
(5) Et	Et	70	14	62	—
(6) Ph	Et	20	10	30	5^b
(7) Ph	Et	70	12	63	—

^{a}NC(Ph)CHCH(Ph)CN.

^{b}NC(Ph)CH—⟨○⟩—C(CN)=C(OR^2)R^1.

of the latter was explained as being a result of an oxidation of the enol tautomer of the ketones (Scheme 34).

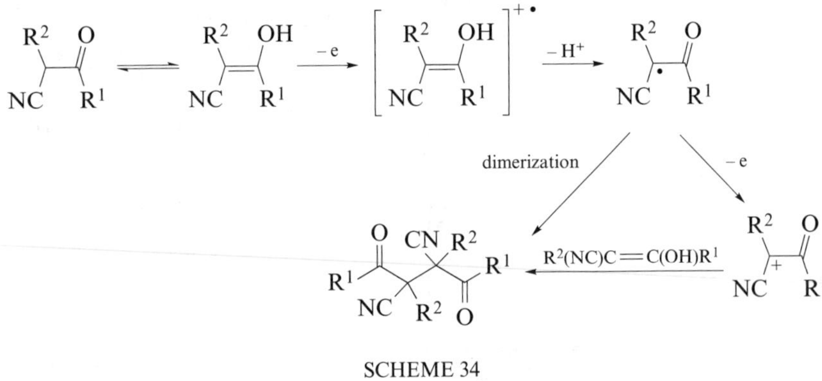

SCHEME 34

B. Thiocyanate Group

Ortho-nitrophenylthiocyanate was studied by polarography over a wide range of pH values in buffered aqueous ethanol solutions[78]. In acidic media three cathodic waves were observed and the ratio of wave heights was 4:0.9:1.3. Over the pH range 1.5–5 only a single cathodic wave was observed and its height was approximately that of the first wave in strong acidic media. The formation of two and then three waves was observed between pH 5–9. Coulometric *n*-values at different pH and reduction potentials were dependent on the duration of the electrolyses, being approximately 4 at the potential of the first cathodic wave and up to 6 at more negative potentials. The non-integer *n*-values point to a complicated reaction course with the participation of different chemical follow-up reactions. Preparative electrolyses of *ortho*-nitrophenylthiocyanate at large-area Hg electrodes led to mixtures of different composition of three products, 2-aminobenzothiazole-3-oxide, 2-aminobenzothiazole and 2,2'-diamino-diphenyl-disulphide, the latter probably formed from the easily oxidized 2-aminothiophenol. The relative amounts of products formed are pH-dependent and, for a given pH, potential-dependent. Thus in 0.5 M HCl solutions at –0.6V (*vs* Hg/Hg$_2$SO$_4$ sat.) 86% of N-oxide was obtained while at –1.6 V the yield dropped to 51%. Also, 44% of the disulphide was formed. At pH 8.0 and –1.6 V the yields for the N-oxide and the disulphide were 22% and 69%, respectively (Scheme 35).

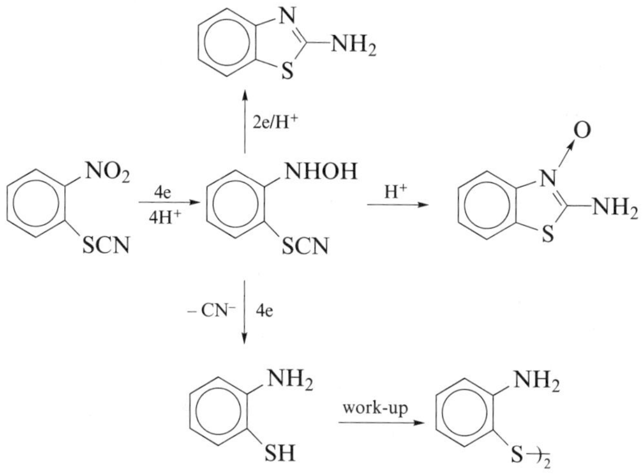

SCHEME 35

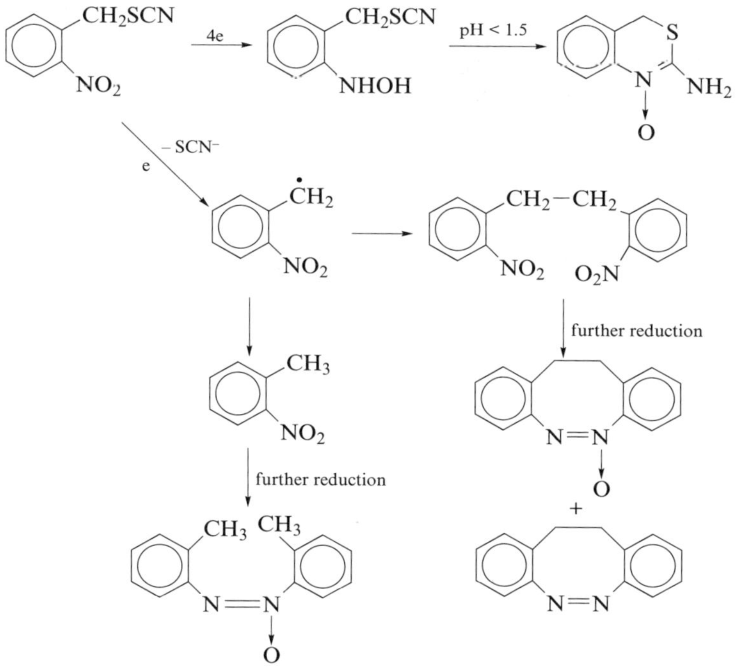

SCHEME 36

Hlavaty[79] studied the *ortho*-nitrobenzylthiocyanate under similar experimental conditions used for the phenyl derivative[78] to assess the influence of the CH_2 group separating the SCN group from the aromatic ring on the course of the electrolytic reduction. The controlled potential electrolysis in strongly acidic solution afforded 2-aminobenzo(1,3)-thiazine-1-oxide in 75% yield, most probably involving a fast reduction of the nitro group to the corresponding hydroxylamine which cyclizes under acidic conditions. In solutions of $1 < pH < 9$ an electroreductive cleavage of the C—S bond occurred via a radical anion leading to the intermediate nitrobenzyl radical, which either forms 1,2-bis(2-nitrophenyl)ethane or *ortho*-nitrotoluene, both being further reduced as shown in Scheme 36.

An interesting application of the electrochemical oxidation of thiocyanate ion is the preparation of alkyl and aryl thiocyanates via anodically generated thiocyanogen[80]. Alcohols have been converted to the corresponding thiocyanates by constant current electrolysis of NaSCN in CH_2Cl_2 containing triphenylphosphite and 2,6-lutidinium perchlorate. The yields were fair to good for the primary and secondary alcohols, but no thiocyanate formation was observed with tertiary ones. Similarly, various aromatic amines and phenols were thiocyanated in a two-step procedure, namely electrochemical preparation of $(SCN)_2$ and subsequent reaction with the substrates[81].

When 3-alkylindoles[82] were electrolysed using Pt electrodes in MeCN solution in the presence of NaSCN, the major products isolated were 2-isothiocyanates with yields ranging from 41 to 79%. An explanation offered for this unexpected result is an *ipso* attack at the 3-position yielding a 3-thiocyano-3*H*-indolium cation, which through a [3,5]-sigmatropic shift followed by [1,5]-hydrogen shift leads to the isothiocyanate.

III. NITROGEN–NITROGEN TRIPLE BOND: DIAZONIUM GROUP

Few developments were made in the electrochemistry of diazonium salts after Fry's review[83]. Most of the studies were concerned with polarographic measurements[84–89] but the main features of the electrochemical process remain the same.

A thorough investigation of the electrochemical reduction of 4-methyl and 4-

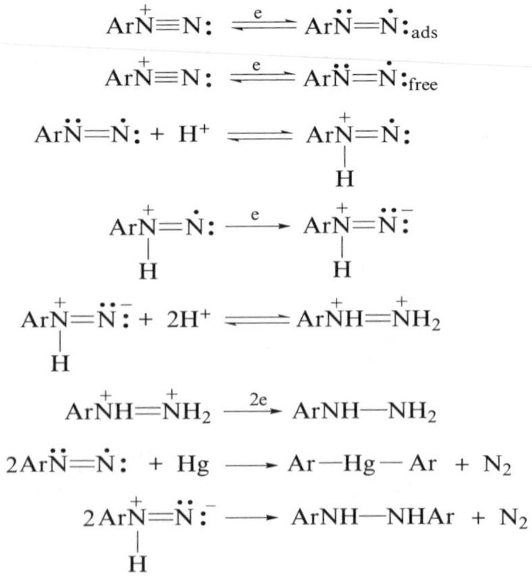

$$Ar\overset{+}{N}{\equiv}N: \; \xrightleftharpoons{e} \; Ar\overset{\cdot\cdot}{N}{=}\overset{\cdot}{N}:_{ads}$$

$$Ar\overset{+}{N}{\equiv}N: \; \xrightleftharpoons{e} \; Ar\overset{\cdot}{N}{=}\overset{\cdot}{N}:_{free}$$

$$Ar\overset{\cdot\cdot}{N}{=}\overset{\cdot}{N}: + H^+ \; \rightleftharpoons \; Ar\overset{+}{N}{=}\overset{\cdot}{N}: \\ \qquad\qquad\qquad\qquad\quad H$$

$$Ar\overset{+}{N}{=}\overset{\cdot}{N}: \; \xrightarrow{e} \; Ar\overset{+}{N}{=}\overset{\cdot\cdot}{N}: \\ \quad H \qquad\qquad\qquad H$$

$$Ar\overset{+}{N}{=}\overset{\cdot\cdot}{N}:^- + 2H^+ \; \rightleftharpoons \; Ar\overset{+}{N}H{=}\overset{+}{N}H_2 \\ \quad H$$

$$Ar\overset{+}{N}H{=}\overset{+}{N}H_2 \; \xrightarrow{2e} \; ArNH{-}NH_2$$

$$2Ar\overset{\cdot\cdot}{N}{=}\overset{\cdot}{N}: \; + \; Hg \; \longrightarrow \; Ar{-}Hg{-}Ar \; + \; N_2$$

$$2Ar\overset{+}{N}{=}\overset{\cdot\cdot}{N}:^- \; \longrightarrow \; ArNH{-}NHAr \; + \; N_2 \\ \quad H$$

SCHEME 37

methoxybenzenediazonium tetrafluoroborates in aqueous media at a mercury cathode was performed using polarography, cyclic voltammetry and controlled potential electrolysis[90]. Four waves were observed in a–c polarography, the first two corresponding to reversible 1-electron transfer reactions, the first being controlled by adsorption. The third wave involved an irreversible 1-electron reduction whereas a 2-electron process was suggested for the fourth. Coulometric measurements at the limiting currents of both the first and second polarographic waves showed a 1-electron transfer. The n-values obtained for the third and fourth waves were smaller than the expected 2- and 4-electrons, indicating that chemical reactions were competing with the electron transfer reactions. These chemical processes were faster for the 4-methyl than for the 4-methoxy derivative and proved to be temperature-dependent. Preparative electrolyses at potentials corresponding to the first two waves afforded a diarylmercury derivative in 90% yield. At the potential of the fourth wave and at low temperatures with moderate stirring, primary arylhydrazines were major products (4-Me: 57%, 4-MeO: 80%) besides the mercury compounds (10%) and the symmetrical hydrazines (80%). The yields of the latter increased up to 30% with non-optimized conditions. These data are summarized in Scheme 37.

The reduction of several 4-substituted benzenediazonium tetrafluoroborates (MeO, Br, H, Me, NO$_2$) were investigated at dropping Hg electrodes both in aqueous and in acetone solutions[89]. Two polarographic waves were observed in aqueous media, in potential regions of +0.05 to –0.02 V (vsAg/AgCl) and –0.97 to –1.03 V with 1:3 ratio of heights and both little dependent on the nature of the substituent. When water was replaced by acetone, the reduction of the diazonium salts was facilitated and an anodic shift of both waves was observed, the $E_{1/2}$ of the first wave being dependent upon the nature of substituent, e.g. 4-methoxy, + 0.19 V and 4-nitro, +0.46 V. The small dependence of $E_{1/2}$ of the first wave on the nature of the substituent was explained by a higher rate of adsorption of the diazonium cation on the electrode in the aqueous media. The mechanism proposed by the authors was essentially the same as depicted in Scheme 37.

The electrochemical reduction of benzenediazonium salts has been shown[83] to be a convenient source of free aryl radicals, which were used in several chemical reactions.

4-Substituted benzenediazonium tetrafluoroborates were reduced on Cu, Fe, Ti, Mo, Al, Zn and Ni electrodes in aqueous solution in the presence of acrylamide, in order to use the free radicals formed by the 1-electron mechanism as a polymerization initiator[91]. Only when Cu cathodes were used was a high degree of conversion of acrylamide achieved. With Fe cathodes, the polymerization occurred to a lesser extent and with the other metals it did not exceed 5–7%. No explanation was offered for this fact and a catalytic effect of Cu or Cu ions on the reduction of the diazonium salts cannot be excluded. The 4-methoxy and 4-methyl derivatives proved to be the most effective initiators with almost no polymerization being observed with the 4-nitro compound. These results were explained by a competitive reaction of the intermediate free radical with H$_2$O. The occurrence of the polymerization by a radical mechanism was proved by using hydroquinone as an inhibitor, and this fact confirms the 1e mechanism for the electrochemical reduction of the diazonium salts.

The electrochemical reduction of benzenediazonium chloride was also studied in the presence of unsaturated compounds like styrene, using Pt as cathode and Cu, Fe or Ti as anode[92]. The main processes observed were addition of a phenyl and a chloro group to the double bond [PhCH(Cl)CH$_2$Ph (I)] and additive dimerization [PhCH$_2$CH(Ph)CH(Ph)-CH$_2$Ph (II)]. These results are due to the catalytic effect of the cations formed by the anodic dissolution of the metal on the reaction between the diazonium salts and the unsaturated compound. The absence of products (I) and (II) when both electrodes were Pt confirms this redox catalysis.

Voltammetric measurements in MeCN at stationary Pt, Au and Hg electrodes showed that the electrodes became deactivated by reaction of radicals formed by reduction of 1-naphthalenediazonium tetrafluoroborate with the metallic electrodes[93]. Similar electrode fouling was observed[94] when cyclic voltammetry of phenyldiazonium tetrafluoroborate was studied in the same solvent. Thus, the reduction potentials measured for diazonium salts have no thermodynamic significance, and must be considered with some reservations when used in correlations with other properties of the ions.

As observed before[83] the first polarographic half-wave potentials determined in aprotic solvents can be correlated with Hammett σ values. This was done[95] for the 4-nitro-, chloro-, methyl-, methoxy- and a series of ring chlorinated dimethylamino-benzene-diazonium salts. A linear correlation was observed when corrected σ_p' constants were used for the dimethylamino group in the 3- and 3,5-halogenated derivatives. The correlation between $E_{1/2}$ values determined in nitromethane solutions for several ortho- and para-substituted diazonium salts was examined[96]. With NO_2, Cl and OMe groups, a linear correlation was obtained but for NMe_2, NEt_2, piperidinyl and pyrrolidinyl substituents, a deviation resulted. This was ascribed to a steric interaction between these ortho-substituents and the diazonium group. This conclusion agreed with the results of the photochemical reactions of these compounds.

Glassy carbon electrode surfaces have been modified by 1-electron electroreductive cleavage of several diazonium salts[97]. During this reaction the C—N bond was cleaved and the resulting aryl radical bound covalently to the carbon surface. The prior adsorption of the diazonium salts was an important feature for the reduction to occur. Functional groups linked to the aromatic ring can react with other chemical reagents or enzymes, allowing further modifications of the electrode surface. An interesting example was described[98] using the diazonium fluoroborate obtained from 4-aminophenylacetic acid. This electrode modified by 4-phenylacetic acid groups, after reaction with a carbondiimide, was used to attach glucose oxidase. The enzymatic activity of the electrode was ascertained by performing enzymatic electrocatalysis.

IV. REFERENCES

1. J. H. P. Utley and R. Lines, in The Chemistry of the Carbon–Carbon Triple Bond (Ed. S. Patai), Wiley, Chichester, 1978, pp. 739–753.
2. J. M. Kern and H. J. Schäfer, Electrochim. Acta, 30, 81 (1985).
3. R. Shao, J. A. Cleary, D. M. La Perriere and D. G. Peters, J. Org. Chem., 48, 3289 (1983).
4. W. M. Moore and D. G. Peters, J. Am. Chem. Soc., 97, 139 (1975).
5. W. M. Moore, A. Salajegheh and D. G. Peters, J. Am. Chem. Soc., 97, 4954 (1975).
6. B. C. Willett, W. M. Moore, A. Salajegheh and D. G. Peters, J. Am. Chem. Soc., 101, 1162 (1979).
7. R. W. Howsam and C. J. M. Stirling, J. Chem. Soc., Perkin Trans. 2, 847 (1972).
8. Yu. M. Kargin, V. Z. Latypova, O. G. Yakovleva, N. N. Khusaenov and A. I. Arkhipov, Zh. Obshch. Khim., 49, 2267 (1979).
9. V. Z. Latypova, O. G. Yakovleva, L. Z. Manapova and Yu. M. Kargin, Zh. Obshch. Khim., 50, 576 (1980).
10. V. Z. Latypova, G. A. Evtyugin, O. G. Yakovleva and Yu. M. Kargin, Zh. Obshch. Khim., 54, 852 (1984).
11. H. Kita and H. Nakajima, J. Chem. Soc., Faraday Trans. 1, 77, 2105 (1981).
12. T. Chiba, M. Okimoto, H. Nagai and Y. Takata, Bull. Chem. Soc. Jpn., 56, 719 (1983).
13. T. Nonaka, M. Takahashi and T. Fuchigami, Bull. Chem. Soc. Jpn., 56, 2584 (1983).
14. T. Shono, I. Nishiguchi and H. Omizu, Chem. Lett., 1233 (1976).
15. E. Kariv-Miller and T. J. Mahachi, J. Org. Chem., 51, 1041 (1986).
16. M. Tokuda and O. Nishio, J. Chem. Soc., Chem. Commun., 188 (1980).
17. M. Tokuda and O. Nishio, J. Org. Chem., 50, 1592 (1985).
18. M. Tokuda, T. Taguchi, O. Nishio and M. Itoh, J. Chem. Soc., Chem. Commun., 606 (1976).
19. G. Le Guillanton, Q. T. Do and J. Simonet, Tetrahedron Lett., 27, 2261 (1986).
20. P. Calas, P. Moreau and A. Commeyras, J. Chem. Soc., Chem. Commun., 433 (1982).
21. P. Calas, C. Amatore, L. Gomez and A. Commeyras, J. Fluorine Chem., 49, 247 (1990).
22. S. Torii, T. Inokuchi and T. Yukawa, J. Org. Chem., 50, 5875 (1985).

23. (a) S. Dérien, E. Duñach and J. Périchon, *J. Am. Chem. Soc.*, **113**, 8447 (1991).
    (b) E. Duñach and J. Périchon, *J. Organomet. Chem.*, **352**, 239 (1988).
    (c) E. Duñach, S. Dérien and J. Périchon, *J. Organomet. Chem.*, **364**, C33 (1989).
    (d) E. Labbé, E. Duñach and J. Périchon, *J. Organomet. Chem.*, **353**, C51 (1988).
24. (a) S. Dérien, J. C. Clinet, E. Duñach and J. Périchon, *J. Org. Chem.*, **58**, 2578 (1993).
    (b) S. Dérien, J. C. Clinet, E. Duñach and J. Périchon, *J. Chem. Soc., Chem. Commun.*, 549 (1991).
    (c) S. Dérien, E. Duñach and J. Périchon, *J. Organomet. Chem.*, **385**, C43 (1990).
25. S. Dérien, J. C. Clinet, E. Duñach and J. Périchon, *J. Organomet. Chem.*, **424**, 213 (1992).
26. (a) L. Coche and J. C. Moutet, *J. Am. Chem. Soc.*, **109**, 6887 (1987).
    (b) A. Deronzier and J. C. Moutet, *Acc. Chem. Res.*, **22**, 249 (1989).
27. C. Cwiklinski and J. Périchon, *Electrochim. Acta*, **19**, 315 (1974).
28. J. W. Johnson, J. L. Reed and W. J. Jones, *J. Electrochem. Soc.*, **114**, 572 (1967).
29. K. Venkateswara Rao and C. B. Roy, *Indian J. Chem.*, **21A**, 34 (1982).
30. (a) M. Katz and H. Wendt, *Electrochim. Acta*, **21**, 215 (1976).
    (b) M. Katz and H. Wendt, *J. Electroanal. Chem.*, **53**, 465 (1974).
31. A. M. Bond, I. D. Heritage and M. H. Briggs, *Langmuir*, **1**, 110 (1985).
32. V. Verniette, C. Daremon and J. Simonet, *Electrochim. Acta*, **23**, 929 (1978).
33. K. Uneyama, K. Takano and S. Torii, *Tetrahedron Lett.*, **23**, 1161 (1982).
34. S. Torii, T. Inokuchi and Y. Hirata, Synthesis, 377 (1987).
35. K. Yoshida, in *The Chemistry of Triple-bonded Functional Groups–Supplement C* (Eds. S. Patai and Z. Rappoport), Wiley, Chichester, 1983, pp. 221–268.
36. T. L. Antonova, L. N. Ivanovskaya, M. Ya. Fioshin and N. N. Savushkina, *Élektrokhimiya*, **23**, 560 (1987).
37. Y. Song and P. N. Pintauro, *J. Appl. Electrochem.*, **21**, 21 (1991).
38. V. Krishnan and A. Muthukumaran, *J. Electrochem. Soc. India*, **32**, 241 (1983).
39. A. Ayyaswami and V. Krishnan, *J. Electrochem. Soc. India*, **32**, 410 (1983).
40. T. Yamada, N. Fujimoto, T. Matsue and T. Osa, *Denki Kagaku Oyobi Kogyo Butsuri Kagaku*, **56**, 175 (1988).
41. T. L. Antonova, L. N. Ivanovskaya, I. A. Avrutskaya, M. Ya. Fioshin and L. I. Gorbacheva, *Élektrokhimiya*, **22**, 546 (1986) and references cited therein.
42. L. Coche, B. Ehui, D. Limosin and J. C. Moutet, *J. Org. Chem.*, **55**, 5905 (1990).
43. C. H. Yang, S. F. Lin, H. L. Chen and C. T. Chang, *Inorg. Chem.*, **19**, 3541 (1980).
44. C. H. Yang and C. T. Chang, *J. Chem. Soc., Dalton Trans.*, 2539 (1982).
45. A. V. Bukhtiarov, V. N. Golyshin, A. P. Tomilov and O. V. Kuź'min, *Zh. Obshch. Khim.*, **58**, 857 (1988).
46. A. P. Tomilov, S. E. Zabusova and N. M. Alpatova, *Élektrokhimiya*, **18**, 504 (1982).
47. A. Gennaro, F. Maran, A. Maye and E. Vianello, *J. Electroanal. Chem.*, **185**, 353 (1985).
48. A. Gennaro, A. M. Romanin, M. G. Severin and E. Vianello, *J. Electroanal. Chem.*, **169**, 279 (1984).
49. M. Sertel and A. Yildiz, *Electrochim. Acta*, **31**, 1287 (1986).
50. L. N. Nekrasov, S. M. Peregudova, L. P. Yur'eva, D. N. Kravtsov, I. A. Uralets and N. N. Zaitseva, *J. Organomet. Chem.*, **365**, 269 (1989).
51. J. E. Swartz and T. T. Stenzel, *J. Am. Chem. Soc.*, **106**, 2520 (1984).
52. A. Webber, E. K. Eisner, J. Osteryoung and J. Hermolin, *J. Electrochem. Soc.*, **129**, 2725 (1982).
53. A. Webber and J. Osteryoung, *J. Electrochem. Soc.*, **129**, 2731 (1982).
54. I. Carelli and M. E. Cardinali, *J. Electroanal. Chem.*, **124**, 147 (1981).
55. (a) M. E. Cardinali and I. Carelli, *J. Electroanal. Chem.*, **125**, 477 (1981).
    (b) I. Carelli, M. E. Cardinali and A. Casini, *J. Electroanal. Chem.*, **105**, 205 (1979).
56. J. Nadra, H. Givadinovitch and M. Devaud, *J. Chem. Res. (M)*, 1831 (1983).
57. O. R. Brown and R. J. Butterfield, *Electrochim. Acta*, **27**, 1647 (1982).
58. M. D. Koppang, N. F. Woolsey and D. E. Bartak, *J. Am. Chem. Soc.*, **106**, 2799 (1984).
59. M D. Koppang, N. F. Woolsey and D. E. Bartak, *J. Am. Chem. Soc.*, **107**, 4692 (1985).
60. C. Amatore, C. Combellas, S. Robveille, J. M. Savéant and A. Thiébault, *J. Am. Chem. Soc.*, **108**, 4754 (1986).
61. G. Le Guillanton and M. Lamant, *Electroorganic Synthesis: Festschrift for Manuel M. Baizer* (Eds. R. D. Little and N. L. Weinberg), Marcel Dekker, New York, 1991, pp. 121–127.
62. T. Shono, J. Terauchi, K. Kitayama, Y. Takeshima and Y. Matsumura, *Tetrahedron*, **48**, 8253 (1992).
63. I. Carelli, A. Curulli and A. Inesi, *Electrochim. Acta*, **30**, 941 (1985).

624   H. Viertler, V. L. Pardini and R. R. Vargas

64. T. Shono, N. Kise, T. Fujimoto, N. Tominaga and H. Morita, *J. Org. Chem.*, **57**, 7175 (1992).
65. T. Shono and N. Kise, *Tetrahedron Lett.*, **31**, 1303 (1990).
66. A. J. Bellamy, J. B. Kerr, C. J. McGregor and I. S. MacKirdy, *J. Chem. Soc., Perkin Trans. 2*, 161 (1982).
67. J. Delaunay, A. Lebouc, G. Le Guillanton, L. M. Gomes and J. Simonet, *Electrochim. Acta*, **27**, 287 (1982).
68. M. Cariou, G. Mabon, G. Le Guillanton and J. Simonet, *Tetrahedron*, **39**, 1551 (1983).
69. T. Inokuchi, M. Kusumoto and S. Torii, *J. Org. Chem.*, **55**, 1548 (1990).
70. V. M. Kazakova, N. E. Minina and V. B. Piskov, *Zh. Obshch. Khim.*, **52**, 961 (1982).
71. M. L. Kaplan, R. C. Haddon, F. B. Bramwell, F. Wudl, J. H. Marshall, D. O. Cowan and S. Gronowitz, *J. Phys. Chem.*, **84**, 427 (1980).
72. (a) H. H. Rüttinger, H. Matschiner and W. Walek, *Z. Chem.*, **21**, 330 (1981).
    (b) H. H. Rüttinger, H. Matschiner, C. Winkelmann and R. Mayer, *Z. Chem.*, **21**, 287 (1981).
73. A. P. Rudenko, M. Ja. Sarubin and F. Pragst, *J. Prakt. Chem.*, **325**, 612 (1983).
74. F. Köleli, C. H. Hamann and J. Martens, *Tetrahedron Lett.*, **30**, 925 (1989).
75. G. Palmisano, B. Danieli, G. Lesma and G. Fiori, *Tetrahedron*, **44**, 1545 (1988).
76. M. A. Le Moing, G. Le Guillanton and J. Simonet, *Electrochim. Acta*, **26**, 139 (1981).
77. M. A. O. Le Moing, G. Le Guillanton and J. Simonet, *Electrochim. Acta*, **27**, 1775, (1982).
78. H. Hlavatý, J. Volke and O. Manousek, *Collect. Czech. Chem. Commun.*, **40**, 3751 (1975).
79. J. Hlavatý, *Collect. Czech. Chem. Commun.*, **50**, 33 (1985).
80. H. Maeda, T. Kawaguchi, M. Masui and H. Ohmori, *Chem. Pharm. Bull.*, **38**, 1389 (1990).
81. P. Krishman and V. G. Gurjar, *Synth. Commun.*, **22**, 2741 (1992).
82. G. Palmisano, E. Brenna, B. Danieli, G. Lesma, B. Vodopivec and G. Fiori, *Tetrahedron Lett.*, **31**, 7229 (1990).
83. A. J. Fry, in *The Chemistry of Diazonium and Diazo Groups* (Ed. S. Patai), Wiley, Chichester, 1978, pp. 489–498.
84. V. Mejstrik, Z. Sagner, L. Drzkova, F. Krampera and M. Matrka, *Chem. Prum.*, **35**, 34 (1985); *Chem. Abstr.*, **102**, 139650e (1985).
85. V. Mejstrik, L. Drzkova, Z. Sagner, F. Krampera and M. Matrka, *Chem. Prum.*, **35**, 83 (1985); *Chem. Abstr.*, **102**, 203470a (1985).
86. V. Mejstrik, L. Drzkova, Z. Sagner, F. Krampera and M. Matrka, *Chem. Prum.*, **36**, 311 (1986); *Chem. Abstr.*, **105**, 87330r (1986).
87. K. Harada, K. Sugita and S. Suzuki, *Nippon Shashin Gakkaishi*, **51**, 275 (1988); *Chem. Abstr.*, **110**, 104760d (1989).
88. N. D. Stepanov, M. P. Noskova, L. S. Litvinova, I. L. Bagal and A. V. El'tsov, *Tezisy Dokl. - Vses. Soveshch. Polyarogr. 7h*, **44** (1978); *Chem. Abstr.*, **93**, 122561z (1980).
89. E. P. Koval'chuk, N. D. Obushak, N. I. Ganushchak and P. Yanderka, *Zh. Obshch. Khim.*, **56**, 1891 (1986).
90. O. Orange, C. E. Hamet and C. Caullet, *J. Electrochem. Soc.*, **128**, 1889 (1981).
91. E. P. Koval'chuk, N. I. Ganushchak, V. I. Kopylets, I. N. Krupak and N. D. Obushak, *Zh. Obshch. Khim.*, **52**, 2540 (1982).
92. N. I. Ganushchak, N. D. Obushak, E. P. Koval'chuk and G. V. Trifonova, *Zh. Obshch. Khim.*, **54**, 2334 (1984).
93. E. Ahlberg, B. Helgée and V. D. Parker, *Acta Chem. Scand.*, **B34**, 181 (1980).
94. A. J. Bard, J. C. Gilbert and R. D. Goodin, *J. Am. Chem. Soc.*, **96**, 620 (1974).
95. E. Fanghänel, J. Kriwanek, W. Ortmann, I. L. Bagal, N. W. Lebedeva and A. V. El'cov, *J. Prakt. Chem.*, **327**, 80 (1985).
96. Von H. Böttcher, A. V. El'cov and N. I. Rtiscev, *J. Prakt. Chem.*, **315**, 725 (1973).
97. M. Delamar, R. Hitmi, J. Pinson and J. M. Savéant, *J. Am. Chem. Soc.*, **114**, 5883 (1992).
98. C. Bourdillon, M. Delamar, C. Demaille, R. Hitmi, J. Moiroux and J. Pinson, *J. Electroanal. Chem.*, **336**, 113 (1992).

CHAPTER 11

Syntheses and uses of isotopically labelled X≡Y groups

KENNETH C. WESTAWAY

Department of Chemistry, Laurentian University, Sudbury Ontario, Canada

and

PETER J. SMITH

Department of Chemistry and Chemical Engineering, University of Saskatchewan, Saskatoon, Saskatchewan, Canada

Supplement C2: The chemistry of triple-bonded functional groups
Edited by S. Patai © 1994 John Wiley & Sons Ltd

I. INTRODUCTION

The Chemistry of Functional Groups series has published three texts dealing with the preparation and uses of isotopically labelled compounds containing triple bonds [1-3]. The most recent chapter dealing with the preparation and uses of isotopically labelled diazonium and diazo compounds was published in 1978 and those discussing the chemistry of alkynes and cyanides date from 1978 and 1970, respectively. In the present chapter, the preparation of isotopically labelled substrates will only be discussed if the procedures are significantly different from those considered in the earlier chapters. The main focus of this chapter will deal with recent investigations where isotopically labelled triply-bonded compounds have been used in novel and informative ways, e.g. in determining reaction mechanisms, etc. It is not the purpose of this chapter to provide a complete literature review.

II. THEORY OF KINETIC ISOTOPE EFFECTS

A. Heavy-atom Kinetic Isotope Effects

Several monographs [4-7] have detailed discussions dealing with heavy-atom and primary and secondary hydrogen–deuterium kinetic isotope effects. The monograph by Melander and Saunders[7] covers the entire area particularly well. For this reason, only a brief summary of the theory of kinetic isotope effects as well as their important uses in the determination of reaction mechanism and transition-state geometry will be presented.

The Bigeleisen treatment[8-10], based on Eyring and coworkers' absolute rate theory[11], assumes that there is a single potential energy surface along which the reaction takes place, and that there is a potential energy barrier separating the reactants from the products. The reaction occurs along the path over the lowest part of the barrier with the transition state at the top of the barrier, i.e. it lies at the energy maximum along the reaction coordinate but at an energy minimum in all other directions. The transition state is assumed to be in equilibrium with the reactants and products and have all the properties of a stable molecule except that one vibrational degree of freedom has been converted into motion along the reaction coordinate.

$$A + B \rightleftharpoons \left[\begin{matrix} \text{Transition} \\ \text{state} \end{matrix} \right]^{\ddagger} \rightleftharpoons \text{Products} \tag{1}$$

The kinetic isotope effect for this

$$\frac{k_1}{k_2} = \frac{\kappa_1}{\kappa_2} \cdot \frac{Q_1^{\ddagger}}{Q_2^{\ddagger}} \cdot \frac{Q_{A_2}}{Q_{A_1}} \cdot \frac{Q_{B_2}}{Q_{B_1}} \tag{2}$$

where the subscripts 1 and 2 refer to the molecules containing the lighter and heavier isotopes, respectively, and the Qs are the complete partition functions for reactants A and

B. Setting $\kappa_1 = \kappa_2$ and applying the harmonic approximation to all non-linear gas molecules leads to an expression for Q_2/Q_1 (equation 3), where S_1 and S_2 are the symmetry numbers of the respective molecules, the Ms are the molecular weights, the Is are the moments of inertia about the three principal axes of the n-atom molecules and the νs are the fundamental vibrational frequencies of the molecules in wave numbers.

$$\frac{Q_2}{Q_1} = \frac{S_1}{S_2}\left(\frac{I_{A_2} \, I_{B_2}}{I_{A_1} \, I_{B_1}}\right)^{1/2}\left(\frac{M_2}{M_1}\right)^{1/2} \prod_i^{3n-6} \exp\left[\frac{(\nu_{1i} - \nu_{2i})hc}{2kT}\right]\left[\frac{1 - \exp\left(-hc\,\nu_{1i}/kT\right)}{1 - \exp\left(-hc\,\nu_{2i}/kT\right)}\right] \quad (3)$$

Using various approximations, a solution to the isotopic rate ratio equation can be obtained. It is found that the isotope rate ratio, k_1/k_2, is dependent on the force constant changes which occur in going from the reactants to the transition state. Consequently, if C—X bond rupture, where the isotopically labelled atom X can be halogen, sulphur, nitrogen, etc., has not progressed at the transition state of the rate-determining step of the overall reaction, there is no change in the force constants involving the isotopic atom and the isotope rate ratio, $k_1 k_2$, will be equal to one. An isotope rate ratio greater than one will be observed if there is a decrease in the force constants at the transition state of the slow step. The greater the decrease in the force constant, the larger the magnitude of the isotope effect.

The observation of a heavy-atom isotope effect, therefore, allows one to determine whether C—X bond weakening (a decrease in force constant) has occurred when the reactant is converted into the transition state of the rate-determining step. Calculations by Saunders[12] and by Sims and coworkers[13] have shown that the magnitude of the leaving-group heavy-atom isotope effect varies linearly with the extent of C—X bond rupture in the transition state for concerted elimination reactions and for nucleophilic substitution reactions, respectively. Since the magnitude of the isotope effect is directly related to the amount of C—X bond rupture in the transition state, these isotope effects provide detailed information about the structure of the transition state.

B. Primary Hydrogen–Deuterium Kinetic Isotope Effects

Although the zero-point energy differences between the isotopic molecules' vibrations are not the only contribution to the isotope effect, they are, however, often the dominant term. This is particularly true for hydrogen–deuterium kinetic isotope effects where the zero-point energy difference is large, and also for large molecules where isotopic substitution does not affect the mass and moment of inertia term significantly. It is usual to assume that the stretching modes are the most important in determining these isotope effects. This is based on the two assumptions: (i) that the bending vibrations are generally of a lower frequency and therefore have smaller zero-point energy differences for isotopic molecules, and (ii) the bending motions in the transition state will be similar to those in the substrates.

Applying these approximations to the rupture of a single C—H bond in a unimolecular process leads to equation 4,

$$\frac{k_H}{k_D} = \exp\left[\left(\frac{hc}{2kT}\right)\left(\nu_H - \nu_D\right)\right] \quad (4)$$

where ν_H and ν_D are the ground-state symmetric stretching frequencies for the C—H and C—D bonds, respectively. Substitution of the appropriate frequencies into equation 4 gives an isotope effect of approximately 7 at 25 °C.

For reactions involving a proton transfer from one molecule to another, however, the situation is more complex. Westheimer[14] and Melander[4] have independently pointed out that, because bond formation and bond breaking are occurring concurrently, new stretching vibrations in the transition state which are not present in the reactants must be considered.

They considered the reaction

$$AH + B^- \longrightarrow [A \cdots H \cdots B]^{\ddagger} \longrightarrow A^- + HB \qquad (5)$$

where $[A \cdots H \cdots B]^{\ddagger}$ is a linear transition state. If this transition state is regarded as a linear molecule, there are two independent stretching vibrational modes which may be illustrated as follows:

$$\longleftarrow \quad ? \quad \longrightarrow \qquad \qquad \longleftarrow \quad \longrightarrow \quad \longleftarrow$$
$$A \cdots H \cdots B \qquad \qquad \quad A \cdots H \cdots B$$
Symmetric   Antisymmetric

Neither of these vibrations corresponds to stretching vibrations of AH or BH. The 'antisymmetric' vibrational mode represents translational motion in the transition state and has an imaginary force constant. The 'symmetric' transition-state vibration has a real force constant but the vibration may or may not involve motion of the central H(D) atom[4,14,15]. If the motion is truly symmetric, the central atom will be motionless in the vibration and thus the frequency of the vibration will not depend on the mass of this atom, i.e. the vibrational frequency will be the same for both isotopically substituted transition states. It is apparent that under such circumstances there will be no zero-point energy differences between the deuterium- and hydrogen-substituted compounds for the symmetric vibration in the transition state. Hence, an isotope effect of 7 at room temperature is expected since the difference in activation energy is the difference between the zero-point energies of the symmetric stretching vibrations of the initial states, i.e., $\frac{1}{2}(hv_H - hv_D)$.

In instances where bond breaking and bond making at the transition state are not equal, i.e. bond breaking is either more or less advanced than bond formation, the 'symmetric' vibration will not be truly symmetric. In these cases, the frequency will have some dependence on the mass of the central atom, there will be a zero-point energy difference for the vibrations of the isotopically substituted molecules at the transition state and k_H/k_D willl have values smaller than 7.

It may be concluded that for reactions where the proton is less or more than one-half transferred in the transition state, i.e. the A—H and H—B force constants are unequal, the primary hydrogen–deuterium kinetic isotope effect will be less than the maximum of 7. The maximum isotope effect will be observed only when the proton is exactly half-way between A and B in the transition state.

C. Secondary α-Hydrogen–Deuterium Kinetic Isotope Effects

In the preceding sections, the bond involving the isotopic atom is broken or formed in the rate-determining step of the reaction. In these cases, the change in rate is referred to as a primary kinetic isotope effect. Isotopic substitution at other sites in the molecule has much smaller effects on the rate. These small isotope effects are collectively referred to as secondary kinetic isotope effects.

As with primary isotope effects, the origin of secondary isotope effects is considered to be mainly due to changes in force constants upon going from reactants to the transition state. For the most part, secondary isotope effects depend on the change in zero-point energy (ZPE). Smaller force constants for the isotopic nuclei in the transition state than in the reactant lead to an isotope effect greater than one (Figure 1a). When the force constants are greater in the transition state than in the reactant, on the other hand, an isotope effect of less than one is observed (Figure 1b).

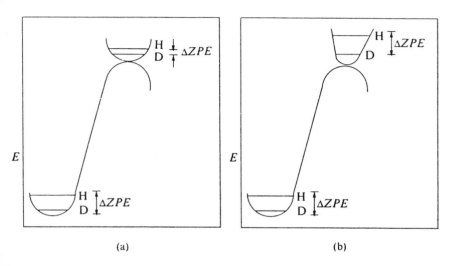

(a)   (b)

FIGURE 1. (a) A reaction where $\Delta ZPE_{(reactant)}$ is greater than $\Delta ZPE_{(transition\ state)}$ and $(k_H k_D)_\alpha > 1.0$. (b) A reaction where $\Delta ZPE_{(reactant)}$ is less than $\Delta ZPE_{(transition\ state)}$ and $(k_H/k_D)_\alpha < 1.0$

Secondary α-hydrogen–deuterium kinetic isotope effects are determined when hydrogen is replaced by deuterium at the α- or reacting carbon. The generally accepted view originally proposed by Streitwieser and coworkers[16] is that the α-deuterium kinetic isotope effects are primarily determined by the changes in the out-of-plane bending vibrations in going from the reactants to the transition state. Solvolysis reactions proceeding via a carbocation are expected to have large normal isotope effects, $(k_H/k_D)_\alpha$, of approximately 1.15. The maximum values expected for various leaving groups are 1.22 for fluoride, 1.15 for chloride, 1.13 for bromide, 1.09 for iodide, 1.19 for ammonia and 1.22 for benzenesulphonate[17,18].

Smaller α-deuterium isotope effects are observed for reactions proceeding via the S_N2 mechanism. This is due to steric interference by the leaving group (LG) and/or the incoming nucleophile (Nu) with the out-of-plane bending vibrations of the Cα—H bonds. This leads to an increased force constant at the S_N2 transition state, **1** (see Figure 1b). In fact, small or inverse isotope effects, $(k_H/k_D)_{\alpha-D} = 0.95 - 1.04$, are observed for the S_N2 reactions of primary substrates[19].

$$\overset{\delta-}{Nu}\text{---}\overset{|\,\,\delta-}{C}\text{---}\overset{\delta-}{LG} \qquad \delta-$$

(1)

α

D. Secondary β-Hydrogen–Deuterium Kinetic Isotope Effects

Secondary β-deuterium kinetic isotope effects arise when the hydrogen(s) on the β-carbon (adjacent to the carbon where the C—X bond rupture is progressing) are replaced by deuterium(s). These isotope effects $(k_H/k_D)_\beta$ are greater than unity for nucleophilic substitution reactions. In addition, the magnitude of the isotope effect increases as the amount of positive charge (carbonium-ion character) on the α-carbon in the transition state **2** is increased. For example, the isotope effect per CD_3 group increases from about 1.03

$$-\underset{\underset{D}{|}}{\overset{|}{C}}_\beta - \underset{\underset{H}{|}}{\overset{|}{C}}_\alpha^{\delta+} \text{-----} \overset{\delta-}{LG}$$

(2)

for ethyl compounds, which undoubtedly react by an S_N2 mechanism, to approximately 1.37 for a t-butyl compound, which reacts by a limiting S_N1 mechanism[20]. A wealth of experimental evidence[21] indicates that these isotope effects are primarily, if not completely, a result of hyperconjugative electron release from the C_β—H bonds[22]. Other studies by Shiner and coworkers:[23, 24] have demonstrated that the magnitude of these isotope effects vary with the dihedral angle between the C_β—H orbital and the developing p-orbital on the α-carbon. The maximum isotope effect in any system is observed when the dihedral angle is either $0°$ or $180°$, i.e. where the overlap between the C_β— H and the p-orbital on the α—carbon is a maximum.

E. Kinetic Isotope Effects Arising from the Difference in Basicity between DO⁻ in D_2O and HO⁻ in H_2O

As already noted in the discussion dealing with the primary hydrogen–deuterium kinetic isotope effect, it is generally agreed that small hydrogen–deuterium isotope effects can arise when the proton is more than or less than one-half transferred to base at the transition state. As a consequence, it is necessary to determine the side of the symmetrical situation on which the transition state lies. This is necessary in order to interpret the magnitude of the primary hydrogen–deuterium isotope effects in terms of the degree of carbon–hydrogen bond rupture at the transition state. Steffa and Thornton[25] approached this problem by comparing the relative reaction rates with DO⁻ in D_2O and HO⁻ in H_2O.

The relative basicity of the hydroxide and deuteroxide ion is determined by the equilibrium (equation 6). The related equilibrium for the conversion of one OD⁻ bond of the solvated deuteroxide ion to one OD bond of heavy water is shown in equation 7 and K must therefore be the direct measure of the relative basicities of OD⁻ and OH⁻. Since OD⁻ is the stronger base, $K = K_B^{1/2} = k^{OD^-}/k^{OH^-} > 1.0$. The magnitude of this secondary isotope effect for complete proton transfer to the base can be calculated using the self-ionization constants of D_2O and H_2O, equations 8 and 9, respectively, and the equilibrium constant, L, defined by equation 10.

$$2OD^- + H_2O \underset{}{\overset{K_B}{\rightleftharpoons}} 2OH^- + D_2O \tag{6}$$

$$OD^- + \tfrac{1}{2}H_2O \underset{}{\overset{K}{\rightleftharpoons}} OH^- + \tfrac{1}{2}D_2O \tag{7}$$

$$2H_2O \xrightleftharpoons{K_H} OH^- + H_3\overset{+}{O} \tag{8}$$

$$2D_2O \xrightleftharpoons{K_D} OD^- + D_3\overset{+}{O} \tag{9}$$

$$2D_3\overset{+}{O} + 3H_2O \xrightleftharpoons{L} 2H_3\overset{+}{O} + 3D_2O \tag{10}$$

From equations 8 and 9 it is seen that

$$\frac{K_H}{K_D} = \frac{\left[OH^-\right]\left[H_3\overset{+}{O}\right]}{\left[H_2O\right]^2} \cdot \frac{\left[D_2O\right]^2}{\left[OD^-\right]\left[D_3\overset{+}{O}\right]} \tag{11}$$

In fact, K_H/K_D is the equilibrium constant, K_{eq}, for the exchange reaction (equation 12).

$$2H_2O + OD^- + D_3\overset{+}{O} \xrightleftharpoons{K_{eq}} 2D_2O + OH^- + H_3\overset{+}{O} \tag{12}$$

From equations 12, 11, 10 and 6 it is evident that

$$K_B = K_{eq}^2 / L \tag{13}$$

Since the equilibrium constants K_{eq} and L can be measured, $k^{OD^-}/k^{OH^-} = K_B^{1/2}$ can be calculated[24] using values of $L = 9.6$, $K_H = 1 \times 10^{-14}$ and $K_D = 1.56 \times 10^{-15}$

The maximum isotope effect, k^{OD^-}/k^{OH^-}, will occur when the proton is completely transferred from H_2O to DO^-, i.e. for the equilibrium reaction shown in equation 6. For reaction at 25°C, $K_B = 2.07$. For a transition state in which the proton is half-transferred between the substrate and base, the isotope effect will be $2.07^{1/2} = 1.44$ at 25 °C. Consequently, the observation of a secondary base isotope effect, k^{OD^-}/k^{OH^-}, which is greater than 1.44 at 25 °C, indicates that the proton is more than one-half transferred to base at the transition state. This allows an interpretation of the primary hydrogen–deuterium isotope effects to be made in terms of the degree of carbon–hydrogen bond rupture in the transition state.

III. THE PREPARATION AND USES OF ISOTOPICALLY LABELLED DIAZO COMPOUNDS

A. Synthesis of Labelled Diazo Compounds

Baldwin and Widdison[26] prepared ethyl α-deuterodiazoacetate by reacting ethyldiazoacetate with two different batches of sodium deuteroxide in deuterium oxide containing hexadecyltrimethylammonium bromide as a phase transfer catalyst. A proton NMR analysis showed the product was more than 98% deuterated at the α-carbon. The deuterated diazoacetate was then converted into two diastereomeric ethyl-(2-bromo-2-methylcyclopropane-1-d)carboxylates by reaction with 2-bromopropene and a rhodium(II) acetate catalyst in dichloromethane at room temperature (equation 14). The two diastereomers, which were obtained in a 1.3:1 ratio, were separated by gas chromatography and

$$N_2=CDCO_2Et + CH_2=C\underset{CH_3}{\overset{Br}{<}} \longrightarrow \underset{D}{\overset{CH_3\quad Br}{\triangle CO_2Et}} + \underset{CO_2Et}{\overset{CH_3\quad Br}{\triangle D}} \quad (14)$$

converted into several α- and β-deuterated (2-methylenecyclopropane acetyl-CoA molecules which were used to measure the isotope effects for the inactivtion of the acyl-coenzyme A dehyrogenase from pig kidney.

Several nitrogen-15 labelled silyl- and organometallic diazomethanes have been prepared for NMR studies[27]. The silyldiazomethanes have been prepared by nitrosation of the product prepared by treating trimethylsilylmethyl chloride with acetamide. The nitrogen-15 label has been incorporated from either nitrogen-15 labelled acetamide and/or nitrogen-15 labelled sodium nitrite. The procedures for preparing these diazomethanes labelled at the β-, at the α-, and at the α- and the β-nitrogens, are presented in equations 15–17, respectively

$$Me_3SiCH_2-N\underset{C=O}{\overset{H}{<}}\underset{CH_3}{} \xrightarrow[H_2SO_4]{Na^{15}NO_2} Me_3SiCH_2-\underset{\underset{CH_3}{|}}{N}-{}^{15}NO \quad (15)$$
$$\underset{C=O}{}$$

$$\xrightarrow[vacuum]{PhCH_2NH_2} Me_3SiCHN^{15}N$$

$$Me_3SiCH_2Cl + Na^{15}NHCOCH_3 \xrightarrow{-NaCl} Me_3SiCH_2{}^{15}NHCOCH_3 \xrightarrow[H_2SO_4]{NaNO_2}$$

$$Me_3SiCH_2{}^{15}N\underset{|}{-}NO \xrightarrow[vacuum]{PhCH_2NH_2} Me_3SiCH^{15}NN \quad (16)$$
$$COCH_3$$

$$Me_3SiCH_2Cl + Na^{15}N\underset{C=O}{\overset{H}{<}}\underset{CH_3}{} \xrightarrow{-NaCl} Me_3SiCH_2-{}^{15}N\underset{C=O}{\overset{H}{<}}\underset{CH_3}{} \xrightarrow[H_2SO_4]{Na^{15}NO_2}$$

$$(17)$$

$$Me_3SiCH_2-{}^{15}N\underset{C=O}{\overset{}{-}}{}^{15}NO \xrightarrow[vacuum]{PhCH_2NH_2} Me_3SiCH^{15}N{}^{15}N$$
$$\underset{CH_3}{}$$

The organometallic diazomethanes were prepared by reacting the appropriately labelled diazomethane precursor with different organometallic amines (equations 18 and 19).

The variously labelled diazomethanes can be prepared from nitrogen-15 labelled methylammonium chloride and sodium nitrite. A synthesis for the formation of diazomethane labelled at both the α- and the β-nitrogens is described in equation 20.

$$L-M-N(CH_3)_2 + (CH_3)_3SiCH = {}^{15}N = N \longrightarrow (L-M)_2C = {}^{15}N = N \tag{18}$$

$$L = (CH_3)_3Ge, (CH_3)_3Sn \text{ and } (CH_3)_2As$$

$$L-M-N(Si(CH_3)_3)_2 + (CH_3)_3Si-CH = N = {}^{15}N \longrightarrow (L-M)_2-C = N = {}^{15}N \tag{19}$$

$$L = (CH_3)_3Pb$$

$$CH_3{}^{15}\overset{+}{N}H_3 \ Cl^- + NH_2\overset{\overset{\displaystyle O}{\parallel}}{C}NH_2 \longrightarrow CH_3{}^{15}NH\overset{\overset{\displaystyle O}{\parallel}}{C}NH_2 \xrightarrow[\text{HCl, 0 °C}]{Na^{15}NO_2}$$

$$CH_3{}^{15}\underset{{}^{15}NO}{N}-\overset{\overset{\displaystyle O}{\parallel}}{C}NH_2 \xrightarrow{KOH} CH_2 = {}^{15}N = {}^{15}N \tag{20}$$

B. Mechanistic Studies Using Diazo Compounds

Bethell and coworkers[28] have used isotope effects to examine the decomposition reaction of alkyl diazo compounds. In particular, these workers wished to determine whether a carbene radical anion, **3**, was generated from an alkyl diazo compound (equation 21).

$$R_2-CN_2 \longrightarrow [R_2-CN_2]^{\bullet-} \longrightarrow [R_2-C\colon]^{\bullet-} + N_2 \tag{21}$$
$$(3)$$

The radical anion has been identified as the initial intermediate formed in the electrochemical reduction of the diazo compound and several authors have suggested, on the basis of product studies of reduction reactions of diazo compounds, that the carbene radical anion is formed from the radical anion by the loss of nitrogen. However, other work suggests that the radical anion reacts with a hydrogen before the loss of nitrogen and that the carbene radical anion is not formed in the reaction (equation 22).

$$R_2-CN_2 \xrightarrow{e^-} [R_2-CN_2]^{\bullet} \xrightarrow{H\bullet} R_2-CH-N_2 \tag{22}$$

Bethell and coworkers[29] reacted 9-diazofluorene with potassium t-butoxide in a 90 volume percent t-butyl alcohol–dimethylsulphoxide mixture. Under these conditions, the diazofluorene decomposes slowly to difluoren-9-ylidenehydrazine. However, it decomposes rapidly in the presence of diazodiphenylmethane or other proton donors to give a high yield of two azines, difluoren-9-ylidenehydrazine and diphenylmethylene (fluoren-9-ylidene) hydrazine (equation 23).

One of the strongest pieces of evidence against the formation of the carbene radical anion comes from an experiment using a nitrogen-15 labelled diazofluorene. The latter was prepared by reacting fluorenone oxime labelled 88% with nitrogen-15 with chloramine (equation 24).

The reaction is initiated when the anion of 2-ethylfluorene transfers an electron to 9-diazofluorene forming the 9-diazofluorene radical anion. The products of the diphenylmethane - 9-diazofluorene radical anion reaction are shown in Scheme 1.

$$\text{Ph}_2\text{C}=\text{N}=\text{N} + \underset{\text{(fluorenylidene)}}{\bigcirc}=\text{N}=\text{N} \xrightarrow{t\text{-BuOK}} \underset{\text{(fluorenylidene)}}{\bigcirc}=\text{N}-\text{N}=\underset{\text{(fluorenylidene)}}{\bigcirc} \qquad (23)$$

$$+ \text{Ph}_2\text{C}=\text{N}-\text{N}=\underset{\text{(fluorenylidene)}}{\bigcirc}$$

$$\underset{\text{(fluorenone)}}{\bigcirc}=\text{O} + {}^{15}\text{NH}_2\text{OH} \rightleftharpoons \underset{\text{(fluorenylidene)}}{\bigcirc}={}^{15}\text{N}_{\text{OH}} \xrightarrow[\substack{\text{OH}^- \\ \text{NH}_3}]{\text{NaOCl}} \underset{\text{(fluorenylidene)}}{\bigcirc}={}^{15}\text{N}=\text{N} \qquad (24)$$

When the nitrogen-15 labelled substrate was used, approximately 85% of the difluoren-9-ylidenehyrazines and 40% of the unsymmetrical azines were labelled with a single nitrogen-15. However, since only 88% of the diazofluorene was labelled with nitrogen-15, virtually all of the difluoren-9-ylidenehydrazine was formed from a diazofluorene radical anion that did not lose nitrogen–15 before it reacted. Moreover, a significant portion of the unsymmetrical azines were also formed from a radical anion that reacted before it lost nitrogen (equation 25). Obviously, the carbene radical anion which forms when the radical anion loses nitrogen was not involved in these reactions.

It is important to note that even though a significant portion of the unsymmetrical azines are not labelled, they may have been formed by attack of carbon-9 of the diazofluorene radical anion on the terminal nitrogen of diazodiphenylmethane with loss of the labelled nitrogen (equation 26). Thus, even the formation of an unlabelled azine does not prove, or even require, that a carbene radical anion is formed in the decomposition of the diazoalkane.

In another study, Bethell and Parker[28] generated the diazodiphenylmethane radical anion electrochemically in CD_3CN containing tetramethylammonium tetrafluoroborate as the supporting electrolyte. They found that the decomposition of the radical anion was first order in the radical anion in both acetonitrile and acetonitrile-d_3. They also showed that the decomposition was not influenced by added proton donors such as diethyl malonate. The kinetic isotope effect found when the solvent was changed from acetonitrile to acetonitrile-d_3 in the decomposition of the radical anion was extremely large at 20.3. The corresponding isotope effect in DMF and DMF-d_7 was approximately 4. These isotope effects were interpreted as primary hydrogen–deuterium kinetic isotope effects associated with the removal of a hydrogen from the solvent in the slow step of the reaction (equation 27). The extremely large isotope effect found when the solvent is acetonitrile was attributed to tunnelling. These isotope effects, coupled with the first-order kinetics, show that the hydrogen transfer and not the loss of nitrogen occurs in the slow step of the reaction. This effectively rules out the formation of the carbene radical anion because the reactions of this high-energy intermediate would be extremely fast.

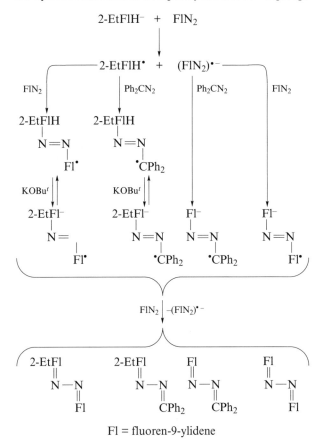

Fl = fluoren-9-ylidene

SCHEME 1

As expected, the product ratios were quite different in acetonitrile and acetonitrile-d_3. For example, the major products were diphenylmethane and benzophenone and only a small amount of benzophenone azine when the reaction was carried out in acetonitrile whereas the percentage of benzophenone azine was ten times larger, i.e. almost equal to the diphenylmethane when the solvent was acetonitrile-d_3. This occurs because the large primary hydrogen–deuterium kinetic isotope effect in acetonitrile-d_3 slows the hydrogen abstraction reaction which leads to diphenylmethane. These product isotope effects are only consistent with the mechanism shown in equation 28 and clearly demonstrate that the radical anion does not decompose by losing nitrogen to form the carbene radical anion when diazoalkanes are reduced.

The amount of azine in the product increases when the initial concentration of diazodiphenylmethane is increased. This occurs because the added diazodiphenylmethane increases the probability of the coupling reaction. Finally, it is probable that the diazodiphenylmethane rather than the diazodiphenylmethane radical anion is the reactant in the formation of the azine because the hydrogen–deuterium isotope effect indicates the primary reaction of the radical anion is with a hydrogen from the solvent.

$$(25)$$

$$(26)$$

Bethell and coworkers[30] have also investigated the oxidation of diazoalkanes. The initial step in the oxidation reaction is the reversible formation of the radical cation, $Ph_2C—N_2^{+\cdot}$, which forms when the diazodiphenylmethane reacts with either copper(II) perchlorate or the stable radical cation salt, tris-(p-bromophenyl)ammoniumyl perchlorate in acetonitrile (equation 29).

$$Ph_2C-N_2{}^{\bullet-} + CH_3CN \longrightarrow (Ph_2C-N_2,H)^- + {}^\bullet CH_2CN \quad (27)$$

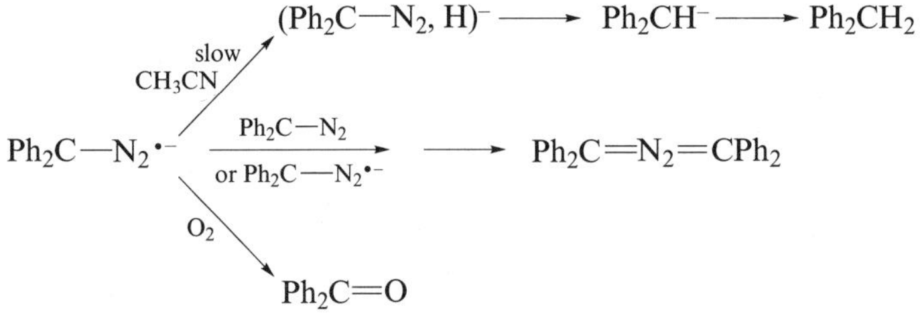

$$Ph_2C-N_2{}^{\bullet-} \begin{array}{c} \overset{\text{slow}}{\underset{CH_3CN}{\nearrow}} (Ph_2C-N_2, H)^- \longrightarrow Ph_2CH^- \longrightarrow Ph_2CH_2 \\ \xrightarrow[\text{or } Ph_2C-N_2{}^{\bullet-}]{Ph_2C-N_2} Ph_2C=N_2=CPh_2 \\ \underset{O_2}{\searrow} \\ Ph_2C=O \end{array} \quad (28)$$

$$Ph_2C=N=N \xrightarrow[\left(Br-\bigcirc-\overset{\bullet+}{N}HClO_4^-\right)_3]{Cu^{II} \text{ or}} Ph_2C=N=\overset{\bullet+}{N} + \begin{array}{c} Cu^I \text{ or} \\ \left(Br-\bigcirc-N\right)_3 \end{array} \quad (29)$$

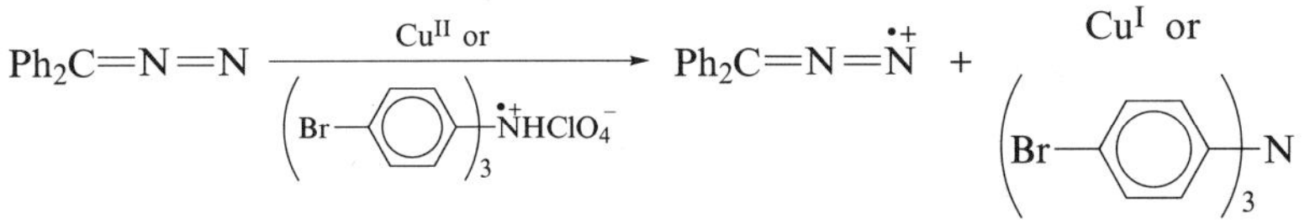

However, it is not known whether the electron is removed from the carbon or a nitrogen. The reaction is first order in the oxidant, so the formation of the radical cation is the rate-determining step of the reaction. The chain process which follows the formation of the radical cation yields the products shown in Scheme 2.

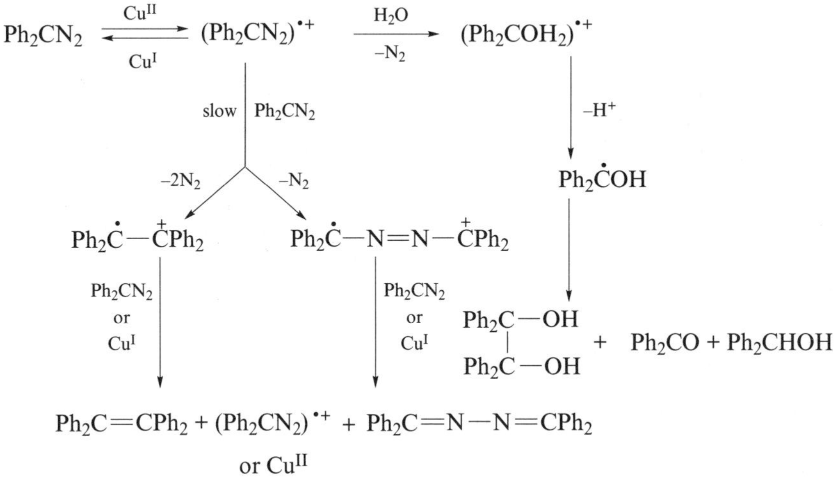

$$Ph_2CN_2 \underset{Cu^I}{\overset{Cu^{II}}{\rightleftharpoons}} (Ph_2CN_2)^{\bullet+} \xrightarrow[-N_2]{H_2O} (Ph_2COH_2)^{\bullet+}$$

$$\text{slow} \Big| Ph_2CN_2 \qquad\qquad \Big| -H^+$$

$$\overset{-2N_2}{\swarrow} \quad \overset{-N_2}{\searrow} \qquad\qquad Ph_2\overset{\bullet}{C}OH$$

$$Ph_2\overset{\bullet}{C}-\overset{+}{C}Ph_2 \qquad Ph_2\overset{\bullet}{C}-N=N-\overset{+}{C}Ph_2$$

$$\begin{array}{cc} Ph_2CN_2 & Ph_2CN_2 \\ \text{or} & \text{or} \quad Ph_2C-OH \\ Cu^I & Cu^I \quad | \qquad\qquad + \quad Ph_2CO + Ph_2CHOH \\ & \qquad\quad Ph_2C-OH \end{array}$$

$$Ph_2C=CPh_2 + (Ph_2CN_2)^{\bullet+} + Ph_2C=N-N=CPh_2$$

$$\text{or } Cu^{II}$$

SCHEME 2

Tetraphenylethylene is the major product because it is produced from the more stable tetraphenylethyl radical cation. Bethell and Parker found that the rate of decomposition of the radical cation was dependent on the concentration of diazodiphenylmethane. This rules out the formation of a carbene radical cation intermediate (equation 30) because the formation of such a high-energy intermediate would be rate-determining and the reaction would be zero-order in diazodiphenylmethane. As the concentration of water in the solvent is increased, the rate increases and then decreases. This is characteristic of a

$$Ph_2C{=}N{=}N \longrightarrow Ph_2C{=}N{=}\overset{\bullet}{\overset{+}{N}} \longrightarrow Ph_2\overset{+}{C}{}^{\bullet} + N{\equiv}N \qquad (30)$$

perchloric acid catalysed reaction. In an effort to confirm this, the authors studied the reaction by adding both H_2O and D_2O to the solvent to show that a proton transfer reaction was occurring. The isotope effect increased as the concentration of water in the solvent increased, i.e. the hydrogen–deuterium kinetic isotope effect increased from 1.5 to 4.5 when the water concentration was increased from 0.046 M to 1.39 M and the radical cation was formed by reaction with copper(II) perchlorate, and from 1.1 to 3.2 when the water concentration was increased from 0.046 M to 0.46 M and the radical cation was formed in the reaction with tris-(p-bromophenyl)ammoniumyl perchlorate. These isotope effects, which are similar to the primary hydrogen–deuterium kinetic isotope effect of 3.25 at 0.66 M water observed for the perchloric acid catalysed reaction of 4,4′-dichlorodiphenyldiazomethane at 29.5 °C[31], were attributed to a rate-determining proton transfer from the solvent to the diazomethane to form a carbocation (equation 31).

$$Ph_2C{=}N_2 \xrightarrow[\text{slow}]{H^+} Ph_2\overset{+}{C}H{-}N_2 \longrightarrow Ph_2CH^+ + N_2 \qquad (31)$$

with Ph_2CHOH (via H_2O) and $Ph_2C{=}CPh_2$ (via $Ph_2C{=}N_2$).

The increase in the isotope effect as the percent water in the solvent increases is accompanied by the expected change in the products of the reaction. As the percent water increases, the amount of the diphenylmethanol increases and the amount of tetraphenylethylene and benzophenone azine decreases. The larger isotope effect is observed when the water concentration is higher, because the slow proton transfer step of the acid catalysed reaction becomes more important as the carbocation pathway accounts for a greater portion of the reaction. Obviously, no isotope effect is observed at low concentrations of water where this reaction is not significant.

Bakke, Bethell and Parker[32] also found that diphenyldiazomethane decomposed by a cation mechanism when the reaction was carried out by anodic oxidation in acetonitrile-water (deuterium oxide) mixtures. The primary hydrogen–deuterium kinetic isotope effect of 3.3 at 17 °C found in this reaction was identical to that found for the acid catalysed reaction above, and confirmed that proton transfer to the diazo carbon occurs in the rate-determining step of the reaction. This is obviously consistent with the formation of a carbocation intermediate; see equation 31. The rate-determining proton transfer reaction has been confirmed by the observation that the reaction ceases when 2,6-lutidine is added.

Yates and coworkers[33,34] have investigated the reactions of syn- and anti-7-substituted 3-diazo-2-norbornanones in acidic medium. The products obtained in D_2O/D_2SO_4 mixtures have indicated that these α-diazo ketones can be converted into exo- and endo-diazonium ions by a proton transfer reaction. The ratio of proton transfer to the exo- and endo-face of the diazo ketone was determined from the deuterated 1-carboxy-cyclohex-2-enes formed when these α-diazo ketones were reacted in D_2O/D_2SO_4 mixtures (equation 32).

The product analysis showed that the protonation was exclusively (>99%) from the exo-face when the unsubstituted, the anti-7-isopropyl- and the anti-7-t-butyl compounds

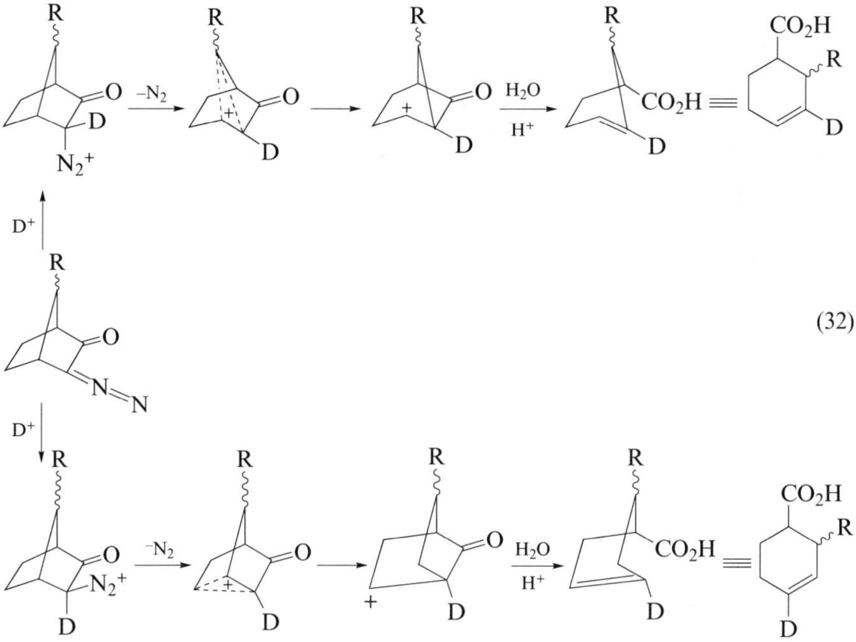

(32)

reacted. However, when a *syn*-7-isopropyl or t-butyl group was present, some protonation occurred from the *endo*-face, presumably because of steric hindrance by the large substituents at C_7. When the *syn*-7 substituent is isopropyl, approximately 6% of the protonation is from the *endo*-face while approximately 18% is from the *endo*-face when the *syn*-substituent is *t*-butyl. A primary hydrogen deuterium kinetic isotope effect of 2.38 was found in the acid catalysed decomposition of *syn*-7-isopropyl-3-diazo-2-norbornanone in acidic medium at 25 °C. This was taken as evidence that these reactions proceeded via an A-S_E2 mechanism where the proton transfer to the carbon of the diazo group is rate-determining (equation 33).

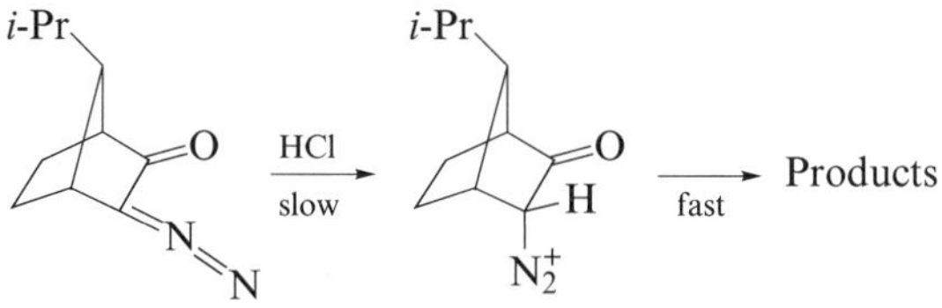

(33)

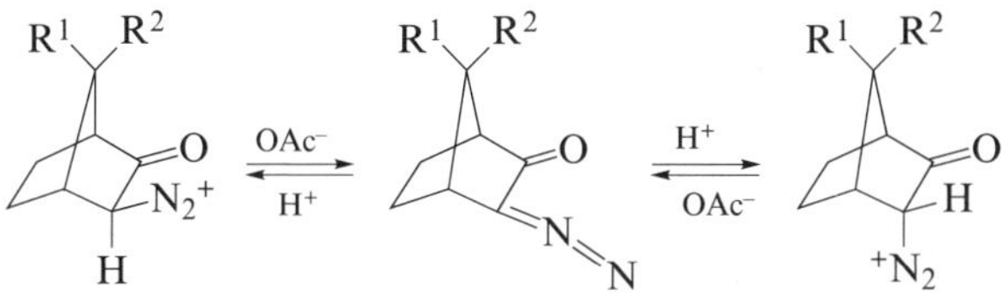

(34)

When these compounds were reacted in acetic acid–sodium acetate buffers on the other hand, the proton transfer reaction was reversible and the diazo compound was in equilibrium with the *endo*-and *exo*-α-keto diazonium ions (equation 34).

The irreversible A-S$_E$2 mechanism was observed in dilute hydrochloric acid solution when there was no substituent at C-7 and when both *syn*- and *anti*-isopropyl and *t*-butyl groups were at carbon-7. The reversible mechanism was only found when an isopropyl or *t*-butyl group was *syn* to the diazo group and the reaction was carried out in the acetic acid–acetate ion buffer.

Thomas and Fry[35] have measured the carbon-14 kinetic isotope effects for the Wolff rearrangement of *para*-substituted diazoacetophenones-1-^{14}C in *t*-butanol containing silver benzoate and triethylamine, and in aniline, in an effort to learn whether the rearrangement occurs by a concerted or a stepwise mechanism. Observing a significant carbon-14 kinetic isotope effect would demonstrate that the reaction was concerted, i.e. that the rearrangement of the alkyl group occurred as the nitrogen was displaced from the substrate rather than in a two-step process involving a carbene intermediate (equation 35).

(35)

The *para*-substituted carbon-14 labelled diazoacetophenones were prepared from the appropriate *para*-substituted benzoic acid-1-^{14}C using conventional methods. A significant carbon isotope effect (k^{12}/k^{14}) of 1.0120 ± 0.0026 was found when the *para*-substituent was methoxy whereas very small isotope effects of between 1.002 and 1.003, i.e. 1.000 considering the experimental error, were found when the *para*-substituents were H and NO$_2$. The same small isotope effect was found when diazoacetophenone reacted in aniline at 120 °C. Because these reactions are first order in nitrogen, the nitrogen is lost in the slow step of the reaction. The primary carbon isotope effect found in the *para*-methoxy-diazoacetophenone reaction indicates that the Wolff rearrangement of this substrate is concerted, i.e. that the nitrogen is eliminated as the phenyl group is transferred and the reaction does not occur via a two-step mechanism and a carbene intermediate. The failure to detect an isotope effect when the *para*-substituent is H or NO$_2$ and in aniline suggests that these reactions occur via a two-step mechanism and the carbene intermediate. However, the authors warn that it is possible that these reactions are concerted but that the carbon-14 kinetic isotope effects for the bond-formation and the bond-rupture processes cancel.

Miller and coworkers[36] have used deuterium labelling to study the mechanism of the electrophilic aromatic substitution reactions of β-aryl-α,β-unsaturated diazoalkanes. It had been proposed that the reaction proceeded via an 8π electrocyclic ring closure followed by a [1,5] sigmatropic hydrogen shift (equation 36).

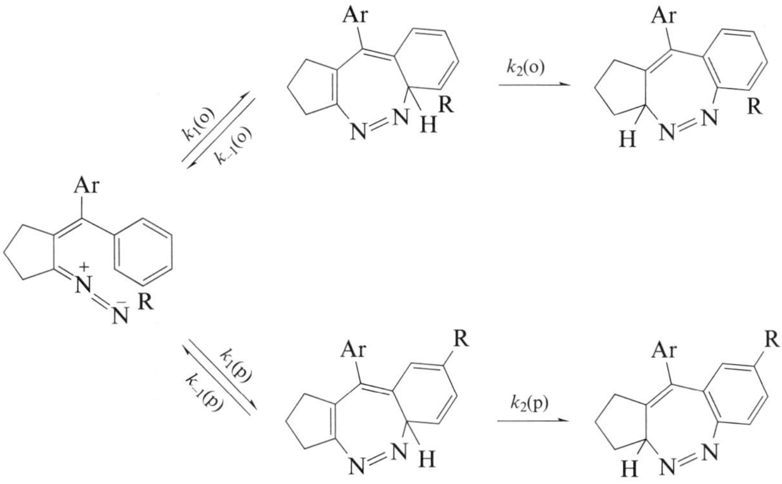

(36)

Since it is highly unlikely that the sigmatropic hydrogen shift is reversible, the reaction either proceeds via a reversible first (k_1) step followed by a rate- determining k_2 step, or via a slow k_1 step followed by a fast k_2 step. The correct mechanism has been determined by reacting the diazo compounds deuterated at the ring closure sites in the aryl rings (equation 37).

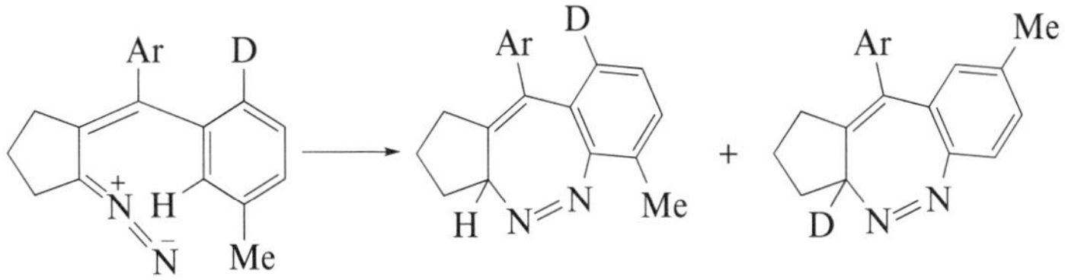

(37)

If the first step is rate-determining, and the product is formed by a rapid migration of a hydrogen or a deuterium (equation 36), the product ratio of **4/5** for attack at the two possible migration sites will not be affected by the presence of deuterium at the migration site. However, if the first step is reversible and the second step is rate-determining, the product ratio will be affected by the primary hydrogen–deuterium kinetic isotope effect associated with the migration of the hydrogen (deuterium) in the second step of the reaction.

Clearly, the product ratio of **4/5** is affected markedly by the presence of deuterium at the migration site. When the undeuterated reactant is used, the product ratio (for migration of the hydrogen adjacent to the methyl group/migration of the hydrogen *para* to the methyl group) is 4.3/1, but it changes to 17.2/1 when a deuterium is *para* to the methyl group, i.e. there is a marked preference for hydrogen migration. When the product ratios were corrected for incomplete deuteration, the product ratio for the hydrogen migration reaction (equation 37) is changed to 20.5/1. Thus, the product ratio has changed by a factor of 20.5/4.3 = 4.8 when a deuterium is present. Obviously, these results indicate the presence of a primary hydrogen deuterium isotope effect on the reaction and are only consistent with a reversible ring closure followed by a rate-determining hydrogen migration.

It is worth noting that the product ratio of **4/5** is also different from the 4.3/1 found for the undeuterated substrate, when the deuterium is adjacent to the methyl group (equation 38). In this case, the corrected product ratio of **4/5** is 1.1/1. The much smaller amount of

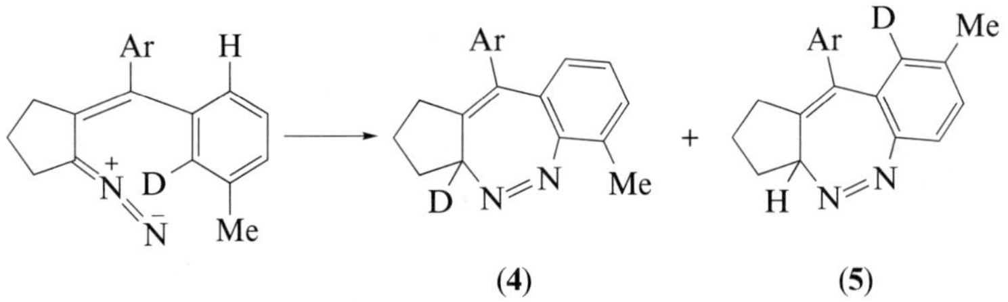

(38)

(4)   **(5)**

product **5** found in the reaction where the deuterium atom migrates is due to the primary hydrogen–deuterium kinetic isotope effect associated with the transfer of the migrating hydrogen. This different product ratio confirms that the hydrogen does not migrate in the fast step of the reaction.

A similar experiment using the diazoalkane with no substituent on the β-aryl group (equation 39) gave similar results, i.e. the ratio

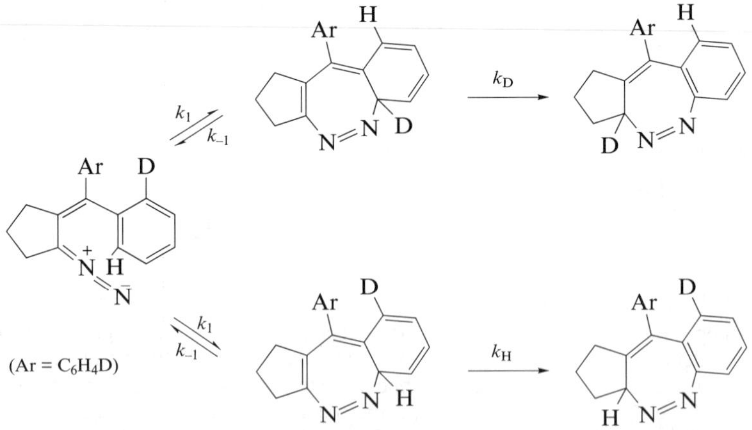

(Ar = C₆H₄D)

(39)

(product formed by hydrogen migration/product formed by deuterium migration) was 4.4/1. It is important to note that this is very close to the value found when the methyl substituent was on the aryl group. Thus, the presence of the substituent on the β-aryl group does not alter the mechanism of the reaction and both compounds react via a reversible first step followed by a rate-determining [1,5] sigmatropic hydrogen shift.

IV. PREPARATION AND USES OF ISOTOPICALLY LABELLED DIAZONIUM SALTS

A. Mechanistic Studies Using Diazonium Salts

The dediazoniation reaction of aryl diazonium salts has been thought to proceed via a phenyl cation that is formed reversibly. The cation has been shown to react with both of the nitrogens in the departing nitrogen molecule and with external nitrogen. These results clearly demonstrate the intermediacy of both a tight phenyl cation–nitrogen ion–molecule pair (**6**) and a free phenyl cation (**7**); see equation 40. Both a large, normal secondary hydrogen-deuterium kinetic isotope effect of 1.52 found for the reaction of 2,4,6-trideutero-benzenediazonium ion and large primary and secondary nitrogen kinetic isotope effects have confirmed the formation of the phenyl cation in these reactions[37].

$$C_6H_5 \overset{+}{-} \overset{+}{N}_\alpha \equiv N_\beta \rightleftharpoons [C_6H_5^+ \overset{N}{\underset{N}{|||}}] \rightleftharpoons C_6H_5^+ + N \equiv N \qquad (40)$$

$$\qquad\qquad\qquad\qquad (6) \qquad\qquad\qquad (7)$$

The secondary deuterium kinetic isotope effect is comprised of two secondary β-deuterium kinetic isotope effects of 1.22/β-D and a secondary δ-deuterium kinetic isotope effect of 1.02[38]. This kinetic isotope effect is identical to those found in several protic and dipolar aprotic solvents such as sulphuric acid–water, trifluoroethanol and methylene chloride. As a result, it has been concluded that the phenyl diazonium salt decomposes by the same mechanism in all solvents.

The primary nitrogen kinetic isotope effect of 1.0384 has been found by Swain and coworkers[38] for rupture of the $C-N_\alpha$ bond. They also measured the secondary nitrogen kinetic isotope effect for this reaction, i.e. for the β-nitrogen, and found a significant isotope effect of 1.0106. The primary nitrogen kinetic isotope effect is almost equal to the theoretical maximum kinetic isotope effect and indicates that almost complete $C-N_\alpha$ bond rupture has occurred in the transition state of the rate-determining step of the reaction. The much smaller β-nitrogen kinetic isotope effect arises because a normal temperature-independent isotope effect and an inverse temperature-dependent kinetic isotope effect combine to give the small normal isotope effect that is observed. The temperature-dependent isotope effect is inverse because the bonding between the two nitrogens increases when the reactants are converted into the transition state. This occurs because the N≡N bond is stronger in nitrogen than in the diazonium ion.

Zollinger and coworkers have continued their pioneering work on the dediazoniation reactions of aryl diazonium salts. For example, Szele, Zollinger and Deshpande[39] have studied the effect of added salt on the dediazoniation reaction of 2,4,6-trimethylben-zenediazonium tetrafluoroborate in 2,2,2-trifluoroethanol. They found that adding 0.365 M potassium thiocyanate to the reaction increased the rate of reaction by 20%. The potassium thiocyanate also reduced the amount of the rearranged and the exchanged diazonium salt found when the β-nitrogen is labelled with nitrogen-15, to 88% and 70%, respectively, of their values in pure 2,2,2-trifluoroethanol (equations 41 and 42). These results are consistent with the proposed mechanism. The added thiocyanate ion can react with both the tight phenyl cation–nitrogen ion-molecule pair, **6** and the free phenyl cation,

$$C_6H_5 \overset{+}{-} \overset{+}{N} \equiv {}^{15}N \longrightarrow [C_6H_5^+ \overset{{}^{15}N}{\underset{N}{|||}}] \overset{k_{-1}}{\longrightarrow} C_6H_5 \overset{+}{-}{}^{15}N \equiv N \qquad (41)$$

$$C_6H_5-\overset{+}{N}\equiv{}^{15}N \longrightarrow [C_6H_5^{+}\overset{{}^{15}N}{\underset{N}{\lVert\lVert}}] \longrightarrow C_6H_5^{+} + N\equiv{}^{15}N \xrightarrow{\overset{N\equiv N}{}}_{k_{-2}}$$ (42)

$$C_6H_5-\overset{+}{N}\equiv N + {}^{15}N\equiv N$$

7, to form phenylthiocyanate and phenylisothiocyanate. This would reduce the probability of the k_{-1} and k_{-2} return reactions occurring (equations 41 and 42), and therefore increase the rate of disappearance of the diazonium ion (rate of reaction). Providing a new reaction pathway for the phenyl cation and the phenyl cation–nitrogen ion–molecule pair, i.e. trapping the intermediates with thiocyanate ion, would reduce the amount of the rearranged diazonium ion by reducing the importance of the k_{-1} step and the exchanged diazonium ion by reducing the return to the tight ion–molecule pair by the k_{-2} step. It is important to note that the different decreases in the percentage of exchange (down to 80% of its previous value) and rearrangement (down to 70% of its previous value) supports the contention that two different carbocation intermediates form in these reactions. Also, the greater decrease in the percent rearrangement (equation 41) than in the exchange reaction (equation 42), i.e. to 70% rather than to 88% of the value in pure 2,2,2-trifluoroethanol, suggests that the tight ion–molecule pair forms first along the reaction coordinate.

Although adding potassium thiocynate affected the exchange and rearrangement reactions significantly, identical secondary deuterium kinetic isotope effects of 1.47 and 1.46 were found for the reaction of the 2,4,6-trideuterobenzenediazonium ion in triflouoroethanol and in the trifluoroethanol–potassium thiocyanate mixtures. Moreover, no unexpected products were found and it was concluded that the potassium thiocyanate did not alter the mechanism of the reaction. This conclusion was supported by the identical, large secondary deuterium kinetic isotope effects which are consistent with the formation of the phenyl cation. Finally, the low (phenylthiocyanate/phenylisothiocyante) ratio of 5/1 is also consistent with the formation of a carbocation intermediate. Thiocyanate/isothiocyanate ratios of between 100 and 1000 have been reported for S_N2 reactions where the nucleophile forms a bond to the α-carbon before the leaving group has departed.

Lewis and coworkers[40,41] have suggested that the benzenespirodiazirine, **8**, rather than the phenyl cation–nitrogen tight ion–molecule pair and the phenyl cation, might be the intermediate in the dediazoniation reaction.

$$\left(\!\!\begin{array}{c}+\end{array}\!\!\right)\!\!\times\!\!\overset{N}{\underset{N}{\lVert}} \longleftrightarrow \left(\!\!\bigcirc\!\!\right) + \overset{N}{\underset{N}{\lVert}}$$

(8)

A comparison of the secondary deuterium kinetic isotope effects found for the dediazoniation and the rearrangement reaction have enabled Szele and Zollinger[42] to conclude that the benzenespirodiazirine intermediate is not formed during the reaction. The large secondary deuterium kinetic isotope effects of 1.46 and 1.51 for the dediazoniation reaction of 2,4,6-trideuterobenzenediazonium tetrafluoroborate in 2,2,2,-trifluoroethanol and in 1,1,1,3,3,3-hexafluoropropanol, respectively, are certainly consistent with the formation of the phenyl cation intermediate in the slow step of the reaction. Szele and Zollinger calculated the isotope effect for the rearrangement reaction (equation 41) from the secondary deuterium kinetic isotope effects and the percent rearrangement for the dediazoniation reactions of β-^{15}N-benzenediazonium ion and the β-^{15}N-2,4,6-trideuterobenzenediazonium ion. They found the secondary deuterium kinetic isotope effect for the rearrangement reaction in trifluoroethanol to be 1.42. This rearrangement isotope effect is almost identical to that found for the ionization reaction and is far too large to indicate the rate-determining formation of a benzenespirodiazirine intermediate. In fact, a small or inverse secondary β-deuterium kinetic isotope effect, like those found in S_N2 reactions[20], would be expected if a benzenespirodiazirine formed in the slow step of the reaction. As a result, it was concluded that phenyl cation–molecule pairs were responsible for the exchange reaction that accompanies the dediazoniation reaction and that the benzenespirodiazirine intermediate is not formed in these reactions.

Zollinger and coworkers[43-46] have also studied the diazo coupling reactions of diazonium salts extensively. The proposed mechanism for the N-coupling reaction between a diazonium ion and an aromatic amine involves formation of an amine–diazonium ion complex, **9**, which is converted into a σ-complex, **10**, which subsequently loses a proton to base forming the product (equation 43), where B is any base, i.e. the amine, the solvent or

$$\text{ArNH}_2 + \text{ArN}_2^+ \underset{\xleftarrow{\hspace{1cm}}}{\overset{K}{\rightleftharpoons}} (\text{ArNH}_2\cdots{}^+\text{N}_2\text{Ar}) \underset{k_{-1}}{\overset{k_1}{\rightleftharpoons}} [\text{ArNH}_2\text{N}_2\text{Ar}]^+$$
$$\qquad\qquad\qquad\qquad\qquad (\mathbf{9})\qquad\qquad\qquad\qquad (\mathbf{10})$$

$$\xrightarrow{k_2[\text{B}]} \text{ArNH}{-}\text{N}_2\text{Ar} + \text{BH}^+ \qquad (43)$$

the anion of the diazonium salt adduct. The $k_2[\text{B}]$ step is fully rate-determining in aqueous medium, i.e. the mechanism is S_E2, whereas in aprotic solvents either the formation of the σ-complex or the $k_2[\text{B}]$ step can be rate-determining[43].

Penton and Zollinger[44] have carried out a detailed investigation of the N-coupling reaction between p-chlorobenzenediazonium tetrafluoroborate and p-chloroaniline in dry, deoxygenated acetonitrile in an effort to confirm that the above mechanism applies to this reaction. This reaction was chosen because it gave the N-coupled product exclusively. The reaction was catalysed by several 'relatively strong' bases such as hexamethylphosphoramide, DMF, DMSO, water, pyridine, chloride ion and even acetone. Small N—H(D) isotope effects ($k_{\text{H}_2\text{O}}/k_{\text{D}_2\text{O}}$) of 1.10 in 5 (v/v)% water and of 1.15 in 30 (v/v)% water were observed for these reactions. These small isotope effects are found because an N—H(D) bond is broken in the rate-determining k_2 step (equation 43). The isotope effects are small because the transition state for this reaction is early and there is only a small amount of N—H bond rupture in the transition state, i.e. the proton is only slightly transferred to solvent in the transition state[1,14]. This conclusion that the proton transfer is rate-determining is supported by the observation that the reactions are strongly catalysed by chloride ion (base) at high percentages of water, so the k_2 step is at least partially rate-determining under these conditions. This conclusion is also supported by the kinetic isotope effects found by changing the solvent from water to deuterated water in the acid catalysed decomposition of the N-coupled diazoamine, 4,4′-dichlorodiazoaminobenzene, i.e. in the reverse of the N-coupling reaction. Here, an inverse primary hydrogen–deuterium kinetic isotope effect, i.e. $k_{\text{H}_2\text{O}}/k_{\text{D}_2\text{O}} = 0.81$, is observed when the percent water in the solvent is 1%. However, the isotope effect increases to 1.37 when the reaction is done in 5% water and is 1.91 at 30% water where the limiting kinetics are observed, i.e. where the rate is independent of the concentration of the basic catalyst. The normal hydrogen–deuterium kinetic isotope effects are consistent with a fast, pre-equilibrium proton transfer at the high-percent water but with a rate-determining proton transfer in 1% water. Thus, when the amount of water in the solvent is low, the N—H(D) bond is forming in the slow step of the microscopic reverse of the N-coupling reaction, i.e. a proton is being transferred from the solvent to the N-coupled diazoamine in the slow step of the reaction. This clearly rules out a mechanism where the σ-complex is formed in the slow step of the N-coupling reaction and the proton transfer from the σ-complex occurs in a fast step. The authors concluded that the N-coupling reaction in acetonitrile at low-percent water occurs via an S_E2 mechanism with a slow proton transfer.

An unusual feature of this N-coupling reaction is that the rate expression varies with the amount of water (base) in the solvent. At low concentrations of water, the rate depends on the concentration of the base (water) but it reaches a limiting rate, independent of the concentration of the base (water), at high concentration of water, i.e. when the water content was $> 30\%$. The limiting kinetics found at high-percent water can be rationalized as explained below. The pseudo-first-order rate constant k_{obs}, for the N-coupling reaction is given by equation 44. In water, $K[ArNH_2] \ll 1$. At low concentrations of water where

$$k_{obs} = (K[ArNH_2]/(1 + K[ArNH_2]))\cdot(k_1 k_2[B]/(k_{-1} + k_2[B])) \tag{44}$$

$k_2[B]/k_{-1} \ll 1$, k_{obs} is equal to $K[ArNH_2]\cdot k_1 k_2[B]/k_{-1}$ and the reaction is first order in B. At high concentrations of water (base), on the other hand, the concentration of B is large and $k_2[B]/k_{-1} \gg 1$. In this case, k_{obs} is equal to $K[ArNH_2]k_1$ and the rate is independent of base, i.e. the k_1 step is rate-determining.

Penton and Zollinger[45] also studied the azo coupling reactions between p-methoxy-benzenediazonium tetrafluoroborate and m-toluidine or N,N-dimethylaniline in dry, deoxygenated acetonitrile. These reactions were of interest because this diazonium ion reacted to form N-coupled and C-coupled products (equation 45). These reactions were

chosen for this study because they gave very high yields of C-coupled products. However, they showed the same base catalysis as the N-coupling reactions discussed above. This led Penton and Zollinger to suspect that the N-coupled product formed initially and that it rearranged to the C-coupled product.

In the reaction with m-toluidine, the kinetic order with respect to the amine and the primary hydrogen–deuterium kinetic isotope effect found when the N—H hydrogens on the amine are replaced with deuterium, decreased as the concentration of the amine in the reaction mixture increased. For example, the isotope effect decreased from 2.06 to 1.04 and the kinetic order with respect to m-toluidine decreased from 1.75 to 1.25, when the concentration of the amine was increased from 0.004 M to 0.256 M at 12.9 °C. Similar changes in the kinetic order and isotope effect were observed at 31.1 and 47.2 °C. Another unusual observation is that, at a particular amine concentration, the kinetic order with respect to the amine and the isotope effect increase with increasing temperature. Also, the enthalpy and entropy of activation for a particular temperature increase with increasing amine concentration. Finally, the reaction is catalysed by the bases, chloride ion and water.

In fact, when either chloride ion or water is added to the reaction mixture, both the rate and the product composition change markedly. For example, the rate increases by fourteen times and the product composition changes from 9% of the N-coupled diazoamino compound to 79% of the diazoamino compound when the chloride ion concentration is increased from zero to 1.4×10^{-3} M. The amount of the N-coupled diazoamino product also increased at all three temperatures used in the study as the amount of amine (base) increased. On the basis of the base catalysis and the hydrogen–deuterium kinetic isotope effects and their studies of other diazo coupling reactions, the authors concluded that the N-coupling and the C-coupling reactions both proceed by an S_E2 mechanism with the proton transfer from the σ-complex rate-determining. The results are best explained by the mechanism presented in Scheme 3.

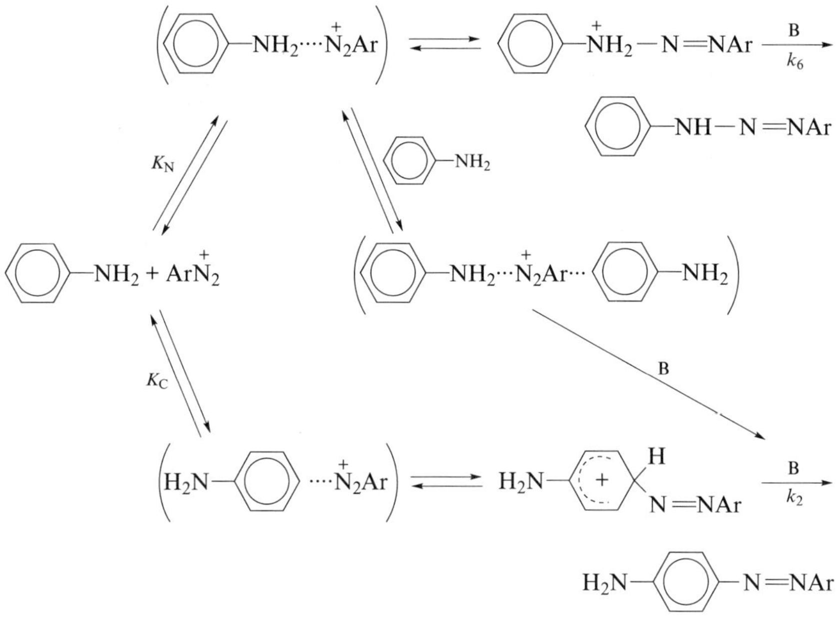

$$\left(\underset{}{\bigcirc}-NH_2 \cdots \overset{+}{N}_2Ar \right) \rightleftharpoons \underset{}{\bigcirc}-\overset{+}{N}H_2-N=NAr \xrightarrow{\underset{k_6}{B}}$$

$$\underset{}{\bigcirc}-NH-N=NAr$$

$$K_N \nearrow \qquad \underset{}{\bigcirc}-NH_2$$

$$\underset{}{\bigcirc}-NH_2 + Ar\overset{+}{N}_2 \qquad \left(\underset{}{\bigcirc}-NH_2 \cdots \overset{+}{N}_2Ar \cdots \underset{}{\bigcirc}-NH_2 \right)$$

$$K_C$$

$$\left(H_2N-\underset{}{\bigcirc} \cdots \overset{+}{N}_2Ar \right) \rightleftharpoons H_2N-\underset{N=NAr}{\overset{H}{\bigcirc{+}}} \xrightarrow{\underset{k_2}{B}}$$

$$H_2N-\underset{}{\bigcirc}-N=NAr$$

SCHEME 3

The isotope effects and kinetic order with respect to the amine decrease with increasing concentration of the amine because the proton transfer to the amine (base) in the k_2 and k_6 steps becomes less rate-determining as the concentration of the amine increases, i.e. the formation of the σ-complexes becomes rate-determining. The much greater amount of diazoamino compound isolated from the reaction at the higher concentrations of base occurs because the N-coupling reaction is faster than the C-coupling reaction. In fact, at 25% water (base), 90% of the product is the diazoamino compound and the N-coupling reaction is 9 times faster than the C-coupling reaction. This conclusion is important because it indicates that the C-coupled product must be formed from the N-coupled product. At low amine concentration, the rearrangement of the N-coupled product to the C-coupled product occurs at almost the same rate as the formation of the C-coupled product.

The unusal temperature dependence of the isotope effect is found because the enthalpy of activation $(\Delta H^{\ddagger})$ for the deuterated compound is less than that for the undeuterated substrate. This means k_H will increase faster with temperature than k_D and the isotope effect will increase with temperature.

Finally, the trends in the enthalpy and entropy of activation are more consistent with the formation of an amine–diazonium ion complex and its reaction with an additional molecule of amine to form the C-coupled product than to have the rearrangement from the N-coupled σ-complex.

Zollinger and coworkers [46] studied the acid catalysed rearrangement of the diazoamino compounds to the azoamino compounds in 20% aqueous acetonitrile (equation 46). Based

$$X-\underset{}{\bigcirc}-N=N-\underset{\underset{CH_3}{|}}{N}-\bigcirc \underset{\longleftarrow}{\overset{H^+}{\rightleftharpoons}} X-\bigcirc-N=N-\bigcirc-NHCH_3 \quad (46) \quad (46)$$

on the secondary deuterium kinetic isotope effects of 1.28 and 1.48 found at high and low buffer concentrations for the reaction between 4′-methoxy-N-methyldiazoaminobenzene and N-methylaniline and N-methyl-2,4,6-trideuteroaniline, and kinetic data, they concluded that the C-coupled compounds are formed from the diazoamino compounds by the Friswell–Green mechanism (equation 47). The large secondary hydrogen–deuterium

$$K_2[N]$$

kinetic isotope effects found in this reaction are consistent with the formation of the positively charged Meisenheimer-type complex in the rate-determining k_3 step of the reaction.

Jackson and Lynch[47] have used deuterium labelling and deuterium kinetic isotope effects to investigate the azo coupling reactions of diazonium ions with 3-substituted indoles in acetonitrile. No primary kinetic isotope effect was observed when 3-deuteroindole reacted with p-nitrobenzenediazonium tetrafluoroborate and it was concluded that the slow step of the reaction was attack of the diazonium ion at the three position of the indole ring (equation 48). The final step in the substitution reaction was the rapid loss of the proton at carbon-3. This conclusion was supported by the observation that there was no base catalysis in this reaction.

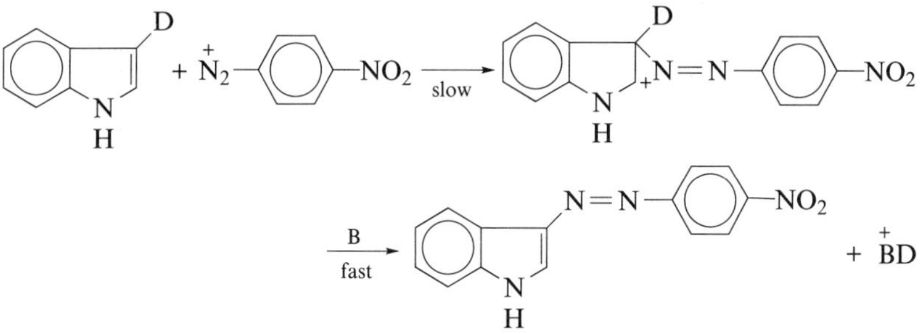

(48)

When 2-deutero-3-methylindole was the substrate, the reaction was base catalysed and a small primary hydrogen–deuterium kinetic isotope effect, which varied between 1.70 and 2.86 and increased as the concentration of the p-nitrobenzenediazonium tetrafluoroborate increased, was observed. These results suggest that the reaction occurs via an *ipso*-substitution-rearrangement mechanism (equation 49), where the *ipso*-substitution is reversible and the removal of the proton at carbon-2 is at least partially rate-determining.

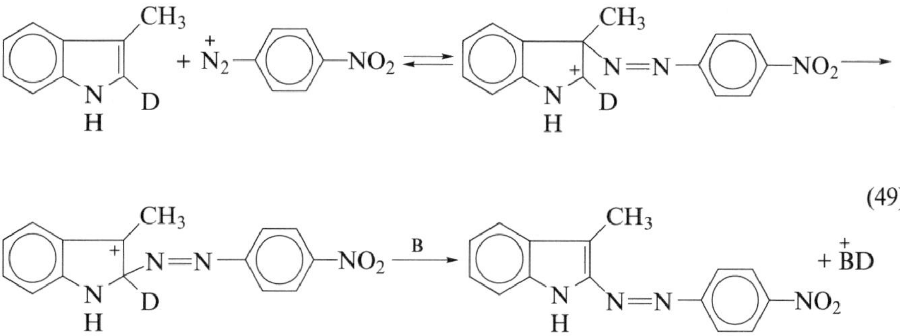

(49)

The behaviour of methyldiazonium ion in water has been investigated in phosphate buffer solutions at a pH of 7.4[48]. The observations indicate that the methyldiazonium ion exchanges protons with deuterated buffers via an equilibrium involving diazomethane (equation 50). The products were recovered as deuterated methanol when the various labelled diazonium ions decomposed to the methyl carbocation, which reacted with deuterated water to form labelled methanol (equation 51).

$$CH_3-\overset{+}{N}\equiv N + OD^- \longrightarrow CH_2=N=N \xrightarrow{D_2O} CH_2D-\overset{+}{N}\equiv N + OD^- \longrightarrow$$

(50)

$$CHD=N=N \xrightarrow{D_2O} CD_2H-\overset{+}{N}\equiv N + OD^- \longrightarrow CD_2=N=N \longrightarrow CD_3-\overset{+}{N}\equiv N$$

$$CD_2H-\overset{+}{N}\equiv N \longrightarrow CD_2H^+ + N\equiv N \xrightarrow{D_2O} CD_2HOD$$

(51)

The product ratios obtained when the methyldiazonium ion generated in four different ways is reacted in a phosphate buffer in D_2O is CH_2DOD = 35%, CHD_2OD = 13% and CD_3OD = 3% with CH_3OD = 48%. This indicates that the exchange reaction with deuterated buffers is fast with respect to the reaction with water to form methanol. The exchange was confirmed by reacting trideuteromethyldiazonium ion with phoshate buffers in water. In this experiment, the product ratio was CD_2HOH = 13.8%, CDH_2OH = 3.0% and CH_3OH = 2.7% with CD_3OH = 80.6%. It was found that these product ratios are consistent with an isotope effect of k_H/k_D = 2.7 for the exchange reaction.

Brown and Doyle[49] have studied the reduction of aryldiazonium ions. Two mechanisms are possible. The reduction could occur via a one-electron transfer 'non-bonded' outer-sphere mechanism where an electron is transferred from a reducing agent to the diazonium ion or via a bonded inner-sphere mechanism, where an intermediate complex undergoes homolytic cleavage (equations 52 and 53), respectively.

$$ArN_2^+ + Red:^- \longrightarrow ArN_2{\cdot} + Red{\cdot} \tag{52}$$

$$Red:^- = \text{Reducing agent}$$

$$ArN_2^+ + Red:^- \longrightarrow Ar{-}N{=}N{-}Red \longrightarrow ArN_2{\cdot} + Red{\cdot} \tag{53}$$

$$Red:^- = \text{Reducing agent}$$

The reaction between p-nitrobenzenediazonium tetrafluoroborate (ArN_2^+) and hydroquinone (H_2Q) in 0.05 M phosphate buffered solutions containing 2% acetonitrile at pH 7.0 was used to determine the mechanism of the reduction reaction. The major products from this reaction are hydroquinone and p-nitrophenyl products, e.g. p-nitrobenzene (equation 54), although other p-nitrophenyl products are formed in low yields. A product analysis

$$HO{-}\left\langle\bigcirc\right\rangle{-}OH + O_2N{-}\left\langle\bigcirc\right\rangle{-}\overset{+}{N_2}\bar{B}F_4 \longrightarrow$$

$$O{=}\left\langle\bigcirc\right\rangle{=}O + O_2N{-}\left\langle\bigcirc\right\rangle{-}H + N_2 \tag{54}$$

showed that the yield of the p-nitro products was twice that of hydroquinone. This suggested that the stoichiometry of the reaction was $2\,ArN_2^+ + H_2Q$. A study of the reaction using UV spectroscopy gave no evidence of an intermediate. Also, the solvent isotope effect k_{H_2O}/k_{D_2O} was equal to 3.6 at pH = 7.0 and 3.9 at pH = 4.5. More surprising was the observation that the p-nitrobenzene formed in the reaction contained no deuterium when the reaction was done in acetonitrile$-D_2O$. This demonstrated that the proton that is added to the benzene ring does not come from either hydroquinone or the solvent. However, 66% of the p-nitrobenzene was deuterated at the $para$-position when $CD_3C{\equiv}N$ rather than acetonitrile was used in H_2O at pH = 7.0. Thus, a large portion of the proton on the benzene ring came from the acetonitrile in the solvent.

The mechanism suggested by these results is

$$H_2Q \overset{K}{\rightleftharpoons} HQ:^- + H^+$$

$$HQ:^- + Ar{-}N_2^+ \overset{slow}{\longrightarrow} Q{\cdot}^- + Ar{-}N_2{\cdot} + H^+ \longrightarrow HQ{\cdot} + Ar{-}N_2{\cdot}$$

$$HQ{\cdot} + Ar{-}N_2^+ \longrightarrow Q + Ar{-}N_2{\cdot} + H^+ \tag{55}$$

and/or

$$Q{\cdot}^- + Ar{-}N_2^+ \longrightarrow Q + Ar{-}N_2{\cdot}$$

The last two reactions, which are extremely fast, account for the stoichiometric relationship $2\,ArN_2^+ + H_2Q$. The reaction is second order kinetically, however, because only one molecule of diazonium ion is consumed up to the rate-determining step of the reaction. The solvent isotope effect is attributed to the fact that the K_a for deuterated hydroquinone, $DO\!-\!C_6H_4\!-\!OD$, is 4.16 times less than that of hydroquinone. The absence of deuterium in the nitrobenzene when the reaction is done in D_2O indicates that the decomposition of the $Ar\!-\!N_2\cdot$ to product does not involve hydroquinone and that hydroquinone is not the source of the hydrogen that is added to the benzene ring in the final step of the reaction. In fact, the partial deuteration of the nitrobenzene suggests that the decomposition of the aryldiazenyl radical occurs by two pathways (equation 56), where SH is the solvent for the reaction.

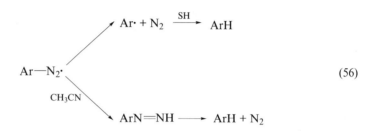

$$(56)$$

Finally, the failure to detect an intermediate, and the observation that the rate constants for the reactions between the hydroquinone monoanion and various *para*-substituted benzenediazonium ions are predicted by the Marcus theory of kinetics, strongly suggest that the reduction occurs by an outer-sphere electron transfer mechanism between the hydroquinone monoanion and the diazonium ion, equation 52.

V. THE PREPARATION AND USES OF ISOTOPICALLY LABELLED CYANO COMPOUNDS

A. Synthesis of Labelled Cyano Compounds

1. Syntheses using carbon-11 labelled cyano groups

Recently, several groups have used the carbon-11 isotope to prepare compounds with the label in the cyano group for medical, biological and mechanistic studies. This radioactive isotope is formed in a cyclotron or linear accelerator by bombarding nitrogen-14 atoms with high-energy (between 18 and 30 MeV) protons[50] (equation 57) ^{11}C decays by emitting a positron and forming ^{11}B (equation 58) with a short half-life of 20.13 minutes. The short

$$^{14}N + p^+ \longrightarrow {}^{11}C + He \tag{57}$$

$$^{11}C \longrightarrow {}^{11}B + e^+ \tag{58}$$

half-life makes this isotope attractive as a tracer in all kinds of medical (pharmaceutical) applications, because all the radioactivity is gone in a few hours and the exposure to radioactivity is reduced. In these applications, a compound that is absorbed in a particular organ in the body is labelled with carbon-11 and injected into the body. Then, the fate of the carbon-11 labelled compound in the body is determined by passing over the body a circular array of detectors that measure the 511-KeV gamma radiation that is released when the positron emitted in the decay of the carbon-11 isotope combines with an electron.

This measurement is called positron emisssion tomography or PET. A computer takes the output from the detectors and provides a picture of a slice through the body, showing the location and the relative concentration of the labelled compound.

The carbon-11 isotope formed in the accelerator reacts with traces of oxygen in the system and is initially recovered as $^{11}CO_2$, which is then converted by reaction with hydrogen into $^{11}CH_4$. The latter is then reacted with ammonia to form $H^{11}CN$ in ten minutes in yields of near 90% (equation 59).

$$^{11}C + O_2 \longrightarrow {}^{11}CO_2 + H_2 \xrightarrow[400\ °C]{Ni} {}^{11}CH_4 \xrightarrow[1000\ °C]{NH_3,\ Pt} H^{11}CN \qquad (59)$$

The $H^{11}CN$ has been converted into cyanogen bromide ($^{11}CNBr$) in high yield by reaction with bromine in a potassium hydroxide–triethylene glycol dimethyl ether solution at 180 °C [51] (equation 60).

$$H^{11}CN + Br_2 \xrightarrow[180\ °C]{KOH} {}^{11}CNBr \qquad (60)$$

These two sources of labelled cyanide, one with a positive cyano group ($^{11}CNBr$) and one with a negative cyano group ($H^{11}CN$), have opened up synthetic routes to a wide variety of labelled compounds. However, an essential requirement with using short-lived isotopes such as carbon-11 is that the synthesis must be completed rapidly. In practice, this means that the total synthesis and purification must be complete within three half-lives, i.e. in the case of carbon-11, within 60 minutes. In spite of this limitation, much work has been done with these synthons.

Although the first use of carbon-11 in a synthesis was reported by Cramer and Kistiakowsky in 1941[52], very little work was done using these synthons until the mid-1970s. Since then, many different compounds with this isotopic label have been synthesized[53–69].

2. Syntheses with carbon-11 labelled cyanide ion or hydrogen cyanide

The $H^{11}CN$ (or $^{11}CN^-$, if the reaction is done under basic conditions) synthon has been mainly used to extend the carbon chain by one carbon. For example, ^{11}C cyanide ion has been used in the synthesis of amino acids labelled in the carboxylate group. This is accomplished using the high pressure–high temperature modification of the Bucherer–Strecker synthesis[70–75]. In this reaction, bisulphite addition complex of an aldehyde reacts with cyanide ion in the presence of ammonium carbonate to form a hydantoin, which is then converted into the amino acid by basic hydrolysis[70, 73–75] (equation 61).

$$(61)$$

An alternative to this method is the acid hydrolysis of the complex formed when cyanide ion reacts with the bisulphite addition complex of an aldehyde in the presence of ammonia[76] (equation 62).

$$R-\overset{\overset{\displaystyle OH}{|}}{\underset{\underset{\displaystyle SO_3^-}{|}}{C}}-H \;+\; {}^{11}CN^- \;\xrightarrow{\;NH_3\;}\; R-\overset{\overset{\displaystyle NH_2}{|}}{\underset{\underset{\displaystyle H}{|}}{C}}-{}^{11}CN \;\xrightarrow[H^+]{\;H_2O\;}\; R-\overset{\overset{\displaystyle NH_2}{|}}{\underset{\underset{\displaystyle H}{|}}{C}}-{}^{11}C\overset{\displaystyle O}{\underset{\displaystyle OH}{\diagup}} \qquad (62)$$

The preparation of the amino acid, alanine, in this way has enabled Langstrom and coworkers to prepare L-tyrosine, L-dopa, L-tryptophan and 5-hydroxy-L-tryptophan labelled in the carboxy group in good yields in less than one hour[77] (equation 63).

$$CH_3CH-{}^{11}C\overset{\displaystyle O}{\underset{\displaystyle OH}{\diagup}} \xrightarrow[\underset{\displaystyle GPT}{\text{Catalase}}]{\text{D-AAO}} CH_3\overset{\overset{\displaystyle O}{\|}}{C}-{}^{11}C\overset{\displaystyle O}{\underset{\displaystyle OH}{\diagup}} \xrightarrow{\beta\text{-tyrosinase}}$$

(with NH_2 below the first carbon)

$$HO-\langle\!\bigcirc\!\rangle\!-CH_2\overset{}{\underset{\underset{\displaystyle NH_2}{|}}{C}}H-{}^{11}C\overset{\displaystyle O}{\underset{\displaystyle OH}{\diagup}}$$

(R substituent at top of ring)

Tryptophanase

$$R\!-\!\left[\text{indole}\right]\!-CH_2\overset{}{\underset{\underset{\displaystyle NH_2}{|}}{C}}H-{}^{11}C\overset{\displaystyle O}{\underset{\displaystyle OH}{\diagup}} \qquad (63)$$

R = H, OH

The addition reaction of carbon-11 labelled cyanide ion to the bisulphite addition adduct of an aldehyde has been extended to prepare carbon-11 labelled amines. Maeda and coworkers[78] prepared both p- and m-octopamine [2-(p-and m-hydroxyphenyl)-2-hydroxyethyl-amine] from the corresponding benzaldehyde by reducing the cyanohydrin formed in the reaction between the appropriate benzaldehyde and cyanide ion both under enzymatic conditions and by the basic modification of the Bucherer–Strecker synthesis, with borane–THF. The synthesis of p-octopamine is presented in equation 64.

$$p\text{-}HOC_6H_4CH{=}O + {}^{11}CN^- \xrightarrow{\;NaHSO_3\;} p\text{-}HOC_6H_4CHOH^{11}CN \xrightarrow{\;BH_3\text{–}THF\;}$$

$$p\text{-}HOC_6H_4CHOH^{11}CH_2NH_2 \qquad (64)$$

Both syntheses were done at 55 °C and the total reaction and purification took approximately 40 minutes. An enzyme catalysed cyanohydrin reduction was also investigated using two different enzymes. Although the enzyme catalysed reactions gave products of high optical purity, they did not improve the final yield, which was between one and two percent. The compounds were, however, formed with radiochemical purities of > 98%.

Both straight-chain and branched fatty acids labelled with carbon-11 at carbon-1 have also been made with $H^{11}CN$. Although these compounds have been made using a Grignard reagent[79, 80], the $H^{11}CN$ method[81] is superior because it can be done on a small scale and without having to use dry solvents. In this method, the $H^{11}CN$ from the linear accelerator was trapped in a KOH/DMSO mixture containing the appropriate alkyl bromide and the mixture was heated for 5 minutes at 120 °C. Then, 6 M sodium hydroxide was added, and the hydrolysis was completed by heating the mixture in a sealed vessel at 180 °C for 10 minutes. Acidification, extraction into ether and chromatography gave the purified fatty acid in approximately one hour (equation 65). Slightly different reaction conditions were required for the 3-methyl fatty acid syntheses.

$$RBr + {}^{11}CN^- \longrightarrow R^{11}CN \xrightarrow{OH^-} \xrightarrow{H^+} R^{11}CO_2H \qquad (65)$$

Spiroperidol, a dopamine-receptor antagonist, has been labelled with carbon-11 at the carbonyl group in a two-step synthesis[82]. The first step involved reacting 8-(3-chloropropyl)-1-phenyl-1,3,8-triazaspiro[4,5]decan-4-one with $H^{11}CN$ in a NaOH/DMSO mixture at 140 °C for 4 minutes. The final product was obtained when the labelled cyanide was reacted with p-fluorophenylmagnesium bromide in xylene for 5 minutes at 140 °C (equation 66). The radiochemical yield was up to 30% and the overall reaction and purification required 40 minutes.

$$(66)$$

The nucleophilic substitution reaction of halides and tosylates with $^{11}CN^-$ has been investigated by Hornfeldt and coworkers[66]. These workers reacted $K^{11}CN$ complexed with Kryptofix 2,2,2 with the diiodide, the ditosylate and the iodotosylate (equation 67), where $n = 3$ or 4. The products were formed in radiochemical yields of between 85 and 95% in 5 minutes. It is interesting that the reaction did not work when $n = 2$ and that the iodide leaving group can be removed exclusively in the presence of the tosylate leaving group.

$$X-(CH_2)_n-Y + K^{11}CN \xrightarrow[70\ °C,\ DMF\ or\ THF]{K\text{-}2,\,2,\,2/K^+,\ KOH} X-(CH_2)_n-{}^{11}CN \qquad (67)$$

$$X = Y = I;\ X = Y = OTs;\ X = OTs,\ Y = I$$

These reactions are useful because the iodides and tosylates formed can react with other nucleophiles to give interesting products. Some examples are given in equation 68.

$$I(CH_2)_n-{}^{11}CN$$

with reagents leading to:

$$\text{(ArO}^-, \text{DMF}, 110\,^{\circ}\text{C)} \longrightarrow \text{Ar}-(CH_2)_n-{}^{11}CN \quad (NO_2)$$

$$\text{NaH, DMSO, } (CH_3CO_2)_2CH_2 \longrightarrow (CH_3CO_2)_2CH-(CH_2)_n-{}^{11}CN$$

(68)

Thorell and coworkers[83] have reacted ${}^{11}CN^-$ with the R–, the S– and the racemic propylene oxide to form the β-hydroxycyanide. A subsequent hydrolysis formed β-hydroxybutyric acids (equation 69).

$$CH_3CH\overset{O}{\underset{}{\diagdown}}CH_2 + {}^{11}CN^- \longrightarrow CH_3\overset{OH}{\underset{|}{C}}HCH_2{}^{11}CN \xrightarrow[H^+]{H_2O} CH_3\overset{OH}{\underset{|}{C}}HCH_2{}^{11}COOH \quad (69)$$

The substitution by ${}^{11}CN^-$ was carried out in an aqueous solution in 10 minutes at 40 °C. This reaction was highly regiospecific with the ${}^{11}CN^-$ attack primarily at the least hindered carbon. Considerable racemization was observed when the hydrolysis was done with sodium hydroxide at 110 °C and almost complete racemization was observed when it was done with sulphuric acid at 150 °C. However, the hydrolysis product had an 87–97% enantiomeric excess when the hydrolysis was done with the weaker acid, hydrochloric acid, at 150 °C. The hydrochloric acid hydrolysis was complete in 10 minutes and the isolated product was obtained in between 45 and 50 minutes. This short reaction time has been reduced further by heating the reaction mixtures in a microwave oven (*vide infra*) but the effect of the microwaves on the chirality has not been determined.

Carbon-11 cyanide has also been used to prepare ${}^{11}C$ busulphan to determine how this drug, which has been used to combat myeloid leukemia, is metabolized in the body. The busulphan was labelled at the terminal carbon of the carbon backbone using the synthetic scheme in equation 70[64].

$$HOCH_2CH_2CH_2Br + {}^{11}CN^- \longrightarrow HOCH_2CH_2CH_2{}^{11}CN \xrightarrow[H^+]{H_2O}$$

$$\xrightarrow[2.\ \ H^+]{1.\ LiAlH_4} HOCH_2CH_2CH_2{}^{11}CH_2OH \xrightarrow{CH_3SO_2Cl}$$

$$CH_3SO_2OCH_2CH_2CH_2{}^{11}CH_2OSO_2CH_3 \quad (70)$$

Busulphan

The reaction between 3-bromo-1-propanol and the aqueous ${}^{11}CN^-$ solution was completed in 7 minutes at 90 °C. The hydrolysis was completed in only 9 minutes at 150 °C giving an 80% yield of the lactone. After a 2 minute reduction at room temperature with lithium aluminium hydride, the product was obtained in 10 minutes by treatment with methanesulphonyl chloride at 110 °C. The overall reaction and purification to pharmaceutical standards was complete in about 70 minutes and the decay-corrected isolated yield was approximately 5%. This was adequate for tracer studies in both monkeys and human subjects[84].

Antoni and Langstrom[65] have recently reported the synthesis of carbon-11 labelled vinyl cyanides from $H^{11}CN$. This provides useful synthons because vinyl chlorides can undergo Michael additions to give a wide variety of carbon-11 labelled compounds. The carbon-11 labelled vinyl and cinnamyl cyanides have been prepared in 35% yield and greater than 99% radiochemical purity in 3 minutes by reacting $^{11}CN^-$ with the appropriate bromide in the presence of a palladium catalyst and 18-crown-6 ether in acetonitrile (equation 71).

$$RCH{=}CHBr + {}^{11}CN^- \xrightarrow[\text{18-crown-6, CH}_3\text{CN}]{\text{Pd(Ph}_3\text{P)}_4} RCH{=}CH^{11}CN \qquad (71)$$

$$R = H, Ph$$

The reaction with vinyl bromide was carried out at 40 °C while the reaction with β-bromostyrene was done at 100 °C.

An example illustrating the synthetic possibilities arising from the Michael addition reaction with these carbon-11 labelled vinyl cyanides is their addition to the carbanion formed from dimethyl malonate[65] (equation 72). These compounds are useful from a

$$(CH_3CO_2)_2CH^- + \underset{{}^{11}CN}{\diagdown} \xrightarrow[\substack{\text{DMSO}\\ \text{80 °C}\\ \text{3 min}}]{\text{NaH}} (CH_3CO_2)_2CHCH_2CH_2{}^{11}CN \qquad (72)$$

synthetic point of view because they can be converted into labelled amino acids. For example, [5-^{11}C]glutamic acid has been prepared in this way (equation 73).

$$CH_2{=}CH^{11}CN + CH_3CONHCH(CO_2Et)_2 \xrightarrow[\text{110 °C, 5 min}]{\text{NaH/DMSO}}$$

$$(73)$$

$$N{\equiv}C^{11}CH_2CH_2$$
$$\underset{CH_3CONH}{\diagup}C(CO_2Et)_2 \xrightarrow{\text{NaOH}} \xrightarrow[\Delta]{\text{HOAc}} HOOC^{11}CH_2CH_2\underset{\underset{NH_2}{|}}{C}HCOOH$$

The [1-^{11}C]D-glucosamine has also been prepared[85] using the aqueous $^{11}CN^-$ solution obtained from the production of the cyanide ion. The N-benzyl imine of D-arabinose was

$$(74)$$

reacted with $^{11}CN^-$ for 5 minutes at 80 °C. This gave an 85% yield of the α-aminonitrile, which was reduced to a mixture of $[1-^{11}C]$D-glucosamine and $[1-^{11}C]$D-mannosamine in 15 minutes at 80 °C with hydrogen over a palladium catalyst (equation 74). The small amount of the $[1-^{11}C]$D-mannosamine was removed by HPLC to give a decay-corrected yield of 10% in 45 minutes.

Recently, microwave radiation has been used to increase the rates of carbon-11 syntheses. This was attempted because microwave radiation had been found to increase the rates of the reactions carried out in sealed teflon containers by up to 1200 times[86-89]. This discovery has been an important addition to the tools available for those using short-lived radioisotopes in synthesis. For example, the time required to carry out the first two steps in the synthesis of the $[^{11}C]$busulphan, i.e. the S_N2 reaction to make the 3-cyano-1-propanol and its acid catalysed hydrolysis and cyclization to the lactone (equation 70), was reduced from 17 minutes to only 1.5 minutes when the reaction was done at even low power levels in a specially designed microwave cavity[90]. As reported[87], the addition of salts to the reaction mixture further increased the rate of reaction.

The formation of $[^{11}C]$D/L-tyrosine using the base catalysed Bucherer–Strecker synthesis (equation 61) was also reduced significantly from 20 minutes to 1 minute using this technique[90].

Carbon-11 can also be inserted into an aromatic ring[91]. Methyl chloroformate has been converted into carbon-11 labelled oxalic acid in a three-step synthesis. The first step (equation 75) was done in a water–dichloromethane mixture using tetrabutylammonium hydroxide and $^{11}CN^-$ under phase transfer catalysis conditions. The conversion to the diethyl ester and to the oxalic acid were done in high yields in 0.5 and 0.25 minute, respectively, in a microwave cavity.

$$CH_3OCCl + {}^{11}CN^- \xrightarrow[\text{H}_2\text{O, CH}_2\text{Cl}_2]{(Bu)_4N^+OH^-} CH_3OC^{11}CN \xrightarrow[\text{EtOH}]{\text{HCl(g)}} EtOC^{11}COEt$$

$$\xrightarrow{\text{HCl}} HO_2C^{11}CO_2H$$

(75)

Both the acid and the ester could be reacted with *ortho*-phenylenediamine under acidic conditions at 150 °C in 5 and 10 minutes, respectively, to give an 87% yield of 2,3-dihydroxyquinoxaline with the carbon-11 label in the aromatic ring (equation 76).

(76)

Recently[50], several new bifunctional, carbon-11 labelled synthons have been prepared from the starting material, chloromethyl pivalate, in at least a 90% yield using the microwave technique (Scheme 4). The comparable reaction times using conventional heating methods are shown for comparison in Table 1. These labelled bifunctional molecules, which can be prepared and purified in minutes, expand markedly the number of molecules that can be prepared with carbon-11 cyanide.

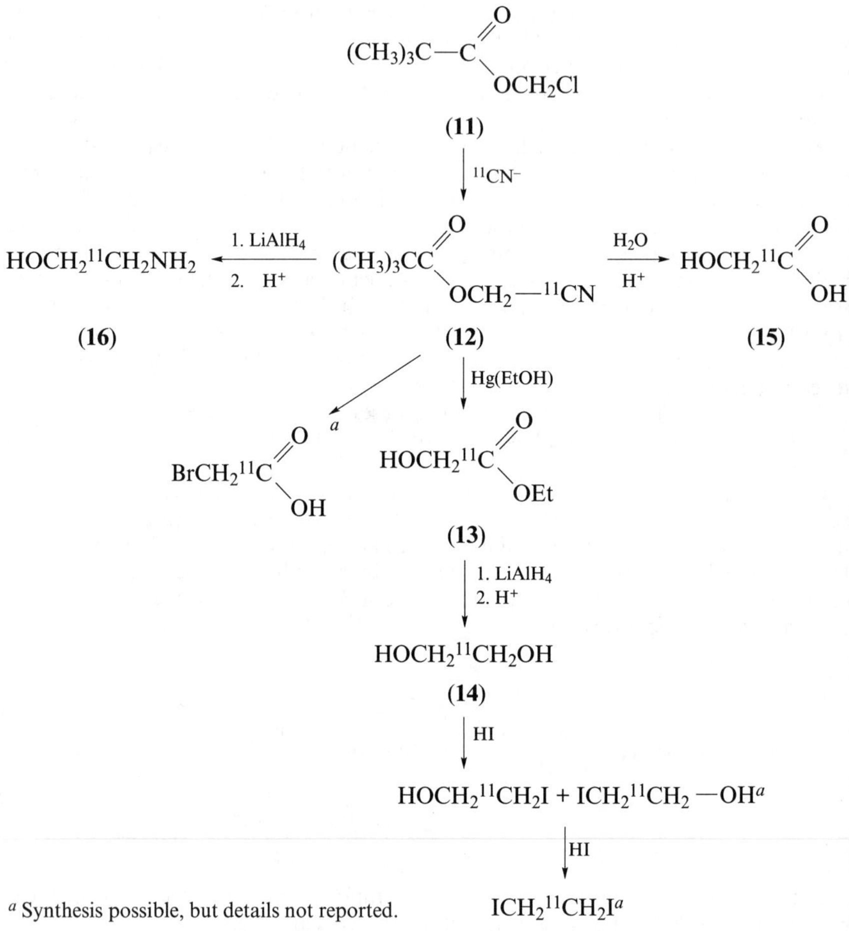

$$(CH_3)_3C-C\underset{OCH_2Cl}{\overset{O}{\diagup}}$$

(11)

$\downarrow$ $^{11}CN^-$

$HOCH_2{}^{11}CH_2NH_2$ $\xleftarrow[\text{2. H}^+]{\text{1. LiAlH}_4}$ $(CH_3)_3CC\underset{OCH_2-{}^{11}CN}{\overset{O}{\diagup}}$ $\xrightarrow[\text{H}^+]{\text{H}_2O}$ $HOCH_2{}^{11}C\underset{OH}{\overset{O}{\diagup}}$

(16) **(12)** **(15)**

$\downarrow$ Hg(EtOH)

$BrCH_2{}^{11}C\underset{OH}{\overset{O}{\diagup}}$ $\overset{a}{\diagup}$ $HOCH_2{}^{11}C\underset{OEt}{\overset{O}{\diagup}}$

(13)

$\downarrow$ 1. LiAlH$_4$
2. H$^+$

$HOCH_2{}^{11}CH_2OH$

(14)

$\downarrow$ HI

$HOCH_2{}^{11}CH_2I + ICH_2{}^{11}CH_2-OH^a$

$\downarrow$ HI

^{a} Synthesis possible, but details not reported. $ICH_2{}^{11}CH_2I^a$

SCHEME 4

TABLE 1. The times required by microwave and by conventional heating for the reactions shown in Scheme 4

Reaction	Microwave heating		Conventional heating	
	time (min)	power (W)	time (min)	temp (°C)
11→12	0.25	100	2	80
12→13	0.50	75	10	95
13→14	—	—	2	r.t.
12→15	0.25	75	5	120
12→16	—	—	2	r.t.

Another novel approach to labelling molecules with carbon-11 has been published by Somawardhana, Sajjad and Lambrecht[92,93]. These workers have absorbed one of the reactants required for the synthesis on a solid support and the reaction has been done by adding the other reagents to the solid support and heating the mixture in a sealed vial. One example illustrating their methodology is the formation of [1-^{11}C]putrescine. The H^{11}CN is trapped on a potassium-hydroxide-coated silica gel. Then, acrylonitrile is added, and the reaction tube sealed at both ends. After the tube had been heated at 75 °C for 5 minutes, the Michael addition product was eluted from the solid support and the solvent evaporated. Finally, the nitrile groups were reduced with a borane-dimethylsulphide complex in THF (equation 77). The total reaction time was less than 40 minutes and the product was obtained in a 53% radiochemical yield.

$$CH_2{=}CHCN + H^{11}CN \xrightarrow[\text{silica gel}]{\text{KOH}} N{\equiv}^{11}CCH_2CH_2C{\equiv}N \xrightarrow[\text{dry THF}]{BH_3, SMe_2}$$

$$\xrightarrow{HCl} H_2N^{11}CH_2CH_2CH_2CH_2NH_2 \tag{77}$$

This technique has also been used to add a cyano group to an aromatic ring[93]. 1-Methoxy-4-methoxycarbonyl pyridinium methyl sulphate was absorbed on silica gel particles. After these particles had been dried under vacuum, they were placed in a column and H^{11}CN was passed through the column. Then the column was sealed and heated at 40 °C for 5 minutes. After the methyl-2[^{11}C]cyano-isonicotinate had been purified by HPLC chromatography, it was converted into 2-[^{11}C]-isonicotinic acid hydrazide in a 30% radiochemical yield by treatment with hydrazine in anhydrous ethanol (equation 78). The synthesis took 30 minutes to complete.

$$\tag{78}$$

3. Syntheses with carbon-11 labelled cyanogen bromide

Labelled cyanogen bromide (^{11}CNBr) has been used by Westerberg and Langstrom[51] to synthesize several aromatic cyanates, and cyanamides, a 1-cyanopyridinium bromide and phenyl and benzyl thiocyanates in good yields in a few minutes (Scheme 5). The yields for each reaction are in the parentheses.

4. Syntheses with carbon-11 labelled diazomethane

Another versatile cyano group precursor is ^{11}C-diazomethane. This synthon can be made in 10 minutes by treating ^{11}CH$_4$ with chlorine and then with hydrazine in ethanolic KOH at 60 °C[94] (equation 79). One of the major uses for diazomethane is the alkylation of carboxylic acid groups. Crouzel and Syrota[95] have used this process to label a calcium channel antagonist PN 200–110 (isrodipine); see equation 80.

$$^{11}CH_4 + Cl_2 \xrightarrow[315\,°C]{CuCl_2} CHCl_3 \xrightarrow[\text{KOH, 60 °C}]{\text{hydrazine, EtOH}} {}^{11}CH_2N_2 \tag{79}$$

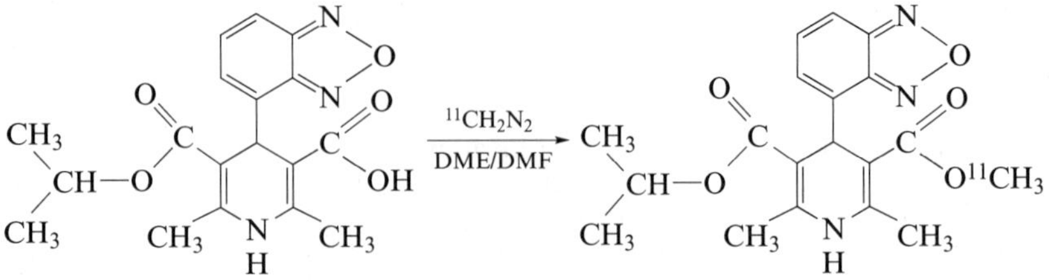

$$\text{(80)}$$

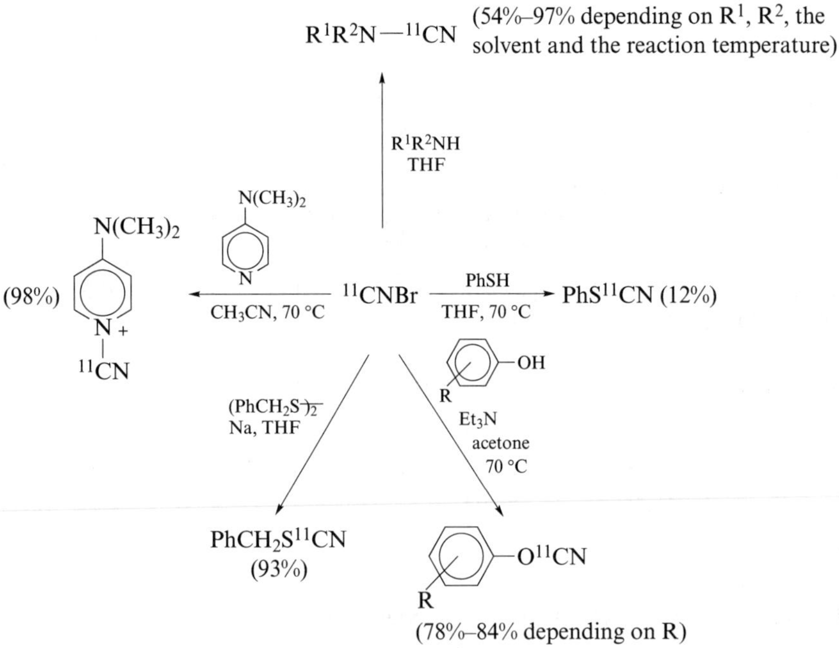

SCHEME 5

5. Other syntheses using labelled cyanides

Chupakhin and coworkers[96] have used doubly labelled ethyl cyanoacetate as a 1,3-C,N-bifunctional nucleophile in an insertion reaction with a pyridine ring to give a doubly labelled 2-triazolylamino-3-carbethoxy-5-nitropyridine. The suggested mechanism for the reaction with the carbon-13, nitrogen-15 labelled ethyl cyanoacetate is presented in Scheme 6.

The first intermediate has been isolated and characterized by NMR. The remaining intermediates are proposed.

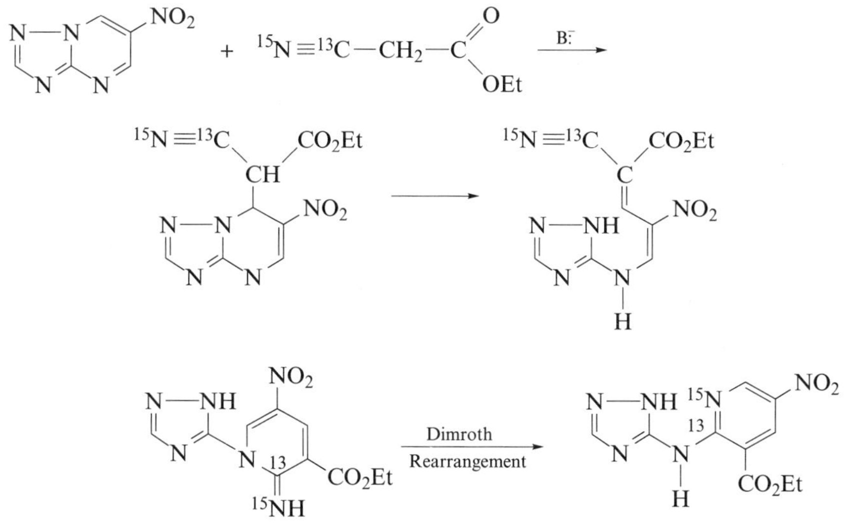

SCHEME 6

B. Mechanistic Studies Using Labelled Cyano Compounds

Togni and Pastor[97] have prepared 2,2-dideutero-α-isocyanoacetate via a hydrogen exchange reaction of α-isocyanoacetate with successive treatments of deuterated water in a deuterochloroform–triethylamine mixture. The deuterated compound was used to measure the hydrogen–deuterium kinetic isotope effect in the addition reaction that occurs when α-isocyanoacetate and an aldehyde react in the presence of a chiral ferrocenylamine–bis(cyclohexyl isocyanide) gold (I) tetrafluoroborate complex to form a mixture of cis- and trans-oxazolinines (equation 81). The hydrogen–deuterium kinetic isotope effect in this

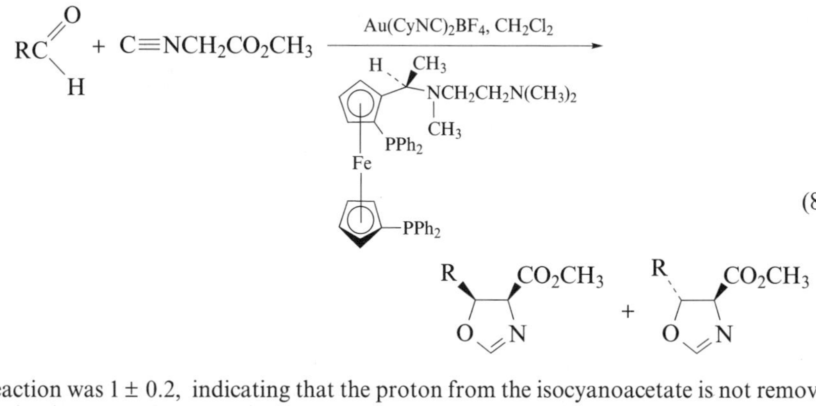

(81)

reaction was 1 ± 0.2, indicating that the proton from the isocyanoacetate is not removed in the slow step of the reaction. This means that formation of the catalyst is not the rate-determining step of the reaction. Other work suggests that the catalyst exists as an equilibrium mixture and that the slow step of the reaction is the attack of the aldehyde on the catalyst (equation 82).

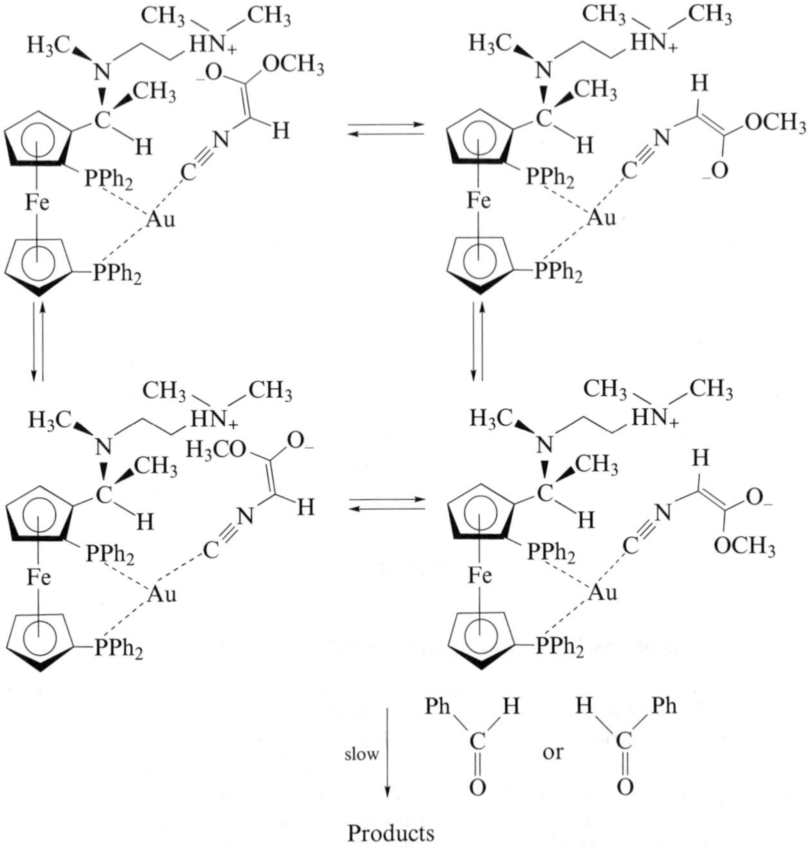

(82)

slow

Products

Tidwell and coworkers[98] have examined the addition of water to 1-cyano-1-ethoxyethene in aqueous sulphuric acid at 25 °C. The primary hydrogen–deuterium kinetic isotope effect found when the reaction is done in deuterated solvent increases from 3.77 to 5.44 as the percent sulphuric acid in the solvent increases from 6.6 M to 10.8 M. These isotope effects demonstrate that the slow step of the reaction is protonation of the vinyl substrate to form the α-cyanocarbocation and that the product is formed in subsequent, fast steps of the Ad-E2 mechanism (equation 83).

$$CH_2=C\diagdown^{CN}_{OEt} \xrightarrow[slow]{H^+} CH_3C+\diagdown^{CN}_{OEt} \xrightarrow[fast]{H_2O} [CH_3\overset{O}{\overset{\|}{C}}-CN] \xrightarrow[fast]{H_2O} CH_3CO_2H$$

(83)

The relatively fast rate (approximately $2 \times 10^{-2}\,S^{-1}$ at 25 °C) for the protonation step is found because of the resonance effect of the cyano group. This resonance interaction was first reported by Olah and coworkers[99], who synthesized the nitrogen-15 labelled cyanodi phenylmethyl cation by treating the nitrogen-15 labelled diphenylmethylcyanohydrin with a FSO$_3$H/SO$_2$ClF solution at −78 °C. The procedure used to prepare the labelled cyanohy drin and cyanodiphenylmethyl cation are shown in equation 84.

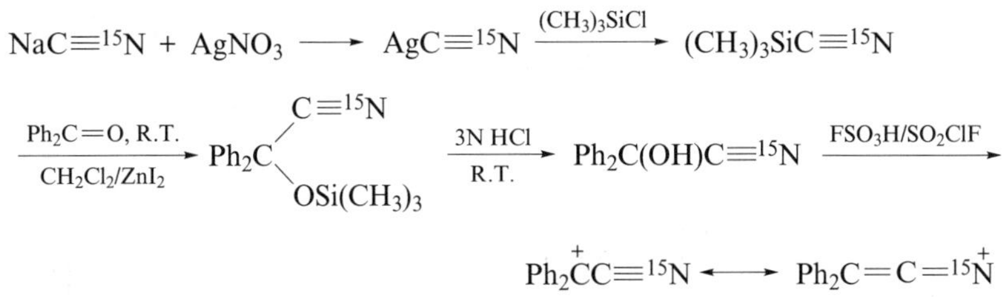

$$NaC\equiv^{15}N + AgNO_3 \longrightarrow AgC\equiv^{15}N \xrightarrow{(CH_3)_3SiCl} (CH_3)_3SiC\equiv^{15}N$$

$$\xrightarrow[CH_2Cl_2/ZnI_2]{Ph_2C=O, R.T.} Ph_2C\begin{array}{c} C\equiv^{15}N \\ \diagup \\ \diagdown \\ OSi(CH_3)_3 \end{array} \xrightarrow[R.T.]{3N\ HCl} Ph_2C(OH)C\equiv^{15}N \xrightarrow{FSO_3H/SO_2ClF} \quad (84)$$

$$Ph_2\overset{+}{C}C\equiv^{15}N \longleftrightarrow Ph_2C=C=^{15}\overset{+}{N}$$

The resonance effect was consistent with the greater chemical shift for the nitrogen in the ^{15}N NMR spectrum of the carbocation, i.e. it increased by 30 ppm from 253 ppm in the cyanohydrin to 283 ppm in the carbocation. This deshielding of only 30 ppm over that of the neutral precursor indicates that there is substantial carbon–nitrogen double-bond character in the carbon–nitrogen bond of the cation because the R—$C\equiv NH^+$ triple bond in the nitrilium cation is shielded by approximately 100 ppm compared to the neutral nitrile. This resonance effect of the cyano group is also supported by the observation that the chemical shift of the terminal nitrogen is close to that of an imine N which has a chemical shift of $\delta = 318$ ppm in the ^{15}N NMR spectrum.

A large primary hydrogen–deuterium kinetic isotope effect has also been reported in the hydrobis(phosphine) platinum(II) complex catalysed hydration of cyano groups[100] (equation 85). The catalyst, *trans*-[PtH(H$_2$O)(PMe$_3$)$_2$][OH], is generated by treating the

$$R-C\equiv N + H_2O \xrightarrow{catalyst} R-\overset{\displaystyle O}{\overset{\|}{C}}-NH_2 \quad (85)$$

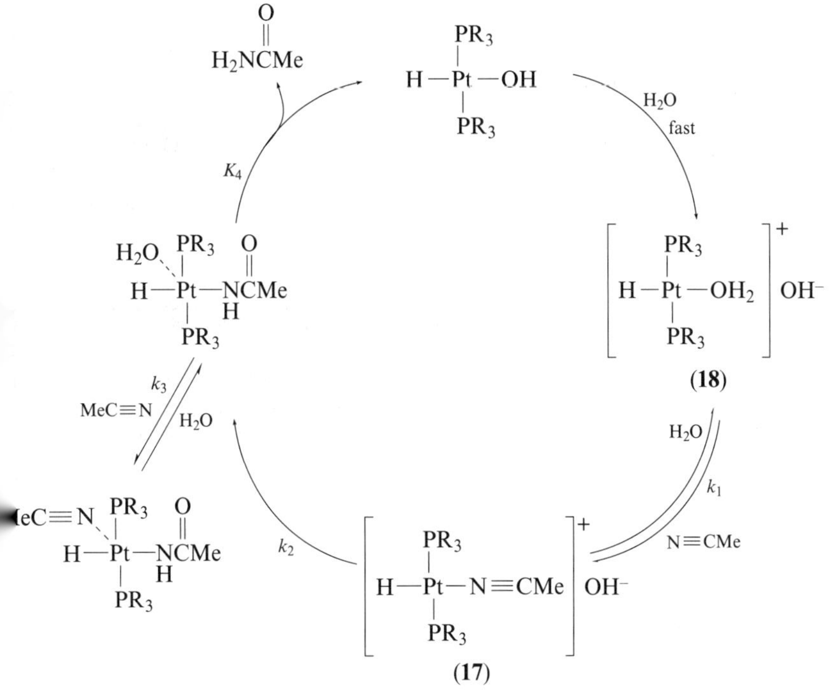

SCHEME 7

corresponding chloride with sodium hydroxide and the reaction is carried out at 25 °C with very little reaction (hydration) of any of the carbon–carbon double bonds. The mechanism that has been suggested for the reaction is shown in Scheme 7.

The second step of the reaction is the replacement of water by the nitrile in a rapid equilibrium that favours the nitrile complex 17. NMR studies of the reaction between the deuterated aqueous form of the catalyst, $trans$-[PtH(D$_2$O)(PEt$_3$)$_2$][PF$_6$], and acetonitrile-d_3 in a 50% mixture of acetone-d_6 and D$_2$O, and of the nitrile form of the catalyst, $trans$-[PtH(CD$_3$C$\equiv$N)(PEt$_3$)$_2$][PF$_6$], with D$_2$O in a 50% acetonitrile-d_3–acetone-d_6 mixture, have confirmed that the nitrile is absorbed more strongly on the catalyst than water, i.e. the NMR studies have shown that the ratio of the nitrile form of the catalyst, 17, to the aqueous form of the catalyst, 18, is 9.8/1.

When the hydroxide ion concentration is normal, the rate constant, k_2, is much greater than the rate constant for the k_4 step, and the k_4 step is rate-determining. The evidence for this conclusion is the primary hydrogen–deuterium kinetic isotope effect of 3.4 that is found when the ratio of water/acetonitrile is less than 30/70 and the solvent is changed from water to deuterated water. Finally, an experiment using the deuterated catalyst, $trans$-[PtD(H$_2$O)(PMe$_3$)$_2$][OH], demonstrated that the deuterium did not exchange out of the catalyst when it was treated with a 10/90 water/acetonitrile mixture at 80 °C for one hour. Since this provides the time for approximately 140 reactions, it was concluded that the acetonitrile is not released in a reductive elimination to form the acetamide and Pt(PMe$_3$)$_2$ (equation 86), but that the proton that is transferred to the leaving acetonitrile comes from the solvent (equation 87).

$$
\begin{array}{c}
\text{H}_2\text{O} \quad \text{PMe}_3 \quad \text{O} \\
\backslash | \| \\
\text{D——Pt——NHCCH}_3 \longrightarrow
\end{array}
\quad
\begin{array}{c}
\text{PMe}_3 \\
| \\
\text{Pt} \\
| \\
\text{PMe}_3
\end{array}
\quad + \quad
\begin{array}{c}
\text{O} \\
\| \\
\text{DHN——CCH}_3
\end{array}
\quad (86)
$$

$$
\begin{array}{c}
\text{H}_2\text{O} \quad \text{PMe}_3 \quad \text{O} \\
\backslash | \| \\
\text{D——Pt——NHCCH}_3 \\
| \\
\text{PMe}_3
\end{array}
\xrightarrow{\text{H}_2\text{O}}
\begin{array}{c}
\text{PMe}_3 \\
| \\
\text{D——Pt——OH} \\
| \\
\text{PMe}_3
\end{array}
+ \text{H}_2\text{NCCH}_3
\quad (87)
$$

In the past only a few studies[101,102] have been reported on base-promoted nitrile forming elimination reactions. Recently, however, a detailed study of the mechanism and the nature of the transition state for this reaction has been carried out. The authors[103] studied the reactions of (E)-O-aryl-$para$-substituted benzaldehyde oximes with hydroxide (deuteroxide) ion in 60% aqueous dimethyl sulphoxide (equation 88). The reactions of compounds 19–21 with hydroxide ion in 60% aq. DMSO were shown to give quantitative yields of the

$$
\begin{array}{c}
p\text{-XC}_6\text{H}_4 \\
\backslash \\
\text{C}=\text{N} \\
/ \backslash \\
\text{H} \text{OAr}
\end{array}
+ \text{OH}^-(\text{OD}^-) \xrightarrow[\text{DMSO}]{60\% \text{ aq.}} p\text{-XC}_6\text{H}_4\text{C}\equiv\text{N} + \text{ArO}^-
$$

(19) Ar = 2, 4-(NO$_2$)$_2$C$_6$H$_3$ (88)

(20) Ar = 4-NO$_2$C$_6$H$_4$ X = (a) MeO, (b) H, (c) Cl,

(21) Ar = C$_6$H$_5$ and (d) NO$_2$

benzonitriles and aryloxides. The ratio of the second-order rate constants, k^{OD^-}/k^{OH^-}, for reaction with OD^- and OH^-, respectively, as well as the calculated values for β and ρ, are given in Table 2.

TABLE 2. The k^{OD^-}/k^{OH^-}, β and ρ values for the cyanide-forming elimination reaction of (E)-O-arylbenzaldehyde oximes promoted by OH^- and OD^- in 60% aq. DMSO at 25 °C

Substrate	k^{OD^-}/k^{OH^-}	β	ρ
19a	1.40 ± 0.14	0.49 ± 0.14	
19b	1.25 ± 0.18	0.32 ± 0.21	0.21± 0.45
19c	1.17 ± 0.17	0.23 ± 0.21	
19d	1.15 ± 0.19	0.20± 0.24	
20a	1.68 ± 0.02	0.75 ± 0.02	
20b	1.49 ± 0.02	0.58 ± 0.02	1.9 ± 0.1
20c	1.64 ± 0.01	0.72 ± 0.01	
20d	1.62 ± 0.01	0.70 ± 0.01	
21a	1.84 ± 0.07	0.88 ± 0.06	
21b	1.71 ± 0.10	0.77 ± 0.02	2.2 ± 0.1
21c	1.75 ± 0.02	0.81 ± 0.02	
21d	1.70 ± 0.02	0.77 ± 0.02	

The β value, which is calculated from the equation $k^{OD^-}/k^{OH^-} = 2.0^{\beta\,104}$, gives the percent transfer (expressed as a fraction) of the hydrogen from the substrate to the base in the E2 transition state. The ρ value, which is found using the rate constants for different aryloxide leaving groups, provides an estimate of the amount of α-carbon–leaving group bond rupture in the transiton state. In addition a β_{lg} value of -0.59 ± 0.01 was calculated from the equation $\log k = \beta_{lg}\, pK^{105}$ using the rate constants for reactions **19a**, **20a** and **21a** and the estimated pK_{lg} values of the aryloxides in 60% aq. DMSO. The significant values β and β_{1g} and product and kinetic studies lead the authors to conclude that the reactions of **19–21** proceed via an E2 elimination pathway. Moreover, the reasonably large β value of 0.49 found for reaction **19a** and the $\beta_{lg} = -0.59$ indicated that the transition state, **22**, had significant cleavage of both the C_β—H bond and the N_α—OAr bonds.

$$
\left[
\begin{array}{c}
\overset{Ar}{\underset{\displaystyle \underset{\delta\text{-OH}}{H}}{C}}{\equiv}N \\
OAr \\
\delta^-
\end{array}
\right]^{\ddagger-}
$$

(22)

The authors concluded that the activated complex had very little carbanionic character at the β-carbon. This presumably occurs because negative charge buildup at C_β is unfavourable since the developing negative charge cannot be stabilized by resonance by the β-aryl group since the C_β—H bond is orthogonal to the π-orbitals of the β-aryl ring. However, stabilization of the negative charge by the leaving group is expected to be appreciable due to resonance delocalization.

For the elimination reactions of **19–21**, the increase in both the k^{OD^-}/k^{OH^-} and the ρ values indicate an increase in the extent of C_β—H bond rupture as the leaving group becomes poorer. This conclusion may be considered in the light of the Jencks–More O'Ferrall energy surface diagram (Figure 2)[106,107]. The authors considered that the reaction of **20** proceeded via a central transition state (Figure 2) with approximately equal

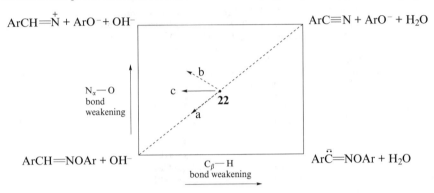

FIGURE 2. The Jencks–More O'Ferrall energy surface diagram for the elimination reaction of (E)-O-aryl benzaldehyde oximes promoted by $^-$OH in 60% aqueous DMSO

extents of both C_β—H and N_α—OAr bond rupture. Changing to a better leaving group would decrease the energy of the upper edge of the diagram, shifting the transition state towards the reactant [parallel motion, dashed arrow (a)] and away from the carbanion corner [perpendicular effect, dashed arrow (b)]. The resultant effect, regardless of the relative magnitudes of (a) and (b), solid arrow (c), indicates that the transition state would shift to the left, i.e. decreasing the extent of C_β—H bond weakening. This is the change in transition state that is observed, i.e. the smaller k^{OD}/k^{OH} and β values found for reaction **19** indicate that there is less C_β—H bond rupture in the transition state with the best leaving group.

The mechanism of the oxidation of primary amines to nitriles with cuprous chloride–dioxygen–pyridine mixtures has recently been investigated by measuring the hydrogen–deuterium kinetic isotope effect in the oxidation of benzylamine to benzonitrile[108]. In earlier work[109], the authors reported that nitriles could be prepared from amines in almost quantitative yield and high purity using this cuprous chloride reagent.

The hydrogen–deuterium kinetic isotope effect was determined by carrying out the reaction using a known mixture of benzylamine and its α, α -d_2 analogue. This gave the ratio of the dideuterated/undeuterated benzylamine in the starting material. The reaction was stopped part way to completion, the unreacted starting materials were acetylated and the ratio of the dideuterated/undeuterated reactants was determined by ^{1}H NMR spectroscopy. The calculated isotope effect was small, i.e. $k_H/k_D=1.24 \pm 0.1$ at 50 °C. This was considered to be a secondary deuterium kinetic isotope effect, indicating that there was no

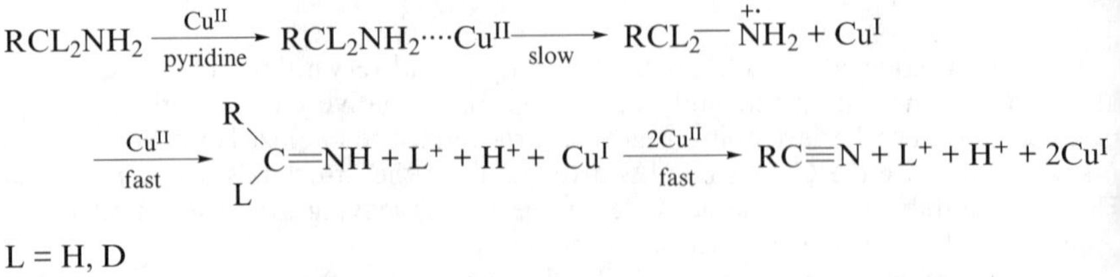

$$RCL_2NH_2 \xrightarrow[\text{pyridine}]{Cu^{II}} RCL_2NH_2\cdots Cu^{II} \xrightarrow[\text{slow}]{} RCL_2\overset{+\cdot}{}NH_2 + Cu^I$$

$$\xrightarrow[\text{fast}]{Cu^{II}} \underset{L}{\overset{R}{\diagdown}}C{=}NH + L^+ + H^+ + Cu^I \xrightarrow[\text{fast}]{2Cu^{II}} RC{\equiv}N + L^+ + H^+ + 2Cu^I$$

$L = H, D$

SCHEME 8

C_α—H (C_α—D) bond rupture at the transition state of the rate-determining step of the reaction.

The mechanism in Scheme 8 was proposed for the oxidation reaction. In the first step, the Cu(II) salt, which is formed in the autooxidation of cuprous chloride, forms a complex with the amine. This is followed by a rate-determining electron transfer from the amine to the Cu(II) species giving Cu(I) and an aminium radical. The subsequent steps were considered to be fast. The authors accounted for the secondary hydrogen–deuterium kinetic isotope effect by suggesting that there was hyperconjugative electron release to the aminium ion nitrogen that forms in the slow step of the reaction.

Recently, Westaway, Matsson and coworkers[110] have measured the incoming nucleophile carbon-11–carbon-14 kinetic isotope effects for the S_N2 reactions between *para*-

TABLE 3. The k^{11}/k^{14} incoming nucleophile kinetic isotope effects for the S_N2 reactions between *para*-substituted benzyl chlorides and labelled cyanide ion in 20% aqueous dimethylsulphoxide at 30 °C

Para-substituent	k^{11}/k^{14}
H	1.0107 ± 0.002
Cl	1.0070 ± 0.0008

substituted benzyl chlorides and labelled cyanide ion in 20% aqueous DMSO at 30 °C. The isotope effects (Table 3) are the largest incoming nucleophile kinetic isotope effects that have been observed. The isotope effects are comprised of a normal temperature-independent term (the ratio of the imaginary frequencies) and a temperature-dependent term arising from changes in the vibrational energy that occur when the reactants are converted into the transition state. In these S_N2 reactions, there is obviously no bond between the nucleophile and the α-carbon (no carbon–carbon bond vibrational energy) in the reactants whereas the carbon–carbon bond is partially formed in the S_N transition state (equation 89). As a result, there is more vibrational energy involving the isotopically labelled carbon in the transition states than in the reactants and therefore the temperature-dependent term is inverse.

$$Z\text{—}\bigcirc\text{—}CH_2Cl + C^*{\equiv}N^- \longrightarrow \left[N{\equiv}C^*\text{----}\overset{\overset{H}{\underset{}{\diagdown}}\ \overset{H}{\diagup}}{C}\text{----}Cl \right]^{\ddagger} \tag{89}$$

$$\longrightarrow Z\text{—}\bigcirc\text{—}CH_2C^*{\equiv}N + Cl^-$$

* = Carbon-11 or carbon-14

The magnitude of these isotope effects can also be used to determine transition state structure. As the amount of bonding between the nucleophile and the α-carbon increases in the transition state, the temperature-dependent term becomes smaller (more inverse). Since the temperature-independent term would be the same for all of the reactions in this study, a smaller isotope effect is indicative of greater (more complete) nucleophile–α-carbon bond formation in the transition state. Thus, these results suggest that the nucleophile–α-carbon transition state bond is slightly shorter (formed more completely) when there is a more electron-withdrawing *para*-substituent in the substrate. This is consistent with what has been postulated for these reactions, because substrates with electron-withdrawing groups are thought to have tighter transition states. However, it is important to finally be able to demonstrate how a change in substituent affects the nucleophile–α-carbon transition state bond in an S_N2 reaction. It is noteworthy that this work has provided the largest incoming nucleophile kinetic isotope effect that has been measured and the first incoming nucleophile isotope effect which is clearly different from 1.00[111,112]. It is important to note that it is only possible to measure these large isotope effects, which are sufficiently sensitive to changes in substituent, because of the large difference in the masses of the isotopes involved.

C. Mass Spectrometry

Several groups have used labelled cyano compounds to study the reactions of ions generated in a mass spectrometer. For example, an extensive investigation was carried out to determine how the loss of hydrogen cyanide from the benzonitrile molecular ion formed upon electron impact, actually occurs. It has generally been assumed that the cyano group remains intact during the electron impact induced fragmentation. Recently, however, Nibbering and coworkers[113] have carried out a detailed study using specifically labelled benzonitriles to investigate whether the cyano group does indeed retain its identity during fragmentation.

Three carbon-13 labelled benzonitriles, **23–25**, were prepared and the relative abundance of the hydrocarbon ions formed in the ion source at 70 °C with masses between $m/z = 74$ and $m/z = 78$ were determined using high-resolution conditions (Table 4).

$$*C\equiv N \qquad C\equiv N \qquad C\equiv N$$

(23)   (24)   (25)

TABLE 4. The relative abundance of the hydrocarbon ions between $m/z = 74$ and $m/z = 78$ ($\Sigma_{74-78} = 100$) in the 70-eV mass spectra of several carbon-13 labelled benzonitriles

m/z	Unlabelled compound	Labelled compound		
		23	24	25
74	8	—	1	3
75	9	12	7	9
76	80	42	12	19
77	3	45	76	68
78	—	1	4	1

The data indicate that the carbon atom of the eliminated HCN molecule is not always the carbon from the original cyano group. This conclusion was reached because there is a partial shift of the $m/z = 76$ (M-HCN or $C_6H_4^{+\cdot}$) peak in the spectrum of the unlabelled benzonitrile to $m/z = 77$ of the ($^{13}C^{12}C_5H_4^{+\cdot}$) peak in the spectrum of the $C_6H_5^{13}CN$ analogue, i.e. some of the carbon-13 in the cyano carbon is not lost with the HCN. Indeed, measures of the metastable peak intensities obtained from the spectrum of the ring carbon-13 labelled substrates

(**24** and **25**) indicate that all the phenyl carbons participate equally in the loss of HCN from the molecular ion decomposition in the first and second field-free regions (longer-lived molecular ions). The authors have suggested a mechanism to account for migration of the cyano group around the aromatic ring in the molecular ion (equation 90).

$$(90)$$

The authors also suggested a possible pathway to account for the involvement of a phenyl ring carbon atom in the loss of HCN from the molecular ion produced from benzonitrile (equation 91).

$$(91)$$

Interesting ion–molecule reactions can be studied using specifically labelled reactant ions. For example, the reactions of the β-distonic ion, $^+CH_2OCH_2CH_2^.$, **26**, and two specifically deuterated analogues, **27** and **28**, with butyronitrile have been investigated[114]. The β-distonic ions **26–28** were generated by electron-impact induced fragmentation of

$$\overset{+}{C}H_2OCH_2\overset{.}{C}H_2 \qquad \overset{+}{C}H_2OCD_2\overset{.}{C}D_2 \quad \text{and} \quad \overset{+}{C}D_2OCH_2\overset{.}{C}H_2$$

(**26**) $m/z = 58$ (**27**) $m/z = 62$ (**28**) $m/z = 60$

1,4-dioxane and 2,2,3,3-tetradeutero-1,4-dioxane, respectively, in the external ion source of a Fourier transform–ion cyclotron resonance spectrometer[115,116]. The isolated ions, **26**, **27** and **28** were allowed to react with $CH_3CH_2CH_2C≡N$ and subsequently, mass spectra were obtained at 0.5-second intervals during a total reaction time of approximately 5 seconds.

A product ion of $m/z = 97$ ($C_6H_{11}N^+$) observed for the reaction of the undeuterated ion, $^+CH_2OCH_2CH_2^.$; with n-C_3H_7CN remained at this m/z value for the reaction of **28** but was found at $m/z = 101$ when the β-distonic ion **27** was the reactant. It was concluded that $C_2H_4^+$ and $C_2D_4^+$ ions are transferred to the n-C_3H_7CN without any isotope scrambling when

$$CH_3CH_2CH_2\overset{+}{C}≡N: + \overset{+}{C}H_2OCD_2\overset{.}{C}D_2 \xrightarrow{-(CH_2O)} CH_3CH_2CH_2\overset{+}{C}≡\overset{.}{N}CD_2\overset{.}{C}D_2 \quad (92)$$

(**27**) $m/z = 62$ $m/z = 101$

either **27** or **28** was the reactant. These reactions are illustrated using **27** as the reacting ion (equation 92).

Subsequent to the formation of the $C_2H_4^{+\cdot}$ ($C_2D_4^{+\cdot}$) adduct ions, i.e. the ions with $m/z = 97$ and 101 (equation 92), the loss of C_2H_4 is noted from the propyl group of n-C_3H_7CN to give an ion which is isomeric to n-butyronitrile, i.e. an ion of $m/z = 69$ in the spectrum of **26** and **28** and $m/z = 73$ in the spectrum of **27** . This process is illustrated (equation 93) for the loss of C_2H_4 from the $m/z = 101$ ion formed in equation 92.

$$\begin{array}{c} \text{C}\equiv\text{N}^+ \\ \text{CH}_2 \qquad\qquad \text{CD}_2 \\ \text{CH}_2 \qquad\qquad\qquad \xrightarrow{-C_2H_4} \quad \overset{\cdot}{\text{C}}\text{H}_2-\text{C}\equiv\overset{+}{\text{N}}-\text{CD}_2\text{CD}_2\text{H} \\ \text{CH}_2-\text{H} \qquad \cdot\text{CD}_2 \end{array} \qquad (93)$$

The ions **26**, **27** and **28** also serve as proton donors to butyronitrile, since a product ion with a $m/z = 70$ ($C_3H_7CNH^+$) was observed in the spectrum for the reaction of **26** while product ions at both $m/z = 70$ and $m/z = 71$ ($C_3H_7CNH^+$ and $C_3H_7CND^+$, respectively), corresponding to the transfer of H^+ and D^+ to n-C_3H_7CN, were noted for the reactions of **27** and **28**. The authors, assuming that all the hydrogens (i.e. the two H and four D for **27** and the four H and two D for **28**) could be transferred to the butyronitrile, calculated identical primary hydrogen—deuterium isotope effects, k_H/k_D, of 2.24 ± 0.10 and 2.33 ± 0.10 for the hydrogen transfer reactions of **27** and **28**, respectively, with n-C_3H_7CN[115].

It was suggested that a reasonable pathway for the protonation of n-C_3H_7CN by the β-distonic ion **26** was reaction via a complex consisting of $C_2H_6^{+\cdot}$, CO and the nitrile molecule. Within such a complex all of the hydrogens (deuterium) of $C_2H_6^{+\cdot}$ ($C_2D_6^{+\cdot}$) are equivalent and protonation (deuteration) can proceed exothermically (equation 94) to give $C_3H_7CNH^+(D^+)$ and $^{\cdot}C_2H_5$. This is consistent with the identical k_H/k_D values found for the reactions of **27** and **28**.

$$\overset{+}{\text{C}}\text{H}_2\text{OCH}_2\overset{\cdot}{\text{C}}\text{H}_2 + C_3H_7CN \longrightarrow C_3H_7\overset{+}{\text{C}}\text{NH} + C_2H_5\cdot + CO \qquad (94)$$

Schwarz and coworkers[117–119] have studied the reactions of the metastable ion produced from Fe^+ and nitriles in the gas phase in a triple-sector mass spectrometer. The Fe^+ is formed when $Fe(CO)_5$ is bombarded with an electron beam of 100-eV kinetic energy. The reaction begins when the bare Fe^+ ion reacts with the neutral nitrile to form an end-on or

$$RCH_2C \equiv N + Fe^+ \longrightarrow RCH_2C \equiv N\cdots Fe^+ \qquad (95)$$

linear metastable ion (equation 95). The $RCH_2C\equiv N\cdots Fe^+$ ions having 8-keV kinetic energy were mass-selected by both magnetic and electric field sectors. Then, either the decomposition of the metastable ion or (by introducing helium gas into the third region of the mass spectrometer) the collision-induced decomposition of the metastable ion was determined by analysing the fragment ions. These mass-selected $RCH_2C\equiv N\cdots Fe^+$ ions decompose by forming an intermediate in which the iron ion activates (coordinates with) C—H bonds in the terminal methyl group of the nitrile in what Breslow has called a remote functionalization reaction. For linear nitriles with between four and seven carbon atoms[117], the linear ion constrains the remainder of the molecule while trying to minimize the strain of the intermediate metallacycles. This means that the Fe^+ atom activates exclusively a C—H bond of the terminal methyl group by oxidative addition, followed by either a β-hydrogen migration to the Fe atom (IV) and reductive elimination of hydrogen (V) or cleavage of a β-C—C bond (II) and reductive elimination of ethylene (III) (Scheme 9).

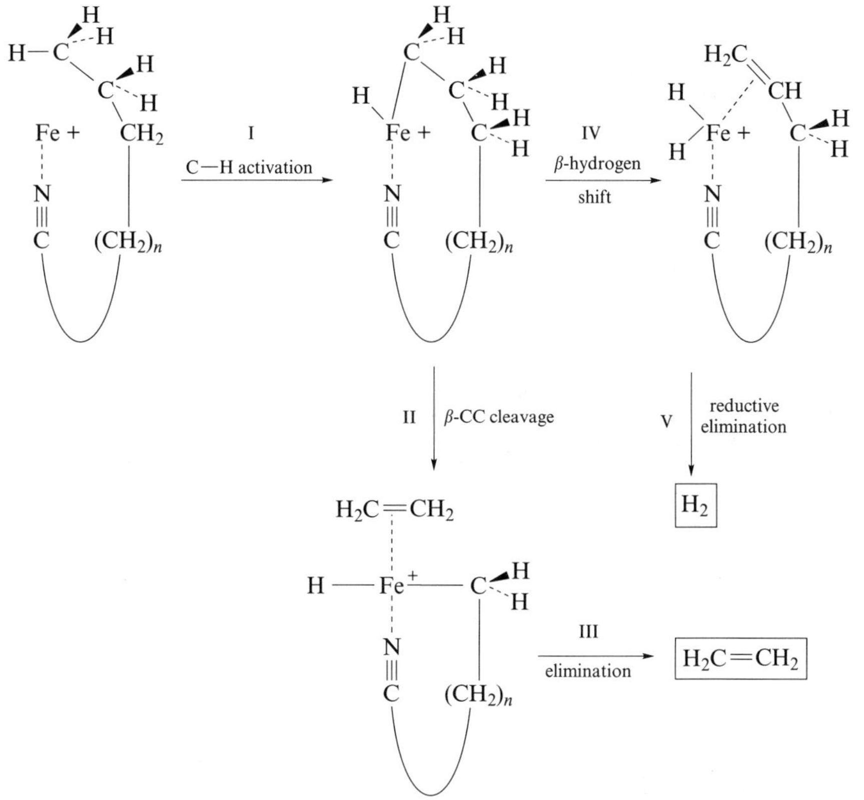

SCHEME 9

The reductive elimination of hydrogen and ethylene accounts for 97% of the products formed from the metastable $RCH_2C\equiv N\cdots Fe^+$ ion although a small amount of alkane is often found from the collision-induced decomposition of the $RCH_2C\equiv N\cdots Fe^+$ ion. The hydrogen atoms for the hydrogen come from the ω and ω-1 carbons of the alkyl group on the nitrile and the carbons for the ethylene come from the ω and ω-1 carbons of the starting material.

Four different isotopomers of 5-cyanononane (**29–31**) were prepared and reacted with Fe^+ ions. The results in Table 5 give the percentages of ethylene and hydrogen that are lost from the $RCH_2C\equiv N\cdots Fe^+$ ions formed from each of the four deuterated cyanononanes.

Because the elimination can occur in either branch of the nonane, isotope effects can be calculated from the product ratios. These deuterium labelling and kinetic isotope effect studies have enabled the authors to determine the rate-determining step of the reaction and to understand the gas-phase decomposition of the metastable ion complex.

The equal amounts of ethylene and ethylene-d_2 formed from the reactions of compounds **29** and **30** show that the formation of the intermediate in the C—H activation step is not rate-determining. If this step were rate-determining, much less ethylene-d_2 would have been formed from compound **30** because there would be a primary hydrogen–deuterium isotope effect associated with the insertion reaction.

$$C\equiv N$$

(29)

$$D_3C \qquad C\equiv N$$

(30)

$$D_3C \qquad C\equiv N$$

(31)

$$D_3C \qquad C\equiv N \qquad CD_3$$

(32)

TABLE 5. The percentage of the various labelled ethylene and hydrogen molecules lost from the $RCH_2C\equiv N\text{---}Fe^+$ ions formed from four different deuterated 5-cyanononanes.

Compound formed	Percent product formed from compound			
	29	**30**	**31**	**32**
C_2H_4	61	61	68	—
$C_2H_2D_2$	39	39	—	60
C_2D_4	—	—	32	40
H_2	73	63	79	—
HD	27	37	—	69
D_2	—	—	21	31

Hydrogen–deuterium kinetic isotope effects have been observed for steps III and IV in Scheme 9, however. The loss of ethylene is 1.56 times greater than the loss of ethylene-d_2 in the reactions of compounds **29** and **30** and the loss of ethylene is 2.13 times greater than the loss of ethylene-d_4 in the reaction of compound **31**. An isotope effect is also observed in the reaction of compound **32**. Here, the loss of the ethylene-d_2 is 1.50 times greater than the loss of ethylene-d_4. These isotope effects clearly demonstrate that the loss of ethylene occurs in a slow step of the reaction, i.e. that step III in Scheme 9 is the rate-determining step in the formation of ethylene.

An examination of the percentages of labelled hydrogen formed in the reactions of the four labelled isotopomers indicates that there is also a hydrogen–deuterium kinetic isotope effect associated with the loss of hydrogen in step IV (Scheme 9). Clearly, the higher percentages of H_2 relative to HD from **29** and **30**, of H_2 relative to D_2 from **31** and of HD relative to D_2 from **32**, indicate that there is an isotope effect associated with the loss of hydrogen from the metastable ion intermediate. In fact, solving the simultaneous equations for the formation of the different labelled hydrogen molecules has shown that there is an isotope effect for both steps IV and V. The isotope effect for step IV, the β-hydrogen shift reaction is 1.59 while the isotope effect for the reductive elimination of hydrogen, step V is 1.70 for k_{H_2}/k_{HD} and 1.44 for k_{HD}/k_{D_2}. Thus both steps IV and V of Scheme 9 are partially rate-determining.

When there are more than seven carbons in the nitrile, products from the reductive elimination of alkanes are found. Studies with deuterated nitriles have shown that these reactions occur when the iron ion inserts into an internal C—C bond (I, equation 96). Then a hydrogen from the internal carbon alpha to the Fe atom, transfers to the iron atom (II) and an alkane is released (equation 96).

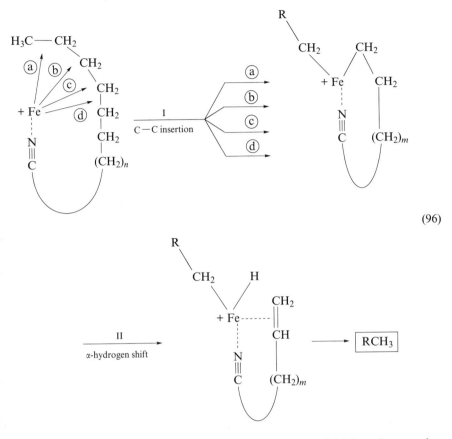

(96)

Two isotopically labelled 8,8-dimethylnonanenitriles, **33** and **34**, have been used to elucidate the mechanism of the reaction with a terminal *t*-butyl group[118]. The major products from both the unimolecular and collision-induced decomposition are hydrogen,

(33)   (34)

methane, butene and C_4H_{10}. The hydrogen eliminated from **33** is almost equally H_2 and HD whereas the hydrogen eliminated from **34** is almost totally H_2. This indicates that the hydrogens come from internal carbons in this long-chain substrate rather than from the terminal methyl groups. Even the small amount of HD found when the iron-nitrile ion formed from **34** reacts is thought to arise from some unimportant deuterium exchange reaction rather than from the transfer of a deuterium from a terminal methyl group.

The formation of methane is thought to occur via one or both of two pathways. In the first step, the iron ion inserts into a terminal carbon–carbon bond, equation 97. Then a β-H from either a terminal methyl group or the internal methylene group is transferred to

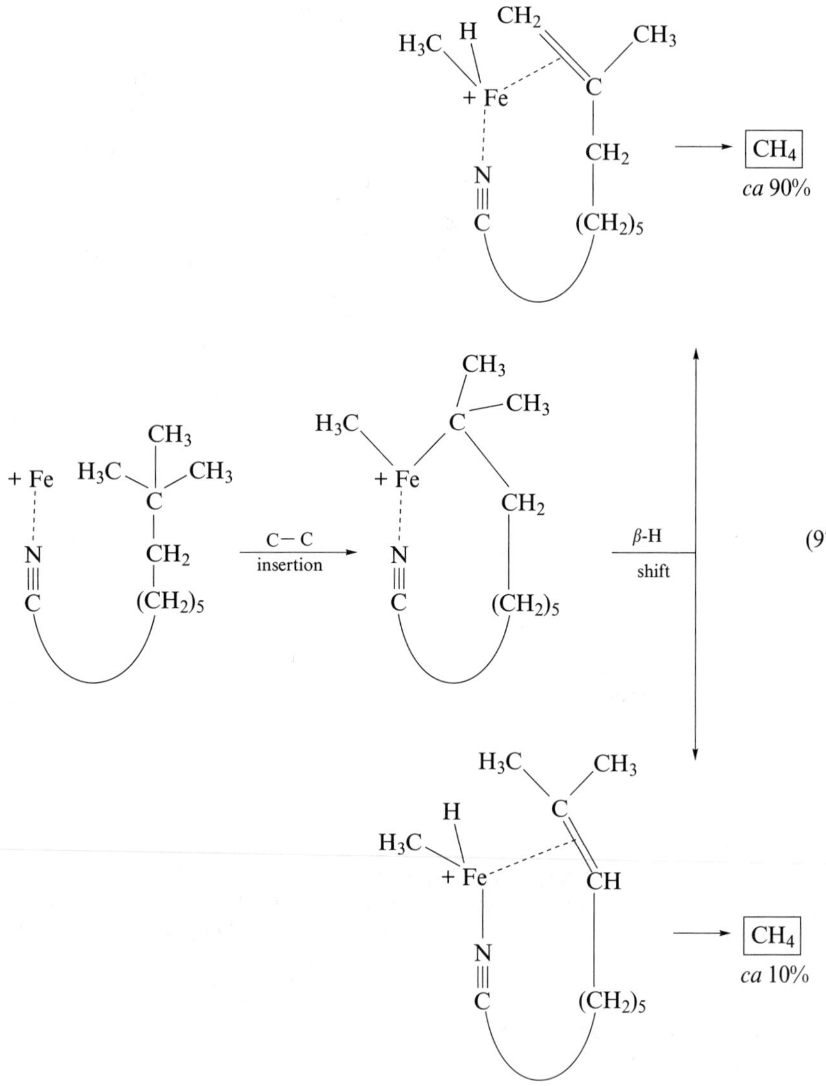

$$
\begin{array}{c}
\text{CH}_2 \\
\text{H}_3\text{C} \quad \text{H} \qquad \text{CH}_3 \\
\diagdown \qquad \diagup \\
+ \text{Fe} \cdots \text{C} \\
\vdots \qquad \text{CH}_2 \longrightarrow \boxed{\text{CH}_4} \\
\text{N} \qquad\qquad ca\ 90\% \\
\parallel \\
\text{C} \qquad (\text{CH}_2)_5
\end{array}
$$

$$
\begin{array}{ccccc}
& \text{CH}_3 & & & \\
& \text{H}_3\text{C} \diagdown | \diagup \text{CH}_3 & \text{H}_3\text{C} \diagdown \diagup \text{CH}_3 & & \\
+\text{Fe} & \text{C} & + \text{Fe} \quad \text{C} & & \\
\vdots & | & \xrightarrow[\text{insertion}]{\text{C-C}} \quad \vdots \quad \text{CH}_2 & \xrightarrow[\text{shift}]{\beta\text{-H}} & (97) \\
\text{N} & \text{CH}_2 & \text{N} & & \\
\parallel & | & \parallel & & \\
\text{C} & (\text{CH}_2)_5 & \text{C} \quad (\text{CH}_2)_5 & &
\end{array}
$$

$$
\begin{array}{c}
\text{H}_3\text{C} \diagdown \diagup \text{CH}_3 \\
\text{H} \qquad \text{C} \\
\text{H}_3\text{C} \diagdown \quad \diagup \\
+ \text{Fe} \cdots \text{CH} \longrightarrow \boxed{\text{CH}_4} \\
| \qquad\qquad ca\ 10\% \\
\text{N} \\
\parallel \\
\text{C} \qquad (\text{CH}_2)_5
\end{array}
$$

the iron atom and methane is eliminated. These two pathways are suggested because the methane produced from the reaction of **33** is almost all CH_4 with only a small amount of CH_3D. In fact, the product ratios suggest that approximately 90% of the methane is formed from a terminal methyl group and a hydrogen from another terminal methyl group. The products found when **34** is the reactant are consistent with this explanation. Here, almost all the product is CD_4 and only a small amount of CD_3H is formed.

Another significant decomposition pathway yields butene. The reaction can occur in two ways (equation 98). In the first step, the iron ion either inserts into a terminal C—H bond (I) or into the β-C—C bond (III). Reaction I is followed by β-C—C bond cleavage (II) while the intermediate formed in III undergoes a β-C—H shift (IV). The reaction is completed when the RCH_2—C≡N···Fe^+ ion eliminates C_4H_8. The butene produced from the reaction of **33** is almost all unlabelled butene while the butene formed from **34** is almost all C_4D_8. This shows that the butene comes primarily from the terminal t-butyl group (equation 98).

(98)

Finally, the RCH_2—C≡N···Fe^+ ion loses C_4H_{10}. The authors suggest that the butane is formed via a mechanism shown in equation 99 in which the iron ion initially inserts into the β-C—C bond. This is followed by the transfer of a β-H to the iron atom and elimination of C_4H_{10}. This mechanism was suggested because the butane from **33** is mainly undeuterated.

Recently, Schwarz and coworkers[119] have used isotopically labelled branched nitriles to demonstrate that the reactions of these iron-nitrile metastable ions involve atoms in both branches of the hydrocarbon chain.

(99)

VI. THE PREPARATION AND USES OF ISOTOPICALLY LABELLED ALKYNES

A. Synthesis of Labelled Alkynes

The general procedures for the incorporation of deuterium into the terminal and propargyl positions of alkynes under basic conditions has been reviewed[120]. Hence, only a brief summary of these procedures will be presented.

Treating a terminal alkyne with equimolar amounts of phenyllithium or n-butyllithium and then with D_2O gives a deuterated alkyne of high isotopic purity (equation 100).

$$RC{\equiv}CH \xrightarrow[\text{2. } D_2O]{\text{1. PhLi or } n\text{-BuLi}} RC{\equiv}CD \qquad (100)$$

Terminal alkynes with > 99% deuterium at the terminal position can also be obtained after several treatments with excess sodium deuteroxide in D_2O (equation 101).

$$RC{\equiv}CH \xrightarrow[D_2O]{\text{NaOD}} RC{\equiv}CD \qquad (101)$$

Small perdeuteroalkynes can be obtained in high purity by adding D_2O to carbides such as magnesium carbide (equation 102).

$$C_3Mg_2 + D_2O \longrightarrow CD_3C{\equiv}CD \qquad (102)$$

Deuterated alkynes have also been prepared by adding specifically deuterated alkyl halides to acetylide salts or deuterated acetylide salts to alkyl halides in liquid ammonia (equations 103 and 104, respectively). Deuterated alkynes can also be prepared by an isomerization reaction with sodium amide in heptane (equation 105). Four separate treatments of terminal alkynes in sodium deuteroxide–D_2O mixtures give isomerized perdeuteroalkynes in greater than 99.5% isotopic purity (equation 106).

$$RC \equiv CNa + CH_3CD_2Br \xrightarrow[NH_3 \, (l)]{} RC \equiv CCD_2CH_3 \qquad (103)$$

$$CD_3C \equiv CNa + CH_3CH_2Br \xrightarrow[NH_3 \, (l)]{} CD_3C \equiv CCH_2CH_3 \qquad (104)$$

$$CH_3(CH_2)_2C \equiv CCD_3 \xrightarrow[\text{2. } H_2O]{\substack{\text{1. excess } NaNH_2, \\ \text{heptane}}} CH_3(CH_2)_2CD_2C \equiv CH \quad (105)$$

$$CH_3CH_2C \equiv CH + NaOD \xrightarrow{D_2O} CD_3C \equiv CCD_3 \qquad (106)$$

Abrams[121,122] has investigated, using various alkali metal amides, the incorporation of deuterium into a series of alkynols. The isomerization and exchange that occurs when the alkynol is treated with the sodium salt of 1,3-diaminopropane-N,N,N^1,N^1-d_4 leads to perdeuteration exclusive of the CH_2OH group of the alkynol. The terminal hydrogen of a 1-alkyne can be either hydrogen or deuterium, depending on whether water or D_2O is used in the quenching step which protonates the terminal acetylide. Some of the results obtained with different alkynols are shown in Table 6.

In all reactions, the triple bond migrates away from the carbon bearing the OH group to give the perdeuterated terminal alkyne. The exchange always occurs in the alkyl group containing the C≡C bond but does not occur at the carbon bearing the hydroxyl group. This presumably is because the acidity of the C—H bond is greatly reduced by the negative oxygen. It can be considered that the formation of the conjugate base of the terminal alkyne essentially traps the final product. A partial mechanism for the isomerization and exchange reactions is illustrated in equation 107.

In a related study, Abrams and Shaw[123] carried out multiple isomerizations of the triple bond in several cycloalkynes producing essentially complete perdeuteration after only two treatments with $LiDN(CH_2)_3ND_2$ in $D_2N(CH_2)_3ND_2$ (equation 108).

TABLE 6. Sites of deuterium incorporation for reactions of several alkynols with the sodium salt of 1,3-diaminopropane-N,N,N^1,N^1-d_4

Substrate	Product	Yield	Percent D-incorporation at the exchangeable hydrogens
$HOCH_2C \equiv C(CH_2)_6CH_3$	$HOCH_2(CD_2)_7C \equiv CD$	48	88
$HO(CH_2)_4C \equiv C(CH_2)_3CH_3$	$HOCH_2(CD_2)_7C \equiv CD$	40	84
$HOCH_2C \equiv C(CH_2)_{14}CH_3$	$HOCH_2(CD_2)_{15}C \equiv CH^a$	18	89
$CH_3CH_2CHOHC \equiv C(CH_2)_5CH_3$	$CH_3CH_2CHOH(CD_2)_6C \equiv CD$	35	88
$CH_3(CH_2)_{12}CHOHC \equiv C(CH_2)_5CH_3$	$CH_3(CH_2)_{12}CHOH(CD_2)_6C \equiv CD$	14	90

^{a}H_2O quench.

$$\underset{CH_3CH_2CH}{\overset{\overset{\displaystyle OH}{|}}{}} -C \equiv C - \underset{CH(CH_2)_4CH_3}{} \xrightarrow{\text{base}} CH_3CH_2CH - C \equiv C - \overset{\overset{\displaystyle O^-}{|}}{\ddot{C}}H(CH_2)_4CH_3$$

$$\xrightarrow{ND_2(CH_2)_3ND_2} \underset{CH_3CH_2CH}{} - \underset{\overset{\displaystyle |}{C}}{\overset{\displaystyle O^-}{|}} = \underset{\overset{\displaystyle |}{C}}{\overset{\displaystyle D}{|}} = CH(CH_2)_4CH_3 \xrightarrow{\text{base}}$$

$$\underset{CH_3CH_2CH}{\overset{\overset{\displaystyle O^-}{|}}{}} - CD = C = \ddot{C} - (CH_2)_4CH_3 \xrightarrow{ND_2(CH_2)_3ND_2}$$

$$\underset{CH_3CH_2CHCD_2C}{\overset{\overset{\displaystyle O^-}{|}}{}} \equiv CCH_2(CH_2)_3CH_3 \xrightarrow{\text{base}} \qquad (107)$$

$$\underset{CH_3CH_2CHCD_2C}{\overset{\overset{\displaystyle O^-}{|}}{}} \equiv C - \ddot{C}H - (CH_2)_3CH_3 \xrightarrow{ND_2(CH_2)_3ND_2} \xrightarrow{\text{base}} \xrightarrow{\text{etc.}} \xrightarrow{\text{etc.}}$$

$$\underset{CH_3CH_2CH(CD_2)_6}{\overset{\overset{\displaystyle O^-}{|}}{}} - C \equiv C \colon^- \xrightarrow{D_2O} \underset{CH_3CH_2CH(CD_2)_6C}{\overset{\overset{\displaystyle OD}{|}}{}} \equiv CD$$

$$\xrightarrow[\text{D}_2\text{N(CH}_2)_3\text{ND}_2]{\text{LiDN(CH}_2)_3\text{ND}_2} \qquad (108)$$

$$n = 10, 11, 13$$

B. Mechanistic Studies Using Labelled Alkynes

A primary hydrogen deuterium kinetic isotope effect of 1.9 was found when Janout and Regen[124] used deuterium oxide in place of water in the hydration of the adduct formed when phenylmercuric hydroxide reacts with terminal alkynes. The first step in this reaction

$$PhHgOH + RC \equiv CH \xrightarrow[\text{CHCl}_3]{60\,°C} PhHgC \equiv CR + H_2O \xrightarrow[\text{CHCl}_3/\text{H}_2\text{O}]{60\,°C} \overset{\overset{\displaystyle O}{\|}}{CH_3CR} \quad (109)$$

$$PhHgC \equiv CR \quad \Longleftrightarrow \quad \underset{\underset{\displaystyle Ph}{|}}{\overset{\underset{\displaystyle Hg}{\overset{\displaystyle +}{}}}{\ddot{C}=C}} \overset{\displaystyle R}{\diagup} \xrightarrow[\text{rate determining}]{H_2O} \qquad (110)$$

is the formation of a σ-bonded mercury acetylide (equation 109), which subsequently is converted into the methyl ketone by hydrolysis in a chloroform–water mixture at 60 °C. The authors suggested that the primary hydrogen–deuterium kinetic isotope effect was found because the proton is transferred from water to the zwitterionic intermediate in the slow, hydration step of the reaction (equation 110).

Although the final yields of ketone are not high, i.e. they range from 50 to 67%, the reaction is very useful because it can be carried out on substrates with several other functional groups. For example, the reaction is successful when acetals, thioacetals, lactones, non-conjugated alkenes, epoxides, alcohols, and secondary bromides are present in the alkyne.

Manceron and Andrews[125] have investigated the alkali metal-induced intermolecular hydrogen transfer to the C≡C bond of an alkyne to form an alkene in solid argon at 15 K (equation 111).

$$3HC \equiv CH + M^0 \longrightarrow CH_2 = CH_2 + 2MC \equiv CH \qquad (111)$$

When acetylene and acetylene-d_2 mixtures are reacted with sodium, potassium or cesium atoms, 77% of the ethylene-d_2 formed when hydrogen from acetylene adds to acetylene-d_2, or deuterium from acetylene-d_2 adds to acetylene, is *trans*. This high *trans/cis* ratio suggests that the first atom adds to the acetylene to give a *trans* vinyl radical (equation 112), which reacts with a second hydrogen to form the *trans*-ethylene-d_2. This mechanism has been proposed because both the cesium acetylide and the cesium–acetylene π-complexes have been detected by infrared spectroscopy of reaction mixtures. Further support for this mechanism is provided by theoretical calculations, which show that the *trans*-vinyl radical formed when a deuterium atom adds to acetylene is more stable than the *cis*-radical by 0.13 kcal mol^{-1}. The authors indicate that this difference in stability would lead to a *trans/cis* ethylene-d_2 ratio of 2.8, which is close to the 3.4 that is found experimentally. They also suggest that the difference between the calculated and the observed *trans/cis* ratio is due to an isotope effect associated with the transfer of deuterium in the first step of the addition reaction. This is reasonable because hydrogen is transferred more readily than deuterium in this reaction, i.e. the amount of $CH_2=CH_2$ formed by the addition of hydrogen to acetylene is greater than the amount of $CD_2=CD_2$ formed by the addition of deuterium to acetylene-d_2.

$$(112)$$

Finally, the *trans/cis* preference found for the formation of ethylene-d_2 has also been observed when deuterium from acetylene-d_2 adds to 2-butyne, when mixtures of acetylene-d_2 and 2-butyne are reacted with alkali metal atoms in solid argon. In this reaction, only the *trans*-2,3-dideutero-2-butene is formed, presumably because the *trans*-butenyl radical is much more stable than the *cis*-butenyl radical.

Lewandos and coworkers[126] proposed the formation of a silver ion–π coordination complex to account for the deuterium exchange that occurs at the terminal acetylenic hydrogen when an alkyne is treated with silver(I) ions and CD_3COOD in nitromethane-d_3 (equation 113). The reaction was carried out in nitromethane-d_3 since this solvent favours the formation of the silver ion–π complexed alkyne. The authors found that the silver π-coordination complex only causes exchange of the terminal acetylenic hydrogen. They were also able to demonstrate that a silver acetylide intermediate, which would also react with CD_3COOD to give deuterium incorporation, was not involved in the reaction.

$$RC\!\!\mathrel{\ooalign{\hss=\hss\cr\hss\equiv\hss}}\!\!CH + CD_3COOD \xrightarrow{\;\;CD_3NO_2\;\;} RC\!\!\mathrel{\ooalign{\hss=\hss\cr\hss\equiv\hss}}\!\!CD + CD_3COOH \qquad (113)$$
$$\quad\;\;Ag^+ \hspace{6.5cm} Ag^+$$

Because terminal alkynes π-coordinated to silver(I) ions were found to undergo deuterium exchange at a rate approximately 10^5 times faster than uncoordinated alkynes[126], it seemed probable that the CD_3COOD was involved in the rate-determining step of the deuterium exchange reaction. This was investigated by determining the dependence of the rate on the concentration of the deuterium donor, CD_3COOD. A partial dependence of the rate on the concentration of the deuterated acetic acid was found for the reactions of three different terminal alkynes[127] (equation 114).

$$\text{rate} = (k_1 + k_2\,[CD_3COOD])[\text{complex}] \qquad (114)$$

Thus, the rate expression suggested that the deuterium exchange occurs by two competing mechanisms. One of these only depends on the concentration of the silver ion–π complex whereas the second depends on the concentrations of both the silver ion–π complex and CD_3COOD, i.e., it has both the silver ion–π complex and CD_3COOD in the rate-determining step.

The authors also measured the hydrogen–deuterium isotope effect associated with the reaction in order to gain insight into the mechanism of the exchange reaction. If cleavage of the terminal acetylenic carbon–hydrogen bond occurs prior to, or during, the rate-determining step, one would observe a normal primary hydrogen–deuterium kinetic isotope effect, k_H/k_D, of at least two. The measured isotope effect, $(k_H/k_D)_{observed}$, however, was found to be inverse, i.e.

$$(k_H/k_D)_{observed} = \left[\frac{\text{rate}_H}{[\text{complex}]_H}\right]\Bigg/\left[\frac{\text{rate}_D}{[\text{complex}]_D}\right] = 0.46 \pm 0.07 \qquad (115)$$

This inverse istope effect is obviously not a primary isotope effect associated with the rupture of the terminal C_{sp}—H (D) bond in the rate-determining step and the authors concluded that it must be a secondary hydrogen–deuterium kinetic isotope effect and that the bond to hydrogen (deuterium) is not broken in the rate-determining step of the reaction. An inverse isotope effect would be observed if the terminal sp-hybridized carbon of the alkyne were converted into an sp^2-hybridized carbon during the rate-determining step of the reaction. If this occurred, the C—H (C—D) bonds would become stronger in the reaction, i.e. the C—H and C—D bonds would have increased force constants and a greater zero-point energy difference at the transition state than in the reactants. As a result, the observed secondary hydrogen–deuterium kinetic isotope effect would be inverse.

A possible rate-determining step for the reaction is illustrated in equation 116.

$$RC{\equiv}CH + Ag^+ \xrightarrow{\ \text{slow}\ } R\overset{+}{C}{=}CH{\cdots}Ag \tag{116}$$

The formation of the vinyl cation–silver ion complex in the slow step of the reaction is consistent with the observation of an inverse secondary deuterium kinetic isotope effect, because the terminal C—H bond undergoes a hybridization change from sp to sp^2 in the rate-determining step of the reaction.

A second mechanism was also proposed to account for the kinetic dependence on the concentration of CD_3COOD (equation 117). In this mechanism, the deuterated acetic acid supplies the D^+, which adds to the silver ion–π complex in the rate-determining step of the reaction. This slow step also has the appropriate hybridization change from sp to sp^2 at the terminal acetylenic carbon and would lead to the inverse isotope effect that is observed.

$$RC{\equiv}C{-}H + D^+ \longrightarrow R\overset{+}{C}{=}C\begin{matrix} H \\ \diagup \\ \diagdown \\ D \end{matrix} \tag{117}$$

with Ag^+ below each structure.

C. Mass Spectrometry

The similarity between the mass spectra of functionalized allenes and acetylenes[128–130] prompted Arseniyadis and coworkers to use several specifically deuterated analogues of β-allenic and γ-acetylenic alcohols, 35–38 and 39–41, respectively, to determine the fragmentation patterns of these alcohols. The relative intensities of some of the important peaks in the 17-eV mass spectra of compounds 35–41 are given in Table 7.

The favourable formation of a common $m/z = 69$ (M-15) ion, $C_4H_5O^+$ and its mono- and di-deuterated analogues is noted in the 17-eV spectra of both the allenes 35–38 and the alkynols 39–41. Collisional activation spectra of the $C_4H_5O^+$ ion and various acylium ions indicate that the structure of this $C_4H_5O^+$ ion is $CH_3CH{=}CH{-}C{\equiv}\overset{+}{O}$[131,132]. The only significant difference in the fragmentation spectra of the allenes and the alkynols is in the percent intensity of the (M-H_2O) ion. For example, the intensity of the M-18 peak in the spectrum of the allene 35 is only 13 whereas the intensity of the M-18 peak is 100 in the spectrum of the alkynol 39. Labelling experiments, using the alkynol 40, indicate that the loss of ·CH_3 to give the $m/z = 69$ peak incorporates carbon-5, whereas an examination of the spectrum of the alkynol 41 shows that extensive hydrogen migration occurs in the molecular ion. The authors suggested that the molecular ion formed from the acetylenic alcohol rearranges via an intermediate 42, where a proton is solvated by a triple bond and a hydroxyl group (Scheme 10). This intermediate can undergo two reactions. In one of these processes, 42 isomerizes by transferring a hydrogen from carbon-3 to carbon-5 giving the

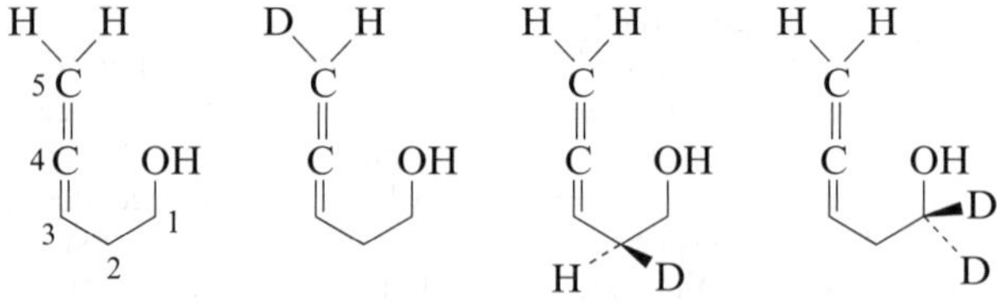

(35) (36) (37) (38)

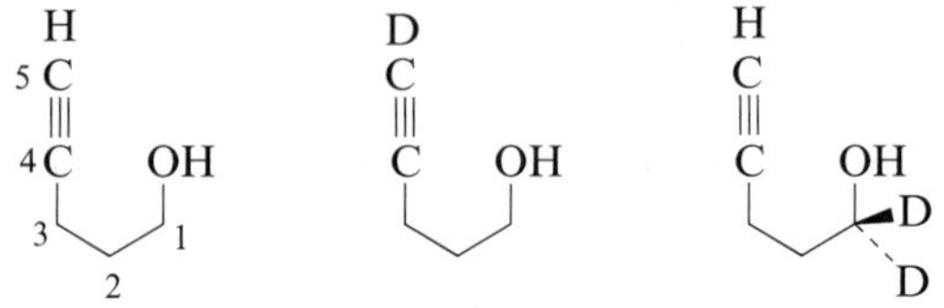

(39) (40) (41)

TABLE 7. The relative intensities[a] of some of the important peaks in the 17-eV mass spectra of compounds 35–41.

m/z	35	36	37	38	39	40	41
86				7(M)			21(M)
85		5(M)	12(M)			16(M)	
84	6(M)				27(M)		
71					100(M-15)		42(M-15)
70			100(M-15)	92(M-16)		8(M-15)	39(M-16)
69	100(M-15)	100(M-16)	29(M-16)	11(M-17)	77(M-15)	65(M-16)	21(M-17)
68				27(M-18)			100(M-18)
67			10(M-18)	11(M-18)		100(M-18)	
66	13(M-18)			6(M-20)	100(M-18)		14(M-20)

[a] The relative intensities of the peaks are expressed relative to a base peak intensity of 100 for the β-allenic alcohols, 35–38, whereas they are relative to a base peak intensity of 77 for the γ-acetylenic alcohols, 39–41.

molecular ion, 43, that would be formed from the β-allenic alcohol. This obviously accounts for the similar mass spectra obtained for the β-allenic and γ-acetylenic alcohols. The second possible reaction for intermediate 42 is to transfer a proton to the oxygen and produce the oxonium ion 44. The latter will readily lose water and transfer a hydrogen to give the M-H$_2$O ion, $[C_5H_6]^+$, $m/z = 66$.

The authors also proposed a mechanism to account for the loss of $\overset{\bullet}{C}H_3$ from the molecular ion formed from the β-allenic alcohol 35. This favourable fragmentation reaction occurs via several steps (Scheme 11).

The distonic ion 45 formed when the $\overset{\bullet}{O}H$ is transferred to carbon-4 of the allenic group undergoes one 1,2-hydrogen shift and two 1,5-hydrogen shifts to give 46, which loses $\overset{\bullet}{C}H_3$ to give the $C_4H_5O^+$ ion (M-15) with a $m/z = 69$.

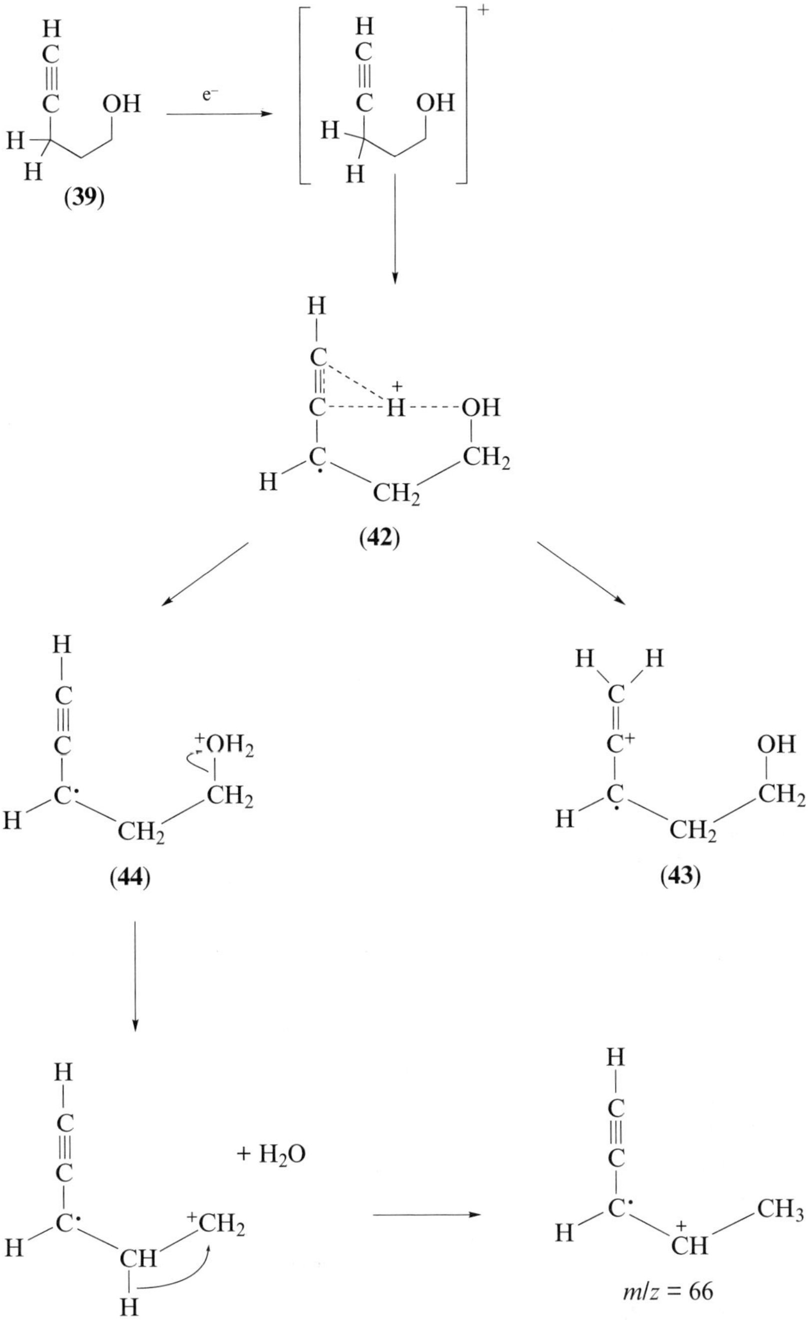

SCHEME 10

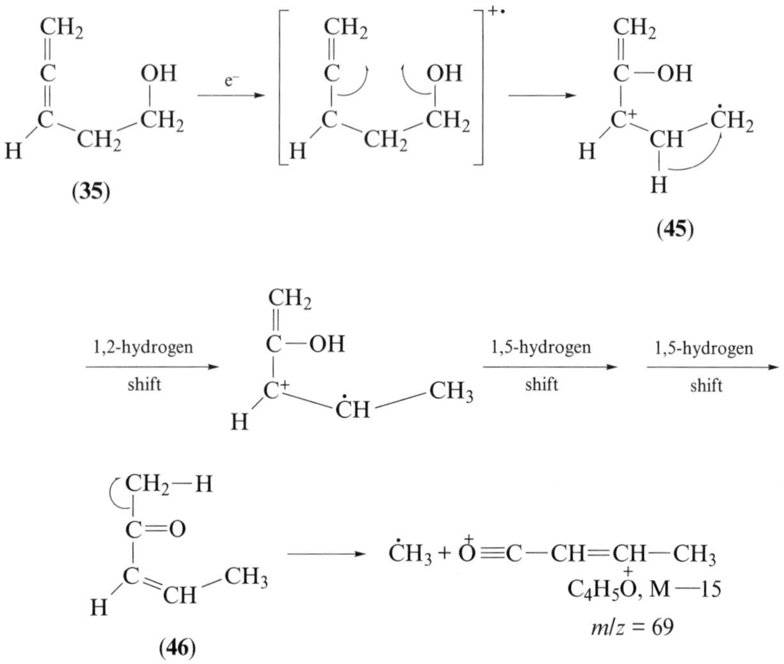

(35)

(45)

$\dot{C}H_3 + \overset{+}{O}\equiv C - CH = CH - CH_3$

$C_4H_5O, M - 15$

$m/z = 69$

(46)

SCHEME 11

VIII. REFERENCES

1. L. Pichat, 'Synthesis and uses of isotopically labelled cyanides', in *The Chemistry of the Cyano Group* (Ed. Z. Rappoport), Wiley, London, 1970, pp. 743–790.
2. J. C. Lavellay and J. Saussey, 'Synthesis and uses of isotopically labelled acetylenes', in *The Chemistry of the Carbon–Carbon Triple Bond* (Ed. S. Patai), Wiley, Chichester, 1978, pp. 957–976.
3. P. J. Smith and K. C. Westaway, 'Preparation and uses of isotopically labelled diazonium and diazo compounds', in *The Chemistry of Diazonium and Diazo Compounds* (Ed. S. Patai), Wiley, Chichester, 1978, pp. 709–749.
4. L. Melander, *Isotope Effects on Reaction Rates*, Ronald Press, New York, 1960.
5. E. Caldin and V. Gold (Eds.), *Proton Transfer Reactions*, Chapman and Hall, London, 1975.
6. W. W. Cleland, M. H. O'Leary and D. B. Northrup (Eds.), *Isotope Effects in Enzyme-catalyzed Reactions*, University Park Press, Baltimore, 1977.
7. L. Melander and W. H. Saunders, Jr., *Reaction Rates of Isotopic Molecules*, Wiley-Interscience, New York, 1980.
8. J. Bigeleisen, *Proceedings International Symposium on Isotope Separation*, North-Holland, Amsterdam, 1958.
9. J. Bigeleisen and M. Wolfsberg, *Adv. Chem. Phys.*, **1**, 15 (1958).
10. J. Bigeleisen, *J. Chem. Phys.*, **17**, 675 (1949).
11. S. Glasstone, K. J. Laidler and H. Eyring, *The Theory of Rate Processes*, McGraw-Hill, New York, 1941.
12. W. H. Saunders, Jr., *Chem. Scr.*, **8**, 27 (1975).
13. L. B. Sims, A. Fry, L. T. Netherton, J. C. Wilson, K. D. Reppond and W. S. Cook, *J. Am. Chem. Soc.*, **94**, 1364 (1972).
14. F. H. Westheimer, *Chem. Rev.*, **61**, 265 (1961).

15. R. A. More O'Ferrall, *J. Chem. Soc. (B)*, 785 (1970).
16. A. Streitwieser, Jr., R. H. Jagow, R. C. Fahey and S. Suzuki, *J. Am. Chem. Soc.*, **80**, 2326 (1958).
17. K. C. Westaway and S. F. Ali, *Can. J. Chem.*, **57**, 1089 (1979).
18. S. R. Hartshorn and V. J. Shiner, Jr., *J. Am. Chem. Soc.*, **94**, 9002 (1972).
19. H. Humski, V. Sendijarevic and V. J. Shiner, Jr., *J. Am. Chem. Soc.*, **96**, 6187 (1974).
20. J. C. Evans and G. Y. S. Lo, *J. Am. Chem. Soc.*, **88**, 2118 (1966).
21. V. J. Shiner, Jr, in *Isotope Effects in Chemical Reactions*, (Eds. C. J. Collins, Jr. and N. S. Bowman), A.C.S. Monograph 167, Van Nostrand-Reinhold, New York, 1970, pp. 122–150.
22. V. J. Shiner, Jr., in Reference 21, p. 138.
23. V. J. Shiner, Jr. and J. S. Humphrey, Jr., *J. Am. Chem. Soc.*, **85**, 2416 (1963).
24. V. J. Shiner, Jr. and J. G. Jewett, *J. Am. Chem. Soc.*, **87**, 1382 (1965).
25. L. Steffa and E. R. Thornton, *J. Am. Chem. Soc.*, **89**, 6149 (1967).
26. J. E. Baldwin and W. C. Widdison, *J. Am. Chem. Soc.*, **114**, 2245 (1992).
27. E. Glozbach, P. Krommes and J. Lorberth, *J. Organomet. Chem.*, **192**, 163 (1980).
28. D. Bethell and V. D. Parker, *J. Chem. Soc., Perkin Trans. 2*, 841 (1982).
29. J. M. Bakke, D. Bethell, P. J. Galsworthy, K. L. Handoo and D. Jackson, *J. Chem. Soc., Chem. Commun.*, 890 (1979).
30. D. Bethell, K. L. Handoo, S. A. Fairhurst and L. H. Sutcliffe, *J. Chem. Soc., Perkin Trans. 2*, 707 (1979).
31. D. Bethell and I. D. Callister, *J. Chem. Soc.*, 3801 (1963).
32. J. M. Bakke, D. Bethell and V. D. Parker, *Acta Chem. Scand.*, **B 41**, 253 (1987).
33. J. D. Kronis and P. Yates, *Tetrahedron Lett.*, **24**, 2419 (1983).
34. J. D. Kronis, M. F. Powell and P. Yates, *Tetrahedron Lett.*, **24**, 2423 (1983).
35. L. Thomas, *Diss. Abstr. Int. B*, **38**, 2187 (1977).
36. T. K. Miller, J. T. Sharp, G. J. Thomas and I. Thompson, *Tetrahedron Lett.*, **22**, 1540 (1981).
37. See Reference 3, pp. 744–747.
38. C. G. Swain, J. E. Sheats and K. G. Harbison, *J. Am. Chem. Soc.*, **97**, 796 (1975).
39. I. Szele, H. Zollinger and A. D. Deshpande, *Helv. Chim. Acta*, **64**, 2721 (1981).
40. E. S. Lewis and J. Insole, *J. Am. Chem. Soc.*, **86**, 34 (1964).
41. E. S. Lewis and R. E. Halliday, *J. Am. Chem. Soc.*, **91**, 426 (1969).
42. I. Szele and H. Zollinger, *Helv. Chim. Acta*, **64**, 2728 (1981).
43. J. R. Penton and H. Zollinger, *Helv. Chim. Acta*, **64**, 1728 (1981).
44. J. R. Penton and H. Zollinger, *Helv. Chim. Acta*, **64**, 1717 (1981).
45. J. R. Penton and H. Zollinger, *Helv. Chim. Acta*, **64**, 1728 (1981).
46. R. P. Kelly, J. R. Penton and H. Zollinger, *Helv. Chim. Acta*, **65**, 122 (1982).
47. A. H. Jackson and P. P. Lynch, *J. Chem. Soc., Perkin Trans. 2*, 1483 (1987).
48. R. H. Smith, S. R. Koepke, Y. Tondeur, C. L. Denlinger and C. J. Michejda, *J. Chem. Soc., Chem. Commun.*, 936 (1985).
49. K. C. Brown and M. P. Doyle, *J. Org. Chem.*, **53**, 3255 (1988).
50. J.-O. Thorell, PhD Dissertation, Karolinska Hospital and Institute, Stockholm, Sweden, 1993.
51. G. Westerberg and B. Langstrom. *Acta Chem. Scand.*, in press.
52. R. D. Cramer and G. B. Kistiakowsky, *J. Biol. Chem.*, **137**, 549 (1941).
53. M. Maeda, S. Nishimura, T. Fukumura and M. Kojima, *J. Labelled Compd. Radioharm.*, **25**, 233 (1988).
54. M. B. Winstead, C.-I. Chern, T.-H. Lin, A. Khentigan, J. F. Lamb and H. S. Winchell, *Int. J. Appl. Radiat. Isot.*, **29**, 443 (1978).
55. M. B. Winstead, D. A. Dougherty, T.-H. Lin, A. Khentigan, J. F. Lamb and H. S. Winchell, *J. Med. Chem.*, **21**, 215 (1978).
56. M. B. Winstead, D. D. Dischino, N. A. Munder and C. Walsh, *Eur. J. Nucl. Med.*, **5**, 165 (1980).
57. A. M. Emran, T. E. Boothe, R. D. Finn, M. M. Vora and P. J. Kothari, *Int. J. Appl. Radiat. Isot.*, **34**, 1013 (1983).
58. R. Amano and C. Crouzel, *Appl. Radiat. Isot.*, **37**, 541 (1986).
59. P. J. Kothari, R. D. Finn, G. W. Kabalka, M. M. Vora, T. E. Boothe, A. M. Emran and M. Mohammadi, *Appl. Radiat. Isot.*, **37**, 471 (1986).
60. B. De Spiegeleer, P. Goethals, G. Slegers, E. Gillis, W. Van den Bossche and P. Moerloose, *J. Nucl. Med.*, **29**, 1107 (1988).
61. S. Stone-Elander, P. Roland, C. Halldin, M. Hassan and R. Seitz, *Nucl. Med. Biol.*, **16**, 741 (1989).

62. J. A. Balatoni, M. J. Adam and L. D. Hall, *J. Labelled Compd. Radiopharm.*, **27**, 1429 (1989).
63. Y. Andersson, P. Malborg and B. Langstrom, *J. Labelled Compd. Radiopharm.*, **30**, 144 (1991).
64. M. Hassan, J.-O. Thorell, N. Warne and S. Stone-Elander, *Appl. Radiat. Isot.*, **42**, 1055 (1991).
65. G. Antoni and B. Langstrom, *Appl. Radiat. Isot.*, **43**, 903 (1992).
66. K. Hornfeldt, G. Antoni and B. Langstrom, *Acta Chem. Scand.*, **46**, 87 (1992).
67. G. Westerberg, P. Malmborg, and B. Langstrom, *J. Labelled Compd. Radiopharm.*, **30**, 126 (1991).
68. C.-Y. Shiue and A. P. Wolf, *J. Labelled Compd. Radiopharm.*, **22**, 171 (1985).
69. M. Maeda, Y. Koga, T. Fukumara and M. Kojima, *Appl. Radiat. Isot.*, **41**, 463 (1990).
70. R. L. Hayes, L. C. Washburn, B. W. Wieland and J. B. Anon, *Int. J. Appl. Radiat. Isot.*, **29**, 186 (1978).
71. L. C. Washburn, B. W. Wieland, T. T. Sun, R. L. Hayes and T. A. Butler, *J. Nucl. Med.*, **19**, 77 (1978).
72. M. R. Zalutsky, J. Wu, P. V. Harper and T. Wickland, *Int. J. Appl. Radiat. Isot.*, **32**, 182 (1981).
73. D. L. Casey, G. A. Digenius, D. A. Wesner, L. C. Washburn, J. E. Chaney, R. L. Hayes and A. P. Callahan, *Int. J. Appl. Radiat. Isot.*, **32**, 325 (1981).
74. J. R. Barrio, R. E. Keen, J. R. Ropchan, N. S. MacDonald, F. J. Baumgartner, H. C. Padgett and M. E. Phelps, *J. Nucl. Med.*, **24**, 515 (1983).
75. C. Halldin, K.-O. Schoeps, S. Stone-Elander and F.-A. Wiesel, *Eur. J. Nucl. Med.*, **13**, 288 (1987).
76. R. Iwata, T. Ido, T. Takahashi, H. Nakanishi and S. Iida, *Appl. Radiat. Isot.*, **38**, 97 (1987).
77. P. Burling, G. Antoni, Y. Watanabe and B. Langstrom, *Acta Chem. Scand.*, **44**, 178 (1990).
78. M. Maeda, Y. Koga, T. Fukumura and M. Kojima, *Appl. Radiat. Isot.*, **41**, 463 (1990).
79. N. D. Poe, G. D. Robinson Jr. and N. S. MacDonald, *Proc. Soc. Exp. Biol. Med.*, **148**, 215 (1975).
80. G. S. Jones, Jr., E. Livin, H. W. Strauss, R. N. Hanson and D. R. Elmaleh, *J. Nucl. Med.*, **29**, 68 (1988).
81. T. Takeahashi, T. Ido, K. Hatano, R. Iwata and H. Nakanishi, *Appl. Radiat. Isot.*, **41**, 649 (1990).
82. J. S. Fowler, C. D. Arnett, A. P. Wolf, R. R. MacGregor, E. F. Norton and A. M. Findley, *J. Nucl. Med.*, **23**, 437 (1982).
83. J.-O. Thorell, S. Stone-Elander, W. A. Konig, C. Halldin and L. Widen, *J. Labelled. Compd. Radiopharm.*, **29**, 709 (1991).
84. M. Hassan, G. Oberg, K. Ericson, H. Ehrsson, L. Eriksson, S. Stone-Elander, J.-O. Thorell, B. Smedmyr, B. Warne and L. Widen, *J. Chemother. Pharmacol.*, **30**, 81 (1992).
85. J. -O. Thorell, S. Stone-Elander and H. Van Holst, *J. Labelled Compd. Radiopharm.*, **32**, 586 (1993).
86. R. Gedye, F. Smith, K. Westaway, H. Ali, L. Baldisera, L. Laberge and J. Rousell.,*Tetrahedron Lett.*, **27**, 279 (1986).
87. R. Gedye, F. Smith and K. Westaway, *J. Microwave Power Electromag. Energy*, **26**, 1 (1991).
88. R. A. Abramovitch, *Org. Prep. Proced. Int.*, **23**, 683 (1991).
89. M. P. Mingos and D. R. Baghurst, *Chem. Soc. Rev.*, **20**, 1 (1991).
90. J.-O. Thorell, S. Stone-Elander and N. Elander, *J. Labelled Compd. Radiopharm.*, **31**, 207 (1992).
91. J.-O. Thorell, S. Stone-Elander, H. von Holst and M. Ingvar, *Appl. Radiat. Isot.*, **44**, 799 (1993).
92. C. W. Somawardhana, M. Sajjad and R. M. Lambrecht, *Appl. Radiat. Isot.*, **42**, 555 (1991).
93. C. W. Somawardhana, M. Sajjad and R. M. Lambrecht, *Appl. Radiat. Isot.*, **42**, 559 (1991).
94. C. Crouzel, R. Amano and D. Fournier *Appl. Radiat. Isot.*, **38**, 669 (1987).
95. C. Crouzel and A. Syrota, *Appl. Radiat. Isot.*, **41**, 241 (1990).
96. O. N. Chupakhin, V. L. Rusinov, A. A. Tumashov, E. O, Sidorov and I. V. Karpin, *Tetrahedron Lett.*, **33**, 3695 (1992).
97. A. Togni and S. D. Pastor, *J. Org. Chem.*, **55**, 1649 (1990).
98. A. D. Allen, F. Shahidi and T. T. Tidwell, *J. Am. Chem. Soc.*, **104**, 2516 (1982).
99. G. A. Olah, K. S. Prakash and M. Arvanaghi, *J. Am. Chem. Soc.*, **102**, 6640 (1980).
100. C. M. Jensen and W. C. Trogler, *J. Am. Chem. Soc.*, **108**, 723 (1986).
101. R. J. Crawford and C. Woo, *Can. J. Chem.*, **43**, 1534 (1965).
102. A. F. Hegarty and P. J. Tuohey, *J. Chem. Soc., Perkin Trans.*, **2**, 1313 (1980).
103. B. R. Cho, J. C. Lee, N. S. Cho and K. D. Kim, *J. Chem. Soc., Perkin Trans. 2*, 489(1989).

104. D. A. Winey and E. R. Thornton, *J. Am. Chem. Soc.*, **97**, 3102 (1975).
105. T. H. Lowry and K. S. Richardson, *Mechanism and Theory in OrganicChemistry*, 3rd edn., Harper and Row, New York, 1987, p. 372.
106. W. P. Jencks, *Chem. Rev.*, **72**, 705 (1972).
107. R. A. More O'Ferrall, *J. Chem. Soc. (B)*, 274 (1970).
108. P. Capdevielle, A. Lavigne, D. Sparfel, J. Baranne-Lafont, N. K. Cuong and M. Maumy, *Tetrahedron Lett.*, **31**, 3305 (1990).
109. P. Capdevielle, A. Lavigne and M. Maumy, *Synthesis*, 453 (1989).
110. K. C. Westaway, O. Matsson, A. S. Axelsson, J. Persson, Y. R. Fang and D. Jobe, unpublished results.
111. J. L. Kurz and M. M. Seif El-Nasr, *J. Am. Chem. Soc.*, **104**, 5823 (1982).
112. P. J. Smith and K. C. Westaway, 'Preparation and uses of isotopically labelled amino, quaternary ammonium and nitro compounds', in *The Chemistry of Amino, Nitroso and Nitro Compounds and Their Derivatives* (Ed. S. Patai), Wiley, Chichester, 1982, p. 1277.
113. T. A. Molenaar-Langeveld, R. H. Fokkens and N. M. M. Nibbering, *Org. MassSpectrom.*, **21**, 15 (1986).
114. D. Wittneben and H.-Fr. Grützmacher, *Org. Mass Spectrom.*, **27**, 533 (1992).
115. D. Wittneben and H.-Fr. Grützmacher, *Int. J. Mass Spectrom. Ion Proc.*, **100**, 543 (1990).
116. B. C. Baumann, J. K. MacLeod and L. Radom, *J. Am. Chem. Soc.*, **102**, 7927 (1980).
117. G. Czekay, T. Drewello and H. Schwarz, *J. Am. Chem. Soc.*, **111**, 4561 (1989).
118. T. Prusse, C. B. Lebrilla, T. Drewello and H. Schwarz, *J. Am. Chem. Soc.*, **110**, 5986 (1988).
119. G. Czekay, T. Drewello, K. Eller, W. Zummack and H. Schwarz, *Organometallics*, **8**, 2439 (1989).
120. J. C. Lavalley and J. Saussey, in *The Chemistry of the Carbon–Carbon Triple Bond* (Ed. S. Patai), Chap. 20, Wiley, Chichester, 1978.
121. S. R. Abrams, *Can. J. Chem.*, **62**, 1333 (1984).
122. S. R. Abrams, *J. Org. Chem.*, **49**, 3587 (1984).
123. S. Abrams and A. Shaw, *Tetrahedron Lett.*, **26**, 3431 (1985).
124. V. Janout and S. L. Regen, *J. Org. Chem.*, **47**, 3331 (1982).
125. L. Manceron and L. Andrews, *J. Phys. Chem.*, **89**, 4094 (1985).
126. J. P. Ginnebaugh, J. W. Maki and G. S. Lewandos, *J. Organomet. Chem.*, **190**, 403 (1980).
127. G. S. Lewandos, J. W. Maki and J. P. Ginnebaugh, *Organometallics*, **190**, 403 (1980).
128. S. Arseniyadis, J. Gore and M. L. Roumestant, *Org. Mass Spectrom.*, **13**, 54 (1978).
129. S. Arseniyadis, P. Guenot, J. Gore and R. Carrie, *Tetrahedron Lett.*, **22**, 2251 (1981).
130. S. Arseniyadis, P. Guenot, J. Gore and R. Carrie, *J. Chem. Soc., Perkin Trans. 2*, 1413 (1985).
131. S. Arseniyadis, A. Maquestiau, R. Flammang, P. Guenot and R. Carrie, *Org. Mass Spectrom.*, **24**, 909 (1989).
132. S. Arseniyadis, A. Maquestiau, R. Flammang, P. Guenot and R. Carrie, *Org. Mass Spectrom.*, **22**, 121 (1987).

CHAPTER 12

Biochemistry of triple-bonded functional groups

DEREK V. BANTHORPE

Department of Chemistry, University College London, 20 Gordon Street, London WC1H 0AH, UK

Supplement C2: The chemistry of triple-bonded functional groups
Edited by S. Patai © 1994 John Wiley & Sons Ltd

I. INTRODUCTION

The various types of natural products that possess C≡C or C≡N groups, the classes of compounds derived therefrom by functionalizations of the groups and what was then known of the biosynthesis of the whole group of compounds has been discussed—in some cases very briefly—in chapters of previous volumes of this Series[1-3]. There are three main types of naturally-occurring compounds that possess triple bonds, the alkynes (in particular polyalkynes), cyanogenic glycosides and isonitriles and these occur together with various minor subsidiary classes. A surprisingly large amount of work has been carried out on the general biochemistry of these at-first-sight obscure metabolites over the last two decades, and here such work is reviewed in outline up to 1992, with concentration on progress since 1980. The physiological, especially toxicological, properties of the group are only touched upon in passing, since they are covered in another chapter.

II. ALKYNES, POLYALKYNES AND DERIVED COMPOUNDS

A. Naturally Occurring Types

This class (≡ acetylenes ≡ polyacetylenes) comprises the most important, largest, most biochemically interesting and best studied group of the compounds under consideration. Tariric acid (1) was characterized in the last century and by 1950 some dozen naturally-occurring alkynes were known. Then came the advent of modern methods of separation and structure determination aided by the ease of detection of the polyalkyne grouping by ultraviolet absorption[4] and by 1985 the number was raised to over the thousand. During this period it was also realized that di-alkyne linkages *in situ* could act as biogenetic precursors for heterocyclic groups such as the furan, thiophene and their reduced ring systems and compounds including such groups—sometimes with no residual triple bond unsaturation—were included in the above count. It is noteworthy that N-heterocyclics with the pyrrole ring are not derived from dialkynes. The alkynes (using the name as generic for both the parent compounds and for the derived heterocyclics) are widespread in nature although they usually occur as a small proportion of the secondary metabolites of a particular species. They have been found in numerous common higher plants—although in fact they only occur in some 15 of the 100 or so families of such recognized by taxonomists; and they also occur in many types of marine organisms (sponges, algae, nudibranches), in *Basidiomycetes* fungi and other microorganisms and even in a few animals where they are accumulated from the diet[5].

TABLE 1. Examples of alkynes

Name	Structure	Source
Crepenynic acid	2	Seed of Compositae spp (daisy)
Dehydromatricaria ester	3	Seed of Compositae spp (daisy)
Falcarinone	4	*Daucus carota* (carrot; Umbellifereae)
Diatrene	5	Fungal spp
–	6	*Carum* spp (caraway; Umbellifereae)
Carlina oxide	7	*Carlina* spp (thistle; Compositae)
Matricaria lactone	8	*Asteraceae* spp (Compositae)
Artemisia lactone	9	*Asteraceae* spp (Compositae)
–	10	*Echinops* spp (thistles; Compositae)
–	11	*Echinops* spp (thistles; Compositae)
α-Terthienyl	12	*Tagetes* spp (marigold; Compositae)
Laurencine	13	Marine algae

The range of types is illustrated in Table 1 and Figure 1; the examples display commonly-occurring functional units but the selection is by no means exhaustive. A particular genus of higher plants, such as *Artemisia* (fam. Compositae), exhibits a wide range of alkynes with types as in Figure 1, supplemented by additional structures exemplified by Figure 2, **14–19**[6]: the spiro-systems in compounds **15–18** are especially characteristic. As examples of numerous phytochemical analyses on higher plants may be cited the characterization of 39 mainly acyclic polyalkynic compounds from *Dahlia* cultivars[7,8], of thiarubrine antibiotics similar to **11** contained in extracts used in African folk medicine and other dithiacyclohexadienes from *Artemisia* spp[9,10]; of spiro-acetals related to **18** from *Chrysanthemum* spp which coexist with novel sulphoxides, cf. **20**[11]; and of alkynic acids unique in inhibiting HMG-Coenzyme A reductase (the enzyme catalysing the rate-determining step

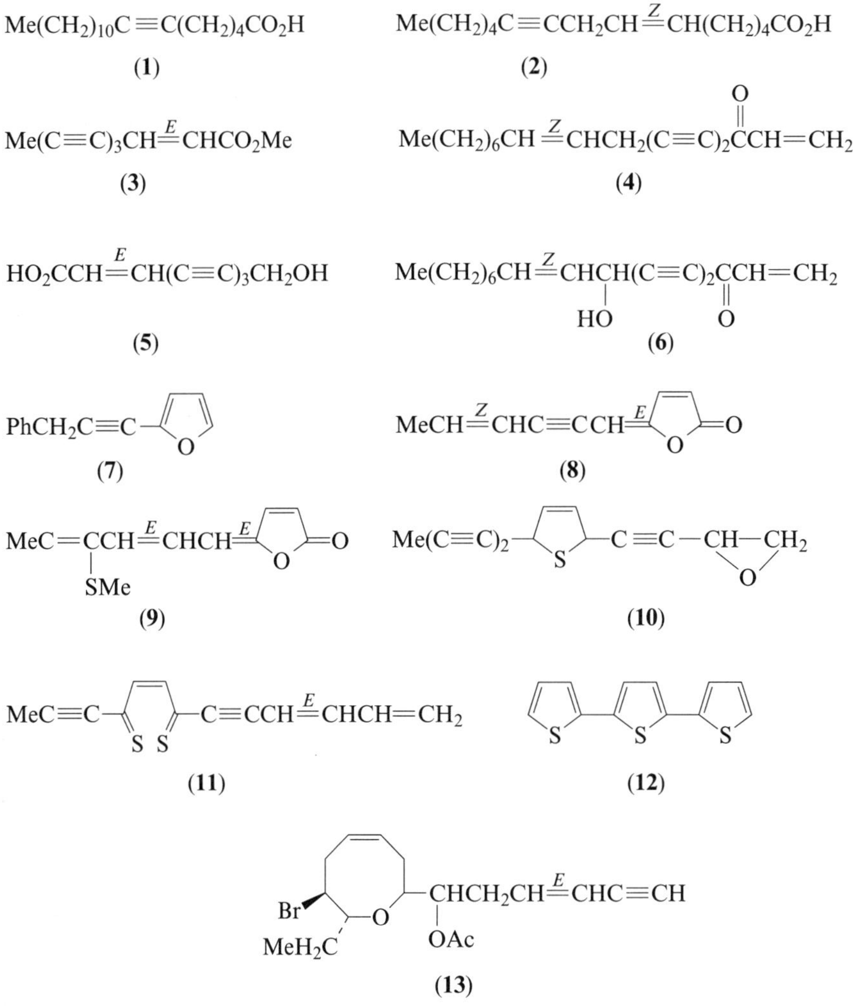

$$Me(CH_2)_{10}C \equiv C(CH_2)_4CO_2H$$

(1)

$$Me(CH_2)_4C \equiv CCH_2CH \overset{Z}{=} CH(CH_2)_4CO_2H$$

(2)

$$Me(C \equiv C)_3CH \overset{E}{=} CHCO_2Me$$

(3)

$$Me(CH_2)_6CH \overset{Z}{=} CHCH_2(C \equiv C)_2 \overset{O}{\overset{\|}{C}} CH = CH_2$$

(4)

$$HO_2CCH \overset{E}{=} CH(C \equiv C)_3CH_2OH$$

(5)

$$Me(CH_2)_6CH \overset{Z}{=} CHCH(C \equiv C)_2CCH = CH_2$$
$$\underset{HO}{|} \qquad \underset{O}{\|}$$

(6)

(7)

(8)

(9)

(10)

(11)

(12)

(13)

FIGURE 1. Some naturally-occurring alkynes

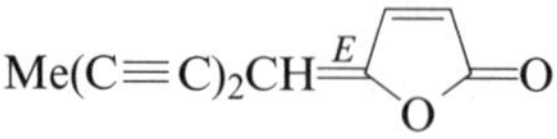

FIGURE 2. Alkynes from *Artemisia* and related species

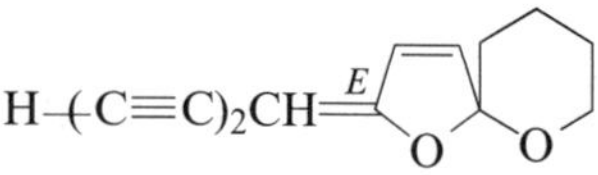

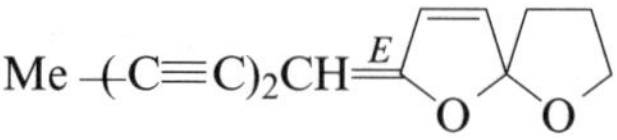

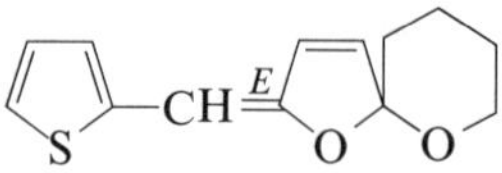

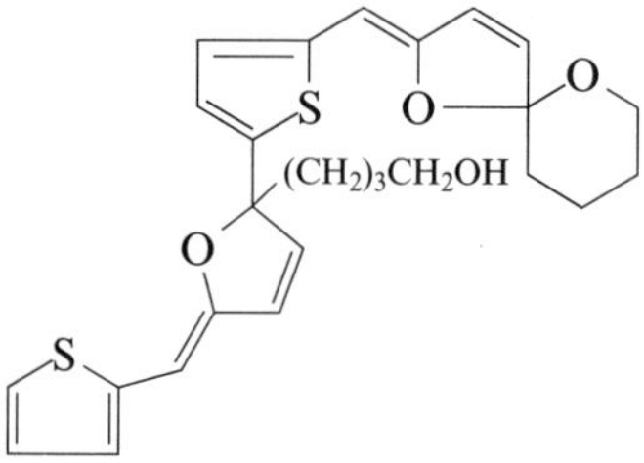

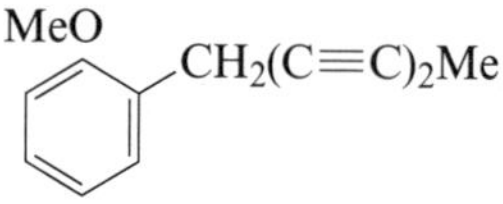

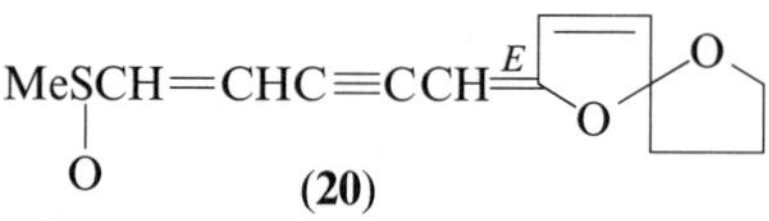

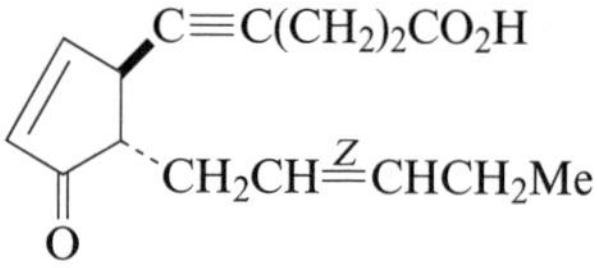

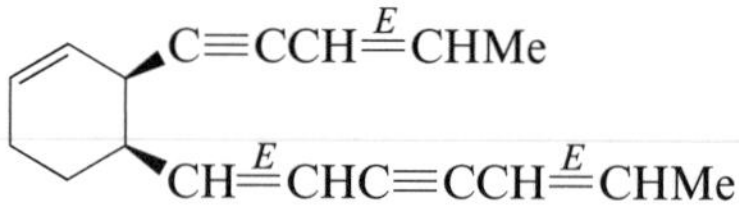

of mevalonoid biosynthesis[12], and numerous thiophenes and polyalkynic glycosides[13-15] from various higher plant species. Unusual compounds are **21** resembling the more familiar icosanoids from a Japanese moss[16] and the Diels–Alder adduct **22** from *Cineraria* spp[17]. The scope and extent of recent work is excellently demonstrated by a series of very detailed reviews on the occurrence, distribution, taxonomy and biosynthesis of polyalkynes in various Families, such as the Araliaceae, Heliantheae, Anthemideae etc. of the Asterales[18-22]. This order contains some 80% of the known polyalkynes—which are usually accumulated in root tissue.

Marine organisms are a rich source of exotic polyalkynes many of which possess structures not found in terrestrial species[23] in addition to conventional acyclic allene-alkyne hydrocarbons and their simple derivatives[24]. For example, red algae (in particular of the *Laurencia* and *Phacelocarpus* genera) accumulate compounds such as **23–26**[25-28] in which bromine derived by enzymic oxidation of Br⁻ from the environment is incorporated into some metabolites. A selection of the numerous compounds in sponges is also in Figure 3. Raspailyne A (**27**) from *Raspailia* spp is unique in that it is an enol ether (often occurring as a glucoside) that is derived from an alkynic aldehyde[29] whilst the acyclics petrosynol (**28**)[30] and siphonadiol (**29**)[31] are an antibiotic and an inhibitor of gastric H^+/K^+ transportation of ATP in *Siphonachalina* spp, respectively. Niphatyne A & B (**30**; $m = 8$, $m = 6$; $m = 10$, $n = 4$) are cytotoxic agents[32,33] and the dibromo-compound **31**[34] shows both anti-tumour and CNS-activity in tests on animals *in vivo*. Marine organisms in

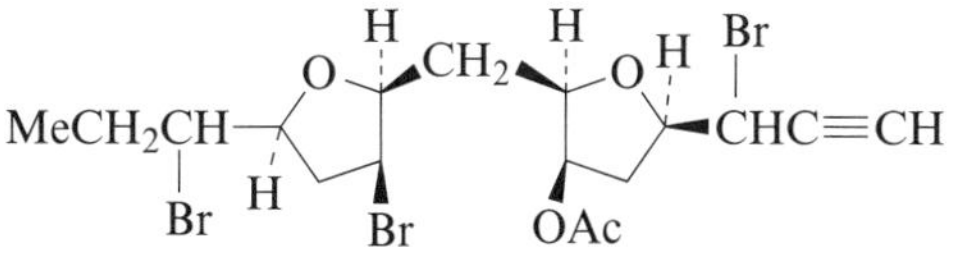

(**23**)

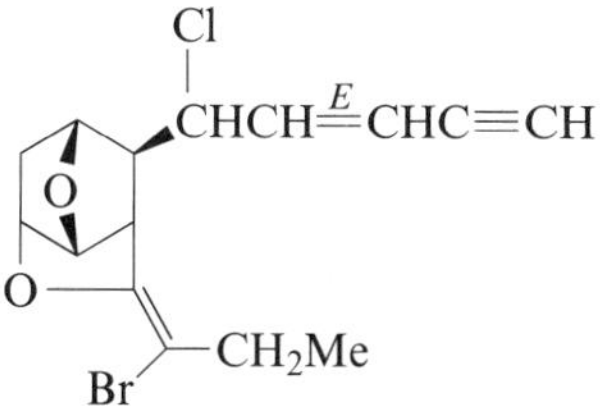

(**24**)

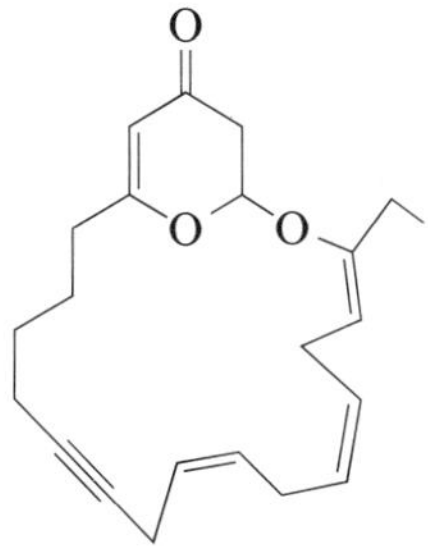

(**25**)

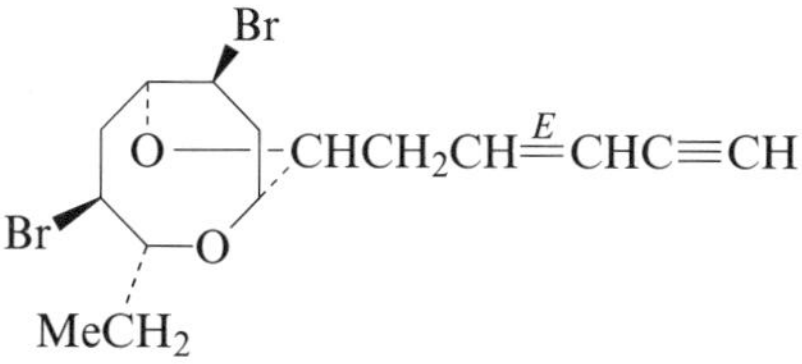

(**26**)

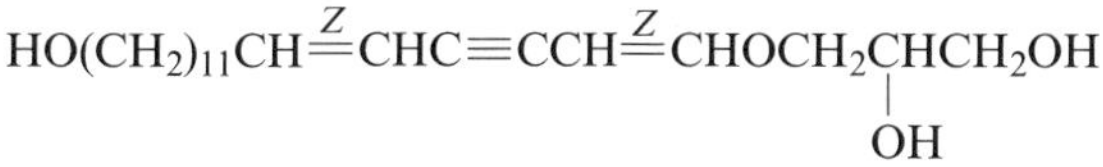

(**27**)

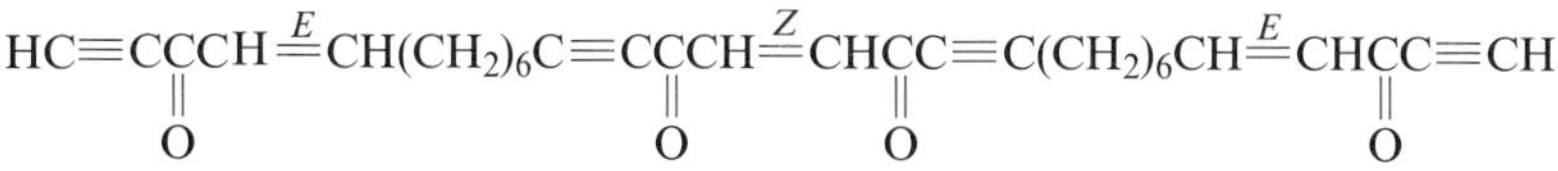

(**28**)

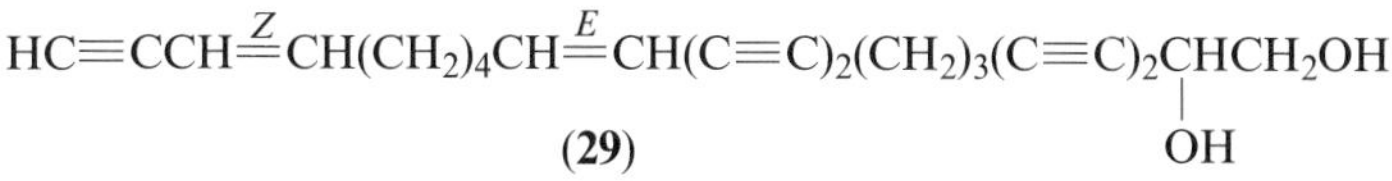

(**29**)

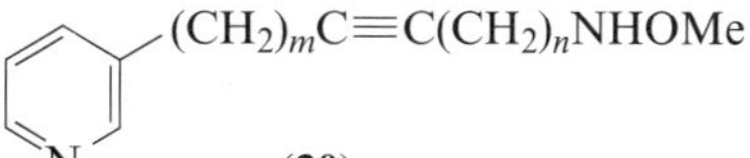

(**30**)

FIGURE 3. (*continued*)

$$\text{BrCH}\overset{E}{=}\text{CHCH}\overset{E_1Z}{=}\text{CH(CH}_2)_4\text{CH}\overset{E}{=}\text{CHC}\equiv\text{C(CH}_2)_3\text{CO}_2\text{H}$$
$$\underset{\text{Br}}{|}$$

(31)

$$\text{HC}\equiv\text{CHC}\overset{Z}{=}\text{CH-(CH}_2)_7\text{HC}$$

$$\text{HC}\equiv\text{CCHCH}\overset{E}{=}\text{CH(CH}_2)_6\text{CH}\overset{Z}{=}\text{CH(CH}_2)_4\text{C}\equiv\text{CCHC}\equiv\text{CCH}\overset{E}{=}\text{CH(CH}_2)_2\overset{\|Z}{\text{CH}}$$
$$\underset{\text{OH}}{|} \qquad\qquad\qquad\qquad\qquad\qquad\qquad \underset{\text{OH}}{|}$$

(32)

FIGURE 3. Alkynes from marine organisms

particular can form long-chain polyalkynes with numbers of carbons in excess of the 18 involved in conventional fatty acid synthesis (see next section). Presumably these are derived from the products of long-chain fatty acid (LCFA) metabolism that are involved in the biosynthesis of plant and animal waxes. Thus the sponges of *Petrosia* spp produce C_{46}-compounds, e.g. **32** that is cytotoxic and inhibits the fertilization of sea urchin eggs[35], and nudibranches of *Deltodoris* sps produce linear polyalkynes with C_{25} to C_{34} chains[36] some of which are chlorinated[37].

Fungi of the Basidomycetes tribe account for about 20% of the known alkynes and the latter occur both in mycelial cultures and in fruiting bodies[38,39]. Compounds from this source are usually straight chain with less oxidative functionalization than those in plants. Common examples are (Figure 4) are hexatriyne (**33**), the aromatic **34** and the thiophene-furan **35**[40], but chloro-compounds (such as **36**) are known[41]. Sometimes sufficient conjugation is present to confer pigmentation on the fungus, in e.g. xerulin (**37**) from *Xerula* spp[42], although some coloured fungi do contain carotenoids solely, or in addition to polyalkynes. Polyalkynes also occur along with other fatty acid metabolites in liverworts and mosses[43].

$$\text{H}\!\!\overset{}{\underset{}{-}}\!\!(\text{C}\equiv\text{C})_3\text{H} \qquad\qquad \text{Ph(C}\equiv\text{C})_2\text{CH}=\text{CHMe} \qquad\qquad$$

(33)   (34)   (35)

$$\text{ClCH}\overset{E}{=}\text{CH(C}\equiv\text{C)}_2\text{CH-CHCH}_2\text{OH}$$
$$\underset{\text{OH}}{|} \quad \underset{\text{OH}}{|}$$

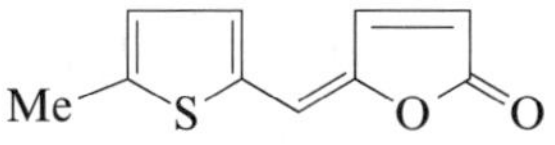

(36)

$$\text{MeCH}\overset{E}{=}\text{CH(C}\equiv\text{C)}_2(\text{CH}\overset{E}{=}\text{CH})_3\text{CH}\overset{E}{=}\!=\!\!O$$

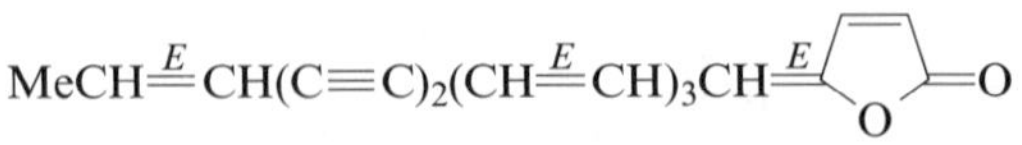

(37)

FIGURE 4. Alkynes from Fungi

B. Biosynthetic Pathways

1. Straight-chain compounds

Crepenynic acid (**2**) was first isolated from seed oil of *Crepsis* spp where it can comprise up to 60% v/v, but it was later found to be widespread in higher plants together with other C_{18} compounds such as stearolic acid (**38**) and some C_{17} compounds (Figure 5). In other early work, ximenynic acid (**39**) and a series of C_{18} alkynic acids (**40–43**; free and esterified) were isolated from *Santalum* spp[44]. It was consequently proposed that straight-chain polyalkynes were formed from oleic acid (a C_{18} compound) by sequential dehydrogenation on the side of the double-bond distal to the carboxyl group[45] such that the Z-unsaturation and the terminal octanoate unit of the parent were retained. This position of attack presumably results from anchoring of the acid group to the enzyme such as to allow desaturation of C_9 onwards. It is now universally accepted that alkynes are of polyketide origin and that oleic acid is indeed the parent on the fatty acid pathway. It has been demonstrated both in microorganisms and in higher plants that most alkynes are formed from one molecule of acetyl–Coenzyme A followed by subsequent condensation of eight malonyl units with loss of CO_2 at each stage: thus the labelling pattern in dehydromatricaria ester biosynthesized from [1-^{14}C]-acetate reveals alternation of tracer as in **44**[46], although as it has a C_{10} chain the original precursor must have been extensively degraded.

$$Me(CH_2)_7C{\equiv}C(CH_2)_7CO_2H \qquad Me(CH_2)_5CH\overset{Z}{=}CHC{\equiv}C(CH_2)_7CO_2H$$

$$\textbf{(38)} \qquad\qquad\qquad \textbf{(39)}$$

$$Me(CH_2)_2(CH{=}CH)_2C{\equiv}C(CH_2)_7CO_2H \quad MeCH_2(CH{=}CH)_2(C{\equiv}C)_2(CH_2)_7CO_2H$$

$$\textbf{(40)} \qquad\qquad\qquad\qquad \textbf{(41)}$$

$$MeCH_2(CH{=}CH)(C{\equiv}C)_3(CH_2)_7CO_2H \qquad H{-}(CH{=}CH)_2(C{\equiv}C)_3(CH_2)_7CO_2H$$

$$\textbf{(42)} \qquad\qquad\qquad\qquad \textbf{(43)}$$

$$Me(\overset{\bullet}{C}{\equiv}C)_3\overset{\bullet}{C}H\overset{E}{=}\overset{\bullet}{C}HCO_2Me$$

$$\textbf{(44)}\ \overset{\bullet}{C} = {}^{14}C$$

FIGURE 5. Related alkynes isolated from higher plants and fungi

Detailed labelling studies using a variety of synthetically-prepared ^{14}C-labelled C_{10} and C_{18} precursors applied to cultures of fungi of *Lepista*, *Polyporus* and *Agrocybe* spp led to the biosynthetic pattern as in Figure 6[47–49]. In particular, incorporation studies showed that the thiophene (**58**), dehydromatricarianol (**52**) and the hydroxy-compound **51** (C_8 and C_{10} compounds) originated from oleic acid via its methyl ester (**45**) and the esters of linoleic acid (**46**) and crepenynic acid (**47**)—all C_{18} compounds. ω-Oxidation (ω-O) can occur at the C_{18} stage (**49→50**) but the efficiencies of incorporation of appropriate precursors indicated that chain-shortening by β-oxidation (β-O, **49→3**) was favoured. **54** to **57** were detected in small amounts by tracer experiments and this showed that the fungi were capable of removing the terminal methyl group by oxidation, reducing the ester function and *trans*-hydroxylation of the double bond via formation of an epoxide. It is noteworthy that very similar metabolites such as matricaria ester (**61**) and dehydromatricaria ester (**3**) are formed by different routes. Such differences are common in secondary metabolism. Only a part of

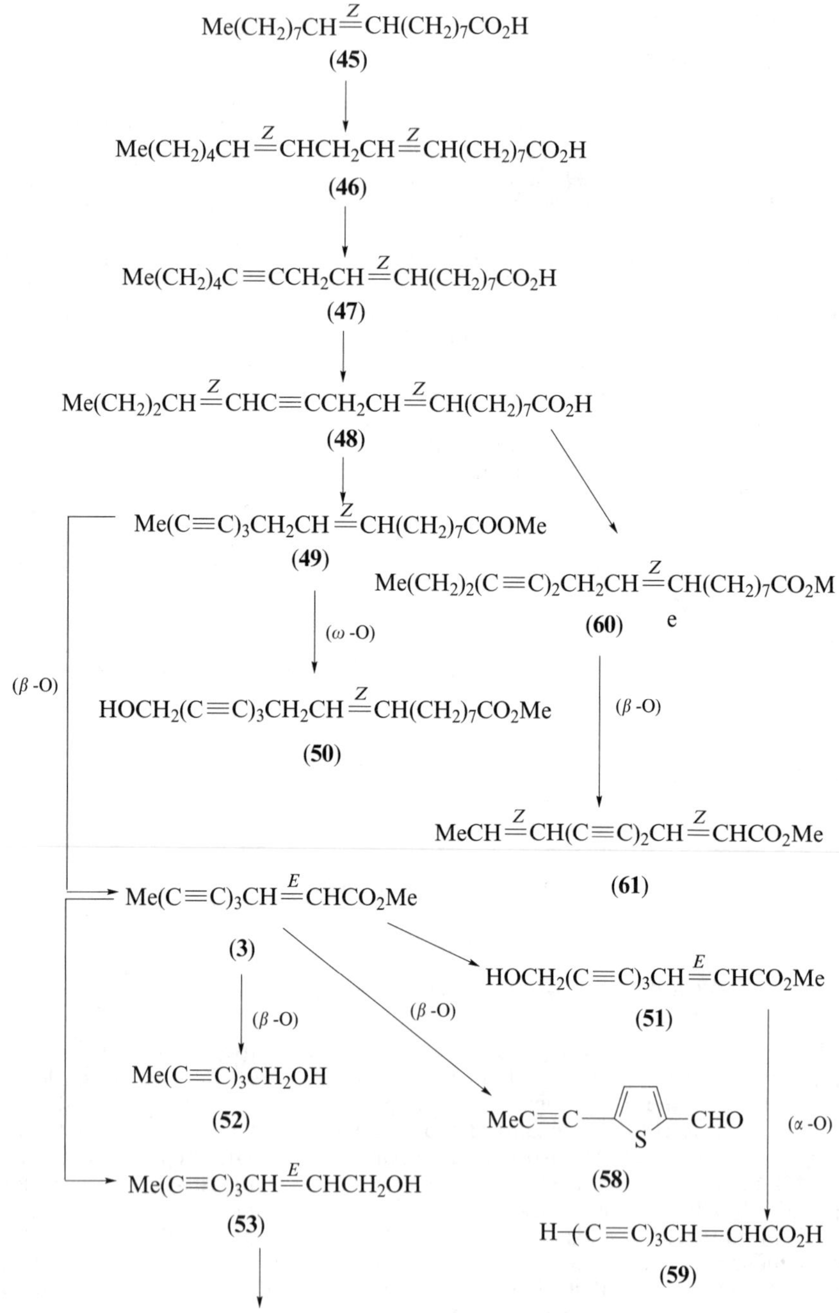

$$Me(CH_2)_7CH \overset{Z}{=} CH(CH_2)_7CO_2H$$
(45)

$$Me(CH_2)_4CH \overset{Z}{=} CHCH_2CH \overset{Z}{=} CH(CH_2)_7CO_2H$$
(46)

$$Me(CH_2)_4C \equiv CCH_2CH \overset{Z}{=} CH(CH_2)_7CO_2H$$
(47)

$$Me(CH_2)_2CH \overset{Z}{=} CHC \equiv CCH_2CH \overset{Z}{=} CH(CH_2)_7CO_2H$$
(48)

$$Me(C \equiv C)_3CH_2CH \overset{Z}{=} CH(CH_2)_7COOMe$$
(49)

$$Me(CH_2)_2(C \equiv C)_2CH_2CH \overset{Z}{=} CH(CH_2)_7CO_2Me$$
(60)

(ω -O)

(β -O)

$$HOCH_2(C \equiv C)_3CH_2CH \overset{Z}{=} CH(CH_2)_7CO_2Me$$
(50)

(β -O)

$$MeCH \overset{Z}{=} CH(C \equiv C)_2CH \overset{Z}{=} CHCO_2Me$$
(61)

$$Me(C \equiv C)_3CH \overset{E}{=} CHCO_2Me$$
(3)

$$HOCH_2(C \equiv C)_3CH \overset{E}{=} CHCO_2Me$$
(51)

(β -O)   (β -O)

$$Me(C \equiv C)_3CH_2OH$$
(52)

$$Me(C \equiv C)_3CH \overset{E}{=} CHCH_2OH$$
(53)

$$MeC \equiv C - \underset{S}{\text{(thiophene)}} - CHO$$
(58)

(α -O)

$$H \text{(} C \equiv C)_3CH = CHCO_2H$$
(59)

FIGURE 6. (*continued*)

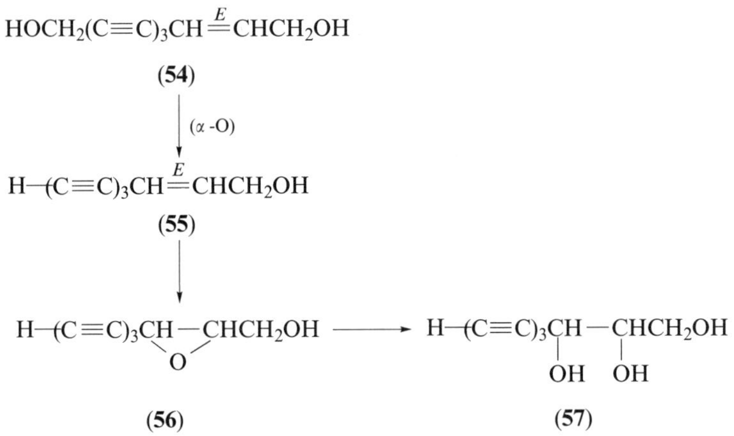

$HOCH_2(C \equiv C)_3CH \overset{E}{=} CHCH_2OH$

(54)

$(\alpha\text{-}O)$

$H-(C \equiv C)_3CH \overset{E}{=} CHCH_2OH$

(55)

$H-(C \equiv C)_3CH-CHCH_2OH \longrightarrow H-(C \equiv C)_3CH-CHCH_2OH$
                        $\underset{O}{\diagdown\diagup}$ $\underset{OH}{|} \quad \underset{OH}{|}$

(56)   (57)

FIGURE 6. Some biosynthetic pathways in fungi:
$(\beta\text{-}O)$, = β-oxidation, i.e. $C_n \rightarrow C_{n-2}$
$(\alpha\text{-}O)$, = α-oxidation, i.e. $C_n \rightarrow C_{n-1}$
$(\omega\text{-}O)$,= ω-oxidation

the deduced pathway is in Figure 6. An example of labelling to give other products is shown by the incorporation of tracer from methyl [1-^{14}C]-dec-2-en-4,6,8-triynoate (62) in *Lepista* spp (Figure 7). Although very reasonable flowcharts can be constructed on the basis of such (albeit limited) tracer studies, it must be appreciated that unique pathways to particular metabolites may not always exist *in vivo* within or between species. The possible presence of batteries of relatively unspecific enzymes could lead to the formation of a

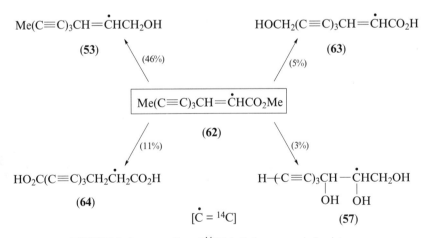

FIGURE 7. Incorporation of ^{14}C-labelled precursor in Lepista sps

metabolic grid whereby alterations in activities of the enzymes under changing conditions or in different periods of the growth cycle could lead to the selection of different pathways to particular metabolites. As no studies have been carried out using purified preparations of the enzymes involved in these reactions, the importance of such grids cannot begin to be assessed.

Figure 6 illustrates many of the common variations of chain-length and functionalization that occur in both fungi and higher plants. Thus:
(a) alkene and alkyne bond systems can both be present;
(b) alkene groups may be oxidized;
(c) an oxygen function at one end of the chain is accompanied by CH_3, CH_2OH, CHO or COOH at the other;
(d) methyls can be removed by ω-oxidation but C_{18} compounds are chain-shortened by β-oxidation or α-oxidation from the carboxyl end of the original oleate precursor (cf Figure 8). Such processes may involve decarboxylation or decarbonylation (of an aldehyde) and can result in the spectrum of odd- and even-numbered carbon chains down to C_6 that are observed. Nothing is known about the details of these processes[50-52].

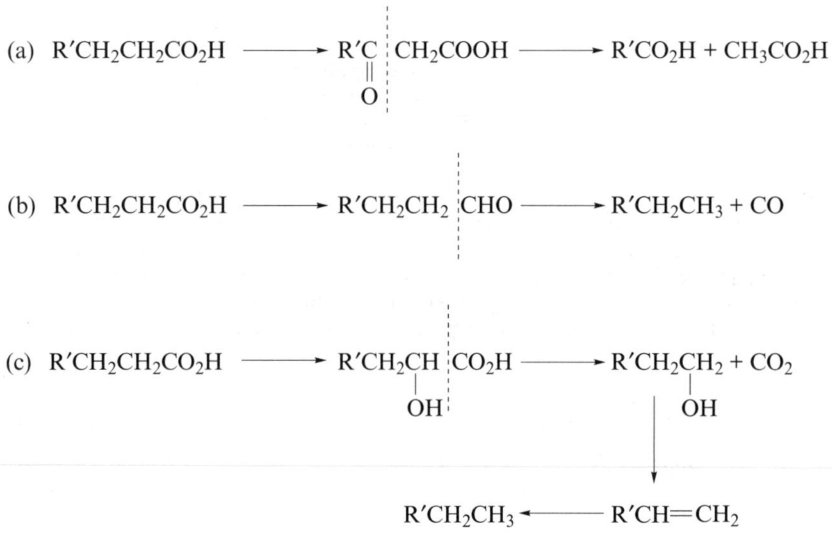

(a) $R'CH_2CH_2CO_2H \longrightarrow R'C \overset{\|}{\underset{O}{}} CH_2COOH \longrightarrow R'CO_2H + CH_3CO_2H$

(b) $R'CH_2CH_2CO_2H \longrightarrow R'CH_2CH_2 \ CHO \longrightarrow R'CH_2CH_3 + CO$

(c) $R'CH_2CH_2CO_2H \longrightarrow R'CH_2\underset{OH}{CH} \ CO_2H \longrightarrow R'CH_2CH_2 + CO_2$
$\underset{OH}{}$

$R'CH_2CH_3 \longleftarrow R'CH = CH_2$

FIGURE 8. Chain-shortening by α- and β-oxidation

A pattern similar to that in Figure 6 is believed to apply in higher plants[47,53] and microorganisms[5,54]. The more effective conversion of the C_{16} compound (65; Figure 9) than the C_{18} compounds (49 and 66) into dehydromatricaria ester (3) in *Artemisia* spp discredits a hypothesis that a direct Baeyer–Villiger type of oxidation of a C_{18}-precursor

$Me(C\equiv C)_3CH_2CH\overset{Z}{=\!=}CH(CH_2)_5CO_2Me$ $Me(C\equiv C)_3CH_2CH\overset{Z}{=\!=}CH(CH_2)_7CO_2Me$
(65)   (49)

$Me(C\equiv C)_3CH_2CH\overset{Z}{=\!=}CHC(CH_2)_7CO_2Me$
$\underset{O}{\|}$
(66)

FIGURE 9. Precursors for polyalkynes in higher plants

is involved in the chain-shortening rather than the presumed β-oxidation. On incorporation of [10-^{14}C; 9,10-^{3}H$_z$]-oleate and crepenylate an unexplained loss of ^{3}H was observed, although the oxidation state of the Z-double bond was apparently unaltered on passage into **3**. A similar loss of tracer occurred in feeding experiments with the fungus *Lepista*[47,53].

Few recent studies have been made on alkyne biosynthesis. Crepenynic acid has been shown to be a precursor of C$_9$ to C$_{14}$ polyalkynes in fungal cultures[55], and was also demonstrated to be a precursor of falcarinol (*cf* **4**, falcarinone) in tissue cultures of *Daucus carota* (carrot)[56]. (14-Z)-14,15-didehydrocrepenynic acid (**67**, Figure 10) is predominately converted into **68** by fungi of *Lepista* spp whereas *Caprinis* spp yield **69**[57].

$$CH_3(CH_2)_2CH \overset{E}{=\!=} CHC \equiv CCH_2CH \overset{Z}{=\!=} CH(CH_2)_7CO_2H$$
$$(67)$$

$$\longrightarrow NC(C \equiv C)_2CH = CHCO_2H$$
$$(68)$$

$$\longrightarrow H + C \equiv C)_3CH - CH_2$$
$$\qquad\qquad\qquad\quad | \quad\ |$$
$$\qquad\qquad\qquad OH \ \ OH$$
$$(69)$$

FIGURE 10. Alternative biotransformations in fungi

Little is known of the desaturation processes whereby the alkene bond is converted into the alkyne grouping—or indeed of the formation of an alkene from a saturated chain in the species studied. An attractive pathway for the former would be via an enol-phosphate in the process of chain extension of the polyketide (*cf* Figure 11 route a), but there is no evidence whatsoever for this—compare the previous discussion of incorporation of crepenyrate and its relatives[58]—and the introduction of unsaturation into a preformed chain by oxidation (O$_2$) in the presence of NADPH as cofactor is universally accepted[59]. The individual hydrogen transfers could be to sulphur (Figure 11, route b) and the intermediacy of the Z-alkene group is favoured by the preferred incorporation of Z-Δ^{14} rather than E-Δ^{14} intermediates (*cf* **48**, Figure 6). A great deal is known of the introduction of Δ^9 in oleic acid and of alkenic unsaturation at other locations in a fatty acid chain in general[59] but how much can be extrapolated to the process of polyalkyne formation is unclear.

(a) $-COCH_2 - SCoA \longrightarrow -C(OH) = CH - SCoA$

$$\longrightarrow -C(OPi) = CH - SCoA \longrightarrow -C \equiv C - SCoA$$

(b)

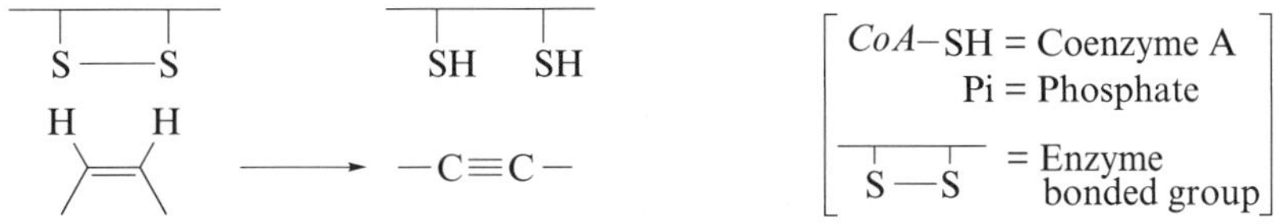

$$\begin{array}{c} S \!-\!\!\!-\! S \\ | \quad | \\ H \quad H \\ \diagdown\!\!\!\diagup\!\!\!=\!\!\!\diagdown\!\!\!\diagup \end{array} \qquad \begin{array}{c} SH \quad SH \end{array} \qquad \left[\begin{array}{l} CoA\!-\!SH = Coenzyme\ A \\ Pi = Phosphate \end{array} \right.$$

$$\longrightarrow -C \equiv C - \qquad \left. \begin{array}{l} S\!-\!\!\!-\! S = Enzyme \\ \qquad\qquad\ bonded\ group \end{array} \right]$$

FIGURE 11. Possible routes for desaturation

Investigation of the desaturase systems has been hampered by the enzymes being tenaciously bonded to membranes and by their substrates typically being micellar at convenient concentrations in buffers. Unlike those unsaturated fatty acids that are components of seed fats and waxes of the plant cuticle, the alkynes—or at best the alkynic acids—seem to occur in most plant tissues. However, tracer studies using homogenates of leaves of *Chrysanthemum* species indicate that the desaturases that accept oleic acid reside in the chloroplasts, whereas the final oxidation of the intermediate C_{13}-tryne and its cyclization to C_{13} spiroketals (*cf* 16) are effected by vacuolar enzymes[60]. This is consistent with the observation that desaturation to form the triple-bond systems of alkynes endogenous to the seed of *Carthamus tinctoria* (safflower) reside in the $10^3 \times g$ fraction after centrifugation[61].

Other functionalizations of straight-chain alkynes often containing alkyne–alkene conjugated systems—such as epoxidation and cyanohydrin formation—are fairly common, and *Z,E*-isomerization of double bonds to give mixtures of alkene isomers is usual; reduction of terminal vinyl groups (such as $—COCH{=}CH_2{\rightarrow}—COCH_2CH_3$) is especially common in the Umbelliferae.

2. Alicylclic and heterocyclic compounds

The variations in chain-lengths, type and position of unsaturation and the nature of functionalization outlined in the previous section leads to a very large number of possible linear alkynes and many hundreds have indeed been isolated. The number of possible compounds is increased enormously when cyclization occurs—and numerous such derivatives have been found and are being regularly routinely discovered. Some types are favoured in higher plants, others in fungi, but the former have the richer variety of structures as there is more opportunity for secondary metabolism in the differentiated tissues of such species than is possible in the more primitive microorganisms. Thus plants produce O- and S-heterocyclic compounds, including epoxides, benzenoids, S-compounds and acetate esters in profusion whereas alkenes and free acids rarely are found. C_9 to C_{18} compounds occur but the main range is C_{13}, C_{14} and C_{17}. In contrast, microorganisms (fungi mainly) contain many free acids and alkenes are frequent, but benzenoids and esters, S-compounds and O-heterocycles are uncommon. C_6 to C_{14} compounds are usual[1] with the main distribution in the range C_8 to C_{11} Marine organisms have a specialized and exotic chemistry which is outlined in Sections II.A and IV.C.

Few studies have been carried out on the biosynthesis of derivatized straight-chain or cyclic compounds, but the possible routes have been codified in a manner that has been generally accepted[62]. Sulphur compounds are believed to result from addition to conjugated dialkyne units (Figure 12a) to yield products such as 9, 10 and 12. The actual sulphur reagent is uncertain, but the biogenetic equivalent of H_2S or MeSH is derived from SO_4^{2-} or cysteine. Biosynthetic studies on the formation of thiophenes in hairy-root tissue cultures of *Tagetes* spp (marigolds) using ^{13}C-labelled substrates support the general validity of these views and demonstrate the utility of such cultures for the study of root products. However, incorporation of S-labelled precursors into dithiacyclohexadienes (such as 70) in similar cultures of *Chaenactis* spp revealed that tracer was incorporated into these and the co-occurring thiophenes (such as 71) at the same rate, and detailed analysis suggested that the previous proposal for sequential formation (*cf* Figure 12c) was unlikely[63]. Furan and pyran rings may be constructed by the routes in (d)–(h) in Figure 12. Some tracer studies favour these processes[64], but incubation of suspension cultures of *Saccharum* cells under [18]O_2 demonstrated that the oxygen atom in the furan ring in certain alkynic acids was derived from air and thus implicated a lipoxygenase system[65]. Larger O-heterocyclic rings occur in many marine organisms and the sequence in Figure 13 has been suggested on the basis of precursor studies to lead to the formation of halogenated C_{15} compounds in red algae of *Laurenica* spp[24]. Aromatic groups are thought to be formed as in Figure 14a and some experimental support for this exists[66]. But the occurrence of alkynes containing an aromatic ring carrying no hydroxy substituents has led to the proposal of an alternative route which, however, involves a rather unlikely methyl shift[67].

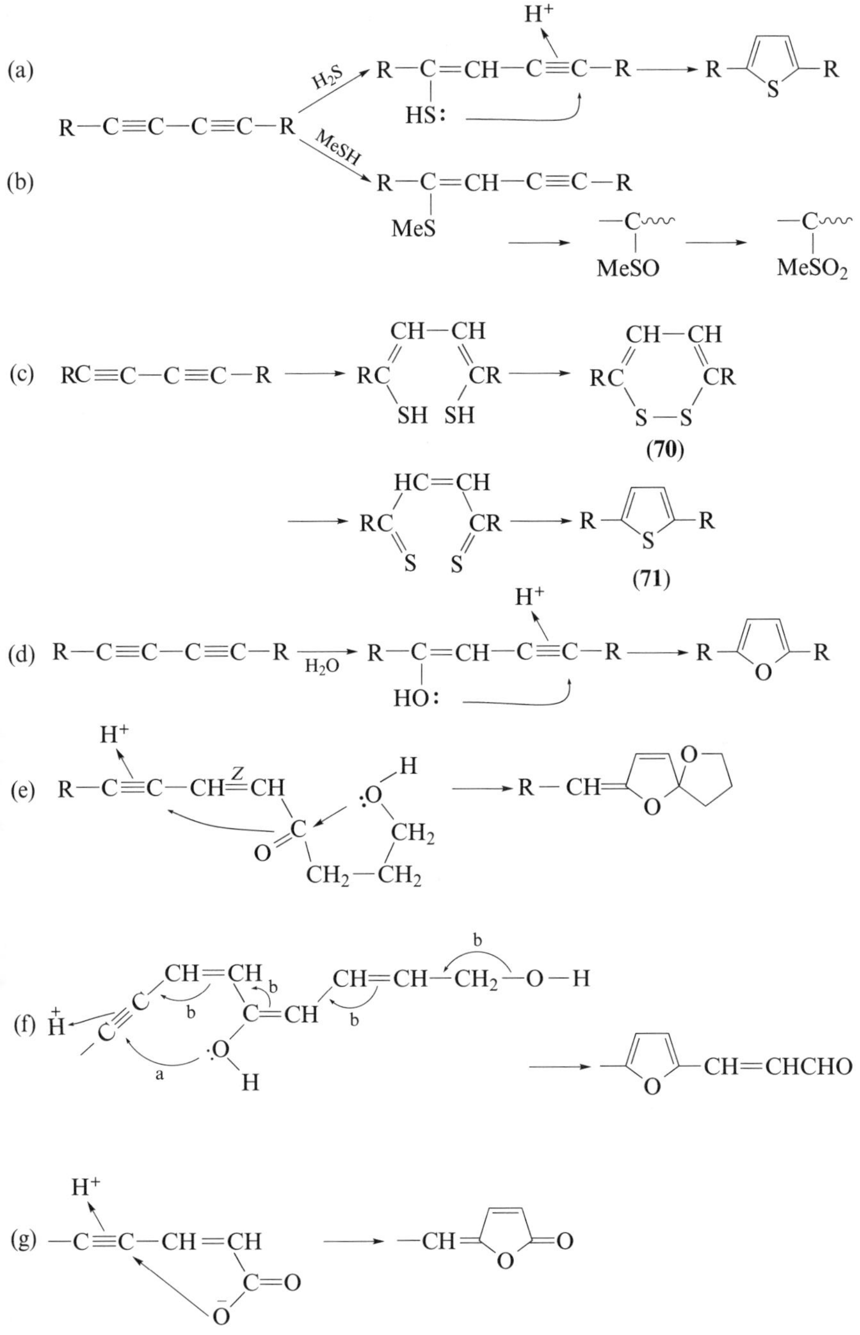

FIGURE 12. (*continued*)

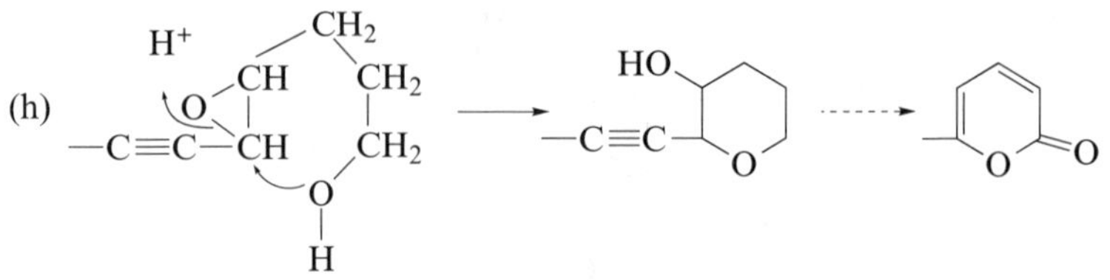

FIGURE 12. Routes to cyclic and heterocyclic alkynes

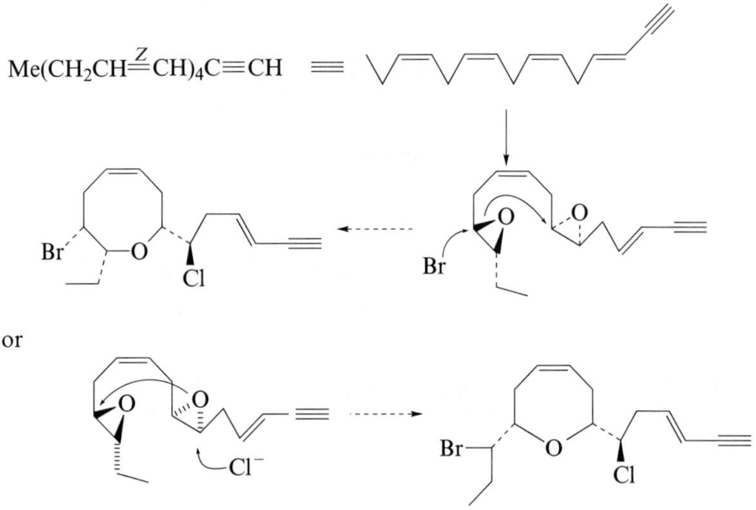

FIGURE 13. O-heterocyclics in red algae

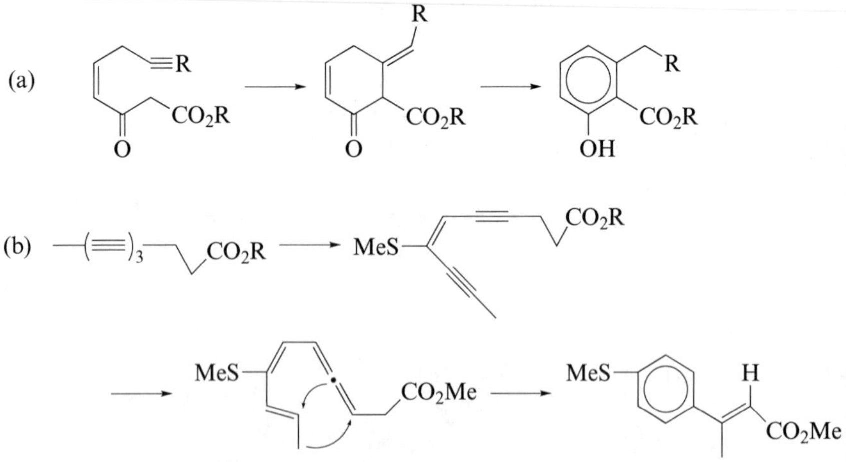

FIGURE 14. Routes to aromatic compounds

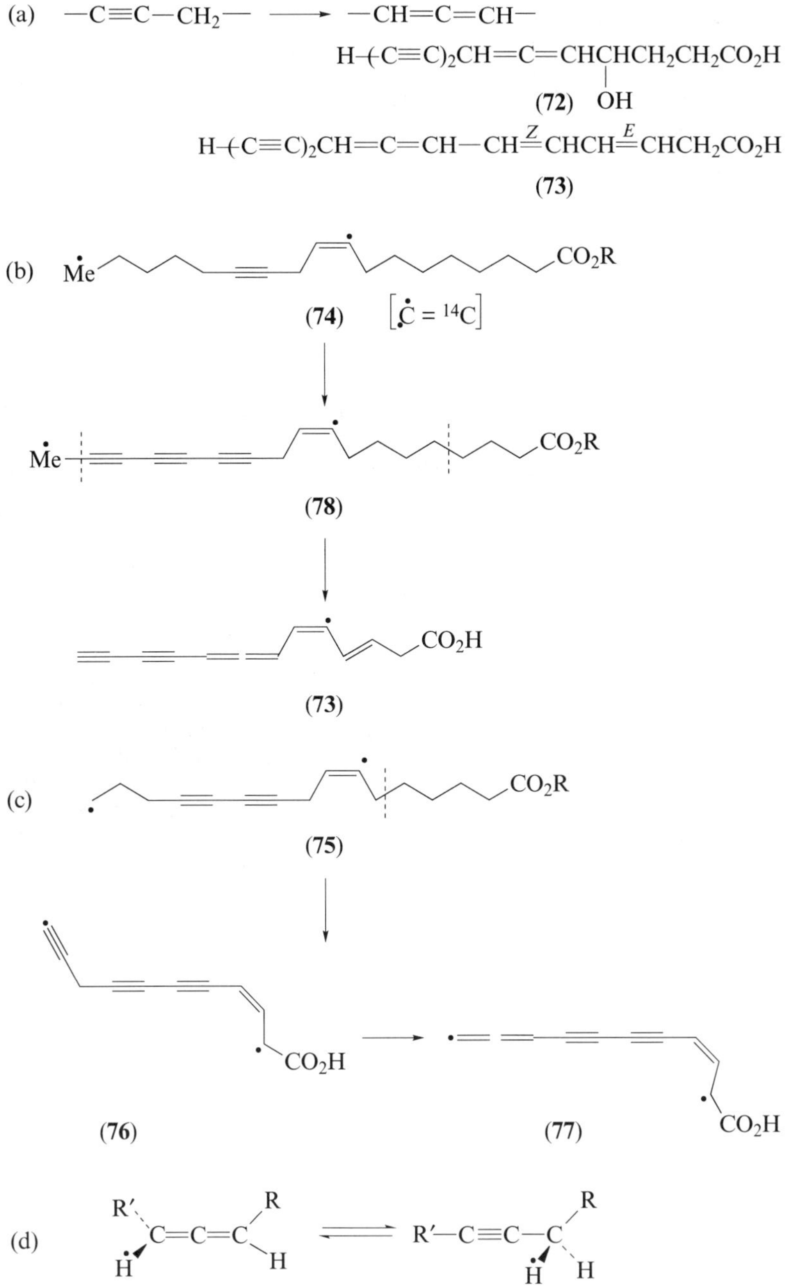

(a) $-C\equiv C-CH_2-$ $\longrightarrow$ $-CH=C=CH-$

$H+C\equiv C)_2CH=C=CHCHCH_2CH_2CO_2H$

(72)   OH

$H+C\equiv C)_2CH=C=CH-CH\overset{Z}{=}CHCH\overset{E}{=}CHCH_2CO_2H$

(73)

(b)

(74) $[\overset{\bullet}{C} = {}^{14}C]$

(78)

(73)

(c)

(75)

(76)   (77)

(d)

FIGURE 15. Propargylic rearrangement and allene formation

Propargylic rearrangement of a alkyne to an allene (Figure 15a) is fairly common in fungi to form, e.g., nemotinic acid (**72**) and mycomycin (**73**). Feeding experiments show that the latter is derived from the C_5 to C_{17} portion of crepenylate (**74**); see Figure 15b. Drosophilin-C (**76**) is found together with its allenic isomer droposphilin D (**77**) in *Drosophila* spp. The trialkyne **78** was not incorporated into **76** but the ^{14}C-labelled dialkyne (**75**: a C_{16}-compound) was an effective precursor[68] with the cleavage as shown demonstrated by tracer-labelling (Figure 15c). The allene–alkyne isomerase from pig liver has been investigated[68] and protonation was found to occur in the Si face at C_3 of the allene moiety: thus the rearrangement is a suprafacial process (Figure 15d) and mimics the stereochemistry of the analogous allylic rearrangement (see Section V.D).

III. CYANOGENIC GLYCOSIDES

A. Types

The most extensively-studied naturally-occurring cyano compounds are the cyanogenic glycosides. These are β-linked derivatives of 2-hydroxynitriles (cyanohydrins; **79**, Figure 16) which are cleaved by endogeneous enzymes to liberate HCN when the parent plant tissue is crushed. Their occurrence, structure, biosynthesis and biological significance has been exhaustively and repetitively reviewed[69–79]. With few exceptions the sugar residue is glucose and, as generally $R^1 \neq R$ (Figure 16), the compounds are chiral. Several examples of epimeric pairs are known, although they do not usually occur in the same plant or even in related species. The common types are exemplified in Figure 17. Linamarin (**80**) and lotaustralin (**81**) are probably the most abundant and widespread; amygladin (**83**) from seed of *Prunus amygladus* (bitter almond); luminin (**84**) and vicianin (**85**) have disaccharide moieties. (*R*)-Prunasin (**86**)—with the same aglycone as **83**—occurs in seed of *Prunus* spp and (*S*)-dhurrin (**88**) in *Sorghum Vulgare*. (*R*)-Dhurrin (**89**) is also known as taxiphyllin. As grouped A to C, the relationships of the aglycones to the amino acids L-valine (**92**), L-isoleucine (**93**), L-leucine (**94**), L-phenylalanine (**95**) and L-tyrosine (**96**) are apparent. There are few members in group B[80] and the derivation of triglochinin (**90**) from phenylalanine (its precursor) is not obvious. Group D also contains few members (*cf* the cyclopentanoid gynocardin, **91**), and is again of obscure parentage at first sight. Unusual glycosides are **97** from *Ilex* spp[81] and the sulphonated compound 98 from *Passiflora* spp[82–86]. Many new cyanogenic glycosides have been recently discovered, including several from the Turneraceae[87] and some with unusual sugar residues[88].

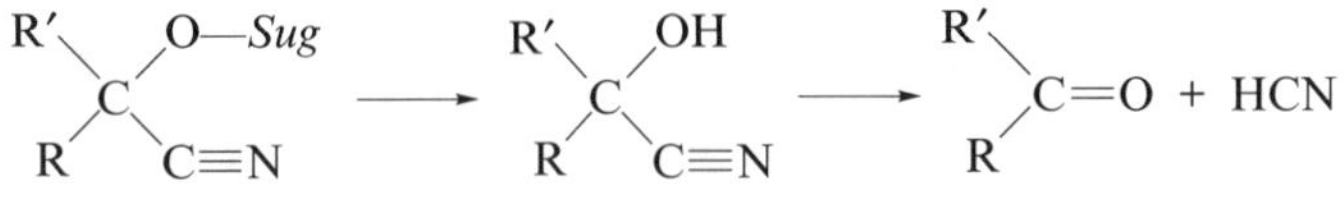

$$\underset{R}{\overset{R'}{\underset{}{\bigg\rangle}}}\underset{C\equiv N}{\overset{O-Sug}{C}} \longrightarrow \underset{R}{\overset{R'}{\underset{}{\bigg\rangle}}}\underset{C\equiv N}{\overset{OH}{C}} \longrightarrow \underset{R}{\overset{R'}{\underset{}{\bigg\rangle}}}C{=}O + HCN$$

(**79**) Sug = sugar residue

FIGURE 16. Breakdown of cyanogenic glycosides

Cyanogenic glycosides occur in several important food and fodder crops, such as *Manihot esculenta* (cassava; tapioca), cereals such as *Sorghum* spp, seeds of *Prunus* spp (apricot; peach; almonds) and leguminous plants such as *Linum* spp (flax) and *Trifolium* (clover) spp. Obvious deleterious effects can occur on ingestion[89,90]. Because of this and the ease of detection (with picric acid paper) of HCN that is evolved by crushed tissue, plant species have been extensively screened for cyanogenic compounds. By 1974, over 1000 cyanogenic species had been detected in 500 genera from 100 families[91] and, although the

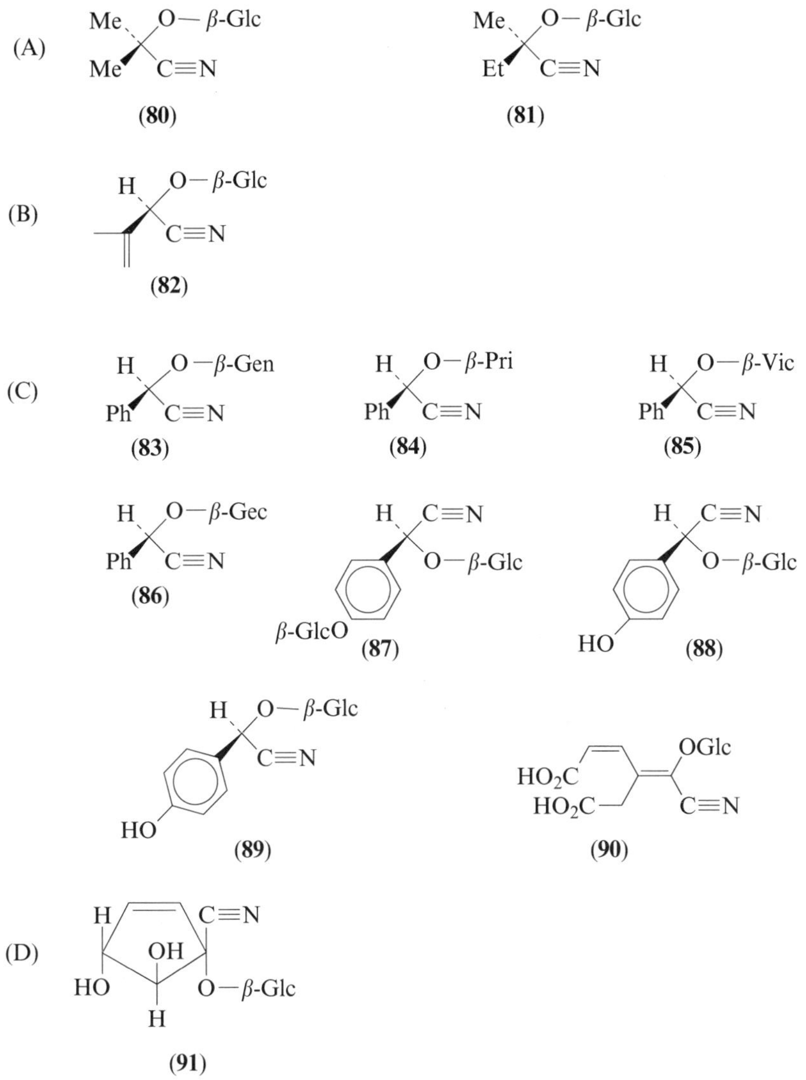

(A)

$$Me \diagdown \diagup O-\beta\text{-Glc}$$
$$Me \diagup \diagdown C{\equiv}N$$
(80)

$$Me \diagdown \diagup O-\beta\text{-Glc}$$
$$Et \diagup \diagdown C{\equiv}N$$
(81)

(B)

$$H \diagdown \diagup O-\beta\text{-Glc}$$
$$C{\equiv}N$$
(82)

(C)

$$H \diagdown \diagup O-\beta\text{-Gen}$$
$$Ph \diagup \diagdown C{\equiv}N$$
(83)

$$H \diagdown \diagup O-\beta\text{-Pri}$$
$$Ph \diagup \diagdown C{\equiv}N$$
(84)

$$H \diagdown \diagup O-\beta\text{-Vic}$$
$$Ph \diagup \diagdown C{\equiv}N$$
(85)

$$H \diagdown \diagup O-\beta\text{-Gec}$$
$$Ph \diagup \diagdown C{\equiv}N$$
(86)

$$H \diagdown \diagup C{\equiv}N$$
$$O-\beta\text{-Glc}$$
β-GlcO
(87)

$$H \diagdown \diagup C{\equiv}N$$
$$O-\beta\text{-Glc}$$
HO
(88)

$$H \diagdown \diagup O-\beta\text{-Glc}$$
$$C{\equiv}N$$
HO
(89)

$$HO_2C$$
$$HO_2C$$ — OGlc, C≡N
(90)

(D)

$$H \diagdown C{\equiv}N$$
OH
HO — O-β-Glc
H
(91)

FIGURE 17. Types of cyanogenic glycosides: Glc = glucose residue; Gen, Pri and Vic = gentiobiose, primaverose (disaccharide) and vicianose residues, respectively

nature of the cyanogens was rarely investigated, the presumption must be that most are glycosides. Excellent sources of linamarin and lotaustralin are *Trifolium repens* (white clover) and *Lotus corniculatus* (birds-foot trefoil), and the main families in which the glycosides accumulate are the Rosaceae, Leguminosae, Gramineae, Araceae, Compositae Euphorbiaceae and Passifloraceae. Usually, aglycones derived from one particular L-amino acid occur in a particular species: the occurrence of valine–isoleucine derivatives together with cyclopentanoid cyanogens in certain *Passiflora* spp is unusual[92]. Cyanogenic

Me　H   Me　H

　　　CO₂H   　CO₂H   CO₂H

Me　　　　　　　　Et

　H　　NH₂   H　　NH₂   H　　NH₂

(92) L-Valine   (93) L-Isoleucine   (94) L-Leucine

　　　　CO₂H   (95) R = H; L-Phenylalanine

　　H　NH₂   (96) R = OH; L-Tyrosine

R

　　　　　H   C≡N

HO

　　　　　C≡N   HO₃SO   O — β-Glc

HO　　O — β-Glc   (98)

　　(97)

glycosides have also been detected in ferns, lichens, mosses and algae[93] and in fungi and other micro-organisms[94–96]. However, certain bacteria and fungi produce HCN by a route that does not involve a glycoside as precursor[97]: rather glycine is broken down to HCN and CO_2 via the intermediate NCCOOH. Some of the CN⁻ thus liberated is converted into β-cyanoalanine (see Section III.D) and a cell-free extract capable of sustaining this degradation has been prepared from *Preudomonas* spp[98].

The location and catabolism of the cyanogenic glycosides have been reviewed[99]. These compounds are situated entirely in the epidermal layers of leaves of *Sorghum bicolor* whereas the glucosyl transferase enzyme that catalyses the last step of their formation is found in both epidermal and mesophyll tissue[100,101].

In contrast, in *Manihot* spp all tissues of the plant contain the glycosides but the microsomal enzyme systems responsible for their biosynthesis are restricted to cotyledons and their petioles: thus the metabolites are actively transported within the plant[102]. At the sub-cellular level the glycosides are stored in the vacuoles[103], whereas the β-glucosidases that cleave them are compartmented away in the cytoplasm[104].

The glycosides are not restricted to the plant kingdom and microorganisms. Butterflies can store up to 95% of the glycosides that they obtain from their diet[105,106] but several *Heliconius* spp (butterflies) and *Zygaena* spp (burnet moths) accumulate linamarin and lotaustralin that do not occur in their food plants and these are synthesized *de novo*[107]. The latter species possess much higher levels of linarin etc. in their tissues than those species that rely on diet to accumulate the glycosides. Radiotracer and ¹³C-labelling demonstrated that valine and isoleucine were incorporated without skeletal degradation[108,109] and the site of biosynthesis and mechanism of transport was elucidated in *Zygaena* spp[110]. The food plants of some species of the moths contained both glycosides—albeit in varying amounts due to the polymorphism that exists in most of the relevant populations—but *de novo* synthesis nevertheless occurred also[111]. The bug *Leptocoris* similarly stores cyanogenic glycosides not present in its food plant, although the latter now contains cyanolipids that may be modified by the insect[112]. Millepede species also synthesize glycosides by the route as in plants[113] and some store mandelonitrile as an alternative defence weapon[114]. The role of cyanogenic compounds in insects[115] and the synthesis and storage of cyanogenic glycosides in butterflies and moths[116] have been reviewed.

B. Biosynthesis

The structure of the common aglycones betray their likely parentage and many experiments have indeed revealed the biogenetic relationships suggested in Figure 17[71,72]. Thus feeding ^{14}C and ^{15}N doubly-labelled amino acids has proved uptake into the target molecules without structural rearrangement or degradation (Figure 18): in particular, the C—N of the precursor remained intact in product. Similar labelling experiments proved intact incorporation into the apparently anomalous cyclopentyl glycosides[117]. The generally-accepted route of biosynthesis is in Figure 18 and there is much evidence for this. Thus, many feeding experiments with expected precursors support the scheme[118], such as L-valine, and the corresponding oxime and nitrile were incorporated into linamarin by microsome fractions of *Trifolium* seedlings that had been grown in the dark[119] (the need to use etiolated material to achieve active cell-free extracts seems common). The postulated intermediate oximes, nitriles etc. could not be isolated and were possibly membrane-bonded within a multi-enzyme complex, but they could be detected by radiochemical trapping experiments. This membrane linkage also severely reduced incorporations of several exogenous precursors, such as the nitrile, by preventing direct access to the biosynthetic conveyor-belt and was made manifest by the observations that the putative intermediates were utilized at different and inconsistent rates in different experiments[120]. The restricted incorporations of obligatory precursors—the so-called 'channelling effect'—was thought to involve the operation of either two multi-enzyme complexes or of two multifunctional proteins. The membrane-bonded nature of the enzyme systems provides a very efficient biosynthetic pathway whereby potentially labile intermediates were protected

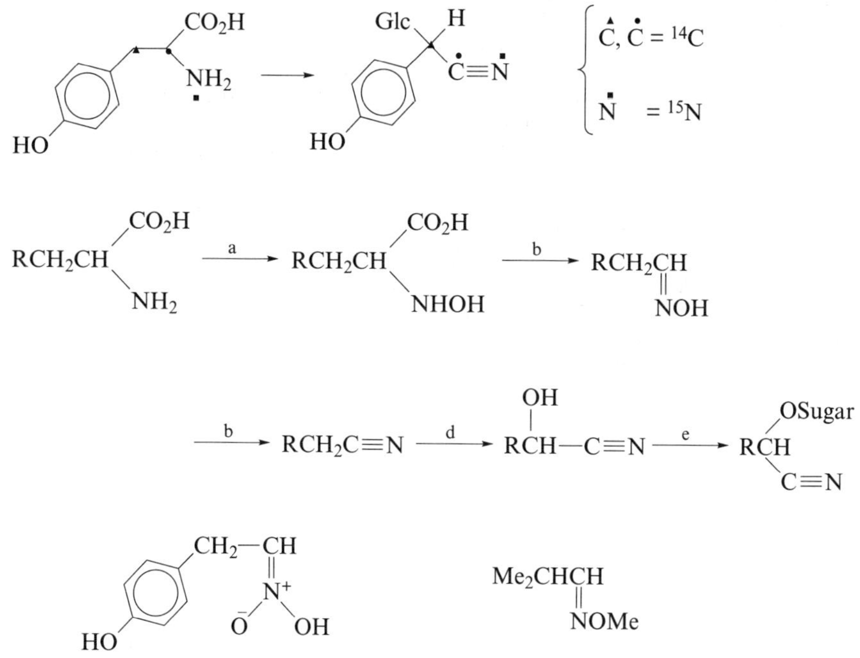

FIGURE 18. Biosynthesis of cyanogenic glycosides

from wasteful side-reactions. As the glucosyltransferase which catalyses the last step of the sequence is a soluble enzyme, microsomal systems must yield the aglycone, not the final glycoside. Similar results were found for the biosynthesis of several target molecules in addition to the aglycone of linamarin[121–123].

Microsome preparations from *Manihot esculenta* converted valine and isoleucine very effectively into linamarin and lotaustralin—but with the exception of 2-cyclopentenylglycine did not accept any of the other amino acids tested. In contrast a broad specificity was shown towards oximes and nitriles as substrates, and *in vitro* biosynthesis was photoreversibly inhibited by CO. This shows the selectivity of the steps a/a+b of Figure 18, in contrast to the later stages, and also implicates cytochrome P-450 in the hydroxylation process(es)[102]. The possibility that the aldoxime be hydroxylated to the 2-hydroxy compound—which is then directly dehydrated to the cyanohydrin, i.e. the hydroxylation in step d preceded the elimination step c—was not supported by precursor or trapping studies[102,124]. Detailed investigations on the biosynthesis of dhurrin (**88**) in microsome systems from *Sorghum* spp proved the intermediacy of N-hydroxytyrosine, (*E*)- and (*Z*)-*p*-hydroxyphenylacetaldoxime, *p*-hydroxyphenylacetonitrile and *p*-hydroxymandelonitrile[125] in agreement with the route in Figure 18. However, studies of consumption of oxygen concomitant with assays of biological activity suggested a hitherto unspecified intermediate between N-hydroxytyrosine and *p*-hydroxyphenylacetaldoxime that was proposed to be 1-nitro-2(*p*-hydroxyphenyl)-ethane probably in its aci-form (**99**)[126,127]. If valid and general, this indicates that step b (Figure 18) is not the simple E2-type decarboxylation that had been implicitly assumed. Cultures of bacteria (*Moraxella* spp) accumulated the O-methyl ether of isobutyraldoxime (**100**): here the pathway to cyanogenic glycosides is trapped by methylation of the oxime stage. On extended incubation methanol is eliminated to yield isobutyronitrile[128].

Feeding of (3*R*)- and (3*S*)-L-[³H₁]-tyrosine alone or in admixture with [U-¹⁴C]-tyrosine gave clear evidence that in *Sorghum* and *Taxus* spp the hydroxylation in step d proceeded with retention of configuration and so indicated the operation of a mixed-function oxidase[129]. Exceptional incorporations (15–72%) were obtained for the uptakes of 2-methylpropanenitrile and the corresponding 2-methylbutane compound into linamarin and lotaustralin by *Lepidoptera* spp: thus steps d and e were clearly demonstrated to occur in insects as well as in plants[130].

The last stage of the pathway (step e) involves a UDP–glucosyl transferase. Glycosylation of mandelonitrile to prunasin (**86**) in extracts from fruit and leaves of black cherry (*Prunus* spp) only formed the (*R*)-epimer, but the preparations showed significant transferase activity towards a number of other substrates, although (*R*)-prunasin was not further glucosylated to amygladin (**83**)[131]. Several glucosyl transferases were purified (up to 200×) from seedlings of *Triglochin* spp and each was specific for a particular cyanohydrin present in the plant. The enzymes were soluble, but although the substrates were membrane-bonded they could equilibrate to bulk solution and be easily isotopically trapped[132]. The transferase from *Sorghum* seedlings has also been partially purified[133].

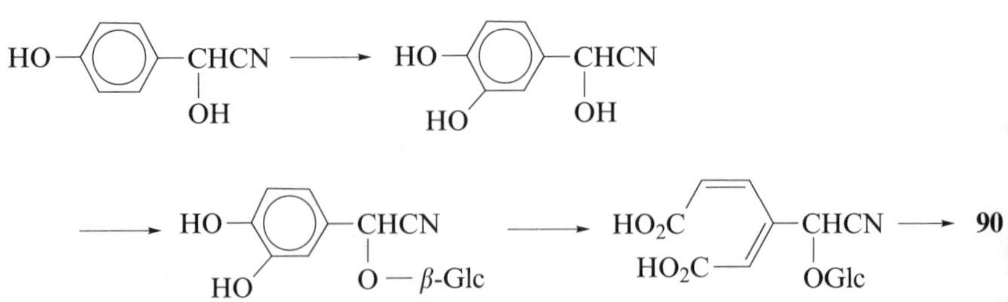

FIGURE 19. Biosynthetic route to triglochinin

The unusual glycoside triglochinin (**90**) co-occurs with (*R*)-dhurrin in *Triglochin* spp and the biosynthetic routes were presumed to be related[134]. This was confirmed by feeding experiments on a monocot (*Triglochin* spp) and a dicot (*Thalictrum* spp), which indicated the scheme in Figure 19 with tyrosine as parent in each example. There was some doubt whether aromatic hydroxylation preceded or followed glycosylation[135].

C. Hydrolysis

Crushing or maceration of the plant (or other) material containing a cyanogenic glycoside causes release of HCN in two stages (Figure 16) catalysed by a β-glycosidase (usually a glucosidase) and an oxynitrilase[136,137]. The best known β-glycosidase is emulsin from *Prunus amygladus* whose target molecule is amygladin but which shows general β-glycosidase activity. All plants that store cyanogenic glycosides apparently contain specific enzymes to effect their cleavage. Thus triglochinin (**90**) and (*R*)-dhurrin (**89**) occur in seeds of *Triglochin* spp and two specific β-glycosidases are present[138], and linamarase from *Trifolium* spp—which is membrane bonded but can be extracted with strong NaCl solutions—is accompanied by at least 3 other β-glucosidases none of which have significant activity towards linamarin[139]. Linamarase has been purified[140] as has vicianin hydrolyase from ferns (*Davallia* sps). The latter cleaves **85** to (*R*)-mandelonitrile and vicianose, has molecular mass 340 k daltons and is comprised of three unequal sub-units. It accepts only a narrow range of cyanogenic glycosides and a few conventional glycosides, requires the (*R*)-epimers as substrates and cleaves disaccharide in preference to monosaccharide linkages[141]. Enzymes catalysing the hydrolysis of dhurrin were exclusively located in the mesophyll tissue-well compartmented from the vacuole-stored glycoside under normal physiological conditions[142].

D. Physiological Role

The function of cyanogenic glycosides in plants has frequently been stated to be that of secondary metabolites generally—that is, as defence compounds of use in repelling, inhibiting and perhaps ultimately poisoning herbivores, pathogens or competitors with subsequent selection for the better protected individuals. Cyanogenic glycosides attracted attention, not so much because of their physiological role but because of their effects on man and his animals—and one, amygladin, has attracted added notice as a possible anti-cancer compound[143]. The role as defence compound is reasonable, but is difficult to accommodate with evidence that the glycosides undergo rapid turn-over in the intact plant[144]. Apparently a β-glucosidase which may not necessarily be the one involved in cyanogenesis in damaged tissue has limited access to the glycoside and triggers the release of HCN. Consequently, the half-life for dhurrin in *Sorghum* sps is only ca 10 h[145].

HCN is toxic to most organisms but it can be removed in higher plants, especially those that accumulate cyanogenic glycosides, by reaction with ʟ-cysteine (**101**) to form β-cyano-ʟ-alanine (**102**), ʟ-asparagine (**103**) and other derivatives (Figure 20). The initial reaction with cysteine is likely to proceed by elimination of H_2S and Michael addition of CN^- to the intermediate α-aminoacrylic acid[146]. This detoxification occurs not only in cyanogenic species (*Linum*; *Trifolium*) but in many noncyanogenic taxa (*Pisum*; *Lupinus*). Usually, asparagine is the main product but β-cyanoalanine (the only cyanoamino acid found in plants) does accumulate in Vicia sps[147], sometimes mainly in the form of the dipeptide γ-glutamyl-β-cyanoalanine (**104**)[148]. **102** and **104** are toxic to animals (LD_{50} 250 mg kg^{-1} for rats) probably owing to inhibition of the action of Vitamin B_6[149]. γ-Glutamyl-β-aminopropionitrile (**105**) is the factor responsible for the toxicity to humans of certain *Lathyrus* sps.

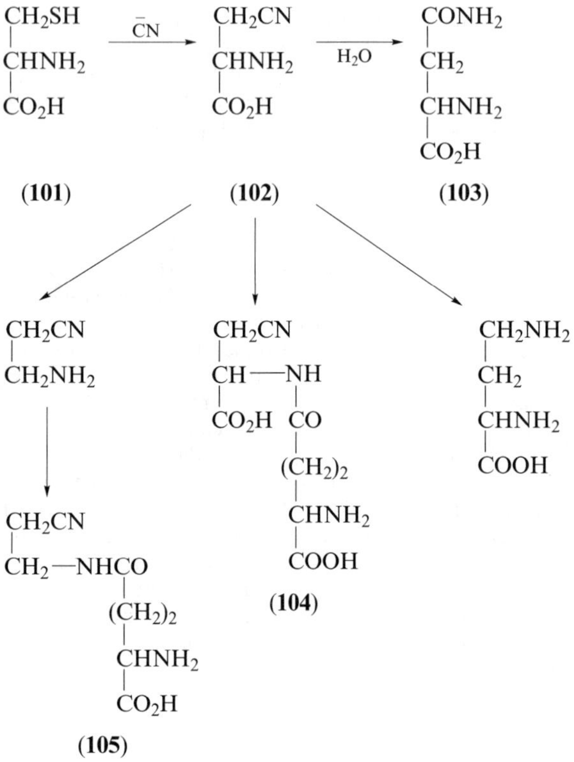

FIGURE 20. Detoxification of CN⁻ by higher plants

Although tracer studies using extracts of several plant species indicated that both L-serine and L-cysteine were substrates for the formation of β-cyano-L-alanine, the use of purified fractions (such as from *Lupinus* spp) revealed that cysteine was the authentic precursor: serine was converted into this by the appropriate enzyme in the crude preparations. The β-cyanoalanine synthase from *Lupinus* was purified to homogeneity and its mechanism of action utilizing pyridoxal phosphate as a coenzyme was elucidated[150]. β-Cyanoalanine hydrolyase which catalyses the hydration of **102→103** has also been purified[151].

In the cyanogenic plants these reactions could detoxify HCN if it should be released in the intact cell, but this begs the question as to why *Lupinus* spp and many others lacking a source of HCN, should possess the appropriate enzymes. A possibility is that cyanogenic glycosides may play a role in essential primary metabolism of plant species that have been selected for their defensive ability[152]. Recently, a 'linostatin pathway' has been proposed to occur in *Hevea* sps, and by extension in others, involving cyclical formation of the glycosides and thence of HCN, and the association of the latter with primary metabolism via β-cyanoalanine synthase[153]: from this viewpoint the glycosides are nitrogen storage compounds and do not exclusively play defensive roles[154,155]. The metabolism of β-cyanoalanine in *Pseudomonas* sps has also been discussed[156], and in some fungi HCN (which may be derived *inter alia* from glycosides) is used to form amino acids in a pathway resembling the Strecker nitrile synthesis, for instance, to form alanine and glutamic acid (Figure 21: R—Me; —CH₂CH₂COOH): apparently this route is not followed in higher plants[157]. Certain insects can detoxify HCN by the β-cyanoalanine pathway, but here it seems likely that the cyanogenic compounds play a defensive rather than a storage role in aposematic species such as *Zygaena* and the heliconid butterflies[157]. Up to 14 families of moths and butterflies of both cryptic and aposematic type contain cyanogens including β-cyanoalanine[158]. Adventitious HCN is usually destroyed in insects by conversion into thiocyanate ($CN^- \rightarrow CNS^-$) by action of the widespread enzyme rhodanese[159].

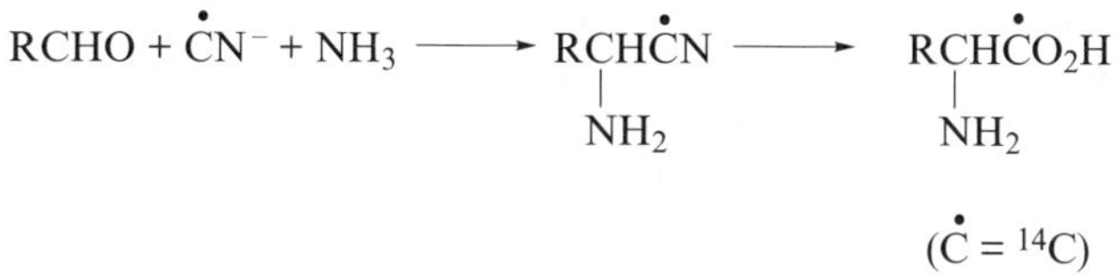

$$RCHO + \overset{\bullet}{C}N^- + NH_3 \longrightarrow RCH\overset{\bullet}{C}N \longrightarrow RCH\overset{\bullet}{C}O_2H$$
$$\qquad\qquad\qquad\qquad\quad | \qquad\qquad\quad |$$
$$\qquad\qquad\qquad\qquad\; NH_2 \qquad\qquad NH_2$$

$$(\overset{\bullet}{C} = {}^{14}C)$$

FIGURE 21. Amino acid production in fungi

IV. OTHER CYANO COMPOUNDS

This section concerns compounds containing triple-bond unsaturation, thus cyanates and thiocyanates ($R-O-C\equiv N$; $R-S-C\equiv N$), but not the iso-compounds ($R-N=C=O$; $R-N=C=S$) are discussed. Early work has been reviewed.

A. Cyanolipids

Since 1970, a new class of cyano compounds has been isolated from certain seed oils that are of interest as some members, like cyanogenic glycosides, liberate HCN on enzymic or mild chemical hydrolysis[160–163]. All the authenticated compounds occur in several genera of the Sapindaceae (soapberry) family[164] and they can comprise up to 50% v/v of the extract, e.g. in kusum seed oil[165]. These cyanolipids are mono- or di-esters of mono- or dihydroxy-nitriles and 4 types are known (Figure 22). The chain length of the fatty acid moiety, which may be saturated or not (e.g. commonly from oleic acid), can be C_{14} to C_{22} with C_{18} and C_{20} predominant and the double bond in type 3 can be Z or E, but structural variations are few. After hydrolysis the α-hydroxynitriles derived from types 1 and 4 are cyanogenic. In many plant species one type occurs to the virtual exclusion of the others: thus type 1 accumulates in *Allophyllus* and *Paullina* spp, whereas type 2 is characteristic of

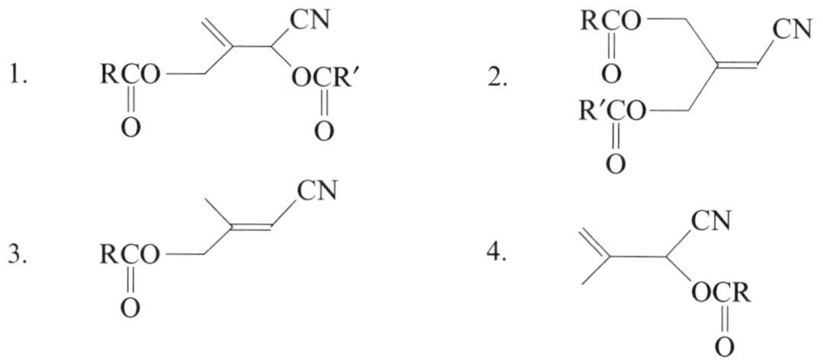

FIGURE 22. Types of cyanolipids

the *Nephelium* and *Sapindus* genera[166]. Few biosynthetic studies are extant. Immature inflorescences of *Koelneuteria* spp incorporated [U-^{14}C]-leucine at up to 0.1% levels into cyanolipids of type 2 [R = Me(CH$_2$)$_{18}$] and type 3 [R = Me(CH$_2$)$_{18}$] and it is likely that this amino acid provides the skeleton for all the class[167].

B. Nitriles

The biochemistry of these compounds (R—C≡N; cyanides) and isonitriles (RN≡C; isocyanides, see Section IV.C) have been jointly reviewed[2,161]. More detail may be obtained in symposia proceedings[168,169] dealing with the metabolism of cyanide in plants, animals and micro-organisms and in reviews on the metabolism and toxicity of aliphatic nitriles[170] and their distribution in living organisms[171]. Pathways involving nitriles in higher plants are neither very extensive nor particularly interesting if the cyanogenic glycosides and cyanolipids are excluded. The compounds are formed usually by amidation of a carboxylic acid followed by elimination of water (RCO$_2$H → RCONH$_2$ → RCN) or by decarboxylation of an amino acid akin to that involved in the formation of cyanogenic glycosides) such as the formation of the auxin indol-3-acetic acid (106) in Crucifereae (Figure 23)[172]. The route can yield unusual products such as 3-cyanopyridines, the best known of which is ricinine (107), a toxic alkaloid from leaves and seed of *Ricinus* spp[161]. *Penicillus* sps and related marine algae convert α-amino acids into nitriles and aldehydes using a bromoperoxidase in the presence of H$_2$O$_2$ and Br$^-$ [173]. Thus valine yields isobutyronitrile together with isobutyraldehyde and alanine forms methyl cyanide and acetic acid. The peroxidases from different sources catalyse similar unspecific reactions. The enzyme from *Penicillus* spp decarboxylates *O*-methyltyrosine which is consistent with the implication of

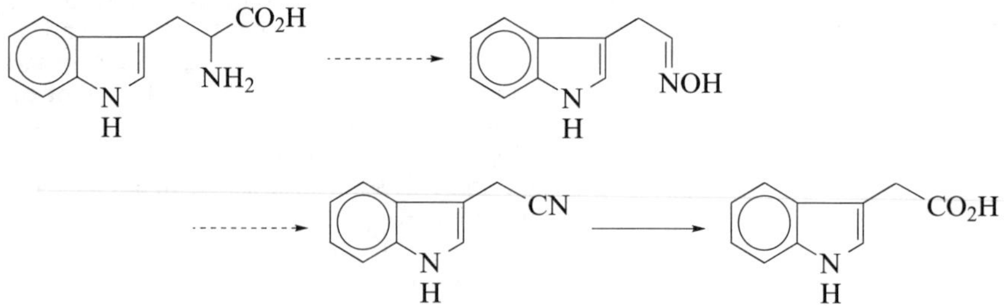

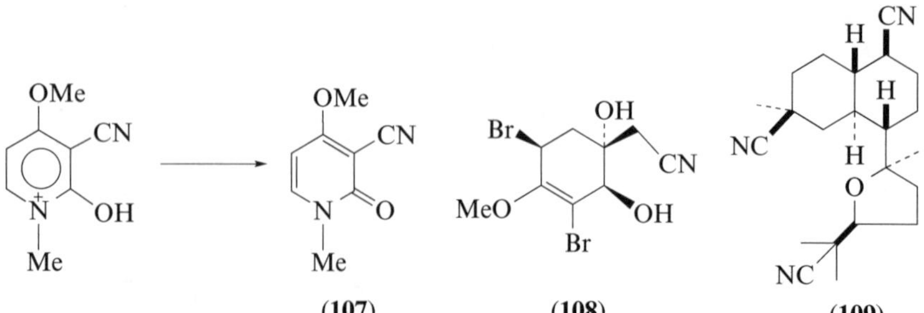

FIGURE 23. Nitriles in higher plants

he system in formation of aeroplysinin I (108) and related compounds in marine sponges ıf the Order Verongida[174]. More complicated are the polyfunctional C_{20} compounds, e.g. 109 in sponges of *Acanthella* spp: these may be of amino acid or diterpenoid origin but letails are unknown[175]. The toxicology and detoxification of nitriles in animals is discussed n another chapter of this volume.

The microbial transformations of nitriles have been extensively studied[176], especially the ;equences in Figure 24 which involve (a) nitrile hydratase (b) amidase and (c) nitrilase as :atalysts. The incentives are the potential use in waste-treatment of toxic residues resulting ïrom the manufacture of acryonitrile or adiponitrile—although at present such biodegra- lation is only partly successful[177]—and also for the conversion of such nitriles into :conomically more attractive compounds[178].

$$\text{(a)} \quad \text{RCN} \xrightarrow[\text{(a)}]{H_2O} \text{RCONH}_2 \xrightarrow[\text{(b)}]{H_2O} \text{RCO}_2\text{H} + \text{NH}_3$$

$$\text{(b)} \quad \text{ArCN} \xrightarrow[\text{(c)}]{2H_2O} \text{ArCO}_2\text{H} + \text{NH}_3$$

FIGURE 24. Microbial transformations of nitriles

Many microorganisms—which should *not* possess nitrile reductase activity—carry out :hese conversions and most are relatively unspecific as to substrate. Bacteria have been ;creened (and taxonomically classified) for acryonitrile hydratase activity[179], amide pro- luction and substrate and inhibition studies have been made for the nitrile hydratase from *Corynebacterium* spp[180–182] and *Rhodococcus* spp[183] and the production of acrylamide has 3een brought to industrial fruition[178]. Bacteria that could utilize acryonitrile as sole nitrogen and carbon source[184] and fungi that accept aliphatic nitriles (C_1 to C_6) as nitrogen ;ource have been studied[185], and fungi of a similar *Fusarium* sps detoxify jojoba toxins :ontaining nitriles[186]. Several enzymes catalysing these processes have been purified. A nitrile hydratase and amidase from *Corynebacterium* sps that convert a dinitrile into a nononitrile (e.g. 1,4-dicyanocyclohexane→4-cyanohexan-1-oic acid) have been purified ıs a dimeric complex[187]. A nitrilase from *Nocardia* spp that accepts *p*-hydroxybenzonitrile ıs sole carbon source was a 12-unit multi-enzyme complex of molecular mass 560 ϰilodaltons[188] and the nitrilase of *Rhodococcus* that accepts aliphatic nitriles was a :omparably large species (15–16 sub-units; 600 kilodaltons)[189]. The gene for nitrile nydratase from *Rhodococcus* spp was cloned, sequenced and expressed in *E. coli* and the .atter used for the conversion of acrylonitrile into acrylamide[190]; similar studies have been :arried out with genes from *Pseudomonas*[191]. Nitrile hydratases from other bacterial ;pecies have been immobilized on solid supports with retention of their activity[192]. An ₁ndustrially-significant process involves using bacteria to convert DL-α-amino nitriles into ɒL-amino acid amides, followed by incubation with a different species that contains both ın amino acid racemase and a L-amino acid amidase to yield the corresponding L-amino ιcid[193].

Cyanide ion in industrial wastes can also be removed by microbial means[194,195] including ₁se of immobilized fungi and bacteria[196]. Cyanide hydrase ($CN^- \rightarrow HCONH_2$) was induced ₁n mycelia of *Stemphylium* and other species by growth in low concentrations of CN^- and the activated strains were immobilized by several methods to yield powerful tools for lecontamination purposes[197].

C. Isonitriles

These compounds ($R-N^+\equiv C^-$; isocyanides) are more widespread in Nature and have a much more varied and interesting biochemistry than their isomers[3,161,198]. Many exhibit antibiotic and other potent biological activities and large numbers have been isolated during screening of bacteria and marine organisms. The compounds from the two latter sources are quite different in type and originate from disparate biosynthetic pathways.

1. Compounds from terrestrial sources

These are all produced by micro-organisms, are derived from amino acids and, with few exceptions, the NC group is conjugated to at least one alkenic residue. They can conveniently be divided into 3 groups, containing in all several dozen examples.

The first group comprises the xanthocillins and their relatives isolated in 1948 from cultures of *Penicillium* spp as the first examples of isonitriles as natural products. Xanthocillin X (**110**, Figure 25) is the parent and the others carry varying substitution patterns and functionalizations of the aromatic rings[199]. These compounds showed a wide spectrum of antibiotic activity against gram-positive and -negative bacteria and pathogenic fungi and yeasts. Other xanthocillins are found in *Aspergillus* and other species; *cf* xanthoascin (**111**) in which the heterocyclic ring is presumably of mevalonoid origin[200] and the recently discovered **112**[201]. Most unusually the isomeric dinitrile emerin (**113**) co-occurs with xanthocillins in an *Aspergillus* sps[202].

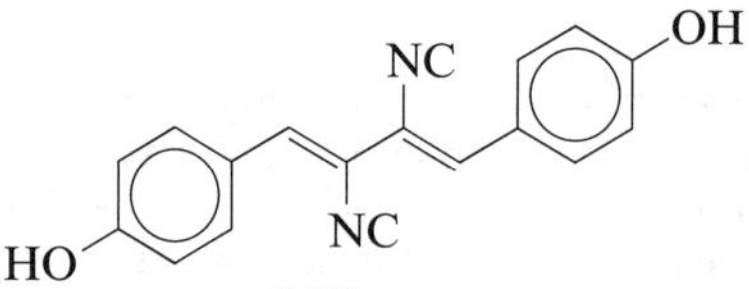

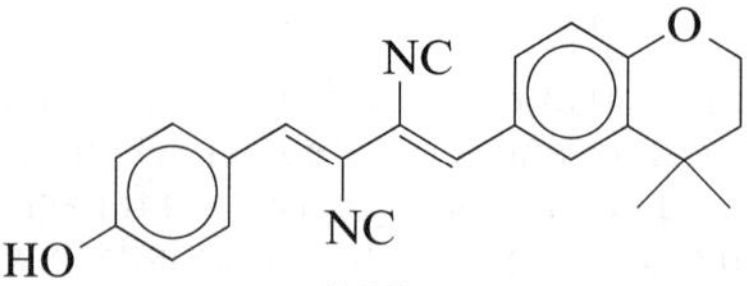

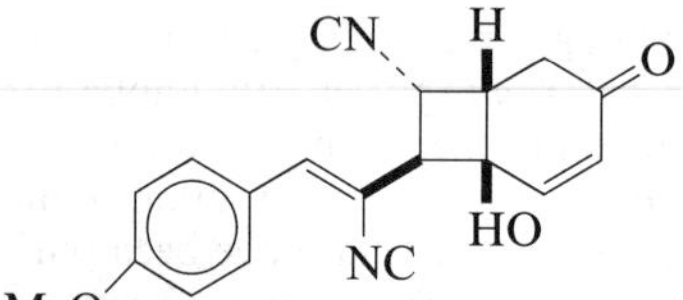

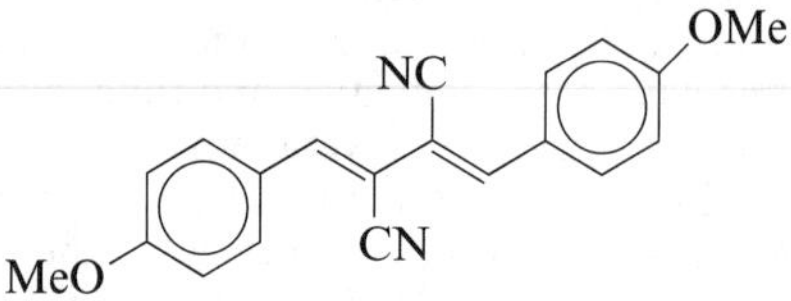

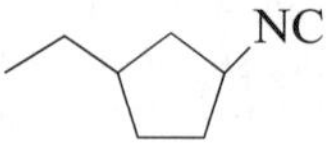

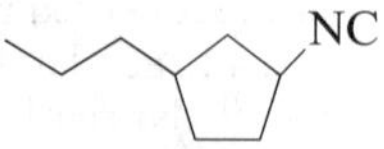

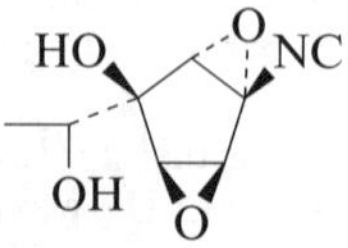

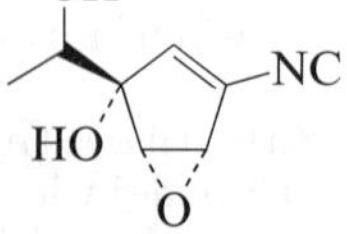

FIGURE 25. (*continued*)

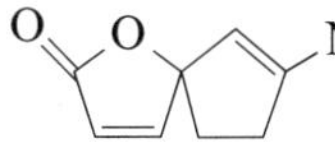

(119) 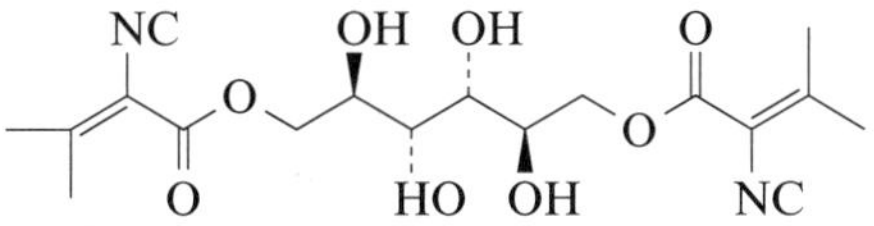   (120)

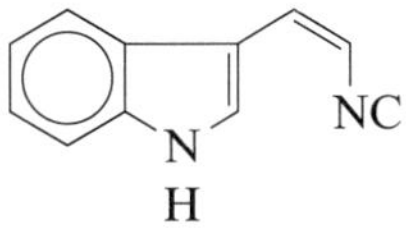

(121) 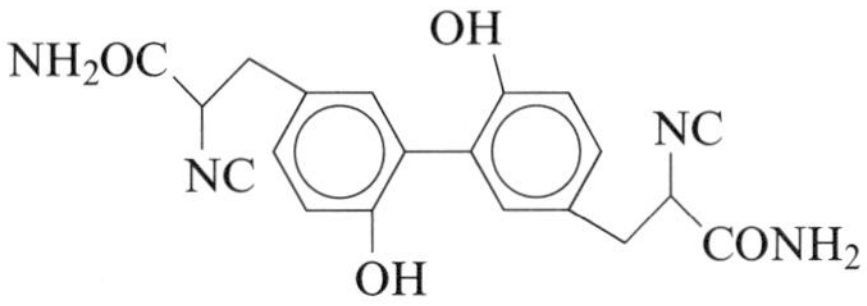   (122)

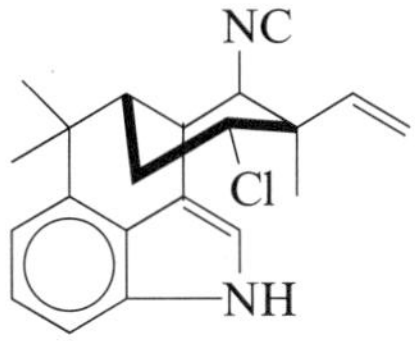

(127)

FIGURE 25. Some terrestrial isonitriles

The second group is based on the cyclopentyl skeletons **114** and **115** with variations in overall oxidation levels and degree of hydration. Dermadin (**116**), an antibiotic from soil fungus of *Trichoderma* sps, was the first discovered[203], and other examples from the same genus are trichovridin **117**[204,205], **118**[206] and the spiro-lactone **119**[207].

The remaining non-marine isonitriles are a motley crew: they include the modified sugar **130** which is a fungistatic antibiotic[208,209]; an indoleacrylisonitrile **121** from *Pseudomonas* spp[210] and the hazimycins which are broad-spectrum antibiotics from *Micromonospora* spp[211,212], e.g. hazimycin factors 5 and 6 which are the (R,S) and $(RR+SS$, i.e. racemate) isomers of **122**. Phenyl isonitrile (PhNC) is not a naturally-occurring compound: it is an artefact generated during distillation of denatured rape oils whereby aniline reacts with organic acids to yield an amide that undergoes radical fragmentation to isocyanate that is in turn subsequently decomposed by unreacted aniline[198].

The biosynthetic routes to these compounds have recently been authoritatively reviewed[161,198]. Inspection suggests that the xanthocillins are derived from tyrosine and indeed [2-^{14}C]-tyrosine was an excellent precursor for **110** in *Penicillium* spp and the radioactivity was confined to the butadiene portion of the molecule. In contrast, negligible incorporation of tracer occurred from the [1-^{14}C]-labelled acid and so decarboxylation of the precursor must have occurred and $C_{(1)}$ of the acid did not become the part of the isonitrile group. Use of tyrosine that was chirally deuteriated at $C_{(3)}$ (**123**) has shown that formation of the double bond in the chain of xanthocillin X (Figure 26) involves loss of the pro-(S) hydrogen in the formal *anti*-periplanar elimination of H^+ and a CO_2^- group[213]. As [^{3}H]-octopamine (**124**), in contrast to tyrosine, was an insignificant precursor it seems unlikely that hydroxylation of the side chain is associated with the double-bond formation[213]. Although it has been established that the nitrogen of the isonitrile group is derived from tyrosine[198,214,215] the origin of the adjacent carbon remains a mystery. The cyanobacteria sps did not utilize glycine, serine, formate, methionine or other compounds associated with tetrahydrofolate metabolism as a C_1 source for synthesis of the NC group, and even though certain of these putative precursors, such as [2-^{14}C]-glycine and [^{14}C-Me]-methionine, could label the methyl group of the O-methyl monoether of **116** that occurs in the fungus *Dichotomyces*, such additives were not involved in the construction of the isonitrile unit[216]. More recent studies showed that [ureido-^{14}C]-citrulline, [^{14}C]-cyanate, [^{14}C]-cyanide or protected cyanide as in [^{14}C]-2-hydroxy-4-methylvaleronitrile also did not supply carbon for the unit[217], despite the fact that cyanide ion provided the entire isonitrile unit in metabolites of marine sponges (see Section IV. C.2). The diformamide **125** has been isolated from cultures of *Penicillium* sps that produce xanthocillin X, but it is unclear if this is an intermediate or a subsequently functionalized metabolite[218]. The isonitrile groups of the xanthocillins may originally have formed part of a much larger functional group situated at the appropriate position on the inter-aromatic chain of carbons. It seems clear that the 'obvious' route to isonitrile via sequential oxidation of N-methylamino-derivatives does not occur.

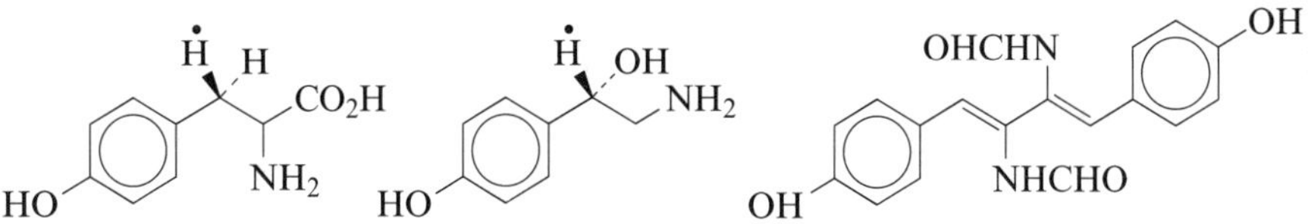

Ḣ = ^{3}H

FIGURE 26. Biosynthesis of xanthocillin

The cyclopentyl isonitriles appear at first sight to be of obscure biogenetic origin but, like the xanthocillins, they are directly derived from tyrosine, *A priori* they could arise from the polyketide pathway by ring-contraction of phenols, or along a route that involved similar contraction of an aromatic acid or by modification of a monoterpenoid. In the event, both [2-^{14}C]- and [1-^{14}C]-acetate were very poorly incorporated into **126** by *Trichoderma* spp whereas L-[U-^{14}C]-tyrosine was an excellent precursor. The similar excellent incorporation of L-[CO$_2$H-^{14}C]-tyrosine revealed that the carboxyl group of the amino acid was retained in the metabolite[219]. Incorporation of ^{13}C-labelled tyrosines then provided the labelling pattern and indicated the pathway (Figure 27a)[220]. The oxidative cleavage of the aromatic ring is well known and occurs in the biosynthesis of certain cyanogenic glycosides (*cf* Section III.B). Studies on the uptake of [^{14}C,^{3}H]-tyrosine in related fungal sps were fully consistent with this route (Figure 27b)[221]. The C$_5$–C$_3$ skeleton of **126** can be converted into compounds with the C$_5$–C$_2$ skeleton **114** by decarboxylation. Alternatively, the loss of a C$_1$ unit may occur following the cleavage of the aromatic ring of tyrosine. Both the C$_5$–C$_3$ and C$_5$–C$_2$ parents can readily be converted into other known metabolites by standard oxidation and hydroxylation steps. The origin of the isonitrile group of **126** has been explored using a series of [^{15}N]-labelled compounds such as tyrosine, its N-formyl derivative, DOPA etc. In no example was significant incorporation observed and, as for xanthocillins, the provenance of this group is quite obscure[198].

The hazimycins appear to be derived by simple coupling of two tyrosine residues *ortho* to the hydroxyl groups[222]. Again the origin of the isonitrile function is unclear. In contrast to the situation for xanthocillins, [Me-^{14}C]-methionine did act as a precursor, albeit considerably less efficiently than tyrosine; and feeding of the ^{13}C-labelled compound proved that the methyl group labelled the isonitrile function. Thus formation of the latter by oxidation of a N-methyl group seemed likely[223] but, inconsistently with this, no label

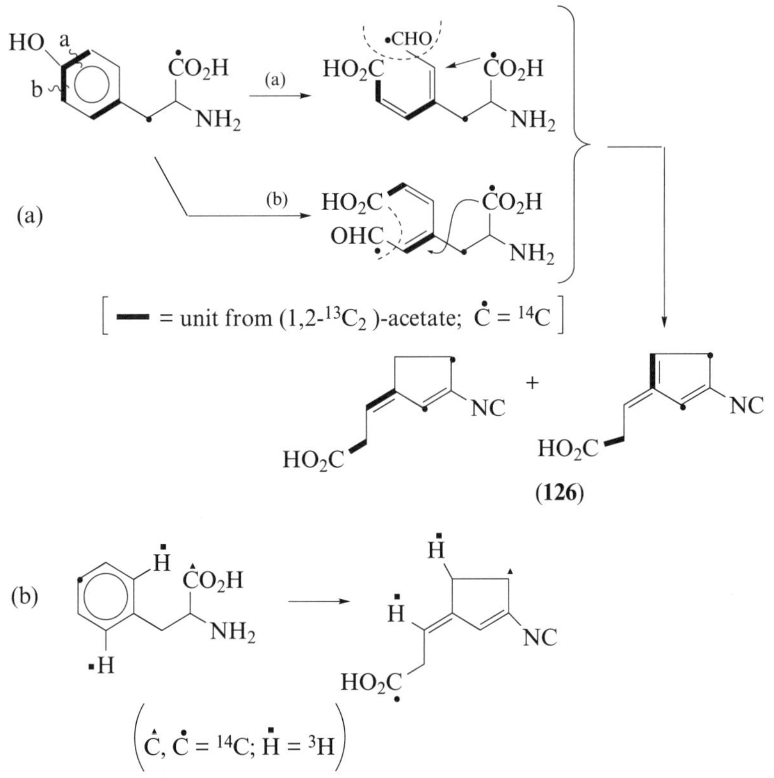

FIGURE 27. Biosynthesis of cyclopentyl isonitriles

from N-[^{132}C]-methyl-DL-tyrosine or [^{14}C]-formaldehyde was transferred to hazimycin[224]. The situation remains confused. The routes of **120** and the indole derivative **121** have not been explored, but by inspection valine together with D-mannitol and tryptophan appear obvious precursors. Hapalindole A (**127**) from blue–green fungi of *Halasiphon* spp is a product of mixed biosynthesis of tryptophan and an acyclic monoterpenoid[225].

2. Compounds from marine sources

In contrast to their terrestrial relatives, those isonitriles of marine origin are terpenoid based, rarely possess conjugated systems and are often accompanied by the corresponding isothiocyanates or formamido compounds, i.e. the triad RNC, RNCS, RNHCHO occurs (Figure 28). Several dozen such compounds are known, mostly from species of tropical sponges[161].

The *Axinella* sponges produce several types of isonitriles based on different sesquiter-penoid skeletons. The main type is exemplified by axisonitrile-1 (**128**), which was the first marine isonitrile to be reported[226] and was also the first example of the axane class of C_{15} compounds. The same sponge also produces axoisonitrile-2 (**129**)[227,228], the spiroxane **130**[229,230] and **131**[231]. *Acanthella* spp in the same Order in contrast yield C_{15}-based iso-nitriles with different skeletons **132**, **133**[232,233] and the richly-functionalized diterpenoid derivatives **134**, **135**[234].

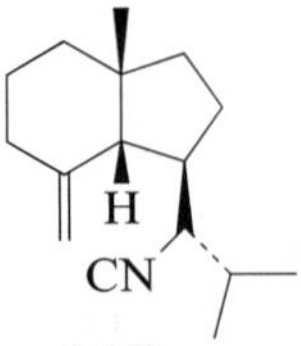

D. V. Banthorpe

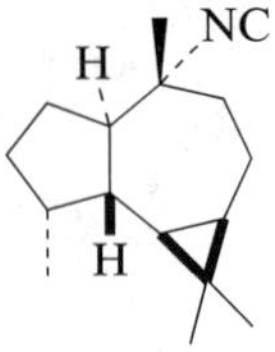

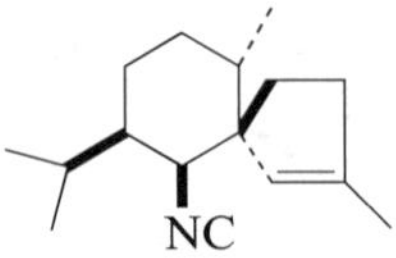

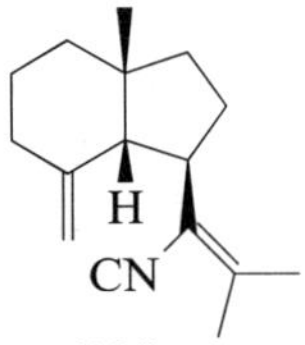

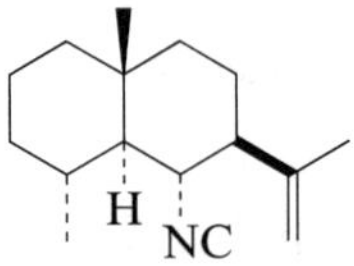

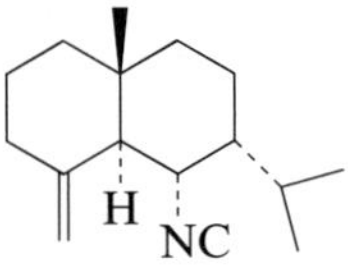

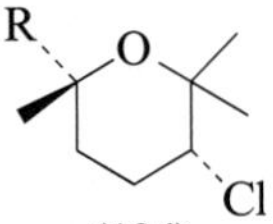

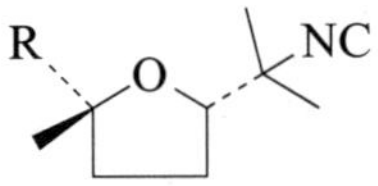

FIGURE 28. (*continued*)

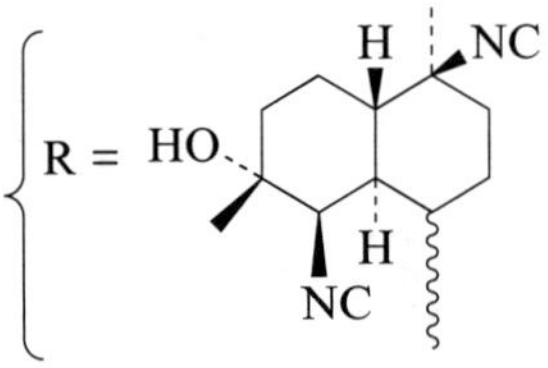

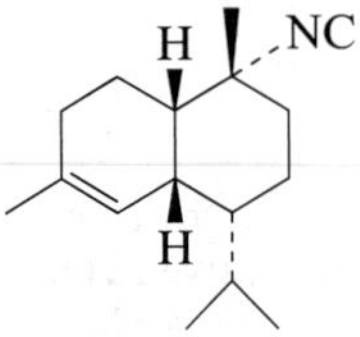

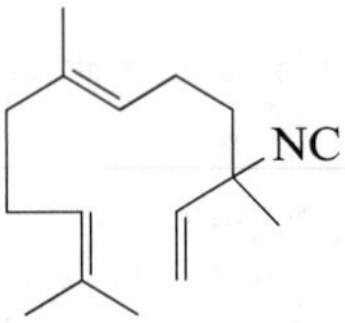

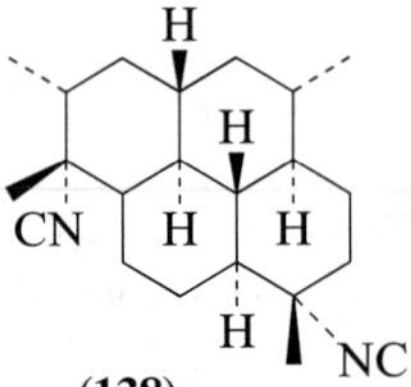

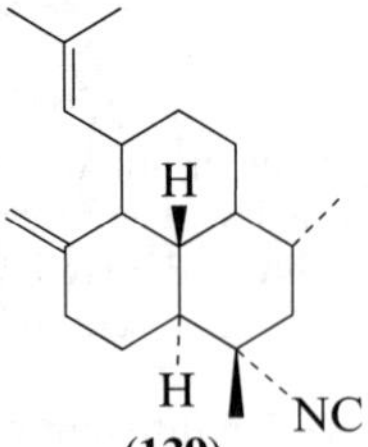

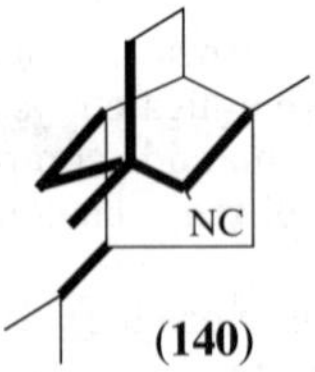

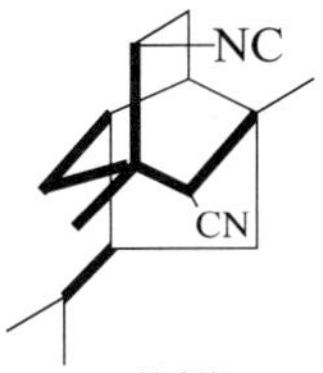

FIGURE 28. Some isonitriles of marine origin

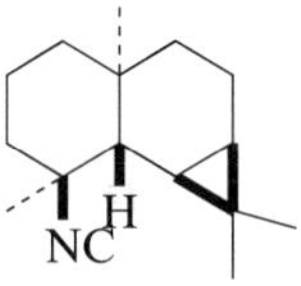

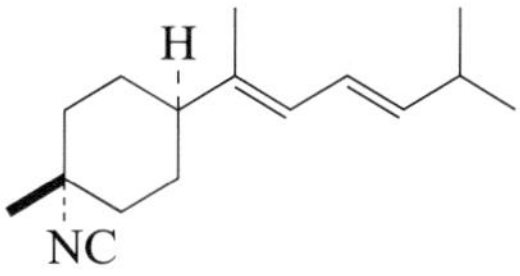

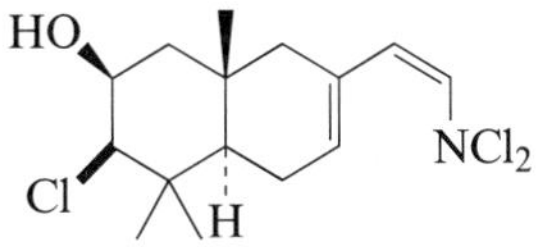

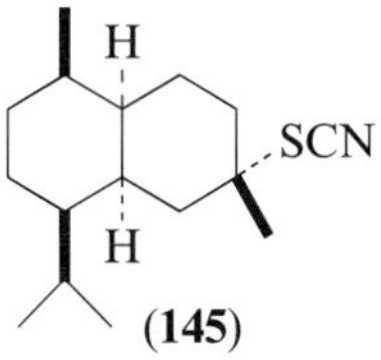

Typical of other sesquiterpenoid and diterpenoid-based isonitriles are **136** and **137** from *Halchondria* spp[235,236] and **138** and **139** from *Adocia* spp[237,238]. Presumably there are chemical defence compounds.

Nudibranches, a type of mollusc, puzzled marine biologists for, although of light colour and lacking an external shell for protection, they were rarely eaten by predatory fish. The reason was that certain species of the creatures accumulated isonitriles from the sponges on which they preyed and these compounds provided valuable weapons in the chemical armoury. 2- and 9-Isocyanopupukeanane (**140, 141**), which possess novel complex sesquiterpenoid rings, were such compounds[239,240] as are **142** and **143**[241,242]. Except for the group of compounds exemplified by **134** and **135**, the marine isonitriles are generally not extensively functionalized.

The biosynthesis of these compounds has recently been summarized together with that of other marine natural products[243]. In comparison with analogous investigations using cultures of micro-organisms, the study of metabolic processes in marine organisms poses severe technical difficulties, and little is known in detail. The occurrence of the aforementioned triads of compounds—indeed sometimes the corresponding amine also co-occurs—suggests a related biosynthetic pattern[244]. It is further reasonable to speculate that the isonitrile group is derived *in vivo* by dehydration of an N-formyl group—but there is clear evidence that this does not occur. All the available information indicates that the isonitrile is the progenitor of the other two members of the triad in an irreversible manner[245]; e.g. the 2-isocyano-, 2-formamido- and 2-isothiocyano-derivatives of **140** were synthesized with ^{13}C in the pendant group and were embedded in gelatine capsules in the sponge *Hymeniacidon* to allow the metabolic interversions in the one-way direction to be ascertained[246]. Many of the marine isonitriles carry the NC functions at secondary or even tertiary centres, which suggests that the group might be incorporated at some step before the elaboration of the skeleton. However, the details are unknown. What is known is that CN^- is the source of the isonitrile group for at least one class of compounds. This, supplied as $Na[^{14}C]$-CN, was incorporated 1.8% in 34 days into di-isocyanadociane (**138**) by a sponge of genus *Amphimedon*[247]. Presumably the CN^- either displaces halogen (common in marine organisms) from a precursor of **138** or neutralizes a carbocationic centre in the same. Attack of the nitrogen atom of the nucleophilic ion is presumably guided by the HSAB principle in either case. [2-^{14}C]-Acetate was not incorporated into **138** in these experiments and this is consistent with the finding that some sponges are not capable of synthesizing terpenoids from acetate[248]. All marine isonitriles are sesqui- or di-terpenoids and many novel structures occur that are unknown in higher plants. However, nothing is known of their origin in sponges, whether synthesized *de novo* or ingested in the diet and subsequently modified by functionalization[249,250].

Several carbonimidic dichlorides, such as **144**, have been isolated from marine sources[251,252]. These compounds unlike their corresponding isonitriles were not biologically active, and the latter may be chlorinated *in vivo* to give protected compounds that could be readily unmasked to give the active species when needed.

D. Thiocyanates

The biochemistry of cyanates and thiocyanates has been reviewed[161,253]. In contrast to the isothiocyanates, the occurrence of thiocyanates is rare in terrestrial organisms[254] although they are gradually being discovered in marine species, e.g. the cadinane sesquiterpenoid (**145**) from sponges[255].

An important source of thiocyanates is as decomposition products of glucosinolates[256]. These mustard oil glycosides are closely related to the cyanogenic glycosides is being formed from amino acids via oximes and can be equally toxic to animals[257]. They occur in the Brassica in most members of the Cruciferae and also in the Caporidaceae, Resedaceae and Moringaceae, and have the general formula **146** (Figure 29). Unlike the limited structural types of the cyanogenic glycosides, the glucosinolates have a wide variation (some 20 types) in side-chain R. On crushing the plant tissue, enzymic hydrolysis occurs by means of thioglucoside glucohydrases (myrosinases) to yield the thiohydroxanthate-*O*-sulphonate (**147**) which rearranges by a Lossen-type process to yield mainly isothiocyanates (**148**) accompanied by nitriles (**149**), thiocyanates (**150**) and lesser products[258]. The relative proportion of products, in particular the yield of **149** and **150**, depends on many factors such as plant species, ambient temperature and the presence of metal ions[73,259–261], and in addition formation of thiocyanate from isothiocyanate may occur during the release of bonded forms. The products are vesicant, toxic mustard oils which occur at high levels in certain wild species of *Brassica* and, when present in minute amounts, are responsible for the distinctive pungent flavour and odour associated with some of the common cultivars of Cruciferae spp used as vegetables and condiments, such as cabbages, turnips, radish, horseradish, mustard and the salad forms of *Lepidium* sps.

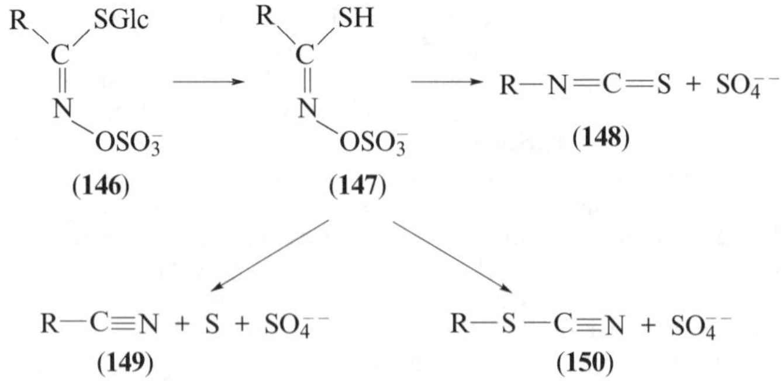

$$
\begin{array}{ccc}
\underset{\substack{\displaystyle \| \\ N \\ | \\ OSO_3^-}}{R\diagdown C \diagup SGlc} & \longrightarrow & \underset{\substack{\displaystyle \| \\ N \\ | \\ OSO_3^-}}{R\diagdown C \diagup SH} & \longrightarrow \quad R{-}N{=}C{=}S \,+\, SO_4^{-\,-} \\
(146) & & (147) & (148)
\end{array}
$$

$$R{-}C{\equiv}N \,+\, S \,+\, SO_4^{-\,-} \qquad\qquad R{-}S{-}C{\equiv}N \,+\, SO_4^{-\,-}$$
$$(149) \qquad\qquad\qquad\qquad (150)$$

FIGURE 29. Degradation of glucosinolates

Relatively high levels of thiocyanates are only found from glucosinates with R as allyl, benzyl or 4-(methylthio)butyl. Thus, a good example is the occurrence of $PhCH_2SCN$ in

Lepidium sativum (salad cress)[262]. It has been proposed that the thiocyanate-enhancing factor of *L. sativum* may be an isomerase that causes *Z,E* interconversion of the aglycone to set up the equilibrium **147** ⇌ **151** (Figure 30). Lossen-type rearrangement of **151** to yield the isothiocyanate **148** is blocked by the sulphate group, and so in this isomer the group R migrates to the S rather than the N atom to yield the thiocyanate **150**. R needs to be capable of generating a stabilized carbanion in order to sustain this process, and this explains the restriction of enhanced thiocyanate formation to compounds with the three groups mentioned above[256]. However, sadly no evidence for the isomerase has been forthcoming: addition of extracts of *L. sativum* to glucosinolates (mainly the 2-phenylethyl compound) from *Nasturtium* spp which did not yield appreciable quantities of thiocyanates did not promote the hoped-for enhancement[261].

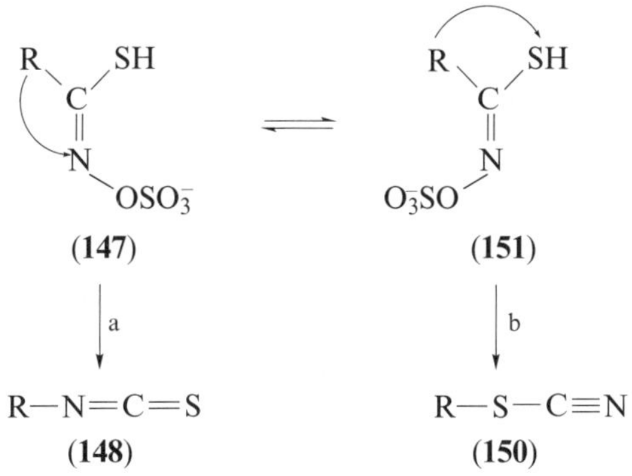

(147)   (151)

R—N=C=S   R—S—C≡N

(148)   (150)

FIGURE 30. Thiocyanate production from glucosinolates

For some years controversy arose whether myrosinase was a single enzyme or a complex of a thiohydrolyase and a synthase, the last species catalysing the rearrangement step. It is now considered to be a single entity that can exist in isoenzymic forms. The enzyme occurs in fungi, bacteria and cabbage aphids, as well as in higher plants, and the enzymes from the *Brassica* are all glycoproteins containing 2 to 4 sub-units and have a molecular mass of 125 to 153 kilodaltons[263-265].

V. BIOLOGICAL TOPICS

A. Biological Activity

Several examples of physiologically-active compounds have been mentioned in pass-ing—especially the cyanogenic glycosides whose major defensive role may be to protect seedlings from slugs and snails[266] and the antibiotics from marine isonitriles. Unfortu-ately, the toxicity of the latter class precludes their adoption in medicine.

Many polyalkynes have also been found to accumulate in plants from the time of initiation of seed germination and this certainly implies some potent biological function. Some are highly toxic and are undoubtedly involved in repelling animal predators[267,268]: thus oenanthetotoxin (**152**) from roots of *Oenanthe crocata* (water dropwort) or *Aethusa ynapium* (fools parsley) exhibit alkaloid-like properties and can cause death in animals and humans[269,270]. Similar compounds (Figure 31) occur in *Pteridium* spp (bracken) and numerous other plants and deter a range of animals from deer to desert rats[268]. Other alkynes are allelopathic agents which inhibit the germination and growth of competing plant species. Phenyl heptariyne (**153**) and α-terthienyl (**12**) are secreted by seedlings of several Compositae[271] (such as *Tagetes* spp) and are potent such inhibitors. They appear to be photosensitized[269], i.e. they are more effective in sunlight or UV than in the dark and so their allelopathic role is influenced by the depth at which they are exuded in the soil. Such compounds are active at levels of 0.4 ppm *in situ*.

$$HOCH_2CH{=}CH(C{\equiv}C)_2(CH{=}CH)_2(CH_2)_2CH(CH_2)_2Me \qquad Ph{-}(C{\equiv}C)_3Me$$

$$\underset{OH}{|}$$

(**152**)   OH   (**153**)

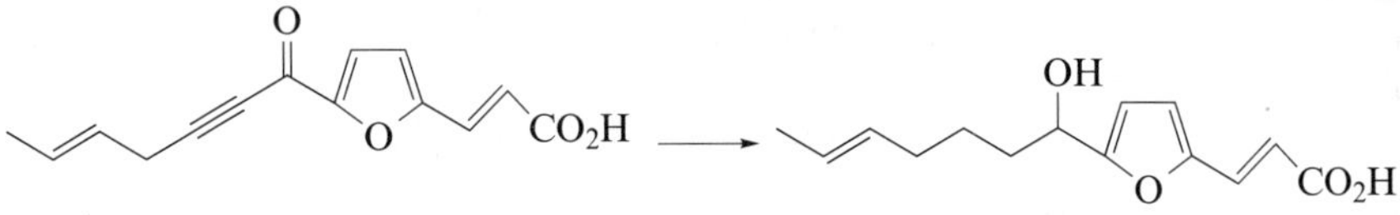

(**154**)   (**155**)

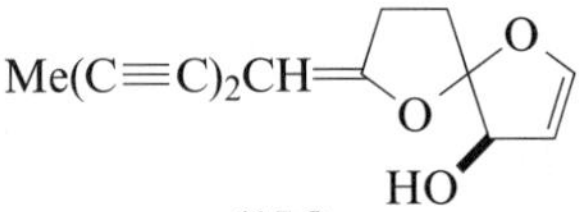

(**156**) 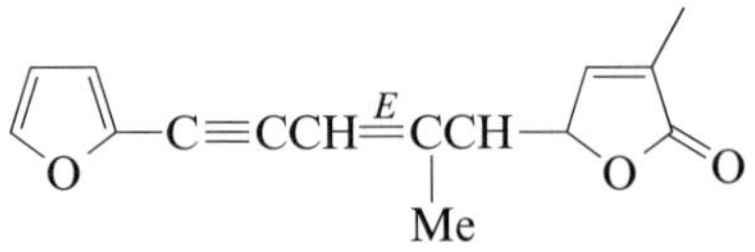   (**157**)

$$MeCH{\overset{E}{=}}CH(C{\equiv}C)_3CH{\overset{E}{=}}CHCH{-}CH_2OH$$

(**158**)   OH

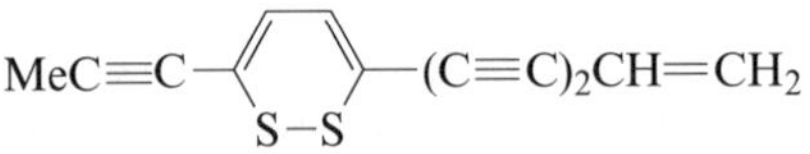

(**159**)

$$CH_2{=}CHCH{-}(C{\equiv}C)_2{-}CH{-}CH{\overset{Z}{=}}CH(CH_2)_6Me$$

$$\underset{OH}{|} \qquad\qquad \underset{OH}{|}$$

(**160**)

FIGURE 31. Some bioactive alkynes

Many alkynes are phytoalexins, that is, they are not constitutive metabolites or, if so, ar present at very low levels, but they are rapidly synthesized *de novo* at locations of tissu infection or damage (caused by pathogens, predators or by simple mechanical wounding and they act to combat especially fungal or bacterial attack. Many furano-alkynes are c this type and a classic example is the formation of wyerone (**154**) following fungal infectio of *Vicia faba* (broad bean). All phytoalexins are short-lived in their source-plants, and th compound is deactivated *in vivo* by reduction to the hexahydro compound **155**. Suc detoxification is also displayed by a *Botrytis* spp—the chocolate-spot fungus of the plant– and so this pathogen is immune to the phytoalexin. Other *Botrytis* spp cannot carry out th reduction and so cannot colonize the bean and are non-pathogenic[272–274]. Other we studied phytoalexins are the spiro-ketal enol ether **156** and freelingyne **157** from *Coleostephe* spp[275] and safynol **158** from *Carthamus tinctora* (safflower)[276]. A powerful technique fc stimulating the production of phytoalexins is to treat cell suspension cultures with th whole pathogens, their cell walls or other fractions derived from the latter. Numero specific phytoalexins of several chemical classes (e.g. sesquiterpenoids) have been th identified without the problems associated with studies on the infected organs of fiel grown plants and large numbers of alkynes have been implicated as defence compounds by those means[277]. Thus thiarubrine A (159) and α-terthienyl (12), the classic nematocides of *Tagetes* and *Bidens* (marigold) species together with a previously unknown relative were elicited when root cultures of the plants were treated with mycelia from the pathogenic *Phytophthora* and *Pythium* spp and the initially low levels of the alkynes were increased up to 30-fold within two days[278]. Another *Bidens* spp that produced no detectable alkynes gave large yields of 153 on similar infection[279]; other thiophens were similarly induced in hairy-root cultures of other *Tagetes* spp[280] whereas straight-chain alkynes resulted from infection of *Arctium* sps[281]. Fourteen alkyne phytoalexins including safynol (158) and its dehydro-derivative were elicited from suspension cultures of *Carthamus* spp by fungal cell-wall preparations[282,283].

As noted above, phytoalexins (i.e. elicitable compounds) may be present at low levels in 'normal' plants (who knows when a plant species is stressed, especially when harvested for photochemical analysis?). Also, a 'normal' constitutive alkyne of one species may not occur unless challenged in another! For example, farcarindiol (160), a characteristic alkyne of the Apiaceae and Araliacea[284,285], was not produced in a *Solanum* spp (eggplant); however, it rapidly accumulated in suspension cultures of the latter after treatment with fungal cell walls[286].

Farcarindiol is a constitutive metabolite of *Daucus carota* (carrot) and is the main component of the mixture of six compounds in the leaf wax that attracts and induces egg-laying by the carrot fly. The compounds all occur in leaves of other Umbellifereae, but the particular profile of components is characteristic of carrot and acts as a marker for the insect[287]. In contrast, other alkynes act as anti-feeding agents for certain *Pieris* (cabbage) butterflies[288].

$$R'C{\equiv}C-CH\overset{Z}{=}CH-C{\equiv}CR$$

(161)

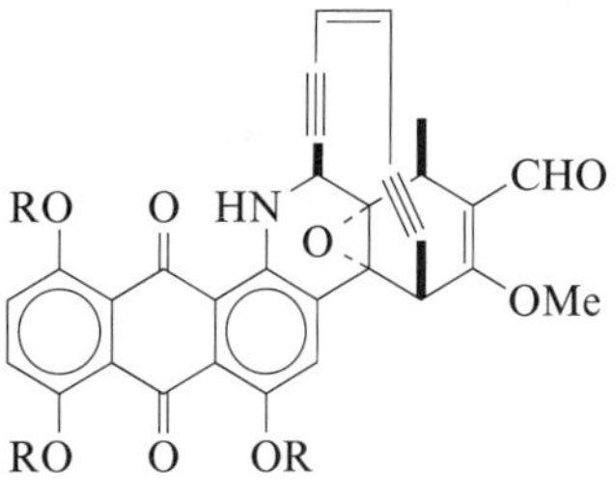

(162) R = H
(163) R = OAc

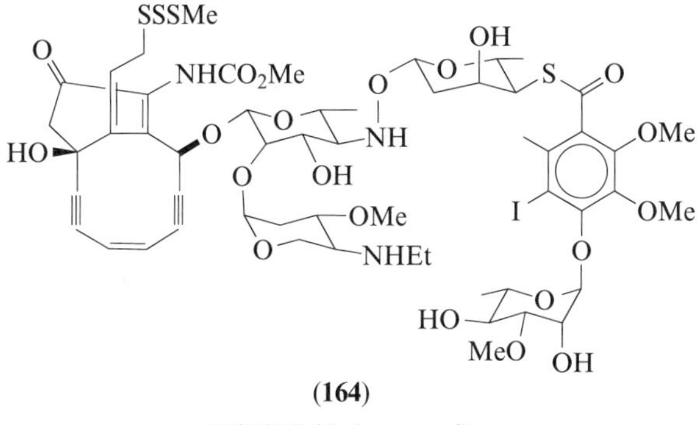

(164)

FIGURE 32. (*continued*)

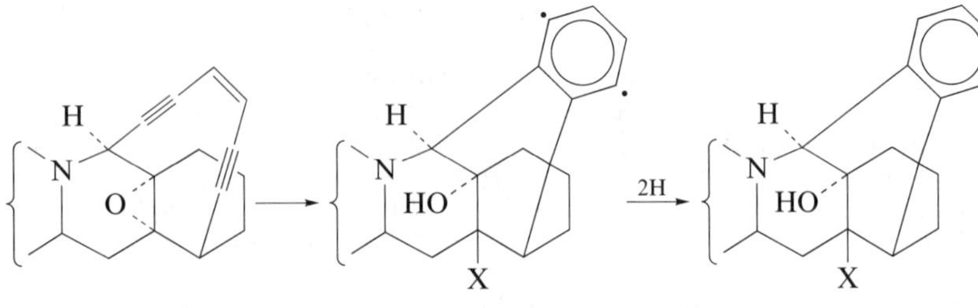

(X = Nucleophilic group)

FIGURE 32. Anti-tumour 1,5-diynes

Several alkynes slow some anti-tumour activity, but recently several bacterial compounds (Figure 32) with especially powerful activity have been discovered and all contain an ene-diyne unit (161).

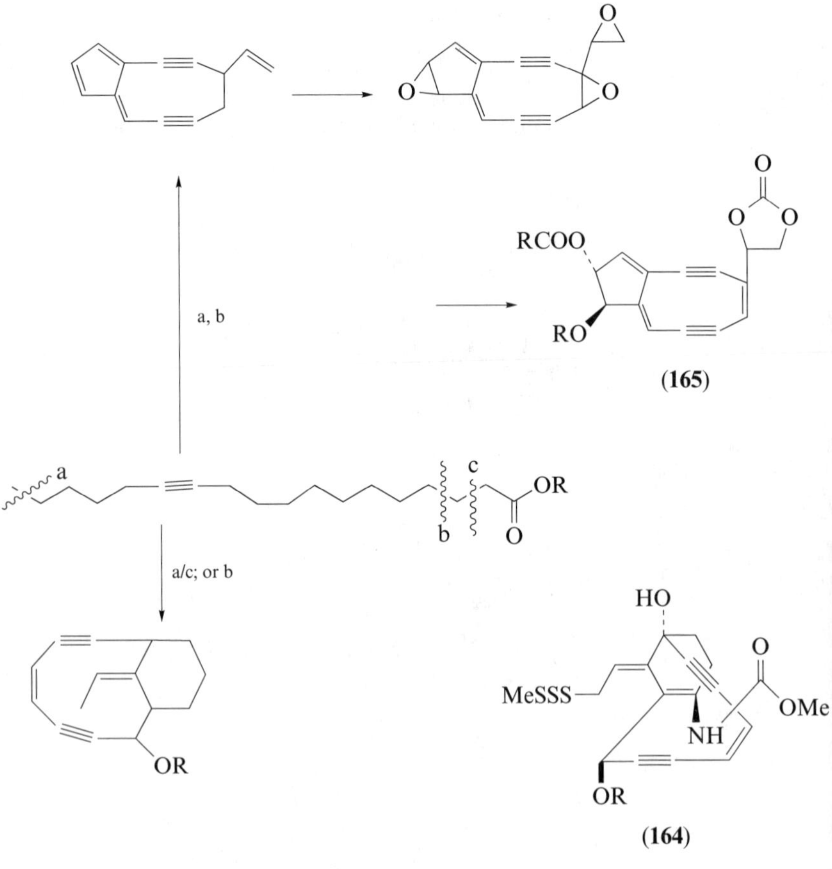

FIGURE 33. Biosynthesis of 1,5-diynes

Dynemicins A and C (**162, 163**) and calicheamicin γ_1 (**164**) are examples[289–291]. The cytotoxic activity depends on the binding of these antibiotics to DNA where the 1,5-diyne ring system acts as a precursor for an aromatic biradical (the Bergman reaction, well known *in vitro*) which in turn abstracts hydrogen from the deoxysugar residues and cleaves the backbone (Figure 33)[292]. Between 1 and 10 molecules of **164** suffice to destroy a tumour cell! The labile methyl trisulphide group in **164** is suspected to play a role (that is not possible, however, in **162** and **163**) as trigger for the cyclization when it is reduced by endogenous thiols; the group also acts as a convenient hook to attach γ-immunoglobulins to target specific tumour cells. Many details of the biosynthesis of the antibiotics have been elucidated. The incorporation of $[1,2-^{13}C_2]$ acetate into the 1,5-diyne 'chromophore' indicated the route to this part of **164** and the related **165** as outlined in Figure 33[293].

The mechanisms of the desaturations and cyclizations remain uninvestigated.

B. Tissue Culture

The use of plant tissue culture as a tool for the elicitation and subsequent characterization of phytoalexins has been outlined (Section V.A); however, the technique has been widely

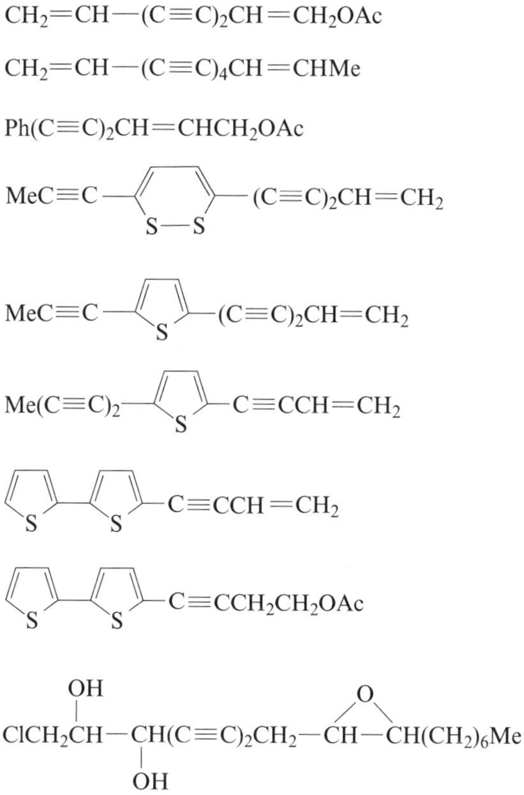

$$CH_2{=}CH{-}(C{\equiv}C)_2CH{=}CH_2OAc$$

$$CH_2{=}CH{-}(C{\equiv}C)_4CH{=}CHMe$$

$$Ph(C{\equiv}C)_2CH{=}CHCH_2OAc$$

$$MeC{\equiv}C{-}\!\!\!\!\!\!\Bigl\langle\!\!\!\Bigr\rangle\!\!\!\!-(C{\equiv}C)_2CH{=}CH_2 \quad (S{-}S)$$

$$MeC{\equiv}C{-}\Bigl\langle_S\Bigr\rangle{-}(C{\equiv}C)_2CH{=}CH_2$$

$$Me(C{\equiv}C)_2{-}\Bigl\langle_S\Bigr\rangle{-}C{\equiv}CCH{=}CH_2$$

$$\Bigl\langle_S\Bigr\rangle\!\!\Bigl\langle_S\Bigr\rangle{-}C{\equiv}CCH{=}CH_2$$

$$\Bigl\langle_S\Bigr\rangle\!\!\Bigl\langle_S\Bigr\rangle{-}C{\equiv}CCH_2CH_2OAc$$

$$\overset{OH}{\underset{\overset{|}{OH}}{ClCH_2CH{-}CH(C{\equiv}C)_2CH_2{-}}}\overset{O}{CH{-}}CH(CH_2)_6Me$$

(**166**)

FIGURE 34. Major alkynes from tissue cultures

used for the study of secondary metabolism in general[277] and many alkynes have been identified in callus, suspension, transformed-root and immobilized cell cultures[294]. Tissue culture also has immense potential as a source of readily available biomass for enzymatic studies on alkynes and other metabolites. A few reports concern the production of glucosinolates in culture[295] but no formation of cyanogenic glycosides or isonitriles in culture has been recorded to my knowledge.

For alkynes, the greatest success has been achieved, firstly, with conventional root cultures and, secondly, with transformed or hairy-root cultures. The former consists of organized root cell-tissue growing in the normal supplemented culture media. The latter result when such cultures are transformed by infection with a consequent introduction of a plasmid from *Agrobacterium* spp: a mass of hair-like threads grow rapidly with fully-retained root morphology and the cultures become independent of the essential plant-growth regulators (auxins and cytokinins) that are essential components of the media for all other types of cell culture. Such transformed cultures often, although not always, biosynthesize and accumulate the pattern of secondary metabolites characteristic of the root of the field-grown plant, and so the technique is well-suited for the study of alkynes as many of the latter are subterranean products. In the event, cultures of the appropriate plants can product 40 to 200% yields (based on dry weight) of that of such metabolites from the root of the parent[294]. Figure 34 lists the major alkynes found in such cultures, but many other types, such as aromatic dialkynes[296], can be formed. Recent leading references give full details as to the methodology and scope of the technique[297–302].

In contrast, yields of alkynes and many other classes of secondary metabolites from cultures of the callus or suspension type that show little or no organogenesis are invariably poor in comparison with those from the intact plant, even when the explant used to establish the culture was from the root. The reason for this is almost certainly due to lack of morphological differentiation as the enzyme systems necessary to produce the derived metabolites are often present and can be extracted in active form. Like many other hydrophobic secondary metabolites, alkynes *in vivo* accumulate in extracellular spaces—ducts, resin canals—or in specialized storage cells such as trichomes[303]. They have also been recorded to form as oily drops in the periderm of roots—possibly organized into pericyclic oil ducts[304] (on the other hand, fungal products are typically excreted directly into the culture medium). When these structures are missing in undifferentiated or very unorganized cultures, any alkyne that is formed must be stored within the cell. Unless there is safe compartmentation in the vacuole the result will be a concentration of toxic metabolite that could prove lethal to the cell population (probably a small proportion of the whole) active in secondary metabolism, or at best could shut down the biosynthetic pathway by feedback inhibition. Alternatively, the levels of the toxic product may be reduced by degradation processes within the cell. In any event, negligible or very low levels of accumulation will be found on harvesting. The levels of thiophene formation in callus of *Tagetes* sps has been shown to be directly coupled to the amount of root regeneration[305,306], and studies with cultures of crown gall tumours show that yields of alkynes correlate well with xylogenesis and that a primitive level of tissue organization may be sufficient to allow the accumulation of these products[294]. Despite such problems some success has been achieved with unorganized tissue[294]. Thus although recent studies using *Panax ginseng* callus did not yield the desired pharmaceutically important ginsenosides, several acyclic alkynes were isolated in good yields including the novel tumour-inhibitor **166**[307–309].

C. Chemotaxonomy

The ease of detection of the conjugated dialkyne chromophore (at 250–350 nm) has resulted in systematic screening of plant families for alkynes on a considerable scale. The target compounds have been found in all plant organs but largely reside in the root (*ca* 85%

TABLE 2. Distribution of alkynes and their derivatives in higher plants

Order	Main families	Main sources
Campanales	Campanulaceae	++
Asterales	Compositae ≡ Asteraceae	+++
	Anthemideae	+
Umbellales	Araliaceae	++
	Umbellifereae ≡ Apiaceae	+++
Rosales	Pittosporaceae	+
Gentianales	Oleaceae	+
Santales	Santalaceae	+

$Me(CH_2)_7C{\equiv}C(CH_2)_7CO_2H$

(167)

$Ph(C{\equiv}C)_3Me$

(168)

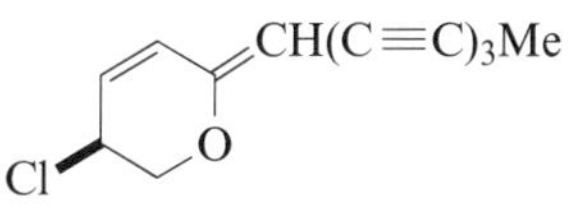

(169)

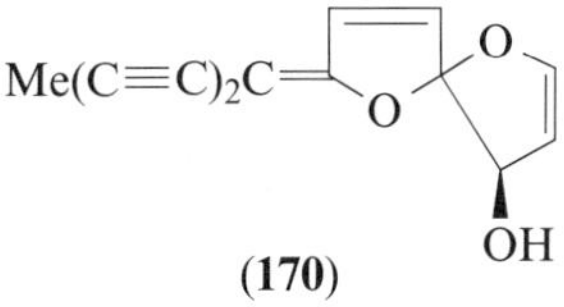

(170)

FIGURE 35. Taxonomically-significant alkynes

of the total) although the patterns are usually highly organ-specific[310,311]. At maximum the recoveries from root tissue are ca 0.2% dry weight.

The alkynes occur almost exclusively in 6 orders of higher plants (Table 2)[5,18,312–314] and the most abundant source is the Compositae. The straight-chain alkynic acids have a rather different distribution from the rest: tarric and stearolic acids (7, 167) occur in the Santales, but also in the Malvales in association with cyclopropene fatty acids. The alkynes have been useful in diverse chemotaxonomic situations[310]. Thus 168 and 169 (Figure 35) are restricted to *Coreopsis* and *Anaphalis* spp, respectively[312], and the spiro-compound 170 is a marker for sub-families of the Compositae[315]. The role of thiophenes in the classification of the Compositae[316], the taxonomy of the Compositae in general[18–22,317] and that of *Thelerperma*[318] and *Argyranthemum* spp[319] have also been discussed.

The cyanogenic glycosides occur in angiosperms, gymnosperms, ferns and fungi[320] and have been detected in numerous genera[321–324] although, as with the alkynes, much confusion and use of differing nomenclature and classifications by different authors over the whole range from super order to family confuse the tissue. The biosynthetic pathway involved is simple and similar in all examples and may represent a relict mechanism that was once more widespread. Nevertheless, the distribution of glycosides has been claimed to be of taxonomic significance[323]. Amygladin (83) only occurs in certain groups of the Rosaceae and is seldom found in the Violaceae, and its relative prunasir (86) has been used as a taxonomic monitor below the level of family in the Rosaceae[324]. The distribution of the unusual cyclopentyl-glucoside gynocardin (91) has also been cited to suggest links between the Flacourtiaceae and the Passifloraceae and to define genera within the families[325,326] and the occurrence of proacacipetalin (82) in the *Acacia* genus was sufficient to revise subgeneric classifications[327–329]. The taxonomic significance of cyanogenic compounds in the Violales has also been assessed[330]. However, problems have arisen: dhurrin (**88**) has been widely accepted as a characteristic marker for the Gramineae but was nevertheless absent from two authenticated bamboo species[331].

Not all the individual specimens in a positively-identified cyanogenic species may contain the appropriate glycosides, i.e. chemical polymorphism may occur. This phenomenon has

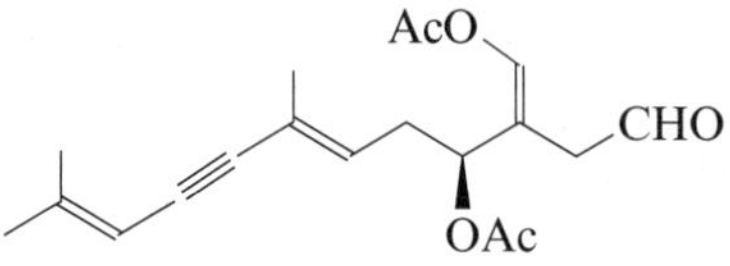

(171) 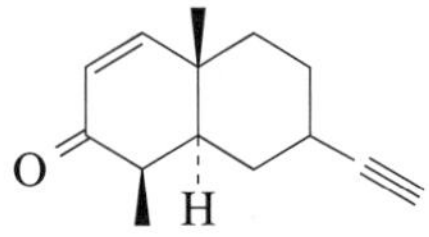   (172)

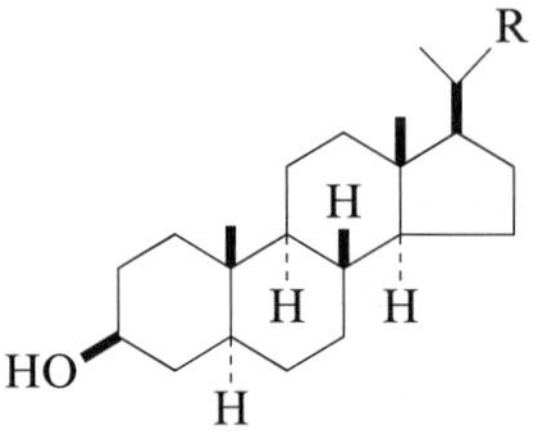

(173) R = ·≡⟨

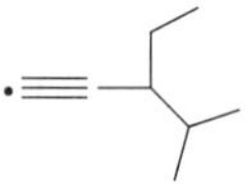

(174) R = ·≡⟨

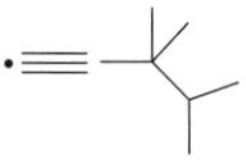

(175) R = ·≡⟨

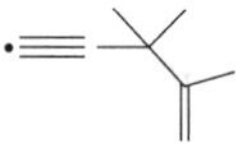

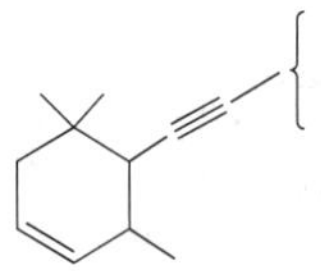

(176) 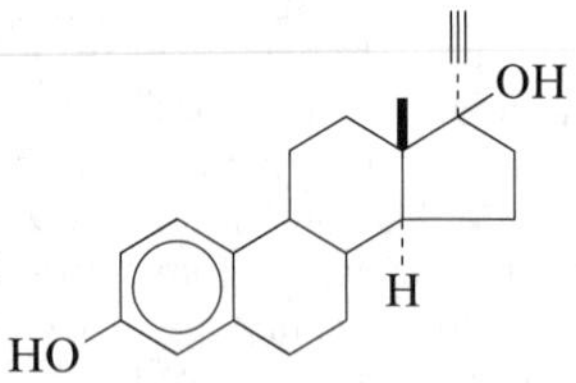   (177)

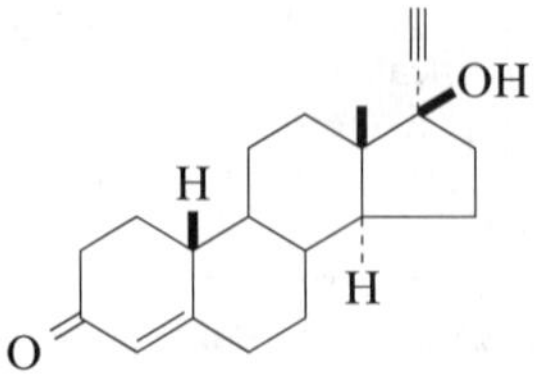

(178)

FIGURE 36. Miscellaneous alkynes

been extensively studied in *Trifolium repens* and *Lotus corniculatus*[332] and also occurs in *Sorghum bicolor*, *Prunus amygladus* and *Manihot esculenta*[333]. Detailed genetic analyses of some species are also available[334,335].

Cyanolipids appear to be a unique taxonomic marker for the Sapindaceae within the Sapindales order. However, the compounds are not found in all the members of the family and doubts have arisen as to their use in analysis at the sub-familial level[326]. The distribution of β-cyano-L-alanine has been used in classification of *Vicia* spp[336].

D. Miscellaneous

Several naturally-occurring terpenoids contain alkyne groups, but nothing is known about the mechanism of insertion of the triple bond which does not occur at a typical position of unsaturation in the parent mevalonoid-derived skeleton. The monoterpenoid oxytoxin-I 171 (Figure 36), which is poisonous to fish, occurs in *Oxynoe* spp of nudibranch and some green algae[337,338]. The nor-sequiterpenoid 172 has been isolated from *Eremophila* spp of plants[339]. Alkynic sterols are found in marine sponges and recently have been discovered in higher plants. Compounds 173–175 were characterized and it was predicted that many sterols now considered to be typically marine will be eventually found in freshwater organisms and terrestrial species[340]. Alkynic carotenoids usually possessing the terminal unit(s) 176 occur in several plants[341] and numerous such compounds have been synthesized in the laboratory for use as possible inhibitors of carotenoid biosynthesis—and hence hoped-for stimulations of steroid formation. In fact, numerous synthetic compounds containing the alkynic, and less often the isonitrile or nitrile groups, have been tested for medicinal or related purposes and several have been adopted. Possibly the most celebrated are the synthetic oestrogens and progesterones, e.g. mestranol (**177**) and norethindrone (**178**), which are components of oral contraceptives.

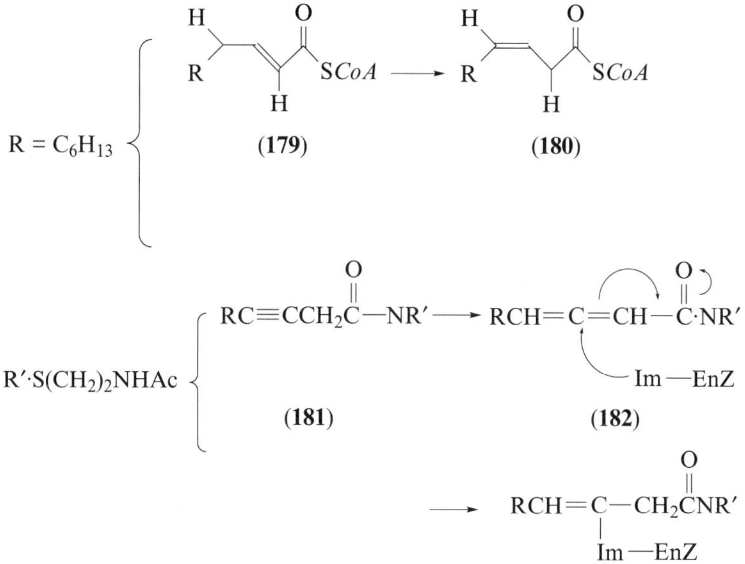

FIGURE 37. Allenes as enzyme inhibitors: Im-Enz represents the imidazole group of a histidine residue at the active site of the enzyme; —CoA represents a Coenzyme A residue

Use of such a synthetic inhibitor is illustrated by studies of the mechanism of the dehydratase-catalysed allylic rearrangement (**179→180**; Figure 37) that is utilized in the direct synthesis of unsaturated fatty acids by anaerobic microorganisms. Evidence for the role of histidine at the active site of the enzyme was provided by incubation with the substrate analogue **181**. This is accepted by the enzyme, but the product of the competing propargylic rearrangement (*cf* Section II.B.2) is the allenic thioester **182**, which is a potent electrophile that alkylates the imidazole group of a histidine residue of the enzyme with resultant irreversible inhibition[342]. Apparently the only enzyme known that produces an allenic product and that is not deactivated by its product is an allene–alkyne isomerase from pig liver[343]. The occurrence of propargylic rearrangement followed by reaction of the allenic compound as above may account for some of the potent biochemical properties of alkynes that have been in previously mentioned sections. In another inhibitor study involving alkynes, cytochrome P-450 from rat liver microsomes was inactivated during the reduction of alkynes ranging from acetylene to octa-1-yne. This was due to N-alkylation of the prosthetic haem group by the substrate: alkynes generally alkylated the pyrrole of ring A although acetylene was exceptional in reacting with more than one nitrogen, whereas those alkenes which were also reduced attacked ring D[344]. Reduction of acetylene by symbiotic plant organisms or intact legume nodules have been much studied[345] as such reactions are widely used to assay the activity of many nitrogen-fixing species by measurement of their capacity as hydrogenases[346–348].

The metabolism of alkynes and alkenes in mammals has been reviewed[349]. The detoxification of alkynes generally follows the well-tried sequence: →oxidation to alcohol→conjugation→excretion. In mammals and other vertebrates the most common conjugates are glucuronides and sulphates but linkage to amino acids, especially glycine and ornithine, is frequently observed. In insects and other invertebrates the formation of a glucoside utilizing a UDP-glucose glucosyl transferase is a major pathway to a soluble excretion product, although routes involving formation of sulphate and phosphate esters can be brought into play[350,351].

One of the most versatile reagents used in biochemistry is cyanogen bromide (BrCN). Its roles in the specific cleavage of peptide chains on the C-terminal side of a seryl residue in order to provide fragments suitable in size for sequencing and also as an activator for the coupling of substrates to resin matrices for use in affinity chromatography are described in any biochemistry text.

VI. REFERENCES

1. E. R. H. Jones and V. Thaller, in *Chemistry of the Carbon–Carbon Triple Bond* (Ed. S. Patai), Part 2, Wiley, Chichester, 1978, p. 621.
2. H. M. Walborsky and M. P. Periasamy, in *Chemistry of Triple Bonded Functional Groups*, Supplement C, (Ed. S. Patai and Z. Rappoport), Part 2, Wiley, Chichester, 1983, p. 835.
3. J. P. Ferris, in *Chemistry of the Cyano Group* (Ed. Z. Rappoport), Wiley, London, 1970, p. 717.
4. J. Lam and L. Hansen, *Methods Plant Biochemistry*, **4**, 159 (1990).
5. F. Bohlmann, T. Burkhardt and C. Zdero, *Naturally-Occurring Acetylenes*, Academic Press, London, 1973.
6. J. M. Marco and O. Barbera, in *Studies in Natural Product Chemistry* (Ed. Atta-ur-Rahmen), Vol. 7, Elsevier, Amsterdam, 1990, p. 201.
7. K. Neuschild, L. P. Christensen and J. Lam, *Phytochemistry*, **31**, 2743 (1992).
8. C. T. Bedford *et al.*, *J. Chem. Soc., Perkin Trans. 1*, 735 (1976).
9. E. Rodriquez, M. Arequillin, T. Nishida and G. H. N. Towers, Experientia, **41**, 419 (1985).
10. F. Balza, I. Lopez, E. Rodriquez and G. H. N. Towers, *Phytochemistry*, **28**, 3523 (1989).
11. J. A. Marco, J. F. Sanz, J. Jakupovic and S. Huneck, *Tetrahedron*, **46**, 6931 (1990).
12. A. D. Patil, J. A. Chan, P. Lois-Fleming, R. J. Moyer and J. W. Westley, *J. Nat. Prod.*, **52**, 15? (1989).

13. J. F. Sanz, E. Falco and J. A. Marco, *Liebigs Justus Ann. Chem.*, 303 (1990).
14. J. Lam, L. P. Christensen and T. Thomasen, *Phytochemistry*, **31**, 2881 (1992).
15. R. Bauer, K. Redl and B. Davies, *Phytochemistry*, **31**, 2035 (1992).
16. J. Ollivier and J. Salaun, *J. Chem. Soc., Chem. Commun.*, 1269 (1985).
17. P. Gonser, J. Jakupovic, G. Grou and L. Vincent, *Phytochemistry*, **29**, 3940 (1990).
18. L. Hansen and P. M. Boll, *Phytochemistry*, **25**, 285 (1986).
19. L. P. Christensen and J. Lam, *Phytochemistry*, **29**, 2753 (1990).
20. L. P. Christensen and J. Lam, *Phytochemistry*, **30**, 11 (1991).
21. L. P. Christensen and J. Lam, *Phytochemistry*, **30**, 2453 (1991).
22. L. P. Christensen, *Phytochemistry*, **31**, 7 (1992).
23. D. J. Faulkner, *Nat. Prod. Reports*, **9**, 323 (1992) and references cited therein.
24. H. Kiyoshi, Y. Shizuri, H. Niwa and Y. Yamada, *Tetrahedron*, **42**, 3781 (1986).
25. A. G. Gonzalez, J. D. Martin, M. Norte, P. Rivera and J. Z. Ruano, *Tetrahedron*, **40**, 3443 (1984).
26. S. M. Waraszkiewicz, H. H. Sun, K. L. Erickson, J. Finer and J. Clardy, *J. Org. Chem.*, **43**, 3194 (1978).
27. R. Kazlauskas, P. J. Murphy, R. J. Wells and A. J. Blackman, *Austr. J. Chem.*, **35**, 113, (1982).
28. A. Fukuzawa, M. Aye, M. Nakamura, N. Tamura and A. Murai, *Tetrahedron Lett.*, **31**, 4895 (1990).
29. G. Guella, I. Mancini and F. Pietra, *J. Chem. Soc., Chem. Commun.*, 77 (1986).
30. N. Fusetani, T. Shiraguki, S. Matsunaga and K. Hashimoto, *Tetrahedron Lett.*, **28**, 4311 (1987).
31. N. Fusetani, M. Sugano, S. Matsunaga and K. Hashimoto, *Tetrahedron Lett.*, **28**, 4311 (1987).
32. E. Quinoa and P. Crews, *Tetrahedron Lett.*, **28**, 2467 (1987).
33. J. Kobayashi *et al.*, *J. Chem. Soc., Perkin Trans. 1*, 3301 (1990).
34. F. J. Schmitz and Y. Gopichard, *Tetrahedron Lett.*, 2637 (1978).
35. G. Cimino, A. De Giulio, S. DeRosa and V. Di Manzo, *Tetrahedron Lett.*, **30**, 3563 (1989).
36. D. Castiello, G. Cimino, S. DeRosa, S. DeStefano and G. Sodano, *Tetrahedron Lett.*, **21**, 5047 (1980).
37. R. P. Walker and D. J. Faulkner, *J. Org. Chem.*, **46**, 1475 (1981).
38. W. B. Turner and D. C. Aldridge, *Fungal Metabolites*, Vol. 2, Academic Press, London 1984.
39. O. G. Yashina and L. I. Vereshchagin, *Usp. Khim.*, **47**, 557, (1978).
40. J. Baeverle, T. Anke, R. Jenke and F. Bosold, *Arch. Microbiol.*, **132**, 194 (1982).
41. J. D. Bu'lock and G. N. Smith, *Biochem. J.*, **87**, 35 (1962).
42. M. Gill and W. Steglich, *Prog. Chem. Org. Nat. Prod.*, **51**, 1 (1987).
43. G. Kohn, O. Vanderkerkhove, E. Hartmann and P. Beutelmann, *Phytochemistry*, **27**, 1049 (1988).
44. D. H. G. Crout, P. J. Fisher, V. S. B. Gaudet, H. Stoeckli-Evans and A. E. Anson, *Phytochemistry*, **25**, 1224 (1986).
45. J. D. Bu'lock, in *Comparative Phytochemistry*, (Ed. T. Swain), Academic Press, London, 1966, p. 79.
46. G. A. Thompson, in *Secondary Plant Products* (Eds. E. A. Bell and B. V. Charlwood), Springer, Berlin, 1980, p. 536.
47. E. R. H. Jones, V. Thaller and J. L. Turner, *J. Chem. Soc. Perkin Trans. 1*, 424 (1975).
48. E. R. H. Jones, C. M. Piggin, V. Thaller and J. L. Turner, *J. Chem. Res. (S)*, 68 (1977).
49. P. Hodge, E. R. H. Jones and G. Lowe, *J. Chem. Soc. (C)*, 1216 (1966).
50. J. D. Bu'lock and H. Gregory, *Biochem. J.*, **72**, 322 (1959).
51. J. D. Bu'lock, D. C. Allport and W. B. Turner, *J. Chem. Soc.*, 1654 (1961).
52. J. D. Bu'lock and H. M. Smalley, *J. Chem. Soc.*, 4663 (1962).
53. R. Jenke and E. Richter, *Phytochemistry*, **15**, 1673 (1976).
54. D. G. Davies, P. Hodge, P. Yates and M. J. Wright, *J. Chem. Soc., Perkin Trans. 1*, 1602 (1978).
55. G. C. Barley *et al.*, *J. Chem. Res. (S)*, 232 (1987).
56. G. C. Barley, E. R. H. Jones and V. Thaller, *Bioact. Mol.*, **7**, 85 (1988).
57. I. W. Farrell, C. A. Higham, E. R. H. Jones and V. Thaller, *J. Chem. Res. (S)*, 234 (1987).
58. G. C. Barley *et al.*, *J. Chem. Soc. (C)*, 3308 (1971).
59. M. I. Gurr and A. J. James, *Lipid Biochemistry*, 3rd edn., Chapman and Hall, London, 1980, p. 90.
60. F. Bohlmann and H. Schulz, *Tetrahedron Lett.*, 4795 (1968).
61. K. I. Ichihara and M. Noda, *Biochim. Biophys. Acta*, **487**, 249 (1977).
62. F. Bohlmann, in *Phytochemistry* (Ed. L. P. Miller), Vol. 3, Van Nostrand, New York, 1973, p. 112.

732   D. V. Banthorpe

63. P. C. Constabel and G. H. N. Towers, *Phytochemistry*, **28**, 93 (1988).
64. F. Bohlmann, R. Jente, W. Lucas, J. Loser and H. Schulz, *Chem. Ber.*, **100**, 3183 (1967).
65. A. Batna and G. Spitellor, *Justus Liebigs Ann. Chem.*, 861 (1991).
66. F. Bohlmann, R. Jente and R. Reinecke, *Chem. Ber.*, **102**, 3283 (1969).
67. M. Luckner, *Secondary Metabolism in Plants and Animals*, Chapman and Hall, London, 1972, p. 100.
68. M. Ahmed, M. T. W. Hearn, E. R. H. Jones and V. Thaller, *J. Chem. Res. (S)*, 125 (1975).
69. E. E. Conn, *Naturwissenschaften*, **66**, 28 (1979).
70. E. E. Conn, *Ann. Rev. Plant Physiol.*, **31**, 433 (1980).
71. E. E. Conn, in *Biochemistry of Plants* (Eds. P. K. Stumpf and E. E. Conn), Vol. 7, Academic Press, New York, 1981, p. 479.
72. A. J. Cutler and E. E. Conn, *Recent Adv. Phytochem*, **16**, 249 (1982).
73. P. M. Dewick, *Nat. Prod. Reports*, **1**, 544 (1984).
74. J. E. Poulton, *Plant Physiol.*, **94**, 401 (1990).
75. A. Nahrstedt, *Proc. Phytochemical Soc. Europe*, **33**, 249 (1992).
76. E. E. Conn, in *Secondary Plant Products* (Eds. E. A. Bell and B. V. Charlwood), Springer, Berlin, 1980, p. 461.
77. D. S. Seigler, *Prog. Phytochem.*, **4**, 83 (1977).
78. A. Kjaer and P. O. Larsen, *Biosynthesis*, **5**, 34 (1977).
79. A. Nahrstedt, P. S. Jensen and V. Wray, *Phytochemistry*, **28**, 623, (1989).
80. C. S. Butterfield, E. E. Conn and D. S. Seigler, *Phytochemistry*, **14**, 993 (1975).
81. K. Ueda, K. Yasutomi and I. Mori, *Chem. Lett.*, 149 (1983).
82. K. C. Spencer and D. S. Seigler, *Phytochemistry*, **24**, 2615 (1985).
83. J. W. Jaroszewski, D. Braun, V. Clausen and C. Cornett, *Planta Med.*, **54**, 333 (1988).
84. E. W. Olafsdottir, J. V. Andersen and J. W. Jaroszewski, *Phytochemistry*, **28**, 127 (1988).
85. J. W. Jaroszewski, J. A. Vanggaard and I. Billeskor, *Tetrahedron*, **43**, 2349 (1987).
86. K. C. Spencer and D. S. Seigler, *Phytochemistry*, **24**, 981 (1985).
87. E. S. Olafsdottir, J. W. Jaroszewski and M. M. Arbo, *Biochem. Syst. Ecol.*, **18**, 435 (1990).
88. E. S. Olafsdottir, C. Cornett and J. W. Jaroszewski, *Acta Chem. Scand.*, **43**, 51 (1989).
89. R. H. Davis, in *Toxic Substances in Crop Plants* (Eds. J. P. F. D'Mello, C. M. Duffus and J. H. Duffus), Royal Soc. Chem., Cambridge, 1991, p. 201.
90. E. E. Conn, in *Herbivores: Their Ingested Secondary Plant Metabolites* (Eds. G. A. Rosenthal and D. H. Jansen), Academic Press, New York, 1979, p. 387.
91. R. D. Gibbs, *Chemotaxonomy of flowering Plants*, Vols. I–IV, McGill-Queens, Montreal, 1974.
92. K. C. Spencer and D. S. Seigler, *Biochem. Syst. Ecol.*, **13**, 303 (1985).
93. V. Plouvier, *C. R. Acad. Sci. Paris*, **282D**, 723 (1976).
94. G. Padmaja and C. Balagopal, *Can. J. Microbiol.*, **31**, 663 (1985).
95. T. M. Flynn and I. A. Southwell, *Phytochemistry*, **26**, 1669 (1987).
96. W. K. Swenson, J. E. Dunn and E. E. Conn, *Phytochemistry*, **28**, 821 (1989).
97. C. J. Knowles, *Bacteriol. Rev.*, **40**, 652 (1976).
98. F. Wissing, *J. Bacteriol.*, **121**, 695 (1975).
99. J. E. Poulton, *Ciba Found. Symp.*, **140**, 67 (1988).
100. M. Kojima, J. E. Poulton, S. S. Thayer and E. E. Conn, *Plant Physiol.*, **63**, 1022 (1979).
101. E. S. Wurtele, S. S. Thayer and E. E. Conn, *Plant Physiol.*, **70**, 1732 (1982).
102. B. Koch, V. S. Nielsen, B. A. Halkier, C. E. Olsen and B. L. Moeller, *Arch. Biochem. Biophys.*, **292**, 141 (1992).
103. J. A. Sanders and E. E. Conn, *Plant Physiol.*, **61**, 154 (1978).
104. H. C. Butcher, G. J. Wagner and H. W. Siegelman, *Plant Physiol.*, **59**, 1098 (1977).
105. A. Nahrstedt and R. H. Davis, *Phytochemistry*, **25**, 2299 (1986).
106. P. C. Constabel, F. Balza and G. H. N. Towers, *Phytochemistry*, **27**, 3533 (1988).
107. A. Nahrstedt and R. H. Davis, *Comp. Biochem. Physiol.*, **68B**, 575 (1981).
108. A. Nahrstedt and R. H. Davis, *Comp. Biochem. Physiol.*, **75B**, 65 (1983).
109. V. Wray, R. H. Davis and A. Nahrstedt, *Z. Naturforsch., Sect. C*, **38**, 538 (1983).
110. S. Franzl, A. Nahrstedt and C. M. Naumann, *J. Insect. Physiol.*, **32**, 705 (1986).
111. T. Eisner, in *Chemical Ecology* (Eds. E. Sondheimer and J. B. Simeone), Academic Press, New York, 1970, p. 157.
112. J. C. Braekman, D. Daloze and J. M. Pasteels, *Biochem. Syst. Ecol.*, **10**, 355 (1980).
113. S. S. Duffey and G. H. N. Towers, *Can. J. Zool*, **56**, 7 (1978).

114. S. S. Duffey, E. W. Underhill and G. H. N. Towers, *Comp. Biochem. Physiol.*, **47B**, 753 (1974).
115. A. Nahrstedt, *Ciba Found Symp.*, **140**, 131 (1988).
116. D. Raubenheimer, *J. Chem. Ecol.*, **15**, 2177 (1989).
117. I. Tober and E. E. Conn, *Phytochemistry*, **24**, 1215 (1985).
118. B. A. Tapper and G. W. Butler, *Phytochemistry*, **11**, 1041 (1972).
119. D. B. Collinge and M. A. Hughes, *Arch. Biochem. Biophys.*, **218**, 38 (1982).
120. A. J. Cutler and E. E. Conn, *Arch. Biochem. Biochem. Biophys.*, **212**, 468 (1981).
121. W. Hoesel and A. Nahrstedt, *Arch. Biochem. Biophys.*, **203**, 753, (1980).
122. A. J. Cutler, W. Hoesel, M. Sternberg and E. E. Conn, *J. Biol. Chem.*, **256**, 4253 (1981).
123. B. L. Moeller and E. E. Conn, *J. Biol. Chem.*, **254**, 8575 (1979).
124. B. L. Moeller, in *Cyanide in Biology* (Eds. B. Vennesland, E. E. Conn, C. J. Knowles, J. Westley and F. Wissing), Academic Press, London, 1981, p. 217.
125. B. A. Halkier, C. E. Olsen and B. L. Moeller, *J. Biol. Chem.*, **264**, 19487 (1989).
126. B. A. Halkier, H. V. Schellar and B. L. Moeller, *Ciba Found. Symp.*, **140**, 49 (1988).
127. B. A. Halkier and B. L. Moeller, *J. Biol. Chem.*, **265**, 21114 (1990).
128. D. B. Harper and J. Nelson, *J. Gen. Microbiol.*, **128**, 1667 (1982).
129. M. A. Rosen, K. J. F. Farnden, E. E. Conn and K. R. Hanson, *J. Biol. Chem.*, **250**, 8302 (1975).
130. R. H. Davis and A. Nahrstedt, *Insect. Biochem.*, **17**, 689 (1987).
131. J. E. Poulton and S. I. Shin, *Z. Natursforsch.*, *Ser. C*, **38**, 369 (1983).
132. W. Hoesel and O. Schiel, *Arch. Biochem. Biophys.*, **229**, 177 (1984).
133. R. F. Reay and E. E. Conn, *J. Biol. Chem.*, **249**, 5826 (1974).
134. A. Nahrstedt, W. Hoesel and A. Walther, *Phytochemistry*, **18**, 1137 (1979).
135. J. W. Jarowszewski and M. G. Ettlinger, *Phytochemistry*, **20**, 819 (1981).
136. W. Hoesel, in *Cyanide in Biology*, see Reference 124, p. 232.
137. W. Hoesel, in *Biochemistry of Plants* (Eds. P. K. Stumpf and E. E. Conn), Vol. 7, Academic Press, Orlando, 1987, p. 725.
138. W. A. Bough and J. E. Gander, *Phytochemistry*, **10**, 77 (1971).
139. P. Boersma, P. Kakej and A. W. Schram, *Acta Bot. Nederl.*, **32**, 39 (1983).
140. M. A. Hughes and A. M. Dunn, *Plant Mol. Biol.*, **1**, 169 (1982).
141. P. A. Lizotta and J. E. Poulton, *Plant Physiol.*, **86**, 322 (1988).
142. S. S. Thayer and E. E. Conn, *Plant Physiol.*, **67**, 617 (1981).
143. J. P. Lewis, *West. Med. J.*, **127**, 55 (1977).
144. E. E. Conn, in *Secondary Plant Products* (Eds. E. A. Bell and B. V. Charlwood), Springer, Berlin, 1980, p. 461.
145. W. A. Bough and J. E. Gander, *Phytochemistry*, **10**, 67 (1971).
146. S. G. Blumenthal, G. W. Butte and E. E. Conn, *Nature (London)*, **197**, 718 (1963).
147. C. Ressler, *J. Biol. Chem.*, **237**, 733 (1962).
148. P. A. Castric and E. E. Conn, *J. Bacteriol.*, **108**, 132 (1971).
149. E. A. Bell, in *Phytochemical Ecology* (Ed. J. B. Harborne), Academic Press, London, 1972, p. 163.
150. T. N. Akopyan, A. E. Braunstein and E. V. Goryachenkova, *Proc. Natl. Acad Sci. U.S.A.*, **72**, 1617 (1975).
151. P. A. Castric, K. J. F. Farnden and E. E. Conn, *Arch. Biochem. Biophys.*, **152**, 62 (1972).
152. D. S. Siegel and P. W. Price, *Am. Naturalist*, **110**, 101 (1976).
153. D. Selmer, R. Lieberei and B. Biehl, *Plant Physiol.*, **86**, 711 (1988).
154. D. Selmer, S. Grocholewski and D. S. Seigler, *Plant Physiol.*, **93**, 631 (1990).
155. A. J. Hruska, *J. Chem. Ecol.*, **14**, 2213 (1988).
156. H. Yanase, T. Sakai and K. Tonomura, *Agric. Biol. Chem.*, **46**, 2925 (1982).
157. A. Nahrstedt, *Plant Syst. Ecol.*, **150**, 35 (1985).
158. J. B. Harborne, *Nat. Prod. Reports*, **6**, 80 (1989).
159. K. Wittholm and C. M. Naumann, *J. Chem. Ecol.*, **13**, 1789 (1987).
160. D. S. Seigler and W. Karahara, *Biochem. Syst. Ecol.*, **4**, 263 (1976).
161. P. J. Scheur, *Acc. Chem. Res.*, **25**, 433 (1992).
162. K. L. Mikolajczak, *Prog. Chem. Fats Other Lipids*, **15**, 97 (1977).
163. A. Ahmad, I. Ahmad and W. M. Osman, *Chem. Ind.* (*London*), 734 (1985).
164. D. S. Seigler, *Biochem. Syst. Ecol.*, **4**, 235 (1976).
165. K. L. Mikolaczak, D. S. Seigler, C. R. Smith, F. A. Wolff and R. B. Bates, *Lipids*, **4**, 617 (1969).
166. K. L. Mikolaczak, C. R. Smith, and L. W. Tjarks, *Lipids*, **5**, 812 (1970).
167. D. S. Seigler and C. S. Butterfield, *Phytochemistry*, **15**, 842 (1976).

734   D. V. Banthorpe

168. B. Vennesland, E. E. Conn, C. J. Knowles, J. Westley and F. Wissing (Eds.), *Cyanide in Biology*, Academic Press, London, 1981.
169. Many authors in *Ciba Found. Symp.*, **140** (1988).
170. A. E. Ahmed and N. M. Trieff, *Prof. Drug Metab.*, **7**, 229 (1983).
171. W. Heubel, *Dtsch. Apoth. Ztg.*, **121**, 863 (1981).
172. T. A. Geissman and D. H. G. Crout, *Organic Chemistry of Secondary Plant Metabolism*, Freeman-Cooper, San Francisco, 1969, p. 205.
173. M. Nieder and L. Hager, *Arch. Biochem. Biophys.*, **240**, 121 (1985).
174. P. R. Berquist and R. J. Wells, in *Marine Natural Products* (Ed. P. J. Scheur), Vol. 5, Academic Press, New York, 1983, p. 80.
175. S. Oman, C. Alberti, T. Fanni and P. Crews, *J. Org. Chem.*, **53**, 5971 (1988).
176. A. E. Ahmed, in *Bioactive Foreign Compounds* (Ed. M. W. Anderson), Academic Press, Orlando, 1985, p. 485.
177. L. A. Thompson, C. J. Knowles, E. A. Linton and J. M. Wyatt, *Chem. Brit.*, **24**, 900 (1988).
178. T. Nagasawa and H. Yamada, *Pure Appl. Chem.*, **62**, 1441 (1990).
179. Y. Satoh and K. Enomoto, *Agric. Biol. Chem.*, **51**, 3193 (1987).
180. Nitto Chem Co. Ltd, *Jap. Patent*, 59, 02693 (1984); *Chem. Abstr.*, **100**, 137399 (1984).
181. T. A. Marant, Y. Vered and Z. Bohak, *Biotechnol. Appl. Biochem.*, **11**, 49 (1989).
182. K. Enomoto and Y. Sato, *Eur. Patent*, 187681 (1986); *Chem. Abstr.*, **105**, 151552 (1986).
183. O. B. Aitaurova, T. E. Pagorelova, O. R. Formina, I. N. Polyakova and A. S. Yanenko, *Biotekhnologiya*, 10 (1991).
184. H. Yamada, Y. Asano, T. Hino and Y. Tani, *J. Ferment. Technol.*, **57**, 8 (1979).
185. M. Kuwahara, H. Yanase, Y. Ishida and Y. Kikuchi, *J. Ferment. Technol.*, **58**, 572 (1980).
186. T. P. Abbott and L. K. Nakamura, *US Patent*, 557827 (1990); *Chem. Abstr.*, **114**, 137922 (1991).
187. Y. Tani, M. Kurihara and H. Nishire, *Agric. Biol. Chem.*, **53**, 3151 (1989).
188. D. B. Harper, *Int. J. Biochem.*, **17**, 677 (1985).
189. M. Kobayashi, N. Yanaka, T. Nagasawa and H. Yamada, *J. Bacteriol.*, **172**, 4807 (1990).
190. P. Beppu, S. Horinouchi, O. Ikehata and T. Endo, *French Patent*, 2633938 (1990); *Chem. Abstr.*, **114**, 18970 (1990).
191. M. Nishiyama, S. Horinouchi, M. Kobayashi and T. Nagasawa, *J. Bacteriol.*, **173**, 2465 (1991).
192. H. Fradet, A. Arnand, G. Rios and P. Galzy, *Biotechnol. Bioeng.*, **27**, 1581 (1985).
193. V. Klages and A. Weber, *PCT Int. Patent*, W089-10969 (1989); *Chem. Abstr.*, **113**, 57454 (1990).
194. R. E. Harris, A. W. Bunch and C. J. Knowles, *Sci. Prog. (Oxford)*, **71**, 293 (1987).
195. M. S. Nawaz, T. M. Heinze and C. E. Cerniglia, *Appl. Environ. Microbiol.*, **58**, 27 (1992).
196. K. Buc and A. Arnand, *Enzyme Microbiol. Technol.*, **4**, 115 (1982).
197. N. Nazly and C. J. Knowles, *Biotechnol. Lett.*, **3**, 368 (1981).
198. M. S. Edenborough and R. B. Herbert, *Nat. Prod. Reports*, **5**, 229 (1988).
199. W. Rothe, *Dtsch. Med. Wochenschr.*, **79**, 1080 (1954).
200. C. Takahashi, S. Sekita, K. Yoshihira and S. Natori, *Chem. Pharm. Bull.*, **24**, 2317 (1976).
201. J. Itoh *et al.*, *J. Antibiot.*, **43**, 456 (1990).
202. M. Ishida, T. Hamasaki and Y. Hatsuda, *Agric. Biol. Chem.*, **39**, 2181 (1975).
203. T. R. Pyke and A. Dietz, *Appl. Microbiol.*, **14**, 506 (1966).
204. M. Nobuhara *et al.*, *Chem. Pharm. Bull.*, **24**, 832 (1976).
205. W. D. Ollis, M. Rey, W. O. Godffredsen, N. Rastrup-Anderson, S. Vangedal and T. J. King, *Tetrahedron*, **36**, 515 (1980).
206. M. Fujiwara, A. Fujiwara and T. Okuda, *Jap. Patent*, 53-15345 (1978); *Chem. Abstr.*, **88**, 26051 (1978).
207. J. E. Baldwin, R. M. Adlington, J. Chandrogiani, M. S. Edenborough, J. W. Keeping and C. B. Ziegler, *J. Chem. Soc., Chem. Commun.*, 816 (1985).
208. L. D. Boeck, M. M. Hoehn, T. H. Sands and R. W. Wetzel, *J. Antibiot.*, **31**, 19 (1978).
209. G. G. Marconi, B. B. Molley, R. Nagarajan, J. W. Martin, J. B. Deeter and J. L. Occolorite, *J. Antibiot.*, **31**, 27 (1978).
210. J. R. Evans, E. J. Napier and P. Yates, *J. Antibiot.*, **29**, 850 (1976).
211. J. A. Marquez *et al.*, *J. Antibiot.*, **36**, 1101 (1983).
212. J. J. Wright *et al.*, *J. Chem. Soc., Chem. Commun.*, 1188 (1982).
213. R. B. Herbert and J. Mann, *Tetrahedron Lett.*, **25**, 4263 (1984).
214. H. Grisebach and F. Konig, *Chem. Ber.*, **105**, 784 (1972).
215. A. Roemer and R. B. Herbert, *Z. Naturforsch., Sect. C.*, **36**, 37 (1982).

216. R. B. Herbert and J. Mann, *Tetrahedron Lett.*, **28**, 3159 (1987).
217. K. M. Cable, R. B. Herbert and J. Mann, *Tetrahedron Lett.*, **28**, 3159 (1987).
218. S. Pfeifer, H. Baer and J. Zarnack, *Pharmazie*, **27**, 536 (1972).
219. J. E. Baldwin *et al.*, *J. Chem. Soc., Chem. Commun.*, 1227 (1981).
220. J. E. Baldwin *et al.*, *Tetrahedron*, **41**, 1931 (1985).
221. R. J. Parry and H. P. Buu, *Tetrahedron Lett.*, **23**, 1435 (1982).
222. J. J. K. Wright, A. B. Cooper, A. T. McPhail, Y. Merrill, T. Nagabhoshan and M. S. Puar, *J. Chem. Soc., Chem. Commun.*, 1188 (1982).
223. P. M. Dewick, *Nat. Prod. Reports*, **3**, 565 (1986).
224. M. S. Puar, H. Munayyer, V. Hedge, B. K. Lee and J. A. Waitz, *J. Antibiot.*, **38**, 350 (1985).
225. V. Bornemann, G. M. L. Patterson and R. E. Moore, *J. Am. Chem. Soc.*, **110**, 2339 (1988).
226. F. Cafieri, E. Fattorusso, S. Magno, C. Santacroce and D. Sica, *Tetrahedron*, **29**, 4259 (1973).
227. E. Fattorusso, S. Magno, L. Mayol, C. Santacroce and D. Sica, *Tetrahedron*, **31**, 369 (1975).
228. E. Fattorusso, S. Magno, L. Mayol, C. Santacroce and D. Sica, *Tetrahedron*, **30**, 3911 (1974).
229. B. Di Blasio *et al.*, *Tetrahedron*, **32**, 473 (1976).
230. D. Caine and H. Deutsch, *J. Am. Chem. Soc.*, **100**, 8030 (1978).
231. B. J. Burreson, C. Christophersen and P. J. Scheuer, *J. Am. Chem. Soc.*, **97**, 201 (1975).
232. L. Minale, R. Riccio and G. Sodano, *Tetrahedron*, **30**, 1341 (1974).
233. P. Ciminiello, E. Fattorusso, S. Magno and L. Mayol, *J. Org. Chem.*, **49**, 3949 (1984).
234. C. W. J. Chang, A. Patna, D. M. Roll, P. J. Scheuer, G. K. Matsumato and J. Clardy, *J. Am. Chem. Soc.*, **106**, 4644 (1984).
235. A. Jengo, L. Mayol and C. Santacroce, *Experientia*, **33**, 11 (1977).
236. B. J. Burreson, C. Christophersen and P. J. Scheuer, *Tetrahedron*, **31**, 2015 (1975).
237. E. J. Corey and R. A. Magriotis, *J. Am. Chem. Soc.*, **109**, 287 (1987).
238. P. Kazlauskas, P. T. Murphy, R. J. Wells and J. F. Blount, *Tetrahedron Lett.*, **21**, 315 (1980).
239. B. J. Burreson, P. J. Scheuer, J. Finer and J. Clardy, *J. Am. Chem. Soc.*, **97**, 4763 (1975).
240. M. R. Hagadone, B. J. Burreson, P. J. Scheuer, J. S. Finer and J. Clardy, *Helv. Chim. Acta*, **62**, 2484 (1979).
241. J. E. Thompson, R. P. Walker, S. J. Wratten and D. J. Faulkner, *Tetrahedron*, **38**, 1865 (1982).
242. N. K. Gulavita *et al.*, *J. Org. Chem.*, **51**, 5136 (1986).
243. M. J. Garson, *Nat. Prod. Reports*, **6**, 143 (1989).
244. D. J. Faulkner, *Nat. Prod. Reports*, **3**, 1 (1986).
245. A. Iengo, S. Santacroce and G. Sodano, *Experientia*, **35**, 10 (1979).
246. M. R. Hagadone, P. J. Scheuer and A. Holm, *J. Am. Chem. Soc.*, **106**, 2447 (1984).
247. M. Garson, *J. Chem. Soc., Chem. Commun.*, 35 (1986).
248. L. Minale, in *Marine Natural Products* (Ed. P. J. Scheuer), Vol. 1, Academic Press, London, 1978, p. 204.
249. J. T. Baker, R. J. Wells, W. E. Oberhaensli and G. B. Hawes, *J. Am. Chem. Soc.*, **98**, 4010 (1976).
250. R. Kazlauskas, P. J. Murphy, R. J. Wells and J. F. Blount, *Tetrahedron Lett.*, **21**, 315 (1980).
251. S. J. Wratten and D. J. Faulkner, *J. Am. Chem. Soc.*, **99**, 7367 (1977).
252. S. J. Wratten, *Tetrahedron Lett.*, 1395 (1978).
253. S. Cohen and E. Oppenheimer, in *Chemistry of Cyanates and Their Thio Derivatives* (Ed. S. Patai), Wiley, London, 1977, p. 923.
254. M. Benn, *Pure Appl. Chem.*, **49**, 197 (1977).
255. H. Y. Ye, D. J. Faulkner, J. S. Shomsky, K. Hong, and J. Clardy, *J. Org. Chem.*, **54**, 2511 (1989).
256. E. W. Underhill, in Reference 46, p. 493.
257. E. E. Conn, *ACS Symp. Ser.*, **380**, 143 (1988).
258. G. F. Spencer and M. E. Dexenbichler, *J. Sci. Food. Agric.*, **31**, 359 (1980).
259. X. Hasapis and A. J. Macleod, *Phytochemistry*, **21**, 291 (1982).
260. X. Hasapis and A. J. Macleod, *Phytochemistry*, **21**, 559 (1982).
261. V. Gil and A. J. Macleod, *Phytochemistry*, **19**, 1657 (1980).
262. V. Gil and A. J. Macleod, *Phytochemistry*, **19**, 1365 (1980).
263. B. Lonnerdal and J. C. Janson, *Biochim. Biophys. Acta*, **315**, 421 (1973).
264. R. Bjorkmann and J. C. Janson, *Biochim. Biophys. Acta*, **276**, 508 (1972).
265. R. Bjorkmann, in *Biochemistry and Chemistry of the Cruciferae* (Eds. J. G. Vaughan, A. J. Macleod and B. M. G. Jones) Academic Press, London, 1976, p. 191.
266. D. A. Jones, in *Phytochemical Ecology* (Ed. J. B. Harborne), Academic Press, London, 1972, p. 103.

267. J. B. Harborne, *Introduction to Ecological Chemistry*, 2nd Edn., Chapman and Hall, London, 1984, p. 212.
268. J. B. Harborne, *Nat. Prod. Reports*, **6**, 302 (1989).
269. G. H. N. Towers, *Prog. Phytochem.*, **6**, 183 (1980).
270. J. B. Harborne, in Reference 267, p. 105.
271. G. Campbell, J. D. H. Campbell, T. Arnason and G. H. N. Towers, *J. Chem. Ecol.*, **8**, 961 (1982).
272. J. Hargreaves, J. W. Mansfield, D. T. Coxon and K. R. Price, *Phytochemistry*, **15**, 1119 (1976).
273. C. J. W. Brooks and D. G. Watson, *Nat. Prod. Reports*, **2**, 427 (1985).
274. J. Hargreaves, J. W. Mansfield, D. T. Coxon and K. R. Price, *Phytochemistry*, **15**, 1126 (1976).
275. P. S. Marshall, J. B. Harborne and G. S. King, *Phytochemistry*, **26**, 2493 (1987).
276. J. B. Harborne, in Reference 270, p. 317.
277. B. E. Ellis, *Nat. Prod. Reports*, **5**, 581 (1988).
278. H. Flores, *Chem. Ind.* (London), 374 (1992).
279. F. DiCosmo, R. Norton and G. H. N. Towers, *Naturwissenschaften*, **69**, 550 (1982).
280. U. Mukundan and M. A. Hjortso, *Appl. Microbiol. Biotechnol.*, **33**, 145 (1990).
281. M. Takasugi, S. Kawashima, N. Katsui and A. Shirata, *Phytochemistry*, **26**, 2957 (1987).
282. H. Greger, *Phytochemistry*, **18**, 1319 (1979).
283. K. G. Tietjen and U. Matern, *Arch. Biochem. Biophys.*, **229**, 136 (1984).
284. P. J. G. M. DeWitt and E. Kodde, *Physiol. Plant*, **81**, 143 (1981).
285. D. M. Elgersma and J. C. Overeem, *J. Plant Physiol.*, **87**, 69 (1981).
286. S. Imoto and Y. Ohta, *Plant Physiol.*, **86**, 176 (1988).
287. E. Staedler and H. R. Buser, *Experientia*, **40**, 1157 (1984).
288. K. Yano, *Insect Biochem.*, **16**, 717 (1986).
289. M. D. Lee *et al.*, *Acc. Chem. Res.*, **24**, 235 (1991).
290. K. Saitoh, T. Miyaki, H. Yamamoto and N. Oda, *Eur. Patent*, 484856 (1992); *Chem. Abstr.*, **117**, 149440 (1992).
291. K. Shiomi *et al.*, *J. Antibiot.*, **43**, 100 (1990).
292. O. D. Hensens and I. H. Goldberg, *J. Antibiot.*, **42**, 761 (1989).
293. O. D. Hensens, J. L. Grincer and I. H. Goldberg, *J. Am. Chem. Soc.*, **111**, 3295 (1989).
294. E. G. Cosio and G. H. N. Towers, in *Cell Culture and Somatic Cell Genetics of Plants* (Eds. I. K. Vasil and F. Constabel), Vol. 5, Wiley, Chichester, 1988, p. 495.
295. G. B. Lockwood, in References 294, p. 467.
296. R. A. Norton and G. H. N. Towers, *J. Plant Physiol.*, **120**, 273 (1985).
297. R. A. Norton, A. J. Finlayson and G. H. N. Towers, *Phytochemistry*, **24**, 719 (1985).
298. K. Ishimaru, H. Yonemitsu and K. Shimomura, *Phytochemistry*, **30**, 2255 (1991).
299. M. Menelaou *et al.*, *Spectrosc. Lett.*, **24**, 1405 (1991).
300. Y. Y. Marchant, *Bioactiv. Mol.*, **7**, 217 (1988).
301. H. Flores, J. J. Pickard and M. W. Hoy, *Bioactiv. Mol.*, **7**, 233 (1988).
302. A. F. Croes, M. Bosveld and G. J. Wullems, *Bioactiv. Mol.*, **7**, 255 (1988).
303. D. A. van Fleet, *Adv. Front. Plant Sci.*, **26**, 109 (1970).
304. B. Garrod and B. G. Lewis, *Trans Brit. Mycol. Soc.*, **72**, 515 (1979).
305. A. F. Croes, A. M. Aarts, M. Bosveld, H. Breteler and C. J. Wullems, *Physiol. Plant.* **76**, 205 (1989).
306. A. F. Croes, A. M. Aarts, M. Bosveld, H. Breteler and C. J. Wullems, *Physiol. Plant.* **76**, 205 (1989).
307. Y. Fujimoto and H. Satah, *Phytochemistry*, **26**, 2850 (1987).
308. S. C. Shim, S. K. Chang, C. W. Hur and C. K. Kim, *Phytochemistry*, **26**, 2849 (1987).
309. Y. Fujimoto and M. Sato, *Chem. Pharm. Bull.*, **36**, 4206 (1988).
310. Y. Y. Marchant, F. R. Ganders, C. K. Wat and G. H. N. Towers, *Biochem. Syst. Ecol.*, **12**, 167 (1984).
311. B. Tosi, G. Lodi, F. Dandi and A. Bruni, *Bioactiv. Mol.*, **7**, 209 (1988).
312. N. A. Sorensen, in *Chemical Plant Taxonomy* (Ed. T. Swain), Academic Press, London, 1963, p. 219.
313. N. A. Sorensen, *Recent Adv. Phytochemistry*, **1**, 187 (1968).
314. P. M. Smith, *Chemotaxonomy of Plants*, Arnold, London, 1976.
315. P. S. Marshall, J. B. Harborne and G. S. King, *Phytochemistry*, **26**, 2493 (1987).
316. K. Downum, D. Provost and L. Swain, *Bioactiv. Mol.*, **7**, 151 (1988).
317. K. L. Stevens, S. C. Witt, and C. E. Turner, *Biochem. Syst. Ecol.*, **18**, 229 (1990).

318. P. Pathak, F. Bohlmann, R. M. King and H. Robinson, *Rev. Latinoam. Quim.*, **18**, 28 (1987).
319. A. G. Gonzalez, B. Bermejo, J. G. Diaz and P. P. Depaz, *Biochem. Syst. Ecol.*, **16**, 17 (1987).
320. P. G. Waterman and A. I. Gray, *Nat. Prod. Reports*, **4**, 175 (1987).
321. R. Hegnauer, *Chemotaxonomie der Pflanzen*, Vol. 1–7, Birkhäuser, Basel, 1962–85.
322. J. B. Harborne, *Phytochemical Methods*, 2nd Edn., Chapman and Hall, London, 1984, p. 142.
323. R. Hegnauer, *Plant System. Ecol.*, Supplement 1, 191 (1977).
324. R. Hegnauer, *Biochem. System.*, **1**, 191 (1973).
325. K. S. Spencer, D. S. Seigler and S. W. Frayley, *Biochem. Syst. Ecol.*, **13**, 433 (1985).
326. D. S. Seigler and W. Kavajara, *Biochem. Syst. Ecol.*, **4**, 263 (1976).
327. B. R. Maslin, E. E. Conn and J. E. Dunn, *Phytochemistry*, **24**, 961 (1985).
328. B. R. Maslin, J. E. Dunn and E. E. Conn, *Phytochemistry*, **27**, 421 (1988).
329. E. E. Conn, D. S. Seigler, B. R. Maslin and J. E. Dunn, *Phytochemistry*, **28**, 817 (1989).
330. S. R. Jensen and B. J. Nielsen, *Phytochemistry*, **25**, 2349 (1986).
331. V. Schwarzmaier, *Chem. Ber.*, **109**, 3379 (1976).
332. H. G. Nass, *Crop. Sci.*, **12**, 503 (1972).
333. D. S. Seigler, *Eco. Bot.*, **30**, 395 (1976).
334. E. E. Conn, in *Secondary Plant Products* (Eds. E. A. Bell and B. V. Charlwood), Springer, Berlin, 1980, p. 461.
335. P. D. Ramani and D. A. Jones, *Pak. J. Bot.*, **16**, 145 (1984).
336. R. Tschiersch and P. Hanelt, *Flora (Jena)*, **157**, 389 (1967).
337. C. Cimino, A. Crispino, V. Dimarzo, M. Govagnir and J. D. Ros, *Experientia*, **46**, 767 (1990).
338. S. W. Ayer and R. J. Anderson, *J. Org. Chem.*, **49**, 3653 (1984).
339. T. Y. S. Nozoe, S. Chang and T. Toda, *Tetrahedron Lett.*, 3663 (1966).
340. T. Akihisa, T. Tamura, T. Matsumato, W. Kokke and T. Yokota, *J. Org. Chem.*, **54**, 606 (1989).
341. O. Straub, *Key to Carotenoids*, Birkhäuser, Basel, 1987.
342. C. R. Kass and K. Bloch, *Proc. Natl. Acad. Sci. U.S.A.*, **58**, 1168 (1967).
343. F. Miesowicz and K. Bloch, *Biochem. Biophys. Res. Commun.*, **65**, 331 (1975).
344. K. L. Kunze, B. L. K. Mangold, C. Wheeler, H. S. Beilan and P. R. Ortiz de Montellano, *J. Biol. Chem.*, **258**, 4202 (1983).
345. M. A. Mascarua-Esparza, R. Villa-Gondalez and J. Mellado-Caballero, *Plant Soil*, **106**, 91 (1988).
346. L. C. Davies, *Ann. Bot.*, **61**, 179 (1988).
347. F. D. H. MacDowall and G. T. Kristjansson, *Can. J. Bot.*, **67**, 360 (1989).
348. A. Sellstedt and L. J. Wirship, *Plant Physiol.*, **94**, 91 (1990).
349. P. R. Ortiz de Montellano, in *Bioactive Foreign Compounds* (Ed. M. W. Anderson), Academic Press, Orlando, 1985, p. 157.
350. D. Deutermann-Hodgeson, in *Biochemistry of Insects* (Ed. M. Rockstein), Academic Press, London, 1978, p. 541.
351. J. B. Harborne (Ed.), *Biochemical Aspects of Plant and Animal Coevolution*, Academic Press, London, 1978.

CHAPTER **13**

Pharmacology of acetylenic derivatives

ZVI BEN-ZVI and ABRAHAM DANON

Department of Clinical Pharmacology, Ben-Gurion University of the Negev and Soroka Medical Center, Beer Sheva, Israel

Supplement C2: The chemistry of triple-bonded functional groups
Edited by S. Patai © 1994 John Wiley & Sons Ltd

I. INTRODUCTION

Naturally occurring alkynes and polyines are relatively common in the plant kingdom. Several hundreds of acetylenic derivatives containing 8–18 carbon atoms with very diversified chemical structures, some with biological activities, have been identified[1-6]. For example, the polyacetylenes isolated from Panax ginseng, panaxynol, and its oxidized derivatives panaxydiol, panaxytriol and chloropanaxydiol exhibit cytotoxic activity[7,8]. Panaxynol is also a potent antiplatelet agent[9]. Several polyacetylenes, especially those containing long conjugated unsaturated bond systems such as phenylheptatriyne, thiophene A and thiarubrine A, possess activities against microorganisms which depend on or are augmented by light of certain wavelengths (UVA, UVB)[6,10].

$$CH_2{=}CHCHC{\equiv}CC{\equiv}CCH_2CH{=}CH(CH_2)_6CH_3$$
$$|$$
$$OH$$

Panaxynol

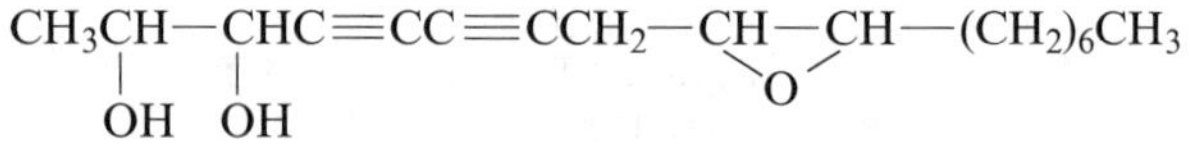

$$CH_3CH{-}CHC{\equiv}CC{\equiv}CCH_2{-}CH{-}CH{-}(CH_2)_6CH_3$$
$$|\quad\ |\qquad\qquad\qquad\quad\ \ \backslash O \diagup$$
$$OH\ \ OH$$

Panaxydiol

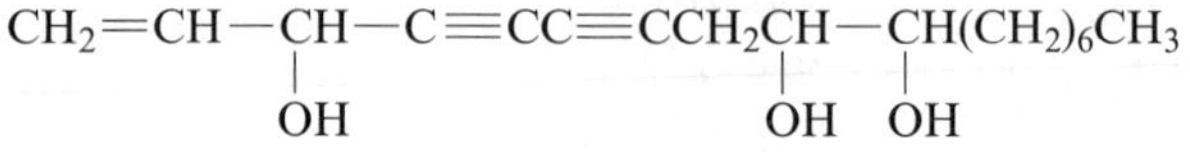

$$CH_2{=}CH{-}CH{-}C{\equiv}CC{\equiv}CCH_2CH{-}CH(CH_2)_6CH_3$$
$$|\qquad\qquad\qquad\quad\ |\quad\ \ |$$
$$OH\qquad\qquad\qquad\ OH\ \ OH$$

Panaxytriol

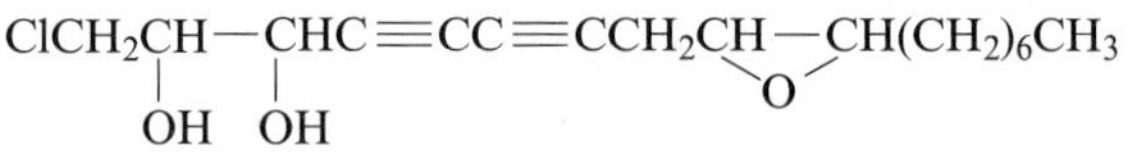

$$ClCH_2CH{-}CHC{\equiv}CC{\equiv}CCH_2CH{-}CH(CH_2)_6CH_3$$
$$|\quad\ \ |\qquad\qquad\qquad\quad\ \backslash O \diagup$$
$$OH\ \ OH$$

Chloropanaxydiol

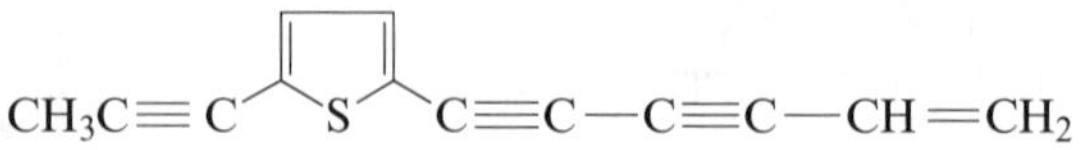

$$CH_3C{\equiv}C{-}\underset{S}{\diagdown}{-}C{\equiv}C{-}C{\equiv}C{-}CH{=}CH_2$$

Thiophene A

$$CH_3C{\equiv}C{-}\underset{S{-}S}{\diagdown}{-}C{\equiv}C{-}C{\equiv}C{-}CH{=}CH_2$$

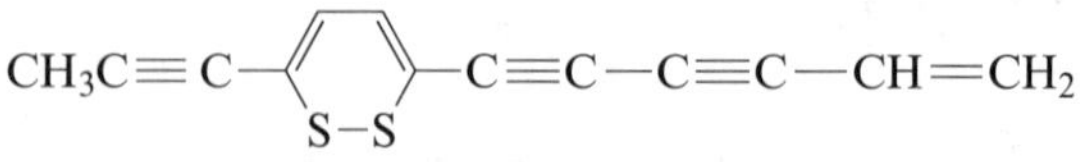

Thiarubrine A

Acetylenes, few of which are biologically active, have also been isolated from micro-organisms and marine organisms, such as the antifungal acetylenic cyclohexene–epoxide derivative asperpentyn isolated from Aspergilus[11] and dactylyne, an acetylenic dibromochloro-ether, isolated from sea hare. Dactylyne was shown to be a very potent inhibitor of drug metabolism[12].

Asperpentyn   Dactylyne

Alkynic drugs are frequently inhibitors of enzyme systems. In some cases acetylenic drugs are competitive inhibitors (see Section VII). In several other cases they exert their biological activities by a special type of irreversible inhibition known as suicide inhibition[13]. The same mechanism has also been referred to as k_{cat} inhibition[14], Trojan horse inhibition[15] and mechanism-based inhibition[16,17]. Suicide inhibitors are relatively inactive molecules but they must have a chemical group that can be activated by the target enzyme, such as a double or triple bond. Initially the suicide inhibitor binds competitively to the target enzyme in the same way as the natural substrate. Subsequently, the inhibitor is converted to an active intermediate, which then binds covalently and irreversibly to the active site of the target enzyme, thus resulting in its deactivation. By contrast, other irreversible inhibitors, such as molecules containing a nitrogen mustard structure, bind covalently to enzymes but are not activated by them. Suicide inhibitors can generally be represented by the following model:

$$E + I \longrightarrow EI \xrightarrow{k_{cat}} EI^{\ddagger} \longrightarrow E—I^{\ddagger} \qquad (1)$$

where E denotes enzyme; I, latent suicide inhibitor; $EI^{\ddagger}$, activated sucide inhibitor–enzyme complex, and $E-I^{\ddagger}$, covalently bound suicide inhibitor. From this model it is evident that the inhibition is a first-order, time-dependent process. Natural substrates can compete with the latent suicide inhibitor before it is activated. However, once the inhibitor is activated, the process cannot be reversed, because the inhibitor does not leave the active site of the enzyme. This intimate association does not even allow nucleophilic scavengers to trap the activated inhibitors.

The major advantage of suicide inhibitors over competitive inhibitors or other types of irreversible inhibitors is their specificity, because mechanism-based inhibitors usually inhibit only the one enzyme or group of enzymes which had activated them[18].

The use of acetylenic compounds as suicide inhibitors can be illustrated in the case of γ-acetylenic GABA. γ-Aminobutyric acid (GABA), an inhibitory neurotransmitter amino acid, modulates various physiological and pathophysiological processes in the central nervous system. Therefore, elevation of brain GABA levels may be beneficial in the treatment of convulsive disease (epilepsy)[19].

GABA is metabolized by the enzyme GABA transaminase (GABA-T) to succinic semialdelhyde[20] (Scheme 1). The first step in the transamination of GABA is the formation of a Schiff base between GABA and pyridoxal phosphate. The Schiff base bond activates the gamma C—H bond of GABA and thus facilitates proton abstraction. Reprotonation leads to the tautomeric form of the Schiff base, which is subsequently hydrolyzed to the products succinic semialdehyde and pyridoxamine phosphate (Scheme 2). Realization of the importance of facilitated deprotonation in this enzymatic pathway has led to the design and synthesis of the tailor-made suicide inhibitor γ-acetylenic GABA.

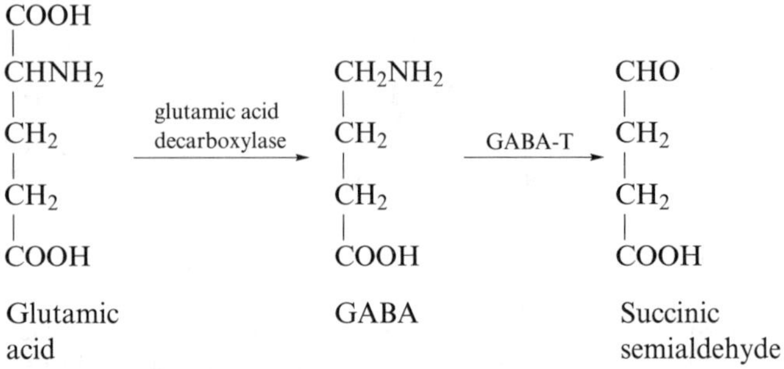

SCHEME 1. Biosynthesis and degradation of GABA

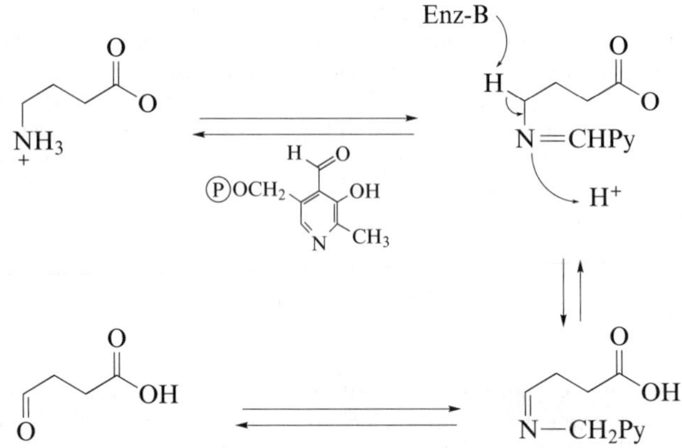

SCHEME 2. Transamination of GABA. Reprinted from Reference 20, 1979, with permission from Pergamon Press Ltd, Headington Hill Hall, Oxford OX3 0BW, UK

Metcalf[20] has suggested two possible mechanistic pathways for the formation of active intermediates of γ-acetylenic GABA following the abstraction of the propargylic proton from the γ-acetylenic GABA-Schiff base (Scheme 3). These activated intermediates could be either an alkyne conjugated to the imine double bond, as illustrated in *path a*, or a conjugated allene (*path b*). Both suggest that the intermediates can alkylate a nucleophilic group in the active site of the enzyme and hence irreversibly inactivate GABA-T.

Indeed, γ-acetylenic GABA was shown to inhibit GABA-T *in vitro* and to elevate the intracellular levels of GABA in rodent brains *in vivo*. Additionally, these experiments indicated that, due to the lipophilic character of γ-acetylenic GABA, its penetration

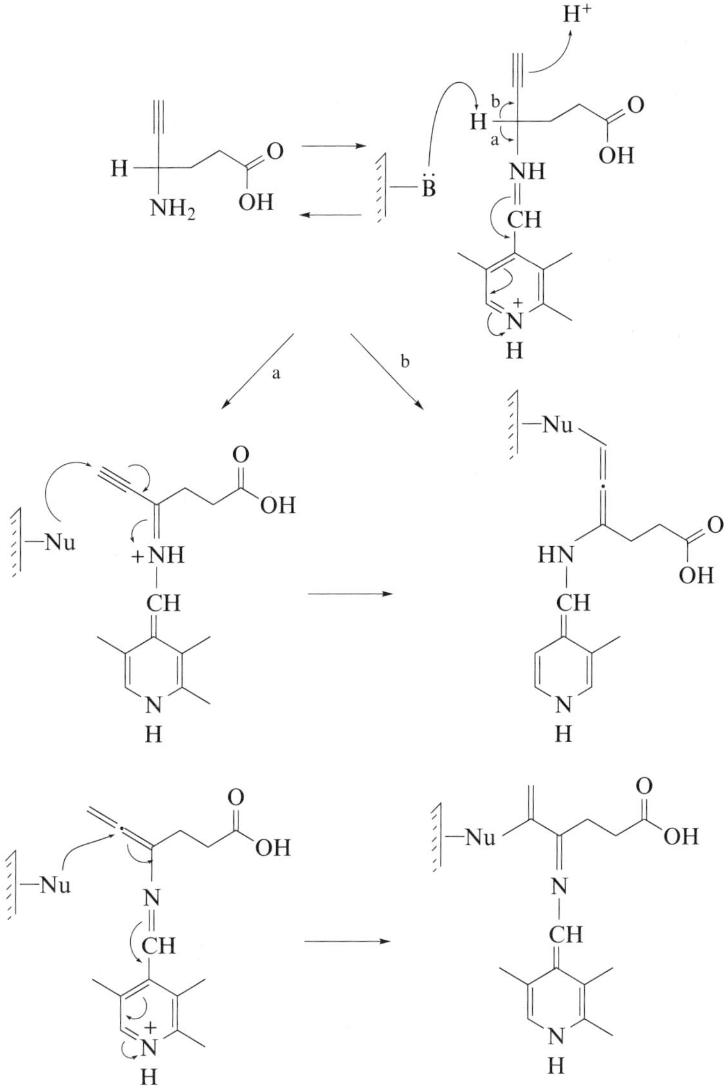

SCHEME 3. Postulated mechanism of GABA-T inhibition by γ-acetylenic GABA. Reprinted from Reference 20, 1979, with permission from Pergamon Press Ltd, Headington Hill Hall, Oxford OX3 OBW, UK

through the blood brain barrier was much better than that of GABA and of another GABA-T suicide inhibitor, vigabatrin (γ-vinyl-GABA)[20], which was recently introduced into clinical use as an antiepileptic drug[21].

Unfortunately, γ-acetylenic GABA also inhibits other pyridoxal phosphate dependent enzymes such as glutamic acid decarboxylase, which catalyzes the formation of GABA from glutamic acid (Scheme 1)[22]. Thus, the inhibition of glutamic acid decarboxylase by γ-acetylenic GABA does not allow the build-up of intracellular GABA levels, making γ-acetylenic GABA an ineffective therapeutic agent.

Other acetylenic derivatives inhibit other enzyme systems by mechanism-based inhibition, for example 3-hydroxydecanoyl thioester dehydratase, lactate dehydrogenase, monoamine oxidase and cytochrome P450[13,14,23,24]. The inhibition of the last two systems by acetylenic compounds will be discussed later in this chapter.

Several alkynic compounds possess specific pharmacological activities by receptor action, although the exact role of the butynyl group in the activity of these drugs is not fully understood. Oxotremorine [1-[4-(1-pyrrolidinyl)-2-butynyl]-2-pyrrolidone], a synthetic drug containing a butynyl group, is a potent muscarinic cholinergic agonist, used only for investigative purposes[25]. The structure–activity relationship of oxotremorine will be discussed in detail in Section VIII. A more selective alkynic muscarinic agonist is McN-A-343 (4-m-chlorphenyl carbamoyloxy)-2-butynyl trimethyl ammonium[26]. Because of its high selectivity for M_1 receptors it is often used for the localization of these receptors in various organs. M_1 receptors are found mainly in the myenteric plexus, peripheral ganglia and in various areas of the brain. Systemic administration of this drug unexpectedly results in an increase in vascular resistance and blood pressure. Due to its high polarity it does not cross the blood brain barrier and is therefore devoid of central activity[27].

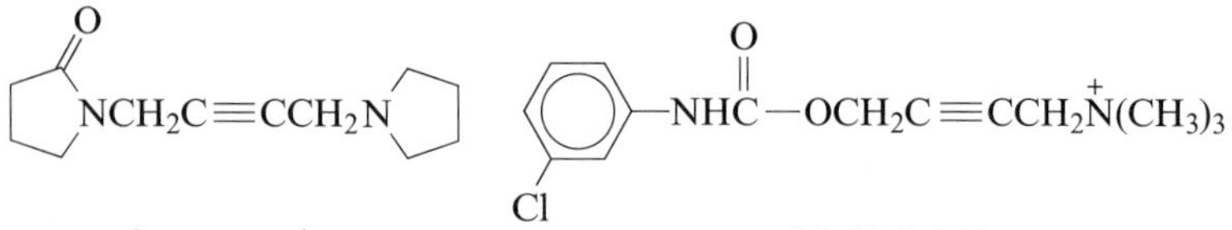

$$\underset{\text{Oxotremorine}}{\text{\Large$\bigcirc$}\!-\!NCH_2C\equiv CCH_2N\!-\!\text{\Large$\bigcirc$}} \qquad \underset{\substack{Cl \\ \text{McN-A-343}}}{\text{\Large$\bigcirc$}\!-\!NHC\!-\!OCH_2C\equiv CCH_2\overset{+}{N}(CH_3)_3}$$

The following pages will dwell on some prominent examples of pharmacologically active acetylenic compounds, and the metabolism of these compounds will also be discussed.

II. METABOLIC DEGRADATION OF ACETYLENIC DRUGS

The major metabolic pathway of terminal acetylenic derivatives is via oxidation to the corresponding acetic acid derivatives. Thus, Sullivan and coworkers[28] and Wade and coworkers[29] showed in 1979 that the major metabolites of ethynylbiphenyls in the rat were biphenyl-4-yl acetic acids. These metabolites are further oxidized to 4'-hydroxybiphenyl-4-yl acetic acids before being excreted in the urine (Scheme 4). Because the *in vitro* metabolism of biphenylacetylenes by rat liver microsomes requires nicotine–adenine–dinucleotide phosphate (NADPH) and molecular oxygen and is inhibited by carbon monoxide, it was concluded that the oxidative metabolism of the acetylenes to the corresponding acetic acid derivatives is mediated by cytochrome P450[28,30]. Acetic acid

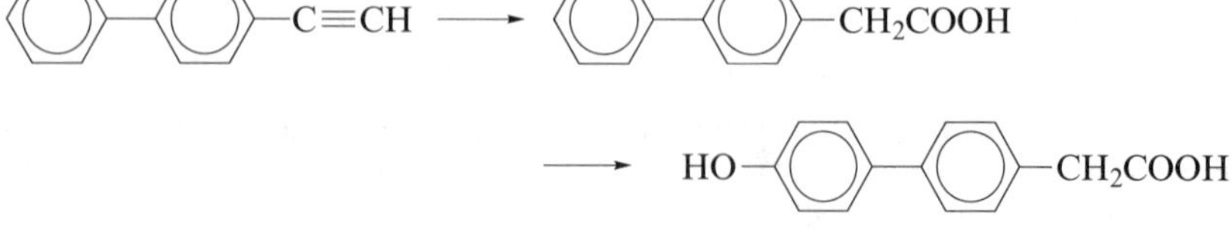

$$\text{\Large$\bigcirc$}\!-\!\text{\Large$\bigcirc$}\!-\!C\equiv CH \longrightarrow \text{\Large$\bigcirc$}\!-\!\text{\Large$\bigcirc$}\!-\!CH_2COOH$$

$$\longrightarrow HO\!-\!\text{\Large$\bigcirc$}\!-\!\text{\Large$\bigcirc$}\!-\!CH_2COOH$$

SCHEME 4. *In vivo* oxidation of acetylenic biphenyl derivatives

metabolites have thus been identified as metabolites of a variety of aromatic acetylenes, such as phenylacetylene[31-33], biphenylacetylene[28-30], acetylenic polycyclic aromatic hydrocarbons, 2-ethynylnaphthalene[34] and 1-ethynylpyrene[33] as well as the aliphatic acetylenes 10-undecynoic acid[35] and 1,2-dichloroacetylene[36].

Since no intermediates could be detected in the *in vitro* studies of the oxidation of acetylenic derivatives, it was assumed that very short-lived intermediates were probably formed. In order to further characterize this pathyway, studies with deuterium and ^{13}C labeled[37-39] biphenylacetylene (acetylenic hydrogen and internal acetylenic carbon, respectively) were initiated. The mass and nuclear magnetic resonance spectra of the resulting biphenyl acetic acid derivative showed that:

(a) The diphenyl moiety was attached to the labeled carbon as in the starting material.
(b) Quantitative 1,2-shift of the deuterium occurred.
(c) There was no oxidation of the nonacetylenic moiety.
(d) Only traces of acetylbiphenyl could be detected.

Parallel chemical oxidation of deuterium labeled biphenylacetylene with metachloroperbenzoic acid yielded the same product as was obtained in the enzymatic reaction[37]. These observations exclude the possible oxidation of the acetylenic terminal C—H bond, because such a reaction would not be compatible with complete conservation of the deuterium atom. Thus, the formation of a ketene derivative was postulated (Scheme 5). However, the nature of the intermediate from which the ketene derivative is formed remains to be elucidated. An oxirene structure was suggested[31,32,37,39] in parallel to alkene epoxidation; however, oxirenes are very unstable species due to electronic and steric factors[40,41]. It is currently believed that a complex between the acetylenic π electrons and cytochrome P450 iron–oxene moiety is formed, leading to the formation of the ketene[32].

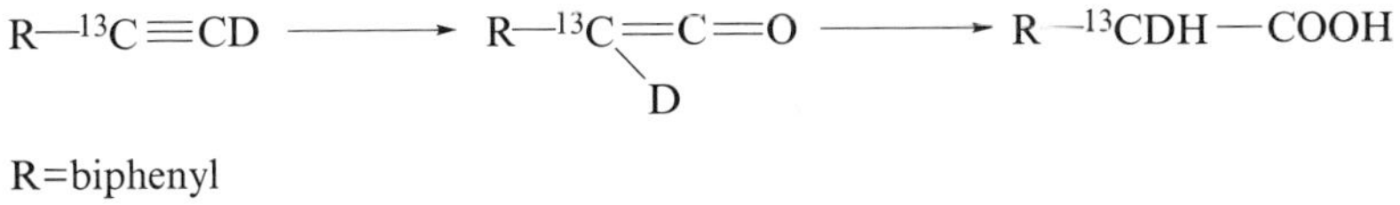

$$R—^{13}C≡CD \longrightarrow R—^{13}\underset{\underset{D}{\diagdown}}{C}=C=O \longrightarrow R —^{13}CDH—COOH$$

R=biphenyl

SCHEME 5. Postulated mechanism of *in vitro* metabolism of acetylenes

White and coworkers[42] showed that 1-octyne is metabolized via a different pathway than the one described above. The intermediary metabolites of 1-octyne bind very quickly to proteins and DNA. However, addition of thiol-containing compounds such as N-acetyl-cysteine diminishes this binding in a concentration-dependent manner. Under these conditions the final metabolite is an N-acetyl-cysteine conjugate of 3-oxo-1-octyne, implicating the latter as an intermediary metabolite (Scheme 6). The high reactivity of the 3-oxo derivative is probably due to its Michael acceptor structure.

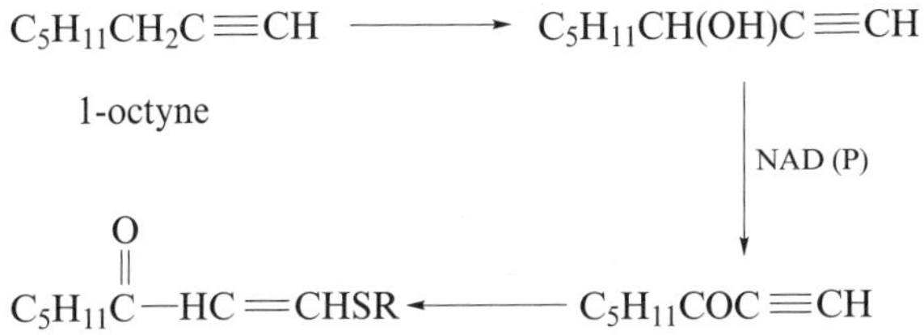

$$C_5H_{11}CH_2C≡CH \longrightarrow C_5H_{11}CH(OH)C≡CH$$

1-octyne

NAD (P)

$$C_5H_{11}\overset{\overset{O}{\|}}{C}—HC=CHSR \longleftarrow C_5H_{11}COC≡CH$$

SCHEME 6. *In vitro* metabolism of 1-octyne

A. Metabolism of 17α-Acetylenic Steroids

Two metabolic pathways have been identified for 17α-acetylenic steroids, namely deethynylation and D-ring homoannulation. Although both detoxification pathways are only of overall minor importance, they seem to be unique for this class of synthetic estrogens and progestins and therefore deserve special consideration.

Studies with ethynyl estrogens in humans *in vivo*[43,44], and with rhesus monkey, baboon and mouse hepatic microsomal preparations[45] showed that small amounts of estrone (and estradiol) are formed via deethynylation. The mechanism of deethynylation is not clear; however, it is known to be mediated by cytochrome P450 (Scheme 7).

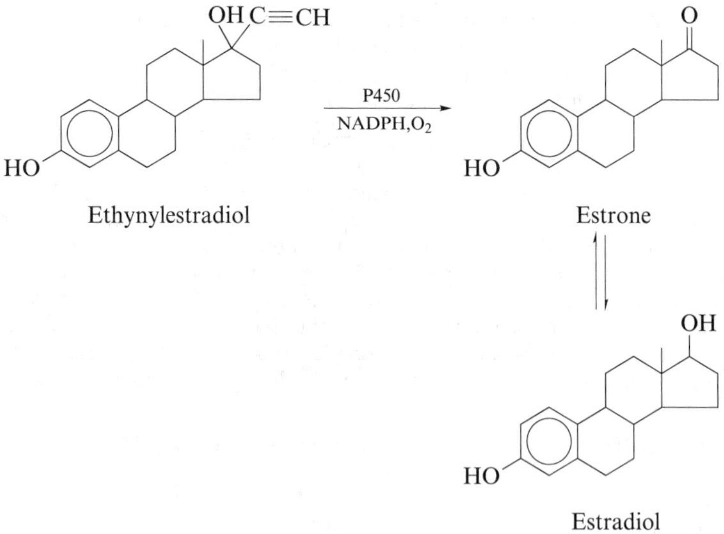

SCHEME 7. Deethynylation of ethynylestradiol

The second pathway, D-ring homoannulation, seems to be of a more general nature than the deethynylation[46–49]. In this pathway, the ethynylcyclopentanol D-ring of the steroid is converted to a cyclohexanone ring, accompanied by formation of carbon dioxide. The latter is derived from the terminal acetylenic carbon, as shown in experiments with labeled steroids. The mechanism suggested for this pathway is depicted in Scheme 8. In the case of norgestrel (for structure, see Table 1A), the acid was isolated under neutral conditions and identified following esterification with diazomethane[50].

B. Conjugation Reactions

Abolin and coworkers[51] showed in 1980 that the hypnotic drug ethchlorvynol undergoes an unusual glucuronidation reaction, namely C-glucuronidation. This type of glucuronidation occurs when a hydrogen attached to a carbon atom has pronounced acidic properties, as in the case of acetylenic hydrogen (Scheme 9). C-glucuronidation has also been observed with few selected nonacetylenic drugs, such as the antiinflammatory drug phenylbutazone and the active constituent of Cannabis, tetrahydrocannabinol[52–54].

Dichloroacetylene, a by-product in the synthesis of trichloroethylene, is conjugated both *in vivo* and *in vitro* with glutathione to yield the 1,2-dichlorovinyl conjugate of glutathione. This conjugate undergoes further hydrolysis to the corresponding cysteine conjugate and

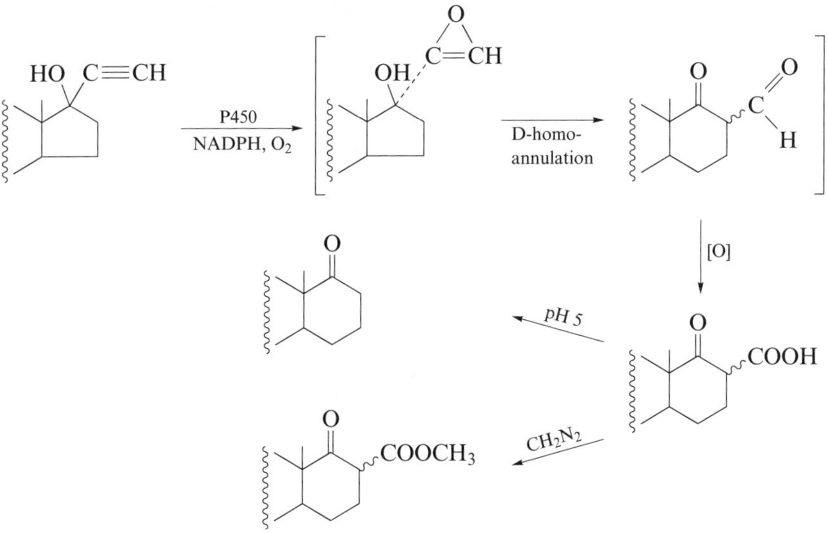

SCHEME 8. Proposed mechanism for D-homoannulation

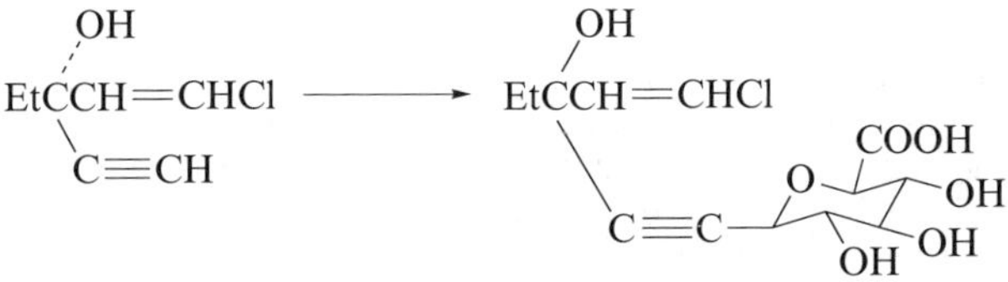

Ethchlorvynol   Ethchlorvynol C-glucuronide

SCHEME 9. Glucuronidation of ethchlorvynol

mercapturic acid[55–57] (Scheme 10). The cysteine as well as the glutathione dichlorovinyl conjugates are of particular toxicological interest, being neuro- and nephrotoxic as well as carcinogenic[56–58].

III. ORAL CONTRACEPTIVE STEROIDS

Estrogens and progesterone are the two female hormones that are synthesized by, and secreted from, the ovary. The natural estrogens, estrone, estradiol and estriol, as well as progesterone, are steroids. They are formed cyclically by the ovary and influence the female reproductive organs, primarily the uterine mucosa, known as the endometrium. Estrogens produce cyclic proliferation of the endometrium during the menstrual cycle, in addition to causing growth and development of the vagina, uterus, Fallopian tubes and breasts and being responsible for other female secondary sex characteristics. Progesterone is secreted from the corpus luteum during the second half of the menstrual cycle and converts the endometrium into its secretory state. It also produces effects on the vagina and endocervical glands. Withdrawal of progesterone levels brings about menstruation. Both estrogens and

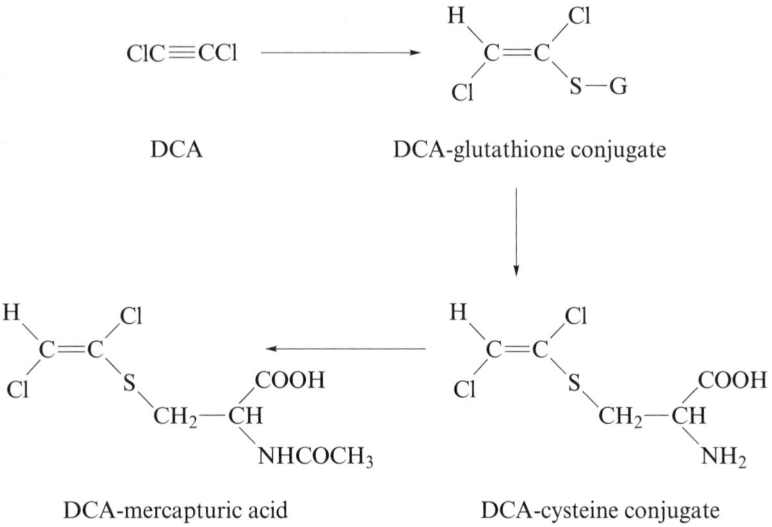

$$ClC\equiv CCl \longrightarrow \overset{H}{\underset{Cl}{\diagdown}}C=C\overset{Cl}{\underset{S-G}{\diagup}}$$

DCA   DCA-glutathione conjugate

$$\overset{H}{\underset{Cl}{\diagdown}}C=C\overset{Cl}{\underset{\underset{\underset{NHCOCH_3}{\diagdown}}{CH_2-CH}}{\underset{S}{\diagup}}\diagdown}\overset{COOH}{\diagup} \longleftarrow \overset{H}{\underset{Cl}{\diagdown}}C=C\overset{Cl}{\underset{\underset{\underset{NH_2}{\diagdown}}{CH_2-CH}}{\underset{S}{\diagup}}\diagdown}\overset{COOH}{\diagup}$$

DCA-mercapturic acid   DCA-cysteine conjugate

SCHEME 10. Metabolism of dichloroacetylene (DCA)

progesterone cause additional metabolic and other actions, some of which will be discussed later as adverse reactions.

Of particular interest is the action of estrogens and progesterone on the hypothalamus and pituitary gland, as this regulates the cyclic secretion of gonadotropins and gonado-tropin-releasing hormone (GnRH) and subsequently of the sex steroids themselves[59]. This action also forms the basis for the development and use of the oral contraceptives.

At puberty, increased secretion of GnRH leads to increased levels of the pituitary gonadotropins follicle stimulating hormone (FSH) and luteinizing hormone (LH)[60]. FSH, in turn, induces growth of ovarian follicles with resultant estrogen secretion. At midcycle the LH secretion peaks and, together with FSH, brings about ovulation. Under the influence of LH the corpus luteum secretes progesterone. The cyclic changes in the secretion of these hormones are regulated by an intricate interplay between the different hormones involving both positive and negative feedback mechanisms.

Although steroidal estrogens and progesterone are relatively well absorbed from the gastrointestinal tract, their oral effectiveness is limited by rapid gut and liver metabolism, the so-called first pass metabolism. The liver is the major site of metabolism of estrogens and progesterone, where they are oxidized to less active or inactive compounds and conjugated with sulfuric and glucuronic acids. Because of rapid metabolism, the half-life of progesterone in plasma is only 5 minutes[61].

A breakthrough in progesterone therapy, and indeed in contraceptive use, occurred with the synthesis in 1951 by Djerassi and coworkers at Syntex of norethindrone which proved an orally effective progestogen[62]. Other 19-nortestosterone compounds followed shortly, such as norethynodrel. The 17-ethynyl substitution provided protection against fast liver metabolism in the same way as in 17α-ethynylestradiol, synthesized in 1938 by Inhoffen and coworkers[63]. The latter compound is also very slowly degraded in the liver, thus providing an orally effective estrogen. The 3-methyl ether of ethynylestradiol, mestranol, is an equally effective oral estrogen.

Following extensive work by Pincus and coworkers on suppression of ovulation in experimental animals by progestational agents, these authors set out in 1955 to carry out

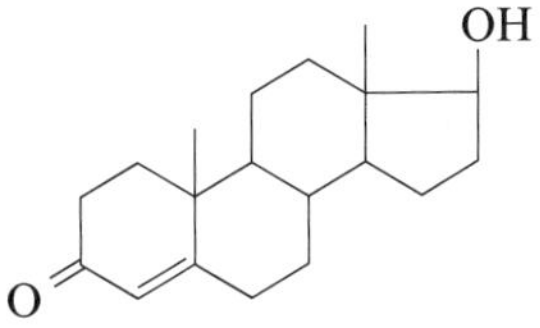

Testosterone

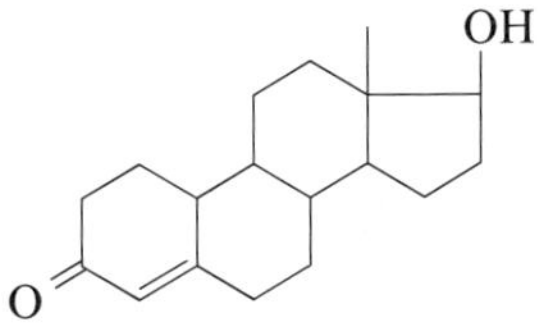

19-Nortestosterone

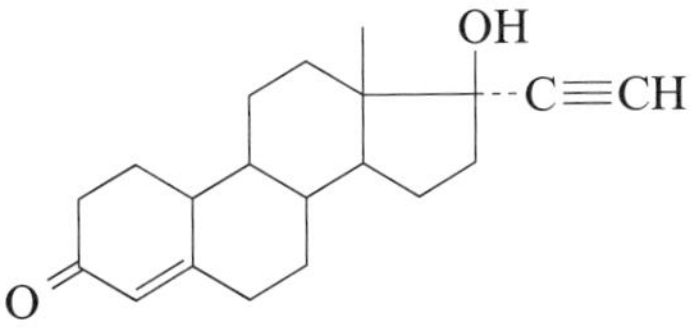

Norethindrone

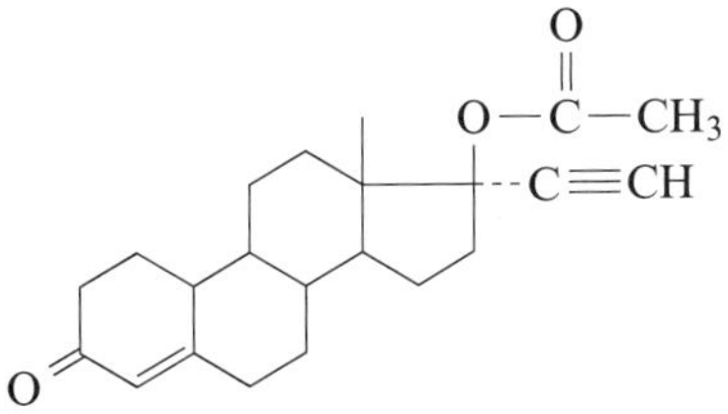

Norethindrone acetate

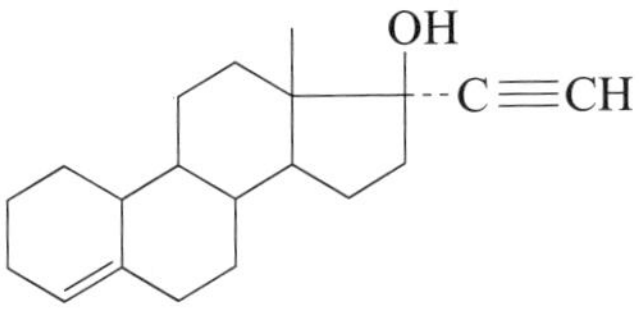

Lynestrenol

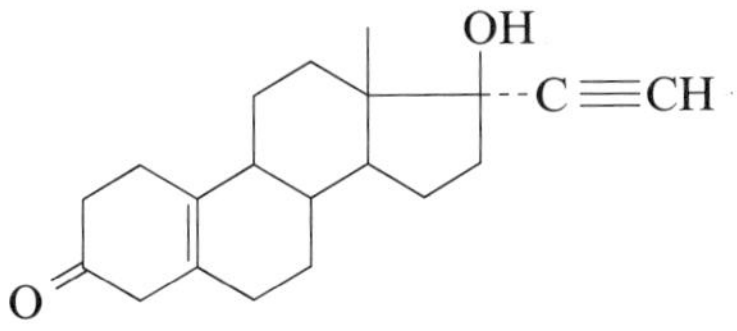

Norethynodrel

the first large scale field studies on oral contraceptives in Puerto Rico[64]. The amazing effectiveness, along with relative safety, of the oral contraceptives paved the way to their

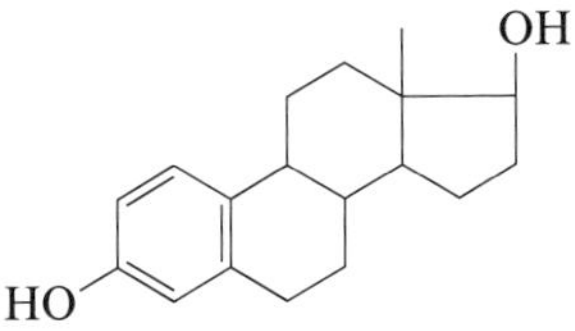

Estradiol

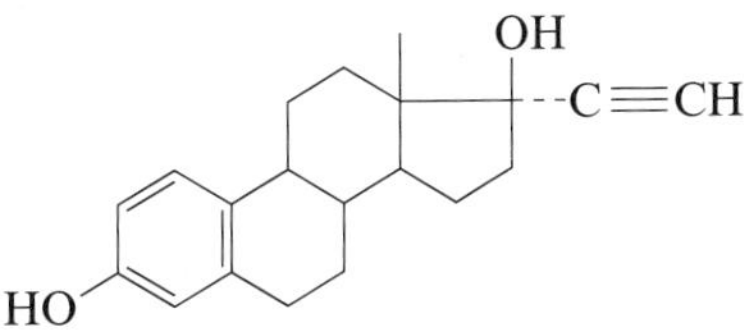

Ethynylestradiol

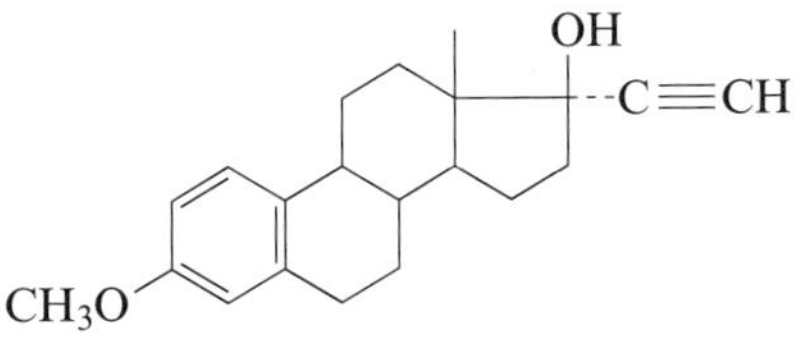

Mestranol

worldwide use over the last 3 decades. The oral contraceptive agents usually comprise an estrogen and a progestogen in different combinations. The most widespread are the combination 'pills', which are taken for 21 days and subsequently omitted for 7 days, before the next cycle starts. In sequential preparations a fixed-dose combination of estrogen and progesterone is taken for several days (usually 10), then a different combination is taken during a second period (biphasic) and posssibly a third combination follows (triphasic). Minipills contain only the progestogen in smaller doses, so as to avoid use of estrogen, which accounts for most of the adverse reactions of the oral contraceptives (see below). Unfortunately, their effectiveness is also reduced.

The estrogenic component of the oral contraceptive agents is almost exclusively ethynylestradiol or mestranol, while currently available progestational steroids are either progesterone derivatives, containing the typical C_{21} structure, or 19-nortestosterone derivatives. Removal of the C_{19} methyl group of testosterone and introduction of the

Levonorgestrel   Norgestimate

Desogestrel   Gestodene

ethynyl group at $C_{17\alpha}$ has resulted in a significant shift of hormonal activity. Thus, norethindrone has a binding affinity to the progesterone receptor of 150% compared with progesterone, while its androgenic activity is reduced to only 15%, compared with metribolone. Norethindrone has practically no affinity for either the estrogenic, glucocorticoid or mineralo-corticoid receptors[61]. A group of 17α-ethynyl-19-nortestosterone progestogens in which the 13-methyl group was replaced by an ethyl group exhibits increased affinities for both the progesterone receptor and the androgen receptor[61]. However, both activities result in suppression of gonadotropin secretion, and the peripheral androgenic activity is counteracted by the estrogen component of the contraceptive pills.

A. Adverse Reactions

Overall, the safety profile of the oral contraceptives proved exceptionally benign. Considering the number of young, healthy persons exposed to 'the pill' over the years, an excellent record is a must. Nevertheless, several adverse effects have been recorded and deserve attention.

In a variety of epidemiologic investigations an association of increased incidence of thrombophlebitis with the use of oral contraceptives has been noted. Earlier studies estimated that the incidence of thrombophlebitis was increased 6–10-fold in contraceptive users, compared with non users[65]. However, subsequent studies reported lower risks, possibly related to both a decrease in the amount of estrogen in the pills and better selection of candidates for oral contraceptive therapy. A similar situation has been reported for coronary and cerebral thrombosis[66–69]. Likewise, the use of oral contraceptives may produce mild to moderate hypertension, some metabolic abnormalities such as increased blood glucose and lipid levels and other effects. Therefore, thorough examination of women before prescribing oral contraceptives and careful monitoring during their use has been encouraged.

IV. DEACTIVATION OF CYTOCHROME P450 BY ACETYLENES

A. Mechanism of Cytochrome P450 Deactivation

White and Muller-Eberhard[70] first showed in 1977 that norethindrone and ethynylestradiol caused time-dependent loss of cytochrome P450. This deactivation of cytochrome P450 was oxygen- and NADPH-dependent. No deactivation occurred in the absence of air or under carbon monoxide, indicating that these oral contraceptives undergo metabolic activation by cytochrome P450, before they finally deactivate the enzyme. This mode of inactivation qualifies the ethynylestrogens as suicide inhibitors of cytochrome P450. Inactivation of P450 could be induced by pretreatment with the enzyme inducer phenobarbital but not with 3-methylcholanthrene or pregnenolone-16α-carbonitrile.

Following *in vivo* administration of the ethynylic estrogen derivatives to rats, green pigments were identified in the liver. Since neither deactivation of cytochrome P450 nor green pigments occurred following administration of estradiol, progesterone or norethandrolone, the ethyl analog of norethindrone, the ethynylic functionality was implicated in the mechanism-based inhibition of cytochrome P450 as well as in green pigment formation[70–71].

Structure–activity relationship studies showed that most drugs containing terminal triple bonds that shared the suicide inhibition property were also lipophilic in nature[72–76] (Table 1). On the other hand, hydrophilic acetylenic drugs are seldom activated by cytochrome P450[75]. The most potent alkylacetylene cytochrome P450 inactivator is 1-decyne, while alkynes with shorter or longer chain lengths are weaker inactivators[75]. In substituted phenylacetylenes the inactivation of cytochrome P450 is insensitive to the electronic nature of the substituents on the aromatic ring[32]. The acetylenic moiety is quite inert and metabolically not easily activated, compared to other functional groups such as olefins. Thus certain acetylenic derivatives are metabolized predominantly via other routes, for example pargyline (see Scheme 18) is mainly metabolized by N-depropargylation and N-demethylation without any activation of the triple bond or formation of green pigment[81]. It is not clear why, for example, hexapropymat is activated by cytochrome P450 while the hypnotic sedative drug ethinamate is not[86]. It is known, however, that the hydrogens on carbon α to the triple bond are not essential for inactivation of cytochrome P450 as shown, for example, by 17α-ethynyl steroids, ethynyl-1-cyclopentanol or ethynyl-1-cyclohexanol. Nor does the acidity of the acetylenic hydrogen play a role in its activation by cytochrome P450[86,87].

The green pigments, isolated by extraction of liver preparations following methylation under mild acidic conditions, were shown to be alkylated derivatives of protoporphyrin IX[73]. The elucidation of their structure was accomplished by UV, NMR and MS. These studies disclosed that the molecular weight of the green pigment was equal to the sum of porphyrin (methyl ester), the acetylenic derivative and one atom of oxygen[74]. The oxygen

TABLE 1. Structure–activity relationship of acetylenic compounds that deactivate cytochrome P450

Compound	Structure	References
A. Acetylenes that destroy cytochrome P450 by heme alkylation		
Acetylene	$HC{\equiv}CH$	72,77,78
Propyne	$CH_3C{\equiv}CH$	79
1-Heptyne	$C_5H_{11}C{\equiv}CH$	75
1-Octyne	$C_6H_{13}C{\equiv}CH$	80,81
1-Decyne	$C_8H_{17}C{\equiv}CH$	75
1-Tridecyne	$C_{11}H_{23}C{\equiv}CH$	75
Phenylacetylene	$C_6H_5C{\equiv}CH$	74
Biphenylacetylene	$4{-}C_6H_5C_6H_4C{\equiv}CH$	37
Ethynylcyclohexane		74
1-Ethynyl-1-cyclohexanol		73,74
1-Ethynyl-1-cyclopentanol		74
4-Phenyl-1-butyne	$C_6H_5CH_2CH_2C{\equiv}CH$	74
3-(2,4-Dichlorophenoxy)-1-propyne	$2,4{-}Cl_2C_6H_3{-}O{-}CH_2C{\equiv}CH$	74
3-Phenoxy-1-propyne	$C_6H_5OCH_2C{\equiv}CH$	74
3-(4-Nitrophenoxy)-1-propyne	$4{-}NO_2C_6H_4OCH_2C{\equiv}CH$	72
2-(2-Propynyl)-4-pentynamide	$(HC{\equiv}CCH_2)_2{-}CHCONH_2$	74
Ethchlorvynol		74,82
Hexapropymat		83
3-Methyl-1-pentyn-3-ol		74
Norethindrone		70
Norgestrel		70
Danazol		73

TABLE 1. (*continued*)

Compound	Structure	References
Ethynylestradiol	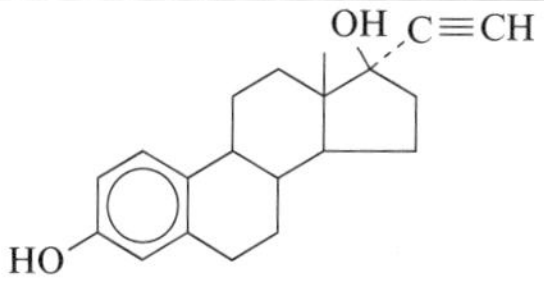	70

B. Acetylenes that destroy cytochrome P450 by other mechanisms

2-Hexyne	$C_3H_7C\equiv CCH_3$	74
2-Decyne	$C_7H_{15}C\equiv CCH_3$	75
4-Methyl-2-octyn-4-ol		74

1-Ethynylpyrene	see Scheme 13	33
2-Ethynylnaphthalene	see Scheme 13	34,95
Norethindrone	see above	84
10-Undecynoic acid	$HC\equiv C(CH_2)_8COOH$	35
Methyl-1-pyrenylacetylene	see Scheme 13	85

was derived from atmospheric oxygen, as indicated by *in vitro* experiments with enriched $^{18}O_2$ atmosphere[80,31]. From NMR data it was concluded that the acetylenes were atttached to the nitrogen of ring A of protoporphyrin IX[79,80], except for acetylene itself[80,83], which bound also to ring D of protoporphyrin, and ethchlorvynol which bound to all four rings of the porphyrin[82]. Moreover, the acetylenic derivatives are always attached to the porphyrin moiety via their terminal acetylenic carbon and the inner acetylenic carbon is

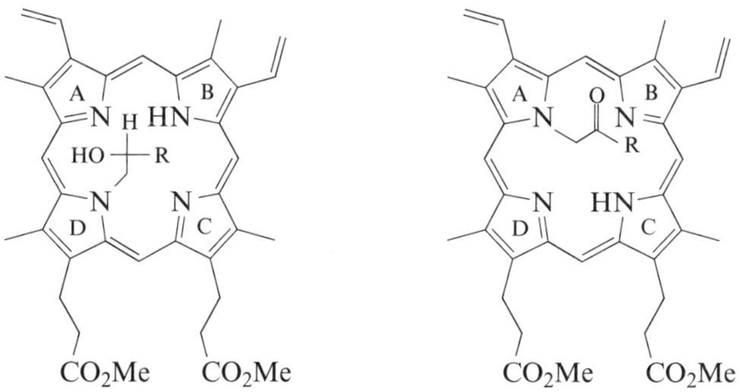

FIGURE 1. Alkylation of protoporphyrin IX by alkenes (left) and alkynes (right)

oxidized to a ketone. The general structure of the green pigments derived from acetylenes is depicted in Figure 1 (right). Although initial experiments of the isolation of the green pigments failed to detect an iron atom, subsequent prelabeling of the cytochrome with[59]Fe and the use of milder isolation procedures revealed that an iron atom was indeed an integral part of the green pigments[71,88].

In vitro, incubation of norethindrone with rat hepatic microsomes resulted in only one norethindrone-derived green pigment. *In vivo*, however, several green pigments associated with this and other acetylenic drugs were observed, depending on the time that elapsed since

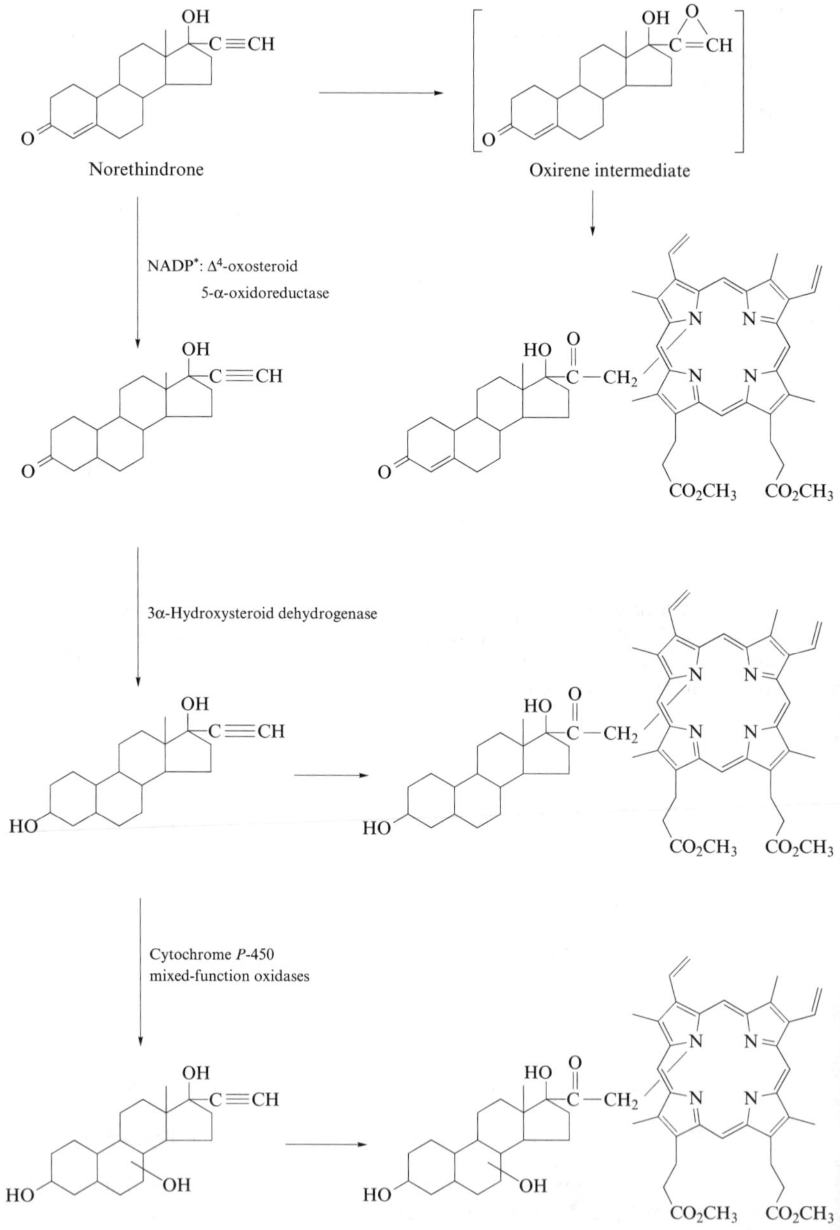

Norethindrone

Oxirene intermediate

NADP$^+$: Δ^4-oxosteroid
5-α-oxidoreductase

3α-Hydroxysteroid dehydrogenase

Cytochrome P-450
mixed-function oxidases

SCHEME 11. *In vivo* metabolism and heme alkylation by norethindrone. Reproduced from Reference 89 by permission of the Biochemical Society and Portland Press

administration[88]. As it turned out, some of the in vivo formed green pigments resulted from initial metabolism of the steroid moiety, mainly in ring A (Scheme 11)[89]. Most of the studies on suicide inhibition of cytochrome P450 by acetylenes have been carried out with rats. Although norethindrone deactivates cytochrome P450 in other species as well, male rats are the most active species in this respect, followed by hamster > rabbit > mouse > marmoset > hen. Human hepatic microsomal activity is very low, exhibiting only 2% of male rat hepatic microsomal activity[90].

Although suicide inhibition of cytochrome P450 accompanied by covalent binding to the heme-moiety has also been observed with drugs containing functional groups like olefins and allenes[80, 91–93], several differences between green pigments formed from olefins and acetylenes have been noted. First, most terminal acetylenes that have been examined formed these pigments. By comparison, only few terminal olefins form such adducts, a typical example being allyl isopropyl acetamide[94]. Second, green pigments formed by activation of olefins contain a hydroxyl group on the inner olefinic carbon, while in acetylenes the functionality on this carbon is always a ketone. Third, while acetylenes are bound to the nitrogen of ring A of protoporphyrin IX (Figure 1, right), the olefins usually bind to ring D[80] (Figure 1, left).

Early studies have postulated oxyrene or cationic carbene intermediates in the formation of either green pigments or the acetic acid metabolites[73,74]. In subsequent investigations, however, these intermediates were ruled out, based on structural and mechanistic consideration[31]. Ortiz de Montellano and coworkers observed that while there was a strong isotope effect in the formation of the acetic acid metabolite, there was no deuterium effect in the formation of the heme-acetylene adduct[37]. The mechanism thus postulated by these authors for the formation of the green pigment as well as the acetic acid metabolites is compatible with the evidence available to date and is shown in Scheme 12. This mechanism

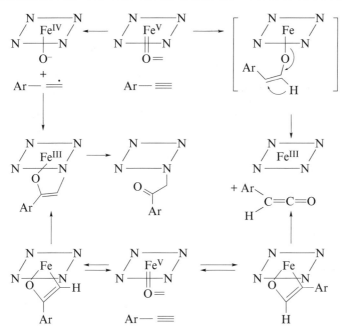

SCHEME 12. Postulated mechanism for heme alkylation by terminal alkynic derivatives. Reproduced with permission from Ref. 74

involves either electron transfer from the heme-oxene complex to the acetylenic derivative or the formation of a ferric cyclooxybutene derivative. Each of these mechanisms can account for both the formation of the metabolites and deactivation of cytochrome P450. Davies and coworkers[84] made the observation that the deactivation of cytochrome P450 by norethindrone accompanied by formation of green pigments did not account for all the cytochrome P450 that was lost. Also, nonterminal acetylenes like 2-hexyne, 2-decyne and 4-methyl-2-octyn-4-ol deactivate cytochrome P450 without formation of green pigments (Table 1B)[89]. Thus, acetylenic derivatives may destroy cytochrome P450 by a different route than heme adduct formation. Osawa and Pohl[95], reviewing the possible mechanisms for cytochrome P450 inactivation, suggested three general mechanisms:

(a) covalent binding of the activated drug to the heme moiety of cytochrome P450,
(b) covalent binding of the activated drug to the protein moiety, and
(c) covalent binding of the heme to the protein moiety mediated by the activated drug.

Several examples that have accumulated in recent years suggest that the deactivation of cytochrome P450 by acetylenic derivatives via the last two suggested pathways are not rare (Table 1B). Thus, CaJacob and coworkers[35] have postulated that the activated intermediate of 10-undecynoic acid deactivated cytochrome P450 by binding to the protein moiety of the cytochrome. It was also shown that while phenylacetylene deactivated the purified cytochrome P450 1A1 via alkylation of the heme moiety, 1-ethynylpyrene and 2-ethynylnaphthalene did so by binding to the protein moiety[33,96]. Norethindrone, for example, in addition to formation of green pigments, is known to induce binding of the heme to the prosthetic protein moiety[95]. The exact structural properties of the various acetylenic derivatives leading to suicidal inactivation of the cytochrome by each of the different possible mechanisms remains to be elucidated.

B. Inhibition of Cytochrome P450 Isozyme Families

Most lipid-soluble drugs, as well as many endogenous compounds, are metabolized to yield more polar compounds which may, in turn, be excreted. Drug oxidation is catalyzed predominantly by a group of enzymes in the smooth endoplasmic reticulum of the liver and other organs, collectively known as cytochrome P450. These monoxygenases are mixed function oxidases, requiring nicotinamide adenine dinucleotide phosphate (NADPH) and molecular oxygen. Investigations over the past decade have revealed that each of the dozens of isozymes of cytochrome P450 is encoded by a separate gene. Most of these isozymes have also quite specific substrates. In addition, hepatic enzyme activity may be modified by environmental factors and drugs that can either induce (i.e. stimulate) or, conversely, inhibit cytochrome P450 activity. The significance of drug metabolism goes far beyond the elimination of potentially toxic compounds. Many harmful effects of foreign compounds, such as specific organ toxicity, carcinogenesis and teratogenesis, may result from the metabolic conversion of xenobiotics to unstable intermediates. Since the rate at which these reactive intermediates form may depend on enzyme induction or inhibition, these processes are of the utmost importance.

1. Inhibition of cytochrome P450 1A

1-Ethynylpyrene (Scheme 13) was the first analog of benzo(a)pyrene that was tested as an inhibitor of benzo(a)pyrene hydroxylase[97]. This ethynyl polycyclic aromatic hydrocarbon (PAH) is a very potent *in vitro* suicide inhibitor of cytochrome P450 1A mediated oxidations of PAH[33,85,97]. The inhibitory property has been attributed to the drug's fitting to the active site of cytochrome P450 1A, as suggested by Jerina and coworkers[98] (see also References 97 and 99). 1-Ethynylpyrene also potently inhibits the 7,12-dimethyl benzo(a

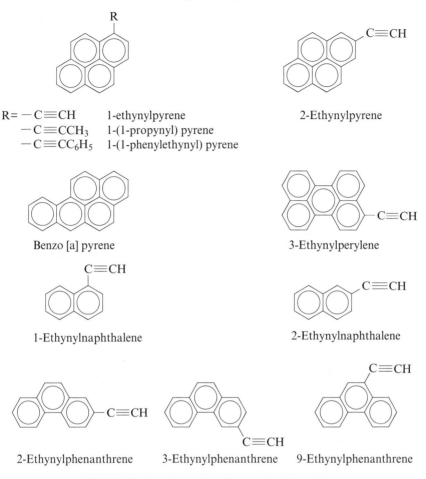

R= —C≡CH   1-ethynylpyrene
   —C≡CCH₃   1-(1-propynyl) pyrene
   —C≡CC₆H₅   1-(1-phenylethynyl) pyrene

2-Ethynylpyrene

Benzo [a] pyrene

3-Ethynylperylene

1-Ethynylnaphthalene

2-Ethynylnaphthalene

2-Ethynylphenanthrene   3-Ethynylphenanthrene   9-Ethynylphenanthrene

SCHEME 13. Acetylenic polycyclic aromatic hydrocarbons

anthracene (DMBA) and benzo(a)pyrene-induced skin tumors in SENCAR mice *in vivo*. This action is related to inhibition of the formation of an oxidized PAH-DNA adduct, which is an essential step in tumor initiation in this model[100]. By contrast, the inhibition of DMBA-induced tumorigenesis by 2-ethynylnaphthalene (Scheme 13) was very weak and the dose–inhibition relationship of this compound was very poor. Thus, it was postulated that 2-ethynylnaphthalene is a weak inhibitor of the skin cytochrome P450 1A isozyme[100].

Structure–activity relationship studies of ethynyl–PAH compounds (Scheme 13) showed that, while all compounds that were tested were competitive inhibitors of cytochrome P450 1A, only very few analogs inhibited the enzyme in a mechanism-based fashion. Thus, while methyl-1-pyrenylacetylene and 3-ethynylperylene are suicide inhibitors of benzo(a)pyrene hydroxylation, phenyl-1-pyrenylacetylene is only a competitive inhibitor[97,99].

Numerous ethynyl PAHs have been tested as potential inhibitors of the *in vitro* de-alkylation of 7-ethoxyphenoxazone[101] and 7-pentoxyphenoxazone[101], specific substrates for cytochrome P450 1A and cytochrome P450 2B1, respectively[102]. The results show that 1-ethynylpyrene was the most potent suicide inhibitor of cytochrome P450 1A while the isomer 2-ethynylpyrene was only a competitive inhibitor of 7-ethoxyphenoxazone deethylase. 1-Ethynylnaphthalene and 2-ethynylnaphthalene inhibited both isozymes in a mechanism-based fashion. Comparison of 3 isomers of ethynylphenanthrene showed that only 2- and 3-ethynylphenanthrene were mechanism-based cytochrome P450 1A inhibitors, while the 9-ethynylphenanthrene was a potent suicide inhibitor of cytochrome P450 2B1 (Scheme 13). These as well as other examples suggest that suicide inhibition of cytochrome P450 1A and cytochrome 2B1 by ethynyl PAHs is probably due to the shape and size of the analogs and to their positioning in the active site of the enzyme in relation to the oxene–iron complex[97,101].

As already mentioned, 2-ethynylnaphthalene was a weak inhibitor of cytochrome P450 1A but exhibited potent inhibitory activity of cytochrome P450 2B1. It was also shown to be a potent inhibitor of 2-naphthylamine N-oxidation *in vitro*, a reaction mediated by cytochrome P450 1A2. Of the two isomers, 2-ethynylnaphthalene was more potent than 1-ethynylnaphthalene. The mechanism of inhibition of cytochrome P450 1A2 proved to be suicide inhibition[34].

Ethynylestradiol was implicated as hepatocarcinogen in the rat[103] and cocarcinogen for DMBA-induced mammary tumors in hamsters[104]. Benrekassa and Decloitre[105] showed that this synthetic estrogen at concentrations of 0.1–1 mM inhibited benzo(a)pyrene metabolism in hepatocytes from naive or phenobarbital-treated rats. Pretreatment of rats with 3-methylcholantrene, an inducer of benzo(a)pyrene metabolism, abolished the inhibitory activity of the synthetic estrogen on benzo(a)pyrene metabolism. This observation suggests that the inhibitory activity of ethynylestradiol is probably not related to the major isozyme that metabolizes benzo(a)pyrene, namely cytochrome P450 1A, but rather to a different isozyme.

2. Inhibition of cytochrome P450 3A

Several 17α-ethynyl steroids inhibit, in a mechanism-based manner, their own major oxidative metabolic pathway, namely hydroxylation at C-2. Guengerich[106] showed, in a large series of 17α-ethynyl steroids, that gestodene was the most potent inhibitor of C-2 hydroxylation of steroids and also of the oxidation of the calcium channel blocking drug nifedipine when studied with purified cytochrome P450 3A4. Although the precise structural requirements for the inhibition of this isozyme are not completely understood, it seems that a Δ15 double bond is an important structural feature. Gestodene also inhibits cytochrome P450 3A5, a minor isozyme of the same subfamily, which mediates the C-2 hydroxylation of estradiol but not of ethynylestradiol[107]. The inhibition by gestodene seems to be specific for the cytochrome P450 3A subfamily, as inhibition of other subfamilies or isozymes of cytochrome P450, namely 2C, 2D6 and 2E1, was not observed.

Gestodene

3. Inhibition of cytochrome P450 22A

Cytochrome P450 22A1 (the trivial name of which is cytochrome P450$_{scc}$) was identified in human and bovine adrenal mitochondria. This enzyme mediates the cleavage of the cholesterol side chain to pregnenolone, which is the first stage in the biosynthesis of several steroidal hormones.

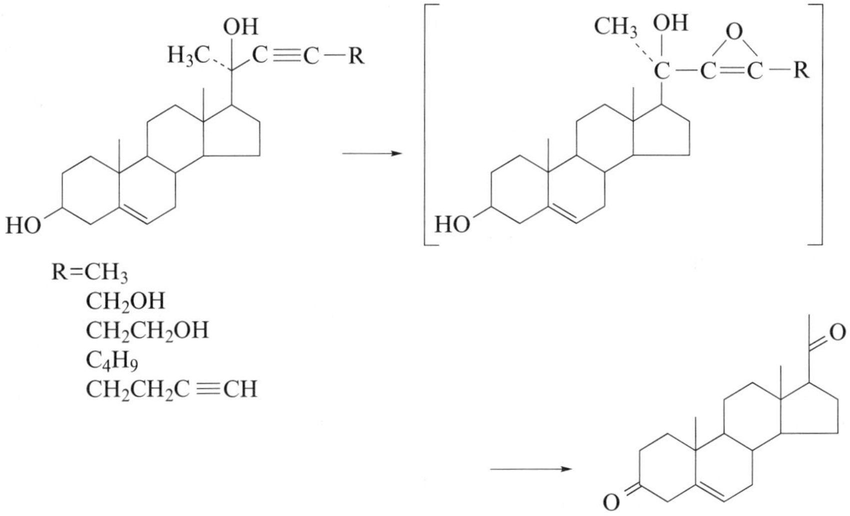

SCHEME 14. Metabolism of pregnenediol derivatives

Acetylenic derivatives of pregnenediol (Scheme 14)[108,109] have been found to be effective mechanism-based inhibitors of cytochrome P450 22A1. Since all the acetylenic derivatives of pregnenediol mentioned are converted to pregnenolones, it is assumed that an active intermediate, possibly an oxirene, might be formed during metabolic oxidation, leading to inactivation of the enzyme.

4. Inhibition of aromatase

The natural estrogens, estradiol and estrone, are synthesized in mammals by aromatization of testosterone and androstenedione, respectively. Estrogens are biosynthesized in

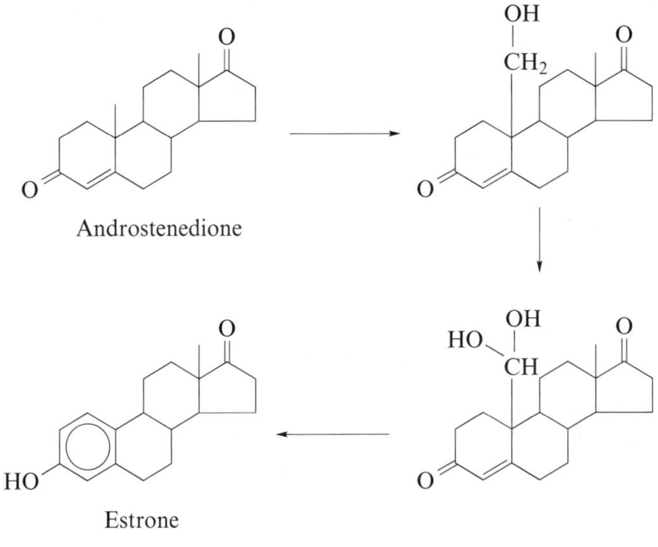

Androstenedione

Estrone

SCHEME 15. Biosynthesis of estrone

fertile females mainly by the ovaries, placenta and breast tissue, whereas in postmenstrual women the major estrogen-producing tissues are the breast, fat and muscles. The biochemical pathway of these reactions is mediated by cytochrome P450 19A1, better known as aromatase. This process (Scheme 15) consists of three NADPH-dependent oxidation steps. The first two are hydroxylation reactions at C-19 to form the gem-dihydroxy derivative. The third step, which is rate-limiting in this pathway, involves hydroxylation at C-2, followed by aromatization of ring A and formation of water and formate[110].

Inhibition of aromatase may have therapeutic implications in diseases that are estrogen-dependent, such as some types of breast and endometrial cancer in women[111–114] and possibly in benign prostatic hyperplasia in men[115]. Several steroidal and nonsteroidal compounds have been screened as aromatase inhibitors. Prominent examples are 4-hydroxyandrost-4-ene-3,17-dione and aminogluthetimide[112,116]. The following discussion will focus primarily on 10-(2-propynyl) estr-4-ene-3,17-dione (PED, also known as MDL 18,962/(Scheme 16). This compound was synthesized and tested as a potential mechanism-based aromatase inhibitor, because of its chemical resemblance to the natural sub-strate, androst-4-ene-3,17 dione, and assuming the inhibitory properties of the acetylenic moiety.

In vitro studies with human term placental microsomes showed that PED displayed the typical properties of a mechanism-based inhibitor of cytochrome P450, namely (a) NADPH-dependent inhibition; (b) time-dependent loss of enzymatic activity, followed by pseudo-first-order kinetics; and (c) natural substrates delayed the irreversible inhibition. The irreversibility of inhibition was further substantiated by dialysis experiments[117,118]. The inhibitory properties of PED were also shown in the pregnant-mare-serum-gonado-tropin-stimulated rat ovarian microsomes (PMSGSROM). Comparison of the inhibitory kinetic parameters of PED in the human placental and PMSGSROM preparations showed close resemblance: time-dependent K_i of 5 nM and 14.5 nM and $t_{1/2}$ of inactivation of 11.8 min and 16.2 min, respectively. The V_{max} of human placental microsomes was, however, 10 times higher than that of PMSGSROM (51.8±9.4 vs 4.2±0.9 pmol/min/mg protein, respectively, with testosterone as substrate). The affinities of aromatase from the two sources were also similar[119,120]. Johnston and coworkers[121] showed that baboon placental aromatase, but not the rhesus placental enzyme, was similar in activity to the human placental preparation.

The inhibitory activity of PED appears to be specific for aromatase. This steroid did not affect the activity of cytochrome P450-mediated ethylmorphine N-deethylation[121], nor did it prolong the pentobarbital sleeping time in rats in vivo, used as an index for in vivo cytochrome P450 activity[122]. Moreover, PED was endocrinologically inactive and had very low affinities for cytosolic estrogen, androgen and progesterone receptors[121,123]. Likewise, PED did not cause any change in serum levels of LH and prolactin[124].

Subcutaneous or oral administration of PED to pregnant-mare-serum-gonadotropin-stimulated rats and to baboons resulted in substantial inhibition of overian estrogen synthesis that lasted several hours[120,124–127]. Similar results were also shown in a peripheral model of aromatization, namely human trophoblast xenografts in athymic mice[128].

While the inhibition of aromatase activity by PED is well established, some controversy exists as to its inhibition of tumor growth in vivo. Zimniski and coworkers[120] and Puett and coworkers[129] showed that PED was a potent inhibitor of dimethylbenzanthracene-induced mammary tumor growth at a dose of 1 mg/kg/day as well as of human ovarian carcinoma in athymic mice. In contrast, di Salle and coworkers[130] and Zaccheo and coworkers[124] found that PED did not affect the growth of established tumors in the same model, nor did it prevent the appearance of new neoplasms. Thus, it appears that more in vivo studies on the inhibition of tumor growth are required before firm conclusions can be drawn on its role in estrogen-dependent tumor biology.

Structure–activity relationship studies have shown that the homolog of PED, 10-(2-butynyl) estr-4-ene-3,17-dione, was a weaker suicide aromatase inhibitor compared to

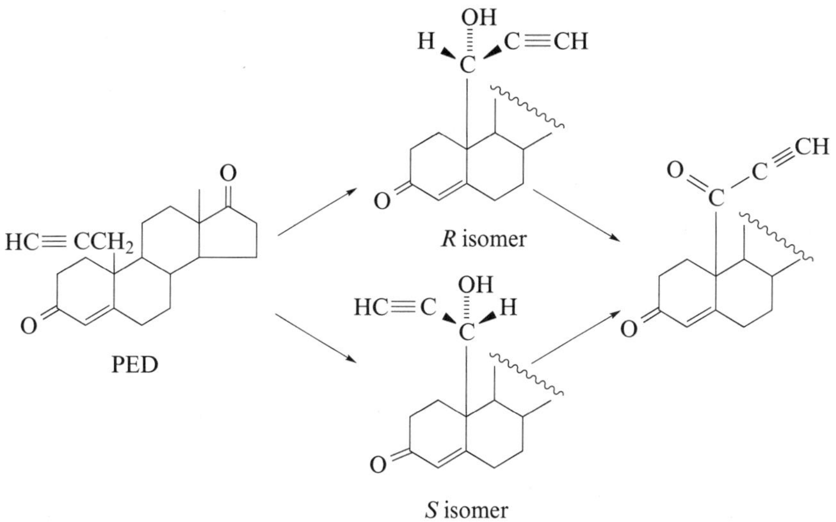

SCHEME 16. Oxidation of PED

PED. The triple-bond positional isomers of PED, 10-(1-propynyl) estr-4-ene-3,17-dione, and the 10-ethynyl, 10-cyano and 1-methyl cyanide analogs were not suicide inhibitors[121,131,132]. The 10-[[1R]-1-hydroxy-2-propynyl], 10-[[1S]-1-hydroxy-2-propynyl] and 10-(1-oxo-2-propynyl) estr-4-ene-3,17-dione were prepared as potential PED metabolites (Scheme 16) and tested for their aromatase inhibitory activity on human placental enzyme. While the [1R]-hydroxy isomer was the only competitive inhibitor of aromatase, the other two oxidized analogs, [1S]-hydroxy and 1-oxo-derivatives, had very weak inhibitory properties compared to PED (Ki 27 μM and 12 μM vs 23 nM for PED and $t_{1/2}$ of 4 min and 2.16 min vs 10.4 min for PED)[117,118]. These observations were in contrast to the *a priori* assumption that 10-(1-oxo-2-propynyl)estr-4-ene-3,17-dione might be the most active derivative in the series, due to its Michael acceptor structure[117,118,121]. Similarly, the 19-hydroxylated propargyl derivatives of 3-deoxyandrost-4-ene-17-one and androst-4-ene-3,6,17-trione were very weak mechanism-based inhibitors of aromatase[133,134].

Johnston and coworkers[135] showed that putative hepatic metabolites of PED may contribute to the *in vivo* inhibitory activity of this compound, although their inhibitory activity is not very strong, compared with PED. Zimniski and coworkers[120] reported that the propyl ester of 10-(2-propynyl)estr-4-ene-17-ol-3-one was inactive as an aromatase inhibitor *in vitro*; however, it was as potent as PED in the rat *in vivo*, suggesting that the ester was hydrolyzed quickly *in vivo*.

Norethindrone (17α-ethynyl-19-nortestosterone) (Table 1) also exhibits properties of a suicide inhibitor of human placental aromatase. It is, however, a very weak inhibitor compared to PED[136,137].

PED was recently shown to inhibit an additional steroid biosynthetic enzyme, namely steroid 19-hydroxylase (aromatase) from bovine adrenals. The observation that steroid 19-hydroxylase is the rate-limiting enzyme in the metabolic formation of 19-nordeoxy-corticosterone, a natural mineralo-corticosteroid that has been implicated in some forms of human and experimental hypertension[138], may bear some implications for the treatment of hypertension. Indeed, PED attenuated hypertension in salt-sensitive hypertensive rats, although the levels of 19-nordeoxycorticosterone were not significantly lowered. Also, the mortality rate was significantly reduced in rats on high salt diet that were treated with PED[139]. Additional work is required before one may assess if this novel approach to the treatment of hypertension can be exploited.

V. PROPARGYLAMINE DERIVATIVES

A. Monoamine Oxidase (MAO)

Monoamine oxidase (MAO) mediates the oxidative deamination of endogenous and exogenous amines as follows:

$$RCH_2NH_2 + H_2O + O_2 \rightarrow RCHO + NH_3 + H_2O_2 \tag{2}$$

Two isozymes, termed MAO A and MAO B, have been described. The early distinction between the two isozymes was based mainly on substrate selectivity and differential inhibition by various inhibitors[140]. Serotonin and noradrenaline are preferentially deaminated by MAO A, while benzylamine and phenylethylamine are better metabolized by MAO B[141–143]. Dopamine and tyramine are equally deaminated by both isozymes. Chlorgyline and selegiline (previously known as deprenyl) are selective inhibitors of MAO A and MAO B, respectively (for review, see Youdim and Finberg[144]). In recent years, human cDNA clones of MAO A and of MAO B were isolated and the primary structures of these isozymes in several species were determined[145–148]. These investigations revealed that the resemblance of the same isozyme in various species seems to be bigger than between the two isozymes in the same species. Human MAO A and MAO B comprise 527 and 520 amino acid residues, respectively, with about 70% homology[145]. There are three polypeptide regions, namely, the adenosine dinucleotide region, the flavin adenine mononucleotide (FAD) region and a third region, the role of which is not clear, in which the homology between the two isozymes is approximately 90%[145]. The two forms of the enzyme are encoded by different genes located on the short arm of chromosome X[149,150]. Both MAO A and B are composed of two subunits, each of which contains an FAD molecule[151].

Monoamine oxidase is located in the outer mitochondrial membrane and is distributed peripherally in neuronal and nonneuronal tissues (liver, intestines) as well as in the central nervous system. Few organs have exclusively one form of the enzyme. Thus, in human placenta over 99% of the enzyme is of the A type[152,153] while platelets[154], lymphocytes[155] and chromaffin cells[144] have only MAO B activity. All other tissues contain both types of the enzyme at varying ratios. The distribution and activity ratio of the two isozymes differ between species. Thus, in rat brain there is appreciable MAO A activity, while in human brain MAO activity is preferentially of the B type. It is intriguing that dopaminergic neurons in the substantia nigra do not contain any MAO activity[156,157]. Instead, the enzyme activity is concentrated in this brain region in glial cells[156,158,159]. This kind of distribution pattern of MAO could be of significance in various diseases of the central nervous system, particularly in Parkinson's disease[160].

B. Propargylamine MAO Inhibitors

MAO inhibitors with a propargylamine moiety inhibit the enzyme in a mechanism-based fashion[142]. They first bind competitively to MAO and then are oxidized by the FAD moiety of the enzyme to active intermediates which bind quantitatively to C-5 of the FAD isoalloxazine ring (Scheme 17)[161,162] (for review see Gerlach and coworkers[163]). Since this deactivation reaction is irreversible and quantitative, chlorgyline and selegiline have been used to map and quantitate MAO A and MAO B activities in the tissues of various animal species[164,165]. Most of the quantitation of MAO A and B in human brains was done

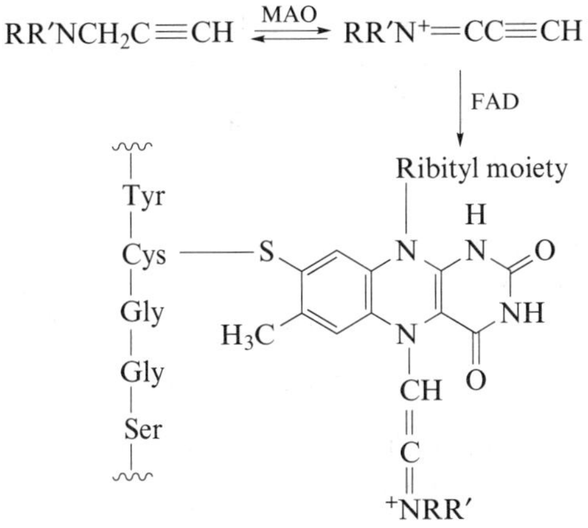

$$RR'NCH_2C\equiv CH \xrightleftharpoons{MAO} RR'N^+=CC\equiv CH$$

$$\downarrow FAD$$

Ribityl moiety

Tyr
|
Cys ——— S
|
Gly H_3C
|
Gly
|
Ser

H
|
N   N   O
|
N   NH

CH   O
‖
C
‖
$^+$NRR'

Selegiline   R=CH$_3$ R' = CHCH$_2$C$_6$H$_5$
                                   |
                                   CH$_3$

Chlorgyline   R=CH$_3$ R' = (CH$_2$)$_3$C$_6$H$_5$

SCHEME 17. Binding of propargylamine derivatives to
FAD moiety of MAO

postmortem[166–168]. The development of the noninvasive technique of positron emission tomography using the ^{11}C labeled MAO inhibitor selegiline has enabled researchers to quantify and map MAO *in vivo*, as shown in monkeys and humans[169–171].

C. Pharmacological Properties of Chlorgyline

Chlorgyline is a selective inhibitor of MAO A *in vitro*[141]. However, *in vivo* some inhibition of MAO B was also observed, especially following acute high doses of chlorgyline or chronic treatment with the drug[172,173]. In the rat, the brain concentrations of cytosolic and granular noradrenaline and serotonin were significantly elevated after treatment with chlorgyline and the concentrations of the deaminated metabolites of these neurotransmitters were decreased[174,175]. In the monkey brain, chlorgyline seemed to inhibit the oxidative deamination of noradrenaline more specifically[176,177]. It was also observed that, concomitantly with the elevation of neurotransmitter levels, the activities of tyrosine hydroxylase and tryptophan hydroxylase were decreased, probably by feedback mechanisms[178]. Chlorgyline also down regulated α and β adrenoceptors, and attenuated blood pressure[178]. This attenuation could be due to either activation of central α_2 adrenoceptors[179] or stimulation of melatonin biosynthesis[180].

Chlorgyline, like many other nonspecific and MAO A inhibitors, potentiates the pressor effect of tyramine on the cardiovascular system in humans and rodents and may result in hypertensive crises. This effect is probably due to inhibition of MAO A activity in the intestines which normally detoxify monoamines in food and beverages such as cheese and wine[181,182]. It is thus believed that only MAO A inhibitors (and mixed inhibitors) have this so-called 'cheese effect'[183,184].

The most common therapeutic use of nonspecific and MAO A inhibitors is as antidepressants. In a few open clinical trials that have been conducted, chlorgyline exhibited anti-depressant and antianxiety effects in depressed patients[185]. It was also effective in hyperactive children with attention deficits[186]. Chlorgyline was, however, not effective in the treatment of obsessive-compulsive disorders[187].

D. Pharmacological Properties of Selegiline

During the early developmental studies of selegiline it was observed that only the L-(–)-enantiomer of the drug had MAO B inhibitory activity, while the D-(+)-enantiomer was practically inactive[188]. Although *in vitro* L-(–)-selegiline is a very potent and selective inhibitor of MAO B, there is some indication of MAO A inhibition *in vivo*. In *ex vivo* experiments in rats the ED50 of MAO B inhibition in the brain was 27 μmol/kg, while the ED50 for MAO A was 1000 μmol/kg, almost 40 times higher[142,189]. The selectivity of L-(–)-selegiline depends on the dose and length of treatment. Analysis of post mortem human Parkinsonian brains treated chronically with selegiline showed complete inhibition of MAO B, while MAO A was only partially inhibited[166]. At high doses selegiline inhibited the metabolism of noradrenaline and increased the occurrence of the 'cheese effect'[190–192]. Inhibition of the oxidative deamination of dopamine in certain brain regions such as the substantia nigra leads to increased dopamine concentrations and thus prolonged dopamine activity. Lauber and Waldmeier[193] showed that in selegiline-treated rats, brain phenylethylamine levels were 60 times higher than in control rat brains. Since this trace amine is a modulator of catecholaminergic transmission, it may play a role in selegiline activity (for review see Boulton[194]). Selegiline has, in addition to its inhibitory activity on MAO B, an inhibitory effect on the neuronal uptake of monoamines[195]. It seems that this effect is due to competition for a neuronal transporter (IC50 = 30 μM)[163]. Selegiline exerts this activity at higher concentrations than needed for MAO B inhibition. Thus, treatment of rats with selegiline for two weeks at a dose of 0.25 mg/kg reduced the uptake of dopamine by striatal slices by 32%, while the same dose inhibits brain MAO B almost completely[195].

The activity of MAO B, but not of MAO A, increases in old age[166,196]. As a result of increased activity of the enzyme, larger amounts of hydrogen peroxide are formed in some brain areas. Hydrogen peroxide and superoxide radicals may be detoxified by three enzymatic pathways: superoxide dismutase, catalase and glutathione peroxidase (equation 3–5). Because these enzymatic pathways are easily saturable, oxidative stress and tissue damage may result. Fortunately, superoxide dismutase and catalase are inducible and

$$2O_2^- + 2H^+ \xrightarrow{\text{superoxide dismutase}} H_2O_2 + O_2 \tag{3}$$

$$2H_2O_2 \xrightarrow{\text{catalase}} H_2O + O_2 \tag{4}$$

$$H_2O_2 + GSH \xrightarrow{\text{glutathione peroxidase}} H_2O + GSSG \tag{5}$$

treatment of rats for three weeks with 2 mg/kg of selegiline produced a 10-fold increase in superoxide dismutase in the striatum[197]. Carrillo and coworkers[198] also showed a threefold increase in the soluble form of this enzyme as well as almost twofold increase in catalase activity. The increased activity of these enzymes may help to quench the harmful effects of hydrogen peroxide and superoxide radicals and prevent the formation of damaging free radicals and lipid peroxidation. Selegiline, however, does not induce any change in glutathione or glutathione peroxidase levels.

Selegiline protects experimental animals against the damaging activity of several neurotoxins such as 1-methyl-4-phenyl-1,2,3,6-tetrahydropyridine (MPTP)[199], 6-hydroxydopamine[200] and DSP-4 [N-(2-chloroethyl)-N-ethyl-2-bromobenzamide][201]. MPTP freely crosses the blood brain barrier in primates and humans and is oxidized in the glial cells of the substantia nigra by MAO B to the ultimate neurotoxin, methyl phenyl pyridinium. The latter destroys dopaminergic neurons in this region, thus resulting in a syndrome similar to Parkinson's disease. Selegiline, as an irreversible MAO B inhibitor, prevents the formation of methyl phenyl pyridinium and hence protects the substantia nigra[202]. The mechanism of protection by selegiline against 6-hydroxydopamine and DSP-4 is not fully understood, but it seems that MAO B is not involved in this mechanism. Recently tetrahydroisoquinoline and N-methylsalsinol were detected in human brain and urine[203,204] and in rat brains[205]. These endobiotics have chemical structures analogous to MPTP and thus, after being activated by MAO B, may cause Parkinsonism. Selegiline may inhibit the formation of the ultimate neurotoxin from these endobiotics, as it protects against MPTP neurotoxicity.

Chronic treatment with selegiline promoted longevity in experimental animals[200] and humans[166]. Knoll[200] showed that rats treated chronically with selegiline had a longer lifespan than saline-treated rats (197.98±2.36 weeks vs 147.05±0.56 weeks, respectively).

The psychostimulant activity of selegiline was initially attributed to the two major metabolites of selegiline, L-amphetamine and L-methamphetamine[206,207]. However, due to their low concentrations in the brain and the low activity of the L-isomers of methamphetamine and amphetamine compared with the psychoactive D-enantiomers, it seems that their contribution cannot be significant[208].

1. Therapeutic uses

Parkinson's disease is a degenerative disease in which a large number of the dopaminergic neurons of the substantia nigra is destroyed. Usually older people are afflicted by this condition. The etiology of the disease is not known; however, several mechanisms have been suggested (for review see Gerlach and coworkers[163]). One such mechanism relates to the observation that exposure to 1-methyl-4-phenyl-1,2,5,6-tetrahydropyridine and endogenous analogs of this environmental pollutant can produce a syndrome similar to Parkinson's disease. Another possible mechanism that can contribute to the etiology of the disease is oxidative damage to the substantia nigra due to the formation of oxygen-containing free radicals. Formation of such radicals was substantiated by the observation of very high concentrations of free Fe^{+2} and Fe^{+3}, accompanied by low levels of ferritin (an iron binding protein) in this region of the brain[209–211]. Ferric and ferrous ions facilitate the formation of toxic oxygen-containing free redicals under physiological conditions (Fenton reactions, equation 6)[212].

$$H_2O_2 + Fe^{+2} \rightarrow OH^- + OH^{\cdot} + Fe^{+3} \tag{6}$$

It was therefore only natural to test the possible beneficial properties of selegiline in Parkinson's disease.

Selegiline as monotherapy for Parkinson's disease is effective when started at the very early stages of the disease[213]. The DATATOP study[214,215] as well as other studies[216,217] showed that treatment with selegiline delayed the time before additional treatment with the standard antiparkinsonian drug L-3,4-dihydroxyphenylalanine (L-DOPA) was necessary by up to a few years. Supplementation of L-DOPA-treated patients with selegiline may produce a slight to moderate improvement, due to attenuation of the on–off effect (end-of-the-dose effect). There may also be favorable effects in the motor disabilities and subjective symptoms. This improvement of symptoms wears off within a few months. It seems that the degree of improvement due to treatment with selegiline in Parkinson's disease depends on the severity of the disease (for a recent review, see Cesura and Pletscher[218]). Vitamin E (tocopherol) has also been suggested as free radical scavenger in the treatment of Parkinson's disease[219]. However, the recent DATATOP report[215] showed that the combination of vitamin E with selegiline had no advantage over selegiline along.

Selegiline was also tested as an antidepressant, despite the belief that the antidepressant property probably resides in MAO A inhibition. Several studies showed, not surprisingly, that selegiline had very little effect, if any, in various types of depression[220–222].

In patients with moderate Alzheimer-type dementia, chronic treatment with selegiline (10 mg daily) had some beneficial effects such as improvements in congnitive functions, learning, alertness and task performance[223–225]. These observations are in line with results obtained in rats chronically treated with selegiline, in which active treatment significantly improved the learning capacity as compared to saline treated controls[200].

2. Adverse reactions

Selegiline seems to be a safe drug with a broad therapeutic index (in mice LD50/ED50 = ca100). It may cause mild gastrointestinal disturbances, bringing about most frequently nausea as well as insomnia and infrequently increases in serum transaminase levels[206,226]. It may exacerbate the side effects of L-DOPA; reduction of the L-DOPA dose eliminates these side effects without affecting the therapeutic benefit. Selegiline does not produce the typical 'cheese effect' generally attributed to MAO inhibitors[188].

E. Structure–Activity Relationship of Propargylamine Derivatives

The identification and characterization of the two isoforms of MAO on the one hand and the synthesis of the specific inhibitors of MAO A and of MAO B chlorgyline and selegiline, respectively, on the other hand, prompted structure–activity relationship studies in order to find more specific and efficacious inhibitors for therapeutic use. These studies revealed that one of the most important determinants of the inhibitory potency of propargylamine analogs as MAO B inhibitors was their pK_a, the optimum value being in the vicinity of 6.2[227].

Further examination of MAO inhibitory activity of selegiline has shown that the levorotatory isomer was 25-fold as potent as the dextrorotatory isomer[228]. The same has been observed with other derivatives having chiral centers[229].

In order to study systematically the structure–activity relationship of propargylamine MAO inhibitors, four major regions of the pargyline molecule, taken as a model, were considered:

$$R \text{---} CH_2 \text{---} \overset{\displaystyle R'}{\underset{\displaystyle }{N}} \text{---} CH_2C\equiv CH$$

$$D \qquad C \qquad B \qquad A$$

Region A—the propargylic moiety, region B—the second substituent on the nitrogen (R′), region C—the distance between the nitrogen atom and the substituent R, and region D—the nature of the substituent R. These studies have revealed that the propargylic group is essential for irreversible inhibition of MAO. This conclusion is compatible with the mechanism of inhibition, which entails activation of the N—CH₂—C≡CH moiety that binds covalently to the FAD component of MAO[161,162]. MAO inhibitors containing the 2-butynyl structure were very active MAO inhibitors[230].

Martin and coworkers[227] showed that one of the essential determinants of activity of propargylamine MAO inhibitors was the size of the R′ substituent. Methyl and hydrogen were shown to be the best in the series. Thus, the inhibitory activity of propargylamines with larger groups at region B such as cyclopropyl, propargyl or 2-butynyl was highly diminished compared with the methyl-containing derivatives[230–232].

The size of region C seems to be crucial in determining whether the derivative in question will be a MAO A, MAO B or mixed inhibitor. Derivatives with a distance of 1–2 carbon atoms between regions B and D have primarily MAO B inhibitory activity, as in selegiline and pargyline (Scheme 18). Region C with a length of 4 carbon atoms has major MAO A inhibitory activity, as in chlorgyline, and very low MAO B activity (U-1520)[231,233–235]. Branching on the benzylic carbon (TZ-996) or alkylation of carbon α to the nitrogen with alkyls larger than methyl (J-502, J-504) produced analogs with very weak MAO B inhibitory activity[235].

Detailed studies of the relationship between the structure of region D and the inhibitory activity on MAO A and MAO B have also been conducted (Scheme 18). These investigations revealed that reduction of the benzene ring to cyclohexyl (J-505) or cyclohexenyl (J-514) in the selegiline series resulted in a strong decline of MAO B activity. Replacement of the benzene ring by furan (U-1424) diminished the inhibitory activity only slightly, while dihydrofuran and tetrahydrofuran derivatives (TZ-1849 and TZ-1037, respectively) were devoid of any MAO B inhibitory activity. Although naphthalene (TZ-1263) and tetraline analogs (J-517, J-512, J-518) were potent MAO B inhibitors, their MAO A inhibitory activity was much more pronounced, such that the inhibitory concentration for MAO A was up to 1000 times smaller than for MAO B. Indanyl derivatives had pronounced inhibitory activity on both MAO A and MAO B (AGN 1135)[231,234,235].

The fact that serotonin (5-hydroxytryptamine), a neurotransmitter and autacoid, is metabolized mainly by MAO A, prompted Fernandez-Alvarez and coworkers to study the structure–activity relationship of propargylamine derivatives in which the substituents in region D were 2-(5-alkoxy-1-methylindolyl) derivatives. All the derivatives that were studied (Table 2) had both MAO A and MAO B inhibitory activities, but some were more selective MAO A inhibitors. Thus, 2-(5-methoxy-1-methylindolyl) and 2-(5-hydroxy-1-methylindolyl) propargylamine derivatives were the most selective MAO A inhibitors, some being 3 times more selective than chlorgyline[230,236–238]. 2-(5-Benzyloxy-1-

TABLE 2. Inhibition of MAO A and B by 2-indolylmethylamine derivatives

R	R′	R″	$\dfrac{\text{IC 50 MAO B}}{\text{IC 50 MAO A}}$		
			5-OH	5-OCH₃	5-OCH₂C₆H₅
H	H	$CH_2C\equiv CH$	> 55	500	1
H	$CH_2C\equiv CH$	$CH_2C\equiv CH$	1	1	1
H	CH_3	$CH_2C\equiv CH$	625	63	1
H	CH_3	$CH_2C\equiv CCH_3$	1925	1400	1
CH_3	H	$CH_2C\equiv CH$	170	2100	1
CH_3	CH_3	$CH_2C\equiv CH$	1540	100	1
CH_3	CH_3	$CH_2C\equiv CCH_3$	174	1	1

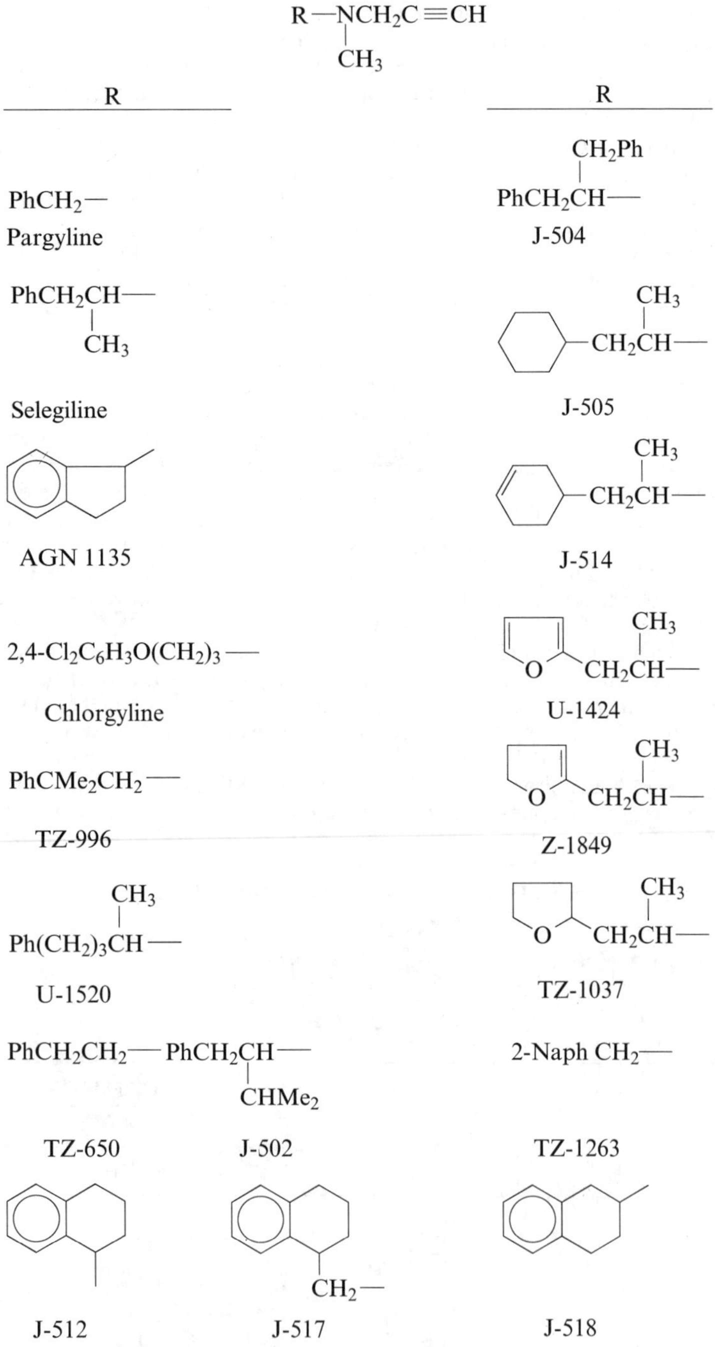
R—NCH2C≡CH
CH3
R
R
PhCH2—
Pargyline
CH2Ph
PhCH2CH—
J-504
PhCH2CH—
CH3
Selegiline
CH3
CH2CH—
J-505
AGN 1135
CH3
CH2CH—
J-514
2,4-Cl2C6H3O(CH2)3—
Chlorgyline
CH3
O
CH2CH—
U-1424
PhCMe2CH2—
TZ-996
CH3
O
CH2CH—
Z-1849
CH3
Ph(CH2)3CH—
U-1520
CH3
O
CH2CH—
TZ-1037
PhCH2CH2—
TZ-650
PhCH2CH—
CHMe2
J-502
2-Naph CH2—
TZ-1263
J-512
CH2—
J-517
J-518

SCHEME 18. *N*-Methyl-*N*-propargylamine MAO inhibitors

methylindolyl) propargylamine derivatives were potent MAO inhibitors, but lacked selectivity[230], as were also the derivatives unsubstituted at C-5 of the indole nucleus[239]. Derivatives of serotonin with branching at region C were completely unselective in spite of being very potent MAO inhibitors[236].

Yu[240] showed that aliphatic amines with chain lengths of 5–10 carbon atoms were good substrates for MAO B. Thus, it was only natural to suspect that propargylamine derivatives containing an alkyl moiety at region D might be potential inhibitors of MAO B. Indeed, N-2-alkyl-N-methyl propargylamines with alkyls 4–7 carbon atoms long proved to be potent MAO B inhibitors[229]. Thus, N-(2-hexyl)-N-methyl propargylamine was more than 1000-fold as active as a MAO B than as a MAO A inhibitor. However, derivatives of this series containing a hydrophilic group (OH, COOH) on their alkyl terminal carnon lost their MAO inhibitory activity. Methylphenyl tetrahydropyridine derivatives containing a propargyl group at C-4 or on the nitrogen were not particularly active suicide inhibitors of MAO B[241].

F. Metabolism of Propargylamine Derivatives

Propargylamine MAO inhibitors are metabolized mainly by N-dealkylation pathways, mediated by cytochrome P450. These pathways are induced by phenobarbital but not by 3-methylcholanthrene and are inhibited by classic cytochrome P450 inhibitors such as SK&F 525A and methyrapone[242–244].

1. Metabolism of selegiline

Incubation of selegiline with rat liver microsomes yielded three metabolites. The two primary metabolites were methamphetamine and norselegiline, and a secondary metabolite, amphetamine, may be derived from either methamphetamine or norselegiline[242] (Scheme 19). The reported rates of formation of these metabolites in rat liver microsomes are 8.91±0.46, 0.055±0.013 and 0.11±0.01 nmol/mg protein/min for methaphetamine, norselegiline and amphetamine, respectively. Metabolism by extrahepatic tissues is very low, compared with the liver[242]. Lung and kidney transform selegiline only to one

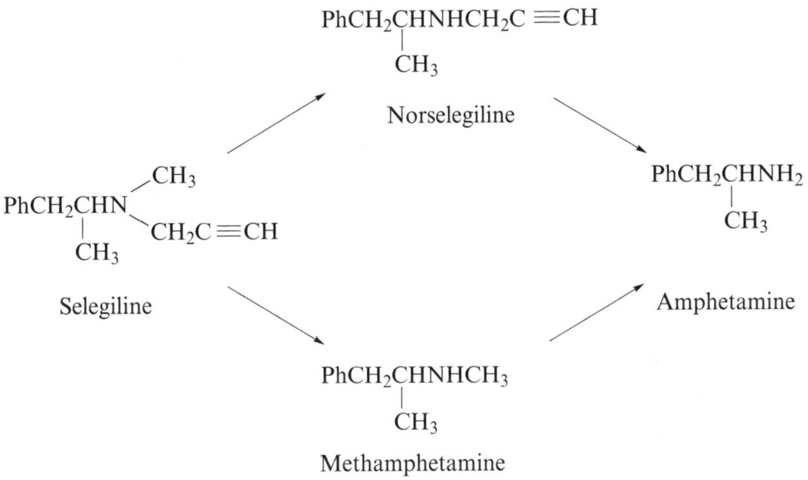

$$PhCH_2\underset{\underset{CH_3}{|}}{C}HNHCH_2C \equiv CH$$

Norselegiline

$$PhCH_2CHN\underset{\underset{CH_3}{|}}{\overset{\diagup CH_3}{\diagdown}}CH_2C \equiv CH$$

Selegiline

$$PhCH_2\underset{\underset{CH_3}{|}}{C}HNH_2$$

Amphetamine

$$PhCH_2\underset{\underset{CH_3}{|}}{C}HNHCH_3$$

Methamphetamine

SCHEME 19. Metabolism of selegiline

metabolite, namely methamphetamine. The latter and amphetamine, as well as their para-hydroxylated derivatives, have been identified as major urinary metabolites in the rat, while norselegiline was a very minor urinary metabolite[242,245] (Scheme 19). A similar metabolic detoxification pattern was also shown for the MAO inhibitors AGN-1135 (N-methyl-N-propargyl indanylamine) and TZ-650 (N-methyl-N-propargyl-2-phenylethylamine) (Scheme 18)[245]. Thus, metabolism of the latter two compounds by either N-depropargylation, N-demethylation or both reactions yielded metabolites corresponding to methamphetamine, norselegiline and amphetamine, respectively.

In man, selegiline at a therapeutic oral dose of 10 mg is metabolized very quickly and is practically undetectable in plasma by currently available methods. Thus, quantitation of metabolites is often used for pharmacokinetic and comparative bioavailabilty studies[246]. Human plasma and urinary metabolites are essentially the same as in rodents, with nor-selegiline being a minor metabolite, which therefore has failed detection in many studies[246–250].

2. Pharmacokinetics of selegiline

L-(–)-Selegiline is readily absorbed from the gastrointestinal tract. In humans maximum plasma levels of 33–45 ng/ml were obtained at 0.5–2 h after oral ingestion of 5 mg[251,252].

L-(–)-Selegiline is fairly liposoluble and thus is readily distributed in all tissues, crosses the blood brain barrier freely and concentrates there in areas rich in MAO B[170]. In the plasma it is 94% bound to proteins, albumin and gamma globulins[253] and it has a volume of distribution of some 300 liters. Its elimination $t_{1/2}$ was estimated at 39.5±23 min[252]. Selegiline undergoes first pass metabolism in the liver after oral administration. After intravenous administration the $t_{1/2}$ of distribution of selegiline was about 9 min[246,254].

Comparison of the disposition of the two enantiomers of selegiline by the novel positron emission tomography method[170] showed that both isomers were distributed very quickly. However, since the D-isomer hardly binds to MAO B, its clearance was much faster than that of the L-isomer. Interestingly, the clearance of the (^{11}C) L-isomer was as fast as the D-isomer in a patient in whom MAO B was already inhibited by previous treatment with selegiline. This observation confirms that covalent binding to the enzyme is the major reason for the slower clearance of the active L-isomer than that of the D-isomer.

3. Metabolism of pargyline

Pargyline is metabolized by rodent liver microsomes to four major metabolites (Scheme 20). Three of these, namely benzylpropargylamine (norpargyline), benzylmethylamine and methylpropargylamine, are products of N-dealkylation reactions. The fourth metabolite is pargyline N-oxide[244]. While the formation of the three N-dealkylated metabolites is mediated by cytochrome P450, there is evidence showing that N-oxidation of pargyline may be mediated by an FAD-containing monooxygenase[244,255,256]. Aromatic hydroxylation of pargyline is secondary to the above-mentioned metabolic pathways[244].

A by-product of depropargylation of simple propargyl drugs such as tripropargylamine and the MAO inhibitors is, of course, the acetylenic aldehyde propiolaldehyde (Scheme 21)[244,257,258]. Therefore, induction of N-depropargylation is expected to augment the formation of propiolaldehyde. Indeed, formation of the latter compound was induced 4.5 times by phenobarbital, compared with control rats (from 0.9 ± 0.2 μmol/30min/g liver to 0.2 ± 0.03 μmol/30 min/g liver, respectively). This observation may be of special importance because propiolaldehyde is an irreversible inhibitor of mitochondrial aldehyde dehydrogenase[257–259] and it may also be responsible for pargyline-induced hepatotoxicity in the rat[260]. Another possible route responsible for pargyline toxicity is the spontaneous decomposition of the metabolite pargyline-N-oxide to acrolein, a well-known hepatotoxin[256].

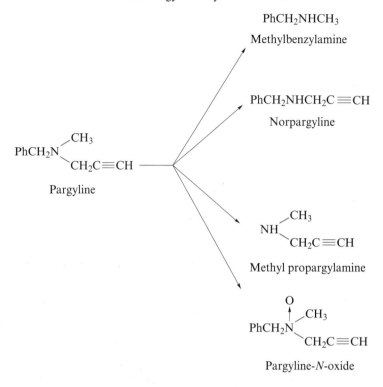

SCHEME 20. Metabolism of pargyline

$$R^1R^2NCH_2C \equiv CH \longrightarrow HC \equiv CHCHO + R^1R^2NH$$

Propiolaldehyde

SCHEME 21. Formation of propiolaldehyde

VI. ACETYLENIC FATTY ACIDS

Cyclooxygenase is an enzyme complex known to catalyze the conversion of arachidonic and related polyunsaturated fatty acids into prostanoids (prostaglandins and thromboxanes). The acetylenic analog of arachidonic acid, eicosatetraynoic acid

$$CH_3CH_2CH_2CH_2CH_2C \equiv CCH_2C \equiv CCH_2C \equiv CCH_2C \equiv CCH_2CH_2CH_2COOH$$

was first synthesized as an intermediate in arachidonic acid synthesis[261]. This compound was later found to inhibit cyclooxygenase activity in an irreversible manner. It inhibited the sheep seminal vesicle preparation cyclooxygenase activity with an IC_{50} of 0.6–2.5 μM[262–264]. However, *in vivo* experiments failed to duplicate this effect. One explanation could be avid incorporation of eicosatetraynoic acid into membrane phospholipids, as does arachidonic acid. Alternatively, the compound may be rapidly oxidized *in vivo*.

Inhibition of cyclooxygenase by eicosatetraynoic acid occurred through a dual mechanism: immediate inhibition was followed by a time-dependent inactivation of the enzyme. The latter could be blocked by α-naphthols, indicating a chain reaction involving a free radical[262,265]. The second stage in cyclooxygenase inhibition by eicosatetraynoic acid is thus mechanism-based.

Twenty carbon fatty acids with 1–3 acetylenic groups also proved to be inhibitors of cyclooxygenase. Triple bonds in positions 9 and 12 resulted in more potent inhibitors than at position 5 of the fatty acid.

The acetylenic analog of Mead Acid, 5,8,11-eicosatriynoic acid, was reported to be a selective inhibitor of platelet 12-lipoxygenase[266]. Acetylenic fatty acids also inhibited 15-lipoxygenases of plant (soybean) as well as of animal (rabbit reticulocyte) origin. The nature of the acetylenic compound significantly affected its activity on the soybean enzyme, but not the rabbit enzyme. For the former, 7,10,13-eicosatriynoic acid was the most powerful inactivator. Addition of a fourth triple bond at position 4 or 5 strongly reduced the rate of inactivation. On the other hand, the rabbit reticulocyte enzyme was inactivated almost equally well by the various acetylenic fatty acids that were tried. The mechanism of inactivation of lipoxygenases was also suicide inhibition[267,268].

VII. THYMIDYLATE SYNTHETASE INHIBITION BY PROPARGYL ANALOGS OF FOLIC ACID

The intracellular synthesis of deoxythymidylate is essential for DNA synthesis in rapidly growing cells. Therefore, fluorouracil and methotrexate exhibit antineoplastic activities by virtue of their inhibition of different enzymes involved in cellular deoxythymidylate biosynthesis (Scheme 22). Recognition of the structure of the active site of thymidylate synthetase has led to the assumption that binding to the cysteine moiety in the active site of this enzyme may inhibit its activity. This approach has thus led to the development of new antineoplastic drugs. A series of quinazoline analogs of folic acid, alkylated at N-10 with propargyl, allyl or propyl groups, has been prepared and tested for thymidylate synthetase inbibition[269]. The most potent analog in this series was the 10-propargyl-5,8-dideazafolic acid (N-[4-[N[(2-amino-4-hydroxy-6-quinazolinyl)methyl]prop-2-ynylamino]-benzoyl]-L-glutamic acid), known as CB 3717 (also PDDF, ICI 155387, NSC 327182). This compound (IC50 = 5 nM) was 15 times as active as the 10-allyl-5,8-dideazafolic acid (CB 3716) and 35 times as active as the 10-propyl-5,8-dideazafolic acid (CB 3715) as a

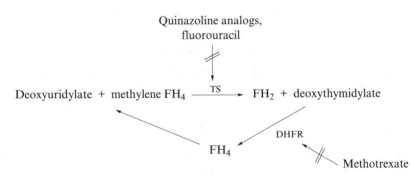

FH_2—dihydrofolate, FH_4—tetrahydrofolate, TS—thymidylate synthetase, DHFR—dihydrofolate reductase

SCHEME 22. Thymidylate cycle and its inhibition

thymidylate synthetase inhibitor (Table 3). CB 3717 was also the most potent of the three compounds in inhibiting thymidylate synthetase in L 1210 cell cultures as well as *in vivo* in mice bearing L 1210 tumors[269]. There was no difference between the three analogs in their inhibitory potential of dihydrofolate reductase, the affinity of CB 3717 towards dihydrofolate reductase being 10 times lower than towards thymidylate synthetase. Thus, the main target of CB 3717 seems to be the latter. Subsequent studies showed that CB 3717 was indeed a potent competitive thymidylate synthetase inhibitor in several human and murine cell lines, irrespective of their sensitivity or resistance to methotrexate[270–275].

TABLE 3. N[10]-substituted 5,8-dideazafolic acid derivatives

Name	R¹	R²
CB 3715	$CH_2CH_2CH_3$	NH_2
CB 3716	$CH_2CH=CH_2$	NH_2
CB 3717	$CH_2C\equiv CH$	NH_2
CB 3717 butynyl analog	$CH_2C\equiv CCH_3$	NH_2
CB 3717 methylcyano analog	$CH_2C\equiv N$	NH_2
CB 3804	$CH_2C\equiv CH$	H
CB 3819	$CH_2C\equiv CH$	CH_3
CB 3828	$CH_2C\equiv CH$	OCH_3

Structure–activity relationship studies testing different substituents on N-10 of 5,8-dideazafolic acid (Table 3), including 2-butyne and methylcyano groups, have shown that the propargyl analog was the best thymidylate synthetase inhibitor. It was also found that the glutamic acid was essential for thymidylate synthetase inhibition[276–279].

Two protein systems are involved in the transport of folates and folate analogs into cells: (1) a high affinity–low capacity transporter for reduced folate, which is also the transporter for methotrexate (RF/MTX carrier), and (2) membrane-associated folate binding proteins (mFBP), which have high affinity for folates[280,281]. CB 3717 is mainly transported by mFBP, to which it binds with very high affinity. However, other 10-propargyl analogs of CB 3717 may have higher affinities for the RF/MTX carrier, as was observed, for example, for 2-desamino-2-methyl-N[10]-propargyl-5,8-dideazafolic acid (CB 3819, ICI-198, 583)[282,283].

In the mammalian cell, CB 3717 is polyglutamated by the enzyme folylpolyglutamyl synthetase (FPGS) to a series of polyglutamates containing 1-5 residues of glutamic acid. These polyglutamate metabolites have two important properties: (1) they are more active (lower Ki) inhibitors of thymidylate synthetase than the parent compound CB 3717, and (2) the activity of the polyglutamates is of longer duration than that of CB 3717. The latter property of the polyglutamates is probably due to their inability to diffuse out of cells. The inhibitory potency of these polyglutamates on thymidylate synthetase activity and duration of their action has been observed in cell cultures from which CB 3717 was washed out. Intracellular CB 3717 diffused out very rapidly, while the intracellular inhibitory activity, presumably related to the polyglutamate derivatives of CB 3717, was maintained for several hours[284–286].

CB 3717 emerged as a very potent drug, inhibiting tumor growth in mice *in vivo*[269]. Likewise, in phase I and II clinical trials CB 3717 exhibited activity against ovarian, breast and liver cancers[287–294]. It produced, however, very serious adverse reactions, namely hepatotoxicity, expressed as raised plasma alanine aminotransferase and nephrotoxicity that resulted in highly elevated levels of plasma urea. Hepato- and nephrotoxicity were also observed in mice, in which CB 3717 precipitated in the renal tubules, causing renal histopathological lesions and significantly reduced glomerular filtration rate[295]. The renal toxicity was attributed mainly to low water solubility of the drug at physiological pH (0.009 and 0.312 mg/ml at pH 5 and 7.4, respectively[296]). Realization of these drawbacks, related to the physicochemical properties of CB 3717, prompted the synthesis of new analogs with higher water solubility at physiological pH. Three new analogs with appreciably higher water solubility resulted from this endeavor: 2-desamino-N^{10}-propargyl-5,8-dideazafolic acid (CB 3804), 2-desamino-2-methyl-N^{10}-propargyl-5,8-dideazafolic acid (CB 3819, ICI 198583) and 2-desamino-2-methoxy-N^{10}-propargyl-5,8-dideazafolic acid (CB 3828) (Table 3). These analogs of CB 3717 also proved to be very potent inhibitors of thymidylate synthetase *in vitro*. The relative potencies of CB 3717, 3804, 3819 and 3828 were 1, 0.125, 0.5 and 1, respectively. Likewise, the IC50 values of these compounds in L1210 cell cultures were 3.4, 0.4, 0.1 and 2 μM, respectively[283,296,297]. However, only CB 3804 and CB 3819 had high affinities for folylpolyglutamyl synthetase, while the affinity of CB 3828 for this enzyme was five times lower than that of CB 3717. This observation can account for the low concentrations of CB 3828 polyglutamates found *in vivo* and in cell cultures treated with the parent compound[296]. As the tetraglutamates of CB 3717, CB 3804 and CB 3819 are 119, 101 and 114 times as active as the parent analogs in inhibiting thymidylate synthetase, it becomes clear that the parent compounds act primarily as prodrugs. Administration of CB 3804 and CB 3819 to mice did not result in elevation of either urea or alanine aminotransferase serum levels, indicating lack of hepato- and nephrotoxicities. The half-lives of elimination of these new analogs were about 5 times shorter than that of CB 3717 and their clearances were 8–15 times faster than that of CB 3717[272,296,297].

Recently it was reported for the first time that a human lymphoblastoid cell line (W1-L1: C2) became resistant to 2-desamino-2-methyl-N^{10}-propargyl-5,8-dideazafolic acid (CB 3819). It was observed that resistance coincided with a 64-fold amplification of the thymidylate synthetase gene expression which led to a 200-fold increase in thymidylate synthetase protein and enzyme activity. Furthermore, CB 3819 was transformed in these cells only to a diglutamate derivative, in contrast to tetra- and pentaglutamate formation in nonresistant cells[298]. These results indicate that although derivatives of CB 3717 show promising therapeutic potential as new antineoplastic drugs, some drawbacks may still occur.

VIII. CHOLINERGIC MUSCARINIC AGENTS

Early studies showed that oxotremorine is a specific muscarinic agonist, later characterized as an M_2 receptor agonist. It was also shown that oxotremorine did not act on nicotinic receptors, as tested on neuromuscular junction preparations[299]. More recent studies, however, showed that some oxotremorine analogs also exhibit agonistic properties on cholinergic nicotinic innervated organs such as the frog rectus abdominis[300]. This activity is more pronounced in quaternary analogs, as compared to tertiary analogs of oxotremorine. Recently, it was also shown that oxotremorine analogs may be weak agonists of other muscarinic receptors as well, most notably the M_4 receptors[301].

The activity of acetylcholine, the neurotransmitter of the parasympathetic autonomic nervous system, is mediated by two types of cholinergic receptors: nicotinic and muscarinic. The former are located mainly in ganglia of the autonomic nervous system and in neuromuscular junctions, while muscarinic receptors prevail in visceral organs that are innervated by the parasympathetic system, such as the gastrointestinal tract and the heart, as well as in the central nervous system. Each of these receptor types has been well characterized pharmacologically by specific agonists, i.e. agents that activate the receptors, and by specific antagonists, i.e. agents that antagonize the effects of agonists on receptors. The alkaloids muscarine and pilocarpine are examples of specific agonists for muscarinic receptors, while atropine is the prototype of specific muscarinic antagonists[27,302].

Because muscarinic receptors are involved in a variety of physiological processes in peripheral organs as well as in the central nervous system, numerous studies have been aimed at understanding the relationship between structure and activity of muscarinic agonists and antagonists. Better understanding of these relationships may eventually lead to the development of new drugs for specific diseases, such as Parkinson's disease and Alzheimer's disease.

In the last decade several muscarinic receptors were characterized by molecular biology methods and were designated as M_1 to M_5. Additionally, new experimental specific agonists and antagonists were developed that further characterize the different muscarinic receptors (for recent review see Hulme and coworkers[303]).

Bioassays have long been used for the screening of central and peripheral activity of muscarinic agents. As the gastrointestinal tract is very rich in M_2 muscarinic receptors, the contraction of the isolated guinea pig ileum has become one of the standard tests for the evaluation of activity mediated by M_2 receptors. The activity of the basal ganglia in the brain is controlled mainly by muscarinic cholinergic nerves and by dopaminergic mechanisms. Increased activity of the muscarinic system in this part of the brain is expressed in tremor and rigidity, which are the major signs of Parkinson's disease. Thus measuring the tremorogenic effect of drugs in mice has become a routine method of screening for central muscarinic activity. Conversely, the abolishment of tremor elicited by a known muscarinic agonist (tremorolytic effect) is a measure of central antimuscarinic activity.

A. Oxotremorine Analogs

The discovery in the early 1960s that oxotremorine is the active metabolite of tremorine, which is inactive by itself[299,304], has since triggered active research on the structural properties and requirements for analogs of oxotremorine containing a butynyl moiety with specific muscarinic activity. Bebbington and coworkers[305] showed that analogs of oxotremorine containing a butenyl, butyl or phenyl group instead of the butynyl group were completely inactive. However, the butynyl group for itself did not impart any muscarinic activity as exemplified in the case of tremorine and N-propargyl trimethylammonium iodide.

Analysis of the structures of the muscarinic agonists muscarine and muscarone has led to the widely held assumption that cholinergic agonists should contain in their structure a cationic group and an anionic group separated optimally by a distance of about 3.5 Å. A third molecular site very vaguely defined was also assumed[306]. Oxotremorine seemed to fulfill all three requirements. It contained pyrrolidine as the cationic site, pyrrolidone as the anionic site and the butynyl group as the required third site. Early studies also showed that analogs of oxotremorine having a ketone or ester functionalities instead of the pyrrolidone were very potent muscarinic agonists[307]. Also, 1-acetoxy-4-trimethyl ammonium-2-butyne, synthesized by Jacob and coworkers, was a very potent muscarinic agonist[308]. On the other hand, oxotremorine analogs containing an ether or succinimide (phthalimide) moieties instead of the pyrrolidone turned out to be muscarinic antagonists[305,309–311].

Resul and coworkers[312] and Ringdahl[313] investigated two series of N-(4-amino-2-butynyl)-2-pyrrilidone (Table 4a) and N-(4-amino-2-butynyl)-N-methylacetamide compounds (Table 4b). In these series, increasing the size of the substituents on the amino group resulted in decreased muscarinic activity *in vitro* in the isolated guinea pig ileum and the *in vivo* tremorogenic activity in mice. Thus, gradually changing the substituents on the amine from pyrrolidine through dimethylamino to methyl butylamino and bigger residues led to

TABLE 4. Muscarinic activity of oxotremorine analogs

(a)

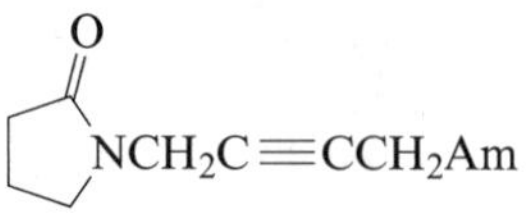

$NCH_2C \equiv CCH_2Am$

Amine	Isolated guinea pig ileum relative activity[a]	Relative tremorogenic activity	Relative affinity	Relative efficacy	% occupancy for 50% response
 (N-piperidine)	1	1	1.00	1.00	3.52
 (N-pyrrolidine)	0.49	1	0.50	4.38	0.79
$N(CH_3)_2$	5.0	6.2	0.03	6.51	0.55
$N(CH_3)(C_2H_5)$	19.5	12	0.045	1.03	3.46
$N(CH_3)(C_3H_7)$	35.8	c			
$N(C_2H_5)_2$	> 4000[b]		0.038	0.055	64.4
$N(CH_3)(C_4H_9)$	antagonist	antagonist			

[a] Activities are presented as equimolar ratios relative to oxotremorine.
[b] Partial agonist.
[c] Tremor not seen at doses below LD50.

(b)

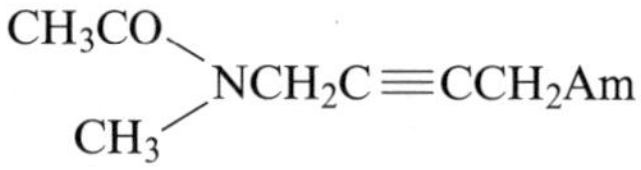

CH_3CO
$NCH_2C \equiv CCH_2Am$
CH_3

Amine	Isolated guinea pig ileum relative activity[a]	Relative tremorogenic activity	Relative affinity	Relative efficacy	% occupancy for 50% response
 (N-piperidine)	2.7	2.5	0.31	1.14	3.1
$N(CH_3)_2$	16.4	30	0.0097	6.46	0.55
$N(CH_3)(C_2H_5)$	58.0	33			
$N(CH_3)(C_3H_7)$	682	c			
$N(CH_3)(C_4H_9)$	b	—			

[a] Activities are presented as equimolar ratios relative to oxotremorine.
[b] Partial agonist.
[c] Tremor not seen at doses below LD50.

the gradual conversion of the activity of the analogs from potent agonists to competitive antagonists (Tables 4a, b). Similar observations were made with secondary or quaternary amines (*in vitro* activity only)[312]. Though the size of the pyrrolidine group seems to be very large, the steric requirements of this group are closer to those of the dimethylamino than to those of the diethylamino group[314,315].

TABLE 5. The muscarinic activity of methyloxotremorine isomers

Position of CH$_3$	Optical isomer	Isolated guinea pig ileum pA$_2$	Relative affinity	Tremorolytic^a potency (μmol/kg)
oxotremorine		5.68^b	1.0	
3'	RS	4.94^b	0.1	
4'	RS	5.00	0.1	37
5'	RS	7.03	12	0.4
	R	7.29		0.19
	S	6.0		4.2
1	RS	7.28	21	0.5
	R	7.55	39	0.26
	S	5.14	0.15	20
4	RS	4.43	0.03	11
2''	RS	5.61	0.45	5
	R	5.89	0.85	2.6
	S	5.17	0.16	56
3''	RS	5.68	0.52	5.2
	R	5.68	0.52	5.8
	S	5.69	0.54	6.5

^{a} Dose required to double the dose of oxotremorine inducing a predetermined intensity in 50% of the mice.
^{b} pK_a.

The series presented in Table 5 is unique in that all the derivatives shown are structural isomers derived from tremorine by introduction of a methyl group. Methylation at carbon 3' of oxotremorine resulted in a potent agonistic activity on the isolated guinea pig ileum and the compound was also a tremorogenic agent. 2''-methyl oxotremorine acts as a partial agonist peripherally in the isolated guinea pig ileum but it antagonizes the tremorogenic activity of oxotremorine. All the other five isomers are competitive antagonists in both test systems[316,317].

Because the introduction of the methyl group on each carbon of oxotremorine forms a chiral center, the separation of the enantiomers made possible the assessment of the effect of chirality on muscarinic activity both *in vitro* and *in vivo*. All the R enantiomers that were tested were more potent than the S enantiomers, except for 3''-methyloxotremorine, in which the R and S enantiomers were equipotent[316,318–321]. The equal activity of the R and S enantiomers of 3''-methyloxotremorine may indicate that introduction of a methyl in this position does not affect binding to the active site. The higher activity of the other R enantiomers did not depend on whether the analog was an agonist, partial agonist or antagonist (Table 5). The same observations were also made for the series of (4-amino-1-methyl-2-butynyl) N-methylacetamide[322–324].

The activity of an agonist depends on its affinity for the specific receptor, which reflects the ability of a drug to bind to a receptor, while its efficacy is a measure of the magnitude of the effect triggered by the drug–receptor complex. Therefore, resolution of the muscarinic receptor activity of oxotremorine analogs into their affinity and efficacy components may give a better insight on the structural requirement for an analog with optimal activity, as either an agonist or antagonist.

Ringdahl[313] observed in a series of tertiary oxotremorine analogs, in which only the amino group was modified (the pyrrolidone and butynyl groups were kept unchanged), that all the analogs examined were full agonists, except for the diethyl analog, which was a partial agonist (70% of maximal response of oxotremorine) (Table 4a). Replacement of the pyrrolidine group by azethidine was accompanied by a 2-fold decrease in affinity but a 4-fold increase in efficacy. The affinity of the dimethylamino analog was 30-fold weaker; however, the efficacy was increased 7-fold compared to oxotremorine. Further methylation led to increased affinity compared to the dimethylamino analog, but the efficacy was not changed. This as well as other studies[316,317,325] revealed lack of correlation between the affinities and efficacies of the analogs in the series that were examined. However, a linear correlation was found between the size of the amine moiety and the relative efficacy. It thus seems that the size of the substituents on the nitrogen and their steric features could affect the efficacy of oxotremorine analogs.

According to the classic receptor occupancy theory, the activity of an agonist is directly proportional to the receptor occupancy. In the simplest system, a 50% receptor occupation should result in 50% of maximal response. However, in many receptor systems 50% of maximal response in achieved at very low receptor occupancy. This phenomenon, referred to as spare receptors, imparts greater sensitivity to the system. In a series of oxotremorine analogs, that were tested in the isolated guinea pig ileum, at 50% of maximal response oxotremorine occupied only 3.5% of the available receptors, and the trimethylammonium derivative occupied only 0.46% of total muscarinic receptors (Table 4a). On the other hand, the partial agonist, i.e. the methyl ethyl analog of oxotremorine, occupied over 50% of the receptors at 50% of maximal response (Table 4a)[313]. The same phenomenon of spare receptors as well as lack of correlation between affinity and efficacy was also found in the acetamide series of oxotremorine analogs (Table 4b)[312].

Oxotremorine has a pK_a of 7.91 and it is mostly protonated at physiological pH (7.4)[326]. The activity of this prototype is believed to be associated with the protonated form. Most oxotremorine analogs do not differ significantly from oxotremorine in their basicity and are also protonated at pH 7.4[313]. Thus, the differences in activity between the analogs of oxotremorine cannot be ascribed to this property. In terms of lipophilicity it seems that secondary and tertiary analogs of oxotremorine are not very different[312,327,328] and are widely distributed in body fluids and tissues including the central nervous system. On the other hand, the quaternary analogs, due to their polarity, do not cross the blood brain barrier and show only peripheral activity. Thus, lipophilicity also cannot account for the differences in activity of secondary and tertiary amines.

In analogy with simple acetylcholine derivatives, where there is delocalization of the positive charge to the substituents on the nitrogen[329], changing the alkyl substituents on the nitrogen of oxotremorine analogs can account in part for the difference in activity with increasing the size of substituents on the nitrogen.

Nordstrom and coworkers[330] showed that the analog of oxotremorine, N-methyl-N-(1-methyl-4-pyrrolidono-2-butynyl) acetamide, acts as a postsynaptic agonist in various muscarinic test models. However, this analog is also an antagonist of presynaptic muscarinic receptors. Such diversity in receptor action may contribute to the differences in overall activity of analogs in this series.

Because small structural or spatial changes in the oxotremorine series could result in conversion of a potent agonist to antagonist, it was argued that muscarinic analogs related to oxotremorine, both agonists and antagonists, bind to the same active site of the receptor[324]. This is in contrast to atropine and other 'classical' antagonists which are believed to bind to an accessory receptor area, different from the binding site of agonists, by a hydrophobic interaction[331-333].

In spite of the plethora of information on the structure–activity relationship in the oxotremorine series, the exact details on its mechanism of action are still unclear.

B. Arecaidine Propargyl Esters

Examination of heterocyclic analogs of acetylcholine with restricted conformational flexibility, such as arecaidine esters, has revealed that the propargyl ester was a more potent M_2 agonist than the allyl ester and the propyl ester, which was in fact an antagonist. The propargyl ester was considerably more active than carbachol (carbamylcholine) and as active as oxotremorine, on the guinea pig ileum and atria as model systems for M_2 receptor-mediated activity[334,335]. A more detailed structure–activity relationship study of arecaidine propargyl esters showed that the 2-butynyl and 2-pentynyl esters were as potent as the propargyl ester; the 2-hexynyl ester was less potent than the propargyl ester 2-heptynyl while the 3-phenyl propargyl esters were competitive antagonists in the guinea pig ileum and atria[336]. The propargyl and 2-butynyl esters were more selective for cardiac M_2 receptors than for ileal M_2 receptors; all the other esters that were screened, including agonists and antagonists, did not show any receptor selectivity.

COOCH$_2$C≡CH

CH$_3$

Shifting of the triple bond of the acetylenic ester from carbons 2–3 to carbons 3–4 resulted in competitive antagonists for the M_2 receptor. This may indicate the importance of the position of the triple bond in relation to the cationic and ester groups. Substitution at C_1 of the propargyl group with a methyl group resulted in competitive antagonist activity. Shifting of the activity from agonist to competitive antagonist by the introduction of a methyl group was also described above in the oxotremorine series[322]. Such observations may suggest that agonists and antagonists in these series bind to the same site on the muscarinic receptor.

Arecaidine propargyl, 2-butynyl and 2-pentynyl esters caused dose-dependent tachycardia in the pithed rat. This activity was blocked by small doses of pirenzepine, indicating that it was mediated by M_1 ganglionic receptors[336].

The N-desmethyl arecaidine propargyl ester was 15 times less active compared with arecaidine propargyl ester on rat atrial and ileal preparations. Substitution of the N-methyl group of arecaidine propargyl ester by larger alkyl groups (ethyl, propyl, butyl and benzyl) resulted in analogs in which the agonistic activity decreased as the bulkiness of the alkyl group increased[337].

The propargyl ester of a bridged derivative of arecaidine, in which there is even lesser conformational flexibility than in arecaidine, was found to be a competitive antagonist in the isolated guinea pig ileum and a partial agonist of cardiac M_2 muscarinic receptors[335].

COOCH$_2$C≡CH

CH$_3$

Recently, it was shown that arecaidine propargyl ester produces relaxation of aortic rings (with intact endothelium) by interaction with M_3 receptors[338]. Also in neuroblastoma

× glioma hybrid cells arecaidine propargyl ester is a partial agonist of M_4 muscarinic receptors[301]. Thus, arecaidine propargyl ester seems to be a rather unselective muscarinic agonist.

IX. ACKNOWLEDGMENTS

We are indebted to Ms Ahuva Selinger for expert assistance in the preparation of this manuscript. Prof. Zvi Ben-Zvi is grateful to the US Environmental Protection Agency (Environmental Toxicology Division) for encouragement in the preparation of this manuscript.

X. REFERENCES

1. F. T. Bohlman, T. Burkhard and C. Zdera, *Naturally Occurring Acetylenes*, Academic Press, New York, 1973.
2. F. Bohlman, in *Chemistry and Biology of Naturally Occurring Acetylenes and Related Compounds* (Eds. J. Lam, H. Breteler, J. T. Arnason and L. Hansen), Elsevier, Amsterdam, 1988, pp. 1–21.
3. G. H. N. Towers and D. Champagne, in *Chemistry and Biology of Naturally Occurring Acetylenes and Related Compounds* (Eds. J. Lam, H. Breteler, J. T. Arnason and L. Hansen), Elsevier, Amsterdam, 1988, pp. 139–150.
4. L. C. Christensen and J. Lam, *Phytochemistry*, **29**, 2753 (1990).
5. J. B. Hudson and G. H. N. Towers, *Pharmacol. Ther.*, **49**, 181 (1991).
6. J. B. Hudson, E. A. Graham, G. Chan, A. J. Finlayson and G. H. N. Towers, *Planta Med.*, **52**, 453 (1986).
7. H. Matsunaga, M. Katano, H. Yamamoto, H. Fujito, M. Mor and K. Takata, *Chem. Pharm. Bull.*, **38**, 3480 (1990).
8. Y. Fujimoto and M. Satoh, *Chem. Pharm. Bull.*, **36**, 4206 (1988).
9. C. M. Teng, S. C. Kuo, F. N. Ko, J. C. Lee, L. G. Lee, S. W. Chen and T. F. Huang, *Biochim. Biophys. Acta.* **990**, 315 (1989).
10. C. P. Constable and G. H. N. Towers, *Planta Med.*, **55**, 35 (1989).
11. A. Muhlenfeld and H. Achenbach, *Phytochemistry*, **27**, 3853 (1988).
12. P. N. Kaul and S. K. Kulkarni, *J. Pharm. Sci.*, **67**, 1293 (1978).
13. R. H. Abeles and A. L. Maycock, *Acc. Chem. Res.*, **9**, 313 (1976).
14. R. R. Rando, *Science*, **185**, 320 (1974).
15. F. M. Miesowicz and K. Block, *J. Biol. Chem.*, **254**, 5868 (1979).
16. D. V. Santi, Y. Wataya and A. Matsuda, In *Enzyme-Activated Irreversible Inhibitors* (Eds. N. Seiler, M. J. Jung and J. Koch-Weser), Elsevier, Amsterdam, 1978, pp. 291–303.
17. C. T. Walsh, *Annu. Rev. Biochem.*, 53, 493 (1984).
18. A. Sjoerdsma, *Clin. Pharmacol. Ther.*, **30**, 3 (1981).
19. M. G. Palfreyman, P. J. Schechter, W. R. Buckett, G. P. Tell and J. Koch-Weser, *Biochem. Pharmacol.*, **30**, 817 (1981).
20. B. W. Metcalf, *Biochem. Pharmacol.*, **28**, 1703 (1979).
21. S. M. Grant and R. C. Heel, *Drugs*, **41**, 889 (1991).
22. M. J. Jung, B. Lippert, B. W. Metcalf, P. J. Schechter, P. Bohlen and A. Sjoerdsma, *J. Neurochem.*, **28**, 717 (1977).
23. T. A. Alston, *Pharmacol. Ther.*, **12**, 1 (1981).
24. P. R. Ortiz de Montellano, in *Bioactivation of Foreign Compounds* (Ed. M. W. Anders), Academic Press, New York, 1985, pp. 121–155.
25. A. Bebbington and R. W. Brimblecombe, *Adv. Drug Res.*, **2**, 143 (1965).
26. S. B. Freedman, E. A. Harley and L. L. Iversen, *Trends Pharmacol. Sci.*, **9**, Suppl. 54 (1988).
27. P. Taylor, in *Goodman and Gilman's The Pharmacological Basis of Therapeutics*, 8th edition (Eds. A. G. Gilman, J. W. Rall, A. S. Nies and P. Taylor), Pergamon Press, New York, 1990, pp. 122–130.
28. H. R. Sullivan, P. Roffey and R. E. McMahon, *Drug Metab. Disp.*, **7**, 76 (1979).
29. A. Wade, A. M. Symons, L. Martin and D. V. Parke, *Biochem. J.*, **184**, 509 (1979).
30. A. Wade, A. M. Symons, L. Martin and D. V. Parke, *Biochem. J.*, **188**, 867 (1980).

31. P. R. Ortiz de Montellano and E. A. Komives, *J. Biol. Chem.*, **260**, 3330 (1985).
32. E. A. Komives, and P. R. Ortiz de Montellano, *J. Biol. Chem.*, **262**, 9793 (1987).
33. W. K. Chan, Z. Sui and P. R. Ortiz de Montellano, *Chem. Res. Toxicol.*, **6**, 38 (1993).
34. G. J. Hammonds, W. L. Alworth, N. E. Hopkins, F. P. Guengerich and F. F. Kadlubar, *Chem. Res. Toxicol.*, **2**, 367 (1989).
35. C. A. C. Jacob, W. K. Chan, E. Shephard and P. R. Ortiz de Montellano, *J. Biol. Chem.*, **263**, 18640 (1988).
36. W. Kanhai, M. Koob, W. Dekant and D. Henschler, *Xenobiotica*, **7**, 905 (1991).
37. P. R. Ortiz de Montellano and K. L. Kunze, *J. Am. Chem. Soc.*, **102**, 7375 (1980).
38. R. E. McMahon, J. C. Turner, G. W. Whitaker and H. R. Sullivan, *Biochem. Biophys. Res. Commun.*, **99**, 662 (1981).
39. P. R. Ortiz de Montellano and K. L. Kunze, *Arch. Biochem. Biophys.*, **209**, 710 (1981).
40. E. G. Lewars, *Chem. Rev.*, **83**, 519 (1983).
41. O. P. Strausz, R. K. Gosavi, A. J. Denes and I. G. Csizmadia, *J. Am. Chem. Soc.*, **98**, 4784 (1976).
42. I. N. H. White, J. B. Campbell, P. B. Farmer, E.Bailey, N. H. Nam and D. C. Thang, *Biochem. J.*, **220**, 85 (1984).
43. F. P. Guengerich, *Life Sci.*, **47**, 1981 (1990).
44. E. D. Helton and J. W. Goldzieher, *J. Toxicol. Environ. Health* **3**, 231 (1977).
45. E. D. Helton, M. C. Williams and J. W. Goldzieher, *Steroids*, **30**, 71 (1977).
46. S. E. Schmid, W. Y. W. Au, D. E. Hill, F. F. Kadlubar and W. Slikker, *Drug Metab. Disp.*, **11**, 531 (1983).
47. J. Salmon, D. Coussediere, C. Cousty and J. P. Raynaud, *J. Steroid Biochem.*, **18**, 565 (1983).
48. J. Salmon, D. Coussediere, C. Cousty and J. P. Raynaud, *J. Steroid Biochem.*, **19**, 1223 (1983).
49. R. E. Ranney, *J. Toxicol Environ. Health*, **3**, 139 (1977).
50. S. F. Sisenwine, H. B. Kimmel, A. L. Liu and H. W. Ruelius, *Drug Metab. Disp.*, **7**, 1 (1979).
51. C. R. Abolin, T. N. Tozer, J. C. Craig and L. C. Gruenke, *Science*, **209**, 703 (1980).
52. W. Dieterle, J. W. Faigle, H. Mory, W. J. Richter and W. Theobald, *Eur. J. Clin. Pharmacol.*, **9**, 135 (1975).
53. S. Levy, B. Yagen and R. Mechoulam, *Science.*, **200**, 1391 (1978).
54. B. Yagen, S. Levy, R. Mechoulam and Z. Ben-Zvi, *J. Am. Chem. Soc.*, **99**, 6444 (1977).
55. W. Kanhai, W. Dekant and D. Henschler, *Chem. Res. Toxicol.*, **2**, 51 (1989).
56. W. Dekant, S. Vamvakas, M. Koob, A. Kochling, W. Kanhai, D. Muller and D. Henschler, *Environ. Health Perspect.*, **88**, 107 (1990).
57. N. J. Patel, J. S. Fullone and M. W. Anders, *Mol. Brain Res.*, **17**, 53 (1993).
58. L. Barret, S. Torch, Y. Usson, B. Gonthier and R. Saxod, *Neurosci. Lett.*, **131**, 141 (1991).
59. F. Naftolin and G. Tolis, *Clin. Obstet. Gynecol.*, **21**, 17 (1978).
60. R. Stanhope and C. G. D. Brook, *J. Endocrinol.*, **318**, 828 (1988).
61. H. Kuhl, *Maturitas*, **12**, 171 (1990).
62. C. Djerassi, L. Miramontes, G. Rosenkranz and F. Sondheimer, *J. Am. Chem. Soc.*, **76**, 4092 (1954).
63. H. H. Inhoffen, W. Longemann, W. Hohlweg and A. Serini, *Chem. Ber.*, **71B**, 1024 (1938).
64. J. Rock, C. M. Garcia and G. Pincus, *Recent Prog. Horm. Res.*, **13**, 323 (1957).
65. M. P. Vessey and R. Doll, *Br. Med. J.*, **2**, 199 (1968).
66. N. M. Kaplan, *Annu. Rev. Med.*, **29**, 31(1978).
67. T. W. Meade, *Am. J. Obstet. Gynecol.*, **158**, 1646 (1988).
68. M. P. Vessey, *Br. J. Fam. Plann.*, **6**, Suppl. 1 (1980).
69. R. W. Shaw, *Br. J. Obstet. Gynecol.*, **94**, 724 (1987).
70. I. N. H. White and U. Muller-Eberhard, *Biochem. J.*, **166**, 57 (1977).
71. I. N. H. White, *Biochem. J.*, **196**, 575 (1981).
72. I. N. H. White, *Biochem. J.*, **174**, 853 (1978).
73. P. R. Ortiz de Montellano, K. L. Kunze, G. S. Yost and B. A. Mico, *Proc. Natl. Acad. Sci., USA*, **76**, 746 (1979).
74. P. R. Ortiz de Montellano and K. L. Kunze, *J. Biol. Chem.*, **255**, 5578 (1980).
75. I. N. H. White, *Biochem. Pharmacol.*, **29**, 3253 (1980).
76. P. R. Ortiz de Montellano and M. A. Correia, *Annu. Rev. Pharmacol. Toxicol.*, **23**, 481 (1983).
77. P. R. Ortiz de Montellano, K. L. Kunze and B. A. Mico, *Mol. Pharmacol.*, **18**, 602 (1980).
78. P. R. Ortiz de Montellano, K. L. Kunze, H. S. Beilan and C. Wheeler, *Biochemistry*, **21**, 1331 (1982).

79. P. R. Ortiz de Montellano and K. L. Kunze, *Biochemistry*, **20**, 7266 (1981).
80. K. L. Kunze, B. L. K. Mangold, C. Wheeler, H. S. Beilan and P. R. Ortiz de Montellano, *J. Biol. Chem.*, **258**, 4202 (1983).
81. I. N. H. White, *Chem. Biol. Interact.*, **39**, 231 (1982).
82. P. R. Ortiz de Montellano, H. S. Beilan and J. M. Mathews, *J. Med. Chem.*, **25**, 1174 (1982).
83. P. R. Ortiz de Montellano, B. A. Mico and K. L. Kunze, in *Molecular Basis of Drug Action* (Eds. T. Singer and R. Ondarza), Elsevier, New York, 1981, pp. 151–166.
84. H. W. Davies, S. G. Britt and L. R. Pohl, *Chem. Biol. Interact.*, **58**, 345 (1986).
85. L.-S. L. Gan, J. Y. L. Lu, D. M. Hershkowitz and W. L. Alworth, *Biochem. Biophys. Res. Commun.*, **129**, 591 (1985).
86. D. K. Lavalle, *The Chemistry and Biochemistry of N-Substituted Porphyrins*, V.C.H. Publishers Inc., New York, 1987, pp. 209–260.
87. I. N. H. White, *Pharm. Res.*, **2**, 141 (1986).
88. I. N. H. White, *Biochem. Pharmacol.*, **31**, 1337 (1982).
89. I. N. H. White, D. C. Blakey, M. L. Green, M. Jarman and M. R. Schulten, *Biochem. J.*, **236**, 379 (1986).
90. I. N. H. White, A. R. Boobis and D. S. Davies, *Biochem. Pharmacol.*, **33**, 459 (1984).
91. P. R. Ortiz de Montellano and K. L. Kunze, *Biochem. Biophys. Res. Commun.*, **94**, 443 (1980).
92. P. R. Ortiz de Montellano and B. A. Mico, *Mol. Pharmacol.*, **18**, 128 (1980).
93. P. R. Ortiz de Montellano, B. A. Mico, J. M. Mathews, K. L. Kunze, G. T. Miwa and A. Y. H. Lu, *Arch. Biochem. Biophys.*, **210**, 717 (1981).
94. P. R. Ortiz de Montellano and B. A. Mico, *Arch. Biochem. Biophys.*, **206**, 43 (1981).
95. Y. Osawa and L. R. Pohl, *Chem. Res. Toxicol.*, **2**, 131 (1989).
96. E. S. Roberts, N. E. Hopkins, W. L. Alworth and P. F. Hollenberg, *Chem. Res. Toxicol.*, **6**, 470 (1993).
97. L. -S. L. Gan, A. L. Acebo and W. L. Alworth, *Biochemistry*, **23**, 3827 (1984).
98. D. M. Jerina, D. P. Michaud, R. J. Feldman, R. N. Armstrong, K. P. Vyas, D. R. Thakker, H. Yagi, R. E. Thomas, D. E. Ryan and W. Levin, in *Microsomes, Drug Oxidation and Drug Toxicity* (Eds. R. Sato and R. Kato), Japan Scientific Societies Press, Tokyo, 1982, pp. 195–201.
99. M. Hall, D. K. Parker, P. L. Grover, J-Y. L. Lu, N. E. Hopkins and W. L. Alworth, *Chem. Biol. Interact.*, **76**, 181 (1990).
100. W. L. Alworth, A. Viaje, A. Sandoval, B. S. Warren and T. J. Slaga, *Carcinogenesis*, **12**, 1209 (1991).
101. N. E. Hopkins, M. K. Foroozesh and W. L. Alworth, *Biochem. Pharmacol.*, **44**, 787 (1992).
102. M. D. Burke, S. Thompson, C. R. Elcombe, J. Helpert, T. Haaparanta and R. T. Mayer, *Biochem. Pharmacol.*, **34**, 3337 (1985).
103. J. D. Yager Jr. and R. Yager, *Cancer Res.*, **40**, 3680 (1980).
104. P. J. Herzog, V. Ilacqua, M. Rustia and R. Gingell, *Toxicol. Appl. Pharmacol.*, **54**, 340 (1980).
105. J. Benrekassa and F. Decloitre, *Biochem. Pharmacol.*, **32**, 347 (1983).
106. F. P. Guengerich, *Chem. Res. Toxicol.*, **3**, 363 (1990).
107. S. A. Wrighton, W. R. Brian, M. A. Sari, M. Iwasaki, F. P. Guengerich, J. L. Raucy, D. T. Molowa and M. van den Branden, *Mol. Pharmacol.*, **38**, 207 (1990).
108. A. Nagahisha, R. W. Spencer and W. H. Orme-Johnson, *J. Biol. Chem.*, **258**, 6721 (1983).
109. O. Olakanmi and D. V. Seybert, *J. Steroid Biochem.*, **36**, 273 (1990).
110. J. Fishman, *Cancer Res.*, **42**, Suppl. 3277 (1982).
111. J. O. Johnston and B. W. Metcalf, in *Novel Approaches to Cancer Therapy* (Ed. P. S. Sunkara), Academic Press, New York, 1986, pp. 307–238.
112. A. M. H. Brodie, *Biochem. Pharmacol.*, **34**, 3213 (1985).
113. J. H. H. Thijssen, M. A. Blankenstein, G. H. Donker and J. Daroszewski, *J. Steroid Biochem.*, **39**, 799 (1991).
114. W. R. Miller, *J. Steroid Biochem.*, **39**, 783 (1991).
115. D. Henderson, U. F. Habenicht, Y. Nishino, U. Kerb and M. F. El Etreby, *J. Steroid Biochem.*, **25**, 867 (1986).
116. A. M. H. Brodie, W. C. Schwarzl, A. A. Shaikh and H. J. Brodie, *Endocrionology*, **100**, 1684 (1977).
117. B. W. Metcalf, C. L. Wright, J. P. Burkhart and J. O. Johnston, *J. Am. Chem. Soc.*, **103**, 3221 (1981).
118. D. F. Covey, W. F. Hood and V. D. Parikh, *J. Biol. Chem.*, **256**, 1076 (1981).

119. M. E. Brandt, D. Puett, D. F. Covey and S. J. Zimniski, *J. Steroid Biochem.*, **31**, 317 (1988).
120. S. J. Zimniski, M. E. Brandt, D. F. Covey and D. Puett, *Steroids*, **50**, 135 (1987).
121. J. O. Johnston, C. L. Wright and B. W. Metcalf, *Endocrinology*, **115**, 776 (1984).
122. J. O. Johnston, C. L. Wright and A. Leeson, *Steroids*, **56**, 180 (1991).
123. M. E. Brandt, D. Puett, R. Garola, K. Fendl, D. F. Covey and S. J. Zimniski, in *Hormone, Cell Biology and Cancer, Perspectives and Potentials* (Eds. W. D. Hankins and D. Puett), Alan R. Liss Inc., New York, 1988, pp. 65–84.
124. T. Zaccheo, D. Guidici, G. Ornati, A. Panzeri and E. di Salle, *Eur. J. Cancer*, **27**, 1145 (1991).
125. E. di Salle, D. Guidici, G. Briatico and G. Ornati, *Ann. N. Y. Acad. Sci.*, **595**, 357 (1990).
126. M. E. Brandt, D. F. Covey and S. J. Zimniski, *J. Enzyme Inhib.*, **4**, 143 (1990).
127. C. Longcope, A. Femino and J. O. Johnston, *Endocrinology*, **122**, 2007 (1988).
128. J. O. Johnston, C. L. Wright and R. C. Shumaker, *J. Steroid Biochem.*, **33**, 521 (1989).
129. D. Puett, M. E. Brandt, D. F. Covey and S. J. Zimniski, in *Endocrinology and Malignancy: Basic and Clinical Issues* (Eds. E. E. Baulieu, S. Jacobelli and W. L. McGuire), Parthenon Publ., Carnforth, England, 1986, pp. 278–289.
130. E. di Salle, G. Briatico, D. Giudici, G. Ornati and T. Zaccheo, *J. Steroid Biochem.*, **34**, 431 (1989).
131. P. A. Marcotte and C. H. Robinson, *Cancer Res.* **42**, Suppl. 3322 (1982).
132. J. O. Johnston, *Steroids*, **50**, 105 (1987).
133. M. Numazawa, A. Mutsumi, K. Hoshi, M. Oshibe, E. Ishikawa and H. Kigawa, *J. Med. Chem.* **34**, 2496 (1991).
134. M. Numazawa. A. Mutsumi, N. Asano and Y. Ito, *Steroids*, **58**, 40 (1993).
135. J. O. Johnston, C. L. Wright, G. W. Holbert and H. D. Benson, *J. Enzyme Inhib.*, **4**, 137 (1990).
136. Y. Osawa, C. Yarborough and Y. Osawa, *Science*, **215**, 1251 (1982).
137. Y. Osawa, Y. Osawa, C. Yarborough and L. Borzynski, *Biochem. Soc. Trans.*, **11**, 656 (1985).
138. G. T. Griffing, M. Holbrook, J. C. Melby, J. Alberta and M. R. Orme-Johnson, *Am. J. Med. Sci.*, **298** 833 (1989).
139. G. T. Griffing, J. C. Melby, M. Holbrook and J. O. Johnston, *Hypertension*, **17**, 771 (1991).
140. J. P. Johnson, *Biochem. Pharmacol.*, **17**, 1285 (1968).
141. C. J. Fowler, B. A. Callingham. T. J. Mantle and K. F. Tipton, *Biochem. Pharmacol.*, **27**, 97 (1978).
142. C. J. Fowler, T. J. Mantle and K. F. Tipton, *Biochem. Pharmacol.*, **31**, 3555 (1982).
143. K. F. Tipton, C. J. Fowler and M. D. Houslay, in *Monoamine Oxidase, Basic and Clinical Frontiers* (Eds. K. Kamijo, E. Usdin and T. Nagatsu), Excerpta Medica, Amsterdam, 1982, pp. 87–99.
144. M. B. H. Youdim and J. P. M. Finberg, *Biochem. Pharmacol.*, **41**, 155 (1991).
145. A. J. W. Bach, N. C. Lan, D. L. Johnson, C. W. Abell, M. E. Bembenek, S. W. Kwan, P. Seeburg and J. C. Shih, *Proc. Natl. Acad. Sci. USA*, **85**, 4934 (1988).
146. Y. P. P. Hsu, W. Weyler, S. Chen, K. B. Sims, W. B. Rinehart, M. Utterback, J. F. Powell and X. O. Breakfield, *J. Neurochem.*, **51**, 1321 (1988).
147. J. F. Powell, Y. P. P. Hsu, W. Weyler, S. Chen, J. I. Salach, K. Andrikopoulus, J. Mallet and X. O. Breakfield, *Biochem. J.*, **259**, 407 (1989).
148. T. Kuwahara, S. Takamoto and A. Ito, *Agr. Biol. Chem.*, **54**, 253 (1990).
149. J. E. Pintar, J. Barbosa, U. Francke, C. M. Castiglione, M. Hawkins Jr. and X. O. Breakfield, *J. Neurosci.*, **1**, 166 (1981).
150. L. M. Kochersperger, E. L. Parker, M. Siciliano, G. J. Darlington and R. M. Denney, *J. Neurosci. Res.*, **16**, 601 (1986).
151. W. Weyler, *Biochem. J.*, **260**, 726 (1989).
152. T. Egashira, *Jap. J. Pharmacol.*, **26**, 493 (1976).
153. W. Weyler and J. I. Salach, *J. Biol. Chem.*, **260**, 13199 (1985).
154. C. H. Donnelly and D. L. Murphy, *Biochem. Pharmacol.*, **26**, 853 (1977).
155. P. A. Bond and R. L. Cundall, *Clin. Chim. Acta*, **80**, 317 (1977).
156. J. Willoughby, V. Glover and M. Sandler, *J. Neural Transm.*, **74**, 29 (1988).
157. K. N. Westlund, R. M. Denney, L. M. Kochersperger, R. M. Rose and C. W. Abell, *Science*, **230**, 181 (1985).
158. C. Konardi, E. Svoma, K. Jellinger, P. Riederer, R. M. Denney and J. Thibault), *Neuroscience*, **26**, 791 (1988).
159. C. Konardi, P. Riederer and H. Henisen, in *Early Diagnosis and Preventative Therapy in Parkinson's Disease* (Eds. H. Przuntek and P. Riederer), Springer-Verlag, Wien, 1989, pp. 243–248.

160. J. G. Richards, J. Saura, J. Ulrich and M. DaPrada, *Psychopharmacology*, **106**, Suppl. 21 (1992).
161. A. L. Maycock, R. H. Abeles, J. I. Salach and T. P. Singer, *Biochemistry*, **15**, 114 (1976).
162. J. I. Salach, K. Detmer and M. B. H. Youdim, *Mol. Pharmacol.*, **16**, 234 (1979).
163. M. Gerlach, P. Riederer and M. B. H. Youdim, *Eur. J. Pharmacol.*, **226**, 97 (1992).
164. N. Gomez, M. Unzeta, K. F. Tipton, M. C. Anderson and A. M. O'Carrol, *Biochem. Pharmacol.*, **35**, 4467 (1986).
165. Y. Arai and H. Kinemuchi, *J. Neural Transm.*, **72**, 99 (1980).
166. P. Riederer and M. B. H. Youdim, *J. Neurochem.*, **46**, 1359 (1986).
167. C. J. Fowler, L. Oreland, J. Marensson and R. Winblad, *Naunyn-Schmiedeberg's Arch. Pharmacol.*, **311**, 263 (1980).
168. S. S. Jossan, P. G. Gillberg, C. G. Gottfries, I. Karlsson and L. Oreland, *Neuroscience*, **45**, 1 (1991).
169. J. S. Fowler, A. P. Wolf, R. R. MacGregor, S. L. Dewey, J. Logan, D. J. Schlyer and B. Langstrom, *J. Neurochem.*, **51**, 1524 (1988).
170. J. S. Fowler, R. R. MacGregor, A. P. Wolf, C. D. Arnett, S. L. Dewey, D. J. Schlyer, D. Christman, J. Logan, M. Smith, H. Sachs, S. M. Aquilonius, P. Bjurling, C. Halldin, P. Hartvig, K. L. Leenders, H. Lundqvist, L. Oreland, C. G. Stalnacke and B. Langstrom, *Science*, **235**, 481 (1987).
171. S. S. Jossan, R. d'Argy, P. G. Gillberg, S. M. Aquilonius, B. Langstrom, C. Halldin, P. Bjurling, C. G. Stalnacke, J. S. Fowler, R. R. MacGregor and L. Oreland, *J. Neural Transm.*, **77**, 55 (1989).
172. D. L. Murphy, J. Sunderland, N. A. Garrick, C. S. Aulakh and R. M. Cohen, in *Clinical Pharmacology in Psychiatry, Selectivity in Psychotropic Drug Action—Promise or Problems* (Eds. S. G. Dahl, L. F. Gram, S. M. Paul and W. Z. Potter), Springer-Verlag, Berlin, 1987, pp. 135–146.
173. E. J. Filinger, *Med. Sci. Res.*, **17**, 547 (1989).
174. E. C. Twist, S. N. Mitchell, T. H. Corn and I. C. Campbell, *Eur. J. Pharmacol.*, **196**, 157 (1991).
175. I. C. Campbell, D. S. Robinson, W. Lovenberg and D. L. Murphy, *J. Neurochem.*, **32**, 49 (1979).
176. N. A. Garrick, M. Scheinin, W. H. Chand, M. Linnoila and D. L. Murphy, *Biochem. Pharmacol.*, **33**, 1423 (1984).
177. J. D. Newman, J. T. Winslow and D. L. Murphy, *Brain Res.*, **538**, 24 (1991).
178. C. J. Fowler and S. B. Ross, *Med. Res. Rev.*, **4**, 323 (1984).
179. J. P. M. Finberg, G. Ari, G. Lavian and D. Hovevey-Sion, *J. Neural Transm.*, Suppl., **32**, 405 (1990).
180. G. F. Oxenkrug, R. B. McCauley, D. J. Fontana, I. M. McIntyre and R. L. Commissaris, *J. Neural Transm.*, **66**, 271 (1986).
181. B. Blackwell and E. Marley, *Br. J. Pharmacol.*, **26**, 120 (1966).
182. B. Blackwell, E. Marley, J. Price and D. Taylor, *Br. J. Psychiat.*, **113**, 349 (1967).
183. J. P. M. Finberg and M. Tenne, *Br. J. Pharmacol.*, **77**, 13 (1982).
184. J. P. M. Finberg and M. B. H. Youdim, *J. Neural Transm. Suppl.* **26**, 11 (1988).
185. S. Lipper, D. L. Murphy, S. Slater and M. Buchsbaum, *Psychopharmacology*, **62**, 123 (1979).
186. A. Zametkin, J. L. Rapoport and D. L. Murphy, *Arch. Gen. Psychiat.*, **42**, 962 (1985).
187. T. P. Zahn, T. R. Insel and D. L. Murphy, *Br. J. Psychiat.*, **145**, 39 (1984).
188. J. Knoll, *Acta Neurol. Scand.*, **95**, 57 (1983).
189. W. E. Haefety, R. Kettler, H. H. Keller and M. DaPrada, *Adv. Neurol.*, **53**, 505 (1990).
190. T. Sunderland, E. A. Mueller, R. M. Cohen, D. C. Jimerson, D. Pickar and D. L. Murphy, *Psychopharmacology*, **86**, 432 (1985).
191. G. Simpson, K. White, E. Pi, J. Razani and R. Sloane, *Psychopharmacol. Bull.*, **19**, 340 (1983).
192. G. Eisenhofer, D. S. Goldstein and R. Stull, *Clin. Chem.*, **32**, 3030 (1986).
193. L. Lauber and P. C. Waldmeier, *J. Neural Transm.*, **60**, 247 (1984).
194. A. A. Boulton, *Prog. Neuro-Psychopharmacol. Biol. Psychiatr.*, **15**, 139 (1991).
195. J. Knoll, *J. Neural Transm., Suppl.*, **25**, 45 (1987).
196. C. G. Gottfries, *Psychopharmacolgy*, **86**, 245 (1985).
197. J. Knoll, *Mech. Ageing Dev.*, **46**, 237 (1988).
198. M. C. Carrillo, S. Kanai, M. Nokubo and K. Kitani, *Life Sci.*, **48**, 517 (1991).
199. R. E. Heikkila, L. Manzino, F. C. Cabbat and R. C. Duvoisin, *Nature*, **311**, 467 (1984).
200. J. Knoll, *Acta Neurol. Scand.*, **126**, 83 (1989).
201. K. T. Finnegan, J. J. Skrat, I. Irwin, L. E. DeLanney and J. W. Langston, *Eur. J. Pharmacol.*, **184**, 119 (1990).

202. M. B. H. Youdim and J. P. M. Finberg, *Adv. Neurol.*, **45**, 127 (1986).
203. P. Dostert, M. Strolin-Benedetti, F. Della Vedova, C. Allievi, R. La Croix, G. Dordain, D. Vernai and F. Durit, *Adv. Neurol.*, **60**, 218 (1993).
204. T. Miwa, N. Takeda, H. Yoshizumi, A. Tatematsu, M. Yoshida, P. Dostert, M. Naoi and T. Nagatsu, *Adv. Neurol.*, **60**, 234 (1993).
205. M. Naoi, P. Dostert, M. Yoshida and T. Nagatsu, *Adv. Neurol.*, **60**, 212 (1993).
206. G. M. Stern. A. J. Lees and M. Sandler, *J. Neural Transm.*, **43**, 245 (1978).
207. G. Engberg, J. Elebring and H. Nissbrandt, *J. Pharmacol. Exp Ther.*, **259**, 841 (1991).
208. S. Yasar, C. W. Schindler, E. B. Thorndike, I. Szelenyi and S. R. Goldberg, *J. Pharmacol. Exp. Ther.*, **265**, 1 (1993).
209. K. A. Jellinger, E. Kienzl, G. Rumpelmaier, W. Paulus, P. Riederer, H. Stachelberger, M. B. H. Youdim and D. Ben-Shachar, *Adv. Neurol.*, **60**, 267 (1993).
210. M. B. H. Youdim, D. Ben-Shachar, G. Eshel, J. P. M. Finberg and P. Riederer, *Adv. Neurol.*, **60**, 259 (1993).
211. E. Sofic, P. Riederer, H. Heinsen, H. Beckmann, G. P. Reynolds, G. Hebenstreit and M. B. H. Youdim, *J. Neural Transm.*, **74**, 199 (1988).
212. C. W. Olanow, *Neurology,* **40**, *Suppl.* 3, 32 (1990).
213. E. Csanda, M. Tarczy, A. Takats, I. Mogyoros, A. Koves and G. Katona, *J. Neural Transm.,* *Suppl.*, **19**, 283 (1983).
214. The Parkinson Study Group (DATATOP), *N. Eng. J. Med.*, **321**, 1364 (1989).
215. The Parkinson Study Group (DATATOP), *N. Eng. J. Med.*, **328**, 176 (1993).
216. P. A. Lewitt and The Parkinson Study Group, *Acta Neurol. Scand.*, **84**, *Suppl.* 79 (1991).
217. J. W. Langston, *Neurology*, **40**, *Suppl.* 3, 61 (1990).
218. A. M. Cesura and A. Pletscher, *Prog. Drug Res.*, **38**, 171 (1992).
219. I. Shoulson, *Ann. N. Y. Acad. Sci.*, **648**, 306 (1992).
220. J. Mann and S. Gershon, *Life Sci.*, **26**, 877 (1980).
221. J. Mandlewicz and M. B. H. Youdim. *Br. J. Psychiat.*, **142**, 508 (1983).
222. P. J. McGrath, J. W. Stewart, W. Harrison, S. Wager, E. N. Nunes and F. M. Quitkin, *Psychopharmacol. Bull.*, **25**, 63 (1989).
223. E. Martini, I. Pataki, K. Szilagyi and V. Vender, *Pharmacopsychiatry*, **20**, 256 (1987).
224. G. L. Piccinin, G. Finaldi and M. Piccirilli, *Clin. Neuropharmacol.*, **13**, 147 (1990).
225. N. Campi, G. P. Todeschini and L. Scarzella, *Clin. Ther.*, **12**, 306 (1990).
226. L. I. Golbe, *Neurology*, **39**, 1109 (1989).
227. Y. C. Martin, W. B. Martin and J. D. Taylor, *J. Med. Chem.*, **18**, 883 (1975).
228. B. J. Robinson, *Biochem. Pharmacol.*, **34**, 4105 (1985).
229. P. H. Yu, B. A. Davis and A. A. Boulton, *J. Med. Chem.*, **35**, 3705 (1992).
230. M. A. Cruces, C. Elorriaga and E. Fernandez–Alvarez, *Eur. J. Med. Chem.*, **26**, 33 (1991).
231. A. S. Kalir, A. Sabbagh and M. B. H. Youdim, *Br. J. Pharmacol.*, **73**, 55 (1981).
232. M. B. H. Youdim, in *Monoamine Oxidase, Basic and Clinical Frontiers* (Eds. K. Kamijo, E. Usdin and T. Nagatsu), Excerpta Medica, Amsterdam, 1982, pp. 171–173.
233. K. F. Tipton, J. M. McCrodden, A. S. Kalir, and M. B. H. Youdim, *Biochem. Pharmacol.*, **31**, 1251 (1982).
234. J. Knoll, in *Monoamine Oxidase: Structure, Function and Altered Functions* (Eds. T. P. Singer, R. W. von Korff and D. L. Murphy), Academic Press, New York, 1979, pp. 431–446.
235. K. Magyar, Z. Ecseri, G. Bernath, E. Satory and J. Knoll, in *Monoamine Oxidases and their Selective Inhibition* (Ed. K. Magyar), Pergamon Press, New York, 1979, pp. 11–21.
236. C. Fernandez Garcia, J. L. Marco and E. Fernandez-Alvarez, *Eur. J. Med. Chem.*, **27**, 909 (1992).
237. M. A. Cruces, C. Elorriaga, E. Fernandez-Alvarez and O. Nieto Lopez, *Eur. J. Med. Chem.*, **25**, 257 (1990).
238. M. A. Cruces, C. Elorriaga and E. Fernandez-Alvarez, *Biochem. Pharmacol.*, **40**, 535 (1990).
239. D. Balsa, E. Fernandez-Alvarez, K. F. Tipton and M. Unzeta, *J. Neural Transm.*, *Suppl.*, **32**, 103 (1990).
240. P. H. Yu, *J. Pharm. Pharmacol.*, **41**, 205 (1989).
241. A. S. Kalgutkar and N. Castagnoli, *J. Med. Chem.*, **35**, 4165 (1992).
242. T. Yoshida, Y. Yamada, T. Yamamoto, and Y. Kuroiwa, *Xenobiotica*, **16**, 129 (1986).
243. T. Yoshida, T. Oguro and Y. Kuroiwa, *Xenobiotica*, **17**, 957 (1987).
244. A. M. Weli and B. Lindeke, *Biochem. Pharmacol.*, **34**, 1993 (1985).
245. H. Kalasz, L. Kerecsen, J. Knoll and J. Pucsok, *J. Chromatogr.*, **499**, 589 (1990).

786 Z. Ben-Zvi and A. Danon

246. E. H. Heinonen, V. Myllyla, K. Sotaniemi, R. Lammintausta, J. S. Salonen, M. Anttila, M. Savijarvi, M. Kotila and U. K. Rinne, *Acta Neurol. Scand.*, **126**, 93 (1989).
247. G. P. Reynolds, J. D. Elsworth, K. Blau, M. Sandler, A. J. Lees and G. M. Stern, *Br. J. Clin. Pharmacol.*, **6**, 542 (1978).
248. M. Schachter, C. D. Marsden, J. D. Parkes, P. Jenner and B. Testa, *J. Neurol. Neurosurg. Psychiat.*, **43**, 1016 (1980).
249. J. D. Elsworth, M. Sandler, A. J. Lees, C. Ward and G. M. Stern, *J. Neural Transm.*, **54**, 105 (1982).
250. S. E. Roselaar, N. Langdon, C. B. Lock, P. Jenner and J. D. Parkes, *Sleep*, **10**, 491 (1987).
251. K. Magyar and L. Tothfalusi, *Pol. J. Pharmacol. Pharm.*, **36**, 373 (1984).
252. A. Benakis, C. Bretton, C. Bouvier and C. Plessas, *Data on file*, Somerset Pharmaceuticals, Inc., Denville, N. J. (1988).
253. E. Szoko, H. Kalasz, L. Kerecsen and L. Magyar, *Pol. J. Pharmacol. Pharm.*, **36**, 413 (1984).
254. E. H. Heinonen and R. Lammintausta, *Acta Neurol. Scand.*, **136**, *Suppl.* 44 (1991).
255. A. M. Weli and B. Lindeke, *Xenobiotica*, **16**, 281 (1986).
256. G. Hellstrom, B. Lindeke, A. H. Khuthier and M. A. Al-Iraqi, *Chem. Biol. Interact.*, **34**, 185 (1981).
257. F. N. Shirota, E. G. DeMaster and H. T. Nagasawa, *J. Med. Chem.*, **22**, 463 (1979).
258. F. N. Shirota, E. G. DeMaster, J. A. Elberling and H. T. Nagasawa, *J. Med. Chem.*, **23**, 669 (1980).
259. E. G. DeMaster, F. N. Shirota and H. T. Nagasawa, *Biochem. Pharmacol.*, **35**, 1481 (1986).
260. E. G. DeMaster, H. W. Summer, E. Kaplan, F. N. Shirota and H. T. Nagasawa, *Toxicol. Appl. Pharmacol.*, **65**, 390 (1982).
261. J. M. Osbond, P. G. Philpott and J. C. Wickens, *J. Chem. Soc.*, 2779 (1961).
262. J. Y. van der Hoek and W. E. M. Lands, *Biochim. Biophys. Acta*, **296**, 374 (1973).
263. D. G. Ahern and D. T. Downing, *Biochim. Biophys. Acta*, **210**, 456 (1970).
264. D. T. Downing, D. G. Ahern and M. L. Bachta, *Biochem. Biophys. Res. Commun.*, **40**, 218 (1970).
265. W. E. M. Lands, P. R. Letellier, L. H. Rome and J. Y. van der Hoek, *Fed. Proc. Fed. Am. Soc. Exp. Biol.*, **31**, 476A (1972).
266. S. Hammarstrom, *Biochim. Biophys. Acta*, **487**, 517 (1977).
267. M. Kulhn, H. G. Holzhutter, T. Schewe, C. Hiebsch and S. M. Rapoport, *Eur. J. Biochem.*, **139**, 577 (1984).
268. H. Kuhn, K. Hayess, H. G. Holzhutter, D. A. Zabolotzski, G. I. Myagkova and T. Schewe, *Biomed. Biochim. Acta*, **50**, 835 (1991).
269. T. R. Jones, A. H. Calvert, A. L. Jackman, S. J. Brown, M. Jones and K. R. Harrap, *Eur. J. Cancer*, **17**, 11 (1981).
270. R. C. Jackson, A. L. Jackman and A. H. Calvert, *Biochem. Pharmacol.*, **32**, 3783 (1983).
271. Y. C. Cheng, G. E. Dutschman, M. C. Starnes, M. H. Fisher, N. T. Nanavati and M. G. Nair, *Cancer Res.*, **45**, 558 (1985).
272. A. L. Jackman, G. A. Taylor, B. M. O'Connor, J. A. Bishop, R. G. Moran and A. H. Calvert, *Cancer Res.*, **50**, 5212 (1990).
273. G. Jansen, J. H. Schornagel, G. R. Westerhof, G. Rijksen, D. R. Newell and A. L. Jackman, *Cancer Res.*, **50**, 7544 (1990).
274. N. J. Curtin, A. L. Harris and G. W. Aherne, *Cancer Res.*, **51**, 2346 (1991).
275. B. F. A. M. van der Laan, G. Jansen, G. A. M. Kathmann, G. R. Westerhof, J. H. Schornagel and G. J. Hordijk, *Int. J. Cancer*, **51**, 909 (1992).
276. T. R. Jones, A. H. Calvert, A. L. Jackman, M. A. Eakin, M. J. Smithers, R. F. Betteridge D. R. Newell, A. J. Hayter, A. Stocker, S. J. Harland, L. C. Davis and K. R. Harrap, *J. Med. Chem.*, **28**, 1468 (1985).
277. M. G. Nair, N. T. Nanavati, I. G. Nair, R. L. Kisliuk, Y. Gaumont, M. C. Hsiao and T. I Kalman, *J. Med. Chem.*, **29**, 1754 (1986).
278. T. R. Jones, R. F. Betteridge, S. Neidle, A. L. Jackman and A. H. Calvert, *Anti-Cancer Drug Design*, **3**, 243 (1989).
279. R. G. Moran, P. D. Coleman and A. Rosowsky, *NCI Monogr.*, **5**, 133 (1987).
280. F. M. Sirotnak, *Pharmacol. Ther.*, **8**, 71 (1980).
281. G. B. Henderson, *Annu. Rev. Nutr.*, **10**, 319 (1990).
282. G. Jansen, J. H. Schornagel, G. R. Westerhof, G. Rijksen, D. R. Newell and A. L. Jackman, *Cancer Res.*, **50**, 7544 (1990).

283. G. R. Westerhof, G. Jansen, N. van Emmerik, I. Kathmann, G. Rijksen, A. L. Jackman and J. H. Schornagel, *Cancer Res.*, **51**, 5507 (1991)
284. E. Sikora, A. L. Jackman, D. R. Newell and A. H. Calvert, *Biochem. Pharmacol.*, **37**, 4047 (1988).
285. R. G. Moran, P. D. Coleman, A. Rosowsky, R. A. Forsch and K. K. Chan, *Mol. Pharmacol.*, **27**, 156 (1985).
286. M. Manteuffel-Cymborowska, E. Sikora and B. Grzelakowska-Sztabert, *Anti-Cancer Res.*, **6**, 807 (1986).
286. B. M. Cantwell, M. Earshaw and A. L. Harris, *Cancer Treat. Rep.*, **70**, 1335 (1986).
288. M. F. Bassendine, N. J. Curtin, H. Loose, A. L. Harris and O. F. W. James, *J. Hepatol.*, **4**, 349 (1987).
289. A. H. Calvert, D. R. Newell, A. L. Jackman, L. A. Gumbrell, E. Sikora, B. Grzelakowska-Sztabert, J. A. M. Bishop, I. R. Judson, S. J. Harland and K. R. Harrap, *NCI Monogr.*, **5**, 213 (1987).
290. S. Vest, E. Bork and H. H. Hansen, *Eur. J. Cancer Clin. Oncol.*, **21**, 201 (1988).
291. B. M. Cantwell, V. Macaulay, A. L. Harris, S. B. Kaye, I. E. Smith, R. A. V. Milsted and A. H. Calvert, *Eur. J. Cancer Clin. Oncol.*, **24**, 733 (1988).
292. C. Sessa, M. Zucchetti, M. Ginier, Y. Willems, M. D'Incalci and F. Cavalli, *Eur. J. Cancer Clin. Oncol.*, **24**, 769 (1988).
293. D. L. Alison, B. Robinson, S. J. Harland, B. D. Evans, and A. H. Calvert, *Br. J. Cancer*, **50**, 242 (1984).
294. A. H. Calvert, D. L. Alison, S. J. Harland, B. A. Robinson, A. L. Jackman, T. R. Jones, D. R. Newell, Z. H. Siddik, E. Wiltshaw, T. J. McElwain, I. E. Smith and K. R. Harrap, *J. Clin. Oncol.*, **4**, 1245 (1986).
295. D. I. Jodrell, D. R. Newell, S. E. Morgan, S. Clinton, J. P. M. Bensted, L. R. Hughes and A. H. Calvert, *Br. J. Cancer*, **64**, 833 (1991).
296. K. R. Harrap, A. L. Jackman, D. R. Newell, G. A. Taylor, L. R. Hughes and A. H. Calvert, *Adv. Enzyme Regul.*, **29**, 161 (1989).
297. A. L. Jackman, D. R. Newell, W. Gibson, D. I. Jodrell, G. A. Taylor, J. A. Bishop, L. R. Hughes and A. H. Calvert, *Biochem. Pharmacol.*, **42**, 1885 (1991).
298. B. M. O'Connor, A. L. Jackman, P. H. Crossley, S. E. Freemantle, J. Lunec and A. H. Calvert, *Cancer Res.*, **52**, 1137 (1992).
299. A. K. Cho, W. L. Haslett and D. J. Jenden, *J. Pharmacol. Exp. Ther.*, **138**, 249 (1962).
300. B. Ringdahl, *Eur. J. Pharmacol.*, **99**, 177 (1984).
301. M. P. Caulfield and D. A. Brown, *Br. J. Pharmacol.*, **104**, 39 (1991).
302. R. L. Lefkowitz, B. B. Hoffman and P. Taylor, in *Goodman and Gilman's The Pharmacological Basis of Therapeutics* (Eds. A. G. Gilman, T. W. Rall, A. S. Nies and P. Taylor), Pergamon Press, New York, 1990, pp. 84–101.
303. E. C. Hulme, N. J. M. Birdsall and N. J. Buckley, *Annu. Rev. Pharmacol. Toxicol.*, **30**, 633 (1990).
304. A. K. Cho, W. L. Haslett and D. J. Jenden, *Biochem. Biophys. Res. Commun.*, **5**, 276 (1961).
305. A. Bebbington, R. W. Brimblecombe and D. Shakeshaft, *Br. J. Pharmacol.*, **26**, 56 (1966).
306. A. H. Beckett, N. J. Harper and J. W. Clitherow, *J. Pharm. Pharmacol.*, **15**, 362 (1963).
307. A. Bebbington, R. W. Brimblecombe and D. G. Rowsell, *Br. J. Pharmacol.*, **26**, 68 (1966).
308. J. Jacob, I. Marszak, L. Bardina, F. A. Marszak and R. Epsztein, *Arch. Int. Pharmacodyn.*, **91**, 303 (1952).
309. D. J. Jenden and A. K. Cho, *Biochem. Pharmacol.*, **12**, Suppl. 38, (1963).
310. J. Levy and E. Michel-Bar, *Therapie*, **22**, 1461 (1967).
311. A. Lindqvist, S. Lindgren, B. Lindeke, B. Karlen, R. Dahlbom and M. R. Blair Jr., *J. Pharm. Pharmacol.*, **22**, 707 (1970).
312. B. Resul, B. Ringdahl, R. Dahlbom and D. J. Jenden, *Eur. J. Pharmacol.*, **87**, 387 (1983).
313. B. Ringdahl, *J. Pharmacol. Exp. Ther.*, **232**, 67 (1985).
314. H. C. Brown and M. D. Taylor, *J. Am. Chem. Soc.*, **69**, 1332 (1947).
315. H. C. Brown and M. Gerstein, *J. Am. Chem. Soc.*, **72**, 2926 (1950).
316. B. Ringdahl and D. J. Jenden, *Mol. Pharmacol.*, **23**, 17 (1983).
317. B. Ringdahl and D. J. Jenden, *Life Sci.*, **32**, 2401 (1983).
318. R. Dahlbom, A. Lindqvist, S. Lindgren, U. Svensson, B. Ringdahl and M. R. Blair Jr., *Experientia*, **30**, 1165 (1975).
319. B. Ringdahl and R. Dahlbom, *Acta Pharm. Suec.*, **15**, 255 (1978).
320. B. Ringdahl, B. Resul and R. Dahlbom, *J. Pharm. Pharmacol.*, **31**, 837 (1979).

788   Z. Ben-Zvi and A. Danon

321. B. Ringdahl and R. Dahlbom, *Acta Pharm. Suec.*, **16**, 13 (1979).
322. R. Amstutz, B. Ringdahl, M. Carlen, M. Roch and D. J. Jenden, *J. Med. Chem.*, 28, 1760 (1985).
323. B. Ringdahl, *Br. J. Pharmacol.*, **82**, 269 (1984).
324. R. Dahlbom, D. J. Jenden, B. Resul and B. Ringdahl, *Br. J. Pharmacol.*, **79**, 299 (1982).
325. B. Ringdahl, *Mol. Pharmacol.*, **31**, 351 (1987).
326. I. Hanin, D. J. Jenden and A. K. Cho, *Mol. Pharmacol.*, **2**, 352 (1966).
327. B. Karlen and D. J. Jenden, *Res. Commun. Chem. Pathol. Pharmacol.*, **1**, 471 (1970).
328. B. B. Brodie, H. Kurz and L. S. Shanker, *J. Pharmacol. Exp. Ther.*, **130**, 20 (1960).
329. B. Pullman, P. H. Courriere and J. L. Coubeils, *Mol. Pharmacol.*, **7**, 397 (1971).
330. O. Nordstrom, P. Alberts, A. Westlind, A. Unden and T. Bartfai, *Mol. Pharmacol.*, **24**, 1 (1983).
331. B. W. J. Ellenbroek, R. J. F. Nivard, J. M. van Rossum and E. J. Ariens, *J. Pharm. Pharmacol.*, **17**, 393 (1965).
332. E. J. Ariens and A. M. Simonis, *Ann. N. Y. Acad. Sci.*, **144**, 842 (1967).
333. E. J. Ariens, A. J. Beld, J. F. Rodrigues de Miranda and A. M. Simonis, in *The Receptors: A Comprehensive Treatise* (Ed. R. D. O'Brien), Plenum Press, New York, 1979, pp. 33–91.
334. E. Mutschler and K. Hultzsch, *Arzneim. Forsch.*, **23**, 732 (1973).
335. E. Mutschler and G. Lambrecht, *Trends Pharmacol. Sci.*, **5**, *Suppl.* 39 (1984).
336. U. Moser, G. Lambrecht, M. Wagner, J. Wess and E. Mutschler, *Br. J. Pharmacol.*, **96**, 319 (1989).
337. M. Wolf-Pflugmann, G. Lambrecht, J. Wess and E. Mutschler, *Arzneim. Forsch.*, **39**, 539 (1989).
338. N. Jaiswal, G. Lambrecht, E. Mutschler, R. Tacke and K. U. Malik, *J. Pharmacol. Exp. Ther.*, **258**, 842 (1991).

CHAPTER **14**

Polycyano compounds

RENATE DWORCZAK and HANS JUNEK

Karl-Franzens University of Graz, Graz, Austria

Supplement C2: The chemistry of triple-bonded functional groups
Edited by S. Patai © 1994 John Wiley & Sons Ltd

I. INTRODUCTION

In the past decade, the chemistry of the cyano group, usually second in abundance and diversity to that of the amino group, has achieved parity and is moving ahead rapidly to new frontiers.

Alexander F. Fatiadi

This statement, made in 1983[1], did not lose its actuality: the field of synthetic applications of cyanocarbons is still expanding. A steadily increasing number of patents on cyano compounds as biologically active materials, dyes, organic metals, etc. indicates the importance of nitrile chemistry for industrial applications.

Malononitrile (MDN) and tetracyanoethylene (TCNE) can be considered the most important starting materials for the synthesis of cyano derivatives. Table 1 gives the number of references in *Chemical Abstracts* concerning these compounds from 1982 to 1992: While the number of citations dealing with TCNE is nearly constant with about 30 publications a year, the number of papers on MDN has reached about 70 to 80 papers. Many of them report on syntheses of compounds containing two or more cyano groups. This is why this chapter will concentrate on malononitrile and tetracyanoethylene. Other polycyano derivatives are discussed more briefly. Certain polycyano compounds (TCNE, tetracyanobenzene, tetracyanoquinodimethane TCNQ) have been used as electron acceptors in charge-transfer complexes. Though complexes cannot be discussed in detail, reaction partners and references are given in tables.

TABLE 1. Number of references in *Chemical Abstracts* 1982–1992 dealing with malononitrile (MDN) and tetracyanoethylene (TCNE)

Year	MDN	TCNE
1982	60	34
1983	64	36
1984	48	29
1985	58	33
1986	63	23
1987	71	36
1988	70	32
1989	80	23
1990	72	38
1991	78	33
1992	77	20

The importance of nitriles in organic chemistry is caused by two facts:

(1) Cyano compounds show typical properties based on the structure–reactivity relationship of the nitrile group(s).

(2) Nitrile groups are reactive moieties for the synthesis of numerous materials, involving additions, ring-closure reactions, etc.

Cyanocarbon chemistry therefore covers a wide field of materials the properties of which depend largely on the presence of one ('nitriles') or more ('polycyano compounds') nitrile groups. The possible role of HCN and polycyano compounds in prebiotic synthesis is one of the most fascinating chapters in the processes that preceded the appearance of life on earth[2].

The aim of this chapter is to present some recent and significant advances in the chemistry of polycyano compounds, especially from the last ten years. Only some patents will be mentioned. Dyes and other special compounds will not be discussed in detail. Literature on the subject that is already covered in reviews which have appeared since 1982 will only be referred to in connection with new results. A list of general reviews, which have been published on cyano compounds between 1981 and 1987, and short abstracts of their contents are given below as well as at the beginning of the sections dealing with malononitrile and with tetracyanoethylene:

Preparation and Synthetic Applications of Cyano Compounds, A. J. Fatiadi, 1983[1]. This comprehensive book chapter (1236 references) deals with synthesis of nitriles and some of their reactions. Synthetic methods and reactions involving cyano substrates are also covered. Various cyano reagents for organic syntheses and reactions of cyanocarbons with electron acceptors are also discussed.

Nitriles, Nitrilium Salts, C. Grundmann, 1985[3]. This book section covers papers on preparations and reactions of mono- and dinitriles and polycyano compounds published from 1951 to approximately 1984 and contains 1518 references.

Utility of α,β-Unsaturated Nitriles in Heterocyclic Synthesis, M. H. Elnagdi and coworkers, 1983[4]. Syntheses of different mono- and polyheterocyclic derivatives via α,β-unsaturated nitriles are surveyed. The scope and limitation of the most important of these approaches are demonstrated. 134 references are given.

Amines, Nitriles, and Other Nitrogen-Containing Functional Groups, G. Kneen, 1981[5], 1982[6], 1983[7], 1984[8], and S. G. Lister, 1986[9]. This series of reviews deals partially with cyanocarbons.

The Synthetic Potentialities of Nitriles in Heterocyclic Synthesis, M. H. Elnagdi and coworkers, 1987[10]. Nitriles have been used as starting materials for a variety of heterocycles: 225 references are given.

Chemistry of 3-Oxonitriles, M. H. Elnagdi and coworkers, 1984[11]. This systematic and comprehensive review (322 references) presents a survey of the methods of preparation, structural studies and chemical reactivities of 3-oxonitriles.

II. CYANOGEN

A. Physicochemical Properties

Physicochemical properties of cyanogen (NC—CN, **1**) were the subject of thorough investigations. Papers published deal with Raman spectra[12], with proton affinity[13], reactions with low-energy electrons[14] and cyanide formation by electron attachment[15]. Kinetics of basic hydrolysis of cyanogen were investigated in detail[16]. Photolysis at 193 nm was studied[17].

B. Reactions

1. Cycloadditions

1 was used for the synthesis of heterocycles. So cycloaddition with oxalyl chloride afforded **3**, which in term gave imidazolecarbonic acid **4** by hydrolysis. As the first step,

(2)   (3)   (4)

(5)   (6)   (7)

formation of ketene **2** is discussed. **4** can be decarboxylated to the corresponding dichloropyrazole. When **1** was reacted with oxalyl bromide, pyrazine **5** was obtained[18]. Aromatic nitrile ylides and **1** gave imidazole carbonitriles **6** in a 1,3-dipolar cyclo-addition[19]. Oxadiazoles **7** were obtained from **1** and oximes[19].

2. Miscellaneous reactions

Cyanogen (**1**) radicals were reacted with alkanes and reactivities were examined for the temperature range from 170 to 740 K[20]. Complex formation with AsF_5 and SbF_5 was observed and discussed on the basis of the HSAB principle[21].

1 reacted with HCN to yield diiminosuccinonitrile (**8**), which formed dicyanopyrazines **9** with inamines. Other ring-closure reactions of **8** were also reported[22].

III. POLYCYANOALKANES

Literature on polycyanoalkanes published in the last decade concentrates on malononitrile (MDN) and tetracyanoethane (TCE). In contrast to earlier years very little has been published on di- and tricyanomethane or on other polycyanoethanes than TCE. A similar situation can be observed concerning higher alkanes. Hence we decided to put our emphasis on malononitrile and tetracyanoethane, to give only a short survey of polycyanopropanes and to disregard other cyanoalkanes.

A. Malononitrile

Though the Knoevenagel reaction may be considered to be typical for malononitrile [MDN, $CH_2(CN)_2$, **10**] this unique CH-acidic synthon has been used in numerous different reactions. A number of reviews has been published on MDN. The most comprehensive ones are listed below:

New Applications of Malononitrile in Organic Chemistry, Parts 1 and 2, A. J. Fatiadi (1978)[23,24]. Applications of malononitrile in organic chemistry are reviewed and 589 references are given. An overview of the chemistry of substituted malononitriles is also included. Mechanisms and synthetic methods employing malononitriles are discussed. Information on malononitrile dimers and trimers, Knoevenagel and Michael condensation reactions, reactions with cyclic polyketones, quinones, sulfur and organometallic compounds is given. Syntheses of new heterocycles, pesticides, charge-transfer complexes and solvatochromic dyes are also covered.

Reactions of Malononitrile Derivatives, F. Freeman (1981)[25]. Preparation, spectral properties, and chemical reactivity of dicyanomethylene compounds halomalononitriles, aminomalononitrile, aminomethylenemalononitrile, alkoxymethylenemalononitriles, hydroxyiminomalononitrile, (*O-p*-tosylnitroso)malononitrile, and arylhydrazonomalono-nitriles are reviewed (367 references). Their use for the preparation of cyanocarbons and heterocyclic systems is described.

Cyclization of Ylidenemalononitriles, E. Campaigne and S. W. Schneller (1976)[26]. The use of malononitrile in annelation reactions (via intermediate dicyanomethylene compounds) is reviewed. Various cyclic compounds like indenones, indanones, naphthalenes, lactones and coumarins are discussed.

Properties and Reactions of Ylidenemalononitriles, F. Freeman (1980)[27]. Dicyanomethylene compounds, their molecular structure and spectral properties, toxicity and analyses are reviewed. Various reactions like hydrolysis, oxidation and reduction, dimerizations, cyclizations, photochemistry and thermolysis are discussed. The review, which contains 380 references, also deals with unsaturated compounds, oxygen, nitrogen, phosphorous and sulfur compounds.

1. Unsubstituted malononitrile

a. Formation and properties. Malononitrile (**10**) is a very powerful synthon in organic synthesis. Therefore the synthesis as well as physicochemical properties of MDN have been of great interest in recent years. A Japanese patent claims the synthesis of **10** from ClCN and acetonitrile by vapor-phase reaction at elevated temperatures[28]. Crystal dynamics and phase transitions of **10** were investigated[29] and the specific heat capacity was measured over a wide range of temperatures[30]. Interactions of **10** with copper electrode surfaces were studied by surface-enhanced Raman spectroscopy[31]. NMR studies deal with solvation phenomena of **10** in aromatic solvents[32] and in dichloromethane[33]. Proton exchange has been investigated by NMR methods[34] as well as ionization of **10** in diluted aqueous HCl[35].

Crown ethers have been used for the complexation of **10** and the resulting complexes were studied. Far-IR studies of **10** complexes of 18-crown-6 have been performed[36]. On the other hand, the complexation of **10** has been used to compare free macrocycles with their complexes with neutral molecules. Information about the conformations of the free crown ethers and their interactions with solvent molecules was obtained[37]. Dicyclohexano-18-crown-6 has shown isomer-dependent complexation of **10**[38].

A quantitative analytical method for the determination of **10** has been developed. Concentrations from 0.0001 to 0.001 mol can be determined by differential polarography[39]. On the other hand, **10** has been used as reactant in the fluorometrical determination of sialic acids[40,41].

b. Reactions.

i. Formation of aliphatic compounds
Reactions of the methylene group: Reactions at the methylene group of the molecule yield dicyanomethylene compounds in many cases. These will be discussed in Section III.A. 3 in detail. Some compounds have been reported, which result from reactions at the methylene group of **10** but are no classical dicyanomethylene compounds. So, for example, nitrosation with nitrous acid in the presence of nucleophilic catalysts yielded the corresponding oxime product $(NC)_2C=NOH$[42].

Benzylidene-cyanothioacetamide (**11**) reacted with **10** to yield **12**, which is a useful intermediate in heterocyclic synthesis[43].

$$\text{(11)} \qquad \text{(10)} \qquad \text{(12)}$$

Dimethylformamide diethyl acetal (**13**) was condensed with **10** to yield **13**. The same reaction was performed with dimethylacetamide diethyl acetal[44]. Reaction of **10** with DMF and POCl$_3$ gave (dimethylaminomethylene)malononitrile (**14**), diazaoctatriene **15** and pyridinedicarbonitrile **16**. **14** was used as synthon for the preparation of N-heterocycles[45].

$$
\begin{array}{ccc}
\text{OEt} & \text{CN} & \\
| & | & \\
\text{HC}-\text{NMe}_2 & + \quad \text{CH}_2 & \longrightarrow \quad \overset{\text{NC}}{\underset{\text{NC}}{\diagup}} \!\!\! \diagdown \text{C}=\text{CH}-\text{NMe}_2 \\
| & | & \\
\text{OEt} & \text{CN} & \\
(\mathbf{13}) & (\mathbf{10}) & (\mathbf{14})
\end{array}
$$

$$
\begin{array}{cc}
& \text{CN} \\
& | \\
\text{Me}_2\text{N}-\text{CH}=\text{N}-\text{C}=\text{CCHC(CN)}_2 \\
& | \\
& \text{Cl} \\
(\mathbf{15}) & (\mathbf{16})
\end{array}
$$

(16) structure: NC and CN on ring, Cl, N, NH$_2$

Three-component coupling of arabinosylureas with triethyl orthoacetate and **10** produced the corresponding dicyanoethylene derivative, which was condensed to arabinosylpyrimidines[46].

Also, a polymer reaction with **10** has been reported. After having cross linked *p*-aminoaryl-cellulose with epichlorohydrin and subsequent diazotation, the reaction with **10** was performed to yield the corresponding arylhydrazonopropanedinitrile derivatives polymeraryl-NHN:C(CN)$_2$[47].

Reactions of the nitrile group(s): In general, these reactions start with a nucleophilic attack at the C atom of the nitrile group. Thus **10** reacted with its anion in basic media to yield the dimerization product 2-amino-1,1,3-tricyanopropene[48,49]. Thiolysis with H$_2$S gave cyanothioacetamide[50]. Acidic alcoholysis in an neutral solvent produced dialkyl propanediimidate dihydrohalides[51]. Metal-catalyzed reactions of **10** with methyl acetoacetate and other β-dicarbonyls have been reported[52].

ii. Formation of carbocycles

Three-membered carbocycles: 3,3-Disubstituted 1,1,2,2-tetracyanocyclopropanes can be obtained by electrolysis of **10** in the presence of ketones and NaBr in ethanol[53].

Six-membered carbocycles: Propenone derivatives react with **10** to yield six-membered cyclic compounds. So 1,3-diarylpropenones yielded 4-cyanocyclohexanols[54] and 4-

$$
\begin{array}{cc}
\text{OEt} & \text{OH} \\
| & \\
\text{CH} & \text{MeCO} \quad \quad \text{R} \\
\| & \\
\text{Me}-\text{C}-\text{C}-\text{C}-\text{CH}_2\text{R} \longrightarrow & \\
\| \quad \quad \| & \text{NH}_2 \\
\text{O} \quad \cdot \quad \text{O} & \text{CN} \\
(\mathbf{17}) & (\mathbf{18})
\end{array}
$$

hydroxycyclohexanedicarbonitriles[55,56]. (Ethoxymethylene)alkanediones **17** afforded benzonitriles **18**[57]. Knoevenagel condensation of levulinaldehyde (**19**) with 3 mol **10** gave pentacyanobicyclooctene **20** while reaction with 2 mol **10** yielded a condensed pyridine derivative[58]. 2,5,5-trimethylhexa-2,3-dien-6-al (**21**) was condensed with two mol **10** to yield **22** in an unexpected reaction[59].

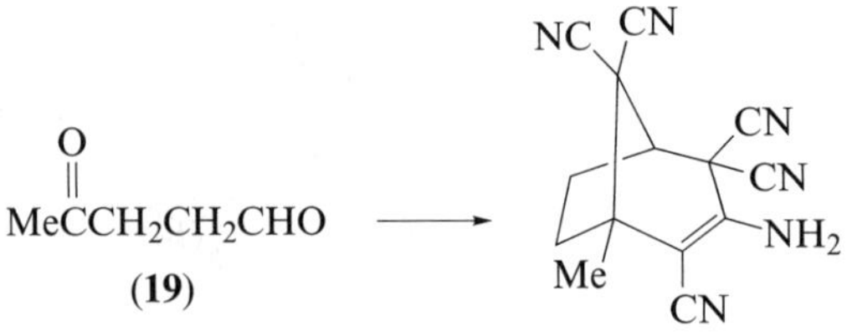
O
MeCCH2CH2CHO
NC
CN
CN
CN
NH2
Me
CN

(**19**) (**20**)

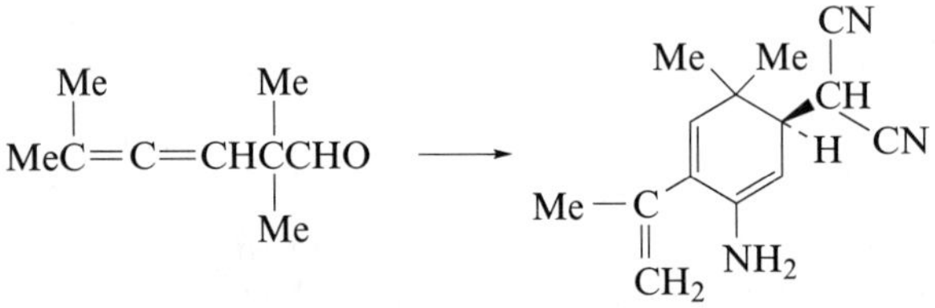
Me
Me
MeC=C=CHCCHO
Me
Me
Me
CN
CH
H
CN
Me—C
CH2
NH2

(**21**) (**22**)

Condensed and other carbocyclic systems: Condensed heterocycles have been prepared either directly or via a condensation product. Acenaphthen (**23**) reacted with **10** in the presence of $AlCl_3$/HCl to yield iminium salt **24**[60]. Tetralones **25** were condensed with **10** and cyclized subsequently in polyphosphoric acid to yield **26**[61].

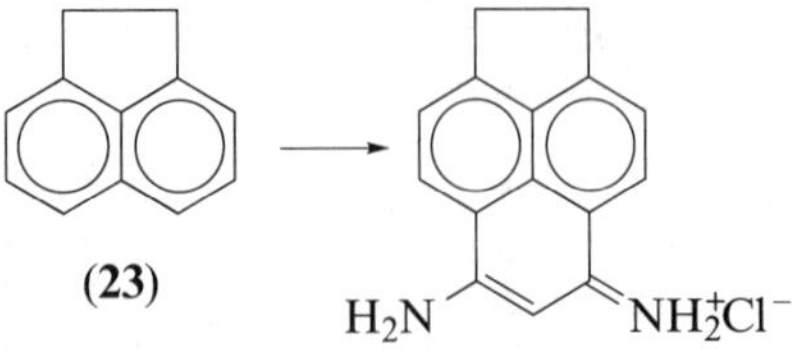
H2N
NH2+Cl−

(**23**) (**24**)

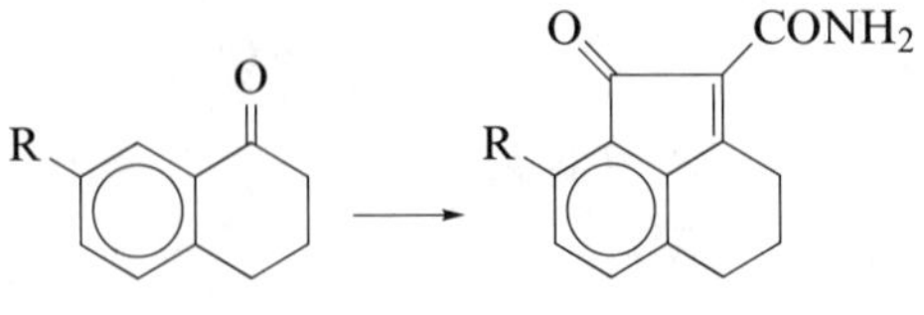
O
R
O
CONH2
R

(**25**) (**26**)

iii. Formation of heterocycles

Five-membered heterocycles: Most papers dealing with the synthesis of cyclic compounds with **10** report on the formation of six-membered or condensed heterocyclic systems. Nevertheless, a number of new five-membered heterocyclic derivatives have been described. So furane, thiophene and pyrrole derivatives were prepared by ring enlargement when oxiranes, thiirane and N-tosylhydrazide, respectively, were reacted with **10** at $-10\,°C$[62,63]. Various other ways to prepare pyrrole derivatives include cyclization of amino acid derivatives with **10** to yield pyrrolecarboxylic esters **27**[64] and cyclocondensation of chloroacetanilide to pyrroline-3-carbonitriles **28**[65].

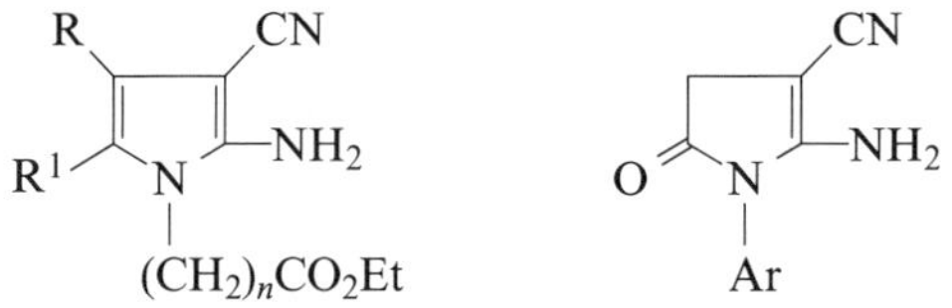

(27)   (28)

Diphenylpyrrole-3-carbonitriles **30** were produced by ammonium acetate-catalyzed cyclization of **10** with a substituted phenylacetophenone **29**[66] and conjugated azoalkenes **31** yielded 1,2-diaminopyrrolo-3-carbonitriles **32**[67].

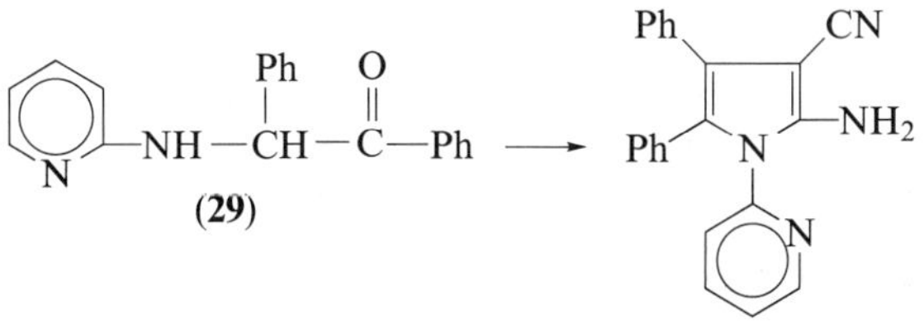

(29)

(30)

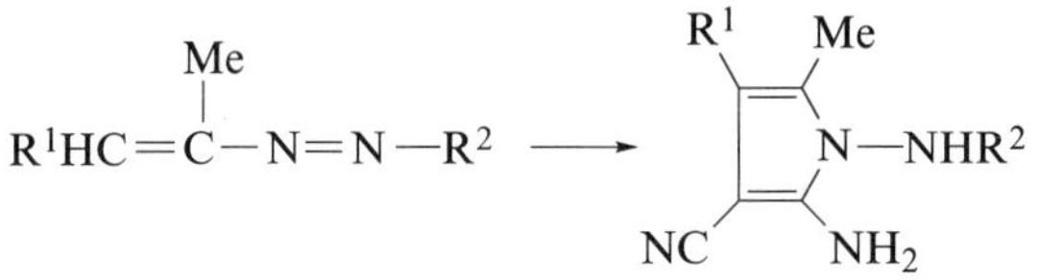

(31)   (32)

Hydrazones have been cyclized with **10** to yield pyrazole derivatives[68,69]. Hydrazines reacted with **10** and orthoformates to produce pyrazole derivatives **33**[70,71]. Tosylazides gave triazolecarbonitrile **34**[72]. Furan derivatives have been prepared by reactions of **10** with halohydrines[63], from **10**, furoin and diethylamine[73]. Reaction of **10** with D-ribose in the presence of $ZnCl_2$ gave furan derivative **35**[74]. The published syntheses of thiophenes include reports on 2-aminothiophene-3-carbonitriles. **10** reacted with ketones and sulfur, catalyzed by aliphatic amines to yield **36**[75,76]. Cyclocondensation of **10** with thioglycolic acid in the presence of a base like sodium ethanolate yielded 2-amino-3-cyano-4-hydroxythiophene (**37**) and its tautomers[77].

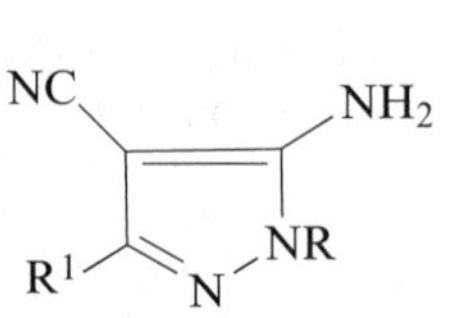

$$\begin{array}{cc} \text{(33)} & \text{(34)} \end{array}$$

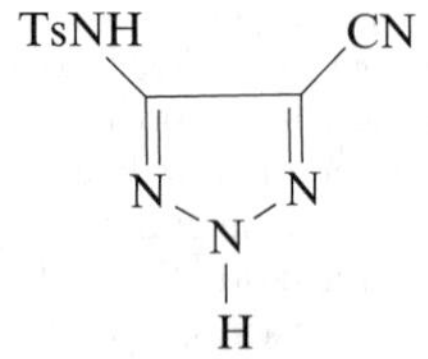

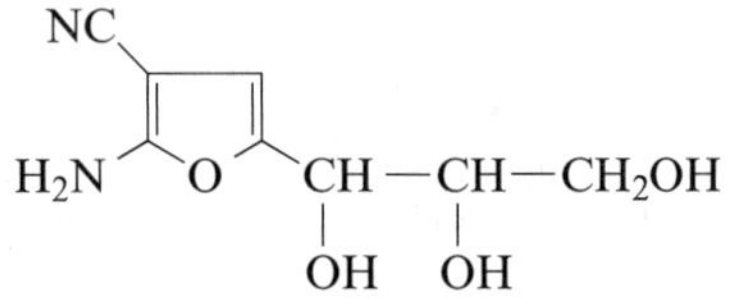

(35)

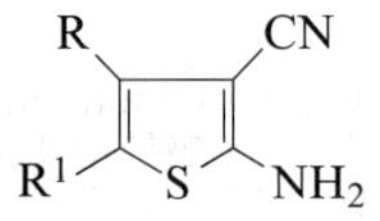

(36) 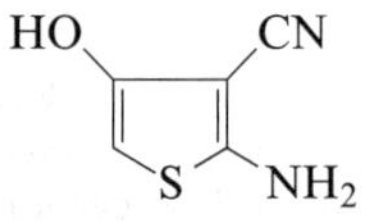   (37)

Chloroxanthione (**38**) reacted with **10** to give dithiole **39** under elimination of nitrogen[78]. Oxadiazole derivatives have been prepared[79,80] as well as isoxazoles[69]. A simple method for the preparation of 3-aminoisothiazole-4-carbonitriles has been described. Thus the sodium salt of **10** was treated with isothiocyanates and then with chloroamine in a one-pot synthesis to give **40** in good yields[81]. When 1,4,2-dithiazolium salts **41** were treated with **10**, the isothiazole derivative **42** was one of the products obtained[82].

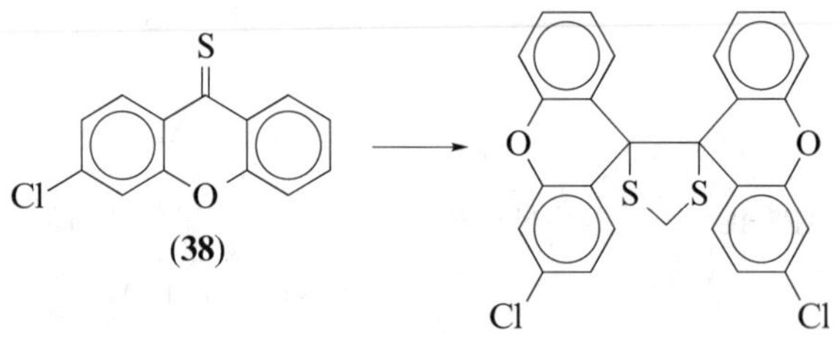

(38)

(39)

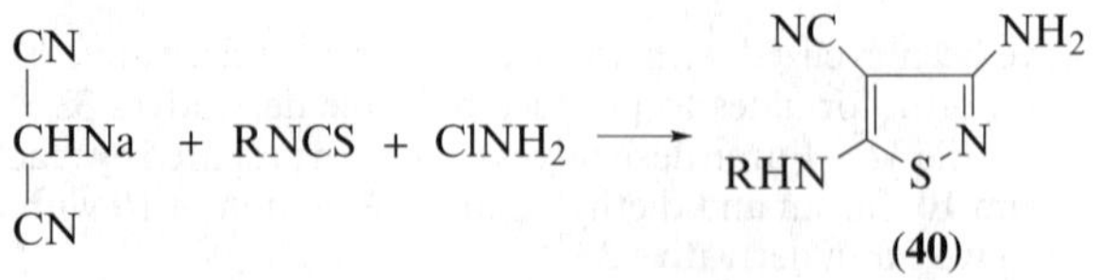

(40)

Six-membered heterocycles: A large number of six-membered heterocycles were pre-pared by condensations of **10**, many of them pyridines with various functional groups. These multifunctional molecules are powerful synthons for condensed heterocyclic ring systems. Therefore, some of the compounds described here will also occur in Section III.A.2.b.

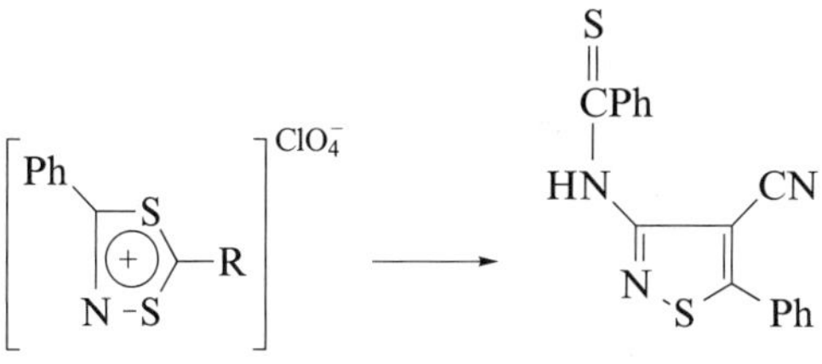

$$\text{(41)} \qquad \longrightarrow \qquad \text{(42)}$$

Nitrogen heterocycles: Various cyano-substituted pyridines have been described. They were prepared from **10** and ethyl cyanoacetic ester[83], and from **10** and acetylenic esters or ketones[84]. Aromatic aldehydes (e.g. 2-furyl, 4-pyridyl, and 2-thienyl aldehydes) have been condensed with **10** and aromatic ketones to yield 2-amino-3-pyridinecarbonitriles[85]. Benzylidenemalononitrile (**43**) reacted with **10** in aniline to yield dicyanopyridine **44**[86].

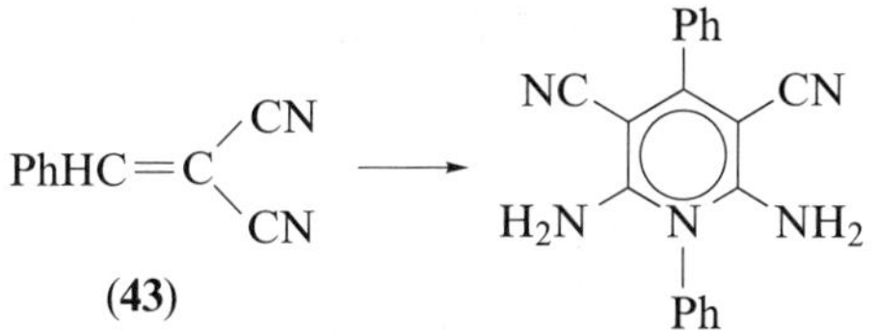

$$\text{PhHC}{=}\text{C}\begin{smallmatrix}\diagup\text{CN}\\ \diagdown\text{CN}\end{smallmatrix} \qquad \longrightarrow$$

(43)

(44)

Chalcones were cyclized with **10** in basic media, e.g. methanol/sodium hydroxide[87] or ammonium acetate[88], to afford pyridines **45** and **46**, respectively.

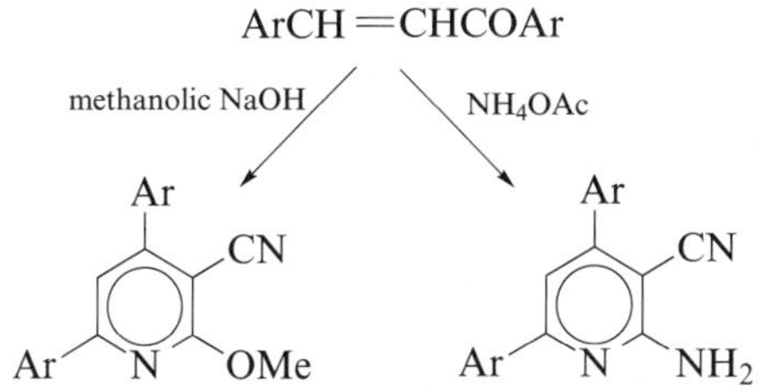

$$\text{ArCH}={=}\text{CHCOAr}$$

methanolic NaOH   NH$_4$OAc

(45)   (46)

Cyclocondensation of aromatic aldehydes with **10** and cyanoselenoacetamide in ethanol/ 4-methylmorpholine gave pyridinedicarbonitrile salts **47**[89]. Alkylation of **47** yielded pyridineselenones **48**. **10** reacted with H$_2$Se in the presence of triethylamine to yield pyridineselenone **49**, which can be used to prepare a number of selenium heterocycles[90]. Cyclization of **10** with α,β-unsaturated carbonyl compounds usually yields 2-oxopyridine derivatives. So 2-oxonicotinonitriles[91–94] and tetrahydropyridine-2-ones[95,96] have been prepared. 6-Oxonicotinonitriles can be prepared from **10** and acetoacetanilide[97]. Some syntheses of pyridinethiones have also been reported.

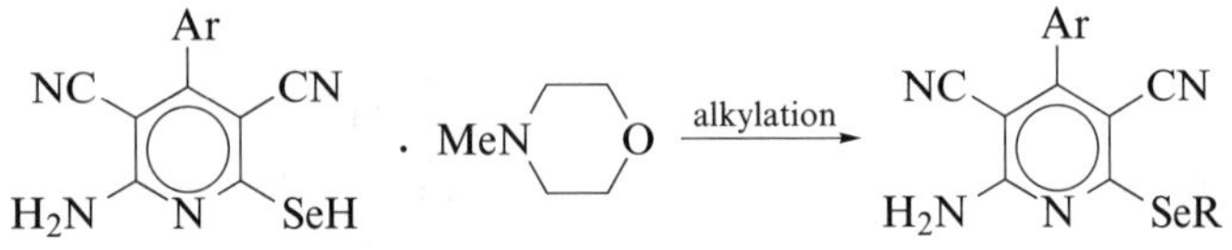

(47)   (48)

(49)   (50)

Cycloaddition of **10** to cyanothioacetamide yielded **50**[98,99]. Pyrimidinethiones **51** have been recyclized by **10** to pyridinethiones **52**[100]. The pyrimidine derivative **54** was prepared by reaction of **10** with the benzamide **53** and acetic anhydride[101].

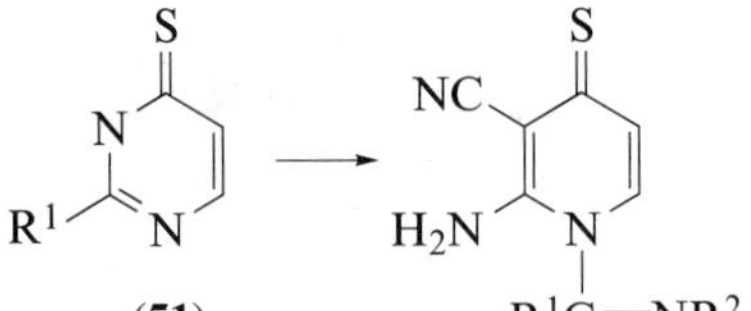

(51)   (52)

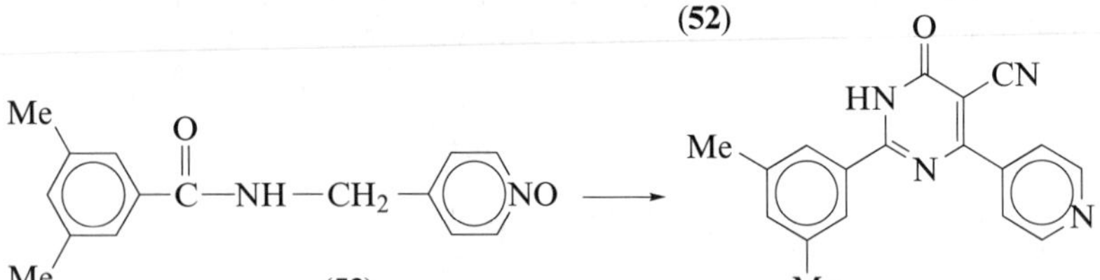

(53)

(54)

Oxygen heterocycles: **10** was reacted with different oxygen-containing condensing reagents to yield 2-amino-3-cyanopyran derivatives. Reaction of arylidenemalonaldehyde **55** gave 2-amino-3-cyano-4*H*-pyran-5-carboxaldehyde **56**[102].

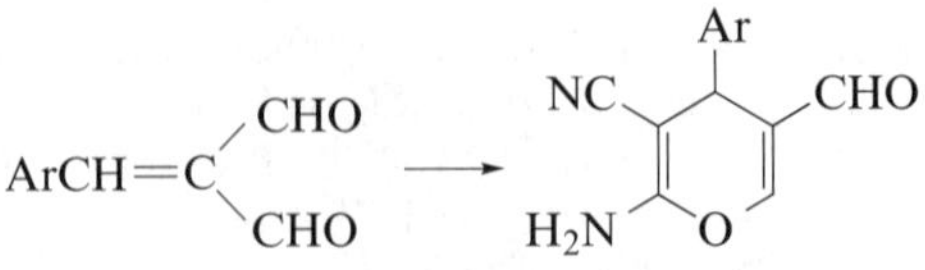

(55)   (56)

Basic ring opening of isoxazoles **57** with an aldehyde yielded **58**, which was cyclized to dicyanopyran **59**[103,104]. 4,5-Unsaturated 3-oxoalkanenitriles can also be condensed with **10** to yield dicyanopyranes **59**[105]. Methiniminium salts **60** have been used to prepare pyrylium salts **61**[106].

Miscellaneous: When anilides **62** were condensed with **10**, 2-aryliminothiopyrans **63** were obtained. **63** can be converted to pyridinethiones **64** under alkaline conditions[107]. **10** reacted with (N-substituted) sulfamides **65** to form 1,2,6-thiadiazines **66**[108].

Condensed and other heterocyclic systems: Nitrogen systems. Pyridazine **67** reacted with **10** and aromatic aldehydes to yield phthalazines **68**[109]. Arylideneindolinones **69** underwent Michael addition with **10**. By subsequent cyclization pyrrolo-indoles **70** were formed[110]. Conjugated azoalkenes **71** and **10** gave pyrrolopyrroles **72**[67].

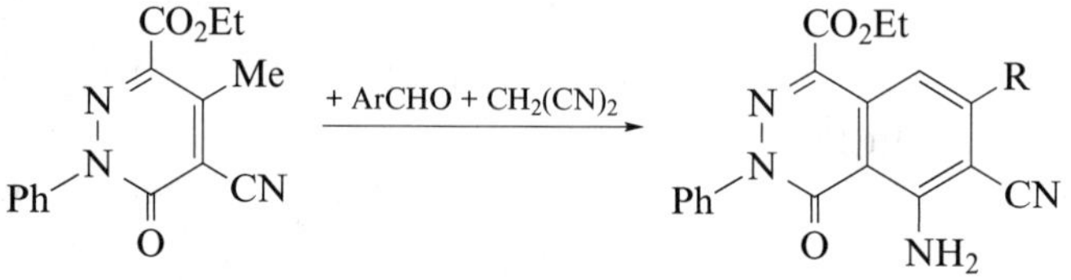

(67)   (68)

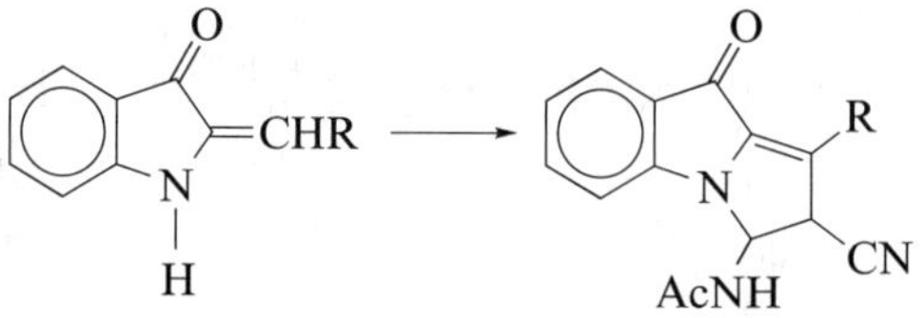

(69)   (70)

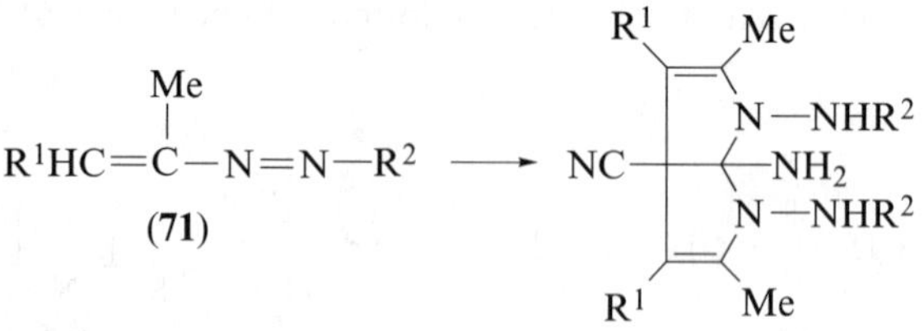

$R^1HC=C-N=N-R^2$ $\longrightarrow$ NC

(71)

(72)

Among the condensed heterocycles prepared by condensation reactions of **10** are many pyridine derivatives. So Knoevenagel condensation of two mol **10** with **19** gave **73**[58].

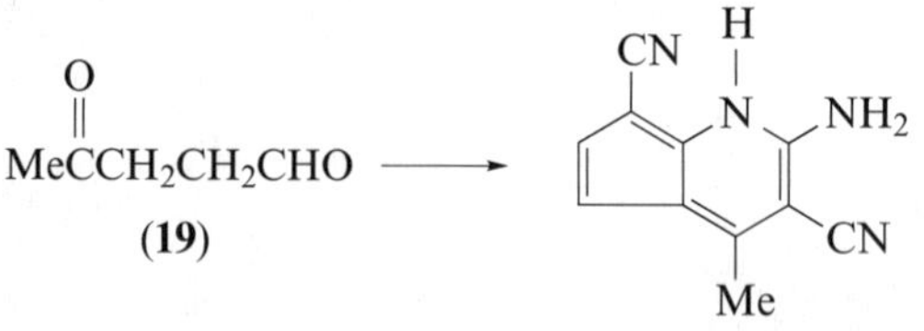

$$\overset{\displaystyle O}{\overset{\displaystyle \|}{MeCCH_2CH_2CHO}} \longrightarrow$$

(19)

(73)

Pyrrolopyridines **75** have been prepared by reaction of **10** with arylpyrrolidinetriones **74**[111]. **10** has been condensed with the hydrazine derivative **76** to yield pyrazoles **77** as well as pyrazolopyridines **78**, depending on the concentration of **10** used. A larger amount of **10** favoured the formation of **78**[69].

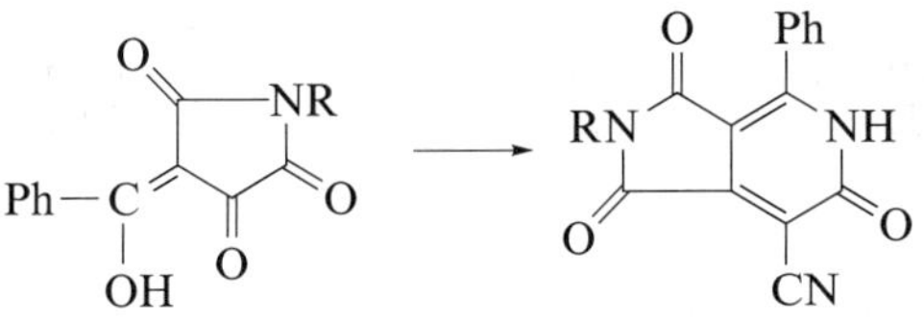

(74)   (75)

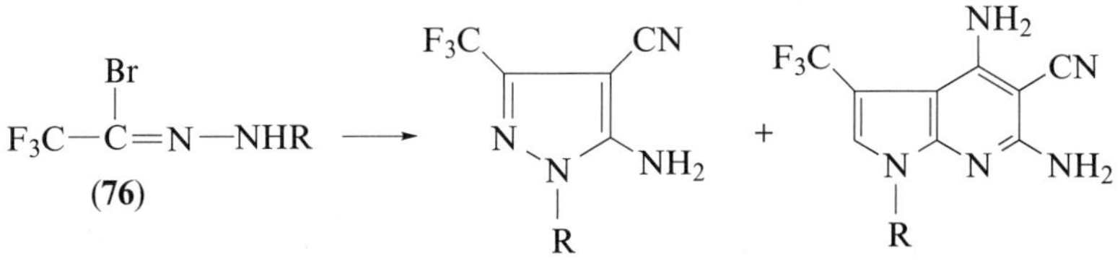

(76)   (77)   +   (78)

Triazolecarboxaldehydes **79** reacted with **10** to give triazolopyridines **80**[112]. Condensation of cyanothioacetamide with two mol **10** yielded **50** and pyridopyridinethione **81** in two steps[98].

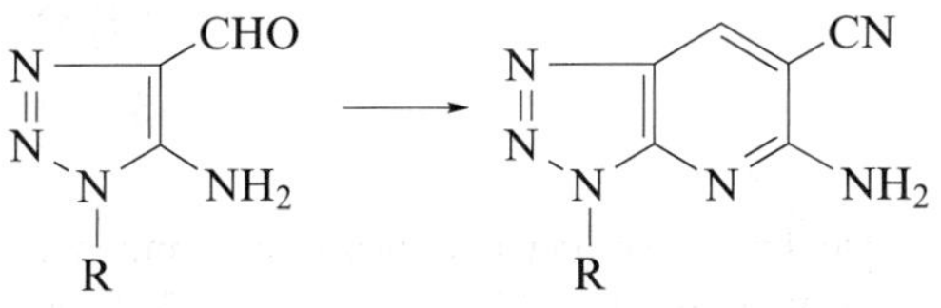

(79)   (80)

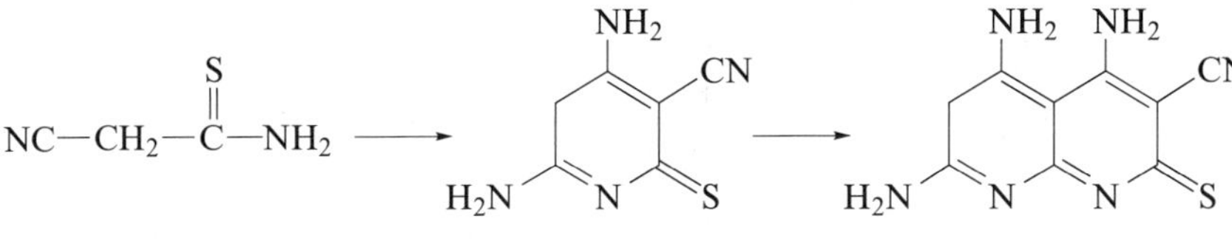

(50)   (81)

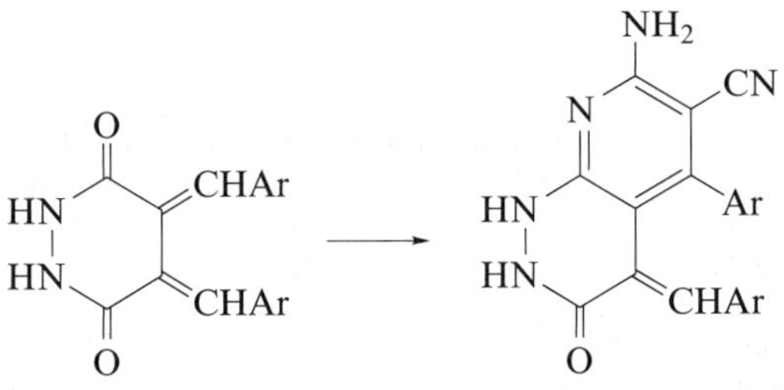

(82)   (83)

Pyridopyridazine derivatives have been prepared either by condensation of **10** with arylidenetetrahydropyridazinedione or from arylhydrazonomalononitriles. So pyridazines **82** gave pyridopyridazinones **83**[113] and hydrazone derivatives **84** afforded **85**[114].

(84)   (85)

Pyrimidine derivatives. With **10**, cyanouracil **86** gave condensation product **87**[115]. The same reaction could also be performed with the corresponding cyanouridine[116].

(86)   (87)

Triazolopyrimidines **88** have been prepared by a three-component reaction of **10** with benzyl azide and an aromatic aldehyde in the presence of sodium ethanolate[117]. Azidotriazoles have been cyclized directly with **10** to yield bis-triazolopyrimidines **89**[118]. Triazolopyrimidines have been transformed into triazolopyridopyrimidines **90**[119].

(88)   (89)   (90)

Transformation of heterocycles has also been used to synthesize condensed nitrogen heterocycles. So *N*-tosylaziridine has been treated with two mol **10** to yield two different bis-adducts depending on the reaction temperature. At 100 °C pyrrolopyridine **91** was formed, while reaction at 10 °C afforded the spiro compound **92**[62]. Condensation of **10** with chloroacetanilides did not only yield pyrrole derivatives like **28** but also pyrroloquinazolines **93**[65].

Oxygen heterocycles. The dicyanocyclopropane derivative **95** has been obtained from **10** and bromofuranone **94**, through Michael addition followed by ring formation[120]. When dithiophenyl-substituted benzophenone **96** was reacted with **10** with catalysis of titanium

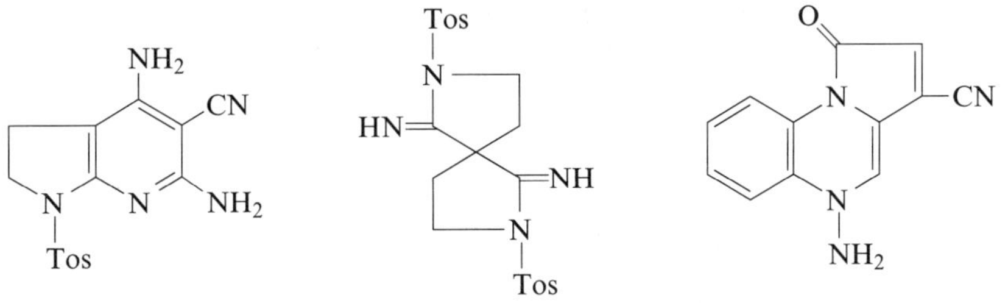

(91) (92) (93)

tetrachloride-pyridine, no Knoevenagel condensation could be observed and, surprisingly, benzofuran **97** was formed[121].

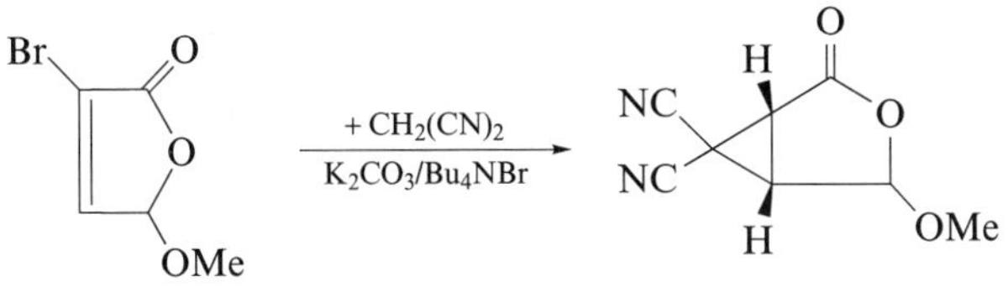

(94) (95)

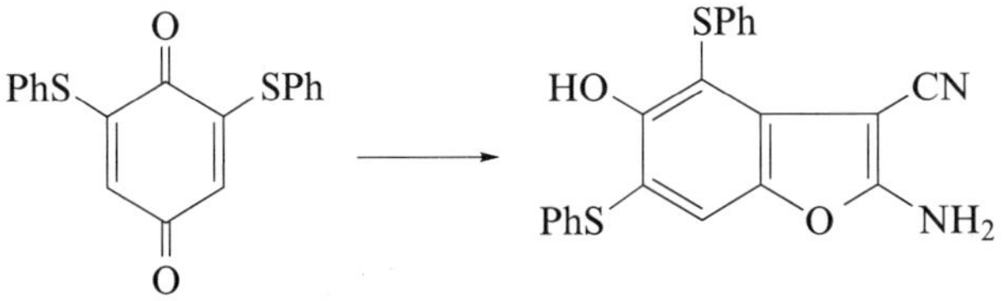

(96) (97)

A number of coumarin derivatives has been reported. For instance, benzocoumarins **99** were produced by condensation of **98** with ArCHO and **10**[122].

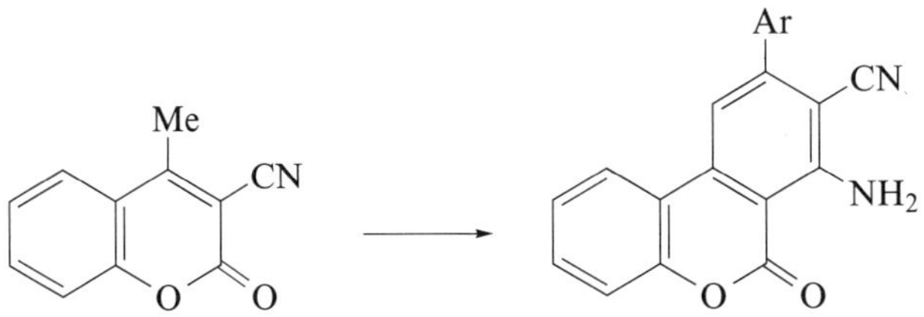

(98) (99)

Benzylidenechromanones **100** reacted with **10** to pyranobenzopyran derivatives **101**, which were recyclized to benzopyranopyridines **102** by sodium ethanolate in ethanol. **100** can be directly reacted to **102** in ethanolic sodium hydroxide[123]. Condensation of **10** with aromatic aldehydes and β-naphthol in ethanolic piperidine gave $4H$-chromenes **103** and **104**[124]; 8-hydroxy-1-naphthalenecarboxaldehyde gave naphthoxepinones **105**[125].

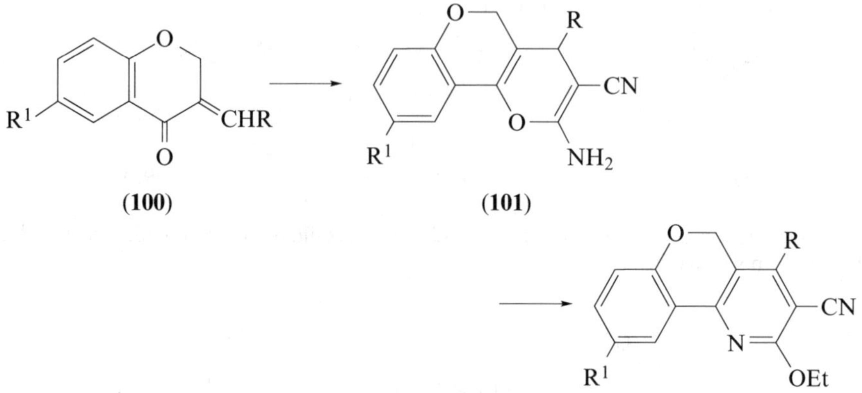

(100)   (101)

(102)

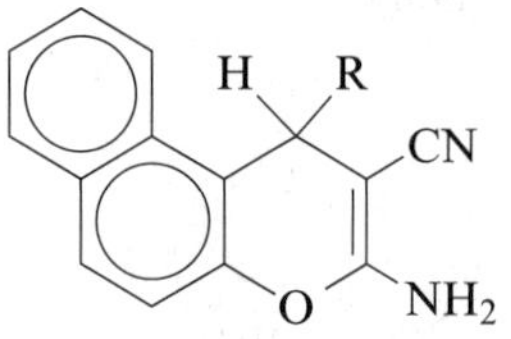

(103) 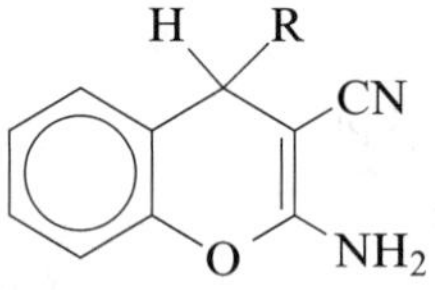   (104) 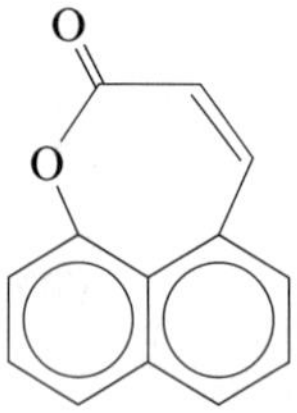   (105)

N,O-systems. Pyrazolones have been condensed with **10** in basic media to yield furopyrazole **106**[126]. Condensation in the presence of a ketone afforded pyranopyrazoles **107**[127].

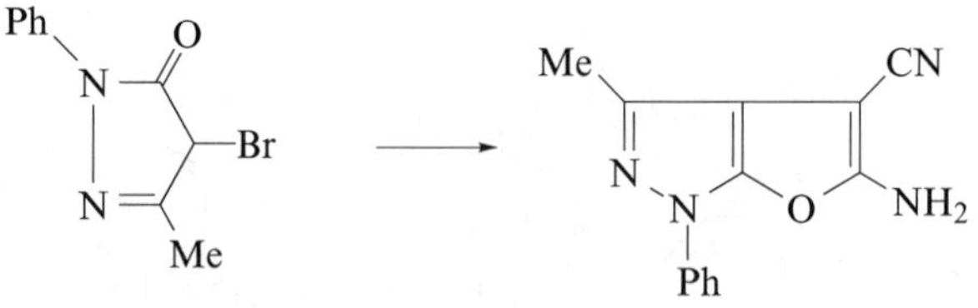

(106)

Pyranopyridines have been obtained by a one-pot reaction of **10** and unsaturated ketones[128]. Benzopyranopyridines **109** have been prepared from isoquinolinediones **108**[129]. Another synthesis was performed by condensation of an o-hydroxyaraldehyde with a cyclic methyl ketone followed by treatment with **10**[130]. Pyranobenzodiazepines **111** have been obtained by condensation of **10** with benzodiazepinones **110**[131]. 4-Dicyanomethylene-1,3-benzoxazines **113** were formed by treating dithiazolidine derivatives **112** with **10**[132].

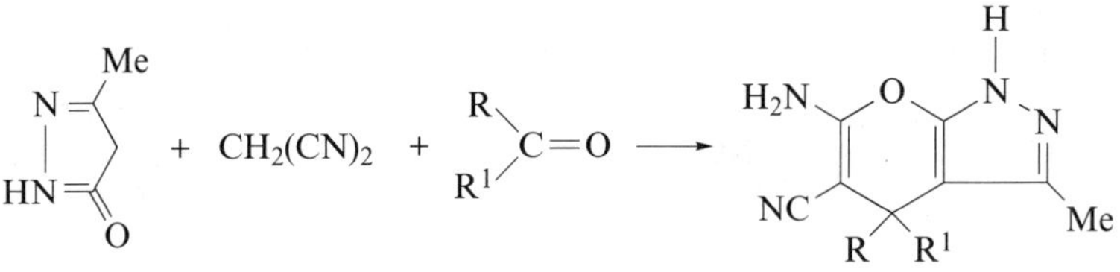

(107)

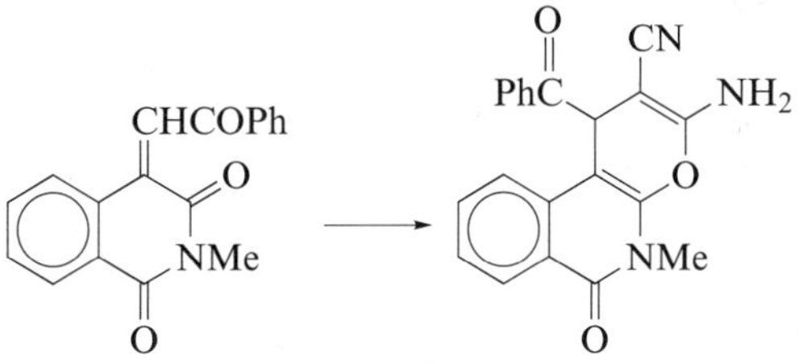

(108)   **(109)**

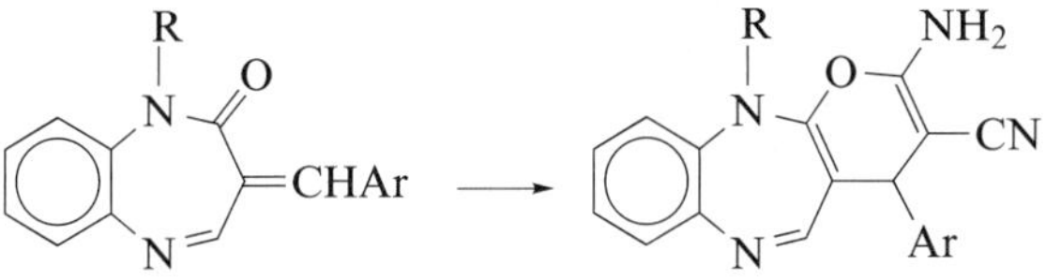

(110)   **(111)**

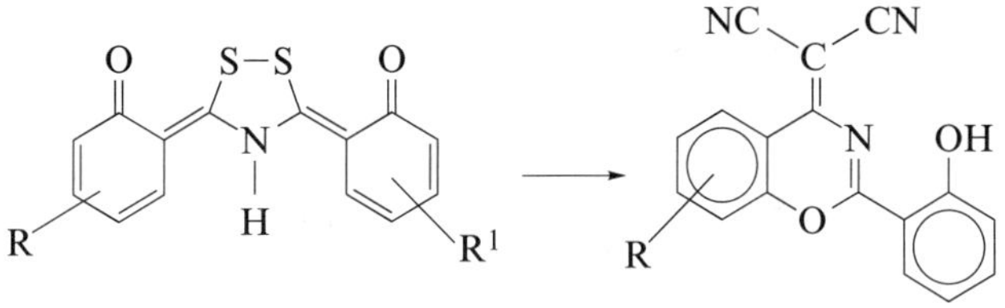

(112)   **(113)**

S,N-systems. When 2-benzothiazoleacetamide **114** reacted with benzaldehydes and **10**, pyridobenzothiazoles **115** were obtained[133]. In a similar procedure, 2-aminothiazoline was condensed with aromatic aldehydes and **10** to yield isomeric thiazolidinopyrimidine derivatives **116** and **117** in a one-spot synthesis[134]. Another one-pot synthesis gave benzo-

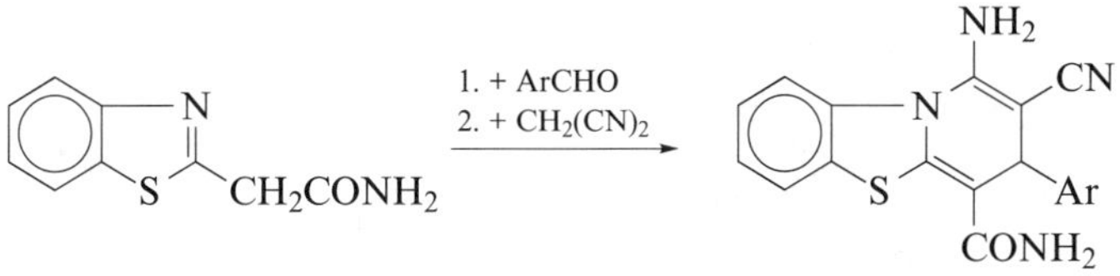

(114)   **(115)**

(116)   (117)   (118)

thiopyranopyridines **118** by condensation of **10** and arylidenethiochromanones and aliphatic alcohols[135].

1,3-Benzothiazin-4-ones were prepared by cyclization of thiosalicyclic acid with **10**[136] and benzylidenebenzothiazinones **119** yielded pyridobenzothiazines **120**[137]. The spiro compound **122** was obtained by reaction of isatins **121** with **10** and cyanothioacetamide[138].

(119)   (120)

(121)   (122)

2. Saturated substituted malononitriles

a. Formation and properties.   Tetracyano compounds can be prepared from **10**. So oxidative dimerization gave 1,1,2,2-tetracyanoethane[139]. Reaction of **10** with a concentrated aqueous solution of formaldehyde in the presence of, e.g., pyridine yielded 1,1,3,3-tetracyanopropane[140]. Arylenedimalononitriles were prepared by the Pd-catalyzed reaction of diiodoarenes with two mol of malononitrile anion[141]. Alkoxymalononitriles were obtained from amino acids and dipeptides under mild conditions in good yields[142].

(123)   (124)

Malononitrile salts. Electrosynthesis of air-stable potassium and ammonium salts of *p*-cycanophenyl malononitrile in liquid ammonia on a platinum grid electrode was reported[143]. Treating **10** with aliphatic alcohols and hydrohalic acids at elevated pressures gave dialkyl propanediimidate dihydrohalides[144]. Acid addition salts (*p*-toluenesulfonate, methanesulfonate and hydrochloride) of aminomalononitrile were prepared[145]. Bis(methylthio)methylene-indanedione **123** reacted with **10** to yield **124**[146].

Catalytic benzylation of **10** was performed under solid–liquid phase-transfer conditions to yield benzylidenemalononitrile **43**[147]. Alkylidenemalononitriles can undergo reduction to yield the corresponding saturated malononitriles[148]. Aryl malononitriles **125** have been obtained from **10** and halobenzenes[149-151] or from aroyl chlorides in three steps[152]. Surprisingly, the aliphatic malononitrile derivative **126** has been obtained when 2-indolone was reacted with **43**. The same reaction was also observed for two other cyclic carbonyl compounds[153].

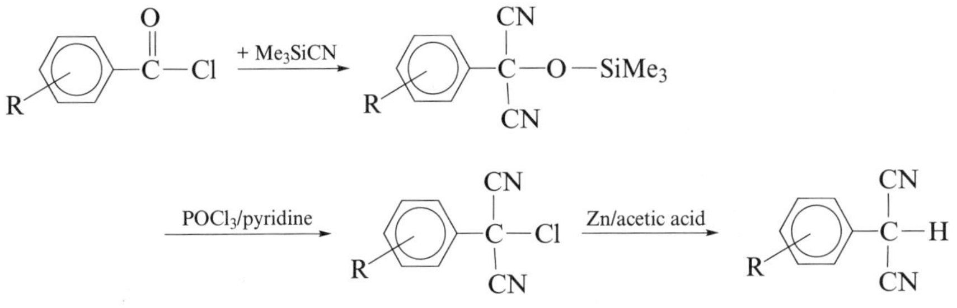

(125)

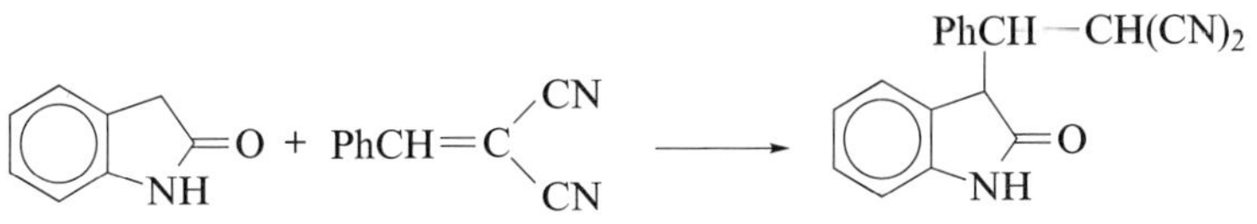

(43)   (126)

Heterocyclic malononitriles. 2-Pyridylmalononitriles have been prepared from pyridinium iodides[154]; 3-pyridylmalononitriles were obtained by reacting 3-halopyridines with malononitrile anion[155]. Other heterocyclic malononitriles like pyrazinemalononitrile[156] and 2-thiazolemalononitrile[157] have been prepared by direct ring-closure reactions from aliphatic precursors.

b. Reactions .
Formation of open-chain derivatives: (Di)alkyl malononitriles **127** were reductively decyanated to the corresponding mononitriles **128** in good to excellent yields[158].

Aryl-*E,E*-azabutadienes were prepared by treatment of ammoniummalononitrile salts with aromatic aldehydes in methanolic sodium acetate solution[159]. The iodine transfer

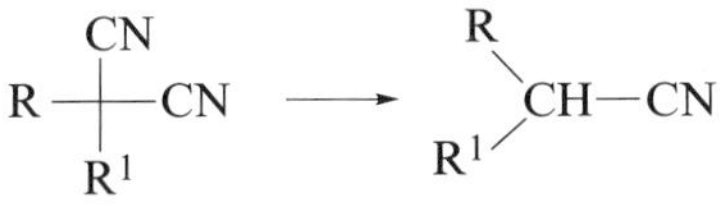

(127)   (128)

reaction and propargylation of alkenes with **129** gave **130**, which is a useful intermediate for cyclopentane derivatives **131**[160].

$$HC\equiv C-CH_2-\underset{\underset{CN}{|}}{\overset{\overset{CN}{|}}{C}}-I \ + \ RHC=CHR'$$

(129)

$$\longrightarrow HC\equiv C-CH_2-\underset{\underset{CN}{|}}{\overset{\overset{CN}{|}}{C}}-CHR-\overset{\overset{I}{|}}{CHR'} \ \xrightarrow{Bu_3SnH}$$

(130)   **(131)**

Vinylsilanes **132** have been reacted with bromomalononitrile to yield the intermediate **133**, which was used for the synthesis of cyclopropane derivatives[161]. Addition reactions of dichloromalononitrile with substituted alkenes and alkadienes can also be used for the preparation of intermediates in carbo- and heterocyclic synthesis[162]. 2-Arylmalononitriles **135** have been produced by coupling malononitriles with aryllead(IV) triacetates like **134**[163].

$$R_3SiCH=CH_2 \ + \ BrCH(CN)_2 \ \longrightarrow \ R_3SiCHBrCH_2CH(CN)_2$$
(132)   **(133)**

$$EtCH(CN)_2 \ + \ p\text{-MeOPhPb(OAc)}_3 \ \longrightarrow \ p\text{-MeOPhC(CN)}_2Et$$
(134)   **(135)**

Formation of heterocycles. Five membered heterocycles: Phenacylmalononitrile underwent cyclization in acidic media to yield substituted furans **136**[164] and **137**[165]. Cyclization with anilines afforded pyrroles **138**[166]. Benzylmalononitriles have been cyclized with hydrazine or hydroxylamine to yield 3,5-diaminopyrazoles or isoxazoles, respectively[167]. Aminomalononitrile tosylate can be reacted with aliphatic or aromatic isothiocyanates to yield 2,5-diaminothiazole-4-carbonitriles[168].

$$PhCOCH_2CH(CN)_2$$

acetic acid/   acetic acid/
hydochloric acid   sulfuric acid

(136)   **(137)**

Six-membered heterocycles: Saturated as well as unsaturated nitrogen heterocycles have been prepared. The malononitrile derivative **139** underwent a ring-closure reaction with

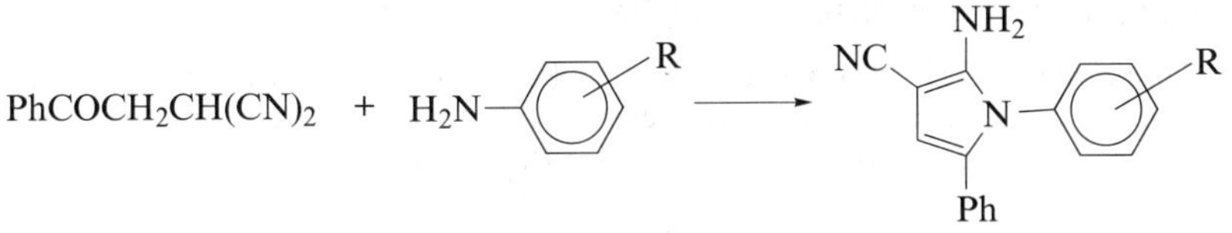

$$PhCOCH_2CH(CN)_2 \quad + \quad H_2N-\bigcirc-R \quad \longrightarrow$$

(138)

P$_4$S$_{10}$ to yield pyridinethione **140**, which by methylation afforded **141**[169]. Dichloromalononitrile was cyclized to give 2,5-dichloronicotinonitrile[162]. Arylazomalononitrile **142** was reacted with guanidinium salt **143** to yield arylazopyrimidine **144**[170,171]. Perhydropyrimidines and -triazines were prepared from (H$_2$N)$_2$C=N—CN 2-ethyl-2-phenylmalononitrile[172]. Tetracyanoperhydropiperazines were obtained from aminomalononitrile salts and aromatic aldehydes[173].

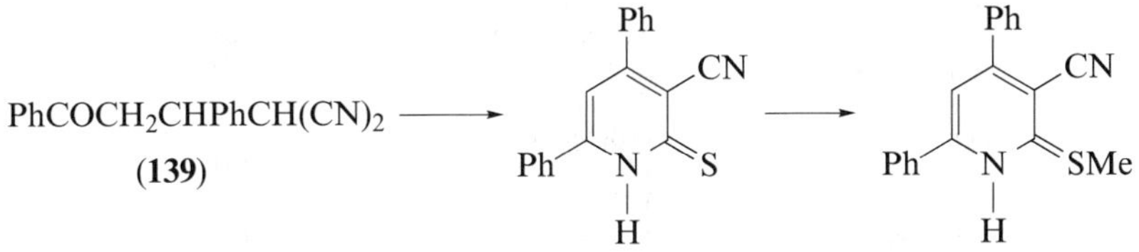

$$PhCOCH_2CHPhCH(CN)_2 \longrightarrow$$

(139)

(140) **(141)**

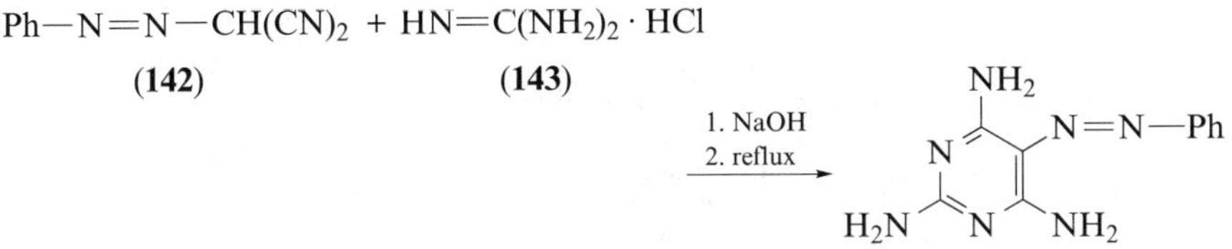

$$Ph-N=N-CH(CN)_2 \quad + \quad HN=C(NH_2)_2 \cdot HCl$$

(142) **(143)**

1. NaOH
2. reflux

(144)

Condensed and other heterocyclic systems: Cyanoacetophenone was condensed with bromomalononitrile to yield the furofuran **145**[174]. Substituted benzylmalononitriles gave **146** by reaction with guanidine[175].

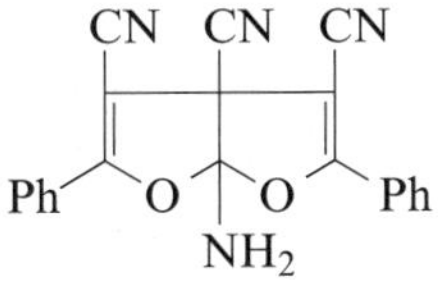

(145) 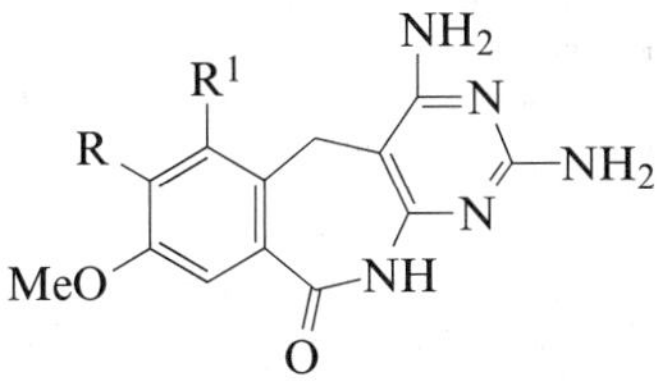 **(146)**

3. Alkylidene- and arylidenemalononitriles

a. Formation and physicochemical data. Papers dealing with the preparation of alkylidene- and arylidenemalononitriles, dicyanomethylene compounds and bis-(dicyanomethylene) derivatives are given in Tables 2, 3 and 4.

TABLE 2. Preparation of alkylidene- and arylidenemalononitriles $R^1R^2C=C(CN)_2$

R^1	R^2	References
Alkyl, phenyl, vinyl	H	180
Amino	Alkyl, phenyl, hetaryl	181
p-Dialkylaminophenyl	H	182
Benzylmercaptyl	Benzylmercaptyl	183
2-Coumarinimino	H	184
Cyanomethyl	Phenyl	185
Methyl	Arylethylidene	186
Methyl	Butadienyl	187
Methyl	2-cyclohexenylethene	188
Methyl	Subst. phenyl	187
N,N-Dimethyl	Aminothiocarbonylmethyl	189
Phenyl	p-B(MeS)$_2$phenyl	190
2-Furyl	H	191–194
Phenyl	H	195
2-Thienyl	H	196

TABLE 3. Preparation of dicyanomethylene derivatives

	References
4-Dicyanomethylene-benzothiopyran	197
Dicyanomethylene-bis(methoxycarbonyl)dithiole	198
Dicyanomethylene-cyclohexadiene, subst.	199
Dicyanomethylene-cyclohexene, subst.	200
Dicyanomethylene-cyclopentene, subst.	200
Dicyanomethylene-dihydropyridine	201
2-Dicyanomethylene-1,3-dioxolane	202
Dicyanomethylene-9-fluorene (subst.)	203–207
Dicyanomethylene-heptadecafulvene	208, 209
2-Dicyanomethylene-imidazole	210
1-Dicyanomethylene-indene	211
4-Dicyanomethylene-nitrobenzene	212
Dicyanomethylene-octadecylcyclohexane	213
Dicyanomethylene-pentadecafulvene	208, 209
3-Dicyanomethylene-phthalide	214
4-Dicyanomethylene-pyran	215
4-Dicyanomethylene-pyrazolone	216
2-Dicyanomethylene-pyridine	217, 218
4-Dicyanomethylene-1,2,3,4-tetrahydro-isoquinoline (subst.)	219,
4-Dicyanomethylene-thiopyran	197, 215
2-Dicyanomethylene-dithiolene	220
Dicyanomethylene-tridecafulvene	208, 209
4-Dicyanomethylene-quinoline	221
4-Dicyanomethylene-thiazolo(5,4-d)pyrimidine	222
3-Dicyanomethylene-thiazolidine	223

TABLE 4. Preparation of bis(dicyanomethylene) compounds (without TCNQ derivatives)

	References
1,4-Bis(dicyanomethylene)cyclohexane	224–226
15,16-Bis(dicyanomethylene)pentacene	
(15,15,16,16-tetracyano-6,13-pentacenequinodimethane)	227
2,4-Bis(dicyanomethylene)-dithietane	220
4,8-Bis(dicyanomethylene)-4,8-dihydrobenzo-(1,2-c:4,5-c')dithiophene	228
1,3-Bis(dicyanomethylene)indan	229

b. Reactions. Ylidenemalononitriles underwent dimerization reactions, the products of which have been brominated[176]. Other electrophilic substitutions (with CF$_3$SCl) gave ((trifluoromethyl)thio)alkyl derivatives[177]. Principal ions in mass spectra of arylidenemalononitriles were determined[178]. An intramolecular pyridinum salt was formed by reactions of ylidenemalononitriles and pyridine *N*-imides[179]. Electrohydrodimerization of substituted benzylidenemalononitriles (**43**) was investigated in detail. Depending on the substituent on the phenyl ring of **43**, either two radical anions couple to give a dimeric anion or one radical anion attacks **43**[230].

i. Formation of open-chain derivatives. Dimerization of **10**[48] or reaction of alkyl cyanoacetates with **10** gave 'dimeric' products **147** and **148**, the reactivity of which was studied in detail[49]. Reaction of **147** and **148** with equimolar amounts of dimethylformamide dimethyl acetal (DMF-DMA) afforded monocondensation products **149** and **150**[231]. Biscondensation product **151** was obtained from **147** with excess DMF-DMA[232].

148, 149 and **151** are highly reactive synthons for various ring-closure reactions and cycloadditions. So **149** gave pyridine **152**[231] and **151** yielded pyrido-pyrimidine **153**[232]. **148** gave halopyridones **A** when cyclized with hydrohalogenic acids; N-aryl-pyridazines **B** were formed with aromatic diazonium salts. With aldehydes **C** were obtained, while salicylic aldehydes afforded benzopyran **D**[233,234]. When 1,1-dicyano-2-ethoxypropene-1 was reacted with DMF-DMA, **E** was obtained. Ring closure with inorganic acids, formic acid and ammonia gave **154, 155** and **156**, respectively[235,236].

(A)   (B)

(C)   (D)   (E)

(154)   (155)   (156)

N,N-Disubstituted 4-amino-1,1-dicyanobuta-1,3-dienes were obtained from ylidene-malononitriles by reactions with triethyl orthoformate and amines[237]. Furfurylidenemalono-nitriles have been subjected to various reactions to form different open-chain derivatives, which are useful intermediates. So **157** was reacted with ketene thioacetal **158** to yield furylhexadienonitriles **159**[238].

(157)   (158)   (159)

Reaction with cyanoethoxyacrylate **160** gave hexadienoates **161**[239].

Phosphonates **162** have been prepared by C-alkylation of furfurylidenemalononitrile with trialkyl phosphites[240].

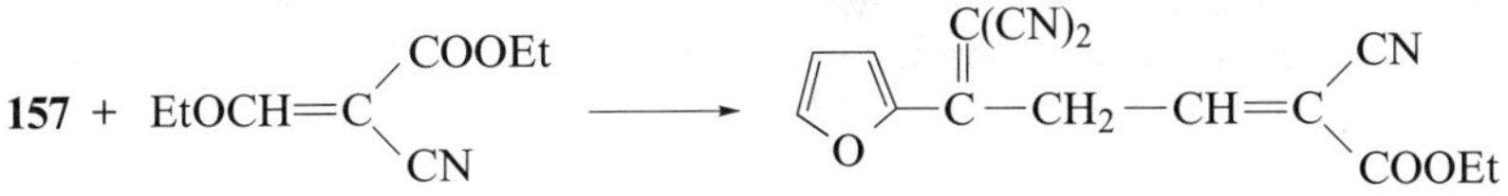

$$157 + EtOCH=C\overset{\diagup COOEt}{\diagdown CN} \longrightarrow$$

(160)

$$\underset{O}{\overset{C(CN)_2}{\diagup}}\parallel C-CH_2-CH=C\overset{\diagup CN}{\diagdown COOEt}$$

(161)

$$\underset{O}{\diagup}CH=C(CN)_2 + \overset{OR}{\underset{OR}{\overset{|}{P}-OR}} \longrightarrow \underset{O}{\diagup}CH\overset{R}{\underset{|}{-}}C(CN)_2$$
$$O=\overset{|}{\underset{OR}{P}}-OR$$

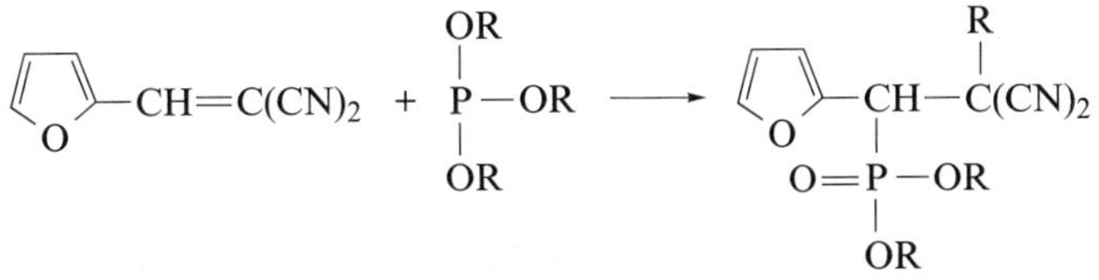

(162)

Multifunctional intermediates have also been prepared from chloromethylene-malononitrile[79,241]. The arylethylidenemalononitrile **163** reacted with benzenediazonium chloride to yield **164**, which cyclized to the pyridazine **165** in dilute acetic acid[242].

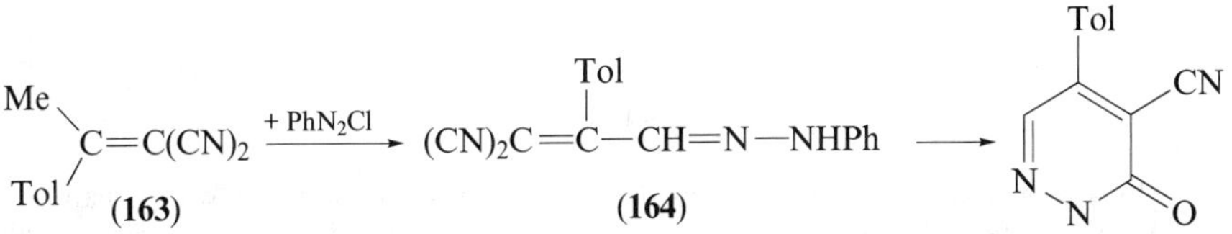

$$\underset{Tol}{\overset{Me}{\diagdown}}C=C(CN)_2 \xrightarrow{+ PhN_2Cl} (CN)_2C=\overset{Tol}{\underset{|}{C}}-CH=N-NHPh \longrightarrow$$

(163)   (164)

(165)

An interesting reaction was observed when **166** was reacted with 2-mercapto-tetrahydropyrimidine **167**. Ring opening yielded isothiocyanate **168**[243].

$$BrCH_2-\overset{Ph}{\underset{|}{C}}=C(CN)_2 +$$

(166)

(167)

$$\longrightarrow SCN \quad NH-CH_2-\overset{Ph}{\underset{|}{C}}=C(CN)_2$$

(168)

ii. Formation of carbocycles.
Five membered carbocycles: Although a number of cyclic compounds have been prepared from ylidenemalononitriles, only a few carbocycles have been discussed. So benzylidenemalononitrile (**43**) was dimerized in DMF/acetic acid to yield *cis*- and *trans*-**169**[244].

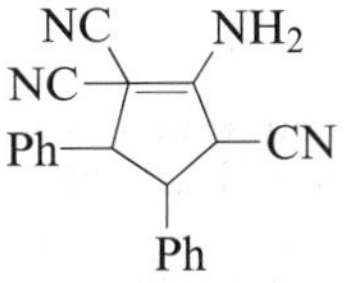

(169)

When benzylidenemalononitrile **43** was reacted with cyclopropenone propanediol ketal **170**, ring enlargement gave cyclopentenone propanediol ketal **171**[245].

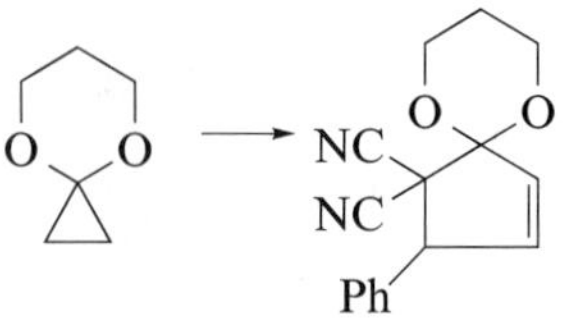

$$
\text{(170)} \longrightarrow \text{(171)}
$$

(170)   (171)

Six-membered carbocycles: A number of 2-amino-1,3-benzodinitriles **172** has been reported[246-250]. R^1 and R^3 generally are (het)aromatic rings; R^2 is, with one exception[250] (NO_2), H. In some cases R^3 is SMe[250,251].

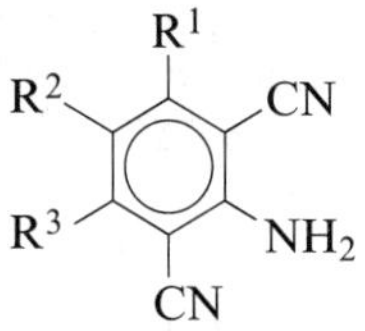

(172)

Arylbenzonitriles can be prepared by condensing arylidenemalononitriles with dienamines. So **43** was reacted with the morpholine derivative **174** to yield biphenyl-2-carbonitrile **175**[252]. When **43** was condensed with 1-phenylethylidenemalononitrile **176**, tricyano-cyclohexadiene **177** was obtained, which in turn afforded the aromatic dinitrile **178** by HCN elimination[253].

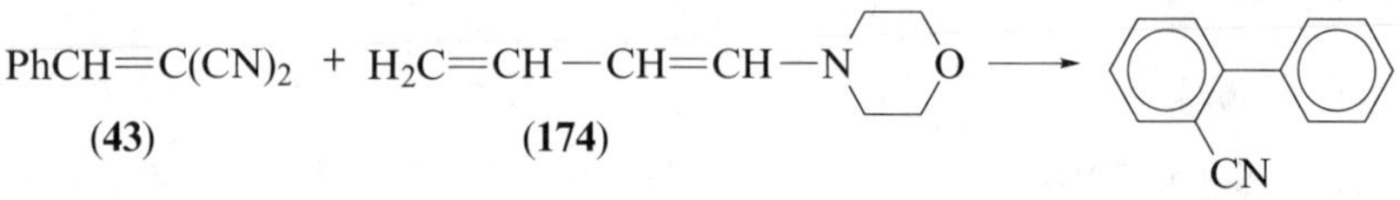

$$
\text{PhCH}{=}\text{C(CN)}_2 \;+\; \text{H}_2\text{C}{=}\text{CH}{-}\text{CH}{=}\text{CH}{-}\text{N}\bigcirc\text{O} \longrightarrow
$$

(43)   (174)

(175)

$$
\text{PhCH}{=}\text{C(CN)}_2 \;+\; \underset{\text{Me}}{\text{PhC}}{=}\text{C(CN)}_2 \longrightarrow
$$

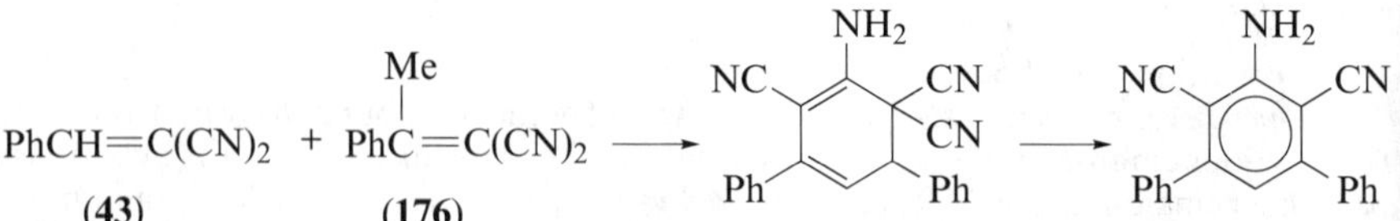

(43)   (176)

(177)   (178)

Polycarbonitriles have been prepared by condensing ylidenemalononitriles with nitriles. So crotonitriles and arylidenemalononitriles gave cyclohexenon imines[254]. Pentacarbonitriles **180** were obtained by reactions of **10** with arylidenemalononitriles **179**[255].

Carboxylates have also been prepared by condensation reactions of ylidenemalononitriles. Thus salicylates **183** and **184** were obtained from (ethoxymethylene)malononitrile **181** and acetoacetates **182**[256].

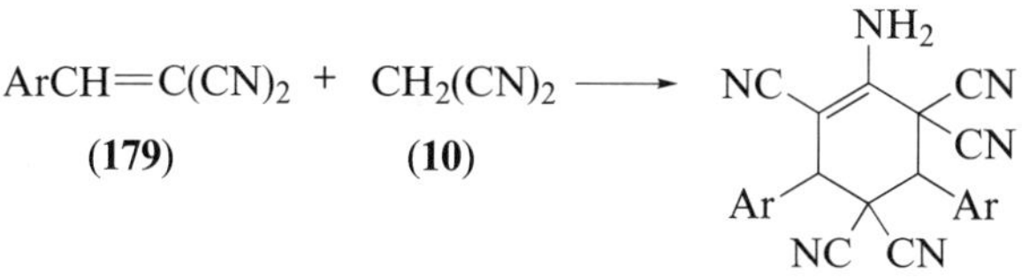

$$\text{ArCH=C(CN)}_2 \;+\; \text{CH}_2\text{(CN)}_2 \longrightarrow$$

(179)   (10)

(180)

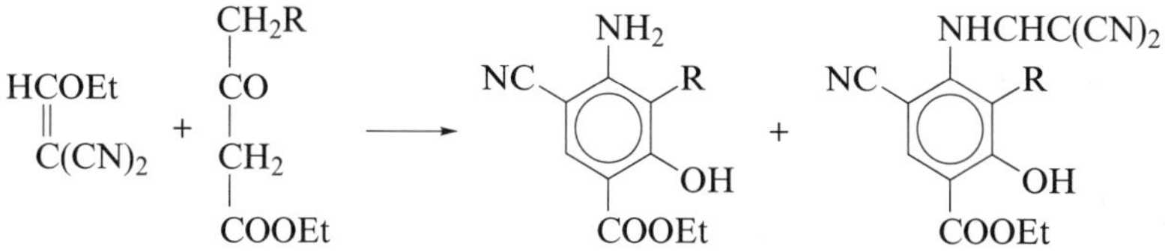

(181)   (182)   (183)   (184)

Tricarboxylate **187** results from potassium carbonate catalyzed condensation of alkylidenemalononitrile **185** with dimethyl acetylenedicarboxylate (**186**)[257].

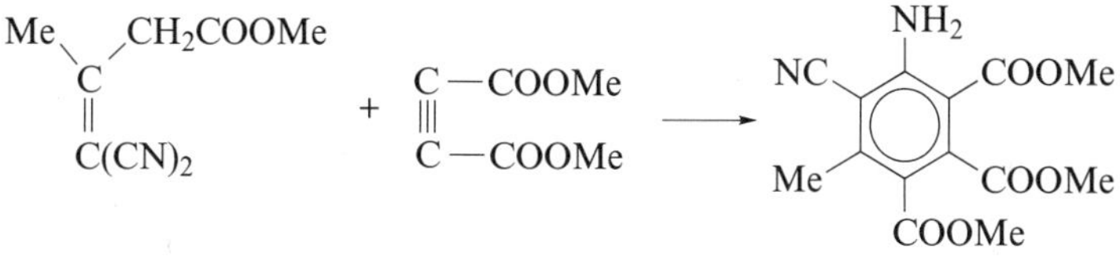

(185)   (186)   (187)

Condensed and other carbocyclic systems: An interesting dimerization reaction was performed with cyclohexylidenemalononitrile **188**. The reaction was performed in methanolic sodium methanolate and yielded the spiro compound **189**[258]. Acidic intramolecular cycloaddition of alkylidenemalononitrile **190** (with concentrated sulfuric acid) gave the naphthalene **191**[259]. 1-Tetraloneanil **192** reacted with **179** to yield phenanthrene **193**[260].

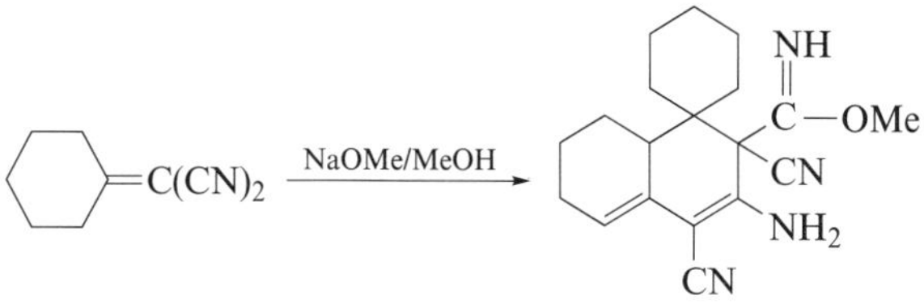

(188)   (189)

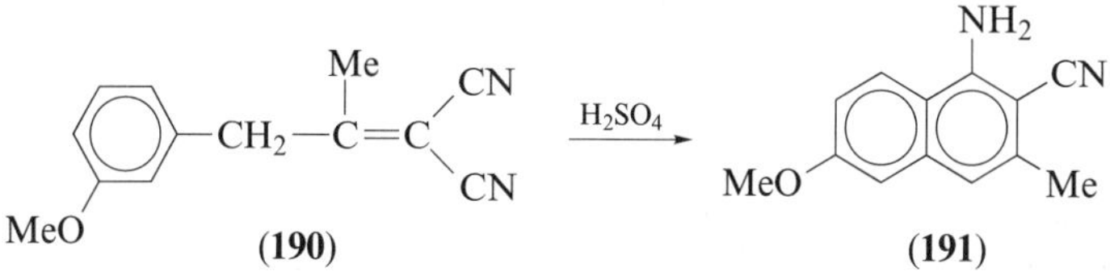

(190)   (191)

$$\text{(192)} \qquad \text{(193)}$$

iii. Formation of heterocycles.

Five-membered heterocycles: Nitrogen heterocycles. Aminomethylenemalononitrile was transformed to azafulvene by lithiation[261]. Cyclohexylidenemalononitrile **188** yielded spiro pyrroline **194** by reaction with trimethylsilyl cyanide. The reaction was catalyzed by KCN/18-crown-6[262].

$$\text{(188)} \qquad \text{(194)}$$

A three-component reaction of the alkylidenemalononitrile **195** with *p*-toluidine and acetone gave pyrroline **196**[263]. Addition of phenylhydrazine to **195** produced the pyrazoline **197**[264].

$$(F_3C)_2C{=}C(CN)_2 \; + \; Me{-}\!\!\!\bigcirc\!\!\!{-}NH_2 \; + \; Me_2CO$$

$$\text{(195)}$$

$$\text{(196)}$$

$$\text{(197)}$$

The nitroethylidene-hydrazine sodium salt **198** was refluxed with the arylidene-malononitrile **179** in acetonitrile to yield the pyrazolidine **199**[265]. Since they are useful as pharmaceuticals and drugs, cyanopyrazoles are of special preparative interest. Additions of hydrazines to ylidenemalononitriles provide a number of substituted pyrazole derivatives **200**. All of them contain one nitrile group, mostly in the position of R^{2} [266-268]. The pyrazoline derivative **202** was obtained from phenylethylidenemalononitrile **201** and cyanacetohydrazide [269].

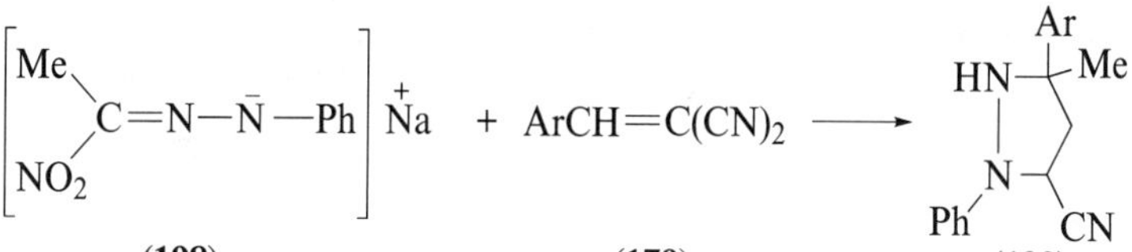

$$
\begin{bmatrix} \underset{NO_2}{\overset{Me}{\diagdown}}C=N-\bar{N}-Ph \end{bmatrix} \overset{+}{Na} \ + \ ArCH=C(CN)_2 \ \longrightarrow
$$

(198)   **(179)**

$$ \text{(199)} $$

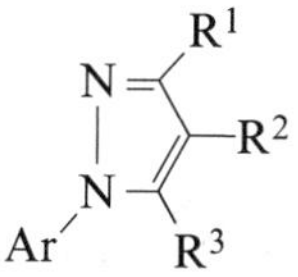

(200)

$$
\underset{Ph}{\overset{Me}{\diagdown}}C=C(CN)_2 \ + \ H_2NNH-COCH_2-CN \ \longrightarrow
$$

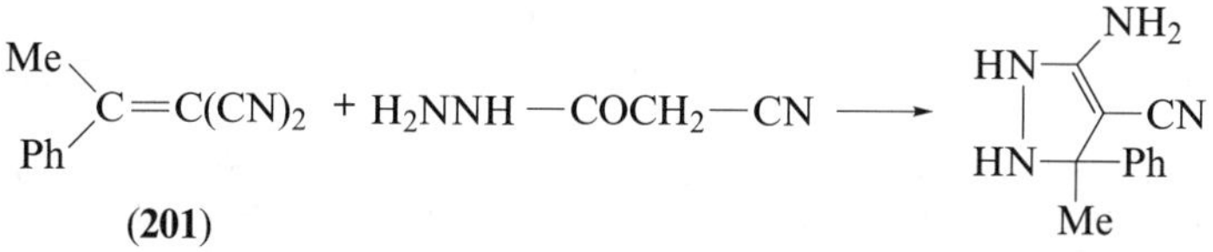

(201)

(202)

Sulfur heterocycles. 3-Amino-4,4-dicyano-3-butenoate was reacted with sulfur to yield a 2,4-diaminothiophene [233]. Reaction of **203** with ylidenemalononitrile **204** gave the heterocyclic ylidenemalononitrile **205** [270]. The dicyano-butene derivative **148** was cyclized with sulfur to yield thiophenecarbonic acid esters **206** [233].

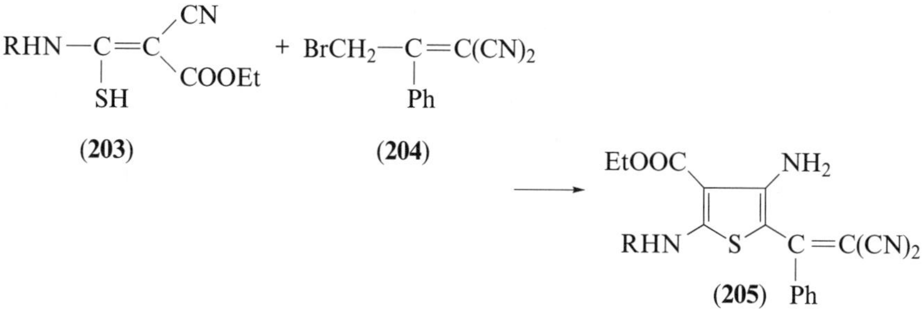

$$
RHN-C=C\underset{COOEt}{\overset{CN}{\diagdown}} \ + \ BrCH_2-\underset{Ph}{\overset{|}{C}}=C(CN)_2
$$

(203)   **(204)**

$\longrightarrow$

(205) Ph

$$
(NC)_2C=\underset{}{\overset{NH_2}{\overset{|}{C}}}-CH_2-CO_2Et \ \xrightarrow{S}
$$

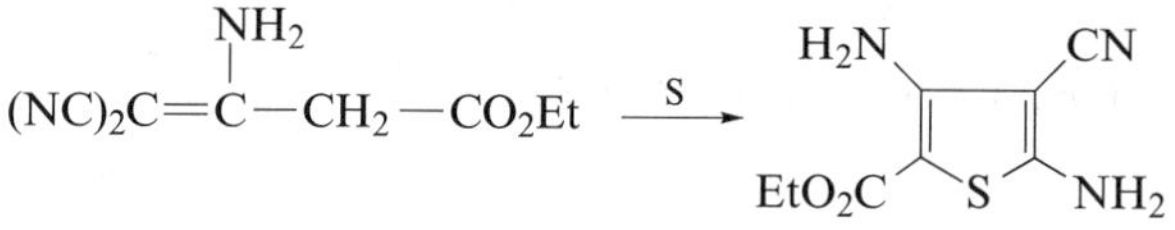

(148)

(206)

Oxygen heterocycles. Arylidenemalononitriles like **43** underwent cycloaddition with aromatic aldehydes (e.g. benzaldehyde) to yield dihydrofuranes like **207**[271].

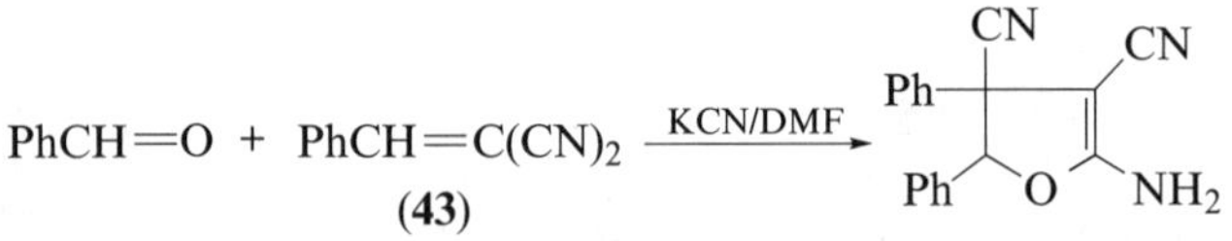

$$PhCH{=}O \ + \ PhCH{=}C(CN)_2 \xrightarrow{\text{KCN/DMF}}$$
(43)

(207)

Six-membered heterocycles: Nitrogen heterocycles. Alkylidenemalononitriles have been cyclodimerized in methanolic sodium methanolate to yield azabicyclooctadienes **208**, which underwent retro-Diels–Alder reaction in xylene or acetic acid to yield the substituted pyridine **209**[272].

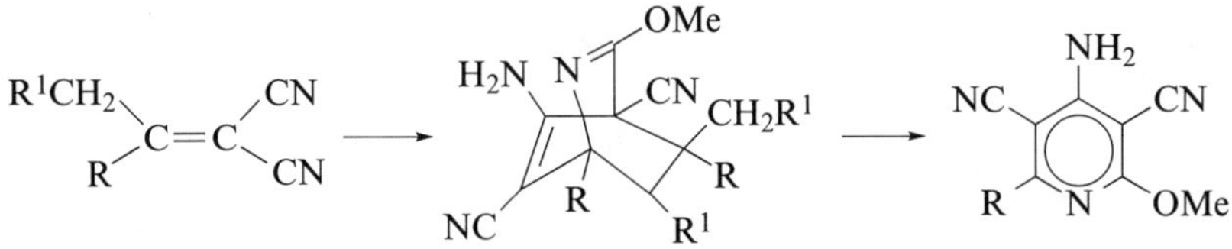

(208)   (209)

Malononitrile dimer (**147**) reacted with dimethylformamide dimethyl acetal **210** to yield **211**, which was cyclized with amines to give aminopyridines **212**[231]. **148**, a mixed dimer prepared from **10** and ethyl cyanoacetate, was cyclized in acidic or basic media to afford cyanopyridones **213** (X = Cl, Br, OMe, OEt, OH)[233]. Reacting **147** with excess **210** gave the product **214**, which was cyclized with ammonia to yield **215**[232].

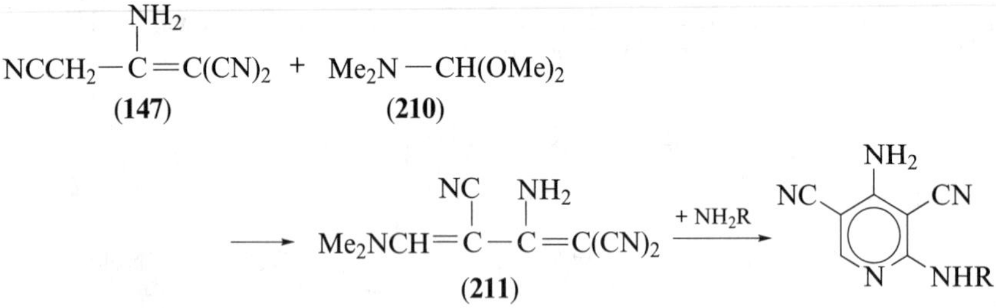

$$NCCH_2{-}\overset{\overset{\displaystyle NH_2}{|}}{C}{=}C(CN)_2 \ + \ Me_2N{-}CH(OMe)_2$$
(147)   (210)

$$\longrightarrow \ Me_2NCH{=}\overset{\overset{\displaystyle NC}{|}}{C}{-}\overset{\overset{\displaystyle NH_2}{|}}{C}{=}C(CN)_2 \xrightarrow{+\ NH_2R}$$
(211)

(212)

$$(NC)_2C{=}\overset{\overset{\displaystyle NH_2}{|}}{C}{-}CH_2{-}CO_2Et \ \longrightarrow$$

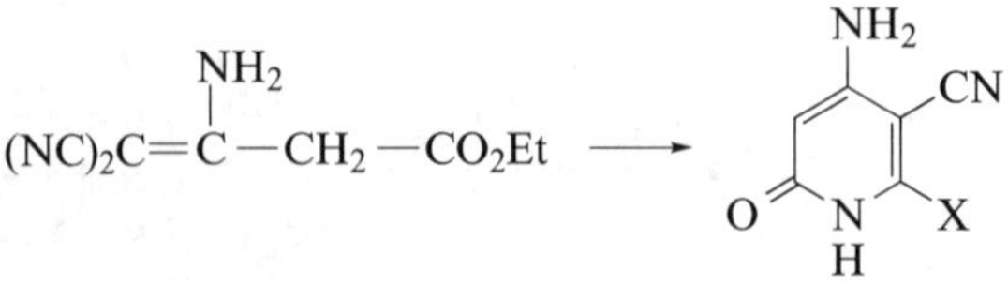

(147)   (213)

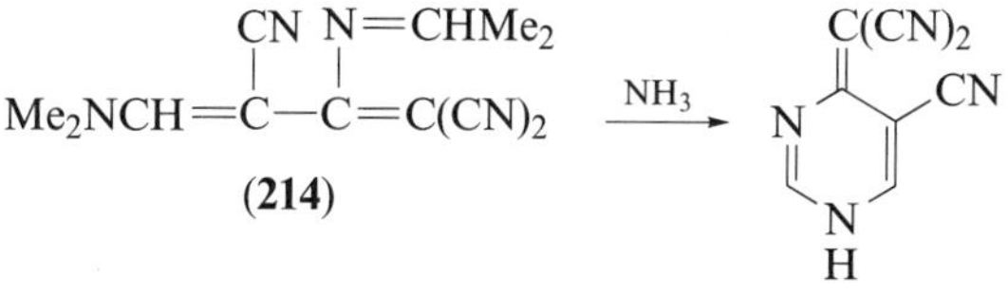

$$Me_2NCH\!=\!\overset{\underset{\displaystyle CN}{|}}{C}\!-\!\overset{\underset{\displaystyle N\!=\!CHMe_2}{|}}{C}\!=\!C(CN)_2 \quad \xrightarrow{\quad NH_3 \quad}$$

(214)

(215)

Pyridines (**218**) can also be obtained, when arylethylidenemalononitriles **216** react with arylmethylidenemalononitriles **217** and **10**[273]. **201** was reacted with **10** and trichloroacetonitrile to yield 2-amino-4-phenyl-6-trichloromethyl-nicotinonitrile[249]. Another three-component reaction yielded ethylthiopyridines **219**. In this case the arylidenemalononitrile (**179**) condensed with **10** and EtSNa[274].

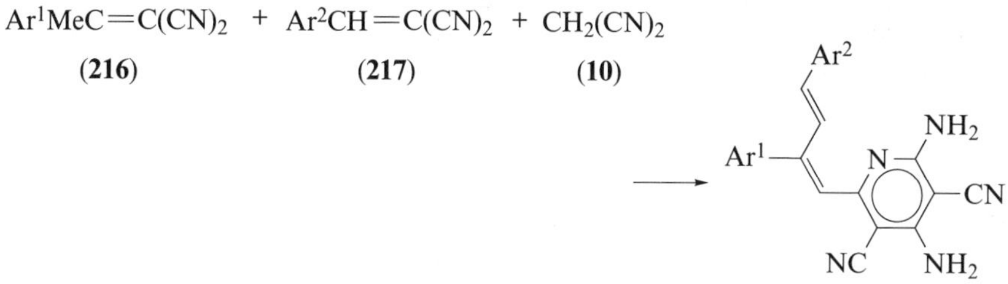

$$Ar^1MeC\!=\!C(CN)_2 \quad + \quad Ar^2CH\!=\!C(CN)_2 \quad + \quad CH_2(CN)_2$$

   (216)   **(217)**   **(10)**

(218)

$$ArCH\!=\!C(CN)_2 \quad + \quad CH_2(CN)_2 + EtSNa \quad \longrightarrow$$

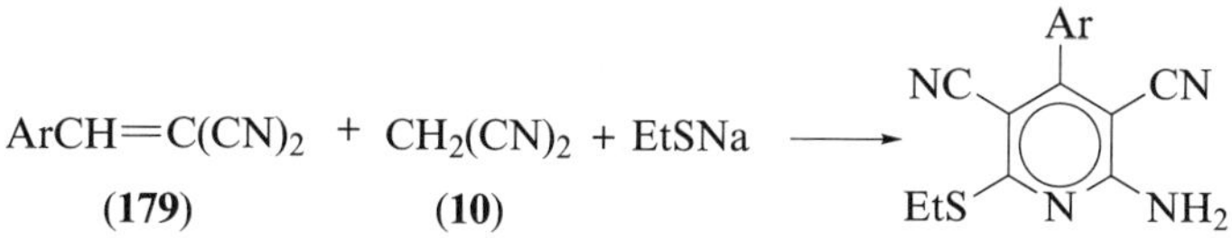

   (179)   **(10)**

(219)

Other pyridines were prepared from tosyloximinomalononitrile and 1-alkoxy-1,3-dienes[275] and by Vilsmeier reaction (DMF, POCl₃) of butylidenemalononitrile[276].

2-Oxo-pyridines were prepared from arylidenemalononitriles and cyanoacetanilide or pyridylcyanoacetamide[277], from arylidenemalononitrile and potassium cyanoacetohydroxamate[278], or arylacetohydrazides[279] and by acidic or basic ring-closure of 3-amino-4,4-dicyano-3-butenoate[233].

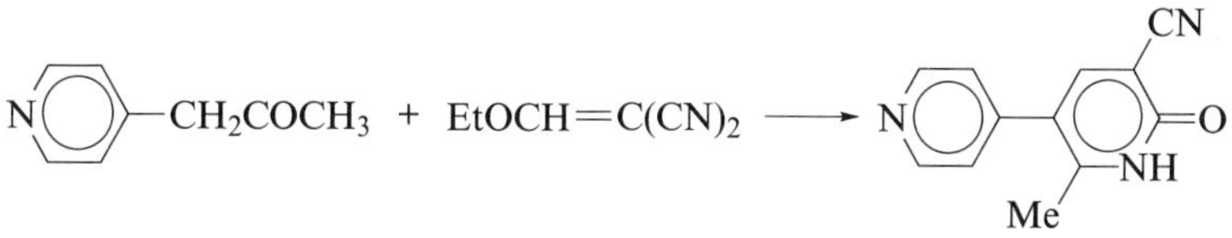

$$N\!\!-\!\!CH_2COCH_3 \quad + \quad EtOCH\!=\!C(CN)_2 \quad \longrightarrow$$

   (F)   **(181)**   **(G)**

Pyridinylpropanone **F** was treated with (ethoxymethylene)malononitrile **181** to yield bipyridyl derivative **G**[280,281]. Pyridinethiones were prepared from β-alkoxy- or β- amino-

malononitriles and cyanothioacetamides. The same reactions have also been performed
with cyanselenoacetamide[282]. Phenylthioacetanilides gave pyridinethiones with various
arylidenemalononitriles[283]. Phenethylidenemalononitriles (201) were reacted with
arylmethylidenecyanothioacetamides H to yield pyridinethiones J[284].

$$ArMeCH{=}C(CN)_2 \;+\; Ar^1CH{=}\underset{\underset{S}{\|}}{\overset{\overset{CN}{|}}{C}}{-}C{-}NH_2 \longrightarrow$$

(201)   (H)   (J)

Substituted 2-furfurylidenemalononitriles K have been reacted with geminal dithiols L
to yield M[285]. Cyclization of enaminonitriles N with sodium ethanolate yielded
pyridinethiones 220[189].

$$R{-}\underset{O}{furyl}{-}CH{=}C(CN)_2 \;+\; R^1CH_2{-}\underset{\underset{SH}{|}}{\overset{\overset{SH}{|}}{C}}R^2 \longrightarrow$$

(K)   (L)   (M)

$$RNH{-}\underset{\underset{S}{\|}}{C}{-}CH_2{-}\underset{\underset{NMe_2}{|}}{C}{=}C(CN)_2 \longrightarrow$$

(N)   (220)

Pyridine spiro compounds 221 were prepared by reacting 188 with cyanothioacetamide[286].
Reaction of 188 with enamines yielded 222[287]. Alkylidenemalononitriles were cyclized with
aromatic diazonium chlorides to yield pyridazines[288]. Cyanoacetamide was dimerized to
223 and reacted with phenyldiazonium chloride to yield the pyridazine 224[289].

(221)   (222)

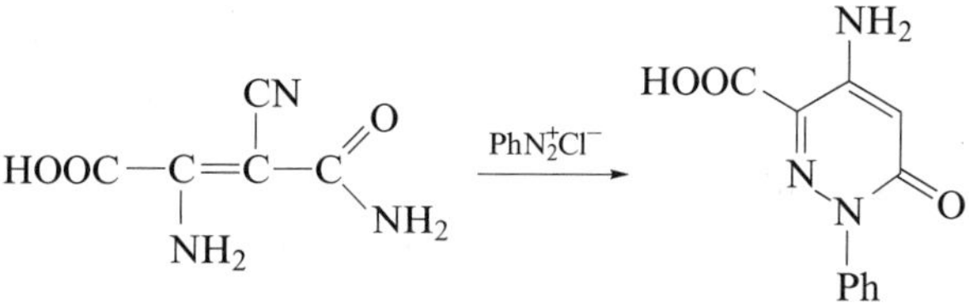

(223)   (224)

Sulfur heterocycles. Some papers report on the synthesis of thiopyrans from cyanothio-acetamides and various ylidenemalononitriles. Subsequent rearrangement yielded pyridinethiones. So arylidenemalononitriles (**179**) yielded **225** in cold ethanol with catalysis of triethyl amine. Refluxing **225** in ethanol/triethylamine gave **226**[290–292].

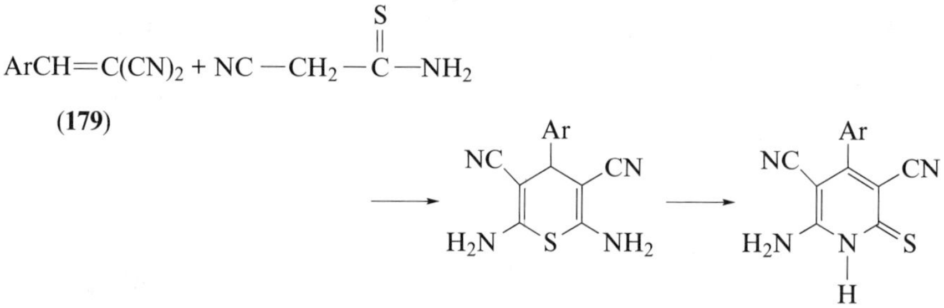

$$ArCH{=}C(CN)_2 + NC-CH_2-\overset{\overset{\displaystyle S}{\|}}{C}-NH_2$$

(179)

(225)   (226)

Reaction of **179** with geminal dithiols **227** yielded dithiane **228**, which was recyclized to pyridinethione **229**[293].

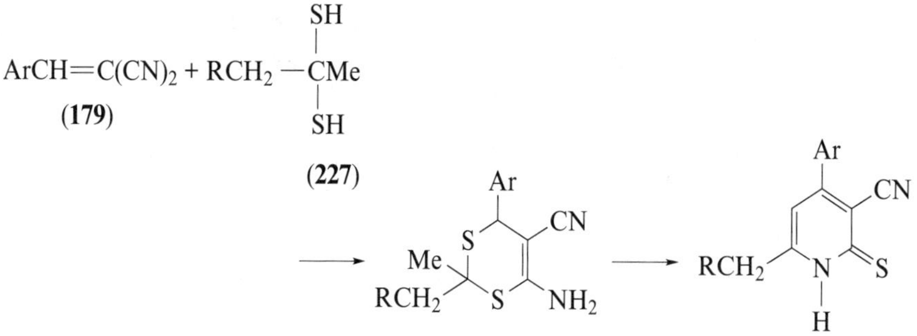

$$ArCH{=}C(CN)_2 + RCH_2-\overset{\overset{\displaystyle SH}{|}}{\underset{\underset{\displaystyle SH}{|}}{C}}Me$$

(179)   (227)

(228)   (229)

Oxygen derivatives. Pyran derivatives **232** were prepared from pyridine-3-ylidene-malononitriles **230** and 1,3-dicarbonyls **231**[294]. Dicyanopyran **234** was obtained from **43** and **233**[295].

Condensed and other heterocyclic systems: Nitrogen heterocycles. Condensed 3-cyanopyridines **236** were prepared from **179** and cycloketones **235** ($n = 1, 2, 3, 5$)[296]. Ring

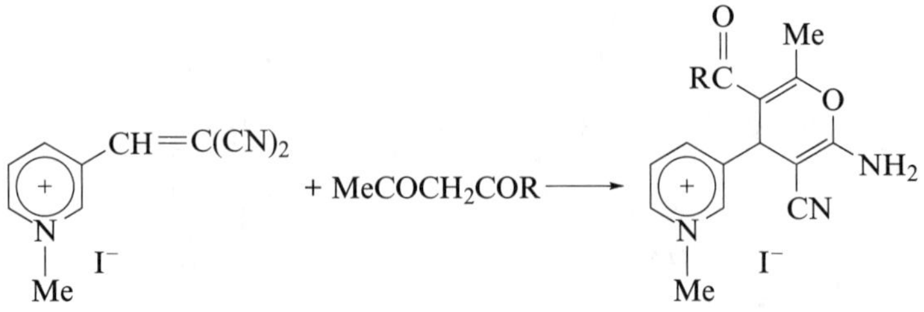

(230)   (231)   (232)

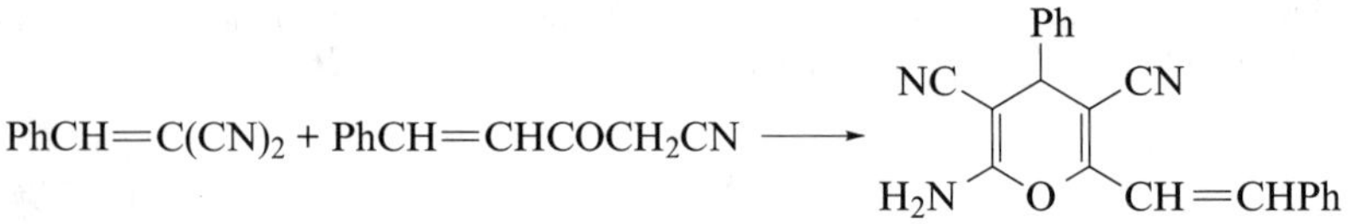

PhCH=C(CN)$_2$ + PhCH=CHCOCH$_2$CN ⟶

(43)   (233)   (234)

ArCH=C(CN)$_2$ +

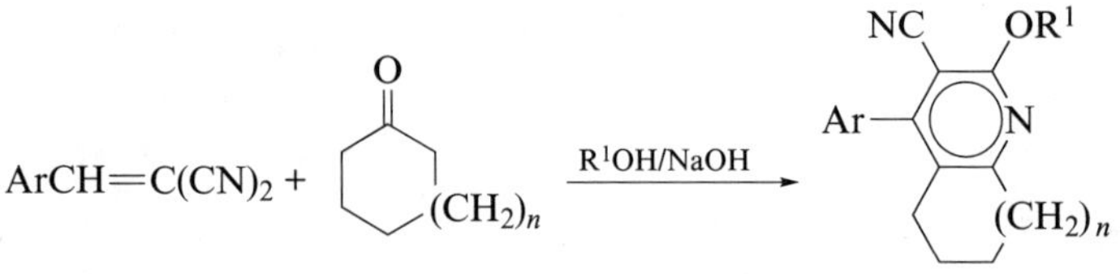

(179)   (235)   (236)

enlargement of indolylidenemalononitrile **237** with pyrazolinone **238** gave the quinoline **239**[297]. Quinoline-N-oxides **241** were obtained from **179** and benzisoxazoles **240**[298].

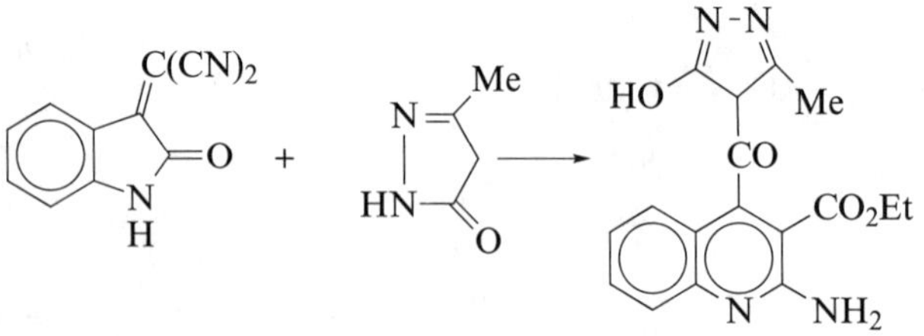

(237)   (238)   (239)

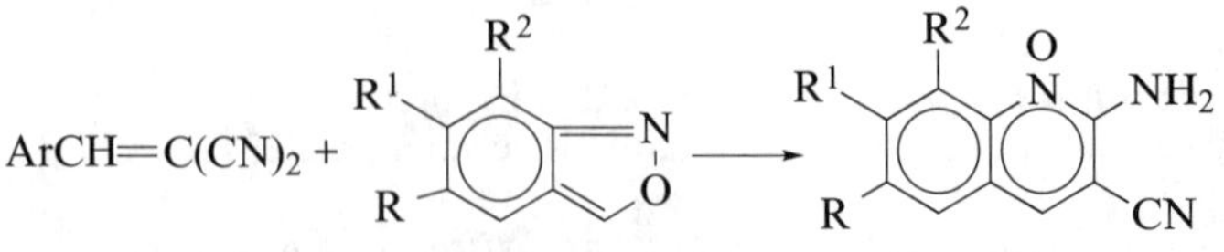

ArCH=C(CN)$_2$ +

(179)   (240)   (241)

Acrylonitriles **242** were cyclized to isoquinoline derivatives with amines like aniline in acetic acid to yield isoquinolines **243**[299]. The dicyanoethenyl derivative **245** was obtained from **147** and *o*-aminobenzaldehyde **244**[233].

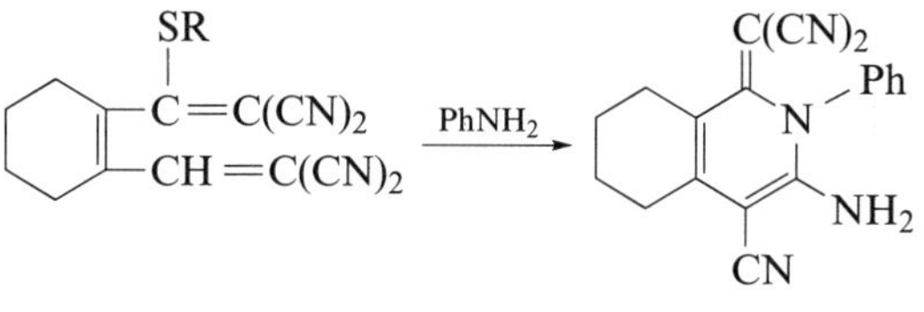

(242)   (243)

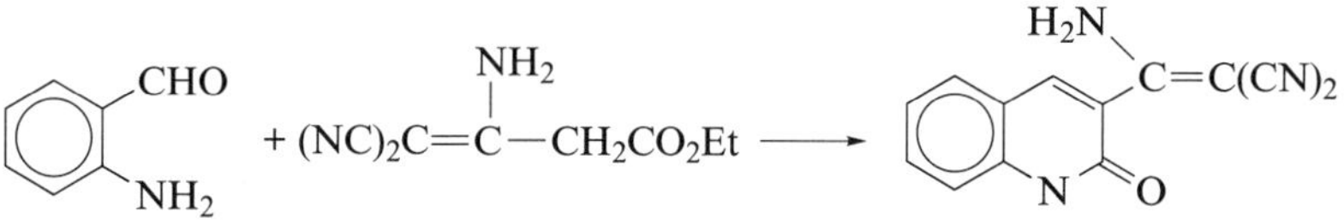

(244)   (147)   (245)

Bis[bis(alkylthio)methylene]hexylidenemalononitriles **246** react with hydrobromic acid to yield **247**[300] and with thiols to **248**[301]. Cyclization with bromine gave **249**[302,303].

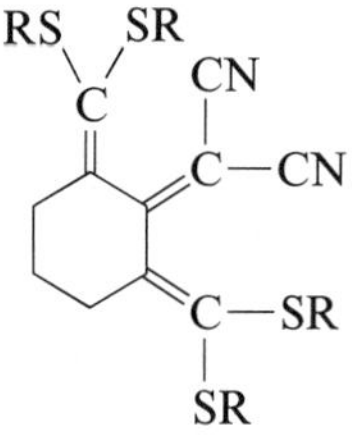

(246)

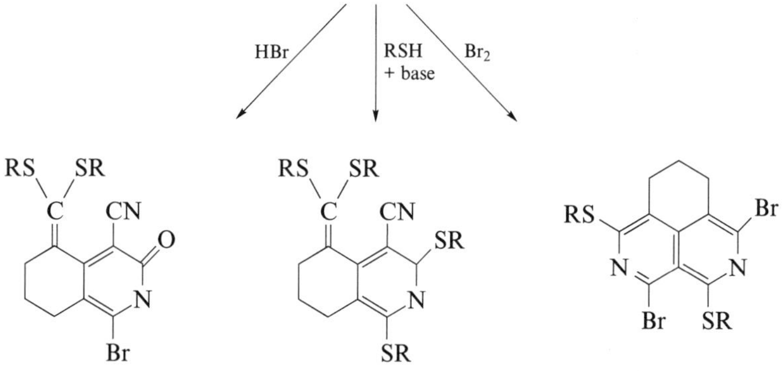

(247)   (248)   (249)

(Ethoxymethylene)malononitrile **181** was condensed with diethyl 3-oxoalkanedioates **250** to compounds **251**, **252** and **253**, depending on the number of methylene groups in **250**[304].

$$
\underset{\textbf{(181)}}{EtOCH{=}C\diagup\!\!\diagdown_{CN}^{CN}} + \underset{\textbf{(250)}}{EtO_2CCH_2\overset{\overset{\displaystyle O}{\|}}{C}CH_2(CH_2)_nCO_2Et}
$$

| (251) | (252) | (253) |

2-(Dicyanomethylene)indandione **254** reacted with α-amino acid esters to yield adducts **255** by Michael addition–elimination. **255** in turn underwent triethylamine-catalyzed cycloaddition with N-methylmaleimide to give **256**[305].

Arylidenemalononitriles (**179**) reacted with benzofuroxanes **257** to yield quinoxaline dioxides **258** by ring enlargement[306]. α-Amino-N-heterocycles were treated with ethoxymethylenemalonester to yield tautomeric products **259**, **260** and **261**[307].

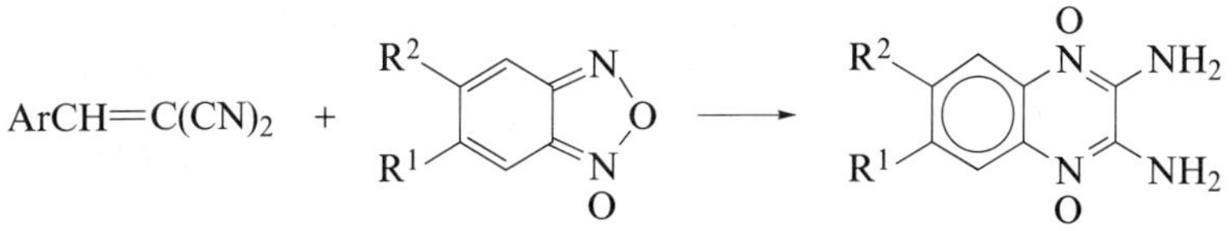

ArCH=C(CN)$_2$ +

(179)   (257)   (258)

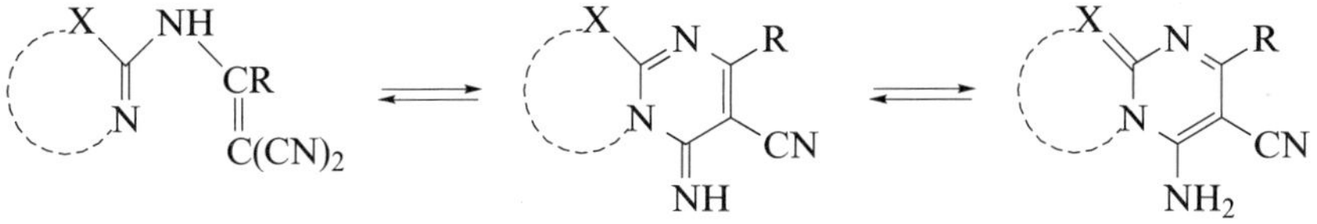

(259)   (260)   (261)

Other N-bridgehead compounds (263) were obtained from pyrazoles 262 and arylidenemalononitriles[308]. Mercaptoimidazole derivatives 264 were reacted with 166 to be transformed to pyrroloimidazoles 265[243].

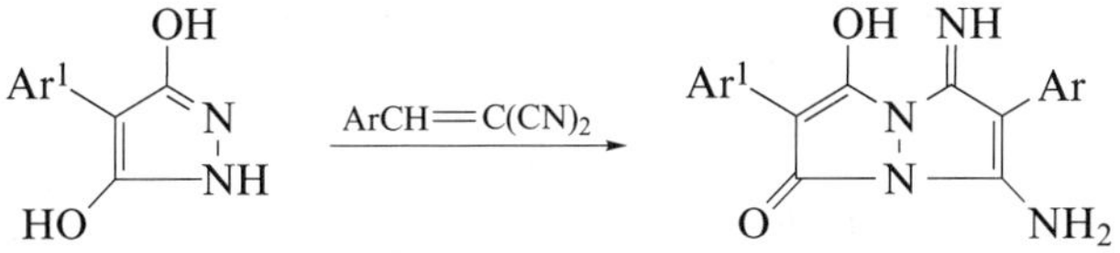

ArCH=C(CN)$_2$

(262)   (263)

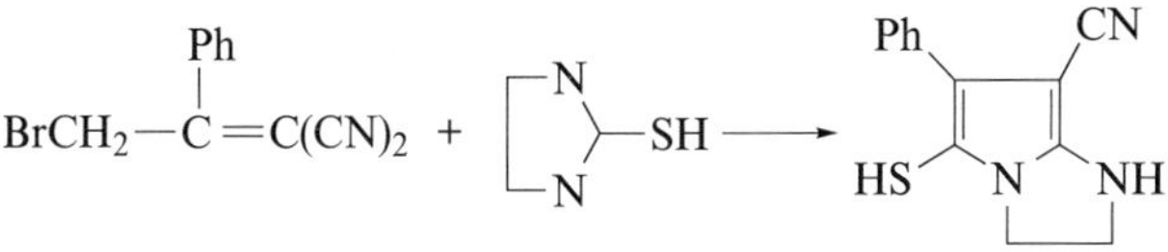

BrCH$_2$—C=C(CN)$_2$ +

(166)   (264)   (265)

ClC$_6$H$_4$NH—N=C—COR^1 + NCCH$_2$—C=C(CN)$_2$ + HN=C(CH$_2$CN)$_2$

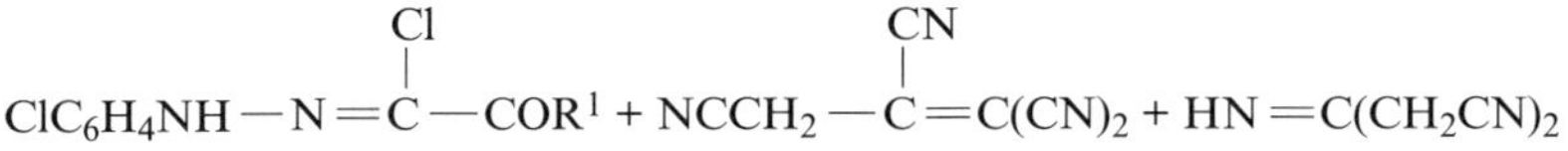

(266)   (147)   (267)

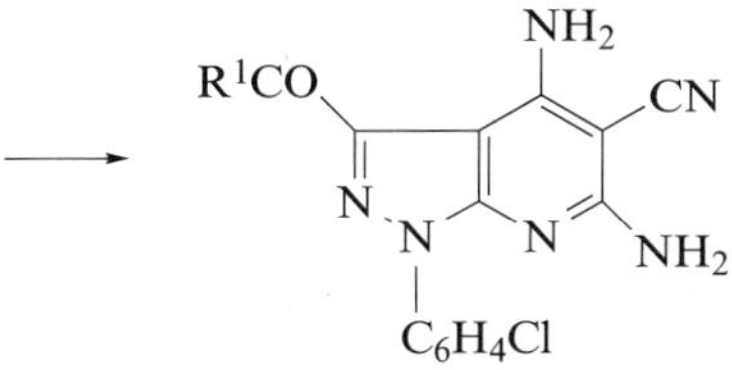

(268)

1,3-Diazepine-2-thione reacted with **166** to yield pyrido(1,2-*a*)(1,3)diazepine[309]. Pyridazinoquinazolines were prepared from ylidenemalononitriles and substituted benzenediazonium chlorides via the corresponding pyridazines[310]. The hydrazinoyl chloride **266** reacted with dimeric malononitrile **147** and iminoglutaronitrile **267** in a one-pot synthesis to yield pyrazolo(3,4-*b*)pyridines **268**[311]. Pyrazolopyridines were also obtained by reactions of chlorophenylpropenylidenemalononitrile and acylhydrazine[312].

Triazolo(1,5-*a*)pyridines were obtained by cyclization of arylidenemalononitriles and 2'-acetyl-2-cyanoacetohydrazide[313] or by reaction of 2-amino-5-methyl-1*H*-1,3,4-triazole with **43**[314]. Arylidenemalononitrile **179** reacted with pyrimidindiones **269** to give pyridopyrimidines **270**[315].

(269)   (179)   (270)

Pyridopyrimidines have also been obtained from **271**, which was cyclized with amines to yield **272**. Acidic hydrolysis gave **273**[232].

(271)   (272)   (273)

Condensation of benzaldehyde 4,4-dimethylaminomethylenehydrazone with (ethoxymethylene)malononitrile gave **274**, which was oxidized to **275**[316,317].

(274)   (275)

Pyridazinecarbonitriles **276** were condensed with benzylidenemalononitrile (**43**) to phthalazinecarbonitriles **277**[318]. **147** reacted with **278** to afford 2-(dicyanomethylene)pyridines **279**, which can be cyclized by sodium methanolate or by hydrobromic acid to form 1,6-naphthyridines **280**[319].

(276)   (277)

(147)   (278)

(279)   (280)

Oxygen heterocycles. Arylidenemalononitriles (**179**) have been cyclized to various condensed oxygen heterocycles. Hence 3-cyano-4-methylcoumarin reacted with **179** to yield benzocoumarins **281** and pyranoquinolines gave benzopyranoquinolines **282**[320]. Reaction of (5,5-dialkyl)-1,3-cyclohexanediones like dimedone afforded benzopyran derivatives **283**[321]. When **179** was reacted with β-tetronic acid (**284**), furopyrans **285** were obtained. Ring transformation of **285** gave difuropyridines **286**[322].

(281)   (282)   (283)

(284)   (285)   (286)

The benzopyran derivative **288** was obtained from **287** and *o*-hydroxybenzaldehyde[233]. 4-Chromanone **289** was condensed with (ethoxymethylene)malononitrile **181** to yield pyranobenzopyrans **290**. The same reaction was performed with 5-homochromanone[323].

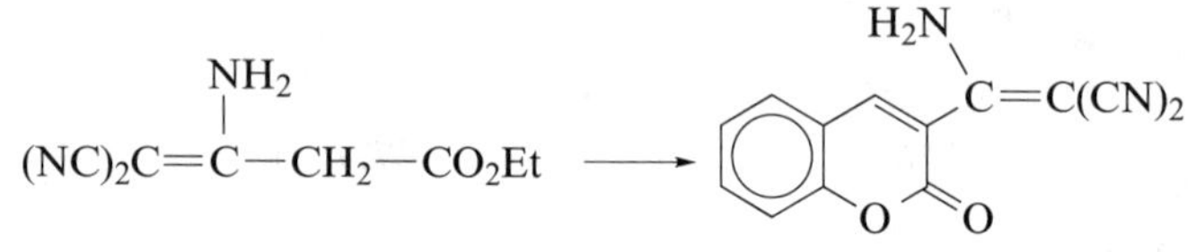

$$(NC)_2C=C-CH_2-CO_2Et \qquad\longrightarrow$$

(287)   (288)

EtOCH=C(CN)$_2$ +

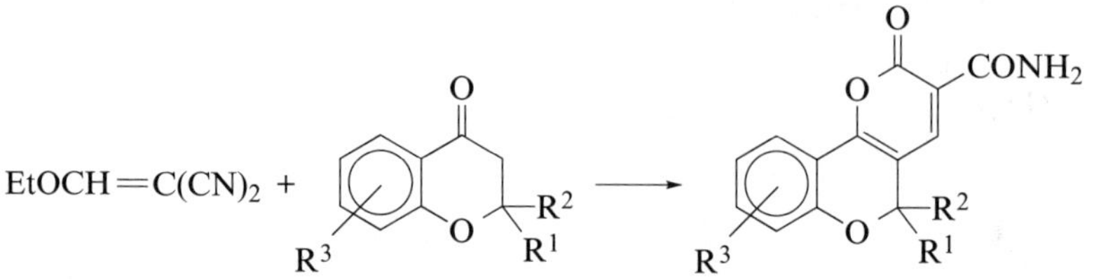

(181)   (289)   (290)

2-(Dicyanomethylene)-1,3-indanedione **254** reacted with Ph$_3$P=C=C=X (**291**), to yield the corresponding pyran derivatives **292**, where X is either O or S[324]. Different products were obtained by reactions of **254** with 1,3-dicarbonyls. Depending on residues R^1 and R^2, spiro compounds **293** (obtained from **254** and dimedone) and **294** or propellanes **295** were isolated[325].

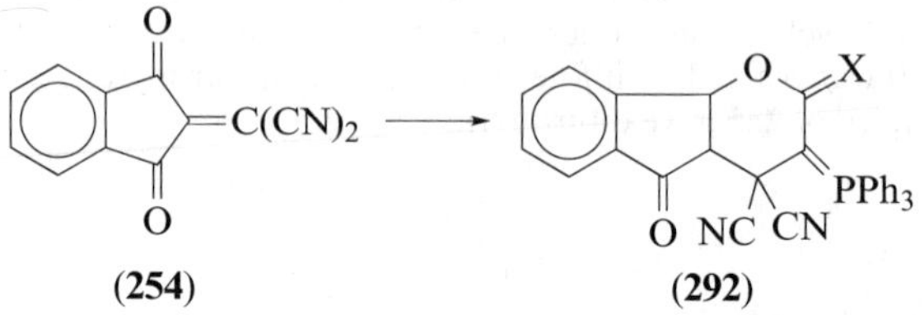

=C(CN)$_2$ $\longrightarrow$

(254)   (292)

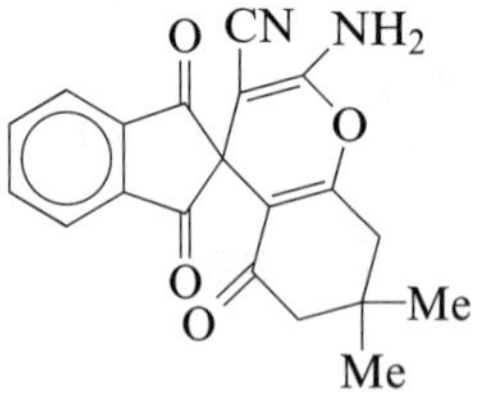

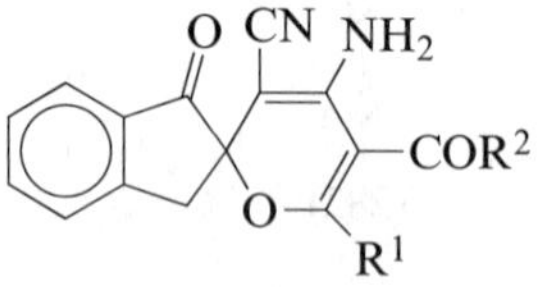

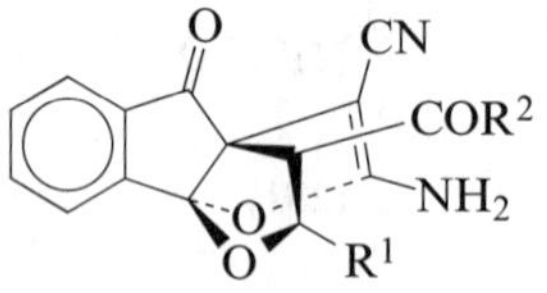

(293)   (294)   (295)

Sulphur heterocycles. α-Haloketones **296** were reacted with (dimercaptomethylene)malononitrile **297** to form thienothiophenes **298**[326]. Cyclocondensation of dicyanomethylene compound **299** with **300** gave benzothiin derivatives **301**[327].

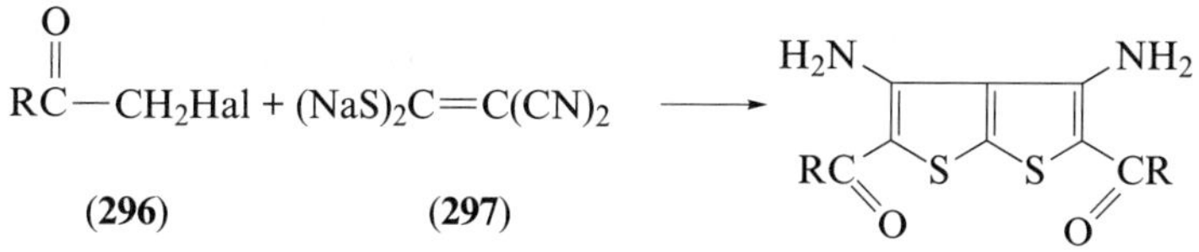

(296) (297) (298)

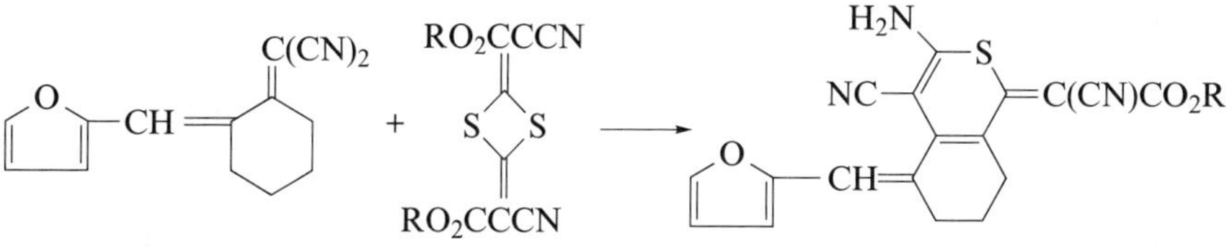

(299) (300) (301)

Nitrogen- and oxygen-containing heterocycles. Benzylidenemalononitrile (**43**) reacted with isoxazole **302** to give isoxazolopyridine **303**[328]. Arylidenemalononitriles (**179**) react with barbituric acid to yield pyranopyrimidinediones **304**[329].

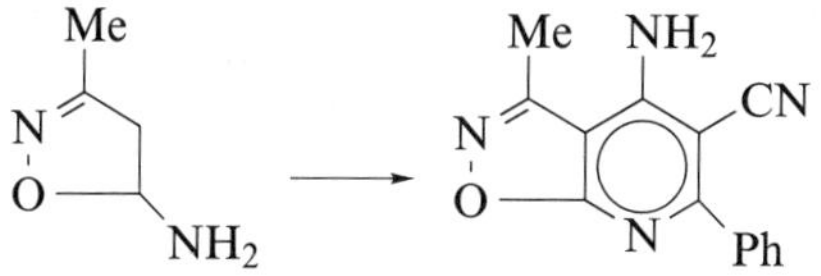

(302) (303)

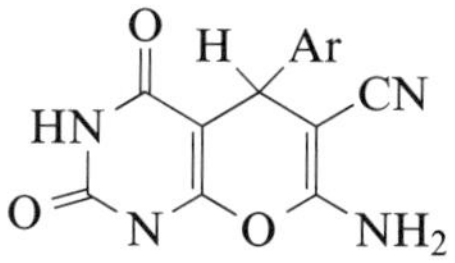

(304)

The ylidenemalononitrile **305**, when refluxed with ammonium acetate in glycol monomethylether for 3 min, gave benzopyranopyrimidines **306** in 90% yield[330].

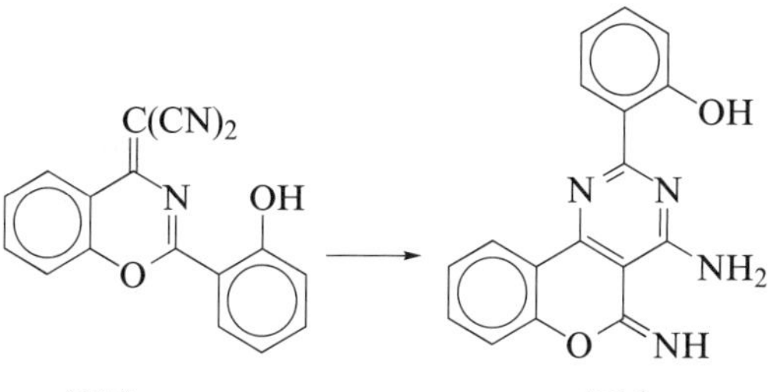

(305) (306)

254 and 3-(dicyanomethylene)-indole-2-one (**307**) reacted with dimethylpyrazolones **308** to afford spiro pyranopyrazoles. **254** and **308** yielded **309** and **310**[331].

(**254**)   (**308**)

(**309**)   (**310**)

Heterocyclic ylidenemalononitriles **311** (X = O, S) were reacted to condensed thienopyrans (X = O) and thienothiopyrans (X = S) **312**[215]. Methylthiomethylenemalononitrile **313** and cyanothioacetamide afforded pyridinethione **314**. After S-alkylation of **314**, cyclization by a base yielded thienopyridines **315**[332].

(**311**)   (**312**)

(**313**)   (**314**)   (**315**)

(**316**)   (**317**)

Benzylidenemalononitrile **43** reacted with pyridinethione **316** to give isoquinoline derivative **317**[99]. Aminothiazines **318** were condensed with methoxyethylidenemalononitrile, after which cyclization with sodium hydride in DMSO yielded pyridothiazines **319**[333].

(318)   (319)

4. Miscellaneous

a. Hydrolysis. Benzylidenemalononitrile (**43**) undergoes significantly reversible hydrolysis in water to yield benzaldehyde and malononitrile. The kinetics and mechanism of this reaction have been investigated thoroughly[334–337]. A review deals with investigations concerning the hydrolysis of *p*-nitrobenzylidene Meldrum's acid and **43**[338].

b. Reduction. As well as the hydrolysis of **43**, also its reduction has been the subject of intensive investigations. Thus the reduction of **43** on a Hg electrode was studied by various electrochemical methods[339,340].

B. Tetracyanoethane (CN)$_2$CHCH(CN)$_2$

1. Formation and properties

Tetracyanoethane (TCE, **320**) was prepared in 70 to 95% yields by reduction of tetracyanoethylene (TCNE) by various hydrides[341]. The configuration[342] and X-ray crystal structure of **320** were determined. The central C—C bond length is significantly longer than in ethane (0.156 nm). All other bond lengths and bond angles have the expected values[343].

2. Cycloaddition reactions

Cycloadditions of **320** have been studied thoroughly. Depending on the reaction partners different carbocyclic and heterocyclic five-membered ring systems were obtained.

a. Formation of five-membered cyano-carbocycles. Tetracyanocyclopentanes **321** were obtained from **320** and 2-alkylidenearylamines[344], with different other nucleophilic reagents[345,346]. Crotonic acid and methyl 1-alkenyl ketones gave tricyanocyclopentenes **322**[347,348] and **323**[349], respectively. Crotonic aldehyde reacted with **320** in alcohols to afford **324**[350].

(321)   (322)   (323)

(324)

b. *Formation of five-membered cyano-heterocycles.*   *N- and O-heterocycles:* Pyrrolines
325 were prepared from **320** and azomethines[351] and transformed to pyrroles **326** by
thermal elimination of HCN[352]. In analogous reactions **320** was reacted with azines[353]. **320**
and ketones afforded dihydrofurans **327** and furans **328**[354,355].

(325)   (326)

(327)   (328)

c. *Condensed and other carbo- and heterocycles.*   Michael reactions of α,β-unsaturated
carbonyl compounds with **320** and subsequent intramolecular cyclization afforded bicyclo
derivatives **329** or tetrahydro-isoindoldiones **330** depending on the substituents R^1–R^4 (in
330 R^2 = H, R^3 = COMe, R^4 = Me)[356,357]. Benzylidene derivatives underwent cycloaddi-

(329)   (330)

(331)   (332)   (333)

tions with **320** to afford spiro compounds **331** (from 5-benzylidenebarbituric acid), **332** (from 2-benzylidene-1,3-indanedione) and **333** (from 3-methyl-4-benzylideneisoxazole)[358]. Spiro compounds were obtained from **320** and benzylideneisoxazolones[359]. Oxabicycloheptanes were prepared from methyl ylidenemalonate[360]. **320** and β-chlorovinyl ketones reacted to yield the polycyclic **334**[361].

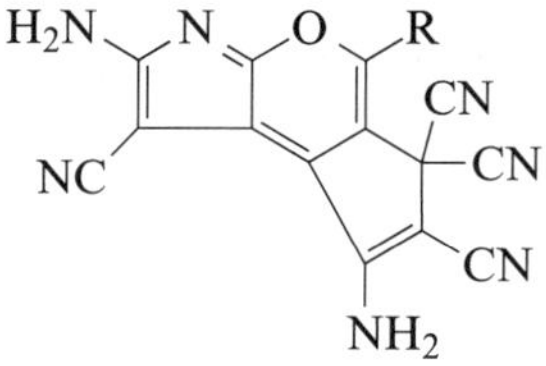

(**334**)

C. Polycyanopropanes

1,1-Dicyanocyclopropane was the subject of thorough structural investigations and compared with cyanocyclopropane and cyclopropane[362]. The vibrational spectra of cis-1,2- and *trans*-1,2-dicyanocyclopropane were studied[363].

1,1,2,2-Tetracyanocyclopropane **335** can easily be reacted with phenylhydrazine to yield the spiro compound **336**. Subsequent ring opening of the cyclopropane afforded pyrazoles **337**[364].

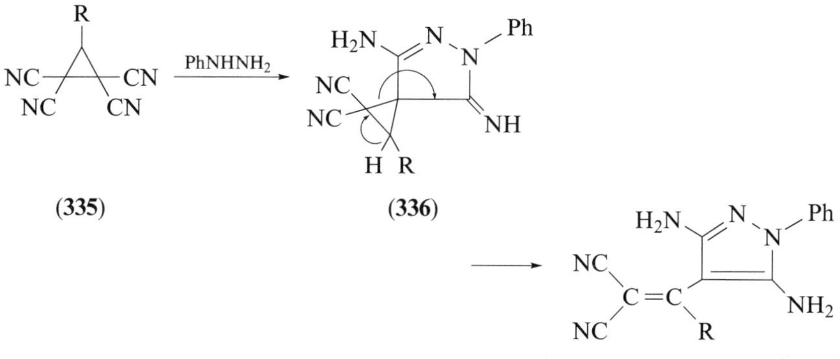

(**335**)   (**336**)

(**337**)

IV. POLYCYANOALKENES

A. Dicyanoethenes

1. Fumaronitriles

a. Synthesis and properties.   Treatment of alkynes with equimolar amounts of bromine produced *trans*-dibromoalkenes, which underwent Rosenmund–von Braun reaction (CuCN/DMF) to afford (substituted) fumaronitriles[365].

b. Reactions.   Fumaronitrile (**338**) gave *N*-tosyl-3,4-dicyanopyrrole in two steps in 70%

overall yield[366]. **338** and nitrilimines, generated by dehydrohalogenation of hydrazinoyl chlorides **339**, underwent cycloaddition with concomitant dehydrocyanation, producing 4-cyanopyrazoles **340**[367].

(338)   (339)   (340)

2. Maleonitriles

a. Synthesis and properties. Aryl and alkyl maleonitriles were obtained by photoisomerization from the corresponding fumaronitriles[365].

b. Reactions. Diaminomaleonitrile **341** can be reacted with orthoformates to afford dicyanoimidazoles **343**. A reaction mechanism via monoimidates **342** was discussed[368].

(341)   (342)

(343)

3. Dicyanoethenes with electron acceptors

a. (Chloromethylene)propanedinitrile. Vinylogous substituted formamidines **346** were obtained by a reaction of (dichloromethylene)propanedinitrile **344** with [bis(dialkyla-mino)methylene]propanedinitriles **345**[79].

(344)   (345)   (346)

4. Dicyanoethenes with electron-donors

a. Acetals. The salts of alkyl dicyanoacetates **347** reacted with chloromethylene ammonium salt **348** in chloroform at room temperature. Oxazinone **349** was obtained in good to excellent yields[369].

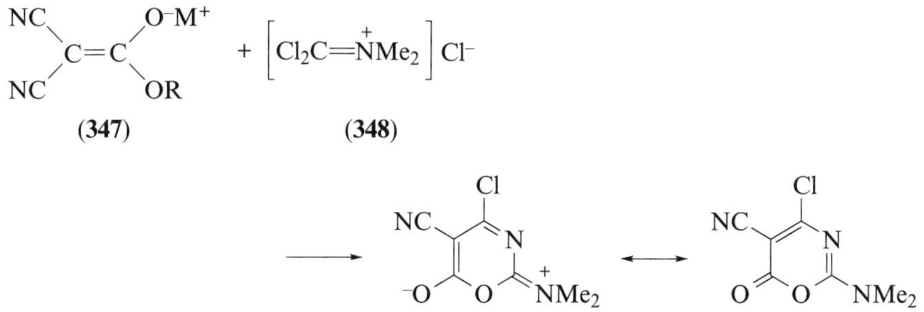

(347)　　　　　(348)

(349)

b. Mercaptals. Thioacetal **313** easily reacts with *o*-amino-thiophenol to yield 2-dicyanomethylene-benzothiazoline **350**, which was used as starting material for electrophilic additions and cyclocondensations to obtain a variety of heterocycles[370]. **313** also reacts with active methylene compounds to yield the corresponding 3-cyano-4-methylthio-2(1*H*)-pyridone derivatives. So **351**, **352** and **353** were obtained from acetone, cyclohexanone and 3-acetylpropionic acid, respectively. Pyridones were obtained from acetophenone and 3-benzoylpropionic acid[371].

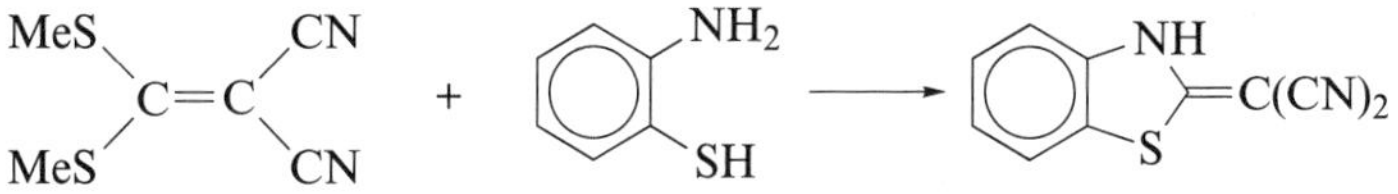

(313)　　　　　　　　　　　　(350)

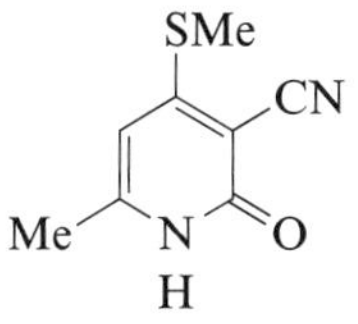

(351) 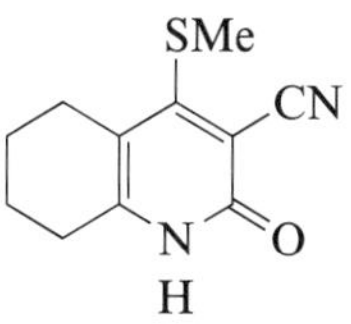　　　　　　(352) 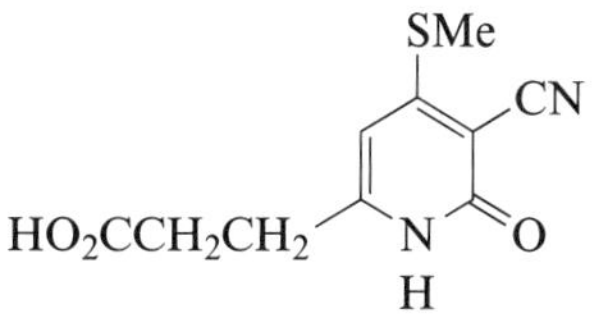　　　　　　(353)

10 reacted with CS_2 to yield disodium dicyandithioacetate **354**, which in turn reacted with γ-bromocrotonic acid derivatives like the nitrile **355** to afford thienothiophene **356**. Further reaction of **356** with DMSO gave a thiophene derivative[372].

Protonation of the potassium salt **357** in water gave the dithiacyclobutane **358**[373]. When **357** was treated with primary alkyl halogenides, the corresponding dialkyl compounds **359** were formed. With RS—COCl dithioketals **360** were obtained[374].

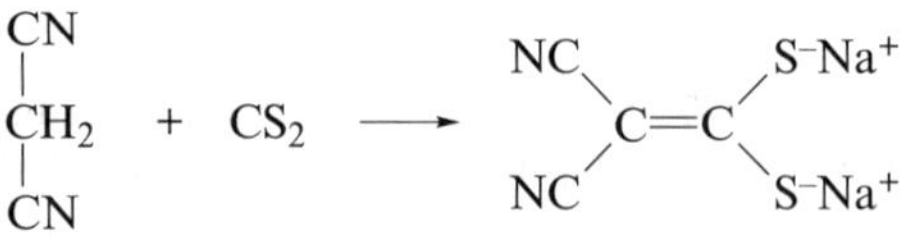

$$
\begin{array}{c}
\text{CN} \\
| \\
\text{CH}_2 \\
| \\
\text{CN}
\end{array}
\quad + \quad \text{CS}_2 \quad \longrightarrow \quad
\underset{\text{NC}}{\overset{\text{NC}}{}}\!\!\diagdown C = C \diagup \!\!\underset{\text{S}^-\text{Na}^+}{\overset{\text{S}^-\text{Na}^+}{}}
$$

(10)   (354)

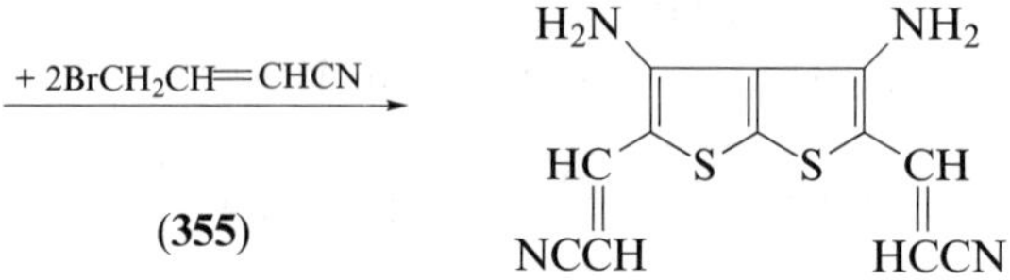

+ 2BrCH$_2$CH=CHCN →

(355)

$$
\underset{\text{NCCH}}{\overset{\text{H}_2\text{N}}{}} \quad \underset{\text{HCCN}}{\overset{\text{NH}_2}{}}
$$

(356)

$$
\underset{\text{NC}}{\overset{\text{NC}}{}}\!\!\diagdown C = C \diagup \!\!\underset{\text{S}^-\text{K}^+}{\overset{\text{S}^-\text{K}^+}{}}
\xrightarrow{\text{H+/H}_2\text{O}}
S = C - C \overset{\text{CN}}{\underset{\text{N}}{\diagdown}} \overset{\text{S}}{\underset{\text{S}}{\diamondsuit}} = C \overset{\text{CN}}{\underset{\text{NH}_2}{\diagup}} - C = S
$$

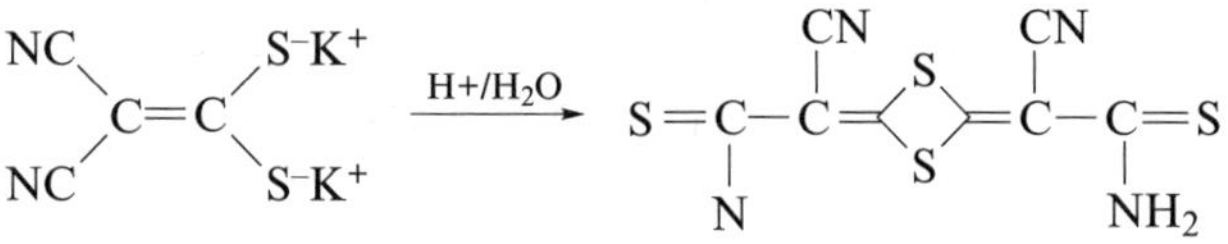

(357)   (358)

(CN)$_2$C=C(CH$_2$R)$_2$ (CN)$_2$C=C(SR)$_2$

(359)   (360)

5. Dicyanoketene

Though dicyanoketene **362** was known about ten years earlier, only in 1980 could its UV photoelectron and mass spectra be taken. As starting molecules 2,6-diazido-3,5-dicyano- or 2,5-diazido-3,6-dicyano-1,4-benzoquinone (**361**) were used[375,376]. The microwave spectrum has also been discussed[377].

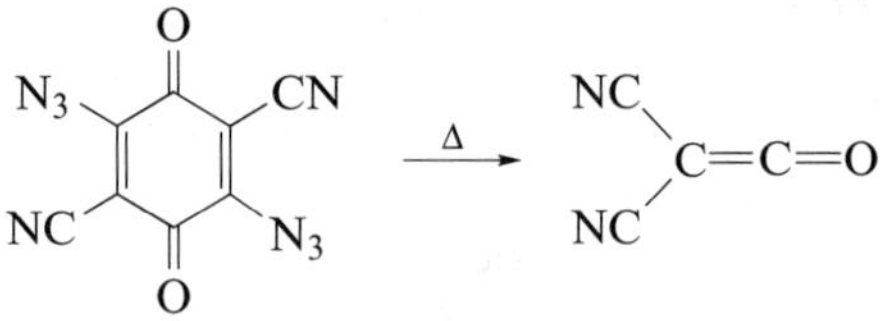

$$
\underset{\text{NC}}{\overset{\text{N}_3}{}}\!\!\diagdown\!\!\underset{\text{O}}{\overset{\text{O}}{\bigcirc}}\!\!\diagup\!\!\underset{\text{N}_3}{\overset{\text{CN}}{}}
\xrightarrow{\Delta}
\underset{\text{NC}}{\overset{\text{NC}}{}}\!\!\diagdown C = C = O
$$

(361)   (362)

B. Tetracyanoethylene

1. Synthesis and properties

Tetracyanoethylene ((NC)$_2$C=C(CN)$_2$, **363**) is the simplest percyanoalkene. The structure makes three types of reactions possible:

   a. Reaction of the double bond. TCNE is a powerful dienophile. The double bond has an extremely low electron density. Thus TCNE reacts easily with dienes and various other electron-rich species to yield a variety of charge-transfer complexes and addition products. Retro-Michael reactions also provide useful synthetic pathways.

b. Substitution of one or two cyano groups. In comparison to reactions of the double bond, reactions of the triple bond or substitution reactions of the cyano group have not been observed very often.

The importance of TCNE in different fields of organic chemistry can be seen in comprehensive reviews which were published in 1986 and 1987 and are given below:

New Applications of Tetracyanoethylene (TCNE) in Organic Chemistry, A. J. Fatiadi (1986)[378]. This review with 501 references deals with reactions of tetracyanoethylene used in organic synthesis. Information on molecular complexes, ozonization of alkenes and acetylenes, dehydrogenation and tricyanovinylation, reactions of TCNE oxide, reactions with ketones and diketones, synthesis of heterocycles and cationic polymerizations are included in this survey. Some industrial and analytical applications are also discussed.

Addition and Cycloaddition Reactions of Tetracyanoethylene in Organic Chemistry, A. J. Fatiadi (1987)[379]. This review focusses on [2 + 2] cycloaddition, the Diels–Alder reaction, the ene reaction and higher-order cycloadditions of TCNE to carbocyclic and heterocyclic substrates. Mechanistic aspects and thermal and photocycloaddition reactions also are discussed. 406 references are given.

New Applications of TCNE in Organometallic Chemistry, A. J. Fatiadi (1987)[380]. Selected reactions used in organometallic synthesis are reviewed. 311 references are given. Structure and bonding of metal–TCNE complexes as well as reactions of TCNE with main-group organometallics, with transition-metal complexes, with metal-coordinated alkenes and alkynes, and reactions of platinum-family complexes are discussed.

Correlation of Charge-transfer Energies of TCNE–Donor Complexes with Ionization Energies of Donor Molecules, J. F. Frey (1987)[381]. 265 TCNE–donor complexes are discussed and 137 references are given.

2. Reactions

a. Reactions of the double bond

i. Formation of charge-transfer complexes. Formation of charge-transfer complexes may be considered a typical reaction for TCNE (**363**). Since the first charge-transfer complex of **363** was reported in 1958[382], many papers have been published on this subject. Many TCNE complexes have also been discussed in the reviews mentioned above[378–381]. Table 5 includes papers on complexes of **363** with organic and inorganic partners which have been published since 1986. Most of them contain conditions of complex formation and detailed spectrochemical characteristics. Formation of charge-transfer complexes takes place so easily that it can also be used for quantitative spectrophotometrical determination of aromatic hydrocarbons[457], glucose and fructose[458], hydrazines and pyrazolones[459], and in pharmaceutical analysis for phenothiazine derivatives[460] and thioxanthenes[461].

ii. Formation of carbocycles. Cycloadditions of TCNE (**363**) with electron-rich olefins or dienes yield a variety of carbocycles, many of them with at least one nitrile group.

Three-membered carbocycles: 1,1,2,2-Tetracyanocyclopropane was prepared in a one-pot synthesis from TCNE and diazomethane. First pyrazoline **364** is formed by cycload-

$$NC\diagup CN \qquad +CH_2N_2 \longrightarrow \qquad -N_2 \longrightarrow$$

(363)   (364)   (365)

TABLE 5. Organic and inorganic components forming charge-transfer complexes with TCNE

Reactant	References
Organic	
Alkaloids, Rauwolfia	383
Anilines, *N*-alkyl	384, 385
Aniline, poly-	386
Anthracene	387
Anions (azide, bromide, chloride, isothiocyanate, iodide, nitrite, sulfite, oxalate anion, Ph$_4$B-anion, thiosulfate)	388
Azomethines	389
Benzene	390
Benzenes, *m*-disubstituted	391
Benzene, hexamethyl-	392
Benzene, nitro	393
Benzene, pentamethyl-	394
Calix(4)arene conformers	395
Carbazole	396
Cyclophanes	397
Cyclophane, 5,8-dimethoxy-2,11-dithia(3.3)*para*-	398
Cyclophane, (2.2)*meta*-	399
Cyclophane, (2.2)*metapara*-	400
Cyclophanes, *para*	401, 402
Cyclophanes, (2.2)*para*-, methylated	403
Cyclophane, (2.2.2)*para*-, triene	402
p-Cymene	404
Durene	394
Ferrocene	405
Furyl derivatives	406
Heterocycles, 5-membered, benzofused	407
Hydrocarbons	408
Hydrocarbons, aromatic	409
Indene	410
Isoindoles, *N*-aryl-	411
Leuco Bindschedler's green	412
Mesitylene	413
Nitriles, aliphatic	414
Nitriles, benzo-	415
Nitriles, naphtho-	415
N-oxides, aromatic	416, 417
N-oxides, imidazole-	418
N-oxides, isoxazole-	418
N-oxides, pyridine-	418
N-oxides, triazene-	419
Olefins	420
Organopolysilanes	421
Peroxides, *endo*	422
Perylene	423
Phenols	424
Phenylhydrazones	425
Picrate	396
p-Quaterphenyl units	426
Poly(azomethines)	427
Quinones, *p*-benzo-	393
Sulfonamide drugs	428
1,3,4-Thiadiazole	429

TABLE 5. (*continued*)

Reactant	References
Thioethers	413
Thiosemicarbazones	430
Vinyl sulfoxides	431
Urea derivatives	432
Metal-organic	
Aluminum, -tris-bipyridyl complexes	433
Borines, mono- and bis(phenaza)- and (phenoxa)	434
Cobaltocene, decamethyl-	435
Cobalt, monomethylcyclopentadienyl derivatives	436
Chromium carbonyls	437
Gold(I), cyclopentadienyl complexes	438
Gold, $(Ph_3P)_2$-	439
Iridium, pentacoordinate alkoxy-alkene	440
Iron, tricarbonylcycloheptatrienone-	441
Iron, -trisbipyridyl complexes	433
Iron, substituted alkylidene-bridged diiron complexes	442
Manganese complexes, allyl-	443
Manganese carbonyls	437
Manganese, cyclopentadienyl	444
Manganese(II) phosphines	445
Molybdenum, carbyne complex	446
Palladium, hydride complex, polynuclear	447
Potassium dimethoxyethane	448
Platinum(II) complexes, allyl-aryl-	449
Platinum(II) complexes, allyl-	443
Platinum, metal dithio acids	450
Ruthenium, di-, diphosphazane-bridged	451
Tin, trialkyl-, halides	452
Tin, trialkyl-, isothiocyanates	453
Tungsten carbonyls	437, 454
Uranium, organic compounds	455
Transition metals, pentadienyl complexes	456
Vanadium, tris-bipyridyl complexes	433

dition in 79% yield. Subsequent slow decomposition in solution with evolution of nitrogen yields the cyclopropane **365**[462].

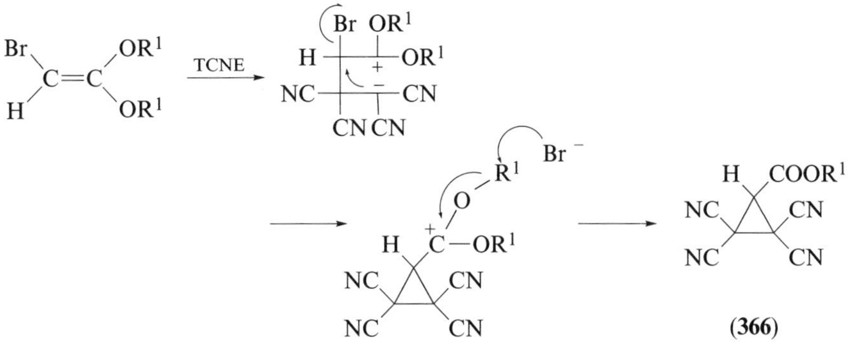

(**366**)

Tetracyanocyclopropane carboxylates **366** have been prepared by reacting dialkoxybromoethylenes with TCNE at low temperatures. The accompanying mechanism has been proposed[463].

Recently, the synthesis of hexacyanocyclopropane **367** has been reported. TCNE was reacted with *t*-butyl hypochlorite in water/dioxane at low temperatures. A 1:3 adduct of **367** with 1,4-dioxane was isolated. Its structure has been determined by single-crystal X-ray diffraction[464].

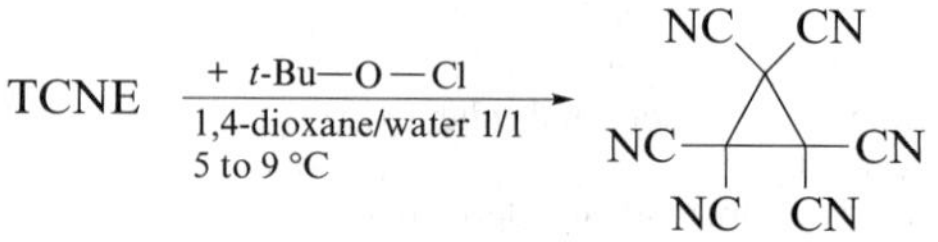

$$\text{TCNE} \quad \xrightarrow[\substack{1,4\text{-dioxane/water } 1/1 \\ 5 \text{ to } 9\,°C}]{+\ t\text{-Bu}-\text{O}-\text{Cl}}$$

(367)

The tetracyanocyclopropane spiro compound **368** was obtained as the sole product from (1.1.1)-propellane and TCNE. Addition of a second TCNE to the double bond of **368** did not occur[465].

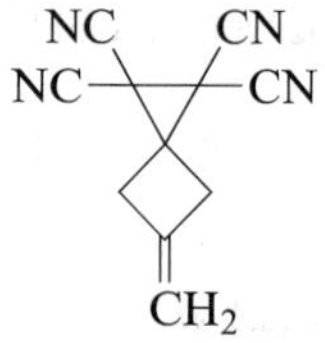

(368)

Four-membered carbocycles: Compounds with sufficiently electron-rich olefinic double bonds react with TCNE to yield tetracyanocyclobutyl derivatives by [2 + 2] cycloadditions. Thus 1-vinylpyrazoles **369**[466] as well as 4-vinyl-pyrazoles **371**[467] react to yield the corresponding substituted tetracyanocyclobutanes **370** and **372**, respectively. The same reaction has been observed with 1-vinylindazoles[466].

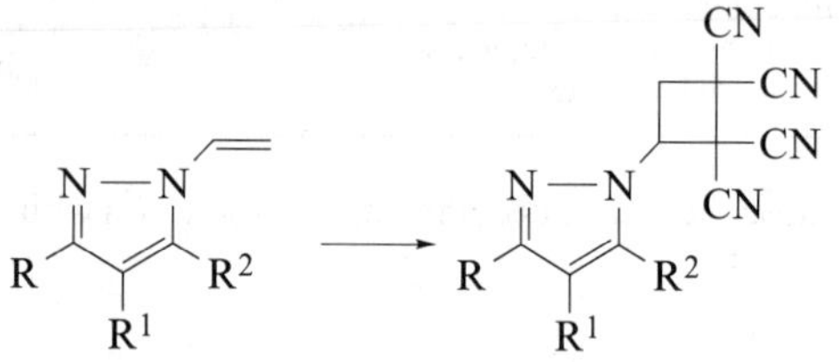

(369)   (370)

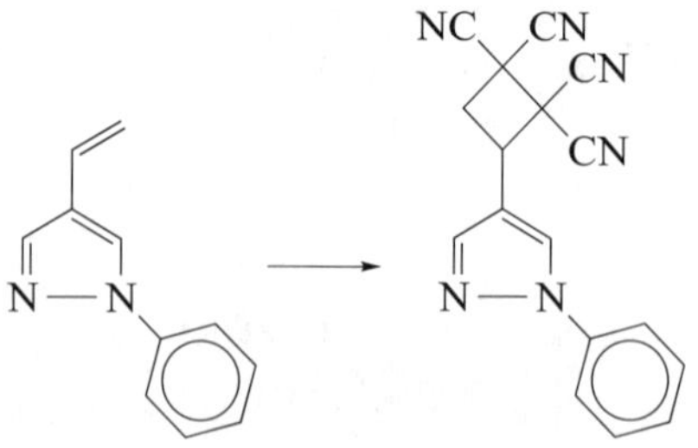

(371)   (372)

4-Benzalamino-3-methyl-5-styrylisoxazoles **373** (R = substituted phenyl groups) and TCNE undergo a photochemical reaction to yield the corresponding substituted cyclobutanes **374**[468]. From reactions of *p*-divinylbenzene with TCNE, a cyclobutane derivative has been obtained, while the reaction of *m*-divinylbenzene yielded a 2:1 Diels–Alder adduct[469].

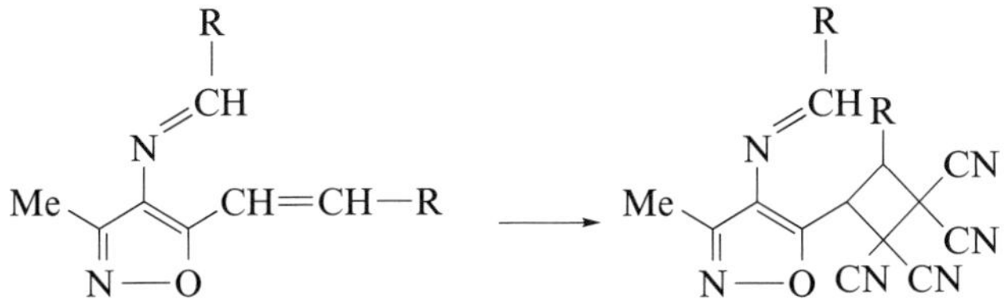

(373)   (374)

Five-membered carbocycles: Ethoxycyclopropane reacted with TCNE in various solvents. In dioxane and acetonitrile 3-ethoxy-1,1,2,2-cyclopentane-tetracarbonitrile **375** was the only product. Performing the reaction in benzene also yields a ring-opened enol ether, which reacts further with TCNE[470].

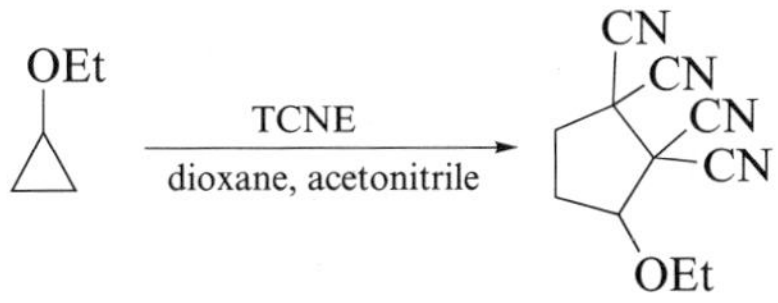

(375)

Spiro(2.4)hept-4-enes with TCNE in methylene chloride or acetonitrile yield various rearrangement products like the tricyanobicyclo(3.3.0)octene derivative **376**, the tetracyanospiro(4.4)nonene **377** and various addition compounds of these derivatives[471].

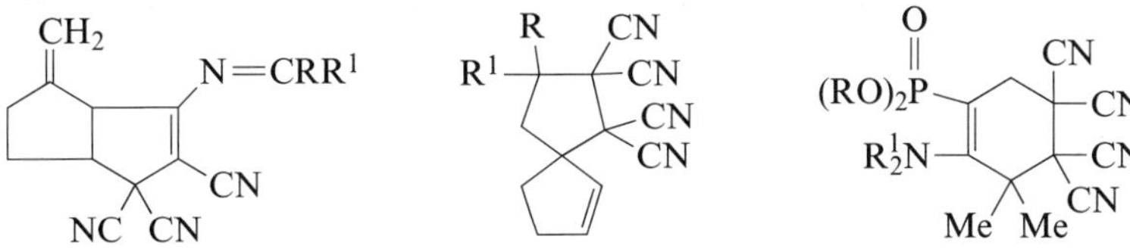

(376)   (377)   (378)

Six-membered carbocycles: Phosphonocyclohexenes (**378**) have been prepared in 22–58% yields by [2 + 4] cycloadditions of TCNE to substituted phosphonic acid esters[472].

Condensed and other multimembered carbocycles: Due to its electronic structure TCNE is an ideal partner for cycloadditions and has also been widely used for trapping of intermediates or as a diagnostic dienophile especially in the last twenty years. Thus four of the six isomeric dihydropentalenes have been trapped in [4 + 2]- and [2 + 2]cycloadditions with TCNE in up to 95% yields[473]. The cyclohepta-trieno(a)phenalene **379** was found to give the [8 + 2]cycloadduct **380**[474]. 1-Benzoxepine and 1-benzothiepine yielded the corresponding dieneadducts[475].

(379)   (380)

When equimolar amounts of TCNE and **381** reacted, a mixture of dihydro-Dewar benzene **382** and dihydrobenzvalene **383** was formed by cycloaddition[476]. Diels–Alder reactions of TCNE are of both preparative and theoretical interest. So recently the reaction with 1,2-dimethylenecyclopentane has been investigated in detail. Kinetical studies by UV-Vis absorption spectroscopy have shown that the [4 + 2]cycloaddition takes place without preorientation by a charge-transfer complex[477].

(381)   (382)   (383)

Direct and reverse Diels–Alder reactions of anthracene and 9,10-dimethylanthracene with TCNE in several solvents have been investigated by measuring kinetic and thermodynamic parameters[478]. 2-Vinylindoles **384** yielded tetrahydrocarbazoles **385** by [4+2]

(384)   (385)

cycloaddition and subsequent [1,3]-H shift[479]. 3-Vinylhydrofurans **386** underwent Diels–Alder reaction with TCNE to generate isobenzofurane derivatives **387** in 60 to 90% yields[480].

Substituted tetramethylidenebicyclo[2.2.2]oct-2-ene (**388**) was shown to have 'tandem Diels–Alder reactivity' toward TCNE and gave the monoadduct **389** and the bisadduct

(386)   (387)

390 depending on the amount of TCNE[481]. A 2:1 Diels–Alder adduct **392** was also obtained from the *m*-divinylbenzene **391**. No 1:2 adduct was isolated[469].

(388)   (389)

(390)

(391)   (392)

Cycloaddition of TCNE to homobenzvalene yields various tetracyanocycloadducts. The photochemical process yields products that differ from the ones obtained by thermal cycloaddition[482,483]. Other polycyclic unsaturated hydrocarbons were also reacted to tetracyano cycloadducts. Hence the dihydrobullvalene **393** yielded **394**. Pyrolysis of **394** gave **393** rather than elimination products[484].

(393)   (394)

Addition of TCNE to tetracyclodecatriene **395** gave two different products **396** and **397**, the ratio of which was dependent on the solvent. When the solvent was changed from benzene to nitromethane, both the rate of reaction and the ratio of **396** to **397** were increased[459].

(395)   (396)   (397)

iii. Formation of heterocycles. Five-membered heterocycles: Cycloadditions of TCNE to heteroaliphatic or heteroaromatic partners yield various heterocycles, most of them with two or more cyano groups. Many of these are useful synthetic intermediates in organic synthesis. So, e.g., 2-amino-5-bromo-3,4-dicyanopyrrole, which can be obtained by acidic cyclization of TCNE with HBr in acetic acid/ethyl acetate/acetone in 66% yield[485], is used in the synthesis of condensed heterocycles.

Two unusual syntheses of thiophene derivatives were performed with small sulfur-containing heterocycles as starting materials. Tetraarylthiiranes **398** and TCNE form a charge-transfer complex, which is transformed on irradiation to the tetracyano-tetrahydrothiophenes **399** by [3 + 2]cycloaddition[486].

(398)   (399)

Cycloaddition of TCNE to thiacyclobutane yielded the tetracyano-2-thia-bicyclo[2.2.0]hexane **400**, which was transformed to 2-dicyanoethenyl-thiophene **401** by thermal rearrangement[487].

Cycloaddition of TCNE to thiobenzophenone afforded 3,4-dicyanothiophene derivatives[488]. A recent paper reported on the synthesis of various heterocycles, such as pyrazoline, oxadiazine, pyrrole and thiazolylpyrrole derivatives by cycloaddition reactions of TCNE[489].

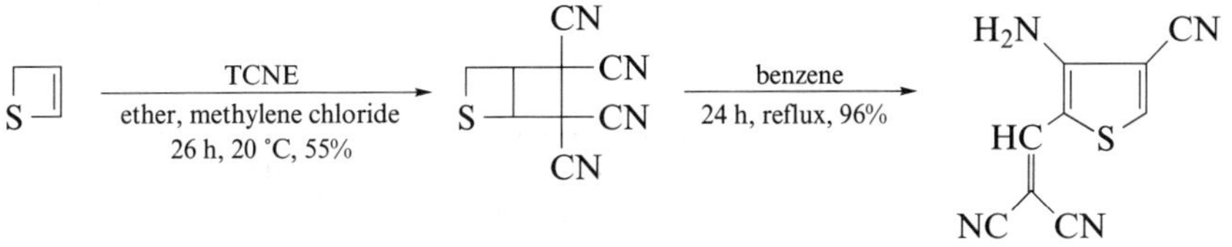

(400)   (401)

Six-membered heterocycles: The Diels–Alder reaction of azadienes with TCNE yielded tetrahydropyridines. Dicyanopyridines **402** have been obtained by cyclization of tetracyanoethylated ketones with hydrochloric (X = Cl) or hydrobromic (X = Br) acid. Yields are between 61 and 84%[490].

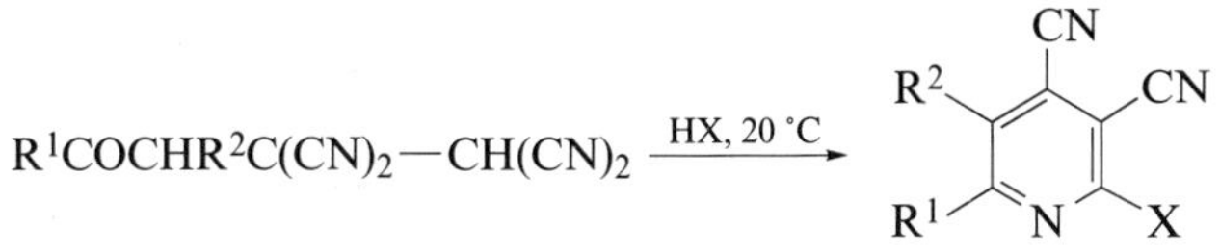

$R^1COCHR^2C(CN)_2$—$CH(CN)_2$ $\xrightarrow{\text{HX, 20 °C}}$

R^1 = Me, Et; R^2 = H   (402)
R^1–R^2 = —(CH$_2$)$_3$—
—(CH$_2$)$_4$—

Condensed and other multimembered heterocycles: Benzoylacetonitrilehydrazones reacted with TCNE to yield pyrazolo[3,4-*b*]pyridines[491]. Pyrazolopyrimidines **403** have been obtained from arylazoaminopyrazoles and TCNE. The first step of the reaction is the formation of a charge-transfer complex[492].

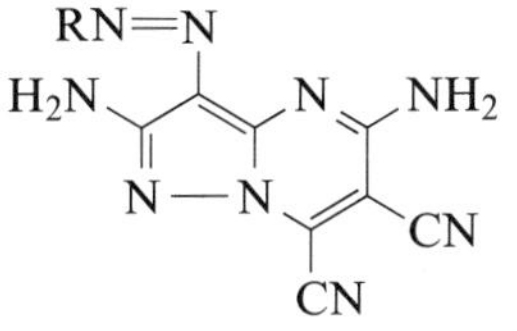

(403)

Some very unusual reactions have also been reported recently. Hence monosaccharides were treated with TCNE in alcoholic hydrofluoric acid. Condensed five-membered hetero-

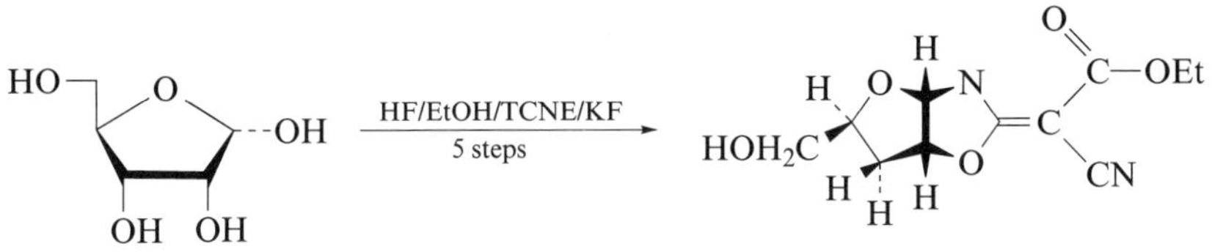

(404)

cycles like furooxazoles and pyranoxazoles were obtained. So furooxazole **404** was prepared from ribose[493].

Diazasemibullvalenes **405** have been prepared from chloro-2-aza-1,3-dienes and TCNE in a simple one-pot synthesis[494]. A remarkable condensation can be observed when TCNE reacts with its monoanion. The reaction is promoted by Lewis acids. The heterocyclic anion was isolated and characterized as the tetraphenylarsonium salt **406**[495].

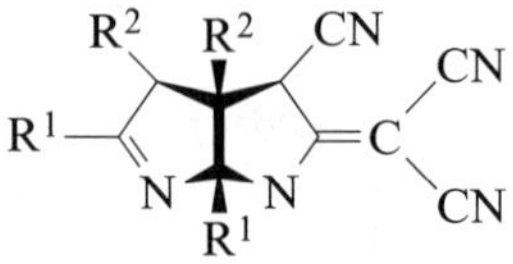

(405)

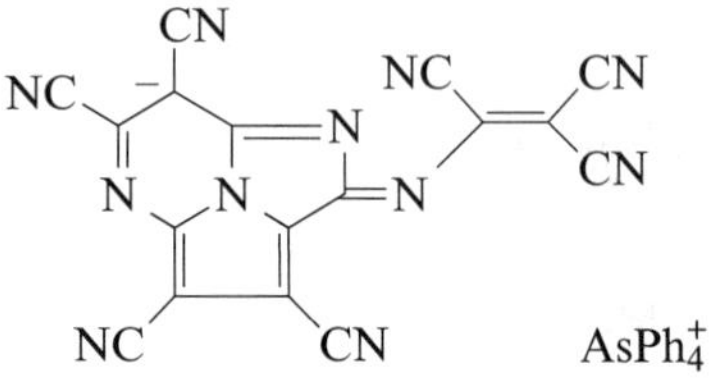

(406)   AsPh$_4^+$

iv. Retro-Michael reactions. Partial retro-Michael reactions of TCNE have been performed to yield a variety of dicyanomethylene compounds. So 1,2,3,4-tetra-chlorocyclopentadiene gave 1,2,3,4-tetrachloropentafulvene-6,6-dicarbonitrile **407**[496], 4,5-dichloro-2-dicyanomethylene-4-cyclopentene-1,3-dione (**408**) was prepared by partial retro-Michael addition of TCNE to 4,5-dichloro-4-cyclopentene-1,3-dione[497] and dicyanomethylenepyrazolones **409** have been prepared from pyrazolones[216].

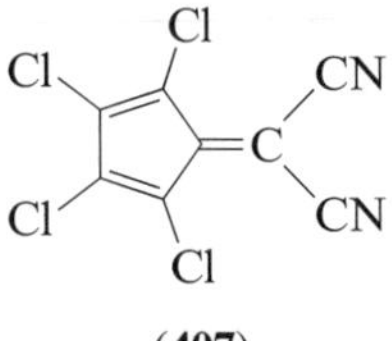

(407) 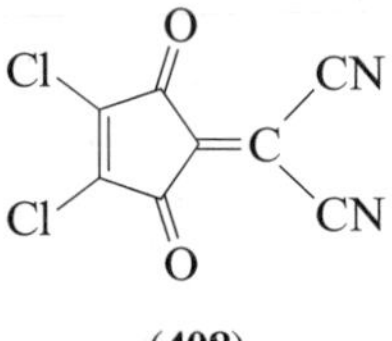   (408) 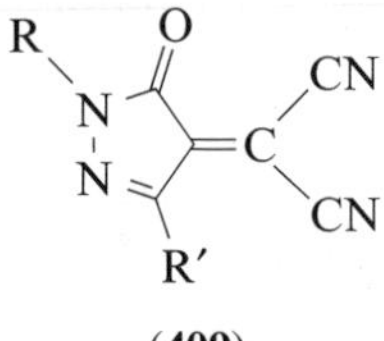   (409)

b. Substitution of one or two nitrile groups

i. Tricyanovinylation. Halogenated primary anilines react with TCNE to yield tricyanovinylated anilino-derivatives **410** in an amine–HCN exchange process[498,499]. *N,N-*

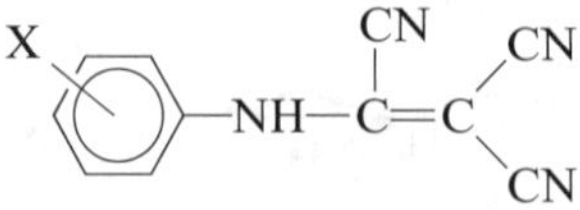

(410) 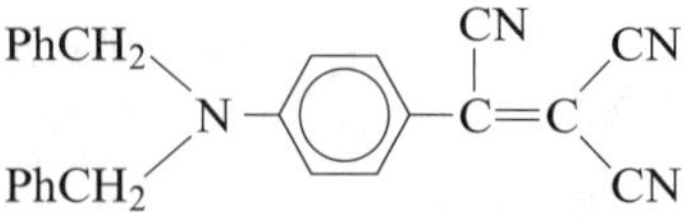   (411)

disubstituted anilines which cannot undergo this type of reaction are tricyanovinylated in the *p*-position of the phenyl group. Thus dibenzylaniline yielded **411**[500].

2-Methyl- and 2-ethyl-1,2,3,4-tetrahydroisoquinoline-1,3-dione were tricyanovinylated in position 4 to yield **412** (R = methyl, ethyl) or the 4-dicyanomethylene derivative **413** (R = methyl)[219].

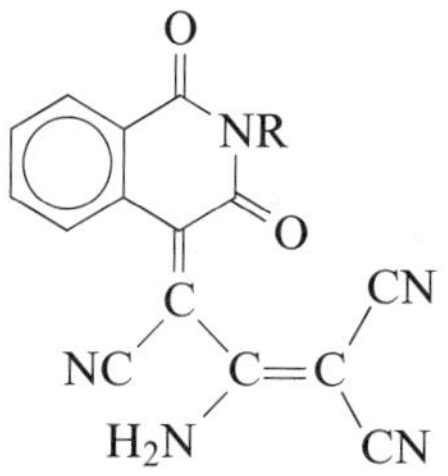

(412) 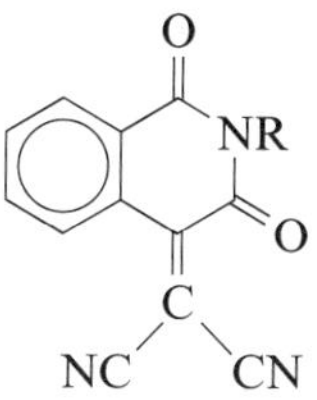   **(413)**

ii. Formation of ketals. Dicyanoketene ketals **414** can be prepared by a simple reaction of TCNE with alcohols in the presence of a base at room temperature[501]. (For other dicyanoethenes with electron donors see also Section IV.A.4.)

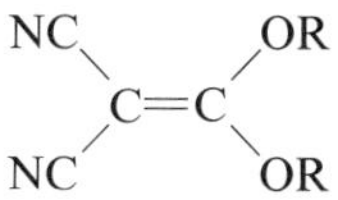

(414)

c. Tetracyanoethylene and polymers. Simple polymers can be transformed to conducting species when they are doped with TCNE. Thus polyacetylene yielded a conducting material (8 Ω cm⁻¹) by formation of an electron-transfer π-complex when it was treated with TCNE/AlCl₃/benzene[502,503]. Analogous experiments have been performed on poly(pyrrole)[503]. Thermally induced currents in poly(*p*-phenylene sulfide) doped with TCNE have been measured[504]. This doping effect has also been investigated for poly(vinylchloride)[505]. TCNE has also been used as a monomer in co-polymerizations with substituted benzylidene anilines[506] or as a sensitizer (with carbazole) on photografting of methacrylic acid in poly(ethylene)[507].

d. Miscellaneous reactions. Symmetric carbonic acid anhydrides have been prepared in 50 to 80% yields by treating aliphatic and aromatic carbonic acids with TCNE. Reactions were performed in benzene at 20 to 80 °C with tertiary amines as catalysts[508].

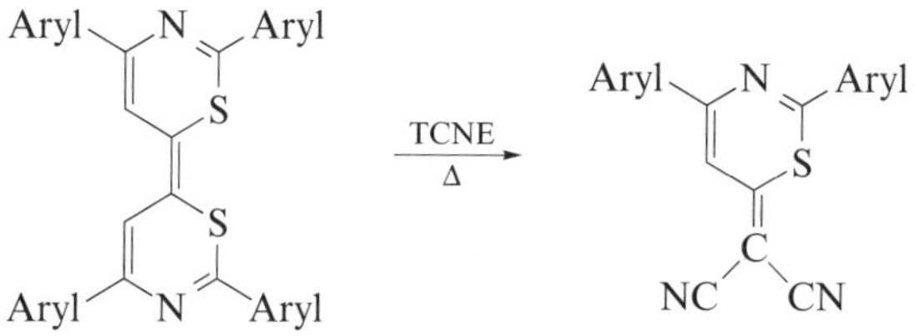

(415)   **(416)**

i. Cleavage reactions. 1,3-Diaryltriazenes reacted with TCNE in the presence of acetic acid to yield schiff's bases and arylhydrazonomalonitriles[509]. 6-Dicyanomethylene-1,3-thiazines **416** were obtained from 6,6-bis-1,3-thiazinylidene derivatives **415**[510].

C. Tetracyanoethylene Oxide

Polarized ethylenes that are both electron donating (an amino or a methylthio group) and electron accepting (two cyano groups) on the adjacent two olefinic carbon atoms were obtained in 70 to 90% yield by reacting thioamides or methyldithiocarboxylates with TCNE oxide **417**. Reactions were performed in benzene at room temperature[511].

From **417** on reaction with the dithiocarboxylate **418**, the dicyanomethylene compound **419** was formed which, when refluxed with a base, afforded **420**[512]. **417** and pyrimidine **421**, when reacted in ether, gave N-dicyanomethylide **422** in 60% yield. **422** has been used for various cycloadditions[513].

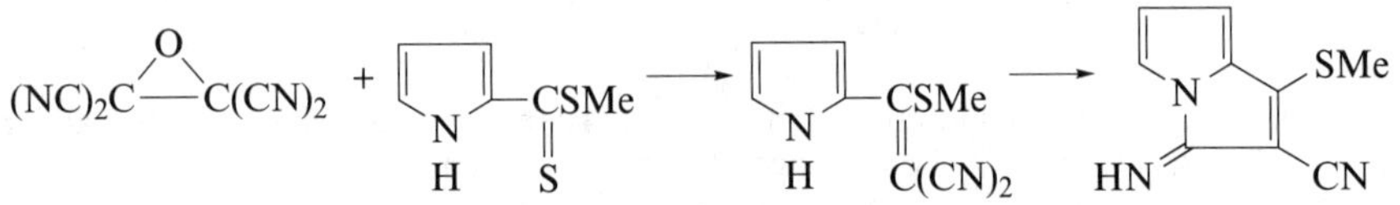

(417)   (418)   (419)   (420)

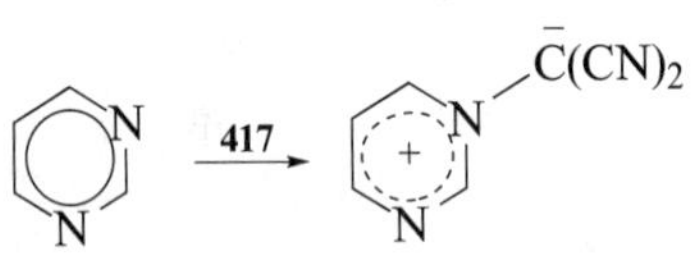

(421)   (422)

An interesting reaction was observed when **417** was combined with (substituted) benzene(s). Mono- (**423a**) or bisadducts (**423b**) were obtained[514].

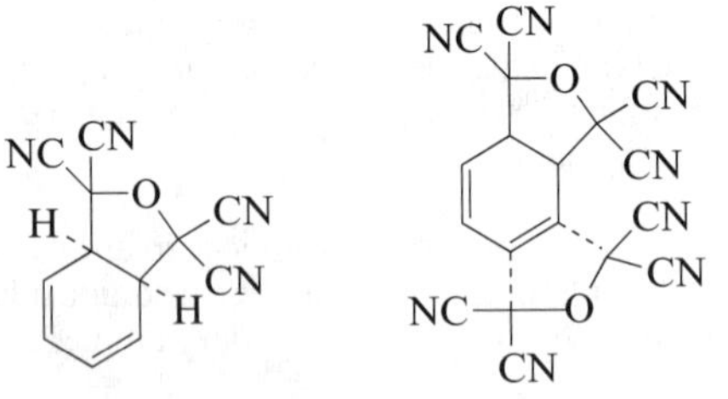

(423a)   (423b)

D. Dicyanopropenes

1,3-Dicyanopropenes provide a valuable source for ring-closure reactions and have been used in a variety of syntheses of N-heterocycles. When α,β-unsaturated esters **424** reacted with **10**, pyridones **425** were obtained and were in turn converted into dinitrile **426** with cyanamide[96]. **426** was cyclized by hydrogen halides and yielded products **427** and **428** (X = Cl, Br, I). Dehalogenation afforded **429** and **430**[515–518]. Ionic species like **431** or

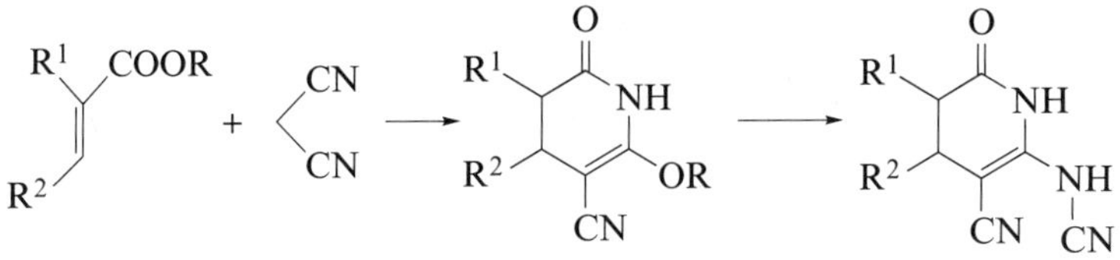

(424) (10) (425) (426)

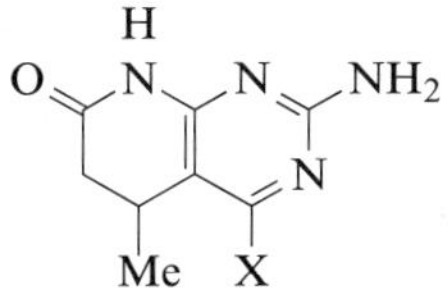

(427) 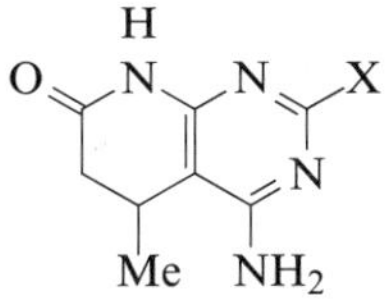 (428)

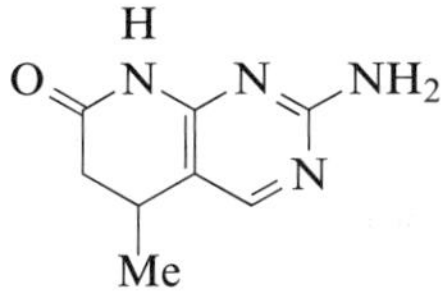

(429) 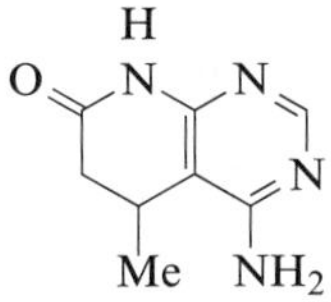 (430)

cyanonitropropenides **433** were cyclized to aminopyrimidine- (**432**) and nitropyridinecarbonitriles (**434**), respectively[519,520].

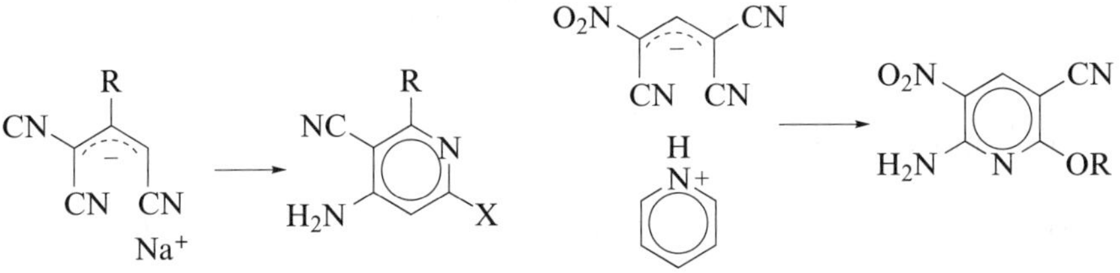

(431) (432) (433) (434)

V. CYANOACETYLENES

Proton affinity of cyanoacetylene was determined and compared to proton affinities of diacetylene and cyanogen[13].

Photochemical cycloadditions of cyanoacetylene and dicyanoacetylene were used to prepare all three isomeric tricyanobenzenes and various tetracyanocyclooctatetraenes **435–439**. Yields were extremely low[521]. Cycloadditions were also performed with [6]paracyclophane to afford a [2 + 2]adduct, and with naphthalenophane and anthracenophane to give [2 + 2]adducts[522].

Isothiazolecarbonitrile **440** was obtained when dicyanoacetylene was cyclized with sulfur dioxide or hydrobromic acid[523].

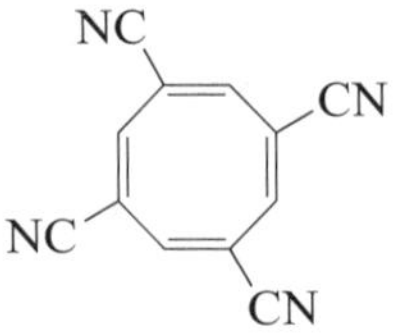

NC
⟩—CN
NC⟨
CN
(435)

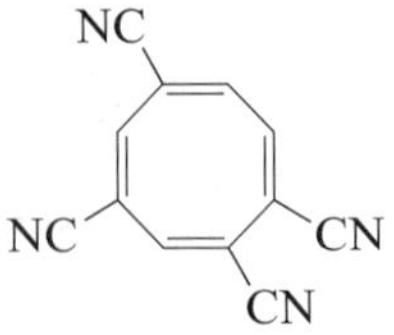

NC
⟩
NC⟨—CN
CN
(436)

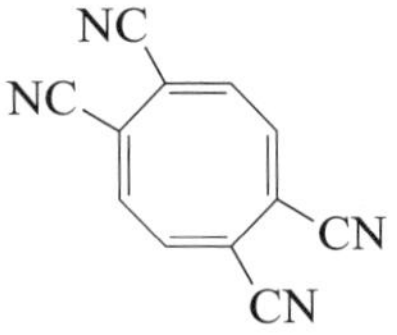

NC
NC⟨
⟩—CN
CN
(437)

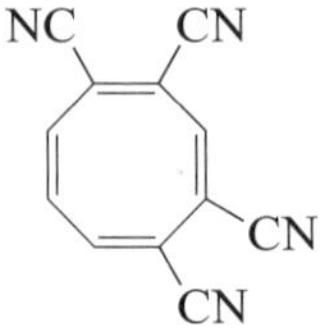

NC   CN
⟩
⟨—CN
CN
(438)

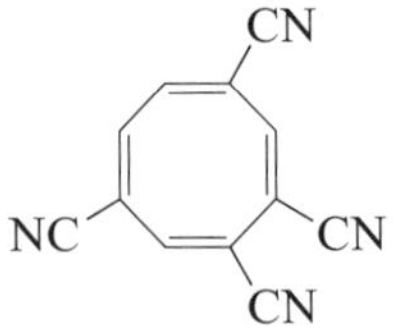

CN
⟩
NC⟨—CN
CN
(439)

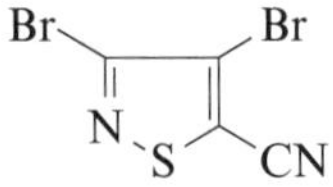

Br   Br
⟨   ⟩
N   S   CN
(440)

VI. POLYCYANOBENZENES

Polycyanobenzenes show an interesting behavior in complex formation and in electroreduction processes. The latter play an important role in various reactions.

A. Dicyanobenzenes

1,3-Dicyanobenzene was converted into 3-cyanobenzoic acid by using the bacterium Rhodococcus rhodochrocus[524]. Intramolecular electron transfer involving isomeric forms of dicyanobenzene was investigated in detail[525] and the mechanisms of electroreduction of all three isomers were studied[526]. When dicyanobenzene was used as catalyst in photoamination reactions of arenes, the first step of the reaction was the formation of an aryl radical cation by electron transfer from the aromatic hydrocarbon to m-dicyanobenzene[527].

B. Tricyanobenzenes

Electroreduction of 1,3,5-tricyanobenzene was studied[526]. When 1,3,5-tricyanobenzene was reacted with silyl and germyl radicals, 1,3,5-tricyanocyclohexadienyl radicals were obtained and detected by ESR spectroscopy[528]. A charge-transfer complex of 1,3,5-tricyanobenzene with 1,3,5-tris(dimethylamino)benzene was reported[529]. Small amounts of isomeric tricyanobenzenes have been obtained by photochemical cycloaddition reactions[521].

C. Tetracyanobenzenes

1. Complex formation

1,2,4,5-Tetracyanobenzene **441** forms charge-transfer complexes with electron donors easily. References dealing with complex formation and properties are given in Table 6. Weak exciplexes were formed with benzene, toluene and xylene. Formation and feedback dissociation of these exciplexes has been found to be strongly dependent on the solvent polarity and the donor ionization potential[530].

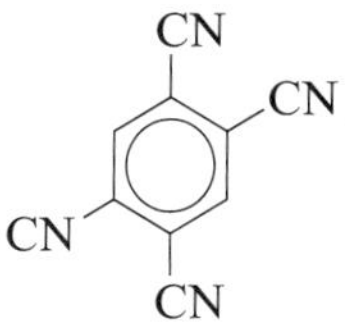

CN

CN

CN

CN

(441)

TABLE 6. Charge-transfer complexes with tetracyanobenzene

Electron donor	References
Anthracene	532–536
Acridine, durene, quinoline	526
Benzene	530, 537
Biphenyl	538, 539
Diphenylacetylene	540
Fluorene	526, 541
Naphthalene	542, 543
tert-Stilbene	544
Thianthrene	545
Toluene	530
Xylene	530, 546

2. Photochemical reactions

441 also underwent different photochemical reactions. When **441** was irradiated in acetonitrile in the presence of *N*-methyl-pyrrolidone, 2,4,5-tricyanoaniline (3%), 1,3,4-tricyanobenzene and 1-(*N*-methyl-2-pyrrolidyl)-benzene-2,4,5-tricarbonitrile were obtained[531]. 5-Tolyl-1,2,4-tricarbonitriles were obtained from **441** and substituted toluenes. Benzene gave no reaction under the same conditions. Benzene, anthracene and phenanthrene reacted with **441** in the presence of methanol to give 1:1:1 addition–elimination products. So **441**, benzene and methanol gave products **442** and **443** together with the thermal reaction product **444**[547].

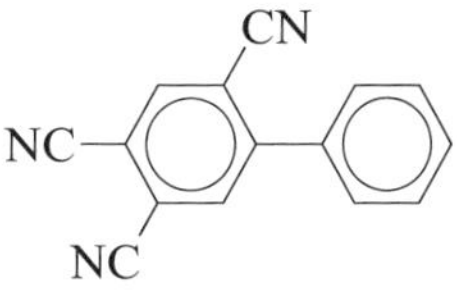

CN

NC—⟨ ⟩—⟨ ⟩

NC

(442)

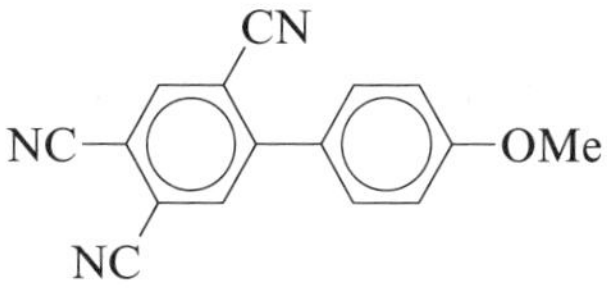

CN

NC—⟨ ⟩—⟨ ⟩—OMe

NC

(443)

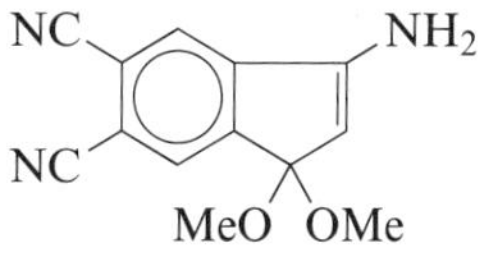

NC⟨ ⟩NH₂

NC

MeO OMe

(444)

VII. POLYCYANOQUINONES

A. Tetracyanobenzoquinone

Though the first papers on cyanil (2,3,5,6-tetracyano-1,4-benzoquinone) were published in the early sixties, this extremely strong electron acceptor did not find much interest later on. But recently, a new method of preparation has been published. Starting from *p*-

bromanil and sodium cyanide, 16% yield was obtained[548]. So work on complex formation and other reactions is likely to follow.

B. Tetracyanoquinodimethane (TCNQ, 445)

1. Formation and properties

a. *Synthesis.* A three-step, one-pot synthesis starting from hydroquinone with an overall yield of 80% was published[549]. A newer synthesis starting from terephthaloyl dichloride and CuCN gave an overall yield of 37% tetracyanoquinodimethane (**445**) in a three-step procedure. The reaction cascade comprised application of ultrasound and electrolysis[550]. In the synthesis of 2,4-disubstituted TCNQ derivatives, trimethylsilyl cyanide was used as cyanide source[551]. In the reaction with **446**, 2-chlorobenzyl thiocyanate **447** was used as cyanating agent to afford substituted TCNQs **448**[552].

$$\begin{array}{ccc} \text{(446)} & \text{(447)} & \text{(448)} \end{array}$$

b. *Complex formation.* **445** has been the subject of thorough investigations in the last ten years. Being an extremely strong electron acceptor, it has been used for complex formation with organic and inorganic electron donors. **445** salts as well as complexes show electric conductivity. Thus **445** is of great importance for the preparation of organic conductors. Most of the papers on **445** published in recent years deals with physicochemical properties of **445**, TCNQ salts and TCNQ complexes. Some reviews have also been published. Electronic and spectroscopic properties of Langmuir–Blodgett films containing TCNQ salts were compared with crystalline molecular conductors in an overview which contains 20 references[581]. A review dealing with conducting organic charge-transfer salts and comprising 114 references was published recently[553]. The synthesis and physicochemical properties of TCNQ compounds were reviewed three years earlier in a comprehensive paper containing 84 references[554].

TCNQ complex formation with electron-rich partners has been used for quantitative spectrophotometric analysis. Thus various pharmaceuticals like benzothiadiazines[555], benzenesulfonamides[556], Terfenadine[557], penicillins[558], antihistamines[559], some MAO inhibitors[560] and procainamide hydrochloride[561] have been determined via their colored charge-transfer complexes. In contrast to these papers a critical evaluation has been published, which pointed out problems occurring in the use of TCNQ for quantitative spectrophotometric analysis. This led the authors[562] to the conclusion that the use of TCNQ should be restricted to qualitative applications.

2. Reactions

a. *Reduction.* TCNQ was reduced electrochemically in aqueous solutions containing lithium, sodium and potassium cations to yield different reduction products, depending on the electrolyte composition[563].

b. Substitution of the nitrile group(s).   The reaction of TCNQ with aliphatic amines has been studied in detail. It is a two-step process proceeding via TCNQ radical anion and afforded 7,7-bis(dialkylamino)-8,8-dicyanoquinodimethane **449**[564]. A similar study has been performed for reactions of TCNQ with anilines[565].

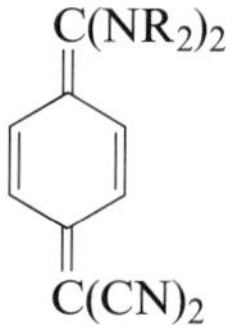

C(NR$_2$)$_2$

C(CN)$_2$

(449)

c. Addition to methylene C atom(s).   Thioaminyl monoradicals reacted with TCNQ to give *p*-phenylenedimalononitriles **450**. Polymerization reactions were also observed[566].

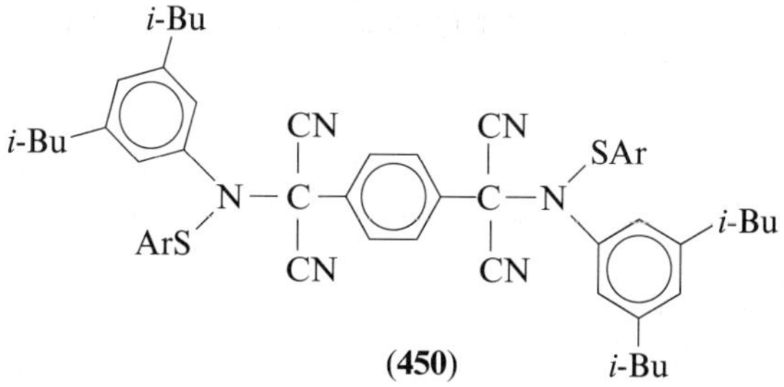

(450)

3. Fused TCNQ derivatives

a. Benzofused systems.   **451**[567], **452** and **453**[227,568] were obtained from the corresponding quinones by titanium tetrachloride-mediated coupling with malononitrile. A soluble tetracyanopentacenequinone derivative was transformed to **454** by base-catalyzed reaction with malononitrile[569]. Octacyano-anthradiquinotetramethane **455** was prepared starting from octahydroanthracene in a multistep synthesis[570].

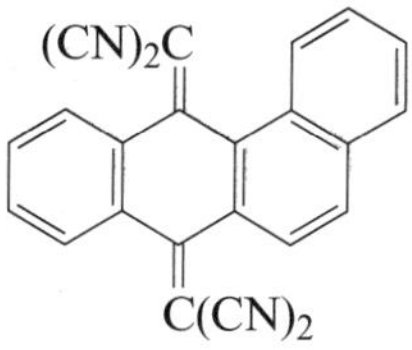

(451) 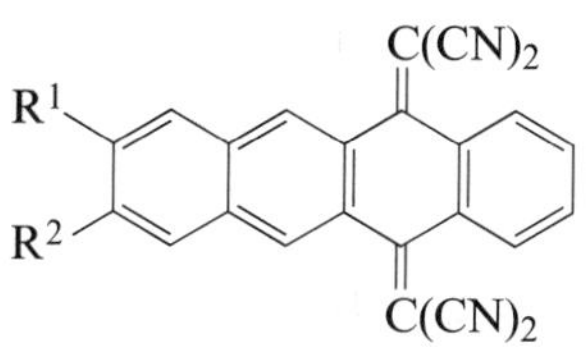   **(452)** 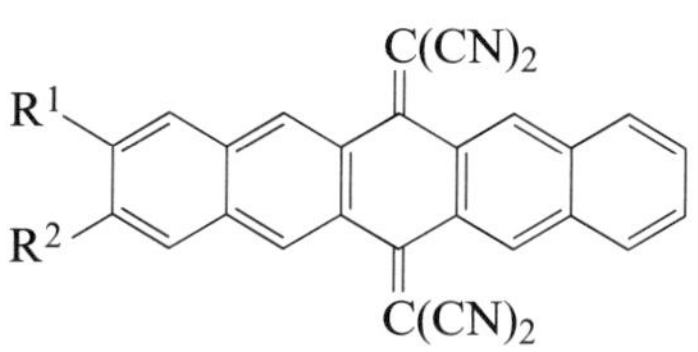   **(453)**

(454)   (455)

b. TCNQ derivatives fused with heterocycles. A variety of such derivatives was synthesized in the last years because of their interesting electrical properties. Hence thiophene- and benzothiophene-fused TCNQs were prepared[571,572]. **456** was obtained from the corresponding quinone and malononitrile[573], and an unsymmetrical product (**457**), formally consisting of TCNQ, thiadiazole and pyrazine, was reported recently[574]. Another unsymmetrically fused TCNQ (**458**) was prepared from the corresponding quinone by Knoevenagel reaction[575].

(456)   (457)   (458)

4. Heteroanalogous TCNQ derivatives

Functionalized TCNQ derivatives with hydroxy or carboxylic acid groups have been synthesized in moderate yields[576]. The pyridine analog of TCNQ, **459**, was found to be an even stronger electron acceptor than TCNQ[577]. Bis(dicyanomethylene)dithiophenes[578], bis(cyanimino)thienothiophenes[579] and bis(dicyanomethylene)biseleno-phenes[580] were reported.

(459)

VIII. REFERENCES

1. A. J. Fatiadi, in *The Chemistry of Functional Groups, Supplement C* (Eds. S. Patai and Z. Rappoport), Wiley, Chichester, 1983, pp. 1057–1303.
2. J. P. Ferris and J. W. J. Hagan, *Tetrahedron*, **40**, 1093 (1984).
3. C. Grundmann, in *Houben-Weyl, Methoden der Organischen Chemie*, Vol. E5, Thieme, Stuttgart, 1985, pp. 1313–1558.
4. M. H. Elnagdi, H. A. Elfahham and G. E. H. Elgemeie, *Heterocycles*, **20**, 519 (1983).
5. G. Kneen, *Gen. Synth. Methods*, **4**, 172 (1981); *Chem. Abstr.*, **96**, 103241 (1982).
6. G. Kneen, *Gen. Synth. Methods*, **5**, 83 (1982); *Chem. Abstr.*, **97**, 109171 (1982).
7. G. Kneen, *Gen. Synth. Methods*, **6**, 193 (1983); *Chem. Abstr.*, **99**, 69683 (1983).
8. G. Kneen, *Gen. Synth. Methods*, **7**, 198 (1985); *Chem. Abstr.*, **103**, 21861 (1985).
9. S. G. Lister, *Gen. Synth. Methods*, **8**, 245 (1986); *Chem. Abstr.*, **105**, 171437 (1986).
10. M. H. Elnagdi, R. M. Mohareb and S. M. Sherif, *Heterocycles*, **26**, 497 (1987).
11. M. H. Elnagdi, M. R. H. Elmoghayer and G. E. H. Elgemeie, *Synthesis*, **1**, 1 (1984).
12. H. Edwards and H. R. Mansour, *J. Mol. Struct.*, **160**, 209 (1987).
13. C. A. Deakyne, T. J. Buckley, M. Meotner and R. Metz, *J. Chem. Phys.*, **86**, 2334 (1987).
14. P. W. Harland and B. J. Mcintosh, *Aust. J. Chem.*, **38**, 967 (1985).
15. A. Kuhn, H. P. Fenzlaff and E. Illenberger, *Chem. Phys. Lett.*, **135**, 335 (1987).
16. Y. L. Wang, M. W. Beach, H. D. Lee and D. W. Margerum, *Inorg. Chem.*, **26**, 2444 (1987).
17. X. Xie, I. A. McLaren, J. B. Halpern and W. M. Jackson, *Zhongguo Jiguang*, **14**, 480 (1987); *Chem. Abstr.*, **108**, 103941 (1988).
18. J. Sundermeyer and H. W. Roesky, *Angew. Chem., Int. Ed. Engl.*, **27**, 1372 (1988).
19. W. Ried and M. Fulde, *Helv. Chim. Acta*, **71**, 1681 (1988).
20. D. L. Yang, T. Yu, N. S. Wang and M. C. Lin, *Chem. Phys.*, **160**, 307 (1992).
21. I. C. Tornieporthoetting, T. M. Klapotke, T. S. Cameron, J. Valkonen, P. Rademachev and K. Kowski, *J. Chem. Soc., Dalton Trans.*, 537 (1992).
22. T. Fukunaga and R. W. Begland, *J. Org. Chem.*, **49**, 813 (1984).
23. A. J. Fatiadi, *Synthesis*, 165 (1978).
24. A. J. Fatiadi, *Synthesis*, 241 (1978).
25. F. Freeman, *Synthesis*, 925 (1981).
26. E. Campaigne and S. W. Schneller, *Synthesis*, 705 (1976).
27. F. Freeman, *Chem. Rev.*, **80**, 329 (1980).
28. N. N., Jpn. Kokai Tokkyo Koho 83185549/29.20.83, Appl. 23.04.82 82/68092, 2 pp./Mitsubishi Gas Chemical Co., Inc. (1983); *Chem. Abstr.*, **100**, 120540 (1984).
29. D. Bougeard, B. Pasquier and N. Lecalve, *Ber. Bunsenges. Phys. Chem.*, **91**, 1273 (1987).
30. T. Wasiutynski, W. Olejarczyk, J. Sciesinski and W. Witko, *J. Phys. C: Solid State Phys.*, **20**, L65 (1987).
31. B. H. Loo, D. O. Frazier and Y. G. Lee, *Chem. Phys. Lett.*, **119**, 312 (1985).
32. M. J. Aroney, E. Patsalides and P. K. Pierens, *Aust. J. Chem.*, **38**, 507 (1985).
33. L. Foucat, M. T. Chenon and L. Werbelow, *J. Phys. Chem.*, **94**, 5791 (1990).
34. J. H. Chen, Thesis, Univ. California, San Diego, CA, USA (1989); *Chem. Abstr.* **114**, 41937 (1991).
35. A. Tsubouchi, N. Hamasaki, N. Matsumura and H. Inoues, *Chem. Express.*, **5**, 565 (1990); *Chem. Abstr.*, **113**, 211879 (1990).
36. W. P. McKenna and E. M. Eyring, *Appl. Spectrosc.*, **40**, 16 (1986).
37. C. J. Van Staveren, V. M. L. J. Aarts, P. D. J. Grootenhuis, J. Van Eerden, S. Harkema and D. N. Reinhoudt, *J. Am. Chem. Soc.*, **108**, 5271 (1986).
38. J. R. Damewood, A. L. Rheingold, T. C. Williamson and J. J. Urban, *J. Org. Chem.*, **53**, 167 (1988).
39. D. P. Zollinger, M. Bos, W. E. Vanderlinden and A. M. W. Vanveenblaauw, *Talanta*, **31**, 723 (1984).
40. S. Honda, S. Iwase, K. Kakehi and S. Suzuki, *Anal. Biochem.*, **160**, 455 (1987).
41. S. Honda, Jpn. Kokai Tokkyo 88 66444/25.03.88, Appl. 09.09.86 86/210550, 7pp/Kanto Chemical Co., Inc. (1988); *Chem. Abstr.*, **110**, 111317 (1989).
42. E. Iglesias and D. L. H. Williams, *J. Chem. Soc., Perkin Trans. 2*, 343 (1989).
43. Y. A. Sharanin, A. M. Shestopalov, V. Y. Mortikov, S. N. Melenchuk, V. K. Promonenkov, B. M. Zolotarev and V. P. Litvinov, *Izv. Akad. Nauk SSSR, Ser. Khim.*, 153 (1986); *Chem. Abstr.*, **106**, 4825 (1987).

44. V. G. Granik, O. S. Anisimova, S. I. Grizik, Y. N. Sheinker and N. P. Soloveva, *Zh. Org. Khim.*, **20**, 673 (1984); *Chem. Abstr.*, **102**, 5674 (1985).
45. M. H. Elnagdi and A. W. M. Erian, *Arch. Pharm.*, **324**, 853 (1991).
46. E. Stankevics, A. Dreimane, E. Liepins, A. Kemme and J. Bleidelis, *Nucleosides Nucleotides*, **2**, 155 (1983).
47. V. Marko and P. Gemeiner, Czech. 229972/01.02.86, Appl. 05.01.82 82/89, 7 pp. (1986); *Chem. Abstr.*, **106**, 5662 (1987).
48. M. Mittelbach, *Monatsh. Chem.*, **116**, 689 (1985).
49. M. Mittelbach, H. Sterk, U. Wagner and H. Junek, *Justus Liebigs Ann. Chem.*, 1131 (1987).
50. G. Gattow and W. Manz, *Z. Anorg. Allg. Chem.*, 149 (1988).
51. J. S. Gramm, U. S. 44122957/01.11.83, Appl. 15.06.82 388625, 3 pp. du Pont de Nemours, e. l., and Co (1983); *Chem. Abstr.*, **100**, 67855 (1984).
52. A. Cesare-Veronese, M. Basato, B. Corain and V. Gandolfi, *J. Mol. Catal.*, **36**, 339 (1986).
53. G. I. Nikishin, M. N. Elinson, T. L. Lizunova and B. I. Ugrak, *Tetrahedron Lett.*, **32**, 2655 (1991).
54. S. K. El-Sadany, S. M. Sharaf, A. I. Darwish and A. A. Youssef, *Indian J. Chem. Soc., Sect. B*, **308**, 567 (1991).
55. M. M. Al-Arab, H. D. Tabba, B. S. Ghanem and M. M. Olmstead, *Synthesis*, 1157 (1990).
56. J. Mirek, *Chem. Scr.*, **28**, 295 (1988); *Chem. Abstr.*, **110**, 114412 (1989).
57. H. W. Schmidt and M. Kores, *Justus Liebigs Ann. Chem.*, 1001 (1988).
58. K. Hartke, M. Fallert and R. Matusch, *Synthesis*, 677 (1986).
59. D. C. Lankin, I. Bernal, N. S. Bhacca, T. J. Delord, G. W. Griffin, J. Korp and S. F. Watkins., *Tetrahedron*, **40**, 2829 (1984).
60. A. G. Kofman, S. A. Podznyakovich, Y. Y. Zinchenko, V. M. Dolmat and V. V. Yadrikhinskii, *Ukr. Khim. Zh.*, **51**, 1213 (1985); *Chem. Abstr.*, **106**, 32515 (1987).
61. E. Campaigne and R. F. Weddleton, *Proc. Indiana Acad. Sci.*, 159 (1985); *Chem. Abstr.*, **108**, 55808 (1988).
62. H. Wamhoff and H. A. Thiemig, *Ber. Dtsch. Chem. Ges.*, **118**, 4473 (1985).
63. T. Matsuda, K. Yamagata, Y. Tomioka and M. Yamazaki, *Chem. Pharm. Bull.*, **33**, 937 (1985).
64. J. M. Sinambela, W. Zimmermann, H. J. Roth and K. Eger, *J. Heterocycl. Chem.*, **23**, 393 (1986).
65. H. Schaefer and K. Gewald, *Monatsh. Chem.*, **120**, 315 (1989).
66. A. Mendel, *Res. Discl.*, 504 (1989); *Chem. Abstr.*, **112**, 138989 (1990).
67. O. A. Attanasi, S. Santeusanio, F. Serra-Zanetti, E. Foresti and A. McKillop, *J. Chem. Soc., Perkin Trans. 1*, 669 (1990).
68. K. Tanaka, S. Maeno and K. Mitsushashi, *Seikei Daigaku Kogakubu Kogaku Hokoka*, 2449 (1984); *Chem. Abstr.*, **101**, 171178 (1984).
69. K. Tanaka, K. Mitsushashi, T. Suzuki and S. Maeno, *Bull. Chem. Soc. Jpn.*, **60**, 4480 (1987).
70. E. A. Jauer, E. Foerster and R. Mayer, Ger. (East) 08.01.86, Appl. 17.08.84 266360, 3 pp./VEB Chemiekombinat Bitterfeld, (1986); *Chem. Abstr.*, **106**, 196441 (1987).
71. M. J. Dooley, R. J. Quinn and P. J. Scammels, *Aust. J. Chem.*, **42**, 747 (1989).
72. S. M. Fahmy and S. O. Abd-Allah, *Egypt. J. Chem.*, **28**, 245 (1985); *Chem. Abstr.*, **107**, 115539 (1987).
73. V. Ertuzun, *Commun. Fac. Sci. Unic. Ankara, Ser. B*, **32**, 41 (1986); *Chem. Abstr.*, **108**, 150180 (1988).
74. K. Eger, T. Storz and S. Spatling, *Justus Liebigs Ann. Chem.*, 1049 (1989).
75. H. Fuerstenwerth, Ger. Offen. 3344294/20.06.85, Appl. 07.12.83 3344294, 22 pp. / Bayer AG, (1985); *Chem. Abstr.*, **103**, 215152 (1985).
76. V. Ertuzun, *Commun. Fac. Sci. Univ. Ankara, Ser. B*, **32**, 1, (1986); *Chem. Abstr.*, **108**, 150194 (1988).
77. N. Ito, H. Aiga, M. Nishihara, N. Yano, T. Nishida and T. Nagayoshi, Jpn. Kokai Tokkyo Koho 85190776/28.09.85, Appl. 09.03.84 84/43850, 3 pp./ Mitsui Toatsu Chemicals, Inc. (1985); *Chem. Abstr.*, **104**, 88420 (1986).
78. I. Zeid, S. Yassin, I. El-Sakka and A. Abass, *Justus Liebigs Ann. Chem.*, 191 (1984).
79. W. Ried and K. Schoepke, *Justus Liebigs Ann. Chem.*, 389 (1986).
80. J. J. Tegeler and C. J. Diamond, *J. Heterocycl. Chem.*, **24**, 697 (1987).
81. C. J. Shishoo, M. B. Devani, S. Ananthan, V. S. Bhadti and G. V. Ullas, *J. Heterocycl. Chem.*, **25**, 759 (1988).
82. K. Yonemoto, I. Shibuya and K. Honda, *Bull. Chem. Soc. Jpn.*, **62**, 1086 (1989).
83. R. M. Mohareb and S. M. Fahmy, *Z. Naturforsch., B: Anorg. Chem., Org. Chem.*, **40B**, 153⁷ (1985).

84. K. A. Kandeel, J. M. Vernon, T. A. Dransfield, F. A. Fouli and A. S. A. Youssef, *J. Chem. Res.* *(S)*, 276 (1990); *Chem. Abstr.*, **113**, 231170 (1990).
85. M. M. El-Kerdawy, H. M. Eisa, A. A. El-Emam, M. A. Massoud and M. N. Nasr, *Orient. J. Chem.*, **6**, 115 (1990); *Chem. Abstr.*, **114**, 42638 (1991).
86. Z. E.-S. Kandeel, K. M. Hassan, N. A. Ismail and M. H. Elnagdi, *J. Prakt. Chem.*, **326**, 248 (1984).
87. G. E. H. Elgemeie, H. F. Zohdi and S. F. Sherif, *Phosphorous, Sulfur Silicon Relat. Elem.*, **54**, 215 (1990).
88. N. Latif, N. Mishriky, B. Haggag and W. Basyouni, *Indian J. Chem., Sect. B*, **24G**, 1230 (1985).
89. Y. A. Sharanin and V. D. Dyachenko, *Ukr. Khim. Zh.*, **56**, 287 (1990); *Chem. Abstr.*, **113**, 131960 (1990).
90. V. D. Dyachenko, Yu. A. Sharonin, V. P. Litvinov, V. P. Nesterov, V. E. Shklover, Yu. T. Struchkov, V. K. Promonenkov and A. V. Turov, *Zh. Obshch. Khim.*, **61**, 747 (1991); *Chem. Abstr.*, **115**, 114386 (1991).
91. B. Singh, U.S. 4413127/01.11.83, Appl. 24.05.82 3810062, 4 pp. / Sterling Drug, Inc. (1983); *Chem. Abstr.*, **100**, 34418 (1984).
92. J. Moragues-Mauri, R. G. W. Spickett, J. Prieto-Soto and A. Vega-Noverola, Span. 537436/16.09.85, Appl. 07.11.84 537426, 7 pp. / Fordonal S. A. (1985); *Chem. Abstr.*, **105**, 226362 (1986).
93. N. S. Ibrahim, R. M. Mohareb and H. Z. Shams, *Z. Naturforsch. B: Chem. Sci.*, **43**, 1351 (1988).
94. M. M. Badran, S. El-Meligie and A. K. El-Ansari, *Rev. Roum. Chim.*, **34**, 2093 (1989); *Chem. Abstr.*, **114**, 6235 (1991).
95. P. Victory, R. Nomen, M. Garriga, X. Tomas and L. G. Sabate, *Afinidad*, **41**, 241 (1984).
96. P. Victory, R. Nomen, O. Colomina, M. Garriga and A. Crespo, *Heterocycles*, **23**, 1135 (1985).
97. A. Habashi, N. S. Ibraheim, R. M. Mohareb and S. M. Fahmy, *Justus Liebigs Ann. Chem.*, 1632 (1986).
98. S. M. Fahmy and R. M. Mohareb, *Tetrahedron*, **42**, 687 (1986).
99. M. H. Elnagdi, S. A. S. Ghozlan, F. M. Adelrazek and M. Ali-Selim, *J. Chem. Res. (S)*, 116 (1991); *Chem. Abstr.*, **115**, 49330 (1991).
100. A. Katoh, C. Kashima and Y. Omote, *Heterocycles*, **22**, 763 (1984).
101. M. F. Brana, J. M. Castellano and M. J. R. Yunta, *Heterocycles*, **22**, 113 (1984).
102. D. Dvorak, D. Saman, J. Hodacova, V. Kral and Z. Arnold, *Collect. Czech. Chem. Commun.*, **52**, 2687 (1987).
103. J. L. Soto, C. Seoane, N. Martin and M. Quinteiro, *Heterocycles*, **22**, 1, (1984).
104. J. A. Ciller, N. Martin, C. Seoane and J. L. Soto, *J. Chem. Soc., Perkin Trans. 1*, 2581 (1985).
105. G. Heinisch, W. Holzer and G. A. M. Nawwar, *Monatsh. Chem.*, **117**, 247 (1986).
106. R. Spitzner and W. Schroth, *J. Prakt. Chem.*, **328**, 314 (1986).
107. E. Augustyn and K. Bogdanowicz-Szwed, *Monatsh. Chem.*, **114**, 1189 (1983).
108. I. Alkorta, A. G. Bielsa, V. J. Aran and M. Stud, *J. Chem. Soc., Perkin Trans. 1*, 1271 (1988).
109. F. F. Abdel-Latif, *Bull. Soc. Chim. Fr.*, 129 (1990).
110. V. S. Velezheva, K. V. Nevskii and N. N. Suvorov, *Khim. Geterotsikl. Soedin.*, 276 (1985); *Chem. Abstr.*, **103**, 22410 (1985).
111. B. Zaleska, *Pol. J. Chem.*, **58**, 469 (1984); *Chem. Abstr.*, **103**, 37382 (1985).
112. G. L'Abbe, A. Vandendriessche and N. Weyns, *Bull. Soc. Chim. Belg.*, **97**, 85 (1988).
113. M. M. Abbasi, A. Akelah, Y. Hafez and E. S. Ismail, *J. Prakt. Chem.*, **329**, 525 (1987).
114. H. Schaefer, K. Gewald and M. Gruner, *J. Prakt. Chem.*, **331**, 878 (1989).
115. T. L. Su, J. T. Huang, J. H. Burchenal, K. A. Watanabe and J. J. Fox, *J. Med. Chem.*, **29**, 709 (1986).
116. T. L. Su, K. Harada and K. A. Watanabe, *Nucleosides Nucleotides*, **3**, 513 (1984).
117. P. L. Barili, G. Biagi, O. Livi, L. Mucci and V. Scartoni, *J. Heterocycl. Chem.*, **24**, 997 (1987).
118. G. L'Abbe, F. Godts, S. Toppet, L. Van Meervelt and G. S. D. King, *Bull. Soc. Chim. Belg.*, **96**, 587 (1987).
119. O. N. Chupakin, V. L. Rusinov, T. N. Pilicheva and A. A. Tumashov, *Synthesis*, 713 (1990).
120. F. Farina, M. C. Maestro, M. V. Martin and M. L. Soria, *Tetrahedron*, **43**, 4007 (1987).
121. J. Y. Becker, J. Bernstein, S. Bittner, E. Harlev and J. A. R. P. Sarma, *J. Chem. Soc., Perkin Trans. 2*, 1157 (1989).
122. F. F. Abdel-Latif, *Gazz. Chim. Ital.*, **121**, 9 (1991).

123. T. Al-Nakib, M. J. Meegan, and D. V. Tyndall, *J. Chem. Res.* (*S*), 322 (1987); *Chem. Abstr.*, **108**, 204521 (1988).
124. F. F. Abdel-Latif, *Indian J. Chem., Sect. B*, **29B**, 664 (1990).
125. V. V. Tkachenko, N. G. Tregub, A. P. Knyazev and V. V. Mezheritskii, *Zh. Org. Khim.*, **26**, 638 (1990); *Chem. Abstr.*, **113**, 152224 (1990).
126. A. O. Abdelhamid and B. Y. Riad, *Arch. Pharm.*, **320**, 1010 (1987).
127. F. F. Abdel-Latif, *Z. Naturforsch., B: Chem. Sci.*, **45**, 1675 (1990).
128. N. Martin, C. Seoane and J. L. Soto, *Tetrahedron*, **44**, 5861 (1988).
129. T. Fujimaki, R. Yamaguchi, H. Otomasu, H. Nagase and K. Kawai, *Chem. Pharm. Bull.*, **33**, 2663 (1985).
130. D. W. Rangnekar and S. V. Dhamnaskar, *J. Heterocycl. Chem.*, **25**, 1767 (1988).
131. H. Abdel-Ghany, A. M. El-Sayed, A. A. Sultan and A. K. El-Shafei, *Synth. Commun.*, **20**, 893 (1990).
132. D. Briel and G. Wagner, Ger (East) 204253/23.11.83, Appl. 07.04.82 238800, 8 pp. / Karl-Marx-Universität, Leipzig (1983); *Chem. Abstr.*, **101**, 23488 (1984).
133. N. M. Fathy and G. F. H. Elgemeie, *Sulfur Lett.*, **7**, 189 (1988).
134. E. Jeanneau-Nicolle, M. Benoit-Guyod and G. Leclerc, *Synth. Commun.*, **21**, 1443 (1991).
135. T. Al-Nakib, D. V. Tyndall and M. J. Meegan, *J. Chem. Res. (S)*, 10 (1988); *Chem. Abstr.*, **109**, 22870 (1988).
136. N. S. Ibrahim, N. M. Abed and Z. E. Kandeel, *Heterocycles*, **22**, 1677 (1984).
137. E. T. McCarthy, D. V. Tyndall and M. J. Meegan, *J. Chem. Res. (S)*, 145 (1988); *Chem. Abstr.*, **110**, 38945 (1989).
138. F. F. Abdel-Latif, *Phosphorous, Sulfur Silicon Relat. Elem.*, **53**, 145 (1990).
139. M. T. Ismail, A. A. Abdel-Wahab and M. T. Makhloof, *J. Electrochem. Soc. India*, **35**, 221 (1986).
140. O. E. Nasakin, M. Yu. Skvortsova, P. M. Lukin, A. Kh. Bulai and A. N. Lyshchikov, U.S.S.R. 1177292/07.09.85, Appl. 23.12.83/3711054/Chuvash State University (1985); *Chem. Abstr.*, **104**, 129514 (1986).
141. M. Uno, K. Seto, M. Masuda, W. Ueda and S. Takahashi, *Tetrahedron Lett.*, **26**, 1553 (1985).
142. H. Nemoto, Y. Kubota and Y. Yamamoto, *J. Org. Chem.*, **55**, 4515 (1990).
143. C. Combellas, M. Lequan, R. M. Lequoin, J. Simon and A. Thiebault, *J. Chem. Soc., Chem. Commun.*, 542 (1990).
144. J. S. Gramm, E. L. Mongan Jr. and P. J. Sheeran, Eur. Pat. Appl. 119799/26.09.84, Appl. 09.03.84 84/301618, Priorities 11.03.83 US 474470, 11 p./du Pont de Nemours, E. I., and Co. (1984); *Chem. Abstr.*, **102**, 24127 (1985).
145. H. Mettler, Eur. Pat. Appl. 298261/11.01.89, Appl. 07.06.88 88/109076, priorities 09.06.87 CH 87/2167, 5 pp/Lonza AG (1987); *Chem. Abstr.*, **111**, 6933 (1989).
146. Y. Tominaga, H. Norisue, Y. Matsuda and G. Kobayashi, *Yukugaku Zasshi*, **104**, 127 (1984); *Chem. Abstr.*, **101**, 90510 (1984).
147. E. Diezbarra, A. Delahoz, P. Sanchezverdu and A. Moreno, *Synthesis*, 391 (1989).
148. H. Chikashita, K. Itoh, M. Miyazaki and S. Nishida, *Synth. Commun.*, **13**, 1033 (1983).
149. S. Takahashi, M. Uno and K. Seto, Jpn. Kokai Tokkyo Koho 85197650/07.10.85, Appl. 16.03.84 84/51749, 6 pp./Nitto Chemical Industry Co., Ltd., (1985); *Chem. Abstr.*, **104**, 88318 (1986).
150. M. Uno, T. Takahashi and S. Takashahi, *J. Chem. Soc., Chem. Commun.*, 785 (1987).
151. M. Uno, K. Seto and S. Takahashi, *J. Chem. Soc., Chem. Commun.*, 932 (1984).
152. S. Yamaguchi, H. Araki and T. Hanafusa, *Chem. Lett.*, 685 (1985).
153. N. Martin-Leon, M. Quinteiro, C. Seoane and J. L. Soto, *Justus Liebigs Ann. Chem.*, 101 (1990).
154. P. Molina and A. Lorenzo, *Tetrahedron Lett.*, **24**, 5805 (1983).
155. T. Sakamoto, E. Kato, Y. Kondo and H. Yamanaka, *Chem. Pharm. Bull.*, **36**, 1664 (1988).
156. M. H. Mohamed, N. S. Ibrahim, M. M. Hussein and M. H. Elnagdi, *Heterocycles*, **27**, 1301 (1988).
157. N. S. Ibrahim, R. M. Mohareb and M. H. Elnagdi, *J. Prakt. Chem.*, **330**, 65 (1988).
158. D. P. Curran and C. M. Seong, *Synlett*, 107 (1991).
159. F. Freeman and D. S. H. L. Kim, *Synthesis*, 698 (1989).
160. D. P. Curran and C. M. Seong, *J. Am. Chem. Soc.*, **112**, 9401 (1990).
161. O. E. Nasakin, P. M. Lukin, A. L. Shevnitsyn, V. D. Sheludyakov and E. A. Chernysov, *Zh. Obshch. Khim.*, **58**, 1168 (1988); *Chem. Abstr.*, **110**, 192906 (1989).
162. M. A. Shvekhgeimer, K. I. Kobrakov, A. G. Pavlov, S. S. Sychev and N. G. Popandopulo, *Dokl. Akad. Nauk SSSR*, **308**, 389 (1989); *Chem. Abstr.*, **112**, 216201 (1990).

163. R. P. Kozyrod, J. Morgan and J. T. Pinhey, *Aust. J. Chem.*, **44**, 369 (1991).
164. A. O. Abdelhamid, A. M. Negm and I. M. Abbas, *Egypt. J. Pharm. Sci.*, **30**, 103 (1989); *Chem. Abstr.*, **112**, 216774 (1990).
165. F. M. Abdelrazek, A. W. Brian and A. M. El-Torgoman, *Chem. Ind. (London)*, 30 (1988).
166. K. M. Hilmy and E. B. Pedersen, *Justus Liebigs Ann. Chem.*, 1145 (1989).
167. J. J. Vaquero, J. C. Delcastillo, L. Fuentes, J. L. Garcia, M. I. Perez and J. L. Soto, *Synthesis*, 33 (1987).
168. F. Freeman and D. S. H. L. Kim, *J. Org. Chem.*, **56**, 4645 (1991).
169. J. L. Soto, C. Seoane, M. J. Rubio and J. M. Botija, *Org. Prep. Proced. Int.*, **16**, 11 (1984).
170. C. O'Murchu, Patentschrift (Switz.) 649292 / 15.05.85, Appl. 22.07.82 82/4474, 3 pp. / Lonza AG, (1985); *Chem. Abstr.*, **103**, 196118 (1985).
171. C. O'Murchu, Patentschrift (Switz.) 651026 / 30.08.85, Appl. 04.11.82 82/6408, 2 pp. / Lonza AG, (1985); *Chem. Abstr.*, **104**, 109674 (1986).
172. Y. V. Erofeev, K. F. Turchin, T. A. Uglova, E. F. Kuleshova and L. S. Gorodetskii, *Khim.-Farm. Zh.*, **22**, 1469 (1988); *Chem. Abstr.*, **111**, 39300 (1989).
173. F. Freeman and D. S. H. L. Kim, *J. Org. Chem.*, **56**, 657 (1991).
174. V. J. Aran, F. Florencio, J. L. Soto, J. Sanzaparicio, C. Seoane and N. Martin, *J. Org. Chem.*, **53**, 5341 (1988).
175. R. Trostschuetz, *Arch. Pharm.*, **324**, 485 (1991).
176. M. J. Mokrosz, *Pol. J. Chem.*, **60**, 631 (1986); *Chem. Abstr.*, **108**, 111840 (1988).
177. M. J. Mokrosz, *J. Fluorine Chem.*, **34**, 201 (1986).
178. M. J. Mokrosz, *Org. Mass Spectrom.*, **19**, 639 (1984).
179. Y. Yamashita, T. Hayashi and M. Masumura, *Tetrahedron Lett.*, **25**, 4429 (1984).
180. A. N. Kasatkin, R. K. Biktimirov, G. A. Tolstikov and L. M. Khalilov, *Zh. Org. Khim.*, **26**, 1191 (1990); *Chem. Abstr.*, **114**, 23392 (1991).
181. E. Schefczik, Eur. Pat. Appl. 327981/16.08.89, Appl. 03.02.89/101853, Priorities 12.02.88 DE 3804394, 13 pp/BASF AG (1989); *Chem. Abstr.*, **112**, 55013 (1990).
182. A. Safarzadeh-Amiri, *Chem. Phys. Lett.*, **129**, 225 (1986).
183. D. H. Suh, J. C. Won, J. C. Jung and D. K. Kim, *J. Polym. Sci., Part C: Polym. Lett.*, **26**, 83 (1988).
184. P. Czerney and H. Hartmann, Ger. (East) 221191/17.04.85, Appl. 01.11.83 256164, 4 pp./VEB Carl Zeiss Jena, (1985); *Chem. Abstr.*, **104**, 111374 (1986).
185. G. E. H. Elgemeie, H. A. Elfahham, S. Elgamal and M. H. Elnagdi, *Heterocycles*, **23**, 1999 (1985).
186. N. Castenedo, K. Peseke, H. Kelling and M. Michalik, *J. Prakt. Chem.*, **325**, 695 (1983).
187. K. Peseke and I. Bohn, Ger. (East) 202696/28.09.83, Appl. 14.07.81 231745, 5 pp./Wilhelm-Pieck-Universität Rostock (1983); *Chem. Abstr.*, **100**, 156375 (1984).
188. Y. P. Yen and M. C. Lew, *J. Chin. Chem. Soc. (Taipei)*, **37**, 591 (1990); *Chem. Abstr.*, **114**, 142719 (1991).
189. H. Takahata, E. C. Wang, T. Nakajima and T. Yamazaki, *Chem. Pharm. Bull.*, **35**, 3139 (1987).
190. Y. Akasaki, K. Sato, H. Tanaka, K. Nukada and H. Sudo, Jpn. Kokai Tokkyo Koho 88174993/ 19.07.88, Appl. 13.01.87 87/4105, 5 pp. / Fuji Xerox Co., Ltd. (1988); *Chem. Abstr.*, **110**, 85426 (1989).
191. R. Kada, V. Knoppova, J. Kovac and P. Cepec, *Collect. Czech. Chem. Commun.*, **49**, 984 (1984).
192. F. Fernandez-Gomez and C. Aguiar-Punal, *Rev. Cubana Quim.*, **1**, 49 (1985); *Chem. Abstr.*, **105**, 133061 (1986).
193. C. Aguiar-Punal and F. Fernandez-Gomez, *Cent. Azucar (Supl. Espec. Deriv. Quim.)*, 21 (1985); *Chem. Abstr.*, **107**, 133641 (1987).
194. H. Yasuda, Y. Idei and Y. Yamada, *Utsunomiya Daigaku Kyoikugakubu Kiyo, Dai-2-bu*, 31 (1987); *Chem. Abstr.*, **107**, 217395 (1987).
195. P. Singh and P. Aggarwal, *Indian J. Chem., Sect. B*, **27**, 1029 (1988).
196. A. K. Mukherjee, S. P. Bhattacharyya, A. De and M. Mukherjee, *Acta Crystallogr., Sect. C: Cryst. Struct. Commun.*, **C40**, 991 (1984).
197. C. H. Chen, G. A. Reynolds, H. R. Luss and J. H. Perlstein, *J. Org. Chem.*, **51**, 3282 (1986).
198. D. S. Yufit, Y. T. Struchkov, E. N. Komarova, A. S. Vyazgin and V. N. Drozd, *Zh. Org. Khim.*, **22**, 2448 (1986); *Chem. Abstr.*, **107**, 58906 (1987).
199. R. Keuhn and H. H. Otto, *Arch. Pharm.*, **319**, 898 (1986).
200. K. Bogdanowicz-Szwed and A. Policht, *J. Prakt. Chem.*, **326**, 721 (1984).
201. K. Peseke, R. Rodriguez-Palacio and J. Gonzales-Feitot, Ger. (East) 244749/15.04.87, Appl. 24.12.85 285356, 4 pp./Wilhelm-Pieck-Universität Rostock (1987); *Chem. Abstr.*, **107**, 217492 (1987).

202. R. Neidlein and D. Kikelj, *Synthesis*, 981 (1988).
203. D. Mysyk, N. M. Sivchenkova, V. Kampars and O. Neilands, *Latv. PSR Zinat. Akad. Vestis, Khim. Ser.*, 621 (1987); *Chem. Abstr.*, **109**, 92406 (1988).
204. Y. Akasaki, K, Nukada and K. Sato, Jpn. Kokai Tokkyo Koho 90134353/23.05.90/Appl. 16.11.88 88/287614, 4pp/Fuji Xerox Co., Ltd. (1990); *Chem. Abstr.*, **113**, 181405 (1990).
205. D. E. Bugner, T. M. Kung and L. J. Rossi, U.S. 4948911/14.08.90, Appl. 18.12.89 452536, 8pp./Eastman Kodak Co. (1990); *Chem. Abstr.*, **114**, 101381 (1991).
206. D. D. Mysyk, I. F. Perepichka, N. M. Sivchenkova, L. I. Kostenko and I. S. Pototskii, *Elektron. Org. Mater.*, 351 (1985); *Chem. Abstr.*, **111**, 47946 (1989) and **107**, 31043 (1987).
207. P. S. Lianis, N. A. Rodios and N. E. Alexandrou, *J. Heterocycl. Chem.*, **25**, 1099 (1988).
208. J. Ojima, S. Ishizaka, Y. Shiraiwa, E. Ejiri, T. Kato, S. Kuroda and H. Takeda, *Chem. Lett.*, 1295 (1986).
209. J. Ojima, S. Ishizaka, Y. Shiraiwa, E. Ejiri, T. Kato and S. Kuroda, *J. Chem. Soc., Perkin Trans. 1*, 1505 (1987).
210. X. Wang and Z. Huang, *Huaxue Xuebao*, **47**, 890 (1989); *Chem. Abstr.*, **112**, 198216 (1990).
211. K. Bogdanowicz-Szwed, H. Feret and M. Lipowska, *Z. Naturforsch., B: Chem. Sci.*, **42**, 623 (1987).
212. H. Suzuki, H. Koide and T. Ogawa, *Bull. Chem. Soc. Jpn.*, **61**, 501 (1988).
213. R. Valters, A. Bace, J. Mednis, V. R. Zin'kovskaya, R. Kampare and O. Neilands, *Latv. PSR Zinat. Akad. Vestis, Kim. Ser.*, 618 (1983); *Chem. Abstr.*, **100**, 67869 (1984).
214. J. A. Moore and J. H. Kim, *Tetrahedron Lett.*, **32**, 3449 (1991).
215. A. P. Mkrtchan, S. G. Kazaryan, A. S. Noravyan and S. A. Vartanyan, *Khim. Geterotsikl. Soedin.*, 395 (1987); *Chem. Abstr.*, **108**, 5880 (1988).
216. H. Junek, M. Klade, H. Sterk and W. Fabian, *Monatsh. Chem.*, **119**, 993 (1988).
217. Y. A. Sharanin, A. M. Shestopalov, G. E. Khoroshilov, V. P. Litvinov and O. M. Nefedov, *Zh. Org. Khim.*, **25**, 1315 (1989); *Chem. Abstr.*, **112**, 76895 (1990).
218. K. Peseke and R. Rodriguez-Palacio, Ger. (East) 250117/30.09.87, Appl. 19.06.86 291443, 3 pp./ Wilhelm-Pieck-Universität Rostock (1987); *Chem. Abstr.*, **109**, 73339 (1988).
219. R. Dworczak, H. Sterk and H. Junek, *Monatsh. Chem.*, **121**, 189 (1990).
220. V. N. Drozd and E. N. Komarova, *Zh. Org. Khim.*, **24**, 1773 (1988).
221. L. A. Sleta and A. K. Sheinkman, *Ukr. Khim. Zh.*, **54**, 190 (1988); *Chem. Abstr.*, **110**, 22923 (1989).
222. K. A. M. El-Bayouki and W. M. Basyouni, *Bull. Chem. Soc. Jpn.*, **61**, 3794 (1988).
223. B. Zaleska, *Pol. J. Chem.*, **62**, 767 (1988); *Chem. Abstr.*, **112**, 138952 (1990).
224. M. Miyashita, Jpn. Kokai Tokkyo Koho 87 26259/04.02.87, Appl. 27.07.85 85/166080, 5 pp./ Nippon Synthetic Chemical Industry Co., Ltd. (1987); *Chem. Abstr.*, **107**, 236134 (1987).
225. T. Kawabata, Jpn. Kokai Tokkyo Koho 88112550/17.05.88, Appl. 27.10.86 86/256254, 6 pp./ Nippon Synthetic Chemical Industry Co., Ltd. (1988); *Chem. Abstr.*, **110**, 7725 (1989).
226. Y. Kawai, A. Mori, H. Higuchi, Y. Takahashi and N. Sakurai, Jpn. Kokai Tokkyo Koho 87145053/29.06.87, Appl. 19.12.85 85/284408, 4 pp./Mitsubishi Gas Chemical Co., Inc. (1987); *Chem. Abstr.*, **109**, 54357 (1988).
227. N. Martin and M. Hanack, *J. Chem. Soc., Chem. Commun.*, 1522 (1988).
228. K. Kobayashi, *Chem. Lett.*, 1511 (1985).
229. T. Hioki, Jpn. Kokai Tokkyo Koho 89 09968/13.01.89, Appl. 30.06.87 87/164804, 4 pp./ Sumitomo Chemical Co., Ltd. (1989); *Chem. Abstr.*, **111**, 41364 (1989).
230. O. Lerflaten, V. D. Parker and P. Margaretha, *Monatsh. Chem.*, **115**, 697 (1984).
231. M. Mittelbach, *Monatsh. Chem.*, **118**, 617 (1987).
232. M. Mittelbach, H. Junek and C. Kratky, *Justus Liebigs Ann. Chem.*, 1107 (1983).
233. M. Mittelbach and H. Junek, *Justus Liebigs Ann. Chem.*, 533 (1986).
234. M. Mittelbach, C. Kratky and U. Wagner, *Justus Liebigs Ann. Chem.*, 889 (1987).
235. M. Mittelbach, G. Kastner and H. Junek, *Arch. Chem.*, **318**, 481 (1985).
236. M. Mittelbach, C. Kratky and H. Junek, *Monatsh. Chem.*, **115**, 1467 (1984).
237. P. Köckritz, R. Sattler and J. Liebscher, *J. Prakt. Chem.*, **327**, 567 (1985).
238. K. Peseke and S. Quinoces J., *Rev. Cubana Quim.*, **1**, 21 (1985); *Chem. Abstr.*, **105**, 190802 (1986)
239. K. Peseke and S. Quinoces J., *Rev. Cubana Quim.*, **1**, 14 (1985); *Chem. Abstr.*, **105**, 190801 (1986)
240. W. M. Abdou, M. D. Khidre and M. R. Mahran, *J. Prakt. Chem.*, **332**, 1029 (1991).
241. E. V. Zheltova, S. G. Churusova, V. A. Kozlov, A. F. Grapov and N. N. Mel'nikov, *Zh. Obshch. Khim.*, **60**, 505 (1990); *Chem. Abstr.*, **113**, 115414 (1990).
242. K. U. Sadek, M. Ali-Selim and R. M. Abdel-Motaleb, *Bull. Chem. Soc. Jpn.*, **63**, 652 (1990).

243. J. Svetlik and F. Turecek, *Tetrahedron Lett.*, **25**, 3901 (1984).
244. S. U. Pedersen, R. G. Hazell and H. Lund, *Acta Chem. Scand., Ser. B*, **B41**, 336 (1987).
245. D. L. Boger, C. E. Brotherton and G. I. Georg, *Org. Synth.*, 32 (1987).
246. P. Milart and J. Sepiol, *Z. Naturforsch. B: Anorg. Chem., Org. Chem.*, **41**, 371 (1986).
247. M. A. El-Maghraby, K. U. Sadek, M. A. Selim and M. H. Elnagdi, *Bull. Chem. Soc. Jpn.*, **61**, 1375 (1988).
248. A. G. A. Elagamey, M. A. Sofan, Z. E. Kandeel and M. H. Elnagdi, *Collect. Czech. Chem. Commun.*, **52**, 1561 (1987).
249. F. M. Abdel-Galil and M. H. Elnagdi, *Justus Liebigs Ann. Chem.*, 477 (1987).
250. H. Schaefer and K. Gewald, *J. Prakt. Chem.*, **327**, 328 (1985).
251. K. Peseke and S. Quinoces J., *Rev. Cubana Quim.*, **1**, 28 (1985); *Chem. Abstr.*, **104**, 190803 (1986).
252. B. Sain and J. S. Sandhu, *J. Org. Chem.*, **55**, 2545 (1990).
253. J. Sepiol and P. Milart, *Tetrahedron*, **41**, 5261 (1985).
254. A. G. Elagamey, *Indian J. Chem., Sect. B*, **27**, 766 (1988).
255. V. N. Nesterov, Y. T. Struchkov, G. E. Khoroshilov, Y. A. Sharanin and V. E. Shklover, *Izv. Akad. Nauk SSSR, Ser. Khim.*, 2771 (1989); *Chem. Abstr.*, **113**, 23157 (1990).
256. H.-W. Schmidt, *Monatsh. Chem.*, **120**, 891 (1989).
257. K. Gewald, M. Gruner and U. Hain, *Z. Chem.*, **27**, 32 (1987).
258. Y. Natano, H. Yokota, M. Igarashi and S. Sato, *Bull. Chem. Soc. Jpn.*, **61**, 3731 (1988).
259. T. Nishiyama and H. Kameoka, *Nippon Kagaku Kaishi*, 699 (1990); *Chem. Abstr.*, **113**, 131693 (1990).
260. K. Bogdanowicz-Szwed and M. Lipowska, *Chem. Scr.*, **28**, 319 (1988).
261. R. R. Schmidt and R. Hirsenkorn, *Tetrahedron*, **39**, 2043 (1983).
262. N. Chatani and T. Hanafusa, *Bull. Chem. Soc. Jpn.*, **63**, 2134 (1990).
263. K. V. Komarov, N. D. Chkanikov, M. V. Galakhov, A. F. Kolomiets and A. V. Fokin, *Izv. Akad. Nauk SSSR, Ser. Khim.*, 1451 (1988); *Chem. Abstr.*, **110**, 57444 (1989).
264. K. V. Komarov, N. D. Chkanikov, S. V. Sereda, M. Y. Antipin, Y. T. Struchkov, A. F. Kolomiets and A. V. Fokin, *Izv. Akad. Nauk SSSR, Ser. Khim.*, 2417 (1988); *Chem. Abstr.*, **110**, 192703 (1989).
265. H. M. Hassaneen and A. S. Shawali, *Indian J. Chem., Sect. B*, **28**, 133 (1989).
266. H. M. Hassaneen, A. S. Shawali, M. S. Algharib and A. M. Farag, *J. Prakt. Chem.*, **330**, 558 (1988).
267. A. S. Tomcufcik, W. E. Meyer and S. S. Tseng, U.S. 4562189/31.12.85, Appl. 09.10.84 659116, 8 pp./American Cyanamid Co., (1985); *Chem. Abstr.*, **105**, 207306 (1986).
268. R. Gehring, E. Klauke, O. Schallner, J. Stetter, H. J. Santel and R. R. Schmidt, Ger. Offen. 3420985/25.04.85, Appl. 06.06.84 3420985, Priorities 15.10.83 DE 3337543, 53 pp./Bayer AG, (1985); *Chem. Abstr.*, **104**, 19599 (1986).
269. F. M. Abdel-Galil, R. M. Abdel-Motaleb and M. H. Elnagdi, *An. Quim., Ser. C*, **84**, 19 (1988); *Chem. Abstr.*, **110**, 75387 (1989).
270. K. Peseke, R. Rodriguez and Y. Rodriguez, *Rev. Cubana Quim.*, **1**, 32 (1985); *Chem. Abstr.*, **105**, 226243 (1986).
271. V. J. Aran and J. L. Soto, *An. Quim., Ser. C*, **79**, 340 (1983); *Chem. Abstr.*, **102**, 24393 (1985).
272. M. Igarashi, Y. Nakano, S. Sato, K. Takezawa and T. Watanabe, *Synthesis*, 68 (1987).
273. Y. T. Abramenko, G. V. Gridunova, A. V. Ivashchenko, K. A. Nogaeva, N. D. Sergeeva, P. A. Sharbatyan, V. E. Shklover and Y. T. Struchkov, *Khim. Geterotsikl. Soedin.*, 1286 (1986); *Chem. Abstr.*, **107**, 7044 (1987).
274. L. Fuentes, J. J. Vaquero, J. C. Del Castillo, M. I. Ardid and J. L. Soto, *Heterocycles*, **23**, 93 (1985).
275. W. Dormagen, K. Rotscheidt and E. Breitmaier, *Synthesis*, 636 (1988).
276. M. Sreenivasulu and G. S. Rao, *Indian J. Chem., Sect. B*, **28B**, 584 (1989).
277. F. F. Abdel-Latif, *Rev. Roum. Chim.*, **35**, 679 (1990); *Chem. Abstr.*, **114**, 228683 (1991).
278. K. U. Sadek, N. S. Ibrahim and M. H. Elnagdi, *Arch. Pharm.*, **321**, 141 (1988).
279. A. K. El-Shafei, A. B. A. G. Ghattas and F. A. Gad, *Rev. Roum. Chim.*, **30**, 817 (1985); *Chem. Abstr.*, **105**, 152895 (1986).
280. B. Singh, *Heterocycles*, **23**, 1479 (1985).
281. S. Julia Arechaga and J. Pi Sallent, Span. 538438/01.11.85, Appl. 10.12.84 538438, 11 pp./Laboratorios Berenguer-Beneyto S. A., (1985); *Chem. Abstr.*, **106**, 4883 (1987).
282. Y. A. Sharanin, A. M. Shestopalov, V. Y. Mortikov, A. S. Demerkov, V. P. Litvinov and G. V. Klokol, *Zh. Org. Khim.*, **24**, 854 (1988); *Chem. Abstr.*, **110**, 94934 (1989).

283. K. Bogdanowicz-Szwed, M. Lipowska and B. Rys, *Justus Liebigs Ann. Chem.*, 1147 (1990).
284. G. E. H. Elgemeie, *Heterocycles*, **31**, 123 (1990).
285. Y. A. Sharanin, A. M. Shestopalov and V. K. Promonenkov, *Zh. Org. Khim.*, **20**, 2002 (1984); *Chem. Abstr.*, **102**, 113330 (1985).
286. F. F. Abdel-Latif, *Pharmazie*, **45**, 283 (1990).
287. F. F. Abdel-Latif, *J. Chin. Chem. Soc. (Taipei)*, **37**, 295 (1990); *Chem. Abstr.*, **113**, 211785 (1990).
288. A. O. Abdelhamid and N. M. Abed, *Rev. Port. Quim.*, **27**, 500 (1985); *Chem. Abstr.*, **107**, 115551 (1987).
289. S. M. Fahmy and R. M. Mohareb, *Synthesis*, 1135 (1985).
290. Y. A. Sharanin, A. M. Shestopalov, V. N. Nesterov, S. N. Melenchuk, V. K. Promonenkov, V. E. Shklover, Y. T. Struchkov and V. P. Litvinov, *Zh. Org. Khim.*, **25**, 1323 (1989); *Chem. Abstr.*, **112**, 76896 (1990).
291. G. E. H. Elgemeie and S. M. Sherif, *Z. Naturforsch., B. Anorg. Chem., Org. Chem.*, **41B**, 781 (1986).
292. F. M. Galil, M. M. Sallam, S. M. Sherif and M. H. Elnagdi, *Justus Liebigs Ann. Chem.*, 1639 (1986).
293. Y. A. Sharanin, A. M. Shestopalov and V. K. Promonenkov, *Zh. Org. Khim.*, **20**, 2012 (1984); *Chem. Abstr.*, **102**, 78794 (1985).
294. A. M. Shestopalov, Y. A. Sharanin, M. R. Khikuba, V. N. Nesterov, V. E. Shklover, Yu. T. Struchkov and V. P. Litvinov, *Khim. Geterotsikl. Soedin*, 205 (1991); *Chem. Abstr.*, **115**, 135873 (1991).
295. G. A. M. Nawwar, S. A. Osman, K. A. M. El-Bayouki, G. E. H. Elgemeie and M. H. Elnagdi, *Heterocycles*, **23**, 2983 (1985).
296. G. E. H. Elgemeie, G. A. Abdelaal and K. Abouhadeed, *J. Chem. Res. (S)*, 128 (1991).
297. A. M. Fahmy, M. Z. A. Badr, Y. S. Mohamed and F. F. Abdel-Latif, *J. Heterocycl. Chem.*, **21**, 1233 (1984).
298. D. Konwar, R. C. Borush and J. S. Sandhu, *Heterocycles*, **23**, 2557 (1985).
299. K. Peseke and G. Heide, Ger. (East) 277675/11.04.90/Appl. 05.12.88 322648, 3 pp./Wilhelm-Pieck-Universität Rostock (1990); *Chem. Abstr.*, **114**, 42589 (1991).
300. K. Peseke and G. Heide, Ger. (East) 277458/04.04.90/Appl. 02.12.88 322530, 3 pp./Wilhelm-Pieck-Universität Rostock (1990); *Chem. Abstr.*, **114**, 42588 (1991).
301. K. Peseke and G. Heide, Ger. (East) 277457/04.04.90/Appl. 02.12.88 322530, 3 pp./Wilhelm-Pieck-Universität Rostock (1990); *Chem. Abstr.*, **114**, 61949 (1991).
302. K. Peseke, G. Heide, H. Fiest and M. Michalik, *J. Prakt. Chem.*, **333**, 119 (1991).
303. K. Peseke and G. Heide, Ger. (East) 277680/11.04.90/Appl. 05.12.88, 3 pp./Wilhelm-Pieck-Universität Rostock (1990); *Chem. Abstr.*, **113**, 231356 (1990).
304. H. W. Schmidt and M. Klade, *Justus Liebigs Ann. Chem.*, 257 (1988).
305. R. Grigg and T. Mongkolaussavaratana, *J. Chem. Soc., Perkin Trans.*, 541 (1988).
306. H. N. Borah, R. C. Boruah and J. S. Sandhu, *Heterocycles*, **22**, 2323 (1984).
307. B. Podanyi, I. Hermecz and A. Horvath, *J. Org. Chem.*, **51**, 2988 (1986).
308. G. Zvilichovsky and M. David, *Synthesis*, 239 (1986).
309. J. Svetlik, F. Turecek and I. Goljer, *J. Org. Chem.*, **55**, 4744 (1990).
310. A. O. Abdelhamid, N. M. Abded and A. M. Farag, *An. Quim., Ser. C Quim. Org. Bio.*, **84**, 22 (1988); *Chem. Abstr.*, **109**, 190360 (1988).
311. M. K. A. Ibrahim, A. M. El-Reedy, M. S. El-Gharib and A. M. Farag, *J. Indian Chem. Soc.*, **64**, 345 (1987).
312. K. Peseke, R. Rodriguez and R. Ramirez, *Rev. Cubana Quim.*, **1**, 37 (1985); *Chem. Abstr.*, **105**, 191004 (1986).
313. M. J. Callejo, P. Lafuente, N. Martin-Leon, M. Quinteiro, C. Seoane and J. L. Soto, *J. Chem. Soc., Perkin Trans. 1*, 1687 (1990).
314. S. M. Hussain, A. S. Ali and A. M. El-Reedy, *Indian J. Chem., Sect. B*, **27**, 421 (1988).
315. M. Gogoi, J. N. Baruah, P. Bhuyan and J. S. Sandhu, *J. Chem. Soc., Chem. Commun.*, 154 (1984).
316. Y. Miyamoto, *Chem. Pharm. Bull.*, **33**, 2678 (1985).
317. Y. Miyamoto, *Nippon Noyaku Gakkaishi*, **11**, 39 (1986); *Chem. Abstr.*, **106**, 312883 (1987).
318. M. H. Elnagdi, N. S. Ibrahim, K. U. Sadek and M. H. Mohamed, *Justus Liebigs Ann. Chem.*, 100 (1988).
319. G. Koitz, H. Junek and B. Thierrichter, *Heterocycles*, **20**, 2405 (1983).

320. E. A. A. Hafez, M. H. Elnagdi, A. G. A. Elgamey and F. M. A. A. El-Taweel, *Heterocycles*, **26**, 903 (1987).
321. L. G. Sharanina, V. N. Nesterov, G. V. Klokol, L. A. Rodinovskaya, V. E. Shklover, Y. A. Sharanin, Y. T. Struchkov and V. K. Promonenkov, *Zh. Org. Khim.*, **22**, 1315 (1986); *Chem. Abstr.*, **106**, 156226 (1987).
322. N. Martin, J. L. Segura, C. Seoane, J. L. Soto, M. Morales and M. Suarez, *Justus Liebigs Ann. Chem.*, 827 (1991).
323. M. C. Sacquet, M. C. Bellassoued-Fargeau, B. Graffe and P. Maitte, *J. Heterocycl. Chem.*, **22**, 757 (1985).
324. F. M. Soliman and M. M. Said, *Phosphorous, Sulfur, Silicon Relat. Elem.*, **61**, 335 (1991).
325. R. Dworczak, H. Sterk, C. Kratky and H. Junek, *Ber. Dtsch. Chem. Ges.*, **122**, 1323 (1989).
326. T. Fuchigami, Z. E.-S. Kandeel, T. Nonaka and H. J. Tien, *J. Chin. Chem. Soc. (Taipei)*, **33**, 241 (1986); *Chem. Abstr.*, **107**, 175909 (1987).
327. K. Peseke, S. J. Quincoces and O. Zayas, Ger. (East) 277461/04.04.90, Appl. 02.12.88 322531, 3pp./Wilhelm-Pieck-Universität Rostock (1990); *Chem. Abstr.*, **114**, 61930 (1991).
328. A. A. Elbannany, L. I. Ibrahiem and S. A. S. Chozlan, *Pharmazie*, **43**, 128 (1988).
329. Y. A. Sharanin and G. V. Klokol, *Zh. Org. Khim.*, **20**, 2448 (1984); *Chem. Abstr.*, **102**, 95608 (1985).
330. D. Briel, S. Leistner and G. Wagner, Ger. (East) 216720/19.12.84. Appl. 01.06.83 251619, 8 pp./ Karl-Marx-Universität Leipzig (1984); *Chem. Abstr.*, **103**, 123503 (1985).
331. R. Dworczak, *Monatsh. Chem.*, **122**, 731 (1991).
332. D. Briel, S. Dumke, G. Wagner and B. Olk, *J. Chem. Res. (S)*, 178 (1991).
333. W. Reid and D. Kuhnt, *Justus Liebigs Ann. Chem.*, 551 (1987).
334. C. F. Bernasconi, K. A. Howard and A. Kanavarioti, *J. Am. Chem. Soc.*, **106**, 6827 (1984).
335. K. A. Howard, Thesis, Univ. California, Santa Cruz, CA, USA (1982); *Chem. Abstr.*, **100**, 102423 (1984).
336. C. F. Bernasconi, A. Kanavarioti and R. B. Killion Jr., *J. Am. Chem. Soc.*, **107**, 3612 (1985).
337. C. F. Bernasconi, J. P. Fox, A. Kanavarioti and M. Panda, *J. Am. Chem. Chem. Soc.*, **108**, 2372 (1986).
338. C. F. Bernasconi, *Tech. Chem. (N. Y.)*, 6 (Invest. Rates Mech. React., 4th Ed. Pt. 1), 425 (1986); *Chem. Abstr.*, **105**, 255469 (1986).
339. G. Abou-Elenien, B. Buecher, N. Ismail, J. Rieser and K. Wallenfels, *Z. Naturforsch., B: Anorg. Chem., Org. Chem.*, **38B**, 1199 (1983).
340. B. Yan, C. Dai, W. Hu and Y. He, *Huaxue Xuebao*, **44**, 1093 (1986); *Chem. Abstr.*, **106**, 127799 (1987).
341. O. E. Nasakin, V. V. Alekseev, G. N. Petrov, V. K. Promonenkov and A. V. Sukhobokov, *J. Appl. Chem. USSR*, **55**, 1286 (1982).
342. Y. S. Chong, H. H. Huang and D. A. Winkler, *J. Mol. Struct.*, **181**, 41 (1988).
343. C. Piccinni-Leopardi, J. P. Declercq, G. Germain and B. Tinant, *Bull. Soc. Chim. Belg.*, **92**, 311 (1983).
344. O. E. Nasakin, P. M. Lukin, P. B. Terent'ev, A. Kh. Bulai, M. I. Shumilin and B. A. Khasakin, *Zh. Org. Khim.*, **20**, 732 (1984); *Chem. Abstr.*, **101**, 151537 (1984).
345. A. B. Zolotoi, S. P. Zilberg, O. E. Nasakin, M. Y. Skvortsova, O. A. Dyachenko, L. O. Atovmyan, A. K. Bulai, A. N. Lyshchikov, P. M. Lukin and S. V. Konovalikhin, *Bull. Acad. Sci. USSR*, **36**, 1685 (1987); *Chem. Abstr.*, **108**, 221535 (1988).
346. A. B. Zolotoi, S. V. Konovalikhin, N. M. Sergeev, A. V. Buevich, P. M. Lukin, A. N. Lyshchikov, N. M. Skvortsova, S. P. Zilberg, P. M. Lukin and O. E. Nasakin, *Zh. Strukt. Khim.*, **31**, 108 (1990); *Chem. Abstr.*, **113**, 190521 (1990).
347. O. E. Nasakin, P. M. Lukin, P. B. Tetrent'ev, A. Kh. Bulai, B. A. Khaskin and V. Ya. Zakharov, *Zh. Org. Khim.*, **21**, 662 (1985); *Chem. Abstr.*, **103**, 104528 (1985).
348. O. E. Nasakin, S. P. Zilberg, A. B. Zolotoi, P. B. Terentev, S. V. Konovalikhin, P. M. Lukin, L. O. Atovmyan, O. A. Dyachenko and A. K. Bulai, *Zh. Org. Khim.*, **24**, 997 (1988).
349. O. E. Nasakin, A. K. Bulai, B. A. Khaskin, P. M. Lukin, M. I. Shumilin and P. B. Terentev, *Zh. Org. Khim.*, **20**, 727 (1984); *Chem. Abstr.*, **101**, 110388 (1984).
350. A. B. Zolotoi, S. P. Zilberg, P. M. Lukin, S. V. Konovalikhin, O. E. Nasakin, A. N. Lyshchikov, O. A. Dyachenko, A. Kh. Bulai and L. O. Atovmyan, *Izv. Akad. Nauk SSSR, Ser. Khim.*, 2561 (1988); *Chem. Abstr.*, **111**, 23105 (1989).
351. O. E. Nasakin, V. V. Alekseev, V. K. Promonenkov, I. A. Abramov, Kh. Bulai and S. Yu. Sil'vestrova, USSR SU 979, 338 (1982); *Chem. Abstr.*, **99**, 5511 (1983).

866 R. Dworczak and H. Junek

352. O. Y. Nasakin, V. V. Alekseyev, A. K. Bulai, P. B. Terentyev and M. Y. Zablotskaya, *Khim. Geterotsikl. Soedin.*, 1062 (1983); *Chem. Abstr.*, **100**, 68112 (1984).
353. O. E. Nasakin, V. V. Alekseev, P. B. Tetrentev, A. K. Bulai and M. Y. Zoblotskaya, *Khim. Geterotsikl. Soedin.*, 1067 (1983); *Chem. Abstr.*, **100**, 103102 (1984).
354. O. E. Nasakin, V. V. Alekseev, I. A. Abramov, P. B. Terentev and A. K. Bulai, USSR SU 1,004,376 15 Mar 1983 (1983); *Chem. Abstr.*, **99**, 105110 (1983).
355. O. E. Nasakin, V. V. Alekseev, P. B. Terentev, A. K. Bulai and V. A. Shmorgunov, *Khim. Geterotsikl. Soedin.*, 1605 (1982); *Chem. Abstr.*, **98**, 107096 (1983).
356. O. E. Nasakin, P. M. Lukin, P. B. Terentev, A. Kh. Bulai, V. D. Sheludyakov, A. B. Zolotoi, A. I. Gusev, O. A. Dyachenko, G. M. Apal'kova and L. O. Atovmyan, *Khim. Geterotsikl. Soedin.*, 1431 (1984); *Chem. Abstr.*, **102**, 24426 (1985).
357. S. P. Zilberg, O. E. Nasakin, A. B. Zolotoi, P. B. Terentev, V. L. Turkhanov, S. V. Konovalikhin, P. M. Lukin, O. A. Dyachenko and A. K. Bulai, *Zh. Org. Khim.*, **24**, 1014 (1988); *Chem. Abstr.*, **110**, 94262 (1989).
358. O. E. Nasakin, P. M. Lukin, P. B. Tetrentev, A. Kh. Bulai, E. J. Zazhivikhina, A. B. Zolotoi, O. A. Dyachenko and L. A. Atovmyan, *Zh. Org. Khim.*, **21**, 916 (1985); *Chem. Abstr.*, **103**, 178218 (1985).
359. A. B. Zolotoi, P. M. Lukin, S. V. Konovalikhin, S. P. Zilberg, O. E. Nasakin and O. A. Dyachenko, *Khim. Geterotsikl. Soedin.*, 1404 (1988); *Chem. Abstr.*, **111**, 115085 (1989).
360. A. B. Zolotoi, P. M. Lukin, S. V. Konovalikhin, N. Yu. Skvortsova, O. E. Nasakin and L. O. Atovmyan, *Khim. Geterotsikl. Soedin.*, 519 (1991); *Chem. Abstr.*, **115**, 158924 (1991).
361. A. B. Zolotoi, P. M. Lukin, A. I. Prokhorov, V. N. Romanov, O. E. Nasakin, O. E. Atovmyan and V. A. Bakhmisov, *Dokl. Akad. Nauk SSSR*, **311**, 122 (1990); *Chem. Abstr.*, **113**, 171917 (1990).
362. T. S. Little, J. R. Durig and W. Y. Zhao, *J. Raman Spectrosc.*, **19**, 479 (1988).
363. G. Schrumpf and H. Dunker, *Spectrochim. Acta, Part A, Mol. Spectrosc.*, **41**, 841 (1985).
364. M. Takahashi, T. Orihara, T. Sasaki, T. Yamatera, K. Yamazaki and A. Yoshida, *Heterocycles*, **24**, 2857 (1986).
365. J. Fitzgerald, W. Taylor and H. Owen, *Synthesis*, 686 (1991).
366. P. Magnus, W. Danikiewicz, T. Katoh, J. C. Huffmann and K. Folting, *J. Am. Chem. Soc.*, **112**, 2465 (1990).
367. H. M. Hassaneen, H. A. Ead, N. M. Elwan and A. S. Shawali, *Heterocycles*, **27**, 2857 (1988).
368. S. J. Johnson, *Synthesis*, 75 (1991).
369. R. Neidlein and Z. Sui, *Synthesis*, 959 (1990).
370. Z.-T. Huang and X. Shi, *Ber. Dtsch. Chem. Ges.*, **123**, 541 (1990).
371. Y. Tominaga, M. Kawabe and A. Hosomi, *J. Heterocycl. Chem.*, **24**, 1325 (1987).
372. D. Wobig, *Justus Liebigs Ann. Chem.*, 115 (1989).
373. H. U. Hummel, W. Forner and H. Procher, *Z. Anorg. Allg. Chem.*, **553**, 95 (1987).
374. J. J. Damico, P. G. Ruminski, L. A. Suba, J. J. Freeman and W. E. Dahl, *Phosphorus Sulfur Silicon Relat. Elem.*, **21**, 307 (1985).
375. A. Hotzel, R. Neidlein, R. E. Schulz and A. Schweig, *Angew. Chem., Int. Ed. Engl.*, **19**, 739 (1980).
376. R. Neidlein and R. Leidholdt, *Ber. Dtsch. Chem. Ges.*, **119**, 844 (1986).
377. R. D. Brown, P. D. Godfrey and T. Sakaizumi, *J. Mol. Spectrosc.*, **129**, 293 (1988).
378. A. J. Fatiadi, *Synthesis*, 249 (1986).
379. A. J. Fatiadi, *Synthesis*, 749 (1987).
380. A. J. Fatiadi, *Synthesis*, 959 (1987).
381. J. E. Frey, *Appl. Spectrosc. Rev.*, **23**, 247 (1987).
382. R. E. Merrifield and W. D. Phillips, *J. Am. Chem. Soc.*, **80**, 2778 (1958).
383. L. Cuervo, M. A. Munoz, P. Guardado, C. Carmona, J. Hidalgo and M. Balon, *J. Phys. Org. Chem.*, **4**, 25 (1991).
384. V. S. Kuts, A. M. Golubenkova and V. D. Pokhodenko, *Teor. Eksp. Khim.*, **23**, 692 (1987); *Chem. Abstr.*, **109**, 72803 (1988).
385. V. S. Kuts and S. A. Biskulova, *Teor. Eksp. Khim.*, **25**, 167 (1989); *Chem. Abstr.*, **111**, 173400 (1989).
386. K. G. Neoh, E. T. Kang, S. H. Khor and K. L. Tan, *J. Polym. Sci., Part A: Polym. Chem.*, **27**, 4365 (1989).
387. A. Takematsu, M. Hoshino, H. Seki and S. Arai, *Chem. Phys. Lett.*, **159**, 282 (1989).
388. S. Spange and G. Heublein, *Z. Chem.*, **28**, 218 (1988).

389. A. M. N. El-Din, Z. *Phys. Chem. (Leipzig)*, **267**, 980 (1986).
390. L. C. Emery and W. D. Edwards, *Int. J. Quantum Chem., Quantum Chem. Symp.*, 347 (1991).
391. B. Uno, Y. Ninomiya, K. Kano and T. Kubota, *Spectrochim. Acta, Part A–Mol. Spectrosc.*, **43**, 955 (1987).
392. B. M. Britt, H. B. Lueck and J. L. Mchale, *Chem. Phys. Lett.*, **190**, 528 (1992).
393. H. Nishihara, Y. Shimano and K. Aramaki, *J. Phys. Chem.*, **91**, 2918 (1987).
394. M. Radomska and R. Radomski, *J. Therm. Anal.*, **37**, 693 (1991).
395. A. Ikeda, T. Nagasaki, K. Araki and S. Shinkai, *Tetrahedron*, **48**, 1059 (1992).
396. J. Masnovi, R. J. Baker, R. L. R. Towns and Z. Chen, *J. Org. Chem.*, **56**, 176 (1991).
397. A. Renault, C. Cohen-Addad, J. Lajzerowicz, E. Canadell and O. Eisenstein, *Mol. Cryst. Liq. Cryst.*, **164**, 179 (1988).
398. C. Cohen-Addad, M. Consigny, G. Dassenza and P. Baret, *Acta Crystallogr. Sect. C: Cryst. Struct.*, **44**, 1924 (1988).
399. C. Cohen-Addad, G. Commandeur, P. Baret and A. Renault, *Acta Crystallogr. Sect. C:Cryst. Struct. Commun.*, **44**, 914 (1988).
400. A. Renault, C. Cohen-Addad, M. J. Crisp, J. P. Dutasta and J. Lajzerowicz-Bonneteau, *Acta Crystallogr. Sect. B: Struct. Sci.* **43**, 480 (1987).
401. A. Renault, C. Cohen-Addad and P. Baret, *Stud. Phys. Theor. Chem.*, **46**, 611 (1987).
402. A. Renault and C. Cohen-Addad, *Acta Crystallogr., Sect. C: Cryst. Struct. Commun.*, **C42**, 1529 (1986).
403. A. E. Mourad, E. H. Eltamany and H. Hopf, *Z. Phys. Chem. (Leipzig)*, **267**, 937 (1986).
404. A. J. Navarro, M. C. Cabeza and J. M. Alvarez, *An. R. Acad. Farm.*, **54**, 110 (1988); *Chem. Abstr.*, **111**, 6859 (1989).
405. T. Matsuura, *Gaodeng Xuexiao Huaxue Xuebao*, **8**, 623 (1987); *Chem. Abstr.*, **108**, 204790 (1988).
406. A. N. Egorochkin, M. A. Lopatin, G. A. Razuvaev, N. P. Erchak, L. M. Ignativich and E. Lukevics, *Dokl. Akad. Nauk SSSR*, **298**, 895 (1988); *Chem. Abstr.*, **110**, 95327 (1989).
407. G. Bocelli, L. Cardellini, G. De Meo, A. Ricci, C. Rizzoli and G. Tosi, *J. Crystallogr. Spectrosc. Res.*, **20**, 561 (1990).
408. T. Asahi and N. Mataga, *J. Phys. Chem.*, **95**, 1956 (1991).
409. N. Mataga, Y. Kanda, T. Asahi, H. Miyasaka, T. Okada and T. Kakitani, *Chem. Phys.*, **127**, 239 (1988).
410. W. D. Edwards, M. Du, J. S. Royal and J. L. Mchale, *J. Phys. Chem.*, **94**, 5748 (1990).
411. A. M. N. El-Din, A. F. E. Mourad, A. A. Hassan and D. Doepp, *Z. Phys. Chem.*, **269**, 832 (1988).
412. N. Nishimura, H. Ninomiya and Y. Osawa, *Chem. Express*, **5**, 945 (1990); *Chem. Abstr.*, **114**, 83875 (1991).
413. D. K. Safin and G. A. Chmutova, *Zh. Org. Khim.*, **23**, 797 (1987); *Chem. Abstr.*, **108**, 130935 (1988).
414. K. Tsujimoto, T. Fujimori and M. Ohashi, *J. Chem. Soc., Chem. Commun.*, 304 (1986).
415. S. Chowdhury and P. Kebarle, *J. Am. Chem. Soc.*, **108**, 5453 (1986).
416. A. V. Ryzhakov and L. L. Rodina, *Khim. Geterotsikl. Soedin.*, 488 (1991); *Chem. Abstr.*, **115**, 182535 (1991).
417. A. V. Ryzhakov, N. E. Polukhina and L. L. Rodina, *Zh. Org. Khim.*, **27**, 219 (1991); *Chem. Abstr.*, **115**, 92037 (1991).
418. V. N. Sheinker, A. D. Garnovsky, E. G. Merinova, O. A. Osipov, A. P. Sadimenko and V. S. Troilina, *J. Mol. Struct.*, **143**, 395 (1986).
419. A. M. N. Eldin, A. A. Hassan, S. K. Mohamed, F. Elatif and H. A. Elfaham, *Bull. Chem. Soc. Jpn.*, **65**, 553 (1992).
420. W. Kaim, B. Olbrich-Deussner and T. Roth, *Organometallics*, **10**, 410 (1991).
421. G. V. Belysheva, V. V. Semenov, A. N. Egorochkin and M. A. Lopatin, *Metalloorg. Chem.*, **3**, 65 (1990).
422. Y. Takahashi, S. Morishima and T. Miyashi, *J. Photochem. Photobiol. A.*, **65**, 157 (1992).
423. T. Goto, T. Kondo and T. Tsunekawa, Japan Kokai Tokkyo Koho JP02 06 399 [9006 399]/ 10.01. 1990, Toray Industries, Inc., (1990); *Chem. Abstr.*, **113**, 124460 (1990).
424. S. Spange, K. Maenz and D. Stadermann, *Justus Liebigs Ann. Chem.*, 1033 (1992).
425. P. Bruni, L. Cardellini, C. Conti, E. Giorgini and G. Tosi, *Gazz. Chim. Ital.*, **120**, 187 (1990).
426. F. Voegtle and K. Kadei, *Ber. Dtsch. Chem. Ges.*, **124**, 903 (1991).
427. H. Yang, A. Jin and A. Natansohn, *J. Polym. Sci., Part A: Polym. Chem.*, **30**, 1953 (1992).
428. A. M. N. El-Din, *Arch. Pharm.*, **319**, 143 (1986).

429. I. Saramet, C. Draghici and M. D. Banciu, *Rev. Roum. Chim.*, **35**, 437 (1990); *Chem. Abstr.*, **114**, 143273 (1991).
430. A. F. E. Mourad and A. R. M. Tawfik, *Bull. Pol. Acad. Sci., Chem.*, **35**, 53 (1987); *Chem. Abstr.*, **108**, 74670 (1988).
431. Y. L. Frolow, L. M. Sinegovskaya, V. V. Keiko, N. K. Gusarova, G. G. Efremova, L. P. Turchaninova and B. A. Trofimov, *Izv. Akad. Nauk SSR, Ser. Kim.*, 1992 (1986); *Chem. Abstr.*, **107**, 39085 (1987).
432. A. M. N. Eldin, *Z. Phys. Chem. (Leipzig)*, **270**, 1217 (1989).
433. A. Flamini and N. Poli, *Inorg. Chim. Acta*, **150**, 149 (1988).
434. S. V. Nesterova, V. A. Kuznetsov, A. Z. Gamzatov and V. O. Reikhsfeld, *Zh. Obshch. Khim.*, **58**, 1296 (1988); *Chem. Abstr.*, **110**, 57719 (1989).
435. D. A. Dixon and J. S. Miller, *J. Am. Chem. Soc.*, **109**, 3656 (1987).
436. M. J. Macazaga, M. S. Delgado and J. R. Masaguer, *J. Organomet. Chem.*, **310**, 249 (1986).
437. B. Olbrich-Deussner, R. Gross and K. Kaim, *J. Organomet. Chem.*, **366**, 155 (1989).
438. T. V. Baukova, E. G. Perevalova and D. N. Kravtsov, *Koord. Khim.*, **15**, 640 (1989); *Chem. Abstr.*, **111**, 233044 (1989).
439. T. V. Baukova, O. G. Ellert, L. G. Kuzmina, N. V. Dvortsova, D. A. Lemenovskii and A. Z. Rubezhov, *Mendeleev Commun.*, 22 (1991); *Chem. Abstr.*, **115**, 293480 (1991).
440. T. S. Janik, K. A. Bernard, M. R. Churchill and J. D. Atwood, *J. Organomet. Chem.*, **23**, 247 (1987).
441. N. Hallinan, P. McArdle, J. Burgess and P. Guardado, *J. Organomet. Chem.*, **333**, 77 (1987).
442. E. Etienne and L. Toupet, *Organometallics*, **9**, 2023 (1990).
443. R. Bertani, G. Carturan, A. M. Maccioni and P. Traldi, *Inorg. Chim. Acta*, **121**, 155 (1986).
444. H. Braunwarth, G. Huttner and L. Zsolnai, *J. Organomet. Chem.*, **372**, C23 (1989).
445. G. A. Gott and C. A. McAuliffe, *J. Chem. Soc., Dalton Trans.*, 1785 (1987).
446. M. Bottrill, M. Green, I. D. Williams, A. G. Orpen and D. R. Saunders, *J. Chem. Soc., Dalton Trans.*, 511 (1989).
447. V. P. Zagorodnikov, S. B. Katser, M. N. Vargaftik, M. A. Poraikoshits and I. I. Moiseev, *Koord. Khim.*, **15**, 1540 (1989; *Chem. Abstr.*, **112**, 198760 (1990).
448. H. Bock, K. Ruppert, D. Fenske and H. Goesmann, *Z. Anorg. Allg. Chem.*, **595**, 275 (1991).
449. H. Kurosawa, K. Shiba, K. Ohkita and I. Ikeda, *Organometallics*, **10**, 3941 (1991).
450. A. E. D. McQueen, A. J. Blake, T. A. Stephenson, M. Schroder and L. J. Yellowlees, *J. Chem. Soc., Chem. Commun.*, 1533 (1988).
451. M. Burdisso, A. Gamba, R. Gandolfi and R. Oberti, *Tetrahedron*, **42**, 923 (1986).
452. M. Slock, S. Hoste, G. P. Vanderkelen and L. Verdonck, *J. Mol. Struct.*, **143**, 389 (1986).
453. P. Verbiest, L. Verdonck and G. P. Vanderkelen, *Spectrochim. Acta, Part A: Mol. Spectrosc.*, **46**, 1097 (1990).
454. T. Roth and W. Kaim, *Inorg. Chem.*, **31**, 1930 (1992).
455. G. Rossetto, N. Brianese, F. Ossola, M. Porchia, S. Sostero and P. Zanella, *Inorg. Chim. Acta*, **132**, 275 (1987).
456. G. H. Lee, S.-M. Peng, S. F. Lush and R. S. Liu, *J. Chem. Soc., Chem. Commun.*, **23**, 1513 (1988).
457. H. J. Petrowitz and M. Wagner, *Fresenius's Z. Anal. Chem.*, **330**, 125 (1988).
458. S. Belal, A. A. Kheir, M. Ayad and S. E. Adl, *Microchem. J.*, **37**, 25 (1988).
459. F. A. Ibrahim, F. Belal and M. S. Rizk, *Analyst*, **111**, 1285 (1986).
460. M. S. Mahrous, *Bull. Fac. Phar. (Cairo Univ.)*, **29**, 37 (1991); *Chem. Abstr.*, **116**, 113652 (1992).
461. M. I. Walash, A. Elbrashy and M. Rizk, *Pharm. Weekbl., Sci. Ed.*, **8**, 234 (1986); *Chem. Abstr.*, **105**, 232530 (1986).
462. U. Eichenauer, R. Huisgen, A. Mitra and J. R. Moran, *Heterocycles*, **25**, 129 (1987).
463. J. Y. Lee and J. H. K. Hall, *J. Org. Chem.*, **55**, 4963 (1990).
464. V. M. Anisimov, A. B. Zolotoi, M. Y. Antipin, P. M. Lukin, O. E. Nasakin and Y. T. Struchkov, *Mendeleev Commun.*, 24 (1992).
465. K. B. Wilberg and S. T. Waddel, *Tetrahedron Lett.*, **28**, 151 (1987).
466. L. A. Eskova, E. V. Petrova, V. K. Turchaninov, E. S. Domnina and A. V. Afonin, *Khim. Geterotsikl. Soedin.*, 924 (1989); *Chem. Abstr.*, **112**, 76093 (1990).
467. M. Medio-Simon and J. Sepulveda-Arques, *Tetrahedron*, **42**, 24 (1986).
468. S. Sailaja, A. K. Murthy and E. Rajanarendar, *Indian J. Chem. Sect. B*, **25**, 191 (1986).
469. A. B. Padias, T. P. Tien and J. H. K. Hall, *J. Org. Chem.*, **56**, 5540 (1991).
470. P. G. Wiering and H. Steinberg, *Recl. Trav. Chim. Pays-Bas*, **105**, 394 (1986).

471. S. Nishida, N. Asanuma, M. Murakami, T. Tsuji and T. Imai, *J. Org. Chem.*, **57**, 4658 (1992).
472. V. K. Brel, E. V. Abramkin and I. V. Martynov, *Izv. Akad. Nauk SSSR, Ser. Khim.*, 2843 (1989); *Chem. Abstr.*, **113**, 40821 (1990).
473. A. Pauli, H. Kolshorn and H. Meier, *Ber. Dtsch. Chem. Ges.*, **120**, 1611 (1987).
474. Y. Sukihara, H. Yamamoto, K. Mizoue and I. Murata, *Angew. Chem.*, **99**, 1283 (1987).
475. H. Hofmann and H. Djafari, *Z. Naturforsch., B: Chem. Sci.*, **44**, 220 (1989).
476. M. Regitz and P. Eisenbarth, *Ber. Dtsch. Chem. Ges.*, **117**, 1991 (1984).
477. R. Sustmann, H. G. Korth, U. Nuchter, I. Siangourifeulner and W. Sicking, *Ber. Dtsch. Chem. Ges.*, **124**, 2811 (1991).
478. V. D. Kiselev, V. B. Malkov and A. I. Konovalov, *Z. Org. Khim.*, **26**, 229 (1990); *Chem. Abstr.*, **113**, 96822 (1990).
479. U. Pindur and M. Eitel, *J. Heterocycl. Chem.*, **28**, 951 (1991).
480. M. H. Cheng, G. M. Yang, J. F. Chow, G. H. Lee, S. M. Peng and R. S. Liu, *J. Chem. Soc., Chem. Commun.*, 934 (1992).
481. G. Burnier and P. Vogel, *Chimia*, **41**, 429 (1987).
482. E. Kim, M. Christl and J. K. Kochi, *Ber. Dtsch. Chem. Ges.*, **123**, 1209 (1990).
483. R. Lang, C. Herzog, R. Stangl, E. Brunn, M. Braun, M. Christl, E. M. Peters, K. Peters and H. G. von Schnering, *Ber. Dtsch. Chem. Ges.*, **123**, 1193 (1990).
484. S. Kirchmeyer and A. D. Meijere, *Ber. Dtsch. Chem. Ges.*, **120**, 2083 (1987).
485. E. E. Swayze, J. M. Hinkley and L. B. Townsend, *Nucleic Acid Chem.*, **4**, 16 (1991).
486. K. Kamata and T. Miyashi, *J. Chem. Soc., Chem. Commun.*, 557 (1989).
487. S. A. Marinuzzi-Brosemer, D. C. Dittmer, M. H. M. Chen and J. Clardy, *J. Org. Chem.*, **50**, 799 (1985).
488. J. R. Moran, R. Huisgen and I. Kalwinsch, *Tetrahedron Lett.*, **26**, 1849 (1985).
489. B. Y. Riad, A. O. Abdelhamid and F. M. A. Galil, *Arab. Gulf J. Sci. Res.*, **8**, 39 (1990); *Chem. Abstr.*, **114**, 185442 (1991).
490. O. E. Nasakin, A. K. Bulai, K. Dager, B. A. Khaskin, Y. G. Nikolayev and P. B. Terentyev, *Khim. Geterotsikl. Soedin.*, 1574 (1984).
491. H.-W. Schmidt, G. Gfrerer and H. Junek, *Z. Naturforsch.*, **37b**, 1327 (1982).
492. A. A. Hassan, Y. R. Ibrahim, N. K. Mohamed and A. F. E. Mourad, *J. Prakt. Chem.*, **332**, 1049 (1990).
493. M. Kohla, A. Klemer, R. Mattes and M. Leimkühler, *Justus Liebigs Ann. Chem.*, 787 (1986).
494. J. Barluenga, F. J. Gonzales, S. Fustero, X. Solans and M. V. Domenech, *J. Chem. Soc., Chem. Commun.*, 1057 (1990).
495. M. Bonamico, V. Fares, A. Flamini and P. Imperatori, *J. Chem. Soc., Perkin Trans. 2*, 121 (1990).
496. H. Junek, G. Uray and G. Zuschnig, *Justus Liebigs Ann. Chem.*, 154 (1983).
497. H. Junek, G. Gfrerer, H. Sterk, R. Thierrichter and G. Zuschnig, *Monatsh. Chem.*, **113**, 1045 (1982).
498. A. Kreutzberger and S. Daus, *Arch. Pharm.*, **320**, 37, 93 (1987).
499. A. Kreutzberger and S. Daus, *Arch. Pharm.*, **319**, 1143 (1986).
500. A. A. Hassan, *Indian J. Chem., Sect. B*, **31B**, 205 (1992).
501. O. E. Nasakin, M. Y. Skvortsova, P. M. Lukin, A. K. Bulai and A. N. Lyshchikov, U.S.S.R. 18.07.85 3930132, Chuvas State University, 1273356, (1985); *Chem. Abstr.*, **106**, 138428 (1987).
502. I. Kulszewicz-Bajer and D. Billaud, *Synth. Met.*, **22**, 239 (1988).
503. I. Kulszewicz-Bajer and D. Billaud, *J. Chem. Soc., Chem. Commun.*, 1720 (1986).
504. T. Takai, M. Ieda, M. Inagaki, M. Inoue and T. Mizutani, *J. Phys. D: Appl. Phys.*, **19**, 1091 (1986).
505. S. V. Bhujle and T. S. Vardrajan, *Polym. Sci. (Symp. Proc. Polym. '91)*, 815 (1991).
506. M. Grigoras, G. David and I. Negulescu, *J. Macromol. Sci., Chem.*, **A28**, 601 (1991).
507. H. Kubota, H. Kozai and Y. Ogiwara, *Eur. Polym. J.*, **26**, 21 (1990).
508. V. Voisin and B. Gastambide, *Tetrahedron Lett.*, **26**, 1503 (1985).
509. T. Mitsuhashi, *J. Chem. Soc., Perkin Trans. 2*, 1495 (1986).
510. W. Schroth, R. Spitzner, S. Freitag, M. Richter and B. Dobner, *Synthesis*, 916 (1986).
511. Y. Tominaga, A. Hosomi, S. Kohra and Y. Matsuoka, *Heterocycles*, **26**, 613 (1987).
512. Y. Tominaga, Y. Matsuoka and A. Hosomi, *Heterocycles*, **27**, 2791 (1988).
513. A. de la Hoz, J. L. G. D. Paz, E. Diez-Barra, J. Elguero and C. Pardo, *Heterocycles*, **24**, 3473 (1986).
514. A. de la Hoz, C. Pardo, J. Elguero and M. L. Jimeno, *Monatsh. Chem.*, **123**, 99 (1992).

870   R. Dworczak and H. Junek

515. P. Victory and M. Garriga, *Heterocycles*, **23**, 1947 (1985).
516. P. Victory and M. Garriga, *Heterocycles*, **23**, 2853 (1985).
517. P. Victory and M. Garriga, *Heterocycles*, **24**, 3053 (1986).
518. P. Victory, A. Crespo, M. Garriga and R. Nomen, *J. Heterocycl. Chem.*, **25**, 245 (1988).
519. H.-W. Schmidt, G. Koitz and H. Junek, *J. Heterocyclic Chem.*, **24**, 1305 (1987).
520. G. H. Reidlinger and H. Junek, *Synthesis*, 835 (1991).
521. J. P. Ferris and J. C. Guillemini, *J. Org. Chem.*, **55**, 5601 (1990).
522. Y. Tobe, A. Takemura, M. Jimbo, T. Takahashi, K. Kobiro and K. Kakiuchi, *J. Am. Chem. Soc.*, **114**, 3479 (1992).
523. Y. L. Zborovskii, I. V. Smirnovzamkov and V. I. Staninets, *Zh. Org. Khim.*, **18**, 675 (1982).
524. J. D. Anunziata, N. S. Galaverna, J. J. Silber and J. O. Singh, *Can. J. Chem.*, **64**, 1491 (1986).
525. L. J. Schaffer and H. Taube, *J. Phys. Chem.*, **90**, 3669 (1986).
526. A. Miniewicz, M. Samoc and D. F. Williams, *Acta Phys. Polonica A*, **74**, 91 (1988); *Chem. Abstr.*, **109**, 202901 (1988).
527. M. Yasuda, Y. Matsuzaki, C. Pac and K. Shima, *J. Chem. Soc., Perkin Trans.*, 745 (1988).
528. A. Alberti and G. F. Pedulli, *Gazz. Chim. Ital.*, **117**, 381 (1987).
529. H. J. Keller, D. Schweitzer, R. Niebl, G. Renner and D. Vonderuhr, *Z. Naturforsch., Sect. B*, **43**, 265 (1988).
530. J. Dresner, W. Ode and J. Prochorow, *J. Phys. Chem.*, **93**, 671 (1989).
531. S. Yamada, Y. Nakagawa, M. Ohashi, S. Suzuki and O. Watabiki, *Chem. Lett.*, 361 (1986).
532. L. Pasimeni and C. Corvaja, *Solid State Commun.*, **53**, 213 (1985).
533. T. Luty and R. W. Munn, *J. Chem. Phys.*, **80**, 3321 (1984).
534. J. C. A. Boeyens and D. C. Levendis, *J. Chem. Phys.*, **80**, 2681 (1984).
535. M. Samoc and D. F. Williams, *J. Chem. Phys.*, **78**, 1924 (1983).
536. F. C. Bos and J. Schmidt, *J. Chem. Phys.*, **84**, 584 (1986).
537. B. B. Craig, A. A. Gorman, I. Hamblett and C. W. Kerr, *Rad. Phys.-Chem. Int.*, **23**, 111 (1984).
538. C. Corvaja and L. Pasimeni, *Chem. Phys. Lett.*, **88**, 347 (1982).
539. L. Pasimeni, D. A. Clemente, C. Corvaja, G. Guella and M. Vicentini, *Mol. Cryst. Liq. Cryst.*, **91**, 25 (1983).
540. L. Pasimeni, D. A. Clemente and C. Corvaja, *Mol. Cryst. Liq. Cryst.*, **104**, 231 (1984).
541. W. Muhle, J. Krzystek, J. J. Stezowski, R. D. Stigler, J. U. Von Schütz and H. C. Wolf, *Chem. Phys.*, **108**, 1 (1986).
542. K. Czarniecka, J. A. Janik, J. M. Janik, R. Kowal, J. Krawczyk, I. Natkaniec, K. Otnes, K. Pigon and J. Wasicki, *J. Chem. Phys.*, **85**, 7289 (1986).
543. C. Corvaja, L. Pasimeni and A. L. Maniero, *Chem. Phys.*, **100**, 265 (1985).
544. G. Agostini, C. Corvaja, G. Giacometti and L. Pasimeni, *Chem. Phys.*, **85**, 421 (1984).
545. A. Miniewicz, D. F. Williams and M. Samoc, *Mol. Cryst. Liq. Cryst.*, **111**, 199 (1984).
546. S. Okajima, E. C. Lim and B. T. Lim, *Chem. Phys. Lett.*, **122**, 82 (1985).
547. S. Yamada, Y. Kimura and M. Ohashi, *Nippon Kagaku Kaishi*, 60 (1984); *Chem. Abstr.*, **100**, 191516 (1984).
548. C. Vazquez, J. C. Calabrese, D. A. Dixon and J. S. Miller, *J. Org. Chem.*, **58**, 65 (1993).
549. R. J. Crawford, *J. Org. Chem.*, **48**, 1366 (1983).
550. S. Yamaguchi, K. Miyamoto and T. Hanafusa, *Bull. Chem. Soc. Jpn.*, **62**, 3036 (1989).
551. S. Yamaguchi and T. Hanafusa, *Chem. Lett.*, 689 (1985).
552. M. R. Bryce, A. M. Grainger, M. Hasan, G. J. Ashwell, P. A. Bales and M. B. Hursthouse, *J. Chem. Soc., Perkin Trans. 1*, 611 (1992).
553. M. R. Bryce, *Chem. Soc. Rev.*, **20**, 355 (1991).
554. K. Kobayashi and Y. Mazaki, *J. Synth. Org., Chem. Jpn.*, **46**, 638 (1988); *Chem. Abstr.*, **110**, 114316 (1988).
555. A. Mohamed, *Talanta*, **35**, 621 (1988).
556. A. Mohamed, *J. Ass. Offi. Anal.*, **72**, 885 (1989).
557. M. E. Abdelhamid and M. A. Abuirjeie, *Talanta*, **35**, 242 (1988).
558. H. F. Askal, G. A. Saleh and N. M. Omar, *Analyst*, **116**, 387 (1991).
559. M. M. Abdel-Khalek, M. E. Abdelhamid and M. S. Mahrous, *J. Ass. Offic. Anal.*, **68**, 1057 (1985); *Chem. Abstr.*, **103**, 147246 (1985).
560. F. Ibrahim, F. Belal, S. M. Hassan and F. A. Aly, *J. Pharm. Biomed. Anal.*, **9**, 101 (1991).
561. A. Mohamed, H. Y. Hassan, H. A. Mohamed and S. A. Hussein, *J. Pharm. Biomed. Anal.*, **9**, 525 (1991).

562. P. P. Brotero and O. A. El-Seoud, *An. Acad. Bras. Cienc.*, **61**, 425 (1989); *Chem. Abstr.*, **115**, 173781 (1991).
563. K. Elkacemi and M. Lamache, *Electrochim. Acta*, **31**, 845 (1986).
564. O. A. El-Seoud, P. P. Brotero, A. Martins and F. P. Ribeiro, *J. Org. Chem.*, **50**, 5099 (1985).
565. O. A. El-Seoud and P. P. Brotero, *An. Acad. Brasileira Cienc.*, **59**, 280 (1987).
566. Y. Miura, T. Ohana and T. Kunishi, *Bull. Chem. Soc. Jpn.*, **63**, 269 (1990).
567. K. Maruyama, K. Nakagawa, N. Tanaka and H. Imahori, *Bull. Chem. Soc. Jpn.*, **62**, 1626 (1989).
568. N. Martin, R. Behnisch and M. Hanack, *J. Org. Chem.*, **54**, 2563 (1989).
569. P. W. Kenny, T. H. Jozefiak and L. L. Miller, *J. Org. Chem.*, **53**, 5007 (1988).
570. U. Muralikrishna and M. Krishnamurthy, *Indian, J. Chem. Sect. A, Inorg.*, **25**, 949 (1986).
571. K. Yui, F. Ogura, T. Otsubo and Y. Aso, *J. Chem. Soc., Chem. Commun.*, 1816 (1987).
572. P. de la Cruz, N. Martin, F. Miguel, C. Seoane, A. Albert, F. H. Cano, A. Gonzalez and J. M. Pingarron, *J. Org. Chem.*, **57**, 6192 (1992).
573. Y. Yamashita, T. Suzuki, G. Saito and T. Mukai, *J. Chem. Soc., Chem. Commun.*, 1044 (1985).
574. Y. Tsubata, T. Suzuki, Y. Yamashita, T. Mukai and T. Miyashi, *Heterocycles*, **33**, 337 (1992).
575. T. Suzuki, C. Kabuto, Y. Yamashita and T. Mukai, *Chem. Lett.*, 1129 (1987).
576. C. A. Panetta, N. E. Heimer, C. L. Hussey and R. M. Metzger, *Synlett*, 301 (1991).
577. Y. Hama, Y. Aso, F. Ogura, Y. Nobuhara and T. Otsubo, *Bull. Chem. Soc. Jpn.*, **61**, 1683 (1988).
578. K. Yui, H. Ishida, F. Ogura, T. Otsubo and Y. Aso, *Chem. Lett.*, 2339 (1987).
579. E. Günther, S. Hünig, K. Peters, H. Rieder, H. G. von Schnering, J.-U. von Schütz, S. Söderholm, H.-P. Werner and H. C. Wolf, *Angew. Chem.*, **102**, 220 (1990).
580. K. Yui, Y. Aso, F. Ogura and T. Otsubo, *Chem. Lett.*, 1179 (1988).
581. J. Richard, P. Delhaes and M. Vandevyver, *New J. Chem.*, **15**, 137 (1991); *Chem. Abstr.*, **114**, 218872 (1991).

CHAPTER **15**

Electrophilic additions to functional groups containing a triple bond

WIENDELT DRENTH

Emeritus Professor, Department of Organic Chemistry, Utrecht University, Utrecht, The Netherlands

Supplement C2: The chemistry of triple-bonded functional groups
Edited by S. Patai   © 1994 John Wiley & Sons Ltd

I. INTRODUCTION AND SCOPE

The chemistry of groups containing a triple bond has been treated in earlier volumes of this series. A volume entitled *The Chemistry of the Cyano Group* was published in 1970[1] and another volume, *The Chemistry of Diazonium and Diazo Groups*, in 1978[2]. *The Chemistry of the Carbon–Carbon Triple Bond* appeared in 1978. Chapter 8 of this volume by G. H. Schmid[3] specifically dealt with electrophilic additions. A supplementary volume was published in 1983[4]. It encompasses chemistry, not only of compounds containing carbon–carbon triple bonds (including arynes), but also of compounds having a functional group with a different type of triple bond, namely a cyano, an isocyano or diazonium function. This supplementary volume, however, did not have a chapter specifically devoted to electrophilic additions. The present chapter reviews the literature of alkynes in the period from 1980 up to 1992. The references on cyanides and isocyanides are from about 1986 and upwards. The coverage is not exhaustive. Arynes are not included in this chapter. Interest in aryne chemistry is mainly in the areas of nucleophilic addition and cycloaddition. Arynes are, for instance, intermediates in natural product synthesis. Electrophilic additions to arynes are quite well possible[5] but in recent years this subject has not been actively investigated. Diazonium compounds, $R-\overset{+}{N}\equiv N\cdot X^-$, are also not included because they, normally, are not subject to electrophilic attack. Diazo compounds, $R^1R^2C=\overset{+}{N}=\overset{-}{N} \leftrightarrow R^1R^2C^-\overset{+}{N}\equiv N$, are also omitted, since electrophilic attack on them occurs exclusively, or almost exclusively, at the carbon atom, not at the triple bond.

Transformation of a carbon–carbon triple bond to a carbon-carbon double bond (equation 1) can proceed through a variety of mechanisms. If the reaction takes place in one step (equation 2), the reaction is called pericyclic and, because in the product the atoms or groups X and Y are always on the same side of the double bond, the addition is a suprafacial one. In principle, it is also possible that X and Y are donated by different molecules (equation 3). Transition state **a** leads to a suprafacial addition product and transition state **b** to an antarafacial one. These termolecular processes would require the simultaneous encounter of three reacting species, which is not likely to happen. In most cases the addition is a stepwise process (equation 4). The intermediate $X-\overset{|}{C}=C-$ will be either a cation, a radical or an anion.

$$X-Y + -C\equiv C- \longrightarrow X-\overset{|}{C}=\overset{|}{C}-Y \tag{1}$$

$$X-Y + -C\equiv C- \left[\begin{array}{c} X^{\cdots}Y \\ C \stackrel{\cdots}{=} C \end{array}\right]^{\ddagger} \longrightarrow \overset{X}{\diagdown}C\equiv C\overset{Y}{\diagup} \tag{2}$$

$$\left[\begin{array}{c} Y^{\cdots}X \quad Y^{\cdots}X \\ C\stackrel{\cdots}{=}C \end{array}\right]^{\ddagger} \qquad \left[\begin{array}{c} Y^{\cdots}X \\ C\stackrel{\cdots}{=}C \\ Y^{\cdots}X \end{array}\right]^{\ddagger} \tag{3}$$

(**a**)   (**b**)

$$X-Y + -C\equiv C- \rightleftharpoons X-\overset{|}{C}=C- \; + Y$$

$$Y + X-\overset{|}{C}=C- \rightleftharpoons X-\overset{|}{C}=\overset{|}{C}-Y \tag{4}$$

$$
\underset{\text{cation}}{X-\overset{|}{C}=\overset{+}{\overset{|}{C}}-} \quad \text{or} \quad \underset{}{\overset{X}{\underset{/}{\overset{\diagup}{C}}} \overset{+}{=} \overset{\diagdown}{C}} \qquad \underset{\text{radical}}{X-\overset{|}{C}=\overset{\bullet}{\overset{|}{C}}-} \qquad \underset{\text{anion}}{X-\overset{|}{C}=\overset{-}{\overset{|}{C}}-}
$$

It should be noted that the ions and radicals will often be part of an ion or radical pair, for example $R^+ \,||\, X^-$ and $R^\bullet X^\bullet$.

Which addition reactions are called electrophilic? In this chapter a definition of electrophilic addition is applied similar to the one used by G. H. Schmid[3]: the transformation of an alkyne to an alkene is said to follow an electrophilic route if, in the transition state of the rate-determining step, the (formerly) acetylenic carbon atoms acquire a partial positive charge or a decrease in negative charge. Many publications on alkyne to alkene transformations, however, do not include mechanistic considerations. It is often not known whether the reaction is concerted or stepwise and, in the latter case, which step is rate-determining. Choices had to be made, admittedly, somewhat arbitrarily. These choices are based on secondary criteria. For instance, sometimes, substituent effects are informative. If a more electron-donating group R in RC≡CH increases the rate, the addition most likely is an electrophilic one. Also, if the attacking reagent is a clearly electrophilic one, the addition probably follows an electrophilic pathway. Similar criteria are applied to the additions to heterocyclic triple bonds.

II. ELECTROPHILIC ADDITIONS TO CARBON–CARBON TRIPLE BONDS

A. Alkynes versus Alkenes

Because of the availability of their π-electrons, alkenes and alkynes are sensitive to electrophilic addition. It is an intriguing question why, with certain reagents and under certain conditions, alkenes are more reactive than alkynes, whereas in other cases the reverse is true. This question has been addressed by Modena and coworkers[6]. They consider protons and carbocations as reagents that react with alkenes at similar or lower rates than with alkynes. Addition of these reagents gives rise to open carbocations in a rate-determining step.

$$
R^1C{\equiv}CR^2 + H_3O^+ \xrightarrow{\ \text{slow}\ } R^1CH{=}\overset{+}{C}R^2 + H_2O
$$

$$
R^1C{\equiv}CR^2 + \overset{\diagdown}{\underset{\diagup}{C}}{}^+ \xrightarrow{\ \text{slow}\ } \overset{\diagdown}{\underset{\diagup}{C}}{-}C(R^1){=}\overset{+}{C}R^2
$$

Generally, halogens and sulphenyl halides add much faster to alkenes than to alkynes. With these reagents the cationic intermediate may have a geometry varying from fully bridged to open, depending on e.g. the halogen and the substituents and on whether the substrate is an alkene or alkyne. Alkenes are more apt to give rise to a cyclic intermediate than are alkynes.

B. Activation of Alkynes in Electrophilic Addition

More electron-releasing substituents at the triple bond facilitate electrophilic attack. A quantitative example, dealing with acid-catalysed hydration, is furnished by Kresge and coworkers[7]; see Section II.C. The rates increase with more negative σ^+ constants of the substituents.

Conversion of a terminal alkyne to a metal acetylide substantially increases the negative charge on the triple-bond carbons and, accordingly, facilitates electrophilic attack.

$$M-C\equiv C-R + E^+ \longrightarrow [M=C=CER]^+$$

Subsequent reaction with a nucleophile affords a metal–vinylidene complex. This subject has been reviewed by Bruce[8]. Reactions with electrophilic alkenes initially lead to a cyclobutenyl complex in a two-step process via a paramagnetic intermediate. Subsequently, the ring opens in a concerted fashion to a butadiene derivative.

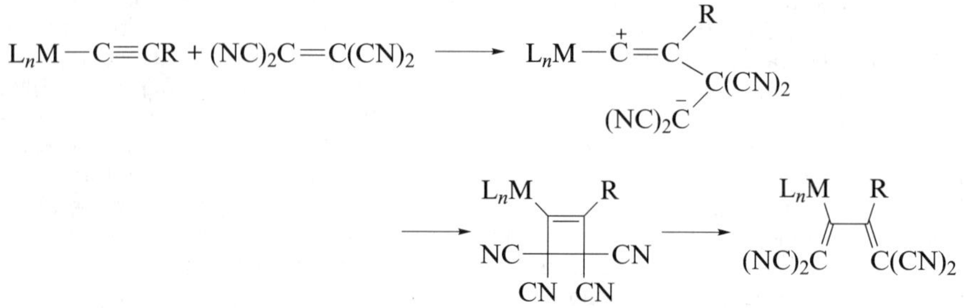

p-Methoxybenzenetellurinic acid anhydride appeared to be a catalyst in the hydration of terminal acetylenes[9]. A tellurium acetylide would be involved.

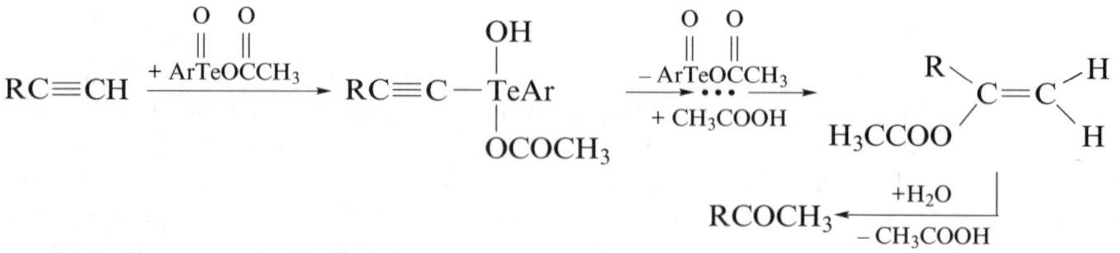

Another example of activation of the triple bond to electrophilic attack is the addition of electrophiles to the triple bond in a propargyl–transition-metal compound[10]. The addition takes the following course:

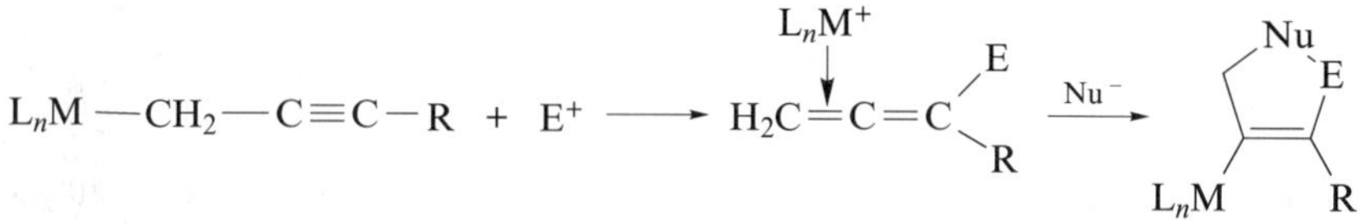

E stands for a host of electrophiles, e.g. SO_3, $(CF_3)_2C=O$ and $(NC)_2C=C(CN)_2$. A photochemical example of activation is mentioned in Section II.C under hydration.

C. Protons

1. Hydration

Acid-catalysed hydration of an unsubstituted acetylene is a slow process. In the industrial practice, metal ions are used as catalysts. If substituents are present that are able to stabilize a neighbouring positive charge, the rate of hydration is much higher.

Lucchini and Modena[11] applied an NMR technique to measure the rate of hydration of $RC\equiv CH$ (R=H, Me and t-Bu) in aqueous sulphuric acid in the absence of metal ions. For instance, at 25 °C in 15.07 mol dm^{-3} sulphuric acid the rate constant of $HC\equiv CH$ amounts to 5.48×10^{-6} s^{-1}, equivalent to a half life-time of 35 h. Similar rates were observed for $MeC\equiv CH$ and t-$BuC\equiv CH$ in 6–7 mol dm^{-3} sulphuric acid. The rate of hydration of $HC\equiv CH$ is lower in deuterated sulphuric acid: $k_H/k_D \approx 2.3$. The authors conclude that a vinyl cation,

$$H_2C = \overset{+}{C}H \quad \text{or} \quad \overset{H}{\underset{HC \overset{+}{=} CH}{}}$$

is formed in the rate-determining step. This species had earlier been detected in the gas phase. According to *ab initio* molecular orbital calculations the cyclic structure would be more stable than the open one by 13 kJ mol^{-1}, anyway if solvent interaction is absent[12].

A rather special substituent that increases the rate of hydration is C_6H_5Hg. Treating a terminal alkyne with phenylmercuric hydroxide affords the acetylide $RC\equiv CHgPh$[13]. By reaction with water the ketone $RCOCH_3$ is obtained in approximately 50% yield if R does not contain an aryl group or carbon–carbon double bond conjugated with the triple bond. For a tellurium-catalysed hydration, see the preceding Section II.B.

Quantitative hydration data have been published by Kresge and coworkers[7] (Table 1). Proton transfer from catalyst to triple bond occurs in the rate-determining step. The observed regioselectivities show that OCH_3 and SCH_3 are better than phenyl at stabilizing the adjacent developing carbocationic centre, consistent with the σ^+-constants of these substitutents, being -0.78, -0.60 and -0.18, respectively. The π-electrons of an alkyne and

TABLE 1. Rate constants of hydration of $R^1C\equiv CR^2$ in dilute aqueous solution at 25 °C

R^1	R^2	k (dm^3 mol^{-1} s^{-1})
C_6H_5	OCH_3	8.81×10^{-2}
C_6H_5	SCH_3	1.21×10^{-3}
C_6H_5	H	2.86×10^{-7}
$3,4,6\text{-}(CH_3)_3C_6H_2$	H	1.21×10^{-4}
$4\text{-}CH_3OC_6H_4$	CH_3	8.45×10^{-6}

an alkene become much more sensitive to proton-catalysed hydration by photo-excitation. Earlier work in this field on aryl-substituted acetylenes has been extended, experimentally as well as theoretically, by Yates and coworkers[14]. The calculations revealed that whereas protonation of the ground state is endothermic, protonation of the S_1 and T_1 states is exothermic, providing a rationale for the greatly enhanced reactivity.

2. Hydrocarboxylation

Addition of carboxylic acids to alkynes generally requires the presence of strong protic acids or other electrophilic catalysts. It was, however, observed that carboxylic acids having pK_a < 4.5 do add to phenylacetylene in the absence of a strong acid catalyst to give enol esters[15]. In case of formic acid, addition occurred by simply heating the alkyne in formic acid at 100 °C. A variety of diaryl, dialkyl, aralkyl and terminal alkynes react, but dimethyl acetylenedicarboxylate was found to be unreactive[16]. The final product is a ketone, while a stoichiometric amount of carbon monoxide is released.

$$R^1C \equiv CR^2 + HCOOH \longrightarrow R^1COCH_2R^2 + CO$$

Kinetics of addition of acetic acid to phenylacetylene, catalysed by strong protic acids, were determined by Monthéard and coworkers[17]. The authors were able to detect the intermediate α-acetoxystyrene. Proton transfer occurs in the rate-determining step. When perchloric instead of sulphuric acid is the catalysing species, this acid slightly decomposes during the reaction and several chlorinated species were detected[18]. The explanation is that the initial cation $Ph\overset{+}{C}=CH_2$ reacts not only with acetic acid, but also with perchlorate ion to $PhC(OClO_3)=CH_2$. This addition product would decompose in a radical fashion to, among other species, chlorine radicals, that give rise to the observed chlorinated products.

$$C_6H_5C \equiv CH + HA \longrightarrow C_6H_5\overset{+}{C}=CH_2 + A^- \qquad \text{slow}$$

$$C_6H_5\overset{+}{C}=CH_2 + CH_3COOH \longrightarrow \underset{\underset{C_6H_5C=CH_2}{|}}{\overset{\overset{OH}{|}}{\underset{|}{\overset{+O=CCH_3}{|}}}} \qquad \text{rapid}$$

$$\downarrow -H^+$$

$$\underset{C_6H_5C-CH_3}{\overset{O}{\|}} + (CH_3CO)_2O \overset{CH_3COOH}{\underset{\longrightarrow}{\longleftarrow}} \underset{C_6H_5C=CH_2}{\overset{\overset{O}{\|}}{\underset{|}{O-CCH_3}}}$$

3. Hydrohalogenation

Hydrogen chloride does not add easily to alkenes at a preparatively useful rate and reacts even slower with alkynes. It appeared that the presence of silicagel or alumina appreciably facilitates hydrohalogenation of alkenes and alkynes when the reaction is performed in methylene chloride[19]. Initially, a product of suprafacial addition is formed, which equilibrates to the thermodynamic E/Z-equilibrium. For instance,

$$C_6H_5C \equiv CCH_3 + HCl \xrightarrow{SiO_2} \underset{E}{\overset{Cl}{\underset{C_6H_5}{>}}C=C\overset{H}{\underset{CH_3}{<}}} \text{ and } \underset{Z}{\overset{Cl}{\underset{C_6H_5}{>}}C=C\overset{CH_3}{\underset{H}{<}}}$$

with isomers E and Z in a 39 : 11 ratio.

Hydrogen bromide adds to 1-phenylpropyne dissolved in methylene chloride mainly in a radical fashion to (Z)-$C_6H_5CH=CBrCH_3$. In the presence of silica or alumina, however, the addition takes an ionic route to (Z)-$C_6H_5CBr=CHCH_3$. Instead of HX, an HX precursor such as $SOCl_2$ or PBr_3 (and moisture or hydroxyl groups of the solid) give comparable results[19].

That surfaces of e.g. alumina play a role in the hydroiodination of alkenes and alkynes had been observed earlier[20]. In refluxing low boiling petroleum ether a mixture of 1-hexyne, iodine and dehydrated alumina afforded in 2 h a 62% yield of the Markovnikov addition product 2-iodo-1-hexene. The authors assume that iodine reacts with surface hydroxyl groups to form HI. The presence of alumina is required for the reaction to occur.

Reddy and Periasamy prepared HI *in situ* from a $BI_3 \cdot$ amine complex and acetic acid in benzene[21]. On addition of 1-decyne or 1-dodecyne, the Markovnikov addition products $RCI\!=\!CH_2$ were obtained in 84% yield.

D. Electrophilic Carbon

1. Alkylation

The rate of reaction between alkyl cations and acetylene and diacetylene has been investigated (Table 2) in order to increase understanding of soot formation in fuel-rich flames. The experiments were performed in a Fourier transform ion cyclotron resonance mass spectrometer. The rate depends on the cation precursor[22,23].

TABLE 2. Rate constants of addition of cations[a] to acetylenes, $k \times 10^{10}$ $(cm^3 s^{-1})$

Cation	Acetylene	Diacetylene
$C_3H_3^+$		14 ± 7
$C_5H_3^+$		5.6 ± 1.7
$C_3H_5^+$	$0.18 - 0.48$	$1.0 - 3.3$

[a]Various structures have been proposed for these ions; see refs. 22 and 23. The reactive isomer of $C_3H_3^+$, is probably linear, while an unreactive isomer would be cyclic.

Several methods of perfluoroalkylation of acetylenes have been investigated for the synthesis of organofluorine compounds. An electrolytically induced addition has been described in Reference 24. For instance, the reaction

$$C_6F_{13}I + HC\!\equiv\!CC(CH_3)_2OH \longrightarrow C_6F_{13}CH\!=\!C(I)C(CH_3)_2OH$$

occurred with 70% yield based on $C_6F_{13}I$ consumed.

It was observed that certain transition metal complexes are active catalysts in perfluoroalkylation of alkynes[25]. For instance, the addition

$$n\text{-}C_8F_{17}I + HC\!\equiv\!CSi(CH_3)_3 \longrightarrow C_8F_{17}CH\!=\!C(I)Si(CH_3)_3, (E)/(Z) = 3$$

occurred at 100 °C in 3 h with a yield of 92% when catalysed by $Fe(CO)_5$. Also, ruthenium and cobalt carbonyls show catalytic activity. Small amounts of amines promote the reaction. The activity decreased in the order $R_FI \gg R_FBr \gg R_FCl$, where R_F = perfluoroalkyl. Similar additions to alkenes were described.

Reaction of alkynes with (perfluoroalkyl)phenyliodonium trifluoromethanesulphonates, $C_nF_{2n+1}IPhOSO_2CF_3$, gives, apart from PhI, products of substitution, addition and reduction[26].

$$C_nF_{2n+1}IPhOSO_2CF_3 + HC\!\equiv\!CR$$

$$\longrightarrow C_nF_{2n+1}C\!\equiv\!CR + C_nF_{2n+1}CH\!=\!C(R)Nu + C_nF_{2n+1}CH\!=\!CHR$$

For $n = 2$ and $R = Ph$ in methanol $C_2F_5C\!\equiv\!CPh$ was obtained exclusively, whereas in formic acid 86% $C_2F_5CH_2C(O)Ph$ was isolated as hydrolysis product of the initially formed $C_2F_5CH\!=\!C(OCHO)Ph$. The first step of the reaction would be electrophilic addition of $C_nF_{2n+1}^+$ to the alkyne.

$$C_nF_{2n+1}I(C_6H_5)OSO_2CF_3 + HC\equiv CR$$

$$\longrightarrow PhI + C_nF_{2n+1}CH = \overset{+}{C}R \cdot \bar{O}SO_2CF_3$$

$$C_nF_{2n+1}C\equiv CR \qquad C_nF_{2n+1}CH = C(R)Nu \qquad C_nF_{2n+1}CH = CHR$$

Nu = OSO_2CF_3, OCHO, OCH_3 or OH.

Electrophilic additions to 1-alkynyltrialkylborates induce a shift of an alkyl group from boron to carbon[27].

$$R_3B^- - C\equiv CR^1 + R^2X \longrightarrow R_2BC(R)=CR^1R^2 + X^-$$

The product of addition can be oxidized to a ketone. When R^2X is (trimethylsilyl)methyl trifluoromethanesulphonate, an alkene is obtained with boron and silicon in the same molecule. Subsequent protolysis or oxidation provided the corresponding allylsilane (1) or ketone (2), respectively[28].

$$CF_3SO_3CH_2SiMe_3 + Li^+[R_3B^- - C\equiv CR^1]$$

Following earlier studies by Kappe and Lube[29] and Potts and Sorm[30], Gotthardt and coworkers[31,32] investigated the cycloaddition of pyrimidiniumolates, which can be regarded as 1,4-dipolar species, to alkynes. Example of addition to an electron-rich alkyne is shown in equation 5. The authors compared their experiments with MO calculations. These calculations indicate that the HOMO of an electron-rich alkyne interacts with the LUMO of the pyrimidiniumolate, whereas the reverse is true for an electron-poor alkyne.

Cyanoketenes, $N{\equiv}C{-}CH{=}C{=}O$, have the tendency to react with alkenes and alkynes to yield a cycloaddition product. Because the ketenic carbonyl carbon has electrophilic character, the reagent will behave as an electrophile. The t-butyl derivative, $N{\equiv}C{-}C(t\text{-Bu}){=}C{=}O$, has received the most attention. Its reactions with alkenes and alkynes have been reviewed by Moore and Gheorghiu[33]. The additions occur either in a $2\pi_a{+}2\pi_s$ concerted fashion or through a zwitterionic intermediate. Moore and coworkers also studied the cycloaddition of chlorocyanoketene to alkynes[34].

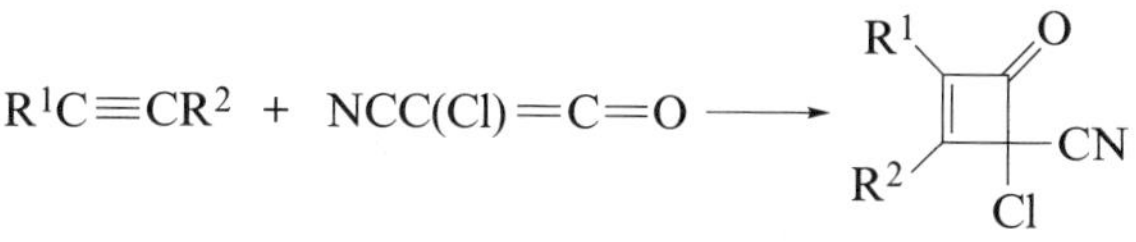

$$R^1C{\equiv}CR^2 \ + \ NCC(Cl){=}C{=}O \longrightarrow$$

For instance, with $R^1{=}Me$ and $R^2{=}Ph$, the yield is 84% when carried out in toluene at 103 °C. Pure products could only be obtained from non-terminal alkynes. 1-Ethoxypropyne behaved differently: cyclohexadienone (**3**) was obtained in 56% yield. 1-Ethoxypropyne also caused enlargement of the cyclobutenones to cyclohexadienones; see, for instance, equation 6. The reaction most likely proceeds through a cyclobutenone $\rightleftharpoons$ vinylketene equilibrium (equation 7).

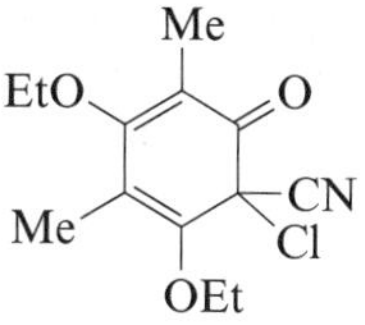

(**3**)

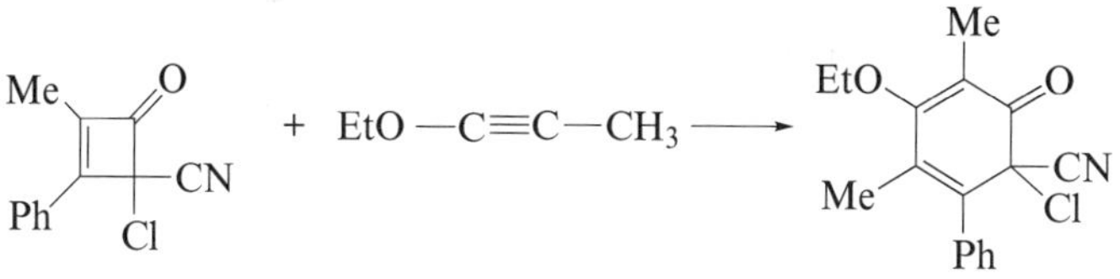

(6)

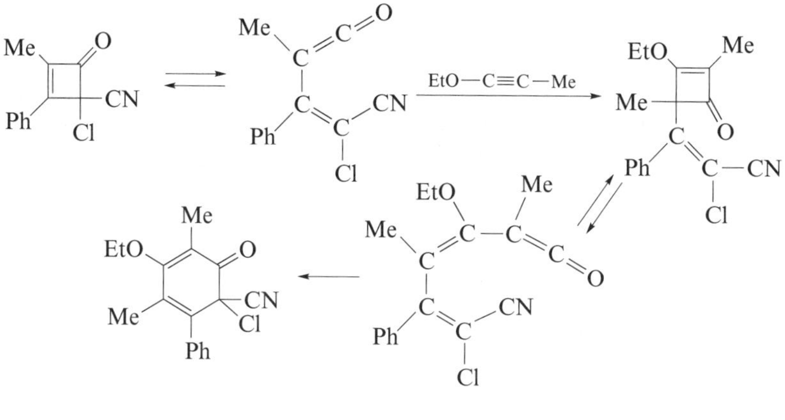

(7)

Reactions between cyclobutenones and alkynes in order to synthesize phenols have also been studied by Danheiser and coworkers[35] [equation 8]. In particular, $R = CH_3O$, CH_3S, $(CH_3)_2N$ and $(CH_3)_3SiO$ are favourable substituents[35,36]. The trimethylsilyloxy-substituted alkyne furnishes a trimethylsilylresorcinol from which the trimethylsilyl group is easily removed. A nickel(0)-catalysed version of this synthesis of phenols from cyclobutenones and alkynes was published by Huffman and Liebeskind[37].

$$(8)$$

As part of their study of the relative reactivity of alkenes and alkynes, Modena and coworkers[38] determined the reactivity ratio toward carbocations. These cations were generated by interaction of either diphenylmethyl chloride or 1-phenylethyl chloride with zinc chloride. Absolute rate constants were not determined. In general, the reactivities of alkynes and alkenes are of the same order of magnitude (see also Section II.A).

Addition of an incipient carbocation to a silylacetylene was applied to functionalize sugars[39]; see, for instance, equation 9. In compound **4**, the positive charge would be stabilized by interaction with the silicon atom.

$$(9)$$

Rodini and Snider demonstrated that dimethylaluminium chloride catalyses an ene reaction between aldehydes and alkenes. As a follow-up they showed that terminal alkynes give a similar reaction with formaldehyde[40].

$$RCH_2C{\equiv}CH + CH_2O$$

$$(5)$$

Moreover, product **5** was formed. In this reaction the vinyl cation

$$RH_2C—\overset{+}{C}=C\overset{\diagup H}{\diagdown CH_2—OAl^-Me_2Cl}$$

would be the common intermediate.

2. Acylation

In general, arylacetyl chlorides react with alkynes $RC{\equiv}CH$ in the presence of aluminium trichloride to chlorovinyl ketones,

$$ArCH_2COCl + HC{\equiv}CR \xrightarrow{\text{AlCl}_3} ArCH_2COCH{=}CRCl$$

The process starts with attack on the alkyne by the carbonyl carbon ($R^1 = ArCH_2$). In either a concerted reaction or through an intermediate carbocation (**6**) the chlorovinyl ketone is formed. Tanabe and Mukaiyama[41] preferred dichlorobis(trifluoromethanesulphonato) titanium(IV) as Friedel-Crafts catalyst over $AlCl_3$, $TiCl_4$ and $Sn(OTf)_2$. After hydrolysis 1,3-diketones were obtained.

$$\left[\begin{array}{c} \overset{\delta-}{O}\text{---}AlCl_3 \\ \| \overset{\delta+}{} \quad \overset{\delta+}{} \\ R^1{—}C\text{-----}C{\equiv}CR \\ | \quad | \\ Cl \quad H \end{array}\right]^{\ddagger} \qquad \begin{array}{c} O \\ \| \\ R^1{—}C{—}C{=}\overset{+}{C}R \\ | \\ H \end{array}$$

(6)

$$HC{\equiv}CR \xrightarrow[\text{2. hydrolysis}]{\substack{\text{1. } (R'CO)_2O,\ TiCl_2(OTf)_2 \\ \text{in } CH_2Cl_2 \text{ under argon}}} \begin{array}{c} O \quad\ O \\ \| \quad\ \| \\ R'CCH_2CR \end{array}$$

It appeared to be important to start the process by adding triflate to the anhydride dissolved in dichloromethane. After stirring for 15 min at 0 °C and standing at ambient temperature for 30 min, the alkyne was added at 0 °C. This one-pot reaction is applicable to a wide variety of terminal alkynes with substituent R containing halogen, ester or carbonyl functional groups. Best results were obtained with 1.7 equivalent of titanium complex and 3.7 equivalent of acetic acid anhydride. With acid chlorides instead of anhydrides, complex mixtures resulted.

Sometimes, the intermediate carbocation behaves differently. For instance, Haack and Beck[42] unexpectedly observed a different behaviour with 4-methoxyphenylacetyl chloride (equation 10). Apparently, the intermediate vinyl cation **8** gives the spiro product **7** by intramolecular attack and concomitant loss of methyl chloride. The reaction is quite general for substituted 4-methoxyphenylacetyl chlorides and substituted acetylenes. This facile formation of a spiro compound was applied by Whiting and coworkers[43] to a larger ring system (equation 11). These spiro compounds show aromatase inhibitor properties.

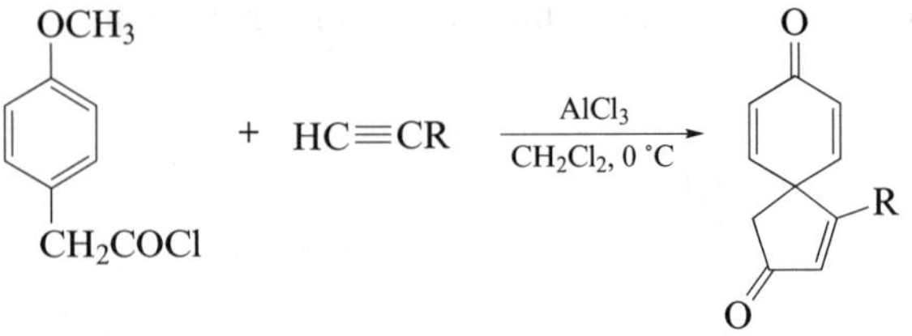

(10)

(7)

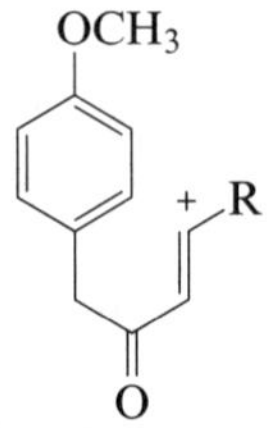

(8)

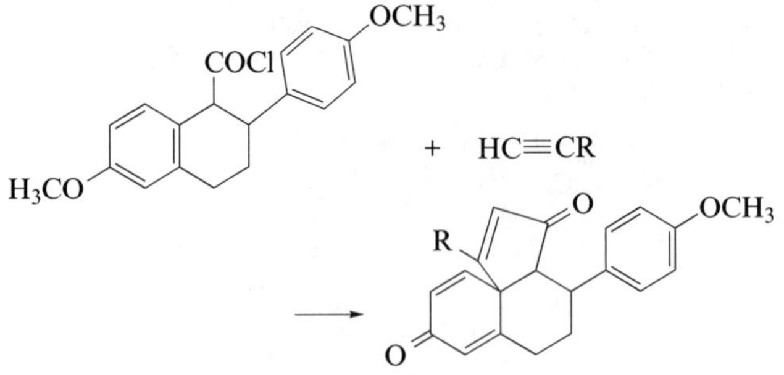

+ CH$_3$Cl   (11)

Another deviation from the expected addition of an acyl chloride to an alkyne under Friedel-Crafts conditions occurs with trimethylsilylalkynes. Instead of or besides addition, substitution takes place[44,45].

$$RCOCl + Me_3SiC{\equiv}CSiMe_3 \xrightarrow{AlCl_3} RCOC{\equiv}CSiMe_3 + Me_3SiCl$$

Apparently, the intermediate carbocation eliminates a trimethylsilyl cation. The cation derived from oxalyl chloride, however, did not eliminate a trimethylsilyl cation and the dihydrofuranone 9 was obtained (equation 12).

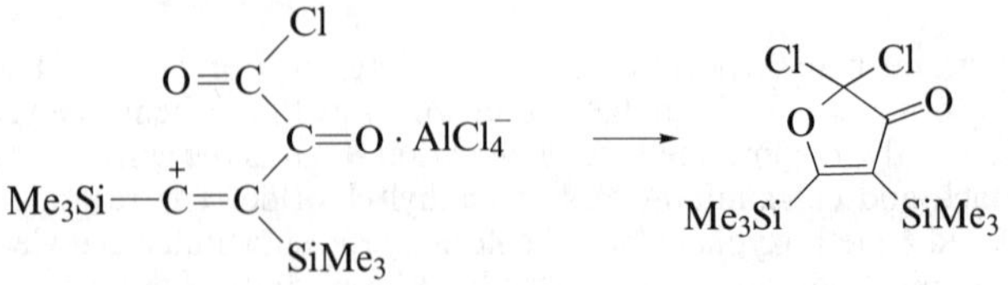

+ AlCl$_3$ (12)

(9)

With mono(trimethylsilyl)acetylene the carbocation has the structure **10** which is stabilized by interaction between C^+ and $SiMe_3$ (cf. structure **4**). If the reaction starts from a 2,2-disubstituted-3-unsaturated acid chloride, cyclopentenones are obtained[46]. (equation 13). In this case the intermediate carbocation is believed to cyclize to a cyclohexene (**15**). A 1,2-acyl shift leads to cation **16**, which is stabilized either by addition of a chloride to give **12** or by loss of a proton to give **13** (equation 14). Similarly, compound **14** is formed.

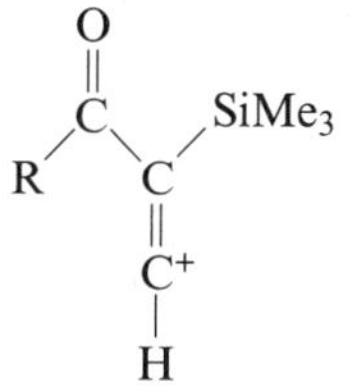

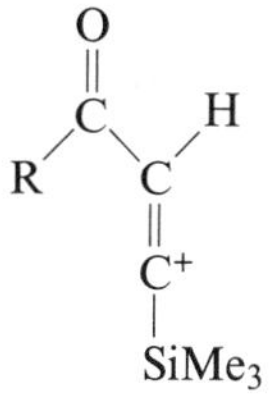

(**10**)   (**11**)

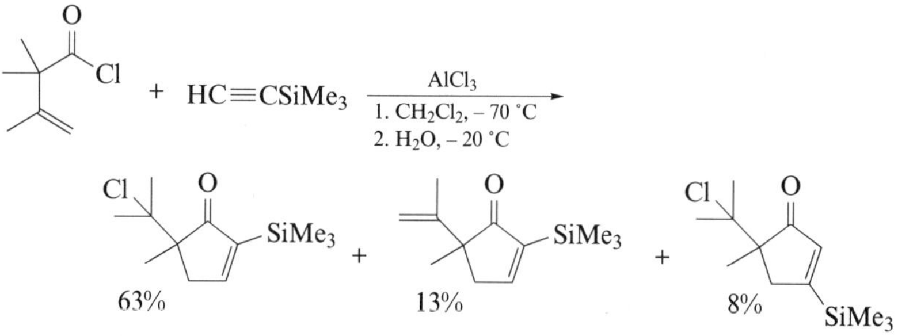

63%   13%   8%

(**12**)   (**13**)   (**14**)

(13)

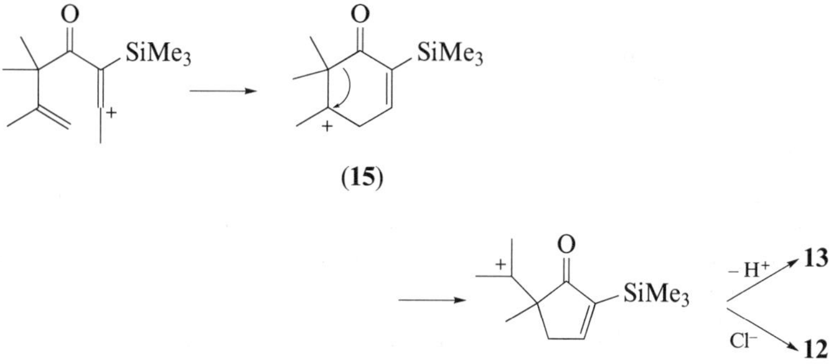

(**15**)

(**16**)

(14)

→ **14**

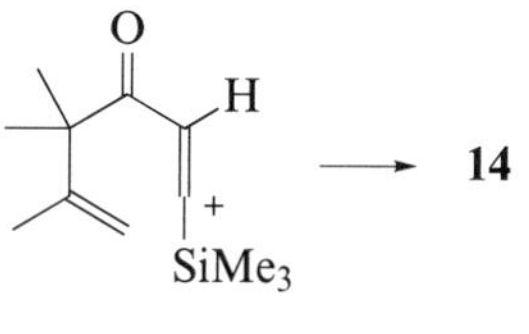

This formation of cyclopentenones from 2,2-disubstituted-3-unsaturated acid chlorides is not restricted to trimethylsilylalkynes. With propyne the cyclopentenones **17** and **18** are obtained. This reaction enables the synthesis of spiro compounds, for instance equation 15. If, however, the acid chloride is monosubstituted at the 2-position the reaction stops at the six-membered ring stage and a phenol results, for instance equation 16.

(17)   **(18)**

$$\text{(15)}$$

R	R^1	R^2	Yield (%)
Me	Me	H	65
SiMe$_3$	H	SiMe$_3$	71
SiMe$_3$	SiMe$_3$	H	17

$$\text{(16)}$$

3. Carbenes

The behaviour of carbenes depends upon whether they are in the singlet or triplet state. Additions of singlet carbenes to unsaturated systems are generally concerted and stereospecific. Additions of triplet carbenes are two-step processes; they are non-stereospecific. The subject of addition of carbenes to alkynes fits into this chapter because most carbenes and carbenoids behave as electrophiles.

Reaction between carbenes or carbenoids and alkynes often leads to cyclopropenes.

$$R^1R^2C: + \quad R^3C{\equiv}CR^4 \longrightarrow$$

The carbenes or carbenoids can be generated in a variety of ways. It is not always clear whether the reaction is concerted or stepwise and whether the carbene behaves as an electrophilic, nucleophilic or radical species. For instance, a carbenoid generated from bismuthonium ylide (19) in the presence of copper(I) chloride would behave as a triplet and add as a radical to a terminal alkyne[47] (equation 17). The dicarbonyl structure and the absence of reaction with methyl propionate to a furane might well indicate electrophilic character of this carbene.

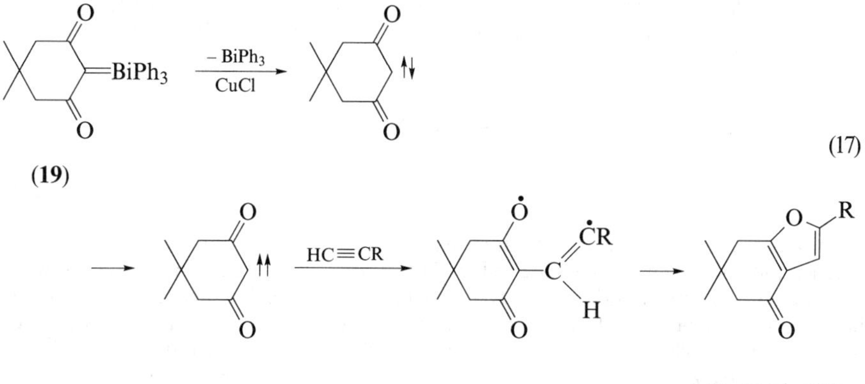

(17)

Yield 21–35%

The reactivity of dichlorocarbene towards carbon–carbon triple bonds was systematically investigated by Dehmlov in the sixties, and this work has been reviewed[48]. Dehmlov generated the carbene e.g. by decomposition of sodium trichloroacetate in 1,2-dimethoxyethane. In additions to ene-yne substrates, sometimes the triple bond reacted, whereas in other cases the double bond accepted the carbene. He observed, for instance, that reaction of the carbene with (E)-1,4-diphenylbutene-1-yne-3 gave the triple-bond addition product 20 exclusively. In reaction with 2-methylpentene-1-yne-3, however, the carbene preferred the double bond, leading to 21. Dichlorocarbene generated from potassium t-butylate and chloroform does not add to electron-poor alkynes, indicating electrophilic behaviour of the carbene.

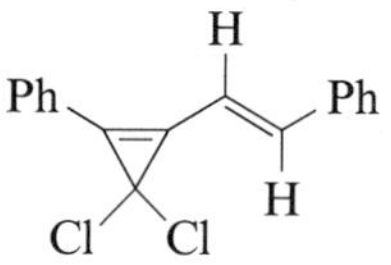

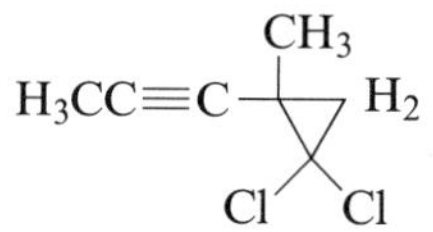

Dehmlov and coworkers[49] extended their investigations to a study of the addition of chlorofluorocarbene to highly substituted alkynes. They synthesized the carbene from dichlorofluoromethane and sodium hydroxide under phase-transfer conditions. Although yields are low, chlorofluorocarbene appeared to be more reactive than dichlorocarbene towards alkynes, especially towards alkynes with bulky substituents.

Moss and coworkers[50] determined the kinetics of this type of reaction. The carbenes were generated by laser-flash photolysis of phenylhalodiazirine (22). The rate constant for addition of PhCCl to $R^3C\equiv CR^4$ varied from 1.8 $dm^3 mol^{-1} s^{-1}$ for R^3 = H and R^4 = COOMe to 22 $dm^3 mol^{-1} s^{-1}$ for R^3 = H and R^4 = Ph. The rate constants for addition of PhCF were slightly lower. The rate constants decreased with increasing π-ionization potential of the alkyne, except for very electron-deficient alkynes such as dimethyl acetylenedicarboxylate. The correlation indicated that in these additions the carbenes generally behave as electrophiles, whereas with the acetylenedicarboxylate carbenic nucleophilicity comes into play. The rate constant of addition of phenylchlorocarbene to 3-hexyne was determined as a function of temperature. The reaction appeared to be entropy controlled: E_a = 8.8 ± 0.4 kJ mol^{-1} and $\Delta S^{\ddagger}$ = –82 J mol^{-1} K^{-1}. The corresponding alkenes have rate constants and activation parameters in the same order of magnitude.

(22)

The reactivity of carbene–metal complexes, amongst others the reactivity with respect to alkenes and alkynes, has been reviewed by Dötz[51]. Just like free carbenes the coordinated carbenes add to triple bonds to give cyclopropene derivatives[52]. Other reaction products, however, are also possible. For instance, the carbene ligand of chromium complex 23 reacts with diphenylacetylene to a mixture of products, including naphthalene derivative 24[53] and furan derivative 25 (equation 18). A carbonyl ligand has participated. Molecular orbital calculations by Hofmann and Hämmerle[54] on this system reveal that the reaction would pass through an η-vinylcarbene type of complex (26) instead of through a planar chromacyclobutene 27. The subsequent steps to yield either phenol or furan could involve vinylketene 28, but this still is a matter of debate[55]. Similar, but more selective, furan syntheses have been observed for carbene complexes based on iron and cobalt[52].

(23)   (24)   (25)

(26)   (27)   (28)

Another route from acetylenes to furans applies a diazocarbonyl as reagent and rhodium(II) acetate as catalyst[56], for instance equation 19. The reaction probably involves a dipolar intermediate 29, and is restricted to terminal acetylenes with positive charge stabilizing substituents and to diazocarbonyls with the reverse substituent effect.

Metal carbene complexes are also active in alkyne polymerization, as shown, for

$$HC \equiv C \text{—} \underset{OCH_3}{\underset{|}{\bigcirc}} \text{—} OCH_3 \quad + \quad EtO \overset{O}{\underset{N_2}{\diagdown}} \underset{}{\diagup} COOEt \tag{19}$$

$$\xrightarrow{Rh_2(OAc)_4} \quad H_3CO \underset{H_3CO}{\bigcirc} \text{—} \underset{O}{\diagup} \overset{COOEt}{\diagdown} OEt$$

$$H_3CO \underset{H_3CO}{\bigcirc} \text{—} \overset{+}{C} \diagup \overset{}{\underset{CH—C}{\diagdown}} \overset{O}{\underset{COOEt}{\diagdown}} C \text{—} OEt$$

(29)

$$n \, HC \equiv CCH_3 \xrightarrow{Ph_2C=W(CO)_5} \left[\underset{H}{\overset{CH_3}{\underset{C}{\overset{|}{\underset{\diagdown}{\overset{C}{\diagup}}}}}} \right]_n \tag{20}$$

instance, by Katz and Lee[57]. An example is the polymerization of propyne, induced by diphenylcarbene pentacarbonyltungsten complex (equation 20). These polymerizations are believed to pass through a metallacyclobutene intermediate. In view of the above-mentioned MO calculations by Hofmann and Hämmerle[54], the structure of the intermediate might need reconsideration. Moreover, labelling experiments by Katz and co-workers[58] seem to indicate that there are different pathways for the reaction of this intermediate to the polymer. With doubly labelled diphenylacetylene, $PhC^* \equiv C^*Ph$, they observed that catalysis by titanium tetrabutoxide results in a polymer in which the labels are separated by a double bond. If, however, molybdenum pentachloride is the catalyst, the labels are separated by a single bond. For an extensive coverage of acetylenic polymers the reader is referred to the relevant chapter of this volume by Masuda, Shirakawa and Takeda.

E. Electrophilic Oxygen, Sulphur and Selenium

1. Oxygen

Oxidation of alkynes by organic peracids has earlier been discussed by Schmid[3]. The mechanism is not always clearly established, but many of these oxidations probably involve transfer of an electrophilic oxygen atom. From a cost-effective point of view hydrogen peroxide has an advantage over organic peracids. Oxygen transfer from hydrogen peroxide is catalysed by metal compounds. Many investigations on this subject have been performed, in particular on the transformation of alkenes to their epoxides. An example with acetylenic substrates will be presented here.

Ishii and Sakata[59] applied as catalyst peroxotungstophosphate with cetylpyridinium counter ions, $\{C_5H_5N^+ \text{—} (CH_2)_{15}CH_3\}_3 (PW_{12}O_{40})^{3-}$. Under two-phase conditions internal alkynes, dissolved in chloroform, were oxidized by aqueous hydrogen peroxide to α,β-epoxy ketones as principal products. For instance, 4-octyne produced 3,4-epoxy-5-octanone (**30**), 5-octen-4-one (**31**) and a small amount of butyric acid (**32**). Although the detailed reaction path is uncertain, the peroxotungstophosphate is believed to be a powerful electrophilic oxygen donor because it epoxidizes an electron-deficient alkene such as Me—CH=CH—C(O)—Et.

$$C_3H_7C\equiv CC_3H_7 \xrightarrow{\text{6 H}_2\text{O}_2,\text{ cat.}}$$

62% (**30**)

$$+ \quad (E)\text{-}H_7C_3\overset{O}{\overset{\|}{C}}CH=CHC_2H_5 \quad + \quad C_3H_7COOH$$

15% (**31**)   5% (**32**)

Epoxidation of alkenes by several single oxygen donors catalysed by metalloporphyrings containing iron or manganese has been widely studied. Oxygen donors are, for instance, iodosylbenzene and sodium hypochlorite. These epoxidations are believed to proceed via an intermediate oxometal complex, the oxo atom having electrophilic character. It was, however, observed that chlorotetraphenylporphirinato iron(III) did not catalyse oxidation of alkynes by iodosylbenzene[60]. Oxidation of disubstituted acetylenes by iodosylbenzene could be achieved in the presence of a ruthenium catalyst, e.g. $RuCl_2(PPh_3)_3$[60]. Under the same conditions terminal acetylenes were cleaved to carboxylic acids. The mechanism has not yet been firmly established.

$$R^1C=CR^2 \longrightarrow \qquad\qquad RC\equiv CH \longrightarrow RCOOH$$

R^1 and R^2 are Ph, alkyl or H

2. Sulphur

Several reactions have been reported involving donation of various types of electrophilic sulphur to a triple bond. Passmore, While and coworkers[61-63] observed 1,3-cycloaddition of the dithionitronium cation, SNS^+, to alkynes, cyanides and thiazyl halides. With alkynes in SO_2 as solvent, the addition products are 1,3,2-dithiazolium ions (**33**); see equation 21. Substituent effects and MO calculations suggest that the LUMO of SNS^+ interacts with the HOMO of the triple bond, even for an electron-poor acetylene such as $CF_3C\equiv CCF_3$. The group of Passmore and White furthermore showed that the 1,3,2-dithiazolium ion, the product of addition of SNS^+ to ethene, has a behaviour similar to that of SNS^+ itself although the former reagent is less reactive[64]. The LUMOs of both reagents match the HOMO of an alkyne. The same applies to the dichlorodithionitronium cation, $\{ClSNSCl\}^+$. With ethyne and propyne in SO_2 as solvent it reacted to the corresponding 1,3,2

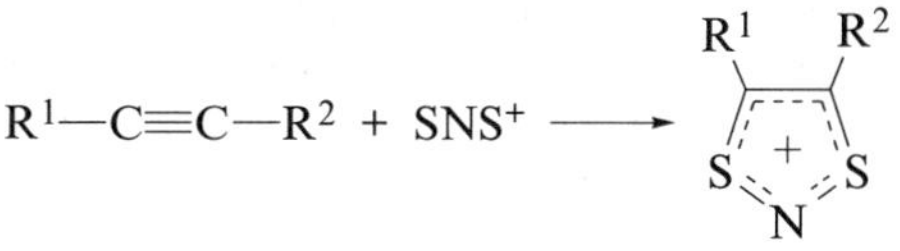

$$R^1-C\equiv C-R^2 + SNS^+ \longrightarrow \begin{array}{c} R^1 \quad R^2 \\ \diagdown \diagup \\ S \overset{+}{\diagdown N \diagup} S \end{array} \tag{21}$$

(33)

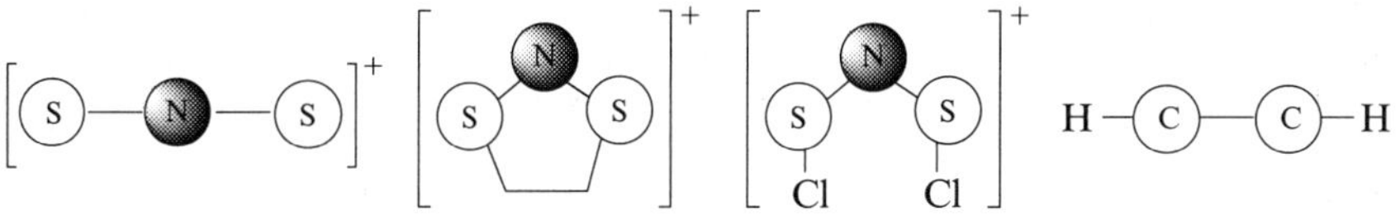

LUMO   LUMO   LUMO   HOMO

dithiazolium cations **(34)** while chlorine was eliminated and trapped as SO_2Cl_2. This reagent was not able to attack $CF_3C\equiv CCF_3$.

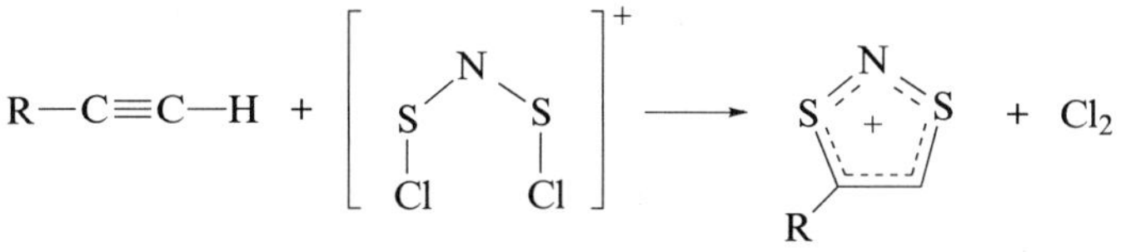

$$R-C\equiv C-H + \begin{bmatrix} N \\ S \diagup \diagdown S \\ | \qquad | \\ Cl \qquad Cl \end{bmatrix}^+ \longrightarrow \begin{array}{c} N \\ S \diagup \overset{+}{\diagdown} S \\ R \end{array} + Cl_2$$

(34)

Many investigations have been performed on the addition of sulphenyl halides, particularly chlorides, to alkenes and alkynes. Thiiranium and thiirenium ions, respectively, would be intermediates. The generality of these ions as intermediates in the addition to alkenes has been questioned in a critical review article[65]. Modena and coworkers[6] listed relative reactivities in the addition of 4-chlorobenzenesulphenyl chloride to alkenes and alkynes. Alkynes appear to be less reactive than alkenes, but the ratio very much depends upon structure. The highest ratio was observed for the parent pair ethene:ethyne, namely 2.82×10^5.

Addition of sulphenyl chlorides to alkynes gives 1:1 adducts in the *trans* configuration. Both Markovnikov and anti-Markovnikov adducts are sometimes obtained. The regio-

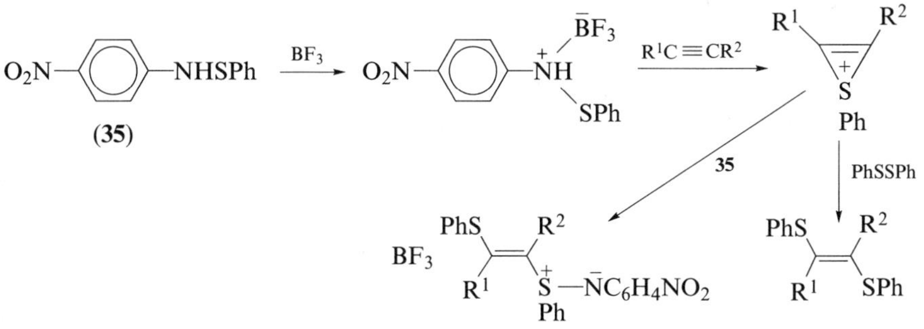

(35)

(37)   **(36)**

orientation is determined by the nucleophilic attack by chloride ion at either carbon of the thiirenium ring, assuming that such a ring is really an intermediate. Product compositions, however, should only be used as a measure of regioselectivity if isomerizations have been excluded. Isomerizations, catalysed by acids and by sulphenyl chlorides themselves have been shown to occur[66,67].

A different source of arenesulphenyl cations has been reported by Montevecchi and coworkers[68,69]. They showed that the BF_3-promoted reaction of 4'-nitrobenzenesulphenanilide (35) with aryl-substituted alkynes in poorly nucleophilic solvents such as chlorobenzene generally led to bissulphides (36) and sulphimides (37) in addition to diphenyl disulphide and 4-nitroaniline. In acetonitrile as solvent, products of capture of the thiirenium ion by the solvent were also observed[68] (equation 22). The thiirenium ion from phenylacetylene reacted even with the poorly nucleophilic solvent chlorobenzene to (E)-$PhSCH=CPhC_6H_4Cl$. With alkyl-substituted alkynes in chlorobenzene as solvent, a small amount of (E)-2-fluorovinyl sulphide $PhSC(R^1)=CF(R^2)$ was also detected. The yield of this sulphide could be increased when the reaction was performed in the presence of tetrabutylammonium tetrafluoroborate. Terminal alkynes gave the corresponding 2-fluorovinyl sulphides in 35–55% and internal alkynes in 65–87% yield. The procedure was unsuccessful for di-*tert*-butylacetylene and gave low yields for arylacetylenes. In acetic acid as solvent the thiirenium ion was captured as (E)-$PhSC(R^1)=CR^2(OAc)$.

A two-sulphur electrophilic addition to alkynes was reported by Bock and coworkers[70]. The reagent is a mixture of the sulphur chloride S_2Cl_2 and aluminium trichloride dissolved in dichloromethane. Initially, 1,2-dithiete 38 is formed, but in the preparative procedure the isolated product is a mixture of 2,6- and 2,5-di-*tert*-butyl-1,4-dithiin (39 and 40). When the reaction is performed under nitrogen in a closed tube, the substituted dithiete is oxidized to its radical cation (equation 23).

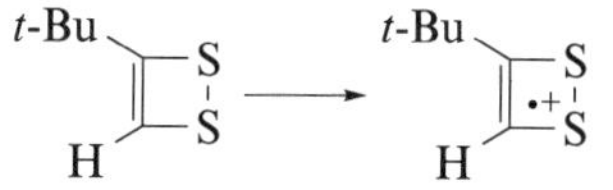

$$\text{(23)}$$

Miller and coworkers[71] treated 1-chloro- and 1-bromo-2-phenylethyne with antimony pentafluoride in liquid sulphur dioxide; benzene was added as a carbocation scavenger. Instead of carbon–halogen heterolysis they unexpectedly observed formation of 2-halo-3-phenylbenzothiophene S-oxide (**41**). On closer examination and depending upon substituent X various other products were observed, e.g. 1-halo-2,2-diphenylvinylsulphinic acids (**42**). The initial step in the formation of these products is probably electrophilic addition of sulphur dioxide, assisted by the Lewis acid. Trapping of the carbon cation by benzene and elimination of antimony pentafluoride explained the various products.

$$\text{Ph}-\text{C}\equiv\text{C}-\text{X} \xrightarrow[\text{C}_6\text{H}_6,\ -78\ ^\circ\text{C}]{\text{SbF}_5,\ \text{SO}_2}$$

(41)

$$\text{Ph}-\text{C}\equiv\text{C}-\text{X} \xrightarrow[\text{SO}_2]{\text{SbF}_5} \text{Ph}\overset{+}{\text{C}}=\text{C} \begin{smallmatrix} \text{X} \\ \\ \text{SO}_2\cdot\text{SbF}_5^- \end{smallmatrix}$$

$$\xrightarrow{\text{C}_6\text{H}_6}$$

$-\text{SbF}_5 + \text{H}^+$ $-\text{SbF}_5\text{OH}^-$

$$\text{Ph}_2\text{C}=\text{C} \begin{smallmatrix} \text{X} \\ \\ \text{SO}_2\text{H} \end{smallmatrix}$$

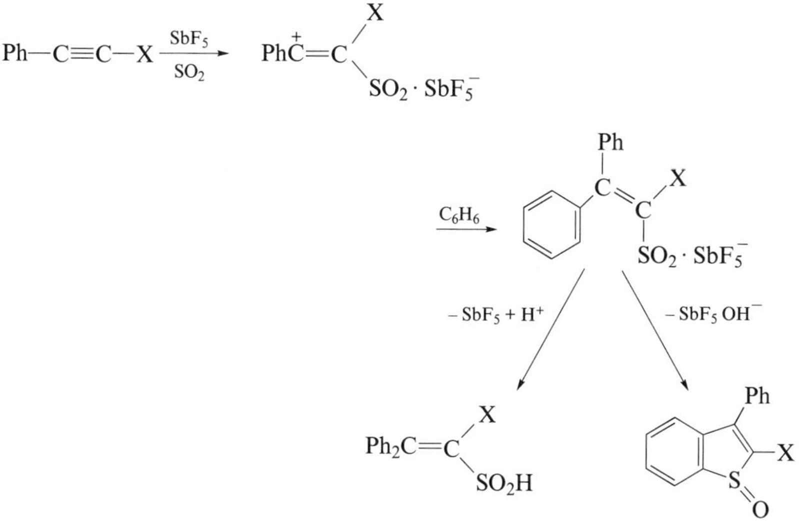

(42)   **(41)**

3. Selenium

Similar to electrophilic sulphur additions, also electrophilic selenium additions to alkynes have been known for a long time. Some more recent examples will be mentioned.

Kataev and coworkers[72] added a number of *para*- and *meta*-substituted benzeneselenenyl chlorides to unsubstituted acetylene in acetic acid solution. The addition products, aryl *trans*-2-chlorovinyl selenides (**43**), were obtained in yields of 69–85% irrespective of the electronic character of the aryl substituent.

$$\text{ArSeCl} + \text{HC}\equiv\text{CH} \longrightarrow \quad \begin{array}{c} \text{ArSe} \\ \\ \text{H} \end{array}\!\!C\!=\!C\!\!\begin{array}{c} \text{H} \\ \\ \text{Cl} \end{array}$$

(43)

By reacting 2-butyne with benzeneselenenyl hexafluoroantimonate, Schmid and Garratt[73] quantitatively isolated 2,3-dimethyl-1-phenylselenirenium hexafluoroantimonate (**44**). This salt could also be synthesized from (E)-2-chloro-3-phenylselenobutene-2 (**45**), but not from its (Z)-isomer, by reaction with AgSbF_6. From reaction of the salt (**44**) with chloride, the (E)- isomer (**45**) could be recovered. These observations support the ring structure of the salt.

$$\text{CH}_3\text{C}\equiv\text{CCH}_3 + \text{C}_6\text{H}_5\text{SeSbF}_6 \longrightarrow \quad \begin{array}{c} \text{C}_6\text{H}_5 \\ | \\ \text{Se} \\ \overset{+}{\triangle} \quad \text{SbF}_6^- \\ \text{H}_3\text{C} \quad \text{CH}_3 \end{array}$$

(44)

$$\underset{\longleftarrow}{\overset{\text{Cl}^-}{\rightleftharpoons}} \quad \begin{array}{c} \text{Cl} \\ \\ \text{H}_3\text{C} \end{array}\!\!C\!=\!C\!\!\begin{array}{c} \text{CH}_3 \\ \\ \text{SeC}_6\text{H}_5 \end{array} + \text{AgSbF}_6$$

(45)

Garratt and coworkers[74] studied the addition of benzeneselenenyl chloride to a large number of propargyl alcohols. The reaction was performed in dichloromethane with the alcohol in excess. Essentially quantitative yields of *anti* addition products were obtained, some of them pure Markovnikov, others pure anti-Markovnikov and a few a mixture of both. The anti-Markovnikov products are formed by kinetic control. They slowly isomerized to the Markovnikov products.

$$\text{PhSeCl} + \text{R}^1\!-\!\text{C}\equiv\text{C}\!-\!\underset{\underset{\text{R}^3}{|}}{\overset{\overset{\text{R}^2}{|}}{\text{C}}}\!-\!\text{OH} \longrightarrow$$

Markovnikov   anti-Markovnikov

Products of addition of benzeneselenenyl chloride to acetylenic alcohols were also reported by Filer and coworkers[75]. The hydroxyl groups were at various positions with respect to the triple bond. The expected cyclizations by reaction of the hydroxyl group with an intermediate selenirenium ion did not occur. An example is given in equation 24.

$$HC\equiv CCH_2CH_2OH + PhSeCl \longrightarrow \underset{Cl}{\overset{H}{}}C=C\underset{CH_2CH_2OH}{\overset{SePh}{}} \quad \begin{array}{l}\text{Yield in } CH_2Cl_2\ 89\%,\\ \text{in AcOH } 84\%\end{array} \quad (24)$$

Cyclization did occur in the reaction between an acetylene containing a carboxylic acid function, and N-phenylselenophtalimide[76] (equation 25). For R = Me the addition stereoselectively afforded the (E)-product, but for R = H a mixture of (Z)- and (E)-isomers (**46** and **47**, respectively) was isolated. It could be shown that the (Z)-isomer was formed via reaction of a second molecule of N-phenylselenophthalimide with the lacton and subsequent elimination. The same reagent, N-phenylselenophthalimide, was applied by Anker and coworkers[77], but now in combination with a fluoride donor, namely the complex of triethylamine and three molecules of hydrogen fluoride (equation 26). The addition is always *anti*. The asymmetric internal alkyne with $R^1 = CH_3$ and $R^2 = n$-C_3H_7 afforded both regio-isomers in approximately equal amounts; for $R^1 = CH_3$ and $R^2 = Ph$, however, the product consisted of (E)-PhFC=C(CH_3)SePh exclusively. When applying terminal acetylenes, complicated product mixtures were obtained, at least partly arising from a second addition of PhSe$^+$ and F$^-$.

$$R-C\equiv C-CH_2CH_2COOH + \text{[phthalimide-N—SePh]} \quad (25)$$

$$\longrightarrow \text{[PhSe—CHR—lactone]} + \text{[phthalimide-NH]}$$

$$(47) \;\rightleftharpoons\; \text{[PhSe}^+\text{]}\; H-\overset{PhSe}{\underset{PhSe}{C}}\text{-lactone}^+ \;\rightleftharpoons\; (46)$$

$$R^1-C\equiv C-R^2 + \text{[phthalimide-N—SePh]} + Et_3N \cdot 3HF$$

$$\longrightarrow \left[\overset{Ph}{\underset{R^1 \quad R^2}{\overset{|}{Se^+}}} \right] \longrightarrow \underset{R^1}{\overset{F}{}}C=C\underset{SePh}{\overset{R^2}{}} \quad (26)$$

Back and Muralidharan[78,79] described the formation of benzeneselenenyl p-toluene-sulphonate (**48**) and its electrophilic addition to alkynes. The benzeneselenenyl p-

toluenesulphonate could not be isolated; it was employed *in situ*. The presence of an intermediate selenirenium ion was suggested by a strong preference for *anti* addition: for $R^1 = R^2 = H$ the $(E):(Z)$ ratio was 90:10. From 1-decyne nearly equal amounts of Markovnikov and anti-Markovnikov addition products were obtained. Addition is possible also to acetylenes with electron-poor triple bonds, albeit in low yield, e.g. to dimethyl acetylenedicarboxylate in a yield of 25%. Formation of the same adducts was observed when the alkyne was heated in benzene with *p*-toluenesulphonic acid, diphenyl diselenide and AIBN or with a sulphinyl sulphon and diphenyl diselenide. In both cases selenenyl sulphonate **48** would be formed through radical reactions, in the former procedure via sulphonate radicals, ArSO$_3^{\cdot}$, and in the latter via a radical induced isomerization.

$$ArSO_3Ag + PhSeCl \longrightarrow Ar\overset{\overset{\displaystyle O}{\|}}{\underset{\underset{\displaystyle O}{\|}}{S}}-OSePh + AgCl$$

(48)

$$48 + R^1-C\equiv C-R^2 \longrightarrow \left[\underset{R^1 \quad R^2}{\overset{\overset{\displaystyle Ph}{\underset{\displaystyle Se}{|}}}{\triangle^{+}}} \quad ArSO_3^{-} \right] \longrightarrow \underset{R^1 \quad\quad O_3SAr}{\overset{PhSe \quad\quad R^2}{C=C}}$$

$$ArSO_3H \xrightarrow{AIBN} ArSO_3^{\cdot} \xrightarrow{PhSeSePh} Ar\overset{\overset{\displaystyle O}{\|}}{\underset{\underset{\displaystyle O}{\|}}{S}}-OSePh$$

$$Ar\overset{\overset{\displaystyle O \quad O}{\| \quad \|}}{\underset{\underset{\displaystyle O}{\|}}{S-S}}-Ar \longrightarrow Ar\overset{\overset{\displaystyle O}{\|}}{\underset{\underset{\displaystyle O}{\|}}{S^{\cdot}}} + {}^{\cdot}O-SAr \longrightarrow Ar\overset{\overset{\displaystyle O}{\|}}{\underset{\underset{\displaystyle O}{\|}}{S}}-OSAr$$

$$Ar\overset{\overset{\displaystyle O}{\|}}{\underset{\underset{\displaystyle O}{\|}}{S}}-OSAr + PhSeSePh \longrightarrow Ar\overset{\overset{\displaystyle O}{\|}}{\underset{\underset{\displaystyle O}{\|}}{S}}-OSePh + PhSeSAr$$

F. Halogens

Addition of halogens to unsaturated carbon–carbon bonds is a classical example of electrophilic addition. Addition to alkynes generally leads to 1,2-dihaloalkenes. Various mechanisms of addition have been discussed in the earlier review by Schmid[3]. No major changes in the mechanistic picture have been published since. The actual addition might be preceded by complex formation. Ault[80] applied the matrix isolation technique to study complex formation between ClF and acetylene, propyne and 2-butyne and also between Cl$_2$ and acetylene and propyne. The data seem to support a T-shaped structure.

Electrophilic addition of halogen passes through either a three-membered cyclic cation (**49**) or an open vinyl cation (**50**). The difference between these cations will not be a distinct one because asymmetric three-membered intermediates (**51**) are quite well possible. If in **51** one of the two bonds is weak, the intermediate resembles an open cation, e.g. **50**.

$$\text{(49)} \qquad \text{(50)} \qquad \text{(51)}$$

One might ask the question whether an electron-poor triple bond will also react via an intermediate cation. Ehrlich and Berliner[81] investigated the bromination of phenylpropiolic acid, its sodium salt and its ethyl ester in aqueous acetic acid solution. Under the reaction conditions the acid actually reacted through its salt. The reaction afforded mainly products of decarboxylation, e.g. bromophenylacetylene.

$$\text{PhC} \equiv \text{C} - \text{COO}^- \xrightarrow{\text{Br}_2} \text{Ph}\overset{+}{\text{C}} = \text{C} \begin{smallmatrix} \diagup \text{COO}^- \\ \diagdown \text{Br} \end{smallmatrix} \xrightarrow{-\text{CO}_2} \text{PhC} \equiv \text{C} - \text{Br}$$

Further reaction with bromine or acetate led to additional products. The ethyl ester reacts approximately 200 times slower than the acid, which reacts through its anion. From this appreciable difference and from the observation that the ethyl ester is two to three times more reactive than the methyl ester, the authors conclude that the addition indeed is an electrophilic one.

Generally, the cations **49**, **50** or **51** are attacked by a halide ion to give the 1,2-dihaloalkene (for instance, see equation 27). The intermediate would be an open rather than a cyclic cation because of appreciable amounts of *cis*-dibromocinnamates in the product mixture.

$$\text{C}_4\text{H}_9\text{C} \equiv \text{CH} + \text{BrCl} \xrightarrow{\text{CCl}_4} \begin{smallmatrix} \text{H}_9\text{C}_4 \diagdown \quad \diagup \text{Br} \\ \text{C} = \text{C} \\ \text{Cl} \diagup \quad \diagdown \text{H} \end{smallmatrix} + \begin{smallmatrix} \text{H}_9\text{C}_4 \diagdown \quad \diagup \text{Cl} \\ \text{C} = \text{C} \\ \text{Br} \diagup \quad \diagdown \text{H} \end{smallmatrix} \qquad (27)$$

$$\qquad\qquad\qquad\qquad\qquad 90\% \qquad\qquad\qquad 10\%$$

Rozen and Brand[82] prepared XF (X = I, Br) by passing fluoride through a suspension of I_2 or Br_2 in CFCl_3 at $-75°\text{C}$. With alkynes, addition of two equivalents of XF occurred, e.g.

$$\text{CH}_3(\text{CH}_2)_3\text{C} \equiv \text{CH} + 2\text{XF} \longrightarrow \text{CH}_3(\text{CH}_2)_3\text{CF}_2\text{CHX}_2$$

With phenylalkynes the reaction took the following course:

$$\text{PhC} \equiv \text{CH} \xrightarrow{\text{XF}} \begin{smallmatrix} \text{Ph} - \text{C} = \text{C} - \text{H} \\ | \quad\quad | \\ \text{F} \quad\; \text{X} \end{smallmatrix} \xrightarrow{\text{XF}} \text{PhCF}_2\text{CX}_2\text{H} + \text{PhCF}_2\text{CFXH}$$

$$\qquad\qquad\qquad\qquad E \text{ and } Z$$

The latter product would be formed in a phenyl-assisted nucleophilic substitution of F for X.

If, apart from the halide anion, other nucleophiles are present, the reaction may take a different course. For instance, in methanol the reaction shown in equation 28 took place[83]. Initially formed 1-bromo-2-methoxyhexene has rapidly reacted further to **52**, which was hydrolysed to ketone **53**. A similar behaviour was observed for bromine instead of chlorobromine[84]. With iodine in methanol a different product composition resulted[85]; terminal alkynes $RC\equiv CH$ (R = Bu, t-Bu and Ph) as well as 3-hexyne almost exclusively gave diiodoalkenes. The authors suggested that this addition of iodine follows a radical course. In the presence of silver nitrate, however, products expected for an ionic process were observed.

$$C_4H_9C\equiv CH \ + \ BrCl$$

$$\xrightarrow{CH_3OH} \quad \underset{31\%}{\overset{\displaystyle H_9C_4}{\underset{\displaystyle Cl}{}}C=C\overset{\displaystyle Br}{\underset{\displaystyle H}{}}} \ + \ \left[\underset{\overset{|}{OCH_3}}{\overset{\overset{|}{OCH_3}}{C_4H_9C}}-CHBr_2 \ + \ \underset{\overset{|}{OCH_3}}{\overset{\overset{|}{Cl}}{C_4H_9C}}-CHBr_2 \right] \qquad (28)$$

$$(52)$$

$$\downarrow$$

$$\underset{69\% \ (53)}{C_4H_9C\overset{\displaystyle O}{\underset{\displaystyle CHBr_2}{\diagdown}}}$$

$$RC\equiv CH \ + \ I_2 \ \xrightarrow{\underset{CH_3OH}{AgNO_3}} \ RC\overset{\displaystyle O}{\underset{\displaystyle CHI_2}{\diagdown}} \ + \ \underset{\displaystyle I}{\overset{\displaystyle R}{}}C=C\overset{\displaystyle I}{\underset{\displaystyle H}{}} \ + \ RC\equiv Cl$$

$$(R = C_4H_9, \ t\text{-}C_4H_9)$$

Instead of silver nitrate, Pagni, Kabalka and coworkers[86-88] used activated and even out-of-the-bottle γ-alumina to promote electrophilic additions of iodine to alkynes (cf. the alumina-catalysed addition of HI). (*E*)-1,2-diiodoalkenes were obtained in 50–98% yield. With terminal alkynes and unactivated alumina, moreover, a minor amount of 1,1,2-triiodo-1-alkene was detected; it was the product of iodine addition to intermediate 1-iodo-1-alkyne. Still another method of electrophilic addition of iodine to alkynes involves mercury(II) salts in dichloromethane at 0 °C[89].

Al-Hassan studied the bromination of alkynylsilanes[90], $RC\equiv CSiMe_3$, R = Ph, $SiMe_3$,

$$R^1C\equiv CR^2 \ + \ HgX_2 \ + \ I_2 \ \longrightarrow \ \underset{\displaystyle X}{\overset{\displaystyle R^1}{}}C=C\overset{\displaystyle I}{\underset{\displaystyle R^2}{}}$$

$$X = Cl, \ OAc, \ \text{and other anions, but not I.}$$

H, Me, c-Hex, Bu and CH_2OH at -12 °C in carbon tetrachloride as solvent; yields of 42–82% of 1,2-dibromo addition products were reported.

The general trend is to increase the selectivity of the addition. Selectivity is often increased by substituting pyridine–halogen charge transfer complexes[91] or tribromides of organic bases[92] for the free halogens. Instead of pyridine an insoluble polymer can be applied. Although the rate of reaction is lower compared to the almost instantaneous reaction of free bromine, this technique facilitates isolation and purification of the product[93].

Zabicky and coworkers[94] applied poly(vinylpyridine) complexes with bromine and bromine chloride: $PVP–Br_2$ and PVP–BrCl. Acetylenic alcohols with the OH group in a propargylic position reacted with $PVP–Br_2$ to addition products of mainly the (E)-configuration, whereas with uncomplexed bromine mixtures of (E)- and (Z)-isomers and by-products resulted. The OH groups did not react when the polymeric reagent was used, whereas they were subject to attack when uncomplexed bromine was the reagent. With propynol pure (E)-2,3-dibromo-2-propen-1-ol was obtained in 22–55% yield in carbon tetrachloride, acetic acid or methanol solution. With free bromine in acetic acid also (E)-2,3-dibromo-2-propen-1-yl acetate was formed together with substantial amounts of (Z)-products. Also PVP–BrCl is more selective than free BrCl; the amounts of dibromo and dichloro adducts are reduced. Bromination of 3-butyn-2-ol and 2-butyne-1,4-diol with $PVP–Br_2$ afforded yields of 42% pure (E)-3,4-dibromo-3-buten-2-ol and 80% crystalline (E)-2,3-dibromo-2-butene-1,4-diol, respectively. On chlorobromination of propynol with PVP–BrCl the crude product consisted of a 1 : 2 mixture of (E)- and (Z)-3-bromo-2-chloro-2-propen-1-ol and a little (E)-2-chloro-1,3-dibromopropene. A side reaction occurring with these PVP complexes is quaternization of the polymer, for instance, with propynol to ${PVP–CH=C(Br)CH_2OH}^+Br^-$ and, in particular with acetylenedicarboxylic acid, probably to $PVP^+–C(COOH)=CHCOO^-$.

Bongini and coworkers[95] converted an Amberlyst bromide into the perbromide by treatment with a carbon tetrachloride solution of bromine. On storage, only a small decrease in active bromine content was observed. This reagent converted phenylacetylene dissolved in dichloromethane, to an approximately 1:3 *cis/trans* mixture of the 1,2-dibromo addition product in 92% yield at room temperature in 1h. A similar result was obtained with 1-hexyne, although the reaction required refluxing for 3h. When the Amberlyst bromide had been treated with a dichloromethane solution of chlorine, the polymer acted as a chlorobromination agent.

$$RC{\equiv}CH + Pol.^+BrCl_2^- \longrightarrow RCCl{=}CHBr + Pol.^+Cl^-$$

For R = Ph, n-Bu and n-C_8H_{17} the yields are 96% (95% Z), 91% (95% E) and 85% (95% E), respectively.

G. Metal-containing Reagents

As part of the enormous expansion of organometallic chemistry in recent years, the number of reports on interactions between alkynes and metal compounds has vastly increased. Often, these interactions are simply of a ligand-coordinated-to-metal nature. In other cases a truly covalent bond is formed, but the nature of the bond-forming process is not often investigated. Therefore, in many reactions the question whether the attack on the triple bond is an electrophilic one, is unanswered.

Many reactions exist in which the metal participates in a catalytic instead of in a stoichiometric amount[96,52]. Although these catalytic transformations often start with an electrophilic addition, they are not considered here.

1. Group 4

Buchwald and Nielsen reviewed interactions between Group 4 metals and alkynes[97]. These authors[98] also described a general method for the preparation of asymmetrically substituted zirconacyclopentadienes **54** applying the so-called Schwartz reagent, Cp_2ZrHCl, that is commercially available. The reaction (equation 29) involved methylation of monoaddition product **55**. With a proton donor or iodine the product was converted to the corresponding diene or 1,4-diiododiene, respectively. Lipshutz and Ellsworth[99] also methylated the monoaddition product **55** (R^1 = H). They transmetallated it with $Me_2Cu(CN)Li_2$ to the corresponding vinylcuprate.

Hydrozirconation of some acetylenes led to zirconacyclopentadienes besides, or instead of, to monoaddition products[100]. Diaddition occurred if, in the monoaddition product, agostic Zr–H interaction existed.

$$(29)$$

$$(54)$$

Instead of the commercially available CpZrHCl, this reagent can be prepared *in situ* from Cp_2ZrCl_2 and lithium triethylborohydride. The reaction mixture hydrozirconates terminal acetylenes without affecting e.g. an ester or trimethylsilyl functionality in the alkyne[101]. An application of the Schwartz reagent in natural product synthesis is the hydrozirconation of the terminal triple bond of $Me_3SiC \equiv C(CH_2)_4C \equiv CH$[102]. A review on hydrozirconation has been published[103].

2. Group 5

Synthesis of an asymmetric metallacyclopentadiene (cf. **54**) could also be achieved with a tantalum reagent[104]. An example is shown in equation 30.

$$Ta(DIPP)_3Cl_2(OEt)_2 \ + \ PhC{\equiv}CPh \xrightarrow{\text{2NaHg}} (DIPP)_3Ta{<}\!\!\!\!\begin{array}{c}Ph\\ \\Ph\end{array} \tag{30}$$

DIPP = 2,6-diisopropylphenoxide

3. Group 12

The mercury(II)-catalysed hydration of acetylene to acetaldehyde was an industrially important process until it was replaced by the process for the direct oxidation of ethylene. In order to understand the process of hydration of acetylenes, oxymercuration of acetylenes has been studied by several groups. For instance, Uemura and coworkers[105] investigated oxymercuration of alkylphenylacetylenes in acetic acid at 20–60 °C. The reaction afforded the two (*E*)-vinylmercury products, **56** and **57**. A bridged mercurinium ion, **58**, was supposed to be an intermediate in this reaction. The ratio of attack by acetate ion on either of the ring carbon atoms varied with substituent R. Electrophilic character of the attacking mercury species was supported by kinetic measurements by Bassetti, Floris and coworkers[106,107] on acetoxymercuration of a variety of substituted acetylenes. Electron-withdrawing substitutents decreased the rate of reaction. In some cases these authors observed *syn*-addition, which would suggest an open instead of a bridged cationic intermediate.

$$PhC{\equiv}CR \xrightarrow[\text{ii. aq. KCl}]{\substack{\text{i. Hg(OAc)}_2,\\ \text{AcOH}}} \quad (56) \quad + \quad (57) \quad (58)$$

4. Group 13

Boron is the prime metal in the area of stoichiometric interactions between metals and unsaturated bonds. Especially, boron hydride additions have been investigated, in particular by H. C. Brown and his students. Nowadays, these addition reactions are well-established text book subjects. A number of reviews on hydroboration have appeared[108–110]. The development of a clear mechanistic picture lagged far behind the applications in synthesis. It was also the group of Brown[111] that contributed to mechanistic understanding by performing careful kinetic measurements using 9-borabicyclo[3.3.1]nonane, abbreviated as 9-BBN-H, as reagent. Reactive alkynes such as 1-hexyne and 3-methyl-1-butyne exhibited first-order kinetics in 9-BBN-H with a rate constant equal to that of reactive alkenes, e.g. 1-hexene. The data indicate that the reaction starts with rate-determining dissociation of the 9-BBN-H dimer. The less reactive 1,2-diphenylethyne, like less reactive alkenes, follows a reaction path in which the dissociation of the diborane is an equilibrium step and the attack of the monomer on the alkyne, rate determining. 9-BBN-H reacts slower with alkynes than with alkenes[112], whereas the reverse is true for e.g. the highly hindered dimesitylborane and the dibromoborane–dimethyl sulfide complex. In addition to an unsaturated bond the boron atom becomes attached to the least substituted carbon centre. This regiochemistry is to be expected if boron acts as an electrophile. It is also favoured by steric effects.

$$\left(\begin{array}{c} H \\ B \end{array}\right)_2 \xrightarrow{\text{slow}} 2\ 9\text{-BBN-H}$$

$$R\text{---}C\equiv CH + 9\text{-BBN-H} \longrightarrow \underset{H}{\overset{R}{>}}C=C\underset{BBN}{\overset{H}{<}} \qquad 46\%$$

$$\underset{H}{\overset{R}{>}}C=C\underset{BBN}{\overset{H}{<}} + 9\text{-BBN-H} \longrightarrow RCH_2CH(BBN)_2 \qquad 27\%$$

Several molecular orbital calculations have been reported on hydroboration, a more recent one by the group of Houk[113]. According to these calculations, in the reaction between BH_3 and ethyne the most favourable transition state has a four-centre structure. The same applies to the transition state of the reaction with propyne; here, boron is indeed linked to the terminal carbon.

In recent years, besides hydroboration also haloboration has been used for the introduction of boron atoms into organic molecules. A short review of this subject has been published[114]. Some examples follow. For instance, Paetzold and coworkers[115] showed that 1-hexyne easily reacts with diphenylboron bromide but not with the chloride. An equilib-

$$H_9C_4\text{---}C\equiv CH + Ph_2BBr \rightleftharpoons \underset{H}{\overset{Ph_2B}{>}}C=C\underset{C_4H_9}{\overset{Br}{<}} \quad \begin{array}{l} \text{equilibrium;} \\ \text{rapid at } -20\ ^\circ C \end{array}$$

$$H_5C_2OOC\text{---}C\equiv CH + Ph_2BBr \longrightarrow \underset{H_5C_2OOC}{\overset{Ph_2B}{>}}C=C\underset{H}{\overset{Br}{<}} + \underset{H_5C_2OOC}{\overset{Ph_2B}{>}}C=C\underset{Br}{\overset{H}{<}}$$

slow at room temp.

$$H_9C_4\text{---}C\equiv CH + Ph_2BBr \xrightarrow{80\text{--}90\ ^\circ C} \underset{Ph}{\overset{PhBBr}{>}}C=C\underset{C_4H_9}{\overset{H}{<}} + \underset{H}{\overset{PhBBr}{>}}C=C\underset{C_4H_9}{\overset{Ph}{<}} \qquad (31)$$

main product; (Z) and (E)

$$\qquad\qquad\qquad (59) \qquad\qquad\qquad\qquad (60)$$

rium is established that shifts to the left at higher temperatures. The addition is regio- and stereospecific. Addition to the ethyl ester of propiolic acid is much slower, has a reverse regiochemistry and is not stereospecific. When the reaction between diphenylboron bromide and 1-hexyne was run at 80–90 °C instead of bromoboration, 1,1-carboboration to **59** was the main reaction (equation 31). Moreover, a side product of 1,2-carboboration (**60**) was observed.

Boron tribromide has also been used as a haloboration reagent, in particular by Suzuki and coworkers[116]. For instance, they reacted it with terminal acetylenes and subsequently converted the monoaddition product to a 1,2-dihalo-1-alkene. Since the addition proceeds regio- and stereoselectively and the halogenation occurs with retention, the 1,2-dihalo products have the (Z)-configuration. Oxidation of the addition product by 30% hydrogen peroxide in a solution buffered at pH 5 and in the presence of potassium acetate afforded a 2-bromoalkanal[117]. A combination of halo- and hydroboration was applied in a one-pot stereoselective synthesis of 1,3-dienes[118]. Diisobutylaluminium hydride (DIBAH) reduces the intermediate divinylboron bromide to the corresponding hydride. For R = Bu and R^1 = Et, 74% pure product could be isolated.

Formal carboboration of a terminal alkyne could be achieved by bromoboration and subsequent palladium-catalysed substitution of alkyl for bromine with alkylzinc chloride.

Treatment of the product with a Brønsted acid afforded a disubstituted, and with an organic halide a trisubstituted, alkene[119].

Bubnov[120] reviewed reactions of allylboranes. In additions of monoallylboranes to terminal acetylenes, ethoxyethyne appeared to give the fastest reaction, supporting electrophilic character of the boron reagent. For example, see equation 32. Reactions between triallylborane and terminal acetylenes give rise to the formation of interesting cage structures. For instance, from **61** and **62** cage **63** is formed in a five-step sequence. 1,1-Carboboration, as observed in the formation of **59**, also occurred in the reaction (equation 33) between triethylboron and bis(trimethylsilylethynyl)dimethylstannane[121]. The authors did not speculate about the mechanism, which should involve shifts of trimethylsilyl and ethyl groups.

$$(32)$$

yield 82 %

(61)   (62)   (63)

$$Et_3B + Me_2Sn(C{\equiv}C{-}SiMe_3)_2 \longrightarrow$$

$$(33)$$

Hydroalumination of alkenes and alkynes has been reviewed[122]. From kinetic data it is concluded that diisobutylaluminium hydride, which is present as its trimer, initially dissociates to the monomer. In attacking the unsaturated bond, the monomer acts as an electrophile. This electrophilic character appears, for instance, from the high rate ratio of addition to $PhC≡CNMe_2$ and $PhC≡CPh$, namely 19,000:1.

Carboalumination of alkynes has also been reviewed[123].

Thallium(III) salts are catalysts in the hydration of acetylenes to ketones. In order to clarify the mechanism of this reaction Uemura and coworkers[124] studied the addition of thallium acetate to acetylenes. The authors suggested that the reaction involves intermolecular acetoxythallation between two initially formed molecules of alkynylthallium diacetate, $RC≡C-Tl(OAc)_2$ (equation 34).

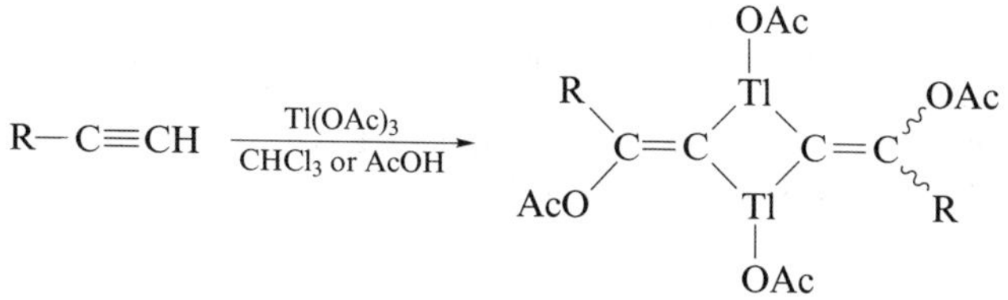

$$R-C≡CH \xrightarrow[\text{CHCl}_3 \text{ or AcOH}]{\text{Tl(OAc)}_3}$$ (34)

5. Group 14

Addition of disilenes to terminal acetylenes gave disilacyclobutenes[125] (see for instance, equation 35). Since only polar alkynes react with **64**, an ionic mechanism, involving intermediate **65**, was considered.

Rates of gas-phase additions of dimethylsilylene, Me_2Si, to acetylene and its mono- and dimethyl derivatives showed an approximate ratio of 1 : 2 : 4. This behaviour could suggest electrophilic character of the silylene[126]. On the other hand, reactivity of dimethylgermylene, Me_2Ge, with acetylenes is high for t-Bu$-C≡C-CN$ and thus not favouring electrophilic character of the germylene[127].

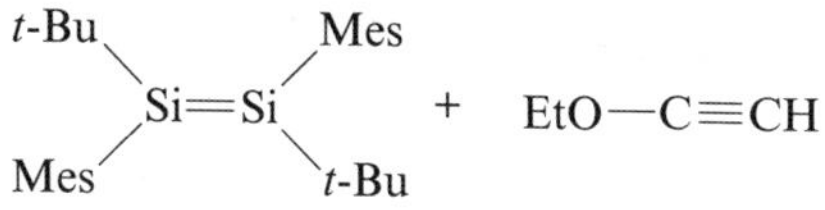

$$\begin{array}{c} t\text{-Bu} \\ \diagdown \\ \text{Mes} \end{array} Si=Si \begin{array}{c} \text{Mes} \\ \diagup \\ t\text{-Bu} \end{array} \quad + \quad EtO-C≡CH \qquad (35)$$

(64)

Mes = 2,4,6-trimethylphenyl

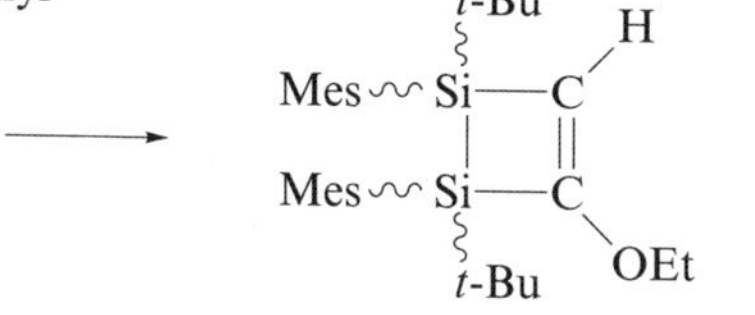

both isomers in
approximately 1:1 ratio

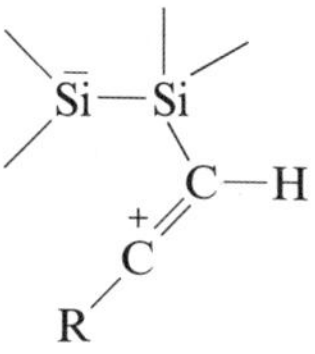

(65)

III. ELECTROPHILIC ADDITIONS TO CYANIDES

Cyanides or nitriles, $RC{\equiv}N$, have two regions for electrophilic attack, namely the triple bond π-electrons and the lone pair on nitrogen. The two possibilities of attack are illustrated by the behaviour of cyanides with respect to transition metal ions. In their gas-phase studies on the interaction between cyanides and transition-metal ions Chen and Miller[128,129] concluded that both side-on and end-on interaction is present. In this respect cyanides resemble the isoelectronic dinitrogen.

$$R{-}C{\equiv}N{:}{\cdots}M^{n+} \qquad\qquad R{-}C{\equiv}N{:}$$

$$\text{end-on} \qquad\qquad\qquad \underset{M^{n+}}{|} \ \text{side-on}$$

A close analogy also exists between the behaviour of dinitrogen, cyanides and isocyanides in their attachment to electron-rich d^6 metal centres, e.g. Mo(0), W(0) and Re(I). This attachment activates the ligand for electrophilic attack. This activation plays a role in the reduction by nitrogenases. For cyanides the following type of reaction has been observed[130]:

$$M^0{-}\overset{\delta+}{C}{\equiv}\overset{\delta-}{N}{-}R \ \xrightarrow{\ H^+\ } \ \left(M{-}C{=}N\begin{array}{l}\nearrow H \\ \searrow R\end{array} \right)^{+} \ \longrightarrow \ \text{products of reduction}$$

Mass spectrometric studies by Schwarz and coworkers on $CH_3(CH_2)_nC^1{\equiv}N{-}M^+$ complexes showed that oxidative addition of $C{-}H$ or $C{-}C$ to the metal ion specifically occurred at an appreciable distance from the metal ion, namely at C^4 or C^5 for M = Fe, Co and Ni[131–133]. This behaviour suggested end-on coordination. A somewhat broader interpretation of this phenomenon was published by Stepnowski and Allison[134]. Copper ions, however, behaved differently and would be coordinated in a side-on fashion[135]. These studies were reviewed and extended to branched chain cyanides[136].

In electrophilic additions to cyanides the electrophile always attacks the nitrogen atom. It is noteworthy to remark that the direction of addition of electrophiles to phosphaalkynes, $R{-}C{\equiv}P$, is generally opposite; attack occurs on carbon[137,138]. Ab initio molecular orbital calculations show that in $R{-}C{\equiv}P$ phosphorous has a net positive and carbon a net negative charge. In $R{-}C{\equiv}N$ nitrogen has a considerable negative charge and carbon has a positive charge[139,140].

In additions to cyanides it is not clear whether the approach is side-on or end-on. It may well be that, often, initial attack is end-on while in a subsequent step the electrophile shifts to a more side-on position. For instance, initial end-on attack seems to occur in gas-phase proton additions. Ab initio molecular orbital calculations have been performed on end-on proton additions to $X{-}C{\equiv}N$ of a broad variety of groups X[141]. The calculated proton affinities are in good agreement with experimental values.

In proton-catalysed additions, proton addition to nitrogen is followed by nucleophilic addition to the cyanide carbon atom. For instance, in hydration the nucleophile is a water molecule. If the substrate does not permit acidic conditions, enzyme-catalysed hydration might be an alternative[142,143].

Additions of electrophilic carbon to cyanides are well known[144]. They occur in e.g. the Ritter reaction which has been reviewed extensively by Bishop[145]. In this reaction a carbenium ion, generated from e.g. t-butanol and sulphuric acid, attacks the nitrogen atom of a cyanide (equation 36). Sometimes, cyanides are applied to trap carbocations. An ex-

TABLE 3. First-order rate constants for reaction of Ph_2CH^+ generated from Ph_2CHCl, with cyanides, $R-C\equiv N$, used as solvents

R	Me	Et	i-Pr	t-Bu
$k \times 10^{-6}$ (s^{-1})	2.5	7.2	6.8	6.1

ample was given by Martínez and coworkers[146] (equation 37). In a study on laser-flash photo-lytically generated carbenium ions with various nucleophiles, it was observed (Table 3) that these ions also attack cyanides used as solvent[147,148].

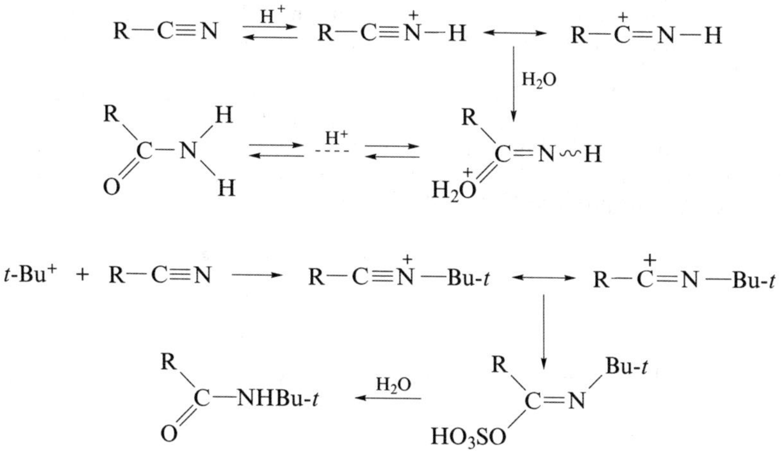

(36)

A *para*-CF$_3$ substituent in one of the phenyl groups increases, whereas a *para*-methyl substituent decreases the rate.

Passmore and coworkers, who studied the addition of (SNS)$^+$ to carbon–carbon triple bonds, also included addition to cyanides[61].

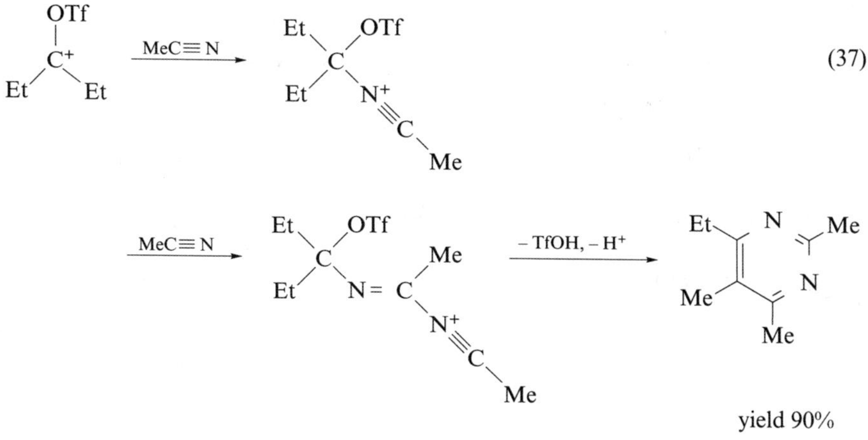

(37)

yield 90%

$$ArAr'CH^+ + N{\equiv}C{-}R \longrightarrow ArAr'CH{-}N^+{\equiv}C{-}R$$

$$(SNS)^+ (AsF_6)^- + R{-}C{\equiv}N \xrightarrow{\ k\ } \underset{N}{\overset{R}{\underset{S\underset{+}{\cdots}S}{\overset{C{\cdots}N}{}}}} (AsF_6)^-$$

A plot of log k versus ionization potential of the substrate is linear; acetylenes and cyanides are on approximately the same line. When reacted with $HC{\equiv}C{-}C{\equiv}N$ the reagent $(SNS)^+$ showed a clear preference for the carbon–carbon triple bond. This result is in agreement with experimentally determined ionization potentials of $C{\equiv}C$ and $C{\equiv}N$ in this molecule of 11.60 and 14.03 eV, respectively.

Reactions of triethylborane with cyanides (equation 38) have been investigated by Yalpani and coworkers[149]. It was assumed that formation of the product was preceded by a dissociation followed by hydroboration of the cyanide. Surprisingly, for R = Me the reaction took a different course. Heterocycle 66 was isolated as the main product (equation 39). With ethyl cyanide products 67 and 68 were observed.

$$Et_3B + R{-}C{\equiv}N \xrightarrow[-C_2H_4]{200\ °C} \quad (38)$$

$$Et_3B \rightleftharpoons Et_2BH + H_2C{=}CH_2$$

$$\xrightarrow[-EtH]{200\ °C} \left[Et_2B{-}N{=}C{=}CH_2 \right]$$

$$(39)$$

(66)

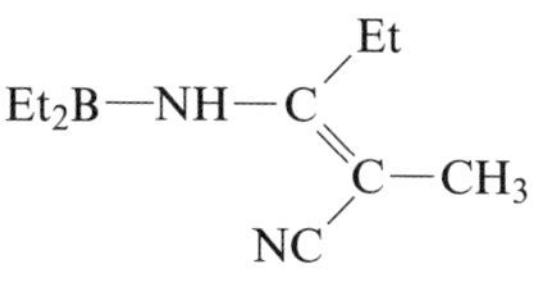

$$Et_2B-NH-C\underset{NC}{\overset{Et}{\underset{}{{\Big\backslash}}}}C-CH_3$$

(E) + (Z)

(67)

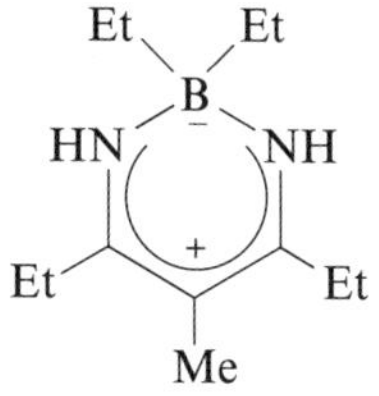

(68)

Electrophilic organometallic complexes are important as Ziegler-Natta catalysts for alkene polymerization. In this connection their behaviour with respect to cyanides has also been investigated. Coordination is often followed by insertion[150]. Only one example is mentioned here, the gas-phase reaction of bis(η^5-cyclopentadienyl)methylzirconium(1+) ion (69) with acetonitrile[150]. The reaction probably took the course shown in equation 40. The system seemed to be in equilibrium since reaction between 69 and deuteriated acetonitrile afforded $Cp_2ZrCD_3^+$. Similarly,

$$Cp_2ZrCH_3^+ + Ph-C\equiv N \longrightarrow Cp_2ZrPh^+ + H_3C-C\equiv N$$

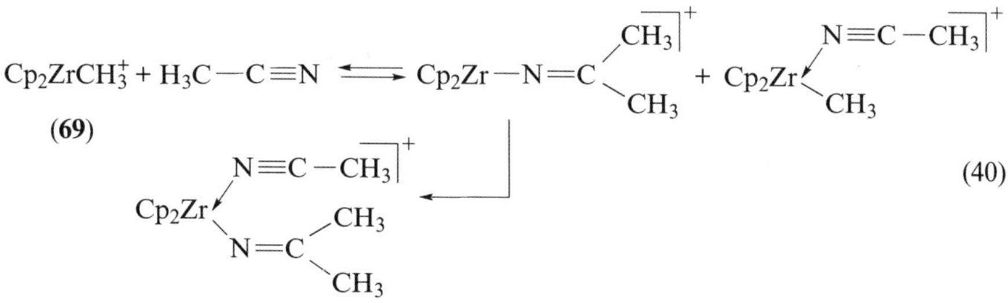

$$Cp_2ZrCH_3^+ + H_3C-C\equiv N \rightleftharpoons Cp_2Zr-N=C\underset{CH_3}{\overset{CH_3}{\underset{}{{\Big\backslash}}}}{}^+ + Cp_2Zr\underset{CH_3}{\overset{N\equiv C-CH_3}{\underset{}{{\Big\backslash}}}}{}^+$$

(69)

$$Cp_2Zr\underset{N=C}{\overset{N\equiv C-CH_3}{\underset{}{{\Big\backslash}}}}\underset{CH_3}{\overset{CH_3}{\underset{}{{\Big\backslash}}}}{}^+$$

(40)

IV. ELECTROPHILIC ADDITIONS TO ISOCYANIDES

Isocyanides, $R-N\equiv C$, are incorporated in the present volume on triple bonds because of their dominant resonance structure $R-\overset{+}{N}\equiv\bar{C}$. This resonance structure suggests a negative charge on the terminal carbon atom. This suggestion is supported by *ab initio* calculations[140], according to which the terminal carbon of $Me-N\equiv C$ has a net negative charge. The nitrogen atom, however, has a negative charge as well. The positive charge resides in the methyl group.

Electrophilic reagents attack uncoordinated isocyanides at their terminal carbon atom. This attack is facile because of the negative charge on this atom or, explained alternatively, because of the resonance stabilization of the intermediate.

$$R-N\equiv C + X^+ \longrightarrow R-N=\overset{+}{C}-X \longleftarrow R-\overset{+}{N}\equiv C-X$$

It should be remarked, however, that nucleophilic attack on the isocyanide carbon atom has also been observed[151].

A quantitative measure of ease of electrophilic attack is the methyl cation affinity as estimated from experimental, mass spectrometric data by Deakyne and Meot-Ner[152].

$$H_3C-N=C + CH_3^+ \longrightarrow H_3C-N=\overset{+}{C}-CH_3 \longleftrightarrow H_3C-\overset{+}{N}\equiv C-CH_3$$

$$H_3C-C\equiv N + CH_3^+ \longrightarrow H_3C-C\equiv\overset{+}{N}-CH_3$$

$$(CH_3)_2O + CH_3^+ \longrightarrow (CH_3)_3O^+$$

For the isocyanide, the cyanide and the ether, these affinities are 510, 410 and 389 kJ mol^{-1}, respectively. These experimental data are in agreement with MP2/6-31G* calculations.

Subsequent to electrophilic attack, a nucleophile is attached to the terminal carbon atom. Thus, 1,1-addition has occurred.

$$R-N=\overset{+}{C}-X \longleftrightarrow R-\overset{+}{N}\equiv C-X \xrightarrow{Nu^-} R-N=C\overset{X}{\underset{Nu}{\diagdown}}$$

Isocyanides are often prepared by dehydration of the corresponding formamide. The reverse transformation occurs with isocyanides in an acidic aqueous medium.

$$R-N=C \xrightarrow{H^+} R-N=\overset{+}{C}-H \longleftrightarrow R-\overset{+}{N}\equiv C-H \xrightarrow{H_2O} R-N=C\begin{smallmatrix}H\\\diagup\\\diagdown\\\overset{+}{O}\\H_2\end{smallmatrix}$$

$$R-N-\overset{H}{C}\overset{\diagup H}{\underset{O}{}} \longleftarrow \cdots \longleftarrow R-N=C\overset{\diagup H}{\underset{OH}{}} \xleftarrow{-H^+}$$

More important than proton additions are additions of electrophilic carbon by which a new carbon–carbon bond is formed. Carbon–carbon bond formation is a major goal in the synthesis of organic compounds. For instance, from a carboxylic acid chloride an α-oxoimidoyl chloride (70) is prepared[153]. Addition of a carbocation to an isocyanide gives a species that is identical to the ion obtained in the Ritter reaction[145]; cf. Section III.

$$R^1-C\overset{O}{\underset{Cl}{\diagup}} + R^2-N=C \longrightarrow R^1-C\overset{O}{\underset{\underset{Cl}{|}}{C}}=N-R^2$$

(70)

$$R-N=C \xrightarrow{R_3^1C^+} R-\overset{+}{N}\equiv C-CR_3^1 \xleftarrow{R^+} R_3^1C-C\equiv N$$

Other reactions, in which carbenium ion addition to an isocyanide is the key step, are the Passerini and Ugi reactions and reactions of similar type. These multicomponent transformations have recently been reviewed[154]. The Passerini reaction starts from an isocyanide, a carbonyl compound and a carboxylic acid.

$$R^1-COOH + R^2R^3C{=}O + R^4-N{=}C \longrightarrow$$

The key step will be attack on the isocyanide by the acid-activated carbonyl compound. Subsequent steps could involve OH-shift and addition of RCOO⁻.

The Brønsted acid could be replaced by the Lewis acid $TiCl_4$. The reaction was believed to have compound **71** as intermediate. Later on it was shown that the Lewis acid simply enhances the electrophilicity of the carbonyl carbon[155].

$$\underset{Cl}{\overset{Cl_3Ti}{\diagdown}}C{=}N-R$$

(71)

Just like the isoelectronic carbon monoxide, an isocyanide is an excellent ligand to metal ions. The chemistry of metal isocyanide complexes has been reviewed by Singleton and Oosthuizen[156]. Only a few examples will be given here. Insertion of an isocyanide into a metal–carbon bond frequently occurs. It is not always clear whether the key step is electrophilic or nucleophilic attack on the coordinated isocyanide or whether the reaction is concerted. Insertion into metal–carbene and metal–carbyne complexes have been reviewed by Aumann[157]. Coordination to the metal considerably affects the chemistry of the isocyanide. If the metal is electron-donating, as in nitrogenase-like centres, the coordinated isocyanide is apt to electrophilic attack at nitrogen[130,158]; cf. Section III.

$$trans\text{-}[ReCl(CNSiMe_3)(dppe)_2] \xrightarrow[-\ Me_3SiX]{\text{weak acid HX}} trans\text{-}[ReCl(CNH)(dppe)_2]$$

dppe = 1,2-(diphenylphosphino)ethane

$$[ReCl(CNR)_3(PMePh_2)_2] \underset{}{\overset{H^+}{\rightleftarrows}} [ReCl(CNHR)(CNR)_2(PMePh_2)_2]^+$$

Sodium isocyanide metallates were attacked by hard electrophiles at the isocyanide nitrogen, whereas soft electrophiles reacted with the metal centre[159].

$$Na[Cp^*W(CO)_n(EtNC)_{3-n}] \xrightarrow{Et_3OBF_4} Cp^*(CO)_n(EtNC)_{2-n}W \equiv C—NEt_2 \quad n = 1, 2$$

$Cp^* = \eta^5$-pentamethylcyclopentadienyl

Most metals are electron-attracting with respect to isocyanide ligands and the coordinated ligand is sensitive to nucleophilic attack. This reversal in behaviour is called 'Umpolung', a term introduced by Seebach[160]. Umpolung is an important feature in organometallic chemistry. An example of such a nucleophilic attack is the initiation step in the polymerization of isocyanides by nickel(II) complexes[161].

$$[Ni(C{=}N—R)_4]^{2+} + X^- \longrightarrow [Ni(C{=}N—R)_3(CX{=}N—R)]^+$$

As an extension of their work on gas-phase decomposition of cyanides coordinated to transition metal ions, Schwarz and coworkers studied similar reactions of isocyanides[162,133]. Unlike the different behaviour of iron and copper complexes of cyanides, the corresponding isocyanide complexes behave similarly.

V. REFERENCES

1. Z. Rappoport (Ed.), *The Chemistry of the Cyano Group*, Wiley, London, 1970.
2. S. Patai (Ed.), *The Chemistry of Diazonium and Diazo Groups*, Wiley, Chichester, 1978, pp. 275–341.
3. G. H. Schmid, in *The Chemistry of the Carbon–Carbon Triple Bond* (Ed. S. Patai), Chapter 8, Wiley, Chichester, 1978, pp. 275–341.
4. S. Patai and Z. Rappoport (Eds.), *The Chemistry of Triple-bonded Functional Groups*, Supplement C, Wiley, Chichester, 1983
5. R. W. Hoffmann, *Dehydrobenzene and Cycloalkynes*, Verlag Chemie, Weinheim, 1967, pp. 182–186.
6. G. Melloni, G. Modena and U. Tonellato, *Acc. Chem. Res.*, **14**, 227 (1981).
7. N. Banait, M. Hojatti, P. Findlay and A. J. Kresge, *Can. J. Chem.*, **65**, 441 (1987).
8. M. I. Bruce, *Pure Appl. Chem.*, **62**, 1021 (1990).
9. N. X. Hu, Y. Aso, T. Otsubo and F. Ogura, *Tetrahedron Lett.*, **27**, 6099 (1986).
10. A. Wojcicki and C. E. Shuchart, *Coord. Chem. Rev.*, **105**, 35 (1990).
11. V. Lucchini and G. Modena, *J. Am. Chem. Soc.*, **112**, 6291 (1990).
12. L. A. Curtiss and J. A. Pople, *J. Chem. Phys.*, **88**, 7405 (1988).
13. V. Janout and S. L. Regen, *J. Org. Chem.*, **47**, 3331 (1982).
14. K. Yates, P. Martin and I. G. Csizmadia, *Pure Appl. Chem.*, **60**, 205 (1988).
15. J.-P. Monthéard, M. Camps and A. Benzaid, *Synth. Commun.*, **13**, 663 (1983); *Chem. Abstr.*, **99**, 194527q (1983).
16. N. Menashe, D. Reshef and Y. Shvo, *J. Org. Chem.*, **56**, 2912 (1991).
17. J.-P. Monthéard, M. Camps and A. Benzaid, *Chem. Lett.*, 523 (1981).
18. J.-P. Monthéard, M. Camps and M. O. Aït-Yahia, *Tetrahedron Lett.*, 1373 (1979).
19. P. J. Kropp, K. A. Daus, S. D. Crawford, M. W. Tubergen, K. D. Kepler, S. L. Craig and V. P. Wilson, *J. Am. Chem. Soc.*, **112**, 7433 (1990).
20. L. J. Stewart, D. Gray, R. M. Pagni and G. W. Kabalka, *Tetrahedron Lett.*, **28**, 4497 (1987).
21. C. K. Reddy and M. Periasamy, *Tetrahedron Lett.*, **31**, 1919 (1990).
22. F. Öztürk, G. Baykut, M. Moini and J. R. Eyler, *J. Phys. Chem.*, **91**, 4360 (1987).
23. F. Öztürk, M. Moini, F. W. Brill, J. R. Eyler, T. J. Buckley, S. G. Lias and P. J. Ausloos, *J. Phys. Chem.*, **93**, 4038 (1989).
24. P. Calas, P. Moreau and A. Commeyras, *J. Chem. Soc., Chem. Commun.*, 433 (1982).

25. T. Fuchikami and I. Ojima, *Tetrahedron Lett.*, **25**, 303 (1984).
26. T. Umemoto, Y. Kuriu and O. Miyano, *Tetrahedron Lett.*, **23**, 3579 (1982).
27. A. Pelter, C. R. Harrison and D. Kirkpatrick, *J. Chem. Soc., Chem. Commun.*, 544 (1973).
28. K. K. Wang and K. E. Yang, *Tetrahedron Lett.*, **28**, 1003 (1987).
29. T. Kappe and W. Lube, *Angew. Chem.*, **83**, 967 (1971).
30. K. T. Potts and M. Sorm, *J. Org. Chem.*, **37**, 1422 (1972).
31. H. Gotthardt and J. Blum, *Chem. Ber.*, **118**, 2079 (1985).
32. H. Gotthardt, J. Blum and K.-H. Schenk, *Chem. Ber.*, **119**, 1315 (1986).
33. H. W. Moore and M. D. Gheorghiu, *Chem. Soc. Rev.*, **10**, 289 (1981).
34. P. L. Fishbein and H. W. Moore, *J. Org. Chem.*, **50**, 3226 (1985).
35. R. L. Danheiser, A. Nishida, S. Savariar and M. P. Trova, *Tetrahedron Lett.*, **29**, 4917 (1988).
36. R. L. Danheiser and S. K. Gee, *J. Org. Chem.*, **49**, 1672 (1984).
37. M. A. Huffman and L. S. Liebeskind, *J. Am. Chem. Soc.*, **113**, 2771 (1991).
38. F. Marcuzzi, G. Melloni and G. Modena, *J. Org. Chem.*, **44**, 3022 (1979).
39. M. Isobe, T. Nishikawa, A. Herunsalee, T. Tsukiyama, Y. Hirose, K. Shimokawa and T. Goto, *Pure Appl. Chem.*, **62**, 2007 (1990).
40. D. J. Rodini and B. B. Snider, *Tetrahedron Lett.*, **21**, 3857 (1980).
41. Y. Tanabe and T. Mukaiyama, *Chem. Lett.*, 673 (1985).
42. R. A. Haack and K. R. Beck, *Tetrahedron Lett.*, **30**, 1605 (1989).
43. F. T. Boyle, Z. S. Matusiak, O. Hares and D. A. Whiting, *J. Chem. Soc., Chem. Commun.*, 518 (1990).
44. L. Birkhofer, A. Ritter and H. Uhlenbrauck, *Chem. Ber.*, **96**, 3280 (1963).
45. D. R. M. Walton and F. Waugh, *J. Organomet. Chem.*, **37**, 45 (1972).
46. M. Karpf, *Tetrahedron Lett.*, **23**, 4923 (1982).
47. T. Ogawa, T. Murafuji, K. Iwata and H. Suzuki, *Chem. Lett.*, 325 (1989).
48. T. Eicher and J. L. Weber, *Top. Curr. Chem.*, **57**, 4 (1975).
49. E. V. Dehmlow and A. Winterfeldt, *Tetrahedron*, **45**, 2925 (1989).
50. R. A. Moss, E. G. Jang and G.-J. Ho, *J. Phys. Org. Chem.*, **3**, 760 (1990).
51. K. H. Dötz, *Angew. Chem.*, **96**, 573 (1984).
52. N. E. Schore, *Chem. Rev.*, **88**, 1081 (1988).
53. K. H. Dötz, *J. Organomet. Chem.*, **140**, 177 (1977).
54. P. Hofmann and M. Hämmerle, *Angew. Chem.*, **101**, 940 (1989); *Angew. Chem., Int. Ed. Engl.*, **28**, 908 (1989).
55. J. S. McCallum, F.-A. Kunng, S. R. Gilbertson and W. D. Wulff, *Organometallics*, **7**, 2346 (1988).
56. H. M. L. Davies and K. R. Romines, *Tetrahedron*, **44**, 3343 (1988).
57. T. J. Katz and S. J. Lee, *J. Am. Chem. Soc.*, **102**, 422 (1980).
58. T. J. Katz, S. M. Hacker, R. D. Kendrick and C. S. Yannoni, *J. Am. Chem. Soc.*, **107**, 2182 (1985).
59. Y. Ishii and Y. Sakata, *J. Org. Chem.*, **55**, 5545 (1990).
60. P. Müller and J. Godoy, *Helv. Chim. Acta*, **64**, 2531 (1981).
61. S. Parsons, J. Passmore, M. J. Schriver and X. Sun, *Inorg. Chem.*, **30**, 3342 (1991).
62. G. K. MacLean, J. Passmore, M. N. S. Rao, M. J. Schriver, P. S. White, D. Bethell, R. S. Pilkington and L. H. Sutcliffe, *J. Chem. Soc., Dalton Trans.*, 1405 (1985).
63. E. G. Awere, N. Burford, C. Mailer, J. Passmore, M. J. Schriver, P. S. White, A. J. Banister, H. Oberhammer and L. H. Sutcliffe, *J. Chem. Soc., Chem. Commun.*, 66 (1987).
64. N. Burford, J. P. Johnson, J. Passmore, M. J. Schriver and P. S. White, *J. Chem. Soc., Chem. Commun.*, 966 (1986).
65. W. A. Smit, N. S. Zefirov, I. V. Bodrikov and M. Z. Krimer, *Acc. Chem. Res.*, **12**, 282 (1979).
66. G. Capozzi, C. Caristi, V. Lucchini and G. Modena, *J. Chem. Soc., Perkin Trans. 1*, 2197 (1982).
67. G. Capozzi, G. Romeo, V. Lucchini and G. Modena, *J. Chem. Soc., Perkin Trans. 1*, 831 (1983).
68. L. Benati, D. Casarini, P. C. Montevecchi and P. Spagnolo, *J. Chem. Soc., Perkin Trans. 1*, 1113 (1989).
69. L. Benati, P. C. Montevecchi and P. Spagnolo, *J. Chem. Soc., Perkin Trans. 1*, 1691 (1990).
70. H. Bock, P. Rittmeyer and U. Stein, *Chem. Ber.*, **119**, 3766 (1986).
71. R.-L. Fan, J. I. Dickstein and S. I. Miller, *J. Org. Chem.*, **47**, 2466 (1982).
72. E. G. Kataev, T. G. Mannafov and Y. Y. Samitov, *J. Org. Chem. USSR (Eng. Transl.).*, **11**, 2366 (1975).
73. G. H. Schmid and D. G. Garratt, *Tetrahedron Lett.*, 3991 (1975).
74. D. G. Garratt, P. L. Beaulieu and V. M. Morisset, *Can. J. Chem.*, **59**, 927 (1981).

75. C. N. Filer, D. Ahern, R. Fazio and E. J. Shelton, *J. Org. Chem.*, **45**, 1313 (1980).
76. T. Toru, S. Fujita and E. Maekawa, *J. Chem. Soc., Chem. Commun.*, 1082 (1985).
77. C. Saluzzo, G. Alvernhe, D. Anker and G. Haufe, *Tetrahedron Lett.*, **31**, 2127 (1990).
78. T. G. Back and K. R. Muralidharan, *Tetrahedron Lett.*, **31**, 1957 (1990).
79. T. G. Back and K. R. Muralidharan, *J. Org. Chem.*, **56**, 2781 (1991).
80. B. S. Ault, *J. Phys. Chem.*, **91**, 4723 (1987).
81. S. J. Ehrlich and E. Berliner, *J. Am. Chem. Soc.*, **100**, 1525 (1978).
82. S. Rozen and M. Brand, *J. Org. Chem.*, **51**, 222 (1986).
83. V. L. Heasley, D. F. Shellhamer, J. A. Iskikian, D. L. Street and G. E. Heasley, *J. Org. Chem.*, **43**, 3139 (1978).
84. S. Uemura, H. Okazaki, M. Okano, S. Sawada, A. Okada and K. Kuwabara, *Bull. Chem. Soc. Jpn.*, **51**, 1911 (1978); *Chem. Abstr.*, **89**, 146527m (1978).
85. V. L. Heasley, D. F. Shellhamer, L. E. Heasley, D. B. Yaeger and G. E. Heasley, *J. Org. Chem.*, **45**, 4649 (1980).
86. R. M. Pagni, G. W. Kabalka, R. Boothe, K. Gaetano, L. J. Stewart, R. Conaway, C. Dial, D. Gray, S. Larson and T. Luidhardt, *J. Org. Chem.*, **53**, 4477 (1988).
87. G. Hondrogiannes, L. C. Lee, G. W. Kabalka and R. M. Pagni, *Tetrahedron Lett.*, **30**, 2069 (1989).
88. S. Larson, T. Luidhardt, G. W. Kabalka and R. M. Pagni, *Tetrahedron Lett.*, **29**, 35 (1988).
89. J. Barluenga, J. M. Martínez-Gallo, C. Nájera and M. Yus, *J. Chem. Soc., Perkin Trans. 1*, 1017 (1987).
90. M. I. Al-Hassan, *J. Organomet. Chem.*, **372**, 183 (1989).
91. G. Bellucci, G. Berti, R. Bianchini, G. Ingrosso and R. Ambrosetti, *J. Am. Chem. Soc.*, **102**, 7480 (1980).
92. J. Berthelot and M. Fournier, *Can. J. Chem.*, **64**, 603 (1986).
93. A. Akelah and D. C. Sherrington, *Chem. Rev.*, **81**, 557 (1981).
94. J. Zabicky, M. Mhasalkar and I. Oren, *Macromolecules*, **23**, 3755 (1990).
95. A. Bongini, G. Cainelli, M. Contento and F. Manescalchi, *Synthesis*, 143 (1980).
96. E. Negishi, *Pure Appl. Chem.*, **53**, 2333 (1981).
97. S. L. Buchwald and R. B. Nielsen, *Chem. Rev.*, **88**, 1047 (1988).
98. S. L. Buchwald and R. B. Nielsen, *J. Am. Chem. Soc.*, **111**, 2870 (1989).
99. B. H. Lipshutz and E. L. Ellsworth, *J. Am. Chem. Soc.*, **112**, 7440 (1990).
100. G. Erker, R. Zwettler, C. Krüger, R. Schlund, I. Hyla-Kryspin and R. Gleiter, *J. Organomet. Chem.*, **346**, C15 (1988).
101. B. H. Lipshutz, R. Keil and E. L. Ellsworth, *Tetrahedron Lett.*, **31**, 7257 (1990).
102. L. Crombie, M. A. Horsham and R. J. Blade, *Tetrahedron Lett.*, **28**, 4879 (1987).
103. J. A. Labinger, in *Comprehensive Organic Synthesis*, Vol. 8 (Eds. B. M. Trost and I. M. Fleming), Pergamon Press, Oxford, 1991, pp. 675–702.
104. J. R. Strickler, P. A. Wexler and D. E. Wigley, *Organometallics*, **10**, 118 (1991).
105. S. Uemura, H. Miyoshi and M. Okano, *J. Chem. Soc., Perkin Trans. 1*, 1098 (1980).
106. M. Bassetti and B. Floris, *J. Org. Chem.*, **51**, 4140 (1986).
107. M. Bassetti, B. Floris and G. Spadafora, *J. Org. Chem.*, **54**, 5934 (1989).
108. P. F. Hudrlik and A. M. Hudrlik, in *The Chemistry of the Carbon–Carbon Triple Bond*, Part 1 (Ed. S. Patai), Wiley, Chichester, 1978, pp. 203–219.
109. A. Pelter, K. Smith and H. C. Brown, in *Best Synthetic Methods, Borane Reagents* (Eds. A. R. Katritzky, O. Meth-Cohn and C. W. Rees), Academic Press, London, 1988.
110. K. Smith and A. Pelter, in *Comprehensive Organic Synthesis*, Vol. 8 (Eds. B. M. Trost and I. Fleming), Pergamon Press, Oxford, 1991, pp. 703–731.
111. H. C. Brown, J. Chandrasekharan and K. K. Wang, *Pure Appl. Chem.*, **55**, 1387 (1983).
112. C. A. Brown and R. A. Coleman, *J. Org. Chem.*, **44**, 2328 (1979).
113. X. Wang, Y. Li, Y.-D. Wu, M. N. Paddon-Row, N. G. Rondan and K. N. Houk, *J. Org. Chem.*, **55**, 2601 (1990).
114. E. Block and A. L. Schwan, in *Comprehensive Organic Synthesis*, Vol. 4 (Eds. B. M. Trost and I. Fleming), Pergamon Press, Oxford, 1991, pp. 357–359.
115. R.-J. Binnewirtz, H. Klingenberger, R. Welte and P. Paetzold, *Chem. Ber.*, **116**, 1271 (1983).
116. S. Hara, T. Kato, H. Shimizu and A. Suzuki, *Tetrahedron Lett.*, **26**, 1065 (1985).
117. Y. Satoh, T. Tayano, H. Koshino, S. Hara and A. Suzuki, *Synthesis*, 406 (1985).
118. S. Hyuga, S. Takinami, S. Hara and A. Suzuki, *Chem. Lett.*, 459 (1986).

119. Y. Satoh, H. Serizawa, N. Miyaura, S. Hara and A. Suzuki, *Tetrahedron Lett.*, **29**, 1811 (1988).
120. Y. N. Bubnov, *Pure Appl. Chem.*, **59**, 895 (1987).
121. B. Wrackmeyer, *J. Organomet. Chem.*, **364**, 331 (1989).
122. J. J. Eisch, in *Comprehensive Organic Synthesis*, Vol. 8 (Eds. B. M. Trost and I. Fleming), Pergamon Press, Oxford, 1991, pp. 733–761.
123. P. Knochel, in *Comprehensive Organic Synthesis*, Vol. 4 (Eds. B. M. Trost and I. Fleming), Pergamon Press, Oxford, 1991, pp. 888–893.
124. S. Uemura, H. Miyoshi, M. Okano and K. Ichikawa, *J. Chem. Soc., Perkin Trans. 1*, 991 (1981).
125. D. J. De Young and R. West, *Chem. Lett.*, 883 (1986).
126. J. E. Baggott, M. A. Blitz, H. M. Frey, P. D. Lightfoot and R. Walsh, *J. Chem. Soc., Faraday Trans. 2*, **84**, 515 (1988).
127. G. Billeb, W. P. Neumann and G. Steinhoff, *Tetrahedron Lett.*, **29**, 5245 (1988).
128. L.-Z. Chen and J. M. Miller, *Can. J. Chem.*, **69**, 2002 (1991).
129. L.-Z. Chen and J. M. Miller, *Org. Mass Spectrom.*, **27**, 19 (1992).
130. A. J. L. Pombeiro and R. L. Richards, *Coord. Chem. Rev.*, **104**, 13 (1990).
131. C. B. Lebrilla, T. Drewello and H. Schwarz, *Int. J. Mass Spectrom. Ion Processes*, **79**, 287 (1987).
132. C. B. Lebrilla, C. Schulze and H. Schwarz, *J. Am. Chem. Soc.*, **109**, 98 (1987).
133. K. Eller and H. Schwarz, *Chem. Rev.*, **91**, 1121 (1991).
134. R. M. Stepnowski and J. Allison, *Organometallics*, **7**, 2097 (1988).
135. C. B. Lebrilla, T. Drewello and H. Schwarz, *Organometallics*, **6**, 2450 (1987).
136. K. Eller, S. Karrass and H. Schwarz, *Ber. Bunsenges. Phys. Chem.*, **94**, 1201 (1990).
137. M. Regitz, *Chem. Rev.*, **90**, 191 (1990).
138. L. N. Markovski and V. D. Romanenko, *Tetrahedron*, **45**, 6019 (1989).
139. S. M. Bachrach, *J. Comput. Chem.*, **10**, 392 (1989).
140. M. H. Palmer, *J. Mol. Struct.*, **200**, (*Theochem.*, **59**), 1 (1989).
141. S. Marriott, R. D. Topsom, C. B. Lebrilla, I. Koppel, M. Mishima and R. W. Taft, *J. Mol. Struct.*, **137**, (*Theochem.*, **30**), 133 (1986).
142. M. A. Cohen, J. Sawden and N. J. Turner, *Tetrahedron Lett.*, **31**, 7223 (1990).
143. N. Klempier, A. de Raadt, K. Faber and H. Griengl, *Tetrahedron Lett.*, **32**, 341 (1991).
144. G. Tennant, in *Comprehensive Organic Chemistry*, Vol. 2 (Ed. I. O. Sutherland), Pergamon Press, Oxford, 1979, p. 540.
145. R. Bishop, in *Comprehensive Organic Synthesis*, Vol. 6 (Eds. B. M. Trost and I. Fleming), Pergamon Press, Oxford, 1991, pp. 261–300.
146. A. G. Martínez, A. H. Fernández and F. M. Jiménez, *J. Org. Chem.*, **57**, 1627 (1992).
147. J. Bartl, S. Steenken, H. Mayr and R. A. McClelland, *J. Am. Chem. Soc.*, **112**, 6918 (1990).
148. J. Bartl, S. Steenken and H. Mayr, *J. Am. Chem. Soc.*, **113**, 7710 (1991).
149. M. Yalpani, R. Köster and R. Boese, *Chem. Ber.*, **125**, 15 (1992).
150. C. S. Christ, J. R. Eyler and D. E. Richardson, *J. Am. Chem. Soc.*, **112**, 4778 (1990).
151. I. D. Cunningham, G. J. Buist and S. R. Arkle, *J. Chem. Soc., Perkin Trans. 2*, 589 (1991).
152. C. A. Deakyne and M. Meot-Ner, *J. Phys. Chem.*, **94**, 232 (1990).
153. W. Kantlehner, in *Comprehensive Organic Synthesis*, Vol. 6 (Eds. B. M. Trost and I. Fleming), Pergamon Press, Oxford, 1991, p. 526.
154. I. Ugi, S. Lohberger and R. Karl, in *Comprehensive Organic Synthesis*, Vol. 2 (Eds. B. M. Trost and I. Fleming), Pergamon Press, Oxford, 1991, pp. 1083–1109.
155. T. Carofiglio, C. Floriani, A. Chiesi-Villa and C. Rizzoli, *Organometallics*, **10**, 1659 (1991) and earlier references cited therein.
156. E. Singleton and H. E. Oosthuizen, *Adv. Organomet. Chem.*, **22**, 209 (1983).
157. R. Aumann, *Angew. Chem.*, **100**, 1512 (1988); *Angew. Chem., Int. Ed. Engl.*, **27**, 1456 (1988).
158. S. Warner and S. J. Lippard, *Organometallics*, **8**, 228 (1989).
159. A. C. Filippou, S. Völkl and P. Kiprof, *J. Organomet. Chem.*, **415**, 375 (1991).
160. D. Seebach, *Angew. Chem.*, **91**, 259 (1979).
161. R. J. M. Nolte and W. Drenth, in *New Methods for Polymer Synthesis* (Ed. W. J. Mijs), Plenum, New York, 1992, pp. 273–310.
162. K. Eller and H. Schwarz, *Chem. Ber.*, **123**, 201 (1990).

CHAPTER **16**

Free-radical addition involving C—C triple bonds

C. CHATGILIALOGLU

I.Co.C.E.A., Consiglio Nazionale delle Ricerche, Via Gobetti 101, 40129 Bologna, Italy

and

C. FERRERI

Dipartimento di Chimica Organica e Biologica, Università di Napoli, Via Mezzocannone 16, 80134 Napoli, Italy

Supplement C2: The chemistry of triple-bonded functional groups
Edited by S. Patai © 1994 John Wiley & Sons Ltd

I. INTRODUCTION

The free-radical addition of molecules to triple bonds as well as their intermediate vinyl radicals has been periodically reviewed. Also, an extensive survey article on this subject has appeared in Patai's series[1]. For this reason this survey, which is not meant to be exhaustive, will reflect the scientific interest of authors, dealing mainly with recent literature.

During the last decade the application of radical addition reactions for solving problems in organic synthesis has grown, as documented by numerous review articles[2,3]. In this survey we will deal mainly with the synthetical aspects of the radical-based addition of a molecule to alkynes, with Section IV dedicated to radical cyclizations and to the preparation of biologically active compounds. In Section II, in order to familiarize the reader with the properties of the vinyl radical, we will briefly examine the structural characteristics of vinyl radicals and make some considerations about the stereochemical outcome of their reactions.

II. VINYL RADICALS

A. Structural Characteristics

Free-radical additions to alkynes generate vinyl radicals (equation 1) and information about the structure of these intermediates has been obtained either by spectroscopic or chemical means[4]. Vinylic intermediates are generally σ-type radicals (1), in which the unpaired electron is in an orbital with substantial s character. The degree of bending and the inversion barrier depends on the α-substituent. That is, for vinyl (R=H) the rate constant for the inversion lies between 3×10^7 and 3×10^9 s^{-1} at -180 °C, whereas 1-methylvinyl inverts somewhat more slowly[5]. Electronegative substituents, such as alkoxy, increase the barrier of inversion[6].

$$R' \!\!=\!\!\!=\!\! R \quad \xrightarrow{\text{X}\bullet} \quad \overset{X}{\underset{R'}{>}}\!\!=\!\!\overset{\bullet}{\sim}R \tag{1}$$

A second π-type structure (2), in which the unpaired electron resides in a p orbital, has often been invoked for vinyls having R groups capable of delocalizing the unpaired electron[4], e.g. R=Ph, C(O)OR, CN. However, EPR data indicate that vinyl radicals with α-C(O)OR substituents are 'bent' and that the 'linear' structure is highly probable for α-phenyl substituted vinyl radicals[7,8]. It is worth pointing out that the 'linear' structure has been identified with certainty for the sterically crowded vinyl 3[9]. Structural information obtained by the muon spin rotation (μSR) technique is consistent with this overall picture[10]. A qualitative explanation of the configurational instability of vinyl radicals (when R = H), compared with that of vinyl anions, is given in a recent paper[11].

(1)   (2)   (3)

The optical absorption spectrum of the $H_2C=CH\bullet$ radical shows two strong bands at the vacuum ultraviolet region[12] and a weak visible absorption attributed to an electronic transition between the ground ($^2A'$) and the first excited state ($^2A''$)[13].

B. Stereochemical Outcome

The stereochemistry of free-radical addition across C—C triple bonds is an intriguing subject. Fragments of homolyzed molecule, such as XY, may be added by either a *syn* or *anti* manner to form two geometrical isomers (equation 2). Although the stereochemical outcome of the addition process can easily be determined, the determination of stereochemical factors that control such a reaction is a difficult task. Two propagation steps are involved in these free-radical chain additions (Scheme 1). In the first step, the X· radical may add reversibly to alkyne to form a vinyl radical. In the absence of directing factors, both *cis* and *trans* radical intermediates can equally be formed as a result of a nonselective free-radical attack on the triple bond. Vinyl radicals can undergo inversion of configuration prior to the atom abstraction (second propagation step). Furthermore, as many radicals add reversibly to olefins, a postisomerization could change the stereochemical outcome. The mechanism depicted in Scheme 1 describes the most complicated situation, which is met more often than one may think. In fact, thiols and tin hydrides are the most popular hydrogen donors in free-radical chemistry and the corresponding thiyl and stannyl radicals are known to add reversibly to C—C triple and double bonds.

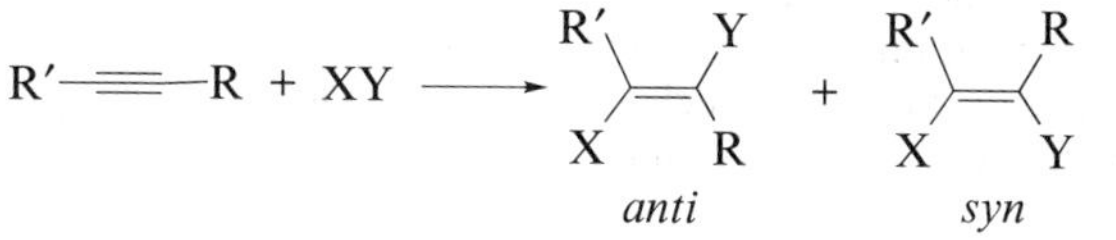

$$R'{=\!=\!=}R + XY \longrightarrow \underset{anti}{\overset{R'\quad Y}{\underset{X\quad R}{>\!=\!<}}} + \underset{syn}{\overset{R'\quad R}{\underset{X\quad Y}{>\!=\!<}}} \tag{2}$$

$$R'{=\!=\!=}R$$

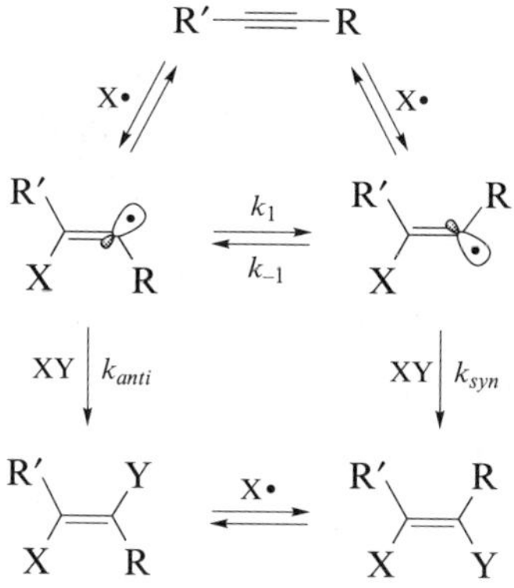

SCHEME 1

The limited number of absolute kinetic data for the individual radical steps in Scheme 1 do not facilitate a desirable synthetic strategy. However, following the fundamental concepts of free-radical chemistry the stereochemical control can be steered, at least in part, provided that postisomerization has been eliminated. Two limiting cases are possible[14,15]: (i) If inversion of the vinyl radical is fast relative to the chain-transfer step, the product ratio is determined by the relative populations of the two 'bent' forms and by stereoselectivity features in the scavenging steps (k_{syn}, k_{anti}). Steric effects have a dominant role in determining which configuration would require less free energy in the transition state. (ii) If the chain-transfer step is comparable to or faster than inversion, the olefin product ratio depends on the stereochemistry of the initial addition step and on the ratios k_1/k_{anti} and k_{-1}/k_{syn}. The former limiting case is also relevant to an addition involving a single, linear vinyl radical.

Giese and coworkers[16,17] have shown that the stereochemistry of vinyl radical **4** can also be guided in a particular direction by changing the abilities of hydrogen donation in the reducing agent and the reaction temperature. Thus, at 0 °C the ratio Z/E is 78:22 with Bu_3SnH as donor, whereas at 260 °C with cyclohexane as H-donor the selectivity (29:71) is reversed. Through the examined series of hydrogen donor, they observed that the compensation of activation entropies and activation entropies leads to an isoselective temperature, which lies between 60 and 80 °C.

$c\text{-}C_6H_{11}$

H

(4)

III. INTERMOLECULAR REACTIONS

A. Addition of Carbon-centered Radicals

Nucleophilic alkyl radicals, i.e. $c\text{-}C_6H_{11}\cdot$ or $t\text{-}Bu\cdot$, add to activated alkynes 3.0–5.2 times slower than the corresponding substituted alkenes[18], whereas nucleophiles having lone pairs of electrons attack alkynes markedly faster than alkenes[19]. Application of Frontier Orbital theory indicates early and late transition states for radical nucleophiles and nonradical nucleophiles, respectively[18].

Gilbert and coworkers employed ESR spectroscopy in studies of radical intermediates formed in the reaction of alkynes[20–23]. In particular, they study the addition of a variety of alkyl, α-hydroxyalkyl and aryl radicals to butynedioic acid. Rate constant at room temperature for the reaction of isopropyl and hydroxymethyl radicals with butynedioic acid are estimated to be 3×10^6 M^{-1} s^{-1} and 1×10^7 M^{-1} s^{-1}, respectively[21,22]. The first-formed vinyl radicals, though not normally detected, react with more alkyne to give a further vinyl radical or abstract a H atom intramolecularly. For example, the addition of a variety of α-alkoxyalkyl radicals to butynedioic acid leads to vinyl radicals most of which undergo a rapid 1,5-shift with a rate constant larger than 10^5 s^{-1} (equation 3)[23].

$$HO_2C \equiv\!\!\!=\!\!\!= CO_2H \ + \ CH_3OCH_2\cdot$$

$$(3)$$

The reaction of a variety of alkynes with alkyl iodides has been the subject of several papers[24–27]. Curran and coworkers have shown that simple secondary and tertiary alkyl iodides add to electron-deficient alkynes under sunlamp photolysis and in the presence of small quantities of $Bu_3SnSnBu_3$ (equation 4) in 70–80% yield[24]. On the other hand, primary alkyl radicals bearing an electron-withdrawing group such as C(O)OR, C(O)R or CN react smoothly with alkyl-substituted alkynes to give the desired product in similar yields[25]. These reactions, which occur by a two-step chain of radical addition and atom transfer (see above), are highly regioselective (*anti* Markovnikov orientation) and with lack of stereoselectivity. Only the phenylsulfonyl acetylene gave exclusively the (Z)-isomer. In order to interpret some unusual stereochemical data of these reactions, Curran and Kim proposed an equilibration between a 'linear' and a 'bent' arrangement of vinyl radical intermediates (equation 5) rather than two 'bent' structures (*cf* Scheme 1)[24].

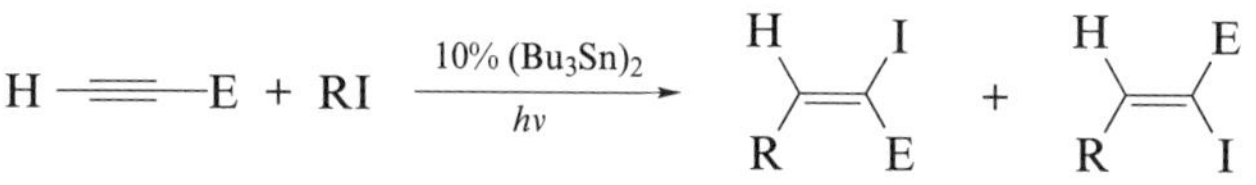

$$H\!\!=\!\!=\!\!E + RI \xrightarrow[h\nu]{10\% \ (Bu_3Sn)_2} \quad \underset{R \quad E}{\overset{H \quad I}{\diagdown\!\!=\!\!\diagup}} \ + \ \underset{R \quad I}{\overset{H \quad E}{\diagdown\!\!=\!\!\diagup}} \tag{4}$$

$$\underset{R}{\overset{H \quad E}{\diagdown\!\!=\!\!\diagup}}\!\!\!\cdot \ \rightleftharpoons \ \underset{H}{\overset{R}{\diagdown\!\!=\!\!\diagup}}\!\!\!\cdot\!\!-E \tag{5}$$

Oshima and coworkers have shown that triethylborane induced radical addition of alkyl iodides to monosubstituted alkynes[26]. With this methodology, the yields are somehow better and the reaction proceeds also with less activated alkynes. As before, these reactions proceeded with high regioselectivity and poor stereoselectivity with the exception of Me$_3$Si- and Ph$_3$Sn-substituted acetylenes where (Z)-isomers are the only observed products. Treatments of terminal or internal acetylenes bearing a variety of substituents with perfluoroalkyl iodides produce the corresponding alkene in good to excellent yields with high regio- and stereoselectivities to give exclusively the *anti*-addition products[27]. An example is reported in equation 6.

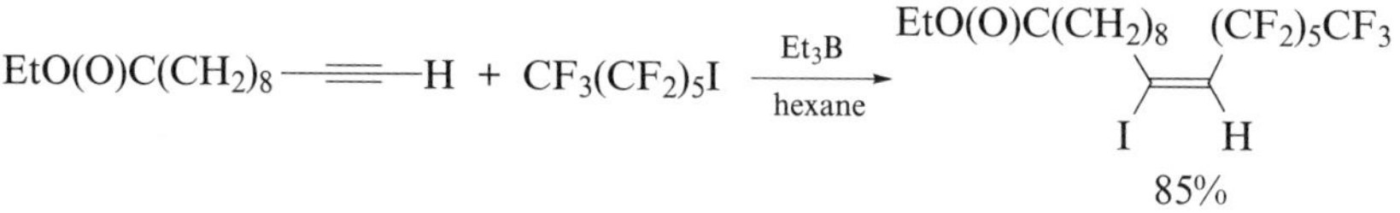

$$EtO(O)C(CH_2)_8\!\!=\!\!=\!\!H + CF_3(CF_2)_5I \xrightarrow[hexane]{Et_3B} \quad \underset{I \quad H}{\overset{EtO(O)C(CH_2)_8 \quad (CF_2)_5CF_3}{\diagdown\!\!=\!\!\diagup}} \tag{6}$$
$$85\%$$

B. Additions of Heteroatom-centered Radicals

1. Sulfur-centered radicals

An extensive survey article on thiyl radicals has recently appeared in Patai's series[28]. Absolute rate constants for the addition of *para*-substituted benzenethiyl radicals to monosubstituted acetylenes has been determined by a flash photolysis technique (Scheme 2)[29,30,30a]. There is a large spread in the reactivities (rate constants vary by 4 orders of magnitude, i.e. 1×10^3 to 1×10^7 M^{-1} s^{-1} at ambient temperature). Evidence that the addition is reversible has also been obtained. The Hammett relations gained by changing the substituents either of the arenethiyl radicals or of the alkynes (Scheme 2) were also investigated to obtain information about the substituent effects. The ρ^+ values character-

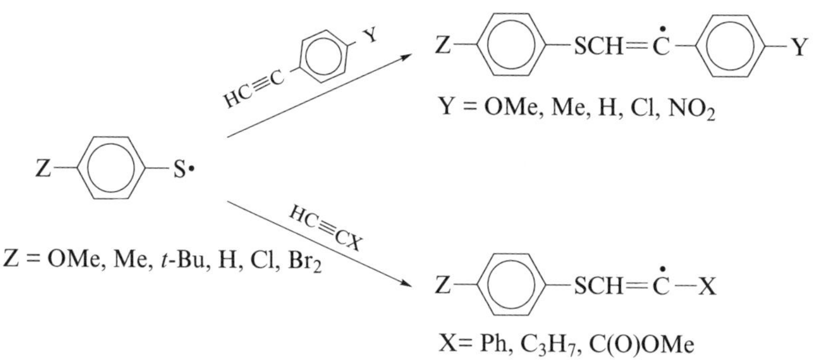

SCHEME 2

izing these reactions indicate an important contribution to the transition state of such polar structures as [p-ZC$_6$H$_6$S$^-$, HC≡CR$^+$]. However, the reactivity of PhC≡CX towards either c-C$_6$H$_{11}$· or PhS· radicals decreases from X = SO$_2$Ph to I to SnR$_3$ or HgR, indicating that both radicals behave as nucleophiles in these reactions, i.e. [PhS$^+$, XC≡CPh$^-$][30b].

The behavior of vinyl radicals (5) generated by addition of a variety of thiyl radicals to butynedioic or propynoic acids has been studied by ESR spectroscopy[31]. The intermolecular abstraction of a thiol hydrogen is in competition with 1,n-hydrogen shifts. An unusual 1,4-shift (equation 7) is shown to occur in cases where the resulting carbon-centered radical bears α-sulfur and α-carboxy substituents, whereas in other examples a 1,5-shift predominates (equation 3).

$$
\begin{array}{ccc}
\underset{\text{(5)}}{
\begin{array}{c}
\text{HO}_2\text{C} \quad\quad \text{CO}_2\text{H} \\
\diagdown\quad\diagup \\
\text{C}=\text{C} \\
\diagup\quad\quad\bullet \\
\text{S} \\
\diagdown \\
\text{CHR} \\
\diagup \\
\text{HO}_2\text{C}
\end{array}}
&
\xrightarrow{\text{1,4-shift}}
&
\begin{array}{c}
\text{HO}_2\text{C} \quad\quad \text{CO}_2\text{H} \\
\diagdown\quad\diagup \\
\text{C}=\text{C} \\
\diagup\quad\quad\diagdown \\
\text{S} \quad\quad \text{H} \\
\diagdown \\
\overset{\bullet}{\text{C}}\text{R} \\
\diagup \\
\text{HO}_2\text{C}
\end{array}
\end{array}
\tag{7}
$$

The addition of thiols to triple bonds is an efficient process leading to β-vinyl sulfides[1]. Although the rate constants for the addition of thiyl radicals to olefins are generally larger than the analogous alkyne reactions[29], in practice, the addition of thiols to acetylenes occurs much more efficiently than to olefins[1]. The main reason for such behavior is probably the higher reversibility of the thiyl radical addition to olefins compared to alkynes. These reactions, which are generally initiated under thermal or photochemical conditions, have recently been performed in the presence of Et$_3$B in good yields (equation 8)[32].

$$
\text{HOCH}_2\text{CH}_2-\!\!\!\equiv\!\!\!-\text{H} + \text{PhSH} \xrightarrow[\text{Et}_3\text{B}]{25\,°\text{C, 4 h}}
\begin{array}{c}
\text{HOCH}_2\text{CH}_2 \quad\quad \text{H} \\
\diagdown\quad\quad\diagup \\
\diagup\quad\quad\diagdown \\
\text{H} \quad\quad \text{SPh}
\end{array}
\tag{8}
$$
$$
91\%\ (Z{:}E = 4{:}6)
$$

Montevecchi and coworkers, in order to investigate the chemical reactivity of various substituted 2-(phenylthio)vinyl radicals without the interference of postisomerization reaction, performed the addition of PhSH in various neat alkynes[33a]. Phenyl-substituted acetylenes gave mainly the (Z)-adduct via *anti* addition (equation 9), whereas with alkyl-substituted acetylenes the stereochemistry was claimed to depend strongly upon the nature of substituents. However, a subsequent reinvestigation showed that *anti* addition occurs predominantly also with alkyl-substituted acetylenes[33b]. To explain their findings, they suggest that in the intermediate vinyl radical there is a bonding interaction between the unpaired electron and the adjacent sulfur, which would essentially prevent attack from the radical scavenger on the side *syn* to PhS.

$$
\text{Ph}-\!\!\!\equiv\!\!\!-\text{X} + \text{PhSH} \xrightarrow{100\,°\text{C}}
\begin{array}{c}
\text{Ph} \quad\quad \text{X} \\
\diagdown\quad\quad\diagup \\
\diagup\quad\quad\diagdown \\
\text{H} \quad\quad \text{SPh}
\end{array}
\tag{9}
$$

$$
\begin{array}{ll}
\text{X = H, Me, Et, Pr} & Z{:}E = 9{:}1 \\
\text{X = } t\text{-Bu} & Z{:}E = 10{:}0
\end{array}
$$

The reaction of diphenyl disulfide with a variety of substituted acetylenes, promoted by thermal decomposition of di-*tert*-butyl peroxide, provides 1,2-bis(phenylthio)ethylene adducts together with the corresponding benzothiophenes in reasonable yield[34]. The mechanism conceived for these transformations is depicted in Scheme 3, where the key step is the intramolecular cyclization of vinyl radicals.

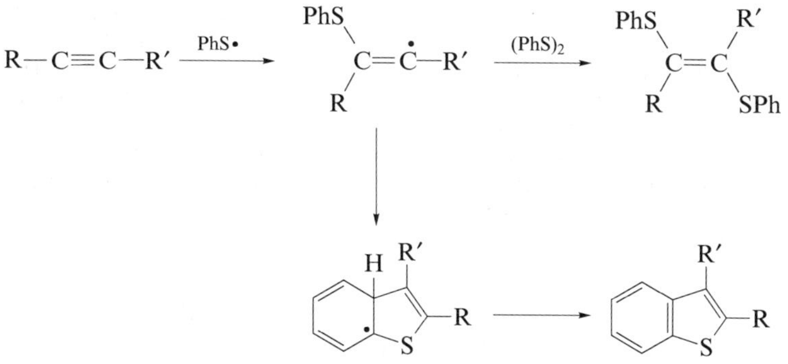

SCHEME 3

Another synthetic route to benzothiophenes has been envisaged[35] in the reaction of o-thioalkyl and o-thioaryl substituted phenyl radicals, generated from the corresponding tetrafluoroborate via electron transfer, with alkynes (equation 10). Yields are good to excellent and some mechanistic considerations have been proposed[35].

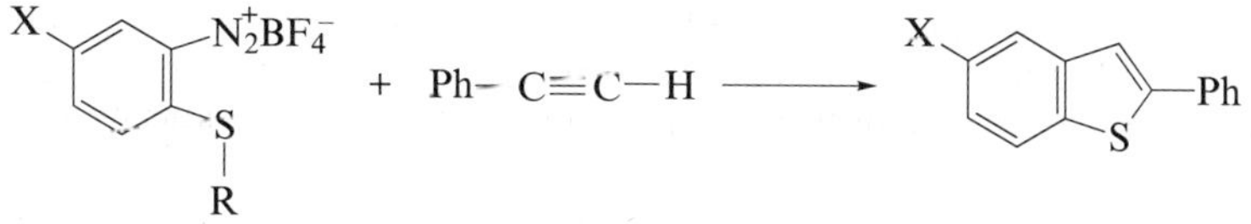

$$\text{(10)}$$

The ability of the β-(phenylthio)vinyl radicals, generated by addition of thiyl radicals to alkynes, to abstract hydrogen atom from appropriately substituted carbon atoms has been studied as a model for the reactions of the deoxyribose units of DNA with similar radicals in vivo[36].

Treatment of monosubstituted acetylenes with alkanethiols under an atmosphere of carbon monoxide gives β-alkyl-α,β-unsaturated aldehydes in reasonable yields (equation 11)[37]. A mechanistic scheme where the alkanethiyl radical adds to the alkyne to give the β-(alkylthio)vinyl radical followed by the reaction with carbon monoxide has been indicated as the key steps in these radical chain reactions. Only the thermodynamically stable (E)-isomer of the carbonylation product was obtained, probably due to a postisomerization process.

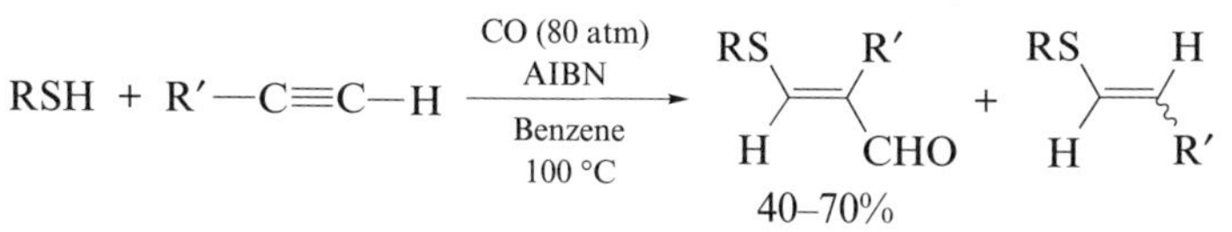

$$\text{(11)}$$

40–70%

The addition of aliphatic and aromatic sulfonyl halides to triple bonds has been accomplished under a variety of experimental conditions[1,38]. Light-catalyzed additions of sulfonyl iodides to acetylenes[39] as well as the thermal addition of sulfonyl bromides to phenylacetylene[40] to form 1:1 adducts have been shown to be stereoselective and to occur in good to excellent yields (equation 12). However, the mechanistic features of these reactions are still open to further investigation. On the other hand, the copper-catalyzed addition of aliphatic and aromatic sulfonyl chlorides[41] or bromides[40] to acetylenes yield mixtures of *trans*- and *cis*-β-halovinyl sulfones. Amiel[40] suggested that the *anti* addition product is a result of a normal radical chain, while the *syn* addition product, which is formed concurrently in the copper-catalyzed reaction, arises presumably from a concerted reaction. Relative reactivies of the addition of *p*-toluenesulfonyl iodide to a variety of substituted phenylacetylenes have been measured and the substituent effect has been investigated using Hammett relations[42]; electro-donating substitutions accelerate the reaction rates which confirm the electrophilic character of the sulfonyl radical and indicate the importance of the polar effects on the transition state.

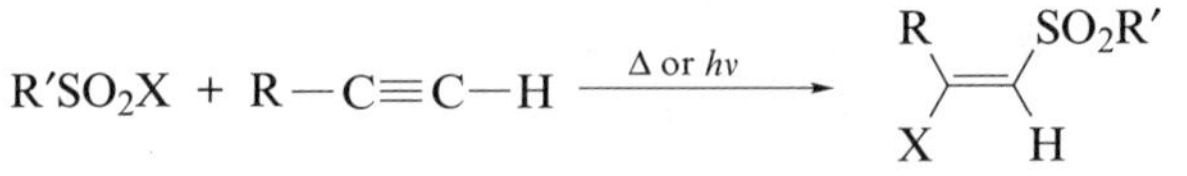

$$R'SO_2X \ + \ R-C{\equiv}C-H \quad \xrightarrow{\Delta \ or \ h\nu} \quad \underset{X \quad\quad H}{\overset{R \quad\quad SO_2R'}{\diagup\diagdown}} \tag{12}$$

A related addition on alkynes has been studied using Se-phenyl areneselenosulfonates ($ArSO_2SePh$)[43,44]. The reaction products and the mechanism are outlined in Scheme 4. The reaction being highly regioselective (*anti*-Markovnikov) and stereoselective (*anti*) is synthetically useful; the β-(phenylseleno)vinyl sulfones can be, in fact, easily transformed by oxidation and elimination in the corresponding acetylenic sulfones. The free-radical selenosulfonation of cyclopropylacetylene has also been studied in order to estimate the rate of chain-transfer relative to ring opening of reaction[45]. Thus, using equimolar quantities of the two reagents, 1,2- and 1,5-adduct are formed in 46% and 24% yield, respectively, indicating that the chain-transfer process competes effectively with the ring-opening (Scheme 5). In agreement with expectation, using an excess of selenosulfonate the

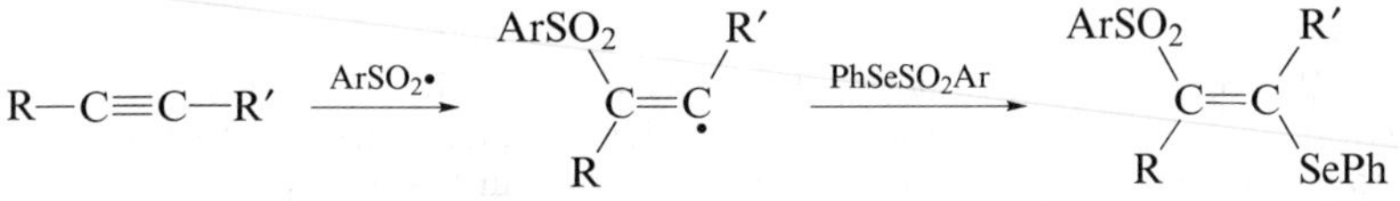

SCHEME 4

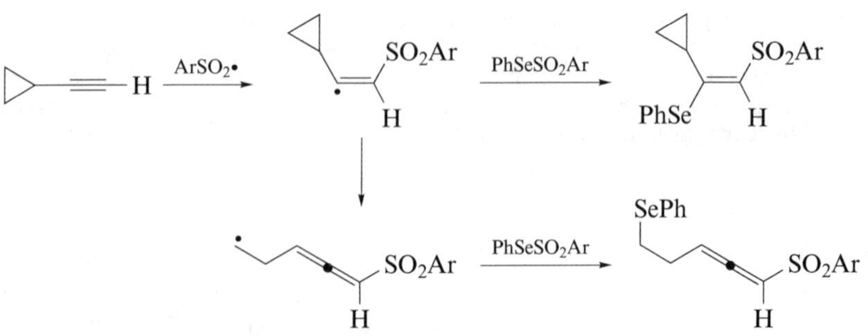

SCHEME 5

bimolecular chain-transfer has been favored at the expense of the unimolecular ring opening, raising the yield of 1,2-adduct up to 76%.

Based on ESR spectroscopy coupled with a rapid-mixing flow technique, it has been found that the reactions of $SO_3^{\cdot-}$ with 3-methylbut-1-yn-3-ol, but-1-yn-3-ol and propy-2-ynyl alcohol lead to vinyl radicals (equation 13)[46]. It has been suggested that the detection of these radicals is due to their stabilization by the SO_3^- ion.

$$SO_3^{\cdot-} + HO\underset{\underset{R}{|}}{\overset{\overset{R}{|}}{C}}-C{\equiv}CH \longrightarrow HO\underset{\underset{R}{|}}{\overset{\overset{R}{|}}{C}}-\overset{\cdot}{C}{=}CHSO_3^- \qquad (13)$$

$$R = H \text{ or } Me$$

2. Selenium- and tellurium-centered radicals

Considering the known addition of selenosulfonates, diphenyl diselenide has also been used in free-radical reactions with acetylenes. The first example was reported in 1987 and describes the photochemical reaction of diphenyl or dimesityl diselenide with dimethyl acetylenedicarboxylate or methyl propiolate to give the corresponding vic-bis(phenylseleno) alkenes in ca 90% yield. The predominant product was the (E)-isomer[47]. To obtain the successful phenylseleno radical additions not only in the case of electron-deficient alkynes, working with higher initial concentrations of the reactants and using efficient means of breaking homolytically the Se—Se bond have been suggested[48]. Under these conditions, phenylacetylene (equation 14) and a variety of other alkynes have been found to react, giving mainly the anti-addition products. The reaction proceeds via a free-radical chain mechanism that involves the formation of a vinyl-type radical by the addition of the phenylseleno radical to the acetylene. Subsequent S_H2 attack of vinyl radical on diphenyl selenide leads to the formation of 1,2-bis(phenylseleno)alkene and regenerates the phenylseleno radical.

$$Ph-C{\equiv}C-H + PhSeSePh \xrightarrow{h\nu} \underset{\underset{Ph}{}}{\overset{\overset{PhSe\qquad H}{}}{C{=}C}}\underset{SePh}{} \qquad (14)$$

$$83\% \ (E/Z = 82/18)$$

Although benzeneselenol adds across the triple bond of phenylacetylene in the presence of molecular oxygen to give the corresponding vinyl selenide (equation 15), inactivated acetylenes proceed very slowly under similar conditions[49a]. However, the addition of PhSeH to inactivated acetylenes has been achieved in the presence of a catalytic amount of diselenide upon irradiation (equation 16)[49a]. Evidence that these reactions proceed via a radical mechanism is also given. Tris(phenylseleno)borane and tris(methylseleno)borane react also with terminal acetylenes to give (Z)-vinyl selenides[49b].

$$Ph{-\!\!\equiv} + PhSeH \xrightarrow{O_2, 25\,°C} \underset{SePh}{\overset{Ph\diagdown}{\diagup}} \qquad (15)$$

$$83\% \ (Z/E > 95/5)$$

$$\text{HOCH}_2\text{—}\!\!\equiv\ +\ \text{PhSeH}\ \xrightarrow[\text{40 °C, } hv]{\text{10 mol\% (PhSe)}_2}\ \begin{array}{c}\text{HOCH}_2\\ \diagdown\!\!=\!\!\diagup\\ \text{SePh}\end{array}\qquad(16)$$

$$62\%\ (Z/E = 64/36)$$

The synthesis of conjugated polymers by the addition polymerization of 1,4-benzenedithiol[50] and 1,4-benzenediselenol[51] to 1,4-diethynylbenzene has been accomplished by means of free-radical reaction. For example, the polymerization of 1,2-benzenediselenol (equation 17) proceeds at such a fast rate as to give 60–70% yield in 6 min. The polymer was isolated and characterized in the *cis/trans* content of double bonds, finding a ratio of 90/10. The isomerization of the *cis* double bond has been obtained by UV irradiation for longer reaction times (18 h).

$$\text{HSe}-\!\!\langle\bigcirc\rangle\!\!-\text{SeH}\ +\ \text{HC}\!\equiv\!\text{C}-\!\!\langle\bigcirc\rangle\!\!-\text{C}\!\equiv\!\text{CH}$$

$$(17)$$

$$\xrightarrow[\text{toluene}]{\text{UV–irr., 60 °C}}\ \left(\!\!\text{HC}\!=\!\text{CH}-\!\!\langle\bigcirc\rangle\!\!-\text{CH}\!=\!\text{CH}-\text{Se}-\!\!\langle\bigcirc\rangle\!\!-\text{Se}\!\!\right)_{\!n}$$

Diphenyl ditelluride has also been found to add efficiently across acetylenic compounds upon irradiation with visible light, providing *vic*-bis(phenyltelluro)alkenes in good yield[52]. PhTeTePh undergoes homolysis upon irradiation with visible light to form *in situ* phenyltelluro radicals (PhTe·); the reactivity of this radical toward carbon–carbon unsaturated bonds is very poor and, in order to succeed in this radical addition, higher concentrations of substrates are essential. Thus, irradiation with wavelengths > 400 nm at 40 °C for several hours of unactivated 1-alkynes, i.e. 1-octyne, gave satisfactory yield of the (*E*)-1,2-bis(phenyltelluro)-1-alkenes. The addition to activated acetylenes, i.e. phenylacetylene, requires lower wavelengths (300 nm) and shorter reaction times.

3. Stannyl-, germyl- and silyl-centered radicals

The hydrostannation of alkenes, which has been known for nearly 30 years, follows a polar or a free radical way depending on substituents and conditions[53,54]. Under free-radical conditions, the stannyl radical generally adds reversibly either to double or triple bonds and this property has been used to control the vinyltin configuration depending on experimental conditions (equation 18). Therefore, with an excess either of acetylenic compound or of tin hydride, it is possible to obtain mainly the (*Z*)-isomer (kinetic control) or the (*E*)-isomer (thermodynamic control)[55].

$$\text{R}\text{—}\!\!\equiv\!\!\text{—H}\ \xrightarrow{\text{Bu}_3\text{SnH}}\ \begin{array}{c}\text{R}\\ \diagdown\!\!=\!\!\diagup\\ \end{array}\!\!\text{SnBu}_3\ \underset{\text{Bu}_3\text{Sn}\cdot}{\rightleftarrows}\ \begin{array}{c}\text{R}\\ \diagdown\!\!=\!\!\diagup\\ \text{SnBu}_3\end{array}\qquad(18)$$

During the last decade, the usefulness of the hydrostannation of acetylenes as the first step in the synthesis of cyclic compounds has been demonstrated; vinyl radical cyclization by this route is a well-established synthetic strategy and this topic will be discussed in some detail in the next section. However, a mild procedure has recently been described, making use of triethylborane as initiator; the Et₃B-induced reaction shows two main features[56,57]

(i) the stannyl radical ($R_3Sn\cdot$) can be generated at low temperature ($-78\ °C$) and the reaction conditions are extremely mild; (ii) the reaction takes place easily in various solvents under high diluted conditions, so it can be applied to cyclization routes.

Allenyltins are conveniently prepared from the terminal alkynes which have a suitable leaving group[58]. An example is shown in equation 19. A two-step one-pot procedure for the synthesis of allene by hydrostannation of alkynes has recently been reported starting from propargylic alcohols; a hydrostannation and subsequent deoxystannylation generates the allenes **6** as shown in equation 20[59]. A chiral version of this procedure has also been described in the paper.

$$Bu_3SnH\ +\ \equiv\!\!-\!\!/^{SR}\ \longrightarrow\ Bu_3Sn\diagdown\!\!=\!\!\bullet\!\!=\quad\quad (19)$$

$$\underset{R^2}{\overset{OH}{R^1\diagup\diagdown\diagup}}\ \xrightarrow[\text{AIBN, heat, 90 °C}]{n\text{-}Bu_3SnH\ (1.1\ eq)}\ \underset{SnBu_3}{\overset{OH}{R^1\diagup\diagdown\!\!=\!\!/^{R^2}}}\ \xrightarrow[\text{CH}_2\text{Cl}_2,\ 0\ °C]{MsCl,\ Et_3N}\ \underset{(6)}{\overset{R^1}{\diagdown\!\!=\!\!\bullet\!\!=\!\!/_{R^2}}}\quad (20)$$

Triphenylgermane has been shown to add to acetylenes in the presence of Et_3B to give alkenyltriphenylgermanes under excellent control of regio- and stereoselectivities[60]. The isomeric Z/E ratios of the products mainly depend on the reaction temperature as well as on the ratio acetylene/Ph_3GeH; at $-78\ °C$ in toluene with a slight excess of 1-dodecyne, (Z)-alkenylgermane has exclusively been isolated, whereas at $60\ °C$ (E)-alkenylgermane is the sole product. The $Ph_3Ge\cdot$ radical is effective to isomerize (Z)-alkenes into their thermo-dynamically more stable (E)-isomers[60].

The addition of silanes to acetylenes is of considerable industrial interest and is catalyzed by transition metal complexes[61]. The hydrosilylation of acetylenes from a free-radical point of view has been studied to a much lesser extent and limited mainly to $Cl_3SiH^{1,62}$. In recent years, tris(trimethylsilyl)silane as an alternative to Bu_3SnH has become more and more popular[63], being a superior reagent from both ecological and practical perspectives. The free-radical addition of tris(trimethylsilyl)silane to a number of mono- and disubstituted acetylenes has been studied[64]. The reaction, which is initiated by either AIBN or Et_3B/O_2 under mild conditions, gives tris(trimethylsilyl)silyl-substituted alkenes in good yields via a radical chain mechanism. The reaction is highly regioselective (*anti*-Markovnikov) and can also show high *cis* or *trans* stereoselectivity depending on the nature of the substituents at the acetylenic moiety. Two examples are given in equations 21 and 22. Although the $(TMS)_3Si\cdot$ radical is shown to isomerize some (Z)-alkenes into their thermodynamically more stable (E)-isomers[65], the postisomerization of the hydrosilylation adduct could not be observed due probably to steric hindrance[66].

$$EtO_2C\!\!=\!\!\!\!=\!\!-H\ +\ (TMS)_3SiH\ \xrightarrow[25\ °C]{Et_3B/O_2}\ \underset{88\%}{\overset{EtO_2C}{\diagdown\!\!=\!\!/^{Si(TMS)_3}}}\quad (21)$$

$$Ph\!\!=\!\!\!\!=\!\!-CO_2Et\ +\ (TMS)_3SiH\ \xrightarrow[80\ °C]{AIBN}\ \underset{85\%}{\overset{Ph}{\diagdown\!\!=\!\!\diagup\!\!\overset{CO_2Et}{\diagdown_{Si(TMS)_3}}}}\quad (22)$$

4. Other heteroatom-centered radicals

Hydroxy radical initiated oxidation of alkynes is important from the point of view of both atmospheric[67] and combustion chemistry[68]. Hatakeyama and coworkers have measured rate constants for the reaction of HO· with acetylene, propyne and 2-butyne under atmospheric conditions[69]. It has been suggested, based on product studies[69], that the β-hydroxyvinyl radicals further react with molecular oxygen to form the corresponding peroxyl radicals[70] and their subsequent reactions give carboxylic acid, α-dicarbonyl compounds and acyl radicals.

ESR results reveal the occurrence of acid-catalyzed conversion of β-hydroxyvinyl radical **7** into carbonyl-conjugated radical **8** in aqueous solution[71]. It is suggested that the reaction involves the rapid protonation of the intermediate enol radical at carbon ($k > 5 \times 10^6$ M^{-1} s^{-1}) followed by deprotonation of the intermediate enol radical cation (equation 23).

$$
\begin{array}{ccc}
\underset{\text{HO}}{\overset{\text{HOCH}_2}{\diagup}}\!\!\diagdown\!\!\overset{\text{CH}_2\text{OH}}{\underset{\bullet}{\diagup}} & \xrightarrow{\;\text{H}^+\;} & \underset{\text{HO}\;\overset{\bullet}{\text{H}}}{\overset{\text{HOCH}_2}{\diagup}}\!\!\diagdown\!\!\overset{\text{CH}_2\text{OH}}{\underset{+\bullet}{\diagup}} \xrightarrow{\;-\text{H}^+\;} \underset{\text{O}\;\;\;\text{H}}{\overset{\text{HOCH}_2}{\diagup}}\!\!\diagdown\!\!\overset{\text{CH}_2\text{OH}}{\underset{\bullet}{\diagup}} \\
(\mathbf{7}) & & (\mathbf{8})
\end{array}
\tag{23}
$$

Diphenylphosphine adds readily to a variety of mono- and disubstituted acetylenes under free-radical conditions[72]. Although (Z)-isomer predominates in the final product mixture, it is clear from a study of the time dependence of the E/Z ratio that a post-isomerization takes place during the reaction.

IV. INTRAMOLECULAR REACTIONS

Both alkyl and vinyl radicals can add to double and triple bonds in an intramolecular fashion, thus generating cyclic products. Since the pioneering work carried out by Stork[73], radical cyclizations have found broad application in organic synthesis and now represent a powerful tool available to synthetic chemists. The chemo-, regio- and stereoselectivity as well as predictability and functional group tolerance of these reactions, together with the entropic advantages, are all important features that justify the great success they have achieved. Several excellent reviews[6,74–76] as well as general guidelines[3] and books[2] have been published. In this section the authors will present some results which appeared in the recent literature, referring to the above-reported references for other meaningful examples as well as for basic information.

A. Addition of Carbon-centered Radicals

The use of carbon-centered radicals in the intramolecular addition to double or triple bonds was introduced by Julia[77]. Cyclization of acetylenic alkyl radicals generated from several bromides was reported by Crandall and Keyton[78], in which the carbon chain separating the reactive center and the acetylenic unit is varied in length (Scheme 6). Competition between the cyclization process to give **12** and the simple hydrogen abstraction to give **11** has been observed; cyclization occurs when the intermediate alkyl radical **9** adds to a triple bond intramolecularly, thus generating the second intermediate vinyl radical **10**. This reaction has to be faster than the reduction step and the authors' suggestion was that, for $n=4$, cyclization resulted fast enough to be efficient and useful in synthesis.

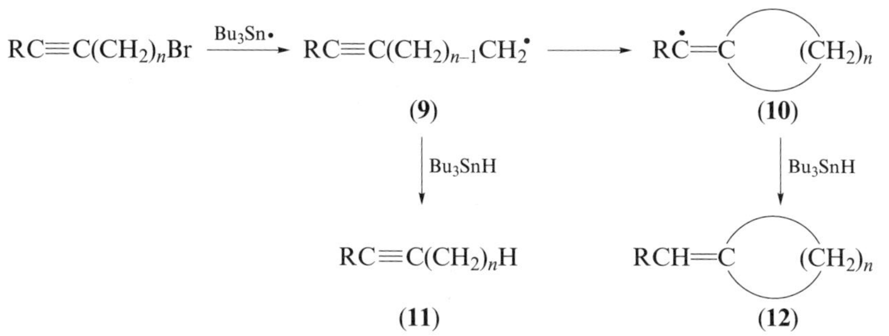

SCHEME 6

Following this strategy, Clive and coworkers constructed a suitable substrate in which an alkyl radical could cyclize onto a triple bond[79,80]. Scheme 7 illustrates the synthesis, starting from a ketone which is first converted to the corresponding enamine. The reaction of the enamine with a Michael acceptor has been carefully evaluated; the Michael acceptor satisfies the requirement of having a group A of such a nature that the C—A bond can be easily homolytically cleaved by a stannyl radical (i.e. SePh or Br). Moreover, it has group B which can be manipulated for further transformations (i.e. SO_2Ph). After the introduction of the α-chain the ketone undergoes the lithium acetylide addition, which suitably locates the triple bond in the molecule for the cyclization. Treatment of the substrate with Ph_3SnH gives the desired methylenecyclopentanone via the predicted cyclization.

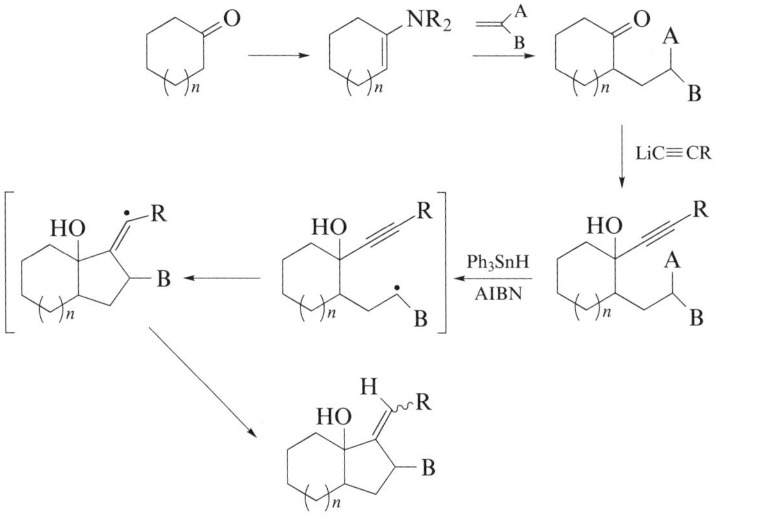

SCHEME 7

A radical-mediated two-step synthesis of spiroketals has recently been reported[81]; the strategy (an example is shown in Scheme 8) involves an intramolecular radical cyclization of a ketal precursor (**14**), which could be easily prepared from 2-methylene-3,4-epoxyoxolanes (**13**). As alternative to tin-hydride mediated cyclization of alkynyl halides, samarium(II) iodide has been used to generate the alkyl radical which adds intramolecularly to the triple bond[82].

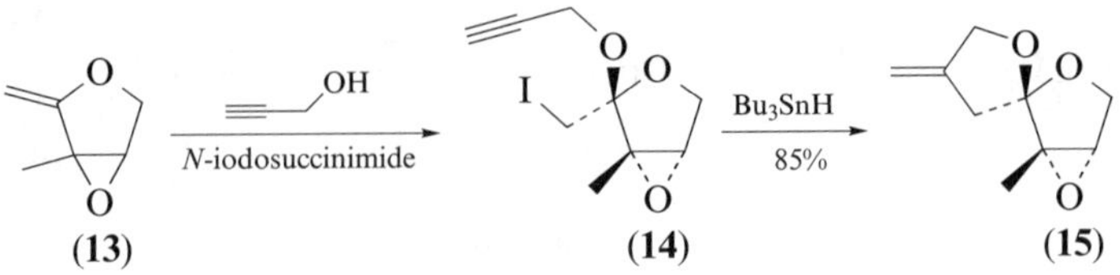

SCHEME 8

A procedure for the synthesis of α-alkylidene-γ-butyrolactones has been conceived and applied by Bachi and Bosch (Scheme 9)[83]. That is, the Bu₃SnH/AIBN-induced carbolactonization of selenocarbonates derived from homopropargylic alcohols proceeds in excellent yield and provides a useful method for the synthesis of monocyclic and bicyclic α-alkylidene-γ-butyrolactones. The free-radical cyclization is characterized by high regioselectivity favoring the five-membered ring. The ratio between *E/Z* stereoisomers about the exocyclic double bond (16) was found to be strongly affected by the nature of the substituents. However, the postisomerization reaction by Bu₃Sn· radical was not taken into account and, therefore, a rationalization of the stereochemical outcome is not possible. In some cases, the feasibility of the 6-*exo*-cyclization for the synthesis of δ-lactones has also been studied[83]. In an analogous route for the formation of lactones, dithiocarbonates of homopropargylic alcohols were shown to be useful as starting materials[84].

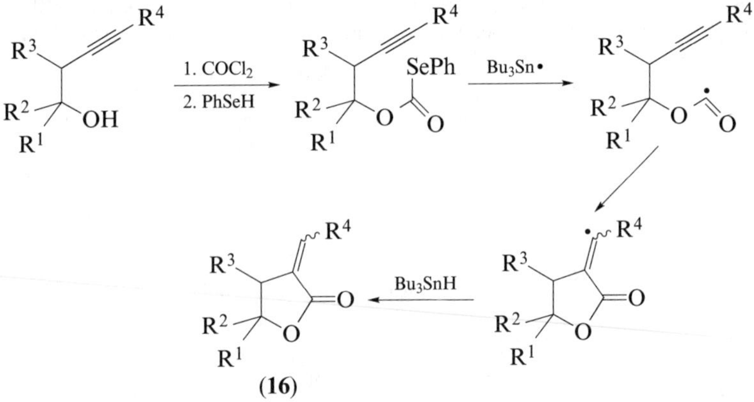

SCHEME 9

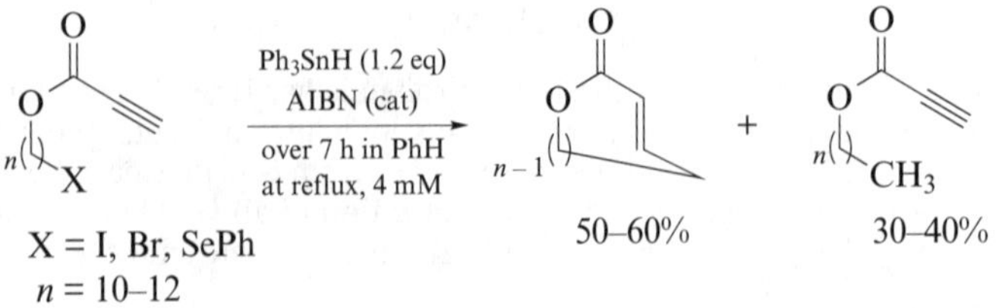

A useful route to 14–16-membered α,β-unsaturated macrocyclic lactones from their corresponding ω-iodoalkyl-propiolate esters under Ph₃SnH/AIBN-mediated conditions has been described by Baldwin's group (equation 24)[85]. As starting materials, bromides and selenides were found to be inferior to the corresponding iodides. Cyclizations were both regio- and stereospecific as only the *trans*-isomers resulting from preferential *endo*-attack were observed. It has been suggested that the initially formed vinyl radical, resulting from a *transoid* orientation, inverts prior to hydrogen abstraction. However, another explanation for the formation of thermodynamically observed *trans*-cyclic product may be the postisomerization which was not taken into consideration for some unclear reasons. Attempts to synthesize analogous 10–13-membered lactones have proven to be unsuccessful, resulting in acyclic products derived from direct reduction of the halide moiety[85].

The extention of radical cyclization of (bromomethyl)dimethylsilyl allyl ethers[86] to propargyl analogs **17** has been studied by Malacria and coworkers[87–89]. The intermediate exocyclic vinyl radical **18** can be either trapped by the hydrogen atom to give, after simple chemical transformations, the trisubstituted alkene **19** (equation 25)[87] or can be added intramolecularly to give cyclic products when suitably located double bonds are present (equations 26 and 27)[88]. An attempt to apply this methodology to the stereoselective synthesis of angular and linear triquinane has also been performed[88]. When R^3 = *tert*-butyl, radical **18** showed an unusual 1,4-hydrogen migration from the R^1 moiety to the vinyl site[89].

Atom transfer cyclization reactions provide a nonreductive complement to the tin-hydride methodology for conducting radical cyclizations. Curran and collaborators have studied the atom transfer cyclization reaction of a variety of substituted iodides (equation 28) and a number of mono- and polycyclic derivatives has been obtained by this route[90]. Two examples of fused and spiro ring formation are reported in equations 29 and 30, respectively. To initiate these chain reactions small amounts of Bu₃SnH/AIBN or Bu₃SnSnBu₃, coupled with heat (85 °C) or light (sunlamp), respectively, are necessary. However, Et₃B as initiator (equation 31) not only increases the overall yield but also avoids the use of toxic organotin compounds[26]. Mechanistic considerations based on the Curtin–Hammett kinetic scheme have been used to analyze the stereochemical trends[90]. It is believed that the inversion of the radical intermediate is more rapid than iodine atom transfer whose rate has been estimated to be 10^8–10^9 M^{-1} s^{-1}.

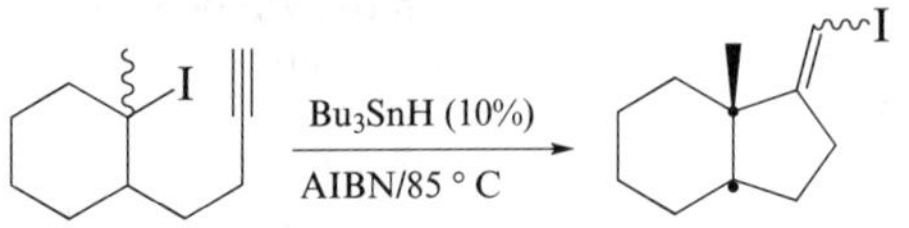

$$(29)$$

40% (E/Z = 95/5)

$$(30)$$

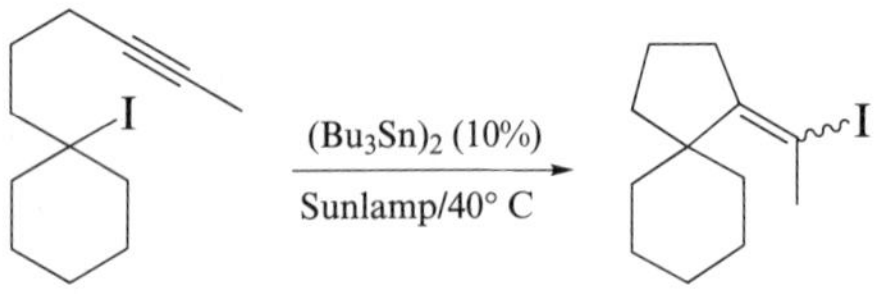

53% (E/Z = 83/17)

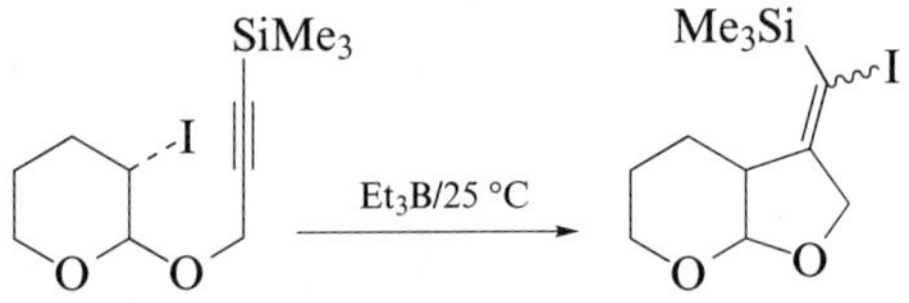

$$(31)$$

94%

In recent years, radical annulation reactions have become of considerably importance in organic synthesis[3]. Curran and coworkers have introduced annulations with alkenes and propargyl iodomalonic esters[90] or iodomalononitriles[91]. Heating of propargyl iodomalononitrile with mono-, di- or trisubstituted alkenes usually produces an intermediate adduct 20 (Scheme 10). This adduct is either reduced with tributyltin hydride to give 21a, or isomerized under iodine transfer conditions to give 21b. These radical annulations are often highly regio- and stereoselective and the overall yields are good to excellent.

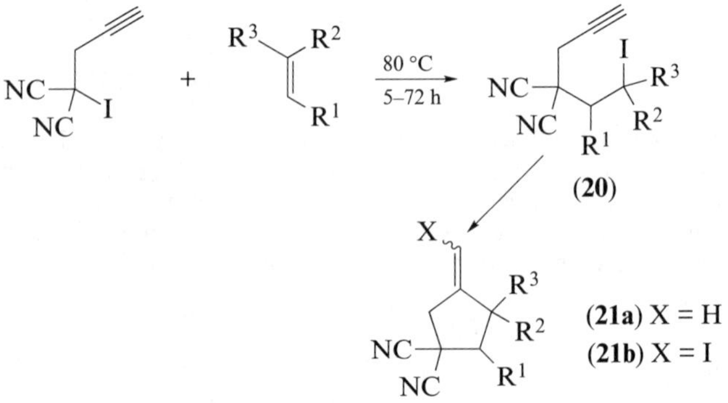

SCHEME 10

The different behavior between malonodinitriles and malonoesters has been discussed and a mechanistic interpretation has been provided.

Arylimidoyl radicals, generated by hydrogen abstraction from N-arylideneanilines (22) with diisopropyl peroxydicarbonate, react with alkynes to give quinolines 23 and 24 in ca 80% overall yield (equation 32)[92]. A mechanistic scheme where arylimidoyl, vinyl and spirocyclohexadienyl radicals are involved as intermediates has been presented and discussed.

$$(22) \quad\quad (23) \quad\quad (24) \tag{32}$$

Oxidation of diethyl α-benzylmalonate (25) by Mn(III) acetate in acetic acid at 70 °C in the presence of mono- or disubstituted alkynes leads to dihydronaphthalene derivatives (26) in moderate to good yields (equation 33)[93]. A mechanistic scheme involving the formation of the corresponding malonyl radical, its addition to a triple bond and intramolecular homolytic aromatic substitution of the vinyl radical adducts is discussed. Absolute rate constants, obtained from competitive studies, for the addition of α-benzylmalonyl radicals to a variety of alkynes cover few orders of magnitude; e.g. the rate constants at 60 °C are 3×10^2 and 1×10^6 M^{-1} s^{-1} for 4-octyne and phenylacetylene respectively. Evidence that the transition state of the addition is characterized by a significant charge transfer from the substrate to the radical is also obtained, in agreement with the electrophilic character of malonyl radical.

$$(25) \quad\quad (26) \tag{33}$$

B. Cyclization of Enynes

The free-radical cyclization of enynes as a synthetic strategy was introduced by Stork and coworkers using $Bu_3SnH/AIBN$-mediated ethynyl cyclization (equation 34)[94]. From a mechanistic point of view, the first step is the selective addition of the stannyl radical to the triple bond of the enyne generating a vinyl radical, which can cyclize onto the double bond[95]. In the absence of the methyl substituents on the olefinic moiety, the groups of Beckwith[96] and Stork[97] have both postulated, and demonstrated with the aid of kinetic studies, that the 6-endo radical 30 arises from a ring expansion of the kinetically favored

$$\tag{34}$$

85%

radical **28** via the cyclopropylcarbinyl intermediate **29** (Scheme 11). The main reason for the selective addition of the stannyl radical to the triple bond is probably the higher reversibility of the stannyl radical addition to olefins compared to alkynes. For practical purposes, it is often possible to steer the cyclization reaction to either the five- or six-membered ring depending on the tin-hydride concentration.

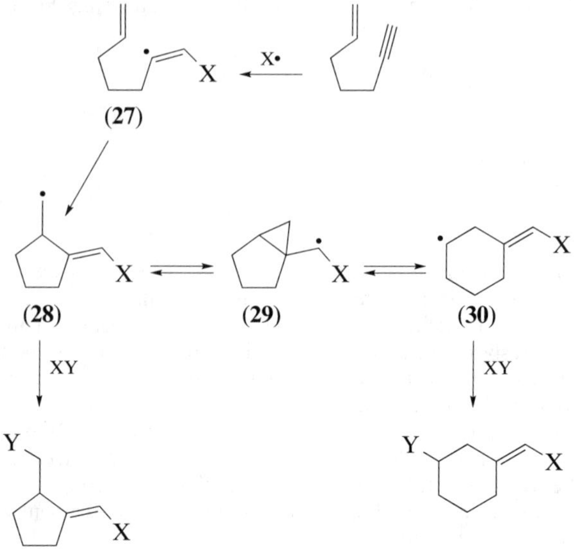

SCHEME 11

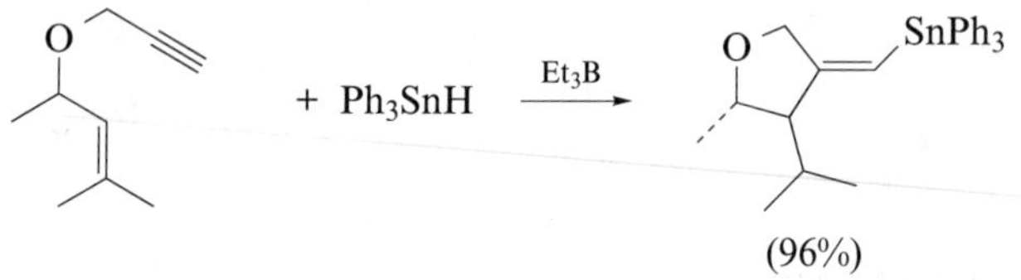

(35)

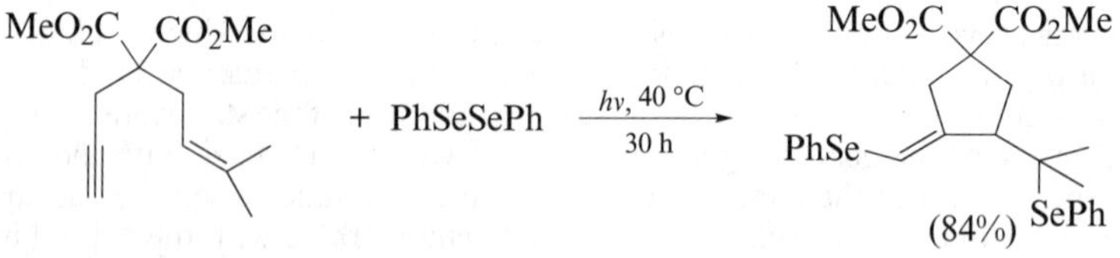

(96%)

(36)

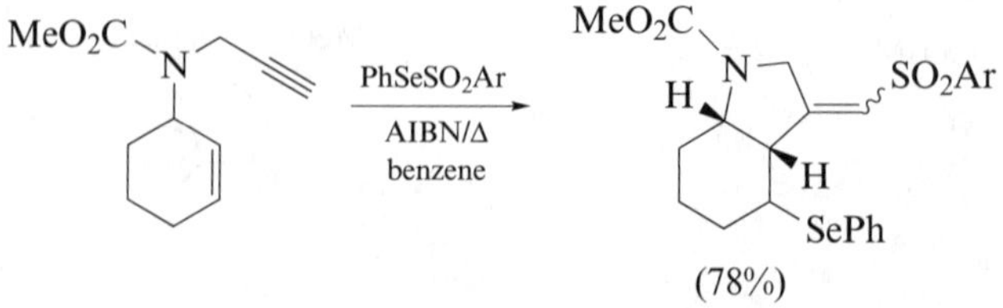

(84%)

(37)

(78%)

The methodology of Et_3B-mediated addition of R_3SnH to an acetylenic bond has also been successfully applied in the cyclization of enynes[56,57]. An example is given in equation 35. The use of triphenylgermanium hydride[60], thiophenol[98], diphenyl diselenide (equation 36)[48], tris(phenylseleno)borane[49b], Se-phenyl areneselenosulfonates (equation 37)[99] and diphenylphoshine[100] under free-radical conditions induces cyclization of enynes like tributyltin hydride. That is, if the chain-transfer step in Scheme 11 is much faster than the ring expansion, the methylenecyclopentane adduct should be the sole product.

The radical cyclization of enynes by the use of tris(phenylseleno)borane has been utilized for the synthesis of α-kainoids, a class of compounds with neuroexcitant properties[49b]. An example is shown in equation 38 where the enyne, prepared by alkylation of 4,4-diethoxycarbonyl-4-trifluoroacetamidobut-2-enoates with prop-2-ynyl bromide, was treated with $(PhSe)_3B$ and AIBN to give in 55% the desired product.

(38)

C. Tandem Cyclizations

The rapid assemblage of complex molecules has been recently obtained using tandem radical reactions; in a tandem sequence, a first-generated radical X˙ undergoes an inter- or (preferably) intramolecular reaction to form Y˙, which is the precursor of the second step of the sequence. The final radical in the sequence, Z˙, has to be converted to a stable product (equation 39).

$$X \longrightarrow X^{\bullet} \xrightarrow{\text{step a}} Y^{\bullet} \xrightarrow{\text{step b}} Z^{\bullet} \xrightarrow[\text{removal}]{\text{radical}} Z \qquad (39)$$

Tandem radical reactions require a very careful evaluation of the reaction rates of each single step involved in the sequence to be successful. Several examples of tandem free-radical processes can be drawn from the synthesis of natural products and have been reviewed in recently published works[2,3,101]; here we would like to report only some of the more recent contributions as examples of the synthesis of cyclic skeletons.

Phenylselenoesters **31** have been reported as precursors to acyl radicals and can be utilized in tandem processes (Scheme 12)[102]. When an alkyl substituent is positioned at the C-5 in the starting selenoester, a clean 6-endo-trig cyclization occurs thus giving the intermediate radical **32**. The subsequent intramolecular addition to triple bond and the final hydrogen trapping lead to the bicyclic product **33**. In model studies, it has been observed that the rates of the addition reactions of substituted acyl radicals (generated from selenoesters) with alkenes bearing electron-withdrawing or radical-stabilizing substituents exceed those of other competitive reactions, such as decarbonylation or reduction.

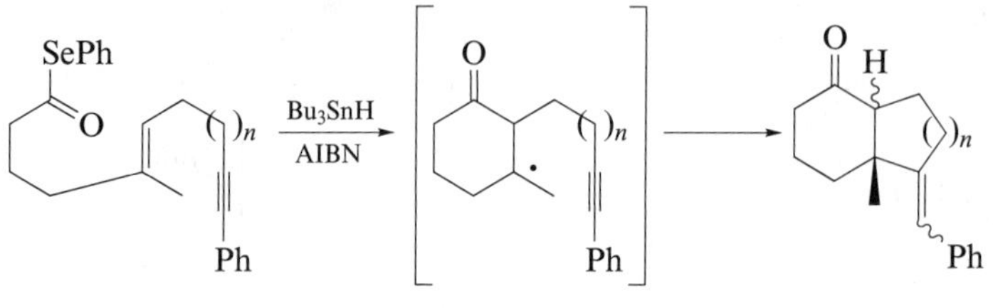

(**31**)   (**32**)   (**33a**) $n = 1$ (>97:3 *cis:trans*)
                                                       (**33b**) $n = 2$ (58:42 *cis:trans*)

SCHEME 12

A tandem 7-*endo*/5-*exo* cyclization route has been tested[103] using the selenoester **34** which enables the construction of [5.3.0]-fused bicyclodecanone systems like **35** in a single step (equation 40). The initial cyclization of the acyl radical is directed to the *endo* mode by the allylic alkoxy substituent at C-5, due to stereoelectronic effects. Model studies were conducted to evaluate the effect of alkyl as well as alkoxy substitutions in this position of the heptenoyl radical system on the cyclization mode. The authors liked to point out that their approach differs from that of Boger because the initial cyclization has been directed to the *endo* mode by the allylic oxygen, whilst in the Boger sequence[102] the *endo* mode was achieved by alkyl substitution at the internal olefin position (steric effects).

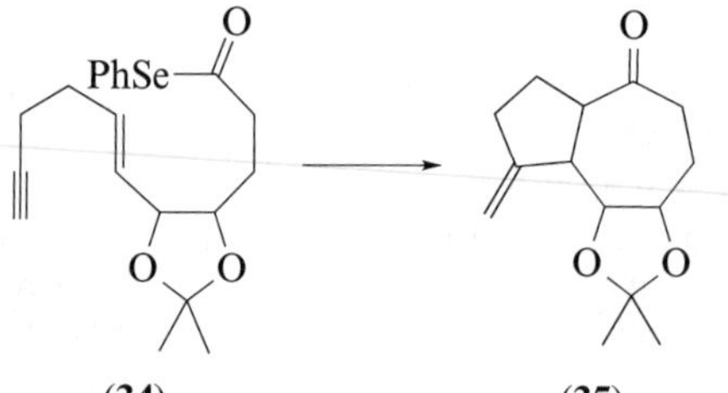

(**34**)   (**35**)   (40)

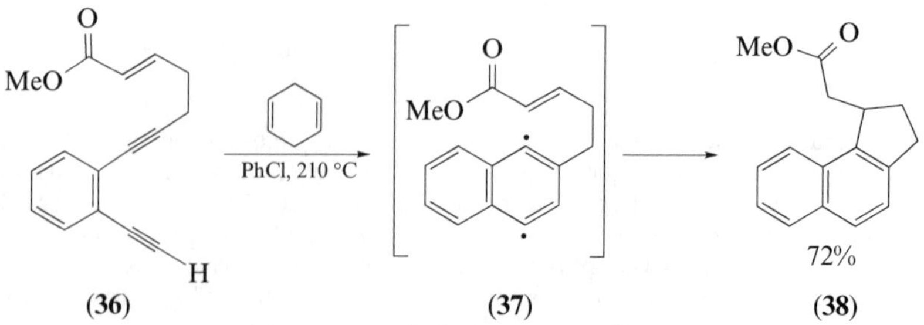

(**36**)   (**37**)   (**38**)

SCHEME 13

Enediynes like **36** were found to be suitable starting materials for tandem processes. A tandem Bergman reaction-radical cyclization has been applied for ring annulation[104] and the sequence for the synthesis of tricyclic compounds **38** is shown in Scheme 13. The yield of the tricyclic product is rather low in the absence of 1,4-cyclohexadiene which acts as hydrogen-atom donor; in fact, a side product can derive from the hydrogen transfer to the intermediate diradical **37** and the rate constant for the radical cyclization of the Bergman diyl intermediate can be considered comparable with the rate of its hydrogen trapping.

Feldman and coworkers have generated radicals from vinylcyclopropanes in the presence of thiyl radicals and studied their reaction with alkynes[105] to give vinylcyclopentenes. A general [3+2] annulation strategy considers vinylcyclopropanes **39** as the three atom components and alkynes **40** as the two atom counterparts (Scheme 14). The vinylcyclopentene products have been obtained as mixtures of *syn-* and *anti-*substituted stereoisomers; the authors considered that product stereochemistry is set during the cyclization of the 5-hexadienyl radical **41** and derives from the two conformations (chair- or boat-like) in the transition state.

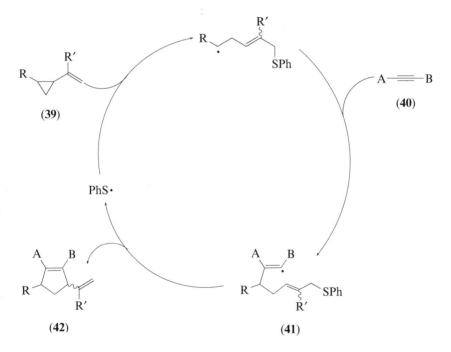

SCHEME 14

2-Vinylidenecyclopentanes **45** have been synthesized by a sequential radical reaction[106]. A careful evaluation of the suitable starting materials led to the choice of thiohydroxamic esters like **43** for starting the radical chain shown in Scheme 15. Photolytical decomposition of **43** in the presence of 5-fold molar excess of a radicophilic olefin generated the 5-hexynyl radical **44** which cyclized onto the triple bond. β-Elimination of a phenylthiyl radical afforded the vinylidenecyclopentane **45** in 60–70% yield. Other insights in the mechanism of these radical chain reactions are given in the paper.

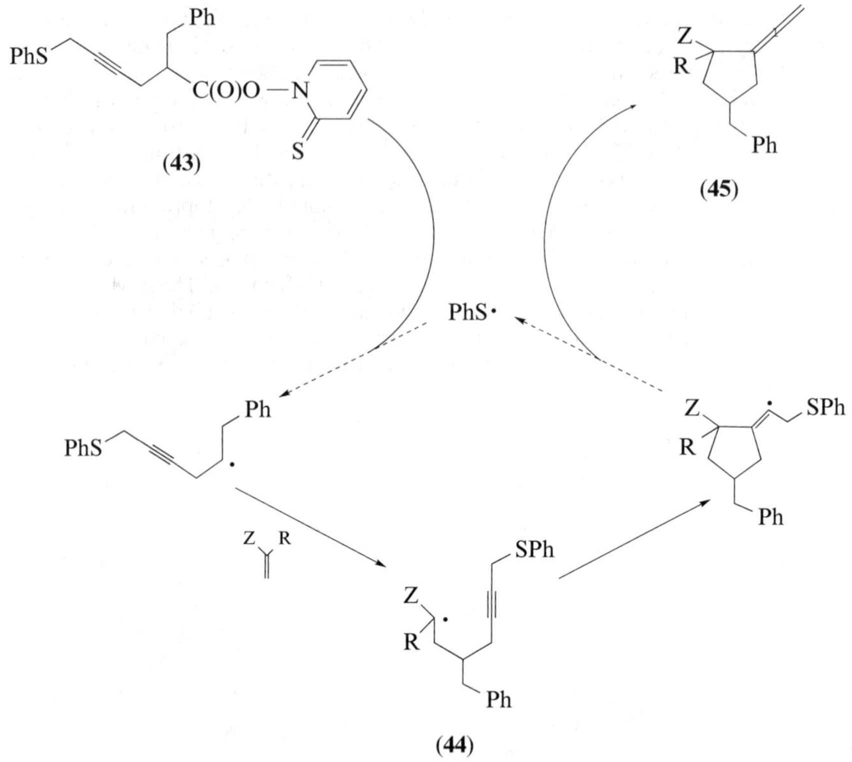

(43)

(45)

(44)

SCHEME 15

D. Synthesis of Natural Products

The field of natural products synthesis has been particularly fruitful in showing the flexibility and power of radical chemistry; several examples from the literature have been collected and commented on in some books[2] and reviews[3,73], therefore in the present chapter we describe only a few examples.

The tricyclic cyclopenta[b]benzofuran skeleton of rocaglamide 47 (an antileukemia agent extracted from *Aglaia elliptifolia*) has been successfully produced by sunlamp irradiation of 46 in the presence of diphenyldisulfide (equation 41)[107]. This work is connected to the

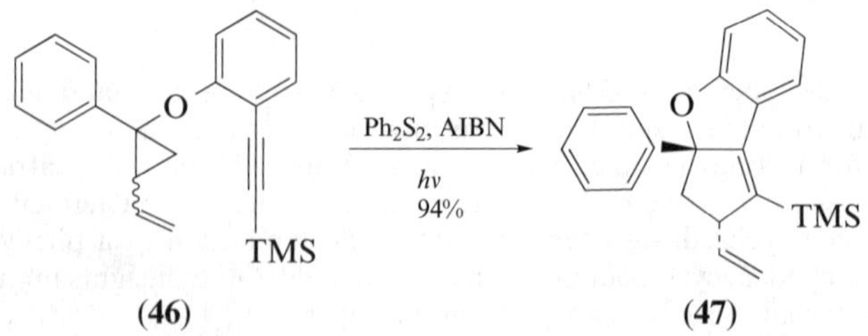

$$\text{Ph}_2\text{S}_2, \text{AIBN} \atop h\nu \atop 94\%$$

(41)

(46)

(47)

studies on the reactivity of vinylcyclopropanes which have been discussed in the previous section.

The volatile monoterpene karahana ether **48** shows an *exo*-methylene function which can be retrosynthetically correlated with a triple bond, that is, it can be generated *via* a radical-mediated 6-*exo-dig* intramolecular cyclization of a suitable alkyl radical onto an alkyne function (equation 42)[108].

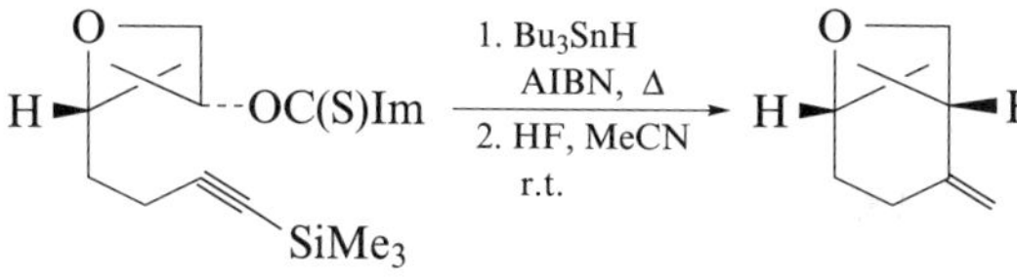

Im = imidazole **(48)**

(42)

(±)-Supinidine **52**, a pyrrolizidine alkaloid with various pharmacologycal activities, has been the target of a synthetic strategy via radical cyclization[109]. The key step of the synthesis of the pyrrolizidine skeleton is the intramolecular addition of an α-sulfonyl radical generated from **49** to the triple bond, followed by the addition of a tin radical to the resulting sulfone **50** (Scheme 16). The resulting product **51** can be further manipulated and easily converted to the target alkaloid.

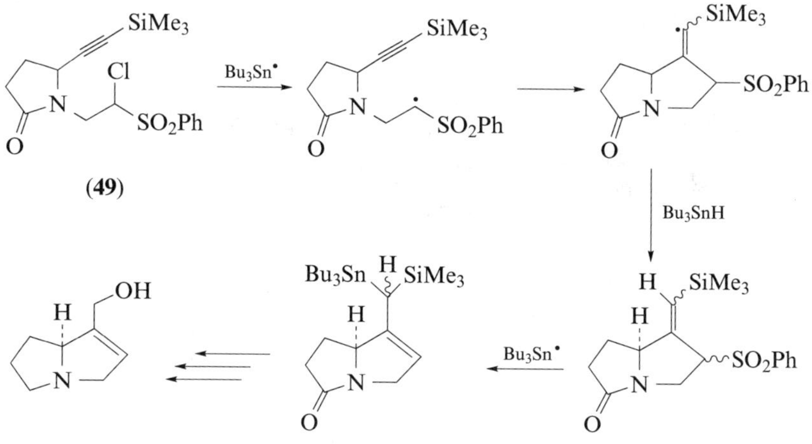

SCHEME 16

An example of cyclization using chiral radicals derived from natural products is given in the 5-*exo-dig* radical cyclization of *N*-alkynyl sulfonamide **53** derived from 2(*S*)-serine[110] to give 2(*S*)-4-exomethyleneproline **54** (equation 43). The exomethylene product has been recovered without any detectable directly reduced, noncyclized product or product derived from an *endo* cyclization.

Tin-mediated tandem cyclization can be useful for the construction of heterocyclic rings as intermediates in natural product synthesis; a series of enynes like **55** has been recently used[111], and the reaction also represents a general route to allyl stannanes (equation 44).

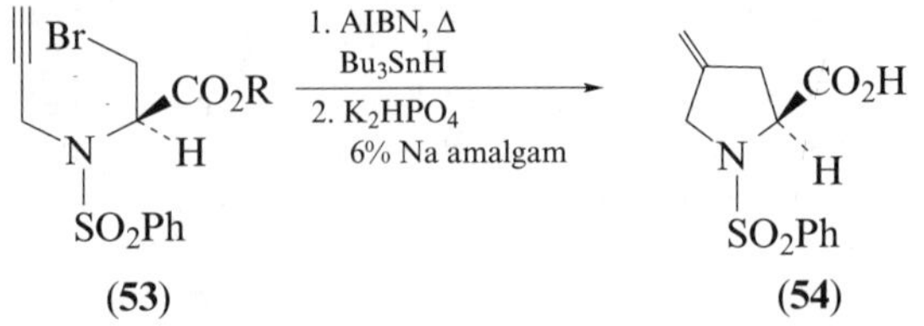

$$
\text{(53)} \quad \xrightarrow[\substack{\text{2. K}_2\text{HPO}_4 \\ \text{6\% Na amalgam}}]{\substack{\text{1. AIBN, }\Delta \\ \text{Bu}_3\text{SnH}}} \quad \text{(54)} \tag{43}
$$

$$
\text{(55) X = O or NPh or C(CO}_2\text{Me)}_2 \quad \xrightarrow[\text{AIBN, }\Delta]{\text{Bu}_3\text{SnH}} \tag{44}
$$

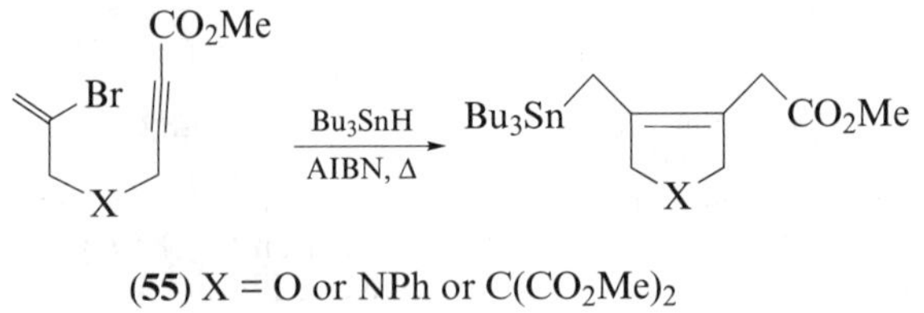

Chiral furo[2,3-*b*]furans can be obtained in a synthetic strategy which utilizes carbohydrates as starting materials[112] and is shown in Scheme 17; D-xylose has been converted to the intermediate **56** which, in the presence of tributyltin hydride, gives the bicyclic product **57** in high yield.

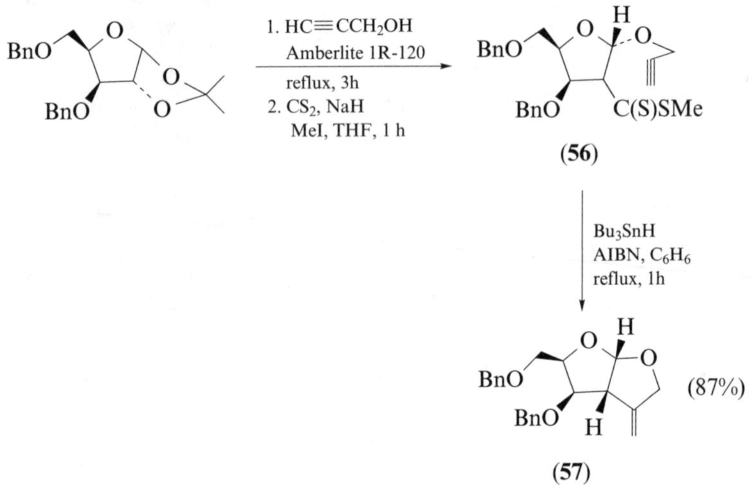

(56)

1. HC≡CCH₂OH
   Amberlite 1R-120
   reflux, 3h
2. CS₂, NaH
   MeI, THF, 1 h

Bu₃SnH
AIBN, C₆H₆
reflux, 1h

(57) (87%)

SCHEME 17

An analogous strategy has been utilized in the total synthesis of (±)-norbakkenolide A (**60**), a sesquiterpene containing an α-spiro lactone fused hydrindane framework[113]. The strategy is illustrated in Scheme 18; the 5-*exo-dig* radical cyclization of the bromoacetal **58** has been performed using *in situ* generated catalytic Bu₃SnH in the presence of AIBN. The resulting spirohemiacetal **59** gave the target in a few steps.

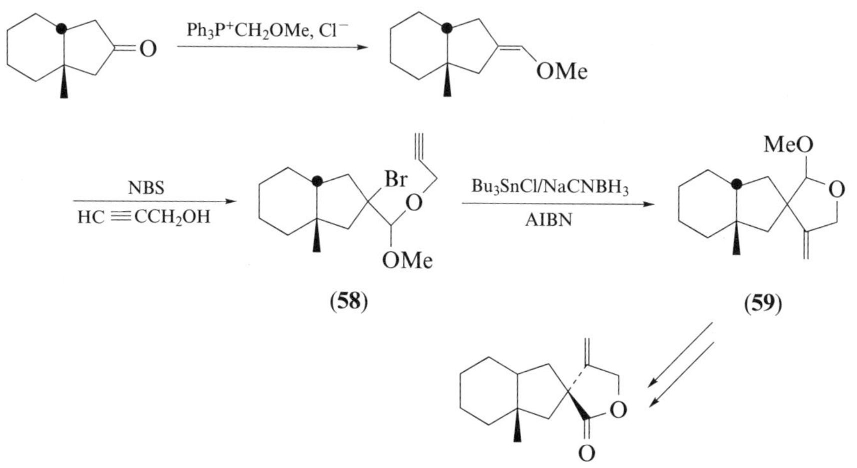

(58) (59)

(60)

SCHEME 18

V. ACKNOWLEDGMENTS

Support from the Progetto Finalizzato Chimica Fine II (CNR, Rome) is gratefully acknowledged.

VI. REFERENCES

1. Y. Amiel, in *The Chemistry of Functional Groups, Supplement C* (Eds. S. Patai and Z. Rappoport), Chap. 10, Wiley, Chichester, 1983, pp. 341–382.
2. For example, see:
   (a) B. Giese, *Radicals in Organic Synthesis: Formation of Carbon–Carbon Bonds*, Pergamon Press, Oxford 1986.
   (b) W. B. Motherwell and D. Crich, *Free Radical Chain Reactions in Organic Synthesis*, Academic Press, London, 1992.
3. For example, see:
   (a) D. P. Curran, in *Comprehensive Organic Synthesis* (Ed. M. F. Semmelhack), Vol. 4, Pergamon Press, Oxford 1991, pp. 715–831.
   (b) D. P. Curran, *Synthesis*, 417 and 489 (1988).
4. L. A. Singer, in *Selective Organic Transformations* (Ed. B. S. Thyagarajan), Vol. 2, Wiley, New York, 1972, pp. 239–268.
5. R. W. Fessenden and R. H. Schuler, *J. Chem. Phys.*, **39**, 2147 (1963).
6. A. L. J. Beckwith and K. U. Ingold, in *Rearrangements in Ground and Excited States* (Ed. P. de Mayo), Vol. 1, Academic Press, New York, 1980, pp. 280–281.
7. H.-G. Korth, J. Lusztyk and K. U. Ingold, *J. Chem. Soc., Perkin Trans. 2*, 1997 (1990).
8. J. E. Bennett and J. A. Howard, *Chem. Phys. Lett.*, **9**, 460 (1971).
9. D. Griller, J. W. Cooper and K. U. Ingold, *J. Am. Chem. Soc.*, **97**, 4269 (1975).
10. (a) D. A. Geeson, M. C. R. Symons, E. Roduner, H. Fischer and S. S. J. Cox, *Chem. Phys. Lett.*, **116**, 186 (1985).
    (b) C. J. Rhodes and M. C. R. Symons, *J. Chem. Soc., Faraday Trans. 1*, **84**, 4495 (1988).
    (c) C. J. Rhodes and E. Roduner, *J. Chem. Soc., Perkin Trans. 2*, 1729 (1990).
11. P. R. Jenkins, M. C. R. Symons, S. E. Booth and C. J. Swain, *Tetrahedron Lett.*, **33**, 3543 (1992).

12. A. Fahr and A. H. Laufer, *J. Phys. Chem.*, **92**, 7229 (1988).
13. H. E. Hunziker, H. Kneppe, A. D. McLean, P. Siegbahn and H. R. Wendt, *Can. J. Chem.*, **61**, 993 (1983).
14. J. A. Kampmeier and G. Chen, *J. Am. Chem. Soc.*, **87**, 2608 (1965).
15. D. K. Wedegaertner, R. M. Kopchik and J. A. Kampmeier, *J. Am. Chem. Soc.*, **93**, 6890 (1971).
16. B. Giese, *Angew. Chem., Int. Ed. Engl.*, **28**, 969 (1989).
17. B. Giese, J. A. Gonzales-Gomez, S. Lachhein and J. O. Metzger, *Angew. Chem., Int. Ed. Engl.*, **26**, 479 (1987).
18. B. Giese and S. Lachhein, *Angew. Chem., Int. Ed. Engl.*, **21**, 768 (1982).
19. J. I. Dickstein and G. I. Miller, in *Chemistry of the Carbon–Carbon Triple Bond* (Ed. S. Patai), Vol. 2, Wiley, New York, 1978.
20. W. T. Dixon, J. Foxall, G. H. Williams, D. J. Edge, B. C. Gilbert, H. Kazarians-Moghaddam and R. O. C. Norman, *J. Chem. Soc., Perkin Trans. 2*, 827 (1977).
21. J. Foxall, B. C. Gilbert, H. Kazarians-Moghaddam, R. O. C. Norman, W. T. Dixon and G. H. Williams, *J. Chem. Soc., Perkin Trans. 2*, 273 (1980).
22. B. C. Gilbert, N. R. McLay and D. J. Parry, *J. Chem. Soc., Perkin Trans. 2*, 329 (1987).
23. B. C. Gilbert and D. J. Parry, *J. Chem. Soc., Perkin Trans. 2*, 875 (1988).
24. D. P. Curran and D. Kim, *Tetrahedron*, **47**, 6171 (1991).
25. D. P. Curran, D. Kim and C. Ziegler, *Tetrahedron*, **47**, 6189 (1991).
26. Y. Ichinose, S. Matsunaga, K. Fugami, K. Oshima and K. Utimoto, *Tetrahedron Lett.*, **30**, 3155 (1989).
27. Y. Takeyama, Y. Ichinose, K. Oshima and K. Utimoto, *Tetrahedron Lett.*, **30**, 3159 (1989).
28. C. Chatgilialoglu and M. Guerra, in *Supplement S: The Chemistry of Sulfur-Containing Functional Groups* (Eds. S. Patai and Z. Rappoport), Chap. 8, Wiley, Chichester, 1993, pp. 363-394.
29. O. Ito, R. Omori and M. Matsuda, *J. Am. Chem. Soc.*, **104**, 3934 (1982).
30. (a) O. Ito and M. D. C. M. Fleming, *J. Chem. Soc., Perkin Trans. 2*, 689 (1989).
    (b) G. A. Russell and P. Ngoviwatchai, in *Sulfur-Centered Reactive Intermediates in Chemistry and Biology* (Eds. C. Chatgilialoglu and K.-D. Asmus), Plenum, New York, 1989, pp. 291–302.
31. B. C. Gilbert, D. J. Parry and L. Grossi, *J. Chem. Soc., Faraday Trans. 1*, **83**, 77 (1987).
32. Y. Ichinose, K. Wakamatsu, K. Nozaki, J.-L. Birbaum, K. Oshima and K. Utimoto, *Chem. Lett.*, 1647 (1987); Y. Ichinose, K. Oshima and K. Utimoto, *Chem. Lett.*, 1437 (1988).
33. (a) L. Benati, P. C. Montevecchi and P. Spagnolo, *J. Chem. Soc., Perkin Trans. 1*, 2103 (1991).
    (b) P. C. Montevecchi, personal communication.
34. L. Benati, P. C. Montevecchi and P. Spagnolo, *J. Chem. Soc., Perkin Trans. 1*, 1659 (1992).
35. R. Leardini, G. F. Pedulli, A. Tundo and G. Zanardi, *J. Chem. Soc., Chem. Commun.*, 1390 (1985).
36. J. Griffiths and J. A. Murphy, *Tetrahedron*, **48**, 5543 (1992).
37. S. Nakatani, J-i. Yoshida and S. Isoe, *J. Chem. Soc., Chem. Commun.*, 880 (1992).
38. C. Chatgilialogu, in *The Chemistry of Sulphones and Sulphoxides* (Eds. S. Patai, Z. Rappoport and C. J. M. Stirling), Chap. 25, Wiley, Chichester, 1988, pp. 1089–1113.
39. W. E. Truce and G. C. Wolf, *J. Org. Chem.*, **36**, 1727 (1971).
40. Y. Amiel, *J. Org. Chem.*, **39**, 3867 (1974).
41. Y. Amiel, *J. Org. Chem.*, **36**, 3691 and 3697 (1971).
42. C. M. M. da Silva Correa and M. D. C. M. Fleming, *J. Chem. Soc., Perkin Trans. 2*, 103 (1987).
43. T. Miura and M. Kobayashi, *J. Chem. Soc., Chem. Commun.*, 438 (1982).
44. T. G. Back, S. Collins and R. G. Kerr, *J. Org. Chem.*, **48**, 3077 (1983).
45. T. G. Back and K. R. Muralidharan, *J. Org. Chem.*, **54**, 121 (1989).
46. T. Ozawa and T. Kwan, *J. Chem. Soc., Chem. Commun.*, 80 (1983).
47. T. G. Back and M. V. Krishna, *J. Org. Chem.*, **53**, 2533 (1988).
48. A. Ogawa, H. Yokoyama, K. Yokoyama, T. Masawaki, N. Kambe and N. Sonoda, *J. Org. Chem.*, **56**, 5721 (1991); A. Ogawa, N. Takami, M. Sekiguchi, H. Yokoyama, M. Kuniyasu, I. Ryu and N. Sonoda, *Chem. Lett.*, 2241 (1991).
49. (a) A. Ogawa, R. Obayashi, M. Sekiguchi, T. Masawaki, N. Kambe and N. Sonoda, *Tetrahedron Lett.*, **33**, 1329 (1992).
    (b) T. Kataoka, M. Yoshimatsu, H. Shimizu and M. Hori, *Tetrahedron Lett.*, **31**, 5927 (1990); T. Kataoka, M. Yoshimatsu, Y. Noda, T. Sato, H. Shimizu and M. Hori, *J. Chem. Soc., Perkin Trans. 1*, 121 (1993).

50. E. Kobayashi, T. Ohashi and J. Furukawa, *Makromol. Chem.*, **187**, 2525 (1986).
51. E. Kobayashi, N. Metaka, S. Aoshima and J. Furukawa, *J. Polym. Sci., Part A: Polym. Chem.*, **30**, 227 (1992).
52. A. Ogawa, K. Yokoyama, R. Obayashi, L.-B. Han, N. Kambe and N. Sonoda, *Tetrahedron*, **49**, 1177 (1993); A. Ogawa, K. Yokoyama, H. Yokoyama, R. Obayashi, N. Kambe and N. Sonoda, *J. Chem. Soc., Chem. Commun.*, 1748 (1991).
53. M. Pereyre, J.-P. Quintard and A. Rahm, *Tin in Organic Synthesis*, Butterworths, London, 1987.
54. W. P. Neumann, *Synthesis*, 665 (1987).
55. M. F. Jung and L. A. Light, *Tetrahedron Lett.*, **23**, 3851 (1982).
56. K. Nozaki, K. Oshima and K. Utimoto, *J. Am. Chem. Soc.*, **109**, 2547 (1987).
57. K. Nozaki, K. Oshima and K. Utimoto, *Tetrahedron*, **45**, 923 (1989).
58. Y. Ueno and M. Okawara, *J. Am. Chem. Soc.*, **101**, 1893 (1979).
59. T. Konoike and Y. Araki, *Tetrahedron Lett.*, **33**, 5093 (1992).
60. K. Nozaki, Y. Ichinose, K. Wakamatsu, K. Oshima and K. Utimoto, *Bull. Chem. Soc. Jpn.*, **63**, 2268 (1990).
61. I. Ojima, in *The Chemistry of Organic Silicon Compounds* (Eds. S. Patai and Z. Rappoport), Vol. 2, Chap. 25, Wiley, Chichester, 1989.
62. H. Sakurai, in *Free Radicals* (Ed. J. K. Kochi), Vol. 2, Chap. 25, Wiley, New York, 1973.
63. C. Chatgilialoglu, *Acc. Chem. Res.*, **25**, 188 (1992).
64. B. Kopping, C. Chatgilialoglu, M. Zehnder and B. Giese, *J. Org. Chem.*, **57**, 3994 (1992).
65. C. Ferreri, M. Ballestri and C. Chatgilialoglu, *Tetrahedron Lett.*, **34**, 5147 (1993).
66. C. Chatgilialoglu, unpublished results.
67. R. Atkinson, K. R. Darnall, A. C. Lloyd, A. M. Winer and J. N. Pitts Jr., *Adv. Photochem.*, **11**, 375 (1979).
68. C. J. Jachimowsky, *Combust. Flame*, **29**, 55 (1977); J. M. Levy, *Combust. Flame*, **46**, 7 (1982).
69. S. Hatakeyama, N. Washida and H. Akimoto, *J. Phys. Chem.*, **90**, 173 (1986).
70. I. R. Slagle, J.-Y. Park, M. C. Heaven and D. Gutman, *J. Am. Chem. Soc.*, **106**, 4356 (1984).
71. B. C. Gilbert and A. C. Whitwood, *J. Chem. Soc., Perkin Trans. 2*, 1921 (1989).
72. T. N. Mitchell and K. Heesche, *J. Organomet. Chem.*, **409**, 163 (1991).
73. G. Stork, in *Current Trends in Organic Synthesis* (Ed. H. Nozaki), Pergamon Press, Oxford, 1983; G. Stork, *Bull. Chem. Soc. Jpn.*, **61**, 149 (1988).
74. D. J. Hart, *Science*, **223**, 883 (1984).
75. A. L. J. Beckwith, *Tetrahedron*, **37**, 3073 (1981).
76. J.-M. Surzur, in *Reactive Intermediates* (Ed. R. A. Abramovitch), Vol. 2, Chap. 3, Plenum Press, New York, 1982.
77. M. Julia, J. M. Surzur and L. Katz, *C.R. Acad. Sci., Ser. C*, **251**, 1030 (1960); M. Julia, *Rec. Chem. Prog.*, **25**, 3 (1964); M. Julia, *Pure Appl. Chem.*, **15**, 167 (1967).
78. J. K. Crandall and D. J. Keyton, *Tetrahedron Lett.*, 1653 (1969).
79. D. L. J. Clive, T. L. B. Boivin and A. G. Angoh, *J. Org. Chem.*, **52**, 4943 (1987).
80. D. L. J. Clive, P. L. Beaulieu and L. Set, *J. Org. Chem.*, **49**, 1314 (1984).
81. V. Dalla and P. Pale, *Tetrahedron Lett.*, **33**, 7875 (1992).
82. S. M. Bennett and D. Larouche, *Synlett*, 805 (1991).
83. M. D. Bachi and E. Bosch, *J. Org. Chem.*, **57**, 4696 (1992).
84. K. Nozaki, K. Oshima and K. Utimoto, *Tetrahedron Lett.*, **29**, 6127 (1988).
85. J. E. Baldwin, R. M. Adlington and S. H. Ramcharitar, *Tetrahedron*, **48**, 3413 (1992).
86. N. Nishiyama, T. Kitajama, M. Matsumoto and K. Itoh, *J. Org. Chem.*, **49**, 2298 (1984); G. Stork and M. Kahn, *J. Am. Chem. Soc.*, **107**, 500 (1985).
87. (a) E. Magnol and M. Malacria, *Tetrahedron Lett.*, **27**, 2255 (1986).
    (b) G. Agnel and M. Malacria, *Synthesis*, 687 (1989); *Synlett*, 58 (1991).
88. (a) M. Journet, E. Magnol, G. Agnel and M. Malacria, *Tetrahedron Lett.*, **31**, 4445 (1990).
    (b) M. Journet and M. Malacria, *J. Org. Chem.*, **57**, 3085 (1992).
89. M. Journet and M. Malacria, *Tetrahedron Lett.*, **33**, 1893 (1992).
90. D. P. Curran, M.-H. Chen and D. Kim, *J. Am. Chem. Soc.*, **111**, 6265 (1989); D. P. Curran and C.-T. Chang, *J. Org. Chem.*, **54**, 3140 (1989).
91. D. P. Curran and C. M. Seong, *J. Am. Chem. Soc.*, **112**, 9401 (1990); D. P. Curran and C. M. Seong, *Tetrahedron*, **48**, 2157 (1992).
92. R. Leardini, A. Tundo and G. Zanardi, *J. Chem. Soc., Chem. Commun.*, 1320 (1984); R. Leardini, D. Nanni, A. Tundo and G. Zanardi, *Gazz. Chim. Ital.*, **119**, 637 (1989).

93. R. Santi, F. Bergamini, A. Citterio, R. Sebastiano and M. Nicolini, *J. Org. Chem.*, **57**, 4250 (1992).
94. G. Stork and R. Mook, Jr., *J. Am. Chem. Soc.*, **109**, 2829 (1987); R. Mook, Jr. and P. M. Sher, *Org. Synth.*, **69**, 75 (1988).
95. For other vinyl radical cyclization, see: G. Stork and P. G. Willard, *J. Am. Chem. Soc.*, **99**, 7067 (1977); G. Stork and N. H. Baine, *J. Am. Chem. Soc.*, **104**, 2321 (1982); G. Stork and R. Mook, Jr., *J. Am. Chem. Soc.*, **105**, 3720 (1983); G. Stork and P. J. Franklin, *Aust. J. Chem.*, **45**, 275 (1992).
96. A. L. J. Beckwith and D. M. O'Shea, *Tetrahedron Lett.*, **27**, 4525 (1986).
97. G. Stock and R. Mook, Jr., *Tetrahedron Lett.*, **27**, 4529 (1986).
98. C. A. Broka and D. E. C. Reichert, *Tetrahedron Lett.*, **28**, 1503 (1987).
99. J. E. Brumwell, N. S. Simpkins and N. K. Terrett, *Tetrahedron Lett.*, **34**, 1219 (1993).
100. J. E. Brumwell, N. S. Simpkins and N. K. Terrett, *Tetrahedron Lett.*, **34**, 1215 (1993).
101. L. F. Tietze and U. Beifuss, *Angew. Chem., Int. Ed. Engl.*, **32**, 131 (1993).
102. D. L. Boger and R. J. Mathvink, *J. Org. Chem.*, **55**, 5442 (1990); D. L. Boger and R. J. Mathvink, *J. Org. Chem.*, **57**, 1429 (1992).
103. D. Batty and D. Crich, *J. Chem. Soc., Perkin Trans. 1*, 3193 (1992).
104. J. Wisniewski, Grissom and T. L. Calkins, *Tetrahedron Lett.*, **33**, 2315 (1992) and references cited therein.
105. (a) K. S. Feldman, R. E. Ruckle and A. L. Romanelli, *Tetrahedron Lett.*, **30**, 5845 (1989).
     (b) K. S. Feldman and R. E. Simpson, *J. Am. Chem. Soc.*, **111**, 4878 (1989) and references cited therein.
106. R. N. Saicic and Z. Cekovic, *Tetrahedron*, **48**, 8975 (1992).
107. K. S. Feldman and C. J. Burns, *J. Org. Chem.*, **56**, 4601 (1991).
108. T. Honda, M. Satoh and Y. Kobayashi, *J. Chem. Soc, Perkin Trans. 1*, 1557 (1992).
109. Y.-M. Tsai, B.-W. Ke, C.-T. Yang and C.-H. Lin, *Tetrahedron Lett.*, **33**, 7895 (1992).
110. R. M. Adlington and S. J. Mantell, *Tetrahedron*, **48**, 6529 (1992).
111. P. J. Parson, M. Stefinovic, P. Willis and F. Meyer, *Synlett*, 864 (1992).
112. G. V. M. Sharma and S. R. Vepachedu, *Carbohydr. Res.*, **226**, 185 (1992).
113. A. Shrikrishna, S. Nagaraju and G. V. R. Sharma, *J. Chem. Soc., Chem. Commun.*, 285 (1993).

CHAPTER **17**

Synthesis and properties of acetylenic polymers

HIDEKI SHIRAKAWA

Institute of Materials Science, University of Tsukuba,Tsukuba, Ibaraki 305, Japan

TOSHIO MASUDA

Department of Polymer Chemistry, Kyoto University, Kyoto 606-01, Japan

and

KENJI TAKEDA

Research Center, Japan Synthetic Rubber Co., Ltd, 25, Miyukigaoka, Tsukuba, Ibaraki 305, Japan

Supplement C2: The chemistry of triple-bonded functional groups
Edited by S. Patai   © 1994 John Wiley & Sons Ltd

I. INTRODUCTION

Acetylenic polymers have alternating single and double bonds along the main chain, which results in a linear arrangement of π-electrons and in characteristic optical, electrical and magnetic properties. It has been known for many years that the catalytic polymerization reactions of acetylene yield cuprene, a dark brown apparently cross-linked amorphous acetylene polymer, as well as oligomers such as benzene, cyclooctatetraene and vinyl-acetylene[1].The polymerization of acetylene to a high molecular weight linear polyene of predominantly *trans* structure was first reported by Natta and coworkers[2] (equation 1).

$$n\mathrm{HC{\equiv}CH} \longrightarrow {-}(\mathrm{CH{=}CH})_{\overline{n}} \qquad (1)$$

However the material obtained was an unprocessable powder. Unlike substituted polyacetylenes, polyacetylene is insoluble, infusible and unstable in air. The discovery of a technique to synthesize the polymer in the form of a free-standing film[3] and the use of electron donors and acceptors to 'dope' it to have metallic conductivity[4] produced intense interest in the polymer in the last two decades. A wide variety of catalyst systems has been described for the polymerization of acetylene. Besides the route via acetylene polymerization, polyacetylene can also be obtained by a two-step route which involves the synthesis of soluble polymer precursors, which are converted to polyacetylene via thermal elimination[5] and transition-metal-catalysed isomerization[6], as well as by polymerization of cyclooctatetraene[7,8], by dehydrochlorination of poly(vinyl chloride)[9] and by dehydration of poly(vinyl alcohol)[10,11].

In general, polymerization of substituted acetylenes provides polymers with alternating double bonds along the main chain and with various pendant groups (equations 2 and 3). Because of the unique properties expected for such polymers, the polymerization of various substituted acetylenes with conventional (radical, ionic, and Ziegler–Natta) catalysts was examined from the late 1950s to the middle of the 1970s. Most of the products, however, were oligomers whose molecular weights were several thousand

$$nHC{\equiv}CR \longrightarrow {-}(CH{=}\underset{\underset{R}{|}}{C}\!)_{\overline{n}} \tag{2}$$

$$nRC{\equiv}CR' \longrightarrow {-}(\underset{\underset{R}{|}}{C}{=}\underset{\underset{R'}{|}}{C}\!)_{\overline{n}} \tag{3}$$

at the highest. One can see the development of such studies in a review by Chauser and coworkers[12]. The successful preparation of polyacetylene membranes, which developed metallic conductivity upon doping raised further interest in substituted polyacetylenes.

In the middle of the 1970s, W and Mo catalysts were found very effective in the polymerization of phenylacetylene. Since then, attention was paid to several transition-metal catalysts, i.e. group 5 and 6 elements (Nb, Ta, Mo and W), Rh and Ziegler–Natta catalysts as effective ones for the polymerization of substituted acetylenes. Though substituted polyacetylenes do not show metallic conductivity on doping, some of them proved to be extremely permeable to gases. More recently, precise synthesis of substituted polyacetylenes by living and stereospecific polymerizations is under intensive research. several review articles are available which deal with the studies on substituted polyacetylenes in the 1970s and 80s[13–20].

Poly(diacetylenes) may be defined, in a broader sense, as the polymers prepared from monomers with two acetylene groups, either conjugated or nonconjugated ones. Depending on the kind of monomers and polymerization conditions, a variety of polymers with different main-chain structures have been described. The nonconjugated diacetylene compounds, especially, α,ω-diynes, are concerned with the various types of polymerization reactions such as cyclopolymerization, oxidative polycondensation, ring-closing poly-addition etc., which can be regarded as the simple application of the conventional reactions concerning the monomeric acetylene group to the bifunctional molecules for polymer synthesis. These reactions usually proceed in solution in the presence of the appropriate catalysts. In contrast, the solid-state polymerization of conjugated diynes was first reported by Wegner[21] in 1969 and is characteristic of the cumulative triple bonds in respect of polymerization mechanism, polymer structure and properties. Polymerization is initiated by applying external stimulation to the monomer crystals such as UV-light or high-energy beam irradiation, thermal annealing, pressure, shear stress etc., without any catalyst. Polymerization proceeds by a 1,4-addition mechanism under the strict control of the monomer crystal lattice to form perfect polymer crystals with nearly the same dimensions and lattice parameters as those of the monomer crystals (equation 4).

$$nRC{\equiv}C{-}C{\equiv}CR' \longrightarrow {=}(\overset{\overset{R}{|}}{C}{-}C{\equiv}C{-}\underset{\underset{R'}{|}}{C}\!)_{\overline{n}} \tag{4}$$

Polymer main chains are fully conjugated and extended in structure, being aligned to compose almost defect-free polymer single crystals with fairly large optical anisotropy. Quasi one-dimensional π-electron systems on the polymer backbones are believed to be the origin of the unique features such as large nonlinear optical susceptibilities, phase transition phenomena etc., which have been attracting much interest from many different scientific areas. Therefore the term 'poly(diacetylenes)' has come to be used in a narrower sense to indicate those conjugated polymers, so-called polydiacetylenes, of which reviews are available[22,23].

Section II of this chapter will discuss the synthesis and properties of polyacetylene, a linear polyene without a substituent. Section III will deal with the synthesis and properties of substituted polyacetylenes and Section IV is going to deal mainly with polydiacetylenes prepared by solid-state polymerization, together with the outline of other polymers formed by the different polymerization mechanisms. This whole chapter discusses the synthesis and properties of various acetylenic polymers, mainly focusing on the developments in the last decade. Two chapters concerning acetylenic polymers have appeared in earlier volumes of the 'Chemistry of Functional Groups' series[23a,b].

II. POLYACETYLENE*

A. Synthesis

A variety of routes have been documented for the synthesis of polyacetylene or linear conjugated polyene molecules and are collected in Table 1.

TABLE 1. Synthetic routes for polyacetylenes

Method	References
A. Polymerization of acetylene monomer	
1. Initiation by Ziegler–Natta catalysts	2, 3, 36–47
2. Initiation by Luttinger catalysts	24–26
3. Initiation by metathesis catalysts	27–29
4. High-pressure polymerization	30, 31
5. Electrochemical polymerization	32
6. Polymerization on metal oxide surfaces	33, 34
7. Initiation by AsF_5	35
B. Polymerization of monomers other than acetylene	
1. Ring-opening polymerization of cyclooctatetraene	7, 8
C. Indirect methods	
1. Durham method	5
2. Isomerization of polybenzvalene	6
3. Dehydrochlorination of poly(vinyl chloride)	9
4. Dehydration of poly(vinyl alcohol)	10, 11

The most widely used catalysts for preparing high-quality, free-standing films of polyacetylene are the Ziegler–Natta catalysts. The Luttinger catalyst is also used occasionally because its preparation does not require rigorous exclusion of moisture and oxygen normally associated with the Ziegler–Natta catalysts. An alternative way other than acetylene polymerization is the Durham method that includes monomer synthesis by thermal cycloaddition between cyclooctatetraene and the appropriate alkynes, the ring-opening metathesis polymerization of the monomers to form soluble prepolymers and their thermal decomposition to give *trans*-polyacetylene.

Among the indirect methods, dehydrochlorination of poly(vinyl chloride) and dehydration of poly(vinyl alcohol) will not be described further in this chapter, because the products may have relatively short conjugated sequences with many irregular structures.

1. Polymerization methods

a. Polymerization of acetylene

i. Polymerization by Ziegler–Natta catalysts. A very wide range of Ziegler–Natta catalysts derived from the transition metals of groups 4–8 in combination with organo-

*By H. Shirakawa.

TABLE 2. Ziegler–Natta catalysts for acetylene polymerization

Transition metal compound	Metal alkyl	References
α-TiCl$_3$	Al(C$_2$H$_5$)$_3$	
VCl$_3$	Al(C$_2$H$_5$)$_3$	
Ti(OC$_3$H$_7$)$_4$	Al(C$_2$H$_5$)$_3$	2
Ti(OC$_4$H$_9$)$_4$	Al(C$_6$H$_{13}$)$_3$	
Ti(OC$_3$H$_7$)$_4$	C$_5$H$_{11}$Li	
TiCl$_4$	Al(i-C$_4$H$_9$)$_3$	
TiCl$_4$	n-C$_4$H$_9$Li	37
TiCl$_4$	Zn(C$_2$H$_5$)$_2$	
TiCl$_4$	Al(C$_2$H$_5$)$_3$	38
Ti(OC$_4$H$_9$)$_4$	Al(C$_2$H$_5$)$_3$	
VCl$_4$	Al(C$_2$H$_5$)$_3$	
VCl$_3$	Al(C$_2$H$_5$)$_3$	
VOCl$_3$	Al(C$_2$H$_5$)$_3$	39
VOCl$_3$	Al(C$_2$H$_5$)$_3$	
VO(acac)$_2{}^a$	Al(C$_2$H$_5$)$_3$	
TiO(acac)$_2$	Al(C$_2$H$_5$)$_3$	
Cr(acac)$_3$	Al(C$_2$H$_5$)$_3$	40
Co(acac)$_3$	Al(C$_2$H$_5$)$_3$	
Ti(acac)$_3$	Al(C$_2$H$_5$)$_2$Cl	36
M(acac)$_3{}^b$	Al(C$_2$H$_5$)$_3$	41
M(naph)$_3{}^c$	Al(C$_2$H$_5$)$_3$	42
Ti(OC$_4$H$_9$)$_4$	[(i-C$_4$H$_9$)$_2$Al]$_2$O	
Ti(OC$_4$H$_9$)$_4$	[(n-C$_6$H$_{13}$)$_2$Al]$_2$O	43
Ti(OC$_4$H$_9$)$_4$	[(C$_2$H$_5$)$_2$Al]$_2$N(n-C$_4$H$_9$)	
Ti(OC$_4$H$_9$)$_4$	C$_2$H$_5$MgBr	44
Ti(OC$_4$H$_9$)$_4$	C$_6$H$_5$MgBr	
Ti(OC$_4$H$_9$)$_4$	n-C$_4$H$_9$Li	45
Ti(O-n-C$_4$H$_9$)$_4$	Al(n-C$_6$H$_{13}$)$_3$	
Ti(O-n-C$_4$H$_9$)$_4$	Al(n-C$_8$H$_{17}$)$_3$	
Ti(O-n-C$_4$H$_9$)$_4$	Al(n-C$_{10}$H$_{21}$)$_3$	46
Ti(O-n-C$_8$H$_{13}$)$_4$	Al(C$_2$H$_5$)$_3$	
Ti(O-n-C$_8$H$_{17}$)$_4$	Al(C$_2$H$_5$)$_3$	
V(acac)$_3$	AlR$_3{}^d$	47
V(mmh)$_3{}^e$	AlR$_3$	

[a] acac = 2,4-pentanedionato (acetylacetonato).
[b] M = Ti, Cr, Fe, Co and Mn.
[c] M = Y, La, Pr,Nd, Gd, Tb and Dy; naph = naphthenate.
[d] R = C$_2$H$_5$, n-C$_3$H$_7$ and i-C$_4$H$_9$.
[e] mmh = 2-methyl-2,3-butanedionato.

metallic compounds of group 1–3a metals are effective for the polymerization of acetylene as shown in Table 2. Unlike ethylene and α-olefins polymerization of acetylene by the Ziegler–Natta catalysts is characterized by formation of cyclic oligomers. For example, Ti(acac)$_3$–Et$_2$AlCl gives cyclic trimer (benzene) as the main product (*ca* 80%) with a trace of ethylbenzene and a small amount of polyacetylene[36]. On the other hand, Ti(OBu)$_4$–Et$_3$Al yields polyacetylene exclusively.

Earlier investigators used a standard polymerization procedure for gaseous monomers that involves bubbling acetylene gas through the catalyst solution with stirring, and produced polyacetylene in the form of a black powder. Since the latter is oxidatively unstable, insoluble in all solvents and infusible, detailed investigation of its properties was unsuccessful.

A major breakthrough occurred in 1971 when Shirakawa and coworkers succeeded in preparing high-quality polyacetylene in the form of a free-standing film with metallic luster[3,41]. They employed a very high concentration of the homogeneous catalyst Ti(OBu)$_4$–Et$_3$Al and allowed acetylene to polymerize on a quiescent surface of the catalyst solution.

The following technique of using Ti(OBu)$_4$–Et$_3$Al that is soluble in common hydro-carbon solvents has been extensively applied and its products were generally used for polyacetylene study: In a typical preparation, 5 mmol of Ti(OBu)$_4$ and 20 mmol of Et$_3$Al, in that order, are added with stirring to 20 ml of toluene in a Schlenk flask. The mixture is left at room temperature for 30 min, when the Schlenk flask is attached to a vacuum line via a flexible joint, degassed at −78 °C and then rotated gently to wet as much of the flask wall with the viscous catalyst solution as possible. Acetylene gas is introduced into the flask at −78 °C. The surface wet with the catalyst solution turns deep red due to the formation of *cis*-polyacetylene, which develops a coppery luster within a few minutes as the thickness of the polymer film increases. The polymerization is interrupted by evacuating the flask. A free-standing film is recovered by washing repeatedly with dry oxygen-free toluene or

trans-polyacetylene (*trans-transoid*) (**1**)

cis-polyacetylene (*cis-transoid*) (**2**)

TABLE 3. *Cis*-contents of polyacetylenes prepared at different temperatures[a]

Temperature (°C)	*Cis*-content (%)
150	0.0
100	7.5
50	32.4
18	59.3
0	78.6
−18	95.4
−78	98.1

[a]Catalyst: Ti(OBu)$_4$–Et$_3$Al; Al/Ti = 4; [Ti] = 10 mmol liter^{-1}.

hexane until the washings become colorless. Finally the film is dried in a stream of dry argon for several minutes when the film is less than *ca* 10 μm, while thicker films are dried by vacuum pumping. The product obtained was *ca* 90–98% *cis*-polyacetylene (**2**)[3]. The polymer has an increasingly higher *trans*-content (**1**) as the polymerization temperature is raised as shown in Table 3.

The thickness of the films can be controlled by varying the acetylene pressure, duration of the polymerization and concentration of the catalyst solution. The thickness increases primarily with increasing acetylene pressure and polymerization time when the same catalyst system is used. Another dominant factor affecting the film thickness may be the concentration of the catalyst. Since this kind of polymerization proceeds by diffusion of acetylene gas through the film initially formed on the gas/catalyst solution interface, the thickness of the growing film may be controlled by the density of the film, which may be governed by the magnitude of the catalyst activity. Thus, use of a highly concentrated catalyst solution results in the formation of a very thin film with high density because the diffusion of acetylene gas into the catalyst solution is suppressed by the initially formed high-density film. In the case of dilute catalyst solutions or less active catalysts, diffusion of acetylene gas into the solution precedes the polymerization and, after several hours, the whole solution becomes a gel-like mass, which can be dried on a Teflon sheet to give a thin film. A low-density foam-like material can be prepared from the gel by freeze-drying it after replacing the solvent by benzene[48].

Since Ziegler–Natta catalysts are thermally unstable, their aging has been carried out at room temperature. Ti(OBu)$_4$–Et$_3$Al, however, is an exception among other Ziegler–Natta catalysts as demonstrated by Naarmann and Theophilou[49]. They have shown that the catalyst keeps its high activity for acetylene polymerization even after aging at a temperature as high as 120 °C. The most salient feature of their method is the high-temperature aging of the catalyst combined with the use of silicon oil as a solvent. The stretch-oriented films prepared by their method showed a conductivity higher than 10^5 S cm^{-1} after iodine doping. The apparent role of the silicon oil is an increase in the viscosity of the catalyst solution, but it is not essential for the production of highly conducting materials. Akagi and coworkers[50], and Tsukamoto and coworkers[51] showed that the use of the high-temperature aged catalyst results in the production of highly stretchable films and, consequently, in increased electrical conductivity of the stretch-oriented and iodine-doped films.

Although the catalyst aged at high temperature keeps its high activity, it is still deactivated to some extent during the aging process. Hence, it is of primary importance to use it at a high concentration. Akagi and coworkers[50] developed a polymerization procedure called the 'solvent evacuation' method, in which the solvent in the aged catalyst is evacuated just before the introduction of the acetylene, leaving a thin layer of the solventless catalyst on the wall of the Schlenk flask. In a modified method called 'Intrinsic Nonsolvent Polymerization'[52], neat Et$_3$Al is added dropwise to neat Ti(OBu)$_4$ in a Schlenk flask keeping the temperature below 0 °C, and the mixture is then subjected to aging at room temperature for 1 h, followed by aging at 150 °C for 1 h. Both methods are found to provide polyacetylene films with high density and high stretchability.

In summary, a homogeneous system is essential for the synthesis of polyacetylene films. Among the homogeneous Ziegler–Natta catalysts, Ti(OBu)$_4$–Et$_3$Al is widely used. The room-temperature aged catalyst gives a film of low density (*ca* 0.4 g cm^{-3} depending mainly on the catalyst concentration) and a texture like an aluminum foil. On the other hand, the high-temperature aged catalyst yields a soft and stretchable film of high density (0.9–1.1 g cm^{-3}).

ii. *Polymerization by single-component catalysts.* Although the standard Ziegler–Natta catalysts are combinations of main catalyst and co-catalyst, several metal complexes alone are reported to be active for acetylene polymerization. Aldissi and coworkers[27] reported that (benzyl)$_4$Ti has an activity without any reducing agent such as R$_3$Al. Hsu and coworkers[53] prepared a metallic gray, bulk polyacetylene by $(C_5H_4)Cp_3Ti_2(Ti-Ti)$[54] [μ-(η^1:η^5-cyclopentadienyl)-tris(η-cyclopentadienyl)dititanium], from acetylene by contacting it at 1 atm with a rapidly stirred solution of 30 mg of the complex in 250 ml of n-hexane. After ca 24 h, a dark reddish voluminous mass of 1.0–1.5 g of solvent-swollen polymer was obtained. At room temperature a polymer having a predominantly $trans$ structure was obtained, while at -80 °C the product was a cis-rich polymer. The highly reactive $Cp_2Ti(PMe_3)_2$ reacts instantly with acetylene in solution forming $Cp_2Ti(C_2H_2)PMe_3$, titanacyclopentadiene [$Cp_2Ti(C_4H_4)$], $trans$-polyacetylene and traces of benzene[55,56]. A very thin film (a few μm thick) was prepared by depositing a layer of this complex on the wall of a Schlenk flask by evaporating the solvent from a saturated solution, followed by polymerization of acetylene[57].

iii. Polymerization by Luttinger catalysts. The Luttinger catalyst consists of a salt or complex of a group VIII metal, nickel chloride for example, and a hydridic reducing agent such as sodium borohydride[24]. Nickel halide–tertiary phosphine complexes themselves are catalysts for the polymerization of acetylene. The catalytic activity is apparently specific for nickel–phosphine complexes, whereas cobalt and palladium complexes are completely inactive. Among the nickel complexes, $NiBr_2 \cdot 2Ph_3P$, $NiI_2 \cdot 2Ph_3P$ and $NiBr_2 \cdot 2(n-Bu_3P)$ are reported to be efficient but the corresponding chloride complexes are not very active[58].

Cataldo reported that chlorine-bridged Rh(I) complexes (e.g. [Rh(1,5-COD)Cl]$_2$ where 1,5-COD is cis,cis-cyclocta-1,5-diene, and [Rh(NBD)Cl]$_2$ where NBD is bicyclo-[2.2.01]hepta-2,5-diene are effective polymerization catalysts of acetylene, particularly in the presence of a base like sodium ethoxide which acts as a co-catalyst[59].

An important characteristic of the Luttinger catalysts is that a hydrophylic solvent such as ethanol, THF or acetonitrile, and even water, can be used in contrast to a requirement of the rigorously dehydrated hydrocarbon solvents for the Ziegler–Natta catalysts. Thus, the Luttinger catalysts have the advantage over the Niegler–Natta catalysts of being stable in air and do not require equipment and handling with the rigid exclusion of oxygen and moisture.

iv. Polymerization by metathesis catalysts. A soluble catalyst was prepared by mixing equimolar amounts of WCl_6 (or $MoCl_5$) and Ph_4Sn in toluene and was aged at 30 °C for 15 min[27]. The formation of uniform films was observed on the quiescent surface of a concentrated solution of WCl_6–Ph_4Sn. A more active catalyst was prepared by using n-BuLi as a co-catalyst instead of Ph_4Sn[28]. The relative activities were in the following order:

$$Ti(OBu)_4-Et_3Al \gg WCl_6-n-BuLi > WCl_6-Ph_4Sn > MoCl_5-Ph_4Sn$$

The metal halides alone are effective without a co-catalyst[60] although the activity is too low to prepare the product in the form of a film.

Combination of WCl_6 and $EtAlCl_2$ (or Et_3Al) is also effective for acetylene polymerization[29]. This catalyst shows high activity towards linear and cyclic alkene metathesis and a metathesis propagation step is to be invoked also in this case. However, this combination is a typical Ziegler–Natta catalyst system, and it may be probable that, in this particular case, the components are acting as a simple coordinated anionic catalyst system.

v. Non-catalytic high-pressure polymerization. A solid-state polymerization of substituted diacetylene has been known for many years and is described in Section IV. Acetylene also undergoes polymerization in the molecular–crystalline orthorhombic phase at room temperature and pressures above 3.5 GPa[30,31,61]. The polymerization was carried out in a diamond anvil that has 0.7-mm-diameter faces. Dominant formation of $trans$-polyacetylene confirmed by Raman spectroscopy suggested that acetylene undergoes $trans$ opening of the triple bond and is polymerized along the diagonal of the bc plane of the unit cell.

vi. Electrochemical polymerization. Polyacetylene was prepared on the surface of a platinum cathode by electrochemical polymerization of acetylene[32]. The electrolytic cell consisted of a platinum cathode, a nickel anode and $NiBr_2$ dissolved in acetonitrile as the electrolyte solution, applying a voltage of 4 to 40 V to start the polymerization for about 50 min. A thin layer of black material was observed on the surface of the platinum cathode, and then grew thicker. During this stage, no precipitation of polyacetylene was observed in the solution.

vii. Polymerization on metal oxide surfaces. Polymerization of acetylene occurs on a highly dehydroxylated surface of rutile $(TiO_2)^{[33]}$, γ-alumina $(Al_2O_3)^{[34,62]}$ and zeolite $KX^{[62]}$ at room temperature to give *trans*-polyacetylene. The results obtained from Raman spectroscopy are analogous with those reported in the literature for plyacetylene prepared by the homogeneous $Ti(OBu)_4$–Et_3Al catalyst. Kurokawa and coworkers observed the formation of a polyene radical on a surface of γ-alumina being fractured in the presence of acetylene in an ESR study[34].

viii. Polymerization by AsF_5. Soga and coworkers observed that acetylene polymerizes to form a thin film on the glass wall when acetylene is exposed to a small amount of AsF_5 in a reaction flask at -75 to $-198\,°C^{[35]}$. Infrared spectra showed that the product is *cis*-polyacetylene doped with AsF_5. The authors suggested that the polymerization proceeds through a charge transfer complex formed between acetylene and AsF_5. In a modification of this method, AsF_5 is introduced into an AsF_3 solution of acetylene[63]. The polymerization occurs instantaneously at temperatures between room temperature and $-90\,°C$. After the reaction, the excess of AsF_5 was removed by pumping, and the solvent was distilled from the reaction mixture to give homogeneous polymeric films.

b. Ring-opening polymerization of 1,3,5,7-cyclooctatetraene. Korshak and coworkers[7] synthesized polyacetylene films by evaporating the solvent from a solution of the catalyst, $W[OCH(CH_2Cl)_2]_nCl_{6-n}$–$Et_2AlCl$, under vacuum by rotating the Schlenk flask to form a solid layer on the surface of the flask, followed by condensing the monomer on the solid catalyst layer for a period of one to several days to provide a solid polyacetylene film.

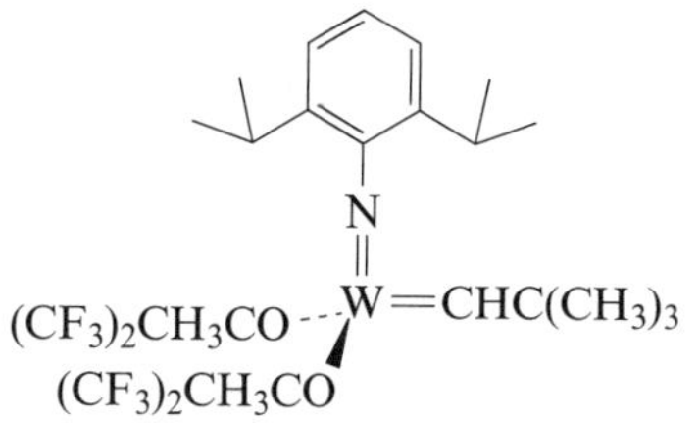

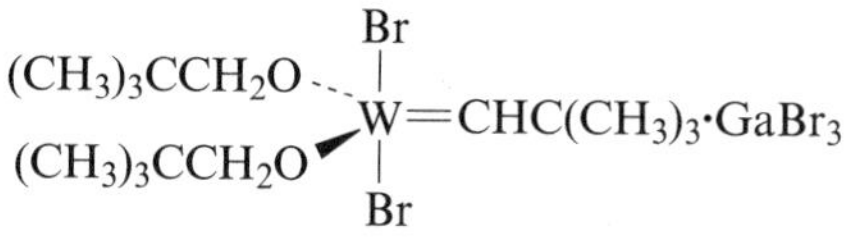

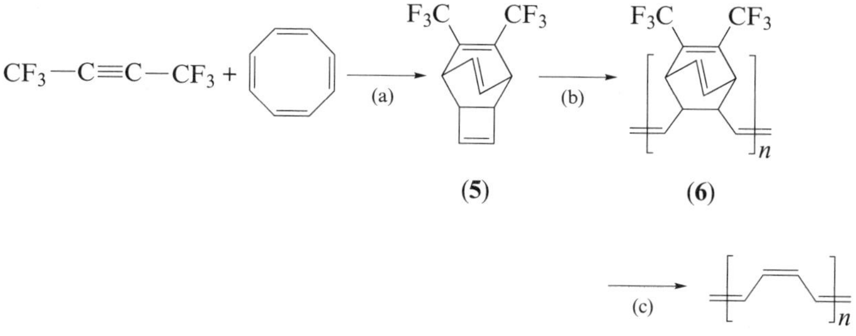

SCHEME 1. The Durham route to polyacetylene

Klavetter and Grubbs presented a versatile and convenient route to polyacetylene films by condensed-phase metathesis polymerization of 1,3,5,7-cyclooctatetraene, using the tungsten-based complexes **3** and **4**[8].

c. Indirect methods

i. Durham method. Edwards and Feast[5] have produced polyacetylene films in a process consisting of three steps. (a) Diels-Alder cycloaddition between 1,3,5,7-cyclooctatetraene and appropriate alkynes, (b) ring-opening metathesis polymerization of the monomers to form soluble prepolymers and (c) thermal decomposition of the prepolymers, as demonstrated in Scheme 1 with hexafluorobut-2-yne as an alkyne. The monomer, 7,8-bis(trifluoromethyltricyclo[4,2,2,0^{2,5}]deca-3,7,9-triene (**5**), is readily prepared in *ca* 80% yield by the thermal reaction between hexafluorobut-2-yne and 1,3,5,7-cyclooctatetraene at 120 °C. The metathesis ring-opening polymerization of **5** occurs selectively at the double bond in the cyclobutene ring to give the prepolymer **6**. The polymerization was carried out using either WCl$_6$–Ph$_4$Sn (W/Sn = 1/2) or TiCl$_4$–Et$_3$Al (Ti/Al = 1/2) catalysts in toluene as a solvent. When the reaction mixture become viscous, the polymerization is terminated by addition of methanol. The prepolymer is soluble in acetone and chloroform, and is thermally unstable and decomposes spontaneously to form polyene and 1,2-bis(trifluoromethyl)benzene on standing in the dark under an atmosphere of dry nitrogen and more rapidly in solution. *Trans*-polyacetylene is obtained by thermal treatment of **6** at 150 °C for 5 h under vacuum (0.01 mmHg) and additional treatment at 210 °C for 3 h. Monomer **3** is widely used for this route, but several other monomers (**7–10**) have also been studied[64].

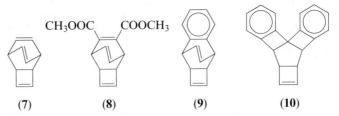

| (**7**) | (**8**) | (**9**) | (**10**) |

The Durham route allows the preparation of polyacetylene in the form of continuous uniform and featureless films, as opposed to the entangled fibrillar morphology obtained by the direct polymerization of acetylene.

ii. Isomerization of polybenzvalene. The highly reactive but readily available monomer benzvalene[65,66] (**11**) can be converted to polybenzvalene (**12**) by ring-opening metathesis polymerization[6]. The polymer is an interesting and unusual one that can be transformed into polyacetylene without elimination of molecular fragments as shown in Scheme 2. The monomer was polymerized by tungsten alkylidene metathesis catalysts[67–69], for example (RO)$_2$W(N(2,6-(*i*-Pr)$_2$Ph))CHCMe$_3$ where R = *t*-butyl, or hexafluoro-*t*-butyl. As a result of its highly strained structure, **12** is a sensitive material that has been observed to undergo spontaneous exothermic decomposition upon the application of severe mechanical stress

(6)

    (**11**)   (**12**)

SCHEME 2. Isomerization of polybenzvalene to polyacetylene

or rapid heating. Isomerization of the polymer can be performed by using transition-metal catalysts. Thus, solutions of $HgCl_2$, $HgBr_2$ and Ag^+ salts in THF transformed films of the polymer into shiny silvery materials resembling polyacetylene in appearance.

d. Synthesis of oriented polyacetylene.   The linear arrangement of π-electrons in the polyacetylene molecule gives rise to anisotropic optical, electrical, magnetic and mechanical properties. To understand these anisotropic properties of polyacetylene, direct syntheses of oriented materials are essential because there are no methods to prepare oriented materials exceptly mehcanical stretching of *cis*-rich polyacetylene films due to their insolubility and infusibility.

i. Epitaxial polymerization on the surfaces of aromatic hydrocarbon crystals. Woerner and coworkers[70,71] have demonstrated that acetylene can be polymerized on the surfaces of biphenyl crystals containing dissolved Ti(OBu)$_4$–Et$_3$Al catalyst to give a highly oriented polyacetylene. The orientation was confirmed through a polarizing optical microscope. Ozaki and coworkers[72] applied the same method using benzene instead of biphenyl and obtained highly oriented thin films. The epitaxial polymerization was observed on the (111) plane of the benzene crystal and arrangements of the polymer chains were found to be along two directions, $\langle 211 \rangle$ and $\langle 112 \rangle$. They explained the epitaxial polymerization with the small mismatching factors between the polyacetylene chain period (*b*-axis) and the lattice periodicities of two directions on the (111) plane of benzene crystal. Yamashita and co-workers[73] reported epitaxial polymerization using naphthalene, anthracene, biphenyl and terphenyl.

ii. Shear flow method. Meyer[74,75] developed a unique procedure to polymerize acetylene under shear conditions and graphoepitaxially in a Couette-type reactor. A glass cylinder as stator was concentrically adjusted around a Teflon drum acting as rotor. The clearance between stator and rotor was 3 mm, in which was filled a catalyst solution of Ti(OBu)$_4$–Et$_3$Al in toluene and held at $-78\,^{\circ}$C. Acetylene gas was introduced into the catalyst solution from the bottom of the reactor, while the Teflon drum was rotated at *ca* 3000 rpm. After the reaction, the films of *ca* 0.5 μm in thickness were lifted off from the glass stator and transferred onto other surfaces. The high degree of orientation was confirmed by crossed polarizers and scanning electron micrograph (SEM) observation.

iii. Polymerization in liquid crystal solvents. Araya and coworkers and Aldissi have developed independently the polymerization of acetylene in liquid crystal solvents. Araya and coworkers[76-78] used an equimolar mixture of 4-(*trans*-4-*n*-propylcyclohexyl)-ethoxybenzene (PCH302) and 4-(*trans*-4-*n*-propylcyclohexyl)-butoxybenzene (PCH304). The mixture was chosen for its chemical stability in the presence of Ziegler–Natta catalysts and for the wide range of temperature in its nematic phase. The catalyst solution was prepared by dissolving Ti(OBu)$_4$ and Et$_3$Al in the liquid crystal mixture. A macroscopic orientation of the solution was attained by letting it flow along the glass wall of the Schlenk flask. Polymerization was carried out by introducing acetylene gas onto the oriented catalyst solution at room temperature (18 $^{\circ}$C). SEM examination of the film showed the high degree of orientation. It was confirmed that the direction of orientation coincides with the flow direction.

Aldissi[79] used the most popular nematic liquid crystal, *p*-methoxybenzylidene-*p*-*n*-butylaniline (MBBA), and aligned the catalyst solution under a magnetic filed of *ca* 4 kG at room temperature. The nematic range of MBBA is 19–30 $^{\circ}$C and it aligns under a magnetic field higher than 2.5 kG. SEM observation of this film revealed that the morphology is not significantly different from the conventional polymer prepared by the usual methods. Akagi and coworkers[80] investigated the behavior of the equimolar mixture of PCH302 and PCH304 and their solutions of Ti(OBu)$_4$ and Et$_3$Al catalyst under a magnetic field by means of optical transmission measurements. The magnetic field required to achieve an ultimate orientation of the liquid crystal was found to be 1 kG, which is shifted to 10 kG upon dissolving the catalyst in a concentration of 0.89 vol% in the mixture. The polymerization was carried out at 10–13 °C in order to keep the nematic phase under a magnetic field of 2–14 kG. Polyacetylene films prepared showed considerable alignment of fibrils in the direction of the applied magnetic field, and the high electrical conductivity of 10^4 S cm^{-1} order upon iodine doping[81,82]. Montaner and coworkers[83] used a modification of Akagi's method by a combination of the gravitational flow and a stronger magnetic field (47 kG) by using a superconducing magnet, achieving a significant improvement in chain orientation.

The equimolar mixture restricted the lower limit of polymerization temperature to near 0 °C, resulting in films having *cis*-contents of 50% at most even though they were as-grown aligned. Low-temperature nematic liquid crystals were developed by mixing several phenylcyclohexane derivatives having different *n*-alkyl and/or *n*-alkoxy groups[84]. Thus, the catalyst solutions using three; five- and six-component systems permitted low-temperature polymerizations ranging from – 20 to – 30 °C, from – 40 to – 60 °C and from –10 to – 45 °C, respectively. Polyacetylene films synthesized by using five- and six-component liquid crystals with an external magnetic force had high *cis*-contents of 85–95% as was to be expected from the low-temperature polymerization.

2. Polymerization mechanisms

Two mechanisms have been proposed for acetylene and substituted acetylene polymerization by transition metal catalysts: one is the metal–alkyl mechanism and the other is the metal–carbene mechanism. In general, it has been proposed that the polymerization of acetylenes by Ziegler–Natta catalysts proceeds by the metal–alkyl mechanism, while the metal–carbene mechanism has been accepted for the polymerization of substituted acetylenes by metathesis catalysts whose main components are halides or complexes of group 5 and 6 transition metals. The latter will be discussed in Section III.

a. The metal–alkyl mechanism in Ziegler–Natta polymerization. We mentioned already that polymerization of acetylene accompanied by some cyclic oligomers, whose amount depends strongly on the catalyst components. Ikeda and Tamaki[85] found that ethylbenzene and toluene form as minor products when acetylene is polymerized with TiCl$_4$–Et$_3$Al and with TiCl$_4$–Me$_3$Al, respectively. Considering also the results obtained by ^{14}C- and ^{2}H-labeled triethylaluminum, they concluded that the formation of cyclic oligomers can be explained by a mechanism that is quite analogous to that of the coordinated anionic polymerization, as shown in Scheme 3.

Permanganate oxidation of polyacetylene prepared by TiCl$_4$–Et$_3$Al, TiCl$_4$–Me$_3$Al and Ti(OBu)$_4$–Et$_3$Al revealed that various carboxylic acids were identified together with the main product, oxalic acid, which originates from the regular repeating unit ($=$CH—CH$=$) in the polymer[86,87]. Since carbon–carbon double bonds are expected to break forming two carboxyl groups in the oxidation reaction, the sources of acetic, propionic and benzoic acid must be the terminal groups in the polyacetylene chains, MeCH$=$CH—, EtCH$=$CH— and PhCH$=$CH—, respectively. Formation of benzoic acid is attributed to phenyl groups formed by a terminal cyclization of the growing chains. Obviously, propionic acid and acetic acid are derived from the catalyst alkyl groups.

As shown in Table 3, *cis*- and *trans*-contents of polyacetylene prepared by the Ziegler–Natta catalysts depends strongly upon the polymerization temperature[3,41,88]. There are two possible explanations for this observation: One is that the fundamental mechanism is the formation of *cis* double bonds by the *cis* insertion of acetylene monomer into the Ti—C bond of the catalyst. This fits the orbital interaction consideration for the role of the catalyst by Fukui and Inagaki[89], according to which the initially formed configuration of the double bond is *cis* as a result of the favorable orbital interaction between the inserting acetylene monomer and the active site of the catalyst. Because the *cis* double bond is thermodynami-

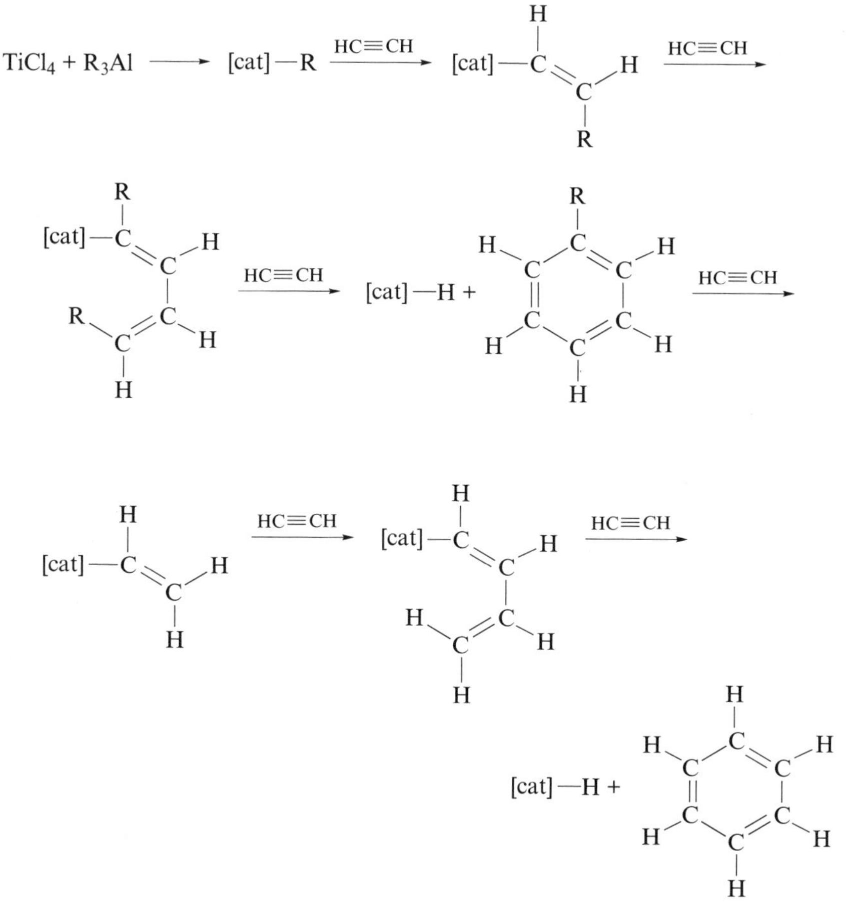

SCHEME 3. Formation of cyclic trimers by the metal–alkyl insertion mechanism

cally less stable than the *trans* one[90], thermal isomerization occurs to decrease the *cis*-content during the chain propagation step, and the extent of thermal isomerization is mainly determined by the polymerization temperature. The second explanation is that there may be different active sites on the catalyst, one of which leads to *trans* opening of the acetylene monomer to give *trans*-polyacetylene directly. This explanation is hard to accept, at least for Ti(OBu)$_4$–Et$_3$Al, because of the successful synthesis of all-*cis* polyacetylene[91].

In summary, the mechanism of acetylene polymerization by the Ziegler–Natta catalyst is quite analogous to that of cyclic trimerization reaction, as shown in Scheme 4[86,87,92]. Thus, both cyclic trimerization and polymerization proceed by the formation of *cis* double bonds by the *cis* insertion of acetylene monomer into the Ti—C bond of the catalyst. Whether the product is a cyclic oligomer or a polymer molecule is dependent on whether the conformation of the growing chain in the vicinity of the active site is *cisoid* or *transoid*, which may be determined by interaction between the growing chain and the catalyst metal. If the growing chain takes *cisoid* conformation, the cyclic trimerization may proceed in preference to the polymerization reaction[87].

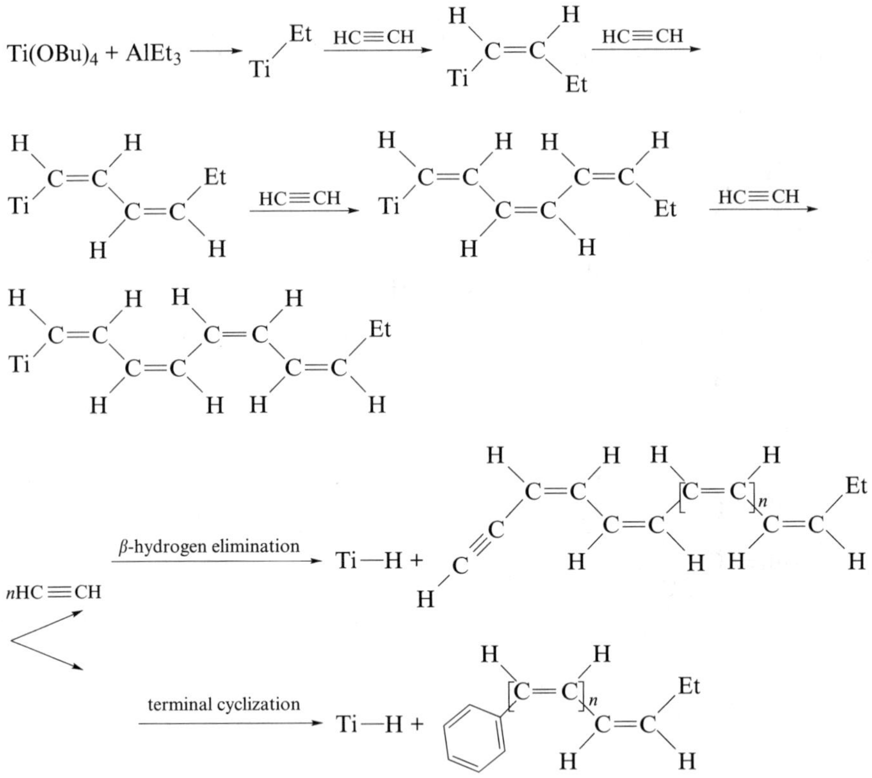

$$\text{Ti(OBu)}_4 + \text{AlEt}_3 \longrightarrow \underset{\text{Ti}}{\overset{\text{Et}}{\diagup}} \xrightarrow{\text{HC}\equiv\text{CH}} \underset{\text{Ti}}{\overset{\text{H}}{\diagup}}\text{C}=\text{C}\overset{\text{H}}{\underset{\text{Et}}{\diagdown}} \xrightarrow{\text{HC}\equiv\text{CH}}$$

SCHEME 4. Formation of polyacetylene by the metal–alkyl insertion mechanism

b. Polymerization mechanism by the Luttinger catalysts. Daniels[58] suggested that the initial step in the polymerization by $\text{NiBr}_2\cdot2\text{R}_3\text{P}$ is replacement of one or more phosphine ligands by acetylene to form a transient π-complexed acetylene intermediate which leads to the polymer by a cycle of ligand insertion and monomer complexation as shown in Scheme 5. The proposed mechanism is based on the fact that the nickel moiety of the product is removed completely by anhydrous HCl in chloroform while halogen is retained, suggesting that nickel and halogen are present as end groups, the halogen as alkyl halide and the nickel as labile organometallic.

c. Mechanism by the metathesis catalysts. Although the metathesis catalysts have been known for many years to act as polymerization catalysts for cycloolefins and substituted acetylenes, not many of the investigations have been extended to the polymerization of acetylene[27-29]. Theophilou and coworkers[28] investigated this reaction catalyzed by WCl_6–BuLi, and proposed the mechanism shown in Scheme 6, which is quite similar to that of the substituted acetylene poplymerization to be described in Section III.A.4. The mechanism involves the replacement of a chlorine atom in WCl_6 by a butyl group, followed by the elimination of HCl to generate an active carbene. This species would then serve as the initiator for the acetylene polymerization.

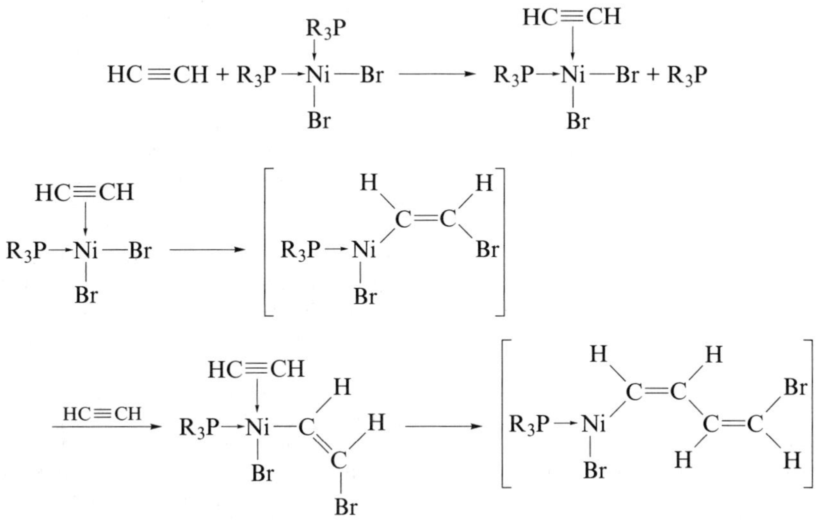

SCHEME 5. Proposed mechanism catalyzed by aNi-complex

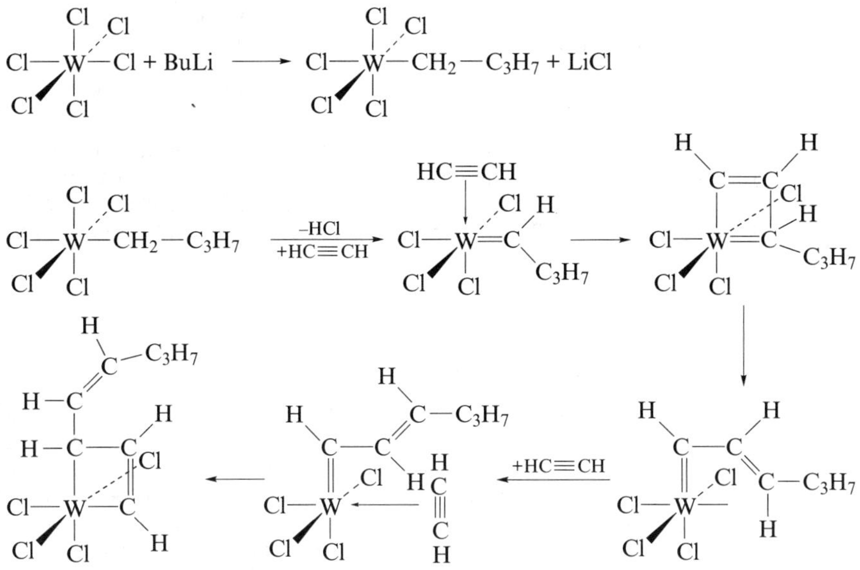

SCHEME 6. Polymerization mechanism by a metathesis catalyst

3. Determination of molecular weight

Insolubility and infusibility of polyacetylene make it impossible to determine its molecular weight by the usual methods based on solutions. Several attempts have been made either to convert polyacetylene to soluble derivatives in order to measure their molecular weight indirectly through that of the derivatives by gel permeation chromatography, or to determine the number-average molecular weight by radioquenching of the polymerization system.

   a. Hydrogenation of alkaline–metal doped polyacetylene to polyethylene.   Shirakawa and coworkers observed that alkaline metal-doped polyacetylene can be hydrogenated at elevated temperature and high hydrogen pressure[93–95] to give polyethylene completely soluble in hot tetraline[94]. Production of completely soluble materials may suggest that the polyacetylene synthesized by Ti(OBu)$_4$–Et$_3$Al has no cross-linkages and that neither degradation of polyacetylene nor of the hydrogenated product takes place during the hydrogenation. However, it is still possible that slight hydrogenolysis occurs during the hydrogenation reaction. The number-average molecular weight ($\bar{M}_n$) determined by this method was *ca* 6000–7000 with a value of $\bar{M}_w/\bar{M}_n$ equal to 2.4.

   b. Chlorination of as-prepared polyacetylene.   Under carefully selected conditions such as reaction temperature –30 °C, low catalyst concentration and the reaction being carried out immediately after the polymerization, polyacetylene prepared by the Luttinger catalyst (Co(NO$_3$)$_2$–NaBH$_4$) at –80 °C can be reacted with chlorine to give completely soluble derivatives[96–98] corresponding to poly(1,2-dichloroethylene). The molecular weight of the soluble chlorinated products was generally of the order of 20,000–50,000 from which the molecular weight of the pristine polyacetylene can be calculated to be 5000–13,000.

   c. Determination of molecular weight by radioquenching.   Radiotagging of the growing chain end with tritiated methanol has been used to obtain the number-average molecular weight of polyacetylene prepared by Ti(OBu)$_4$–Et$_3$Al[99,100]. The reaction involved is

$$Ti—P + CH_3O^3H \longrightarrow TiOCH_3 + P^3H \qquad (7)$$

where ^3H is tritium, P is a polyacetylene chain and P^{3}H is the radiotagged polyacetylene. Depending upon the catalyst concentration, catalyst to activator ratio, monomer pressure and temperature, the values of $\bar{M}_n$ were found to range from 500–120,000[100]. The free-standing *cis*-rich polyacetylene film made by a standard procedure had a quite uniform $\bar{M}_n$ value of 10,500 ± 1500[100].

B. Reactions of Polyacetylene

1. Chemical and electrochemical doping

   Among the properties of polyacetylene, the most striking features must be oxidation and reduction accompanying charge transfer between the polymer and electron acceptors or electron donors. The charge transfer reactions convert the insulating or semiconducting polyacetylene to a conducting polymer with electrical conductivity of over 10^5 S cm^{-1} [49,51,101]. The reaction is widely referred to as 'doping' by analogy with the doping of inorganic semiconductors such as Si, Ge and GaAs in contrast to the oxidation and reduction, because a small amount of the reagents causes a drastic change in electrical, optical, magnetic and even chemical properties. Electron acceptors and electron donors that are effective for the doping of conjugated polymers are called 'dopants'.

   As a result of very long π-conjugation along its backbone, polyacetylene has a small ionization potential and a large electron affinity. The small value for the ionization potential indicates that the polymer is easily oxidized by electron-accepting reagents, and the large value for the electron affinity indicates that the polymer is easily reduced with electron donors. The calculated[102] and experimentally determined[103] values of ionization potential (4.7 eV for *trans*-polyacetylene) are much smaller than those of aromatic hydrocarbons, such as benzene (9.24 eV), naphthalene (8.12 eV) and anthracene (7.55 eV)[104]. Thus, polyacetylene is easily oxidized by a large variety of electron acceptors such as halogens, protonic acids and Lewis acids to give *p*-type doped polyacetylenes bearing positively-charged radical ions along its conjugated chain and anions as the counter ion, as shown in equation 8, where A and A$^-$ denote the electron acceptor and the acceptor

$$\text{(polyene chain)} + A \longrightarrow \overset{\overset{A^-}{+}}{\text{(polyene chain)}}{}_{\bullet} \tag{8}$$

$$\overset{\overset{A^-}{+}}{\underset{\bullet}{\text{(polyene chain)}}} + A \longrightarrow \overset{\overset{A^-}{+}}{\underset{\bullet}{\text{(polyene chain)}}}\,\overset{\bullet}{\underset{\underset{A^-}{+}}{\text{(chain)}}} \tag{9}$$

$$\overset{\overset{A^-}{+}}{\underset{\underset{A^-}{+}}{\underset{\bullet}{\text{(polyene chain)}}}} \longrightarrow \overset{\overset{A^-}{+}}{\underset{\underset{A^-}{+}}{\text{(polyene chain)}}} \tag{10}$$

anion, respectively. Subsequent charge transfer on the same chain forms a pair of radical cations (equation 9). A dication forms as a result of diffusion of the radicals by changing the bond alternation along the conjugation (equation 10). The overall reaction can be written as equation 11, where $(CH)_x$ denotes polyacetylene.

$$(CH)_x + xyA \longrightarrow (CH^{y+}A^-_y)_x \tag{11}$$

Polyacetylene can also be reduced by alkaline metals to give *n*-type doped polyacetylene bearing dianions and alkaline–metal cations as the counter ion as shown in equation 12, where D and D$^+$ denote the electron donor and the donor cation, respectively.

$$(CH)_x + xyD \longrightarrow (D^+_y CH^{y-})_x \tag{12}$$

The most investigated acceptor (*p*-type dopant) is iodine. The doping is carried out either in gas phase or in chloroform or carbon tetrachloride solution. Other halogen derivatives, such as Cl_2, Br_2, ICl, ICl_3, IBr may cause an electrophilic addition reaction to give halogenated polyacetylenes with no conjugated double bonds, although there are some doping effects at the initial stage of the reaction. Raman scattering from the conducting polyacetylene doped by iodine indicates that the doped iodine takes the form of either I_3^- or I_5^- [53,105–110]. ^{129}I Mössbauer spectroscopic studies[111–113] have revealed that the iodine–doped polyacetylene contains anion species of I^-, I_3^- and I_5^-, where the structural arrangement of I_3^- and I_5^- is linear with symmetrical charge population. The content of each iodide anion depends on the dopant concentration y due to the chemical reactions 13–15, where

$$\tfrac{1}{2} I_2 + e^- \longrightarrow I^- \tag{13}$$

$$I^- + I_2 \longrightarrow I_3^- \tag{14}$$

$$I_3^- + I_2 \longrightarrow I_5^- \tag{15}$$

the mono-iodide anion I^- is created by charge transfer from the polymer and consumed by the formation of I_3^- and I_5^- during the doping.

Doping by $FeCl_3$ is usually carried out in nitromethane solution. The anion species is identified by ^{57}Fe Mössbauer spectroscopy[114,115] and the extended X-ray absorption fine structure (EXAFS)[116,117] as $FeCl_4^-$. The doping reaction involves dissociation of $FeCl_3$ in nitromethane (**16a**) followed by oxidation of polyacetylene by $FeCl_2^+$ (**16b**). Thus, the overall doping reaction can be written as equation (**16c**).

$$2FeCl_3 \rightleftharpoons FeCl_2^+ + FeCl_4^- \qquad (16a)$$

$$FeCl_2^+ + e^- \longrightarrow FeCl_2 \qquad (16b)$$

$$(CH)_x + 2xyFeCl_3 \longrightarrow [(CH(FeCl_4^-)_y)_y]_x + xyFeCl_2 \qquad (16c)$$

Doping of cycloacetylene with AsF_5 was found to involve oxidation of the polymer according to the following reaction[118]:

$$3AsF_5 \longrightarrow 2AsF_6^- + AsF_3 \qquad (17)$$

Various acceptor and donor dopants are summarized in Table 4.

TABLE 4. Various dopants for polyacetylene

	Acceptor dopants
Halogens	Cl_2, Br_2, I_2, ICl, ICl_3, IBr, IF_5
Lewis acids	PF_5, AsF_5, SbF_5, BF_3, BCl_3, BBr_3, SO_3, $AlCl_3$
Protonic acids	HF, HCl, HNO_3, H_2SO_4, $HClO_4$, FSO_3H, $ClSO_3H$, CH_3SO_3H
Transition-metal halides	$TiCl_4$, $ZrCl_4$, $HfCl_4$, $NbCl_5$, $TaCl_5$, $MoCl_5$
	WCl_6, $FeCl_3$, UF_6, $LnCl_3$ (Ln = La, Ce, Pr, Nd, Sm)
Organic acceptors	$TCNE^a$, $TCNQ^b$, chloranil, DDQ^C
Electrolyte anions	Cl^-, Br^-, I^-, ClO_4^-. PF_6^-, AsF_6^-, SbF_6^-, BF_4^-
	various sulfonic acid anions
Miscellaneous	$XeOF_4$, XeF, $(NO_2)^+(PF_6^-)$, $(NO_2)^+(SbF_6^-)$,
	$(NO_2)^+(BF_4)^-$, $(NO)^+(SbCl_6)^-$, $(Me_3O^+)(SbCl_6^-)$
	FSO_2OOSO_2F, $AgClO_4$, $AgBF_4$, H_2IrCl_6

	Donor dopants
Alkaline metal	Li, Na, K, Rb, Cs
Alkal-earth metal	Ca, Sr, Ba
Lanthanide metal	Eu
Miscellaneous	R_4N^+, R_4P^+, R_4As^+, R_3S^+

^{a}Tetracyanoethylene.
^{b}7,7,8,8-Tetracyanoquinodimethane.
^{c}2,3-Dichloro-5,6-dicyano-p-benzoquinone.

Polyacetylene can be controllably oxidized or reduced by simple electrochemical procedures[119]. Cyclic voltammetry studies on free-standing films of *cis*-polyacetylene show that they can be reversibly oxidized at *ca* +3.6 V *vs* Li and reversibly reduced at *ca* +1.4 V *vs* Li. The reactions occurring at the cathode (equation 18a) and at the anode (equation 18b) are given below.

$$(CH)_x + xyA^- \longrightarrow (CH^{y+}A^-_y)_x + xye^- \qquad (18a)$$

$$(CH)_x + xyD^+ + xye^- \longrightarrow (D^+_yCH^{y-})_x \qquad (18b)$$

Electrochemical doping provides some advantages over chemical doping. Thus both p-type and n-type dopings can be achieved simultaneously, the doping level can be precisely controlled by the applied voltage and, 'undoping' (reverse reaction of doping) proceeds smoothly by changing the applied voltage.

2. Stereospecific chlorination of p-doped polyacetylene

Polyacetylene reacts with chlorine rapidly to give a white polymer that is equivalent to poly(1,2-dichloroethylene) and no stereospecificity has been reported on the chlorinated polymer. The addition of halogens is an important general reaction of carbon–carbon double bonds in a stereospecific and regiospecific sense. For instance, the electrophilic additions of chlorine to ethylene, buta-1,3-diene and hexa-1,3,5-triene have been shown to proceed by 1,2-*trans* (*anti*), 1,4-*cis* (*syn*) and 1,6-*trans* (*anti*) attack, respectively. This stereospecificity has been rationalized with a mixing rule of $\sigma-\pi$ orbital interaction[120–122]. The reaction of a long conjugated polyene like polyacetylene with chlorine may produce an atactic chlorinated polyene, because random 1,2n additions occur to result in a random addition product[123].

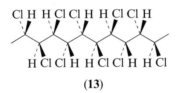

(13)

Akagi and coworkers[123–125] have found that doping by $FeCl_3$ or I_2 followed by addition of chlorine to the doped polyacetylene produces a stereoregular chlorinated polyacetylene, i.e. di-syndiotactic poly(1,2-dichloroethylene) (13). In this case, tacticity of the polymer, defined as the ratio of stereoregular to atactic segments in the chlorinated polymer, strictly depends upon the dopant concentration of iodine before the chlorine addition. This means that the stereospecific chlorination depends on how many positively charged polyene segments are generated by the chemical doping, which thus plays an essential role in governing the stereospecific chlorination. This also means that undoped polyene segments still remaining in partially doped polyacetylene will suffer random chlorination by 1,2n additions, yielding atactic segments in the chlorinated product. Thus, the chemical doping is a promising method to control the stereospecificity of the addition reaction.

C. Properties of Polyacetylene

1. Electrical properties

Pristine polyacetylene is a typical organic semiconductor irrespective of its *cis–trans* content. The room-temperature conductivity measured by a conventional two-probe method and the energy gap determined by equation 19 are 1×10^{-4} S cm^{-1} and 0.56 eV for 92.5% *trans*-polymer, and 4×10^{-9} S cm^{-1} and 0.93 eV for 80.0% *cis*-polymer, respectively[126]. In equation 19 σ is the electrical conductivity at T and ΔE_G is the energy gap

$$\sigma = \sigma_0 \exp[-\Delta E_G/(2kT)] \qquad (19)$$

between the valence and conduction bands. If the electronic conduction of the pristine polyacetylene can be explained by a band model, the energy gap should coincide with the photoconductive threshold. The energies corresponding to absorption edges at longer wavelengths for *cis*- and *trans*-polyacetylene were found from the visible spectra to be 1.78 eV (695 nm) and 1.35 eV (920 mm), respectively. The smaller values of the band gap obtained by the temperature-dependent conductivity are due to inpurity levels in the band gap. Park and coworkers[127] observed that the sign of thermoelectric power of the pristine polyacetylene is positive and its value is *ca* +900 μV K^{-1}, indicating *p*-type behavior consistent with an unexpected *p*-type doping of the polymer by electron-accepting catalyst residues.

As described in Section II.B.1 above, doping causes a drastic change in the electrical properties of polyacetylene. The initial values of electrical conductivity were of the order of 10^2 S cm^{-1} for unoriented materials[4,124–130] and, when doped by iodine and AsF$_5$, were enhanced to the order of 10^3 S cm^{-1}, which was obtained in the parallel direction of the doped films oriented by mechanical stretching[127,131]. Improvements in polymerization methods and in the catalyst systems also enhanced the electrical conductivity. Highly oriented films prepared in liquid crystal solvents (Section II.A.1.d.iii) exhibited a conductivity higher than 10^4 S cm^{-1}, as did also a well stretch-oriented film prepared by Ti(OBu)$_4$– Et$_3$Al dissolved in silicon oil and aged at 120°C[132]. In further studies Naarmann and Theophilou[49] and Tsukamoto and coworkers[51] attained a conductivity of *ca* 10^5 S cm^{-1}.

2. Magnetic properties

Pristine *trans*-polyacetylene has a narrow spin resonance with a *g* value (2.0026 ± 0.005) close to the free-electron value and an intensity corresponding to a Curie-law susceptibility from approximately one unpaired spin per 3000 carbon atoms[126]. When acetylene is polymerized directly in an ESR sample tube at –78 °C, there is no signal around the *g* value suggesting that *cis*-polyacetylene has no unpaired electrons. ESR measurements on stretched films suggested strongly that the signal has its origin in the π-electrons in *trans*-polyacetylene[133]. Thus, it has been suggested that the signal has its origin in the bond alternation kink in the conjugated *trans* sequence as shown in **14**. Note that the unpaired

(14)

electron is located at the site where two *trans* conjugated sequences having opposite bond alternation merge to form the kink. The unpaired electorn on the kink is often referred to as a netural soliton[134,135]. Upon doping, the Curie-law susceptibility decreases to a level lower than 1 ppm in the highly conducting regime[136]. The neutral soliton is either oxidized by an acceptor or reduced by a donor, to form either a positively charged soliton that is equivalent to a delocalized carbonium ion or a negatively charged soliton that is equivalent to a stabilized carbanion. These charged solitons are believed to be the carriers for electrical conduction. The temperature-independent Pauli susceptibility shows an abrupt increase by heavy doping (7%).

3. Optical properties

Polyacetylene has a strong absorption due to the lowest $\pi \rightarrow \pi^*$ transition in the visible region with λ_{max} at 594 nm for the all-*cis* polymer and at 700 nm for the all-*trans* polymer[137]. The ε_{max} of these absorptions is *ca* 4 ×10^5 cm^{-1}. Thus, very thin films of *trans*- and *cis*-

polymers show deep blue and clear red colors, respectively. As the thickness of the polymer films increases, reflection becomes predominant to give a coppery luster in the *cis* and a more silvery shine in the *trans*-polymers.

Photoconductivity has been observed in *trans*-polyacetylene, but found to be absent in *cis*-polymer[138,139]. On the other hand, photoluminescence has been observed in the *cis*- but not in the *trans*-polymer[139,140]. Lauchlan and coworkers[139] interpreted these results in terms of photogeneration of a soliton–antisoliton pair in *trans*-polyacetylene followed by its separation which induces photoconduction, but there is no recombination and hence no luminescence. In *cis*-polyacetylene, the creation of a soliton and its migration are considered to be energetically unfavorable because the ground state of the *cis*-polymer (*cis-transoid*) (**2**) is more stable than the *trans-cisoid* structure. Consequently, the photoexcited state in *cis*-polymer recombines immediately after the excitation to release luminescence.

Sinclair and coworkers[141] measured the third-order nonlinear optical susceptibility of *trans*-polyacetylene. The measured susceptibility $\chi^{(3)}$ was 5×10^{-10} esu, which is comparable to the magnitude of the large nonlinear susceptibilities measured in the polydiacetylenes.

4. Mechanical properties

Young's modulus (E) of an infinitely long linear polymer can be roughly estimated by equation 20, where k, r_0 and S are the force constant of the C—C bond stretching in the

$$E = k \cdot r_0/S \tag{20}$$

repeating unit of the polymer, the equilibrium bond length of the bond and the cross section of the molecule, respectively. Taking into account the simple structure without any pendant groups, the bond multiplicity and the planarity of both *cis*- and *trans*-polyacetylene, one might expect that polyacetylene has a very large Young's modulus. In fact, Young's modulus and the tensile strength of *trans*-polyacetylene films prepared by the nonsolvent method and stretch-oriented by 7–8 times were 100 GPa and 0.9 GPa, respectively. Those of *cis*-polymer films with the same draw ratio were 25–30 GPa and 0.6 GPa, respectively[50]. The result that Young's modulus of the *trans*-polymer is 3–4 times larger than that of the *cis*-polymer is consistent with values calculated by Treloar's method[142] of *trans*- and *cis*-isomers being *ca* 300–400 Gpa and 100 GPa, respectively. Cao and coworkers[143,144] reported that Young's modulus and the tensile strength of *trans*-polyacetylene prepared by the nonsolvent method and by modified stretch-orientation up to 15 times, were 50 GPa and 0.9 GPa, respectively, and that the electrical conductivity increases linearly with Young's modulus and with the tensile strength, the polymer chain orientation being the implicit variable.

III. SUBSTITUTED POLYACETYLENES*

A. Synthesis

1. Polymerization catalysts

a. Metal chloride-based catalysts. Among various transition metal chlorides, those of group 5 and 6 transition metals (**15**) and their mixtures with suitable organometallic cocatalysts (**16**) are specifically effective in the polymerization of substituted acetylenes[15,16]. The catalyst components are commercially available, the catalysts are easy to

*By T. Masuda.

$$MCl_n \ (M = Nb, Ta, Mo, W; n = 5, 6)$$

$$(15)$$

$$MCl_n\text{--cocatalyst (1:1) (cocatalyst: } Ph_4Sn, Bu_4Sn, Ph_3SiH, Et_3SiH, Ph_3Sb, Ph_3Bi)$$

$$(16)$$

prepare and highly active, and hence they are eminently suitable for the synthesis of substituted polyacetylenes.

MoCl$_5$ and WCl$_6$ alone can induce the polymerization of various monosubstituted acetylenes. The use of a suitable organometallic cocatalyst enhances the catalytic activity. With these catalysts, polymer molecular weight is low or medium ($<10^5$) for 1-hexyne and phenylacetylene but reaches one million for sterically crowded monomers like t-butylacetylene and $ortho$-substituted phenylacetylenes.

For the polymerization of disubstituted acetylenes, MoCl$_5$ and WCl$_6$ alone are inactive, and it is necessary to use the catalyst/cocatalyst mixtures (16), which are active for sterically less crowded monomers (e.g., 2-octyne and 1-chloro-1-octyne). In contrast, NbCl$_5$ and TaCl$_5$ by themselves polymerize disubstituted acetylenes with bulky substituents such as 1-(trimethylsilyl)-1-propyne. Diphenylacetylene and its derivatives, however, are polymerizable only with the TaCl$_5$–cocatalyst systems. The Nb and Ta catalysts selectively afford cyclotrimers from most monosubstituted acetylenes.

b. Metal carbonyl-based catalysts. The catalysts (17) obtained by ultraviolet irradiation of CCl$_4$ solutions of Mo(CO)$_6$ and W(CO)$_6$ are effective in the polymerization of substituted acetylenes[145,146]. Other metal carbonyls such as Cr(CO)$_6$, Mn$_2$(CO)$_{10}$ and Re$_2$(CO)$_{10}$ do not yield any active species under similar conditions. Carbon tetrachloride solvent takes part in the formation of the active species and, therefore, cannot be replaced by toluene which is often used for metal chloride-based catalysts. As compared with catalysts 15 and 16, catalyst 17 shows lower activity (slower polymerizations) and higher

$$M(CO)_6\text{--}CCl_4\text{--}hv(M = Mo, W)$$

$$(17)$$

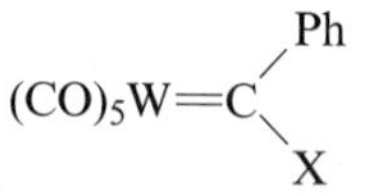

(CO)$_5$W=C(Ph)(X)

(18a) X = OMe
(18b) X = Ph

(CO)$_4$W=C(OMe)

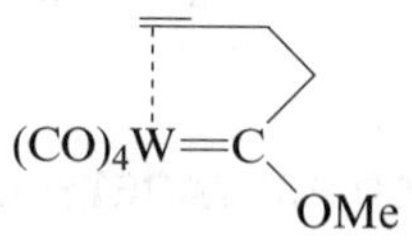

(19)

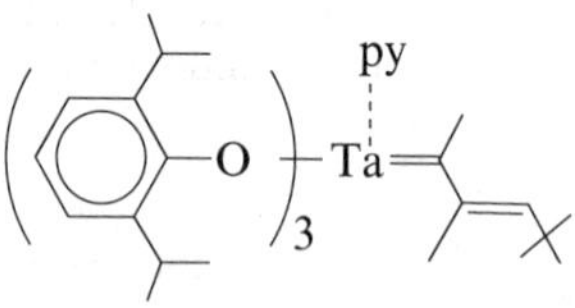

(20)

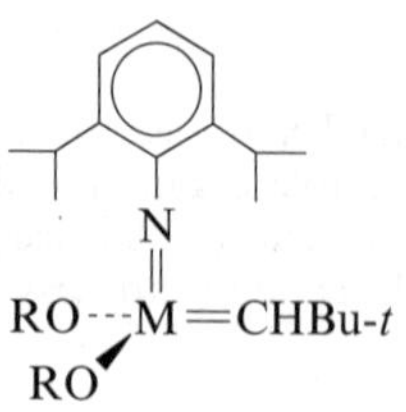

(21) M = Mo, W;
     R = t-Bu, CF$_3$Me$_2$C, (CF$_3$)$_2$MeC
(21a) M = Mo; R = (CF$_3$)$_2$MeC

RO---M=CHBu-t
RO

molecular weights of the formed polymers. It is useful also in olefin metathesis and metathesis polymerization of cycloolefins[146].

c. Metal-carbene catalysts. Katz and Lee found that the Fischer carbene **18a** and the Casey carbene **18b** polymerize phenylacetylene and *tert*-butylacetylene[147], but **18b** is less stable and more active than **18a**. These metal carbenes effect also metathesis polymerization of cycloolefins. These results strongly support the idea that the polymerization of acetylenes by group 5 and 6 transition metal catalysts proceeds by a metal–carbene mechanism.

Metal carbenes **19** and **20** have been reported to be effective in the polymerization of substituted acetylenes. Since **19** has an olefin ligand that can be removed when an acetylene monomer approaches, it is more active than **18**[148]. Metal carbenes **20** and **21a** induce living polymerizations of substituted acetylenes[149,150] (see below). In general, metal carbene catalysts are not very active, but the initiation reaction thereby is simple, and hence they are useful for the investigation of kinetics, etc.

$$\text{mesitylene·M(CO)}_3 \qquad \text{(MeCN)}_3\text{M(CO)}_3$$

$$\textbf{(22)} \text{ M = Mo, W} \qquad \textbf{(23)} \text{ M = Mo, W}$$

$$\text{(CO)}_4\text{BrW}\equiv\text{CPh}$$

$$\textbf{(24)}$$

(25) M = Nb, Ta

d. Organometallic complex catalysts. Several complexes involving group 5 and 6 transition metals (**22–25**) serve as catalysts for the polymerization of substituted acetylenes. It is assumed that these complexes react with monomers to form metal carbenes. Unlike **17**, activation by UV light is not required for **22**[151] and **23**[152]. Complex **24** is thought to isomerize into a metal carbene, $(CO)_4W=CBrPh$[153]. Binuclear complexes **25** polymerize disubstituted acetylenes as do $NbCl_5$ and $TaCl_5$[154]. These complexes are more difficult to prepare, their catalytic activities are diverse and their initiation mechanism is not clear. Hence, if one uses them, it is important to determine their characteristics.

e. Living polymerization catalysts. Catalyst systems **26** and **27** effect living polymerization of various substituted acetylenes. In this process, which is a polymerization without

$$\text{MoOCl}_4\text{–Bu}_4\text{Sn–EtOH (1:1:1)} \qquad \text{MoCl}_5\text{–Bu}_4\text{Sn–EtOH (1:1:1)}$$

$$\textbf{(26)}^{155} \qquad \textbf{(27)}^{156}$$

termination and chain transfer, the molecular weight of the formed polymer increases in direct proportion to monomer conversion, and the molecular weight distribution is usually narrow. **26** is superior to **27** as living polymerization catalyst, since the former gives polymers with narrower molecular weight distributions. The monomers that undergo living polymerization with **26** include 1-chloro-1-octyne, *o*-[(trifluoromethyl)-phenyl]acetylene and *t*-butylacetylene. See Section III A.5 for the details of the living polymerizations.

f. Stereospecific polymerization catalysts. The poly(phenylacetylene) prepared with Ziegler catalyst **28** possesses mainly the *cis-cisoidal* structure (as evidenced by the C—H out-of-plane deformation at 740 cm^{-1} in the IR spectrum), and is insoluble in all solvents owing to its high crystallinity[13]. Rhodium catalysts such as **29** and **30** provide a soluble, *cis–transoidal* poly(phenylacetylene)[157,158]. This polymer exhibits a sharp peak due to the olefinic proton at δ 5.8 in the ^{1}H NMR spectrum.

Fe(acac)$_3$–Et$_3$Al (1:3)

(28)   **(29)**   **(30)**

g. Catalysts effective in polar solvents. Ziegler-type catalysts (e.g., **28**) tend to decompose in polar solvents or in the presence of polar monomers, and hence are not applicable to such systems. In contrast, some group 8 transition metal catalysts like **29** and **30** work even in alcohols, and others such as PdCl$_2$ polymerize oxygen-containing monomers (e.g. HC≡CCH$_2$OH[159], HC≡CCO$_2$H[160]). MoCl$_5$ polymerizes an OH-containing monomer[161]. Study of such catalysts and monomers will enable the synthesis of new interesting polymers.

h. Ionic polymerization catalysts. Acetylenes having electron-withdrawing groups can be polymerized by anionic initiators. For instance, hexafluoro-2-butyne polymerizes with CsOBu-*t* and RbOBu-*t* in THF or toluene to give a colorless, insoluble, thermally very stable polymer[162]. Another example is the polymerization of 9-anthrylmethyl propiolate by Et$_3$N in DMF ($\bar{M}_n$ ~2 × 103)[163]. The spontaneous polymerization of quarternized ethynylpyridines is thought to proceed via an ionic or zwitter-ionic mechanism[164,165]. The polymers formed are highly charged and their backbones are extensively conjugated.

2. Polymerization of monosubstituted acetylenes

Typical examples of the polymerization of monosubstituted acetylenes are shown in Table 5. As seen, Fe, Mo, W and Rh catalysts, all of which involve transition metals, are particularly effective. It is noted that not only sterically unhindered monomers but also very crowded ones afford high molecular weight polymers.

a. Hydrocarbon acetylenes. Aliphatic terminal acetylenes with *prim-* and *sec*-alkyl groups provide orange to yellow, high molecular weight polymers, when polymerized with iron alkanoate-organoaluminum catalysts[13,193], but yield no high polymers with Mo or W chloride-based catalysts. Recently it has been reported that phenoxy complexes of Nb and W polymerize 1-alkynes[166]. On the other hand, *t*-alkylacetylenes, which are sterically very crowded, can be polymerized by Mo and W catalysts, and the $\bar{M}_w$ of the polymer reaches several hundred thousand[168].

Typical catalysts for the polymerization of phenylacetylene contain W, Rh and Fe. W catalysts produce an auburn polymer; WCl$_6$–Ph$_4$Sn is highly active, while W(CO)$_6$–CCl$_4$–hν is useful to achieve high molecular weight ($\bar{M}_n$ ~1 × 10^5)[16]. The polymerization by Rh catalysts proceeds in alcohols and amines to form a yellow polymer[157,158]. When Fe(acac)$_3$–Et$_3$Al is used, the poly(phenylacetylene) formed is insoluble in any solvent[194].

TABLE 5. Polymerization of monosubstituted acetylenes

Monomer	Catalyst	$\dfrac{\text{MW}}{10^3}$, $[\eta]$	Ref.
a) Hydrocarbon acetylenes			
HC≡CBu-n	WCl$_2$(OC$_6$H$_4$-o,o-Me$_2$)$_4$	170($\overline{M}_n$)	166
HC≡CBu-s	Fe(acac)$_3$–i-Bu$_3$Al	1.0($[\eta]$)	167
HC≡CBu-t	MoCl$_5$	330($\overline{M}_n$)	147, 168
HC≡CCH$_2$CH$_2$Ph	Fe(acac)$_3$–Et$_3$Al	97($\overline{M}_n$)	169
HC≡CPh	WCl$_6$–Ph$_4$Sn	15($\overline{M}_n$)	170
HC≡CPh	W(CO)$_6$–CCl$_4$–hν	80($\overline{M}_n$)	171
HC≡CPh	(COD·RhCl)$_2$	10–1000($\overline{M}_w$)	172
HC≡C-Naph-1	(WCl$_4$·SiMe$_3$)$_2$	95(M_n)	173
b) Si-containing acetylenes			
HC≡CCH(SiMe$_3$)C$_5$H$_{11}$-n	MoCl$_5$–Et$_3$SiH	450($\overline{M}_w$)	174
HC≡CCH(SiMe$_2$-n-C$_6$H$_{13}$)Pr-n	MoCl$_5$–Ph$_3$Sb	410($\overline{M}_w$)	175
HC≡CCH(SiMe$_2$Ph)Pr-n	WCl$_6$–Ph$_3$Sb	210($\overline{M}_w$)	175
HC≡CC≡CCHSiPh$_2$Bu-t	MoCl$_5$–Et$_3$SiH	31($\overline{M}_n$)	176
HC≡CC$_6$H$_4$-p-SiMe$_3$	WCl$_6$–Bu$_4$Sn	108($\overline{M}_n$)	177
c) Acetylenes with other heteroatoms			
HC≡CCH$_2$OH	PdCl$_2$	insol.	159
HC≡CCO$_2$-n-Bu	(NBD·RhCl)$_2$	84($\overline{M}_w$)	178
HC≡C-α-thiophene	WCl$_6$–Bu$_4$Sn	20($\overline{M}_n$)	179
HC≡CCH$_2$NH$_2$	Mo(OEt)$_5$–EtAlCl$_2$	insol.	180
HC≡CCH$_2$NHCH$_2$Ph	[Rh·COD·chel]$^+$X$^-$	4($\overline{M}_n$)	181
HC≡CCN	Ti(OBu)$_4$–Et$_3$Al	0.11($[\eta]$)	182
HC≡C(CF$_2$)$_5$CF$_3$	WCl$_6$–Ph$_4$Sn	0.047($[\eta]$)	183
HC≡CC$_{10}$H$_6$-4-CF$_3$	W(CO$_6$)–CCl$_4$–hν	110($\overline{M}_w$)	184
HC≡CCH$_2$Cl	MoCl$_5$	—	185
d) Ortho-substituted phenylacetylenes			
HC≡CC$_6$H$_2$-o,o-Me$_2$-p-t-Bu	W(CO)$_6$–CCl$_4$–hν	2600($\overline{M}_w$)	186
HC≡CC$_6$H$_4$-o-SiMe$_3$	W(CO)$_6$–CCl$_4$–hν	3400($\overline{M}_w$)	187
HC≡CC$_6$H$_4$-o-GeMe$_3$	WCl$_6$	690($\overline{M}_w$)	188
HC≡CC$_6$H$_4$-o-CF$_3$	W(CO)$_6$–CCl$_4$–hν	1600($\overline{M}_w$)	189, 190
HC≡CC$_6$H$_4$-2,4,5-(CF$_3$)$_3$	WCl$_6$–Ph$_4$Sn	94($\overline{M}_n$)	191
HC≡CC$_6$F$_4$-Bu-p	W(CO)$_6$–CCl$_4$–hν	900($\overline{M}_w$)	192
HC≡CC$_6$F$_5$	WCl$_6$–Ph$_4$Sn	0.61($[\eta]$)	192

b. Heteroatom-containing acetylenes. Examples of the polymerizations of heteroatom-containing acetylenes have been increasing (Table 5). The heteroatoms include Si, halogens, O, S and N. Especially, Si and F endow the polymers with unique properties and functions and are unlikely to deactivate polymerization catalysts. Hence the synthesis of Si- and F-containing polyacetylenes has been examined particularly extensively.

(Trimethylsilyl)acetylene is polymerizable with W catalysts, but the product polymer is partly insoluble in any solvent[195]. 3-(Trimethylsilyl)-1-alkynes, in which Si is not directly bonded to the acetylenic carbon, provide yellow, high molecular weight polymers in the presence of Mo catalysts[174]. (Perfluoroalkyl)acetylenes yield white polymers soluble only in fluorine-containing solvents[183].

Though it has been reported that some acetylenes having OH and COOH groups can be polymerized by group 6 and 8 transition-metal catalysts, it is rather difficult to obtain soluble, high molecular weight polymers from them[159,160]. Ethynylthiophenes show a polymerizability similar to that of phenylacetylene[179,196]. A few studies on the polymerization of N-containing acetylenes have appeared[180-182].

c. Ortho-substituted phenylacetylenes. As mentioned above, an interesting tendency has been observed in the polymerization of monosubstituted acetylenes by W and Mo catalysts. That is, sterically uncrowded acetylenes like 1-hexyne do not produce high molecular weight polymers, whereas monomers with bulky groups such as *tert*-butylacetylene do so, while the molecular weight of poly(phenylacetylene) is in between. In this respect the polymerizations of *ortho*-substituted phenylacetylenes are interesting, and in fact high molecular weight polymers have been obtained from them, as discussed below.

o-(Methylphenyl)acetylene, one of the simplest *ortho*-substituted phenylacetylenes, polymerizes virtually quantitatively not only with WCl$_6$-based catalysts but also with MoCl$_5$-based counterparts[197]. The highest $\overline{M}_w$ of poly[*o*-(methylphenyl)acetylene] reaches *ca.* 8×10^5, which is attained with the W(CO)$_6$–CCl$_4$–hν catalyst. Quite interestingly, (*p-t*-butyl-*o,o*-dimethylphenyl)acetylene, an *ortho*-dimethyl substituted phenylacetylene, polymerizes with W(CO)$_6$–CCl$_4$–hν to provide in high yield a totally soluble high molecular weight polymer ($\overline{M}_w > 2 \times 10^6$)[186]. Both trifluoromethyl and trimethylsilyl groups are very bulky, and the former is electron-withdrawing while the latter is electron-donating. The two phenylacetylenes having these groups at the *ortho* position polymerize in high yields with W and Mo catalysts. The $\overline{M}_w$ of the formed polymers exceeds one million. Thus, the steric effect of the *ortho* substituents greatly affects the polymerizability and the polymer molecular weight of phenylacetylenes, while the electronic effect hardly influences them.

3. Polymerization of disubstituted acetylenes

In general, disubstituted acetylenes are sterically more crowded than their monosubstituted counterparts and, consequently, their effective polymerization catalysts are restricted virtually to group 5 and 6 transition-metal catalysts. Among disubstituted acetylenes,

TABLE 6. Polymerization of disubstituted acetylenes

Monomer	Catalyst	$\dfrac{\text{MW}}{10^3}$,[η]	Ref.
a) Hydrocarbon acetylenes			
MeC≡CPr	MoCl$_5$	1100($\overline{M}_w$)	198
PrC≡CPr	NbCl$_5$	insol.	199
PhC≡CMe	TaCl$_5$–Bu$_4$Sn	1500($\overline{M}_w$)	200
PhC≡CC$_6$H$_{13}$	TaCl$_5$–Bu$_4$Sn	1100($\overline{M}_w$)	201
b) Si-, Ge-containing acetylenes			
MeC≡CSiMe$_3$	TaCl$_5$	730($\overline{M}_w$)	202
MeC≡CSiMe$_3$	NbCl$_5$	220($\overline{M}_w$)	202
MeC≡CSiMe$_3$	TaCl$_5$–Ph$_3$Bi	4000($\overline{M}_w$)	203
MeC≡CSiMe$_2$C$_6$H$_{13}$	TaCl$_5$–Ph$_3$Bi	1400($\overline{M}_w$)	204
MeC≡CGeMe$_3$	TaCl$_5$	>100($\overline{M}_w$)	205
c) Acetylenes with other heteroatoms			
MeC≡CSBu	MoCl$_5$–Ph$_3$SiH	180($\overline{M}_w$)	206
MeSC≡CC$_6$H$_{13}$	MoCl$_5$–Ph$_3$SiH	130($\overline{M}_w$)	207
ClC≡CC$_6$H$_{13}$	MoCl$_5$–Bu$_4$Sn	1100($\overline{M}_w$)	208
ClC≡CPh	MoCl$_5$–Bu$_4$Sn	690($\overline{M}_w$)	209,210
BrC≡CPh	MoCl$_5$–Bu$_4$Sn	~20($\overline{M}_w$)	211
d) Diphenylacetylenes			
PhC≡CPh	TaCl$_5$–Bu$_4$Sn	insol.	201
PhC≡CC$_6$H$_4$-*p-t*-Bu	TaCl$_5$–Bu$_4$Sn	3600($\overline{M}_w$)	212
PhC≡CC$_6$H$_4$-*p-n*-Bu	TaCl$_5$–Bu$_4$Sn	1100($\overline{M}_w$)	212
PhC≡CC$_6$H$_4$-*p*-SiMe$_3$	TaCl$_5$–Bu$_4$Sn	2200($\overline{M}_w$)	213,214
PhC≡CC$_6$H$_4$-*p*-SiMe$_2$Pr-*i*	TaCl$_5$–Bu$_4$Sn	1600($\overline{M}_w$)	215

those with less steric hindrance polymerize with Mo and W catalysts and tend to give cyclotrimers with Nb and Ta catalysts. On the other hand, sterically more crowded disubstituted acetylenes do not polymerize with Mo or W catalysts, but do so with Nb and Ta catalysts. The polymers from disubstituted acetylenes having two identical groups or two groups of similar sizes are generally insoluble in any solvent. Most polymers from disubstituted acetylenes are colorless, though some aromatic polymers are yellow. Table 6 lists typical examples of the polymerizations of disubstituted acetylenes.

a. Hydrocarbon acetylenes. 2-Alkynes (e.g. 2-octyne), which are sterically not very crowded, polymerize with Mo catalysts to give polymers with $\bar{M}_w$ over one million[198]. For these monomers, W and Nb catalysts are less active, and Ta catalysts yield only cyclotrimers. Symmetrical dialkylacetylenes (e.g. 4-octyne) are slightly more crowded, and consequently Nb, Ta and W catalysts exhibit high activity, while Mo catalysts are not active[199]. Since 1-phenyl-1-alkynes (e.g., 1-phenyl-1-propyne) involve even larger steric effects, Nb and Ta catalysts[200,201] produce polymers having $\bar{M}_w$ M of 1×10^5–1×10^6. In contrast, W catalysts yield only oligomers of $\bar{M}_n$ lower than 1×10^4, and Mo catalysts are inactive.

b. Heteroatom-containing acetylenes. 1-Trimethylsilyl-1-propyne (TMSP), a sterically highly crowded Si-containing acetylene, polymerizes with Nb and Ta catalysts, but does not with Mo or W catalysts[202]. The $\bar{M}_w$ of the polymer obtained with $TaCl_5$–Ph_3Bi reaches four million[203], which is among the highest for all the substituted polyacetylenes. The monomers in which one of the methyl groups on the Si of 1-trimethylsilyl-1-propyne is replaced either by a *n*-hexyl or a phenyl group are polymerizable[204], while those in which the methyl group bonded to the acetylenic carbon is replaced by ethyl or higher alkyl groups are not polymerizable because of steric hindrance. It has been revealed in a patent[205] that 1-(trimethylgermyl)-1-propyne polymerizes much faster than TMSP, but the details including the polymer molecular weight are not clear.

Mo catalysts are uniquely effective in the polymerization of S-containing disubstituted acetylenes like 1-(*n*-butylthio)-1-propyne and 1-(methylthio)-1-octyne[206,207]. Though there is the possibility that the S as well as the O in the monomer deactivates group 5 and 6 transition-metal catalysts, the basicity of the S is weakened by the conjugation with the triple bond, resulting in the lower coordinating ability to the propagating species.

Cl-containing monomers afford high molecular weight polymers. For instance, the polymerization of 1-chloro-1-octyne and 1-chloro-2-phenylacetylene is catalyzed by $MoCl_5$–Bu_4Sn and $Mo(CO)_6$–CCl_4–*h*ν to give polymers whose $\bar{M}_w$ values reach about one million[208–210]. It appears that the electron-withdrawing chlorine atom plays some role in the inertness of these monomers to Nb, Ta and W catalysts.

c. Diphenylacetylenes. Diphenylacetylene itself forms a polymer in the presence of $TaCl_5$–Bu_4Sn[201]. The polymer possesses a very high thermal stability, but is insoluble in any solvent. Regarding polymer solubility, there is a tendency that polyacetylenes having two identical alkyl groups in the repeating unit are insoluble in any solvent, whereas polyacetylenes having both a methyl and a long alkyl group are soluble in various solvents. By analogy, one can hypothesize that *para*- or *meta*-substituted diphenylacetylenes provide soluble polymers.

Recently, soluble, high molecular weight polymers have indeed been obtained from several diphenylacetylenes with substituents. For instance, 1-phenyl-2-[(*p*-trimethylsilyl)-phenyl]acetylene polymerizes with $TaCl_5$–cocatalyst in *ca.* 80% yield[213,214]. The polymer is soluble in toluene and in $CHCl_3$, and its $\bar{M}_w$ value is as high as about two million. In contrast, $TaCl_5$ alone $NbCl_5$–cocatalyst are ineffective with this monomer, unlike with 1-(trimethylsilyl)-1-propyne. The diphenylacetylenes with *m*-Me_3Si, *m*-Me_3Ge, *p*-*t*-Bu and *p*-*n*-Bu groups polymerize similarly, leading to totally soluble, high molecular weight polymers[212–215].

4. Polymerization mechanisms

a. Metal–carbene and metal–alkyl mechanisms. At least two mechanisms are possible for the propagation in the polymerization of acetylenes by transition metal catalysts: one is the metal–carbene mechanism (equation 21) and the other is the metal–alkyl mechanism (equation 22). Though completely different from each other, both lead to a main chain composed of alternating carbon–carbon double bonds and hence are difficult to distinguish on the basis of the polymer structure.

$$\text{\textasciitilde\textasciitilde\textasciitilde} C{=}M \quad \xrightarrow{\;C{\equiv}C\;} \quad \text{\textasciitilde\textasciitilde\textasciitilde} \begin{array}{c} C{-}M \\ | \quad | \\ C{=}C \end{array} \quad \longrightarrow \quad \text{\textasciitilde\textasciitilde\textasciitilde} \begin{array}{c} C \quad M \\ \| \quad \| \\ C{-}C \end{array} \tag{21}$$

(M: metal)

$$\text{\textasciitilde\textasciitilde\textasciitilde} C{=}C{-}M \quad \xrightarrow{\;C{\equiv}C\;} \quad \text{\textasciitilde\textasciitilde\textasciitilde} C{=}C{-}C{=}C{-}M \tag{22}$$

The metal–carbene mechanism is now being accepted for the polymerization of substituted acetylenes by group 5 and 6 transition–metal (Nb, Ta, Mo, and W) catalysts. This is based on the following observations: (i) some of the catalysts are known to generate metal carbenes; (ii) isolated metal carbenes initiate polymerization; (iii) acetylenes copolymerize with cycloolefins for which similar metal–carbene mechanisms have been established; and (iv) certain linear olefins act as chain transfer agents. Apart from group 5 and 6 transition-metal catalysts, both Ziegler-type and Rh catalysts act as polymerization catalysts. The polymerization of acetylenes by Ziegler-type catalysts is thought to proceed by the metal–alkyl mechanism[36,87] (see Section II.A.2). It is assumed that the polymerization of phenylacetylene by Rh catalysts also proceeds by the metal–alkyl mechanism[216]. However, more convincing data are necessary before the metal–alkyl mechanism is accepted for the Rh-catalyzed polymerization. In the case of Nb, Ta, Mo and W catalysts, detailed mechanistic information has been accumulated.

b. Formation of metal carbenes. It has been confirmed by NMR that a W carbene is formed in the WCl_6–Me_4Sn catalyst system[217] (equation 23). The reaction of $MoCl_5$ with MeLi provides a Mo carbene, which can be employed to olefination of carbonyl compounds[218] (equation 24). The formation of a W carbene in the $W(CO)_6$–CCl_4–$h\nu$ system is indirectly proved by the fact that it gives 1,1-dichloro-1-propene from 1-propene[219] (equation 25).

$$WCl_6 + 2Me_4Sn \quad \xrightarrow{\;-\,2Me_3SnCl\;} \quad Me_2WCl_4 \quad \xrightarrow{\;-\,CH_4\;} \quad H_2C{=}WCl_4 \tag{23}$$

$$MoCl_5 + 2MeLi \quad \longrightarrow \quad CH_2{=}MoCl_3 + CH_4 \tag{24}$$

$$W(CO)_6 + CCl_4 \quad \xrightarrow{\;h\nu\;} \quad \left[\begin{array}{c} Cl \\ \diagdown \\ \diagup \quad C{=}W \\ Cl \end{array}\right] \quad \xrightarrow{\;C{=}C-\;} \quad \begin{array}{c} Cl \\ \diagdown \\ \diagup \quad C{=}C- \\ Cl \end{array} \tag{25}$$

Though $MoCl_5$ and WCl_6 alone also polymerize monosubstituted acetylenes, how the metal–carbene intermediates are formed in these systems is not clear. Various suggestions have been reported[220,221] (equations 26 and 27).

$$W(CO)_3(dppe)\ (THF) \xrightarrow{\quad HC\equiv CPh \quad} (HC\equiv CPh)\ W\ (CO)_3\ (dppe)$$

$$(dppe = Ph_2PCH_2CH_2PPh_2)$$

$$\xrightarrow{\qquad} \begin{array}{c} Ph \\ \diagdown \\ \diagup \\ H \end{array} C{=}C{=}W(CO)_3(dppe) \quad (26)$$

$$W(CO)_6 \xrightarrow[\ HC\equiv CPh\ ]{\ hv\ } (HC\equiv CPh)\ W\ (CO)_5 \xrightarrow{\qquad}$$

$$\begin{array}{c} Ph \\ \diagdown \\ \diagup \\ H \end{array} C{=}C{=}W(CO)_5 \xrightarrow[nHC\equiv CPh]{-CO} \begin{array}{c} {-}(CH{=}C)_n{-} \\ | \\ Ph \end{array} \quad (27)$$

c. Polymerization by isolated metal carbenes. Katz and Lee found that the Fischer carbene (**18a**) and the Casey carbene (**18b**), both involving a W center, polymerize various substituted acetylenes[147]. Earlier, the present author and coworkers pointed out that the polymerization of substituted acetylenes by Mo and W catalysts would proceed most probably by the metal–carbene mechanism[222], and the results of Katz and Lee strongly supported our proposal. Schrock and collaborators have reported that isolable Ta (**20**) and Mo (**21a**) carbenes also polymerize substituted acetylenes[149,150] (see Section III.A.5.c). These results provide direct evidence for the metal-carbene mechanism.

d. Copolymerization with cycloolefins. It is known that ring-opening metathesis polymerization of cycloolefins proceeds via metal carbenes. Hence, if a substituted acetylene copolymerizes with a cycloolefin, this will suggest that the metal–carbene mechanism is valid for the polymerization of substituted acetylenes (equation 28). In fact, copolymer formation between norbornene (NBE) and phenylacetylene (PA) has been confirmed[223].

Both PA and NBE react simultaneously in their copolymerization by WCl$_6$; the reactivity ratio of PA to NBE is *ca* 5:1. The product is a copolymer according to GPC, UV and NMR analyses. The PA–NBE dyads can be directly observed by correlation peaks a and b in H–H COSY NMR[223] (Figure 1), and the formation of a random copolymer is suggested on the basis of the copolymer composition curves. When an electron-withdrawing substituent is introduced into the phenyl ring of PA, the monomer becomes less reactive, probably owing to its weaker coordinating ability[224].

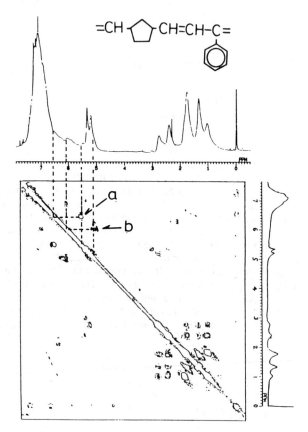

FIGURE 1. H–H COSY NMR spectrum of copoly(phenyl-acetylene/norborene)

e. Chain transfer to olefins.   The nature of effective transfer agents is closely related to the polymerization mechanism. For example, in the ring-opening metathesis polymerization of cycloolefins, which proceeds via metal carbenes, acyclic olefins serve as transfer agents[225]. If the metal–carbene mechanism applies also to the polymerization of substituted acetylenes, suitable olefins are expected to effect chain-transfer reaction (equation 29).

Considering to this proposal, the effects of various olefins have been examined in the polymerization of phenylacetylenes catalyzed by WCl_6–Ph_4Sn. Si-containing olefins, especially trimethylvinylsilane, were found to work as effective chain-transfer agents[226] (Table 7), causing remarkable molecular-weight decreases with phenylacetylenes having electron-withdrawing ring substituents. This is because such monomers have weak coordinating ability to the propagating end. Similarly, disubstituted acetylenes containing an electron-withdrawing chlorine readily undergo chain transfer to trimethylvinylsilane[227].

$$\text{\tiny www}C{=}M \xrightarrow{\;C{=}C\;} \begin{array}{cc} \text{\tiny www}C-M \\ |\quad\ | \\ C-C \end{array} \longrightarrow \begin{array}{cc} \text{\tiny www}C \quad M \\ \|\quad\ \| \\ C \quad C \end{array}$$

$$\xrightarrow{\;C{\equiv}C\;} \begin{array}{c} \text{\tiny www}C \\ \| \\ C \end{array} + \begin{array}{c} C\text{\tiny wwwwwwwwwww}C{=}M \\ \| \\ C \end{array} \tag{29}$$

TABLE 7. Effect of trimethylvinylsilane (TMVS) as transfer agent on the polymerization of phenylacetylenes by WCl$_6$–Ph$_4$Sn(1:1) [a]

Monomer	$\overline{M}_n/10^3$		$\overline{M}_w/10^3$	
	None	TMVS	None	TMVS
HC≡CC$_6$F$_4$Bu-p	125	6.0	293	14
HC≡CC$_6$H$_4$CF$_3$-o	182	29	491	61
HC≡CC$_6$H$_4$SiMe$_3$-o	180	56	367	150
HC≡CC$_6$H$_2$-o,o-Me$_2$-p-t-Bu	221	127	499	696

[a] Polymerized in toluene at 30 °C for 24 h; [M]$_0$ = 0.50 M, [WCl$_6$] =10 mM. [TMVS] = 0.10 M.

5. Control of polymerization

An important aspect of polymer chemistry is the precise synthesis of polymers, which can be achieved most typically by living polymerization and by stereospecific polymerization. In particular, the former provides also methods of synthesizing end-functionalized

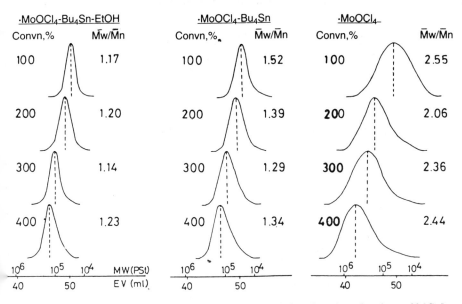

FIGURE 2. Polymerization of 1-chloro-1-octyne by MoOCl$_4$-based catalysts (in toluene, 30 °C, 5 min each, [M]$_0$ = [M]$_{added}$ = 0.10 M, [Cat] = 20mM)

polymers, block copolymers etc. Thus, this section specifically discusses the living polymerization catalyzed by **26** and other catalysts and the stereospecific polymerization by **29**, **30**, etc. for substituted acetylenes.

a. Living polymerization of 1-chloro-1-octyne by 26 and 27. The $MoOCl_4$–Bu_4Sn–EtOH catalyst, **26**, induces living polymerization of 1-chloro-1-octyne in toluene at 30 °C[155] (Figure 2). The $\bar{M}_n$ of the polymer increases progressively with repeated additions of monomer feeds. Meanwhile, the $\bar{M}_w/\bar{M}_n$ ratio, which is a measure of the molecular weight distribution (MWD), is maintained at about 1.15. The estimated mole ratio of the propagating species to the Mo catalyst, however, is rather small (*ca* 0.025). These results manifest the occurrence of a living polymerization. In contrast, when $MoOCl_4$ alone and $MoOCl_4$–Bu_4Sn are used as catalysts, the $\bar{M}_w/\bar{M}_n$ ratio is not so close to unity, although the polymer molecular weight increases with increasing monomer consumption.

With catalyst **26**, 1-chloro-1-alkynes with different alkyl lengths also undergo living polymerizations[155]. Consequently, the sequential addition of 1-chloro-1-hexadecyne (A), 1-chloro-1-hexyne (B) and 1-chloro-1-hexadecyne (A) in this order provides an A–B–A-type triblock copolymer. Similarly, one can obtain a B–A–B-type triblock copolymer. These are the first examples of block copolymers from substituted acetylenes.

The polymerization of 1-chloro-1-octyne by $MoCl_5$–Bu_4Sn–EtOH (**27**) also produces polymers with narrow MWDs ($\bar{M}_w/\bar{M}_n$ 1.2–1.3)[156]. The $\bar{M}_n$ increases in proportion to the monomer conversion. Hence this polymerization is concluded to be living, although **26** (with $MoOCl_4$) is superior to **27** (with $MoCl_5$) as a living polymerization catalyst.

b. Living polymerization of other acetylenes by 26. Table 8 summarizes the living polymerization of various substituted acetylenes by $MoOCl_4$–Bu_4Sn–EtOH, **26**. Not only 1-chloro-1-alkynes (disubstituted acetylenes) but also several *ortho*-substituted phenylacetylenes and *t*-butylacetylene (monosubstituted acetylenes) undergo living polymerization in toluene in the temperature range – 30 to + 30 °C. In all cases, the $\bar{M}_w/\bar{M}_n$ ratios are

TABLE 8. Living polymerization of substituted acetylenes

Monomer	Catalyst	$\dfrac{M_w}{M_n}$	Ref.
$ClC\equiv CC_6H_{13}$	**26**[a]	1.13	155,156
$HC\equiv CC_6H_4Me$-*o*	**26**	1.30	228
$HC\equiv CC_6H_4(i\text{-}Pr)$-*o*	**26**	1.23	228
$HC\equiv CC_6H_4SiMe_3$-*o*	**26**	1.07	229
$HC\equiv CC_6H_4GeMe_3$-*o*	**26**	1.08	188
$HC\equiv CC_6H_4CF_3$-*o*	**26**	1.06	230
$HC\equiv CC_6F_4Bu$-*p*	**26**	1.16	231
$HC\equiv CBu$-*t*[b]	**26**	1.12	232
$MeC\equiv CSiMe_3$	$NbCl_5$	1.17	233
$MeC\equiv CMe$	**20**[c]	1.03	149
$(HC\equiv CCH_2)_2C(CO_2Et)_2$	**21a**[d]	~1.20	150
$HC\equiv CC_6H_4Cl$-*m*[e]	**30**[f]	~2	234
$HC\equiv CBu$-*t*[e]	$MoCl_5$	~2	235

[a]Catalyst **26**: $MoOCl_4$–Bu_4Sn–EtOH (1:1:1).
[b]A stereoregular (*cis* 97%) and living polymer formed.
[c]Catalyst **20**: $[o,o\text{-}(i\text{-}Pr)_2C_6H_3O]_3(py)Ta = CRR'$.
[d]Catalyst **21a**: $[(CF_3)_2MeCO]_2(ArN=)Mo=CHBu\text{-}t$.
[e]The propagating species involved is not living but long-lived.
[f]Catalyst **30**: $[(1,5\text{-cyclooctadiene)·RhCl}]_2$.

fairly close to unity (1.05–1.30), and the $\bar{M}_n$ increases in direct proportion to monomer conversion.

t-Butylacetylene undergoes stereospecific living polymerization with $\mathbf{26}^{232}$. The methyl–carbon signal in the ^{13}C NMR of the polymer obtained with MoOCl$_4$ alone splits into two peaks, a$_1$ and a$_2$, indicating the presence of both cis and trans structures (Figure 3). In

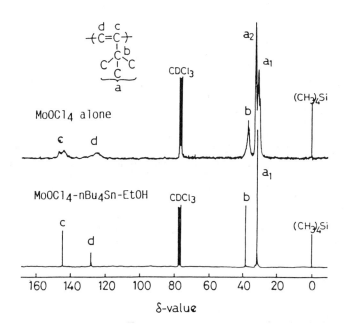

FIGURE 3. ^{13}C-NMR spectra of poly(t-butylacetylene) obtained in toluene at 0 °C

contrast, the polymer obtained with $\mathbf{26}$ shows virtually only peak a$_1$ for the methyl carbons, and hence it is thought to consist exclusively of the *cis* structure. It is noteworthy that the other peaks of this polymer are also very sharp owing to its high stereoregularity. So far, only a few examples of stereospecific and living polymerizations have been known, such as anionic and coordination polymerizations of methyl methacrylate and the metathesis polymerization of an F-containing norbornadiene.

c. Living polymerization by isolated metal carbenes. 2-Butyne polymerizes in a living manner with a Ta carbene, $\mathbf{20}$, in toluene at -30 °C to give a polymer having narrow MWD $(\bar{M}_w/\bar{M}_n 1.03)^{149}$. The M_n increases in proportion to the monomer consumption. 1-Pentyne did not polymerize in a living fashion with $\mathbf{20}$, which was attributed to a metathesis reaction of the polymer formed.

The cyclopolymerization of diethyl dipropargylmalonate by [(CF$_3$)$_2$MeCO]$_2$-ArN=)Mo=CH Bu–t ($\mathbf{21a}$) in 1,2-dimethoxyethane at ambient temperature also proceeds in a living fashion to provide a conjugated polymer having both five- and six-membered rings (see Table 8)^{150}. The living nature of this polymerization has also been demonstrated by the synthesis of a block copolymer with 2,3-dicarbomethoxynorbornadiene.

d. Stereospecific polymerization. Four steric structures are theoretically possible for substituted polyacetylenes:

| *cis-cisoid* | *cis-transoid* | *trans-cisoid* | *trans-transoid* |

The geometric structure of poly(phenylacetylene) has been studied extensively but mostly only qualitatively[13]. Ziegler-type catalysts like Fe(acac)$_3$–Et$_3$Al provide principally the *cis-cisoidal* form, which is crimson, crystalline, and insoluble in toluene[194]. Rh catalysts [e.g., (norbonadiene·RhCl)$_2$] yield a yellow, toluene-soluble poly(phenylacetylene) exclusively composed of the *cis-transoidal* structure according to ^{1}H-NMR spectroscopy[157,158,172]. It is known that Mo catalysts give *cis*-rich yellow polymers, while W catalysts provide *trans*-rich, auburn polymers[222].

The geometric structure of poly(*t*-butylacetylene) can easily be determined by ^{13}C NMR[236] and ^{1}H NMR[153]. MoCl$_5$ gives a higher *cis* content than WCl$_6$ does, and the *cis* content is higher in oxygen-containing polymerization solvents like anisole than in conventional solvents like toluene[236]. Eventually, all-*cis* poly(*t*-butylacetylene) can be obtained by MoCl$_5$ in anisole, while the *cis* content for the WCl$_6$–toluene system is the lowest and about 50%.

The geometric structure of poly[1-(trimethylsilyl)-1-propyne] is being elucidated[237,238]. Preliminary results indicate that the *cis* contents are *ca* 40 and 60% with TaCl$_5$ and NbCl$_5$, respectively; however, no stereoregular polymer has yet been obtained.

B. Properties and Functions

1. Properties of polymers

One might think that the properties of substituted polyacetylenes resemble those of the unsubstituted polyacetylene. As seen in Table 9, however, they are quite different from each other particularly when the former carry bulky substituents. Further, the properties of substituted polyacetylenes greatly depend on the kind of substituents.

TABLE 9. Comparison of substituted polyacetylenes with polyacetylene

	Polyacetylene	Substituted polyacetylene
Structure	$+C=C+_n$ H H	$+C=C+_n$ R R'
Color	Black (powder)	Various colors
Solubility	Insoluble	Soluble
Crystallinity	Crystalline	Amorphous
Air stability	Unstable	Stable
Typical function	Electrical Conductivity	Gas permeability

Polyacetylene is black in the powdery state and shows a metallic luster in the filmy state. On the other hand, substituted polyacetylenes are variously colored depending on the number and kind of substituents (Table 10). *Ortho*-substituted poly(phenylacetylene)s have deep colors, i.e. brown to purple. The color of poly(phenylacetylene) depends on the catalysts, and it is auburn with W catalysts. Polymers from aromatic disubstituted

TABLE 10. Color and UV–visible spectral data of substituted polyacetylenes[a]

$+C=C+_n$ R R'		Color	λ_{max} (nm)	ε_{max} $(M^{-1}cm^{-1})$	$\lambda_\varepsilon < 50$[b] (nm)	Ref.
R	R'					
H	$C_6H_4SiMe_3$-o	Purple	542	6400	700	187
H	C_6H_4Me-o	Brown	466	4300	650	197
H	$C_6H_4CF_3$-o	Brown	458	5300	620	190
H	Ph	Auburn[c]	~260	5000	560	222
Ph	$C_6H_4SiMe_3$-p	Orange	430	4800	500	214
H	$SiMe_3$	Yellow	297	2760	460	195
H	n-Bu	Orange	285	1600	450	239
Cl	Ph	Yellow	315	2500	420	209
Me	Ph	Colorless	283	2500	400	240
H	t-Bu	Colorless	283	1500	350	168
Me	C_5H_{11}	Colorless	280	700	350	198
Cl	C_6H_{13}	Colorless	280	720	320	208
Me	$SiMe_3$	Colorless	234	3400	290	20

[a] The data arranged according to the wavelength of $\lambda_\varepsilon < 50$.
[b] Virtual cutoff wavelength of absorption.
[c] With W catalysts (cf. yellow with Mo catalysts).

acetylenes and aliphatic monosubstituted acetylenes with less bulky groups are usually lightly colored. In contrast, polymers from aliphatic disubstituted acetylenes, 1-phenyl-1-alkynes and t-butylacetylene are colorless. It is not clear at present why poly(phenyl-acetylenes) with *ortho*-substituents are so deeply colored. However, the color of substituted polyacetylenes is usually explicable in terms of both steric effects and the contribution of substituents to the conjugation. When the cutoff wavelength in the UV-visible spectrum is shorter than 400 nm, the polymer is colorless, and otherwise it is colored.

Most substituted polyacetylenes obtained with group 5 and 6 transition–metal catalysts are soluble in common organic solvents with relatively low polarities such as toluene, chloroform, and tetrahydrofuran (Table 11). Aliphatic polyacetylenes are also soluble in

TABLE 11. Solubility properties of substituted polyacetylenes[a,b]

$+C=C+_n$ R R'		Hexane	Toluene, $CHCl_3$	Anisole, $(CH_2Cl)_2$	acetone MeOH
R	R'				
H	t-Bu				
Me	C_5H_{11}	Soluble	Soluble	Insoluble	Insoluble
Me	$SiMe_3$				
Me	Ph				
Cl	Ph	Insoluble	Soluble	Soluble	Insoluble
Ph	$C_6H_4SiMe_3$-p				
Et	Et				
Ph	Ph	Insoluble	Insoluble	Insoluble	Insoluble
CF_3	CF_3				

[a] See Tables 5 and 6 for references.

n-alkanes, while they are insoluble in more polar solvents like anisole and 1,2-dichloroethane. On the other hand, an opposite tendency is seen with aromatic polyacetylenes. This agrees with the principle that like dissolves like. An exception is that the polymers from symmetrically disubstituted acetylenes [e.g. poly(diphenylacetylene)] are usually insoluble in any solvent. This can be explained by assuming that these polymers are densely packed and have less surface, resulting in weaker interaction with a solvent.

It is well known that polyacetylene is oxidized in air even at room temperature. Substitituted polyacetylenes are generally more stable, but their stability varies considerably, depending on the nature of the substituent[242]. One of the methods to evaluate polymer stability is to measure the onset temperature (T_0) of weight loss in the thermogravimetric analysis in air (Table 12). The T_0 values of the polymers from *prim*- or *sec*-

TABLE 12. Thermal stability of substituted polyacetylenes[a]

$$-(C=C)_n-$$
$$\quad |\ \ |$$
$$\quad R\ \ R'$$
$T_0,^b$ (°C)

R	R'	
H	Bu	150
H	*i*-Pr	150
H	Ph	200
H	*t*-Bu	200
Me	SiMe$_3$	280
Me	Ph	280
Cl	*n*-C$_6$H$_{13}$	300
Cl	Ph	310
Ph	C$_6$H$_4$SiMe$_3$-*m*	400
Ph	C$_6$H$_4$SiMe$_3$-*p*	420
Ph	Ph	500
CF$_3$	CF$_3$	525

[a] Data from References. 16, 201, 214, 241 and 242.
[b] T_0: the onset temperature of weight loss in the thermogravimetric analysis performed in air.

alkylacetylenes are *ca* 150 °C, and they decompose rather fast even at room temperature. The values for poly(2-alkynes), poly(*t*-butylacetylene) and poly(phenylacetylene) are around 200 °C; these polymers degrade gradually at room temperature. The polymers from MeC≡CSiMe$_3$, MeC≡CPh, ClC≡CC$_6$H$_{13}$ and ClC≡CPh have T_0 values of approximately 300 °C, and do not decompose at room temperature even over long periods of time. Interestingly, poly(diphenylacetylene) and poly(hexafluoro-2-butyne) have the highest thermal stabilities among all the known substituted polyacetylenes. The excellent thermal stability of the former is virtually retained in the *para*- and meta-silylated derivatives. Recently the stability of poly(phenylacetylene) in solution has been discussed[243].

The mechanical properties of substituted polyacetylenes also depend strongly on substituents[244,245]. Thus, polyacetylenes with long *n*-alkyl groups are generally ductile and can be elongated by several tens to several hundred percent. They are rather soft and show smaller tensile moduli. On the contrary, aromatic polyacetylenes are generally hard and brittle, and their films break even when elongated only by a few percent. Most substituted polyacetylenes are amorphous, being glassy below the glass transition temperature (T_g), and rubbery above T_g. Unless long *n*-alkyl groups are present, substituted polyacetylenes have T_g values around 200 °C that do not depend strongly on the kind of substituents. This is quite different from the case of vinyl polymers, and attributable to the fairly stiff alternating double-bond structure of the main chain.

While polyacetylene is an electrical semiconductor, substituted polyacetylenes are virtually insulators (Table 13). Among them, the aliphatic polymers possess specific conductivities as low as 10^{-18} S cm^{-1}, while aromatic monosubstituted acetylene polymers have values in the order of *ca* 10^{-15} S cm^{-1}. The low conductivities of substituted polyacetylenes are attributable to their twisted main-chain conformation induced by the bulky substituents. The unpaired-electron densities of most disubstituted acetylene polymers are below a detection limit of 1×10^{15} spin g^{-1}. In contrast, poly(phenylacetylene) and its ring-substituted derivatives have appreciable amounts of unpaired electrons, and the electrical conductivity of substituted polyacetylenes is roughly in parallel with their unpaired-electron density. This is reasonable because both are governed by the extent of conjugation of the main chain.

TABLE 13. Electrical conductivity (σ) and unpaired-electron density of substituted polyacetylenes[a]

$-(C=C)_n-$
   $\vert$ $\vert$
  R R′
		σ, (25°C)
		 (S·cm^{-1})
			Unpaired electron,
			 (spin·g^{-1})

R	R′		
H	*t*-Bu	$<1 \times 10^{-18}$	$<1 \times 10^{15}$
Me	C_5H_{11}	$<1 \times 10^{-18}$	$<1 \times 10^{15}$
Cl	Ph	$<1 \times 10^{-18}$	$<1 \times 10^{15}$
Cl	C_6H_{13}	4×10^{-18}	$<1 \times 10^{15}$
Me	Ph	7×10^{-18}	$<1 \times 10^{15}$
Me	SBu	3×10^{-17}	$<1 \times 10^{15}$
Ph	$C_6H_4SiMe_3$-*m*	3×10^{-16}	—
H	$C_6H_4CF_3$-*o*	2×10^{-15}	6×10^{17}
H	$C_6H_4SiMe_3$-*o*	4×10^{-15}	7×10^{17}
H	Ph	4×10^{-15}	8×10^{16}
H	C_6H_2-*o*,*o*-Me$_2$-*p*-*t*-Bu	1×10^{-13}	2×10^{18}
H	H(*cis*)	2×10^{-9}	2×10^{18}
H	H (*trans*)	4×10^{-5}	3×10^{19}

[a] See Tables 5 and 6 for references.

2. Functions of polymers

Typical functions of substituted polyacetylenes are based on their (i) high gas permeability and (ii) electronic and photonic properties. The former originates from the rigid main chain and bulky substituents. Though electrical insulators, substituted polyacetylenes are more or less conjugated polymers, and this feature has been utilized to develop their electronic and photonic functions such as photoconductivity, electrochromism, optical nonlinearity and ferromagnetism.

a. Gas and liquid permeation. Polyacetylenes with bulky groups are, in general, highly permeable to gases[246–248]. Table 14 lists the oxygen permeability coefficients (P_{O2}) and the ratios of permeability coefficients for oxygen and nitrogen (P_{O2}/P_{N2}) for substituted polyacetylenes at 25 °C. The P_{O2} values are in the range *ca* 6000–5 barrers.

It was previously known that poly(dimethylsiloxane) (P_{O2} 600 barrer, P_{O2}/P_{N2} 2) is the most permeable to oxygen among all the conventional polymers. However, fairly many substituted polyacetylenes, e.g. those from MeC≡CSiMe$_3$, MeC≡CGeMe$_3$, PhC≡C$_6$H$_4$SiMe$_3$-*m*, PhC≡CC$_6$H$_4$SiMe$_3$-*p*, and MeC≡CSiEt$_3$, are more oxygen-permeable. They possess bulky and rigid substituents such as Me$_3$Si, Me$_3$Ge and *t*-Bu. The *n*-alkyl and phenyl derivatives, on the other hand, are less permeable than poly(dimethylsiloxane).

TABLE 14. Oxygen permeability coefficients (P_{O_2}) and P_{O_2}/P_{N_2} of substituted polyacetylenes [a]

$+(C=C)_n$ R R'		$P_{O_2,}$[b] barrer	$\dfrac{P_{O_2}}{P_{N_2}}$	Ref
R	R'			
Me	SiMe₃	3000 -6000	1.7	248 - 253
Me	GeMe₃	1800[c]	1.5	205
Ph	C₆H₄SiMe₃-m	1200	2.0	213, 214
Ph	C₆H₄SiMe₃-p	1100	2.1	213, 214
Me	SiEt₃	860	2.0	254
H	C₆H₂-2,4,5-(CF₃)₃	780	2.1	191
Ph	C₆H₄(t-Bu)-p	780	2.3	212
Me	SiMe₂Et	500	2.2	250
H	C₆H₃-2,5-(CF₃)₂	450	2.3	191
H	t-Bu	130	3.0	248
H	C₆H₄GeMe₃-o	110	2.0	188
Me	SiMe₂Pr	100	2.8	250
Ph	C₆H₄Bu-p	100	1.7	212
Me	SBu	79	4.4	206
H	C₆H₄SiMe₃-o	78	3.3	248
Me	SiMe₂CH₂SiMe₃	75	3.6	248
Me	SiMe₂CH₂CH₂SiMe₃	50	3.6	248
Me	SC₈H₁₇	50	2.6	206
Cl	C₈H₁₇	47	2.9	248
Me	C₆H₄SiMe₃-m	38	2.7	255
Me	C₇H₁₅	35	2.5	248
Me	C₅H₁₁	34	2.5	256
Cl	C₆H₁₃	32	2.9	248
H	CH(n-C₅H₁₁)SiMe₃	27	3.1	248
H	C₆H₄CF₃-o	25	3.4	248
Me	SiMe₂C₆H₁₃	18	4.2	248
Ph	C₆H₁₃	14	2.5	248
Et	Ph	12	2.7	248
H	C₆H₄Me-o	8.1	2.7	248
Me	Ph	6.3	2.9	248
Cl	Ph	5.1	5.1	248

[a]Measured at 25 °C.
[b]1 barrer = 1×10^{-10} cm³(STP)·cm/(cm²· s· cmHg).
[c]Estimated as 0.4 times an average value, 4500, of poly(TMSP); cf reference 205.

Though poly[1-(trimethylsilyl)-1-propyne] [poly(TMSP)] and poly(dimethylsiloxane) are typical gas-permeable polymers, their properties are quite different from each other. For instance, the T_g of the former polymer is >200 °C[244], while that of the latter is –127 °C. Further, gases permeate through poly(TMSP) by the dual mode mechanism involving Langmuir adsorption[251–253], whereas the permeation through poly(dimethylsiloxane) is simply explained in terms of the solution-diffusion mechanism governed by Henry's law.

There are two properties to be improved in poly(TMSP) as oxygen-enrichment membrane; i.e. the small P_{O_2}/P_{N_2} and the decrease of P_{O_2} with time. To raise the P_{O_2}/P_{N_2} ratio, modifications with F₂ gas[257] and CF₄ plasma[258] have been examined. Poly(dimethylsiloxane) grafts have been introduced onto poly(TMSP) in an attempt to suppress the decrease in P_{O_2}[259].

Pervaporation is a method of separating liquid mixtures, especially ethanol–water mixtures, with the use of a polymer membrane. Poly(TMSP) membrane has proved highly permselective for ethanol in the ethanol-water pervaporation[260-263]. Thus, the separation factor, α(EtOH/ H_2O), reaches 17 at 10 wt% ethanol in the feed. This value is similar to that for poly(dimethylsiloxane), a well-known ethanol-permselective membrane. In contrast, other substituted polyacetylenes are nonpermselective or rather water-permselective. On the basis of its excellent adsorptivity for organic compounds, poly(TMSP) has been applied to a regenerable column packing which removes trace organic solvents from water[264].

   b. Electronic and photonic functions.   Poly{[(o-trimethylsilyl)phenyl]acetylene} shows photoconductivity when irradiated with visible light[265]. The photocurrent is about 100 times larger than the dark current, and a mechanism in which positive holes are the main carriers has been proposed. Photoconductivity of a poly(1,6-heptadiyne) derivative containing a carbazole moiety has recently been studied[266]. The electrochromism of poly{[(o-trimethylsilyl)phenyl]acetylene} has been studied[267]. This polymer is originally colored purple but becomes colorless when doped electrochemically into a cationic species. The reason for the color change is that the absorption shifts from the visible region to the infrared region on electrical oxidation.

   Laser lights with short wavelengths are important in optical data processing, and conjugated polymers are expected to be useful materials for their generation. Optical third-harmonic generation measurements have been performed for poly(phenylacetylenes)[268-270]. The magnitudes of the third-order susceptibility $\chi^{(3)}$ were 7×10^{-12} esu for poly(phenylacetylene)[268] and in the order of 10^{-11}–10^{-13} esu for its derivatives[269,270].

   The possibility that poly(phenylacetylenes) with a radical on each pendant group become polymer ferromagnets was suggested by Ovchinnikov and coworkers[271]. To examine this proposal, several research groups have synthesized such polymers[272-276], some of which indeed contain considerable amounts of radicals. Ferromagnetism, however, has not been observed.

   A few polyacetylenes with liquid crystalline side groups have been reported. For instance, a polymer ($\bar{M}_w$ 16×10^4) is obtained in high yield with WCl_6 from cholesteryl 4-pentynoate[277]. The polymer shows a transition to a thermotropic liquid-crystalline phase at 89 °C and a transition to the isotropic phase at 189 °C. Other examples are poly(1,6-heptadiynes) with mesogenic side groups[278]. Studies on their applications are under way.

IV. POLY(DIACETYLENES)*

A. Synthesis and Properties of Polymers from Nonconjugated Diynes

   Diacetylene compounds with two nonconjugated acetylene groups, especially α,ω-diynes may undergo various types of polymerization to provide polymers with different main chain structures.

1. Cyclopolymers from α, ω-diynes

   Cyclopolymerization of an α, ω-diyne was first reported by Stille[279] on 1,6-heptadyne by use of a Ziegler–Natta catalyst [$TiCl_4$/i-Bu_3Al]. The conjugated polyene structure including a six-membered ring was proposed for the soluble polymers with conductivities of 10^{-10}–10^{-13} S cm^{-1}. Conductivities of free-standing poly(1,6-heptadiyne) films[280] were improved up to 1 S cm^{-1} by treating with acceptors such as I_2.

*By K.Takeda

The cyclopolymerization of α,ω-diynes (equation 30) has been intensively studied by Choi and coworkers with a variety of dipropargyl compounds containing different substituents (31)[266,278,281–288].

$$(30)$$

monomers (31) catalysts (32)

$XR^1R^2 = CH_2$, CMe_2, $CMePh$, $C(OH)H$, $MoCl_5$, $MoCl_5$–Bu_4Sn, $MoCl_5$–$EtAlCl_2$
$C(OH)Ph$, CPh_2, $C(CO_2Me)_2$, $C(CO_2Et)_2$, WCl_6, WCl_6–Bu_4Sn, WCl_6–$EtAlCl_2$,
$C(CO_2CH_2CF_3)_2$, $C[CO_2CH(CF_3)_2]_2$, $TiCl_4$–$EtAlCl_2$, $TaCl_5$, $PdCl_2$
$C[CO_2(CH_2)_6C_{Vz}]_2$, $C(OC_6H_5)_2$, O, S,
NPh, $SiMe_2$, $SiMePh$, $SiPh_2$, $GeMe_2$,
$GePh_2$

Among the transition-metal catalyst systems (32) examined, $MoCl_5$ and WCl_6 catalysts were found to be highly effective for the cyclopolymerization of dipropargyl compounds. In some cases the combinations of $MoCl_5$ with $EtAlCl_2$ or Bu_4Sn were more effective than $MoCl_5$ alone for the specific monomers. The polyene structures with cyclic recurring units in the polymer main chains were confirmed by IR, NMR (^1H and ^{13}C) and UV-visible spectroscopies. The resultant polymers with molecular weight (M_n) $\approx 1.27 \times 10^5$ were soluble in common organic solvents such as benzene, chloroform and tetrahydrofuran. Other characteristics of the cyclopolymers were their excellent thermal and oxidative stabilities. The conductivities of polymers were increased from 10^{-11} to 10^{-4}–10^{-2} S cm^{-1} on I_2 doping. Functionalization of polymers by the substitution with the photoconductive[266] or mesogenic[278] side groups is a recent topic in the study of the cyclopolymers.

2. Polymers with conjugated diacetylene units

Oxidative polycondensation of α,ω-diynes, the applications of the Glaser or the Eglinton coupling reaction, are synthetic routes to obtain polymers with conjugated diacetylene units in the main chain structure (33, equation 31) which may be cross-linked intermolecularly by UV-light irradiation or thermal annealing. Since Hay reported the oxidative coupling of m-diethynylbenzene[289], oxidative polycondensation has been frequently employed to prepare polymers with conjugated diyne units including polyethers[290], polyesters[291] etc. from α,ω-diynes including the corresponding moieties.

$$H-C\equiv C-R-C\equiv C-H \xrightarrow{Cu^+,\ O_2} -(C\equiv C-R-C\equiv C)_n \qquad (31)$$

$$(33)$$

The similar main chain structures with conjugated diacetylene units (34, 35) can be prepared (equations 32 and 33) by polyaddition or polycondensation using components containing the conjugated diyne units[292,293]. The processable polymers become insoluble on photoirradiation or by thermal annealing via intermolecular cross-linking.

$$HO-R-C\equiv C-C\equiv C-R-OH \xrightarrow{Cl-OC-R'-CO-Cl}$$

$$-\!\!\left(O-R-C\equiv C-C\equiv C-R-O-\overset{\overset{\displaystyle O}{\|}}{C}-R'-\overset{\overset{\displaystyle O}{\|}}{C}\right)_{\!\!n}^{} \quad (32)$$

(34)

$$HO-R-C\equiv C-C\equiv C-R-OH \xrightarrow{OCN-R'-NCO}$$

$$-\!\!\left(O-R-C\equiv C-C\equiv C-R-O-\overset{\overset{\displaystyle O}{\|}}{C}-NH-R'-NH-\overset{\overset{\displaystyle O}{\|}}{C}\right)_{\!\!n}^{} \quad (33)$$

3. Polyphenylenes

Soluble polyphenylenes with a branched structure (**36**) were prepared (equation 34) from diethynylbenzene and phenylacetylene by trimerization of the acetylene group in the presence of Ziegler catalysts[294,295]. The oligomers (called 'H-resin') can be cross-linked on heating to yield thermally and chemically stable resins[296].

$$R'C\equiv CH + R(C\equiv CH)_2 \longrightarrow \left(\!\!\left(\!\!\bigcirc\!\!\right)\!\!-R\!\!\right)_{\!\!n} \quad (34)$$

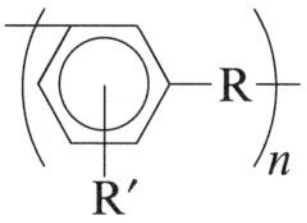

(36)

Polyphenylenes with excellent thermal stability can be prepared (equations 35 and 36) by Diels–Alder reactions using diethynylbenzene as a component. Phenylated polyphenylenes

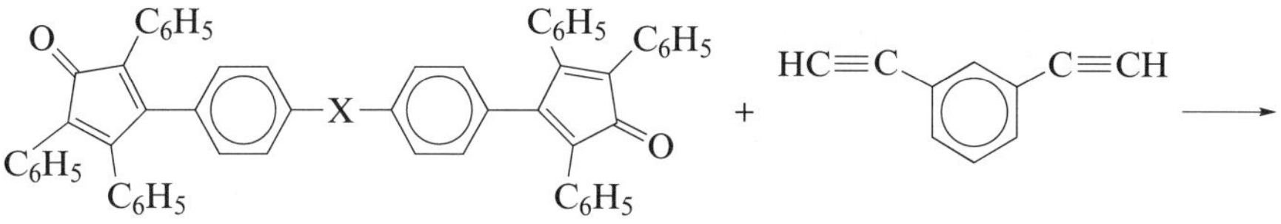

(a) X = nil
(b) X = O
(c) X = S

$$(35)$$

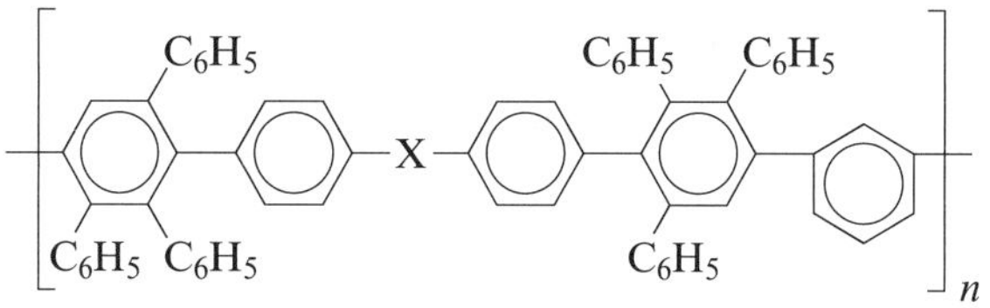

(a) X = nil
(b) X = O
(c) X = S

(37)

(37) with molecualr weight (M_n) of several ten thousands were colorless and soluble in common organic solvents[297,298], while poly(p-phenylene) (38) was yellow in color, highly crystalline and insoluble[299].

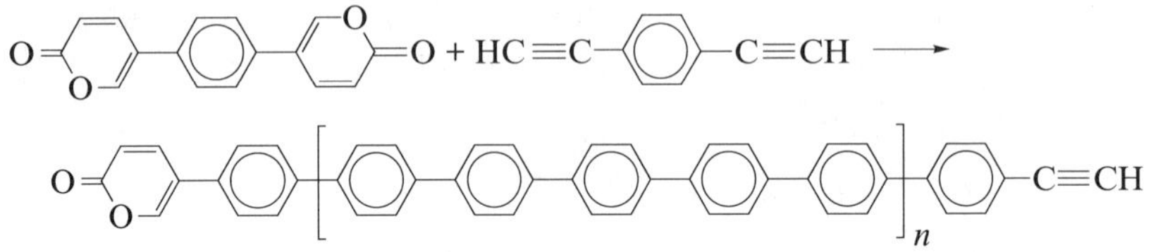

(38)

B. Synthesis and Properties of Copolymers Composed of Nonconjugated Diyne Components

Some of the copolymers composed of diacetylene should be mentioned, even though they may not fit exactly the definition of polydiacetylenes.

1. Metal containing poly-yne polymers

Some acetylide bonds with transition metals are stable enough to form organometallic polymer chains, which are usually insoluble in organic solvents due to the high rigidity of the polymer chains. Hay prepared polymers containing silicon, arsenic or mercury in the polymer backbones[300]. A tin containing poly-yne polymer showed limited delocalization of electrons on the polymer chains interrupted at the metal moieties[301]. Sonogashira and coworkers succeeded in the preparation of soluble poly-yne polymers containing transition metals[302–304]. The key factors for successful results were the choice of transition metals (Pt, Pd and Ni) and the use of bulky tributylphosphine ligand providing the solubility and the

$$trans\text{-}(PBu_3)_2MCl_2 + trans\text{-}(PBu_3)M(C\equiv C-C\equiv CH)_2 \xrightarrow[\text{CuI, amine}]{-HCl}$$

$$-[trans\text{-}M(PBu_3)_2-C\equiv C-C\equiv C]_n \quad (37)$$

$$(M = Pt, Pd; R = H, Me, Et, iso\text{-}Pr, Et-O-)$$

(39)

$$trans\text{-}(PBu_3)_2M(C\equiv C-Y-C\equiv CH)_2 \xrightarrow[\text{CuCl, O}_2\text{, Me}_2\text{NCH}_2\text{CH}_2\text{NMe}_2]{-H}$$

$$-[\,trans\text{-}M(PBu_3)_2-C\equiv C-Y-C\equiv C-C\equiv C-Y-C\equiv C\,]_n \quad (38)$$

$$(M = Pt, Pd; Y = nil, -C_6H_4-)$$

(40)

$$trans\text{-}(PBu_3)_2Ni(C\equiv CH)_2 + HC\equiv C-Y-C\equiv CH \xrightarrow[\text{CuI, amine}]{-HC\equiv CH}$$

$$-[\,trans\text{-}Ni(PBu_3)_2-C\equiv C-Y-C\equiv C\,]_n \quad (39)$$

$$(Y = nil, -C_6H_4-) \quad (41)$$

987

TABLE 15. Representative metal-containing poly-yne polymers

Polymer	$[\eta]^a$	M_w	Color
$+(PBu_3)_2Pt-C\equiv C-C\equiv C+_n$	2.11	120,000	yellow
$+(PBu_3)_2Pt-C\equiv C-\langle benzene\rangle-C\equiv C+_n$	2.76	69,000	yellow
$+(PBu_3)_2Pt-C\equiv C-\langle benzene\rangle-C\equiv C+_n$	2.64	78,000	yellow
$+(PBu_3)_2Pt-C\equiv C-\langle benzene\rangle-C\equiv C+_n$	0.43	63,000	pale yellow
$+(PBu_3)_2Pd-C\equiv C-\langle benzene\rangle-C\equiv C+_n$	0.46	19,000	pale yellow
$+(PBu_3)_2Ni-C\equiv C-\langle benzene\rangle-C\equiv C+_n$	0.19	15,000	orange
$+(PBu_3)_2Ni-C\equiv C-\langle benzene\rangle-C\equiv C-C\equiv C-\langle benzene\rangle-C\equiv C+_n$	0.20	13,000	orange
$+(PBu_3)_2Pt-C\equiv C-\langle benzene\rangle-C\equiv C-C\equiv C-\langle benzene\rangle-C\equiv C+_n$	5.01	95,000	yellow
$+(PBu_3)_2Pd-C\equiv C-\langle benzene\rangle-C\equiv C-C\equiv C-\langle benzene\rangle-C\equiv C+_n$	1.01	26,000	yellow

^{a}In THF, 25 °C.

stability of the complexes. Poly-yne polymers containing Pt and Pd (**39, 40**) were synthesized either by polycondensation (equation 37) accompanied by dehydrohalogenation in the presence of a cupric salt, or by oxidative coupling (equation 38). Ni-containing poly-yne polymers (**41**) were synthesized by alkynyl-exchange reaction (equation 39).

Metal-containing poly-yne polymers with molecular weight (M_w) of ca 1.2×10^5 were soluble in usual organic solvents such as benzene, methylene chloride, tetrahydrofuran etc. The thermal stability of the polymers was increased with the metals, in the order of Pt > Pd > Ni. Films with a thickness of several microns and with a tensile strength as large as 9.0×10^3 g/mm^2 were fabricated from the Pt-containing polymer. One of the characteristics of these polymers is their lyotropic liquid crystal behavior in solution[305]. Some representative polymers prepared are listed in Table 15.

Poly-yne polymers composed of phthalocyanine (Pc) complexes (**42**) with silicon were synthesized (equation 40) by Mitulla and Hanack[306]. The conductivities of the polymers were up to 10^{-12} S cm^{-1}, and increased to 10^{-5} S cm^{-1} by I$_2$ doping.

$$Si(Pc)Cl_2 + BrMgC{\equiv}C{-}(\!Y{-}C{\equiv}C)_{\overline{m}}MgBr$$

$$\xrightarrow{\text{THF}} {-}[\,Si(Pc)C{\equiv}C{-}(\!Y{-}C{\equiv}C)_{\overline{m}}\,]_n \qquad (40)$$

(**42**)

$$m = 0$$
$$m = 1, Y = {-\!-}$$
$$m = 1, Y = {-}\!\langle\bigcirc\rangle\!{-}$$

2. Metal containing poly-ene polymers

Poly-ene polymers including metal moieties, for example selenium (**43**), were prepared (equation 41) from 1,4-benzenediselenol and diethynylbenzene in the presence of a radical initiator or by UV-light irradiation[307]. The polymer was insoluble in organic solvents and the conductivity after I$_2$ doping was 10^{-5} S cm^{-1}.

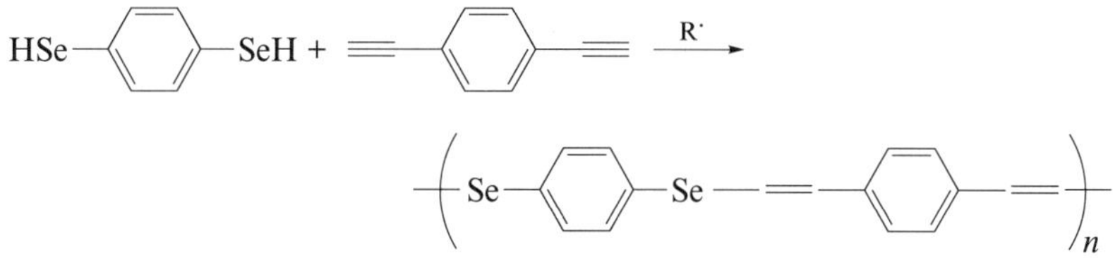

$$HSe-\langle\bigcirc\rangle-SeH + \equiv\!\!-\langle\bigcirc\rangle-\!\!\equiv \xrightarrow{R^{\cdot}}$$

$$\left(\!\!-Se-\langle\bigcirc\rangle-Se-\!\!=\!\!-\langle\bigcirc\rangle-\!\!=\!\!-\right)_n \quad (41)$$

(**43**)

3. Organosilicon polymers from diethynyl silanes

Recently, three types of novel poly(disilanylenebutenyne-1,4-diyls) (**44**) with molecular weights (M_w) of $1.0 \times 10^4 - 1.0 \times 10^5$ were synthesized (equation 42) from diethynyldisilanes by use of rhodium(I) catalysts[308]. The scission of Si—Si bonds by UV-light irradiation took place easily for poly(disilanylenebutenye-1,4-diyls) in solution and in a solid film, while poly[(methylphenylsilylene)butenyne-1,4-diyl] was found to be photochemically stable. Conductivities of poly(disilanylbutenyne-1,4-diyls) doped with SbF_5 were $10^{-3}-10^0$ S cm^{-1} in air and $10^{-4}-10^{-3}$ S cm^{-1} in vacuo.

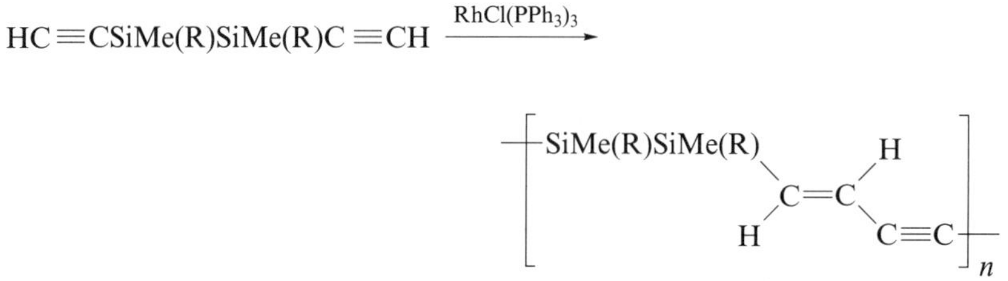

$$HC\equiv CSiMe(R)SiMe(R)C\equiv CH \xrightarrow{RhCl(PPh_3)_3}$$

$$\left[\begin{array}{c} -SiMe(R)SiMe(R) \qquad\qquad H \\ \diagdown \qquad\qquad \diagup \\ C=C \\ \diagup \qquad\qquad \diagdown \\ H \qquad\qquad C\equiv C- \end{array}\right]_n \quad (42)$$

(**44**)

4. Boron containing polymers composed of α,ω-diynes

Hydroboration of α,ω-diynes with boron tribromide yielded (equation 43) a novel type of organoborane polymer (**45**) with reactive B—Br groups, which is a poly-Lewis acid[309].

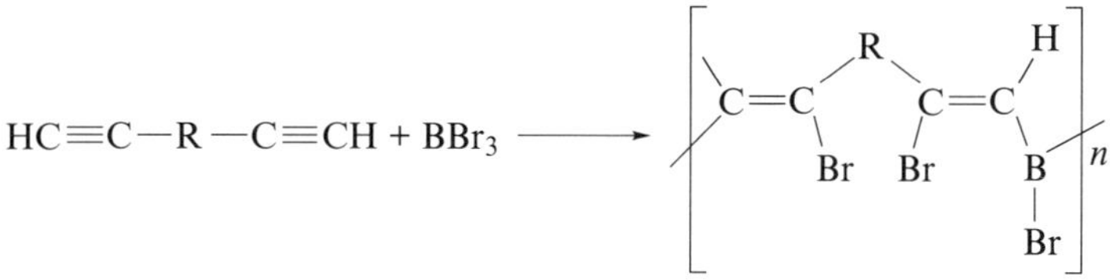

$$HC\equiv C-R-C\equiv CH + BBr_3 \longrightarrow \left[\begin{array}{c} \qquad R \qquad H \\ \diagdown \diagup \quad \diagdown \diagup \\ C=C \qquad C=C \\ \diagup \quad \diagdown \quad \diagdown \\ Br \quad Br \qquad B \\ \qquad\qquad | \\ \qquad\qquad Br \end{array}\right]_n \quad (43)$$

(**45**)

5. Polypyrones, copolymers composed of diynes and CO₂

Cycloaddition copolymerization of diynes (Equation 44, product 46) such as 3,11-tetradecadiyne, 3,9-dodecadiyne, 1,3- or 1,4-di(2-hexynyl)benzene with CO_2 was successfully carried out in the presence of zero-valent nickel catalyst [NiCOD)₂ 2P(n-C₈H₁₇)₃] prepared from bis(1,5-cyclooctadiene)nickel and tri-n-octylphosphine ligand[310,311]. Synthesis of soluble ladder-type polymers (**47**) was carried out by cycloaddition copolymerization of 1,7-cyclotridecadiyne (**48**) and CO_2 (equation 45).

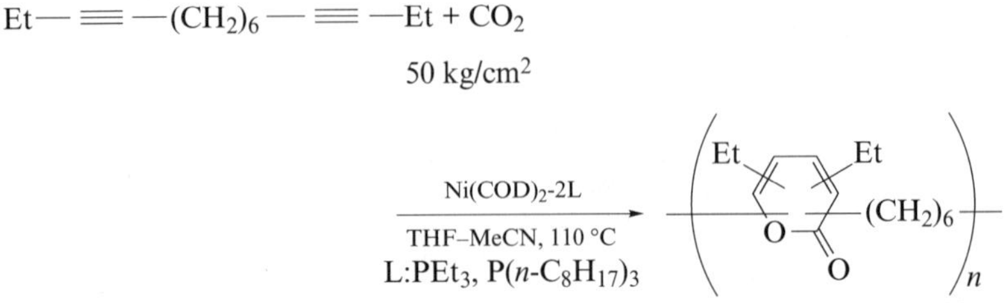

$$\text{Et} \underline{\quad} \equiv \underline{\quad} (CH_2)_6 \underline{\quad} \equiv \underline{\quad} \text{Et} + CO_2$$

$$50 \text{ kg/cm}^2$$

$$\xrightarrow[\substack{\text{THF–MeCN, 110 °C} \\ \text{L:PEt}_3, \text{P}(n\text{-}C_8H_{17})_3}]{\text{Ni(COD)}_2\text{-2L}}$$

(44)

$$\sim 90\%, \; M_n = \sim 18{,}000$$

(46)

$$\underset{\textbf{(48)}}{\left[\substack{(CH_2)_4 \\ (CH_2)_5} \right]} + \underset{\substack{20 \text{ kg/cm}^2}}{CO_2} \xrightarrow[\text{THF–MeCN, 60 °C}]{\text{Ni(COD)}_2\text{-2P}(n\text{-}C_8H_{17})_3}$$

(45)

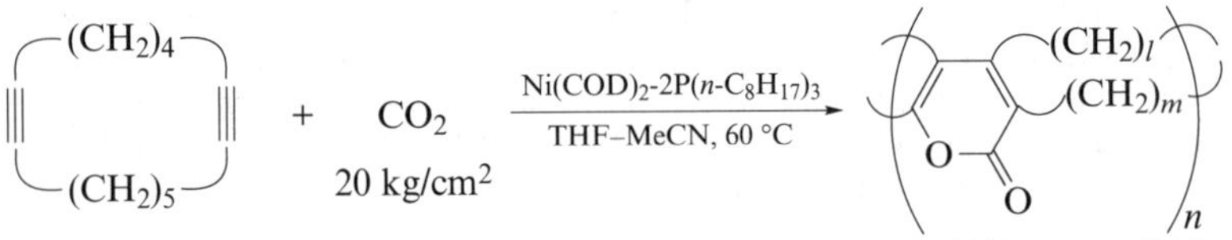

$$\sim 100\%, \; M_n = 6000$$

(47)

C. Polymers Prepared by Solid State Polymerization from Conjugated Diynes

1. Synthesis

a. Conjugated diynes for solid state polymerization.   Since the reports on the solid state polymerization of diacetylene compounds were presented by Wegner[21,312–314], hundreds of conjugated diynes have been synthesized and screened in order to obtain polydiacetyl-enes with unique characteristics. Symmetrical conjugated diacetylene compounds ($R\text{—}C\equiv C\text{—}C\equiv C\text{—}R$) are commonly synthesized by the oxidative coupling of mono-substituted acetylenes developed by Glaser[315] (equation 46) and by Eglinton and McCrae[316] (equation 47). The Cadiot–Chodkiewicz[317] coupling reaction (equation 48) an effective synthetic route to unsymmetrical diacetylene compounds ($R\text{—}C\equiv C\text{—}C\equiv C\text{—}R'$). Chemi-cal modification of the substituents may be carried out, if necessary, for obtaining further

$$2R\text{—}C\equiv CH \xrightarrow{\text{Cu}^+, \text{O}_2} R\text{—}C\equiv C\text{—}C\equiv C\text{—}R \tag{46}$$

$$2R\text{—}C\equiv CH \xrightarrow[\text{pyridine}]{\text{Cu(OAc)}_2, \text{O}_2} R\text{—}C\equiv C\text{—}C\equiv C\text{—}R \tag{47}$$

$$R\text{—}C\equiv CH + BrC\equiv C\text{—}R' \xrightarrow[\text{amine}]{\text{Cu}^+, \text{O}_2} R\text{—}C\equiv C\text{—}C\equiv C\text{—}R' \tag{48}$$

molecular designs.
   In Table 16 are listed some representative conjugated diynes containing various substitu-ents.

b. Phenomenological aspects of solid state polymerization.   Polymerization is usually initiated by UV-light or high-energy beam irradiation or by thermal annealing of monomer crystals. As polymerization proceeds, a drastic color change from colorless to either blue or red is observed, depending on the substituents. The polymerization mechanism is a 1,4-addition, as is shown schematically in Figure 4.

TABLE 16. List of abbreviations and structures of the substituents for diacetylene monomers ($R - C \equiv C - C \equiv C - R'$)

Abbreviation		Substituent(s)
PTS	(49)	$-CH_2 - O - SO_2 - \langle C_6H_4 \rangle - CH_3$ (R = R')
PTS-12	(50)	$-(CH_2)_4 - O - SO_2 - \langle C_6H_4 \rangle - CH_3$ (R = R')
TCDU	(51)	$-(CH_2)_4 - O - CO - NH - \langle C_6H_5 \rangle$ (R = R')
nBCMU	(52)	$-(CH_2)_n - O - CO - NH - CH_2 - CO - O - NH - C_4H_9$ (R = R')
C$_4$UC$_n$	(53)	$-(CH_2)_4 - O - CO - NH - C_nH_{2n+1}$ (R = R')
DCH	(54)	$-CH_2 - N \langle \text{carbazole} \rangle$ (R = R')
BIPO	(55)	$\text{HO} - \begin{array}{c} H_3C \quad CH_3 \\ \diagup \\ \diagdown \\ H_3C \quad CH_3 \end{array} N - \dot{O}$ (R = R')
(m, n)	(56)	$R = -(CH_2)_{m-1} - CH_3$ $\qquad R' = -(CH_2)_n - COOH$

The conversion from monomer to polymer in the crystal proceeds via a diffusionless phase transition to form perfect polymer single crystals with the lattice parameters close to those of the original monomer crystals[318]. During the course of the polymerization, the polymer chains and the monomer matrix form a homogeneous solid solution without any nucleation of the product phases, thus neither destruction nor fibrillation can be observed in the resultant polymer single crystals. Solid state polymerization of conjugated diynes is known as a typical and rare example of topochemical polymerization, which implies that the polymerization proceeds under strict control of the monomer crystal lattice and that the crystal structure of the polymer is essentially related to the original monomer crystal structure. As–polymerized polymer main chains are fully conjugated and extended in structure, being aligned to a definite direction to compose nearly defect-free polymer single crystals with fairly large optical anisotropy. Polymer single crystals with such unique features can never be prepared by conventional crystallization from solution or from a melt. Polydiacetylene single crystals thus prepared have been studied as ideal models for quasi-one-dimensional π-electron systems, especially of electrooptical interest. However, the extended polymer chain structure with high rigidity is also the reason why most polydiacetylenes are insoluble, which is the main bottleneck for the fundamental characterization and the application of polydiacetylenes. Various attempts to obtain polydiacetylenes of desirable forms have been made, e.g. growth of thin single crystals by a melt-shear method[319], Langmuir–Brodgett film fabrication[320,321], fabrication of oriented thin films[5] by vacuum deposition[322], epitaxial crystal growth on an alkali–halide crystal surface[323] and frozen-state polymerization[324].

   c. Molecular design and crystal engineering.   Solid state polymerization of diacetylene compounds proceeds by 1,4-addition between adjacent monomer molecules along to the stacking axis. Interactions such as hydrogen bonds, dipole–dipole interactions through the substiuents sometimes assist the molecular packing in the crystal, so that the substituents may enhance the polymerizability. On the other hand, diphenyldiacetylene is hardly polymerizable in the crystalline state due to the steric hindrance of the phenyl group next to the triple bond. While this indicates the great importance of the molecular design, it is to be noted that the solid-state polymerizability is essentially dependent on the packing properties of the monomer molecules in the crystal lattice, i.e. whether or not the monomer molecules are arranged in appropriate positions for the 1,4-addition reaction. The idea is supported by the information that a given diacetylene compound does exibit different polymerizabilities, depending on different crystal modifications[313]. The change in the polymerizabilities due to the solvato-chromism, as is the case of 2,4-hexadiyne-1,6-bis(*m*-tolylurethane)[325], is another basis for the argument.

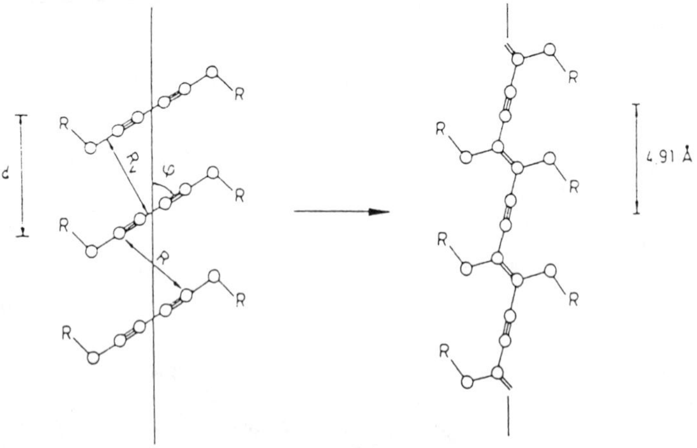

FIGURE 4. Packing model for the diacetylene monomers[22]. Reproduced by permission of Springer-Verlag

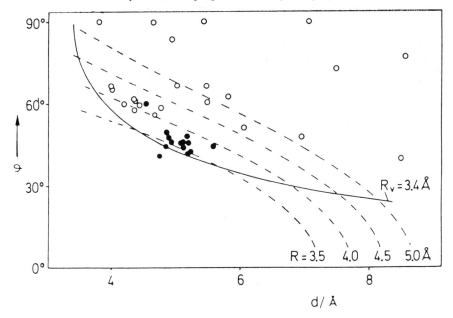

FIGURE 5. Relation between d (distance between the centers of the adjacent stacking monomer molecules) and φ (stacking angle). The broken lines indicate the distance between the reaction sites[22]. ● refers to polymerizable monomer crystals and ○ to unpolymerizable monomer crystals. Reproduced by permission of Springer-Verlag

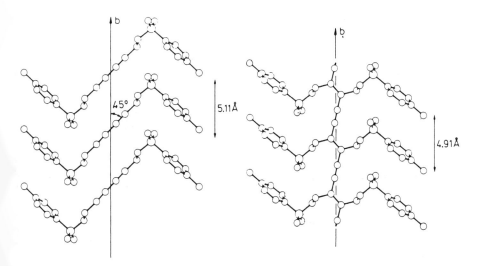

FIGURE 6. Projection of PTS monomer amd polymer crystal structures on the plane of the polymer backbone[22]. Reproduced by permission of Springer-Verlag

The relationship between polymerizability and monomer crystal structure was demonstrated by Baughman and coworkers[326–328], based on crystallographic data of both polymerizable and unpolymerizable diynes. Figure 4 is the schematic packing model of monomer molecules along the stacking axis, where φ and d are, respectively, the stacking angle and the distance between the centers of adjacent monomer molecules. As indicated by solid circles in Figure 5, the monomers with high poymerizability are plotted in the limited region with the values of $d \sim 5$ Å, and $\varphi \cong 45°$, i.e. almost all highly reactive monomers are stacked with reaction sites distances R between 3.5 and 4.0 Å. One exception is the special case of DCH (**54**), where the reaction is connected with a phase transition phenomenon[329].

As is seen in the projection of the PTS (**49**) monomer and polymer crystal structures on a common plane given in Figure 6, the distance d between the centers of adjacent monomer units along the b-axis decrease by 0.2 Å from monomer to polymer, while the substituents of p-toluenesulfonate groups are kept in nearly the same positions[330,331]. The important role of the flexible side groups is suggested to cause the high polymerizability of PTS by the relaxation of the stress in the crystal during the polymerization process.

d. Polymerization mechanism.   Solid state polymerization of diacetylene compounds proceeds by a chain reaction mechanism via radical or carbene intemediates. Characteriza-

TABLE 17. The structures of the oligomeric reaction intermediates and the stable products[12]

Symbols	Structures		
Spin S		Simplified structures	Chains
— DR — diradicals $S = 0, S^* = 1$			butatriene $7 > n \geq 2$
— DC — dicarbenes $S = 0, S^* = 1, 2$			acetylene $n \geq 7$
— AC — asymmetric carbenes $S = 1$			acetylene $n \geq 2$
— SO — stable oligomer $S = 0$			acetylene $n \geq 3$

tion of the reaction intermediates at the initial reaction stage was unsuccessful until Sixl and coworkers made precise experiments on photoinduced polymerization at very low temperature[332–340]. Initiation was carried out by the irradiation of either a xenon high pressure arc at 310 nm or the 308 nm line of an excimer laser (ArF) to PTS single crystal at 10 K in order to reduce the reaction rates. The individual steps of the polymerization were investigated by subsequent photobleaching at low temperture or by thermal annealing up to 100 K. From spectroscopic data, it was concluded that the mechanism of the solid state polymerization could be described by three series of intermediates with active chain ends and a series of stable products, which were assigned to the diradicals (DR), the dicarbenes (DC), the asymmetric carbenes (AC) and the stable oligomers (SO) by correlation to the ESR spectra and by agreement with the data calculated from a one-dimensional electron gas model (see Table 17).

The longest DR oligomer experimentally confirmed was the hexamer molecule ($n = 6$). Longer DR oligomers ($n \geq 7$) with butatriene structure were energetically unstable and transfomed to dicarbene (DC) oligomers with acetylene structure. Because of the high efficiency of the thermal addition process, low molecular weight SO molecules could be confirmed only at very low temperature, but not at the usual polymerization conditions.

As almost all common polydiacetylenes are insoluble due to the high rigidity of polymer main chains, the molecular weight and molecular weight distribution have been measured only for soluble polydiacetylenes from special monomers such as nBCMU ($n = 3,4$) (52) and PTS-12 (50). Poly(3BCMU) and poly(4BCMU) were the first soluble polydiacetylenes with substituents composed of flexible and solvating sites[341–344]. The molecular weight of poly(3BCMU) was estimated by the light scattering method to be 4.8×10^5.

Figure 7 shows the weight distribution of poly(PTS-12) at the different stages of polymerization[345]. An average degree of polymerization (P_n) was 60 in the initial stage at conversion below 1% and the fraction peaks shifted to the high molecular weight polymers with $P_n = 800$ at conversions above 20%. The molecular weight (M_w) of poly(PTS-12)[346] was measured to be 7.4×10^5. The kinetic chain length as a function of conversion for PTS-12 is available in the literature[347].

The time-dependent conversion curves for solid state polymerization display a sigmoidal shape with a steep increase in the polymerization rate with an onset at ca 10% conversion to polymer. The observed autocatalytic effect[312,348,349] can be explained with a mechanical model based on strain-energy considerations. The activation energies for the thermal, UV-light and γ-ray polymerization of PTS were estimated to be ca 21–24, 2–3 and 2–3 kcal mol^{-1}, respectively[349].

2. Properties

a. Polymer chain structure.   Main chain structures of polydiacetylenes may be depicted (Figure 8) either as acetylene structures (i), or as a butatriene structure (ii), however no polymer with butatriene structure has been confirmed as yet by X-ray analysis[350–352]. This is consistent with the results about the structures of low molecular weight oligomers of Sixl, namely that the butatriene structure is energetically favorable only for oligomers with chain lengths shorter than hexamer. Therefore the main chain structure of polydiacetylenes is reasonably assumed to be an acetylene, ene-yne structure.

b. Conformational transition in solution.   Extensive studies on the solution properties of polydiacetylenes were made for poly(3BCMU) and poly(4BCMU). Figure 9 shows the phase diagram of poly(3BCMU) in relation to the solvent–nonsolvent ratio and polymer concentration[341]. On addition of hexane (nonsolvent) to poly(3BCMU) solution in chloroform, a drastic color change from yellow to blue takes place at the chloroform mole fraction $X_C = 0.71$. The blue solution is homogeneous and stable enough not to precipitate over a long period of storage and even by the ultra-centrifugation. Further addition of

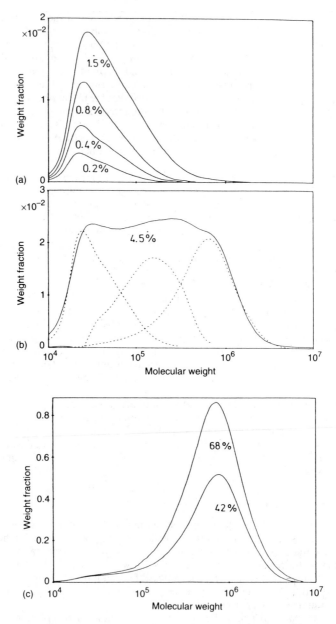

FIGURE 7. Molecular weight distribution of PTS-12 at different conversions[22]. Reproduced by permission of Springer-Verlag

excess hexane results in the precipitated polymers colored blue. These color changes are reversible and independent of the polymer concentration in the region below 0.003

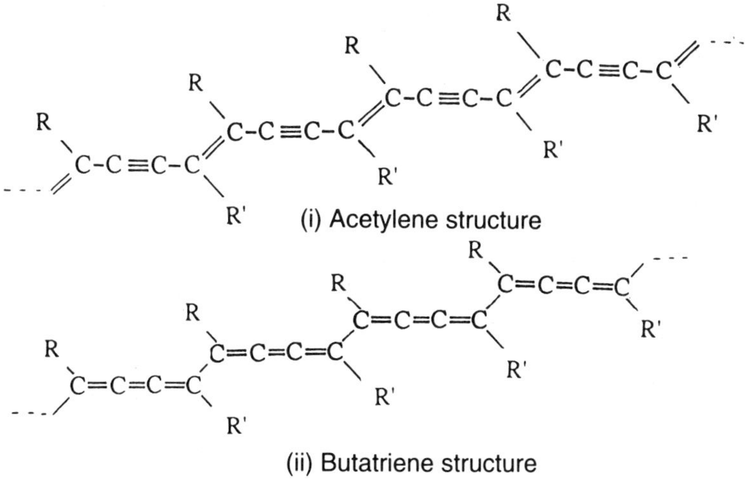

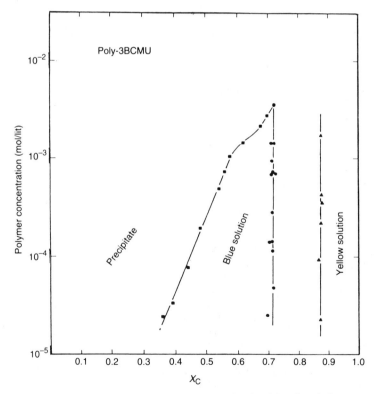

FIGURE 8. Chain structures of polydiacetylene: (i) acetylene structure, (ii) butatriene structure.

FIGURE 9. Phase diagram for poly(3BCMU) in chloroform/n-hexane mixtures[341]. Reproduced by permission of the American Institute of Physics

mol/l, which suggests that the color transitio n is due to the conformational change of the individual polymer chains. Visible absorption and FT-IR spectra show that the polymer chains in the blue solution were planar and extended in structure, being stabilized by intramolecular hydrogen bonds between the substituents via urethane linkages. The effective conjugation length, i.e. the planarity of the conjugated units, of the polymers in blue solution was evaluated to be >30 monomer repeat units. In the yellow solution the polymer structure became less planar by interruption of the hydrogen bonds and the conjugation length was estimated to be 7 monomer repeat units. Models of distortion structures (helical or wormlike)[353,354] and of structural defects[355] were proposed to explain the respective conformations with the different effective conjugation lengths. Similar information is available for poly(4BCMU) in the literature[356–358].

   c. Phase transition phenomena in crystals.   Characteristic features of polydiacetylenes in the solid state are the phase transition phenomena between two phases, the so-called blue-phase (A-phase) and red-phase (B-phase). The color transition usually occurs from the blue to the red phase on thermal or photo excitation. The irreversibility of the phase transition can be explained by the limited conformational freedom in the crystal in contrast to that of single polymer chains in solution.

   (i) Reversible thermochromism.   The first thermochromic reversibility of polydiacetylene was reported by Rubner and coworkers[359] for poly(ETCD) [= poly(C_4UC_2)], whose mechanism was proposed by Tanaka and collaborators from NMR spectroscopy[360,361]. Koshihara and coworkers[362,363] made detailed measurements on the thermochromic behavior of a series of poly(C_4UC_n) [$n = 1–10$] (53) with alkylurethane substituents and confirmed that the thermochromic reversibilities were characteristic of the whole poly(C_4UC_n) family. As-polymerized poly(C_4UC_n)s are spectoscopically blue (A-phase) with reflectance maxima at about 1.9 eV (640 nm) at 290 K due to the ^1B_u exciton transition. On heating up to 420 K, the polymers turn red (B-phase) with a peak at about 2.3 eV (560 nm). Figure 10 shows the correspondence of the reflectance intensity at 1.94 eV (A-phase peak) and DSC thermogram. T_2 and T_1 are the critical temperatures for the A-to-B transition in the heating process and for the B-to-A transition in the cooling process, respectively. As shown in Fig. 10a, a typical hysteresis loop due to the reversible A–B transition can be traced in the region below the polymer melting point (T_3). When the temperature is increased beyond T_3, the reflectivity at 1.94 eV can not be recovered anymore on cooling, and the A–B transition becomes irreversible. T'_3 is the temperature of solidification of once molten polymer. When the sample is heated up to 500 K, no distinct hysteresis loop can be traced, as shown in Fig. 10b.

   Figure 11 shows the phase diagram in terms of critical phase transition temperatures depending on the terminal alkyl chain length n in the substituents. Poly(C_4UC_n) exists in the A-phase at the temperature below T_1 and, between T_1 and T_2, is the region where A- and B-phases are co-existing. Polymers in the B-phase kept at a temperature below T_3 can give again the A-phase on cooling. The respective electronic structures and the stretching modes of —C=C— and —C≡C— in the polymer chains in both A- and B-phases were characterized by reflectance and Raman spectroscopies. The amount of hydrogen bonds via urethane moieties was estimated by the IR band intensity of the N—H sretching mode at 3300 cm^{-1}. It was confirmed that the breaking and formation of the hydrogen bonds were entirely reversible unless the samples were heated beyond T_3 in heating and cooling cycles, while only one-half of the hydrogen bonds broken by heating beyond T_3 could be recovered on cooling to T'_3. The polymers once heated above T_3 (B'-phase) could never give again the A-phase on cooling, though the B'-phase was similar to the B-phase in reflectance and Raman spectra. As a result, it is concluded that the interactions between the substituents like hydrogen bonds play an important role in the thermochromic reversibility by stabilizing the conformation in the B-phase which can yield again the A-phase on cooling.

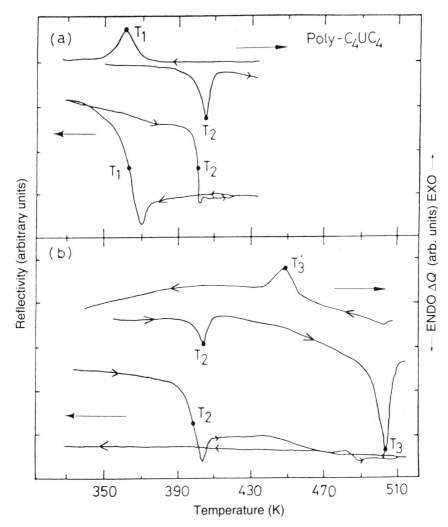

FIGURE 10. Temperature dependences of reflectivity for poly(C_4UC_4) crystal at 1.94 eV and DSC thermograms[362]. Reproduced by permission of the American Institute of Physics

(ii) *Reversible photochromism*. It was believed that the intense photoirradiation of polydiacetylene crystals in the blue phase brings about an irreversible phase transition to the red phase (B-phase in Figure 9) by direct excitation of the π-electron systems on the polymer backbone accompanied by a drastic conformational change. The first successful reversible photoinduced phase transition[364] was presented for poly(C_4UC_3) single crystals as recently as 1992. The idea for the investigation was based on the thermodynamical consideration that the free energy of a polymer crystal could be expressed by a potential curve with double local minima corresponding to A- and B-phases separated by an energy barrier and that the B-phase should be photoexcitated to the A-phase at a temperature

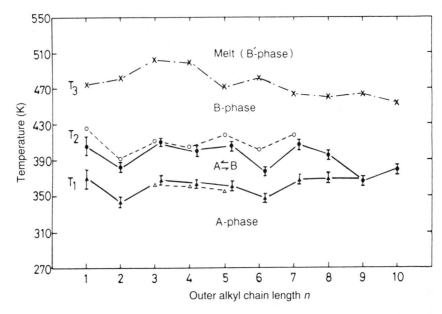

FIGURE 11. Dependency of the transition temperatures (T_1, T_2 and T_3) on *n in the side groups of poly*(C_4UC_n)[362]. Reproduced by permission of the American Institute of Physics

within the hysteresis region, where two potential minima were close in depth each other. A poly(C_4UC_3) single crystal was irradiated with a single shot (2.81 eV) from a pulse dye laser pumped by an XeCl excimer laser at the temperature corresponding to point 2 (390 K) in Figure 12. The electronic absorption (ε_2) and Raman spectra displayed almost full A-to-B conversion in the heating cycle shown by dashed curves in Figure 13. Similarly, the sample was excited by a single pulse (3.18 eV) at point 4 (370 K) in the cooling cycle with *ca* 50% conversion from the B- to the A-phase. The threshold values were determined to be *ca* 2.4 eV for the A-to-B and *ca* 2.7 eV for the B-to-A transitions, respectively. Care was taken to check the heating effects by laser irradiation in pump and probe spectroscopy. The domain size of photoinduced phase transition per photon was evaluated to be 140 monomer repeat units, which gave the first evidence of a photoinduced cooperative phase-transition phenomena on polymer chains.

Figure 14 shows the phase conversion fraction as a function of the exciting photon energy for both A-to-B and B-to-A phase changes. The excitation spectra for the converted fraction showed the threshold located 0.3–0.5 eV higher than the exciton absorption peak for both the A-to-B and B-to-A phase changes and the extremely low efficiency of photoinduced phase changes at the exciton absorption peak. The similarity between the excitation spectra for photoconversion and the photoconductivity spectra strongly suggested that the photoinduced phase transition was triggered by the photogenerated charge carriers (polaron or bipolarons) and not by excitons.

d. Optical properties

(i) Linear optical properties. The spectra of as-polymerized polydiacetylene single crystals show intense absorption peaks either at *ca* 1.6 eV (640 nm) for the blue phase or at *ca* 1.9 eV (560 nm) for the red phase, depending on the kind of substituents. Any other

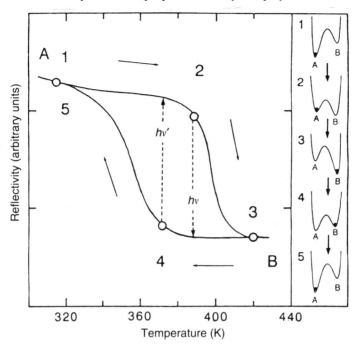

FIGURE 12. Temperature dependence of the reflectivity of C_4UC_3 crystals at 1.95 eV (left-hand side) and schematic diagrams of free energy (right-hand side)[364]. Reproduced by permission of the American Physical Society

distinct spectral peaks corresponding to the polymer main chain structure cannot be observed in the region up to 10 eV, except the phonon side bands. The origins of the main peaks in the spectra are attributed to the ^1B_u state excitons, the lowest allowed excited state excitons, based on results of photoconductivity measurements[365]. The linear optical properties of polydiacetylene single chains are discussed in terms of the effective conjugation lengths for poly(3BCMU) and poly(4BCMU) solutions colored blue ($E_0 = 1.6 \times 10^4$ cm^{-1}), red ($E_0 = 1.9 \times 10^4$ cm^{-1}) and yellow ($E_0 = 2.1 \times 10^4$ cm^{-1})[366]. In the exciton absorption region, polydiacetylenes in the red phase show luminescence spectra with the mirror-image of the exciton absorption curve, whereas the blue phase does not show observable luminescence. The luminescence spectra are preferentially polarized along to the polymer main chains. It can be concluded that the excitons in the blue phase decay to a self-trapped state before undergoing radiative recombination, whereas the excitons in the red phase undergo direct radiative recombination[367].

(ii) Nonlinear optical properties. Polydiacetylenes with fully conjugated and extended structures have been regarded as one of the most promising nonlinear optical materials with extra large susceptibilities[368], short response time[369,370] and high optical damage threshold[371,372,381]. As for the second-order nonlinear optical properties, many unsymmetric diacetylenes synthesized with expectation of the second harmonic generation (SHG) have failed to give satisfactory results[371–380]. This may be due to the polymer crystal structure with centrosymmetry or to the negligible extent of electronic coupling between the substituents and the polymer backbone. High potentiality of the extremely large third-

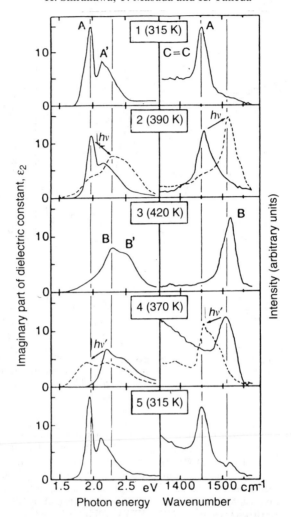

FIGURE 13. Spectra of the imaginary part of the dielectric
constant (ε_2) for the exciton absorption (left part) and Raman
spectra of the C=C stretching mode (right part) at various
temperatures[364]. Reproduced by permission of the American
Physical Society

order optical nonlinearities of conjugated polymers was demonstrated for the first time by
the $\chi^{(3)}$ measurement of polydiacetylenes[381]. This group[381] observed a 600-fold increase in
$\chi^{(3)}(3\omega)$ intensities of TCDU crystals during the conversion from monomer to polymer,
accompanied by the generation of a polarized harmonic field along the polymer chain axis.
The $\chi^{(3)}(3\omega)$ component along the poly(PTS) chain was measured to be (8.5 ± 0.5)
$\times 10^{-10}$ esu at 1.89 μm, which is comparative to the value of $(4.0 \pm 2.0) \times 10^{-10}$ esu at 10.6
μm for germanium. The values of n_2 related to the refractive index variation Δn to the light

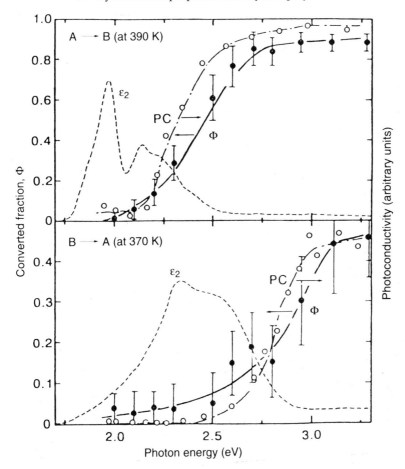

FIGURE 14. Dependence of the converted fraction (Φ) on the exciting photon energy (closed circles) for the phase transitions. Open circles and dashed lines show the action spectra of photoconductivity[364]. Reproduced by permission of the American Physical Society

intensity I ($\Delta n = n_2 I$) were estimated at 1.8×10^{-6} MW^{-1}cm^2 for poly(PTS) and 3×10^{-8} MW^{-1} cm^2 for CS$_2$. With a 1.06 μm, 1 GW/cm^2 pump, the gain for four-wave parametric amplification in the near-infrared in poly(PTS) was about ten times larger than the gain for the three-wave parametric amplification in LiNbO$_3$. The measurement of nonresonant third-order nonlinear optical susceptibility $\chi^{(3)}(\omega)$ proportional to the intensity-dependent index of reflection was made on poly(PTS) at 1.9 μm to obtain the largest measured nonresonant $\chi^{(3)}$ value of 3×10^{-9} esu[382]. Since the first proposals for the use of conjugated polymers in nonlinear optics were made, extensive studies have been continuing on various polydiacetylenes; however, no consistent picture of the nonlinear optical properties of polydiacetylenes in terms of excitonic structures has as yet been established. For example, there are the arguments whether the position of the dominant two-photon accessible state responsible for most of the two-photon absorption oscillator strength lies below[383] or above[384,385] the energy gap (E_g). Recently, Hasegawa and coworkers investigated the optical nonlinearities of oriented poly(C_4UC_4) thin films in the blue and red phases by means of third-order harmonic generation (THG) and electromodulation spectra measurements[386,387]. Based on the results from the third-order optical susceptibility spectra $|\chi^{(3)}(-3\omega;\omega,\omega,\omega)|$ indicating prominent three-photon resonance at the respective excitons (Figure 15) and the electro-modulation spectra Im $\chi^{(3)}(-\omega;0,0,\omega)$ together with the energy derivatives of Im$\chi^{(1)}$ spectra (Figure 16), the microscopic description of the nonlinear optical processes for poly(C_4UC_4) in both phases were expressed by the three-level models

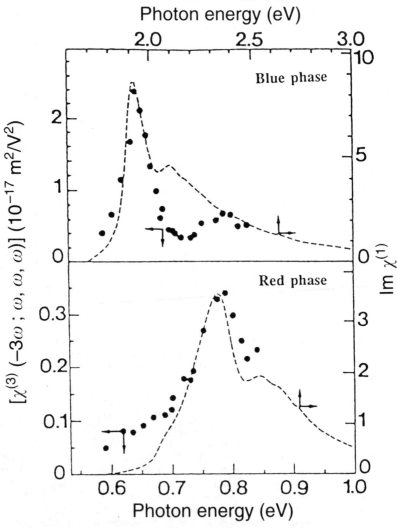

FIGURE 15. Third-order nonlinear optical susceptibility spectra Im$[\chi^{(3)}(-3\omega;\omega,\omega,\omega)]$ in comparison with the exciton absorption spectra $[Im\chi^{(1)}]^{387}$. Reproduced by permission of Elsevier Sequoia

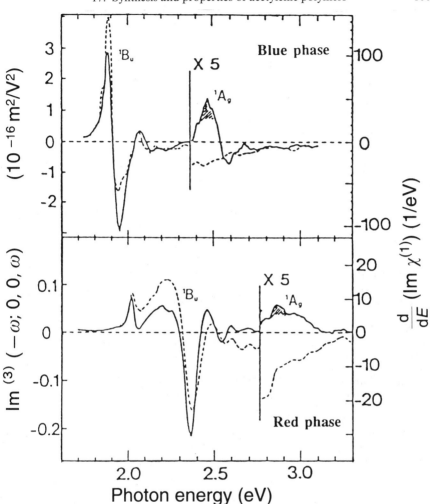

FIGURE 16. Third-order nonlinear optical susceptibility spectra $[I_m\chi^{(3)}(-\omega;0,0,\omega)]$ in comparison with the energy derivative absorption spectra[387]. Reproduced by permission of Elsevier Sequoia

given in Figure 17. The difference of optical nonlinearities between the two phases were explained by the important contribution of the forbidden (two-photon-allowed) ^1A_g excitons lying above ^1B_u excitons lying above ^1B_u excitons, in addition to the dominant role of ^1B_u excitons. The experimentally observed exciton specra were in good agreement with the results calculated by Abe[388,389] using a simple tight-binding model with long-range electron–electron interactions.

The temporal response of the third-order nonlinear optical susceptibility in poly(PTS) crystal was determined by time-resolved degenerate four-wave mixing (DFWM) by Carter[390] and coworkers. The excited lifetime T_1 ws measured to be 1.8 ± 0.5 ps at 652 nm in the resonant region. The values of phase relxation time T_2 measures by Hattori and Kobayashi[391] for a cast film of poly(3BCMU) were 30 and 90 fs for the respective incident

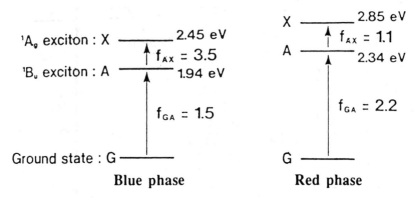

FIGURE 17. Three-level models for poly(C_4UC_4) in blue and red phases[387]. Reproduced by permission of Elsevier Sequoia

light wavelength of 648 and 582 nm. The accumulated data on the fundamental characteristics of polydiacetylenes lead several groups to the expectation of optical device applications such as ultrafast optical switching[392-395], the study of which is now in progress.

   e. Electrical properties.   In spite of the metallic brilliance of polymer single crystals, the measured conductivities of as-polymerized polydiacetylenes are less than 10^{-10} S cm^{-1}, due to the finite energy gaps between the highest filled and the lowest empty molecular orbitals that originated from the limited bond-alternation in the polymer chains[396,397]. The difficulty in doping polydiacetylene single crystals is related to the crystal rigidity, for the conductivities of less rigid states in LB-film or cast film were improved up to 10^{-6} S cm^{-1} by I_2 doping[398,399]. Nakanishi and coworkers[400] developed a doping method to realize conductivities of 10^{-4} S cm^{-1} for poly(PTS) single crystals by polymerizing monomer crystals grown in the presence of dopants such as SbF$_5$, AgClO$_4$ and FeCl$_3$. γ-ray irradiation of PTS crystals in SO$_3$ atmosphere is another method to achieve conductivities of 10^{-4} S cm^{-1}.

   The possibility of ultrahigh electron mobility on polydiacetylene chains was discussed by Wilson[401]. The carrier velocity was estimated to be 2.2×10^3 m s^{-1}, which was greater than the mobility of any conventional semiconductor. The results were explained in relation to a solitary wave acoustic polaron characteristic of a one dimensional π-electron system.

   f. Magnetic properties.   Ovchinnikov's group presented sensational reports on polydiacetylenes from 1,4-bis-(2,2,6,6,-tetramethyl-4-oxy-4-piperidyl-1-oxyl)butadiyne (BIPO) (**55**), which exhibited ferromagnetic properties ($M_c = 0.3$ G/g, $H_c = 100$ Oe) at 273 K[402,271]. Mechanisms of 1,4- and 1,2-addition reactions were proposed, taking account of the polymer structures capable of parallel spin orientation and of the drastic polymerization conditions[271]. The Curie temperature of the polymer was estimated to be 150–190 °C, however the spontaneous magnetization of the products was only 0.1% of the theoretical value for one-dimensional organic ferromagnets. The experimental results were hard to reproduce[403] so that these remain many unresolved question.

   g. Mechanical properties.   Polydiacetylenes have been regarded as ideal models for studing the mechanical properties of polymer chains, for only solid state polymerization of conjugated diynes can afford polymer single crystals composed of highly oriented polymer chains with extended and planar structures. Very large mechanical strength were measured for the polydiacetylene single crystals. For example, the fiber-like poly(TCDU) (**51**) single crystals **404** were deformed elastically by strains of over 3%. Young's modulus in the chain direction was measured to be 45 GPa for poly(TCDU) and poly(DCH), 62 GPa for poly(ETCD)[405], which agreed well with a Young's modulus of 50 GPa for poly(PTS) estimated from the force constants measured by Raman spectroscopy[406]. Creep or time-dependent deformation could not be detected during the deformation of the crystals in tension parallel to the chain direction[405], which was a remarkable and characteristic mechanical property of polydiacetylenes. The deformation of poly(TCDU) crystals on compression parallel to the chain direction provided the first evidence for chain twinning in polymer crystals[406–409]. Considering the excellent mechanical properties of polydiacetylene single crystals, reinforcement with polydiacetyles is expected to produce promising composites with high stiffness and strength, low creep, good thermal stability and low density[410].

V. REFERENCES

1. J. A. Nieuwland and R. R. Vogt, in *The Chemistry of Acetylene*, Reinhold, 1945, p. 138.
2. G. Natta, G, Mazzanti and R. Corradini, *Atti. Acad. Naz. Lincei, Rend. Classe sri. fis. mat. e nat.* [8] **25**, 3 (1958).
3. T. Ito, H. Shirakawa and S. Ikeda, *J. Polym. Sci., Polym. Chem. Ed.*, **12**, 11 (1974).
4. H. Shirakawa, E. J. Louis, A. G. MacDiarmid, C. K. Chiang and A. J. Heeger, *J. Chem. Soc., Chem. Commun.*, 578 (1977).
5. J. H. Edwards and W. J. Feast, *Polym. Commun.*, **21**, 595 (1980).
6. T. M. Swager, D. A. Dougherty and R. H. Grubbs, *J. Am. Chem. Soc.*, **110**, 2973 (1988).
7. Y.V. Korshak, V. V. Korshak, G. Kanischka and H. Höcker, *Makromol. Chem., Rapid Commun.*, **6**, 685 (1985).
8. F. L. Klavetter and R. H. Grubbs, *J. Am. Chem. Soc.*, **110**, 7807 (1988).
9. C. S. Marvel, J. H. Sample and M. F. Roy, *J. Am. Chem. Soc.*, **61**, 3241 (1939).
10. C. A. Finch, in *Polyvinyl Alcohol*, Wiley, New York, 1973, p. 477.
11. K. Maruyama, M. Take, N. Fujii and Y. Tanizaki, *Bull. Chem. Soc. Jpn.*, **59**, 13 (1986).
12. M. G. Chauser, Yu. M. Rodionov, V. M. Misin and M. I. Cherkashin, *Russ. Chem. Rev. (Engl. Transl.)*, **45**, 348 (1976); *Usp. Khim.*, **45**, 695 (1976).
13. C. I. Simionescu and V. Percec, *Prog. Polym. Sci.*, **8**, 133 (1982).
14. H. W. Gibson and J. M. Porchan, in *Encyclopedia of Polymer Science and Engineering* (Ed. J. I. Kvoschwitz), 2nd ed., Vol. 1, Wiley, New York, 1984, p. 87.
15. T. Masuda and T. Higashimura, *Acc. Chem. Res.*, **17**, 51 (1984).
16. T. Masuda and T. Higashimura, *Adv. Polym. Sci.*, **81**, 121 (1986).
17. H. W. Gibson, in *Handbook of Conducting Polymers* (Ed. T. A. Skotheim), Chap. 11, M. Dekker, New York, 1986.
18. A. A. Matnishyan, *Russ. Chem. Rev. (Engl. Transl.)*, **57**, 367 (1988); *Usp. Khim.*, **57**, 656 (1988).
19. G. Costa, in *Comprehensive Polymer Science* (Ed. G. Allen). Vol. 4, Chap. 9, Pergamon, Oxford, 1989.
20. T. Masuda and T. Higashimura, *ACS Adv. Chem. Ser.*, NO. 224, Chap. 35, 1990.
21. G. Wegner, *Z. Naturforsch.*, **24b**, 824 (1969).
22. H.-J. Cantow (Ed.), *Polydiacetylenes*, Springer-Verlag, Berlin, 1984.
23. D. Bloor and R. R. Chance (Eds.), *Polydiacetylenes*, NATO ASI Series, Dodrecht, 1985.
    (a) W. D. Huntsman, in *The Chemistry of the Carbon–Carbon Triple Bond*, Part 2 (Ed. S. Patai), Wiley, Chichester, 1978, Ch. 13.
    (b) W. D. Huntsman, in *The Chemistry of the Carbon–Carbon Triple Bond*, Supplement B, Part 2 (Eds S. Patai and Z. Rappoport), Wiley, Chichester, 1983, Ch.22.
24. L. B. Luttinger, *Chem. Ind. (London)*, 1135 (1960).
25. M. Monkenbusch, B. S. Morra and G. Wegner, *Makromol. Chem., Rapid Commun.*, **3**, 69 (1982).
26. L. Terlemezyan and M. Mihailov, *Makromol. Chem., Rapid Commun.*, **3**, 613 (1982).
27. M. Aldissi, C. Linaya, J. Sledz, F. Schué, L. Giral, J. M. Fabre and M. Rolland, *Polymer*, **23**, 243 (1982).

28. N. Theophilou, A. Munardi, R. Azna, J. Sledz, F. Schué and H. Naarmann, *Eur. Polym. J.*, **23**, 15 (1987).
29. A. J. Amass, M. S. Beevers, T. R. Farren and J. A. Stowell, *Makromol. Chem., Rapid Commun.*, **8**, 119 (1987).
30. K. Aoki, Y. Kakudate, S. Usuba, M. Yoshida, K. Tanaka and S. Fujiwara, *Solid State Commun.*, **64**, 1329 (1987).
31. A. Aoki, Y. Kakudate, S. Usuba, M. Yoshida, K. Tanaka and S. Fujiwara, *Synth. Met.*, **28**, D91 (1989).
32. S.-A. Chen and H.-J. Shy, *J. Polym. Sci., Polym. Chem. Ed.*, **23**, 2441 (1981).
33. V. Rives-Arnau and N. Sheppard, *J. Chem. Soc., Faraday Trans. 1*, **76**, 394 (1980).
34. N. Kurokawa, M. Tabata and J. Soma, *J. Polym. Sci., Polym. Lett. Ed.*, **19**, 355 (1981).
35. K. Soga, Y. Kobayashi, S. Ibeda and S. Kawakami, *J. Chem. Soc., Chem. Commun.*, 931 (1980).
36. H. Shirakawa and S. Ikeda, *J. Polym. Sci., Polym. Chem. Ed.*, **12**, 929 (1974).
37. W. H. Watson, Jr., W. C. McMordie, Jr. and L. G. Lands, *J. Polym. Sci.*, **55**, 137 (1961).
38. M. Hatano, S. Kambara and S. Okamoto, *J. Polym. Sci.*, **51**, s26 (1961).
39. S. Kambara, M. Hatano and T. Hosoe, *Kogyo Kagaku Zasshi*, **65**, 720 (1962).
40. E. Angelescu and I. V. Nicolescu, *J. Polym. Sci. Part C*, **22**, 203 (1968).
41. H. Shirakawa and S. Ikeda, *Polym. J.*, **2**, 231 (1971).
42. Z.-Q. Shen, M.-J. Yang, M.-X. Shi and Y.-P. Cai, *J. Polym. Sci., Polym. Lett. Ed.*, **20**, 411 (1982).
43. M. Catellani, S. Destri, A. Bolognesi and E. Albizzati, *Makromol. Chem.*, **187**, 1345 (1986).
44. I. Kmínek and J. Trekoval, *Makromol. Chem., Rapid Commun.*, **7**, 53 (1986).
45. A. Munardi, N. Theophilou, R. Aznar, J. Sledz, F. Schué and H. Naarmann, *Makromol. Chem.*, **188**, 395 (1987).
46. M. Soga, S. Hotta and N. Sonoda, *Synth. Met.*, **30**, 251 (1989).
47. S. Y. Oh, K. Akagi and H. Shirakawa, *Synth. Met.*, **32**, 245 (1989).
48. H. Shirakawa and S. Ikeda, *Synth. Met.*, **1**, 175 (1979/80).
49. H. Naarmann and N. Theophilou, *Synth. Met.*, **22**, 1 (1987).
50. K. Akagi, M. Suezaki, H. Shirakawa, H. Kyotani, M. Shimomura and Y. Tanabe, *Synth. Met.*, **28**, D1 (1989).
51. J. Tsukamoto, A. Takashashi and K. Kawasaki, *Jpn. J. Appl. Phys.*, **29**, 125 (1990).
52. K. Akagi, K. Sakamaki and H. Shirakawa, *Macromolecules*, **25**, 6725 (1993).
53. S. L. Hsu, A. J. Signorelli, G. P. Pez and T. H. Baughman, *J. Chem. Phys.*, **69**, 106 (1978).
54. G. P. Pez, *J. Am. Chem. Soc.*, **98**, 8072 (1976).
55. H. G. Alt, H. E. Engelhardt, M. D. Rusch and L. B. Kool, *J. Am. Chem. Soc.*, **107**, 3717 (1985).
56. H. G. Alt, H. E. Engelhardt, M. D. Rausch and L. B. Kool, *J. Organomet. Chem.*, **329**, 61 (1987).
57. J. R. Martinez, M. D. Rausch, J. C. W. Chien and H. G. Alt, *Makromol. Chem.*, **190**, 1309 (1989).
58. W. E. Daniels, *J. Org. Chem.*, **29**, 2936 (1964).
59. F. Cataldo, *Polymer*, **33**, 3073 (1992).
60. M. G. Voronkov, V. B. Pukhnarevich, S. P. Sushchinskaya, V. Z. Annenkova, V. M. Annenkova and N. J. Andreeva, *J. Polym. Sci., Polym. Lett. Ed.*, **18**, 53 (1980).
61. K. Aoki, S. Usuba, M. Yoshida, Y. Kakudate, K. Tanaka and S. Fujiwara, *J. Chem. Phys.*, **89**, 529 (1988).
62. J. Heaviside, P. J. Hendra, P. Tsai and R. P. Cooney, *J. Chem. Soc., Faraday Trans. 1*, **74**, 2542 (1980).
63. M. Aldissi and R. Liepins, *J. Chem. Soc., Chem. Commun.*, 255 (1984).
64. J. H. Edwards, W. J. Feast and D. C. Bott, *Polymer*, **25**, 395 (1984).
65. T. J. Katz, E. J. Wang and N. Acton, *J. Am. Chem. Soc.*, **93**, 3783 (1971).
66. T. J. Katz J. Roth, N. Acton and E. Carnahan, *Org. Synth.*, **53**, 157 (1973).
67. C. J. Schaverien, J. C. Dewan and R. R. Schrock, *J. Am. Chem. Soc.*, **108**, 2771 (1986).
68. R. R. Schrock, J. Feldman, L. F. Cannizzo and R. H. Grubbs, *Macromolecules*, **20**, 1169 (1987).
69. J. Kress, A. Aguero and J. A. Osborn, *J. Mol. Catal.*, **36**, 1 (1986).
70. T. Woerner, A. G. MacDiarmid and A. J. Heeger, *J. Polym. Chem., Polym. Lett. Ed.*, **20**, 305 (1982).
71. T. Woerner, A. G. MacDiarmid A. Feldblum and A. J. Heeger, *J. Polym. Chem., Polym. Lett. Ed.*, **22**, 119 (1984).

72. M. Ozaki, Y. Ikeda and T. Akakawa, *J. Polym. Chem., Polym. Lett. Ed.*, **21**, 989 (1983).
73. Y. Yamashita, S. Nishimura, K. Shimamura and K. Monobe, *Makromol. Chem.*, **187**, 1757 (1986).
74. W. H. Meyer, *Synth. Met.*, **4**, 81 (1981).
75. W. H. Meyer, *Mol. Cryst. Liq. Cryst.*, **77**, 137 (1981).
76. K. Araya, A. Mukoh, T. Narahara and H. Shirakawa, *Chem. Lett.*, 1141 (1984).
77. K. Araya, A. Mukoh, T. Narahara and H. Shirakawa, *Synth. Met.*, **14**, 199 (1984).
78. K. Araya, A. Mukoh, T. Narahara, K. Akagi and H. Shirakawa, *Synth. Met.*, **17**, 247 (1984).
79. M. Aldissi, *J. Polym. Chem., Polym. Lett. Ed.*, **23**, 167 (1985).
80. K. Akagi, M. Ito, S. Katayama, H. Shirakawa and K. Araya, *Mol. Cryst. Liq. Cryst.*, **172**, 115 (1989).
81. K. Akagi, S. Katayama, H. Shirakawa, K. Araya, A. Mukoh and T. Narahara, *Synth. Met.*, **17**, 241 (1987).
82. H. Shirakawa, K. Akagi, S. Katayama, K. Araya, A. Mukoh and T. Narahara, *J. Macromol Sci.-Chem.*, **A25**, 643 (1988).
83. A. Montaner, M. Rolland, J. L. Sauvajol, M. Galtier, R. Almairac and J. L. Ribet, *Polymer*, **29**, 1101 (1988).
84. K. Akagi, S. Katayama, M. Ito, H. Shirakawa and K. Araya, *Synth. Met.*, **28**, D51 (1989).
85. S. Ikeda and A. Tamaki, *J. Polym. Sci. B*, **4**, 605 (1966).
86. S. Ikeda and A. Tamaki, *Int. Symp. Macromol. Chem., Tokyo and Kyoto*, I-124 (1966).
87. S. Ikeda, *Kogyo Kagaku Zasshi*, **70**, 1880 (1967).
88. T. Ito, H. Shirakawa and S. Ideka, *Kubunshi Ronbunshu*, **33**, 339 (1976).
89. K. Fukui and S. Inagaki, *J. Am. Chem. Soc.*, **97**, 4445 (1975).
90. T. Ito, H. Shirakawa and S. Ikeda, *J. Polym. Sci., Polym. Chem. Ed.*, **13**, 1943 (1975).
91. H. Shirakawa, Y.-X. Zhang and K. Akagi, in *Conjugated Polymers and Related Materials: The interconnection of Chemical and Electronic Structure, Proceedings of the Eighty-first Nobel Symposium* (Eds. W. R. Salaneck *et al.*) Chap. 5, Oxford University Press, 1993, p. 65.
92. M. A. Schen, F. E. Karasz and J. C. W. Chien, *J. Polym. Sci., Polym. Chem. Ed.*, **21**, 2787 (1983).
93. H. Shirakawa, A. Hamano, S. Kawakami, M. Sato, K. Soga and S. Ikeda, *Z. Phys. Chem. Neue Folge*, **120**, 235 (1980).
94. K. Soga, S. Kawakami, H. Shirakawa and S. Ikeda, *Makromol. Chem., Rapid Commun.*, **1**, 523 (1980).
95. K. Soga, S. Kawakami, H. Shirakawa and S. Ikeda, *Makromol. Chem., Rapid Commun.*, **1**, 643 (1980).
96. G. Lieser, G. Wegner, W. Müller and V. Enkelmann, *Makromol. Chem., Rapid Commun.*, **1**, 621 (1980).
97. V. Enkelmann, G. Lieser, W. Müller and G. Wegner, *Chem. Scr.*, **17**, 141 (1981).
98. V. Enkelmann, G. Lieser, M. Monkenbusch, W. Müller and G. Wegner, *Mol. Cryst. Liq. Cryst.*, **77**, 111 (1981).
99. J. C. W. Chien, J. D. Capistran, F. E. Karasz, L. C. Dickinson and M. A. Schen, *J. Polym. Sci., Polym. Lett. Ed.*, **21**, 93 (1983).
100. J. C. W. Chien, F. E. Karasz, M. A. Schen and J. A. Hirsch, *Macromolecules*, **16**, 1694 (1983).
101. J. Tsukamoto, *Adv. Phys.*, **41**, 509 (1992).
102. J. L. Brédas, R. R. Chance and R. H. Baughman, *J. Chem. Phys.*, **76**, 3673 (1982).
103. W. R. Salaneck, H. R. Thomas, C. B. Duke, A. Paton, E. W. Plummer, A. J. Heeger and A. G. MacDiarmid, *J. Chem. Phys.*, **71**, 2044 (1979).
104. CRC *Handbook of Chemistry and Physics*, 67th ed. (Ed. R. C. Weast), CRC Press, Florida, 1986, p. E-83.
105. I. Harada, M. Tasumi, H. Shirakawa and S. Ikeda, *Chem. Lett.*, 1411 (1978).
106. S. Lefrant, L. S. Lichtmann, H. Temkin, D. C. Fitchen, D. C. Miller, G. E. Whitwell and J. M. Burlitch, *Solid State Commun.*, **29**, 191 (1979).
107. R. H. Baughman, S. L. Hsu, G. P. Pez and A. J. Signorelli, *J. Chem. Phys.*, **68**, 5405 (1978).
108. J. C. W. Chien, F. E. Karasz and K. Shimamura, *Macromolecules*, **15**, 1012 (1982).
109. N. S. Murthy, G. G. Miller and R. H. Baughman, *J. Chem. Phys.*, **89**, 2523 (1988).
110. G. Perego, G. Lugli, U. Pedretti and G. Allegra, *Makromol. Chem.*, **189**, 2787 (1988).
111. T. Matsuyama, H. Sakai, H. Yamaoka, Y. Maeda and H. Shirakawa, *Solid, State Commun.*, **40**, 563 (1981).

1010   H. Shirakawa, T. Masuda and K. Takeda

112. G. Kaindl, G. Wortmann, S. Roth and K. Menke, *Solid State Commun.*, **41**, 75 (1982).
113. T. Matsuyama, H. Sakai, H. Yamaoka, Y. Maeda and H. Shirakawa, *J. Phys. Soc. Jpn.*, **52**, 2238 (1983).
114. H. Sakai, Y. Maeda, T. Kobayashi and H. Shirakawa, *Bull. Chem. Soc. Jpn.*, **56**, 1616 (1983).
115. Z. Kucharski, M. Lukasiak, J. Suwalski and A. Pron, *J. Phys. Paris*, **C3**, 321 (1983).
116. H. Kuroda, I. Ikemoto, K. Asakura, H. Ishii, H. Shirakawa, T. Kobayashi, H. Oyanagi and T. Matsushita, *Solid State Commun.*, **46**, 235 (1983).
117. K. Asakura, I. Ikemoto, H. Kuroda, T. Kobayashi and H. Shirakawa, *Bull. Chem. Soc. Jpn.*, **58**, 2113 (1985).
118. T. C. Clarke, R. H. Geiss, W. D. Gill, P. M. Grant, H. Morawitz and G. B. Street, *Synth. Met.*, **1**, 21 (1979/80).
119. P. J. Nigrey, A. G. MacDiarmid and A. J. Heeger, *Mol. Cryst. Liq. Cryst.*, **83**, 309 (1982).
120. K. Fukui and H. Fujimoto, *Bull. Chem. Soc. Jpn.*, **39**, 2116 (1966).
121. K. Fukui, *Tetrahedron Lett.*, 2427 (1965).
122. N. T. Anh, *J. Chem. Soc., Chem. Commun.*, 1089 (1968).
123. K. Akagi, T. Kadokura and H. Shirakawa, *Int. J. Quantum Chem., Quantum Chem. Symp.*, **24**, 41 (1990).
124. K. Akagi, T. Kadokura and H. Shirakawa, *J. Mol. Struct. (Theochem)*, **232**, 211 (1991).
125. K. Akagi, T. Kadokura and H. Shirakawa, *Polymer*, **33**, 4058 (1992).
126. H. Shirakawa, T. Ito and S. Ikeda, *Makromol. Chem.*, **179**, 1565 (1978).
127. Y.-W. Park, A. J. Heeger, M. A. Druy and A. G. MacDiarmid, *J. Chem. Phys.*, **73**, 946 (1980).
128. C. K. Chiang, C. R. Fincher, Jr., Y. W. Park, A. J. Heeger, H. Shirakawa, E. J. Louis, S. C. Gau and A. G. MacDiarmid, *Phys. Rev. Lett.*, **39**, 1098 (1977).
129. C. K. Chiang, M. A. Druy, S. C. Gau, A. J. Heeger, E. J. Louis, A. G. MacDiarmid, Y. W. Park and H. Shirakawa, *J. Am. Chem. Soc.*, **100**, 1013 (1978).
130. C. K. Chiang, Y. W. Park, A. J. Heeger, H. Shirakawa, E. J. Louis and A. G. MacDiarmid, *J. Chem. Phys.*, **69**, 5098 (1978).
131. Y. W. Park, M. A. Druy, C. K. Chiang, A. G. MacDiarmid, A. J. Heeger, H. Shirakawa and S. Ikeda, *J. Polym. Sci., Polym. Lett. Ed.*, **17**, 195 (1979).
132. H. Naarmann, *Synth. Met.*, **17**, 223 (1989).
133. S. Kuroda, M. Tokumoto, N. Kinoshita and H. Shirakawa, *J. Phys. Soc. Jpn.*, **51**, 693 (1982).
134. W.P. Su, J. R. Schrieffer and A. J. Heeger, *Phys. Rev. Lett.*, **42**, 1698 (1979).
135. M. J. Rice, *Phys. Lett.*, **71A**, 152 (1979).
136. S. Ikehata, J. Kaufer, T. Woerner, A. Pron, M. A. Druy, A. Sivak, A. J. Heeger and A. G. MacDiarmid, *Phys. Rev. Lett.*, **45**, 1123 (1980).
137. H. Shirakawa, T. Ito and S. Ikeda, *Polym. J.*, **4**, 460 (1973).
138. S. Etemad, T. Mitani, M. Ozaki, T. C. Chung, A. J. Heeger and A. G. MacDiarmid, *Solid State Commun.*, **40**, 75 (1981).
139. L. Lauchlan, S. Etemad, T. C. Cung, A. J. Heeger and A. G. MacDiarmid, *Phys. Rev.*, **B24**, 3701 (1981).
140. K. Yoshino, S. Hayashi, Y. Inuishi, H. Kato and Y. Watanabe, *Jpn. J. Appl.Phys.*, **21**, L653 (1982).
141. M. Sinclair, D. Moses, A. J. Heeger, K. Vilhelmsson, B. Vakl and M. Salour, *Solid State Commun.*, **61**, 221 (1987).
142. L. R. Treloar, *Polymer*, **1**, 95 (1960).
143. Y. Cao, P. Smith and A. J. Heeger, *Polymer*, **32**, 1210 (1991).
144. Y. Cao, P. Smith and A. J. Heeger, *Synth. Met.*, **41**, 181 (1991).
145. T. Masuda, K. Yamamoto, and T. Higashimura, *Polymer*, **23**, 1663 (1982).
146. T. Szymanska-Buzar, *J. Mol. Cat.*, **48**, 43 (1988).
147. T. J. Katz and S. J. Lee, *J. Am. Chem. Soc.*, **102**, 422 (1980).
148. D. J. Liaw, A. Soum, M. Fontanille, A. Parlier, and H. Rudler, *Makromol. Chem., Rapid Commun.*, 6, 309 (1985)
149. K. C. Wallace, A. H. Liu, W. M. Davis, and R. R. Schrock, *Organometallics*, **8**, 644 (1989).
150. H. H. Fox and R. R. Schrock, *Organometallics*, **11**, 2763 (1992).
151. K. Tamura, T. Masuda, and T. Higashimura, *Polym. Bull.*, **30**, 537 (1993).
152. B.-Z. Tang and N. Kotera, *Macromolecules*, **22**, 4388 (1989).
153. T. J. Katz, T. H. Ho, N.-Y. Shih, Y.-C. Ying, and V. I. W. Stuart, *J. Am. Chem. Soc.*, **106**, 2659 (1984).

154. F. A. Cotton, W. T. Hall, K. J. Kann, and F. J. Karol, *Macromolecules*, **14**, 233 (1981).
155. T. Masuda, T. Yoshimura, and T. Higashimura, *Macromolecules*, **22**, 3804 (1989).
156. T. Yoshimura, T. Masuda, and T. Higashimura, *Macromolecules*, **21**, 1899 (1988).
157. A. Furlani, C. Napoletano, M. V. Russo, A. Camus, and N. Marsich, *J. Polym. Sci., Part A, Polym. Chem.*, **27**, 75 (1989).
158. W. Yang, M. Tabata, S. Kobayashi, K. Yokota, and A. Shimizu, *Polymer* **23**,1135 (1991).
159. I. A. Akopyan, S. G. Grigoryan, G. A. Zhamkochyan, and S. G. Matsoyan, *Vysokomol. Soed.*, **A17**, 2517 (1975).
160. T. Masuda, M. Kawai, and T. Higashimura, *Polymer*, **23**, 744 (1982).
161. Y.-H. Kim, K.-Y. Choi, and S.-K. Choi, *J. Polym. Sci., Part C, Polym. Lett.*, **27**, 443 (1989).
162. T. Narita, T. Hagihara, H. Hamana, M. Yoshizawa and S. Nishimura, *Makromol. Chem., Rapid Commun.*, **13**, 189 (1992)
163. C. I. Simionescu and M. Grigoras, *J. Polym. Sci., Part A, Polym. Chem.* **27**, 3201 (1989).
164. S. Subramanyam, A. Blumstein, and K.-P. Li, *Macromolecules*, **25**, 2065 (1992).
165. S. Subramanyam and A. Blumstein, *Macromolecules*, **25**, 4058 (1992).
166. Y. Nakayama, K. Mashima, and A. Nakamura, *Macromolecules*, **26**, 6267 (1993).
167. T. Ciardelli, S. Lanzillo, and O. Pieroni, *Macromolecules*, **7**, 175 (1974).
168. T. Masuda, Y. Okano, Y. Kuwane and T. Higashimura, *Polym. J.*, **12**, 907 (1980).
169. S. Y. Oh, F. Oguri, K. Akagi and H. Shirakawa, *J. Polym. Sci., Part A, Polym. Chem.*, **31**, 781 (1993).
170. T. Masuda, K.-Q. Thieu, N. Sasaki, and T. Higashimura, *Macromolecules*, **9**, 661 (1976).
171. T. Masuda, K. Yamamoto and T. Higashimura, *Polymer*, **23**, 1663 (1982).
172. A. Furlani, C. Napoletano, M. V. Russo, and W. J. Feast, *Polym. Bull.*, **16**, 311 (1986).
173. M. Yamaguchi, M. Hirama, and H. Nishihara, *Chem.,Lett.*, 1667 (1992).
174. T. Masuda, H. Tajima, T. Yoshimura, and T. Higashimura, *Macromolecules*, **20**, 1467 (1987).
175. T. Masuda, K. Tsuchihara, K. Ohmameuda, and T. Higashimura, *Macromolecules*, **22**, 1036 (1989).
176. M. Yamaguchi, K. Torisu, K. Hiraki, and T. Minami, *Chem. Lett.*, 2221 (1990).
177. T.-L. Chang, L. J. Holzknecht, H. B. Mark, Jr., T. H. Ridgway and H. Zimmer, *J. Polym. Sci., Part A, Polym. Chem.*, **27**, 989 (1989).
178. T. Hosina, Y. Morimoto, M. Tabata, and K. Yokota, *Polymer Preprints, Japan*, **41**(2), 295 (1992).
179. Y.-S. Gal, H.-N. Cho, and S.-K. Choi, *J. Polym. Sci., Part A, Polym. Chem.*, **24**, 2021 (1986).
180. Y.-S. Gal, B. Jung, W.-C. Lee and S.-K. Choi, *J. Polym. Sci., Part A, Polym. Chem.*, **30**, 2657 (1992).
181. A. Furlani, R. Paolesse, M. V. Russo, A. Camus and N. Marsich, *Polymer*, **28**, 1221 (1987).
182. C. Carlini and J. C. W. Chien, *J. Polym. Sci., Polym. Chem. Ed.*, **22**, 2749 (1984).
183. K. Tsuchihara, T. Masuda and T. Higashimura, *Polym. Bull.*, **20**, 343 (1988).
184. T. Okano, K. Ito, K. Kodaira, K. Hosokawa, M. Nishida, T. Ueda and H. Muramatsu, *J. Fluorine Chem.*, **38**, 139 (1988).
185. J. Kunzler and V. Percec, *J. Polym. Sci., Part A, Polym. Chem.*, **28**, 1043 (1990).
186. T. Yoshida, Y. Abe, T. Masuda and T. Higashimura, *Polymer Preprints, Japan*, **37**(2), 144 (1988).
187. T. Masuda, T. Hamano, K. Tsuchihara and T. Higashimura, *Macromolecules*, **23**,1374 (1990).
188. T. Mizumoto, T. Masuda and T. Higashimura, *J. Polym. Sci., Part A, Polym. Chem.*, **31**, 2555 (1993).
189. H. Muramatsu, T. Ueda and K. Ito, *Macromolecules*, **18**, 1634 (1985).
190. T. Masuda, T. Hamano, T. Higashimura, T. Ueda and H. Muramatsu, *Macromolecules*, **21**, 281 (1988).
191. Y. Hayakawa, M. Nishida, T. Aoki and H. Muramatsu, *J. Polym. Sci., Part A, Polym. Chem.*, **30**, 873 (1992).
192. T. Yoshimura, T. Masuda, T. Higashimura, K. Okuhara and T. Ueda,*Macromolecules*, **24**, 6053 (1991).
193. W. J. Trepka and R. J. Sonnenfeld, *J. Polym. Sci., Part A-1*, **8**, 2721 (1970).
194. R. J. Kern, *J. Polym. Sci., Part A-1*, **7**, 621 (1969).
195. Y. Okano, T. Masuda, and T. Higashimura, *J. Polym. Sci., Polym. Chem. Ed.*, **2**, 1603 (1984).
196. S. Tanaka, K. Okuhara, and K. Kaeriyama, *Makromol. Chem.*, **187**, 2793 (1986).
197. Y. Abe, T. Masuda and T. Higashimura, *J. Polym. Sci., Part A, Polym. Chem.*, **27**, 4267 (189).
198. T. Higashimura, Y.-X. Deng and T. Masuda, *Macromolecules*, **15**, 234 (1982).

199. T. Masuda, T. Takahashi, A. Niki and T. Higashimura, *J. Polym. Sci., Polym. Chem. Ed.*, **24**, 809 (1986).
200. T. Masuda, A. Niki, E. Isobe and T. Higashimura, *Macromolecules*, **18**, 2109 (1985).
201. A. Niki, T. Masuda and T. Higashimura, *J. Polym. Sci., Part A, Polym. Chem.*, **25**, 1553 (1987).
202. T. Masuda, E. Isobe and T. Higashimura, *Macromolecules*, **18**, 841 (1985).
203. T. Masuda, E. Isobe T. Hamano and T. Higashimura, *Macromolecules*, **19**, 2448 (1986).
204. T. Masuda, E. Isobe T. Hamano and T. Higashimura, *J. Polym. Sci., Part A, Polym. Chem.*, **25**, 1353 (1987).
205. Air Products, US Patent 4759776 (1988).
206. T. Masuda, T. Matsumoto, T. Yoshimura and T. Higashimura, *Macromolecules*, **23**, 4902 (1990).
207. T. Matsumoto, T. Masuda and T. Higashimura, *J. Polym. Sci., Part A, Polym. Chem.*, **29**, 295 (1991).
208. T. Masuda, T. Yoshimura, K. Tamura and T. Higashimura, *Macromolecules*, **20**, 1734 (1987).
209. T. Masuda, Y. Kuwane and T. Higashimura, *J. Polym. Sci., Polym. Chem. Ed.*, **20**, 1043 (1982).
210. T. Masuda, M. Yamagata and T. Higashimura, *Macromolecules*, **17**, 126 (1984).
211. M. Yamagata, T. Masuda and T. Higashimura, *J. Polym. Sci., Polym. Chem. Ed.*, **22**, 2275 (1984).
212. H. Kouzai, T. Masuda and T. Higashimura, *Polymer Preprints, Japan*, **42**(2), 275 (1993).
213. K. Tsuchihara, T. Masuda and T. Higashimura, *J. Am. Chem. Soc.*, **113**, 8548 (1991).
214. K. Tsuchihara, T. Masuda and T. Higashimura, *Macromolecules*, **25**, 5816 (1992).
215. K. Tsuchihara, T. Masuda and T. Higashimura, *J. Polym. Sci., Part A, Polym. Chem.*, **31**, 547 (1993).
216. M. Tabata, W. Yang, K. Yokota and Y. Nozaki, *Polymer Preprints, Japan*, **41**(6), 1986 (1992).
217. K. J. Ivin and B. D. Milligan, *Makromol. Chem. Rapid. Commun.*, **8**, 269 (1987).
218. T. Kaufmann, P. Fiegenbaum and R. Wieschollek, *Angew. Chem., Int. Ed. Engl.*, **23**, 531, 532 (1984).
219. F. Garnier, P. Krausz and H. Rudler, *J. Organomet. Chem.*, **186**, 77 (1980).
220. K. R. Birdwhistell, S. J. N. Burgmayer and J. L. Templeton, *J. Am. Chem. Soc.*, **105**, 7789 (1983).
221. S. J. Landon, P. M. Shulman and G. L. Geoffroy, *J. Am. Chem. Soc.*, **107**, 6739 (1985).
222. T. Masuda, N. Sasaki and T. Higashimura, *Macromolecules*, **8**, 717 (1975).
223. H. Makio, T. Masuda and T. Higashimura, *Polymer*, **34**, 1490 (1993).
224. H. Makio, T. Masuda and T. Higashimura, *Polymer*, **34**, 2218 (1993).
225. K. J. Ivin, *Olefin Metathesis*, Chap. 15, Academic Press, London, 1983, .
226. H. Kouzai, T. Masuda and T. Higashimura, *Macromolecules*, **25**, 7096 (1992).
227. H. Kouzai, T. Masuda and T. Higashimura, *J. Polym. Sci., Part A, Polym. Chem.*, **31**, 1887 (1993).
228. T. Mizumoto, T. Masuda and T. Higashimura *Polymer Preprints, Japan*, **41**(6), 1980 (1992).
229. T. Masuda, J. Fujimori, M. Z. Ab. Rahman and T. Higashimura, *Polym. J.*, **25**, 535 (1993).
230. T. Masuda, K. Mishima, J. Fujimori, M. Nishida, H. Muramatsu and T. Higashimura, *Macromolecules*, **25**, 1401 (1992).
231. T. Masuda, K. Mishima, H. Seki and T. Higashimura, *Poly. Bull.*, in press.
232. M. Nakano, T. Masuda and T. Higashimura, *Macromolecules*, **27**, No. 5 (1994).
233. J. Fujimori, T. Masuda and T. Higashimura, *Polym. Bull*, **20**, 1 (1988).
234. M. Tabata, W. Yang and K. Yokota, *Polym. J.*, **22**, 1105 (1990).
235. J. F. Kunzler and V. Percec, *Polym. Bull.*, **18**, 303 (1987).
236. Y. Okano, T. Masuda and T. Higashimura, *Polym. J.*, **14**, 477 (1982).
237. G. Costa, A. Grosso, M. C. Sacchi, P. C. Stein and L. Zetta, *Macromolecules*, **24**, 2858 (1991).
238. H. Izumikawa, T. Masuda and T. Higashimura, *Polym. Bull.*, **27**, 193 (1991).
239. T. Masuda, Y.-X. Deng and T. Higashimura, *Bull. Chem. Soc. Jpn.*, **56**, 2798 (1983).
240. T. Masuda, T. Takahashi and T. Higashimura, *Macromolecules*, **18**, 311 (1985).
241. T. Masuda, B.-Z. Tang, T. Higashimura and H. Yamaoka, *Macromolecules*, **18**, 2369 (1985).
242. J. A. Jackson, *J. Polym. Sci., Polym. Chem. Ed.*, **10**, 2935 (1972).
243. J. Sedlacek, J. Vohlidal and Z. Grubisic-Gallot, *Makromol. Chem., Rapid Commun.*, **14**, 51 (1993).
244. T. Masuda, B.-Z. Tang, A. Tanaka, and T. Higashimura, *Macromolecules*, **19**, 1459 (1986).
245. H. Seki, T. Masuda, B.-Z. Tang, A. Tanaka, and T. Higashimura, *Polymer*, in press.
246. H. Odani and T. Masuda, in *Polymers for Gas Separation* (Ed. N. Toshima), Chap. 4, VCH, New York, 1992.
247. T. Masuda, *Maku (Membrane)*, **13**, 195 (1988).
248. T. Masuda, Y. Iguchi, B.-Z. Tang and T. Higashimura, *Polymer*, **29**, 2041 (1988).

249. T. Masuda, E. Isobe, T. Higashimura and K. Takada, *J. Am. Chem. Soc.*, **105**, 7473 (1983).
250. K. Takada, H. Matsuya, T. Masuda and T. Higashimura, *J. Appl. Polym. Sci.*, **30**, 1605 (1985).
251. Y. Ichiraku, S. A. Stern and T. Nakagawa, *J. Membrane Sci.*, **34**, 5 (1987).
252. L. C. Witchey-Lakshmanan, H. B. Hopfenberg and R. T. Chern, *J. Membrane Sci.*, **48**, 321 (1990).
253. N. A. Plate, A. K. Bokarev, N. E. Kalieuzhnyi, E.G. Litvinova, V. S. Khotimskii, V. V. Volkov and Yu. P. Yampolskii, *J. Membrane Sci.*, **60**, 13 (1991).
254. E. Isobe, T. Masuda, T. Higashimura and A. Yamamoto, *J. Polym. Sci. Polym. Chem. Ed.*, **24**, 1839 (1986).
255. K. Tsuchihara, T. Oshita, T. Mausda and T. Higashimura, *Polym. J.*, **23**, 1273 (1991).
256. T. Higashimura, T. Masuda and M. Okada, *Polym. Bull.*, **10**, 114 (1983).
257. M. Langsam, M. Anand and E. J. Karwacki, *Gas Sep. Purifi.*, **2**, 162 (1988).
258. X. Lin, J. Xiao, Y.-L. Yu, J. Chen, G.-D. Zheng and J.-P. Xu, *J. Appl. Polym. Sci.*, **48**, 231 (1993).
259. Y. Nagase, T. Ueda, K. Matsui and M. Uchikura, *J. Polym. Sci., Part B, Polym. Phys.*, **29**, 171 (1991).
260. T. Masuda, B.-Z. Tang and T. Higashimura, *Polym. J.*, **18**, 565 (1986).
261. K. Ishihara, Y. Nagase and K. Matsui, *Makromol. Chem., Rapid Commun.*, **7**, 43 (1986).
262. T. Masuda, M. Takatsuka, B.-Z. Tang and T. Higashimura, *J. Membrane Sci.*, **49**, 69 (1990).
263. Y. Nagase, K. Ishihara and K. Matsui, *J. Polym. Sci., Part B, Polym. Phys.*, **28**, 377 (1990).
264. M. Langsam and L. M. Robeson, *ACS Polymer Preprints*, **32**(2), 87 (1991).
265. E. T. Kang, K. G. Neoh, T. Masuda, T. Higashimura, and M. Yamamoto, *Polymer*, **30**, 1328 (1989).
266. J.-W. Park, J.-H. Lee, H.-N. Cho and S.K. Choi, *Macromolecules*, **26**, 1191 (1993).
267. T. Fujisaka, M. Suezaki, T. Koremoto, T. Inoue, T. Masuda and T. Higashimura, *Polymer Preprints, Japan*, **38** (3), 797 (1989).
268. D. Neher, A. Wolf, C. Bubeck and G. Wegner, *Chem. Phys. Lett.*, **163**, 116 (1989).
269. D. Neher, A. Wolf, M. Leclerc, A. Kaltbeitzel, C. Bubeck and G. Wegner, *Synth. Met.*, **37**, 249 (1990).
270. J. Le. Moigne, A. Hilberer, and C. Strazielle, *Macromolecules*, **25**, 6705 (1992).
271. Yu. V. Korshak, T. V. Medvedeva, A. A. Ovchinnikov and V. N. Spector, *Nature*, **326**, 370 (1987).
272. H. Iwamura and S. Murata, *Mol. Cryst., Liq. Cryst., Lett. Sect.*, **176**, 33 (1989).
273. H. Nishide, N. Yoshioka, K. Inagaki, T. Kaku and E. Tsuchida, *Macromolecules*, **25**, 569 (1992).
274. N. Yoshioka, H. Nishide, T. Kaneko, H. Yoshiki and E. Tsuchida, *Macromolecules*, **25**, 3838 (1992).
275. Y. Miura, K. Inui, F. Yamaguchi, M. Inoue, Y. Teki, T. Takui and K. Itoh, *J. Polym. Sci., Part A, Polym. Chem.*, **30**, 959 (1992).
276. L. Dulog and S. Lutz, *Makromol. Chem., Rapid Commun.*, **14**, 147 (1993).
277. J. L. Moigne, A. Hilberer and F. Kajzar, *Makromol. Chem.*, **193**, 515, (1992).
278. S.-H. Jin, S.-J. Choi, W.-S. Ahn, H.-N. Cho, and S.-K. Choi, *Macromolecules*, **26**, 1487 (1993).
279. J. K. Stille and D. A. Frey, *J. Am. Chem. Soc.*, **83**, 1697 (1961).
280. H. W. Gibson, F. C. Bailey, A. J. Epstein, H. Rommelmann, S. Kaplan, J. Harbour, X. -Q. Yang, D. B. Tanner and J. M. Pochan, *J. Am. Chem. Soc.*, **105**, 4417 (1983)
281. Y.-H. Kim, Y.-S. Gal, U.-Y. Kim and S.-K. Choi, *Macromolecules*, **21** 1991 (1988).
282. Y.-H. Kim, K.-Y. Choi and S.-K. Choi, *J. Polym. Sci., Part C: Polym. Lett.*, **27**, 443 (1989).
283. O.-K. Cho, Y.-H. Kim, K.-Y. Choi and S.-K. Choi, *Macromolecules*, **23**, 12 (1990).
284. M.-S. Ryoo, W.-C. Lee and S.-K. Choi, *Macromolecules*, **23**, 3029 (1990).
285. M.-S. Jang, S.-K. Kwon and S.-K. Choi, *Macromolecules*, **23**, 4135 (1990).
286. S.-H. Han, U.-Y. Kim, Y.-S. Kang and S.-K. Choi, *Macromolecules*, **24**, 973 (1991).
287. S.-H. Jin, H.-N. Cho and S.-K. Choi, *J. Polym. Sci., Part A, Polym. Chem.*, **31**, 69 (1993).
288. S.-H. Jin, S.-H. Kim, H.-N. Cho and S.-K. Choi, *Macromolecules*, **24**, 6050 (1991).
289. A. S. Hay, *J. Org. Chem.*, **25**, 1275 (1960).
290. A. S. Hay, *J. Polym. Sci., Polym. Lett. Ed.*, **19**, 227 (1981).
291. S. Kuehling, H. Kuel, and H. Hoecker, *Macromolecules*, **23**, 4192 (1990).
292. G. Wegner, *Makromol. Chem.*, **134**, 219 (1970).
293. Y. Ozcayir, and A. Blumstein, *J. Polym. Sci., Part A: Polym. Chem. Ed.*, **24**, 1217 (1986).
294. A. L. Chalk and A. R. Gilbert, *J. Polym. Sci., Part A-1*, **10**, 2033 (1972).

295. W. Bracke, *J. Polym. Sci., Part A-1*, **10**, 2097 (1972).
296. P. M. Hergenrother, *J. Macromol., Sci.-Rev. Macromol. Chem.*, **C19**, 1 (1980).
297. H. Mukamal, F. W. Harris and J. K. Stille, *J. Polym. Sci., Part A-1*, **5**, 2721 (1967).
298. G. K. Noren and J. K. Stille, *J. Polym. Sci., Part D*, **5**, 395 (1971).
299. H. F. Vankerckhoven, Y. K. Gilliams and J. K. Stille, *Macromolecules*, **5**, 541 (1972).
300. A. S. Hay, *J. Polym. Sci., Part A-1*, **7**, 1625 (1969).
301. T. Tahara, K. Seto and S. Takahashi, *Polym. J.*, **19**, 301 (1987).
302. K. Sonogashira, S. Takahashi and N. Hagiwara, *Macromolecules*, **10**, 879 (1977).
303. S. Takahashi, M. Kariya, T. Yatake, K. Sonogashira and N. Hagiwara, *Macromolecules*, **11**, 1064 (1978).
304. N. Hagiwara, K. Sonogashira and S. Takahashi, *Adv. Polym. Sci.*, **41**, 149 (1981).
305. S. Takahashi, E. Murata, M. Kariya, K. Sonogashira and N. Hagiwara, *Macromolecules*, **12**, 1016 (1979).
306. K. Mitulla and M. Hanack, *Z. Naturforsch.*, **35b**, 1111 (1980).
307. E. Kobayashi, N. Metaka, S. Aoshima and J. Furukawa, *J. Polym. Sci., Part A, Polym. Chem.*, **30**, 227 (1992).
308. J. Ohshita, A. Matsuguchi, K. Furumori, R.-F. Hong, M. Ishikawa, T. Yamanaka, T. Koike and J. Shioya, *Macromolecules*, **25**, 2134 (1992).
309. Y. Chujo, I. Tomita and T. Saegusa, *Macromolecules*, **23**, 687 (1990).
310. T. Tsuda, K. Murata and Y. Kitaike, *J. Am. Chem. Soc.*, **114**, 1498 (1992).
311. T. Tsuda and K. Murata, *Macromolecules*, **25**, 6120 (1992).
312. G. Wegner, *Makromol. Chem.*, **145**, 85 (1971).
313. J. Kaiser, G. Wegner and E. W. Fisher, *Israel. J. Chem.*, **10**, 157 (1972).
314. G. Wegner, *Makromol. Chem.*, **154**, 35 (1972).
315. C. Glaser, *Chem. Ber.*, **2**, 522 (1869).
316. G. Eglinton and W. McCrae, *Adv. Org. Chem.*, **4**, 225 (1963).
317. W. Chodkiewicz and P. Cadiot, *Compt. Rend.*, **241**, 1055 (1955)
318. V. Enkelmann, R. J. Leyrer and G. Wegner, *Makromol. Chem.*, **180**, 1787 (1979).
319. M. Thakur and S. Meyler, *Macromolecules*, **18**, 2341 (1985).
320. B. Tieke, G. Wegner, B. Naegele and H. Ringsdorf, *Angew. Chem., Int. Ed. Engl.*, **15**, 764 (1976).
321. B. Tieke, H. J. Graf, G. Wegner, B. Naegele, H. Ringsdorf, S. Banerjiie, D. Day and J. B. Lando, *Colloid Polym. Sci.*, **225**, 521 (1977).
322. T. Kanetake, K. Ishikawa, T. Koda, Y. Tokura and K. Takeda, *Appl. Phys. Lett.*, **51**, 1957 (1987).
323. J. L. Chollet, F. Kajzar and J. Messies, *J. Chem. Phys.*, **88**, 6647 (1988).
324. K. Takeda and M. Hasegawa, *Synthetic Metals*, **18**, 413 (1987).
325. G. N. Patel, E. N. Duesler, D. Y. Curtin and I. C. Paul, *J. Am. Chem. Soc.*, **16**, 461 (1980).
326. R. H. Baughman, *J. Polym. Sci., Polym. Phys. Ed.*, **12**, 1511 (1974).
327. R. H. Baughman and K. C. Yee, *J. Polym. Sci., Macromol. Rev.*, **13**, 219 (1978).
328. R. H. Baughman and R. R. Chance, *Ann. Acad. Sci. N.Y.* **313**, 705 (1978).
329. V. Enkelmann, R. J. Leyrer, G. Schleier and G. Wegner, *J. Mater. Sci.*, **15**, 168 (1980).
330. D. Kobelt and E. F. Paulus, *Acta Crystallogr.*, **B30**, 232 (1974).
331. V. Enkelmann, *Acta Crystallogr.*, **B33**, 2842 (1977).
332. C. Bubeck, H. Sixl and H. C. Wolf, *Chem. Phys.*, **32**, 231 (1978).
333. C. Bubeck, W. Neumann and H. Sixl, *Chem. Phys.*, **48**, 269 (1980).
334. H. Sixl, W. Hersel and H. C. Wolf, *Chem. Phys. Lett.*, **53**, 39 (1978).
335. W. Hersel, H. Sixl and G. Wegner, *Chem. Phys. Lett.*, **73**, 288 (1980).
336. W. Neumann and H. Sixl, *Chem. Phys.*, **50**, 273 (1980).
337. W. Neumann and H. Sixl, *Chem. Phys.*, **58**, 303 (1981).
338. C. Bubeck, W. Hersel, H. Sixl and J. Waldmann, *J. Chem. Phys.*, **51**, 1 (1980).
339. R. A. Huber, M. Schwoerer, H. Benk and H. Sixl, *Chem. Phys. Lett.*, **78**, 416 (1981).
340. H. Gross and H. Sixl, *Mol. Chem. Phys. Lett.*, **91**, 262 (1982).
341. G. N. Patel, R. R. Chance and J. D. Witt, *J. Chem. Phys.*, **70**, 4387 (1979).
342. G. N. Patel and E. K. Walsh, *J. Polym. Sci., Polym. Lett. Ed.*, **17**, 203 (1979).
343. R. R. Chance, M. L. Shand, C. Hogg and R. Silbey, *Phys. Rev.*, **B22**, 3540 (1980).
344. R. R. Chance, G. N. Patel and J. D. Witt, *J. Chem. Phys.*, **71**, 206 (1979).
345. G. Wenz, Dissertation, Freiburg (1983).
346. G. Wenz and G. Wegner, *Makromol. Chem., Rapid Commun.*, **3**, 231 (1982).

347. V. Enkelmann, G. Wenz, H. A. Muller, M. Schmidt and G. Wegner, *Mol. Cryst. Liq. Cryst.*, **105**, 11 (1984).
348. D. Bloor, L. Koski, G. C. Stevens, F. H. Preston and D. J. Ando, *J. Mater. Sci.*, **10**, 1678 (1975).
349. R. R. Chance and G. N. Patel, *J. Polym. Sci., Polym. Phys. Ed.*, **16**, 859 (1978).
350. G. Wegner and V. Enkelmann, *Angew. Chem.*, **89**, 432 (1977).
351. V. Enkelmann, R. J. Leyrer and G. Wegner, *Makromol. Chem.*, **180**, 1787 (1979).
352. A. Kobayashi, H. Kobayashi, Y. Tokura, T. Kanetake and T. Koda, *J. Chem. Phys.*, **87**, 4962 (1987).
353. G. Porod, *Monatsh. Chem.*, **80**, 251 (1949).
354. O. Kratky and G. Porod, *Recl. Trav. Chim. Pays-Bas*, **68**, 1106 (1949).
355. H. Kuhn, *Fortschr. Chem. Org. Naturst.*, **16**, 169 (1958); **17**, 404 (1959).
356. G. Wenz and G. Wegner, *Makromol. Chem., Rapid Commun.*, **3**, 206 (1982).
357. C. Plachetta, N. O. Rau, A. Hauck and R. C. Schulz, *Makromol. Chem., Rapid Commun.*, **3**, 249 (1982).
358. C. Plachetta and R. C. Schultz, *Makromol. Chem., Rapid Commun.*, **3** 815 (1982).
359. M. F. Rubner, D. J. Sandman and C. Velazquez, *Macromolecules*, **20**, 1296 (1987).
360. H. Tanaka, M. Thakur, M. A. Gomez and A. E. Tonelli, *Macromolecules*, **20**, 3094 (1987).
361. H. Tanaka, M. A. Gomez, A. E. Tonelli and M. Thakur, *Macromolecules*, **22**, 1208 (1989).
362. S. Koshihara, Y. Tokura, K. Takeda, T. Koda and A. Kobayashi, *J. Chem. Phys.*, **92**, 7581 (1990).
363. K. Takeda, T. Koda, S. Koshihara and Y. Tokura, *Synthetic Metals*, **41–43**, 231 (1991).
364. S. Koshihara, Y. Tokura, K. Takeda and T. Koda, *Phys. Rev. Lett.*, **68**, 1148 (1992).
365. D. Bloor and F. H. Preston, *Phys. Status Solidi*, **A39**, 607 (1977).
366. M. L. Shand, R. R. Chance and R. Silbey, *Chem. Phys. Lett.*, **64**, 448 (1979).
367. K. Koda, T. Hasegawa, K. Ishikawa, K. Kubodera, H. Kobatashi and K. Takeda, *J. Lumin.*, **48–49**, 321 (1991).
368. J. Ducuing, in *Nonlinear Spectroscopy* (Ed. N. Bloembergen), Enrico Fermi Course **44**, Academic Press, New York, 1975.
369. P. W. Smith and W. J. Tomlinson, *IEEE Spectrum, June*, **26** (1981).
370. P. W. Smith, *Bell. Syst. Tech. J.*, **61**, 1975 (1982).
371. A. F. Garito, *Polym. Prepr.*, **23**, 155 (1982).
372. A. F. Garito, K. D. Sing and C. C. Teng, *ACS Symp. Ser.*, **233**, 1 (1983).
373. A. F. Garito, C. C. Teng, K. Y. Wong and O. Zammani Khamiri, *Mol. Cryst. Liq. Cryst.*, **106**, 219 (1984).
374. P. S. Kalyanaraman, A. F. Garito, A. R. McGhie and K. N. Desai, *Makromol. Chem.*, **180**, 1393 (1979).
375. J. E. Sohn, A. F. Garito, K. N. Desai, R. S. Narang and M. Kuzyk, *Makromol Chem.*, **180**, 2975 (1979).
376. A. R. McGhie, C. F. Lipscomb, A. F. Garito, K. N. Desai and P. S. Kalyanaraman, *Makromol. Chem.*, **182**, 965 (1981).
377. A. F. Garito, C. F. Horner, P. S. Kalyanaraman and K. N. Desai, *Makromol. Chem.*, **181**, 1605 (1980).
378. A. F. Garito, K. D. Singer, K. Hayes, G. F. Lipscomb, S. J. Lalama and K. N. Desai, *J. Opt. Soc. Am.*, **70**, 1399 (1980).
379. S. J. Lalama, K. D. Singer, A. F. Garito and K. N. Desai, *Appl. Phys. Lett.*, **39**, 940 (1981).
380. K. J. Wong, C. C. Teng and A. F. Garito, *J. Opt. Soc. Am.*, **B1**, 434 (1984).
381. C. Sauteret, J. -P. Hermann, R. Frey, F. Praedere, J. Ducuing, R. H. Baughman and R. R. Chance, *Phys. Rev. Lett.*, **36**, 956 (1976).
382. J. P. Hermann and P. W. Smith, *Dig. Tech. Pap. -Int. Quantum Electron. Conf., 11 th, 1980*, p.656.
383. S. Etemad, P. W. Townsend, G. L. Baker and Z. Soos, in *Organic Molecules for Nonlinear Optics and Photonics* (Ed. J. Messier), Kluwer Academic Publishers, 1991, p. 489.
384. G. P. Agrawal, C. Cojan and C. Flytzanis, *Phys. Rev.*, **B17**, 776 (1985).
385. W. Weikang, *Phys. Rev. Lett.*, **61**, 1119 (1988).
386. T. Hasegawa, K. Ishikawa, K. Takeda, H. Kobayashi and K. Kubodera, *Chem. Phys. Lett.*, **171**, 239 (1990).
387. T. Hasegawa, K. Ishikawa, T. Koda, K. Takeda, H. Kobayashi and K. Kubodera, *Synthetic Metals*, **41–43**, 3151 (1991).
388. S. Abe, J. Yu and W. P. Su, *Phys. Rev.*, **B45**, 8264 (1992).

389. S. Abe, *J. Photo. Sci. Tech.*, **6**, 247 (1993).
390. G. M. Carter, J. V. Hryniewicz, M. K. Thakur, Y. J. Chen and S. E. Meyler, *Appl. Phys. Lett.*, **49**, 998 (1986).
391. T. Hattori and T. Kobayashi, *Chem. Phys. Lett.*, **133**, 230 (1987).
392. P. W. Townsend, G. L. Baker, N. E. Schlotter, C. F. Klausner and S. Etemad, *Appl. Phys. Lett.*, **53**, 1782 (1988).
393. P. W. Townsend, J. L. Jackel, G. L. Baker, J. A. Shelburne and S. Etemad, *Appl. Phys. Lett.*, **55**, 1829 (1989).
394. M. Thakur and D. M. Krol, *Phys. Lett.*, **56**, 1213 (1990).
395. D. M. Krol and M. Thakur, *Appl. Phys. Lett.*, **56**, 1406 (1990).
396. W. Schermann and G. Wegner, *Makromol. Chem.*, **175**, 667 (1974).
397. G. Wegner and W. Schermann, *Colloid Polym. Sci.*, **252**, 655 (1974).
398. D. Day and J. B. Lando, *J. Appl. Polym. Sci.*, **26**, 1605 (1981).
399. K. Se, H. Ohnuma and T. Kotaka, *Polymer Preprints, Japan*, **31**(4), 710 (1982).
400. H. Nakanishi, F. Mizutani, F. Nakanishi, M. Kato and K. Hasumi, *J. Polym. Sci., Polym. Lett.*, **21**, 983 (1983).
401. E. G. Wilson, in *Polydiacetylenes* (Eds. D. Bloor and R. R. Chance), 155, NATO ASI Series, 1985.
402. Yu. V. Korshak, A. A. Ovchinnikov, A. M. Shapiro, T. V. Medvedeva and V. N. Spector, *JETP Lett.*, **43**, 399 (1986).
403. J. S. Miller, J. C. Calabrese, D. T. Glatzhofer and A. J. Epstein, *J. Appl. Phys.*, **63**, 2949 (1988).
404. R. H. Baughman, H. Gleiter and N. Sendfeld, *J. Polym. Sci., Polym. Phys. Ed.*, **13**, 1871 (1975).
405. C. Galiotis and R. J. Young, *Polymer*, **24**, 1023 (1983).
406. D. N. Batchelder and D. Bloor, *J. Polym. Sci., Polym. Phys. Ed.*, **17**, 569 (1979).
407. R. J. Young, D. Bloor, D. N. Batchelder and C. L. Hubble, *J. Mater. Sci.*, **13**, 62 (1978).
408. R. J. Young, R. Dulniak, D. N. Batchelder and D. Bloor, *J. Polym. Sci., Polym. Phys. Eds.*, **17**, 1325 (1979).
409. R. J. Young and J. Petermann, *Makromol. Chem.*, **182**, 621 (1981).
410. C. Galiotis, P. H. J. Yeung, R. J. Young and D. N. Batchelder, *J. Mater. Sci.*, **19**, 3640 (1984).

CHAPTER **18**

Arynes and heteroarynes

HAROLD HART

Department of Chemistry, Michigan State University, East Lansing, Michigan 48824, USA

Supplement C2: The chemistry of triple-bonded functional groups
Edited by S. Patai © 1994 John Wiley & Sons Ltd

I. INTRODUCTION

Arynes and heteroarynes are reactive intermediates derived formally by the removal of two adjacent hydrogens from, respectively, a carbocyclic or a heterocyclic aromatic ring. Prototypical examples are benzyne (**1**) and 3,4-didehydropyridine (**2**). Since these molecules contain a partial triple bond, they are properly regarded as -*ynes* and, as will be seen, undergo many reactions typical of that functionality, modified by the inherent strain caused by their nonlinearity.

(1)   **(2)**

    This review is an update of the aryne chapter that appeared in an earlier volume of this series[1]. That chapter covered mainly the literature from 1970–1979, earlier work having already been reviewed extensively[2,3]. Here, the literature is covered from 1980–1992. Some of this literature has been periodically and very briefly reviewed[4].

    Unlike its predecessor, this review also includes heteroarynes, though the treatment is less exhaustive than previous comprehensive reviews of six-membered[5,6] and five-membered[5,7] heteroarynes.

    Although most chemists refer to **1** as benzyne and to **2** as 3-pyridyne, one will not find these names in *Chemical Abstracts* (CA). Benzyne is listed there as 1,3-cyclohexadiene-5-yne, and under this heading one can also find various substituted benzynes (alkyl, halo, etc.). Although this name emphasizes the appropriateness of benzyne to this volume on the triple bond, other names are used for other carbocyclic or heterocyclic arynes. There, CA uses the *dehydro* nomenclature (benzyne was called *dehydrobenzene* by its discoverer, G. Wittig[2]). Thus to locate naphthyne references, one must search under naphthalene, 1,2- or 2,3-didehydro; similarly, one must search every aromatic or heteroaromatic ring system to determine whether its various dehydro derivatives have been studied. Since many aromatic systems also have *di*hydro derivatives, and since errors regarding dehydro and dihydro creep into the indices, it is difficult to be certain that a search is complete. Some references can also be found under the general topic *arynes* in the General Subject Index of CA, but this listing is quite incomplete. It is hoped that, through the pursuit of cross references, this review is reasonably complete, but apologies are extended to authors whose work may somehow have been missed.

II. THE STRUCTURE OF BENZYNE

A. Infrared Spectrum

The structure of benzyne, which was originally based on pioneering studies of its infrared spectrum determined in a nitrogen matrix at low temperature[8], has been the subject of considerable recent controversy. At the heart of this controversy was the original assignment of a fairly intense IR band at 2085 cm^{-1} to the —C≡C— stretching vibration. This assignment was at first confirmed by several experimental[9] and normal coordinate analysis[10] studies. In 1986, however, this seemingly established picture of benzyne was challenged by gas-phase photodetachment studies[11] of the benzyne anion $C_6H_4^{\cdot-}$. This ion was

$$O^- + C_6H_6 \longrightarrow C_6H_4^- + H_2O$$

generated from benzene (or benzene-d_6) using chemical ionization mass spectrometry. O^-, produced from N_2O, removed H_2^+ mainly from *ortho* positions, generating the *o*-benzyne anion. The relationship between the electronic structure of *o*-$C_6H_4^-$ and neutral

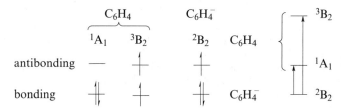

FIGURE 1. Electronic structure of singlet and triplet neutral *o*-benzyne and doublet *o*-benzyne anion

o-C_6H_4 is illustrated qualitatively in Figure 1. Photoelectron detachment from the anion leads directly to either the singlet or triplet state of *o*-benzyne (see the right part of Figure 1), thus yielding a value for the singlet–triplet splitting (37.7 kcal mol^{-1}, in reasonable agreement with a theoretical calculation[12]), as well as for the electron affinity of neutral benzyne (0.56 eV). Careful vibrational analysis of the $^1A_1 \leftarrow {}^2B_2$ photodetachment spectra led to a revised value of 1860 ± 15 cm^{-1} for the —C≡C— bond stretch in neutral *o*-benzyne.

This assignment, which is well outside the normal carbon–carbon triple-bond range, provoked both theoretical[12–14] and experimental[15] attention. The latter study, which involved irradiation of phthalic anhydride (3) in Ne, Ar, Xe, N_2 and CO matrices using high resolution, single-site infrared spectra with isotopic shifts (spectra of C_6H_4, 1,2-$^{13}C_2C_4H_4$ and C_6D_4 were observed), calculated frequencies (MP2/6-31G) and experimental symmetries as well as line shapes and intensities, led to a definitive assignment of the IR spectrum

of benzyne (20 bands were observed and assigned; confirmation of some weak band assignments is still required, through determining the Raman spectrum of benzyne). A low-intensity band at 1846 cm^{-1} in a Ne matrix (1844 cm^{-1} in C_6D_4 and 1793 cm^{-1} in 1,2-$^{13}C_2C_4H_4$) is assigned to the C≡C stretch, consistent with the photodetachment assignment[11] and recent theoretical calculations[13,14]. The previously observed band at 2085 cm^{-1} is now assigned to cyclopentadieneylidene ketene (5) which is simultaneously formed during the photolysis[15,16]. The full photochemical scheme is shown, and the benzyne–

cyclopentadienylideneketene interconversion ($1 \underset{-CO}{\overset{CO}{\rightleftarrows}} 5$) is discussed in Section IV.

The low frequency for the —C≡C— bond is thought to indicate a substantial loss of alkyne character in benzyne[15], and the absence of 'cyclohexatriene'-type bonds suggests that the benzene π-electron delocalization is largely preserved; that is, localization to a cumulene-type structure[12] such as 8 is considered unimportant. The most recently calculated geometry for benzyne (9) using MP2(TZ + 2P)[14] indicates a somewhat longer 'triple-bond' length (1.259 Å) than is common in most acetylenes (1.20 Å). That is, the bond is a weak triple bond.

B. Other Spectra

The UV/VIS absorption spectrum of benzyne has been reported[16,17]. The benzyne was produced in an argon matrix by irradiation of **10**[17] or **11**[16,17]. The full scheme from **11** is

(10)   **(11)**

shown. Two paths to benzyne were proposed, each leading to a different mixture of benzyne and cyclopentadienylideneketene. Depending on the radiation energy, reaction proceeds via either ketenecarbene **12** or bis-ketene **13**, and in each case benzocyclopropenone **4**

(identified by an IR band at 1857 cm^{-1}). Photoequilibration of the two product mixtures of **1** + **5** (+ CO) was achieved, and since nearly pure **1** was present in one of these mixtures, it was possible to deduce the UV/VIS spectra of each component. Benzyne has a low-intensity broad σ–σ* transition in the region of 3.3 eV and two more intense higher energy bands at 5.08 and 6.25 eV corresponding to the ^1L_b and ^1L_a bands in the spectrum of benzene.

Although the two studies[15,16] establish quite well the IR and UV/VIS spectra of benzyne, there are some differences in the two proposed photochemical schemes (from **3**[15] and from **11**[16]); for example, in the former[15], **4** is not on a direct route to benzyne, whereas in the latter[16] it is. Resolution of these differences awaits further study.

A weak microwave spectrum of benzyne, generated by pyrolysis of **11** (and confirmed by pyrolysis of **3** or **14**), was observed[18]. Assignment of the spectrum to other possible pyrolysis products was carefully excluded. The observed rotational constants ($A_0 = 6990$, $B_0 = 5707$ and $C_0 = 3140$) agree very well (< 1%) with most recent theoretical calculations[14].

(14)

Photoelectron spectra of benzyne have been reported[19,20]. Most recent data[20] lead to two ionization potentials for o-benzyne: 9.03 eV for removal of an out-of-plane (π) electron, and 9.77 eV for removal of an in-plane (σ) electron. A thermochemical cycle using these data and ionization potentials and bond dissociation energies (BDE) for phenyl radical and cation led to a value for the singlet–triplet gap of o-benzyne that is in good agreement (38.5 ± 2.4 kcal mol^{-1}) with the independently determined[11] value (37.7 ± 0.7).

C. Theoretical Calculations

Benzyne has been the subject of numerous theoretical calculations[1]. In addition to those already mentioned[12–15] others of varying degrees of sophistication have been reported[21–25]. Each yields its own optimum geometry, valence ionization energies[25] and other structural parameters. The most recent[14] is probably the most reliable.

D. Heat of Formation

The original report[26] of an experimental heat of formation of o-benzyne, based on mass spectrometric studies, gave a value of 118 ± 5 kcal mol^{-1}. Other values (107[27], 118[28], 100[29]) were subsequently reported. Each of these reports suffered from considerable experimental uncertainty. In 1991, three independent studies[30–32], each based on somewhat different methodology, gave a consistent value of 105 ± 3 kcal mol^{-1} for $\Delta H_f^{\circ}[o\text{-}C_6H_4]$, with higher values[32] of 116 ± 3 and 128 ± 3 kcal mol^{-1} respectively for m- and p-benzyne.

III. GENERATION OF ARYNES

A. Benzyne

In general, three major methods have been used in the past to generate benzyne[1,2], the particular choice of method depending on the medium (solvent, gas phase, matrix) and the subsequent chemistry to be carried out. These methods involve as precursors (A) 2-halogenophenyl anions, (B) benzenediazonium or related carboxylates and (C) benzo-fused heterocyclic systems that easily fragment thermally or photochemically with the elimination of small stable molecules. Typical examples of the three methods are illustrated in the accompanying equations. Each methodology is now so well established[1,2] that it requires no further discussion here. However, since the previous review[1] several novel improved methods or variations on these methods have been described.

Type A 1.

2.

Type B 1.

2.

Type C 1.

2.

1. Type A1

Benzyne can be produced by the reaction of *o*-chloro- or *o*-bromosulfoxides (**15**) with Grignard reagents[33]. When X = Br or Cl, Grignard exchange occurs mainly at the sulfoxide

to give the *o*-halo-Grignard **16**. When X = Cl, **16** can be trapped at room temperature with electrophiles (i.e. PhCHO) but at 70 °C it gives benzyne (trapped by furan). When X = Br, elimination of **16** to benzyne occurs at room temperature. When X = I, attack at I competes with attack at S and renders the method ineffective. The starting sulfoxides **15** were prepared from the corresponding sulfides by *m*-CPBA oxidation.

Tosylate can function in place of halogen as the leaving group in aryne generation. The method has been most successfully applied to naphthynes[34] as in the following example[35]:

(17)   **(18)** 51%

The method has the advantage that the *o*-bromotosylate can usually be prepared by bromination and tosylation of the corresponding phenol.

(19)   **(20)**   **(21)** 90%

Metal–halogen exchange in *o*-haloaryltriflates (**19**, X = Br, I) occurs at –78 °C with BuLi to produce arynes[36]. Mesylate or tosylate in place of triflate gave unsatisfactory results. Addition of alkoxyarynes produced this way to 2-methoxyfuran (**23**) is regioselective and proceeds head-to-head.

(22)   **(23)**   **(24)**   **(25)**

Aryl triflates have been used to generate arynes via other routes than metal–halogen exchange. For example, fluoride ion displacement of the trimethylsilyl group in **26** provides a convenient route to benzyne under mild conditions[37]. The required benzyne precursor **26** was prepared from *o*-trimethylsilylphenoxytrimethylsilane by sequential treatment with BuLi and (Tf)$_2$O. Analogous aryne precursors have also been prepared less directly, as in the case of amide benzyne **31**[38]. Examples of **32** include various methylfurans, diphenylisobenzofuran and tetraphenylcyclopentadienone. Examples of nucleophiles include methanol, phenol, lithium thiophenoxide and others.

(26)   **(27)** 61%

(28)   **(29)**   **(30)**

(33)   **(32)**   **(31)**   **(34)**

Aryl triflates **35**, generated from the corresponding phenols, react with lithium diisopropylamide (LDA) in diisopropylamine (DIA) to give **36** in good yield[39]. The reaction was demonstrated (via substitution pattern in the products) to proceed via an aryne intermediate. The choice of LDA as the base is critical; BuLi, NaNH₂, sodium acetylide and 2-lithiofuran all failed to generate an aryne from **35** (R = *p*-Ph). Application of the method to other nucleophiles than DIA, or to cycloadditions, has not yet been demonstrated.

(35)   **(36)**

Whereas *o*-halolithium or magnesium arenes readily undergo elimination to arynes, *o*-Cl- and *o*-F-copper reagents do not, and can be used in nucleophilic displacements, as in these examples[40]. An activated form of copper is used to directly form the organocopper reagent; the more common route via organolithium or Grignard precursors requires temperatures so low that the nucleophilic displacements cannot proceed.

(37)   **(38)**   **(39)** R = Me, Et, PhCO, MeCO

2. Type A2

These reactions are based primarily on the idea that the acidity of hydrogens *ortho* to a halogen is enhanced by the electron-withdrawing effect of the halogen. If a second electron-withdrawing group (EWG) is located *meta* to the halogen, the acidity of the hydrogen between the two substituents will be further enhanced. The 2-oxazolyl group has been

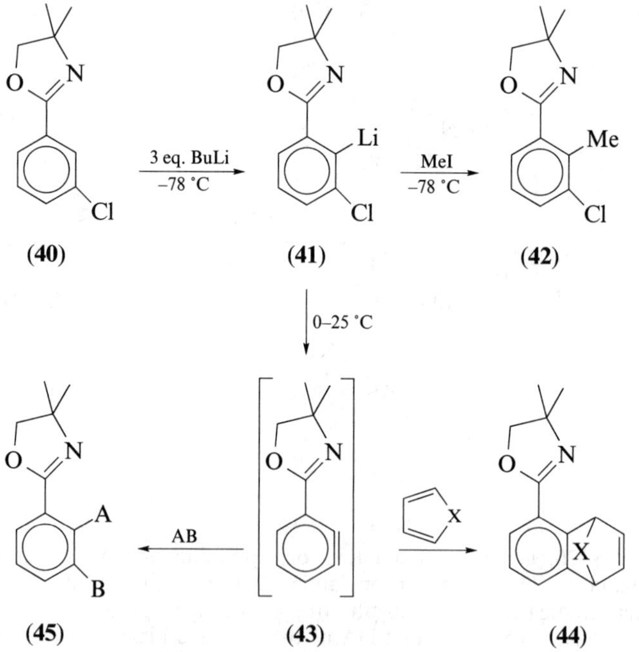

(40)   3 eq. BuLi   **(41)**   MeI   **(42)**
                     −78 °C   −78 °C

0–25 °C

(45)   AB   **(43)**   **(44)**

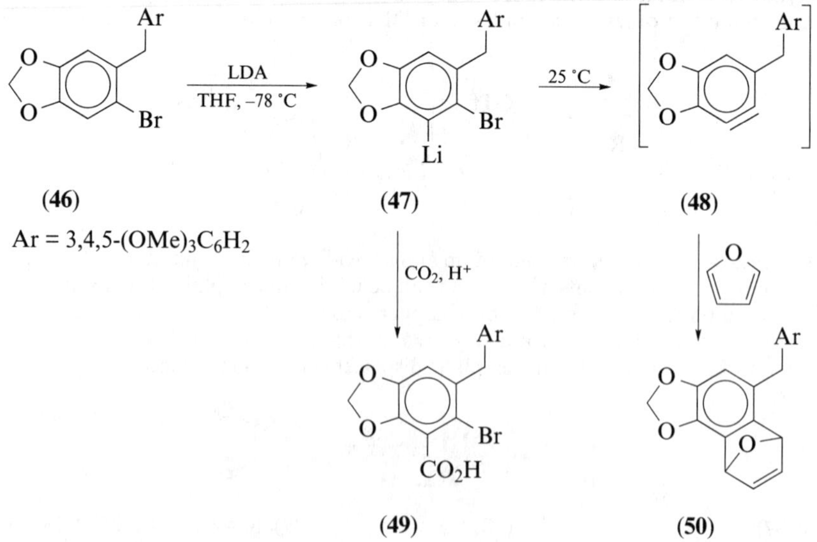

Ar = 3,4,5-(OMe)$_3$C$_6$H$_2$

(46)   LDA, THF, −78 °C   **(47)**   25 °C   **(48)**

CO$_2$, H$^+$

(49)   **(50)**

used in this way, to provide a useful route to 1,2,3-trisubstituted benzenes[41,42]. Thus the (*m*-chlorophenyl)oxazoline **40** (readily prepared from *m*-chlorobenzoic acid) is lithiated exclusively between the two substituents to give **41**, which is stable at –78 °C and can be trapped with electrophiles (i.e. MeI). On warming, however, LiCl is lost to give benzyne **43**, which may be trapped with various nucleophiles AB or with dienes, as shown. Since the oxazolyl group in the products is readily converted to a carboxyl or other functions, the method has considerable utility. Details as to the regiochemistry of the nucleophilic addition to **40** are discussed in Section VI.B.

LDA is a useful base for generating a benzyne from *m*-alkoxyaryl bromides[43]. Thus the aryl and not the benzylic proton is removed from **46** to give **47** which, on warming, produces aryne **48**. The *m*-alkoxy substituent is essential, since *m*-bromoanisole, but not the *o*- or *p*-isomers, undergoes analogous chemistry.

3. Type B

An X-ray crystal structure has been reported for 2-diazoniobenzenecarboxylate hydrate (explosive!), the common benzyne precursor produced through diazotization of anthranilic acid[44]. The aryl carbon–carboxylate carbon bond length is 1.526 Å, significantly longer than the usual bond length between sp^2-hybridized carbons (1.46 Å), suggesting little conjugation of the CO_2 group with the phenyl ring. Also, the carboxylate group is rotated 25.9° from the benzene ring plane. However, this structure is calculated to be only 1.5 kcal mol^{-1} less stable than the planar conformer. The rotational barrier is estimated at 9.9 kcal mol^{-1}, larger by about 5 kcal mol^{-1} than that for the *p*-isomer of **51**, suggesting an attractive interaction between the two *ortho* substituents.

(51)

Perdeuterobenzyne has been prepared by the diazonium carboxylate route[45]. The required precursor **53** was obtained from potassium 2-aminobenzoate by exchange with D_2O over a Pt catalyst and, on diazotization with isoamyl nitrite in the presence of tetraphenylcyclopentadienone or 6,6-dimethylfulvene, gave the corresponding deuterated cycloadducts.

(52)   (53)

An alkoxy group *ortho* to the diazonium moiety in **51** may interfere with the normal decomposition pathway. For example, the 3-alkoxybenzyne **54** could not be prepared from **55**, but was easily prepared from its regioisomer **56**[46,47]. This may be a consequence of the alternate decomposition mode, to give diazoquinone **57**. This path may be avoided by using the diazonium carboxylate hydrochloride[47].

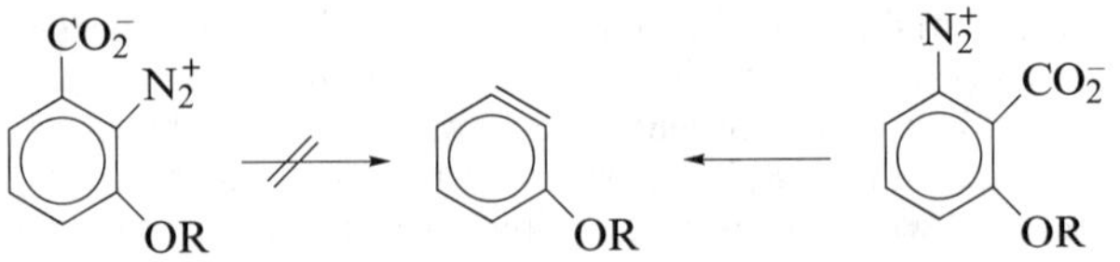

(55)   (54)   (56)

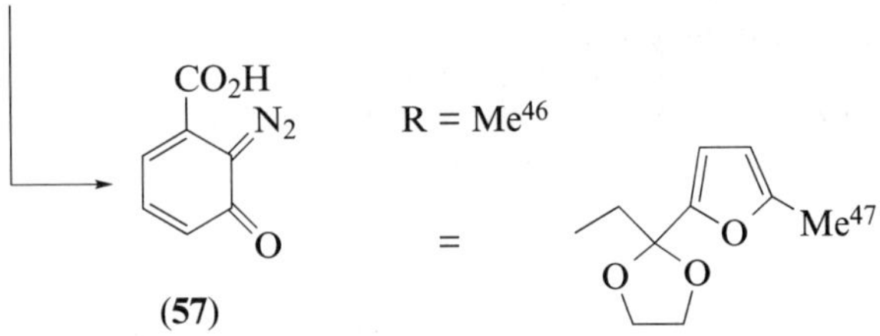

R = Me[46]

(57)   =

The diaryliodonium carboxylate route to benzynes (Type B2) has been adapted to the polymer-bound precursor **58**, prepared from Merrifield's resin by standard methods[48]. (An earlier example of a *polymer-bound* benzyne has been described[49].) The polymer afforded benzyne **59** when heated above 200 °C, as demonstrated by trapping with 2-furoic acid; the initial cycloadduct decarboxylates spontaneously at that temperature, affording naphthols **60** in about 20% yield. The three-phase test[50] involving transfer of aryne from **58** to a

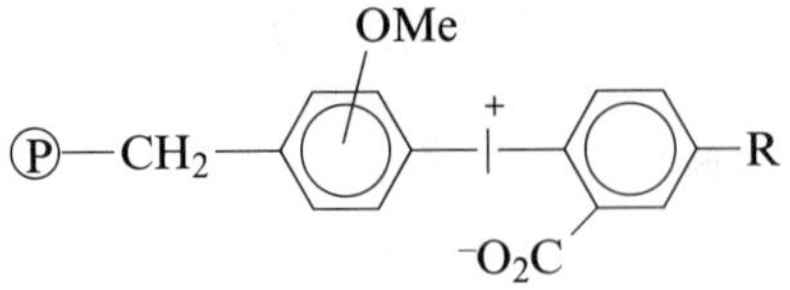

(58)

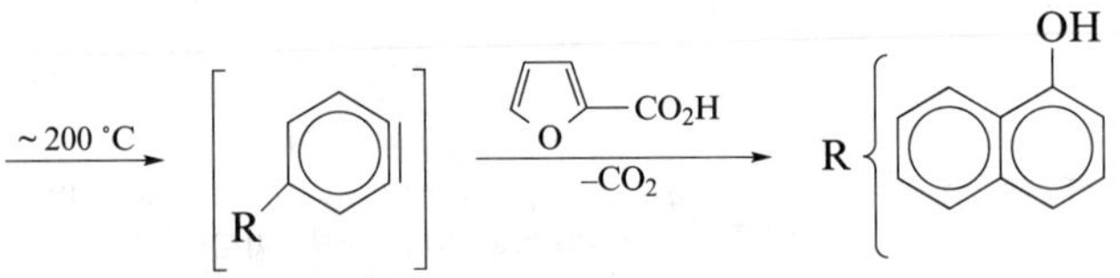

(59)   (60)

polymeric 2-furoic ester via diethylbenzene as solvent gave a lifetime for benzyne of 5s, slightly longer when R was an EWG. Similar results were obtained with polymer-bound 1(2-carboxyaryl)triazenes **61** and 2-carboxyarylsulfonates **62**, but the aryne yields were lower than with **58**[51].

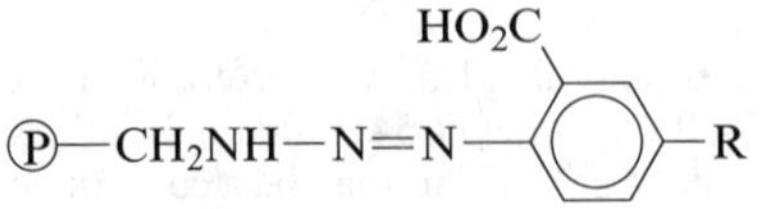

(61) 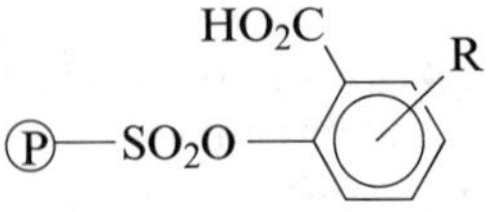   (62)

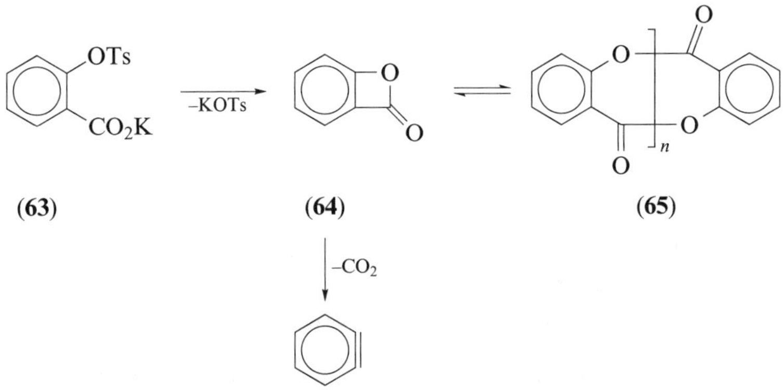

(63)   **(64)**   **(65)**

(1)

There is growing evidence that some decompositions of this type, those that involve decarboxylation, may not be concerted but involve discrete intermediates[52,53]. For example, thermal decomposition of potassium 2-carboxyphenyl p-toluenesulfonate **63** at 170 °C occurs without weight loss; carboxylate bands disappear and ester bands are formed, and di-, tri- and polysalicylides **65** are formed. At higher temperatures (300 °C), however, products from **1** (for example, cycloadducts with tetraphenylcyclone present as a trap, or products derived from nucleophilic attack) are formed in low to modest yield. These results require a $C_6H_4CO_2$ intermediate, most likely benzoxet-2-one **64** (or valence tautomers)[53]. Similar results were obtained with polymer-bound diaryliodonium carboxylates[52].

4. Type C

The lead tetraacetate oxidation of 1-aminobenzotriazole (ABT) over a wide temperature range gives a higher yield of biphenylene **67** than most other benzyne precursors[54]. With the thought that the high propensity for benzyne dimerizations in these reactions might be due to a high local non-steady-state concentration of benzyne being formed in the vicinity

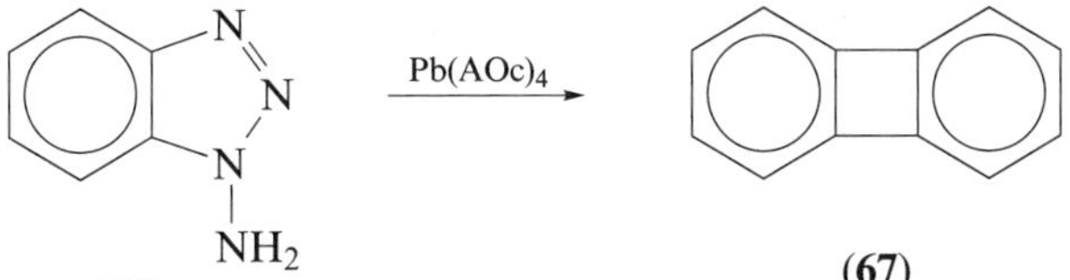

(66)   **(67)**

of each drop of added reagent, the procedure was modified[55] by using two syringe pumps to add lead tetracetate and **66** simultaneously from opposing ports of a reaction flask. When this technique was applied to 4-phenyloxazole **68** at 0 °C, essentially quantitative yields of cycloadduct **69** were formed; the usual method, in contrast, gave mainly biphenylene and other products, and little if any **69**. Through retro-Diels–Alder reaction, **69** serves as an excellent source of isobenzofuran[55].

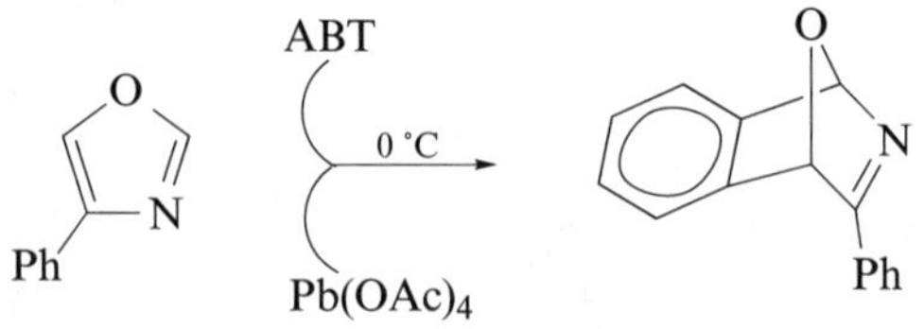

(68)   **(69)**

B. Other Benzynes

The same methods used for benzynes are generally applied to other carbocyclic rings. Type A methods tend to be most commonly used because of the accessibility of the haloarenes.

Two discrete cyclopropabenzynes **71** and **74** are intermediates in the reactions of 2- and 3-bromocyclopropabenzenes[56] with the complex base formed from *t*-BuOH and sodamide. That **73** mainly gives **74** and not **71** suggests that **74** is the more stable of the two arynes, and theoretical calculations support this conclusion.

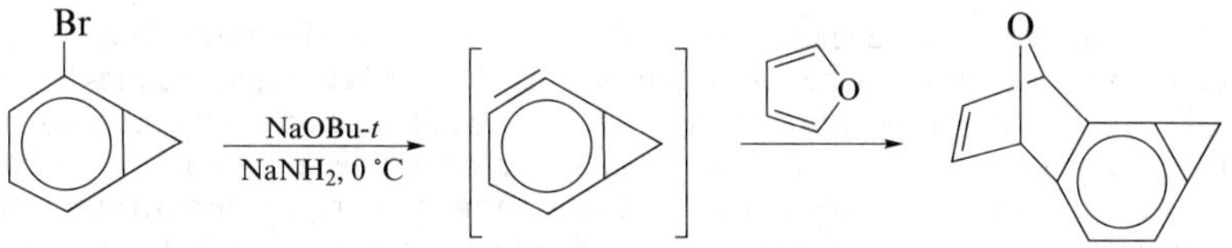

(70)   **(71)**   **(72)** 10%

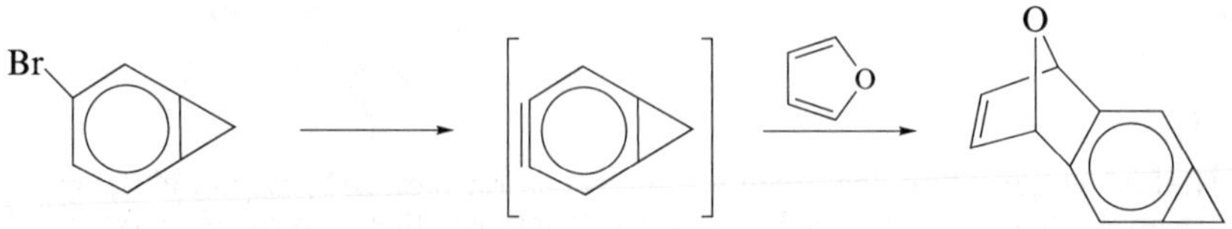

(73)   **(74)**   **(75)** 51%

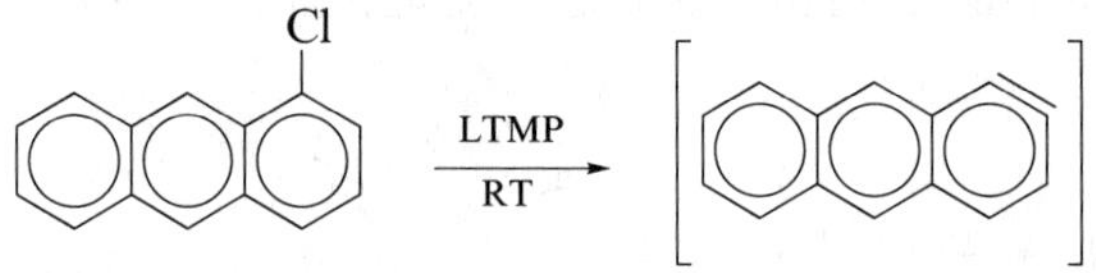

(76)   **(77)**

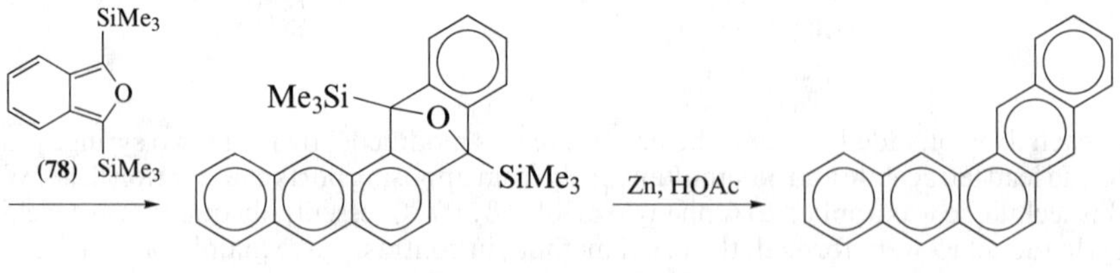

(79)   **(80)**

1,2-Anthracyne **77**, generated from 1-chloroanthracene and lithium tetramethylpiperidide (LTMP), is an intermediate in a new route to pentaphene **80**[57]. The required 1-chloroanthracene was also prepared via aryne routes[58].

Annulyne **82** has been generated from bromide **81** and sodamide, and has been trapped by various furans; removal of the oxygen bridge leads to the annelated annulenes[59].

Dehydrocyclophanes have been reported[60]. Evidence has been presented recently that in such species, the aryne ring can rotate 180° through the cyclophane ring before trapping[61]. Thus whether one starts with *anti*-bromide **84a** or *syn*-bromide **84s**, a mixture of *syn* and *anti* adducts **86** is obtained (Nu = OBu-*t* and OH), although *anti* adducts predominate from *anti*-bromide, and *syn* adducts from *syn*-bromide. Rotation interconverts the *anti*- and *syn*-arynes **85a** and **85s**, presumably through perpendicular aryne **87**, some of which gives the intramolecular cycloaddition product **88**. Clearly the initially formed arynes are sufficiently long-lived to undergo this rotation despite a 20-fold molar excess of *t*-butoxide.

2,3-Naphthyne is produced in good yield via a Type B method, from (4'-methoxyphenyl)-2-naphthyliodonium 3-carboxylate (**89**), and from an analogous polymer-bond precursor[52].

(**89**)   (**90**)

(**91**) 55%

Phenanthryne (and methyl and benzo analogs) are produced in good yield by a Type C method, very low pressure pyrolysis at 700–900 °C[62] of the 9,10-dicarboxylic anhydrides. The arynes were identified by mass spectrometry and by products derived from reaction with benzene or 1,3-butadiene.

(**92**)   (**93**)

C. Benzynes as Intermediates

Benzynes have been proposed as intermediates in a number of reactions which, however, are not useful as general aryne sources for synthesis.

Reactions of aryl halides with metals often produce arynes. Tetrafluorobenzyne **95** is thought to be an intermediate in the thermal decomposition of vanadium–THF complex

94. When the solvent is dibutyl ether, perfluorotriphenylene **96** is formed, whereas in benzene as solvent, the product is cycloadduct **97**[63].

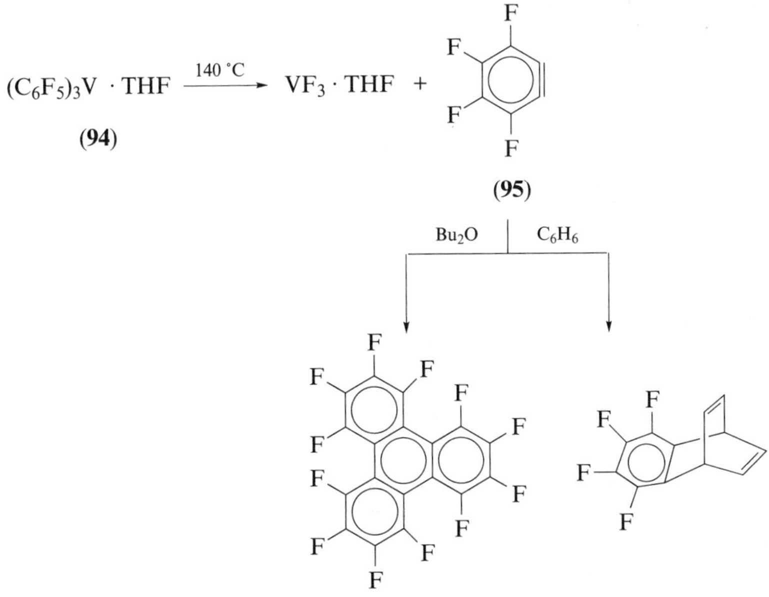

$$(C_6F_5)_3V \cdot THF \xrightarrow{140\ °C} VF_3 \cdot THF +$$

(94)

(95)

Bu$_2$O   C$_6$H$_6$

(96)   **(97)**

An especially efficient triphenylene synthesis involves *o*-sodiofluorobenzene as an intermediate[64]. Temperature control is critical. When the temperature reaches about –30 °C in the second reaction stage (**99 → 1**), the mixture is cooled in a liquid nitrogen bath. Even so, the reaction is so exothermic that the temperature rises to 20 °C. The fast decomposition of **99** may give rise to a high benzyne concentration, hence a high yield of **100**. Analogous other metal derivatives (i.e. Li) give only minor amounts of **100**.

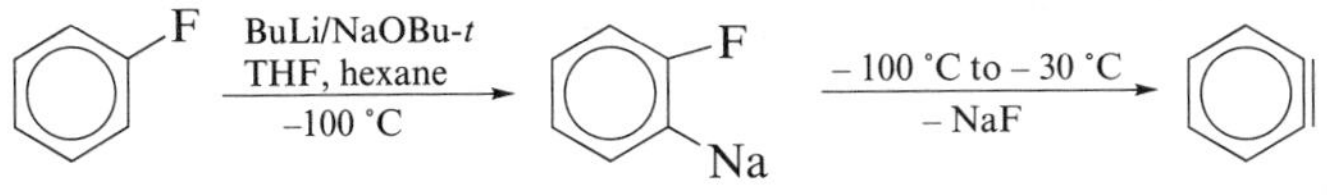

BuLi/NaOBu-*t*
THF, hexane
–100 °C

– 100 °C to – 30 °C
– NaF

(98)   **(99)**   **(1)**

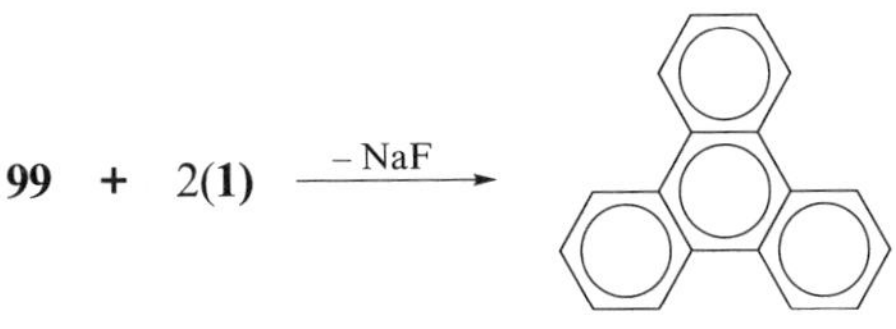

$$99 \ + \ 2(1) \xrightarrow{- NaF}$$

(100) 66%

Aromatic 1,2-dihalides (I, Br) gave only traces of benzyne on gas-phase reaction with alkali metal vapors under microwave or ultrasound excitation[65]. Electroreduction of such halides in the presence of tertiary amines can, however, generate benzyne; trapping products are isolated in low (10–20%) yield[66].

Biphenylene was observed among the products of the reaction of o-chloronitrobenzene with disodium ditelluride, and is thought to arise from benzyne dimerization[67]. The major product, however was phenazine (102).

CO_2-laser-induced photolysis of chloropentafluorobenzene produces $C_6F_5^{\cdot}$ which then loses fluorine atoms to produce C_6F_4, thought to be tetrafluorobenzyne[68].

$$C_6F_5Cl \xrightarrow{h\nu} Cl + C_6F_5 \xrightarrow{h\nu} F + C_6F_4$$
$$(95)$$

Decomposition of benzenethiol on a molybdenum surface (Mo 110) produces chemisorbed benzyne[69]. Benzene absorbs on the same surface in a π-bound geometry parallel to the surface, and at 400 K dehydrogenation to a stable surface intermediate, proposed to be tilted benzyne, occurs[70].

A remarkable and so far unique generation of a benzyne intermediate has been proposed in the thermal decomposition of azidoquinone **103** which, in refluxing benzene, produces cycloadduct **104** in 38% yield[71]. Loss of nitrogen and CO with re-bonding is thought to give diradical **105**[72], which suffers a trimethylsilyl shift from carbon to oxygen to give benzyne **106**. Diels–Alder addition to the solvent accounts for the observed product **104**.

Benzyne is a necessary intermediate in the diazotization of 2-[(2-acetoxyethyl)sulfinyl(and sulfonyl)]anilines **107** and **116**, but a detailed study of the way it is formed gave some surprises[73,74]. Products from the aprotic diazotization of **107** included biphenylene **67** and dibenzo-1,4-thioxin **112**. EPR evidence for radical intermediates was obtained. Formation of **112** requires cleavage of the S—O bond, most probably via benzoxathiete **110** as shown. Although yields of **67** and **112** were low, when the diazotization was carried out in the presence of benzyne traps such as diphenylisobenzofuran or anthracene, respectable yields (30–40%) of the corresponding benzyne cycloadducts were obtained.

An alternative route to benzyne from **107** might involve formation and decomposition of 1,2,3-benzthiadiazole-1-oxide **114**. However, independent synthesis of **114**[74] showed that it is remarkably stable; its thermal decomposition requires a temperature of 135 °C! Consequently **114** cannot be an intermediate in the room-temperature diazotizations. This is in sharp contrast with 1,2,3-benzothiazole-1,1-dioxide **115**, which is a well-known benzyne precursor at 0 °C and which explodes spontaneously in the solid state[75]. It is a plausible intermediate in the diazotization of **116**, which affords biphenylene and dibenzo-

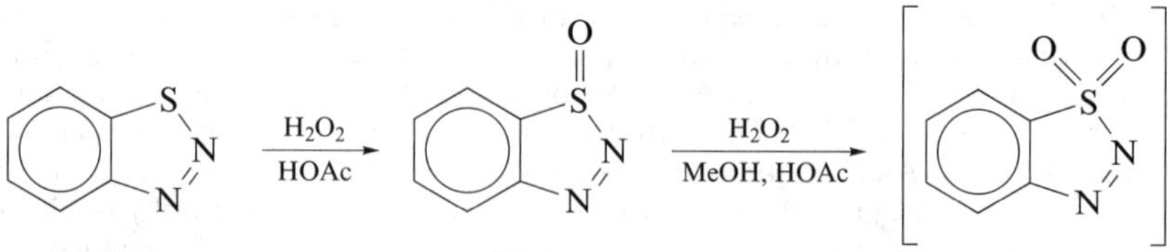

(113)   (114)   (115)

1,4-thioxin-S-oxide **117**. The same products were obtained by hydrogen peroxide oxidation of **114**.

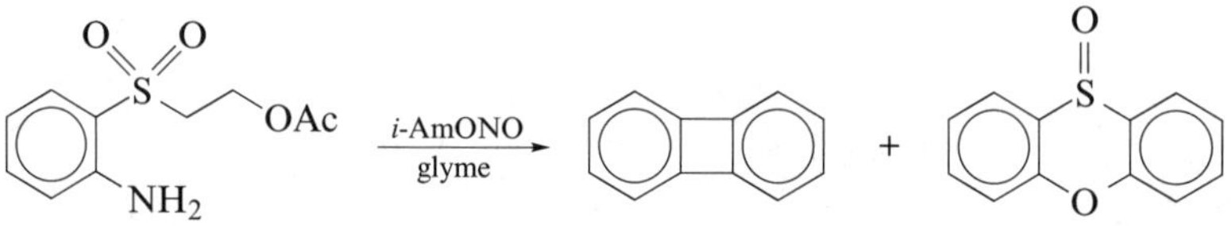

(116)   (67)   (117)

Benzyne is formed when 1-chlorobenzvalene **118** is treated with *n*-BuLi at –105 °C, as shown by trapping with furans or BuLi[76]. Apparently 1,6-dehydrobenzvalene **119**, a presumed intermediate, is unstable with respect to benzyne even at that low temperature. 2,3-Naphthyne was similarly formed from the benzo analog of **118**.

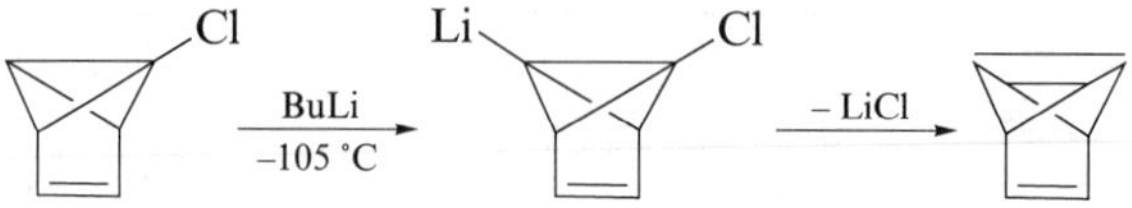

(118)   (119)

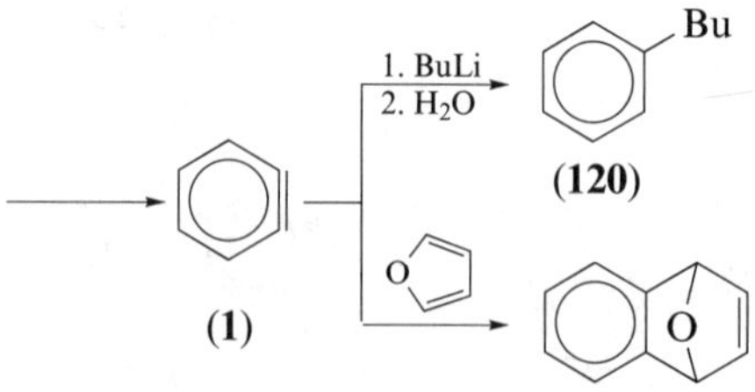

(120)

(1)

(27)

1-Aminobenzotriazole and related triazoles, all of which are aryne precursors, are autocatalytic inactivators of cytochrome P-450 and related oxidases[77]. Evidence has been presented that the enzyme oxidizes these substrates to benzyne which then binds covalently to the heme prosthetic group of the enzyme, possibly as in **120**, thus deactivating it.

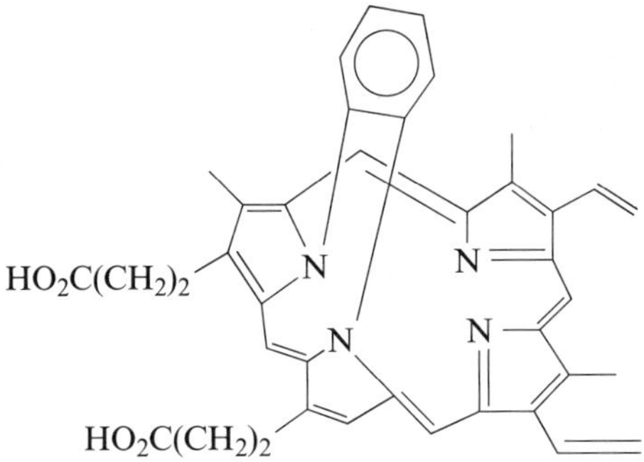

(120)

IV. THE BENZYNE–CYCLOPENTYLIDENECARBENE REARRANGEMENT

Benzyne (**1**) and cyclopentylidenecarbene (**121**) are isoelectronic. The formation of benzyne from **121** in the gas phase was first proposed in 1979[78], and the ensuing studies have recently been reviewed[79]. The rearrangement of **121** → **1** was suggested from pyrolysis results with the alkylidene Meldrum's acid **122**. Besides high yields of the expected carbon dioxide, acetone and cyclopentadiene, substantial amounts of aromatic products such as biphenylene (27%) and triphenylene (12%) were obtained that were most easily accounted for as a consequence of rearrangement of carbene **121** to benzyne and subsequent di- or trimerization.

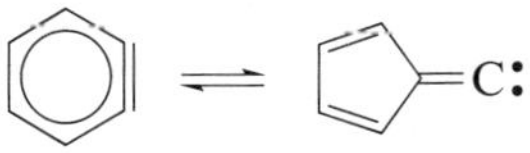

(1)   **(121)**

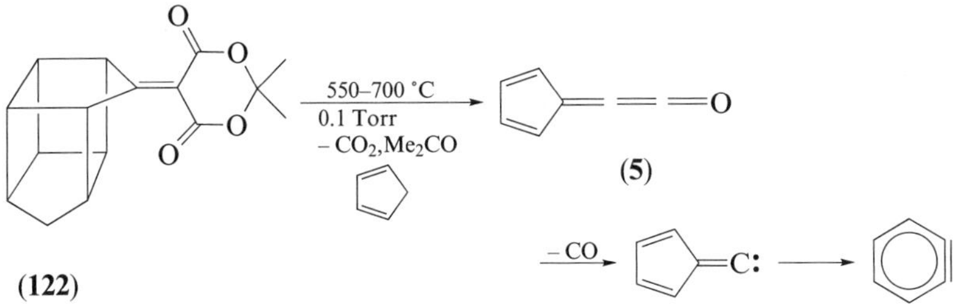

(122)   **(121)**   **(1)**

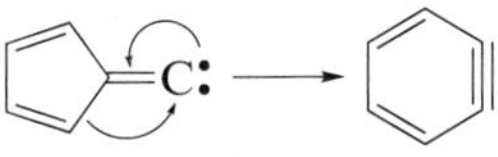

(122)   **(1)**

It should be recalled that matrix photochemical studies[15,16] discussed in Sections II.A and II.B require the interconversion of **1** and **121**. That this rearrangement may occur in solution at ordinary temperatures was suggested by the reactions of 6-halopentafulvene **123**[80]. With lithium piperidide as the base, **123** gave the piperidinopentafulvene **125** at – 70 °C, but the piperidinobenzene **126** at 25 °C. The authors suggest that carbene **128** is captured at low temperature to give **125** but rearranges to 3,5-di-*t*-butylbenzyne **129** in the room-temperature experiment. Additional evidence for the intermediacy of **129** was provided by trapping with diphenylisobenzofuran to give **127**. However, attempts to trap carbene **128** (by [2+1] cycloadditions) were unsuccessful, suggesting the possibility that carbenoid **124** may rearrange directly to **129**.

Benzyne is calculated to be appreciably more stable than cyclopentylidenecarbene[25,81,82], and indeed the latest calculations[82] suggest an energy difference of 31 kcal mol^{-1} with **121** as a shallow energy minimum, bound by less than 5 kcal mol^{-1}. Nevertheless, the question of the reversibility of this rearrangement (i.e. $\mathbf{1} \rightarrow \mathbf{121}$) at the very high temperatures used to generate benzyne from gas-phase precursors such as phthalic anhydride has been raised[83].

Whereas diazotization of labelled anthranilic acid **130** at 84 °C (in the presence of excess unlabelled acid) gave biphenylene **131** without label rearrangement, pyrolysis of labelled phthalic anhydride **132** (again, with excess unlabelled anhydride, so that the presence of tetra-labelled biphenylene in the product is virtually nil) at 830 °C gave approximately a 1:1 mixture of **131** and **133**[83]. Similar labelling results were obtained in the pyrolysis of **134** cf **11**) at 650–830 °C.

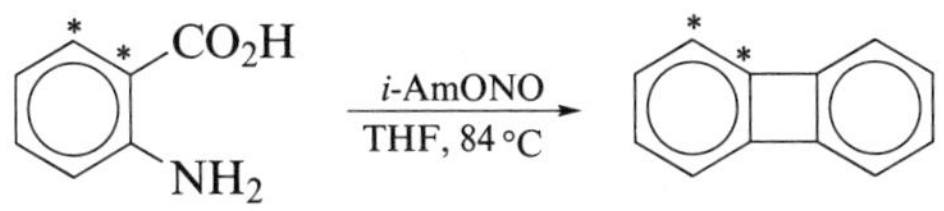

(130)   (131)

$* = {}^{13}\text{C}$

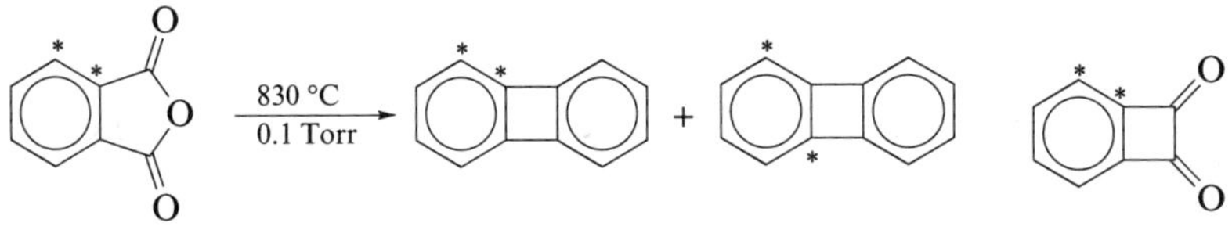

(132)   (131)   (133)   (134)

$* = {}^{13}\text{C}$

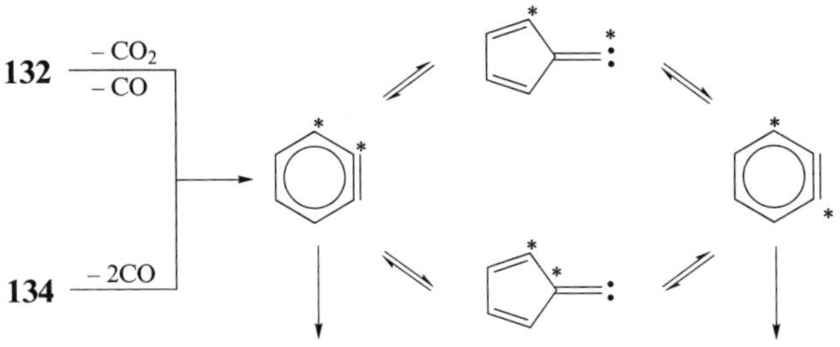

(131)   (133)

Two explanations have been offered to account for this label scrambling. In the first[83], benzyne–cyclopentylidenecarbene interconversion rationalizes the result. In the second[9b], scrambling is proposed to occur *prior* to benzyne formation, as a consequence of the symmetry of **132** (and **134**), and the two ways that CO_2 (or CO) might be lost in the first step to give ketenecarbenes **135** and **136**. Wolff rearrangement to **137** and **138** respectively followed by loss of CO and rearrangement to labelled benzynes (thermodynamically favorable) leads to the final product.

A choice between the two explanations remains somewhat elusive. Although the *photochemical* interconversion of benzyne, carbon monoxide and cyclopentylideneketene **5** has been demonstrated at low temperatures in neon and argon matrices[15], there is as yet no clearcut evidence for **5** (i.e. **137** + **138**) in the pyrolysis of phthalic anhydride[84], although it has been detected (IR band at 2089 cm^{-1}) in the pyrolysis of other precursors such as **139** and **140** (600–700 °C, then trap in Ar matrix at 10 K); biphenylene is also formed, from benzyne generated via **121** by CO loss from **5**[85].

Pyrolytic studies with **141**[86], which produces benzyne and biphenylene at lower temperatures than phthalic anhydride, showed increasing scramble of label in the biphenylene with increasing temperatures (400–850 °C). This result is thought to support the thermal conversion of benzyne to cyclopentylidenecarbene mechanism.

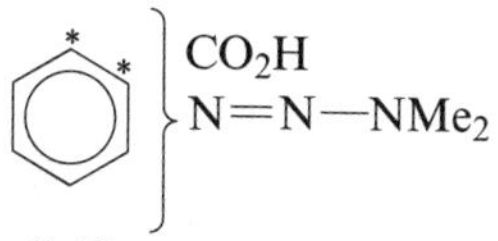

(141)

Pyrolytic studies on several other aromatic anhydrides proceed with ring contraction, presumably via aryne–cyclopentylidenecarbene interconversion. Tetraphenylphthalic anhydride **142** gave the green hydrocarbon **143**, presumably via insertion of the intermediate carbene into the C—H bond of an adjacent phenyl group[87]. 1,2-Naphthynes appear

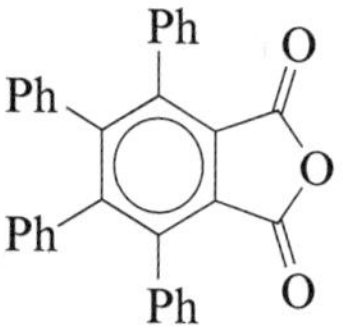

(142)

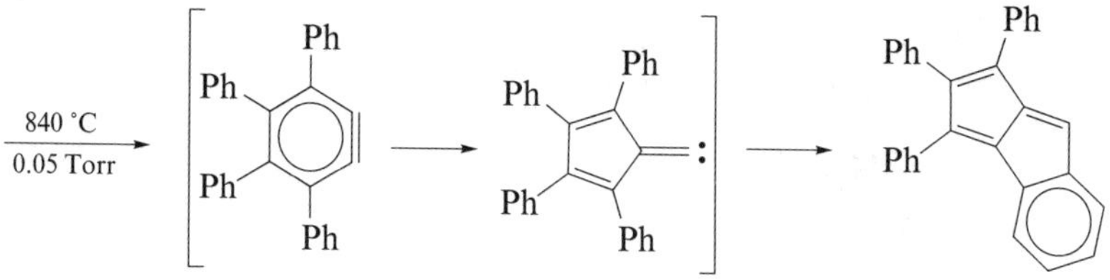

(143) 74%

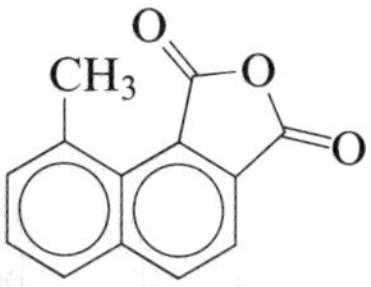

(144)

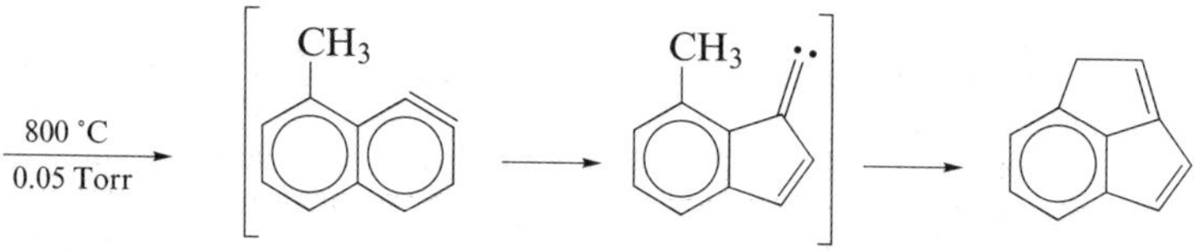

(145) 85%

to undergo ring contraction, but 2,3-naphthynes (which would have to disturb the aromaticity of the adjacent arene ring) do not. For example, pyrolysis of **144** gave **145**[88] and **146** gave **147**[89]. On the other hand, **148**, which should generate the aryne **149**, gave fluoranthene, most likely via a radical mechanism[89]; ring contraction to **151**, in which the aromaticity of the benzenoid ring is perturbed, is not observed. There is no evidence that 2,3-didehydronaphthalene can rearrange, via hydrogen migration, to the more stable 1,2-isomer[90].

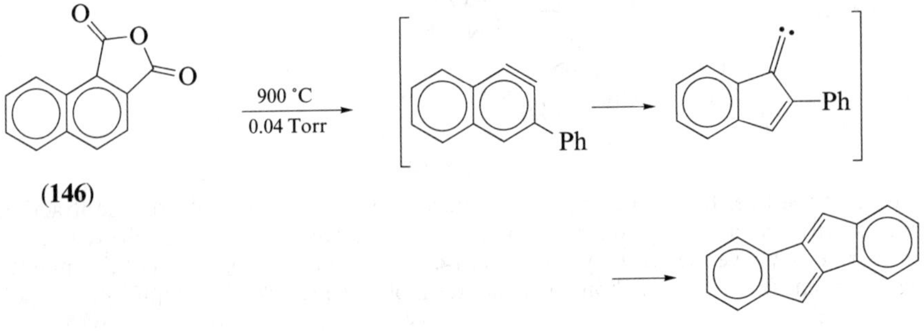

(146)

900 °C
0.04 Torr

Ph

Ph

(147) 95%

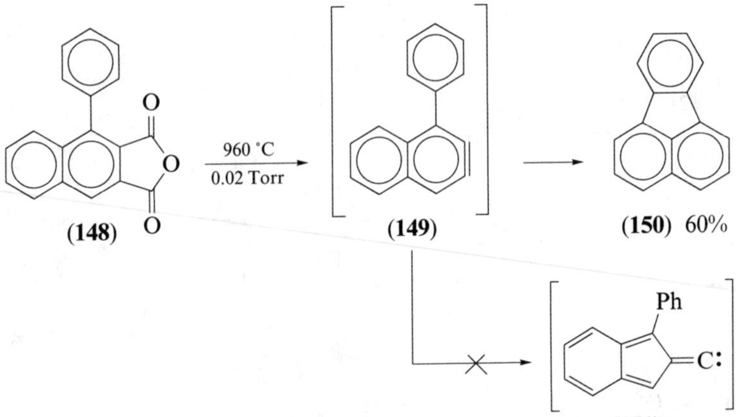

(148)

960 °C
0.02 Torr

(149)

(150) 60%

Ph

C:

(151)

3,4-Didehydrophenanthrene **154** and the ring-contracted carbene **155** are thought to be intermediates in the pyrolysis of anhydride **152** and dione **153**[91]. The observed products were phenanthrene **156** and the two ethynylacenaphthylenes **160** and **161**, which may be formed as indicated.

The conversions of **144** → **145** and **146** → **147** involve carbene insertion into a methyl and phenyl group, respectively. In the *o*-tolyl anhydride **162** both options are possible, but in fact its pyrolysis gave only products derived from interaction with adjacent methyl group[9]. The products were fluorene **168** (49%) and the two benz[*e*]indenes **169** and **170** (11% each). The initial intermediate is aryne **163**, possibly the source of fluorene as shown, though other routes to this product cannot be ruled out. Ring-contraction of **163** gives carbene **164** which, through insertion into the methyl group, leads to the other two products.

As can be seen from several of these reactions, what began as an interesting mechanistic question, the interconversion of arynes and cyclopentylidenecarbenes, has evolved into a useful synthetic method for certain polycyclic aromatic compounds.

V. CYCLOADDITION REACTIONS OF ARYNES

A. General Remarks

One of the two reaction types which dominate the synthetic utility of arynes as reactive intermediates is cycloaddition. Cycloadditions can be subdivided into several categories. For example, benzynes undergo [4+2] and [2+2] cycloadditions as well as 1,3-dipolar cycloadditions, the 'ene' reaction, and miscellaneous others. These reactions may occur in an inter- or intramolecular mode. Further, multiple and tandem aryne reactions can be used for multiple-bond construction in a single step. Each of these reaction types, as well as a few miscellaneous reactions, is discussed in the following sections.

B. Diels–Alder Cycloadditions

The general features of Diels–Alder reactions of arynes are thoroughly discussed in an earlier review[1] and are only briefly summarized here. Because of their considerable electrophilicity and high reactivity, arynes are particularly aggressive but short-lived dienophiles. Practical consequences of this behavior are that cycloaddition proceeds best with cyclic dienes whose cisoid conformation is consistent with the Diels–Alder transition state for good orbital overlap with the aryne 'triple' bond. Thus five-membered cyclic dienes, including five-membered aromatic heterocycles, are particularly successful diene partners for arynes. However, acyclic dienes and larger ring cyclic dienes may also be successful provided that the cisoid and nearly planar conformation is readily attainable. Also, arynes are sufficiently potent electrophiles that they can cycloadd to aromatic systems. The less disruption of aromaticity, the more facile are the reactions. The most common aromatic partner of arynes is the central ring of anthracenes, though naphthalenes and even benzenes may also cycloadd, especially to the more reactive arynes i.e. to those with electron-withdrawing substituents.

It should be noted that because of the extreme reactivity of arynes, other reactions such as 2+2 cycloaddition or the ene reaction may complete with the Diels–Alder mode of addition more frequently than is the case with less reactive dienophiles. This is especially true if the diene component is not conformationally predisposed to the 4+2 cycloaddition mode. Also, as with other dienophiles, Diels–Alder reactions of arynes are stereoselective.

This section is organized according to type of diene partner for arynes.

1. Acyclic dienes

2,3-Disubstituted butadienes are good Diels–Alder partners for benzyne. The method affords, after dehydrogenation, a 6+4 carbon assembly route to naphthalenes. Addition of benzyne to **173** affords naphthalene containing 98±2% D, whereas the resulting naphthalene from **174** contained only 2±2% D[95]. Thus not only is the cycloaddition stereoselective, but the elimination from dihydronaphthalene **175** is almost exclusively *syn*.

(171)   (172)

R = CH$_3$[93] (14%) R = CO$_2$Me[94] (59%)

(173) R^1 = D, R^2 = H   (175)   (176)
(174) R^1 = H, R^2 = D

Yields of benzyne cycloadducts with 1-vinylcyclohexene or 1-vinylcyclopentene are quite low[93] but with 1,2-bis-methylenecycloalkanes the yields can be quite good (**177** → **178**).

Reaction of bis-diene **179** with benzyne gives both the mono-adduct **180** and the bis-adduct **181**[97]. Cycloaddition of 3-methoxybenzyne to diene **182** was regioselective, giving, after dehydrogenation, **183** and **184** in a 4:1 ratio.

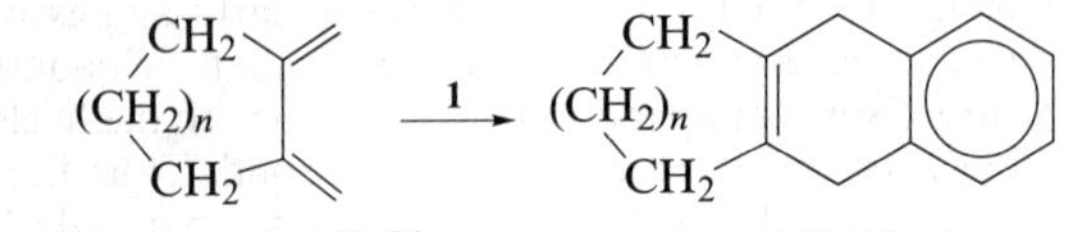

(**177**) $n = 0, 1, 2$[93, 96]   (**178**)

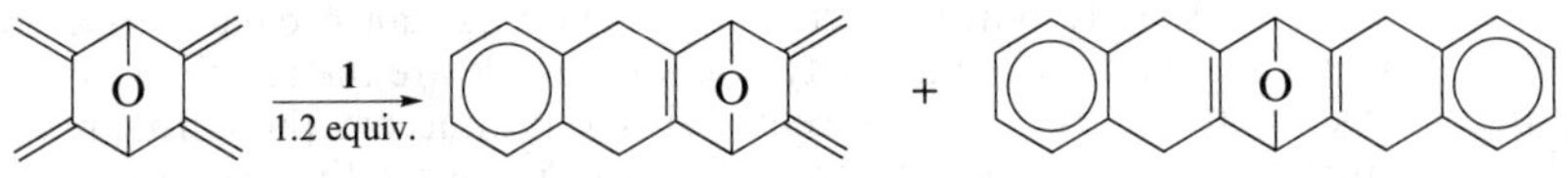

(**179**)   (**180**) 20%   (**181**) 5%

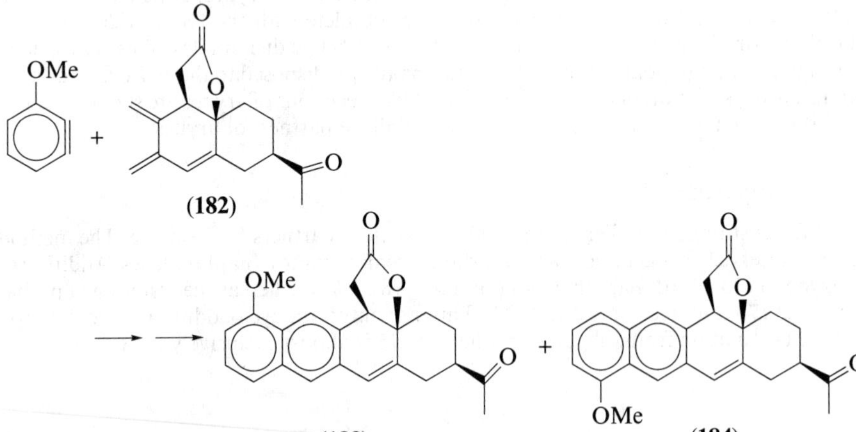

(**182**)

(**183**)   (**184**)

2. Cyclic dienes

The reaction of benzynes with simple cyclopentadienes[99,100] and fulvenes[101] continues to provide the most direct route to benzonorbornadienes. Addition of benzyne (and other dienophiles) to fused ring cyclopentadienes is π-facially selective[102]. For example, cycloaddition to **185** gives *endo* adduct **186** (71%) which, in a separate step, adds a second equivalent of benzyne [2+2] fashion to give **187**. With **188**, the monoadduct **189** could not be isolated since the 2+2 cycloaddition at the very strained central double bond of **189** is faster than the first step, giving only **190** (84%) regardless of the benzyne/**188** mol ratio. The relationship between **187** and **190** was established by hydrogenation. The exclusive *endo* attack on **185** and **188** has been rationalized theoretically in two different ways[102,103].

Benzyne addition to fulvenes **191** also occurs exclusively from the *endo* face[104], but when the 1-carbon bridge in **188** is expanded to a 2-carbon bridge, as in **193**, cycloaddition of benzyne is less stereoselective and *exo* attack predominates (**195**:**194** = 81:19), a result which is contrary to what might have been predicted on steric grounds[102,103]. In all of these studies on π-facial selectivity, benzyne was generated via the benzenediazonium carboxylate route.

(185)   **(186)**   **(187)**

(188)   **(189)**   **(190)**

(192)

(191) X = H, Me, OMe, Cl, F, NO$_2$

(193)   **(194)**   **(195)**

Cyclohexadiene **196** also reacts stereoselectively with benzyne, attack at the carbonyl face being preferred (**197:198** = 9:1)[105]. Cyclohexadiene **199** gave cycloadduct **200**, but in only 4.3% yield; benzyne in this example was generated from *o*-fluorobromobenzene and magnesium[106].

Azulene (**201**) reacts with benzyne to give cycloadduct **202**, the yield being best (33%) when **1** was generated via Pb(OAc)$_4$ oxidation of 1-aminobenzotriazole[107]. The product is derived from addition across the cyclopentadiene moiety of **201**. Although a concerted cycloaddition is possible, reaction may take place via electrophilic attack of benzyne at C1 of azulene (known to be prone to electrophilic attack) to form zwitterionic intermediate **203**, followed by ring closure at C3a.

(196)   **(197)**   **(198)**

(199)   **(200)**   **(201)**   **(202)**

1,6-Methano[10]annulene **204** adds benzyne with ring closure to give **205** (18–45%)[108,109].

(203)   **(204)**   **(205)**

Benzyne cycloadditions have been useful in synthesizing sterically crowded naphthalenes[110]. Recent examples include 1,2,4-tri-*t*-butylnaphthalene **207** from cyclopentadienone **206**[111]. Tetra-*t*-butylcyclopentadienone, however, turned out to be inert toward benzyne. The synthesis of **210** could be accomplished by adding benzyne to tetra-*t*-

(206)   **(207)**

(208)   **(4 equiv)**   **(209)**   **(210)**

butylcyclobutadiene. Adduct **209**, obtained in 49% yield, rearranged thermally to the desired **210**[112].

Benzynes add to cyclopentadienones; sometimes the carbonyl group is retained, as in **212**[113], but more often it is lost[114] during the cycloaddition to give aromatic product, as in the conversion of **214** to **215**[115].

(211)   (212)   (213)

(214)   (215) 92%

3. Benzenoid aromatics

Arynes are among the few dienophiles that react sufficiently exothermically to be capable of adding to and hence disrupting the aromaticity of a benzene ring. With simple benzynes a large excess of the arene is usually required for synthetic utility[116], but electron-withdrawing substituents on the aryne or electron donors on the benzenoid ring enhance the reaction, as expected. Also, intramolecular attack may be more favored than inter-molecular attack on a simple benzenoid ring. Thus cyclophane **216** gave internal cycloadduct **218** in 66% yield via aryne **217**[117] (cf **84** → **88**). More spectacular, because the molecule is neither strained nor conformationally constrained, is the intramolecular cycloaddition that **219** undergoes when treated with excess n-BuLi. Adduct **220** was obtained in 25% yield, possibly via a lithiated intermediate such as **221**[118].

(216)   (217)   (218)

The susceptibility of a benzenoid ring to dienophilic attack may be enhanced through metallation. For example, **225** was one of two products obtained from the reaction of hexabromobenzene with mesitylmagnesium bromide[119]. This product clearly arises from cycloaddition of arynes derived from **222** to a mesitylene ring (a small amount of the C_{2v}

regioisomer of **225** is also formed). Deuterium labeling quenching and competition experiments with added mesitylene showed that aryne addition occurs across the 2,5-positions of **223** and that the reaction of the Grignard reagent **223** with benzyne interme-diates is at least 27× faster than is cycloaddition to the corresponding hydrocarbon. (Evidence for intermediates such as **226** and the mechanism by which **224** is formed in this reaction is discussed in Section VI.D.)

(**219**)   (**220**)

(**221**)

(**222**)   (Mes-MgBr)
                  (**223**)

(**224**) 30%   (**225**) 20%

(**226**)   (**223**)   (**227**)   **225**

Styrenes may give Diels–Alder adducts with arynes, the exo-cyclic double bond being part of the diene moiety[1]. Yields are generally rather low. The competition between [4+2] cycloaddition and Michael addition of benzyne (generated from chlorobenzene and sodium amide) to α-dimethylaminostyrene **228** had been investigated experimentally and theoretically[120]. The yield of **229** (isolated as phenanthrene after aromatization on chromatography) was 10%; the dimethylaminostilbene **230** (approx. 30%) was identified by mass spectrometry.

(**228**)   (**229**)   (**230**)

Polynuclear aromatics such as naphthalenes and anthracenes, whose aromaticity is less disrupted by cycloadditions, are common substrates for reaction with arynes. Tetrafluorobenzyne cycloadds only to the unsubstituted ring of 1,2,3,4-tetrafluoronaphthalene (**231**) to give **233** in low yield[121]. The aryne was presumably generated from Grignard reagent **232** at a rather high temperature and over an extended time. If the corresponding lithio analog was used at –15 to 0 °C, no **233** was formed, the only isolable product being nonafluorobiphenyl. In contrast, the lithio analog reacted with naphthalene itself at 0 °C to give cycloadduct in 27% yield.

(**231**)   (**232**)   (**233**) 12%

Ever since Wittig's one-step synthesis of triptycene (**234**) from benzyne and anthracene[122], aryne cycloaddition to anthracenes constitutes a preferred route to a variety of substituted triptycenes. The rigid framework of triptycenes is attractive for studying many phenomena in organic chemistry, including restricted rotation at tetrahedral carbon, hydrogen bonding, molecular gearing, intramolecular rearrangements and so on. A complete list is beyond the scope of this review, but Table 1 lists some of the more interesting examples of triptycenes synthesized via the aryne–anthracene route. Most commonly the diazonium carboxylate precursor to the aryne is used, but other methods have also been employed.

(**234**)

TABLE 1. Triptycenes via aryne–anthracene cycloadditions

Aryne	Anthracene	Reference
Benzyne	1-OMe-9-Et	123
Benzyne	1-F(Cl,Br)-9-i-Pr	124
Benzyne	1-F-9-(1,1-Me$_2$-2-phenylethyl)	125
Benzyne	9-(3,5-Me$_2$benzyl)	
	(and 1-OMe; 2-OMe; 1,8-Cl$_2$ analogs)	126
Benzyne	1,8-Cl$_2$-9-benzyl	126
Benzyne	9-[1-(3,5-Me$_2$phenyl)ethyl]	126
Benzyne	9-allyl (and 1,4-Me$_2$; 2,3-Cl$_2$ analogs)	127
Benzyne	1,8-bis(aza)-2,7-di-t-Bu	
	(and 2,7-di-Ph analog)	128
Benzyne	bis(9-anthryl)methane	129
3-F; 3-Cl; 3-Br; 3-CF$_3$	9(1,1-Me$_2$-2-phenylethyl)	125
4,5-Br$_2$	2,6-di-t-Bu	130
3,6-Me$_2$	1-Cl-9-(2-Me benzyl)	131
3,6-Me$_2$	1-Br-9-(2-Me benzyl)	123
3,6-Me$_2$	9-aryloxymethyl [Ar=p-OMe-C$_6$H$_4$; C$_6$H$_5$;	132
	p-NO$_2$-C$_6$H$_4$; 3,4-(NO$_2$)$_2$-C$_6$H$_3$]	
3,6-Me$_2$	ethylene glycol acetal of 9-formyl	133
3,6-Me$_2$	9-(1,1-Me$_2$-3-butenyl)	134
3,6-Me$_2$	9-(1,1-Me$_2$-2-phenylethyl)	125
3,6-Me$_2$	1,8-Cl$_2$-9-(3,5-Me$_2$ benzyl)	
	(and 1,8-F$_2$ analog)	126
3,6-Me$_2$	1,4-Me$_2$-9-allyl	127
3,5-di-t-Bu	1-F(Cl, Br)-9-i-Pr	124
3,6-Cl$_2$	9-(1,1-Me$_2$-2-phenylethyl)	125, 135
3,6-(OMe)$_2$	9-(1,1-Me$_2$-2-phenylethyl)	125
Cl$_4$	9-(1,1-Me$_2$-2-phenylethyl)	135, 136
Cl$_4$	cis- and $trans$-1,2-Me$_2$-aceanthrene	137
Cl$_4$	Z- and E-9-(1-Me-1-propenyl)	137
Cl$_4$	9-allyl	127
Cl$_4$	1,2,3,4-Cl$_4$-9-allyl	127
Br$_4$; F$_4$	9-(1,1-Me$_2$-2-phenylethyl)	136

(235)   **(236)**   T$_p$ =

A novel example of this reaction is the double benzyne addition to cyclopropenone **235** to give **236**, an intermediate en route to the novel tritriptycyl cyclopropenium cation[138].

A series of 1,8-disubstituted anthracenes **237** was treated with 3-substituted benzynes **238** to give *syn-* and *anti-*triptycenes **239** and **240**[139] (8 examples; no product was obtained when X = CN, Y = Cl). Substantial regioselectivity was apparent in most cases (*syn* when Y = Me, *anti* in most other examples), and was rationalized in terms of the electron-donating or-releasing properties of the substituents.

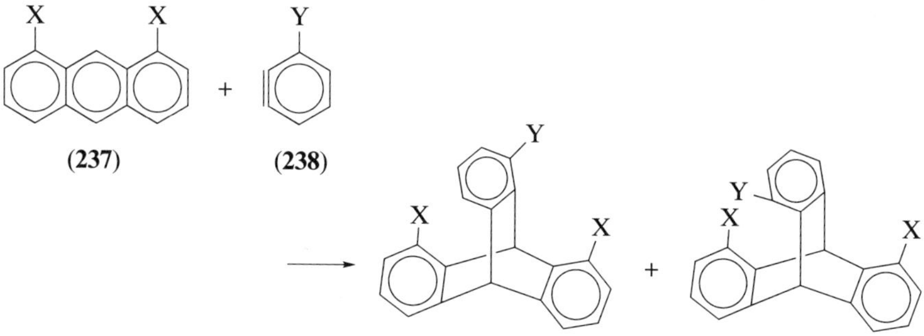

(237) (238) (239) (240)

(X = Cl, CN, CO₂Me)   (Y = Me, Cl, CO₂Me)

4. Five-membered aromatic heterocycles

The reactions of benzyne with various kinds of heterocyclic compounds has been reviewed[140]. Aromatic five-membered heterocycles **241** (X = O, N, S, Si, Ge, etc.) react with benzynes to give [4 + 2] cycloadducts (and sometimes other products) **242**, from which the X-bridge can be eliminated or rearranged in various ways, often with aromatization of the bicyclic ring. The methodology permits a new aromatic ring to be fused to an already existing one (in the aryne). Some recent examples are described here.

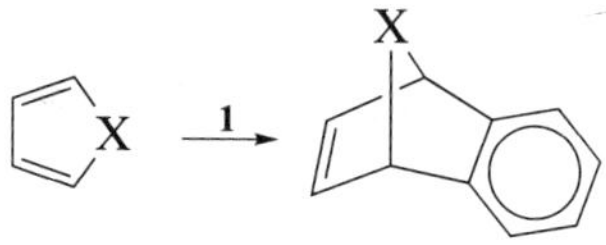

(241)   (242)

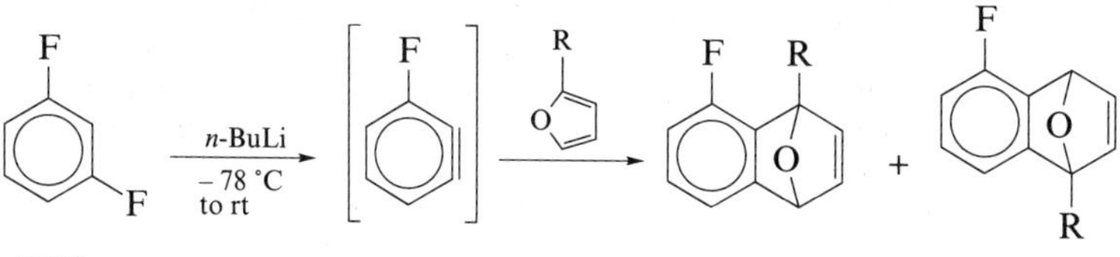

(243)   (244)   (245)

(R = Me, Et, *i*-Pr, *t*-Bu)

The cycloaddition of 3-fluorobenzyne to 2-alkylfurans shows considerable regioselectivity in favor of the *syn* isomer **244**[141]. The range is from 2:1 (R = Me) to 10:1 (R = *t*-Bu) and is rationalized in terms of a nonsynchronous Diels–Alder transition state that reflects the considerable polarization of 3-fluorobenzyne and the slight polarization of the 2-alkylfuran, coupled with a steric effect of the alkyl substituent.

3-Methoxybenzyne and 2-acetoxyfuran gave a mixture of both regioisomeric adducts in unspecified yield or ratio[142]. On the other hand, benzyne cycloaddition to **246** occurs exclusively *syn* to the spirocyclopropane functionality[143], a result that has been rationalized theoretically.

(246)   (247) 26%

2-Furancarbonamides **248** give good yields of cycloadducts with benzyne[144]. Reduction and flash vacuum pyrolysis of **249** led to isobenzofurans, which were again subjected to benzyne cycloaddition to form **250**, of potential interest as CNS-active agents.

(248)   (249)   (250)

Furanophanes give modest yields of cycloadducts with benzyne or 2,3-naphthyne[145].

(251)   (252) 25%

With macrocycle **253**, readily synthesized from furan and acetone, benzyne added to all four furan rings to give a single cycloadduct **254** in high yield[146]. Attempts to remove all four oxygen bridges from **254** failed. Benzyne also gave bis-adduct **256** in good yield from **255**, a partially ring-opened precursor of **253**. However bis-furan **257** failed to give a cycloadduct with benzyne.

Isobenzofurans are especially useful dienes toward arynes. 1,3-Diphenylisobenzofuran has long been used as a trap for benzynes[2], but only recently has the utility of this reaction been demonstrated through the generation of simpler isobenzofurans and their trimethylsilyl derivatives[58,147]. Treatment of acetal **258** with one equivalent of MeLi and a catalytic amount of diisopropylamine in ether gave solutions of **259**, and subsequent addition of *o*-

(253)   (254) 84%

(255)   (256) 97%   (257) 92%

(258)   (259)   (260)

(261)   (262)   (263)

bromochlorobenzene and one more equivalent of methyllithium generated benzyne, which added to 259 to give cycloadduct 260 (54%). Alternatively, use of three equivalents of BuLi with 258 generated dilithio derivative 261, which was trapped with trimethylsilyl chloride to give **262**. Subsequent addition of one equivalent of bromobenzene and LTMP afforded crystalline benzyne adduct **263** (59%), which could be converted to **260** by displacement at silicon. The advantage of the second route is that a variety of simple haloarenes can be used as sources of a variety of arynes in the second stage of the sequence. Two examples are shown.

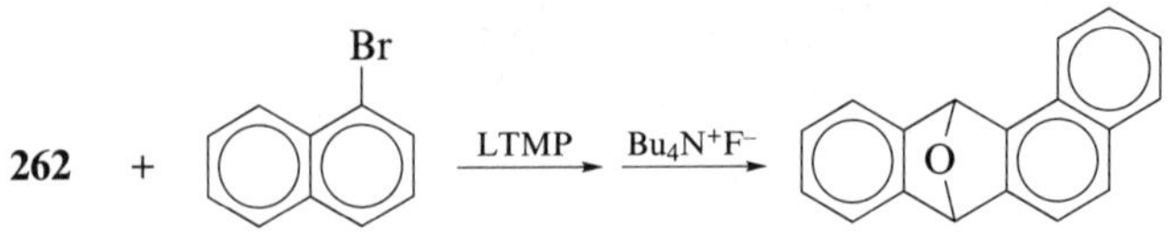

262 + [structure with Br] $\xrightarrow{\text{LTMP}}$ $\xrightarrow{\text{Bu}_4\text{N}^+\text{F}^-}$ [structure]

(**264**) 70%

262 + [structure with Br and CH$_3$] $\xrightarrow{\text{BuLi}}{\text{LTMP}}$ $\xrightarrow{\text{KOH}}{\text{Me}_2\text{SO}}$ [structure with CH$_3$]

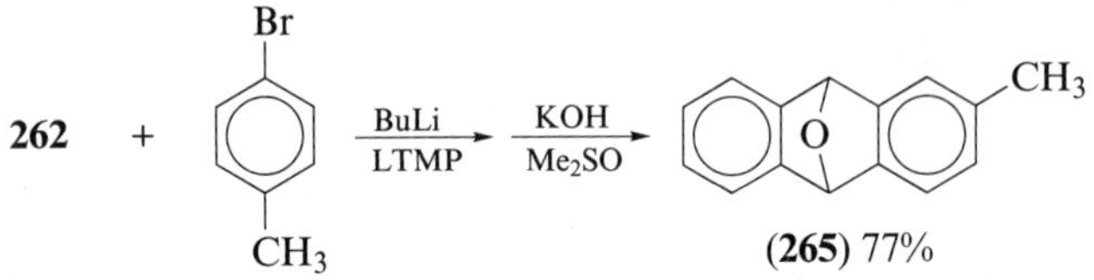

(**265**) 77%

Similar transformations can be carried out with *ortho* ester **266** which, in a one-pot procedure, was converted to **268** via isobenzofuran **267** in 89% yield[148]. The regioselectivity of these cycloadditions with substituted arynes has been investigated[149] and found to be slightly in favor of that isomer (**269**) in which the EtO and R groups are *syn*.

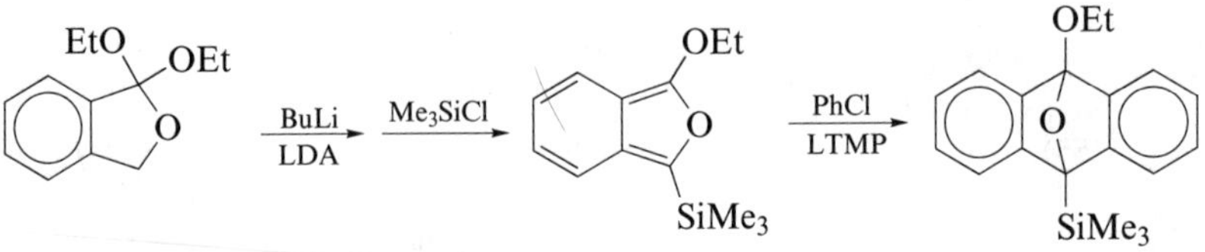

[structure EtO, OEt, O] $\xrightarrow{\text{BuLi}}{\text{LDA}}$ $\xrightarrow{\text{Me}_3\text{SiCl}}$ [structure OEt, O, SiMe$_3$] $\xrightarrow{\text{PhCl}}{\text{LTMP}}$ [structure OEt, O, SiMe$_3$]

(**266**)   (**267**)   (**268**) 89%

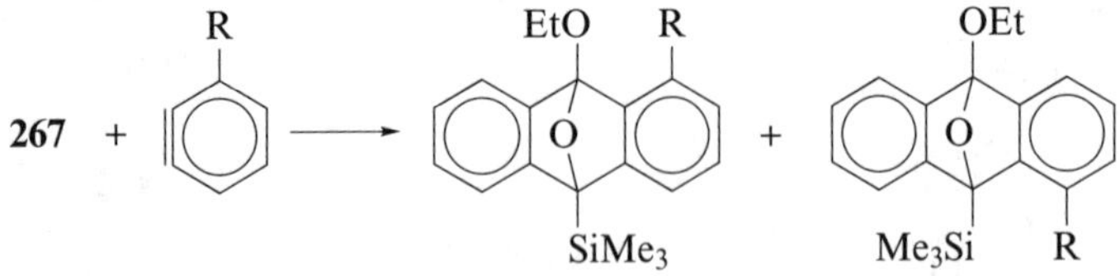

267 + [structure with R] $\longrightarrow$ [structure EtO, R, O, SiMe$_3$] + [structure OEt, O, Me$_3$Si, R]

(**269**)   (**270**)

R = Me   Cl   Br   OMe
269/270 = 1.5   1.1   1.7   4.0

Aryne cycloadditions to furans have been employed in the synthesis of bio-related molecules. For example, adduct **272** served as the starting point for the preparation of the aromatic segment of rifamycin W (an antibiotic)[150]. Previously unreported benzyne **274** (its precursor **273** was prepared in several steps from piperonal) readily gave cycloadduct **275** with furan, but unfortunately for a proposed route to podophyllotoxin, did not undergo

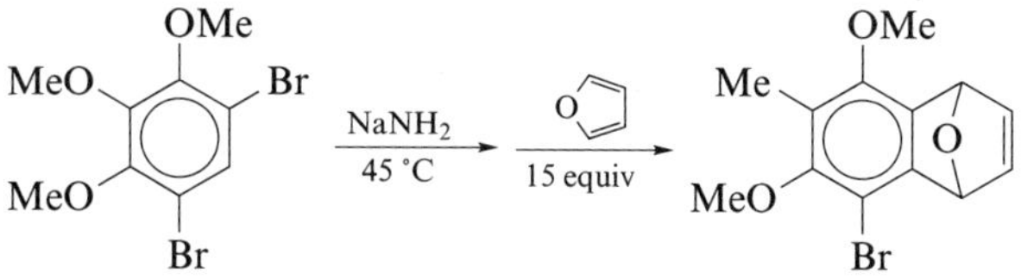

(271)

(272) 94%

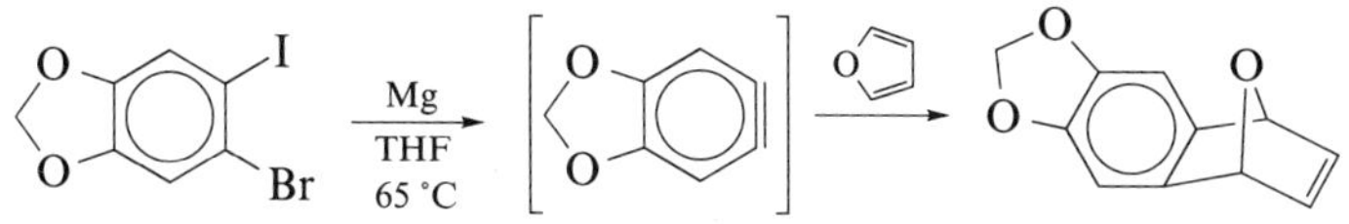

(273)

(274)

(275) 88%

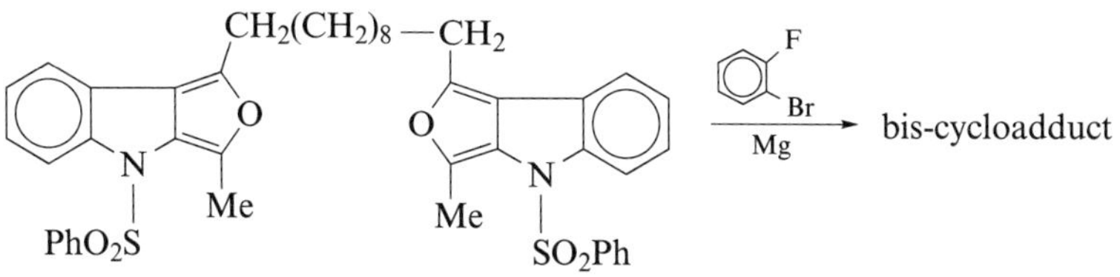

(276)

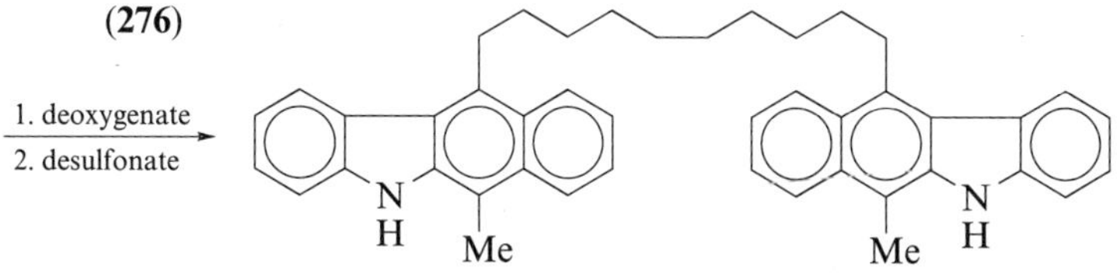

bis-cycloadduct

(277) 55% overall

[2+2] cycloadditions with a variety of olefins[151]. The proposed DNA intercalator **277** was prepared by a double benzyne addition to bis-furan **276**[152].

Aryne cycloadditions have been coupled with radical cyclizations to rapidly assemble multiring systems[153]. For example, benzyne addition to furan **278** followed by radical cyclization onto the newly formed double bond gave **280** in two steps and fair yield.

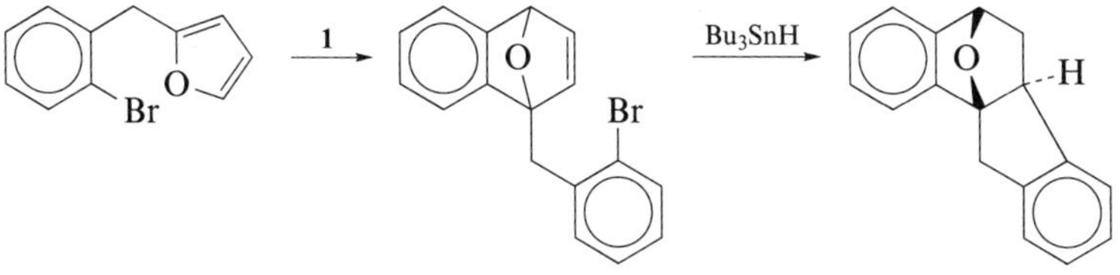

(278)

(279) 50%

(280) 67%

Pyrroles, like furans, undergo [4+2] cycloadditions with arynes. For good yields, however, it is important to have some functionality other than hydrogen attached to the nitrogen. A variety of 7-azabenzonorbornenes have been prepared in this way[2,154]. The reaction has been used to prepare and study the dynamics of nitrogen inversion in such compounds as **282**.

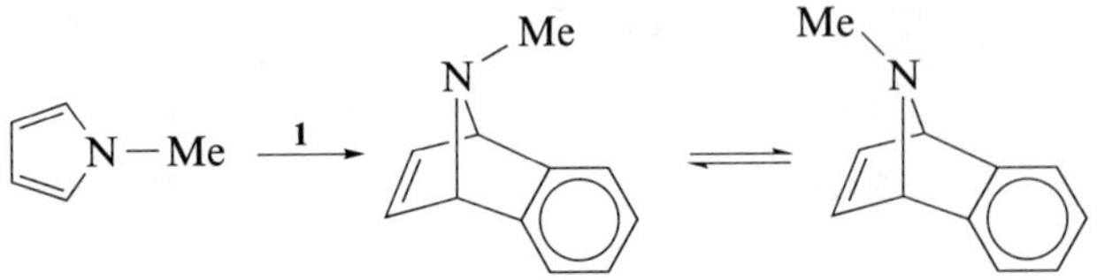

(281)   (282) *syn*   (282) *anti*

Isoindoles react to give good yields of cycloadducts. Interesting examples include the bis- and anti-cycloadditions of tetramethylbenzyne to bis-isoindoles **283** and **285**[155]. Pyrroloindole **287** gave benzyne adduct **288** in good yield[156], but the corresponding free pyrrole (H in place of CO$_2$Me) gave no adduct, emphasizing the need for protecting the pyrrole N—H.

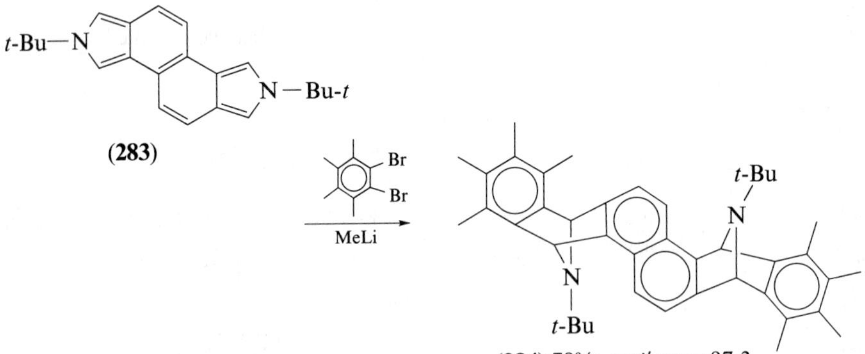

(283)

(284) 78%; *anti/syn* = 97:3

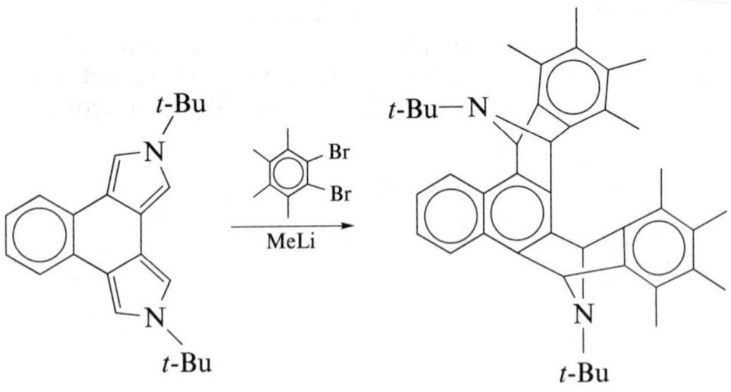

(285)   (286) 76%

The aromaticity of 3-vinylindoles is easily disrupted by cycloaddition to dienophiles such as benzyne[157,158]. For example, the 1,1-bis(3-indolylethene) **289** gave a mixture of cycloadduct **290** and its aromatic oxidation product **291**[158]. Reaction of **292** with benzyne gave the

(287) →[1] **(288)** 75%

(289) →[1, THF, 45 °C] **(290)** + **(291)**

(292) →[Excess 1] **(293)** 15%

benzo[a]carbazole **293** in low yield[157], through cycloaddition and further oxidation. The absence of a protecting group in the indole nitrogen probably contributes to the low yield in this example.

Thiophene is generally considered the most aromatic of the five-membered heterocycles, hence its reluctance to function as a diene in cycloadditions. A detailed study[159] of the reaction of thiophene with six common benzyne precursors showed that side reactions with intermediates en route to benzyne often dominate the reaction course. Best results were obtained with diphenyliodonium-2-carboxylate **294**, which gave up to 30% yields of naphthalene, presumably mainly through intermediate **295**. Other benzyne-derived products [benzo(b)thiophene, 0.5%; 2-phenylthiophene, trace; diphenyl sulfide, trace; 1- and 2-naphthyl phenyl sulfides, 9 and 6%, respectively; o-(2-thienyl)biphenyl, 4%] bring the

(294) →[220 °C] **(295)** →[− S]

aryne-derived yield to nearly 50%. This is larger than obtained from benzene and benzyne generated in the same way (10%), showing that thiophene is more reactive than benzene as a diene, but neither is particularly useful in this regard.

Decomposition of **294** in eleven substituted thiophenes established that [4+2] and [2+2] benzyne cycloaddition occur, but that the former predominates[160]. For example, 2-methylthiophene gave 24% of 1-methylnaphthalene [4+2] and only 2% of 2-methylnaphthalene [2+2], as shown. Small yields (< 4%) of benzo[*b*]thiophenes formed in these reactions may arise from a 1,3-dipolar addition mechanism initiated by benzyne attack on sulfur[159–161].

Although the reaction of arynes with thiophenes as a route to naphthalenes leaves much to be desired, the reaction with thiophene-1,1-dioxides is synthetically useful[162]. An excess of benzyne precursor (benzenediazonium carboxylate) is used, and the solvent was refluxing 1,2-dichloroethane. Yields of naphthalenes are moderate and conversions are good. Thus 3-methylbenzyne and tetrachlorothiophene dioxide **296** gives naphthalene **297** in 65% yield (96% conversion). The thiophene dioxides may be prepared by *m*-CPBA oxidation of the corresponding thiophenes.

Tin, germanium and silicon five-membered heterocycles have been treated with arynes. Stannole **298** (X = Sn) gave naphthalene **300**, presumably via the norbornadiene **299**

(X = Sn), which, however, could not be isolated, and aromatized (with elimination of [SnMe$_2$]) even at –20 °C[163]. On the other hand, the germanium analog **299** (X = Ge) could be obtained (71%) as a crystalline material[164], and underwent bridge elimination only at 70 °C in CCl$_4$ with $t_{1/2}$ = 40 min. Silicon analogs of **299** are considerably more stable. Thus **301**, prepared via the silacyclopentadiene and benzyne, melts at 226.5–227 °C[165].

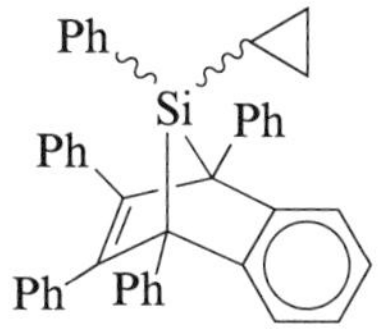

(301)

Of the five-membered heterocycles with more than one heteroatom, cycloadditions are most useful with oxazoles. When the benzyne is generated from anthranilic acid and isoamyl nitrite in refluxing dioxane, the initial adduct **303** eliminates nitrile to give isobenzofuran **304**, which then adds a second equivalent of benzyne to give **305**[166]. As already described (Section III. A. 4), if the benzyne is generated from 1-aminobenzotriazole and Pb(OAc)$_4$ using a high dilution technique, the intermediate **303** can be isolated[55]. Examples include R = Ph, p-NO$_2$C$_6$H$_4$ and p-An with R^2 = R^5 = H. The cycloaddition reaction of 2,5-diphenyloxazole (**302**, R^2 = R^5 = Ph, R = H) with benzyne is more complex than that of 4-phenyloxazole (**302**, R^2 = R^5 = H, R = Ph) because of subsequent reactions at the unsubstituted C=N bond of intermediate **303** (R^2 = R^5 = Ph, R = H) prior to its elimination of HCN to form diphenylisobenzofuran[55].

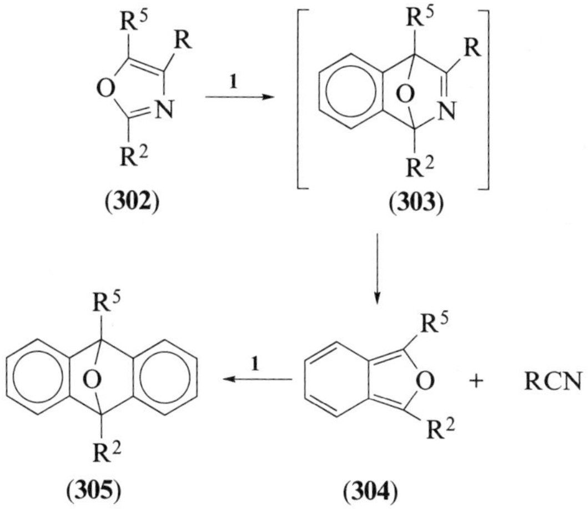

(302)   **(303)**

(305)   **(304)**   +   RCN

2,5-Bis-(trifluoromethyl)-1,3,4-oxadiazole reacts with benzyne in a similar manner. The initial adduct eliminates nitrogen to give an isobenzofuran, which then adds a second equivalent of benzyne to give the observed product **309**[167]. Interestingly, the sulfur analog of **306** did not react with benzyne.

$$\underset{(306)}{\text{[CF}_3 \text{ oxadiazole]}} \xrightarrow[\text{reflux THF}]{1} \underset{(307)}{\text{[intermediate]}}$$

$$\Big\downarrow - N_2$$

$$\underset{(309)}{\text{[CF}_3 \text{ epoxide tricycle]}} \xleftarrow{1} \underset{(308)}{\text{[CF}_3 \text{ isobenzofuran]}}$$

5. Six-membered heterocycles

The early literature has been thoroughly reviewed[140]. α-Pyrones are known to add benzyne, often with elimination of CO_2 from the resulting adduct. The method has been applied to pyranoindolones **310** to provide a short general route to benzo[*b*]carbazoles **311**[168].

$$\underset{(310)\ R = H,\ Me}{\text{[pyranoindolone]}} \xrightarrow[\substack{\text{reflux} \\ -CO_2}]{1,\ \text{ClCH}_2\text{CH}_2\text{Cl}} \underset{(311)\ 31-44\%}{\text{[benzocarbazole]}}$$

2-Pyridones react similarly, but without elimination of the heteroatom bridge. Adducts **313** were obtained in modest yield from *N*-methyl-2-pyridones **312**[169] and other *N*-alkyl derivatives[170]. With **312** ($R^1 = H$), however, the major products resulted from benzyne attack at oxygen (i.e. 2-phenoxypyridines) although some **313** ($R^1 = H$) was formed and, in a few instances, *N*-phenylation also occurred in low yield[169].

$$\underset{(312)}{\text{[pyridone with } R^3, R^4, R^5, R^6, R^1\text{]}} \xrightarrow{1} \underset{(313)}{\text{[bicyclic adduct]}}$$

A route to isoquinolines **315** with electron-withdrawing substituents has been developed through aryne cycloaddition to 1,2,4-triazines (**314**)[171]. Although 4-methylbenzyne reacted unselectively with triazines (to give mixtures of 6- and 7-methylisoquinolines), the reactions of 3-methylbenzyne with triazines **316** and **317** were highly regioselective in a strange way that as yet defies explanation; **316** gave the 5-methylisoquinoline **318** (30%) whereas **317** gave the 8-methylisoquinoline **319** (61%).

(**314**)   (**315**) 77%

(**316**) R^5 = Ph, R^6 = H (**318**) R = Me, R' = H; R^5 = Ph, R^6 = H
(**317**) R^5 = R^6 = Ph   (**319**) R = H,R' = Me; R^5 = R^6 =Ph

Benzyne (from anthranilic acid) cycloadds to mesoionic pyrimidines **320**; the adducts **321** were obtained in 42–62% yields[172]. Examples include R^1 = Ph or CH_2Ph, R^2 = Ph or CH_3 and R^5 = alkyl or CH_2Ph. Thermal decomposition of **321** (70–210 °C depending on substitution) gave near-quantitative yields of isoquinolones **322**. If the initial benzyne addition is conducted at *elevated* temperatures and if phenyl isocyanate elimination from the initial adduct **321** occurs at relatively *low* temperatures, then benzyne cycloadducts of the isoquinolones may also be formed (cf additions to **312**).

(**320**)   (**321**)   (**322**)

C. [2 + 2] Cycloadditions

In general, [2 + 2] cycloadditions to simple alkenes (to form benzocyclobutenes) proceed in only moderate to low yields, and by a stepwise rather than a concerted mechanism[1,173]. As expected, because of the electrophilic nature of benzyne, the reactions proceed best with alkenes bearing electron-donating substituents. Despite these limitations, the reaction sometimes offers a simple, direct, one-step route to useful synthetic intermediates.

1,1-Dimethoxyethene (**323**) is reported to be one of the most successful alkene partners for [2 + 2] cycloaddition to arynes[174]. The aryne is generated from a bromoarene precursor using sodium amide as the base, in either refluxing THF or no solvent. It is important that the sodamide be either fresh, or always opened under nitrogen in a glovebag; otherwise the reaction may be erratic, with an induction period[175]. Reactions of **326**, **328** and **332** were regiospecific; none of the regioisomer **334** was detected. 2-Bromoanisole gave the same product as 3-bromoanisole (**326**), implying a common benzyne intermediate (3-methoxybenzyne), and the same was true for **328** and 3-bromoveratrole **335**, the

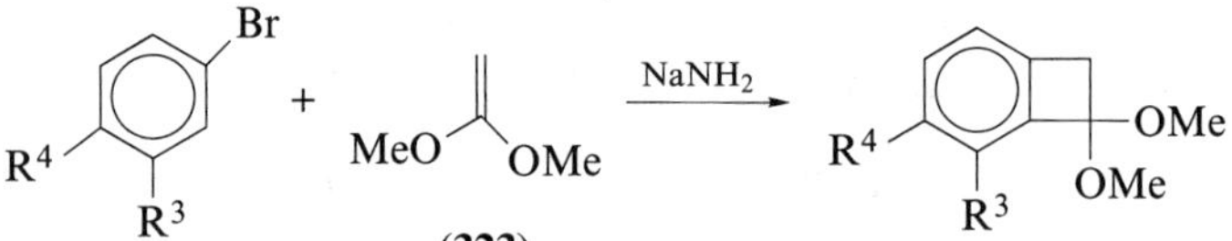

(**324**) $R^3 = R^4 = H$ (**325**) 63%
(**326**) $R^3 = OMe$, $R^4 = H$ (**327**) 76%
(**328**) $R^3 = R^4 = OMe$ (**329**) 68%
(**330**) $R^3 = H$, $R^4 = OMe$ (**331**) 56% + 8% **334**
(**332**) $R^3 = Cl$, $R^4 = H$ (**333**) 76%

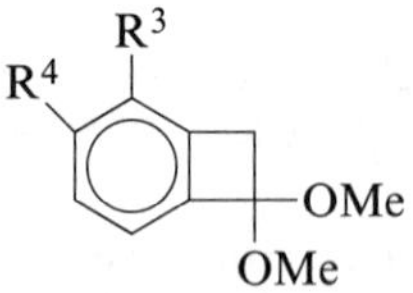

common aryne in this case being 3,4-dimethoxybenzyne. The regioselectivity can be rationalized if one considers the reaction as analogous to that of nucleophiles with arynes containing inductively electron-withdrawing substituents:

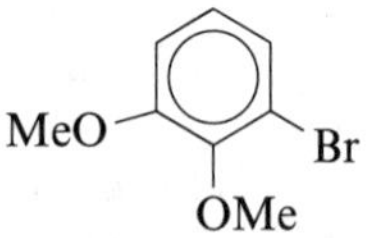

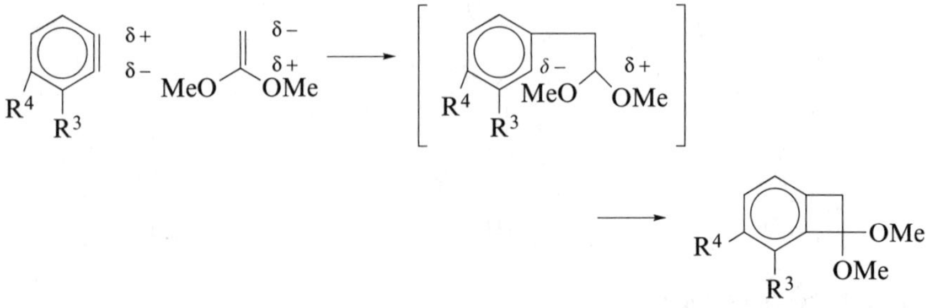

The dimethoxybenzocyclobutenes can be quantitatively hydrolyzed to benzo-cyclobutenones, providing useful functionality in the strained ring. Conversion of **336** to **337** in 79% overall yield provided a key step in the total synthesis of taxodione **338**[176]. The synthetic utility of the methodology was considerably extended by providing a good route from the initial cycloadducts to the corresponding benzocyclobutenediones[175].

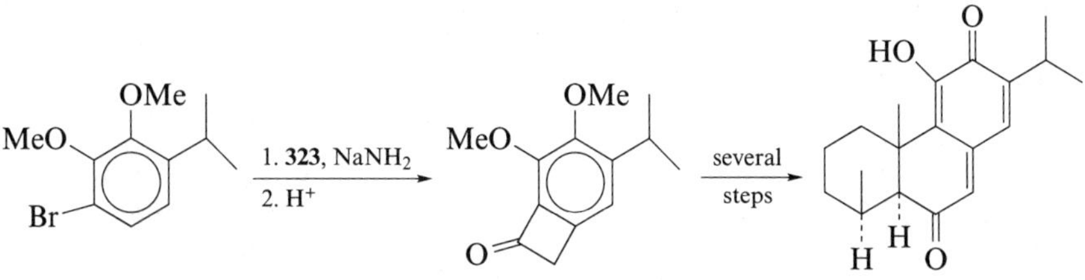

(336)   (337)   (338)

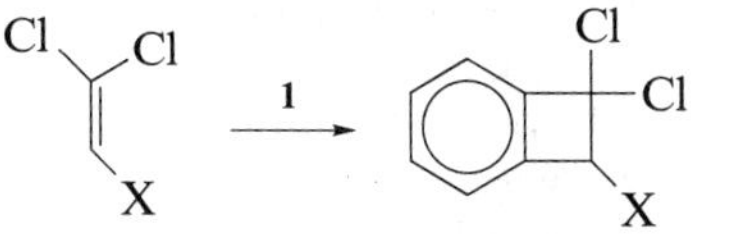

(339) X = H, Cl

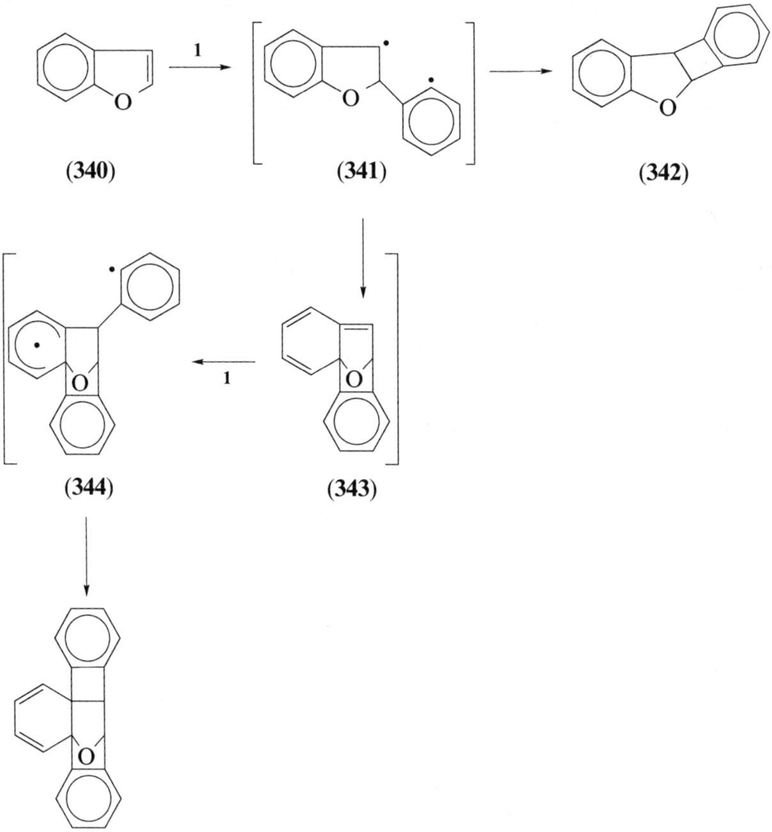

(340)   (341)   (342)

(344)   (343)

(345)

1,1-Dichloro- and trichloroethene give cycloadducts with benzyne (from anthranilic acid)[177]. Yields are modest (X=H, 27–61%; X=Cl, 18–42%). Tetrachloroethene failed to give a cycloadduct.

Benzofuran **340** can be regarded as alkoxyalkene (analogous to **323**) or as a furan, with regard to benzyne cycloaddition. When treated with benzyne (from *o*-benzenediazonium carboxylate in refluxing ClCH$_2$CH$_2$Cl) **340** (4-fold excess) gave **342** (4%) and **345** (32%)[178]. Despite the large excess of **340**, the major product (**345**) resulted from the addition of two equivalents of benzyne. A stepwise mechanism via **341** was proposed, although an initial [4+2] cycloaddition to the furan moiety to give **343** directly cannot be ruled out. This curious formation of the novel **345** should attract further study.

The strained double bond in **27** reacts with benzyne (from benzenediazonium carboxylate) to give the *exo*-[2+2] adduct **346**[178]. Analogous cycloadditions to norbornene and norbornadiene are known[1]. It was recently reported that 7-*t*-butoxynorbornadiene gives both [2+2] regioisomers **348** and **349**, as well as the homo[4+2] cycloadduct **350**[179]. The predominance of *syn* (**348**) over *anti* (**349**) adduct is thought to be a consequence of stabilization of the diradical intermediate by the *t*-butoxy oxygen atom.

(**27**)   (**346**) 39%

(**347**)   (**348**) 14%   (**349**) 6%   (**350**) 18%

(**351**) R = Me, H   (**352**) 43%

(**353**)   (**354**) 8%

Cyclic conjugated dienes could give [4+2] or [2+2] cycloadducts. Although in general [4+2] cycloaddition predominates, **351**[180] and **353**[109] have structural features which retard this mode of addition, and only the [2+2] mode was observed. [4+2] Cycloaddition across the diene system in **351** or **353** would result in adjacent double 2-carbon bridging across a cycloheptene (1,7 and 3,6) or a cyclooctadiene (3,8 and 4,7), respectively. Also, although the diene moieties in **351** and **353** are nearly planar, [4+2] cycloaddition would require bringing the terminal atoms of that moiety closer together, thus increasing strain. Instead, [2+2] cycloaddition occurs to give **352** (43%) or **354** (8%). It is thought that benzyne adds *syn* to the oxygen in **354** and *anti*, for steric reasons, to the *gem*-dimethyl group in **352** (R = Me).

Initial [2+2] cycloaddition of benzyne to homoazulene **355** is followed by ring opening to give benzo[*b*]homoheptalene **357** in good yield[181]. The alternative [4+2] cycloaddition is probably disfavored for the same reasons as with **351** and **353**.

(**355**)   (**356**)   (**357**) 51%

Benzazetidines are conceivable products from the [2+2] cycloaddition of arynes across the C=N bond. Evidence for the transient formation of these intermediates has been presented[182]. However, yields are very low and the azetidines have never been isolated, only products from their supposed intermediacy. For example, benzyne and amidine **358** gave *N*-phenylanthranaldehyde **360** (8%) and acridine **361** (17%), presumably via azetidine **359**.

(**358**)   (**359**)   (**360**)

(**361**)

The reaction of benzyne with enamines[183] has been reinvestigated[184]. In the original study[183] it was proposed that the two observed products from the reaction of benzyne with 1-pyrrolidinocyclohexene **362**, (i.e. **364** and, after hydrolysis, **366**) arose from a common dipolar intermediate **363** as shown. An alternative explanation is that **364** arises from a stepwise [2 + 2] cycloaddition as shown, but that **365** arises independently from a concerted 'ene' reaction. If this hypothesis is correct, then there should in fact be *two* possible 'ene' products, **367** and **368**. Careful product analysis showed the presence of about 1% of previously unobserved **368**, together with 15–25% of **367** and 15–25% of [2 + 2] product

364. These data, together with observations on the reaction of 1,2-didehydro-*o*-carborane (an analog of *o*-benzyne) with enamines, tend to support the two independent mechanisms proposal. Competition between [2+2] cycloadditions and 'ene' reactions in the case of cyclic alkenes is well known[1,2,185].

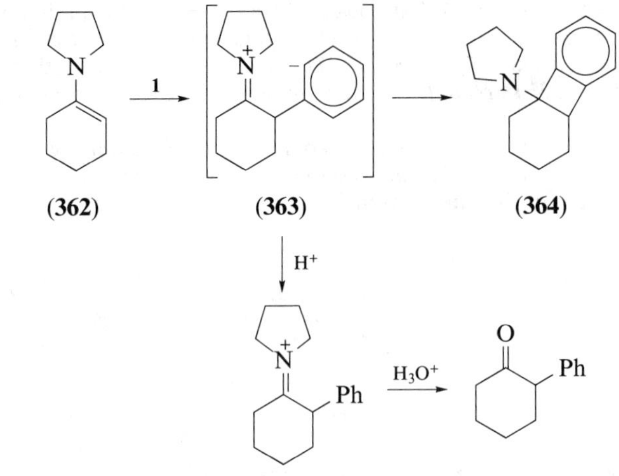

(362)   (363)   (364)

(365)   (366)

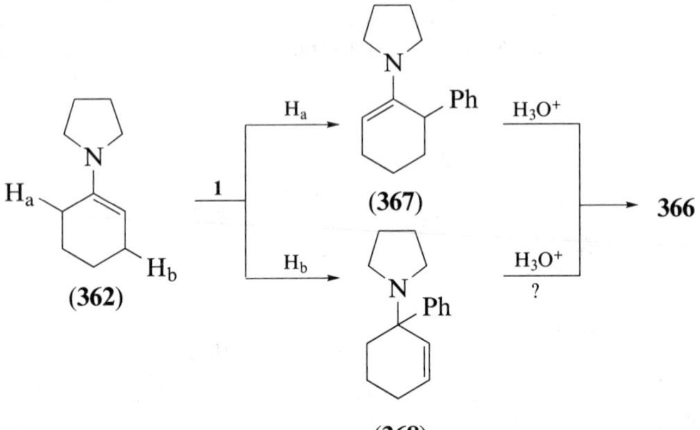

(367)   366

(368)

The reaction of benzyne (generated from chloro- or bromobenzene and sodamide) with O-silylated enolates of carboxylic esters **369** provides a convenient route to *ortho*-alkylbenzoic

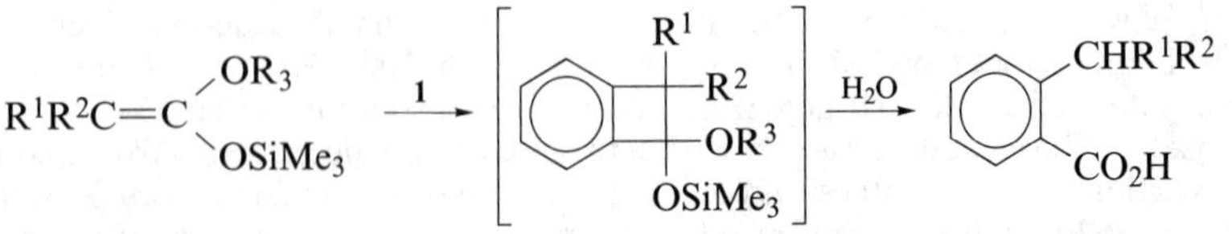

$R^1R^2C=C$ with OR_3 and $OSiMe_3$

(369)   (370)   (371)

acids **371**[186]. The reaction probably proceeds via a [2 + 2] cycloaddition to give **370** which is hydrolyzed during work-up.

D. 1,3-Dipolar Cycloadditions

A wide variety of stable 1,3-dipoles (azides, nitrones, diazo compounds and many others) undergo cycloadditions with arynes[1,2,140]. In this section, a few recent examples that are synthetically useful are described.

Pyridinium dicyanomethides **372** react with benzyne (from anthranilic acid) to give, after HCN loss, cyanobenzoindolizines **373** in 10–35% yield[187]. Similarly, isoquinolinium dicyanomethide **375** gave **376** in 74% yield[188]. The products (i.e. **373**) are also 1,3-dipoles and can react with a second equivalent of benzyne (again, with loss of HCN) to give benzo[a]isoindolo[1,2,3-cd]indolizines **374**, which are 18π annulenes[187].

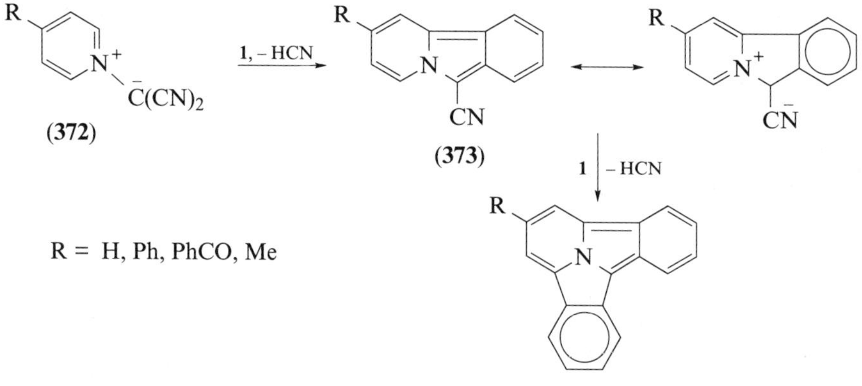

(372)

(373)

(374)

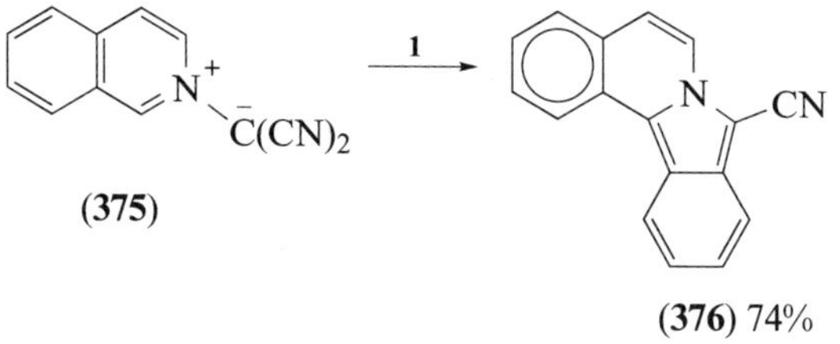

R = H, Ph, PhCO, Me

(375)

(376) 74%

Diazoketones are well known to cycloadd arynes[189]. With 1,2-naphthyne, diazoketone **377** gave adducts in which **378** predominates[190], and with 4-methylbenzyne gave more **380** than **381**[191]. Thus the carbon (δ-) of the diazoketone preferentially attacks C2 of each aryne.

Acyclic diazoketones **382** (n = 1, 2, 3, 5) gave benzyne cycloadducts **383** in 45–75% yields[192]. A large number of spiro[fluorene-9,3'-indazoles] **384** with various substituents in the positions shown have been prepared by 1,3-cycloaddition of benzynes to the corresponding 9-diazofluorenes[193]. The inner salts of 3-oxo-pyrazolidinium and 3-oxo-1,2-diazetidinium hydroxides react with benzyne generated from benzenediazonium carboxylate to give bicyclic adducts **386** and **388**, respectively, containing a N—N bond[194]. Yields were mainly in the 50–75% range, depending on substituents.

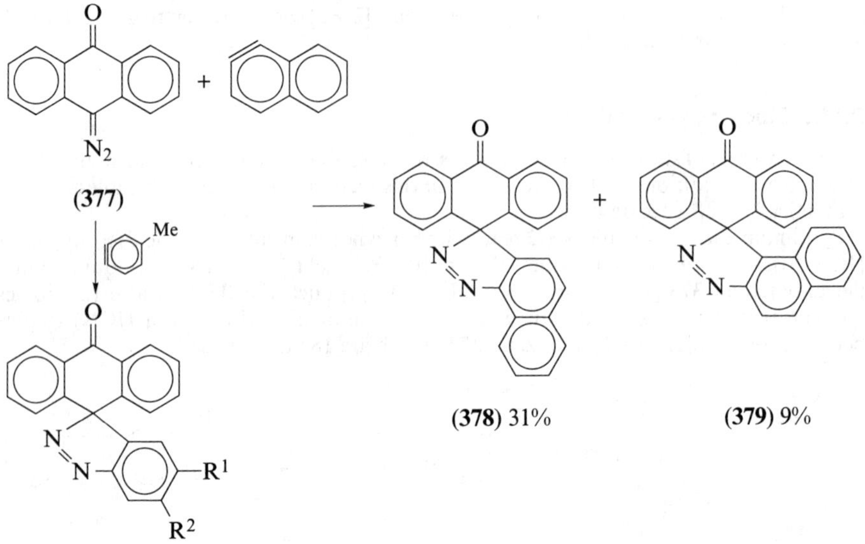

(377)

(378) 31%   **(379)** 9%

(380) R^1 = Me, R^2 = H; 53%
(381) R^1 = H, R^2 = Me; 29%

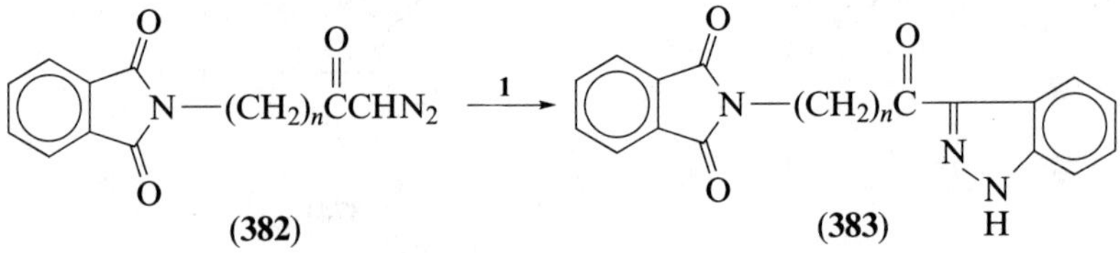

(382)   **(383)**

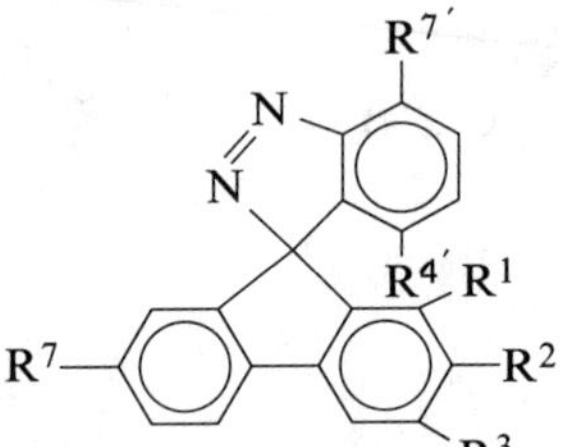

(384)

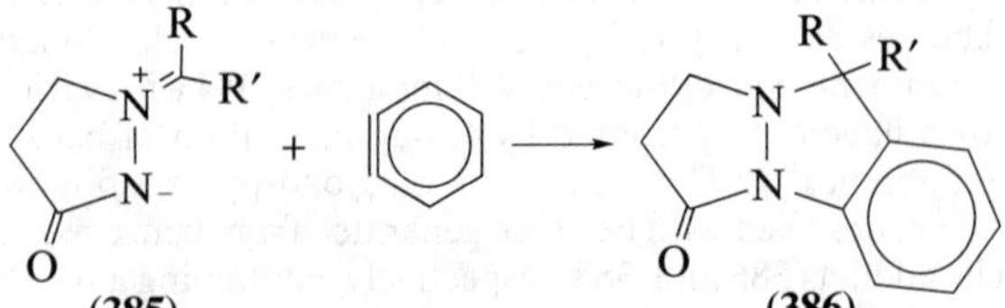

(385)   **(386)**

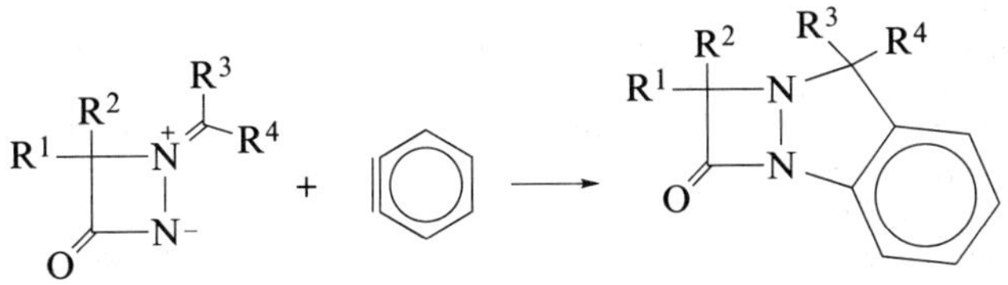

(387)   (388)

Benzoylnitrile oxide reacted with benzyne to give cycloadduct benzisoxazole **390** in 45% yield[195]. The product originally isolated from the reaction of benzyne with dehydro-dithizone **391** was **393**, presumably formed by loss of phenylnitrene from intermediate **392**. By carrying out the reaction at 70 °C instead of the original 150 °C, it was possible to isolate cycloadduct **392** as a stable orange-red solid in 39% yield[196].

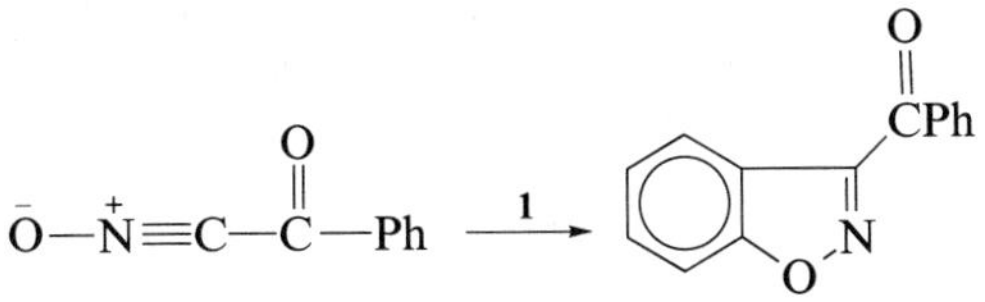

(389)   (390)

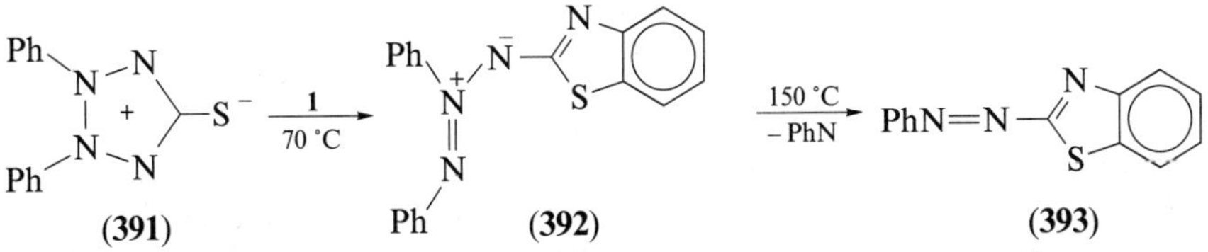

(391)   (392)   (393)

The reaction of benzyne with certain selena- and thiadiazoles has been studied exten-sively and revealed some interesting mechanistic twists and turns. With benzoselenadiazole

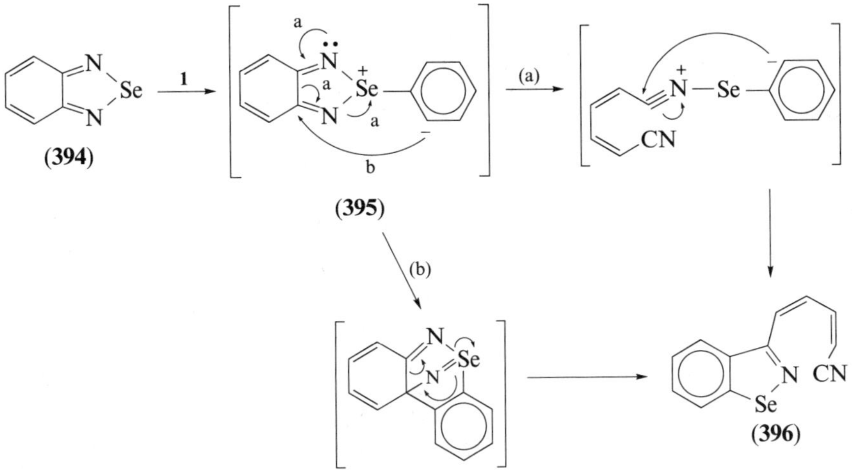

(394)   (395)   (396)

394, the major product (88%) was **396**, with *cis,cis* geometry of the side chain; very minor amounts of product with the *cis,trans* side chain were also formed[197]. A possible mechanism is shown. 1,4-Cycloaddition of benzyne across the N=C—C=N moiety of **394** does *not* occur. Instead, the product **396** arises from formal 1,3-cycloaddition across the C=N—Se moiety, though reaction probably proceeds via **395**, formed by nucleophilic attack of Se on benzyne. Two paths (*a* or *b*) are possible from **395** to **396**. Numerous substituted analogs of **394** react similarly[198]. Under some conditions, the product **396** may undergo further rearrangement, as in the conversion of **398** (R = Me) to **399** at 140 °C or photochemically. This reaction is facilitated when R = Cl. Thus no **398** could be isolated from the reaction of benzyne with **397** (R = Cl); the only product isolated was **399** (R = Cl).

The reaction with naphthoselenadiazole **400** took still a different course[199]. Benzyne (from benzenediazonium carboxylate) gave **403** in very low yield (3%). The initial product is thought to be **401** (that is, **400** reacts with benzyne in the same manner as anthracene, to give a heterotriptycene). Although **401** could not be isolated, **400** did give an analogous adduct **404** (26%) with dimethyl acetylenedicarboxylate. Presumably **401** reacts with a second equivalent of benzyne in a manner analogous to **394** to generate **402**, which is then aromatized (by benzyne?) to **403**. In support of this mechanism, **404** reacted with benzyne to give **405** in 16% yield.

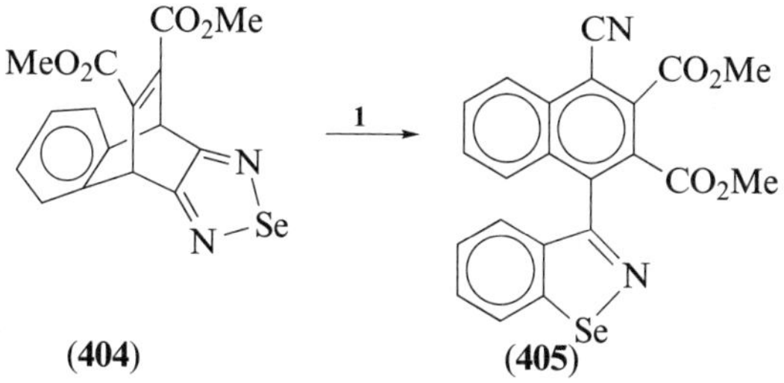

(**404**)   (**405**)

The sulfur analogs of these selenium compounds react less efficiently with benzyne. Thus the yield of **396** (S in place of Se) was only 5% from the benzothiadiazole[197]. However, with the parent thiadiazoles **406** the yields are improved. Thus 1,2-benzoisothiazoles **409** are formed in 25–75% yield, depending on R[200], probably by the mechanism shown. Selenium analogs behaved analogously[200]. In contrast with **394** or **406** and **409**, the iso-analog of **409** (i.e. **410**) does react with benzyne in a [4+2] cycloaddition mode to give acridine **411**, but the yield is very low (5%).

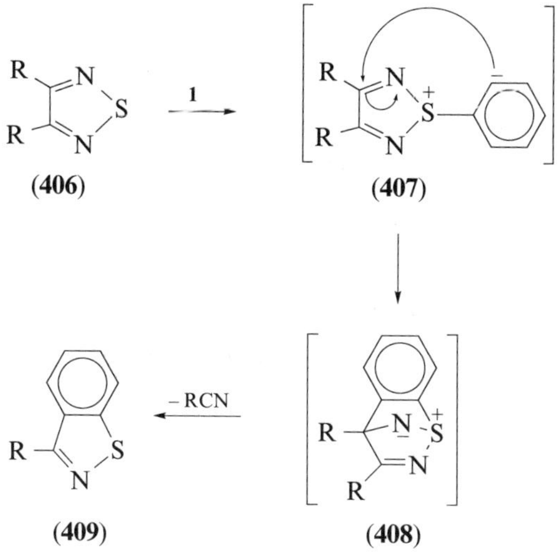

(**406**)   (**407**)

(**409**)   (**408**)

Azides add to benzynes to give benzotriazoles[1,2]. Two recent examples include the reaction of 1-azido-2-methylnaphthalene **412** with 3-methylbenzyne to give **413** and its regioisomer[201], and the addition of benzyne to an (azide)indium(III) porphyrin to give the corresponding (benzotriazolato)indium(III) porphyrin[202].

Finally, 1,3-cycloadditions have been coupled with radical cyclizations to quickly assemble polycyclic systems, as in the conversion of betaine **415** to benzyne cycloadduct **416** and thence to aminoketone **417**[153].

(410) → [ ] → (411)

(412) + → (413) + (414)

(415) → (416) $\xrightarrow{\text{Bu}_3\text{SnH}}$ (417)

E. Miscellaneous Cycloadditions

Cyclopropanes normally do not give 1,3-addition products with benzyne, although the highly strained bicyclobutane **418** gave a low yield of **419**, as well as 'ene' product **420**[203].

(418) $\xrightarrow{1}$ (419) 6–10%   +   (420) 35–50%

(421) $\xrightarrow{1}$ (422) 18%

Consequently the insertion of benzyne into the C—Si bond of silacyclopropane **421** is novel[204]. The benzyne was generated *in situ* from *o*-bromofluorobenzene and magnesium in refluxing THF. Under similar conditions, benzyne also inserted into the C—Si bond of silacyclopropene **423** to give silaindene **424** in 37% yield[205]. This result is to be contrasted with the reaction of dimethylcyclopropene with benzyne, which gives **426** in low yield, presumably via an initial [2 + 2] cycloaddition[206], and of triphenylcyclopropene **427** which gives the 'ene' product **428**, and tetraphenylcyclopropene, which does not react[206]. It is thought that **424** is formed via diradical intermediates[205].

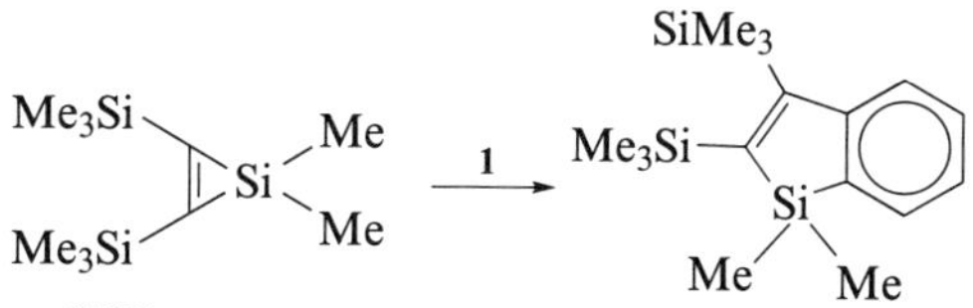

(423)   (424) 37%

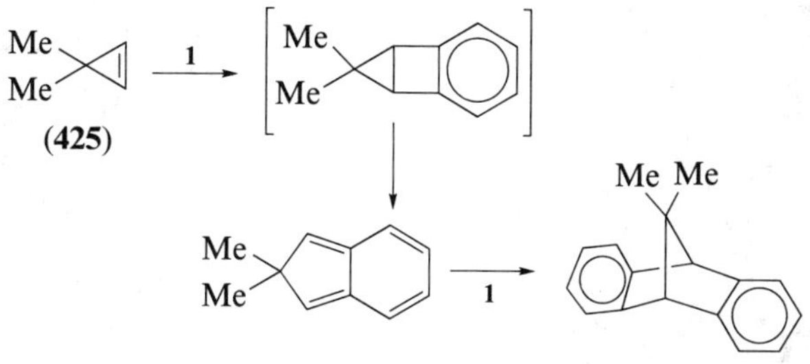

(425)

(426) 5%

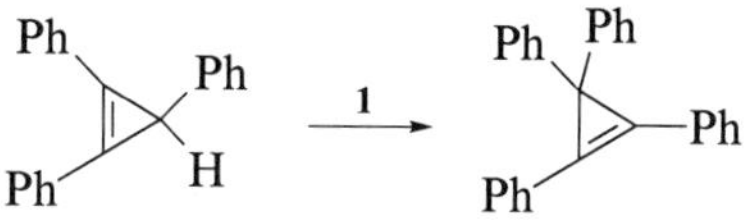

(427)   (428) 60–70%

A net [4+2] cycloaddition of benzyne to vinyl isocyanates allows one to construct phenanthridinones and benzphenanthradinones in one step and 40–50% yield[207]. Thus lead tetraacetate oxidation of aminobenztriazole **66** in the presence of **429** gave **430** in good yield. Whether the mechanism involves a concerted [4+2] addition to the 2-azadiene moiety of the isocyanate or a polar stepwise addition remains for further study.

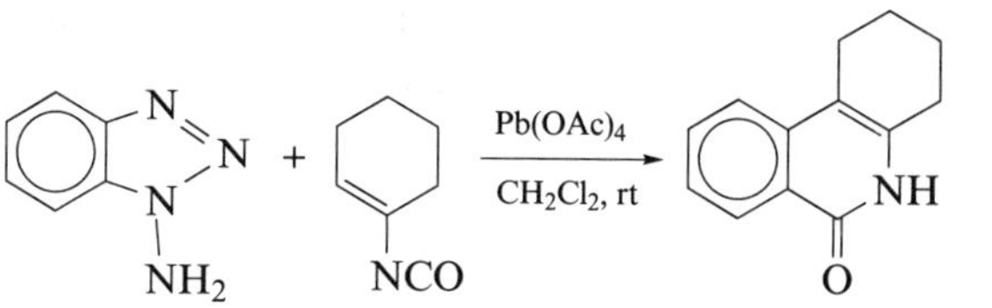

(66)   (429)   (430) 58%

Soon after its availability, C_{60} was subjected to reaction with benzyne. A series of adducts with the formulas $[C_{60} + (C_6H_4)_n]$, where $n = 1$–4, was detected by mass spectrometry[208,209]. The monoadduct was isolated via high performance liquid chromatography and characterized by its ^1H and ^{13}C NMR spectra[208]. These data show that benzyne adds across the pyracyclene bond (i.e. across the bond between two fused six-membered rings, as shown in **432**). Consequently the ^1H NMR spectrum appears as an AA'BB' multiplet for the benzenoid ring protons, and the ^{13}C spectrum shows 19 of 20 peaks expected for the symmetry of **432**.

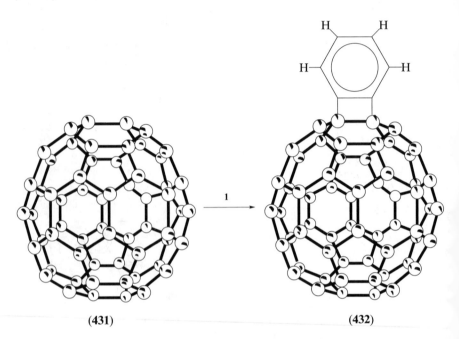

(431)   (432)

A monoadduct of C_{70} was also formed, but its structures as well as those of bis- and higher benzyne adducts of C_{60} remain to be determined. Also, there were indications that another monoadduct of C_{60} may also be formed under certain benzyne addition conditions[208].

F. Cycloadditions with Two or More Aryne Intermediates

Theoretical calculations show that 1,3- and 1,4-benzadiyne **433** and **434** are each considerably higher in energy than benzyne, though it is not clear which of the two diynes is the more stable[24]. Furthermore, their isomerization to 1,3,5-hexatriyne is predicted to be highly exothermic[24,210]. Consequently, even though $C_6H_2^+$ ions are the dominant ions in the mass spectra of all three tetrabromobenzenes[211], it seems unlikely that these diynes are involved in the solution chemistry of various precursors. Instead, it is more likely that two

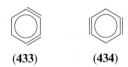

(433)   (434)

monoarynes are generated in sequence. For example, Wittig found that **435**, treated with Mg in THF containing excess furan, gave mainly monoadduct **436** and only 5% of bis-adduct **437**[212]. With BuLi in place of Mg, the yield of **437** increased to 15% but no **436** was isolated. He concluded that the reaction proceeded stepwise via monoarynes **438** and **439**, and since yields were low and precursor **435** required a multistep synthesis, no further work was done.

By switching to the much more accessible tetrabromoarenes **440**, the reaction became synthetically useful[213]. Yields of **441**, for example, were in the 60–80% range, even for rather hindered cases ($R^1 = R^2 = OMe$; $R^3 = R^4 = Me$). From 1,2,4,5-tetrabromobenzene and furan, an easily separable mixture of *syn-* and *anti-***441** ($R^1 = R^2 = R^3 = R^4 = H$) was obtained, and these bis-endoxides have been useful in complex iptycene synthesis[214] and in the synthesis of molecular belts and waves[215].

$$R^1 = R^2 = H, Me, OMe;$$
$$R^1 = Me, R^2 = OMe, \text{etc.}$$

Pyrroles, cyclopentadienes and fulvenes have been used with **440** to make a large array of bis-adducts[216]. By using only one equivalent of BuLi, the monoadducts can be isolated in good yield. These can subsequently be treated with a second diene to give nonsymmetric bis-adducts, as in the two syntheses of **447**[216] shown.

Anthracenes react to give iptycenes **448**[217]. In all of these reactions, the 1,2,4,5-tetrahalobenzene acts as the synthetic equivalent of 1,4-benzadiyne **434**.

The isolation of mono-cycloadducts in high yield from **440** suggests that metal–halogen exchange occurs stepwise. However, when $R^1 = R^2 = $ an EWG such as chlorine, double exchange at *para*-bromines occurs even at −78 °C[218]. Thus di-lithio derivative **449** is formed exclusively, and can be trapped as **450**. If warmed with furan, it eliminates only LiBr (*not* LiCl), to give bis-adduct **451**. This elimination is probably stepwise, and not via a 1,4-benzadiyne.

By using a 1,2,3,4-tetrahalobenzene in which metal–halogen exchange occurs preferentially at the 1 and 4 halogens, one has a synthetic equivalent of 1,3-benzadiyne **433**[217,219]. For example, the highly strained permethylphenanthrene **454** can be synthesized in this way by a double cycloaddition to **452**[219]. Initial exchange occurs primarily at the iodine atoms of **452**.

The synthesis of benzo[1,2-*c*:3,4-*c'*:5,6-*c''*]trifuran **459** (the furan analog of triphenylene) was accomplished by three aryne cycloadditions onto a single benzene ring[220]. The arynes were generated by two different methods. The dibromodiazonium carboxylate hydrochloride **455** and furan gave adduct **456**, which was then treated with excess potassium amide.

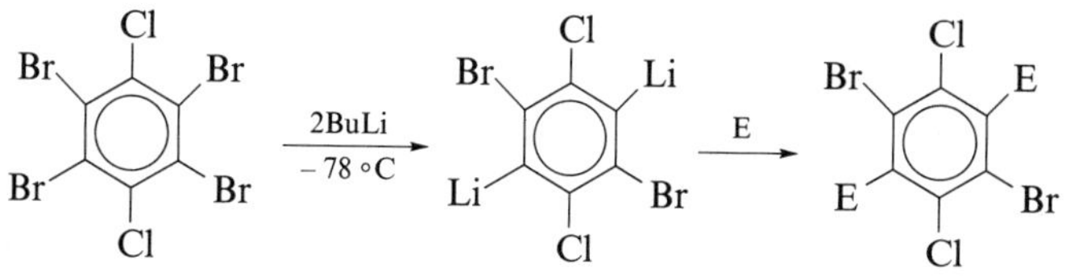

(440) R$_1$ = R$_2$ = Cl   **(449)**   **(450)**

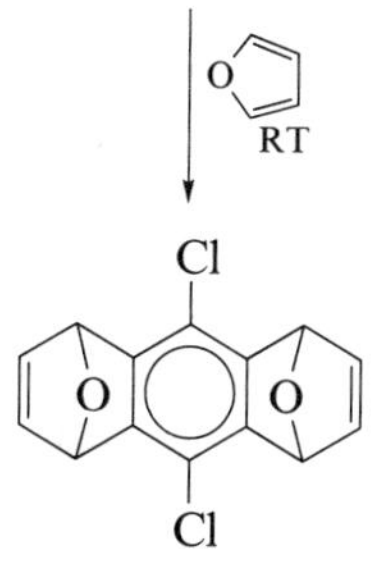

(451)

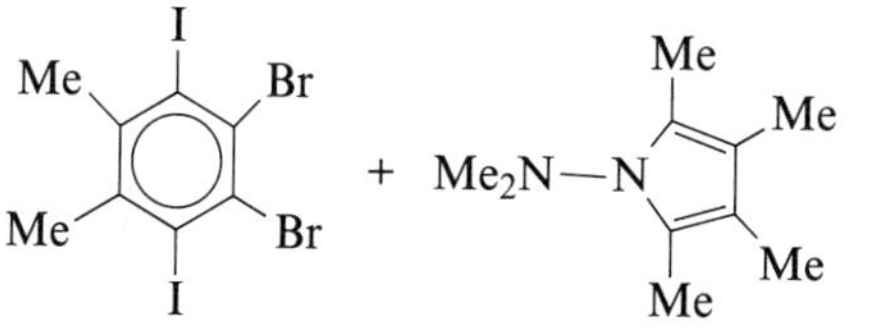

(452)

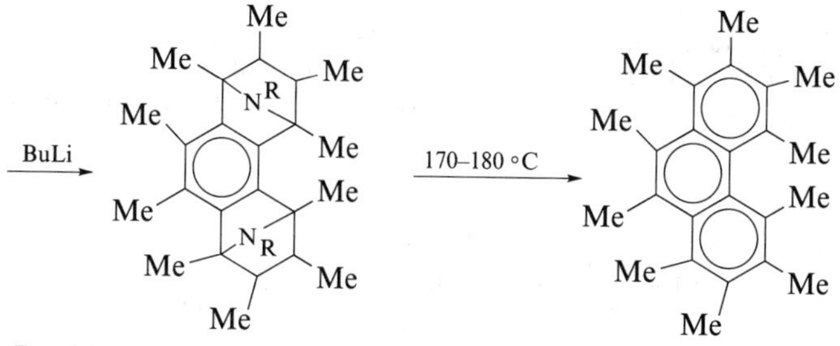

R = NMe$_2$ **(453)** 16%   **(454)** 97%

In addition to recovered **456** (38%) there was obtained monoadduct **457** and, in very low yield, bis-adduct **458**. Additional **458** (11%) could be obtained by resubjecting **457** to the aryne-generating conditions. Hydrogenation and thermolysis converted **458** to **459** in essentially quantitative yield. In these reactions, the 3,6-dibromoanthranilic acid functions as a synthetic equivalent for 1,3,5-benzatriyne, albeit the overall yield of **458** is very low.

An attempt to develop a multigram synthesis of **458** provoked a reinvestigation of aryne generation from polyhalobenzenes[221]. The main results are the following. Hexabromobenzene, BuLi (1 equiv) and furan give monoadduct **460** via 3,4,5,6-tetrabromobenzyne[222]; further metallation occurs at the 2-Br (and with 2 equiv of BuLi, at the 2,3-bromines, as established by low-temperature proton quenching) and elimination

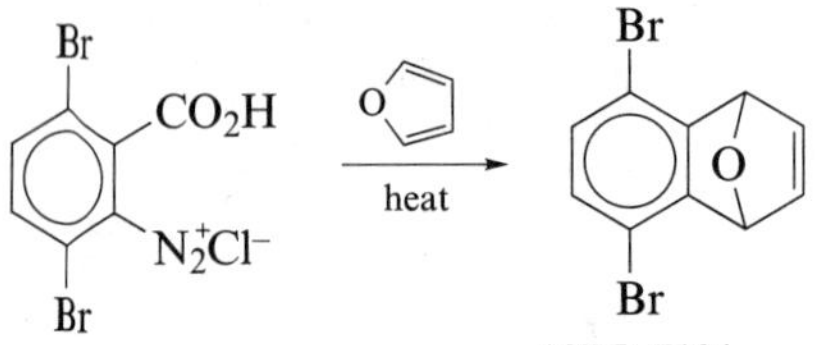

(455) → (456) 70%

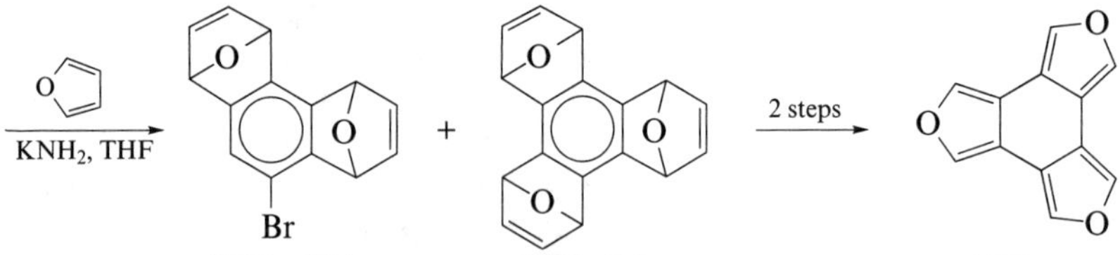

(457) 42%   (458) 3%   (459)

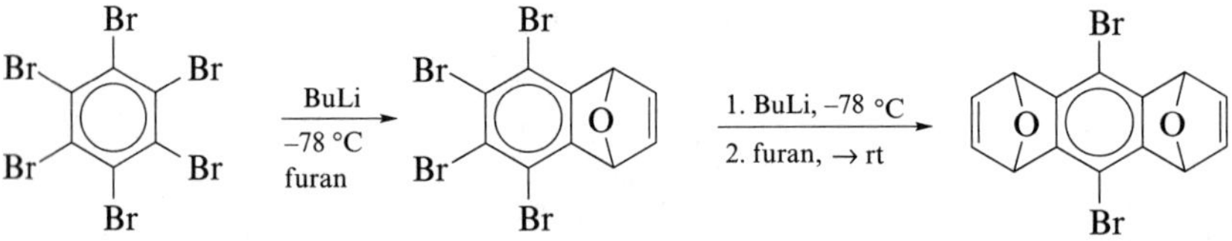

(460) 50%   (461) 43%

occurs at the 3-Br to give 'linear' bis-adduct **461**. Similar results were obtained with **440** (R$_1$ = R$_2$ = Cl).

The product from monolithiation and proton quench of **460** (i.e. **462**) also gave **461** with sodamide, but with BuLi gave the angular bis-adduct **463**, presumably as shown. Better yields of **463** were obtained[221] from **464**, the monoadduct of furan and **440** (R^1 = R^2 = H)[223].

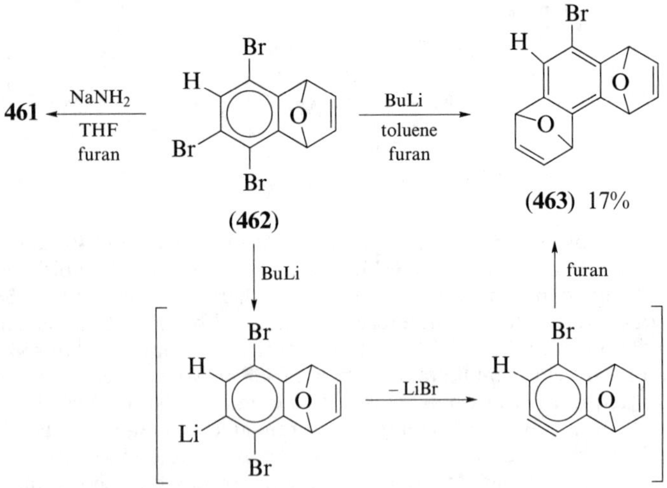

(462)   (463) 17%

Elimination of HBr from **463** (or double elimination from **464**) with sodamide (or potassium amide) to give the desired **458** proceeds in only low yield (*ca* 10%). Arynes **465** (*syn* and *anti*) resist formation, probably as a consequence of deformation of the benzene ring by the two fused oxanorbornene rings. This may cause some bond fixation as shown in **463**, thus retarding HBr elimination.

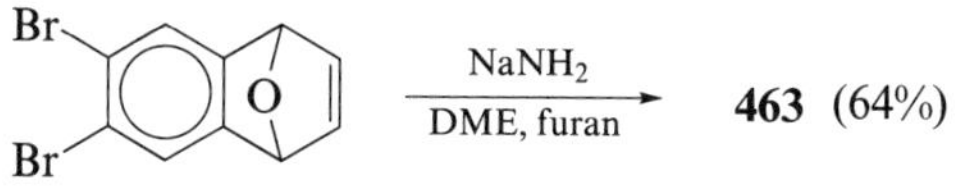

(464) $\xrightarrow[\text{DME, furan}]{\text{NaNH}_2}$ **463** (64%)

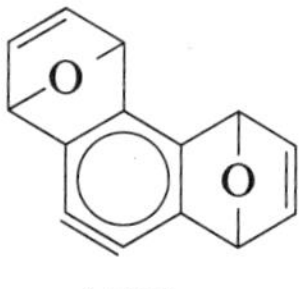

(465)

Di- and tri-aryne equivalents other than polyhalobenzenes have only been briefly explored. Bistriazole **466** was synthesized as a 1,4-benzadiyne equivalent[224]. Treatment with lead tetraacetate (LTA) in the presence of various dienes give bis-adducts in good yield (mainly 70–90%, 17 examples). The main advantages over tetrahalobenzenes **440** is that dienes with functionality that might not survive BuLi can be tolerated, as shown in examples **467–469** (the latter a bis-1,3-dipolar cycloaddition). The regioselectivity in these cycloadditions is also striking. The major disadvantage is that precursor **466** requires a multistep synthesis.

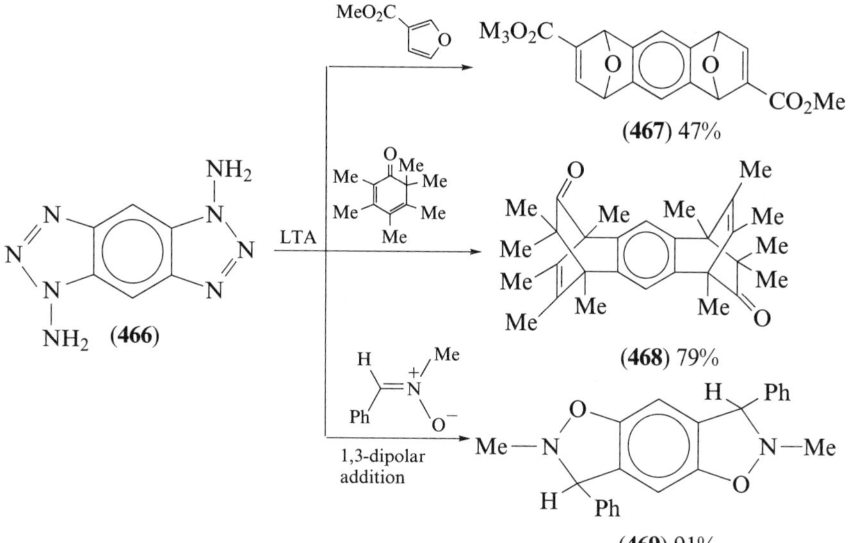

(467) 47%

(468) 79%

(469) 91%

Bis-(*o*-bromotosylates) of naphthalene have been used as equivalents of di-arynes **470–473**, useful in the synthesis of polynuclear aromatic hydrocarbons[225]. The bromotosylates, prepared by brominating and tosylating the corresponding dihydroxynaphthalenes, were treated with phenyllithium and furan to give bis-adducts, which were then aromatized by hydrogenation and acid-catalyzed dehydration in standard fashion. In this way, chrysene was prepared from **474** (41% overall), triphenylene from **475** (16% overall), benz[*a*]anthracene from **476** (29% overall) and naphthacene from **477** (25% overall). Substituted analogs (for example, dimethyl derivatives) could be obtained in comparable yields using substituted furans.

Bis-arynic reagents in which the two arynes are in separate, insulated arene rings have also been developed[226]. For example, bis-anthranilic acid **478** is a synthetic equivalent for diaryne **479**; aprotic diazotization of **478** in the presence of various dienes (i.e., tetraphenylcyclone, 2,5-dimethylfuran, anthracene) gave the expected bis-adducts in 40–

60% yield. In a similar fashion, **480** has been condensed with bis-tetracyclones and bis-α-pyrones to produce polymers (i.e. **482** and regioisomers) derived from bis-aryne **481**[227]. Examples of Y include *p*-phenylene, *p,p'*-diphenyl ether and 8 others.

Finally, a double [2+2] cycloaddition with *p*-dibromobenzene and 1,1-dimethoxyethylene gave, after acid hydrolysis, a 20% yield of the novel bis-benzocyclobutenone **483**[175]. Thus *p*-dibromobenzene acts as a synthetic equivalent for diyne **434**, and cycloaddition to the presumed intermediate aryne **484** is regioselective.

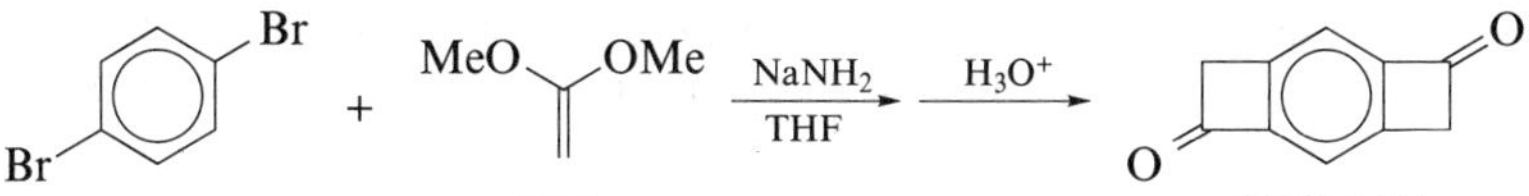

(323)　　　　　　　　　　　　　　　　　(483) 20%

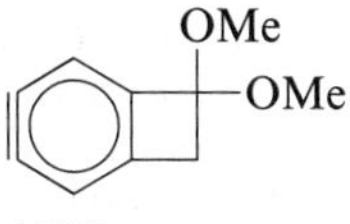

(484)

Although diynes **433** and **434** have not been established as intermediates in solution (their ions may exist in the gas phase, as indicated by mass spectra), metal complexes of **434** have been described, prepared in a stepwise manner[228].

VI. NUCLEOPHILIC ADDITION TO ARYNES

A. General Remarks

This subject has been thoroughly reviewed[1,2], and only the main features will be briefly summarized here in order to set the following sections in context. Nucleophiles generally exhibit a smaller range of reactivities toward arynes than toward other substrates. This is because of the extreme reactivity and electrophilicity of arynes. Even so, selectivities are evident; with benzyne in liquid ammonia, for example, the nucleophilicity order is $PhS^- > PhO^- > RO^-$ and with benzyne in alcohol the order is $I^- > Br^- > Cl^- \gg EtOH$[2]. In water at 318 °C the relative reactivities toward 4-methylbenzyne are PhS^- (46), I^- (6.2), piperidine (3.0), Br^- (1.7), $PhNH_2$ (1.3), PhO^- (1.0), Cl^- (1.0), NH_3 (0.5), F^- (0.2)[229]; this compressed

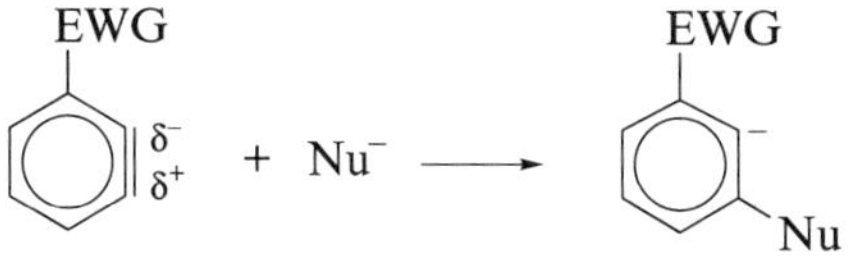

(485)　　　　　　　　　　　　(486)

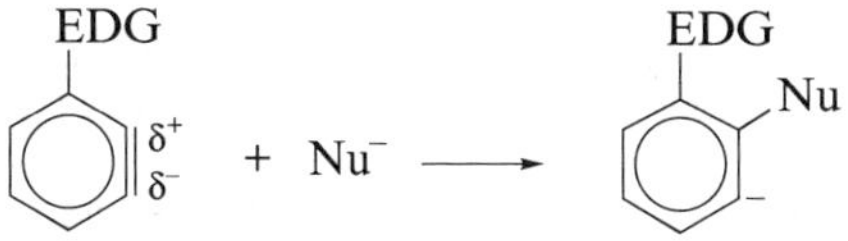

(487)　　　　　　　　　　　　(488)

reactivity range is no doubt considerably broader, however, at the lower temperatures at which most reactions of arynes are carried out.

Substituents influence the position of attack by nucleophiles. Electron-withdrawing groups (EWG) in the 3-position polarize the triple bond as shown (**485**) and stabilize an adjacent negative charge, hence favor '*meta*' addition of the nucleophile (**486**). Electron-donating groups have the opposite effect (**487** → **488**). These effects are diminished if the substituent is more remote from the 'triple' bond, and are mitigated by steric factors in the substituent and nucleophile. Even so, as will be seen in the following sections, nucleophilic attack on substituted arynes is often highly regioselective, hence synthetically useful.

A computer-assisted mechanistic evaluation that includes nucleophilic aromatic substitution via aryne intermediates has been developed[230].

B. Carbon Nucleophiles

Benzyne intermediates can be used to introduce two adjacent carbon chains on a benzene ring. For example, treatment of *o*-bromoiodobenzene **489** with slightly over 2 equivalents of a vinyl Grignard reagent results in metal–halogen exchange followed by elimination and addition of the reagent to benzyne, giving *o*-vinyl Grignard reagent **490**. Addition of an alkylating agent such as an allylic bromide introduces the second side-chain[231]. A specific example is shown. Yields are good (60–65%). Trimethylsilyl vinyl Grignard reagents react similarly[232].

$$E = H, I, CHO, CO_2H$$

In analogous fashion, aryl Grignard reagents can be used in place of the vinyl Grignard reagent, to synthesize unsymmetrically substituted biaryls **494** (yields mainly 60–70% for 10 examples)[233].

As mentioned in Section III.A.2, 3-(2-oxazolyl)arynes **43** react with nucleophiles to give 1,2,3-trisubstituted benzenes **45**[41,42]. For example, reaction of **43** (prepared as already described) with *n*-butyllithium followed by a wide variety of electrophiles and subsequent

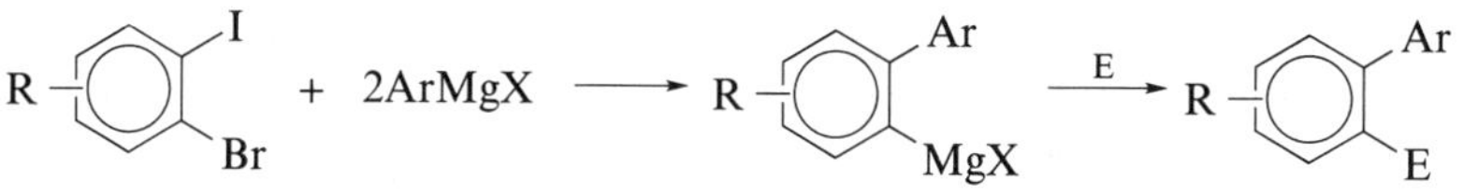

(494)

acid-catalyzed hydrolysis of the oxazoline moiety gave the 2,3-disubstituted benzoic acids **496** in 53–70% yields (R = *n*-Bu; E = H, Me, CH_2OH, CHO, PhCHOH, PhC=O). The regiochemistry of the RLi addition (**495**), which is contrary to what might be expected because the 2-oxazolyl substituent is electron-withdrawing (see **485** → **486**), is thought to be a consequence of initial coordination of the lithium with the substituent, as in **497**, thus introducing the R group to the 2-position. Nonpolar solvents favor this regiochemistry, but attack of R at C1 (instead of C2) can occur in solvents that themselves can coordinate the Li (i.e. Et_2O, THF) and also with R groups in which the lithium is already internally coordinated, such as *o*-lithioanisole.

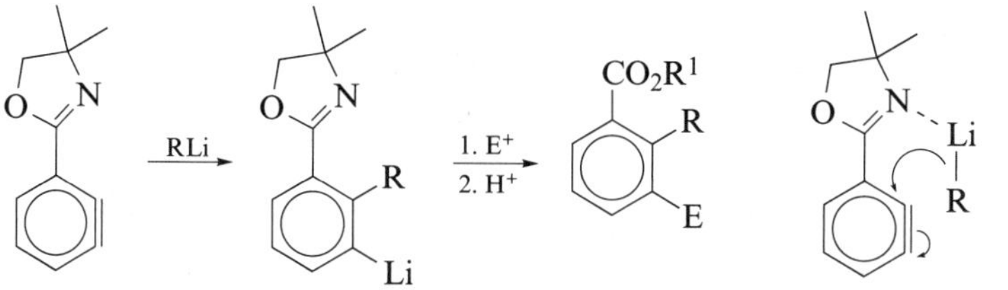

(43) (495) (496) (497)

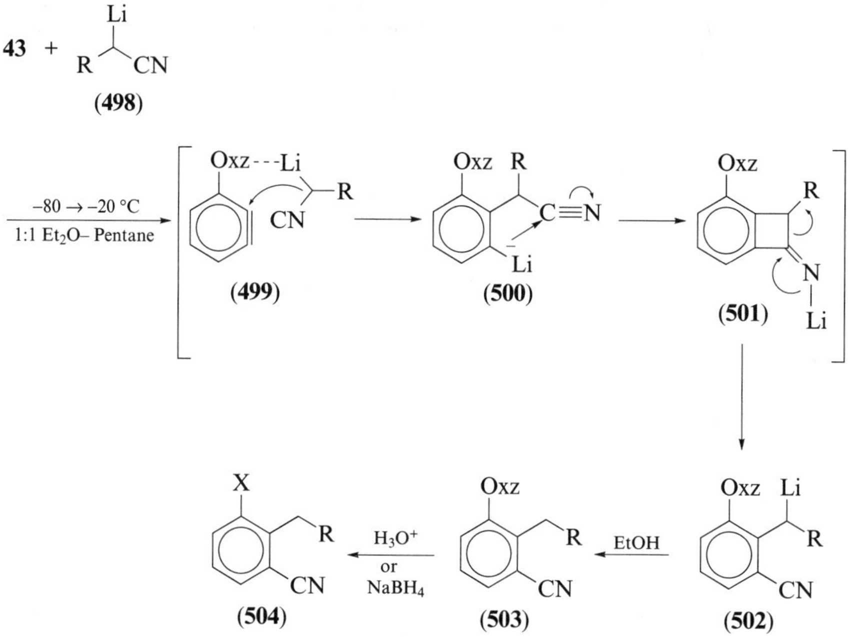

(498) (499) (500) (501) (502) (503) (504)

X = CO_2Me or CHO

Attack at C2 was also observed with α-lithionitriles **498**, which gave, after hydrolysis, **504**. The mechanism involves a rearrangement; after initial attack at C2 to give **500**, intramolecular nucleophilic attack on the nitrile function gives **501**, then **502**. Yields of **503** are mainly in the 40–70% range, and subsequent hydrolysis or reduction gives 2,3-disubstituted benzoates or benzaldehydes **504** in high yield. If desired, intermediate **502** can be alkylated prior to the final steps.

The reverse regiochemistry in additions to **43** was accomplished with cuprate nucleophiles. Thus **506**, the regioisomer of **496**, could be obtained in good yield as shown (examples include R = Me, Bu, Ph; E = H, allyl, acyl).

$$
\begin{array}{ccc}
\textbf{(40)} & \textbf{(505)} & \textbf{(506)}
\end{array}
$$

1. RLi, – 78 °C
2. R$_2$CuLi, rt

1. E
2. H$_3$O$^+$

Alkylation of arynes generated from aryl bromides and KNH$_2$/NH$_3$ with α-nitrile anions has also been observed. Yields of **508** were in the 50–80% range[234–236]. Aryne intermediacy (rather than direct displacement of bromine) was evident from results with **509** or analogs (2,4-dimethoxy- or 2,3,4-trimethoxy bromobenzenes, 2-bromo-4-methylanisole) which gave mixtures **512** and **513** (ratio approx. 1:5)[237,238]. The predominance of **513** is not surprising, since it is the sole product from **511** and also can be formed from **510**. Yields are generally good, though with some substrates amination competes with alkylation. Amination can be avoided by using a complex base[239].

(**507**) Y = Me, OMe, CH$_2$OMe   (**508**) R = alkyl

$$ + \; RCH_2CN + KNH_2 \xrightarrow{NH_3} $$

$$
\begin{array}{ccccc}
\textbf{(509)} & \textbf{(510)} & \textbf{(511)} & \textbf{(512)} & \textbf{(513)}
\end{array}
$$

The same type of rearrangement described for 3-(2-oxazolybenzyne) **43** and α-lithionitriles (see **503**) has also been observed in some instances when the arynes were generated from substituted aryl bromides using LDA in THF. For example, **515** was obtained in 23–83%

yield (depending on R and Ar)[240]. The mechanism is similar to that proposed for **503**; however, nucleophilic attack occurs *meta* to the electron-withdrawing methoxy substituent to give **516**. Interestingly, changing the 4-substituent in **514** from —CH$_2$R to —OMe or using an alkane-nitrile (alkyl in place of Ar) gave mainly the product of normal nucleophilic attack on the aryne (i.e. analogs of **512**, **513**). One factor is the nucleophilicity vis-a-vis basicity of intermediate **516**, which may either be protonated to give 'normal' alkylation product, or react intramolecularly with the nitrile function.

These studies have been extended to a variety of other aryne precursors and nitriles[241–245]. One of the more interesting examples involves the one-step synthesis of aminoisoquinoline **519** from precursors **517** and **518**, thought to occur by the mechanism shown[243]. Some of this work has been reviewed[246].

Anions derived from unsaturated amides or acids have recently been used as nucleophiles toward arynes[247,248]. The products are 4-aryl unsaturated amides (**520**, **521**) or acids thought to arise as shown. A variety of alkoxy substituted arynes were used. With methylsubstituted arynes, rearranged structures appeared in the products. For example, with **522** (R = H) indan **528** was obtained (51%) together with a small amount of **526** (R = H). With **522** (R = Me), however, the major product (55%) was **526** (R = Me), and no indan derivative was produced. A possible mechanism is shown.

Anions of phthalides react with benzynes in a useful way, leading to a short general synthesis of anthraquinones[249]. The initial product is **532**, but by allowing the reaction to stand in air for 20 h before workup, the anthraquinone is obtained. Yields are moderate to good, mainly 40–60%. With unsymmetric reagents the reaction can be quite regioselective; thus 4-methoxyphthalide and 3-methoxybenzyne gave mainly (39%) 1,5-dimethoxy-anthraquinone and only 2% of the regio (1,8) isomer.

Two modifications which lead directly to anthraquinones without the need for the final oxidation step have been described. In one[250], the phthalide carries a phenylsulfonyl substituent in the 3-position, whereas in the other[251] a 3-cyanophthalide is used. In this way the intermediate **535** formed by attack of the phthalide on the benzyne can collapse directly to an anthraquinone. A number of naturally occurring anthraquinones have been prepared this way[251,252].

In somewhat related chemistry, anions derived from benzocyclobutenols react with arynes to produce anthracenes[253] (the benzocyclobutenols were obtained by reducing benzocyclobutenones, themselves prepared via aryne [2+2] reactions with ketene acetals[174,175]). The mechanism is illustrated with a specific example. Alkoxide **538** derived from benzocyclobutenol **536** undergoes ring opening and the resulting benzyl anion attacks aryne **539** regioselectivity at the 1-position; the phenyl anion then closes back on the

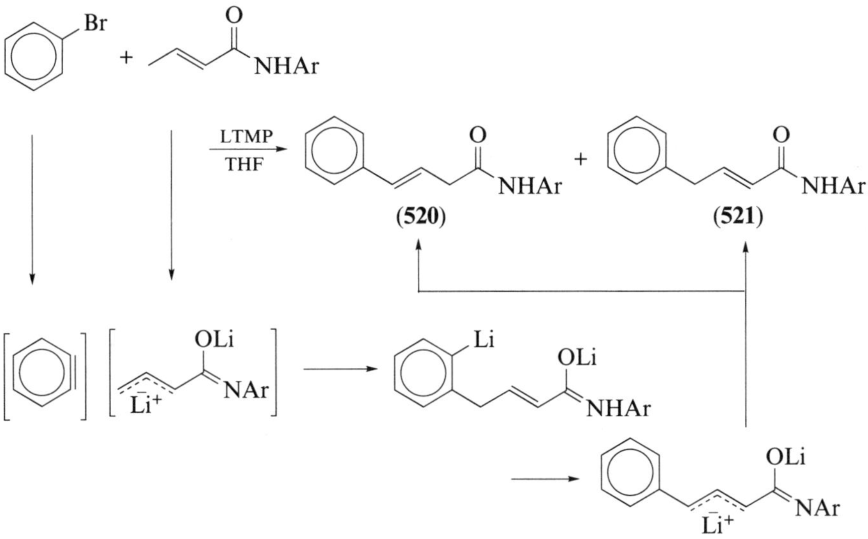

(520) (521)

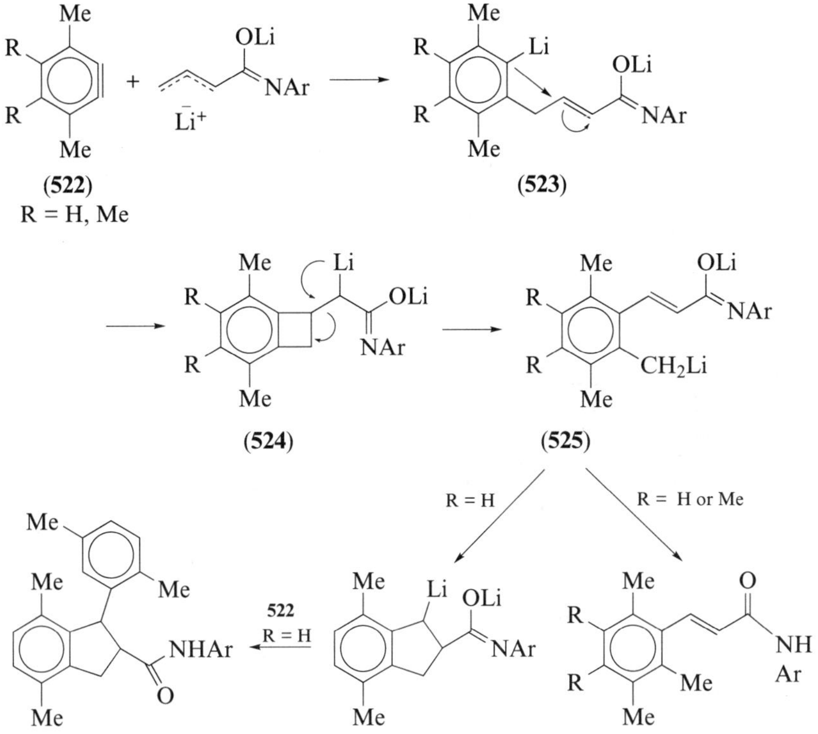

(522) R = H, Me

(523)

(524) (525)

(526) (527) (528)

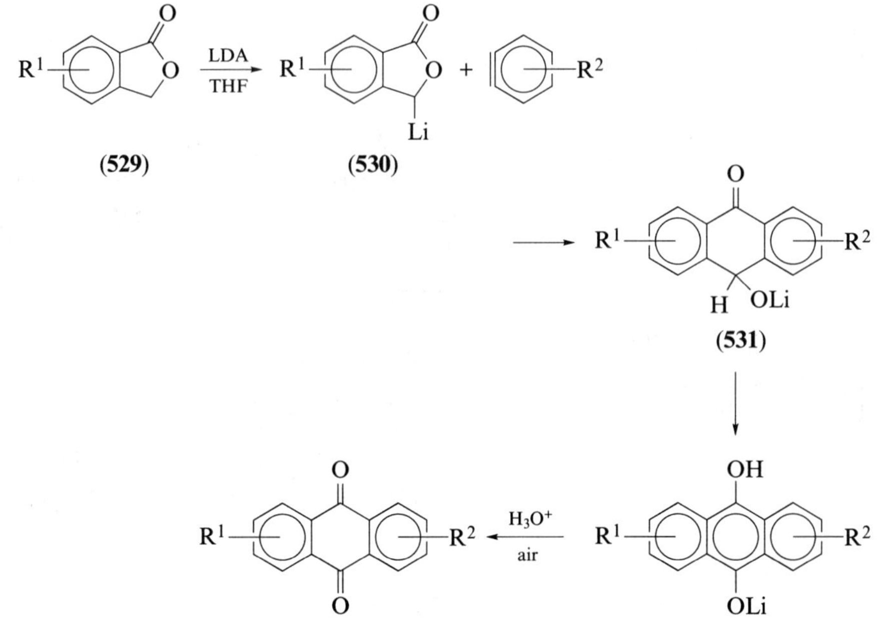

(529) (530)

(531)

(533) (532)

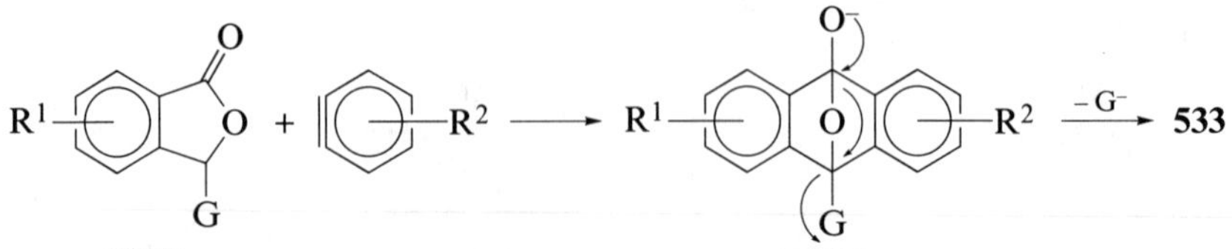

(534) (535)

G = PhSO$_2^-$, CN

carbonyl group to give **540** which aromatizes on workup. Yields are moderate to good (40–68%) for 11 examples.

Many of the reactions just described, that involve benzocyclobutene intermediates, are reminiscent of earlier work on arynic condensations of ketone enolates[1,254]. In general, the initial product **543** from cyclic enolates **542** can give benzocyclobutenols **544**, ring-expanded benzocyclenones **545** or phenylated ketones **546** depending on n, the base and other substituents that may be present. One limitation was that success in isolating cyclobutanols **544** was restricted mainly to enolates with $n = 1$–3. This restriction has now been overcome by using enolates derived from monoketals of 1,2-diketones[255]. The mixed base sodium amide–sodium t-butoxide is essential, and yields of **548** are mainly in the 65–93% range. The ring juncture is *cis* for 5–8-membered rings, but *trans* for 9–12. This methodology also works with acyclic analogs of **547**. It is thought that the adjacent oxygens of the ketal function coordinate with the sodium ion in intermediates such as **543**, stabilizing it and hence slowing down the ring-opening mechanisms that normally lead to products such as **545** and **546**.

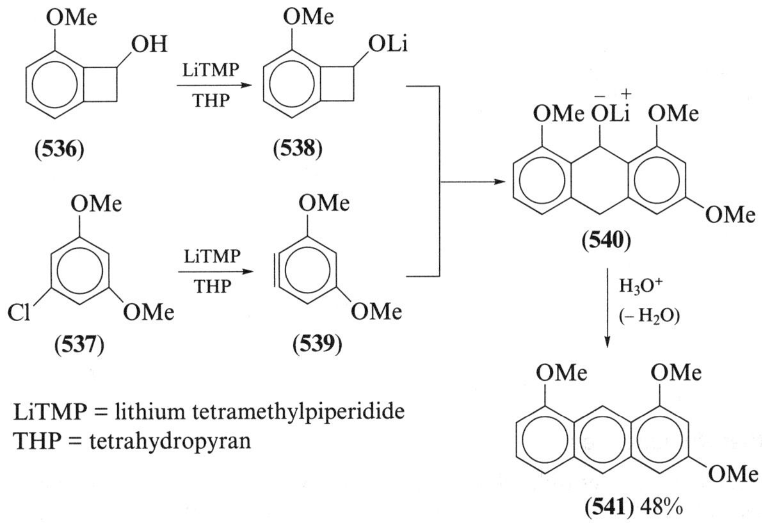

(536) (538) (540)

(537) (539) (541) 48%

LiTMP = lithium tetramethylpiperidide
THP = tetrahydropyran

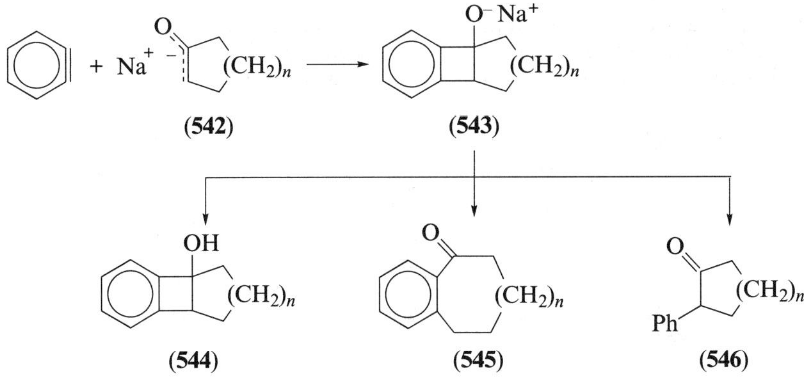

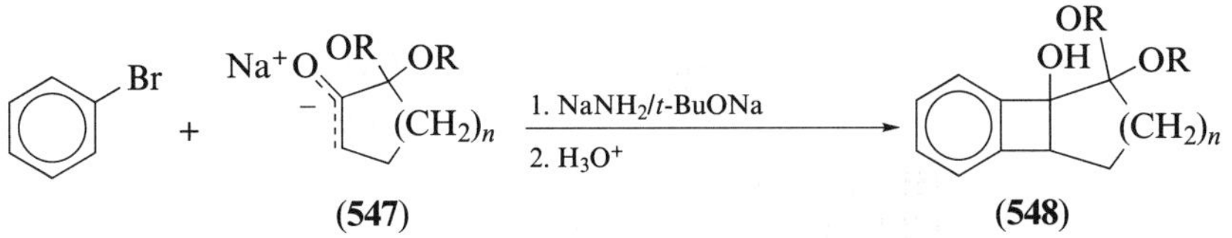

(542) (543)

(544) (545) (546)

(547) (548)

The ketal function in the product of these condensations allows the process to be continued, as shown with one example (**549** → **551** → **552**)[256] leading to reduced forms of polybiphenylenes. In view of the many transformations possible with these cyclobutenols[257] the methodology is attractive for the synthesis of polycyclic compounds.

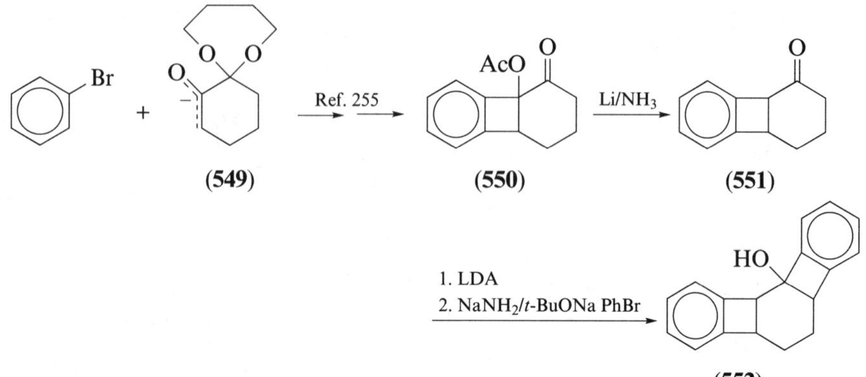

(549)   (550)   (551)

1. LDA
2. NaNH$_2$/t-BuONa PhBr

(552)

C. Other Nucleophiles

The reaction of arynes with primary and secondary amines provides a convenient route to alkylated anilines[1,2]. This reaction has now been extended to several di- and tri-substituted arynes. With symmetric arynes (from **553** and **554**) only one product is possible. With the aryne from **555**, substitution occurs *meta* to the methoxyl substituent to give **556** (R^1= Me, R^2 = OMe), as expected. With **557** the most acidic proton R^3 is eliminated and the resulting 3,4-dimethoxybenzyne is attacked regioselectively. With **558** and **559** substitution occurs adjacent to the bromine, clearly demonstrating the arynic nature of the reaction. In all these cases, a single amine is formed in good to excellent yield, the yields at the low end tending to occur with bulky amine nucleophiles.

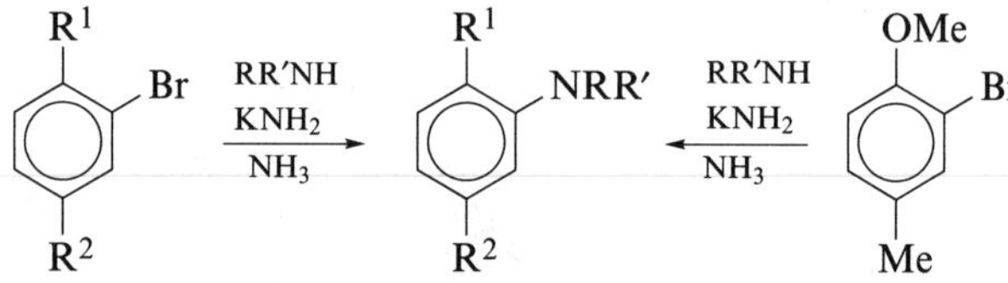

(553) R^1 = R^2 = Me[234]   (556)
(554) R^1 = R^2 = OMe[235]

(555)[238]

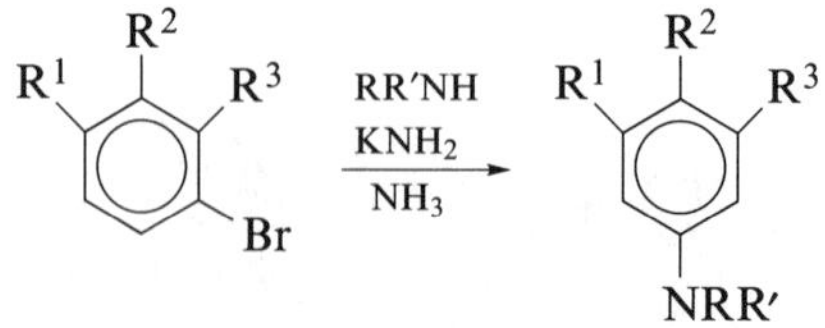

(557) (R^1 = R^2 = OMe, R^3 = H)[234]   (560)
(558) (R^1 = R^2 = OMe, R^2 = H)[238]
(559) (R^1 = R^2 = R^3 = OMe)[238]

Only with **561** were mixtures obtained, due to the weaker directing effect of Me vis-a-vis MeO (compare with **557**). Even so, **564** predominated over **565** by anywhere from 6–9:1, suggesting that most of the reaction proceeds via aryne **562**.

Poly(ethylene glycol)s (PEG) and their dimethyl ethers have been used as phase-transfer catalysts for the reaction of aryl halides with diphenylamine[258]. Since mixtures of triarylamines were obtained from a single aryl bromide (for example, o-bromotoluene gave both o- and m-tolyldiphenylamine), arynes are intermediates. A kinetic study led the authors to propose the following mechanism:

$$Ph_2NH \xrightarrow[PEG]{KOH} Ph_2\bar{N}$$

$$ArBr + Ph_2N^- \xrightarrow{rds} aryne$$

$$aryne + Ph_2NH \longrightarrow ArNPh_2$$

The PEG functions to complex potassium ion.

N-methyl-4-piperidone **568** acted as a nitrogen nucleophile, *not* a carbon enolate nucleophile, when treated with PhBr and sodium amide in THF[259]. A possible mechanism for the formation of unexpected product **569**, whose structure was established by an X-ray structure, is shown. Initial attack by nitrogen is followed by intramolecular attack at the carbonyl carbon and a Hofmann elimination.

A one-step synthesis of acridones takes advantage of somewhat similar chemistry[260]. Treatment of methyl N-methylanthranilate **570** with lithium N-isopropylcyclohexylamide (LilCA) in THF generated N-lithio salt **571**; addition of a halobenzene as the aryne source then produced the acridone **572** as shown, in good yield (40–60%). Substituted haloarenes gave substituted acridones. Utility of the method was demonstrated with a short synthesis of the alkaloid acronyeine **573**.

A high yield synthesis of diaryl ethers involves nucleophilic attack by phenoxide on arynes generated from iodobenzenes[261]. The arynic nature was established with substituted iodobenzenes (o-MePhI and phenol, for example, gave 35% m-tolyl phenyl ether as well as 25% of o-tolyl phenyl ether). Diaryl sulfides were similarly prepared from thiophenols and iodoarenes.

(574) 80%

A ketones enolate functions as an oxygen-rather than a carbon-nucleophile in the intramolecular capture of an aryne (575 → 576)[43].

A number of reactions have been developed involving sulfur ylids formed by attack of sulfur nucleophiles on arynes. For example, alkyl o- and p-(phenylthio)benzoates are obtained in modest yield by reaction of benzyne (from anthranilic acid) with the corre-

(575)   **(576)** 61%

$$Ar = 3,4,5\text{-}(OMe)_3C_6H_2$$

sponding (alkylthio)benzoic acids[262]. Since the reaction works with the *p*-isomer **579**, the alkyl transfer in the final stage can occur either inter- or intramolecularly.

Phenyl aryl sulfides are prepared in excellent yield from benzyne and ethyl aryl sulfides by a similar mechanism (10 examples, 87–97% yields)[263]. Even *o*-bis(phenylthio)benzene could be obtained (90%) from the corresponding bis(ethylthio)benzene and excess benzyne. When Ar in these reactions was a 9-anthryl group similar reactions occurred except for 9-*t*-butylthioanthracene **583**, which gave Diels–Alder adducts **584** and **585** instead[264]. Apparently, steric hindrance at the sulfur is sufficient to negate its effectiveness as a nucleophile, even toward the highly reactive and sterically unencumbered benzyne.

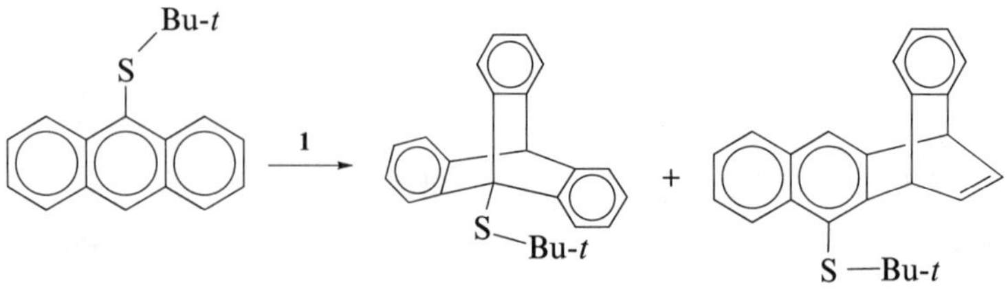

(583)   (584) 36%   (585) 13%

The reaction of benzyne (from 2-carboxybenzenediazonium chloride) with 2,5-dihydrothiophenes **586** affords 1-phenylthio-1,3-butadienes **589** in good yield[265]. The reaction probably involves betaine **587** and ylid **588** which then undergoes electrocyclic ring opening. The Z-dienes predominate (approx. 9:1) although minor amounts of the E-isomer are formed (H[+]-catalyzed isomerization), and minor amounts of the corresponding thiophenes are also formed, probably through dehydrogenation of **586** by benzyne. That the ring-opening is stereospecific is shown by the conversion of **590** to **591** (90% yield).

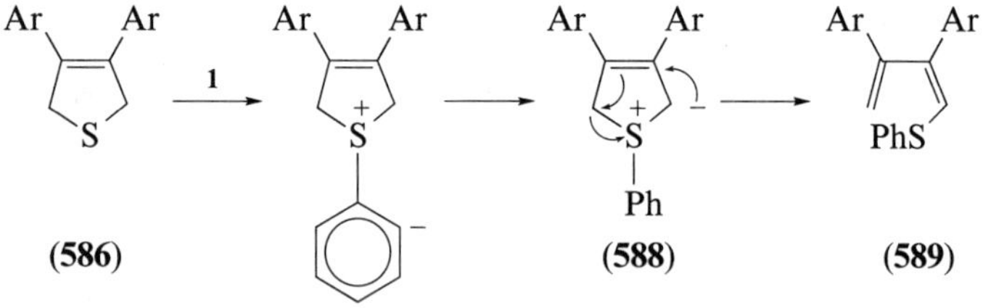

(586)   (588)   (589)

(587)

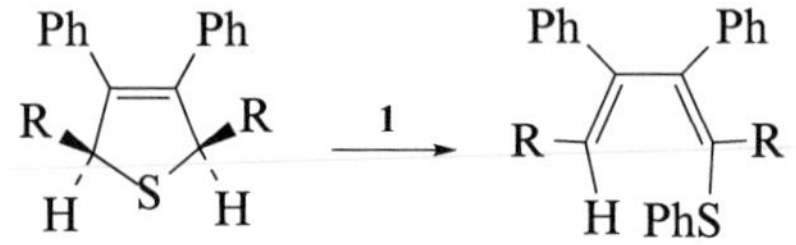

(590)   (591)

R = Me, Et

1,3-Dithiolanes **592** (X = S) are converted by benzyne to phenyl dithiocarboxylates **596** and/or to phenyl vinyl sulfide **597** depending on which ylid, **594** or **595**, is formed from initial intermediate **593**[266]. The reaction provides a two-step route from aromatic aldehydes to dithioesters in moderate (20–35%) yield; aliphatic aldehyde-derived dithiolanes give mainly **597**.

More useful is the reaction with 1,3-oxathiolanes **592** (X = O), which proceeds only via **595**[267]. If $R^1 = R^2 = R^3$, the reaction provides a good route to **597** (71%), since the other product, acetone, is volatile and easily removed. In other cases (for example, R^1 and/or R^2 is aryl), the reaction provides a good method for deprotection of carbonyl compounds from 1,3-oxathiolanes. The reaction also provides a good route to various phenyl vinyl sulfides that are otherwise difficult to obtain. Two examples are shown (**598** → **599** and **600** → **601**). Finally, since phenyl vinyl sulfides can be hydrolyzed to carbonyl compounds, the methodology provides a new 1,2-carbonyl transposition route. For example, 3-cholestanone

602 can be converted in five steps and 25% overall yield to the 2-isomer 605 via oxathiolane 603 and vinyl sulfide 604.

Reaction of benzyne with thiiranes gives phenyl vinyl sulfides in good yields (9 examples, 64–89% yields). The reaction is stereospecific, as shown by the formation of **608** from *cis*-thiirane **606**, and the corresponding *trans* vinyl sulfide from the *trans*-thiirane[268].

Benzyne generated from 2-carboxybenzenediazonium chloride cleaves cyclic and acyclic sulfides, as in the conversion of thiolane **609** to **613**[269]. Apparently protonation to the sulfonium salt **612** is faster than cleavage of intermediates **610** and/or **611**.

Contrary to an earlier report[270], arynes can insert into the S—S bond of diaryl disulfides, albeit in low (15–30%) yields, to give *o*-bis(arylthio)benzenes[271]. Somewhat higher yields can be obtained with diselenides[272].

The benzyne–Stevens rearrangement[1,2] continues to find utility as a carbon–carbon bond-forming reaction, especially in the synthesis of cyclophanes[273,274]. One example, the synthesis of strongly bent **617**[274] from bis(sulfide) **616** via a double Stevens rearrangement, is shown.

The reaction of benzyne with 1,4,2-benzodithiazine **618** to give a low yield of thianthrene **619** is probably initiated by nucleophilic attack of sulfur[275], as shown.

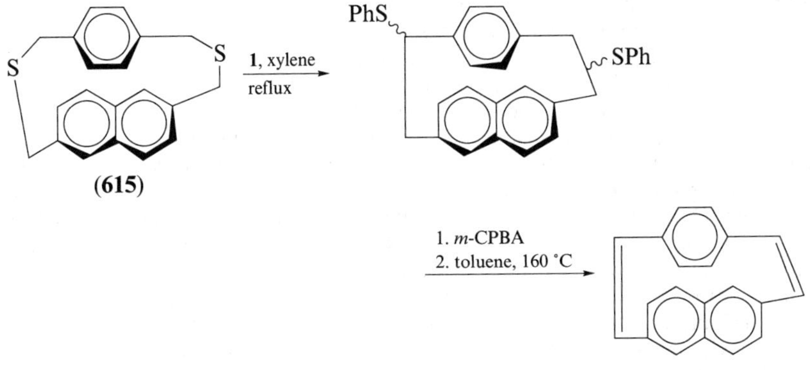
S
S
1, xylene
reflux
PhS
SPh
1. m-CPBA
2. toluene, 160 °C

(615) (617)

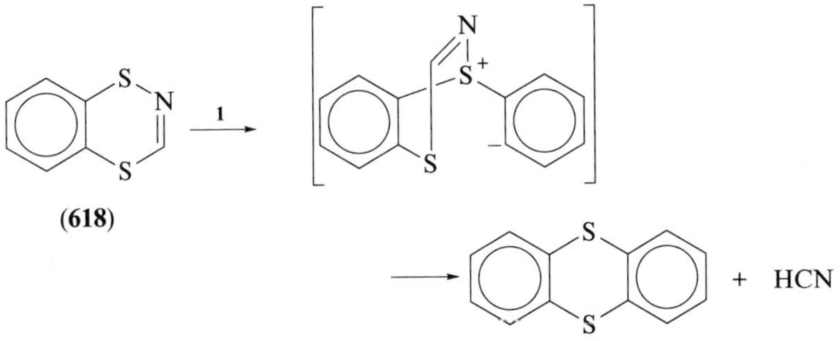
S
N
S
1
N
S
+
S
S
S
+ HCN

(618) (619)

D. Multiple and Tandem Aryne Reactions via Nucleophiles

Arenes with more than one halogen atom have been used to generate two or more aryne intermediates, each of which can be captured by nucleophiles. For example, a recent one-step pentaphene (**621**) synthesis uses *o*-dichlorobenzene in this way (*cf* **536** + **537** → **541**)[253].

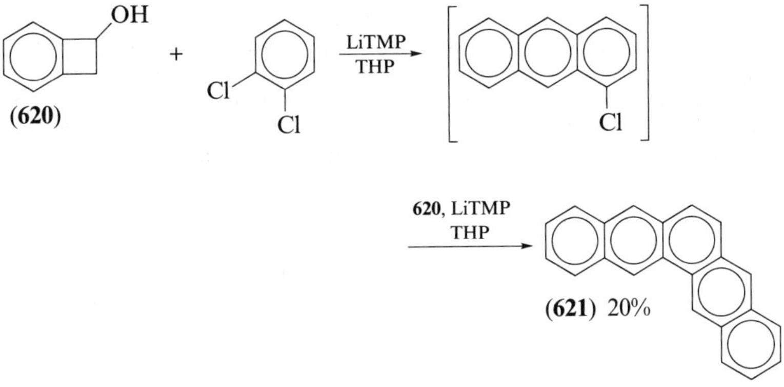
OH
+
Cl
Cl
LiTMP
THP
Cl
620, LiTMP
THP

(620) (621) 20%

With few exceptions[210–212] arynes used in the nucleophilic additions discussed so far are generated either by removing with base an aryl proton adjacent to a leaving group (Type A2, Section III.A.2), or by decomposing an anthranilic acid derivative (Type B1, Section III.A.3). Reactions described in the remainder of this section are initiated instead by metal–halogen exchange (Type A1, Section III.A.1).

1,2,3-Trihalobenzenes **622** (readily prepared from 2,6-dihaloanilines) undergo metal–halogen exchange mainly at the central (C2) halogen. This happens because the negative charge on the resulting organometallic is stabilized by two adjacent electron-withdrawing substituents. The resulting **623** can then eliminate MX to generate 3-haloaryne **624** which should react regioselectively with nucleophiles at C1 because the substituent is an EWG. If the nucleophile is an organometallic RM, the expected product **625** is set up for a second MX elimination to generate a new aryne **626**, which can again add the nucleophile to produce **627** and/or **628**. If the R group is an electron-withdrawing group (such as aryl, vinyl, ethynyl) then one would expect addition to **626** to be also regioselective and furnish only **627**. Finally, the product of this tandem aryne sequence is an organometallic reagent which might then be treated with various electrophiles to give an isolable product. If a

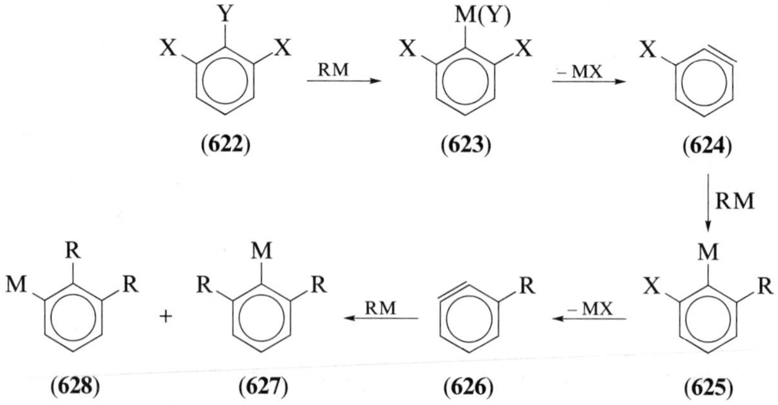

(622)   (623)   (624)

(628)   +   (627)   (626)   (625)

carbon electrophile were used, one would have replaced all three carbon–halogen bonds of **622** with carbon–carbon bonds in a single reaction.

In the event, treatment of **622** (X = Br or Cl, Y = Br or I) with 3 equivalents of an aryl Grignard reagent followed by an electrophilic quench gave *m*-terphenyls **629** in good yield (60–85%)[276]. The reaction is not subject to steric hindrance (for example, Ar = 2,6-dimethylphenyl, mesityl, 1-naphthyl or 2-anisyl). It provides a one-pot route to *m*-terphenyls substituted at the 2′ position (or, looked at another way, 2,6-diaryl benzene derivatives), difficult to obtain in other ways. This *m*-terphenyl synthesis has been used to create two new classes of cyclophanes (cupped- and cappedophanes)[277,278].

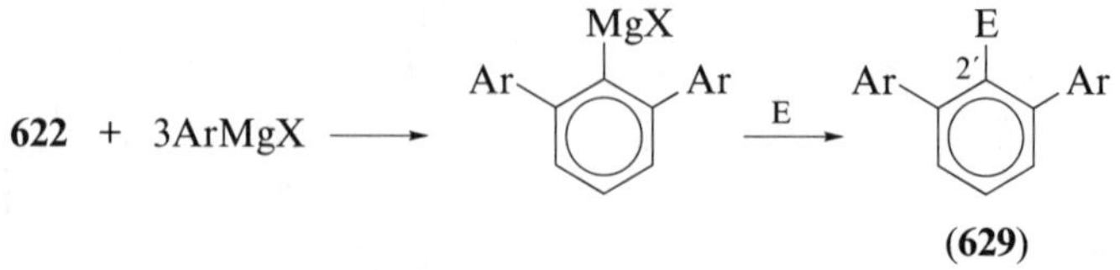

622   +   3ArMgX   ⟶

(629)

Aryllithiums give much lower yields of *m*-terphenyls. This is probably because aryllithiums are more basic than the corresponding Grignard reagents, and remove aryl protons from the trihalobenzenes, leading to other products.

Only two of the three equivalents of aryl Grignard used in the transformation of **622** to **629** appear in the product; the first equivalent is used for metal–halogen exchange (**622** → **623**). To conserve a rare or expensive aryl Grignard reagent, one can first treat **622** with one equivalent of vinylmagnesium bromide to bring about the exchange, then add just two equivalents of the aryl Grignard reagent[277].

Vinyl Grignard reagents can also be used as the nucleophiles. For example, a series 1,3-bis(1'-cycloalkenyl)benzenes **630** have been prepared via the tandem aryne sequence in reasonable yield[279]. Similarly, trimethylsilyl vinyl Grignard **492** converts **622**, via **631**, to indenone **632** in which the three halogens have been replaced by three functionally differentiated adjacent carbons[232].

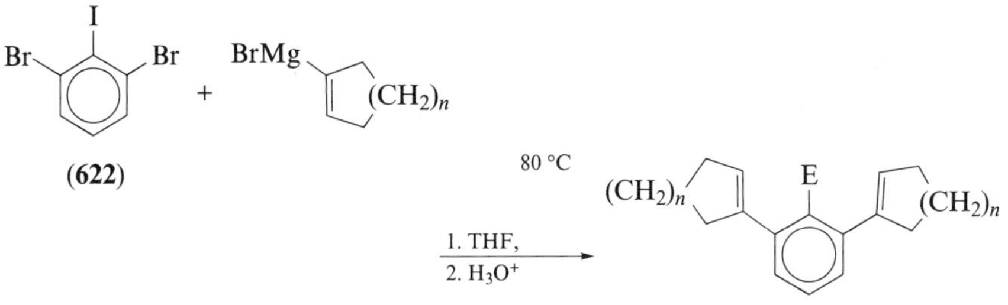

(622)

(630) n = 1–4
(E = H, Me, I)

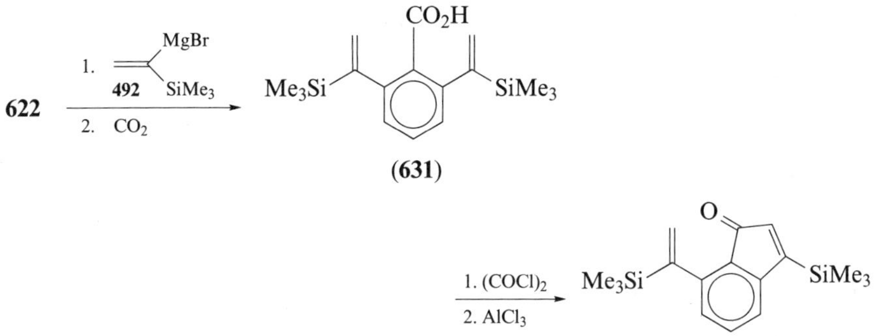

(631)

(632) 40%

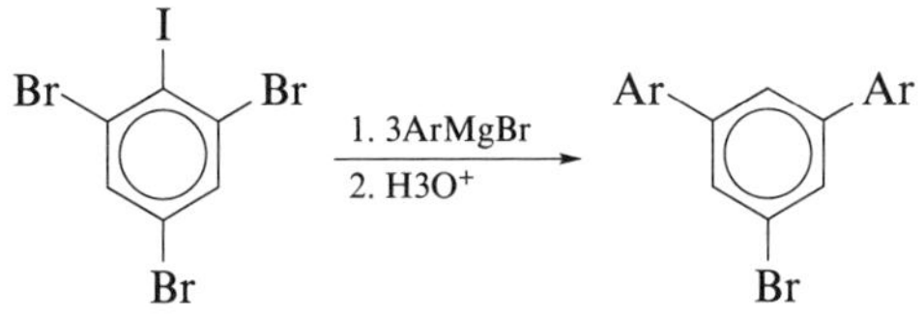

(633)

(634)

*Tetra*halobenzenes also undergo regioselective metal–halogen exchange and aryne generation, the particulars of which depend on the location of the halogens. 1,2,3,5-tetrahalobenzene **633** reacts analogously to 1,2,3-trihalobenzenes, with exchange occurring at the 2-halogen. Thus **633** gave 5'-bromo-*m*-terphenyls **634** (80%)[276].

1,2,4,5-Tetrahalobenzenes such as **635** react with aryl Grignard reagents to give *p*-terphenyls **636**[233]. Four equivalents of the Grignard reagent are required, two for metal–halogen exchange and two as nucleophiles. Yields are in the 30–55% range (10 examples).

(635)   (636)

(635)   (637)

(639)   (640)

(641)

Although there are some mechanistic uncertainties, there is evidence that the reaction occurs via two successively formed arynes produced after metal–halogen exchange at *both* iodines. If arynes were formed independently at each halogen pair, one would expect a mixture of *m*- and *p*-terphenyls as products. It is thought that the regioselectivity in the nucleophilic additions to aryne intermediates **638** and **639** is a consequence of the need to keep the two carbanionic sites on the benzene ring remote (i.e. *para*) to each other.

This methodology was used recently to synthesize cyclophane **641**, with three face-to-face benzene rings[280].

The situation with 1,2,3,4-tetrahalobenzenes and Grignard reagents is somewhat more complex. The major products (*ca* 50% yield) from **642** and unhindered aryl Grignards were 1,2,3-triarylbenzenes **643**, with minor amounts of diarylbenzene **644**[281]. For example, the novel polyphenyl **645** was synthesized in one step and 30% overall yield from **642** (X = Br, Y = Cl) and 4-biphenylylmagnesium bromide, with a methyl iodide quench. However, with the hindered mesitylmagnesium bromide, the product was the 1,2,4-trimesitylbenzene, *not* the 1,2,3-isomer.

(642) X = Br, Y = Cl   **(643)** 40–50%   **(644)** 5%
or X = I, Y = Br

(645) 30%

A mechanism involving three aryne intermediates formed in tandem was proposed to account for the products, with initial metal–halogen exchange occurring at bromine in preference to chlorine, or at iodine in preference to bromine. Low-temperature quenching experiments established **646** as the first exchange product. Aryne **647** adds the nucleophile *meta* to Y as expected, to give **648** which is set up for the second aryne formation. Again, nucleophilic addition occurs mainly *meta* to X to give **650**. The third aryne **651** can then give either 1,2,3-product **652** or 1,2,4-product **653**. The predominant formation of **652**, except when Ar is the bulky mesityl group, is a somewhat surprising result and perhaps a weak point in the mechanism. It would be important to generate **651** directly by another route and establish whether or not it does react with ArMgX to give **652**. The formation of **654** as well as **650** from **649** is not too surprising since both substituents are electron-withdrawing, and minor product **644** could arise from **654** after another metal–halogen exchange. Also, some product may arise via a di-Grignard derived from a second exchange at X in **646**. Clearly, further work is required to establish the mechanism more firmly.

The reaction of pentahalobenzenes with Grignard reagents has not been extensively studied, but hexahalobenzenes react with excess Grignard reagent (6–8 equivalents) in THF as solvent in a very useful way, to give exclusively 1,2,4,5-tetraarylbenzenes[282] (early experiments using ether as solvent gave similar products but in < 10% yield[283]). For example, hexabromobenzene (or 1,2,4,5-tetrabromo-3,6-dichlorobenzene) give, after proton quench, 1,2,4,5-tetraarylbenzenes in 50–75% yield (11 examples, including phenyl-, mesityl-[119], biphenylyl-, 1- and 2-naphthyl-). The actual reaction product is the 1,2,4,5-tetraaryl-3,6-di-Grignard reagent **662**, as shown by quenching with electrophiles other than a proton. A mechanism from **655** to **662** is shown. There are at least four (and possibly as many as eight) tandemly-formed aryne intermediates, only the first (**658**) and last (**661**) of which are shown. Quenching evidence for mono-Grignard **656** and di-Grignard **657** intermediates was obtained. The controlling factor in nucleophilic addition to all the aryne intermediates (for example, **658** → **659** or **661** → **662**) is thought to be the need to keep anionic centres in the di-Grignards as remote from one another as possible. In a sense, one can regard a hexahalobenzene as made up of two 1,2,3-trihalobenzenes, in which case treatment with an aryl Grignard gives a *m*-terphenyl at each end of the molecule (*cf* **622** → **629**).

As discussed above with 1,2,3-trihalobenzenes, vinyl Grignard reagents also undergo metal–halogen exchange and nucleophilic aryne capture, and can replace aryl Grignard reagents in all of these reactions. For example, **635** → **664** and **665** → **666**[284]. Acetylenic Grignard reagents, however, in most cases fail to undergo the metal–halogen exchange required to initiate the reactions. This difficulty can be overcome by first adding one equivalent (in the case of 1,2,3-trihalobenzenes) or two equivalents (in the case of 1,2,4,5-

tetrahalo- or hexahalobenzenes) of an alkyl or vinyl Grignard to effect the exchange, then adding the resulting halogenated aryl Grignard to an excess of the desired alkynyl Grignard, as in the conversion of **667** to **668**.

The scope of multiple and tandem aryne additions, which as we have seen are useful for forming two or more carbon–carbon bonds in a one-pot reaction, remains to be further explored.

VII. NATURAL PRODUCTS VIA ARYNES

There has been considerable progress in this area since the previous review[1], particularly in the application of aryne chemistry to alkaloid synthesis, and a thorough review with over 100 references has appeared[285] covering the literature through 1986. Earlier work focussed mainly on ring construction via intramolecular nucleophilic attack on an aryne generated by dehydrohalogenation of an aryl halide[1]. Examples include benzylisoquinoline-derived alkaloids (an aryne is generated on the arene ring of the benzyl group and cyclizes onto the isoquinoline nitrogen or onto a suitably located phenolic oxygen in the isoquinoline moiety), benzophenanthridines (aryne cyclization onto carbon *ortho* to the amide ion of an aniline), and of course the considerable body of work based on benzocyclobutenes (prepared via intramolecular aryne cyclization on the α-anions of nitriles) and their isomeric *o*-quinonemethides. More recently emphasis has shifted to cycloadditions, particularly *inter*molecular aryne reactions of the Diels–Alder type. Classes of natural products synthesized in this way include the aporphinoids, protoberberines, benzophenanthridines, ellipticines, anthracyclines and *o*-naphthoquinones.

With a few exceptions, results will be described here that have appeared since the comprehensive review[285] (i.e. references of 1987 and later), although some reference to earlier work is essential for background information.

A. Intramolecular Cycloadditions

One of the early examples of an intramolecular Diels–Alder reaction of benzynes involved cycloaddition to a furan that was tethered to the benzyne moiety[47,286]. Diazotization of anthranilic acid **669** gave in 86% yield the cycloadduct **670**, which was subsequently converted to the *o*-naphthoquinone **671** (mansonone E; 7 steps) as well as to the structurally related mansonones F and I and biflorin[287].

(**669**)   (**670**)   (**671**)

(**672**)   (**673**)   (**674**) 63%

In **669** a 3-atom chain links the aryne and furan moieties. The reaction was also successful with a 4-atom link (**672** → **674**)[47] but, with an analogous 2-atom link, fragmentation to diazoquinone **676** occurred[47]. The analogous chemistry, but with an all-carbon link (that would be less prone to fragmentation), has not been explored.

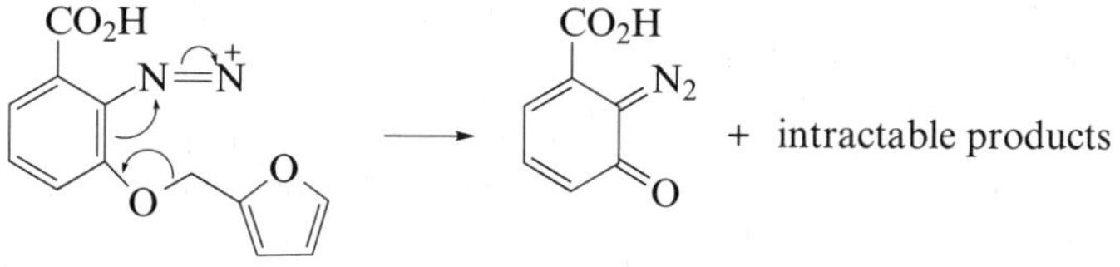

(**675**)　　　　　　　　　　(**676**)

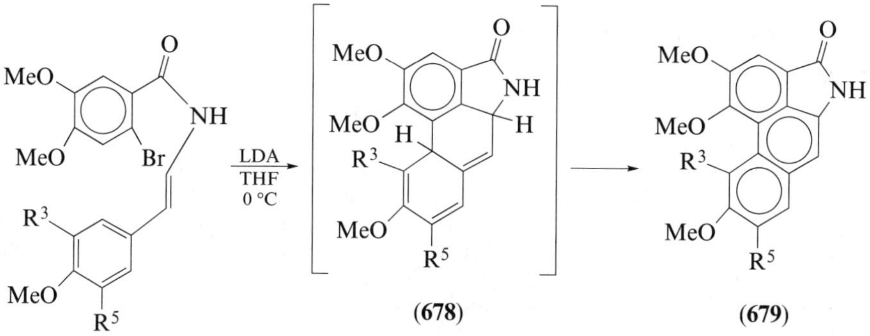

(**677**) R^3 = R^5 = H; R^3 = OMe, R^5 = H; R^3 = H, R^5 = OMe

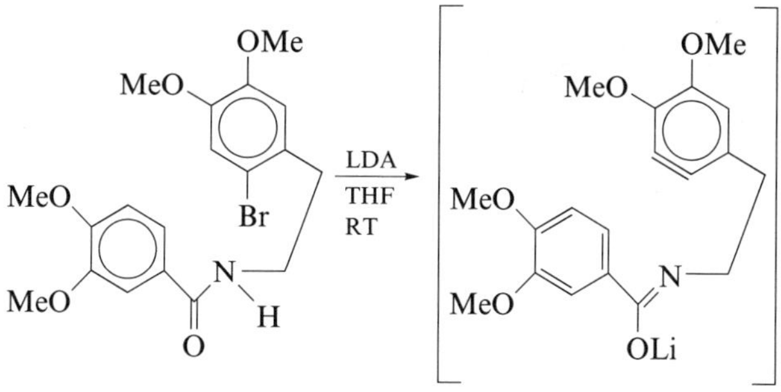

(**678**)　　　　　　　　(**679**)

(**680**)　　　　　　　　(**681**)

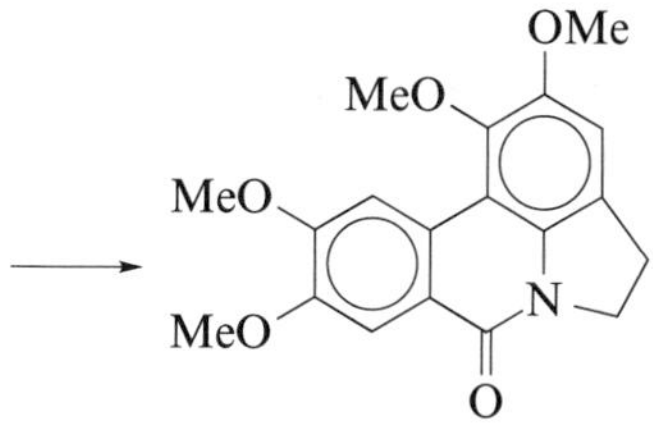

(**682**) 74%

An intramolecular benzyne cycloaddition to a styrenic diene has been used to simultaneously construct two carbon–carbon bonds of the central ring of a phenanthrene moiety, in a new approach to aristolactam alkaloids **679**[288]. Treatment of *E*-alkene **677** ($R^3 = R^5$ = H) with LDA generated the aryne which added to the styrene moiety, presumably to give **678** which was oxidized to the natural product **679** under the reaction conditions (yield 35%). The mechanism of the oxidation step is discussed briefly below. Analogs with a methoxy substituent at R^3 or R^5 were also prepared.

A similar strategy has been used to prepare **682**, a compound closely related to *Amaryllidaceae* alkaloids[289]. In this case, the aryne presumably adds to a 1-azadiene, formed from the amide and base. With 2 equivalents of LDA the yield of **682** was only 25%, but a large excess (16 equiv) raised the yield to 74%. The reaction may involve a nonsynchronous cycloaddition or a nucleophilic attack on the aryne; a loss of 'hydride' is necessary to complete the reaction.

B. Intermolecular Cycloadditions

1. Aporphinoids

The aporphinoid skeleton **683** could be constructed via an *inter*molecular aryne cycloaddition to the styrene moiety of an alkylidenetetrahydroisoquinoline, thus simultaneously creating bonds **b** and **c** of the C-ring (*cf* the *intramolecular* aryne approach in **677** → **679**, which creates the **a** and **c** bonds). This approach has met with considerable success[290]. The following examples illustrate some of the structural variations that are possible. Best yields were obtained when preformed benzenediazonium-2-carboxylate was used as the aryne source. The synthesis of dehydroaporphines **685** (12 examples), 4,5-dioxoaporphines **687** (4 examples) and aristolactams **689** proceeds in modest to good yields[290g]; in the latter case, substantial amounts of [2+2] adduct **690** is also formed.

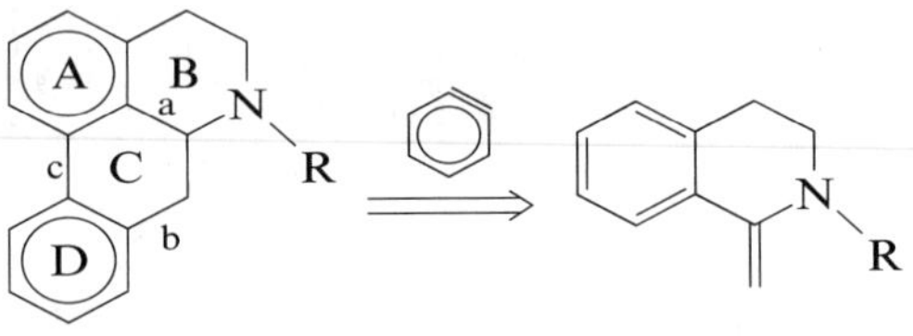

(683)

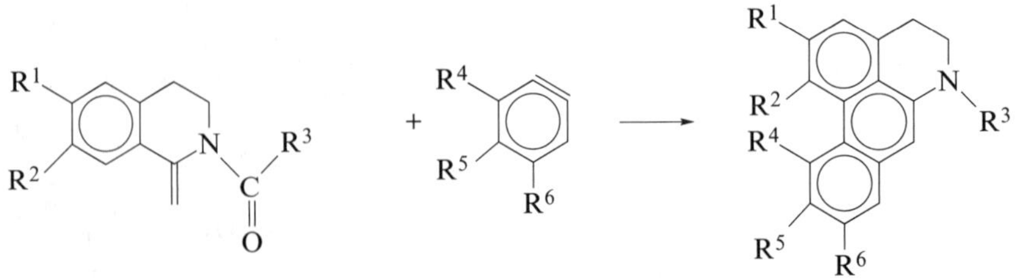

(684)

(685)

(686)   (687)

(688)   (689)   (690)

These cycloadditions can be highly regioselective[290b], as in a recent synthesis of apomorphine analog **693** from **691** and 3,4-dimethoxybenzyne; **693** was isolated in 34% yield, and none of its regioisomer could be detected in the reaction mixture. Thus initial attack is by the more nucleophilic site of the diene on the more electrophilic (and less hindered) site of the aryne.

(691)   (692)   (693)

As to the final oxidation (aromatization) step that follows these cycloadditions in the same reaction mixture and prior to workup, some results suggest a stepwise process[290f]. Treatment of either Z-**694** or E-**694** (or a mixture of the two) gives **697** in approximately identical yield (52%). Since only one of the two initial cycloadducts, cis-**695**, can undergo a benzyne-promoted 1,4-syn elimination of hydrogen, it is suggested that aromatization proceeds by hydride abstraction by benzyne to generate carbocation **696**, which then loses a proton.

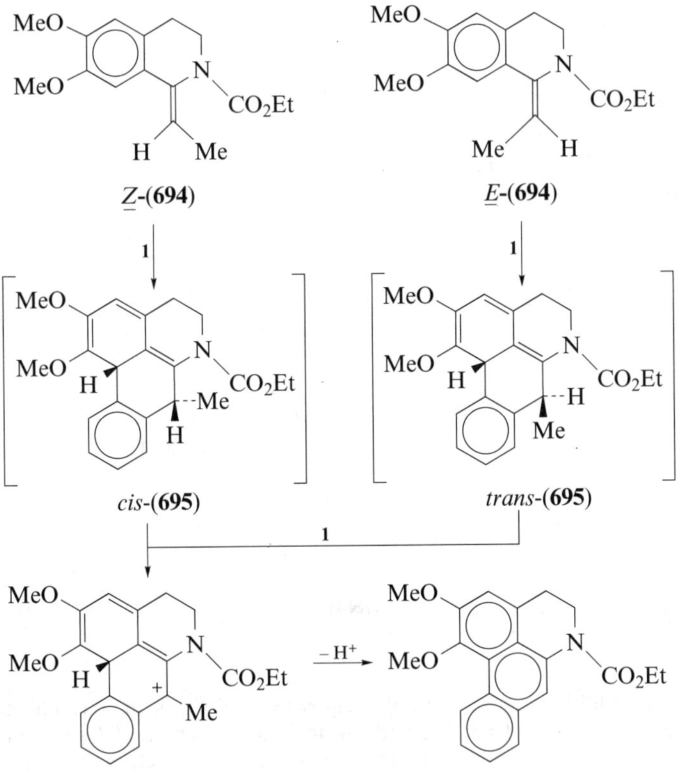

Z-(694)

E-(694)

cis-(695)

trans-(695)

(696)

(697)

2. Protoberberines

These isoquinoline alkaloids are characterized by the tetracyclic structure present in **699**, and have been synthesized via pyrrolinediones **698** and arynes[291]. For example, **698** (R^1 = OMe, R^2 = Br) and 4,5-dimethoxybenzyne (generated via aprotic diazotization of the anthranilic acid) in refluxing DME gave **699** (R^1 = R^3 = OMe, R^2 = Br) in 44% yield. Hydrogenolysis of the bromine gave 8-oxypseudopalmatine **699** (R^1 = R^3 = OMe, R^2 = H)[291b,c]. It is important that R^2 not be H; otherwise side reactions, including arylation

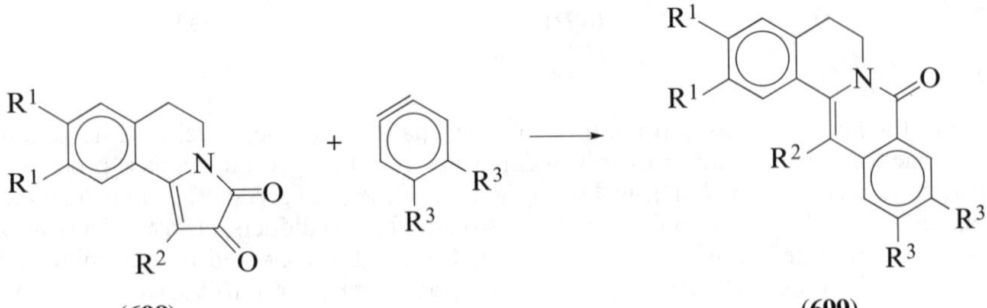

(698)

(699)

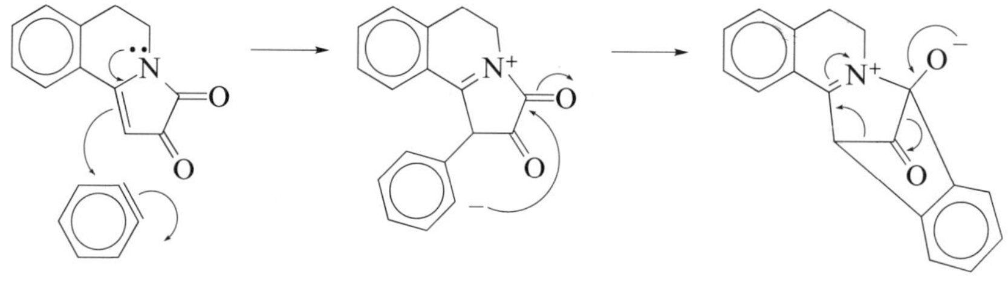

(700)

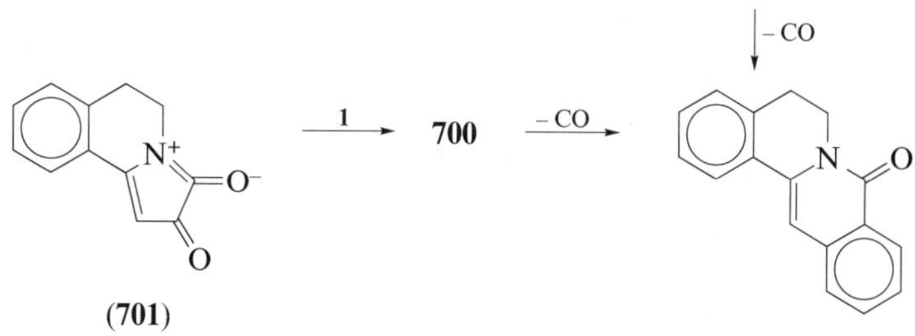

(701)

at that position, can occur. As seen in this example, halogen is an effective blocking group at R^2, and can later be removed if necessary. Although the mechanism has not been studied, the reaction may well proceed through attack of the aryne on a nucleophilic site on the pyrrolinedione followed by cyclization to **700**. Elimination of CO from **700** may be assisted by a second equivalent of aryne. Another alternative is cycloaddition to the 2-azadiene **701**.

The methodology has been extended to some indole alkaloids **702**[291c].

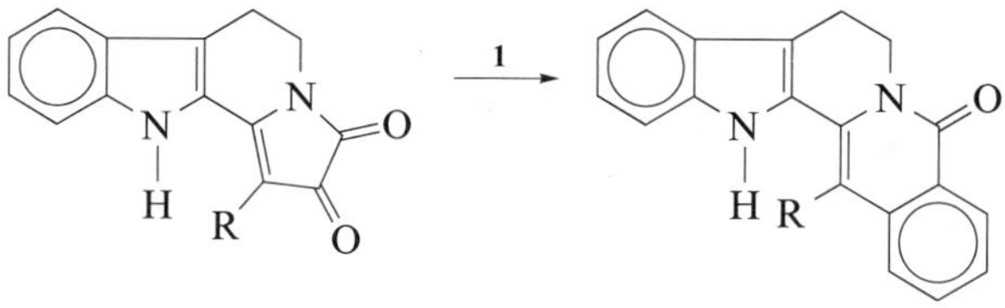

(702)
R = Et, Br, Cl

3. Benzo[c]phenanthridines

Three different intermolecular benzyne cycloaddition approaches to these alkaloids have been developed. One[292] is analogous to the protoberberine route (Section VII.B.2), and constructs the B-ring. The nitrogen in **703** was protected (R = Me, CH₂Ph) and the remaining R groups were usually H or alkoxy. Yields were mainly in the 30–50% range. The reaction of **703** (R^4, R^5 = OCH₂O) with 3,4-dimethoxybenzyne (R^1 = H, R^2 = R^3 = OMe) was regioselective, the only product being **704** (R^1 = H, R^2 = R^3 = OMe, R^4, R^5 = OCH₂O, R = Me) in 29% yield.

The second approach constructs the C-ring, and uses a benzyne–pyrone cycloaddition[293]. Yields were good, usually in the 70–80% range (6 examples). The nitrogen in **705**

(703)   (704)

(705)   (706)

must be protected, or O-phenylation of the amide occurs. The product **706** is easily saponified and decarboxylated to give the benzo[*c*]phenanthridine.

The third approach, like the first, constructs the B-ring. The key step, applied to the synthesis of *N*-nornitidine **710**[294], involved aryne cycloaddition to a vinyl isocyanate[207]. 4,5-Dimethoxybenzyne, generated from aminobenzotriazole **707** by Pb(OAc)$_4$ oxidation, added to vinyl isocyanate **708** to give **709**, which was not purified but converted directly to **710** (overall yield in the 30–35% range).

(707)   (708)   (709)

(710)

4. Aryl C-glycosides

A key step in the synthesis[295] of Gilvocarcin M **713**, an aryl *C*-glycoside antibiotic, is the regioselective cycloaddition of the aryne derived from **711** to 2-methoxyfuran, to give

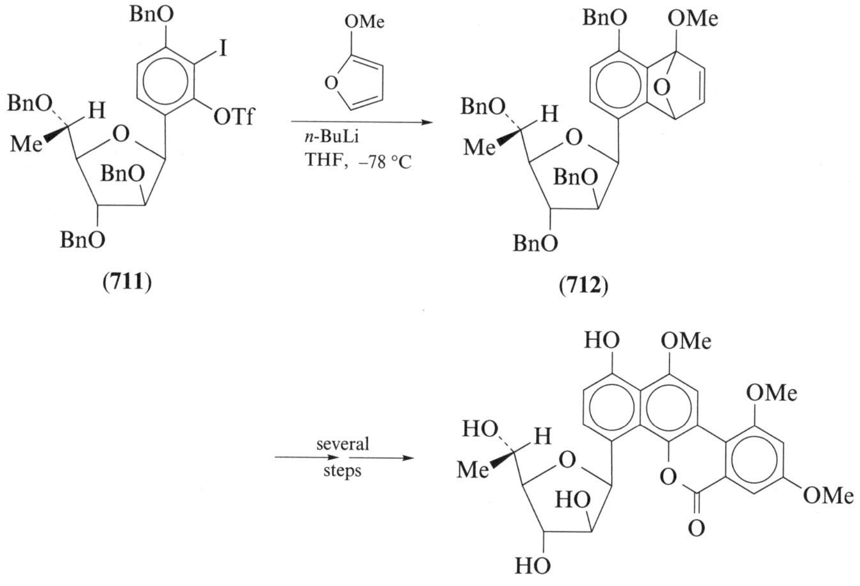

(711)

(712)

(713)

adduct **712** in 85% yield (9% of the regioisomer was also isolated). The observed regioselectivity is attributed to the polar effect of the benzyloxy substituent on the aryne, which controls the mode of cycloaddition.

VIII. HETEROARYNES

The literature prior to approximately 1981–2 has been thoroughly and exhaustively reviewed[5–7]; consequently, this section will deal only with literature since then, and only with the major types of six- and five-membered heteroarynes.

A. Six-membered Heteroarynes

1. Didehydropyridines

a. Structure. These are by far the most studied and best authenticated of the heteroarynes[5]. Two recent *ab initio* calculations[24,296] conclude that 3,4-didehydropyridine (**2**) is more stable than the 2,3-isomer by 7.4[296] or 13.9[24] kcal mol^{-1}. Both are predicted to be ground state singlets, the singlet–triplet gap for **2** being 30.7 kcal mol^{-1} (similar to that for *o*-benzyne), and for the 2,3-isomer, 17.5 kcal mol^{-1}. Optimum geometries for all six possible didehydropyridines have been calculated[296]. The 'triple' bond length in **2** is predicted to be about 1.26 Å[296].

The infrared spectrum of matrix-isolated 3,4-didehydropyridine **2** has been reported[297]. This is the first direct observation of a heteroaryne. 3,4-Pyridinedicarboxylic anhydride **714** was deposited in an Ar or N_2 matrix and irradiated at 13 K; with light > 340 nm **714** gave CO, CO_2 and peaks assigned to **2**, whereas subsequent irradiation with higher-energy light caused the disappearance of those peaks and the appearance of HCN, HC≡CH, HC≡C—C≡CH and HC≡C—CN bands, presumably due to fragmentation of **2**. A key feature of the IR spectrum was a band at 2085 cm^{-1}, attributed to the C≡C stretch of the C3—C4 bond, by analogy with a band at that frequency that was originally attributed to the triple bond of o-benzyne. However, since the original assignment of this band in benzyne is now attributed to cyclopentylidene ketene (see Section II.A), it is possible that this band is due to the analogous ketene **715**. The C≡C stretch in benzyne give rise to a weak band at 1860 cm^{-1}, and since no band for **2** in the 1860 cm^{-1} range was noted, one must view the pyridyne assignment[297] with some skepticism until the results are either confirmed or refuted.

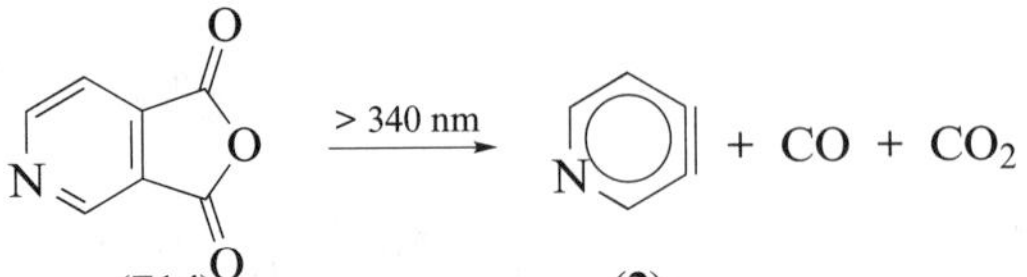

(714) → (2) + CO + CO₂

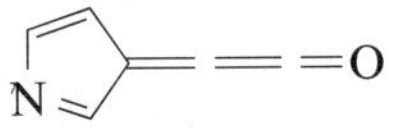

(715)

Similar irradiation experiments on 2,3-pyridinedicarboxylic anhydride **716** resulted in ring-opened products, and gave no evidence for the formation of **717**[298].

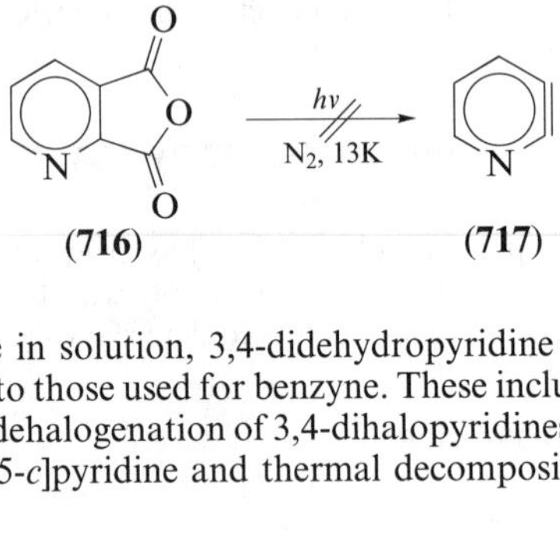

(716)   (717)

b. Generation. For use in solution, 3,4-didehydropyridine has been generated in a variety of ways analogous to those used for benzyne. These include dehydrohalogenation of 3- and 4-halopyridines, dehalogenation of 3,4-dihalopyridines, lead tetraacetate oxidation of 1-aminotriazolo[4,5-c]pyridine and thermal decomposition of pyridine-3-diazonium-4-carboxylate[5,6].

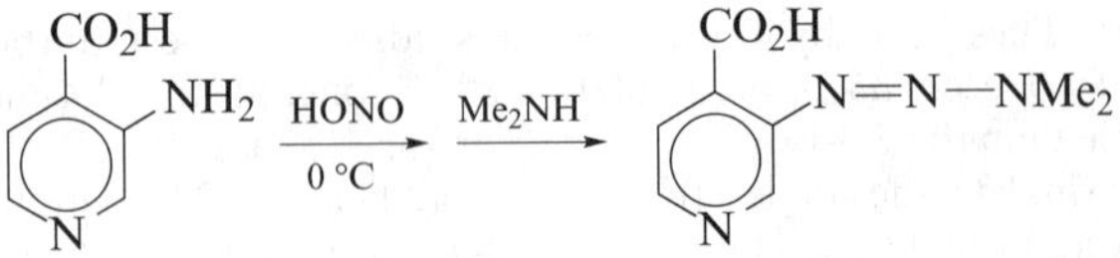

(718) 73%

The latter method, so commonly used for benzyne, is less used with pyridyne because the precursor is difficult to prepare and somewhat unstable. As an alternative, the use of 3-(3,3-dimethyltriazen-1-yl)pyridine-4-carboxylic acid **718** has been recommended[299]. It is readily prepared as shown, and is a shelf-stable crystalline solid. It decomposes in refluxing

(719) 40%

acetonitrile containing trifluoracetic acid to generate the heteroaryne which, for example, can be trapped by furan.

Pyridyne is formed from 3-bromopyridyl sulfoxide **720** and phenylmagnesium bromide at elevated temperatures[33] (*cf* **15** → **16** → **1**).

For reactions with nucleophiles, 3,4-didehydropyridine is effectively generated from 3-bromopyridine using the complex base NaNH$_2$/*t*-BuONa in THF[300]. For example, 3- and 4-aminopyridines **721** and **722** were obtained in 72–95% yield (approx. 1:1 ratio; usually a slight excess of **722**), via nucleophilic addition to the intermediate heteroaryne.

The most recent successful method for generating 3-pyridynes uses 4-trialkylsilyl-3-pyridyl triflates as precursors[301] (a similar approach but with a bromine in place of the triflate group[302] is considerably less successful). They were synthesized by *o*-metallation and silylation of carbamate **723** to give **724**, which was then converted to triflate **725**. Treatment of **725** (R = Et) with fluoride (either tetrabutylammonium fluoride or CsF + 18-crown-6) generated the pyridyne which was trapped with several dienes, such as furan.

 c. Thermal rearrangements in the gas phase.   The aryne–cyclopentylidene carbene rearrangement (Section IV) has been examined briefly in the pyridyne series. Flash vacuum pyrolysis of 2-phenylquinoline-3,4-dicarboxylic anhydride **726**[303] at 800 °C/0.06 Torr gave indeno[1,2-*b*]indole **729** in 30–40% yield. The product (red) is a dibenzoazapentalene, hence reactive and was isolated in a dihydro form after addition of various nucleophiles across the 1-azadiene moiety. The proposed mechanism is analogous to that for **146** → **147**, and involves rearrangement of quinolyne **727** to carbene **728** as the key step.

The analogous *o*-tolyl heteroaryne precursor **730** gave mainly 11*H*-benzo[*a*]carbazole **733** through insertion into the methyl C—H bond[303b].

In contrast with the 3,4-didehydroquinoline intermediates, the 3,4-pyridyne intermediate obtained on pyrolysis of anhydride **734** (*cf* **726**) did not rearrange to a carbene or otherwise maintain its structural identity, but instead fragmented mainly to 1,4-diphenylbutadiyne **735** and benzonitrile, with minor amounts (< 4%) of 1- and 2-phenylnaphthalenes[304]. Deuterium labeling established that the phenyl groups in **735** come from the C2 and C5 positions in **734**, and that the benzonitrile comes mainly (85%) from the C6 phenyl group (and 15% from the C2 phenyl group). Hence the major fragmentation path is as shown in **737**.

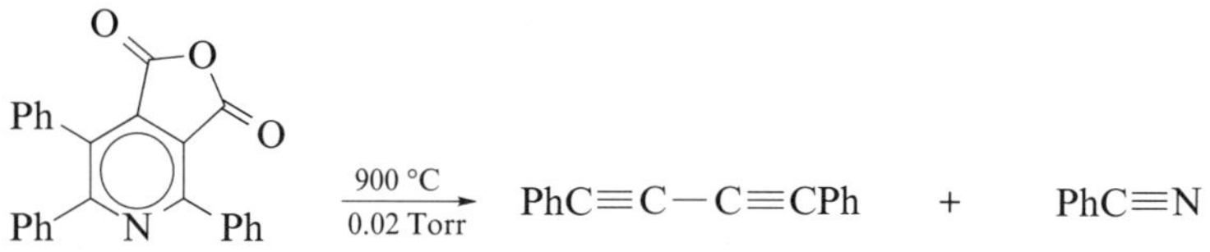

$$\text{(734)} \xrightarrow[\text{0.02 Torr}]{900\,°C} \text{PhC}\equiv\text{C}-\text{C}\equiv\text{CPh} + \text{PhC}\equiv\text{N}$$

(735) 25%   (736) 54%

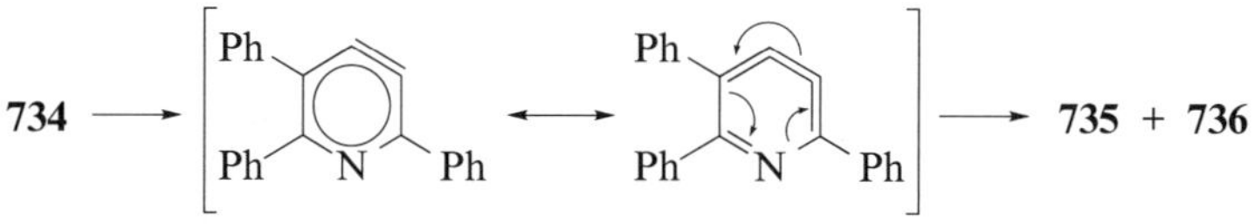

$$734 \longrightarrow [\text{(737)}] \longrightarrow 735 + 736$$

d. Cycloadditions. Although 3,4-didehydropyridine (**2**) gives cycloadducts with furans and tetraphenylcyclopentadienone[6], it is reported not to give an adduct with anthracene[299,305] and either no[299] or a [2+2] adduct[306] with cyclopentadiene. However, it is now thought that the [2+2] product **738** is in fact the [4+2] adduct **739**[307], formed in low yield from 3,4-didehydropyridine generated through aprotic diazotization of 3-aminoisonicotinic acid or, more conveniently, from **718**.

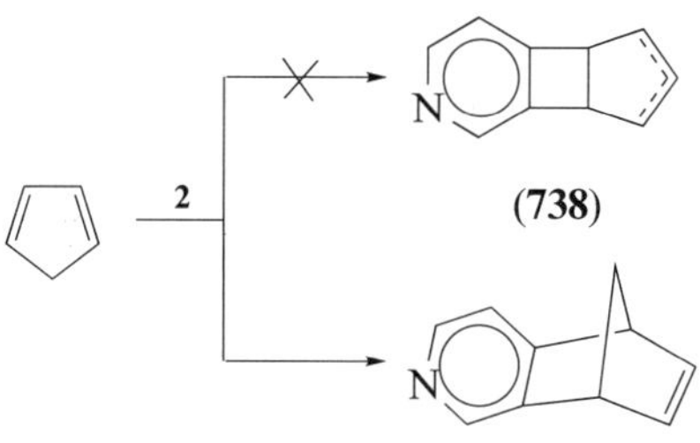

(738)

(739) 16%

A three-step synthesis of the antitumor alkaloid ellipticine **741** involves, as a key step, the cycloaddition of 3,4-didehydropyridine generated from **718** to the diene moiety in

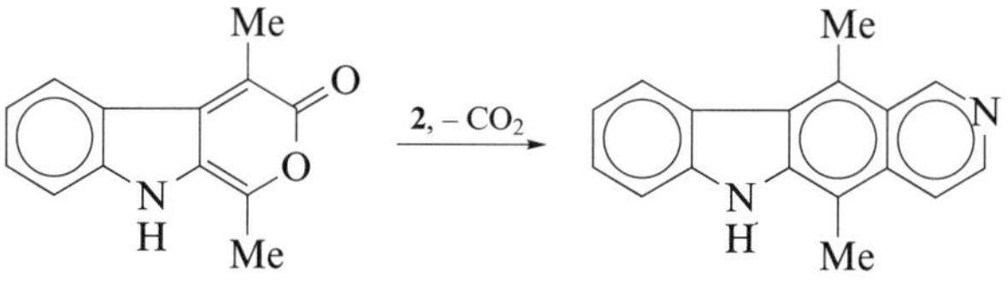

(740)   (741) 20%

pyranoindolone **740**[308]. The cycloaddition is *not* regioselective, an equal amount of isoellipticine, (the regioisomer of **741**) also being formed. An alternative approach[309] to ellipticine, using isoindole **742** and 3-pyridyne generated from lead tetraacetate oxidation of the appropriate aminotriazole, gave adduct **743** and its regioisomer in 62% yield (ratio 55:45). These were converted in 3 steps to **741** and its regioisomer. 5-Amino analogs of ellipticine and isoellipticine were similarly prepared.

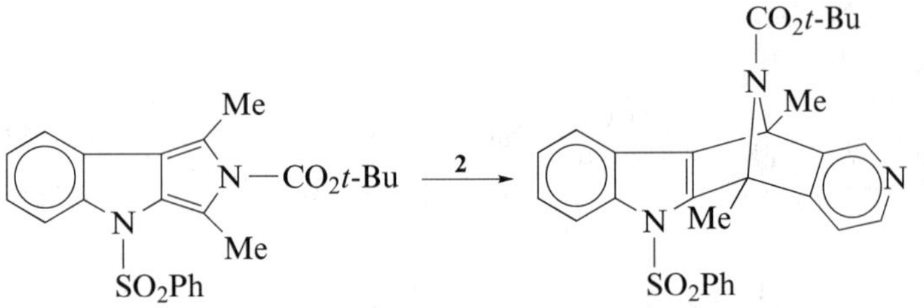

(**742**)

(**743**) 62%
+ regioisomer

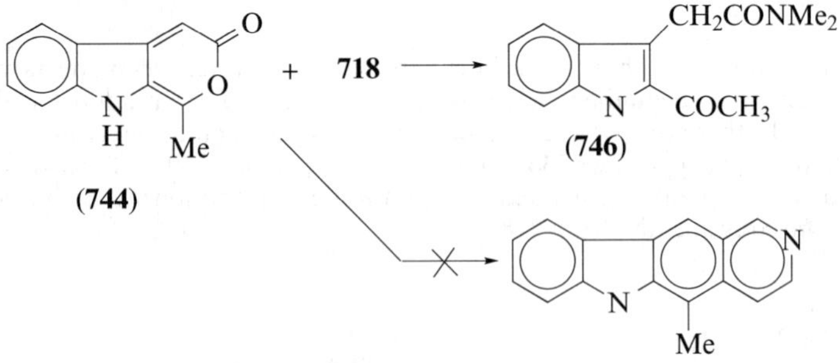

(**744**) + **718** →

(**746**)

(**745**)

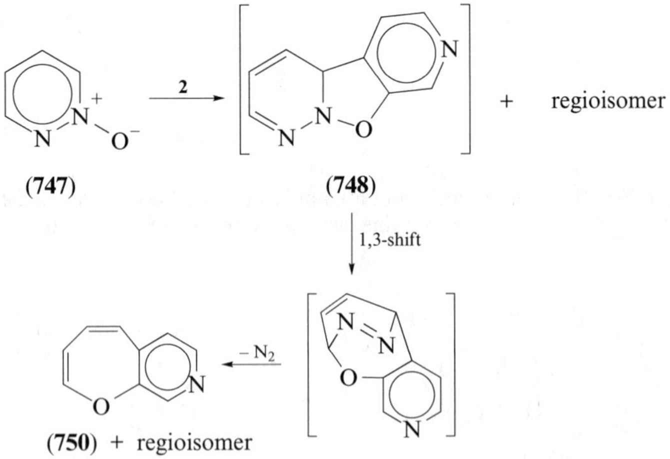

(**747**)

(**748**) + regioisomer

1,3-shift

(**749**)

(**750**) + regioisomer

An attempt to prepare olivacine **745** from **744** (cf **740**) failed; instead, amide **746** was obtained in 40% yield[310], derived from nucleophilic attack of dimethylamine on **744**; no pyridyne adduct was formed. Apparently the 1-methyl substituent in **740** is sufficient to slow down nucleophilic attack by dimethylamine and allow the pyridyne cycloaddition, which is rather sluggish, to occur. With the methyl group adjacent to the carbonyl group absent in **744**, nucleophilic attack is faster than the cycloaddition.

Finally, an example of a 1,3-dipolar cycloaddition has appeared[311]. Reaction of pyridynes and quinolynes with pyridazine-N-oxides **747** gave pyrido-oxepins **750** in 20–30% yields. The reaction is thought to involve cycloaddition to give **748** which, after a 1,3-shift of the oxygen from N to C and nitrogen loss, gives the product. The pyridynes were produced by lead tetraacetate oxidation of 1-aminotriazolopyridines.

e. Nucleophilic additions. Carbanions derived from nitriles add to 3-pyridyne to give 3- and 4-substituted pyridines **751** and **752** in nearly equal amounts and 40–50% total yield[300]. In a similar manner, enolates derived from 1,2-diketone monoketals (cf **547** → **548**) give [2 + 2] adducts **754** and **755** and their ring-opened isomers **756** and **757** in approximately 50% total yield[300].

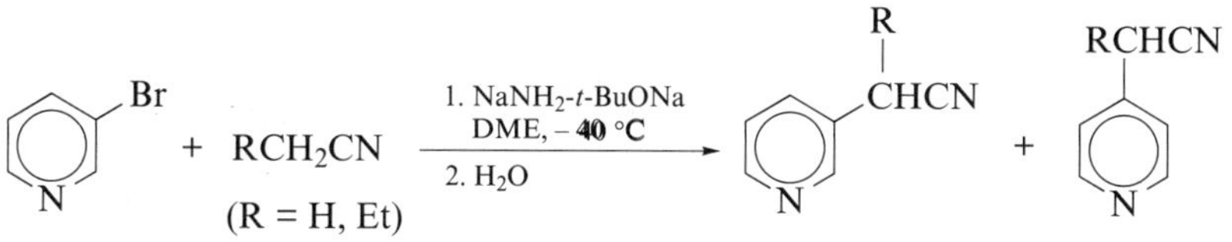

(751)   (752)

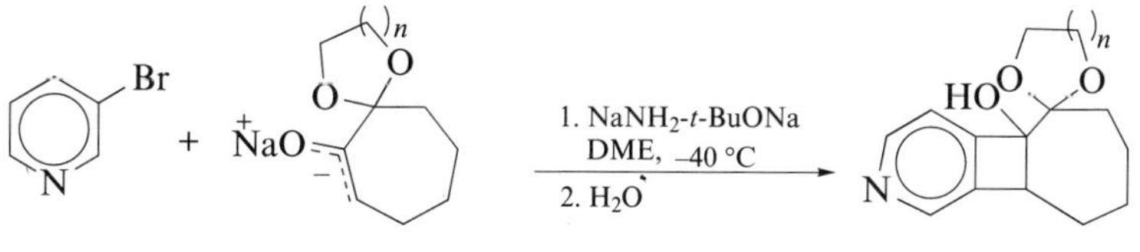

(753) n = 1,2   (754)

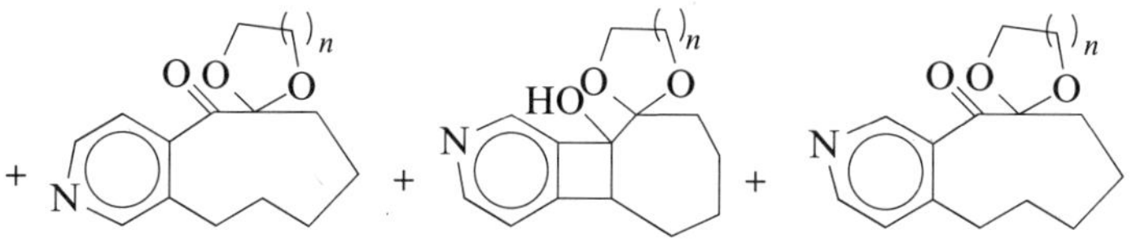

(756)   (755)   (757)

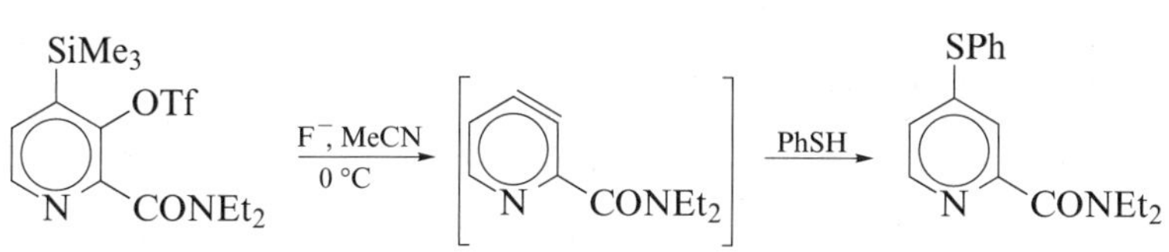

(758)   (759) 55%

Thiophenol adds regioselectively to substituted pyridyne **758** to give the 4-substituted pyridine **759**[301]; the 3-isomer was not detected (< 2%). Since the substituent on **758** is electron-withdrawing, the regioselectivity is expected. With 3-pyridyne itself (generated in the same way) the 3- and 4-phenylthiopyridines were formed in equal amounts (42% each).

The carbanion of 3-cyanophthalide (*cf* **534**)[251] cycloadds to 3-pyridynes to give 2-azaanthraquinones in good yield[312]. The mechanism is as shown for **535** → **533**. With 7-methoxy-3-cyanophthalide **762** and 2,6-dimethoxypyridyne **761**, the cycloaddition was regioselective, giving mainly **763** as expected for nucleophilic attack at C4 of **761**[312]. In a somewhat similar manner, the nitrogen nucleophiles **764** added regioselectively to **761** to give di- and triazaacridones **765** (21–27%)[313].

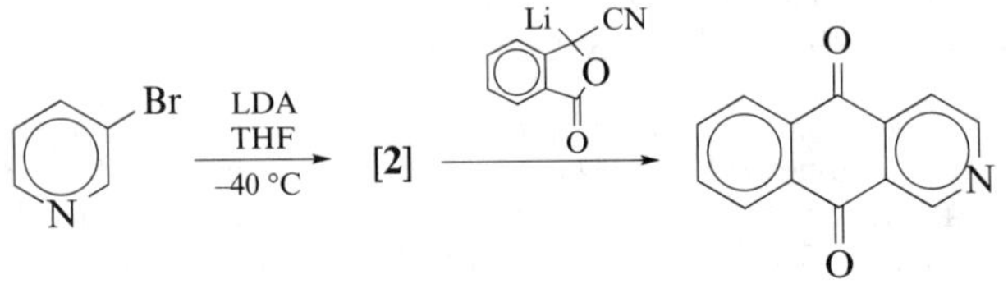

(**760**) 65%

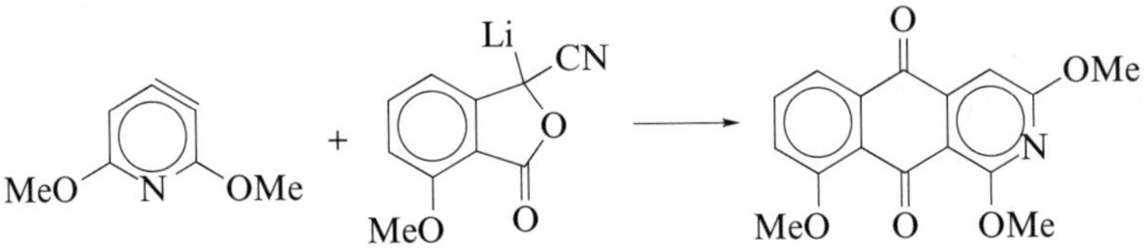

(**761**)   (**762**)   (**763**) 32%

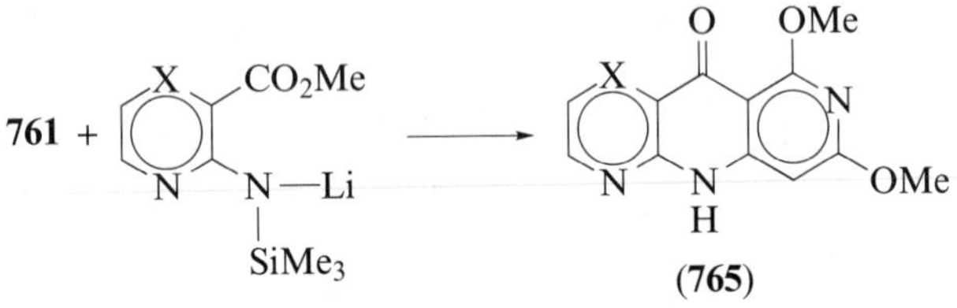

(**764**) X = CH, N

(**765**)

2. Didehydrodiazines

Semiempirical[314] and *ab initio*[24] calculations have been carried out on didehydro-pyridazines **766** and **767**, didehydropyrimidine **768**[315] and didehydropyrazine **769**. These calculations predict that **766** is more stable than **767** by 11–12 kcal mol^{-1} for reasons similar to those for the greater stability of **2** over **717**. This result is in accord with the considerably greater experimental evidence for **766** compared with **767**[5,6]. Although a large diradical character was claimed for **769**[314], this conclusion was not substantiated by the *ab initio* calculations, nor by a recalculation of the semiempirical result[24].

The mass spectra and pyrolysis of anhydrides **770** and **771** have been interpreted in terms of heteroarynes **766** and **767**, respectively[316], but the evidence is weak and led the authors to conclude that **767** is more stable than **766**, whereas the opposite would be expected[24,314].

Convincing experimental evidence for 2-*t*-butyl-4,5-didehydropyrimidine **773** has been presented[315]. Aminotriazole **772** was used as a source for **773**, which was trapped with dienes (furan, tetraphenylcyclone), 1,3-dipoles, a nucleophile and an electrophile. Several

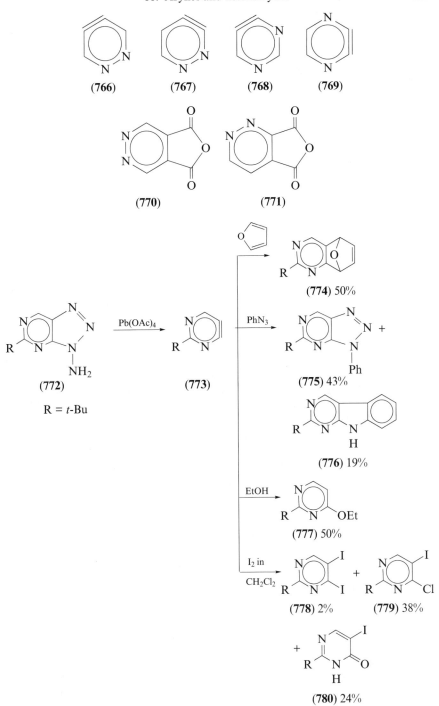

(766) **(767)** **(768)** **(769)**

(770) **(771)**

(772) **(773)**

R = *t*-Bu

Pb(OAc)₄

(774) 50%

PhN₃

(775) 43% +

(776) 19%

EtOH

(777) 50%

I₂ in
CH₂Cl₂

(778) 2% + **(779)** 38%

+

(780) 24%

features of these reactions support calculations[24,315] that show the 'triple' bond in **773** to be polarized with C4 positive and C5 negative, due to electron release of the unshared electron pair from the adjacent nitrogen. For example, only **775** (and not its regioisomer) was formed, due to attack by the negative α-nitrogen of phenyl azide **781** on C4 of **773**. Minor product **776** is thought to arise by loss of nitrogen from intermediate **782** prior to ring closure.

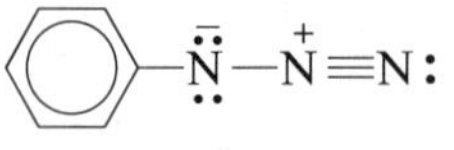

(781)

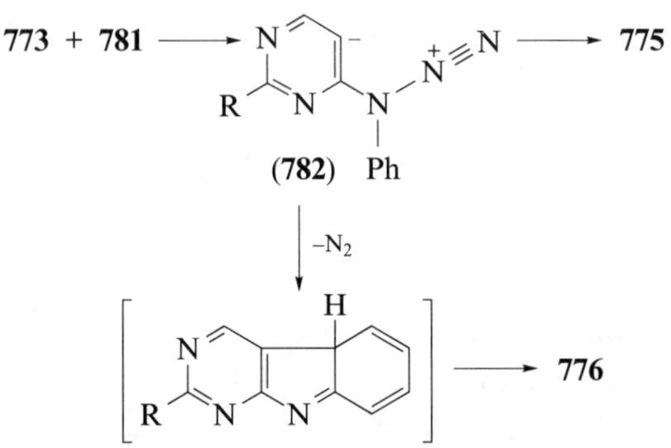

(782) Ph

$-N_2$

776

Only **777** (and not its regioisomer) was formed by attack of nucleophile EtOH on the positive end of the triple bond in **773**. On the other hand, electrophile I_2 attacks the more negative end (C5) to give intermediate **783**, which may then react with nucleophiles present to give the observed products.

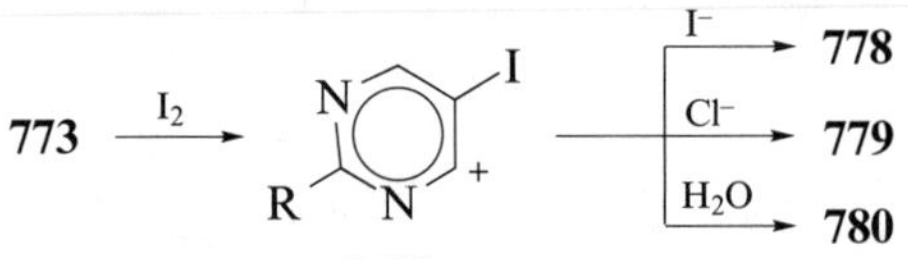

(783)

B. Five-membered Heteroarynes

Ab initio calculations have been performed on the didehydrofurans, thiophenes and pyrroles, pyrazoles and imidazoles, as well as on tetradehydro analogs[24]. These calculations predict that 3,4-didehydrofuran **784** is more stable than its 2,3-isomer **785** by 15 kcal mol^{-1}, but that in the thiophene series the two isomers have nearly the same energy. The

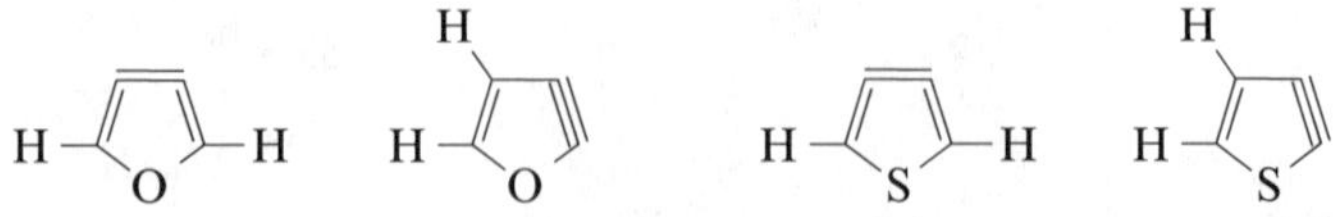

(784)　　　　**(785)**　　　　**(786)**　　　　**(787)**

relative energies in the pyrrole series are 0, 7.9 and 9.6 kcal mol^{-1}, **788** being the most stable. All of these species with one heteroatom are predicted to have barriers to ring opening, though the 3,4- and O—C5 bonds in **785** are unusually long.

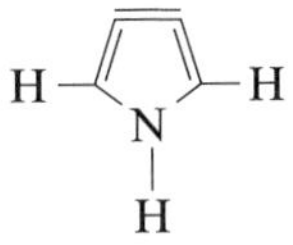

(788) 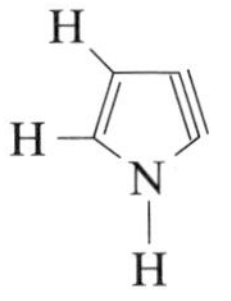   (789) 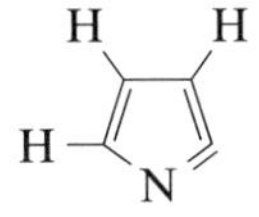   (790)

Although **791** has the lowest energy of the three didehydropyrazoles, it is predicted to ring open without a barrier, through rupture of the N—N bond. Isomer **792** is slightly lower in energy than **793**. Isomer **794** has the lowest energy of the three didehydroimidazoles, with **795** and **796** lying 7.6 and 29 kcal mol^{-1} above it, respectively.

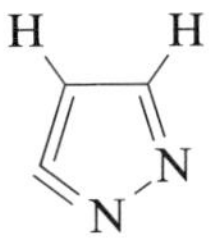

(791) 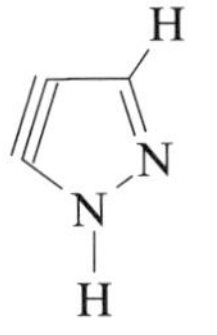   (792) 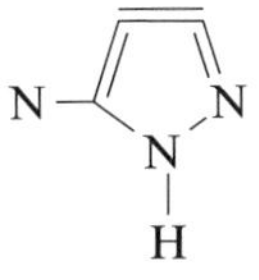   (793)

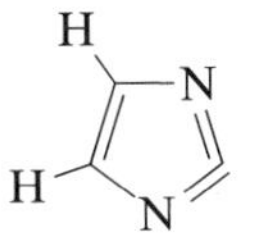

(794) 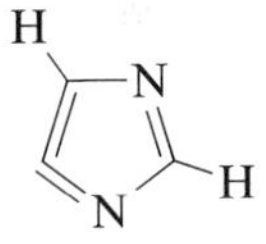   (795) 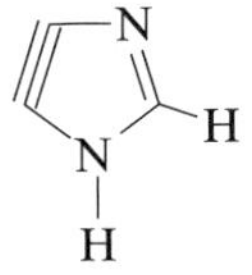   (796)

Tetradehydrofuran, pyrazole and imidazole were all unstable, but tetradehydrothiophene and pyrrole possessed bound structures.

Most experimental evidence for the existence of five-membered heteroarynes is questionable; indeed it was stated in 1967[317] that 'no unambiguous evidence is yet available for the formation of dehydro derivatives of five-membered aromatic heterocycles', and the picture has not changed greatly since then, despite a major (with 591 references) and excellent review on the subject[7]. The only possible exception is for 2,3-didehydrothiophene **787** proposed as an intermediate in the pyrolysis of 2,3-thiophenedicarboxylic anhydride[318–321], and a recent re-investigation using matrix isolation techniques found no direct evidence for **787**, though two ring-opened isomers were identified[322].

In the earlier work[318–320], **797** was subjected to flash vacuum pyrolysis (500 °C) in the presence of various dienes as traps. For example, with thiophene as a trap[159] the identifiable products were thianaphthene **799** (59%) and sulfur, and with furan, hydroxythianaphthenes **801** (1%) and cyclopentenothiophenes **802** (8%) were identified. Adducts **798** and **800** could not be detected, and alternative routes to the observed products, as well as to those obtained with other traps (1,3-cyclohexadiene, benzene, cyclopentadiene, 2,3-dimethyl-1,3-buta-diene and propyne) were presented. Pyrolysis of **797** in the presence of hexafluoroacetone

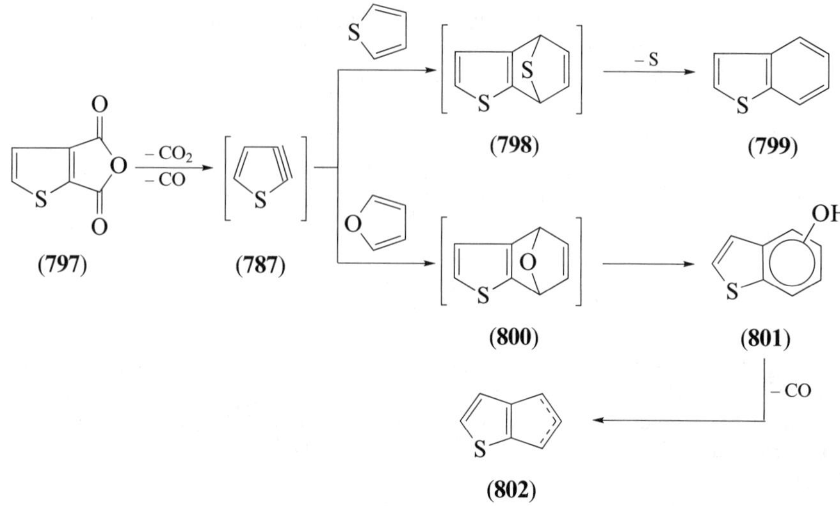
O
O
O
S
– CO2
– CO
(797)
(787)
S
S
(798)
– S
(799)
O
(800)
OH
(801)
– CO
(802)

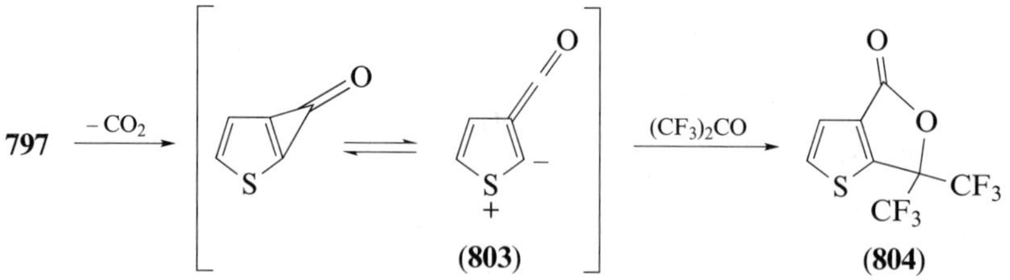
797
– CO2
O
S
(803)
(CF3)2CO
CF3
CF3
(804)

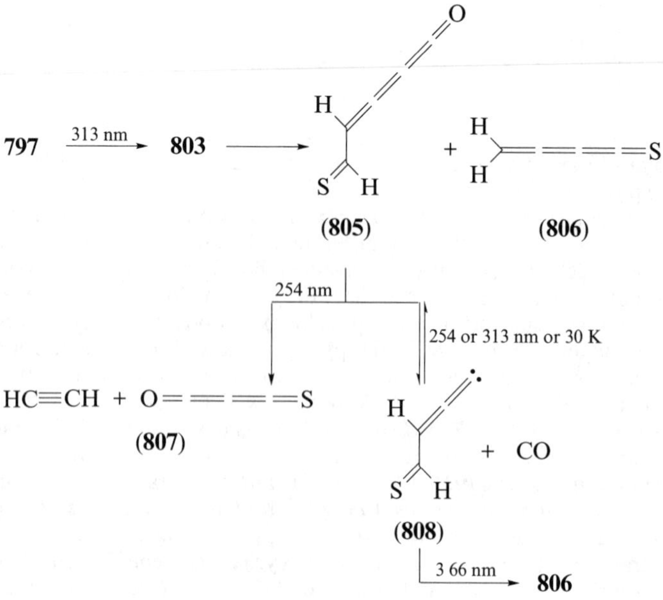
797
313 nm
803
H
S
H
(805)
+
H
H
S
(806)
254 nm
254 or 313 nm or 30 K
HC≡CH + O=====S
(807)
+ CO
(808)
3 66 nm
806

gave lactone **804** in 35% yield[321], showing that (as with other anhydrides) loss of CO_2 occurs first to give **803**. (Calculations show[322] that **803** is 13 kcal more stable than its C2 isomer, perhaps accounting for the loss of CO_2 from the C2 and not C3 position of **797**). It is possible that the various trapping products also arise from **803**, not **787**.

When the pyrolysis of **797** was repeated[322] (a temperature of 850 °C was required for complete conversion) and the pyrolyzate was isolated in an argon matrix at 12 K, the volatile products were CO_2, CO, CS_2, HC≡CH, OCS, thioketene and CS, identified by IR. Most of the material, however, remained in the quartz pyrolysis tube as a shiny black solid.

More interesting results were obtained from *photolysis* of **797** and its various ^{13}C and ^{2}H isotopomers in an argon matrix at 23 K. The main products were **805** and **806**; carbene **808**, and **807** were also identified (IR). Two of these substances (**806** and **808**) are previously unknown isomers of 2,3-didehydrothiophene. It is possible that the heteroaryne is an intermediate in some of these conversions (**803** → **806** and **808** → **806**) but it could not be detected.

A recent attempt to generate 2,3-indolyne **810** from triflate **809** failed, and instead gave indigo (51%), possibly via indoxyl **812**[323].

To date, then, the unequivocal identification of a five-membered heteroaryne remains a challenge.

IX. OMISSIONS

Two areas of aryne chemistry, aryne–metal complexes and aryne radical ions, have been omitted from this review.

Although a large number of aryne–metal complexes have been prepared and character-ized, and they have an interesting chemistry of their own, that chemistry is in general quite different from the types of reactions usually associated with arynes. One might have hoped that aryne–metal complexes could act as shelf-stable storehouses for arynes, that could then be released on demand to do their usual chemistry (as in the manner, for example, of cyclobutadiene complexes), but to date this has not proved to be the case. The synthesis and chemistry of aryne–metal complexes has been reviewed[324], and the interested reader is also provided with a few references to work that has appeared since those reviews[325].

Aryne radical ions have been used mainly to deduce thermochemical information about neutral arynes[11,29,31,326], although they also exhibit their own chemistry (H abstraction, fragmentation). The chemistry is insufficiently developed systematically to be ripe for review at this time. Metal complexes of aryne ions have also been observed[327], and it will be interesting to follow their chemistry as it unfolds.

X. REFERENCES

1. T. L. Gilchrist, in *The Chemistry of Functional Groups, Supplement C* (Eds. S. Patai and Z. Rappoport) Wiley, Chichester, 1983, pp. 383–419.
2. R. W. Hoffmann, *Dehydrobenzene and Cycloalkynes*, Academic Press, New York, 1967.
3. R. W. Hoffmann, in *Chemistry of Acetylenes* (Ed. H. G. Viehe), Dekker, New York, 1969, p. 1063; W. Pritzkow, *Z. Chem.*, **10**, 330 (1970); E. K. Fields, in *Organic Reactive Intermediates* (Ed. S. P. McManus), Academic Press, New York, 1973, pp. 449–508; W. J. le Noble, *Highlights of Organic Chemistry*, Chap. 19; Dekker, New York, 1974; J. T. Sharp, in *Comprehensive Organic Chemistry*, Vol. 1 (Ed. J. F. Stoddard), Pergamon Press, Oxford, 1979, pp. 477–490. C. Grundmann in *Methoden der Organischen Chemie (Houben-Weyl)*, Vol. 5/2b, G. Thieme Verlag, Stuttgart, 1981, pp. 613–648.
4. R. H. Levin, in *Reactive Intermediates*, Vol. 3 (Eds. M. Jones, Jr. and R. A. Moss), Wiley, New York, 1985, pp. 1–18; M. R. Crampton, in *Organic Reaction Mechanisms* (Eds. A. C. Knipe and W. E. Watts), Wiley, New York, 1986, pp. 263–264; *ibid.*, 1987, p. 259; *ibid.*, 1988, pp. 258–259; *ibid.*, 1989, pp. 281–282; *ibid.*, 1990, p. 318; *ibid.*, 1991, p. 266; *ibid.*, 1992, pp. 272–273; M. S. Baird, in *Annual Reports on the Progress of Chemistry B*, Vol. 81, Royal Society of Chemistry, Burlington House, London, 1984, pp. 79–81; *ibid.*, Vol. 80, 1983, pp. 101–102; *ibid.*, Vol. 79, 1982, pp. 83–84; D. W. Knight, *ibid.*, Vol. 78, 1981, pp. 97–99; *ibid.*, Vol. 77, 1980, pp. 67–68.
5. M. G. Reinecke, *Tetrahedron*, **38**, 427 (1982).
6. H. C. Van Der Plas and F. Roeterdink, in *The Chemistry of Functional Groups, Supplement C* (Eds. S. Patai and Z. Rappoport), Wiley, Chichester, 1983, pp. 421–511.
7. M. G. Reinecke, in *Reactive Intermediates*, Vol. 2 (Ed. R. A. Abramovitch), Plenum, New York, 1982, pp. 367–526.
8. O. L. Chapman, K. Mattes, C. L. McIntosh, J. Pacansky, G. V. Calder and G. Orr, *J. Am. Chem. Soc.*, **95**, 6134 (1973); O. L. Chapman, C.-C. Chang, J. Kolc, N. R. Rosenquist and H. Tomioka, *J. Am. Chem. Soc.*, **97**, 6586 (1975); I. R. Dunkin and J. G. MacDonald, *J. Chem. Soc., Chem. Commun.*, 772 (1979).
9. (a) H. H. Nam and G. E. Leroi, *J. Mol. Struct.*, **157**, 301 (1987).
   (b) C. Wentrup, R. Blanch, H. Briehl and G. Gross, *J. Am. Chem. Soc.*, **110**, 1874 (1988).
10. J. W. Laing and R. S. Berry, *J. Am. Chem. Soc.*, **98**, 660 (1976); H.-H. Nam and G. E. Leroi, *Spectrochim. Acta*, **41A**, 67 (1985).
11. D. G. Leopold, A. E. S. Miller and W. C. Lineberger, *J. Am. Chem. Soc.*, **108**, 1379 (1986).
12. K. Rigby, I. H. Hillier and M. Vincent, *J. Chem. Soc., Perkin Trans. 2*, 117 (1987).
13. A. C. Scheiner, H. F. Schaefer III and B. Liu, *J. Am. Chem. Soc.*, **111**, 3118 (1989).
14. A. C. Scheiner and H. F. Schaefer III, *Chem. Phys. Lett.*, **177**, 471 (1991).
15. J. G. Radziszewski, B. A. Hess, Jr. and R. Zahradnik *J. Am. Chem. Soc.*, **114**, 52 (1992).
16. J. G. Simon, N. Münzel and A. Schweig, *Chem. Phys. Lett.*, **170**, 187 (1990).
17. N. Münzel and A. Schweig, *Chem. Phys. Lett.*, **147**, 192 (1988).
18. R. D. Brown, P. D. Godfrey and M. Rodler, *J. Am. Chem. Soc.*, **108**, 1296 (1986).
19. M. J. S. Dewar and T.-P. Tien, *J. Chem. Soc., Chem. Commun.*, 1243 (1985).
20. X. Zhang and P. Chen, *J. Am. Chem. Soc.*, **114**, 3147 (1992); J. A. Blush, H. Clauberg, D. W. Kohn, D. W. Minsek, X. Zhang and P. Chen, *Acc. Chem. Res.*, **25**, 385 (1992).
21. W. Thiel, *J. Am. Chem. Soc.*, **103**, 1420 (1981).
22. M. J. S. Dewar, G. P. Ford and C. H. Reynolds, *J. Am. Chem. Soc.*, **105**, 3162 (1983).
23. C. W. Bock, P. George and M. Trachtman, *J. Phys. Chem.*, **88**, 1467 (1984).
24. L. Radom, R. H. Nobes, D. J. Underwood and W.-K. Li, *Pure Appl. Chem.*, **58**, 75 (1986).
25. I. H. Hillier, M. A. Vincent, M. F. Guest and W. VonNiessen, *Chem. Phys. Lett.*, **134**, 403 (1987).
26. H.-F. Grützmacher and J. Lohmann, *Justus Liebigs Ann. Chem.*, **705**, 81 (1967).
27. H. M. Rosenstock, R. Stockbauer and A. C. Parr, *J. Chim. Phys.*, **77**, 745 (1980).
28. S. K. Pollack and W. J. Hehre, *Tetrahedron Lett.*, **21**, 2483 (1980).
29. M. Moini and G. E. Leroi, *J. Phys. Chem.*, **90**, 4002 (1986).
30. J. M. Riveros, S. Ingemann and N. M. M. Nibbering, *J. Am. Chem. Soc.*, **113**, 1053 (1991).
31. Y. Guo and J. J. Grabowski, *J. Am. Chem. Soc.*, **113**, 5923 (1991).

32. P. G. Wenthold, J. A. Paulino and R. R. Squires, *J. Am. Chem. Soc.*, **113**, 7414 (1991).
33. N. Furukawa, T. Shibutani and H. Fujihara, *Tetrahedron Lett.*, **28**, 2727 (1987).
34. W. Tochtermann, G. Stubenrauch, K. Reiff and U. Schumacher, *Chem. Ber.*, **107**, 3340 (1974).
35. C. S. Le Houillier and G. W. Gribble, *J. Org. Chem.*, **48**, 1682 (1983).
36. T. Matsumoto, T. Hosoya, M. Katsuki and K. Suzuki, *Tetrahedron Lett.*, **32**, 6735 (1991).
37. Y. Himeshima, T. Sonoda and H. Kobayashi, *Chem. Lett.*, 1211 (1983).
38. K. Shankaran and V. Snieckus, *Tetrahedron Lett.*, **25**, 2827 (1984).
39. P. P. Wickham, K. H. Hazen, H. Guo, G. Jones, K. H. Reuter and W. J. Scott, *J. Org. Chem.*, **56**, 2045 (1991).
40. G. W. Ebert, D. R. Pfennig, S. D. Suchan and T. A. Donovan, Jr., *Tetrahedron Lett.*, **34**, 2279 (1993).
41. A. I. Meyers and W. Rieker, *Tetrahedron Lett.*, **23**, 2091 (1982); A. I. Meyers and P. D. Pansegrau, *Tetrahedron Lett.*, **24**, 4935 (1983); A. I. Meyers and P. D. Pansegrau, *Tetrahedron Lett.*, **25**, 2941 (1984); A. I. Meyers and P. D. Pansegrau, *J. Chem. Soc., Chem. Commun.*, 690 (1985).
42. P. D. Pansegrau, W. F. Rieker and A. I. Meyers, *J. Am. Chem. Soc.*, **110**, 7178 (1988).
43. M. E. Jung and G. T. Lowen, *Tetrahedron Lett.*, **27**, 5319 (1986).
44. C. J. Horan, C. L. Barnes and R. Glaser, *Chem. Ber.*, **126**, 243 (1993).
45. R. A. Russell and R. N. Warrener, *J. Lab. Comp. Radiopharm.*, **14**, 239 (1978).
46. R. A. Snow, D. M. Cottrell and L. A. Paquette, *J. Am. Chem. Soc.*, **99**, 3734 (1977).
47. W. M. West and D. Wege, *Aust. J. Chem.*, **39**, 635 (1986).
48. F. Gavina, S. V. Luis, P. Gil and A. M. Costero, *Tetrahedron Lett.*, **25**, 779 (1984); F. Gavina, S. V. Luis, A. M. Costero and P. Gil, *Tetrahedron*, **42**, 155 (1986).
49. S. Mazur and P. Jayalekshmy, *J. Am. Chem. Soc.*, **101**, 677 (1979).
50. J. Rebek, Jr. and F. Gavina, *J. Am. Chem. Soc.*, **96**, 7112 (1974); J. Rebek, Jr., *Tetrahedron*, **35**, 723 (1979).
51. F. Gaviña, S. V. Luis, P. Ferrer, A. M. Costero and P. Gil, *Tetrahedron*, **42**, 5641 (1986).
52. S. V. Luis, F. Gaviña, P. Ferrer, V. S. Safont, M. C. Torres and M. I. Burguete, *Tetrahedron*, **45**, 6281 (1989).
53. S. V. Luis, P. Ferrer and M. I. Burguete, *J. Org. Chem.*, **55**, 3808 (1990).
54. C. D. Campbell and C. W. Rees, *Proc. Chem. Soc.*, 296 (1964); *J. Chem. Soc. (C)*, 742 (1969).
55. S. E. Whitney and B. Rickborn, *J. Org. Chem.*, **53**, 5595 (1988); S. E. Whitney, M. Winters and B. Rickborn, *J. Org. Chem.*, **55**, 929 (1990).
56. Y. Apeloig, D. Arad, B. Hilton and C. J. Randall, *J. Am. Chem. Soc.*, **108**, 4932 (1986).
57. R. Camenzind and B. Rickborn, *J. Org. Chem.*, **51**, 1914 (1986).
58. S. L. Crump, J. Netka and B. Rickborn, *J. Org. Chem.*, **50**, 2746 (1985); J. Netka, S. L. Crump and B. Rickborn, *J. Org. Chem.*, **51**, 1189 (1986).
59. R. H. Mitchell and P. Zhou, *Tetrahedron Lett.*, **31**, 5277 (1990).
60. D. T. Longone and J. A. Gladysz, *Tetrahedron Lett.*, 4559 (1976); D. T. Longone and G. R. Chipman, *J. Chem. Soc., Chem. Commun.*, 1358 (1969); D. J. Cram and A. C. Day, *J. Org. Chem.*, **31**, 1227 (1966); H. J. Reich and D. J. Cram, *J. Am. Chem. Soc.*, **91**, 3527 (1969).
61. N. Mori, M. Horiki and H. Akimoto, *J. Am. Chem. Soc.*, **114**, 7927 (1992).
62. H.-F. Grützmacher and U. Straetmans, *Tetrahedron*, **36**, 807 (1980).
63. G. A. Razuvaev, L. I. Vyshinskaya, V. V. Drobotenko and V. N. Latyaeva, *Dokl. Akad. Nauk SSSR (Engl. Transl.)*, 304 (1982).
64. M. Fossatelli and L. Brandsma, *Synthesis*, 756 (1992).
65. D. Otteson and J. Michl, *J. Org. Chem.*, **49**, 866 (1984).
66. N. Egashira, J. Takenaga and F. Hori, *Bull. Chem. Soc. Jpn.*, **60**, 2671 (1987).
67. R. A. Zingaro and C. Herrera, *Bull. Chem. Soc. Jpn.*, **62**, 1382 (1989).
68. I. R. Slagle and D. Gutman, *J. Phys. Chem.* **87**, 1818 (1983).
69. J. T. Roberts and C. M. Friend, *J. Chem. Phys.*, **88**, 7172 (1988).
70. A. C. Liu and C. M. Friend, *J. Chem. Phys.*, **89**, 4396 (1988).
71. K. Chow and H. W. Moore, *J. Org. Chem.*, **55**, 370 (1990).
72. L. D. Foland, J. O. Karlsson, S. T. Perri, R. Schwabe, S. L. Xu, S. Patil and H. W. Moore, *J. Am. Chem. Soc.*, **111**, 975 (1989).
73. A. Naghipur, K. Reszka, A.-M. Sapse and J. W. Lown, *J. Am. Chem. Soc.*, **111**, 258 (1989); A. Naghipur, J. W. Lown, D. C. Jain and A.-M. Sapse, *Can. J. Chem.*, **66**, 1890 (1988).
74. A. Naghipur, K. Reszka, J. W. Lown and A.-M. Sapse, *Can. J. Chem.*, **68**, 1950 (1990).
75. G. Wittig and R. W. Hoffmann, *Org. Synth.*, **47**, 4 (1962); *Org. Synth. Coll. Vol. V*, 60 (1973).

76. A. D. Schlüter, J. Belzner, U. Heywang and G. Szeimies, *Tetrahedron Lett.*, **24**, 891 (1983).
77. P. R. Ortiz de Montelano and J. M. Mathews, *Biochem. J.*, **195**, 761 (1981); P. R. Ortiz de Montellano, B. A. Mico, J. M. Mathews, K. L. Kunze, G. T. Miwa and A. Y. H. Lu, *Arch. Biochem. Biophys.*, **210**, 718 (1981); P. R. Ortiz de Montellano, J. M. Mathews and K. C. Langry, *Tetrahedron*, **40**, 511 (1984).
78. R. J. Armstrong, R. F. C. Brown, F. W. Eastwood and M. E. Romyn, *Aust. J. Chem.*, **32**, 1767 (1979).
79. R. F. C. Brown, *Recl. Trav. Chim. Pays-Bas*, **107**, 655 (1988); R. F. C. Brown and F. W. Eastwood, *Synlett.*, 9 (1993).
80. K. Hafner, H.-P. Krimmer and B. Stowasser, *Angew. Chem., Int. Ed. Engl.*, **22**, 490 (1983).
81. Y. Apeloig, R. Schreiber and P. J. Stang, *Tetrahedron Lett.*, **21**, 411 (1980).
82. N. A. Burton, G. E. Quelch, M. M. Gallo and H. F. Schaefer III, *J. Am. Chem. Soc.*, **113**, 764 (1991).
83. M. Barry, R. F. C. Brown, F. W. Eastwood, D. A. Gunawardana and C. Vogel, *Aust. J. Chem.*, **37**, 1643 (1984).
84. R. F. C. Brown, N. R. Browne, K. J. Coulston, L. B. Danen, F. W. Eastwood, M. J. Irvine and A. D. E. Pullin, *Tetrahedron Lett.*, **27**, 1075 (1986).
85. R. F. C. Brown, N. R. Browne, K. J. Coulston, F. W. Eastwood, M. J. Irvine, A. D. E. Pullin and U. E. Wiersum, *Aust. J. Chem.*, **42**, 1321 (1989).
86. R. F. C. Brown, K. J. Coulston, F. W. Eastwood and C. Vogel, *Aust. J. Chem.*, **41**, 1687 (1988).
87. R. F. C. Brown, K. J. Coulston, F. W. Eastwood, T. Korakis, *Tetrahedron Lett.*, **29**, 6791 (1988).
88. R. F. C. Brown, K. J. Coulston, F. W. Eastwood and S. Saminathan, *Aust. J. Chem.*, **40**, 107 (1987).
89. M. R. Anderson, R. F. C. Brown, K. J. Coulston, F. W. Eastwood and A. Ward, *Aust. J. Chem.*, **43**, 1137 (1990).
90. R. F. C. Brown, N. R. Browne, K. J. Coulston and F. W. Eastwood, *Aust. J. Chem.*, **43**, 1935 (1990).
91. M. Adeney, R. F. C. Brown, K. J. Coulston, F. W. Eastwood and I. W. James, *Aust. J. Chem.*, **44**, 967 (1991).
92. R. F. C. Brown, F. W. Eastwood and C. J. Smith, *Aust. J. Chem.*, **45**, 1315 (1992).
93. R. P. Thummel, W. E. Cravey and D. B. Cantu, *J. Org. Chem.*, **45**, 1633 (1980).
94. B. Tarnchompoo, C. Thebtaranonth and Y. Thebtaranonth, *Tetrahedron Lett.*, **28**, 6671 (1987).
95. R. K. Hill and M. G. Bock, *J. Am. Chem. Soc.*, **100**, 637 (1978).
96. R. P. Thummel, W. E. Cravey and W. Nutakul, *J. Org. Chem.*, **43**, 2473 (1978); R. P. Thummel, *Croat. Chem. Acta*, **53**, 659 (1980).
97. Y. Bessière and P. Vogel, *Helv. Chim. Acta*, **63**, 232 (1980).
98. B. Demarchi and P. Vogel, *Tetrahedron Lett.*, **28**, 2239 (1987).
99. A. Padwa and W. F. Rieker, *Tetrahedron Lett.*, **22**, 1487 (1981); A. Padwa, W. F. Rieker and R. J. Rosenthal, *J. Am. Chem. Soc.*, **105**, 4446 (1983).
100. G. L. Grunewald, V. M. Paradkar, B. Pazhenchevsky, M. A. Pleiss, D. J. Sall, W. L. Seibel and T. J. Reitz, *J. Org. Chem.*, **48**, 2321 (1983); G. L. Grunewald, H. S. Arrington, W. J. Bartlett, T. J. Reitz and D. J. Sall, *J. Med. Chem.*, **29**, 1972 (1986); G. L. Grunewald, K. M. Markovich and D. J. Sall, *J. Med. Chem.*, **30**, 2191 (1987); G. L. Grunewald, A. J. Kolar, M. S. S. Palanki, V. M. Paradkar, T. J. Reitz and D. J. Sall, *Org. Prep. Proc.*, **22**, 747 (1990).
101. W. Adam, V. Lucchini, E.-M. Peters, K. Peters, L. Pasquato, H. G. von Schnering, K. Seguchi, H. Walter and B. Will, *Chem. Ber.*, **122**, 133 (1989).
102. L. A. Paquette, R. V. C. Carr, M. C. Böhm and R. Gleiter, *J. Am. Chem. Soc.*, **102**, 1186 (1980); M. C. Böhm, R. V. C. Carr, R. Gleiter and L. A. Paquette, *J. Am. Chem. Soc.*, **102**, 7218 (1980).
103. F. K. Brown and K. N. Houk, *J. Am. Chem. Soc.*, **107**, 1971 (1985).
104. L. A. Paquette and M. Gugelchuk, *J. Org. Chem.*, **53**, 1835 (1988); M. Gugelchuk and L. A. Paquette, *J. Am. Chem. Soc.*, **113**, 246 (1991).
105. J. M. Coxon, M. J. O'Connell and P. J. Steel, *J. Org. Chem.*, **52**, 4726 (1987).
106. D. Kaufmann and A. de Maijere, *Chem. Ber.*, **116**, 1897 (1983).
107. T. M. Cresp and D. Wege, *Tetrahedron*, **42**, 6713 (1986).
108. G. Fischer, E. Beckmann, H. Prinzbach, G. Rihs and J. Wirz, *Tetrahedron Lett.*, **27**, 1273 (1986).
109. I. J. Anthony, L. T. Byrne, R. K. McCulloch and D. Wege, *J. Org. Chem.*, **53**, 4123 (1988), cf. A. Menzek and M. Balci, *Tetrahedron*, **49**, 6071 (1993).
110. A. Oku, T. Kakihana and H. Hart, *J. Am. Chem. Soc.*, **89**, 4554 (1967); H. Hart and A. Oku, *J. Org. Chem.*, **37**, 4269 (1972).

111. Z. Yoshida, F. Kawamoto, H. Miyoshi and H. Ikikoshi, to Sekisui Chemical Co., Ltd., *Jpn. Kokai Tokkyo Koho*, **1979**, 138, 549; *Chem. Abstr.*, **92**, P163763q (1980); S. Miki and Z. Yoshida, *Kokagaku*, **11**, 58 (1987).
112. S. Miki, T. Ema, R. Shimizu, H. Nakatsuji and Z. Yoshida, *Tetrahedron Lett.*, **33**, 1619 (1992).
113. S. Mondal, T. K. Bandyopadhyay and A. J. Battacharya, *Ind. J. Chem.*, **22B**, 448 (1983).
114. T. K. Bandyopadhyay and A. J. Battacharya, *Ind. J. Chem.*, **21B**, 91 (1982).
115. M. Hori, T. Kataoka, H. Shimizu and M. Okitsu, *Heterocycles*, **15**, 1061 (1981); M. Hori, T. Kataoka, H. Shimzu and M. Okitsu, *Chem. Pharm. Bull.*, **29**, 1244 (1981).
116. I. Tabushi, H. Yamada, Z. Yoshida and R. Oda, *Bull. Chem. Soc. Jpn.*, **50**, 285 (1977).
117. D. T. Longone and J. A. Gladysz, *Tetrahedron Lett.*, 4559 (1976).
118. W. J. Houlihan, Y. Uike and V. A. Parrino, *J. Org. Chem.*, **46**, 4515 (1981).
119. H. Hart, C.-J. F. Du and J. Mohebelian, *J. Org. Chem.*, **53**, 2720 (1988).
120. L. N. Koikov, P. B. Terent'ev and Yu. G. Bundel', *J. Org. Chem. USSR (Engl. Transl.)*, 1397 (1983); L. N. Koikov, P. B. Terent'ev, I. O. Gloriozov and Yu. G. Bundel', *J. Org. Chem. USSR (Engl. Transl.)*, 1483 (1985).
121. G. L. Cantrell and R. Filler, *J. Fluorine Chem.*, **29**, 417 (1985).
122. G. Wittig and R. Ludwig, *Angew. Chem.*, **68**, 40 (1956).
123. G. Yamamoto, *Chem. Lett.*, 1373 (1990).
124. G. Yamamoto and M. Ōki, *Bull. Chem. Soc. Jpn.*, **56**, 2082 (1983).
125. G. Yamamoto, M. Suzuki and M. Ōki, *Bull. Chem. Soc. Jpn.*, **56**, 306 (1983).
126. G. Yamamoto and M. Ōki, *Bull. Chem. Soc. Jpn.*, **54**, 473 (1981).
127. S. Hatakeyama, T. Mitsuhashi and M. Ōki, *Bull. Chem. Soc. Jpn.*, **53**, 731 (1980).
128. H. Quast and N. Schon, *Justus Liebigs Ann. Chem.*, 381 (1984).
129. Y. Kawada and H. Iwamura, *J. Org. Chem.*, **45**, 2547 (1980).
130. T. A. Shatskaya, M. G. Gal'pern, V. R. Skvarchenko and E. A. Luk'yanets, *J. Org. Chem. USSR (Engl. Transl.)*, **56**, 341 (1986).
131. G. Yamamoto, *Chem. Lett.*, 741 (1991).
132. Y. Tamura, G. Yamamoto and M. Ōki, *Bull. Chem. Soc. Jpn.*, **60**, 1781 (1987).
133. G. Yamamoto and M. Ōki, *Chem. Lett.*, 1163 (1987).
134. K. Yonemoto, Y. Nakai, G. Yamamoto and M. Ōki, *Chem. Lett.*, 1739 (1985).
135. G. Yamamoto and M. Ōki, *Chem. Lett.*, 1523 (1980).
136. G. Yamamoto M. Suzuki and M. Ōki, *Bull. Chem. Soc. Jpn.*, **56**, 809 (1983).
137. H. Kikuchi, S. Seki, G. Yamamoto, T. Mitsuhashi, N. Nakamura and M. Ōki, *Bull. Chem. Soc. Jpn.*, **55**, 1514 (1982).
138. J. M. Chance, J. H. Geiger and K. Mislow, *J. Am. Chem. Soc.*, **111**, 2326 (1989).
139. M. E. Rogers and B. A. Averill, *J. Org. Chem.*, **51**, 3308 (1986).
140. M. R. Bryce and J. M. Vernon, in *Advances in Heterocyclic Chemistry*, Vol. 28 (Eds. A. R. Katritzky and A. J. Boulton), Academic Press, New York, 1981, pp. 183–229.
141. G. W. Gribble, D. J. Keavy, S. E. Branz, W. J. Kelly and M. A. Pals, *Tetrahedron Lett.*, **29**, 6227 (1988); G. W. Gribble and W. J. Kelly, *Tetrahedron Lett.*, **22**, 2475 (1981).
142. R. A. Russell, D. A. C. Evans and R. N. Warrener, *Aust. J. Chem.*, **37**, 1699 (1984).
143. L. A. Paquette, K. E. Green, R. Gleiter, W. Schäfer and J. C. Gallucci, *J. Am. Chem. Soc.*, **106**, 8232 (1984).
144. H. F. G. Linde and G. Cramer, *Arch. Pharm. (Weinheim)*, **322**, 565 (1989).
145. W. Tochtermann and K. Luttman, *Tetrahedron Lett.*, **28**, 2521 (1987).
146. H. Hart and Y. Takehira, *J. Org. Chem.*, **47**, 4370 (1982).
147. D. J. Pollart and B. Rickborn, *J. Org. Chem.*, **51**, 3155 (1986).
148. S. Mirsadeghi and B. Rickborn, *J. Org. Chem.*, **52**, 787 (1987).
149. D. J. Pollart and B. Rickborn, *J. Org. Chem.*, **52**, 792 (1987).
150. M. Nakata, M. Kinoshita, S. Ohba and Y. Saito, *Tetrahedron Lett.*, **25**, 1373 (1984).
151. M. E. Jung, P. Y.-S. Lam, M. M. Mansuri and L. M. Speltz, *J. Org. Chem.*, **50**, 1087 (1985).
152. G. W. Gribble and M. G. Saulnier, *J. Chem. Soc., Chem. Commun.*, 168 (1984).
153. T. Ghosh and H. Hart, *J. Org. Chem.*, **53**, 2396 (1988); T. Ghosh and H. Hart, *J. Org. Chem.*, **54**, 5073 (1989); T. Ghosh and H. Hart, *J. Org. Chem.*, **53**, 5192 (1988).
154. J. W. Davies, M. L. Durrant, M. P. Walker, D. Belkacemi and J. R. Malpass, *Tetrahedron*, **48**, 861 (1992); J. W. Davies, M. L. Durrant, M. P. Walker and J. R. Malpass, *Tetrahedron*, **48**, 4379 (1992); D. Belkacemi, J. W. Davies, J. R. Malpass, A. Naylor and C. R. Smith, *Tetrahedron*, **48**, 10161 (1992); J. W. Davies, J. R. Malpass, J. Fawcett, L. J. S. Prouse, R. Lindsay and D. R.

Russell, *J. Chem. Soc., Chem. Commun.*, 1135 (1986); J. W. Davies, J. R. Malpass and R. E. Moss, *Tetrahedron Lett.*, **27**, 4071 (1986).

155. R. P. Kreher and T. Hildebrand, *Angew. Chem., Int. Ed. Engl.*, **26**, 1262 (1987); H. Preut, Th. Hildebrand and R. P. Kreher, *Acta Crystallogr.*, **C44**, 203 (1988).
156. C.-K. Sha, K.-S. Chuang and J.-J. Young, *J. Chem. Soc., Chem. Commun.*, 1552 (1984).
157. S. Brooks, M. Sainsbury and D. K. Weerasinge, *Tetrahedron*, **38**, 3019 (1982).
158. L. Pfeuffer and U. Pindur, *Chimia*, **40**, 124 (1986) and earlier references cited therein.
159. D. Del Mazza and M. G. Reinecke, *Heterocycles*, **14**, 647 (1980); D. Del Mazza and M. G. Reinecke, *J. Org. Chem.*, **53**, 5799 (1988).
160. M. G. Reinecke and D. Del Mazza, *J. Org. Chem.*, **54**, 2142 (1989).
161. D. Del Mazza and M. G. Reinecke, *J. Chem. Soc., Chem. Commun.*, 124 (1981).
162. J. Nakayama, M. Kuroda and M. Hoshino, *Heterocycles*, **24**, 1233 (1986).
163. C. Grugel, W. P. Neumann and M. Schriewer, *Angew. Chem., Int. Ed. Engl.*, **18**, 543 (1979).
164. W. P. Neumann and M. Schriewer, *Tetrahedron Lett.*, **21**, 3273 (1980); M. Schriewer and W. P. Neumann, *J. Am. Chem. Soc.*, **105**, 897 (1983); W. Ando and T. Tsumuraya, *Organometallics*, **8**, 1467 (1989).
165. W. Ando, Y. Hamada and A. Sekiguchi, *J. Chem. Soc., Chem. Commun.*, 787 (1982).
166. G. S. Reddy and M. V. Bhatt, *Tetrahedron Lett.*, **21**, 3627 (1980).
167. V. G. Seitz and H. Wassmuth, *Chem-Zeit*, **112**, 80 (1988).
168. C. J. Moody, *J. Chem. Soc., Chem. Commun.*, 925 (1984); C. J. Moody, *J. Chem. Soc., Perkin Trans. 1*, 2505 (1985).
169. M. Kuzuya, E. Mano, M. Adachi, A. Noguchi and T. Okuda, *Chem. Lett.*, 475 (1982); M. Kuzuya, A. Noguchi, E. Mano and T. Okuda, *Bull. Chem. Soc. Jpn.*, **58**, 1149 (1985); M. Kuzuya, A. Noguchi, S. Kamiya and T. Okuda, *Chem. Pharm. Bull.*, **33**, 2313 (1985).
170. F. W. Muellner and L. Bauer, *J. Heterocycl. Chem.*, **20**, 1581 (1983).
171. A. M. d'A. Rocha Gonsalves, T. M. V. D. Pinho e Melo and T. L. Gilchrist, *Tetrahedron*, **48**, 6821 (1992).
172. T. Kappe and D. Pocivalnik, *Heterocycles*, **20**, 1367 (1983).
173. W. F. Maier, G. C. Lau and A. B. McEwen, *J. Am. Chem. Soc.*, **107**, 4724 (1985).
174. R. V. Stevens and G. S. Bisacchi, *J. Org. Chem.*, **47**, 2393 (1982).
175. L. S. Liebeskind, L. J. Lescosky and C. M. McSwain, Jr., *J. Org. Chem.*, **54**, 1435 (1989).
176. R. V. Stevens and G. S. Bisacchi, *J. Org. Chem.*, **47**, 2396 (1982).
177. O. Abou-Teim, M. C. Goodland and J. F. W. McOmie, *J. Chem. Soc., Perkin Trans. 1*, 2659 (1983); M. S. South and L. S. Liebeskind, *J. Org. Chem.*, **47**, 3815 (1982); M. A. O'Leary, M. B. Stringer and D. Wege, *Aust. J. Chem.*, **31**, 2003 (1978); H. Dürr, H. Nickels, L. A. Pacala and M. Jones, *J. Org. Chem.*, **45**, 973 (1980).
178. I. J. Anthony and D. Wege, *Aust. J. Chem.*, **37**, 1283 (1984).
179. L. Enescu, A. Ghenciulescu, F. Chiraleu and I. G. Dinulescu, *Rev. Roum. Chim.*, **34**, 765 (1989).
180. T.-K. Yin, J. G. Lee and W. T. Borden, *J. Org. Chem.*, **50**, 531 (1985).
181. L. T. Scott, M. A. Kirms, H. Günther and H. von Puttkamer, *J. Am. Chem. Soc.*, **105**, 1372 (1983).
182. C. W. G. Fishwick, R. C. Gupta and R. C. Storr, *J. Chem. Soc., Perkin Trans. 1*, 2827 (1984).
183. M. E. Kuehne, *J. Am. Chem. Soc.*, **84**, 837 (1962).
184. H. L. Gingrich, Q. Huang, A. L. Morales and M. Jones, Jr., *J. Org. Chem.*, **57**, 3803 (1992).
185. M. Kato, Y. Okamoto, T. Chikamoto and T. Miwa, *Bull. Chem. Soc. Jpn.*, **51**, 1163 (1978).
186. S. M. Ali and S. Tanimoto, *J. Chem. Soc., Chem. Commun.*, 1465 (1988).
187. K. Matsumoto, T. Uchida, T. Sugi and Y. Yagi, *Chem. Lett.*, 869 (1982); K. Matsumoto, T. Uchida, T. Sugi and T. Kobayashi, *Heterocycles*, **20**, 1525 (1983); K. Matsumoto, T. Uchida, K. Aoyama, M. Nishikawa, T. Kuroda and T. Okamoto, *J. Heterocycl. Chem.*, **25**, 1793 (1988).
188. Y. Tominaga, Y. Shiroshita, Y. Matsuda and A. Hosomi, *Heterocycles*, **26**, 2073 (1987).
189. J. C. Fleming and H. Shechter, *J. Org. Chem.*, **34**, 3962 (1969); G. Baum, R. Bernard and H. Shechter, *J. Am. Chem. Soc.*, **89**, 5307 (1967); G. Baum and H. Shechter, *J. Org. Chem.*, **41**, 2120 (1976).
190. K. Hirakawa, T. Toki, K. Yamazaki and S. Nakazawa, *J. Chem. Soc., Perkin Trans. 1*, 1944 (1980).
191. K. Hirakawa, Y. Minami and S. Hayashi, *J. Chem. Soc., Perkin Trans. 1*, 577 (1982).
192. V. G. Kartsev, O. V. Isakova and R. X. Vikeneev, *Khim. Geterots. Soedin.*, 78 (1984).
193. W. Burgert, M. Grosse and D. Rewicki, *Chem. Ber.*, **115**, 309 (1982).
194. E. C. Taylor and D. M. Sobieray, *Tetrahedron*, **47**, 9599 (1991).

195. T. Yamamori, Y. Hiramatsu and I. Adachi, *J. Heterocycl. Chem.*, **18**, 347 (1981).
196. K. T. Potts, A. J. Elliott, G. R. Titus, D. Al-Hilal, P. F. Lindley, G. V. Boyd and T. Norris, *J. Chem. Soc., Perkin Trans. 1*, 2692 (1981); K. T. Potts, A. J. Elliott, G. R. Titus, D. Al-Hilal, P. F. Lindley, G. V. Boyd and T. Norris, *Acta Crystallogr.*, **B38**, 682 (1982).
197. C. D. Campbell, C. W. Rees, M. R. Bryce, M. D. Cooke, P. Hanson and J. M. Vernon, *J. Chem. Soc., Perkin Trans. 1*, 1006 (1978).
198. M. R. Bryce, C. D. Reynolds, P. Hanson and J. M. Vernon, *J. Chem. Soc., Perkin Trans. 1*, 607 (1981).
199. J. M. Vernon, M. R. Bryce and T. A. Dransfield, *Tetrahedron*, **39**, 835 (1983).
200. M. R. Bryce, T. A. Dransfield, K. A. Kandeel and J. M. Vernon, *J. Chem. Soc., Perkin Trans. 1*, 2141 (1988); *cf* M. R. Bryce, P. Hanson and J. M. Vernon, *J. Chem. Soc., Chem. Commun.*, 299 (1982).
201. J. J. Kulagowski, G. Mitchell, C. J. Moody and C. W. Rees, *J. Chem. Soc., Chem. Commun.*, 652 (1985).
202. R. Guilard, S. S. Gerges, A. Tabard, P. Richard, M. A. El Borai and C. Lecomte, *J. Am. Chem. Soc.*, **109**, 7228 (1987).
203. M. Pomerantz, *J. Am. Chem. Soc.*, **88**, 5349 (1966).
204. D. Seyferth, D. P. Duncan, M. L. Shannon and E. W. Goldman, *Organometallics*, **3**, 574 (1984).
205. D. Seyferth, S. C. Vick and M. L. Shannon, *Organometallics*, **3**, 1897 (1984).
206. J. A. Berson and M. Pomerantz, *J. Am. Chem. Soc.*, **86**, 3896 (1964).
207. J. H. Rigby, D. D. Holsworth and K. James, *J. Org. Chem.*, **54**, 4019 (1989).
208. S. H. Hoke II, J. Molstad, D. Dilettato, M. J. Jay, D. Carlson, B. Kahr and R. G. Cooks, *J. Org. Chem.*, **57**, 5069 (1992).
209. M. Tsuda, T. Ishida, T. Nogami, S. Kurono and M. Ohashi, *Chem. Lett.*, 2333 (1992).
210. S. E. Stein and A. Fahr, *J. Phys. Chem.*, **89**, 3714 (1985).
211. H. Hart and B. Osterby, unpublished observations.
212. G. Wittig and H. Härle, *Justus Liebigs Ann. Chem.*, **623**, 17 (1959).
213. H. Hart, C.-Y. Lai, G. Nwokogu, S. Shamouilian, A. Teuerstein and C. Zlotogorski, *J. Am. Chem. Soc.*, **102**, 6649 (1980).
214. H. Hart, N. Raju, M. A. Meador and D. L. Ward, *J. Org. Chem.*, **48**, 4357 (1983).
215. J. P. Mathias and J. F. Stoddart, *Chem. Soc. Rev.*, 215 (1992) and earlier references cited therein.
216. H. Hart, C.-Y. Lai, G. C. Nwokogu and S. Shamouilian, *Tetrahedron*, **43**, 5203 (1987).
217. H. Hart, S. Shamouilian and Y. Takehira, *J. Org. Chem.*, **46**, 4427 (1981).
218. H. Hart and G. C. Nwokogu, *Tetrahedron Lett.*, **24**, 5721 (1983).
219. H. Hart and S. Shamouilian, *J. Org. Chem.*, **46**, 4874 (1981).
220. M. B. Stringer and D. Wege, *Tetrahedron Lett.*, **21**, 3831 (1980).
221. F. Raymo, F. H. Kohnke, F. Cardullo, U. Girreser and J. F. Stoddart, *Tetrahedron*, **48**, 6827 (1992).
222. P. R. Ashton, N. S. Isaacs, F. H. Kohnke, G. S. d'Alcontres and J. F. Stoddart, *Angew. Chem., Int. Ed. Engl.*, **28**, 1261 (1989).
223. H. Hart, A. Bashir-Hashemi, J. Luo and M. A. Meador, *Tetrahedron*, **42**, 1641 (1986).
224. H. Hart and D. Ok, *J. Org. Chem.*, **51**, 979 (1986); D. Ok and H. Hart, *J. Org. Chem.*, **52**, 3835 (1987).
225. C. S. LeHoullier and G. W. Gribble, *J. Org. Chem.*, **48**, 1682 (1983); G. W. Gribble, R. B. Perni and K. D. Onan, *J. Org. Chem.*, **50**, 2934 (1985).
226. J. Nakayama, A. Sakari and M. Hoshino, *J. Org. Chem.*, **49**, 5084 (1984); J. Nakayama, K. Matsuzaki, M. Tanuma, K. Saito and M. Hoshino, *J. Chem. Soc., Chem. Commun.*, 974 (1986).
227. J. M. Dineen, E. E. Howell, Jr. and A. A. Volpe, *Polymer Preprints*, 282 (1983).
228. S. L. Buchwald, E. A. Lucas and J. C. Dewan, *J. Am. Chem. Soc.*, **109**, 4396 (1987); M. A. Bennett, J. S. Drage, K. D. Griffiths, N. K. Roberts, G. B. Robertson and W. A. Wickramasinghe, *Angew. Chem., Int. Ed. Engl.*, **27**, 941 (1988); D. P. Hsu, E. A. Lucas and S. L. Buchwald, *Tetrahedron Lett.*, **31**, 5563 (1990).
229. M. Zoratti and J. F. Bunnett, *J. Org. Chem.*, **45**, 1776 (1980).
230. C. E. Peishoff and W. L. Jorgensen, *J. Org. Chem.*, **50**, 1056 (1985).
231. H. Hart and A. Saednya, *Synth. Commun.*, **18**, 1749 (1988).
232. T. K. Vinod and H. Hart, *Tetrahedron Lett.*, **29**, 885 (1988).
233. H. Hart, K. Harada and C.-J. F. Du, *J. Org. Chem.*, **50**, 3104 (1985); see also A. Adejare and D. D. Miller, *Tetrahedron Lett.*, **25**, 5597 (1984).

234. H. Y. Xin and E. R. Biehl, *J. Org. Chem.*, **48**, 4397 (1983).
235. Y. X. Han, M. V. Jovanovic and E. R. Biehl, *J. Org. Chem.*, **50**, 1334 (1985).
236. S. P. Khanapure and E. R. Biehl, *J. Org. Chem.*, **52**, 1333 (1987).
237. E. R. Biehl, A. Razzuk, M. V. Jovanovic and S. P. Khanapure, *J. Org. Chem.*, **51**, 5157 (1986).
238. A. Razzuk and E. R. Biehl, *J. Org. Chem.*, **52**, 2619 (1987).
239. M. C. Carre, A. S. Ezzinadi, M. A. Zouaoui, P. Geoffroy and P. Caubère, *Synth. Commun.*, **19**, 3323 (1989).
240. S. P. Khanapure, L. Crenshaw, R. T. Reddy and E. R. Biehl, *J. Org. Chem.*, **53**, 4915 (1988).
241. E. R. Biehl, S. P. Khanapure, U. Siriwardane and M. Tschantz, *Heterocycles*, **29**, 485 (1989).
242. J. L. Self, S. P. Khanapure and E. R. Biehl, *Heterocycles*, **32**, 311 (1991).
243. A. R. Deshmukh and E. R. Biehl, *Heterocycles*, **34**, 99 (1992).
244. A. R. Deshmukh, M. Morgan, L. Tran, H. Zhang, M. Dutt and E. R. Biehl, *Heterocycles*, **34**, 1239 (1992).
245. J. H. Waggenspack, L. Tran, S. Taylor, L. K. Yeung, M. Morgan, A. R. Deshmukh, S. P. Khanapure and E. R. Biehl, *Synthesis*, 765 (1992).
246. E. R. Biehl and S. P. Khanapure, *Acc. Chem. Res.*, **22**, 275 (1989).
247. A. R. Deshmukh, L. Tran and E. R. Biehl, *J. Org. Chem.*, **57**, 667 (1992).
248. A. R. Deshmukh, H. Zhang, L. ducTran and E. Biehl, *J. Org. Chem.*, **57**, 2485 (1992).
249. P. G. Sammes and D. J. Dodsworth, *J. Chem. Soc., Chem. Commun.*, 33 (1979); D. J. Dodsworth, M.-P. Calcagno, E. U. Ehrmann, B. Devadas and P. G. Sammes, *J. Chem. Soc., Perkin Trans. 1*, 2120 (1981).
250. R. A. Russell and R. N. Warrener, *J. Chem. Soc., Chem. Commun.*, 108 (1981).
251. S. P. Khanapure, R. T. Reddy and E. R. Biehl, *J. Org. Chem.*, **52**, 5685 (1987).
252. S. P. Khanapure and E. R. Biehl, *J. Nat. Prod.*, **52**, 1357 (1989); S. P. Khanapure and E. R. Biehl, *Synthesis*, 33 (1991).
253. J. J. Fitzgerald, N. E. Drysdale and R. A. Olofson, *Synth. Commun.*, **22**, 1807 (1992); J. J. Fitzgerald, N. E. Drysdale and R. A. Olofson, *J. Org. Chem.*, **57**, 7122 (1992); I. Fleming and T. Mah, *J. Chem. Soc., Perkin Trans. 1*, 964 (1975).
254. P. Caubère, *Top. Curr. Chem.*, **73**, 49 (1978), but see especially pp. 72–84.
255. M.-C. Carré, B. Gregoire and P. Caubère, *J. Org. Chem.*, **49**, 2050 (1984); B. Gregoire, M.-C. Carré and P. Caubère, *J. Org. Chem.*, **51**, 1419 (1986).
256. M. A. Zouaoui, A. Mouaddib, B. Jamart-Gregoire, S. Ianelli, M. Nardelli and P. Caubère, *J. Org. Chem.*, **56**, 4078 (1991).
257. M-C. Carré, B. Jamart-Gregoire, P. Geoffroy and P. Caubère, *Tetrahedron*, **44**, 127 (1988).
258. K. Sukata and T. Akagawa, *J. Org. Chem.*, **54**, 1476 (1989).
259. N. Rodier, L. Pinelli, G. Adam, J. Andrieux and M. Plat, *Bull. Soc. Chim. France*, I-217 (1984).
260. M. Watanabe, A. Kurosaki and S. Furukawa, *Chem. Pharm. Bull.*, **32**, 1264 (1984).
261. R. B. Bates and K. D. Janda, *J. Org. Chem.*, **47**, 4374 (1982).
262. J. Nakayama, T. Fujita and M. Hoshino, *Chem. Lett.*, 1777 (1982).
263. J. Nakayama, T. Fujita and M. Hoshino, *Chem. Lett.*, 249 (1982).
264. N. Nakamura, *Chem. Lett.*, 1795 (1983); N. Nakamura, *J. Am. Chem. Soc.*, **105**, 7172 (1983).
265. J. Nakayama, Y. Kumano and M. Hoshino, *Tetrahedron Lett.*, **30**, 847 (1989).
266. J. Nakayama, H. Ozasa and M. Hoshino, *Heterocycles*, **22**, 1053 (1984).
267. J. Nakayama, H. Sugiura, A. Shiotsuki and M. Hoshino, *Tetrahedron Lett.*, **26**, 2195 (1985).
268. J. Nakayama, S. Takeue and M. Hoshino, *Tetrahedron Lett.*, **25**, 2679 (1984).
269. J. Nakayama, K. Hoshino and M. Hoshino, *Chem. Lett.*, 677 (1985).
270. N. Petragnani and V. G. Toscano, *Chem. Ber.*, **103**, 1652 (1970).
271. J. Nakayama, T. Tajiro and M. Hoshino, *Bull. Chem. Soc. Jpn.*, **59**, 2907 (1986).
272. D. J. Guliver, E. G. Hope, W. Levason, S. G. Murray, D. M. Potter and G. L. Marshall, *J. Chem. Soc., Perkin Trans. 2*, 429 (1984); C. Lambert and L. Christiaens, *Tetrahedron Lett.*, **25**, 833 (1984).
273. J. Bruhin, F. Gerson, R. Möckel and G. Plattner, *Helv. Chim. Acta*, **68**, 377 (1985).
274. N. E. Blank, M. W. Haenel, C. Krüger, Y.-H. Tsay and H. Wientges, *Angew. Chem., Int. Ed. Engl.*, **27**, 1064 (1988).
275. J. Nakayama, H. Fukushima, R. Hashimoto and M. Hoshino, *J. Chem. Soc., Chem. Commun.*, 612 (1982).
276. C.-J. F. Du, H. Hart and K.-K. D. Ng, *J. Org. Chem.*, **51**, 3162 (1986).

277. T. K. Vinod and H. Hart, *J. Am. Chem. Soc.*, **110**, 6574 (1988); T. Vinod and H. Hart, *J. Org. Chem.*, **55**, 881 (1990).
278. T. K. Vinod and H. Hart, *J. Am. Chem. Soc.*, **112**, 3250 (1990); T. K. Vinod and H. Hart, *J. Org. Chem.*, **55**, 5461 (1990); R. S. Grewal and H. Hart, *Tetrahedron Lett.*, **30**, 4271 (1990); T. K. Vinod and H. Hart, *J. Org. Chem.*, **56**, 5630 (1991); R. S. Grewal, H. Hart and T. K. Vinod, *J. Org. Chem.*, **57**, 2721 (1992); U. Lüning, C. Wangnick, K. Peters and H. G. v. Schnering, *Chem. Ber.*, **124**, 397 (1991); U. Lüning and C. Wangnick, *Justus Leibigs Ann. Chem.*, 481 (1992).
279. H. Hart and T. Ghosh, *Tetrahedron Lett.*, **29**, 881 (1988).
280. J.-J. Chiu, H. Hart and D. L. Ward, *J. Org. Chem.* **58**, 964 (1993).
281. T. Ghosh and H. Hart, *J. Org. Chem.*, **53**, 3555 (1988).
282. K. Harada, H. Hart and C.-J. Du, *J. Org. Chem.*, **50**, 5524 (1985).
283. W. Dilthey and G. Hurtig, *Chem. Ber.*, **67**, 2004 (1934); T. A. Geissman and R. C. Mallett, *J. Am. Chem. Soc.*, **61**, 1788 (1939).
284. C.-J. F. Du and H. Hart, *J. Org. Chem.*, **52**, 4311 (1987).
285. L. Castedo and E. Guitian, in *Studies in Natural Products Chemistry*, Vol. 3 (Ed. Atta-ur-Rahman), Elsevier, Amsterdam, 1989, pp. 417–454; see also L. Castedo and E. Guitian, in *Natural Products Chemistry 3* (Eds. Att-ur-Rahman and P. W. LeQuesne), Springer, Berlin, 1988, pp. 235–245.
286. W. M. Best and D. Wege, *Tetrahedron Lett.*, **22**, 4877 (1981).
287. W. M. Best and D. Wege, *Aust. J. Chem.*, **39**, 647 (1986).
288. J. C. Estévez, R. J. Estévez, E. Guitián, M. C. Villaverde and L. Castedo, *Tetrahedron Lett.*, **30**, 5785 (1989).
289. D. P. Meiras, E. Guitián and L. Castedo, *Tetrahedron Lett.*, **31**, 2331 (1990).
290. (a) L. Castedo, E. Guitián, J. M. Saá and R. Suau, *Tetrahedron Lett.*, **23**, 457 (1982).
    (b) L. Castedo, E. Guitián, C. Saá, R. Suau and J. M. Saá, *Tetrahedron Lett.*, **24**, 2107 1983).
    (c) C. Saa, E. Guitián, L. Castedo and J. M. Saá, *Tetrahedron Lett.*, **26**, 4559 (1985).
    (d) N. Atanes, E. Guitián, C. Saá, L. Castedo and J. M. Saá, *Tetrahedron Lett.*, **28**, 817 (1987).
    (e) N. Atanes, L. Castedo, E. Guitián and J. M. Saá, *Heterocycles*, **26**, 1183 (1987).
    (f) N. Atanes, L. Castedo, A. Cobas, E. Guitián, C. Saá and J. M. Saá, *Tetrahedron*, **45**, 7947 (1989).
    (g) N. Atancs, L. Castedo, E. Guitián, C. Saá, J. M. Saá and R. Suau, *J. Org. Chem.*, **56**, 2984 (1991).
    (h) B. Gómez, G. Martin, E. Guitián, L. Castedo and J. M. Saá, *Tetrahedron*, **49**, 1251 (1993).
291. (a) C. Saá, E. Guitián, L. Castedo, R. Suau and J. M. Saá, *J. Org. Chem.*, **51**, 2781 (1986).
    (b) A. Cobas, E. Guitián, L. Castedo and J. M. Saá, *Tetrahedron Lett.*, **29**, 2491 (1988).
    (c) A. Cobas, E. Guitián and L. Castedo, *J. Org. Chem.*, **57**, 6765 (1992).
292. G. Martin, E. Guitián, L. Castedo and J. M. Saá, *Tetrahedron Lett.*, **28**, 2407 (1987); G. Martin, E. Guitián, L. Castedo and J. M. Saá, *J. Org. Chem.*, **57**, 5907 (1992).
293. D. P. Meirás, E. Guitián and L. Castedo, *Tetrahedron Lett.*, **31**, 143 (1990); D. Pérez, E. Guitián and L. Castedo, *J. Org. Chem.*, **57**, 5911 (1992).
294. J. H. Rigby and D. D. Holsworth, *Tetrahedron Lett.*, **32**, 5757 (1991).
295. T. Matsumoto, T. Hosoya and K. Suzuki, *J. Am. Chem. Soc.*, **114**, 3568 (1992).
296. H. H. Nam, G. E. Leroi and J. F. Harrison, *J. Phys. Chem.*, **95**, 6514 (1991).
297. H. H. Nam and G. E. Leroi, *J. Am. Chem. Soc.*, **110**, 4096 (1988).
298. H. H. Nam and G. E. Leroi, *Tetrahedron Lett.*, **31**, 4837 (1990).
299. C. May and C. J. Moody, *Tetrahedron Lett.*, **26**, 2123 (1985); *cf* J. Nakayama, O. Simamura and M. Yoshida, *J. Chem. Soc., Chem. Commun.*, 1222 (1970).
300. B. Jamart-Gregoire, C. Leger and P. Caubère, *Tetrahedron Lett.*, **31**, 7599 (1990).
301. M. Tsukazaki and V. Snieckus, *Heterocycles*, **33**, 533 (1992).
302. F. Effenberger and W. Daub, *Chem. Ber.*, **124**, 2119 (1991).
303. (a) R. F. C. Brown, K. J. Coulston, F. W. Eastwood and M. R. Moffatt, *Tetrahedron Lett.*, **32**, 801 (1991).
    (b) R. F. C. Brown, K. J. Coulston, F. W. Eastwood and M. R. Moffatt, *Tetrahedron*, **48**, 7763 (1992).
304. R. F. C. Brown, N. Choi and F. W. Eastwood, *Tetrahedron Lett.*, **33**, 3787 (1992).
305. G. Wittig and E. Knaus, *Chem. Ber.*, **91**, 895 (1958).
306. T. Kauffmann, J. Hansen, K. Udluft and R. Wirthwein, *Angew. Chem., Int. Ed. Engl.*, **3**, 650 (1964); T. Kauffmann, *Angew. Chem. Int. Ed. Engl.*, **4**, 543 (1965).

307. M. G. Reinecke, E. S. Brown, B. P. Capehart, D. E. Minter and R. K. Freeman, *Tetrahedron*, **44**, 5675 (1988).
308. C. May and C. J. Moody, *J. Chem. Soc., Chem. Commun.*, 926 (1984); C. May and C. J. Moody, *J. Chem. Soc., Perkin Trans. 1*, 247 (1988).
309. C.-K. Sha and J.-F. Yang, *Tetrahedron*, **48**, 10645 (1992).
310. S. P. Modi and S. Archer, *J. Org. Chem.*, **54**, 5189 (1989).
311. J. Kurita, N. Kakusawa, S. Yasuike and T. Tsuchiya, *Heterocycles*, **31**, 1937 (1990).
312. S. P. Khanapure and E. R. Biehl, *Heterocycles*, **27**, 2643 (1988).
313. S. P. Khanapure, B. M. Bhawal and E. R. Biehl, *Heterocycles*, **32**, 1773 (1991).
314. M. J. S. Dewar and D. R. Kuhn, *J. Am. Chem. Soc.*, **106**, 5256 (1984).
315. M. Tielemans, R. Promel and P. Geerlings, *Tetrahedron Lett.*, **29**, 1687 (1988); M. Tielemans, V. Areschka, J. Colomer, R. Promel, W. Langenaeker and P. Geerlings, *Tetrahedron*, **48**, 10575 (1992).
316. F. De Sio, S. Chimichi, R. Nesi and L. Cecchi, *Heterocycles*, **19**, 1427 (1982).
317. See Reference 2, p. 293.
318. M. G. Reinecke and J. G. Newsom, *J. Am. Chem. Soc.*, **98**, 3021 (1976).
319. M. G. Reinecke, J. G. Newsom and L.-J. Chen, *J. Am. Chem. Soc.*, **103**, 2760 (1981).
320. M. G. Reinecke, J. G. Newsom and K. A. Almqvist, *Tetrahedron*, **37**, 4151 (1981).
321. M. G. Reinecke, L. J. Chen and K. A. Almqvist, *J. Chem. Soc., Chem. Commun.*, 585 (1980).
322. J. H. Teles, B. A. Hess, Jr. and L. J. Schaad, *Chem. Ber.*, **125**, 423 (1992).
323. G. W. Gribble and S. C. Conway, *Synth. Commun.*, **22**, 2129 (1992).
324. M. A. Bennett and H. P. Schwemlein, *Angew. Chem., Int. Ed. Engl.*, **28**, 1296 (1989); S. L. Buchwald and R. B. Nielsen, *Chem. Rev.*, **88**, 1047 (1988); S. L. Buchwald and R. A. Fisher, *Chem. Scr.*, **29** 417 (1989).
325. J. F. Hartwig, R. G. Bergman and R. A. Anderson *J. Am. Chem. Soc.*, **112**, 3234 (1990); M. A. Bennett, K. D. Griffiths, T. Okano, V. Parthasarathi and G. B. Robertson, *J. Am. Chem. Soc.*, **112**, 7047 (1990); G. Erker, U. Korek, R. Petrenz and A. L. Rheingold, *J. Organometal. Chem.*, **421**, 215 (1991); S. L. Buchwald and S. M. King, *J. Am. Chem. Soc.*, **113**, 258 (1991); J. Bodiguel, P. Meunier, M. M. Kubicki, P. Richard, B. Gautheron, G. Dousse, H. Lavayssiere and J. Satge, *Organometallics*, **11**, 1423 (1992); J. K. Cockcroft, V. C. Gibson, J. A. K. Howard, A. D. Poole, U. Siemeling and C. Wilson, *J. Chem. Soc., Chem. Commun.*, 1668 (1992); Z. Lu, K. A. Abboud and W. M. Jones, *J. Am. Chem. Soc.*, **114**, 10991 (1992).
326. H. M. Rosenstock, R. Stockbauer and A. C. Parr, *J. Chim. Phys.*, **77**, 745 (1980); W. J. van der Hart, E. Oosterveld, T. A. Molenaar-Langeveld and N. M. M. Nibbering, *Org. Mass Spectrom.*, **24**, 59 (1989); S. Gronert and C. H. DePuy, *J. Am. Chem. Soc.*, **111**, 9253 (1989).
327. Y. Huang and B. S. Freiser, *J. Am. Chem. Soc.*, **111**, 2387 (1989); Y. Huang and B. S. Freiser, *J. Am. Chem. Soc.*, **112**, 1682 (1990); J. R. Gord, B. S. Freiser and S. W. Buckner, *J. Chem. Phys.*, **94**, 4282 (1991).

CHAPTER **19**

Ynol ethers and esters

PETER J. STANG

Department of Chemistry, The University of Utah, Salt Lake City, Utah 84112, USA

and

VIKTOR V. ZHDANKIN

Department of Chemistry, University of Minnesota-Duluth, Minnesota 55812, USA

Supplement C2: The chemistry of triple-bonded functional groups
Edited by S. Patai © 1994 John Wiley & Sons Ltd

I. INTRODUCTION

Acetylenic ethers and esters represent an important class of functionalized acetylene derivatives of the hypothetical alkynols, $RC\equiv COH$. The chemistry of acetylenic ethers has been well developed since their first preparation about 100 years ago[1]. Several exhaustive reviews[1-3] covering the literature on acetylenic ethers and their analogs up to 1985 have been published in the last 30 years. In contrast to acetylenic ethers, esters of alkynols were unknown until the mid-1980s when the first preparation of alkynyl tosylates was reported[4]. In the following years a wide variety of alkynyl carboxylate, phosphate and sulfonate esters has been prepared from alkynyl iodonium salts. The chemistry of these novel derivatives of alkynols has been summarized in a recent review[5]. In the last 10 years considerable interest and research activity has arisen toward alkynols themselves and such derivatives as alkynolate salts and silyl ynol ethers. The present chapter will cover the chemistry of acetylenic ethers and esters as well as related derivatives of ynols with emphasis on new developments in this subject during the last 5–10 years.

II. YNOLS AND YNOLATES

A. Ynols

Ynols or hydroxyacetylenes (1) are the triple bond analogs of enols. Like enols they are tautomers of the corresponding carbonyl species–ketenes (2) (equation 1)[6]. This tautomeric equilibrium is almost completely on the side of ketenes (2), which according to high-level *ab initio* calculations are some 37 kcal mol^{-1} more stable than the corresponding ynols[7,8]. However, in the gas phase alkynols may exist as relatively stable molecules due to a very high energy barrier for the isomerization $1 \rightarrow 2$. For example, the activation energy

$$RC\equiv COH \quad \xrightarrow{\qquad\qquad} \quad RHC=C=O \qquad (1)$$

$$\textbf{(1)} \qquad\qquad\qquad\qquad \textbf{(2)}$$

for the isomerization of hydroxyacetylene to ketene is calculated[9] to be 73 kcal mol^{-1}. Due to its stability in the form of isolated molecules, hydroxyacetylene has been considered as a possible constituent of interstellar clouds, planetary atmospheres and flames[10]. Moreover, two direct observations of hydroxyacetylene have been reported recently[9,11]. In 1986 Schwarz and coworkers reported the generation and identification of hydroxyacetylene in a tandem mass spectrometer[9]. More recently, hydroxyacetylene (4) has been generated photochemically from semisquaric acid (3) in an argon matrix at a temperature of 12 K (equation 2) and characterized by IR spectroscopy[11].

$$\xrightarrow[-2CO \text{ in 3 steps}]{hv} \quad HC\equiv COH \qquad (2)$$

$$\textbf{(3)} \qquad\qquad\qquad\qquad\qquad \textbf{(4)}$$

Interestingly, the IR spectrum of **4** is typical of any unsymmetrically substituted alkyne with the stretching band of the triple bond at 2198 cm^{-1}, the terminal C—H at 3340 and the O—H at 3501 cm^{-1}.

In a series of studies by Kresge and coworkers the chemistry of ynols was examined in aqueous solution[6,12–14]. In this research phenylhydroxyacetylene (**6**) was generated as a transient intermediate by flash photolysis of phenylhydroxycyclopropenone (**5**) (equation 3) and its subsequent transformations were investigated kinetically[12,13].

$$
\underset{\textbf{(5)}}{\underset{\text{Ph} \quad \text{OH}}{\overset{\text{O}}{\triangle}}}
\xrightarrow[-\text{CO}]{hv}
\underset{\textbf{(6)}}{\text{PhC}\equiv\text{COH}}
\longrightarrow
\text{PhHC}=\text{C}=\text{O}
\quad (3)
$$

$$
\downarrow \text{H}_2\text{O}
$$

$$
\text{PhCH}_2\text{CO}_2\text{H}
$$

One of the most remarkable results of this research[6] was the determination of the unusually high acidity of ynol (**6**) with an estimated pK_a of less than 2.8. This indicates that ynols are at least 7 orders of magnitude more acidic than their double bond analogs– enols, and more acidic than even carboxylic acids. The results of *ab initio* calculations[8] confirm and provide insight into the high acidity of hydroxyacetylenes.

B. Ynolate Anions

1. Preparation

Acetylene diolate salts, MOC≡COM (M = alkali or alkali earth metal), are well known products of oxo-carbon chemistry which can be prepared by the interaction of the corresponding metal with carbon monoxide[15–18]. For example, the colorless potassium ethynediolate salt (**7**) can be obtained as the principal product by passing carbon monoxide into a solution of potassium in liquid ammonia (equation 4)[15].

$$
2\text{K} + 2\text{CO} \xrightarrow{\text{liquid NH}_3} \underset{\textbf{(7)}}{\text{KOC}\equiv\text{COK}} \quad (4)
$$

The structure of ethynediolate salts was investigated by X-ray diffraction[15] and matrix IR spectroscopy[18]. The CC bond length in compound **7** is 1.21 Å, which is typical of the acetylenic triple bond. The CO distance of 1.28 Å has an intermediate value between a single and a double bond. However, IR spectroscopy indicates a single bond character for the CO bond in the ethynediolate anion[18]. Further support of structure **7** as the ynolate salt was obtained from its chemistry: one of the products of the protonolysis of this salt with anhydrous hydrogen chloride is hexahydroxybenzene, which may result from the trimerization of the initially formed dihydroxyacetylene, HOC≡COH[17].

$$
\underset{\textbf{(8)}}{\overset{\text{O}}{\underset{}{\text{RCCHBr}_2}}}
\xrightarrow{\text{Base}}
\underset{\textbf{(9)}}{\overset{\text{O}^-}{\underset{}{\text{RC}=\text{CBr}_2}}}
\xrightarrow{t\text{-BuLi}}
\underset{\textbf{(10)}}{\overset{\text{O}^-}{\underset{}{\text{RC}=\text{CBrLi}}}}
\xrightarrow[-\text{LiBr}]{}
\underset{\textbf{(11)}}{\text{RC}\equiv\text{CO}^-}
\quad (5)
$$

There is substantial recent interest in the preparation and chemistry of alkynolate anions. In 1982 Kowalski and coworkers found a general approach to alkynolate anions **11** starting from α-haloketones **8** by a sequence of reactions represented in equation 5[19]. The key step in this scheme (equation 5) is the rearrangement of the carbene generated from the intermediate **10**, obtained by metal–halogen exchange in the haloenolate anion **9**. The reaction usually is performed in a THF solution at low temperature, however, the resulting solutions of lithium alkynolate salts are relatively stable even at room temperature. Instead of α-dibromoketones (**8**) monobromo- or monochloroketones can be used as the starting material; such bases as lithium hexamethyldisilazide or methyllithium are applied in the enolization step **8** → **9**[19]. In the following papers[20,21] Kowalski and coauthors further modified this preparation of alkynolates (equation 5) using the ethyl esters of carboxylic acids and dibromomethyllithium as the starting materials for the generation of enolates **9**.

Several less general approaches to alkynolate anions have also been reported in the literature[22–25]. Trimethylsilylethynolate (**13**) can be generated by deprotonation of ketene **12** in THF solution (equation 6)[22]. Phenylethynolate (**15**) has been prepared via isoxazole **14** (equation 7)[23], and lithiumethynolate (**17**) has been reported as an intermediate product in the reaction of dihydrofurane **16** with butyllithium (equation 8)[24].

$$Me_3SiCH{=}C{=}O \xrightarrow[-100\,°C]{BuLi,\ THF} Me_3SiC{\equiv}COLi \qquad (6)$$

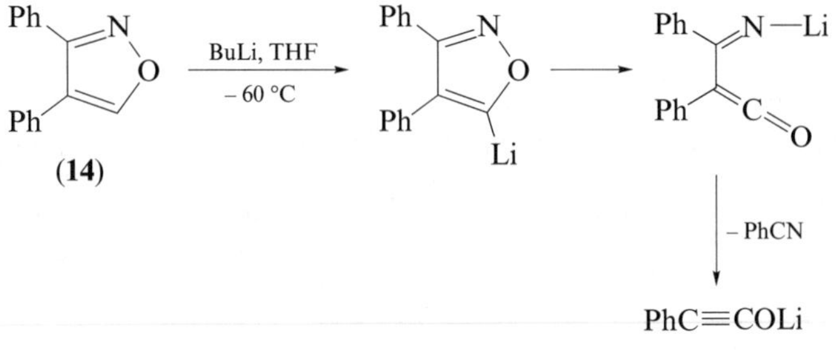

$$\text{(12)} \qquad\qquad\qquad\qquad \text{(13)}$$

(7)

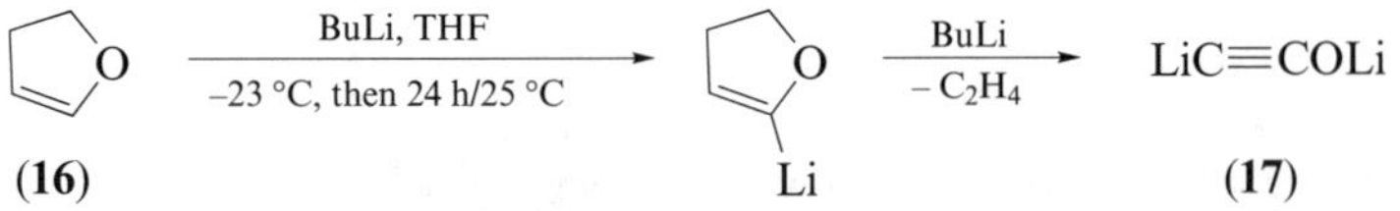

$$- PhCN$$

$$PhC{\equiv}COLi$$

$$\text{(15)}$$

(8)

(**16**)   Li   (**17**)

A recent approach to lithium alkynolates involves cleavage of some ynol ethers and esters with methyllithium[21,25]. Desilylation of silyl ynol ethers with methyllithium at room temperature in THF solution leads to the selective formation of the corresponding alkynolates (equation 9)[21]. Similarly, alkynyl tosylates can be selectively cleaved with methyllithium at low temperatures (equation 10)[25].

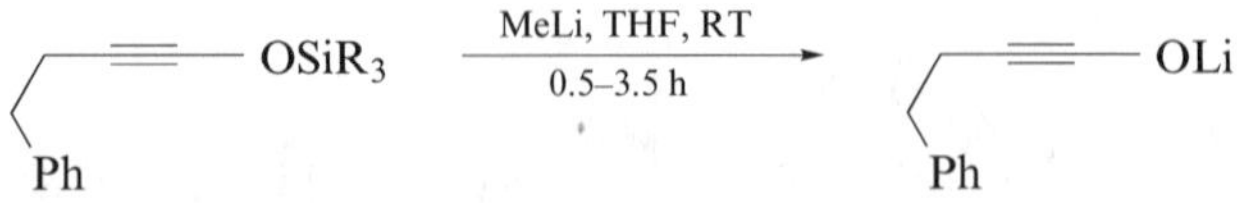

(9)

$$SiR_3 = Si(i\text{-}Pr)_3 \text{ or } Si(t\text{-}Bu)Me_2 \qquad\qquad 92\text{–}93\%$$

$$RC\equiv COTs \xrightarrow[-20\ °C]{MeLi,\ THF\ or\ DME} RC\equiv COLi \qquad (10)$$

$$R = t\text{-Bu or } s\text{-Bu}$$

2. Reactions

According to its structure, which can be represented by two resonance contributors (equation 11), an ynolate anion has two nucleophilic centers. In principle, attack of an electrophilic reagent on the oxygen atom of an ynolate anion leads to the formation of an ynol ether or ester, whereas C-attack gives a ketene molecule.

$$RC\equiv CO^- \longleftrightarrow R\bar{C}=C=O \qquad (11)$$

Both types of reactivity of ynolate anion have been reported in the literature. The O-attack is typical for the reactions of lithium ynolates with trialkylchlorosilanes[21,24,25] and dialkylchlorophosphates[26]. Lithium ynolates, generated as shown in equations 5–10, react with sterically hindered trialkylchlorosilanes in THF affording silyl ynol ethers as primary products (equation 12)[21,25]. However, in some cases the silyl ynol ethers are unstable at room temperature and isomerization to the more stable ketenes, or decomposition, occurs[21–24]. The ketene rearrangement usually occurs in reactions of lithium alkynolates with methyl substituted silyl chlorides; a typical example of such a rearrangement is represented by reaction 13[24].

$$RC\equiv COLi + R'_3SiCl \xrightarrow{THF,\ -78\ °C\ to\ RT} RC\equiv COSiR'_3 \qquad (12)$$

$$50\text{–}93\%$$

$$R = t\text{-Bu, } s\text{-Bu, cyclohexyl, PhCH}_2\text{CH}_2\text{, Ph, etc.}$$
$$SiR'_3 = Si(i\text{-Pr})_3 \text{ or } Si(t\text{-Bu})Me_2$$

$$LiC\equiv COLi \xrightarrow[THF,\ -60\ °C\ to\ RT]{2RMe_2SiCl} \left[RMe_2SiC\equiv COSiMe_2R \right] \qquad (13)$$

$$R = H,\ Me,\ t\text{-Bu} \qquad\qquad (RMe_2Si)_2C=C=O$$

$$40\text{–}56\%$$

The analogous reaction of lithium alkynolates with trialkylgermanium and trialkyltin chlorides results in the exclusive formation of the corresponding ketenes[25]. However, a mechanism implying the intermediate formation of the ynol derivatives and their subsequent rearrangement to ketenes cannot be ruled out.

Reactions of lithium alkynolates with dialkylchlorophosphates afford alkynyl phosphate esters as the major products (equation 14)[26]. However, reactions with acyl chlorides under similar conditions results in both O-acylation (product **20**) and C-acylation (product **21**) (equation 15)[26].

$$RC\equiv COLi + ClPO(OEt)_2 \xrightarrow{THF,\ -20\ °C} RC\equiv COPO(OEt)_2 \qquad (14)$$

$$R = t\text{-Bu, } n\text{-Bu, Me, Ph} \qquad 30\text{–}56\%$$

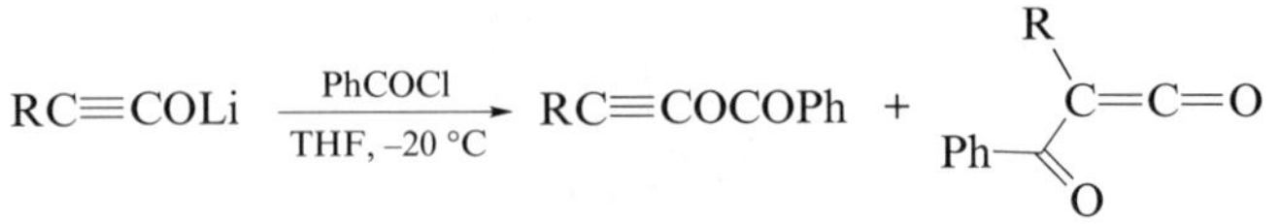

$$RC \equiv COLi \xrightarrow[\text{THF, } -20\ °C]{\text{PhCOCl}} RC \equiv COCOPh \ + \ \underset{Ph}{\overset{R}{\diagdown}}C = C = O \qquad (15)$$

R = t-Bu, Ph   (20)   (21)

Carbon electrophiles such as ketones and aldehydes react with ynolates, forming exclusively products of C-attack. Addition of phenylethynolate (15) to cyclohexanone affords spirolactone 23, the formation of which can be explained by the cyclization of the originally formed ketene 22 (equation 16)[19,23]. Similarly, the reaction of lithium alkynolates with benzaldehyde can be rationalized by the intermediate formation of ketene 24 (equation 17), however, the final product in this case is the unsaturated acid 25 but not a lactone[19].

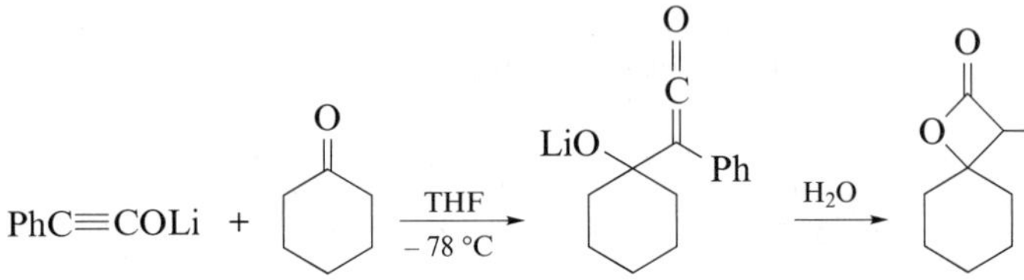

$$PhC \equiv COLi \ + \ \text{cyclohexanone} \xrightarrow[-78\ °C]{\text{THF}} \ \textbf{(22)} \xrightarrow{H_2O} \ \textbf{(23) 43\%} \qquad (16)$$

(15)   (22)   (23) 43%

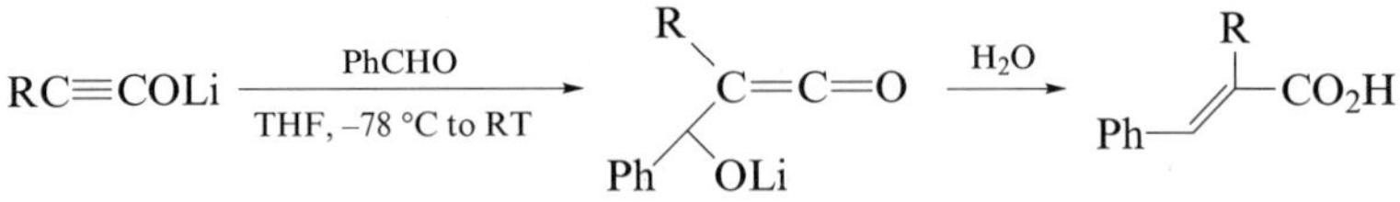

$$RC \equiv COLi \xrightarrow[\text{THF, } -78\ °C \text{ to RT}]{\text{PhCHO}} \underset{Ph \quad OLi}{\overset{R}{\diagdown}}C = C = O \xrightarrow{H_2O} \underset{Ph}{\overset{R}{\diagup}}-CO_2H \qquad (17)$$

(24)   (25) 75–85%

R = t-Bu, Ph

The reactions of alkynolates with alcohols leading to esters 26 also imply the intermediate formation of ketenes (equation 18)[19]. However, initial formation of ynols resulting from the O-protonation of the alkynolate anion and subsequent rearrangement to ketenes cannot be excluded.

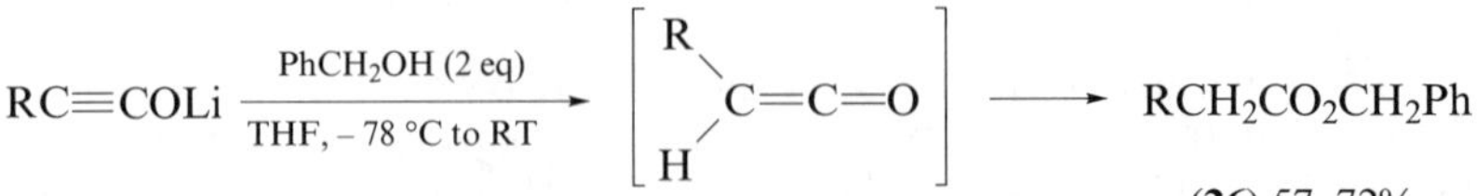

$$RC \equiv COLi \xrightarrow[\text{THF, } -78\ °C \text{ to RT}]{\text{PhCH}_2\text{OH (2 eq)}} \left[\underset{H}{\overset{R}{\diagdown}}C = C = O \right] \longrightarrow RCH_2CO_2CH_2Ph \qquad (18)$$

(26) 57–72%

R = t-Bu, Ph

Reaction of lithium alkynolates with alcohols is especially useful for organic synthesis as a method of ester homologation[20]. In this case ethyl esters of carboxylic acids 27 are used for the generation of lithium alkynolates in a multistep one-pot procedure (equation 19). Subsequent ethanolysis of lithium alkynolate in the presence of HCl gives homologated esters 28 in good yield.

A modified homologation reaction[20b] can be used for the stereoselective one-pot synthesis of functionalized 1,3-dienes 30 from esters of α,β-unsaturated acids 29 according to equation 20.

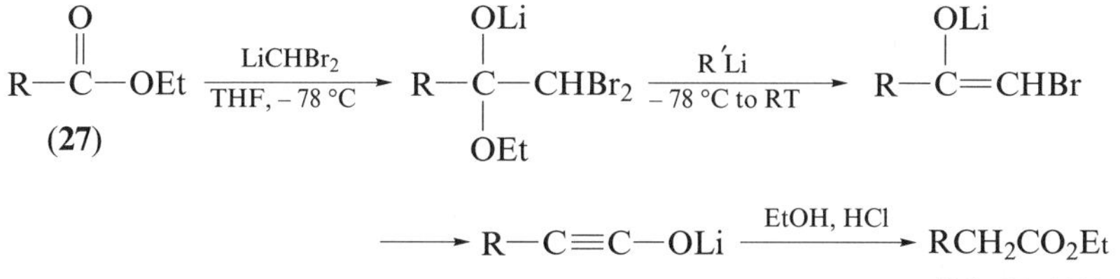

$$R—C\equiv C—OLi \xrightarrow{\text{EtOH, HCl}} RCH_2CO_2Et \quad (19)$$

(28) 67–90%

R = Ph, CH_2CH_2Ph, CMe_2CH_2Ph, cyclohexyl, 1-naphthyl, etc.
R′ = tetramethylpiperidine, hexamethyldisylazane, Bu, s-Bu

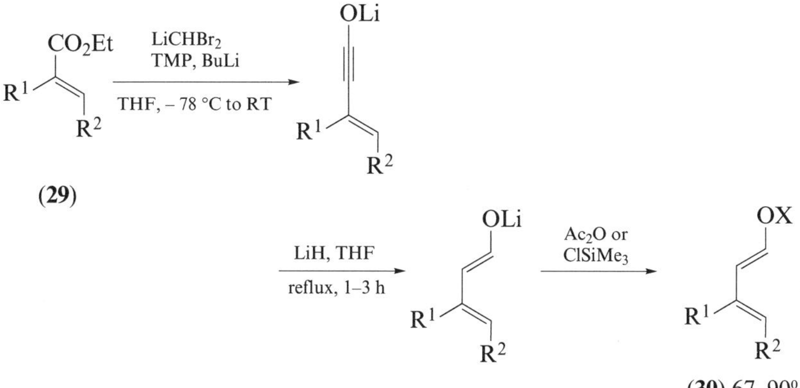

(29)

(20)

(30) 67–90%

R^1, R^2 = H, Me, Ph, C_5H_{11}, —$(CH_2)_4$—
X = Ac or Me_3Si; TMP = tetramethylpiperidine

C. Ynolate Complexes of Transition Metals

Ynolate (or ketenyl) ligands RCCO⁻ in organometallic chemistry have become easily accessible by the coupling of alkylidyne and carbonyl ligands[27–30]. The first coupling of this type was reported in 1976 in the reaction of tungsten alkylidyne–carbonyl complex **31** with trimethylphosphine leading to ynolate complex **32** (equation 21)[31].

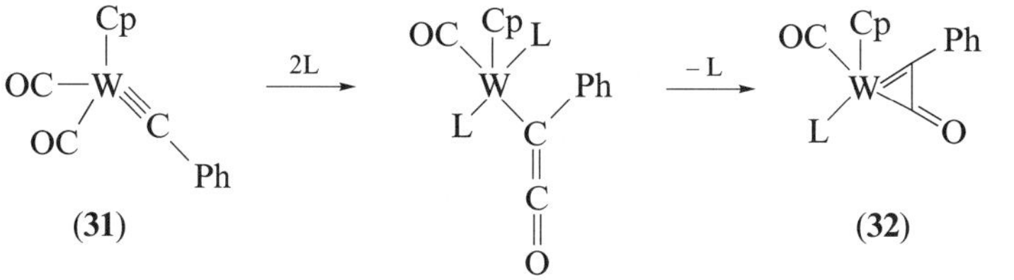

L = Me_3P, Cp = cyclopentadienyl

Additional examples of nucleophile-induced alkylidyne–carbonyl coupling reactions have been described in recent reviews[27–30]; photoinduction of such coupling has also been reported[32].

Ynolate metal complexes can be selectively alkylated or acylated to form the corresponding ynol ether complexes **33** (equation 22) or ynol ester complexes **34** (equation 23)[33].

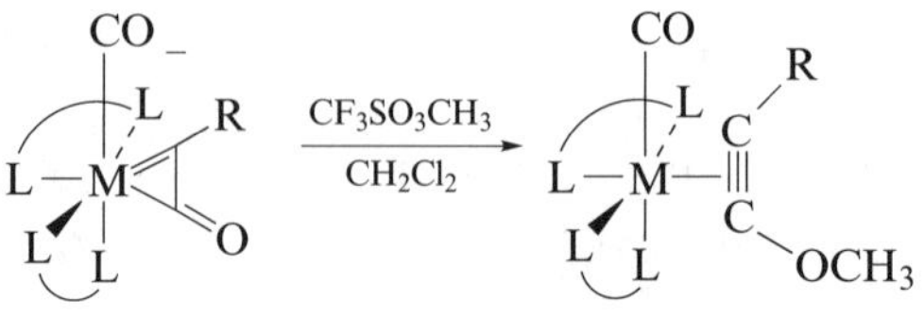

(33) (22)

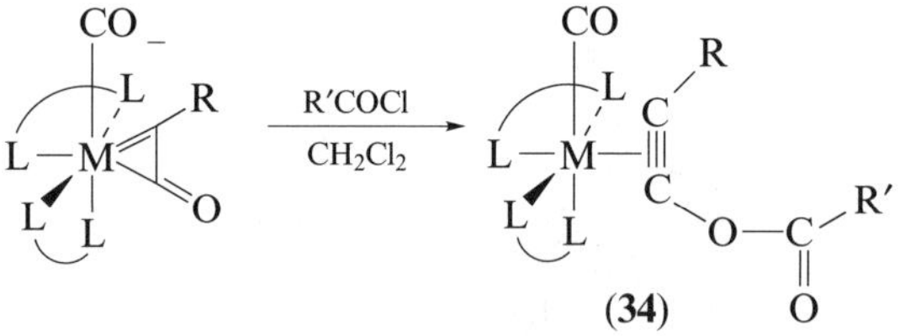

(34) (23)

M = Mo, W; R = Ph, p-$Me_3CC_6H_4$; R′ = Me, CMe_3, CH_2Ph, Ph;
L_2 = dithiocarbamate or pyrrolecarboxaldehyde methylimine

D. Silyl Ynol Ethers

1. Preparation and properties

The first reliable synthesis of silyl ynol ethers **37** was reported in 1985 by Maas and Brückmann[34]. This method employs α-diazocarbonyl compounds **35** and their silylated derivatives **36** as starting material (equation 24) and is based on a carbene rearrangement[34].

$$\underset{(\mathbf{35})}{RC(=O)-CH(=N_2)} \xrightarrow[\text{EtN}(i\text{-Pr})_2]{\text{R}'\text{SiOTf}} \underset{(\mathbf{36})}{RC(=O)-CSiR'_3(=N_2)} \xrightarrow[\text{2 h, reflux}]{\text{benzene}} \underset{(\mathbf{37})\ 39\text{–}87\%}{RC{\equiv}COSiR'} \qquad (24)$$

R = aryl, heteroaryl
SiR'_3 = $SiEt_3$, $Si(i\text{-}Pr)_3$, $SiMe_2(t\text{-}Bu)$

To date the most simple and general procedure for the preparation of silyl ynol ethers is the reaction of lithium alkynolates, generated *in situ* from ethyl esters, with silyl chlorides, according to equation 25[21]. However, this method (equation 25) does not work for the preparation of siloxyalkynes in which R is a lower alkyl group or hydrogen[35]. Danheiser

$$RCO_2Et \xrightarrow[\text{5 equiv } (i\text{-Pr})_3\text{SiCl, THF, }-78 \text{ to } 0\,^\circ\text{C}]{\substack{\text{2.2 equiv LiTMP, 2.2 equiv CH}_2\text{Br}_2\\ \text{5 equiv BuLi, THF, }-78 \text{ to } 30\,^\circ\text{C}}} \underset{50\text{–}75\%}{RC{\equiv}COSi(i\text{-}Pr)_3} \qquad (25)$$

R = cyclohexyl, $PhCH_2CH_2$, Ph, etc.
TMP = 2,2,6,6-tetramethylpiperidine

and coworkers[35] have found a more conventional approach to silyl ynol ethers by the base-promoted dehydrohalogenation of (Z)-2-bromovinyl ethers **38** (equations 26 and 27). This method can be applied even for the synthesis of the parent siloxyacetylene **39**, as well as other silyl ynol ethers.

$$Br{\overset{\displaystyle\diagup\!\!=\!\!\diagdown}{}}OSiR_3 \quad \xrightarrow[\text{then 1.5 equiv EtOH at } -78\,°C]{\text{2.0–2.1 equiv LDA, THF, } 0\,°C} \quad HC{\equiv}COSiR_3 \qquad (26)$$
$$\textbf{(38)} \qquad\qquad\qquad\qquad\qquad\qquad \textbf{(39)} \ 88\text{–}96\%$$

$$Br{\overset{\displaystyle\diagup\!\!=\!\!\diagdown}{}}OSiR_3 \quad \xrightarrow[\text{then 2.2–2.5 equiv MeI at } -78 \text{ to } 25\,°C]{\text{2.0–2.2 equiv LDA, THF, } 0\,°C} \quad MeC{\equiv}COSiR_3 \qquad (27)$$
$$\textbf{(38)} \qquad\qquad\qquad\qquad\qquad\qquad \textbf{(40)} \ 32\text{–}69\%$$

$$SiR_3 = Si(i\text{-}Pr)_3,\ SiMe_2(t\text{-}Bu),\ SiPh_2(t\text{-}Bu)$$
$$LDA = \text{lithium diisopropylamide}$$

Silyl ynol ethers bearing bulky silyl groups (i-Pr$_3$Si, t-BuMe$_2$Si) are thermally stable oils which can be distilled without decomposition or purified by chromatography[21,25,34,35]. Products containing small groups are less stable: triethylsiloxyacetylenes partially decompose on column chromatography or on distillation[34], and trimethylsilyl derivatives are unstable even at room temperature[21]. All siloxyalkynes are moisture sensitive; their hydrolysis affords the corresponding carboxylic acids[34].

Silyl ynol ethers have highly characteristic spectral properties. The triple bond stretching in the IR of siloxyalkynes is at 2260–2270 cm^{-1}, which compares with that of ynol esters and alkoxyacetylenes (see Sections III and IV of the present review)[25,34]. Particularly noteworthy and characteristic are the acetylenic carbon signals in the ^{13}C NMR. The signals of the α-carbons are in the normal alkyne region of 85–89 ppm whereas the β-carbons are at 27–40 ppm[34,35]. This considerable upfield shift of the β-carbons can be explained by the electron-rich nature of the siloxyacetylenes due to resonance donation according to equation 28.

$$R\overset{\beta}{C}{\equiv}\overset{\alpha}{C}OSiR_3' \quad\longleftrightarrow\quad R\bar{C}{=}C{=}\overset{+}{O}SiR_3' \qquad (28)$$

2. Chemistry

Silyl ynol ethers have found some practical application in the synthesis of trisubstituted olefins[36] and as partners in [2+2] cycloaddition reactions[35,37,38]. Kowalski and coworkers[36] reported that reaction of siloxyacetylenes with aldehydes in the presence of TiCl$_4$ affords trisubstituted olefins with very high E/Z stereoselectivity under mild conditions (equation 29).

$$RC{\equiv}COSi(i\text{-}Pr)_3 \quad \xrightarrow[\text{MeOH}, -78\,°C \text{ to RT}]{R'CHO,\ TiCl_3,\ CH_2Cl_2,\ -78\,°C} \quad \underset{H}{\overset{R'}{\diagdown}}{C}{=}{C}\underset{CO_2Me}{\overset{R}{\diagup}} \qquad (29)$$

$$R,\ R' = \text{Alkyl or Aryl} \qquad 60\text{–}65\%$$

Similarly to alkoxyacetylenes, siloxyacetylenes are good partners in [2+2] cycloaddition reactions with ketenes[35,37,38]. Kowalski and coworkers have found that bubbling of ketene through solutions of siloxyalkynes at 0 °C affords silyloxycyclobutenones **41** in

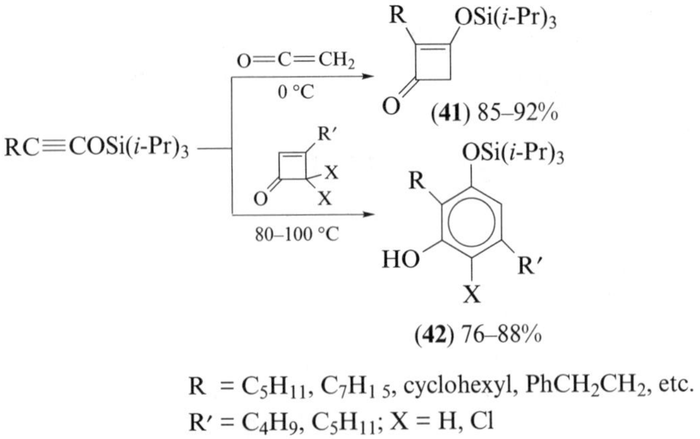

$$RC{\equiv}COSi(i\text{-}Pr)_3 \quad
\begin{array}{c}
\xrightarrow[0\,°C]{O{=}C{=}CH_2} \\[2mm]
\xrightarrow[80{-}100\,°C]{}
\end{array}$$

(41) 85–92%

(42) 76–88%

(30)

R = C$_5$H$_{11}$, C$_7$H$_{15}$, cyclohexyl, PhCH$_2$CH$_2$, etc.
R' = C$_4$H$_9$, C$_5$H$_{11}$; X = H, Cl

good yield. The same authors[38] also reported the reaction of siloxyalkynes with cyclobutenones leading to resorcinols **42** (equation 30).

In independent research Danheiser and coworkers[35,37] applied the reaction of siloxyalkynes with cyclobutenones to the synthesis of various highly substituted aromatic compounds **47** (equation 31).

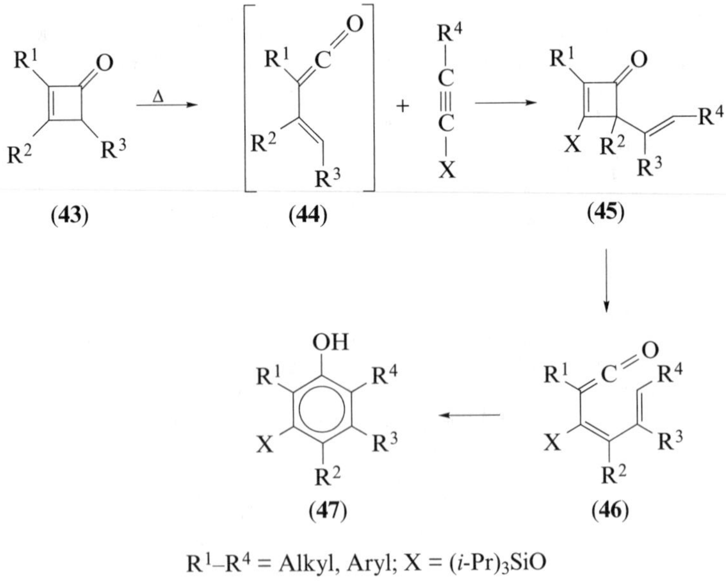

(43) (44) (45)

(31)

(47) (46)

R^1–R^4 = Alkyl, Aryl; X = (i-Pr)$_3$SiO

According to the authors[35,37], the driving force of this annulation reaction (equation 31) is the high reactivity of siloxyacetylenes with vinylketene **44**, which is thermally generated by the 4π electrocyclic opening of cyclobutenone **43**. The [2+2] cycloaddition of siloxyacetylene and ketene **44** affords 4-vinylcyclobutenone derivative **45**. Subsequent electrocyclic cleavage of this intermediate leads to dienylketene **46**, which undergoes a 6π electrocyclic closure to furnish the final product **47**. Further development of this strategy by Danheiser[37] expanded the scope of the method and provided access to a variety of important polycyclic aromatic and heteroaromatic systems. Particularly interesting is the application of an analogous annulation strategy in the total syntheses of the biologically active natural quinones *Dan Shen* **49** (equation 32)[39] and *Maesanin* **51** (equation 33)[40]. In these reactions (equations 32 and 33) the key aromatic annulation step is accomplished by irradiation to generate a ketene intermediate (similar to **44**, equation 31) from diazo ketones **48**, **50** by a photochemical Wolf rearrangement[39,40].

$$(32)$$

$$(33)$$

III. ACETYLENIC ETHERS

A. Preparation

1. Dehydrohalogenation of haloacetals and haloalkenyl ethers

The most general and well-developed procedures for the preparation of acetylenic ethers are based on dehydrohalogenation or dehalogenation of various haloacetals (for example, acetals of chloroacetaldehyde **52**) or haloalkenyl ethers **53** (equation 34)[1-3]. The first product in these eliminations usually is alkoxyacetylide **54** which can be subsequently quenched with a variety of electrophiles such as water, alkyl halides, ketones, etc to give the final product **55**.

$$R = \text{Alkyl}; X = \text{Cl, Br or H}; El = \text{electrophile}$$

Numerous examples of such reactions (equation 34) have been reported in earlier reviews[1-3]. In the well-known standard synthesis of alkoxyalkynes[41,42] reaction of acetals of chloroacetaldehyde **52** with sodium amide in liquid ammonia has been used to generate sodium alkoxyacetylide **54**. Recently, this elimination procedure has been further improved by using lithium diethylamide in THF solution (equation 35)[43]. In this work[43] *in situ* generated lithium alkoxyacetylide **56** was subsequently converted to either ethoxyacetylene **57** by quenching with saturated aqueous sodium chloride, or ethoxyethynyl carbinols **58** by subsequent reaction with ketones or aldehydes.

$$R = \text{Alkyl, Aryl}; R = \text{H, Alkyl} \qquad \text{(58) 30–90\%}$$

Dehydrohalogenation or dehalogenation of haloalkenyl ethers is probably the most universal approach to alkoxyacetylenes[1-3]. During the last 10 years this approach has been significantly improved and applied to the preparation of various acetylenic ethers. In contrast to older procedures, in which NaOH or NaNH$_2$ are used as bases, these new methods usually employ alkyllithium or potassium hydride. In 1985 Smithers[44] developed a procedure based on the reaction of α-chloro-β,β-dibromovinyl ethers **59** with butyllithium (equation 36). In contrast to previous methods, the first step of this reaction is metallation of the vinylic bromine bond and subsequent elimination of LiBr resulting in bromoalkyne **60**. The consequent halogen–metal exchange leads to lithium alkoxyacetylide **61**, which can be further functionalized *in situ* by reaction with such electrophiles as H_2O, ketones and alkyl bromides[44].

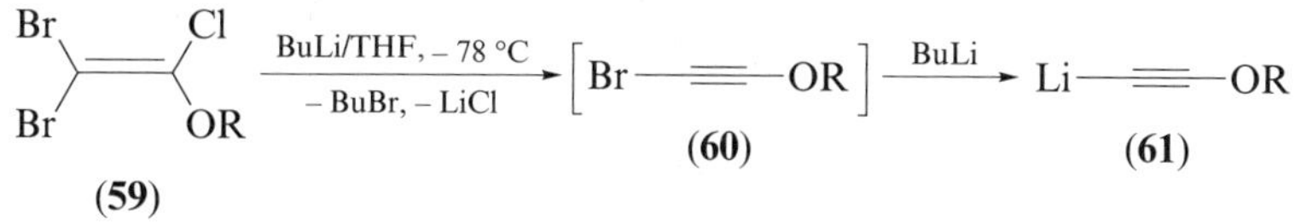

$$
\textbf{(59)} \xrightarrow[- \text{BuBr}, - \text{LiCl}]{\text{BuLi/THF}, - 78\,°C} \left[\text{Br} \!=\!\!=\! \text{OR} \right] \xrightarrow{\text{BuLi}} \text{Li} \!=\!\!=\! \text{OR} \quad (36)
$$

$$
\textbf{(60)} \qquad\qquad \textbf{(61)}
$$

$$
R = Me, Et, C_8H_{17}, \text{etc.}
$$

Recently, the dehalogenation method was further modified by using 1,2-dichlorovinyl ethers **62** as starting compounds[45,46]. Löffler and Himbert applied the reaction of 1,2-dichlorovinyl ethers **62** with butyllithium to generate lithium alkoxyacetylides **61**, which were subsequently quenched with ketones or arylated by the palladium-catalyzed cross-coupling with aryl halides to yield alkoxyarylacetylenes **63** (equation 37)[45].

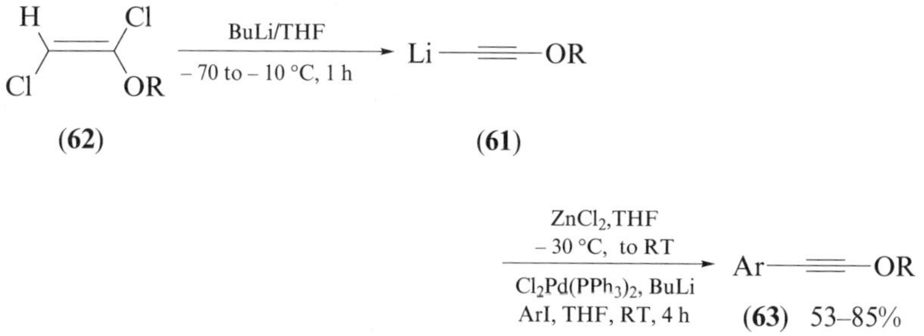

$$
\textbf{(62)} \xrightarrow[- 70 \text{ to } - 10\,°C, 1\,h]{\text{BuLi/THF}} \text{Li} \!=\!\!=\! \text{OR}
$$

$$
\textbf{(61)}
$$

$$
\xrightarrow[\substack{\text{Cl}_2\text{Pd(PPh}_3)_2, \text{BuLi} \\ \text{ArI, THF, RT, 4 h}}]{\substack{\text{ZnCl}_2,\text{THF} \\ - 30\,°C, \text{ to RT}}} \text{Ar} \!=\!\!=\! \text{OR} \quad (37)
$$

$$
\textbf{(63)} \quad 53\text{–}85\%
$$

$$
R = Et, i\text{-Pr}, Bu; \\
Ar = Ph, 4\text{-}ClC_6H_4, 4\text{-}MeC_6H_4, 3\text{-}ClC_6H_4
$$

This procedure (equation 37) is especially convenient, since 1,2-dichlorovinyl ethers **62** can be easily prepared from readily available and inexpensive trichloroethylene and sodium alkoxides[45]. A similar method has been used by Greene and coworkers[46] for the

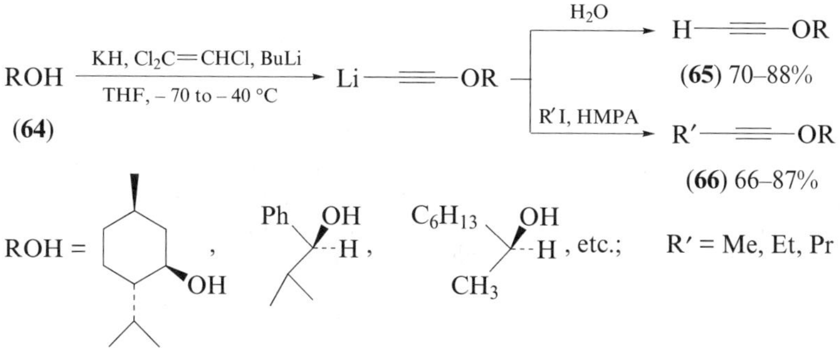

$$
\textbf{ROH} \xrightarrow[\text{THF}, - 70 \text{ to } - 40\,°C]{\text{KH, Cl}_2\text{C}=\text{CHCl, BuLi}} \text{Li} \!=\!\!=\! \text{OR}
$$

$$
\textbf{(64)}
$$

$$
\begin{cases} \xrightarrow{H_2O} H \!=\!\!=\! \text{OR} \quad \textbf{(65)}\ 70\text{–}88\% \\[2mm] \xrightarrow{R'I, HMPA} R' \!=\!\!=\! \text{OR} \quad \textbf{(66)}\ 66\text{–}87\% \end{cases} \quad (38)
$$

$$
R' = Me, Et, Pr
$$

$$
\text{ROH} =
$$

Ph $>$ OH --H , C_6H_{13} $>$ OH --H , etc.; with CH_3 groups shown preparation of chiral acetylenic ethers **65**, **66** from the corresponding alcohols **64** and trichloroethylene (equation 38).

Based on the elimination reaction, Pericàs and coauthors[47] have developed a method for the synthesis of acetylenic ethers, derived from tertiary alcohols with a bulky alkyl group such as *tert*-butyl and adamantyl. The key step in this synthesis is the dehydrobromination of 1-bromo-2-alkoxyethylene **67** with sodium amide in ammonia or lithium diisopropylamide in a hexane–THF solution (equation 39)[47].

$$
\underset{\substack{\text{Br} \qquad \text{OR} \\ (\mathbf{67})}}{\overset{\text{H} \qquad \text{H}}{\diagup\!\!=\!\!\diagdown}} \quad \xrightarrow[\text{2. H}_2\text{O}]{\text{1. NaNH}_2/\text{NH}_3 \text{ or LDA/THF}} \quad \underset{59\text{–}75\%}{\text{H}\!\!=\!\!\!=\!\!\!=\!\!\text{OR}} \tag{39}
$$

R = *t*-Bu, 1-adamantyl

A similar elimination procedure has been used for the preparation of relatively stable, sterically hindered dialkoxyacetylenes **69** (equation 40)[48,49]. In this case the precursors are 1,2-diethers **68** which can be prepared from 1,4-dioxane in several steps.

$$
\underset{\substack{\text{Br} \qquad \text{H} \\ (\mathbf{68}),\, Z+E\text{ mixture}}}{\overset{\text{RO} \qquad \text{OR}}{\diagup\!\!=\!\!\diagdown}} \quad \xrightarrow{\text{NaNH}_2/\text{NH}_3} \quad \underset{(\mathbf{69})\ 46\text{–}92\%}{\text{RO}\!\!=\!\!\!=\!\!\!=\!\!\text{OR}} \tag{40}
$$

R = *i*-Pr, *t*-Bu, $-$CH$_2$$-$(*t*-Bu)

Recently, a dehydrohalogenation approach was applied to the generation of 1,4-dioxacyclohexyne (*p*-dioxyne) **71** (equation 41)[50]. This highly unstable cyclic acetylenic diether was formed as a transient molecule in the reaction of dibromide **70** with *t*-BuLi at −78 °C. Consequent cyclotrimerization of **71** gives the final isolated product **72**.

$$\tag{41}$$

(**70**)   (**71**)   (**72**)

2. Functionalization of terminal alkoxyacetylenes

The parent alkoxyacetylenes, HC≡COR, are the most readily available representatives of acetylenic ethers. Moreover, ethoxyacetylene is even commercially available, and can generally be easily prepared by standard, well developed, procedures[1–3]. These compounds can usually be further functionalized by standard methods employed for terminal alkynes.

The most general method is the reaction of alkoxyacetylides **73** with electrophiles according to equation 42.

$$HC{\equiv}COR \xrightarrow{\text{Base}} MC{\equiv}COR \xrightarrow{\text{El}} EIC{\equiv}COR \qquad (42)$$

$$(73), M = Li, Na, etc.$$

Several examples of such reactions have already been shown in Section III.A.1 of the present review. Recently, this method has been widely used for the preparation of various element-containing alkoxyacetylenes. For example, Lukashev and Kazankova[51] synthesized a number of useful phosphorylated alkoxyacetylenes by the reaction of lithium alkoxyacetylides with phosphinehalides (equation 43).

$$X_2PCl + LiC{\equiv}COR \longrightarrow X_2PC{\equiv}COR \qquad (43)$$

$$X = i\text{-}Pr, t\text{-}Bu, Ph, C_6F_5, OAlk, N(i\text{-}Pr)_2; R = Me, Et, Bu, Ph$$

A similar approach has been employed in the syntheses of silicon[52], germanium[53], tin[53] and boron[54] substituted alkoxyacetylenes.

Another well-known transformation of terminal acetylenes is an oxidative dimerization leading to butadiynes. Recently Pericàs and coauthors[55] have found that alkoxyacetylenes also can be oxidatively dimerized in the presence of a copper(I) catalyst yielding relatively stable 1,4-dialkoxy-1,3-butadienes **73** (equation 44).

$$2H{-}C{\equiv}C{-}OR \xrightarrow[\text{CuI·2TMEDA (cat.)}]{\text{O}_2\text{, acetone}} RO{-}C{\equiv}C{-}C{\equiv}C{-}OR \qquad (44)$$

$$(73) \; 65\text{--}95\%$$

$$R = t\text{-}Bu, 1\text{-}adamantyl, cyclohexyl, etc.$$

3. Reactions of haloacetylenes with alkoxides

Nucleophilic substitution of halogen at an acetylenic carbon in principle represents another general approach to acetylenic ethers (equation 45).

$$RC{\equiv}CX \xrightarrow{\text{R'ONa}} RC{\equiv}COR' \qquad (45)$$

$$X = Cl, Br, I$$

A study of the reaction of phenylethynyl halides with methoxide anion revealed a complex addition–elimination mechanism for this substitution[56]. Such reactions require application of DMSO as a solvent and afford alkynyl ethers only in 42–46% yield[57]. Reactions of phosphorylated chloroacetylenes **74** with alkoxide anions are even less selective and afford products of nucleophilic addition to the triple bond as major products[58]. However, analogous reaction of **74** with phenoxide anion gives phenoxyacetylenes **75** in moderate yield (equation 46)[58].

$$(RO)_2P(O)C{\equiv}CCl \xrightarrow[\text{either, RT, 2 h}]{\text{PhONa}} (RO)_2P(O)C{\equiv}COPh \qquad (46)$$

$$(74) \qquad\qquad\qquad (75) \; 28\text{--}32\%$$

$$R = Me, Et$$

Use of a better leaving group in the starting acetylene in principle should make this reaction more selective and improve the yield of the product. Indeed, alkynyl iodonium salts which have a very good leaving group, iodobenzene, react with alkoxide anion under mild conditions affording alkynyl ethers as major products[59]. For example, bis-iodonium acetylene **76** reacts with sodium phenolate even at low temperatures to give diphenoxyacetylene **77** in 57% isolated yield (equation 47)[59].

$$\overset{+}{Ph\overset{}{I}}C\equiv C\overset{+}{\overset{}{I}}Ph\cdot 2TfO^- + 2PhOLi \quad \xrightarrow[-78\,^\circ C\text{ to RT, 15 min}]{CH_2Cl_2} \quad PhOC\equiv COPh \quad (47)$$

$$(74) \qquad\qquad\qquad\qquad\qquad\qquad\qquad\qquad\qquad (74)\ 57\%$$

B. Properties and Characterization

The stability of acetylenic ethers depends on the size and electronic properties of the substituents. Aliphatic acetylenic monoethers are usually stable at room temperature and can be purified by distillation under reduced pressure at temperatures below 75 °C[41,44]. Phenoxyacetylenes are even more stable and can survive heating above 150 °C[58]. The most stable representative of dialkoxyacetylenes, di-t-butoxyethyne, decomposes at room temperature in a few days, however, it can be stored in a refrigerator as a crystalline solid (mp 8.5 °C) for months[48]. In general, the thermal stability of dialkoxyacetylenes correlates well with the effective van der Waals radii for the corresponding alkoxy groups and decreases in the following order: $(t\text{-Bu})OC\equiv CO(t\text{-Bu}) >> (i\text{-Pr})OC\equiv CO(i\text{-Pr}) \approx (t\text{-Bu})CH_2OC\equiv COCH_2(t\text{-Bu}) > EtOC\equiv COEt > MeOC\equiv COMe$[49]. The least stable, dimethoxyethyne, decomposes even at low temperature; it has been detected only by NMR spectroscopy at $-40\,^\circ C$[60]. Similar dependence of the thermal stability on the size of alkoxy groups has been found for 1,4-dialkoxy-1,3-butadiynes (**73**, equation 44): the most stable one is a sterically hindered bis(1-adamantyloxy) derivative, which can be heated to 150 °C without apparent decomposition[55]. The only known cyclic acetylenic diether-p-dioxyne (**71**)–is unstable even at $-78\,^\circ C$ and can be observed only as a transient intermediate[50]. An *ab initio* MO study predicts that the electronic ground state of p-dioxyne is a singlet state which has a nonplanar acetylenic structure with a considerable amount of diradical character[61].

Two different types of chemical transformation are relevant to the thermal decomposition of alkoxyacetylenes: polymerization and fragmentation[49]. The thermal polymerization is typical for methoxy- and neopentyloxyalkynes which do not bear hydrogen atoms β to oxygen in the alkyl substituent[49]. The fragmentation of alkynyl ethers into olefins and ketenes due to a concerted process involving an intramolecular H-shift is common in the case of alkyl substituents bearing a β hydrogen (equation 48)[47].

$$\underset{\underset{|}{-C\equiv C}}{\underset{H}{\overset{\underset{\beta}{}\diagup}{\underset{}{\bigg)}}}\underset{}{\overset{\alpha}{\diagup}}\overset{O}{}\quad\xrightarrow{\Delta}\quad \underset{}{\overset{H}{\diagdown}}C=C=O\ +\ \underset{}{\overset{\diagdown}{\underset{\diagup}{}}}\underset{\beta}{C}=\underset{\alpha}{C}\underset{}{\overset{\diagup}{\diagdown}}\quad (48)$$

The ease of this fragmentation (equation 48) is roughly proportional to the number of β hydrogen atoms. Thus, the elimination in ethoxyethyne takes place only at 120 °C, but in t-butoxyethyne the process occurs with appreciable rate already at 40 °C[47]. According to semiempirical MO calculations, this fragmentation is concerted and exhibits a highly synchronous character[62]. The predicted order of reactivity $-C\equiv CO-$

(t-Bu) > —C≡CO(i-Pr) > —C≡COEt in this study[62] completely coincides with the experimental observations. The particular ease of fragmentation in t-butoxyacetylenes can be used for synthetic purposes[63–65] (see also Section III.C).

Alkoxyalkynes have characteristic spectral properties[44–49,51–55]. The triple bond stretching in the IR of alkoxyacetylenes varies from 2290 cm^{-1} in 1-ethoxyhexyne[41] to 2130–2160 cm^{-1} in the parent[44,47] and silylated[45] alkoxyethynes. The signals of both acetylenic carbons in the ^{13}C NMR are located in the normal alkyne region of 80–90 ppm[41,47,55], which is distinctly different from siloxyacetylenes (see Section II.D.1) and ynol esters (Section IV.B). The mass spectra of alkoxyacetylenes almost always have a peak for the molecular ion and easily identifiable, characteristic fragmentation[46,47,55].

The electronic structure of alkoxyacetylenes has been investigated by various physical methods. A study of the ^{13}C NMR spectra of ethoxyacetylene and silicon, germanium and tin containing ethynyl ethers showed that the alkoxy group is an effective p-donor in relation to the triple bond[66]. Some structural data about geometry and dipole moments in methoxy and ethoxy ethynes have been established by microwave spectroscopy[67,68]. The dipole moment of 1.94D of ethoxyethyne is substantially greater than of diethyl ether (1.25D), which can be attributed to an appreciable contribution of the resonance structure EtO$^+$=C=$\bar{\text{C}}$H[68].

C. Chemistry

1. Addition of electrophiles and nucleophiles

The high reactivity of acetylenic ethers toward both electrophilic and nucleophilic reagents is determined by the highly polarized character of the triple bond due to the contribution of the resonance structure RO$^+$=C=C$^-$R. The carbon atom β to oxygen is the usual site of electrophilic attack, while the α-carbon is reactive toward nucleophiles (equation 49).

$$R\bar{C}=C-OR' \xleftarrow{\ Nu^-\ } \left[RC\equiv C-OR' \longleftrightarrow R\bar{C}=C=\overset{+}{O}R' \right]$$

$$\underset{Nu}{|}$$

$$\xrightarrow{\ El^+\ } RC=\overset{+}{C}-OR' \quad (49)$$

$$\underset{El}{|}$$

Both directions of the addition to the triple bond in alkoxyacetylenes have been well documented in previous comprehensive reviews[1–3]. A typical reaction with an electrophilic reagent is acidic hydrolysis which, in the case of monoalkoxyacetylenes, may proceed with explosive violence[1]. However, sterically hindered dialkoxyacetylenes are less reactive, for example, di-t-butoxyethyne is quantitatively recovered unchanged after treatment with 2N H$_2$SO$_4$ for 45 min and can be hydrated only under more severe conditions in the presence of mercury sulfate as a catalyst (equation 50)[48a].

$$(t\text{-Bu})OC\equiv CO(t\text{-Bu}) \xrightarrow[\text{RT, 165 min}]{\text{6N H}_2\text{SO}_4,\ \text{HgSO}_4\text{(cat.)}} (t\text{-Bu})OCH_2COOH \quad (50)$$

$$\textbf{(78)}\ 59\%$$

Less sterically hindered dialkoxyacetylenes, such as diisopropylethyne, react with acids under milder conditions[48a].

The reactions of alkoxyacetylenes with such electrophiles as halogens, carbonyl compounds, derivatives of boron, mercury, sulfur, silicon and germanium usually proceed regioselectively according to equation 49[1-3].

Recently, Lukashev and Kazankova[51] reported reactions of phosphorylated alkoxyacetylenes with phosphine halides (equation 51) and halides of silicon or germanium (equation 52). Usually the final products in these reactions are ketenes (**79, 81**), however, an intermediate formation of vinyl ethers **80** has been detected in reaction 52.

$$X_2PC\equiv COR + R_2'PHal \xrightarrow[-RHal]{} \begin{matrix} X_2P \\ \diagdown \\ R_2P \diagup \end{matrix} C=C=O \quad \textbf{(79)} \quad (51)$$

$$X_2PC\equiv COR + R_3'EHal \longrightarrow \begin{matrix} X_2P \diagdown \quad \diagup Hal \\ C=C \\ R_3'E \diagup \quad \diagdown OR \end{matrix} \quad \textbf{(80)}$$

$$\xrightarrow[-RHal]{} \begin{matrix} X_2P \\ \diagdown \\ R_3'E \diagup \end{matrix} C=C=O \quad \textbf{(81)} \quad (52)$$

E = Si, Ge; X and R = Alk, Ar, C_6F_5; R = Me, Et, Bu

Nucleophilic reagents usually attack the α-carbon of alkoxyacetylenes (equation 49). A large number of reactions of acetylenic ethers with such nucleophiles as amines, water, alcohols, thiols and organolithium compounds have been reported in previous reviews[1-3]. More recent examples are reactions of phosphorylated alkoxyacetylenes with amines or alcohols (equation 53)[51] and the reduction of chiral acetylenic ethers with $LiAlH_4$ to enol ethers[46].

$$R_2P(S)C\equiv COR' + XH \longrightarrow R_2P(S)CH=C(OR')X \quad (53)$$

R, R' = i-Pr, t-Bu; X = Alk_2N, AlkO

2. Reactions with metal carbonyls

In recent reactions of dialkoxyacetylenes with a variety of metal carbonyls, such as $Co_2(CO)_8$, $Co(CO)_2(C_5H_5)$, $Ni(CO)_4$, $Fe_2(CO)_9$, have attracted special attention for two reasons: firstly, complexation with some metal carbonyls may lead to substantial stabilization of the usually unstable dialkoxyacetylenes[48,49,60] and, secondly, metal cabonyls can catalyze cyclooligomerization of dialkoxyacetylenes to give some important products of oxocarbon chemistry[69,70]. Serratosa and coworkers have found that unstable dialkoxyacetylenes can be trapped in the form of stable crystalline hexacarbonyl dicobalt complexes **82** by reaction with excess $Co_2(CO)_8$ in pentane (equation 54)[48,49,60].

On the other hand, the reaction of acetylenic diethers with a catalytic amount of $Co_2(CO)_8$ leads to the corresponding aromatic cyclic trimers in 5% yield[48,49]. The yield of the trimerization product **83** can be substantially improved to 60% by using $Ni(CO)_4$ as a catalyst (equation 55)[69-71].

Acid solvolysis of cycloadduct **83** with trifluoroacetic acid affords a quantitative yield of hexahydroxybenzene, which is an important precursor in the synthesis of rhodizonic acid

$$\text{ROC}\equiv\text{COR} \xrightarrow[\text{pentane, } -78\ ^\circ\text{C to RT}]{\text{CO}_2(\text{CO})_8\ (\text{excess})} \quad \underset{\textbf{(82)}\ 2.2–35\%}{\text{RO}-\text{C}\overset{\overset{\displaystyle \text{OR}}{|}}{\underset{}{\text{C}}}\underset{\text{Co(CO)}_3}{\text{Co(CO)}_3}} \tag{54}$$

R = Me, Et, *i*-Pr, *t*-Bu, CH$_2$—*t*-Bu

$$(t\text{-Bu})\text{OC}\equiv\text{CO}(t\text{-Bu}) \xrightarrow[\text{pentane, } -78\ ^\circ\text{C to RT}]{\text{Ni(CO)}_4} \quad \underset{\textbf{(83)}\ 60\%}{\text{C}_6(\text{O}(t\text{-Bu}))_6} \tag{55}$$

and some other products of oxocarbon chemistry[69,70]. Another representative of oxocarbons–the croconate dianion **86**–can be easily prepared in two steps from compound **84**, which is formed in the reaction of di-*t*-butoxyacetylene with Fe$_2$(CO)$_9$ (equation 56)[72].

$$(t\text{-Bu})\text{OC}\equiv\text{CO}(t\text{-Bu}) \xrightarrow[\text{benzene, RT, 6–7 h}]{\text{Fe}_2(\text{CO})_9} \quad \textbf{(84)}\ 77\%$$

$$\xrightarrow[\text{RT, 15 h}]{\text{CF}_3\text{CO}_2\text{H}} \quad \textbf{(85)}\ 80\% \xrightarrow[\text{H}_2\text{O, RT, 5 h}]{\text{O}_2,\ \text{KOH}} \quad \textbf{(86)}\ 63\% \tag{56}$$

3. Fragmentation

In principle, the thermal decomposition of *t*-butoxyacetylenes can provide an efficient route to ketenes by a fragmentation according to equation 48 (Section III.B). However, only silylketenes possess sufficient stability to be isolated as individual compounds[63]. A number of silylketenes **88** have been prepared by Pericàs and coauthors by a gentle

$$R_3SiC\equiv CO(t\text{-}Bu) \xrightarrow[-CH_2=C(CH_3)_2]{80-110\ °C} \begin{array}{c} H \\ \diagdown \\ R_3Si \end{array} C=C=O \qquad (57)$$

(87)

(88) 63–100%

$$R_3Si = Me_3Si,\ t\text{-}BuMe_2Si,\ t\text{-}BuPh_2Si$$

heating of 1-t-butoxy-2-silylethynes 87 with simultaneous distillation of the desired product (equation 57).

Less stable ketenes can be generated from the corresponding t-butoxyacetylenes *in situ* and subsequently trapped. So, the pyrolysis of the parent t-butoxyacetylene under mild conditions in the presence of amines affords the corresponding pure acetamides 89 in essentially quantitative yield (equation 58)[65]. This result can be explained by the intermediate formation of the unstable parent ketene which subsequently acetylates amines *in situ*.

$$HC\equiv CO(t\text{-}Bu) \xrightarrow[-CH_2=C(CH_3)_2]{CHCl_3,\ reflux} \left[\begin{array}{c} H \\ \diagdown \\ H \diagup \end{array} C=C=O \right] \xrightarrow{R^1R^2NH} MeCONR^1R^2 \qquad (58)$$

(89)

$$R^1 = H,\ i\text{-}Pr;\ R^2 = i\text{-}Bu,\ t\text{-}Bu,\ C_6H_{13},\ cyclo\text{-}C_6H_{11},\ Ph,\ PhCH_2,\ etc.$$

Ketene generated by this method can also be trapped by a [2+2] cycloaddition reaction with a second molecule of t-butoxyacetylene. Such a sequence of reactions (equation 59)[44] can provide an efficient approach to 1,3-cyclobutanedione (90).

$$HC\equiv CO(t\text{-}Bu) \xrightarrow[30\ °C,\ 86\ h]{CH_2Cl_2} \left[\begin{array}{c} H \\ \diagdown \\ H \diagup \\ (t\text{-}Bu)OC\equiv CH \end{array} C=C=O \right] \longrightarrow \underset{(t\text{-}Bu)O}{\overset{O}{\diamondsuit}} \xrightarrow[-10\ to\ 15\ °C]{CF_3CO_2H} \underset{O}{\overset{O}{\diamondsuit}} \qquad (59)$$

70%   (90) 88%

4. Cycloaddition

In general, alkoxyacetylenes are very active in [2+2] cycloaddition reactions. The most useful and best investigated are reactions with isocyanates[73,74] and cyclobutenone derivatives[75,76]. Usually, the original four-membered ring products resulting from such cycloadditions are not stable enough to be isolated as individual compounds. Serratosa and coworkers attempted to prepare *N*-acylazetone 91 by the reaction of di-t-butoxyacetylene with benzoyl isocyanate. However, the major isolated product in this reaction was the six-membered nitrogen heterocycle 92 arising from a rearrangement of the primary [2+2] adduct 91 (equation 60)[73].

Danheiser and coworkers have developed a convenient synthetic approach to useful, highly substituted, aromatic compounds by the reaction of alkoxyacetylenes with cyclobutenone derivatives (equation 61)[75,76]. The mechanism of this annulation is similar to the analogous reaction of siloxyacetylenes (equation 31, Section II.D.2) and the key step is a [2+2] cycloaddition of alkoxyacetylene with a vinylketene intermediate[75].

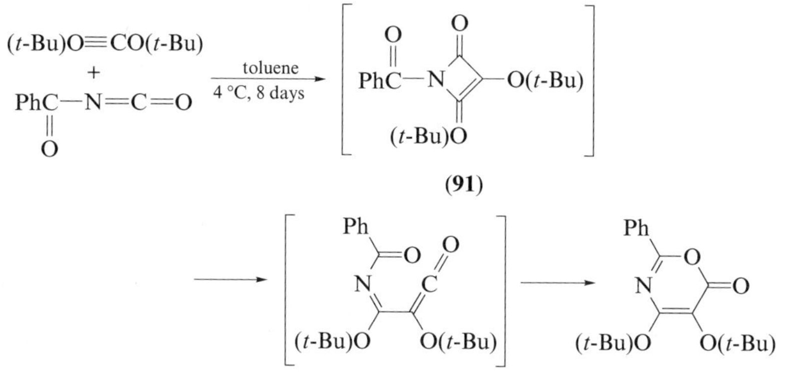

$(t\text{-Bu})O\equiv CO(t\text{-Bu})$
+
$\underset{\underset{O}{\|}}{PhC}-N=C=O$ $\xrightarrow[4\,^\circ C,\,8\,\text{days}]{\text{toluene}}$

$$\left[\begin{array}{c} \text{PhC}-N-\text{O}(t\text{-Bu}) \\ (t\text{-Bu})O \end{array} \right]$$

(91)

$$\longrightarrow \left[\begin{array}{c} Ph \\ (t\text{-Bu})O \qquad O(t\text{-Bu}) \end{array} \right] \longrightarrow \begin{array}{c} Ph \\ (t\text{-Bu})O \qquad O(t\text{-Bu}) \end{array} \qquad (60)$$

(92) 16%

$$\begin{array}{c} R^1 \quad O \\ R^2 \quad R^3 \end{array} + \begin{array}{c} R^4 \\ | \\ C \\ ||| \\ C \\ | \\ OMe \end{array} \xrightarrow[80\,^\circ C,\,4\text{--}26\,h]{\text{benzene}} \begin{array}{c} OH \\ R^1 \qquad R^4 \\ OMe \qquad R^3 \\ R^2 \end{array} \qquad (61)$$

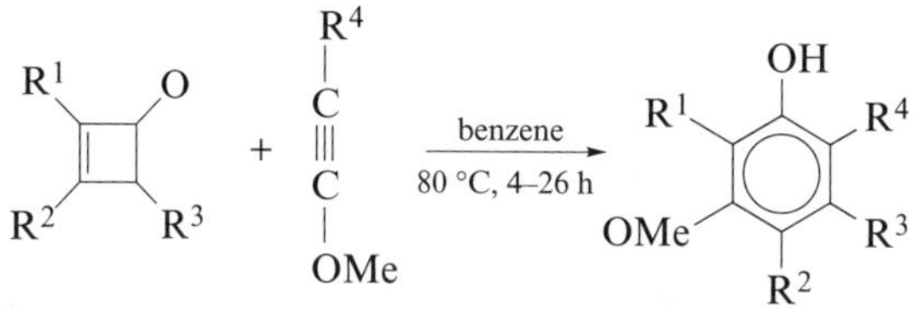

(93) 65–92%

$R^1\text{--}R^4 = \text{H, alkyl, O-Alkyl}$

This aromatic annulation strategy has been successfully applied in an efficient total synthesis of the antitumor antibiotic mycophenolic acid[76].

IV. ACETYLENIC ESTERS

A. Preparation

The most general approach to the preparation of functionalized acetylenes including acetylenic ethers involves elimination techniques, mostly dehydrohalogenation of appropriate olefin precursors (see Sections II.D.1 and III.A.1). However, these and related procedures do not work for the preparation of alkynol esters 94–96, and this undoubtedly is one of the major reasons why these compounds remained unknown until the mid-1980s[4,5].

$$RC\equiv COSOR' \qquad RC\equiv COP(O)(OR')_2 \qquad RC\equiv COC(O)R' \qquad (62)$$
$$(94) \qquad\qquad (95) \qquad\qquad\qquad (96)$$

The first successful preparation of acetylenic esters involved the use of alkynyl (phenyl)iodonium salts 97–99, which are now readily available by a variety of methods[77]. The first reported members of acetylenic esters, alkynyl sulfonates 94, have been prepared by treatment of alkynyl(phenyl)iodonium tosylates or mesylates (97) with catalytic CuOTf or AgOTf in dry acetonitrile (equation 63)[4,78].

$$RC{\equiv}\overset{+}{C}IPh \cdot \overset{-}{O}SO_2R' \xrightarrow[\text{25 °C, 2–8 h}]{\text{CH}_2\text{CN, 0.05–0.1 CuOTf}} RC{\equiv}COSOR' \qquad (63)$$

$$(97) \qquad\qquad\qquad\qquad (94) \ 54–72\%$$

$$R = Me, Bu, s\text{-}Bu, t\text{-}Bu, Ph$$
$$R' = Me, p\text{-}MeC_6H_4$$

In a similar manner, decomposition of alkynyl(phenyl)iodonium phosphates **98** affords acetylenic phosphate esters (equation 64)[79,80]. This reaction does not require any catalyst, since alkynyl(phenyl)iodonium phosphates **98** are substantially less stable than tosylates **97**, and smoothly decompose upon standing at room temperature for a few hours in CH$_2$Cl$_2$ or CHCl$_3$ yielding alkynyl phosphates **95** and iodobenzene as the by-product.

$$RC{\equiv}\overset{+}{C}IPh \cdot \overset{-}{O}P(O)(OR')_2 \xrightarrow[\text{25 °C}]{\text{CHCl}_3 \text{ or CH}_2\text{Cl}_2} RC{\equiv}COP(O)(OR') \qquad (64)$$

$$(98) \qquad\qquad\qquad\qquad (95) \ 31–58\%$$

$$R = Me, Bu, s\text{-}Bu, t\text{-}Bu, Pr, C_6H_{13}$$
$$R' = Me, CH_2Ph$$

Alkynyl(phenyl)iodonium carboxylates **99** are even less stable than phosphates **98** and decompose immediately upon generation from iodonium triflates **100** or benzoate **101** with the formation of the desired alkynyl carboxylates **96** (equation 65)[81].

$$PhI(O_2CR')_2 + RC{\equiv}CLi \xrightarrow[\text{78 °C to RT}]{\text{THF}}$$

$$(101)$$

$$\xrightarrow[\text{CH}_2\text{Cl}_2, \text{ H}_2\text{O}]{\text{NaO}_2CR'} RC{\equiv}\overset{+}{C}IPh \cdot \overset{-}{O}C(O)R'$$

$$RC{\equiv}\overset{+}{C}IPh \cdot \overset{-}{O}Tf \qquad\qquad (99)$$

$$(100) \qquad\qquad \longrightarrow RC{\equiv}COC(O)R' \quad (65)$$

$$(96) \ 2–57\%$$

$$R = Me, Bu, s\text{-}Bu, t\text{-}Bu, t\text{-}Pr, MeOCMe_2$$
$$R' = Me, CMe_3, Ph, p\text{-}MeOC_6H_4, p\text{-}NO_2C_6H_4,$$
$$3,5\text{-}(MeO)_2C_6H_3, CH_2CH_2Ph, CHPh_2$$

$$(66)$$

Formalistically, reactions 63 and 64 are nucleophilic acetylenic substitutions (S_N-A) with the corresponding anions (sulfonate, phosphate or carboxylate) acting as nucleophiles and alkynyliodonium species 97–99 as the electrophilic substrates. However, the actual details of the mechanism are considerably more complex (equation 66).

The first step in this scheme is a Michael addition of the nucleophile to the β-carbon of the alkynyliodonium salt to give the ylide 102. Loss of iodobenzene from 102 gives alkylidenecarbene 103, which rearranges to alkyne 104 in the absence of external traps. This mechanism is experimentally supported by the isolation of cyclic by-products 108 besides the major products, the alkynyl esters 107 in the reaction of alkynyliodonium salt 105 with nucleophiles (equation 67)[82]. These cyclic enol ethers are the result of the insertion of the intermediate carbene 106 into the tertiary-ε-carbon–hydrogen bond.

$$(CH_3)_2CH(CH_2)_2C\overset{+}{\equiv}\overset{-}{C}IPh \cdot \bar{O}Tf \; + \; Nu^- \; \xrightarrow[-PhI]{} \; \left[\begin{array}{c} Nu \\ C{=}C{:} \quad H \\ CH_2 \qquad C(CH_3)_2 \\ CH_2 \end{array} \right]$$

(105)   (106)

$$(CH_3)_2CH(CH_2)_2C\equiv CNu$$

Nu = $PhCO_2$, $(EtO)_2PO_2$, etc.   (107)   (108)

(67)

Another experimental result, the isolation of vinyliodonium salt 110 in the reaction of alkynyliodonium salt 109 with trimethylsilyl azide in wet CH_2Cl_2 (equation 68)[83], supports the intermediate formation of iodonium ylide 102 (Nu = N_3) in the mechanism shown in equation 66.

$$RC\equiv \overset{-}{C}IPh \cdot \overset{-}{B}F_4 \; + \; Me_3SiN_3$$

(109)

$$\xrightarrow{CH_2Cl_2} \left[\begin{array}{c} N_3 \\ C{=}\overset{-}{C}{-}\overset{+}{I}Ph \\ R \end{array} \right] \xrightarrow{H_2O} \begin{array}{c} N_3 \qquad H \\ C{=}C \\ R \qquad \overset{+}{I}Ph \cdot \overset{-}{B}F_4 \end{array}$$

(68)

(110)

To date, reaction of alkynyliodonium salts with the corresponding anionic nucleophiles is the most general and selective approach to acetylenic esters. However, two more conventional but less general syntheses have also been reported in the literature[26,84]. Hoppe and Gonschorrek[84] have prepared enynyl carbamates 112, which also are members of the family of alkynyl esters, by treatment of allene 111 with butyllithium (equation 69).

More recently we have reported the preparation of alkynyl esters by the reaction of lithium alkynolates with the corresponding acid chlorides (equations 14 and 15, Section II.B.2)[26]. This method works well for the synthesis of alkynyl phosphates, however, it has only limited applicability in the preparation of alkynyl carboxylates[26].

R\C=C=C/H ... (structure)

$$\text{(111)} \xrightarrow[\text{THF}]{\text{BuLi}} \text{(112)} \quad (69)$$

B. Properties, Characterization and Structure

Pure alkynyl esters **94–96** are colorless or pale yellow liquids or solids with a low melting point. Most of them are relatively stable and can be stored in a refrigerator for extended periods. Of the three classes of alkynyl esters the sulfonates **94** are the most thermally stable; some of them may even be distilled in vacuum at temperatures of 100–130 °C[78]. The carboxylate esters **96** are the least stable, in fact, so far only alkynyl benzoates (**96**, R′ = Ar) and pivalates (**96**, R′ = t-Bu) have been isolated in pure form. Alkynyl esters of such acids as formic and acetic are too unstable to isolate[81]. All three classes of alkynyl esters and, especially, carboxylates are water sensitive and rapidly decompose by hydrolysis in the presence of moisture.

Alkynyl esters have highly characteristic spectral properties[78–80]. The triple bond stretching in the IR of all three classes **94–96** is in the region of 2260–2290 cm^{-1}. In addition, the carboxylate esters **96** have an intense carbonyl stretch around 1760–1770 cm^{-1}, whereas the phosphates **95** show a very strong P=O signal at 1280–1300 cm^{-1} and the sulfonates **94** have their characteristic SO$_2$ vibrations around 1390 and 1185 cm^{-1}. The mass spectra of alkynyl esters almost always have a peak for the molecular ions and characteristic further fragmentations. The chemical shifts of acetylenic carbons in the ^{13}C NMR are in the same region as for siloxyacetylenes which is consistent with the electron-donating resonance effect of the ester moieties (see Section II.D.1). Hence, these alkynyl esters **94–96** are in fact electron-rich acetylenes despite the electron-withdrawing inductive effect of all three ester moieties.

X-ray structural data have been reported for three alkynyl esters: ethynyl benzoate[85] **113**, propynyl p-nitrobenzoate[86] **114** and propynyl tosylate[85] **115**.

$$\text{HC}{\equiv}\text{COC(O)PPH} \quad \text{CH}_3\text{C}{\equiv}\text{COC(O)C}_6\text{H}_4\text{NO}_2\text{-}p \quad \text{CH}_3\text{C}{\equiv}\text{COSO}_2\text{C}_6\text{H}_4\text{CH}_3\text{-}p \quad (70)$$

$$\text{(113)} \qquad\qquad\qquad \text{(114)} \qquad\qquad\qquad \text{(115)}$$

The most important structural features of acetylenic esters **113–115** are the following[85,86]: (i) the acetylenic fragment, as expected, is essentially linear with C≡C—O bond angles in the range of 175–178°; (ii) the general structural features of the ester moiety of both alkynyl carboxylates and sulfonates closely resemble those of their saturated and enol analogs; (iii) the C$_{sp}$—O bond length is about 1.33 Å, which is significantly shorter than C—O bonds in the corresponding enol and saturated esters. Theoretical *ab initio* calculations on carboxylate and sulfonate esters are in agreement with the observed structural features and, particularly, confirm the shortening of the C—O bond for the sp-hybridized carbon[85,86].

C. Chemistry

1. Addition of electrophiles and nucleophiles

The chemical reactivity of acetylenic esters is similar to the behaviour of the other ynol derivatives (see Sections II.B.2 and III.C.1) and is determined by the polar character of the ynolic triple bond. All three types of esters **94–96**, as expected, hydrolyze rapidly under acidic as well as basic conditions[87]. However, in contrast to ynol ethers (Section III.C.1) esters also react under *neutral* conditions.

Hydrolysis of alkynyl esters **116–118** under acidic conditions results in the formation of the carboxylic acids derived from the acetylenic moiety of the esters (equations 71–73).

$$CH_3C \equiv COC(O)Ph \xrightarrow[H_2SO_4]{H_2O} CH_3CH_2CO_2H + PhCO_2H \tag{71}$$
$$(116)$$

$$BuC \equiv COP(O)Et_2 \xrightarrow[H_2SO_4]{H_2O} C_5H_{11}CO_2H + (EtO)_2PO_2H \tag{72}$$
$$(117)$$

$$BuC \equiv COSO_2Ar \xrightarrow[H_2SO_4]{H_2O} C_5H_{11}CO_2H + ArSO_3H \tag{73}$$
$$(118)$$

These reactions proceed via an Ad_E2 mechanism involving a rate-limiting proton transfer to the β-carbon and formation of vinyl cation **119**, which reacts rapidly with H_2O to give consequently enol **120**, mixed anhydride **121** and the final products (equation 74)[87].

$$RC \equiv COR' \xrightarrow[slow]{H^+} RCH = \overset{+}{C}OR' \xrightarrow[fast]{H_2O} RCH = \overset{\overset{+}{O}H_2}{\underset{|}{C}}OR' \tag{74}$$
$$(120)$$

$$RCO_2H + R'OH \longleftarrow RCH_2\overset{O}{\overset{\|}{C}}OR'$$
$$(121)$$

$$R = Alk, R' = C(O)Ph, ArSO_2, (EtO)_2P(O)$$

When reactions of alkynyl tosylates with acids are carried out under anhydrous conditions, 1,1-vinylic diesters **122** can be isolated (equation 75)[88]. The regio- and stereoselectivity of this reaction (equation 75) is consistent with an Ad_E mechanism, which was proposed for the first step of the acidic hydrolysis of alkynyl esters (equation 74). Moreover, the consequent hydrolysis of **122** yields the same products as the hydrolysis of alkynyl tosylates[89].

$$RC{\equiv}COTs \xrightarrow[\text{CH}_2\text{Cl}_2 - 20 \text{ to } 25\,°\text{C}]{\text{HX}} \begin{array}{c} R \\ \\ H \end{array}C{=}C\begin{array}{c} OTs \\ \\ X \end{array} \qquad (75)$$

$$\textbf{(122)}$$

$$R = t\text{-Bu, } s\text{-Bu}; \; X = Cl, OTf, OTs, OC(O)CF_3$$

The mechanism of the neutral reaction is more unusual, and was investigated in detail for alkynyl benzoates[90]. Neutral hydrolysis of propynyl benzoate (116) in aqueous acetonitrile leads to keto-ester 124 in 46% yield along with the expected propanoic and benzoic acids. The mechanism of this reaction involves an unusual cyclization pathway including the intermediate formation of dioxalene 123 (equation 76). This mechanism was unveiled by labeling studies in $H_2\,^{18}O$ as well as careful NMR studies. Further evidence for this mechanism comes from the isolation and characterization dioxalenes 125 from the reaction of alkynyl benzoates with anhydrous methanol (equation 77)[90].

$$\begin{array}{c} CH_3C{\equiv}C{-}O \end{array} \cdots \longrightarrow \left[\begin{array}{c} Ph \quad Ph \\ O \quad O \\ Me \end{array} \right] \longrightarrow \begin{array}{c} O \qquad O \\ \| \qquad \| \\ CH_3CCH_2OCOPh \end{array} \quad (76)$$

$$\textbf{(116)} \qquad\qquad \textbf{(123)} \qquad\qquad \textbf{(124)}$$

$$RC{\equiv}COC(O)Ph \xrightarrow[60\,°\text{C}]{\text{MeOH}} \begin{array}{c} He \quad OMe \\ O \quad O \\ R \end{array} \qquad (77)$$

$$\textbf{(96)} \qquad\qquad \textbf{(125)}$$

$$R = H, Me, t\text{-Bu}$$

Neutral hydrolysis of alkynyl tosylates and phosphates proceeds similarly to carboxylates to give the corresponding products, i.e. $RC(O)CH_2OP(O)(OEt)_2$ and $RC(O)CH_2OTs$, analogously to equation 76.

Reactions of alkynyl esters with nucleophilic reagents usually involve the attack of the nucleophile on the electrophilic acyl moiety (i.e. C=O, P=O, SO$_2$) but not the triple bond. A typical example of such a process is the formation of lithium ynolates in the reaction of alkynyl tosylates with methyl lithium[25] (see equation 10 in Section II.B.1). Similarly, the base-catalyzed hydrolysis of alkynyl esters most likely proceeds via $^-$OH attack on the acyl moiety and the subsequent standard mechanistic steps.

2. Cycloaddition

Analogously to siloxy- and alkoxyacetylenes (see Sections II.D.2 and III.C.4), alkynyl esters are good partners in [2+2] cycloaddition reactions. A typical example of such a reaction is the cycloaddition of benzoate 116 and tosylate 128 with azete 126 (equations 78 and 79). Both reactions gave the novel Dewar pyridines **127**, **129** as the major products (equations 78 and 79)[91].

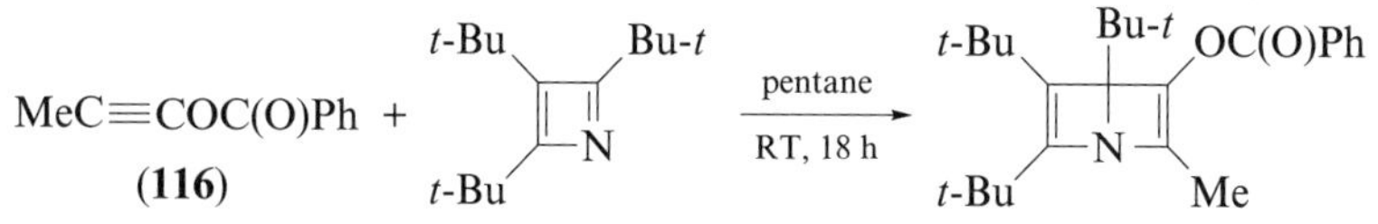

$$\text{MeC}\equiv\text{COC(O)Ph} \quad + \qquad\qquad\qquad \xrightarrow[\text{RT, 18 h}]{\text{pentane}} \qquad\qquad\qquad\qquad (78)$$

(**116**)   (**126**)   (**127**) 83%

$$t\text{-BuC}\equiv\text{COTs} \quad + \qquad\qquad\qquad \xrightarrow[\text{RT, 22 h}]{\text{pentane}} \qquad\qquad\qquad\qquad (79)$$

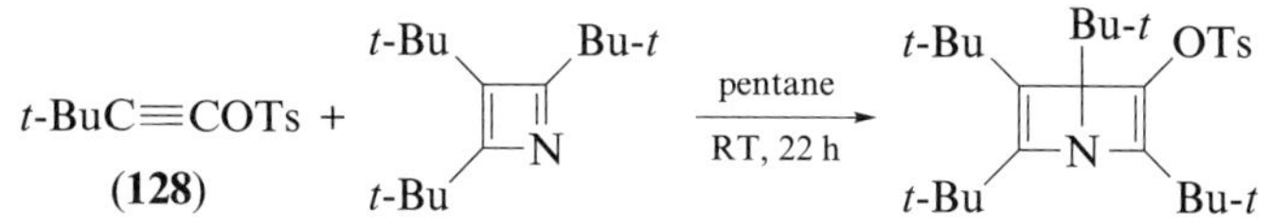

(**128**)   (**126**)   (**129**) 79%

Analogous reaction of the parent ethynyl benzoate **113** with **126** yields two regioisomers **130**, **131** (equation 80)[91].

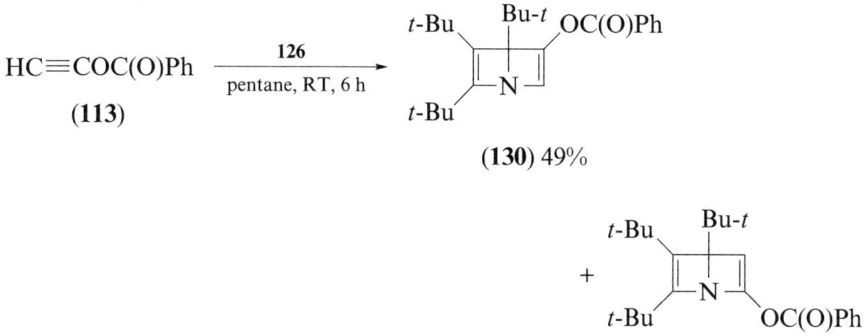

$$\text{HC}\equiv\text{COC(O)Ph} \quad \xrightarrow[\text{pentane, RT, 6 h}]{\textbf{126}}$$

(**113**)

(**130**) 49%

$$+ \qquad\qquad\qquad\qquad\qquad\qquad\qquad\qquad (80)$$

(**131**) 34%

As expected, similar products of [2+2] cycloaddition are formed in the reaction of azete **126** with siloxy and ethoxyacetylene[91].

D. Biochemistry

Acetylenic esters show promising biological activity as enzyme inhibitors[92,93]. They belong to a recently discovered class of enzyme-activated inhibitors, also referred to as 'suicide substrates', which are structural analogs of the normal physiological substrate of the target enzyme, with a built-in latently reactive functional group activated during normal catalytic action by the enzyme. Specifically, propynyl benzoates **132** are efficient inhibitors of various serine proteases[92]. For example, benzoates **132** inhibited α-chymotrypsin, the prototypical serine protease, 20–50-fold more effectively than conventional powerful inhibitors of chymotrypsin such as diphenylcarbamoyl chloride, phenylmethanesulfonyl fluoride (PMSF) and diisopropyl fluorophosphate (DFP). Hexynyl diethyl phosphate (**117**) effectively inhibits phosphotriesterase from *Pseudomonas diminuta* with < 1% residual activity in less than 1 min and the partitioning ratio, the number of inhibitor molecules hydrolyzed per enzyme inactivated, of approximately 1200[93].

$$
\text{MeC} \equiv \text{C} - \text{O} - \overset{\text{O}}{\underset{\|}{\text{C}}} - \underset{}{\langle\quad\rangle} - \text{R} \qquad \text{BuC} \equiv \text{C} - \text{O} - \overset{\text{O}}{\underset{\|}{\text{P}}}(\text{OEt})_2 \qquad (81)
$$

$$
\textbf{(132)} \quad \text{R = H, OMe} \qquad\qquad\qquad \textbf{(117)}
$$

Some preliminary data also indicate substantial antitumor activity of all three types of alkynyl esters (**94–96**) towards murine leukemia cells (L1210) and human T-lymphoblast cells[94].

V. ACKNOWLEDGMENT

Financial support of our own work, described herein, by the National Institute of Health Research Grant 2ROCA16903 to PJS at Utah is gratefully acknowledged. We also thank our colleagues for stimulating discussions and productive collaborations as herein described and cited.

VI. REFERENCES

1. L. Brandsma, H. J. T. Bos and J. F. Arens, in *Chemistry of Acetylenes* (Ed. H. G. Viehe), Chap.11, Marcel Dekker, New York, 1969, pp. 751–860; J. F. Arens, in *Advances in Organic Chemistry*, Vol. 2 (Eds. R. A. Rafael, E. C. Taylor and H. Wynberg), Interscience, New York, 1960, pp. 117–212.
2. H. Meerwein, in *Methoden der Organischen Chemie (Houben-Weil)*, 4th ed., Vol. 6/3, Chap. 1, Georg Thieme Verlag, Stuttgart, 1965, pp. 116–118; D. A. Ben-Efraim, in *The Chemistry of the Carbon–Carbon Triple Bond*, Vol. 2, (Ed. S. Patai), Chap. 18, Wiley-Interscience, Chichester, 1978.
3. S. I. Radchenko and A. A. Petrov, *Russ. Chem. Rev.*, **58**, 948 (1989).
4. P. J. Stang and B. W. Surber, *J. Am. Chem. Soc.*, **107**, 1452 (1985).
5. P. J. Stang, *Acc. Chem. Res.*, **24**, 304 (1991).
6. A. J. Kresge, *Acc. Chem. Res.*, **23**, 43 (1990).
7. W. J. Bouma, R. H. Nobes, L. Radom and C. E. Woodward, *J. Org. Chem.*, **47**, 1869 (1982).
8. B. J. Smith, L. Radom and A. J. Kresge, *J. Am. Chem. Soc.*, **111**, 8297 (1989).
9. B. von Baar, T. Weiske, J. K. Terlouw and H. Schwarz, *Angew. Chem., Int. Ed. Engl.*, **25**, 282 (1986).
10. D. J. DeFrees and A. D. McLean, *J. Phys. Chem.*, **86**, 2835 (1982).
11. R. Hochstrasser and J. Wirz, *Angew. Chem., Int. Ed. Engl.*, **28**, 181 (1989).
12. Y. Chiang, A. J. Kresge, R. Hochstrasser and J. Wirz, *J. Am. Chem. Soc.*, **111**, 2355 (1989).
13. A. D. Allen, A. J. Kresge, N. P. Schepp and T. T. Tidwell, *Can. J. Chem.*, **65**, 1719 (1987).
14. N. Banait, M. Hojatti, P. Findlay and A. J. Kresge, *Can. J. Chem.*, **65**, 441 (1987).
15. E. Weiss and W. Büchner, *Helv. Chim. Acta*, **46**, 1121 (1963) and references cited therein.
16. W. Büchner, *Helv. Chim. Acta*, **49**, 907 (1966).
17. W. Büchner, *Helv. Chim. Acta*, **48**, 1229 (1965).
18. A. Ayed, L. Manceron and B. Silvi, *J. Phys. Chem.*, **92**, 37 (1988).
19. C. J. Kowalski and K. W. Fields, *J. Am. Chem. Soc.*, **104**, 321 (1982).
20. (a) C. J. Kowalski and R. E. Reddy, *J. Org. Chem.*, **57**, 7194 (1992); C. J. Kowalski, M. S. Haque and K. W. Fields, *J. Am. Chem. Soc.*, **107**, 1429 (1985).
    (b) C. J. Kowalski and G. S. Lal, *Tetrahedron Lett.*, **28**, 2463 (1987).
21. C. J. Kowalski, G. S. Lal and M. S. Haque, *J. Am. Chem. Soc.*, **108**, 7127 (1986).
22. R. P. Woodbury, N. R. Long and M. W. Rathke, *J. Org. Chem.*, **43**, 376 (1978).

23. I. Hoppe and U. Schöllkopf, *Justus Liebigs Ann. Chem.*, 219 (1979).
24. B. L. Groh, G. R. Magrum and T. J. Barton, *J. Am. Chem. Soc.*, **109**, 7568 (1987).
25. P. J. Stang and K. A. Roberts, *J. Am. Chem. Soc.*, **108**, 7125 (1986).
26. V. V. Zhdankin and P. J. Stang, *Tetrahedron Lett.*, **34**, 1461 (1993).
27. F. R. Kreissl, in *Organometallics in Organic Synthesis* (Eds. A. de Meijere and H. tom Dieck), Springer-Verlag, Berlin, 1987, p. 105.
28. H. Fischer, P. Hoffmann, F. R. Kreissl, R. R. Schrock, U. Schubert and K. Weiss, *Carbyne Complexes*, VCH, Weinheim, 1988.
29. A. Mayr and H. Hoffmeister, *Adv. Organomet. Chem.*, **32**, 227 (1991).
30. H. P. Kim and R. J. Angelici, *Adv. Organomet. Chem.*, **27**, 51 (1987).
31. F. R. Kreissl, A. Frank, U. Schuberat, T. L. Lindner and G. Huttner, *Angew. Chem., Int. Ed. Engl.*, **15**, 632 (1976).
32. J. B. Sheridan, D. B. Pourreau, G. L. Geoffroy and A. L. Rheingold, *Organometallics*, **7**, 289 (1988).
33. K. A. Belsky, M. F. Asaro, S. Y. Chen and A. Mayr, *Organometallics*, **11**, 1926 (1992).
34. G. Mass and R. Brückmann, *J. Org. Chem.*, **50**, 2801 (1985).
35. R. L. Danheiser, A. Nishida, S. Savariar and M. Trova, *Tetrahedron Lett.*, **29**, 4917 (1988).
36. C. J. Kowalski and S. Sakdart, *J. Org. Chem.*, **55**, 1977 (1990).
37. R. L. Danheiser, R. G. Brisbois, J. J. Kowalczyk and R. F. Miller, *J. Am. Chem. Soc.*, **112**, 3093 (1990).
38. C. J. Kowalski and G. S. Lal, *J. Am. Chem. Soc.*, **110**, 3693 (1988).
39. R. L. Danheiser, D. S. Casebier and J. L. Loebach, *Tetrahedron Lett.*, **33**, 1149 (1992).
40. R. L. Danheiser and D. D. Cha, *Tetrahedron Lett.*, **31**, 1527 (1990).
41. M. S. Newman, J. R. Geib and W. M. Stalick, *Org. Prep.*, **4**, 89 (1972); M. S. Newman and W. M. Stalick, *Org. Synth.*, **57**, 65 (1977); W. M. Stalick, R. N. Hazlett and R. E. Morris, *Synthesis*, 287 (1988).
42. L. Brandsma, in *Preparative Acetylenic Chemistry*, 2nd ed., Elsevier, Amsterdam, 1988, p. 174.
43. S. Raucher and B. L. Bray, *J. Org. Chem.*, **52**, 2332 (1987).
44. R. H. Smithers, *Synthesis*, 556 (1985).
45. A. Löffler and G. Himbert, *Synthesis*, 495 (1992).
46. A. Moyano, F. Charbonnier and A. E. Greene, *J. Org. Chem.*, **52**, 2919 (1987).
47. M. A. Pericàs, F. Serratosa and E. Valenti, *Tetrahedron*, **43**, 2311 (1987).
48. (a) A. Bou, M. A. Pericàs and F. Serratosa, *Tetrahedron*, **37**, 1441 (1981).
    (b) A. Bou, M. A. Pericàs, A. Riera and F. Serratosa, *Org. Synth.*, **65**, 68 (1987).
49. M. A. Pericàs, A. Riera and F. Serratosa, *Tetrahedron*, **38**, 1505 (1982).
50. M. A. Pericàs, A. Riera, O. Rossell, F. Serratosa and M. Seco, *J. Chem. Soc., Chem. Commun.*, 942 (1988).
51. N. V. Lukashev and M. A. Kazankova, *Phosphorus, Sulfur, and Silicon*, **49/50**, 179 (1990).
52. M. A. Kazankova and I. F. Lutsenko, *Vestn. Mosk. Univ., Ser. 2: Khim.*, **24**, 315 (1983); G. Himbert and L. Henn, *Justus Liebigs Ann. Chem.*, 1358 (1984).
53. S. V. Ponomarev and I. F. Lutsenko, *Vestn. Mosk. Univ., Ser. 2: Khim.*, **28**, 3 (1987).
54. B. M. Mikhailov, M. E. Gurskii and M. G. Gverdtsiteli, *Izv. Akad. Nauk SSSR, Ser. Khim.*, 1456 (1977).
54. E. Valenti, M. A. Pericàs and F. Serratosa, *J. Am. Chem. Soc.*, **112**, 7405 (1990).
56. R. Tanaka, M. Rodgers, R. Simonaitis and S. I. Miller, *Tetrahedron*, **27**, 2651 (1971).
57. R. Tanaka and S. I. Miller, *Tetrahedron Lett.*, 1753 (1971).
58. V. A. Garibina, A. A. Leonov, A. V. Dogadina, B. I. Ionin and A. A. Petrov, *J. Gen. Chem. USSR*, **55**, 1771 (1985).
59. P. J. Stang and V. V. Zhdankin, *J. Am. Chem. Soc.*, **113**, 4571 (1991).
60. A. Messeguer, F. Serratosa, and J. Rivera, *Tetrahedron Lett.*, 2895 (1973).
61. S. Olivella, A. Pericàs, A. Riera, F. Serratosa and A. Solè, *J. Am. Chem. Soc.*, **109**, 5600 (1987).
62. A. Moyano, M. A. Pericàs, F. Serratosa and E. Valenti, *J. Org. Chem.*, **52**, 5532 (1987).
63. E. Valenti, M. A. Pericàs and F. Serratosa, *J. Org. Chem.*, **55**, 395 (1990).
64. M. A. Pericàs, F. Serratosa and E. Valenti, *Synthesis*, 1118 (1985).
65. E. Valenti, M. A. Pericàs, F. Serratosa and D. Maña *J. Chem. Res (S)*, 118 (1990).
66. G. A. Kalabin, D. F. Kushnarev, S. M. Shostakovskii and T. K. Voropaev, *Izv. Akad. Nauk SSSR, Ser. Khim.*, 2459 (1975); Yu. K. Grishin, S. V. Ponomarev and S. A. Lebedev, *Zh. Org.*

Khim., **10**, 404 (1974); B. Wrackmeyer, *J. Organomet. Chem.*, **166**, 353 (1979); A. Sebald and B. Wrackmeyer, *Spectrochim. Acta, Part A*, **37**, 365 (1981).

67. D. Engelsen, H. A. Dijkerman and J. Kerssen, *Recueil*, **84**, 1357 (1965); D. Engelsen, *J. Mol. Spectrosc.*, **30**, 466 (1969); B. P. Van Eijck, A. Dubrulle, J. Demaison and J. L. Ripoll, *J. Mol. Spectrosc.*, **112**, 95 (1985).
68. A. Bjørseth, *J. Mol. Struct.*, **20**, 61 (1974).
69. F. Serratosa, *Acc. Chem. Res.*, **16**, 170 (1983).
70. F. Serratosa, *J. Organomet. Chem.*, **413**, 445 (1991).
71. F. Camps, J. Coll, J. M. Moretò and J. Torras, *J. Org. Chem.*, **54**, 1969 (1989).
72. D. Fornals, M. A. Pericàs, F. Serratosa, J. Vinaixa, M. Font-Altaba and X. Solans, *J. Chem. Soc., Perkin Trans. 1*, 2749 (1987).
73. M. A. Pericàs, F. Serratosa, E. Valenti, M. Font-Altaba and X. Solans, *J. Chem. Soc., Perkin Trans. 2*, 961 (1986).
74. M. A. Pericàs, F. Serratosa and E. Valenti, *J. Chem. Soc., Perkin Trans. 2*, 151 (1987).
75. R. L. Dahneiser and S. K. Gee, *J. Org. Chem.*, **49**, 1672 (1984).
76. R. L. Daheiser, S. K. Gee and J. J. Perez, *J. Am. Chem. Soc.*, **108**, 806 (1986).
77. P. J. Stang, *Angew. Chem., Int. Ed. Engl.*, **31**, 274 (1992).
78. P. J. Stang, B. W. Surber, Z. C. Chen, K. A. Roberts and A. G. Anderson, *J. Am. Chem., Soc.*, **109**, 228 (1987).
79. P. J. Stang, M. Boehshar and J. Lin, *J. Am. Chem. Soc.*, **108**, 7832 (1986).
80. P. J. Stang, T. Kitamura, M. Boehshar and H. Wingert, *J. Am. Chem. Soc.*, **111**, 2225 (1989).
81. P. J. Stang, M. Boehshar, H. Wingert and T. Kitamura, *J. Am. Chem. Soc.*, **110**, 3272 (1988).
82. P. J. Stang and D. R. Fischer, unpublished observations.
83. M. Ochiai, M. Kunishima, K. Fuji and Y. Nagao, *J. Org. Chem.*, **53**, 6144 (1988).
84. D. Hoppe and C. Gonschorrek, *Tetrahedron Lett.*, **28**, 785 (1987).
85. P. J. Stang, C. M. Crittell, A. M. Arif, M. Karni and Y. Apeloig, *J. Am. Chem. Soc.*, **113**, 7461 (1991).
86. P. J. Stang, T. Kitamura, A. M. Arif, M. Karni and Y. Apeloig, *J. Am. Chem. Soc.*, **112**, 374 (1990).
87. A. D. Allen, K. A. Roberts, T. Kitamura, P. J. Stang and T. T. Tidwell, *J. Am. Chem. Soc.*, **110**, 622 (1988).
88. P. J. Stang and K. A. Roberts, *J. Org. Chem.*, **52**, 5213 (1987).
89. R. A. Cox, M. McAllister, K. A. Roberts, P. J. Stang and T. T. Tidwell, *J. Org. Chem.*, **54**, 4899 (1989).
90. A. D. Allen, T. Kitamura, R. A. McClelland, P. J. Stang and T. T. Tidwell, *J. Am. Chem. Soc.*, **112**, 8873 (1990).
91. G. Maas, M. Regitz, R. Rahm, H. Wingert, J. Schneider, P. J. Stang and C. M. Crittell, *J. Chem. Soc., Chem. Commun.*, 1456 (1990).
92. D. Segal, Y. Shalitin H. Wingert, T. Kitamura and P. J. Stang, *FEBS Lett.*, **247**, 217 (1989).
93. J. N. Blankenship, H. Abu-Soud, F. M. Raushel, D. R. Fischer and P. J. Stang, *J. Am. Chem. Soc.*, **113**, 8560 (1991).
94. E. De Clercq, D. R. Fischer and P. J. Stang, to appear.

CHAPTER **20**

Alkynyl(phenyl)iodonium and related species

PETER J. STANG

Department of Chemistry, The University of Utah, Salt Lake City, Utah 84112, USA

I. INTRODUCTION

The triple bond of alkynes constitutes one of the oldest, simplest and most valuable functional groups in organic chemistry. Besides the common hydrocarbon acetylenes, a wide variety of functionalized alkynes (**1**) are known that play an important role in numerous organic transformations. The latest members of the family of functionalized alkynes are the alkynyl iodonium salts (**2**). These compounds are a subgroup of the dicoordinated positively-charged 8-I-2 iodine species L_2I^+ (**3**)[1]. The first example of this type of molecules (**4**) was initially reported exactly one hundred years ago in 1894 by the German chemists Hartman and Meyer[2]. To date, the best known and most widely used members of this class of polycoordinated iodine compounds are the diaryliodonium salts (**5**)[1].

Supplement C2: The chemistry of triple-bonded functional groups
Edited by S. Patai   © 1994 John Wiley & Sons Ltd

$$RC\equiv CY$$

(1) Y = CN, NO$_2$, NR$_2$, SiR$_3$, PR$_2$, OR, ROC(O), halogens, etc.

$$RC\equiv CI^+PhX^- \quad L_2I^+ \quad IC_6H_4I^+C_6H_5 \quad HSO_4^- \quad Ar_2I^+ \ X^-$$

$$(2) \qquad\qquad (3) \qquad (4) \qquad\qquad\quad (5)$$

    The first alkynyl(phenyl)iodonium salt, the rather unstable chloride (8), was reported in low yield, by the interaction of β-phenylethynyl lithium (6) with (dichloroiodo)benzene (7) (equation 1)[3]. The Russian chemist Merkushev and his coworkers[4] next reported the isolation of 9 in 64% yield (equation 2). However, this interesting but hygroscopic product was only sparsely characterized. It was not until the 1980s, and in particular the last half-dozen years, that these novel, uniquely functionalized acetylenes became readily available. Since then they have gained considerable importance and widespread use as synthons for the electrophilic acetylene species 'RC≡C$^+$' due to the versatility of their reactions with a wide variety of nucleophiles. This chapter will cover the preparation, characterization and chemistry of alkynyl(phenyl)iodonium salts (2) and related species with emphasis on our own recent contributions to this rapidly evolving area of acetylene chemistry. Two previous reviews[1,5] describe the early developments and chemistry in this new field.

$$PhC\equiv CLi + PhICl_2 \xrightarrow[0-5\,°C]{Et_2O} PhC\equiv CI^+PhCl^- \qquad (1)$$

$$(6) \qquad\quad (7) \qquad\qquad\qquad\qquad (8)$$

$$HC\equiv C-\!\!\bigcirc\!\!-C\equiv CH + PhI(OCOCF_3)_2$$

$$\xrightarrow[20\,°C]{CHCl_3} Ph^+IC\equiv C-\!\!\bigcirc\!\!-C\equiv CI^+Ph \ 2CF_3CO_2^- \qquad (2)$$

$$(9)$$

II. PREPARATION, CHARACTERIZATION AND STRUCTURE

Initial methods of preparation[6] of alkynyl(phenyl)iodonium compounds, as the tosylate salts (11), involved the interaction of terminal alkynes with [hydroxy-(tosyloxy)iodo]benzene[7] (10) (equation 3). This method, however, suffers from lack of generality, low product yields, and separation problems from the concomitantly formed vinyl species (12) (equation 3). Recent modifications[8] and improvements[9] in this procedure have resulted in wider applicability and product yields of 60–90% of iodonium tosylates (11).

$$RC\equiv CH + PhI(OH)OTs \xrightarrow[reflux]{CHCl_3} RC\equiv CI^+Ph\bar{O}Ts + \overset{R}{\underset{TsO}{\diagdown}}C=CHI^+Ph\bar{O}Ts \qquad (3)$$

$$(10) \qquad\qquad\qquad\qquad\quad (11) \qquad\qquad\qquad (12)$$

    Likewise, interaction of commercially available iodosobenzene (13) with silylacetylenes (14) results in alkynyl(phenyl)iodonium tetrafluoroborates (15) in 50–85% isolated yields (equation 4)[10].

$$\text{PhIO} + \text{RC}\equiv\text{CSiMe}_3 \xrightarrow[\text{CH}_2\text{Cl}_2,\ 20\ °\text{C}]{\text{Et}_2\text{O}\cdot\text{BF}_3} \xrightarrow[\text{H}_2\text{O}]{\text{NaBF}_4} \text{RC}\equiv\text{CI}^+\text{PhBF}_4^- \qquad (4)$$

$$\textbf{(13)} \qquad\quad \textbf{(14)} \qquad\qquad\qquad\qquad\qquad\qquad\quad \textbf{(15)}$$

The most general and efficient methods of preparation involve iodonium trifluoromethanesulfonates (triflates). Treatment of a silylalkyne or stannylalkyne[17] with *in situ* generated Zefirov's reagent[11,12] **(16)** affords the stable, microcrystalline alkynyl(phenyl)iodonium triflates **(18)** in good to excellent yields (equation 5)[13]. The mechanism of this process involves the electrophilic addition of **16** to alkyne **17** to give the bridged iodonium and/or vinyl cation intermediates **19** and subsequent conversion to products (equation 6). Less reactive substrates, such as the parent system **(17, R=H)**, require the use of tin alkynes due to the considerably greater stabilizing effect[14] of a β-Sn compared to a β-Si group on cations.

$$\text{PhIO} + (\text{CF}_3\text{SO}_2)_2\text{O} \xrightarrow[0\ °\text{C}]{\text{CH}_2\text{Cl}_2} \left[\ \underset{\text{PhI}}{\overset{\text{OTf}}{|}}\!\!-\text{O}-\!\!\underset{\text{IPh}}{\overset{\text{OTf}}{|}}\ \right]$$

$$\textbf{(13)} \qquad\qquad\qquad\qquad\qquad\qquad\qquad \textbf{(16)}$$
$$(5)$$

$$\xrightarrow[\text{CH}_2\text{Cl}_2,\ 0\ °\text{C to }20\ °\text{C}]{\textbf{17}:\ \text{RC}\equiv\text{CMR}_3':\ \text{M} = \text{Si, Sn}} \text{RC}\equiv\text{CI}^+\text{Ph}\bar{\text{O}}\text{S}_2\text{CF}_3 + (\text{R}_3'\text{M})_2\text{O}$$

$$\textbf{(18)}\ 45\text{--}95\%$$

$$\text{R}_3'\text{MC}\equiv\text{CR} + \underset{\text{PhI}}{\overset{\text{OTf}}{|}}\!\!-\text{O}-\!\!\underset{\text{IPh}}{\overset{\text{OTf}}{|}}$$

$$\textbf{(17)}:\ \text{M} = \text{Si, Sn} \qquad \textbf{(16)} \qquad\qquad \textbf{(19a)} \qquad\qquad\qquad \textbf{(19b)}$$

$$\text{RC}\equiv\text{CI}^+\text{Ph}\bar{\text{O}}\text{Tf} + \text{R}_3'\text{MOIPh}$$

$$\textbf{(18)} \qquad\qquad\qquad \textbf{(20)} \qquad\qquad\qquad (6)$$

$$\textbf{17} + \textbf{20} \longrightarrow$$

$$\textbf{18} + (\text{R}_3'\text{M})_2\text{O}$$

The most versatile method of preparation uses the readily available alkynylstannanes[15] **(21)** and the mixed phenyliodonium triflate[16] **(22)** as the iodonium transfer agent (equation 7). This procedure works very well for the synthesis of a wide variety of hydrocarbon as well as β-functionalized[17] alkynyl(phenyl)iodonium triflates [**23**: Y=CN, Cl, RC(O), $(CH_3)_2NC(O)$, $ArSO_2$, etc.].

Likewise, this methodology is well suited for the preparation of novel bis(iodonium)ethyne[18] **(24)** (equation 8) as well as bis(iodonium)diynes **(25–27)**[19]. This procedure is also applicable[20] to the preparation of buta-1,3-diynyl(phenyl)iodonium triflates **(28)** (equation 9). The use of iodosyl triflate[21] **(29)** and two equivalents of silylacetylene affords the dialkynyliodonium salts **(30)** (equation 10)[22].

$$R_3SnC{\equiv}CY + PhI^+CN\bar{O}Tf \xrightarrow[-40\ °C\ to\ -25\ °C]{CH_2Cl_2} YC{\equiv}CI^+Ph\bar{O}Tf \qquad (7)$$

$$\textbf{(21)} \qquad\qquad \textbf{(22)} \qquad\qquad\qquad\qquad \textbf{(23)}\ 42\text{–}89\%$$

$$R_3SnC{\equiv}CSnR_3 + 2PhI^+CN\bar{O}Tf \longrightarrow PhI^+C{\equiv}CI^+Ph\ 2\bar{O}Tf \qquad (8)$$

$$\textbf{(22)} \qquad\qquad\qquad\qquad\qquad \textbf{(24)}\ 80\%$$

$$PhI^+C{\equiv}C{-}(CH_2)_{\overline{n}}C{\equiv}CI^+Ph\ 2\bar{O}Tf \qquad\qquad PhI^+C{\equiv}C{-}\langle\bigcirc\rangle{-}C{\equiv}CI^+Ph\ 2\bar{O}Tf$$

$$\textbf{(25)}\ n = 2,\ 4,\ 5\ (90\text{–}93\%) \qquad\qquad\qquad\qquad \textbf{(26)}\ 82\%$$

$$PhI^+C{\equiv}C{-}\langle\bigcirc\rangle{-}\langle\bigcirc\rangle{-}C{\equiv}CI^+Ph\ 2\bar{O}Tf$$

$$\textbf{(27)}\ 92\%$$

$$RC{\equiv}C{-}C{\equiv}CSnR_3' + PhI^+CN\ \bar{O}Tf$$

$$\textbf{(22)}$$

$$\xrightarrow[-40\ °C]{CH_2Cl_2} RC{\equiv}C{-}C{\equiv}CI^+Ph\bar{O}Tf + R_3'SnCN \qquad (9)$$

$$\textbf{(28)}\ 72\text{–}96\%$$

$$2RC{\equiv}CSiMe_3 + O{=}IOTf \xrightarrow[-78\ °C\ to\ rt]{CH_2Cl_2} (RC{\equiv}C{\rightarrow}_2 I^+\ \bar{O}Tf + (Me_3Si{\rightarrow}_2O \qquad (10)$$

$$\textbf{(22)} \qquad\qquad\qquad\qquad\qquad \textbf{(30)}\ 42\text{–}83\%$$

The vast majority of alkynyl(phenyl)iodonium salts are stable microcrystalline solids. Their exact stability depends upon both the counter-ion and the substituents on the alkyne. Stability decreases with the increasing nucleophilicity of the counter-ion. Hence, the most stable, and therefore widely used, counter-ions are $CF_3SO_3^-$ and BF_4^- with somewhat lower stability for $ArSO_3^-$ and $MeSO_3^-$ as well as $CF_3CO_2^-$, whereas halides as anions are rather unstable and decompose in a matter of hours or less.

Alkynyl(phenyl)iodonium species are readily characterized by spectroscopic means. The infrared spectra show a weak, but clearly discernible, signal for the $C\equiv C$ absorption between 2120–2190 cm^{-1} along with signals characteristic of the counter-ions; broad, intense bands between 1100–1000 cm^{-1} for BF$_4^-$ and two strong absorptions around 1270 and 1000 cm^{-1} for CF$_3$SO$_3^-$. The FAB mass spectra usually exhibit a peak for the *intact* cationic portion RC≡CİPh, (M-OTf)$^+$ or (M-BF$_4$)$^+$, with reasonable intensity, along with characteristic, readily identifiable fragmentation patterns.

Most valuable and characteristic, however, are the ^{1}H and ^{13}C NMR data. In the ^{13}C spectra the C$_\beta$ and C$_\alpha$ signals are generally between 110–120 ppm and 20–40 ppm, respectively. The downfield shift of the β-carbon signals, relative to the common acetylenic carbon signals[23] between 60–90 ppm, is due to the electron deficiency on this carbon as a consequence of resonance:

$$RC\equiv C-I^+ArX^- \longleftrightarrow R\overset{+}{C}=C=IArX^-$$

The upfield shift of the α-carbon resonance is due to the spin–orbital effects[24] of the heavy iodine atom. For alkynyl(phenyl)iodonium triflates there is an additional signal, centred at 121 ppm (q, $J_{C-F} \approx 320$ Hz), due to the carbon of the CF$_3$SO$_3$ moiety.

The ^{1}H spectra exhibit three highly diagnostic signals between 8.25–7.50 ppm in a 2:1:2 ratio for the *o:p:m* hydrogens of the phenyl group. The low-field nature of these resonance is in accord with the strong electron-withdrawing and concomitant deshielding effect of the iodonium moiety in these salts: σ_I for iodine is only 0.45 whereas it is 0.85 for —I(OAc)$_2$, 1.17 for —ICl$_2$ and 1.24 for PhĪBF$_4^-$ (with PhI$^+$OTf comparable)[25].

Single-crystal X-ray structures have been determined to date for four alkynyl-(phenyl)iodonium salts (**31–34**). The X-ray data unambiguously establish the structures of these species as a pseudo-trigonal-bipyramid (**35**) with a T-shaped geometry. The equatorial positions L$_e$ are occupied by the phenyl group and two sets of nonbonding lone-pair electrons, whereas the axial positions, L$_a$, are taken up by the alkyne and the counter-ion, respectively. The I—C(Ph) bond distance is 2.1 Å, the I—C(alkyne) 2.0 Å, with a nearly 90° bond angle between these two carbon ligands. The I—O distances (to the respective anions) are generally greater than 2.5 Å and well outside the sum of the covalent radii of I and O and the I—O single-bond distance of 1.99 Å. However, secondary interactions generally result in geometries distorted from the above ideal trigonal-bipyramid[15].

HC≡CI$^+$PhŌSO$_2$CF$_3$ NCC≡CI$^+$PhŌSO$_2$CF$_3$

(**31**)[26] (**32**)[17]

PhC≡CI$^+$PhŌSO$_2$C$_6$H$_4$CH$_3$ *t*-BuC≡CI$^+$PhIŌ$_4$

(**33**)[8] (**34**)[27] (**35**)

III. CHEMISTRY

A. Reaction with Nucleophiles

The great majority of known[28] reactions of acetylenes are with electrophiles either via the acetylide ions RC≡C$^-$, or by way of electrophilic addition reactions to the triple bond. Nucleophilic acetylenic substitutions (S$_N$-A) are generally unfavorable[29]. Alkynyl(phenyl)iodonium species may serve as synthons for the electrophilic alkynyl cation 'RC≡C⁺', and thereby achieve an umpolung of the more common, normal, electrophilic acetylene reactivity. This formally S_N-A process with alkynyliodonium species is due to the excellent leaving-group ability of the iodonium moiety[30], via loss of neutral iodobenzene, in a manner analogous to the loss of N_2 from diazonium salts. As indicated below, the actual mechanism of reaction of alkynyl(phenyl)iodonium species with nucleophiles is considerably more complicated than a simple, direct S_N-A process analogous to the well known S_N-2 or S_N-1 reactions of saturated substrates. Nevertheless, interaction of a wide variety of nucleophiles with alkynyliodonium species have been reported to yield diverse functionalized alkynes.

1. Organic nucleophiles

Interaction of alkynyl(phenyl)iodonium triflates **18** with NaO_2CPh (equation 11), $NaOSO_2Ar$ (equation 12) and $NaOP(O)(OR')$(equation 13) results in the corresponding alkynyl benzoate[31] (**36**), alkynyl sulfonate[8] (**37**) and alkynyl phosphate[32] (**38**) esters, respectively. These novel compounds represent a new class of hitherto unknown esters that combine two of the simplest, most common and valuable organic functionalities, the carbon–carbon triple bond and an ester, into a single molecular framework[33]. These alkynyliodonium-derived acetylenic esters have interesting properties and chemistry[33] including enzyme inhibition[34], cycloadditions[35], ynolate generation[36], an unusual mechanism of hydration[37] and electrophilic additions[38].

$$RC{\equiv}CI^{+}Ph\bar{O}Tf + NaOC(O)Ph \longrightarrow RC{\equiv}CO\overset{\overset{O}{\|}}{C}Ph \qquad (11)$$

$$\textbf{(18)} \qquad\qquad\qquad\qquad\qquad \textbf{(36)}$$

$$RC{\equiv}CI^{+}Ph\bar{O}Tf + NaOSO_2Ar \longrightarrow RC{\equiv}COSO_2Ar \qquad (12)$$

$$\textbf{(18)} \qquad\qquad\qquad\qquad\qquad \textbf{(37)}$$

$$RC{\equiv}CI^{+}Ph\bar{O}Tf + NaOP(O)(OR')_2 \longrightarrow RC{\equiv}CO\overset{\overset{O}{\|}}{P}(OR')_2 \qquad (13)$$

$$\textbf{(18)} \qquad\qquad\qquad\qquad\qquad \textbf{(38)}$$

Similarly, reaction of **18** with NaSCN affords the corresponding alkynyl thiocyanates (**39**) in good yields (equation 14)[39]. Interaction of **18** with Ph_3P results in essentially quantitative yields of the respective alkynyl phosphonium salts **40** (equation 15)[40,41]. The similar reaction of alkynyliodonium tosylates **11** with $(RO)_3P$ results in **41** (equation 15)[40,41]. The similar reaction of alkynyliodonium tosylates **11** with $(RO')_2PS_2K$ gives the corresponding alkynyl thiophosphate esters[42] **42** (equation 17) that are also biologically active and related to phosphate esters **38**[43]. Similar products, although generally obtained in lower yields, are observed in reactions of the diiodonium salts **24**–**26** with nucleophiles. For example, interaction of PhS^- and Ph_3P with **24** gives **43** and **44**, respectively[18]. Reaction of **25** and **26** with Ph_3P and NaOTs results in the corresponding bis-phosphonium salts **45** and **47** and bis-sulfonates **46** and **48**, respectively[44]. Advantage was taken of the reaction of phosphines with alkynyliodonium species in the synthesis of new $1\lambda^5,3\lambda^5$-diphospholium ions (**49**) by interaction of **11** with $Ph_2PCH_2PPh_2$ (equation 18)[45].

$$RC{\equiv}CI^+Ph\bar{O}Tf + NaSCN \longrightarrow RC{\equiv}CSCN \quad (14)$$

$$\textbf{(18)} \qquad\qquad\qquad\qquad \textbf{(39)}$$

$$RC{\equiv}CI^+Ph\bar{O}Tf + Ph_3P \xrightarrow[-PhI]{CH_2Cl_2} RC{\equiv}CP^+Ph_3\bar{O}Tf \quad (15)$$

$$\textbf{(18)} \qquad\qquad\qquad\qquad\qquad \textbf{(40)}$$

$$RC{\equiv}CI^+Ph\bar{O}Ts + (RO)_3P \longrightarrow RC{\equiv}C\overset{\displaystyle O}{P}(OR)_2 \quad (16)$$

$$\textbf{(11)} \qquad\qquad\qquad\qquad \textbf{(41)}$$

$$RC{\equiv}CI^+Ph\bar{O}Ts + (R'O)_2PS_2K \longrightarrow RC{\equiv}C\overset{\displaystyle S}{P}(OR')_2 \quad (17)$$

$$\textbf{(11)} \qquad\qquad\qquad\qquad\qquad \textbf{(42)}$$

$$PhSC{\equiv}CSPh \qquad\qquad Ph_3P^+C{\equiv}CP^+Ph_32\bar{O}Tf$$

$$\textbf{(43)} \qquad\qquad\qquad\qquad \textbf{(44)}$$

$$Ph_3P^+C{\equiv}C{-}(CH_2)_{\overline{n}}\,C{\equiv}CP^+Ph_32\bar{O}Tf \qquad TsOC{\equiv}C{-}(CH_2)_{\overline{n}}\,C{\equiv}COTs$$

$$\textbf{(45)} \qquad\qquad\qquad\qquad\qquad\qquad \textbf{(46)}$$

$$Ph_3P^+C{\equiv}C{-}\langle\bigcirc\rangle{-}C{\equiv}CP^+Ph_32\bar{O}Tf$$

$$\textbf{(47)}$$

$$ArO_2SOC{\equiv}C{-}\langle\bigcirc\rangle{-}C{\equiv}COSO_2Ar$$

$$\textbf{(48)}$$

More complex nucleophiles can be reacted with alkynyl(phenyl)iodonium salts as well. For example, protected aminomalonates (50) and 18 give the corresponding alkynylmalonates 51 in 30–90% isolated yields (equation 19)[13]. Likewise, a variety of β-dicarbonyl enolates 52–54 react with 55 to give alkynyl products 56–58 (equations 20–22)[46]. These reactions may be looked upon as alkynylations or, in other words, the triple-bond analogs of the well-established alkylation reactions. Unfortunately, they only work with 'soft' nucleophiles such as 52–54 ($\bar{O}$Ts, PhCO$_2^-$ Ph$_3$P, etc.). Nucleophiles such as R$\bar{O}$, or simple enolates, do not seem to work.

$$RC{\equiv}CI^+Ph\bar{O}Ts + Ph_2PCH_2PPh_2 \longrightarrow \left[\begin{array}{c} R \\ C \\ \| \\ C \\ | \\ Ph_2PCH_2PPh_2\bar{O}Ts \\ + \end{array}\right]$$

$$\textbf{(11)}$$

$$\textbf{(49)}$$

$$\longrightarrow Ph_2P{\diagup}{\diagdown}P^+Ph_2 \quad (18)$$

$$\overset{R}{}$$

$$\bar{O}Ts$$

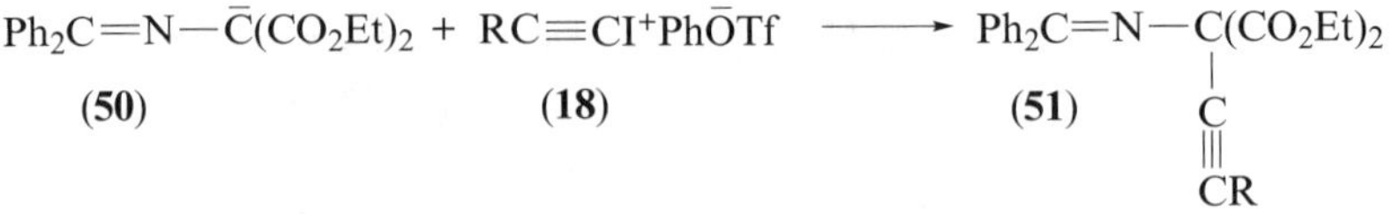

$$Ph_2C{=}N{-}\bar{C}(CO_2Et)_2 \; + \; RC{\equiv}CI^+Ph\bar{O}Tf \longrightarrow Ph_2C{=}N{-}C(CO_2Et)_2 \quad (19)$$

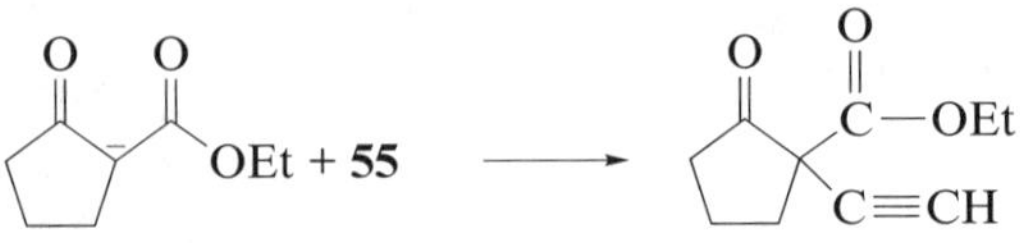

(50) (18) (51)

with substituent:
$$\underset{\displaystyle CR}{\overset{\displaystyle C}{\underset{\|}{\,}}}$$

$$\left(EtO\overset{O}{\underset{\|}{C}}\right)_2\bar{C}Ph \; + \; HC{\equiv}CI^+Ph\bar{B}F_4 \longrightarrow \left(EtO\overset{O}{\underset{\|}{C}}\right)_2\overset{Ph}{\underset{|}{C}}{-}C{\equiv}CH \quad (20)$$

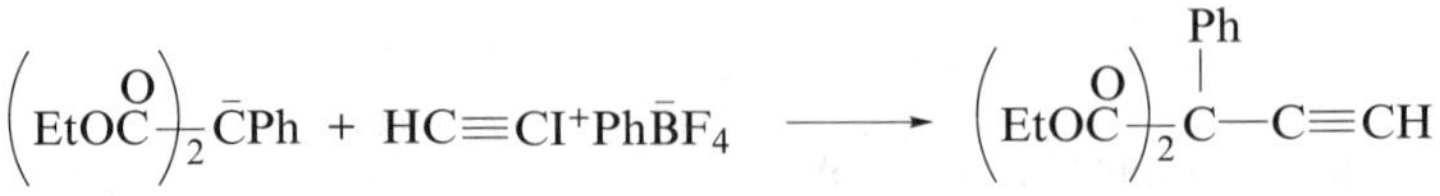

(52) (55) (56)

$$(53)\;{+}\;55 \longrightarrow (57) \quad (21)$$

(53) (57)

$$(54){-}R\;{+}\;55 \longrightarrow (58) \quad (22)$$

(54) (58)

2. Organometallic nucleophiles

Alkynyl(phenyl)iodonium salts are also excellent substrates for interaction with organometallic nucleophiles. Reaction of Vaska's complex (59) and its Rh-analog (60) with 18 results in σ-alkynyl complexes 61 and 62, respectively, in nearly quantitative yields (equations 23 and 24)[47]. These reactions are oxidative additions to Ir and Rh, respectively, and represent one of the best ways to introduce a σ-acetylide ligand into metal complexes.

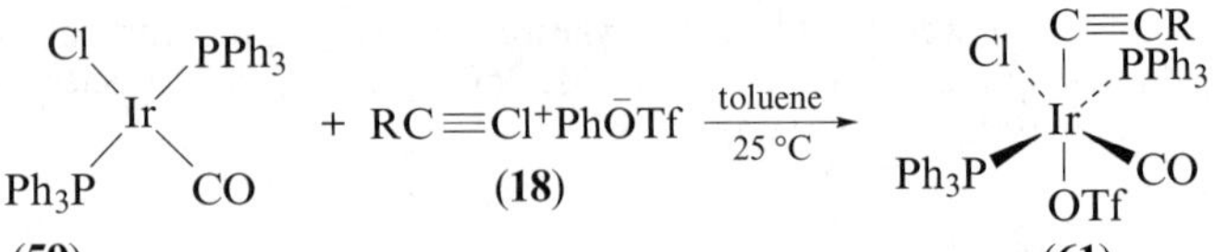

$$(59) \; + \; RC{\equiv}CI^+Ph\bar{O}Tf \; \xrightarrow[25\,°C]{toluene} \; (61) \quad (23)$$

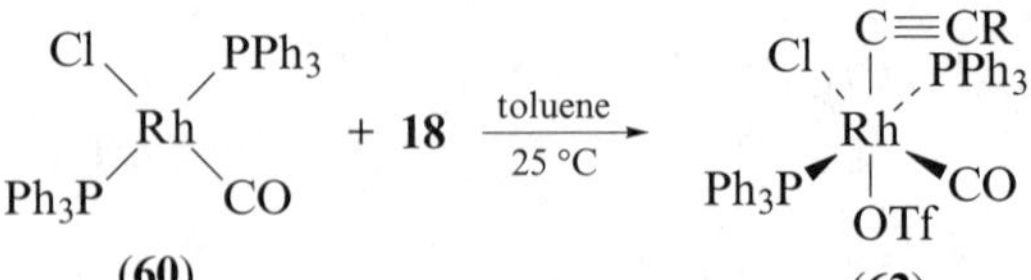

(59) (18) (61)

$$(60) \; + \; 18 \; \xrightarrow[25\,°C]{toluene} \; (62) \quad (24)$$

(60) (62)

Similarly, reaction of 26 with 59 and 60 results in novel rigid-rod, cationic, bimetallic σ-complexes 63 and 64 (equations 25 and 26)[48]. Complexes 63 and 64 are of some interest, both as potential catalysts and as possible nonlinear optical materials[48].

$$26 + 59 \xrightarrow[25\,°C]{CH_3CN} \left[CH_3CN-\underset{\underset{PPh_3}{|}}{\overset{\overset{OC}{|}\,PPh_3}{Ir}}\overset{Cl}{\diagdown}-C\equiv C-\!\!\!\!\!\left\langle\!\!\!\bigcirc\!\!\!\right\rangle\!\!\!\!\!-C\equiv C-\underset{\underset{PPh_3}{|}}{\overset{\overset{PPh_3}{|}\,CO}{Ir}}\overset{Cl}{\diagup}-NCCH_3 \right]^{2+} 2\bar{O}Tf \quad (25)$$

(63)

$$26 + 60 \xrightarrow[25\,°C]{CH_3CN} \left[CH_3CN-\underset{\underset{PPh_3}{|}}{\overset{\overset{OC}{|}\,PPh_3}{Rh}}\overset{Cl}{\diagdown}-C\equiv C-\!\!\!\!\!\left\langle\!\!\!\bigcirc\!\!\!\right\rangle\!\!\!\!\!-C\equiv C-\underset{\underset{PPh_3}{|}}{\overset{\overset{OC}{|}\,PPh_3}{Rh}}\overset{Cl}{\diagdown}-NCCH_3 \right]^{2+} 2\bar{O}Tf \quad (26)$$

(64)

3. Mechanism of reaction

Although all of the reactions of alkynyliodonium species involving nucleophiles are formally S_N-A processes, the actual mechanistic process is more complex. Experimental evidence indicates that the reaction involves Michael addition, unsaturated carbene formation and rearrangement (equation 27). The nucleophile, Nu^-, adds in a Michael fashion to the β-carbon of the alkynyliodonium species to give the I-ylide **65**. Evidence[49] for this step is provided by the trapping of ylide **65** to give the stable, isolated, vinyliodonium salt **67** when **15** is reacted with Me_3SiN_3 in wet CH_2Cl_2 (equation 28). Loss of iodobenzene from ylide **65** results in the known[50] alkylidenecarbene (**66**) that, in the absence of a suitable trap, rearranges to the alkyne (equation 27). Supporting data[51] for the carbene **66** include trapping via insertion into Si—H and O—H bonds (equation 29) when **11** is reacted with N_3^- in the presence of Et_3SiH or in methanol, respectively. Moreover, carbene **66** may also be trapped intramolecularly via insertion into a 1,5-C—H bond (equation 30) to give the cyclic vinyl azide **68**[51].

$$RC\equiv CI^+PhX^- + Nu^- \xrightarrow{-X^-} \left[\underset{R}{\overset{Nu}{\diagdown}}C=C=IPh \longleftrightarrow \underset{R}{\overset{Nu}{\diagdown}}C=\bar{C}=\overset{+}{I}Ph \right] \quad (27)$$

(65a)   (65b)

$$\Big\downarrow {-PhI}$$

$$RC\equiv CNu \xleftarrow{\;O\;} \left[\underset{R}{\overset{Nu}{\diagdown}}C=C\colon \right]$$

(66)

$$RC\equiv CI^+PhBF_4^- + Me_3SiN_3 \xrightarrow{aq.CH_2Cl_2} \underset{R}{\overset{N_3}{\diagdown}}C=C\underset{I^+PhBF_4^-}{\overset{H}{\diagup}} \quad (28)$$

(15) R = Me, t-Bu, n-C_8H_{13}   (67) 50–90%

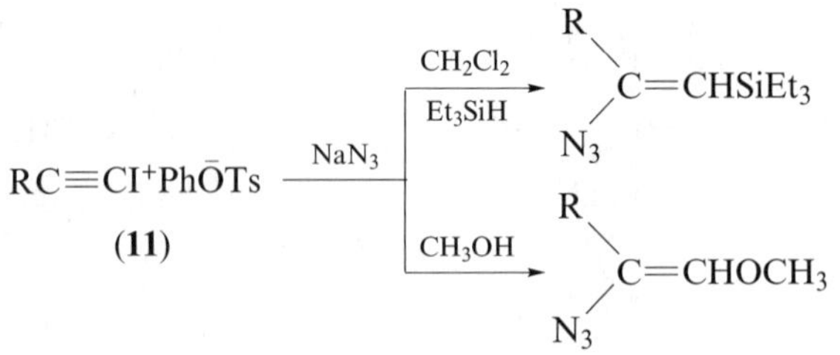

$$RC\equiv CI^+Ph\bar{O}Ts \xrightarrow[NaN_3]{}
\begin{cases}
\xrightarrow[Et_3SiH]{CH_2Cl_2} & \underset{N_3}{\overset{R}{C}}=CHSiEt_3 \\[2em]
\xrightarrow{CH_3OH} & \underset{N_3}{\overset{R}{C}}=CHOCH_3
\end{cases} \tag{29}$$

(11)

Ochiai and coworkers[52] have used these tandem Michael-carbene-insertion process in a nice synthesis of cyclopentene derivatives as illustrated by **69** and **70** using alkynyl(phenyl)iodonium salts (equation 31). Similarly, $PhSO_2H$ addition to alkynyl(phenyl)iodonium salts in methanol results[53] in the formation of e.g., **71** that, upon treatment with Et_3N, gives both rearranged sulfone **72** and the cyclopentene **73** via the unsaturated carbene, as illustrated by the example in equation 32.

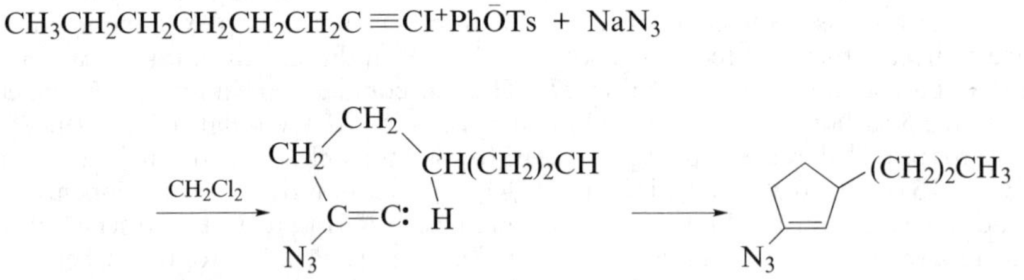

$$CH_3CH_2CH_2CH_2CH_2CH_2C\equiv CI^+Ph\bar{O}Ts \; + \; NaN_3$$

(30)

(68)

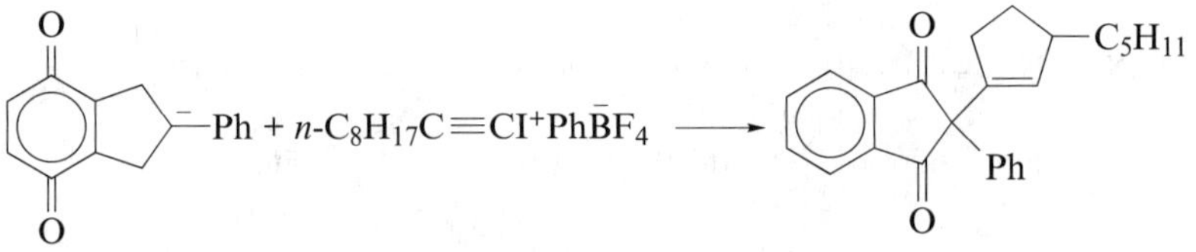

(31)

(69)

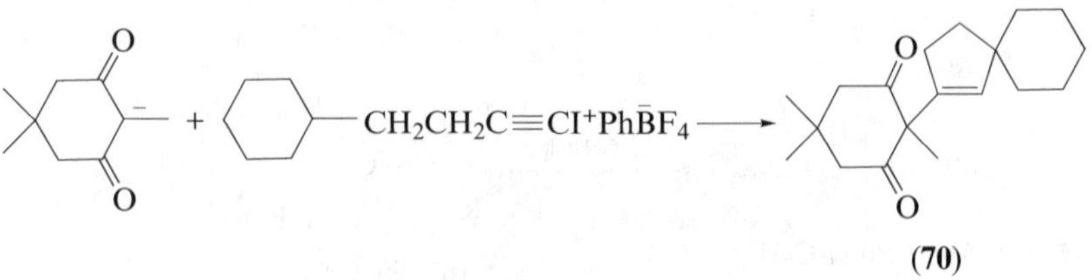

(70)

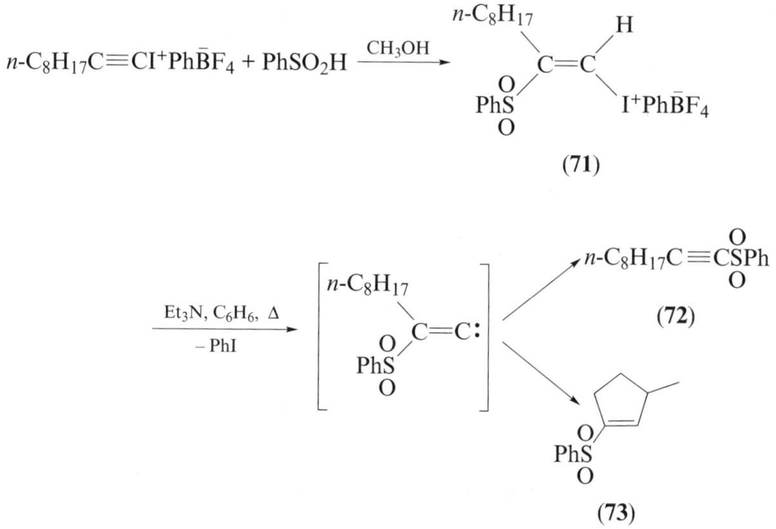

$$n\text{-}C_8H_{17}C\equiv CI^+Ph\bar{B}F_4 + PhSO_2H \xrightarrow{CH_3OH}$$

(71)

(32)

$$\xrightarrow[-\,PhI]{Et_3N,\ C_6H_6,\ \Delta}$$

$n\text{-}C_8H_{17}C\equiv CSPh$ (with O, O groups)

(72)

(73)

B. Cycloadditions

Acetylenes, particularly ones with electron-withdrawing substituents such as $MeO_2CC\equiv CCO_2Me$, are known to have a rich cycloaddition chemistry. Since alkynyl(phenyl)iodonium species are highly electron-deficient alkynes, they are expected to be excellent cycloaddition partners in a variety of electrocyclic processes.

1. 1,3-Dipolar cycloadditions

The first reported [2+3]-cycloaddition of alkynyliodonium salts by Varvoglis and coworkers[54] involved benzonitrile oxides 74 as 1,3-dipoles (equation 33). The predominant regioisomer observed was 75 but, in some cases, 76 was also seen[54]. α-Diazocarbonyl compounds 77 also behave as 1,3-dipoles towards $Me_3SiC\equiv C\overset{+}{I}Ph\bar{O}Tf$ (equation 34)[55]. The final products are the stable microcrystalline heterocyclic (4-pyrazolyl)phenyliodonium triflates (79), the result of a rapid hydrogen shift in the initial cycloadduct 78. No other regioisomer was detected in these reactions. Likewise, only regioisomer 81 was observed in the reaction of 77 with 80 (equation 35). Similarly, reaction of methyl azide 82 with 81 gave only the regioisomer 83 in a [2+3] cycloaddition (equation 36)[55]. The observed regioisomers are in agreement with analogous [2+3] cycloaddition reactions of polar alkynes such as $HC\equiv CC(O)R$ in 1,3-dipolar reactions[56] and with FMO theory[57] of such reactions.

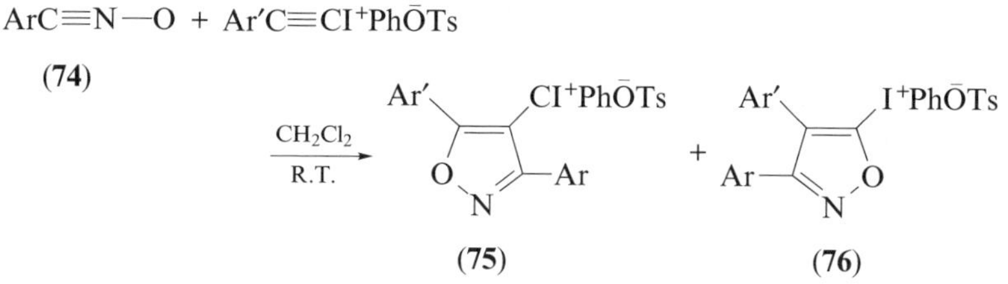

$$ArC\equiv N\!-\!O + Ar'C\equiv CI^+Ph\bar{O}Ts$$

(74)

$$\xrightarrow[R.T.]{CH_2Cl_2}$$

(75)   +   (76)   (33)

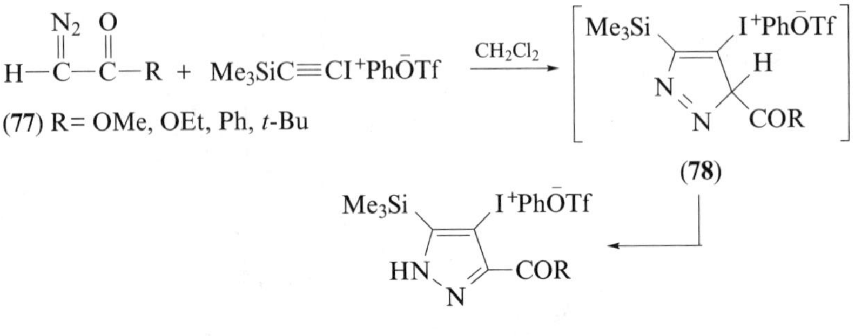

$$
\begin{array}{cc}
\underset{(77)\ R=\text{OMe, OEt, Ph, }t\text{-Bu}}{\overset{\overset{N_2}{\|}\ \overset{O}{\|}}{H-C-C-R}} + Me_3SiC\!\equiv\!CI^+Ph\bar{O}Tf & \xrightarrow{CH_2Cl_2}
\end{array}
\left[
\begin{array}{c}
Me_3Si \quad\quad I^+Ph\bar{O}Tf \\
\quad\quad\quad H \\
N \\
N \quad COR \\
\mathbf{(78)}
\end{array}
\right]
\tag{34}
$$

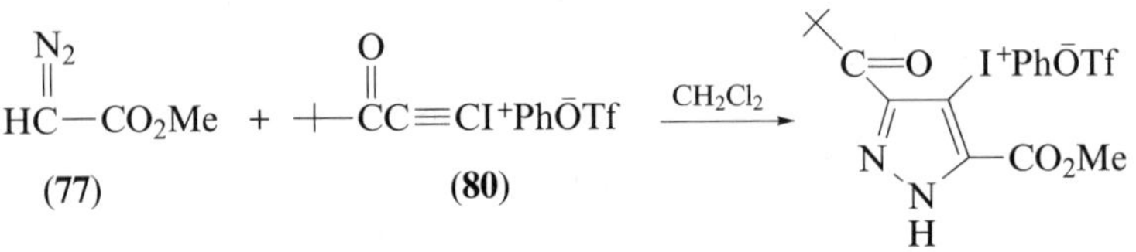

$$
\begin{array}{c}
Me_3Si \quad\quad I^+Ph\bar{O}Tf \\
HN \quad\quad COR \\
N \\
\mathbf{(79)}
\end{array}
\longleftarrow
$$

$$
\underset{\mathbf{(77)}}{\overset{\overset{N_2}{\|}}{HC-CO_2Me}} + \underset{\mathbf{(80)}}{\overset{\overset{O}{\|}}{+\!-CC\!\equiv\!CI^+Ph\bar{O}Tf}} \xrightarrow{CH_2Cl_2}
\begin{array}{c}
C\!=\!O \quad I^+Ph\bar{O}Tf \\
N \quad\quad CO_2Me \\
N \\
H \\
\mathbf{(81)}
\end{array}
\tag{35}
$$

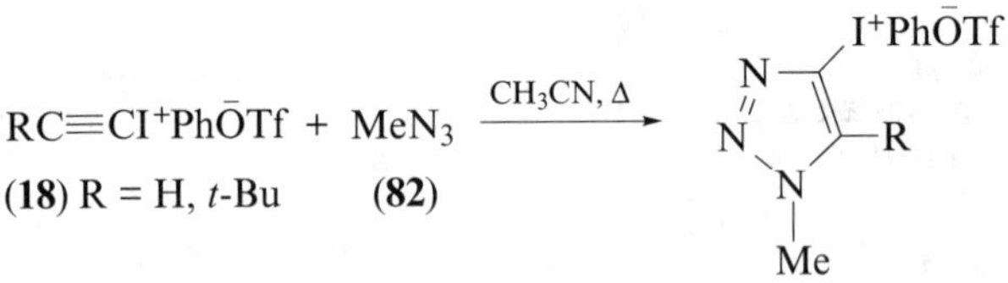

$$
\underset{\mathbf{(18)}\ R=H,\ t\text{-Bu}}{RC\!\equiv\!CI^+Ph\bar{O}Tf} + \underset{\mathbf{(82)}}{MeN_3} \xrightarrow{CH_3CN,\ \Delta}
\begin{array}{c}
N \quad I^+Ph\bar{O}Tf \\
N \\
N \quad R \\
N \\
Me \\
\mathbf{(83)}
\end{array}
\tag{36}
$$

2. Diels–Alder reactions

Electron-deficient acetylenes are superb [2+4] cycloaddition partners. The β-substituted alkynyliodonium salts **23** are highly electron deficient and hence readily undergo Diels–

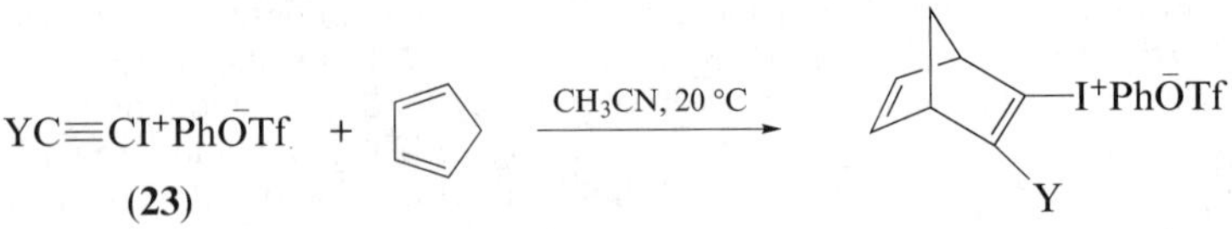

$$
\underset{\mathbf{(23)}}{YC\!\equiv\!CI^+Ph\bar{O}Tf} + \text{(cyclopentadiene)} \xrightarrow{CH_3CN,\ 20\ ^\circ C}
\begin{array}{c}
I^+Ph\bar{O}Tf \\
Y \\
\mathbf{(84)}
\end{array}
\tag{37}
$$

$$
Y = CN,\ ArSO_2,\ RC(O)
$$

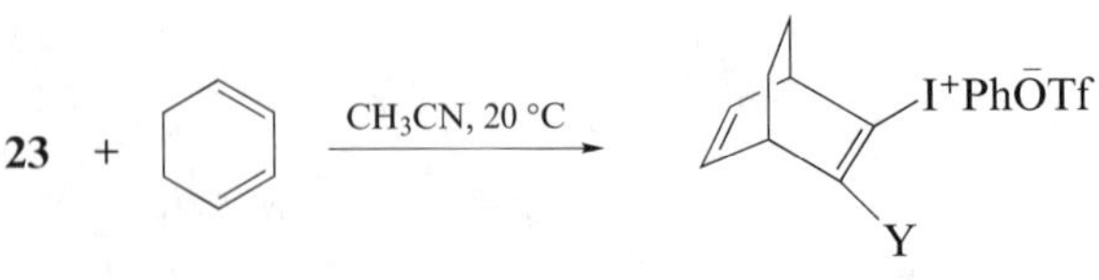

$$
\mathbf{23} + \text{(cyclohexadiene)} \xrightarrow{CH_3CN,\ 20\ ^\circ C}
\begin{array}{c}
I^+Ph\bar{O}Tf \\
Y \\
\mathbf{(85)}
\end{array}
\tag{38}
$$

$$(39)$$

$$(40)$$

Alder reactions with a variety of dienes as illustrated in equations 37–40, and the resultant cycloadducts **84–87** are stable, microcrystalline vinyliodonium salts. These cycloadducts are set up for further synthetic elaboration by virtue of the two functionalities Y and the iodonium moiety. Vinyliodonium salts, like alkynyliodonium species, are known[58] to react with a wide variety of nucleophiles.

The bisiodonium ethyne **24** is even more reactive than **18** and readily undergoes Diels–Alder cycloaddition with cyclopentadiene and furans as shown in equation 41. The remarkable fact of the cycloaddition of **24** is that, unlike the usual Diels–Alder reactions that require extensive heating and/or pressure, these reactions occur in a matter of minutes at –35 °C.

$$(41)$$

Furthermore, cycloadducts **88** and **89** may be reacted further with $RC{\equiv}CLi$ to give ene-diynes **91** (equation 42)[59] or with a variety of nucleophiles resulting in **92** (equation 43)[60].

C. Miscellaneous Reactions

The reactions discussed so far involved either reaction with nucleophiles or cycloadditions whose mechanisms are reasonably well understood. However, reactions at the I or the ipso-carbon of the phenyl as well as perhaps the α-carbon of the alkyne are possible, as are other reactions whose mechanisms are not yet known. For example, either thermal or photochemical decomposition of **93** results in iodoalkyne (**94**) and phenyl tosylate (**95**) (equation 44)[8]. Reaction of **96** with 2-lithiofuran (**97**) resulted in an exchange of carbon ligands and formation of **98**; the analogous 2-lithiothiophene (**99**) gave **100** (equation 45).

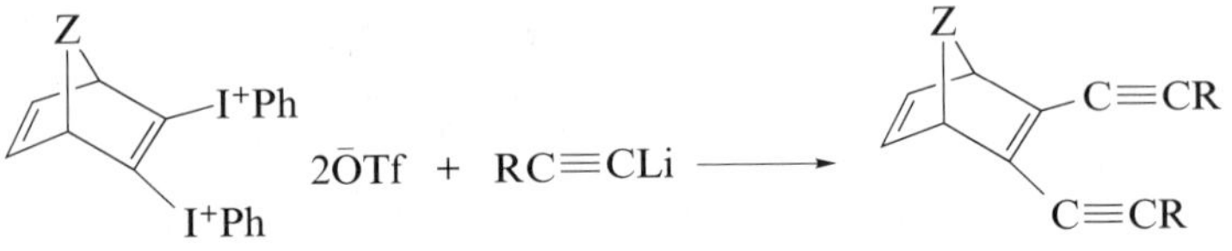

$$ \text{(42)} $$

(88) Z = CH$_2$
(89) Z = O

(91)

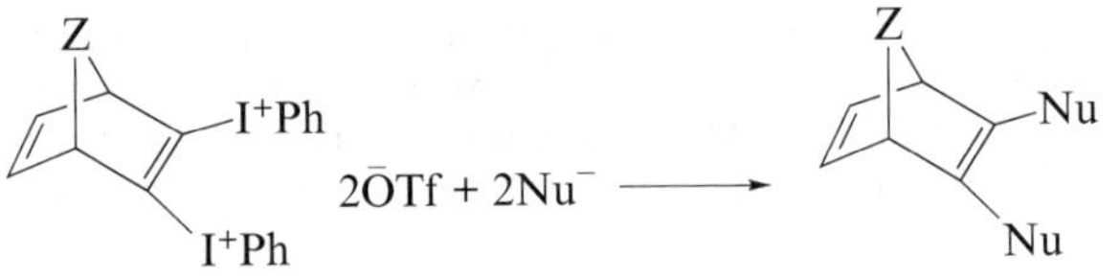

$$ \text{(43)} $$

(88) Z = CH$_2$
(89) Z = O

(92) Nu = Br, I, CN

$$ \text{PhC}\equiv\text{CI}^+\text{Ph}\bar{\text{O}}\text{Ts} \xrightarrow{\Delta \text{ or } h\nu} \text{PhC}\equiv\text{CI} + \text{PhOTs} \qquad (44) $$

(93)   **(94)**   **(95)**

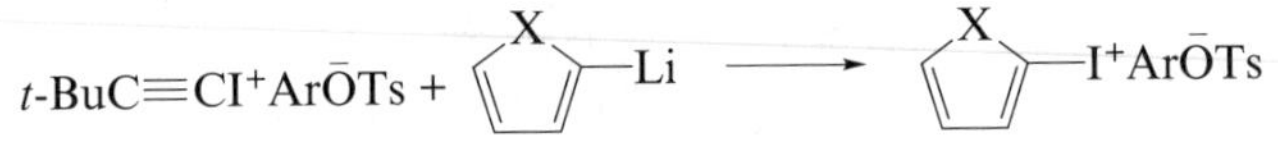

$$ t\text{-BuC}\equiv\text{CI}^+\text{Ar}\bar{\text{O}}\text{Ts} + \qquad \text{(45)} $$

(96)   **(97)** X = O   **(98)** X = O
                (99) X = S   **(100)** X = S

$$ \text{RC}\equiv\text{CI}^+\text{Ph}\bar{\text{B}}\text{F}_4 + \text{R}'\text{CO}_2\text{Na} \xrightarrow{\text{THF/H}_2\text{O}} \text{RCCH}_2\text{OCR}' \qquad (46) $$

(15)   **(101)**   **(102)**

$$ \text{RC}\equiv\text{CI}^+\text{Ph}\bar{\text{O}}\text{Ts} + \text{R}'\text{OH} \xrightarrow[\text{CO, base}]{\text{Pd(OAc)}_2} \text{RC}\equiv\text{CCOR}' \qquad (47) $$

(11)   **(103)**

Treatment of **15** with sodium carboxylates (**101**) in aqueous THF gave keto esters (**102**) as products (equation 46)[62]. Carbonylation of **11** in alcohol solvents, catalyzed by Pd(OAc)$_2$ results in alkynyl carboxylates (**103**) (equation 47)[63]. Coupling of alkynyl(phenyl)iodonium

$$R^2Cu(SMe_2)MgBr_2 + R^1C\equiv CH \longrightarrow \underset{(104a)}{\overset{R^1}{\underset{R^2}{\diagdown}}C=C\overset{H}{\underset{Cu}{\diagup}}}$$

$$\xrightarrow[\text{Et}_2\text{O},\, -78\text{ to }25\,°\text{C},\,12\text{ h}]{11} \underset{(105a)}{\overset{R^1}{\underset{R^2}{\diagdown}}C=C\overset{H}{\underset{C\equiv CR}{\diagup}}} \tag{48}$$

$$R^1Cu(SMe_2)MgBr_2 + R^2C\equiv CH \longrightarrow \underset{(104b)}{\overset{R^2}{\underset{R^1}{\diagdown}}C=C\overset{H}{\underset{Cu}{\diagup}}}$$

$$\xrightarrow[\text{Et}_2\text{O},\, -78\text{ to }25\,°\text{C},\,12\text{ h}]{11} \underset{(105b)}{\overset{R^2}{\underset{R^1}{\diagdown}}C=C\overset{H}{\underset{C\equiv CR}{\diagup}}}$$

tosylates (11) with vinylcopper reagents (104) results in conjugated enynes 105 (equation 48)[64].

Control of olefin geometry is achieved by the mode of addition of the alkylcopper reagent as outlined in equation 48. The reaction is completely stereospecific[64] and nicely complements the widely employed Pd-catalyzed cross-coupling processes[65] for the formation of these valuable, conjugated, aliphatic molecules.

A similar coupling is possible between 11 and alkynylcopper reagents (106) resulting in conjugated diynes (107) (equation 49)[66].

$$11 + (R'C\equiv C)_2 Cu(CN)Li_2 \xrightarrow[-78\text{ to }25\,°\text{C}]{\text{THF}} RC\equiv C-C\equiv CR' \tag{49}$$
$$\quad\quad\quad (106) \quad\quad\quad\quad\quad\quad\quad\quad\quad (107)$$

IV. ALKYNYL XENON SPECIES

Xe(II) species are isoelectronic with I(III) species. Hence, it was of interest to see if analogous alkynyl xenon species could be made and if they might be stable to isolation. Indeed reaction of silylalkynes with XeF_2 under appropriate conditions results in 108, the xenon analogs of alkynyliodonium salts (equation 50)[67]. Unfortunately, these novel alkynyl xenonium salts are only stable in solution below $-30\,°$C. They were identified by spectral means including ^{129}Xe NMR. For example, the t-Bu compound 109 shows a strong ^{129}Xe signal at -1818 ppm (relative to external XeF_2 in $CDCl_3$ solution) as well as a $C\equiv C$ signal in the IR at 2200 cm^{-1}, data in accord with the proposed structure[67]. Reaction of 109 with Ph_3P results in the known[46] phosphonium salt 110 (equation 51). It is likely that this reaction with Ph_3P proceeds via loss of Xe gas and formation of the very high energy alkynyl cation[68], $t\text{-BuC}{\equiv}C^+$, followed by capture with Ph_3P. The analogous He species generated via a tritium recoil process is also postulated to decompose via a high energy alkynyl cation[69].

$$RC{\equiv}CSiMe_3 + XeF_2 \xrightarrow[CD_2Cl_2, -78\ to -45\ °C]{BF_3 \cdot OEt_2} RC{\equiv}CXe^+\bar{B}F_4 \qquad (50)$$

$$(108)\ R = t\text{-Bu, Me}_3Si$$

$$t\text{-BuC}{\equiv}CXe^+BF_4 + Ph_3P \xrightarrow[-78\ °C]{CH_2Cl_2} t\text{-BuC}{\equiv}CP^+Ph_3\bar{B}F_4 + Xe \qquad (51)$$

$$(109) \qquad\qquad\qquad\qquad (110)$$

V. CONCLUSIONS

It is evident from this brief chapter that alkynyliodonium species represent a new class of versatile and valuable functionalized alkynes. Although they have only been available and known for a scant decade, they have already emerged as one of the most useful reagent in acetylene and related chemistry. They serve as highly reactive substrates for nucleophilic acetylenic substitutions, S_N-A, with diverse nucleophiles, thereby allowing the preparation of hitherto unknown or not easily available functionalized acetylenes, such as the new alkynyl esters. They are excellent cycloaddition partners in a variety of electrocyclic reactions. However, this is just the beginning of the many possible uses and applications of these novel, easily prepared, functionalized acetylenes. Their electrophilic character, in contrast to the more common nucleophilic acetylenic reactivity, allows an umpolung of the chemistry of alkynes, and hence alkynyliodonium species serve as synthons for the high energy, generally inaccessible, alkynyl cation, $RC{\equiv}C^+$. Their ready availability from commercial precursors, general stability, and ease of handling and versatility should stimulate further imaginative uses and greatly enhance the continued development of acetylene chemistry.

VI. ACKNOWLEDGMENTS

Financial support of our own work, described herein, by the National Institute of Health Research Grant 2ROCA 16903 is gratefully acknowledged. We also thank many colleagues for stimulating discussions and productive collaborations as herein described and cited.

VII. REFERENCES

1. A. Varvoglis, *The Chemistry of Polycoordinated Iodine*, VCH Publishers, Inc., New York, 1992.
2. C. Hartman and V. Meyer, *Ber. Dtsch. Chem. Ges.*, **27**, 426 (1894).
3. F. M. Beringer and S. Galton, *J. Org. Chem.*, **30**, 1930 (1965).
4. E. B. Merkushev, L.-G. Karpitskaya and G. I. Novosel'tseva, *Dolk. Akad. Nauk SSSR*, **245**, 607 (1979).
5. P. J. Stang, *Angew. Chem., Int. Ed. Engl.*, **31**, 274 (1992).
6. G. F. Koser, L. Rebrovic and R. H. Wettach, *J. Org. Chem.*, **46**, 4324 (1981); L. Rebrovic and G. F. Koser, *J. Org. Chem.*, **49**, 4700 (1984).
7. For a review on **10** and its chemistry, see: R. M. Moriarty, R. K. Vaid and G. F. Koser, *Synlett*, 365 (1990).
8. (a) P. J. Stang and B. W. Surber, *J. Am. Soc. Chem.*, **107**, 1452 (1985).

(b) P. J. Stang, B. W. Surber, Z. C. Chen, K. A. Roberts and A. G. Anderson, *J. Am. Chem. Soc.*, **109**, 228 (1987).
9. T. Kitamura and P. J. Stang, *J. Org. Chem.*, **53**, 4105 (1988); P. J. Stang and T. Kitamura, *Org. Synth.*, **70**, 215 (1991).
10. M. Ochiai, M. Kunishima, K. Sumi, Y. Nagao, E. Fujita, M. Arimoto and H. Yamaguchi, *Tetrahedron Lett.*, **26**, 4501 (1985).
11. N. S. Zefirov, V. V. Zhdankin and A. S. Koz'min, *Izv. Akad. Nauk SSSR, Ser Khim.*, 1682 (1988); N. S. Zefirov, V. V. Zhdankin, Y. V. Dan'kov and A. S. Koz'min, *Zh. Org. Khim.*, **20**, 446 (1984).
12. J. Gallos, A. Varvoglis and N. W. Alcock, *J. Chem. Soc., Perkin Trans.* 1, 757 (1985). T. R. Hembre, C. P. Scott and J. R. Norton, *J. Org. Chem.*, **52**, 3650 (1987).
13. M. D. Bachi, N. Bar-Ner, C. M. Crittell, P. J. Stang and B. L. Williamson, *J. Org. Chem.*, **56**, 3912 (1991).
14. K. A. Nguyen, M. S. Gorgen, G.-T. Wang and J. B. Lambert, *Organometallics*, **10**, 2798 (1991); J. B. Lambert and G. T. Wang, *Tetrahedron Lett.*, **29**, 2551 (1988); J. B. Lambert, G. T. Wang and D. H. Teramura, *J. Org. Chem.*, **53**, 5422 (1988); J. B. Lambert and E. C. J. Chelius, *J. Am. Chem. Soc.*, **112**, 8120 (1990); M. R. Ibrahim and W. L. J. Jorgensen, *J. Am. Chem. Soc.*, **111**, 819 (1989). Review. J. B. Lambert, *Tetrahedron*, **46**, 2677 (1990).
15. C. Cauletti, Furlani and A. Sebald, *Gazz. Chim. Ital.*, **118**, 1 (1988); F. G. Kleiner and W. P. Neumann, *Justus Liebigs Ann. Chem.*, **716**, 19 (1968); W. E. Davidsohn and M. C. Henry, *Chem. Rev.*, **67**, 73 (1967).
16. V. V. Zhdankin, C. M. Crittell, P. J. Stang and N. S. Zefirov, *Tetrahedron Lett.*, 4828 (1990).
17. B. L. Williamson, P. J. Stang and A. Arif, *J. Am. Chem. Soc.*, **115**, 2590 (1993); P. J. Stang, V. V. Zhdankin and B. L. Williamson, *J. Am. Chem. Soc.*, **113**, 5870 (1991).
18. P. J. Stang and V. V. Zhdankin, *J. Am. Chem. Soc.*, **113**, 4571 (1991); P. J. Stang and V. V. Zhdankin, *J. Am. Chem. Soc.*, **112**, 6437 (1990).
19. P. J. Stang, R. Tykwinski and V. V. Zhdankin, *J. Org. Chem.*, **57**, 1861 (1992).
20. P. J. Stang and Jörg Ullman, *Synthesis*, 1073 (1991).
21. J. R. Dalziel, H. A. Carter and F. Aubke, *Inorg. Chem.*, **15**, 1247 (1976).
22. P. J. Stang, V. V. Zhdankin and A. M. Arif, *J. Am. Chem. Soc.*, **113**, 8997 (1991).
23. G. C. Levy, R. L. Lichter and G. L. Nelson, *Carbon-13 Nuclear Magnetic Resonance Spectroscopy*, 2nd ed., Wiley-Interscience, New York, 1980.
24. P. Pyykkö, A. Görling and N. Rösch, *Mol. Phys.*, **61**, 195 (1987); Y. Yomura, Y. Takeuchi and N. Nakawa, *Tetrahedron Lett.*, 639 (1969); A. A. Cheremisin and P. Schastnev, *J. Magn. Reson.*, **40**, 459 (1980).
25. A. A. Mironova, I. I. Maletina, S. V. Iksanova, V. V. Orda and L. M. Yagupolskii, *Zh. Org. Khim.*, **25**, 306 (1989).
26. P. J. Stang, A. M. Arif and C. M. Crittell, *Angew. Chem., Int. Ed. Engl.*, **29**, 287 (1990).
27. M. Ochiai, M. Kunishima, K. Fuji, Y. Nagao and M. Shiro, *Chem. Pharm. Bull.*, **37**, 1948 (1989).
28. H. G. Viehe, *Chemistry of Acetylenes*, Dekker, New York, 1969.
29. J. I. Dickstein and S. I. Miller, in *The Chemistry of the Carbon–Carbon Triple Bond* (Ed. S. Patai), Chap. 19, Wiley-Interscience, London, 1978, pp. 813–955.
30. K. B. Wiberg, W. E. Pratt and M. G. Matturro, *J. Org. Chem.*, **47**, 2720 (1982).
31. P. J. Stang, M. Boehshar, H. Wingert and T. Kitamura, *J. Am. Chem. Soc.*, **110**, 3272 (1988); P. J. Stang, M. Boehshar and J. Lin, *J. Am. Chem. Soc.*, **108**, 7832 (1986).
32. P. J. Stang, T. Kitamura, M. Boehshar and H. Wingert, *J. Am. Chem. Soc.*, **111**, 2225 (1989).
33. For a review see: P. J. Stang, *Acc. Chem. Res.*, **24**, 304 (1991).
34. J. N. Blenkenship, H. Abu-Soud, W. A. Francisco, F. M. Raushel, D. R. Fischer and P. J. Stang, *J. Am. Chem. Soc.*, **113**, 8560 (1991); D. Segal, Y. Shalitin, H. Wingert, T. Kitamura and P. J. Stang, *FEBS. Lett.*, **247**, 217 (1989).
35. G. Maas, R. Rahm, F. Krebs, M. Regitz, P. J. Stang and C. M. Crittell, *Chem. Ber.*, **124**, 1661 (1991); G. Maas, M. Regitz, R. Rahm, H. Wingert, J. Schneider, P. J. Stang and C. M. Crittell, *J. Chem. Soc., Chem. Commun.*, 1456 (1990).
36. P. J. Stang and K. A. Roberts, *J. Am. Chem. Soc.*, **108**, 7125 (1986).
37. A. D. Allen, T. Kitamura, R. A. McClelland, P. J. Stang and T. T. Tidwell, *J. Am. Chem. Soc.*, **112**, 8873 (1990).
38. P. J. Stang and K. A. Roberts, *J. Org. Chem.*, **52**, 5213 (1987).

39. D. R. Fischer, B. L. Williamson and P. J. Stang, *Synlett.*, 535 (1992).
40. P. J. Stang and C. M. Crittell, *J. Org. Chem.*, **57**, 4305 (1992).
41. M. Ochiai, M. Kunishima, Y. Nagao, K. Fuji and E. Fujita, *J. Chem. Soc., Chem. Commun.*, 1708 (1987).
42. J. S. Lodoya and G. F. Koser, *J. Org. Chem.*, **55**, 1513 (1990).
43. Z. D. Lu and Z. C. Chen, *J. Org. Chem.*, **58**, 1924 (1993).
44. R. Tykwinski and P. J. Stang, *Tetrahedron*, **49**, 3043 (1993).
45. A. Schmidpeter, P. Mayer, J. Stocker, K. A. Roberts and P. J. Stang, *Heteroatom Chem.*, **2**, 569 (1991).
46. M. Ochiai, T. Ho, Y. Takaoka, Y. Masaki, M. Kunishima, S. Tani and Y. Nagao, *J. Chem. Soc., Chem. Commun.*, 118 (1990).
47. P. J. Stang and C. M. Crittell, *Organometallics*, **9**, 3191 (1990).
48. P. J. Stang and R. Tykwinski, *J. Am. Chem. Soc.*, **114**, 4411 (1992).
49. M. Ochiai, M. Kunishima, K. Fuji and Y. Nagao, *J. Org. Chem.*, **53**, 6144 (1988).
50. P. J. Stang, in *Methoden Org. Chem.*, 4th ed. Vol. E19b, Part I (Houben-Weyl), 1992, pp. 84–165; *Acc. Chem. Res.*, **15**, 348 (1982); *Chem. Rev.*, **78**, 383 (1978).
51. T. Kitamura and P. J. Stang, *Tetrahedron Lett.*, **29**, 1887 (1988).
52. M. Ochiai, M. Kunishima, Y. Nagao, K. Fuji, M. Shiro and E. Fujita, *J. Am. Chem. Soc.*, **108**, 8281 (1986).
53. M. Ochiai, M. Kunishima, S. Tani and Y. Nagao, *J. Am. Chem. Soc.*, **113**, 3135 (1991).
54. E. Kotali, A. Varvoglis and A. Bozopoulos, *J. Chem. Soc., Perkin Trans. 1*, 827 (1989).
55. G. Maas, M. Regitz, U. Moll, R. Rahm, F. Krebs, R. Hector, P. J. Stang, C. M. Crittell and B. L. Williamson, *Tetrahedron*, **48**, 3527 (1992).
56. M. Regitz and M. Heydt, in *1,3-Dipolar Cycloaddition Chemistry*, Vol. 1 (Ed. A. Padwa), Chap. 4, Wiley, New york, 1984.
57. R. Huisgen Vol. 1, Chap. 1 and K. Houk and K. Yamaguchi, Vol. 2, Chap. 13, in *1,3-Dipolar Cycloaddition Chemistry* (Ed. A. Padwa), Wiley, New York, 1984.
58. (a) M. Ochiai, K. Sumi, Y. Takaoka, M. Kunishima, Y. Nagao, M. Shiro and E. Fujita, *Tetrahedron*, **44**, 4095 (1988).
    (b) M. Ochiai, K. Sumi, Y. Nagao and E. Fujita, *Tetrahedron Lett.*, **26**, 2351 (1985).
59. P. J. Stang, T. Blume and V. V. Zhdankin, *Synthesis*, **1**, 35 (1993).
60. P. J. Stang, A. Schwatz, T. Blume and V. V. Zhdankin, *Tetrahedron Lett.*, **33**, 6759 (1992).
61. A. J. Margida and G. F. Koser, *J. Org. Chem.*, **49**, 4703 (1984.
62. M. Ochiai, M. Kunishima, K. Fuji and Y. Nagao, *J. Org. Chem.*, **54**, 4038 (1989).
63. T. Kitamura, I. Mihara, H. Taniguchi and P. J. Stang, *J. Chem. Soc., Chem. Commun.*, 614 (1990).
64. P. J. Stang and T. Kitamura, *J. Am. Chem. Soc.*, **109**, 7561 (1987).
65. Reviews: W. J. Scott and J. E. McMurry, *Acc. Chem. Res.*, **21**, 47 (1988); E. Negishi, *Acc. Chem. Res.*, **15**, 340 (1982); A. Suzuki, *Acc. Chem. Res.*, **25**, 178 (1982); J. K. Stille, *Angew. Chem.*, **98**, 504 (1986); *Angew. Chem., Int. Ed. Engl.*, **25**, 508 (1986); *Pure Appl. Chem.*, **57**, 1771 (1985); A. Suzuki, *Pure Appl. Chem.*, **57**, 1749 (1985).
66. T. Kitamura, T. Tanaka, H. Taniguchi and P. J. Stang, *J. Chem. Soc., Perkin, Trans. 1*, 2892 (1991).
67. V. V. Zhdankin, P. J. Stang and N. S. Zefirov, *Chem. Commun.*, 578 (1992).
68. P. J. Stang, Z. Rappoport, M. Hanack and L. R. Subramanian, *Vinyl Cations*, Academic Press, New York, 1979.
69. G. Angelini, M. Hanack, J. Vermehren and M. Speranza, *J. Am. Chem. Soc.*, **110**, 1298 (1988).

CHAPTER **21**

Mass spectrometry studies on molecules and ions with C≡C bonds

CHRYS WESDEMIOTIS, MICHAEL J. POLCE and ŠÁRKA
BERANOVÁ

Department of Chemistry, The University of Akron, Akron, Ohio 44325-3601, USA

Supplement C2: The chemistry of triple-bonded functional groups
Edited by S. Patai © 1994 John Wiley & Sons Ltd

1183

LIST OF ABBREVIATIONS

AE	appearance energy
CAD	collisionally activated dissociation
EI	electron ionization
FTICR	Fourier-transform ion cyclotron resonance
ICR	ion cyclotron resonance
KERD	kinetic energy release distribution
MI	metastable ion
NR	neutralization–reionization
NRMS	neutralization–reionization mass spectrometry
PIPECO	photoion–photoelectron coincidence
QIT-MS	quadrupole ion trap mass spectrometer
TOF	time of flight
TPEPICO	threshold photoelectron-photoion coincidence

This chapter is divided into three sections. Section 1 covers the electron ionization (EI) mass spectra of acetylenic compounds and discusses the types of singly and doubly charged cations formed on electron impact. Section 2 concerns the unimolecular chemistry of ions with C≡C bonds. Finally, Section 3 is devoted to the ion-molecule reactions of acetylenic ions and acetylenes. The material mainly originates from articles that have been published within the last decade and is presented with considerable detail. This review does not claim to be exhaustive; nevertheless, it provides examples from all areas of active mass spectrometry research in C≡C bonded molecules, so that the reader can learn about the important and novel developments in this area.

I. THE ELECTRON IONIZATION (EI) MASS SPECTRA OF ALKYNES

A. 70 eV EI and Charge Exchange Mass Spectra

The principal fragments in the 70 eV electron ionization (EI) mass spectra of *n*-alkynes containing 7–10 C-atoms allow for a ready distinction between isomers[1]. The most

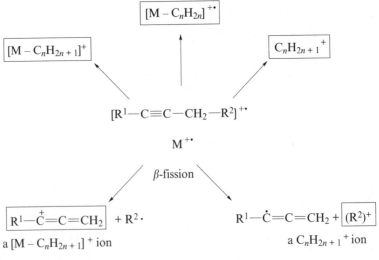

$$[M - C_nH_{2n}]^{+\bullet}$$

$$[M - C_nH_{2n+1}]^+ \qquad C_nH_{2n+1}^+$$

$$[R^1-C\equiv C-CH_2-R^2]^{+\bullet}$$

$$M^{+\bullet}$$

β-fission

$$R^1-\overset{+}{C}=C=CH_2 \quad + R^2\bullet \qquad\qquad R^1-\overset{\bullet}{C}=C=CH_2 + (R^2)^+$$

a $[M - C_nH_{2n+1}]^+$ ion a $C_nH_{2n+1}^+$ ion

SCHEME 1

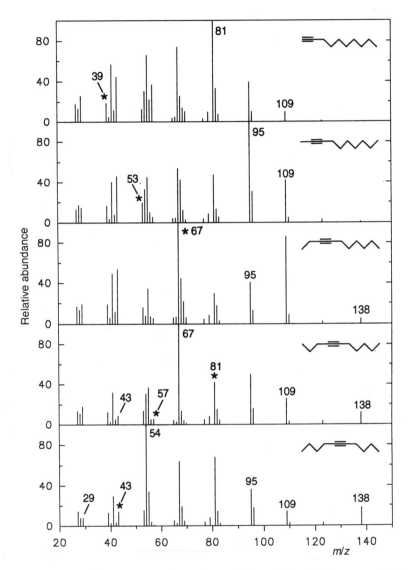

FIGURE 1. 70 eV EI mass spectra of linear decynes. The products of β-cleavage are indicated by *. Adapted from reference 1

pronounced differences are observed for the losses of $C_nH_{2n+1}\cdot$ radicals and C_nH_{2n} alkenes and in the abundances of $C_nH_{2n+1}^+$ and $M^{+\cdot}$ (Scheme 1). Generally, the terminal acetylenes show negligible $M^{+\cdot}$ intensities. The most abundant fragment for all 3-alkynes is the $[M-C_nH_{2n+1}]^+$ cation of m/z 67, generated by β-fission (see Scheme 1). However, for all other alkynes, β-fission is not a very important dissociation channel; this is illustrated in Figure 1 for the isomeric decynes.

The distinctive features of the 70 eV EI mass spectra are preserved in the $CS_2^{+\cdot}$ charge exchange mass spectra of the n-alkynes[1], in which the molecular ions are formed with < 1 eV internal energy[2]. On the other hand, the metastable ion (MI) spectra of $M^{+\cdot}$, which probe alkyne ions of threshold energies, provide only a limited differentiation among isomers[1].

The structure characteristic signals of the isomeric heptynes are enhanced in the $N_2O^{+\cdot}$ charge exchange mass spectra in which a relatively high internal energy (*ca* 4 eV)[2] is deposited upon ionization[1]. In contrast, there is no such enhancement for the octynes, while for the nonynes the spectral differences are reduced in the $N_2O^{+\cdot}$ charge exchange vis-à-vis the EI spectra. The latter result suggests that isomer interconversion becomes more facile as the size of the alkyne increases. For all alkynes studied, the $N_2O^{+\cdot}$ charge exchange spectra do not show the enhancement of β-fission fragments observed upon field ionization, despite the fact that both ionization methods produce alkyne molecular ions with similar internal energy contents (*ca* 4 eV)[1]. Consequently, field dissociation must be responsible for the abundant β-fission fragments that had been observed in field ionization mass spectra of alkynes[3].

B. 12 eV EI Mass Spectra

The 12 eV, low-temperature EI mass spectra of *n*-alkynes (Figure 2) show much fewer fragments than the above-discussed 70 eV EI spectra[4]. At low energy, alkynes essentially dissociate by cleavages of C_nH_{2n+1} and C_nH_{2n} (Scheme 1) giving rise to sizable $C_mH_{2m-3}^+$ and $C_mH_{2m-2}^{+\cdot}$ products, respectively. $M^{+\cdot}$ is the base peak (100% relative intensity) for ethyne, propyne and 1-butyne but drops to 54% for 1-pentyne and to <10% for the larger 1-alkynes whose spectra are dominated by the $C_mH_{2m-3}^+$ fragments $C_5H_7^+$ (*m/z* 67) or $C_6H_9^+$ (*m/z* 81). Apparently, $C_mH_{2m-3}^+$ ions of very low heat of formation can be formed

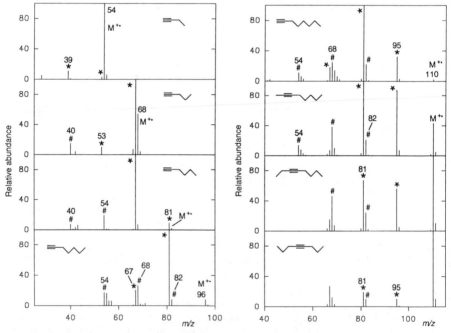

FIGURE 2. 12 eV, low-temperature (350 K) EI mass spectra of linear alkynes. The $C_mH_{2m-3}^+$ (∗) and $C_mH_{2m-2}^{+\cdot}$ (#) series are labeled. The spectrum of acetylene consists of $C_2H_2^{+\cdot}$ (100%) and the spectrum of propyne of $C_3H_4^{+\cdot}$ (100%) and $C_3H_3^{+\cdot}$ (9%). Adapted from Reference 4 .

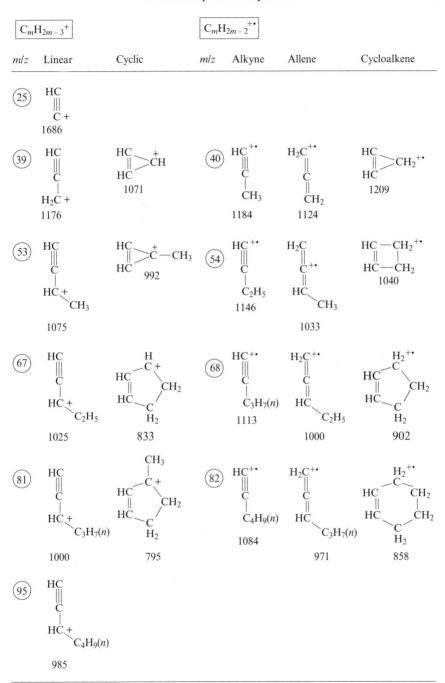

SCHEME 2. ΔH°_f in kJ mol^{-1} of $C_mH_{2m-3}^+$ and $C_mH_{2m-2}^{+\bullet}$ cations. All values from Reference 4

from precursors with >5 carbons. Thermochemical data (Scheme 2) suggest that these ions must have cyclic structures, as cyclic $C_mH_{2m-3}{}^+$, m >5, exhibit exceptionally high thermodynamic stabilities.

Moving the triple bond towards the center of the alkyne substantially increases the relative abundance and hence stability of the corresponding molecular ion (Figure 2)[4]. With the isomeric octynes the fragmentation yields (total fragment ion current divided by [M^+]) are 125 for 1-, 6.3 for 2-, 2.1 for 3- and 0.77 for 4-octyne. Analogous, albeit not as profound, trends were observed at 70 eV[1]. Thus, determination of the fragmentation yield of isomeric alkynes suffices to ascertain the position of the triple bond[4].

Appearance energy (AE) measurements on $C_mH_{2m-3}{}^+$ indicate that its formation requires a considerable activation energy[5,6]. For example, 1-hexyne$^{+\cdot}$ ($C_6H_{10}{}^+$) → cyclopentenium ($C_5H_7{}^+$) + $\cdot CH_3$ (Scheme 3) necessitates 117 kJ mol^{-1} more than the corresponding reaction enthalpy[6,7]. The presence of a substantial barrier is also reflected by the large kinetic energy release observed upon production of $C_5H_7{}^+$ from several C_6H_{10} alkynes[6]. The mechanism postulated for the formation of $C_5H_7{}^+$ from 1-hexyne is illustrated in Scheme 3[6]. Similar reaction sequences can operate for homologous $C_mH_{2m-3}{}^+$ ions.

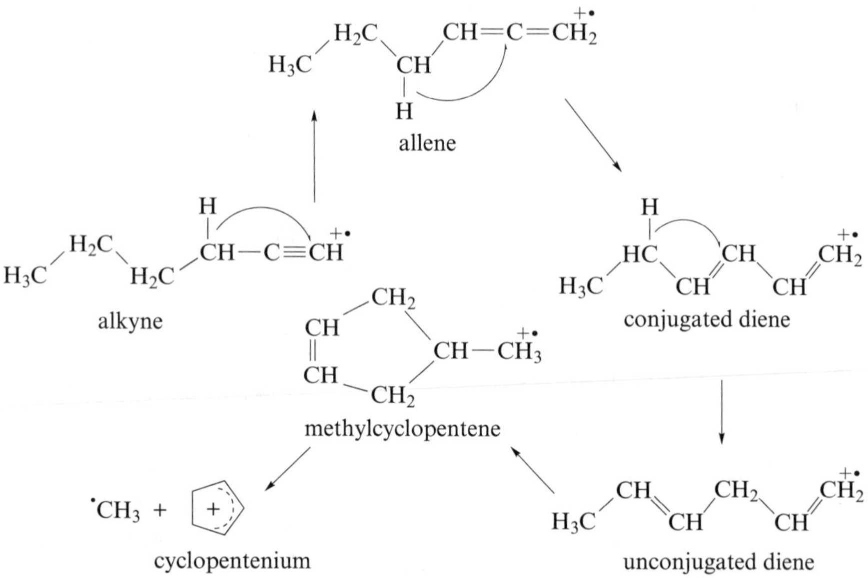

SCHEME 3

Concerning the formation of $C_mH_{2m-2}{}^{+\cdot}$, it has been found that the acetylenic and most of the terminal H atoms are not retained in the alkene molecule being eliminated[8]. Further, from the possible $C_mH_{2m-2}{}^{+\cdot}$ products (viz. Scheme 2), the allenic and cyclic (>C_4) ions are

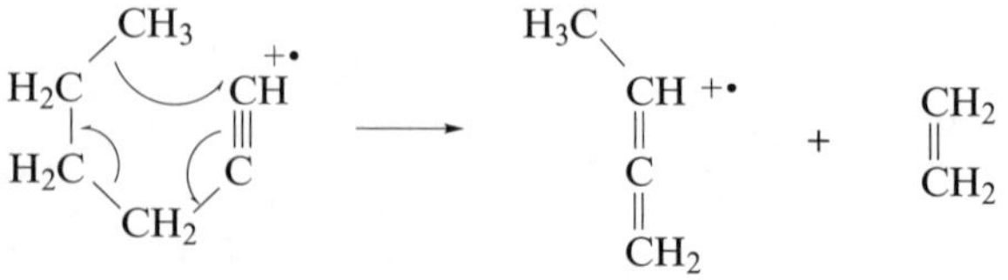

SCHEME 4

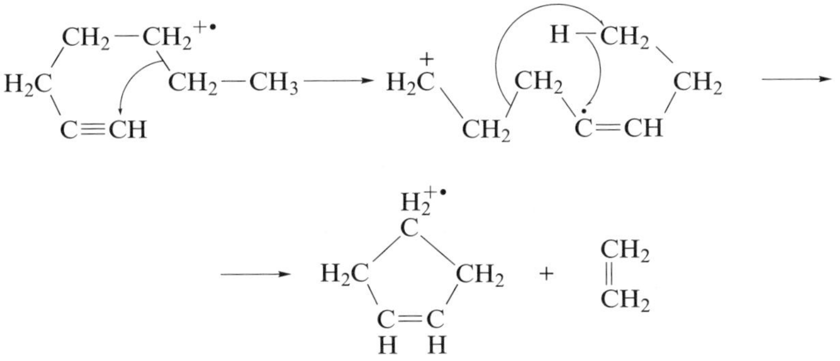

SCHEME 5

substantially more stable than their acetylenic isomers[2,4] and, therefore, more likely to be formed at the thermodynamic threshold. Schemes 4 and 5 rationalize the generation of such allenic and cyclic $C_mH_{2m-2}^{+\cdot}$ ions from 1-hexene and 1-heptene, respectively[4].

C. Doubly Charged Ion Mass Spectra

Electron ionization of alkynes also produces doubly charged ions[9], which can be separated from the simultaneously generated singly charged ions by using doubly charged ion ('2E') mass spectrometry[9,10]. Acetylene and phenyl substituted alkynes yield abundant M^{2+} (base peaks) but aliphatic molecules give no observable molecular dications. As illustrated for C_5H_8 isomers in Figure 3, the 2E mass spectra of linear alkynes differing in the position of the triple bond are indistinguishable; however, branched isomers show some differences, originating from enhanced fragmentation at the substitution center[9].

The 2E mass spectra of all aliphatic acetylenes contain intense $C_nH_2^{2+}$ product ions. For the smaller alkynes (propyne, butynes, pentynes, unbranched hexynes and heptynes), base

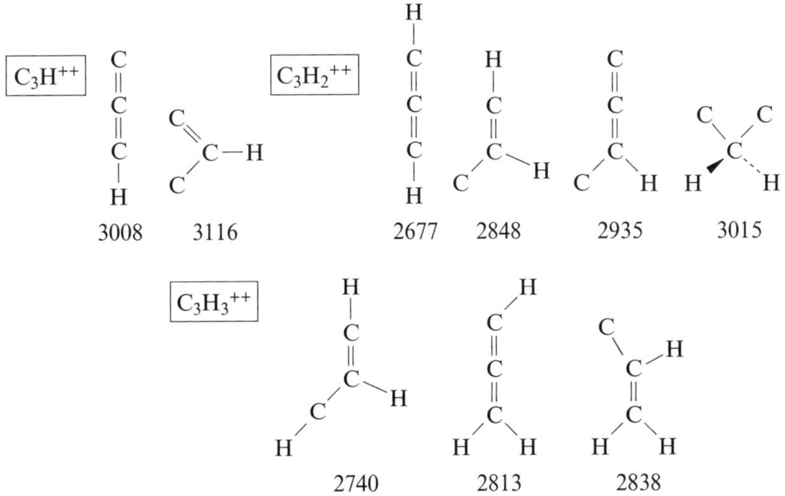

SCHEME 6. Calculated (MINDO/3) ΔH°_f (in kJ mol^{-1}) of $C_3H_n^{++}$ dications ($n = 1$–3). All values from Reference 9

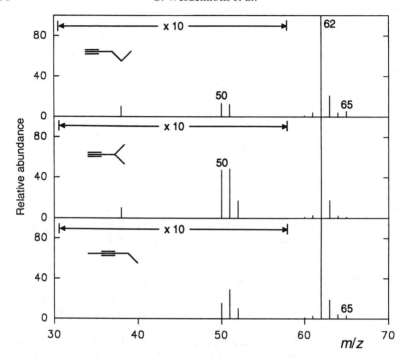

FIGURE 3. 2E mass spectra of C_5H_8 isomers. The center spectrum contains larger m/z 50–51 fragments than other two spectra. Adapted from Reference 9

peak is the $C_nH_2^{2+}$ fragment resulting by hydrogen losses from M^{2+}, i.e. $C_nH_2^{2+}$ with n equal to the number of carbon atoms in the molecular ion (see; e.g., Figure 3). As the size of the alkyne increases, C—C bond scission accompanies the H eliminations; thus, both 1-octyne (C_8H_{14}) and 1-nonyne (C_9H_{16}) give rise to $C_7H_2^{2+}$ base peaks. Branching may also promote C—C cleavage, as in the case of 3,3-dimethyl-1-butyne (C_6H_8) whose spectrum is dominated by $C_5H_2^{2+}$ (100% vis-à-vis 16% for $C_6H_2^{2+}$)[9]. In contrast to the aliphatic acetylenes, aromatic acetylenes yield much weaker $C_nH_2^{2+}$; as mentioned above, the latter species mainly lead to M^{2+} [9].

Besides $C_nH_2^{2+}$, other important fragments in the 2E spectra of aliphatic alkynes are C_nH^{2+} and $C_nH_3^{2+}$. Further, 2- and 3-hexyne also show abundant $C_6H_6^{2+}$ products (35% and 20%, respectively), while 1-nonyne forms sizable $C_7H_6^{2+}$ (52%) and $C_8H_6^{2+}$ (58%). It has been suggested that the $C_nH_6^{2+}$ dications have the structure $CH_3C_{4-6}CH_3^{2+}$ [9].

Semiempirical MINDO/3 theory[11] predicts that the lowest-energy $C_nH_2^{2+}$ ions have linear structures with the H atoms on the terminal carbons (Scheme 6). Such linear structures must indeed be formed because their ΔH°_f values lead to calculated appearance energies for $C_nH_{2n-2} \rightarrow C_nH_2^{2+} + (2n-4)H$ that agree reasonably with experimentally determined AEs (24–47 eV)[9]. Also for C_nH^{2+} and $C_nH_3^{2+}$, theory suggests that the lowest-energy isomers are linear and contain single hydrogens on the end carbons (Scheme 6); evidently, this connectivity minimizes the coulombic repulsion between the two positive charges, thus providing a higher thermodynamic stability.

D. Heteroatom Substituted Alkynes

The γ-acetylenic alcohol **1** and the isomeric β-allenic alcohol **2** bear no methyl substitu-ents but produce abundant fragments by ˙CH_3 loss in both their 17 eV EI mass spectra and

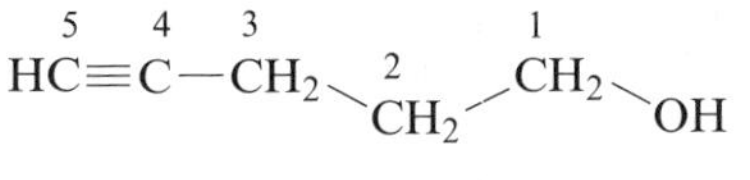

$$\overset{5}{HC}\equiv\overset{4}{C}-\overset{3}{CH_2}\diagdown\overset{2}{CH_2}\diagup\overset{1}{CH_2}\diagdown OH$$

(1)

$$\overset{5}{H_2C}=\overset{4}{C}=\overset{3}{CH}\diagdown\overset{2}{CH_2}\diagup\overset{1}{CH_2}\diagdown OH$$

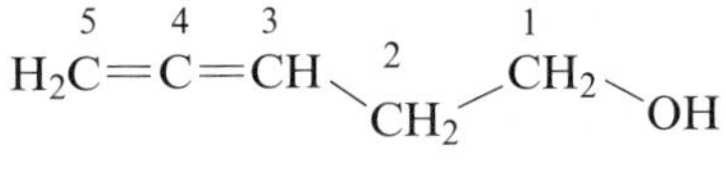

(2)

the MI spectra of the corresponding molecular cations[12]. Based on collisionally activated dissociation (CAD)[13], the structure of the resulting $C_4H_5O^+$ product (m/z 69) is $CH_3CH=CHC\equiv O^+$. Hence, complicated rearrangements precede the $\cdot CH_3$ elimination. Deuterated isotopomers of **2** show that the C(5) carbon is lost in this fragmentation. Further, one H atom is transferred from either C(1) or C(2) to an almost equal extent, while the hydroxylic H atom of **2** does not participate in the methyl loss[12]. Very similar results are obtained for labeled $\mathbf{1^{+\cdot}}$. The common formation of $CH_3CH=CHC\equiv O^+$ from both $\mathbf{1^{+\cdot}}$ and $\mathbf{2^{+\cdot}}$ implies that similar intermediates are traversed before the cleavage of a methyl radical. A plausible mechanism, supported by the labeling data, is illustrated in Scheme 7 and involves isomerization of $\mathbf{1^{+\cdot}}$ to $\mathbf{2^{+\cdot}}$ by a hydrogen transfer that is 'catalyzed' by the hydroxyl group[12,14]. The intermediate $\mathbf{1a^{+\cdot}}$ corresponds to a proton solvated by a triple bond and a hydroxyl group and should be prone to H_2O loss. Indeed, water elimination is an important dissociation of $\mathbf{1^{+\cdot}}$, with $[\mathbf{1^{+\cdot}}-H_2O]/[\mathbf{1^{+\cdot}}-\cdot CH_3]$ equal to 1.3 in the EI and 0.24 in the MI spectrum. For comparison, $[\mathbf{2^{+\cdot}}-H_2O]/[\mathbf{2^{+\cdot}}-\cdot CH_3] = 0.13$ and <0.01, respectively, indicating that the isomerization $\mathbf{1^{+\cdot}} \to \mathbf{1a^{+\cdot}} \to \mathbf{2^{+\cdot}}$ is irreversible. From $\mathbf{2^{+\cdot}}$, $\cdot CH_3$ can be cleaved after a multistep pathway (Scheme 7) beginning with migration of the OH group

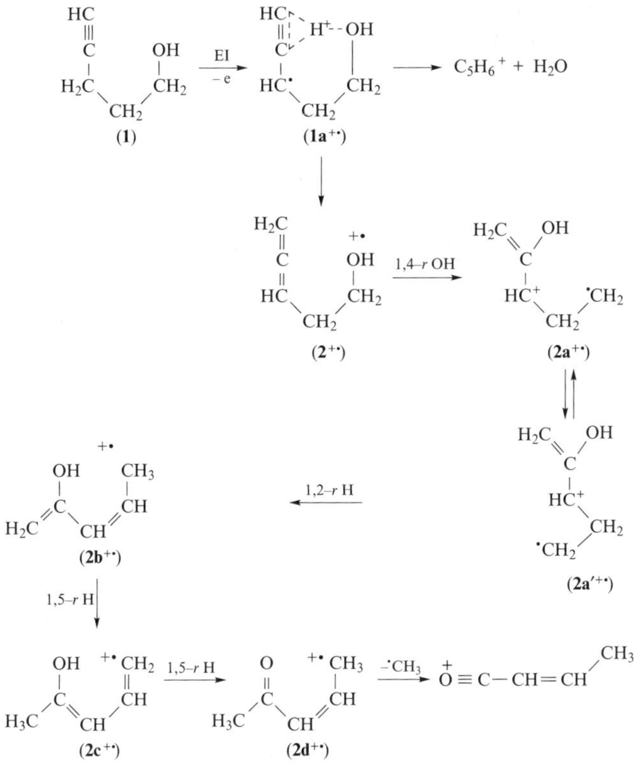

SCHEME 7

onto C(4). This rearrangement gives rise to the distonic ion $2a^{+\cdot}$ in which C(1) and C(2) can lose their positional identity by a fast 1,2-migration of the cationic center to the radical site[12,15]. Subsequent H-rearrangements ultimately lead to the keto ion $2d^{+\cdot}$ from which the observed $CH_3CH=CHC\equiv O^+$ fragment can be formed by a simple C—C bond rupture.

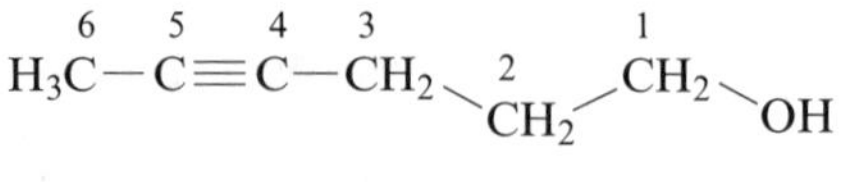

$$
\overset{6}{H_3C}-\overset{5}{C}\equiv\overset{4}{C}-\overset{3}{CH_2}\diagdown\underset{CH_2}{\overset{2}{}}\diagup\overset{1}{CH_2}\diagdown OH
\qquad\qquad
\overset{6}{H_3C}-\overset{5}{CH}=\overset{4}{C}=\overset{3}{CH}\diagdown\underset{CH_2}{\overset{2}{}}\diagup\overset{1}{CH_2}\diagdown OH
$$

$$
(3)\qquad\qquad\qquad\qquad\qquad\qquad\qquad(4)
$$

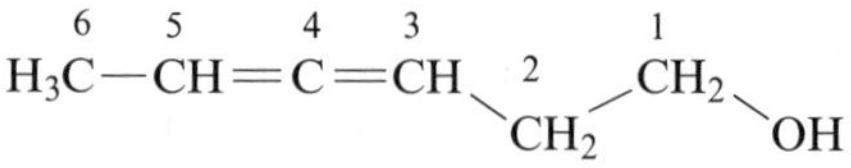

The homologous ions $3^{+\cdot}$ and $4^{+\cdot}$ behave in a parallel manner[12]. They eliminate $^\cdot C_2H_5$ [incorporating C(5) and C(6)] to predominantly yield the acylium cation $CH_3CH=CHC\equiv O^+$, presumably via a reaction sequence similar to Scheme 7. In addition, these heavier homologs also lose $^\cdot CH_3$. In the EI spectra, the cleaved methyl originates from C(6) and also C(1). However, metastable $3^{+\cdot}$ and $4^{+\cdot}$ primarily lose C(1); a mechanism conforming to this finding is shown in Scheme 8[12].

The neighboring group effects encountered upon the decomposition of the aliphatic alkynes 1 and 3 are quite common with aromatic acetylenes[16–18]. For example, base peak in the EI spectrum of *ortho*-methoxyphenylacetylene (5) is the ([M–H]+ fragment (*m/z* 131). This product is not noticed in the spectra of the *para* isomer and of phenylacetylene itself, inferring that the observed H$^\cdot$ loss is the result of an *ortho* effect. The mechanism postulated for this reaction is depicted in Scheme 9 and involves isomerization to ionized benzopyran (6^+) before H$^\cdot$ elimination[17]. In line with this proposal, the CAD spectra of [M–H]$^+$ from 5 and 6 are indistinguishable.

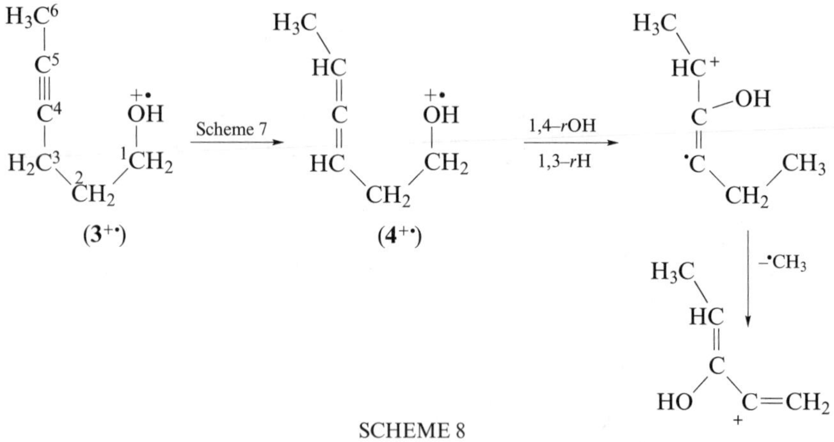

SCHEME 8

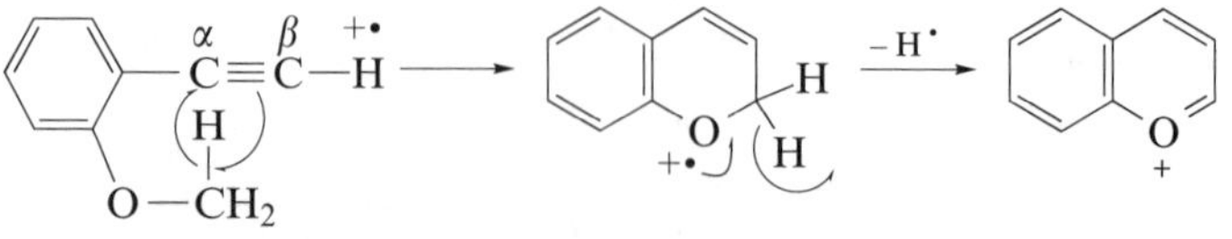

SCHEME 9

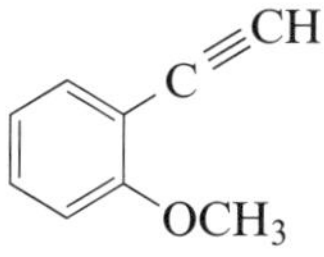

(5) 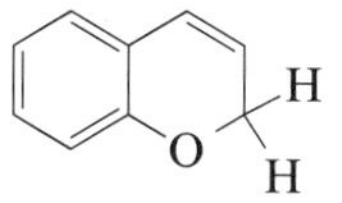 (6) 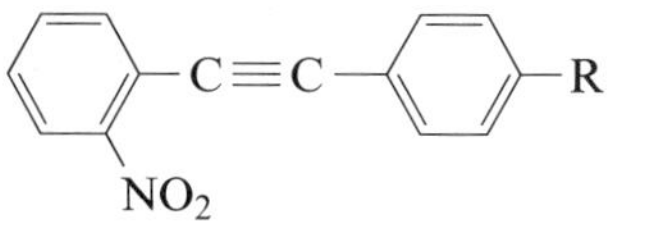 (7) R = H
     (8) R = CH$_3$

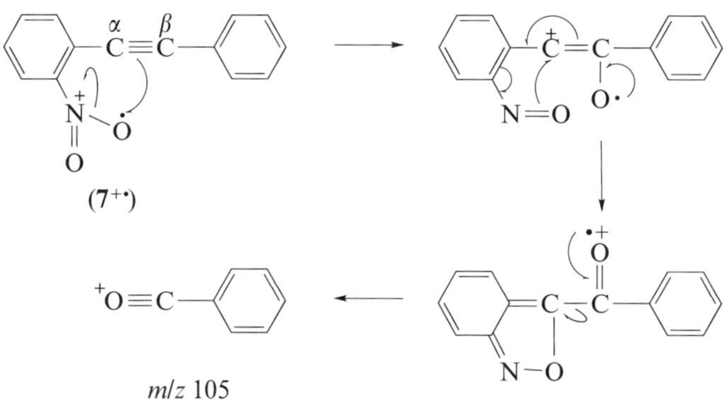

SCHEME 10

Similar *ortho* effects have been reported upon the EI-induced dissociation of nitro-substituted aromatic acetylenes[16,18], such as **7** and **8**. The major product in the mass spectrum of **7** has the composition C$_7$H$_5$O and is, according to CAD data, the benzoyl cation PhC≡O$^+$ (*m/z* 105)[16]. Formation of this ion necessitates O-transfer to one of the acetylenic carbons. Since **8** gives rise to a *p*-CH$_3$—C$_6$H$_4$C≡O$^+$ base peak (*m/z* 119) the oxygen must be transferred to the β-carbon, as rationalized in Scheme 10[16]. It is noteworthy that a hydrogen atom preferentially rearranges to the α-acetylenic carbon (Scheme 9). Thus, the most favorable transition state for these *ortho* effects depends significantly on the nature of the migrating moiety[16–18].

II. UNIMOLECULAR CHEMISTRY OF ACETYLENIC IONS

A. Metal Ion (M$^+$) -C$_2$H$_2$ Bond Strengths

Reaction of the transition metal ions Sc$^+$, Y$^+$ and La$^+$ (formed by laser desorption[19]) with C$_2$H$_4$ in a Fourier-transform ion cyclotron resonance (FTICR) mass spectrometer mainly yields MC$_2$H$_2^+$ + H$_2$[20]. Photodissociation of MC$_2$H$_2^+$ with a tunable laser enables one to monitor the threshold for the process MC$_2$H$_2^+$ → M$^+$ + C$_2$H$_2$ and thus determine the bond energies of the metal ion–acetylene bonds. The so-obtained D(M$^+$–C$_2$H$_2$) values are 218 kJmol^{-1} for all three metals[20]. This result is in excellent agreement with the energies predicted by theory for Sc$^+$ (222 kJmol^{-1}) and Y$^+$ (222 kJmol^{-1}), but not for La$^+$ (276 kJmol^{-1})[21]. The discrepancy with La$^+$ could be due to the fact that LaC$_2$H$_2^+$ → La$^+$ + C$_2$H$_2$ occurs via multiphoton and not, as expected, single-photon dissociation; in such a case the experimental bond energy could be a lower limit[20].

B. Dissociation Dynamics of the Cluster Ion (OCS:C₂H₂)⁺·

Chemical ionization of a *ca* 2% C_2H_2 mixture in OCS (0.04–0.10 Torr) leads *inter alia* to formation of the cluster ion $(OCS:C_2H_2)^{+\cdot}$.[22] Expectedly, the CAD spectrum of this cation contains abundant OCS^+ (100%) and $C_2H_2^{+\cdot}$ (42%) fragments; however, $^{+\cdot}SC_2H_2$ (65%) is also produced, suggesting chemical bonding in the $(OCS:C_2H_2)^{+\cdot}$ cluster as opposed to a mere electrostatic interaction. The detailed dynamics of unimolecular dissociation of $(OCS:C_2H_2)^{+\cdot}$ have been assessed by studying its metastable dissociation and photodissociation[22].

The MI spectrum of $(OCS:C_2H_2)^{+\cdot}$ shows two products, namely $OCS^{+\cdot}$ (*ca* 10% at the lowest source pressure) and $^{+\cdot}SC_2H_2$ (100%). Increasing the ion source pressure reduces the internal energy of the complex by collisional cooling; this in turn raises $[^{+\cdot}SC_2H_2]/[OCS^+]$, revealing that formation of $^{+\cdot}SC_2H_2$ requires a lower critical energy than formation of $OCS^{+\cdot}$. Further, the peak for $OCS^{+\cdot}$ is narrow and near-gaussian, with a kinetic energy release distribution (KERD)[23,24] peaking near zero and an average kinetic energy release ($\langle T \rangle$) of only 6 kJ mol⁻¹. In sharp contrast, the $^{+\cdot}SC_2H_2$ peak is flat-topped and yields a triangular KERD that is strongly shifted from zero and extends up to $T_{max} = 92$ kJ mol⁻¹ ($\langle T \rangle = 41$ kJ mol⁻¹)[22]. These data clearly demonstrate that the dissociation $(OCS:C_2H_2)^{+\cdot} \rightarrow OCS^{+\cdot} + C_2H_2$ is continuously endothermic and has no reverse activation energy (Figure 4) while $(OCS:C_2H_2)^{+\cdot} \rightarrow CO + {}^{+\cdot}SC_2H_2$ must be associated with a considerable reverse activation energy, which causes the observed peak broadening[25]. Based on T_{max} of the latter reaction, $\Delta H_f^\circ(CO + {}^{+\cdot}SC_2H_2)$ must lie at least 92 kJ mol⁻¹ below ΔH°_f $(OCS^{+\cdot} + C_2H_2)$, i.e. the overall process $OCS^{+\cdot} + C_2H_2 \rightarrow CO + {}^{+\cdot}SC_2H_2$ must be >92 kJ mol⁻¹ exothermic (see Figure 4). *Ab initio* calculations show that this is possible if $^{+\cdot}SC_2H_2$ has the thioketene or thiirene structures; the ethynethiol isomer is ruled out. The experimental data do not allow one to distinguish between a thioketene or thiirene product ion. Most likely only one of them is formed because the KERD for $^{+\cdot}SC_2H_2$ is not bimodal[22].

Photodissociation of $(OCS:C_2H_2)^{+\cdot}$ with a tunable laser (2.1–3.5 eV) yields solely $OCS^{+\cdot}$ and $C_2H_2^{+\cdot}$ in a branching ratio of 6:1 which remains independent of the photon energy[22]. The peak shapes of the $OCS^{+\cdot}$ and $C_2H_2^{+\cdot}$ photoproducts change substantially upon

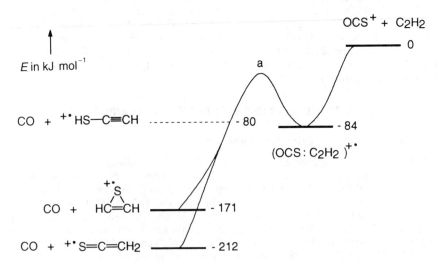

FIGURE 4. Potential energy diagram of the metastable reactions of the cluster ion $(OCS:C_2H_2)^{+\cdot}$. All energies shown are *ab initio* values from Reference 22. [a]CO loss requires less critical energy than C_2H_2 loss (see text). Since metastable ions have narrow internal energy distributions (<0.5 eV)[25], this transition state lies somewhere between 0 and >–50 kJ mol⁻¹

varying the laser polarization angle from 0° to 90°. This strong anisotropy of the angular distributions of the photoproducts suggests that dissociation of $\{(OCS:C_2H_2)^{+\cdot}\}^*$ occurs much faster than rotation of the cluster and therefore from a repulsive excited state [22]. The average kinetic energy releases of the photoproducts increase linearly with photon energy. Modeling the photodissociation as an impulsive half-collision[26] allows one to calculate the reduced mass of the atoms connected in the breaking bond. The obtained value (9.4 amu) is very close to the reduced mass for a carbon–sulfur bond (8.7 amu); apparently, the $(OCS:C_2H_2)^{+\cdot}$ cluster is a sulfur–carbon bonded species with the connectivity $(O—C—S—CH—CH)^{+\cdot}$. Such a structure also is the most stable form found by *ab initio* theory, which predicts a cluster bond energy of 84 kJ mol^{-1} (Figure 4). This value agrees well with the experimental bond dissociation energy (<111 kJ mol^{-1})[22] provided by analysis of the photodissociation KERDs using impulsive theory.

C. Photoionization and Dissociation of Acetylene Oligomers

Mass spectrometry is increasingly used for the detection of clusters. In order to unequivocally characterize the size of a neutral cluster, it is important to know whether the

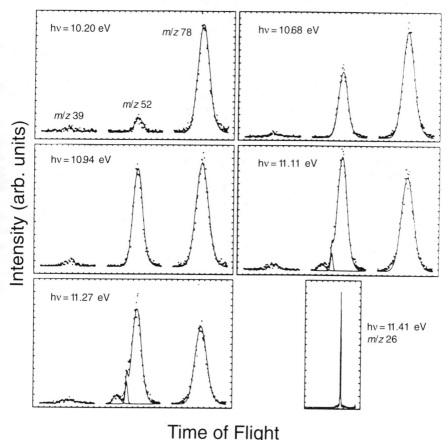

Time of Flight

FIGURE 5. TPEPICO time-of-flight mass spectra of supersonically expanded acetylene at five photon energies. Lower right: HC≡CH (monomer) mass peak indicating narrow shape of translationally cold parent ions. Reproduced from Reference 28 by permission of the American Institute of Physics

ions observed in the spectrum arise by direct ionization or by dissociative ionization of larger clusters[27]. Differentiation of these processes can be achieved by threshold photoelectron photoion coincidence (TPEPICO) spectroscopy of cluster beams produced by supersonic expansion[27]. In supersonic beams, the translational temperatures perpendicular to the beam axis are very small (*ca* 5 K); therefore; ions formed directly from such beams have exceptionally narrow dispersion in their time of flight (TOF) in a TOF mass spectrometer. In sharp contrast, ions from dissociative photoionization give rise to substantially broader peaks owing to the kinetic energy released during fragmentation[27].

The TPEPICO mass spectra of the acetylene clusters arising upon supersonic expansion of a 1:3 acetylene:argon mixture are shown in Figure 5[28]. At all photon energies used (ranging from 10.20 to 11.27 eV) the dominant peaks correspond to $C_{2n}H_{2n}^{+\cdot}$ ions ($n = 1-3$, i.e. *m/z* 26, 52, 78). The absence of narrow TOF signals for masses larger than the monomer confirms that acetylene clusters ionize only by dissociative pathways. Further, the spectra of Figure 5 differ significantly from the TPEPICO spectrum of benzene, indicating that the *m/z* 78 ion from acetylene clusters does not represent benzene [28].

Ab initio calculations on neutral and ionized $(C_2H_2)_2$ and $(C_2H_2)_3$ show that the equilibrium geometries of the neutral dimer and trimer are markedly different from those of the respective molecular cations[28]. As a consequence, adiabatic ionization of the clusters has very poor Franck–Condon factors. Vertical ionization, on the other hand, produces $(C_2H_2)_2^{+\cdot}$ and $(C_2H_2)_3^{+\cdot}$ well above their dissociation limits to $C_3H_3^+ + \cdot H$ (or $C_4H_4^{+\cdot} + H_2$) and $C_4H_4^{+\cdot} + C_2H_2$, respectively, thus justifying the lack of narrow $C_4H_4^{+\cdot}$ and $C_6H_6^{+\cdot}$ peaks in Figure 5.

The precursors of the wide $C_4H_4^{+\cdot}$ (*m/z* 52) and $C_6H_6^{+\cdot}$ (*m/z* 78) fragments in Figure 5 can be found by statistical modeling of the KERDs of their gaussian peaks. The procedure is explained in detail in Reference 28. Statistical analysis of the peak widths observed for $C_4H_4^{+\cdot}$ reveals that formation of $C_4H_4^{+\cdot}$ from the neutral trimer, via $(C_2H_2)_3 + h\nu \rightarrow C_4H_4^{+\cdot} + C_2H_2$, would lead to a $C_4H_4^{+\cdot}$ isomer with $\Delta H^\circ_{f,0K} = 1240$ kJmol⁻¹. This value is in excellent agreement with the 0 K heats of formation of ionized vinylacetylene (1240), butatriene (1240) and cyclobutadiene (1226 kJmol⁻¹). On the other hand, $C_4H_4^{+\cdot}$ generation from the tetramer, via $(C_2H_2)_4 + h\nu \rightarrow C_4H_4^{+\cdot} + 2C_2H_2$, would lead to $\Delta H^\circ_{f,0K}(C_4H_4^{+\cdot}) = 1010$ kJmol⁻¹. The latter value is >170 kJmol⁻¹ lower than $\Delta H^\circ_{f,0K}$ of any known $C_4H_4^{+\cdot}$ ion. Consequently, $C_4H_4^{+\cdot}$ must be formed by dissociative ionization of $(C_2H_2)_3$, not $(C_2H_2)_4$ [28]. Similar analysis of the peak widths of $C_6H_6^{+\cdot}$ (*m/z* 78) suggests that the most likely route to this product is the dissociative ionization $(C_2H_2)_4 + h\nu \rightarrow C_6H_6^{+\cdot} + C_2H_2$; the heat of formation predicted for $C_6H_6^{+\cdot}$ is 1160 kJmol⁻¹, which agrees very well with $\Delta H^\circ_{f,0K}$(benzvalene ion) = 1163 kJmol⁻¹ [28,29]. The $C_4H_4^{+\cdot}$ and $C_6H_6^{+\cdot}$ products most likely arise by C_2H_2 evaporation from the incipient

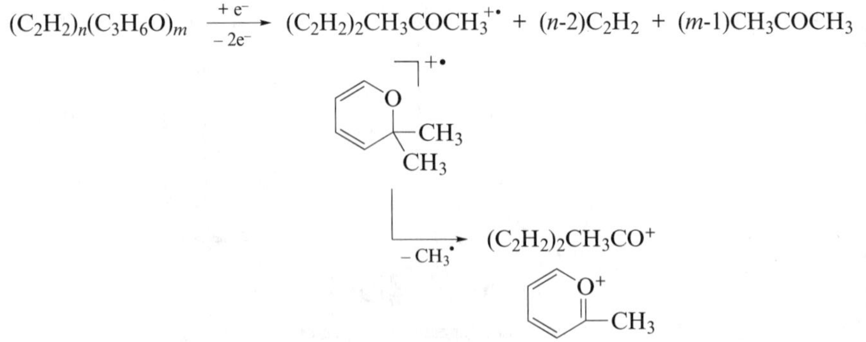

$$(C_2H_2)_n(C_3H_6O)_m \xrightarrow[-2e^-]{+e^-} (C_2H_2)_2CH_3COCH_3^{+\cdot} + (n-2)C_2H_2 + (m-1)CH_3COCH_3$$

SCHEME 11

complexes $C_4H_4^{+\cdot}$:C_2H_2 and $C_6H_6^{+\cdot}$:C_2H_2, respectively, which are formed upon photo-ionization of $(C_2H_2)_3$ and $(C_2H_2)_4$ with sufficient excitation (*vide supra*) for immediate C_2H_2 elimination.

Elimination of monomer moieties from a transient cluster molecular ion has also been observed for mixed clusters[31]. For example, the two dominant heterocluster ions arising upon electron ionization of a supersonically expanded mixture of *ca* 3% $HC{\equiv}CH$ in acetone are $(C_2H_2)_2(CH_3COCH_3)_1^{+\cdot}$ and $(C_2H_2)_2(CH_3CO)_1^{+}$. The magic number 2 for acetylene has been proposed to result from the production of covalently bonded cyclic ions by intercluster ion–molecule polymerization reactions (Scheme 11)[31]. Formation of such stable ions should be exothermic, the reaction heat being dissipated by evaporating the remaining monomer molecules.

D. Novel Heteroatom Substituted Acetylenes and Their Cations

Ionized hydroxyacetylene ($HC{\equiv}C{-}OH^{+\cdot}$, **9**$^{+\cdot}$), the enol isomer of the ketene radical cation ($H_2C{=}C{=}O^{+\cdot}$, **10**$^{+\cdot}$), cannot be formed directly because **9** does not exist in condensed phases[32]. It is, however, available by dissociative ionization of propynoic acid (Scheme 12)[33]. The CAD spectrum of **9**$^{+\cdot}$ is completely different from that of **10**$^{+\cdot}$, indicating that the exothermic[33,34] isomerization **9**$^{+\cdot}$ → **10**$^{+\cdot}$ does not occur in the time scale of the mass spectrometer ($<10\mu s$)[33].

Metastable **10**$^{+\cdot}$ dissociates exclusively into $CH_2^{+\cdot}$ + CO (Figure 6)[33]. The small kinetic energy release accompanying this reaction ($T_{0.5} = 0.2$ kJ mol^{-1}) is in keeping with a direct cleavage with no reverse activation energy[13,25]. Metastable **9**$^{+\cdot}$ yields the same products, $CH_2^{+\cdot}$ + CO, but with a substantially larger kinetic energy release ($T_{0.5} = 2.5$ kJ mol^{-1})[33]. This result implies that **9**$^{+\cdot}$ undergoes a rate-determining isomerization to **10**$^{+\cdot}$ prior to CO loss (Figure 6)[25]. The theoretical value for the isomerization barrier is 331 kJ mol^{-1}[33]. Since **9**$^{+\cdot}$ → $^+HC{=}C{=}O$ + $\cdot$H is not observed, it must involve a significantly higher critical energy[33].

From the stable cation **9**$^{+\cdot}$, it is possible to generate and study the elusive hydroxyacetylene (**9**) by neutralization–reionization mass spectrometry (NRMS)[35–39]. In a NRMS experiment (Scheme 13), mass-selected $HC{\equiv}C{-}OH^{+\cdot}$ (keV domain) is first neutralized, e.g. by collisions with Xe; any remaining ions are removed from the beam path by electrostatic

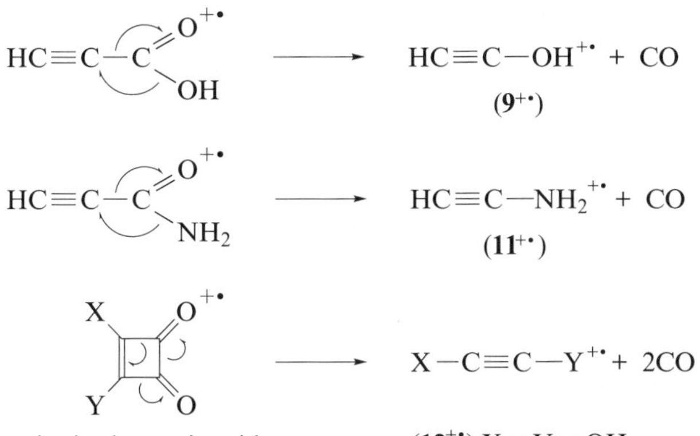

$$HC{\equiv}C{-}C{\overset{O^{+\cdot}}{\underset{OH}{\diagdown}}} \longrightarrow HC{\equiv}C{-}OH^{+\cdot} + CO$$
$$(\mathbf{9}^{+\cdot})$$

$$HC{\equiv}C{-}C{\overset{O^{+\cdot}}{\underset{NH_2}{\diagdown}}} \longrightarrow HC{\equiv}C{-}NH_2^{+\cdot} + CO$$
$$(\mathbf{11}^{+\cdot})$$

$$\overset{X}{\underset{Y}{\diagdown}}{\overset{O}{\underset{O}{\square}}}^{+\cdot} \longrightarrow X{-}C{\equiv}C{-}Y^{+\cdot} + 2CO$$

ionized squaric acid
(X = Y = OH) or derivatives

(**12**$^{+\cdot}$) X = Y = OH
(**13**$^{+\cdot}$) X = Y = NH$_2$
(**14**$^{+\cdot}$) X = Y = OCH$_3$
(**15**$^{+\cdot}$) X = OH, Y = NH$_2$

SCHEME 12

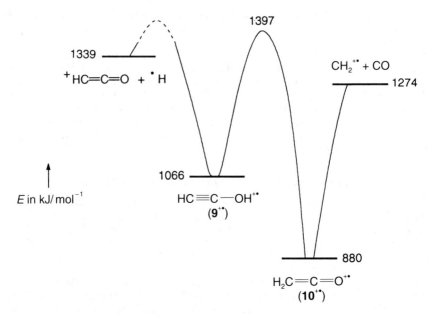

FIGURE 6. Potential energy diagram for the metastable reactions of ionized hydroxy-acetylene, based on data from References 2 and 33

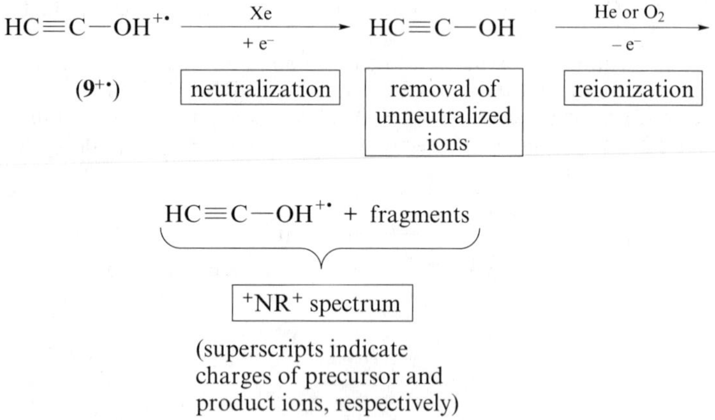

$$\text{HC} \equiv \text{C} - \text{OH}^{+\bullet} \xrightarrow[\substack{+ e^-}]{\text{Xe}} \text{HC} \equiv \text{C} - \text{OH} \xrightarrow[\substack{- e^-}]{\text{He or O}_2}$$

$$(\mathbf{9^{+\bullet}}) \qquad \boxed{\text{neutralization}} \qquad \begin{array}{c}\text{removal of}\\ \text{unneutralized}\\ \text{ions}\end{array} \qquad \boxed{\text{reionization}}$$

$$\text{HC} \equiv \text{C} - \text{OH}^{+\bullet} + \text{ fragments}$$

$$\boxed{^{+}\text{NR}^{+} \text{ spectrum}}$$

(superscripts indicate
charges of precursor and
product ions, respectively)

SCHEME 13

deflection, and the residual neutral beam is reionized by collisions with, e.g., He or O_2. The so-resulting neutralization–reionization ($^{+}NR^{+}$ Xe/He) spectrum of $\mathbf{9^{+\bullet}}$ contains all structurally diagnostic features present in the CAD spectrum of $\mathbf{9^{+\bullet}}$, indicating that the acetylenic structure is preserved upon the $^{+}NR^{+}$ sequence[33]. The $^{+}NR^{+}$ spectrum of $\mathbf{10^{+\bullet}}$ is characteristically different[33]. Consequently, HC≡C—OH ($\mathbf{9}$) does not isomerize to the thermodynamically more stable H_2C=C=O ($\mathbf{10}$). This experimental finding is in consent with theoretical calculations, which predict that the exothermic reaction $\mathbf{9} \rightarrow \mathbf{10}$ ($\Delta H^{\circ} = -151$ kJmol^{-1}) must surmount an appreciable barrier ($E^{\ddagger} = 305$ kJmol^{-1})[32].

Similar studies have been reported for many other, previously unknown acetylenes, including aminoacetylene (**11**) as well as the difunctional alkynes $X—C\equiv C—Y$ (X,Y = OH, NH_2, OCH_3)[40–42]. Ion **11**$^{+\cdot}$ is accessed by decarbonylation of the amide of propynoic acid (Scheme 12); ions $X—C\equiv C—Y^{+\cdot}$ (**12**$^{+\cdot}$–**15**$^{+\cdot}$) are available from appropriately substituted squaric acids (Scheme 12). All these ions produce structure characteristic CAD and $^+NR^+$ spectra, demonstrating that both the acetylenic ions and the corresponding neutral molecules reside in deep potential wells.

It is noteworthy that CAD of **9**$^{+\cdot}$, **11**$^{+\cdot}$ and **12**$^{+\cdot}$–**15**$^{+\cdot}$ generates (*inter alia*) prominent fragments by cleavage of the triple bond[33,40–42]. This decomposition yields long-sought organic cations, such as $C=O^+H$ (isoformyl cation) from **9**$^{+\cdot}$, $C=N^+H_2$ from **11**$^{+\cdot}$ and $C=O^+CH_3$ (O-methylated CO) from **14**$^{+\cdot}$, all of which are of potential interest in interstellar chemistry[42,43]. MS/MS/MS experiments[13] confirm that these ions exist as distinct species; hence, they must be separated by considerable barriers from the more stable isomers $HC\equiv O^+$, $HC\equiv NH^+$ and $CH_3C\equiv O^+$, respectively.

E. Vinylacetylene and Isomeric Radical Cations

Ionized vinylacetylene (**16**$^{+\cdot}$) and the $C_4H_4^{+\cdot}$ ions from pyridine, benzene and 1,5-hexadiyne produce identical MI spectra containing $C_4H_3^+$ and $C_4H_2^{+\cdot}$ fragments in the abundance ratio of *ca* 1.2:1[30]. All $C_4H_4^{+\cdot}$ precursors give rise to gaussian signals for $C_4H_3^+$ ($T_{0.5} = 4\,kJ\,mol^{-1}$) and dish-topped signals for $C_4H_2^{+\cdot}$ ($T_{0.5} = 40\,kJ\,mol^{-1}$)[30]. These MI data suggest that the isomerization barriers between $C_4H_4^{+\cdot}$ isomers are smaller than the corresponding dissociation thresholds, so that a common structure is reached before dissociation. Upon CAD, $C_4H_4^{+\cdot}$ isomers yield overall similar spectra, too, but some reproducible differences are now observed in the $[C_2H_3^+]$ / $[C_2H_2^+]$ ratio[44]. Ion **16**$^{+\cdot}$, in which a vinyl group is present, consistently generates more $C_2H_3^+$ than $C_4H_4^{+\cdot}$ ions from other sources, e.g., benzene[44]. Evidently, a second stable $C_4H_4^{+\cdot}$ isomer must exist. CAD does not allow one, however, to pinpoint unequivocally whether the other $C_4H_4^{+\cdot}$ structure (from benzene) is ionized butatriene (**17**$^{+\cdot}$), methylenecyclopropene (**18**$^+$) or cyclobutadiene (**19**$^{+\cdot}$). Appropriate precursors for the latter three ions are butatriene and the tetracyclic compounds **18a** and **19a** (Scheme 14)[45], respectively.

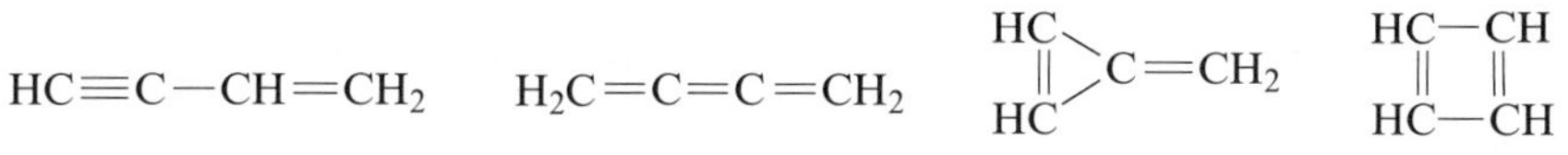

$$HC\equiv C—CH=CH_2 \qquad H_2C=C=C=CH_2$$

(**16**)   (**17**)   (**18**)   (**19**)

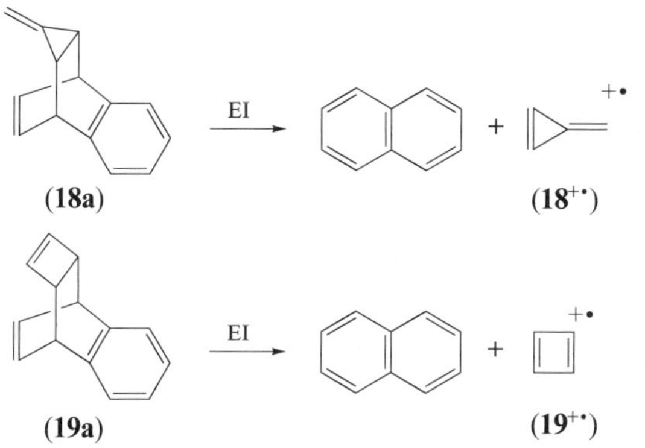

SCHEME 14

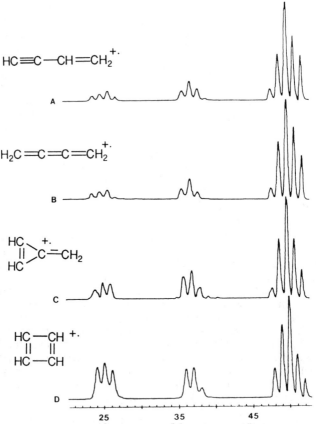

FIGURE 7. $^+NR^+$ (Hg/He) mass spectra of (A) $16^{+\cdot}$, (B) $17^{+\cdot}$, (C)$18^{+\cdot}$, and (D) $19^{+\cdot}$. Ions $16^{+\cdot}$ and $17^{+\cdot}$ were produced by EI of vinylacetylene and butatriene, respectively. Ions $18^{+\cdot}$ and $19^{+\cdot}$ were generated as shown in Scheme 14. Reprinted with permission from Reference 46. Copyright (1989) American Chemical Society

Isomers $16^{+\cdot}$–$19^{+\cdot}$ are best distinguished by NRMS experiments[46] (Scheme 13). Figure 7 shows their $^+NR^+$Hg/He spectra, obtained by Hg neutralization of $C_4H_4^{+\cdot}$ and He reionization of the immediate neutrals. Usually, NR spectra are acquired under *ca* single-collision conditions, i.e. at target pressures corresponding to $\geq70\%$ transmittance[36–38]. Here, the pressure was increased to 30% transmittance (*ca* 2 collisions per affected species)[25b,36] to dissociate *and* reionize the neutrals formed in the neutralization step[46]. The $^+NR^+$ spectra of $16^{+\cdot}$–$19^{+\cdot}$ are characteristically distinct from each other, particularly in the abundances of $C_4H_4^{+\cdot}$ (*m/z* 52, recovered molecular ion), $C_3^{+\cdot}$ (*m/z* 36), $C_2H_3^{+\cdot}$ (*m/z* 27) and $C_2H_2^{+\cdot}$ (*m/z* 26). Expectedly, ion $16^{+\cdot}$ shows the largest $[C_2H_3^+]$. Structure indicative differences are also observed in the charge-reversal ($^+NR^-$ C_6H_6) spectra of $16^{+\cdot}$–$19^{+\cdot}$ (Figure 8), in which the $C_4H_4^{+\cdot}$ precursor cations are first neutralized and then reionized to anions. Such a reaction sequence can be effected by multiple collisions with the same target (e.g. C_6H_6 or Xe)[36,47]. The $^+NR^-$ spectra permit isomer differentiation

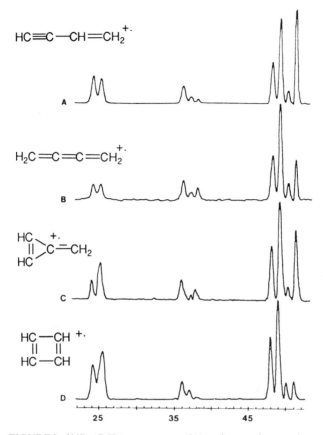

FIGURE 8. $^+NR^-$ (C_6H_6) mass spectra of (A) **16$^{+\cdot}$**, (B) **17$^{+\cdot}$**, (C) **18$^{+\cdot}$** and (D) **19$^{+\cdot}$**. Reprinted with permission from Reference 46. Copyright (1989) American Chemical Society

on the basis of the relative intensities of m/z 51, 38 and 26/25. It is noteworthy that ion **16$^{+\cdot}$** produces the greatest [m/z 51], consistent with formation of the stable acetylide anion $^-C{\equiv}C{-}CH{=}CH_2$.

The superiority of $^+NR^{+,-}$ spectra in $C_4H_4^{+\cdot}$ isomer differentiation results from the fact that the isomerization barriers between neutrals are generally higher than those between the corresponding ions. Therefore, any dissociation between the neutralization and reionization events proceeds without much interconversion, leading to unique fragments and hence diagnostic spectra after reionization. Using the spectra of Figures 7 and 8 as reference for structures **16$^{+\cdot}$–19$^{+\cdot}$**, it is possible to determine the precise composition of $C_4H_4^{+\cdot}$ precursors from various samples[48]. For example, $C_4H_4^{+\cdot}$ from benzene (70 eV EI) gives $^+NR^{+,-}$ spectra that are consistent with a mixture of *ca* 1/3 **16$^{+\cdot}$** plus *ca* 2/3 **18$^{+\cdot}$**, 3-butyn-1-ol exclusively forms **16$^{+\cdot}$**, presumably by 1,2-H_2O elimination ($HC{\equiv}C{-}CH_2{-}CH_2{-}OH^+ \rightarrow HC{\equiv}C{-}CH{=}CH_2^{+\cdot} + OH_2$); and 1,4-benzoquinone yields *ca* 1/10 **16$^{+\cdot}$** plus *ca* 9/10 **19$^{+\cdot}$**[48].

C. Wesdemiotis *et al.*

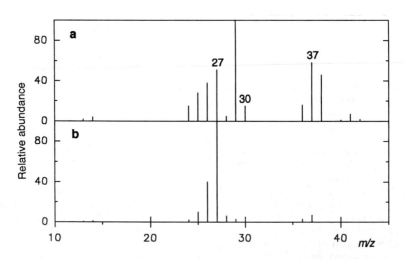

$$HC\equiv C-CH=OH^+ \qquad H_2C=HC-C\equiv O^+$$

$$(20^+) \qquad\qquad (21^+)$$

$$HC\equiv CH + CHO^+$$

$$(320)$$

$$\overset{+}{HC}=CH-CH=O \xrightarrow{(379)} H_2C=\overset{+}{C}-CH=O$$

$$(301) \qquad\qquad (285)$$

$$HC\equiv C-CH=OH^+ \qquad\qquad H_2C=CH-C\equiv O^+$$

$$(171) \qquad\qquad\qquad (0)$$

$$(20^+) \qquad\qquad\qquad (21^+)$$

(471)

(357)

The numbers in
parentheses are *ab initio*
relative energies in kJ mol^{-1}

$$H_2C=CH^+ + CO$$

$$(237)$$

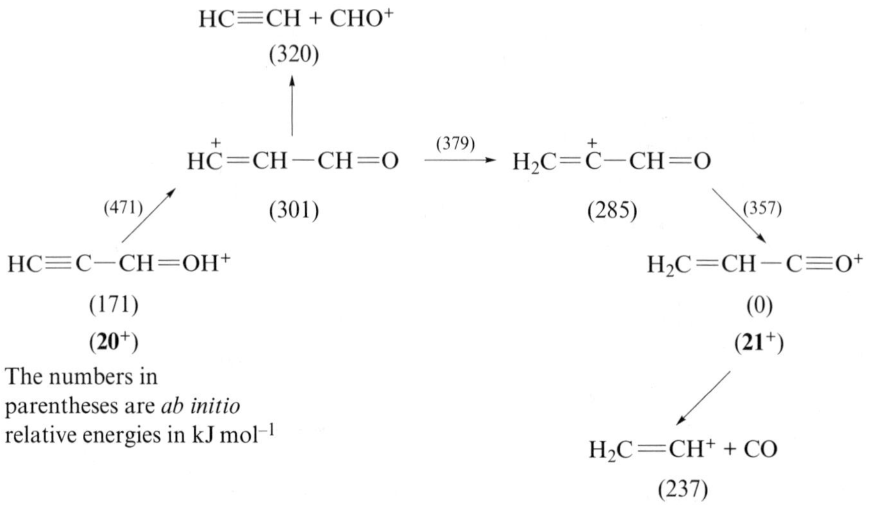

SCHEME 15

FIGURE 9. CAD mass spectra of the $C_3H_3O^+$ cations generated by (a) protonation of propynal and (b) dissociative EI of methyl vinyl ketone. The accepted structures of these $C_3H_3O^+$ cations are (a) $HC\equiv C-CH=OH^+$ (**20$^+$**) and (b) $H_2C=CH-C\equiv O^+$ (**21$^+$**), respectively. Metastable **20$^+$** loses approximately equal amounts of CO ($T_{0.5}$ = 6.3 kJ mol^{-1}) and C_2H_2 ($T_{0.5}$ = 5.3 kJ mol^{-1}) whereas metastable **21$^+$** only cleaves CO ($T_{0.5}$ = 0.45 kJ mol^{-1}); all peak shapes are gaussian. Adapted from Reference 50

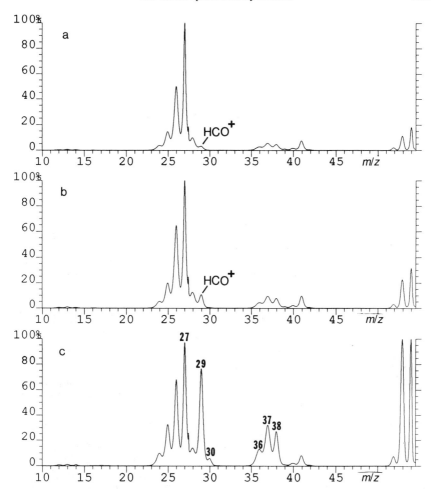

FIGURE 10. CAD mass spectra of $C_3H_3O^+$ cations. (a) $[M-CH_3]^+$ from $H_2C=CH-CO-CH_3$; (b) $[M-H]^+$ from $HC\equiv C-CH_2-OH$; (c) $[M-CH_3]^+$ from $HC\equiv C-CH(CH_3)-OH$. The $C_3H_3O^+$ cations have the structures (a) $H_2C=CH-C\equiv O^+$ (21^+), (b) mainly 21^+ and (c) mainly $HC\equiv C-CH=OH^+$ (20^+)[51]

F. The Hydroxypropynyl Cation $HC\equiv C-CH=OH^+$ (Protonated Propynal)

Ab initio calculations find the $C_3H_3O^+$ ion $HC\equiv C-CH=OH^+$ (20^+) to lie 171 kJ mol^{-1} above the propenoyl cation $H_2C=CH-C\equiv O^+$ (21^+)[49] whose experimental heat of formation is 751 kJ mol^{-1}. Despite such thermodynamic instability, 20^+ should be detectable because its isomerization to 21^+ requires overcoming large barriers (Scheme 15)[49,50]. This is indeed confirmed by the distinctively different MI and CAD spectra of 20^+ and 21^+ generated in pure form by dissociative electron ionization of methyl vinyl ketone and protonation of propynal, respectively (Figure 9)[50]. Note that the dominant CAD fragment from 21^+ (at *m/z* 27) arises by the structure-diagnostic decomposition $CH_2=CH-C\equiv O^+ \rightarrow C_2H_3^+$ (*m/z* 27) + CO. This fragment is weaker for isomer 20^+

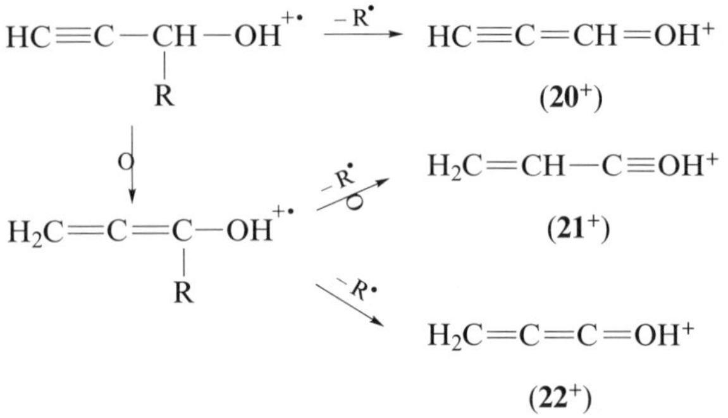

$$HC\equiv C-\underset{\underset{R}{|}}{C}H-OH^{+\cdot} \xrightarrow{-R^{\cdot}} HC\equiv C=CH=OH^+$$

$$(20^+)$$

$$\downarrow^{\curvearrowleft}$$

$$H_2C=C=\underset{\underset{R}{|}}{C}-OH^{+\cdot} \xrightarrow{-R^{\cdot}} H_2C=CH-C\equiv OH^+$$

$$(21^+)$$

$$\xrightarrow{-R^{\cdot}} H_2C=C=C=OH^+$$

$$(22^+)$$

SCHEME 16

which in turn is characterized by the unique fragments of m/z 29 ($HC\equiv O^+$) and 30 (presumably $HC=OH^{+\cdot}$) as well as by relatively intense $C_3H_{0-2}^+$ peaks (m/z 36–38).

$C_3H_3O^+$ (m/z 55) is an important fragment in the EI mass spectra of acetylenic alcohols[52]. It is the base peak for 2-propyn-1-ol ($HC\equiv C-CH_2-OH$) and 3-butyn-2-ol ($HC\equiv C-CH(CH_3)-OH$), from which $C_3H_3O^+$ can arise by elimination of ˙H and ˙CH₃, respectively. If these dissociations were simple α-cleavages[53], then the $C_3H_3O^+$ product would be cation **20⁺**. This is clearly not the case for propynol; its $[M-H]^+$ fragment yields a CAD spectrum (Figure 10b) that is fairly similar to that of ion **21⁺** (Figure 10a), indicating that ˙H loss from $HC\equiv C-CH_2-OH^{+\cdot}$ mainly leads to the acylium cation **21⁺**[50,51]. A tiny amount of **20⁺** is also formed, as $[HCO^+]$ and $[C_3H_{0-2}^+]$ are visibly larger in Figure 10b than in the reference CAD spectrum of **21⁺** (Figure 10a)[51]. A completely different result is obtained for $C_3H_3O^+$ from butynol. Now, the CAD spectrum (Figure 10c) includes all products diagnostic of **20⁺** (e.g. m/z 29, 30, 36–38) with respectable abundances. Evidently, ˙CH₃ elimination from $HC\equiv C-CH(CH_3)-OH^{+\cdot}$ produces a substantial amount of ion **20⁺**. Nonetheless, some **21⁺** is coproduced judging from the relatively high intensity of $C_2H_3^+$ (m/z 27) in Figure 10c vis-à-vis the $[m/z$ 27] value for pure **20⁺** (Figure 9a)[50,51].

The CAD spectra of $C_3H_3O^+$ from $HC\equiv C-CH(R)-OH$ (R = H, CH₃) cannot be reproduced by linear superposition of the reference spectra of pure **20⁺** and **21⁺**[50]. Thus, an additional isomer, **22⁺**, must be contained in these $C_3H_3O^+$ cations. The proportions of **21⁺** and **22⁺** *increase* upon reducing the ionizing electron energy[50]. This result precludes the direct interconversions **20⁺** → **21⁺** and **20⁺** → **22⁺**, as the extent of such post-isomerizations of initially formed **20⁺** ought to *decrease* at the lower internal energy[13,25].

The observed results can be understood if one keeps in mind that the precursor ions $HC\equiv C-CH_2-OH^{+\cdot}$ and $HC\equiv C-CH(CH_3)-OH^{+\cdot}$ can undergo exothermic rearrangement to form more stable allenic ions (see Section I and Scheme 3). For example, ΔH°_f ($HC\equiv C-CH_2-OH^{+\cdot}$) = 1046 kJ mol^{-1} and ΔH°_f ($H_2C=C=CH-OH^{+\cdot}$) = 879 kJmol^{-1}[50]; thus, all propynol ions with sufficient energy to isomerize should readily do so near threshold. The allenic precursor ions could then be the source of **21⁺** and **22⁺** (Scheme 16)[50]. At higher internal energies, the acetylene ion → allene ion interconversion becomes less competitive compared to direct cleavages, so that now the relative fraction of **20⁺** type products increases.

According to the CAD data of Figures 9 and 10, the isomerization proclivity of $HC\equiv C-CH_2-OH^{+\cdot}$ is much larger than that of its methyl homolog $HC\equiv C-CH(CH_3)-OH^{+\cdot}$, as the former generates a minuscule amount of **20⁺** while the latter yields almost exclusively **20⁺**. These trends are verified by neutralization–reionization spectra (Figure 11)[51]. The $^+NR^+$ spectra of $[M-H]^+$ from $HC\equiv C-CH_2-OH$ and of pure

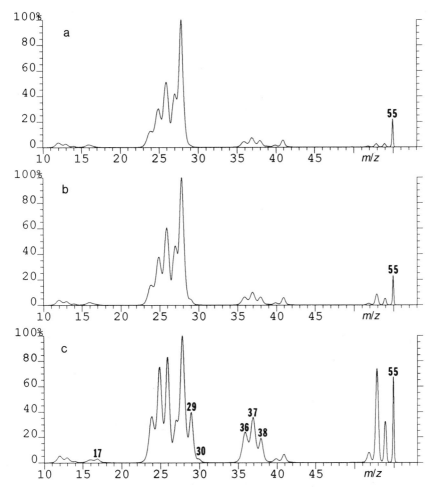

FIGURE 11. $^+NR^+$ (Xe/O$_2$) mass spectra of (a) [M–CH$_3$]$^+$ from H$_2$C=CH—CO—CH$_3$, (b) [M–H]$^+$ from HC≡C—CH$_2$—OH and (c) [M–CH$_3$]$^+$ from HC≡C—CH(CH$_3$)—OH[51]

21$^+$ are very similar, in keeping with the predominant formation of H$_2$C=CH—C≡O$^+$ (**21**$^+$) upon 'H loss from ionized 2-propyn-1-ol. In contrast, the $^+NR^+$ spectrum of [M–CH$_3$]$^+$ from HC≡C—CH(CH$_3$)—OH is distinctively different and contains markedly stronger ions at m/z 17, 29, 30 and 36–38, which are indicative for the presence of a substantial amount of isomer **20**$^+$ (see Figure 9a).

Finally, it is important to notice the appearance of intact precursor ions (m/z 55) in the $^+NR^+$ spectra; their observation provides strong evidence that the radicals [HC≡C—CH—OH]' (**20**) and [H$_2$C=CH—C=O]' (**21**) exist as bound species (lifetime $>ca$ 1 μs). The abundance of the m/z 55 peak is smaller for **21**, which can dissociate to C$_2$H$_3$' + CO by a simple bond cleavage. Isomer **20** cannot yield similarly stable products without rearrangement and, therefore, leads to a more intense molecular ion after reionization[51].

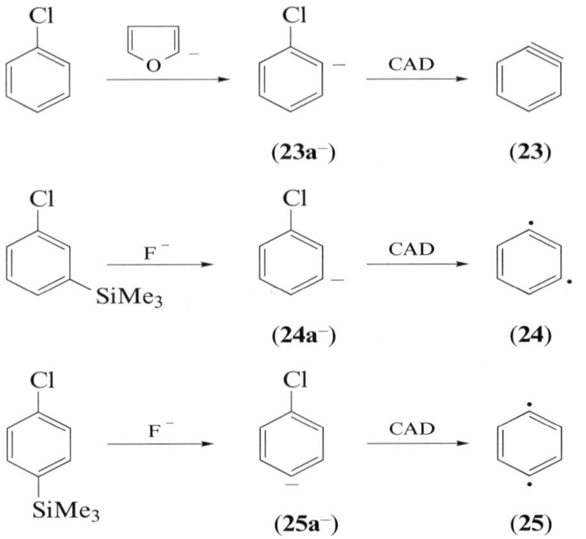

(23a⁻)   (23)

(24a⁻)   (24)

(25a⁻)   (25)

SCHEME 17

G. Benzynes and Related Anions

The elusive *o*-, *m*- and *p*-benzynes (23–25) can be prepared in the gas phase by CAD of *o*-, *m*- and *p*-chlorophenyl anions, respectively (Scheme 17)[54]. These anions are readily accessible in a flowing afterglow ion source by appropriate ion–molecule reactions [54]. The *o*-anion (23a⁻) is formed in isomerically pure form by proton abstraction from chlorobenzene by furanide anion; the *m*- and *p*-anions (24a⁻, 25a⁻) are generated from the reactions between F⁻ and the corresponding (chlorophenyl)trimethylsilanes (Scheme 17).

Mass selection of the phenyl anions in a triple quadrupole mass spectrometer[55] followed by CAD yields exclusively Cl⁻ + C₆H₄ (Scheme 17). Monitoring [Cl⁻] as a function of the center-of-mass collision energy[13] allows one to determine the critical energies for Cl⁻ appearance $(E^{\ddagger})$[56]. These are 67 ± 4, 122 ± 7 and 173 ± 3 kJ mol⁻¹ for the perdeutero isotopomers of 548 23a⁻, 24a⁻ and 25a⁻, respectively. From these energies $(E^{\ddagger})$, the heats of formation of the sought benzynes can be calculated according to the equation.

$$\Delta H°_f(C_6H_4) = \Delta H°_f(C_6H_4Cl^-)^{54} - \Delta H°_f(Cl^-)^2 + (E^{\ddagger} + 2.5)$$

in which 2.5 kJ mol⁻¹ is an expansion-work term for converting the activation energies $E^{\ddagger}$ to enthalpies[56]. The $\Delta H°_f(C_6H_4)$ values determined in this manner are 446 kJ mol⁻¹ for *o*-benzyne (23), 510 kJ mol⁻¹ for *m*-benzyne (24) and 574 kJ mol⁻¹ for *p*-benzyne (25)[54].

That C₆H₄Cl⁻ → Cl⁻ + C₆H₄ liberates a complete C₆H₄ unit has been established by a NRMS experiment[57]. Here, chlorophenyl anion is produced in a negative chemical ionization ion source (using N₂O/CH₄ reagents), mass selected and subjected to keV CAD with He. Under these high-energy conditions, three fragmentations are observed (Figure 12a), namely H˙ loss (giving rise to *m/z* 110), H˙+HCl loss (*m/z* 74) and C₆H₄ loss (*m/z* 35). Post-ionization of the neutral fragments to cations by collisions with O₂ leads to the neutral fragment-reionization (⁻N_fR⁺) spectrum[58,59] of Figure 12b; it contains an abundant

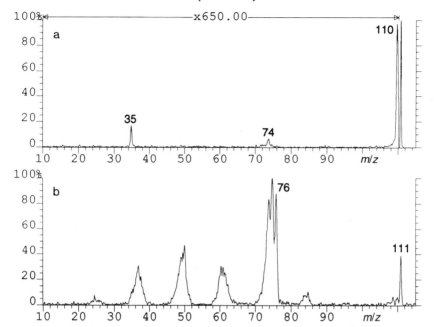

FIGURE 12. (a) CAD spectrum of $C_6H_4Cl^-$ anions. (b) Neutral fragment-reionization, $^-N_fR^+$ (He/O$_2$), spectrum of $C_6H_4Cl^-$, obtained by ionizing to cations (with O$_2$) all neutral losses liberated from $C_6H_4Cl^-$ upon CAD with He[57]

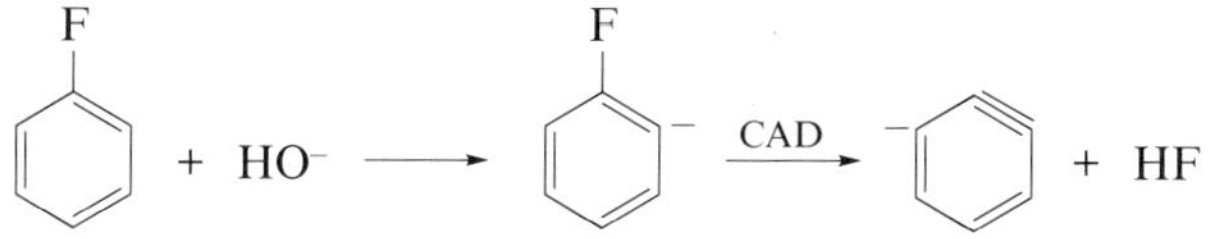

SCHEME 18

$C_6H_4^{+\cdot}$ product (at m/z 76), consistent with the presence of intact C_6H_4 moieties in the neutral loss mixture from CAD of $C_6H_4Cl^-$ [57]. Some neutralization of the $C_6H_4Cl^-$ precursor ion also takes place as attested by the appearance of a reionized $C_6H_4Cl^+$ peak (m/z 111) in the spectrum. It is unlikely that the intense $C_6H_4^{+\cdot}$ ion originates (entirely) from neutralized–reionized $C_6H_4Cl^+$ (i.e. via $C_6H_4Cl^+ \rightarrow C_6H_4^{+\cdot} + \cdot Cl$), because the even-electron closed-shell cation $C_6H_4Cl^+$ should mainly eliminate HCl, not $\cdot Cl$[53]. Indeed, the more saturated analogon $C_6H_6Cl^+$ (protonated chlorobenzene) shows $C_6H_5^+$ but *no* $C_6H_6^{+\cdot}$ in its neutralization–reionization mass spectrum[60].

Bromophenyl anions parallel the behavior of chlorophenyl anions, yielding $C_6H_4 + Br^-$ on CAD[61]. In sharp contrast, the o-fluorophenyl anion, ($C_6H_4F^-$), produced in a flowing afterglow source according to Scheme 18, forms exclusively $C_6H_3^- + HF$ upon low-energy CAD[61]. This result is surprising, as o-benzyne is considerably less acidic than HF[2]. The reason of $C_6H_4 F^-$'s deviation is not known[61].

H. The Structures of Fe$^+$ – Acetylene Complexes

Reaction of C_2H_2 with Fe(CO)$_5$ in a CI ion source leads to ligand exchange generating abundant Fe(C_2H_2)$_x^+$ ions ($x = 2$–4)[62]. Whether these products remain as ion–molecule

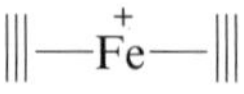

C. Wesdemiotis *et al.*

(26a⁺) 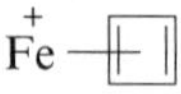   (26b⁺) 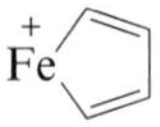   (26c⁺)

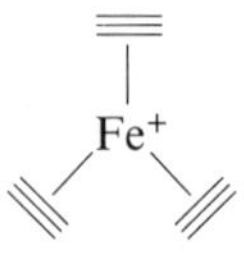

(27a⁺) 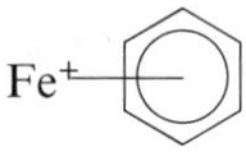   (27b⁺)

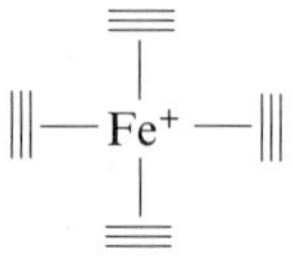

(28a⁺) 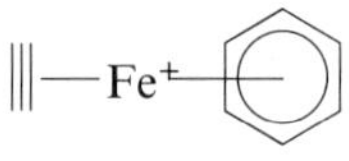   (28b⁺) 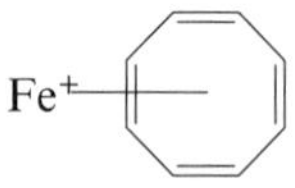   (28c⁺)

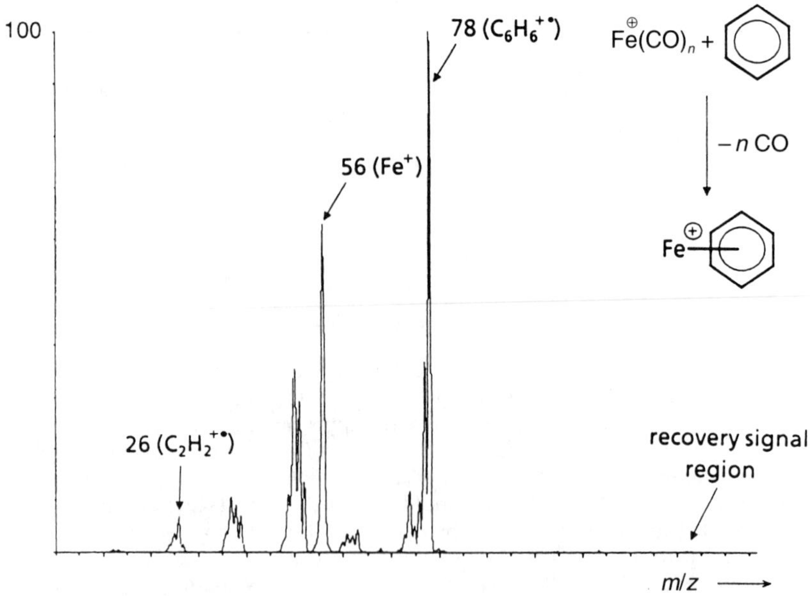

FIGURE 13. $^{+}NR^{+}$ (Xe/O$_2$) mass spectrum of Fe(C$_6$H$_6$)$^{+}$ (ion **27b⁺**) . Reproduced from Reference 62 by permission of Elsevier Science

complexes (**26a⁺** – **28a⁺**) or whether Fe^{+}-catalyzed fusion to form cyclic isomers (**26b⁺**–**28b⁺**, **26c⁺**, **28c⁺**) has taken place can best be answered by neutralization–reionization experiments (see Scheme 13)[35–40, 46–48].

This is first illustrated for Fe(C$_2$H$_2$)$_3^{+}$, **27a⁺**. Fe^{+}-mediated cyclotrimerization of the acetylene units to benzene would produce Fe(C$_6$H$_6$)$^{+}$, **27b⁺**. This ion can be formed

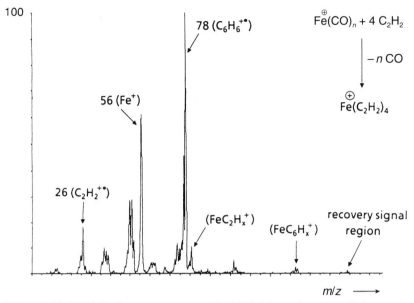

FIGURE 14. $^+NR^+$ (Xe/O$_2$) mass spectrum of Fe(C$_2$H$_2$)$_3^+$. Reproduced from Reference 62 by permission of Elsevier Publishers BV

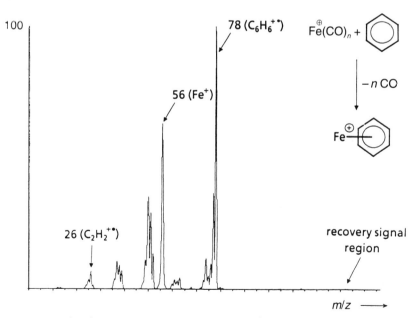

FIGURE 15. $^+NR^+$ (Xe/O$_2$) mass spectrum of Fe(C$_2$H$_2$)$_4^+$. Reproduced from Reference 62 by permission of Elsevier Publishers BV

independently by reacting $Fe(CO)_5$ with C_6H_6 in the CI ion source. Neutralization of $Fe(C_6H_6)^+$ with Xe and reionization of the emerging products with O_2 leads to the $^+NR^+$ Xe/O_2 spectrum of Figure 13. Before proceeding with the interpretation of this spectrum, it is important to consider what happens upon adding an e^- to $27b^+$. Removing the charge from this cation significantly reduces the binding energy between the metal and the organic ligand: $D(Fe^+-C_6H_6) = 230$ kJmol^{-1} [63] compared to $D(Fe-C_6H_6) < 125$ kJmol^{-1} [62]. Therefore, a substantial fraction of $27b$ should fall apart to $Fe + C_6H_6$ upon neutralization. This expectation is indeed confirmed by the $^+NR^+$ spectrum of $27b^+$ which is dominated by signals at m/z 56 (Fe^+) and 78 ($C_6H_6^{+\cdot}$), originating from the aforementioned dissociation. Further decomposition of $C_6H_6^{+\cdot}$ can account for the other peaks observed in the spectrum. The near absence of ions above m/z 78 also reveals that the fragmentation $27b \rightarrow Fe + C_6H_6$ is nearly complete. Apparently, the transition $27b^+ \rightarrow 27b$ yields the weakly bound $Fe(C_6H_6)$ above its dissociation threshold, presumably due to an unfavorable Franck–Condon factor caused by severe geometry differences between $27b^+$ and $27b^{36-38}$.

Subjecting $Fe(C_2H_2)_3^+$, $27a^+$, to $^+NR^+$ leads to a characteristically different spectrum (Figure 14)[62]. The appearance of $C_6H_6^{+\cdot}$ (m/z 78 in this spectrum clearly shows that some $27a^+$ has isomerized to $Fe(C_6H_6)^+$, $27b^+$. Notice, however, that $[C_2H_2^+]$, $[Fe^+]$ and $[Fe(C_xH_y)^+]$ are markedly larger in Figure 14 than in Figure 13, consistent with partial retention of the original unfused structure $Fe(C_2H_2)_3^+$.

The $^+NR^+$ spectrum of $Fe(C_2H_2)_2^+$, $26a^+$, is dominated by peaks for Fe^+, $C_2H_2^{+\cdot}$ and $Fe(C_2H_x)^+$ ($x = 0$–2), in keeping with the bis-acetylene structure $26a^+$ [62]. The intensities of $C_4H_4^{+\cdot}/C_4H_3^{++}$ are negligible, indicating that isomerization to the cyclobutadiene complex $26b^+$ or to the metallacycle $26c^+$ has not occurred to a measurable extent. Similarly, the $^+NR^+$ spectrum of $Fe(C_2H_2)_4^+$, $28a^+$, contains no detectable $C_8H_{7-8}^+$ (Figure 15)[62], excluding $28a^+ \rightarrow 28c^+$. However, the $C_6H_6^{+\cdot}$ base peak and the presence of weak yet diagnostic $Fe(C_2H_x)^+$ and $Fe(C_6H_x)^+$ signals in the $^+NR^+$ spectrum point out that $28a^+$ partially rearranges to $28b^+$. It is noteworthy that reionized precursor ions ('recovery signals') are practically absent from all $^+NR^+$ mass spectra, in keeping with very weak bonding in the neutrals 26–28[62].

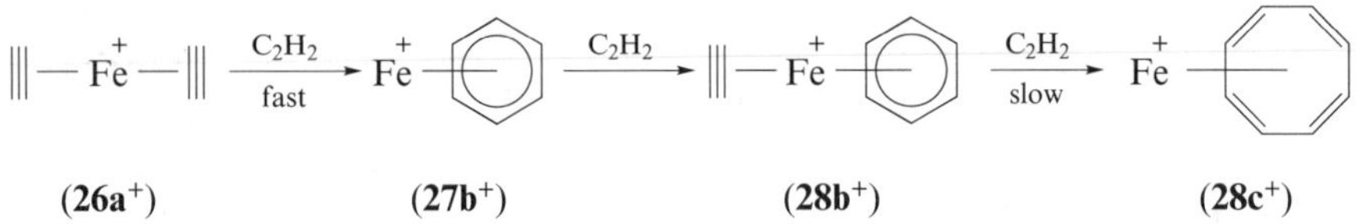

(26a⁺)   (27b⁺)   (28b⁺)   (28c⁺)

SCHEME 19

The experimental observations are summarized in Scheme 19: the Fe^+-catalyzed oligomerization of acetylene becomes rapid once three C_2H_2 units have assembled around the metal ion but slows down upon addition of a fourth molecule, showing that further polymerization requires a substantial barrier[62].

III. ION-MOLECULE REACTIONS WITH ACETYLENES

A. Chemical Ionization of Alkynes with Metal Ions

An elegant method for locating the position of a triple bond of an alkyne is *in situ* derivatization in a chemical ionization source, followed by MS/MS[13] of the adduct. The most successful approach has involved reaction of the alkyne with $Fe(CO)_x^+$ [from

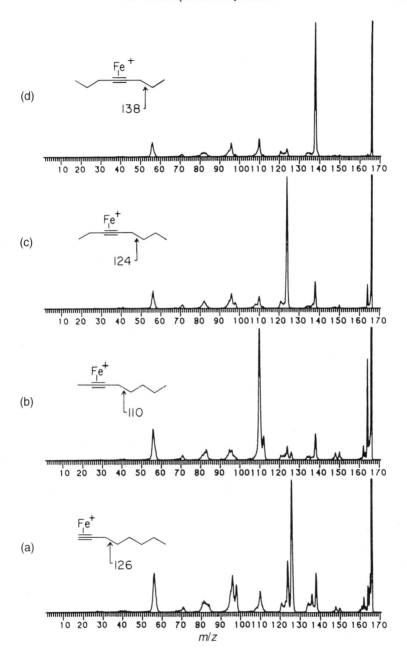

FIGURE 16. CAD mass spectra of $Fe(C_8H_{14})^+$ (m/z 166) from (bottom to top) (a) 1-octyne, (b) 2-octyne, (c) 3-octyne and (d) 4-octyne. Reprinted with permission from Reference 64. Copyright (1985) American Chemical Society

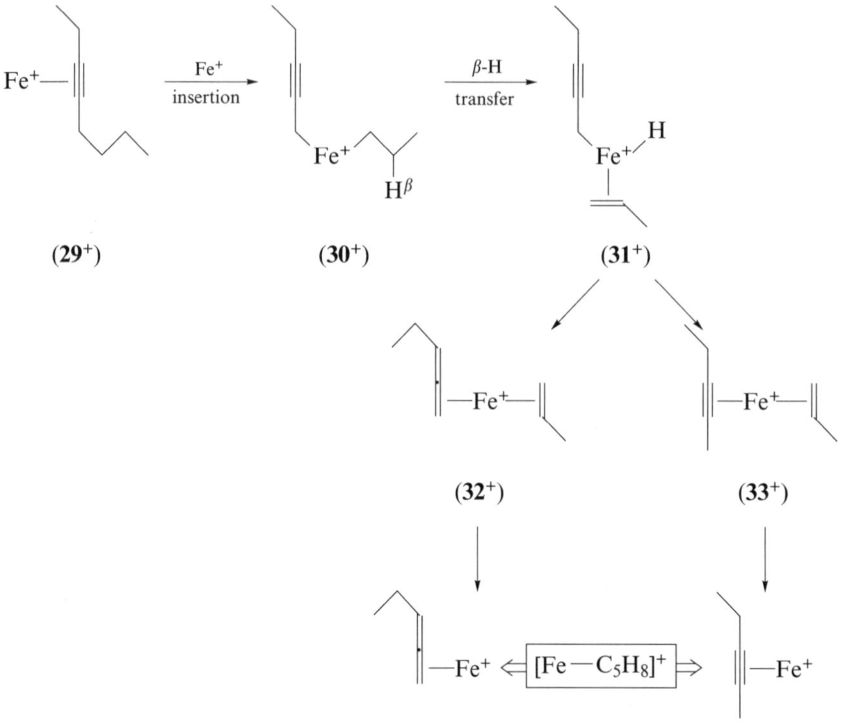

C. Wesdemiotis *et al.*

Fe⁺—‖ →(Fe⁺ insertion)→ ‖ →(β-H transfer)→ ‖

Fe⁺ insertion

Fe⁺ Hᵝ

Fe⁺ H

(29⁺)   (30⁺)   (31⁺)

—Fe⁺—‖   ‖—Fe⁺—‖

(32⁺)   (33⁺)

—Fe⁺ ⇐ [Fe—C₅H₈]⁺ ⇒ ‖—Fe⁺

(34⁺)   (35⁺)

SCHEME 20

$Fe(CO)_5$] to form a Fe(alkyne)⁺ complex; on CAD, the latter produces structure-indicative fragments, based on which the position of the acetylenic bond can be determined[64–67]. This is illustrated for isomeric octynes in Figure 16[64].

The principal fragments indicated in Figure 16 arise by cleavage of the propargylic bond. This reaction has been proposed to commence by Fe^+ insertion into the propargylic bond ($29^+ \rightarrow 30^+$ in Scheme 20)[65]. Subsequent β-H transfer by the metal ion to the propargyl moiety produces a Fe^+ complex in which two unsaturated ligands are bound to the metal ion (32^+ or 33^+); the H atom primarily moves to C(3) of the acetylenic unit to yield an allene ligand (32^+) and to a lesser extent to C(1) to yield a 2-alkyne ligand (33^+). Expulsion of the smaller ligand from the Fe^+ complex gives rise to the most abundant fragment[65]. For example, 3-octyne (Scheme 20) leads to a $Fe(C_5H_8)^+$ base peak (m/z 124 in Figure 16c), consisting of *ca* 75% 34^+ (allene product) and *ca* 25% 35^+ (2-alkyne product). This composition was ascertained by comparing the CAD spectrum of $Fe(C_5H_8)^+$, formed in the ion source from 3-octyne, to the CAD spectra of authentic 34^+ and 35^+, generated from 1,2-pentadiene and 2-pentyne, respectively[65].

Deuterium labeling corroborates the migration of a β-H atom[67]. If no β-hydrogen is available after propargylic insertion (as, e.g., in 3-hexyne or 2-pentyne), the Fe^+ ion either catalyzes interconversion to an isomer which possesses a β-H atom (3-hexyne → 2-hexyne), or promotes rearrangement to dienes (2-pentyne → pentadienes) and then inserts into the vinylic bonds[65].

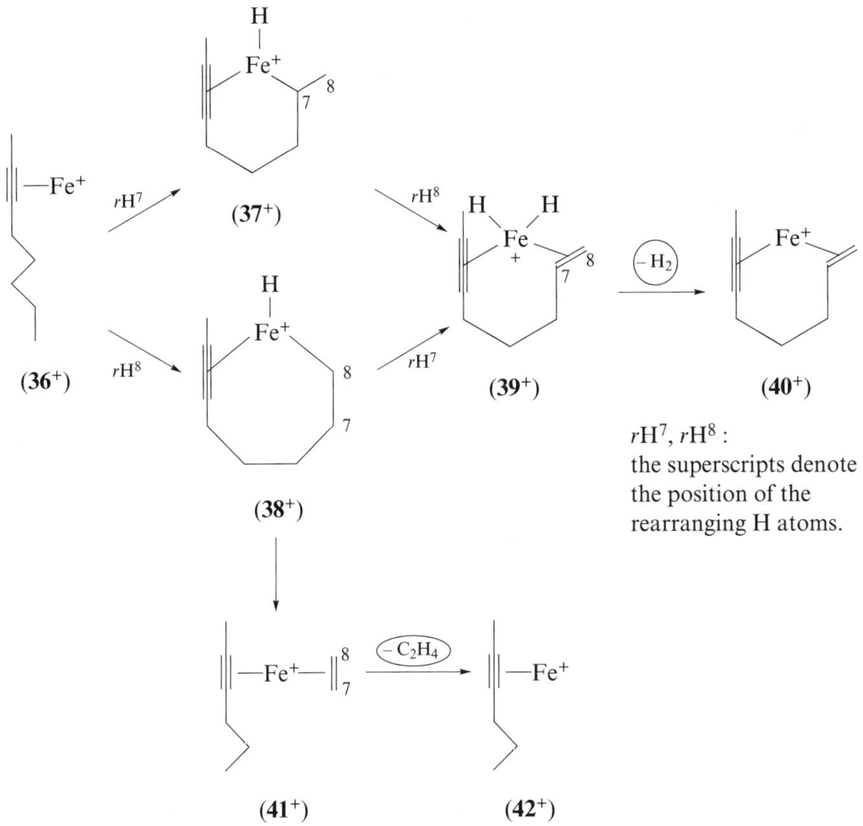

(37^+)

(36^+)

rH^7

rH^8

rH^8

rH^7

(38^+)

(39^+)

$(-H_2)$

(40^+)

rH^7, rH^8 :
the superscripts denote
the position of the
rearranging H atoms.

$(-C_2H_4)$

(41^+) (42^+)

SCHEME 21

B. Remote Functionalization of Alkynes by Metal Ions

Although the structurally diagnostic fragments generated from the reaction of Fe^+ with alkynes arise as shown in Scheme 20, i.e. by oxidative addition of the metal ion to a C—C bond followed by β-H transfer, other important products call for an additional pathway involving functionalization of remote C—H bonds[68]. This reaction channel is illustrated for 2-octyne in Scheme 21 and explains the relatively high abundance of H_2 and C_2H_4 losses from $Fe(2\text{-octyne})^+$ in Figure 16b.

The reaction sequences of Scheme 21 are supported by the following facts:(i) the CAD spectrum of the product of H_2 loss (40^+) is identical to that of authentic $Fe(1\text{-octene-6-yne})^+$; (ii) deuterated 2-octynes undergo regiospecific H_2 and C_2H_4 losses, largely involving the triple bond-remote positions $C(7)$ and $C(8)$[68].

The above studies have been conducted with a beam instrument, where the ion–molecule adduct $Fe(alkyne)^+$ is first isolated before its spontaneous or collisionally activated dissociation products are assessed. It has been shown, however[69], that reaction of bare Fe^+ with alkynes in an ICR mass spectrometer leads to similar products; now, the intermediate $Fe(alkyne)^+$ adduct may not be observed due to the absence of stabilizing collisions in the ICR environment[70].

Besides Fe^+, many other metal ions yield analogous results. The nature of the metal ion influences the critical energies of the insertion and abstraction steps in Schemes 20 and 21 and, thus, determines the overall reaction rate[71]. The detailed gas-phase chemistry between bare or ligated metal ions and a large variety of organic substrates, including alkynes, is thoroughly reviewed in Reference 19.

C. Reaction of Acetylene with $C_3H_3^+$

The isomeric structures of $C_3H_3^+$ have been studied extensively by theory[72,73]. The most stable isomers are the aromatic cyclopropenylium ion 43^+ and the propargyl ion 44^+. Ion 43^+ is 105 kJmol^{-1} more stable and is accessed in relatively pure form upon electron ionization of allene. Ion 44^+ is formed with high purity (*ca* 90%) by dissociative ionization of propargyl iodide. Other unsaturated organic compounds, such as propyne and propargyl chloride or bromide, lead to mixtures of both isomers[74].

$$(43^+) \qquad\qquad (44^+)$$

Ions 43^+ and 44^+ are characterized by substantially different bimolecular reactivities[74-79]. While 43^+ does generally not react, 44^+ undergoes a variety of addition and condensation reactions. Considerable attention has been devoted to the reactions of ion 44^+ with its precursor, with acetylene and with diacetylene, as such processes possibly initiate soot formation in flames[74,76-78].

In an FTICR instrument, 44^+ from propargyl iodide (C_3H_3I) reacts with C_3H_3I to yield, *inter alia*, $C_6H_6^{+\cdot}$, $C_7H_7^+$ and $C_7H_9^{+}$[78]. Addition of acetylene causes partial isomerization of 44^+ to more stable 43^+ thus markedly reducing the overall $C_3H_3^+$ reactivity; a stable $C_5H_5^+$ adduct is not observed[74]. When C_2D_2 is reacted with 44^+, isotope exchange takes place generating nonreactive (i.e. with structure 43^+) $C_3H_2D^+$, $C_3HD_2^+$ and $C_3D_3^+$. These results suggest that an intermediate, long-lived $C_5H_5^+$ collision complex is formed, in which H/D exchange and the isomerization $44^+ \rightarrow 43^+$ can occur. The low pressures prevailing in the FTICR-MS prohibit collisional stabilization of the $C_5H_5^+$ complex, so that it dissociates back to $C_3H_3^+ + HC{\equiv}CH$. At the higher pressures of the quadrupole ion trap mass spectrometer (QIT-MS), $C_5H_5^+$ has indeed been observed[80]. The reactions of 44^+ (from propargyl bromide) with C_2H_2 and $^{13}C_2H_2$ have been monitored in the QIT-MS as a function of time and reveal that the nascent $C_5H_5^+$ adduct decomposes to yield *ca* 1/3 unreactive $C_3H_3^+$ (43^+) and *ca* 2/3 reactive $C_3H_3^+$ (44^+); the latter reacts anew with acetylene at longer delay times[80].

The reaction of 44^+ with diacetylene ($HC{\equiv}C{-}C{\equiv}CH$) causes some $44^+ \rightarrow 43^+$ interconversion but also extensive addition/condensation sequences[74]:

$$C_3H_3^+ + C_4H_2 \rightarrow C_7H_5^+$$

$$C_3H_3^+ + C_4H_2 \rightarrow C_5H_3^+ + C_2H_2$$

$$C_5H_3^+ + C_4H_2 \rightarrow C_9H_5^+$$

$$C_5H_3^+ + C_4H_2 \rightarrow C_7H_3^+ + C_2H_2$$

$$C_7H_3^+ + C_4H_2 \rightarrow C_{11}H_5^+$$

etc.

The formation of such products strongly supports the supposition that ion–molecule reactions of $C_3H_3^+$ play an important role in soot formation.

D. Reaction of Acetylene with Other Small Hydrocarbon Ions

The reaction of $C_xH_y^+$ cations ($x = 1-4$; $y = 0-4$) with $HC\equiv CH$ has been investigated by ICR[81]. CH_y^+ (e.g. from CH_4), $C_2H_y^+$ (e.g. from C_2H_2, C_2H_4 or C_2H_6) and $C_3H_y^+$ (from allene or propyne) exclusively undergo condensation reactions in which most commonly $\cdot$H or H_2, and to a lesser extent $\cdot CH_3$ or CH_4, are eliminated to form product ions with a larger carbon skeleton than the reactant ions. A notable exception is $C_3H_3^+$, which does not generate any condensates in an ICR instrument (*vide supra*).

The situation is quite different with $C_4H_y^+$ reactants. $C_4^{+\cdot}$ and C_4H^+ (both from diacetylene) condense with $HC\equiv CH$ to yield $C_6H^+ + \cdot H$ and $C_6H_2^{+\cdot} + \cdot H$, respectively. $C_4H_2^{+\cdot}$ (from diacetylene) leads to some condensation (5% $C_6H_3^+ + \cdot H$) but mainly yields an association product (95% $C_6H_4^{+\cdot}$). $C_4H_3^+$ (from vinylacetylene) solely undergoes association with C_2H_2, producing $C_6H_5^+$. Finally, $C_4H_4^{+\cdot}$ (from vinylacetylene) + C_2H_2 gives rise to 10% $C_6H_4^{+\cdot} + H_2$, 75% $C_6H_5^+ + \cdot H$ and 15% $C_6H_6^{+\cdot}$ (association product).

The association reactions observed are strongly exothermic[2,81]. So that the adducts can survive intact, the reaction heat must be dissipated by collisions[82] or photon emission[83]. Because association takes place at pressures $\leq 1\times10^{-6}$ Torr, photon emission is the most likely means of stabilization[81]. Further, the stability of the association products suggests that these species are covalently bound rather than loosely held ion–molecule complexes[81]. The addition and condensation pathways occurring in the ICR cell could be the source of the large unsaturated hydrocarbons detected in flames and astrochemical environments.

E. The Bimolecular Reactions $C_2H_2^{+\cdot} + C_2H_4$ and $C_2H_2 + C_2H_4^{+\cdot}$

The reaction of $C_2H_4^{+\cdot}$ with C_2H_2 in an ICR cell produces $C_3H_3^+ + \cdot CH_3$ and $C_4H_5^+ + \cdot H$ (Table 1)[84,85]; with $C_2D_4^{+\cdot}$ or C_2D_2, complete scrambling is observed for both reaction channels[84]. The product distribution changes drastically if $C_2H_2^{+\cdot}$ is reacted with C_2H_4 (Table 1). Now, much less $C_3H_3^+$ and $C_4H_5^+$ are formed, and the major product becomes $C_2H_4^{+\cdot}$ arising by charge exchange in the reactants[84]. Labeling shows that, also in the $C_2H_2^{+\cdot}/C_2H_4$ reaction, $C_3H_3^+$ and $C_4H_5^+$ are generated after complete H-atom scrambling. In contrast, the distribution of labels in the charge exchange channel yielding $C_2H_4^{+\cdot} + C_2H_2$ is not statistical (<20% scrambling)[84].

The above data suggest that $C_3H_3^+$ and $C_4H_5^+$ are produced in both ion–molecule reactions through an intermediate in which all six hydrogen atoms become equivalent. This intermediate could be ionized 1,3-butadiene (Figure 17)[84] or any other $C_4H_6^{+\cdot}$ ion, all of which have been found to undergo rapid hydrogen and carbon interchange prior to unimolecular decay[86]; as the most stable $C_4H_6^{+\cdot}$ isomer[2], $CH_2=CH-CH=CH_2^{+\cdot}$ is a

TABLE 1

	Ion–Molecule Reactions		Unimolecular dissociations of $C_4H_6^{+\cdot}$ at the same total energy as the bimolecular reactions	
	$C_2H_4^{+\cdot} + C_2H_2$	$C_2H_2^{+\cdot} + C_2H_4$	$C_2H_4^{+\cdot} + C_2H_2$	$C_2H_2^{+\cdot} + C_2H_4$
Product	(%)	(%)	(%)	(%)
$C_3H_3^+$	74	20	50	25
$C_4H_5^+$	26	16	43	56
$C_2H_4^{+\cdot}$		64	6	19

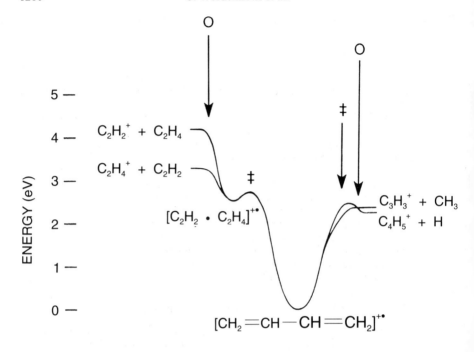

FIGURE 17. Potential energy diagram for the $C_4H_6^{+\cdot}$ system. Tight and orbiting transition states are indicated by '‡' and 'O', respectively. Reproduced from Reference 84 by permission of the American Institute of Physics

reasonable candidate. For the charge exchange channel $C_2H_2^{+\cdot} + C_2H_4 \rightarrow C_2H_2 + C_2H_4^{+\cdot}$, the $C_4H_6^{+\cdot}$ intermediate does not need to be sampled. This reaction can also occur in a loose $(C_2H_2 \cdots C_2H_4)^{+\cdot}$ ion–molecule complex (Figure 17), in which long-range electron jump is possible but isotope exchange not easy.

It is interesting to compare the branching ratios from the bimolecular reactants $C_2H_4^{+\cdot}/$ C_2H_2 and $C_2H_2^{+\cdot}/C_2H_4$ to those from the unimolecular dissociation of $C_4H_6^{+\cdot}$ ions of the same internal energy (Table 1). Based on photoion-photoelectron coincidence (PIPECO) experiments[87], $C_4H_6^{+\cdot}$ ions of internal energy equal to the $C_2H_4^{+\cdot} + C_2H_2$ energy level lead to a smaller $[C_3H_3^+]/[C_4H_5^+]$ ratio than the ion–molecule reaction itself. Similarly, $C_4H_6^{+\cdot}$ ions with energy equal to that of $C_2H_2^{+\cdot} + C_2H_4$ yield more $C_4H_5^+$ than the respective bimolecular reaction; in addition, the charge exchange channel is suppressed (from 64% to 19%).

These seemingly contradictory results are consistent with the transition-state switching model[88]. According to this theory, the reaction coordinate for a unimolecular dissociation contains two transition states: an orbiting transition state near the products and a tight transition state near the reactant (Figure 17). The tight transition state is the saddle point for reactions with a reverse activation energy. For continuously endothermic systems, a minimum flux is expected to occur between the orbiting transition state and the reactant because of the competition between the following two factor: (i) a decrease in local flux due to the conversion of nonfixed energy into potential energy and (ii) an increase in local flux due to conversion of vibrational degrees of freedom of the reactant ion into translational and rotational degrees of freedom of the products[84,88].

The dissociation $C_4H_6^{+\cdot} \rightarrow C_3H_3^+ + {}^\cdot CH_3$ has no saddle point (Figure 17)[84]. The absence of a barrier for the reverse reaction is indicated by the narrow, gaussian peak shape of $C_3H_3^+$ from metastable $C_4H_6^{+\cdot}$. For the reasons outlined above, this reaction has a tight *and* an orbiting transition state; the tight transition state (i.e. the place of minimum flux) has an activation energy that is smaller than the difference in zero point energies between $C_4H_6^{+\cdot}$ and $C_3H_3^+ + {}^\cdot CH_3$. The $C_4H_5^+ + {}^\cdot H$ channel does have a substantial reverse activation energy (Figure 17), reflected by the flat-topped peak shape in the corresponding metastable transition. The maximum (saddle point) in the $C_4H_5^+ + {}^\cdot H$ reaction coordinate corresponds to the tight transition state; an orbiting transition state may also exist, but the flux through it will always be greater than through the saddle point due to the presence of an appreciable reverse barrier[84]. The $C_2H_4^{+\cdot} + C_2H_2$ and $C_2H_2^{+\cdot} + C_2H_4$ channels share a common tight transition state through individual orbiting transition states (Figure 17). The potential well between these two types of transition states corresponds to the loosely bound complex $(C_2H_4 \cdots C_2H_2)^{+\cdot}$ [84].

As mentioned above, the charge exchange process in the bimolecular reaction $C_2H_2^{+\cdot} + C_2H_4$ should mainly proceed via $(C_2H_4 \cdots C_2H_2)^{+\cdot}$; after the reactants enter the shallow well, they prefer exiting through the orbiting transition state (to $C_2H_4^{+\cdot} + C_2H$) and not through the tight transition state (to $C_4H_6^{+\cdot}$) over which the flux is minimal. Such a charge exchange sequence cannot be associated with any significant scrambling. The *ca* 20% scrambling observed upon $C_2H_2^{+\cdot} + C_2H_4 \rightarrow C_2H_2 + C_2H_4^{+\cdot}$ must result from the small amount of $(C_2H_4 \cdots C_2H_2)^{+\cdot}$ that samples $C_4H_6^{+\cdot}$ over the tight transition state (Figure 17); it cannot be excluded, however, that some isotope exchange occurs in the shallow well, too.

Based on the transition state switching model, bimolecular and unimolecular branching ratios at the same energy can differ substantially because of the strong dependence of branching rations on angular momentum, *J*. The model[88], when applied to the system represented by Figure 17, predicts that the high-*J* (bimolecular) reaction should lead to a much greater $[C_3H_3^+]/[C_4H_5^+]$ value than the low-*J* (unimolecular) reaction[84]. This fully agrees with the experimental data (Table 1)[84,87]. The preference of $C_3H_3^+$ formation in the bimolecular reaction most probably results from the fact that the rotational constant of the tight transition state to $C_3H_3^+ + {}^\cdot CH_3$ is smaller than the constant of the tight transition state to $C_4H_5^+ + {}^\cdot H$, thus facilitating passage to $C_3H_3^+$ for higher-*J* systems[84].

F. Reactions of Carbon Cluster Ions with Acetylene

Bombardment of graphite with a Nd:YAG laser leads to desorption of abundant C_n^+ cluster ions ($n = 3$–20)[89,90]. The reactivity of C_n^+ with D_2, O_2 and CH_4 decreases with n and stops at $C_{>9}^+$ [91,92]. This result is consistent with the existence of two distinct C_n^+ structures, namely a linear ($n = 3$–9) which is reactive because it possesses terminal carbene sites, and a cyclic ($n \geq 10$) which is unreactive due to the absence of carbenic termini[91,92]. The reactions with D_2, O_2 and CH_4 proceed to completion for all linear ions besides C_7^+, which remains partially unreactive. This deviating behavior suggests the presence of two C_7^+ isomers, having a reactive linear and a nonreactive cyclic structure, respectively.

In contrast to D_2, O_2 and CH_4, which do not react with $C_{\geq 10}^+$, acetylene does react with the larger, presumably monocyclic cluster ions [92]. C_n^+, $n = 10$–14, exclusively associate with C_2H_2 to form $C_{n+2}H_2^+$ adducts. No condensation products are formed. The excess energy of the adducts is most likely dissipated by radiation, not collisions which would require much higher pressures than those attainable in an FTICR-MS ($<2.0\times10^{-7}$ Torr in these experiments)[93,94].

For C_n^+, $n = 3$–9, the major reaction with C_2H_2 is condensation with loss of $H^\cdot$, viz. $C_n^+ + C_2H_2 \rightarrow C_{n+2}H^+ + H^\cdot$ (illustrated in Scheme 22 for C_4^+)[92]. C_4^+, C_6^+ and C_8^+, which contain an even number of carbons, also undergo some condensation with loss of C_3 ($\leq 40\%$), viz.($C_n^+ + C_2H_2 \rightarrow C_{n-1}H_2^+ + C_3$ (Scheme 22). In addition, C_7^+, C_8^+ and C_9^+ form association adducts, viz. $C_n^+ + C_2H_2 \rightarrow C_{n+2}H_2^+$. This latter reaction is characteristic for cyclic C_n^+ ions (*vide supra*); thus, C_7^+, C_8^+ and C_9^+ must contain both linear and cyclic structures, in agreement with the earlier evidence on C_7^+ [91].

For C_7^+, the presence of two isomers has also been verified by MS/MS[92]: here, the linear fraction is first depleted by reacting C_7^+ with an excess of D_2[91]. Mass-selection of the unreacted C_7^+ followed by reaction with HC≡CH leads to one product only, the $C_9H_2^+$ adduct. Exclusive association is diagnostic for the larger ($n \geq 10$) cluster ions which are cyclic. Such a result provides positive and unequivocal evidence for the presence of two isomers for C_7^+ [92].

The mechanism proposed for the condensation reactions of the linear $C_{\leq 9}^+$ is presented in Scheme 22 for the case of C_4^+ (**45$^+$**). Insertion of a carbene end of **45$^+$** into the triple bond of acetylene yields intermediate **46$^+$** whose excess energy can be disposed by cleavages of either H˙ (to give a $C_{n+2}H^+$ product) or C_3 (to give a $C_{n-1}H_2^+$ product)[92]. Expulsion of C_3 takes place only for even cluster ions, revealing a preference to form product ions with an odd number of C atoms.

The primary condensation products $C_{n+2}H^+$ ($n = 3$–9) and $C_{n-1}H_2^+$ ($n = 4, 6, 8$) still contain an unreacted carbene site (see ions **47$^+$** and **48$^+$** in Scheme 22); therefore, they can undergo further ('secondary') reactions with acetylene[92]. Observed are the secondary condensations $C_{n+2}H^+ + C_2H_2 \rightarrow C_{n+4}H_2^+ + H˙$ and $C_{n-1}H_2^+ + H_2 \rightarrow C_{n+1}H_3^+ + H˙$, as well as (for $n \geq 5$) secondary associations to $C_{n+4}H_3^+$ and $C_{n+1}H_4^+$. Association is preferred by the odd cluster ions and condensation by the even cluster ions.

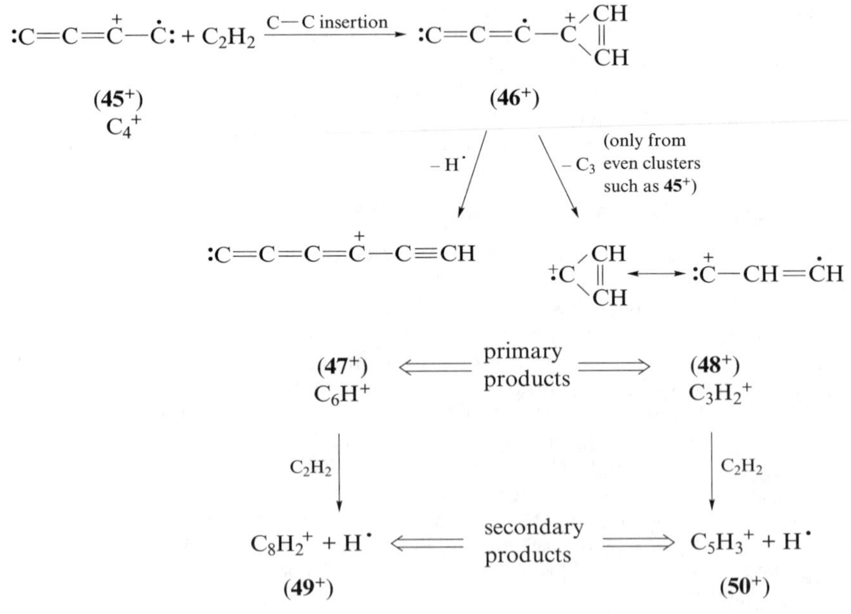

SCHEME 22

IV. SUMMARY

(1) Isomeric alkynes can be distinguished from their EI mass spectra . Although partial isomerizations take place during ionization and/or decomposition, differences remain and allow for definitive structural assignments in combination with reference spectra.

(2) Alkynes and isomeric allenes also undergo partial equilibration after ionization. Generally, the smaller ions exhibit the largest rearrangement proclivities.

(3) The most facile means of C≡C-isomer distinction is by MS/MS of metal ion adducts. The major fragments observed on MS/MS arise by insertion of the metal ion into the propargylic bond, ultimately leading to rupture of this bond. The second most important fragmentation pathway involves functionalization by the metal ion of bonds remote to the C≡C-center.

(4) Chemical reactions take place in ionized clusters containing acetylene or other triple-bonded moieties. The reaction heat liberated upon formation of the new bond(s) is dissipated by condensation (i.e. elimination of a neutral residue from the cluster), unless collisional or radiative stabilization is possible.

(5) Unimolecular dissociation of appropriately structured molecular ions can yield elusive C≡C-bonded products (e.g. ionized ethynol or the benzynes). These novel acetylenes can conveniently be studied in the gas phase by neutralization–reionization mass spectrometry techniques.

V. REFERENCES

1. R. S. Mercer and A. G. Harrison, *Org. Mass Spectrom.*, **21**, 717 (1986).
2. S. G. Lias, J. E. Bartmess, J. F. Liebman, J. L. Holmes, R. D. Levin and W. G. Mallard, *J. Phys. Chem. Ref. Data*, **17**, Suppl. No.1 (1988).
3. W. Wagner and K. Levsen, *Adv. Mass Spectrom.*, **14**, 222 (1979).
4. J. L. Courtneidge, A. G. Davis and A. Maccoll, *Int. J. Mass Spectrum. Ion Processes*, **101**, 167 (1990).
5. W. Wagner-Redeker, K. Levsen, H. Schwarz and W. Zummack, *Org. Mass Spectrom.*, **16**, 361 (1981).
6. P. Wolkoff, J. L. Holmes and F. P. Lossing, *Adv. Mass Spectrom.*, **8A**, 743 (1980).
7. J. L. Holmes, M. Fingas, and F. P. Lossing, *Can. J. Chem.*, **59**, 80 (1981).
8. Z. Dolejsek, V. Hanus and K. Vokac, *Adv. Mass Spectrom.*, **3**, 503 (1966).
9. Z. R. Appling, B. E. Jones, L. E. Abbey, D. E. Bostwick and T. F. Moran, *Org. Mass Spectrom.*, **18**, 282 (1983).
10. D. L. Kemp and R. G. Cooks, in *Collision Spectroscopy* (Ed. R. G. Cooks), Plenum Press, New York, 1978, p. 260.
11. R. C. Bingham, M. J. S. Dewar and D. J. Lo, *J. Am. Chem. Soc.*, **97**, 1285 (1975).
12. S. Arseniyadis, A. Maquestiau, R. Flammang, P. Guenot and R. Carrie, *Org. Mass Spectrom.*, **24**, 909 (1989).
13. K. L. Busch, G. L. Glish and S. A. McLuckey, *Mass Spectrometry/Mass Spectrometry*, VCH Publishers, Inc., New York, 1988.
14. (a) A. Maquestiau, C. Jartay, D. Beugnies, R. Flammang, R. Houriet, E. Rolli and G. Bouchoux, *Int. J. Mass Spectrum. Ion Processes*, **82**, 33 (1988).
    (b) W. H. Jones, R. D. Mariani and M. L. Lively, *Chem. Phys. Lett.*, **108**, 602 (1984).
15. (a) H. Schwarz, T. Weiske, K. Levsen, A. Maquestiau and R. Flammang, *Int. J. Mass Spectrom. Ion Phys.*, **45**, 367 (1982).
    (b) G. Bouchoux and Y. Hoppilliard, *Int. J. Mass Spectrom. Ion Phys.*, **47**, 55 (1983).
    (c) G. Bouchoux, R. Flammang and A. Maquestiau, *Org. Mass Spectrom.*, **20**, 154 (1985).
16. D. V. Ramana and N. V. S. Rama Krishna, *Org. Mass Spectrom.*, **24**, 66 (1989).
17. D. V. Ramana and N. V. S. Rama Krishna, *Org. Mass Spectrom.*, **24**, 317 (1989).
18. D. V. Ramana and N. V. S. Rama Krishna, *Org. Mass Spectrom.*, **24**, 485 (1989).
19. K. Eller and H. Schwarz, *Chem. Rev.*, **91**, 1121 (1991).
20. Y. A. Ranasinghe and B. S. Freiser, *Chem. Phys. Lett.*, **200**, 135 (1992).

21. C. W. Bauschlicher, Jr. and S. R. Langhof, *J. Phys. Chem.*, **95**, 2278 (1991).
22. S. T. Graul and M. T. Bowers, *J. Phys. Chem.*, **95**, 8328 (1991).
23. D. T. Terwilliger, J. F. Elder, Jr., J. H. Beynon and R. G. Cooks, *Int. J. Mass Spectrom. Ion Phys.*, **16**, 225 (1975).
24. B. A. Rumpf and P. J. Derrick, *Int. J. Mass Spectrom. Ion Processes*, **82**, 239 (1988).
25. (a) J. L. Holmes and J. T. Terlouw, *Org. Mass Spectrom.* **15**, 383 (1980).
    (b) J. L. Holmes, *Org. Mass Spectrom.*, **20**, 169 (1985).
26. G. E. Busch and K. R. Wilson, *J. Chem. Phys.*, **56**, 3638 (1972).
27. J. A. Booze and T. Baer, *J. Chem. Phys.*, **96**, 5541 (1992).
28. J. A. Booze and T. Baer, *J. Chem. Phys.*, **98**, 186 (1993).
29. The most probable precursors of the minor products $C_4H_3^+/C_4H_2^{+\cdot}$ (at m/z 51/50 in Figure 5) are the dissociative ionizations $(C_2H_2)_2 + h\nu \rightarrow C_4H_2^{+\cdot} + H_2$ and $(C_2H_2)_2 + h\nu \rightarrow C_4H_3^+ + \cdot H$. These ions cannot be secondary fragments of $C_4H_4^{+\cdot}$ because their peaks are noticeably narrower than the $C_4H_4^{+\cdot}$ peak[28]. $\langle T \rangle$ for $C_4H_2^{+\cdot}$ and $C_4H_3^+$ are very similar to the values observed for $C_4H_2^{+\cdot}$ and $C_4H_3^+$ from metastable vinylacetylene ions [30].
30. C. Lifshitz, D. Gibson, K. Levsen and I. Dotan, *Int. J. Mass Spectrom Ion Processes*, **40**, 157 (1981).
31. S. G. Whitney, M. T. Coolbaugh, G. Vaidyanathan and J. F. Garvey, *J. Phys. Chem.*, **95**, 9625 (1991).
32. W. J. Bouma, R. H. Nobes, L. Radom and C. E. Woodward, *J. Org. Chem.*, **47**, 1869 (1982).
33. B. v. Baar, T. Weiske, J. K. Terlouw and H. Schwarz, *Angew. Chem., Int. Ed. Engl.*, **25**, 282 (1986).
34. W. J. Bouma, P. M. W. Gill and L. Radom, *Org. Mass Spectrom.*, **19**, 610 (1984).
35. P. O. Danis, C. Wesdemiotis and F.W. McLafferty, *J. Am. Chem. Soc.*, **105**, 7454 (1983).
36. C. Wesdemiotis and F. W. Mc Lafferty, *Chem. Rev.*, **87**, 485 (1987).
37. J. K. Terlouw and H. Schwarz, *Angew, Chem. Int. Ed. Engl.*, **26**, 805 (1987).
38. J. L. Holmes, *Mass Spectrom. Rev.*, **8**, 513 (1989).
39. M. J. Polce and C. Wesdemiotis, *Mass Spectrometry in the Biological Sciences: A Tutorial* (Ed. M. L. Gross), Kluwer Academic Publishers, Dordrecht, 1992, p. 303.
40. B. v. Baar, W. Koch, C. Lebrilla, J. K. Terlouw, T. Weiske and H. Schwarz, *Angew. Chem., Int. Ed. Engl.*, **25**, 827 (1986).
41. J. M. Buschek, J. L. Holmes and F. P. Lossing, *Org. Mass Spectrom.*, **21**, 729 (1986).
42. J. K. Terlouw, P. C. Burgers, B. L. M. v. Baar, T. Weiske and H. Schwarz, *Chimia*, **40**, 357 (1986).
43. D. Smith, *Chem. Rev.*, **92**, 1473 (1992).
44. W. Wagner-Redeker, A. J. Illies, P. R. Kemper and M. T. Bowers, *J. Am. Chem. Soc.*, **105**, 5719 (1983).
45. J. C. Ray, Jr. P. O. Danis, F. W. Mclafferty and B. K. Carpenter, *J. Am. Chem. Soc.*, **109**, 4408 (1987).
46. M. Y. Zhang, C. Wesdemiotis, M. Marchetti, P. O. Danis, J. C. Ray, Jr. B. K. Carpenter and F. W. McLafferty, *J. Am. Chem. Soc.*, **111**, 8341 (1989).
47. M. J. Polce and C. Wesdemiotis, *J. Am. Chem. Soc.*, **115**, 10849 (1993).
48. M. Y. Zhang, B. K. Carpenter and F. W. McLafferty, *J. Am. Chem. Soc.*, **113**, 9499 (1991).
49. G. Bouchoux, Y. Hoppilliard and J.-P. Flament, *Org. Mass Spectrom.*, **20**, 560 (1985).
50. J. L. Holmes, J. K. Terlouw and P. C. Burgers, *Org. Mass Spectrom.*, **15**, 140 (1980).
51. M. J. Polce and C. Wesdemiotis, unpublished results.
52. F. W. McLafferty and D. B. Stauffer, *Wiley/NBS Registry of Mass Spectral Data*, Wiley, New York, 1989.
53. F. W. McLafferty and F. Turecek, *Interpretation of Mass Spectra*, 4th edition, University Science Books, Mill Valley, CA, 1993.
54. (a) P. G. Wenthold, J. A. Paulino and R. R. Squires, *J. Am. Chem. Soc.*, **113**, 74514 (1991).
    (b) P. G. Wenthold and R. R. Squires, *J. Am. Chem. Soc.*, **116** to appear (1994).
55. S. T. Graul and R. R. Squires, *Mass Spectrom. Rev.*, **7**, 263 (1988).
56. S. T. Graul and R. R. Squires, *J. Am. Chem. Soc.*, **112**, 2517 (1990).
57. S. Beranová and C. Wesdemiotis, unpublished results.
58. M. M. Cordero, J. J. Houser and C. Wesdemiotis, *Anal. Chem.*, **65**, 1594 (1993).
59. M. M. Cordero and C. Wesdemiotis, *Anal. Chem.*, **66, 861** to appear (1994).
60. A. W. McMahon, F. Chadikun and A. G. Harrison, *Int. J. Mass Spectrom. Ion Processes*, **87**, 275 (1989).
61. S. Gronert and C. H. DePuy, *J. Am. Chem. Soc.*, **111**, 9253 (1989).

62. D. Schröder, D. Sülzle, J. Hrusak, D. K. Böhme and H. Schwarz, *Int. J. Mass Spectrom. Ion Processes* **110**, 145 (1991).
63. R. L. Hettich, T. C. Jackson, E, M. Stanko and B. S. Freiser, *J. Am. Chem. Soc.*, **108**, 5086 (1986).
64. D. A. Peake and M. L. Gross, *Anal. Chem.*, **57**, 115 (1985).
65. D. A. Peake and M. L. Gross, *Organometallics*, **5**, 1236 (1986).
66. D. A. Peake, S.-K. Huang and M. L. Gross. *Anal. Chem.*, **59**, 1557 (1987).
67. C. Schulze, H. Schwarz, D. A. Peake and M. L. Gross, *J. Am. Chem. Soc.*, **109**, 2368 (1987).
68. C. Schulze and H. Schwarz, *Chimia*, **41**, 29 (1987).
69. K. Eller and H. Schwarz, *Int. J. Mass Spectrom. Ion Processes*, **93**, 243 (1989).
70. For organometallic chemistry with ICR see:
    (a) B. S. Freiser, *Talanta*, **32**, 697 (1985).
    (b) J. L. Beauchamp, in High Energy Processes in Organometallic Chemistry (Ed.R.E. Suslick), Symposium Series 333, American Chemical Society, 1987.
71. C. Schulze and H. Schwarz, *Organometallics* **9**, 2164 (1990).
72. L. Radom, P. C. Hariharan, J. A. Pople and P. v. R. Schleyer, *J. Am. Chem. Soc.*, **98**, 19 (1976).
73. J. Leszczynski, B. Weiner and M. C. Zerner, *J. Phys. Chem.*, **93**, 139 (1989).
74. F. Ozturk, G. Baykut, M. Moini and J. R. Eyler, *J. Phys. Chem.*, **91**, 4360 (1987).
75. J. L. Holmes and F. P. Lossing, *Can, J. Chem.*, **57**, 249 (1979).
76. K. C. Smyth, S. G. Lias and P. Ausloos, *Combust. Sci. Tech.*, **28**, 147 (1982).
77. P. Ausloos and S. G. Lias, *J. Am. Chem. Soc.*, **103**, 6506 (1981).
78. G. Baykut, F. W. Brill and J. R. Eyler, *Combust, Sci. Tech.*, **45**, 233 (1986).
79. D. Smith and N. G. Adams, *Int. J. Mass Spectrom. Ion Processes*, **76**, 307 (1987).
80. M. Moini, *J. Am. Soc. Mass Spectrom*, **3**. 631 (1992).
81. V. G. Anicich, W. T. Huntress, Jr. and M. J. McEwan, *J. Phys. Chem.*, **90**, 2446 (1986).
82. F. W. Brill and J. R. Eyler, *J. Phys. Chem.*, **85**, 1091 (1981).
83. P. G. Miasek and J. L. Beauchamp, *Int. J. Mass Spectrom. Ion Phys.*, **15**, 49 (1979).
84. M. F. Jarrold, L. M. Bass, P. R. Kemper, P. A. M. van Koppen and M. T. Bowers, *J. Chem. Phys.*, **78**, 3756 (1983).
85. M. T. Bowers, D. D. Elleman and J. L. Beauchamp, *J. Chem. Phys.*, **72**, 3599 (1968).
86. D. H. Rusell, M. L. Gross, J. van der Greef and N. M. M. Nibbering, *J. Am. Chem. Soc.*, **101**, 2086 (1979).
87. S. A. Werner and T. Baer, *J. Chem, Phys.*, **62**, 2900 (1975).
88. W. J. Chesnavich, L. Bass, T. Su and M. T. Bowers, *J. Chem. Phys.*, **74**, 2228 (1981).
89. S. W. McElvany, W. R. Greasy and A. O'Keefe, *J. Chem. Phys.*, **85**, 632 (1986).
90. S. W. McElvany, H. H. Nelson, A. P. Barqanovski, C. H. Watson and J. R.Eyler, *Chem. Phys, Lett.*, **134**, 214 (1987).
91. S. W. McElvany, B. I. Dunlap and A. O'Keefe, *J. Chem. Phys.*, **86**, 715 (1987).
92. S. W. McElvany, *J. Chem. Phys.*, **89**, 2063 (1988).
93. V. G. Anicich, W. T. Huntress Jr. and M. J. McEwan, *J. Phys Chem.*, **90**, 2446 (1986).
94. V. G. Anicich, G. A. Blake, J. K. Kim, M. J. McEwan and W. T. Huntress Jr. *J. Phys.; Chem.*, **88**. 4608 (1984).

Author index

This author index is designed to enable the reader to locate an author's name and work with the aid of the reference numbers appearing in the text. The page numbers are printed in normal type in ascending numerical order, followed by the reference numbers in parentheses. The numbers in *italics* refer to the pages on which the references are actually listed.

Konovalikhin, S.V. 833(345, 346, 348, 350), 834(357), 835(359, 360), 865, 866
Konovalov, A.I. 844(478), 869
Konwar, D. 824(298), 864
Koob, M. 745(36), 747(56), 781
Kool, L.B. 952(55, 56), 1008
Koopmans, T. 22(108), 105, 155(6), 186
Kopchik, R.M. 919(15), 942
Koppang, M.D. 607(58, 59), 623
Koppel, H. 28(117), 105, 185(108), 189
Koppel, I. 44(158), 106, 278(250), 285, 906(141), 915
Koppel, I.A. 271(221), 285
Koppen, P.A.M.van 1215–1217(84), 1221
Kopping, B. 526(203), 576, 927(64), 943
Kopylets, V.I. 621(91), 624
Kopylova, L.I. 321(204), 370
Korakis, T. 1041(87), 1128
Korbalai, M.H. 480(36), 483
Korek, U. 1125(325), 1134
Koremoto, T. 983, 984(267), 1013
Korenowski, G.M. 547(314), 578
Kores, M. 796(57), 858
Korkowski, P.F. 333(266), 372
Kornberg, B.E. 213(117), 228
Korobitsyna, I.K. 426(265), 465
Korolev, V.A. 171, 177–179(52), 187
Korp, J. 796(59), 858
Korshak, V.V. 946, 948, 953(7), 1007
Korshak, Yu.V. 983, 984(271), 1006(271, 402), 1013, 1016
Korshak, Y.V. 946, 948, 953(7), 1007
Korth, H.-G. 844(477), 869, 918(7), 941
Kortüm, G. 241–243, 245(77a), 281
Koschinsky, R. 507(105), 574
Koschmieder, S.U. 139(108), 150
Kosemura, S. 223(177), 229
Koser, G.F. 1166(6, 7), 1170(42), 1178(61), 1180, 1182
Koshihara, S. 998(362, 363), 999(362, 364), 1000(362), 1001–1003(364), 1015
Koshino, H. 903(117), 914
Koski, H.K. 115, 116, 146(26), 148
Koski, L. 995(348), 1015
Kosmynina, A.S. 260(164), 284
Kostenko, L.I. 812(206), 862
Köster, R. 908(149), 915
Kosugi, M. 491(19), 572
Kotaka, T. 1006(399), 1016
Kotake, H. 328(240), 371
Kotali, E. 1175(54), 1182
Kotera, M. 416(206), 463
Kotera, N. 967(152), 1010
Koteswari Prasad, A. 473, 480(9), 483
Kothari, P.J. 652(57, 59), 685
Kotila, M. 770(246), 786
Kottirsch, G. 300(71), 367

Kotzybahibert, F. 215(132), 229
Koutsantonis, G.A. 496(43), 572
Kouwenhoven, A.P. 514(144), 574
Kouzai, H. 970, 971(212), 974(226, 227), 982(212), 1012
Kovac, B. 183(94), 189
Kovac, J. 812(191), 861
Kovacic, P. 442(394, 396, 397), 468
Koval'chuk, E.P. 620(89), 621(89, 91, 92), 624
Kovalev, I.P. 493(33), 572
Kovalevskaya, A.M. 458(541), 471
Koves, A. 765(213), 785
Kowal, R. 853(542), 870
Kowalczyk, J.J. 354(346), 373, 564(421), 580, 1143, 1144(37), 1163
Kowalski, C.J. 1138(19, 20a, 20b, 21), 1139(21), 1140(19, 20a, 20b), 1142(21), 1143(21, 36, 38), 1144(38), 1162, 1163
Kowski, K. 793(21), 857
Koyabu, Y. 421(233), 464
Koyama, K. 444(413), 445(417), 468
Kozai, H. 849(507), 869
Kozhikhova, I.N. 458(544), 471
Kozhushkov, S.I. 162, 178(34), 187
Kozima, S. 289(8), 366
Kozlov, V.A. 815(241), 862
Kozlowska-Gramsz, E. 445(420), 468
Koz'min, A.S. 1167(11), 1181
Kozuka, T. 448(440), 468
Kozyrod, R.P. 810(163), 860
Kraebel, C.M. 117(32), 148
Kraemer, W.P. 29, 30(121), 105, 181(74), 188
Krafft, M.E. 343(306), 373, 508(111), 574
Kral, T. 223(179), 230
Kral, V. 800(102), 859
Kramer, J.B. 427, 428(274), 465
Kramer, T. 478(25), 483
Kramer, W. 447(433), 468
Krampera, F. 620(84–86), 624
Krasnaya, Zh.A. 408(168), 462
Kratky, C. 813(232), 814(232, 234), 820, 828(232), 830(325), 862, 865
Kratky, O. 998(354), 1015
Kratz, D. 165(45–47), 169, 178(45), 187, 308(107), 368
Krause, J. 265(201, 202), 284
Krauss, M. 3, 4(29, 30), 103
Krausz, P. 972(219), 1012
Kravtsov, D.N. 603(50), 623, 841(438), 868
Krawczyk, J. 853(542), 870
Krebs, A. 158(20, 22), 178(20), 187
Krebs, F. 571(481), 581, 1170(35), 1175(55), 1181, 1182
Kreeger, R.L. 547(324), 578
Kreher, R.P. 1058(155), 1130

Index compiled by K. Raven

Subject index

1323

Index compiled by P. Raven